Neuroanatomy Guidance to Successful Neurosurgical Interventions

Anatomy guide for Safe Neurosurgical Intervention
The Torch of Successful Neurosurgery

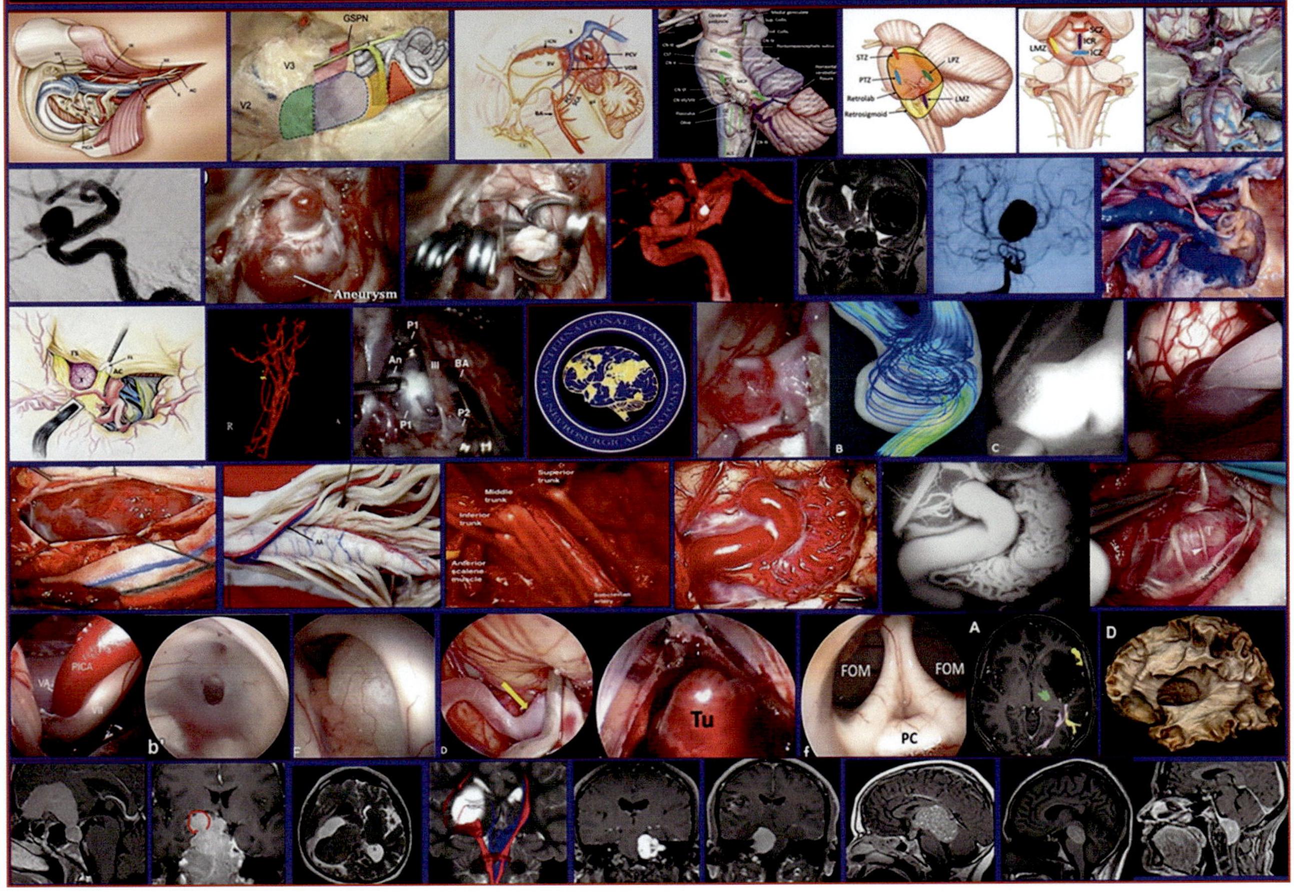

Imad N. Kanaan • Vladimír Beneš
Editors

Neuroanatomy Guidance to Successful Neurosurgical Interventions

The Torch of Neurosurgical Excellence

Volume II

Springer

Editors
Imad N. Kanaan
Neurosciences Center, Section of Neurosurgery
King Faisal Specialist Hospital & Research
Center Institution, Alfaisal University
College of Medicine
Riyadh, Kingdom of Saudi Arabia

Vladimír Beneš
Dept. of Neurosurgery
Central Military Hospital Prague
Praha 2, Czech Republic

ISBN 978-3-031-59837-1 ISBN 978-3-031-59838-8 (eBook)
https://doi.org/10.1007/978-3-031-59838-8

This Springer imprint is published by the registered company Springer Nature Switzerland AG
The registered company address is: Gewerbestrasse 11, 6330 Cham, Switzerland

Foreword

My personal congratulation to Professor Imad N. Kanaan and Professor Vladimír Beneš for publishing such a special neurosurgical book, which is the result of scientific programs of the last 2 years of webinars organized by both authors.

This was a very important neurosurgical educational program for many young and practicing neurosurgeons from all over the world, who had the opportunity to follow all these exciting lectures.

During the 2 years, the most outstanding and experienced neurosurgeons have delivered important insights from their expertise for the state-of-the-art of neurosurgery.

The publication of these important lectures on different aspects of neurosurgery such as "Management of intracranial skull base and intraspinal tumors," "Current technologies for treatment of neurovascular pathologies," as well as "Functional neurosurgery and peripheral nerve injury" is another step towards spreading this spirit of education to all other neurosurgeons who have not had the opportunity to attend live neurosurgical webinars in the last 2 years.

I hope that this important book will reach all neurosurgeons, particularly the new generation of neurosurgeons in every country.

My personal thanks go to both editors, Dr. Imad N. Kanaan and Dr. Vladimír Beneš, for promoting neurosurgical training and education worldwide and for publishing these important contributions to the world of neurosurgery.

Hannover, Germany

Madjid Samii

Preface 1

The successful global venture into neurosurgical education online had its genesis in navigating the constrictions imposed by the late stages of the coronavirus pandemic. This phenomenon is reminiscent of the proverbial maxim, "every cloud has a silver lining," underscoring positive outcomes born out of adversity. This laudable initiative owes its realization to the confluence of advancements in communication and digital technologies, spurred onward by the fervor of our eminent mentors and the aspirational zeal of international neurosurgeons and trainees.

This reference textbook of neuroanatomy guidance for successful neurosurgical interventions stands as a seminal academic publication, a "beacon to neurosurgical excellence," adeptly bridging the chasm between applied anatomy and the intricacies of neurosurgical intervention. Its 75 chapters authored by elite academic neurosurgeons cover the wide spectrum of neurosurgical practice. Adopting a stylistic and conceptual approach evocative of the visual artistry of Leonardo Da Vinci and the balance between art and science, the narrative intertwines vibrant illustrations with anatomical highlights and expertly selected surgical pearls, thereby offering profound guidance for the pursuit of safe neurosurgical interventions.

The expeditious responses of the devoted members of the "IANA" and our friends and esteemed colleagues within the international neurosurgical community were integral to the realization of this monumental endeavor. Their willingness to share their knowledge and experience with their younger international peers and trainees exemplifies the collaborative spirit within this field, and indeed was fundamental to accomplish this publication. Most importantly, I am really blessed to have the greatest support, inspiration, and understanding of my beloved family—my wife Huda and my two precious daughters Faten and Nadia—without whose encouragement this achievement would not have been possible.

Riyadh, Saudi Arabia — Imad N. Kanaan

Preface 2

Neuroanatomy Guidance to Successful Neurosurgical Interventions: The Torch of Neurosurgical Excellence is the book which is exactly what the title declares. The authors of 75 chapters are world known and respected leaders in their field. They have spent a considerable amount of time and effort to put together chapters which would be educational, very much pictorial and which would cover the recent state-of-the-art in respected areas. All areas of neurosurgery are covered with emphasis on educational value.

Starting with history with special emphasis on neuroanatomy and a few general chapters dealing with evolution, plasticity, and neurosurgical facilities requirements the following pages are devoted to individual subspecialties of neurosurgery. Part II of the book covers the whole vascular neurosurgery, aneurysms, AVMs, cavernomas, and ischemia. Part III is devoted to skull base tumors. All aspects are discussed, again with the special emphasis on neuroanatomy and surgical approaches. Part IV is dedicated to ventricular tumors which form the specific group and specific management requirements like hydrocephalus. Part V, gliomas, deals with the intrinsic tumors on the basis of white matter anatomy. One chapter deals with brainstem tumors and another with pediatric tumors which by all means form a specific and special group with problems different from that in adults. Part VI starts with an excellent anatomy of spinal cord vasculature followed by spinal tumors and peripheral nerves where the anatomical knowledge is of extreme importance. Part VII, functional, in its epilepsy chapters is again based on anatomy of white matter tracts. In this part, the movement disorders, trigeminal neuralgia, and normal pressure hydrocephalus are also included. Quite fittingly, the book is concluded by a special article on complications. Complications appear, despite the fact that generally all the preceding chapters do have one common feature—how to avoid the complications. The other common feature is the fact that all the authors have special interest in neuroanatomy and are either members or friends of International Academy of Neurosurgical Anatomy. The special interest of the individual authors, neuroanatomy, ensures the high number of high-quality images. These by themselves do have very high educational value.

We hope and believe the book will be of extreme help and reference, especially to young neurosurgeons, to those who want to excel in their chosen discipline. We believe the book will serve all neurosurgeons as well as others for a good orientation on what the recent neurosurgery may offer to our patients.

I would like to appreciate and thank all those who helped me put the book together, without such a help the task would be impossible. I also would like to thank my wife Eva for her understanding and never ending support.

Prague, Czech Republic　　Vladimír Beneš

Contents

Part IV Ventricular Tumors

Part V Gliomas

Part VI Spinal Tumors and Peripheral Nerves

Part VII Functional

Part VIII Complication in Neurosurgery

Contributors

Javier Abarca-Olivas Department of Neurosurgery, Hospital General Universitario de Alicante, Alicante, Spain

Hadie Adams Imperial College Healthcare NHS Trust, London, UK

Guilherme Agnoletto Department of Neurosurgery, Clinical Neurosciences Center, University of Utah, Salt Lake City, UT, USA

Mohamed Helmy Neurosurgery Department, Huashan Hospital Fudan University, China National Neurological & Neurosurgical Center, Shanghai, People's Republic of China

Neurosurgery Department, Shanghai Donglei Brain Hospital, Shanghai, People's Republic of China

Filippo Flavio Angileri Department of Neurosurgery, AOU Policlinico "G. Martino", University of Messina, Messina, Italy

Miguel A. Arraez Carlos Haya University Hospital, University of Malaga, Malaga, Spain

Cinta Arraez-Manrique Carlos Haya University Hospital, University of Malaga, Malaga, Spain

Heba Azooz Department of Neurosurgery, Children Cancer Hospital Egypt, Cairo, Egypt

Matias Baldoncini Laboratory of Microsurgical Neuroanatomy, Second Chair of Gross Anatomy, School of Medicine, University of Buenos Aires, Buenos Aires, Argentina

Elena Baño-Ruiz Department of Neurosurgery, Hospital General Universitario de Alicante, Alicante, Spain

Mustafa K. Başkaya Department of Neurological Surgery, University of Wisconsin-Madison, Madison, WI, USA

Jaafar Basma Arkansas Neuroscience Institute, CHI Saint-Vincent, Sherwood, AR, USA

Vladimír Beneš Institute of Clinical Neurodisciplines, Department of Neurosurgery and Neurooncology, First Faculty of Medicinefavo, Charles University and Military University Hospital, Prague, Czech Republic

Department of Neurosurgery and Neurooncology, First Faculty of Medicine, Charles University and Military University Hospital, Prague, Czech Republic

Giacomo Bertolini Neurosurgery Division, Head and Neck Department, Azienda Ospedaliero-Universitaria di Parma, Parma, Italy

Xu Bin Neurosurgery Department, Huashan Hospital Fudan University, China National Neurological & Neurosurgical Center, Shanghai, People's Republic of China

Luis A. B. Borba Department of Neurosurgery, Mackenzie Evangelical University Hospital, Curitiba, PR, Brazil

Department of Neurosurgery, Federal University of Paraná, Curitiba, PR, Brazil

Roman Bošnjak University Medical Center Ljubljana, Ljubljana, Slovenia

University of Ljubljana, Medical faculty, Ljubljana, Slovenia

Ondřej Bradáč Institute of Clinical Neurodisciplines, Department of Neurosurgery and Neurooncology, First Faculty of Medicine, Charles University and Military University Hospital, Prague, Czech Republic

Department of Neurosurgery and Neurooncology, First Faculty of Medicine, Charles University and Military University Hospital, Prague, Czech Republic

Department of Neurosurgery, Second Faculty of Medicine, Charles University and Motol University Hospital, Prague, Czech Republic

David P. Bray Department of Neurosurgery, Emory University School of Medicine, Atlanta, GA, USA

Giovanni Broggi Department of Neurosurgery, Fondazione IRCCS Istituto Neurologico Carlo Besta, Milan, Italy

Adéla Bubeníková Department of Neurosurgery and Neurooncology, First Faculty of Medicine, Charles University and Military University Hospital, Prague, Czech Republic

Institute of Clinical Neurodisciplines, Department of Neurosurgery and Neurooncology, First Faculty of Medicinefavo, Charles University and Military University Hospital, Prague, Czech Republic

Department of Neurosurgery, Second Faculty of Medicine, Charles University and Motol University Hospital, Prague, Czech Republic

Institute of Clinical Neurodisciplines, Department of Neurosurgery and Neurooncology, First Faculty of Medicine, Charles University and Military University Hospital, Prague, Czech Republic

Department of Neurosurgery and Neurooncology, First Faculty of Medicine, Charles University and Military University Hospital, Prague, Czech Republic

Department of Neurosurgery, Second Faculty of Medicine, Charles University and Motol University Hospital, Prague, Czech Republic

Alvaro Campero Department of Neurological Surgery, Hospital Padilla, Tucumán, Argentina

Alberto Campione Department of Neurosurgery, San Filippo Neri Hospital/ASLRoma1, Rome, Italy

Paolo Cappabianca Division of Neurosurgery, Department of Neurosciences, Reproductive and Odontostomatological Sciences, Università degli Studi di Napoli "Federico II", Naples, Italy

Salvatore Cardali Department of Neurosurgery, "Papardo" General Hospital, University of Messina, Messina, Italy

Andrea L. Castillo Department of Neurosurgery, Antonio Lenin Fonseca Hospital, Managua, Nicaragua

Guilherme H. W. Ceccato Department of Neurosurgery, Mackenzie Evangelical University Hospital, Curitiba, PR, Brazil

Fady T. Charbel Department of Neurosurgery, University of Illinois, Chicago, IL, USA

Dimitris Charitos Department of Neurosurgery, Lariboisière Hospital, APHP Nord, Assistance Publique – Hôpitaux de Paris, Paris Cité University, Paris, France

Experimental Neurosurgery Laboratory, Department of Neurosurgery, Lariboisière Hospital, APHP Nord, Assistance Publique – Hôpitaux de Paris, Paris Cité University, Paris, France

Iype Cherian Department of Neurosurgery, Krishna Vishwa Vidyapeeth, Karad, Maharashtra, India

Alfredo Conti Neurosurgery, IRCCS Istituto delle Scienze Neurologiche di Bologna, Bologna, Italy

Dipartimento di Scienze Biomediche e NeuroMotorie (DIBINEM), Alma Mater Studiorum University of Bologna, Bologna, Italy

Giulia Cossu Department of Neurosurgery, University Hospital of Lausanne, Lausanne, Switzerland

William T. Couldwell Department of Neurosurgery, Clinical Neurosciences Center, University of Utah, Salt Lake City, UT, USA

Antonello Curcio Department of Neurosurgery, AOU Policlinico "G. Martino", University of Messina, Messina, Italy

Roy T. Daniel Department of Neurosurgery, University-Hospital Lausanne, Lausanne, Switzerland

Chandrashekhar E. Deopujari Department of Neurosurgery, Bombay Hospital Institute of Medical Sciences, Mumbai, Maharashtra, India

Yücel Doğruel Department of Neurosurgery, Health Sciences University, Tepecik Training and Research Hospital, İzmir, Turkey

Víctor Fernández-Cornejo Department of Neurosurgery, Hospital General Universitario de Alicante, Alicante, Spain

Ioan Alexandru Florian Department of Neuroscience, University of Medicine and Pharmacy Iuliu Hatieganu, Cluj-Napoca, Romania

Department of Neurosurgery, Cluj County Emergency Clinical Hospital, Cluj-Napoca, Romania

Ioan Stefan Florian Department of Neuroscience, University of Medicine and Pharmacy Iuliu Hatieganu, Cluj-Napoca, Romania

Department of Neurosurgery, Cluj County Emergency Clinical Hospital, Cluj-Napoca, Romania

Sébastien Froelich Department of Neurosurgery, Lariboisière Hospital, APHP Nord, Assistance Publique – Hôpitaux de Paris, Paris Cité University, Paris, France

Experimental Neurosurgery Laboratory, Department of Neurosurgery, Lariboisière Hospital, APHP Nord, Assistance Publique – Hôpitaux de Paris, Paris Cité University, Paris, France

Giada Garufi Department of Neurosurgery, "Papardo" General Hospital, Messina, Italy

Turab Gasimov Department of Neurological Surgery, University of Wisconsin-Madison, Madison, WI, USA

Mario Giordano Department of Neurosurgery, INI – International Neuroscience Institute, Hannover, Germany

Atul Goel Department of Neurosurgery, Seth G.S. Medical College and K.E.M Hospital, Parel, Mumbai, Maharashtra, India

Department of Neurosurgery, Lilavati Hospital and Research Center, Mumbai, Maharashtra, India

Juan L. Gómez-Amador Department of Neurosurgery, National Institute of Neurology and Neurosurgery "Manuel Velasco Suárez", Mexico City, Mexico

Cristina Gómez-Revuelta Department of Neurosurgery, Hospital General Universitario de Alicante, Alicante, Spain

Pablo González-Lopéz Alicant General Hospital, University of Alicant, Alicant, Spain

Department of Neurosurgery, Hospital General Universitario de Alicante, Alicante, Spain

Takeo Goto Department of Neurosurgery, Osaka Metropolitan University Graduate School of Medicine, Osaka, Japan

J. André Grotenhuis Emeritus Professor of Neurosurgery, Radboud University Medical Centre, Nijmegen, The Netherlands

Michelle Grouls Department of Neurosurgery, Lariboisière Hospital, APHP Nord, Assistance Publique – Hôpitaux de Paris, Paris Cité University, Paris, France

Experimental Neurosurgery Laboratory, Department of Neurosurgery, Lariboisière Hospital, APHP Nord, Assistance Publique – Hôpitaux de Paris, Paris Cité University, Paris, France

Abuzer Güngör Faculty of Medicine Department of Neurosurgery, İstinye University, Istanbul, Turkey

Liv Hospital Vadistanbul, Neurosurgery Clinic, Istanbul, Turkey

Arthur Hosmann Department of Neurosurgery & Gamma Knife Surgery, P.D. Hinduja Hospital & MRC, Mumbai, Maharashtra, India

Department of Neurosurgery, Medical University of Vienna, Vienna, Austria

Timothée Jacquesson Skull Base and Pituitary Unit, Department of Neurosurgery B, Neurological Hospital Pierre Wertheimer, Lyon, France

Dragan Jankovic Department of Neurosurgery, University Medical Center, Johannes Gutenberg University Mainz, Mainz, Germany

DieuMerci Kantenga Kabulo Department of Neurosurgery, University of Zimbabwe, Harare, Zimbabwe

Kazadi K. N. Kalangu University of Zimbabwe, Department of Neurosurgery, Harare, Zimbabwe

Imad N. Kanaan Neurosciences Center, Section of Neurosurgery, King Faisal Specialist Hospital & Research Center Institution, Alfaisal University, College of Medicine, Riyadh, Kingdom of Saudi Arabia

Ekkehard M. Kasper Boston University School of Medicine, Boston, MA, USA

Department of Neurosurgery, St. Elizabeth Medical Center, Boston, MA, USA

Faculty of Health Sciences, McMaster University, Hamilton, ON, Canada

Yoko Kato Department of Neurosurgery, Fujita Health University Bantane Hospital, Nagoya, Japan

Abdel Raouf Kayssi Arkansas Neuroscience Institute, CHI Saint-Vincent, Sherwood, AR, USA

Mehdi Khaleghi Department of Neurosurgery, The James Cancer Hospital and Wexner Medical Center, The Ohio State University, Columbus, OH, USA

Ramez W. Kirollos Department of Neurosurgery, National Neuroscience Institute, Singapore, Singapore

Duke-NUS Graduate Medical School, Singapore, Singapore

Ali F. Krisht Arkansas Neuroscience Institute, CHI Saint-Vincent, Sherwood, AR, USA

Artem Kuptsov Department of Neurosurgery, Hospital General Universitario de Alicante, Alicante, Spain

José Alberto Landeiro Department of Neurosurgery, Antonio Pedro University Hospital, Fluminense Federal University, Niterói, Brazil

Yujun Liao Fudan University, Shanghai, China

James K. Liu, MD, FACS, FAANS Skull Base Institute of New Jersey, Neurosurgeons of New Jersey, NYU Langone Neurosurgery Network, Livingston, NJ, USA

Department of Neurosurgery, Cooperman Barnabas Medical Center, RWJ Barnabas Health, Livingston, NJ, USA

Martin Májovský Institute of Clinical Neurodisciplines, Department of Neurosurgery and Neurooncology, First Faculty of Medicine, Charles University and Military University Hospital, Prague, Czech Republic

Igor L. Maldonado CHRU de Tours, Tours, France

Laboratoire d'Anatomie, Faculté de Médecine de Tours, Tours, France

INSERM, Imaging Brain & Neuropsychiatry iBraiN U1253, University of Tours, Tours, France

Sai Mannam Perelman School of Medicine, University of Pennsylvania, Philadelphia, PA, USA

Mario Ammirati College of Science and Technology, Temple University, Philadelphia, PA, USA

Jaime L. Martínez Santos Department of Neurological Surgery, Medical University of South Carolina, Charleston, SC, USA

Luciano Mastronardi Department of Neurosurgery, San Filippo Neri Hospital/ASLRoma1, Rome, Italy

Laura Stone McGuire Department of Neurosurgery, University of Illinois, Chicago, IL, USA

José Orlando de Melo Junior Department of Neurosurgery, Paulo Niemeyer State Brain Institute, Rio de Janeiro, Brazil

Ahmed Al Menabbawy Department of Neurosurgery, University Medicine Greifswald, Greifswald, Germany

Ali Tayebi Meybodi Department of Neurological Surgery, New Jersey Medical School, Newark, NJ, USA

Basant K. Misra Department of Neurosurgery & Gamma Knife Surgery, P.D. Hinduja Hospital & MRC, Mumbai, Maharashtra, India

Jacques J. Morcos, MD Department of Neurosurgery, The University of Miami, Miami, FL, USA

Department of Neurosurgery, UT Houston, Houston, TX, USA

Hiroki Morisako Department of Neurosurgery, Osaka Metropolitan University Graduate School of Medicine, Osaka, Japan

Ramesh Nair Imperial College Healthcare NHS Trust, London, UK

Farshad Nassiri Department of Neurosurgery, Clinical Neurosciences Center, University of Utah, Salt Lake City, UT, USA

Juan Nieto-Navarro Department of Neurosurgery, Hospital General Universitario de Alicante, Alicante, Spain

Kenji Ohata Department of Neurosurgery, Osaka Metropolitan University Graduate School of Medicine, Osaka, Japan

Soichi Oya Department of Neurosurgery, Gunma University Graduate School of Medicine, Maebashi, Gunma, Japan

Nelson M. Oyesiku Department of Neurosurgery, University of North Carolina School of Medicine, Chapel Hill, NC, USA

Dhiraj Parihar Department of Neurosurgery, Seth G.S. Medical College and K.E.M Hospital, Parel, Mumbai, Maharashtra, India

Victor Hugo Pérez Forensic Sciences Institute, TSJ, Autonomous University of Mexico State, Mexico City, Mexico

Daniel M. Prevedello Department of Neurosurgery, The James Cancer Hospital and Wexner Medical Center, The Ohio State University, Columbus, OH, USA

Mubashir Mahmood Qureshi Aga Khan University Hospital, Nairobi, Kenya

Manuel de Jesús Encarnacion Ramírez Neurosurgery Department, RUDN University, Russian Federation, Moscow, Russia

Manuel Encarnacion Ramirez Laboratory of Microsurgical Neuroanatomy, Second Chair of Gross Anatomy, School of Medicine, University of Buenos Aires, Autónoma de Buenos Aires, Argentina

Lukas Rasulić Faculty of Medicine, University of Belgrade, Belgrade, Serbia

Department for Peripheral Nerve Surgery, Functional Neurosurgery, and Pain Management Surgery, Clinic for Neurosurgery, University Clinical Centre of Serbia, Belgrade, Serbia

Ehab El Refaee Department of Neurosurgery, University Medicine Greifswald, Greifswald, Germany

Department of Neurosurgery, Cairo University, Cairo, Egypt

Roberto Colasanti Department of Neurosurgery, Maurizio Bufalini Hospital, AUSL della Romagna, Cesena, Italy

Andrea De Rosa Division of Neurosurgery, Department of Neurosciences, Reproductive and Odontostomatological Sciences, Università degli Studi di Napoli "Federico II", Naples, Italy

Miroslav Samardžić Faculty of Medicine, University of Belgrade, Belgrade, Serbia

Department for Peripheral Nerve Surgery, Functional Neurosurgery, and Pain Management Surgery, Clinic for Neurosurgery, University Clinical Centre of Serbia, Belgrade, Serbia

Amir Samii Department of Neurosurgery, INI – International Neuroscience Institute, Hannover, Germany

Marcos V. Sangrador-Deitos Department of Neurosurgery, National Institute of Neurology and Neurosurgery "Manuel Velasco Suárez", Mexico City, Mexico

Thomas Santarius Department of Neurosurgery, Cambridge University Hospitals NHS Trust, Cambridge, UK

Department of Clinical Neurosciences, University of Cambridge, Cambridge, UK

Henry W. S. Schroeder Department of Neurosurgery, University Medicine Greifswald, Greifswald, Germany

Abhidha Shah Department of Neurosurgery, Seth G.S. Medical College and K.E.M Hospital, Parel, Mumbai, Maharashtra, India

Nishit J. Shah Department of ENT, Bombay Hospital Institute of Medical Sciences, Mumbai, Maharashtra, India

Muhammad Yousuf Shaikh Department of Neurosurgery, LNH, Karachi, Pakistan

Salman Sharif Department of Neurosurgery, LNH, Karachi, Pakistan

Puneet Sharma Imperial College Healthcare NHS Trust, London, UK

Rajeev Sharma Department of Neurosurgery, Osaka Metropolitan University Graduate School of Medicine, Osaka, Japan

All India Institute of Medical Sciences, New Delhi, India

Uzair Ahmed Siddiqui LNH, Karachi, Pakistan

Juan E. Small Lahey Hospital and Medical Center, Burlington, MA, USA

Anna Štekláčová Institute of Clinical Neurodisciplines, Department of Neurosurgery and Neurooncology, First Faculty of Medicinefavo, Charles University and Military University Hospital, Prague, Czech Republic

Zhangjie Su Department of Neurosurgery, Cambridge University Hospitals NHS Trust, Cambridge, UK

Department of Clinical Neurosciences, University of Cambridge, Cambridge, UK

Albert Sufianov Federal Centre of Neurosurgery (Tyumen) of Ministry of Health of Russian Federation, Tyumen, Russia

Rinat Sufianov I.M.Sechenov First Moscow State Medical University, Moscow, Russia

Matthew Z. Sun Department of Neurosurgery, UT Southwestern, Dallas, TX, USA

Francesco Tomasello Humanitas Clinical Institute of Catania, Misterbianco, Catania, Italy

Keki E. Turel Department of Neurosurgery, Bombay Hospital Institute of Medical Sciences, Mumbai, India

Sara Venturini Department of Neurosurgery, Cambridge University Hospitals NHS Trust, Cambridge, UK

Rodolfo Villalobos-Díaz Department of Neurosurgery, National Institute of Neurology and Neurosurgery "Manuel Velasco Suárez", Mexico City, Mexico

Robert Walker Department of Neurosurgery, St. Elizabeth Medical Center, Boston, MA, USA

Abhijit G. Warade Department of Neurosurgery & Gamma Knife Surgery, P.D. Hinduja Hospital & MRC, Mumbai, Maharashtra, India

Eva M. Wu, MD Department of Neurosurgery, The University of Miami, Miami, FL, USA

Anterior Petrosal Approach

41

Sébastien Froelich, Michelle Grouls, and Dimitris Charitos

Learning Objectives

- Knowledge of the key anatomical landmarks and their variations to perform an anterior petrosal approach
- Knowledge of the step-by-step technique of the anterior petrosal approach
- Knowledge of indication for use and limitations
- Knowledge of the potential complications and tips and tricks to avoid them

41.1 Introduction

The anterior petrosal approach (APA) offers access to the posterior fossa through the temporal fossa floor. It is a "looking-down" approach contrary to the posterior petrosal approach, which is a "looking-up" approach. It exposes the petroclival region, including upper clivus and mid-clivus, and the anterolateral part of the pons. Compared to the retrosigmoid approach, the APA also provides direct access to the temporal fossa floor, Meckel's cave, and the petrous apex (PA). It offers exposure of the trigeminal nerve from the foramen ovale (FO) to the root entry point. In the case of petroclival meningiomas, which is one of the most common indications for this approach, it exposes the tumor base including its feeders first, transforming a skull base meningioma into a convexity meningioma. However, one of the main issues of this approach is temporal lobe (TL) retraction, which increases when looking further upwards. It remains a challenging procedure due to the complexity of the anatomy, the narrow working space below the TL that must be preserved, and the deep location of the targeted area. Therefore, indications for this approach must be carefully evaluated. In this chapter, an overview of the key anatomical landmarks and variations, step-by-step operative techniques, and tips and tricks to avoid complications is presented to make the best use of this approach.

S. Froelich (✉) · M. Grouls · D. Charitos
Department of Neurosurgery, Lariboisière Hospital, APHP Nord, Assistance Publique – Hôpitaux de Paris, Paris Cité University, Paris, France

Experimental Neurosurgery Laboratory, Department of Neurosurgery, Lariboisière Hospital, APHP Nord, Assistance Publique – Hôpitaux de Paris, Paris Cité University, Paris, France
e-mail: Sebastien.froelich@aphp.fr; M.Grouls@umcutrecht.nl

41.2 History and Background

Exposing the internal auditory canal (IAC) and posterior cranial fossa by a middle cranial fossa approach to section the vestibulocochlear nerve in Meniere's disease was first described by Parry in 1904 [1]. However, it took decades before House first precisely described the "middle fossa approach" to access the IAC and its contents through the middle cranial fossa for vertigo, tinnitus, and vestibular schwannomas [2]. Later, the extended middle fossa approach was described to increase the exposure of the cerebellopontine angle [3]. Kawase, focusing on the drilling of the PA anterior to the IAC, popularized this now called "Kawase approach" for the resection of petroclival meningiomas and surgical treatment of lower basilar artery aneurysms [4, 5]. Over the years, the approach was mainly described by neurosurgeons, as the "anterior petrosal approach" and "anterior petrosectomy" represent only the bone drilling part of the approach [6–10].

I. N. Kanaan, V. Beneš (eds.), *Neuroanatomy Guidance to Successful Neurosurgical Interventions*,
https://doi.org/10.1007/978-3-031-59838-8_41

41.3 Anatomy Highlights

The most important landmarks for the anterior petrosal approach that need to be identified to delineate the area of drilling and to avoid injury to surrounding critical structures are:

- Foramen spinosum
- Foramen ovale
- Arcuate eminence
- Greater petrosal nerve
- Petrous ridge

The temporal fossa floor is made of elevations and prominences (Fig. 41.1), and only a few landmarks can help surgeon to envision the structures at risk under the surface of the bone and delineate the bone corridor to drill (Fig. 41.2).

- Foramen Spinosum (FS): FS is the most lateral key landmark (Fig. 41.2). It can be easily identified by following the middle meningeal artery (MMA) branches in a retrograde manner toward the artery trunk. In rare occasion, FS may be duplicated [11].
- Foramen Ovale (FO): FO can be identified immediately anteromedial (from 2 to 7.5 mm) to FS, the mandibular nerve (V3), entering FO [12] (Fig. 41.2). FO may be hidden behind a prominence of bone. In rare occasion, two FOs can be found as well as merged FO and FS [11].
- Arcuate Eminence (AE): The most prominent elevation in the petrous bone is the AE, which provides an indication of the anatomical position of the superior semicircular canal (SSC) (Fig. 41.2). Nevertheless, there is not always an exact correspondence between the AE and the superior semicircular canal [13].
- Greater Petrosal Nerve (GSPN): Slightly lateral and anterior to the AE lies the hiatus of the GSPN, from where the nerve emerges from the bone and runs anteriorly and slightly medially toward the trigeminal nerve (Fig. 41.2). GSPN runs into a small inconsistent groove and the sphenopetrosal groove and is embedded meningeal connective tissue [14]. Once the temporal fossa dura is detached from the floor of the temporal fossa, several bundles of connective tissue may remain stuck to the bone and may be mistaken for the GSN. The bone covering the geniculate ganglion (GG) and the proximal part of GSPN is dehiscent in 14.5% of cases [15]. Below the GPSN lies the petrous segment of the internal carotid artery. Predominantly, a thin wall of bone is found between the internal carotid artery (ICA) and the GSPN before it runs below the trigeminal nerve [16].
- Internal Carotid Artery (ICA): The petrous segment of the ICA lies immediately below GSPN (Fig. 41.2). Both run anteriorly and slightly medially with a 7° angle between both [12, 17]. The length of the petrous ICA is variable ranging from 4.5 to 19 mm with a mean value of approximately 10 mm. The carotid canal is dehiscent in more than 80% of cases [18], and the ICA is only covered by temporal dura (TD) and connective tissue [19]. In some cases, the dehiscence of the CC starts before the ICA reaches the lateral aspect of V3. The distance from the GSPN hiatus to the mandibular nerve (V3) is approximately 1 cm.
- Internal Auditory Canal (IAC): There are several techniques to localize the IAC that all carry a degree of imprecision due to variable anatomy and the influence of head rotation. Fisch described the SSC as the unique essential landmark for the identification of the IAC. He reports that the IAC can be located by dissecting within a 60° angle anterior to the superior semicircular canal [20]. Because

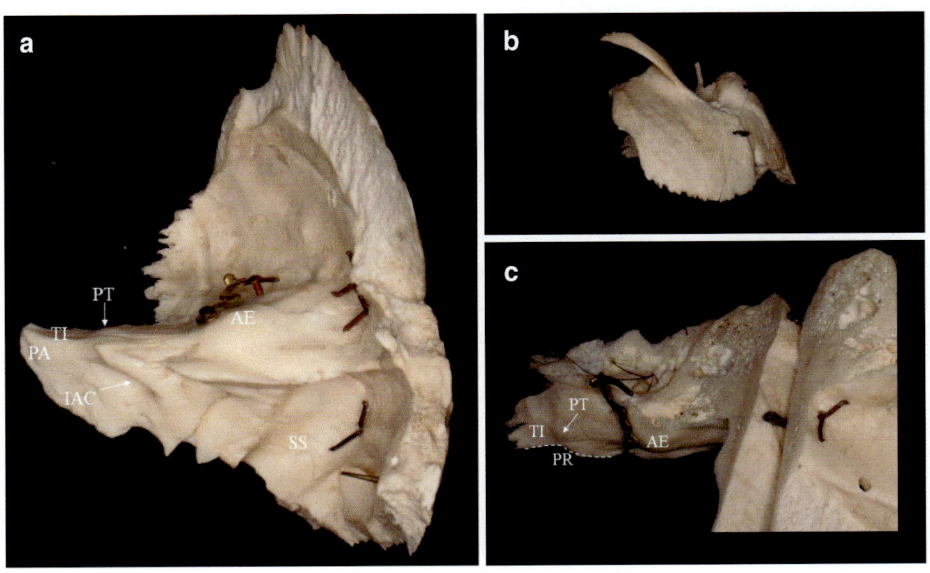

Fig. 41.1 Right temporal bone. (**a**) Posteromedial view of the petrous bone. (**b**) Superior posterolateral view, perspective before removing the squamous part. (**c**) operative view. *PA* petrous apex, *TI* trigeminal impression, *PT* princeteau tubercle, *AE* arcuate eminence, *IAC* internal auditory canal, *SS* sigmoid sulcus, *PR* petrous ridge

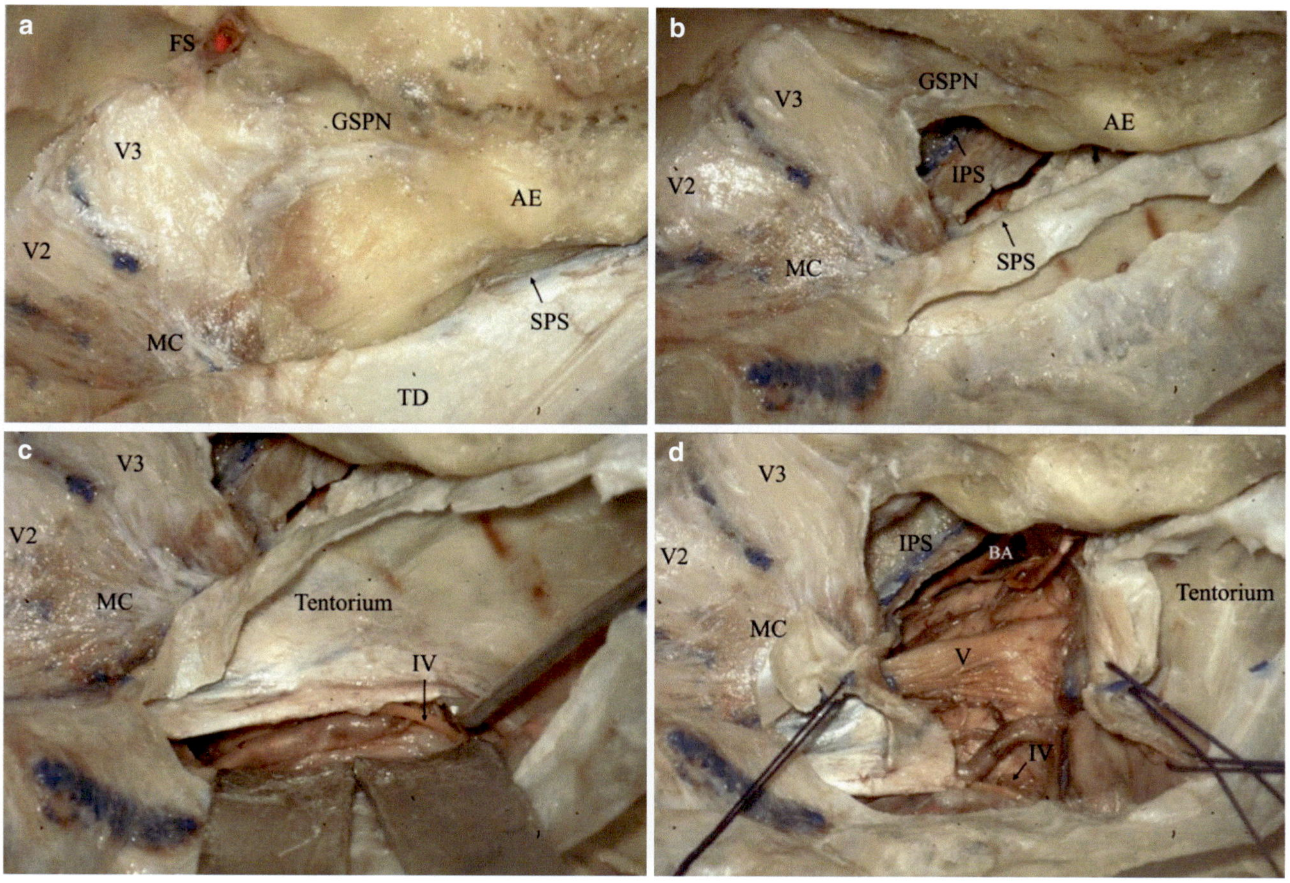

Fig. 41.2 Anatomical dissections of the anterior petrosal approach. (**a**) Anatomical landmarks on the temporal fossa floor after temporal dura peeling. (**b**) Posterior fossa dura and temporal dura incision after petrous bone drilling. (**c**) Exposure of the tentorium and IV CN entry point below the free edge. (**d**) Final exposure. *FS* foramen spinosum, *AE* arcuate eminence, *SPS* superior petrosal sinus, *IPS* inferior petrosal sinus, *GSPN* greater superior petrosal nerve, *MC* Meckel's cave, *BA* basilar artery

intraoperative angle measurement is challenging, Garcia-Ibanez's technique is used frequently. With this technique, the orientation of the IAC is described as a line bisecting the main axes of GSPN and the AE [21]. Most important to be aware of is that the thickness of bone above the IAC diminished as drilling progresses laterally [22].

- Cochlea: The cochlea is the most difficult structure to precisely locate in the petrous bone. The basal turn of the cochlea is parallel to GSPN and about 60° angle with the IAC. With a variable depth to the surface of the temporal fossa floor, it is located somewhere below GG between the IAC and the ICA. It should be noted that the cochlea will be located deeper than the IAC [23]. In a cadaveric study, the average distance between cochlea and ICA was between 2.1 and 4.3 mm [24, 25]. Following the line of sight of the approach, the cochlea is therefore most often hidden in a dark corner below GG; however, there are significant variations from one patient to the other. Therefore, precise evaluation on a thin-cut bone CT scanner of the cochlea's depth and its relationship with the GG and ICA is key information before surgery and should be correlated with the line of sight of the approach (Fig. 41.2).
- Superior Petrosal Sinus (SPS): The SPS lies along the petrous ridge in a groove, the superior petrosal sulcus, which is more pronounced in its posterolateral part [7, 26] (Fig. 41.2). It originates in the cavernous sinus and drains into the transverse-sigmoid junction (TSJ). Most often, the sinus passes above the trigeminal porus. However, in some cases, the sinus runs below or divides above and below. The superior petrosal venous complex drains in the SPS either above or lateral to the level of IAC (type I), between MC and IAC (type II, most frequent), or above and medial to MC (type III) [26, 27]. The sinus may drain to both CS and TSJ (65%) only to the TSJ (37%) or only to the CS (3%) [28].
- Inferior Petrosal Sinus (IPS): IPS is draining venous blood from the CS to the junction of the sigmoid sinus (SS) and superior jugular bulb (JB) (Fig. 41.2). It is the main venous drainage of the CS. It lies in the inferior petrosal sinus groove along the petro-occipital fissure. This groove is absent in 10% of cases and forms a bone tunnel in 3.5% [29]. The size and drainage of the IPS are variable, and multiple classifications have been described. It most often drains into the jugular bulb or jugular vein

but may connect directly to the deep cervical plexus [30]. It is a classic target for venous catheterization of the CS or venous sampling and may be embolized before the anterior petrosal approach to avoid bleeding [31].

41.4 Preoperative Evaluation of Imaging

Since transpetrosal approaches are technically complex, a precise understanding of surgical and radiological anatomy is mandatory. Beside the complexity of anatomy of petrous bone and related neurovascular structures, variations from one patient to another are numeous starting with the pneumatization of the mastoid and PA. The reported prevalence of PA pneumatization in the literature ranges from 5% to 33% [32]. The degree of pneumatization is highly variable and can range from no air cells at all to a fully pneumatized PA. It is therefore mandatory to carefully study individual anatomy prior to surgery on the preoperative workup which should systematically include thin-cut CT scan and MRI including MR-v.

The following characteristics of the petrous bone and venous anatomy should be studied before surgery to avoid unexpected findings and complications [33].

- Pneumatization of the zygomatic root, mastoid, and PA
- Clival depression evaluated with the angle between petrous ridge and clivus
- FS and FO position and relationships
- Presence of a retro-gasserian tubercle of princeteau along the petrous ridge, lateral to the trigeminal depression
- Distance between cochlea and middle fossa floor
- Distance between SSC and middle fossa floor
- Distance between cochlea and ampulla of the SSC
- Distance between cochlea and ICA
- Dehiscence of middle fossa floor over petrous ICA
- Dehiscence of middle fossa floor over GG
- Venous anatomy of the temporal veins (including Labbe's vein), SPS, and IPS

When studying the petrous bone, evaluation of its shape, the air cells, the relationship between cancellous bone and cortical bone, and the thickness of the cortical bone over critical structures can be used as a road map during this approach. Not recognizing a bone dehiscence over the petrous ICA or over the GG before the operation can easily lead to unforgivable surgical complications. Rarely, the FS and FO are fused, increasing the likelihood of V3 injury due to coagulation and sectioning of the MMA (Fig. 41.3).

The location and drainage pattern of the venous system can be a contraindication for a transpetrosal approach. It is important to examine the course of the Labbe vein, as well as the course of other temporal veins. Labbe vein may be absent or represented by more than one vessel [34]. In our experience, a long low horizontal pathway along the temporal lobe increases the risk of injury during craniotomy. When not already occluded by the tumor, sacrificing the superior petrosal veins can lead to venous complications. In cases lacking either the medial part or the lateral part of the SPS, as found in 40% of cerebellopontine angles, the superior petrosal veins may lose their drainage pathway if the remaining segment of the SPS is sectioned [35]. The surgeon should assess anatomical variations before surgery such as a connection between the SPS and the cavernous sinus or drainage of the middle cerebral veins in SPS or through FO into pterygoid plexus [34, 36].

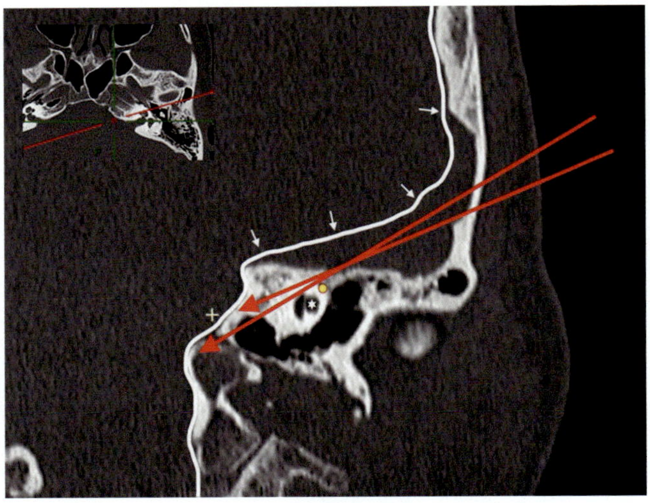

Fig. 41.3 Coronal CT showing the cochlea (white star) and the GG (yellow). The white line represents the elevated dura mater (white arrows). The red arrows show the different lines of sight when drilling more laterally toward the cochlea and with additional retraction

41.5 Approach/Surgical Intervention

41.5.1 Positioning and Setting

The patient is positioned supine, with the head fixed in 3-pin Mayfield head clamp with rotation of 70°, slight extension, and inclination downward. The shoulder is slightly elevated if needed. The thorax is elevated to position the head above the level of the heart. Neuronavigation is prepared with merged CT (bone reconstruction kernel, thin slices) and MRI sequences. Neuromonitoring is prepared for at least the sixth and seventh cranial nerves. Facial nerve monitoring is mandatory to identify greater superficial petrosal nerve and GG and to monitor facial nerve function during the approach [37]. Abducens nerve monitoring is mandatory for petroclival lesions and/or if there is a need to drill the true PA below Meckel's cave. An external lumbar drain is not routinely used to keep CSF in the dural sac, which also serves as a protection of the brain.

41.5.2 Soft Tissue Dissection and Craniotomy

Figure 41.4 gives an overview of the key surgical steps. A frontotemporal question mark incision is made starting immediately anterior to the tragus. The skin is retracted anteriorly, until the fat pad comes into view. An interfacial dissection is performed to further separate the skin flap from the temporalis muscle (TM) and to allow additional retraction of both to further expose the anterior aspect of the temporal fossa floor. The posterior root of the zygomatic arch is exposed. The pericranium is incised all along the superior temporal line and a retrograde dissection of the TM is performed preserving the deep periosteal layer of the muscle, which protects the vessels and nerves that vascularize and innervate it. The TM is retracted anteriorly, and a temporal craniotomy is performed. The lower margin of the temporal craniotomy is drilled until the level of the temporal fossa floor is reached. Mastoid air cells extending toward the zygomatic arch may be opened at this level and need to be closed with bone wax. The anterior border of the craniotomy below the retracted TM is drilled using a 4-mm coarse diamond drill, to gain additional exposure of the anterior aspect of the temporal fossa floor and FO.

41.5.3 Kawase Triangle Exposure

The TD is then progressively elevated from posterior to anterior from the temporal fossa floor using a sharp dissector. Shrinking the dura with low-power coagulation can help at the beginning of dural peeling from the middle fossa floor. To identify FS, the MMA running along the dura is followed toward the foramen. The artery is then coagulated and cut 2 to 3 mm distal to the foramen. Immediately anterior and slightly medial to FS, the FO is

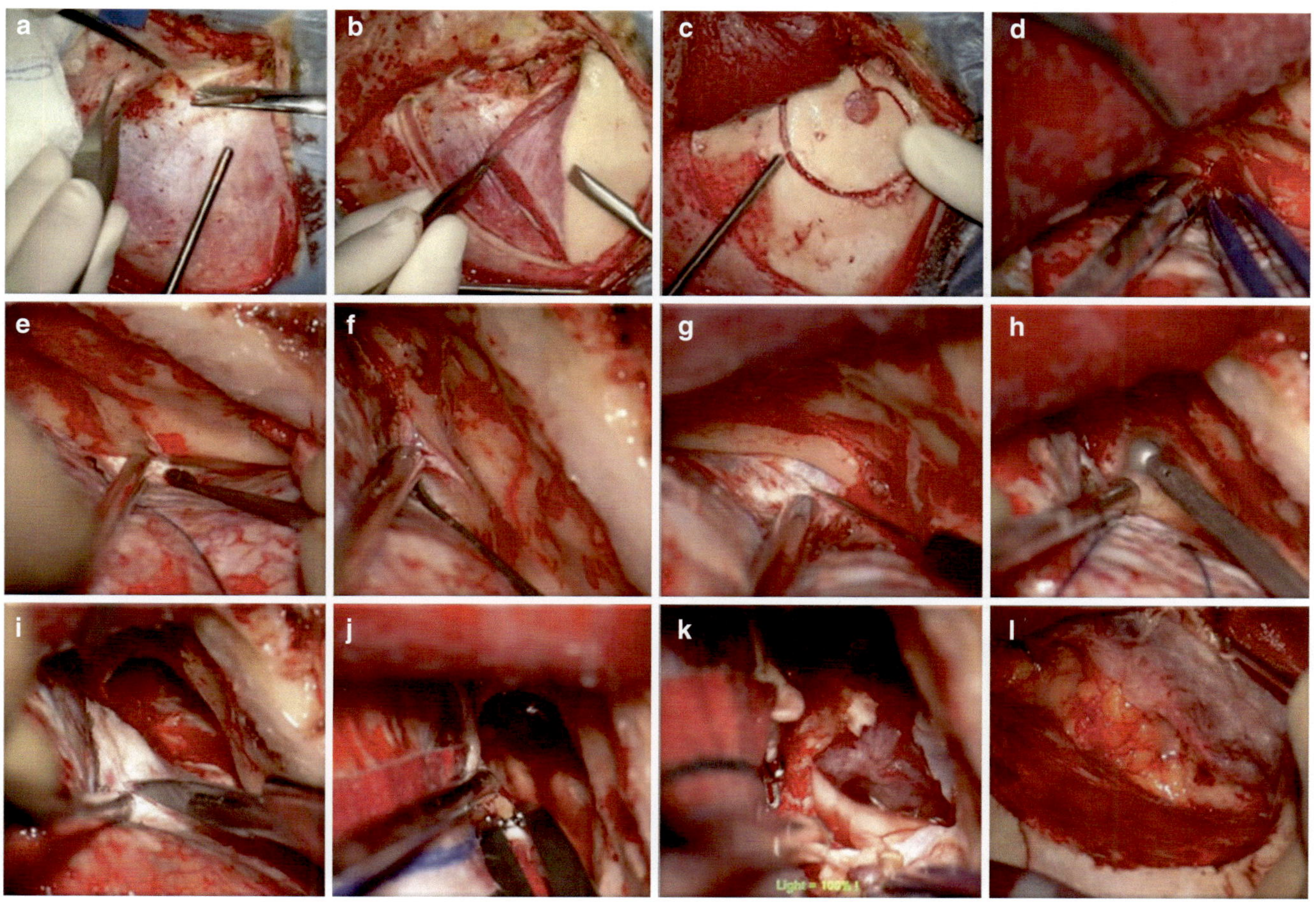

Fig. 41.4 Anterior petrosal approach. (**a**) Interfacial dissection. (**b**) Retrograde temporalis muscle dissection. (**c**) Temporal craniotomy. (**d**) Coagulation of the MMA distal to FS. (**e**) Temporal dura peeling. (**f**) Identification of GSPN. (**g**) Incision periosteum FO. (**h**) Premeatal drilling. (**i**) Opening temporal dura. (**j**) Cutting SPS. (**k**) Opening tentorium. (**l**) Closure with autologous abdominal fat graft and fibrin glue

identified. A bump of bone may obscure the exposure of FO and can be drilled with a 3-mm diamond drill before incising the periosteum along the V3. The periosteal layer along FO is incised, and the dura propria is then peeled from V3. The dura is further peeled anterior to V3 until FR is reached, to allow a more homogeneous retraction of the TD. In the case of venous bleeding, hemostasis is obtained with hemostatic matric and tamponade to avoid coagulation and thermal injury to the nerves. Subsequently, the AE is exposed posteriorly. Retraction of the temporal lobe is achieved by using dural sutures. Slightly anterior to the AE, GSPN is identified exiting a small bone canal (hiatus of the greater superficial petrosal hiatus). Neurostimulation is used to confirm the position of GSPN. The dura is further peeled from the GSPN from back to front using a sharp periosteal elevator and microscissor, until the previously partially exposed V3 is reached [14]. The dura is then elevated more medially toward the petrous ridge and the PA. Along the lateral aspect of V3, the periosteal layer of the dura follows the bone anteriorly below V3. This dural layer should then be sectioned along V3 to further elevate the dura propria from V3 toward Meckel's cave and to expose the PA (Fig. 41.5). For most cases, there is no need to peel the dura from V2. Trigeminal nerve and Meckel's cave itself can be further elevated from the trigeminal impression and ICA exposing the true PA. At last, the petrous ridge is exposed. One should be aware of the "false petrous ridge," which is the lateral border of the SPS groove. The medial border is the true petrous ridge that needs to be exposed. Care should be taken not to enter the sinus while peeling its lateral wall from the groove.

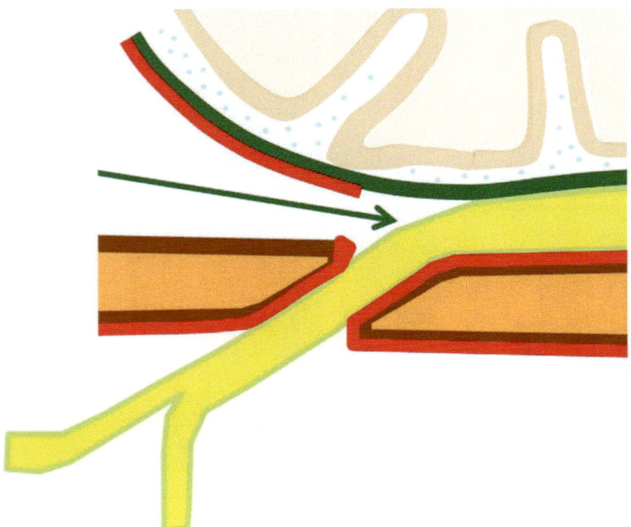

Fig. 41.5 Schematic overview of cutting the periosteal layer (in red) of dura over nerve. Keeping the CSF in the dural sac also serves as a protection for the TL. (Illustration from S. Froelich)

41.5.4 Anterior Petrosectomy

For the anterior petrosectomy, four drilling zones can be identified (Fig. 41.6):

- Premeatal zone: between lateral margin of V3 and anterior border of IAC
- Supra-meatal zone: above the IAC
- Post-meatal zone: behind the IAC
- True PA: below Meckel's cave

Anterior petrous bone drilling is performed with a 3-mm coarse diamond or pure diamond under continuous irrigation. Drilling starts in the safe medial corner of the premeatal space at the level of Princeteau tubercle lateral to the trigeminal impression (Fig. 41.7). Drilling progressively extends laterally toward the ICA and GSPN, within the cancellous bone, until the cortical bone covering the ICA is identified. The carotid canal is opened only if petrous ICA control is needed. The cortical bone of the posterior surface of the petrous bone and dura is progressively exposed, using a pure diamond drill. The dura of the posterior surface is then followed laterally until the dura of the anterior aspect of the internal auditory canal is reached. Drilling then continues anteriorly and inferiorly toward the IPS. In the case of profuse venous bleeding hemostatic matrix and tamponade can be used.

The most challenging part is enlarging laterally toward the cochlea since there are no ideal landmarks to precisely locate the cochlea. However, lateral drilling extension is critical as the more the drilling is pushed laterally toward the cochlea, the more exposure of the lower clivus and brainstem is obtained. Therefore, the use of neuronavigation in this part is recommended to help locate the cochlea (Fig. 41.3). If

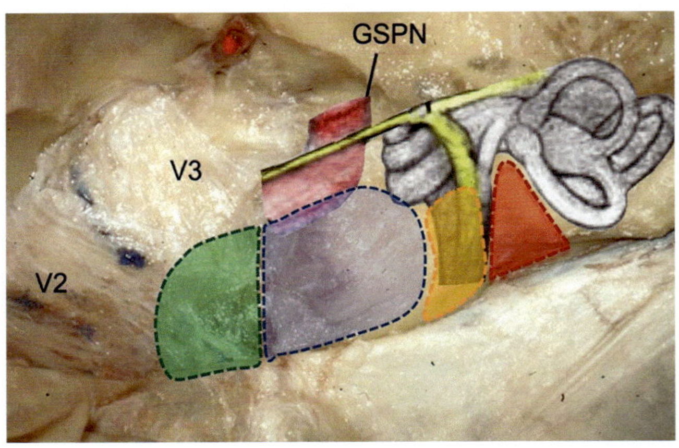

Fig. 41.6 Right-sided cadaveric dissection with delineated areas for anterior petrosectomy: (red) postmeatal, delineated by the IAC and SSC; (yellow) suprameatal, superiorly to the IAC; (blue) premeatal, delineated by IAC, GSPN, V3, ICA, and IPS; (green) true PA, under Meckel's cave. (Illustration from S. Froelich)

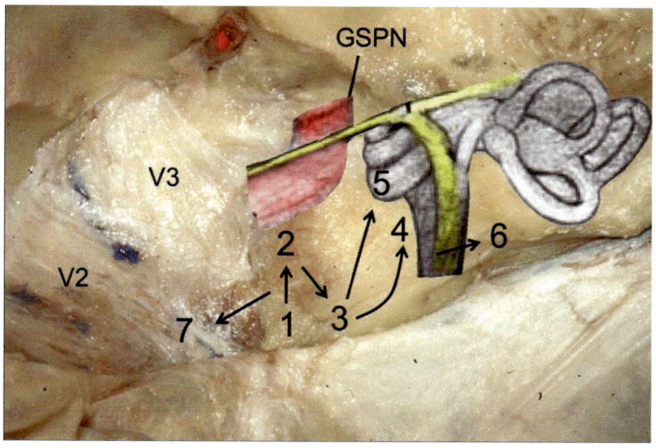

Fig. 41.7 Right-sided cadaveric dissection indicating the drilling steps: Drilling starts in safe zone number 1, medial to the premeatal zone, and immediately lateral to V3. It then proceeds toward the ICA (2) before going posteriorly again along the posterior fossa dura (3) and then following this dura, along the IAC (4). The most challenging part is to drill toward the cochlea laterally (5). The post-meatal space can be drilled (6) and the true PA (7). (Illustration from S. Froelich)

necessary, post-meatal drilling area can add providing a better exposure of the middle cerebellar peduncle and cisternal segment of the VII and VIII cranial nerves. The last part of the anterior petrosectomy comprises drilling of the true PA below Meckel's cave, just behind the Princeteau tubercle. Drilling of this part of the petrous bone will allow the downward transposition of V3.

41.5.5 Dura Opening

The posterior fossa dura is opened first starting inferiorly, allowing additional CSF release, and extending toward the SPS. Care is taken not to injure the superior petrosal venous complex which joins the SPS close to the trigeminal porus. Care is taken also not to injure the trigeminal nerve that may be pushed laterally against the posterior fossa dura in the case of petroclival tumor, especially meningiomas.

The TD is then opened along approximately between 1 and 2 cm parallel to the SPS. Doing so, the TD is used to protect the temporal lobe. The incision extends anteriorly on dura that was peeled from the roof of Meckel's cave. From this temporal incision, a counter incision is made toward the SPS. The SPS is then ligated, close to the trigeminal porus to preserve the venous drainage of the superior petrosal venous complex and is sectioned between both ligations. The use of a micro-Doppler can be helpful to evaluate the blood flow into the SPS and petrosal veins. Until the trigeminal nerve is located, excessive coagulation of the SPS should be avoided. The tentorium is then sectioned toward the free edge. The trochlear nerve should be first located on the other side of the arachnoid covering the tentorium and protected with a cottonoid before coagulation and the final section of the free edge. Once the tentorium is sectioned, the margins of the incision are slightly coagulated to allow retraction of the dura. To open Meckel's cave, a counter incision is made below the SPS, along the lateral aspect of Meckel's cave and V3. In the case of tumor extending into Meckel's cave medial to the trigeminal nerve, the nerve is usually pushed against the lateral aspect of the trigeminal porus and is extremely thin.

41.5.6 Closure

In the case of pneumatized petrous apex, a thin layer of bone wax is placed. The dura mater is reapproximated and pieces of abdominal fat graft are placed into the drilled bone volume to fill the dead space in between the dural margins that could not be sutured. Fibrin glue is used to seal the dura and fat. The bone flap is fixed with plates.

41.6 Surgical Pearls

- Optimize the position to reduce ICP
- Avoid self-retaining retractor
- Optimize the line of sight to reduce temporal lobe retraction
- Avoid tearing the vascular bundles between the temporal fossa floor and the dura
- Avoid using coagulation in proximity to the cranial nerves or only at the lowest intensity
- Control extradural venous bleeding with the hemostatic agent and tamponade
- Peel the dura mater from posterior to anterior avoiding excessive traction
- VII cranial nerve monitoring is mandatory

One of the most challenging parts of this approach is the preservation of the temporal lobe. Firstly, to reduce the retraction, it is mandatory to constantly adjust the position of the microscope to use the optimum line of sight for the target and use the least possible amount of retraction (Fig. 41.8).

A permanent interaction with the anesthesiologist is needed to obtain brain relaxation. Mannitol and hyperventilation can be used if needed. A lumbar drainage is not routinely placed. During the initial dura peeling, the CSF protects the brain (Fig. 41.5). When more relaxation is needed, once the bone flap is removed, a 2- to 3-mm incision of the temporal convexity dura can be made to allow the CSF to slowly bead up while peeling the dura from the temporal fossa floor.

Self-retaining retractor can and should be avoided as much as possible. Dynamic retraction is achieved with two

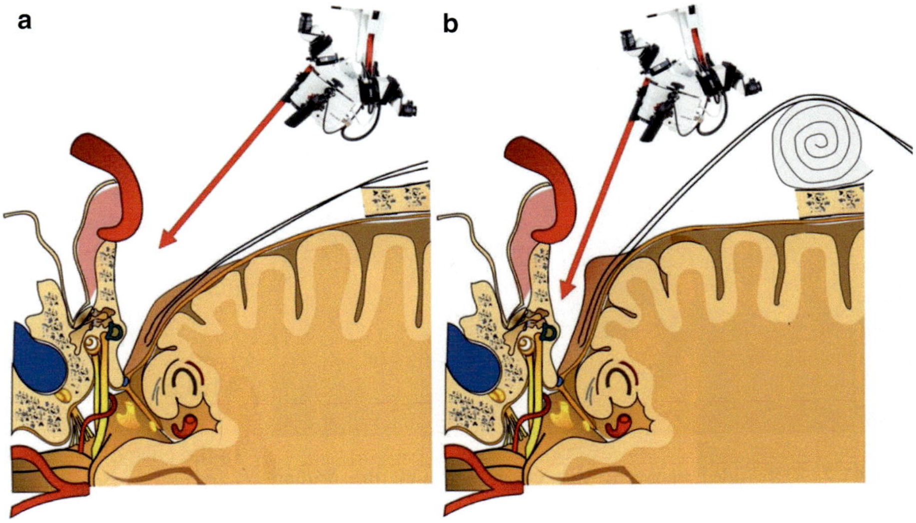

Fig. 41.8 Retraction of the TD and temporal lobe using stitches and a sponge roll. (**a**) Retraction without the sponge roll. (**b**) The sponge roll is used to avoid lateral to medial compression of the temporal by the stiches. (Illustration from K. Watanabe & S Froelich)

or three dural stitches placed on the TD (5-0 monofilament) to keep the dura elevated from the floor (Fig. 41.8). To avoid those stitches also push the dura and the temporal lobe downward toward the central cisterns while running along its outer surface, the three stitches run over a 3-cm diameter roll of sponges positioned along the superior margin of the craniotomy (Fig. 41.8).

Exposed small vascular bundles, exiting bone foramina along the temporal fossa floor, joining the TD, should be coagulated and cut with micro-scissors to constant bleeding from the bone during the rest of the operation. Coagulation close to the nerves should be avoided. Venous bleeding can be stopped with tamponade or hemostatic matrix. In the case of arterial bleeding of the sectioned dural layers along FO, only precise very low-power coagulation under continuous irrigation should be done.

Peeling the dura from GSPN should be done gently from posterior to anterior under continuous EMG monitoring to avoid traction injury to the GG. In the case of significant adherence, sharp dissection using a micro-scissors or a micro-blade is recommended to avoid significant traction and injury to the nerve.

41.7 Risk of Complications

- Temporal lobe injury
- Venous infarct
- Hearing loss
- Facial nerve palsy
- Trochlear nerve palsy
- Abducens nerve palsy
- CSF leakage
- Meningitis
- Wound infection
- Pseudomeningocele
- Carotid artery injury
- Temporalis muscle atrophy
- Seizures

41.8 Discussion

The APA is one of the pillars of skull surgery. However, it remains one of the most challenging approaches consisting of drilling a deep volume of bone, located immediately below the TL, and crossed and surrounded with critical neurovascular structures that can only be imagined in the mind's eye with the help of anatomical landmarks. A precise knowledge of the anatomy and variations of the petrous bone is therefore mandatory and cannot be hardly replaced using neuronavigation.

Indications for APA are numerous including (Fig. 41.9):

- Intrapetrous lesions (e.g., chondrosarcoma and chordoma), mainly chondrosarcoma because of its more lateral position in comparison with chordoma
- Petroclival lesions (e.g., meningioma, schwannoma, and epidermoid cyst), especially the petroclival meningiomas located at the level of the mid- and upper clivus
- Meckel's cave lesions (e.g., trigeminal schwannoma)
- Pons lesions (e.g., cavernoma and tumor)

One should keep in mind that the APA is a looking-down approach, and the more the lesion is extending superiorly above the level of the free edge, the more TL retraction will be needed to look up. There is no need for APA for lesions located below the free edge, but which do not extend down below the level of the petrous ridge as Kawase triangle drilling is made to remove the ridge and expose lesions behind it. Posterior petrosal, a looking-up approach, retrosigmoid, or transsylvian approaches may be more appropriate for lesions extending above the free edge avoiding significant retraction of the TL.

The limits of this approach are lesions extending laterally in the CPA displacing VII and VIII as APA does not provide adequate exposure of the cisternal segment of VII and VIII cranial nerves. In such a situation, post-meatal drilling can help provide additional exposure.

Another limitation is lesions extending inferiorly or located into the lower part of the pons. To increase exposure inferiorly, drilling can be further extended laterally toward the cochlea and ICA to the cost of additional risk for those structures. The use of 0 and 30° endoscopes can also greatly enhance the visualization around the corner and overcome some of those limitations (Fig. 41.10).

The comparison between the open and endonasal endoscopic techniques is well described by Van Gompel [38]. With the open APA, a larger portion of the PA can be drilled with adequate visualization of more lateral structures, such as internal auditory canal and cochlea. However, endonasal endoscopic approach is more appropriate for more inferomedial lesions, such as cholesterol granuloma or chordoma arising from the clivus and extending into the PA. The consistency of the lesion is also an important matter to consider, with EEA being more appropriate for soft lesions.

The APA can be combined with a posterior petrosal approach as a so-called combined petrosal approach. The combination of a "looking-down" approach, the APA, and a "looking-up" approach to the retroinfundibular space, the posterior petrosal approach, provides the advantage of multiple optimal lines of sight to look at a large deep-seated lesion and its relationship with critical related neurovascular structures.

In the case of combined petrosal approach, the anterior petrosectomy drilling is the final step of the petrous ridge drilling like the clinoidectomy is the final step of a frontotemporal approach with sphenoid ridge drilling (Fig. 41.11) (Table 41.1).

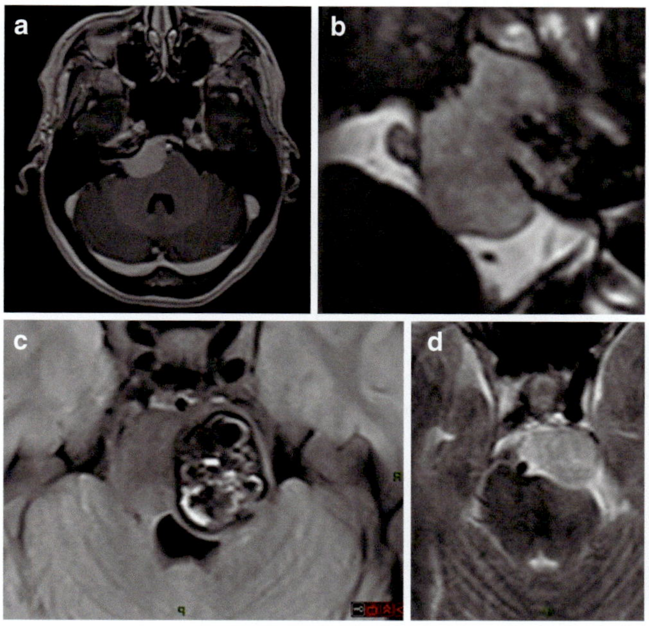

Fig. 41.9 (**a**) Petroclival meningioma. (**b**) Chondrosarcoma. (**c**) Cavernoma. (**d**) Chordoma

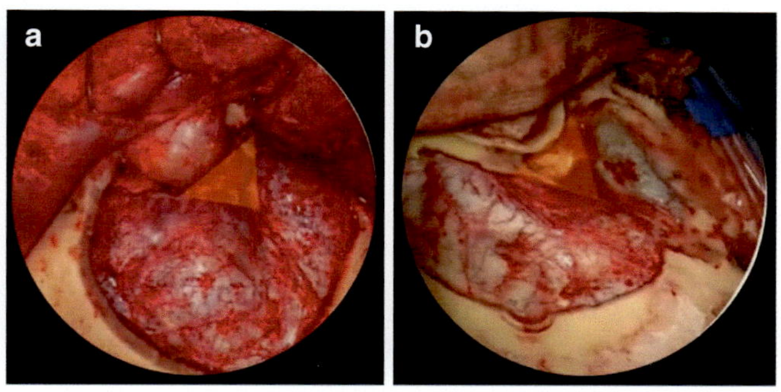

Fig. 41.11 (**a**) Frontotemporal approach. (**b**) Combined petrosal approach

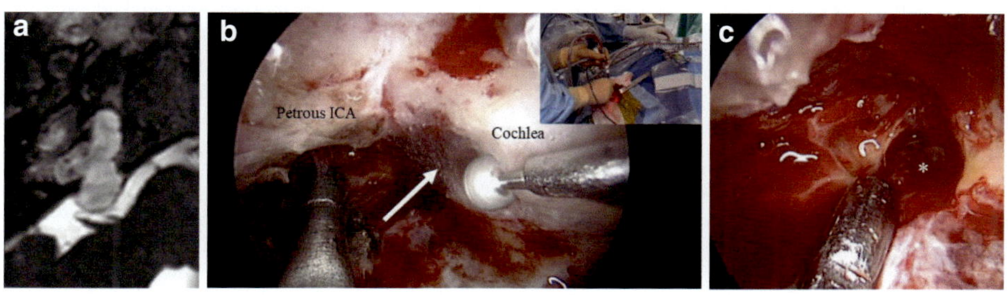

Fig. 41.10 Petrous apex chondrosarcoma resected through an endoscopic-assisted APA. (**a**) Axial T2-weighted MRI showing the lesion. (**b**) Endoscopic angled view (30° angled scope) with chopsticks technique, white arrow pointing to the tumor inferior to the cochlea (**c**) with 70° angled endoscope, and the tumor (white asterisk) is visualized inferior to the cochlea

Table 41.1 Comparison between the frontotemporal approach and the combined petrosal approach

	Frontotemporal approach	Combined petrosal approach
Craniotomy	Frontotemporal	Temporo-occipital
Corridor	Between the frontal and temporal lobes	Between the temporal lobe and the cerebellum
Drilling	Sphenoid ridge	Petrous ridge
Line of sight	Along the orbit	Along the labyrinth
Line of sight	Drilling the anterior clinoid process	Drilling the PA

41.9 Conclusion

The APA is a challenging approach because of the complexity of petrous bone anatomy, the fragility of the temporal lobe that needs to be elevated, the variable venous anatomy, and the depth of the corridor. However, it provides a direct and optimal line of sight for some deep-seated lesions especially those located in the petroclival region and in the anterolateral aspect of the pons. The use of this approach, instead of other approaches such as retrosigmoid, posterior petrosal, or endoscopic endonasal, has to be carefully considered and should be the result of a thoughtful process and careful analysis of anatomical variations. Considering the limits of the corridor provided by the APA, one should be aware that the use of the endoscope can also significantly increase the area of exposure.

References

1. Parry RH. A case of tinnitus and vertigo treated by division of the auditory nerve. 1904. J Laryngol Otol. 1991;105(12):1099–100.
2. House WF. Middle cranial fossa approach to the petrous pyramid. Report of 50 cases. Arch Otolaryngol. 1963;78:460–9.
3. Bochenek Z, Kukwa A. An extended approach through the middle cranial fossa to the internal auditory meatus and the cerebello-pontine angle. Acta Otolaryngol. 1975;80(5–6):410–4.
4. Kawase T, Toya S, Shiobara R, Mine T. Transpetrosal approach for aneurysms of the lower basilar artery. J Neurosurg. 1985;63(6):857–61.
5. Aziz KM, van Loveren HR, Tew JM Jr, Chicoine MR. The Kawase approach to retrosellar and upper clival basilar aneurysms. Neurosurgery. 1999;44(6):1225–34; discussion 34–6.
6. Harsh GR, Sekhar LN. The subtemporal, transcavernous, anterior transpetrosal approach to the upper brain stem and clivus. J Neurosurg. 1992;77(5):709–17.
7. Fournier HD, Mercier P, Velut S, Reigner B, Cronier P, Pillet J. Surgical anatomy and dissection of the petrous and peripetrous area. Anatomic basis of the lateral approaches to the skull base. Surg Radiol Anat. 1994;16(2):143–8.
8. Aristegui M, Cokkeser Y, Saleh E, Naguib M, Landolfi M, Taibah A, et al. Surgical anatomy of the extended middle cranial fossa approach. Skull Base Surg. 1994;4(4):181–8.
9. Wigand ME. Enlarged middle fossa approach to the cerebello-pontine angle. Technique and indications. Rev Laryngol Otol Rhinol (Bord). 1998;119(3):159–62.
10. Al-Mefty O, Fox JL, Smith RR. Petrosal approach for petroclival meningiomas. Neurosurgery. 1988;22(3):510–7.
11. Khairnar KB, Bhusari PA. An anatomical study on the foramen ovale and the foramen spinosum. J Clin Diagn Res. 2013;7(3):427–9.
12. Krayenbuhl N, Isolan GR, Al-Mefty O. The foramen spinosum: a landmark in middle fossa surgery. Neurosurg Rev. 2008;31(4):397–401; discussion 2.
13. Santos FP, Longo MG, May GG, Isolan GR. Computed tomography evaluation of the correspondence between the arcuate eminence and the superior semicircular canal. World Neurosurg. 2018;111:e261–e6.
14. Jittapiromsak P, Sabuncuoglu H, Deshmukh P, Nakaji P, Spetzler RF, Preul MC. Greater superficial petrosal nerve dissection: back to front or front to back? Neurosurgery. 2009;64(5 Suppl 2):253–8; discussion 8–9.
15. Isaacson B, Vrabec JT. The radiographic prevalence of geniculate ganglion dehiscence in normal and congenitally thin temporal bones. Otol Neurotol. 2007;28(1):107–10.
16. Paullus WS, Pait TG, Rhoton AI Jr. Microsurgical exposure of the petrous portion of the carotid artery. J Neurosurg. 1977;47(5):713–26.
17. Spiessberger A, Baumann F, Kothbauer KF, Aref M, Marbacher S, Fandino J, et al. Bony dehiscence of the horizontal petrous internal carotid artery canal: an anatomic study with surgical implications. World Neurosurg. 2018;114:e1174–e9.
18. Hearst MJ, Kadar A, Keller JT, Choo DI, Pensak ML, Samy RN. Petrous carotid canal dehiscence: an anatomic and radiographic study. Otol Neurotol. 2008;29(7):1001–4.
19. Osawa S, Rhoton AL Jr, Tanriover N, Shimizu S, Fujii K. Microsurgical anatomy and surgical exposure of the petrous segment of the internal carotid artery. Neurosurgery. 2008;63(4 Suppl 2):210–38; discussion 39.
20. Fisch U. Transtemporal surgery of the internal auditory canal. Report of 92 cases, technique, indications and results. Adv Otorhinolaryngol. 1970;17:203–40.
21. Garcia-Ibanez E, Garcia-Ibanez JL. Middle fossa vestibular neurectomy: a report of 373 cases. Otolaryngol Head Neck Surg. 1980;88(4):486–90.
22. Tayebi Meybodi A, Mignucci-Jimenez G, Lawton MT, Liu JK, Preul MC, Sun H. Comprehensive microsurgical anatomy of the middle cranial fossa: part I-osseous and meningeal anatomy. Front Surg. 2023;10:1132774.
23. Mastronardi L, Sameshima T, Ducati A, De Waele LF, Ferrante L, Fukushima T. Extradural middle fossa approach. Proposal of a learning method: the "rule of two fans". Technical note. Skull Base. 2006;16(3):181–4.
24. Rhoton AL Jr. The temporal bone and transtemporal approaches. Neurosurgery. 2000;47(3 Suppl):S211–65.
25. Dew LA, Shelton C, Harnsberger HR, Thompson BG Jr. Surgical exposure of the petrous internal carotid artery: practical application for skull base surgery. Laryngoscope. 1997;107(7):967–76.
26. Tanriover N, Abe H, Rhoton AL Jr, Kawashima M, Sanus GZ, Akar Z. Microsurgical anatomy of the superior petrosal venous complex: new classifications and implications for subtemporal transtentorial and retrosigmoid suprameatal approaches. J Neurosurg. 2007;106(6):1041–50.
27. Mortazavi MM, Cox MA, Saker E, Krishnamurthy S, Verma K, Griessenauer CJ, et al. The superior petrosal sinus: a review of anatomy, embryology, pathology, and neurosurgical relevance. Neurosurg Rev. 2018;41(3):713–8.
28. Tayebi Meybodi A, Mignucci-Jimenez G, Lawton MT, Liu JK, Preul MC, Sun H. Comprehensive microsurgical anatomy of the

middle cranial fossa: part II-neurovascular anatomy. Front Surg. 2023;10:1132784.
29. Ekanem UI, Olewnik L, Porzionato A, Macchi V, Iwanaga J, Loukas M, et al. Morphology of the groove of the inferior petrosal sinus: application to better understanding variations and surgery of the skull base. Anat Cell Biol. 2022;55(2):135–41.
30. Lv X, Wu Z. Anatomic variations of internal jugular vein, inferior petrosal sinus and its confluence pattern: implications in inferior petrosal sinus catheterization. Interv Neuroradiol. 2015;21(6):769–73.
31. Fournier HD, Pasco-Papon A, Mercier P, Menei P, Hayek G, Guy G. The anterolateral transpetrosal approach to the clivus and the pons. Preoperative embolisation of the inferior petrosal sinus. Interv Neuroradiol. 1997;3(3):255–60.
32. Adams Perez J, Rassier Isolan G, Pires de Aguiar PH, Antunes AM. Volumetry and analysis of anatomical variants of the anterior portion of the petrous apex outlined by the kawase triangle using computed tomography. J Neurol Surg B Skull Base. 2014;75(3):147–51.
33. di Russo P, Giammattei L, Passeri T, Fava A, Voormolen E, Bernat AL, et al. Lariboisiere hospital pre-operative surgical checklist to improve safety during transpetrosal approaches. Acta Neurochir. 2022;164(11):2819–32.
34. Padget DH. The cranial venous system in man in reference to development, adult configuration, and relation to the arteries. Am J Anat. 1956;98(3):307–55.
35. Matsushima K, Matsushima T, Kuga Y, Kodama Y, Inoue K, Ohnishi H, et al. Classification of the superior petrosal veins and sinus based on drainage pattern. Neurosurgery. 2014;10(Suppl 2):357–67; discussion 67.
36. Suzuki Y, Matsumoto K. Variations of the superficial middle cerebral vein: classification using three-dimensional CT angiography. AJNR Am J Neuroradiol. 2000;21(5):932–8.
37. Tomio R, Akiyama T, Ohira T, Horikoshi T, Yoshida K. Usefulness of facial nerve monitoring for confirmation of greater superficial petrosal nerve in anterior transpetrosal approach. Acta Neurochir. 2014;156(10):1847–52.
38. Van Gompel JJ, Alikhani P, Tabor MH, van Loveren HR, Agazzi S, Froelich S, et al. Anterior inferior petrosectomy: defining the role of endonasal endoscopic techniques for petrous apex approaches. J Neurosurg. 2014;120(6):1321–5.

Anatomy and Surgical Aspects of Supracerebellar Infratentorial Approach

42

Vladimír Beneš, Martin Májovský, and Adéla Bubeníková

Learning Objectives
- Understanding the anatomy and surgical corridor to the pineal region is vital for every neurosurgical resident due to its broad application in many pathologies
- There is important variability and flexibility of the approach (medial and lateral unilateral or bilateral medial) which can be well studied in order to deliver the best both surgical and treatment outcomes
- SCIT can be combined with several other approaches, bearing in mind the pathology dictates the approach and not vice versa
- Introduction of endoscopic assistance is extremely helpful in this approach, mainly in the subtemporal area and in the upper midline region, anterolateral aspect of the brainstem, cranial nerves, and basilar artery with its tributaries that are easy to distinguish

42.1 Introduction

A pineal region and cerebellopontine angle-petrous apex-petroclival regions are deep-seated areas of the brain surrounded by eloquent structures. Differential diagnosis of space-occupying lesions in this region encompasses pineal gland cysts, pineal gland tumors, metastases, germ cell tumors, meningiomas, gliomas, ependymomas, hemangioblastomas, and neuroectodermal tumors. The treatment strategy is based mainly on the anatomical characteristics of the tumor and histological type. Except for radiosensitive germinomas, complete surgical excision is the treatment of choice.

A microsurgical supracerebellar infratentorial approach, pioneered by Fedor Krause in 1913, recently frequently with endoscopic assistance (SCIT) is the essential approach to the upper part of the posterior fossa. The SCIT is a straightforward extracerebral approach via the natural corridor between the superior aspect of the cerebellum and the tentorium [1]. Lately, the unilateral SCIT (either medial or lateral) is employed, while the midline bilateral approach is nearly abandoned. Since SCIT is an excellent approach to the petroclival area, it may be combined with retrosigmoid and even as far caudally as far-lateral approach. Conversely, the SCIT may be combined with a supratentorial falcotentorial route in certain lesions. Quite frequent is the tentorial opening in all aforementioned variants.

In general, all lesions located below the vein of Galen are indicated for resection using SCIT [2, 3]. Possible alternatives to the SCIT in selected patients are the occipital transtentorial approach or an interhemispheric transcallosal approach [4]. Lateral SCIT used for the petroclival region mostly deals with meningiomas, schwannomas (V nerve), and cholesteatomas. Lateral SCIT with transtentorial opening is usually indicated for middle fossa meningiomas and medial temporal gliomas [5]. In certain cases, the approach can be used for amygdalohippocampectomy in epilepsy patients [6]. Nontumorous lesions are mostly cavernomas. See Table 42.1 for an overview of our experience with SCIT in various pathologies during the 20-year period.

Magnetic resonance imaging is not always sufficient for definitive diagnosis, and hence, biopsy is recommended in some cases for further histological evaluation. Before surgery, every patient with a tumor in the pineal region should be examined on human chorionic gonadotropin (hCG) and alpha-fetoprotein (AFP) in serum and cerebrospinal fluid, which may diagnose germinomas even without surgery.

V. Beneš (✉) · M. Májovský
Institute of Clinical Neurodisciplines, Department of Neurosurgery and Neurooncology, First Faculty of Medicine, Charles University and Military University Hospital, Prague, Czech Republic
e-mail: vladimir.benes@uvn.cz

A. Bubeníková
Institute of Clinical Neurodisciplines, Department of Neurosurgery and Neurooncology, First Faculty of Medicine, Charles University and Military University Hospital, Prague, Czech Republic

Department of Neurosurgery, Second Faculty of Medicine, Charles University and Motol University Hospital, Prague, Czech Republic

I. N. Kanaan, V. Beneš (eds.), *Neuroanatomy Guidance to Successful Neurosurgical Interventions*,
https://doi.org/10.1007/978-3-031-59838-8_42

Table 42.1 Overview of our experience with SCIT during a 22-year period. The group "other" consists of arteriovenous malformation, teratoma, hemangioblastoma, plexus papilloma, and colloid cyst one each

SCIT during 2000–2022	
Pathology	n
Total	*179*
Meningiomas total	74
Lateral (petrous apex, CPA, PC)	64
Medial (pineal, rectus sinus)	10
Pineal cysts	28
Cavernomas total	21
Mesencephalon	9
Pons	8
Thalamus	4
BG and thalamic gliomas	15
Pineal tumors	10
Cholesteatomas	8
Metastases	7
BG cysts	3
Ependymomas	3
PNET	3
V nerve schwannoma	2
Other	5

Abbreviations: *CPA* cerebellopontine angle, *BG* basal ganglia, *PC* petroclival, *PNET* primitive neuro-ectodermal tumor

When germinoma is ruled out, radical resection of the tumor is the goal of the surgery.

For patients with preoperative hydrocephalus, external ventricular drainage, endoscopic third ventriculostomy, or shunting procedure might be performed as a first step before tumor resection. At the beginning of surgery, the lumbar drain is introduced to allow for some cerebrospinal fluid (CSF) drainage, which affords the cerebellum some relaxation and opens space between the cerebellum and tentorium [5].

42.2 Anatomical Highlights

- Via the midline approach, the bridging veins often block the surgical corridor in the midline and some of these might be sacrificed in order to create a safe surgical corridor
- Thickened and frequently nontransparent arachnoid of the ambient and pineal cisterns needs to be sharply dissected, starting with ambient cistern laterally to the Rosenthal vein where arachnoid is usually transparent
- The ambient cistern as an extension of the quadrigeminal cistern contains the trochlear nerve, posterior cerebral artery, superior cerebellar artery, and the basal vein of Rosenthal
- Then, the arachnoid of the pineal cistern overlying the pineal gland is sharply dissected in order to expose deep cerebral veins
- The upper vermian veins can be sacrificed during the approach if necessary, except for the most anterior one: the cerebellomesencephalic vein
- In unilateral approaches, usually one ascending vein from the upper cerebellar hemisphere entering the tentorium is present, it is coagulated and cut
- In the paramedian approach, targeted on the pineal region, no other vein is traversing the surgical corridor
- In the lateral approach, the petrosal vein is preserved laterally, except for cases of meningiomas in which the superior petrosal sinus represents the tumor origin, and the sinus is thus occluded. In the lateral approach, the IV nerve is exposed very early
- The potential of all three approaches may be enhanced by unilateral tentorium incision. Extreme cautiousness must be enhanced during the incision in order to be prepared for potential so-called tentorial lakes. If possible, no harm to these veins should be made in order to prevent difficult bleeding control.

42.3 Positioning and Monitoring of the Patient

A sitting position is preferred when using the SCIT approach. It allows caudal displacement of the cerebellum by gravity without the need of additional retraction and prevents the pooling of blood in the surgical field. A main drawback of the sitting position is the risk of venous air embolism, which might be a potentially fatal complication. Preoperative echocardiography is highly recommended to rule out patent foramen ovale, which represents a contraindication for the sitting position. Placement of a central venous catheter before surgery is mandatory. Echocardiographic monitoring is used along with the electrophysiological monitoring according to the lesion location. Legs should be approximately the same height as the heart to balance central venous pressure. The patient's head in the Mayfield head holder is flexed to place the tentorium parallel to the floor but still allowing two fingerbreadths between the chin and sternum to avoid jugular vein compression. The most comfortable position is then achieved by surgical table inclination, or rotation, or both. A microscope or exoscope is introduced prior to dura opening (Fig. 42.1).

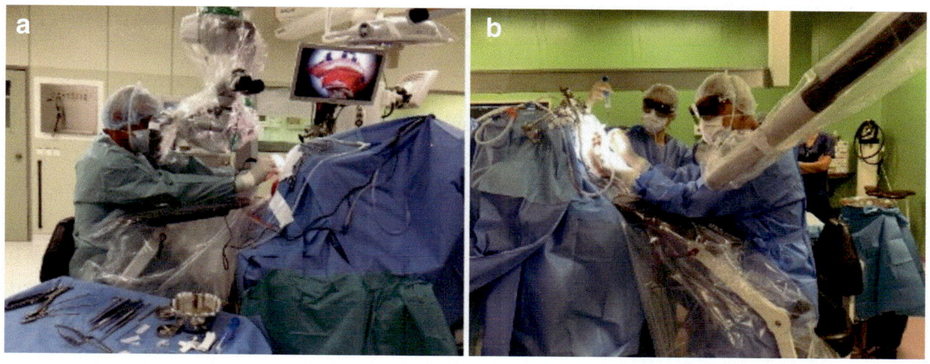

Fig. 42.1 (**a**) Using microscope, the arms of the surgeon are stretched, but the line of vision is straight. (**b**) Using exoscope is more ergonomic, but the screen is out of direct axis

42.4 Skin Incision

For midline SCIT, a linear skin incision is made from approximately 2 cm above the inion to the level of C2. Muscles are split with monopolar electrocautery in the midline along the ligamentum nuchae to avoid unnecessary muscle damage. When performed properly, minimal bleeding occurs with this procedure. Bleeding from emissary veins is stopped using bone wax. For paramedian and lateral SCIT (see below), the vertical skin incision runs approximately in the middle between the occipital protuberance and external auditory meatus.

42.5 Three Types of Craniotomy and Target Areas

Generally, three types of SCIT are available: midline SCIT, paramedian SCIT, and lateral SCIT.

42.5.1 Midline SCIT (mSCIT)

Burr holes are placed on both sides of the transverse sinuses, and the dura is carefully detached. A craniotome is then used to connect burr holes creating bone flap. After removing the bone flap, the torcular Herophili and both transverse sinuses are exposed in the superior aspect of craniotomy. Caudally, craniotomy extends some 3–4 cm below the transverse sinuses. Epidural bleeding is stopped using a gelatin sponge and tack-up sutures. mSCIT is reserved for complex midline lesions. This approach is also recommended to those who are not well familiar with the region. In the last 10 years, we have used the midline approach in only one case.

42.5.2 Paramedian SCIT (pSCIT)

A craniotomy is situated on one side, immediately lateral to the torcular Herophili, and some 4 cm of the transverse sinus is exposed (Fig. 42.2). This approach allows for the protection of a dominant transverse sinus and to spare bridging veins from and opposite side of the cerebellar hemisphere and posterior ascending veins from the vermis. pSCIT is suitable for pineal tumors, dorsal mesencephalic and posterior thalamic lesions. The side of the approach depends on the localization of the lesion. However, whenever possible, the side of nondominant a higher positioned transverse sinus is chosen.

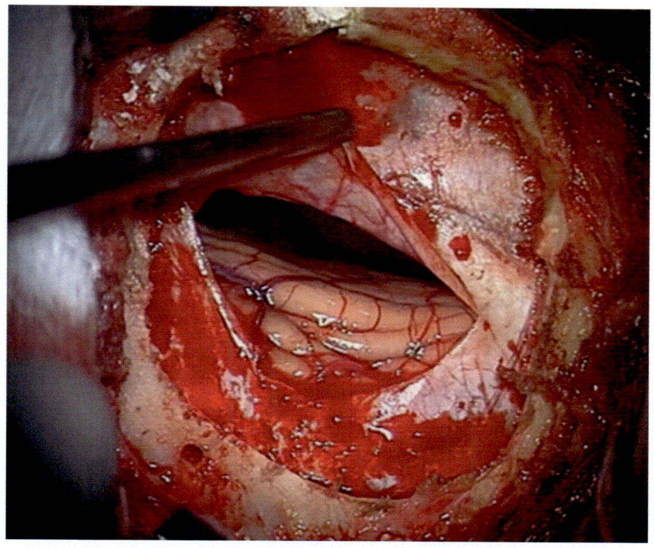

Fig. 42.2 Lateral paramedian SCIT. Dura opening after the 20 cc of CSF was drained via lumbar drain. The cerebellar hemisphere descended, and the space of approach already opened. Note the modest dura opening, which prevents cerebellar prolapse

42.5.3 Lateral SCIT (lSCIT)

Craniotomy exposes the lateral portion of the transverse sinus and the upper segment of the sigmoid sinus. This approach is used for cerebellopontine angle: the superior petrosal sinus, petroclival area, and lateral mesencephalic lesions. Inferiorly, the limit of lSCIT is the internal acoustic meatus. If necessary, cavum Meckeli can be entered and the petrous apex can be resected [6]. This route is also used for supratentorial medial temporal lesions.

A close cooperation with an anesthesiologist is crucial throughout the entire craniotomy in all three cases. Moreover, if an air embolism occurs (detected by a decrease in end-tidal CO_2), the neurosurgeon needs to react promptly. Steps to be taken include wound irrigation with saline, waxing of bone edges, lowering of the patient's head, compression of the jugular veins, and aspiration of air from the right atrium via a central venous catheter.

42.6 Dural Opening

The dura is opened only after some 20 cc of CSF is drained via the lumbar drain (Fig. 42.3). Dura is usually incised in curvilinear fashion, starting laterally on each side toward the midline. The occipital sinus is ligated and then cut, and the resulting dural flap is reflected upward exposing the superior aspect of the cerebellum. For lesions in the petroclival area and cerebellopontine angle, in some cases additional dural incision in the caudolateral direction is performed to expose also the upper lateral aspect of the cerebellum allowing to reach the level of internal acoustic meatus (IAM) (Fig. 42.2, dashed line). In all approaches, the dura is opened 15–20 mm below the transverse sinuses to prevent prolapse of the cerebellum through the dural opening (Fig. 42.3).

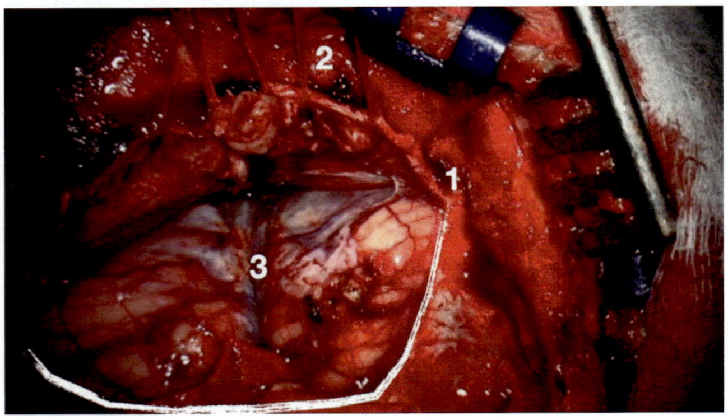

Fig. 42.3 Midline approach. (1) Shape of dural opening, (2) dural stitches to lift both transverse sinuses and torcular, and (3) cerebellar vermis

42.7 Approach to the Pineal Region

In the midline approach, the bridging veins are often blocking the surgical corridor in the midline and some of these might be sacrificed (Fig. 42.4). Thickened and frequently nontransparent arachnoid of the ambient and pineal cisterns is opened sharply starting with ambient cistern laterally to the Rosenthal vein where arachnoid is usually transparent. The dissection proceeds to the midline. The same opening is done on the opposite side. Then, the arachnoid of the pineal cistern overlying the pineal gland is sharply dissected and deep cerebral veins are exposed (Fig. 42.5). The lesion is identified and dealt with accordingly. Extreme care must then be taken in preserving the integrity of all the vessels. Superior cerebellar arteries and numerous perforators are encountered on the arterial side. On the venous side, the following veins are exposed: basal veins of Rosenthal, internal cerebral veins, tectal and pineal veins, internal occipital veins, and the vein of Galen. The upper vermian veins can be sacrificed, except for the most anterior one, namely the cerebellomesencephalic vein (Fig. 42.5).

In unilateral approaches, usually one ascending vein from the upper cerebellar hemisphere entering the tentorium is present. The vein is coagulated and cut. In the paramedian approach targeted on the pineal region, no other vein is traversing the surgical corridor. In the lateral approach, the petrosal vein is preserved laterally, except for cases of meningiomas in which the superior petrosal sinus represents the tumor origin and the sinus is thus occluded. In the lateral approach, the IV nerve is exposed very early. In meningioma cases, this nerve is traversing the dorsal aspect of the tumor and it is difficult to dissect the nerve free and preserve it.

The upper limit of SCIT is the basal veins, while the lowermost limit is the inferior colliculi. The IV nerve origin is generally below the line of vision; however, the nerve is well exposed on the lateral aspect of the mesencephalon (Fig. 42.6). The upper portion of the IV ventricle can easily be approached using a short paravermian incision. Relevant anatomy of the petrous apex one should be well familiar with is depicted in Fig. 42.7.

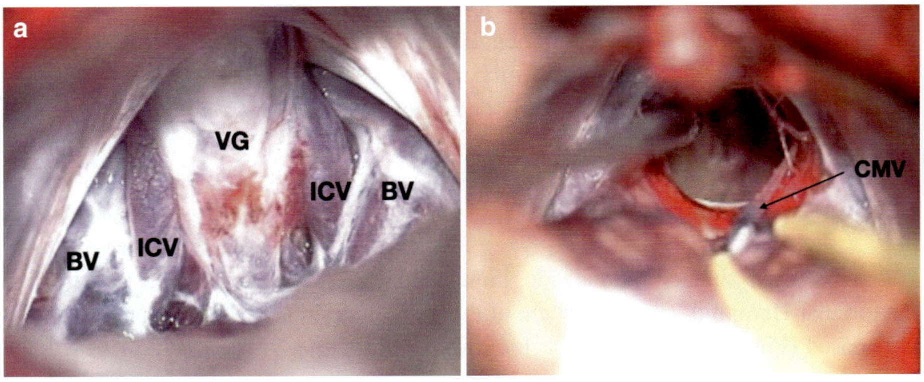

Fig. 42.4 Medial approach for the pineal cyst. Major veins dissected (*BV* basal vein of Rosenthal, *ICV* internal cerebral veins (left and right), *VG* vein of Galen). (**a**) After the resection, the III ventricle is opened. (**b**) Note the preserved cerebellomesencephalic vein (CMV)

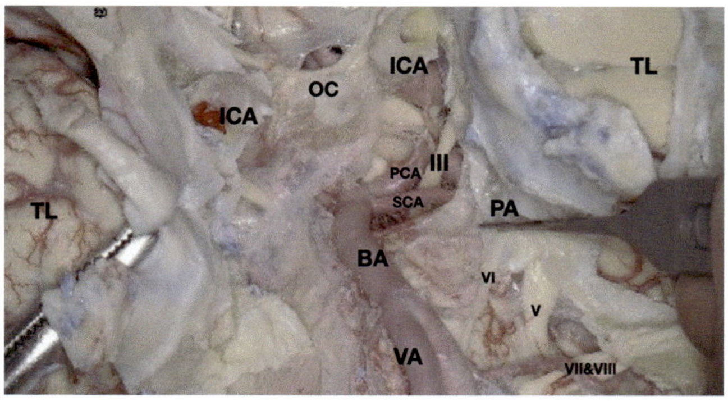

Fig. 42.5 Petrous apex region from below. Oculomotor (III), trigeminal (V), and abducens (VI), with facial and vestibulocochlear (VII and VIII) cranial nerves, brainstem, temporal lobe, and associated relevant structures. The outer arachnoid of cerebellopontine and interpeduncular cisterns is well seen. Abbreviations used: *BA* basilar artery, *ICA* internal carotid artery, *PA* petrous apex, *SCA* superior cerebellar artery, *VA* vertebral artery, *OC* optic chiasm, *TL* temporal lobe

The potential of all three approaches may be enhanced by unilateral tentorium incision. The tentorial incision is customized according to the lesion. It usually is triangular with the top anterior to the rectus sinus and medial cut running toward the incisura, the lateral one running toward the superior petrosal sinus dorsal to the IV nerve entry into the cavernous sinus. However, ordinarily only a limited tentorial resection is sufficient—a "V"-shaped incision with the apex approximately in the middle of the tentorium and the anterolateral extension tailored according to the target lesion. Tentorium resection allows attacking much more superiorly localized brainstem and posterior thalamus lesions and resection of the lesions located in the basal medial parts of the occipital and temporal lobes. Anteriorly, it allows an approach as far as the Liliequist membrane and posterior clinoid area. In petroclival meningiomas, the tentorium opening permits very early control of the brainstem/tumor

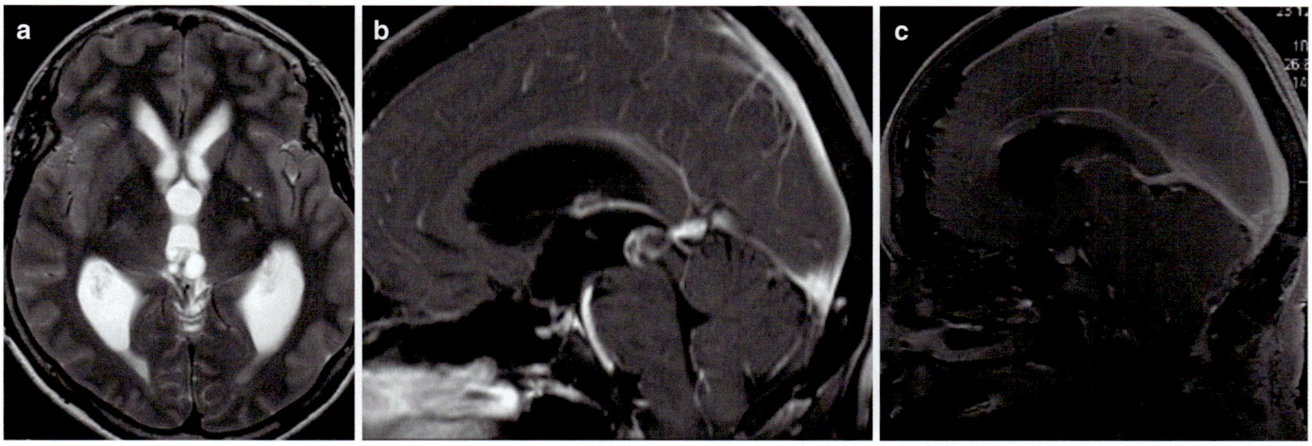

Fig. 42.6 Medial approach. Ependymoma of the III ventricle (**a**, **b**) pre- and (**c**) postoperative images

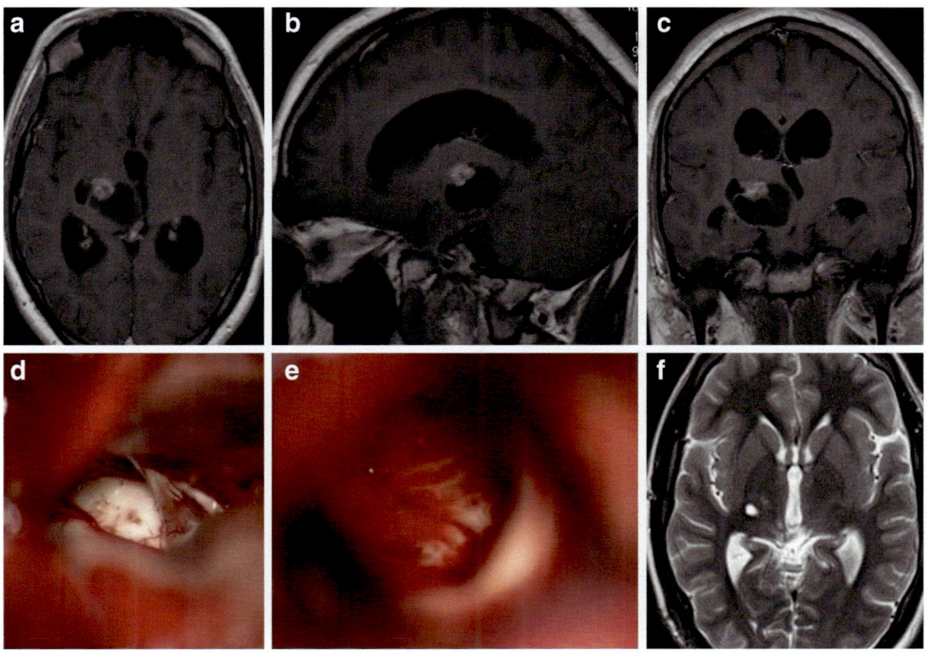

Fig. 42.7 Medial superior approach. Thalamic pilocytic astrocytoma. (**a**, **b**, **c**) Preoperative MR. (**d**) After the minimal tentorium opening, the entry point into the thalamus is chosen. (**e**) The tumor bed after the resection and (**f**) six-months postoperative MR

interface at the tumor upper medial limit. The most extensive resections are done in the anteriorly located rectus sinus meningiomas, where the rectus sinus is occluded. The tentorium is resected on both sides along with the rectus sinus, and then, the falx is cut anterior to the lesion and down to the inferior sagittal sinus. It is important to note that reversed flow in the inferior sagittal sinus may be the only deep venous drainage. The danger of tentorial opening lies with the presence of so-called tentorial lakes—the venous spaces. The bleeding is difficult to control, and it is best to check for the presence of these lakes in advance and avoid them if possible. If this is not possible, the opening should be stepwise while applying the hemoclips to prevent bleeding.

42.8 Surgical Pearls

- Introduce more stitches into the dura flap and lift the dura properly along with the transverse sinus. Such a maneuver opens the surgical corridor considerably.
- A tackling suture introduced into the opened tentorium edge will allow some more additional space in the upward direction.
- Start the cisternal dissection lateral at the ambient cistern well away from the midline and continue in an upward direction toward the pineal cistern in the midline. This will allow some CSF drainage, and also, the anatomy of pineal region will be much better displayed.
- Cut the arachnoid sharply, do not tear it.
- Restrict your coagulation and coagulate only vessels belonging to the tumor.
- Deploy navigation for lesions below the surface.
- Be prepared to use endoscopic assistance, customarily for inferior parts of the lesion hidden by the cerebellum and for the lesion in the petroclival area.
- The IV nerve is dorsal to the petroclival tumors. In sphenopetroclival meningiomas, it is difficult to maintain nerve integrity.
- Endoscopic assistance is extremely helpful in the subtemporal area, and in the upper midline region, anterolateral aspect of the brainstem, cranial nerves, and basilar artery with its tributaries are easy to distinguish.

42.9 Discussion

42.9.1 Summary of Approaches and Target Lesions

In Table 42.2, the potential of SCIT is shown, and Figs. 42.8, 42.9, 42.10, 42.11, 42.12, 42.13, 42.14, and 42.15 document all these possibilities. In lesions reaching more downward, like some clival meningiomas, the SCIT can be combined with retrosig and even as far caudal as the far-lateral approach (Fig. 42.14). Conversely, in tumors extending supratentorially like some rectus sinus meningiomas the SCIT can be combined with interhemispheric or falcotentorial access (Fig. 42.15).

42.9.2 Duraplasty and Wound Closure

The dura is closed in watertight fashion using a 3-0 suture and reinforced with a fibrin glue or a dural graft. The bone flap is fixed with bone sutures or miniplates. The wound is then closed in anatomic layers, commonly without a drain. After surgery, the patient stays at the intensive care unit, where a slow awakening from anesthesia is ensured before an extubation attempt. A lumbar drain is usually left in place until the first postoperative day.

Table 42.2 Summary of surgical corridors alternatives of the SCIT approach

Medial
1. Anterior (pineal, posterior III ventricle)
2. Superior (posterior thalamus)
3. Inferior (mesencephalon, upper IV ventricle)
Lateral
1. Anterior (petrous apex)
2. Superior (upper tentorial edge)
3. Inferior (pons)
Superolateral (medial temporal and occipital)
Combined
1. Retrosigmoid-far lateral
2. Falcotentorial-interhemispheric

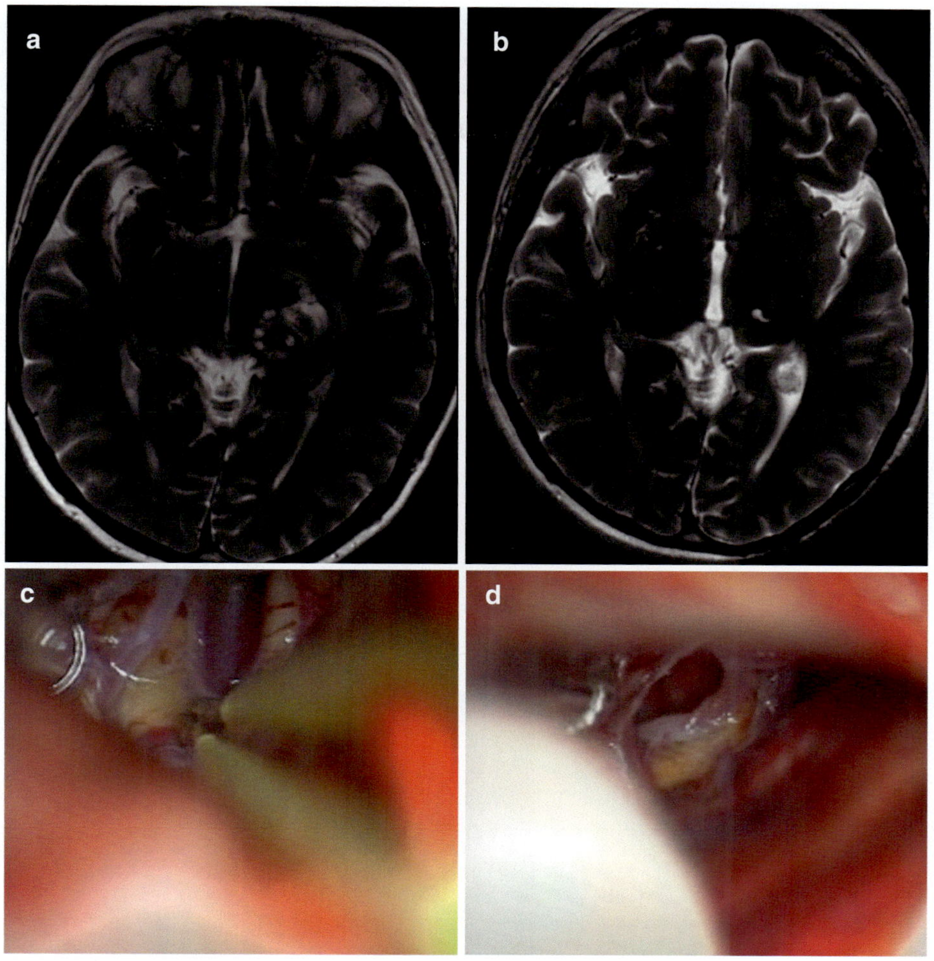

Fig. 42.8 Medial superior approach. (**a**) Thalamic cavernoma on preoperative MR and (**b**) postoperative MR. (**c, d**) After the limited tentorium section, the posterior aspect of the thalamus is exposed. The cavity after the cavernoma resection will be checked using the endoscope

42.9.3 Pitfalls

Obviously, the most important pitfall is insufficient knowledge of regional anatomy. Technical errors are not that frequent, largely because SCIT normally remains in the hands of experienced neurosurgeons. At this point, only simple errors are to be noted: too low dural opening and ensuing cerebellar herniation, venous injury with ensuing venous infarction, and perforator injury on the mesencephalic surface. In older patients, a sitting position and thorough CSF drainage may lead to massive pneumocephalus. In some patients, CSF flow disturbances may be seen and dealt with postoperatively. It is not uncommon that hydrocephalus develops even after mass lesion resection.

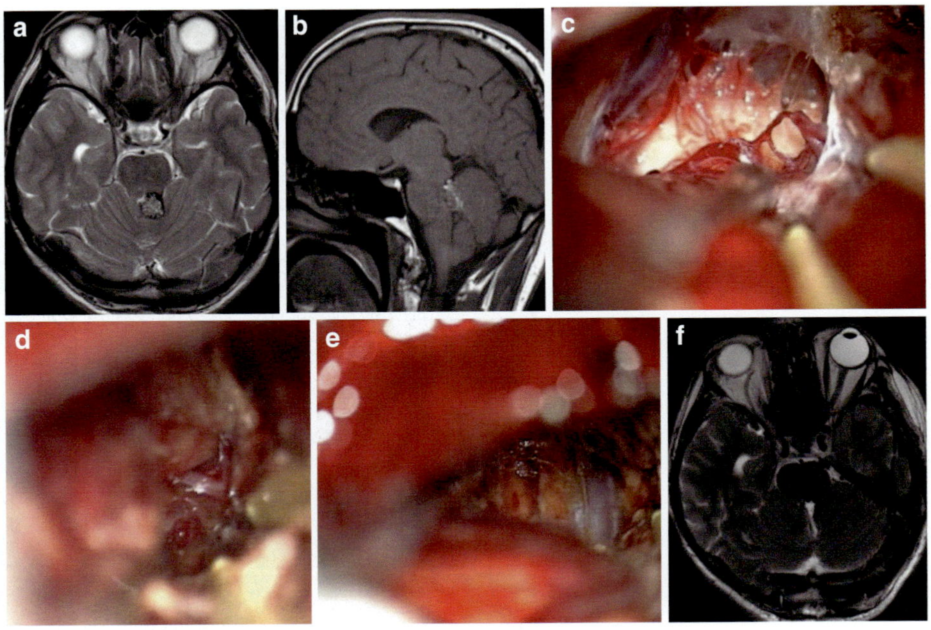

Fig. 42.9 Medial inferior approach for the IV ventricle cavernoma. (**a**, **b**) Preoperative MR (**c**) approach. (**d**, **e**) Via short paravermian cerebellotomy, the cavernoma is exposed; there is also successful preservation of three veins on the floor of the IV ventricle after cavernoma resection. (**f**) Postoperative MR showing the limited cerebellar section

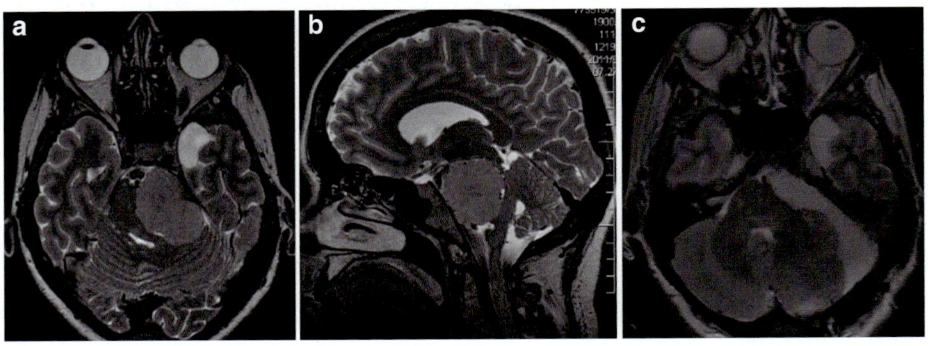

Fig. 42.10 Lateral approach. Petrous apex meningioma. (**a**, **b**) Preoperative MR and (**c**) postoperative MR

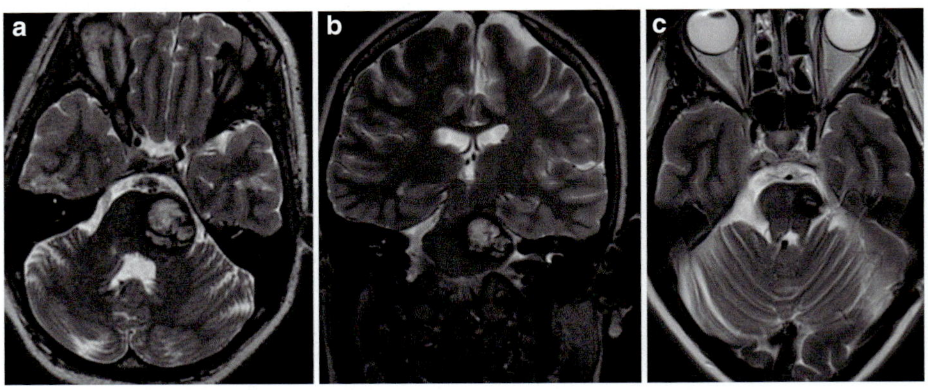

Fig. 42.11 Lateral inferior approach. Pontine cavernoma. (**a**, **b**) Preoperative MR and (**c**) postoperative MR

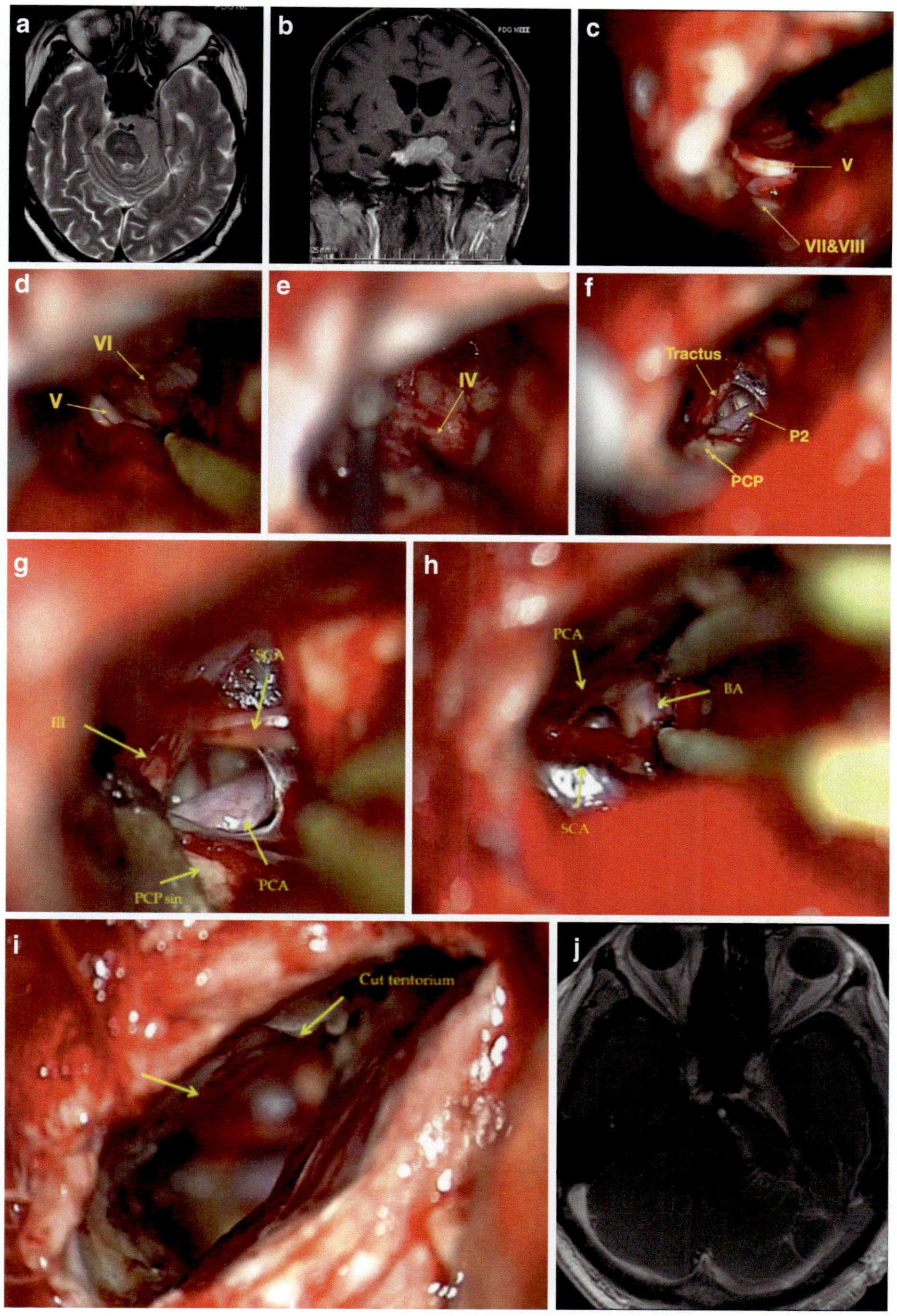

Fig. 42.12 Lateral superior approach. Sphenopetroclival meningioma. (**a**, **b**) Preoperative MR. (**c**) VII, VIII, and V cranial nerves; (**d**) V and VI cranial nerves; (**e**) the IV nerve traversing the tumor; (**f**) opposite side PCP, P2, tractus, and pituitary stalk. (**g**) Ipsilateral oculomotor nerve, superior cerebellar artery (SCA) and posterior cerebral artery (PCA), posterior clinoid process (PCP), (**h**) basilar artery (BA), PCA and SCA, and Liliequist membrane (shiny structure). (**i**) The approach and the triangular opening of tentorium. (**j**) postoperative MR, remnant in the cavernous sinus

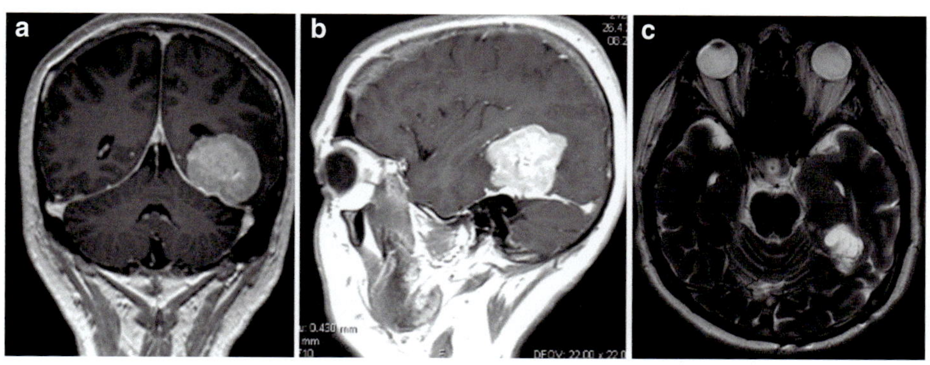

Fig. 42.13 Superolateral approach. Superior petrosal sinus meningioma. (**a**, **b**) Preoperative MR and (**c**) postoperative MR

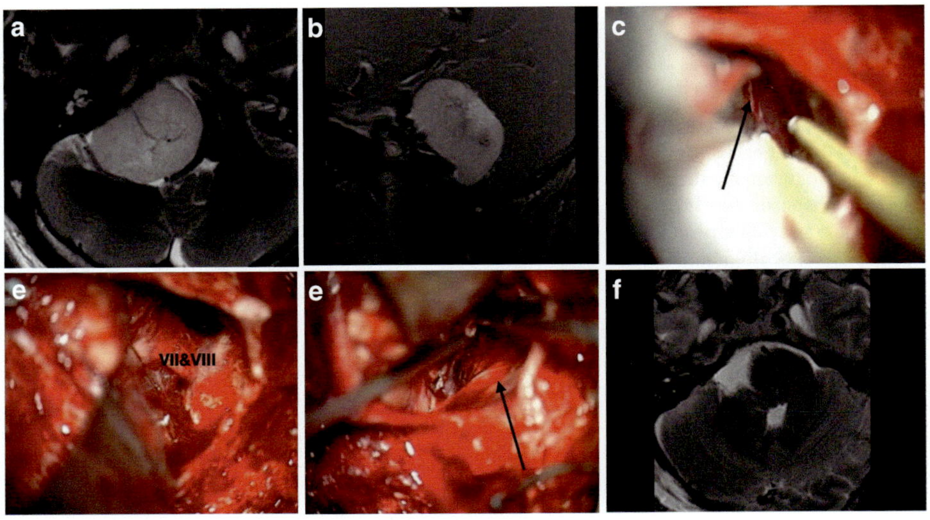

Fig. 42.14 Combined approach: SCIT-retrosigmoid-far lateral. Clival meningioma. (**a**, **b**) Preoperative MR; (**c**) SCIT phase; the mesencephalon is seen (*arrow*); (**d**) retrosigmoid phase with VII and VIII nerves, and (**e**) far-lateral phase. (**e**) Exposure of the jugular foramen (*arrow*). (**f**) Postoperative MR

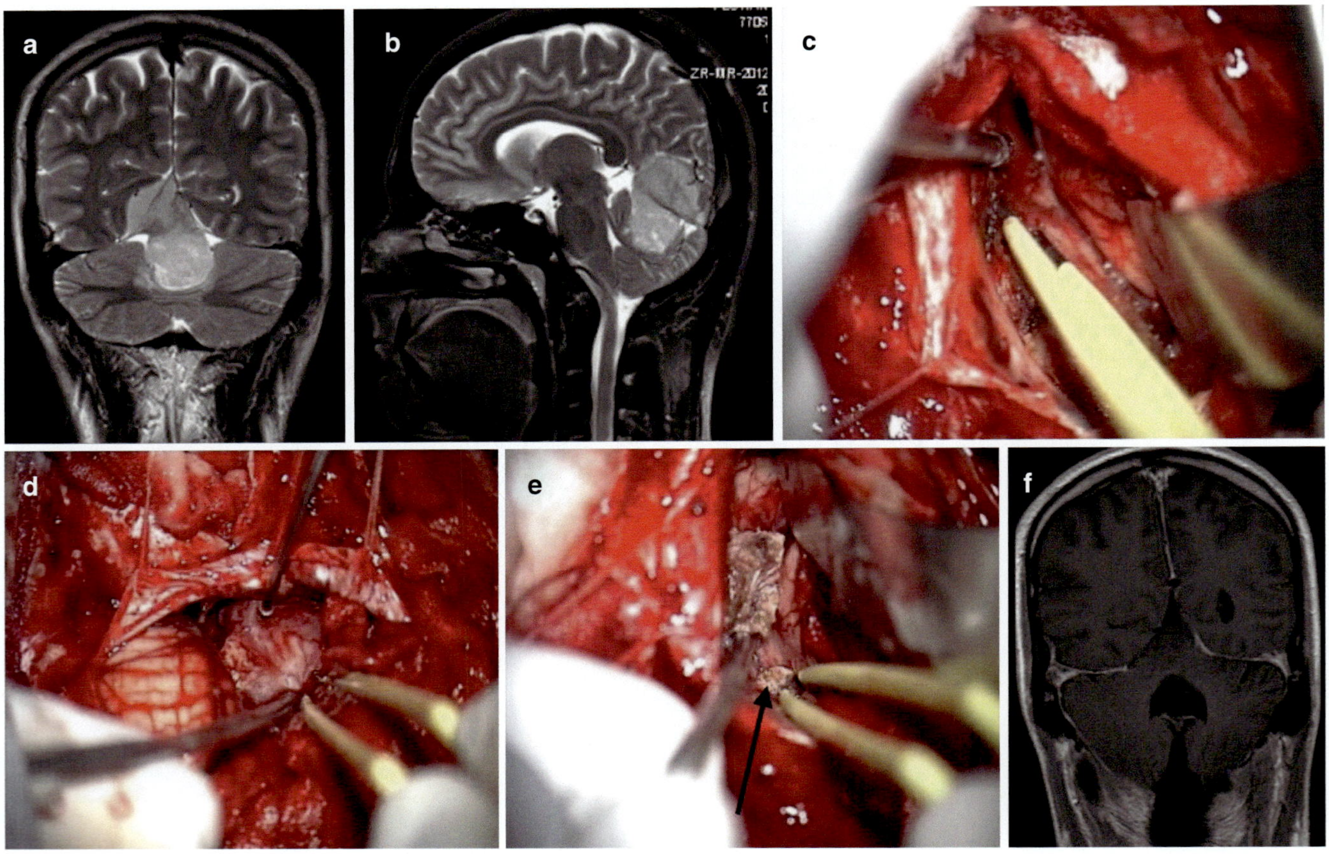

Fig. 42.15 Combined approach: SCIT-interhemispheric-falcotentorial. Sinus rectus meningioma. (**a**, **b**) Preoperative MR. (**c**) Resection of the right falcotentorial angle part of the meningioma and left falcotentorial angle after the tumor resection is exposed. (**d**) The main bulk of the tumor resected via SCIT. (**e**) Finally, the tentorium on both sides, sinus rectus and falx, are cut and resected (*arrow*). (**f**) Postoperative MR

42.10 Conclusion

Midline SCIT is reserved for the most complex lesions and it is used only rarely. Paramedian SCIT is a very useful and versatile approach. The pineal region represents a strict extracerebral route to the lesion and good anatomical orientation, and it allows the treatment of a plethora of various lesions with different locations. The lateral SCIT for the petroclival area allows to deal with meningiomas and other lesions in a very straightforward manner. As opposed to various middle fossa approaches, the craniotomy is easy and fast, the tumor is easily approached, and origin along the superior petrosal sinus and anteriorly to the clival area is dealt with first. However, preservation of the IV nerve, which is directly posterior to the tumor, is difficult. Routine, safety, simplicity, and speed are the best attributes of the approach.

This study was supported by AZV grant no. NV19-04-00272.

References

1. Kodera T, Bozinov O, Sürücü O, et al. Neurosurgical venous considerations for tumors of the pineal region resected using the infratentorial supracerebellar approach. J Clin Neurosci. 2011;18(11):1481–5.
2. Konovalov AN, Pitskhelauri DI. Principles of treatment of the pineal region tumors. Surg Neurol. 2003;59(4):250–68.
3. Sonabend AM, Bowden S, Bruce JN. Microsurgical resection of pineal region tumors. J Neuro-Oncol. 2016;130(2):351–66.
4. Oliveira J, Cerejo A, Silva PS, Polónia P, Pereira J, Vaz R. The infratentorial supracerebellar approach in surgery of lesions of the pineal region. Surg Neurol Int. 2013;4:154.
5. Porter JM, Pidgeon C, Cunningham AJ. The sitting position in neurosurgery: a critical appraisal. Br J Anaesth. 1999;82(1):117–28.
6. Giammattei L, Starnoni D, Benes V, et al. Extreme lateral supracerebellar infratentorial approach: surgical anatomy and review of the literature. World Neurosurg. 2021;147:89–104.

43 Neurosurgical Approach to Vestibular Schwannoma

Luciano Mastronardi and Alberto Campione

Learning Objectives

- Distinguish indications of surgical approaches to vestibular schwannomas
- Describe the relevant anatomical structures involved
- Summarize the steps of each procedure

43.1 Introduction

Vestibular schwannomas (VS) account for 5–10% of all intracranial tumors [1, 2]. The current incidence stands at about 34 cases/million/year, an increase compared to the past due to the extension of general population life expectancy and easier access to second-level diagnostic aids, such as magnetic resonance (MR) imaging. The mean age at diagnosis is 60 years, and the mean size at presentation is about 13.4 mm; therefore, Koos grades III and IV are a small minority [1]. Treatment options consist of watchful waiting, stereotactic radiosurgery (SRS), and microsurgery. The neurosurgical approach to these tumors, with the objective of curing the disease through a gross total resection (GTR), must guarantee anatomical and functional preservation of the facial nerve (CN7), of the cochlear nerve (CN8), where hearing preservation (HP) is deemed feasible, and a good quality of life for patients with a long survival expectancy.

Surgery is indicated in case of:

- Preoperative socially useful hearing (AAO-HNS Class A-B) in Koos grades I–II
- In case of preoperative unserviceable hearing, Koos grades III–IV
- In case of preoperative unserviceable hearing, Koos grade II showing progression (>2 mm/year) or causing disabling symptoms

43.2 Background

Three surgical approaches are indicated for the management of VS:

- Retrosigmoid Approach (RS): Used in the majority of cases for its versatility, it is indicated for small and large tumors and when HP is a goal of the operation
- Translabyrinthine Approach (TL): Indicated for larger tumors, when HP is not feasible
- Middle Fossa Approach (MF): Less commonly used, indicated for small intracanalicular (IC) tumors for the sake of HP.

43.3 Anatomy Highlights

VS surgery requires an in-depth knowledge of the anatomy of the petrosal bone. If we consider it as a pyramid with a lateral base and an anterior-medial apex, the surgical approaches described below allow the manipulation of VS by crossing one of the sides of the pyramid. TL traverses the lateral base via mastoidectomy and drilling of the semicircular canals. RS exposes the posterior wall of the petrous bone, which forms the anterior limit of the cerebellopontine angle (CPA). Finally, MF exposes the floor of the middle cranial fossa, which—in correspondence with the rhomboid fossa—constitutes the roof of the petrous bone and of the internal acoustic canal (IAC) that runs through it. It follows that, in order to understand the surgical procedure described in the individual approaches, it is necessary to master the stratification of the mastoid anatomical structures, the projection of

L. Mastronardi (✉) · A. Campione
Department of Neurosurgery, San Filippo Neri Hospital/ASLRoma1, Rome, Italy
e-mail: luciano.mastronardi@aslroma1.it

I. N. Kanaan, V. Beneš (eds.), *Neuroanatomy Guidance to Successful Neurosurgical Interventions*,
https://doi.org/10.1007/978-3-031-59838-8_43

the IAC as it appears in the latero-medullary and cerebello-pontine cisterns, and the limits of the boid fossa.

43.3.1 Mastoid Anatomy

The mastoid anatomy can be well represented in a diagram with three triangles of progressively smaller size, useful for identifying the landmarks necessary for surgical exposure (Fig. 43.1). Fukushima's outer triangle has as vertices the asterion (posterior), the root of zygoma (RZ), and the mastoid tip MT (inferior) and corresponds to the most superficial layer of the mastoid process, immediately accessible after dissection of the soft tissues. Fukushima's inner triangle has the sinodural angle (posterior), the mastoid antrum (anterior), and the digastric groove (DG) (inferior) as vertices. It corresponds to the second layer of the mastoid process, consisting of compact bone exposed after the drilling of the mastoid cells. The contents of Fukushima's inner triangle include the tegmen tympani, i.e., the basal temporal dura mater superiorly, the surface of the sigmoid sinus (SS) posteriorly, and the bony labyrinth anteriorly. The anterior-superior limit of the latter is the mastoid antrum, which allows to locate the lateral semicircular canal (LSC) in the depth. DG is an important landmark as it is always posterior and lateral to the stylomastoid foramen, the exit point of CN7 from the fallopian canal. Once the mastoid antrum and DG are exposed, CN7 will follow an intraosseous course along the segment that connects these landmarks. Egg-shelling of the anterior-superior structures of Fukushima's inner triangle exposes the contents of MacEwen's (or suprameatal) triangle. The latter has, as sides, half the distance between the mastoid antrum and the sinodural angle (superiorly) and half the distance between the mastoid antrum and DG (anteriorly). The resulting posterior limit is an oblique line passing through the major axis of the posterior semicircular canal (PSC). The vestibule is located on a deeper plane than the bony labyrinth and in particular in correspondence with the inferior limb of the PSC and the common crus with the superior semicircular canal (SSC) (Fig. 43.2) [3]. The medial wall of the vestibule corresponds to the posterior border of the IAC. At the fundus of the latter, there are two small perpendicularly oriented bony prominences: the transverse crest, which separates the superior (SVN) and inferior vestibular nerves (IVN), and the Bill's bar, which separates SVN, dorsally, from CN7, ventrally (Fig. 43.3).

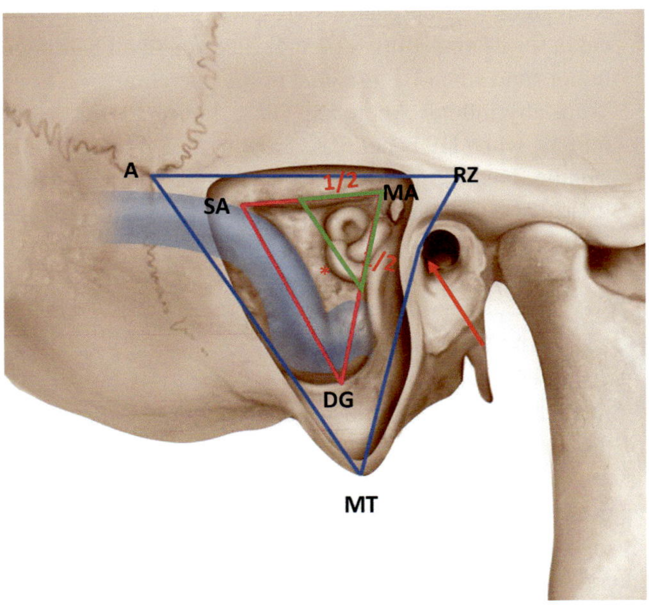

Fig. 43.1 Anatomical triangles of mastoid process. The three triangles of mastoid process are outlines (see text for discussion and anatomical implications). Fukushima's outer triangle outlined in blue, *A* asterion, *RZ* root of zygoma, *DG* digastric groove, and red arrow pointing to the spine of Henle as a superficial landmark of external auditory meatus during surgical dissection. Fukushima's inner triangle outlined in red, *SA* sinodural angle, *MA* mastoid antrum, *DG* digastric groove. MacEwen's triangle outlined in green, superior and anterior borders marked as one-half of the respective one within Fukushima's inner triangle (1/2, in red), the posterior border coursed parallel to the lateral semicircular canal (asterisk)

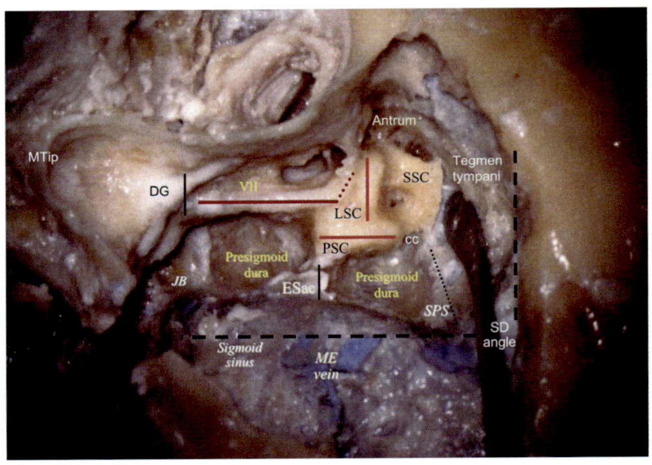

Fig. 43.2 Mastoid regional anatomy scheme. Schematic representation of regional anatomy arranged as multiple parallel and perpendicular lines at the end of a mastoidectomy, left side. The main landmark is the antrum, the lateral semicircular canal (LSC) being exposed once the antrum is opened. Posterior and perpendicular to the LSC is the posterior semicircular canal (PSC). The superior semicircular canal (SSC) shares the common crus (cc) with the PSC. Further superiorly, the superior petrosal sinus (SPS) is encountered as an oblique line leading to middle fossa dura or tegmen tympani, which extends posteriorly until covering the sigmoid sinus and thus yielding the sinodural angle (SD angle). The sigmoid sinus is crossed perpendicularly both anteriorly and posteriorly by two structures, the endolymphatic sac (EnSac) and the mastoid emissary vein (ME vein), respectively. As regards the facial nerve (7cn), its tympanic segment is anteroinferior and approximately parallel to LSC. Running inferiorly, the 7cn is parallel to the sigmoid sinus and exits at the level of the digastric ridge (DG). As a result of progressive egg-shelling, presigmoid dura has become apparent and the sigmoid sinus has been skeletonized down to the jugular bulb (JB)

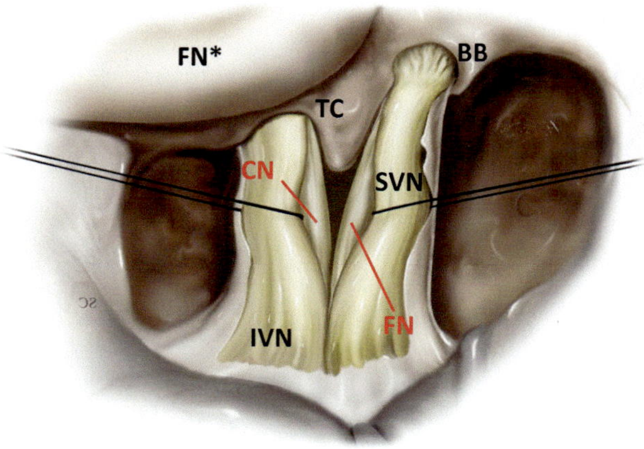

Fig. 43.3 Contents of the vestibulofacial bundle at the fundus of internal auditory canal. The transverse crest interposes between the two vestibular nerves. The Bill's bar serves as a landmark to recognize the surgical plane of dissection between the superior vestibular nerve and the facial nerve, here emphasized in red as the cochlear nerve. *TC* transverse crest, *BB* Bill's bar, *CN* cochlear nerve, *SVN* superior vestibular nerve, *IVN* inferior vestibular nerve, *FN* facial nerve, *FN** facial nerve within fallopian canal

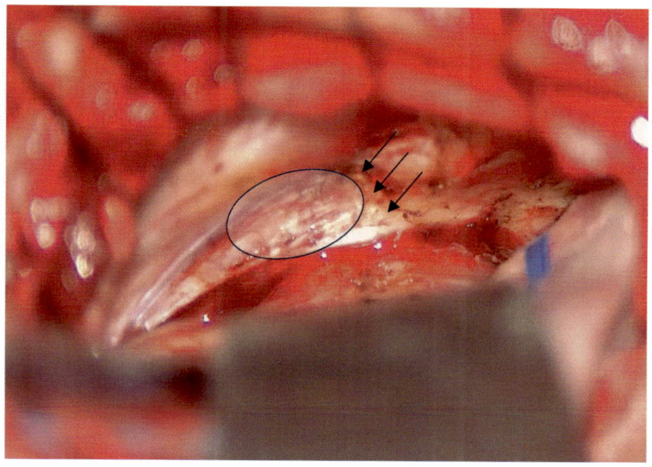

Fig. 43.4 Tübingen line. Caudal-to-cranial folds are emphasized within the oval-shaded area; their ends follow a line along the tip of the black arrows

43.3.2 Internal Auditory Canal Landmark in Retrosigmoid Exposure

The surgical technique of RS will be discussed in the following section. In this paragraph, we illustrate which anatomical landmark to follow to identify the porus acusticus and the course of the IAC. In the presence of voluminous VS, the posterior wall of the petrous bone is obscured by the tumor bulk. Several dural folds extend from the jugular foramen and follow a cranial course for about 5–7 mm. The level at which these folds end and the dura closely adheres to the posterior wall of the petrosal bone is the Tübingen line [4] (Fig. 43.4), which forms the projection of the posterior-inferior limit of the IAC. It is a reliable landmark for locating the porus acusticus and for retrolabyrinthine meatotomy. The suprameatal tubercle represents the superior-anterior limit of the porus.

43.3.3 Middle Cranial Fossa and Rhomboid Fossa Anatomy

It is useful to present the anatomy of the middle cranial fossa in a lateral-to-medial fashion, as it is exposed in a subtemporal approach. The first anatomical landmark to look for is the middle meningeal artery, which runs through the foramen spinosum which is the epicenter of the early stage of surgical exposure. Anterior to it are the three branches of the trigeminal nerve (CN5); posteriorly, the contents of the petrosal bone. Posterior-lateral and running anterior-medially along the major axis of the petrosal bone is the greater superficial petrosal nerve (GSPN), which arises from the geniculate ganglion (GG) and courses anteriorly, deep to V3 branch of CN5. The GSPN runs on a bone layer superficial to the petrosal segment of the internal carotid artery (ICA). Posterior-lateral to the posterior end of the GSPN, i.e., the GG, oriented more or less perpendicular to it, is the arcuate eminence (AE), which is the bony prominence corresponding to the underlying SSC. Following the course of the GSPN anteriorly, it ends perpendicular to the posterior border of V3. The AE posteriorly, the GSPN laterally, and V3 anteriorly constitute the sides of the rhomboid fossa, bordered medially by the petrous ridge. The IAC runs approximately along the bisector of the angle formed by the AE and the GSPN, which has the GG as its vertex. Consequently, the IAC divides the rhomboid fossa into two triangles, namely premeatal and postmeatal. The cochlea is located in the premeatal triangle, in the angle between GSPN and IAC, medial to the course of the petrous ICA. The course of the anatomical structures of middle cranial fossa can be schematized in a diagram [5], as in Fig. 43.5.

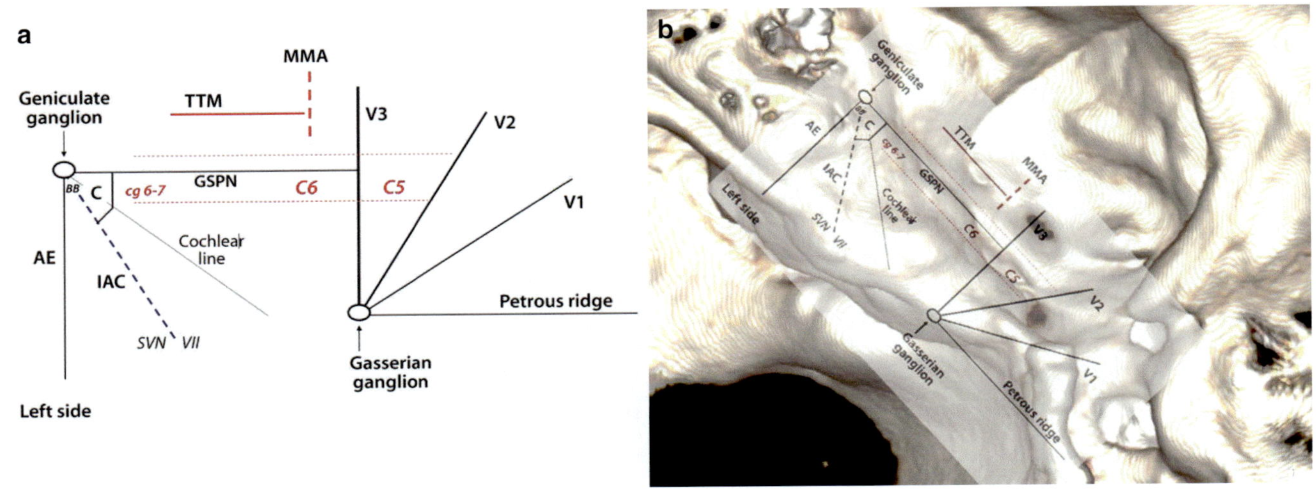

Fig. 43.5 Middle cranial fossa: anatomical scheme. Anatomical structures contained in the middle cranial fossa can be schematized according to the "rule of two fans." Two bordering fans (posterior and anterior) are represented, each shifted at 90° as compared to the other one. Posterior fan: geniculate ganglion as the vertex, the rays being the arcuate eminence (AE) posteriorly, and the greater superficial petrosal nerve (GSPN) laterally. The inner auditory canal (IAC) approximately marks the bisector of the angle in between (*SVN* superior vestibular nerve, *VII* facial nerve, *C* cochlea). Anterior fan: The gasserian ganglion is the vertex. The rays of the fan are V3 posteriorly, V2, V1, and the petrous ridge medially. *MMA* middle meningeal artery, *TTM* tensor tympani muscle, C5 and C6, segments of ICA according to Fukushima's classification

43.4 Approaches

In our experience, the most useful position to operate on VS is Fukushima's modified lateral position, which combines satisfactory surgical exposure with anesthesiological safety.

The patient is placed on the side contralateral to the lesion. This reduces the neck rotation required for adequate surgical exposure and preserves the venous outflow through the contralateral jugular vein. The patient's back is brought close to the edge of the operating table, and the shoulders are brought to the superior border. To prevent pressure injuries, bony prominences are padded. In positioning the nondependent arm, the shoulder is brought anteriorly and gently pulled caudally to allow for broader exposure. The head is placed in a three-point fixation head holder so that the two posterior pins are at the inion and at the contralateral mastoid process, while the frontal one is positioned behind the hairline for a better cosmetic result. The head is rotated contralaterally to the lesion so that the petrous ridge is perpendicular, the mastoid is parallel to the floor, and the porus acusticus is in line of sight. It is flexed so that the chin is at two fingers breadth from the sternum—this maneuver brings the mastoid process further away from the shoulder. Then, the head is elevated and the vertex is tilted down so that the mastoid process is at the highest point of the operative field (Fig. 43.6). The intraoperative monitoring devices such as somatosensory and motor evoked potentials, CN7 online EMG, and brainstem evoked auditory potentials (BAEPs)—if hearing preservation (HP) is attempted—are activated and positioned at the end of patient positioning.

43.4.1 Retrosigmoid Approach

The root of the zygoma is palpated and marked with a horizontal line, whose posterior prosecution to the inion is the projection of the superior nuchal line (SNL). The MT is palpated and marked as well; the horizontal line that runs posteriorly from the digastric point parallel to the SNL is the projection of the inferior nuchal line (INL). SNL and INL serve as superior and inferior borders of the projection of the posterior cranial fossa. A retroauricular straight or C-shaped incision of about 5 cm is performed, running from the supramastoid crest to the level of the mastoid tip, 2 cm posterior to the mastoid process. The intersection between the incision line and the SNL locates the asterion and the genu between transverse sinus and sigmoid sinus (Fig. 43.6d), which constitute the superior and anterior limits of the craniotomy, respectively.

After Incision

- Scalp elevation in two layers to expose the underlying fascia.
- Fascial graft (about 3 × 3 cm) harvested for dural closure.
- Neck muscles are incised and elevated from the suboccipital surface; inferiorly, dissection is carried out until the digastric groove (DG) is identified.
- Soft tissues retracted with blunt-tip hooks to expose the bone and allow for asterion identification.
- A 4-mm or 5-mm extra-coarse diamond bar is used for a 3 × 3 cm craniotomy (Fig. 43.7a).

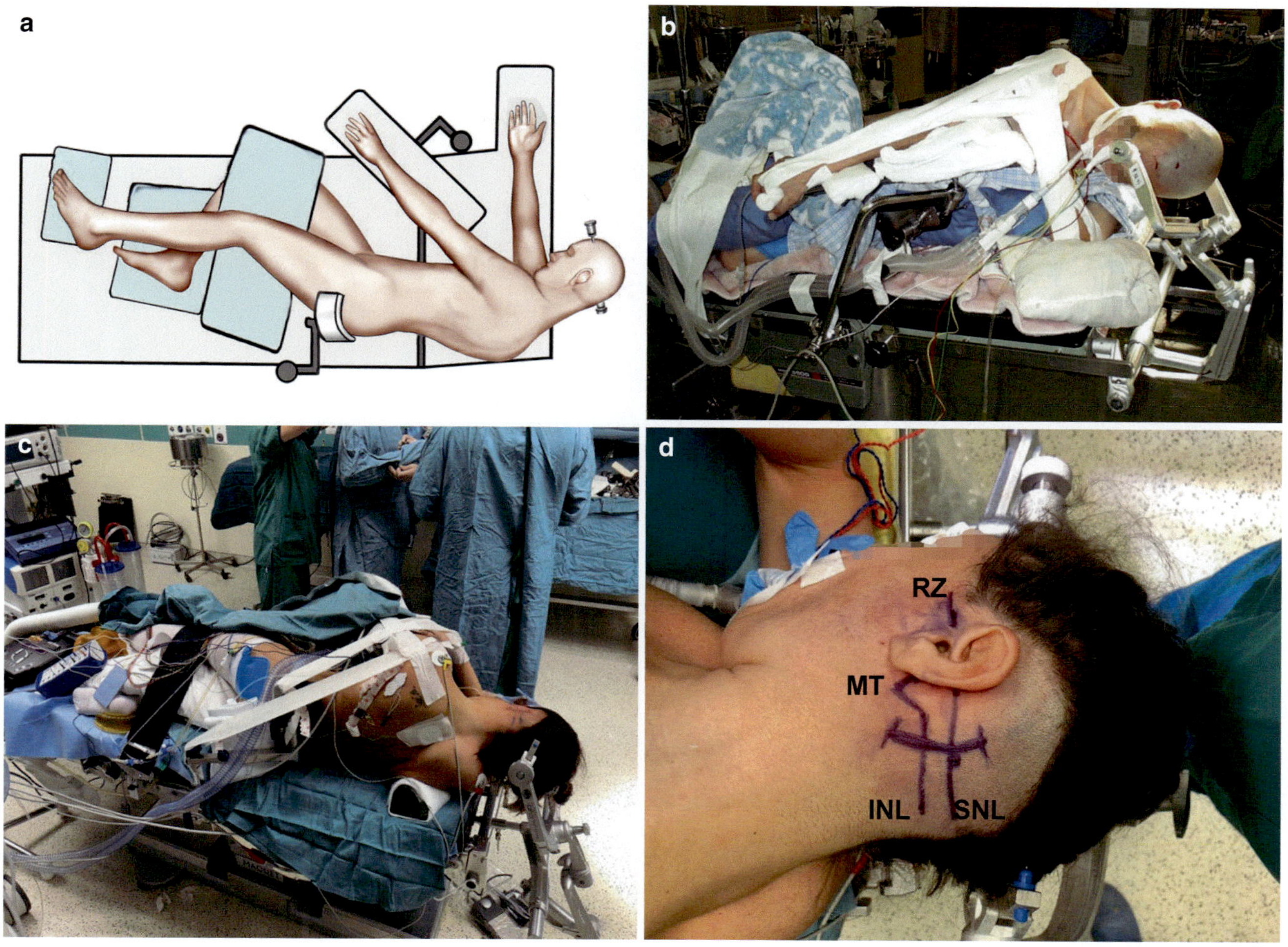

Fig. 43.6 Fukushima's lateral position. (**a**) Diagram of a patient in Fukushima position viewed from above. (**b**) The nondependent shoulder is pulled anteriorly and caudally and then secured by tape to the arm board. The vertex is tilted down to raise the mastoid process as much as possible. (**c**) Note the roll under the axilla to prevent any compression of the brachial plexus and elevate the should for improving neck positioning. The head is rotated contralaterally. (**d**) Superficial landmarks of the region. The incision curves posterior to the pinna. It intersects with SNL, thus marking the projection of the asterion and (deep) the genu between transverse sinus and sigmoid sinus. *RZ* root of zygoma, *MT* mastoid tip, *SNL* superior nuchal line, *INL* inferior nuchal line

- Its anterior margin runs from asterion to DG; superior margin from asterion along the SNL for about 3 cm; and posterior and inferior sides complete the roughly square flap.
- Drilling continues until an eggshell of bone is reached, first at the anterior-inferior corner, then continuing superiorly along the sigmoid sinus margin to expose it safely, and then at the transverse-sigmoid junction and finally along the course of the transverse sinus.
- Once sinuses are exposed, egg-shelling of the posterior and inferior edges of the craniotomy is completed and the bone operculum is elevated.
- Dura mater is opened in a semicircular shape, covering part of the cerebellar hemisphere to act as a protective sheath during retraction (Fig. 43.7b).
- The arachnoid membrane of the lateral medullary cistern is opened, and the CSF is aspirated for cerebellar relaxation.
- A 2-mm tip brain spatula is inserted to hold (not retract) cerebellum and gain insight into the CPA.
- At this point, especially if in case of large VS with posterior cerebellar displacement, it is necessary to stimulate—with monopolar probe—the surface of the tumor to exclude that facial nerve (FN) runs dorsally.
- Tumor capsule is incised with laser thulium or scissors, and debulking is performed, which is of paramount importance in order to proceed with safe dissection of the neurovascular structure.
- Tumor should be removed piecemeal from the inside, with gradual thinning (up to 2–3 mm thickness) of the

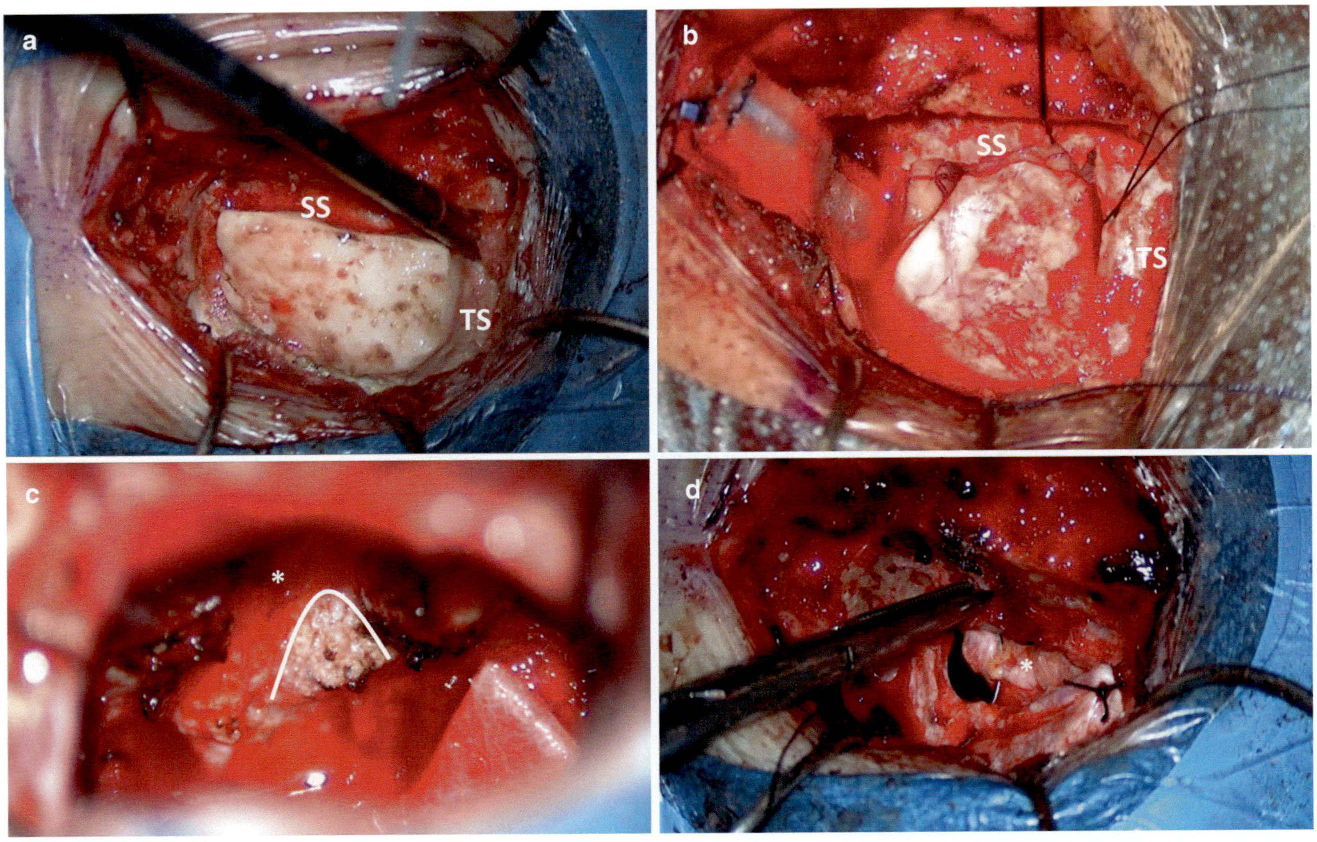

Fig. 43.7 Retrosigmoid approach. (**a**) Retrosigmoid craniotomy performed with a high-speed drill with 4-mm extra-coarse diamond burr. *SS* sigmoid sinus, *TS* transverse sinus. (**b**) Semicircular incision of dura mater along the borders of the sinuses so as to protect the cerebellum beneath. *SS* sigmoid sinus, *TS* transverse sinus. (**c**) The internal auditory canal has been opened (asterisk), and its contents are exposed. A V cut (curved line) has been performed by microscissors (or, alternatively, with thulium laser hand-held fiber) on the dorsal surface of the intrameatal tumor to expose the vestibular nerves. The intracisternal portion of the tumor has been already excised. (**d**) Dural closure: The fascial graft (asterisk) is sutured underlay to provide a "reverse umbrella" mechanism of CSF containment

intracapsular portion in a circumferential fashion to be able to elevate the capsule.

- Ultrasonic aspirator can be used for large, soft, or fibrous tumors as well.
- Maintain the arachnoid plane around the tumor for successive steps of capsule dissection from surrounding neurovascular structures.
- Proximal FN can be identified in the lateral aspect of the pontomedullary junction, just close to the choroid plexus of the lateral recess.
- Intracisternal FN is shifted either straight upward, obliquely, or caudally along the ventral aspect of the tumor capsule. At this level, the displacement, splitting, or fanning of the nerve is extremely variable and requires extreme care with dissection, particularly in the last 10-mm segment near the inferior margin of the enlarged IAC.
- The FN can be identified and dissected as the whitish band while confirming its response by monopolar stimulus, starting from 2 mA or more (on the capsule for nerve course localization) to 0.3 to 0.05 mA (directly on the nerve for functional verification).
- If the capsule thickness is down to 1 cm, a stimulation of 0.7–1.2 mA is feasible. If the response is 0.07 or 0.5 mA, the tumor capsule becomes very thin, less than 5 mm of thickness. If FN responds to 0.05–0.2 mA stimuli, the probe is already on the nerve.
- FN could be ventral (31–52%), anterosuperior (38.5–48%), anteroinferior (5.3–21%), or, rarely, dorsal (0.3–3.8%) to tumor capsule.
- Meatus of IAC should be identified—in the first instance in Koos I-II VS or after due debulking for larger tumors—following the Tübingen line.
- The dura posterior to the meatus is incised or removed by laser in an inverted "U" shape, with the base approximately at the fovea, which corresponds to the apex of the endolymphatic sac.
- The inverted U-shaped dural incision exposes the bone that has to be removed and extends about 6–8 mm toward the fovea and its base extends 2 mm above and 2 mm below the IAC.
- The meatotomy can be performed by ultrasonic aspirator (bone tip) or drill; in this case, a 4-mm extra-coarse dia-

mond burr is used to unroof the IAC (Fig. 43.7c) until the transverse crest can be appreciated.

- It is important to remove the bone from the upper and lower corners of the IAC to expose FN and CN8, which are usually compressed by the tumor as it bulges out of the meatus.
- According to Fukushima's technique, a V cut is performed with microscissors or thulium laser on the dorsal surface of tumor effacing porus acusticus (Fig. 43.7c) and debulking is obtained with microinstruments.
- This results in exposure of SVN and IVN; the lateral edge of the latter should be kept intact in order to protect CN8 if HP is attempted.
- As both the nerves lie in the same plane, the edge of IVN also helps to establish the dissection plane between tumor and CN8.
- The distal part of FN can be identified as the whitish thin band when IVN and SNV are elevated at the fundus of IAC.
- The portion of the tumor within the IAC must be separated from adjacent nerves, cut from its nerve of origin, and fragmented to allow for piecemeal removal. The McElveen-Hitzelberger dissector is the most appropriate instrument for the task.
- The use of bipolar coagulation must be minimized so as not to damage FN, CN8, or the internal auditory artery (IAA), essential for HP.
- GTR of the portion of VS penetrating deeply into the fundus of IAC is difficult under a straight microscopic view. This problem can be overcome with endoscopic devices (***endoscopic-assisted microneurosurgery***).
- The 30° or 70° angled rigid endoscopes or the flexible endoscope allow for complete IAC visualization, down to its fundus, and residual tumor identification. As a result, the lesion can be removed completely under endoscopic guidance. Kumon et al. [6] found the extent of resection to be significantly higher in case of endoscope-assisted procedures, especially for VS located beyond the mid-portion of the IAC, with no augmented risks of FN or CN8 complications.
- Flexible endoscopes represent a step forward in endoscopic-assisted microsurgery. In our experience, a 4-mm Flexible Video Endoscope (4-mm × 65-cm, Karl Storz, Inc.) is introduced—under both microscopic and endoscopic visualization, to prevent injury to CPA structures—at the end of microsurgical resection, and the endoscopic tip is oriented into the IAC to detect tumor remnants. If residual tumor is identified, microsurgical resection is pursued and further endoscopic controls are repeated until GTR is achieved [7].
- The remaining tumor capsule is removed with standard microsurgical tools from the brainstem and cranial nerves, in a medial-to-lateral direction using thin-blade microscissors or a sickle knife to avoid putting traction on the nerves. All strands of the arachnoid tethering the tumor should be cut sharply with microscissors rather than pulled. A firmly adherent tumor capsule should be thinned until it is transparent and supple rather than peeled away. In cases of tight adhesion, the surgeon should consider leaving 1–2 mm thickness of thinned tumor capsule on the FN.
- Dissection of functioning CN8 requires an extremely gentle microneurosurgical technique. It appears as the yellowish-whitish band at the caudal aspect of the tumor. It is very important to identify the true tumor capsule and elevate it gently and sharply while protecting the flattened cochlear nerve with Surgicel, Gelfoam, and microcottonoids (1–2 mm in size). If any decrease of the BAER wave V is detected, the surgeon should stop dissection, irrigate with diluted steroids, and wait for recovery.
- Brief spurts of current rather than prolonged cauterization should be used for hemostasis to avoid current spread jeopardizing surrounding nerves. It is a good practice to use a cotton patty to isolate the nerves from the tips of the bipolar cautery forceps.
- Bone wall of IAC is covered with wax to occlude air cells and avoid CSF leaks.
- Autologous fascial graft is inserted into the dural defect as an underlay hourglass-shaped plug [8] (Fig. 43.7d). It is then fixed under operative microscope magnification with separated 3–0 silk stitches (with an "inside-to-outside" direction) to the dura mater.
- The patch is then augmented with one layer of absorbable hemostats (TachoSil®, Takeda, Japan) and with a dural sealant (DuraSeal, Covidien LLC, Mansfield, Massachusetts or Tisseel, Baxter, Deerfield, Illinois, USA).
- In all cases, the autologous bone is put back in place and fixed with miniscrews, the bone defect around being filled with bone cement.

43.4.2 Translabyrinthine Approach [9]

The same position is used, and the same incision as for RS is performed.

- Soft tissues are elevated in two layers, so as to expose temporalis fascia and muscle, periosteum, and sternocleidomastoid fascia to fashion a muscular-fascial flap that is important in obtaining a watertight, cosmetic closure.
- Temporalis muscle graft is harvested from underneath temporalis fascia and saved for packing of the eustachian tube and middle ear in later stages of the procedure.
- Soft tissues are elevated anteriorly to reveal the posterior lip of the external auditory canal, the spine of Henle, and

the root of the zygoma. At this point, Fukushima's outer triangle is exposed.

- Using the high-speed drill with a large cutting burr (5–6 mm) and continuous irrigation, the cortex over the mastoid bone is removed. It is helpful to first outline the boundaries of bone to be removed: Tracing the anterior and superior borders of Fukushima's outer triangle marks a skewed "T" shape.
- The angle in between marks the surface projection of the mastoid antrum and the LSC. Cortical bone is removed within these boundaries, working anterior to posterior and superior to inferior. As the cortical bone is removed, air cells are encountered.
- Posteriorly, over the sigmoid sinus, the bone remains compact. Bone removal proceeds 1 cm behind the sigmoid sinus, maintaining a uniform depth as it is exposed.
- A dominant sigmoid sinus may be the source of dramatic bleeding during the early stages of drilling; a preoperative angiographic study is therefore advisable so as to prevent excessively aggressive skeletonization.
- The mastoid air cells are removed anteriorly and superiorly to expose the middle fossa dura (tegmen tympani) and sinodural angle posteriorly.
- Moving anteriorly, air cell removal exposes the compact bone of the bony labyrinth and the mastoid antrum. Maintaining the same relative depth, air cell removal proceeds inferiorly to reveal the DG and eventually outline Fukushima's inner triangle.
- For maximal exposure in the case of a retrolabyrinthine approach, the posterolateral portion of the bony labyrinth must be completely defined.
- With a medium-coarse diamond burr (2–3 mm), the small air cells surrounding the semicircular canals are removed.
- The ridge covering the LSC is first identified as the antrum is opened.
- The tympanic segment of FN is parallel and 1–2 mm anterior to LSC. Proceeding inferiorly, the descending segment of FN can be traced as a pink line running from the LSC to the DG.
- Anteriorly, approximately 12–15 mm deep to the external auditory meatus and the spine of Henle, lies the "genu" of the fallopian canal. Therefore, bone removal in the anterior direction at this level must be carried through with extreme care.
- FN is egg-shelled using the diamond burr from the genu to the stylomastoid foramen under constant.
- At this stage, the contents of MacEwen's (or suprameatal) triangle are exposed (Fig. 43.8a).
- Exposure of the semicircular canals may be cumbersome in that the labyrinthine structures are fragile and may be damaged by excessive drilling: Gentle egg-shelling should be performed until a transparent-blue line (Fig. 43.8a, lower left) can be seen in each of the canals.
- Moving posteriorly, as the PSC is defined, the retrofacial air cells are located just inferior to it and toward the jugular dome. These air cells are removed to skeletonize the jugular bulb.
- For a retrolabyrinthine approach [10], after presigmoid dural exposure, the dura mater is incised parallel to the sigmoid sinus. The dura is retracted anteriorly exposing VS, FN, and CN8 within CPA.
- Locate and preserve the arachnoid sheath that invests the tumor capsule and makes up for an optimal cleavage surface.
- The tumor mass can then be debulked using ring curettes or ultrasonic aspiration. Within CPA, FN is dissected from the tumor capsule in the medial-to-lateral direction. The continuous use of the stimulator helps to determine the trajectory of the nerve over the tumor periphery.
- For translabyrinthine approach, the LSC and PSC are first opened with the drill. The anterior end of the LSC is carefully removed: Preservation of its anterior wall protects the tympanic segment of FN.
- Removal of the superior segment of the PSC exposes the common crus which it shares with the SSC. The inferior limb of the PSC is followed by the vestibule. The medial wall of the vestibule corresponds to the posterior border of the IAC.
- The cortical bone surrounding IAC is defined by egg-shelling superior and inferior to the LSC so that 180°–270° of its circumference is skeletonized (Fig. 43.8b).
- The floor of IAC is uncovered first by removing the bone that separates it from the jugular bulb inferiorly. If a high-riding jugular bulb is encountered, it is first skeletonized and then gently pushed inferiorly with the periosteum and held in place with bone wax.
- Beginning in the region of the porus acusticus, the compact bone surrounding the LSC is egg-shelled with small coarse diamond burrs. As the bone is thinned at the lateral end of the canal, the transverse crest becomes apparent.
- The paper-thin shell of bone in the region of the porus acusticus is elevated first with a fine dissector, with the bone over the fundus of IAC saved for last. The superior lip of the porus is generally the most difficult to manage because of the proximity of FN.
- The dura of the IAC is then sharply opened using a #11 blade.
- The transverse crest and the Bill's bar are palpated in the fundus with a right-angle pick. While palpating Bill's bar, SVN is gently displaced and divided with a small right-angle hook to allow for visualization of FN. IVN and CN8 are then separated laterally (Fig. 43.8d).
- The contents of the IAC are dissected from FN in a lateral-to-medial fashion and then removed; in selected cases, CN8 may be preserved and this can be useful for a later stage of cochlear implantation.

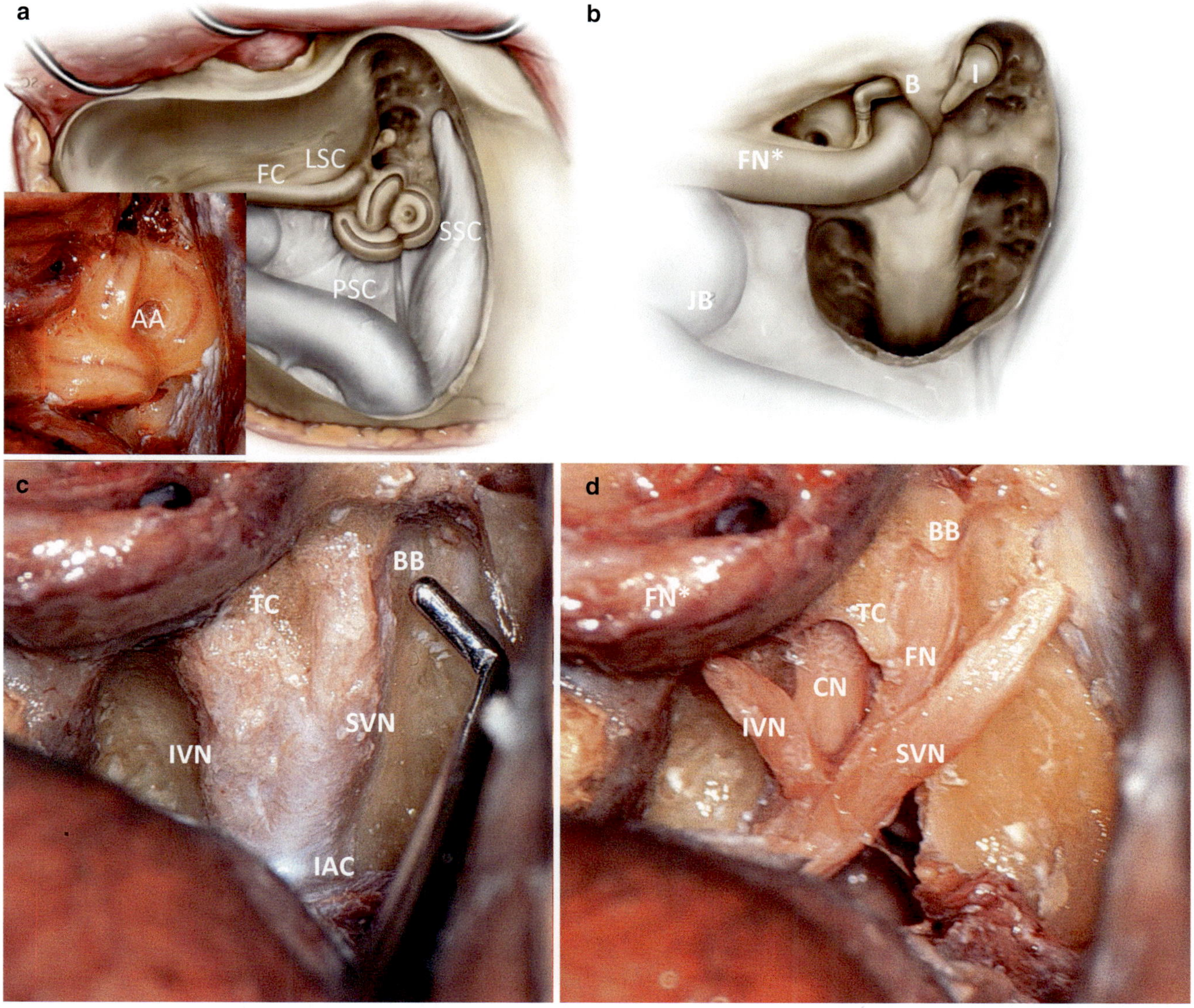

Fig. 43.8 Translabyrinthine approach. (**a**) Contents of MacEwen's triangle are exposed. In the lower left, insight of bony labyrinth drilling and exposure of the blue line along each of the canals (see the text for details). *FC* fallopian canal, *LSC, SSC, and PSC* lateral, superior, and posterior semicircular canal, respectively; and *AA* arcuate artery. (**b**) Surgical field after complete bony labyrinth drilling. *B* buttress, *I* incus, *FN** facial nerve inside fallopian canal, *JB* jugular bulb. (**c**) Real-life correlate of Figure b. *TC* transverse crest, *BB* Bill's bar, *SVN and IVN* superior and inferior vestibular nerve, respectively; and *IAC* inner auditory canal. (**d**) The contents of the internal auditory canal have been dissected from the facial nerve, which runs laterally and vertically in the fallopian canal

- The incus is removed, and the graft of temporalis muscle previously harvested is packed carefully through the epitympanum, occluding the origin of the Eustachian tube to reduce the risk of CSF leakage.
- The dural incision is closed up to the internal auditory canal, and strips of autologous (abdominal) fat are placed in the gaps of the dura so as to seal the CSF space.
- The previously fashioned musculofascial flap is closed tightly over the adipose graft, and the postauricular incision is closed in two layers.

43.4.3 Middle Fossa Approach [11]

MF consists of an anterior petrosectomy via an extradural subtemporal trajectory. A preauricular linear or "lazy-S" skin incision is performed (Fig. 43.9a), followed by and a 4 × 4 cm temporal craniotomy, extended inferiorly with drilling to expose the floor of the middle cranial fossa. The trajectory of the approach is therefore parallel to the lateral skull base. Epidural exposure of the middle fossa begins with the elevation of the "dura propria," i.e., the dura mater of the

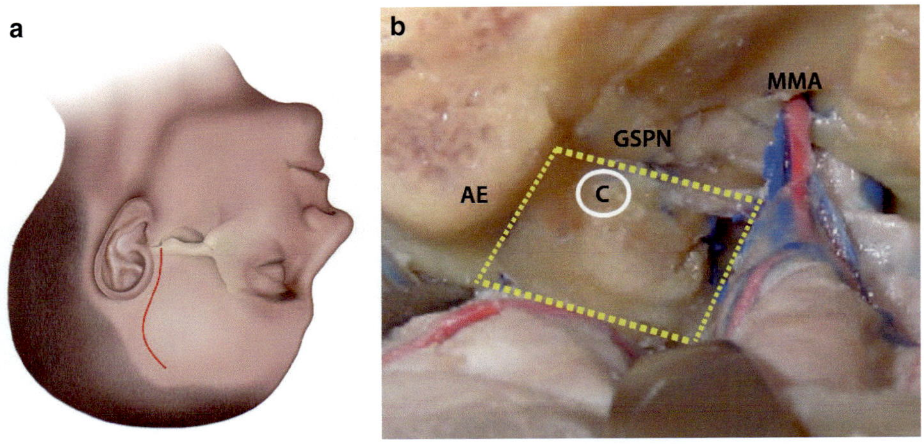

Fig. 43.9 Middle fossa approach. (**a**) Skin incision for a middle fossa approach. (**b**) Rhomboid fossa in a dissection specimen. *MMA* middle meningeal artery, *GSPN* greater superficial petrosal nerve, *AE* arcuate eminence, *C* cochlea

temporal lobe, which is separated during the procedure from the dura mater covering the trigeminal nerve and gasserian ganglion. For this reason, this approach is also known as the "interdural approach."

- The first landmark to identify is the middle meningeal artery, which must be coagulated and divided about 1 cm distal to the foramen spinosum, to avoid its elastic retraction inside the foramen and subsequent bleeding.
- Proceeding with the elevation of the dura propria, posteriorly and along the major axis of the petrous bone, the GSPN is identified, which runs superficially to the bone covering petrous ICA. However, bony atresia of the canal that normally intervenes between GG and the GSPN, or between GSPN and ICA is not uncommon: Therefore, it is necessary to proceed with caution with the dissection, using monopolar and laser probes as little as possible to avoid catastrophic vascular lesions.
- The safest spot to search for the GSPN is posterior and deep to the posterior border of V3, roughly halfway between the foramen ovale and the gasserian ganglion.
- Continuing along the course of GSPN up to GG, AE is exposed posteromedially and the rhomboid fossa is identified (Fig. 43.9b).
- Although IAC runs in it following the bisector of the angle between AE and GSPN, many anatomical variants exist.
- It is advisable to expose IAC starting the drilling of the rhomboid fossa medially, along the petrous ridge, continuing laterally in the direction of GG. This allows the roof of IAC to be removed and its contents to be exposed without directly manipulating FN.
- In order to preserve hearing, it is necessary to respect the anatomical integrity of the cochlea, which is located at the apex of the angle between IAC and GSPN, 4–5 mm anteromedial to GG and posteromedial to the petrous ICA.
- Once the contents of IAC are exposed, FN occupies an anterosuperior position in it, while CN8 is anteroinferior.
- Dissection of the intracanalicular VS proceeds with microsurgical instruments lateral to medial to prevent traction on the nerves at the level of the porus acusticus.
- When the removal is complete, haemostasis is perfected with fibrin glue, the bone operculum is repositioned, and the incision is closed in layers.

43.5 Surgical Pearls

43.5.1 Retrosigmoid Approach

- Fukushima lateral position
- 3 × 3 cm keyhole opening and microneurosurgical technique
- 2-mm space CSF aspiration (from lateral medullary cistern)
- Clean, bloodless, and sharp dissection
- Maintain arachnoid plane
- Sufficient internal tumor debulking
- Identify facial nerve (monopolar stimulator)
- Final capsule elevation: V-cut technique
- Preserve AICA, PICA, perforators, and petrosal veins

43.5.2 Translabyrinthine Approach

- Fukushima lateral position
- Knowledge of mastoid anatomy and triangles
- Identification of antrum, semicircular canals, sigmoid sinus, and tegmen tympani
- Drilling of lateral and posterior semicircular canals for IAC exposure
- Clean, bloodless, and sharp dissection
- Maintain arachnoid plane and microneurosurgical technique

- Sufficient internal tumor debulking for safe final capsule elevation
- Identify facial nerve (pink line, monopolar stimulator)
- Preserve AICA, PICA, perforators, and petrosal veins

43.6 Discussion

Comparative data among the different approaches are sparse in the literature. Ansar et al. [12], in a systematic review of 35 studies involving a total of 5064 patients, analyzed the results of every approach stratified by a single outcome. With regard to HP, MF was superior to RS for IC tumors (<1.5 cm), but not in the case of larger size. On the contrary, considering the anatomo-functional preservation of CN7, RS was superior to MF and TL, all the more significantly as tumor size increased. The risk of CSF fistula was greatest with RS. On the other hand, there were no significant differences regarding the extent of resection, major complications not related to cranial nerves, and mortality, as also confirmed by the results of Tolisano et al. [13]. Contrasting data derive from the systematic review by Jun et al. [14] (9 studies, 2429 patients), from which it would emerge that TL is more effective than RS in GTR, but not in near-total or subtotal resection; moreover, they would not have found significant differences in the incidence of CSF fistulas.

The HP outcome with RS, in recent years, stands at 57%, 37%, and 12% for IC tumors, Koos II, and Koos III-IV, respectively, as stated by a systematic review of 26 studies (2034 patients) by Preet et al. [15]. However, it should be considered that these data are burdened by considerable heterogeneity of the single reviewed studies. Regarding the hearing outcome of patients operated with TL, Thompson et al. [16] report a favorable outcome—i.e., open-set speech perception—in 85% of patients in case of cochlear implant at the time of VS excision.

43.7 Conclusion

The surgical treatment of vestibular schwannomas is a complex and extensive topic that requires constant practice in recognizing and manipulating delicate anatomical structures. The three approaches illustrated must form part of the baggage of a skull base surgeon and remain current and necessary, even in the context of an increasingly integrated multidisciplinary approach to this pathology.

References

1. Reznitsky M, Petersen M, West N, Stangerup SE, Caye-Thomasen P. Epidemiology of vestibular Schwannomas - prospective 40-year data from an unselected National Cohort. Clin Epidemiol. 2019;11:981–6.
2. Reznitsky M, Petersen M, West N, Stangerup SE, Caye-Thomasen P. The natural history of vestibular schwannoma growth-prospective 40-year data from an unselected national cohort. Neuro-Oncology. 2021;23(5):827–36.
3. Mastronardi L, Campione A, Cacciotti G, Roperto R, Giacobbo Scavo C, Zomorodi A, et al. Translabyrinthine approach. In: Mastronardi L, Fukushima T, Campione A, editors. Advances in vestibular Schwannoma microneurosurgery. Cham: Springer International Publishing; 2019. p. 53–67.
4. Campero A, Martins C, Rhoton A, Tatagiba M. Dural landmark to locate the internal auditory canal in large and giant vestibular schwannomas: the Tübingen line. Neurosurgery. 2011;69(1 Suppl Operative):ons99–102; discussions.
5. Mastronardi L, Sameshima T, Ducati A, De Waele LF, Ferrante L, Fukushima T. Extradural middle fossa approach. Proposal of a learning method: the "rule of two fans." Technical note. Skull Base. 2006;16(3):181–4.
6. Kumon Y, Kohno S, Ohue S, Watanabe H, Inoue A, Iwata S, et al. Usefulness of endoscope-assisted microsurgery for removal of vestibular schwannomas. J Neurol Surg B Skull Base. 2012;73(1):42–7.
7. Corrivetti F, Cacciotti G, Scavo CG, Roperto R, Stati G, et al. Flexible endoscopic assistance in the surgical management of vestibular schwannomas. Neurosurg Rev. 2021;44(1):363–71.
8. Mastronardi L, Cacciotti G, Caputi F, Roperto R, Tonelli MP, Carpineta E, et al. Underlay hourglass-shaped autologous pericranium duraplasty in "key-hole" retrosigmoid approach surgery: technical report. Surg Neurol Int. 2016;7:25.
9. Sameshima T, editor. Fukushima's microanatomy and dissection of the temporal bone. 2nd ed. Raleigh: AF-Neurovideo; 2007. p. 115.
10. Sameshima T, Mastronardi L, Friedman AH, Fukushima T. Microanatomy and dissection of temporal bone for surgery of acoustic neuroma and petroclival meningioma. 2nd ed. Raleigh: AF Neurovideo; 2007.
11. Mastronardi L, Corrivetti F, Scavo CG, Cacciotti G, Roperto R, et al. Extended middle cranial fossa approach for lesions invading infratemporal fossa: anatomic study and clinical application. World Neurosurg. 2020;138:83.
12. Ansari SF, Terry C, Cohen-Gadol AA. Surgery for vestibular schwannomas: a systematic review of complications by approach. Neurosurg Focus. 2012;33(3):E14.
13. Tolisano AM, Littlefield PD. Adverse events following vestibular schwannoma surgery: a comparison of surgical approach. Otol Neurotol. 2017;38(4):551–4.
14. Jun W, Gao YL, Yu HG, Huang QL, Long XQ, Liu GH, et al. Comparison of translabyrinthine and retrosigmoid approach for treating vestibular schwannoma: a meta-analysis. Clin Neurol Neurosurg. 2020;196:105994.
15. Preet K, Ong V, Sheppard JP, Udawatta M, Duong C, Romiyo P, et al. Postoperative hearing preservation in patients undergoing retrosigmoid craniotomy for resection of vestibular schwannomas: a systematic review of 2034 patients. Neurosurgery. 2020;86(3):332–42.
16. Thompson NJ, O'Connell BP, Brown KD. Translabyrinthine excision of vestibular schwannoma with concurrent cochlear implantation: systematic review. J Neurol Surg B Skull Base. 2019;80(2):187–95.

44 Retrosigmoid Microsurgery of Giant Vestibular Schwannoma

Mario Ammirati and Roberto Colasanti

Learning Objectives

- Know the epidemiology and the clinical presentation of giant vestibular schwannomas.
- Learn about the microsurgical strategy for the removal of such lesions.
- Understanding the importance of intraoperative monitoring, as well as other technological tools, for safe microsurgical resection.
- Know the pros and cons of the different surgical approaches to vestibular schwannomas.
- Learn about alternative treatment strategies to reduce surgical risks and improve the functional outcome.

44.1 Introduction

Schwannomas arise from the nerve sheath's Schwann cells and account for about 8% of all intracranial tumors. Vestibular schwannomas originate from the inferior vestibular nerve in most cases, and constitute 90% of all intracranial schwannoma, with an estimated incidence of approximately 1 per 100,000 per year [1–4].

In the last decades, the technological and surgical techniques advancements, together with a better understanding of the natural history of the disease, have progressively improved the functional outcomes of patients harboring vestibular schwannomas [1–4]. However, when dealing with large and especially giant vestibular schwannomas (defined by a maximum extrameatal diameter greater than 4 cm), it is more difficult to achieve good postoperative outcomes. Giant tumors usually represent up to 2% of all vestibular schwannomas in Western countries, but they may be the majority in developing countries [1, 3].

Because of their size, these lesions cause significant brainstem compression and compress multiple cranial nerves. As a consequence, their clinical presentation varies from that of patients with smaller tumors. The cochlear nerve is the most commonly clinically involved cranial nerve (hearing loss in 96–100%, and tinnitus in 42–46%), followed by the trigeminal nerve (in 40–78%), and by the lower cranial nerves (in 10–14%). Diplopia may be due to sixth-nerve palsy. The compression of the middle cerebellar peduncle and cerebellum causes gait instability in 44–88%, while long tract signs (present in 14–25% of the cases) are due to brainstem compression. In more severe cases, signs of intracranial hypertension (e.g., mental confusion, papilledema with visual deterioration, etc.) may be observed. The obstruction of the cerebrospinal fluid pathways may cause hydrocephalus [1, 3, 5–9].

Hence, the microsurgical removal of the lesion, with consequent decompression of the brainstem and of the surrounding neurovascular structures, is the cornerstone of giant vestibular schwannomas management. However, their size and their close relationship to important neurovascular structures may make the microsurgical resection very difficult, thus increasing the risk of postoperative morbidity and mortality. In particular, the considerable compression and stretching of cranial nerves complicate facial nerve preservation with the chance of functional hearing preservation even questioned by various authors [1, 6, 7]. As a consequence, there is still no clear consensus regarding the best treatment option: some authors consider the complete surgical removal as the final goal to be reached in 1, 2, or more stages; others believe that the so-called "functional sparing surgery policy" based on a subtotal resection followed by stereotactic radiosurgery of the tumor remnant either upfront or when tumor progression is demonstrated could be helpful to preserve facial nerve function [1, 3, 5–9].

M. Ammirati (✉)
College of Science and Technology, Temple University, Philadelphia, PA, USA

R. Colasanti
Department of Neurosurgery, Maurizio Bufalini Hospital, AUSL della Romagna, Cesena, Italy

I. N. Kanaan, V. Beneš (eds.), *Neuroanatomy Guidance to Successful Neurosurgical Interventions*,
https://doi.org/10.1007/978-3-031-59838-8_44

Here we present our technique for the microsurgical removal of giant vestibular schwannomas through the retrosigmoid approach, and discuss the therapeutic strategy for achieving good functional outcomes when dealing with such difficult lesions.

44.2 Anatomy Highlights

The asterion marks the junction of the transverse and the sigmoid sinuses [4, 10].

The vestibulocochlear nerve and facial nerve go through the cerebellopontine angle, from the pons to the internal auditory canal. The nerves have a precise location at the lateral end of the meatus, which is divided into four parts by a horizontal ridge ("transverse" or "falciform" crest) and a vertical crest ("Bill's bar"). The facial nerve occupies the anterosuperior quadrant of the lateral meatus, the cochlear nerve the anteroinferior portion, the superior and inferior vestibular nerves pass through the respective posterior foramina. The anterior inferior cerebellar artery (AICA) is the main artery of the middle neurovascular complex [4, 10].

During their growth, vestibular schwannomas progressively involve various cranial nerves and cerebellar arteries. On the lateral side, they frequently determine an enlargement of the meatus, while less commonly invade the vestibule and cochlea. On the medial side, they compress the brainstem and cerebellum. Large and giant vestibular schwannomas commonly compress structures also of the upper and lower neurovascular complexes [1, 4, 10].

Vestibular schwannomas usually push forward the facial and the cochlear nerves (anterosuperiorly or anteroinferiorly in relation to the direction of the growth of the tumor), as they normally originate from the posteriorly placed vestibular nerves. Being stretched around the anterior half of the tumor capsule, the facial and the cochlear nerves are more easily identified laterally, at the level of the meatal fundus, where the degree of displacement is less conspicuous [1, 4, 10]. After the intracapsular debulking of the tumor, the facial and vestibulocochlear nerves may be more safely located on the brainstem side of the lesion at the lateral end of the pontomedullary sulcus, just rostral to the glossopharyngeal nerve and just anterosuperior to the foramen of Luschka, the flocculus, and the choroid plexus protruding from the foramen of Luschka [1, 4, 10].

44.3 Surgical Approach

We use the retrosigmoid approach for the excision of vestibular schwannomas (Figs. 44.1, and 44.2).

Continuous intraoperative neurophysiological monitoring includes brainstem auditory evoked potentials, somatosensory evoked potentials, and facial nerve electromyography. Direct intraoperative facial nerve stimulation is carried out liberally and repeatedly, as it is useful in locating the facial nerve and evaluating its functional continuity at the end of the operation.

Transesophageal or transthoracic echocardiography is employed for immediate detection of air embolism.

Flexion-extension x-rays of the cervical spine are obtained to rule out the rare latent instability.

Under somatosensory evoked potentials, we position the patient semisitting with the head slightly flexed (leaving a room of about two fingers between the chin and the sternum) and rotated 30° toward the side of the tumor.

The operating table is tilted elevating the legs of the patient above the level of the heart, at least as high as the transverse sinus, in order to decrease the risk of air embolism [1, 2, 4, 11, 12].

Neuronavigation is used with a fused CT-MRI dataset and the microscope focal point is used as the virtual navigation pointer. The navigation data are used to trace on the patient's skin the distal transverse sinus, its junction with the sigmoid sinus, the sigmoid sinus, and the jugular bulb. The skin incision is positioned about 1 cm medial to the sigmoid sinus (about two fingers behind the ear) extending superiorly to 1 cm above the transverse sinus and inferiorly about 2 cm inferior to the jugular bulb (approximately from the tip of the ear to 1 cm below the mastoid tip). After standard exposure of the lateral occipital bone, a small 2- to 3-cm craniotomy is carried out by placing a single bur hole very close to the junction of the transverse-sigmoid sinus. The retrosigmoid craniotomy has to be large enough to expose the borders of the transverse sinus, the transverse-sigmoid junction, the sigmoid sinus, and also the jugular bulb in case of large lesions with significant extension into the lateral cerebellomedullary cistern. Then, the horizontal portion of the occipital squama is progressively removed with a rongeur.

The dura is opened with a C-shaped incision that runs parallel to the course of the transverse sinus and sigmoid sinus, with the concavity facing medially. At this point, under the operating microscope we expose the inferior portion of the lateral cerebellomedullary cistern by gently supporting the cerebellar tonsil; this maneuver is greatly facilitated by extensive removal of the horizontal portion of the occipital squama. The lateral portion of the cisterna magna is also opened. The drainage of cerebrospinal fluid from these cisterns is a paramount step. Obviously, in cases of florid hydrocephalus, a ventriculostomy catheter is placed immediately prior to the retrosigmoid incision.

At this point, the cerebellar hemisphere is gently supported using a self-retaining retractor in order to expose the tumor. At this stage, the lateral portion of the tumor is progressively exposed by a selective drilling of the posterior lip of the porus acusticus and the internal auditory canal. In

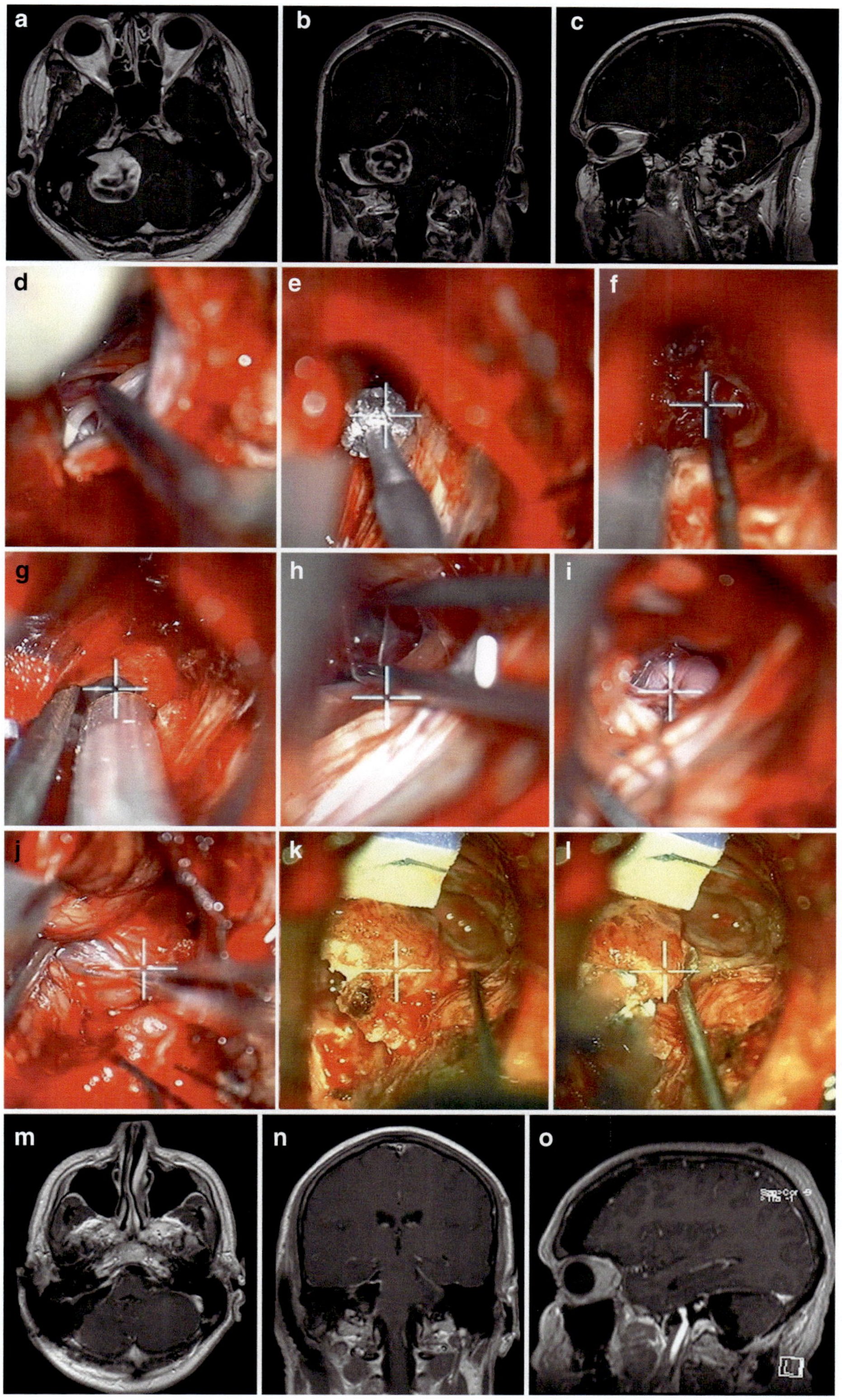

Fig. 44.1 Right-sided giant vestibular schwannoma in a 55-year-old man with complete loss of hearing, and imbalance leading to falls. Tumor was completely removed using the retrosigmoid approach while the patient was in the semisitting position. (**a–c**) Preoperative MRI images showing the giant tumor compressing the brainstem. (**d–l**) Intraoperative pictures showing the 11th nerve visualized after the draining of the lateral cerebellomedullary cistern (**d**), an important preresection step; the drilling of the posterior wall of the internal auditory canal (**e**); the removal of the intrameatal portion of the lesion (**f**); the intratumoral debulking dissection technique (**g**); view of the inferior pole of the surgical field after the intratumoral debulking (**h**, **i**); separation of the tumor-arachnoid plane at the level of the brainstem (**j**); stimulation of the facial nerve (**k**); careful dissection of the tumor from the facial nerve (**l**). (**m–o**) Postoperative MRI images showing total tumor removal and decompression of the brainstem

patients with preoperative serviceable hearing, it is crucial to preserve the posterior semicircular canal and the common crus during the drilling of the canal. The high-definition CT navigation dataset is employed to make this drilling easier and safer by precisely defining its limits [13–15]. Particular attention has to be paid if a high jugular bulb is detected on the preoperative imaging [16]. The Tübingen line, which is identified by dural wrinkles on the posterior surface of the petrous bone, is a projection of the internal auditory canal and is especially useful when drilling the canal in large ves-

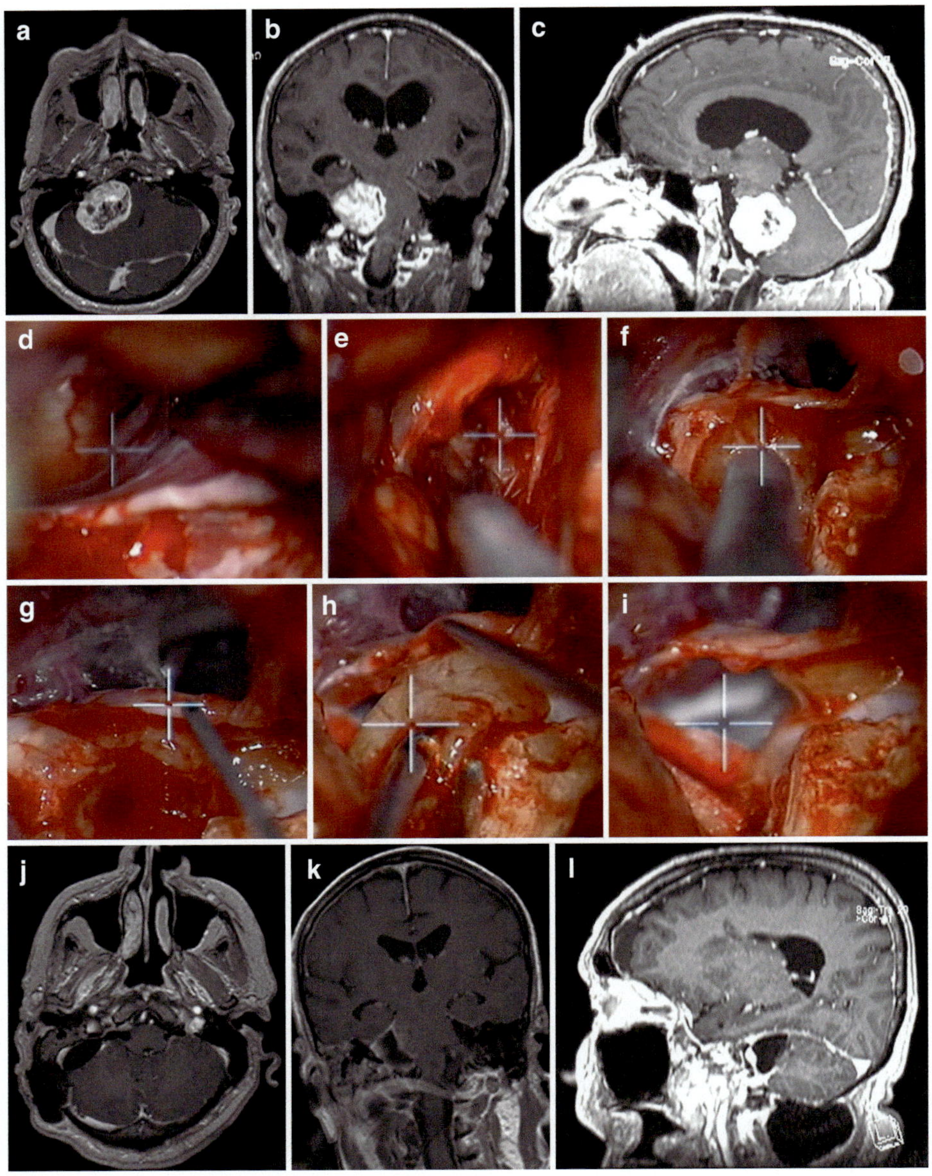

Fig. 44.2 Right-sided giant vestibular schwannoma in a 61-year-old man with complete loss of hearing, confusion, and imbalance leading to falls. Tumor was completely removed using the retrosigmoid approach with the patient in the semisitting position. (**a**–**c**) Preoperative MRI images showing the giant tumor compressing the brainstem and determining obstructive hydrocephalus. Intraoperatively, a ventricular drainage was placed. (**d**–**l**) Intraoperative pictures showing the 11th nerve visualized after the draining of the lateral cerebellomedullary cistern (**d**), an important preresection step; the progressive intratumoral debulking (**e**, **f**); stimulation of the facial nerve (**g**); careful dissection of the tumor from the facial nerve (**h**); and facial nerve free of tumor (**i**). (**j**–**l**) Postoperative MRI images showing total tumor removal and decompression of the brainstem

tibular schwannomas [4]. The endoscopic assistance may be helpful to widen the exposure while minimizing the retraction/manipulation of the neurovascular structures [1, 2, 4, 17].

After the completion of the drilling, intratumoral debulking is accomplished alternating the use of microsurgical instruments and the ultrasonic aspirator. Then, after a generous intratumoral debulking, separation of the tumor-arachnoid plane is carried out by gently grasping the tumor or, occasionally, the arachnoid with bayonet forceps and peeling off one structure away from the other. Since vestibular schwannomas are extra-arachnoidal tumors, in most cases it is possible to separate the tumor and the arachnoid. We work on the lesion in a nonsequential manner by going back and forth to different regions. We favor to not coagulate and instead use temporary packing with topical hemostatic agents because coagulation tends to destroy the tumor-arachnoid interface, thus making dissection more difficult and risky. In addition, an excessive coagulation before vascular structures are completely dissected may interrupt the vascular supply to the small but vital brainstem perforators.

After the identification of the lower cranial nerves, they are dissected from the inferior pole of the tumor, and protected with a small collagen-based sponge; next, the AICA is gently detached from the tumor, and the superior pole is carefully separated from the trigeminal nerve and the surrounding neurovascular structures. At this point, once an effective intracapsular debulking of the tumor has been carried out, the vestibulocochlear and facial nerves should be identified at the brainstem.

If the patient has serviceable hearing preoperatively, an effort to preserve the cochlear nerve should always be made even though serviceable hearing will be maintained postoperatively at most in only about 10% of patients with a large tumor [1, 6, 7]. After identification of the facial nerve at the brainstem, we start carefully dissecting the tumor from the facial nerve by progressively debulking the lesion so that the capsule collapses on the facial nerve. Finally, the capsule is detached from the facial nerve by sharp dissection.

Since there is often a quite firm adhesion between the tumor and the facial nerve just a few millimeters medial to the porus acusticus, a careful and gentle dissection is required in this region in order to preserve the facial nerve function. In addition, sometimes, the capsule of the tumor tenaciously adheres to the brainstem, usually to the pons, and the arachnoidal plane at that level is practically nonexistent. In this situation, it is preferable to attempt the dissection of this tumor portion when most of the lesion has been resected, and greater room and more operative trajectories are available. After removal of lesion in the cerebellopontine angle is accomplished, tumor is resected from the internal auditory canal under combined direct microscopic and endoscopic vision [1, 2, 4, 17].

Indeed, while the tumor size is clearly relevant, the tumor consistency and the adhesions between tumor/brainstem/cranial nerves and vessels are extremely relevant as well. Patience in performing tumor debulking from different angles, facilitated by the retrosigmoid approach in the semisitting position, as well as early recognition of significant adhesion between the tumor and relevant neurovascular structures, is paramount to achieving a good outcome [1, 2, 4, 17].

If a small layer of tumor is very adherent to the brainstem or to the facial or cochlear nerve and the relative intraoperative potentials worsen during manipulation, it is advisable to leave a residual of the tumor in order to preserve function [1–4, 6, 7, 18].

Pieces of muscle or fat mixed with fibrin glue are employed to close the opened air cells in the region of the internal auditory canal, as well as opened mastoid cells. At the end of the operation, jugular venous compression is used to discover any open or torn veins for final hemostasis. An accurate water-tight closure of the dura is performed. When a craniectomy is carried out, a large piece of gelfoam sponge is placed over the dura mater, and the bone defect is closed with a methylmetacrylate plastic or with other synthetic preformed material [4].

44.4 Special Recommendations

1. A careful preoperative review of neuroimaging (brain MRI and thin-slice CT scan of the skull base) is paramount to evaluate the relationships of the vestibular schwannoma with surrounding neurovascular structures, the location of the emissary veins, the location and height of the jugular bulb, and the topographical relationship between inner ear structures and the widened internal auditory canal.
2. When semisitting position is planned, a transthoracic echocardiography is performed for the detection of a patent foramen ovale determining a right-to-left cardiac shunt. In addition, a preoperative flexion-extension X-ray of the cervical spine is recommended to rule out severe degenerative disease or instability. In patients with neurofibromatosis type II, cervical spine MRI scans are evaluated to exclude intraspinal tumors.
3. In the semisitting position, the legs of the patient are elevated above the level of the heart, at least as high as the transverse sinus, in order to increase venous pressure and ultimately achieve a continuous positive venous pressure, thus reducing the risk of air embolism. This goal is reached thanks also to other measures such as a correct fluid input and positive end-expiratory pressure (PEEP) ventilation. Continuous intraoperative end-tidal CO_2 monitoring, in combination with transesophageal or transthoracic echocardiography, is employed for immediate detection of air embolism.
4. The retrosigmoid craniotomy has to expose the transverse and sigmoid sinus, and the jugular bulb in case of large lesions. The horizontal portion of the occipital squama has to be removed.
5. The drainage of cerebrospinal fluid from the lateral cerebellomedullary cistern (as well as from the cisterna magna, in case of tumor extension) makes easier and safer the following tumor removal.
6. Intraoperative navigation may precisely define the limits of petrous pyramid drilling, thus making the exposure of the intrameatal tumor portion easier and safer.
7. Intraoperative monitoring is paramount to guide microsurgical removal and decrease postoperative complications.
8. A generous intracapsular debulking of the cisternal part of the tumor facilitates the exposure of the facial and vestibulocochlear nerves on the brainstem side.
9. Since vestibular schwannomas are extra-arachnoidal tumors, in most cases it is possible to separate the tumor and the arachnoid by a gentle and careful bimanual dissection. An excessive coagulation should be avoided as it tends to destroy the tumor-arachnoid interface, while potentially interrupting important vascular supply.
10. Muscle or fat are used to close the opened air cells.

44.5 Surgical Pearls

- Complete exposure of the transverse and sigmoid sinus through a careful removal of the horizontal portion of the occipital squama.

- The drainage of cerebrospinal fluid facilitates the following tumor removal.
- Intraoperative navigation to guide petrous pyramid drilling.
- Intraoperative monitoring.
- A generous intracapsular debulking should precede the exposure of the facial and vestibulocochlear nerves on the brainstem side.
- An excessive coagulation could destroy the tumor-arachnoid interface, while potentially interrupting important vascular supply.
- Careful closure of the opened air cells.

44.6 Discussion

Giant vestibular schwannomas may cause considerable compression of the brainstem, as well as of multiple cranial nerves and vascular structures. As a consequence, the microsurgical removal plays a key role in their management in order to decompress the various neurovascular structures and relieve patients' symptoms. However, the adherence of vestibular schwannoma's capsule to the surrounding vascular and nervous structures (in particular to the brainstem and to the facial nerve) may make their microsurgical resection challenging and increase postoperative morbidity and mortality [1, 3, 5–9, 18].

44.6.1 Microsurgical Approaches for Giant Vestibular Schwannomas

The retrosigmoid approach is our favored surgical route for the removal of vestibular schwannomas. Indeed, this simple and well-known approach allows the microsurgical resection of lesions of different size and is particularly helpful when dealing with large and giant tumors. The retrosigmoid approach provides an effective exposure of the internal auditory canal and of the cerebellopontine angle, thus setting advantageous conditions for preservation of the vestibulocochlear and facial nerves. In case of significant preoperative hydrocephalus, a frontal or occipital ventriculostomy is carried out before we position the patient for surgery [1, 2, 4].

The retrosigmoid approach can be carried out in the supine, prone, or lateral positions. However, we prefer the semisitting position because it improves the drainage of blood and cerebrospinal fluid out of the surgical field, thus reducing the need for coagulation and suctioning during tumor removal. The decreased need for coagulation at the tumor-arachnoid interface may avoid the binding of these two structures, hence facilitating the dissection. The decreased need for suctioning may prevent the risk of trauma to neurovascular structures from the suction tip. Furthermore, the semisitting position may increase the venous drainage, thus allowing to reduce the venous bleeding and intracranial pressure, as well as the amount of cerebellar retraction. In addition, the semisitting position is dynamic and by changing the table height and the surgeon's position, a multitude of views/surgical portals become apparent [1, 2, 4].

Nowadays, the sitting position has lost favor among neurosurgeons partly due to assumptions of increased complications. However, the literature is inconclusive on this matter, and several protocols, modified sitting positions, the use of antigravity trousers, and optimal anesthetic considerations, may be effectively used in the clinical setting in order to prevent its major complications. In addition, as abovementioned, the sitting position offers various advantages and, thanks to the appropriate adjustment of the operating table, may allow different viewing/working angles difficult to achieve in other positions [1, 2, 4, 11, 12, 19–21].

However, even if we favor the semisitting position, the pros of the prone position or any of its variations (e.g., low risk of air emboli, more comfort for the surgeon who does not operate with outstretched arms) should be recognized [2, 11, 19].

Neuronavigation may be effectively used to create a topographical framework that may help to maximize the drilling of the posterior wall of the internal auditory canal via the retrosigmoid route without violating the integrity of the labyrinth, the endolymphatic sac, and the vestibular aqueduct or injuring the cochlear nerve [1, 13, 14].

The endoscopic assistance may be very useful to better visualize the deepest extensions of the lesions, as well as to bring into the surgical field some blind regions that may not be adequately and safely visualized with the microscope alone (e.g., the intrameatal portion of the tumor), and to provide (like microscopic high magnification does) a clear distinction of the interface between the tumor and the surrounding neurovascular structures. Hence, endoscopic assistance may allow an enhanced exposure, while minimizing the retraction/manipulation of the neurovascular structures [2, 17]. We recommend the introduction of the endoscope under microscopic control in order to decrease the risk of accidental mechanical injury. In this regard, the picture-in-picture technology may permit an effective integration between the endoscope and the microscope by efficiently displaying the endoscopic image in the microscope oculars, therefore reducing the operative time [2, 17].

The "extended retrosigmoid approach," which is characterized by a bony skeletonization of the sigmoid/transverse sinuses with an additional mastoidectomy, may provide (thanks to the mobilization of the sinuses themselves) a better exposure of lesions deeply extending in the cerebellopontine angle, while reducing retraction of the cerebellum [22]. However, the downward relaxation of the cerebellar hemisphere, which is determined by the drainage of cerebrospinal fluid from the cisterna magna, is fully sufficient to prevent significant cerebellar retraction in the majority of the cases. Therefore, this allows to avoid the higher risk of cerebrospinal fluid leakage and venous sinus thrombosis, which is associated with the "extended retrosigmoid approach" [22].

Various suprameatal and inframeatal/infralabyrinthine extensions of the retrosigmoid approach have been proposed for the removal of giant lesions extending from Meckel's cave to the petrous internal carotid artery with a variable grade of petrous bone involvement [15, 23–30].

In general, in giant vestibular schwannomas, the simple retrosigmoid approach is the most useful; indeed, if patient intratumoral debulking is performed, the tumor will be progressively shrunk and will deliver itself into the simple retrosigmoid window. It is not necessary to expose the totality of the tumor at the beginning of the operation [1, 2, 4].

Discussion of transtemporal approaches to giant vestibular schwannomas is outside the scope of this contribution. Suffice to say that many variations of the transtemporal approaches have been reported for the removal of giant vestibular schwannomas. These approaches provide variable degrees of exposure with minimal cerebellar retraction, but they are time-consuming and are associated with a high risk of approach-related complications that are proportionate to the extent of bony drilling and the neural and vascular manipulations [1, 9, 31, 32].

44.6.2 Facial Nerve Outcome

The main predictor of facial nerve anatomical and functional preservation is represented by the size of the vestibular schwannoma. Indeed, patients with a vestibular schwannoma >3 cm have a sixfold greater risk of facial nerve dysfunction than patients with small lesions [1, 6, 7].

Tumor size being comparable, higher postoperative rates of good facial nerve function are observed in solid rather than in cystic vestibular schwannomas, due to the usual strict adherence of cystic lesions to the brainstem and/or cranial nerves [18].

The microsurgical removal of large vestibular schwannomas through the translabyrinthine approach or the retrosigmoid approach is associated with an 80–93% rate of anatomical facial nerve preservation. The long-term postoperative facial nerve outcome is worse in patients with giant vestibular schwannomas than in those with tumors <3.9 cm. In cases of giant vestibular schwannomas, the facial nerve is anatomically preserved in 78–94%, with a good postoperative facial function (House-Brackmann Grade I–III) in 38–62% [1, 6, 7].

Samii et al. evaluated the postoperative functional outcome in 50 patients with giant vestibular schwannomas versus 167 patients with lesions <3.9 cm. They reported excellent or good facial nerve function in 75% of the patients with giant tumors and 91% of those with smaller lesions. However, a statistically significant difference between the two groups was observed for the rate of patients with normal function (25% compared with 63%, respectively; $p < 0.05$) [1].

This is because the facial nerve is gradually stretched by the tumor growth, becoming progressively thinner and thinner and more vulnerable during surgery [1, 6, 7].

Diffusion tensor tractography may be a useful tool for preoperatively predicting facial nerve location and displacement in vestibular schwannoma (success rate 62.5–100%) [7], although intraoperatively the facial nerve position may change as the tumor is debulked.

An accurate surgical strategy, as above described, is paramount to safely identify the facial nerve along its course (a generous intracapsular debulking of the cisternal part of the tumor should always precede the exposure of the facial nerve on the brainstem side) [1, 2, 4, 6, 7].

Intraoperative monitoring/stimulation is paramount to determine the location and function of the facial nerve during surgery [1, 2, 4, 6, 7].

44.6.3 Hearing Preservation

Many authors prefer the translabyrinthine approach, as they think that is practically not possible to preserve a serviceable hearing in patients with large and giant vestibular schwannomas. However, as pointed by other authors, we believe that a hearing-preservation technique should always be offered to the patients unless this implicates an excessive risk of postoperative morbidity [1, 6, 7]. As a consequence, we favor the use of the retrosigmoid approach for the removal of large and giant vestibular schwannomas, as it is a simple and well-known approach, associated with low-complication rates, and it allows to anatomically preserve the cochlear nerve in the majority of the cases [1, 6, 7].

The tumor size is the most important predictive factor for hearing preservation, with larger tumors showing the lower postoperative hearing-preservation rates [1, 6, 7]. In addition, only about 10% of patients with large vestibular schwannomas have preoperatively serviceable hearing function [7]. However, even in patients harboring giant lesions but with useful hearing before surgery, Samii et al. observed that there is still an 11% chance of postoperative hearing preservation [1].

In order to achieve these results, it is paramount to an accurate intraoperative safeguard of the integrity of the cochlear nerve, cochlea, and internal auditory artery (the labyrinthine artery, which supplies blood to the vestibulocochlear nerve and the inner ear). Intraoperative monitoring of the cochlear nerve may be helpful for its preservation [1, 7]. In addition, as abovementioned, neuronavigation is a useful tool for inner ear preservation during internal auditory canal drilling [13, 14]. Finally, the endoscopic assistance may make easier and safer the removal of the intrameatal portions of the tumor [2, 17].

44.6.4 Postoperative Complications

The tumor growth determines a long-standing compression of surrounding neurovascular structures resulting in progressive and significant structural modifications that raise the risk of postoperative complications, in particular in large and giant vestibular schwannomas [1, 6, 7]. Higher neurological complication rates in large and giant tumors have been reported by various authors [1, 6–9].

Roland et al. reported a series of 54 patients who underwent surgical resection of vestibular schwannomas >3 cm. All tumors were removed through a translabyrinthine (82%) or retrosigmoid/transmeatal (18%) approach, and an overall total tumor removal rate of 73% was achieved. Postoperatively, abducens nerve palsies were observed in 7% of the patients, lower cranial nerve dysfunction in 4%, cerebrospinal fluid leaks in 5%, and pseudomeningocele in 6%. The major determinant of significant morbidity in this series was postoperative intracranial hemorrhage that required surgical evacuation in 4% [32]. In general series, postoperative symptomatic intracranial hemorrhage has an incidence of about 1–2.4%, but it rises up to 3–4% in large and giant lesions [1, 6, 7, 32].

Yamakami et al. reported excellent functional outcomes in a series of 50 consecutive patients who underwent microsurgical resection of large vestibular schwannomas through the retrosigmoid suboccipital approach, but cerebellar hemorrhage determined persistent ataxia in two patients (4%) [1].

An accurate hemostasis, as well as the intraoperative preservation of the arteries and veins that do not supply the tumor, is paramount to reduce the risk of postoperative intracranial hemorrhage [1, 6, 7, 27].

The most dreadful morbidity that can occur in a vestibular schwannoma case is a vascular complication. Sughrue et al. evaluated 32,870 cases of vestibular schwannomas from 100 articles. They observed an incidence of vascular complications of about 1% (337 patients, hemorrhagic injuries in more than 80% of the cases), far below the rate of 7% reported by other authors. Vascular complications were associated with a mortality rate of 19%, and AICA hemorrhage was the leading cause of death [7, 33]. Vascular complications had the highest incidence in patients who underwent resection through the retrosigmoid route (2.6%), followed by the middle cranial fossa (1%), and the translabyrinthine (0.9%) approaches [33]. This may be related to the fact that the tumor lies between the surgeon and the AICA when operating via a translabyrinthine route, while the surgeon must work close and past the AICA through a retrosigmoid approach [9, 33]. On the other hand, the simple and well-known retrosigmoid route allows earlier visualization of the various vascular structures and cranial nerves compared with the more complex lateral transtemporal or presigmoid approaches, thus potentially reducing the risk of injury [1, 4, 15, 27]. In addition, the rate of vascular complications was significantly higher in tumors >25 mm (1.3%) than in those ≤25 mm (0.7%) ($p < 0.0001$) [33].

Various authors consider the tumor size as a major predictor of postoperative cerebrospinal fluid complications, with a 4–10% incidence in patients with large or giant vestibular schwannomas [1, 6, 7, 9, 33]. Samii et al. registered a higher rate of cerebrospinal fluid leak in patients with giant lesions (6%) than in those with smaller tumors (2.4%), even though the difference was not significant ($p > 0.05$) [1]. According to other authors, the major determinant of postoperative cerebrospinal fluid complications is the surgical approach and not the tumor size [7, 33]. Sughrue et al. registered the highest incidence in patients who underwent resection through the translabyrinthine route (27%), followed by the middle cranial fossa (7%), and the retrosigmoid (6%) approaches [33].

Infections have been considered as a major determinant of surgical morbidity and death. Sughrue et al. observed a statistically significant higher infection rate in patients with tumors >25 mm (3%) than in those ≤25 mm (1%) ($p < 0.0001$) [33].

44.6.5 Alternative Treatment Strategies to Decrease Risks and to Improve the Functional Outcome

Various treatment strategies have been proposed to reduce postoperative complications in giant vestibular schwannomas and to enhance the functional outcome. Subtotal or radical intracapsular tumor resection has been advocated by some authors to decrease the risk of injury to the different neurovascular structures. Other groups have reported staged removal via a single approach or both the retrosigmoid and the translabyrinthine approaches with excellent tumor resection rates and without mortality and major morbidity [1, 5–7, 32].

Radiosurgery is not an alternative to microsurgical resection for giant vestibular schwannomas causing severe compression of the brainstem. However, various authors have proposed the so-called "functional sparing surgery policy" (based on a subtotal resection followed by immediate stereotactic radiosurgery or delayed stereotactic radiosurgery only when tumor progression is demonstrated) to improve functional outcomes [1, 3, 5–9].

Bretonnier et al. evaluated the postoperative functional outcomes of 60 patients with giant vestibular schwannomas. Radiotherapy was performed in 12 patients for a growing residual lesion. The tumor control rate was better in the gross total resection group, but there was no tumor recurrence after radiotherapy in the subtotal resection group. On the other hand, the facial nerve outcome was significantly better in the subtotal resection group [3]. Other studies reported similar data both for large and giant tumors [1, 3, 7, 8]. Hence, a "functional nerve-sparing resection" could represent a reasonable and effective strategy to achieve satisfactory outcomes, in particular in low-volume centers [1, 3, 7, 8].

44.7 Conclusions

An accurate and meticulous microsurgical technique, together with the use of intraoperative monitoring, may allow to achieve good tumor removal and good functional outcome (in particular facial nerve function) in patients with giant vestibular schwannomas, with low postoperative morbidity.

On the other hand, when the lesion is very adherent to the facial or cochlear nerve, as well as to the brainstem, and a clear arachnoid plane may not be identified, a near total/subtotal resection (followed by radiosurgery in case of tumor progression) should be advocated in order to preserve function.

References

1. Samii M, Gerganov VM, Samii A. Functional outcome after complete surgical removal of giant vestibular schwannomas. J Neurosurg. 2010;112(4):860–7.
2. Ammirati M, Colasanti R. Overview of skull base tumors. In: Youmans and Winn neurological surgery. 8th ed. Elsevier; 2023. p. 1341–64.e3.
3. Bretonnier M, Bernard F, Tinois J, Troude L, Cebula H, Godey B, et al. Functional sparing surgery policy for giant vestibular schwannomas. Clin Otolaryngol. 2020;45(5):762–7.
4. Tatagiba M, Roser F, Schuhmann MU, Ebner FH. Vestibular schwannoma surgery via the retrosigmoid transmeatal approach. Acta Neurochir. 2014;156(2):421–5; discussion 425.
5. Silva J, Cerejo A, Duarte F, Silveira F, Vaz R. Surgical removal of giant acoustic neuromas. World Neurosurg. 2012;77(5–6):731–5.
6. Zou P, Zhao L, Chen P, Xu H, Liu N, Zhao P, et al. Functional outcome and postoperative complications after the microsurgical removal of large vestibular schwannomas via the retrosigmoid approach: a meta-analysis. Neurosurg Rev. 2014;37(1):15–21.
7. Huang X, Xu J, Xu M, Chen M, Ji K, Ren J, et al. Functional outcome and complications after the microsurgical removal of giant vestibular schwannomas via the retrosigmoid approach: a retrospective review of 16-year experience in a single hospital. BMC Neurol. 2017;17(1):18.
8. Zhang S, Liu W, Hui X, You C. Surgical treatment of Giant vestibular schwannomas: facial nerve outcome and tumor control. World Neurosurg. 2016;94:137–44.
9. Grinblat G, Dandinarasaiah M, Braverman I, Taibah A, Lisma DG, Sanna M. Large and giant vestibular schwannomas: overall outcomes and the factors influencing facial nerve function. Neurosurg Rev. 2021;44(4):2119–31.
10. Rhoton AL. The cerebellopontine angle and posterior fossa cranial nerves by the retrosigmoid approach. Neurosurgery. 2000;47(3 Suppl):S93–129.
11. Jadik S, Wissing H, Friedrich K, Beck J, Seifert V, Raabe A. A standardized protocol for the prevention of clinically relevant venous air embolism during neurosurgical interventions in the semisitting position. Neurosurgery. 2009;64(3):533–8; discussion 538–9.
12. Ammirati M, Theeb Lamki T, Brian Shaw A, Forde B, Nakano I, Mani M. A streamlined protocol for the use of the semi-sitting position in neurosurgery: a report on 48 consecutive procedures. J Clin Neurosci. 2013;20(1):32–4.
13. Pillai P, Sammet S, Ammirati M. Application accuracy of computed tomography-based, image-guided navigation of temporal bone. Neurosurgery. 2008;63(4 Suppl 2):326–32; discussion 332–3.
14. Colasanti R, Tailor ARA, Zhang J, Ammirati M. Image-guided, microsurgical topographic anatomy of the endolymphatic sac and vestibular aqueduct via a suboccipital retrosigmoid approach. Neurosurg Rev. 2015;38(4):715–21.
15. Colasanti R, Tailor ARA, Zhang J, Ammirati M. Functional Petrosectomy via a suboccipital Retrosigmoid approach: guidelines and topography. World Neurosurg. 2016;87:143–54.
16. Shao KN, Tatagiba M, Samii M. Surgical management of high jugular bulb in acoustic neurinoma via retrosigmoid approach. Neurosurgery. 1993;32(1):32–6; discussion 36–7.
17. Salma A, Ammirati M. Real time parallel intraoperative integration of endoscopic, microscopic, and navigation images: a proof of concept based on laboratory dissections. J Neurol Surg B Skull Base. 2012;73(1):36–41.
18. Mastronardi L, Campione A, Boccacci F, Scavo CG, Carpineta E, Cacciotti G, et al. Koos grade IV vestibular schwannomas: considerations on a consecutive series of 60 cases-searching for the balance between preservation of function and maximal tumor removal. Neurosurg Rev. 2021;44(6):3349–58.
19. Feigl GC, Decker K, Wurms M, Krischek B, Ritz R, Unertl K, et al. Neurosurgical procedures in the semisitting position: evaluation of the risk of paradoxical venous air embolism in patients with a patent foramen ovale. World Neurosurg. 2014;81(1):159–64.
20. Choque-Velasquez J, Colasanti R, Resendiz-Nieves JC, Gonzáles-Echevarría KE, Raj R, Jahromi BR, et al. Praying sitting position for pineal region surgery: an efficient variant of a classic position in neurosurgery. World Neurosurg. 2018;113:e604–11.
21. Choque-Velasquez J, Colasanti R, Resendiz-Nieves JC, Raj R, Lindroos AC, Jahromi BR, et al. Venous air embolisms and sitting position in Helsinki pineal region surgery. Surg Neurol Int. 2018;9:160.
22. Raza SM, Quinones-Hinojosa A. The extended retrosigmoid approach for neoplastic lesions in the posterior fossa: technique modification. Neurosurg Rev. 2011;34(1):123–9.
23. Matsushima K, Kohno M, Komune N, Miki K, Matsushima T, Rhoton AL. Suprajugular extension of the retrosigmoid approach: microsurgical anatomy. J Neurosurg. 2014;121(2):397–407.
24. Seoane E, Rhoton AL. Suprameatal extension of the retrosigmoid approach: microsurgical anatomy. Neurosurgery. 1999;44(3):553–60.

25. Samii M, Metwali H, Samii A, Gerganov V. Retrosigmoid intradural inframeatal approach: indications and technique. Neurosurgery. 2013;73(1 Suppl Operative):ons53–9; discussion ons60.
26. Samii M, Tatagiba M, Carvalho GA. Retrosigmoid intradural suprameatal approach to Meckel's cave and the middle fossa: surgical technique and outcome. J Neurosurg. 2000;92(2):235–41.
27. Sato Y, Mizutani T, Shimizu K, Freund HJ, Samii M. Retrosigmoid intradural suprameatal-inframeatal approach for complete surgical removal of a Giant recurrent vestibular schwannoma with severe petrous bone involvement: technical case report. World Neurosurg. 2018;110:93–8.
28. Colasanti R, Tailor ARA, Gorjian M, Zhang J, Ammirati M. Microsurgical and endoscopic anatomy of the extended retrosigmoid inframeatal infratemporal approach. Neurosurgery. 2015;11(Suppl 2):181–9; discussion 189.
29. Colasanti R, Tailor ARA, Lamki T, Zhang J, Ammirati M. Maximizing the petroclival region exposure via a suboccipital retrosigmoid approach: where is the intrapetrous internal carotid artery? Neurosurgery. 2015;11(Suppl 2):329–36; discussion 336–7.
30. Colasanti R, Tailor ARA, Zhang J, Ammirati M. Expanding the horizon of the suboccipital retrosigmoid approach to the middle incisural space by cutting the tentorium cerebelli: anatomic study and illustration of 2 cases. World Neurosurg. 2016;92:303–12.
31. Rhoton AL. The temporal bone and transtemporal approaches. Neurosurgery. 2000;47(3 Suppl):S211–65.
32. Roland JT, Fishman AJ, Golfinos JG, Cohen N, Alexiades G, Jackman AH. Cranial nerve preservation in surgery for large acoustic neuromas. Skull Base. 2004;14(2):85–90; discussion 90–1.
33. Sughrue ME, Yang I, Aranda D, Rutkowski MJ, Fang S, Cheung SW, et al. Beyond audiofacial morbidity after vestibular schwannoma surgery. J Neurosurg. 2011;114(2):367–74.

45 Retrosigmoid Approach and Hybrid Use of Endoscopy in CPA Tumors

Imad N. Kanaan

Learning Objectives
- Describe modified surgical position and the proper placement of the craniotomy.
- Highlight important anatomical landmarks & strategies to guide a safe tumor resection.
- Tailor surgical approach to fit the corresponding pathology.
- Mark value of hybrid use of endoscopy to trace & treat residual or multicompartment tumors.
- Provide measures to maintain the integrity of vital structures & score excellent outcomes.
- Review potential indications for retrosigmoid approach and hybrid use of endoscopy.

45.1 Introduction

The retrosigmoid (RS) approach is a versatile corridor frequently employed for the management of a variety of neoplastic and vascular pathologies that are confined to the cerebellopontine angle (CPA) and its surrounding structures. This approach is particularly well-suited for the treatment of vestibular or trigeminal schwannomas, meningiomas, and epidermoid tumors, which are among the most common pathologies encountered in this region. Additionally, vascular pathologies or exophytic tumor growth arising from the cerebellum, brainstem, or clivus can also be effectively managed with the RS approach.

The complex regional anatomy of the CPA presents a significant challenge to skull base surgeons, owing to the presence of vital and delicate functional structures within the petrous bone. Cranial nerves, brainstem, and the cerebellum are all tightly packed into this area, along with a dense network of arteries and veins. However, progress in modern imaging techniques and neuronavigation, recent development of novel microsurgical equipment, and surgical optics coupled with implementation of intraoperative neurophysiology monitoring (IOM) have prompted a resurgence of interest in the RS approach as a safe and highly effective intervention.

Moreover, the use of endoscopy in a hybrid fashion has further expanded the visual reach of the neurosurgeon, allowing for better visualization and access to obscured lesions, without the need for complex skullbase exposure.

45.2 History and Background

The term "cerebellopontine angle" (CPA) was coined by Hellenberger and Kock in 1800 and later became known as the "bloody angle" due to the high rate of surgical complications associated with intervention on this region, as noted by Harvey Cushing [1]. The retrosigmoid (RS) approach to the CPA evolved from the unilateral suboccipital approach, which was first credited to Ballance in 1894 and Thomas Annandale in 1895 for the removal of a vestibular schwannoma [2].

The RS approach was successfully performed by Krause using an osteoblastic flap in 1903 [3] and later popularized by Dandy in 1934, which led to a reduction in its associated mortality rate [4, 5]. This approach was subsequently refined by McKenzie and Alexander [6] who introduced the seated position, and eventually perfected by Kunze and Rand in 1957 with the use of the microscope, complemented by the transmeatal approach in 1965 [7, 8].

The presigmoid-retrolabyrinthine approach was first described by H. Silverstein in 1980 [9] and later modified

I. N. Kanaan (✉)
Neurosciences Center, Section of Neurosurgery, King Faisal Specialist Hospital & Research Center Institution, Alfaisal University, College of Medicine, Riyadh, Kingdom of Saudi Arabia

I. N. Kanaan, V. Beneš (eds.), *Neuroanatomy Guidance to Successful Neurosurgical Interventions*,
https://doi.org/10.1007/978-3-031-59838-8_45

and adopted by pioneer skull base surgeons for direct access to CPA tumors, Meniere's disease, and the placement of auditory brainstem implants.

The retrosigmoid (RS) approach can be used either alone or in combination with other established approaches, such as the combined presigmoid petrosal approach described by Hakuba [10] and later refined and popularized by Fukushima, Almefty, and others [11–14]. In 1988, Samii published a tailored version of retrosigmoid suprameatal approach known as the "RIZA approach," which involves limited drilling of the petrous apex and inferior incision of the tentorium [15]. These approaches provide wider exposure to Meckel's cave, the tentorial notch, the petroclival region, the anterolateral aspect of the brain stem, and the basilar artery [11–17]. Recently, the use of modern neuronavigation and the resurgence of hybrid endoscopic techniques have optimized visualization of regional neurovascular structures, including the internal acoustic meatus (IAM), brainstem, and the aforementioned anatomical locations. This strategy facilitates successful interventions for vascular decompression [18–20], microsurgical resection of residual or expanding tumors, and helps achieve better surgical outcomes [21–26]. This innovative approach represents a promising future direction for the management of lesions in the CPA and its surrounding regions.

45.3 Anatomical Highlights (Figs. 45.1 and 45.2)

(a) *External anatomical landmarks (variable):*
  - Asterion: A bony landmark corresponding to the confluent of transverse with sigmoid sinuses.
  - The axis of a line between the root of the zygoma and the inion lies above the lateral sinus.
  - DP (Digastric Point) corresponds to sigmoid sinus, above the posterior digastric belly.

(b) *Internal anatomical landmarks:*
  - Tentorium, the petrosal vein rostral position, and CN 4 is at the tentorium free edge.
  - Meckel's cave at petrous apex, Gasserian ganglion of CN 5.
  - Dorello' canal includes CN 6 & IPS at the most anteromedial portion of the petrous ridge.
  - IAC: Average 8.5 mm in length and about 4 mm in diameter (CN 7 & CN 8).
  - Labyrinthine A. originated from AICA enters IAC inferior anterior.
  - The fundus separates the internal auditory canal from the adjacent cochlea and vestibule.
  - Flocculus next to middle cerebellar peduncle covers the Pontomedullary sulcus (exist of 6 & 7 CN).
  - Jugular foramina (9, 10, 11 nerves, jugular vein).

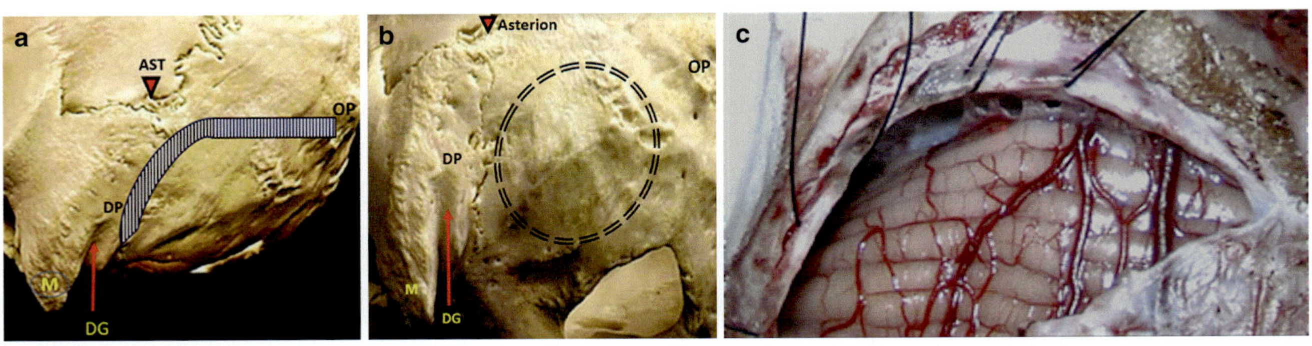

Fig. 45.1 Retrosigmoid approach left-side "antomical highlights" (**a**) Posterior view: red arrow head (asterion), *DP* diagastric protuberance, *OP* occipital protuberance, *M* mastoid tip, Red arrow-DG: digastric groove, double-dotted circle: craniotomy line. (**b**) Lateral view: striped blue lines: lateral and sigmoid sinus markings. (**c**) Y-shape opening of the dura demonstrating the effect of gravity on the cerebellum

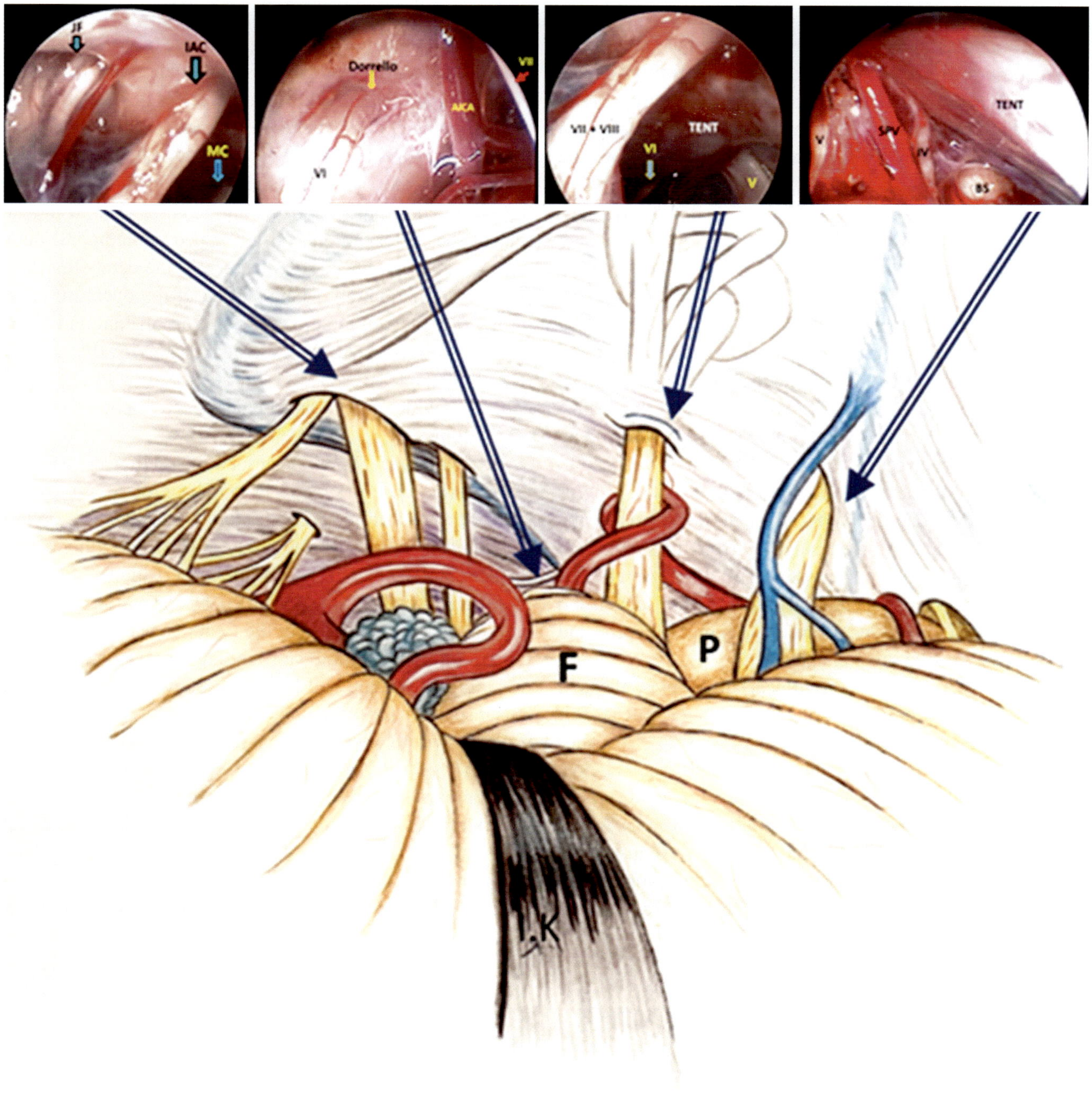

Fig. 45.2 Retrosigmoid approach left-side "antomical highlights" Artistic drawing showing cranial nerves 4th, 5th, 6th, 7th, 8th, and the caudal cranial nerves going through their corresponding skull base channel. *MC* Meckel's cave, *IAC* internal auditory canal, *DC* Dorello's canal, *JF* jugular foramina and the 4th cranial nerve at the tentorial edge, *AICA* anterior inferior cerebellar artery, *F* flocculus, *P* pons. (Note: The top row of the picture includes endoscopic views of anatomical structures seen at the different endoscopic access levels)

45.4 Approach Selection and Indication

The choice of neurosurgical approach to CPA lesions depends on several factors, including the nature of the lesion, its location, size and extent, the presence of intracanalicular growth, baseline hearing function, as well as the patient's and surgeon's preference and experience. Variation in temporal bone anatomy, aeration, Trautmann's triangle measurement, petrous slope, and position of the sigmoid sinus may also affect proper selection [27, 28]. The tailored retrosigmoid approach with the recent hybrid use of endoscopy as an extended visual arm remains the most commonly used avenue for the successful treatment of a wide range of clinical conditions including schwannoma, meningioma, epidermoid, aneurysm, vascular compression, brainstem cavernoma, exophytic or extension of adjacent neoplasm such as chordoma, brainstem or cerebellar tumors. However, each surgical intervention has its indications, merits and limitations, and the neurosurgeon should be acquainted with the different approaches in order to select the proper surgical strategy and achieve the best outcome.

45.5 Retrosigmoid Approach (RS)

Position: For the retrosigmoid approach (RS), the patient is placed in a modified park-bench position, with the upper trunk elevated 25° to facilitate venous outflow, and the head secured in a Mayfield clamp. A 3-axis rotation of the head is performed, similar to the semisitting position popularized by Yaşargil [29], which includes anterior flexion of the head to bring the tent into the field, rotation to the ipsilateral shoulder to adjust the line of vision along the long axis of the petrous, and a side tilt of the head away from the shoulder to provide space for bimanual freedom (Fig. 45.3). An alternative position that is frequently advocated is the semisitting position as popularized by Yaşargil and Samii; the pros and cons are described in other chapters.

Exposure: A standard curvilinear incision is elevated approximately 2–3 cm behind the postauricular sulcus, followed by incision of the soft tissue and exposure of the suboccipital region. A bur-hole created a few millimeters below the asterion, which is a landmark corresponding to the underlying transverse sinus junction with the sigmoid sinus. A second bur-hole drilled 3–4 cm medial to the first one just under the axis line from root of the zygoma to the inion, corresponding to the lateral sinus. The third bur-hole placed in the lower part of the posterior fossa about 1 cm medial to digastric point (DP) as shown in Fig. 45.1. The use of neuronavigation is recommended for safe placement of the craniotomy. The initial dura opening pertains to a small incision over the vicinity of the cisterna magna to allow for controlled egress of cerebrospinal fluid and hence relaxation of the underlying cerebellum. Subsequently, two separate leaflets, a superior and a lateral one, are fashioned to raise to the edge of the transverse and sigmoid sinuses, allowing for clear visualization of the tentorium and posterior petrous temporal dura. The medial dural sheet is left to cradle the cerebellum and prevent out herniation. Manipulation of the operating table during the intervention aids the washout of fluid and blood from the surgical field. With the help of gravity, the cerebellopontine angle (CPA) and infratentorial working space enhance with the least dynamic retraction of the cerebellum.

Intervention: The microsurgical dissection of the cerebellopontine cistern superiorly and lysis of the arachnoid toward the tentorium reveal the tumor capsule. The double arachnoid membrane envelops the tumor, providing a safe dissection plane that allows for the preservation of fine neurovascular structures, as described by Yaşargil. The membrane is sharply opened over the tumor capsule with respect for its integrity toward surrounding vital structures. For resection of medium and large-size Schwannomas, meticulous inspection for the anomalous position of the facial nerve is performed, followed by a small cruciate dorsal incision of the tumor capsule, complemented with progressive internal decompression of the tumor using the Ultrasonic Surgical Aspirator and onion-shell microsurgical techniques under continuous irrigation. The least use of bipolar coagulation for point hemostasis is preferred. A 2–3 mm diamond drill is used to open the internal auditory canal (IAC) and resect small primary or residual tumors in the porus acusticus under

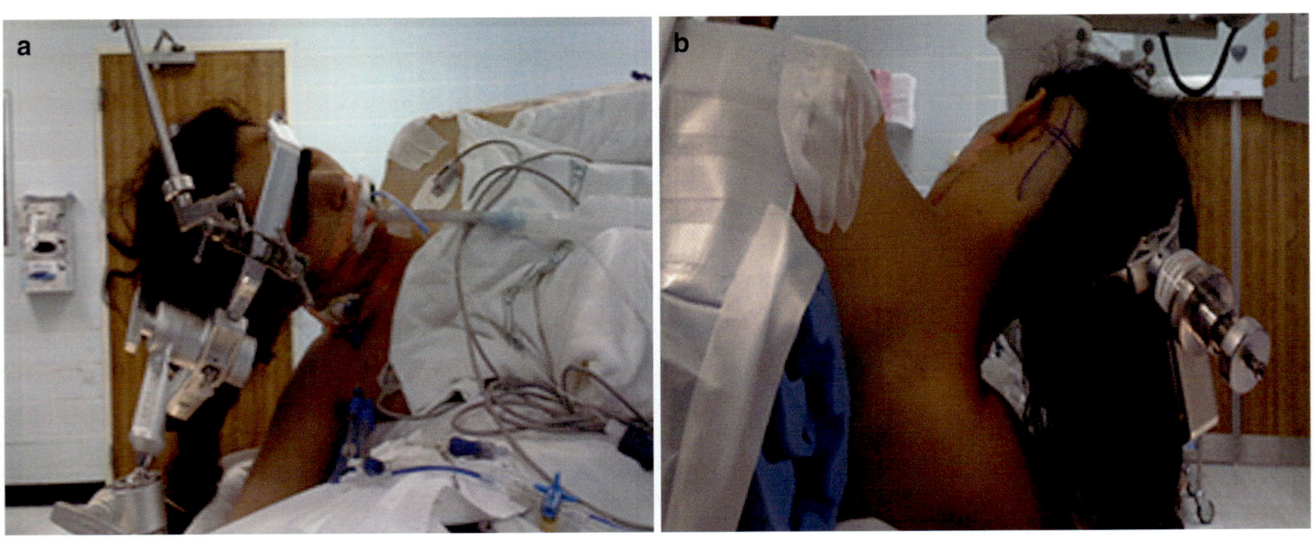

Fig. 45.3 Modified park-bench position by the author, utilizing the advantage of gravity yet minimizing the risk of air embolism

neurophysiology monitoring and endoscopic view (Figs. 45.4 and 45.5). During tumor capsule resection, the stimulation probe and the panoramic view of the endoscopy are also used to identify the facial nerve.

It is advisable to leave severely adherent capsule residual on the facial nerve or brain stem to prevent potential neurovascular injury and catastrophic outcome. For trigeminal schwannomas, epidermoid, and some meningiomas with

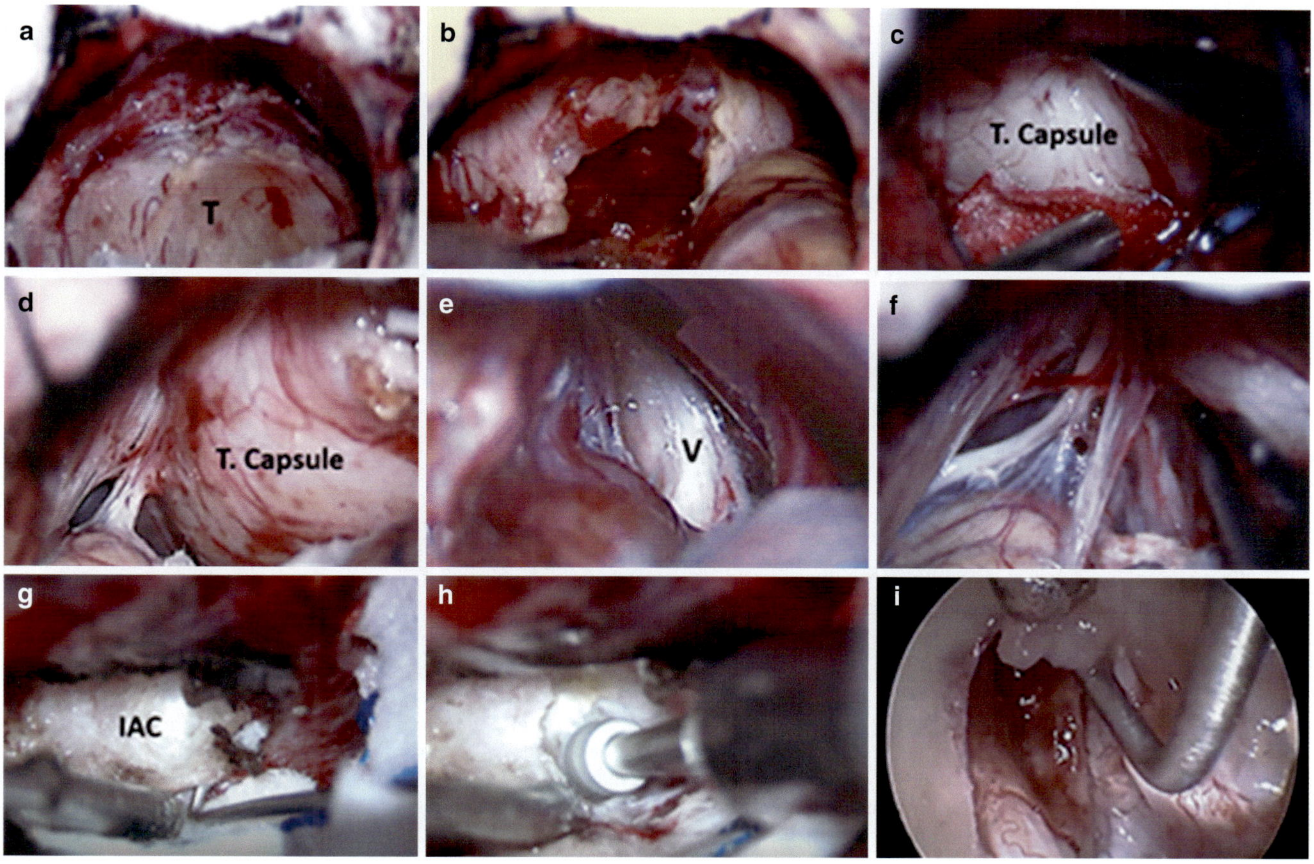

Fig. 45.4 Left retrosigmoid craniotomy for resection of a vestibular schwannoma (**a**) Exposure of the tumor, (**b**) central decompression, (**c–f**) Microsurgical dissection of the tumor capsule while maintaining the integrity of the surrounding cranial nerves using the stimulation probe and the two-forceps technique under copious irrigation, (**g** and **h**) Identifying and drilling the posterior lip of IAC. (**i**) Using the endoscope to retrieve any residual tumor

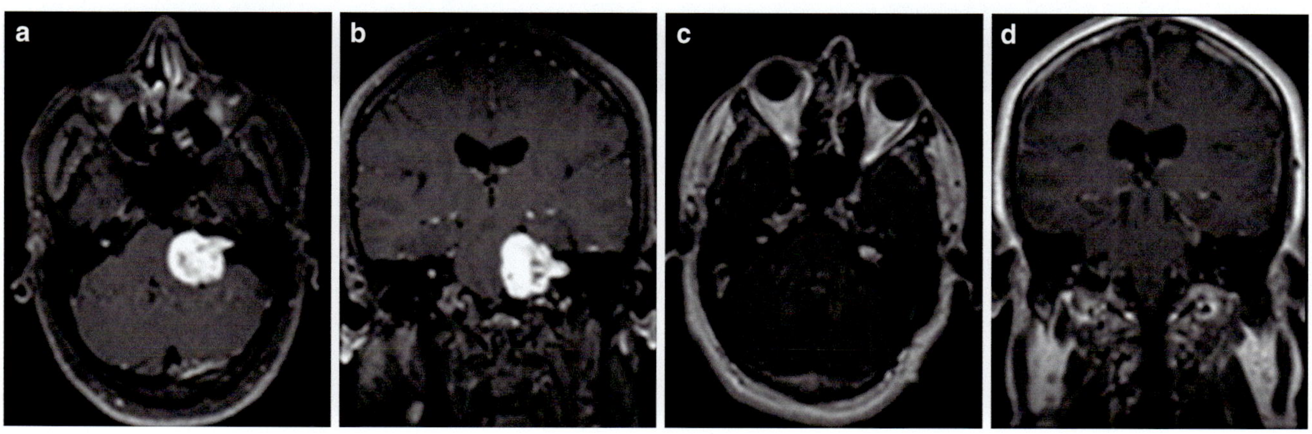

Fig. 45.5 (**a–d**) MRI with contrast of a patient with left CPA vestibular schwannoma in axial and coronal projections pre- and postoperatively confirmed gross resection of the schwannoma

superior and medial extension along the tentorial notch or petroclival region, satisfactory removal of the tumor is achieved using modified suprameatal approach with inferior medial incision of the tentorium, as described by Samii and more recently supported by the hybrid use of endoscopy (see representative cases in Figs. 45.6, 45.7, 45.8, 45.9, 45.10, and 45.11).

Closure: Following resection, primary dura closure or duraplasty using autologous or synthetic dural graft is perfomed to provide a watertight closure. Any exposed bony air cells are tamponaded using bone wax and fat graft with tissue glue to prevent CSF leak. The craniotomy bone flap is fixed using a low-profile titanium-plating system. The procedure is concluded with a multilayer closure of the surgical site

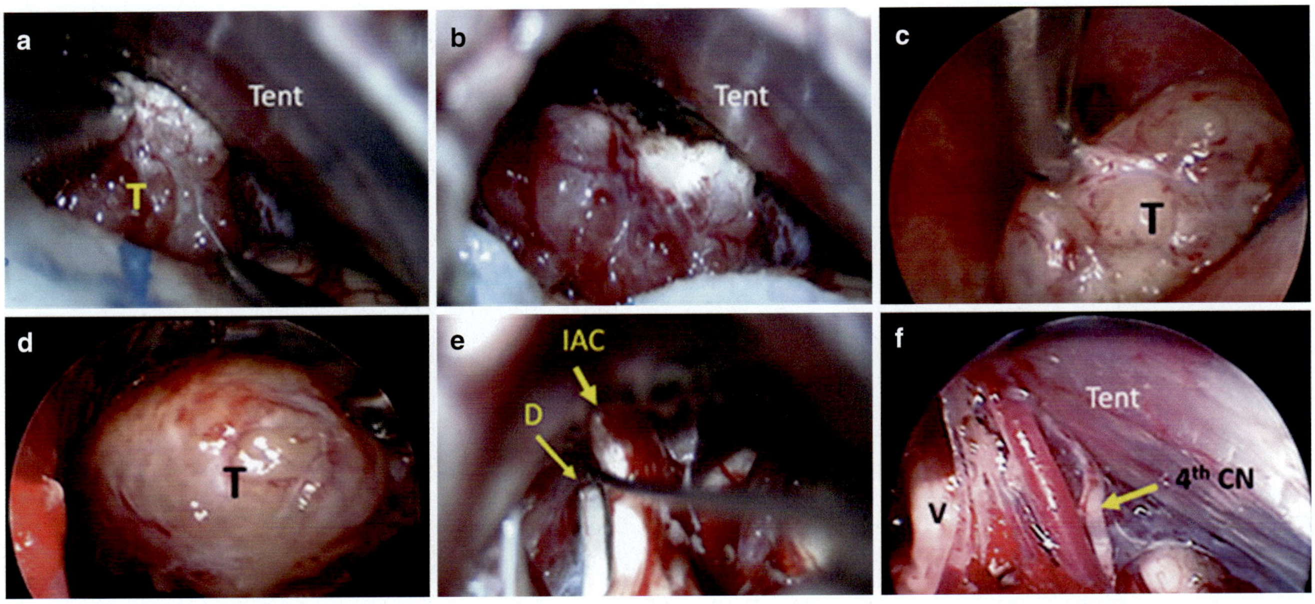

Fig. 45.6 Endoscopic-assisted resection of left CPA meningioma (**a**) Devascularization of tumor at the tentorial attachment. (**b**) Segmental basal resection and central decompression of the meningioma. (**c**) Peeling the arachnoid and dissection of the tumor capsule. (**d**) Gross removal of the meningioma. (**e**) Showing preservation of regional cranial nerves 6, 7, 8 postresection (Thin yellow arrow points to Dorello-canal, short thick arrow to IAC). (**f**) Endoscopic view of trigeminal and 4th CNs at the tentorial edge

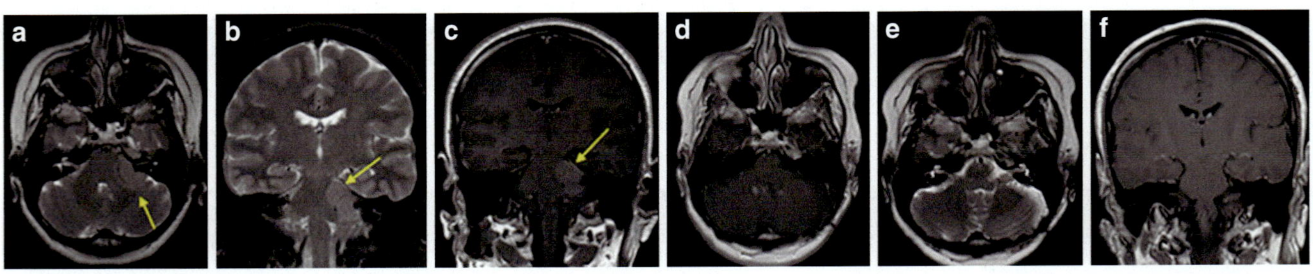

Fig. 45.7 MRI studies of the patient with left CPA Meningioma. (**a–c**) Preoperative MRI depicting left posterior petrosal meningioma with infratentorial superior extension. (**d–f**) Postoperative MRI confirmed gross resection of the meningioma

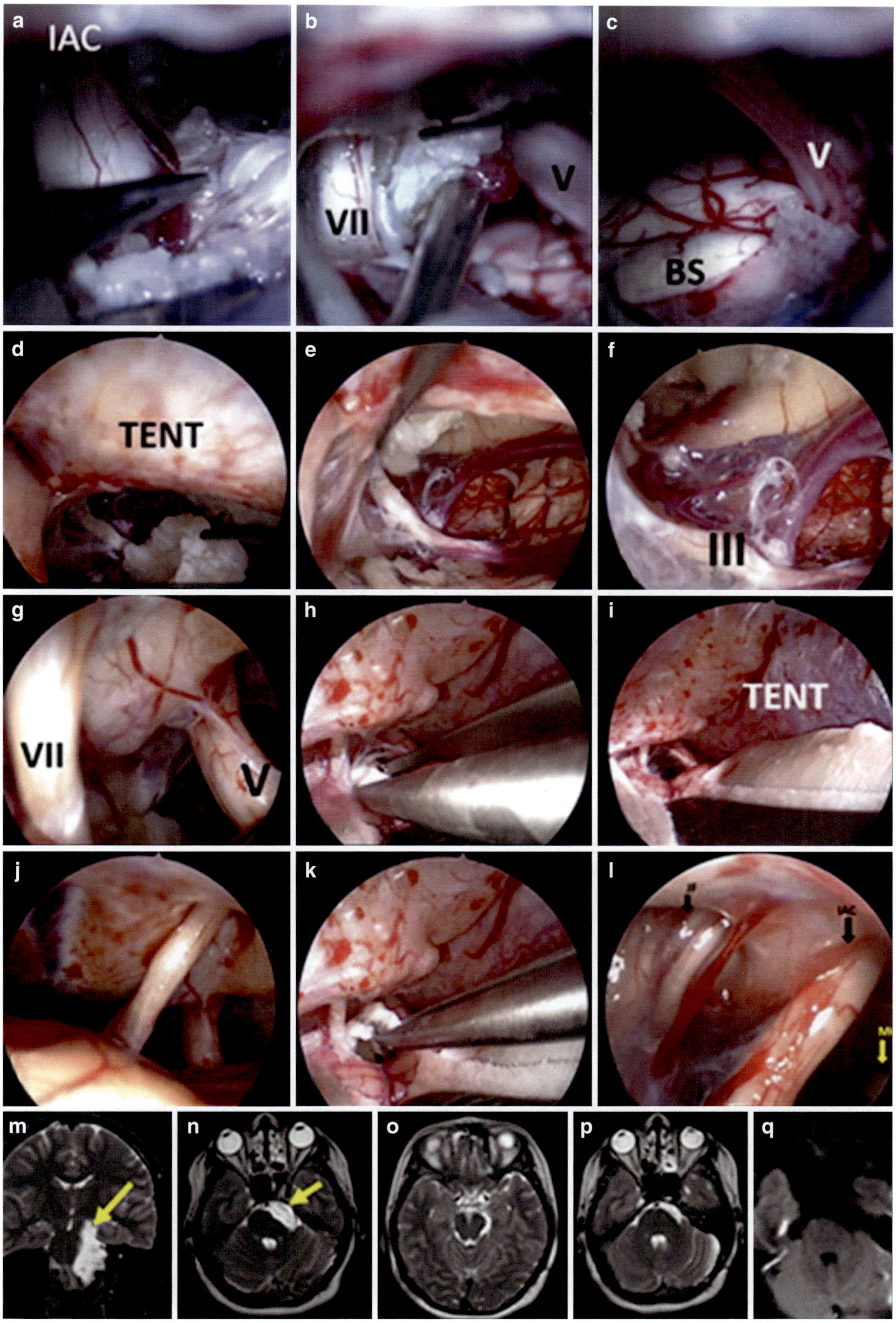

Fig. 45.8 Hybrid microsurgical resection of CPA epidermoid with superior extension. (**a–c**) Microscopic removal of the epidermoid between the 5th, 7th, and 8th CNs. (**d–f**) Endoscopic resection of the superior extension of epidermoid and liberating 3rd CN. (**g–l**) Progressive endoscopic gross removal of the epidermoid while maintaining the integrity of the CNs. (**m–q**) MRI preoperative and postoperative images in T2 and DWI confirming the gross tumor resection

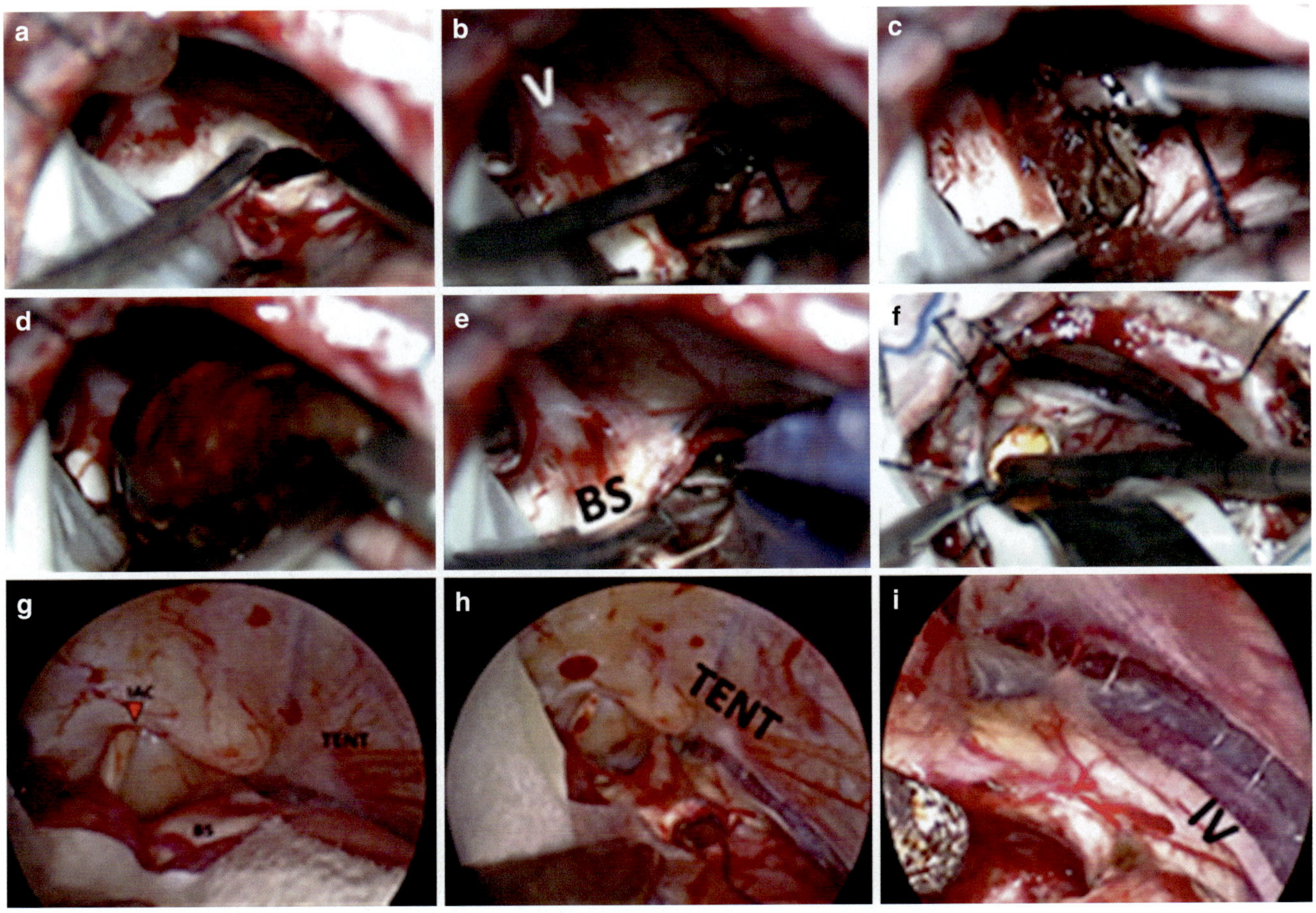

Fig. 45.9 Left retrosigmoid approach for resection of brainstem cavernoma. (**a**) Incision of the thin discolored spot of the brainstem just above the trigeminal nerve exit (**b**) Exposure & circumferential dissection of the cavernoma using the stimulator probe (**c**) Mobilization of the lesion (**d**) Total inclusive excision of the cavernoma (**e**) Meticulous hemostasis (**f**) Introduction of the endoscope (**g–i**) Endoscopic visualization of the CPA, showing 4th, 5th, 7th, 8th CNs, and inspection of the resection cavity for any residual lesion

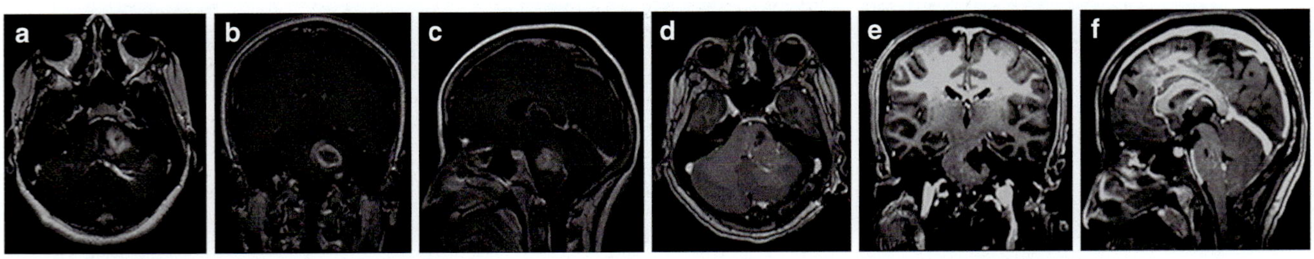

Fig. 45.10 MRI study of the patient with a left brainstem cavernoma. (**a–c**) Preoperative MRI depicting left pontine cavernoma with surrounded hematoma. (**d–f**) Postoperative MRI confirming complete resection of the cavernoma

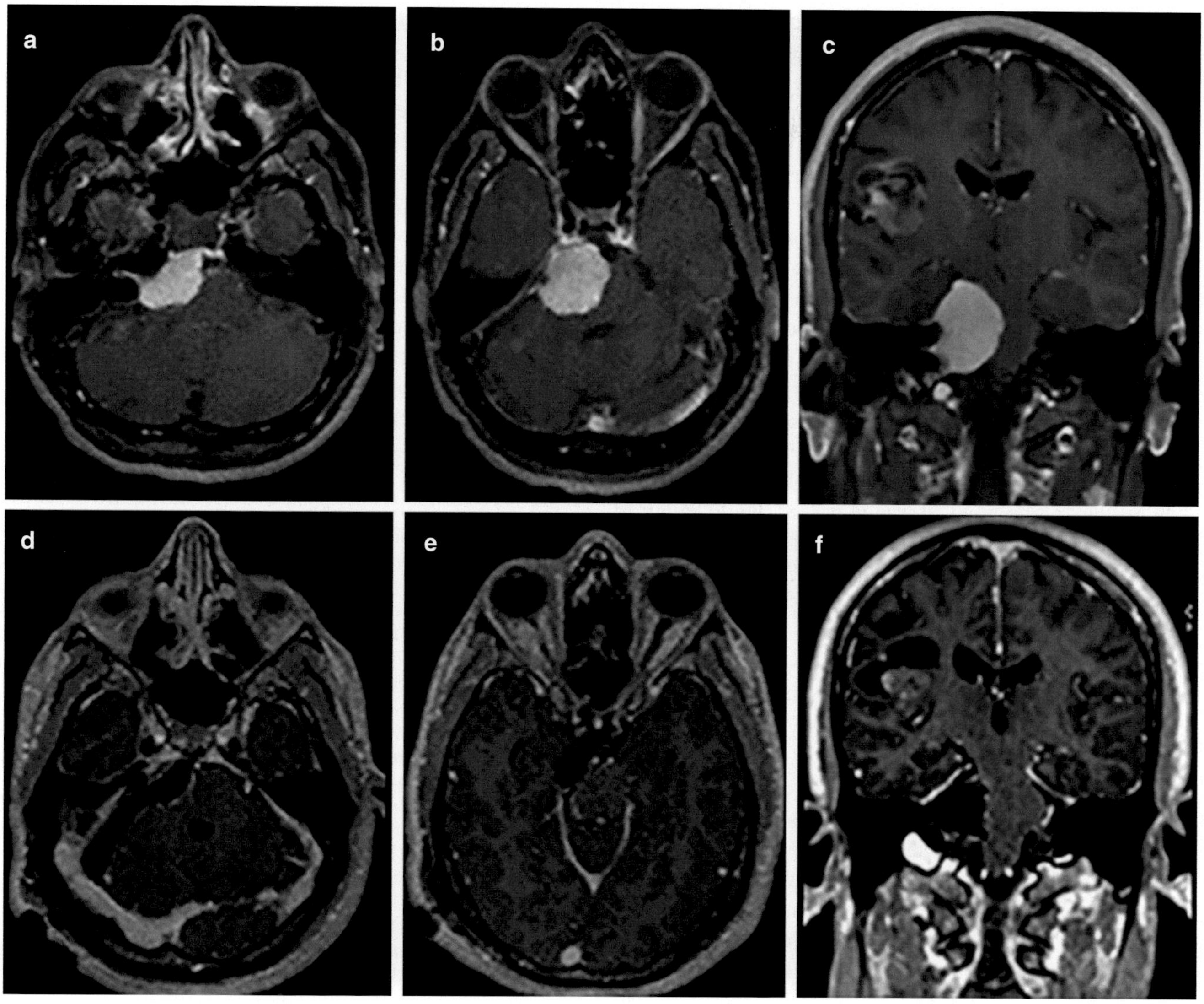

Fig. 45.11 MRI right petroclival meningioma with superior extension into Meckel's cave Microsurgical total resection via retrosigmoid suprameatal approach facilitated by the hybrid use of endoscopy, paying attention to the prominent venous sinuses on the right side. (**a–c**) Preoperative MRI-T1 with contrast depicting a right petroclival meningioma. (**d–f**) Postoperative MRI-T1 with contrast confirmed total resection of the tumor

45.6 Special Considerations

In patients with large tumors and risk of associated hydrocephalus, a temporary simple CSF diversion procedure may be performed to reduce increased intracranial pressure, minimize cerebellar retraction, and prevent resultant acute hydrocephalus.

During the surgical intervention, dural sinus thrombosis and venous injury are of concern. Special attention must be paid to protect the dural sinuses and their patency during exposure, limit the number of cerebellar bridging veins sacrificed to minimum, and maintain the integrity of the petrosal vein and venous network over the brainstem.

45.7 Surgical Pearls

- Early release of CSF from the basal cisterns or lumbar drain relaxes the cerebellum.
- The petrosal vein (Dandy's vein) coursing toward its junction with the superior petrosal sinus (SPS) at the rostral aspect of the cerebellopontine angle (CPA) should be respected.
- During tumor resection, the concept of "D" should be applied, which refers to **d**evascularization, **d**ecompression, and **d**issection (peeling), the arachnoid off the tumor capsule.

- Tumor removal should be performed under copious irrigation with minimal use of thermocoagulation and minimal or dynamic retraction.
- Intraoperative neurophysiological monitoring (IOM) should be employed to ensure the integrity of cranial nerves and the brain stem during tumor resection or management of vascular lesions.
- Sites to trace the facial nerve are: IAC, at the pontomedullary sulcus under the flocculus, and on the surface of the tumor capsule by using the stimulation probe.
- Hybrid use of endoscopy can be considered to trace tumor extension into the internal acoustic meatus (IAM) or along the tentorial notch.
- A stimulating probe can be used as a dissector during chopstick resection of epidermoid.
- Small residual tumors that are severely adherent to the brainstem, perforators, or cranial nerves should be left in place to prevent complications.
- Suspected mastoid air cells inadvertently exposed should be obliterated with bone wax or a piece of muscle & complemented with watertight closure.

45.8 Discussion

The retrosigmoid approach is a safe and commonly utilized multichannel surgical approach for managing a variety of pathologies affecting the cerebellopontine angle (CPA). To ensure successful management, familiarity with relevant radiological images, knowledge of anatomical landmarks, and mastery of the different surgical approaches are necessary prerequisites. Hakuba's innovative approach, which combines both retrosigmoid and presigmoid avenues, has undergone several recent modifications to reduce potential pitfalls and provide greater anterolateral access with minimal cerebellar retraction [30]. These features make it easier to perform microsurgical intervention for lesions situated within the petroclival region, the anterolateral aspect of the brain stem, the tentorial hiatus, or Meckel's cave. However, this approach still involves significant complex bone dissection, requires more time, longer hospital stays, and is more costly. The recent introduction of modern neuronavigation and endoscopic-assisted microsurgery, combined with the promotion of adjuvant treatment options, has prompted a shift toward simpler retrosigmoid approaches to achieve comparable success rates for treating residual lesions or trans-compartment tumor extensions, excluding giant petroclival meningiomas or extensive chordomas. This strategy may negate the need for such complex skull base bone exposure. Many scientific publications support the greater role of endoscopic-assisted approaches in inspecting displaced and hidden position of cranial nerves, retrieving residual tumors from hidden sites, and identifying offending vessels during microvascular decompression surgery.

45.9 Conclusion

The retrosigmoid (RS) approach has become a cornerstone of surgical management for CPA pathologies. Its use has evolved with advances in technology and surgical techniques, including the suprameatal route with inferior tentorial incision or hybrid endoscopic approach, which has allowed for successful microsurgical management of a variety of neoplastic or vascular lesions with minimal need for complex bone exposure. Emerging adjuvant treatments in interventional neuroradiology, stereotactic radiation, and medical oncology can provide numerous benefits and should be considered as part of a comprehensive management plan.

References

1. Cushing H. Tumours of the nervous acusticus and the syndrome of the cerebellopontine angle. Philadelphia: W.B. Saunders; 1917. 296 p.
2. Ballance C. Some points in the surgery of the brain and its membranes. London: Macmillan and Co; 1907. https://doi.org/10.1097/00007611-190803000-00007.
3. Krause F. Zur Freilegung der hinteren felsenbeingflache und des kleinhirns. Beitrage Klin Chir. 1903;37:728–64.
4. Dandy WE. Removal of cerebellopontile (acoustic) tumors through a unilateral approach. Arch Surg. 1934;29(3):337–44. https://doi.org/10.1001/archsurg.1934.01180030002001.
5. Dandy WE. Results of removal of acoustic tumors by the unilateral approach. Arch Surg. 1941;42(6):1026–33. https://doi.org/10.1001/archsurg.1941.01210120061006.
6. McKenzie KG, Alexander E Jr. Acoustic neuroma. Clin Neurosurg. 1954;2:21–36.
7. Rand RW, Kurze T. Micro-neurosurgical resection of acoustic tumors by a transmeatal posterior fossa approach. Bull Los Angel Neurol Soc. 1965;30:17–20.
8. Rand RW, Kurze T. Preservation of vestibular, cochlear, and facial nerves during microsurgical removal of acoustic tumors: report of two cases. J Neurosurg. 1968;28(2):158–61. https://doi.org/10.3171/jns.1968.28.2.0158.
9. Silverstein H, Norrell H. Retrolabyrinthine surgery: a direct approach to the cerebellopontine angle. Otolaryngol Head Neck Surg. 1980;88(4):462–9. https://doi.org/10.1177/019459988008800425.
10. Hakuba A, Nishimura S, Jang BJ. A combined retroauricular and preauricular transpetrosal-transtentorial approach to clivus meningiomas. Surg Neurol. 1988;30(2):108–16. https://doi.org/10.1016/0090-3019(88)90095-X.
11. Fukushima T. Combined supra- and infra-parapetrosal approach for petroclival lesions. In: Sekhar LN, Janecka IP, editors. Surgery of cranial base tumors. New York: Raven Press; 1993. p. 661–9.
12. Al-Mefty O, Fox JL Sr, Smith RR. Petrosal approach for petroclival meningiomas. Neurosurgery. 1988;22(3):510–7.
13. Samii M, Ammirati M. The combined supra-infratentorial presigmoid sinus avenue to the petro-clival region. Surgical technique and clinical applications. Acta Neurochir. 1988;95(1):6–12. https://doi.org/10.1007/BF01793075.
14. Spetzler RF, Daspit CP, Pappas CTE. The combined supra- and infratentorial approach for lesions of the petrous and clival regions; experience with 46 cases. J Neurosurg. 1992;76(4):588–99. https://doi.org/10.3171/jns.1992.76.4.0588.
15. Samii M, Tatagiba M, Carvalho GA. Retrosigmoid intradural suprameatal approach to Meckels cave and the middle fossa: surgical

technique and outcome. J Neurosurg. 2000;92(2):235–41. https://doi.org/10.3171/jns.2000.92.2.0235.
16. Ambekar S, Amene C, Sonig A, Guthikonda B, Nanda A. Quantitative comparison of retrosigmoid intradural suprameatal approach and retrosigmoid transtentorial approach: implications for tumors in the petroclival region. J Neurol Surg B Skull Base. 2013;74(5):300–4. https://doi.org/10.1055/s-0033-1348025.
17. Lynch JC, Pereira C, Welling L, Gonçalves M, Zanon N. Extended retrosigmoid approach for cerebellopontine angle meningiomas: operative technique and results-a series of 28 patients. J Neurol Surg B Skull Base. 2018;79(5):458–65. https://doi.org/10.1055/s-0037-1620278.
18. Magnan JGH. Atlas of neurosurgical techniques. In: Sekhar LN, editor. Endoscope-assisted microvascular decompression. Stuttgart: Thieme; 2006.
19. Teo C, Nakaji P, Mobbs RJ. Endoscope-assisted microvascular decompression for trigeminal neuralgia: technical case report. Neurosurgery. 2006;59(suppl_4):ONS-E489–90.
20. Jarrahy R, Eby JB, Cha ST, Shahinian HK. Fully endoscopic vascular decompression of the trigeminal nerve. Minim Invasive Neurosurg. 2002;45(1):32–5. https://doi.org/10.1055/s-2002-23586.
21. de Divitiis O, Cavallo LM, Dal Fabbro M, Elefante A, Cappabianca P. Freehand dynamic endoscopic resection of an epidermoid tumor of the cerebellopontine angle: technical case report. Operative. Neurosurgery. 2007;61(5):E239–40. https://doi.org/10.1227/01.neu.0000303222.40145.1a.
22. Shahinain HK, Eby JB, Ocon M. Fully endoscopic excision of vestibular schwannomas. Minim Invasive Neurosurg. 2004;47(6):329–32. https://doi.org/10.1055/s-2004-830122.
23. Vaz-Guimaraes F, Gardner PA, Fernandez-Miranda JC. Fully endoscopic retrosigmoid approach for posterior petrous meningioma and trigeminal microvascular decompression. Acta Neurochir. 2015;157(4):611–5. https://doi.org/10.1007/s00701-014-2332-1.
24. Tatagiba M, Matthies C, Samii M. Microendoscopy of the internal auditory canal in vestibular schwannoma surgery. Neurosurgery. 1996;38(4):737–40.
25. Schroeder HW, Oertel J, Gaab MR. Endoscope-assisted microsurgical resection of epidermoid tumors of the cerebellopontine angle. J Neurosurg. 2004;101(2):227–32. https://doi.org/10.3171/jns.2004.101.2.0227.
26. Kanaan IN. Hybrid use of endoscopy in skull base surgery. In: Deopujari C, Quereshi M, editors. WFNS guide for neuroendoscopy. Noida: Salubris Medical Publishers; 2022. p. 155–62. www.Sularis.in
27. Erkmen K, Pravdenkova S, Al-Mefty O. Surgical management of petroclival meningiomas: factors determining the choice of approach. Neurosurg Focus. 2005;19(2):1–12. https://doi.org/10.3171/foc.2005.19.2.8.
28. Wong AK, Stamates MM, Bhansali AP, Shinners M, Wong RH. Radiographic assessment of the presigmoid retrolabyrinthine approach. Surg Neurol Int. 2017;8(1):129. https://doi.org/10.4103/sni.sni_243_16.
29. Yaşargil M. Acoustic neuromas. In: Yaşargil MG, Abernathey CD, editors. Microneurosurgery vol. IV B: microsurgery of CNS tumors. New York: Thieme; 1996.
30. Fava A, di Russo P, Passeri T, Camara B, Paglia F, Matano F, Okano A, Giammattei L, Froelich S. The mini-combined transpetrosal approach: an anatomical study and comparison with the combined transpetrosal approach. Acta Neurochir. 2022;164(4):1079–93. https://doi.org/10.1007/s00701-022-05124-x.

46 Neuroanatomy for Neurobionic Hearing Restoration

Amir Samii and Mario Giordano

Learning Objectives
- Anatomy and physiology of the auditory system
- Overview of the modern technology for hearing restoration
- Surgical approaches and their variants to different treatment target structures along the auditory pathway

46.1 Introduction

The sensory system plays a crucial role in receiving and processing information enabling individuals to be aware of their surrounding space. Among the special sense organs, the auditory system is responsible for perceiving and comprehending sounds in our environment. It comprises various peripheral structures, including the outer, middle, and inner ear, as well as central components within the brain such as the cochlear nuclei, superior olivary nuclei, inferior colliculus, medial geniculate nuclei, and auditory cortex. These are connected by multiple parallel and overlapping pathways with interrelated functions working together to process and interpret auditory stimuli [1].

Within the auditory brain circuits, several important features of sounds are encoded, including frequency, attenuation, and spatial location. The processing of auditory information is not fixed but rather undergoes continuous changes facilitated by descending feedback circuits. These changes are depending on attentional focus and the perceived significance of specific environmental cues.

Neural prostheses are the focus of biomedical scientists since the end of last century. The aim of these devices is to restore lost neural functions as a result of congenital or acquired disease. The first results for the restoration of auditory function were encouraging [2] and pushed the researcher to continue the development of progressively more complex devices many of which are currently used in the medical field for earing restoration in a large number of patients all over the world.

In this chapter, we present an overview on the anatomical basis of hearing restoration describing the actually available technology and methods of implant.

46.2 Anatomy Highlights of the Auditory System

- The cochlea is the first important target for hearing restoration.
- The round window of the cochlea is the point of electrode insertion for the cochlear implant (CI).
- Choroid plexus and the foramen of Luschka are related to the cochlear nuclei: these are used as landmarks for the positioning of auditory brainstem implants (ABI).
- The inferior colliculus (IC) is a bilateral, symmetrical structure of the midbrain. It is the target of a novel implant for hearing restoration: the auditory midbrain implant (AMI).

The process of hearing involves the complex transformation of sound vibrations from the external environment into electrical signals. By processing the one-dimensional stimuli received by each ear, the auditory system ingeniously constructs the so-called auditory objects. Thus, it performs the impressive task of both the analysis and synthesis of sound [1].

Vibrating objects produce energy waves that cause variations of the air pressure in the surrounding environment. This pressure variation in the time unit is defined as frequency of the sound measured in hertz (Hz): the hearing range for human ear is 20 to 20,000 Hz. Another important characteristic of the sound is its intensity that may be defined as the sound power per unit area and is usually measured using the

A. Samii (✉) · M. Giordano
Department of Neurosurgery, INI – International Neuroscience Institute, Hannover, Germany
e-mail: a.samii@ini-hannover.de

I. N. Kanaan, V. Beneš (eds.), *Neuroanatomy Guidance to Successful Neurosurgical Interventions*,
https://doi.org/10.1007/978-3-031-59838-8_46

decibel scale (dB): the healthy human ear has a range from 0 to 130 dB. It is important to notice that the intensity of the sound precepted by the ear differs in dependence of the frequency of the sound wave. It is clear that all these physical features of the sound have to undergo to a conversion process in order to create the acoustic sensation in our organism. The ear organ is deputed to this transformation and it is classically divided into external, middle, and inner ear.

The external ear is simply composed by the pinna and the external auditory meatus, a 2.5 cm canal that enters the temporal bone ending at the tympanic membrane. At this level, the first transformation of the stimulus happens with conversion of air vibrations into tympanic membrane vibrations transmitted into the middle ear.

The middle ear is a cavity of the temporal bone filled with air that enters and exits via the eustachian tube that reaches the pharynx allowing compensation of air pressure on both sides of the membrane. This cavity contains the 3 ossicles: malleus, incus, and stapes which amplify the vibration transferring it to the oval window and consequently to the cochlea. After the tympanic membrane, the middle ear is another important structure to amplify the transmission: it has been calculated that the input impedance at the cochlea is 200-fold greater than at tympanic membrane [3].

46.3 Cochlea and Round Window

The cochlea is the organ of hearing and its name derives from the Greek word shell of snail due to its particular form. It was described by Eustachi in 1564 but only in 1761 the Neapolitan anatomist Cotugno are described in detail its anatomy using corrosion casts in "*De aqueductibus Auris Humana Interna*" [4]. Its morphology resembles a fluid-filled spiral tube composed by 2.5–3 turns [5] of 1 × 0.5 cm with a length of 3.2 to 4.2 cm. The bone overlying on its basal turn protrudes into the middle ear cavity and is called Promontory an important structure for functional auditory tests and as landmark for the cochlear implant. The promontory separates the oval and the round window that is situated posteroinferiorly and is covered by the round window membrane. It has a key role as access point during cochlear implant surgery. In fact, the cochlea is the first important target for the hearing restoration implants that will be described in this chapter. The tubular structure of the cochlea has 3 parallel scale: the scala tympani, scala vestibule, and scala media. In the scala media, the basilar membrane contains the organ of Corti that encloses the mechanical receptor cells constituted by 3 rows of outer hair cells and one row of inner hair cells. The cochlea, thanks to the organ of Corti (first described by the Italian anatomist Alfonso Corti in 1851) that act as a regenerative mechanical amplifier, plays the important role of detection and analysis of the sound. The pressure changes at the oval window initiates waves through the scala tympani and scala vestibule of the cochlea. These waves exert pressure on the basilar membrane, which responds differently depending on the frequency of the stimulus with a mapping inside the cochlea itself. Higher frequencies cause the basilar membrane to peak closer to the base and vice versa lower frequencies closer to the apex. When the hair cells are stimulated, changes in electric potential occur within the outer hair cells, leading to a mechanical response that alters their length. Thus, the signal is perceived by hair cells which receive two different types of innervation: afferent innervation from spiral ganglion neurons and efferent innervation from the superior olivary complex. Inner hair cells have a more direct function representing the encoder of the auditory signal, and each inner hair cell is connected by multiple type I spiral ganglion neurons that represent the majority. The outer hair cells, on the other hand, have a much smaller afferent innervation connecting with type II spiral ganglion neurons. This last kind of interaction has been seen to be fundamental in the modulation of cochlear sensitivity driving the olivocochlear efferent reflex [6] and in nociception after noise-induced damage [7]. The outer cells have a mechanical role in the reverse transduction: changing the length of their cell body with consequent variation of the stiffness of the basilar membrane can modify the signal transmitted to the sensory inner hair cells. In fact, when they are in an elongated state, stimulation of inner hair cells is largely inhibited: it has been proposed that this mechanism is at the base of our ability to selectively attend auditory signals in noisy environments [8]. The output of the cochlea is achieved by the auditory nerve fibers. The spiral ganglion neurons are bipolar cell and their body is located in the modiolus of the cochlea. Their distal process passes through the bony lamina and the so-called habenula perforata innervating the inner hair cells and, in small percentage (type II), the outer hair cells [9]. The central processes of the spiral ganglion neurons form the auditory nerve entering into the brainstem. There is a frequency selectivity of individual fibers: the innervation at the base of the cochlea is tuned to high frequencies, while at the apex is tuned to low frequencies [1].

46.4 Cochlear Nuclei

The central auditory system has a high grade of complexity due to its multiple parallel and partially interrelating connections. This sophisticate system has not only to provide the extraction of the stimulus but also to deal with its fast temporal fluctuations. The subcortical pathways have been the target of growing interest due to the fact that they have proven to be the site of several auditory processing functions previously attributed to the cortex [10]. They include the cochlear nucleus, superior olivary nuclei, lateral lemniscus, inferior colliculus, and medial geniculate nuclei. The central auditory system at the level of the brainstem can be then schematized in an anterior stream mainly responsible for sound localization and a posterior stream involved in the complex stimulus analysis [1]. The anterior pathway is started by the anterior branch of the auditory nerve that innervate the anterior cochlear nucleus while the posterior branch terminates in the dorsal cochlear nucleus (DCN) and (PVCN) creating the posterior pathway. It is very important to notice that the nerve fibers maintain the configuration of their origin in the cochlea creating a tonotopical organization in the cochlear nuclei with separate frequencies [11]. It should be also emphasized, due to the topic of this chapter, that the topographical anatomy and localization of these nuclei are of pivotal importance because this anatomical region is the target of auditory brainstem implants. It has been shown that structures on the surface of the medulla oblongata such as choroid plexus, the angle between the vestibulocochlear and glossopharyngeal nerves, and the foramen of Luschka are related to the cochlear nuclei [12]. Such structures are considered landmarks for the morphological guidance to the surgeon in the positioning of auditory brainstem implants.

In the ventral auditory stream, the anteroventral cochlear nucleus (AVCN) project fibers to the superior olivary complex where there is a localization of the sound comparing the inputs from both ears. The two populations of neurons contained in the AVCN project in different areas [10]. The spherical cells connect to the medial superior olivary nuclei (MSO) bilaterally and to the ipsilateral lateral superior olive (LSO) while the globular bushy cells project to the contralateral medial nucleus of the trapezoid body participating in the comparison of the stimuli at the two ears [1, 13]. The MSO and LSO project to the inferior colliculus (IC) respectively to the ipsilateral and contralateral side through the lateral lemniscus encountering its two intrinsic nuclei: dorsal and ventral nuclei of the lateral lemniscus that enhance the accuracy in sound localization [14].

The dorsal auditory pathway has an important role in the complex stimulus analysis. The DCN contributes to the "non-lemniscal" system that is involved in the multimodal integration and reflexes. On the other hand, the PVCN cells project mainly to the ventral nucleus of the lateral lemniscus on the opposite side participating in the extraction of the temporal fluctuations of the stimuli due to the accuracy of the response [15].

46.5 Inferior Colliculus

The anterior and posterior auditory pathways of the brainstem converge to the IC. This is another fundamental anatomical structure, not only for its physiological importance but also as target of a novel implant for the hearing restoration: the auditory midbrain implant (AMI).

The IC is a bilateral, symmetrical structure of the midbrain. The two ICs are situated caudal to the superior colliculi and together with them form the eminences of the quadrigeminal body, part of the tectal region. Caudal to the IC emerges the trochlear nerve, another important anatomical landmark of this region. The IC can be schematically divided into three areas: the central nucleus, the dorsal cortex, and the external nucleus [16]. The central nucleus is the largest constituent of the IC: it receives afferences from the lemniscal pathway and is tonotopically organized due to its laminar anatomic structure [17]. This organization pattern is particularly important as target in the electrode insertion of AMI. The dorsal cortex, on the other hand, does not receive inputs from the olivary system but projections from the contralateral IC and from the auditory cortex: it may have a role in the adaptation to new sounds [18]. The external nucleus has been hypothesized to have a more integrative function with multisensory interactions due to its connection also with dorsal columns and trigeminal nuclei.

Thus, the IC not only enables the localization of sound by integrating input from both ears but also integrates inputs from other sensory modalities. Some cells in the IC respond to visual stimuli in the opposite visual field due to connections with the superior colliculi aiding in mapping physical space using both auditory and visual information. The somatosensory inputs from the spinal cord and cranial nerves contribute to the body's ability to orient itself toward the source of a sound [19].

The medial geniculate body (MGB) is the auditory centrum of the thalamus and a fundamental multifunctional relay of the auditory system that receives stimuli from the IC of both sides and project to the auditory cortex. It is located at medial and caudal to the lateral geniculate nucleus and classically divided ventral, dorsal, and medial nuclei [20].

The dorsal and medial nuclei receive projection mainly from the dorsal cortex and external nucleus of the IC: they are part of the extra-lemniscal pathway involved in the integrative role and complex somatosensory integration of the auditory system [1]. On the contrary, the ventral nucleus is part of the specific (lemniscal) pathway. It projects to the primary auditory cortex with the adjacent cortical fields and has a tonotopic organization resembling the central nucleus of the IC: efferent fibers to the cortex maintain the frequency-specific and temporal properties gathered in the ascending pathway of the brainstem.

The auditory cortex is the final target of the ascendant auditory system. The knowledge about this anatomical region has been acquired mainly from the research studies in the cat [21]. In humans, it has been more difficult to identify this area in the brain: it can be located in the temporal lobe, within the upper part of the superior temporal gyrus (Brodmann area 41). It has to be considered that the primary auditory cortex (AI), the first cortical area that receives and processes auditory inputs with from the MGB, is surrounded by the so-called "belt" region that has connections mainly with dorsal and medial nuclei of MGB. Neurons in AI are tonotopically organized, meaning that different frequencies are represented in an ordered manner across the cortical surface. Functionally, the AI has tasks of processing the basic auditory features such as frequency, intensity, and timing. The belt region, due to its different connections and structure, is responsible of integrating data from multiple auditory sources and contributes to higher-level processing, for example sound localization or extrapolation of important auditory information. Parabelt regions, that surround the belt area, are also interconnected with the belt and frontal lobe areas [22] thus are supposed to have the function of auditory information association with other sensory inputs.

For the complete overview about the normal functional anatomy of the auditory system, it is important to mention the centrifugal pathways. This system starts from the cortex and reaches the IC through the MCG. It is a sort of control mechanism of the cortex itself on the response to the different inputs at the level of the brainstem. The centrifugal pathway continues to the olivary complex and with its last tract: the so-called olivocochlear system. It has generated great interests due to its inhibitory function on the cochlear response to different stimuli [23].

46.6 Approaches and Surgical Technique of Hearing Restoration

The complexity of the auditory system with its parallel and overlapping pathways can appear overwhelming and difficult to restore artificially. Of course, replicate the complex spatial and temporal pattern of neural firing of this structure is a challenging task that has been approached in a systematic and segmental fashion by contemporary medical technology. Modern devices with multiple electrodes are able to produce high levels of speech comprehension and are successfully implanted in hundred thousand of patients. In this chapter, we focus on three types of devices and their surgical implant: cochlear implant (CI), auditory brainstem implant (ABI), and auditory midbrain implant (AMI).

46.6.1 Cochlear Implant

Cochlear implants are electronic devices used as standard therapy to restore hearing function in individuals with sensory deafness. The implant consists of an electrode array that is positioned within the cochlea, allowing for selective stimulation of the auditory nerves and transmission of distinct pitch perceptions. By mimicking the frequency organization of the cochlea, CIs can generate a diverse pattern of neuronal stimulation in the auditory nerve in response to complex sound signals. This pattern forms the basis for speech understanding with the implant [24]. These devices replace the function of the inner hair cells in the organ of Corti (see anatomy highlights section). Thus, a technical simulation of the natural hearing process is performed with tonotopic presentation of the frequencies along the basilar membrane. Of course, this mechanism cannot be compared with the complexity of the physiological hearing, in fact only a limited number of electrically separated channels is available for signal transmission.

Indications for CI have changed during the years mainly due to the evolution of the devices and progressive successful results. The main candidates are patients with bilateral high-grade hearing impairment or sensory hearing loss approaching deafness, unilateral sensory deafness, and high-frequency hearing loss with residual hearing in the low frequencies [24]. In patients with residual hearing, it is crucial to finely tune the cochlear implant to complement the existing acoustic function, and by achieving this synergy between the two modalities, significant improvements in hearing outcomes can be achieved. While relying solely on residual acoustic hearing may be inadequate for clear speech understanding, when it is combined with a cochlear implant, the results are exceptionally positive [25].

In children, the age of implant has been reduced during last decade to take advantage of the plasticity of the developing brain and the surgery is indicated when objective measurements fail to detect any reliable stimulus responses and alternative therapeutic options do not offer the potential for normal hearing or language development. Typically, the procedure can be done after the sixth month of life within the first year, and in cases of bilateral deafness, simultaneous bilateral implantation is preferred [26].

Preoperative diagnostic includes functional tests and imaging studies. Among the functional tests, the promontory test is pivotal to verify the function of hearing nerve: an electrode is inserted onto the promontory of the cochlea and the subjects are then instructed to provide feedback regarding the sensations they experience, including whether they perceive sound, sensation, or pain. If patients can identify the electrical stimulation between 50 and 1600 Hz as a recognizable sound, it suggests that the cochlear nerve is functional [27]. Other preoperative examinations include ear inspection, objective, and subjective audiometry.

The imaging studies are necessary to exclude malformations of the inner ear and other intracranial pathologies. Routinely, thin-slice high-field MRI and thin-slice (0.5 mm thickness) CT scan are performed.

The CI device is constituted by various parts. A microphone with a speech processor which transforms of the acoustic information into a logical sequence of electrical impulses, the transmitter for a transcutaneous transmission

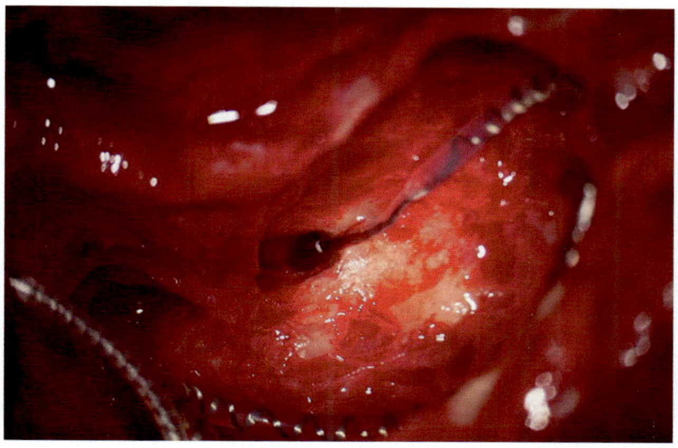

Fig. 46.1 Intraoperative image of a cochlear implant: After the posterior tympanostomy, exposure of the round window and incision of its membrane, the electrode is inserted in the cochlea

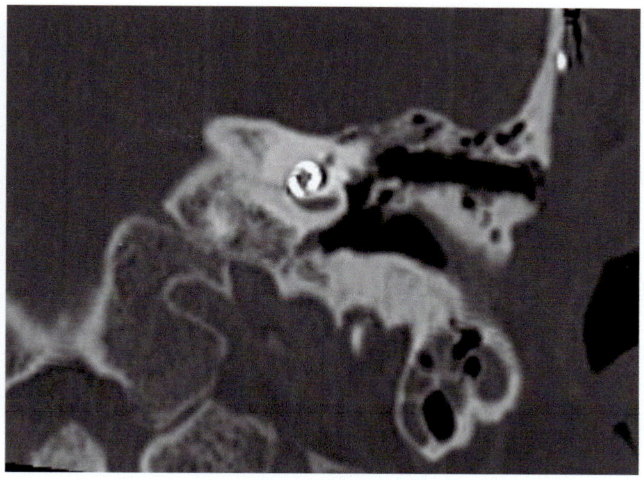

Fig. 46.2 Postoperative thin-slice CT scan showing the correct positioning of the electrode array into the cochlear canal

and power supply of the implant per induction, subcutaneous receiver coil for reception of the signal, a demodulator and an electrode array with different intracochlear electrode contacts for transmission of the impulses to the hearing nerve. The surgical technique for CI has been refined during the years and is actually very well-known and standardized as described by *Lenarz* [24] consisting of a transmastoid approach with posterior tympanostomy. A retroauricular C-shaped incision is performed and a partial drilling of the mastoid is achieved for the exposure of the antrum with incus, the mastoid course of the facial nerve, and the canal of the chorda tympani. A bone groove for the implant is then prepared circa 1 cm above to the sinus-dura angle and is connected with the mastoidectomy area. Further bone is drilled around the chorda tympani and facial nerve canal to have access to the promontory, the round and oval windows (Posterior tympanostomy). Exposure of the round window and incision of its membrane are then performed obtaining the opening of the cochlea. The electrode is inserted in the cochlea through this opening at a depth calculated on the basis of its anatomy and, eventually, residual hearing (Fig. 46.1). After a closure of the cochlea, the electrode carrier is fixed in the mastoid with fibrin glue or clips to avoid its dislocation. The wound is then closed in a layered fashion.

Imaging studies are performed intraoperatively with X-ray to verify the correct position of the implant and postoperatively with thin-slice CT scan to assess the course of the electrode in the cochlea (Fig. 46.2). The neurophysiological monitoring during the surgery is mandatory to achieve excellent postoperative results: electrode impedances, electrically triggered stapedius reflex, triggered action potential of the hearing nerve, and cochlear monitoring are routinely performed. The application of robotics with preoperatively planned trajectories for the drilling from the mastoid surface to the cochlea [28] could lead to shortening of surgical time and increasing of patients' safety. Apart from device failure (2–4% of the cases), potential postoperative complications can be divided into early or late. The first are mainly represented by infections, bleeding, or inner ear damage, while in the second category migration of the electrode, thinning and necrosis of the skin covering the implant and late labyrinthitis or meningitis are seen. The procedure has proved to be extremely safe with high successful rate especially in average word perception.

46.6.2 Auditory Brainstem Implant (ABI)

Cochlear implant is effective for hearing restoration in subjects with congenital and acquired sensorineural hearing loss. However, some patients harbor conditions that involve the cochlea itself or the auditory nerve which makes peripheral cochlear implant ineffective. These include bilateral vestibular schwannomas (neurofibromatosis type 2, NF2) (Fig. 46.3), trauma with consequent bilateral temporal fracture, or severe ossification of the cochlea [29, 30]. In order to bypass the unserviceable hearing nerve and cochlea in the auditory circuit, another type of implant, the ABI has been developed for the direct stimulation of the cochlear nuclei at the brainstem. It is, of course, mandatory the correct selection of the patients for both the tumor and non-tumor (i.e., Cochlear ossification or aplasia) pathologies. In fact, it has to be noticed that some patients affected by NF2 and already treated for microsurgical schwannoma resection could still profit from CI if the cochlear nerve is preserved and its functional status is verified by the promontory test [31].

The components of the ABI device are very similar to CI with external (microphone, battery, speech processor, external magnet, and antenna) and internal parts (internal magnet, receiver, and electrode array). In selected patients, the sur-

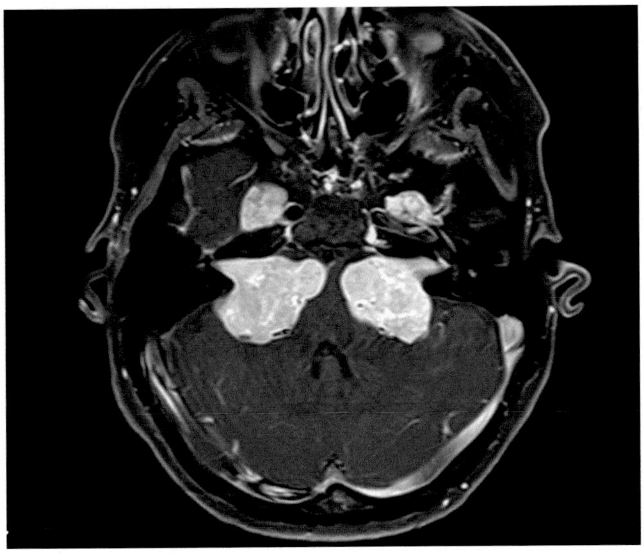

Fig. 46.3 Magnetic resonance imaging T1-weighted sequence with contrast showing a case of neurofibromatosis 2 with the typical bilateral vestibular schwannomas compressing the brainstem

gery for ABI can also be performed during the same session of the vestibular schwannoma removal.

As for CI, a complete preoperative evaluation of the candidates should be performed including not only the above-mentioned auditory tests but also a high-quality neuroimaging. The imaging protocol is usually the same performed for the preoperative planning of vestibular schwannoma resection and includes thin-slice CT scan and MRI with high-resolution T2 and eventually diffusion tensor imaging sequences to better identify the nerves of the cerebellopontine angle (CPA) [32]. There are two possible access to the cochlear nucleus: translabyrinthine or retrosigmoid approach. In the experience of our institute, the retrosigmoid approach is the technique of choice for both schwannoma removal and ABI [30, 33] and the surgery is performed under general anesthesia in semi-sitting position. After an extended C-shaped incision for the housing of the implant, a classical retrosigmoid craniotomy is performed exposing the margin of the sigmoid and transverse sinus with extension inferiorly to the floor of posterior fossa. As for the CI, an extra bony cavity contoured to the shape of the device is drilled superiorly to the lateral sinus to contain the implant. After dura incision and delicate cerebellar retraction, the lateral cerebellomedullary cistern is opened to allow cerebrospinal fluid drainage. In large schwannomas that restrict the access to the internal acoustic canal, initial debulking is performed before the drilling of the canal. The intrameatal tumor portion is then removed followed by the complete schwannoma resection in the CPA using a bimanual technique under constant irrigation to allow the preservation of the surrounding neurovascular structures [34]. After the vestibular schwannoma removal, the structures in the CPA angle are clearly visible: in particular the CN VII and VIII (if anatomically preserved) nerve complex, the brainstem, and the lower cranial nerves.

At this point, the important landmarks for the superficial identifications of the cochlear nucleus are identified: the foramen of Luschka and the lateral recess that represent the lateral opening of the fourth ventricle. These structures can be visualized by gentle retraction of the cerebellum, following the ninth nerve and the choroid plexus. A test probe is then inserted at the level of the lateral recess and evoked auditory brainstem potentials are recorded. In case of positive response, the multichannel electrode contact paddle (dimension: 3 mm × 8–10 mm) is placed in situ. The dacron mesh of the paddle promotes fibrous tissue growth securing it in position. Analogous to the CI, the receiver is then placed in the previously prepared bony groove and all the layers, starting from the dura, are progressively closed. The patients are monitored in the intensive care unit for 24 h and a CT scan is performed to verify the correct position of the implant and the absence of surgical complications such as hydrocephalus or bleeding. The implant is activated some days after the surgery: there is a first fitting with a basic training period of some weeks including daily speech and hearing therapy [35]. The evaluation of auditory performance is then obtained after this period and at 3–6 months interval. During this initial activation period, the patient is monitored for non-auditory side effects such as bradycardia, vertigo, and motor long tract stimulation. Auditory outcome after ABI is variable and mainly dependent by surgeon experience, post-implant programming, and motivation of the subjects during the rehabilitation phase. The patients should be fully aware before the surgical procedure that the implant does not achieve normal sound quality and that sound awareness is a reasonable goal [29]. In fact, the most important benefit of ABI is, in more than 90% of the patients, an improvement of sentences understanding with ABI combined with lip-reading [36].

46.6.3 Auditory Midbrain Implant (AMI)

The studies regarding the outcome of ABI have demonstrated that the results in NF2 patients are inferior if compared with non-tumor patients (ie. Cochlear aplasia) [29]. Various hypothesis has been considered to explain such results and the most probable reason could be a damage of the auditory nuclei at the level of the brainstem mainly due to tumor compression or its surgical removal. Thus, a new stimulation target of the auditory pathway distal to the cochlear nuclei has been searched. Of course, this area must be surgically accessible and have a defined neuronal organization to permit systematic spatial stimulation of different functional regions. The central nucleus of the IC satisfies these criteria especially for its well-defined tonotopic organization. Thus, the hypothesis to obtain a certain amount of hearing restoration

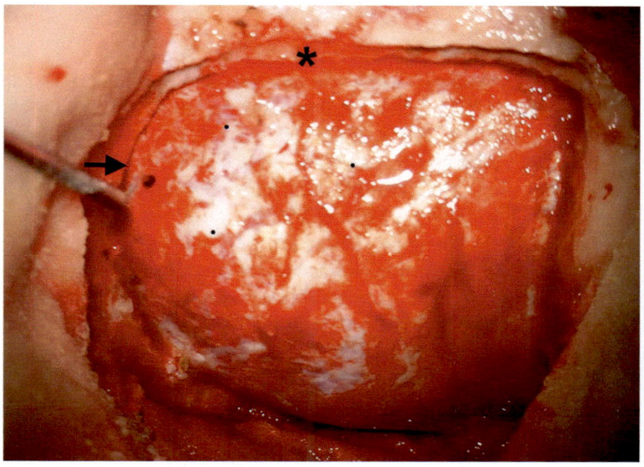

Fig. 46.4 Extended retrosigmoid craniotomy on the left side: the medial enlargement of the craniotomy allows the removal of the vestibular schwannoma and a supracerebellar infratentorial approach for the access to the inferior colliculus for the auditory midbrain implant. *: transverse sinus. Arrow: sigmoid sinus

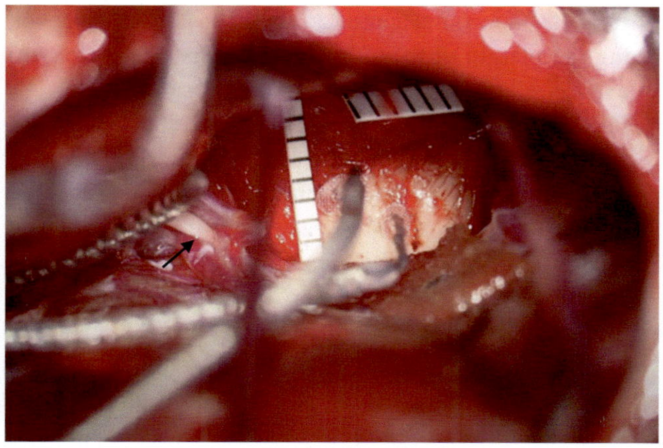

Fig. 46.5 Auditory midbrain implant with double array: via supracerebellar infratentorial approach the IC is visualized and both arrays are inserted preserving the vascular structures and the trochlear nerve (arrow)

with the direct electrical stimulation of the IC has become the basis of the AMI. Two study groups have started the experience of IC stimulation. The group of Colletti et al. [37] performed surface stimulation with a Med-El ABI array and our group using penetrating stimulation (termed AMI) with a new Cochlear DBS array [38, 39]. In our first clinical trial, we placed the AMI in five NF2 patients: the implant was performed in the same surgical session of vestibular schwannoma removal. We used the retrosigmoid approach with the same principles described in the ABI section. This approach has the advantage that can be easily extended to a lateral supracerebellar infratentorial approach (Fig. 46.4) to access the quadrigeminal cistern and consequently allowing an optimal insertion of the array at the level of the IC along the tonotopic gradient of the central nucleus. After the tumor removal in semi-sitting position, the access to the quadrigeminal cistern via supracerebellar approach is performed with the aid of neuronavigation for the exact localization of the IC and precise trajectory of insertion for the electrode. The semi-sitting position allows the cerebellum to relax downwards away from the tentorium facilitating the access to the supracerebellar infratentorial space. At the tentorial notch, the arachnoid is carefully dissected, and the quadrigeminal cistern is exposed from the posterior-lateral direction. Important landmarks of this region are the trochlear nerve, superior cerebellar artery, and deep veins that have to be clearly preserved to avoid postoperative complications. The array is then implanted posterolateral-to-anteromedial direction with a slight superior-to-inferior deflection. This single-shank multisite array was designed according to the dimensions of the human IC and consists of 20 linearly spaced platinum ring electrodes. The dacron mash at the basis of the electrode provides additional fixation. As for the ABI, the receiver-stimulator is fixed within the previously prepared bony bed, the sutured dura is covered with a piece of gel foam, and the craniotomy and the skin layer are closed systematically. The location of the electrode array in the IC is postoperatively controlled using thin-slice CT scan: one of the challenges of this procedure during the first clinical trial was the precise placement of the AMI array into the ICC in the correct direction along the tonotopic gradient of the ICC. Based on encouraging experimental results, we have started a second AMI clinical trial in 2017 that was funded by the National Institutes of Health and includes the implant of a new two-shank AMI array (performed in 7 cases) (Fig. 46.5). Each shank was designed with a site spacing of 300 μm for 10 of the 11 sites to sufficiently span the tonotopic axis of the ICC [40]. In our second trial, the AMI has proven to be a safe procedure with no postoperative surgical complication in the treated subjects. It is a feasible alternative when ABI failed or cannot be performed and has achieved encouraging results in terms of safety and improvements in lip-reading and environmental awareness but has not yet provided sufficient speech perception. The technique requires a learning curve: it is important to underline that the results in the recently treated patients are better in term of auditory function in comparison with the first cases.

46.7 Surgical Pearls

- Mastoidectomy allows the exposure of the antrum with incus, the mastoid course of the facial nerve, and the canal of the chorda tympani. Further bone is drilled around the chorda tympani and facial nerve canal to have access to the promontory, the round, and oval windows.

- Lateral suboccipital approach should be performed exposing the margin of the sigmoid and transverse sinus with extension inferiorly to the floor of posterior fossa.
- Foramen of Luschka and lateral recess represent the lateral opening of the fourth ventricle and they are the landmarks for the superficial identifications of the cochlear nucleus. These structures can be visualized by gentle retraction of the cerebellum, following the ninth nerve and the choroid plexus.
- Supracerebellar infratentorial approach: the arachnoid is carefully dissected, and the quadrigeminal cistern is exposed from the posterior-lateral direction. Important landmarks of this region are the trochlear nerve, superior cerebellar artery, and deep veins that have to be clearly preserved to avoid postoperative complications.

46.8 Discussion

It is then clear that the comprehension of the auditory pathways is much more complex than other systems, such as the visual circuit, because the stages of signal analysis are not clearly separated. Consequently, it is challenging to employ a simple functional framework to understand the anatomical and physiological findings. Even if, in the complexity of the auditory sensory system seemed impossible to replicate in a prosthetic device, the actual technology has allowed to develop tailored implants for different types of pathology. Of course, the CI is nowadays a well-established technique with a high successful rate and very low complications due to its great development during decades. However, newer and even experimental techniques such as ABI and AMI are showing extremely promising results for patients that are not suitable for CI.

46.9 Conclusion

During last decades, many neural prostheses to restore sensory function have been introduced and among these, the auditory implants are the most effective. In order to understand the principles at the basis of these technologies, the deep knowledge of the topographic and functional anatomy is pivotal.

Despite the complexity of the human auditory system, the devices implanted at different stages of the hearing pathways have proved their efficacy for the treatment of hearing deficits with different etiology. Near the excellent results of the CI, more complex and pioneering approaches, such as ABI and AMI, have demonstrated encouraging results and great potential for future development.

References

1. Pickles JO. Auditory pathways: anatomy and physiology. Handb Clin Neurol. 2015;129:3–25. https://doi.org/10.1016/B978-0-444-62630-1.00001-9. PMID: 25726260.
2. Bilger RC. Auditory results with single-channel implants. Ann N Y Acad Sci. 1983;405:337–42. https://doi.org/10.1111/j.1749-6632.1983.tb31647.x. PMID: 6575656.
3. Nakajima HH, Dong W, Olson ES, Merchant SN, Ravicz ME, Rosowski JJ. Differential intracochlear sound pressure measurements in normal human temporal bones. J Assoc Res Otolaryngol. 2009;10(1):23–36. https://doi.org/10.1007/s10162-008-0150-y. Epub 2008 Dec 9. PMID: 19067078; PMCID: PMC2644388.
4. Rask-Andersen H, Liu W, Erixon E, Kinnefors A, Pfaller K, Schrott-Fischer A, Glueckert R. Human cochlea: anatomical characteristics and their relevance for cochlear implantation. Anat Rec (Hoboken). 2012;295(11):1791–811. https://doi.org/10.1002/ar.22599. Epub 2012 Oct 8. PMID: 23044521.
5. Tian Q, Linthicum FH Jr, Fayad JN. Human cochleae with three turns: an unreported malformation. Laryngoscope. 2006;116(5):800–3. https://doi.org/10.1097/01.mlg.0000209097.95444.59. PMID: 16652091.
6. Froud KE, Wong AC, Cederholm JM, Klugmann M, Sandow SL, Julien JP, Ryan AF, Housley GD. Type II spiral ganglion afferent neurons drive medial olivocochlear reflex suppression of the cochlear amplifier. Nat Commun. 2015;6:7115. https://doi.org/10.1038/ncomms8115. PMID: 25965946; PMCID: PMC4432632.
7. Delacroix L, Malgrange B. Cochlear afferent innervation development. Hear Res. 2015;330(Pt B):157–69. https://doi.org/10.1016/j.heares.2015.07.015. Epub 2015 Jul 29. PMID: 26231304.
8. Ciganović N, Warren RL, Keçeli B, Jacob S, Fridberger A, Reichenbach T. Static length changes of cochlear outer hair cells can tune low-frequency hearing. PLoS Comput Biol. 2018;14(1):e1005936. https://doi.org/10.1371/journal.pcbi.1005936. PMID: 29351276; PMCID: PMC5792030.
9. Spoendlin H, Schrott A. The spiral ganglion and the innervation of the human organ of Corti. Acta Otolaryngol. 1988;105(5–6):403–10. https://doi.org/10.3109/00016488809119493. PMID: 3400442.
10. Felix RA 2nd, Gourévitch B, Portfors CV. Subcortical pathways: towards a better understanding of auditory disorders. Hear Res. 2018;362:48–60. https://doi.org/10.1016/j.heares.2018.01.008. Epub 2018 Jan 31. PMID: 29395615; PMCID: PMC5911198.
11. Osen KK, Roth K. Histochemical localization of cholinesterases in the cochlear nuclei of the cat, with notes on the origin of acetylcholinesterase-positive afferents and the superior olive. Brain Res. 1969;16(1):165–85. https://doi.org/10.1016/0006-8993(69)90092-4. PMID: 5348847.
12. Rosahl SK, Rosahl S, Walter GF, Hussein S, Matthies C, Samii M. Cochlear region of the brainstem. J Neurosurg. 2000;93(4):724–9. https://doi.org/10.3171/jns.2000.93.4.0724. PMID: 11014562.
13. Grothe B, Pecka M, McAlpine D. Mechanisms of sound localization in mammals. Physiol Rev. 2010;90(3):983–1012. https://doi.org/10.1152/physrev.00026.2009. PMID: 20664077.
14. Pecka M, Siveke I, Grothe B, Lesica NA. Enhancement of ITD coding within the initial stages of the auditory pathway. J Neurophysiol. 2010;103(1):38–46. https://doi.org/10.1152/jn.00628.2009. Epub 2009 Oct 21. PMID: 19846624.
15. Golding NL, Oertel D. Synaptic integration in dendrites: exceptional need for speed. J Physiol. 2012;590(22):5563–9. https://doi.org/10.1113/jphysiol.2012.229328. Epub 2012 Aug 28. PMID: 22930273; PMCID: PMC3528977.
16. Yang Y, Lee J, Kim G. Integration of locomotion and auditory signals in the mouse inferior colliculus. elife. 2020;9:e52228.

https://doi.org/10.7554/eLife.52228. PMID: 31987070; PMCID: PMC7004561.

17. Straka MM, Hughes R, Lee P, Lim HH. Descending and tonotopic projection patterns from the auditory cortex to the inferior colliculus. Neuroscience. 2015;300:325–37. https://doi.org/10.1016/j.neuroscience.2015.05.032. Epub 2015 May 19. PMID: 26002315.
18. Lumani A, Zhang H. Responses of neurons in the rat's dorsal cortex of the inferior colliculus to monaural tone bursts. Brain Res. 2010;1351:115–29. https://doi.org/10.1016/j.brainres.2010.06.066. Epub 2010 Jul 23. PMID: 20615398.
19. Gruters KG, Groh JM. Sounds and beyond: multisensory and other non-auditory signals in the inferior colliculus. Front Neural Circuits. 2012;6:96. https://doi.org/10.3389/fncir.2012.00096. PMID: 23248584; PMCID: PMC3518932.
20. Jones EG. Chemically defined parallel pathways in the monkey auditory system. Ann N Y Acad Sci. 2003;999:218–33. https://doi.org/10.1196/annals.1284.033. PMID: 14681146.
21. Reale RA, Imig TJ. Tonotopic organization in auditory cortex of the cat. J Comp Neurol. 1980;192(2):265–91. https://doi.org/10.1002/cne.901920207. PMID: 7400399.
22. Hackett TA, Stepniewska I, Kaas JH. Subdivisions of auditory cortex and ipsilateral cortical connections of the parabelt auditory cortex in macaque monkeys. J Comp Neurol. 1998;394(4):475–95. https://doi.org/10.1002/(sici)1096-9861(19980518)394:4<475::aid-cne6>3.0.co;2-z. PMID: 9590556.
23. Guinan JJ Jr. Cochlear efferent innervation and function. Curr Opin Otolaryngol Head Neck Surg. 2010;18(5):447–53. https://doi.org/10.1097/MOO.0b013e32833e05d6. PMID: 20717032; PMCID: PMC3075443.
24. Lenarz T. Cochlear implant – state of the art. GMS Curr Top Otorhinolaryngol Head Neck Surg. 2018;16:Doc04. https://doi.org/10.3205/cto000143. PMID: 29503669; PMCID: PMC5818683.
25. Shannon RV. Advances in auditory prostheses. Curr Opin Neurol. 2012;25(1):61–6. https://doi.org/10.1097/WCO.0b013e32834ef878. PMID: 22157109; PMCID: PMC4123811.
26. Illg A, Giourgas A, Kral A, Büchner A, Lesinski-Schiedat A, Lenarz T. Speech comprehension in children and adolescents after sequential bilateral cochlear implantation with long interimplant interval. Otol Neurotol. 2013;34(4):682–9. https://doi.org/10.1097/MAO.0b013e31828bb75e. PMID: 23640090.
27. Ito J, Tsuji J, Sakakihara J. Reliability of the promontory stimulation test for the preoperative evaluation of cochlear implants: a comparison with the round window stimulation test. Auris Nasus Larynx. 1994;21(1):13–6. https://doi.org/10.1016/s0385-8146(12)80004-6. PMID: 7980189.
28. Labadie RF, Balachandran R, Noble JH, Blachon GS, Mitchell JE, Reda FA, Dawant BM, Fitzpatrick JM. Minimally invasive image-guided cochlear implantation surgery: first report of clinical implementation. Laryngoscope. 2014;124(8):1915–22. https://doi.org/10.1002/lary.24520. Epub 2014 Jan 6. PMID: 24272427; PMCID: PMC4453761.
29. Deep NL, Choudhury B, Roland JT Jr. Auditory brainstem implantation: an overview. J Neurol Surg B Skull Base. 2019;80(2):203–8. https://doi.org/10.1055/s-0039-1679891. Epub 2019 Feb 14. PMID: 30931229; PMCID: PMC6438789.
30. Lesinski-Schiedat A, Frohne C, Illg A, Rost U, Matthies C, Battmer RD, Samii M, Lenarz T. Auditory brainstem implant in auditory rehabilitation of patients with neurofibromatosis type 2: Hannover programme. J Laryngol Otol Suppl. 2000;27:15–7. https://doi.org/10.1258/0022215001904806. PMID: 11211430.
31. Roehm PC, Mallen-St Clair J, Jethanamest D, Golfinos JG, Shapiro W, Waltzman S, Roland JT Jr. Auditory rehabilitation of patients with neurofibromatosis Type 2 by using cochlear implants. J Neurosurg. 2011;115(4):827–34. https://doi.org/10.3171/2011.5.JNS101929. Epub 2011 Jul 15. PMID: 21761973; PMCID: PMC3590004.
32. Gerganov VM, Giordano M, Samii M, Samii A. Diffusion tensor imaging-based fiber tracking for prediction of the position of the facial nerve in relation to large vestibular schwannomas. J Neurosurg. 2011;115(6):1087–93. https://doi.org/10.3171/2011.7.JNS11495. Epub 2011 Aug 26. PMID: 21962081.
33. Samii M, Matthies C, Tatagiba M. Management of vestibular schwannomas (acoustic neuromas): auditory and facial nerve function after resection of 120 vestibular schwannomas in patients with neurofibromatosis 2. Neurosurgery. 1997;40(4):696–705; discussion 705–6. https://doi.org/10.1097/00006123-199704000-00007. PMID: 9092842.
34. Samii M, Gerganov V, Samii A. Improved preservation of hearing and facial nerve function in vestibular schwannoma surgery via the retrosigmoid approach in a series of 200 patients. J Neurosurg. 2006;105(4):527–35. https://doi.org/10.3171/jns.2006.105.4.527. PMID: 17044553.
35. Lenarz T, Moshrefi M, Matthies C, Frohne C, Lesinski-Schiedat A, Illg A, Rost U, Battmer RD, Samii M. Auditory brainstem implant: part I. Auditory performance and its evolution over time. Otol Neurotol. 2001;22(6):823–33. https://doi.org/10.1097/00129492-200111000-00019. PMID: 11698803.
36. Ebinger K, Otto S, Arcaroli J, Staller S, Arndt P. Multichannel auditory brainstem implant: US clinical trial results. J Laryngol Otol Suppl. 2000;27:50–3. https://doi.org/10.1258/0022215001904743. PMID: 11211440.
37. Colletti V, Shannon R, Carner M, Sacchetto L, Turazzi S, Masotto B, Colletti L. The first successful case of hearing produced by electrical stimulation of the human midbrain. Otol Neurotol. 2007;28(1):39–43. https://doi.org/10.1097/01.mao.0000247808.47712.02. PMID: 17195744.
38. Samii A, Lenarz M, Majdani O, Lim HH, Samii M, Lenarz T. Auditory midbrain implant: a combined approach for vestibular schwannoma surgery and device implantation. Otol Neurotol. 2007;28(1):31–8. https://doi.org/10.1097/01.mao.0000247819.16325.7d. PMID: 17195743.
39. Lim HH, Lenarz T, Joseph G, Battmer RD, Samii A, Samii M, Patrick JF, Lenarz M. Electrical stimulation of the midbrain for hearing restoration: insight into the functional organization of the human central auditory system. J Neurosci. 2007;27(49):13541–51. https://doi.org/10.1523/JNEUROSCI.3123-07.2007. PMID: 18057212; PMCID: PMC6673101.
40. Lim HH, Lenarz T. Auditory midbrain implant: research and development towards a second clinical trial. Hear Res. 2015;322:212–23. https://doi.org/10.1016/j.heares.2015.01.006. Epub 2015 Jan 20. PMID: 25613994; PMCID: PMC4417550.

47 Anatomy and Variation for Combined Transpetrosal Approach

Kenji Ohata and Hiroki Morisako

Learning Objectives

- Clarify the most versatile basic procedure for the anterior and posterior combined transpetrosal approach by preserving the labyrinthine bone capsule.
- Outline the basic procedural steps, including exposure of the sigmoid sinus, peeling off of the dura mater, removal of the petrosal bone, identification of the semicircular canals, dural incision, and cutting of the cerebellar tent.
- Detail subtypes in the combined transpetrosal approach, depending on the range of removal of the petrous apex, removal of the semicircular canals and cochlea.
- Summarize various variations of combined transpetrosal approach in actual cases.

47.1 Introduction

The resection of skull base lesions has traditionally been associated with significant morbidity and mortality. Advances in microsurgical techniques and skull base approaches in combination have significantly improved surgical outcomes and facilitated radical resection.

Surgical excision of tumors involving the petroclival region has a technical challenge, mainly due to access, especially for their relationship with the brain stem, cranial nerve, temporal lobe, and blood vessels.

The anterior and posterior combined transpetrosal approach is usually advantageous for minimizing temporal lobe retraction, thereby allowing access to the ventral brainstem and permitting the removal of petroclival lesions.

K. Ohata (✉) · H. Morisako
Department of Neurosurgery, Osaka Metropolitan University Graduate School of Medicine, Osaka, Japan
e-mail: hmorisako@oum.ac.jp

47.2 History/Background

The anterior and posterior combined transpetrosal approach is a method to reach the petroclival area through the presigmoid space with a temporal craniotomy and a small suboccipital craniotomy, followed by mastoidotomy and petrosectomy [1–3]. The areas that can be reached are the inferior optic chiasm anteriorly, the third ventricle and hypothalamic region parietally, the jugular foramen caudally, and the sigmoid sinus laterally [4]. In cases where hearing is not preserved preoperatively, the surgical field can be expanded by removing the entire labyrinth and the vestibule to expose the dura mater of the internal auditory canal and the posterior cranial fossa.

47.3 Anatomy Highlights

- The important structures during the anterior and posterior combined transpetrosal approach include the base of the middle cranial fossa, the crest of the petrosal bone, the cerebellar tent, the superior petrosal sinus (SPS), the arcuate eminence, the great superficial petrosal nerve (GSPN), the endolymphatic sac, the internal auditory canal, and the sigmoid sinus (Fig. 47.1).
- The position of the posterior middle cranial fossa almost coincides with the supramastoid crest on the lateral surface of the temporal bone. The crest of the petrosal bone runs anteromedially at an angle of approximately 53° to the midline anteroposterior axis. The arcuate eminence is a bony protuberance found almost in the center of the petrosal bone, and directly below this is the superior semicircular canal. The arc of the superior semicircular canal faces the petrous apex. When exposing the membranous labyrinth of the superior semicircular canal, the membranous labyrinth can be confirmed as a so-called blue line by carefully excising the arcuate eminence in a direction almost perpendicular to the crest of the petrosal

I. N. Kanaan, V. Beneš (eds.), *Neuroanatomy Guidance to Successful Neurosurgical Interventions*,
https://doi.org/10.1007/978-3-031-59838-8_47

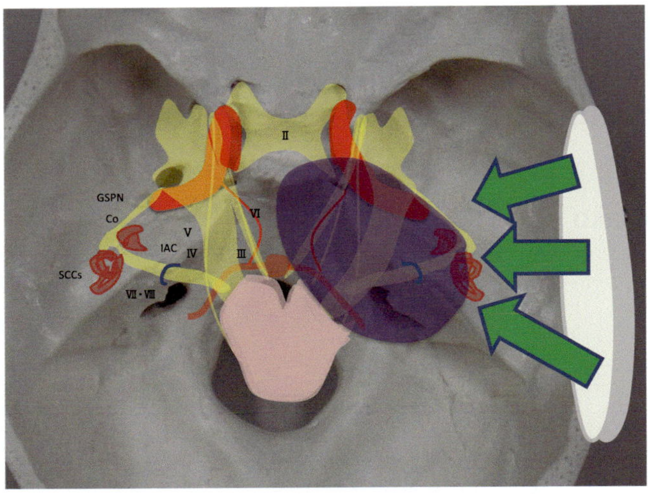

Fig. 47.1 Schematic approach of the combined transpetrosal approach for petroclival lesions. *GSPN* greater superficial petrosal nerve, *IAC* internal auditory canal, *Co* cochlea, *SCCs* semicircular canals

bone. The endolymphatic sac exists as a cul-de-sac between the inner and outer lobes of the posterior fossa, 12.2 mm anterior to the anterior margin of the sigmoid sinus, and the posterior semicircular canal lies 4.2 mm deep.

- Anteriorly lateral to the arcuate eminence is the hiatus facialis, from which emerges the GSPN, which runs along the inferior surface of the Gasserian ganglion in the direction of the foramen lacerum. This nerve branches from the geniculate ganglion. The geniculate ganglion is only covered with an extremely thin bone cortex, and great care must be taken when exposing this part. The area surrounded by the GSPN, trigeminal nerve impression, arcuate eminence and the crest of the petrosal bone is the posteromedial triangle (Kawase's triangle).
- When drilling out the petrosal bone while preserving hearing, the position of the internal auditory canal is estimated from the upper surface of the petrosal bone. As for this estimation method, Fisch's method assumes that the angle formed by the arcuate eminence and the internal auditory canal is 60° [5]. In addition, Garcia-Ibanez et al. reported that the internal auditory canal lies on the line that bisects the angle formed by the extension of the GSPN and the extension of the arcuate eminence [6]. Using this presumed area as a guide, excise the inner side of the arcuate eminence anteriorly to expose the dura mater of the upper wall of the internal auditory canal.
- In order not to damage the facial nerve in the internal auditory canal, it is important not to damage the dura mater of the internal auditory canal. The base of the internal auditory canal is the shallowest part of the upper surface of the petrosal bone, and the internal auditory foramen is the deepest part. For this reason, at first, the dura mater on the posterior wall of the inner auditory foramen is exposed at the ridge of the petrosal bone, and then the surface of the petrosal bone is carefully excised in the distal direction.
- The entrance to the tentorium of the trochlear nerve is located 9 mm posterior to the posterior end of the oculomotor triangle and 2 mm below the superior edge of the notch of the tentorium [7].

47.4 Approach (Surgical Intervention)

The basic procedural steps of the anterior and posterior combined transpetrosal approach are exposure of the sigmoid sinus, peeling off of the dura mater, removal of the petrosal bone, identification of the semicircular canals, dural incision, and cutting of the cerebellar tent.

47.4.1 Patient Positioning, Skin Incision, and Craniotomy

After induction of general anesthesia, a spinal drainage is placed. The patient is placed in a semi-prone park bench position. The head is fixed using 3-point fixation, with the head rotated and the vertex down to keep the temporal side of the head in the horizontal plane (Fig. 47.2a). The skin incision starts at the upper margin of the zygomatic arch anterior to the tragus, turns 2–3 cm above the ear, and then descends behind the posterior margin of the mastoid process. After raising the skin flap, a temporo-fascial pericranial flap is harvested based on the sternocleidomastoid muscle pedicle, which is used for the dural reconstruction at the end; to prevent postoperative CSF leakage (Fig. 47.2b). A temporal craniotomy is performed prior to mastoidotomy. Burr holes are placed at six points, including anatomical landmarks to avoid injury to the sigmoid sinus. The first burr hole is made at the asterion, the second at the intersection of the supramastoid crest with the squamous suture, the third at the mastoid emissary foramen, the fourth at the root of the zygoma, and the fifth and sixth at the anterior and posterior aspects of the squamous suture on the temporal bone, respectively (Fig. 47.2c).

47.4.2 Splitting Mastoidotomy and Exposure of the Sigmoid Sinus

Following the standard temporal craniotomy, the outer table of the mastoid process is split as a thin triangular plate to prevent post-operative cosmetic deformity. The sigmoid sinus is carefully exposed from the transverse sinus–sigmoid sinus junction down to its short horizontal segment [8] (Fig. 47.2c).

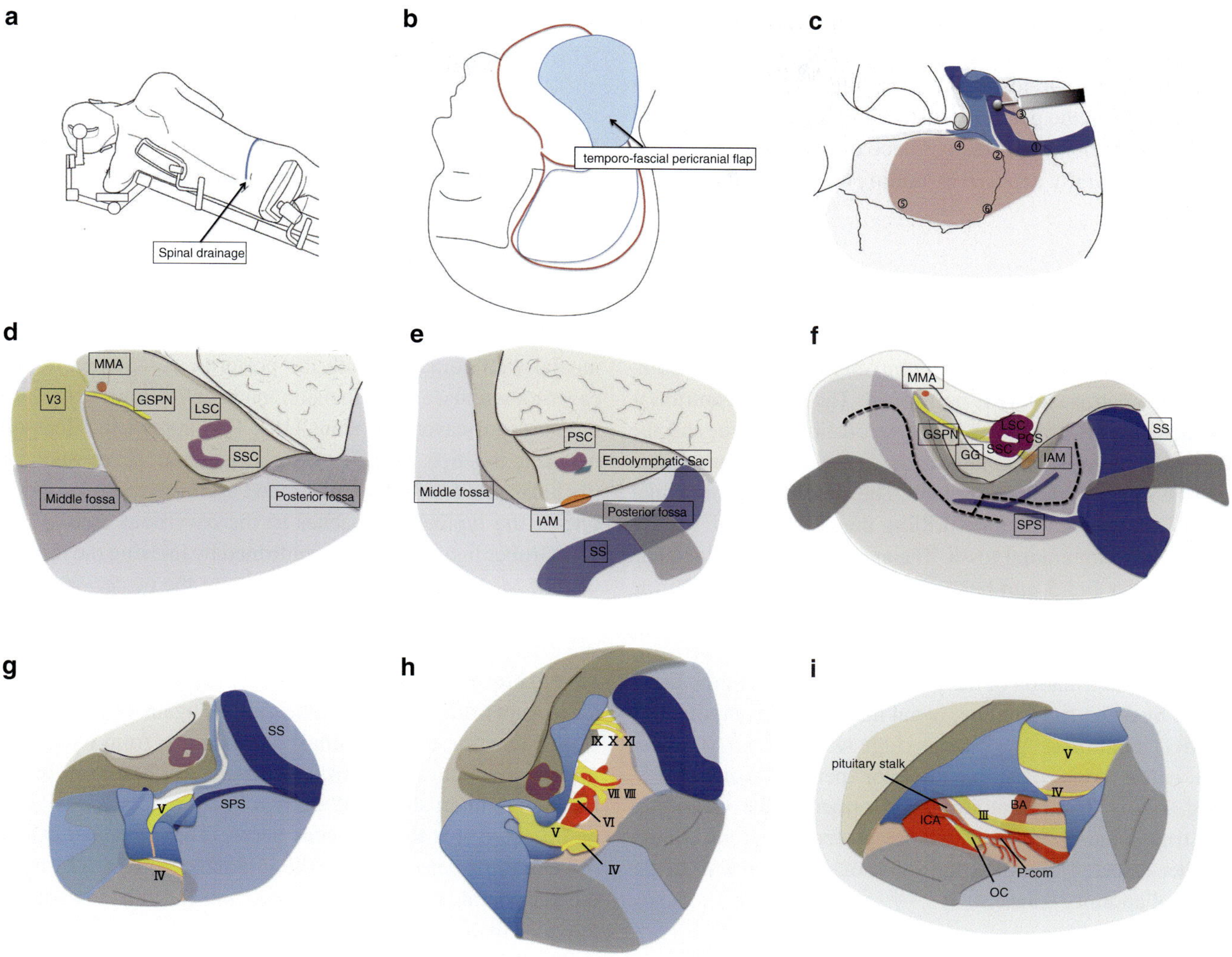

Fig. 47.2 Illustrations demonstrating surgical steps in the combined transpetrosal approach. (**a**) Patient positioning (**b**) Skin incision and a temporo-fascial pericranial flap based on the sternocleidomastoid muscle pedicle (**c**) Craniotomy (brown area) and splitting mastoidotomy (blue area) (**d**) Dural dissection on the middle skull base (**e**) Dural dissection from the petrous bone at the posterior cranial fossa (**f**) After a small amount of bone at the petrous ridge is drilled out to just expose the entire course of the SPS, the subtemporal dural incision is made as far anteriorly as possible, and its medial extent runs along the lateral margin of Meckel's cave. The presigmoid dura is opened along the drilled petrosal portion of the temporal bone as far anteriorly as possible while the drainage of the SPV into the SPS is inspected, and the SPS is divided anterior to it. (**g**) The tentorium is cut behind the dural entrance of the trochlear nerve (orange line), and the trochlear nerve is exposed. (**h**) Observation of cranial nerves and blood vessels in the posterior cranial fossa. (**i**) Observation of cranial nerves and blood vessels in the supratentorial space. *GSPN* greater superficial petrosal nerve, *MMA* middle meningeal artery, *LSC* lateral semicircular canal, *SSC* superior semicircular canal, *V* trigeminal nerve, *PSC* posterior semicircular canal, *IAM* internal auditory meatus, *SS* sigmoid sinus, *GG* geniculate ganglion of facial nerve, *SPS* superior petrosal sinus, *IV* trochlear nerve, *VI* abducens nerve, *VII* facial nerve, *VIII* acoustic nerve, *IX* glossopharyngeal nerve, *X* vagus nerve, *XI* accessory nerve, *III* oculomotor nerve, *ICA* internal carotid artery, *BA* basilar artery, *P-com* posterior communicating artery, *OC* optic chiasm

47.4.3 Dural Dissection and Exposure of the Posteromedial Triangle (Kawase's Triangle)

Dural dissection from the petrous and mastoid portions of the temporal bone is then done for a safe and swift petrosectomy. In the middle fossa, the dura matter over the temporal base is gently reflected to fully expose the entire course of the petrous ridge; by cutting the middle meningeal artery and dissecting the GSPN. The GSPN, which exits the hiatus facialis anteriorly to the arcuate eminence and heads for the foramen lacerum, runs with the lesser superficial petrosal nerve and the superior tympanic artery. These two nerves and one artery run between the two dura maters of the middle cranial fossa. The GSPN can be preserved by carefully dissecting the meningeal dura and this nerve after sharply incising the outer periosteal dura. The petrous bony surface is confirmed until the trigeminal impression, and the posteromedial triangle (Kawase's triangle) is exposed (Fig. 47.2d).

In the posterior fossa, the presigmoid dura is carefully dissected from the posterior surface of the petrous portion of the temporal bone, and the entrance of the internal auditory canal is epidurally exposed by cutting the endolymphatic sac (Fig. 47.2e).

47.4.4 Petrosectomy

After the dural dissection is complete, bony work encompassing a petrosectomy is performed. The mastoid antrum is opened deep behind the external acoustic aperture, and the labyrinthine capsule containing the lateral semicircular canal can be confirmed at its bottom. The posterior semicircular canal can be confirmed by drilling the petrous bone surface of the posterior cranial fossa. The anterior semicircular canal can be seen 1–3.5 mm below the arcuate eminence. During drilling, care is taken to keep the membranous labyrinths of the semicircular canals intact for hearing preservation.

After confirmation of the geniculate ganglion of the facial nerve by using free-running facial muscle electromyography (EMG) monitoring, the petrous ridge was resected to expose the entire course of the superior petrosal sinus (SPS) (Fig. 47.2f). The osteotomy is performed in the vicinity of the trigeminal impression, and the posterior petrous bone is further resected to aim at the superior posterior wall of the internal auditory canal.

47.4.5 Dural Opening and Cutting of the Tentorium

The middle fossa dura is opened along the inferior temporal lobe toward the SPS. The subtemporal dural incision is made as far anteriorly as possible, and its medial extent runs along the lateral margin of Meckel's cave. The presigmoid dura is opened along the drilled petrosal portion of the temporal bone as far anteriorly as possible, and the drainage of the superior petrosal vein (SPV) into the SPS is inspected. The SPS was transected at a point anterior to the drainage point of the SPV into the SPS to preserve its venous drainage (Fig. 47.2f). After opening Meckel's cave and identifying the running course of the trigeminal nerve, the tentorium was divided from the point of the transection of the SPS to its hiatus 5 mm behind the dural entrance of the trochlear nerve (Fig. 47.2g).

47.4.6 Opening of Meckel's Cave and the Posterior Component of the Cavernous Sinus

The superior wall of Meckel's cave is the dura propria of the middle cranial fossa and the tentorium cerebellum, which is continuous with the dura mater. The cerebellar tentorium continues from the middle cranial fossa dura and folds back at the cerebellar tentorial notch, ending in the trigeminal Gasserian ganglion. The lateral and inferior walls are extensions of the dura mater of the posterior fossa [9]. The anteromedial wall coincides with the posterior component of the cavernous sinus, and an incision made here allows entry into the posterior cavernous sinus [10]. This approach is located between the first branch of the trigeminal nerve and the trochlear nerve, which is called the Parkinson triangle, and is one of the typical approaches to the cavernous sinus [11]. The Gruber ligament can be confirmed by incising the medial wall of Meckel's cave posteriorly along the posterior petroclinoid fold. This ligament is a tough fibrous connective tissue that connects the tip of the petrous bone and the posterior clinoid process and is often calcified [12]. Observation of the anterosuperiorlateral part of the Gruber ligament reveals the C5-C4 junction of the internal carotid artery, from which the meningohypophyseal trunk originated. Furthermore, when tracing the internal carotid artery to the proximal side, it can be observed that the abducens nerve, which has just exited the Dorello's canal, crosses the internal carotid artery from the peripheral side to the proximal side and runs toward the lateral wall of the cavernous sinus. Incision of the dura proximally along the abducens nerve opens the inferior petrosal sinus, which continues from the cavernous sinus and surrounds the Dorello's canal.

47.4.7 Observation of Cranial Nerves and Blood Vessels in the Posterior Cranial Fossa

In the posterior cranial fossa, the trigeminal nerve and cranial nerves VII and VIII can first be identified. The IX-XI cranial nerves can be seen on the caudal side, and the basilar artery, anterior inferior cerebellar artery, superior cerebellar artery, and abducens nerve on the brainstem side. In addition, opening Meckel's cave creates mobility in the trigeminal nerve, making it easier to observe the median side of the posterior cranial fossa (Fig. 47.2h).

47.4.8 Observation of Cranial Nerves and Blood Vessels in the Supratentorial Space

The supratentorial region can be observed by moving the visual axis of the microscope from the caudal posterior to the cranial anterior and retracting the temporal lobe. The top of the basilar artery, the superior cerebellar artery, the posterior cerebral artery, the posterior communicating artery, and the oculomotor nerve can be confirmed. In front, the posterior part of the optic nerve, the C2 part of the internal carotid artery, and the dura entrance of the oculomotor nerve can be observed. On the medial side, the lower surface of the optic chiasm, the pituitary stalk, the floor of the third ventricle, and the posterior clinoid process can be observed (Fig. 47.2i).

47.5 Representative Cases

47.5.1 Anterior and Posterior Combined Transpetrosal Approach

Case 1: 40 years old female, petroclival meningioma

A workup of light-headedness and right hearing loss indicated a large petroclival meningioma (Fig. 47.3a–c).

After making a fascia-periosteal flap, a temporal suboccipital craniotomy was performed. Since the sigmoid sinus had well developed, the bone was drilled out caudally from the junction of the transverse sinus and the sigmoid sinus to expose the sigmoid sinus, and then a splitting mastoidectomy was performed (Fig. 47.3f).

The mastoid antrum was opened, the MMA was coagulated, and the dura mater of the middle skull base was peeled along the petrous rim to expose the trigeminal impression (Fig. 47.3g). Posteriorly, the endolymphatic sac was coagulated and cut at the presigmoid on the posterior surface of the petrous bone, and the dura mater of the posterior cranial fossa was peeled (Fig. 47.3h). The petrous rim was then removed to expose the SPS (Fig. 47.3i).

A dural incision was made along the superior border of the SPS at the inferior aspect of the temporal lobe and a presigmoid dural incision was also performed on the posterior fossa side. After removing the tumor in Meckel's cave, a partial tentorium incision was made along the tumor attachment (Fig. 47.3j). The trigeminal nerve was compressed caudally by the tumor (Fig. 47.3k). Next, a tentorium incision was also made posteriorly along the posterior margin of the tumor. The trochlear nerve was preserved to reach the medial edge of the tentorium. The anterior petroclinoid ligament was resected at the superior margin of Meckel's cave (Fig. 47.3l). Next, after the trigeminal nerve was displaced upward and bone removal at the petrous apex was added using a diamond drill. After detachment of the trigeminal nerve and the tumor while reducing the tumor volume on the posterior fossa side, coagulation of the tumor attachment on the midline side of the clival bone resulted in tumor mobility. The trochlear nerve and the tumor capsule were dissected from the posterior to the anterior side to confirm the tentorial entrance of the trochlear nerve (Fig. 47.3m). A superior petrosal vein (SPV) was found to be occluded by tumor infiltration. The posterior portion of the tent, which was the remaining tumor attachment, was resected.

The visual field was shifted forward, and the trigeminal nerve V1, V2, and the lateral wall of the cavernous sinus were detached while pulling the temporal lobe. The superior wall of Meckel's cavity was resected from anterior to posterior. At this time, the trochlear nerve was confirmed on the lateral wall of the cavernous sinus about 5 mm anterior to the tentorium entrance, so the Parkinson's triangle was released lateral side of the trochlear nerve (Fig. 47.3n). When the posterior petroclinoid ligament was excised between the trigeminal nerve and the trochlear nerve, the posterior clinoid process was confirmed to be deep. The oculomotor nerve was confirmed when the tumor on the tentorium was removed while pulling the tentorium to the caudal side (Fig. 47.3o). The oculomotor cistern was opened, and the tumor on the medial side of the oculomotor nerve was removed from between the oculomotor nerve and the trochlear nerve, and the pituitary stalk was confirmed in the deep part.

Next, we moved to the resection of the residual tumor on the caudal side of the cerebellar pontine angle. Tumors around the internal auditory canal were excised after additional drilling of the petrous bone (Fig. 47.3p). After removing the petrous apex, the residual tumor around the abducens nerve was removed, and the dura mater at the attachment of the tumor was coagulated (Fig. 47.3q).

The patient showed no neurological worsening (Fig. 47.3d, e).

47.5.2 Anterior and Posterior Combined Transpetrosal Approach with Labyrinthectomy

Case 2: 34 years old female, recurrent petroclival meningioma

A workup of hearing loss and facial paralysis indicated a left petroclival meningioma. Partial tumor resection was performed by the previous doctors using a combined presigmoid approach. Two years after the operation, the residual lesion increased, and the patient was referred to our hospital (Fig. 47.4a–c). There was no useful hearing before surgery, and facial paralysis (House-Brackmann grade 3) was confirmed.

A temporooccipital suboccipital craniotomy was performed to confirm the petrous ridge, which had been removed

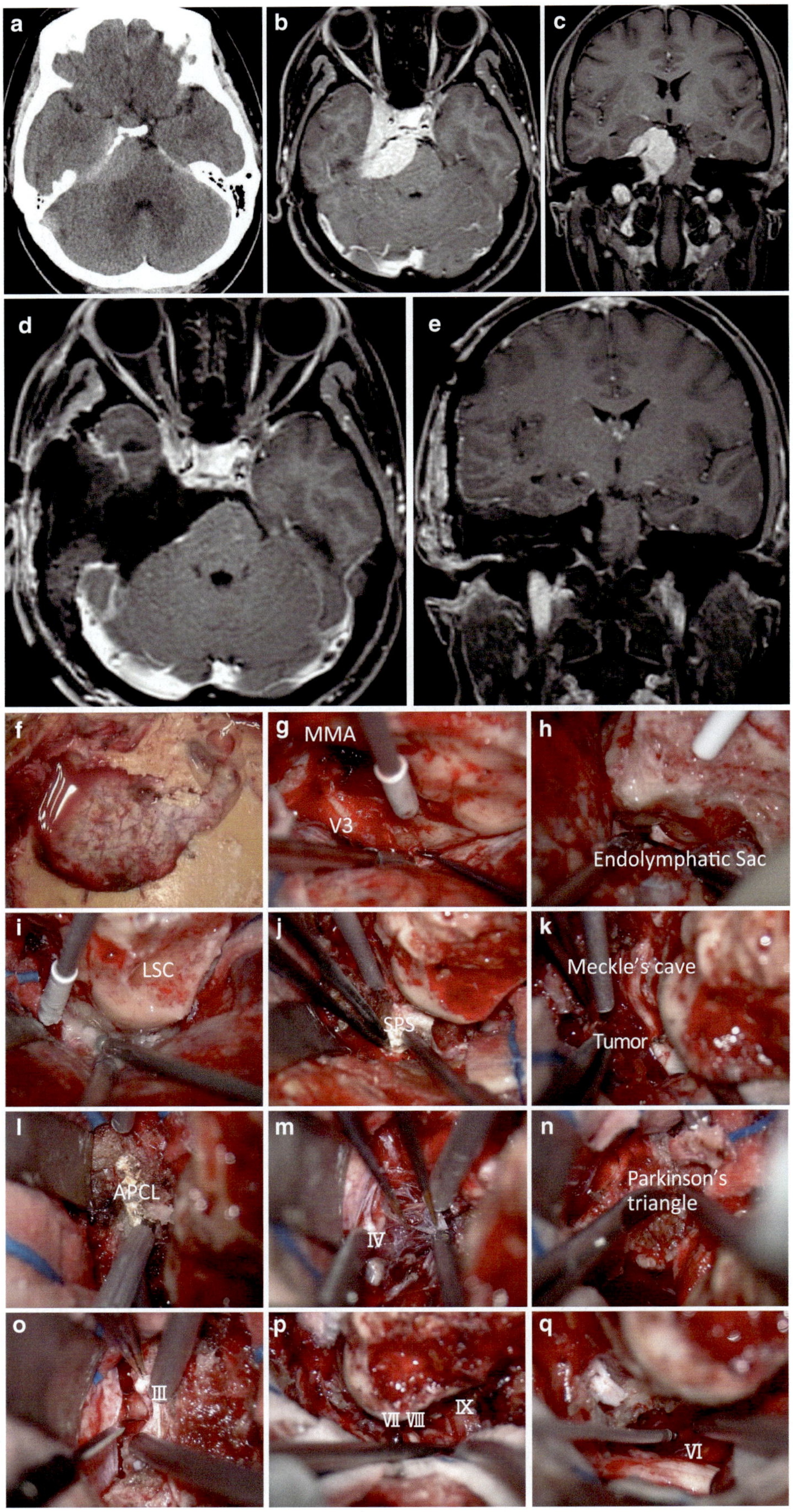

Fig. 47.3 Case 1 (**a**) Preoperative axial CT image. (**b**, **c**) Preoperative axial and coronal T1-weighted MR images with gadolinium enhancement. (**d**, **e**) Postoperative axial and coronal T1-weighted MR images with gadolinium enhancement. (**f–q**) Intraoperative photographs from Case 1. A temporal suboccipital craniotomy was performed (**f**). The trigeminal impression was exposed at the middle skull base (**g**). The dura mater of the posterior cranial fossa was peeled (**h**). The petrous rim was drilled to expose the SPS (**i**). A tentorium incision was made along the tumor attachment (**j**). The trigeminal nerve was compressed caudally by the tumor (**k**). The anterior petroclinoid ligament was resected at the superior margin of Meckel's cave (**l**). The trochlear nerve and the tumor capsule were dissected from the posterior to the anterior side (**m**). The Parkinson's triangle was released on the lateral side of the trochlear nerve (**n**). The oculomotor cistern was opened, and the tumor on the medial side of the oculomotor nerve was removed (**o**). Tumors around the internal auditory canal were excised after additional drilling of the petrous bone (**p**). The residual tumor around the abducens nerve was removed (**q**). *MMA* middle meningeal artery, *V* trigeminal nerve, *LSC* lateral semicircular canal, *SPS* superior petrosal sinus, *APCL* anterior petroclinoid ligament, *IV* trochlear nerve, *III* oculomotor nerve, *VII* facial nerve, *VIII* acoustic nerve, *IX* glossopharyngeal nerve, *VI* abducens nerve

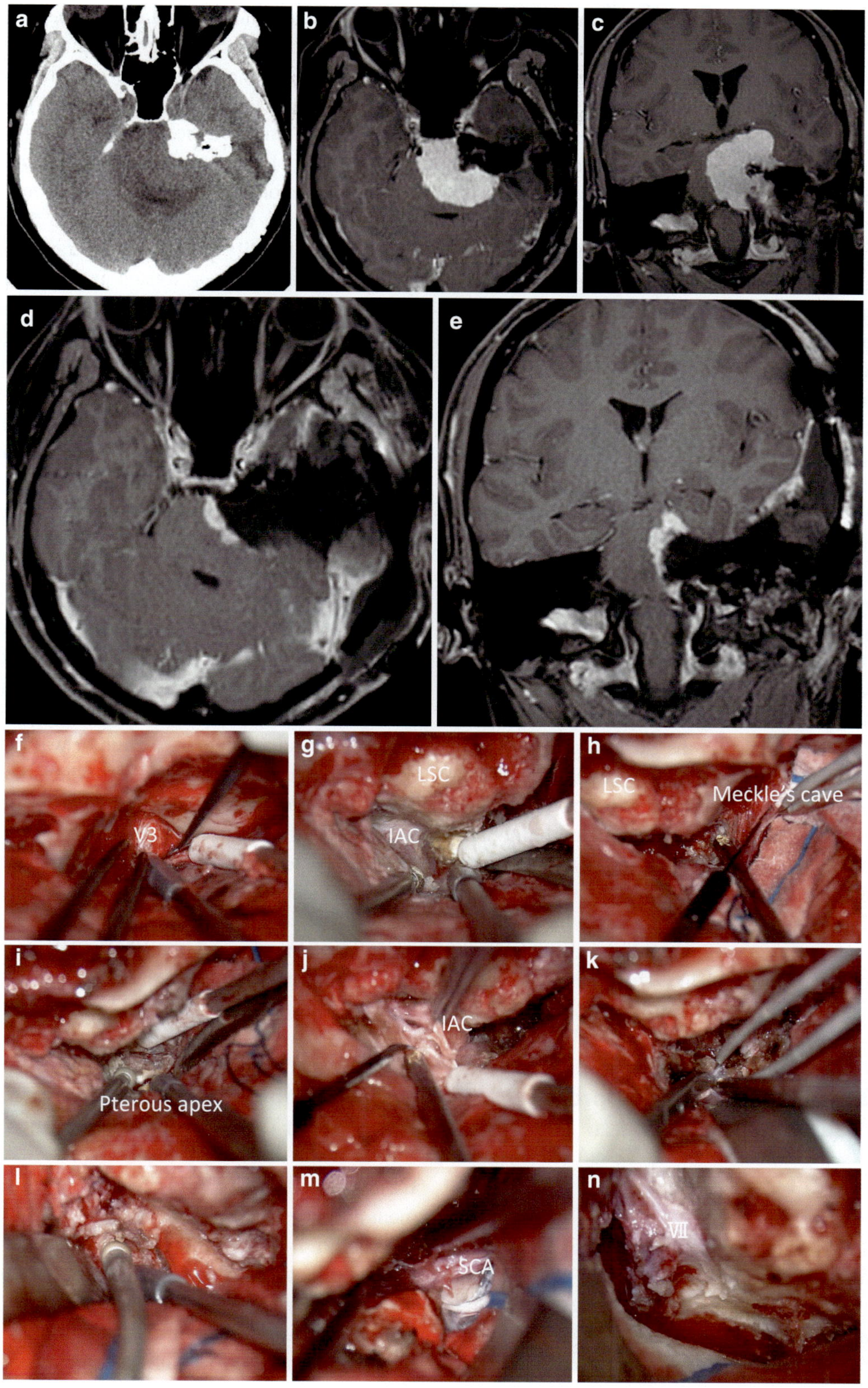

Fig. 47.4 Case 2 (**a**) Preoperative axial CT image. (**b**, **c**) Preoperative axial and coronal T1-weighted MR images with gadolinium enhancement. (**d**, **e**) Postoperative axial and coronal T1-weighted MR images with gadolinium enhancement. (**f–n**) Intraoperative photographs from Case 2. The trigeminal nerve was detached from the dura of the temporal lobe to expose the trigeminal impression (**f**). The superior and posterior semicircular canals were removed to open the posterior and superior walls of the IAC (**g**). Meckle's cave was opened (**h**). The petrous apex was removed (**i**). The presigmoid dura was incised (**j**). The SPS was cut at the posterior margin of the tentorial attachment of the tumor (**k**). Additional bone was removed using a diamond drill from the upper wall of the IAC to the petrous apex (**l**). Internal decompression of the tumor and retraction of the temporal lobe allowed identification of the superior margin of the tumor, the trochlear nerve, the oculomotor nerve, and the SCA (**m**). The tumor was removed as much as possible from the lower wall of the internal auditory canal to the upper wall of the jugular foramen (**n**). *V* trigeminal nerve, *LSC* lateral semicircular canal, *IAC* internal auditory canal, *SCA* superior cerebellar artery, *VII* facial nerve

in the previous presigmoid approach, exposing the semicircular canals. The trigeminal nerve was detached from the dura of the temporal lobe to expose the trigeminal impression (Fig. 47.4f). The posterior surface of the petrous bone and the dura mater of the posterior fossa were exposed, and a diamond drill was used to remove the bone at the rim of the petrous bone. Since the patient had already lost hearing, the superior and posterior semicircular canals were removed to open the posterior and superior walls of the IAC (Fig. 47.4g).

Next, we gradually removed the calcified part located in the center of the tumor attachment. A dura incision was made on the inferior aspect of the temporal lobe, and the trigeminal nerve fibers were identified by extending the dura incision posteriorly along the lateral V3 (Fig. 47.4h). Then, bone removal at the petrous apex and tumor removal with the calcified part were added (Fig. 47.4i). Subsequently, the dura of the presigmoid was incised (Fig. 47.4j). SPS was coagulated at the posterior margin of the tentorial attachment of the tumor (Fig. 47.4k). Internal decompression of the tumor revealed the trochlear nerve at the superior margin of the tumor. Additional bone was removed using a diamond drill from the upper wall of the IAC to the petrous apex (Fig. 47.4l). An incision was made at the superior wall of Meckle's cave while exposing the trigeminal nerve. Retraction of the temporal lobe allowed identification of the superior margin of the tumor, the trochlear nerve, the oculomotor nerve, and the SCA (Fig. 47.4m). The basilar artery and abducens nerve were exposed when the tumor was resected on the median side at the lower margin of the tumor. Since the tumor capsule was adhered tightly with the brainstem and abducens nerve, a portion of the tumor capsule was remained. Finally, the tumor was removed as much as possible from the lower wall of the internal auditory canal to the upper wall of the jugular foramen, and the tumor was removed at about 95% (Fig. 47.4n). The patient showed no neurological worsening (Fig. 47.4d, e).

47.5.3 Combined Transpetrosal Approach with Retrosigmoid Approach

Case 3: 30 years old male, cavernous petrotentorial meningioma

Examination of facial pain indicated a large cavernous petrotentorial meningioma (Fig. 47.5a, b).

The temporalis muscle was then retracted forward to expose the temporal bone (Fig. 47.5e). After a temporooccipital suboccipital craniotomy, the sigmoid sinus was exposed from the junction with the transverse sinus. After exposure of the rim of the petrous, bone was partially removed to expose the SPS (Fig. 47.5f).

A temporal dural incision and dura of the presigmoid were incised. The SPS was amputated posteriorly because the tumor had spread along the SPS to the junction of the transverse sinus and sigmoid sinus. The Meckel's cave was opened, and the tumor within it was removed (Fig. 47.5g). While incising and removing the calcified anterior petroclinoid ligament in the tumor, the temporal lobe was retracted to expose the tumor on the middle cranial fossa side.

The dura of the retrosigmoid was incised along the transverse-sigmoid sinus, and a tentorial incision was added posteriorly along the tumor attachment. After retracting the cerebellum to identify the 7th and 8th cranial nerves, the posterior part of the tumor was rotated gradually with a dural incision at the posterior margin of the petrous bone (Fig. 47.5h). Tumor excision proceeded by extending the tentorium incision along the posterior margin of the tumor while moving this posterior part of the tumor forward (Fig. 47.5i). Next, the trochlear nerve could be identified by additional removal of the tip of the petrous bone. An incision was made in the lateral wall of the cavernous sinus along the trochlear nerve (Fig. 47.5j). When the tumor located in the posterior component of the CS was excised, the tumor was pulled caudally, and the oculomotor nerve was identified at the superior margin of the tumor. The oculomotor cistern was released to follow the oculomotor nerve from the back to the front, and the tumor was detached from the oculomotor nerve (Fig. 47.5k). The tumor was excised anterior to the oculomotor nerve, and the internal carotid artery was confirmed (Fig. 47.5l). After decompressing the tumor above the trigeminal nerve, the tumor was detached from the brainstem to expose the SCA. Then, when the trigeminal nerve was moved upward and the tumor on the posterior caudal side was excised, the abducens nerve encased in the tumor was confirmed. The tumor was carefully removed separately on the cranial and caudal sides of the abducens nerve, and the tumor removal was completed (Fig. 47.5m).

The patient showed transient oculomotor nerve palsy, which improved completely 3 months after surgery (Fig. 47.5c, d).

47.5.4 Combined Presigmoid Approach with Transcondylar Approach

Case 4: 60 years old female, recurrent petroclival meningioma

A workup of gait disturbance indicated a left CS-petroclival meningioma about 7 years ago. Partial tumor resection was performed by the previous hospital using a subtemporal approach. Four years after the operation, the residual lesion increased, and partial tumor resection via suboccipital approach and cyberknife (28 Gy/7 Fr) was performed. Two years later, the residual tumor increased, and she was referred to our hospital (Fig. 47.6a–c). The patient showed complete oculomotor and abducens nerve palsy before surgery.

After the patient was placed in the right-side down prone-park bench position, the head rotated in the opposite direc-

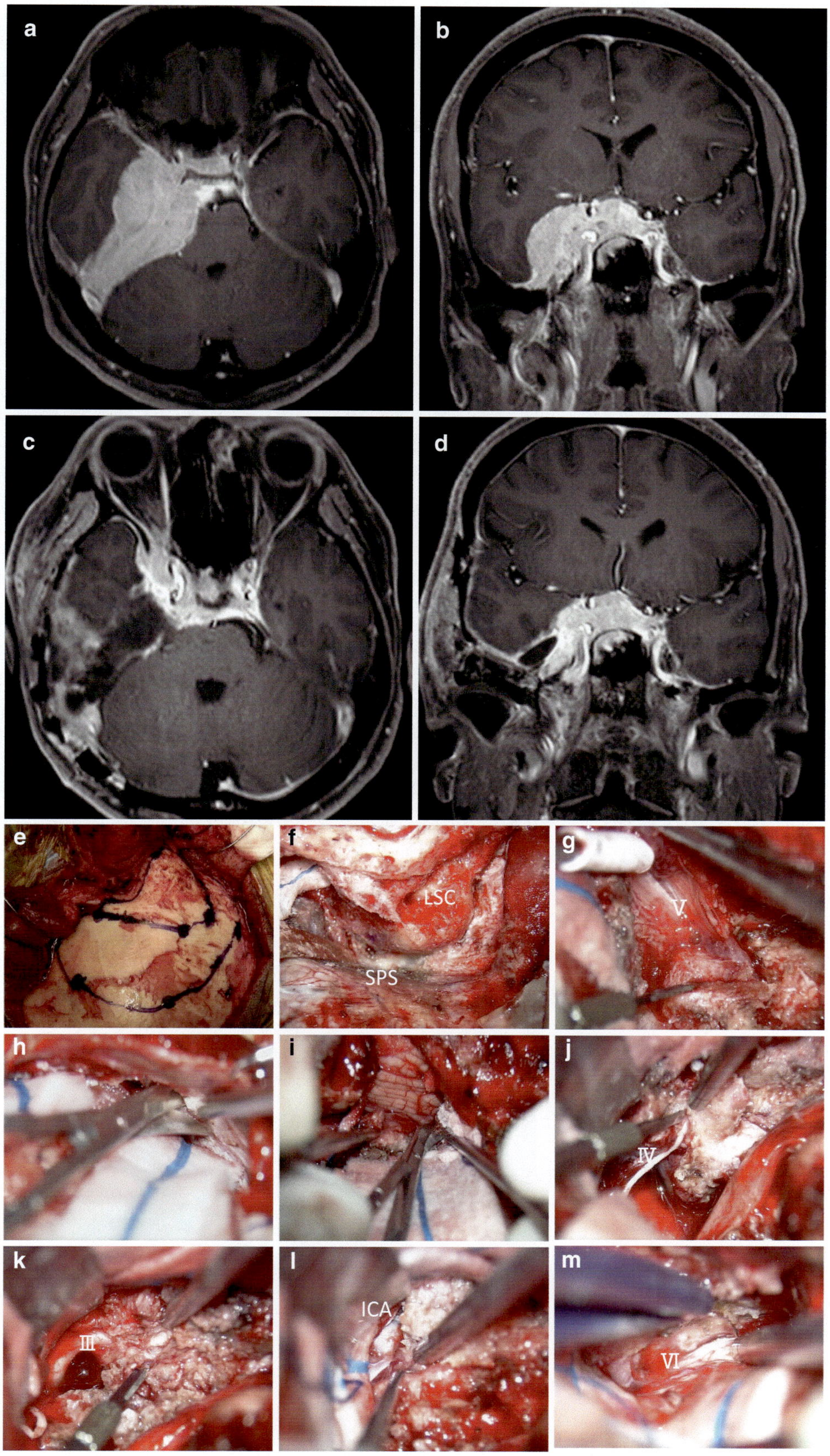

Fig. 47.5 Case 3. (**a**, **b**) Preoperative axial and coronal T1-weighted MR images with gadolinium enhancement. (**c**, **d**) Postoperative axial and coronal T1-weighted MR images with gadolinium enhancement. (**e**–**m**) Intraoperative photographs from Case 3. A temporooccipital suboccipital craniotomy was performed (**e**). The rim of the petrous bone was partially removed to expose the SPS (**f**). The tumor that occupied Meckel's cave was removed (**g**). The tumor attachment was incised via a retrosigmoid approach (**h**). The tentorium incision along the posterior margin of the tumor was made (**i**). An incision was made in the lateral wall of the cavernous sinus along the trochlear nerve (**j**). The oculomotor cistern was released to follow the oculomotor nerve from the back to the front (**k**). The internal carotid artery was confirmed (**l**). The tumor was carefully removed separately on the cranial and caudal sides of the abducens nerve (**m**). *LSC* lateral semicircular canal, *SPS* superior petrosal sinus, *V* trigeminal nerve, *IV* trochlear nerve, *III* oculomotor nerve, *ICA* internal carotid artery, *VI* abducens nerve

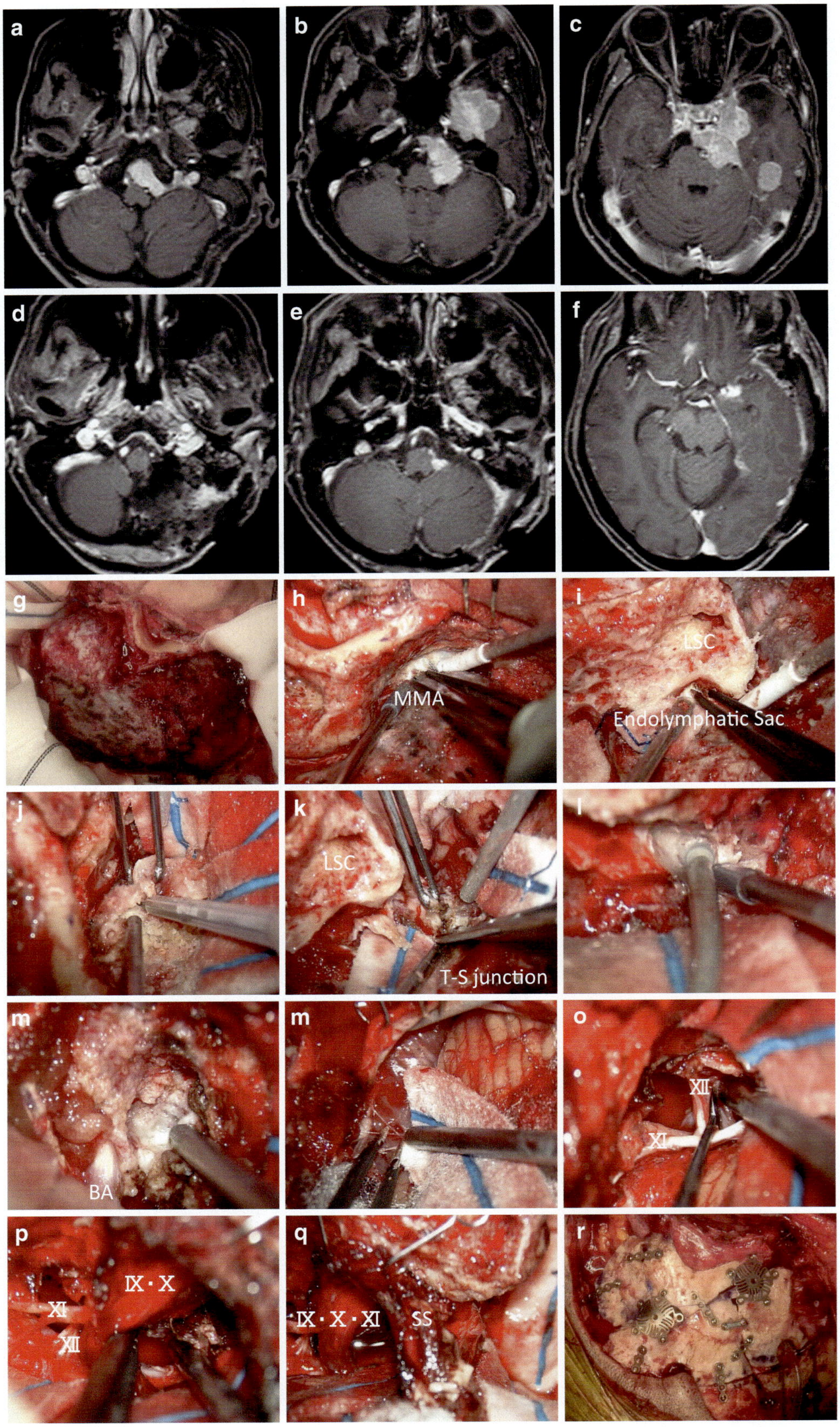

Fig. 47.6 Case 4. (**a–b**) Preoperative axial T1-weighted MR images with gadolinium enhancement. (**d–f**) Postoperative axial T1-weighted MR images with gadolinium enhancement. (**g–r**) Intraoperative photographs from Case 4. A temporooccipital suboccipital craniotomy with opening the foramen magnum was performed (**g**). The MMA was coagulated and cut at the foramen spinosum (**h**). The endolymphatic sac was coagulated and cut (**i**). The tumor in the middle cranial fossa was first excised (**j**). The tumor located on the posterior lateral side of the tentorium was excised while making a tentorium incision (**k**). Additional bone removal at the petrous apex was performed (**l**). The basilar artery and brainstem were highly adhered to the tumor capsule (**m**). After incision of the retrosigmoid dura, a left lower petroclival lesion was revealed (**n**). Tumor around the hypoglossal nerve was removed (**o**). Subcapsular tumor resection was performed between the facial nerve and the lower cranial nerve (**p**). Subtotal resection of the tumor was achieved (**q**). The bone fragments were fixed with miniplates (**r**). *MMA* middle meningeal artery, *LSC* lateral semicircular canal, *T-S junction* transverse sinus-sigmoid sinus junction, *BA* basilar artery, *XI* accessory nerve, *XII* hypoglossal nerve, *IX* glossopharyngeal nerve, *X* vagus nerve, *SS* sigmoid sinus

tion, and the mastoid process raised. After a temporooccipital suboccipital craniotomy, opening the foramen magnum and drilling of transcondylar fossa were added (Fig. 47.6g).

Next, the MMA was coagulated and cut at the foramen spinosum (Fig. 47.6h), and the trigeminal impression was exposed. The endolymphatic sac was coagulated and cut (Fig. 47.6i), and the rim of the petrous bone was removed to expose the SPS. After a dural incision, the tumor in the middle cranial fossa was first excised. After debulking the tumor and detaching the temporal lobe from the tumor (Fig. 47.6j), the medial border of the tentorium was confirmed. A dural incision was then added posteriorly on the tentorium along the SPS, a presigmoid dural incision was also made on the posterior fossa side, and the SPS was coagulated just before the junction of the transverse sinus and the sigmoid sinus. The tumor located on the posterior lateral side of the tentorium was excised while making a tentorium incision (Fig. 47.6k). The tumor in Meckel's cave was resected. The internal auditory canal was opened with additional bone removal at the petrous bone (Fig. 47.6l). The left oculomotor nerve was cut with priority given to the excision of the tumor. After confirmation of the optic nerve and internal carotid artery, the posterior communicating artery and tumor were dissected, and the lesion extending to the medial temporal lobe was excised. The basilar artery was confirmed when the tumor was removed from the prepontine cistern above the internal auditory canal. The basilar artery and brainstem were highly adhered to the tumor capsule, some of the tumor capsule remained with the surrounding scar tissue (Fig. 47.6m).

Next, we moved to tumor resection via a transcondylar approach (Fig. 47.6n). Incision of the retrosigmoid dura, with additional opening of the hypoglossal canal and removal of the cortical bone of the jugular tubercule. We proceeded to reduce the tumor volume by pulling it out from the caudal side of the hypoglossal nerve into the hypoglossal canal (Fig. 47.6o). Subcapsular tumor resection was performed between the lower cranial nerve and the hypoglossal nerve and between the facial nerve and the lower cranial nerve to decompress the brainstem (Fig. 47.6p). The tumor was resected while coagulating and exposing the dura mater at the tumor attachment in the lower part of the petroclival area as much as possible, and subtotal resection of the tumor was performed (Fig. 47.6q).

The dural defect around the foramen magnum was sutured with subcutaneous fat. The bone removal around the hypoglossal nerve canal and opened mastoid air cells were filled with subcutaneous fat tissues. After covering the posterior side of the petrous bon by using a temporal fascia periosteal flap, the bone fragments were fixed with miniplates (Fig. 47.6r). Subtotal removal of the tumor was achieved (Fig. 47.6d–f).

47.6 Surgical Pearls

- *Splitting mastoidotomy and exposure of the sigmoid sinus*

  In cases with a well-developed sigmoid sinus, the distance from the outer table of the mastoid process to the sigmoid sinus is very short, so there is no need to insist on splitting the mastoid process to avoid sinus injury.
- *Petrosectomy*

  In drilling the petrous apex, care should be taken not to damage the geniculate ganglion and cochlea. The geniculate ganglion is located in the petrous bone about 3–4 mm deep in the hiatus facialis, and this position roughly coincides with the intersection of the extension line of the GSPN and the line connecting the external and internal auditory foramen. Since the cochlea is also medial and anterior to this junction, cochlear injury is avoided by not removing the anteromedial apex of the arcuate eminence. In addition, as long as the GSPN is preserved anteromedially and the petrous bone is removed in the visible range, the carotid canal is not damaged. After confirmation of the geniculate ganglion of the facial nerve by using free-running facial muscle EMG monitoring, the petrous ridge was resected to expose the entire course of the SPS.

  In order to remove the petrous apex, it is necessary to lift the dura mater of Meckel's cave, but damage to the dura mater on the petrous surface may cause bleeding from the cavernous sinus or inferior petrosal sinus, so a blind procedure should be avoided. For this reason, bone removal at the petrous apex can be safely performed by obtaining mobility of the trigeminal nerve after dural incision and then removing additional bone according to the extent of lesion progression [3].
- *Dural opening and cutting of the tentorium*

  The SPS was transected at a point anterior to the drainage point of the SPV into the SPS to preserve its venous drainage.

47.7 Discussion

Surgical excision of tumors involving the petroclival area is still a significant technical challenge, mainly due to access, especially for medially located lesions, as well as their relationship with the brain stem, cranial nerves, temporal lobe, blood vessels, and venous sinuses. Traditionally, transcranial approaches to such lesions include the retrosigmoid, anterior transpetrosal, and combined presigmoid approach.

Retrosigmoid approach is simple and quickly performed with no risk of temporal retraction. Disadvantage is limited access to the prepontine region and Meckel's cave. If it is necessary to gain more access to the cerebellopontine angle, careful removal of the petrous apex with a high-speed drill is required and when the petrosal vein is well developed, surgical corridor is narrowed.

The anterior transpetrosal approach is an essential technique for a skull base neurosurgeon. This approach is well indicated for small or middle-sized tumors located around petrous apex.

The anterior and posterior combined presigmoid approach is reserved for relatively large meningiomas located around the petroclival region. By using combined presigmoid approach, the surgeon's operative distance to these corridors is shorter than the retrosigmoid approach, and such extensive access leads to better control of the basilar artery and perforator vessels with minimal retraction of the cerebellum and temporal lobe.

In recent years, the evolution of neuroendoscopic techniques has markedly improved the instrumentation and neurosurgical facility of endoscope-assisted procedures. An endoscopic endonasal approach has emerged as an alternative to the classic surgical approaches, providing a direct route to the clival bone. However, such an endonasal approach still has a limited role for skull base lesions involving the lateral posterior cranial fossa. In contrast, an endoscopic transcranial transpetrosal approach has the potential to provide direct access to petrous apex with minimal petrosectomy in near future.

47.8 Conclusion

We have explained the combined transpetrosal approach and its variation for skull base lesions. Complications associated with the approach itself can be avoided by observing the precautions mentioned above. Tumor removal of skull base lesions via combined transpetrosal approaches requires familiarity with the microsurgical anatomy, and training using cadaveric dissections is also useful. We hope that this chapter will prove helpful to surgeons performing a combined transpetrosal approach.

References

1. Hakuba A, Nishimura S, Inoue Y. Transpetrosal-transtentorial approach and its application in the therapy of retrochiasmatic craniopharyngiomas. Surg Neurol. 1985;24(4):405–15.
2. Morisako H, Goto T, Ohata K. Petroclival meningiomas resected via a combined transpetrosal approach: surgical outcomes in 60 cases and a new scoring system for clinical evaluation. J Neurosurg. 2015;122(2):373–80.
3. Morisako H, Ohata H, Shinde B, Nagahama A, Watanabe Y, Goto T. Minimal anterior and posterior combined transpetrosal approach for large petroclival meningiomas. J Neurosurg. 2021;135(4):1180–9.
4. Kunihiro N, Goto T, Ishibashi K, Ohata K. Surgical outcomes of the minimum anterior and posterior combined transpetrosal approach for resection of retrochiasmatic craniopharyngiomas with complicated conditions. J Neurosurg. 2014;120(1):1–11.
5. Fisch U. Transtemporal surgery of the internal auditory canal. Report of 92 cases, technique, indications and results. Adv Otorhinolaryngol. 1970;17:203–40.
6. Garcia-Ibanez E, Garcia-Ibanez JL. Middle fossa vestibular neurectomy: a report of 373 cases. Otolaryngol Head Neck Surg. 1980;88(4):486–90.
7. Inoue T, Rhoton AL Jr, Theele D, Barry MES. Surgical approaches to the cavernous sinus: a microsurgical study. Neurosurgery. 1990;26(6):903–32.
8. Goto T, Ishibashi K, Morisako H, Nagata T, Kunihiro N, Ikeda H, Ohata K. Simple and safe exposure of the sigmoid sinus with presigmoid approaches. Neurosurg Rev. 2013;36(3):477–82.
9. Taptas JN. The so-called cavernous sinus: a review of the controversy and its implications for neurosurgeons. Neurosurgery. 1982;11(5):712–7.
10. Morisako H, Goto T, Ohata H, Goudihalli SR, Shirosaka K, Ohata K. Safe maximal resection of primary cavernous sinus meningiomas via a minimal anterior and posterior combined transpetrosal approach. Neurosurg Focus. 2018;44(4):E11.
11. Parkinson D. A surgical approach to the cavernous portion of the carotid artery. Anatomical studies and case report. J Neurosurg. 1965;23(5):474–83.
12. Umansky F, Elidan J, Valarezo A. Dorello's canal: a microanatomical study. J Neurosurg. 1991;75(2):294–8.

Anatomy and Surgical Approaches to Pineal Region Tumors (PRTs)

48

Imad N. Kanaan

Learning Objectives

- Identify strategies and treatment options for pineal region tumors.
- Highlight vital anatomical landmarks to guide the safe resection of pineal region tumors.
- Underscore the value of hybrid use of endoscopy in surgery of pineal region tumors.
- Discern merits and limitations of each surgical approaches to the pineal region.
- Share recommendations to improve outcome and minimize complications.

48.1 Introduction

Traditionally, the pineal gland was often referred to as the "third eye" due to its deep location within the brain and its association with light. It holds significance in mystic and esoteric spiritual traditions, wherein it is believed to serve as a gateway controlling the realms of spirit, soul & thought, and serving as a metaphysical bridge between physical and spiritual dimensions.

Embryonically, the pineal gland originates from the ectodermal tissue around the seventh week of development. It was first described by Hirophilos (335–280 BC), and later recognized as a gland by Galen (circa 130–210). In 1543, Vesalius likened its shape to that of "a cone from a pine tree", giving rise to its current name, the "pineal gland". In the seventeenth century, Descartes theorized that it housed the spirit and soul, proposing the term "epiphysis cerebri" for it [1–4]. The first documented report of a pineal tumor can be attributed to Derlingcort in 1717.

For several decades, surgical intervention in the pineal region remained stagnant, being considered inoperable. However, a turning point emerged at the beginning of the nineteenth century when Cushing reported the first clinical-autopsy case in 1904. This case pertained to a 28-year-old-man afflicted by a tectal plate glioma. Cushing's frustration became evident as he continued to work to achieve sufficient exposure to justify a viable attempt at pineal tumor removal. Horsley's attempt at the first direct intervention in 1905 marked a crucial milestone. Subsequently, Oppenheim and Krause achieved the landmark feat of successfully excising a tumor from the pineal region in 1913 [5]. Their groundbreaking accomplishment set the stage for further advancements. Krause described his standard Midline Supracerebellar Infratentorial Approach (M-SCITA) in 1926 [6], a framework that would undergo fine-tuning and gain prominence, ultimately being popularized by Stein in 1971 [7]. More recently, the Unilateral Paramedian Approach (PM-SCITA) was introduced by Kulwin et al. in 2016 [8], following a cadaver research study of the supracerebellar venous anatomy by Ueyama et al. [9]. This approach demonstrated superiority over the conventional midline version (M-SCITA). In 1921, Dandy proposed the Parieto-occipital Transcallosal Approach [10], followed by Horrax's performance of the first successful occipital transtentorial resection in 1937 [11]. Poppen later modified this approach [12] in 1960, and Ausman further refined it by introducing the ¾ prone position in 1988 [13]. In 1997, Ellenbogen and Moores [14] described the Endoscopic Third Ventriculostomy (ETV) with simultaneous biopsy (Bx) for management of pineal region tumor- associated hydrocephalus. This has evolved into the primary procedure of choice for germ cell line tumors and hydrocephalus in contemporary practice [15–17].

I. N. Kanaan (✉)
Neurosciences Center, Section of Neurosurgery,
King Faisal Specialist Hospital & Research Center Institution,
Alfaisal University, College of Medicine,
Riyadh, Kingdom of Saudi Arabia

I. N. Kanaan, V. Beneš (eds.), *Neuroanatomy Guidance to Successful Neurosurgical Interventions*,
https://doi.org/10.1007/978-3-031-59838-8_48

48.2 Classification and Clinical Aspects

The classification of pineal region tumors is dynamic in nature, encompassing a diverse array of tumor types. Germ-cell tumors (GCTs) are most prevalent, accounting for 50% of pineal tumors, and are characterized by their proclivity for dissemination through the cerebrospinal fluid (CSF). Among these, the more aggressive variants are categorized as non-germinomateous germ cell tumor (NGGCTs), while immature teratomas are associated with positive oncoproteins marker in both serum and CSF. These markers, including alpha-fetoprotein (AFP), beta human chorionic gonadotropin (BhCG), lactate dehydrogenase (LDH), and placental alkaline phosphatase (PAP) [6], support the diagnosis, can guide therapy, and are utilized in monitoring therapy response and long-term follow-up [18–24].

Germinomas are highly sensitive to radiation and specific chemotherapy treatment. Bifocal germinomas—those with simultaneous involvement of pineal and suprasellar regions—represent 8% of CNS germinoma cases [25, 26]. Tumors arising from the pineal parenchyma are rare and are negative for the aforementioned markers. Such tumors include pineocytomas as well as the more aggressive pineoblastomas, which are classified as grade-IV tumors. Pineoctoma, mature teratomas, and tumors originating from adjacent structures like meningiomas, ependymomas, or epidermoids are predominantly managed by surgery, whereas pineoblastomas require resection and adjuvant treatment akin to treatment applied for yolk-sac tumors and embryonal cell carcinomas.

Clinically, a triad of symptoms are indicative of pineal region tumors, namely: (i) signs of increased intracranial pressure due to obstructive triventricular hydrocephalus, (ii) Parinuad's syndrome related to tectum/brain stem compression and 4th cranial nerves, and (iii) hormonal manifestation due to potential suprasellar region involvement in bifocal germinomas, leading to precocious puberty and diabetes insipidus. In cases where pineal region tumor is suspected, patients should undergo tumor markers studies, in addition to magnetic resonance (MRI) and magnetic resonance venography (MRV) of the brain. These serve to support the initial diagnosis, and facilitate appropriate treatment strategy planning, and selection of the safest surgical approach. Furthermore, spinal MRI should be considered, to exclude potential (to rule out the possibility of) drop metastasis.

48.3 Anatomical Highlights (Figs. 48.1 and 48.2)

- The virtual anatomical boundaries of the pineal region include: Galen venous system and splenium of corpus callosum superiorly, quadrigeminal plate inferiorly, posterior wall of the third ventricle anteriorly, quadrigeminal cistern and cerebellar vermis posteriorly, and thalamic pulvinars laterally.
- The pineal gland is supplied by the pineal artery, which originates from the medial posterior choroidal artery, arising from the posterior cerebral artery.
- Rich in situ vascular anastomosis explains the rarity of infarcts in the mesencephalon and the relative tolerance of the midbrain to direct surgical intervention.
- The great vein of Galen forms a roof-like, dense venous network above the pineal body and the quadrigeminal plate. It is formed from where the internal cerebral veins unite with the basal veins of Rosenthal. Other tributaries of the vein of Galen include the precentral cerebellar vein, internal occipital vein, posterior mesencephalic vein, and posterior ventricular vein.
- The joining of the great vein of Galen with the inferior sagittal sinus forms the straight sinus.

The anatomical landscape of the pineal region resembles the one for the pituitary region in several aspects. Both areas are midline and centrally located regions and both can be accessed via an extra-axial route. These regions embrace a small gland with a deep surrounding vascular network (galenic confluent venous system vs arterial circle of Willis). The pituitary region lies within the anterior incisura and is in close proximity to the chiasmatic cistern, the third ventricle infundibulum and anterior commissure, whereas the pineal region lies within the posterior incisura and is in close prox-

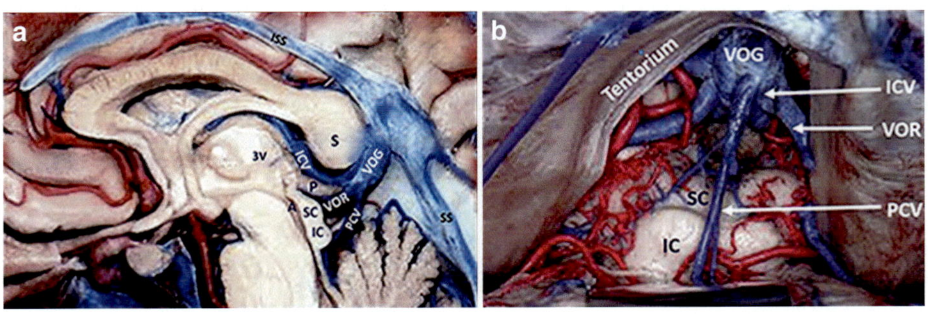

Fig. 48.1 Anatomical landmarks of the pineal region (**a**) Highlight important neurovascular structures on sagittal projection. (**b**) Highlight important neurovascular structures on coronal projection. (Courtesy of "Rhoton Collection ®", AANS, Research and Education Foundation (NERF))

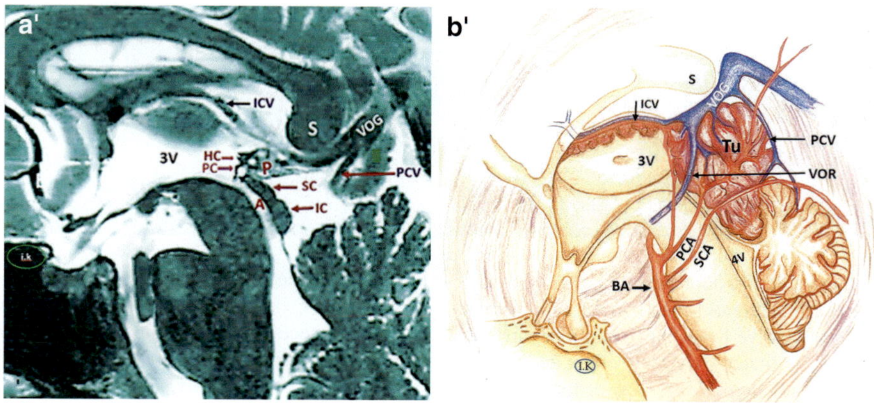

Fig. 48.2 Imaging and schematic drawing of Pineal region. (**a′**) Localization of the anatomical landmarks on an artistic sagittal MRI image. (**b′**) Artist drawing of a pineal region tumor showing its intimate relationship to the surrounding neurovascular structures. KEY: *A* aqueduct, *HC* habenular commissure, *IC* inferior colliculus, *ICV* Internal cerebral vein, *P* pineal gland, *PC* posterior commissure, *PCA* posterior cerebral artery, *PCV* precentral vein, *S* splenium, *SC* superior colliculus, *SCA* superior cerebellar artery, *3V* third ventricle, *4V* fourth ventricle, *VOG* vein of Galen, *VOR* vein of Rosenthal, *Tu* tumor

imity to the quadrigeminal cistern, the posterior wall of the third ventricle and posterior commissure.

48.4 Surgical Approaches

The two most recommended approaches to pineal region tumors are the Paramedian Supra-Cerebellar Infra-Tentorial Approach (PM-SCITA) and the Occipital Trans-tentorial Approach (OTA), whereas, the standard approach (M-SCITA) and the Posterior Inter-hemispheric Trans-Callosal Approach (PITCA) are considered less favorable due to the potential risks and complications associated with them. Yet they may be considered for some selected indications. Management of multi-compartmental tumors may necessitate combined or staged interventions, or considering wide fenestration of the tentorium and the falx cerebri to widen the exposure and enhance the extent of resection [27].

48.4.1 Occipital Transtentorial Approach (OTA): (Figs. 48.3 and 48.4)

The occipital transtentorial approach provides a surgical exposure that is short, safe, and wide, and prevents the inherit complications associated with the sitting position. This approach is frequently selected for management of large and vascular pineal region tumors with superior, inferior, anterior, or dorsal extensions, but with a limited view into the contralateral field.

It is preferable to place an external ventricular drain and select the non-dominant site for this approach, in which the patient is placed in a 3/4 prone position with the head slightly rotated toward the contralateral side, in which position gravity enables natural displacement of the occipital lobe. A u-shaped skin incision is created, followed by 4 × 5 cm occipital craniotomy, Exposing the margins of the transverse and superior sagittal sinus through the bone window. The dura mater is incised in a semicircular fashion and reflected medially over the superior sagittal sinus. Under microscopic view, the falx cerebri, straight sinus, and tentorium cerebelli are identified. A linear incision of the tentorium is made parallel to the straight sinus, up to the posterior tentorial edge.

Extent of mass lesion exposure and resection is dictated by its pathology. Pineal region meningiomas mostly originate from the falcotentorial junction. Their safe resection proceeds by interruption of their meningeal supply from the falx and tentorium, followed by internal debulking, and is completed by tumor capsule dissection and resection using the two forceps pealing technique while maintaining the integrity of the arachnoid membrane and the neurovascular structures. Special attention must be paid not to injure the vein of Galen and its tributaries. The use of endoscope surveillance to enhance surgical visualization and achieve better exposure and safer resection has been adopted in current practice. These steps are also applied for piecemeal removal of other types of tumors after sharp opening of the thickened arachnoid membrane of the pineal/quadrigeminal cisterna.

48.4.2 The Midline Supracerebellar-Infratentorial Approach (M-SCITA). (Figs. 48.5 and 48.6)

The standard Midline Supracerebellar-Infratentorial Approach (M-SCITA) is commonly selected as a midline approach with a direct view of the pineal region tumor via an extra-axial corridor without transgressing the Galenic sys-

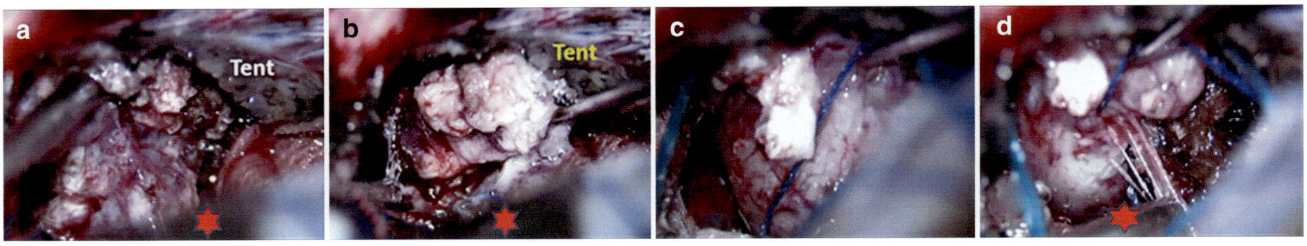

Fig. 48.3 Right occipital trans-tentorial approach for gross microsurgical resection of pineal region meningioma in a 46-year-old female with a six-month history of progressive headache. The procedure was performed in ¾ prone position, facilitated by gravity effect and temporary insertion of lumbar drain. (**a**–**d**) steps of tumor resection following long parafalcine tentorial incision parallel to and 1 cm away from straight sinus; devascularization, internal decompression, dissection, and gross resection of the meningioma, maintaining cisterna arachnoid membrane integrity (*) soft spatula and cotton patty covering the right occipital lobe, applied in dynamic fashion

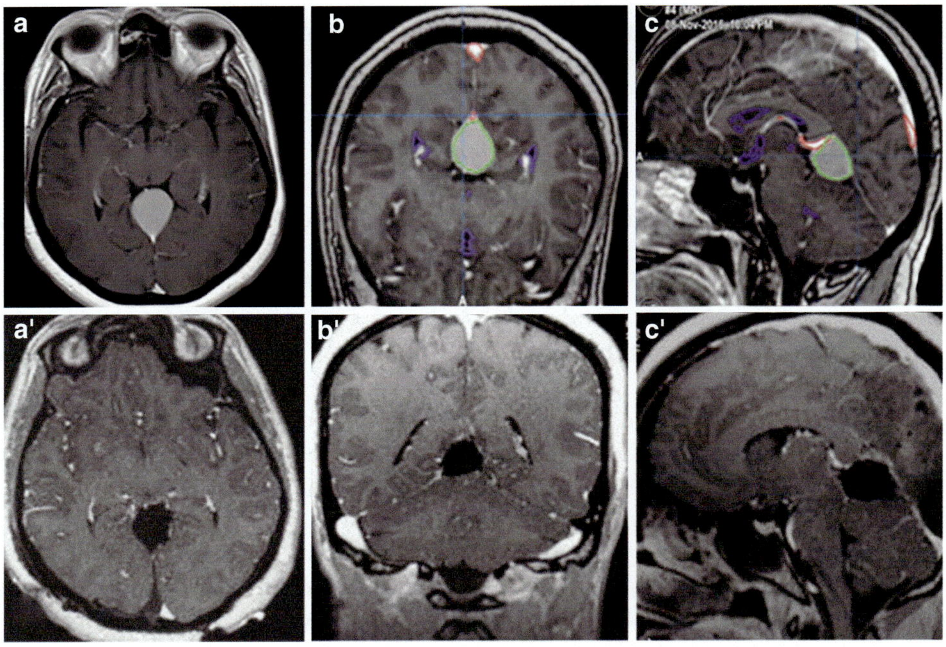

Fig. 48.4 Magnetic Resonance Imaging (MRI) of the same patient. (**a**–**c**) preoperative MRI in axial, coronal and sagittal views depict the homogeneously enhanced pineal region meningioma. (**a′**–**c′**) Postoperative MRI control confirmed the gross total resection of the meningioma. Patient has full clinical recovery and no signs of recurrence after 10 years follow up

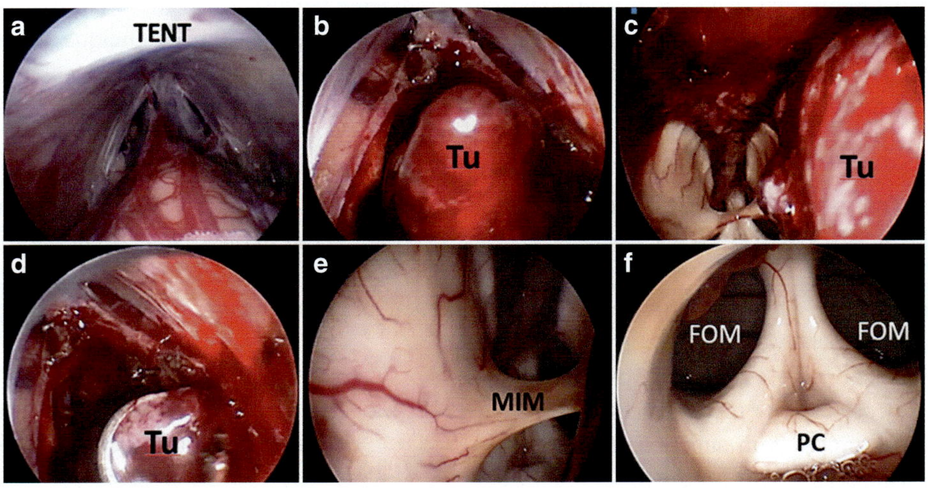

Fig. 48.5 Supracerebellar infratentorial approach (SCITA) with hybrid use of endoscopy. Stepwise endoscopic-assisted near total resection of pineal parenchymal tumor grade II-WHO. 49-year-old female presented with 5-month history of refractory headache with negative tumor markers. Procedure performed via SCITA approach in semi-sitting position. (**a**) Endoscopic view of SCIT-region and central bridging vein. (**b**) Tumor mass identify under the tentorial notch after release of bridging vein (**c**, **d**) Progressive resection of tumor using cavitron surgical aspirator and ring curettes (**e**, **f**) Endoscopic panoramic view of 3rd ventricle and its anatomical landmarks including interthalamic adhesion (massa intermedia) **MIM**, Foramina of Monro **FOM** and posterior commissure **PC**. Patient remained clinically intact, received postoperative fractionated radiotherapy, and no signs of recurrence on updated MRI 8 years later

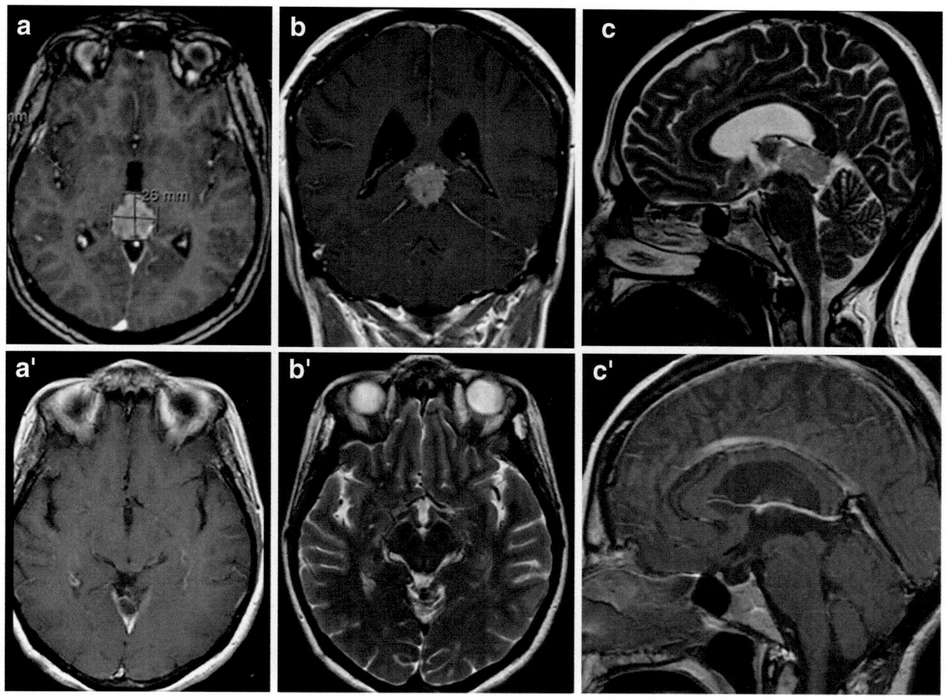

Fig. 48.6 (**a–c**) Preoperative MRI in axial, coronal, and T2 sagittal projection demonstrates the homogeneously enhanced tumor with extension into the posterior third ventricle. (**a′–c′**) Latest follow up MRI 8 years post-treatment showed no signs of residual/recurrent tumor

tem located superior to the tumor. It is preferably performed with the patient in a sitting position, with the main advantages of gravity-assisted cerebellar retraction away from the tentorium that helps easier dissection of adherent veins off the tumor surface and decreased pooling of blood and cerebrospinal fluid (CSF) in the surgical field. Once the corridor between the cerebellum and inferior surface of the tentorium is developed, the precentral cerebellar vein and/or the superior vermian vein are identified and in most cases cut before the tumor is dissected. Rarely, the tectal veins or the superior and inferior quadrigeminal veins are very well developed, hindering the approach to the tumor and requiring an oblique trajectory. Dividing the precentral cerebellar vein is generally not advised; but, it may be performed without major sequalae. Low-seated torcula and steep tentorial angle may limit this approach; however, this limitation can be overcome by the recently introduced unilateral paramedian version of this approach, described below, and the hybrid use of endoscope.

Inherit potential complications associated with sitting position include air embolism, pneumocephaly, and hypotension; these can be decreased or avoided by considering a semi-sitting position, the regular use of doppler sonography monitor, copious irrigation, and meticulous hemostasis. Alternatively, some neurosurgeons may prefer the use of the Concorde position to avoid these potential complications and mostly in pediatrics or elderly [28].

48.4.3 Endoscopic-Assisted Paramedian SCIT Approach (PM-SCITA): (Fig. 48.7)

Among the merits of this approach are that it provides a simple and less invasive functional exposure, maintaining the integrity of the torcula, contralateral dominant transverse sinus, and the vermian bridging veins. This approach overcomes the limitation of the steep tentorial angle and the obstruction of the culmen while facilitating the "cross-court view" of the contralateral tumor extension, hence enhancing the extent of tumor resection with better surgical outcomes [8, 16, 29–34].

The patient is preferably placed either in a semi-sitting position or in a lateral decubitus position. A paramedian skin incision of 8 cm length is placed over the non-dominant transverse sinus (left side) level under neuronavigation guidance. One third of the incision is then extended above the sinus level and two thirds below it. A unilateral 4 × 5 suboccipital craniotomy is performed using one or two burr holes over the lower edge of the transverse sinus and about 1.5–2 cm lateral to the torcula. The exposure should include the lower margin of the transverse sinus. A curved dura incision is made and reflected upward over the transverse sinus. One or two encountered bridging veins on the cerebellar surface can be coagulated and sacrificed without major sequelae, leaving the medial prominent bridging veins intact. However, the tension on such bridging vein during en-route manipula-

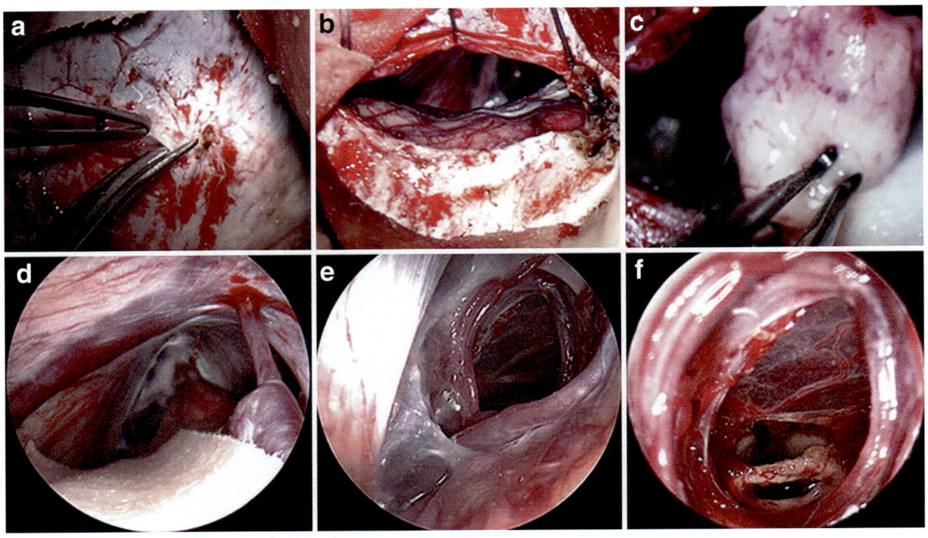

Fig. 48.7 Paramedian SCITA left side with hybrid use of endoscopy. (**a**) Left suboccipital craniotomy in semi-sitting position, semi-elliptical incision of dura under exposed edge of transverse sinus. (**b**) Absence of major bridging veins on para median surface of cerebellum. Relaxed and inferiorly displaced cerebellum by gravity with satisfactory exposure of infra-tentorial space. (**c**) Microsurgical dissection and resection of tumor mass after central decompression with special care not to injure Galen venous system displaced anteriorly and superiorly. (**d**–**f**) Endoscopic visualization of surrounding vascular structures and surveillance for any residual lesion extension (Courtesy of H. Schroeder)

tion can be minimized by performing a small semilunar dural patch incision around its junction with one of the remnant embryonic tentorial sinuses. The surgical microscope is introduced and two retraction sutures are placed along the posterior aspect of the tentorium to uplift the transverse sinus and enhance the supracerebellar infratentorial avenue. This is followed by a sharp incision of the thickened arachnoid membrane and entering both the quadrigeminal and ambient cisterns, paying special attention to maintain the integrity of the neurovascular structure, providing arachnoid protection to the trochlear nerve. Progressive dissection is continued medially to adequately visualize the midline large tumor at the posterior incisura from an oblique trajectory. The complex venous system is usually displaced superiorly and medially. At this stage, a zero degree endoscope may be introduced for surveillance and visualization of blind surgical points, before mounting the scope on a pneumatic scope-holder to complement the progressive microsurgical tumor resection under copious irrigation. This approach concludes with water-tight dura closure, fixation of the bone flap using micro-plate system, and wound closure in layers [35–41].

48.4.4 Posterior Interhemispheric Trans-callosal Approach: (PITA)

This approach, with the patient in a prone or lateral position, is favored by pediatric neurosurgeons for resection of posterior third ventricular tumors, thalamic tumors, and some large pineal region tumors with superior and anterior extension into the third ventricles. This approach may result in either transient or permanent surgical morbidities such as visual impairment resulting from calcarine lobe venous infarction. The occurrence of such infarction is precipitated by the retraction of the confined occipital lobe adjacent to the falx cerebi and the tentorium, or the sacrifice of several bridging veins during the exposure. Other potential morbidities include short-term memory deficit during en-route injury of both fornices, and disconnection syndrome because of incising the posterior corpus callosum [42–44].

48.4.5 Endoscopic Third Ventriculostomy (ETV) & Biopsy (BX) (Figs. 48.8 and 48.9)

Simultaneous ETV and tumor biopsy is a safe, minimally invasive procedure and is currently the primary intervention of choice for pineal region tumors with positive tumor markers, or for pineal region tumors with negative marker but radiological indications of germinoma associated with hydrocephalus.

This approach can be performed safely and effectively using a single burr hole technique with a 0° rigid endoscope, providing better visibility and no additional risk of memory loss, and can be enhanced by the use of 30° endoscope or placing a second burr hole more anteriorly.

The procedure starts with CSF collection for cytology and tumor markers, followed by third ventriculostomy and finally biopsy of the tumor. Taking a biopsy after (rather than before) the ETV prevents tumor biopsy-associated bleeding visually obscuring third ventricular floor landmark, which may necessitate abandoning the ETV procedure [14–18].

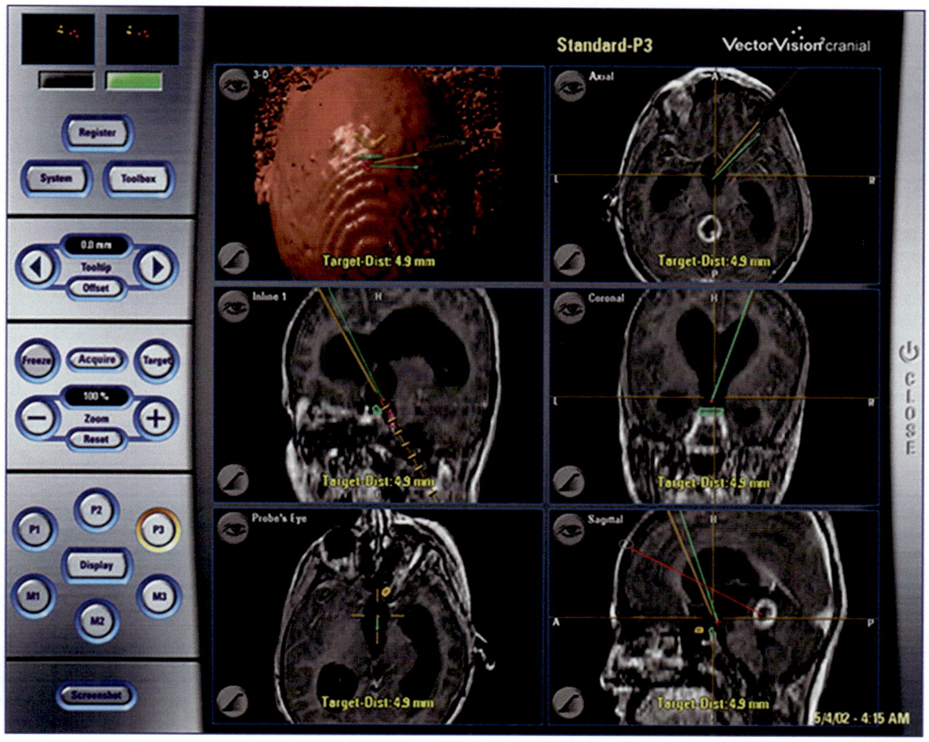

Fig. 48.8 Endoscopic third ventriculostomy (ETV) and simultaneous PRTs biopsy. MRI showed intraoperative guidance for endoscopic 3rd ventriculostomy and simultaneous pineal region tumor biopsy, facilitated by use of neuronavigation. The final pathology confirmed a germ cell tumor. The patient had negative tumor markers and negative spinal MRI preoperatively. He underwent adjuvant treatment with radiation and chemotherapy and achieved excellent response and long-term outcome with a follow up of 14 years

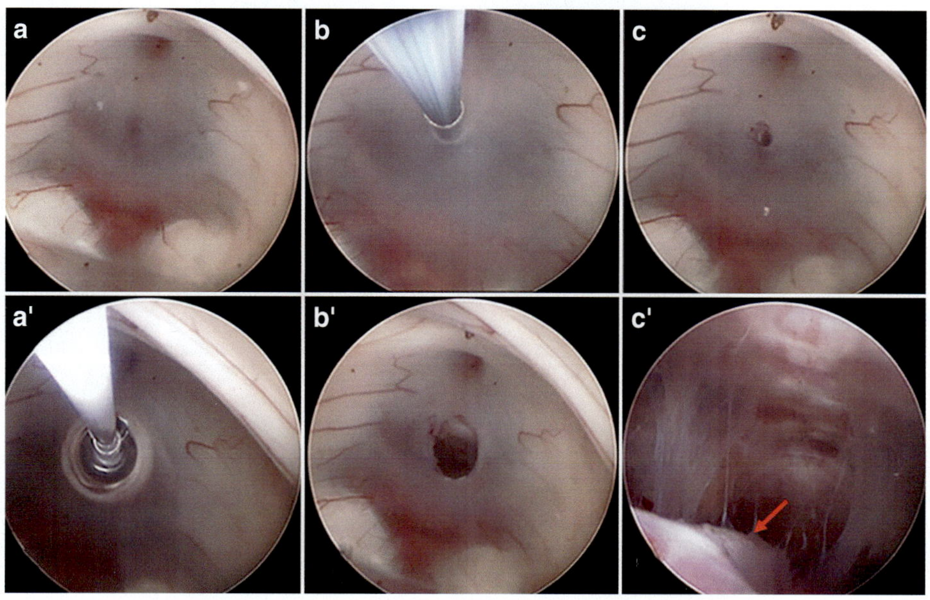

Fig. 48.9 Endoscopic third ventriculostomy (ETV). (**a**) Identifying the most transparent area of the floor of third ventricle between mammillary bodies posteriorly, and dorsum sellae and infundibular recess anteriorly. (**b**, **c**) Fenestration of the 3V floor using blunt probe, or laser fiber just behind the dorsum Sella. (**a′**, **b′**) Dilatation of the fenestration by using French Fogarty catheter or alternative small balloon catheter. (**c′**) Endoscopic-assisted release of any thickened membrane or trabeculations and visualization of the basilar artery (red arrow)

Upon visualization of the tumor after ETV completion, the 30° angled lens is rotated to achieve a posterior direction of view in cases where a single entry site is used. If a separate anterior entry for endoscopic biopsy is used, a 0° lens allows visualization of the posterior third ventricle. A single intermediately positioned burr hole is preferred for this approach. Upon tumor identification, biopsy forceps are used to obtain tumor tissue samples from areas on the surface that are relatively avascular, most likely represent pathological tissue, and need the least torque. Varying degrees of hemorrhage may occur following the biopsy, the majority of which may be controlled with continuous irrigation, balloon tamponade, or electrocautery.

All cases with pineal region tumors should be discussed within a multi-disciplinary team that includes an experienced neurosurgeon, medical and radiation oncologists in order to select the proper management plan and to achieve good outcome (Figs. 48.10 and 48.11).

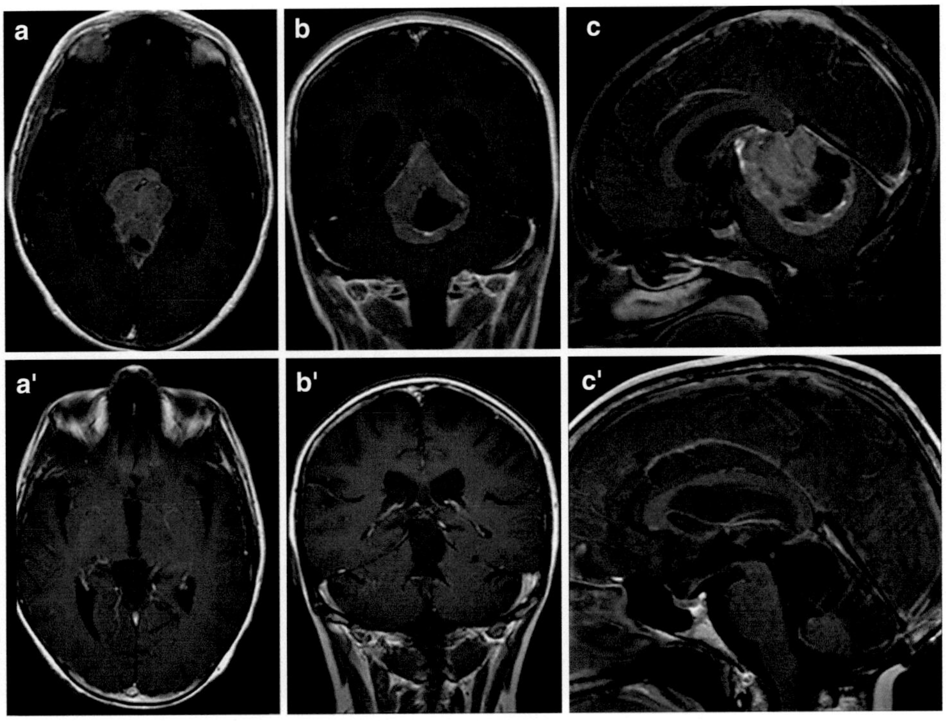

Fig. 48.10 Pinealoblastoma in 29-year-old patient with negative tumor markers, treated with partial tumor resection, chemotherapy, and radiation therapy. The patient has been in remission for 15 years but remains on treatment for epilepsy. (**a**–**c**) Pre-treatment MRI demonstrates the giant pineal region tumor with caudal extension and anterior superior extension into the posterior 3rd ventricle. (**a′**–**c′**) Recent MRI after 15 years showed excellent treatment response with no signs of recurrence

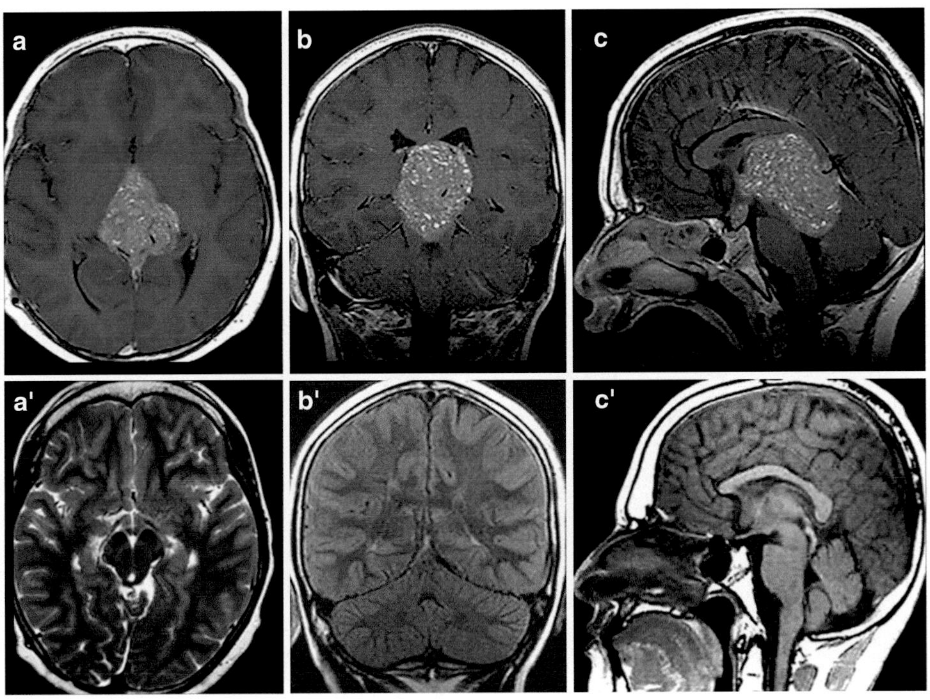

Fig. 48.11 A case of giant pineal region germ cell tumor treated with chemotherapy and fractionated radiotherapy. Patient required transient VP-shunt for ensuing hydrocephalus upon completing his radiation course. The patient has been in remission for 12 years with no neurological deficits. (**a**–**c**) Pre-treatment MRI depicted a large pineal region tumor with extension into the posterior third ventricle, caudal extension, and compression of the tectal plate. (**a′**–**c′**) Post-treatment MRI in axial T2, coronal, and sagittal T1 views demonstrated no signs of residual or major tumor recurrence. This recent follow-up was performed 12 years after the initial treatment and intentionally done without contrast due to the partially impaired kidney function provoked by the side effect of chemotherapy

48.5 Surgical Pearls

- Tumor marker studies and complete whole brain and spinal MRI and MRV should be conducted to establish a diagnosis and plan the intervention, as well as to save the patient potentially unnecessary major surgery.
- Endoscopic Third Ventriculostomy (ETV) with concomitant biopsy (BX) is considered the primary treatment of choice for pineal region germ-cell tumors with hydrocephalus. ETV should proceed ahead of the BX.
- Primary VP-shunt placement should be avoided in treatment of germ cell tumors with hydrocephalus in order to prevent peritoneal metastasis.
- Use of neuronavigation is of great value for preoperative planning, localization, and safe surgical guidance.
- Unilateral PM-SCITA overcomes the limitations caused by the steep tentorial angle or other anatomical obstruction of the standard M-SCITA, making it a safer and less invasive functional approach.
- The Occipital Trans-tentorial Approach (OTA) is recommended for resection of larger PRTs with tumor extensions.
- Hybrid use of endoscopy promotes less invasive intervention and enhances surgical field visualization and extent of tumor resection.
- Integrity of the vein of Galen and its tributaries with maintenance of the arachnoid membrane toward adjacent neurovascular structures is of paramount importance for safe tumor resection.
- It is important to avoid coagulating and dividing the precentral cerebellar vein near the confluence of the basal veins of Rosenthal, as this leads to progression of venous thrombosis of the collateral venous circulation and results in hemorrhagic infarction [45].
- Stereotactic biopsy is less favorable intervention for pineal region tumor due to the risk of bleeding and limited tissue sampling and interpretation. However, it may be considered useful in some eccentric pineal region tumors with smaller ventricles [46, 47].
- A multidisciplinary approach is important for optimal treatment strategy selection. The role of radiation therapy and chemotherapy should not be underestimated as both stand at the forefront for treatment of germ cell tumors, its variants, and other malignant pineal region tumors [18–20, 48].

48.6 Discussion

The incidence of pineal region tumors (PRTs) is rare and encompasses a wide range of pathologies. The complex neurovascular anatomy of the pineal region poses a major challenge during their surgical treatment. Surgical excision is the treatment of choice for most PRTs with excellent clinical outcomes; whereas, the management of germ cell-line tumors and other radiosensitive malignancies exclusively require radiation therapy, chemotherapy, and ETV to treat the associated hydrocephalus. The most frequently used open surgical approaches are the PM-SCITA or its standard version (M-SCITA), the OTA and PITA. Selection of the appropriate intervention is dictated by the tumor pathology, location and direction of growth, its relations to the adjacent neurovascular structures and tentorial angle, patient age, as well as the surgeon's preference and experience. Each of these approaches has its own merits, indications, limitations, and potential complications. The mainstay for successful management rests on individualizing treatment plans and having a multidisciplinary treating team.

48.7 Conclusion

The intricacy of the pineal region anatomy and complex surgical approaches to pineal region tumors pose a challenging endeavor. The pathology and extension of the tumor, presence of tumor markers or hydrocephalus, the availability of a multi-disciplinary approach team, and surgical experience all contribute towards the selection of a proper management strategy. Variations of potential surgical interventions include the paramedian SCITA or its standard midline SCITA, the occipital transtentorial approach, or the posterior inter hemispheric transcallosal approach for selective indications. Each approach can be enhanced by the hybrid use of endoscopy. Furthermore, endoscopic third ventriculostomy with concomitant biopsy is the best intervention for management of germinomas, immature teratomas, and other similar cancers complemented with radiation therapy and chemotherapy.

References

1. Lokhorst GJ. Descartes and the pineal gland. Stanford Encyclopedia of Philosophy; 2008. https://plato.stanford.edu/entries/pineal-gland/. Accessed 13 July 2010.
2. Choudhry O, Gupta G, Prestigiacomo CJ. On the surgery of the seat of the soul: the pineal gland and the history of its surgical approaches. Neurosurg Clin. 2011;22(3):321–33.
3. Pendl G. The surgery of pineal lesions – historical perspective. In: Neuwelt AE, editor. Diagnosis and treatment of pineal region tumors. Baltimore: Williams and Wilkins; 1984. p. 139–54.
4. Matula C. Tumors of the pineal region. In: Principles of neurological surgery. 2012; W.B. Saunders. p. 565–84. https://clinicalgate.com/tumors-of-the-pineal-region/
5. Oppenheim H, Krause F. Operative Erfolge bei Geschwülsten der Sehhügel und Vierhügelgegend, vol. 50. Berlin Klin Wschr; 1913. p. 2316–22. [Google Scholar].
6. Krause F. Operative Frielegung der Vierhugel, nebst Beobachtungen uber Hirndruck und Dekompression. Zentrabl Chir. 1926;53:2812–9. [Google Scholar].

7. Stein BM. The infratentorial supracerebellar approach to pineal lesions. J Neurosurg. 1971;35:197–202. [PubMed] [Google Scholar].
8. Kulwin C, Matsushima K, Malekpour M, Cohen-Gadol AA. Lateral supracerebellar infratentorial approach for microsurgical resection of large midline pineal region tumors: techniques to expand the operative corridor. J Neurosurg. 2016;124(1):269–76. https://doi.org/10.3171/2015.2.JNS142088.
9. Ueyama T, Al-Mefty O, Tamaki N. Bridging veins on the tentorial surface of the cerebellum: a microsurgical anatomic study and operative considerations. Neurosurgery. 1998;43(5):1137–45.
10. Dandy WE. An operation for the removal of pineal tumors. Surg Gynecol Obstet. 1921;33:113–9. [Google Scholar].
11. Horrax G. Extirpation of a huge pinealoma from a patient with pubertas praecox. Arch Neurol Psychiatr. 1937;37:385–97. [Google Scholar].
12. Poppen JL. The right occipital approach to a pinealoma. J Neurosurg. 1966;25:706–10. [PubMed] [Google Scholar].
13. Ausman JI, Malik GM, Dujovny M, Mann R. Three-quarter prone approach to the pineal-tentorial region. Surg Neurol. 1988;29:298–306. [PubMed] [Google Scholar].
14. Ellenbogen RG, Moores LE. Endoscopic management of a pineal and suprasellar germinoma with associated hydrocephalus: technical case report. Minim Invasive Neurosurg. 1997;40:13–5. [PubMed] [Google Scholar].
15. Morgenstern PF, Souweidane MM. Pineal region tumors: simultaneous endoscopic third ventriculostomy and tumor biopsy. World Neurosurg. 2013;79(2 Suppl):S18.e9–13. [PubMed] [Google Scholar].
16. Azab WA, Nasim K, Salaheddin W. An overview of the current surgical options for pineal region tumors. Surg Neurol Int. 2014;5:39. https://doi.org/10.4103/2152-7806.129430.
17. O'Brien DF, Hayhurst C, Pizer B, Mallucci CL. Outcomes in patients undergoing single-trajectory endoscopic third ventriculostomy and endoscopic biopsy for midline tumors presenting with obstructive hydrocephalus. J Neurosurg. 2006;105:219–26.
18. Zaazoue MA, Goumnerova LC. Pineal region tumors: a simplified management scheme. Childs Nerv Syst. 2016;32:2041–5.
19. Tamrazi B, Nelson M, Blüml S. Pineal region masses in pediatric patients. Neuroimaging Clin N Am. 2017;27:85–97. https://doi.org/10.1016/j.nic.2016.08.002. [PubMed] [CrossRef] [Google Scholar].
20. Favero G, Bonomini F, Rezzani R. Pineal gland tumors: a review. Cancers. 2021;13(7):1547. https://doi.org/10.3390/cancers13071547. Published online 2021 Mar 27. PMCID: PMC8036741PMID: 33801639.
21. Carr C, O'Neill BE, Hochhalter CB, Strong MJ, Ware ML. Biomarkers of pineal region tumors: a review. Ochsner J. 2019;19:26–31. https://doi.org/10.31486/toj.18.0110. [PMC free article] [PubMed] [CrossRef] [Google Scholar].
22. Wang KY, Chen MM, Malayil Lincoln CM. Adult primary brain neoplasm, including 2016 World Health Organization classification. Neuroimaging Clin N Am. 2021;31:121–38. https://doi.org/10.1016/j.nic.2020.09.011. [PubMed] [CrossRef] [Google Scholar].
23. Baranzelli MC, Kramar A, Bouffet E, Quintana E, Rubie H, Edan C, Patte C. Prognostic factors in children with localized malignant nonseminomatous germ cell tumors. J Clin Oncol. 1999;17(4):1212. CAS [PubMed] [Google Scholar].
24. Lee D, Suh YL. Histologically confirmed intracranial germ cell tumors; an analysis of 62 patients in a single institute. Virchows Arch. 2010;457:347–57.
25. Jennings MT, Gelman R, Hochberg F. Intracranial germ-cell tumors: natural history and pathogenesis. J Neurosurg. 1985;63(2):155–67. [PubMed] [Google Scholar].
26. Dhall G, Khatua S, Finlay JL. Pineal region tumors in children. Curr Opin Neurol. 2010;23:576–82. [PubMed] [Google Scholar].
27. Ziyal IM, Sekhar LN, Salas E, Olan WJ. Combined supra/infratentorial-transsinus approach to large pineal region tumors. J Neurosurg. 1998;88(6):1050–7. https://doi.org/10.3171/jns.1998.88.6.1050. PMID: 9609300.
28. Kobayashi S, Sugita K, Tanaka Y, Kyoshima K. Infratentorial approach to the pineal region in the prone position: Concorde position. Technical note. J Neurosurg. 1983;58:141–3. [PubMed] [Google Scholar].
29. Yaşargil MG. Microneurosurgery, vol. 1. New York: Thieme; 1987.
30. Thaher F, Kurucz P, Fuellbier L, Bittl M, Hopf NJ. Endoscopic surgery for tumors of the pineal region via a paramedian infratentorial supracerebellar keyhole approach (PISKA). Neurosurg Rev. 2014;37(4):677–84. https://doi.org/10.1007/s10143-014-0567-1.
31. Van den Bergh R. Lateral-paramedian infratentorial approach in lateral decubitus for pineal tumours. Clin Neurol Neurosurg. 1990;92(4):311–6.
32. Choque-Velasquez J, Resendiz-Nieves J, Jahromi BR, Colasanti R, Baluszek S, Muhammad S, Hernesniemi J. Midline and paramedian supracerebellar infratentorial approach to the pineal region: a comparative clinical study in 112 patients. World Neurosurg. 2020;137:e194–207.
33. Ogata N, Yonekawa Y. Paramedian supracerebellar approach to the upper brain stem and peduncular lesions. Neurosurgery. 1997;40(1):101–5.
34. Zaidi HA, Elhadi AM, Lei T, Preul MC, Little AS, Nakaji P. Minimally invasive endoscopic supracerebellar-infratentorial surgery of the pineal region: anatomical comparison of four variant approaches. World Neurosurg. 2015;84(2):257–66.
35. Gore PA, Gonzalez LF, Rekate HL, Nakaji P. Endoscopic supracerebellar infratentorial approach for pineal cyst resection: technical case report. Oper Neurosurg. 2008;62(3 Suppl 1):108–9. discussion 109.
36. Broggi M, Darbar A, Teo C. The value of endoscopy in the total resection of pineocytomas. Oper Neurosurg. 2010;67(3 Supp l):ons159–65.
37. Sood S, Hoeprich M, Ham SD. Pure endoscopic removal of pineal region tumors. Childs Nerv Syst. 2011;27:1489–92. Technical Note Published: 24 May 2011.
38. Shahinian H, Ra Y. Fully endoscopic resection of pineal region tumors. J Neurol Surg B Skull Base. 2013;74:114–7. https://doi.org/10.1055/s-0033-1338165. PMCID: PMC3712663. PMID: 24436899.
39. Gu Y, Hu F, Zhang X. Purely endoscopic resection of pineal region tumors using infratentorial supracerebellar approach: how I do it. Acta Neurochir. 2016;158(11):2155–8. https://doi.org/10.1007/s00701-016-2895-0. Epub 2016 Aug 9. PMID: 27506850.
40. Hua W, Xu H, Zhang X, Yu G, Wang X, Zhang J, Pan Z, Zhu W. Pure endoscopic resection of pineal region tumors through supracerebellar infratentorial approach with 'head-up' park-bench position. Neurol Res. 2023;45(4):354–62. https://doi.org/10.1080/01616412.2022.2146266. Epub 2022 Dec 12. PMID: 36509700 Review.
41. Ding Y, Sun L, Hu Y, Zhai W, Zhang L, Yu Z, Wu J, Chen G. Combined microscopic and endoscopic surgery for pineal region meningiomas using the occipital-parietal transtentorial approach. Front Oncol. 2022;12:828361. https://doi.org/10.3389/fonc.2022.828361.
42. Rhoton AL Jr. Cranial anatomy and surgical approaches. Neurosurgery. Lippincott: Williams and Wilkins; 2003.
43. Liu JK. Endoscopic-assisted interhemispheric parieto-occipital transtentorial approach for microsurgical resection of a pineal region tumor: operative video and technical nuances. Neurosurg Focus. 2016;40(video suppl 1):1. https://doi.org/10.3171/2016.1.FocusVid.15450. PMID: 26722692.

44. Jia W, Ma Z, Liu IY, Zhang Y, Jia G, Wan W. Transcallosal interforniceal approach to pineal region tumors in 150 children. J Neurosurg Pediatr. 2011;7(1):98–103.
45. Kodera T, Bozinov O, Sürücü O, Ulrich NH, Burkhardt JK, Bertalanffy H. Neurosurgical venous considerations for tumors of the pineal region resected using the infratentorial supracerebellar approach. J Clin Neurosci. 2011;18(11):1481–5.
46. Zacharia BE, Bruce JN. Stereotactic biopsy considerations for pineal tumors. Neurosurg Clin. 2011;22(3):359–66. [PubMed] [Google Scholar].
47. Regis J, Bouillot P, Rouby-Volot F, Figarella-Branger D, Dufour H, Peragut JC. Pineal region tumors and the role of stereotactic biopsy: review of the mortality, morbidity, and diagnostic rates in 370 cases. Neurosurgery. 1996;39:907–12. https://doi.org/10.1097/00006123-199611000-00003. discussion 912–4. [PubMed] [CrossRef] [Google Scholar].
48. Iorio-Morin C, Kano H, Huang M, Lunsford LD, Simonová G, Liscak R, Cohen-Inbar O, Sheehan J, Lee CC, Wu HM, et al. Histology-stratified tumor control and patient survival after stereotactic radiosurgery for pineal region tumors: a report from the international gamma knife research foundation. World Neurosurg. 2017;107:974–82. https://doi.org/10.1016/j.wneu.2017.07.097. [PubMed] [CrossRef] [Google Scholar].

49 Surgical Strategies for Jugular Foramen Tumors

James K. Liu, Andrea L. Castillo, and Ali Tayebi Meybodi

Learning Objectives

At the end of this chapter, you will be able to

- Understand the anatomy of the jugular foramen.
- Describe the different approaches to access lesion around the jugular foramen.
- Know clinical and surgical pearls for avoiding complications and maximizing cranial nerve preservation.

49.1 Introduction

The jugular foramen is a complex anatomical structure that can vary considerably in size, shape. and dimensions. It has an intracranial and extracranial compartment that makes its understanding very challenging, creating the approach for lesions around this area difficult to access due to its complex anatomy, deep location, and associated delicate neurovascular structures, that if damaged, can pose a great morbidity to patients [1–3].

The most common tumors found within the jugular foramen are paragangliomas, followed by schwannomas, and less commonly, meningiomas, chordomas, chondrosarcomas, and plasmacytomas. Metastases and malignant tumors arising in adjacent anatomic structures (nasopharynx, parotid, and temporal bone), in advanced stages, may also spread to the jugular foramen secondarily. Lastly, endolymphatic sac tumors arise at the posterior medial aspect of the petrous bone and can potentially extend to the jugular foramen [4, 5].

It is important to have an extended knowledge of the anatomy of the jugular foramen and its surroundings, to understand each type of tumor growth pattern, and how it is related to the neurovascular structures that encloses them. Having these tools will help the neurosurgeon develop the appropriate surgical techniques to remove these tumors with greater confidence, better extent of volumetric tumor removal, and improved surgical outcomes while avoiding cranial nerve morbidities.

49.2 History/Background

The surgical removal of these tumors remains a challenge. Advances in neuroimaging, microsurgery, and modern cranial base surgery have allowed safe resection of these tumors with lower rates of morbidity and mortality [6]. Total removal of these tumors increases the likelihood of achieving surgical cure and many approaches have been described to access this region.

Fisch [7] has described several lateral infratemporal fossa approaches for resection of large foramen jugular tumors and emphasizes transection and blind sac closure of the external ear canal, permanent rerouting of the facial nerve, and anterior displacement of the mandible for exposure of the vertical infratemporal C7 segment of the internal carotid artery. These maneuvers may result in conductive hearing loss, facial nerve palsy, and problems with jaw opening or mastication, respectively. A combined two-stage resection has also been advocated for larger Fisch Type D2 tumors (tumors with intracranial extension greater than 2 cm in diameter)

J. K. Liu (✉)
Skull Base Institute of New Jersey, Neurosurgeons of New Jersey, NYU Langone Neurosurgery Network, Livingston, NJ, USA

Department of Neurosurgery, Cooperman Barnabas Medical Center, RWJ Barnabas Health, Livingston, NJ, USA
e-mail: jliu@neurosurgerynj.com

A. L. Castillo
Department of Neurosurgery, Antonio Lenin Fonseca Hospital, Managua, Nicaragua

A. T. Meybodi
Department of Neurological Surgery, New Jersey Medical School, Newark, NJ, USA
e-mail: at1085@njms.rutgers.edu

I. N. Kanaan, V. Beneš (eds.), *Neuroanatomy Guidance to Successful Neurosurgical Interventions*,
https://doi.org/10.1007/978-3-031-59838-8_49

[8]. In these cases, the extracranial portion of tumor is removed during the first stage, and the intracranial tumor is removed during a second stage at a later date to minimize the complications of excessive blood loss when dealing with hypervascular paragangliomas.

Subsequently, Al-Mefty [9] described a combined infratemporal and posterior fossa approach for the removal of giant tumors of this region. He emphasized a single-stage surgery where both a temporal and posterior fossa craniotomy is done, having the advantage of less chances of anesthesia complications and infections due to second-stage surgery. Nevertheless, the anterior transposition of the facial nerve is also described in this technique, which can carry a potential risk of postoperative facial paresis or paralysis [10].

Most recently we described a one-stage transjugular posterior infratemporal fossa approach (that is a variation of Fukushima's [11] postauricular transmastoid technique) that allows radical resection of tumors that are located around the jugular foramen, the lower clivus, and the high cervical region from an anterolateral direction. This approach is a combination of the transmastoid, retro- and infralabyrinthine, transjugular, infrajugular transcondylar transtubercular, and high cervical approaches. Total exposure of the jugular foramen can be achieved, and multidirectional approaches can be performed, including suprajugular (infralabyrinthine), transjugular, and infrajugular (retrosigmoid/transcondylar) exposures. Both intracranial and extracranial tumor can be removed in a one-stage procedure without the transection of the external ear canal and permanent rerouting of the facial nerve, having less chances of facial nerve paresis or paralysis [6, 12]. Instead, a fallopian bridge technique is used where the facial nerve remains protected in its bony canal. Nevertheless, in select cases, where there is extension into the middle ear, such as jugulotympanic paragangliomas, transection and blind sac closure of the external ear canal may be necessary to provide adequate exposure at the cost of conductive hearing loss.

Although there are a variety of surgical approaches to the jugular foramen, it is important to individualize each case and tailor the approach for the patient to provide the appropriate necessary exposure to remove the most amount of tumor with the least amount of neurovascular morbidity.

49.3 Anatomy Highlights

- The jugular foramen has several components, which can be described according to osseous landmarks and its adjacent bony structures, dural configuration, neurovascular components that traverse through and around it, and muscular relationships. From a general perspective, the jugular foramen is bordered anteriorly by the carotid artery, laterally by the facial nerve, medially by the hypoglossal nerve, and inferiorly by the vertebral artery [3, 13] (Fig. 49.1). The structures that go through the foramen are the sigmoid sinus and jugular bulb, the inferior petrosal sinus, meningeal branches of the ascending pharyngeal and occipital arteries, the glossopharyngeal, vagus, and accessory nerves, the tympanic branch of the glossopharyngeal nerve (Jacob's nerve), the auricular branch of the vagus nerve (Arnold's nerve), and the cochlear aqueduct [1, 14].
- From the osseous perspective, the foramen is located between the temporal bone and the occipital bone and is configured around the sigmoid and inferior petrosal sinus. These sinuses serve as landmark and divide the foramen into two compartments: a larger lateral sigmoid part through which the sigmoid sinus empties and a smaller anteromedial petrosal part through which the inferior petrosal sinus empties [1, 12]. The junction of the sigmoid and petrosal parts is the site of bony prominences on the opposing surfaces of the temporal and occipital bones, called the intrajugular processes, which are joined by a fibrous bridge called the intrajugular septum, that separates the sigmoid and petrosal part of the foramen [15]. The jugular process extends behind the jugular foramen and connects the clival and squamosal parts of the occipital bone, and also forms the posteromedial wall of the foramen. Its upper surface presents an oval prominence, the jugular tubercle, that is located above the hypoglossal canal and below the glossopharyngeal, vagus, and accessory nerves which pass over it [16].
- The jugular foramen from the intracranial perspective is located approximately 5 mm below the porus of the internal auditory canal and 5 mm above the intracranial orifice of the hypoglossal canal. The lateral edge is located below the lateral end of the internal acoustic meatus. The medial edge faces the jugular tubercle, which is the junction of the basal and condylar parts of the occipital bone. From the extracranial perspective, the anterior margin is formed by the carotid canal and the lateral margin is formed by the styloid process and stylomastoid foramen (Fig. 49.1). The tympanic cavity is located medial to the tympanic membrane, and the endolymphatic sac is situated on the posterior surface of the petrous bone [1, 14, 15].
- The dural architecture from the intracranial orifice can be divided into three compartments: the petrosal compartment situated anteromedially, the sigmoid compartment situated posteriorly, and the intrajugular or neural compartment situated between the petrosal and sigmoid compartments at the site of the intrajugular processes [1]. This intrajugular compartment is especially important since it has a glossopharyngeal meatus through which the glossopharyngeal nerve passes, and a vagal meatus, through which the vagus and accessory nerves pass, separated only by a dura fold. The only intradural place at which we

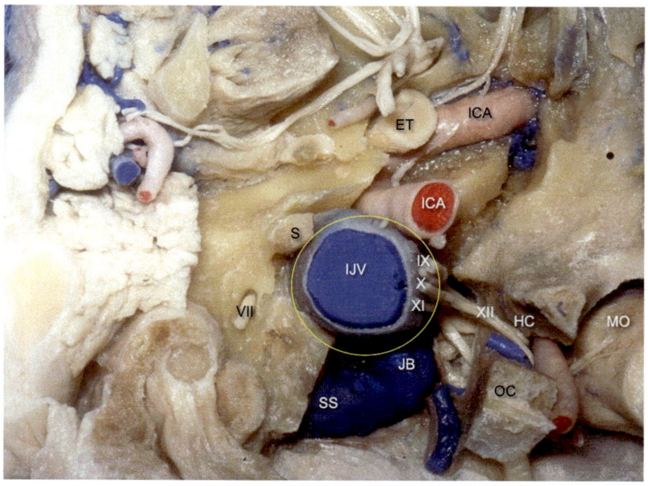

Fig. 49.1 Inferior view of the jugular foramen. From a general perspective, the JF (which is the area in the yellow circle) is bordered anteriorly by the ICA, laterally by the VII nerve, medially by the XII nerve, and inferiorly by the VA. Facial nerve passing the stylomastoid foramen on the lateral side of the IJV. The hypoglossal nerve from the medulla oblongata going through the hypoglossal canal above the OC and exiting the jugular foramen on the medial side of the IJV. *ET* Eustachian tube, *HC* hypoglossal canal, *ICA* internal carotid artery, *IJV* internal jugular vein, *JB* jugular bulb, *MO* medulla oblongata, *OC* occipital condyle, *S* styloid process, *SS* sigmoid sinus, *VII* facial nerve, *IX* glossopharyngeal nerve, *X* vagus nerve, *XI* accessory nerve, *XII* hypoglossal nerve. (Courtesy of the Rhoton Collection)

can differentiate the glossopharyngeal nerve consistently form the vagus nerve is just proximal to this dural septum [17, 18].

- From an extracranial perspective the glossopharyngeal, vagus, accessory, and hypoglossal nerves exit the skull on the medial side of the internal carotid artery and jugular vein. The glossopharyngeal and hypoglossal nerves pass anteriorly along the lateral surface of the internal carotid artery, and the accessory nerve descends posteriorly across the lateral surface of the internal jugular vein. The vagus nerve descends inferiorly within the carotid sheath [1, 19].
- The vascular relationships are highly important for surgical approaches to the jugular foramen. The internal carotid artery courses superiorly, posterior to the external carotid artery and anteromedial to the internal jugular vein to enter into the carotid canal at the skull base. At the level of the skull base, the internal jugular vein courses posterior to the internal carotid artery. The vertical segment of the infratemporal internal carotid artery enters the carotid canal with the carotid sympathetic nerves and ascends superiorly, and then turns anteromedially, where it then courses horizontally towards the petrous apex. The external carotid artery ascends anterior to the internal carotid artery, and gives rise to six branches, which can be divided into anterior and posterior groups. The posterior group (ascending pharyngeal artery, occipital artery, and posterior auricular artery) can be involved in jugular foramen pathologies. The vertebral artery ascends superiorly, passing through the transverse foramen of the atlas that is located posteroinferior to the jugular foramen. Branches of the vertebral artery, such as the meningeal, posterior spinal, and posteroinferior cerebellar artery, can be can be encountered in some jugular foramen pathologies [1, 14].
- The sigmoid sinus empties into the jugular foramen. After coursing down the sigmoid sulcus, the sinus angles anteriorly to the jugular foramen, crossing the occipitomastoid suture, then is directed forward below the petrous temporal bone at the site of the jugular bulb. The dome of the jugular bulb extends upward in the posterior wall of the internal auditory canal, and it is usually larger on the right side, matching the larger diameter of the sigmoid sinus on that side. From the jugular bulb, flow is directed downward and behind the tympanic bone into the internal jugular vein [1, 14].

49.4 Approach/Surgical Intervention

The jugular foramen can be exposed by several approaches depending on the type, size, and configuration of the lesion. Each case must be carefully assessed before selecting the adequate approach. Frequently, subtle nuances distinguish each approach from another adding complexity for proper understanding. To simplify this, we can divide the approaches based on anterolateral and posterolateral perspectives. The anterolateral approaches include the dissection of the structures located in front and lateral to the sigmoid sinus and jugular foramen, and posterolateral approaches area are accessed via dissection of structures behind and lateral to the sigmoid sinus and jugular foramen (Table 49.1) [20]. Every approach has its advantages and disadvantages as well as their specific indications for each type of tumor.

Table 49.1 Surgical approaches classification for jugular foramen tumors

Anterolateral	Posterolateral	Combined
Postauricular transtemporal approach	Retrosigmoid approach	Combined transmastoid retro- and infralabyrinthine transjugular transcondylar trans tubercular high cervical approach
Preauricular subtemporal infratemporal approach	Far lateral approach Transcondylar Supracondylar Paracondylar	

49.4.1 Posterolateral Approaches

49.4.1.1 Retrosigmoid Approach

This approach is probably the most common and well-known approach for jugular foramen tumors in neurosurgery. It is indicated for tumors that are predominantly intradurally with little or no extension to the extradural compartment. It is performed by a C-shaped retroauricular skin incision posterior and parallel to the outline of the pinna and a lateral, suboccipital craniotomy exposing the dura inferior and posterior to the transverse and sigmoid sinuses. The dura is opened and the cerebellum is gently displaced medially away from the posterior petrous surface of the temporal to expose the lateral aspect of the brain stem and intracranial portions of the cranial nerves exiting through the internal acoustic meatus and jugular foramen [1, 20]. Even though it is a simple and straightforward approach, this is not suitable for large tumors that goes out the extracranial compartment of the jugular foramen [21]. There is an extension of the retrosigmoid approach described by ken Matsushima et al. in 2014 where the area inferior to the acoustic meatus, medial to the endolymphatic depression and surrounding the superior half of the glossopharyngeal dural fold is drilled to access the suprajugular aspect of the foramen jugular. This is called the suprajugular approach and is suitable for tumors that extend into the upper part of the jugular foramen and above the jugular bulb [22].

49.4.1.2 Far Lateral Approach

The far lateral approach is a more inferior extension of the retrosigmoid approach that better access the foramen magnum and craniocervical junction, providing a corridor inferior to the lower cranial nerves. Its major indication is for tumors that extends down to the foramen magnum in front or lateral to the lower brainstem at the craniocervical junction. The jugular foramen is exposed posteriorly [1]. The skin incision can be performed using a retroauricular C-shaped incision or by a classic hockey stick incision starting at the mastoid tip, running posteriorly to the inion, and inferiorly to the spinous process of C2. The muscles attached to the occipital bone are detached en block and reflected inferiorly. A retrosigmoid suboccipital craniotomy that unroofs the foramen magnum and C1 hemi-laminectomy are performed. It has three variations depending on which area is desired for access: the jugular foramen, the lower clivus, and the premedullary area [20]. The basic far lateral approach without drilling of the condyle may be all that is needed to reach some lesions located along the anterolateral margin of the foramen magnum, nevertheless it can provide a route through which the transcondylar, supracondylar, and paracondylar approaches can be completed [13, 23].

The transcondylar approach is performed by drilling the posterior portion of the condyle and allows a progressive increase in the working space to the anterior part of the brainstem. The percentage of condyle to be removed may range from only the posterior third with no instability of the craniovertebral junction, to as far as the posterior half, which will require fixation of the craniovertebral junction due to instability. The complete transcondylar approach will have the hypoglossal canal as the anterior limit. This approach is generally used to reach the premedullary area [24].

The supracondylar approach is the one that is used to reach the jugular tubercle and is accomplished by drilling the occipital bone above and behind the occipital condyle by completely sparing the occipital condyle and including the drilling of the condylar fossa. The condylar fossa contains the posterior condylar canal at the bottom, and drilling it results in a defect in the posterior part of the jugular tubercle [13, 20, 24].

The paracondylar extension is the one considered as a possible route for those lesions that involve the posterior part of the jugular foramen, which can include paragangliomas of the jugular foramen and dumbbell schwannomas of the lower cranial nerves. The approach is directed through the jugular process of the occipital bone and involves the skeletonization and opening of the hypoglossal canal, partial drilling of the lateral portion of the occipital condyle and the mastoid tip [13, 20, 24].

49.4.2 Anterolateral Approaches

49.4.2.1 Postauricular Transtemporal Approach

This approach can provide excellent exposure of the jugular foramen, intrapetrous carotid artery, and lateral skull base. It is accessed from a lateral approach through the mastoid and through the neck (mastoid-neck approach) [20]. A postauricular C-shaped skin incision provides exposure for a mastoidectomy and neck dissection. The external auditory canal can be either preserved or transected with blind sac closure, depending on the anterior extension of the tumor [1]. The mastoidectomy primarily involves the infralabyrinthine region with exposure of the sigmoid sinus, jugular bulb, and mastoid segment of the facial nerve. Hearing does not have to be sacrificed and it can be preserved by maintaining the foot plate of the stapes, but to fully expose the lateral half of the jugular foramen, the mastoid portion of the facial nerve is mobilized anteriorly (which can cause facial nerve palsy), the styloid process must be resected and de-attachment of the rectus capitis lateralis muscle from the jugular process of the occipital bone is performed [11, 20, 25]. Both middle and posterior cranial fossa dura in front (Trautman's triangle) and behind of the sigmoid sinus are exposed [26].

49.4.2.2 Preauricular Subtemporal Infratemporal Approach

This approach exposes the jugular foramen anteriorly and it may be suitable for select tumors that extend along the petrous portion of the internal carotid artery, through the eustachian tube or through the cancellous portion of the petrous apex [1]. A preauricular skin incision extending across the zygomatic process of the temporal bone into the cervical region is made. A fronto-temporal craniotomy with or without removal of the superior and lateral orbital rim is performed. The middle cranial fossa floor is removed from lateral to medial until the carotid canal is seen. Both the eustachian tube and tensor tympani muscle are sacrificed. Removal of the styloid process allows anterior mobilization of the internal carotid artery and access the clival region. Drilling of Kawase's triangle allows resection of the petrous apex and provides a corridor to the posterior fossa [9, 20, 25]. Fisch in 1978 described three types of infratemporal fossa approaches to the lateral skull base. Type A approach allows access to the temporal bone in its infralabyrinthine component and is suitable for tumors of the jugular foramen. The external auditory canal is transected at the bone-cartilage junction [7, 8]. However, this approach is most often combined with a lateral approach to access tumors that have more anterior extension, type B or C Fisch's classification of paraganglioma tumors, requiring two-staged surgeries, a bigger chance of auditory loss and facial nerve damage. Type A tumors are jugulotympanic paragangliomas. Type B tumors are jugular paragangliomas with no destruction of the bone and type C tumors are jugular paragangliomas with destruction of the infralabyrinthine compartment of the temporal bone [8].

49.4.3 Combined Approach

49.4.3.1 Combined Transmastoid Retro- and Infralabyrinthine Transjugular Transcondylar Transtubercular Transcervical Approach

A combined approach can be used to gain total exposure of the jugular foramen in a single-stage surgery from an antero-lateral route. Allows radical resection of tumors that are located around the jugular foramen, the lower clivus, and the high cervical region. This approach is a combination of the transmastoid, retro- and infralabyrinthine transcondylar transtubercular, and transcervical approaches. Multidirectional angles and working corridors can be performed including suprajugular, transjugular, and infrajugular exposures maximizing the advantages of all the approaches mentioned above. Both intracranial and extracranial tumors can be removed in a one-stage procedure. Blind sac closure of the external ear canal and permanent routing of the facial nerve is generally not necessary, making conductive hearing loss and facial nerve palsy less common. Nevertheless, these maneuvers can be performed selectively based on the indicated pathology. Furthermore, access to the lower clivus is facilitated by anterior translocation of the vertical portion of the internal carotid artery and inferior translocation of the lower cranial nerves, if needed [6, 12].

This complex approach for total jugular foramen exposure can be simplified in a stepwise fashion: (1) Postauricular C-shaped infratemporal incision (2) Retrolabyrinthine mastoidectomy (3) High cervical exposure (4) Skeletonization and anterior translocation of the facial nerve (5) Lateral suboccipital craniotomy and transcondylar-transtubercular exposure. (6) Exposure of the internal jugular vein, jugular bulb, and sigmoid sinus and (7) Intradural exposure (if needed) [6, 12] (Figs. 49.2, 49.3, 49.4, 49.5, 49.6, and 49.7).

The patient is placed in the supine position with the head turned laterally away from the side of the lesion. A retroauricular curvilinear C-shaped skin incision is started 2 to 3 cm posterior to the upper border of the ear (Fig. 49.2). It continues posteroinferiorly into the neck over the anterior border of the sternocleidomastoid muscle and under the mandibular angle. The entire body and tip of the mastoid, the spine of Henle, the posterior point of the root of the zygoma, the supramastoid crest, and the asterion must be exposed before the mastoidectomy if performed [6, 12].

A retro and infralabyrinthine mastoidectomy is then performed, skeletonizing semicircular canals, fallopian canal, sigmoid sinus, and jugular bulb (Fig. 49.3). Then a high cervical exposure is done to identify the extracranial portions of the lower cranial nerves, the internal carotid artery, and the internal jugular vein (Fig. 49.4). After the

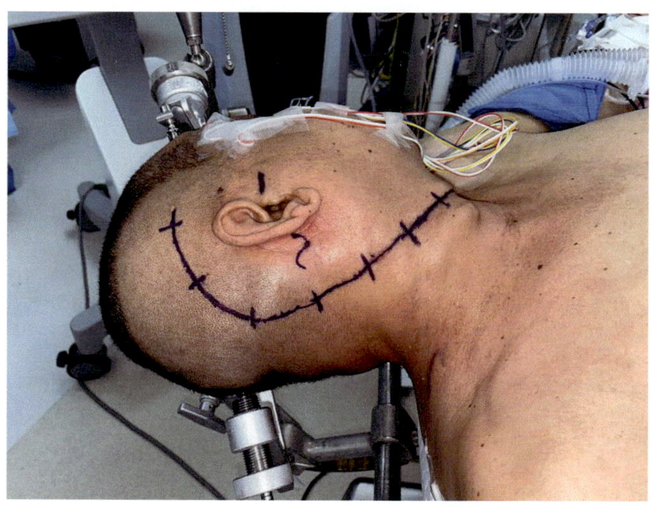

Fig. 49.2 Right-sided C-shaped retro-auricular incision. The incision is started approximately 2 to 3 cm posterior to the upper border of the ear. It continues posteroinferiorly into the neck over the anterior border of the sternocleidomastoid muscle and under the mandibular angle

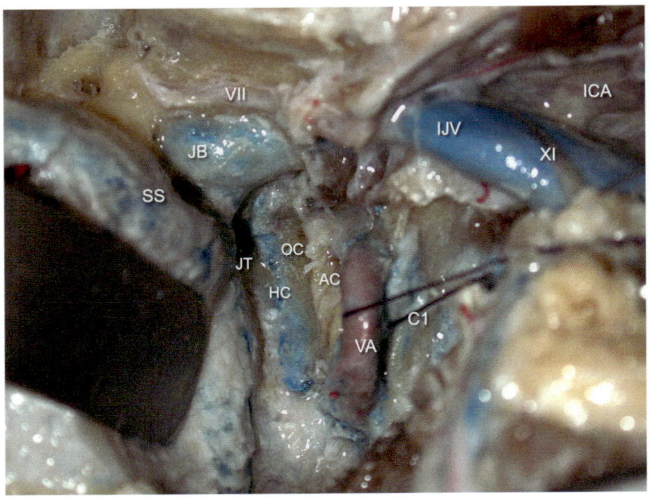

Fig. 49.3 Lateral view of the combined approach with a high cervical exposure, reflecting the posterior fossa dura. The JB passing through the jugular foramen. The IJV descends along the ACI with the lower cranial nerves. The vertebral artery ascends through the transverse process of C1 and usually passes behind the atlantal condyle. *AC* Atlantal Condyle, *ICA, C1* cervical spine vertebrae 1 internal carotid artery, *HC* hypoglossal canal, *IJV* internal jugular vein, *JB* jugular bulb, *JT* jugular tubercule, *OC* occipital condyle, *PFD* posterior fossa dura, *SS* sigmoid sinus, *VA* vertebral artery, *VII* facial nerve, *XI* accessory nerve

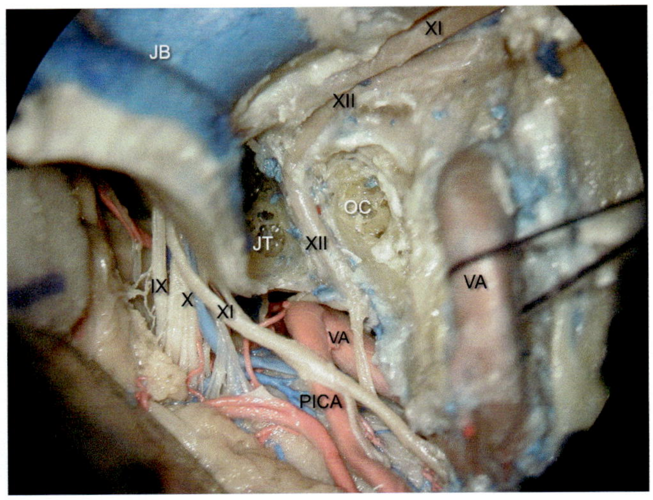

Fig. 49.5 Intradural and extradural views of the lower cranial nerves passing through the jugular foramen. The glossopharyngeal, vagus, and accessory nerves arise from the medulla in the postolivary sulcus and pierce the dural roof of the jugular foramen to pass through it. The IX nerve enters the jugular foramen through the glossopharyngeal meatus, and the X and XI nerves through the vagus meatus. The PICA arises from the posterior or lateral surfaces of the VA. The XII exiting through the hypoglossal canal above the OC. *JB* jugular bulb, *JT* jugular tubercule, *OC* occipital condyle, *PICA* postero-inferior cerebellar artery, *VA* vertebral artery, *IX* glossopharyngeal nerve, *X* vagus nerve, *XI* accessory nerve, *XII* hypoglossal nerve

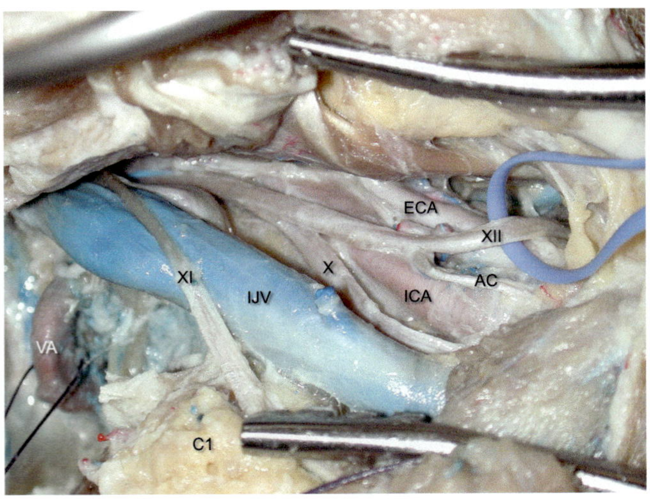

Fig. 49.4 Extracranial perspective of the glossopharyngeal, vagus, accessory, and hypoglossal nerves. The IX and XII nerve pass anteriorly along the lateral surface of the ICA. The XI nerve descends posteriorly across the lateral surface of the IJV. The vagus descends inferiorly within the carotid sheath. *AC* ancestor colous of the hypoglossal nerve, *C1* cervical spine vertebrae 1, *ECA* external carotid artery, *ICA* internal carotid artery, *IJV* internal jugular vein, *VA* vertebral artery, *IX* glossopharyngeal nerve, *X* vagus nerve, *XI* accessory nerve, *XII* hypoglossal nerve

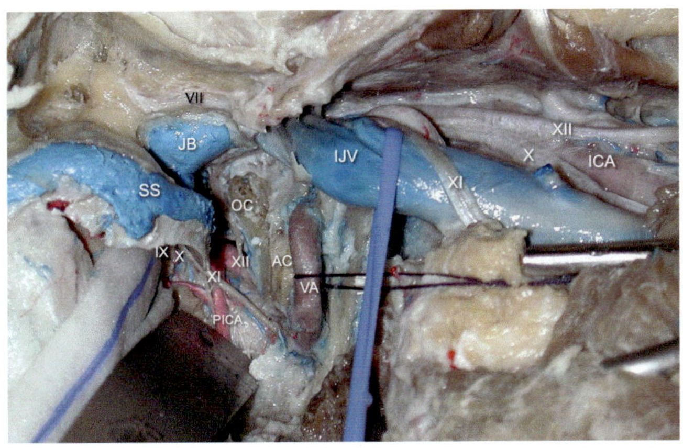

Fig. 49.6 Final view of the combined approach with high cervical exposure. From the intradural perspective the lower cranial nerves leaving the medulla and entering the jugular foramen. Hypoglossal nerve exiting through the hypoglossal canal. Vertebral artery below the OC. Sigmoid sinus empties into the jugular foramen after coursing down the sigmoid sulcus, crosses the occipitomastoid suture at the site of the jugular bulb. From the jugular bulb the flow is directed downward into the IJV. *AC* atlantal condyle, *ICA* internal carotid artery, *IJV* internal jugular vein, *JB* jugular bulb, *OC* occipital condyle, *PICA* postero-inferior cerebellar artery, *SS* sigmoid sinus, *VA* vertebral artery, *VII* facial nerve, *IX* glossopharyngeal nerve, *X* vagus nerve, *XI* accessory nerve, *XII* hypoglossal nerve

subcutaneous tissue and platysma muscle are divided, blunt dissection is used to identify the posterior angle of the mandible, as well as the anterior border of the sternocleidomastoid muscle. Later, the facial nerve in the fallopian canal is fully skeletonized from the genu to the stylomastoid foramen with a diamond burr. The mastoid tip is removed with a high-speed drill to decompress the facial nerve from the stylomastoid foramen [6, 12].

The suboccipital triangle, which is bound by the deep layer of muscles, is the key anatomic landmark for this portion of

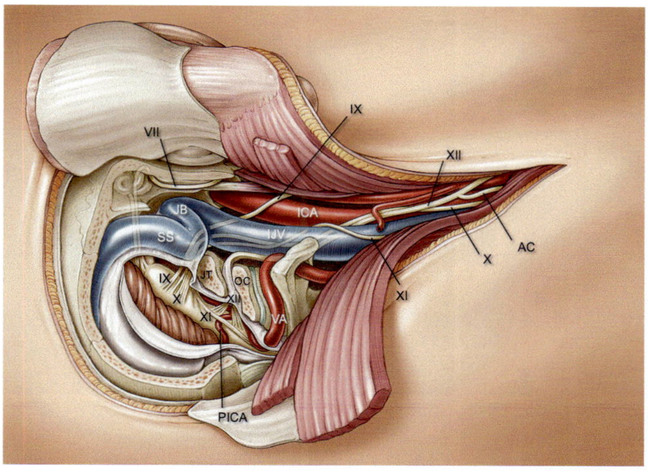

Fig. 49.7 Final view of the combined approach with high cervical exposure with its neurovascular relationships. *AC* ancestor colous of the hypoglossal nerve, *ICA* internal carotid artery, *IJV* internal jugular vein, JB jugular bulb, *JT* jugular tubercule, *OC* occipital condyle, *PICA* postero-inferior cerebellar artery, *SS* sigmoid sinus, *VA* vertebral artery, *VII* facial nerve, *IX* glossopharyngeal nerve, *X* vagus nerve, *XI* accessory nerve, *XII* hypoglossal nerve. (Copyright Dr. James K, Liu MD)

the exposure, and it can be opened by detaching the insertions of the superior and inferior oblique muscles from the transverse process of C1 and reflecting them posteriorly. This triangle involves the dorsal ramus of the C1 nerve root and the V3 horizontal segment of the vertebral artery. A lateral suboccipital craniectomy is then performed with a craniotome and roungers. Extradural reduction of the occipital condyle and jugular tubercle are the key maneuvers of this step (Figs. 49.5 and 49.6). Removal of the posterior and medial one third of the occipital condyle is generally adequate to increase the surgical corridor to the ventral foramen magnum [6, 12].

After complete exposure of the sigmoid sinus, jugular bulb, and internal jugular vein, the tumor mass can be palpated within these venous structures. After all arterial feeders to the tumor are coagulated, the internal jugular vein is then ligated just inferior to the tumor mass. The sigmoid sinus is occluded just above the tumor mass with a suture ligature. Alternatively, an endoluminal occlusion using gelfoam packing can be used also to avoid any durotomies. The lateral wall of the internal jugular vein is incised and removed with the tumor up to the jugular bulb and sigmoid sinus. If needed, for intradural pathology, a retrosigmoid dural incision can be made to access the intradural portion of the tumor [6, 12] (Fig. 49.7).

49.4.4 Representative Cases

The approach to the jugular foramen can vary significantly depending on the type, size, and configuration of the tumor. Here, we present 4 cases in which the combined approach can be used to treat large jugular foramen tumors.

49.4.4.1 Case 1

A 24-year-old female patient presented with progressive headaches, dysphagia, and right-sided weakness from a left jugular foramen meningioma invading the jugular bulb with compression of the brainstem (Fig. 49.8a, b). The tumor was resected via a combined transmastoid infralabyrinthine transjugular transcervical approach to remove this tumor. In this case, blind sac closure was not necessary since there was no tumor extension into the middle ear. The retrosigmoid corridor was opened to remove the intradural portion of the tumor. Then, the internal jugular vein was ligated and the sigmoid sinus was occluded endoluminally (Fig. 49.9a–d). The jugular bulb and vein were opened to remove the intraluminal invasion by the tumor. Postoperatively, the patient was neurologically intact with no cranial nerve deficits. Postoperative MRI showed gross total resection of the tumor. (Fig. 49.8c, d)

49.4.4.2 Case 2

A 51-year-old-male presented with progressive dysphagia and significant weight loss with 9th through 12th cranial nerve palsies. MRI demonstrated a giant craniocervical junction chordoma with jugular foramen invasion, extension into the right parapharyngeal space and intradural cervicomedullary junction, and near-complete erosion of the right occipital condyle (Fig. 49.10a, b). The tumor was resected via a right combined infralabyrinthine transjugular transclival transcervical approach. The internal jugular vein was ligated and the sigmoid sinus was occluded endoluminally with gelfoam. The jugular bulb was opened to reveal intrabulbar chordoma. A gross total resection of the tumor in all the invaded spaces was achieved and the pre-clival dura was resected (Fig. 49.10c, d). Multi-layered reconstruction with an alloderm graft and fat graft was performed. A second-stage occipital cervical fusion was performed because of the craniocervical instability caused by tumor invasion into the right occipital condyle. Postoperatively, the patient remained at his preoperative neurological baseline without any new neurological deficits.

49.4.4.3 Case 3

A 34-year-old-female presented with left-sided hearing loss, left vocal cord paralysis, and severe gait ataxia from a left dumbbell jugular foramen schwannoma involving the cerebellopontine angle (Fig. 49.11a, b). A gross total resection was achieved via a combined infralabyrinthine, transsigmoid transjugular approach (Fig. 49.11c, d). Postoperatively, the patient improved gait function and incurred no new cranial nerve deficits.

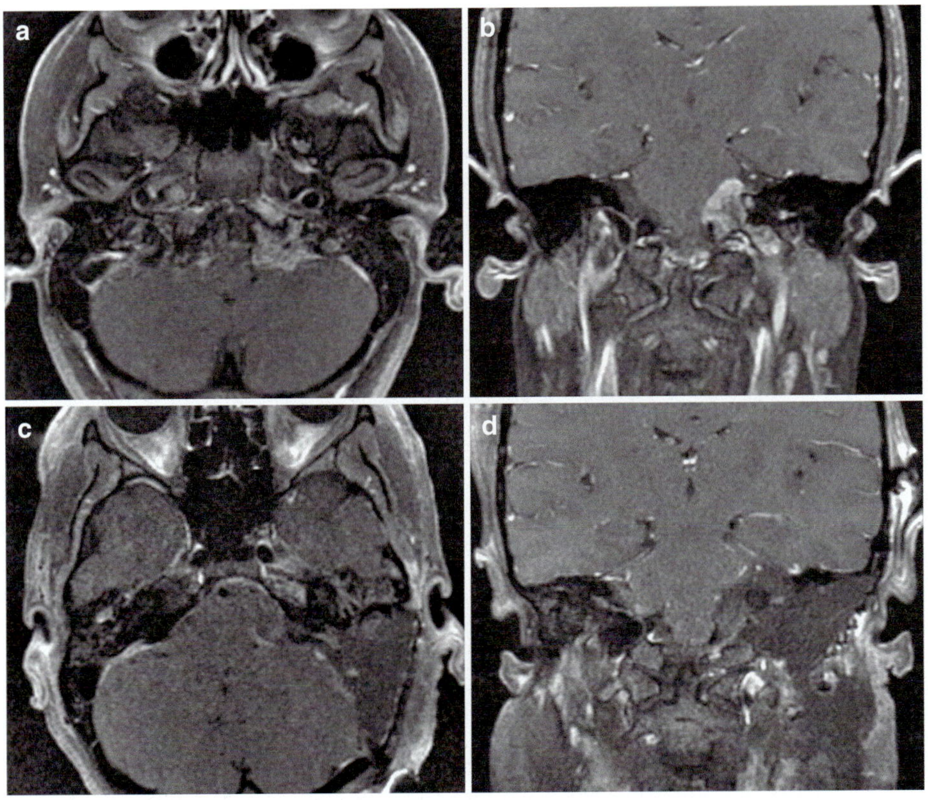

Fig. 49.8 Pre-op and post-po MRI views of a left jugular foramen meningioma. (**a** and **b**) Shows T1 gadolinium-enhanced images of a homogenous mass in the jugular foramen. (**c** and **d**) shows gross total resection of the tumor

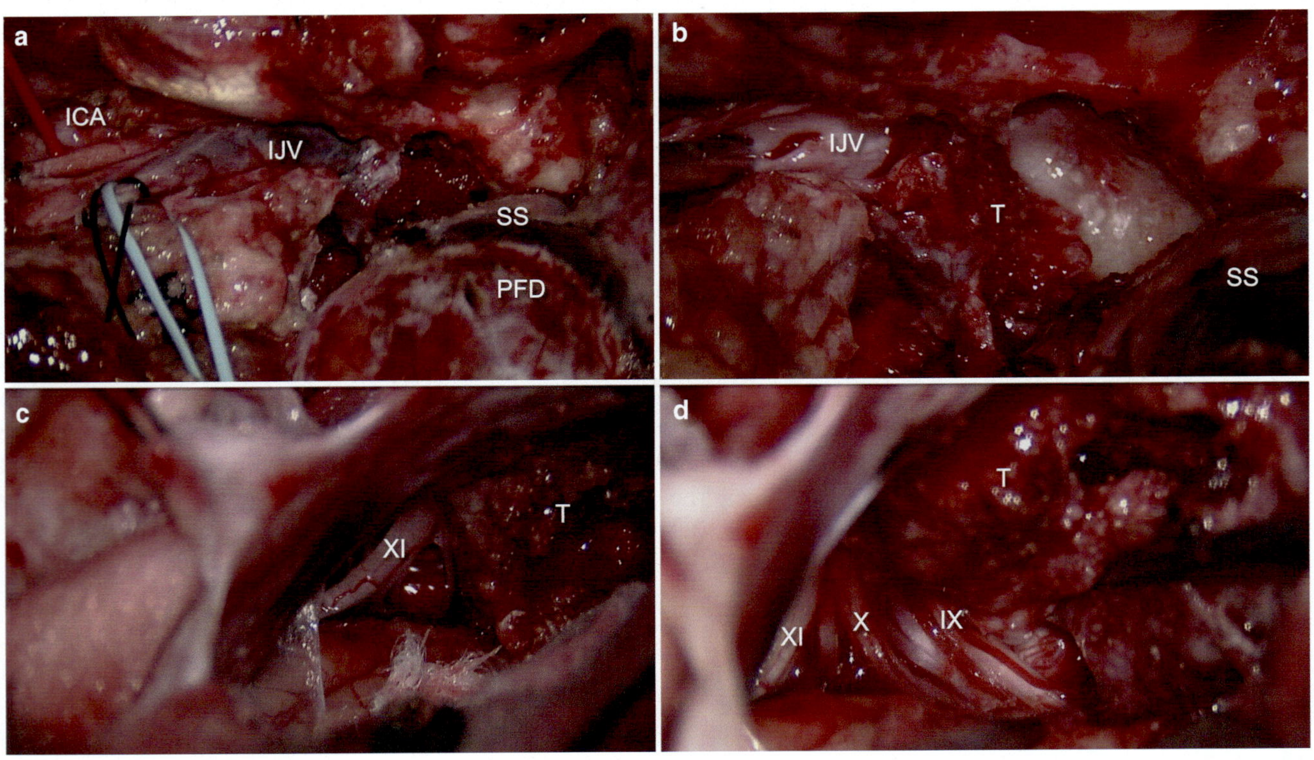

Fig. 49.9 (**a**) Left-sided exposure of jugular foramen via extended anterolateral infralabyrinthine transjugular approach for resection of jugular foramen meningioma. (**b**) After tying off the IJV and endoluminal occlusion of the sigmoid sinus, the lateral wall of the sigmoid sinus, jugular bulb, and internal jugular vein are excised to expose the intraluminal tumor in the jugular bulb. (**c** and **d**) Retrosigmoid exposure of the intradural portion of the tumor at the jugular fossa. The tumor is carefully dissected off of the lower cranial nerves. *ICA* internal carotid artery, *IJV* internal jugular vein, *PFD* posterior fossa dura, *SS* sigmoid sinus, *T* tumor

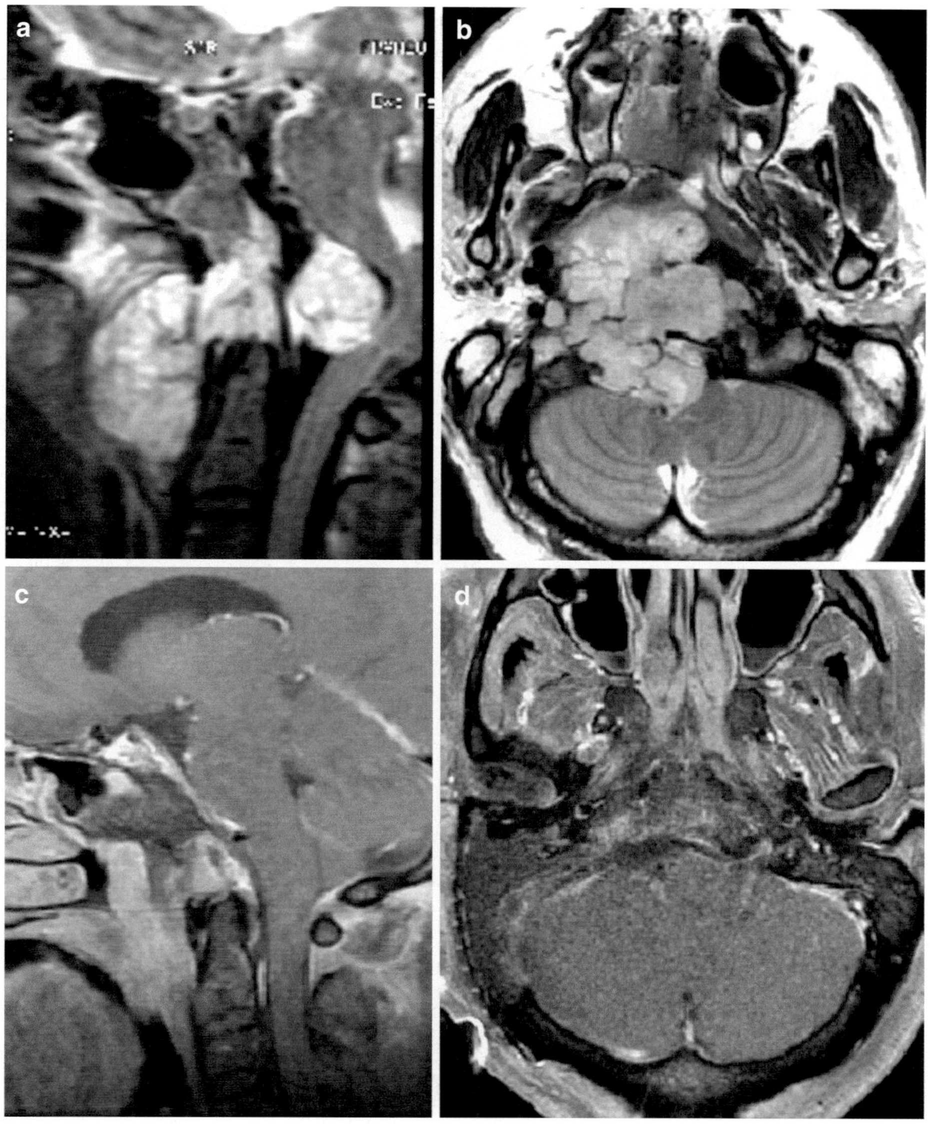

Fig. 49.10 Pre-op and Post-op MRI views of an extensive cranio-cervical chordoma invading the jugular foramen, cerebello-medullary cistern and parapharyngeal space. (**a** and **b**) shows a T2 MRI signal with a large lobulated mass centered on the right parapharyngeal space with intradural and extradural extension. The tumor is compressing the anterior pons and the airway deviated to the left side. (**c** and **d**) show a gross total resection of the tumor

49.4.4.4 Case 4

A 78-year-old-female patient presented with progressive left-sided hearing loss, left vocal cord paralysis, and dysphagia. MRI showed a left jugular paraganglioma (glomus jugulare) invading the jugular bulb and vein (Fig. 49.12a, b). After preoperative embolization, resection was performed via a left infralabyrinthine transjugular transcervical approach. The internal jugular vein was ligated, and the sigmoid sinus was occluded endoluminally with gelfoam. The jugular venous system was opened up to reveal the intraluminal tumor. The tumor was entirely extradural and was entirely resected from the intraluminal jugular bulb and vein (Fig. 49.12c, d). Postoperatively, she remained at her neurological baseline without new neurological deficits.

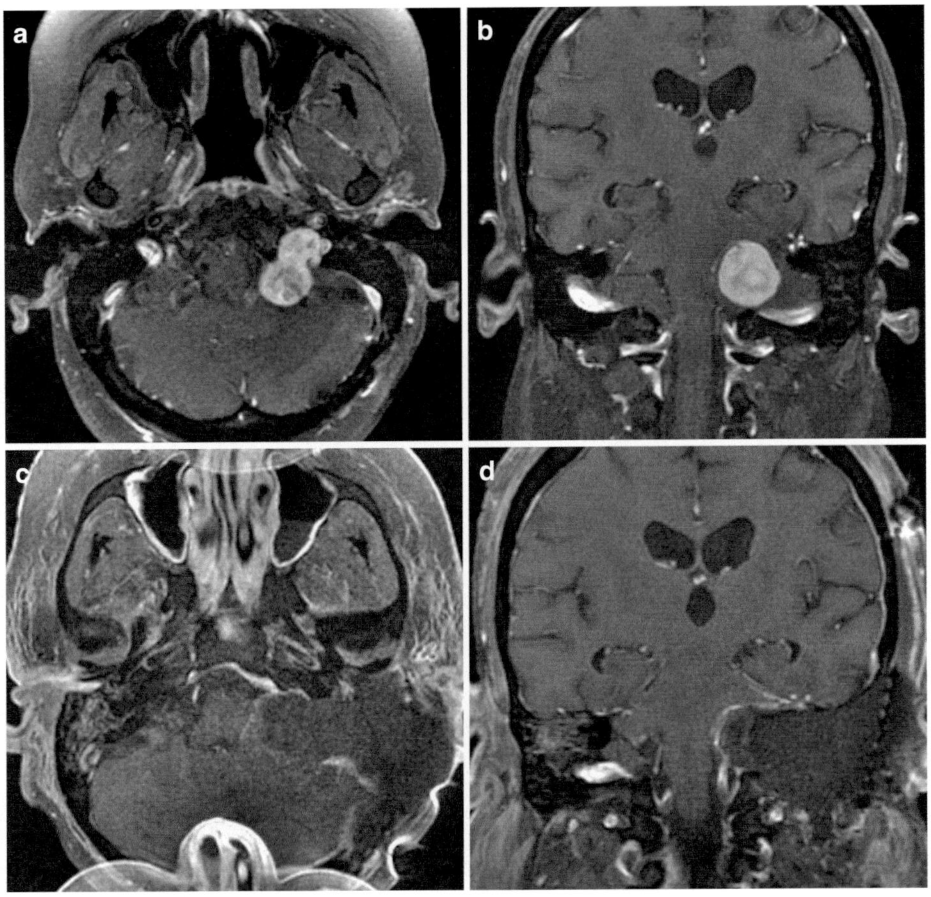

Fig. 49.11 Pre-op and post-op MRI views of a jugular foramen Schwannoma. (**a** and **b**) shows a gadolinium-enhanced T1 weighted MR images with a well- circumscribed dumbbell-shaped Schwannoma invading the jugular foramen. (**c** and **d**) shows gross total resection of the tumor

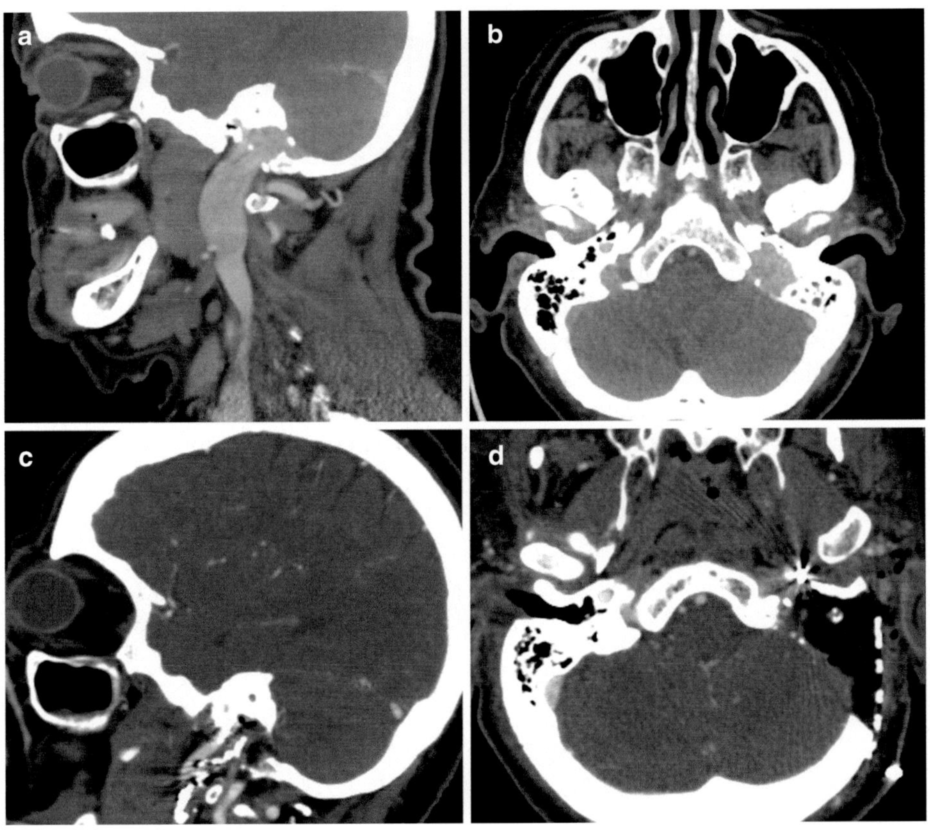

Fig. 49.12 Pre-op and Post-op CT angiogram demonstrating a jugular paraganglioma (glomus jugulare). (**a** and **b**) shows a large jugular paraganglioma invading the internal jugular vein and jugular bulb within the temporal bone. (**c** and **d**) shows gross total resection of the tumor

49.5 Surgical Pearls

- The combined transmastoid retro- and infralabyrinthine transjugular transcondylar transtubercular transcervical approach described above with a C shape retroauricular incision allows for single-staged radical resection of large complex jugular foramen tumors. This approach has the advantage of providing total exposure of the jugular foramen with multidirectional angles of attack. The infratemporal carotid artery can be exposed without transection of the external ear canal, permanent rerouting of the facial nerve, or mandibular translocation.
- The anterolateral approaches can provide an excellent choice of approach when the tumor is not extending to the infratemporal ICA or when middle fossa approach is needed with tumor that extended to the petrous apex.
- The posterolateral approaches, especially the retrosigmoid approach, are indicated for tumors that are located predominantly intradurally with little or no extension to the extradural compartment.
- Surgery of jugular foramen tumors (JFTs) often requires vascular control of the jugular venous system by ligating the internal jugular vein (IJV) and sigmoid sinus (SS) to allow intrabulbar access. Occlusion of the SS traditionally involves pre- and retrosigmoid durotomies to allow application of vascular clips or suture ligatures, which can risk postoperative cerebrospinal fluid (CSF) leakage and pseudomeningoceles. An endoluminal sigmoid sinus occlusion with gelfoam technique that is entirely extradural can avoid any durotomies that can result in postoperative CSF leaks. This is performed by placing pieces of gelfoam endoluminally into the proximal sigmoid sinus after the jugular vein is tied off.
- Endoscopic assistant can be performed to assess the degree of tumor resection and improve visualization of the lower cranial nerves after the removal of the tumor.

49.6 Discussion

The approach to the jugular foramen can vary significantly depending on the type, size, and configuration of the tumor. Also, experience and extensive knowledge of the anatomy of the region is imperative for achieving the best results for the patient. We aim to have a philosophy of maximal safe resection with avoidance of neurovascular injury, and optimal preservation of functioning cranial nerves. In cases where there is pre-existing irreversible cranial nerve palsies, a more aggressive approach can be considered.

Selecting the approach to the jugular foramen must be based on surgical experience, the relationship of the tumor to the surrounding neurovascular structures, the extend of pre-existing cranial nerve damage, and the goals of the surgery (either maximal removal with cranial nerve preservation or a more aggressive approach when there is pre-existing irreversible cranial nerve palsies). Also having a multidisciplinary team that have the experience of managing this type of tumors is mandatory. We encourage all neurosurgeons to first have an extensive anatomical knowledge of the jugular foramen through cadaveric dissections or 3D simulations before attempting on performing these types of approaches.

49.7 Conclusion

The combined transmastoid retro- and infralabyrinthine transjugular transcondylar transtubercular high cervical approach can be performed in a stepwise fashion for resection of complex jugular foramen tumors including (meningiomas, schwannomas, paragangliomas, and chordomas). Total exposure of the jugular foramen can be achieved, and multidirectional approaches can be performed, including suprajugular, infrajugular, and transjugular exposures. Each case should be carefully assessed to find the most suitable approach for the patient that will allow the surgeon to remove the tumor with minimal neurovascular damage. Extensive knowledge of the anatomy and a multidisciplinary team is needed (neuro-otologist, head and neck surgeon, laryngologist, neuro-anesthesiologist, neurointerventionalist, radiation oncologist) to provide the patient the best optimal postoperative outcomes.

Acknowledgments The authors wish to thank the Rhoton Collection for the illustrations used in Fig. 49.1.

References

1. Rhoton AL. Jugular foramen. In: Rhoton's cranial anatomy and surgical approaches. Oxford: Oxford University Press; 2019.
2. Shapiro R. Compartmentation of the jugular foramen. J Neurosurg. 1972;36:340–3.
3. Tekdemir I, Tuccar E, Aslan A, Elhan A, Deda H, Ciftci E, Akyar S. The jugular foramen: a comparative radioanatomic study. Surg Neurol. 1998;50:557–62.
4. Lo WW, Solti-Bohman LG. High-resolution CT of the jugular foramen: anatomy and vascular variants and anomalies. Radiology. 1984;150:743–7.
5. Ramina R, Maniglia JJ, Fernandes YB, Paschoal JR, Pfeilsticker LN, Neto MC, Borges G. Jugular foramen tumors: diagnosis and treatment. FOC. 2004;17:31–40.
6. Liu JK, Sameshima T, Gottfried ON, Couldwell WT, Fukushima T. The combined transmastoid retro- and infralabyrinthine transjugular transcondylar transtubercular high cervical approach for resection of glomus Jugulare tumors. Operat Neurosurg. 2006;59:ONS-115–25.
7. Fisch U. Infratemporal fossa approach to tumours of the temporal bone and base of the skull. J Laryngol Otol. 1978;92:949–67.

8. Fisch U, Pillsbury HC. Infratemporal fossa approach to lesions in the temporal bone and base of the skull. Arch Otolaryngol Head Neck Surg. 1979;105:99–107.
9. Al-Mefty O, Fox JL, Rifai A, Smith RR. A combined infratemporal and posterior fossa approach for the removal of giant glomus tumors and chondrosarcomas. Surg Neurol. 1987;28:423–31.
10. Al-Mefty O, Teixeira A. Complex tumors of the glomus jugulare: criteria, treatment, and outcome. J Neurosurg. 2002;97:1356–66.
11. Fukushima T. Manual of skull base dissection. Raleigh: AF-Neuro Video; 2010.
12. Liu JK, Zhao K, Baredes S, Jyung RW. Extended anterolateral infralabyrinthine transjugular approach for microsurgical resection of giant glomus vagale tumor: operative video and technical nuances. J Neurol Surg B Skull Base. 2021;82:S59–60.
13. Wen HT, Rhoton AL, Katsuta T, de Oliveira E. Microsurgical anatomy of the transcondylar, supracondylar, and paracondylar extensions of the far-lateral approach. J Neurosurg. 1997;87:555–85.
14. Ayeni SA, Ohata K, Tanaka K, Hakuba A. The microsurgical anatomy of the jugular foramen. J Neurosurg. 1995;83:903–9.
15. Inserra MM, Pfister M, Jackler RK. Anatomy involved in the jugular foramen approach for jugulotympanic paraganglioma resection. FOC. 2004;17:41–4.
16. Mintelis A, Sameshima T, Bulsara KR, Gray L, Friedman AH, Fukushima T. Jugular tubercle: morphometric analysis and surgical significance. JNS. 2006;105:753–7.
17. Tubbs R, Griessenauer C, Bilal M, Raborn J, Loukas M, Cohen-Gadol A. Dural septation on the inner surface of the jugular foramen: an anatomical study. J Neurol Surg B. 2015;76:214–7.
18. Silverstein H, Willcox TO, Rosenberg SI, Seidman MD. The jugular Dural fold – helpful skull base landmark to the cranial nerves. Skull Base. 1995;5:57–61.
19. Rubinstein D, Burton BS, Walker AL. The anatomy of the inferior petrosal sinus, glossopharyngeal nerve, vagus nerve, and accessory nerve in the jugular foramen. Am J Neuroradiol. 1995;16:185–94.
20. McGrew B, Matusz P, De Caro R, Loukas M, Tubbs R, Griessenauer C. Surgical approaches to the jugular foramen: a comprehensive review. J Neurol Surg B. 2015;77:260–4.
21. Jackson CG, Kaylie DM, Coppit G, Gardner EK. Glomus jugulare tumors with intracranial extension. FOC. 2004;17:45–50.
22. Matsushima K, Kohno M, Komune N, Miki K, Matsushima T, Rhoton AL. Suprajugular extension of the retrosigmoid approach: microsurgical anatomy. JNS. 2014;121:397–407.
23. Sen CN, Sekhar LN. Surgical management of anteriorly placed lesions at the craniocervical junction? An alternative approach. Acta Neurochir. 1991;108:70–7.
24. Luzzi S, Lucifero AG, Bruno N, Baldoncini M, Campero A, Galio R. Far lateral approach. Acta Biomed. 2021;92:e2021352.
25. Sanna M. Atlas of microsurgery of the lateral Skull Base. New York: Thieme; 2011.
26. Shane Tubbs R, Griessenauer C, Loukas M, Ansari SF, Fritsch MH, Cohen-Gadol AA. Trautmann's triangle anatomy with application to posterior transpetrosal and other related skull base procedures. Clin Anat. 2014;27:994–8.

50 Merits of Endoscopic Approach to Clivus Tumors and Challenges

Takeo Goto and Rajeev Sharma

Learning Objectives
- Assess the safety of the extended endoscopic endonasal transclival approach.
- Learn the steps and technical nuances of the extended endoscopic endonasal transclival approach.
- Learn the likely complications of the extended endoscopic endonasal transclival approach along with their avoidance and management.

50.1 Introduction

Midline posterior skull base lesions extending from retrosellar area down till CV Junction displacing the neurovascular structures dorsally and laterally can be managed from anterior by extended endoscopic endonasal transclival approaches. Though it is relatively easier to manage extradural pathologies, intradural lesions can also be resected with experience. High-resolution CT with reconstruction (both soft tissue and bone windows) helps in detailed understanding of the anatomy like sphenoid sinus and clivus pneumatization, bony dehiscence over paraclival and parasellar carotid prominences and optic nerves, relationship of vidian canal to foramen lacerum, configuration of dorsum sellae and posterior clinoids, caroticoclinoid foramen, ossification of interclinoid ligaments, etc. Besides the routine MRI sequences in the axial, sagittal, and coronal planes, fat-saturation T1 MRI sequence without and with contrast is essential for differentiating pathology from the fatty bone marrow present in normal clivus. CT angiography is important for study of vascular anatomy, and "kissing parasellar carotid arteries" is an absolute contraindication for endonasal approach especially to upper clivus [1]. State of the art facilities in the operation room like dedicated extended endoscopic set and long monoshaft microinstruments, image guidance system, arterial microdoppler, and an experienced skull base surgical team with equal competence in transcranial approaches are essential.

T. Goto (✉)
Department of Neurosurgery, Osaka Metropolitan University Graduate School of Medicine, Osaka, Japan
e-mail: gotot@omu.ac.jp

R. Sharma
All India Institute of Medical Sciences, New Delhi, India

50.2 History and Background

The first attempt to remove a central skull base lesion by transcranial approach was done by Caton and Paul in 1893 [2]. Since the use of endoscope for sellar surgery by nasal route as early as in 1970s, the extent of skull base access by this corridor have progressively advanced with time with the improvement in expertise, evolution of intraoperative gadgets, and development of intensive care, thus making endoscopic endonasal approach as the standard of care for midline skull base tumors since 1990s. Many endonasal approaches have been developed by the combined team efforts of Neurosurgeons and Otorhinolaryngological surgeons. University of Pittsburgh Medical Center Team comprising Neurosurgeon Amin Kassam and Otorhinolaryngologists Ricardo L. Carrau and Carl Snydermann have contributed a lot in expanding the access of endoscopic endonasal approach from sella to the remaining skull base [3]. Since their first publication of the experience of the extended endoscopic endonasal approach from posterior clinoids to foramen magnum as early as 2005, it has progressively improved with time [4].

I. N. Kanaan, V. Beneš (eds.), *Neuroanatomy Guidance to Successful Neurosurgical Interventions*,
https://doi.org/10.1007/978-3-031-59838-8_50

50.3 Anatomy Highlights

- Upper, middle, and lower transclival approaches may need mobilization of normal anatomic structures (pituitary gland, paraclival ICA, and eustachian tubes, respectively) for adequate exposure to the lesion [2].
- Extended endoscopic endonasal petrosectomy is of three types: medial, inferior, and inferomedial [5–7].
- Unfavorable anatomical features like chonchal type of sphenoid sinus and reduced intercarotid distance are its absolute contraindications.

50.4 Approach (Surgical Intervention)

The step-by-step description of extended endonasal approach (binostril, 4-hands technique) to midline posterior skull base lesions is as follows: supine position with neutral head or slight head extension (to place dorsum sellae lower down and parallel to the OT floor) and neck slightly rotated to right side and tilt to left side, unilateral middle turbinectomy with lateralization of contralateral middle turbinate, middle meatal antrostomy on the side of turbinectomy (inferior extent as per extent of pathology), Hadad-Bussagasteguy flap harvest and placement in maxillary sinus cavity, detachment of nasal septum from keel of sphenoid, posterior septectomy (up to 2 cm) for binostril access, bilateral ethmoidectomy (posterior and/or anterior as per need), ipsilateral superior turbinate excision, wide anterior sphenoidotomy till medial pterygoid plates, drilling of the intrasphenoid septae, and identification of intrasphenoidal bony landmarks like tubercular recess, MOCR, MOCR, clival recess, parasellar carotid prominence, paraclival carotid prominence, etc. Sphenoidotomy and posterior septectomy are done only if upper and middle clival regions need to be accessed [8].

The further approach proceeds either above (superior corridor) or below (inferior corridor) the inferior wall (floor) of sphenoid sinus using angled (45° or 70°) endoscopes. The superior corridor leads to sellar and retrosellar areas, whereas the inferior corridor leads to the rhinopharyngeal segment of the clivus, foramen magnum, and CV junction (Fig. 50.1). As the posterior skull base is divided into three (cranial, middle, and caudal) levels, it will be discussed under these headings:

1. Cranial level of posterior skull base (sellar clivus): It extends from the posterior interclinoid line above to the floor of sella turcica below. Some classifications consider Dorello's canal as the lower limit [9]. Access to this retrosellar area requires bone removal of sellar face, sellar floor, tubercular recess, planum, MOCR, parasellar bone (to expose anterior wall of cavernous sinus), uni/bi-lateral pituitary transposition, thus exposing dorsum sellae for drilling followed by uni/bi-lateral posterior clinoids dissection and removal using angled (45° or 70°) endoscope [1]. Blood loss during anterior transcavernous approach, interdural pituitary transposition, and posterior clinoid removal may be troublesome and needs to be controlled by hemostats (like flowseal, Surgiflo) [10]. Retroinfundibular fossa or interpeduncular cistern and its contents become accessible by this approach [9].
2. Middle level of posterior skull base (sphenoidal clivus): It extends from sellar floor above to the line connecting hypoglossal canals below. Some classifications consider foramen lacerum or choana as the lower limit [9]. Access to this rhinopharyngeal segment of the clivus is below the inferior wall (floor) of sphenoid sinus. After identification of vomer-sphenoid junction in midline, expose the bone

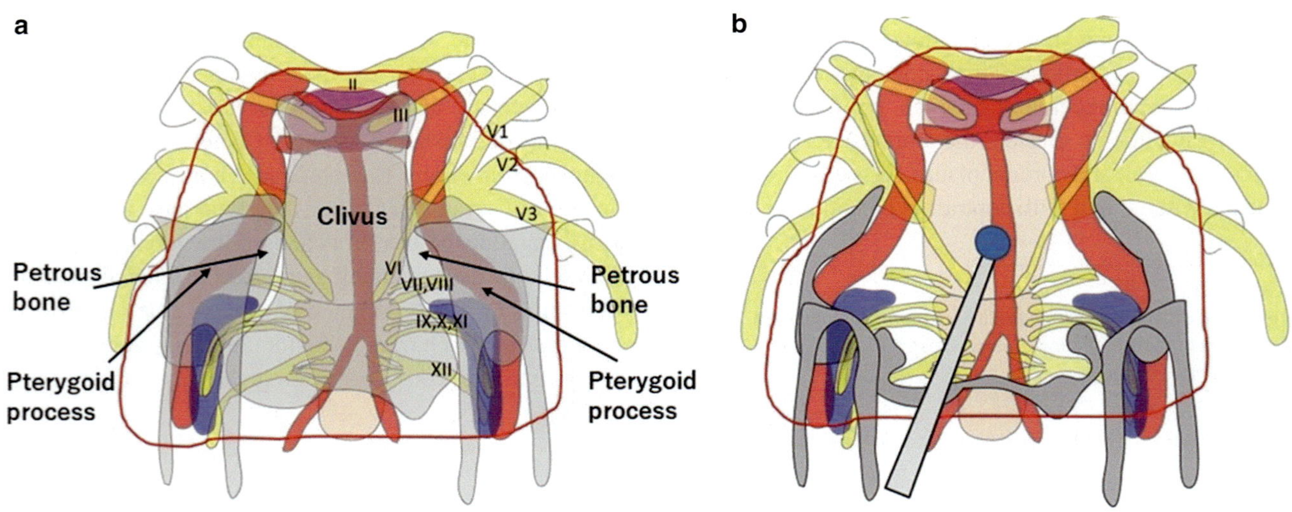

Fig. 50.1 Animated diagram showing the three-dimensional extent of bone drilling. (**a**) Before bone drilling; (**b**) After bone drilling

laterally by submucosal dissection, and thus palatovaginal canal and pterygoid canal are identified on either side. Then, the exposed vomer and inferior wall (floor) of the sphenoid sinus are drilled out. This bone drilling is continued laterally till pterygoid canal (0.5 cm lateral to vomer-sphenoid junction) on either side. This dissection of the rhinopharyngeal mucosa and lateralization of longus capitis and longus colli muscles is continued inferiorly till hypoglossal canals on either side. The lateral boundary of this exposure is represented by the paraclival ICA, foramen lacerum, abducent nerve, medial pterygoid plate, eustachian tube, and hypoglossal canal on both sides. The exposed clivus bone is drilled out to thin egg shell using coarse diamond drill bit under copious water irrigation and then removed using dissector, thus freeing the paraclival ICA. The free paraclival ICA can be mobilized laterally, thus providing access to petrous apex, free edge of tentorium, trochlear nerve, and posterior root of trigeminal nerve. Prepontine cistern and its contents get accessible by this approach [9].

3. Caudal level of posterior skull base (nasopharyngeal clivus): It extends from the line connecting hypoglossal canals above up to CV junction. Premedullary cistern and its contents get accessible by this approach [9].

Extradural tumor is removed after completing bone work. For intradural tumors, T-shaped durotomy is done and bleeding from the basilar dural venous plexus is controlled using appropriate hemostats (flowseal, Surgiflo). During Durotomy, care should be taken to avoid injury to the structures underneath. So, dural incision is started in midline higher up at the sellar floor and then extended inferiorly. Intradural dissection is done with the same old microneurosurgical principles. After completing the intradural tumor removal, multilayered skull base reconstruction is done using dural closure (approximation suture fixed by Osaka knot), intradural fat graft without or with subdural inlay graft placement, onlay fascial graft, autologous fat graft to fill the clival bone dead space, reinforced with pedicled vascularized nasoseptal flap and sealent [1]. Even in cases with intact dura (like after extradural tumor removal), vascularized nasoseptal flap needs to be placed to cover and protect the exposed internal carotid arteries.

Transclival approach is required for retroinfundibular tumors with or without extension into interpeduncular and prepontine cisterns like retrochiasmatic craniopharyngiomas, clival chordomas, chondrosarcomas, meningiomas, and others like germinomas, and teratomas, etc. [11] Advantage of this approach is lack of any brain retraction without any neurovascular restraint along the path, thus excluding injury to the cranial nerves and vessels. This advantage is secondary to the wide panoramic view provided by the angled endoscopes. This improves total resection rates with reduced duration of hospitalization and morbidity rates [12].

A 45-year-old gentleman presented with chronic headache and recent onset left abducens palsy. Preoperative CEMRI brain showed a heterogeneously enhancing midline lesion involving the whole of clivus with extension into interpeduncular cistern. This lesion was excised nearly completely by extended transclival approach and histopathologic assessment was suggestive of chordoma

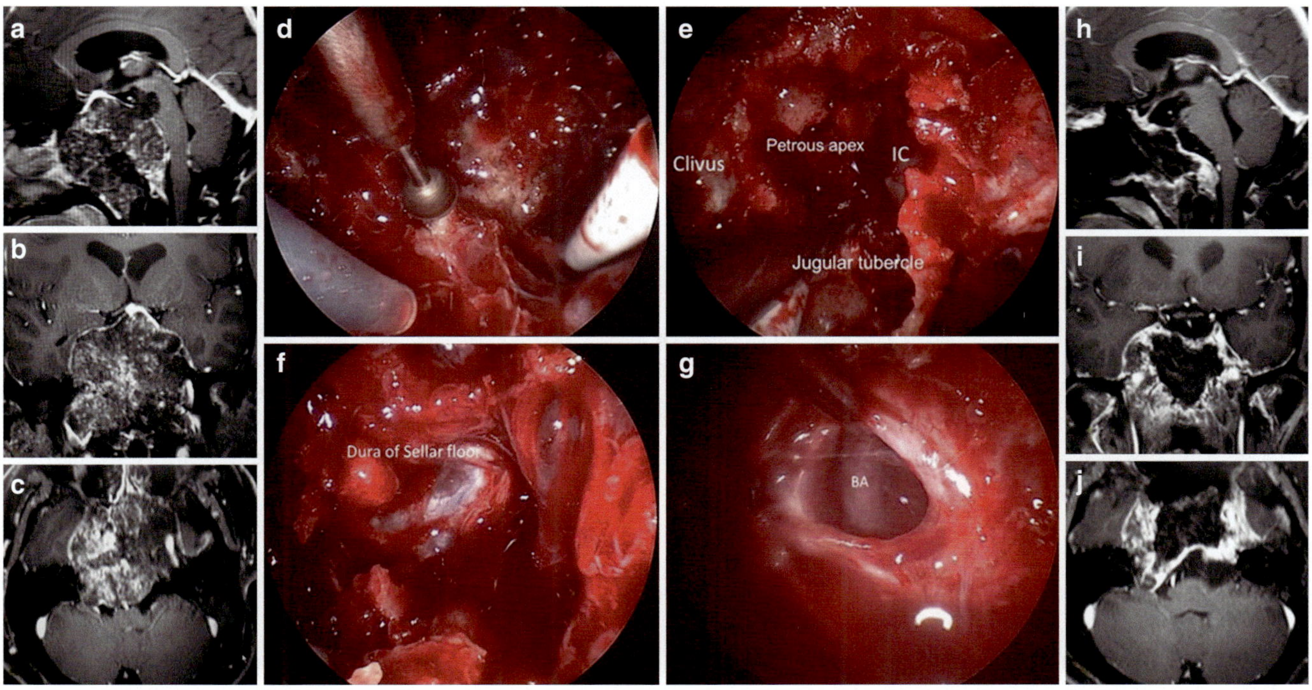

Fig. 50.2 (**a–c**) Preoperative CEMRI brain showing a heterogeneously enhancing midline lesion involving the whole of clivus with extension into interpeduncular cistern; (**d–g**) intraoperative images showing bone drilling, the exposed intradural structures after tumor excision including the intradural part through the dural defect; (**h–j**) postoperative MRI showing nearly complete excision of chordoma with the enhancing Hadad flap

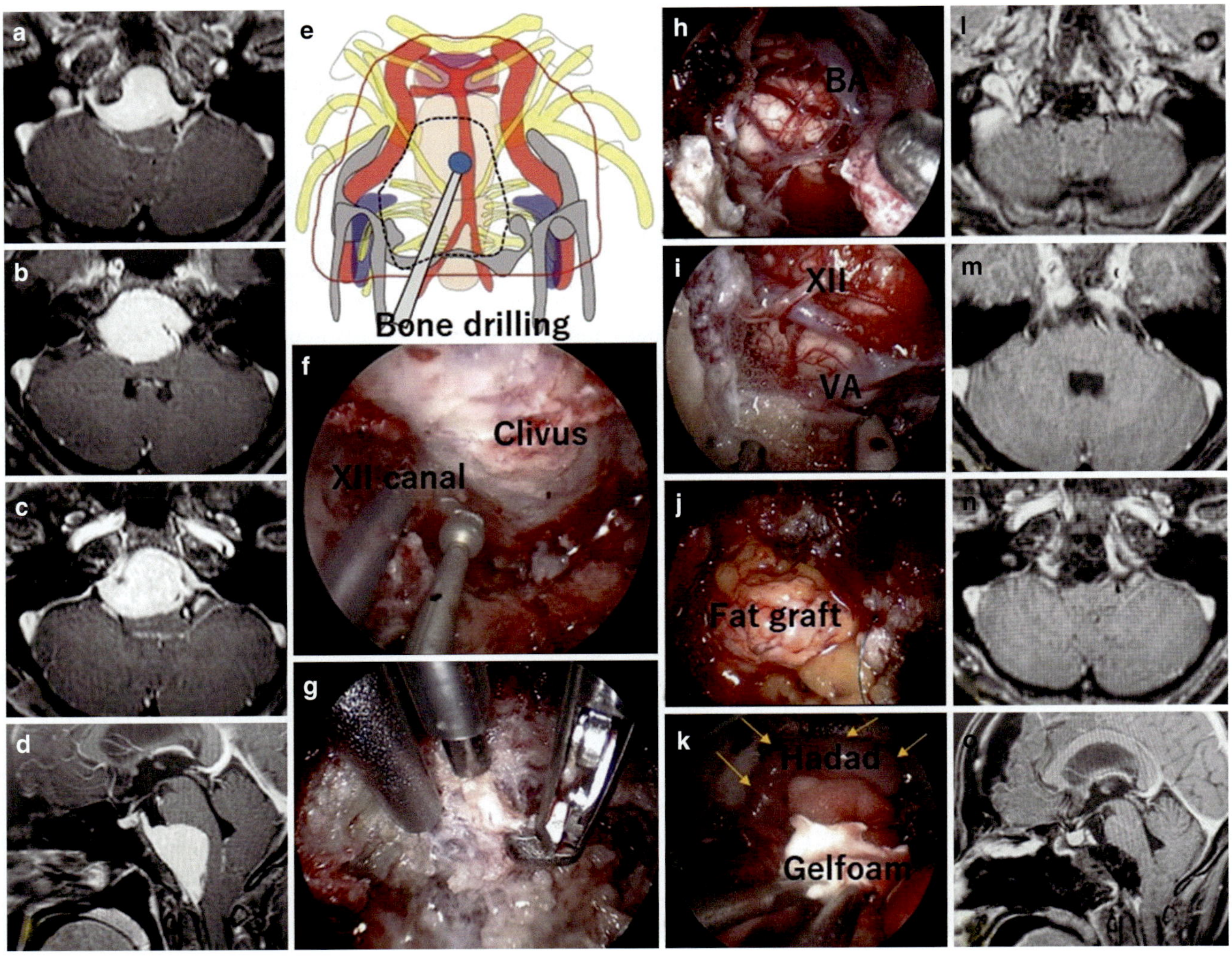

Fig. 50.3 (**a**–**d**) Preoperative CEMRI brain showing a homogeneously enhancing midline lesion; (**e**) three-dimensional extent of bone drilling; (**f**–**k**) intraoperative images showing bone drilling, intradural structures after tumor removal, and reconstruction; (**l**–**o**) postoperative MRI showing complete excision of meningioma

(Fig. 50.2). Another 50-year-old lady presented with wasting of right half of tongue since last 6 months. Preoperative CEMRI brain showed a homogeneously enhancing midline lesion involving the middle and lower third of clivus up to C2 with dural tail. This lesion was excised nearly completely by extended transclival approach and histopathologic assessment was suggestive of transitional meningioma (Fig. 50.3).

50.5 Surgical Pearls

- Midline of the clival bone is a safe spot to start drilling.
- Interdural pituitary transposition provides direct access to the interpeduncular cistern without any manipulation of the adjacent neurovascular vital structures.
- During en-bloc interdural pituitary transposition, the jelly-like neurohypophysis which is more densely attached posteriorly needs to be detached sharply without any undue overpressure by dissector.
- Rule out the presence of sellar bridges laterally (ossified interclinoid ligament) on CT axial bone windows. If present, they should be fractured meticulously before removing posterior clinoids.
- Vidian nerve - Eustachian tube junction (VEJ) is a very reliable landmark to locate the second genu of the petrous ICA.
- Avoid stretching of the pituitary stalk during dorsum sella drilling and posterior clinoid removal.
- Be aware of the persistent trigeminal artery located just behind the neurohypophysis.

50.6 Discussion

PCP and dorsum sella removal is a very useful adjunct to extended endonasal transsphenoidal transclival approach to the retrosellar space, thus providing access to the interpeduncular and prepontine cisterns. It can be done in an intradural, extradural, or interdural manner (Table 50.1).

Table 50.1 Comparison of various techniques of pituitary transposition for dorsectomy and posterior clinoidectomy [10, 13, 14]

	Intradural (Kassam, 2008) [15]	Extradural (Silva, 2012) [13]	Interdural (Fernandez-Miranda, 2014) [10]
Approach	Transdorsum sellae approach to interpeduncular cistern	Transclival approach to interpeduncular cistern	Transclival transcavernous approach to interpeduncular cistern
Corridor	Intradural	Pituitary mobilized superiorly still covered by both dural layers (meningeal and periosteal)- periosteal layer gets disrupted later	Natural corridor provided by the separation of the inner and outer dural layers at the lateral margin of the sellae
Opening of outer dural layer of the cavernous sinus	No	Uncontrolled and slightly restrictive	Controlled and wide open
Likely avulsion of IHA from ICA wall during posterior clinoidectomy	No	Slightly more likely	Less likely
Coagulation and cutting of IHA (U/L, B/L)	Yes	No	Yes
Intraoperative blood loss	Less	Slightly less	More (transcavernous)
Completeness of excision of PCP	May be incomplete	May be incomplete sometimes (posterolateral tip of posterior clinoid may not be exposed)	Complete (even prominent posterior clinoids are removed)- provides a more superior exposure of the posterior clinoids
Intraoperative pituitary handling	Manipulation and bilateral disruption of its venous drainage (superior and inferior intercavernous sinuses are coagulated and cut)	Less disruption of the venous drainage pathways	Preservation of the attachment of the medial (sellar) wall of the cavernous sinus to the pituitary gland during medial and superior mobilization provides additional protection to the glandular tissue and allows for preservation of the remaining venous drainage pathways, namely the superior intercavernous sinus and paraclinoid venous channels.
Risk of intraoperative pituitary gland damage	High	Low	Lowest
Risk of postop pituitary gland dysfunction	High	Low	Lowest

The biggest challenge during dorsectomy and posterior clinoidectomy is the profuse venous bleeding encountered while dissecting and cutting the overlying dura due to the presence of intercavernous sinuses and the basilar venous plexus. It needs to be controlled diligently using hemostatic agents (flowseal, Surgiflo, etc.) as bipolar coagulation is usually ineffective [15].

Drill-related parasellar ICA injury can happen during drilling of dorsum sellae in the deep and narrow corridor as the sellar intercarotid distance is around 20 mm and posterior interclinoid distance is around 15 mm [1]. So, whole rotating length of the drill except its tip should be covered (Fig. 50.2d). Left ICA injuries are more common than right ICA. Intraoperative or immediate postoperative angiogram is recommended [2]. Paraclival ICA can get injured during bone drilling or removing egg-shell of bone. Coarse-diamond 1 mm drill bits are preferred for bony bleeding (drilling-associated hemostasis).

The high possibility of CSF rhinorrhea following transclival approaches is greatly reduced by multilayered skull base reconstruction reenforced with vascularized nasoseptal flaps [2]. Clival defects may require extended nasoseptal flaps incorporating the nasal cavity floor to provide the extra length required for reconstruction in these cases. Filling the clival dead space with the autologous fat graft provides a smooth bed for the placement of the pedicled flap. Flap margins should be in direct contact with the bone. Synthetic tissue glue and gelfoam packing over the strategically placed nasoseptal flap followed by nasal tampons helps in maintaining flap position, thus preventing its migration. Lumbar drain may be placed in select cases for 2–3 days for draining CSF at the rate of 5–10 ml per hour if there is no extensive pneumocephalus on postoperative NCCT head [8]. These vascularized flaps show linear enhancement on postoperative contrast MRI (Fig. 50.2h–j); absence of enhancement indicates flap failure. Postoperative CSF rhinorrhea is usually managed by endoscopic re-exploration [2].

Possibility of iatrogenic craniovertebral instability needs to be considered in the laterally extended "far-medial" transclival approaches to lower clivus requiring extensive condylar compartment drilling. Less than 50% condylar drilling poses minimal risk of craniovertebral junction instability [8].

Preoperative cranial neuropathies may get exacerbated due to intraoperative manipulation and are likely to improve gradually. Tumors like meningiomas displace adjacent cranial nerves in the direction away from its epicenter. Cranial nerves anatomy can be delineated accurately on preoperative MRIs. Middle and lower petroclival tumors displacing lower cranial nerves anteriorly have high risk of iatrogenic lower cranial nerve injury by endonasal approach and thus should not be operated by endonasal approach [8]. Iatrogenic cranial nerve injury essp abducens palsy is seen following transclival approaches due to its interdural course. Abducens nerve enters the clival dura at Dorello's point which is situated 1 cm off midline on either side and approx. 20 mm below posterior clinoid process. Dorsal meningeal artery, a branch of meningohypophyseal trunk, is a useful landmark for its identification [1].

Residual tumor on postoperative imaging may be present posterior and lateral to the critical structures like ICA. Residual posterior clinoids may be seen. As posterior clinoids are attached to ligaments, these ligaments need to be cut sharply to ensure complete delivery of lateralmost portion of the posterior clinoid tip [15].

50.7 Conclusion

A naïve endoscopic skull base surgeon should master the transclival bony work by operating an adequate number of purely extradural clival lesions (less demanding) before putting his/her hands on clival lesions with intradural extension (more demanding). Good skull base reconstruction with/without the support of lumbar drain is needed for clival lesions with intradural extension to prevent high pressure CSF leaks.

References

1. Cavallo LM, Esposito I, Notaris MD, Esposito F, Tschabitscher M, Cappabianca P. Endoscopic anatomy of the clivus and posterior fossa. In: Stamm AC, editor. Transnasal endoscopic skull base and brain surgery. Tips and pearls. 1st ed. New York: Thieme; 2011. p. 265–74. https://doi.org/10.1055/b-0034-81706.
2. Rao D, Stein R, Jenson M, Patel J, Fiester P. Imaging review of the endoscopic operative corridor for clivus lesions. Curr Probl Diagn Radiol. 2023;52(1):66–76. https://doi.org/10.1067/j.cpradiol.2022.07.005.
3. Prevedello DM, Doglietto F, Jane JA, Jagannathan J, Han J, Laws ER. History of endoscopic skull base surgery: its evolution and current reality. J Neurosurg. 2007;107(1):206–13. https://doi.org/10.3171/JNS-07/07/0206.
4. Kassam A, Snyderman CH, Mintz A, Gardner P, Carrau RL. Expanded endonasal approach: the rostrocaudal axis. Part II. Posterior clinoids to the foramen magnum. Neurosurg Focus. 2005;19(1):E4.
5. Loymak T, Belykh E, Abramov I, et al. Comparative analysis of surgical exposure among endoscopic endonasal approaches to petrosectomy: an experimental study in cadavers. J Neurol Surg B Skull Base. 2022;83(5):526–35. https://doi.org/10.1055/s-0041-1741067.
6. Loymak T, Tungsanga S, Abramov I, Sarris CE, Little AS, Preul MC. Comparison of anatomic exposure after petrosectomy using anterior transpetrosal and endoscopic endonasal approaches: experimental cadaveric study. World Neurosurg. 2022;161:e642–53. https://doi.org/10.1016/j.wneu.2022.02.076.
7. Borghei-Razavi H, Truong HQ, Fernandes Cabral DT, et al. Endoscopic endonasal petrosectomy: anatomical investigation, limitations, and surgical relevance. Oper Neurosurg Hagerstown Md. 2019;16(5):557–70. https://doi.org/10.1093/ons/opy195.
8. Vaz-Guimaraes Filho F, Wang EW, Snyderman CH, Gardner PA, Fernandez-Miranda JC. Endoscopic endonasal "far-medial" transclival approach: surgical anatomy and technique. Oper Tech Otolaryngol Head Neck Surg. 2013;24(4):222–8. https://doi.org/10.1016/j.otot.2013.09.003.
9. Patel CR, Fernandez-Miranda JC, Wang WH, Wang EW. Skull base anatomy. Otolaryngol Clin N Am. 2016;49(1):9–20. https://doi.org/10.1016/j.otc.2015.09.001.
10. Fernandez-Miranda JC, Gardner PA, Rastelli MM, et al. Endoscopic endonasal transcavernous posterior clinoidectomy with interdural pituitary transposition. J Neurosurg. 2014;121(1):91–9. https://doi.org/10.3171/2014.3.JNS131865.
11. Prevedello DM, Kassam AB, Fernandez-Miranda JC, Gardner PA, Carrau RL, Snyderman CH. Transsellar/Transdorsal approach via a pituitary transposition to the interpeduncular cistern. In: Stamm AC, editor. Transnasal endoscopic skull base and brain surgery. Tips and pearls. 1st ed. New York: Thieme; 2011. p. 256–62. https://doi.org/10.1055/b-0034-81705.
12. Stamm AC, Pignatari SSN, Vellutini E, Hermann D, Timperley D. Endoscopic transnasal craniectomy: approaches to the clivus and posterior fossa. In: Stamm AC, editor. Transnasal endoscopic skull base and brain surgery. Tips and pearls. 1st ed. New York: Thieme; 2011. p. 275–82. https://doi.org/10.1055/b-0034-81707.
13. Silva D, Attia M, Kandasamy J, Alimi M, Anand VK, Schwartz TH. Endoscopic endonasal posterior clinoidectomy. Surg Neurol Int. 2012;3:64. https://doi.org/10.4103/2152-7806.97008.
14. Silva D, Attia M, Schwartz TH. Endoscopic endonasal posterior clinoidectomy LETTER TO EDITOR 2015. J Neurosurg. 2015;122(2):478–9. https://doi.org/10.3171/2014.8.JNS141783.
15. Kassam AB, Prevedello DM, Thomas A, et al. Endoscopic endonasal pituitary transposition for a transdorsum sellae approach to the interpeduncular cistern. Neurosurgery. 2008;62(3 Suppl 1):57–72; discussion 72–4. https://doi.org/10.1227/01.neu.0000317374.30443.23.

Anatomy and Tailored Surgical Approach to Foramina Magnum Meningioma

51

José Alberto Landeiro and José Orlando de Melo Junior

Learning Objectives

- Identify the main anatomical landmarks of the foramen magnum (FM) region.
- Distinguish the types of FM meningioma according to their classification.
- Critically analyze the need for condylectomy in FM meningioma.
- Select and tailor the best approach for FM meningioma.
- Apply the main concepts and techniques for surgical resection of FM meningioma.

51.1 Introduction

Foramen magnum meningiomas (FMMs) are defined as tumors whose insertion zones are mainly located in the foramen magnum (FM) region [1]. They represent about 2–3% of all intracranial meningiomas and the most frequent FM tumors, accounting for about 70% of all cases, followed by schwannomas [1, 2]. FMMs are typically slow-growing and clinical symptoms are insidious [3, 4]. Thus, the clinical pattern often delays diagnosis and reveals large tumors at presentation.

The most frequent initial symptoms are occipital headache and neck pain, usually exacerbated by Valsalva maneuvers that increase intracranial pressure. Muscle weakness, sensory and gait disturbances, dysphagia, dysarthria, dyspnea, and sphincter dysfunction are other progressive symptoms [5]. Classic FM syndrome is defined by the development of unilateral arm sensory and motor deficits, which progress to the ipsilateral leg, then the contralateral leg, and finally contralateral upper extremity [6].

J. A. Landeiro (✉)
Department of Neurosurgery, Antonio Pedro University Hospital, Fluminense Federal University, Niterói, Brazil

J. O. de Melo Junior
Department of Neurosurgery, Paulo Niemeyer State Brain Institute, Rio de Janeiro, Brazil

FMMs are located anterior and anterolateral to the spinal dentate ligament in about 90% of the cases, followed by posterolateral, posterior, and more rarely, purely anterior [6, 7]. Microsurgical resection is the treatment of choice for the majority of these lesions [4]. Posteriorly situated tumors are usually approached through a median suboccipital approach, while far-lateral (FL) and extreme-lateral approaches are used to reach ventral and ventrolateral lesions [7]. Condylectomy, vertebral artery (VA) transposition/mobilization, extent of cervical laminectomy, and mastoidectomy are steps that may be performed and tailored to the individual case [8, 9]. Intraoperative neurophysiological monitoring of the lower cranial nerves (IX, X, XI, and XII) and motor and somatosensory-evoked potentials are helpful to detect early signs of dysfunction during surgery [1, 10].

Surgical morbidity and mortality are considerable and have an estimated incidence of 7.5–32% and 0–7.5%, respectively [11–14]. Some factors that affect the prognosis are location, tumor size (larger lesions naturally widen the surgical corridor and favor microsurgical resection), tumor invasiveness, extradural extension, VA entrapment, absence of intraoperative arachnoid plane, and adhesions in cases of reoperation [3].

51.2 History

The first report of FMM was announced in 1872 after an autopsy by Hallopeau at Lariboisière Hospital in Paris [15]. The first described surgical attempt was performed by Frazier and Spiller in 1922 [16]. In 1938, Cushing and Eisenhardt classified FMMs into a craniospinal group that arises above and grows downward through the FM and a spinocranial group that arises below and grows upward through the FM [3]. In 1988, George et al. applied the lateral suboccipital approach for resection of anteriorly located FM tumors, including meningiomas; it combined the exposure and control of the VA and the sigmoid sinus and, for further lateral

I. N. Kanaan, V. Beneš (eds.), *Neuroanatomy Guidance to Successful Neurosurgical Interventions*,
https://doi.org/10.1007/978-3-031-59838-8_51

enlargement, medial transposition of the VA and section of the SS were performed [17].

An important contribution was made by Bruneau and George in 2008, who proposed a classification system that helps select the best surgical approach and anticipate the position of lower cranial nerves in relation to the tumor [1]. Based on preoperative imaging characteristics of the lesion, three parameters are used: (1) the compartment of development (intradural; extradural; or intra and extradural); (2) dural insertion in the anteroposterior plane (anterior, if insertion is on both sides of the anterior midline; lateral, if insertion is between the midline and the dentate ligament; or posterior, if insertion is posterior to the dentate ligament); and (3) relation to the VA (above; below; or on both sides).

Recently, Yamahata et al. studied the space between the condyle and neuraxis created by the mass effect of FMMs and classified them into clival (lower third of clivus), foraminal (at the level of FM), and atlantal (below the hypoglossal canal) subtypes [18]. The clival was associated with the narrower surgical space compared to foraminal and atlantal subtypes. Another contribution was made by Silveira et al. who revisited Cushing's classification and proposed the Leal-Landeiro classification of the craniospinal FMM into type 1 (the tumor grows up to the pontomedullary sulcus) and type 2 (the tumor grows above the pontomedullary sulcus) [19]. Craniospinal type 1 can be reached through the posterior midline subtonsillar approach if it displaces the brainstem to the opposite side. Craniospinal type 2 needs more cerebellar retraction and the FL with or without the transcondylar approach is a better option than the subtonsillar approach.

51.3 Anatomy Highlights

- The concept of FM region is defined according to three anatomical borders: anterior, lateral, and posterior [11]. It is delimited at the anterior border from the lower third of the clivus to the upper edge of the body of C2, at the lateral border from the jugular tubercle to the upper aspect of C2 lamina, and at the posterior border from the anterior edge of the squamous occipital bone and C2 spinous process.
- The anatomy of the craniovertebral junction is complex and the relevant structures are the atlantoaxial joint, caudal portion of the medulla, cerebellar tonsils, inferior vermis, fourth ventricle, rostral part of the spinal cord, upper cervical and lower cranial nerves, and suboccipital (V3) and intradural (V4) segments of the VA with their branches [10] (Fig. 51.1).

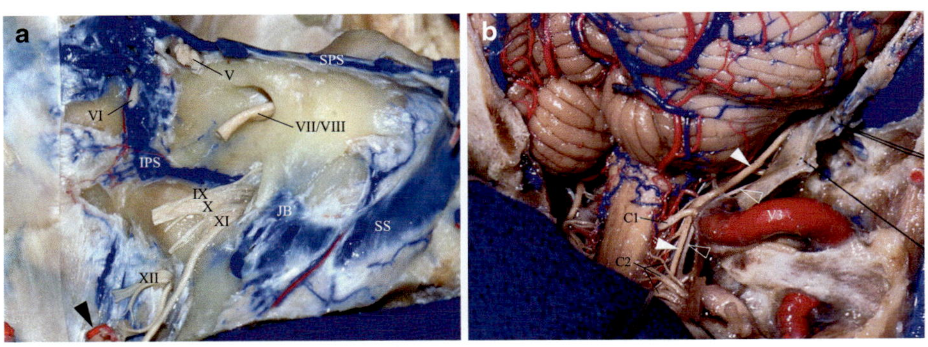

Fig. 51.1 Craniovertebral junction anatomy. (**a**) Posterior view of the clivus, petrous bone, and craniovertebral junction on the right side. Superior petrosal sinus (*SPS*), inferior petrosal sinus (*IPS*), sigmoid sinus (*SS*), jugular bulb (*JB*), V4 segment of the VA (*black closed arrowhead*), trigeminal (*V*), abducens (*VI*), vestibulocochlear and facial (*VII/VIII*), glossopharyngeal (*IX*), vagus (*X*), accessory (*XI*), and hypoglossal (*XII*) cranial nerves. (**b**) Posterolateral view of the craniovertebral junction through the FL approach on the right side. Dentate ligament (*white opened arrowhead*), accessory cranial nerve (*white closed arrowhead*), C1 (*C1*) and C2 (*C2*) posterior nerve rootlets, and V3 segment of the VA (*V3*)

51.4 Approaches

51.4.1 Far-Lateral Approach

FL approach combines a lateral suboccipital craniotomy with hemilaminectomy of C1 and is the basis for the supracondylar, paracondylar, and transcondylar approaches [20]. It is indicated for anterior and lateral FMMs [1] (Fig. 51.2).

The patient is placed in lateral decubitus position. Semi-sitting, park-bench, or prone positions can also be used [10]. The head is fixed on a three-point fixation device and subjected to three movements: (1) flexed so as to distance two fingers from the sternal notch; (2) contralateral lateralization towards the shoulder to increase the angle between the FM and the arch of C1 (distraction maneuver); (3) contralateral rotation by 15° (Fig. 51.3a).

We use the traditional hockey-stick incision. It begins at the tip of the mastoid process ipsilateral to the tumor and posterior to the sternocleidomastoid muscle, runs superiorly and vertically towards the asterion and then curves towards the inion, at the level of the superior nuchal line, then curves downward in the occipitocervical midline to the level of the spinous process of C3 [21] (Fig. 51.3b). Dissection of the muscular plane is carried out through the nuchal ligament and followed by subperiosteal muscle dissection to expose the suboccipital and the posterior arch of the C1, with attention to the course of the VA (Fig. 51.3c).

After adequate bone exposure and identification of the VA in the sulcus arteriosus of the posterior arch of C1, an ipsilateral hemilaminectomy of C1 is performed with drill and Kerrison. Then, lateral suboccipital craniotomy is performed and includes opening of the ipsilateral half of the FM and maximal lateral retrosigmoid enlargement (Fig. 51.3d). The most anterior and lateral suboccipital retrocondylar bone of the FM must be removed, usually by drilling with a cutting and/or diamond bits, towards the condylar fossa. Venous bleeding is observed from the condylar canal containing the condylar emissary vein and denotes the most anterior limit of the approach and the anatomical reference of the posterior aspect of the occipital condyle. If necessary, the transcondylar approach should be performed to provide safe working space. Dura mater is opened in C-shape and laterally anchored with stitches. Tumor can be already visualized (Fig. 51.4a). Every effort should be made to prevent dura shrinkage due to the drying effect of microscope lighting and exposure to air by covering the dura with pieces of hydrated absorbable gelatin sponge (e.g., Gelfoam®—Pfizer) and frequent irrigation, also avoiding dura coagulation, for primary closure at the end of the procedure.

Microsurgical steps for tumor resection consist of the following (Fig. 51.4): (1) Identification of interposed neurovascular structures; (2) Release and mobilization of the spinal cord through the section of the first dentate ligament in order to expand the surgical corridor and to reduce the risk of injury to neural structures. If necessary, the C1 posterior nerve rootlet can also be sectioned distally to its connection with the accessory nerve, without major clinical repercussions [1, 10, 22]; (3) Early tumor devascularization through coagulation and section of the dural implant; (4) Tumor debulking (internal decompression) with ultrasonic aspirator, bipolar coagulation and microscissors; (5) Microdissection of the surrounding neurovascular structures interspersed with debulking, always maintaining the arachnoid plane. Subcapsular dissection may improve resection and prevent injury to neurovascular structures in cases where the tumor is extremely adherent without a defined arachnoid plane. It is acceptable to leave behind small pieces to preserve function. Hemostasis is performed with hemostatic agent and bipolar coagulation. Primary dural closure is performed in a watertight fashion. Suboccipital bone flap is fixed with miniplates and screws. Closed-suction submuscular drain is used. Wound closure is carried out in a standard fashion.

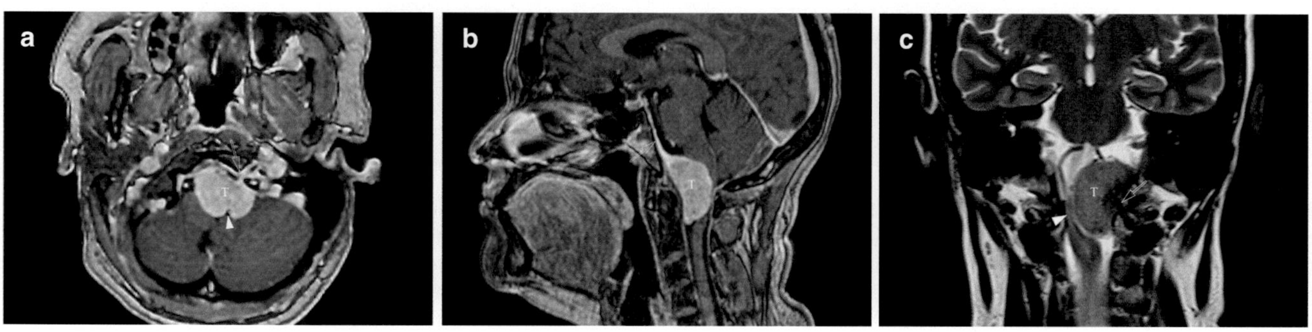

Fig. 51.2 Left lateral FMM. (**a**) Contrast-enhanced axial T1-weighted MRI scan shows a large left intradural lateral FMM (*T*) extending into the hypoglossal canal (*white opened arrow*) and displacing the VA posteriorly (*white closed arrowhead*). (**b**) Contrast-enhanced sagittal T1-weighted MRI scan shows a dural tail (*white opened arrow*) of the meningioma (*T*) extending to the middle clivus. **c** Coronal T2-weighted MRI scan shows the tumor (*T*) involving the VA below, above and around the VA (*white opened arrow*), and causing severe compression of the spinal cord (*white closed arrowhead*)

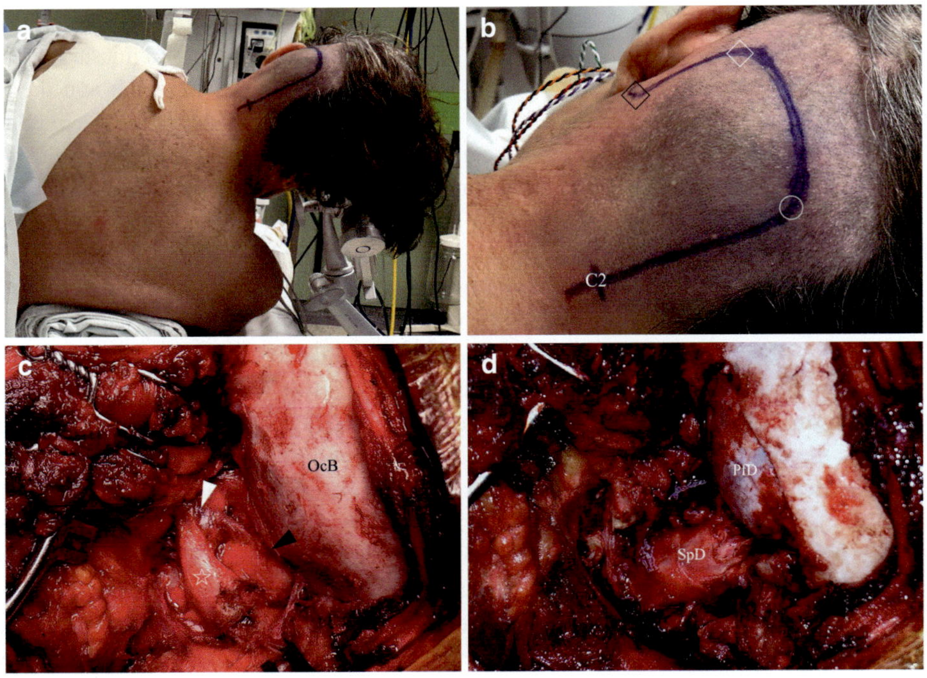

Fig. 51.3 FL approach for a left lateral FMM. (**a**) Right lateral decubitus position. (**b**) Left-sided hockey-stick incision. Mastoid tip (*black opened diamond*), asterion (*white opened diamond*), inion (*white opened circle*), and spinal process of C2 (*C2*). (**c**) Bone exposure. V3 segment of the VA (*white closed arrowhead*), posterior border of the FM (*black closed arrowhead*), posterior arch of C1 (*white opened star*), and occipital bone (*OcB*). (**d**) Posterolateral suboccipital craniotomy and hemilaminectomy of the posterior arch of C1 on the left side were performed exposing the dura mater of the spinal cord (*SpD*) and posterior fossa (*PfD*)

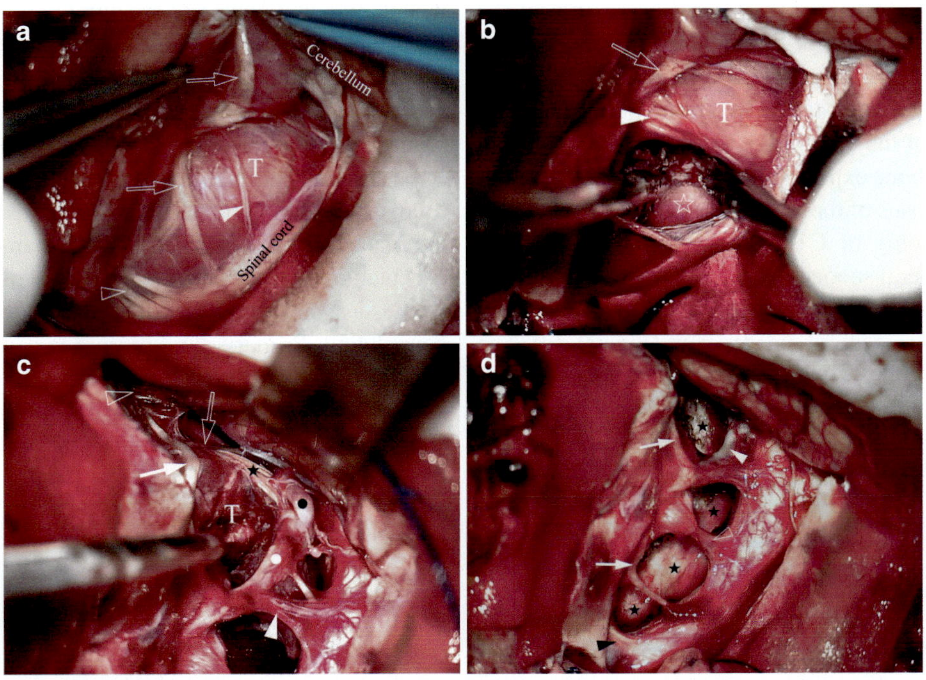

Fig. 51.4 Microsurgical steps for resection of a left lateral FMM through a FL approach. (**a**) Tumor (*T*) exposed. Accessory cranial nerve (*white opened arrow*), C1 (*white closed arrowhead*) and C2 (*white opened arrowhead*) nerve rootlets, cerebellum (*Cerebellum*), and spinal cord (*Spinal cord*). (**b**) Resection is started at the inferior pole of the tumor (*T*) working between the nerve rootlets. Once the tumor is debulked, its capsule can be mobilized laterally. Medially mobilization should be avoided to prevent injury to the neuroaxis. Accessory cranial nerve (*white opened arrow*), C1 (*white closed arrowhead*), and anterior spinal cord dura (*white opened star*). (**c**) Resection of the superior pole of the tumor (*T*) is the most dangerous part for the lower cranial nerves. The position of the lower cranial nerves in this case is difficult to predict. Careful dissection must be performed in order to localize them. Accessory (*white closed arrow*), glossopharyngeal and vagus (*white opened arrowhead*), and hypoglossal (*black closed star*) cranial nerves, hypoglossal canal (*white opened arrow*), C1 posterior nerve rootlets (*white closed arrowhead*), VA (*white closed circle*), and posterior inferior cerebellar artery (*black closed circle*). (**d**) Gross total resection of the tumor. Note the surgical corridors created between the nerves. Dura (*black closed stars*), accessory cranial nerve (*white closed arrow*), VA (*white closed arrowhead*), and C1 (*white opened arrowhead*) and C2 (*black closed arrowhead*) posterior nerve rootlets

The relationship of the tumor to the V4 segment of the VA plays an important role in predicting the position of the lower cranial nerves [1] (Fig. 51.5). For meningiomas located exclusively inferior to the VA, microsurgical resection is started at the most caudal portion of the tumor and the lower cranial nerves can be found displaced superiorly and posteriorly. Meningiomas located superiorly or circumferentially involving the VA are technically more complex, the position of the lower cranial nerves cannot be anticipated, and a more careful dissection must be performed in order to localize them. The nerves in this case can be displaced in any direction and microsurgical removal is performed through intratumoral debulking with the aid of an ultrasonic aspirator, interposing an effort to identify and protect the nerves in its periphery. The most consistent anatomical position of the lower cranial nerves is in the jugular foramen, so it is useful to first identify them distally. Dura mater is inseparable from the adventitia of the VA and the attempt to completely remove tumors with infiltration in this region can cause vascular rupture [23].

51.4.2 Posterior Midline Approach

Posterior midline approach combines a median suboccipital craniotomy with C1 laminectomy [1]. It is indicated for posterior FMM and for craniospinal Leal-Landeiro type 1 anterolateral FMM [1, 19] (Fig. 51.6).

The patient can be placed in semi-sitting, lateral decubitus, park-bench, or prone positions [1, 6, 24]. Incision is outlined in the midline from the external occipital protuberance to the spinous process of C3. Dissection of the muscular plane is performed through the nuchal ligament, followed by subperiosteal exposure of the bilateral suboccipital bone and the posterior arches of C1 and C2. Median suboccipital craniotomy is followed with wide opening of the FM combined with C1 laminectomy. Dura mater is opened in a "Y" fashion and laterally anchored with sutures. The neuroaxis is anteriorly displaced in cases of posterior FMM and tumor resection is performed meticulously with ultrasonic aspirator and microscissors without direct manipulation of neural structures. Hemostasis is ensured. Dura mater is closed in a watertight fashion. Autologous cranioplasty is fixed with titanium miniplates and screws. Closed-suction submuscular drain is used. Wound closure is carried out in a standard fashion.

51.4.3 Anterior Midline Approach (Endoscopic Endonasal Approach)

The main indication is for small anterior midline FMM with no vascular encasements and with limited inferior extension, i.e., not below the odontoid tip [25] (Fig. 51.7).

The patient is placed in supine position with the head fixed in neutral position. Transnasal binostril approach with inferior turbinate out-fracture, wide sphenoidotomy, and vascularized nasoseptal flap preparation is followed by clivus drilling. The limits of the bone removal are superiorly by the floor of the sella, inferiorly by the FM, and laterally by the internal carotid artery, hypoglossal canal, and occipital condyle. Bleeding of the basilar plexus can be controlled with gentle pressure with cotton pads and hemostatic agents. Dura mater is opened after extensive coagulation to devascularize the tumor. Following internal tumor debulking with ultrasonic aspirator, microsurgical extracapsular dissection is performed to separate the tumor from the lateralized lower cranial nerves and VAs. Dural defect is reconstructed with autologous free graft (or dural substitute) and covered with vascularized nasoseptal flap. Dural sealant or fibrin glue can be used to reinforce it. The reconstruction is held in place with foley balloon catheter. Postoperative CSF lumbar drainage may be applied to aid healing.

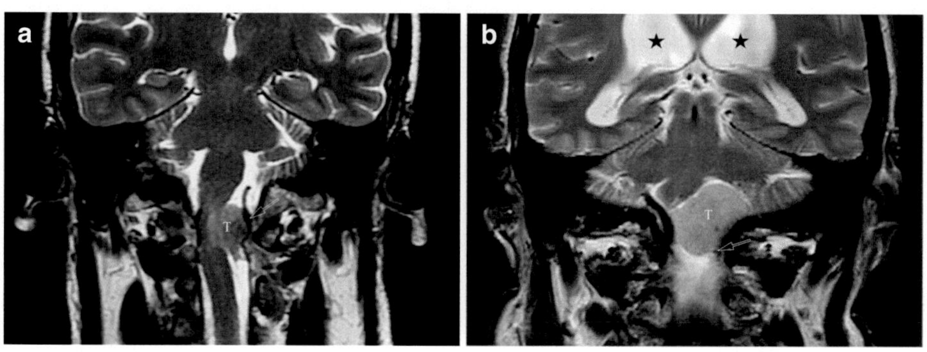

Fig. 51.5 Classification of Bruneau and George of the FMMs according to their relation to the VA [1]. (**a**) Tumor (*T*) located below the intradural segment of the VA (*white opened arrow*). The lower cranial nerves can be found displaced superiorly and posteriorly. (**b**) Tumor (*T*) located above the hypoplastic intradural segment of the VA (*white opened arrow*). The lower cranial nerves can be displaced in any direction. This patient also had enlarged ventricles and hydrocephalus (*black closed stars*) secondary to severe brainstem compression

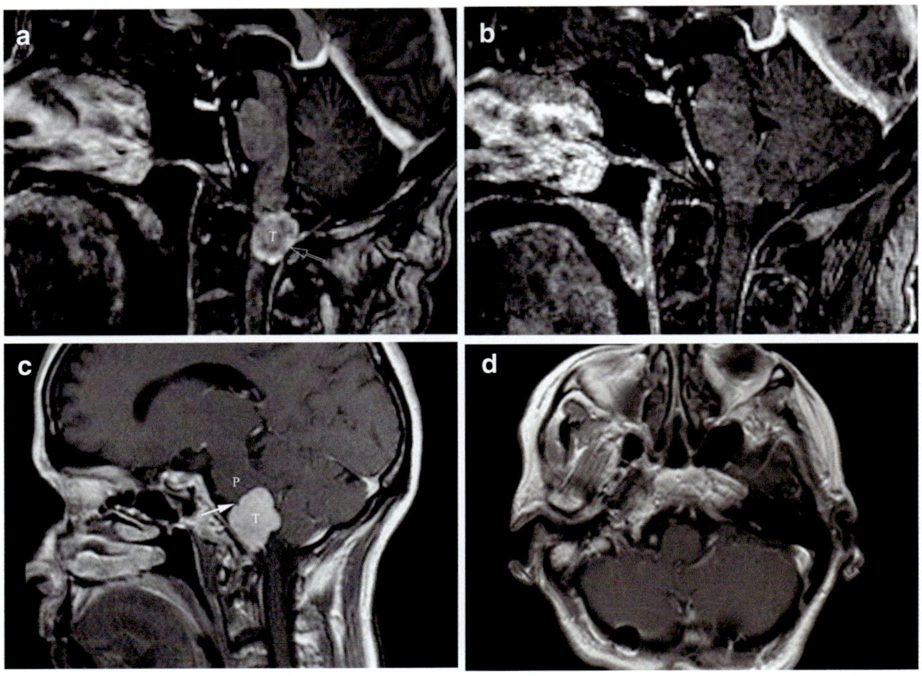

Fig. 51.6 Posterior midline approach for FMM. (**a**) Contrast-enhanced sagittal T1-weighted MRI scan shows a posterior FMM (*T*) and its dural attachment (*white opened arrow*). (**b**) Postoperative contrast-enhanced sagittal T1-weighted MRI scan shows complete tumor resection through a traditional posterior midline approach. (**c**) Contrast-enhanced sagittal T1-weighted MRI scan shows a large lateral FMM (*T*) displacing superiorly the pons (*P*) and below the pontomedullary sulcus (*white closed arrow*), indicating a craniospinal Leal-Landeiro type 1 anterolateral FMM. (**d**) Postoperative contrast-enhanced axial T1-weighted MRI scan shows complete tumor resection through a posterior midline subtonsillar approach

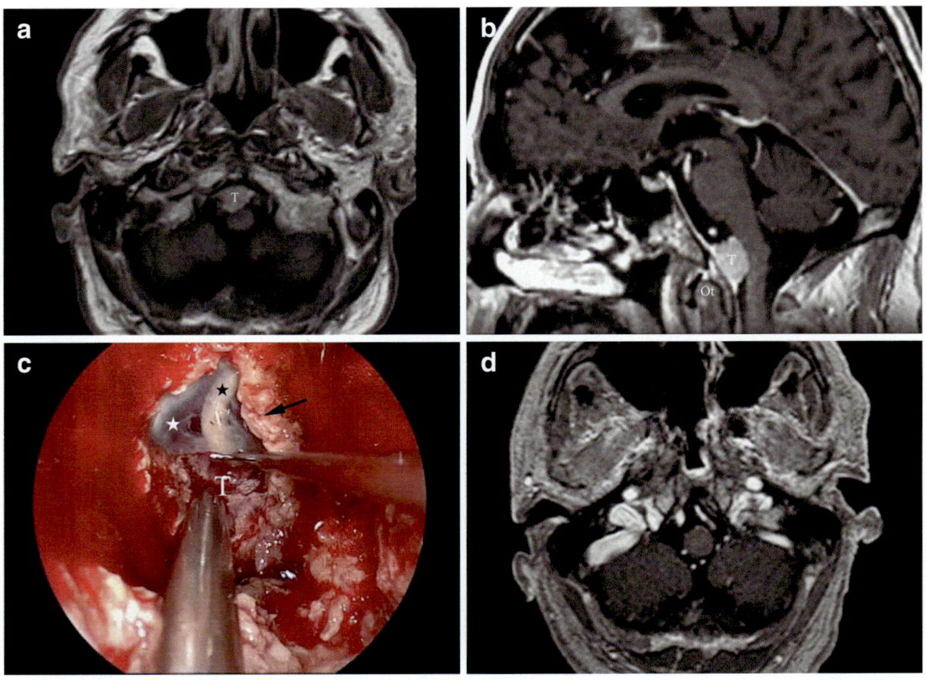

Fig. 51.7 Endoscopic endonasal transclival approach for small anterior FMM. (**a**, **b**) Contrast-enhanced axial and sagittal T1-weighted MRI scan, respectively, shows a small strict anterior midline meningioma without neurovascular involvement and above the odontoid tip (*Ot*). (**c**) Intraoperative view. Following tumor (*T*) debulking with ultrasonic aspirator, extracapsular dissection is performed from the VA (*black closed star*) preserving the arachnoid plane (*white closed star*). Dura infiltrated by the tumor (*black closed arrow*). (**d**) Postoperative contrast-enhanced axial T1-weighted MRI scan shows complete tumor resection. (Reprinted with permission from the Elsevier Inc. Schmidek & Sweet Operative Neurosurgical Techniques: Indications, Methods, and Results, 7th Edition, Volume 1, José Alberto Landeiro, Roberto Leal Silveira, Cassius Vinícius Corrêa dos Reis and Gustavo Augusto Porto Sereno Cabral, chapter 31, Surgical Management of Tumors of the Foramen Magnum, p. 367, Copyright Elsevier, 2022)

51.5 Surgical Pearls

- Appropriate classification of the FM meningioma is crucial for surgical strategy and outcome.
- FL approach without condylectomy is adequate for resection of almost all anterolateral and anterior FMMs.
- Extensive tumor debulking and preservation of the arachnoid plane are critical surgical steps for safe tumor mobilization and resection while preserving the neurovascular structures.
- Subcapsular dissection of the tumor is an alternative technique to improve resection and prevent damage to neurovascular structures in cases where the tumor is adherent without a defined arachnoid plane.
- The main objective of surgery is to preserve the function. Acute postoperative lower cranial deficit can be catastrophic and fatal. For this reason, it is acceptable to leave behind small pieces of tumor adherent to the lower cranial nerves, brainstem, or infiltrating into the jugular foramen. Ultimately, dissection of the functioning lower cranial nerves encased by the tumor is not safe.

51.6 Discussion

Microsurgical resection is the treatment of choice for the majority of FMMs [4]. Subtotal resection should be considered in fibrous and calcified tumors with VA entrapment or adhered to lower nerves, due to high surgical morbidity [26]. Extradural involvement is also often associated with a more invasive and infiltrative behavior, including the adventitia of the VA, which may preclude total resection [27]. Ultimately, the microsurgical technique must focus on a functional preservation surgery. In contrast, patients with chronic deficit of the lower cranial nerves are more able to tolerate a worsening of the deficit in the postoperative period due to previous compensation and adaptation to the dysfunction [3].

Different routes have been used to approach these lesions and can be grouped into anterior, lateral, and posterior approaches. Tumor location, extension and size, neurovascular encasement, and surgeon's experience are some factors that guide the decision making in selecting and tailoring the best approach. The posterior midline approach is mainly indicated for posterior FMMs, and the FL approach, also known as posterolateral approach or lateral suboccipital approach, is indicated for anterior and lateral FMMs [1]. The posterior midline subtonsillar approach has been used to approach the lateral FMMs with acceptable exposure due to the surgical space created by the tumor [19, 28]. The pontomedullary sulcus plays an important role in defining the need for a more lateral approach, according to the Leal-Landeiro classification [19]. The transoral approach has limited indication due to the higher risk of meningitis, neurologic morbidity, and mechanical instability, since it requires an extensive odontoid and C1 arch resection for adequate surgical exposure of the anterior brainstem [29]. The anterior extended endoscopic endonasal approach is promising with the main advantage of reduced manipulation of the lower cranial nerves, particularly for strict midline tumor with no vascular encasements and no inferior extension below the odontoid tip [25]. Extensive drilling of the odontoid process could produce cervical instability, notably when transverse ligament and its attachments are disrupted [25].

Transcondylar approach and VA transposition/mobilization are still matters of debate and controversial techniques [30–32]. They aim to expand the surgical corridor, which is laterally delimited by the occipital condyle and medially by the lateral margin of the bulbocervical junction [6]. VA transposition is routinely performed in the extreme-lateral approach and a rare step included in the FL approach [2]. Some authors suggest that VA transposition/mobilization may be useful for FMMs with extradural extension and dural insertion on both sides of the anterior midline, and for the small clival type, as reported by Yamahata [20]. In a cadaveric study, condylectomy can increase visibility to the clivus region by 15.9° (removal of the posterior third of the condyle) and 19.9° (removal of the posterior half of the condyle) [32]. However, according to the natural history of these tumors, symptoms and diagnosis are usually late, at a time when the tumor reaches sufficient proportions to displace the nervous structures, naturally expanding the surgical corridor and visibility to the clivus [6, 13]. In our experience, drilling the occipital condyle, VA transposition/mobilization, or extreme-lateral approach are not necessary in most cases and should be tailored on an individual basis.

Transcondylar approach may be necessary in strictly anterior and small meningiomas and in those involving the VA circumferentially [31, 32]. There is a consensus in the literature that removal of the posterior third of the occipital condyle does not lead to craniocervical instability [10]. Classically, the intracranial orifice of the hypoglossal canal approximately demarcates the union of the posterior and middle thirds of the occipital condyle, being considered the most anterior intraoperative limit when removing the posterior third [20]. However, considerable anatomical variability of the hypoglossal canal is described; in up to 22% of individuals, the hypoglossal canal may be located in the anterior third of the occipital condyle [33]. Condylar removal down to the hypoglossal canal in these patients may increase the risk of craniocervical instability [34].

51.7 Conclusion

Favorable surgical outcome of the FMM can be achieved through adequate anatomical knowledge of the craniocervical junction, precise classification of the tumor, selecting and tailoring the best approach, and applying the concepts of the microsurgical technique with a focus on functional preservation surgery.

Acknowledgments The authors would like to thank Prof. Marcos Tatagiba, MD, PhD, and Daniel M Prevedello, MD, for their contributions to this chapter.

References

1. Bruneau M, George B. Foramen magnum meningiomas: detailed surgical approaches and technical aspects at Lariboisière Hospital and review of the literature. Neurosurg Rev. 2008;31(1):19–32.
2. Arnautović KI, Al-Mefty O, Husain M, Burton CR, Payne S, Addington-Hall J, et al. Ventral foramen magnum meninigiomas. J Neurosurg. 2000;92:71–80.
3. Colli BO, Carlotti-Junior CG, Assirati-Junior JA, Borba LAB, Coelho-Junior V de PM, Neder L. Foramen magnum meningiomas: surgical treatment in a single public institution in a developing country. Arq Neuropsiquiatr. 2014;72:528–37.
4. Bocchetti A, Cioffi V, Gragnaniello C, Falco R de. Versatility of sub-occipital approach for foramen magnum meningiomas: a single centre experience. J Spine Surg. 2017;3:411.
5. Leon-Ariza DS, Campero A, Romero Chaparro RJ, Prada DG, Vargas Grau G, Rhoton AL. Key aspects in foramen magnum meningiomas: from old neuroanatomical conceptions to current far lateral neurosurgical intervention. World Neurosurg. 2017;106:477–83.
6. Boulton MR, Cusimano MD. Foramen magnum meningiomas: concepts, classifications, and nuances. Neurosurg Focus. 2003;14(6):e10.
7. Talacchi A, Biroli A, Soda C, Masotto B, Bricolo A. Surgical management of ventral and ventrolateral foramen magnum meningiomas: report on a 64-case series and review of the literature. Neurosurg Rev. 2012;35:359–68.
8. Sen CN, Sekhar LN. Surgical management of anteriorly placed lesions at the craniocervical junction-an alternative approach. Acta Neurochir. 1991;108:70–7.
9. Bertalanffy H, Gilsbach JM, Mayfrank L, Klein HM, Kawase T, Seeger W. Microsurgical management of ventral and ventrolateral foramen magnum meningiomas. Acta Neurochir Suppl. 1996;65:82–5.
10. Flores BC, Boudreaux BP, Klinger DR, Mickey BE, Barnett SL. The far-lateral approach for foramen magnum meningiomas. Neurosurg Focus. 2013;35:E12.
11. George B, Lot G, Boissonnet H. Meningioma of the foramen magnum: a series of 40 cases. Surg Neurol. 1997;47:371–9.
12. Pamir MN, Kiliç T, Özduman K, Türe U. Experience of a single institution treating foramen magnum meningiomas. J Clin Neurosci. 2004;11:863–7.
13. Bassiouni H, Ntoukas V, Asgari S, Sandalcioglu EI, Stolke D, Seifert V. Foramen magnum meningiomas: clinical outcome after microsurgical resection via a posterolateral suboccipital retrocondylar approach. Neurosurgery. 2006;59:1177–85.
14. Samii M, Klekamp J, Carvalho G. Surgical results for meningiomas of the craniocervical junction. Neurosurgery. 1996;39:1086–95.
15. Hallopeau H. Note sur deux faits de tumeur de mesocephale. Gaz Med Paris. 1874;3:2.
16. Frazier CH, Spiller WG. An analysis of fourteen consecutive cases of spinal cord tumor. JAMA. 1922;79:1024–8.
17. George B, Dematons C, Cophignon J. Lateral approach to the anterior portion of the foramen magnum. Application to surgical removal of 14 benign tumors: technical note. Surg Neurol. 1988;29:484–90.
18. Yamahata H, Yamaguchi S, Takayasu M, Takasaki K, Osuka K, Aoyama M, et al. Exploitation of simple classification and space created by the tumor for the treatment of foramen magnum meningiomas. World Neurosurg. 2016;87:1–7.
19. da Silveira RL, Von Zuben D, de O, Bertani R, Landeiro JA. Foramen magnum meningioma approached by the midline subtonsilar approach – revisiting Cushing's classification of craniospinal meningiomas. Surg Neurol Int. 2023;14:56.
20. Rhoton J. The far-lateral approach and its transcondylar, supracondylar, and paracondylar extensions. Neurosurgery. 2000;47:S195–209.
21. Lanzino G, Paolini S, Spetzler RF. Far-lateral approach to the craniocervical junction. Neurosurgery. 2005;57:367–71.
22. Komotar RJ, Zacharia BE, McGovern RA, Sisti MB, Bruce JN, D'Ambrosio AL. Approaches to anterior and anterolateral foramen magnum lesions: a critical review. J Craniovertebr Junction Spine. 2010;1:86–99.
23. Bruneau M, Cornelius JF, George B. Antero-lateral approach to the V3 segment of the vertebral artery. Neurosurgery. 2006;58:ONS29-35.
24. Wu Z, Hao S, Zhang J, Zhang L, Jia G, Tang J, et al. Foramen magnum meningiomas: experiences in 114 patients at a single institute over 15 years. Surg Neurol. 2009;72:376–82.
25. Khattar N, Koutourousiou M, Chabot JD, Wang EW, Cohen-Gadol AA, Snyderman CH, et al. Endoscopic endonasal and transcranial surgery for microsurgical resection of ventral foramen magnum meningiomas: a preliminary experience. Oper Neurosurg (Hagerstown, Md). 2018;14:503–14.
26. Pirotte BJM, Brotchi J, Dewitte O. Management of anterolateral foramen magnum meningiomas: surgical vs conservative decision making. Neurosurgery. 2010;67:58–70.
27. Wu P, Guan Y, Wang M, Zhang L, Zhao D, Cui X, et al. Classification and microsurgical treatment of foramen magnum meningioma. Chinese Neurosurg J. 2023;9:3.
28. Herlan S, Ebner FH, Nitz A, Hirt B, Tatagiba M, Roser F. The midline suboccipital subtonsillar approach to the cerebellomedullary cistern and its structures: anatomical considerations, surgical technique and clinical application. Clin Neurol Neurosurg. 2014;125:98–105.
29. Paun L, Gondar R, Borrelli P, Meling TR. Foramen magnum meningiomas: a systematic review and meta-analysis. Neurosurg Rev. 2021;44:2583–96.
30. Mohammad HUR, Vooturi S, Panigrahi M. Far lateral approach: is condylar resection required? Neurol India. 2016;64:455–61.
31. Wu B, Chen LY, Huang GF, Liu WD. Disposal of occipital condyle in far lateral approach for ventrolateral foramen magnum meningiomas. World Neurosurg. 2016;93:29–37.
32. Nanda A, Vincent DA, Vannemreddy PSSV, Baskaya MK, Chanda A. Far-lateral approach to intradural lesions of the foramen magnum without resection of the occipital condyle. J Neurosurg. 2002;96:302–9.
33. Naderi S, Korman E, Çitak G, Güvençer M, Arman C, Şenoğlu M, et al. Morphometric analysis of human occipital condyle. Clin Neurol Neurosurg. 2005;107:191–9.
34. Mazur MD, Couldwell WT, Cutler A, Shah LM, Brodke DS, Bachus K, et al. Occipitocervical instability after far-lateral Transcondylar surgery: a biomechanical analysis. Neurosurgery. 2017;80:140–5.

52 Anatomy for Microsurgery of Paragangliomas

Luis A. B. Borba and Guilherme H. W. Ceccato

Learning Objectives
- Anatomical features of jugular foramen related to key surgical steps
- Description of skull base approaches towards jugular foramen
- To differentiate the surgical approaches regarding lesion features

52.1 Introduction

The most common tumor of jugular foramen (JF) is paraganglioma in 56.8–57.1% of cases [1]. However, considering nonglomic tumors of JF, the most common is schwannoma in 38.6–60% of cases, followed by meningiomas in 20–22.7% of cases, chordomas/chondrosarcomas in 13.6–16.7%, and malignant carcinomas in 3.3–9%, with differential rare cases of aneurysmal bone cysts, cholesteatomas, chondroma, lymphangioma, endolymphatic sac tumors, or inflammatory granulomas [1]. Nonglomic tumors may be classified as intrinsic or extrinsic, if arising from the structures located in the jugular foramen or from adjacent locations extending to the foramen, respectively [1] (Fig. 52.1).

The symptomatology among jugular foramen masses is similar considering the local anatomical relationships [1]. Regarding nonglomic tumors of JF, for example, about 67% of patients present impairment of CN VIII, 56% of CN X, 36.6% of CN IX, and 36.6% of CN XI, with lateropulsion present in 56% of cases and facial nerve affected in 20% of patients [1]. Collet-Sicard syndrome, a combination of impairment of CNs IX to XII, occurs in 1/3 of patients [1].

L. A. B. Borba (✉)
Department of Neurosurgery, Mackenzie Evangelical University Hospital, Curitiba, PR, Brazil

Department of Neurosurgery, Federal University of Paraná, Curitiba, PR, Brazil

G. H. W. Ceccato
Department of Neurosurgery, Mackenzie Evangelical University Hospital, Curitiba, PR, Brazil

I. N. Kanaan, V. Beneš (eds.), *Neuroanatomy Guidance to Successful Neurosurgical Interventions*,
https://doi.org/10.1007/978-3-031-59838-8_52

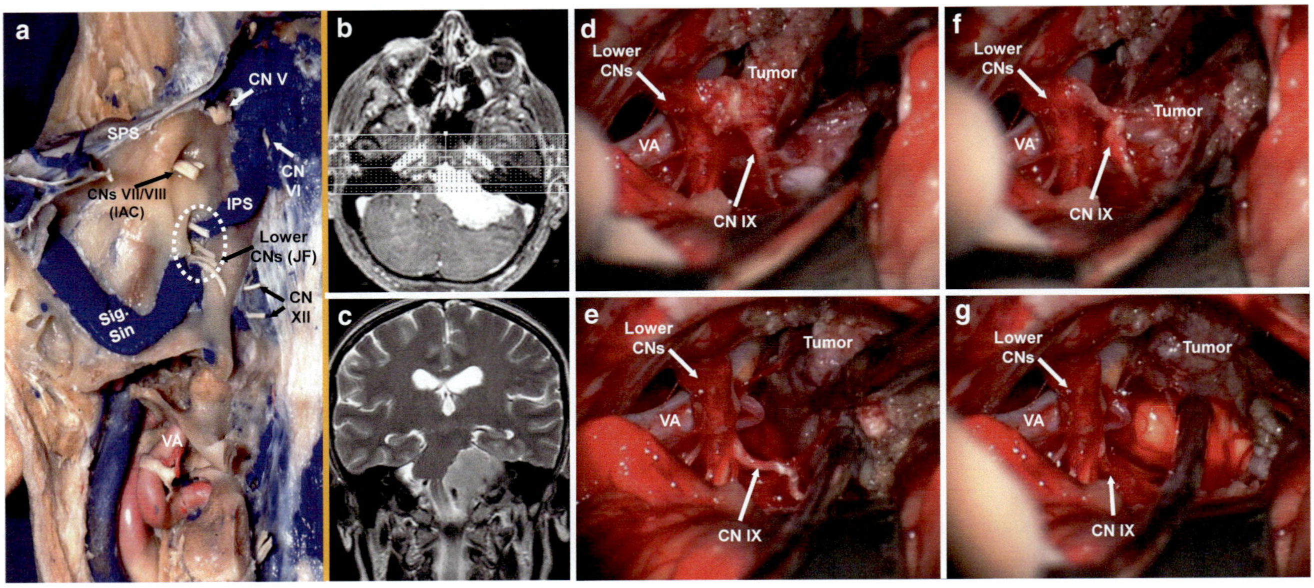

Fig. 52.1 Extension towards jugular foramen of a petrosal meningioma. (**a**) Global view of intracranial relationships of petrosal part of temporal bone, highlighting topography of jugular foramen. (**b**) Axial T1WI (**c**) Coronal T2WI depicting a petrosal meningioma. (**d–g**) Step-by-step of dissection of glossopharyngeal nerve from the meningioma, demonstrating the progressive extension of the mass towards jugular foramen area. *CN* cranial nerve, *IAC* internal auditory canal, *IPS* inferior petrosal sinus, *Sig. Sin.* sigmoid sinus, *SPS* superior petrosal sinus, *VA* vertebral artery

52.2 History

The first report of a glomus jugular tumor resection goes back to 1945 by the otologic surgeon Rosenwasser; however, the first surgical exploration of jugular foramen dates from 1934 by another otologic surgeon named Seiffert [2, 3]. As JF masses present commonly with hearing disturbances, they were initially treated by otorhinolaryngologists. The first reports of neurosurgical intervention to jugular foramen tumors date from 1951, and since the 70s, there has been a more intense development of skull base approaches and multidisciplinary teams [4].

In 1952, it was first described by Capps the anterior transposition of facial nerve associated with ligation of sigmoid sinus and jugular vein. However, due to significant complications of surgical treatment, radiotherapy started to be preferred to manage these lesions for many years, with the first paper advocating radiation published next year to Capps [4]. In 1953, Guild published findings of glomus formations along the Jacobson's and Arnold's nerves, and thanks to this description, it became possible to differentiate between glomus tympanicum and jugular [3, 4]. Arteriography was introduced for investigation in 1953, considering the feature of these tumors to be highly vascularized [4]. Microsurgical refinements progressed and reports of jugular vein and sigmoid sinus ligation, anterior facial nerve transposition, and arterial devascularization of the tumor surfaced in 1964. In 1965, Gejrot already highlighted the importance of preserving the medial wall of internal jugular vein and bulb to avoid injury to lower cranial nerves [4]. In 1968, Rosenwasser reviewed his almost 25 years of experience with 36 glomus tumors, reporting 17% rate of overall mortality; however, considering just 22 cases of glomus tympanicum, there was a 27% recurrence rate and 5% mortality, while the 10 lesions with skull base extensions presented 80% rate of recurrence and 50% of mortality [4]. In 1969, CT scans started to spread and became of general use in 1974; during this period, House suggested avoiding the transposition of facial nerve and instead using the extended facial recess corridor [4]. In 1971, Kempe et al. reported the addition of a suboccipital craniotomy to the standard mastoidectomy, removing tumor extensions to the posterior fossa.

Initially, in the first cases, intradural extensions were left to be treated by radiation or in a next procedure; however, gradually it was attempted and became possible to resect both intra and extradural extensions of jugular foramen tumors [2]. By the end of 70s, Fisch described a systematic scheme for classification and management of glomus jugular tumors and detailed the infratemporal fossa approach with its variations according to the extensions of the lesion [4, 5]. Balloon occlusion testing was introduced in the mid-70s and preoperative embolization in the 80s. In 1985, Brown depicted a 10-year follow-up review of 127 patients with glomus jugular tumors, reporting surgical control in 64% of patients and a mortality rate of 12% [4]. In 1987, Al-Mefty and Teixeira detailed approach refinements to resect giant glomus jugular tumors and temporal bone chondrosarcomas [2]. In 1994, a review by Green et al. of 52 patients with glomus jugular tumors reported an average follow-up of 3.4 years with 85% of cases with total resection and no mortality, detailing improvements in the technique of rerouting facial nerve [4]. In 1997, Pensack and Jackler described the fallopian bridge technique, advocating avoiding rerouting facial nerve and working around the facial canal [6]. In 1999, 136 cases of temporal bone paragangliomas from Fisch's group were reviewed, reporting 83% of total resection and 98% disease-free survival, with 88% of patients with rerouting of facial nerve presenting House-Brackmann I-II postoperatively [4]. In 2004 and 2010, Borba et al. advocated no need for facial nerve rerouting to resect large glomus jugular tumors [7, 8]. In 2013, Lope Ahmad et al. reported 121 cases of jugular foramen paragangliomas, with about 55.4% presenting intracranial extension. Surgical control in 96% of cases and a low rate of perioperative complications in 1.6% of cases (as CSF leak) were achieved, besides preservation of lower cranial nerves in 63% of cases, with no mortality [4, 9].

52.3 Anatomy Highlights

- Jugular foramen is located in an oblique orientation between temporal and occipital bones
- Glossopharyngeal nerve usually is a superior single nerve bundle; vagus nerve presents as multiple rootlets and accessory nerve consists of two fascicles, one being spinal
- Lower cranial nerves are covered by the anteromedial wall of internal jugular vein/bulb through a lateral surgical pathway
- The roof of a domed jugular fossa contributes to the floor of tympanic cavity
- Facial recess is a pathway to tympanic cavity from the mastoid
- Deep into facial recess rests internal carotid artery
- Incus is a landmark to help identify mastoid segment of facial nerve
- Mobilization of digastric muscle helps exposure of extracranial segment of facial nerve
- Arterial blood supply to jugular foramen area comes mainly from ascending pharyngeal or occipital arteries, as well as can have some contribution from caroticotympanic branches
- Inferior petrosal sinus drains into jugular bulb, just as other local emissary veins do

Jugular foramen (JF) is in an oblique orientation between temporal and occipital bones (Fig. 52.2). In general, related structures with jugular foramen are the internal carotid artery anteriorly; styloid process anterolaterally; hypoglossal nerve and occipital condyle medially; mastoid tip, facial nerve, and digastric muscle laterally; rectus capitis lateralis muscle posteriorly; vertebral artery inferiorly, and labyrinth superiorly [10, 11] (Fig. 52.3). JF is commonly divided into a *pars nervosa* anteromedially and a *pars venosa* posterolaterally; however, different classifications have already been proposed. The first one harbors a meningeal branch of ascending pharyngeal artery, inferior petrosal sinus, and CN IX, and the last one contains jugular bulb and CNs X and XI. It is possible that there exists a fibrous septum separating these two compartments connecting intrajugular process of occipital and temporal bones, or even no clear division [12] (Fig. 52.4).

CN IX usually consists of a unique bundle traveling to jugular foramen, positioned anteromedial to the other nerves and closest to internal carotid artery when emerges from the foramen. CN X usually consists of multiple rootlets traveling to jugular foramen, while CN XI usually of one or two fascicles (cranial and spinal) [13, 14]. Tympanic branch of CN IX (Jacobson's nerve) emerges from inferior ganglion and travels to tympanic canaliculus to constitute tympanic plexus

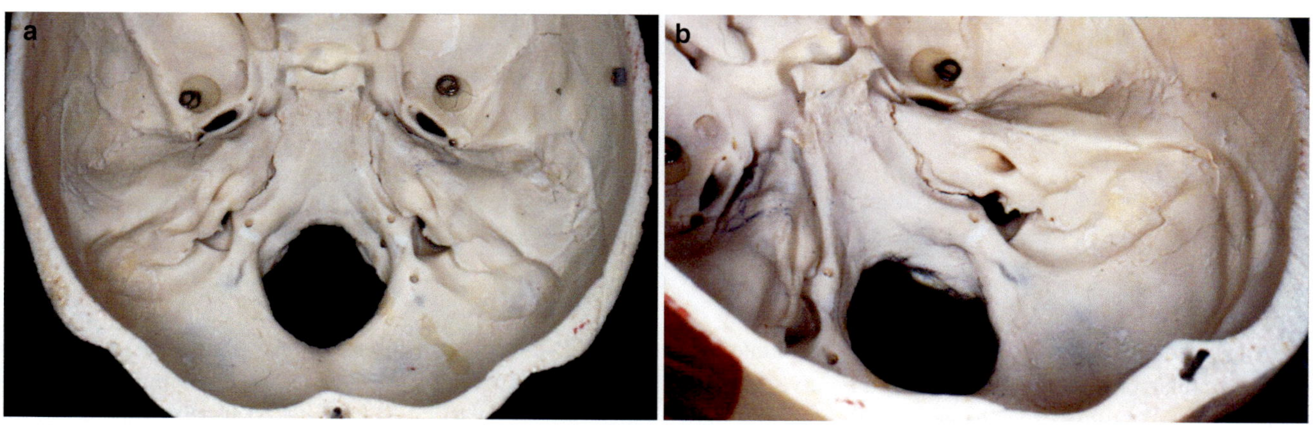

Fig. 52.2 Osseous relationships of jugular foramen, which is located between the petrosal part of temporal bone and occipital bone. Its proximity with hypoglossal canal and occipital condyle, as well as with internal auditory canal, is highlighted. An intrajugular process of temporal bone is observed, as well as the sigmoid groove

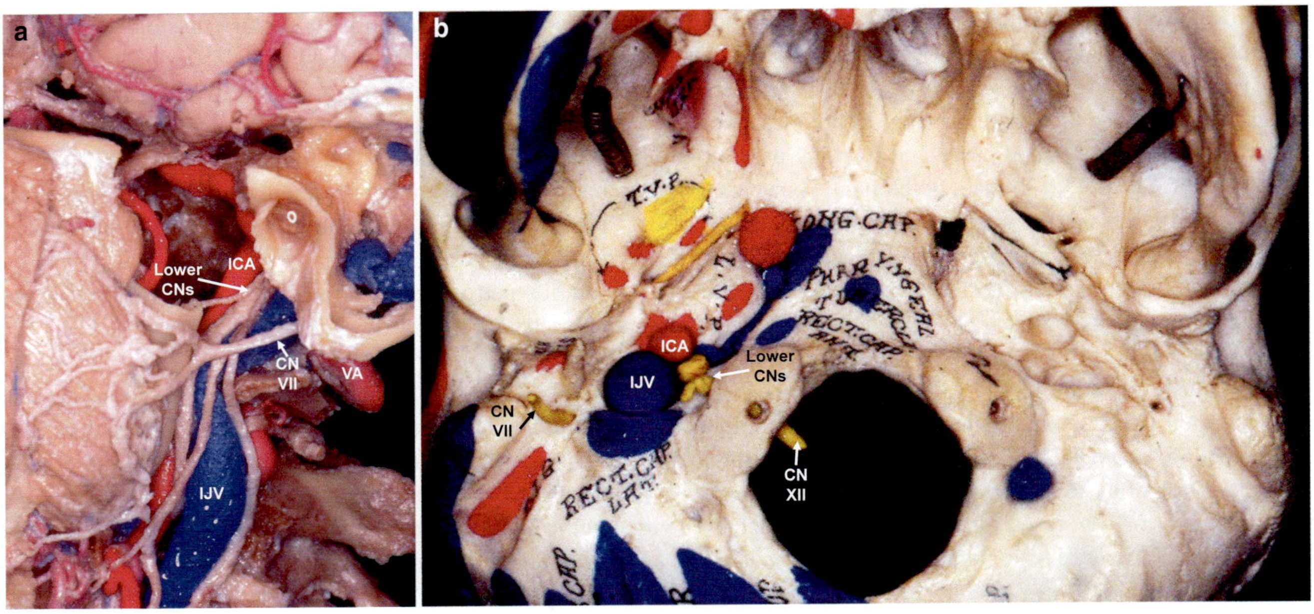

Fig. 52.3 Anatomical boundaries of jugular foramen, by lateral (**a**) and inferior (**b**) views. It demonstrated the internal carotid artery anteriorly; hypoglossal nerve and occipital condyle medially; mastoid tip, styloid process, facial nerve, and digastric muscle laterally; rectus capitis lateralis muscle posteriorly; vertebral artery inferiorly; and labyrinth superiorly. CNs IX-XI are seen running medially to internal jugular vein. *CN* cranial nerve, *ICA* internal carotid artery, *IJV* internal jugular vein, *VA* vertebral artery

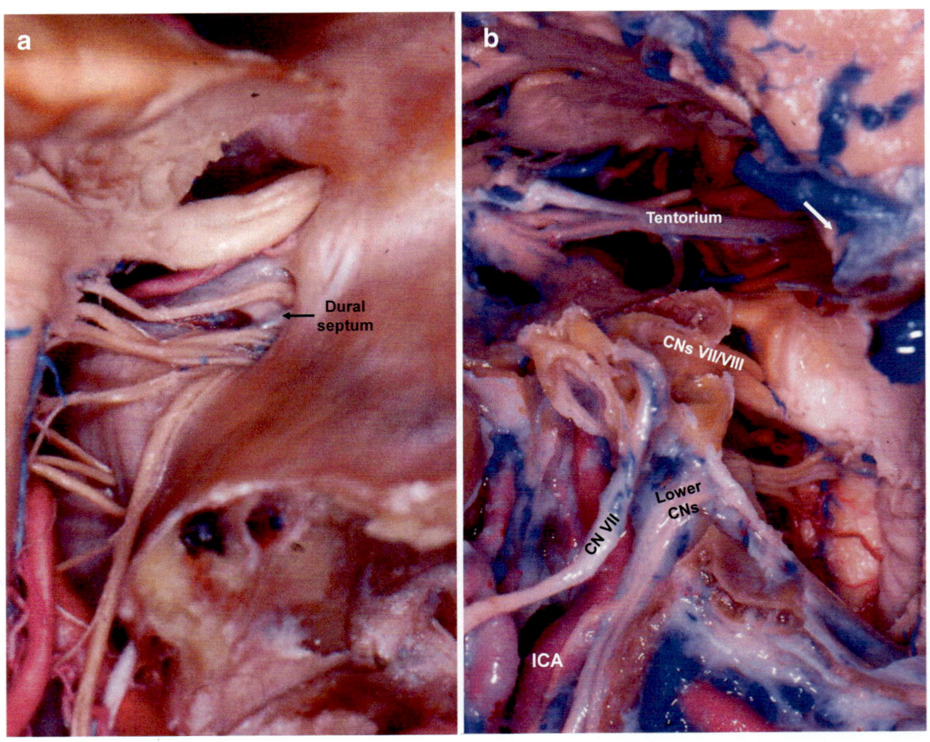

Fig. 52.4 (**a**) Anatomy of the nerves in the jugular foramen, highlighting the dural septum inside the foramen dividing it into pars nervosa and pars venosa. (**b**) Lateral overview of posterior fossa structures. *CN* cranial nerve, *ICA* internal carotid artery

to contribute then to the lesser petrosal nerve in middle fossa [13]. Auricular branch of CN X (Arnold's nerve) emerges from superior ganglion of vagus nerve with possible contribution of inferior ganglion of CN IX traveling to mastoid canaliculus towards external auditory canal, passing close to mastoid segment of facial nerve [13]. It is important to comment on the proximity of hypoglossal nerve, consisting of some rootlets grouping towards hypoglossal canal.

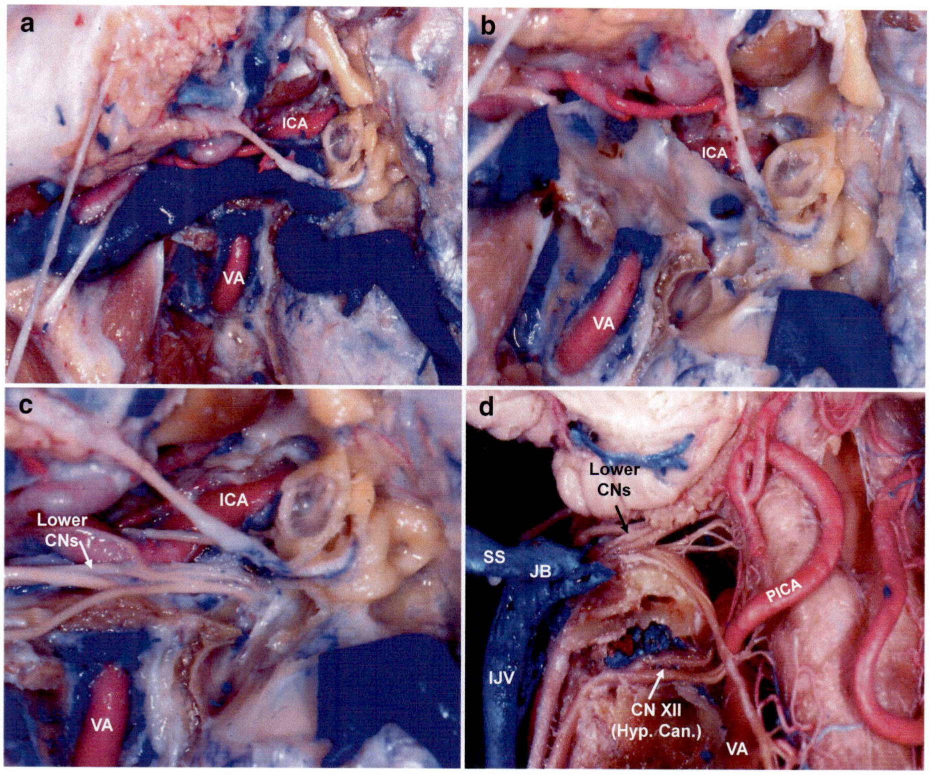

Fig. 52.5 View highlighting the medial trajectory of lower cranial nerves related to internal jugular vein and jugular bulb, as they are covered by the medial wall of IJV through a lateral corridor. (**a**–**c**) Stepwise removal of internal jugular vein exposing underlying lower cranial nerves. (**d**) Posteroanterior view depicting lower CNs medial to jugular bulb. *CN* cranial nerve, *Hyp. Can.* hypoglossal canal, *ICA* internal carotid artery, *IJV* internal jugular vein, *JB* jugular bulb, *PICA* posterior inferior cerebellar artery, *SS* sigmoid sinus, *VA* vertebral artery

Lower cranial nerves travel in the jugular foramen medial to jugular bulb and internal jugular vein, and the medial wall of the vein protects the nerves during microsurgical resection as surgical corridor is from lateral to medial. By this way, the medial wall should be spared if there is preoperative lower cranial nerves' functional preservation. Lesions arising laterally in the foramen displace the nerves medially; however, lesions arising medially as from foramen magnum area displace nerves laterally [10] (Figs. 52.5).

Tumors extending to middle ear may be reached through the facial recess corridor, keeping facial nerve inside its canal, which is just skeletonized (Fig. 52.6). During mastoidectomy, a helpful landmark is to identify incus, which points to the position of facial canal. Jugular fossa contains jugular bulb, and they form the floor of middle ear when in a domed fashion with a well-developed jugular bulb. Jugular foramen tumors may spread towards middle ear eroding through roof of jugular foramen/floor middle ear [15] (Fig. 52.7). The mean depth of right jugular fossa is 12.54 mm and that of the left one is 8.72 mm; higher jugular fossa may have a significative projection over internal auditory canal/middle ear, and mean width of right jugular fossa is 7.2 mm and that of left is 5.8 mm [15]. Mean extracranial width of jugular foramen is around 13 mm, with no significative difference between sides. Interestingly, the distance from jugular foramen and lateral stylomastoid foramen is 4–5 mm [15].

Arterial supply to jugular foramen comes mainly from ascending pharyngeal or occipital arteries and may also have contributions from caroticotympanic branches of internal carotid artery. Depending on the case, involvement of vertebral artery may also be seen, which is in close contact with lower cranial nerves intradurally [13]. Carotid canal lies just anteromedial to jugular foramen and highlights the proximity of petrosal segment of internal carotid artery inside temporal bone with middle ear and jugular foramen. It is important to mention the stylomastoid artery, branching from posterior or occipital artery and running towards stylomastoid foramen to supply CN VII and tympanic cavity.

From a surgical point of view, drilling infralabyrinthine mastoid and mobilizing digastric muscle exposes an arc length of 4.1 mm and provide an angle of attack of 86.4° at the lateral aspect of jugular vein leaving jugular foramen. Removal of rectus capitis lateralis muscle provides an additional arc length of 5.2 mm and a posterior angle of exposure of 108.4°. Removal of styloid process and muscle adds an anterior arc length of 4.1 mm and an 85.5° angle of attack, as well as better access to internal carotid artery entering petrous part of temporal bone (Fig. 52.3). Combining these steps, a total 280.3° of exposure around jugular foramen is obtained [11].

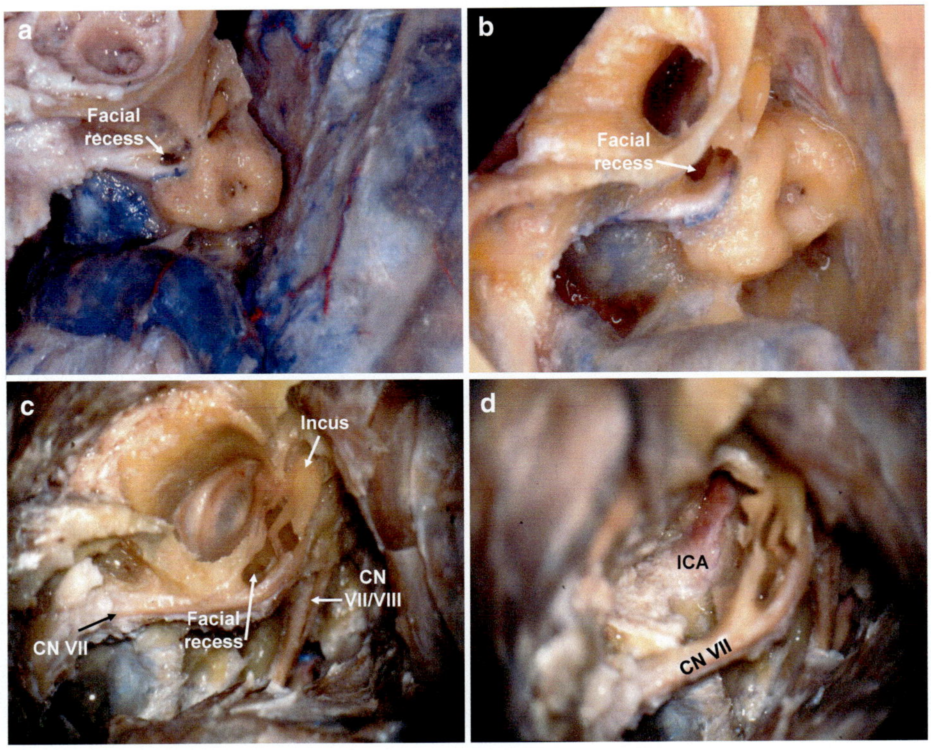

Fig. 52.6 Overview highlighting the anatomical relationships of facial recess. *CN* cranial nerve, *ICA* internal carotid artery

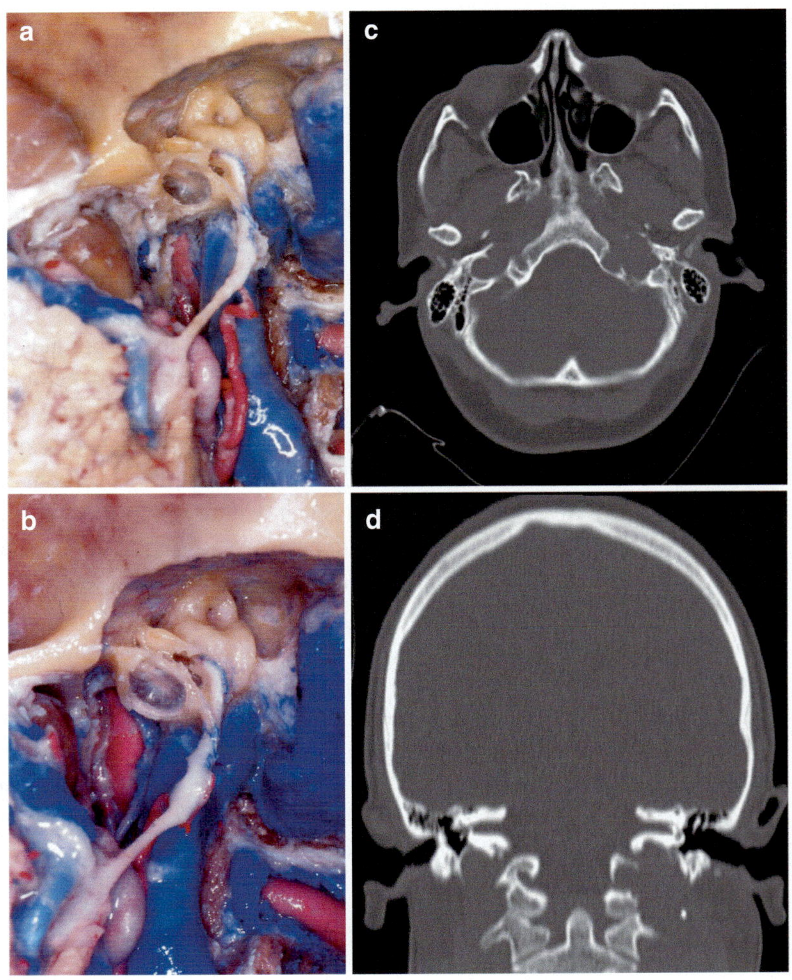

Fig. 52.7 (**a**, **b**) Proximity and relationship of jugular bulb with middle ear, especially tympanic cavity. (**c**, **d**) Proximity or the roof of jugular fossa with internal auditory canal

52.4 Clinical Aspects and Intervention

The most common lesions involving jugular foramen are the paraganglioma, followed by schwannomas and meningiomas.

52.4.1 Paragangliomas

Jugulotympanic paragangliomas present an incidence of about 0.07 per 100.000 persons per year [16, 17]. They are more common between age 50 and 60 and more frequently affect female patients. Despite containing secretory granules histologically, just 1–3% of tumors are functional. Less than 5% have malignancy features with usual metastases to regional lymph nodes; however, despite being usually benign, they can be locally aggressive and infiltrate temporal bone [17–19]. Common initial symptoms are hearing loss in 61.7% of cases and tinnitus in 56.1%; also, specifically neck/ear pain occurs in 27.2%, hoarseness in 23%, and dysphagia in 16.5% [17]. Patients usually present with cranial nerve deficits in 29.7% of cases, usually with CN IX compromise in 43.3%, CN X in 41.7%, CN XI in 25%, and CN XII in 26.7% [17]. A retrotympanic tumor may be identified in around 90% of cases [20] (Fig. 52.8). Tumor growth rate is usually less than 1 mm per year, and diagnosis may be late, only when tumor presents with big size, and by this way it is important to identify growing lesions [17].

Microsurgical resection is preferred in young and healthy patients, with small and more accessible tumors, usually Fisch grade A and B. Surgical decision should be individualized in greater and more complex tumors considering the patient's functional status and surgical risks, as some degree of postoperative cranial nerve neuropathy may be not tolerated in older and more debilitated patients. Main surgical benefit over radiotherapy is greater chance of cure and cytoreduction and relief mass effect. In cases not suitable for surgical intervention, or residual lesions, radiosurgery may be indicated, considering preoperative neurological deficits and performance status, as radiosurgery yields a less rate of morbidity [17]. Use of preoperative embolization may be done in cases of larger tumors usually greater than 3 cm, with multiple feeders and presenting preoperative lower cranial nerves deficits, as there is the theoretical risk of accidental embolization of vasa nervorum.

Microsurgical resection presents a 91% rate of radical resection. Facial nerve function was normal in the postoperative period in 85% of cases and kept in its position in 95% of cases. In 17.6% of patients, there was a new postoperative lower cranial nerve deficit, but swallowing function recovered adequately in follow-up; however, vocal disturbance presented worst recovery [10].

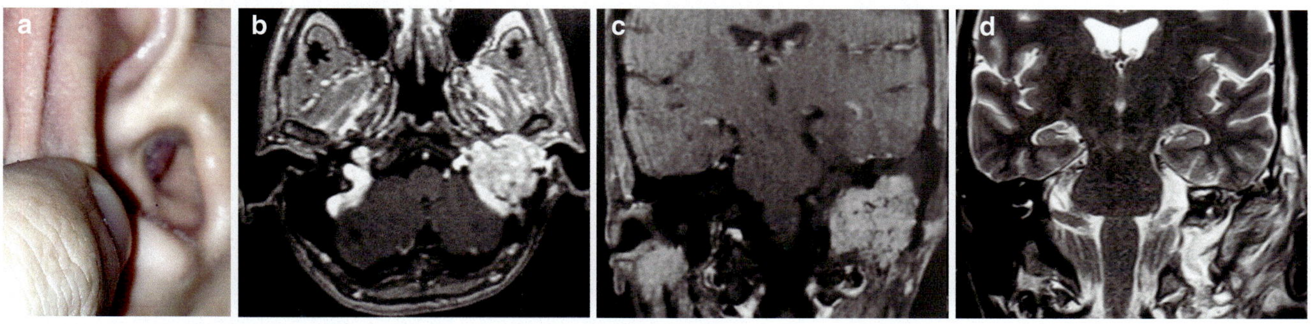

Fig. 52.8 31-year-old female patient harboring a tympanojugular paraganglioma presenting with facial palsy, dizziness, and deafness. (**a**) Mass extending into external auditory canal. (**b**, **c**) Preoperative MRI. (**d**) Postoperative MRI

52.5 Approach/Surgical Intervention

Most lesions at jugular foramen need microsurgical resection as the primary treatment, with radiation reserved for residual cases, with few reports of chemotherapy; embolization is another tool used commonly as a preoperative adjunct to microsurgical resection. Different approaches can be employed, regarding the extension of the lesions, and a careful evaluation of preoperative neurological deficits associated with patient performance status should be done [21].

52.5.1 Positioning and Skin Incision

Patient is placed supine with the head rotated to the contralateral side in 45–60° and slightly extended up to the malar eminence be the highest point in the surgical field to open the space between angle of mandible and neck; ipsilateral shoulder is also slightly elevated and care must be taken to avoid compression of internal jugular vein. It is important to prepare an area of abdomen and/or thigh if necessary for fat or fascia harvesting. An arciform incision is initiated 2 cm above the pinna, running ~3 cm behind the ear progressing up to ~3 cm below mastoid tip, extending anteriorly over the anterior border of sternocleidomastoid muscle for ~3 cm.

Skin is reflected anteriorly, and it is important to identify the great auricular nerve crossing around sternocleidomastoid muscle. Cartilaginous part of external acoustic canal may or may not be sectioned depending on the type of approach. In the surgical field, the posterior aspect of temporalis muscle and insertion point/proximal part of sternocleidomastoid muscle are exposed. Temporalis fascia is dissected from underlying muscle and progressively sternocleidomastoid muscle is also dissected and reflected posteriorly by keeping attached to temporalis fascia; some possible exposed lateral edge of deep muscle layer as from splenius capitis may be also reflected together with sternocleidomastoid muscle. Temporalis muscle is mobilized anteriorly (Fig. 52.9). Digastric muscle is then detached from digastric groove and mobilized anteriorly.

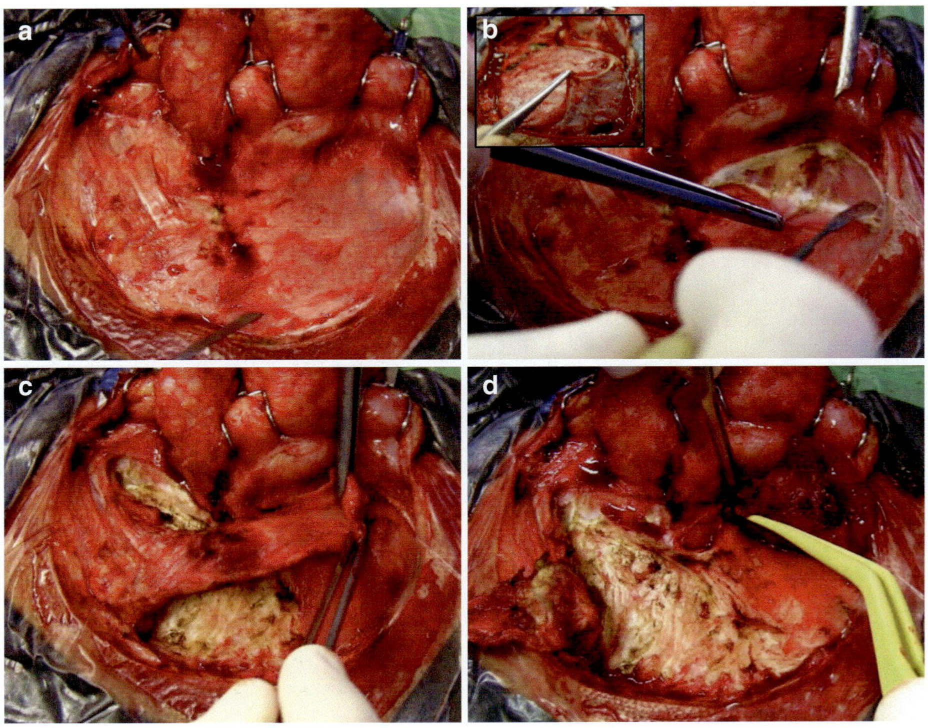

Fig. 52.9 Intraoperative detail of preparation from the beginning of pedicled flaps to be used in the closure. A key point is to dissect temporalis fascia from underlying temporalis muscle and keep it attached to sternocleidomastoid muscle while reflecting this muscle

52.6 Dissection of Cervical Neurovascular Structures

In the neck, it is important to dissect and identify the major neurovascular structures. Exposure and control of common, internal, and external carotid arteries may be performed, and any blood supply for tumor is ligated/coagulated when identified, as branches of ascending pharyngeal/occipital arteries. Internal jugular vein should also be properly dissected and identified, as may be ligated if necessary. Also, lower cranial nerves running to the neck should be adequately identified: CN IX running over lateral aspect of distal internal carotid artery entering petrous bone, CN X between ICA and IJV, CN XI running posteriorly over IJV, and CN XII performing a loop over carotid bifurcation and projecting anteriorly, giving a branch to the ansa cervicalis.

52.6.1 Mastoidectomy

Mastoidectomy is then performed, initially creating a bony shell of cortical bone which is removed to help in reconstruction later. Next, mastoid air cells are carefully removed, skeletonizing underlying structures: bony labyrinth, facial canal, sigmoid sinus, and jugular bulb. Facial canal is not opened exposing facial nerve, just in case of preoperative facial nerve palsy or signs of facial canal infiltration/destruction (Fig. 52.10). In this way, tumor extensions to middle ear may be reached through the facial recess, a corridor between mastoid segment of facial nerve and chorda tympani. An important point is that the mastoid segment of facial nerve is usually pointed by incus, helping in its identification. Just skeletonizing facial canal then allows working around in the way of fallopian bridge technique (Fig. 52.11). Attention should be paid working through facial recess due the proximity with internal carotid artery [6] (Figs. 52.12 and 52.13). If there is facial nerve compromise, the nerve may be either decompressed or the part affected resected and grafted, usually employing great auricular nerve. Nerve graft may be just positioned covering a thin nerve or performing end-to-end neurorrhaphy (Fig. 52.10). Finally, mastoid tip is removed, as well as rectus capitis lateralis muscle, exposing the whole venous trajectory including sigmoid sinus, jugular bulb and internal jugular vein.

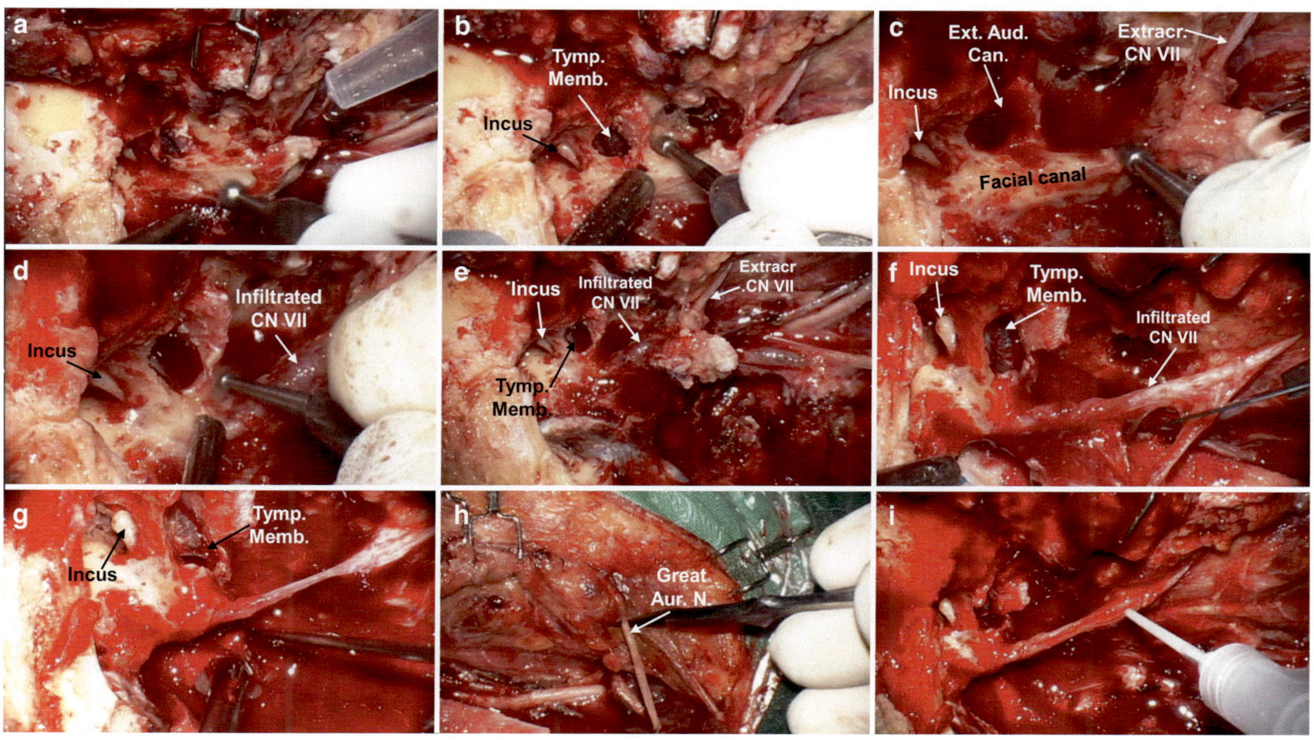

Fig. 52.10 Intraoperative details of the progressive exposure of a mastoid segment of facial nerve infiltrated by tumor. It was needed to dissect and remove the infiltrating tumor and graft facial nerve with great auricular nerve. *CN* cranial nerve; *Ext. Aud. Can.* external auditory canal, *Extracr.* extracranial, *Great Aur. N.* great auricular nerve, *Tymp. Memb.* tympanic membrane

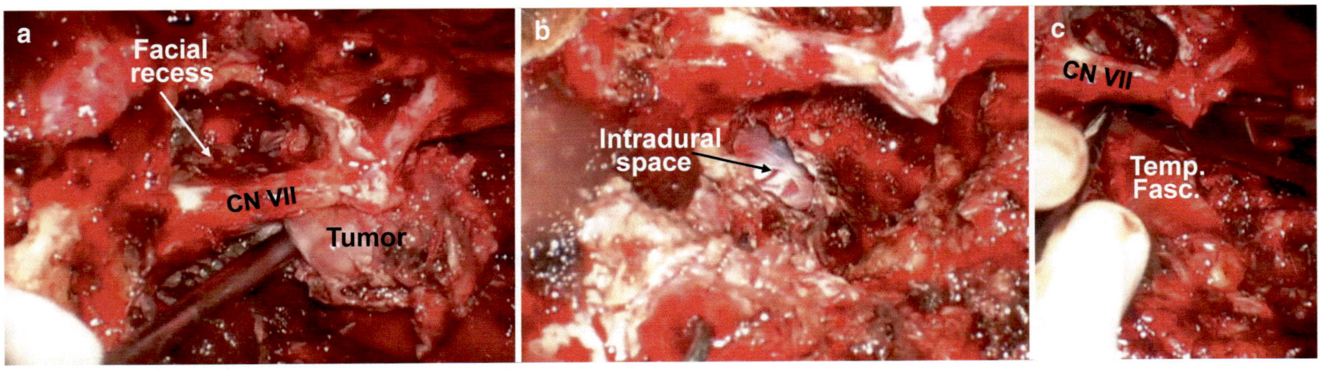

Fig. 52.11 (**a**) Removal of the extradural part of a jugular foramen paraganglioma after dissecting and working around facial canal. (**b**) Dural defect exposing intradural compartment. (**c**) Covering of dural defect with pedicled flap of temporalis fascia. *CN* cranial nerve, *Temp. Fasc.* temporalis fascia

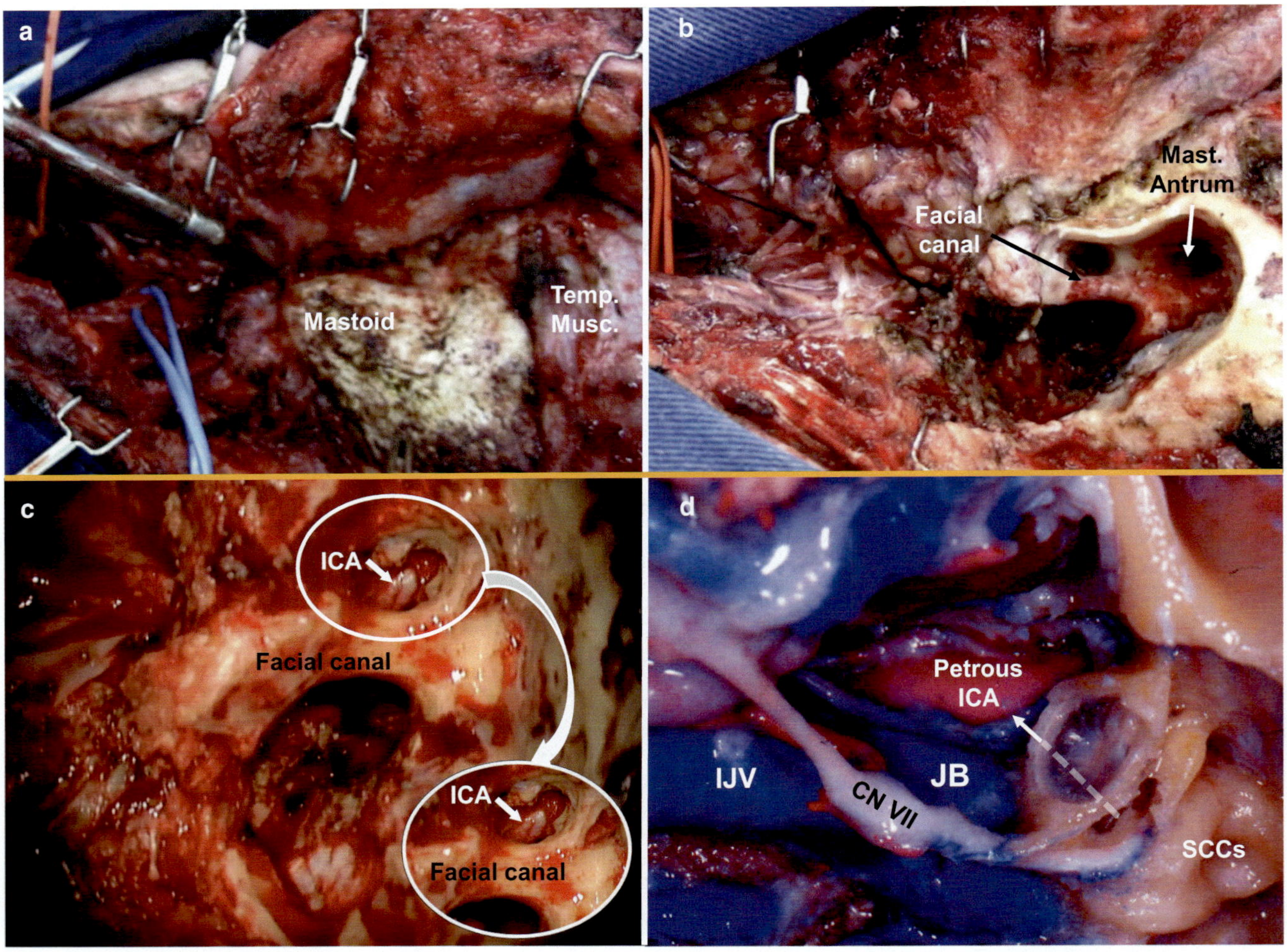

Fig. 52.12 (**a**, **b**) Demonstration of skeletonized facial canal allowing to work around it in the way of a fallopian bridge technique. (**c**, **d**) Proximity of the internal carotid artery when accessing middle ear through facial recess is demonstrated. *CN* cranial nerve, *ICA* internal carotid artery, *IJV* internal jugular vein, *J* jugular bulb, *Mast. Antrum* mastoid antrum, *Mid. Ear Ossic.* middle ear ossicle, *SCCs* semicircular canals, *Temp. Musc.* temporalis muscle

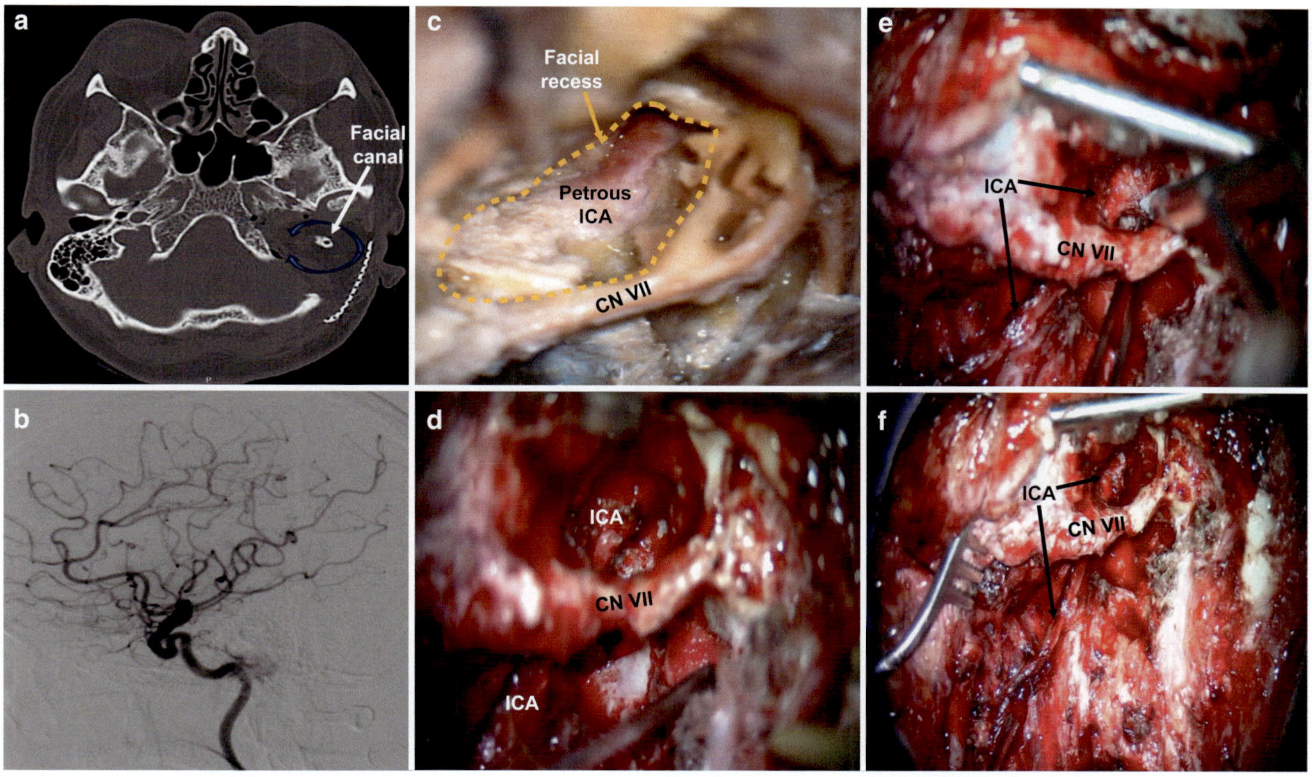

Fig. 52.13 (**a**) Postoperative CT scan following resection of a jugular foramen paraganglioma. Preservation of facial canal highlights its use in the fallopian bridge technique. (**b**) Preoperative angiography depicting tumor blood supply from internal carotid artery. (**c**) Demonstration of proximity of ICA through facial recess, which is more precisely defined as a small space between chorda tympani and facial nerve. (**d**) Intraoperative close view through facial recess highlighting internal carotid artery during resection of a left jugulotympanic paraganglioma. (**e**) Tumor resection close to ICA through facial recess. (**f**) Overview of surgical exposure. *CN* cranial nerve, *ICA* internal carotid artery

52.7 Venous Structures Management

Mainly paragangliomas at jugular foramen, as well some other tumors, may infiltrate into lumen of jugular vein/bulb and sigmoid sinus, leading to occlusion of these venous outflow, which highlights importance of preoperative venous study (Fig. 52.14). To completely resect the lesion, sigmoid sinus is first ligated proximally above the extension of the tumor, and internal jugular vein is ligated below tumor projection. Lateral wall of sigmoid sinus is incised, removing intraluminal tumor, and especial care must be taken in this step as profuse venous bleeding may occur, associated with risk of air embolism. After complete proximal intraluminal tumor removal, sigmoid sinus is packed with hemostatic material and ligated. Also, in the neck the lateral wall of internal jugular vein is incised and intraluminal tumor is removed. A key point is that lower cranial nerves run medial to IJV and are protected by the anteromedial wall of the vein from a lateral approach. In this way, if the patient presents preoperative preservation of the function of lower cranial nerves, the anteromedial wall of internal jugular vein and jugular bulb should be preserved and excessive coagulation in this region avoided. An intrabulbar tumor removal is performed at the level of jugular bulb preserving the medial wall and it is important to pay attention to venous bleeding coming from emissary veins at this region, as well as from inferior petrosal sinus (Fig. 52.15). In case of proven ipsilateral lower cranial nerves' preoperative deficit, the medial wall of internal jugular vein may be resected (Fig. 52.16).

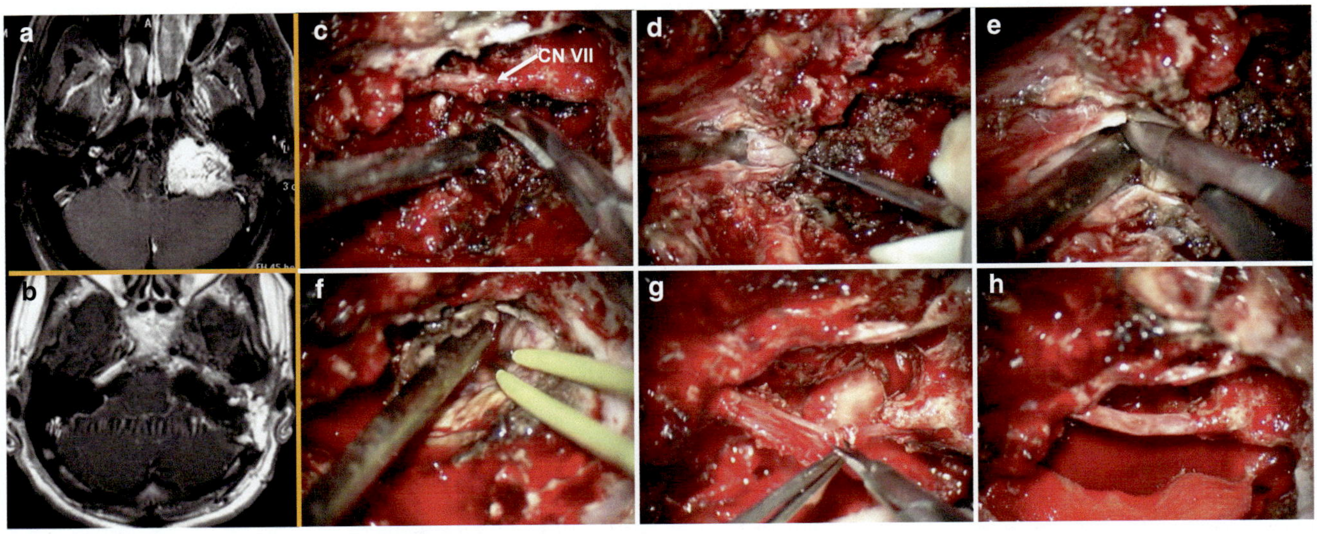

Fig. 52.14 (**a**) Preoperative (**b**) Postoperative MRI following resection of a jugular foramen paraganglioma. (**c**) Dissection and resection of extradural tumor. (**d**) Exposure of intraluminal tumor inside internal jugular vein. (**e**) Section of compromised dura. (**f**) Progressive removal of compromised dura. (**g**) Dissection and removal of infiltrating tumor of facial nerve. (**h**) Facial nerve graft with great auricular nerve. *CN* cranial nerve

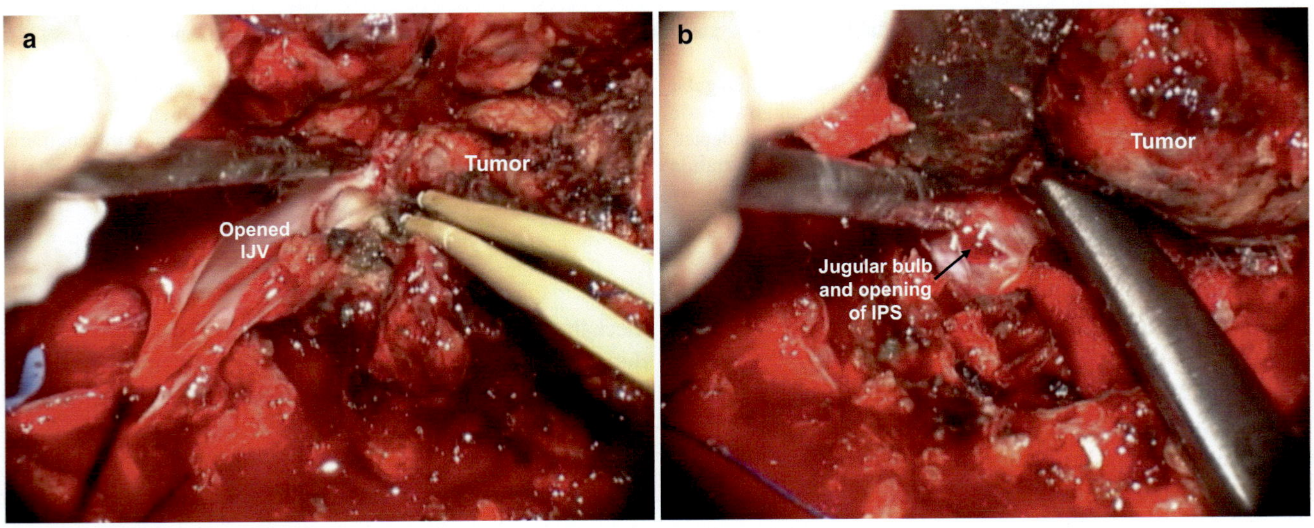

Fig. 52.15 (**a**) Demonstration of tumor projected inside internal jugular vein. (**b**) Exposure of opened jugular bulb highlighting orifice of inferior petrosal sinus. *IPS* inferior petrosal sinus

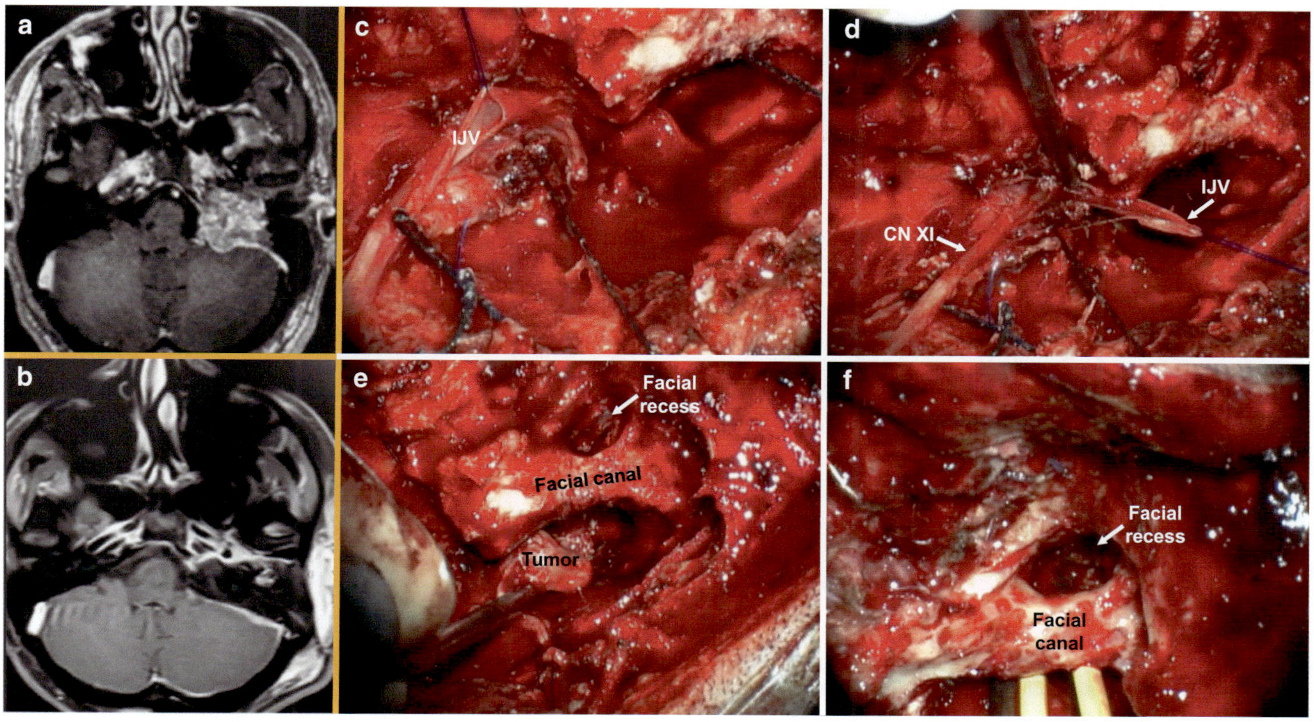

Fig. 52.16 Left-sided jugular foramen paraganglioma in a patient with proven ipsilateral lower cranial nerves deficits. (**a**) Preoperative (**b**) Postoperative MRI. (**c**, **d**) Incision of internal jugular vein and removal of intraluminal tumor. Next, removal of medial wall of internal jugular vein is performed. (**e**, **f**) Tumor removal around facial canal

52.8 Closure

Closure should be performed carefully. Eustachian tube should be packed as it is involved when opening middle ear, as well as mastoid air cells should be closed. Dural defects must be watertight-sutured, and in cases of compromised dura, pedicled flaps should be advanced to close it, as temporalis fascia attached to sternocleidomastoid muscle, temporalis muscle, or even abdominal fat or fascia lata (Figs. 52.9 and 52.11c).

52.9 Surgical Pearls

We employ variations of the infratemporal fossa (ITF) approach described by Fisch [5, 8, 10] and refined by Al-Mefty [2], considering the following features: (Fig. 52.17)

- Tumor blood supply
- Tumor extensions along internal carotid artery
- Middle ear involvement
- Facial nerve function

ITFA: retrofacial infralabyrinthine approach. In cases of tumor with blood supply mainly from external carotid artery, without anterior/superior extension around internal carotid artery (Fig. 52.18).

ITFB: infralabyrinthine pre- and retrofacial approach (fallopian bridge technique), without occlusion of external acoustic meatus and without resection of middle ear structures. Indicated in cases involving ICA and with blood supply by caroticotympanic branches of ICA, as it can be reached anterior to facial nerve. However, this approach is not intended for too much superior and anterior projections (Figs. 52.12 and 52.13).

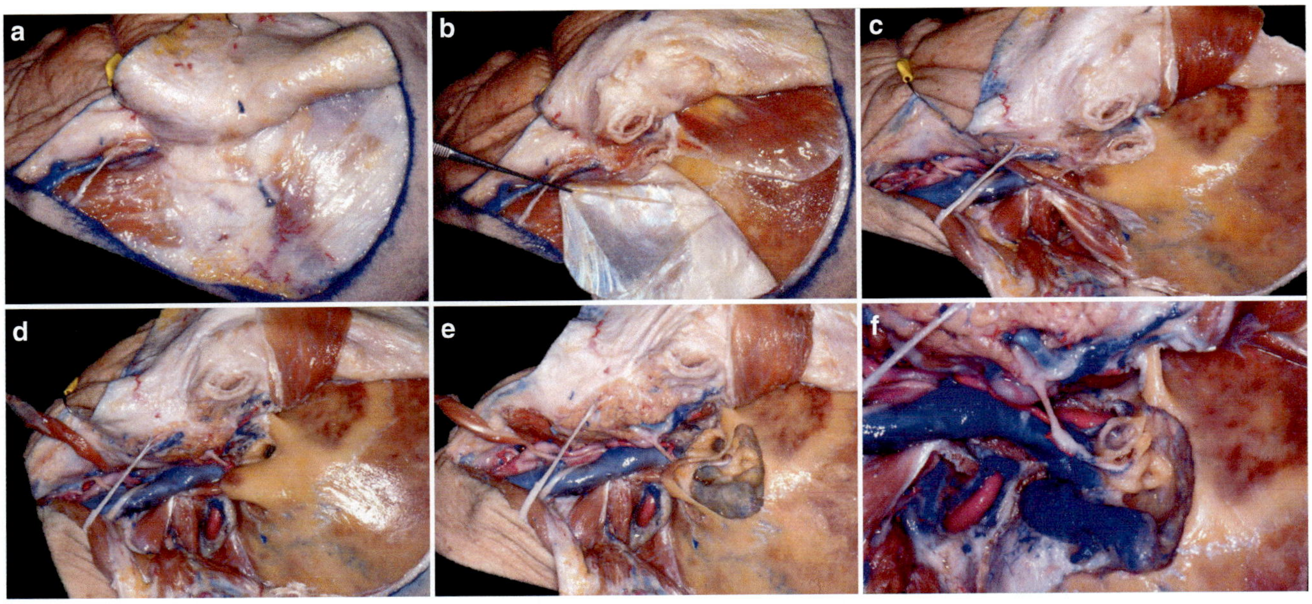

Fig. 52.17 Step-by-step anatomical dissection of infratemporal fossa approach. (**a**) Great auricular nerve is identified running over sternocleidomastoid muscle and should be preserved. (**b**) Sternocleidomastoid muscle is reflected posteriorly attached to temporalis muscle fascia to build a large pedicled flap to be used in closure. (**c**) Neck dissection should be performed to obtain vascular control of internal jugular vein and internal carotid artery. (**d**) Digastric muscle is mobilized from digastric groove, exposing facial nerve leaving stylomastoid foramen. (**e**, **f**) A mastoidectomy is performed to expose the whole venous course of sigmoid sinus, jugular bulb, and internal jugular vein

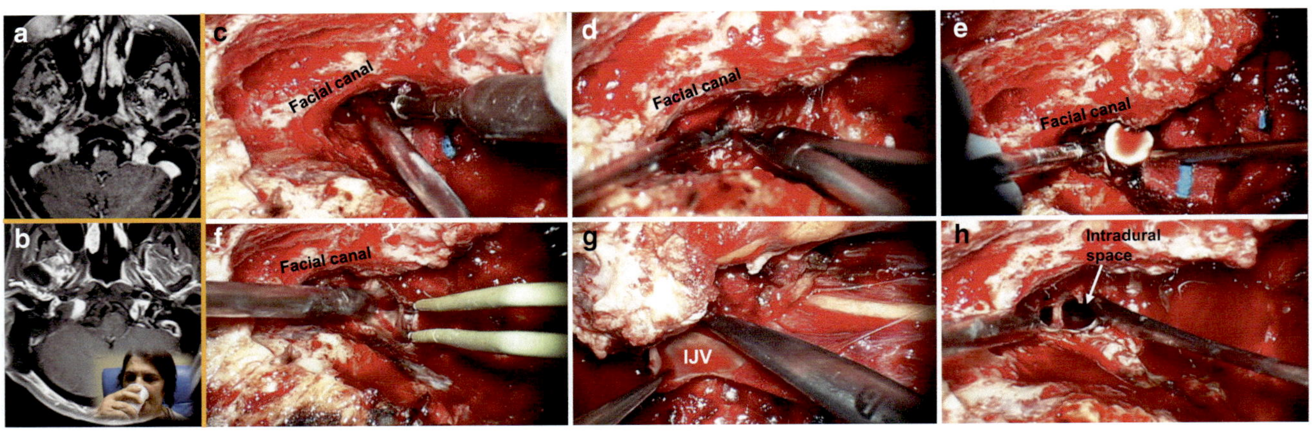

Fig. 52.18 Microsurgical resection of a jugular foramen paraganglioma employing a retrofacial approach (Infratemporal fossa approach type A). (**a**) Preoperative and (**b**) Postoperative MRI, inset image demonstrating preservation of lower cranial nerves function. (**c**–**h**) Intraoperative details demonstrating resection of extra and intradural tumor extensions, as well as intraluminal growth into internal jugular vein. *IJV* internal jugular vein

ITFC: infralabyrinthine pre- and retrofacial approach with occlusion of external acoustic meatus and resection of middle ear structures. Facial nerve may be preserved within the facial canal. Tumors involving carotid artery running towards cavernous sinus with blood supply from ICA and anterior/superior extensions can be reached (Fig. 52.19).

ITFD: includes anterior transposition of facial nerve (Fig. 52.17h) Preferred for cases where facial canal is destroyed, and the nerve may be fixed to a new position, as well as presenting hearing loss. Indicated in cases with great anterior and superior extensions involving completely the petrous segment of carotid artery and extending towards cavernous sinus.

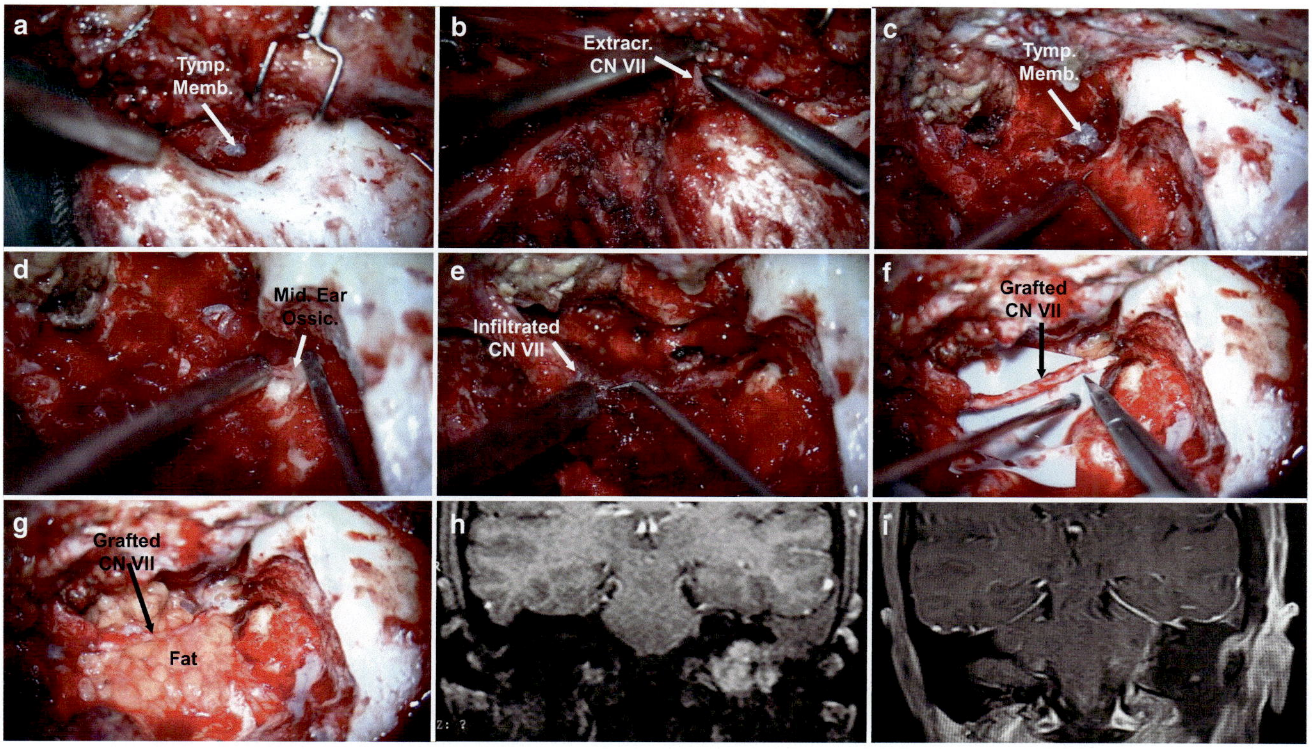

Fig. 52.19 Microsurgical resection of a jugular foramen paraganglioma. (**a**) Mastoid exposure depicting tympanic membrane. (**b**) Identification of extracranial segment of facial nerve. (**c**) Progressive drilling of mastoid and walls of ear canal. (**d**) Removal of middle ear ossicle. (**e**) Dissection and removal of infiltrated segment of facial nerve. (**f**) Nerve graft with great auricular nerve. (**g**) Fat covering empty space. (**h**) Preoperative MRI. (**i**) Postoperative MRI. *CN* cranial nerve, *Extracr.* extracranial, *Tymp. Memb.* tympanic membrane

52.10 Discussion

52.10.1 Paragangliomas

Head and neck paragangliomas (HNPGLs) arise from neural crest remnants along skull and neck accounting for 0.03% of all neoplasms and 0.6% of all head and neck tumors, with up to 20 anatomical locations described [22, 23]. Most common site of origin is from carotid body in 60% of cases, followed by jugulotympanic lesions ~30% and vagal tumors in ~10% [23, 24].

Hereditary patterns of HNPGLs account for about 40% of cases, usually associated with mutations of SDHx genes (more commonly subunit D), enzyme involved in Krebs cycle [25]. SDHx mutations are involved with 5 paraganglioma syndromes, varying tumor sites, multifocality, functionality, associated tumor as renal cell carcinomas, and malignancy [26]. Moreover, there are other familial syndromes involved with paragangliomas as Neurofibromatosis type 1, Multiple endocrine neoplasia type 2, Von-Hippel Lindau syndromes, and Carney-Stratakis dyad [25]. Around 40% of HNPGLs are associated with germline mutations, increasing for 70% of cases when multifocal, as well as higher rate of germline mutations is found in familial cases [22]. However, it is common to patients who harbor a germline mutation without a familial history, which could be explained by an incomplete penetrance, and regarding SDHD mutation, a paternal transmission usually leads to development of paragangliomas, while maternal transmission is rarely associated with tumor development [27].

Jugular foramen paragangliomas are highly vascularized lesions, usually with blood supply from external carotid artery through ascending pharyngeal or occipital arteries, as well as possibly with blood supply from caroticotympanic branches of internal carotid artery, or contribution of vertebrobasilar system [27]. An important point is that usually paraganglioma grows into lumen of internal jugular vein/jugular bulb or sigmoid sinus causing occlusion of venous outflow. Preoperative angiogram is essential to carefully study arterial blood supply and venous drainage (Fig. 52.20).

Histologically, paragangliomas present a *zellballen* pattern, composed of nests of chiefs cells containing neurosecretory granules, surrounded by fibrovascular sustentacular cells [28]. Immunohistochemistry shows positive staining for chromogranin and synaptophysin neuroendocrine markers, as well as S100 positivity in sustentacular cells [28, 29].

Preoperative workup should include careful clinical history and examination, as well as laboratory investigation to identify potential secreting tumors, despite uncommon. Adrenergic lesions may manifest with increased blood pres-

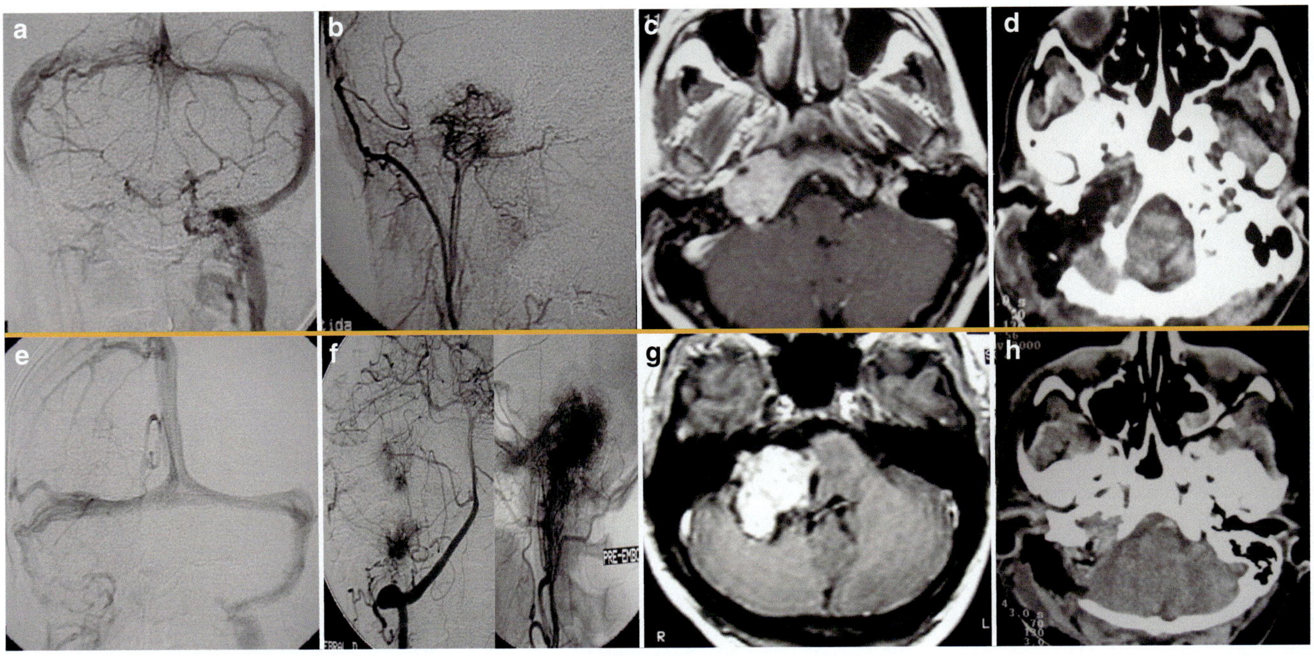

Fig. 52.20 Angiographic features of two cases of jugular foramen paragangliomas (**a–d** and **e–h**). (**a**, **e**) Venous phase of angiogram depicting occlusion of ipsilateral sigmoid sinus. (**b**) Tumor blood supply coming from external carotid artery. (**f**) Angiogram demonstrating blood supply from vertebrobasilar system beside external carotid artery. (**c**, **g**) Preoperative MRI. (**d**, **h**) Postoperative CT scan

sure, palpitation, tachycardia, or sweating, and especially in these situations screening for synchronous tumors is important, as usually head and neck paragangliomas are more associated with parasympathetic system, while thoracic/abdominal/pelvic tumors with sympathetic system. Diagnosis is performed with dosage of plasmatic and 24-h urine catecholamines and metanephrines, as well as plasmatic 3-methoxytyramine in cases of dopamine lesions [25]. Preoperative alpha-blockage is advisable in norepinephrine or epinephrine-secreting lesions to avoid intraoperative hypertensive crisis; however, postoperative hypotension may occur due to sudden withdrawal of catecholamine source. Beta-blockage may be employed to manage reflex tachycardia due to complete alpha-block, and beta-blocker should never be administered alone as an unopposed alpha effect may lead to hypertensive crisis. Exclusively, dopamine-secreting tumors are rare and usually are not associated with adrenergic symptoms, delaying diagnosis, and in these specific cases preoperative alpha and beta-blockage is usually not indicated.

CT scan better demonstrates bone involvement, especially erosion around jugular foramen in a "moth-eaten" pattern. MRI depicts a hypointense in T1WI and hyperintense in T2WI, usually with flow voids and presenting with intense heterogeneous contrast enhancement, with a salt-and-pepper appearance [28]. Jugulotympanic paragangliomas may be classified either by Fisch [5] or Jackson-Glasscock [30] system. Fisch classification is more widely used and helps to tailor surgical approaches, which are usually variants of infratemporal fossa approach described by him. Jugular foramen paragangliomas may extend between intra and extracranial spaces, as well as between intra and extradural compartments; extensions towards middle ear may be due to the path along Jacobson's nerve [27].

Functional imaging is important to complement whole-body disease staging; SDHD-mutated paragangliomas exhibit strong somatostatin receptor (SSTR) subtype 2 expression and may be evaluated by ^{68}Ga-radiolabelled somatostatin analogues. Also, 6-[^{18}F] fluoro-3,4-dihydroxyphenylalanine ([^{18}F]-FDOPA)-PET–CT is a good alternative for detecting and staging HNPGLs, until recently the gold standard, but with less sensitivity than SSTR PET-CT [25, 31].

52.11 Conclusion

Jugular foramen lesions require extensive anatomical knowledge for surgical management. Different types of tumors may be found at this location, most commonly paragangliomas. Careful preoperative evaluation is essential to define the best treatment. Laboratory training is crucial to achieve the best surgical results performing complex approaches towards jugular foramen.

References

1. Guinto G, Kageyama M, Trujillo-Luarca VH, Abdo M, Ruiz-Than A, Romero-Rangel A. Nonglomic tumors of the jugular foramen: differential diagnosis and prognostic implications. World Neurosurg. 2014;82(6):1283–90.
2. Al-Mefty O, Fox JL, Rifai A, Smith RR. A combined infratemporal and posterior fossa approach for the removal of giant glomus tumors and chondrosarcomas. Surg Neurol. 1987;28(6):423–31.
3. Michael LM, Robertson JH. Glomus jugulare tumors: historical overview of the management of this disease. Neurosurg Focus. 2004;17(2):1–5.
4. Constanzo F, Nogueira GF, Neto MC, Ramina R. Historical aspects and surgical nuances on the craniocervical approach to the jugular foramen. Arq Bras Neurocir. 2022;41(2):e174–9.
5. Fisch U. Infratemporal fossa approach to tumours of the temporal bone and base of the skull. J Laryngol Otol. 1978;92(11):949–67.
6. Pensak ML, Jackler RK. Removal of jugular foramen tumors: the fallopian bridge technique. Otolaryngol Head Neck Surg. 1997;117:586–91.
7. Borba LA, Ale-Bark S, London C. Surgical treatment of glomus jugulare tumors without rerouting of the facial nerve: an infralabyrinthine approach. Neurosurg Focus. 2004;17(2):51–5.
8. Borba LA, Araújo JC, de Oliveira JG, Giudicissi Filho M, Moro MS, Tirapelli LF, Colli BO. Surgical management of glomus jugulare tumors: a proposal for approach selection based on tumor relationships with the facial nerve. J Neurosurg. 2010;112(1):88–98.
9. Lope Ahmad RAR, Sivalingam S, Konishi M, De Donato G, Sanna M. Oncologic outcome in surgical management of jugular paraganglioma and factors influencing outcomes. Head Neck. 2013;35(4):527–34.
10. Zottis-Grapiglia C, Díaz-Castillejos A, Rehder R, Luis AB. Glomus yugular: Revisión de la literatura y técnica quirúrgica. Rev Mex Neurosci. 2015;16(1):73–89.
11. Basma J, Michael LM II, Sorenson JM, Robertson JH. Deconstruction of the surgical approach to the jugular foramen region: anatomical study. J Neurol Surg B Skull Base. 2019;80(05):518–26.
12. Inserra MM, Pfister M, Jackler RK. Anatomy involved in the jugular foramen approach for jugulotympanic paraganglioma resection. Neurosurg Focus. 2004;17(2):41–4.
13. Constanzo F, Coelho Neto M, Nogueira GF, Ramina R. Microsurgical anatomy of the jugular foramen applied to surgery of glomus jugulare via craniocervical approach. Front Surg. 2020;7:27.
14. Sen C, Hague K, Kacchara R, Jenkins A, Das S, Catalano P. Jugular foramen: microscopic anatomic features and implications for neural preservation with reference to glomus tumors involving the temporal bone. Neurosurgery. 2001;48(4):838–48.
15. Das SS, Saluja S, Vasudeva N. Complete morphometric analysis of jugular foramen and its clinical implications. J Craniovertebr Junction Spine. 2016;7(4):257.
16. Wang Z, Zhang Z, Huang Q, Yang J, Wu H. Surgical management of extensive jugular paragangliomas: 10-year-experience with a large cohort of patients in China. Int J Surg. 2013;11(9):853–7.
17. Dharnipragada R, Butterfield JT, Dhawan S, Adams ME, Venteicher AS. Modern management of complex tympanojugular paragangliomas: a systematic review and meta-analysis. World Neurosurg. 2022;170:149.
18. Sanna M, Shin SH, De Donato G, Sivalingam S, Lauda L, Vitullo F, Piazza P. Management of complex tympanojugular paragangliomas including endovascular intervention. Laryngoscope. 2011;121(7):1372–82.
19. Szyfter W, Kopeć T, Kawczyński M. Glomus caroticum, jugulare and vagale – problems in diagnosis and treatment. Otolaryngologia Polska = Polish Otolaryngol. 2006;60(3):305–12.
20. Fayad JN, Keles B, Brackmann DE. Jugular foramen tumors: clinical characteristics and treatment outcomes. Otol Neurotol. 2010;31(2):299–305.
21. David CA. Preoperative planning and surgical approaches to tumors of the jugular foramen. Oper Tech Neurosurg. 2005;8(1):19–24.
22. Cleere EF, Martin-Grace J, Gendre A, Sherlock M, O'Neill JP. Contemporary management of paragangliomas of the head and neck. Laryngoscope Investig Otolaryngol. 2022;7(1):93–107.
23. Williams MD. Paragangliomas of the head and neck: an overview from diagnosis to genetics. Head Neck Pathol. 2017;11:278–87.
24. Rana MU, Østhus AA, Heimdal K, Jebsen P, Revheim MER, Osnes TA. Head and neck paragangliomas in Norway, importance of genetics, updated diagnostic workup and treatment. Acta Otolaryngol. 2021;141(3):303–8.
25. Taïeb D, Wanna GB, Ahmad M, Lussey-Lepoutre C, Perrier ND, Nölting S, … Pacak K. Clinical consensus guideline on the management of phaeochromocytoma and paraganglioma in patients harbouring germline SDHD pathogenic variants. Lancet Diabetes Endocrinol. 2023;11:345.
26. Offergeld C, Brase C, Yaremchuk S, Mader I, Rischke HC, Gläsker S, … Neumann HP. Head and neck paragangliomas: clinical and molecular genetic classification. Clinics. 2012;67:19–28.
27. Lee KY, Oh YW, Noh HJ, Lee YJ, Yong HS, Kang EY, … Lee NJ. Extraadrenal paragangliomas of the body: imaging features. AJR Am J Roentgenol. 2006;187(2):492–504.
28. Breton JM, Arkun K, Tischler AS, Qamar AS, Sillman JS, Heilman CB. Clinical and histopathological principles for the diagnosis of a recurrent paraganglioma of the jugular foramen initially diagnosed as a middle ear adenoma: illustrative case. J Neurosurg Case Lessons. 2021;2(5):CASE21307.
29. Lin EP, Chin BB, Fishbein L, Moritani T, Montoya SP, Ellika S, Newlands S. Head and neck paragangliomas: an update on the molecular classification, state-of-the-art imaging, and management recommendations. Radiol Imaging Cancer. 2022;4(3):e210088.
30. Jackson CG, Glasscock ME, Harris PF. Glomus tumors: diagnosis, classification, and management of large lesions. Arch Otolaryngol. 1982;108(7):401–6.
31. Janssen I, Chen CC, Taieb D, Patronas NJ, Millo CM, Adams KT, … Pacak K. 68Ga-DOTATATE PET/CT in the localization of head and neck paragangliomas compared with other functional imaging modalities and CT/MRI. J Nucl Med. 2016;57(2):186–191.

Part IV

Ventricular Tumors

Intraventricular Endoscopic Anatomy

53

Henry W. S. Schroeder and Ehab El Refaee

Learning Objectives

- Characterize the surgical anatomy of the lateral and third ventricles.
- Obtain the surgical information needed for endoscopic third ventriculostomy procedures.
- Understand the importance of the choroid plexus for orientation in distorted anatomy.

53.1 Introduction

Endoscopic third ventriculostomy (ETV) is a well-established procedure that targets the diversion of cerebrospinal fluid (CSF) from the third ventricle directly to the subarachnoid space bypassing any obstruction of the CSF pathway distal to the middle of the third ventricle. The reputation of this procedure evolved due to the fact that valve-regulated shunts are still associated with high failure rates and related complications [1]. Performing an endoscopic third ventriculostomy needs proficient orientation with the neuroanatomy of the lateral and third ventricles together with adroit handling of neuroendoscopic instruments [2–4].

53.2 History and Background

The idea of combining a lens endoscope with an electric light source by Carl Friedrich Nitze in the nineteenth century initiated the successful application of endoscopy in medicine [5]. Initially, endoscopy was predominantly used in urology. However, the endoscope was later adopted for neurosurgical procedures as well. The urologist Victor Darwin Lespinasse (1878–1946) from Chicago is credited with the first endoscopic surgery within the brain. In 1910, he coagulated the choroid plexus with the aid of a cystoscope in two infants suffering from hydrocephalus. One child died, the other survived 5 years. Lespinasse did not publish it but reported it at a local medical society according to a footnote in the book "The Principles of Neurological Surgery" by Loyal Edward Davis published in 1936 [6]. Encouraged by Dandy (1886–1946), who reported an open ventriculostomy via a subfrontal route with intentional cutting of one optic nerve in 1922 [7], Mixter (1880–1958) performed the first endoscopic third ventriculostomy on the 6th of February 1923, in a 9-months-old infant. Using a urethroscope, he perforated the floor of the third ventricle with a flexible sound [8]. However, the endoscopic techniques did not get widely accepted at that time because of poor illumination and insufficient instruments. Furthermore, with the development of valve-regulated shunts, a technically easier procedure with less complications was available since the beginning of the 1950s [9]. Since the beginning of the 90s, sophisticated ventriculoscopes specifically designed for intraventricular procedures have been developed, and endoscopic techniques were established as the procedure of choice for several types of CSF pathway obstructions [10, 11].

53.3 Anatomical Highlights

53.3.1 Anatomy of the Lateral Ventricles

The lateral ventricles are bilateral C-shaped structures that follow the contour of the choroidal fissure. They are segmented into five parts: frontal horn, body, atrium, occipital horn, and temporal horn. The frontal horn has no choroid plexus. The floor is formed by the corpus callosum. The head

H. W. S. Schroeder (✉)
Department of Neurosurgery, University Medicine Greifswald, Greifswald, Germany
e-mail: henry.schroeder@uni-greifswald.de; http://www.medizin.uni-greifswald.de/neuro_ch

E. El Refaee
Department of Neurosurgery, University Medicine Greifswald, Greifswald, Germany

Department of Neurosurgery, Cairo University, Cairo, Egypt
e-mail: Ehab.ElRefaee@med.uni-greifswald.de

I. N. Kanaan, V. Beneš (eds.), *Neuroanatomy Guidance to Successful Neurosurgical Interventions*,
https://doi.org/10.1007/978-3-031-59838-8_53

of the caudate nucleus is the lateral border (Fig. 53.1a) and the septum pellucidum is the medial one (Fig. 53.1b). In chronic hydrocephalus, the septum is usually atrophic and perforated (Fig. 53.2a). Sometimes, only the vessels remain. The corpus callosum is forming most of the walls of the lateral ventricle. The rostrum is forming the floor of the frontal horn, the genu forming the anterior boundary, the genu and body form the roof, and the splenium covers the walls of the posterior compartment [12]. The choroid plexus also follows the choroidal fissure in a C-shaped manner from the third ventricle back to the atrium and then turns forward to the temporal horn (Fig. 53.2b). The foramen of Monro is a crescent-shaped structure situated at the floor of the lateral ventricle on both sides, and it connects the lateral ventricles to the third ventricle. The column of the fornix is the main border of the foramen (Fig. 53.3). At the foramen of Monro, thalamostriate, septal, and choroidal veins merge together to form the internal cerebral veins, which run at the roof of the third ventricle in the tela choroidea backwards to the large vein of Galen [12]. At the edge of the foramen of Monro, the choroid plexus is situated posteriorly. The size and shape of the foramen of Monro are very variable (Fig. 53.4). In case of hydrocephalus, the foramen becomes circular and larger. However, there are numerous variations in this area in cases of hydrocephalus. Sometimes, the anatomy is completely distorted, and the orientation becomes very difficult (Fig. 53.5). Then the choroid plexus and the main draining veins remain the essential landmarks [13, 14]. The choroid plexus guides the surgeon to the foramen of Monro and third ventricle. For the endoscopic access to the third ventricle via the foramen of Monro, an accurate approach planning is required to avoid injuring the fornix anteriorly, the septal vein medially, or the thalamostriate vein posterolaterally. The appropriate location of the burr hole has to be determined according to the individual anatomic situation of each patient after carefully studying the MR imaging. Neuronavigation is helpful in cases of distorted anatomy [15].

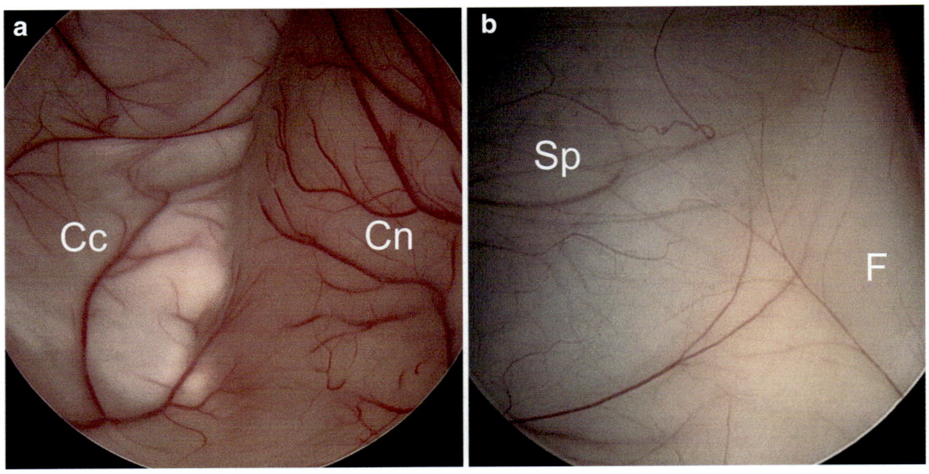

Fig. 53.1 (**a**) Anterolateral view with a 45° endoscope within the right lateral ventricle showing the head of the caudate nucleus (Cn) and the corpus callosum (Cc). (**b**) Medial view with 45° endoscope shows the septum pellucidum (Sp) and the fornix (F)

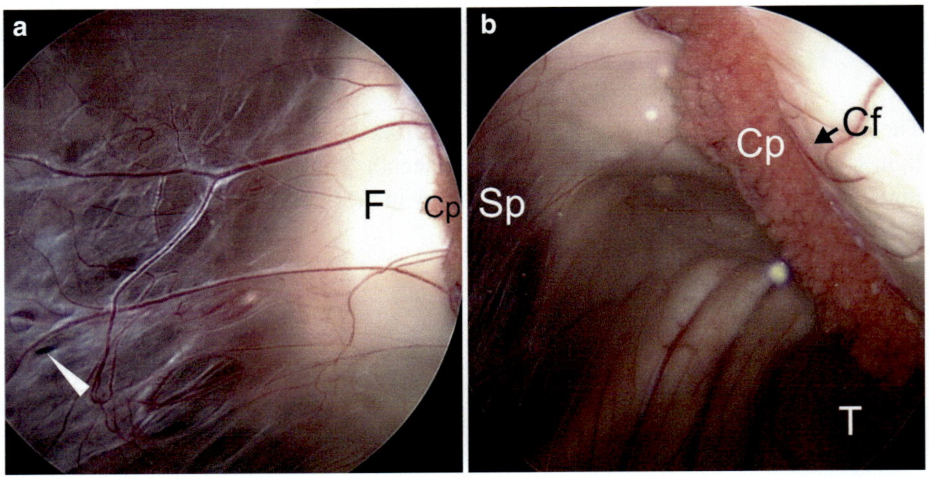

Fig. 53.2 (**a**) Medial view with a 45° endoscope in the right lateral ventricle showing the fenestrated septum pellucidum (arrowhead) with fornix (F) and choroid plexus (Cp), as a sign of chronic hydrocephalus. (**b**) Posteromedial view with 45° endoscope to the cella media and trigone (T) of the lateral ventricle showing the choroidal fissure (arrow, Cf) covered by choroid plexus (Cp) and bounded medially with the fenestrated septum pellucidum (Sp)

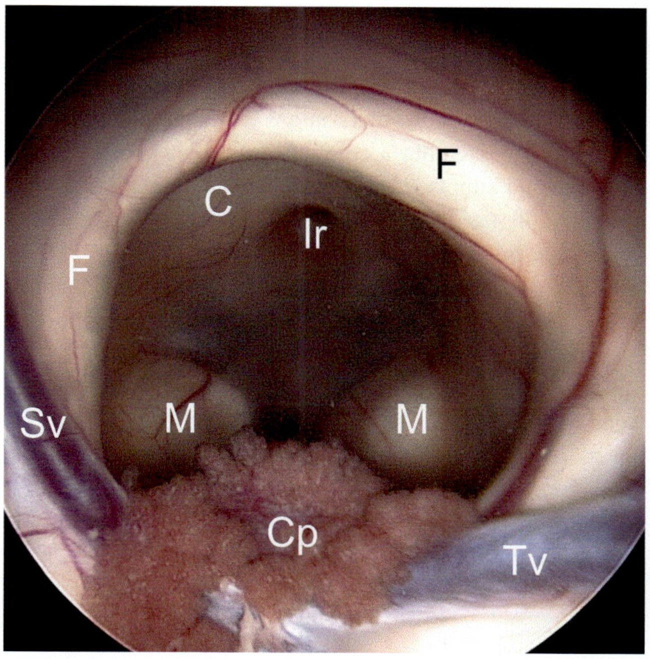

Fig. 53.3 Endoscopic view with a 0° endoscope within the right lateral ventricle showing the foramen of Monro bounded by the septal vein (Sv), thalamostriate vein (Tv), choroid plexus (Cp), and fornix (F). The floor of the third ventricle with mammillary bodies (M), infundibular recess (Ir), and chiasm (C) is visible

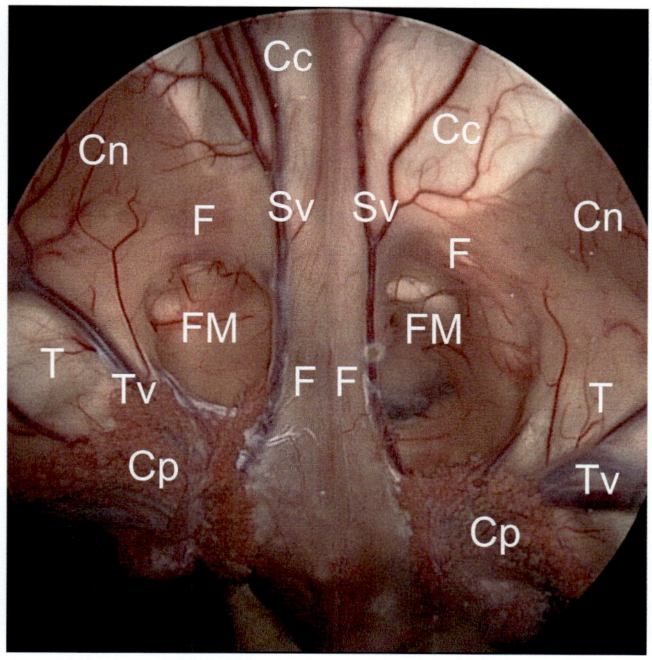

Fig. 53.5 View with a 0° endoscope into a cyclops ventricle without septum pellucidum showing corpus callosum (Cc), head of caudate nucleus (Cn), septal veins (Sv), thalamostriate veins (Tv), fornix (F), foramina of Monro (FM), choroid plexus (Cp), and thalamus (T)

Fig. 53.4 (**a–d**) Endoscopic view with a 0° endoscope within the right lateral ventricle showing four different foramina of Monro. The appearance is very variable (septal vein (Sv), thalamostriate vein (Tv), choroid plexus (Cp), and fornix (F))

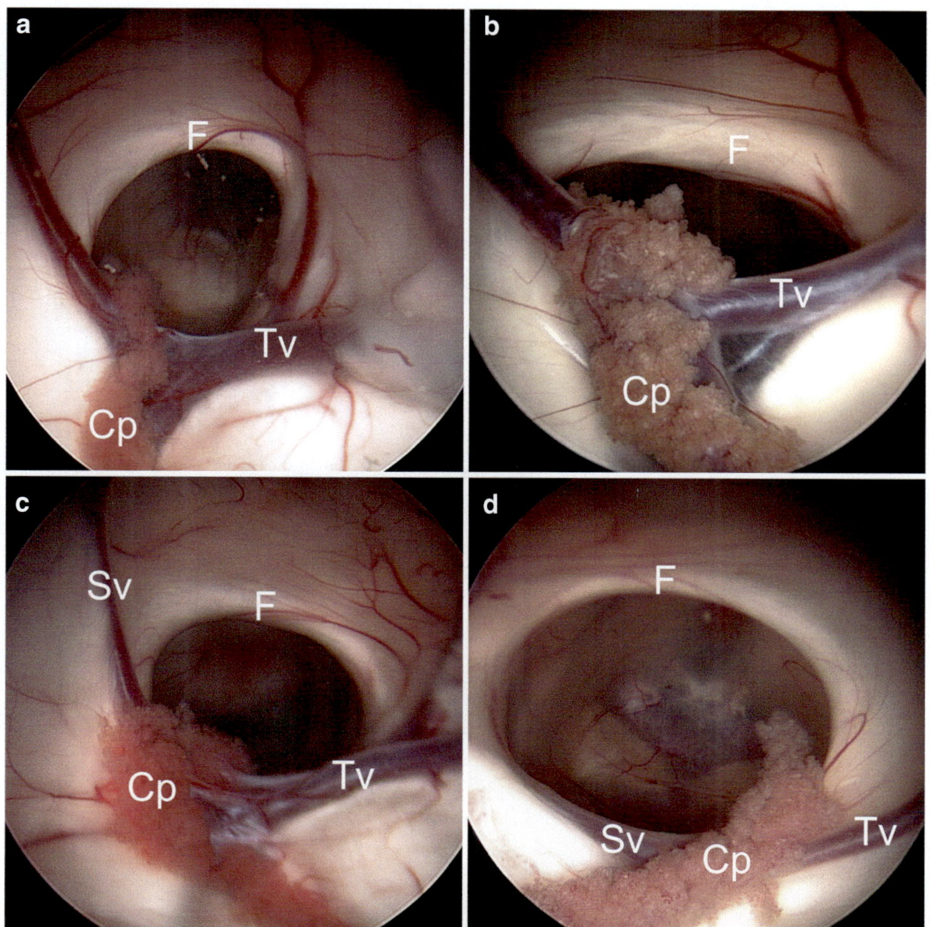

53.3.2 Anatomy of the Third Ventricle

The third ventricle is a funnel-shaped space in the midline connected to the two lateral ventricles via the foramina of Monro and enlarges in case of hydrocephalus [14]. The lateral walls are made by hypothalamus and thalamus. The floor of the third ventricle is formed by the tuber cinereum, infundibular recess, and mammillary bodies (Fig. 53.6). In cases of hydrocephalus, the tuber cinereum might become paper-thin and translucent, providing ideal conditions for a third ventriculostomy.

The most anterior part of the third ventricle can be visualized by 30° and 45° endoscopes, where the anterior commissure, lamina terminalis, which covers the anterior cerebral arteries complex, and the optic chiasm are identified (Fig. 53.7a). The anterior commissure is composed of a tight bundle of fibers that crosses the midline in front of the columns of the fornix. The lamina terminalis covers the interval between the anterior commissure and the optic chiasm [16]. Looking with a 45° endoscope medially shows the contralateral foramen of Monro with fornix and anterior commissure (Fig. 53.7b). The subfrontal route was used to open the lamina terminalis creating a communication between the third ventricle and suprasellar cistern [17]. This ventriculostomy route is still used for CSF release in various transcranial suprasellar procedures [18, 19].

The posterior part of the third ventricle can be visualized from a coronal burr hole with a 45° endoscope. The first structure recognized is the massa intermedia which connects both lateral walls of the third ventricle (Fig. 53.8a). It is not a commissure and can be cut. The posterior wall includes the aqueduct of Sylvius, posterior commissure, pineal recess, habenular commissure, and suprapineal recess (Fig. 53.8b).

The roof is formed by the tela choroidea containing the internal cerebral veins and the choroid plexus.

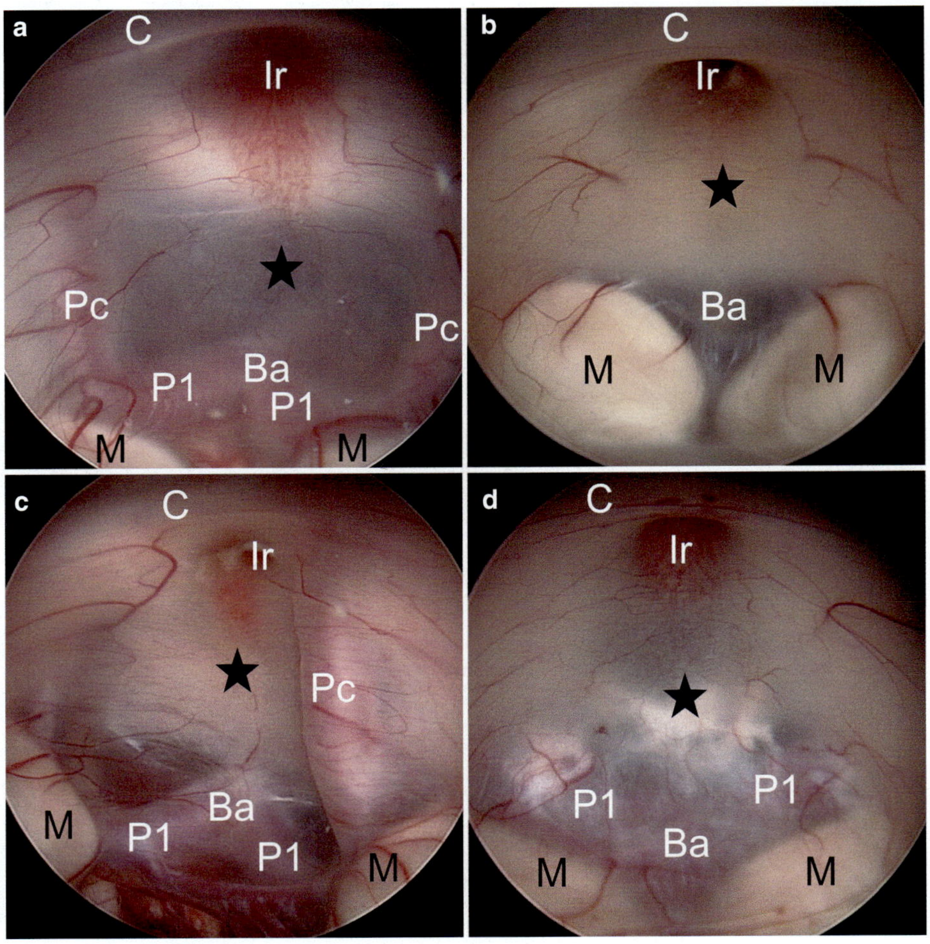

Fig. 53.6 (**a–d**) View with a 0° endoscope to the floor of the third ventricle in four different patients showing chiasm (C), infundibular recess (Ir), mammillary bodies (M), posterior communicating artery (Pc), P1 segment of posterior cerebral artery (P1), and basilar artery (Ba). The black star indicates the ideal fenestration site. The appearance is very variable

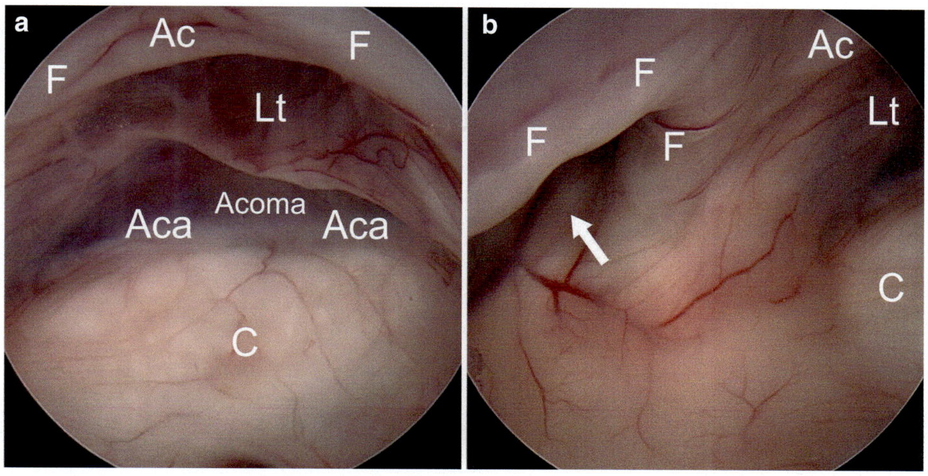

Fig. 53.7 (**a**) Anterior view with a 45° endoscope showing the anterior third ventricle with lamina terminalis (Lt) covering the anterior cerebral arteries (Aca) and anterior communicating artery (Acoma), chiasm (C), and the anterior commissure (Ac) between the fornices (F). (**b**) View with 45° endoscope to the contralateral left foramen of Monro (arrow) showing the contralateral fornix (F), anterior commissure (Ac), lamina terminalis (Lt), and chiasm (C)

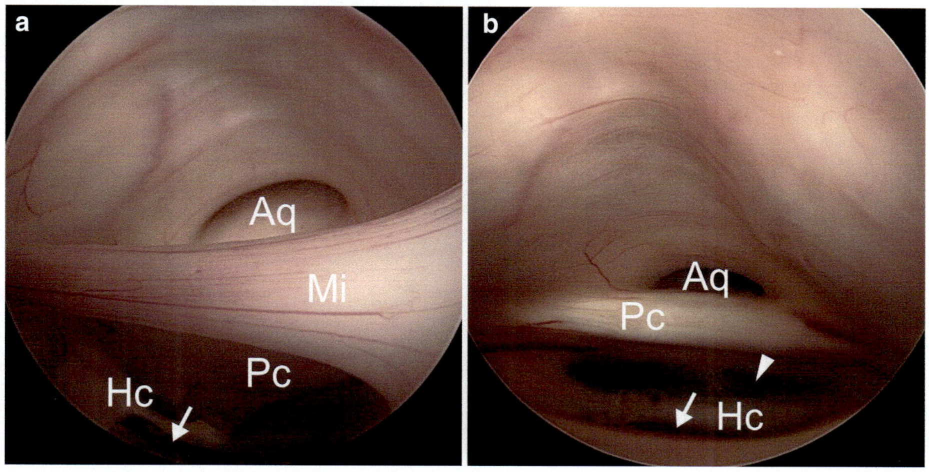

Fig. 53.8 (**a**) Posterior view with a 45° endoscope showing the posterior third ventricle with massa intermedia (Mi), entrance to the aqueduct (Aq), suprapineal recess (arrow), and posterior (Pc) and habenular commissure (Hc). (**b**) Posterior view with a 45° endoscope after passing the massa intermedia showing the posterior third ventricle with entrance to the aqueduct (Aq), posterior (Pc) and habenular commissure (Hc), pineal (arrowhead), and suprapineal recess (arrow)

53.3.3 Surgical Technique of Endoscopic Third Ventriculostomy

This procedure is performed under general anesthesia and in the supine position. The head is fixed in a Mayfield clamp in a 30° anteflexed position. The ideal trajectory towards the floor of the third ventricle via the foramen of Monro is planned according to the preoperative MRI. Neuronavigation is not routinely used, except in cases with distorted anatomy or small ventricles. The entry point is usually located at the coronal suture, 2.5 cm lateral to the midline on the right side. The left side is only selected when there is a significant asymmetry with a larger foramen of Monro or ventricle size. After local shaving, disinfection, and draping, a 3 cm skin incision and a 1 cm burr hole are made. After cruciate opening of the dura, the cortex is coagulated. The ventricle is tapped with a Cushing cannula and CSF is obtained for lab investigations. Then the endoscopic sheath containing the trocar is inserted perpendicular to the burr hole into the frontal horn of the right lateral ventricle. Thereafter, the trocar is removed and replaced with the ventriculoscope. After endoscopic orientation with the regional anatomy, the operating sheath is advanced through the foramen of Monro under direct visualization. After identification of the floor with infundibular recess and mammillary bodies, the ideal fenestration site is recognized in front of the basilar artery just behind the dorsum sellae (Fig. 53.9a). A blunt perforation of the floor of the third ventricle using a Decq forceps is performed (Fig. 53.9b). The fenestration is enlarged by spreading the jaws of the forceps (Fig. 53.9c) and then by inflating the balloon of a Fogarty balloon catheter (Fig. 53.9d). The size of the ventriculostomy depends on the width of the floor, but should be approximately 4 mm in general. It is of utmost importance to open not only the floor of the third ventricle but also the Liliequist's membrane to restore the pulsatile CSF flow in the prepontine cistern (Fig. 53.9e). To confirm that the subarachnoid space has been reached, endoscopic inspection with a 30° endoscope should show the basilar artery (Fig. 53.9f). Looking with a 45° endoscope to the left shows the oculomotor nerve running in between the posterior cerebral artery and superior cerebellar artery (Fig. 53.9g). After making sure that no active bleeding exists, the endo-

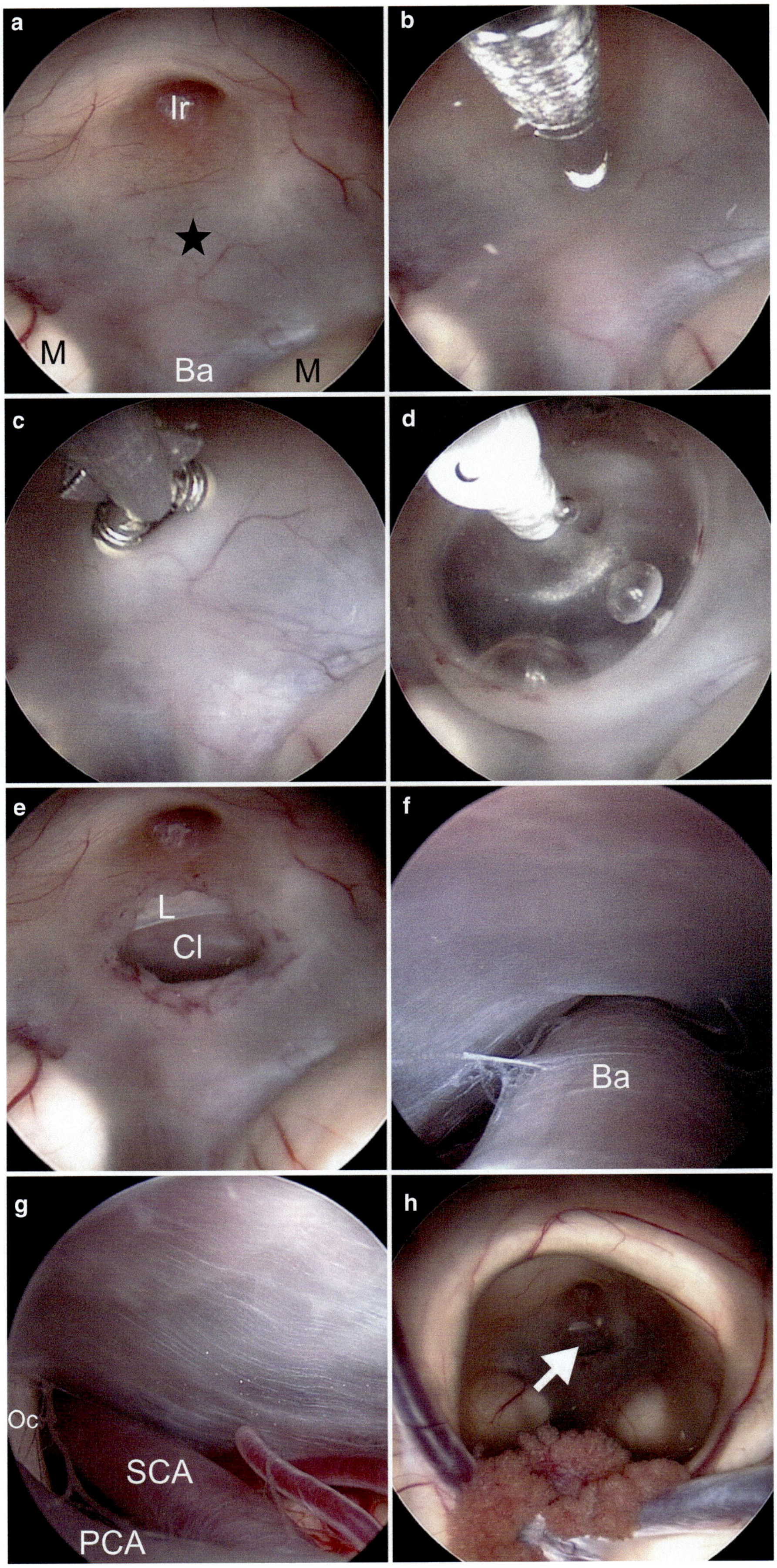

Fig. 53.9 Steps of endoscopic third ventriculostomy (**a**) Identification of the ideal ventriculostomy site (black star) halfway between infundibular recess (Ir) and mammillary bodies (M) in the midline in front of the basilar artery (Ba). (**b**) Blunt puncture of the floor with closed Decq forceps. (**c**) Widening of the perforation by spreading the jaws of the Decq forceps. (**d**) Enlarging the fenestration by inflation of the balloon of a 3 French Fogarty catheter. (**e**) Accurate ventriculostomy behind the clivus (Cl). The Liliequist membrane (L) is fenestrated as well. (**f**) Visual confirmation of reaching the prepontine cistern by seeing the basilar artery (Ba) with a 30° endoscope looking posteriorly. (**g**) View with a 45° endoscope to the left showing the oculomotor nerve (Oc) running between superior cerebellar artery (SCA) and posterior cerebral artery (PCA) (**h**) Final endoscopic view showing the intact foramen of Monro after completing the third ventriculostomy (arrow)

scope is removed together with the endoscopic sheath from the third and lateral ventricles. The final view of the foramen of Monro shows no damage of fornix, veins, or choroid plexus, indicating that a correct entry point was chosen (Fig. 53.9h). The burr hole is closed with a piece of gel foam, and the skin is closed in layers.

In case of hydrocephalus caused by a tumor in the posterior third ventricle, an ETV and a tumor biopsy are frequently required. Whether these procedures can be done via one burr hole depends on the individual anatomy of the patient. If the foramen of Monro is large, both procedures can be done via one burr hole. If the foramen is small, two burr holes are required if no flexible scope is available.

53.4 Surgical Pearls

- A precise analysis of the preoperative MRI for choosing the ideal approach is mandatory. The size of the ventricles and foramen of Monro, as well as vascular anomalies, should be taken into account [20].
- In case of small foramina of Monro, a smaller ventriculoscope should be used to prevent forniceal injury.
- Blunt perforation of the floor is preferred.
- The optimal perforation site is usually in the midline, halfway between the infundibular recess and mammillary bodies. However, if a vascular anomaly exists, the perforation site must be adapted accordingly.
- Sharp dissection for a thick Liliequist's membrane is preferred than blunt dissection to prevent arterial injury.
- Although holding the endoscope by an experienced assistant is an option, we prefer using an endoscope-holding arm to fix the endoscopic sheath, which enables bimanual handling of the instruments and balloon catheter independently of the skills of the assistant.

53.5 Discussion

Distinct developments in neuroendoscopy have been encountered during the last 30 years allowing for better visualization and more precise surgical steps. This improved perception served for a better understanding of the ventricular anatomy, especially of the lateral and third ventricles. The morbid anatomy of the ventricles varies according to the cause of hydrocephalus and the associated pathologies. Identification of the neurovascular structures is crucial for performing a safe and successful endoscopic surgery. Therefore, it is strongly encouraged to train neuroendoscopy in the lab before doing surgical procedures in the OR.

Several endoscopic techniques have been developed to treat obstructed hydrocephalus [11]. The workhorse for most types of obstructed hydrocephalus is still ETV [21–23]. It is indicated for all obstructions distal to the middle of the third ventricle. ETV can also be combined with other endoscopic procedures such as septostomy, tumor biopsy, or choroid plexus coagulation according to the pathological features of the disease that caused hydrocephalus [24–26]. Despite being a straightforward procedure, cautious respect for the anatomical structures and meticulous handling of instruments are important to prevent major intraoperative complications [27, 28].

53.6 Conclusion

Profound knowledge of the intraventricular anatomy is essential for a successful endoscopic surgery. Knowing the normal anatomy is a prerequisite for recognizing the distorted anatomy in hydrocephalic conditions. ETV remains the procedure of choice for most types of obstructed hydrocephalus.

Acknowledgments We thank Marc Matthes, MSc, for his help in preparing the illustrations.

References

1. Hasanain AA, Abdullah A, Alsawy MFM, Soliman MAR, Ghaleb AA, Elwy R, et al. Incidence of and causes for ventriculoperitoneal shunt failure in children younger than 2 years: a systematic review. J Neurol Surg A Cent Eur Neurosurg. 2019;80(1):26–33.
2. Schroeder HWS, Niendorf WR, Gaab MR. Complications of endoscopic third ventriculostomy. J Neurosurg. 2002;96(6):1032–40.
3. El Refaee E, Zohdi A, Schroeder HWS. In reply: a new trend of blended learning in neurosurgical training: fellowship of neuroendoscopy. Neurosurgery. 2022;90(2):e57.
4. Dezena RA. Atlas of endoscopic neurosurgery of the third ventricle: basic principles for ventricular approaches and essential intraoperative anatomy. Cham: Springer; 2017. 271 p.
5. Reuter HJ, Reuter MA. Philipp Bozzini und die Endoskopie des 19. JH. Stuttgart: Max-Nitze-Museum; 1988.
6. Davis LE. Principles of neurological surgery. Philadelphia: Lea & Febiger; 1936. p. 405.
7. Dandy WE. An operative procedure for hydrocephalus. Bull Johns Hopkins Hosp. 1922;33:189–90.
8. Mixter WJ. Ventriculoscopy and puncture of the floor of the third ventricle. Boston Med Surg J. 1923;188:277–8.
9. Nulsen FE, Spitz EB. Treatment of hydrocephalus by direct shunt from ventricle to jugular vein. Surg Forum. 1952;2:399–403.
10. Schroeder HWS. A new multipurpose ventriculoscope. Neurosurgery. 2008;62(2):489–91; discussion 491–2.
11. Schroeder HWS, Oertel J, Gaab MR. Endoscopic treatment of cerebrospinal fluid pathway obstructions. Neurosurgery. 2008;62(6 Suppl 3):1084–92.

12. van Aalst J, Beuls EAM, van Nie FA, Vles JSH, Cornips EMJ. Acute distortion of the anatomy of the third ventricle during third ventriculostomy. Report of four cases. J Neurosurg. 2002;96(3):597–9.
13. Morota N, Watabe T, Inukai T, Hongo K, Nakagawa H. Anatomical variants in the floor of the third ventricle; implications for endoscopic third ventriculostomy. J Neurol Neurosurg Psychiatry. 2000;69(4):531–4.
14. Salamon G, Huang YP, Michotey P, Moscow N, Raybaud C, Farnarier P, et al. Third ventricle and suprasellar cisterns. In: Radiologic anatomy of the brain. Berlin/Heidelberg: Springer; 1976. p. 176–91.
15. Schroeder HW, Wagner W, Tschiltschke W, Gaab MR. Frameless neuronavigation in intracranial endoscopic neurosurgery. J Neurosurg. 2001;94(1):72–9.
16. Rhoton AL Jr. The lateral and third ventricles. Neurosurgery. 2002;51(4 Suppl):S207–71.
17. Morello G, Migliavacca F. Third ventriculostomy (Stookey and Scarff operation) in the treatment of benign aqueductal stenosis. Acta Neurochir. 1959;7:417–24.
18. Andaluz N, Zuccarello M. Fenestration of the lamina terminalis as a valuable adjunct in aneurysm surgery. Neurosurgery. 2004;55(5):1050–9.
19. Giussani C, Di Cristofori A. Lamina terminalis fenestration: an important neurosurgical corridor. Handb Clin Neurol. 2021;180:217–26.
20. El Refaee E, Baldauf J, Schroeder HWS. Bilateral occlusion of the foramina of Monro after third ventriculostomy. J Neurosurg. 2012;116(6):1333–6.
21. Baldauf J, Oertel J, Gaab MR, Schroeder HWS. Endoscopic third ventriculostomy for occlusive hydrocephalus caused by cerebellar infarction. Neurosurgery. 2006;59(3):539–44; discussion 539–44.
22. Baldauf J, Oertel J, Gaab MR, Schroeder HWS. Endoscopic third ventriculostomy in children younger than 2 years of age. Childs Nerv Syst. 2007;23(6):623–6.
23. Oertel JMK, Baldauf J, Schroeder HWS, Gaab MR. Endoscopic options in children: experience with 134 procedures. J Neurosurg Pediatr. 2009;3(2):81–9.
24. Warf BC. The impact of combined endoscopic third ventriculostomy and choroid plexus cauterization on the management of pediatric hydrocephalus in developing countries. World Neurosurg. 2013;79(2 Suppl):S23.e13–5.
25. Fabbro S, Tuniz F, Piccolo D, Cramaro A. Late-onset occlusion of the Monro foramina after endoscopic third ventriculostomy in adults: case discussion and review of the literature. Surg Neurol Int. 2020;11:326.
26. O'Brien DF, Hayhurst C, Pizer B, Mallucci CL. Outcomes in patients undergoing single-trajectory endoscopic third ventriculostomy and endoscopic biopsy for midline tumors presenting with obstructive hydrocephalus. J Neurosurg. 2006;105(3 Suppl):219–26.
27. Schroeder HWS, Warzok RW, Jamal JA, Gaab MR. Fatal subarachnoid hemorrhage after endoscopic third ventriculostomy – case report. J Neurosurg. 1999;90:153–5.
28. Bernard R, Vallee F, Mateo J, Marsella M, George B, Payen D, et al. Uncontrollable high-frequency tachypnea: a rare and nearly fatal complication of endoscopic third ventriculostomy: case report and literature review. Minim Invasive Neurosurg. 2010;53(5–6):270–2.

Anatomy and Surgical Approaches to Lateral Ventricle Tumors

54

Ekkehard M. Kasper and Robert Walker

Learning Objectives

- To understand epidemiology and pathogenesis of intraventricular tumors (IVT)
- To review clinical features and radiographic characteristics of selected tumors
- To employ available imaging techniques for presurgical planning for these lesions
- To understand perioperative clinical management

Table 54.1 Common intraventricular pathologies encountered in adult patients

Neoplastic lesions	Nonneoplastic lesions
Ependymoma, subependymoma, SEGA, neurocytoma, choroid plexus papilloma, choroid plexus carcinoma, meningioma, glioma, lymphoma, medulloblastoma, PNET, teratoma and germ cell tumors, pinealoma, sarcoma, metastasis	Colloid cyst Neuroepithelial cyst Choroid plexus cyst Tuberculoma Infection/abscess Parasites

54.1 Introduction

Tumors located within the lateral ventricles of the brain (IVT) make up a very small portion of primary brain tumors and can also impose as metastasis from cancer. Among primary CNS tumors, these IVT can represent the entire spectrum of CNS neoplasms. Subtypes that have been reported include those related to the ventricular lining (such as subependymoma, ependymoma, and subependymal giant cell astrocytoma) as well as all forms of glioma (e.g., chordoid glioma, oligodendroglioma, and astrocytoma) or true choroid plexus-derived tumors (meningioma, plexus papilloma, and plexus carcinoma). On rare occasions, one can also encounter lymphoma, germ cell tumors, central neurocytoma, and metastatic disease among other entities (Table 54.1).

Each tumor subtype has unique pathological characteristics with distinct epidemiology, radiographic and pathophysiological features, and molecular determinants. Each will require specific management considerations directly affecting prognosis and outcomes. However, there are striking commonalities that apply to the management of all lesions. Most of these tumors which fill and expand the ventricular space will present with a vague initial clinical presentation of mostly headache, nausea/vomiting, dizziness, cognitive impairment, and blurred vision due to mass effect and possibly hydrocephalus. Only rarely, seizures are seen.

Other patients will present with secondary symptoms in late stages when mass effect or infiltration dominates the clinical picture or when spontaneous hemorrhage occurs. At this point, patients may present with focal symptoms such as weakness, sensory changes, apraxia, aphasia, tinnitus, or visual field cuts caused by compression of nearby fiber tracts. IVT can cause almost any neurological symptom, depending on the growth direction and its primary property of either displacing or infiltrating the surrounding tissue. Seizures may occur if IVT reach the cortical surface.

E. M. Kasper (✉)
Boston University School of Medicine, Boston, MA, USA

Department of Neurosurgery, St. Elizabeth Medical Center, Boston, MA, USA

Faculty of Health Sciences, McMaster University, Hamilton, ON, Canada
e-mail: Ekkehard.kasper@steward.org

R. Walker
Department of Neurosurgery, St. Elizabeth Medical Center, Boston, MA, USA

I. N. Kanaan, V. Beneš (eds.), *Neuroanatomy Guidance to Successful Neurosurgical Interventions*,
https://doi.org/10.1007/978-3-031-59838-8_54

Table 54.2 Neoplasm classification

Tumors of the ventricular wall and septum	Ependymoma, subependymoma, neurocytoma, subependymal giant cell astrocytoma
Tumors of the choroid plexus	Papilloma, plexus carcinoma, meningioma, solitary fibrous tumor, metastasis
Mixed tumors	Lymphoma, glioma, PNET, medulloblastoma, teratoma, germ cell tumors, sarcoma, craniopharyngioma, metastasis

When approaching a patient with IVY, the most useful step to create a working diagnosis is based on the presumed site of origin: ventricular wall, choroid plexus proper, or involvement of adjacent structures (Table 54.2).

We shall discuss the advantages and risks of each surgical approach to IVT by using the most frequent intraventricular tumor, meningioma, as an example.

54.2 Anatomical Highlights

- Identify the most suitable sulcus of the lobar surface nearest the lesion.
- Minimize white-matter dissection via transsulcal dissection.
- Perform white matter incision in parallel to the optic radiation to minimize compromise to Meyer's loop.
- Understand macroscopic anatomy of the lateral ventricles and related structures.
- Know vascular anatomy: Supply and drainage to the structures in the lateral ventricles.

54.2.1 Detailed Surgical Anatomy of the Lateral Ventricle [30, 31]

- Understanding the surgical anatomy of the lateral ventricle is crucial because it can help select the best surgical approach and often predict the achievable EOR.
- The lateral ventricles (LV) are paired, C-shaped structures that wrap around the thalamus. Each LV has a frontal (anterior) horn, body (central part), atrium, temporal (inferior) horn, and posterior (occipital) horn. Each of these parts has a distinct roof, floor, and medial and lateral walls. We recommend standard anatomy textbooks to review cross-sectional illustrations.
- The anterior horn is bordered medially by the septum pellucidum, foramen Monroi, and column of the fornix and by the head of the caudate nucleus laterally. The corpus callosum forms the roof, with the genu being the anterior wall and the rostrum, the floor.
- The body boundaries consist of the septum pellucidum and the fornix medially as well as the medial part of the caudate nucleus laterally. The roof is composed of the inferior part of the corpus callosum. The superior aspect of the thalamus forms the floor.
- The atrium of the lateral ventricle communicates with the frontal horn, the occipital horn, and the temporal horn laterally. It is surrounded by various structures that include visual projection fibers laterally inferiorly and the choroidal fissure medially. The anterior wall is formed by the pulvinar and the posterior pole of the thalamus. The roof is composed of splenial fibers.
- The occipital horn is surrounded medially by the bulbar prominence and calcar avis. Laterally, it is bordered by the splenium of the corpus callosum, optic radiation, and the inferior longitudinal fasciculus.
- The temporal horn contains the choroid plexus and is bordered by the inferior thalamus, the tail of the caudate nucleus at the roof, and the hippocampus as the floor.
- The blood supply to the LV arises from the choroidal arteries, which originate from the anterior circulation (AchA) and the posterior circulation (PchA). AchA originates directly from the internal carotid artery and supplies the choroid plexus. The PchA arises from the posterior cerebral artery, fed by the basilar artery. It is then divided into lateral and medial divisions: the lateral division forms 1–9 branches that may arise from P1, whereas the medial division forms 1–3 branches arising from the BA or P1.
- The venous drainage system of the LV consists of medial and lateral ventricular wall veins. The medial veins include the anterior and posterior septal veins, medial atrial veins, transversal veins of the hippocampus, and veins of the amygdaloid body (see Fig. 54.8). The lateral wall veins include the anterior caudate, thalamostriate, posterior caudate, thalamocaudate, lateral atrial, and inferior ventricular veins. In addition to the veins that drain the choroid plexus, the superior choroid vein, inferior choroid vein, and medial choroid vein are distinguished. These veins drain into the internal cerebral veins and the basal vein of Rosenthal.

54.3 Intraventricular Meningioma (IVM)

IVM are rare in neurosurgical practice and are covered in most textbooks in a minimalistic fashion [1–3]. IVMs account for 0.5–4.5% of all meningiomas in adults and 1–5% of intraventricular tumors [4–6]. According to larger series, most are found in the lateral ventricle, followed by a smaller number in the third ventricle and rare occurrences in the fourth ventricle (78%, 16%, and 6%, respectively) [7, 8]. Among the LV, the atrium is the most common location (44.4%) with the left side being more frequently affected [8]. IVM have female-to-male ratio of 2:1 [9, 10] and a median age of 40.6 years at onset [11].

54.3.1 Clinical Features

The presentation of IVM depends on size: small tumors (<1 cm) usually show no symptoms, in contrast to large IVM which can produce vague and only later specific symptoms based on growth direction [12]. In 1938, Cushing and Eisenhardt distinguished 5 key symptoms: (1) headaches related to increased intracranial pressure; (2) contralateral homonymous hemianopia; (3) contralateral sensory and motor deficits; (4) cerebellar symptoms; and (5) paralexia when located on the dominant side [13]. The combination of weakness, hemianopia, and dysphasia has been termed: "Syndrome of the Trigone". In 2003, Nakamura reviewed 532 cases of IVM and found that IVM usually have a slow growth pattern before causing symptoms [8, 12].

The most frequently reported IVM symptoms include headache, nausea, vomiting, and visual disturbances [4, 14–16], due to compromise of the geniculo-calcarine tract, which runs laterally and inferiorly to the atrium [15]. Visual symptoms include homonymous hemianopia, diplopia, or blurred vision [11, 16]. Other clinical features include mental status impairment, aphasia, hemiparesis, numbness, gait disturbance, and epilepsy [11, 12, 16]. If the mass extends downwards, clinical symptoms may include ipsilateral cerebellar peduncle signs. Some patients have been reported to present with intraventricular hemorrhage [17, 18].

54.3.2 Radiological Features: (See Figs. 54.1, 54.2 and 54.3 Below)

Radiological features of IVM are similar to other meningiomas [4]. CT usually reveals a hyperdense mass with some calcification (32.6%) and homogeneous enhancement. Hydrocephalus (see Figs. 54.3 and 54.7) and peritumoral edema can be present [16, 19–23].

CTA will demonstrate arterial supply and venous drainage, often from the choroid plexus of the body of the lateral ventricle (superiorly) and the choroid plexus of the temporal horn (inferiorly) (see Fig. 54.3). Similarly, MRA may show feeders from the posterior choroidal artery (PChA) and anterior choroidal artery (AchA) [22].

On MRI, IVM appear iso- or hypointense on T1 images and iso- or hyperintense on T2- images, with homogenous enhancement after gadolinium administration [7, 15]. Functional MRI and (DTI) tractography can further assist in optimizing the strategies for approaches [23–25]. If a MRP is obtained, strong perfusion can be demonstrated in some IVM.

Beyond this, digital subtraction angiography (DSA) can identify feeders, usually from medial and lateral PchA and branches of AchA. DSA also shows venous drainage usually into the vein of Galen, or towards the basal vein of Rosenthal (see Fig. 54.4) [27, 28].

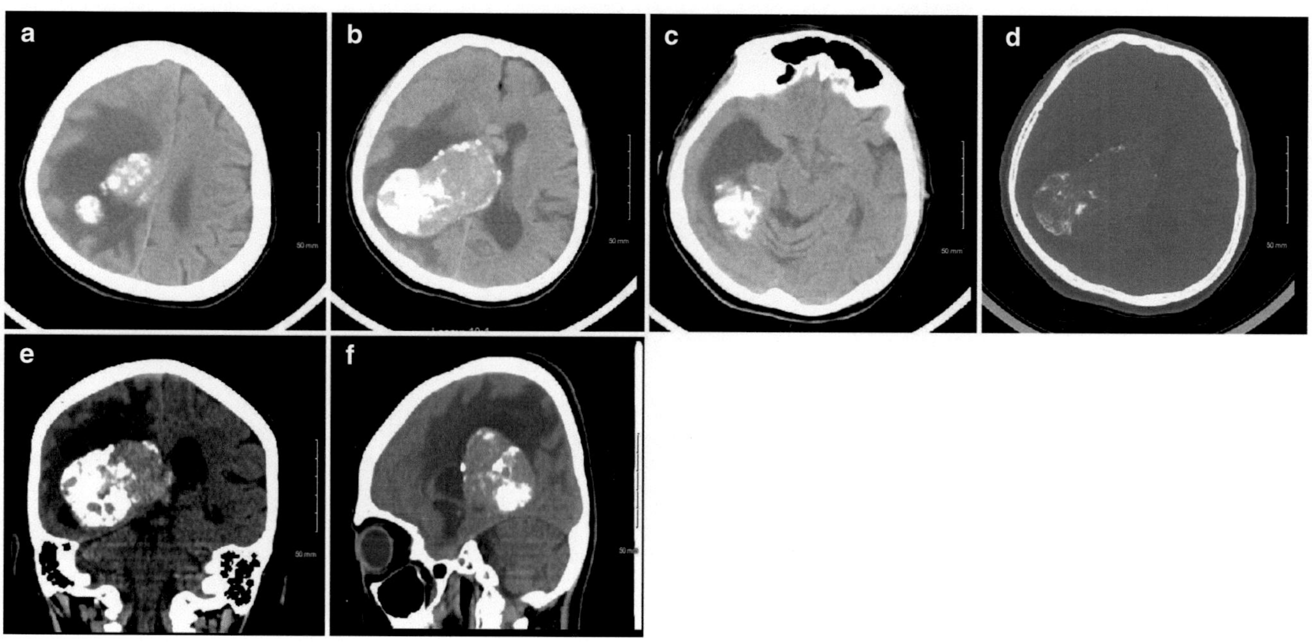

Fig. 54.1 (**a**): Basic radiographic characteristics of IVM on CT. Native CT in A,B,C: Axial cut soft tissue window; D: Bone window; E,F: Coronal and sagittal soft tissue window. This large IVM displays native hyperdensity and intrinsic calcifications. There is local mass effect, midline shift, sulcal effacement, and peritumoral edema (venous outflow obstruction + ipsilateral entrapment hydrocephalus). (**b**): Typical CTA for a large IVM. This large IVM shows hypervascularity and contrast enhancement. There are two vascular pedicles identified from the Anterior and Posterior Choroidal Arteries with medial and lateral contributories. Axial cuts A, Coronal B, Sagittal C. The venous outflow was complex and created a large draining vein superiorly medially with dilatation and tortuosity of the venous triangle (formed by the septal/caudate and thalamostriate veins). The internal cerebral veins were enlarged and displaced to the contralateral side

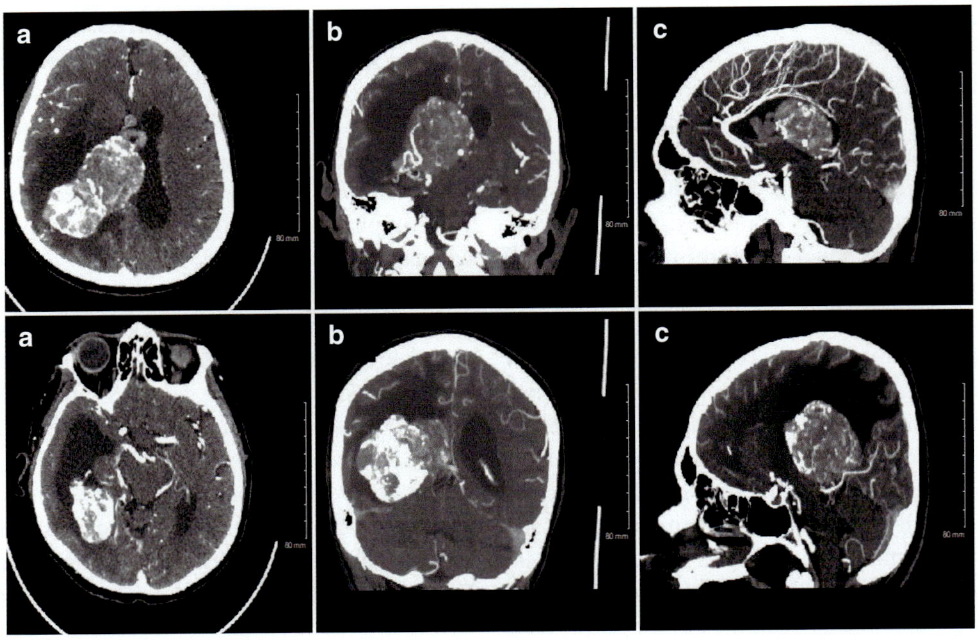

Fig. 54.1 (continued)

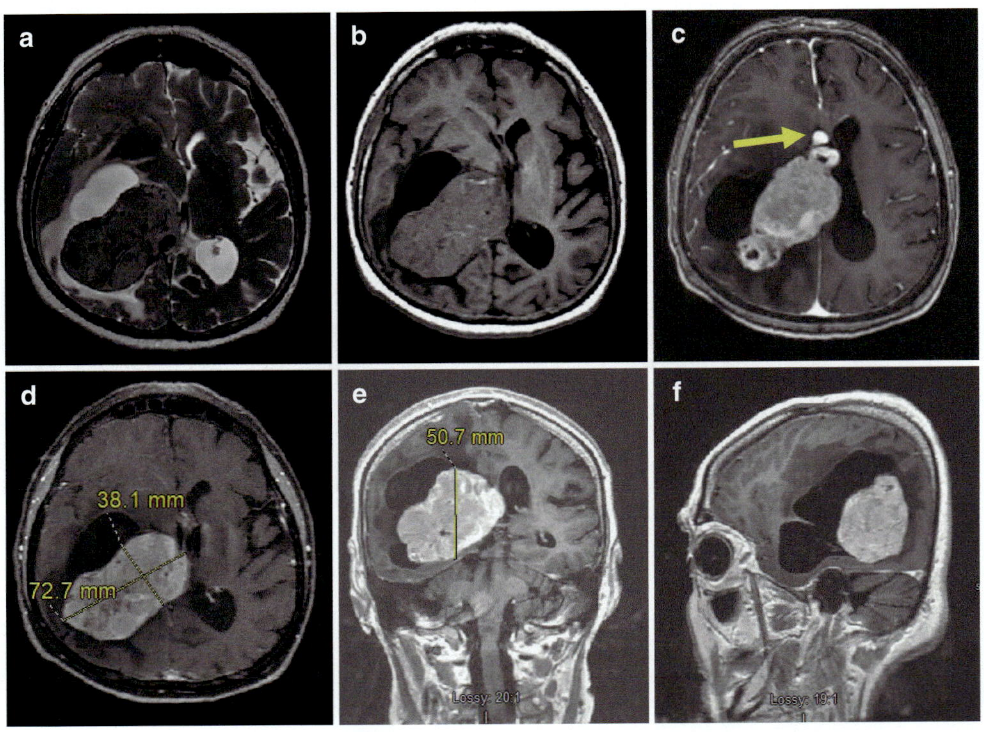

Fig. 54.2 Basic radiographic characteristics of IVM on MRI. MRI sequences reveal an intraventricular mass that is surrounded by CSF (best seen on T2 imagines in **a**); This lesion is T1-isodense to cortex on pre-contrast studies (**b**) and shows clear hypervascularity (**c**). **d–f**: After application of I/V contrast Gadolinium, the tumor shows avid enhancement

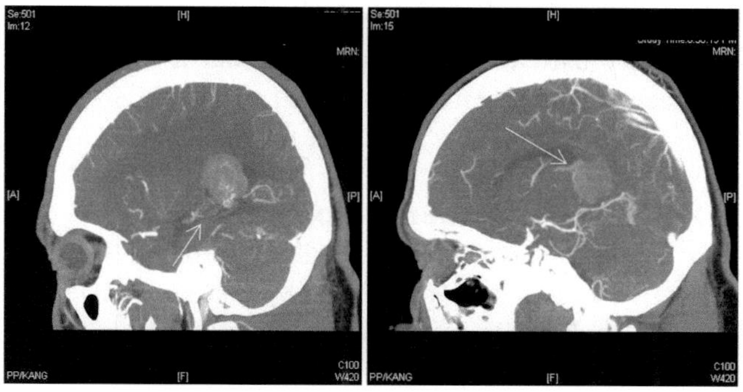

Fig. 54.3 Sagittal CTA/CTV reveals an inferior-anterior vascular pedicle and an anterior-superior supply

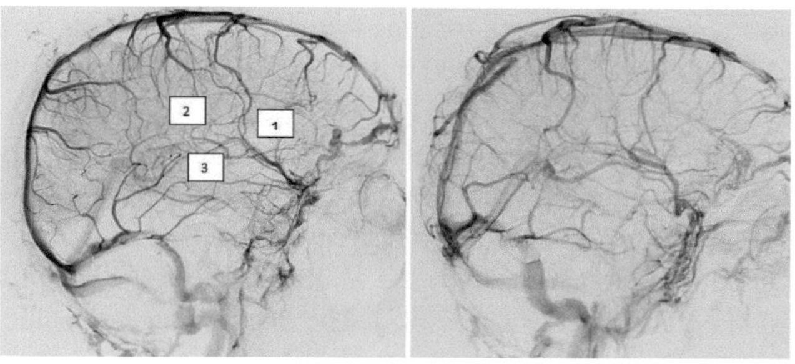

Fig. 54.4 The venous anatomy of the ventricle and surrounding structures is very variable. The main draining veins are flowing towards the internal cerebral veins. Anteriorly = tributaries to the septal vein (1) and head of caudate vein and a sup. Choroidal vein. Medial = medial atrial veins and inferior choroidal vein; thalamostriate vein (2). Posterior = veins of the occipital horn. Inferior = inferior ventricular veins (3). (Angiograms: courtesy of Prof. Thanh Nguyen of Boston Medical Center, Boston University). For further details, the interested reader is referred to www. Internal Cerebral Vein | neuroangio.org [26]

54.3.3 Histology and Pathology

Bertalanffy et al. reported that the majority of IVM were of meningothelial or mixed type [5] with Zhang et al. showing that 47.62% accounted for meningothelial tumors, followed by 28.57% for transitional (mixed) and 19.05% for fibroblastic IVM [22].

54.3.4 Conservative Management of Trigone Meningiomas

Observation is suggested for some patients, especially older people with small tumors on the left. Here, the slow growth rate of tumors outweighs the risk for intervention [11, 15]. If patients are not suitable candidates for microsurgical resection, radiation therapy (SRS or SRT) is an alternative, allowing reasonable local control rates as demonstrated by this author's team [29].

54.3.5 Open Surgical Approaches and Complication Avoidance for Lesions of the Atrium

Surgical approaches to IVM are challenging and pose significant risks. These can be divided into risks specific to the type and configuration of the lesion, and risks related to the approach (e.g., visual field deficits due to damage to projection fibers) [30–35]. In addition, on the dominant side, there is a risk of speech disturbances and cognitive impairment in language areas [30–32]. There is some risk of seizures due to cortical dissection and a risk of damage to AchA and PchA [30–32, 35]. Lastly, there is potential damage to the thalamus and limbic system, located anteriorly and medially [30, 31].

54.4 Surgical Intervention

The choice of surgical strategy for IVM depends on (A) patient factors (e.g., clinical status and presence of preoperative deficits, comorbidities, and goals of care) and (B) specific tumor factors (e.g., precise location and 3D-configuration of the lesion, size, growth pattern, pathology, vascular anatomy, surrounding structures, and whether or not the dominant hemisphere is involved) [11, 36–38]. The ideal surgical approach is the one that provides "the shortest distance to the lesion with a perpendicular field of view requiring minimal retraction avoiding trajectory through important structures" [39]. Key elements of resection of IVM were summarized by Piepmeier as: "gentle retraction, early control of feeding arteries, and careful piecemeal resection" [40].

54.4.1 Overview of Surgical Approaches

These can generally be divided into anterior, posterior, and lateral corridors.

54.4.2 Anterior Approaches

54.4.2.1 Distal Transsylvian Approach for Small IVM

A wide Sylvian fissure opening distinguishes Heschl's gyri [33]. Nagata et al. used the lateral transsylvian approach to resect IVM through the posterior Sylvian fissure and superior temporal sulcus for patients with wide and narrow Sylvian fissures, respectively [41]. This technique provides direct, short-distance access with little brain retraction allowing GTR without mortality [41]. However, patients might develop transient aphasia and dyscalculia due to damage to

Heschl's gyri, optic radiation, association fibers, and the parietal lobe [33, 41]. This approach is unsuitable for large tumors for which the sulcus may be tight and not accessible [36].

54.4.3 Lateral Approaches

54.4.3.1 Superior Parietal Lobule Approach (Fig. 54.5)

Here, a cortical incision is made via the superior parietal lobule to expose the calcar avis, corpus callosum, pulvinar, and trigone [33]. It can be used for large IVM in the dominant hemisphere extending superiorly [22, 32, 35, 42], providing good surgical outcomes with low complication rates [5, 11, 22].

Bertalanffy reported a series of 14 IVM approached via the superior parietal lobule [5] and Menon et al. employed it in the left hemisphere providing good results and possible complete resections with little risk of speech or motor deficits [11]. Zhang et al. performed a retrospective study of 21 IVM patients with superior parietal lobule access in nine: all achieved good surgical outcomes [22]. However, this approach carries the risk of seizures and visual deficit [5, 11, 22].

54.4.3.2 Intraparietal Sulcus Approach

Bertalanffy used this approach and found it to be a safe approach for IVM tumor resection [5] and so did Silva et al. [43]

54.4.3.3 Transtemporal Approach

This approach was described by DeLaTorre in 1963, followed by Olivecrona, Criscuolo and Symon, and McDermott [15, 44–46]. It is performed through an incision in the middle or inferior temporal gyrus in the nondominant hemisphere or the lower surface of the temporal lobe in the dominant hemisphere [37]. It can be used for IVM without upward extension or those with temporal horn extensions [11, 22]. Wang et al. reported on 35 IVM and used a transtemporal approach for nine lesions in the temporal horn [37]. Menon et al. also used this approach in 4/13 IVM with temporal horn extensions [11].

The advantages include short access to the atrium, early access to anterior choroidal artery, decompression of a trapped temporal horn, and a decrease in the risk to the optic radiation and speech, with the ability to obtain GTR with good outcomes [11, 35, 37, 47]. The disadvantages include: injury to bridging veins and the inferior optic radiation, causing contralateral quadrantanopia, as well as Wernicke's aphasia in the dominant hemisphere [11, 22, 37].

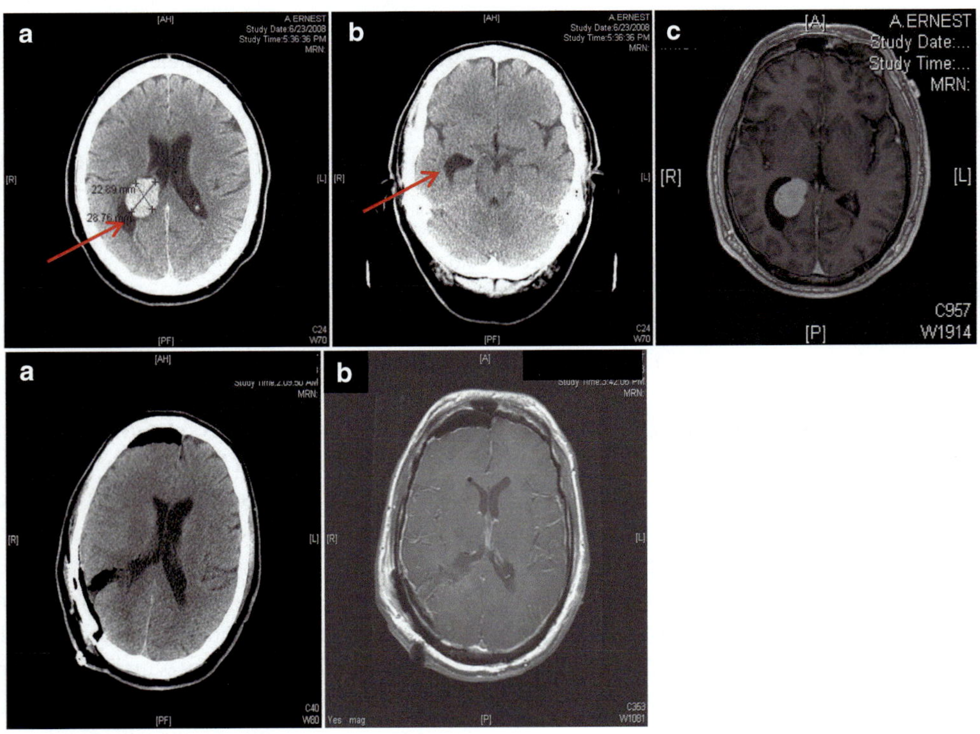

Fig. 54.5 Presurgical CT (native) and MRI (T1 Gado) for R-sided IVM. Preoperative CT axial cuts post-contrast in **a**, **b**; Note Hydrocephalus. MRI T1 post-contrast in **c**. Postsurgical CT (**a**: native) and MRI (**b**: T1 post-contrast) after parietal approach to IVM on the right side

54.4.3.4 Temporoparietal Approach (Fig. 54.7)

This approach was described by Cushing et al. in 1938, specifying a cortical incision at the temporoparietal junction [13]. Grujicic et al. published a series of 40 patients with lateral IVM, of which 12 were operated this way with good surgical outcomes [48]. However, this approach is not widely accepted because of possible optic radiation injury causing significant field deficits [34].

54.4.3.5 Combined Temporal and Parietal Approach

A combined temporal and parietal approach for IVM was reported by Couillard et al. achieving GTR with excellent outcome [49].

54.4.4 Posterior Approaches

54.4.4.1 Parieto-Occipital Approach (Figs. 54.8 and 54.9)

First described by Cramer et al. in 1960, and later by Fornari et al. [42, 50], this approach can be used on both the dominant and nondominant sides [51, 52]. It is suitable to resect large lesions and Zhang et al. used it to resect 16 tumors with diameters of >5 cm [22].

This technique has advantages and Fornari et al. achieved GTR in 18 IVM cases [42]. Nakamura et al. used this parieto-occipital approach in 10/11 patients with IVM reporting no mortality [8]. Furthermore, Liu et al. used this approach in 20/24 cases of trigonal meningioma and GTR was achieved in all but one patient with significant improvement in symptoms. These authors considered it safe when performed with piecemeal resection [4]. This approach can also provide good control of choroidal vessels [22].

In 2015, Sun et al. used this approach for 60 IVM employing fiber-tracking in 30. He accomplished GTR in all [53]. Wang et al. also used this approach in 38/51 patients with lateral IVM, with GTR in most patients [37], as did Gelabert-González et al. [28]

However, this approach has not been widely accepted because of the risk of optic radiation injury [51] as seen in a study by Ødegaard et al. who reported no mortality in 20 patients, with GTR and resolution of most symptoms. However, their most common complication was contralateral visual field defects [51].

To this end, Saleem et al. studied 16 cadavers to define the course of the optic radiation. They found that the optic radiation covered the lateral ventricle wall, whereas the medial wall was completely void of such fibers [34], concluding that a medial approach can avoid optic radiation injury [34]. This is supported by other studies reporting that the optic radiation passes inferolaterally [8, 54]. Furthermore, preoperative DTI-based tractography can help identify the visual pathways [24].

54.4.4.2 Occipital Corticotomy or Lobectomy

This approach can provide good exposure to the trigone and minimize retraction of brain, but provides poor access to blood supply and often inflicts homonymous hemianopia [46, 47, 55]. For these reasons, this approach has been largely abandoned.

54.4.4.3 Posterior Interhemispheric Parieto-Occipital Precuneal (Para-Splenial) Approach

This interhemispheric approach was initially described by Yasargil [56] and is used for small- and medium-sized lesions of the medial trigonal wall [57]. It can provide access to the trigone without causing optic radiation injury, with optimum surgical outcomes [58]. Nishizaki has used it in the lateral semi-prone position accomplishing GTR [58]. The disadvantages of this approach include a narrow and deep surgical corridor, poor control of vascular supply, and risk of venous air embolism from the sitting position [4, 58].

54.4.4.4 Posterior Transcallosal Approach

This approach was proposed by Kempe and Blaylock for IVM in the dominant hemisphere: [59] The incision is made through the splenium of the corpus callosum to resect IVM that extend upwards and medially [8, 33]. It can also be used for small IVM near the midline limiting postoperative deficits [22]. Advantages include preservation of the optic radiation [34]and good exposure of PchA [8]. However, it carries some risk for a hemispheric disconnection syndrome (alexia without agraphia) and mutism [60–63]. Jun et al. emphasized that this approach is not appropriate for paratrigonal lesions [64].

In our own practice, we recommend to simplify the choices and stick to two general recommendations:

(A) Superior parietal lobule or trans-parietal sulcus approach for right-sided lesions.
(B) Parieto-occipital intraparietal sulcus approach for left-sided lesions.

My recommendation is to keep the head position of the patient orthogonal whenever possible. That means: head straight up and without tilt. This will help to understand topography and maintain intraaxial orientation. Once the surgeon has gained more experience, adjustments can be made to positioning to optimize certain strategic steps (in particular: gaining control over vasculatures)—though the resulting intraoperative exposure of lesion and adjacent eloquent areas will be less familiar.

Presurgical management is not different from other brain tumors: In the operating room, adequate line-up is established. To gain additional working space intracranially, we give Mannitol (1–1.5 g/kg) and Furosemide (20 mg) as well as a single dose of steroids (Dexamethasone 10 mg). The patient is slightly hyperventilated (etPO2 = 30 mm Hg) and the bed is in reverse Trendelenburg position to facilitate venous outflow. This accomplishes good brain relaxation.

Right-sided lesions (see Figs. 54.1 and 54.5):

(A) The superior parietal lobule/transparietal sulcus approach is excellent for IVM on the nondominant side. In these cases, a standard supine position is chosen with the head slightly tilted to the contralateral side and the chin tucked in.

This provides short working distances from the surface of the brain, familiar viewing angles, and good access to its choroid plexus. Given that the geniculo-calcarine tract is inferior, this access carries low risk for visual field defects (quadrantanopia) and limits traction to motor fibers with little risk for postoperative hemiparesis. If the choroidal blood supply is mostly inferior, a right middle temporal lobe access may be preferred, which facilitates early control over feeders, yet at the expense of a larger visual field cut.

Whenever dissecting white matter, we recommend making the incision in parallel to the tracts to splay the respective fibers rather than cutting them orthogonally.

Left-Sided Lesions (Figs. 54.6 and 54.7):

(B) The pareto-occipital route or trans-intraparietal sulcus approach is suitable for complex IVM on the dominant side.

We position these patients ¾ prone with the lesional side up (Fig. 54.6). Again, this will give comparatively short, safe, and very familiar working distances to the lesion with little retraction. The skin incision is similar to a Frazier burr hole (ca. 5 cm above the inion, about 3.5 cm lateral) with the precise entry point guided by the individual anatomy (best visualized on T2 sequences). Dissection into the depth of the sulcus will provide about 2 cm additional distance from the brain surface towards the lesion, followed by about 2 cm of white matter dissection in parallel to the optic radiation, keeping the postoperative visual field defect to a minimum.

To avoid traumatizing the parenchyma from repetitive moves during tumor mobilization, we employ three stationary retractors to create an oval cross-sectional diameter of the working channel with little pressure points on the cortex. Once the ventricle is reached, the ependyma is exposed and coagulated with minimal bipolar intensity. In large lesions, one often encounters the tumor mass plastered against the surgical entry point. However, diligent study of presurgical MRI T2 sequences will reveal a shallow CSF rim surrounding most IVM which will help to pick the optimal starting point for resection.

The resection strategy remains simple: Study the vessels (veins!) on the surface of the tumor carefully and preserve them until the arterial supply is taken down. Create an incision into the tumor to allow internal debulking in a piecemeal fashion. This can be done with conventional bipolar coagulation and microscissors. I personally do not use the CUSA in the ventricle since the device is bulky and restrains visualization during dissection. Seek optimal control over any potential bleeders from within the tumor and minimize spillage of blood into the CSF which can result in post-op hydrocephalus.

After decoring the tumor, take the capsule out in segments, with careful coagulation of the attachment to the choroid plexus. One needs to pay great attention to a low bipolar setting when dissecting the choroid plexus itself, to avoid inflicting thermal damage to the adjacent ventricular wall (thalamus and internal capsule!). Once resection is complete, irrigate the ventricle copiously to get rid of any heme or debris, which helps to avoid postoperative hydrocephalus. If in doubt, leave a postoperative drain (e.g., a standard EVD) in place for 48 h. The ventricular system is filled with fluid and the wound is closed in layers. We extubate the patient in the OR under strict BP control (SBP < 140 and minimal to no PEEP) and with deep mental status to avoid coughing, which can create venous pressure spikes leading to bleeding in the ventricle. High inspiratory FiO2 is maintained for the first 12–24 h to help reabsorb pneumo-

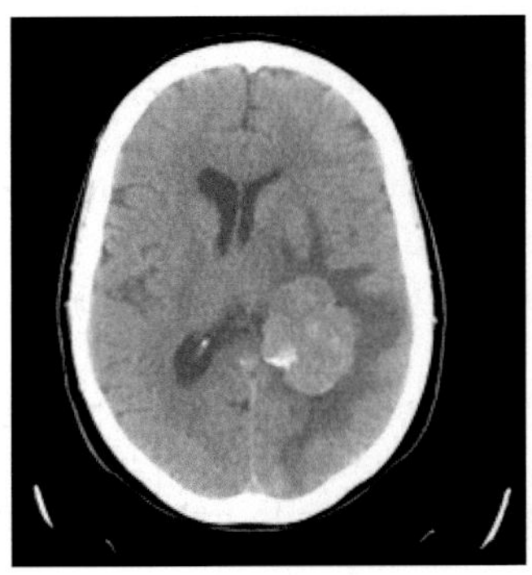

Fig. 54.6 LEFT-sided supratentorial approach to the lateral ventricle. CT w/o contrast shows a 4 × 5cm hyperdense lesion in the atrium of the left ventricle. CTA and MRI demonstrate a large 5 cm left-sided trigonal mass. The lesion shows hyperdense contrast enhancement on CTA (**a–c**) and the typical signal characteristics and irregular delineation on T1 post-contrast sequences (**d–f**). Note the very favorable fluid cleft surrounding the lesion as seen on T2 images (**g**). The DTI images (**h**) show the splaying of the optic radiation around the atrial tumor. H2: Approach to surgery on the left side: The patient's head is positioned in ¾ prone in this case and a paramedian incision is made approximately at Frazier's point (here modified to 3.5 cm from the midline and 5 cm superior to the inion to go transsulcal)

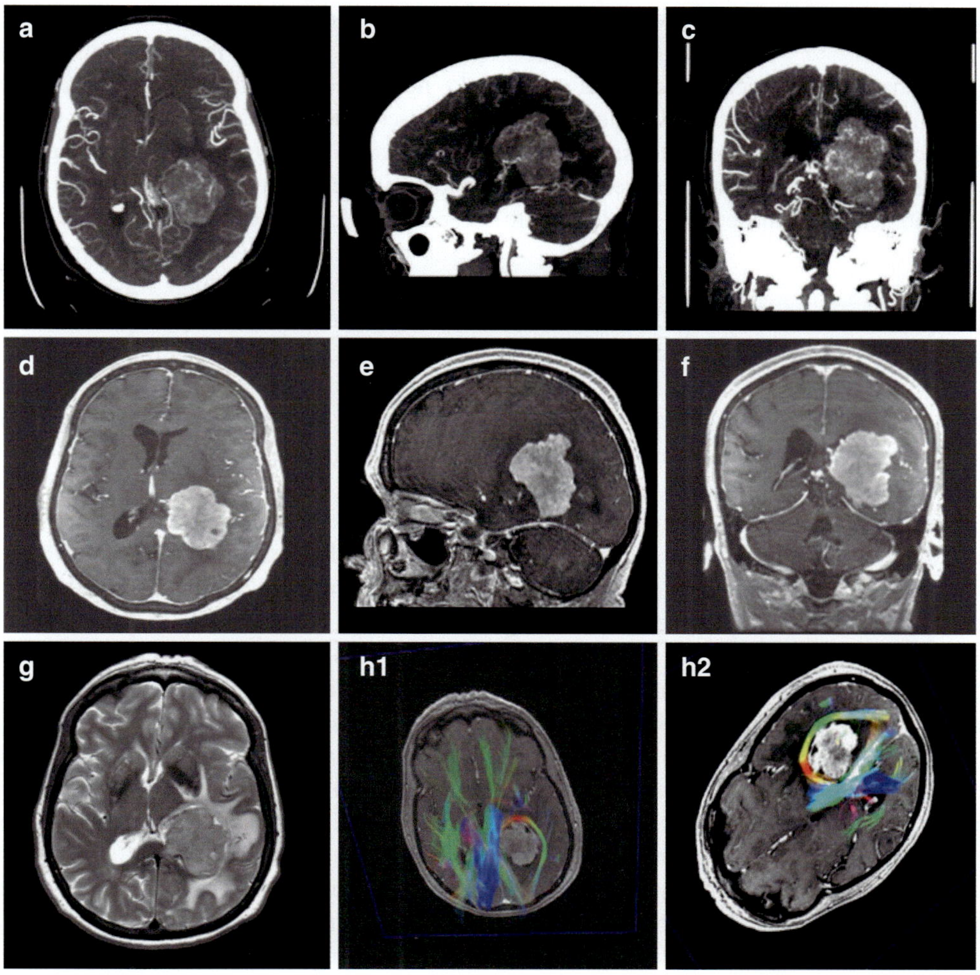

Fig. 54.6 (continued)

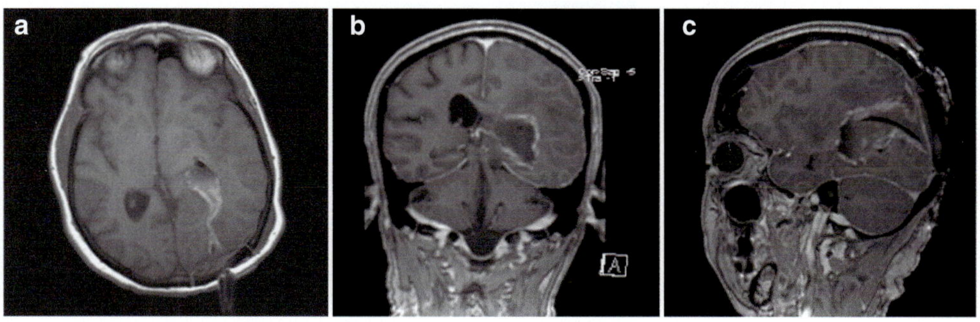

Fig. 54.7 Postsurgical MRI scan after resection of a left-sided IVM. T1 post-contrast studies show a complete removal of the tumor and sparing of the calcarine fissure. Views: Axial **a**, Coronal **b**, and Sagittal **c**

cephalus. The patient is placed in the ICU for 1–2 days with a native CT at ca. 4 h. Post-op to exclude bleeding or hydrocephalus. An MRI is obtained the day after to assess EOR and any potential strokes. The remainder of our postoperative regimen for these patients does not differ from other brain tumors.

If the patient shows any intraventricular heme on post-op CT (Fig. 54.5), we consider it safest to simply take the patient back to the OR for a second look, in case one of the choroidal vessels may continue to bleed. Dexamethasone is weaned off within 4 days and the patient can be mobilized on POD#1 to sit in a chair and can be moved to a regular floor.

54.5 Intraventricular Subependymoma

This tumor originates from the ventricular lining and subsequently protrudes into the ventricle. Similar to the surgical decision-making described above, the characteristic features of this tumor can guide us in its surgical management. The available surgical approaches to choose from are identical to those presented before.

54.5.1 Epidemiology and Background [65]

Ventricular subependymomas (VSE) are rare and slow growing glial tumors arising under the ventricular lining (ependyma) anywhere along the neuraxis. It is thought that they arise from Bergmann's glia during corticogenesis after neural tube closure, but the precise histogenesis remains to be determined.

VSE can occur anywhere in the lateral ventricles, along the septum pellucidum, in the fourth ventricle and even in the spine. VSE are mostly well-circumscribed nodular tumors, and due to their very slow progression, they frequently remain asymptomatic only to be found at autopsy. However, if enlarging, these IVT most frequently present with obstructive hydrocephalus causing headache and nausea from increased ICP. VSE very infrequently can hemorrhage into the tumor bed and present with ICH, although this remains clinically inapparent and is mostly found in pathological specimen.

Macroscopically, these tumors often have a rubbery consistency, and they may have a fibrous texture, although they do not present as firm and cohesive as meningiomas. Microscopically, cystic degeneration is often seen. According to the latest WHO classification, subependymomas are classified as WHO grade I tumors. Many lesions present with ependymal features of pseudorosettes/rosettes or with astrocytic characteristics (Rosenthal fibers). The appearance is very bland and hypocellular with minimal atypia and very little, if any, mitotic activity.

Tumor location and extent of resection remain the most relevant predictors for prognosis and outcome [65]. Surgery is the mainstay of therapy, and only in very rare cases, adjuvant radiation is needed.

54.5.2 Radiographic Features [66] (See Figs. 54.8 and 54.9)

VSE are usually well-demarcated hypo- or isodense lesions on CT imaging and appear hypo- or isointense on T1-MRI and mildly hyperintense on T2. Contrast enhancement has been described occasionally. There may be microcalcifications which can be seen on CT images. VSE may impose as single or multiple lesions with a pedicle arising from the ventricular wall and protruding into the ventricle itself. On CTA, SE usually show minimal perfusion and no hypervascularity. These tumors are not supplied by the choroidal vessels, but by very small vessels that provide terminal branches to the site of parenchymal origin.

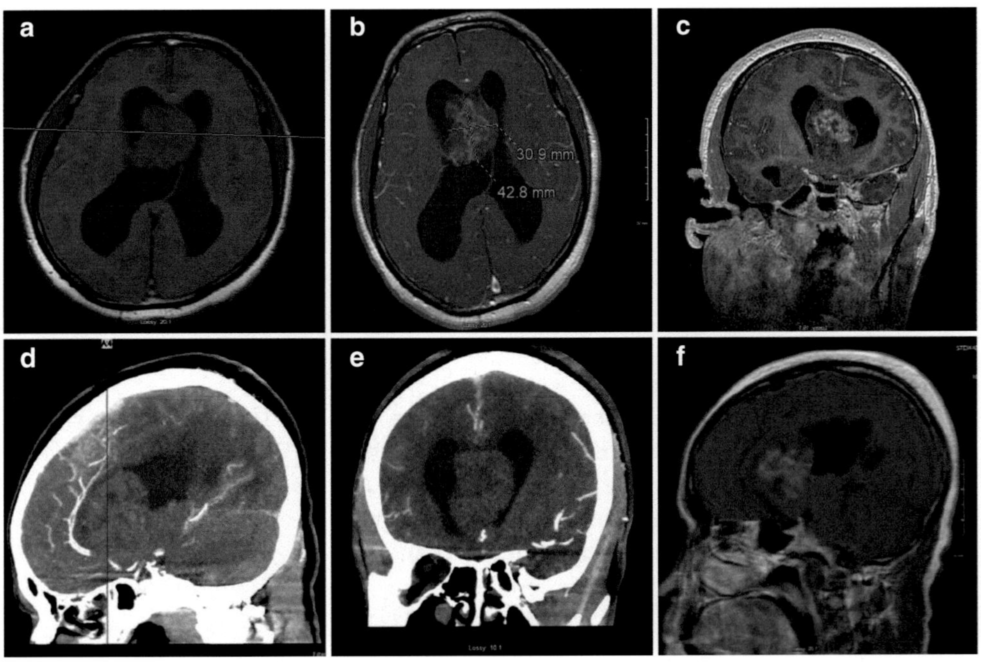

Fig. 54.8 Preoperative MRI scans for intraventricular subependymoma. First row: (**a**) Precontrast axial T1, (**b**) Post-contrast axial T1/MPRAGE (**c**) Coronal cut; Second row: (**d**) Post-contrast sagittal CTA MIB and (**e**) coronal CTA MIB; (**f**) Post-contrast sagittal T1/MPRAGE

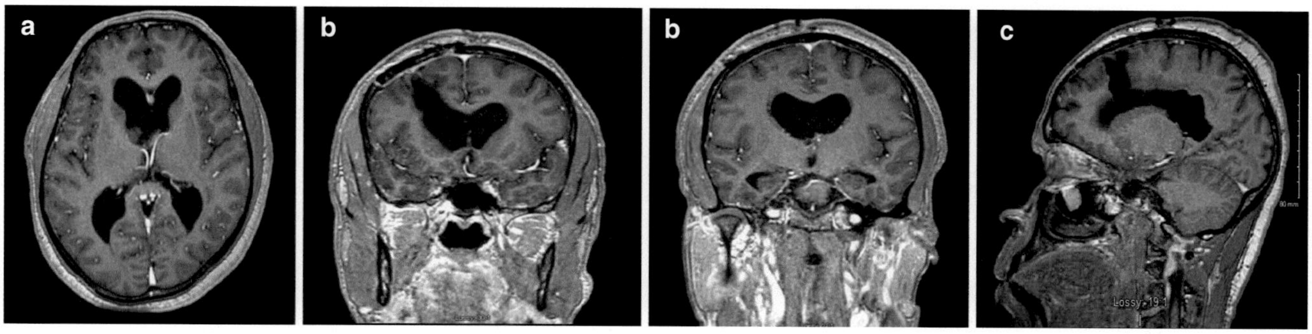

Fig. 54.9 Postoperative MRI scans for intraventricular subependymoma after resection via a right frontal paramedian transsulcal approach. (**a**) Post-contrast axial T1/MPRAGE, (**b**) Post-contrast coronal T1/MPRAGE, (**c**) Post-contrast sagittal T1/MPRAGE. See the excellent near total removal including resection of the septum anteriorly. Only minimal residual disease is visible on the axial cuts, with a nodule at the junction of the head of the caudate and internal capsule and a small residual above the bilateral fornices on coronal cuts. The sagittal cut illustrates the access corridor

54.5.2.1 Surgical Decision Making

VSE can create hydrocephalus providing enlarged working spaces for the surgeon. Surgery should be diligently planned and utilize navigation. It is helpful to approach these tumors via a corridor that puts the lesion closest to the cortical surface. In the example presented here, a VSE arising from the septum is showing a significant tumor volume near the midline and should be approached from the nondominant side.

In our example, the tumor is attached to anterior portion of the septum, separating the lateral ventricles with more bulk on the right. Since the VSE obstructs the Foramen Monroi, the most suitable approach mimics the trajectory of a standard EVD with entry at Kocher's point. We recommend seeking out the deepest sulcus between the superior and medial frontal gyrus aiming for a transsulcal corridor. This can be appreciated on the coronal projections in the postoperative MRI scan (Fig. 54.9). After entering the ventricle, a fresh frozen pathology specimen should corroborate the working diagnosis. If confirmed, piecemeal resection follows. I do not use the CUSA inside the ventricle. Instead, I piecemeal the mass to dissect it off its attachment to the surrounding structures. Only medially, I proceed with a more aggressive resection strategy and resect the septum itself to gain control over the attachment of the VSE.

The goal of surgery is a maximal safe resection, which does not always represent a gross total resection since this tumor type has a parenchymal origin. In our case, there was a clear second attachment to the head of the ipsilateral caudate, as seen on MRI. We most definitely wanted to preserve the neurological status of the patient and hence diligently resected the VSE under the microscope until the surface of the caudate head was identified abutting the tumor. We then followed the tumor margin, staying medially in the ventricle on the side of the VSE. A safety zone was created inferior medially and centrally where the tumor on the septum touched the bilateral fornices. Given the significant morbidity that comes with injury to the fornix, we decided to leave a tiny remnant inferiorly in order to avoid compromise to cognitive skills and memory. Both these goals were accomplished as can be seen on the postoperative images (Fig. 54.9). As a result of this strategy, we were able to accomplish a near total resection in a patient who remained neurologically intact.

54.6 Surgical Pearls

- Let the shortest distance from the tumor to the cortical surface guide your approach to IVT on the nondominant side.
- Right lateral superior lobule transsulcal approach is the preferred access to right IVM.
- Right anterior frontal transsulcal approach near Kocher's point is recommended for VSE.
- Left paramedian parieto-occipital transsulcal approach is chosen by us for left IVM to protect the language function on the dominant side.
- Each approach carries specific complications which may be prominent, if the tumor grows in a lobulated fashion protruding into eloquent areas or tracts.
- Study presurgical T2 MR-images in great detail preoperatively. The fluid cleft around the tumor surface will help the surgeon to mobilize tumor parts under eloquent cortex.
- If available, use Diffusion Tensor Tractography to plan white matter dissection trajectories as to avoid compromising nearby tracts.

54.7 Discussion

As can be seen from the synopsis above, there are many suitable resection strategies for IVTs and each of the surgical corridors comes with clear pros and cons. Many of the approaches can be used for various pathologies, and the pertinent literature is steadily growing to optimize our management of these challenging lesions. Better imaging modalities (e.g., DTI) make these cases more predictable and help to decrease iatrogenic morbidity.

For planning purposes, we always obtain presurgical MRI and CTA, but we almost always supplement these with fMRI and DTI as well as a formal DSA. This contributes to an optimal choice of the surgical corridor and has clearly improved patient outcomes by minimizing surgical morbidity.

Age and location of the lesion as well as radiographic characteristics remain the best predictor of the underlying histology and a thorough knowledge of the expected pathology aids in defining the goals of surgery.

In our opinion, all symptomatic large IVM should be approached via open microsurgical resection or port-based systems. Control over vasculature is superior in this setting and bleeding into the ventricle can best be minimized under direct visual control.

More recent developments of modern endoscopes with multiple working channels have opened-up the opportunity to address some IVTs differently: e.g., posterior third ventricular lesions in the pineal region can be well addressed via minimal access endoscopic surgery from anterior entry point trajectories, allowing the eloquent areas or tracts such as the corpus callosum to be left uncompromised by access corridors. Furthermore, the development of transnasal-transphenoidal surgery techniques and better endoscopic instruments has facilitated the management of sellar lesions with intraventricular extension as well as IVM of the inferior-anterior third ventricle (such as type IV craniopharyngiomas or prolapsing pituitary adenomas). With the establishment of true centers of excellence in endoscopic neurosurgery, we shall see further expansion of the application of these techniques in the near future.

54.8 Conclusion

Resection of IVTs is challenging because such tumors are surrounded by essential anatomical structures. The surgical approaches for resections differ from case to case, and the selection of the optimal approach depends on various factors related to the patient and tumor. Each approach has associated risks and complications. Surgical resection can be facilitated by the use of functional MRI and neuronavigation systems. However, if diligent dissection is performed with standard approaches, these tumors can be managed very well with little associated morbidity.

References

1. Piepmeier JM, Spencer DD, Sass KJ, George TM. Lateral ventricular masses: chapter 19. In: Apuzzo M, editor. Brain surgery, vol. 1. New York: Churchill Livingstone; 1993. p. 581ff.
2. Ojemann RR. Supratentorial meningiomas: clinical features and surgical management. Chapter 89. In: Wilkins R, Rengachary S, editors. Neurosurgery, vol. 1. 2nd ed. New York: McGraw Hill; 1996. p. 873ff.
3. Burger P, Scheithauer B. Tumors of the central nervous system. In: Atlas of tumor pathology, vol. 10. Washington, DC: Armed Forces Institute of Pathology; 1994. p. 261.
4. Liu M, Wei Y, Liu Y, Zhu S, Li X. Intraventricular meningiomas: a report of 25 cases. Neurosurg Rev. 2006;29(1):36–40.
5. Bertalanffy A, Roessler K, Koperek O, et al. Intraventricular meningiomas: a report of 16 cases. Neurosurg Rev. 2006;29:30–5.
6. Lyngdoh BT, Giri PJ, Behari S, Banerji D, Chhabra DK, Jain VK. Intraventricular meningiomas: a surgical challenge. J Clin Neurosci. 2007;14:442–8.
7. Bhatoe HS, Singh P, Dutta V. Intraventricular meningiomas: a clinicopathological study and review of the literature. Neurosug Focus. 2006;20:1–6.
8. Nakamura M, Roser F, Bundschuh O, Vorkapic P, Samii M. Intraventricular meningiomas: a review of 16 cases with reference to the literature. Surg Neurol. 2003;59:491–503; discussion 503–504.
9. Nanda A, Bir SC, Maiti T, Konar S. Intraventricular meningioma: technical nuances in surgical management. World Neurosurg. 2016;88:526–37.
10. Erman T, Göcer AI, Erdogan E, et al. Intraventricular meningiomas: a review and report of eight cases. Neurosurg Quart. 2004;14:154–60.
11. Menon G, Nair S, Sudhir J, Rao R, Easwer HV, Krishnakumar K. Meningiomas of the lateral ventricle – a report of 15 cases. Br J Neurosurg. 2009;23:297–303.
12. Winkler PA, Buhl R, Tonn JC. Intraventricular meningiomas. In: Lee JH, editor. Meningiomas: diagnosis, treatment and outcome. London: Springer; 2009. p. 491–514.
13. Cushing H, Eisenhardt L. Meningiomas: their classification, regional behaviour, life history and surgical end results, vol. III. Springfield: Charles C Thomas; 1938. p. 139–49.
14. Guidetti B, Delfini R, Gagliardi FM, Vagnozzi R. Meningiomas of the lateral ventricles. Clinical, neuroradiologic, and surgical considerations in 19 cases. Surg Neurol. 1985;24(4):364–70.
15. McDermott MW. Intraventricular meningiomas. Neurosurg Clin N Am. 2003;14:559–69.
16. Ma J, Cheng L, Wang G, Lin S. Surgical management of meningioma of the trigone area of the lateral ventricle. World Neurosurg. 2014;82:757–69.
17. Romeike BFM, Joellenbeck B, Skalej M, Scherlach C, Kirches E, Mawrin C. Intraventricular meningioma with fatal haemorrhage: clinical and autopsy findings. Clin Neurol Neurosurg. 2007;109:884–7.
18. Bosnjak R, Derham C, Popovic M, Ravnik J. Spontaneous intracranial meningioma bleeding: clinicopathological features and outcome. J Neurosurg. 2005;103:473–84.
19. Kendall B, Reider-Grosswasser I, Valentine A. Diagnosis of masses presenting within the ventricles on computed tomography. Neuroradiology. 1983;25:11–22.
20. Kloc W, Imielinski BL, Wasilewski W, Stempniewicz M, Jende P, Karwacki Z. Meningiomas of the lateral ventricles of the brain in children. Childs Nerv Syst. 1998;14:350–3.

21. Lang I, Jackson A, Strang FA. Intraventricular hemorrhage caused by intraventricular meningioma: CT appearance. AJNR Am J Neuroradiol. 1995;16:1378–81.
22. Zhang WH, Xie M, Liu H, Wang X, Lin MH. Surgical challenges for lateral ventricle meningiomas: a consecutive series of 21 patients. J Huazhong Univ Sci Technol Med Sci. 2015;35:742–6.
23. Staempfli P, Reischauer C, Jaermann T, Valavanis A, Kollias S, Boesiger P. Combining fMRI and DTI: a framework for exploring the limits of fMRI-guided DTI fiber tracking and for verifying DTI-based fiber tractography results. NeuroImage. 2008;39(1): 119–26.
24. Kamada K, Todo T, Morita A, Masutani Y, Aoki S, Ino K, Kawai K, Kirino T. Functional monitoring for visual pathway using real-time visual evoked potentials and optic-radiation tractography. Neurosurgery. 2005;57(1 Suppl):121–7; discussion 121–127.
25. Li F, Lin J, Zhu G, Meng H, Wu N, Hu R, Feng H. Neuroimaging and functional navigation as potential tools to reduce the incidence of surgical complications of lateral ventricular meningiomas. Clin Neurol Neurosurg. 2011;113(7):564–9.
26. Internal Cerebral Vein | as seen on www.neuroangio.org. Accessed 17 May 2023.
27. Guidetti B, Delfini R. Lateral and fourth ventricle meningiomas. In: Al-Mefty O, editor. Meningiomas. New York: Raven Press; 1991. p. 569–87.
28. Gelabert-González M, García-Allut A, Bandín-Diéguez J, Serramito-García R, Martínez-Rumbo R. Meningiomas of the lateral ventricles. A review of 10 cases. Neurocirugia (Astur). 2008;19:427–33.
29. Christ SM, Farhadi DS, Junzhao S, Mahadevan A, Thormann M, Lam FC, Yu X, Kasper EM. Efficacy and safety of primary stereotactic radiosurgery in patients with intraventricular meningiomas. Adv Radiat Oncol. 2022;8(1):101098. https://doi.org/10.1016/j.adro.2022.101098. PMID: 36407683; PMCID: PMC9666502.
30. Pendl G, Öztürk E, Haselsberger K. Surgery of tumours of the lateral ventricle. Acta Neurochir. 1992;116:128–36.
31. Le Gars D, Lejeune JP, Peltier J. Surgical anatomy and surgical approaches to the lateral ventricles. Adv Tech Stand Neurosurg. 2009;34:147–87.
32. Ellenbogen RG. Transcortical surgery for lateral ventricular tumors. Neurosurg Focus. 2001;10:1–13.
33. Kawashima M, Li X, Rhoton AL Jr, Ulm AJ, Oka H, Fujii K. Surgical approaches to the atrium of the lateral ventricle: microsurgical anatomy. Surg Neurol. 2006;65(5):436–45.
34. Mahaney KB, Abdulrauf SI. Anatomic relationship of the optic radiations to the atrium of the lateral ventricle: description of a novel entry point to the trigone. Neurosurgery. 2008;63(Suppl 2):S195–203.
35. D'Angelo VA, Galarza M, Catapano D, Monte V, Bisceglia M, Carosi I. Lateral ventricle tumors: surgical strategies according to tumor origin and development – a series of 72 cases. Neurosurgery. 2005;56(Suppl1):S36–45.
36. Zanini MA, Faleiros AT, Almeida CR, Clara CA, Gabarra RC. Trigone ventricular meningiomas: surgical approaches. Arq Neuropsiquiatr. 2011;69:670–5.
37. Wang X, Cai BW, You C, He M. Microsurgical management of lateral ventricular meningiomas: a report of 51 cases. Minim Invasive Neurosurg. 2007;50(6):346–9.
38. Seçer HI, Düz B, Izci Y, Tehli O, Solmaz I, Gönül E. Tumors of the lateral ventricle: the factors that affected the preference of the surgical approach in 46 patients. Turk Neurosurg. 2008;18(4):345–55.
39. Tew JM, Larson JJ Jr. Intraventricular meningioma. In: Kaye AH, Black PML, editors. Operative neurosurgery. London: Harcourt Publishers; 2000. p. 575–85.
40. Piepmeier JM. Tumors and approaches to the lateral ventricles. J Neuro-Oncol. 1996;30:267–74.
41. Nagata S, Sasaki T. Lateral trans-sulcal approach to asymptomatic trigonal meningiomas with correlative microsurgical anatomy: technical case report. Neurosurgery. 2005;56:438.
42. Fornari M, Savoiardo G, Morello G, Solero CL. Meningiomas of the lateral ventricles. Neuroradiological and surgical considerations in 18 cases. J Neurosurg. 1981;54:64–74.
43. Silva DO, Matis GK, Costa LF, Kitamura MA, Birbilis TA, Azevedo Filho HR. Intraventricular trigonal meningioma: neuronavigation? No, thanks! Surg Neurol Int. 2011;2:113.
44. De La Torre E, Alexander E, Davis CH, Crandell DL. Tumours of the lateral ventricles of the brain. J Neurosurg. 1963;20:461–70.
45. Olivecrona H, Tönnis W. Handbuch der Neurochirurgie, Berlin Heidelberg, vol. 4. New York: Springer; 1974. p. 175–7.
46. Criscuolo GR, Symon L. Intraventricular meningioma. A review of 10 cases of the National Hospital, Queen Square (1974–1985) with reference to the literature. Acta Neurochir. 1986;83:83–91.
47. Rubino PA, Rhoton AL Jr, Tong X, Oliveira E. Three-dimensional relationships of the optic radiation. Neurosurgery. 2005;57:219–27.
48. Grujicic D, Cavallo LM, Somma T, Illic R, Milicevic M, Raicevic S, Gazibara MS, Villa A, Savic D, Solari D, Cappabianca P. Intraventricular meningiomas: a series of 42 patients at a single institution and literature review. World Neurosurg. 2017;97:178–88.
49. Couillard P, Karmi MZ, Abdelkader AM. Microsurgical removal of an intraventricular meningioma with ultrasound guidance, and balloon dilatation of operative corridors: case report and technical note. Surg Neurol. 1996;45:155–60.
50. Cramer F. The intraventricular meningiomas: a note of the neurologic determinants governing the surgical approach. Arch Neurol. 1960;3:98.
51. Ødegaard KM, Helseth E, Meling TR. Intraventricular meningiomas: a consecutive series of 22 patients and literature review. Neurosurg Rev. 2013;36:57–64.
52. Zhao X, Shen X, Chen X, et al. Integrated functional neuronavigation-guided resection of small meningiomas of the atrium via the paramedian parieto-occipital approach. Clin Neurol Neurosurg. 2015;128:47–52.
53. Sun GC, Chen XL, Yu XG, Zhang M, Liu G, Hou BK, Ma XD. Functional neuronavigation-guided transparieto-occipital cortical resection of meningiomas in trigone of lateral ventricle. World Neurosurg. 2015;84:756–65.
54. Ebeling U, Reulen HJ. Neurosurgical topography of the optic radiation in the temporal lobe. Acta Neurochir. 1988;92(1–4):29–36.
55. Sarubbo S, De Benedictis A, Milani P, Paradiso B, Barbareschi M, Rozzanigo U, Colarusso E, Tugnoli V, Farneti M, Granieri E, Duffau H, Chioffi F. The course and the anatomo-functional relationships of the optic radiation: a combined study with 'post mortem' dissections and 'in vivo' direct electrical mapping. J Anat. 2015;226(1):47–59.
56. Yasargil MG. Parieto-occipital inter-hemispheric approach. In: Yasargil MG, editor. Microneurosurgery, vol. IVB. New York: Thieme; 1996. p. 56–7.
57. Faquini I, Fonseca RB, Vale de Melo SL, Negri H, Vieira E, Saboia T, Azevedo-Filho H. Trigone ventricular meningiomas: is it possible to achieve good results even in the absence of high tech tools? Surg Neurol Int. 2015;6:180.

58. Nishizaki T, Ikeda N, Nakano S, Okamura T, Abiko S. Occipital inter-hemispheric approach for lateral ventricular trigone meningioma. Acta Neurochir. 2009;151:1717–21.
59. Kempe LG, Blaylock R. Lateral-trigonal intraventricular tumors. A new operative approach. Acta Neurochir (Wien). 1976;35(4):233–42.
60. Geffen G, Walsh A, Simpson D. Comparison to the effects of transcortical and transcallosal removal of intraventricular tumors. Brain. 1980;103:773–88.
61. Jeeves MA, Simpson DA, Geffen G. Functional consequences of the transcallosal removal of intraventricular tumors. J Neurol Neurosurg Psychiatry. 1970;42:134–42.
62. Nakasu Y, Isozumi T, Nioka H, Handa J. Mechanism of mutism following the transcallosal approach to the ventricles. Acta Neurochir. 1991;110:146–53.
63. Levin HS, Rose JE. Alexia without agraphia in a musician after transcallosal removal of a left intraventricular meningioma. Neurosurgery. 1979;4(2):168–74.
64. Jun CL, Nutik SL. Surgical approaches to intraventricular meningiomas of the trigone. Neurosurgery. 1985;13:416–20.
65. Siffert J, Jallo G. Ependymomas and intraventricular tumors. Chapter 21. In: Principles of neurooncology. New York: McGraw Hill; 2005. p. 432–45.
66. Sattouf M, Intraventricular meningioma. Case study, Radiopaedia.org. Accessed on 15 Jul 2024. https://doi.org/10.53347/rID-186623.

55 Microsurgical Anatomy of Third Ventricle Approaches

Francesco Tomasello and Alfredo Conti

Learning Objectives

- Third ventricle microsurgical anatomy
- Different lesions affecting third ventricle dictate the surgical strategy according to the growth pattern inside the ventricular cavity.
- Microsurgical tips in anterior approaches and posterior approaches.
- Venous meticulous preservation as main goal.
- Minimal callosotomy for transcallosal approaches. Transphoraminal approach widened by transchoroidal route.
- CSF patency should be achieved as possible.

55.1 Introduction

In the realm of neurosurgery, few endeavors demand the utmost precision, intricate knowledge, and unwavering expertise like the exploration of the third ventricular area [1]. The third ventricle is a vital anatomical structure nestled deep within the cerebral hemisphere. Unveiling the complexity of this intricate surgical anatomy requires a comprehensive grasp of its boundaries, connections, and functional relationships. The surgical anatomy of the third ventricle [2–16] represents an indispensable cornerstone of neurosurgical practice. It is through this labyrinthine landscape that surgeons navigate to access and treat a myriad of complex pathologies, such as tumors, cysts, and vascular malformations. Thus, a thorough comprehension of its anatomical intricacies paves the way for successful surgical interventions and optimal patient outcomes. There are several surgical approaches used to access and remove third ventricle tumors. Here, we summarize the main microsurgical approaches to anterior and posterior parts of the third ventricle. Beyond the tumor characteristics and location, the choice of approach depends on various factors, like patient's clinical condition and surgeon's experience. The selection of the surgical approach and the overall treatment plan should be personalized and discussed thoroughly between the patient, their family, and the medical team.

This chapter embarks on a journey, aiming to elucidate the surgical anatomy of the third ventricle, shedding light on its structural features, topographical relationships, and clinical implications. By exploring the neuroanatomical foundation that underlies this delicate region, we aspire to equip surgeons with sufficient understanding of the surgical approaches, challenges, and potential complications encountered within this demanding realm.

55.2 Anatomical Highlights

- The third ventricle is a median sagittal slit that connects the lateral ventricles with the fourth ventricle.
- The third ventricle extends forward to the lamina terminalis stretched between the optic chiasm and the rostrum of the corpus callosum.
- The surgical access to this cavity needs at least small neural incision.
- The lateral wall is formed by the medial aspect of the thalamus, the hypothalamus anteriorly and the subthalamus posteriorly.
- The third ventricle extends forward to the lamina terminalis stretched between the optic chiasm and the rostrum of the corpus callosum.

F. Tomasello (✉)
Humanitas Clinical Institute of Catania, Misterbianco, Catania, Italy
e-mail: ftomasel@unime.it

A. Conti
Neurosurgery, IRCCS Istituto delle Scienze Neurologiche di Bologna, Bologna, Italy

Dipartimento di Scienze Biomediche e NeuroMotorie (DIBINEM), Alma Mater Studiorum University of Bologna, Bologna, Italy
e-mail: alfredo.conti2@unibo.it

I. N. Kanaan, V. Beneš (eds.), *Neuroanatomy Guidance to Successful Neurosurgical Interventions*, https://doi.org/10.1007/978-3-031-59838-8_55

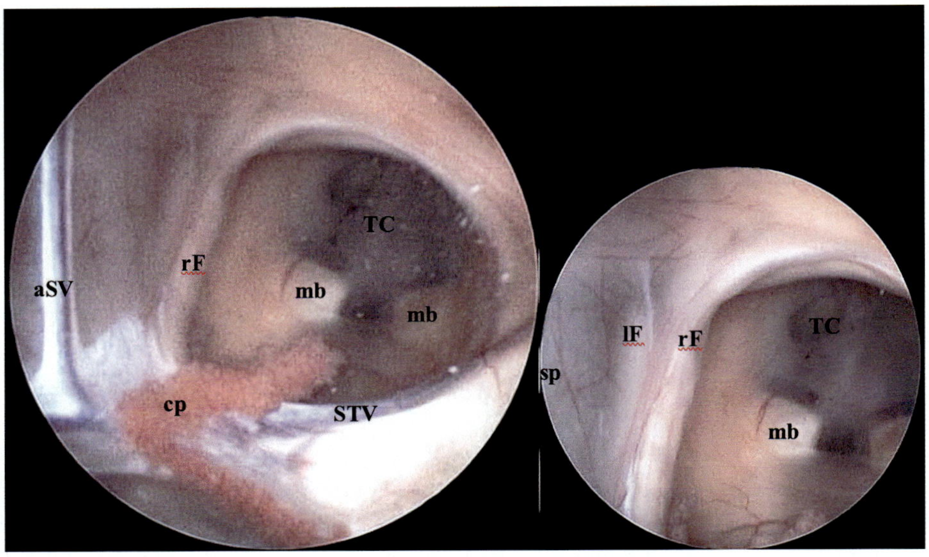

Fig. 55.1 Endoscopic view of the Monro foramen(left). An enlarged detail (right). *rF* right fornix column, *lF* left fornix column (visible in transparency beyond the pellucid septum), *aSV* anterior septal vein, *STV* striated thalamic vein, *cp* choroid plexus, *mb* mammillary body, *TC* tuber cinereum, *sp* septum pellucidum

- Above the lamina terminalis, the anterior wall of the ventricle is formed by the diverging columns of the fornix and the anterior commissure.
- At the base of the lamina terminalis, immediately above the optic chiasm, the ventricle has a small angular diverticulum which takes the name of the optic recess.
- Just behind the optic recess, the anterior part of the ventricular floor consists mainly of hypothalamic structures.
- The floor of the third ventricle is formed by the tuber cinereum and the mammillary bodies (Figs. 55.1 and 55.2).
- The roof of the third ventricle is a thin ependymal membrane stretched between the lateral walls and the choroid plexus, occupying the choroidal fissure. The body of the fornix forms the roof of the third ventricle.

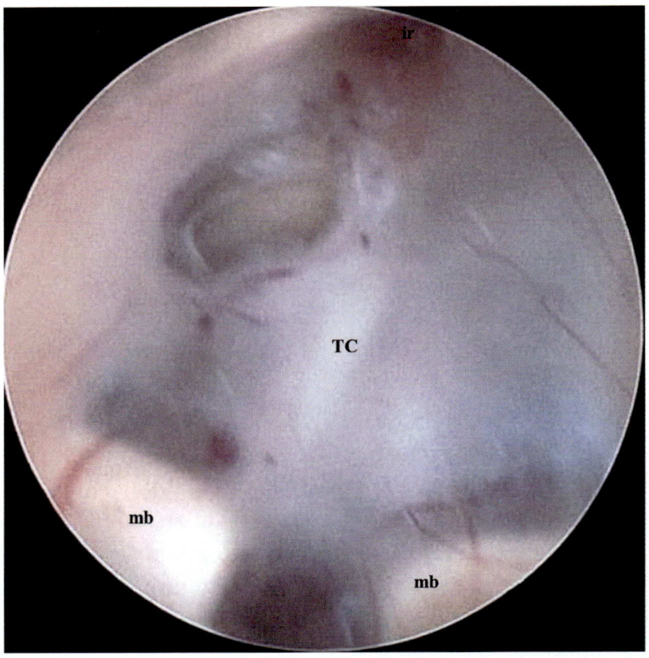

Fig. 55.2 Endoscopic view of the floor of the third ventricle (after third ventriculostomy). *mb* mammillary body, *TC* tuber cinereum, *ir* infundibular recess

55.3 History/Background

Several types of lesions can affect the third ventricle [17–23]. These lesions can be both benign or malignant, and their impact on the function of the third ventricle and surrounding structures can vary and influence the treatment modality.

Craniopharyngiomas and Neuroendocrine pituitary tumors typically develop in the anterior pituitary gland and stalk, but sometimes extend into the third ventricle. Colloid Cysts typically tend to develop within the third ventricle near the foramen of Monro. Pineal Region Tumors, including pineal parenchymal tumors, germ cell tumors, or gliomas, originate from the pineal gland or in the surrounding area and can extend into the posterior third ventricle. Gliomas, including pilocytic astrocytomas, high-grade gliomas, and ependymomas, may originate from periventricular glia. Neurocytomas typically arise in the lateral ventricles, but can extend into the third ventricle. Metastasis may indirectly involve this area as well. Nonetheless, this is not an exhaustive list and other less common lesions or malformations can also affect the third ventricle requiring surgery. Various surgical approaches have been described over time, building upon the pioneering work of Dandy. The diagnosis, treatment, and prognosis of these lesions depend on several factors, including their specific characteristics, location, and individual patient considerations.

55.4 Microsurgical Anatomy of Third Ventricle Approaches

Tumors located in the vicinity of the third ventricle pose considerable challenges in terms of exposure and removal. The optimal surgical approach is determined by the tumor's location in relation to neighboring structures such as the lateral and third ventricle, the foramen of Monro, aqueduct of Sylvius, optic nerves and chiasm, pineal gland, sella turcica, pituitary gland, fornix, midbrain, and thalamus. These operative approaches can be categorized based on whether they are suitable for accessing the anterior or posterior half of the third ventricle.

55.5 Anterior Approaches

55.5.1 Transylvian Approach

The transylvian translamina terminalis approach [2, 11, 24–27] involves a surgical corridor created by dissection through the sylvian fissure, allowing direct access to the third ventricle. This technique provides a favorable trajectory for resection of the anterior ventricle tumors while providing early visualization and minimizing manipulation of critical neurovascular structures.

Before splitting the sylvian fissure itself, the basal cisterns (i.e., the optic-carotid cistern) are opened to allow the cerebrospinal fluid to drain and relax the brain, which facilitates the splitting process. Typically, the splitting of the sylvian fissure begins at the level of the pars triangularis, where the space between the frontal and temporal lobes is wider due to the natural retraction of the pars triangularis of the inferior frontal gyrus. The superficial part of the sylvian fissure is opened using a scalpel with a no. 11 blade and microscissors. The dissection is usually performed on the frontal side of the superficial sylvian vein, which courses laterally toward the tip of the temporal lobe to empty into the sphenoparietal sinus. Sometimes, the frontal and temporal lobes can be strongly adherent, making it challenging to identify the dissection plane and split the superficial part of the sylvian fissure. In such situations, it is advisable to initially deepen the dissection into the operculoinsular compartment of the sylvian fissure to identify the branches of the middle cerebral artery and create a tunnel along the deep part of the fissure. Once the optic and carotid cisterns are opened, together with the proximal sylvian cistern, the internal carotid artery is dissected free until its bifurcation to identify the A1 segment of the anterior cerebral artery.

The A1 segment originates from the carotid bifurcation within the carotid cistern and typically courses above either the optic chiasm or the optic nerve before entering the lamina terminalis cistern. In around 70% of cases, the junction of the anterior communicating artery with the right and left A1 segments occurs above the chiasm, while in 30% of cases, it happens above the nerve. Shorter A1 segments are often tightly stretched over the chiasm, whereas longer segments pass anteriorly over the optic nerve and can be elongated and tortuous, reaching either the tuberculum sellae or the planum sphenoidale. The medial lenticulostriate perforators, ranging from 1 to 11 branches (average 6.4), arise from the superior, posterior, or posterosuperior aspect of the proximal half of the A1 segment. They take a direct posterior and superior course to the anterior perforated substance, supplying the optic chiasm, optic tract, genu of the internal capsule, and the anterior part of the globus pallidus. They may also extend to the adjacent part of the posterior limb of the internal capsule and, less commonly, to the thalamus.

The anterior communicating artery unites the paired anterior cerebral arteries in the lamina terminalis cistern. Only in 20% of cases does the anterior communicating artery communicate with two A1 segments of equal size. In the remaining cases, it is associated with the hypoplasia of one A1 segment. In about 75% of cases, the anterior communicating artery exists as a single channel. However, a spectrum of anomalies exists in other cases, ranging from duplications and triplications to fenestrations, reticular patterns, loops, and bridges. The perforators from the anterior communicating artery, ranging from 0 to 4 (average 1.6), usually arise from its posteroinferior aspect to supply the infundibulum, the anterior perforated substance, the optic chiasm, the subcallosal area, and the preoptic areas of the hypothalamus.

The recurrent artery of Heubner, arising from the anterior cerebral artery in 78% of cases, typically originates from the proximal A2 segment [28, 29]. It loops back on its parent vessel and courses anteriorly to the A1 segment in 60% of cases. It can be visualized before the A1 segment when the frontal lobe is elevated. This artery is the largest and longest branch directed to the anterior perforated substance. After its origin, it passes above the carotid bifurcation and accompanies the middle cerebral artery into the medial part of the sylvian fissure before entering the anterior and middle portions (93%) of the full mediolateral extent of the anterior perforated substance. It supplies the most anterior and inferior part of the head of the caudate nucleus and putamen, as well as the adjacent part of the anterior limb of the internal capsule.

The A2 segment is also the source of the central or basal perforating arteries, which pass posteriorly to enter the optic chiasm, the lamina terminalis, and the anterior forebrain below the corpus callosum. They supply the anterior hypothalamus, septum pellucidum, medial portion of the anterior commissure, columns of the fornix, and the anteroinferior part of the striatum.

Fig. 55.3 Intraoperative visualization of the lamina terminalis through a transylvian route

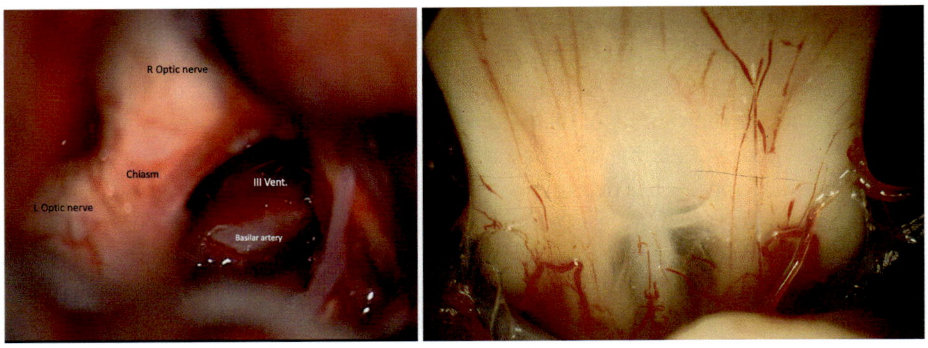

Roughly two third of the external surface of the anterior wall of the third ventricle is exposed behind this arterial complex, while the remaining upper one-third is concealed behind the rostrum of the corpus callosum. The visible portion of the anterior wall consists of the optic chiasm and the lamina terminalis (LT) (Fig. 55.3). The LT is then closely related to various structures, including both A1 tracts of the anterior cerebral artery (ACA), the complex formed by the anterior communicating artery (AComA), both recurrent Heubner arteries, both fronto-orbital arteries, arteries supplying the hypothalamus, proximal A2 segments of the ACA, and the anterior communicating and anterior cerebral veins. These structures reside within the LT cistern, where the LT forms the posterior boundary. The LT is a thin layer of gray matter covered by a pial layer and attaches as a delicate membrane to the upper surface of the chiasm. The LT forms a distinct angle with the broad, flattened surface of the chiasm, creating a recess known as the optic recess. This recess is bounded below by the upper surface of the chiasm and terminates in bays on both sides, leaving depressions on the surface of the two optic nerves. The LT, although varying in development, constitutes the widest segment of the third ventricle. It extends posteriorly and superiorly, filling the space between the chiasm and the rostrum of the corpus callosum. The LT gradually thickens in its upper part, detaches from the optic tracts, expands to its maximum width, and then tapers again to reach the anterior commissure.

In opening the LT, caution must be paid to preserving the perforating branches penetrating the anterior wall of the third ventricle during retraction of the ACA-AComA complex to expose the LT. With respect to the neural relationships of the LT, it is recommended that meticulous attention be paid in opening the thinned LT, to prevent damage to the anterior commissure and the rostrum of the corpus callosum above, the optic chiasm below, and the optic tracts, the columns of the fornix, and the hypothalamic walls laterally. Regarding the LT approach to the anterior third ventricle as a portion of the subfrontal exposure, Stein [30, 31] called attention to the vascular supply of the dorsum of the optic chiasm in this region. There are important anastomotic branches from the AComA and ACA that might be interrupted or injured by such an approach.

55.5.2 Interhemispheric Trans Lamina Terminalis Approach

To gain a wide access corridor to the anterior part of the third ventricle with minimal brain retraction and without incision of neural structures, the trans-LT route via a subfrontal [23] or interhemispheric approach [19, 32–34] can be used as well. This route has been described more frequently for craniopharyngiomas [32, 35, 36] and for a wide spectrum of pathological entities located within the anterior third ventricle or invading or compressing the walls of the chamber, but intrinsic lesions of the hypothalamus [2, 37–39] can also be successfully removed via this route. A discussion of the relative advantages and disadvantages of the different surgical approaches to the LT region is, however, beyond the scope of this chapter.

The interhemispheric approach to the LT is obtained through bifrontal craniotomy. The craniotomy should be performed as low as possible on the orbital roofs. In all cases, the frontal sinus is opened. The mucous membrane within the sinus is not dissected or cauterized to avoid inflammation, but the internal bone lamina of the sinus is removed to reduce dead space. The crista galli and residual bone are removed using a high-speed drill.

The dura is transversely opened along the anterior orbital bone edge, extending as far forward as possible to minimize damage to the frontal bridging veins. The self-retaining retractor blades are positioned to provide traction across the arachnoid enveloping the frontal lobes. The falx cerebri is then cut as far forward as possible, and bleeding from the severed end of the superior and/or inferior sagittal sinus is stopped using electrocoagulation. With the aid of the operating microscope, the arachnoid is sharply divided with microscissors, carefully suctioning the cerebrospinal fluid to tear

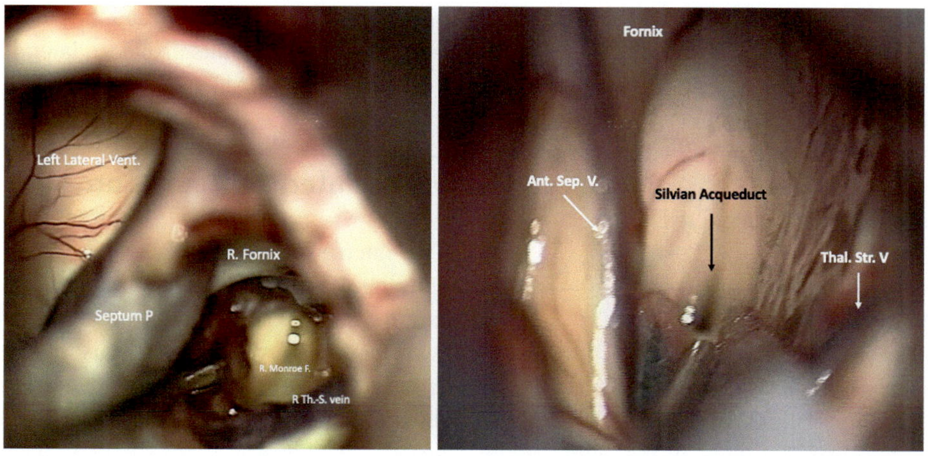

Fig. 55.4 *Left image*: Transcallosal exposure of the third ventricle with visualization of the foramen of Monro, fornix, thalamo-striate vein, choroidal plexus, and opened septum pellucidum. Intraoperative visualization of the right Monro foramen after septal opening. *Right Image*: Transcallosal exposure of the third ventricle with visualization of the foramen of Monro, fornix, thalamo-striate vein, anterior septal vein, and the third ventricular access to sylvan aqueduct

the olfactory cistern. The olfactory bulb and olfactory tract are preserved from the orbital surface of the frontal lobes bilaterally up to the olfactory trigonal region. The next step involves entering the basal interhemispheric fissure. The floor of the tuberculum sellae and optic chiasm are observed, noting the extent of exposure and the degree of frontal lobe retraction required.

55.5.3 Trancallosal Approach

The transcallosal approach provides a direct and adequate pathway to the lateral ventricles (40), and the transforaminal, transchoroidal, and interforniceal approaches provide access to the third ventricle through its roof, granting optimal tumor resection with minimal manipulation and hazard for neurovascular structures [41].

The patient is placed in a standard supine position, with the head in neutral position, slightly flexed and fixed in a three pin Mayfield-Kees headholder. A horseshoe anteriorly based skin incision of 5 cm is usually performed. The approach is performed on the nondominant side. The bone flap should be designed to be 2/3 anterior to the coronal suture and 1/3 posterior. The position of the bridging veins, possibly identified by neuronavigation, can be considered for minimal modifications of the bone flap design. Dural opening should be tailored according to the bridging veins' position. Neuronavigation can be used to their location. Careful microdissection of the arachnoid is required to improve mobilization of the veins and access to the interhemispheric fissure. In some cases, the veins penetrate a dural duplication before reaching the superior sagittal sinus that requires multiple dural microincision of these dural layers to improve vein mobilization. Interhemispheric fissure dissection is performed with sharp instruments. Initial opening of the fissure produces CSF outflow and brain relaxation. Two cottonoids, rolled and saturated with saline, are placed at the two extremities of the cisternal opening to achieve a 2-cm lateral retraction along 4 cm of the fissure. The interhemispheric opening proceeds along the falx to identify the cingulate gyrus, both pericallosal and callosomarginal arteries. The corpus callosum is distinguished by its white color as opposed to the grey-cream color of the cingulate cortex.

A sagittal callosotomy of 1.5–2 cm in length is usually performed with bipolar coagulation and suction at the ventricular entrance. Ependymal arteries should be coagulated. Some authors propose a transversal incision to improve white matter fibers' preservation. The retractor is then advanced paying attention not to injure the basal ganglia. A wide opening of the septum pellucidum allows a quick and accurate identification of anatomical landmarks and the orientation of the entry with respect to both Monro foramina [42, 43] (Fig. 55.4). It also facilitates postoperative CSF circulation.

55.6 Transforaminal Approach

The exploration of the third ventricle should always be attempted through Monro foramina, especially when they are enlarged by a tumor. The exposure of both Monro foramina allows bilateral access to the third ventricle and tumor resection. Care must be taken to avoid manipulation of the fornix in the superomedial portion of the foramen and in the anterior pillar. The posterior portion of the foramen is usually limited by the confluence of anterior septal and thalamostriate veins where they reach the internal cerebral vein. When the Monro foramen is not dilated by tumor, however, access to the third ventricle is limited. In such case, different strategies, described below, can be adopted.

55.7 Transchoroidal Approach

The transchoroidal approach is performed by opening the choroidal fissure, which is located between the fornix and the thalamus, through the tenia fornicis or the tenia thalami (Fig. 55.5), to expose the roof of the third ventricle by gently mobilizing the fornix to the contralateral side. The access between the choroid plexus and the fornix avoids damage to the thalamus that results displaced laterally and protected by the choroid plexus [44].

55.8 Interforniceal Approach

The interforniceal approach (Fig. 55.6) is performed by splitting the midline forniceal raphe with subsequent perforation of the diencephalic roof along the plane between the two fornix bodies. We suggest this route in the presence of a "cavum vergae," creating a natural corridor to the roof of the third ventricle. It is very important to stay on the midline because all neurovascular structures, such as body of the fornix, internal cerebral vein, and choroidal arteries, are laterally placed.

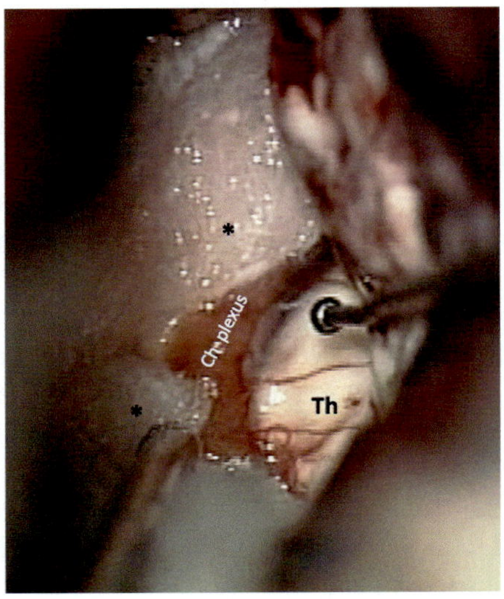

Fig. 55.5 Intraoperative picture of the trans-choroidal approach. The choroidal fissure between the thalamus and the choroidal plexus is opened and the tenia thalami is divided to improve access to the third ventricle chamber. * On the left, cottonoids protect the fornix

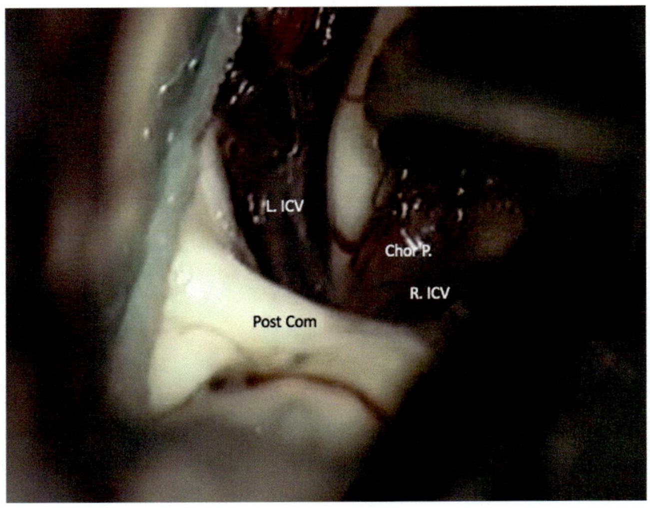

Fig. 55.6 Intraoperative photograph demonstrating exposition of the whole chamber of the third ventricle with an interforniceal corridor exposed through a cavum vergae. Both cerebral veins, posterior commissure, and pineal recess are recognized

55.9 Posterior Approaches

55.9.1 Infratentorial-Supracerebellar Approach

Krause initially documented the infratentorial-supracerebellar approach in 1926 [44]. However, due to high morbidity rates, the technique fell out of favor for several years. It was reintroduced successfully by Stein [30] in 1971, taking advantage of the advancements in microsurgery during that time.

Typically, the infratentorial-supracerebellar [45] approach is performed with the patient positioned in a sitting position [46, 47]. To perform the procedure, the incision is made starting from approximately 3/4 cm above the external occipital protuberance; a midline incision extends to the level of the C3 spinous process. A proper craniotomy is performed below the torcula, providing sufficient room and access for surgical instruments. Craniotomy is preferred over craniectomy to reduce the risk of complications. Care is taken to ensure a clear view along the tentorium by exposing the transverse sinus. The use of a microvascular Doppler can be useful during bone drilling to identify the sinus. The bony edges are carefully waxed, and venous bleeding is controlled to prevent air embolism. The intracranial pressure is assessed by palpating the dura in the posterior fossa, and if necessary, cerebrospinal fluid (CSF) is released through a ventricular drain or opening the cisterna magna through a small dural incision. A semilunar dura mater opening is made by starting at the lateral aspects of the exposure and moving towards the falx cerebelli and occipital sinus. The occipital sinus may be divided if patent, while the falx cerebelli is cut and the occipital sinus is secured. The dural flap is reflected upwards and tensioned with sutures attached to the retractor frame. Care must be taken to avoid excessive retraction that could obstruct the transverse sinuses. Arachnoid adhesions and bridging veins between the cerebellum and tentorium are cauterized and divided, opening the infratentorial corridor to the pineal region. The posterior fossa has extensive collateral circulation, reducing the risk of complications from venous sacrifice. However, caution is necessary when dividing bridging veins close to the tentorial surface to control subsequent dural bleeding. With the attachments divided, the cerebellum descends away from the tentorium, creating a favorable corridor with minimal retraction. The dorsal surface of the cerebellum is protected with padding material.

Arachnoid dissection is generally sufficient to obtain cerebellar mobilization without retractors, revealing additional adhesions and bridging veins near the anterior vermis, which are cauterized and retracted. Deeper dissection is then performed with the use of an armrest to support the surgeon's arms. The opalescent arachnoid of the quadrigeminal cistern is opened, and the precentral cerebellar vein is cauterized and divided. Pineal region tumors may be covered by small branches of the choroidal and superior cerebellar arteries, which can be safely coagulated, but care must be taken not to damage the deep venous system. The trajectory is redirected inferiorly to avoid damaging the vein of Galen and the confluence of the deep venous system. With the quadrigeminal arachnoid opened, cerebellar retraction is maximized, and the tumor's posterior surface is initially debulked. Further debulking is performed using various instruments, and the tumor's capsule is separated from the surrounding thalamus. Vessels along the capsule wall, mostly of choroidal origin, do not need to be preserved. Dissection progresses laterally or centrally to reach the third ventricle, and the tumor is dissected off the midbrain. The most challenging part of the dissection is often separating the tumor from the collicular plate (Fig. 55.7). The final attachment of the tumor is along the velum interpositum and deep venous system, which must be carefully cauterized and divided to avoid a rupture. An experienced surgeon can determine the extent of tumor invasion and whether it can be completely resected without significant morbidity. Radical removal is beneficial for tumors with pseudoencapsulation, reducing the risk of postoperative hemorrhage and improving the efficacy of adjuvant therapy.

Achieving meticulous hemostasis is crucial since there is limited tissue turgor to control bleeding in the tumor bed. Low-intensity cautery is the preferred method for direct hemostasis. The excessive use of hemostatic agents should be avoided as they may enter the ventricle and lead to obstructive hydrocephalus, shunt malfunction, or chemical meningitis. If necessary, long strips of Surgicel can be draped

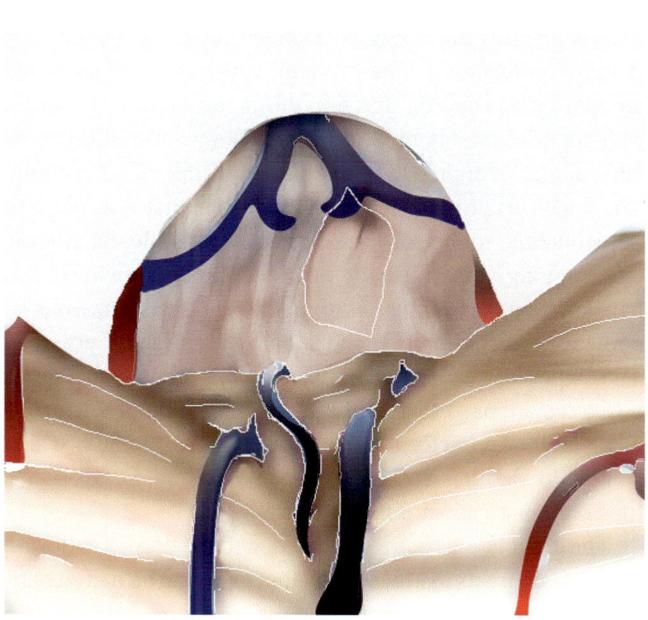

Fig. 55.7 Schematic drawing depicting the vision of the pineal region through the infratentorial-supracerebellar approach

over the cerebellum and tumor bed to promote hemostasis with minimal risk. Once hemostasis is achieved, the dura is closed meticulously to ensure a watertight seal. The craniotomy flap is secured using titanium miniplates. Generally, a more comprehensive closure of the dura and bone flap reduces the risk of postoperative headaches and wound complications.

55.9.2 Interhemispheric-Transcallosal Approach

The interhemispheric-transcallosal approach, first described by Dandy in 1921, creates a corridor between the falx and the hemisphere along the parieto-occipital junction [44, 46–48]. The preferred patient position for this approach is typically the lateral or three-quarter prone position.

The positioning of the bone flap depends on the tumor's location within the third ventricle. A wide craniotomy allows for flexibility in identifying a corridor that minimizes the sacrifice of bridging veins. Different skin incisions can be used, but a U-shaped flap extending across the midline and reflected laterally provides sufficient exposure. The craniotomy is centered over the vertex, with the lower margin positioned 2 cm above the apex of the lambdoid suture. Burr holes are drilled over the superior sagittal sinus at the front and back, and a craniotome is used to create a generous biparietal craniotomy, usually eccentric to the nondominant hemisphere. It is important to remove 1–2 cm of bone on the dominant side to enhance operative illumination. Bleeding from the sinus can be significant, but it is manageable with hemostatic agents since it occurs at low pressure.

A U-shaped durotomy is reflected medially, positioning its base on the sinus. The bridging veins are examined, and an approach is chosen that minimizes the sacrifice of veins. The depth of the operative corridor allows for multiple possible angles of approach through a small superficial opening. The ideal trajectory for each patient is one that is as close to perpendicular to the corpus callosum as possible. Sacrificing one bridging vein is usually necessary to achieve sufficient exposure, but sacrificing more than one increases the risk of cortical venous infarction and should be avoided if possible. The exposed hemisphere is protected, and two malleable retractors are used to gently mobilize the parietal lobe laterally. Another retractor may be utilized along the falx, which can be divided inferiorly to enhance exposure. Dissection is then continued under a microscope.

The cingulate gyri are separated, and the corpus callosum is identified by its distinct white appearance. The paired pericallosal arteries are observed and may be retracted to one side or each side using separate retractors, depending on their location. The corpus callosum may be thinner over the tumor, and a bulge on its dorsal surface may indicate the tumor's position. MRI-based frameless stereotactic guidance can aid in positioning the callosotomy. A 2-cm anterior-posterior opening is typically sufficient for most tumors, while the lateral extent of the opening is limited to ensure tumor resection without harming the pericallosal arteries.

The retractors are then shifted to retract the medial and lateral margins of the callosotomy, allowing visualization of the tumor's dorsal surface through the tela choroidea. The layer is opened, and a small biopsy is taken for frozen tissue diagnosis. It's crucial to identify and preserve the internal cerebral veins early in the dissection. Lesions in this region often displace the internal cerebral veins off the midline, but occasionally they can be found overlying the tumor in the midline. Once the deep venous system is recognized, internal debulking can proceed similarly to the supracerebellar-infratentorial approach. The degree of dissection required to enter the third ventricle depends on the size of the tumor and the presence of hydrocephalus. Care must be taken to avoid damaging the anterior roof of the third ventricle where the fornices are located. The difficulty of achieving complete resection is influenced by the tumor's point of origin, with caution needed for exophytic thalamic tumors, and more aggressive resection is possible for tumors originating from the pineal gland or velum interpositum.

55.9.3 Occipital-Transtentorial Approach

The occipital-transtentorial approach initially described by Jamieson in 1971 represents an alternative to supracerebellar route for posterior third ventricle [12, 45, 46, 49–52]. The three-quarter prone position is generally preferred due to the gravitational assistance in retracting the occipital lobe. In most cases, the exposure occurs between the right occipital lobe and the falx cerebri. Opting for a right-sided approach safeguards the dominant visual cortex against potential retraction injury. The primary anatomical advantage of this approach is the absence of bridging veins that extend from the occipital lobe into the superior sagittal sinus. This significantly reduces the risk of cortical venous infarction typically associated with the interhemispheric approach, as long as the inferior cerebral vein remains intact. Although the inferior cerebral vein drains a significant portion of the occipital lobe, it usually lies in a protected area lateral to the operative exposure. Division of the tentorium provides excellent visualization of the collicular plate, making this approach well-suited for tumors that require substantial access. One possible drawback of this approach is its oblique trajectory for midline lesions, which can potentially disorient inexperienced surgeons. Like other approaches to the pineal region, ventricular drainage can be beneficial in achieving brain relaxation.

To initiate the procedure, a U-shaped scalp flap is created on the right occipital region and reflected downwards. The inner part of the incision is positioned slightly left of the midline, starting at the level of the torcula. A burr hole is drilled across the superior sagittal sinus just above the torcula, followed by another one drilled 6–10 cm higher along the midline. A craniotomy extending along the midline is created using a turning maneuver. The dura can be opened either as a U-shaped flap based on the sagittal sinus or as triangular leaves based on the sagittal sinus and the lower edge of the craniotomy.

The dissection within the occipital corridor is performed under a microscope, taking advantage of gravity to retract the occipital lobe. The straight sinus is identified, and the tentorium is divided adjacent to it. The initial incision starts near the junction of the straight sinus and the torcula extends forward towards the incisura. Venous channels within the tentorium can be controlled using bipolar cautery or hemoclips. For additional exposure, a retractor can be placed on the falx, and in some cases, the falx and inferior sagittal sinus may be divided. Retention sutures can be placed on the cut tentorial edge to retract it laterally. At this stage, the arachnoid covering the tumor, the galenic system, and the quadrigeminal cistern comes into view.

Caution must be exercised during the dissection of the dense arachnoid over the pineal region to avoid injury to the deep venous system. The dissection is carried out using an arachnoid knife and microscissors. The vein of Galen (Fig. 55.8) is typically encountered first, followed by the right basal vein of Rosenthal, the internal cerebral veins, and the precentral cerebellar vein. Right-sided arachnoid dissection is often sufficient for tumor resection, but for a more extensive exposure, left-sided dissection and division of the precentral cerebellar vein may be performed. Once the arachnoid dissection is complete, lesions of the superior vermis, collicular plate, posterior third ventricle, pineal gland, and splenium can be easily accessed.

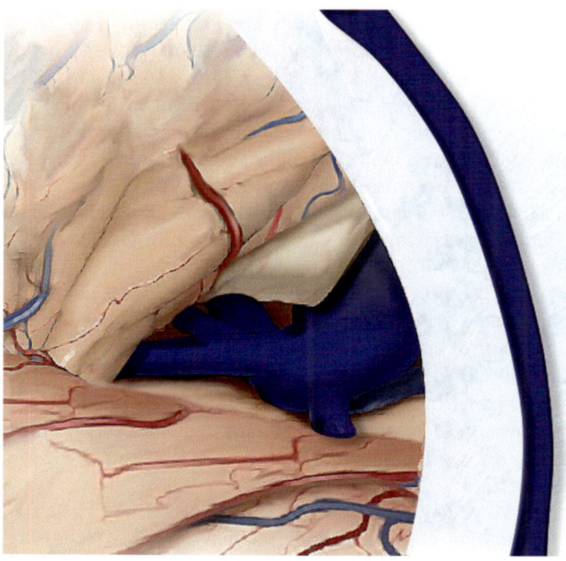

Fig. 55.8 Schematic drawing depicting the vision of the pineal region and internal venous system confluence through the occipital-transtentorial approach

Most of the mentioned lesions displace the deep venous system towards the posterior and superior regions, making it easier to identify the deep veins and necessitating dissection between them. However, lesions originating from the posterior leaf of the tela choroidea, the free edge of the tentorium, or the falcotentorial junction displace the galenic system anteriorly or inferiorly, making it invisible to the surgeon. In such cases, the tumor needs to be internally debulked, keeping in mind that the critical veins are located just beyond the deep capsule. Once adequate decompression is achieved, the tumor capsule can be cautiously dissected, and the underlying veins can be identified. Generally, capsular dissection should begin laterally and inferiorly and proceed medially and superiorly. The extent of resection that can be achieved depends on the tumor type, specific location, and invasiveness.

55.10 Surgical Pearls

- Various surgical approaches to third ventricle have been described. Choice of the strategy mostly depends on tumor pathology, extension, size, and relationship with crucial neurovascular structures.
- Venous preservation is crucial. It begins from interhemispheric dissection, but deep venous structures sparing is vital.
- Midline identification through all the surgical step is a major issue.
- Relax the brain by means of ventricular drainage before starting dissection and tumor exposure.
- Drain cysts and debulk tumor before attempting tumor capsule dissection.

55.11 Discussion

Accurate knowledge of the microsurgical anatomy of the third ventricle is indispensable for the successful management of various pathological conditions, including tumors, cysts, hydrocephalus, and vascular malformations. Understanding the precise locations of critical structures, including the fornix, thalami, hypothalamus, and vascular supply, enables surgeons to navigate these intricate regions with precision and minimize the risk of complications. The course of the choroid plexus and the thalamostriate vein to the foramen of Monro, where it joins the anterior septal vein to form the internal cerebral vein, can be used to guide the surgeon for localization of the fornix, thalamus, and septum pellucidum. Advancements in imaging techniques, three-dimensional navigation, and intraoperative imaging, such as ultrasounds and indocyanine-green videoangiography, have greatly facilitated the visualization and comprehension of the third ventricle's anatomical intricacies. These tools provide invaluable preoperative planning assistance, enabling surgeons to tailor their approaches based on the individual patient's anatomy.

However, it is important to recognize the dynamic nature of the surgical anatomy of the third ventricle. Each patient may present with unique variations, necessitating adaptability and intraoperative decision-making. One key point for a successful surgery in the third ventricle is the possible achievement of multiple points of access to the lesion. Control of bleeding is a major concern because high-power coagulation may lead to damage of relevant neurovascular structures, and thus, should be avoided. Gentle compression with the suction probe over cottonoids may help to reach hemostasis. In cystic lesions, aspiration is followed by wall dissection. In solid tumors, initial central debulking followed by circumferential dissection must be attempted. To prevent mistakes, it is necessary to orient the dissection according to reliable anatomical landmarks. The plane between the tumor and ependymal surface should be maintained. Surgical expertise, coupled with a comprehensive understanding of the anatomical nuances, remains the key for successful outcomes.

55.12 Conclusions

Surgical access to the third ventricle is one of the most challenging neurosurgical procedures. Detailed anatomical knowledge and tumor characteristics are crucial. The major challenge is to dissect lesions in a deep and narrow cavity. Different surgical approaches are available according to the exact location of tumors and their pathological features. Venous preservation and CSF patency should be achieved in order to avoid complications. It is mandatory that surgical manipulation avoids damage to important anatomical structures such as fornix, thalamus, and hypothalamus.

References

1. Chakravarthi SS, Kassam AB, Fukui MB, Monroy-Sosa A, Rothong N, Cunningham J, et al. Awake surgical management of third ventricular tumors: a preliminary safety, feasibility, and clinical applications study. Oper Neurosurg (Hagerstown). 2019;17(2):208–26.
2. Choudhri O, Chang SD. Subfrontal trans-lamina terminalis approach to a third ventricular craniopharyngioma. Neurosurg Focus. 2016;40 Video Suppl 1:2016.1.FocusVid.15416.
3. Erturk M, Kayalioglu G, Ozer MA. Morphometry of the anterior third ventricle region as a guide for the subfrontal (translaminaterminalis) approach. Neurosurg Rev. 2003;26(4):249–52.
4. Fernandes-Silva J, Silva SM, Alves H, Andrade JP, Arantes M. Neurosurgical anatomy of the floor of the third ventricle and related vascular structures. Surg Radiol Anat. 2021;43(12):1915–25.
5. Graziano F, Ganau M, Meccio F, Iacopino DG, Ulm AJ. The transcallosal anterior interfoniceal approach: a microsurgical anatomy study. J Neurol Surg B Skull Base. 2015;76(3):183–8.
6. Hendricks BK, Spetzler RF. Contralateral interhemispheric transcallosal transchoroidal approach for resection of a tumor in the lateral wall of the third ventricle: 2-dimensional operative video. Oper Neurosurg (Hagerstown). 2019;17(5):E197.
7. Hernesniemi J, Romani R, Dashti R, Albayrak BS, Savolainen S, Ramsey C, et al. Microsurgical treatment of third ventricular colloid cysts by interhemispheric far lateral transcallosal approach—experience of 134 patients. Surg Neurol. 2008;69(5):447–53; discussion 53–6
8. Jean WC, Tai AX, Hogan E, Herur-Raman A, Felbaum DR, Leonardo J, et al. An anatomical study of the foramen of Monro: implications in management of pineal tumors presenting with hydrocephalus. Acta Neurochir. 2019;161(5):975–83.
9. Ma Y, Lan Q. An anatomic study of the occipital transtentorial keyhole approach. World Neurosurg. 2013;80(1–2):183–9.
10. Meyer FB, Bruce JN. Introduction to microsurgery of the third ventricle, pineal region, and tentorial incisura. Neurosurg Focus. 2016;40 Video Suppl 1:2016.1.FocusVid.Intro.
11. Mortini P, Gagliardi F, Boari N, Roberti F, Caputy AJ. The combined interhemispheric subcommissural translaminaterminalis approach for large craniopharyngiomas. World Neurosurg. 2013;80(1–2):160–6.
12. Muscas G, Battista F, Lastrucci G, Della PA. An alternative route to the posterior half of the third ventricle: the transoccipital horn approach. Technical note. World Neurosurg. 2019;132:223–9.
13. Nanda A, Konar S, Kalakoti P, Maiti T. Posterior interhemispheric approach and microsurgical resection of a pineal parenchymal neoplasm of intermediate differentiation. Neurosurg Focus. 2016;40 Video Suppl 1:2016.1.FocusVid.15439.
14. Savu R, Mohan D, Mohan A, Moisa H, Ciurea AV. A modern minimally invasive transcallosal approach to the third ventricle. A personal experience of 30 patients. Chirurgia (Bucur). 2014;109(1):80–9.
15. Spina A, Gagliardi F, Bailo M, Boari N, Caputy AJ, Mortini P. Comparative anatomical study on operability in surgical approaches to the anterior part of the third ventricle. World Neurosurg. 2016;95:457–63.
16. Symss NP, Ramamurthi R, Kapu R, Rao SM, Vasudevan MC, Pande A, et al. Complication avoidance in transcallosal transforaminal approach to colloid cysts of the anterior third ventriclen: an analysis of 80 cases. Asian J Neurosurg. 2014;9(2):51–7.
17. Abrahams JM, Forman MS, Lavi E, Goldberg H, Flamm ES. Hemangiopericytoma of the third ventricle. Case report. J Neurosurg. 1999;90(2):359–62.
18. Akyüz C, Köseoğlu V, Bertan V, Söylemezoğlu F, Kutluk MT, Büyükpamukçu M. Primary intracranial germ cell tumors in children: a report of eight cases and review of the literature. Turk J Pediatr. 1999;41(2):161–72.
19. Behari S, Banerji D, Mishra A, Sharma S, Chhabra DK, Jain VK. Intrinsic third ventricular craniopharyngiomas: report on six cases and a review of the literature. Surg Neurol. 2003;60(3):245–52; discussion 52–3
20. Ben Nsir A, Gdoura Y, Thai QA, Zhani Kassar A, Hattab N, Jemel H. Intraventricular glioblastomas. World Neurosurg. 2016;88:126–31.
21. Bhatoe HS, Singh P, Dutta V. Intraventricular meningiomas: a clinicopathological study and review. Neurosurg Focus. 2006;20(3):E9.

22. Cai Q, Wang J, Wang L, Deng G, Chen Q, Chen Z. A classification of lesions around interventricular foramen and its clinical value. Int J Clin Exp Pathol. 2015;8(9):9950–7.
23. Carmel PW. Tumours of the third ventricle. Acta Neurochir. 1985;75(1–4):136–46.
24. Adib SD, Herlan S, Ebner FH, Hirt B, Tatagiba M, Honegger J. Interoptic, trans-lamina terminalis, opticocarotid triangle, and caroticosylvian windows from mini-supraorbital, frontomedial, and pterional perspectives: a comparative cadaver study with artificial lesions. Front Surg. 2019;6:40.
25. Cai M, Ye Z, Ling C, Zhang B, Hou B. Trans-eyebrow supraorbital keyhole approach in suprasellar and third ventricular craniopharyngioma surgery: the experience of 27 cases and a literature review. J Neuro-Oncol. 2019;141(2):363–71.
26. Maira G, Anile C, Colosimo C, Cabezas D. Craniopharyngiomas of the third ventricle: trans-lamina terminalis approach. Neurosurgery. 2000;47(4):857–63; discussion 63–5
27. Silva PS, Cerejo A, Polónia P, Pereira J, Vaz R. Trans-lamina terminalis approach for third ventricle and suprasellar tumours. Clin Neurol Neurosurg. 2013;115(9):1745–52.
28. Campero A, Ajler P, Goldschmidt E, Rica C, Martins C, Rhoton A. Three-step anterolateral approaches to the skull base. J Clin Neurosci. 2014;21(10):1803–7.
29. Lejeune JP, Toussaint P. Surgical anatomy and surgical approaches of the third ventricle. Neurochirurgie. 2000;46(3):188–202.
30. Stein BM, Fraser RA, Tenner MS. Tumours of the third ventricle in children. J Neurol Neurosurg Psychiatry. 1972;35(6):776–88.
31. Stein BM. Third ventricular tumors. Clin Neurosurg. 1980;27:315–31.
32. Shi XE, Wu B, Fan T, Zhou ZQ, Zhang YL. Craniopharyngioma: surgical experience of 309 cases in China. Clin Neurol Neurosurg. 2008;110(2):151–9.
33. Amagasa M, Ishibashi Y, Kayama T, Suzuki J. A total removal case of cavernous angioma at the lateral wall of the third ventricle with interhemispheric trans-lamina terminalis approach. No Shinkei Geka. 1984;12(4):517–22.
34. Han S, Tie X, Qin X, Wang Y, Wu A. Frontobasal interhemispheric approach for large superasellar craniopharyngiomas: do the benefits outweigh the risks? Acta Neurochir. 2014;156(1):123–30. discussion 30–1
35. Hori T, Kawamata T, Amano K, Aihara Y, Ono M, Miki N. Anterior interhemispheric approach for 100 tumors in and around the anterior third ventricle. Neurosurgery. 2010;66(3 Suppl Operative):65–74.
36. Zhang YQ, Ma ZY, Wu ZB, Luo SQ, Wang ZC. Radical resection of 202 pediatric craniopharyngiomas with special reference to the surgical approaches and hypothalamic protection. Pediatr Neurosurg. 2008;44(6):435–43.
37. Carrasco R, Pascual JM, Reina T, Nieto S, Linera J, Sola RG. Chordoid glioma of the third ventricle attached to the optic chiasm. Successful removal through a trans-lamina terminalis approach. Clin Neurol Neurosurg. 2008;110(8):828–33.
38. Kobayashi T, Tsugawa T, Hashizume C, Arita N, Hatano H, Iwami K, et al. Therapeutic approach to chordoid glioma of the third ventricle. Neurol Med Chir (Tokyo). 2013;53(4):249–55.
39. Liu JK. Modified one-piece extended transbasal approach for translamina terminalis resection of retrochiasmatic third ventricular craniopharyngioma. Neurosurg Focus. 2013;34(1 Suppl):Video 1.
40. Apuzzo ML, Chikovani OK, Gott PS, Teng EL, Zee CS, Giannotta SL, et al. Transcallosal, interfornicial approaches for lesions affecting the third ventricle: surgical considerations and consequences. Neurosurgery. 1982;10(5):547–54.
41. Ulm AJ, Russo A, Albanese E, Tanriover N, Martins C, Mericle RM, et al. Limitations of the transcallosal transchoroidal approach to the third ventricle. J Neurosurg. 2009;111(3):600–9.
42. Nagata S, Rhoton AL, Barry M. Microsurgical anatomy of the choroidal fissure. Surg Neurol. 1988;30(1):3–59.
43. Rhoton AL. The lateral and third ventricles. Neurosurgery. 2002;51(4 Suppl):S207–71.
44. Hu X, Ren YM, Yang X, Liu XD, Huang BW, Chen TY, et al. Surgical treatment of pineal region tumors: an 18 year-experience at a single institution. World Neurosurg. 2023;172:e1–e11.
45. Konovalov AN, Pitskhelauri DI. Principles of treatment of the pineal region tumors. Surg Neurol. 2003;59(4):250–68.
46. Baumgartner J, Tournade A, Plas JY. Surgical and pathological study of the approach to the superior choroid plexus of the 3 rd ventricle. One case of posterior choroid arteriovenous aneurysm (author's transl). Neurochirurgie. 1981;27(2):97–102.
47. Lozier AP, Bruce JN. Surgical approaches to posterior third ventricular tumors. Neurosurg Clin N Am. 2003;14(4):527–45.
48. Patel PG, Cohen-Gadol AA, Mercier P, Boop FA, Klimo P. The posterior transcallosal approach to the pineal region and posterior third ventricle: intervenous and paravenous variants. Oper Neurosurg (Hagerstown). 2017;13(1):77–88.
49. Danaila L, Radoi M. Surgery of tumors of the third ventricle region. Chirurgia (Bucur). 2013;108(4):456–62.
50. Moshel YA, Parker EC, Kelly PJ. Occipital transtentorial approach to the precentral cerebellar fissure and posterior incisural space. Neurosurgery. 2009;65(3):554–64; discussion 64
51. Sai Kiran NA, Vidyasagar K, Srinivasa R, Sivaraju L, Raj V, Furtado SV, et al. Occipital interhemispheric transtentorial approach for tumors of posterior third ventricular region: review of surgical results. Neurol India. 2022;70(4):1417–26.
52. Sardhara J, Kumar A, Srivastava AK, Nangarwal B, Behari S. Gravity assisted retraction less occipito transtentorial and transsplenial approach for posterior third ventricular dermoid. Neurol India. 2021;69(6):1557–9.

56 Endoscopic Resection of Colloid Cysts

Henry W. S. Schroeder

Learning Objectives

- Identify the anatomical intraventricular structures which can be visualized using an endoscopic approach to colloid cysts of the third ventricle.
- Define the ideal entry point for endoscopic resection of colloid cysts.
- Discuss the surgical steps for endoscopic resection of colloid cysts.
- Describe the stepwise use of different hemostasis techniques.

56.1 Introduction

Colloid cysts are rare lesions of endodermal origin which most frequently occur within the third ventricle. Although benign in nature, they may lead to sudden death, especially in younger patients. Therefore, careful attention is required. However, not all colloid cysts need immediate surgery. Many cysts do not grow over a long period of time and do not obstruct the CSF pathways. MR imaging surveillance is sufficient under these circumstances. There is still a debate over which surgical treatment is better for removal of colloid cysts. Most studies have shown that the microsurgical resection has a higher total resection rate, but more complications than the endoscopic approach [1]. This chapter demonstrates important factors for a successful endoscopic cyst resection.

56.2 History and Background

Michael Powell is credited with the first publication of a series of neuroendoscopic procedures for colloid cysts with the goal to fenestrate the membrane and aspirate the cyst content in 1983 [2]. In 1988, Auer et al. reported one colloid cyst within their series of 133 neuroendoscopic procedures. However, they did not succeed to fenestrate the colloid cyst membrane and had to switch to a microsurgical procedure [3]. The endoscopic series of the 1980s and early 1990s described mainly a cyst fenestration, aspiration of the cyst content, and shrinking of the cyst capsule by coagulation. This resulted in certain percentage of cyst recurrences [4, 5]. The last 30 years faced a rapid evolution of neuroendoscopic approaches to colloid cysts of the third ventricle, resulting in a procedure with comparable efficacy but reduced invasiveness compared to microsurgical procedures.

56.3 Anatomy Highlights

- The appearance of a colloid cyst in the foramen of Monro is very variable (Fig. 56.1). The cyst may dilate the foramen or compress it. Sometimes the cyst is not visible and covered by a thick choroid plexus.
- Critical structures at the foramen of Monro are the fornix, thalamostriate vein, septal vein, and choroid plexus (Fig. 56.2).

H. W. S. Schroeder (✉)
Department of Neurosurgery, University Medicine Greifswald, Greifswald, Germany
e-mail: Henry.Schroeder@uni-greifswald.de

I. N. Kanaan, V. Beneš (eds.), *Neuroanatomy Guidance to Successful Neurosurgical Interventions*, https://doi.org/10.1007/978-3-031-59838-8_56

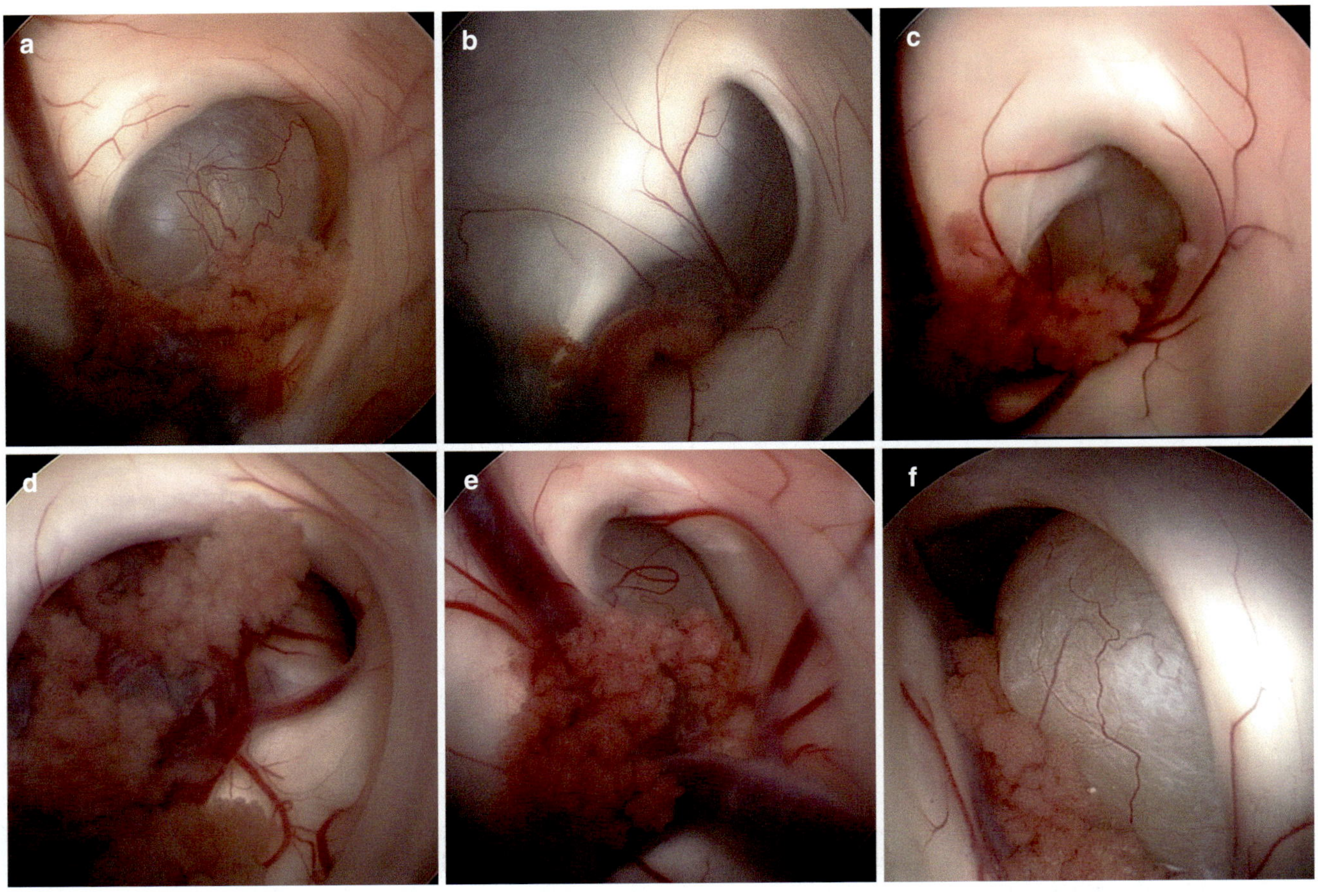

Fig. 56.1 (**a–f**) Endoscopic views demonstrating the high variety of the appearance of colloid cysts at the foramen of Monro

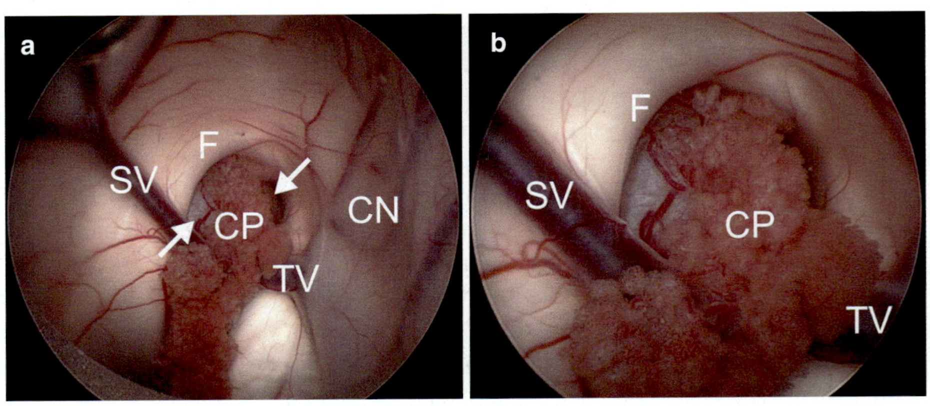

Fig. 56.2 Endoscopic image of a colloid cyst showing the relation to adjacent neurovascular structures. (**a**) The colloid cyst (arrows) is covered by choroid plexus (CP) and very closely related to the fornix (F). The septal (SV) and thalamostriata (TV) veins are at the dorsal aspect of the foramen. Laterally is the head of the caudate nucleus (CN). (**b**) Close-up of the same cyst

56.4 Approach/Surgical Intervention

56.4.1 Preoperative Considerations

Not all colloid cysts are good candidates for an endoscopic resection. Careful evaluation of the preoperative MR imaging is crucial for a successful surgery. Ventricle size, the appearance of the cyst at the foramen of Monro, the size of the foramen, and most importantly, the colloid cyst content have to be taken into consideration. If the foramen of Monro is small and the cyst large with a black signal in the T2-weighted images, an endoscope-assisted microsurgical approach might be preferable. In the presented case example, MR imaging revealed a biventricular dilation caused by a colloid cyst of the third ventricle (Fig. 56.3). The colloid cyst is equally apparent in both foramina of Monro. Therefore, a right frontal approach is chosen. The cyst content is isointense in the T2-weighted MR images, indicating an intermediate consistency. We always aim for a gross total resection of the cyst. If we would fail purely endoscopically, we would switch to an endoscope-assisted microsurgical procedure.

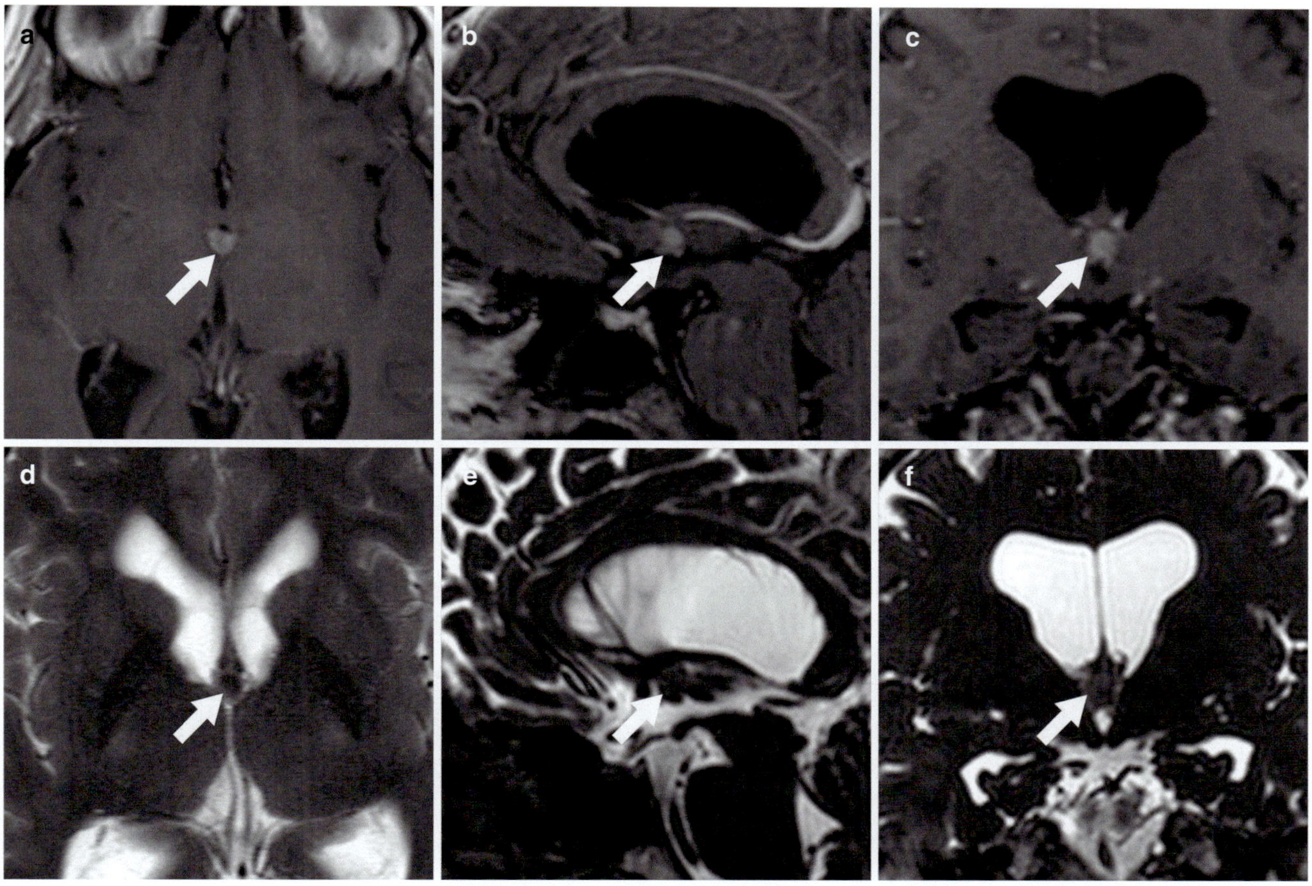

Fig. 56.3 The 58-year-old female presented with headache and nausea. (**a–c**) T1-weighted MR imaging shows a small hyperintense colloid cyst (arrow) within the third ventricle causing bilateral dilation of the lateral ventricles. (**d–f**) In the T2-weighted images, the cyst (arrow) is isointense indicating a medium consistency

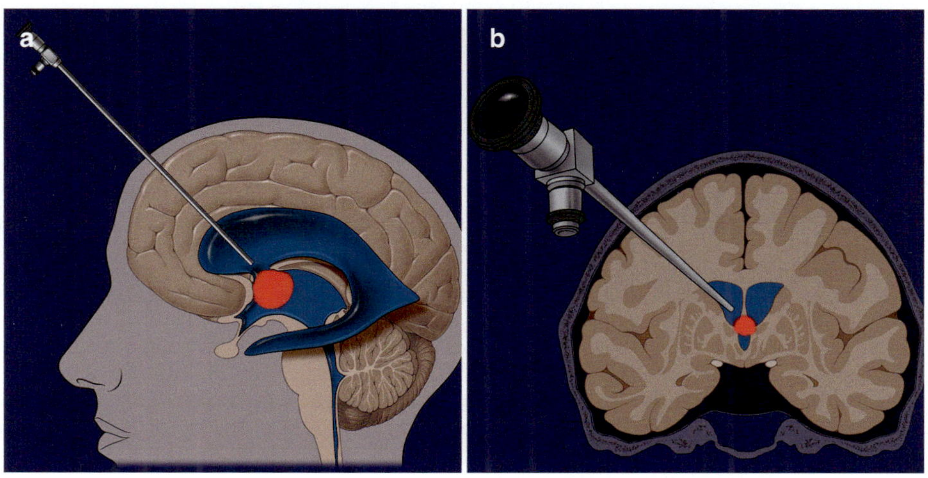

Fig. 56.4 (**a** and **b**) Schematic drawing showing the anterolateral approach trajectory to a colloid cyst of the third ventricle (copyright Henry W. S. Schroeder)

56.4.2 Surgical Procedure

We use a navigated, single portal, transcortical, transventricular, and transforaminal approach with an anterolateral approach trajectory (Fig. 56.4). The patient is positioned supine with the head in a sharp three-point head fixation after general anesthesia was induced. The neck is in slight anteflexion to have the entry point at the highest level to prevent excessive CSF loss (Fig. 56.5). Cefuroxime is administered intravenously as a single shot antibiotic prophylaxis. Neuronavigation (BrainLab, Munich, Germany) is used for planning of the ideal approach trajectory which slides just over the head of the caudate nucleus (Fig. 56.6) [6]. The burr hole should be placed far anteriorly to the coronal suture, but

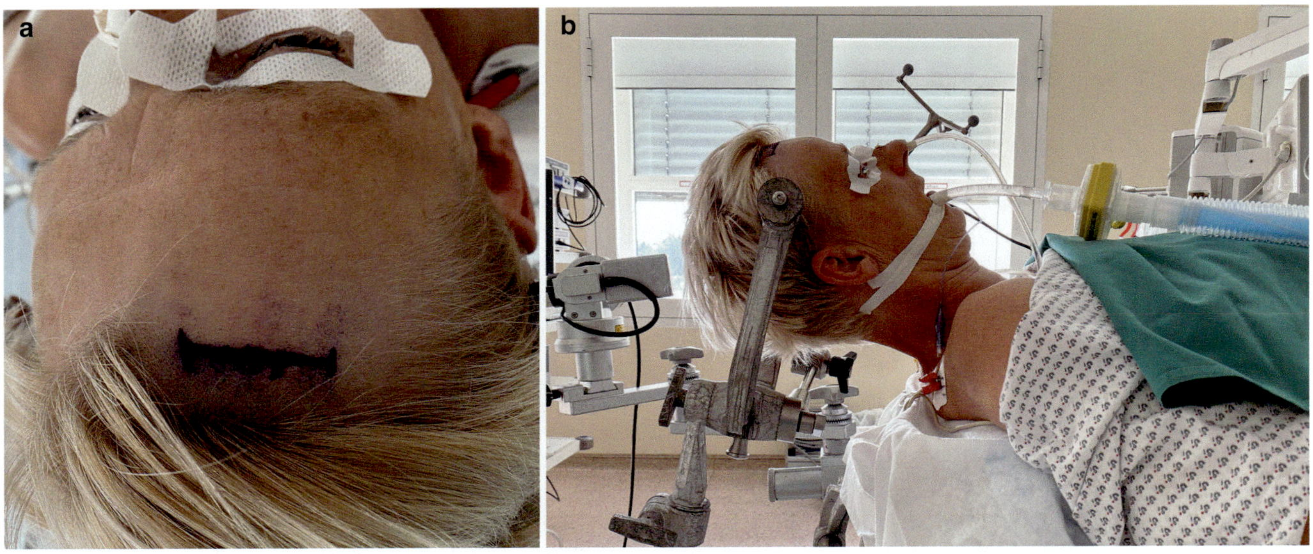

Fig. 56.5 (**a**) For cosmetic reasons skin incision just behind the hairline. (**b**) The patient is positioned supine with sharp head fixation in a Mayfield clamp. The head is in neutral position and slightly anteflexed. The burr hole is at the highest point to avoid excessive loss of CSF

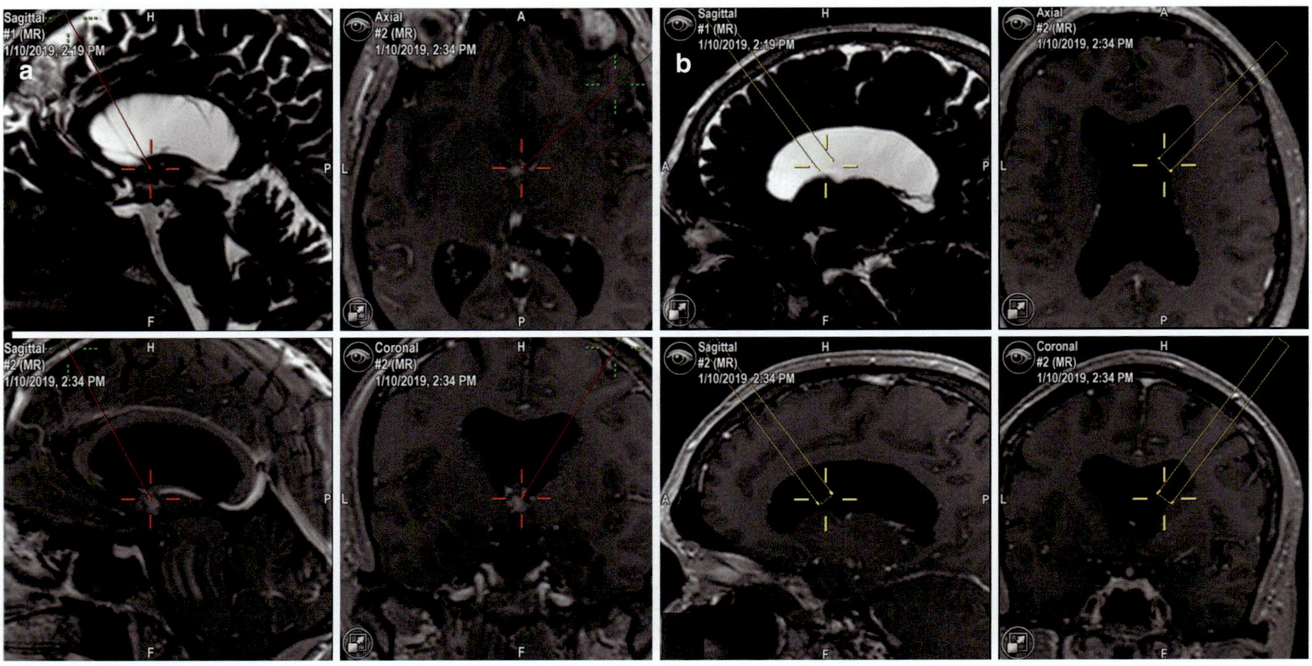

Fig. 56.6 Screenshots of the approach planning and insertion of the endoscope into the lateral ventricle. (**a**) Planning of a right anterolateral approach. The entry point is behind the hairline far lateral. The trajectory goes just over the head of the caudate nucleus. The approach is from the right side because both the ventricles are dilated. (**b**) Insertion of the endoscopic sheath under navigational guidance into the lateral ventricle. The endoscopic sheath is used to retract the head of the caudate nucleus a little bit laterally to give the optimal approach to the cyst

for cosmetic reasons the skin incision is usually made behind the hairline. In bald heads, we place the incision in a forehead wrinkle. After disinfection and draping, a 3-cm skin incision is made and a 1-cm burr hole is placed. The dura is incised in a cruciate fashion, followed by coagulation of the cortex. Then, the endoscopic sheath of the LOTTA ventriculoscope [7] with its trocar inside is introduced under navigational guidance into the frontal horn of the lateral ventricle. In order to get the optimal trajectory to the cyst, the head of the caudate nucleus is gently pushed laterally with the endoscopic sheath once the ventricle is reached.

After removal of the trocar, the ventricle is inspected with the ventriculoscope. The endoscopic image shows the ideal trajectory just sliding over the head of the caudate nucleus (Fig. 56.7a). The colloid cyst obstructing the foramen on Monro is visualized (Fig. 56.7b). The choroid plexus

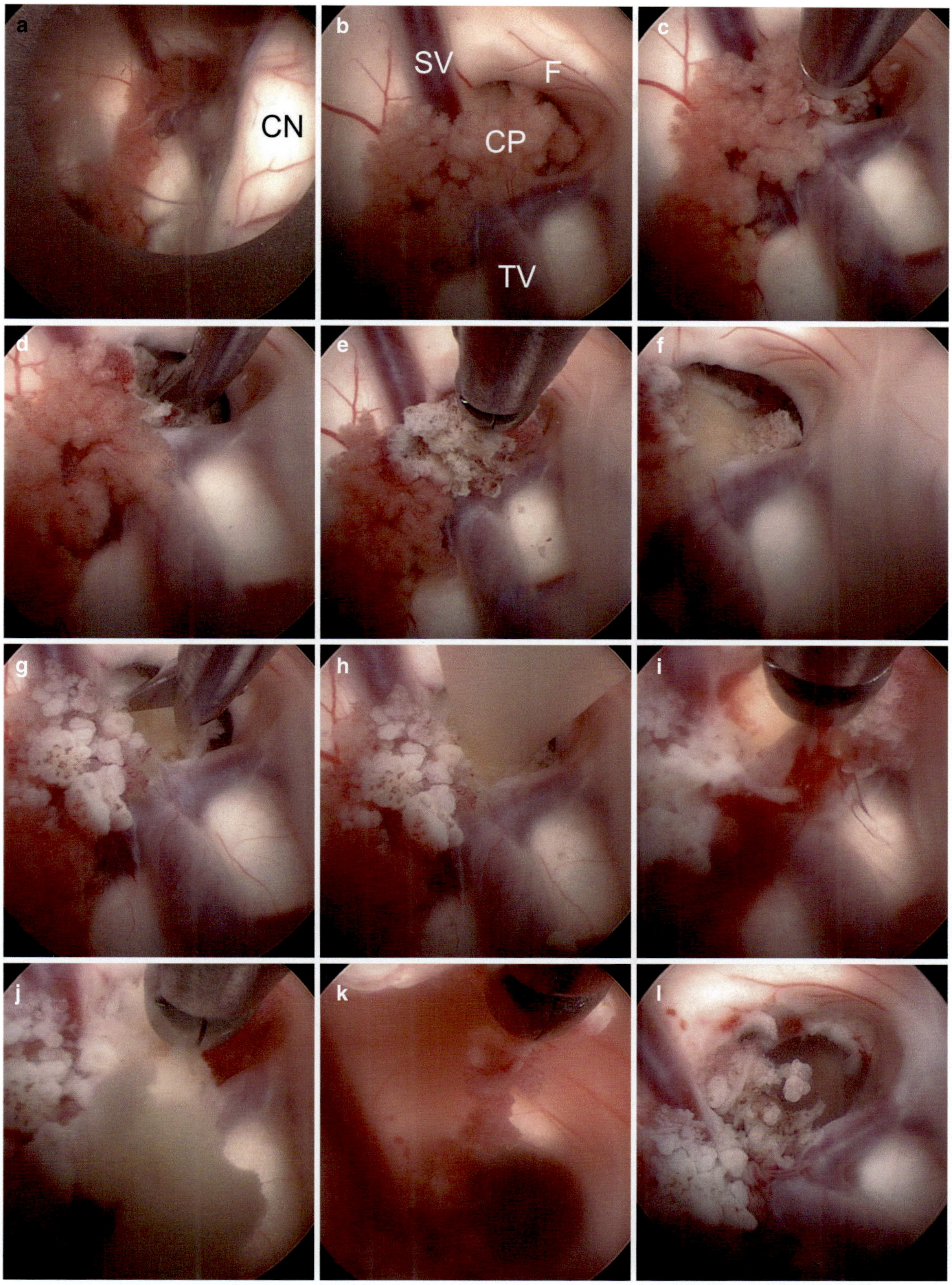

Fig. 56.7 Endoscopic images showing the resection of the colloid cyst. (**a**) View after entering the right lateral ventricle shows that the approach trajectory is just sliding over the head of the caudate nucleus (CN). (**b**) Colloid cyst covered by choroid plexus (CP) occluding the foramen of Monro with fornix (F), septal vein (SV), and thalamostriate vein (TV). (**c**) Coagulation of the choroid plexus with a bipolar diathermy probe. (**d**) Cutting of the coagulated choroid plexus. (**e**) Removal of the choroid plexus with grasping forceps. (**f**) Exposed colloid cyst within the third ventricle. (**g**) Incision of the cyst capsule with scissors. (**h**) Aspiration of the cyst content with a suction tube. (**i**) Mobilization of the cyst into the lateral ventricle. (**j**) Removal of the cyst and yellowish content with a grasping forceps. (**k**) Resection of choroid plexus to increase the passage through the small foramen of Monro. (**l**) Free foramen of Monro after cyst removal with intact fornix and veins

is covering the cyst. Fornix, septal vein, and thalamostriate veins are identified. The choroid plexus covering the cyst is coagulated (Fig. 56.7c). The choroid plexus is cut and removed (Fig. 56.7d and e). Thereafter, the cyst is clearly apparent in the foramen (Fig. 56.7f). The cyst capsule is incised with scissors (Fig. 56.7g) and the colloid content of the cyst is aspirated with a pediatric 2.7-mm endotracheal suction tube and a syringe (Fig. 56.7h). Then the cyst is mobilized into the lateral ventricle (Fig. 56.7i). The cyst wall and residual content are removed with a grasping forceps (Fig. 56.7j). Finally, parts of the choroid plexus are excised because the choroid plexus itself is obstructing the CSF flow in the very small foramen of Monro (Fig. 56.7k). The bimanual dissection technique was not used in this case because the cyst wall was very fragile and ruptured easily. Therefore, it was removed in a piece-meal fashion. Care has to be taken to grasp only the cyst wall and not the tela choroidea and choroid plexus. After cyst resection, a vigorous irrigation of the ventricles is done to remove all spilled cyst content. The third ventricle is inspected to make sure the cyst is totally removed and the aqueduct is not blocked by cyst debris. The final inspection shows the patent foramen of Monro with intact fornix and veins (Fig. 56.7l). The ventriculoscope is removed under endoscopic view from the lateral ventricle and cortical puncture channel to rule out active bleeding. No external ventricular drainage is inserted. A gelatin sponge is placed in the burr hole and the burr hole covered with a titanium plate. Galea and skin are sutured. The patient is observed overnight in the intermediate care unit. MR imaging is performed next morning. The postoperative course was uneventful. The symptoms resolved immediately after surgery. The postoperative MR imaging obtained 4 years after surgery revealed no cyst recurrence (Fig. 56.8).

When the cyst wall is tough, a bimanual dissection technique can be used to deal with the vascular pedicle of the cyst. With a flexible grasping forceps which is inserted via the left side-channel of the ventriculoscope, the cyst capsule is held up by gentle traction, allowing visualization of the cyst pedicle. With a bipolar diathermy probe which is introduced via the main working channel, the vessels are coagulated and then cut with microscissors [8].

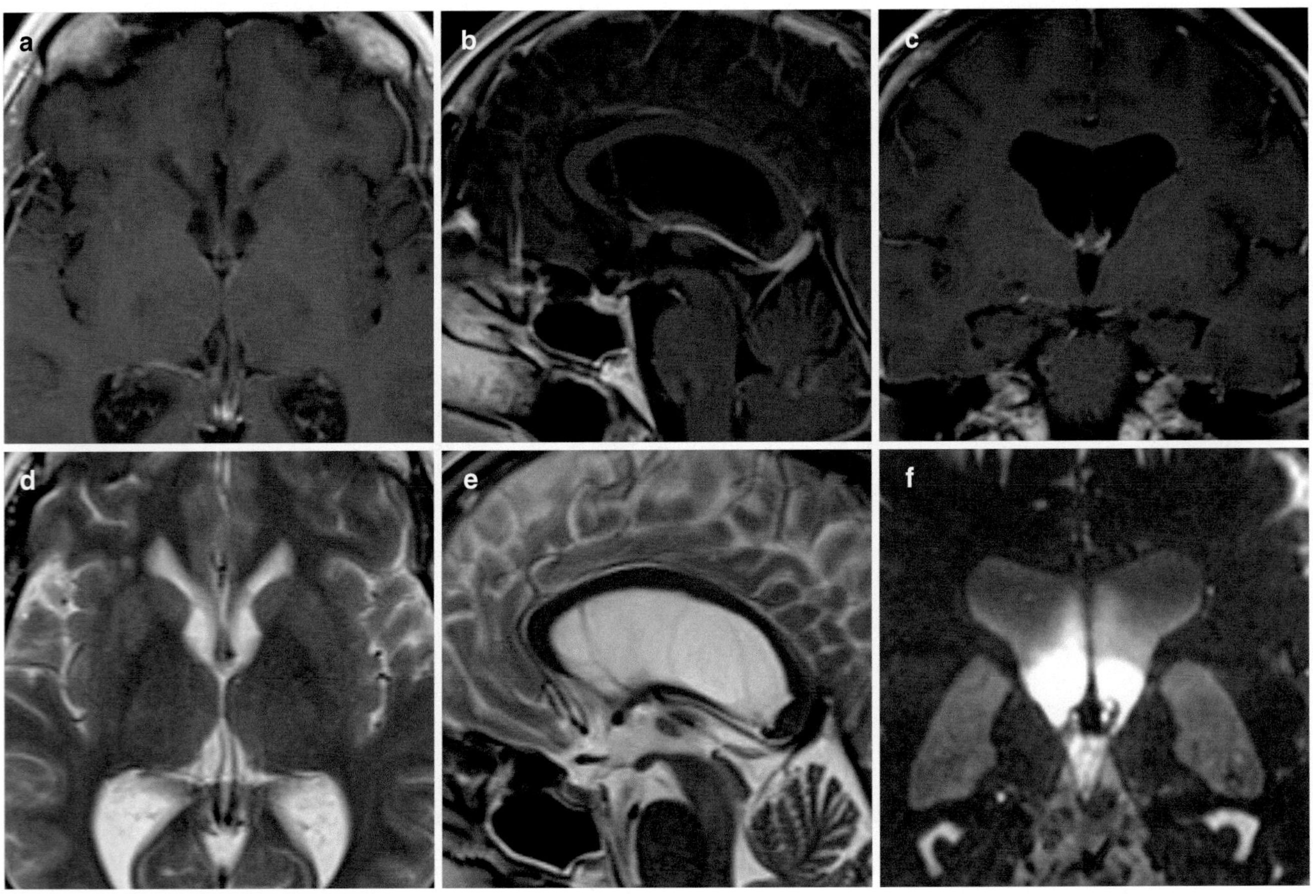

Fig. 56.8 MR imaging obtained 4 years after surgery shows no cyst recurrence. The ventricles decreased in size. The patient has no complaints

56.4.3 Hemostasis Techniques

In case of minor hemorrhage, the small chamber irrigation technique is used to control the hemorrhage [9]. The endoscope is withdrawn 3–4 mm into the endoscopic sheath to form a small chamber at the tip of the endoscopic sheath which can easily be cleared from blood by irrigation. With this good visualization, the bleeding vessel is easily identified and coagulated using the bipolar rod.

In severe hemorrhage with complete red-out, the small chamber irrigation technique will fail. In this scenario, the dry-field technique is used [10]. The CSF is aspirated with the aid of a pediatric endotracheal suction tube and a syringe. Usually, the bleeding stops spontaneously without the need for surgical hemostasis when the hemorrhage area is free of CSF. A ventricular collapse is prevented by filling the ventricle with ringer solution after hemostasis has been achieved and the dissection is continued.

56.5 Surgical Pearls

- Choose the ideal anterolateral entry point and approach trajectory using neuronavigation.
- Use an endoscope with a large working channel allowing aspiration of the colloid content of the cyst.
- Use a ventriculoscope with at least 2 working channels enabling bimanual dissection.
- Use a bimanual sharp dissection technique of the cyst pedicle.
- In case of moderate hemorrhage, the small chamber irrigation technique is used for hemostasis.
- In severe hemorrhage, the dryfield technique should be used.
- Irrigate abundantly to remove all spilled cyst content.
- Make sure the cyst is completely removed.
- Fornix and thalamostriate vein must be preserved.
- Check the posterior aspect of the third ventricle to exclude a blocked aqueduct by debris or blood clot.

56.6 Discussion

Although meta-analyses show that the microsurgical approach has a higher gross total resection rate, less recurrences, and fewer reoperations [11], endoscopic procedures have less complications and are therefore preferred by many surgeons [12–14]. In a more recent review which only included studies published after 2000, the outcome differences between microsurgical and endoscopic surgeries were less obvious [1].

Profound knowledge of the intraventricular anatomy is required for a successful endoscopic colloid cyst removal. Details of the anatomy have to be obtained from the preoperative MR imaging in order to select appropriate candidates for an endoscopic approach. Size and location of the cyst, size of the lateral ventricles and foramen of Monro, and the cyst content have to be considered. Once a patient is selected for an endoscopic approach, accurate approach planning is of utmost importance. The approach is usually via a far anterolateral entry point, sliding just over the head of the caudate nucleus.

A significant limitation of the endoscopic surgical procedure was the initially performed one-hand technique. With a one-handed technique, the surgeon can only pull, coagulate, or cut. When the cyst is released, it is falling back to the foramen, obstructing the vision to the cyst pedicle. In the bimanual resection technique, the cyst wall is held under tension with the aid of a grasping forceps which is introduced via a side-channel of the endoscope. Under good vision to the pedicle, the vessels of the pedicle are coagulated and finally cut with scissors. Sharp dissection is crucial to avoid excessive venous bleeding. If the foramen of Monro is large, the endoscopic sheath can be used to retract the fornix a little bit cranially to get better access to the tela choroidea where the cyst is attached to. If necessary, the foramen of Monro may be enlarged by opening of the choroidal fissure [15].

Since recurrences are common when cyst capsule remains behind after colloid cyst resection [16], we usually aim for a gross total resection of the cyst. If via the endoscopic approach a gross total resection cannot be achieved, we switch to a microsurgical tube-assisted technique where the tube is inserted along the endoscope track.

56.7 Conclusion

The majority of colloid cysts can be totally removed using an endoscopic bimanual sharp dissection technique. However, patient selection and individual approach planning are crucial for a successful surgery.

References

1. Sethi A, Cavalcante D, Ormond DR. Endoscopic versus microscopic transcallosal excision of colloid cysts: a systematic review in the era of complete endoscopic excision. World Neurosurg. 2019;132:e53–e8.
2. Powell MP, Torrens MJ, Thomson JL, Horgan JG. Isodense colloid cysts of the third ventricle: a diagnostic and therapeutic problem resolved by ventriculoscopy. Neurosurgery. 1983;13(3):234–7.
3. Auer LM, Holzer P, Ascher PW, Heppner F. Endoscopic neurosurgery. Acta Neurochir. 1988;90(1–2):1–14.
4. Abdou MS, Cohen AR. Endoscopic treatment of colloid cysts of the third ventricle. Technical note and review of the literature. J Neurosurg. 1998;89(6):1062–8.

5. Lewis AI, Crone KR, Taha J, van Loveren HR, Yeh HS, Tew JM Jr. Surgical resection of third ventricle colloid cysts. Preliminary results comparing transcallosal microsurgery with endoscopy. J Neurosurg. 1994;81(2):174–8.
6. Schroeder HW, Wagner W, Tschiltschke W, Gaab MR. Frameless neuronavigation in intracranial endoscopic neurosurgery. J Neurosurg. 2001;94(1):72–9.
7. Schroeder HW. A new multipurpose ventriculoscope. Neurosurgery. 2008;62(2):489–91; discussion 91–2
8. Marx S, Schroeder HWS. Endoscopic bimanual sharp dissection technique for gross-total resection of colloid cysts: technical note. J Neurosurg. 2020;135(1):318–26.
9. Manwaring JC, El Damaty A, Baldauf J, Schroeder HW. The small-chamber irrigation technique (SCIT): a simple maneuver for managing intraoperative hemorrhage during endoscopic intraventricular surgery. Neurosurgery. 2014;10(Suppl 3):375–9; discussion 9
10. Oertel J, Linsler S, Csokonay A, Schroeder HWS, Senger S. Management of severe intraoperative hemorrhage during intraventricular neuroendoscopic procedures: the dry field technique. J Neurosurg. 2018;131(3):931–5.
11. Sheikh AB, Mendelson ZS, Liu JK. Endoscopic versus microsurgical resection of colloid cysts: a systematic review and meta-analysis of 1,278 patients. World Neurosurg. 2014;82(6):1187–97.
12. Boogaarts HD, Decq P, Grotenhuis JA, Le Guerinel C, Nseir R, Jarraya B, et al. Long-term results of the neuroendoscopic management of colloid cysts of the third ventricle: a series of 90 cases. Neurosurgery. 2011;68(1):179–87.
13. Gaab MR, Schroeder HW. Neuroendoscopic approach to intraventricular lesions. J Neurosurg. 1998;88(3):496–505.
14. Schroeder HW, Gaab MR. Endoscopic resection of colloid cysts. Neurosurgery. 2002;51(6):1441–4; discussion 4–5
15. Tawk RG, Akinduro OO, Grewal SS, Brasiliense L, Grand W, Grotenhuis A. Endoscopic transforaminal transchoroidal approach to the third ventricle for cystic and solid tumors. World Neurosurg. 2020;134:e453–e9.
16. Vorbau C, Baldauf J, Oertel J, Gaab MR, Schroeder HWS. Long-term results after endoscopic resection of colloid cysts. World Neurosurg. 2019;122:e176–e85.

Endoscopic Transchoroidal Fissure Approach to the Posterior Third Ventricle

57

Roman Bošnjak

Learning Objectives

- Endoscopic anatomy of the lateral and third ventricle and suprapineal region.
- Indications for the transchoroidal fissure approach.
- Surgical technique of purely endoscopic transtubular approach to the lat. Ventricle.
- Limitations of the transchoroidal fissure approach.
- Tips and tricks.

57.1 Introduction

Transforaminal approach allows access to anterior half of the third ventricle. Dissection of the choroidal fissure enlarges foramen of Monro posteriorly and enables exposure of the posterior half of the third ventricle behind the adhesio interthalamica, and with further more posterior dissection, also directs view into the aqueduct, suprapineal, and tectal areas [1–3].

Choroidal fissure is the thinnest part at the medial wall of the lateral ventricle between the fornix and thalamus, overlaid by the choroidal plexus.

From superior transventricular perspective, separation of the choroidal plexus away from the fornix opens the fissure and exposes the velum interpositum first. Here, medial or lateral to the ipsilateral great intracerebral vein, the third ventricle can be entered through its roof. Alternatively, the choroid plexus can be turned medially and splitting is performerd from underaspect of choroid plexus and lateral to internal cerebral vein (subchoroidal), again away from fornix to minimize surgical injury [4]. Interforniceal (strictly midline splitting of the fornices) puts in risk both fornices from midline cut itself, traction, and manipulative trauma and has no advantage in comparison to less invasive unilateral transchoroidal or subchoroidal approach. It can be testified only in the presence of cavum septi pellucidi and cystic lesion in the third ventricle.

Endoscopic technique improves visualization and enables close and angled views of deep neurovascular structures and hidden remnants.

Some endoscopic techniques provide relatively unrestricted bimanual dissection under superior close endoscopic visualization and illumination and thus possibility of microsurgical mass removal (e.g., through tubular retractor, endonasally) [5, 6]. In full endoscopic techniques, a special probe is applied through the working channel (ultrasonic, probe for continuous aspiration, and diode laser).

This chapter only explains the transtubular transcortical transventricular transchoroidal approach to the posterior third ventricle and suprapineal-tectal region in more detail.

57.2 History/Background

Intraventricular lesions present a challenge of modern neurosurgery. Third ventricle lesions can be reached from above (transcortical or interhemispheric transcallosal), from the front (translamina terminalis), or from below and behind (suprapineal transvelum interpositum) [7]. Transcallosal interhemispheric and transcortical approaches are the mainstream approaches to the lateral and third ventricle, but may be related to significant morbidity (injury of Papez circuit, injury of thalamostriate and internal cerebral veins, fornix injury, haemathocephalus and clogging of aqueduct, direct thalamic injury, hemiparesis, and ischemic injury from retraction blade) [8]. Main microsurgical and endoscopic approaches to the third ventricle are listed in Table 57.1.

The introduction of endoscopic techniques into cranial neurosurgery significantly reduced local trauma of brain parenchyma along the trajectory of endoscope insertion. Endoscopic transcortical transventricular approach is most often employed in endoscopic third ventriculostomy (ETV)

R. Bošnjak (✉)
University Medical Center Ljubljana, Ljubljana, Slovenia

University of Ljubljana, Medical faculty, Ljubljana, Slovenia

I. N. Kanaan, V. Beneš (eds.), *Neuroanatomy Guidance to Successful Neurosurgical Interventions*,
https://doi.org/10.1007/978-3-031-59838-8_57

Table 57.1 Main microsurgical and endoscopic approaches to the third ventricle

Microsurgical
Transcallosal interhemispheric
Anterior transcallosal-transchoroidal approach [4]
Transcallosal-subchoroidal approach [4]
Transcallosal-transforniceal-transchoroidal
Contralateral transcallosal approach
Posterior transcallosal
Transcortical transventricular
Anterior transchoroidal/subchoroidal
Posterior transchoroidal/subchoroidal
Trans-lamina terminalis (pterional, supraorbital [14]
Supracerebellar-infratentorial (SCIT)
(Sub)occipital transtentorial (OTA)
Purely endoscopic
Transtubular transcortical transventricular transforaminal
Transtubular transcortical transventricular transchoroidal/subchoroidal [2]
Purely endoscopic SCIT [15]
Supraorbital trans-lamina terminalis
Full endoscopic
Frontal transforaminal transchoroidal [3]
Combined (rigid shaft and steerable-flexible endoscope) [1]

[9, 10]. Although full endoscopic technique may be the most atraumatic for the brain, it offers very limited or no maneuverability of probes within single or dual working channels. Biopsy with endoscopic (rhino-mouth like) biopsy forceps is possible in nonvascular lesions. New technologies enable tumor debulking or gross total removal in full endoscopy using specialized probes for tissue destruction (ultrasonic probe, probe for automated continuous aspiration of tumor tissue, and diode laser probe) [11].

Endoscopic biopsy of posterior third ventricle and tectal area lesions can be performed during ETV using the same trajectory, combining two different trajectories [12] or using a flexible endoscope (alone or pushed through the endoscopic shaft) [1].

There are a very few reports on full endoscopic choroidal fissure splitting using endoscopic bipolar forceps and dissectors and endoscopic ultrasonic aspirator—through the forehead burr hole only [3]. However, this frontal transfroaminal approach includes only a limited splitting of choroidal fissure anteriorly at the venous angle.

Tubular retractors, introduced first in microsurgery by Kelly more than 30 years ago, enable in combination with endoscopic visualization and illumination sufficient surgical exposure and permit 2-hand surgery and use of microsurgical technique [6]. Tubular retractors disperse retraction forces evenly on the surrounding tissues with retraction pressure of <10 mmHg (below ischemia treshold of ≥ 20 mmHg).

By tilting the tube and using transhenoidal instruments, further increase of exposed surgical field area and operability is achieved. Transforaminal approach exposes only anterior part of the third ventricle. Transchoroidal fissure splitting through a tubular retractor expands the foramen of Monro posteriorly from the ETV trajectory over and behind adhesio interthalamica to the posterior third ventricle roof and floor, suprapineal and the tectal area [2]. However, angled view may give us a look into a blind spots, but this does not increase operability as they can't be reached safely. This is especially true in trigonal approach where we can see far into occipital and temporal horns, but can't do nothing else [13]. Steerable flexible endoscope technology in combination with rigid shaft may solve this »angled & far distant approach« in the future. Oi et al. [1] has combined a flexible endoscope through rigid shaft for transforaminal biopsy of posterior third ventricle lesion, entering from Kocher point.

57.3 Anatomy Highlights

- Foramen of Monro can be enlarged by posteriorly splitting the choroidal fissure.
- Choroidal fissure is a slit between fornix medially and thalamus laterally.
- Choroid plexus is attached gently to the choroid fissure and covers it from direct vision.
- Junction of the thalamostriate, anterior septal, and internal cerebral vein at posterior rim of foramen of Monro (venous angle) may have variations.
- Velum interpositum is a space between two ultrathin horizontal membranes between thalami (sup. and inf. choroidal tenias attached to striae medullaris sup. and inf.) and contains intracerebral veins.
- Fornix is responsible for a short-term spatial and working memory.

In extremely simplified shema, the third ventricle is located in the diencephalon (as a cavity between two thalami), the floor is mesencephalon and the roof are fornices. Adhesion interthalamica divides the third ventricle into anterior and posterior parts. The wall of the third ventricle is surrounded by eloquent neurovascular structures, such as the thalamus, hypothalamus, optic chiasm, mamillary bodies, midbrain, fornix, pineal gland, and internal cerebral veins. The tegmentum (the majority of the floor of the third ventricle) houses motor nucleus of oculomotor nerve,

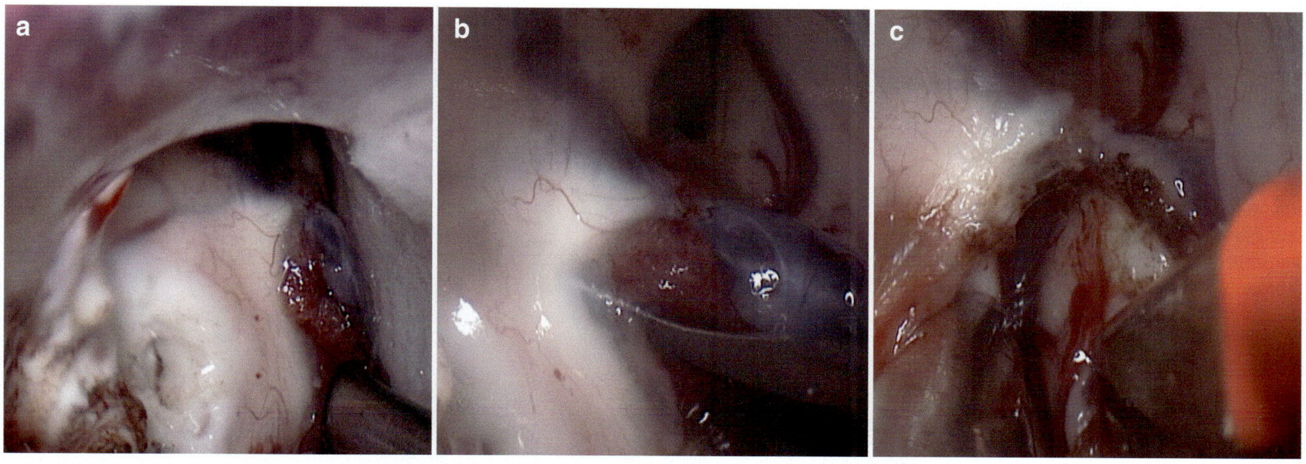

Fig. 57.1 Endoscopic transtubular view of the venous angle behind the right foramen of Monro, lateralization of choroidal plexus, and the entry point for choroidal fissure splitting. In transchoroidal approach, the choroidal plexus is pushed lateraly to expose the choroidal fissure (**a**). The tenia choroidea superior between the fornix and plexus might need a sharp dissection with sickle knife or scissors to enter a velum interpositum space (**b**). In its deeper aspect, an ipsilateral internal cerebral vein will be easily found and followed backward during splitting of the fissure (**c**). Note oblique plane of the fissure from laterosuperior to inferomedial direction. Avoid vertical dissection which will mislead you directly into the thalamus, go underneath the fornix

accessory motor nucleus of oculomotor nerve (Edinger-Westphal), nucleus of trochlear nerve, and mesencephalic nucleus of trigeminal nerve; red nucleus; substantia nigra; and peraqueductal gray matter, ascending and descending tracts.

From the superior ventricular view, the choroidal fissure is a slit between the fornix and thalamus. It is also a gentle attachment of the choroid plexus, which covers and hides fissure from direct view (Figs. 57.1a and 57.2b, c).

Splitting the fissure is done by detaching the choroid plexus away from the fornix from medial aspect of plexus (transchoroidal) or from lateral aspect (subchoroidal). By doing so, the tenia choroidea superior is opened along superior medullary line (stria). Parallel and inferior to stria medullaris superior is a stria medullaris inferior to which tenia choroidea inferior is attached. These two choroidal tenias comprise a space between two ultrathin membranes—a velum interpositum space, which contains intracerebral veins and branches of medial posterior choroidal arteries. When there is a dilatation and bulging of this space, this is called cavum veli interpositi.

Türe et al. [16] reported 4 variations at the junction of the thalamostriate vein (TSV), anterior septal vein (ASV), and internal cerebral veins (ICVs). The true "venous angle" was situated in the posterior margin of the foramen of Monro (Figs. 57.1 and 57.2) and a "false angle" was located 3–7 mm posterior to the foramen of Monro.

The fornix connects the hippocampus and other limbic structures and is responsible for spatial and working memory [17]. At the level of the temporal horn, choroidal fissure is situated between the stria terminalis superomedially and the fimbria of the hippocampus inferolaterally. The choroidal point is the most inferior point of the fissure where a choroidea anterior enters the temporal horn and supplies choroid plexus. The

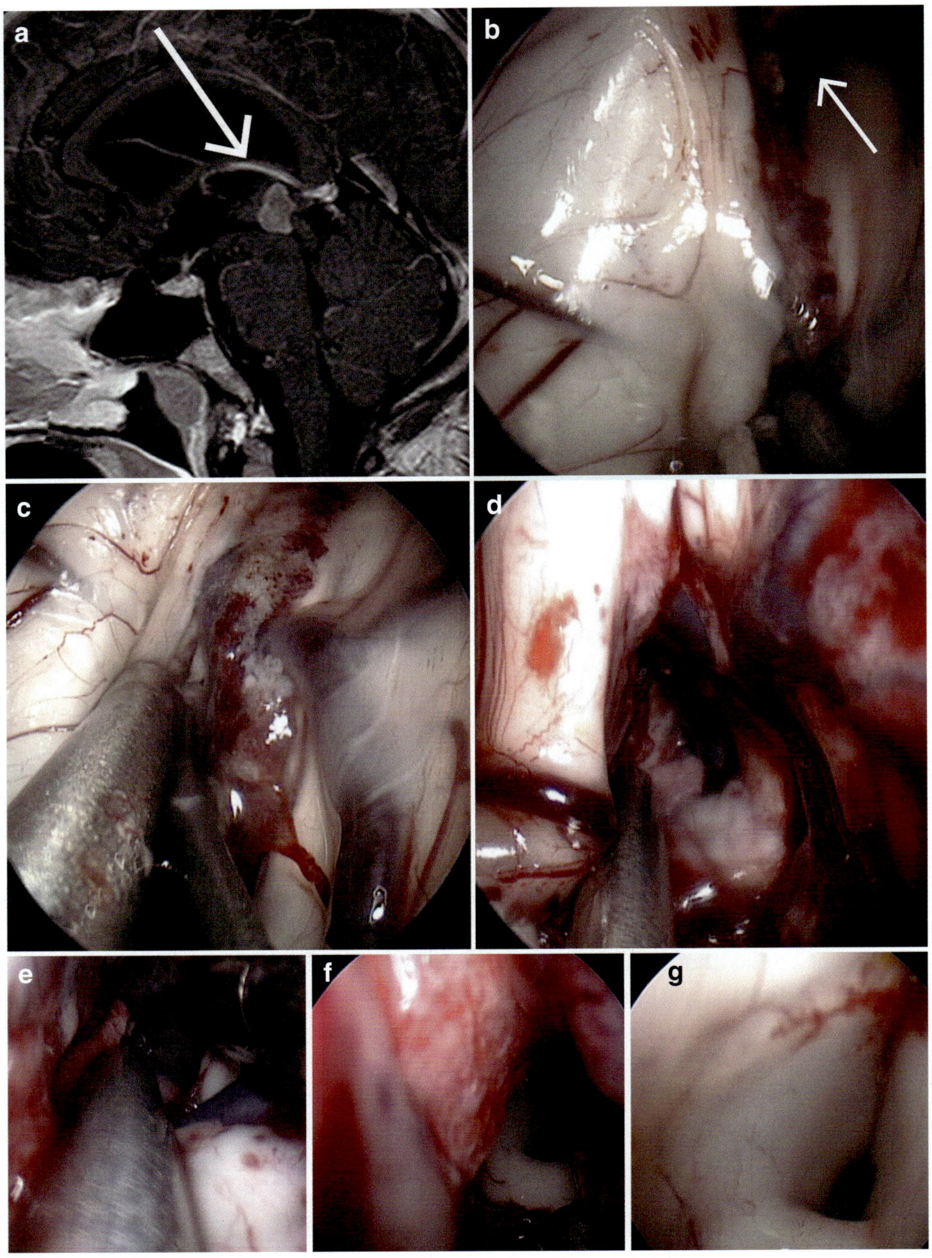

Fig. 57.2 Purely endoscopic transtubular transchoroidal removal of papillary tumor of pineal region (PTPR, WHO 2–3) from the posterior third ventricle. This approach applies microsurgical bimanual technique to split the fissure and remove the tumor. It releases CSF circulation and enables a direct view into the aqueduct (**e–g**). Note that this is a dry field technique. The arrow (**a**) shows a trajectory from right coronary suture 3 cm lateral from midline toward aqueduct. Note very small foramen of Monro and unremarkable thalamostriatal vein in this case (arrow, **b**). Do not start splitting the choroidal fissure in the middle, but at foramen of Monro, just behind the septal vein (may be sacrified). Preserve the fornix untouched (here under aspirator's tip) (**c**). Study the vein pattern of choroidal veins in the choroid plexus, which may be sometimes very large (as large as internal cerebral vein), but appear immediately under the plexus (**c**, **d**). Internal cerebral vein(s) are located much *deeper* (**d**, **e**)

whole course of the choroidal fissure incl. atrial and temporal parts is described in detail in the study of Nagata et al. [18].

57.4 Approach/Surgical Intervention

This section reports on surgical technique for purely endoscopic transtubular-transcortical-transventricular-transchoroidal approach to the third ventricle.

Transchoroidal fissure splitting through a tubular retractor, inserted transcortically at post-Kocher point, expands a single trajectory for ETV to move posteriorly along the fornix from foramen Monro to the posterior third ventricle roof and floor and the tectal area by a simple tilting of the tube. Transtubular endoscope-controlled approach enables microsurgical bimanual manipulation with anatomical structures (e.g., dissection of fissure itself) and tumor removal or debulking. Before applying this approach, consider ventricular size and the size of foramen Monro (not advisable to use this approach in very small ventricular size). Dedicated bayonette shaft instruments and key hole aspirators are preconditions for a successful work.

The optimal entrance point for ETV is best established with the neuronavigational device. Kanner et al. [9] proposed that the optimal trephination point was 30 mm lateral to the midline and 10 mm anterior to the coronal suture. In contrast, Chen and Nakaji [10]. found that a highly variable optimal point for ETV entry is 30.1 ± 7 mm perpendicular to the midline and 8.9 ± 14.1 mm behind the coronal suture. As we want to work over the central part of the choroidal fissure (behind adhesio interthalamica), a "3 cm lateral, 1 cm behind" coronal suture is a shema for empirical use. The right side is selected as default.

As the tube diameter is 13–15 mm, a simple burr hole is not enough. Small circular/oval trephination of at least 25 mm is adviced. This size enables unobstructed tube tilting.

There are only a few commercial brain tubes. We explained our improvisation technique with expandable PVC sheet elsewhere [19]. A transparent PVC sheet of 7 × 5 cm can be cut out from a sterilization bag. It is wrapped around a navigation probe 3 times or around a navigated endoscope, and after insertion and probe withdraw, it is expanded with Fogarty baloon catheter to diameter of 11–13 mm; (see video in [20]). In case no navigation is available, insert external ventricular catheter first and use it as a guide for your tube. You may also use peel-away sheat and magnetic navigation probe to cannulate the ventricle. Avoid injury to cortical (bridging) veins at the tube insertion site. Transcortical insertion of the port is more safe than transulcal.

Upon insertion of the tube, the CSF pours out and the ventricles collapse. The microsurgical work through the tube and under endoscopic visualization continues in a dry field.

The frontal horn and body of lateral ventricle are carefully inspected from the entry point in the roof of the ventricle. The bird view reveals main anatomical points: foramen of Monro, thalamostriate vein, anterior septal vein, venous angle, body of fornix, and choroidal plexus with superior choroidal vein (Fig. 57.1).

The splitting starts with posterior enlargement of foramen Monro. Abundant choroid plexus can be coagulated superficially with shaft bayonetted bipolar forceps to shrink. An anterior septal vein may be necessary to coagulate and separate from main trunk. This is the critical point to start splitting (just behind ASV and medial to internal cerebral-thalamostriate venous loop). A detachment of plexus away from fornix is done in direction from foramen Monro to posterior direction. Bear in mind that in small lesions around the aqueduct, the third ventricle cavity may collapse and both thalami come closer to midline. An ipsilateral thalamus thus slides more under the fornix and splitting in the central part of the fissure should not be directed into vertically deep as the thalamus will be entered (Fig. 57.2c). It is critical to start just behind the ASV where the cavity is preserved and easiest to enter and ICV can be easiest to find by following the medial aspect of anterior septal vein toward internal cerebral vein (Fig. 57.1). CSF from unobstucted anterior third ventricle will pour out at the moment ICV is found and lateralized. The part II of splitting goes then along internal cerebral vein backward. The third ventricle is entered from the roof between both ICVs. Some cottonoid strips may be necessary to insert between ICVs to keep the working space opened. Avoid standard size of cottonoid (too big). Cut it to appropriate narrow strip. Fogarty catheter may be also useful as a spacer. Otherwise, refill with water. Cottonoid may stick to the fornix and later tear the fornix when cottonoid is carefully removed at the end of surgery.

The most posterior part of the third ventricle gives a beautiful view directly into aqueduct and habenula and disrupting the tenia of the posterior wall of the third ventricle between pulvinars reveals suprapineal area and enters the cistern of lamina quadrigemina (Fig. 57.2f, g). At the end of surgery, aqueduct patency is checked by direct view, eventual clot aspirated from it, and ETV is performed (if not already done as emergent surgery to solve the acute hydrocephalus).

Alternatively, it is possible to enter posterior third ventricle from endoscopic supracerebellar-infratentorial approach via transvenous-suprapineal corridor. This is advisable only in pineal tumors extending into the posterior third ventricle, where the veins are already distended and displaced apart by the tumor [15].

57.5 Surgical Pearls

- The tip of navigated endoscope may hit the septum (Fig. 57.1a) or fornix upon too vigorous perforation of the roof of the lateral ventricle. Approach slowly under endoscopic visualization.
- Start splitting the choroidal fissure anteriorly (not in the middle) just behind the anterior septal vein and follow its medial aspect deep to find internal cerebral vein under the fornix (Fig. 57.1). Consider that you are working in dry field and that both thalami came more together to the midline (»kissing thalami«) as the third ventricle collapsed when CSF was removed. When you would start splitting in the middle, you may easily enter into ipsilateral thalamus directly and miss the fissure (Fig. 57.2c).
- Dissect away from fornix, possibly without touching it directly, stand lateral to the head (not behind), so the fornix will be seen in horizontal position.
- Keep working space between internal cerebral veins opened by using cottonoid strips or re-expand the collapsed cavity by filling it with water and continue working in wet field.
- If your trajectory is too lateral, the thalamus belly will be in the way towards the foramen of Monro and the fissure. This may result in thalamic contusion, subependimal bleeding, and laceration from repetitive use of instruments. The genu of internal capsule is positioned very superficial at this point and prone to contact injury.
- Do frozen section in germinoma suspicious tumors (radiosensitive).
- Perform blood and CSF tumor markers tests (α-fetoprotein, β-hCG) in emergency ETV or EVD; try biopsy with ETV when feasible.

- At the end, check for a clot in deep aspect of posterior third ventricle under adhesio interthalamica and above the aqueduct. You will have to move apart a little the thalami to reach and aspirate the clot.

57.6 Discussion

Third ventricle is the deepest and one of the most difficult areas of the brain to access. Access through a deep narrow corridor from any direction is highly related to retractor injury and manipulation with critical structures causing cortical, subcortical, hypothalamic, thalamic, mesencephalic, forniceal, and optic path damage, especially in infiltrative lesions. The interhemispheric transcallosal approach [4] with interforniceal, transchoroidal, and subchoroidal variants, a workhorse of open neurosurgical approaches, is related to significant morbidity and mortality [8]. Interforniceal approach bears additional risk for bilateral fornix injury with permanent memory impairment.

The combined use of tubular retractor and endoscope enabled sufficient corridor for applying bimanual instrument manipulation under endoscopic visualization and illumination. Microsurgical removal or significant debulking of tumors in the posterior third ventricle was possible in minimally invasive way. Image-guided navigation in such procedures is an indispensible tool not only intraoperatively, but also for preoperative planning the optimal trajectory and access area. The later may be nearly doubled by tilting of the tubular retractor 20 ° [13].

Benign and malign lesions of third ventricle are very diverse and may include colloid cysts, germinomas, gliomas (pilocytic astrocytoma, diffuse midline glioma with mutation H3 K27), ependymoma, subependymoma, craniopharyngioma, meningioma, CP papilloma, pineal parenchymal tumor of intermediate differentiation, other pineal tumors, papillary tumor of pineal region, metastases, and vascular lesions. Neuroendoscopic procedures have a great advantage in the management of chemo- or radiosensitive tumors, such as germinoma, pineoblastoma, or primitive neuroectodermal tumor [1]. Vascular lesions should be avoided in transtubular endoscopic surgery.

Risk of damage to the ipsilateral fornix and deep venous system should be avoided. Bilateral rupture anterior to the thalamic level may lead to lasting anterograde amnesia [17]. Unilateral lesioning of the fornix (contusion, laceration) produces usually transient memory problems for few months. Left fornix dominance has been observed. Thalamostriate vein has a large drain area of deep white matter and ganglia and its injury results in hemiplegia, mutism, drowsiness, hemorrhagic infarction of the basal ganglia, or death. Adhesio interthalamica may be absent in many people. Injury limited strictly to it is usually asymptomatic.

57.7 Conclusion

Purely endoscopic transtubular-transcortical-transventricular-transchoroidal approach to the third ventricle provides a stable and reliable surgical access to the third ventricle while minimizing traction damage. Many small- to middle-sized and well-encapsulated solid lesions or cysts (e.g., colloid cysts) can be completely removed using bimanual instrumental manipulation of the tissue in the same way as in open microsurgery. Careful planning of the optimal trajectory and surgery simulation and intraoperative use of image-guided navigation is mandatory, especially in infiltrative tumors. Here, maximally safe image-guided debulking means going as close as possible, but not to far to avoid disastrous mesencephalic, hypothalamic, thalamic, forniceal, and optic path damage. A deep contralateral side is better controlled than ipsilateral one [4].

The procedure of purely endoscopic transtubular microsurgery is very static and a demanding one even for an experienced neuroendoscopist with all equipment available.

It seems that close endoscopic view and bimanual tissue manipulation through the neuroport give at least same or better surgical results in comparison to open microsurgery in well-selected lesions of maximal size 2.5–3 cm, but further studies are necessary [1–3, 6, 12, 19].

References

1. Oi S, Kamio M, Joki T, Abe T. Neuroendoscopic anatomy and surgery in pineal region tumors role of neuroendoscopic procedure in the "minimally-invasive preferential" management. J Neuro-Oncol. 2001;54(3):277–86.
2. Tawk RG, Akinduro OO, Grewal SS, Brasiliense L, Grand W, Grotenhuis A. Endoscopic transforaminal transchoroidal approach to the third ventricle for cystic and solid tumors. World Neurosurg. 2020;134:e453–9.
3. Tirado-Caballero J, Rivero-Garvía M, González-Pombo M, Cárdenas-Ruiz-Valdepeñas E, Kaen A, Márquez-Rivas J. Fully endoscopic transforaminal–transchoroidal approach for tectal area tumor removal. World Neurosurg. 2020;137:164–72.
4. Bozkurt B, Yağmurlu K, Belykh E, Tayebi Meybodi A, Staren MS, Aklinski JL, et al. Quantitative anatomic analysis of the transcallosal-transchoroidal approach and the transcallosal-

subchoroidal approach to the floor of the third ventricle: an anatomic study. World Neurosurg. 2018;118:219–29.
5. Kelly PJ. Future perspectives in stereotactic neurosurgery: stereotactic microsurgical removal of deep brain tumors. J Neurosurg Sci. 1989;33(1):149–54.
6. Ogura K, Tachibana E, Aoshima C, Sumitomo M. New microsurgical technique for intraparenchymal lesions of the brain: transcylinder approach. Acta Neurochir. 2006;148(7):779–85.
7. Lozier AP, Bruce JN. Surgical approaches to posterior third ventricular tumors. Neurosurg Clin N Am. 2003;14(4):527–45.
8. Hassaneen W, Suki D, Salaskar AL, Levine NB, DeMonte F, Lang FF, et al. Immediate morbidity and mortality associated with transcallosal resection of tumors of the third ventricle. J Clin Neurosci. 2010;17(7):830–6.
9. Kanner A, Hopf NJ, Grunert P. The "optimal" burr hole position for endoscopic third ventriculostomy: Results from 31 stereotactically guided procedures. Minim Invasive Neurosurg. 2000;43(4):187–9.
10. Chen F, Nakaji P. Optimal entry point and trajectory for endoscopic third ventriculostomy: evaluation of 53 patients with volumetric imaging guidance. J Neurosurg. 2012;116(5):1153–7.
11. Bosnjak R. Advanced neuroendoscopy [Internet]. www.youtube.com. 2022 [cited 2023 Apr 10]. https://www.youtube.com/watch?v=oJ3-KwrUjcE
12. Chibbaro S, Di Rocco F, Makiese O, Reiss A, Poczos P, Mirone G, et al. Neuroendoscopic management of posterior third ventricle and pineal region tumors: technique, limitation, and possible complication avoidance. Neurosurg Rev. 2012;35(3):331–40.
13. Gagliardi F, Piloni M, Bailo M, Gragnaniello C, Boari N, Donofrio CA, et al. Anatomic study on neuroendoportal transcortical approach to lateral ventricles. World Neurosurg. 2019;131:e255–64.
14. Krishna V, Blaker B, Kosnik L, Patel S, Vandergrift W. Translamina terminalis approach to third ventricle using supraorbital craniotomy: technique description and literature review for outcome comparison with anterior, lateral and trans-sphenoidal corridors. Minim Invasive Neurosurg. 2011;54(05/06):236–42.
15. Spazzapan P, Velnar T, Bosnjak R. Endoscopic supracerebellar infratentorial approach to pineal and posterior third ventricle lesions in prone position with head extension: a technical note. Neurol Res. 2020;42(12):1070–3.
16. Türe U, Yaşargil MG, Al-Mefty O. The transcallosal—transforaminal approach to the third ventricle with regard to the venous variations in this region. J Neurosurg. 1997;87(5):706–15.
17. Calabrese P, Markowitsch HJ, Harders AG, Scholz M, Gehlen W. Fornix damage and memory. Cortex. 1995;31(3):555–64.
18. Nagata S, Rhoton AL, Barry M. Microsurgical anatomy of the choroidal fissure. Surg Neurol. 1988;30(1):3–59.
19. Bosnjak R, Antolin A, Jeglic A, Felbabic T, Velnar T. Navigated endoport in the purely endoscopic microsurgery of intraventricular and other deep-seated brain lesions: a case report. Acta Med Acad. 2021;49(1, 2, 3):70. https://www.ama.ba/index.php/ama/article/view/427. Accessed 10 Apr 2023
20. Bosnjak R. Neurosurgery Endoport-guided endoscopic removal of intraventricular tumor—Dr. Bosnjak [Internet] wwwyoutubecom. 2013 [cited 2023 Apr 9]. https://www.youtube.com/watch?v=zMno8C07txk

Anatomy and Telovelar Approach to the Fourth Ventricular Pathology

58

Thomas Santarius, Heba Azooz, and Sara Venturini

Comfort of the surgeon is safety for the patient.

David Hardy, Cambridge.

Learning Objectives

- To understand surgical anatomy relevant to the telovelar approach to know when to use it and when alternative approaches may be more appropriate.
- To realise what one needs to learn and practise to be able to use this approach safely.
- To learn about possible complications and how to avoid them.

58.1 Introduction

The telovelar approach is an anatomical approach to tumours in and around the fourth ventricle and brainstem. Its key benefits are that it follows 'natural corridors', thus avoiding significant disruption of parenchymal structures. It preserves pial surfaces and thus attenuates the effects of retraction. Importantly, it allows early exposure and control of blood supply to the majority of tumours in and around the fourth ventricle. It can be tailored to the lesion and the objective of the surgery, from partial unilateral, through bilateral to combined telovelar-retrosigmoid (TV-R) [1], and supplemented with tailored or partial laminectomy or laminoplasty of the posterior arch of atlas, axis, or even lower cervical vertebrae, as required. These can be done in one sitting or as a staged procedure [1].

T. Santarius (✉)
Department of Neurosurgery, Cambridge University Hospitals NHS Trust, Cambridge, UK

Department of Clinical Neurosciences, University of Cambridge, Cambridge, UK
e-mail: ts381@cam.ac.uk

H. Azooz
Department of Neurosurgery, Children Cancer Hospital Egypt, Cairo, Egypt

S. Venturini
Department of Neurosurgery, Cambridge University Hospitals NHS Trust, Cambridge, UK
e-mail: sara.venturini@nhs.net

58.1.1 History/Background

Access to the fourth ventricle and brainstem was limited until the early twentieth century, when Walter Dandy described gaining access to the fourth ventricle via the transvermian approach in 1945 [2]. As anatomical descriptions of structures in this region evolved, neurosurgeons started to describe the use of dissection of the cerebromedullary fissure as a way of accessing lesion in the fourth ventricle and brainstem. Mussi and Rhoton in 2000 were the first to use the term *telovelar* (TV) to describe their approach where the velum was opened to expose the fourth ventricular roof and lateral recess [3]. Since then, further modifications and extensions to this approach have been described, including removal of the C1 posterior arch and combining the TV with other surgical approaches [4].

58.1.2 Indications

The TV approach is suitable for access and resection of lesions in and around the fourth ventricle, i.e. the cerebellum and the brainstem. Symptoms are often related to hydrocephalus as well as compression or invasion of surrounding structures, including brainstem and cranial nerves. Such lesions include ependymomas, medulloblastoma, pilocytic astrocytomas, haemangioblastomas as well as cavernomas, aneu-

I. N. Kanaan, V. Beneš (eds.), *Neuroanatomy Guidance to Successful Neurosurgical Interventions*,
https://doi.org/10.1007/978-3-031-59838-8_58

rysms, and arteriovenous malformations. The fourth ventricular floor is highly eloquent and cranial nerve palsies can occur due to tumour or other lesional effects as well as post-operatively due to intraoperative manipulation and ischaemia. Detailed knowledge of the relevant local anatomy as well as the nature of the interaction of the lesion with the anatomy is crucial for planning and execution of any surgical intervention in this area.

58.2 Anatomy Highlights

- The essence of the telovelar approach is opening of the cerebello-medullary fissure in an analogous way to opening of the Sylvian fissure.
- This is possible because inferior half of the roof of the fourth ventricle can be disconnected from the brainstem by cutting the tela choroidea and the inferior medullar vellum.
- By doing this, the surgeon is connecting the foramina of Magendie and Lushka.
- For practical purposes, the entire floor of the fourth ventricle ought to be considered 'eloquent' and there is always some price to pay, in terms of neurological deficit, if this is transgressed.
- Each branch of the first 3 segments of PICA (anterior medullary, lateral medullary, and tonsillo-medullary) branches supplies the brainstem or other highly eloquent structures until proven otherwise. Although branches of the telovelo-tonsillar segment may also supply brainstem, this is not likely and most will be choroidal vessels.
- PICAs are highly variable and some (up to 20%) of PICAs may have extradural origin.

58.3 Anatomy

The fourth ventricle is a tent-shaped space with the floor facing dorsally and the rostral and caudal roof plates meeting at the apex called fastigium (Fig. 58.1). The rostral plate (superior half) of the roof is formed of the superior medullary velum stretched between the left and right superior cerebellar peduncles heading towards their decussation in the tegmentum of the midbrain. The caudal plate (inferior half) is formed by the noduloflocular lobe superiorly and the tela choroidea inferiorly. The noduloflocular lobe consists of nodulus medially and flocullus laterally, the most caudal segments of the vermis and cerebellar hemisphere, respectively. Stretched between them is the inferior medullary velum. Like the superior medullary velum, the inferior medullary velum contains neural tissue, but this is thought to be non-functional. As the inferior medullary velum gradually thins caudally, ultimately it loses its neural tissue altogether resulting in the pia coming in direct contact with the ependyma (Figs. 58.1 and 58.2). From this point on, it is the tela choroidea. The pia with its blood vessels invaginates into the fourth ventricle, forming the choroid plexus. The gradual atresia of the tissues of the roof of the fourth ventricle culminates in one medial and two lateral perforations of its roof, the foramina of Magendie and Luschka, respectively (Fig. 58.1). The lateral walls of the upper half of fourth ventricle are formed by the superior cerebellar peduncles, flanked by the middle cerebellar peduncle laterally (Fig. 58.1). The lateral wall of the lower half is formed by the inferior cerebellar peduncle rostrally, and more caudally, it opens widely into the lateral recess communicating with the cerebellomedullary cistern via the foramen of Luschka. The attachment of the tela choroidea to the inferior cerebellar peduncle is called the taenia (of the fourth ventricle) (Fig. 58.1).

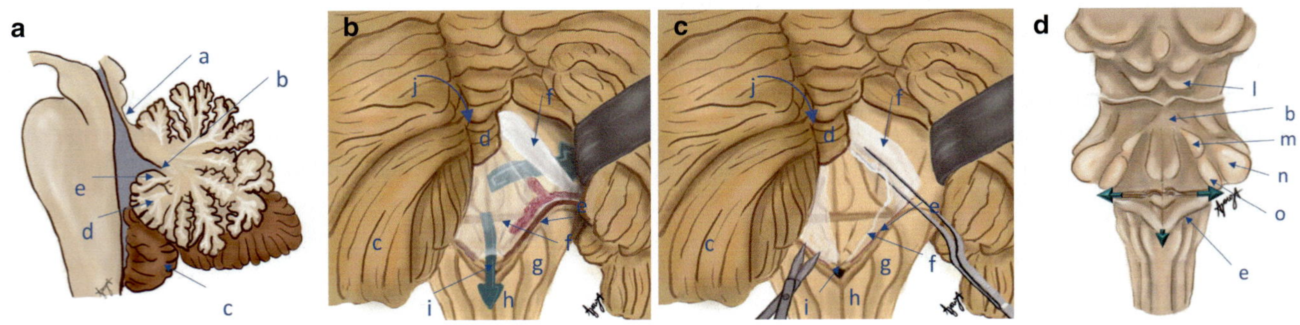

Fig. 58.1 Basic anatomy of the fourth ventricle relevant to the telovelar approach. (**a**) The roof of the fourth ventricle—midline sagittal view of the fourth ventricle. (**b**) Inferior half of the fourth ventricle as seen after the right tonsil has been retracted laterally exposing the tela choroidea and the inferior medullary vellum. (**c**) The tela choroidea has been cut to allow direct inspection of the floor of the fourth ventricle. (**d**) Posterior view of the hindbrain (rhombencephalon) with the cerebellum removed by sectioning the cerebellar peduncles, which allowed viewing of the entire fourth ventricle. *a*—superior medullary velum, *b*—fastigium, *c*—tonsil, *d*—uvula, *e*—nodulus, *f*—inferior medullary velum; *e*—taenia (a line of attachment of the tela choroidea to the brainstem), *f*—translucent tela choroidea with choroid plexus of the fourth ventricle seen through it, *g*—tubercle of the cuneate nucleus, *h*—tubercle of the gracile nucleus, *i*—foramen of Magendie, *j*—left tonsilo-uvular fissure, the right tonsilo-uvular fissure has been opened, *k*—right trochlear nerve emerging from underneath the inferior colliculus (l) and rostral to the superior medullary vellum (b), *m*—superior cerebellar peduncle, *n*—middle cerebellar peduncle, *o*—inferior cerebellar peduncle, *f*—taenia (a line of attachment of the tela choroidea to the brainstem); red arrows—foramina of Luschka, turquois arrow—foramen of Magendie

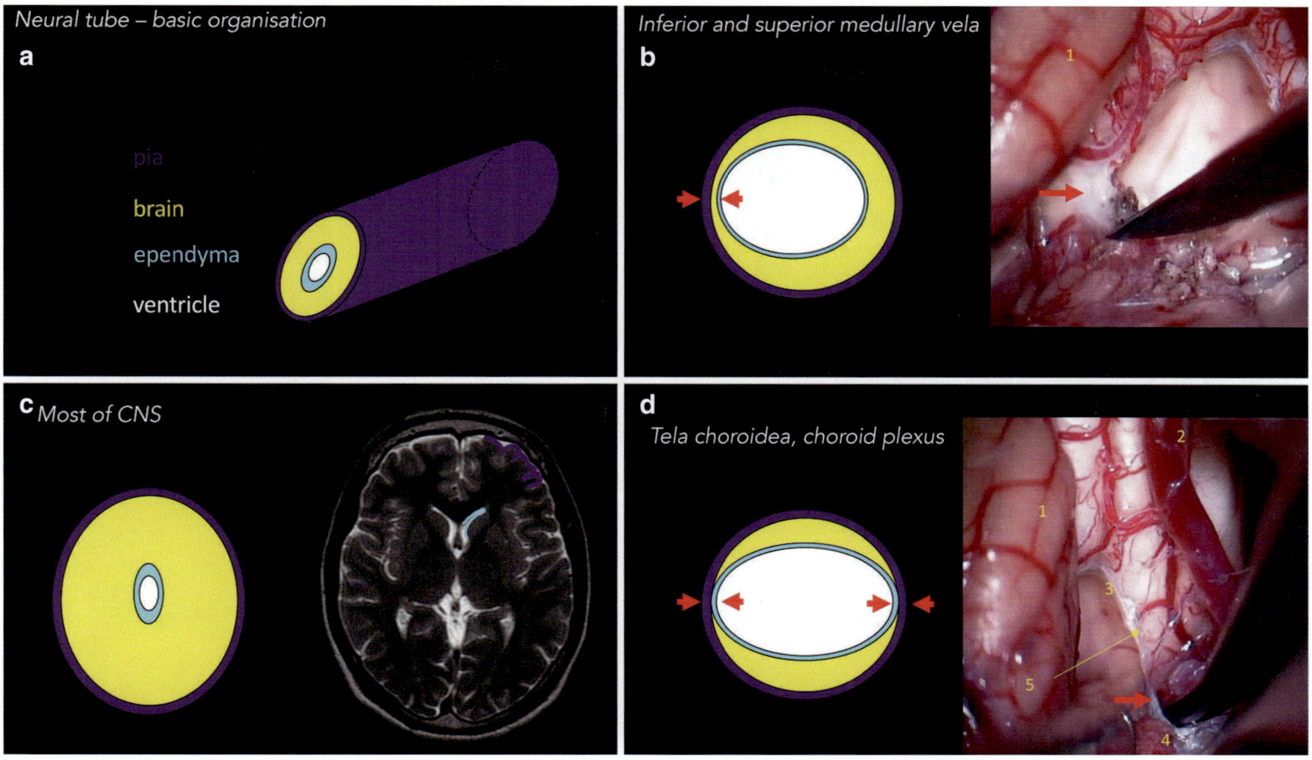

Fig. 58.2 The relevance of the basic organisation of the CNS to the understanding of the tela choroidea and medullary vela. (**a**) Neural tube—basic organisation, (**b**) Brain parenchyma forms a thick layer throughout most of CNS. (**c**) In some parts of the brain, there is a major thinning of the neuronal layer with only a small amount of brain tissue between the ependyma and pia, such as in the case of the inferior and superior medullary velum. Intra-operative photograph: the tella choroidea has been divided on both sides with the right-sided inferior medullary vellum (red arrow) to about to be cut (1—right tonsil retracted with a sucker). (**d**) With a complete atresia of the neuronal tissue, the ependyma and pia come to direct contact forming the tela choroidea. The pial arteries and veins invaginate into the ependyma of the ventricle, forming the choroid plexus. Intra-operative photograph: the tella choroidea is being divided (red arrow). (1—right tonsil retracted with a sucker, 2—left PICA, 3—obex seen via the foramen of Magendie which has been enlarged by partially cutting the tella choroidea from the foramen of Magendie towards the inferior medullary vellum, 4—choroid plexus on the ventricular side of the tela choroidea, 5—taenia, a line of attachment of the tela choroidea to the brainstem)

The floor of the fourth ventricle is highly eloquent and, under normal circumstances, no entry to deeper lesions in the brainstem can be carried out without causing some neurological deficit. Superior cerebellar peduncles form the roof and lateral walls of the upper half of the fourth ventricle and are the main cerebellar output (Fig. 58.1). It is likely that their disruption, which can occur during resection of medulloblastomas, for example, is thought to be the main cause of the cerebellar mutism syndrome (CMS) [5]. The inferior cerebellar peduncles constitute the lateral boundary of the fourth ventricle inferiorly, and as they turn dorsally towards the cerebellum, they then form a part of the upper border of the lateral recess (Fig. 58.1). The inferior cerebellar peduncles constitute the main direct cerebellar input from the spinal cord, mostly the dorsal columns. The tela choroidea and inferior medullary velum are devoid of functional neural tissue and can be cut all the way from the foramen of Magendie to the foramina of Luschka. This allows the inferior half of the fourth ventricular roof to be elevated, granting unobstructed access to the entire fourth ventricle (Fig. 58.1). When viewed from behind, as during the midline suboccipital approach, these structures can't be seen at first as they are covered by the cerebellar tonsils. Lateral retraction of the tonsils and widening of the vallecula, the space between the tonsils, reveal the uvula, the penultimate caudal segment of the vermis, located in the depth of the vallecula.

The posterior inferior cerebellar arteries (PICAs) are closely related to structures encountered during the telovelar approach (Figs. 58.3 and 58.9). The PICA is a paired artery arising from the vertebral arteries intradurally, although rarely it can arise extradurally or from the basilar artery. It usually has the anterior medullary, lateral medullary and tonsilo-medullary, and telovelo-tonsillar and cortical branches. It is a highly variable artery in terms of its origin, course, and its supply territory, but, as a rule of thumb, the branches of the first three segments supply brainstem, while the latter two do not. The majority of the branches of the telovelo-tonsillar segment can be divided during the T-V approach as they will supply the choroid plexus of the fourth ventricle and the tumour. The cortical segment begins when PICA emerges in the groove

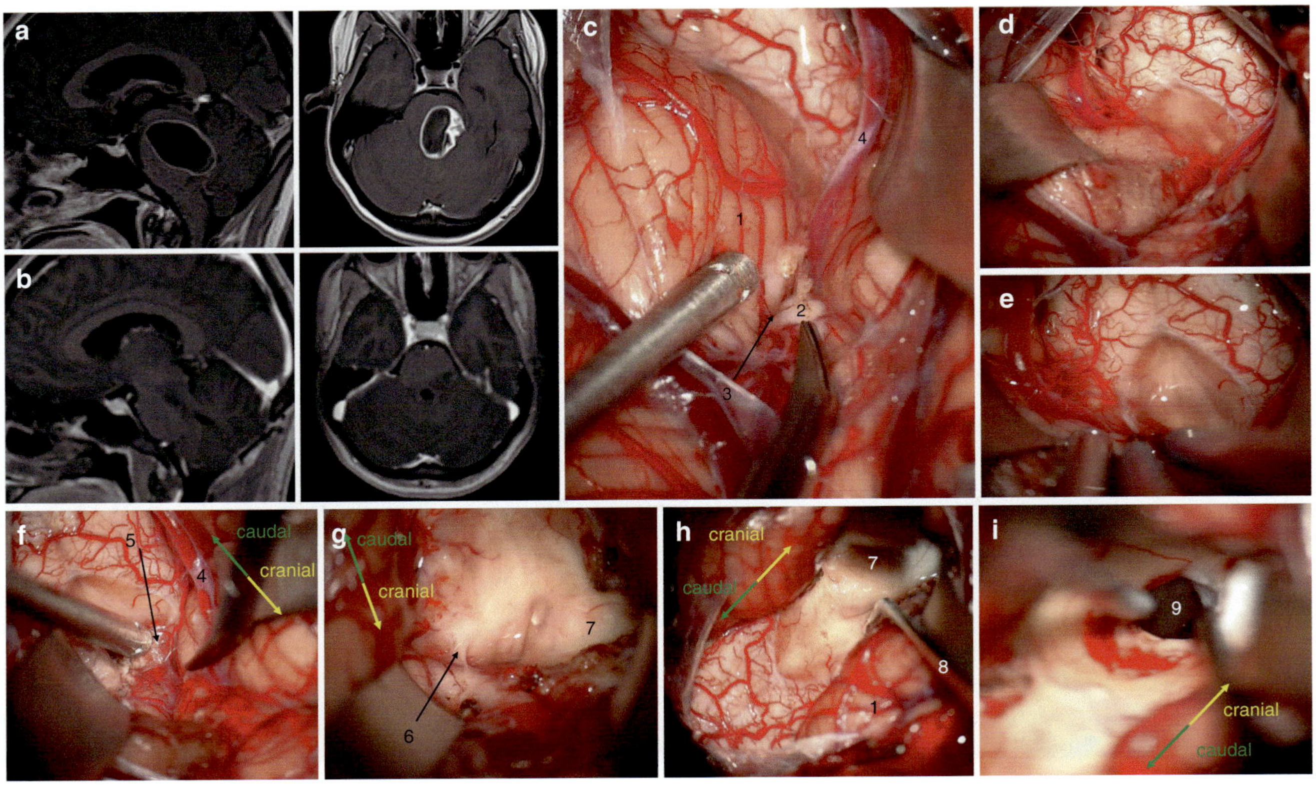

Fig. 58.3 Telovelar approach to a brainstem pilocytic astrocytoma projecting to the floor of the fourth ventricle. 14-year old with progressive 'balance issues' over 5 months, followed by right-sided numbness and diplopia. On examination, there was bilateral INO, loss of all sensory modalities on the right, mild right-sided weakness of the upper and lower extremities. Face was symmetrical. (**a**) Pre-operative MRI (T1 with contrast). (**b**) Post-operative MRI (T1 with contrast). (**c**) Opening of the tonsillo-uvular fissure. As the entry is relatively high and the tumour extends deep into the brainstem, every bit of widening of the natural corridor will be useful, including the tonsillo-uvular fissure. It is usually shallow, but sometimes it can contribute significantly to widening of the gateway to the fourth ventricle. (**d**) Gradual opening of the cerebellomedullary fissure by coagulating choroidal branches of bilateral PICAs and cutting tela choroidea and inferior medullary velum from the foramen of Magendie to the foramina of Luschka. (**e**, **f**) Cutting tela choroidea [5]. (**g**) Cutting inferior medullary velum [6] and exposing the tumour covered by ependyma and a very thin layer of brain tissue bulging to the fourth ventricle. (**h**) Mapping of the floor of the fourth ventricle with a bipolar stimulator [8]. In addition, continuous MEPs and SSEPs were run to monitor the integrity of the motor and sensory pathways, respectively. (**i**) The tumour was entered via functionally silent and thinnest portion of brain parenchyma and debulked. 1. Right tonsil, 2. Uvula, 3. Tonsilo-uvular fissure, and 4. Left PICA

between the cerebellar hemispheres and tonsils laterally and the vermis medially. Blood supply to the pons and medulla, including the floor of the fourth ventricle, is derived from the perforating branches of the basilar, vertebral, posterior inferior, and anterior inferior cerebellar arteries entering the brainstem from its anterior and lateral surfaces and not directly from PICAs branches within the fourth ventricle.

The embryology of the region can be useful to understand important anatomical relations. When the neural tube forms, the pia is on the outside and the ependyma is on the inside, with parenchyma in between these. Differential proliferation and atresia result in distinct amount of parenchyma according to the location (Fig. 58.2). The inferior medullary velum, which is part of the nodulofloccular lobe, contains only a thin layer of non-functional brain tissue. At the tela choroidea, there is no parenchyma at all and the pia and ependyma are at a direct apposition. Invagination of the pia into the ependyma gives rise to the choroid plexus.

58.4 Approach/Surgical Intervention

58.4.1 Pre-operative Considerations

Detailed history, examination, and review of all the relevant imaging and laboratory investigations are essential when planning surgery. History and examination can inform us about the directionality and growth dynamics of the lesion as well its relationship with the relevant anatomical structures. Imaging will add further anatomical detail which must be assessed together with the clinical information. The likely tumour diagnosis and alternative and complementary therapies are also highly relevant when defining the aims of surgery. Planning of the optimum approach then follows. Considerations of whether CSF diversion is required prior to craniotomy should be made. If so, the choice usually lies between a temporary external ventricular drain (EVD) and an endoscopic third ventriculostomy (ETV). There is growing body of evidence, suggesting that pre-operative endoscopic third ventriculostomy may reduce the need of extracranial CSF diversion [6, 7].

58.4.2 Positioning

The patient's head should ideally be immobilised using a rigid head clamp. Especially in the paediatric population, care should be taken to avoid possible complications such as depressed skull fractures and extradural haematoma, which have been reported for posterior fossa craniotomies [8, 9]. A range of positions can be used depending on the patient's anatomy and physiology, lesion, and surgeon preference. All require a degree of neck flexion to create a wider access (longer distance between the spinous process of C2 and lower suboccital region, a greater distance between C1and foramen magnum) and place the floor of the fourth ventricle closer to the vertical plane, which is ergonomically advantageous. In the first instance, maximum head flexion is achieved in the atlanto-occipital joint and only then some additional cervical spine flexion is applied. The patient should be strapped to ensure safe operating table manipulation, especially head elevation and lateral tilt (Fig. 58.4). The following patient positions can be used:

1. Concorde/prone: achieved by flexing at the occipito-cervical joint, adding neck flexion and elevation. This is thought to be optimal for young children. It has the disadvantage of venous congestion, which can be mitigated by elevating the head above the level of the heart (Fig. 58.4).
2. Lateral/oblique (modified park bench): This position can be used if the pathology extends laterally to the CP angle and anteriorly in front of the brainstem. Disadvantages include that the anatomy is not centred, there is a need to support the upper cerebellar hemisphere, and there is a risk of brachial plexus injury if the bottom arm is not adequately supported [10].
3. Sitting: It provides a very clear operative field, as blood and CSF drain out of operative site. Its disadvantages are higher risk of cardiovascular instability and hypotension and air embolism. Unless the patient's trunk is forward, the surgical corridor may be heading uncomfortably upwards.

We use versions of the Concorde position pretty much exclusively, as with modern operative tables and flexible operating theatre setup, it is feasible to operate from behind the vertex as well as from either side of the patient (Fig. 58.5). If the majority of the fourth ventricular lesion is extending to one lateral recess/foramen of Lushka/CP angle, the cerebello-medullary, and/or -pontine cisterns, the head can be tilted to the that side and the surgeon can follow the tumour comfortably by standing on the opposite side of the patient.

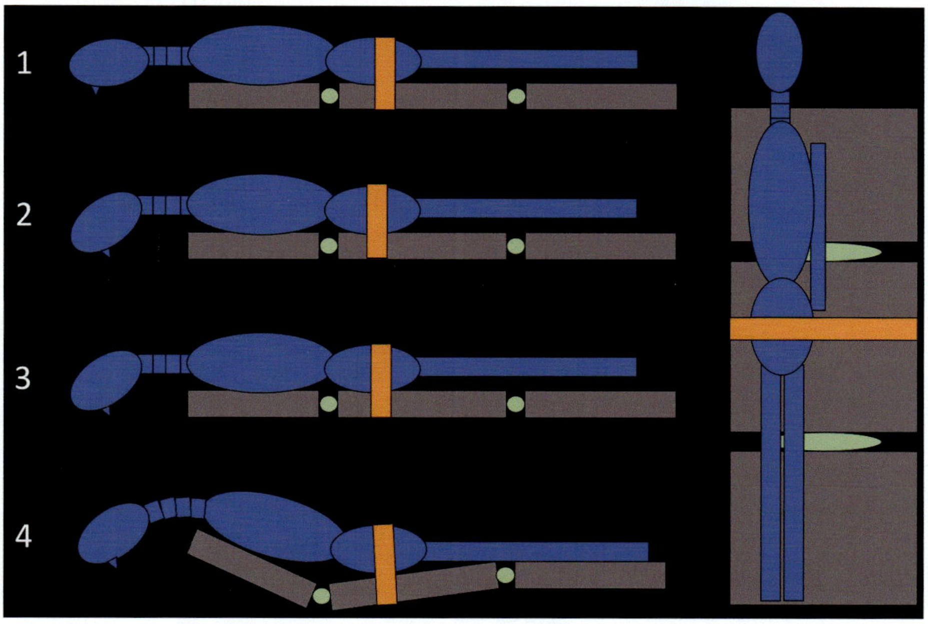

Fig. 58.4 Positioning for telovelar approach—stepwise diagram

58.4.3 Surgical Technique

This can vary depending on exact tumour location. Here we describe the basic approach which can be modified as required.

Usually, a linear midline incision starting 1–2 cm above the external occipital protuberance down to C2 is performed. The exact extension depends on several factors, which include the caudal extent of the lesion, the need for lateral bone exposure, whether this is symmetrical, and the depth from skin to bony structures, especially at the level of the foramen magnum. The latter will influence the ease and comfort with which the extremes, especially the rostral aspect of the lesion, will be handled. Transverse incisions have also been described, but these are not really suitable for most indications of T-V approach [11].

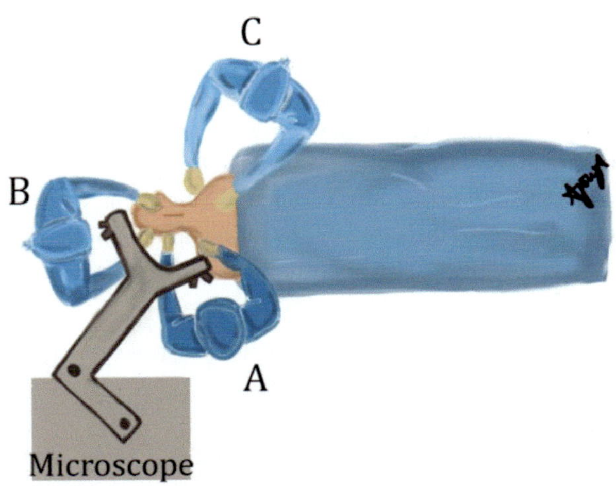

Fig. 58.5 Operating theatre setup for telovelar approach allowing the surgeon to operate from the top (**b**) and either side of the patient (**a** and **c**)

Midline suboccipital craniotomy is performed, typically starting with burr holes at either side of the midline in the safe distance caudally to the transverse sinus (Fig. 58.6). The exact distance from the foramen magnum depends on the size and location of the lesion and how crowded is the posterior fossa, whether CSF can be released from the cisterna magna, how high does the inferior half of the roof of the fourth ventricle need to be elevated, etc. The craniotomy is usually narrower than that required for a hemispheric lesions, for which we generally favour unilateral approaches. However, for most lesions it is critical that the craniotomy is suitably wide at the level of the foramen magnum as this is the main bony access port. If required, the craniotomy is extended with rongeurs and drill as far laterally as the medial limit of the occipital condyles. Caudally, the access can be further extended as required, especially in the case of caudally spreading ependymomas, by undercutting the posterior arch of C1, and performing laminectomy of C1, C2 or, rarely, even lower. The course of the vertebral artery surrounded by the vertebral venous plexus as well as anatomical variations, such as *ponticulus posticus* and extradural origin of PICA, need to be kept in mind. Sticking strictly to the subperiostal plane using suitable periosteal elevator when exposing bone will protect neuro-vascular structures, including the marginal and vertebral venous plexus.

Dural opening should be performed with care, as this usually requires crossing the occipital sinus which, especially in younger patients, can transmit relatively large volume of blood (Fig. 58.6). Liga haemostatic clips can be useful to stop bleeding from the sinus. The dura can be opened in Y-shaped fashion, although we prefer a 'red wine glass' shape as this facilitates better closure and the flap can be partially 'rotated' as required. At least some CSF can usually be released from the cisterna magna which results in proportionate relaxation of the contents. Gentle separation of the cerebellar tonsils allows to expose the cerebellomedullary fissure. The uvula is

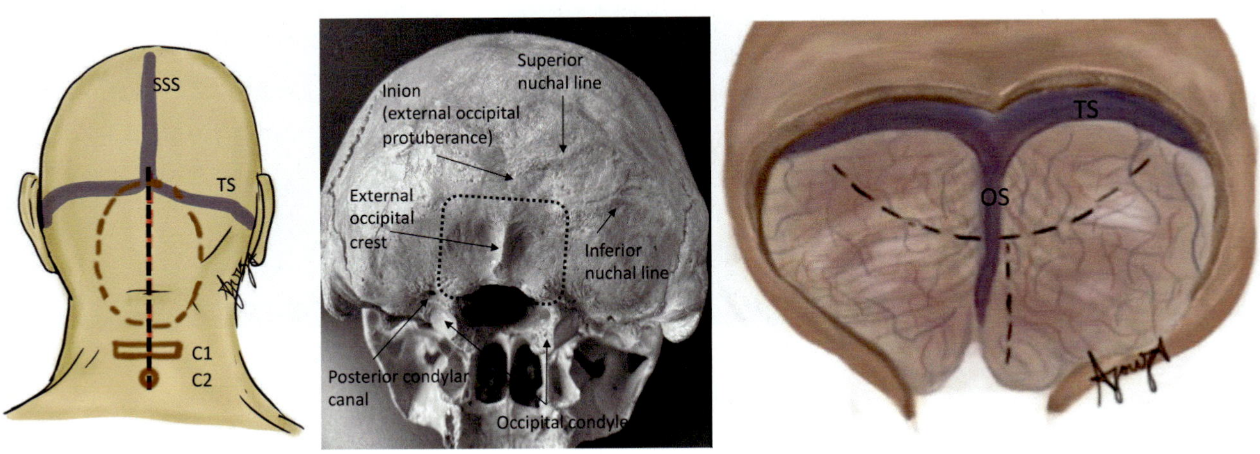

Fig. 58.6 Anatomy relevant to suboccipital craniotomy for the telovelar approach—from skin to dural incision. *TS* transverse sinus, *OS* occipital sinus, *SSS* superior sagittal sinus

visualised, and it is usually possible to open at least a portion of the tonsillo-uvular fissure. The PICAs are identified and usually left with the tonsils. The posterior columns, foramen of Magendie, and obex are identified. The translucent tela choroidea is cut in parallel with the inferior cerebellar peduncle (continuation of the posterior columns), while coagulating and dividing the choroidal branches of the PICA. Soon the milky semi-translucent inferior medullary velum is encountered and divided in the same fashion as the tela choroidea. Once most of all of the lateral attachment of the roof of the inferior half of the fourth ventricle is cut, the tonsils and the rest of the inferior cerebellum can be elevated, exposing the entire fourth ventricular cavity (Fig. 58.7) [11]. Most often than not, especially in the case of large tumours filling in most of the inferior fourth ventricle (e.g. ependymomas and medulloblastomas), the tela choroidea and the inferior medullary velum are fused with the tumour and, essentially, form the capsule of the tumour with the choroidal arteries serving as the main feeders of the tumour. The draining veins are the ependymal/subependymal veins and large tumour veins may be seen draining towards the Galenic (upper half of the roof via the pre-central vein) or via the lateral recess to the petrosal group of veins (lower half of the roof via the vein of the cerebello-medullary fissure*)*, Figs. 58.8 and 58.9. High-resolution T2 or T1 + C MRI scans are usually sufficient to disclose their calibre and location.

Dura should be closed in watertight fashion to reduce the risk of pseudomeningocoele and CSF leak post-operatively. Dural substitutes can be used if the dural leaflets cannot be approximated. Bone flap is replaced and fixed with cranial plates. Fascial layers are closed, again, aiming at watertight seal. Skin is sutured with preferred absorbable or removable material. We use continuous absorbable sutures in children and staples in adults.

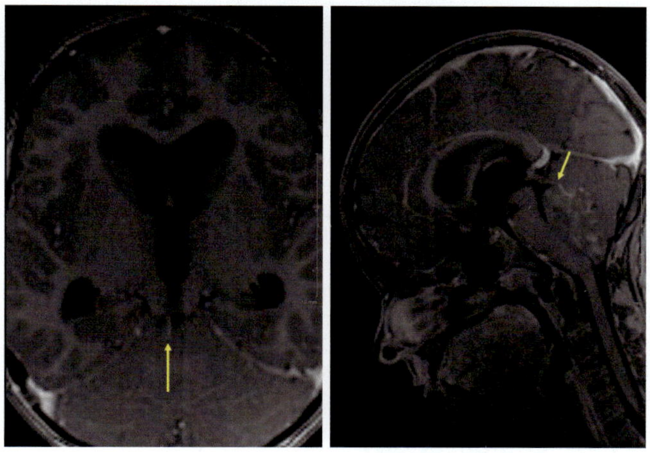

Fig. 58.8 Tumour-draining subependymal vein (yellow arrow) via the superior medullary velum to the pre-central vein in a case of a medulloblastoma

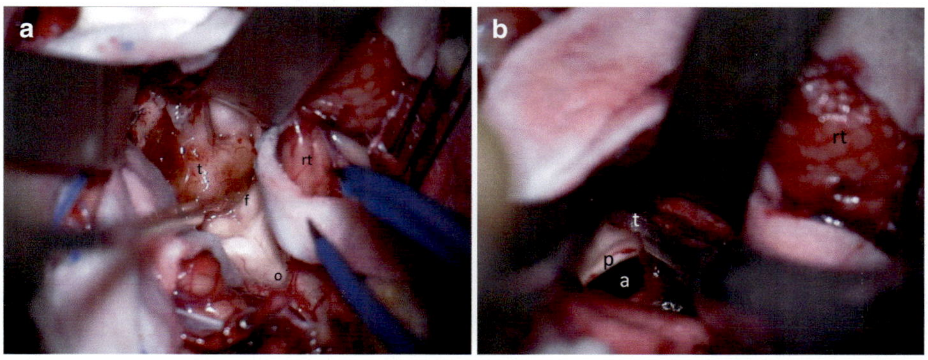

Fig. 58.7 Telovelar approach to the tumour in the floor of the upper fourth ventricle (subependymoma). *a* aqueduct of Sylvius, *f* right facial colliculus, *o* obex, *p* posterior commissure forming the roof of the cerebral aqueduct (**a**) Demonstration of the tumour in the upper half of the fourth ventricle. *rt* right tonsil, *t* tumour (**b**) View into the third ventricle via the cerebral aqueduct after resection of the tumour

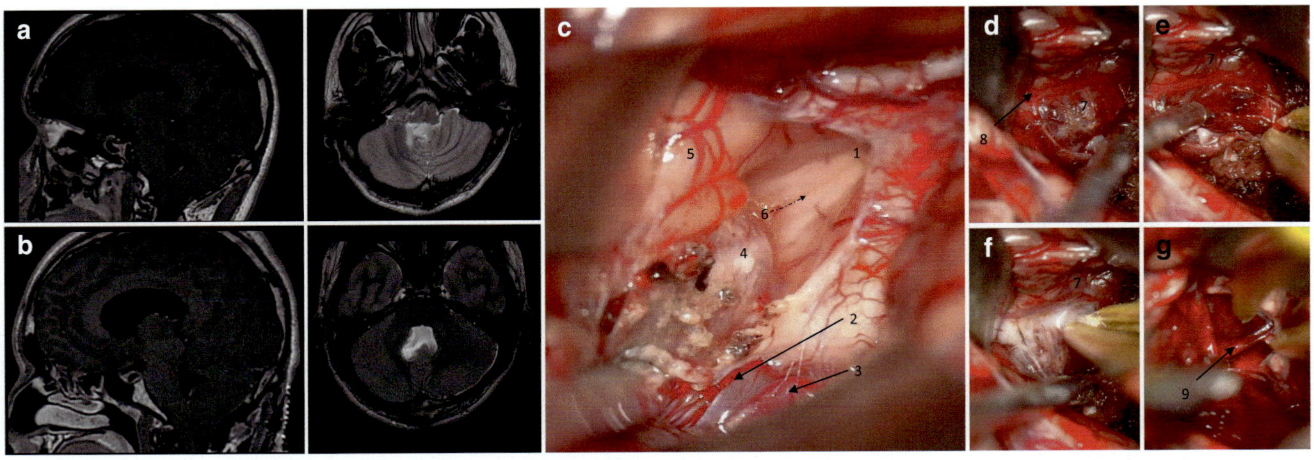

Fig. 58.9 Telovelar approach to a medulloblastoma. Similarly to tumours that are more obviously accessed via the telovelar approach, we use this approach also in medulloblastoma. It allows early access to feeding arteries, resulting surgery being highly controlled. The surgeon needs to know where the tap is to turn it off before entering the tumour. This is especially relevant in small children. 14-year old with headache, papilloedema, and worsening gait. (**a**) Pre-operative scans. (**b**) Post-operative scans. (**c**) Left tonsil retracted together with PICA. (**d–g**) Telovelo-tonsillar segment of right PICA directly supplying tumour. 1. Obex. 2. Tela choroidea with choroidal branches from PICA. 3. Left PICA. 4. Exposed tumour. 5. Tumour expanding lower vermis covered by a thin layer of brain and pia. 6. Median sulcus of the floor of the fourth ventricle. 7. Tumour. 8. Telovelo-tonsilar segment of PICA. 9. A large choroidal vein draining tumour blood towards the petrosal graining group via the foramen of Luschka

58.4.4 Complications

Because of the anatomical location of fourth ventricular lesions and their close relationships to surrounding structures, one must be mindful and aware of potential complications with this approach.

Hydrocephalus is a common complication, with 10–20% patients with post-fossa tumour requiring permanent shunting [12, 13]. Similarly, pseudomeningocoele can occur post-operatively and is more prevalent in the paediatric population, with an incidence of 10–15% of children [14]. Occasionally, this can result in CSF leak from the surgical wound, requiring further treatment. Aseptic meningitis and "posterior fossa fever" can occur [15]. Steroids are used to treat this condition. It is likely that minimising intra-operative blood spillage into the subarachnoid space and minimising the amount of post-operative blood within the CSF circulation can help reducing post-operative hydrocephalus and 'posterior fossa fever'.

Transient or permanent cranial nerve palsies can complicate the post-operative period and the relative likelihood depends on the location and type of pathology. INO, VI, and VII nerve palsies are most common in lesions in the fourth ventricle, especially if these involve its floor; although, strictly speaking, INO is not a nerve palsy, but rather result of disruption of the medial longitudinal fasciculus (MLF) connecting VI (and the horizontal integration gaze centre on the floor of the fourth ventricular floor) and VI nuclei in the midbrain. The MLF runs superficially and close to the median sulcus. Palsy of the lower cranial nerves, especially IX and XI, can result in dysarthria, dysphonia, and impairment of swallowing and airway protection and tracheostomy and/or gastrostomy may be required. This is more commonly a result of direct nerve and vascular damage rather than injury of the nucleus in the medulla.

A rarer complication, more common in children, is posterior fossa syndrome (mutism/pseudobulbar palsy) [16]. This is defined as post-operative ataxia, hypotonia and speech, and cognitive and emotional impairment, especially irritability. The symptoms often develop and peak with some delay (1–2 days after surgery) and last for weeks and months. In approximately 10–20%, there is mild to moderate permanent deficit. Its aetiology remains unclear, but it is thought to result from disruption of the cerebellar output (superior cerebellar peduncles and cerebellar nuclei) either through direct damage, oedema, or vascular insufficiency.

Injury to the inferior cerebellar peduncle can result in equilibrium disturbance, truncal ataxia, and staggering gait oscillation of head.

58.5 Surgical Pearls

- Consider first maximum safety of each part of each operation and only then try to minimise the access route.
- Achieve maximum head flexion in the atlanto-occipital join and only then in the cervical spine, if/when required.
- Subperioesteal dissection. Get on the bone and stay on the bone. This will not only reduce bleeding because you keep marginal sinus and vertebral venous plexus intact, but also keep other neurovascular structures safe.
- Brisk bleeding can occur from occipital, marginal, and other venous sinuses. These can be immediately and relatively easily be controlled by grasping the bleeding dural edges with a forceps elevating them under tension. This can be supplemented by elevating the head before liga-clips, dural tension sutures, are applied.
- Anticipate possibility of air embolism when venous sinuses, veins, and even intra-osseous venous channels are open, especially when the site of opening is high above the atrium of the heart. Irrigate with saline, alert the anaesthetist, lower the head using operating table.
- Keep PICA on the tonsils.
- Determine the destination of each branch of PICA before you divide it.
- Don't use any fixed retraction until you feel it is really required—with time you will surprise yourself by how rarely you will actually need it.
- Keep pia intact—it is better in protecting brain and vascular structures than any cottonoid.
- You may be able to open the tonsillo-uvular fissure. It is not always possible without neurovascular trauma, in which case it should not be pursued. However, when it is possible, opening of the tonsillo-uvular fissure can considerably widen the access corridor to your lesion and reduce the amount of retraction.
- Most tumours in and around the fourth ventricle, especially ependymomas, medulloblastomas, pilocytic astrocytomas, and haemangioblastomas, will derive some if not most of their blood supply from the choroidal branches from one or both PICAs.
- Consider (within reason) each tumour as an AVM; take feeding vessels early, veins late (it is better to drain blood into the vein than into the sucker), and consider using temporary aneurysm clips in case of brisk bleeding. Coagulate bleeding point of a vessel, not (necessarily) the vessel or blood.
- Good closure is important in this operation and good closure starts with good opening, both in terms of planning and execution. It is important to prevent tissue loss and devascularisation. Sufficient access to dural edges is important to comfortably accomplish watertight dural closure as this reduces the incidence of wound-related complications, including CSF leak and infection as well as pseudomeningocoele.
- Tumour draining veins. T2 and T1 + C MRI scans are usually sufficient to disclose their calibre and location—look towards the lateral recess and superior medullary vellum. Coagulation is not always the best method of haemostasis. Gentle packing with Surgicel and/or haemostatic gel matrix and patience can be more effective and less likely to result in focal brain damage.

58.6 Discussion

Discussions regarding the telovelar approach often relvolve around its advantage of reducing the likelihood of the cerebellar mutism syndrome (CMS). It has become increasingly clear that CMS is a result of the disruption of the cerebellar outflow tracts, namely the deep cerebellar nuclei and superior cerebellar peduncles. Telovelar approach is often contrasted to the transvermian approach as an approach that is meant to spare patients CMS. Although it is always better to avoid transgressing neurovascular structures by following natural corridors, it is unlikely that CMS is resulting from disruption of the vermian structures. It is more likely that the surgeon loses orientation and strays to deep nuclei or superior cerebellar peduncles.

We now know that most medulloblastomas arise from the rhombic lip, the inferior aspect of the ventricular roof corresponding to the ventricular aspect of the nodulus. Therefore, even medulloblastoma is best considered an intra-ventricular tumour, thus best approached from the ventricle, i.e. via the telovelar approach. Hence, the discussion as to whether to use the telovelar of transversion approach does not hinge on whether damage to the cerebellar vermis on approach results in cerebellar mutism. Instead, the key question is why to use the transvermian approach at all.

58.7 Conclusion

The telovelar approach is an anatomical approach which, by using mostly natural corridors and thus avoiding traversing any neural tissue, minimises approach-related morbidity. For tumours in fourth ventricle, most of their blood supply comes from PICA and its choroidal branches. Identifying the PICA course, coagulation, and division of the choroidal vessels helps devascularising the tumour early during the approach. This is particularly relevant to young children. The tela choroidea often forms the capsule of tumours in the fourth ventricle and can be useful as a dissection plane. The telovelar approach can be modified to suit each patient's anatomy and pathology and it can be combined with other approaches for more extensive tumours.

References

1. Budohoski KP, Guilfoyle MR, Barone DG, Kirollos RW, Trivedi RA, Santarius T. Radical, staged approach to extensive posterior fossa pediatric ependymoma: 3-dimensional operative video. Oper Neurosurg. 2018;14(6):705.
2. Dandy WE. Brain tumors: general diagnosis and treatment. Sugery of the Brain; 1945.
3. Mussi AC, Rhoton AL. Telovelar approach to the fourth ventricle: microsurgical anatomy. J Neurosurg. 2000;92(5):812–23.
4. Patra DP, Turcotte EL, Stonnington HO, Green DL, Batjer HH, Bendok BR. The telovelar approach: part 1—historical perspectives and anatomic considerations. Contemp Neurosurg. 2023;45(1):1.
5. Gudrunardottir T, Sehested A, Juhler M, Schmiegelow K. Cerebellar mutism. Childs Nerv Syst. 2011;27(3):355–63.
6. Sainte-Rose C, Cinalli G, Roux FE, Maixner W, Chumas PD, Mansour M, et al. Management of hydrocephalus in pediatric patients with posterior fossa tumors: the role of endoscopic third ventriculostomy. J Neurosurg. 2001;95(5):791–7.
7. Bhatia R, Tahir M, Chandler CL. The management of hydrocephalus in children with posterior fossa tumours: the role of pre-resectional endoscopic third ventriculostomy. Pediatr Neurosurg. 2009;45(3):186–91.
8. Vitali AM, Steinbok P. Depressed skull fracture and epidural hematoma from head fixation with pins for craniotomy in children. Childs Nerv Syst. 2008;24(8):917–23. discussion 925
9. Di Rocco C. Depressed skull fractures and epidural hematoma from head fixation with pins for craniotomy in children. Childs Nerv Syst. 2008;24(8):925.
10. Velho V, Naik H, Bhide A, Bhople L, Gade P. Lateral semi-sitting position: a novel method of patient's head positioning in suboccipital retrosigmoid approaches. Asian J Neurosurg. 2019;14(1):82–6.
11. Naylor RM, Rotter J, Proctor MR, Bite U, Ahn ES. Above-the-hairline suboccipital transverse approach for the treatment of Chiari I malformation. Oper Neurosurg. 2023;24(5):542–7.
12. Chen CC, Wei ST, Tsaia SC, Chen XX, Cho DY. Cerebrolysin enhances cognitive recovery of mild traumatic brain injury patients: double-blind, placebo-controlled, randomized study. Br J Neurosurg. 2013;27(6):803–7.
13. Onorini N, Spennato P, Orlando V, Savoia F, Calì C, Russo C, et al. The clinical and prognostic impact of the choice of surgical approach to fourth ventricular tumors in a single-center, single-surgeon cohort of 92 consecutive pediatric patients. Front Oncol. 2022;12:821738.
14. Sastry RA, Walek K, Leary OP, Rex N, Shaaya EA, Poggi JA, et al. Incidence, characteristics, and outcomes of Pseudomeningocele and cerebrospinal fluid fistula after posterior fossa surgery. World Neurosurg. 2022;164:e1094–102.
15. Carmel PW, Greif LK. The aseptic meningitis syndrome: a complication of posterior fossa surgery. Pediatr Neurosurg. 1993;19(5):276–80.
16. Ashida R, Nazar N, Edwards R, Teo M. Cerebellar mutism syndrome: an overview of the pathophysiology in relation to the cerebrocerebellar anatomy, risk factors, potential treatments, and outcomes. World Neurosurg. 2021;153:63–74.

Endonasal Endoscopic Skull Base Surgery in Children: Anatomical and Technical Considerations

59

Chandrashekhar E. Deopujari and Nishit J. Shah

Learning Objectives

- To understand the developmental anatomy of the skull base through embryology and imaging at various stages of growth in a child.
- Help in selection of proper cases suitable for the approach.
- To choose proper instruments for endonasal endoscopy in children.
- To learn variations in technical steps suitable to children.
- Importance of various closure techniques during endonasal surgery for intradural pathologies.

59.1 Introduction

Skull base surgery in children is not practised as often as in adults. Apart from the fact that pathologies in this age group are different than adults, developmental issues of the anatomy of skull base produce various challenges in surgery. This has not been elaborated or critically looked at either by the paediatric or skull base surgery groups in the past and has only received due recognition recently through efforts of the neuro-oncology and craniofacial plastic surgeons [1–3].

A proper understanding of embryology and anatomy of the skull base through radiological and cadaveric studies has encouraged the application of open as well as endonasal endoscopic surgeries in this area. We have presented here the relevant embryological and anatomical facts and evaluated the surgical experience for endonasal endoscopic approach in children and described our team's strategy in management of various lesions encountered in our practice.

C. E. Deopujari (✉)
Department of Neurosurgery, Bombay Hospital Institute of Medical Sciences, Mumbai, Maharashtra, India

N. J. Shah
Department of ENT, Bombay Hospital Institute of Medical Sciences, Mumbai, Maharashtra, India

59.2 Development of the Skull Base

The skeleton of the head may be divided into the facial skeleton—viscerocranium, and the head proper—the neurocranium. The neurocranium is further divided into the skull base—chondro/desmocranium, and the skull vault—calvarium.

The embryological origin of tissues is postulated by the "New Head Hypothesis" of Gans and Northcutt [4, 5]. The CNC, unlike the Neural Crest (NC) in the rest of the body, forms and migrates before the fusion of the cranial neural folds. Through a process of epithelial to mesenchymal transformation (EMT), it gives rise to connective tissue of the head and pharyngeal arches and the anterior half of the skull base. The face and thereby the viscerocranium is developed from the first two pharyngeal arches. The coronal suture is the junction between the tissues derived from rostral mesenchyme (posteriorly) and neural crest (anteriorly) [6–8].

The desmocranium, also referred to as the chondrocranium, is formed majorly through endochondral ossification. The notochord plays an important role in the induction and organisation of chondrogenesis. The rostral end of the notochord lies just below the region of the sella turcica within the body of the future sphenoid bone and is called 'Chorda dorsalis' [8]. This is thus a common region for the occurrence of chordomas. The tract of the notochord passes just outside and anterior to the sphenoid, comes in contact with the endoderm of the pharynx, and gives rise to the 'Tornwaldt's bursa' [9] and the corresponding depression on the anterior surface of the clivus—the 'Fossa navicularis'. From here, it re-enters the clivus and passes through the apical ligament of the dens, through the vertebral column.

The notochordal tissue at the chorda dorsalis is rich in sulphated glycosaminoglycans and is able to cause the mesenchymal cells in the area to become densely packed, to form cartilage [8]. The chordal cells also produce 'Chordin,' a key regulator of BMP-4, which controls ossification of the entire skull base and produces symmetry. The chondrifica-

I. N. Kanaan, V. Beneš (eds.), *Neuroanatomy Guidance to Successful Neurosurgical Interventions*,
https://doi.org/10.1007/978-3-031-59838-8_59

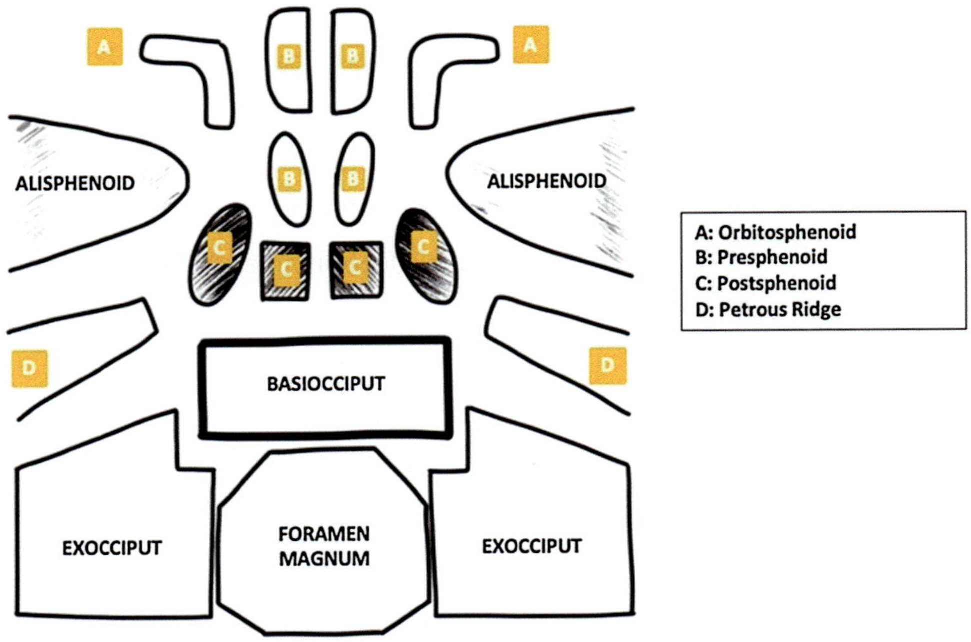

Fig. 59.1 The embryological progenitor cartilages of the sphenoid bone

tion process begins around week seven and produces a continuous cartilaginous framework of the skull base, encircling the pre-existing neurovascular structures. The first cartilage to be formed is the parachordal cartilage, which contributes to the formation of the basiocciput [7]. By birth, most of the skull base has ossified and growth occurs through the sphenooccipital and sphenopetrosal synchondroses [6–8].

The sphenoid develops through numerous cartilages, namely—alisphenoid (greater wing), orbitosphenoid (lesser wing), postsphenoid (dorsum sellae and clivus), presphenoid, and basisphenoid (body) [10] (Fig. 59.1). The tract of the adenohypophysis passes through the body of the sphenoid and may aberrantly persist to give rise to a patent craniopharyngeal canal.

The ethmoid contributes to the cribriform part of the anterior midline skull base. There are three connections of importance between the frontal and ethmoidal bones, namely—the foramen caecum, prenasal space, and fonticulus nasofrontalis. These may persist beyond the embryonal period and give rise to encephalocoeles [8].

59.3 Radiological Insights into Developmental Anatomy

The skull base in children is constantly changing as regards to size and pneumatisation. This is equally true of their facial morphology. The surgeon must thus consider the progression of such changes that affect the surgical corridor and manoeuvrability of the endoscope and instruments, so as to obtain safe access and exposure, while limiting damage and unnecessary resection of normal structures. The earliest report of endoscopic skull base surgeries in paediatric population was in 1990s for repair of congenital defects [11].

There are typically three bottlenecks along the corridor—the external nares/ pyriform aperture, the nasal passage at the level of the middle meatus, and the sellar floor itself, at the level of the inter-carotid space [12–15]. It is generally understood that while bony apertures are fixed, soft tissues can be stretched and manipulated to navigate instruments past them. Hence, it is useful to take measurements in both CT and MR images, in addition to clinical inspection, to get a complete idea about the anatomy in every patient.

The crucial challenges faced by the surgeon in paediatric skull base surgery are discussed:

1. *Narrow Pyriform aperture*: The first challenge is at the pyriform aperture, which forms the bony entrance of the nasal cavity. The aperture is made of the convergence of three bones: the vomer, the palatine bone, and the maxilla. The lateral bony limitation may hamper the manoeuvrability of instruments in children. The external nasal apertures are smaller in neonates and infants and show a very slow rate of growth (Fig. 59.2). The lowest admissible diameter of the nares is 3 mm [15], below which a paediatric endoscope of 2.7 mm diameter ought not to be forced through. A smaller size of the nares may prohibit the introduction of more than one instrument per side, impeding the four-hand technique used for complex procedures. The width measurements of the pyriform aperture, as assessed from pre-op CT scan, from the lateral borders, at various ages are: 15 mm at birth, 18 mm around 2 years, and 18 mm and 20 mm by 5 years.

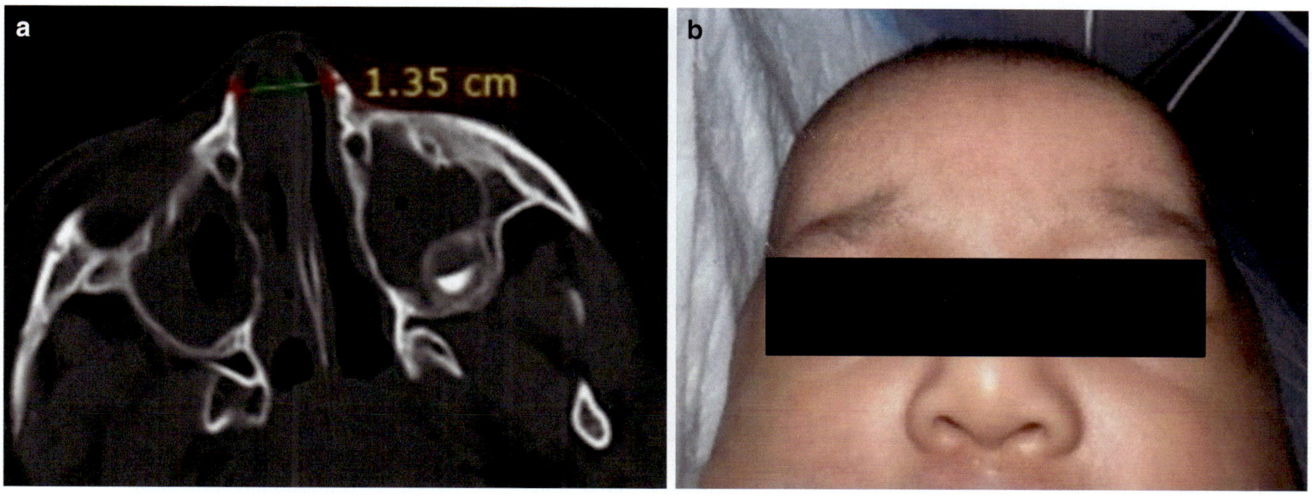

Fig. 59.2 Small pyriform aperture in children. (**a**) CT axial bone window showing the narrow size of the pyriform aperture in an infant. (**b**) Very small size of external nares in an infant

With small changes of 5 mm in the diameter, the cross-sectional dimension increases twofold, affording a four-handed instrumentation. The ease of access improves from 2 years to 6 years of age with gradual increase in the pyriform aperture diameter from 15 mm to 20 mm. A pyriform aperture of 20 mm provides an unrestricted path for four-handed surgical access [16]. If the diameter of the pyriform aperture is less than 15 mm, it is considered a relative contraindication for endoscopic surgeries.

2. *Inter-turbinate distance*: The distance between the inferior or middle meati of the two sides has not been shown to vary significantly with age [14, 15], therefore we may expect the same ease or difficulty navigating across them as we experience in adult surgery. The average inter-turbinate distance in children is 6–10 mm. This can easily be increased by lateralisation of the turbinates. In case of severely narrowed nasal passages, resection of a combination of middle turbinate and the bony septum is usually sufficient to create a roomy corridor.
3. *Sinus pneumatisation*: The sphenoid sinus pneumatisation is important for all trans-sphenoid surgeries. The visualisation of all the important neurovascular landmarks is important not only to keep the surgeon oriented, but also to avoid any potential complications. The sphenoid sinus is not pneumatised at birth. The pneumatisation of the sphenoid begins as early as 2–3 years of age and progresses very slowly, from anterior to posterior and inferior to superior [15, 17]. The pneumatisation occurs at the planum and the sella and moves towards the clivus after 10–15 years of age. The sinus is pneumatised to its full potential by adolescence [17].

   While a lack of pneumatisation may necessitate additional drilling, the bone being much softer in children does not significantly increase the difficulty of the approach. It does, however, mask most of the bony landmarks such as the optic canals and the carotid impressions, and thus, compel the use of adjuncts such as neuronavigation and Doppler. The patterns of sphenoid pneumatisation are classified based on the depth of pneumatisation with respect to the anterior wall of the sella.

   (a) Conchal sphenoid: it is a barely pneumatised sphenoid, where none of the posterior landmarks are clearly visible, a challenge for endonasal procedure (Fig. 59.3).
   (b) Presellar sphenoid: the pneumatisation occurs till the anterior face of the sella (Fig. 59.4). A trans-sphenoid approach may be challenging due to the relative non-visualisation of all landmarks, but can be overcome by sphenoid bone drilling with the aid of neuro-navigation.
   (c) Sellar sphenoid: The pneumatisation goes beyond the anterior wall of the sella. When it approaches the clivus, it is known as complete sellar pneumatisation (Fig. 59.5). This type is the most suitable for trans-sphenoid surgeries.
4. *Intercarotid distance* is one of the most important bottlenecks in the surgical approach to the midline skull base, significantly for trans-sellar, trans-tubercular, and trans-clival approaches. This parameter may change with age and sphenoid pneumatisation; however, it is generally agreed that it is smallest in children from birth to 4 years of age, and the lowest acceptable ICD for negotiating instruments is 10 mm [14]. Interestingly, the shorter length of the nasal cavity in children compensates for the smaller width, thus maintaining a similar angle of exposure as we see in adult patients. The lower limit of ICD should be 10 mm at the sella and 15 mm at the clivus for one to attempt an endonasal approach. ICD less than 8 mm is an absolute contraindication. In children less

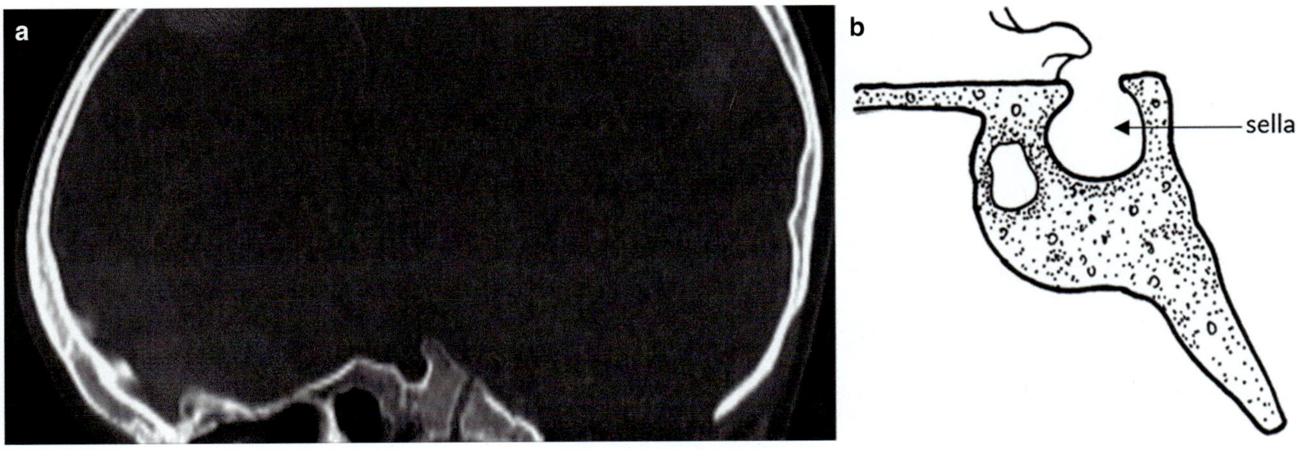

Fig. 59.3 Conchal pattern of pneumatisation of the sphenoid sinus. (**a**) CT bone window sagittal view. (**b**) Illustration of the same

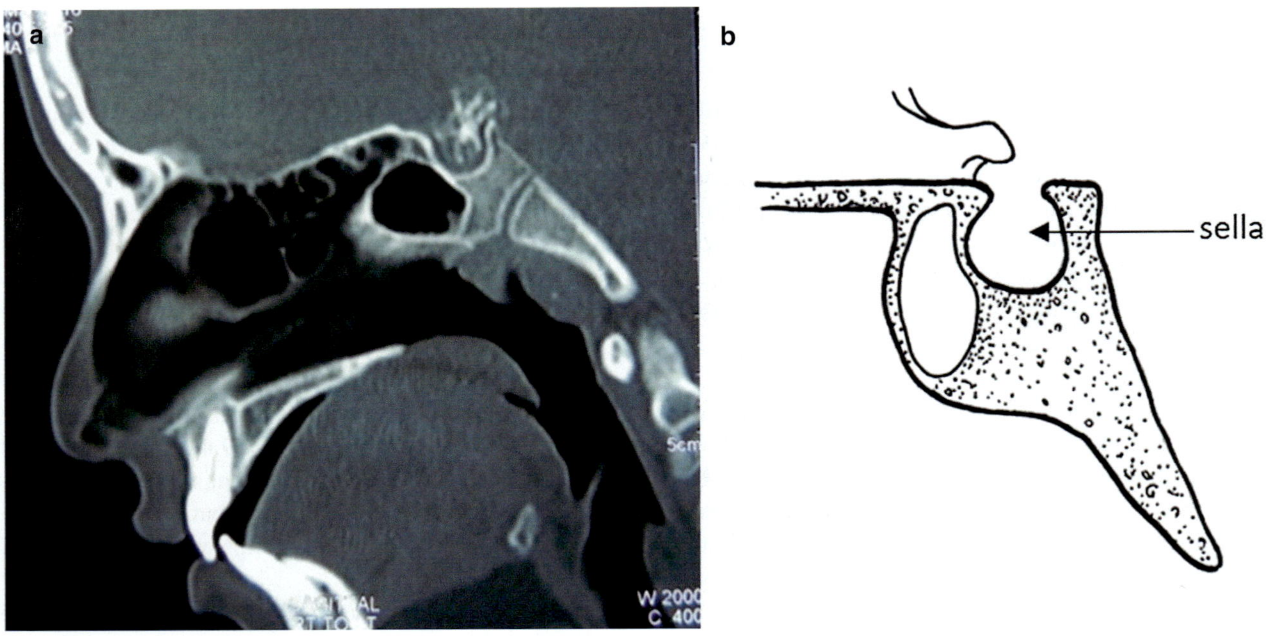

Fig. 59.4 Presellar pattern of pneumatisation of the sphenoid sinus. (**a**) CT bone window sagittal view (**b**) Illustration of the same

than 2 years of age, the ICA distance is less than 10 mm in children aged less than 2 years, making them a high-risk group for endonasal transsphenoidal surgeries [18]. This parameter must be regularly measured in pre-op CT Angiography or MRI T2 images (Fig. 59.6).

We have rarely seen a case where a skull base approach was not possible simply due to the age of the patient, and in fact with the use of properly sized scope and instruments, it is quite straightforward to perform a regular surgery with standard skull base principles [19, 20].

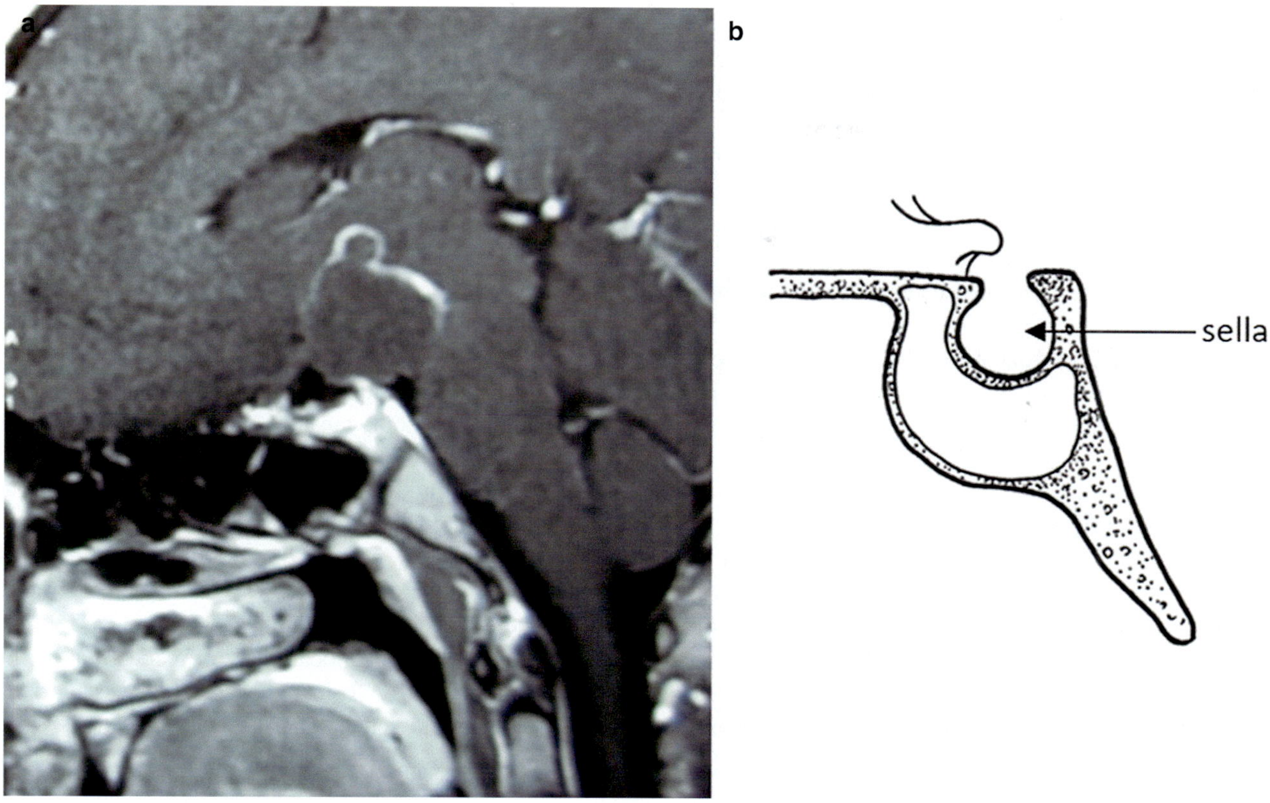

Fig. 59.5 Sellar pattern of pneumatisation of the sphenoid sinus. (**a**) CT bone window sagittal view (**b**) Illustration of the same

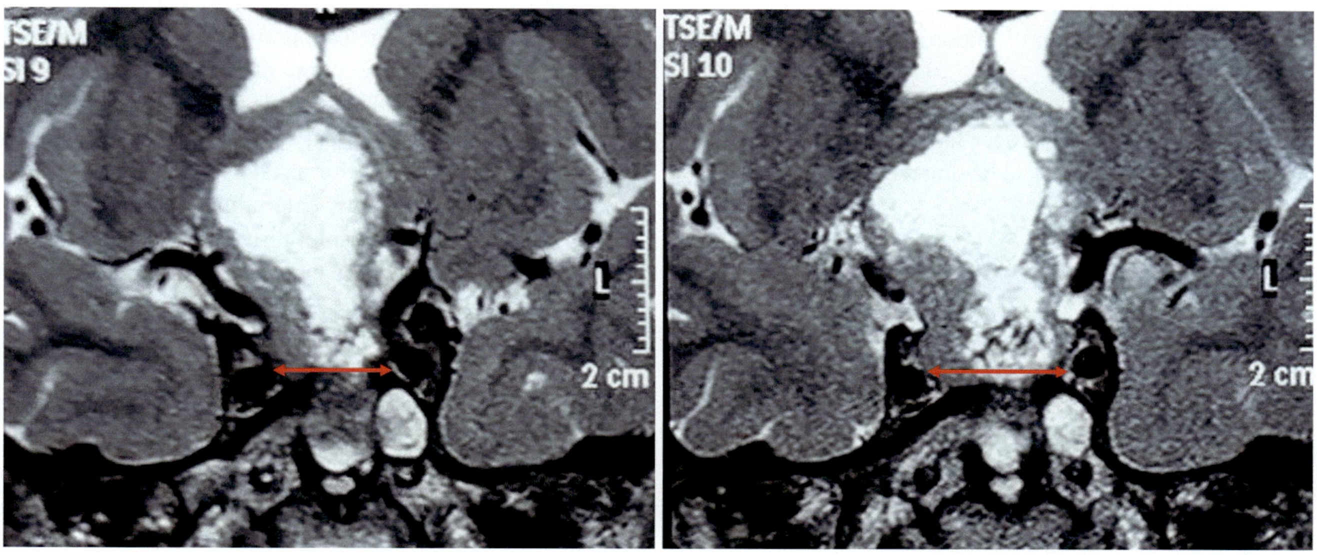

Fig. 59.6 Intercarotid distance in a case of sellar craniopharyngioma, with measurements showing the distance to be 12 mm

59.3.1 Anatomical Highlights

- The corridor for the endonasal approach in children has several bony and cartilaginous constraints.
- Anatomical landmarks are not completely formed at birth and develop over time.
- Small nasal aperture and narrow Intercarotid space are most important.
- Pneumatisation of sinuses proceeds in an age-dependent manner.
- In congenital defects, contents of the sac need to be carefully evaluated specially to rule out incorporation of neurovascular structures.

59.4 Surgical Approach

The surgical goal is to completely remove the lesion while minimizing collateral damage. This is particularly important in growing children to avoid disrupting overall craniofacial development. The preferred endonasal skull base approaches for paediatric patients include transethmoid, transsphenoid, and transpterygoid.

The surgical approach is determined by the site and extent of the lesion. Uninostril approach is less destructive and preserves the nasal septum, but restricts instrument manipulation. Binostril approach eases instrument manoeuvrability, but may require resection of important growth centre at the posterior bony septum.

The preservation of normal nasal anatomy is of utmost importance, not only for proper facial development but also to save important anatomical landmarks for the future.

59.4.1 Steps of Surgery

Patient positioning and equipment: The patient is placed supine with the head on a horse-shoe head holder, with 15 degrees of head elevation and 15 degrees of rotation to face the operating surgeon. The thigh may be prepared for harvesting a fascia lata graft and fat. The degree of neck extension is adjusted to adequately visualise the pathology. A lumbar drain is inserted only in cases where CSF leak is anticipated. We generally avoid it for fear of post-op meningitis.

We use a paediatric nasal endoscope which is of 2.7 mm diameter. Though the reduction in the scope diameter reduces the illumination transmitted via the light cable, vision of the surgical field cannot be compromised in an already narrow nasal cavity and 4 K HD camera system is preferred as it gives a crisp high-definition image.

With repeated introduction of the scope and instruments, the nasal mucosa may suffer injury which may later translate into nasal crusts and fibrin exudates, increasing chances of synechiae formation. Hence, all instrument manoeuvring must be done with utmost dexterity and fine movements. We prefer using otology micro-instruments in the nose to raise the nasoseptal flaps and for tumour dissection.

Electromagnetic Neuronavigation is set up for all extended procedures and if the sphenoid is inadequately pneumatised. A microvascular Doppler with nasal probe is frequently used.

Nasal stage: The initial nasal packing is done using large cottonoids, soaked and squeezed dry, using 4% xylocaine solution with 1:10000 adrenaline. Nasal packing must be done diligently, without causing mucosal trauma, lest it causes bleeding and hinders the scope vision. The inferior turbinates and the middle turbinates are lateralised to improve the nasosphenoid corridor and increase the working space. The middle turbinate is preserved, if possible, as it serves as an important landmark for the future, but may be sacrificed in cases where space is a constraint.

The next step is to raise local flaps for reconstruction. In case of an anticipated CSF leak, a nasoseptal vascularised Haddad Bassagasteguy flap (HBF) is raised, which is the workhorse flap for skull base reconstruction [21]. It is based on the septal branches of sphenopalatine artery. The superior incision is started from the sphenoid ostium and brought along the septum, staying under the olfactory groove markings, until the bony-cartilaginous junction of the septum. From this junction, the incision is swung as high as possible along the cartilaginous septum until the mucocutaneous junction of the septum. The inferior incision is started from above the eustachian tube and curved towards the septum along the choana. The inferior incision can go onto the floor or even the inferior meatus, determining the width of the HBF. The superior and the inferior incisions are then joined with a vertical incision at the mucocutaneous junction of the septum anteriorly.

The harvesting of an HBF renders the ipsilateral septal cartilage bare. In order to decrease nasal morbidity and crust formation and hasten septal healing, the opposite septal mucosa is harvested for a reverse flap. The superior and the inferior incisions remain similar as in HBF, except that they are taken only until the bony-cartilaginous junction of the septum. The vertical incision is taken on the anterior face of sphenoid, thus sacrificing the septal branches of the sphenopalatine arteries. In order to gain a binostril access to the sella, a posterior septectomy is done, which entails removal of vomer and the perpendicular plates of ethmoid. While the HBF is parked in the nasopharynx, the reverse flap is brought on to the opposite side through the posterior septectomy and used to cover the bare septal cartilage on the opposite side.

In some cases where a leak is not expected, one can harvest rescue flaps. Rescue flaps are essentially incomplete Haddad flaps that are based both anteriorly and posteriorly.

The superior and the inferior incisions are made exactly as in HBF except that the incisions are stopped at the bony cartilaginous junction on both sides. The posterior septectomy is done and the flaps are dropped down to gain access to the sphenoid sinus. This is a more conservative approach which preserves the septal mucosa on both sides and gives a feel of a normal septum once reposited. In an event of a CSF leak, either of the rescue flaps may be converted to a HBF and used for reconstruction.

Sphenoid stage: The classic endonasal trans-sphenoidal approach suffices for sphenoidal encephaloceles, pituitary tumours, and purely infradiaphragmatic Rathke cleft cysts (RCC) and craniopharyngiomas of the sellar variety, while extended trans-tubercular and trans-planar approaches are required for suprasellar craniopharyngiomas and RCCs, epidermoids, and hypothalamic-optic pathway lesions. Trans-ethmoid and cribriform approaches are required for mainly CSF leaks and occasionally to augment repairs of congenital defects. Inferior trans-clival extensions are required for odontectomy and chordomas.

Once the anterior wall of the sphenoid sinus is exposed, the rostrum and the keel of the sphenoid are removed using Luc's forceps. In a paediatric sphenoid sinus, which is not fully pneumatised, a neuronavigation system is an absolute must to double check on the landmarks before drilling. Opening of the posterior ethmoids in all cases would help not only to create more space, but also to identify the optic nerves by tracing them from the orbital apex.

The sphenoid sinus is stripped bare off its mucosa. This helps not only in better identification of the landmarks, but is also essential for proper flap uptake should we use one. All the sphenoidal septations are drilled and flattened. The exposure is said to be complete once optic nerves, lateral Optico-Carotid Recesses (OCR), and cavernous carotid bulges are identified and visualised properly. A straight suction should be able to get lateral to the ipsilateral carotid to confirm that enough space is available for manipulation.

Intracranial stage: For all cases, the dura is exposed and incised. We prefer an inverted U-shaped incision to lift up a flap of dura for pituitary surgery, since this is more convenient during closure. For other sellar-suprasellar pathologies, appropriate extensions to the bony exposure are made. The intercavernous sinuses may need to be coagulated in midline to prevent any injuries to carotid perforators before making necessary dural extensions. The position of the cavernous carotid arteries must also be periodically reconfirmed, using a combination of neuronavigation and intra-op Doppler probe.

Sellar and suprasellar pathologies are approached through a standard sellar approach, with dissection proceeding from posterior to lateral, anterior, and lastly superior, to prevent premature descent of the arachnoid.

Anterior extensions necessitate removal of the bone of the tuberculum and, at times, the planum. The anterior intercavernous sinus is coagulated and cut to extend the dural incision. One needs to extend more anteriorly for a suprachiasmatic pathology than a retrochiasmatic one, and a craniopharyngioma may necessitate multiple extensions. Although infrachiasmatic craniopharyngiomas are preferentially managed through the endonasal route [22], we practice a combined approach for larger retrochiasmal craniopharyngiomas extending into the third ventricle, which entail the addition of a transcranial approach (usually endoscopic, through a single burr hole) to target the suprachiasmatic/intraventricular component, followed by endonasal surgery after a short interval [23].

Lateral extensions, although rarely used, may be required in specific instances for specific tumour pathologies. We have used a corridor between the terminal ICA and M1 segment of the MCA to reach the lateral extension of a craniopharyngioma, to reach as far laterally as the sylvian cistern. A standard sellar approach with bony exposure extended laterally over the cavernous sinuses was used, coupled with a 30° endoscope and angled instruments. The surgeon must demonstrate a high degree of flexibility and adaptability in their approach, to match the variability of the tumour itself, so as to be able to target it successfully [24].

Pituitary transposition may be required to approach a retroinfundibular craniopharyngioma or more posterior pathology such as in the interpeduncular cistern, viz. epidermoid or dorsum sellae meningioma. We do not perform this in children, since in our experience in adults, it produces pan hypopituitarism in almost all cases.

A trans-arachnoidal plane of dissection is created between the lesion and important neurological structures, to avoid injuring small feeders. Sharp dissection is preferred during extratumoural cisternal dissection, to avoid avulsion injuries. The infundibulum may be sharply divided if inseparably involved by the lesion and the patient already has hypopituitarism.

In case of an encephalocoeles, the bony edges of the defect must be well-demarcated and the neck of the dural sac visualised. The herniated brain may be shrunk using cautery after making sure of its contents and excised if too bulky. The arachnoid defect is carefully exposed to achieve closure. Any remaining defect is plugged using fat globules. This is followed by multi-layered closure using fat, fascia, tissue glue, pedicled mucoperiosteal flap, and a variety of packing materials such as gelatin sponge or collagen sheets.

59.5 Illustrative Cases

Case 1 A 4-month-old infant was investigated by a paediatrician for difficulty in breathing and observed to have a right nasal mass. Imaging revealed the mass to be a fronto-ethmoidal encephalocele (Fig. 59.7a). The child was referred to our institute for further management. It was deemed fit to be treated by endoscopic trans-nasal surgery (Fig. 59.7b). The overlying mucosa was first coagulated (Fig. 59.7c) and excised to reveal the dural sac. The bony defect was delineated all around (Fig. 59.7d). The dura was coagulated to shrink it. After confirming the absence of any CSF leak, the defect was reconstructed using a piece of septal cartilage (Fig. 59.7e). The repair was then completed using fat, naso-septal flap, fibrin glue (Fig. 59.7f), and gelfoam (Fig. 59.7g). The infant recovered without complication.

Case 2 A 13-year-old girl, known case of Prolactinoma, on pharmacotherapy, presented with worsening of visual function after initial improvement. Imaging revealed a large pituitary tumour with multiple extensions, Hardy-Wilson Grade III C, Knosp Grade I—II (Fig. 59.8a, b). It was decided to operate it through a trans-sphenoidal route. Intraoperatively, the tumour was found to be soft and easily suckable (Fig. 59.8c, d). A complete excision could be achieved. The anterior third ventricular cavity was completely exposed by the end of the surgery (Fig. 59.8e) and the sellar defect was closed in a robust multi-layered manner to prevent CSF leak. Post-operatively, the patient had transient hypocortisolism and DI, both of which improved over a course of weeks.

Case 3 An eight-year-old boy was brought with complaints of headache and diminution of vision for a couple of months. Imaging advised by his family physician revealed a sellar cystic mass, supposedly a Rathke's cleft cyst (Fig. 59.9a). The patient was treated with trans-sphenoidal surgery (Fig. 59.9b). We drained the cyst (Fig. 59.9c, d) and lined it with nasal mucosa (Fig. 59.9e), to prevent recurrence, as described in literature [25, 26]. The patient had an uneventful post-op course and remained recurrence-free after 4 years (Fig. 59.9f).

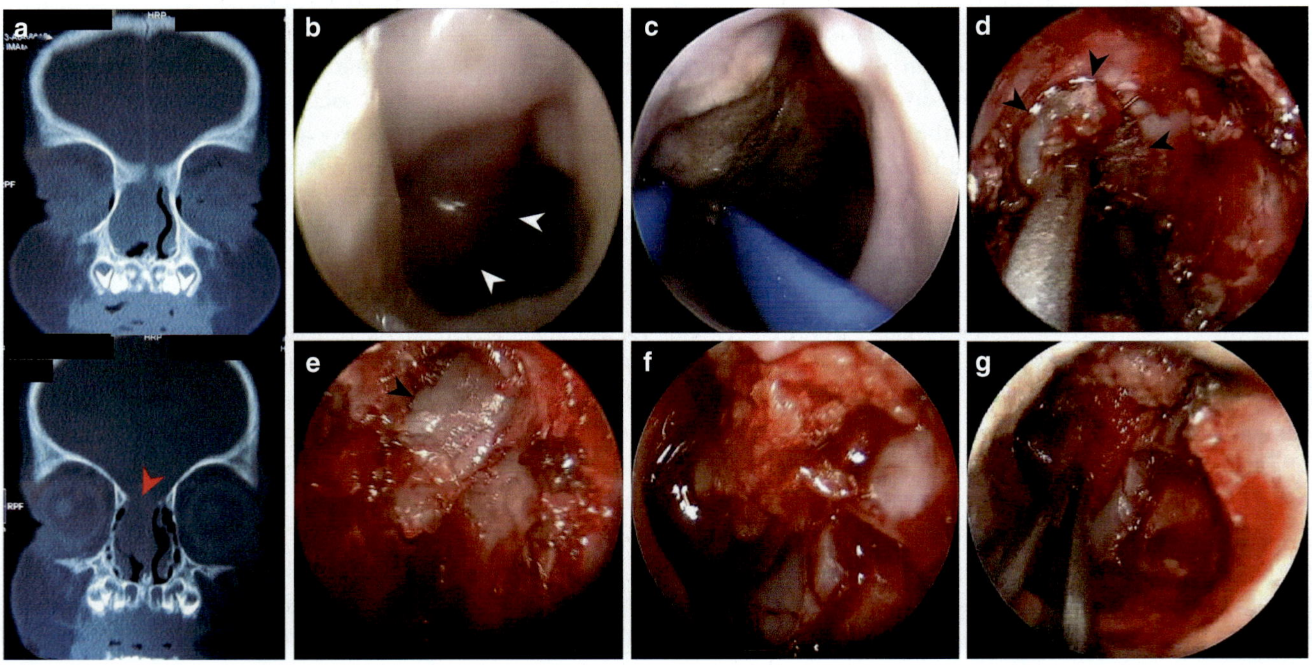

Fig. 59.7 Surgery for fronto-ethmoidal encephalocele. (**a**) CT bone window coronal views showing the defect at the level of the frontoethmoidal junction, just right of the midline (red arrowhead). (**b**) Endoscopic view of the encephalocele sac (white arrowheads). (**c**) Overlying mucosa being coagulated. (**d**) Dural sac of the encephalocele after excision of overlying mucosa. Bony defect visible all around (black arrowheads). (**e**) Bony defect bridged (black arrowhead) with piece of septal cartilage. (**f**) Repair with fat and nasoseptal flap. (**g**) Final appearance of the repair

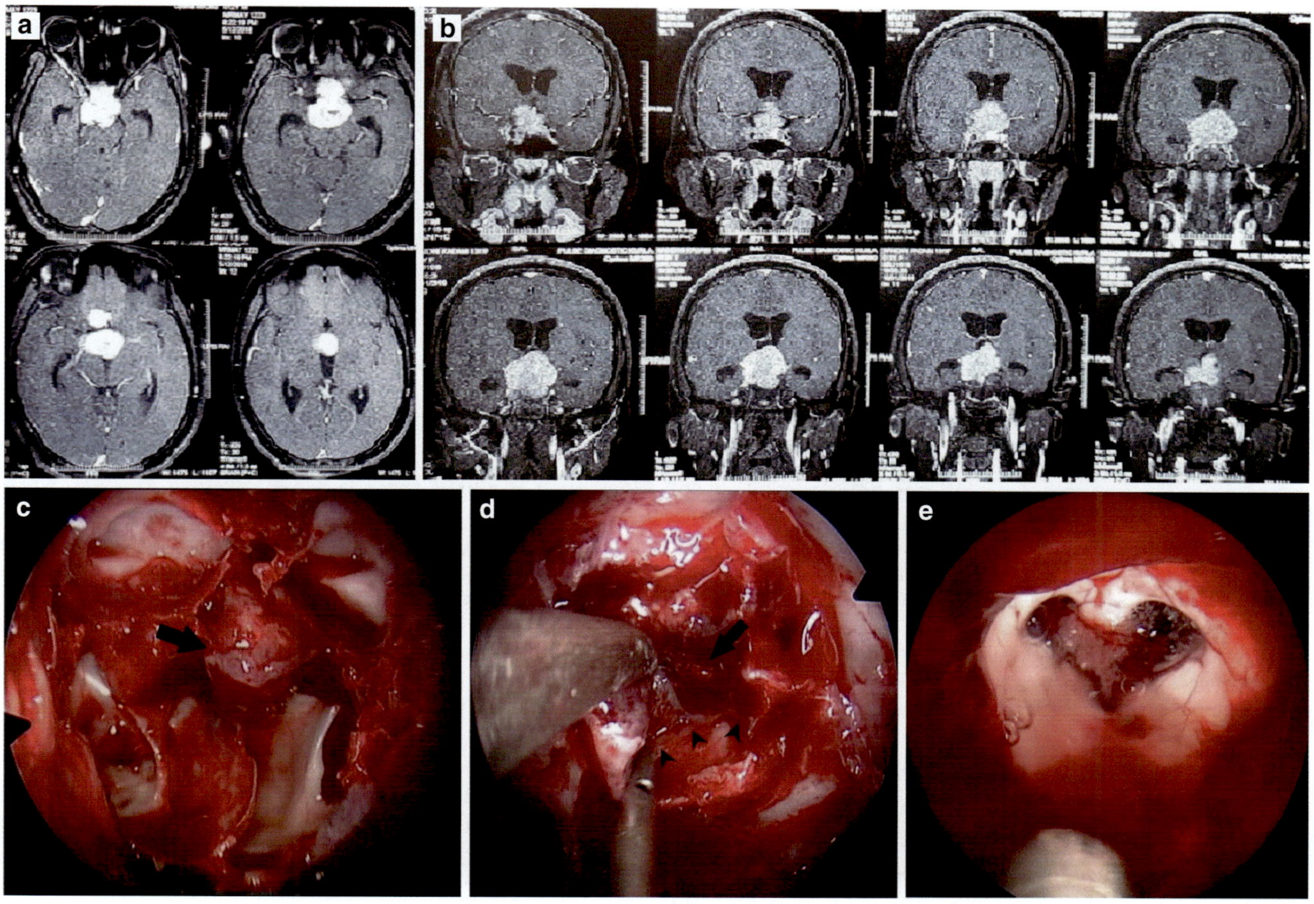

Fig. 59.8 Giant Prolactinoma unresponsive to Medical Treatment. (**a**) Axial CEMRI showing an intensely enhancing sellar-suprasellar mass. (**b**) Coronal CEMRI showing the relationship of the mass with the cavernous sinuses and the anterior third ventricle. (**c**) Endoscopic view of the sellar floor with focal invasion of the bone and tumour mass seen prolapsing through the defect (black arrow). (**d**) Soft, vascular tumour (black arrow) being resected with a suction cannula. Normal pituitary seen posteriorly with the plane of separation (black arrowheads). (**e**) Anterior third ventricular cavity seen at the end of tumour excision

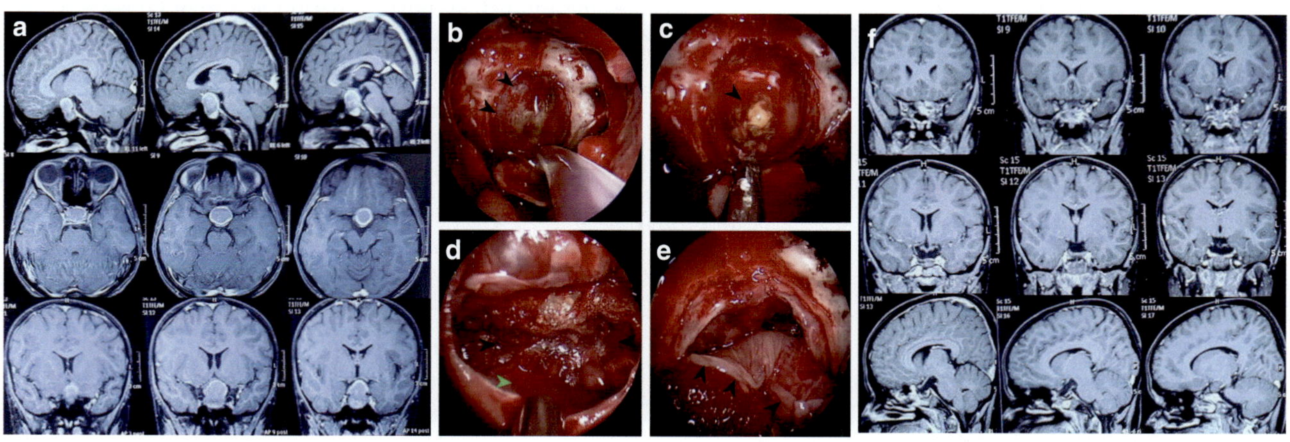

Fig. 59.9 Rathke's Cleft Cyst. (**a**) CEMRI in sagittal, axial, and coronal orientations showing a cystic sellar lesion. (**b**) Tense bulge of cystic lesion (black arrowhead). (**c**) Cyst puncture with extrusion of epithelial debris (black arrowhead). (**d**) Cavity post-cyst excision (black arrowhead). Normal pituitary seen in the background (green arrowhead). (**e**) Nasal mucosal flap being placed into the cyst cavity, coated over a piece of gelfoam (black arrowheads). (**f**) Post-op MRI showing complete excision of the cyst

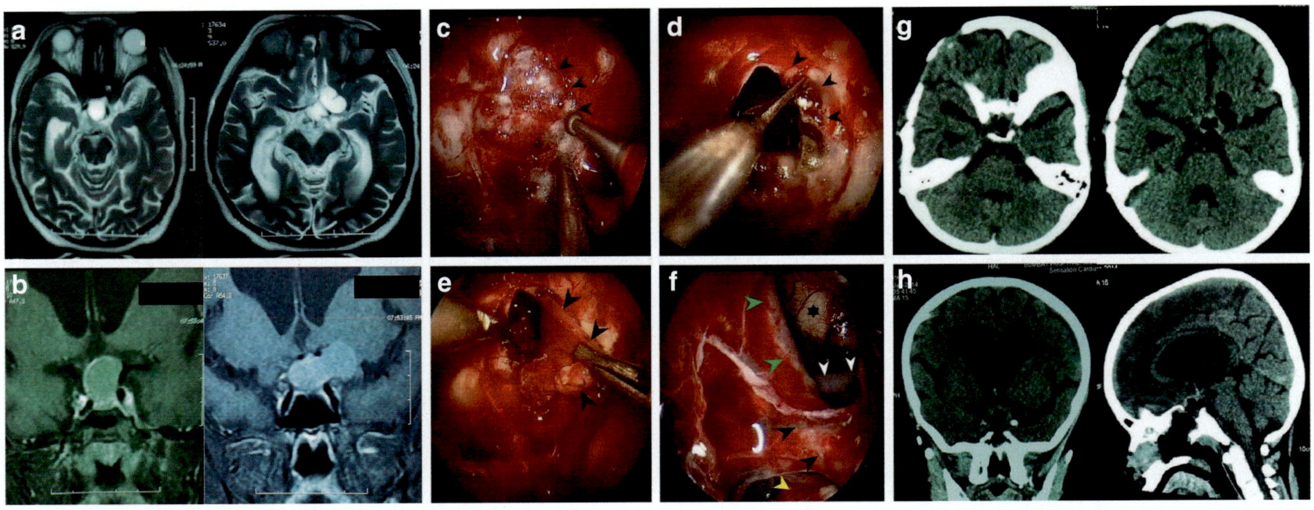

Fig. 59.10 Recurrent Sellar-suprasellar Craniopharyngioma with left parasellar extension. (**a**) Axial T2W image showing a sellar craniopharyngioma with left parasellar extension. (**b**) Coronal T1W MRI image showing the neck of the parasellar extension in the angle between the left ICA, MCA and ACA. (**c**) Sella being drilled over the region of the left cavernous sinus (black arrowheads). (**d**) Durotomy being extended on the left (black arrowheads) to facilitate parasellar tumour excision. (**e**) Traction being applied to the tumour capsule (black arrowheads). (**f**) Cavity after tumour excision showing the corridor into the left sylvian cistern between the left ICA (black arrowheads), ACA (green arrowheads), and the left MCA (white arrowheads). The left optic nerve (yellow arrowhead is seen below the ICA. The inferior cortex of the left frontal lobe is seen in the background (black asterix). (**g**, **h**) Post op CT scan showing complete tumour excision

Case 4 A 9-year-old boy was brought by his parents with complaints of progressive diminution of vision, noticed 4 weeks before. On examination, he was found to have severe visual impairment, worse in the left eye—only perception of light, while he was able to count fingers at 3 ft. with his right eye. Fundoscopy revealed optic atrophy in both eyes. Imaging showed a sellar craniopharyngioma with a large left parasellar extension (Fig. 59.10a, b). We decided to operate the patient by a trans-sphenoidal approach. The parasellar extension was considered amenable to this approach due to the presence of a wide neck connecting it to the midline component. During surgery, the sellar exposure (Fig. 59.10c) and durotomy were extended slightly more on the left (Fig. 59.10d) and the lateral extent of the cyst was excised with gentle traction (Fig. 59.10e, f). The child recovered without any complication, and the post-op scan showed complete removal of the tumour (Fig. 59.10g, h).

59.6 Complications

- CSF leak: This is the most common and feared complication of trans-nasal skull base surgery. Its occurrence can be predicted intraoperatively, based on the size of the skull base defect and the amount of intra-arachnoid dissection performed [27]. Pre-emptive steps should be taken to ensure adequate watertight closure, including the use of gasket seal or bath plug repair techniques, reinforced with fibrin glue, and vascularised pedicled mucoperichondrial grafts. The repair is endoscopically inspected on the fourth or fifth day. If a CSF leak develops at any time during the post-operative period, empirical antibiotics must be instituted immediately and the patient taken up for endoscopic CSF leak repair surgery at the earliest.
- Infundibular injury: The infundibulum being a thin and flimsy structure may be very difficult to protect. The key principle is to avoid blunt dissection or traction on the delicate structures and avoid drilling of calcified fragments. Diabetes insipidus (DI), often alternating with SIADH, is the most common hormonal abnormality in the immediate post-op period and may be seen in up to 18% of cases. Most cases recover over time, while about 2% patients experience permanent DI. Regimens of hormone replacement therapy are now widely available, albeit expensive to maintain.
- Vascular injury: Several structures such as the cavernous or the intercavernous sinuses may get injured during trans-sphenoidal surgery. Perforators from the circle of Willis may also get injured during the intracranial phase of the surgery, especially due to the narrow corridors in children. As children are unable to tolerate blood loss well, care must be taken to go slow and follow optimal microsurgical techniques to avoid injury to vessels, keeping a low threshold for transfusion.
- Neural injury: Occasionally, pressure or traction on cavernous sinus can produce III, IV, or VI nerve paresis. Hypothalamic injury should be avoided with gentle dissection technique.
- Endonasal trauma: This is a result of overcrowding of instruments in a narrow corridor, with frequent frictional

injury due to the passage of the instruments as well as bright light of the endoscope may result in thermal injury to the mucosa with complications like crusting and discharge, parosmia, formation of synechiae, and development of external nasal deformity. It is imperative to involve an ENT colleague for adequate management of the nasal phase of the surgery and post-op endoscopic nasal inspection and cleaning.

59.7 Surgical Pearls

- Endonasal endoscopic approach can be safely practised in children over 3 years.
- Surgery can still be performed safely with a focussed study of pre-operative imaging and the use of adjuncts such as neuronavigation and Doppler.
- Most important limitations to instrument manoeuvrability are constituted by a narrow pyriform aperture and small intercarotid distance.
- Two-nostril-four-hand approach can be successfully done for children.
- Ideally suited for pituitary tumours, as proved by a large worldwide experience.
- Also recommended for craniopharyngiomas, Rathke's cleft cysts, and chordomas.
- Various vascularised mucosal flaps are available for closure even in children.
- Importance of postoperative nasal care for hygiene to prevent synechiae.
- It is a two-surgeon interdisciplinary surgery and a familiar team is of utmost importance.

59.8 Discussion

Endonasal endoscopic surgery for skull base lesions is being increasingly practiced. Though first described for young children in 1999 for repair of congenital defects by Van Den Abeele T et al. in craniofacial literature [11], it has come to the attention of neurosurgeons after Locatelli advocated the endoscopic approach for congenital skull base defects [28]. The role of this technique for repair of congenital defects is invaluable [21]. Improvement in endoscopy equipment, better understanding of CT/ MR imaging during growth, and surgical adjuncts, viz. navigation have contributed to more frequent use of this approach [14, 29]. Numerous studies and systematic reviews reporting the results of endoscopic endonasal surgery in children are now available [23, 30–32]. Jane et al. reported this technique to be better than conventional open surgery for selected cases [33].

Pituitary surgery and sellar craniopharyngiomas are best suited for this approach in presence of good pneumatisation of sphenoid [32]. However, suprasellar lesions, viz. Craniopharyngiomas, Rathke's cleft cysts, and chordomas, need more skill and experience and a team approach with a familiar ENT and neurological surgeon. The binostril approach can be carried out for radical tumour excision and also allows opportunity for proper reconstruction with vascularised flaps [20].

The extension to endonasal endoscopic odontoidectomy and anteriorly placed lesions up to craniovertebral junction has also been successfully done in a series of 133 patients [34]. The anatomical studies done in other continents have also shown it to be feasible across the globe [15]. We have not encountered any case where it was not possible due to the young age of the patient.

Feasibility of procedure and good cosmetic outcome are being increasingly reported [35]. Retrospective review of 11 years' experience in 2 centres has recently been reported for 92 patients undergoing 130 procedures with good results and complications equivalent to the adult cohort [36]. A recent systematic review recommends more consistent reporting for assessment of outcome after analysing 12 studies discussing a total of 399 patients aged 0–18 years which met inclusion criteria [37].

59.9 Conclusion

The endonasal endoscopic approach has added a new and important dimension to the management of skull base lesions. After analysing the various pathologies treated across all age categories, we feel that it is a safe and efficacious approach in appropriately selected cases over 3 years of age. The replicable results from specialised teams at various centres should reassure new surgeons with further interest in this area. The philosophy of radical vs. conservative approach for craniopharyngioma or optochiasmal-hypothalamic gliomas remains with the treating group. However, better clearance of the tumour and better visual outcome have been shown in most studies.

The role of team approach is most crucial in our opinion for proper selection of cases, instrumentation, closure of the defects as well as post-op follow-up for avoiding short-term complications and obtaining good results in long term.

In very young children, especially below 3 years of age, highly specialised care is required. Though good reconstruction with vascularised flaps has been obtained even in infants, possibility of need for transcranial repair should be kept in mind. With more use of this approach, it will soon be established as a standard approach in children for many skull base pathologies.

We thank Aniruddha Bhagwat, Suhas Malineni, and Sai Spoorthi Naik for their help with the manuscript, and members of our neurosurgical skull base team for their role in the surgical management of the patients in our institutional series. We thank Amol Mittal for his contributions to the artwork in this chapter.

References

1. Hayhurst C, Williams D, Yousaf J, Ichardson DR, Pizer B, Mallucci C. Skull base surgery for tumors in children: long-term clinical and functional outcome. J Neurosurg Pediatr. 2013;11:496–503.
2. Tsai EC, Hawryluk G, Rutka JT. Tumors of the skull base in children. In: Neuro-oncology of CNS tumors. Berlin Heidelberg: Springer; 2006. p. 555–65.
3. Gruber DP, Brockmeyer D. Pediatric skull base surgery. 1. Embryology and developmental anatomy. Pediatr Neurosurg. 2003;38:2–8.
4. Gans C, Northcutt RG. Neural crest and the origin of vertebrates: a new head. Science (New York, NY). 1983;220:268–74.
5. Northcutt RG. The new head hypothesis revisited. J Exp Zool B Mol Dev Evol. 2005;304:274–97.
6. Jin SW, Sim KB, Kim SD. Development and growth of the normal cranial vault: an embryologic review. J Korean Neurosurg Soc. 2016;59:192–6.
7. Kerr DA, Rosenberg AE. Embryology of the skull base and vertebral column. In: Chordomas and chondrosarcomas of the skull base and spine. Academic Press; 2018. p. 3–9.
8. Di Ieva A, Bruner E, Haider T, Rodella LF, Lee JM, Cusimano MD, Tschabitscher M. Skull base embryology: a multidisciplinary review. Childs Nerv Syst. 2014;30:991–1000.
9. Miller RH, Sneed WF. Tornwaldt's bursa. Clin Otolaryngol Allied Sci. 1985;10:21–5.
10. Yamamoto M, Abe H, Hirouchi H, Sato M, Murakami G, Rodríguez-Vázquez JF, Abe S. Development of the cartilaginous connecting apparatuses in the fetal sphenoid, with a focus on the alar process. PLoS One. 2021;16:e0251068. https://doi.org/10.1371/JOURNAL.PONE.0251068.
11. Van Den Abbeele T, Elmaleh M, Herman P, François M, Narcy P. Transnasal endoscopic repair of congenital defects of the skull base in children. Arch Otolaryngol Head Neck Surg. 1999;125:580–4.
12. Nemzek WR, Brodie HA, Hecht ST, Chong BW, Babcook CJ, Seibert JA. MR, CT, and plain film imaging of the developing skull base in fetal specimens. AJNR Am J Neuroradiol. 2000;21:1699.
13. Hughes DC, Kaduthodil MJ, Connolly DJA, Griffiths PD. Dimensions and ossification of the normal anterior cranial fossa in children. Am J Neuroradiol. 2010;31:1268–72.
14. Banu MA, Guerrero-Maldonado A, Mccrea HJ, Garcia-Navarro V, Souweidane MM, Anand VK, Heier L, Schwartz TH, Greenfield JP. Impact of skull base development on endonasal endoscopic surgical corridors. J Neurosurg Pediatr. 2014;13:155–69.
15. Sawant N, Sabarish SS, George T, Shivhare P, Sudhir BJ, Kesavadas C, Abraham M, Jissa VT, Nair P. A study of the developing paediatric skullbase anatomy and its application to endoscopic endonasal approaches in children. Neurol India. 2020;68:1065–72.
16. London NR, Rangel GG, Walz PC. The expanded endonasal approach in pediatric skull base surgery: a review. Laryngoscope Investig Otolaryngol. 2020;5:313.
17. Rastatter JC, Snyderman CH, Gardner PA, Alden TD, Tyler-Kabara E. Endoscopic endonasal surgery for sinonasal and skull base lesions in the pediatric population. Otolaryngol Clin N Am. 2015;48:79–99.
18. Tatreau JR, Patel MR, Shah RN, McKinney KA, Wheless SA, Senior BA, Ewend MG, Germanwala AV, Ebert CS, Zanation AM. Anatomical considerations for endoscopic endonasal skull base surgery in pediatric patients. Laryngoscope. 2010;120:1730–7.
19. Na MK, Jang B, Choi KS, et al. Craniopharyngioma resection by endoscopic endonasal approach versus transcranial approach: a systematic review and meta-analysis of comparative studies. Front Oncol. 2022;12:6518.
20. Deopujari CE, Shah NJ, Shaikh ST, Karmarkar VS, Mohanty CB. Endonasal endoscopic skullbase surgery in children. Childs Nerv Syst. 2019;35:2091–8.
21. Shah N, Deopujari C, Bommakanti V. The reconstruction of skull base defects in infants using pedicled nasoseptal flap-a review of four cases. Childs Nerv Syst. 2019;35:2157–62.
22. Jane JA, Prevedello DM, Alden TD, Laws ER. The transsphenoidal resection of pediatric craniopharyngiomas: a case series. J Neurosurg Pediatr. 2010;5:49–60.
23. Deopujari CE, Karmarkar VS, Shah N, Vashu R, Patil R, Mohanty C, Shaikh S. Combined endoscopic approach in the management of suprasellar craniopharyngioma. Childs Nerv Syst. 2018;34:871–6.
24. Koutourousiou M, Fernandez-Miranda JC, Wang EW, Snyderman CH, Gardner PA. The limits of transsellar/transtuberculum surgery for craniopharyngioma. J Neurosurg Sci. 2018;62:301–9.
25. Kino H, Akutsu H, Tanaka S, et al. Endoscopic endonasal cyst fenestration into the sphenoid sinus using the mucosa coupling method for symptomatic Rathke's cleft cyst: a novel method for maintaining cyst drainage to prevent recurrence. J Neurosurg. 2019;133:1710–20.
26. Kuan EC, Yoo F, Chyu J, Bergsneider M, Wang MB. Treatment outcomes of Rathke's cleft cysts managed with marsupialization. J Neurol Surg B Skull Base. 2017;78:112–5.
27. Stapleton AL, Tyler-Kabara EC, Gardner PA, Snyderman CH, Wang EW. Risk factors for cerebrospinal fluid leak in pediatric patients undergoing endoscopic endonasal skull base surgery. Int J Pediatr Otorhinolaryngol. 2017;93:163–6.
28. Locatelli D, Rampa F, Acchiardi I, Bignami M, Pistochini A, Castelnuovo P. Endoscopic endonasal approaches to anterior skull base defects in pediatric patients. Childs Nerv Syst. 2006;22:1411–8.
29. Banu MA, Rathman A, Patel KS, Souweidane MM, Anand VK, Greenfield JP, Schwartz TH. Corridor-based endonasal endoscopic surgery for pediatric skull base pathology with detailed radioanatomic measurements. Operat Neurosurg. 2014;10:273–93.
30. Chen W, Gardner PA, Branstetter BF, Liu SD, Chang YF, Snyderman CH, Goldstein JA, Tyler-Kabara EC, Schuster LA. Long-term impact of pediatric endoscopic endonasal skull base surgery on midface growth. J Neurosurg Pediatr. 2019;23:523–30.
31. McDowell MM, Chiang M, Abou-Al-Shaar H, Zenonos GA, Wang EW, Snyderman CH, Gardner PA. Applications of endoscopic endonasal surgery in early childhood: a case series. Pediatr Neurosurg. 2021;56:519–28.
32. Kim YH, Lee JY, Phi JH, Wang KC, Kim SK. Endoscopic endonasal skull base surgery for pediatric brain tumors. Childs Nerv Syst. 2019;35:2081–90.
33. Soldozy S, Yeghyayan M, Yağmurlu K, Norat P, Taylor DG, Kalani MYS, Jane JA, Syed HR. Endoscopic endonasal surgery outcomes for pediatric craniopharyngioma: a systematic review. Neurosurg Focus. 2020;48:E6. https://doi.org/10.3171/2019.10.FOCUS19728.
34. Chivukula S, Koutourousiou M, Snyderman CH, Fernandez-Miranda JC, Gardner PA, Tyler-Kabara EC. Endoscopic endonasal skull base surgery in the pediatric population: clinical article. PED. 2013;11:227–41.
35. Low CM, Hwang PH. Endoscopic skull base surgery in children. Curr Otorhinolaryngol Rep. 2022;10:440–6.
36. Behbahani M, Rastatter JC, Eide J, Karras C, Walz P, Suresh K, Leonard JR, Alden TD. Pediatric endoscopic endonasal skull base surgery: a retrospective review over 11 years. World Neurosurg. 2023;170:e70–8.
37. Lopez EM, Farzal Z, Dean KM, Miller C, Morse JC, Ebert CS, Kimple AJ, Thorp BD, Zanation AM. Outcomes in pediatric endoscopic skull base surgery: a systematic review. J Neurol Surg B Skull Base. 2023;84:024–37.

Part V

Gliomas

Anatomy and White Matter Dissection for Surgery of Low-grade Gliomas

60

Abuzer Güngör and Yücel Doğruel

Learning Objectives

- Understanding organization and anatomy of white matter.
- Understanding the basic principles of white matter dissection.
- Surgical planning and approach selection in management of low-grade gliomas.
- Learning the role of imaging techniques in preoperative planning for low-grade gliomas.

60.1 Introduction

Understanding white matter is crucial for various aspects such as brain function, individual differences, diagnosis and treatment, imaging advancements, and surgical approaches. Researchers utilize various techniques to identify and avoid damaging white matter tracts, including white matter dissection, white matter tractography, quantitative analysis of white matter tracts, and structural and functional imaging [1–5].

Fiber dissection is a time-consuming and complex procedure that significantly contributes to understanding brain anatomy, particularly in microneurosurgery. It enhances surgical orientation, reduces complications, and enables examination of the three-dimensional structure of white matter tracts. Neurosurgeons benefit from fiber dissection by improving their understanding of spatial relationships and aiding surgical approach selection [6, 7].

60.2 Background

The Klingler's technique is the most widely utilized approach for white matter dissection [1, 8]. The procedure involves several steps, including brain preparation, leptomeningeal removal, cortical sparing, dissection, and visualization. Leptomeningeal removal involves the meticulous elimination of the dura mater vessels and the arachnoid covering the brain to expose the white matter tracts. Dissection is performed using specialized tools in a systematic manner, progressing from lateral to medial or vice versa. The dissected white matter tracts are visualized using staining, microscopy, or imaging techniques to enable a detailed examination of their structure. Integration of fiber dissection with other methodologies like tractography, augmented reality, and virtual reality simulations has shown promise in improving surgical orientation and reducing complications in neurosurgery [6, 7, 9–14].

Different researchers have used Klingler's technique with minor modifications. The suggested time interval for initiating fixation after death and brain removal ranged from 8 h to 18 h, with a preference for completion within 12 h. Fixation involves immersing specimens in a formalin solution, with the most common concentration being 10%. However, concentrations ranging from 10% to 15% or below 10% (typically 4% to 5%) have also been employed. The duration of fixation varies from 24 h to 3 months [1]. For brain preservation, freezing times can range from 8 h to 1 month, while temperatures can vary from −5 °C to −80 °C. Thawing can be done at room temperature or with the use of water. After thawing, specimens can be preserved in either formalin or alcohol solution during the dissection process [1].

A. Güngör (✉)
Faculty of Medicine Department of Neurosurgery, İstinye University, Istanbul, Turkey

Liv Hospital Vadistanbul, Neurosurgery Clinic, Istanbul, Turkey

Y. Doğruel
Department of Neurosurgery, Health Sciences University, Tepecik Training and Research Hospital, İzmir, Turkey

I. N. Kanaan, V. Beneš (eds.), *Neuroanatomy Guidance to Successful Neurosurgical Interventions*,
https://doi.org/10.1007/978-3-031-59838-8_60

60.3 Anatomy Highlights

- Association Fibers (Short Association fibers, superior longitudinal fasciculus/arcuate (SLF/AF) complex, extreme capsule, uncinate fasciculus, fronto-occipital fasciculus, cingulum, inferior longitudinal fasciculus, fornix).
- Commissural Fibers (Corpus callosum, anterior commissure, hippocampal commissure).
- Projection Fibers (Internal capsule - corona radiata).

60.4 Association Fibers

60.4.1 Short Association Fibers (U Fibers)

Before initiating fiber dissection, the sulcal and gyral anatomy should be studied (Fig. 60.1a, b). Upon removing the gray matter, the U fibers (short association fibers) can be observed spanning the brain's surface, connecting neighboring cortices across the hemisphere (Fig. 60.1c). These fibers, extending vertically to a depth of approximately 2–2.5 cm, should be considered when planning a transcortical incision to minimize injury. The trajectory of these fibers can be determined by visualizing the adjacent sulcus [15].

60.4.2 Superior Longitudinal Fasciculus/Arcuate (SLF/AF) Complex

The SLF/AF complex, a crucial collection of association fibers within the cerebral hemispheres' white matter, serves to interconnect the frontal lobe with the parietal, occipital, and temporal lobes on the same side (Fig. 60.1d). The AF and SLF were previously considered synonymous; however, this outdated nomenclature should be discarded, as numerous studies have clearly established the two bundles as distinct entities, both anatomically and functionally. Therefore, the AF and SLF should be regarded as separate structures, and efforts should be made to differentiate them accurately in research and clinical settings [16]. The pathways of the SLF/AF are integral to fundamental cognitive functions such as language, memory, attention, and emotions [17].

60.4.3 Extreme Capsule

The extreme capsule (Fig. 60.1e), a prominent white matter pathway situated beneath the insular surface, comprises a collection of short association fibers that traverse the central core of the brain [18]. It encompasses all the opercula, forming connections between the medial aspect of these opercula and the insula by way of the peri-insular sulcus [19]. The extreme capsule plays a crucial role in facilitating higher-level language comprehension, as it serves as a vital pathway connecting the middle temporal lobe and the ventrolateral prefrontal cortex [20, 21].

60.4.4 Uncinate Fasciculus

The uncinate fasciculus, a white matter pathway, serves as a neural conduit that establishes connections between the prefrontal cortex and the temporal lobe, encompassing vital structures like the amygdala and hippocampus [22]. Its fibers emerge from the anterior temporal lobe and amygdala, then curve in an upward trajectory behind the external capsule and inward of the insular cortex, before continuing upward into the posterior portion of the orbital gyrus within the temporal lobe (Fig. 60.1f). Bilateral damage to the uncinate fasciculus is hypothesized to affect memory-related behavior, including reward-based decision-making and memory expression [22].

60.4.5 Fronto-Occipital Fasciculus

The inferior fronto-occipital fasciculus (IFOF), a prominent white matter tract, originates in the occipital and parietal lobes and culminates in the inferior frontal lobe (Fig. 60.1f). It is situated alongside the uncinate fasciculus and courses through the infero-lateral insula, utilizing the extreme and external capsules as its adjacent structures [23]. While the primary function of the IFOF is associated with semantic language processing and transmission, emerging evidence from various studies suggests its involvement in connecting the salience network with the executive control network [23].

60.4.6 Cingulum

The cingulum bundle, a notable white matter tract, establishes connections among frontal, parietal, and medial temporal regions, facilitating the integration of subcortical nuclei with the cingulate gyrus (Fig. 60.2). Although it exhibits a visually uninterrupted structure, the composition of the cingulum constantly evolves due to the continuous addition and removal of fibers within the bundle. The cingulum bundle has important role in various cognitive functions such as executive control, emotion regulation, and pain processing (specifically in the dorsal cingulum). Additionally, the parahippocampal cingulum has been associated with episodic memory [24].

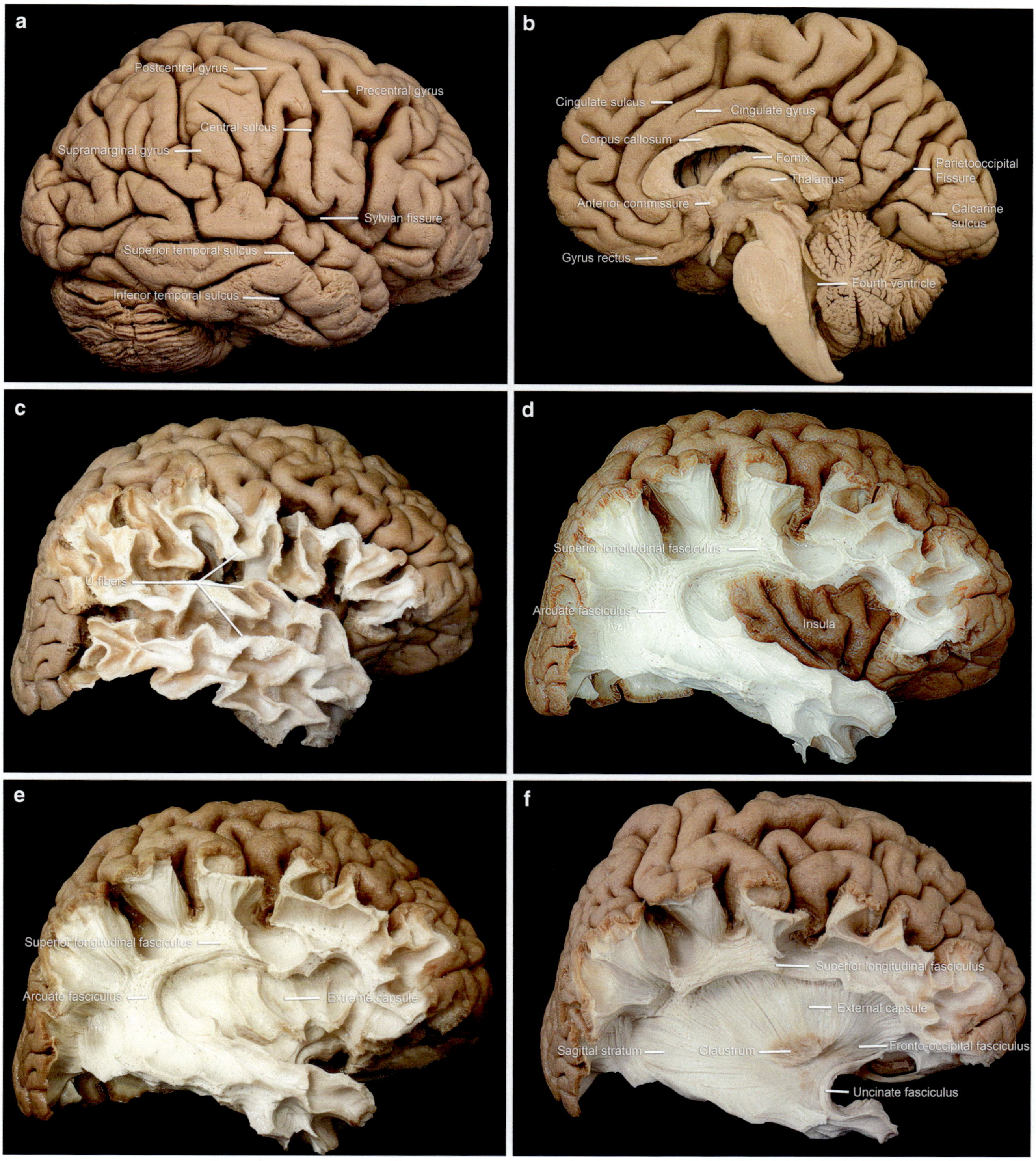

Fig. 60.1 Fiber dissection of the lateral aspect of the brain. (**a**) Lateral cortical surface of a right cerebral hemisphere after removal of arachnoid and vessels. (**b**) Medial cortical surface of a right cerebral hemisphere after removal of arachnoid and vessels. (**c**) Removal of cortical grey matter exposes U fibers. (**d**) SLF and AF are demonstrated after removing the U fibers. (**e**) Removing the insular cortex reveals the extreme capsule. (**f**) Fronto-occipital and uncinate fasciculus are exposed after dissection of the insular frontal and temporal lobes

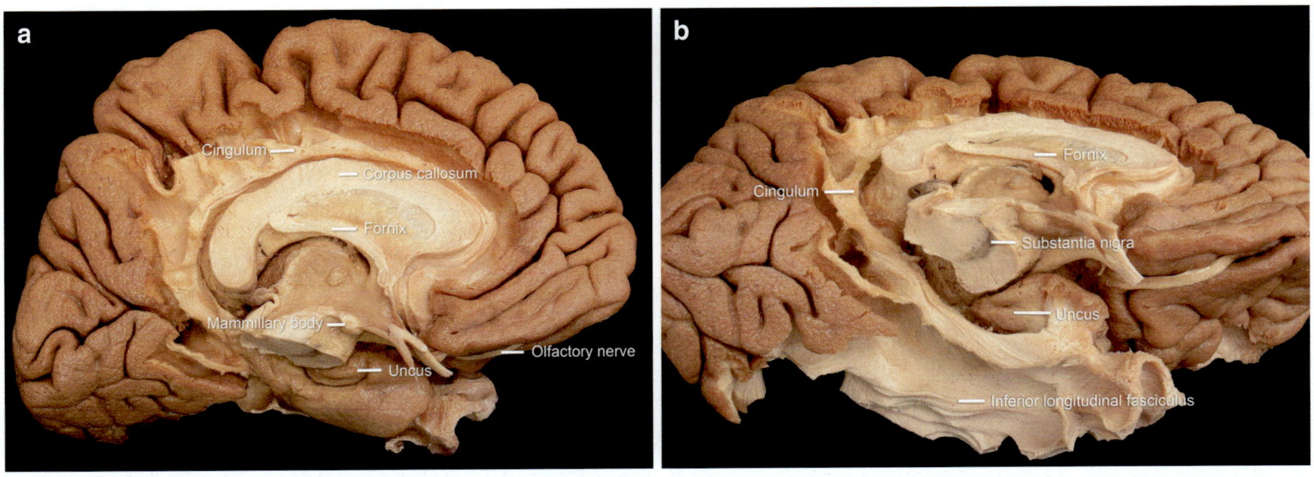

Fig. 60.2 Fiber dissection of medial and inferior surfaces of the brain. (**a**) Cingulum is seen after the removal of the cingulate cortex. (**b**) Inferior longitudinal fasciculus is demonstrated in the inferior surface of the hemisphere

60.4.7 Inferior Longitudinal Fasciculus

The inferior longitudinal fasciculus is a white matter pathway that predominantly establishes connections between the occipital and the anterior temporal lobes (Fig. 60.2). It plays a vital role in various functions. It is involved in visual functions such as object and face processing, reading, lexical and semantic processing, emotion processing, and visual memory [25].

60.4.8 Fornix

The fornix is a white matter bundle located within the mesial aspect of the cerebral hemispheres, serving as a crucial connector in the limbic circuitry. It extends longitudinally from the mesial temporal lobe to the diencephalon and basal forebrain (Fig. 60.2). Originating from the subiculum, its fibers combine with the alveus to form the fimbria of the fornix, which then expands and becomes the crura of the fornix. The crura arch beneath the splenium of the corpus callosum and project contralaterally through the forniceal commissure. Further rostrally, the fornix divides into left and right columns that descend into the basal forebrain, positioned anterior to the interventricular foramina. At the level of the anterior commissure, the columns further divide, forming the pre-commissural fornix and the postcommissural fornix [26].

60.5 Commissural Fibers

60.5.1 Corpus Callosum

The corpus callosum, the largest commissural tract in the forebrain, serves as a vital white matter structure that establishes connections between the two cerebral hemispheres of the brain (Fig. 60.3a, b) [27]. The corpus callosum consists of different parts: rostrum, genu, body, isthmus, and splenium. Its white fiber connections, also known as radiations, can be categorized into four types: superior/dorsal connections, ventral/inferior connections, anterior connections, and posterior connections. The forceps minor, originating from the genu and rostrum, curves anteriorly into the medial and basal frontal lobes. The forceps major extends posteriorly from the splenium into the medial aspect of the occipital lobes. The dorsal radiations are bounded by vertical fibers of the corona radiata, while the ventral radiations closely follow the lateral ventricle [28].

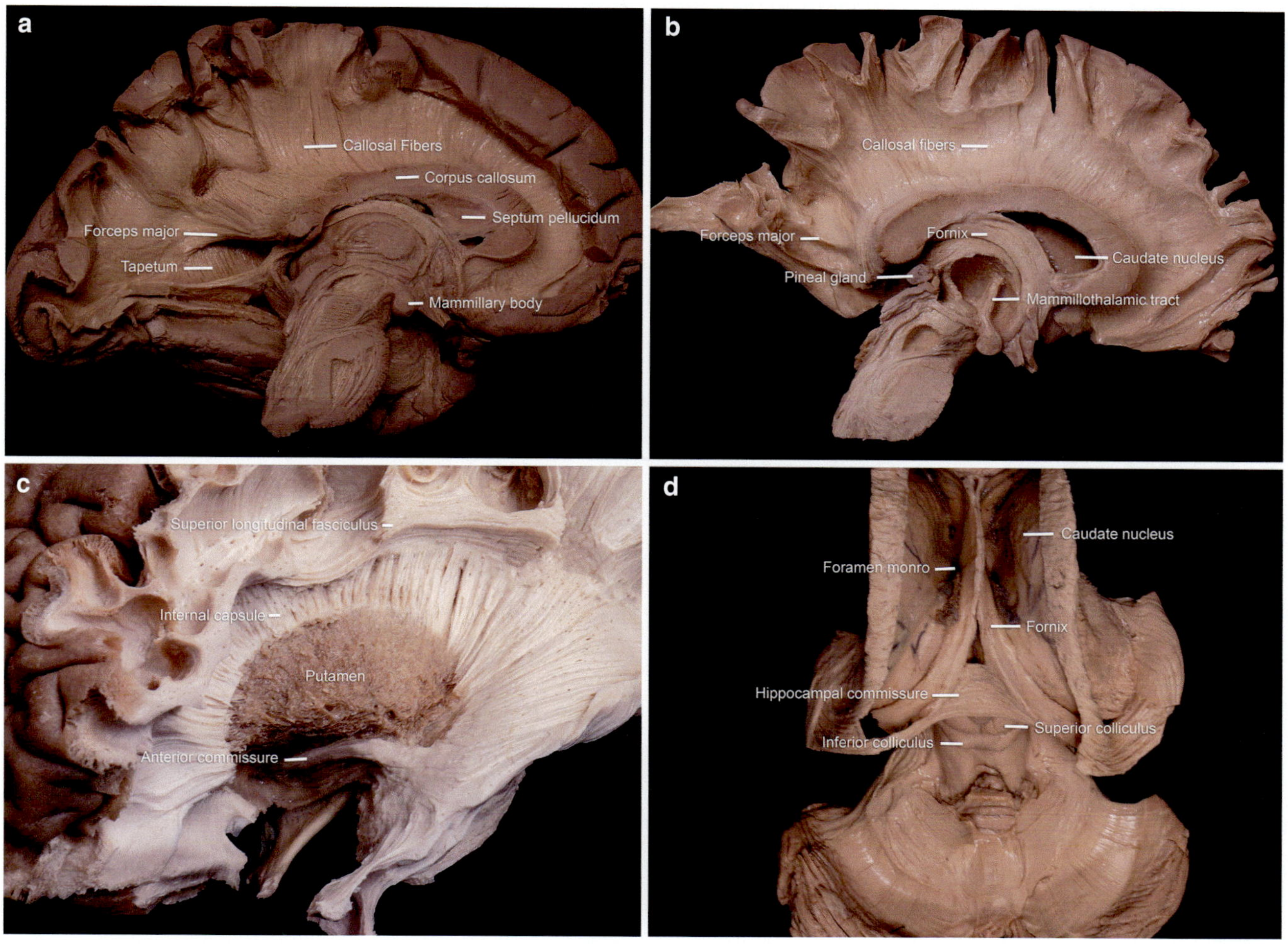

Fig. 60.3 Fiber dissection of commissural fibers. (**a**) Callosal fibers are demonstrated after the removal of the cingulum. (**b**) Callosal fibers are shown along their course. (**c**) Anterior commissure is revealed after dissecting the uncinate and fronto-occipital fascicles. (**d**) Hippocampal (Forniceal) commissure is demonstrated

60.5.2 Anterior Commissure

The anterior commissure is a white matter tract that establishes a connection between the temporal lobes across the midline and olfactive areas (Fig. 60.3c). The anterior commissure serves as a conduit for interhemispheric transmission of visual, auditory, olfactory, and gustatory stimuli and has been implicated in color perception, fear- or stress-driven behaviors, and aspects of sexuality in medical contexts [29].

60.6 Hippocampal (Forniceal) Commissure

The hippocampal commissure is a small bundle of nerve fibers positioned ventral to the splenium of the corpus callosum (Fig. 60.3d) [30]. The hippocampal commissure has been investigated in animal studies, and it has been proposed to contribute to the secondary generalization of focal epileptiform discharges. The precise developmental mechanisms underlying the anomalies of the hippocampal commissure in mice remain unclear [31].

60.7 Projection Fibers

60.7.1 Internal Capsule—Corona Radiata

The internal capsule (IC) is a subcortical white matter structure located in the inferomedial region of each cerebral hemisphere (Fig. 60.4a). It consists of myelinated fiber tracts that ascend and descend, forming essential connections between the cerebral hemispheres and subcortical structures, the brainstem, and the spinal cord. These fiber tracts traverse the basal ganglia as they course through the internal capsule, facilitating communication between different regions of the brain [32]. At the upper edge of the putamen, the external and internal capsule fibers converge to form the corona radiata. This radiating fiber bundle courses through the superior

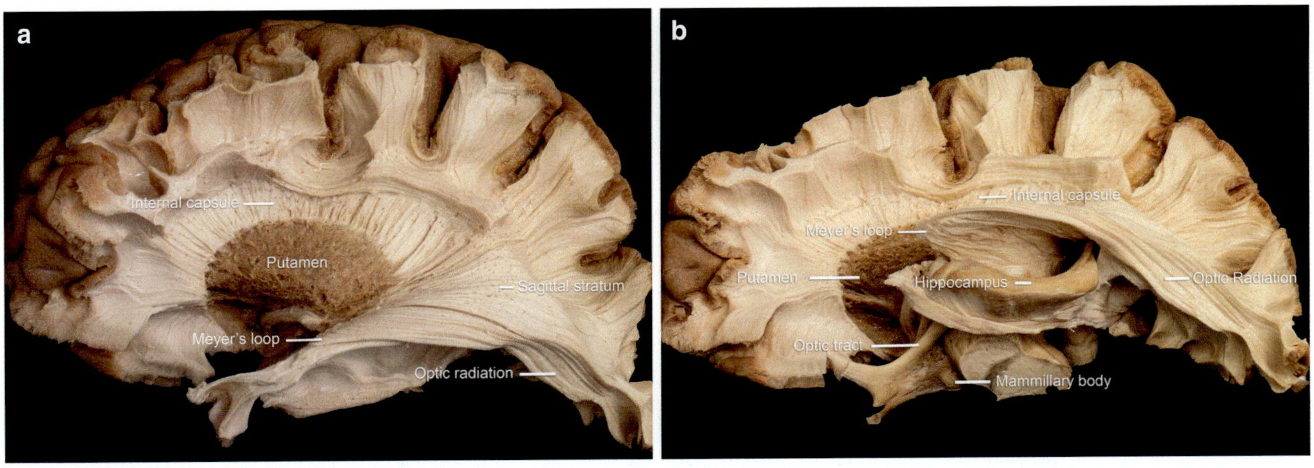

Fig. 60.4 Demonstration of internal capsule fibers, Meyer's loop, and optic radiation. (**a**) Fibers of the internal capsule and their relations with surrounding structures are demonstrated. (**b**) Meyer's loop and optic radiation fibers are shown

regions of the frontal and parietal lobes. In this region, the corona radiata joins with the callosal fibers, collectively forming the centrum semiovale [33].

During its posterior extension, the sublenticular part of the internal capsule forms a distinct loop known as Meyer's loop, which arises during the embryological development of the temporal lobe (Fig. 60.4b). Referring to this loop as the posterior thalamic peduncle, instead of the optic radiation alone, provides a more precise designation, as it encompasses not only optic fibers (Fig. 60.4b), but also serves as a bidirectional pathway for information exchange between the cerebral cortex and other regions [34].

60.7.2 Approach

Illustrative Case 1 10-year-old female patient presented with left-sided hemiparesis. MRI demonstrated a mass in the right globus pallidus (Fig. 60.5).

Tumors originating primarily in the globus pallidus are infrequent occurrences. Without infiltration, they may displace neighboring tissues, such as the internal capsule or hypothalamus. The challenge lies in approaching these tumors without causing harm to vital neighboring structures, including the putamen and internal capsule. Nonetheless, the tumors often facilitate the creation of a pathway that allows access without compromising the integrity of the surrounding functional parenchyma. Consequently, upon tumor removal, the putamen, internal capsule, and unaffected region of the globus pallidus can be preserved without damage.

The patient is operated via anterior interhemispheric transcallosal approach. Histopathologic examination revealed a pylocitic astrocytoma. The patients' symptoms are improved immediately after the surgery and discharged without complications.

Illustrative Case 2 22-year-old male patient presented with headache, dizziness, and generalized tonic-clonic seizure. Subsequent MRI revealed T2 hyperintense, T1 hypointense insular tumor (Fig. 60.6).

In situations where DTI is not accessible, the selection of the approach should be based on a careful assessment of the white matter anatomy, utilizing trans-sulcal and trans-cisternal corridors. The primary focus in insular tumor surgery is to minimize the use of bipolar coagulation and ensure the protection of encountered lenticulostriate arteries in order to mitigate the risk of neurological deficits. Cavitron Ultrasonic Surgical Aspirator (CUSA) should be used in low-power settings during insular surgery, if needed. Papaverine can be administered to address vasospasm in middle cerebral artery branches.

The present patient was operated via pterional transsylvian approach. Histopathologic examination revealed diffuse glioma grade II. The patient was discharged without complications.

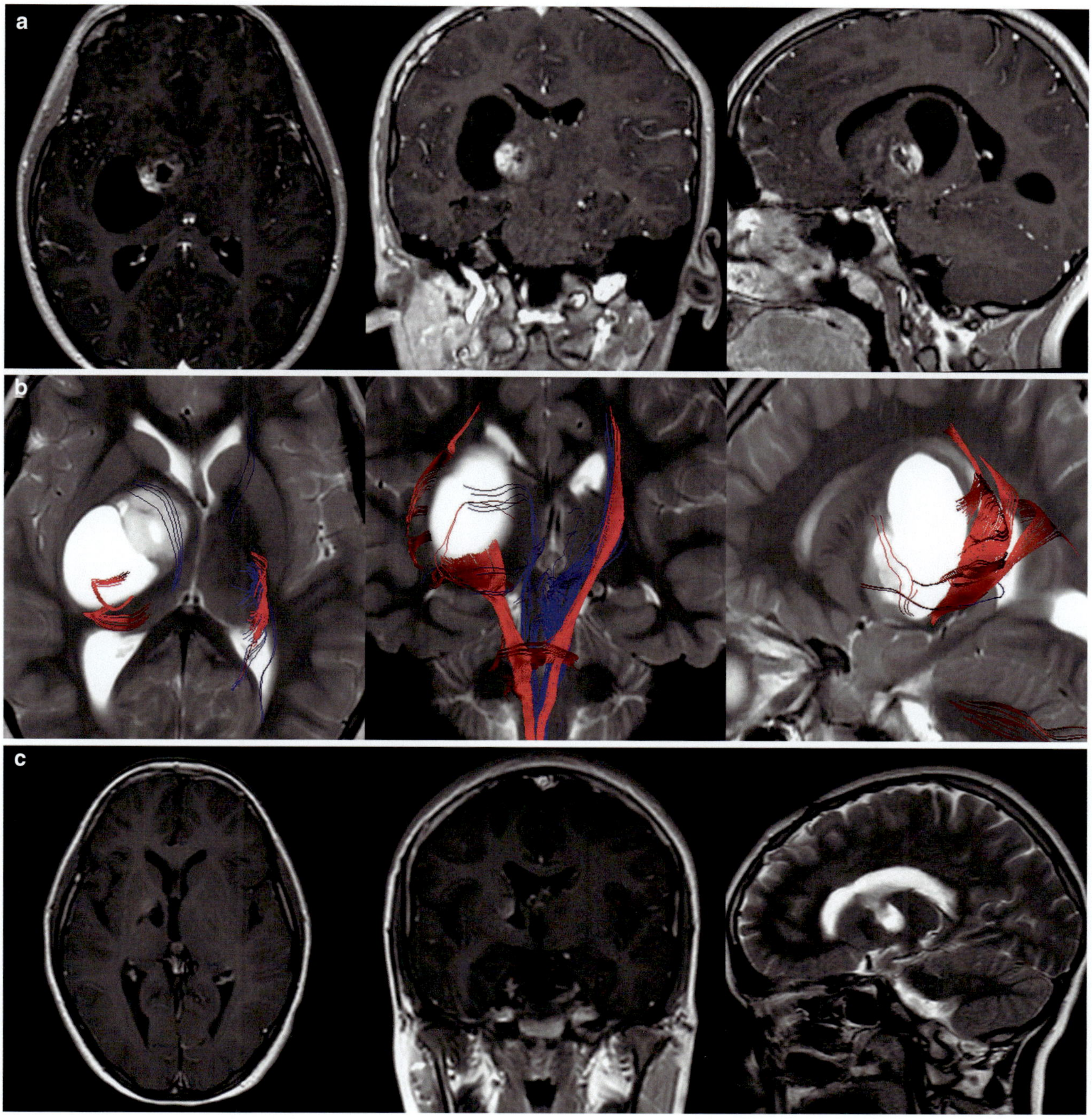

Fig. 60.5 Preoperative and postoperative radiological images of a 10-year-old female patient with a mass in the right globus pallidus with a large cystic component. (**a**) Preoperative MRI imaging of the patient demonstrates the lesion. (**b**) Preoperative DTI imaging shows the motor and sensory fibers are displaced posteriorly because of the mass effect of the lesion. (**c**) Postoperative MRIs demonstrate gross total resection of the tumor after the resection through the anterior interhemispheric approach

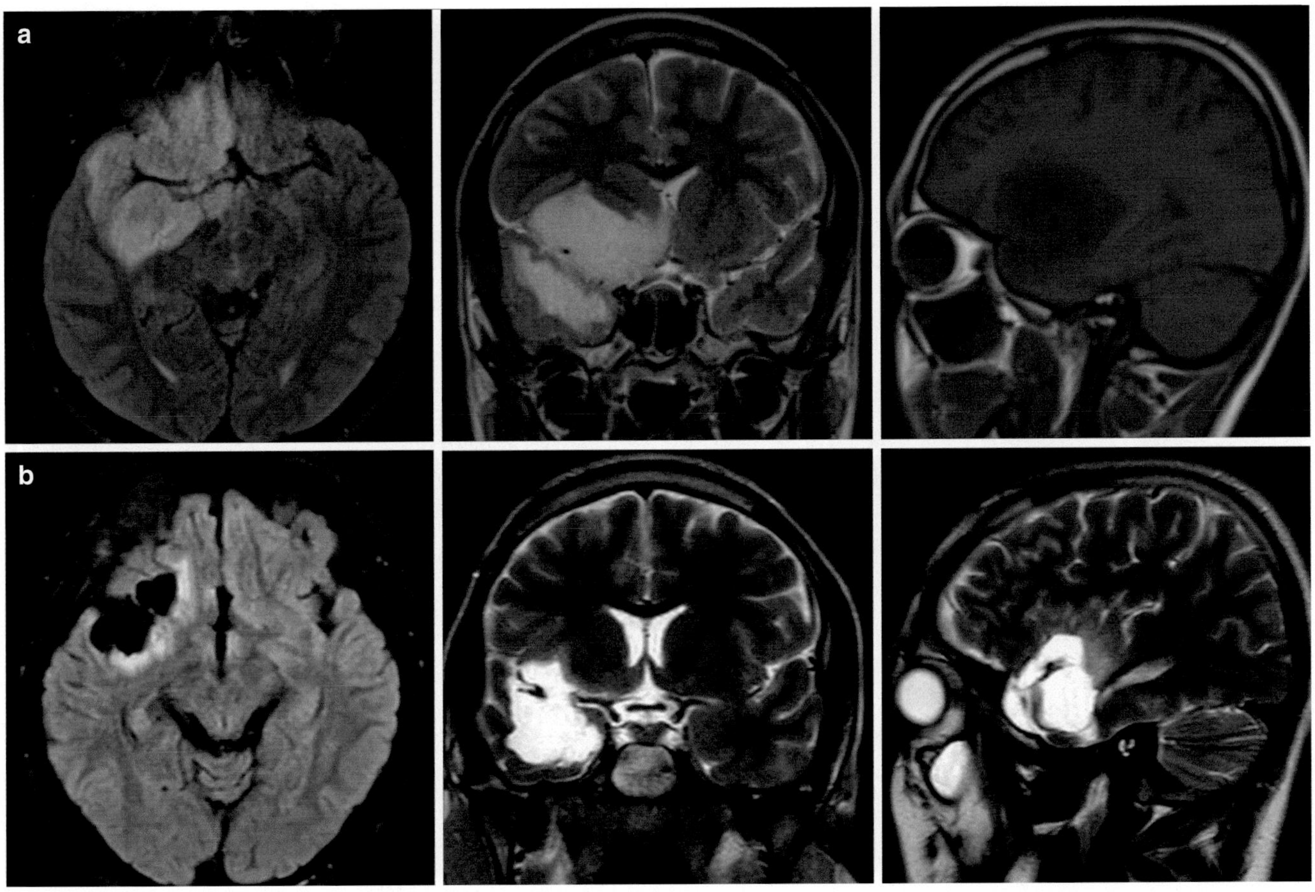

Fig. 60.6 Preoperative and postoperative radiological images of a 22-year-old male patient with an insular tumor. (**a**) Preoperative MRI demonstrates the lesion. (**b**) The surgical cavity and extent of the resection are demonstrated

60.8 Surgical Pearls

- Studying fiber dissection improves knowledge of white matter anatomy and surgical orientation, leading to reduced complications and enhanced outcomes.
- Combining the knowledge of fiber anatomy gained from fiber dissection with advanced neuroimaging techniques provides valuable insights for refining surgical methods.
- Identifying and individualizing white matter tracts is important for determining the appropriate surgical approach, especially in deep-seated tumors.
- Surgical planning should be based on routes that pass through sulci and fissures, which serve as natural corridors, while preserving white matter structures as much as possible.
- Preservation of arteries and veins is essential for achieving optimal postoperative outcomes.
- Minimizing the use of bipolar coagulation during resection is recommended. After the resection is completed, low-power settings should be used sparingly.

60.9 Discussion

The precise identification of white matter tracts is significant when determining the appropriate surgical approach. Initially, the primary consideration lies in the tumor's point of origin, as its growth inevitably exerts pressure on the adjacent white matter fibers, causing them to displace outward. During surgical planning, tractography can be employed to map the altered positions of these fibers, aiding in surgical decision-making.

A comprehensive understanding of the anatomical structure of white matter holds significant importance in the realm of tumor surgery, particularly when dealing with patients afflicted by gliomas. Gliomas can interact with adjacent white matter tracts through displacement, infiltration, or disruption. Notably, while gray matter demonstrates a higher capacity for plasticity, the plasticity of white matter is relatively constrained. Even small tumors that invade white matter pathways may not possess compensatory mechanisms for functional reorganization, potentially resulting in neurological impairments evident from the onset of the disease [35]. Thorough preoperative evaluation of the tumor, arteries, veins, and white matter through detailed imaging analysis plays a crucial role in formulating an effective surgical strategy. In glioma surgery, the utilization of advanced surgical support technologies, including navigation systems, electrophysiological monitoring, awake craniotomy, and fluorescence-guided surgery, becomes indispensable. Neurosurgeons must adeptly develop surgical techniques to accurately identify and delineate the boundaries of the tumor, preserve critical arteries and veins, safeguard the eloquent cortex and neural fibers, and manage any bleeding that may occur during the tumor resection process [36].

60.10 Conclusion

Fiber dissection, coupled with modern neuroimaging, offers valuable neuroanatomical insights. It refines surgical methods, fosters innovation, and enhances our comprehension of the brain's intricate structure. Precise identification of white matter tracts guides surgical approaches, particularly when dealing with tumors that displace adjacent fibers.

References

1. Dziedzic TA, Balasa A, Jeżewski MP, et al. White matter dissection with the Klingler technique: a literature review. Brain Struct Funct. 2021;226:13–47.
2. Essayed WI, Zhang F, Unadkat P, et al. White matter tractography for neurosurgical planning: a topography-based review of the current state of the art. Neuroimage Clin. 2017;15:659–72.
3. Alms C, Eseonu CI. Comparative quantification of diffusion tensor tractography using automated whole brain MRI tractography for intracranial tumor surgery: technical note. Cureus. 2022;14:e25546.
4. Li M, Song L, Zhang Y, et al. White matter network of oral word reading identified by network-based lesion-symptom mapping. iScience. 2021;24:102862.
5. Aabedi AA, Young JS, Chang EF, et al. Involvement of white matter language tracts in glioma: clinical implications, operative management, and functional recovery after injury. Front Neurosci. 2022;16:932478. Epub ahead of print 11 July 2022
6. Türe U, Yaşargil MG, Friedman HS, et al. Fiber dissection technique: lateral aspect of the brain. Neurosurgery. 2000;47:417–26.
7. Egemen E, Celtikci P, Doğruel Y, et al. Microsurgical and tractographic anatomical study of transtemporal-transchoroidal fissure approaches to the ambient cistern. Oper Neurosurg. 2020;20:189–97.
8. Klingler J. Erleichterung der makrokopischen Präparation des Gehirns durch den Gefrierprozess. Orell Füssli; 1935.
9. Yaşargil MG, Türe U, Yaşargil DCH. Surgical anatomy of supratentorial midline lesions. Neurosurg Focus. 2005;18:E1.
10. Cristina G, Klara B, Florian IS, et al. The three-dimensional architecture of the internal capsule of the human brain demonstrated by fiber dissection technique. Ars Medica Tomitana. 2015;20:115–22.
11. Komaitis S, Kalamatianos T, Drosos E, et al. A stepwise laboratory manual for the dissection and illustration of major limbic structures. Evidence from the Klingler's Technique. 2021. Epub ahead of print
12. Çavdar S, Aydin AE, Algin O, et al. Fiber dissection and 3-tesla diffusion tensor tractography of the superior cerebellar peduncle in the human brain: emphasize on the cerebello-hypthalamic fibers. Brain Struct Funct. 2020;225(1):121–8.
13. Gurses ME, Gungor A, Gökalp E, et al. Three-dimensional modeling and augmented and virtual reality simulations of the white matter anatomy of the cerebrum. Oper Neurosurg. 2022;23:355–66.
14. Martino J, Hamer PCDW, Vergani F, et al. Cortex-sparing fiber dissection: an improved method for the study of white matter anatomy in the human brain. J Anat. 2011;219:531–41.
15. Shah A, Goel A, Jhawar SS, et al. Neural circuitry: architecture and function—a fiber dissection study. World Neurosurg. 2019;125:e620–38.
16. Janelle F, Iorio-Morin C, D'amour S, et al. Superior longitudinal fasciculus: a review of the anatomical descriptions with functional correlates. Front Neurol. 2022;13:794618.
17. Kamali A, Sair HI, Radmanesh A, et al. Decoding the superior parietal lobule connections of the superior longitudinal fasciculus/arcuate fasciculus in the human brain. Neuroscience. 2014;277:577–83.
18. Ribas EC, Yağmurlu K, de Oliveira E, et al. Microsurgical anatomy of the central core of the brain. J Neurosurg. 2018;129:752–69.
19. Demirtaş OK, Güngör A, Çeltikçi P, et al. Microsurgical anatomy and insular connectivity of the cerebral opercula. J Neurosurg. 2022;137:1509–23.
20. Saur D, Kreher BW, Schnell S, et al. Ventral and dorsal pathways for language. Proc Natl Acad Sci. 2008;105:18035–40.
21. Makris N, Pandya DN. The extreme capsule in humans and rethinking of the language circuitry. Brain Struct Funct. 2009;213:343–58.
22. Von Der Heide RJ, Skipper LM, Klobusicky E, et al. Dissecting the uncinate fasciculus: disorders, controversies and a hypothesis. Brain. 2013;136:1692–707.
23. Conner AK, Briggs RG, Sali G, et al. A connectomic atlas of the human cerebrum—chapter 13: tractographic description of the inferior fronto-occipital fasciculus. Oper Neurosurg. 2018;15:S436–43.
24. Bubb EJ, Metzler-Baddeley C, Aggleton JP. The cingulum bundle: anatomy, function, and dysfunction. Neurosci Biobehav Rev. 2018;92:104–27.
25. Herbet G, Zemmoura I, Duffau H. Functional anatomy of the inferior longitudinal fasciculus: from historical reports to current hypotheses. Front Neuroanat. 2018;12:77.
26. Senova S, Fomenko A, Gondard E, et al. Anatomy and function of the fornix in the context of its potential as a therapeutic target. J Neurol Neurosurg Psychiatry. 2020;91:547–59.
27. Wang P, Meng C, Yuan R, et al. The organization of the human corpus callosum estimated by intrinsic functional connectivity with white-matter functional networks. Cereb Cortex. 2020;30:3313–24.
28. Shah A, Jhawar S, Goel A, et al. Corpus callosum and its connections: a fiber dissection study. World Neurosurg. 2021;151:e1024–35.
29. Liu X, Kinoshita M, Shinohara H, et al. A fiber dissection study of the anterior commissure: correlations with diffusion Spectrum imaging tractography and clinical relevance in gliomas. Brain Topogr. 2022;35:232–40.
30. Raybaud C. The corpus callosum, the other great forebrain commissures, and the septum pellucidum: anatomy, development, and malformation. Neuroradiology. 2010;52:447–77.
31. Magara F, Müller U, Li Z-W, et al. Genetic background changes the pattern of forebrain commissure defects in transgenic mice under-

expressing the β-amyloid-precursor protein. Proc Natl Acad Sci. 1999;96:4656–61.
32. Costa M, Braga VL, Yagmurlu K, et al. A technical guide for fiber tract dissection of the internal capsule. Turk Neurosurg. 2018;28(6):934–9.
33. Güngör A, Baydin S, Middlebrooks EH, et al. The white matter tracts of the cerebrum in ventricular surgery and hydrocephalus. J Neurosurg. 2017;126:945–71.
34. Goga C, Türe U. The anatomy of Meyer's loop revisited: changing the anatomical paradigm of the temporal loop based on evidence from fiber microdissection. J Neurosurg. 2015;122:1253–62.
35. Smits A, Zetterling M, Lundin M, et al. Neurological impairment linked with cortico-subcortical infiltration of diffuse low-grade gliomas at initial diagnosis supports early brain plasticity. Front Neurol. 2015;6:137.
36. Saito K. Glioma surgery and surgical supporting technology. No Shinkei Geka. 2021;49:536–48.

Neuroanatomical Landmarks to Approach Insular Tumors

61

Pablo González-Lopéz, Javier Abarca-Olivas, and Juan Nieto-Navarro

Learning Objectives

- Understand anatomical and phylogenetic insular development.
- Understand anatomical, cortical, and subcortical insular relationships.
- Acquire the ability to choose the correct approach for tumors around the insular cortex.

61.1 Introduction

Intrinsic brain tumor surgery has evolved tremendously in the past two decades due to an avalanche of technological advances in the field of diagnostic techniques, planning tools, intraoperative visualization, neuromonitoring, awake surgery, and intraoperative neuroimaging. However, none of them has replaced the knowledge of brain anatomy. A detailed study of the brain cortical and subcortical organization should drive us to a better understanding of how the intrinsic gliomas invade the gyri and even how they grow through the subcortical avenues that the white matter tracts represent [1, 2]. But, if there is an area in the brain that requires a specific mental preparation before performing a surgical procedure, that is the insula.

Even though the insular cortex constitutes only 2% of the total cortical surface area, it owns a wide range of functions due to its extensive connections with adjacent and distant areas of the brain. The insula presents some special anatomical features compared to the rest of the human brain. First, this lobe is obscured by the frontal, parietal, and temporal opercula. The tremendous development of the telencephalic neocortex has pushed the paralimbic cortical areas of the insula to a deeper location, while the temporal lobe neocortex was moved anterolaterally till contacting with the frontal and parietal lobes and closing the sylvian fissure. This movement is known as the maturation of telencephalic flexure and represents a specific feature of the human brain development [3].

These intrinsic developmental brain forces have pushed the insula to the floor of the sylvian fissure, being in close relationship with relevant vascular and highly eloquent cortical and subcortical regions. Understanding this *flexure* is key to create a mental idea of the insular lobe's surrounding structures as the opercular speech-related cortical regions. The rest of the frontoparietal operculum covering this hidden lobe is constituted by the primary auditory area represented by Heschl gyrus, as well as the primary motor and somatosensory areas. Regarding the subcortical anatomy, the superior longitudinal and arcuate fasciculus, as well as the temporal stem, sagittal stratum, and corticospinal tract, are the white matter fibers around and closely related to the insula.

The insular lobe is considered as the external layer of the so-called central core, a key part of the human brain, where the basal ganglia, internal capsule, and diencephalon are lying. This core of highly relevant structures is shielded by two arterial systems: the neocortical branches of the middle cerebral artery on the insular surface and the proximal perforating arterial system formed by the lenticulostriate arteries.

Having said this, it is understandable how relevant the study of the general and individual anatomy of each case must be to achieve the most efficient surgical result in terms of maximizing the tumor resection and minimizing the postoperative neurological deficits. The aim of this chapter is no other than highlighting some key anatomical landmarks to

P. González-Lopéz (✉) · J. Abarca-Olivas · J. Nieto-Navarro
Department of Neurosurgery, Hospital General Universitario de Alicante, Alicante, Spain
e-mail: gonzalez_pab@gva.es; nieto_juanav@gva.es

I. N. Kanaan, V. Beneš (eds.), *Neuroanatomy Guidance to Successful Neurosurgical Interventions*,
https://doi.org/10.1007/978-3-031-59838-8_61

consider during the preoperative process of surgical planning to understand the location of the tumor, how this knowledge can help us to decide our approach, and finally how to take intraoperative decisions based on the previously designed surgical map.

61.2 History/Background

The first reported insular successful resections were reported by Penfield and Faulk in the 50 s [4]. These surgical explorations were carried out in epilepsy cases in awake patients. The technique consisted of an anterior temporal lobe resection to widely expose the insula. Direct surgical stimulation was performed on the insular cortex, describing the clinical findings and trying to demonstrate the epileptic foci. Various degrees of surgical resections were carried out on the insular surface. Parallel to Penfield's efforts, Guillaume pioneered the first transopercular approaches to the insular lobe based on electrocorticographic findings [5].

Classic and pioneer series in insular tumor resections revealed frequent postoperative neurological deficits as hemiparesis, speech disturbances, and different impairments as apraxia and mutism. Nevertheless, some of those cases were posteriorly examined showing that the damage seemed to be caused due to a direct injury over the opercular cortex and corona radiata. These lesions could have been caused by direct invasion of the lesion and removal, excessive retraction, and ischemic injuries due to direct manipulation and/or coagulation of the middle cerebral artery branches. After these very first efforts to successfully operate into and around the insula, it was Professor Yasargil who opened the door to a new era in neurosurgical care with the introduction of microsurgical techniques such as splitting the sylvian fissure to access the insular lobe. He published his first long series of 57 purely insular and 93 operculoinsular lesions with good results [6, 7]. This fact provoked a renewed interest in insular tumor surgery.

61.3 Anatomical Highlights

61.3.1 The Telencephalic Flexure

One of the most fascinating characteristics of the insula lies in the fact that the neuronal network that makes up its cortex has diverse phylogenetic origins, and therefore owns very different functions. The phylogenetic origin of the insula has stored many cortical evolutive changes. This is why the insula is described as a paralimbic structure, as it serves as a pivotal cortex in between the paleo−/archicortical limbic regions and the newly developed neocortex in the dorsal telencephalon. Developmental embryology studies have shown that the insular cortex seems to be the first cortical structure developed in the fetus and appears to run from allo- to isocortex. Then, the disproportional development of the neocortical telencephalon pushes the insula to a central position in the evolving brain, storing part of the paleo- and archicortex within its surface. This is how the insula 'sequesters' and keeps limbic and paralimbic rudiments on it [8].

A cytoarchitectonic analysis of the insular cortex in primates has shown that it is composed of three clearly differentiated areas: agranular (anterior and ventral), nongranular (pericentral), and granular (posterior and dorsal) [9]. Furthermore, if we analyze the insular connections with the rest of the primate brain, we can state that it relates to neighboring but also very distant areas. Thus, the posterior regions (dorsal—granular) present evident connections with the retroinsular region, supplementary motor area, as well as the temporal, pre-, and postcentral neocortical areas, presenting mainly motor and auditory functions. On the other hand, more anterior regions (ventral—agranular and nongranular) are connected to the orbitofrontal, entorhinal and periamygdaline cortex, specializing in functions related to olfaction, emotional memory, and visceral and autonomic reactions Fig. 61.1.

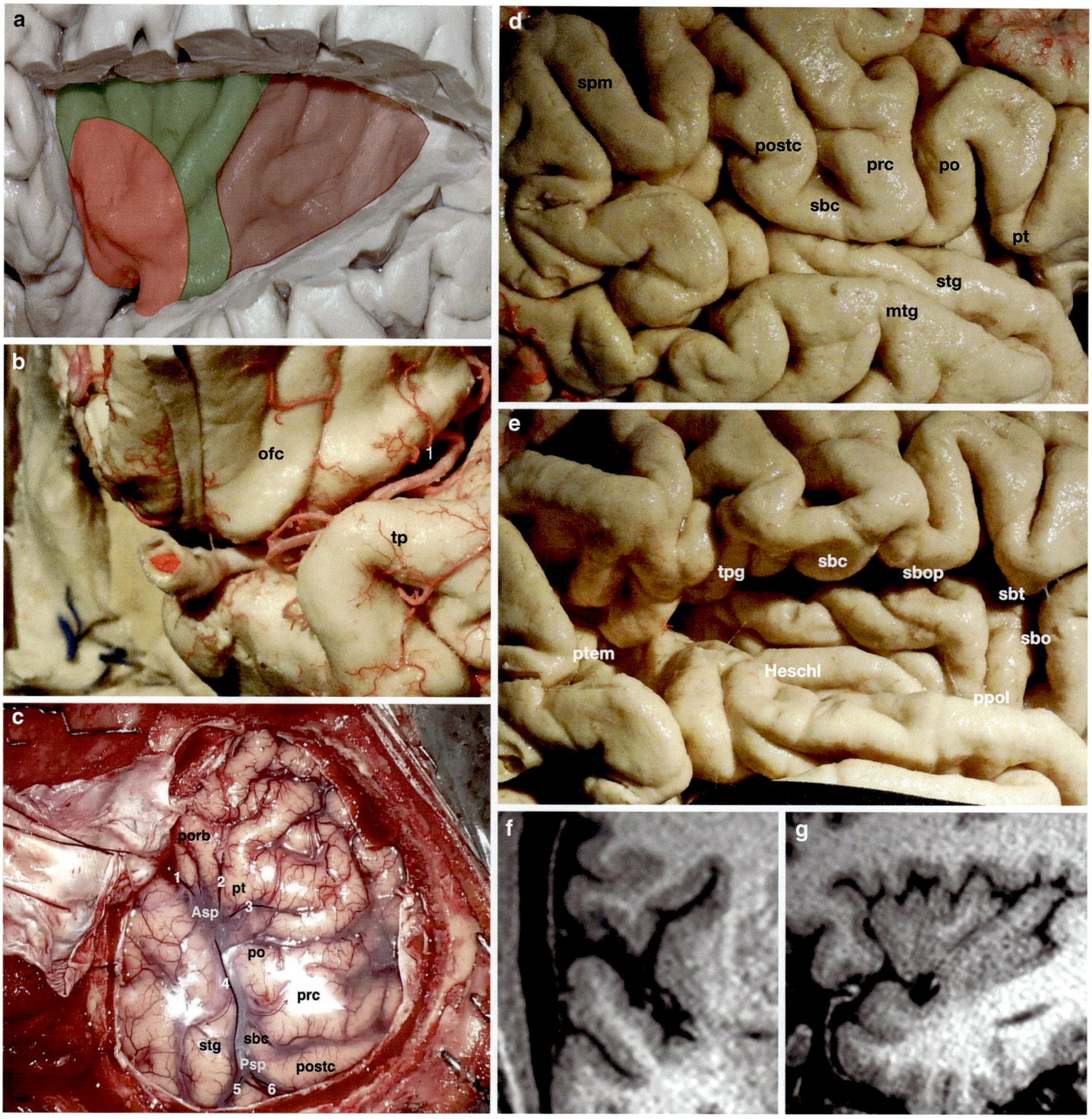

Fig. 61.1 (**a**) Lateral view of the left insular lobe on an anatomical specimen after having removed the orbitofrontal, frontoparietal, and temporal opercula. The different cortical areas are illustrated: paleocortical (red), archicortical (green), and neocortical (brown). (**b**) Basal view of the frontobasal (ofc) and temporomesial (tp) regions, and its close relationship with the proximal branch of the sylvian fissure (1). (**c**) Intraoperative view of a left side human brain after opening the dura on a pterional craniotomy. The different parts of the sylvian fissure and opercula are highlighted: proximal (1), horizontal (2), ascending (3), distal (4), superior (6), and inferior terminal (5) branches of the sylvian fissure; pars orbitalis (porb), pars triangularis (pt), pars opercularis (po), precentral (prc), subcentral (sbc), postcentral (postc) regions, superior temporal gyrus (stg), as well as the anterior (Asp) and posterior (Psp) sylvian points. Two red arrows are shown exiting from the fissure on opposite frontal and temporal directions. (**d**, **e**) Right side brain hemisphere showing the opercula before (**d**) and after (**e**) opening the Sylvian fissure: pars triangularis (pt), pars opercularis (po), precentral (prc), subcentral (sbc), postcentral (postc) regions, supramarginal gyrus (spm), superior (stg) and middle temporal gyri (mtg), suborbital (sbo), subtriangular (sbt), subopercular (sbop), subcentral (sbc), transverse parietal (tpg) gyri, planum temporale (ptem), transverse Heschl gyri (Heschl), and planum polare (ppol). (**f**, **g**) Coronal and sagittal T1-weighted MRI slices showing the location of the insult and anatomical relationships

Each of these areas is related to different anatomical cortical landmarks and deeply related and connected through various white matter structures. The common origin, close relationship and underlying fiber tracts connecting adjacent areas, might serve as the substrate through which gliomas grow in and around the insula.

61.3.2 The Sylvian Fissure and Opercula

The insular lobe represents the floor of the lateral or sylvian fissure. This is the first landmark we can identify after opening the dura through a pterional craniotomy. Despite the anatomical variations that can appear, the sylvian fissure is a constant structure that provides the surgeon the necessary information about the relative location and anteroposterior extension of the insula. The superficial sylvian veins are easily recognized, as they generally run superficially into the fissure. They usually receive tributaries from the parietal, frontal, and/or temporal sides. This variability makes it mandatory to study the main pattern on each case to decide from which side the splitting of the fissure should start. Sometimes, these veins are inconstant on its pathway or even can be absent, making the fissure identification difficult. This is especially common when we need to approach the posterior insula through the fissure, as it is quite common that the sylvian vein does not reach the distal segments of the fissure (posterior sylvian point). In those cases, we recommend looking for the M4 branches of the middle cerebral artery that will arise from the fissure in two directions, superiorly and inferiorly. The lateral fissure is the only sulcus in the convexity of the brain where we can find this pattern. Other adjuncts, such as intraoperative neuronavigation, ultrasound, or indocyanine green, can help us to identify the fissure Fig. 61.1.

The sylvian fissure shows a hook shape from its proximal part in the internal carotid artery bifurcation zone, towards its distal part in the supramarginal neocortex. This looped trajectory is divided into two segments. The anterior segment or sylvian stem starts when the internal carotid gives rise to the middle cerebral artery and runs posteriorly and laterally following the shape of the lesser sphenoid wing, with which it is in close relationship. When reaching the lateral aspect of the brain, this anterior segment finalizes at the so-called anterior sylvian point. This is an important landmark as it usually serves as a safe area to start splitting the fissure proximally and distally. The posterior segment of the fissure starts at this level and trifurcates in horizontal, ascending, and posterior rami. The horizontal runs anteriorly and separates the pars orbitalis from triangularis, the ascending is located in between the pars triangularis and opercularis, while the posterior ramus keeps running posteriorly in between the frontoparietal and temporal opercula. The supramarginal gyrus is defined at the sylvian fissure end by the ascending or terminal ramus of the posterior branch Fig. 61.1.

The posterior segment of the fissure marks the anteroposterior extension of the insular lobe and separates two sides of neocortical regions: the superior and inferior opercula. The superior side is formed by the fronto-orbital and frontoparietal. The fronto-orbital refers to the lateral aspect of the posterolateral fronto-orbital gyrus, limited by the posterolateral part of the transverse orbital sulcus and the horizontal ramus of the sylvian fissure. This horizonal ramus represents the border between the fronto-orbital and frontoparietal opercula. Therefore, from anterior to posterior, the pars triangularis, ascending ramus of the lateral fissure, pars opercularis (generally named Broca's area in the dominant hemisphere), interrupted by the inconstant diagonal sulcus, anterior subcentral sulcus, pre-, sub-, and postcentral gyrus, posterior subcentral sulcus, and dorsal aspect of the supramarginal gyrus are found. It is relevant to highlight that the subcentral gyrus represents a u-shaped junction between the pre- and postcentral gyri and only in 8% of the cases it is hidden inside the fissure. The inferior or temporal operculum follows a straightforward direction from the ventral part of the supramarginal gyrus towards the temporal pole. The main interruption of its natural course is the junction of the transverse gyrus of Heschl to this superior temporal gyrus. Heschl gyrus represents the primary auditory area and runs from the depth of the fissure to its surface following an oblique anterior and lateral direction. These gyri generally join just below the subcentral gyrus on the posterior third of the superior temporal gyrus. The area in between this anatomical junction and the dorsal aspect of the supramarginal gyrus is supposed to have highly eloquent functions mainly related with speech and has been classically described as Wernicke's area in the dominant hemisphere. Therefore, it seems reasonable to respect this posterior third of the temporal operculum if a temporal lobectomy is needed [10–12] Fig. 61.1.

A wide opening of the arachnoid adjacent to the sylvian vein and splitting the sylvian fissure will provide access to the sylvian fossa. This is a cisternal space through which surgical exploration is feasible as it is protected by superior and inferior subopercular arachnoid layers. This space has two limbs: the superior, situated between the insular cortex and frontoparietal operculum; and the inferior, in between the insula and temporal operculum. Therefore, this sylvian cistern is faced by the insular surface and the superior and inferior opercular extensions which go deep into the fissure and are hidden from a direct surgical view. A third limb can be described following a perpendicular direction to the previous one (represented in the coronal plane) and is located between the anterior insular surface and the fronto-orbital operculum. Being aware of this fronto-orbital, frontoparietal, and temporal subopercular extensions is mandatory as they keep quite constant relationships among them and with

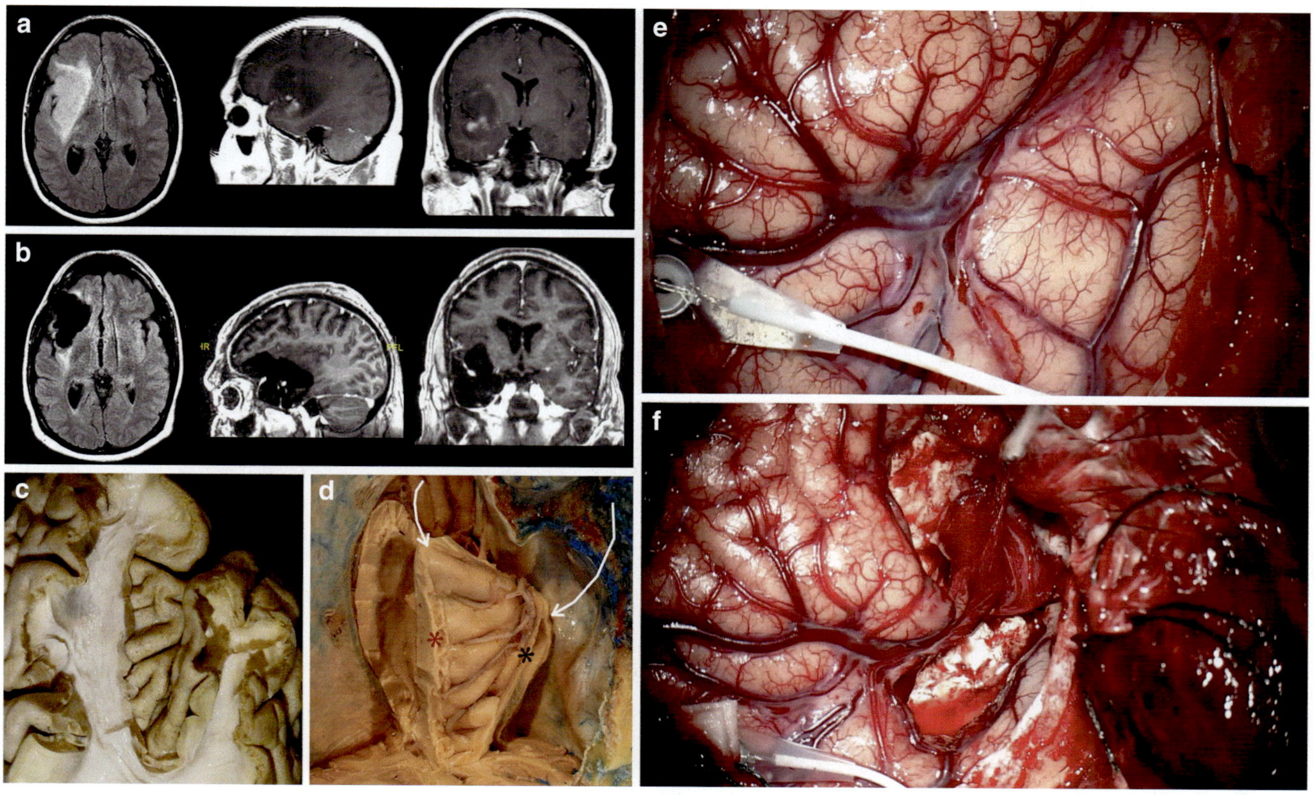

Fig. 61.2 (**a**) Preoperative axial FLAIR, sagittal T1-weighted GD MRI, and axial T1-weighted GD MRI of a right insular lobe tumor extended to the orbitofrontal and temporal opercula. (**b**) Postoperative axial FLAIR, sagittal T1-weighted GD MRI, and axial T1-weighted GD MRI of the same case showing a posteromedial small remnant close to the corticospinal tract after having performed a transopercular resection. (**c**) Right side hemisphere of a human brain specimen showing the close relationship to the insular lobe with the arcuate fascicle in surgical position. (**d**) Right side hemisphere of a human brain specimen showing the frontal and temporal transopercular trajectories (white arrows), as well as the relative location of the superior (red asterisk) and inferior (black asterisk) peninsular sulci. (**e**, **f**) Intraoperative pre- and post-resection views

the insular lobe. The fronto-orbital subopercular extensions (suborbital gyri) are vertically oriented and face the anterior insular surface, separated by the anterior periinsular sulcus; the frontoparietal (subtriangular, subopercular, subcentral, and transverse parietal gyri) faces the superior insular surface separated by the superior periinsular sulcus; finally, the temporal operculum inner extension (planum temporale, transverse gyri of Heschl, and planum polare) faces the inferior part of the insula with the inferior periinsular sulcus among them [10, 13] Figs. 61.1 and 61.2.

61.3.3 Topographic Anatomy of the Insular Surface

The insula is classically described as a pyramid with three walls, a base, and an apex. The three walls (anterior, superior, and inferior) are related with three subopercular surfaces, three sulci, and two junction points, all converging into the insular apex. Deep to the apex the base lies at the level of the limen insula. This base is a 'flat' surface formed by the claustrum and the external capsule, which separate the insular lobe from the deeper basal ganglia. All these structures together will give rise to the so-called central core, which is crossed and related to highly functional white matter tracts that will be discussed in next sections.

The insular surface is delineated by the anterior, superior, and inferior periinsular sulcus. The anterior and superior converge at the anterior insular point, which is an anatomical landmark to the superior extension of the anterior limb of the internal capsule, head of the caudate nucleus, and the frontal horn of the lateral ventricle. The superior and inferior sulci join at the posterior insular point. Underlying this region, the posterior limb and retrolenticular component of the internal capsule can be found. Deeper to the latter structures, the posterior insular point gives access to the atrium of the lateral ventricle. The transverse gyrus of Heschl arises exactly from this point and runs anterolaterally to join the superior temporal gyrus as previously described. The anterior and inferior sulci finally converge in the limen insula, which is an area covered by olfactory cortex and serves as an anatomical junction in between the entorhinal, paraolfactory, fronto-orbital, and temporomesial cortices, with the paleo- and archicortex of the ventral insula. Medially, the limen insula is

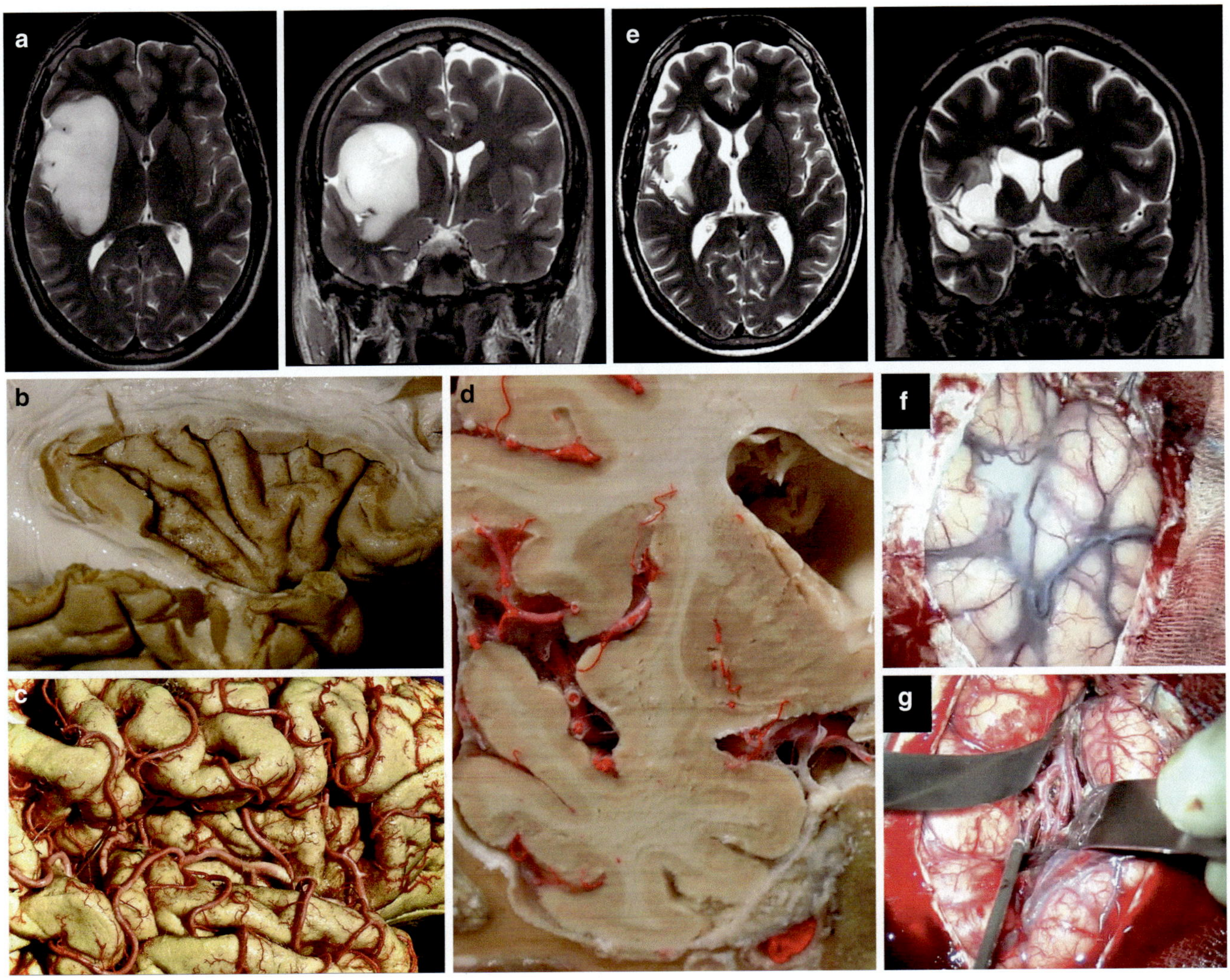

Fig. 61.3 (**a**) Preoperative axial and coronal T2-weighted MRI showing an insular tumor confined to the insular lobe. (**b**) Right side hemisphere of a human brain specimen showing the close relationship to the insular lobe with the arcuate fascicle. (**c**) Right side hemisphere of a human brain specimen showing the MCA branches after opening the opercula. (**d**) Right side coronal on a human brain specimen at the level of the insular apex, showing the sylvian fissure depth anatomy, MCA branches and short and long perforators, as well as some of the lenticulostriate arteries. (**e**) Postoperative axial and coronal T2-weighted MRI with a small remnant in the superior, posterior, and medial aspect of the insular lobe. (**f**, **g**) Intraoperative pre- and post-resection views after a transylvian approach

near the anterior perforated substance, which should be a surgical limit during insular tumors resection. From a sulcogyral point of view, the insula can be divided into anterior and posterior lobes. The anterior is mostly related to the limbic areas in terms of connections and cytoarchitecture, while the posterior lobe is closer to the neocortex. These lobes are divided by the central sulcus of the insula. Three to five short anterior gyri and generally two long posterior gyri are found on both sides of this sulcus Fig. 61.3.

The fact that these gyri present different phylogenetic origins and connections to the surrounding brain seems to influence the growing patterns of tumors in the insular lobe. Therefore, the most common patterns are purely insular tumors, operculoinsular and insular lesions invading the fronto-orbital and temporomesial paraolfactory regions. A careful preoperative analysis of the images should help us to identify these different growing patterns and trace an optimal surgical plan, deciding before going to the operating room where to enter the tumor from and how much to resect. This is, indeed, what we call a surgical strategy.

61.3.4 Arteries of the Insula

An arterial tree branching from the M1 segment of the middle cerebral artery feeds the insular cortex covering its lateral surface. In addition, a group of thinner arteries, the lenticulostriate, arise from the M1 trunk, internal carotid artery bifurcation, or even from the proximal segments of the M2 inferior and/or superior trunks. This group of vessels ascends vertically into the basal ganglia and the internal capsule to irrigate them after traversing the anterior perforated substance. The M1 segment turns 90 degrees around the limen insulae and forms the genu, where it generally bifurcates into the superior and inferior trunks. The olfactory cortex covering the limen as well as the underlying tracts is fed by small perforators coming from these early segments of the middle cerebral artery. Along its cortical course through the insular surface, the main superior and inferior trunks give rise to 8–12 branches which will continue their path towards the anterior, superior, and inferior periinsular sulci. Once at the level of the sulci, this M2 distal branches turn again almost 180 degrees and follow the shape of the subopercular surface till reaching the sylvian fissure. The segment in between the periinsular sulci and the sylvian fissure is known as M3, while the M4 branches start once the lateral aspect of the hemisphere is reached Fig. 61.3.

One of the most constant M3 branches is the prefrontal artery, which usually arises from the anterior insular point and runs laterally to emerge into the anterior sylvian point. Therefore, this is a good anatomical landmark when these areas need to be approached. The prefrontal, precentral, and central arteries give rise to small arteries which feed the anterior insular lobe. The central ascends in the central insular sulcus. The anterior and posterior parietal arteries generally branch into small perforators to feed the anterior long gyrus; the temporo-occipital, angular, and posterior temporal arteries supply the posterior long gyrus [14, 15] Fig. 61.3.

All these small perforators also feed the insula in pathological conditions. Therefore, if a transylvian approach is performed, they must be identified, carefully coagulated, and properly cut. However, when performing a transopercular approach, it is less technically demanding to deal with these small perforators, as during the subpial resection, just a controlled irrigation is enough to control their bleeding in most of the cases. However, special attention must be given to the long perforators, which are located more dorsally around the posterior central sulcus of the insula and posterior lobe. These long perforators run through the dorsal insular cortex and some of them reach and feed the corona radiata. Thus, an accidental damage of these vessels can cause a small but clinically very relevant infarct [14] Fig. 61.3.

61.3.5 Subcortical Anatomy

The same forces that occur during embryological development to place the insular lobe in a central position of the human brain are also responsible for the special relationship that this structure has with the white matter tracts it shelters. Therefore, the subcortical organization of the insula is arranged in consecutive layers of grey and white matter and circumferentially surrounded by longitudinal, curvilinear, and curved long association fiber tracts.

Just underlying the cortical grey matter layers, the extreme capsule appears as the first association fiber system. This layer of fibers represents the short ‘u’-fibers that interconnect adjacent gyri into the neocortical dorsal telencephalon. The extreme capsule represents the connection between the different gyri of the insular cortex and also presents longer branches that turn around the anterior, superior, and inferior periinsular sulci connecting the insula with the adjacent fronto-orbital, frontoparietal, and temporal opercula, respectively. Certain opercular and insular tumors use this superficial layer to invade the adjacent homonymous regions. Moreover, it is postulated that operculoinsular neocortical lesions travel through the extreme capsule opercular projections Fig. 61.4.

The next subinsular layer is made of grey matter and located underlying the extreme capsule into its most ventral position, mostly related with the anterior insular lobe. This is the claustrum, which continues at the level of the limen insulae in close relationship with a couple of association fiber systems (uncinate and inferior frontooccipital). Before describing these association fiber systems, it is relevant to remind how the insula, on its trip to the central core during the embryological development, and the anterolateral direction of the temporal lobe, has forced the fibers connecting the frontal with the parietal, occipital, and temporal lobe to a specific location. Therefore, the pathways in between the frontal and parietal lobes (superior longitudinal fascicle) remain dorsal and run in a longitudinal way as both lobes are in the same axial axis. The main fiber system connecting the frontal and occipital lobes (inferior frontooccipital fascicle) is located ventral to the insula and runs again in a longitudinal way but shows an italic ‘s’ shape as with a curve into the limen of the insula, as the occipital pole is a bit inferiorly located compared to the frontal lobe. Finally, the uncinate (connecting the fronto-orbital and temporomesial structures) and the arcuate (connecting frontal and posterior temporal areas) are forced and curved in a ventral and dorsal location to the insula, respectively. Summarizing all these association fiber tracts, we can divide a dorsal and ventral stream Fig. 61.4.

The ventral stream is composed of the uncinate and the inferior frontooccipital fasciculus, which are long association fiber systems that expand and open like a fan close to their targets, but are compressed and tight at the level of the limen insula and keep a close relationship with the claustrum. Those insular tumors which remain close to the limen might use the uncinate fascicle to invade the fronto-orbital and/or the temporomesial region, or most likely, paleopallial origin tumors could use this fiber system to grow, and therefore partially invade the insula. The inferior frontooccipital

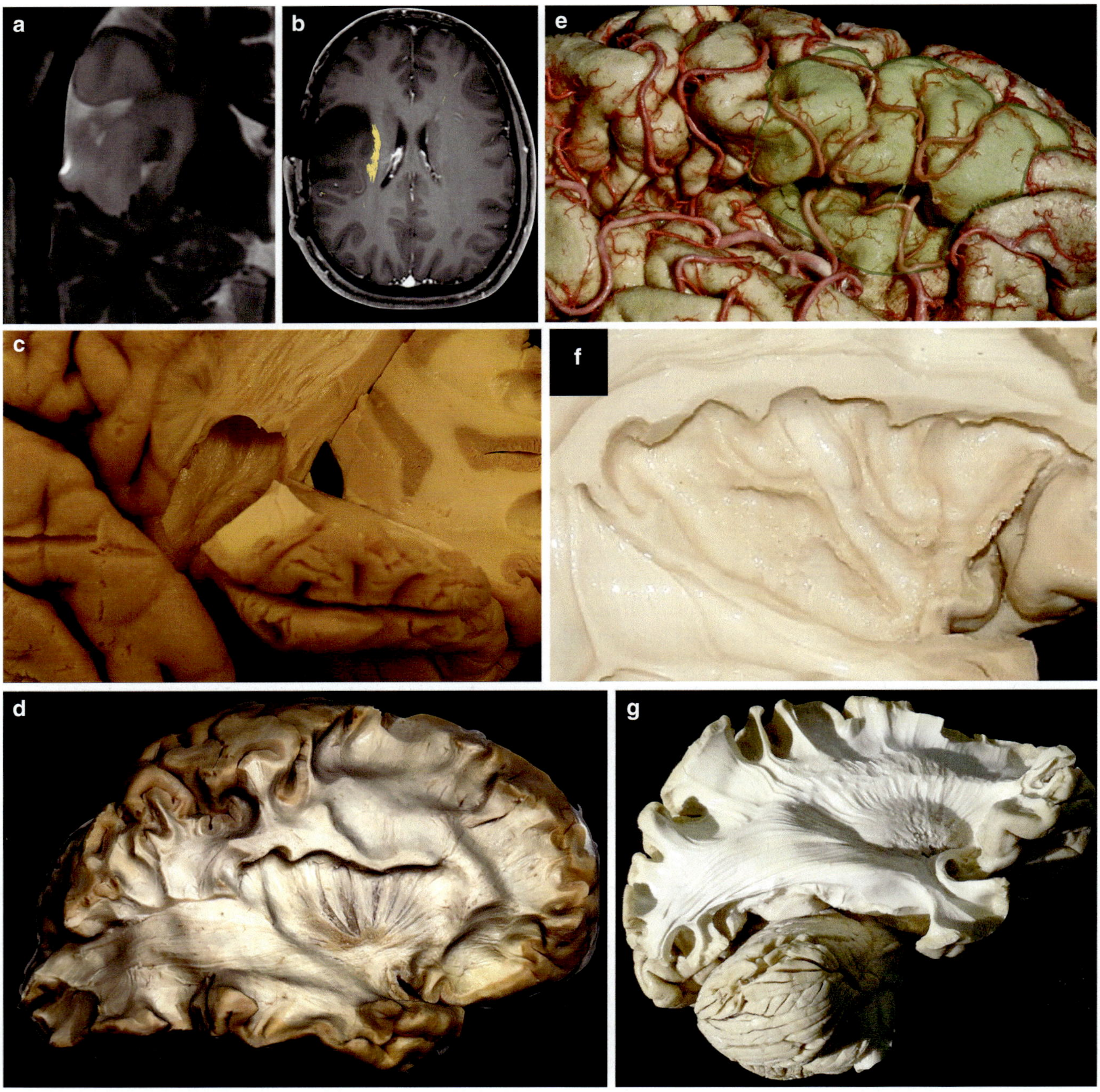

Fig. 61.4 (**a**) Preoperative coronal T2-weighted MRI and (**b**) axial T1-weighted GD MRI sequence with the DTI pyramidal tract (yellow) of a frontoparietal operculum tumor invading the insula through the extreme capsule. (**c**) Right side hemisphere of a human brain after partially dissecting the frontal and insular lobes showing the connection of the frontal operculum and the insular cortex through the extreme capsule. (**d**) Right hemisphere of a human brain after a white matter dissection showing the inferior frontooccipital, arcuate, and uncinate fascicles, as well as the claustrum, external capsule, and putamen, all related to the insular lobe. (**e**) Basal view of the subopercular space and its close relationship to the insula and M3 branches. The area the tumor in A-B is occupying is highlighted in *green*. (**f**) Extreme capsule dissection on a right insular lobe. (**g**) The corona radiata is exposed in this right hemisphere dissection

is involved in semantic language processing, visual switching tasks, and goal-oriented behavior, while the uncinate is supposed to allow mnemonic associations in the temporal lobe that might modify responses through direct interactions with the orbitofrontal cortex [16, 17].

Moving to the opposite side of the insula, around the posterior insular point, the arcuate fascicle turns more than 90° on its way from the pars opercularis to the superior temporal gyrus. This anatomical detail is very important, as it is one of the main limitations to perform transopercular approaches to the insula in this region where the dorsal stream remains in the subcortical layers Fig. 61.4.

If we continue our dissection one layer deeper and remove the extreme capsule and claustrum, the first subinsular projection fiber system is represented by the external capsule, which represents a theoretical boundary that most of the low-grade insular gliomas will not cross spontaneously, thus respecting the putamen that underlies the external capsule. The location of the putamen matches quite accurately with the deep projection of the insular cortex. Therefore, the peri-insular sulci are appropriate landmarks to predict the location of the basal ganglia Fig. 61.4.

Removing the putamen gives us an overview of the next and deepest capsule, the internal capsule. All these structures (extreme, external, and internal) are known as capsules as they encapsulate the lateral basal ganglia. Thus, the external and internal capsule join on the top of the putamen. This junction happens in a very thin space due to the narrow aspect of the external capsule, and one of the key points is the one in which the corticospinal tract runs down from the corona radiata into the internal capsule. This junction takes place deep into the superior periinsular sulcus and quite close to the posterior insular point. This is a very delicate area during surgical resection of posteriorly growing insular tumors Fig. 61.5.

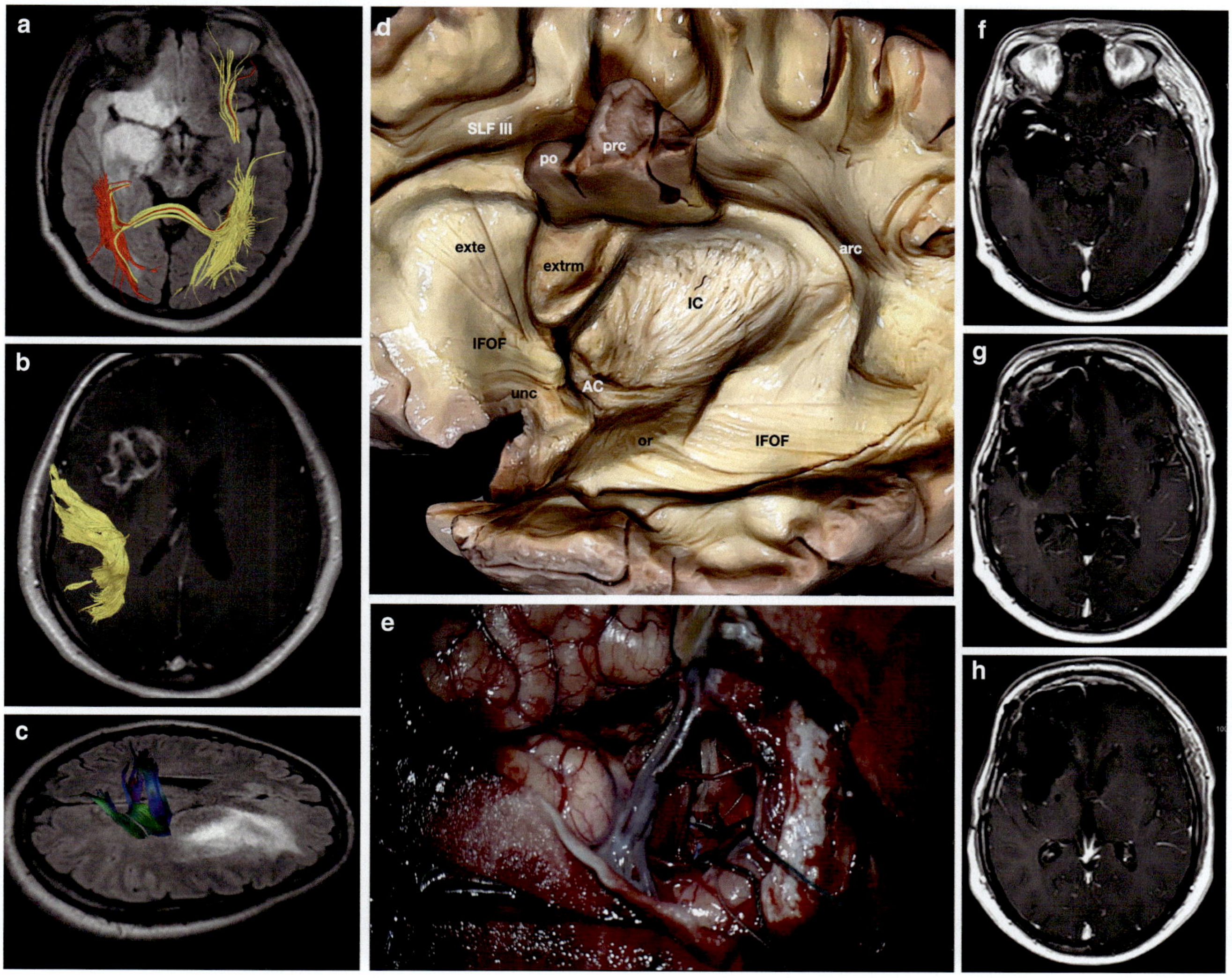

Fig. 61.5 (**a**) Preoperative axial FLAIR MRI sequence and the right (red) and left side inferior frontooccipital (yellow) fascicles related to a operculoinsular tumor in the right hemisphere. (**b**) Preoperative axial T1-weighted GD MRI showing the tumor and the right-side arcuate fascicle (yellow). (**c**) Preoperative axial FLAIR MRI sequence and the corticospinal tract (blue and green) related to the operculoinsular tumor. (**d**) Left side hemisphere human brain dissection showing the most relevant structures related to the insular tumors: superior longitudinal fascicle III (SLF III), pars opercularis (po), precentral gyrus (prc), arcuate fascicle (arc), inferior frontooccipital fascicle (IFOF), external capsule (exte), uncinate fascicle (unc), extreme capsule (extrm), anterior commissure (AC), internal capsule (IC), and optic radiations (or). (**e**) Intraoperative view after a transopercular approach for the operculoinsular tumor. (**f–h**) Postoperative T1-weighted GD MRI after a transopercular resection

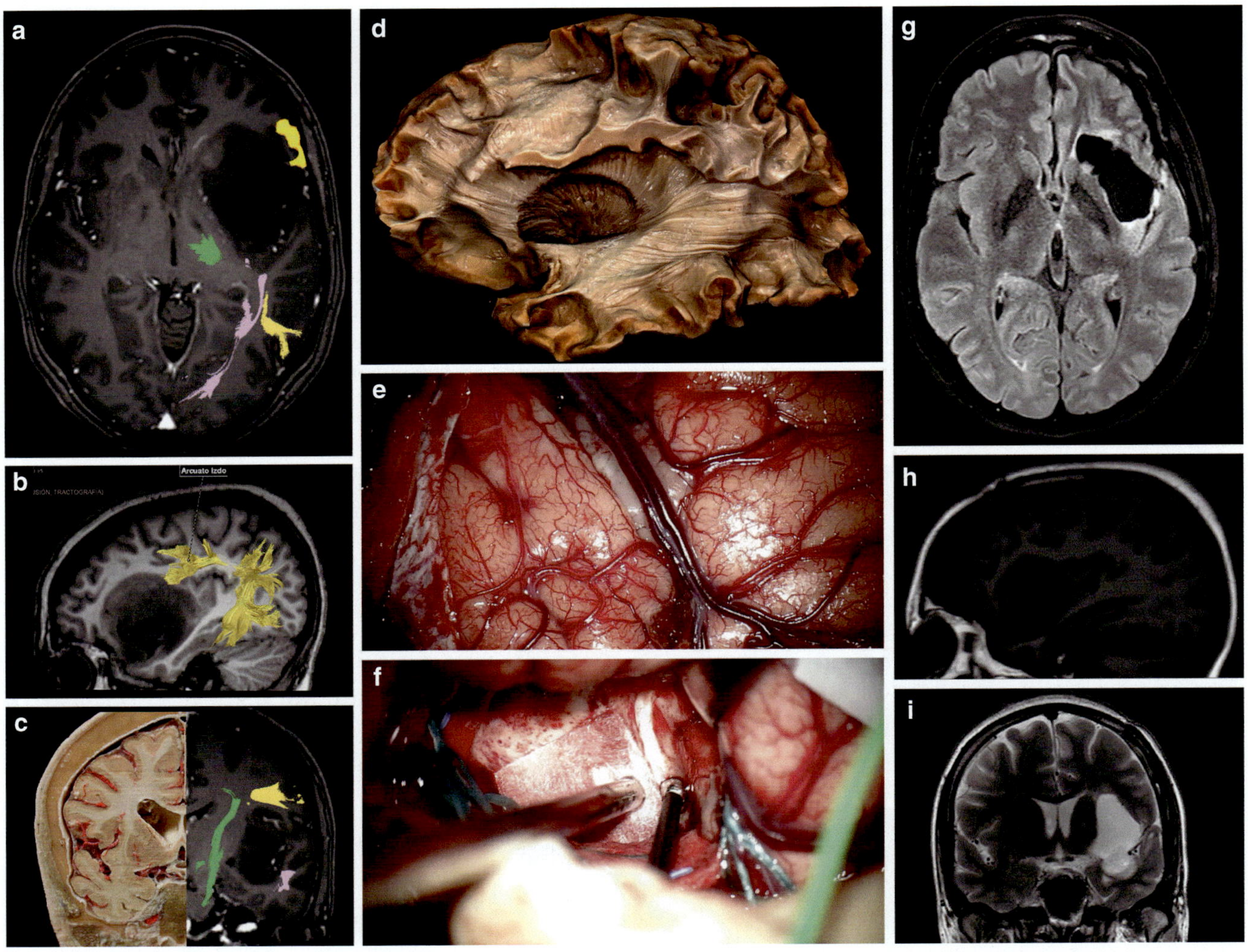

Fig. 61.6 (**a**) Preoperative axial T1-weighted GD MRI with tractography of the corticospinal tract (green), arcuate (yellow) and inferior frontooccipital (pink) fascicles, as well as speech function fMRI (orange) in a left-sided insular glioma case. (**b**) Preoperative sagittal T1-weighted MRI with the arcuate fascicle (yellow) and this anterior insular lobe tumor. (**c**) Coronal cut combining the anatomic view in the right, and a T1-weighted GD MRI with tractography of the corticospinal tract (green), arcuate (yellow), and inferior frontooccipital (pink) fascicles. (**d**) Left hemisphere of a human brain after a white matter dissection showing the inferior frontooccipital, arcuate, and uncinate fascicles, as well as the internal capsule after having removed the basal ganglia, all related to the insular lobe. (**e**) Intraoperative view after opening the sylvian fissure arachnoid membrane showing the tumor underlying the sylvian veins. (**f**) Intraoperative view at the end of the resection showing the subcortical stimulation near the posterior insular point with positive motor response. (**g–i**) Postoperative axial FLAIR, sagittal T1-weighted, and coronal T2-weighted after the transylvian resection where a small posterior tumor remnant is seen

Once we have created a mental image of all these white matter structures, we can correlate the surface topography of the periinsular sulci with the underlying fascicles. Thus, deep into the anterior periinsular sulcus, the anterior limb of the internal capsule appears running to the prefrontal cortex; the superior periinsular sulcus is situated just ventral to the frontal projections of the superior longitudinal and arcuate fascicles and lies superficial to corona radiata projections of the genu and posterior limb of the internal capsule; and the inferior periinsular sulcus has a special feature as, together with the limen insulae cover, the so-called temporal stem represents the only junction through which the temporal lobe is secured ventrally to the rest of the hemisphere. It consists of a bundle of fibers which separate different areas inside the temporal lobe, acting sometimes as edges. This bundle of fibers enters the temporal lobe between the anterior limit of the insula and the temporal horn of the lateral ventricle. It is composed of the uncinate, inferior frontooccipital, anterior commissure, and optic radiations Fig. 61.6.

61.4 Surgical Pearls

- Choose the transopercular or transylvian approach based on the location of the tumor and its extension to adjacent gyri.
- Motor Evoked Potentials (MEPs) are highly recommended when approaching this area to control possible reversible

damage to the lenticulostriate arteries, distal branches of the MCA, and corticospinal tract projections.

- When a transopercular approach is intended, venous superficial anatomy is mandatory to be considered to plan where to start the sylvian fissure splitting.
- Subpial resection must be intended to perform, especially when working near small arteries to reduce the possibility of their damage.

61.5 Discussion

Insular tumors can be divided and named according to different classifications. All of them add singular and relevant information. However, one of the most common ways to classify these challenging lesions is the one based on their infiltrative diffuse pattern. There is no way to understand and classify these lesions on this anatomical basis without understanding the topographic anatomy of the insula.

It was Professor Yasargil who first proposed this simple but extremely efficient classification from a surgical perspective. This is very important as it not only provides anatomical and functional information, but also focuses on the surgical planning to the most adequate approach [1, 2, 7].

Therefore, tumors can be classified as insular (the ones restricted to the insular lobe itself—*3a*), operculoinsular (invading the adjacent perisylvian operculum—*3b*), and insular lesions invading the fronto-orbital and temporomesial paraolfactory regions (5). This classification fits perfectly with the three different cytoarchitectonic areas found in the insula: agranular (ventral), nongranular (pericentral), and granular (dorsal). The ventral shows more connections to the orbitofrontal, entorhinal, and periamygdaline cortex and is near the uncinate fascicle, which indeed connects these regions, while the dorsal is mostly connected to the retroinsular areas and neocortical opercula through the extreme capsule [7].

Being aware of these growing patterns offers a rich information about the ideal approach. Therefore, we mainly use transylvian approaches for lesions which are restricted to the insular lobe. If the vascular anatomy covering the insular surface is challenging to deal with, a transopercular approach could be an option, especially in the nondominant hemisphere. Transopercular approaches are without any doubt ideal for operculoinsular tumors as the tumor itself is related to the opercular units. In these cases, the corticotomy is carried out through the affected operculum (frontal or temporal) or through both of them in those cases where both sides are infiltrated. The paleopallial tumors (insular and regions related to the paleo- and archipalium as the orbitofrontal and entorhinal cortices) are probably the most challenging ones to be approached. Different options are: transylvian and access to the adjacent orbitofrontal and temporomesial through the fissure itself, transplanum polare approach for those ones restricted to the insula and the temporomesial region, or even transopercular under awake conditions when the operculum is free of tumor, and we need to create a corridor through it [18] Figs. 61.2, 61.3, 61.5, and 61.6.

61.6 Conclusion

The best tool and most constant tool that the neurosurgeon owns to distinguish the tumor from the healthy brain tissue is not other than a strong anatomical orientation acquired by training in the laboratory performing brain dissections under the microscope view and years of surgical experience.

Any brain surgery requires a profound mental process through which the neurosurgeon must go before planning and performing the surgery itself, and the most successful ingredient of this mental process is the tumor anatomy. Pathology follows anatomy, and understanding the way in which tumors develop, grow, and travel through the brain avenues will provide us the necessary information to understand the clinical symptoms, perform safer and more effective surgeries, and even predict potential postoperative deficits our patients might present.

References

1. Yasargil MG. Microneurosurgery. CNS tumors: surgical anatomy, neuropathology, neuroradiology, neurophysiology, clinical considerations, operability, treatment options, vol. VIA. Stuttgart: Thieme; 1994.
2. Yasargil MG. Microneurosurgery. CNS tumors: surgical anatomy, neuropathology, neuroradiology, neurophysiology, clinical considerations, operability, treatment options, vol. IVB. Stuttgart: Thieme; 1994.
3. González-López P, Cossu G, Thomas CM, Marston JS, Gómez C, Pralong E, et al. Cadaveric white matter dissection study of the telencephalic flexure: surgical implications. Front Neurol. 2022;13:757757.
4. Wilder P, Faulk ME. The insula; further observations on its function. Brain. 1955;78(4):445–70.
5. Guillaume MMJ, Mazars G. Technique de résection de l'insula dans les épilepsies insulaires. Rev Neurol. 1949;81:900–3.
6. Yaşargil MG, von Ammon K, Cavazos E, Doczi T, Reeves JD, Roth P. Tumours of the limbic and paralimbic systems. Acta Neurochir. 1992;118(1–2):40–52.
7. Yaşargil MG, Reeves JD. Tumours of the limbic and paralimbic system. Acta Neurochir. 1992;116(2–4):147–9.
8. Kalani MY, Kalani MA, Gwinn R, Keogh B, Tse VC. Embryological development of the human insula and its implications for the spread and resection of insular gliomas. Neurosurg Focus. 2009;27(2):E2.
9. Mesulam M-M, Mufson EJ. The insula of Reil in man and monkey. Association and Auditory Cortices. 1985:179–226.
10. González-López P, Gunness N, Olivas JA, Elbabaa SK, Caffo M, López PM. Insular brain tumor surgery: part 1—insular lobe anatomy. Contempor Neurosurg. 2019;41(5):1–7.
11. Ribas GC, Ribas EC, Rodrigues CJ. The anterior sylvian point and the suprasylvian operculum. Neurosurg Focus. 2005;18(6):1–6.

12. Tanriover N, Rhoton AL, Kawashima M, Ulm AJ, Yasuda A. Microsurgical anatomy of the insula and the sylvian fissure. J Neurosurg. 2004;100(5):891–922.
13. Türe U, Yaşargil DC, Al-Mefty O, Yaşargil MG. Topographic anatomy of the insular region. J Neurosurg. 1999;90(4):720–33.
14. Türe U, Yaşargil MG, Al-Mefty O, Yaşargil DC. Arteries of the insula. J Neurosurg. 2000;92(4):676–87.
15. Delion M, Mercier P, Brassier G. Arteries and veins of the sylvian fissure and insula: microsurgical anatomy. Adv Techn Stand Neurosurg. 2016;43:185–216.
16. Conner AK, Briggs RG, Sali G, Rahimi M, Baker CM, Burks JD, et al. A connectomic Atlas of the human cerebrum—Chapter 13: Tractographic Description of the inferior fronto-occipital fasciculus. Operat Neurosurg. 2018;15(suppl_1):S436–43.
17. Von Der Heide RJ, Skipper LM, Klobusicky E, Olson IR. Dissecting the uncinate fasciculus: disorders, controversies and a hypothesis. Brain. 2013;136(6):1692–707.
18. González-López P, Gunnes N, Abarca Olivas J, Elbabaa S, Caffo M, Moreno LP. Insular brain tumor surgery: part 4—surgical approaches. Contempor Neurosurg. 2019;41(8):8.

The Left Inferior Parietal Lobule: Anatomy, Function, and Surgery

62

Igor L. Maldonado

Learning Objectives

- Cortical anatomy related to parenchymal approaches to the inferior parietal lobule.
- Morphology of the main white matter fasciculi of the inferior parietal lobule and neighboring areas.
- Current concepts of cortical and subcortical functional anatomy evidenced during electrical mapping in awake surgery.
- Recent updates concerning the implication of the inferior parietal lobe connectivity in language and consequences for surgery.
- Anatomical mental picture for planning surgery of the dominant temporoparietal junction due to previous awareness of the eloquent structures present at the resection cavity boundaries.

62.1 Introduction

In the exploration of the intricate interplay between brain anatomy and functional considerations in surgical scenarios, the forthcoming sections embark on a journey into the surgical anatomy of the left Inferior Parietal Lobule (IPL). The morphological aspects are delineated according to the prevailing observations in surgical settings, whereas the functional ones are elucidated through monitoring conducted during awake procedures [1, 2].

Surgery in or near the left dominant IPL is often complex, given the density of functional sites at both the cortical and subcortical levels, particularly for language. A detailed study of the results of neuroimaging, intraoperative ultrasound, and intraoperative electrical mapping can establish the safe limits of resection. As the use of direct cerebral stimulation in awake conditions enables the identification of functional boundaries to optimize tumor removal, the exeresis can be continued until the functional zones have been encountered [3, 4].

This chapter's primary goal is to present the anatomical basis of the mental image needed for surgical planning when the objective is to achieve the maximum possible resection with the preservation of eloquent brain areas. First, a concise review of gross anatomy is provided. Second, the main clinical manifestations generally observed in each mapping stage (cortical or subcortical) are detailed and analyzed in conjunction with the relevant structures that need to be considered when interpreting these findings. This chapter does not intend to exhaustively present the microanatomical details and theories about brain circuitry, which are permanently evolving. In addition, variants in which linguistic functions predominate in the right cerebral hemisphere are out of scope and will not be discussed here.

62.2 History and Background

Until the late twentieth century, there were few reported studies on surgery inside the IPL, especially when dealing with intraparenchymal tumors in the so-called dominant hemisphere. Early attempts were largely constrained by limited anatomical knowledge and the absence of adequate imaging technologies. These endeavors were characterized by cautious and often empirical approaches, often driven by the pressing need to alleviate symptoms. Pioneers like Wilder Penfield, through their groundbreaking work in functional brain mapping, unveiled the critical role of the IPL in complex cognitive functions. These advancements enabled a more nuanced understanding of the region's functional significance, prompting a paradigm shift in surgical strategy from the earlier cautious stance to one that embraced strategic resections while preserving vital functions.

I. L. Maldonado (✉)
CHRU de Tours, Tours, France

Laboratoire d'Anatomie, Faculté de Médecine de Tours, Tours, France

INSERM, Imaging Brain & Neuropsychiatry iBraiN U1253, University of Tours, Tours, France
e-mail: limamaldonado@univ-tours.fr

I. N. Kanaan, V. Beneš (eds.), *Neuroanatomy Guidance to Successful Neurosurgical Interventions*, https://doi.org/10.1007/978-3-031-59838-8_62

Based on the analysis of 28 patients who underwent parenchymal surgery within the parietal lobe, Russell et al. (2005) observed an association between tumor size, patient age, or the presence of a glioblastoma, and the occurrence of neurological deficits [5]. Notably, only patients with gliomas excised from the dominant hemisphere exhibited new postoperative association deficits. The authors highlighted that, as low-grade gliomas exhibit gradual growth and their infiltrative nature into eloquent parenchyma allows for partial functional preservation, a heightened incidence of new neurological deficits is observed in comparison to glioblastoma cases.

The most typical situation in which the functional anatomy of the IPL demonstrates maximum exuberance is that of a right-handed patient diagnosed with an infiltrating cortico-subcortical tumor involving the left temporoparietal junction. In the case of low-grade glioma, it is very often a relatively young working-age patient. Often, the first symptom is an episode of seizures or sometimes language disorders. In other situations, such as metastases or high-grade tumors, the symptomatology can be more exuberant and rapidly evolving. Nevertheless, in almost all situations, the tumor site, as well as the surrounding parenchyma, is eloquent.

Studies focused on surgery of the left dominant IPL were rarely published up to the early 2010s, particularly in the context of gliomas. Postoperative complications have, however, been reported, notably encompassing sensory and language impairments. In light of this, awake surgery has been progressively proposed. Southwell et al. (2017) emphasized that in a limited subset of patients presenting with preoperative language deficits, the resection of IPL gliomas can lead to enhanced language functionality [6]. Nevertheless, in patients with intact preoperative language capabilities, the removal of IPL gliomas may precipitate new language deficits. Such morbidity can be reduced through the strategic implementation of cortical and subcortical stimulation mapping techniques. This methodology preserves or enhances the quality of life while maximizing resection extent.

62.3 Anatomical Highlights

- The inferior parietal lobe comprises two horseshoe-shaped gyri: the supramarginal gyrus circumventing the posterior end of the lateral sulcus and the angular gyrus circumventing the posterior end of the superior temporal sulcus. These gyri collectively correspond roughly to the parietal prominence of the skull.
- Intercalated between the supramarginal and angular gyrus is the sulcus intermedius primus, also known as Jensen's sulcus. This sulcus exhibits frequent variations in both depth and length.
- Following the removal of the superficial white matter, the inferior parietal lobe becomes a convergence point for two associated fasciculi from the superior longitudinal system: the ventral component of the superior longitudinal fasciculus (SLF-III) and the posterior (or vertical) component.
- Traveling through the deep white matter beneath the inferior parietal lobe is the arcuate fasciculus (AF), an association fasciculus closely related to the anterior limiting sulcus of the insula and the corona radiata.

62.4 Gross Anatomy

Knowledge of the three-dimensional arrangement of the main cortical and subcortical elements inside and around a parenchymal approach is a sine qua non for surgery of the temporoparietal junction.

Figure 62.1 shows a postmortem anatomical specimen. The meninges have been removed. We note the relevant gyri and sulci (annotated) on the lateral aspect of the cerebral hemisphere. The anterior portion of the IPL is bounded anteriorly by the postcentral sulcus. The supramarginal gyrus (SMG) begins with a narrow gyral fold connecting it to the lower end of the postcentral gyrus. Taking a slightly irregular trajectory, it circumvents the posterior end (posterior ramus) of the lateral sulcus (of Sylvius). A second, often discontinuous, horseshoe-shaped convolution, the angular gyrus (AG), circumvents the posterior end of the superior temporal sulcus. The sulcus intermedius primus (Jensen's sulcus) is intercalated between these two convolutions (SMG and AG) and varies in both depth and length.

These elements must be precisely identified during the patient's MRI examination. On the preoperative MRI, it is important to determine whether the SMG, the AG, or both are affected by the tumor, as well as the presence of extensions towards the superior parietal lobule (beyond the intraparietal sulcus), temporal lobe (T1, T2), occipital lobe, or postcentral gyrus. It is also important to be aware of any anatomical variation in this area, which is very frequent. The so-called *pre-AG variant* consists of a narrow convolution between the AG and the SMG [7], while the *pre-SMG variant* exhibits a narrow convolution between the postcentral gyrus and the SMG.

For the purpose of anatomical demonstration, the specimens in Figs. 62.2, 62.3, 62.4, and 62.5 were prepared with the Klingler technique (for fiber dissection). The gray matter, made darker and more brittle by the anatomical preparation, could be removed to expose the cerebral white matter. In the white matter, the operator must be aware of the presence of three essential elements of the SLF/AF system and their topography [8]: the ventral (or opercular) component of the SLF-III, the AF, and the posterior (or vertical) segment.

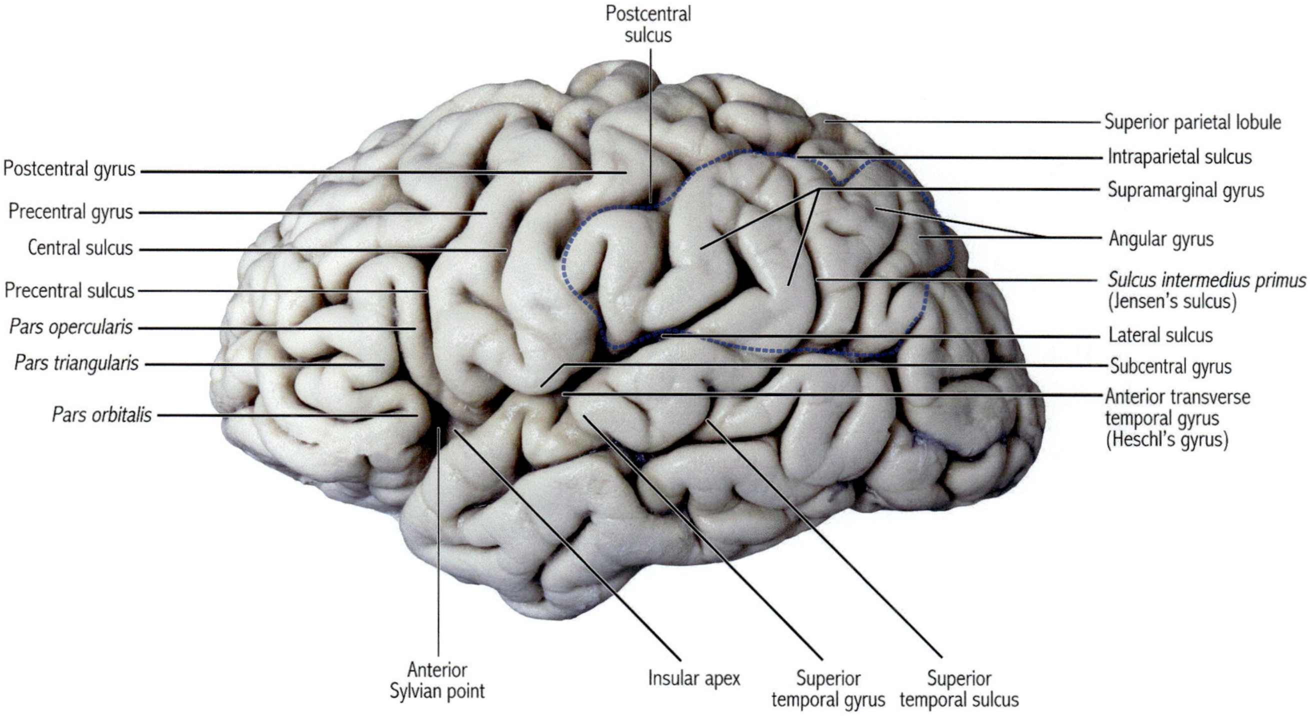

Fig. 62.1 Critical anatomical elements of the sulcogyral anatomy of the inferior parietal lobule (dotted blue line) and neighboring regions

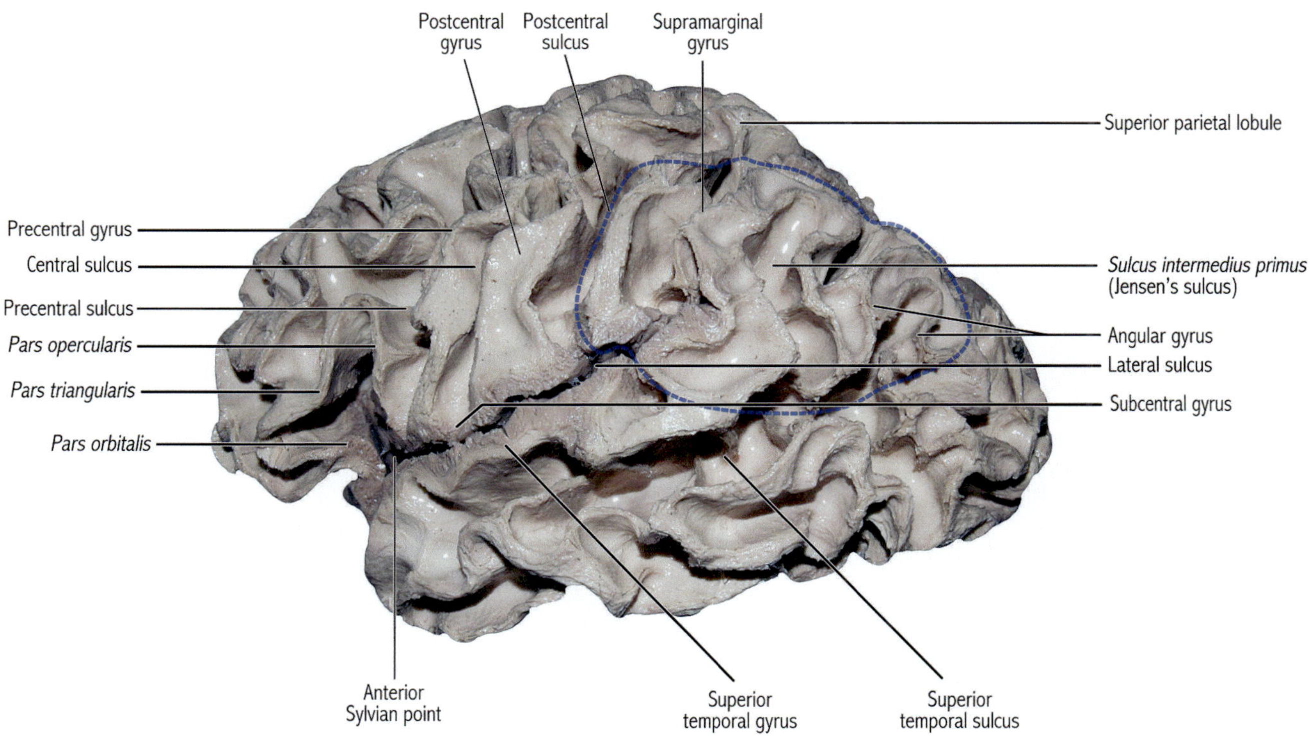

Fig. 62.2 Superficial (subcortical) white matter, critical anatomical accidents, and their correspondence with the sulcogyral anatomy (a different specimen). A laboratory dissection according to the Klingler fiber dissection technique. Blue dotted line: inferior parietal lobule

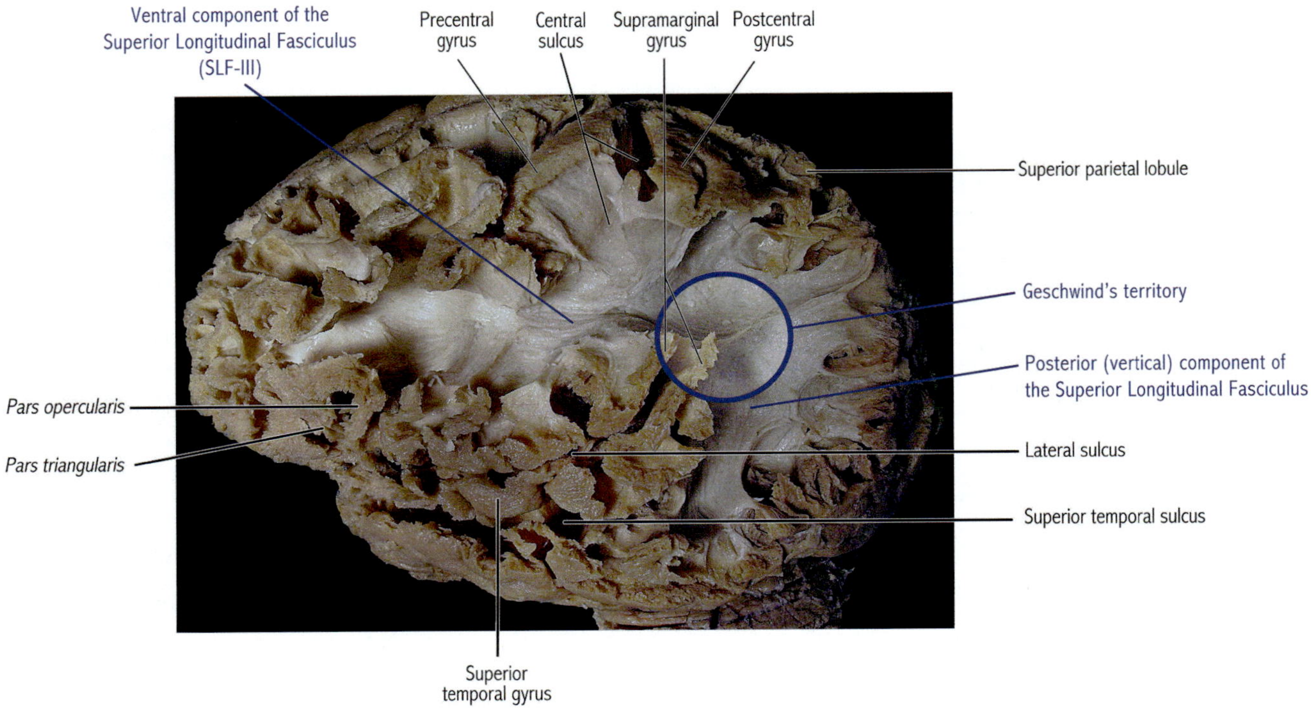

Fig. 62.3 Intermediate stage of fiber dissection. Two components of the superior longitudinal system are exposed: the ventral (opercular) component or SLF-III and the posterior or vertical component. They converge on the so-called Geshwind's territory (blue circle)

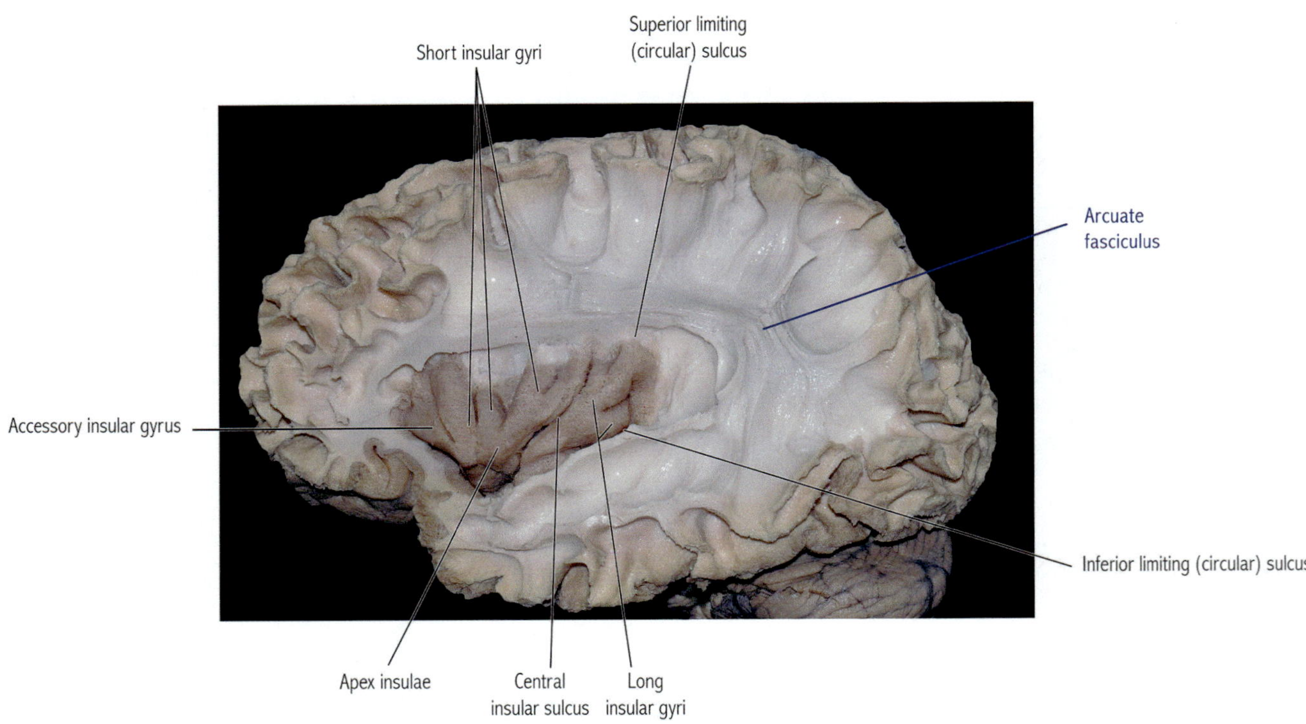

Fig. 62.4 Deep stage of fiber dissection. The deepest component of the SLF/AF system is exposed, which is the arcuate fasciculus. Note the close relationship of this fasciculus with the superior limiting sulcus of the insula. Adapted with permission from Maldonado et al. 2018 [38]

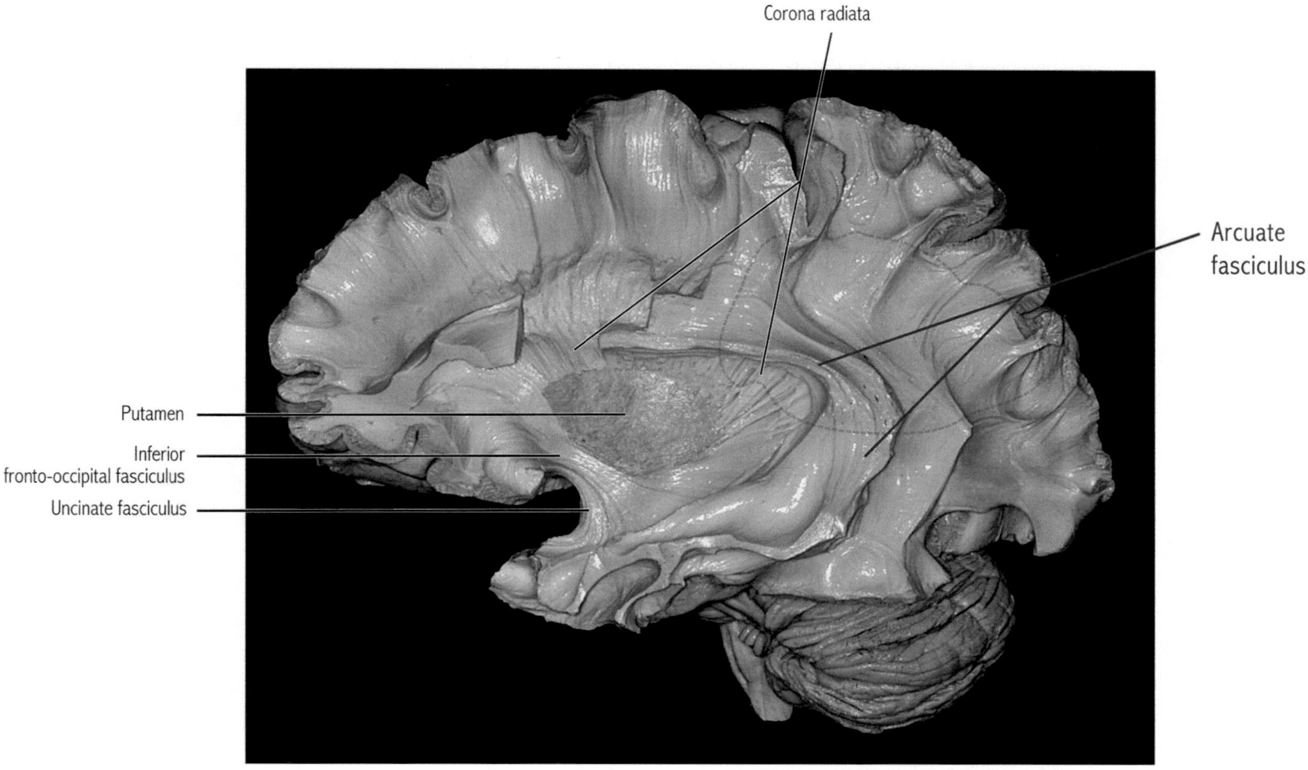

Fig. 62.5 Anatomical relationships in the deep white matter to the inferior parietal lobe (blue dotted line). Note the proximity between the arcuate fasciculus and the fibers of the corona radiata, exposed through the superior limiting sulcus of the insula

62.5 Cortical Functional Anatomy

As mentioned above, the cortical functional anatomy is ideally explored by direct brain stimulation in an awake patient. Resection and electrical stimulation must be performed simultaneously or in close alternation to ensure the identification of functional sites and the safety of the procedure.

It is desirable to expose the entire cortical surface of the IPL. A craniotomy that extends beyond the limits of the tumor is preferable to a limited opening that requires the surgeon to work under the bony and dural edges. The primary sensory cortex and language areas should be identified without this technical inconvenience. A sufficiently wide craniotomy provides a rich functional map in which the main eloquent areas can be identified, allowing the neurosurgeon to navigate safely.

62.5.1 Superiorly

The anterosuperior limit of the IPL relates to the posteroinferior border of the postcentral gyrus. Electrical stimulation typically causes dysesthesia of the contralateral hand and fingers (primary sensory area). These sites are usually identified without difficulty and tagged. Depending on the case, degree of infiltration, and mapping results, the limit of resection may or may not correspond to the postcentral sulcus or tumoral limits (as revealed by intraoperative ultrasound). In the postero-superior direction, it is often possible to advance the resection to the tumor's boundaries (or beyond, in selected cases).

62.5.2 Inferiorly

Cortical language sites are found in the inferior and lateral limits of the operative approach. In these regions, functional mapping is positive in virtually all cases. During electrostimulation of the ventral boundaries of the resection, anomia is usually observed in the posterior part of the superior temporal gyrus, anteroinferior portion of the SMG, or posterior region of the middle temporal gyrus. The functional sites of language most often detected are represented in Fig. 62.6. Other possible observations include semantic disorders, hesitation, and perseveration.

These lower limits of resection correspond to language sites in the posterosuperior portions of the left temporal lobe, especially Wernicke's area [9]. Interindividual variability in this region is much greater than in sensory or motor areas. The operator should also be aware that significant reorganization can be induced by slow-growing lesions, such as low-grade gliomas [10–13]. Consequently, one should never

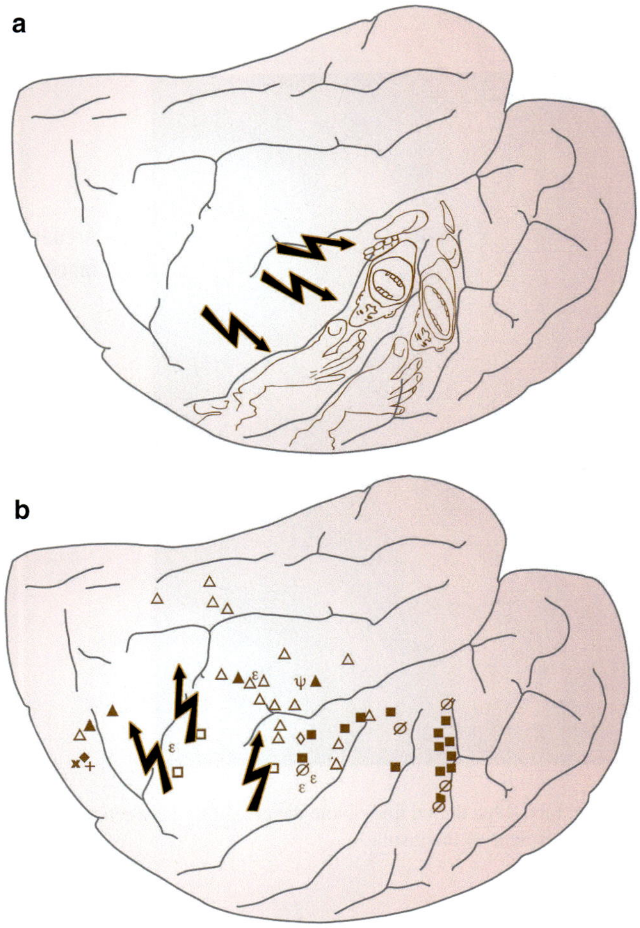

Fig. 62.6 Summary of sites where clinically evident manifestations were induced by cortical electrostimulation. The maps are oriented as in the operative position. The patient's head is turned to the right. (**a**) Sensorimotor map. Stimulation typically induces contraction, involuntary movements, or dysesthesias in specific sites of the corresponding homunculi. (**b**) Language map during counting, picture naming, and spontaneous speech. ∅ Speech arrest; Δ Anomia; ▲ Semantic paraphasia; □ Phonological paraphasia; ■ Articulatory troubles; ◇ Word telescoping; ◆ Gender error; ✕ Increased Latency; ✚ Syntactic problems; Ψ Perseveration; ε Hesitation. Based on the maps of 14 awake patients [4, 38]

expect the distribution of language sites in the dominant temporoparietal junction to follow a rigid pattern. Morphological anatomy alone is not reliable enough to predict function [14]. Intra-axial procedures in this topography are currently recommended to be performed in an awake condition whenever possible. This approach permits a better understanding of the language structures involved at the individual level both before and during resection.

62.5.3 Anteriorly

The postcentral sulcus and the primary somatosensory cortex (in the postcentral gyrus) border the IPL anteriorly. Here mapping is also positive in virtually all cases. The most frequent manifestations are dysesthesias of the hand or fingers, followed by dysesthesias of the forearm, face, tongue, and lips. When dysesthesias are present, they show a somatotopic organization that is consistent across individuals. This pattern recalls the classic descriptions of Penfield's sensory homunculus and involves, from the upper to the lower zones: (i) the forearm, hand, and/or fingers for the upper limb and (ii) the face and/or tongue for the cephalic segment.

In most patients undergoing surgery of the temporoparietal junction, the primary motor area in the precentral gyrus is not part of the anterior boundaries of resection. It is relevant, however, in cases of tumors that invade the postcentral gyrus and (more rarely) the precentral gyrus. In this situation, the resection may progress anteriorly, and electrostimulation should map contralateral motor responses.

Stimulation of the tumor's anteroinferior borders is also positive for language in many cases. It is not uncommon for patients to experience articulation disturbances during stimulation of the parietal operculum. It is worth noting that this manifestation is quite reproducible during stimulation of the (more anterior) ventral premotor cortex. Its use as a reference for the initial electrical stimulation parameters has been proposed.

The anterior functional limit of an IPL resection cavity is thus mostly determined by somatosensory manifestations, although some variation in the exact location of each response has been observed, probably due to interindividual differences or modifications induced by slow-growing lesions [10, 15–18]. Additional teams have reported this phenomenon since Penfield and Boldrey's pioneering work [19, 20].

Traditionally, awake surgery is not considered essential for identifying sensorimotor areas, as somatosensory-evoked potential techniques can provide useful information [21, 22]. However, awake protocols have the great advantages of allowing for tests of precision of movement and measuring more than muscle contraction by assessing spatial cognition, somatosensory feedback, the relationship with vestibular function, and the appearance of hemineglect [17]. Other authors have reported that the excision of part of the parietal cortex can cause severe deficits, particularly disorders of sensory function [5, 23]. Damage to the primary sensory cortex has been associated with Verger-Déjerine cortical sensory syndrome, which involves loss of light touch sensation, decreased two-point discrimination, joint position, and astereognosis.

Most information on functional recovery after sensory cortex injury comes from studies of stroke or noninvasive functional imaging techniques. The latter has recently enabled the study of compensation mechanisms such as the activation of the contralateral somatosensory cortex, reorganization of cortical mapping, displacement of the cortical functional site, amplification of a preserved functional area, or activation of the secondary somatosensory cortex [24–28]. These mechanisms cause significant or partial recovery, depending on the extent of the damaged area. Despite the possibilities of functional reorganization, the importance of preserving sensory functions during the rehabilitation process has been widely reported [29, 30], with sensory deficits being associated with a significant delay in restoring normal motricity and walking.

62.5.4 Posteriorly

Given that cortical mapping is generally less rich in the posterior limits of the IPL, the posterior boundaries of resection are often determined by the tumor morphology (and can extend beyond tumor boundaries in select cases). As mentioned above, these boundaries are identified, among other things, using intraoperative ultrasound, a practice that enables overcoming problems of brain shift in neuronavigation.

Thus, in the posterior limits of the IPL, stimulation usually does not induce sensorimotor responses or speech disturbance, with some infrequent exceptions. The exceptions mainly involve language manifestations: reproducible hesitation or anomie could be observed during electrical stimulation of the posterior part of the AG or the parieto-occipital junction in a limited number of cases, as could an increase in response latency and syntactic errors. When this occurs, stopping the resection in that direction is recommended. Although it is difficult to draw definitive conclusions about these less typical locations, we can hypothesize that they are part of a functional reorganization induced by slow tumor growth. They are, a priori, much less likely to be observed in rapidly growing lesions, such as brain metastases and low-grade gliomas.

62.6 Subcortical Functional Anatomy

Clinical responses observed during electrical stimulation should be considered the result of the disturbance of functional networks. Direct stimulation of white matter in the depth of the IPL elicits clinically evident language manifestations in the vast majority of patients and sensorimotor manifestations in a few, depending mainly on the anterior extension of the resection cavity. The correct interpretation of these findings requires considering the anatomy of the fiber bundles that travel through or end in this region of the cerebral hemisphere (Fig. 62.7).

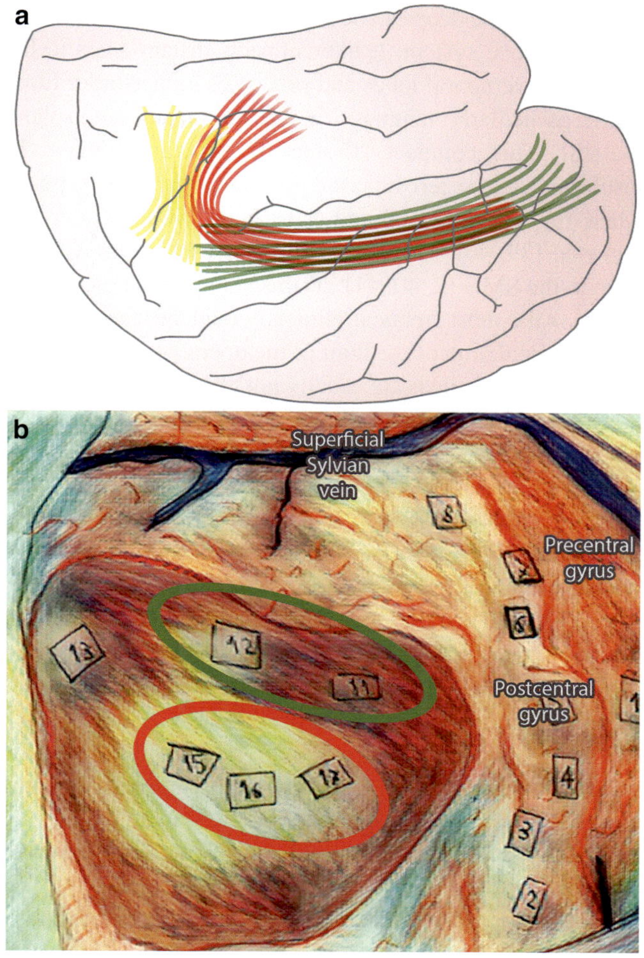

Fig. 62.7 Anatomical location of major association white matter fasciculi encountered during exeresis in left IPL: the arcuate fasciculus (red), the ventral, opercular component (SLF-III, green), and the vertical posterior component (yellow) of the SLF. The drawing is oriented as in the operative position. The patient's head is turned to the right. (**a**) Major association fasciculi in the IPL parenchyma. (**b**) Corresponding subcortical mapping. Stimulation in the topography of the SLF III elicits articulatory disturbances during stimulation. In the deep white matter, stimulation in the topography of the arcuate fascicle elicits very reproducible phonological paraphasias

62.6.1 Medially (To the Depth)

Growing evidence shows that language connectivity is organized along at least two broad streams [31, 32]: the dorsal stream, which is primarily involved in the processing of language for motor functions, including speech production and articulation, and the ventral stream, which is involved in the comprehension and semantic processing of language. Using cortical and subcortical electrostimulation during surgery of the IPL, the operator can map elements of the dorsal stream, of which SLF/AF fibers are critical.

The AF is the deepest and longest component of the SLF/AF associative system. It interconnects distant areas in the frontal and temporal lobes and processes higher-order functions, particularly language in the dominant hemisphere [33–38]. This compact "C-shaped" structure frames the insula and its posterosuperior portion is partially exposed when the IPL's cortex and superficial white matter are removed [39, 40]. The posterosuperior part of the AF runs deep to the SMG [2, 40, 41]. At this site, phonological paraphasia is the most frequent clinically evident response elicited by stimulation of the floor of the operative cavity, where semantic disturbances are usually not observed.

Different components of the SLF/AF system are known to relate to different aspects of language processing. Damage to the AF is an acknowledged cause of conduction aphasia, which combines repetition and phonological disorders. In accordance with Wernicke-Geschwind's classic model, paraphasic speech in patients with this syndrome can be interpreted as the inability of temporal sites of language to monitor speech production by Broca's area [42]. Conduction aphasia has also been reported following pure cortical lesions or mixed cortical and subcortical lesions. However, AF lesion may occur without repetition deficits [31, 35]. As a result of the hodological organization of language networks [33, 38, 42, 43], phonological paraphasia can be induced by electrical stimulation at different sites along the path of the AF [41, 44, 45].

Articulatory disorders with or without a complete cessation of speech are also among the most frequent findings during subcortical awake mapping of the SMG superficial white matter. During naming or counting tasks, dysarthria or anarthria may be induced by stimulation of the white matter area from the Rolandic operculum to the posterior end of the Sylvian fissure in the SMG. This topography corresponds to the ventral component of the horizontal portion of the SLF, the SLF-III, according to experimental [46], tractographic [37], and laboratory anatomical findings [47]. Sites of pure articulatory disturbances are typically deeper than those of phonological paraphasias, most likely due to the course of SLF-III and AF fibers [2, 40].

Importantly, in his theoretical working memory model, Baddeley [48–50] describes an articulatory loop comprising a short-term phonological stock and an articulatory component of repetition, which revitalizes the memorized information. Functional neuroimaging studies suggest that the phonological stock involves the left SMG and that the subvocal repetition system is associated with the left inferior frontal cortex and the ventral premotor cortex [51–53]. The fact that electrical stimulation of these cortical centers, along with their interconnecting white matter pathway, can elicit articulatory disturbances suggests the presence of at least some anatomical correlates of the articulatory loop within the surgical field [54].

Fiber dissection and Diffusion Tensor Imaging (DTI) studies [38, 55–59] agree that, in addition to being interconnected by the long arcuate fibers, the frontal and temporal lobes are interconnected by shorter fasciculi that converge to the so-called Geschwind's territory in the IPL [38, 42, 60, 61]. As part of this indirect pathway, vertically oriented fibers from the posterior component of the SLF/AF system arise from the posterior portions of T1 and T2. This vertical, posterior, component seems to play an important role in reading processes, especially those involving spelling–sound inconsistency, pseudowords, and learning. Electrical stimulation of its topography can generate transient reading difficulties, such as "letter-by-letter" reading or difficulties in correlating between print characters and sounds. Interruption of this fasciculus may result in difficulties in reading irregular words and pseudowords, providing evidence suggesting parallel feedback connections linking the phonological and lexical networks [62, 63].

In cases of anterior tumor extension, sensory manifestations may be observed during subcortical mapping due to the presence of thalamocortical projections. Contralateral upper extremity dysesthesias, as well as movement disturbance and upper limb and facial dysesthesias, may be induced during stimulation of the anterior wall of the operative cavity. In the case of a large anterior invasion crossing the postcentral gyrus, it is possible to observe facial movements during stimulation at an intermediate depth. Vertigo and nystagmus are other less frequent but possible manifestations during stimulation of the insulo-opercular junction.

In the specific case of large temporal extension, it is not unusual to observe semantic paraphasias or syntactic errors close to the cortical temporal sites at which anomia or semantic disorders were observed. At the subcortical level, a critical element of the ventral stream is the inferior fronto-occipital fasciculus [64, 65], which is strongly involved in semantic processing. Its electrical stimulation is, however, more frequently performed during fronto-opercular, insular, or temporal approaches and elicits reproducible transient semantic disturbances [15, 44].

Direct cerebral stimulation is considered a powerful tool for revealing functional anatomy. It offers an exceptional opportunity to better understand language, cognitive networks, and the patient's individual anatomy [16, 41, 44, 53, 54, 66, 67]. Despite plasticity mechanisms that might induce the rearrangement of language sites in cases of slow-growing lesions, [68] this phenomenon does not seem to impact the responsiveness of the SLF/AF system to functional mapping [69]. Consequently, checking for the presence of phonological and articulatory disturbances in the wall operative cavity remains mandatory.

While DTI images can be integrated with the neuronavigation system, it is essential to consider that brain shift during tumor resection may introduce navigation imprecisions, posing a risk to white matter tracts. Therefore, when available, DTI should not be used as a standalone technique for guiding resection near eloquent tracts. The repetitive use of electrical stimulation, however, proves valuable in identifying major pathways before any potential interruptions occur.

62.7 Surgical Pearls

- Functional electrical mapping plays a critical role in handling eloquent areas. In this context, possessing a comprehensive mental picture of the anatomy is fundamental for effective surgical planning. This mental map enables the surgeon to anticipate procedural steps and accurately interpret the sensorimotor and language mapping results, ensuring a safer and more successful procedure.
- Given the procedure's goal of achieving a maximum resection while minimizing permanent neurological deficits, it is crucial to adhere to the mapping results when defining the boundaries of tumor resection in highly eloquent areas rather than relying solely on the sulco-gyral anatomy.
- For the resection of an infiltrating lesion (e.g., low-grade glioma) in the IPL, anterosuperior resection boundaries are often determined by the primary sensory cortex, while inferior boundaries are constrained by the presence of temporal language areas.
- Within the depths of the supramarginal gyrus, the occurrence of phonological paraphasias during electrical stimulation serves as a valuable alert for the surgeon, indicating the presence of the AF, which is often part of the deep limits of resection. More superficially, articulatory disturbances may be elicited during stimulation of the topography of SLF-III.
- Postoperative transient sensorimotor and language neurological deficits are common when the resection extends to contact eloquent structures. As part of the initial stages of preoperative planning, patients must be appropriately warned and reassured about these potential outcomes.

62.8 Discussion

In the present chapter, we revisit the surgical anatomy of the left dominant IPL, from morphology to functional aspects, as revealed by direct cerebral stimulation. It is important to note that a certain degree of interindividual variability exists in sulco-gyral anatomy and cortical maps, particularly those for language. We suggest that no rigid model can be proposed for resections in this region, but certain observations are reproducible at sites that frequently correspond to resection boundaries, namely: the primary sensory cortex, anterosuperiorly; language areas in the posterosuperior portion of the temporal lobe, inferiorly; and SLF/AF white matter pathways, in deeper portions of the operating cavity.

Surgery in the IPL in the dominant hemisphere for language can be performed safely in awake patients by adapting tumor excision to individual functional boundaries. Identifying these elements is very important for optimizing the surgical results of the functional plan. Nevertheless, it is important to note that the operator encounters highly eloquent areas, and complications are still possible even when procedures follow standardized protocols. Complications may be related to the functional roles mentioned earlier in this chapter or to nearby vascular (e.g., angular artery) or neural (e.g., optic radiations in tumors extending posteriorly) structures.

In the immediate postoperative period, transient language disorders are very often reported. The patient must be fully informed of this postoperative evolution—with probable transient worsening—during the earliest planning stages. Such disorders manifest most frequently as speech apraxia, difficulty in naming or calculation, phonological disorder, or slowness. When the temporal language areas are respected, comprehension is generally well preserved. When connectivity has also been preserved, and in the absence of vascular complications, most recovery occurs within 2–3 weeks. Long-term postoperative functional evolution is strongly influenced by tumor histology and the patient's general state.

62.9 Conclusion

A clear and accurate mental picture of anatomy is surely one of the most valuable skills in the brain surgeon's arsenal. This ability to visualize and understand intrinsic brain anatomy, even when some structures are hidden in the parenchyma, is fundamental to guiding our actions safely and efficiently. The surgeon can then anticipate structures and individual variations of which they were previously aware. This allows the operator to avoid damage to critical areas, minimize the risk of complications, and ensure a more precise approach.

References

1. Deras P, Moulinié G, Maldonado IL, Moritz-Gasser S, Duffau H, Bertram L. Intermittent general anesthesia with controlled ventilation for "asleep-awake-asleep" brain surgery: a prospective series of 140 gliomas in eloquent areas. Neurosurgery. 2012;71:764–72.
2. Maldonado IL, Moritz-Gasser S, Duffau H. Does the left superior longitudinal fascicle subserve language semantics? A brain electrostimulation study. Brain Struct Funct. 2011;216:263–74.
3. Duffau H, Thiebaut de Schotten M, Mandonnet E. White matter functional connectivity as an additional landmark for dominant temporal lobectomy. J Neurol Neurosurg Psychiatry. 2008;79:492–5.

4. Berger MS, Deliganis AV, Dobbins JD. The effect of extent of resection on recurrence in patients with low-grade cerebral hemisphere gliomas. Cancer. 1994;74:1784–91.
5. Russell SM, Elliott R, Forshaw D, Kelly PJ, Golfinos JG. Resection of parietal lobe gliomas: incidence and evolution of neurological deficits in 28 consecutive patients correlated to the location and morphological characteristics of the tumor. J Neurosurg. 2005;103:1010–7.
6. Southwell DG, Riva M, Jordan K, Caverzasi E, Li J, Perry DW, et al. Language outcomes after resection of dominant inferior parietal lobule gliomas. J Neurosurg. 2017;127:781–9.
7. Kiriyama I, Miki H, Kikuchi K, Ohue S, Matsuda S, Mochizuki T. Topographic analysis of the inferior parietal lobule in high-resolution 3D MR imaging. AJNR Am J Neuroradiol. 2009;30:520–4.
8. Porto de Oliveira JVM, Raquelo-Menegássio AF, Maldonado IL. What's your name again? A review of the Superior Longitudinal and Arcuate Fasciculus evolving nomenclature. Clin Anat. 2021;34(7):1101–10.
9. Vigneau M, Beaucousin V, Herve PY, Duffau H, Crivello F, Houde O, et al. Meta-analyzing left hemisphere language areas: phonology, semantics, and sentence processing. NeuroImage. 2006;30:1414–32.
10. Duffau H. Brain plasticity and tumors. Adv Tech Stand Neurosurg. 2008;33:3–33.
11. Duffau H. New concepts in surgery of WHO grade II gliomas: functional brain mapping, connectionism and plasticity-a review. J Neuro-Oncol. 2006;79:77–115.
12. Desmurget M, Bonnetblanc F, Duffau H. Contrasting acute and slow-growing lesions: a new door to brain plasticity. Brain. 2007;130:898–914.
13. Duffau H. Lessons from brain mapping in surgery for low-grade glioma: insights into associations between tumour and brain plasticity. Lancet Neurol. 2005;4:476–86.
14. Pouratian N, Bookheimer SY. The reliability of neuroanatomy as a predictor of eloquence: a review. Neurosurg Focus. 2010;28:E3.
15. Duffau H. Intraoperative cortico-subcortical stimulations in surgery of low-grade gliomas. Expert Rev Neurother. 2005;5:473–85.
16. De Witt Hamer PC, Moritz-Gasser S, Gatignol P, Duffau H. Is the human left middle longitudinal fascicle essential for language? A brain electrostimulation study. Hum Brain Mapp. 2011;32(6):962–73.
17. Duffau H. Awake surgery for nonlanguage mapping. Neurosurgery. 2010;66:523–8; discussion 528–9
18. Maldonado IL. Letter to the editor regarding "clinical application of brain plasticity in neurosurgery". World Neurosurg. 2021;146:399.
19. Penfield W, Boldrey E. Somatic motor and sensory representation in the cerebral cortex of man as studied by electrical stimulation. Brain. 1937;60:389–443.
20. Penfield W, Welch K. Instability of response to stimulation of the sensorimotor cortex of man. J Physiol. 1949;109:358–65.
21. Grant GA, Farrell D, Silbergeld DL. Continuous somatosensory evoked potential monitoring during brain tumor resection. Report of four cases and review of the litterature. J Neurosurg. 2002;97:709–13.
22. Szelényi A, Bello L, Duffau H, Fava E, Feigi GC, Galanda M, et al. Intraoperative electrical stimulation in awake craniotomy: methodological aspects of current practice. Neurosurg Focus. 2010;28:E7.
23. Hommet C, Bardet F, De Toffol B, Perrier D, Biraben A, Vignal JP, et al. Unilateral spatial neglect following right inferior parietal cortectomy. Epilepsy Behav. 2004;5:416–9.
24. Jang SH. Contra-lesional somatosensory cortex activity and somatosensory recovery in two stroke patients. J Rehabil Med. 2011;43:268–70.
25. Jang SH, Ahn SH, Lee J, Cho YW. Cortical reorganization of sensori-motor function in a patient with cortical infarct. NeuroRehabilitation. 2010;26:163–6.
26. Rossini PM, Tecchio F, Pizzella V, Lupoi D, Cassetta E, Pasqualetti P, et al. On the reorganization of sensory hand areas after monohemispheric lesion: a functional (MEG)/anatomical (MRI) integrative study. Brain Res. 1998;782:153–66.
27. Rossini PM, Tecchio F, Pizzella V, Lupoi D, Cassetta E, Pasqualetti P. Interhemispheric differences of sensory hand areas after monohemispheric stroke: MEG/MRI integrative study. NeuroImage. 2001;14:474–85.
28. Cramer SC, Moore CI, Finklestein SP, Rosen BR. A pilot study of somatotopic mapping after cortical infarct. Stroke. 2000;31:668–71.
29. Hurwitz LJ, Adams GF. Rehabilitation of hemiplegia: indices of assessment and prognosis. Br Med J. 1972;1:94–8.
30. Van Buskirk C, Webster D. Prognostic value of sensory defect in rehabilitation of hemiplegics. Neurology. 1955;5:407–11.
31. Hickok G, Poeppel D. Dorsal and ventral streams: a framework for understanding aspects of the functional anatomy of language. Cognition. 2004;92:67–99.
32. Hickok G, Osaka K, Serences JT. Area Spt in the human planum temporale supports sensory-motor integration for speech processing. J Neurophysiol. 2009;101:2725–32.
33. Catani M, ffytche DH. The rises and falls of disconnection syndromes. Brain. 2005;128:2224–39.
34. Geschwind N. Disconnexion syndromes in animals and man: part I. Neuropsychol Rev. 1965;20:128–57.
35. Anderson JM, Gilmore R, Roper S, Crosson B, Bauer RM, Nadeau S, et al. Conduction aphasia and the arcuate fasciculus: a reexamination of the Wernicke-Geschwind model. Brain Lang. 1999; 70:1–12.
36. Bernal B, Altman N. The connectivity of the superior longitudinal fasciculus: a tractography DTI study. Magn Reson Imaging. 2010;28(2):217–25. http://www.ncbi.nlm.nih.gov/entrez/query.fcgi?cmd=Retrieve&db=PubMed&dopt=Citation&list_uids=19695825
37. Makris N, Kennedy DN, McInerney S, Sorensen AG, Wang R, Caviness VS Jr, et al. Segmentation of subcomponents within the superior longitudinal fascicle in humans: a quantitative, in vivo, DT-MRI study. Cereb Cortex. 2005;15:854–69.
38. Catani M, Jones DK, ffytche DH. Perisylvian language networks of the human brain. Ann Neurol. 2005;57:8–16.
39. Maldonado IL, Zemmoura I, Destrieux C. Gross anatomy of the human insula. In: Turgut M, Yurttaş C, Tubbs RS, editors. Isl Reil insula hum brain. Cham: Springer International Publishing; 2018. [cité 15 août 2023]. p. 15–22. https://doi.org/10.1007/978-3-319-75468-0_2.
40. Maldonado IL, Moritz-Gasser S, de Champfleur NM, Bertram L, Moulinie G, Duffau H. Surgery for gliomas involving the left inferior parietal lobule: new insights into the functional anatomy provided by stimulation mapping in awake patients. J Neurosurg. 2011;115:770–9.
41. Duffau H, Capelle L, Sichez N, Denvil D, Lopes M, Sichez JP, et al. Intraoperative mapping of the subcortical language pathways using direct stimulations. An anatomo-functional study. Brain. 2002;125:199–214.
42. Catani M, Mesulam M. The arcuate fasciculus and the disconnection theme in language and aphasia: history and current state. Cortex. 2008;44:953–61.
43. Catani M, Mesulam M. What is a disconnection syndrome? Cortex. 2008;44:911–3.
44. Duffau H, Peggy Gatignol ST, Mandonnet E, Capelle L, Taillandier L. Intraoperative subcortical stimulation mapping of language pathways in a consecutive series of 115 patients with Grade II glioma in the left dominant hemisphere. J Neurosurg. 2008;109:461–71.

45. Duffau H. The anatomo-functional connectivity of language revisited. New insights provided by electrostimulation and tractography. Neuropsychologia. 2008;46:927–34.
46. Petrides M, Pandya DN. Projections to the frontal cortex from the posterior parietal region in the rhesus monkey. J Comp Neurol. 1984;228:105–16.
47. Schmahmann JD, Pandya DN. Fiber pathways of the brain. Oxford: Oxford University Press, Inc.; 2006.
48. Baddeley AD, Hitch G. Working memory. In: Bower GA, editor. The psychology of learning and motivation: advances in research and theory. New York, NY: Academic Press; 1974. p. 47–89.
49. Baddeley A. Working memory. Science. 1992;255:556–9.
50. Baddeley A. The episodic buffer: a new component of working memory? Trends Cogn Sci. 2000;4:417–23.
51. Zatorre RJ, Meyer E, Gjedde A, Evans AC. PET studies of phonetic processing of speech: review, replication, and reanalysis. Cereb Cortex. 1996;6:21–30.
52. Paulesu E, Frith CD, Frackowiak RS. The neural correlates of the verbal component of working memory. Nature. 1993;362:342–5.
53. Duffau H, Capelle L, Denvil D, Gatignol P, Sichez N, Lopes M, et al. The role of dominant premotor cortex in language: a study using intraoperative functional mapping in awake patients. NeuroImage. 2003;20:1903–14.
54. Duffau H, Gatignol P, Denvil D, Lopes M, Capelle L. The articulatory loop: study of the subcortical connectivity by electrostimulation. Neuroreport. 2003;14:2005–8.
55. Dejerine J. Anatomie des Centres Nerveux. Tome 1. Paris: Rueff et Cie; 1895.
56. Dejerine J. Anatomie des Centres Nerveux. Tome 2. Paris: Rueff et Cie; 1901.
57. Ludwig E, Klingler J. Atlas cerebri humani. Basel: S. Karger A.G; 1956.
58. Catani M, Thiebaut de Schotten M. A diffusion tensor imaging tractography atlas for virtual in vivo dissections. Cortex. 2008;44:1105–32.
59. Mori S. MRI atlas of human white matter. 1st éd ed. Amsterdam/Boston, MA: Elsevier; 2005.
60. Fernandez-Miranda JC, Rhoton AL Jr, Alvarez-Linera J, Kakizawa Y, Choi C, de Oliveira EP. Three-dimensional microsurgical and tractographic anatomy of the white matter of the human brain. Neurosurgery. 2008;62:SHC989–1026. discussion SHC1026-8.
61. Catani M, Howard RJ, Pajevic S, Jones DK. Virtual in vivo interactive dissection of white matter fasciculi in the human brain. NeuroImage. 2002;17:77–94.
62. Zemmoura I, Herbet G, Moritz-Gasser S, Duffau H. New insights into the neural network mediating reading processes provided by cortico-subcortical electrical mapping: neural basis of reading. Hum Brain Mapp. 2015;36:2215–30.
63. Young JS, Lee AT, Chang EF. A review of cortical and subcortical stimulation mapping for language. Neurosurgery. 2021;89:331–42.
64. Maldonado IL, Destrieux C, Ribas EC, de Abreu Brito Guimarães BS, Cruz PP, Duffau H. Composition and organization of the sagittal stratum in the human brain: a fiber dissection study. J Neurosurg. 2021;135(4):1214–22. https://doi.org/10.3171/2020.7.JNS192846.
65. Sarubbo S, De Benedictis A, Maldonado IL, Basso G, Duffau H. Frontal terminations for the inferior fronto-occipital fascicle: anatomical dissection, DTI study and functional considerations on a multi-component bundle. Brain Struct Funct. 2013;218:21–37.
66. Mandonnet E, Nouet A, Gatignol P, Capelle L, Duffau H. Does the left inferior longitudinal fasciculus play a role in language? A brain stimulation study. Brain. 2007;130:623–9.
67. Thiebaut de Schotten M, Urbanski M, Duffau H, Volle E, Levy R, Dubois B, et al. Direct evidence for a parietal-frontal pathway subserving spatial awareness in humans. Science. 2005;309:2226–8.
68. Duffau H, Lopes M, Arthuis F, Bitar A, Sichez JP, Van Effenterre R, et al. Contribution of intraoperative electrical stimulations in surgery of low grade gliomas: a comparative study between two series without (1985-96) and with (1996-2003) functional mapping in the same institution. J Neurol Neurosurg Psychiatry. 2005;76:845–51.
69. Duffau H. Does post-lesional subcortical plasticity exist in the human brain? Neurosci Res. 2009;65:131–5.

White Fibers Anatomy for Approaches to Limbic Tumors

63

Pablo González-Lopéz, Artem Kuptsov, and Víctor Fernández-Cornejo

Learning Objectives

- Analyze the phylogenetic evolution of the limbic structures and the way in which they occupy a mediobasal position into the human brain.
- Understand the limbic system components and anatomical relationships among them, as well as the white matter connections.
- Identify the anatomical location of the limbic lesions as well as their invasion patterns based on the understanding of the white fibers anatomy.
- Choose the most appropriate approach based on clinical presentation, tumor location, and invasion patterns.

63.1 Introduction

The limbic system has been classically understood as the phylogenetically old limbic lobe, some other subcortical structures, and their connections. As a result of the phylogenetic evolution, the limbic system reached a medial and basal location in the hemisphere in between the diencephalon and the massively enlarged cerebral neocortical hemispheres [1]. The tremendous evolution of the telencephalon and central pivotal position of the diencephalon and ventricles around it are responsible for the special shape of the white matter fibers that form the limbic system. Thus, most of these bundles are arranged in a '*belt-shape*' wrapping around the midline structures as the corpus callosum, thalamus, and choroidal fissure, following a parallel path to the lateral ventricles [2].

This capricious architecture and shape in the mediobasal aspect of the hemispheres are responsible for its hidden location, covered by relevant cortical and subcortical structures that makes it difficult to approach it surgically.

P. González-Lopéz (✉) · A. Kuptsov · V. Fernández-Cornejo
Department of Neurosurgery, Hospital General Universitario de Alicante, Alicante, Spain
e-mail: gonzalez_pab@gva.es

The limbic system components can be summarized as follows:

1. Orbitofrontal gyri (also considered as paralimbic).
2. Septal area.
3. Limbic cortex.
   (a) Cingulate gyrus.
   (b) Isthmus cinguli.
   (c) Parahippocampal gyrus.
4. Hippocampal formation.
   (a) Dentate gyrus.
   (b) Hippocampus.
   (c) Subicular Complex.
5. Amygdala.

All these structures show relevant interconnections in between them as well as with the surrounding association cortices. Therefore, a deep understanding of all those cortical and subcortical structures remains of great importance when approaching lesions in these areas. The main goal of this chapter is to provide a general overview of the anatomy of the limbic region with a surgical perspective to summarize the different surgical routes to the limbic tumors.

63.2 History/Background

It was Thomas Willis who first described the anatomical features of the cortical region in the medial edge of the telencephalon, just in the border between the cerebral cortex and the thalamus (1664), and named this structure as the *limbus*, a word derived from the Latin language, meaning "the border" [3]. More than two centuries later, Paul Broca described that this circle on the medial edge of each hemisphere was made of the grey matter regions that form a transition between neocortex and the diencephalon and called it the

I. N. Kanaan, V. Beneš (eds.), *Neuroanatomy Guidance to Successful Neurosurgical Interventions*,
https://doi.org/10.1007/978-3-031-59838-8_63

Broca's Great Limbic Lobe as the curved rim of the cortex which includes the cingulate and the parahippocampal gyri [4]. As this great limbic lobe showed an anterior connection with the olfactory system, at the beginning, it was thought to be involved in the neural processes of the sense of olfaction. It was not till 1937 that James Papez, an American physician, introduced other kinds of sensory afferent fiber systems to reach the limbic cortex, and due to this fact, this system was probably also related with functions as emotion (Papez's Circuit) [5].

63.2.1 The Phylogenetic Evolution of the Limbic System

An important fact to understand the human brain development stages is to reach a comprehensive understanding of the central nervous system's phylogenetic development [1, 6]. The easiest way to get it is reviewing the most important stages of the brain evolution in different species, as it was shown in previous chapters [1].

- The very first brains of *fish* developed with purely olfactory neurons (rhinencephalon) 500 million years ago. This rudimentary brain was named paleopallium (olfactory brain).
- The first *amphibia* developed 100 years later the archipallium and started leaving space between paleo and archipallial structures, for the future appearance of basal ganglia.
- *Reptiles* were the next animals showing some evolutive changes in their brains (350 million years ago). The subpallium appeared between paleo and archipallium. This reptilian's brain was mainly composed of brainstem and these newly appeared striatal nuclei (subpallium).
- 250 million years ago, the *paleomammals* started developing the neopallium as an expansion from the rudimentary cerebral cortex of reptiles, differentiating into the limbic lobe.
- *Neomammals* continued the development of neopallium (65–180 million years ago), and the mass of cerebral cortex increased progressively to become the neocortex.

As just mentioned, the result of this great telencephalic expansion mainly based on neopallial structures, the older limbic system took a position on the medial and basal surface of the hemisphere. As the most complex central nervous system, the human brain, somehow 'summarizes' the phylogenetic evolution during the embryological stages. Therefore, the paleopallium (root of the future hippocampal formation) is one of the first structures developing to a highly organized corticosubcortical system.

63.2.2 An Overview of the Limbic System Embryology

The first appearance of the nervous system during the fetal development takes place approximately during the day 18, with the beginning of neurogenesis. The *neural plate* invaginates to form the neural groove. An important fact in this stage is the neural induction produced by different substances, which in this first period induce the ectoderm to form that initial neural structure (neuroblasts, neuroglial, and neural crest-derived cells), rather than epidermis. During the third week, development of the major divisions in the brain takes place: prosencephalon, mesencephalon, and rhombencephalon.

Till the sixth week, the rhinencephalon forms the main structure derived from the prosencephalon, but in human primates, and during the 6–7th week (42–45 days), the cerebral lobes begin their shape development from the telencephalon (Frontal, Parietal, Occipital, and Temporal). These first cerebral lobes start growing then, displacing the latest structures as the rhinencephalon to a hidden position in rostral and ventral locations. The rhinencephalon will follow a parallel development leading to some limbic structures.

The *cortical plate* appears on the day 52 and is the landmark for the cortical layers' development. This initial cerebral cortex is derived from the pallium, which has three differentiated regions: the *paleopallium*, *archipallium* (these latter are sometimes considered as the same), and the *neopallium*, which is initially located between the archipallium (hippocampal formation) and the paleopallium (cingulate). The paleo and archipallium will form together the limbic lobe, which is mainly derived from the old rhinencephalon, having a different cortical pattern (allocortical and mesocortical cerebrum). The hippocampal formation (dentate gyrus and cornu ammonis) is the first cortical area in acquiring a pattern of higher differentiation [7]. Simultaneously, waves of neuroblasts migrate to a subpial position, differentiating into fully mature neurons in the telencephalic structures (main cerebral lobes). The frontal, central, parietal, occipital, part of temporal and insula, and claustrum belong to this neocortical cerebrum [8].

63.2.3 An Overview of the Limbic System Functions

A large amount of experimental and nonexperimental studies have been performed to elucidate the limbic system functions. In 1954, Pool published some hypothetical functions of different parts of the limbic lobe, based on the observation and experimentation over some of his patients [9]. Eight patients diagnosed with brain tumors in some cortical representations of the limbic lobe were included in his study, in which he analyzed the different symptomatology presented at admission and recorded the effects caused by electrical stimulation in different limbic regions. He concluded that specific visceral phenomena are processed and controlled by the limbic system or *visceral brain*. Under the term visceral brain, Pool included the anterior cingulate gyrus, anterior temporal and posterior orbital cortex, as well as the insula, uncus, amygdaloid nuclei, and parahippocampal gyrus. Some months later, Scoville reviewed different works involving the limbic system, especially the studies developed by MacLean and Pool [10]. Scoville asserted that all the major structures belonging to the limbic lobe (orbitofrontal, cingulum, hippocampus, and amygdala) were related to the emotional and motivational functions. More recently, Reep et al. [11] reviewed the mammalian brain evolution and the changes that have occurred in their limbic structures. They agreed on a reduction of the olfactory structures and neighbor areas in the limbic system (entorhinal cortex, amygdala, hippocampus, and septal regions), while the neocortical areas increased in size in primates. The authors hypothesized that these changes could be due to adaptation to diurnality, demanding a higher visual development, instead of olfaction. Some modern and old studies have deeply dealt with the interhemispheric connections between different limbic areas, concluding the importance of the anterior commissure as a limbic commissural system showing connections among bilateral septal areas, amygdala, as well as temporooccipital neocortices [12–16].

The most generally accepted and classically studied functions of the limbic system have been the control of emotional behavior and visceral responses [9, 17, 18]. The emotion perception could be a highly processed role, in which the amygdala receives different inputs, mainly from hypothalamus and cerebral cortex (cingulum gyrus and prefrontal cortex), generating the conscious feeling of them [19]. The aggressive reactions in which the limbic lobe also plays an important role are probably remnants of the survival instinct and the reaction to fear [20]. Regarding the capacity of learning and memory, it is fully accepted that most of these circuits are responsible for mnemonic paradigms [17, 21]. In this sense, the mesial region of the temporal lobe highlights as a capital center for memory and learning processing [4].

63.3 Anatomical Highlights

63.3.1 Orbitofrontal Gyri and Olfactory Cortex

The orbitofrontal cortex is located ventrally into the frontal lobe, representing one of its three surfaces. It is sitting above the anterior cranial fossa in close relationship with the orbital roof. It is divided into four segments: anterior, posterior, lateral, and medial through the H-shaped orbitofrontal sulcus. The medial orbitofrontal gyrus and the gyrus rectus are separated by the olfactory sulcus, on which the olfactory bulb and tract run posteriorly to reach the olfactory cortex where it is divided into the medial and lateral olfactory striae that delineate the so-called olfactory trigone, located just anterior and medial to the anterior perforated substance.

This area of the brain is thought to play a role in motivation, emotions, and social behavior modulation. It processes rewards and punishments, acting as an integration cortical region through its connections to the temporomesial regions, amygdala, entorhinal and cingulate cortices, hippocampal formation, basal ganglia, and hypothalamus.

The main white matter fiber tracts involved on its connections are the uncinate fascicle, the inferior frontooccipital fascicle, and anterior limb of the internal capsule via the thalamocortical and corticostriatal fibers [22] (Fig. 63.1).

63.3.2 Septal Area

It is considered an anterior part of the limbic system. The septal area is in very close relationship with the posterior part of the gyrus rectus, medial olfactory stria, anterior perforated substance, rostrum of the corpus callosum, and anterior cingulum [23]. It is a small area formed by the paraterminal gyrus and subcallosal area, separated by two sulci, which in most of the cases are oriented vertically: anterior and posterior paraolfactory. The posterior paraolfactory sulcus separates the paraterminal gyrus from the lamina terminalis and the anterior aspect of the third ventricle [24].

The septal region is a central station with direct connections to the extraventricular inner and outer limbic belts. Therefore, the paraterminal gyrus is connected to the so-called prehippocampal rudiment or indusium griseum, which is an extension of the fasciola cinerea coming posterosuperior from the dentate gyrus. The paraterminal gyrus projects laterally and basally through the diagonal band of Broca [25]. These structures form the inner extraventricular limbic belt. On the other hand, the subcallosal area is connected to the anterior segment of the cingulate gyrus, which wraps around the corpus callosum. The subcallosal area projects to the gyrus rectus and orbitofrontal cortex in close relationship with the medial olfactory stria (Fig. 63.1).

63.3.3 Cingulate Gyrus

The cingulate gyrus is connected to the subcallosal region adjacent to the rostrum of the corpus callosum. It then wraps

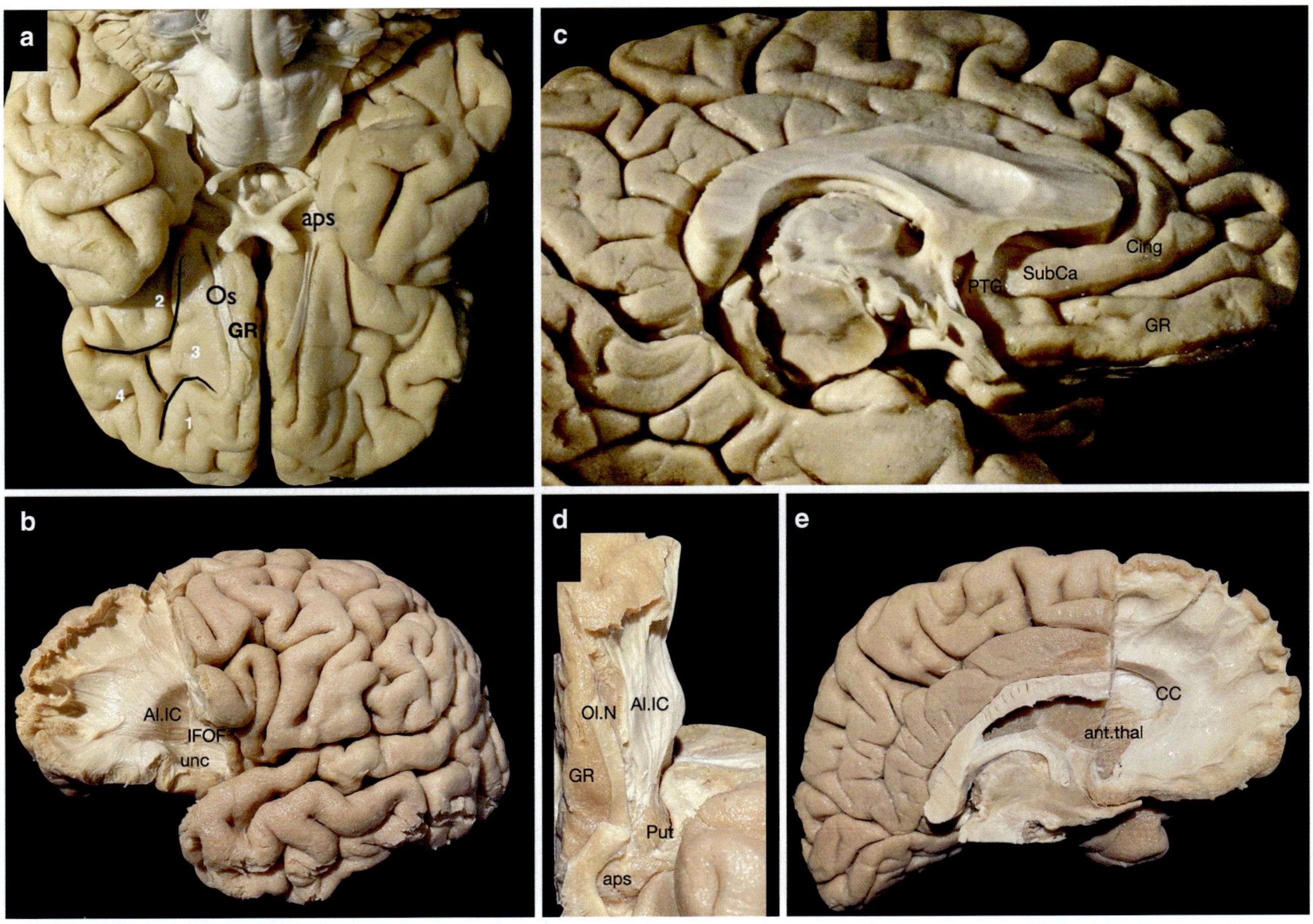

Fig. 63.1 (**a**) Basal view of a brain specimen. The 'H-shaped' orbitofrontal sulcus is shown in between the anterior (1), posterior (2), medial (3), and lateral (4) orbitofrontal gyri. The olfactory tract is covered by the olfactory sulcus (Os). The gyrus rectus (GR) lies medial to the olfactory tract. The anterior perforated substance (aps) is situated between the olfactory striae. (**b**) Lateral view of a left-brain specimen. The frontal lobe has been dissected partially showing the frontobasal connections through the anterior limb of the internal capsule (Al.IC), inferior frontooccipital (IFOF), and the uncinate (unc) fascicles. (**c**) Mediobasal view of a left frontal lobe. The close relationship and anatomical connections among the cingulate gyrus (Cing), gyrus rectus (GR), subcallosal area (SubCa), and paraterminal gyrus (PTG) are demonstrated. (**d**) Basal view of a left-brain specimen on its frontal aspect. The lateral part has been dissected by Klingler's method showing the close relationship of the orbitofrontal region with the ventral putamen (Put), anterior perforated substance (aps), anterior limb of the internal capsule (Al.IC), olfactory nerve (Ol.N), and the gyrus rectus (GR). (**e**) Medial aspect of a left brain hemisphere showing the medial aspect of the internal capsule—anterior thalamic radiations (ant.thal) and corpus callosum (CC)

around the genu and continues posteriorly above the body of the corpus callosum. Once it reaches the level of the splenium, it turns inferiorly and connects with the parahippocampal gyrus through the isthmus cinguli, which is the region where the direction of the limbic cortex (cingulum runs posteriorly and at the level of the retrosplenial area-isthmus-becomes the parahippocampal gyrus, which moves laterally 'avoiding' the diencephalon. The Cingulate gyrus is a single structure with no evident subgyri on it and is limited by the supracallosal and cingulate sulci, which separate the cingulum from the corpus callosum below and the neocortex of different association areas above (gyrus rectus, prefrontal cortex, medial frontal gyrus, paracentral lobe, and precuneus). The cingulate sulcus shows some constant sulcal branches that can help us to identify its different divisions. Therefore, two common sulcal branches are the paracentral and marginal sulci, which ascend to the edge of the interhemispheric fissure joining the lateral convexity of the neocortex. These two sulci define the paracentral lobe, which represents the midline projection of the primary motor and somatosensory areas. The marginal or ascending ramus separates the paracentral lobe from the cingulo-parietal connection that takes place posteriorly at the level of the precuneus. Here, a variable shape subparietal sulcus separates the posterior cingulate from the precuneus.

Despite its length and shape, it is difficult to find anatomical divisions of the cingulate gyrus. However, there are many divisions based on the cytoarchitecture of the cingulum and functions. The most accepted was proposed by Vogt et al. in 2004 [26], which divides the cingulate gyrus in four parts:

- Anterior: found connected to the subcallosal area (subgenual) and turning around the genu of the corpus callosum (perigenual). It shows strong connections to the limbic and paralimbic structures through the gyrus rectus and septal region.
- Middle: the so-called midcingulate cortex approximately starts when the cingulate cortex becomes straight after turning around the genu of the callosum and reaches the area of the marginal sulcus.
- Posterior: starting below the marginal sulcus ascending branching and reaching the isthmus. Limited by the cingulate, subparietal, and parietooccipital sulci.
- Retrosplenial: cytoarchitectonic different structures within the ventral part of the posterior cingulum just adjacent to the splenium.

This regional classification is based on the different neuronal layers' organization and studied functions. However, we need a more anatomical and radiological classification that helps us to stratify differently located tumors inside the cingulum. Thus, we use the classification proposed by Gong, which considers the AC-PC line as a reference. A couple of vertical perpendicular lines are traced superiorly from the AC and PC. These two lines divide the cingulum in anterior, middle, and posterior cingulum and can be easily traced in a midsagittal plane MRI.

The cingulum bundle that is underlying the cortex follows a direction parallel to the gyrus shape and is shown to connect the orbitofrontal cortex to the temporomesial region. Therefore, it is considered a long association tract with a very characteristic belt shape. On its way, it projects a few connections to the adjacent neocortical areas, as it happens to the prefrontal cortex, and mainly to the precuneal region [27]. The cingulum connections are also part of the output of the so-called Papez's circuit, receiving fibers from the anterior thalamic nuclei which project to the anterior cingulate gyrus and finally to the prefrontal cortices [28] (Fig. 63.2).

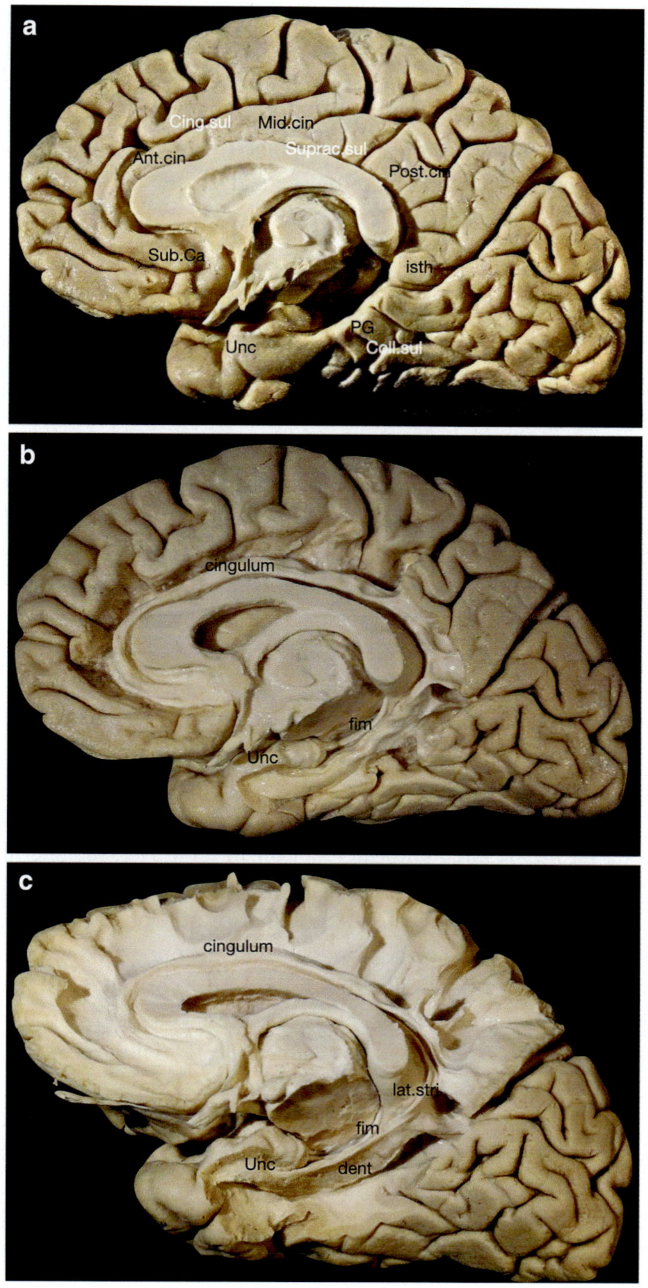

Fig. 63.2 (**a**) Medial view of the right hemisphere. The main limbic cortical areas are highlighted: subcallosal area (Sub.Ca), anterior cingulum (Ant.cin), cingulate sulcus (Cing.sul), middle cingulum (Mid.cin), supracallosal sulcus (Suprac.sul), posterior cingulum (Post.cin), isthmus cinguli (isth), parahippocampal gyrus (PG), collateral sulcus (Coll.sul), and uncus (Unc). (**b**) The cortex covering the limbic areas has been carefully removed showing the underlying white matter fibers and connections; mainly the belt-shaped cingulum bundle wrapping around the corpus callosum and running to the uncus (Unc). The intraventricular fimbria (fim) can be seen. (**c**) The parahippocampal component of the cingulum has been scrapped away showing the basal aspect of the dentate gyrus (dent). Intraventricularly, the fimbria (fim) continues towards the crus fornicis, and extraventricularly, the dentate continues with the fasciola cinerea to the lateral stria (lat.stri)

63.3.4 Isthmus Cinguli

There are a few and quite superficial anatomical descriptions of the isthmus of the cingulate gyrus. Generally, it is described as the most posterior part of the cingulate gyrus. From an anatomical and surgical perspective, the isthmus is a relevant structure located in the midline inferior and posterior to the splenium of the corpus callosum, just above the terminal branch of the parietooccipital sulcus. This is the point where the limbic cortex of the cingulate continues into the parahippocampal gyrus that moves forward following an anterolateral direction. This is the pivotal point where the limbic system moves from medial to basal. The isthmus is in close relationship with the parasplenial, ambiens, and quadrigeminal cisterns, but also limits medially the atrium of the lateral ventricle (Fig. 63.2).

63.3.5 Parahippocampal Gyrus

The temporal is the only lobe of the brain that owns four surfaces: the lateral, superior, basal, and mesial. The parahippocampal gyrus forms the limit in between the basal and the mesial aspects of the temporal lobe. Considering the temporal uncus as an anteromedial extension of the parahippocampal gyrus, the latter shows a hockey-stick shape. Running posteriorly, the parahippocampal gyrus is somehow limited by the junction of the parietooccipital sulcus and the anterior extension of the calcarine fissure. This is considered as the anatomical limit in between the parahippocampal gyrus and isthmus cinguli. At this point, the parahippocampal gyrus is directly connected cortically and subcortically with the lingual gyrus.

The parahippocampal gyrus is a long curved limbic structure which embraces the midbrain following the tentorial edge from the falcotentorial junction all the way to the lateral wall of the cavernous sinus, passing through the tentorial incisura, where its most anteromedial projection, the uncus, is found bulging. The anatomical limits of this gyrus are the anterior limb of the calcarine fissure posterosuperiorly, the collateral sulcus posterolaterally, the rhinal sulcus anterolaterally, the hippocampal sulcus superiorly, and the free tentorial edge and contents of the ambient cistern medially.

The collateral sulcus is an important anatomical landmark as it separates the parahippocampal from the fusiform gyrus and marks the collateral eminence at the temporal horn floor. The collateral sulcus is commonly interrupted anteriorly and followed by the rhinal sulcus, which represents the lateral limit of the so-called entorhinal area. This rhinal sulcus, indeed, represents the anterolateral limit of the uncus. More medially, the parahippocampal gyrus and uncus can be clearly differentiated as the uncal notch appears as a structure on the cortical surface creating this boundary. The hippocampal sulcus represents the anatomical limit in between the parahippocampal and the dentate gyri [29] (Fig. 63.2).

The cingulum itself and the inferior longitudinal fasciculus have been proved to be the most relevant connecting fibers related to the parahippocampal gyrus.

The cingulum belt connects to the isthmus cinguli and anterior cingulate cortex. It spreads from the anteromedial aspect of the parahippocampal cortex and runs posteromedially to the retro-subsplenial area where it branches in a Y-shape to enter the lingual gyrus longitudinally and the posterior cingulum through the isthmus in a curved way around the splenium of the corpus callosum. These fibers run forward to terminate into the subcallosal and septal areas.

The inferior longitudinal fasciculus runs a bit more lateral and shows a more longitudinally oriented direction, terminating into the pericalcarine regions of the cuneus and lingual gyrus. This fascicle also shows associative connections to the hippocampus, fusiform, inferior, and superior gyri of the temporal lobe on some DTI-tractography studies (Fig. 63.2).

63.3.6 Hippocampal Formation

The hippocampus is hidden inside the temporal horn of the lateral ventricle and anatomically related to the dentate and parahippocampal gyri, as its shape and location have been also influenced by the telencephalic flexure. It is part of an intraventricular limbic belt. For surgical and anatomical purposes, it can be divided into head, body, and tail, the head being the largest and more anterior part of it. The head is an intraventricular structure, facing anteriorly and superiorly to the extraventricular amygdala, which are separated by the uncal recess in the anterior tip of the temporal horn and frequently fuse together in the medial part (tip of the uncal pyramid). The posterior limit of the head is the point where the anterior choroidal artery enters the ventricle through the inferior choroidal point, and where the choroid plexus and fimbria start running posteriorly. This is the beginning of the body of the hippocampus, which occupies the medial aspect of the floor of the temporal horn. The medial aspect of the hippocampus is limited by the choroidal fissure, which gives access to the ambient cistern. Once the body reaches the entry to the atrium of the lateral ventricle, the hippocampus wraps around the pulvinar of the thalamus and runs superomedially as the tail of the hippocampus, which is thinner and fuses the calcar avis into the medial wall of the atrium, and the subsplenial region.

The main hippocampal white matter connections are covering the hippocampus itself. Therefore, the alveus is a white matter structure that covers the body and forms the fimbria. This fimbria is separated from the dentate gyrus inferiorly by the fimbrodentate sulcus. Posteriorly, the fimbria gets a more defined shape and wraps around the thalamus as the crus fornicis which joins the contralateral one as the hippocampal commissure below the anterior splenial region, finally giving rise to the body of the fornix that travels forward into the floor of the lateral ventricle. When the body reaches the foramen of Monro, it splits again into two columns, forming the superior and anterior limits of each foramen, and runs down-

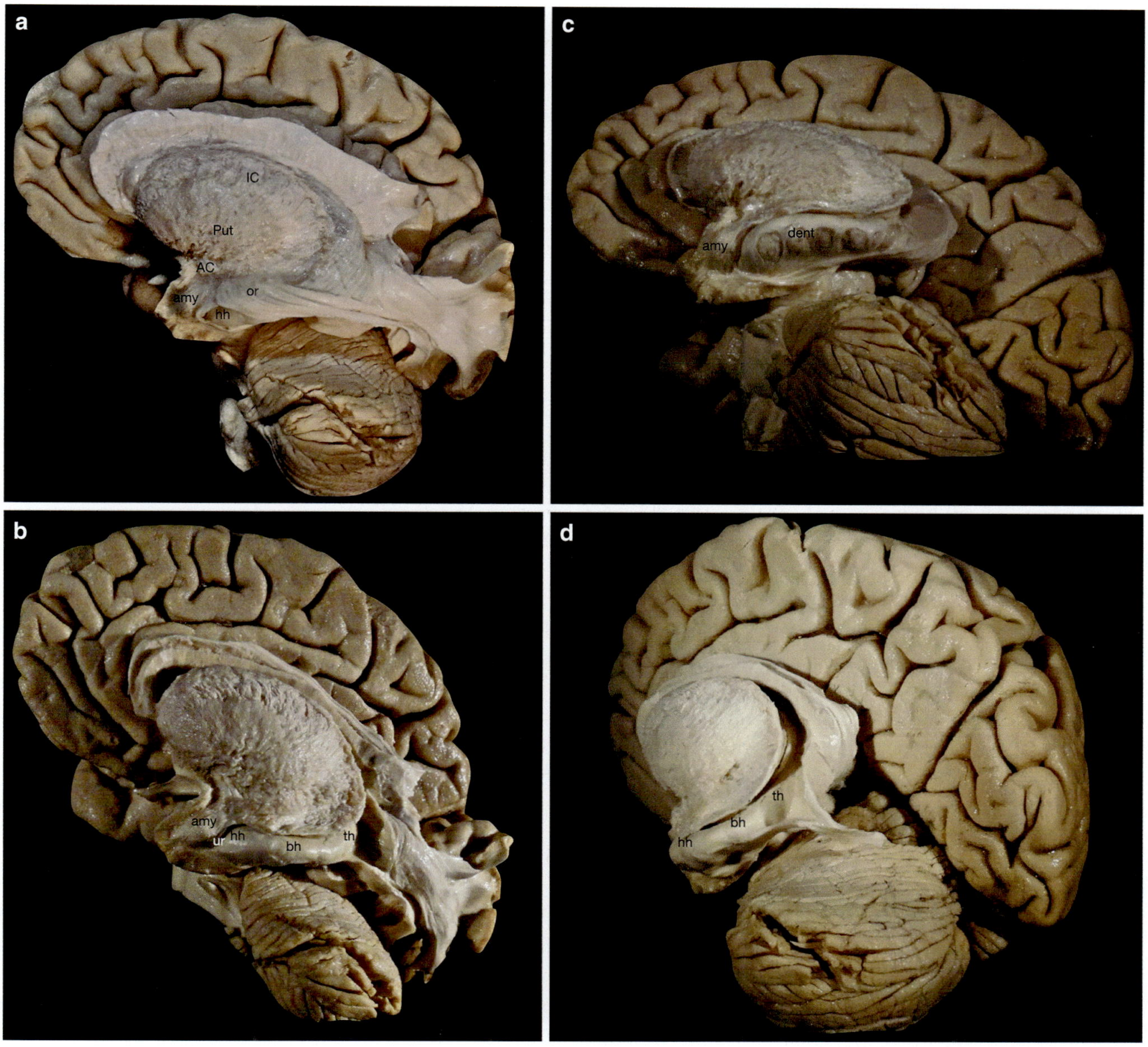

Fig. 63.3 (**a**) Lateral view of a human brain, in which most of the lateral aspect tracts have been removed. The mediobasal temporal structures are shown in close relationship with the ventral putamen (Put), anterior commissure (AC), and different internal capsule (IC) components. Therefore, the amygdala (amy) is located caudal and dorsal to the limen insulae on an extraventricular position. The head of the hippocampus (hh) is in an intraventricular position on the most rostral part of the temporal horn covered by the Meyer's loop part of optic radiations (or). (**b**) The optic radiations have been peeled away showing the relationship of the amygdala (amy) and head of hippocampus (hh). The uncal recess is shown in between these two structures. The body (bh) and tail (th) of the hippocampus are shown inside the temporal horn. (**c**) Basal view of the structures showing the dentate gyrus (dent). (**d**) The head (hh), body (bh), and tail (th) of the hippocampus are shown in this posterolateral view in close relationship with the temporal horn and atrium

wards into the hypothalamus at the lateral walls of the third ventricle to reach the mammillary bodies. This is an important part of the so-called Papez's circuit, which continues through the mammillothalamic tract, anterior thalamic nuclei, and anterior thalamic radiations to the anterior cingulate and prefrontal cortices (Fig. 63.3).

The hidden location of the hippocampal structures inside the temporal horn and atrium is covered by the wall and roof of this part of the lateral ventricle. The neocortex, as well as the relevant association, commissural and projection fibers cover in different ways and segments this part of the limbic system. The superior and inferior longitudinal, inferior frontooccipital and arcuate fascicles, as well as the optic radiations, retro- and sublenticular projections of the internal capsule, and the anterior commissure and tapetum cover these deep structures. Therefore, approaches to the hippocampal formation are challenging and require an understanding of the 3D architecture of all these fibers (Fig. 63.3).

63.3.7 Amygdala and Entorhinal Cortex

The amygdala is an area located in a pivotal and central part of the limbic system in close relationship with the uncus and head of the hippocampus. This junction of grey matter nuclei is located into the depth of the anterior part of the uncus.

Superiorly, it is related to the ventral aspect of the basal ganglia (most likely the globus pallidus), and this is probably the less defined boundary and subsequently the most difficult to be identified during surgery. Inferiorly, the amygdala protrudes into the roof of the anterior temporal horn, in close relationship with the location of the hippocampal head. Medially, it contacts the crural cistern through the anteromedial uncus, and anteriorly and laterally, it continues with the entorhinal region.

The amygdala receives different connections, among which the anterior commissure, stria terminalis, caudate nucleus, and amygdaloseptal and amygdolahypothalamic fibers are the most relevant ones.

This deep location makes accessing the amygdala a challenging task, as it is also covered by cerebral cortex and relevant subcortical white matter fascicles, as the uncinate and inferior frontooccipital (Fig. 63.3).

The anatomy of the limbic system can be challenging to summarize in a general way. Therefore, we propose to go through each of its different components separately, but with a special emphasis on the morphologic cortical relationship and the subcortical connections between them. The specific anatomy of these mediobasal structures shows specific surgical approaches that are summarized in the discussion.

63.4 Surgical Pearls

- Study the axial, sagittal, and coronal cuts on MRI looking for landmarks that might give you an understanding of the anatomical location of the tumor.
- Analyze the invasion pattern of the tumor and try to correlate its potential growing directions with the white matter limbic belts.
- Choose an approach that you feel comfortable with.
- Respect the surrounding tissue anatomy and eloquence by mapping techniques or choosing an anatomical route (midline and/or basal).
- Use the gravity as an adjunct to your resection.
  - Supine position with head lateral rotation for midline tumors (sick hemisphere falling in ipsilateral, and healthy hemisphere falling in contralateral approaches).
  - Lateral position with the head extended to the contralateral side in subtemporal approaches will help to split the subtemporal space and keep the temporal lobe less retracted.
  - Sitting position for basal supracerebellar transtentorial approaches will help to release pressure on the cerebellum and will aim the tumor to come to surgical view after opening the tentorium and starting the corticotomy.
- Always look for cisterns to release CSF and keep good access during the whole procedure in case you need extra relaxation.

63.5 Discussion

The mediobasal location of the limbic lobe, as well as the special belt-shape white matter structures which connect its containing structures themselves, is responsible for the specific recommended surgical approaches to reach these regions (orbitofrontal, septal, cingulum, isthmus cinguli, parahippocampus, hippocampal formation, amygdala—temporomesial pole).

63.5.1 Orbitofrontal and Septal

Orbitofrontal and septal areas are anatomically connected through the posterior aspect of the gyrus rectus. These lesions can be approached subfrontally in case of a more orbitofrontal and basal extension, or anterior interhemispheric in case of a more septal invasion pattern (Figs. 63.4 and 63.5).

63.5.2 Cingulate Tumors

These can be divided into anterior, central, and posterior, rest in the cerebral hemispheres' midline, and covered by relevant cortical and subcortical structures. Awake surgery and intraoperative mapping techniques have allowed surgeons to approach these lesions through 'non-eloquent' cortical and subcortical regions. Despite this new trend, we prefer to approach these tumors following classic anatomical pathways as the ones described decades ago by Professor Yasargil. Therefore, the interhemispheric ipsi- or even contralateral approach is usually our preferred route as they allow us to follow anatomical cisternal corridors to enter the sick functional unit directly without disrupting the healthy tissue [1] (Figs. 63.4 and 63.5).

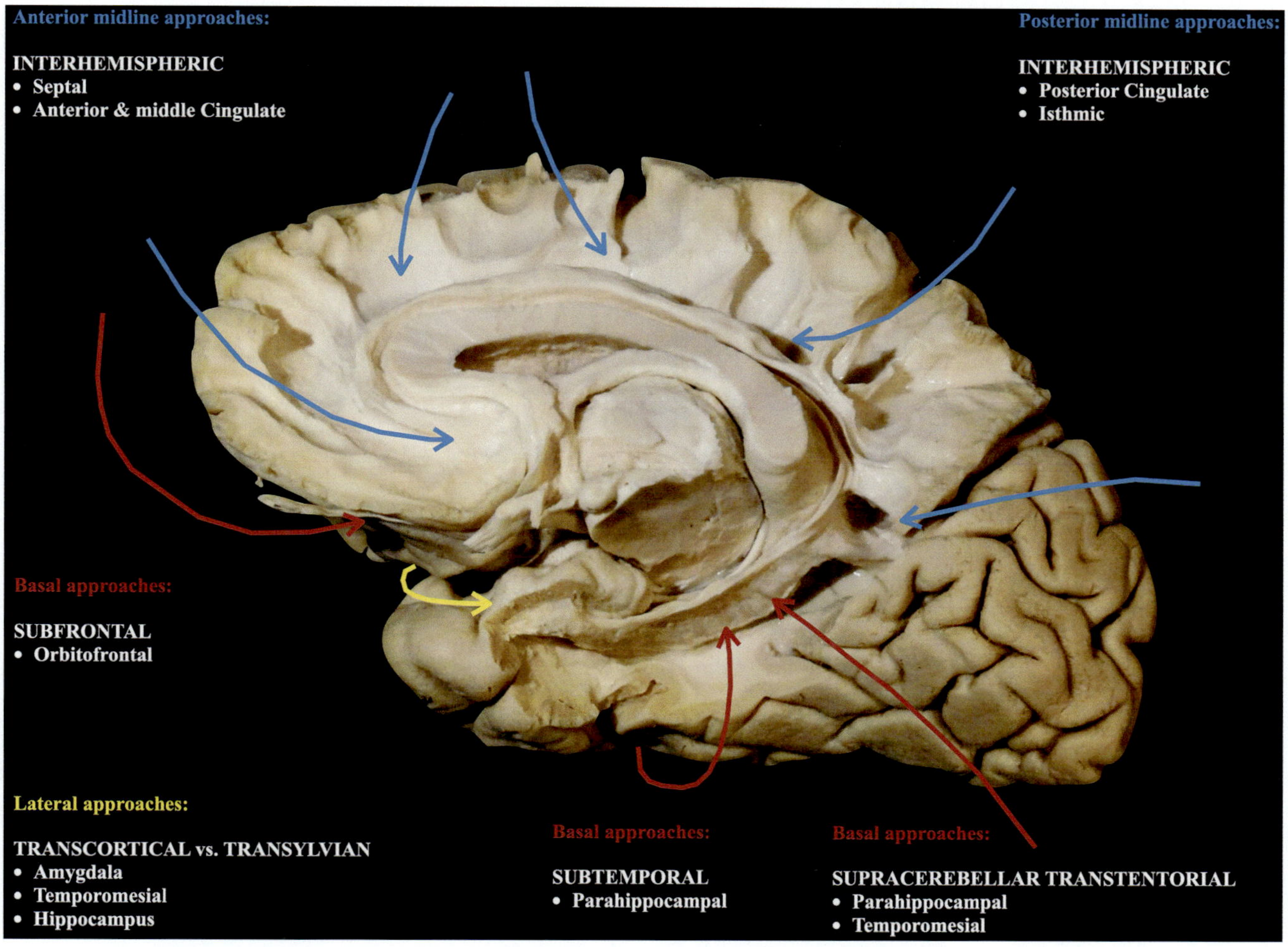

Fig. 63.4 Review of the most common limbic tumor approaches summarized in midline, basal, and lateral

63.5.3 Isthmic Tumors

These are rarely isolated tumors, as they usually appear invading the cingulum and/or parahippocampal gyri as well as other adjacent areas such as the lingual gyrus and the precuneus/cuneus. Depending on the anatomical invasion pattern, individual anatomic features, and clinical presentation, surgical approaches to the isthmus vary from posterior interhemispheric (transisthmic) to transcortical through the superior parietal lobe and/or precuneus. Our preference is more to the anatomic interhemispheric routes. These routes can be ipsi- or contralateral depending on the invasion pattern. Isthmic tumors invading the posterior cingulum can be approached through the ipsi- or contralateral side of the interhemispheric fissure. However, the height of the falcotentorial junction and straight sinus related to the isthmus must be carefully studied in those cases where the parahippocampal gyrus is mostly invaded (Figs. 63.4 and 63.5).

63.5.4 Parahippocampal Tumors

The parahippocampal gyrus is covered by relevant neocortex of the temporal lobe in the lateral aspect, the fusiform gyrus basally, and highly eloquent longitudinal white fibers shelling its deep location. We are prone to anatomical resections respecting the eloquence of the surrounding tissue. Therefore, in our practice, the transcortical approach through the temporal pole is an option for anteriorly located lesions into the parahippocampal gyrus invading the uncus. Theoretically, the subtemporal route respects the anatomy; however, it requires

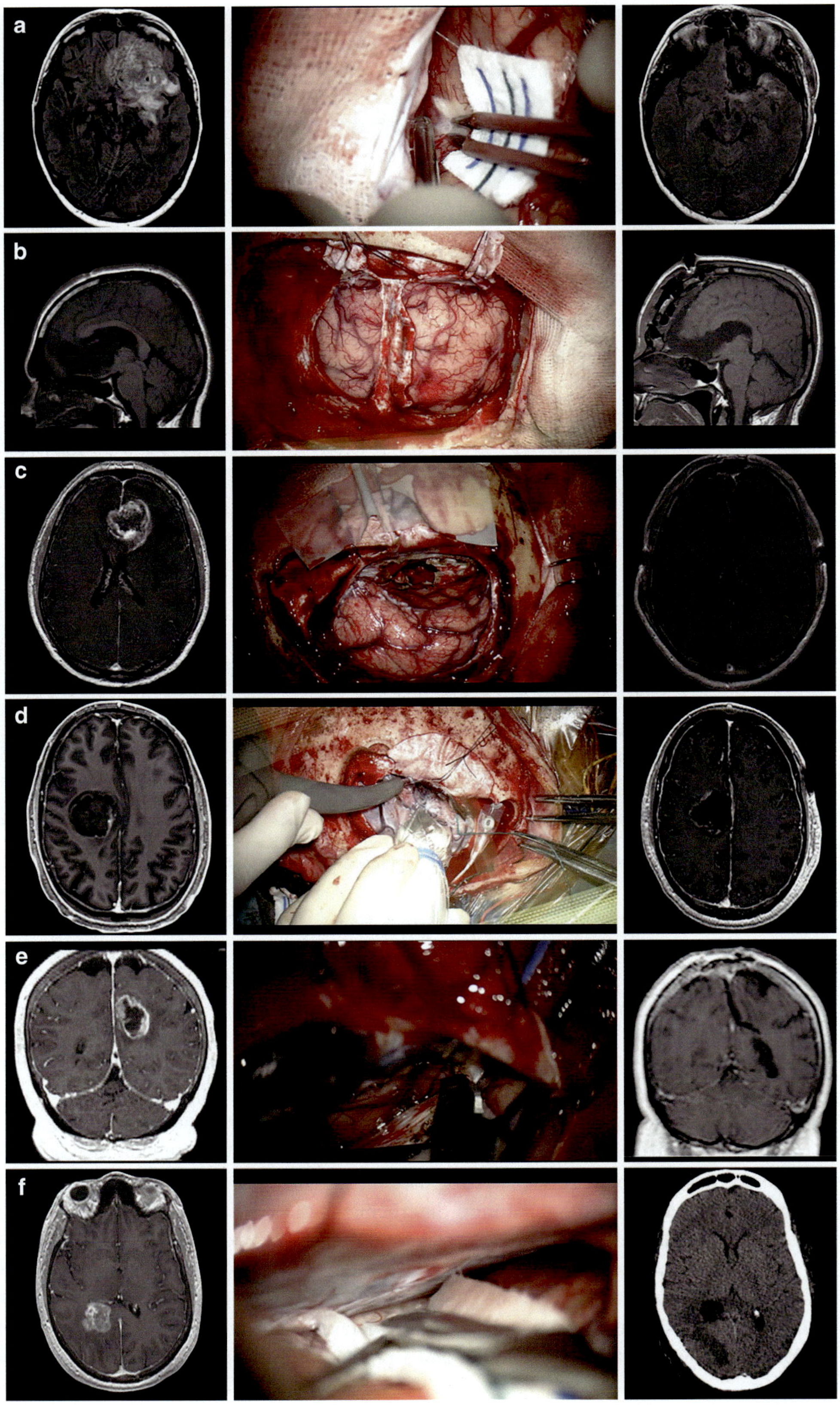

Fig. 63.5 Different midline limbic tumors approached through basal and midline approaches. (**a**) Left orbitofrontal GBM removed through a basal subfrontal approach. Preoperative axial FLAIR MRI sequence, intraoperative view of the left subfrontal approach, and postoperative axial FLAIR MRI sequence. (**b**) Right septal grade II diffuse astrocytoma operated through a midline anterior interhemispheric approach with head rotation to the right side. Preoperative sagittal T1-weighted MRI sequence, intraoperative view of the right anterior interhemispheric approach, and postoperative sagittal T1-weighted MRI sequence. (**c**) Left anterior cingulum GBM operated through a right-side midline contralateral anterior interhemispheric transfalcine approach with head rotation to the right side. Preoperative axial T1-weighted GD MRI sequence, intraoperative view of the approach, and postoperative axial T1-weighted GD MRI sequence. (**d**) Right middle cingulum GBM operated through a left side midline contralateral anterior interhemispheric transfalcine approach with head rotation to the left side. Preoperative axial T1-weighted GD MRI sequence, intraoperative view of the approach, and postoperative axial T1-weighted GD MRI sequence. (**e**) Left posterior cingulum GBM operated through a right-side midline contralateral posterior interhemispheric transfalcine approach with head rotation to the right side. Preoperative coronal T1-weighted GD MRI sequence, intraoperative view of the approach, and postoperative coronal T1-weighted GD MRI sequence. (**f**) Right isthmus GBM operated through a right-side midline posterior interhemispheric transfalcine approach with head rotation to the right side. Preoperative axial T1-weighted GD MRI sequence, intraoperative view of the approach, and postoperative axial T1-weighted GD MRI sequence

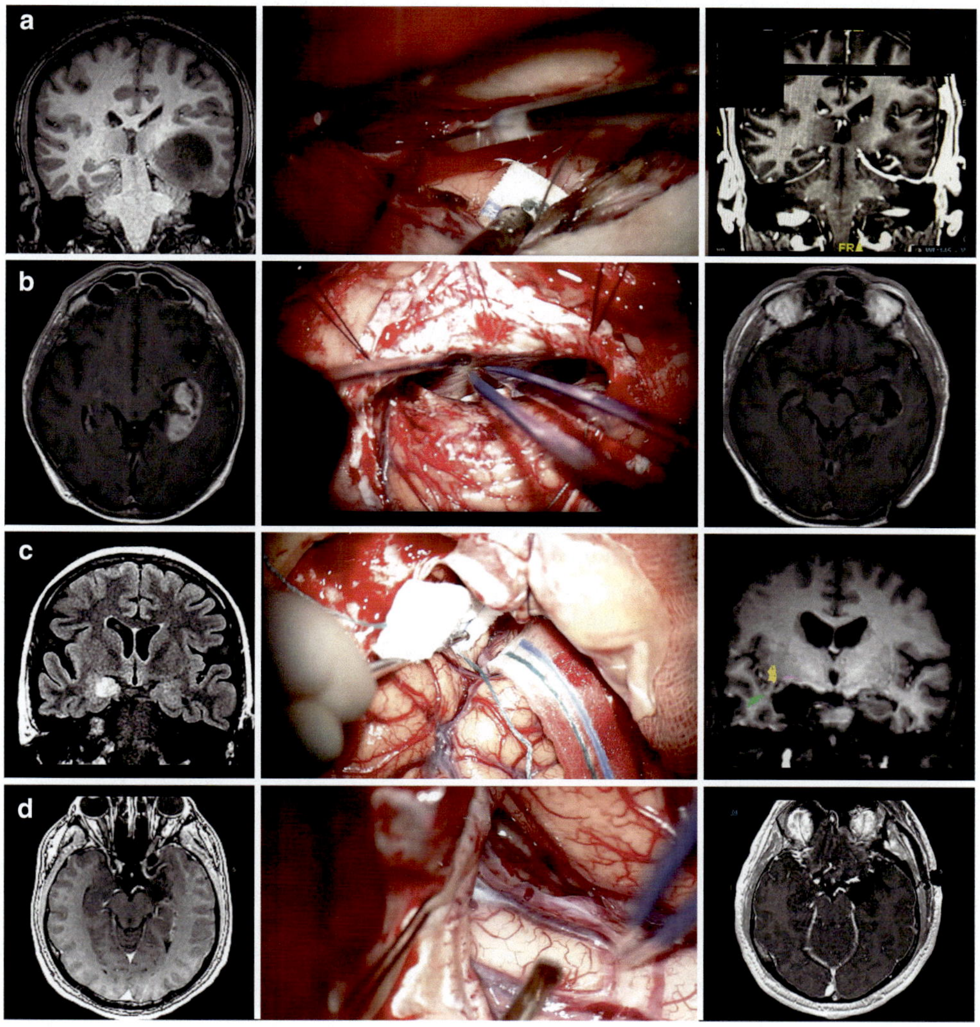

Fig. 63.6 Different mediobasal temporal limbic tumors approached through basal and lateral approaches. (**a**) Left parahippocampal Grade II diffuse astrocytoma operated through a basal subtemporal approach in lateral position. Preoperative coronal T1-weighted MRI sequence, intraoperative view of the approach, and postoperative coronal T1-weighted GD MRI sequence. (**b**) Left hippocampal GBM operated through a basal supracerebellar transtentorial approach in a semi-sitting position. Preoperative axial T1-weighted GD MRI sequence, intraoperative view of the approach, and postoperative axial T1-weighted GD MRI sequence. (**c**) Right amygdalar DNET operated through a lateral transylvian approach. Preoperative coronal FLAIR MRI sequence, intraoperative view of the approach, and postoperative coronal T1-weighted MRI sequence with implementation of tractography: inferior frontooccipital (yellow), uncinate (pink), and optic radiations (green). (**d**) Left amygdalar DNET operated through a lateral transcortical transplanum polare approach. Preoperative axial T1-weighted GD MRI sequence, intraoperative view of the approach, and postoperative coronal T1-weighted GD MRI sequence

different degrees of temporal lobe retraction, the reason why we generally try to avoid this maneuver. In cases with a posterior parahippocampal extension, we recommend the supracerebellar transtentorial approach as it provides enough working basal field after opening the tent [30] (Figs. 63.4 and 63.6).

63.5.5 Hippocampal and Amygdala Tumors

Purely hippocampal tumors are rare as they tend to include other adjacent regions such as the amygdala and/or the dentate and parahippocampal gyri. Depending on the invasion pattern and clinical presentation, we aim for transylvian, transplanum polare, or supracerebellar transtentorial approaches. The transylvian and transplanum polare are our preference in those cases where tumors invade the temporomesial pole, amygdala, and anterior hippocampal structures. When the tumors are located more posterior reaching the tail, the supracerebellar transtentorial is a good option. Both anterior and posterior routes respect the temporal stem, which is our goal when approaching these areas (Figs. 63.4 and 63.6).

63.5.6 Diffuse Limbic Tumors

Different combinations of approaches in two or even three stages are good options when tumors invade different limbic structures.

63.6 Conclusion

The capricious shape of the different limbic belts is inducing the way in which we approach tumors in these areas. A deep study of the phylogenetic evolution of the mediobasal limbic areas will lead us to understand the different eloquent cortical and subcortical structures covering the limbic system when approached from the lateral and superior convexity. This is the reason why we aim for anatomical approaches trying to enter more cortical tumor extensions. Midline and basal approaches are key to accomplish this goal. The tumor invasion pattern is also important to analyze as it can lead us to choose anterior/posterior midline vs. anterior/posterior basal approaches.

References

1. Yaşargil MG, von Ammon K, Cavazos E, Doczi T, Reeves JD, Roth P. Tumours of the limbic and paralimbic systems. Acta Neurochir. 1992;118(1–2):40–52.
2. González-López P, Cossu G, Thomas CM, Marston JS, Gómez C, Pralong E, et al. Cadaveric white matter dissection study of the telencephalic flexure: surgical implications. Front Neurol. 2022;13:757757.
3. Nosair E. Limbic System: when emotions control us, we remain little different from animals.
4. Van Groen T, Kadish I, Hoef V, Wyss L. Neuroscience in medicine. Neurosci Med Third Ed Chap. 2008;16:379–96.
5. Papez JW. A proposed mechanism of emotion. Arch Neurol Psychiatr. 1937;38(4):725.
6. Yaşargil MG. Microneurosurgery: microneurosurgery of CNS tumors. Stuttgart: Georg Thieme Verlag; 1996. p. 1894–996.
7. Kier EL, Fulbright RK, Bronen RA. Limbic lobe embryology and anatomy: dissection and MR of the medial surface of the fetal cerebral hemisphere. AJNR Am J Neuroradiol. 1995;16(9):1847–53.
8. Sadler TW. Langman's medical embryology. 12th ed. Philadelphia, PA: Lippincott Williams and Wilkins; 2011.
9. Pool JL. The visceral brain of man. J Neurosurg. 1954;11(1):45–63.
10. Scoville WB. The limbic lobe in man. J Neurosurg. 1954;11(1):64–6.
11. Reep RL, Finlay BL, Darlington RB. The limbic system in mammalian brain evolution. Brain Behav Evol. 2007;70(1):57–70.
12. Colnat-Coulbois S, Mok K, Klein D, Pénicaud S, Tanriverdi T, Olivier A. Tractography of the amygdala and hippocampus: anatomical study and application to selective amygdalohippocampectomy: laboratory investigation. J Neurosurg. 2010;113(6):1135–43.
13. Condés-Lara M, Paz C, Jiménez JR, Martínez G, Martínez-Lorenzana G, Larriva-Sahd J. Electrophysiological responses of interfascicular neurons of the rat anterior commissure to activation from the anterior olfactory nucleus, medial frontal cortex, and posterior nucleus of the amygdala. Brain Res. 2003;982(2):288–92.
14. Gurdjian ES. Olfactory connections in the albino rat, with special reference to the stria medullaris and the anterior commissure. J Comp Neurol. 1925;38(2):127–63.
15. Lohman AH, Mentink GM. The lateral olfactory tract, the anterior commissure and the cells of the olfactory bulb. Brain Res. 1969;12(2):396–413.
16. Martínez-Lorenzana G, Jiménez JR, Condés-Lara M. Interamygdaloid connection of basolateral nucleus through the anterior commissure in the rat. Neurosci Lett. 2004;366(2):154–7.
17. Duvernoy HM. The human hippocampus: functional anatomy, vascularization and serial sections with MRI. 3rd ed. Berlin,Heidelberg/New York, NY: Springer-Verlag; 2005.
18. Heimer L, Van Hoesen GW. The limbic lobe and its output channels: implications for emotional functions and adaptive behavior. Neurosci Biobehav Rev. 2006;30(2):126–47.
19. Sánchez-Navarro JP, Román F. Amigdala, corteza prefrontal y especializacion hemisferica en la experiencia y expresion emocional. Annals of Psychology. 2004;20(2),223–40. https://revistas.um.es/analesps/article/view/27351.
20. Jacobson S, Marcus EM, Pugsley S. The limbic system, temporal lobe, and prefrontal cortex. In: Neuroanatomy for the neuroscientist. Cham: Springer International Publishing; 2018. p. 477–529.
21. Tamraz JC, Comair YC, Lüders HO. Atlas of regional anatomy of the brain using MRI. With functional correlations. With Funct Correlat First Ed Chap. 2005;6:161–83.
22. Burks JD, Conner AK, Bonney PA, Glenn CA, Baker CM, Boettcher LB, et al. Anatomy and white matter connections of the orbitofrontal gyrus. J Neurosurg. 2018;128(6):1865–72.
23. Naidich TP, Daniels DL, Haughton VM, Pech P, Williams A, Pojunas K, et al. Hippocampal formation and related structures of the limbic lobe: anatomic-MR correlation. Part II Sagittal sections. Radiology. 1987;162(3):755–61.
24. Mark LP, Daniels DL, Naidich TP, Hendrix LE, Maas E, Anatomic moment. The septal area. AJNR Am J Neuroradiol. 1994;15(2):273–6.
25. Williams PL, Warwick R, Dyson M, Bannister LH. Gray's anatomy. 37th ed. Edinburgh: Churchill Livingstone; 1989.
26. Vogt BA, Hof PR, Vogt LJ. Cingulate gyrus. In: Paxinos G, Mai JK, editors. The human nervous system. 2nd ed. San Diego, CA: Academic Press; 2004. p. 915–49.
27. Wu Y, Sun D, Wang Y, Wang Y, Ou S. Segmentation of the cingulum bundle in the human brain: a new perspective based on DSI tractography and fiber dissection study. Front Neuroanat. 2016;10:84.
28. Weininger J, Roman E, Tierney P, Barry D, Gallagher H, Murphy P, et al. Papez's forgotten tract: 80 years of unreconciled findings concerning the thalamocingulate tract. Front Neuroanat. 2019;13:14.
29. Wen HT, Rhoton AL Jr, de Oliveira E, Cardoso AC, Tedeschi H, Baccanelli M, et al. Microsurgical anatomy of the temporal lobe: part 1: mesial temporal lobe anatomy and its vascular relationships as applied to amygdalohippocampectomy. Neurosurgery. 1999;45(3):549–91.
30. González-López P, Luna E, Abarca-Olivas J, Daniel RT. How I do it: paramedian supracerebellar transtentorial approach for a temporomesial glioma. Acta Neurochir. 2021;163(5):1311–6.

Anatomy and Challenges for Surgical Removal of Brainstem Tumors: The Mesencephalic Region

64

Miguel A. Arraez, Pablo González-Lopéz, and Cinta Arraez-Manrique

Learning Objectives
- Indication of surgery at the brainstem
- Knowledge of safe entry zones of the Mesencephalon: anatomical basis
- The choice of the approach for brainstem surgery
- Operative nuances for microsurgical removal

64.1 Introduction

Brainstem lesions had been traditionally considered unamenable for surgery due to high morbidity and mortality. The earliest operations on the brainstem aimed at sectioning fiber pathways for the treatment of pain. Lesions displacing rather than invading the brainstem (some neoplasms, vascular lesions, and infectious diseases) started to be surgically treated few decades ago [1]. The current knowledge of neuroanatomy, physiology, and refinement of intraoperative technologies (among them intraoperative neurophysiological monitoring) allows for safe and successful removal of brainstem tumors.

Generally speaking, we consider a group of infiltrating glial tumors that are more prevalent in younger patients and also a different group of benign non-infiltrating lesions more frequent in adults. Among them, cavernous angioma is one of the most emblematic lesions in order to consider the usefulness of the big effort of undertaking surgery at the brainstem. Cavernomatous lesions are prone to bleed with disastrous consequences. Surgery may cure the patient, but the risks are obvious. Hence, surgical indication is the first step. Good indications are lesions with repeated bleedings, symptomatic tumors at the medullary region, growing tumors, and altered level of consciousness due to size or hemorrhage. On the contrary, bad indications for surgery are small and/or inaccessible lesions, rapid resolution of neurological deficit, and medullary/asymptomatic tumors [2].

M. A. Arraez (✉) · C. Arraez-Manrique
Department of Neurosurgery, Carlos Haya University Hospital, University of Malaga, Malaga, Spain
e-mail: marraezs@uma.es

P. González-Lopéz
Department of Neurosurgery, Alicant General Hospital, University of Alicant, Alicant, Spain
e-mail: gonzalez_pab@gva.es

64.2 Anatomy Highlights

Neurovascular and intrinsic anatomy

- Intercrural, parapeduncular, quadrigeminal, and velum interpositum cisterns.
- Periocullomotor area.
- Supra and infracollicular area.
- Lateral mesencephalic sulcus.

Brainstem tumors' surgery has historically represented one of the most challenging procedures. The brainstem is a neural structure located in the tentorial incisura and the posterior fossa, through which the long pathways travel, connecting the telencephalon with the vital centers of the brainstem itself, the spinal cord, and the cerebellum. The cranial nerves nuclei, as well as their emergence through its surface, are located into the brainstem thickness. The external surface of the trunk is in close relationship with important posterior circulation vascular structures. Therefore, surgically approaching its intrinsic pathology requires a strong anatomical knowledge, in addition to a refined microsurgical technique.

The brainstem is divided into three parts: the midbrain, protuberance, and medulla oblongata, from cranial to caudal. The midbrain represents the smaller part in volume and length, but probably the most challenging one in terms of surgical approaches due to its relative location in the center of the tentorial incisura and superior part of the posterior fossa, connecting the supra- and infratentorial central nervous system areas.

I. N. Kanaan, V. Beneš (eds.), *Neuroanatomy Guidance to Successful Neurosurgical Interventions*,
https://doi.org/10.1007/978-3-031-59838-8_64

64.3 Mesencephalon. Surface Anatomy. Limits and Divisions

The mesencephalon (midbrain) connects the metencephalon (protuberance-pons and cerebellum) with the diencephalon and upper telencephalon through the cerebral peduncles.

As every brainstem region, the intrinsic anatomy of the midbrain is divided in rostral, central and dorsal areas. The midbrain anterolateral face shows two clearly identifiable grooves such as the ventrolateral fissure or lateral mesencephalic groove, which separates the rostral midbrain base from the central region or tegmentum, and the dorsolateral fissure that indicates the limit between the tegmentum and the dorsal tectum. The cerebral Aqueduct of Sylvius is located in between the tectum and the tegmentum (Fig. 64.1).

The most rostral area of the midbrain, or base, is formed by the cerebral peduncles on both sides and the interpeduncular fossa in between them, from which the oculomotor nerves (III CN) emerge, and in which roof the posterior thalamoperforating arteries penetrate. These are branches of the basilar top and P1 segment of the posterior cerebral artery (PCA). On the posterior face, the mesencephalic lamina quadrigemina stands out being occupied by four small eminences known as the superior and inferior colliculi. Both on its anterior and lateral faces, the anatomical limit between the pons and the midbrain is clearly identifiable at the level of the pontomesencephalic sulcus. The upper limit that separates the midbrain from the diencephalon is given by the optic tract on its anterior face. On the lateral aspect, this upper limit is defined by a plane caudal to the lateral and medial geniculate bodies as well as to the pulvinar nucleus. On the posterior face, the tectal plate clearly defines the mesencephalic tectum on the surface and is delimited above by the pineal gland and the brachium of the superior colliculus, and below by the infracolicular recess from which the trochlear nerve (IV CN) arises laterally (Fig. 64.2).

If we analyze a midline sagittal section from its medial aspect, the lower limit of the midbrain is seen passing through a plane caudal to the inferior colliculus dorsally and through the rostral border of the fibers of the pons ventrally. The upper limit, which separates it from the diencephalon, passes through the posterior commissure dorsally and the caudal portion of the mammillary body ventrally (Fig. 64.1).

64.3.1 Perimesencephalic Cisterns

The arachnoid cisterns represent natural pathways where the cranial nerves and vascular structures arise and travel. These are key highways through which surgical approaches are designed to reach the different areas of the central nervous system. Despite its small size and volume, but due to its central position into the tentorial incisura, the midbrain is 360° surrounded by cisterns. These are generally named perimesencephalic cisterns, and accurately understanding their anatomy is key to perform safe surgery in and around the midbrain. These cisterns are divided into anterior, lateral, and posterior (Fig. 64.3).

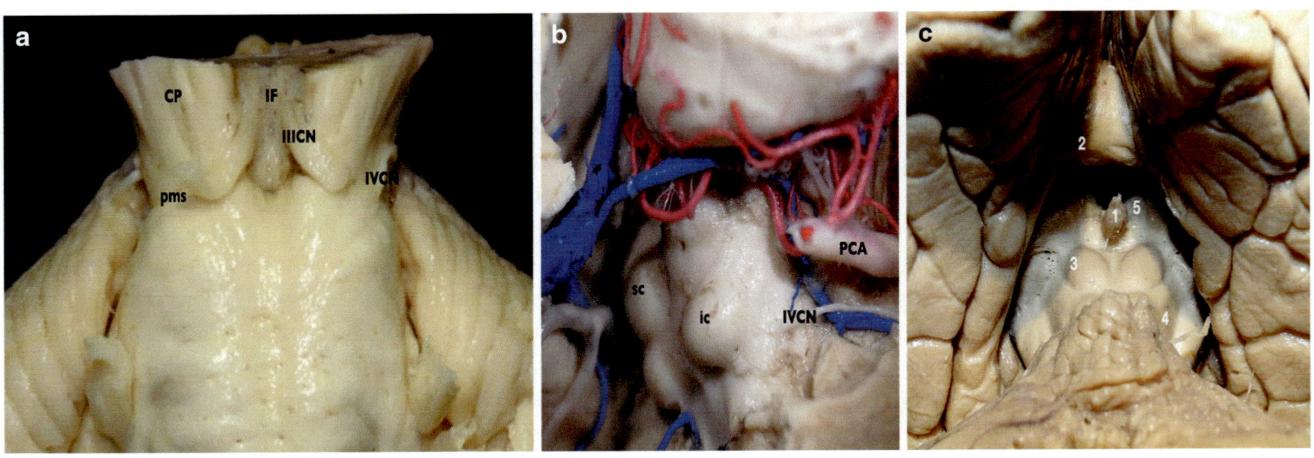

Fig. 64.1 Cadaveric anatomical dissections showing the anterior, posterolateral, and posterior views of the upper brainstem. (**a**) Midbrain ventral surface where the pontomesencephalic sulcus represents the inferior limit. Both cerebral peduncles and oculomotor nerves are showed on both sides of the interpeduncular fossa. (**b**) Dorsolateral supracerebellar exposure of the tectal plate and lateral aspect of the tegmentum and its close relationship to the trochlear nerve, posterior cerebral artery, and the lateral mesencephalic sulcus and vein. (**c**) Midline posterior supracerebellar view of the tectal plate and pineal region. The splenium represents the quadrigeminal cistern roof. 1. Pineal gland; 2. Splenium; 3. Superior colliculus; 4 Inferior colliculus; 5. Habenula; *CP* cerebral peduncle, *ic* inferior colliculus, *IF* interpeduncular fossa, *IIICN* oculomotor nerve, *IVCN* trochlear nerve, *PCA* posterior cerebral artery, *pms* pontomesencephalic sulcus, *sc* superior colliculus

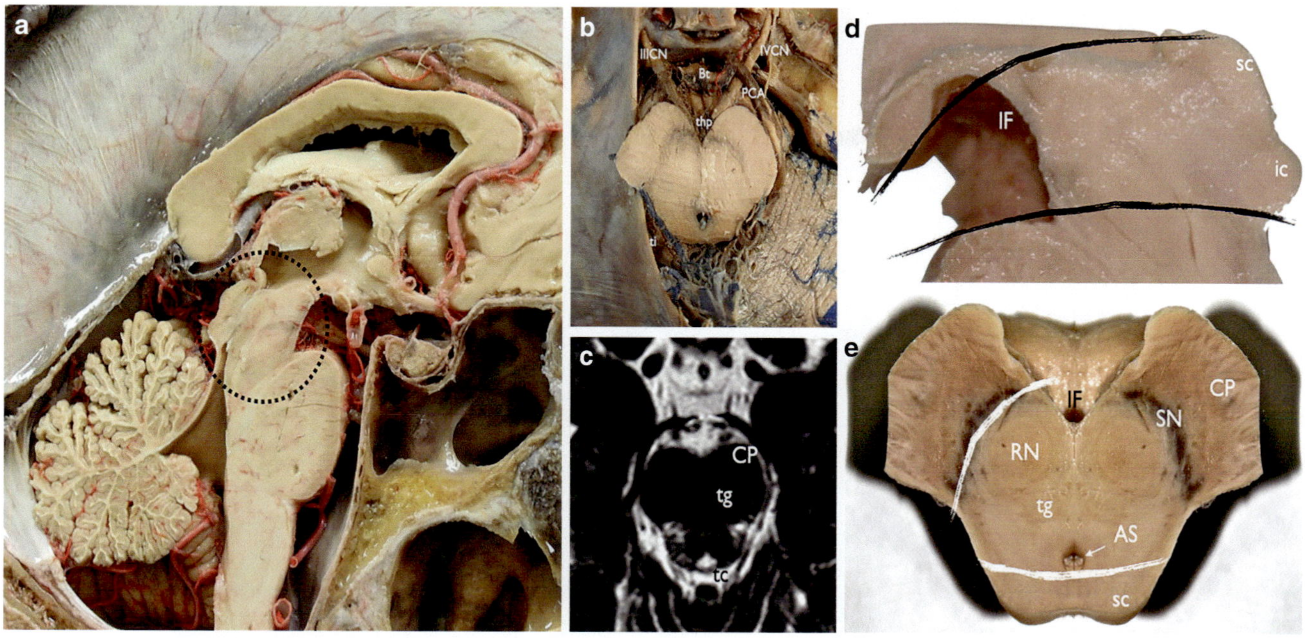

Fig. 64.2 (**a**) Midsagittal view of a human cadaveric piece, where the relative location of the upper brainstem—midbrain is located (black line circle) connecting the telencephalon and diencephalon with the rest of the stem. (**b**) Superior view of the tentorial incisura. Notice that the right-sided tentorial layer has been removed so as to expose the tentorial cerebellar surface and its close relationship to the lateral and posterior aspects of the mesencephalon. Anteriorly, the midbrain is closely related to the interpeduncular fossa and its contents. (**c**) Axial FIESTA MRI sequence showing the midbrain and its internal divisions. (**d**) Magnified detail of a sagittal cut showing the upper and lower limits of the midbrain, tracing two lines, one from the upper limit of the superior colliculus to the inferior aspect of the mammillary bodies, and the second one from the inferior border of the inferior colliculus to the pontomesencephalic sulcus, respectively. (**e**) Internal midbrain divisions and structures in an axial cut through the superior colliculus. *AS* aqueduct of sylvius, *Bt* basilar artery tip, *CP* cerebral peduncle, *IF* interpeduncular fossa, *IIICN* oculomotor nerve, *IVCN* trochlear nerve, *PCA* posterior cerebral artery, *RN* red nucleus, *SN* substantia nigra, *tc* tectum, *tg* tegmentum, *thp* thalamoperforators

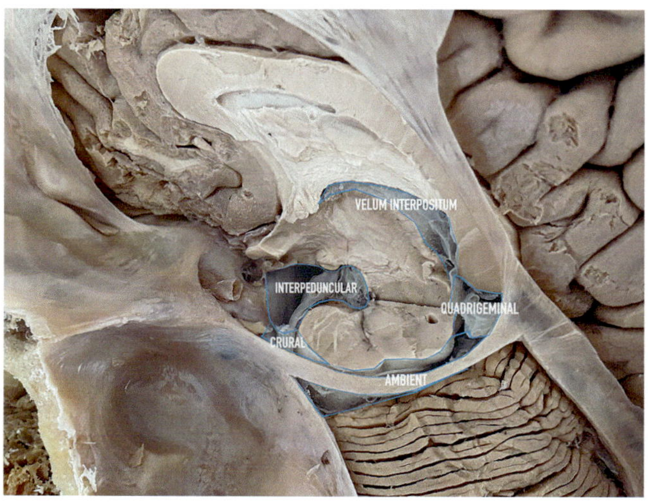

Fig. 64.3 Anatomical specimen where the left hemisphere has been removed after performing a sagittal midline cut till the superior collicular level, where a left axial cut has been carried out too. The fall and tentorium have been partially opened showing the midbrain anatomical relationships and the location of the perimesencephalic cisterns

64.3.1.1 Anterior

Interpeduncular (intercrural) cistern. It forms the confluence of the supratentorial and infratentorial subarachnoid spaces. It is in close contact superiorly with the inferior surface of the diencephalon. The posterior wall is formed by the cerebral peduncles, the posterior perforated substance, and the ventral surface of the upper portion on the pons. The anterior and lateral walls are formed by the diencephalic and mesencephalic parts of Liliequist membrane. The inferior surface is composed of the medial pontomesencephalic and two lateral pontomesencephalic membranes. Laterally, it joins the ambient cistern inferiorly and is limited superiorly by the carotid and crural cisterns and the mesial temporal lobe. It contains the bifurcation of the basilar artery, PCA, and SCA peduncular segments, perforating branches of the basilar artery, PCA, and posterior communicating arteries, the basal vein of Rosenthal, interpeduncular and posterior communicating veins, as well as the third cranial nerve arising from the midbrain surface (Fig. 64.4a, b).

64.3.1.2 Lateral (Parapeduncular)

Crural cistern. The crural cistern represents a contiguous space from the interpeduncular cistern and follows along the inferior and lateral surface of both crus cerebri, between the peduncle and the temporal uncus posterior aspect. The crural cistern mainly contains the anterior and medial posterior choroidal arteries and the basal vein of Rosenthal (Fig. 64.4a, c).

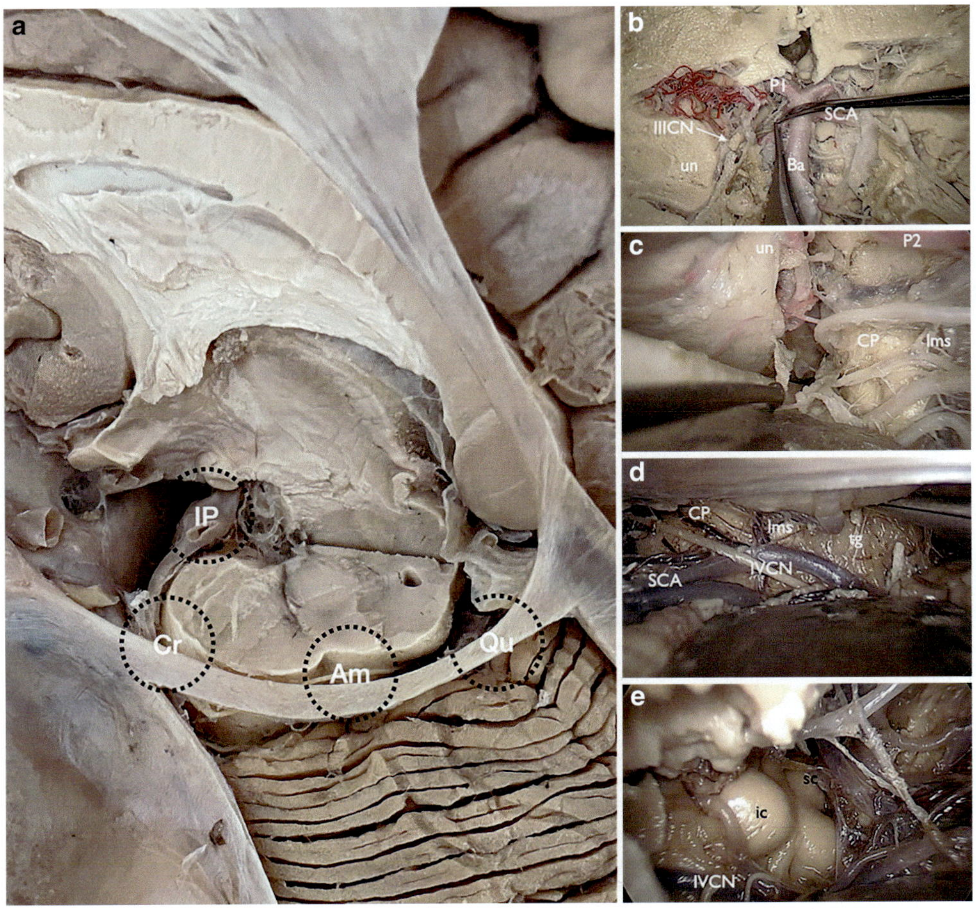

Fig. 64.4 (**a**) Magnified anatomical dissection highlighting the interpeduncular, crural, ambient, and quadrigeminal cisterns. (**b**) Anterior coronal cut through the anterior third ventricle, showing the basilar artery in the center of the image, entering the interpeduncular cistern. Both oculomotor nerves are leaving the stem between the posterior cerebral and superior cerebellar arteries, surrounded by a thick layer of arachnoid (lateral leaves of Liliequist membranes). (**c**) Left lateral subtemporal view of the crural cistern posterior aspect. The P2 segment of the posterior cerebral, as well as the anterior choroidal arteries are sown in between the temporal uncus and the cerebral peduncle. The lateral mesencephalic sulcus is found laterally. (**d**) Left lateral supracerebellar infratentorial view of the ambient cistern crossed by the trochlear nerve and superior cerebellar artery. The lateral mesencephalic sulcus is highlighted in between the anterior and posterior parts of the midbrain. (**e**) Left paramedial supracerebellar infratentorial view of the dorsolateral access to the quadrigeminal cistern. The tectal plate is limiting this cistern anteriorly. *Am* ambient cistern, *Ba* basilar artery, *CP* cerebral peduncle, *Cr* crural cistern, *IIICN* oculomotor nerve, *IP* interpeduncular cistern, *IVCN* trochlear nerve, *P1* P1 segment of the posterior cerebral artery, *P2* P2 segment of the posterior cerebral artery, *Qu* quadrigeminal cistern, *SCA* superior cerebellar artery, *ic* inferior colliculus, *lms* lateral mesencephalic sulcus, *sc* superior colliculus, *tg* tegmentum, *un* temporal uncus

Ambient cistern. It somehow connects with the posteriorly located quadrigeminal cistern, wraps around the midbrain, and communicates anteriorly with the prepontine and crural cisterns. Laterally, it extends to the attachment of the choroid plexus. The two wings communicate above the roof of the third ventricle. It can be divided into supratentorial and infratentorial compartments. The supratentorial compartment contains the basal vein and the posterior cerebral artery. The infratentorial compartment contains the superior cerebellar artery and the fourth cranial nerve (Fig. 64.4a, d).

64.3.1.3 Posterior

Quadrigeminal cistern. The quadrigeminal cistern is located on the so-called posterior tentorial incisural space. It is anteriorly facing the quadrigeminal plate and the pineal gland, while the vermis lies posteriorly. The callosal splenium represents the roof while the anterior medullary velum and the lingual cerebelli form the floor. It connects laterally with the ambient and superiorly with the velum interpositum cisterns. It contains the vein of Galen, posterior pericallosal arteries, third portion of the SCA, perforating branches of the PCA and SCA, and the third segment of the PCA (Fig. 64.4a, e).

Velum interpositum cistern. It extends from the habenular commissure to the foramen of Monro. The roof is formed by the ventral splenium and the floor is formed by the roof of the third ventricle. Its anterior wall converges below the fornix into the foramen of Monro, while its posterior face has no clear limit with the quadrigeminal cistern. This cistern contains the medial posterior choroidal arteries, splenothalamic branches, and the internal cerebral veins (Fig. 64.3).

64.4 Intrinsic Anatomy

The midbrain axial contents can be divided from anterior to posterior into base, tegmentum, cerebral aqueduct, and tectum. The midbrain houses some important gray matter structures and several "en route" white matter fiber bundles that traverse its thickness.

The ventral midbrain is known as the cerebral peduncle or crus cerebri and is composed of cortical projections to the stem and spinal cord. The pontine projections of the parietal, occipital, and temporal lobes are lateral, the frontopontine projections are central, and the corticospinal and corticobulbar tracts occupy a medial position in the base of the midbrain through its course just dorsal to the interpeduncular fossa. Behind the crus cerebri, the substantia nigra forms the limit between the anterior base and the central mesencephalic tegmentum. Dorsally, the red nucleus is found in a medial position, while the lateral lemniscus is located lateral. The medial lemniscus lies in the thickness of the tegmentum between the posterior part of the red nucleus and the substantia nigra. In a somewhat more dorsal position, we will find the Edinger-Westphal and oculomotor nuclei for the III CN at the level of the superior colliculus and medial to the spinothalamic bundle. The trochlear nucleus is located caudal to the III CN nucleus and at the level of the inferior colliculus. Both nuclei are interconnected by the medial longitudinal fasciculus (MLF), which is placed lateral to these nuclei, and its damage involves the so-called internuclear ophthalmoplegia. Behind these nuclei, the cerebral aqueduct appears connecting the third and fourth ventricles. It is surrounded by the periaqueductal gray matter, which projects to the spinal cord and inhibits upward transmission of pain.

The quadrigeminal colliculi are located on the dorsal part of the midbrain forming what we know as the mesencephalic tectum. The superior colliculi receive sensory information, which projects to the lateral geniculate body to direct gaze toward sensory stimuli. The inferior colliculus receives auditory information primarily through the lateral lemniscus and it is projected to the primary auditory cortical area through the auditory radiations (Fig. 64.5).

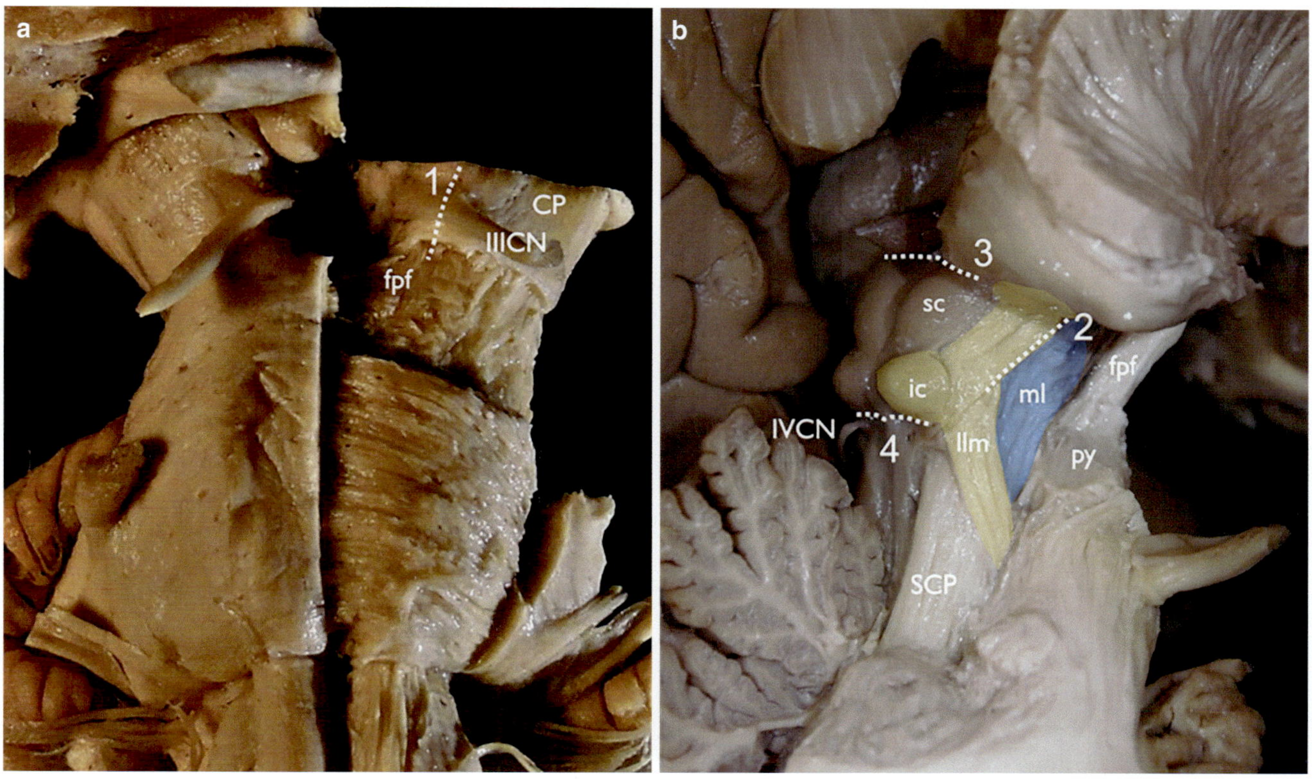

Fig. 64.5 (**a**) Anterior aspect of the brainstem. The left hemiprotuberance has been dissected showing the transverse-oriented middle cerebellar peduncle fibers. The descending cerebral peduncle fibers are shown in the mesencephalic thickness running downwards just lateral to the oculomotor nerve exit zone. The perioculomotor area is shown as a vertically oriented brainstem incision just lateral to the oculomotor nerve exit zone and in close relationship with the frontopontine fibers. (**b**) Dorsolateral aspect of the right brainstem. The lateral (yellow) and medial (blue) lemnisci are shown into the tegmental space. The lateral mesencephalic sulcus represents a common lateral midbrain entry zone, running in an obliquely oriented vertical incision as shown. The supra- and infracollicular incisions are shown above and below the collicular plate. 1. Perioculomotor area; 2. Lateral mesencephalic sulcus; 3. Supracollicular area; 4. Infracollicular area; *CP* cerebral peduncle, *IIICN* oculomotor nerve, *IVCN* trochlear nerve, *SCP* superior cerebellar peduncle, *fpf* frontopontine fibers, *ic* inferior colliculus, *llm* lateral lemniscus, *ml* medial lemniscus, *py* pyramidal tract, *sc* superior colliculus

64.5 'Safe Entry Zones'

Due to the large number of highly functional gray and white matter structures located in the midbrain thickness as well as the relevant neurovascular components hosted by the perimesencephalic cisterns, surgical routes to its intrinsic pathology require a carefully planned craniotomy and surgical strategy to choose the safest brainstem entry zone and draw an optimized trajectory. Literature is extensive remarking the importance of the anatomical aspects of the safe entry zones to the brainstem and related surgical approaches [3–11].

Based on its location in the midbrain, intrinsic pathology can be divided mainly into ventral, ventrolateral, and dorsal. The craniotomy and bone work at the base of the skull will depend mainly on the chosen entry point and the direction of growth of the midbrain lesion (Fig. 64.5).

- The *perioculomotor area* is an anterior region located in the posterior surface of the interpeduncular cistern between the PCA and SCA, lateral to the exit of the III CN and medial to the pyramidal tract. It is positioned between the frontopontine, corticobulbar, and corticospinal pathways laterally and the oculomotor nerve exit point medially. The immediately lateral to the III CN cerebral peduncle fibers are the frontopontine, whose injury is highly expected during this entry to the midbrain anterior aspect. The first structures found in the dorsal midbrain are the red nucleus and the intramesencephalic segment of the oculomotor nerve that traverses it (Fig. 64.5a).
- The *lateral mesencephalic sulcus* is a groove that runs on the lateral surface of the midbrain between the cerebral peduncle and the lateral lemniscus of the tegmentum. It is also known as the ventrolateral fissure and extends in the sagittal plane from the pontomesencephalic sulcus to the medial geniculate body. This entry zone is mainly described to deal with ventrolateral midbrain intrinsic lesions. Depending on the surgical trajectory, we will be closer to the red nucleus, decussation of the superior cerebellar peduncle and substantia nigra (if the direction is ventral), or to the spinothalamic tracts, medial lemniscus, and periaqueductal structures (if the direction is more dorsal) (Fig. 64.5b).
- An entry through the *supracollicular area* requires an incision in the axial plane above the superior colliculus. It is important to limit this cut to the tectal area (posterior to the cerebral aqueduct), to avoid an intramesencephalic segment of the III CN and the MLF damage. Directing the trajectory laterally, the habenula, the trigeminal mesencephalic tract, the central tegmental tract, and finally the red nucleus stand out (Fig. 64.5b)
- The *infracollicular area* incision is made between the IV CN exit zone fibers and the lower border of the inferior colliculus. The entrance to the tectum in a lateral direction is related, from surface to depth, with the superior cerebellar peduncle and the lateral lemniscus (Fig. 64.5b)

64.6 Intervention

The technique for the surgical removal is of paramount importance in brainstem surgery and of particular interest in case of cavernoma excision to avoid postoperative morbidity in such a benign lesion. Morbidity may come from the dissection between the lesion and the surrounding normal brainstem fibers and nuclei. The other aspect of concern is the entrance inside the brainstem [2–4]. One tenet is just entering brainstem where the lesion abuts the surface. For deeper lesions, the rule of the two points can be used: one point in the center of the lesion, another one in the closest vicinity to the pial surface and then an imaginary line would depict the trajectory of our approach, and hence, the choice of the surgical approach. But the latest incorporation in the last two decades has been the concept of safe entry zone, or how to get inside the brainstem without adding morbidity (see "Anatomy highlights"). Figures 64.6, 64.7, 64.8, and 64.9 represent different approaches for resection of brainstem lesions at the Mesencephalon.

When dealing with brainstem cavernoma, another aspect to be considered is the timing for the operation. Surgery in the acute stage is riskier and has more probability of leaving behind some of the tumor in certain cases. Waiting too long potentially allows for the creation of a gliotic plane that hampers the safe resection of the tumor. Some authors do consider a timeframe of around 6 weeks after the bleeding as appropriate to undertake the operation [11].

Intraoperative neurophysiological monitoring (IOMN) deserves special consideration. Among the intraoperative technologies, IONM yields the highest possible safety in the resection of brainstem tumors [12]. Main targets at the mesencephalon are ascendent somatosensory and descendent and motor pathways. Electromyography of the third nerve is also desirable. Conjugate gaze pathways cannot be monitored being one of the challenges of IONM at the mesencephalon.

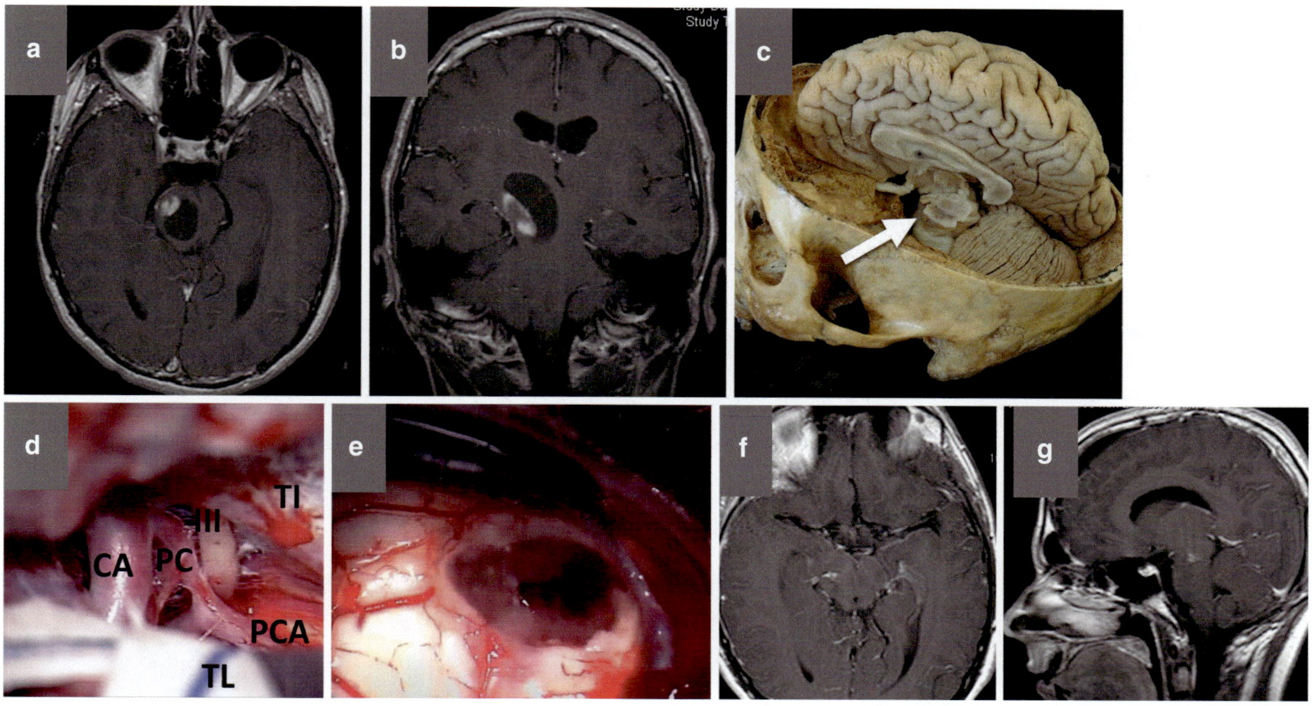

Fig. 64.6 Thalamo-mesecephalic solid-cystic pilocytic astrocytoma in a 16 years old boy (**a** and **b**). (**c**) Scope of the approach and anatomical target of the approach (anterolateral mesencephalon; arrow). (**d**) Surgical field after exposition of the tentorial incisura. *CA* carotid artery, *PC* posterior communicating artery, *PCA* posterior cerebral artery, *III* third nerve, *TI* tentorial incisura. (**e**) Exposure of the tumor at the right peduncle. (**f** and **g**) postoperative MRI

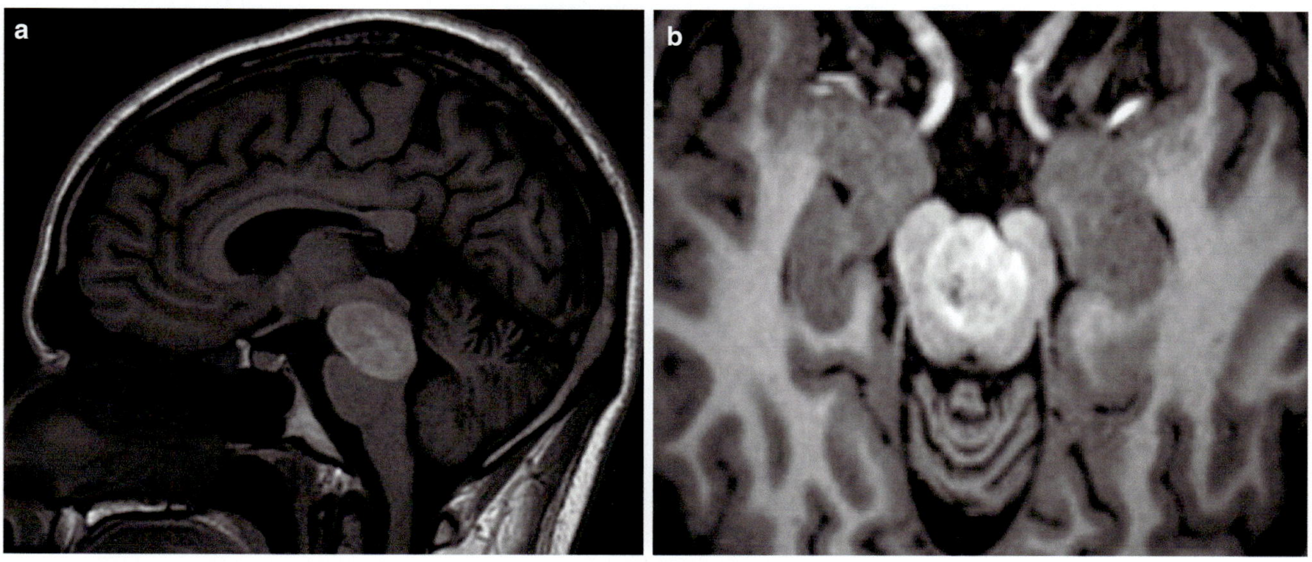

Fig. 64.7 Mesencephalic melanocytic tumor (**a**, **b**). (**c**) subtemporal suboccipital approach through the infracollicular safe entry zone. *T* tumor, *TE* transected tentorium, *IV* fourth nerve exiting brainstem, *OL* occipital lobe. (**d** and **e**) postoperative MRI

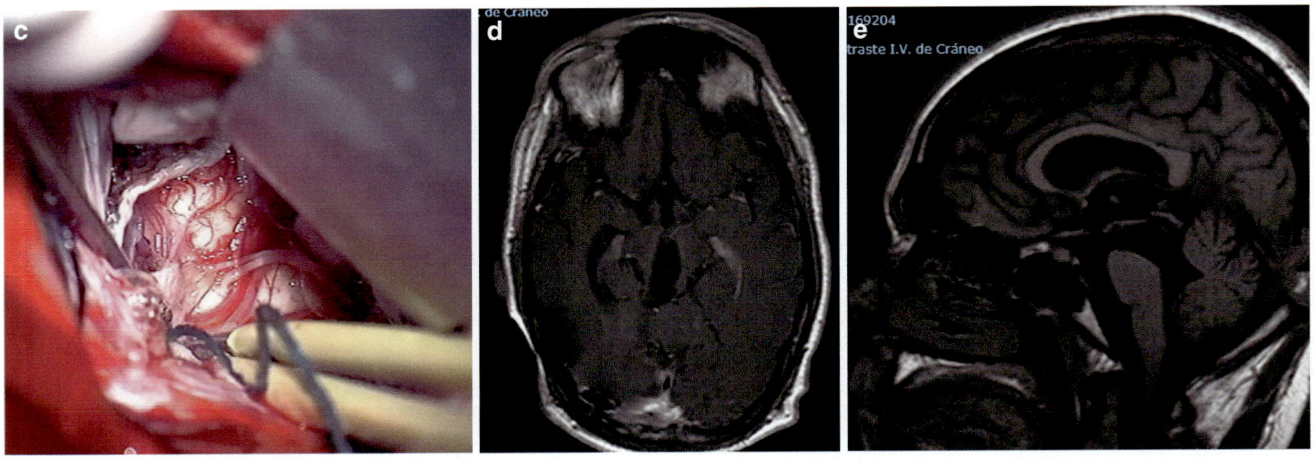

Fig. 64.7 (continued)

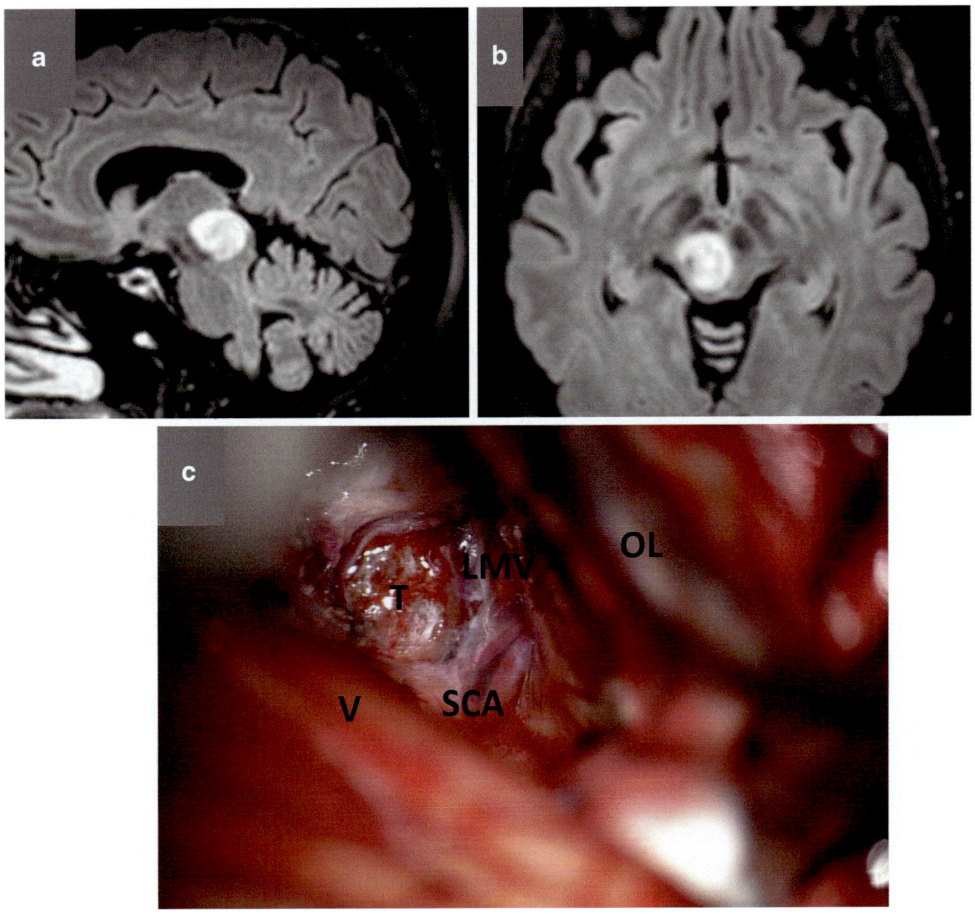

Fig. 64.8 (**a** and **b**) Tegmental pilocytic astrocytoma. Right supracerebellar infratentorial approach (**c**) after section of tentorium and right occipital retraction. LMV lateral mesencephalic vein, *T* tumor, *V* vermian region, *LMV* lateral mesencephalic vein. *SCA* superior cerebellar artery, *Ol* occipital lobe after tentorial section

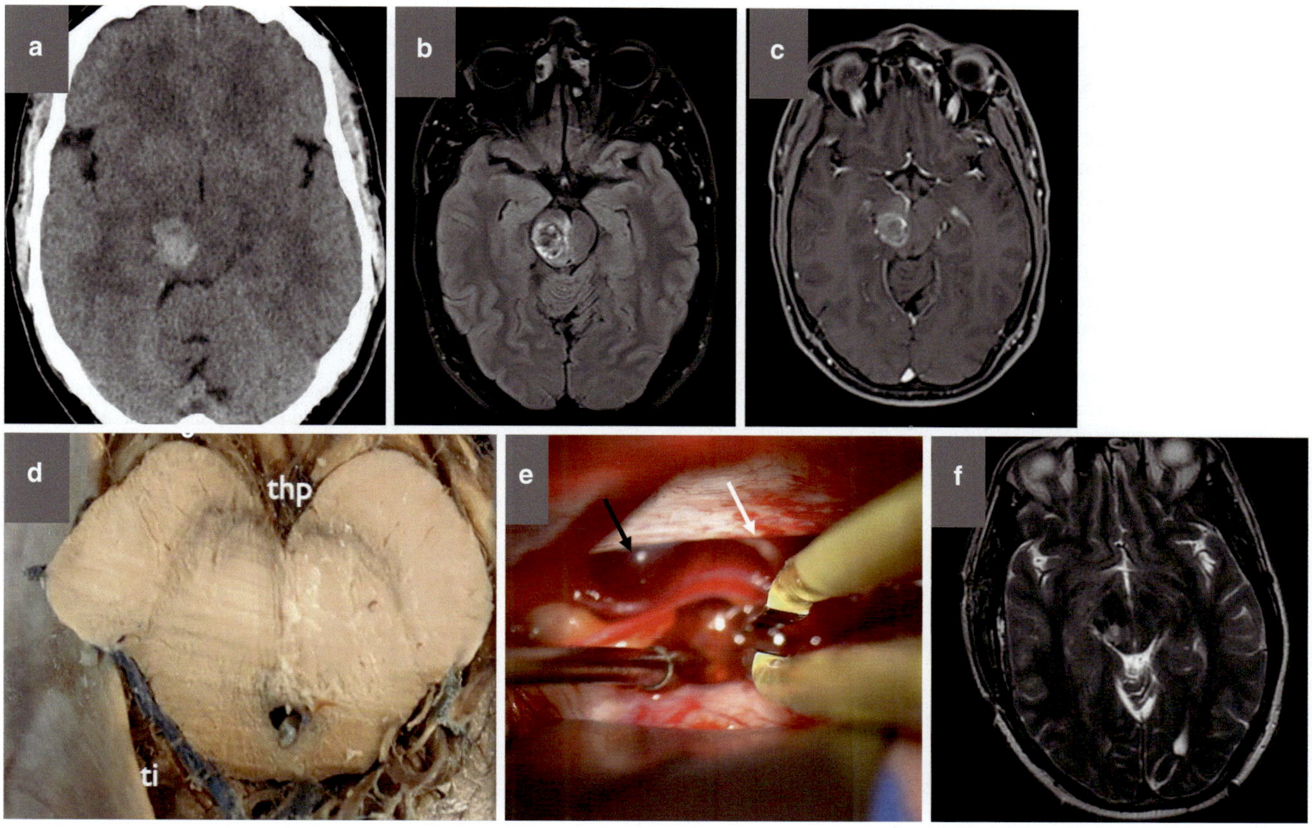

Fig. 64.9 Hemorrhagic cavernous malformation at the right mesencephalon (**a–c**). (**d**) anatomical relationship of the lateral mesencephalon with the tentorium. (**e**) right subtemporal approach to the lateral mesencephalon. Exposure of the lesion and removal. White arrow is pointing out the IV nerve and its relationship with tentorium. Black arrow, accompanying venous anomaly. (**f**) postoperative MRI

64.7 Surgical Pearls

- Entrance through safe entry zones or where the lesion abuts the surface
- Careful dissection between the tumor and normal brainstem
- Timing of the operation to avoid gliotic tissue around cavernomas
- Intraoperative neurophysiological monitoring

64.8 Discussion

Tumors at the brainstem represent a formidable challenge in neurosurgery. Mesencephalon represents a very difficult area to deal with due to the difficulty of the anatomy, difficult surgical access, and functional and eloquent structures. The current sophisticated intraoperative technologies cannot overcome the necessity of a thorough knowledge of the mesencephalic anatomy and its safe entry zones. Many factors have to be taken into account, including the timing for the operation in cavernoma cases. The dissecting plane between the lesion and normal brainstem has to be very carefully managed to avoid morbidity. The safe resection of tumors at this level of the anatomy is now possible due to the combination of these strategies, Intraoperative Neurophysiological Monitoring being one of the most crucial refinements among the different tools of the neurosurgical armamentarium.

64.9 Conclusion

The accurate knowledge of the mesencephalic anatomy is crucial for successful tumor removal. The concept of safe entry zone and careful dissection of the tumor avoids postoperative morbidity. Intraoperative neurophysiological monitoring is mandatory in mesencephalic surgery.

References

1. Morcos JJ, Haines SJ. History of brain stem surgery. Neurosurg Clin N Am. 1993;4(3):357–65.
2. Ohue S, Fukushima T, Kumon Y, Ohnishi T, Friedman AH. Surgical management of brainstem cavernomas: selection of approaches and microsurgical techniques. Neurosurg Rev. 2010;33(3):315–22.
3. Cavalcanti DD, Preul MC, Kalani MY, Spetzler RF. Microsurgical anatomy of safe entry zones to the brainstem. J Neurosurg. 2016;124(5):1359–76.
4. Recalde RJ, Figueiredo EG, de Oliveira E. Microsurgical anatomy of the safe entry zones on the anterolateral brainstem related to surgical approaches to cavernous malformations. Neurosurgery. 2008;62(3 Suppl 1):9–15.
5. Yagmurlu K, Rhoton AL Jr, Tanriover N, Bennett JA. Three-dimensional microsurgical anatomy and the safe entry zones of the brainstem. Neurosurgery. 2014;10 Suppl 4:602–19; discussion 619–20
6. Zhong J, Li ST, Xu SQ, Wan L. Surgical treatment of ventral mesencephalic cavernoma. Acta Neurochir. 2007;149(10):1057–61.
7. Abou-Al-Shaar H, White TG, Peto I, Dehdashti AR. Left suboccipital supracerebellar transtentorial approach for resection of tectal cavernous malformation. Neurosurg Focus Video. 2019;1(2):V19.
8. Kalani MYS, Yağmurlu K, Spetzler RF. The interpeduncular fossa approach for resection of ventromedial midbrain lesions. J Neurosurg. 2018;128(3):834–9.
9. Graffeo CS, Scherschinski L, Baranoski JF, Srinivasan VM, Lawton MT. Resection of quadrigeminal midbrain cavernous malformation using the supracollicular safe entry zone. World Neurosurg. 2023;169:51.
10. Delaunois J, Vaz G, Raftopoulos C. Transsylvian transuncal approach for an anterior midbrain cavernous malformation resection: a case report. Oper Neurosurg (Hagerstown). 2018;14(3):E38–43.
11. Rajagopal N, Kawase T, Mohammad AA, Seng LB, Yamada Y, Kato Y. Timing of surgery and surgical strategies in symptomatic brainstem cavernomas: review of the literature. Asian J Neurosurg. 2019;14(1):15–27.
12. Rauschenbach L, Santos AN, Dinger TF, et al. Predictive value of intraoperative neuromonitoring in brainstem cavernous malformation surgery. World Neurosurg. 2021;156:e359–73.

Neuroanatomy and Surgical Approaches to Paediatric Posterior Fossa Tumours

65

Thomas Santarius, Heba Azooz, and Zhangjie Su

Learning Objectives

- To outline the specific surgical challenges in managing common posterior fossa tumours in the paediatric cohort.
- To highlight the importance of a surgical goal and holistic consideration when dealing with various types of tumours in the posterior fossa.
- To provide insights into different surgical approaches, their relevant anatomical considerations, technical nuances, and applications.
- To discuss the prevention and management of complications associated with resection of paediatric posterior fossa tumours.

65.1 Introduction of Paediatric Posterior Fossa Tumours

Posterior fossa tumours in the paediatric population represent a surgical challenge and require multidisciplinary management to deliver optimum outcomes. Most common paediatric tumours in this region are medulloblastoma, pilocytic astrocytoma (PA), and ependymoma. Less common tumours include brainstem glioma (PA and diffuse), germ cell tumours, haemangioblastoma, atypical teratoid rhabdoid tumour (ATRT), choroid plexus papilloma/carcinoma, and tumours in the cerebellopontine angle (CPA, e.g. epidermoid, vestibular schwannoma, meningioma). These tumours range in their invasiveness to the surrounding neural structures and present specific surgical challenges, warranting individualised surgical planning and adjuvant oncological treatments.

65.2 Anatomy Highlights

- It is possible to approach posterior fossa from any direction. The limits are formed by venous channels (superior sagittal, transverse and sigmoid sinuses), cranial nerves, and most structures contained in brainstem and some of the cerebellum, the ICA, the vertebral artery as well as the middle and inner ears.
- There are several natural corridors that allow access to the target lesion without the damage of the neurovascular tissue, e.g. the cerebellomedullary fissure and tonsillomedullary fissure used in the telovelar approach.
- The trochlear nerve can be an important craniocaudal landmark (Figs. 65.1 and 65.2) for surgery in and around the midbrain tumours, including the tumours of the pineal region. Trochlear nerves can also be used as pointers to the junction of the lower tectum and the superior medullary velum as they emerge just inferior to the inferior colliculi.
- The accessory nerve is a useful and dependable guide to the jugular foramen which, in turn, can help identifying the vagal and glossopharyngeal nerves. Both of these nerves, together with the cranial accessory rootlets, arise from the retroolivary sulcus, while the hypoglossal nerve arises from the preolivary sulcus.
- The anatomy and physiology of the venous system in the posterior fossa are highly variable and it is crucial to only divide veins and venous channels (e.g. the cerebellar draining veins, the occipital sinus, etc.) after careful consideration.
- The bridging veins on the occipital lobe are directed forward so that the superior sagittal sinus downstream of the lambdoid suture is often completely devoid of bridging veins, thus enabling relatively safe interhemispheric approach to posterior tentorial incisura (Fig. 65.3).

T. Santarius (✉) · Z. Su
Department of Neurosurgery, Cambridge University Hospitals NHS Trust, Cambridge, UK

Department of Clinical Neurosciences, University of Cambridge, Cambridge, UK
e-mail: ts381@cam.ac.uk; zhangjie.su1@nhs.net

H. Azooz
Department of Neurosurgery, Children Cancer Hospital Egypt, Cairo, Egypt

I. N. Kanaan, V. Beneš (eds.), *Neuroanatomy Guidance to Successful Neurosurgical Interventions*,
https://doi.org/10.1007/978-3-031-59838-8_65

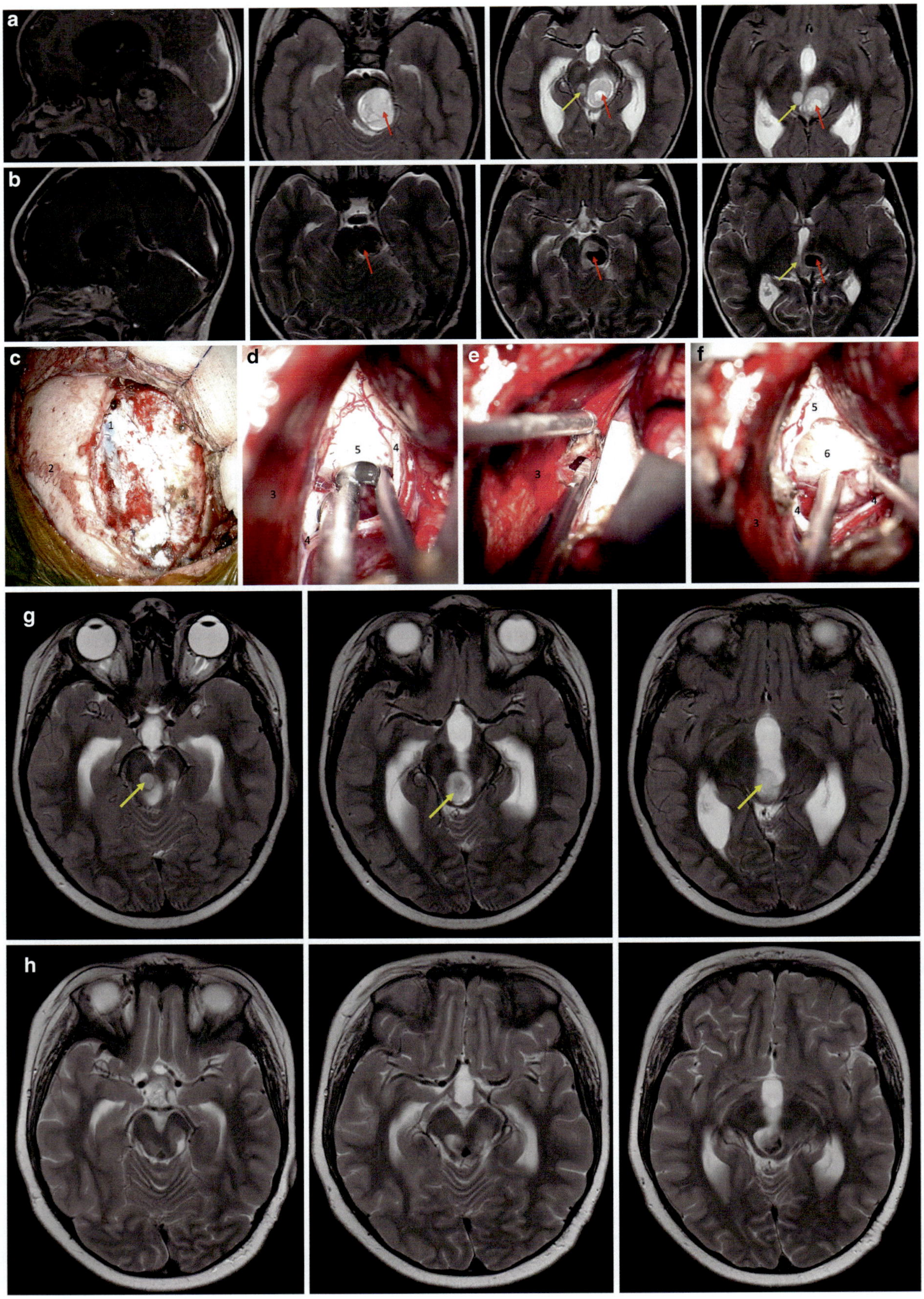
a
b
c
1
2
d
5
4
3
4
e
3
f
5
6
3
4
4
g
h

Fig. 65.1 SCIT with division of the tentorium to resect an extensive pilocytic astrocytoma of the brainstem. (**a**) Four-year old boy presented with 6-week history of ophthalmoplegia and vomiting. (**b**) Post-operative imaging revealed complete resection of the left-sided tumour. The right-sided tumour (yellow arrow) was not in continuity with the left-sided tumour (red arrow). (**c**) Craniotomy, as it appeared at the end of resection and following watertight closure of the dura. (**d**) Supracerebellar infratentorial approach. (**e**) Tentorium being cut. (**f**) Tentorium was cut allowing unrestricted access to the entire tumour. (**g**) A T2 axial image 2 years later, showing no recurrence of the left-sided tumour, but progression of the right-sided tumour (yellow arrow). This was completely resected via the left-sided interhemispheric trans-callosal approach (**h**). 1. Left transverse sinus. 2. Lambdoid suture. 3. Intact tentorium. 4. Left trochlear nerve. 5. Midbrain. 6. Tumour

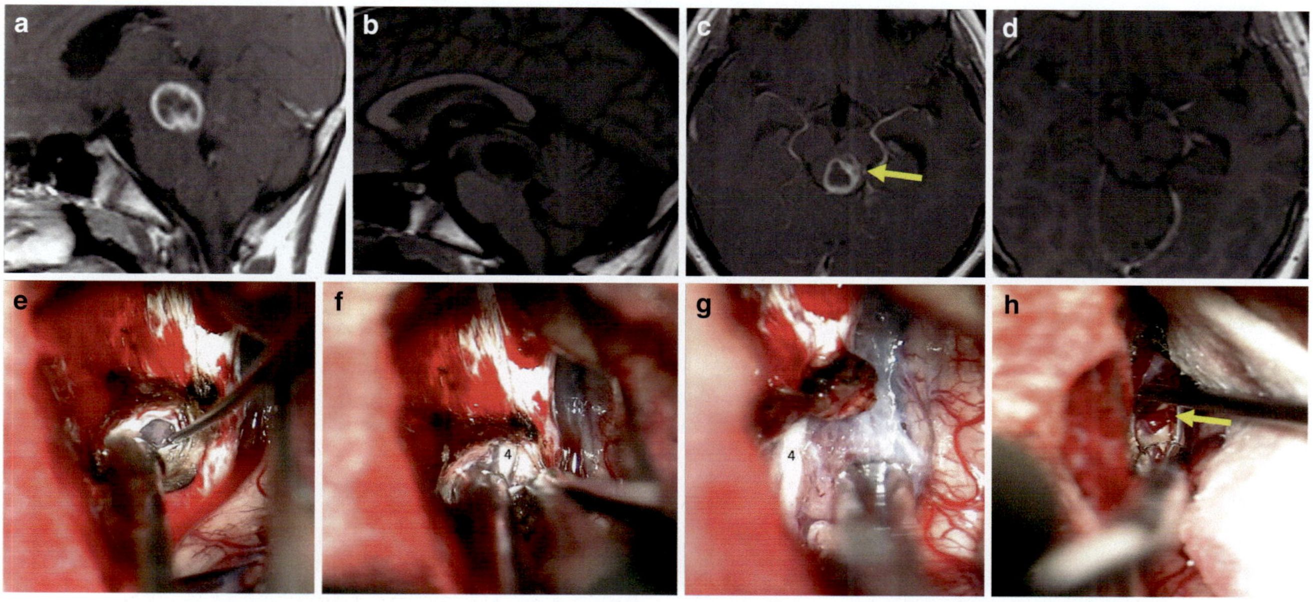

Fig. 65.2 Subtemporal transtentorial approach. A 6-year old girl presented with a few week history of ophthalmoplegia, progressive gait disturbance, 'imbalance,' and right hemiparesis. (**a–c**). Preoperative imaging showing a contrast-enhancing lesion in the left lower tegmentum. (**d**) Post-operative image. (**e–h**). Left subtemporal approach dividing the tentorium while taking care not to injure the trochlear nerve (4). (**g**). The entry to the midbrain was made in front of the vein of the lateral mesencephalus sulcus (yellow arrow—see also **c**)

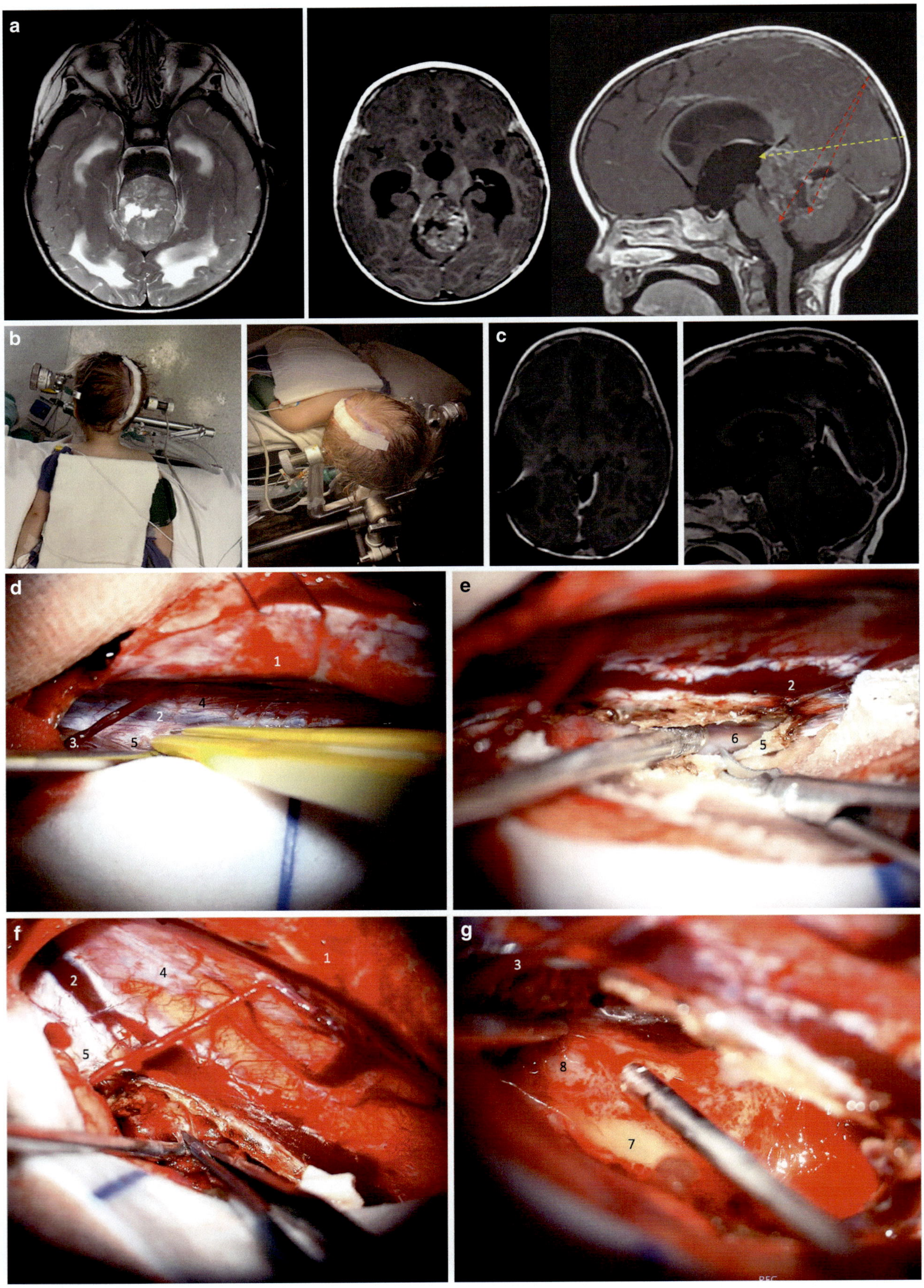
a
b
c
d
1
4
2
3
5
e
2
6
5
f
1
4
2
5
g
3
8
7

Fig. 65.3 Interhemispheric transtentorial approach. Interhemispheric transtentorial approach is preferable to the supracerebellar infratentorial approach for lesion in the posterior incisural space (pineal, tectal, cerebellar) with substantial ventro-dorsal extension. Reaching and gently resecting these tumour margins (red arrows) would not be possible without substantial and almost certainly harmful venous sacrifice and retraction. This is example of an 18-month old with an atypical teratoid rhabdoid tumour. (**a**) Preoperative images. (**b**) Intra-operative positioning. (**c**) Post-operative images. (**d**) Parietal interhemispheric approach. The bone flap straddles the midline and dural flap is reflected over the superior sagittal sinus. The bipolar is at the falcotentorial junction, just inferior to the straight sinus. (**e**) The tentorium is cut along the straight sinus. The location size of the incision depends on the target lesion. (**f**) Resection of the tumour. (**g**) Final stages of tumour resection with some flakes of the tumour seen on the walls of the upper fourth ventricle (superior cerebellar peduole). The tumour was probably arising from the anterior cerebellar vermis of the superior medullary velum (now missing and so providing view of the fourth ventricle). 1. Dura reflected over the superior sagittal sinus. 2. Straight sinus. 3. Occipital hemispheric draining vein. 4. Falx. 5. Tentorium. 6. Tumour. 7. Ependyma of the fourth ventricle. 8. Remnants of the tumour on the left superior cerebellar peduncle

65.3 Surgical Approaches

In general, surgical approaches to posterior fossa lesion in paediatric patients include posterior midline, suboccipital paramedian, lateral and far-lateral, and anterior approaches. Optimal approach is the one tailored to the specific case, based on aims of the surgery, tumour location, size, consistency, and shape. Other factors such as patient's body habitus and surgeon's preferences also dictate the choice operative approach. An overview of surgical approaches is provided in Fig. 65.4.

65.3.1 Midline Suboccipital Approaches

Telovelar approach. It uses natural corridors, the cerebellomedullary and tonsillouvular fissures, to gain access to the fourth ventricle. Given that most lesions located in and around the fourth ventricle are supplied by the branches of the PICA, this approach allows their early identification and control of the tumour blood supply. Furthermore, this anatomically based approach facilitates orientation and so minimises risks resulting from vascular and neural injury, especially the cerebellofugal pathways such as the superior cerebellar peduncles.

Transvermian approach. This is typically carried out by transecting the inferior one third of the vermis. It is used for large lesions occupying the rostral fourth ventricle without extension to the lateral recesses. However, the surgical exposure is limited to midline lesions and the vermis must be divided, leading to vermis split syndrome manifested by nystagmus, gait disturbance, head/neck oscillation, and truncal ataxia. In addition, paediatric patients are prone to develop cerebellar mutism with this approach, probably by surgeons losing their orientation and damaging cerebellar nuclei, especially the dentate nucleus and cerebellar centrifugal pathways. For these reasons, we do not use this approach and have very comfortably managed all fourth-ventricular lesions via the telovelar approach, both in children and in adults.

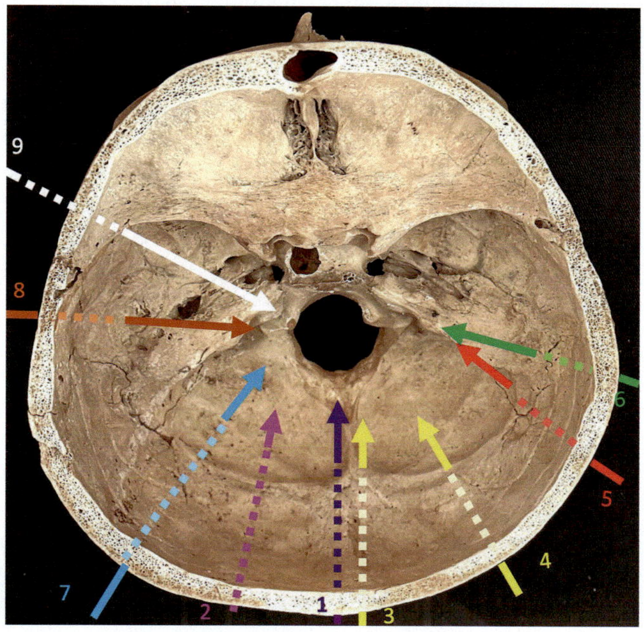

Fig. 65.4 Approached to the posterior fossa. 1. Midline telovelar/transvermian approach. 2. Direct transquamous to target a cerebellar hemisphere lesion. 3.Midline infratentorial supracerebellar approach. 4. Paramedian supracerebellar infratentorial approach. 5. Retrosigmoid approach. 6. Transpetrosal approach. 7. Far-lateral (transcondylar) approach. 8. Anterior transtentorial (subtemporal) approach. 9. Transsylvian pre-temporal/subtemporal transpetrosal

65.3.2 Supracerebellar Infratentorial Approach (SCIT)

This approach (via midline or paramedian route) exploits the natural supracerebellar infratentorial space to access lesions in the tentorial surface of the cerebellum (especially those located anteriorly), the pineal region, posterolateral mesencephalon and diencephalon, posterior third ventricle, and mesial temporal and occipital lobes. It is important to study the relationship between the tumour and the vein of Galen as well as other surrounding vascular structures in preoperative images as the anatomy could be significantly distorted. The narrow and deep operative corridor of this approach presents certain technical challenges, some of which could be mitigated by the sitting position. Lateral visualisation from this approach is limited by the angle of the tentorium, and the inferior visualisation obstructed by the apex of the vermian culmen. Another drawback of this approach via the midline route is invariable sacrifice of the midline bridging vermian veins as well as the occipital venous sinus and, in some cases, important emissary veins around the torcular. The midline approach also exposes and, quite unnecessarily, puts at risk both transverse sinuses and the torcular. The advantages of paramedian supracerebellar infratentorial approach over its midline counterpart include: (1) manipulation of one cerebellar hemisphere only, and (2) improved trajectory to the inferior pole of the tumour with the caudally sloping lateral cerebellum. We therefore generally prefer the paramedian route. The higher-positioned and non-dominant (usually the left) transverse sinus dictates the side of the approach. However, other factors, such as the location and direction of extensions of the target lesion, need also to be considered. Retraction sutures can be anchored to the dural flap and used to superiorly retract the transverse sinus together with the posterior tentorium (Fig. 65.5). However, this needs to be performed judiciously as this manoeuvre can impair venous drainage (causing congestion of the operative field) and result in thrombosis of the sinus. Of course, there are cases where the midline approach is appropriate (Figs. 65.5 and 65.6). If a lesion extends out of the posterior fossa, the tentorium can be incised to gain comfortable access to the lesion (Fig. 65.1).

65.3.3 Paramedian Approaches to Cerebellum

This approach can be applied to reach unilateral cerebellar lesions. We prefer these to midline approaches as they are simpler, faster, and result in less compromise to venous outflow (by not sacrificing the occipital and marginal venous sinuses).

65.3.4 Lateral (Upper and Lower) Approaches

These approaches have been developed for resection of pathologies in the CPA, with the retrosigmoid approach being the most commonly used. The versatile exposure of this lateral approach can reach almost all lesions within the CPA and ventrolateral region of the brainstem, and its adaptability allows combination with various transpetrosal approaches, such as retrolabyrinthine, translabyrinthine, and transcochlear approaches, or even the midline suboccipital approach. The first key step of this approach is identification of the transverse sinus, with the incision usually marked 1/3 above

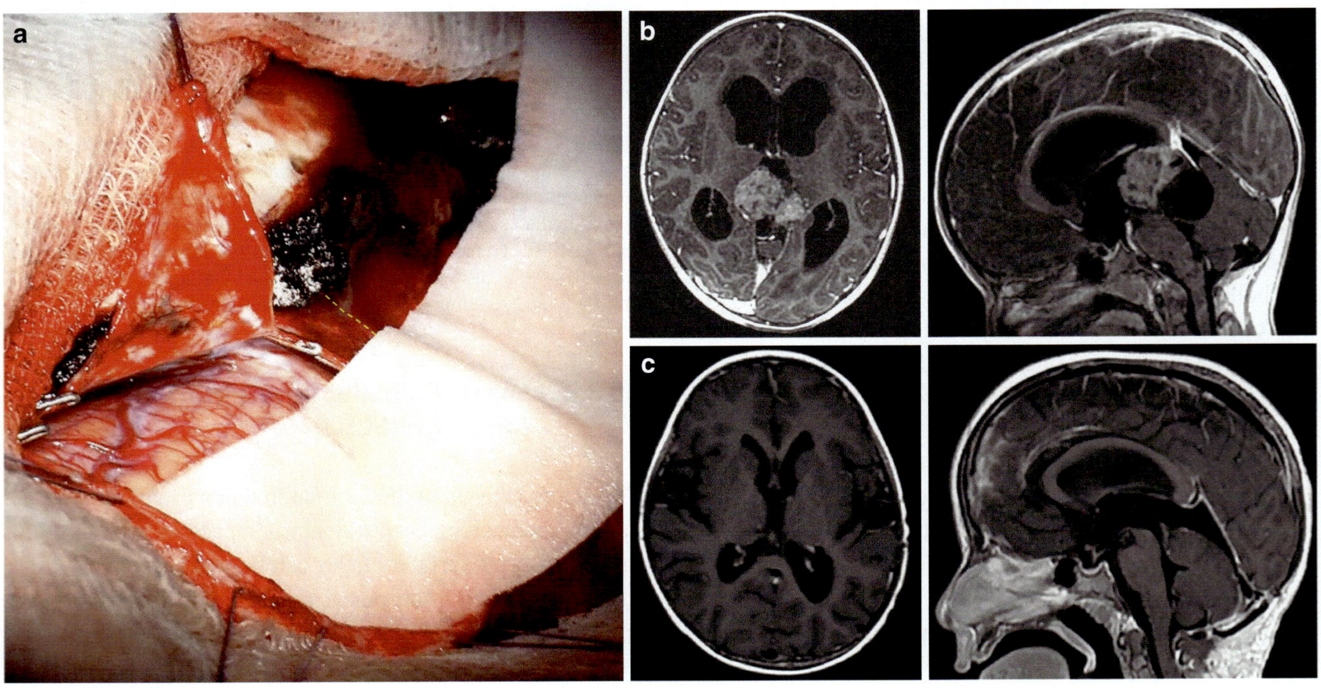

Fig. 65.5 Midline suboccipital craniotomy with unilateral supracerebellar infratentorial approach (SCIT). This pineoablastoma in the 4-year old patient was extending laterally, more so on the right. Therefore, a contralateral SCIT was chosen as the main approach. However, to be able to reach all corners of the tumour bed, if required, midline suboccipital craniotomy was fashioned with the dural flap raised only on the left side. In this case, the right dural window was not required. This strategy allows maximum flexibility while preserving normal anatomy; in this case suboccipital sinus is an important venous draining channel of the posterior fossa. (**a**) Intra-operative image. Interrupted yellow line marks the midline. The dural flap is reflected over the left transverse sinus exposing the junction of the suboccipital and tentorial cerebellar surface. (**b**) Preoperative MRI (T1 + C). (**c**) Post-operative MRI (T1 + C) showed no residual tumour

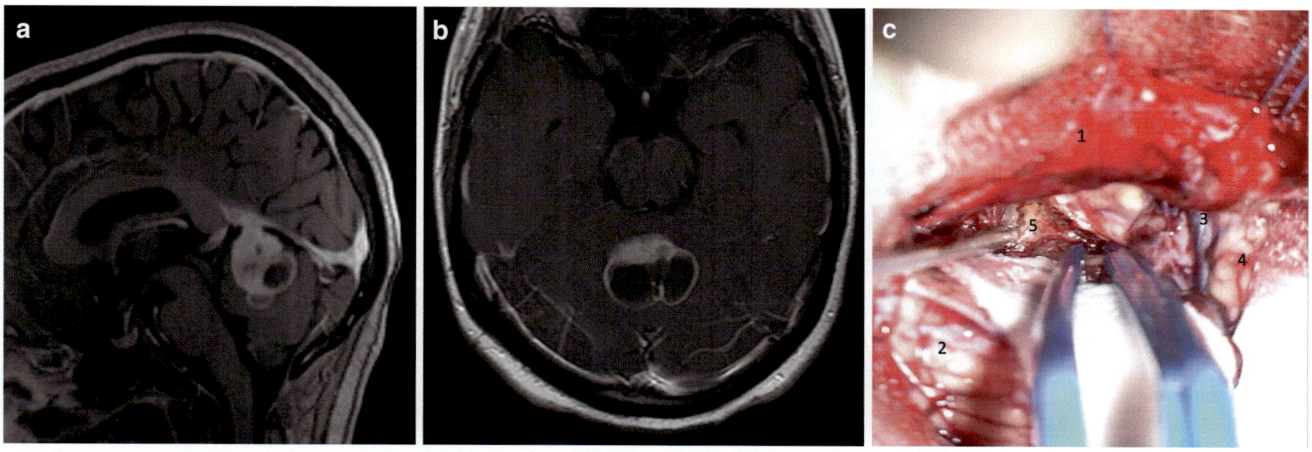

Fig. 65.6 Midline supracerebellar infratentorial approach to a midline anterior vermian pilocytic astrocytoma. Preoperative MRI with contrast sagittal (**a**) and axial (**b**). Intra-operative image (**c**). 1. Dural flap. 2. Left cerebellar hemisphere. 3. Inferior vermian vein. 4. Vermis. 5. Tumour

and 2/3 below it. The initial burr hole should be placed with caution and ideally guided by neuronavigation, as the exact location of the transverse and sigmoid sinuses could be variable. Instead of burr holes, we fashion a ~3 mm trench perpendicular to the sinus, starting inferiorly to the sinus and working our way to the sinus. We use a combination of a high-speed drill and Kerrison's rongeur. The craniotomy is carried out inferior to the transverse sinus and posterior to the sigmoid sinus or it can cross the transverse sinus, if required. The roof of the sigmoid (and the transverse, if required) sinus may be further 'egg-shelled" and removed after elevation of the bone flap. Exposing the sinus is the cardinal step to expand the operative corridor through this approach. The retraction sutures along the dura of sigmoid sinus allow its lateral and anterior mobilisation, expanding the lateral aspect of the operative corridor, while reducing the retraction on the cerebellar hemisphere. Again, caution needs to be exercised to minimise the risk of sinus thrombosis.

The upper trajectory of this approach allows the surgeon to reach lesions superior and posterior to the CN VII/VIII complex (supralateral cerebellar approach) and the lower trajectory allows resection of lesions inferior and anterior to the CN VII/VIII complex (infralateral cerebellar or infrafloccular approach). In the supralateral route, identification of the petro-tentorial junction is a crucial step to avoid unintentional exposure/tearing of the supracerebellar bridging veins and controlling the retraction vector to prevent damage to the CN VII/VIII complex, especially its cochlear component. The retromastoid supracerebellar routes can also be used to expose the tentorium, allowing access to the ventrolateral petroclival area, posterior basal temporal lobe, and floor of the middle fossa (detailed discussion of such approaches is beyond the scope of this chapter). If the craniotomy is extended caudally, this route may also be suitable to lesions around the jugular foramen. In some cases of extensive tumours, various combined approaches can be used [1]. Strategic movements of the suction device allow dynamic retraction for selective exposure of the surgical corridor, obviating the need to employ fixed rigid brain retractors. Together with generous sharp dissection of arachnoid bands/membranes within the surgical field, the dynamic retraction minimises the risk of cerebellar and CN injury.

65.3.5 Far-lateral Approach

The far-lateral suboccipital approach was designed to expose lesions at the ventral and ventrolateral craniovertebral junction extending from the jugular bulb to anterolateral aspect of the upper cervical spine. There are variants of this approach including supra-, para-, and trans-condylar approaches to tackle different pathologies in this region such as neoplastic (e.g. meningiomas or lower CN schwannomas) or vascular lesions (e.g. ventral aneurysms of the distal vertebral arteries and anterior lower brainstem cavernous malformations). The far-lateral transcondylar approach entails a suboccipital craniotomy posterior to the sigmoid sinus and C1 hemilaminectomy followed by partial resection of the ipsilateral occipital condyle. In practice, this is rarely required.

65.3.6 Transtentorial Approaches

We use anterior and posterior transtentorial approaches to reach lesion around the anterior and posterior tentorial incisura, respectively. To access the anterior and lateral incisura, we typically use either the transsylvian pretemporal or the subtemporal approaches.

The transsylvian pretemporal approach is for more anterior (Fig. 65.7) and the latter for more laterally located lesions (Fig. 65.2). In addition to dividing the tentorial edge,

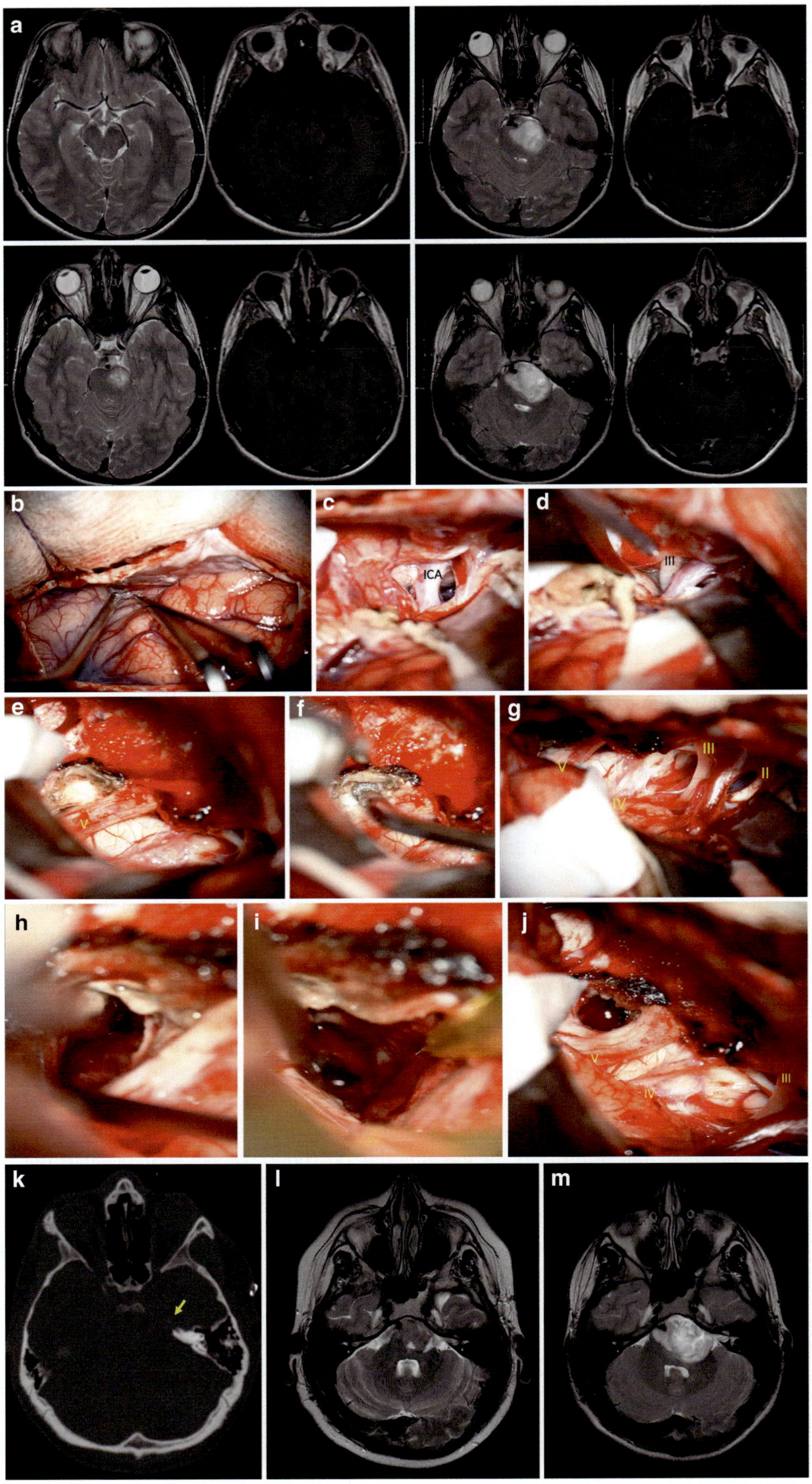

Fig. 65.7 Trans-sylvian pre-subtemporal approach. 16-year old male woke up with severe headache associated with photophobia; later that day developed dysarthria and right hemiparesis. He was alert and orientated with left trochlear and abducens palsy and right upper motor neuron weakness of face and IX to XI nerve-supplied muscle weakness, including poor gag reflex. There was right upper extremity monoplegia, severe right lower extremity hemiparesis, and grossly intact sensation. The aim of surgery was to obtain tissue for diagnosis and decompress brainstem tracts and nerves. The secondary objective was to remove as much of the tumour volume as possible without compromising patient's functional outcome. The diagnosis came back as diffuse midline glioma WHO grade IV with H3K27 mutation. The patient's clinical condition has improved a great deal following surgery and he was able to mobilise independently with only mild hemiparesis. He underwent chemo-radiotherapy. (**a**) Pre-op scans. (**b–d**) Left subfrontal approach with splitting of the Sylvian fissure and exposing the ICA and oculomotor nerve. (**e**, **f**) Division of the superior petrosal sinus and drilling of the petrous apex. (**g**) Wider anatomical context. (**h–j**) Working corridor underneath the trigeminal nerve. (**k**) Post-operative CT showing small, tailored anterior petrusectomy. (**l**) Post-operative T2 MRI showing considerable decrease of volume of the tumour following surgery and chemo-radiotherapy when compared to the preoperative scan (**m**). Nevertheless, the patient succumbed to the disease just over a year after surgery

this approach is often supplemented by a degree of anterior petrusectomy. The transsylvian pretemporal approach also allows a more caudal reach.

We use the posterior transtentorial approach to access pineal region lesions when these are located primarily above the great veins or extend caudally, deep into the cerebello-mesencephalic fissure (Fig. 65.3). In this case, great deal of retraction would be required to access the most caudal part of the tumour. Moreover, a great deal of finesse is required to dissect such lesions off the superior cerebellar peduncles, tectum, and numerous small perforators and venous structures in the quadrigeminal cistern if significant neurological injury is to be prevented. Generally, the posterior approach is interhemispheric or suboccipital. We rarely use the latter as the subhemispheric (suboccipital) approach is largely equivalent to the supracerebellar infratentorial approach (SCIT) [2], which we would use in preference to the subhemispheric approach. In some rare cases, however, the transverse sinus can be positioned so caudally that SCIT would be difficult and the suboccipital transtentorial approach is then more appropriate.

65.3.7 Anterior Approaches (Transnasal or Transoral)

The aforementioned lateral/far-lateral approaches are preferred for intradural lesions. The anterior routes, such as the transnasal endoscopic and transoral approaches, are reserved primarily for extradural pathologies, such as chordomas.

In summary, selection of a surgical approach should aim for minimally invasive exposure but maximally safe resection. Optimum positioning will facilitate comfortable working conditions for the surgeon. The surgeons should aim to perform the most effective and safest surgeries, not 'spectacular' surgeries.

65.4 Surgical Pearls

- The surgical plan should not only be minimally invasive but also, and more importantly, maximally safe.
- Cerebellar mutism. Cerebellar mutism results from impairment of the output pathways of the cerebellum. At surgery, the most commonly injured cerebellar output pathways are those contained in the superior cerebellar peduncles, forming the lateral and dorsal walls of the upper fourth ventricle.
- Cranial nerves injury: When exposing CPA tumour, retraction of the cerebellum in a purely medial direction must be avoided since the vector of retractor will be parallel to the sensitive cochlear nerve fibres, increasing the risk of hearing loss.
- Good closure begins at the opening. A carefully planned incision, sharp dissection, meticulous dural opening with an adequate margin, and zero or minimal coagulation of the dura, all contribute to a good closure. Post-operative CSF leak or fistulae is a significant risk after resection of posterior fossa tumour. To minimise these, a watertight dural closure is mandatory. Primary closure is preferred over free-flap auto- or allografts.
- Always think of air embolism, especially when the head is elevated above the level of the heart and when planned or accidental opening of veins and venous sinuses occurs. Never forget emissary veins when doing bone work. Preoperative oesophageal doppler to rule out presence of cardiac right-to-left shunt is mandatory.
- Aseptic meningitis, arachnoid adhesions, and post-operative hydrocephalus: To minimise these complications, spillage of blood into the arachnoid space should be minimised during the opening, resection, and closure by covering the intradural field with lintine strips.
- Retraction injury: The best retractor is the gravity. Appropriate head positioning, arachnoid dissection, and CSF egress all facilitate opening and maintenance of the surgical corridor with no or minimal retraction. When applying fixed retraction, this should be used more as a "holder" or "protector" rather than a traction instrument. If prolonged retraction is absolutely required, transient relaxation will facilitate reperfusion and reduce the risk of ischaemia.

65.5 Discussion

65.5.1 Surgery vs. Natural History and Alternative Treatments

The natural history and alternative treatment options should always be taken into consideration when developing a management plan. For example, in the case of most hemispheric cerebellar PA, surgery can be curative and total resection ought to be the goal of surgery. However, in a more eloquent location, when the interface of tumour and brain is less clear and firm, partial resection even in the case of PA is preferable as the tumour can usually be at least partially controlled with chemotherapy and many PAs are likely to cease growing after puberty [3]. The effectiveness and availability of chemo- and radiotherapy could influence how aggressive the tumour resection needs to be around the critical neural structures of the posterior fossa. Even in high and average risk groups of paediatric medulloblastoma patients, the chemo- and radiotherapy regimens have improved the 5-year overall and event-free survival to greater than 80% and resulted in 70–75% cure rate in children aged 3 years or older [4].

65.5.2 Vascular Principles

Some tumours, such as choroid plexus papilloma and haemangioblastoma, as well as some meningiomas, ependymomas, and medulloblastomas, may have very high blood volume through flow (especially when related to patient's circulating blood volume) and it is helpful to employ principles and tricks used in surgery for AVMs, including consideration of preoperative embolisation, early identification and control of feeding arteries, and preservation of venous outflow (it is better to send blood back to the circulation rather than to the collection bottle). The use of temporary clips, both at the arterial and venous ends, may be helpful.

65.5.3 Surgical Adjuncts

Electrophysiology is used for mapping and monitoring of ascending and descending projection fibres and CNs (Fig. 65.8). It can not only be informative of the resection limits, but also facilitate early identification of nerves, their nuclei, and functional pathways. They are also useful in guiding the resection of lesions that involve dynamic retraction, such as auditory-evoked potentials during surgery in the CPA.

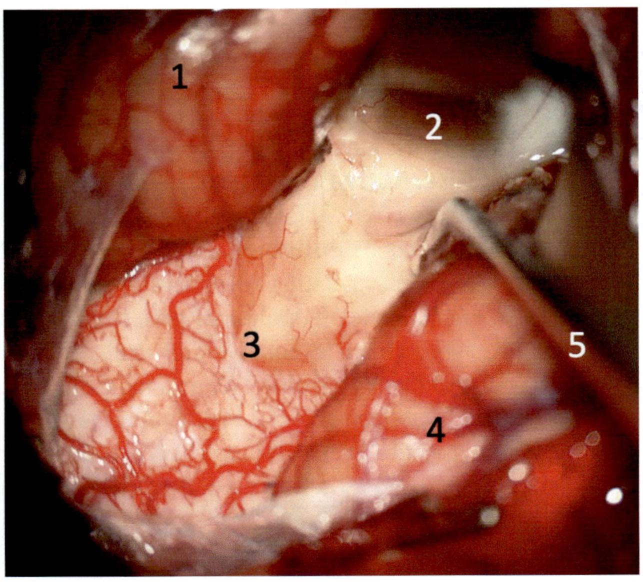

Fig. 65.8 The use of neurophysiology is essential for maximising safety and the degree of resection of tumour in and around brainstem. 1. Left tonsil, 2. Brainstem tumour covered by ependyma bulging into the fourth ventricle, 3. Obex, 4. Right tonsil, 5. Concentric bipolar stimulator

Intraoperative MRI can be useful to evaluate residual and its relationship to the surrounding anatomical structures and it may help avoiding additional general anaesthesia for post-operative imaging, especially in younger paediatric patients.

Endoscope and curved instruments can also be applied to visualise the regions where microscopic vision is blocked by neural structures, enabling the surgeon to reach distant parts of the tumour while avoiding direct (e.g. retraction) and indirect (e.g. division of a vein) injury to the brain (Fig. 65.9).

65.5.4 Minimally Invasive and Maximally Safe Approaches

The knowledge of anatomy allows the tailoring of optimum approaches to each individual case. Although midline craniotomy is commonly used for hemispheric cerebellar lesions (such as PA), unilateral craniotomy is usually all that is required. It is more direct, smaller approach and it does not involve division of the suboccipital sinus. Similarly, most pineal region tumours can be approached from a unilateral approach, thus avoiding division of the suboccipital sinus and peritorcular emissary veins. Sometimes, when a tumour extends laterally on either side, it may be prudent to keep an option of approach via both, the left and right supracerebellar approaches. In these cases, we chose to carry out midline incision and midline suboccipital craniotomy, but open dura only on one side. The other side is used if it is not feasible to reach the tumour safely from one approach (Fig. 65.6). The telovelar approach is probably the best example of utilising natural anatomical corridors to approach posterior fossa lesions. Even more important than how minimally invasive is the approach is to make sure that it is maximally safe. That said, these two goals are not and should not be thought as mutually exclusive.

65.5.5 *En Bloc* Resection Versus Piecemeal Resection

The surgical technique with either *en bloc* or piecemeal resection (with internal debulking) depends on the presumed pathology of the tumour and the pre- and intraoperative assessment of the tumoural consistency. As a rule, the vast majority of tumours are resected piecemeal, following the principle of devascularise, debulk, and dissect off neurovascular structures. However, in highly vascular tumours, such as haemangioblastoma and choroid plexus papilloma, piecemeal resection is hazardous and *en bloc* resection should be carried out (Fig. 65.10).

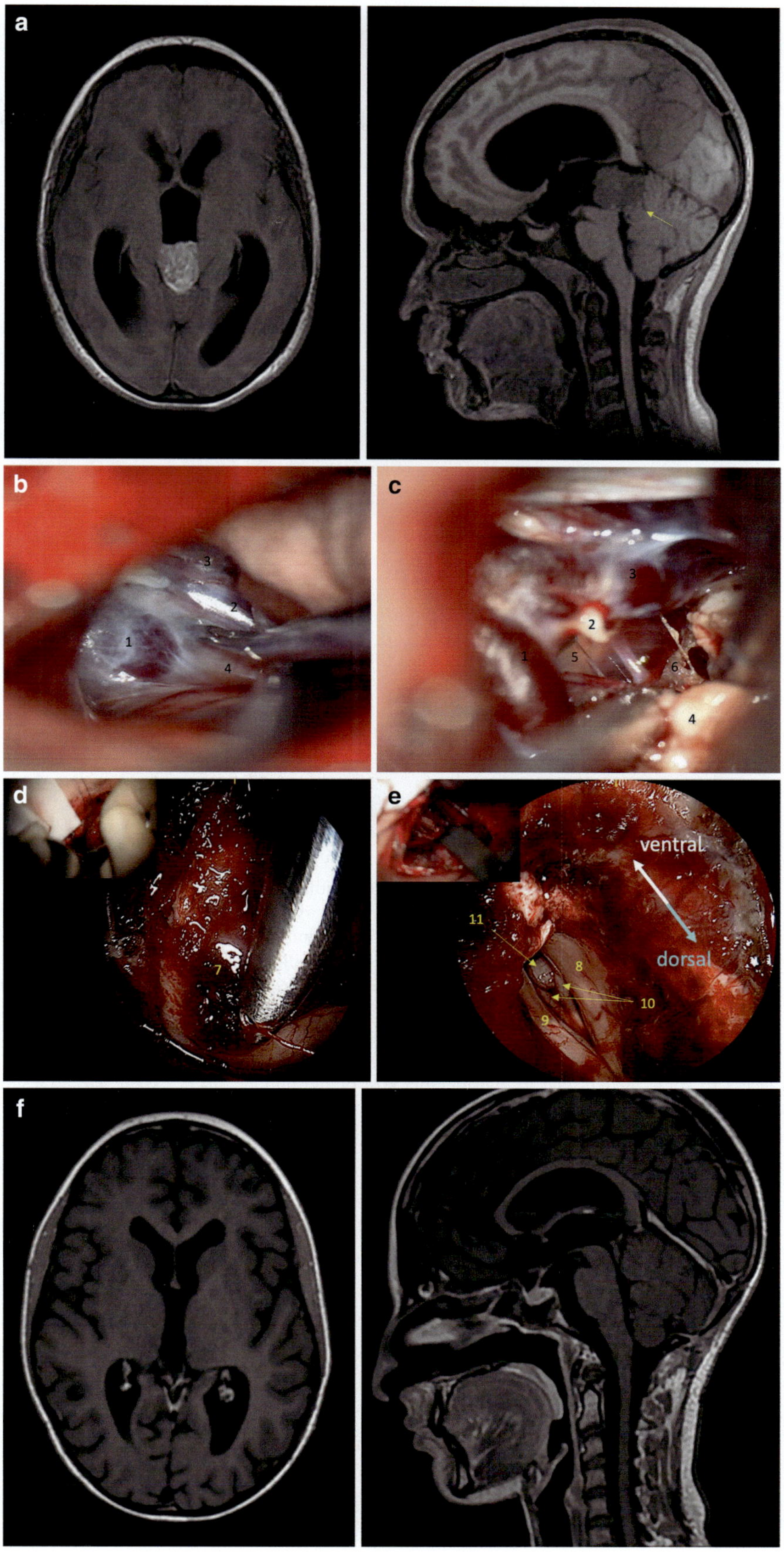

Fig. 65.9 Supracerebellar infratentorial approach with endoscopic assistance to resect a pineal tumour. (**a**) 12-year old girl developed progressive headaches over 9 months. More recently, she developed 'blurred vision'. Eventually, she suffered a seizure and imaging revealed hydrocephalus secondary to a pineal tumour. Despite the low-laying tumour, we judged that with endoscopic assistance we will be able to reach all aspects of the tumour safely, especially as the tectum has been reflected posteriorly. As a result, we will not be working 'over' or 'through' the tectum, but rather 'above' it. An angled endoscope and instruments will facilitate the resection, especially its final stages. (**b**) Arachnoid dissection posteriorly to the great veins. (**c**) Tumour resection. (**d**) Resection of remnants of the tumour with a 45° endoscope and curved suction. (**e**) View of the third ventricle at the end of resection. (**f**) Post-operative MRI with contrast showing complete resection of tumour and much more obvious tectal plate (yellow arrow). 1. Left internal cerebral vein (ICV). 2. Precentral vein. 3. Right internal cerebral vein (ICV). 4. Tumour. 5. Upper layer of tela choidea (= upper wall of the velum interpositum). This is attached to striae medullaris thalami of the medial wall of the thalami anteriorly and the habenular commissure and the pineal gland posteriorly. 6. Lower layer of tela choidea (= upper wall of the velum interpositum) with choroid plexus fused with the capsule of the tumour. This is attached to the body of the fornix (seen as a white structure through the tela choroidea) and hippocampal commissure. 7. Remnant of the tumour on the medial inferior wall of the thalamus, which was subsequently removed with gentle suction under visual control via a 45° endoscope. 8. Medial wall of the left thalamus. 9. Medial wall of the right thalamus. 10. Columns of the fornix. 11. Anterior commissure

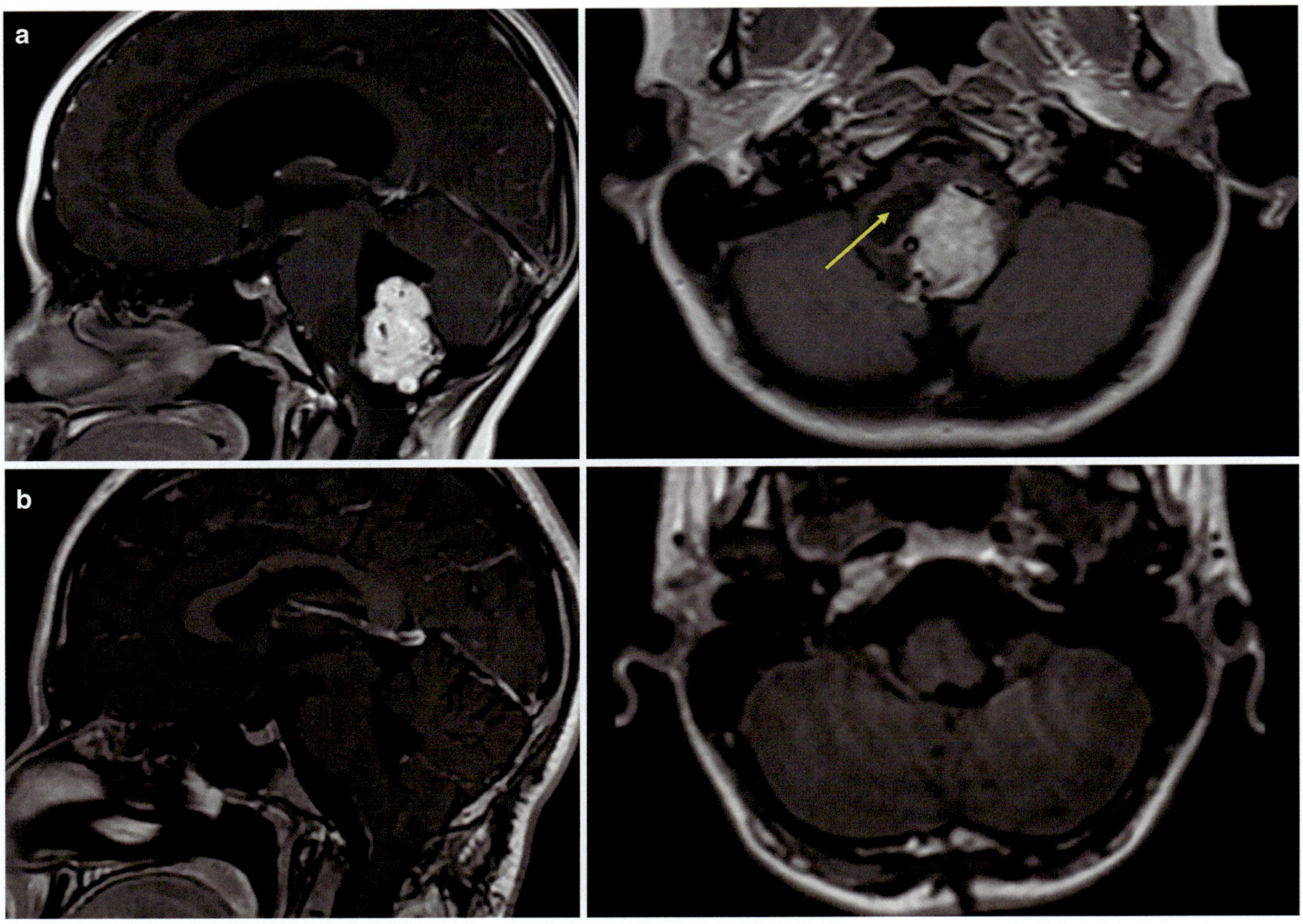

Fig. 65.10 Haemangioblastoma. 13-year old boy with two-year history of morning nausea and vomiting, 3–4-week history of pulsatile tinnitus, left-sided diplopia, and transient right upper limb weakness. Examination revealed left trochlear nerve palsy and papilloedema on fundoscopy. MRI head (**a**) showed avidly enhancing midline posterior fossa tumour with extensive flow voids, cyst (yellow arrow), and associated hydrocephalus. Spine MRI was negative for drop metastases. The challenges that need to be anticipated in this case include difficult access to some feeders due to considerable size of the tumour, firm consistency, high-vascularity (debulking is not really an option), and limited working space around the tumour due to presence of highly eloquent neurovascular structures. Preoperative embolisation, but this was not feasible due to numerous small feeders coming from both PICAs (especially the left). Despite careful preoperative planning and temporary clipping of the left PICA, 1 L of blood loss occurred during surgery. However, the patient had excellent functional outcome with no new post-operative and recovery of preoperative neurological deficits. There was no residual on post-operative imaging (**b**)

65.5.6 Staged Surgery

Excessive blood loss, long duration, and the need for updated imaging and/or functional assessment are some reasons why an operation may have to be staged. Usually, the same access corridor is used in the second stage. As illustrated in the case of a large teratoma (Fig. 65.11) or a large ependymoma [1], an operation may be staged to functionally assess patients and employ a separate approach using a secondary corridor to reach the residual tumour more gently and safely, leading to less damage and better surgical outcome for the patients.

65.6 Conclusion

Management of posterior fossa tumours in paediatric patients requires multi-disciplinary team collaboration, meticulous preoperative work up, and surgical planning. Understanding of the pathologies and relevant neuroanatomy is fundamental to the selection of an approach for minimally invasive exposure and maximally safe resection to achieve optimal outcomes and avoid complications. This requires knowledge, skill, focus, tenacity, and last but not the least, kindness.

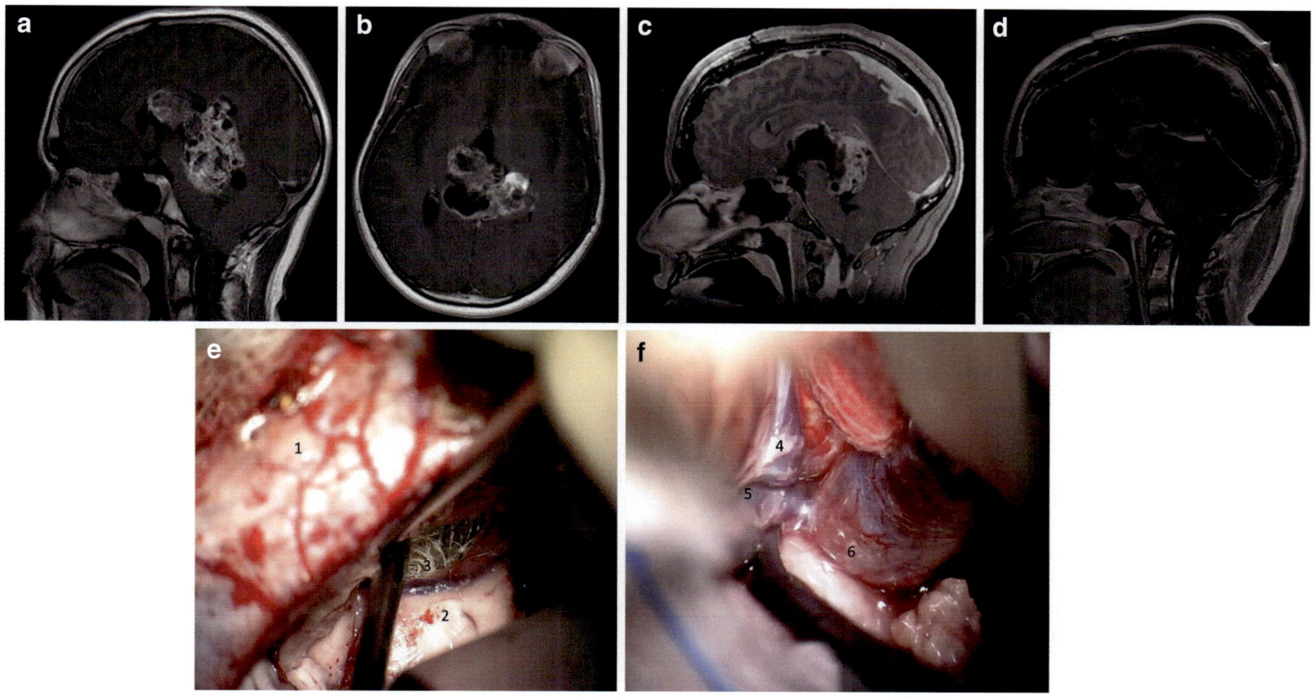

Fig. 65.11 Staged approach to a rapidly decompensating pineal region teratoma. 16-year old male patient rapidly deteriorated and had to be operated on emergency basis. Two separate approaches were used, interhemispheric transcallosal subforniceal (**e**) and interhemispheric transtentorial (**f**) in two separate operative sessions with 1 day in between. An MRI was obtained in between (**c**). (**a**, **b**). Preoperative MRI. (**c**) MRI between the two stages of the resection. (**d**) Final post-operative result. 1. Dural flap reflected over the frontal section of the superior sagittal sinus. 2. Right thalamus. 3. Hair and keratin seen through the foramen of Monroe. 4. Posterior pericallosal vein. 5. Anterior calcarine vein. 6. Vein of Galen

References

1. Budohoski KP, Guilfoyle MR, Barone DG, et al. Radical, staged approach to extensive posterior fossa pediatric ependymoma: 3-dimensional operative video. Oper Neurosurg (Hagerstown). 2018;14(6):705.
2. Hart MG, Santarius T, Kirollos RW. How I do it—pineal surgery: supracerebellar infratentorial versus occipital transtentorial. Acta Neurochir. 2013;155(3):463–7.
3. Greuter L, Guzman R, Soleman. Typical pediatric brain tumors occurring in adults-differences in management and outcome. Biomedicines. 2021;9(4):356.
4. Gajjar AJ, Robinson GW. Medulloblastoma-translating discoveries from the bench to the bedside. Nat Rev Clin Oncol. 2014;11(12):714–22.

Part VI

Spinal Tumors and Peripheral Nerves

Spinal Cord Vascularity

66

Victor Hugo Pérez and Juan E. Small

Learning Objectives
After reviewing this chapter, the reader will be able to perform the following features:

- Identify the major arterial supply to the anterior spinal cord.
- Identify the major arterial supply to the dorsal spinal cord.
- Compare and contrast the differences and morphometric characteristics of the fine and complex anatomy of the spinal anatomy.

66.1 Introduction

A thorough understanding of the anatomy of the spinal cord vascularity is of crucial importance for both surgical and endovascular treatment. The details and intricacies of the cervical spinal artery, the dominant anterior thoracic artery (DATA), the great anterior radiculomedullary artery, and the great posterior radiculomedullary artery are therefore of great interest to the specialized personnel endeavoring in this field.

66.2 Anatomical Highlights

- Anterior spinal artery and a small branch from the great anterior radiculomedullary artery supply the anterosuperior two-thirds of the spinal cord.
- The artery of Adamkiewicz (anterior radiculomedullary artery) arises from the left side of the aorta between T8 and T12.
- The midline sulcus is an important anatomical landmark for surgical orientation.
- Importance of dentate ligament for bilateral fixation if the spinal cord and boundary between anterior and posterior is half of the canal.

66.3 History and Background

Centuries prior, relatively easy access and identification of arterial spinal cord structures led to several important discoveries regarding the major radiculomedullary branches supplying the spinal cord. Although today, the majority of healthcare professionals have a good understanding of major branches such as the artery of Adamkiewicz as well as the artery of cervical enlargement, other branches such as the DATA or the great posterior radiculomedullary artery have received much less attention and are not well known [1–5].

66.4 Vascular Supply of the Spinal Cord

The study of the vascular anatomy of the spinal cord is essential to understand and interpret imaging studies and the surgical and endovascular management of tumors and vascular malformations.

In this study of 50 spinal cords in cadaveric specimens from the Institute of Forensic Science in Mexico City whose average age was 45 years, the spinal cord was removed after performing a longitudinal dissection in the midline from C1 to L5. All cadavers underwent laminectomy, and resection of the spinal cord by cutting the anterior and posterior spinal roots in their portion closest to the intervertebral foramen. Subsequently, the main radicular arteries were identified. With a 5× surgical magnifying lens, the large arteries were injected with red latex and those smaller arteries with diluted industrial paint, subsequently fixed in a 5% formaldehyde solution (Fig. 66.1).

V. H. Pérez (✉)
Forensic Sciences Institute, Autonomous University of Mexico State, TSJ, Mexico City, Mexico

J. E. Small
Lahey Hospital and Medical Center, Burlington, MA, USA
e-mail: Juan.E.small@Lahey.org

I. N. Kanaan, V. Beneš (eds.), *Neuroanatomy Guidance to Successful Neurosurgical Interventions*,
https://doi.org/10.1007/978-3-031-59838-8_66

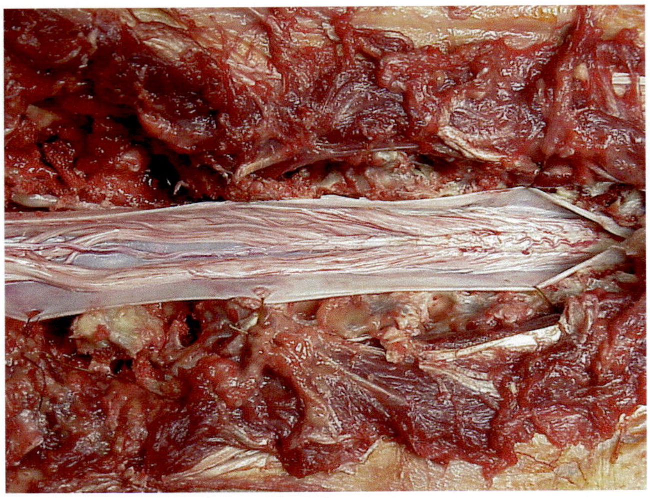

Fig. 66.1 Image of the dissection of the spinal cord after performing laminectomies, the dura has been opened to visualize the cauda equina

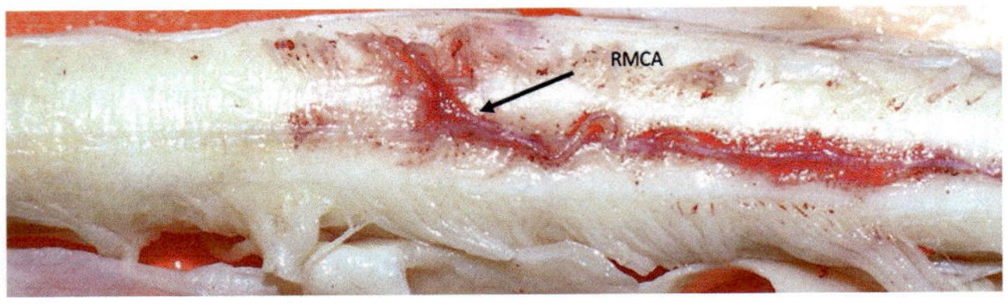

Fig. 66.2 This image shows the cervical radiculo-medullary artery (RMCA) in C5 with a diameter of 730 μm

The vascular supply of the spinal cord is basically made up of three arteries that run longitudinally along it: the anterior spinal artery that has an average diameter of 500 μm and two posterior spinal arteries with an average diameter of 250 μm. The anterior spinal artery originates from the union of two branches of the vertebral artery which join at the level of the decussation of the pyramids, then descend on the anterior surface of the spinal cord in a small groove that forms the anterior median fissure. The anterior spinal artery gives off branches that supply the anterior portion of the medulla oblongata, the hypoglossal nerve, and the anterior two-thirds of the medulla oblongata. It anastomoses with cervical spinal arteries throughout the spinal cord, and at the thoracolumbar level, it joins with the artery of Adamkiewicz. The posterior spinal arteries arise directly from the vertebral arteries or indirectly from the posteroinferior cerebellar arteries and then descend on the posterolateral surface of the spinal cord close to the posterior roots, one on each side. These arteries supply the posterior third of the spinal cord.

The longitudinal system joins a highly variable series of transverse tributary vessels that enter the medullary canal through the intervertebral foramina, accompanying the nerves and spinal roots called radicular arteries. The anastomoses between the longitudinal vessels and the segmental vessels occur on the surface of the spinal cord. The segmental vessels correspond to approximately 31 vessels that accompany the same number of nerve roots, the vast majority of which do not reach the main arteries of the spinal cord, such as the posterior spinal arteries and the anterior spinal artery. They are called radicular arteries because they only supply the spinal nerves and the sensory ganglion; the radiculopial arteries reach only the pia mater. The radiculomedullary arteries are found in a group of 8–10 arteries that reach the spinal cord and anastomose with the arteries of the longitudinal system, enter the sides of the spinal cord and divide into two branches, one anterior and one posterior, which accompany the respective nerves.

In a study of 50 spinal cords, we found at least one artery in the cervical segment in 44 specimens, the smallest measuring 130 μm and the largest measuring 730 μm. Up to four radiculomedullary cervical arteries were observed in four spinal cords. The site where it was most frequently observed was in the C7 root in a total of 14 arteries, nine on the right side and five on the left side (Fig. 66.2).

66.5 Vascular Supply of the Thoracic Segment of the Spinal Cord

Regarding the anterior thoracic arteries of the spinal cord, we found the DATA which was identified between T1 and T11 in 94% of the specimens, being the most frequent site at the level of T4 on the left side, the average diameter of these arteries was 446 μm with a range of 270 to 759 μm. At least 1 DATA was evident in 94% of spinal cord specimens (47/50). A single DATA was present in 66% of the specimens (33/50). Two DATAs were identified in 26% (13/50). Three DATAs were identified in a single (1/50) specimen (2%), all arising on the left at the level of T3, T7, and T9 (Figs. 66.3, 66.4, and 66.5).

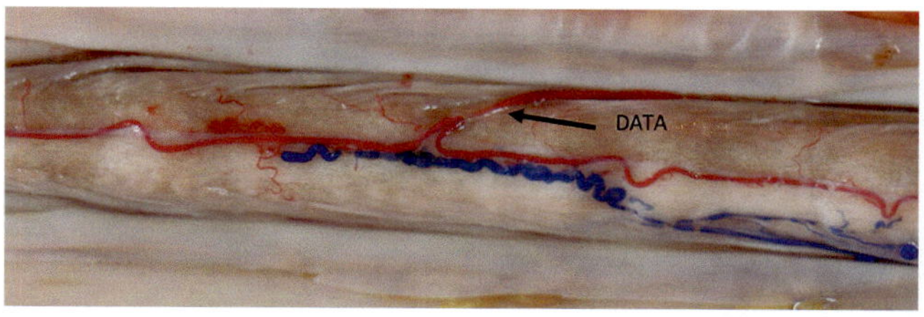

Fig. 66.3 Image of the largest DATA (759 μm) in left T3

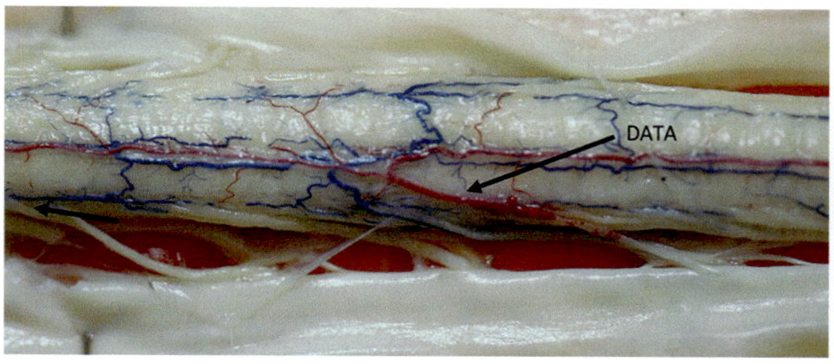

Fig. 66.4 DATA image (500 μm) in right T3

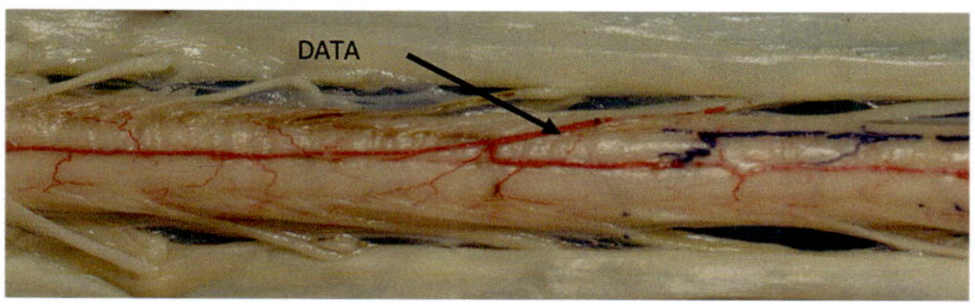

Fig. 66.5 DATA image (460 μm) in left T3. Dominant vessels such as the one shown in this image are easily distinguished from very small caliber radicular arteries

66.6 Adamkiewicz Artery

In relation to the supply of the anterior portion of the lumbosacral segment, the great anterior radiculomedullary artery, also known as the artery of Adamkiewicz, plays a determining role. It is considered the most important artery due to its size and its importance in the supply of the thoracolumbosacral segment as well as its anastomosis with the anterior spinal artery. The artery of Adamkiewicz provides the major supply to the upper and middle thoracic segment of the spinal cord. In our study, the great anterior spinal radiculomedullary artery had a diameter ranging from 0.490 mm to 1.214 mm. It was present on the left side in 45 specimens and on the right side in five specimens. It occurred more frequently on the left side at the level of T9 in 13 specimens and at T8 in 10 specimens. In six specimens, the artery of Adamkiewicz was found to be duplicated (Figs. 66.6, 66.7, 66.8, 66.9, 66.10, and 66.11).

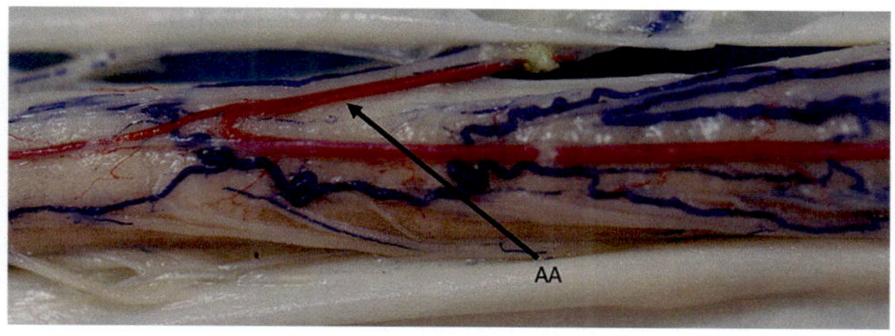

Fig. 66.6 Artery of Adamkiewicz (AA) at left side, T8, diameter: 0.640 mm

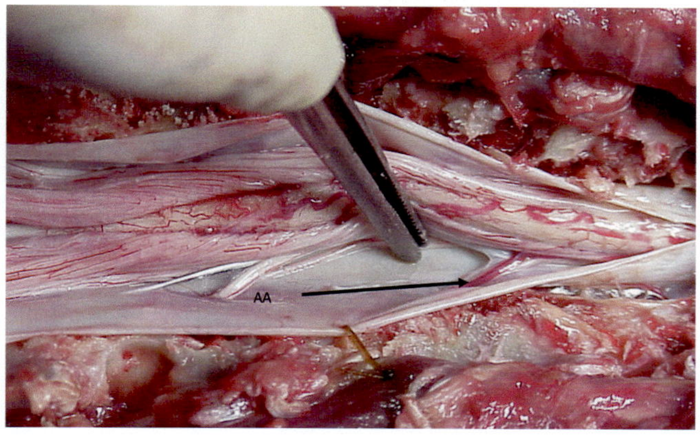

Fig. 66.7 Artery of Adamkiewicz dissected in situ

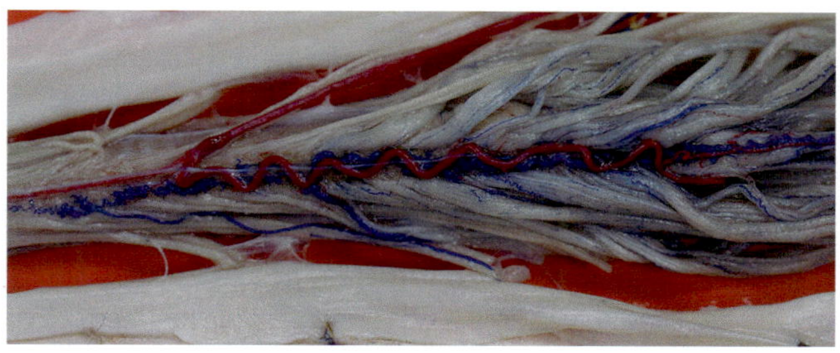

Fig. 66.8 In this image, the Adamkiewicz artery (AA) demonstrates a sinuous course as an anatomical variant. The anterior spinal ligament (AL) that fixes the artery to the spinal cord on its anterior face is clearly observed

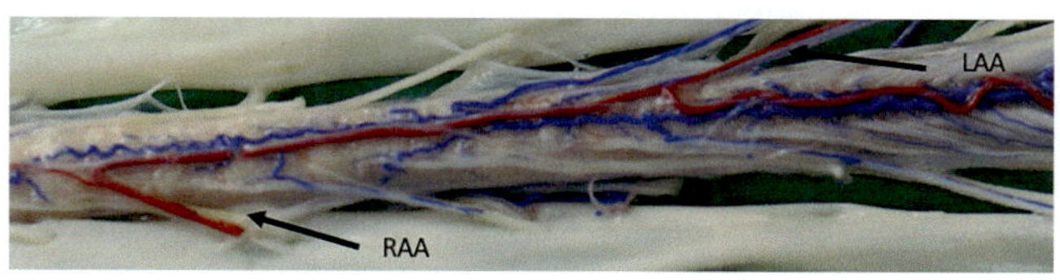

Fig. 66.9 Spinal cord with two arteries of Adamkiewicz: the left (LAA), at T11 with a diameter of 0.47 mm, and the right (RAA), at right T8 with a diameter of 0.74 mm

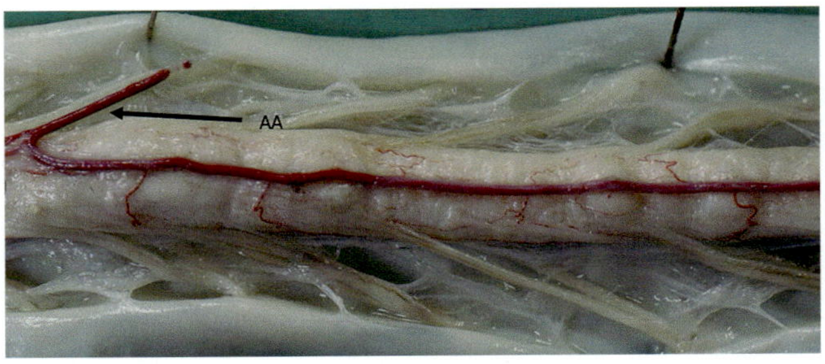

Fig. 66.10 A close-up of the Adamkiewicz artery originating from left T3, the highest level in our study

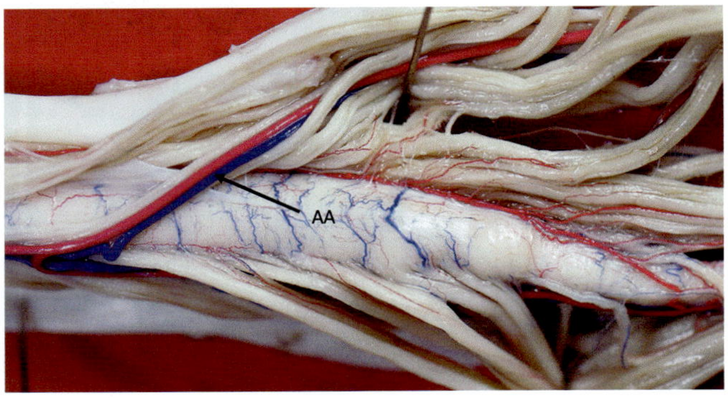

Fig. 66.11 Artery of Adamkiewicz rising from left L2, the lowest level

66.7 Great Posterior Radiculomedullary Artery

The existence of a great posterior radiculomedullary artery has attracted little attention and has been a matter of debate [4]. Its site is more caudal than that of the anterior radicularis magna artery. In our study, a GPRA was identified in 72% of spinal cord specimens (36/50). A unilateral left-sided GPRA was present (13/50). A unilateral right-sided GPRA was present in 22% of the specimens (11/50).

The average GPRA size was 0.44 mm with a range of 0.120–0.678 mm on the left and 0.260–0.635 mm on the right. Although the caliber of the GPRAs did not approximate that of the AKA, the vessels were truly dominant and easily distinguished from the tiny caliber of the radicular arteries, which accompany posterior spinal nerves. In the specimen that demonstrated three right-sided GPRAs, they were seen at T11 (0.371 mm), T12 (0.319 mm), and L1 (0.240 mm) (Fig. 66.12).

66.8 Surgical Pearls

- Proper position of the patient on the surgical table is of paramount importance to improve venous return and avoid related nerve's injuries.

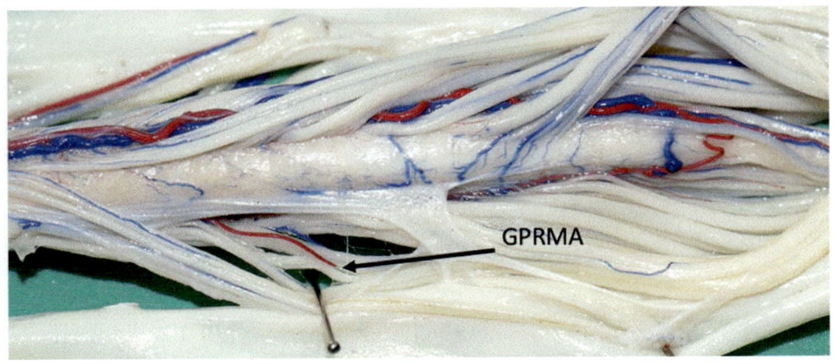

Fig. 66.12 GPRMA at level T12 left with a diameter of 0.120 mm

- Avoid excessive cord manipulation or traction leading to cord ischemia and poor surgical outcome.
- Use intraoperative neurophysiology monitoring (IOM) to prevent complications and enhance surgical outcome.
- Release of dental ligaments facilitates mobility of the spinal cord and exposure of anteriorly located lesions and adjacent vascular structure such as the vertebral artery in FM-meningioma.
- Special respect to the integrity of anterior spinal artery and the artery of Adamkiewicz.
- Minimal use of traction or thermocoagulation to avoid ischemic injury to the vulnerable spinal cord.
- Watertight dura closure to avoid CSF leak and its related complication.

66.9 Conclusion

With respect to the frequency and size of the dominant cervical radiculomedullary arteries, it is important to take into account that most of them are located in the C6 and C7 cervical nerves; therefore, the spine surgeon must take the above into account, since performing spinal surgery at these levels has a greater possibility of injuring these arteries.

The DATA is observed in the majority of people, and this contradicts the notion that the thoracic spinal segment represents an ischemia risk area due to the false belief that it has a very poor blood supply. However, in the studies we have conducted, the DATA is very constant and contributes decisively to the blood supply of this segment. In this study, we have confirmed the existence of the great posterior radiculomedullary artery, which some decades ago was denied by some researchers. The anastomosis of the great posterior radicular artery with the posterior spinal arteries that originate from the bifuracation of the Adamkiewicz artery makes it possible to face and tolerate situations of ischemia in surgical procedures.

References

1. Djindjian R, editor. Spinal angiomas: advances in diagnosis and therapy. Springer-Verlag; 1978. p. 1–17.
2. Gailloud P. The artery of von Haller: a constant anterior radiculome-dullary artery at the upper thoracic level. Neurosurgery. 2013;73:1034–43.
3. Lazorthes G. Blood supply and vascular pathology of the spinal cord. In: Pia HW, Djindjian R, editors. Spinal angiomas. Berlin, Heidelberg: Springer; 1978.
4. Pérez PV, Small JE. Anatomy of the great posterior radiculomedullary artery. AJNR Am J Neuroradiol. 2019;40(12):2010–5.
5. Small JE, Pérez PV. The dominant anterior thoracic artery of the spinal cord. AJNR Am J Neuroradiol. 2023;44:111–4.

Microsurgical Removal of Spinal Tumors

67

Francesco Tomasello, Giada Garufi, and Salvatore Cardali

Learning Objectives
- Diagnosis and compartmentalization of spine tumors.
- Crucial knowledge of neurosurgical anatomy.
- Surgical approaches to spine tumors.
- Preoperative planning.
- Intraoperative tools.
- Functional outcome.

67.1 Introduction

Surgical treatment of spinal tumors is a challenging topic: despite the prevalence of spinal tumors is relatively low in comparison to lesions of the CNS, they represent a complex pathology which is often difficult to manage.

Microsurgical resection remains the primary treatment for most spinal tumors.

Tumors of the spine can be divided into primary and metastatic tumors, with the latter being the most common presentation. The standard classification divides spinal tumors, based on their anatomical location, in relation to the spinal cord into extradural and intradural tumors, and then into extramedullary spinal cord tumors (EMSCTs), and intramedullary spinal cord tumors (IMSCTs). This classification dramatically influences the surgical approach to the lesion in terms of surgical removal, vertebral stability, and final outcome of the patients.

F. Tomasello (✉)
Humanitas Clinical Institute of Catania, Misterbianco, Catania, Italy
e-mail: ftomasel@unime.it

G. Garufi
Department of Neurosurgery, "Papardo" General Hospital, Messina, Italy

S. Cardali
Department of Neurosurgery, "Papardo" General Hospital, University of Messina, Messina, Italy
e-mail: salvatore.cardali@unime.it

Among spinal tumors which occur extradurally, schwannomas (29%), meningiomas (25%), and exophytic ependymomas (13%) are the most common neoplasms, while ependymomas (55%) and astrocytomas (30%) are the most common intradural intramedullary tumors statistically reported [1]. Metastases are the most common extradural extramedullary tumors, and they typically involve bony structures. MRI is considered the state of the art to diagnose intradural spinal tumors whereas in some cases, CT scans, angiography, CSF analyses, and neurophysiological examination could add a valuable help in diagnosis. The preoperative management of these lesions is focused either on the suspected histopathological diagnosis and on the clinical presentation and the anatomical location. Although intramedullary lesions are rarer and harder to manage, a well-planned treatment could lead to an excellent outcome without treatment-related deficits. Technical advances in imaging, neuromonitoring, minimally invasive approaches, and radiotherapy have improved the outcome of intradural spinal tumors. However, the outcome remains poor in malignant intramedullary tumors.

Gross total resection is the gold standard therapy for spinal tumors and may be curative for well-circumscribed lesions (this is particularly true for EMSCTs and low-grade IMSCTs). However, adjuvant radiation and chemotherapy can be indicated for patients that would experience increased morbidity from an iatrogenic damage. Intraoperative tools are systematically used in spinal tumor surgery in order to improve the surgical removal and to increase the neurological outcome. As well as with lesions occurring in eloquent regions of the brain, intraoperative neurophysiological monitoring (IONM) of spinal cord lesions has been proven to be safe and effective: SSEP, MEP, and D-wave monitoring are the most used and useful tools for this kind of surgery. While surgery is the mainstay treatment for many of these lesions, radiation and chemotherapy are of growing importance in recurrent and multifocal disease.

I. N. Kanaan, V. Beneš (eds.), *Neuroanatomy Guidance to Successful Neurosurgical Interventions*, https://doi.org/10.1007/978-3-031-59838-8_67

67.2 Anatomical Highlights

- Anatomical location related to spinal cord and dural plane dictates surgical strategy, pathology, and prognosis.
- The role of dentate ligament section to improve spinal cord mobility.
- Posterior median sulcus is the main landmark for myelotomy.
- Posterolateral sulcus is an adjunctive landmark for intramedullary lateral tumors.
- Pial retraction of myelotomic borders.

67.3 History/Background

67.3.1 EMSCTs

Extramedullary intradural spinal cord tumors are rare variants of SCTs, characterized by their extraaxial location, accounting for about 40–45% of all spinal tumors [2]. The first case of a resected EMSCT in the literature was reported by Sir Victor Alexander Haden Horsley in 1888 which described the resection of a "fibromyxoma" in a 42-year-old patient [3]. Meningiomas are the most common primary extramedullary tumors (25%), most likely to occur in the thoracic spine, in the posterolateral side of the spinal canal and in female patients. Nerve sheath tumors, spinal schwannomas, or neurofibromas account for 10–15% of all the EMSCTs, and they usually originate from the sensory dorsal rootlets: typical schwannomas are intradural (10–15%) but it is not unusual to find "dumbbell" tumors with an intraextradural growth (10%). Dumbbell schwannomas were considered in some series, to be at an higher risk to be malignant, particularly in young patients [4]. Among ependymomas, about 40% occurs within the filum terminalis being mostly myxopapillary, WHO grade I. In filum terminale ependymomas surgery, the role of the tumor capsule plays a fundamental role to reduce the risk of local recurrence: an en bloc resection of both the tumor and the capsule itself is a predictor of the best surgical outcome [5]. Intradural extramedullary tumors typically compress and displace the spinal cord, nerve roots and, occasionally, surrounding structures.

The main goal of EMSCTs tumors is to control the origin and the extent of the lesion: a posterior durotomy could give a direct control of posterior or posterolateral located lesion. For highly vascularized tumors of those that are partly calcified (e.g., meningioma), the debulking could be more challenging and a preoperative surgical planning is mandatory to achieve the GTR without any functional damage (Case 1) [6]. Schwannomas nearly exclusively originate from the posterior sensory root. A T-shaped dural opening with a root-sparing surgery is safe and recommended.

Metastases are the most frequent extradural tumors and most commonly affect the thoracic vertebrae. Goals of surgery are usually palliative and consist of decompression of neural elements and instrumentation for stabilization. The use of radiofrequency thermal ablation (RFA) with vertebroplasty or kyphoplasty represents a minimally invasive, safe, and effective strategy for the palliative treatment of multimetastatic spine disease, particularly in fragile patients (Case 4).

67.3.2 IMSCTs

Intradural spinal cord tumors account for 10–20% of primary CNS tumors, with extramedullary tumors (EMSCT) accounting for up to two-thirds of this group and one-third are intramedullary tumors (IMSCTs). Among them, ependymomas are considered the most common, representing about 40–60% of all intramedullary tumors, where astrocytomas being the second in-line accounting for about 30%. Hemangioblastoma is the third reported lesion, representing 3–8% of IMSCTs [1, 7, 8]. Surgery for IMSCTs has been introduced in 1911 with the Elsberg and Beer's two-stage procedure which revolutionized the management of this surgery. From then, evidence-based guidelines recommend some technical tips to achieve the best surgical outcome with the best functional and neurological outcome. The combination of a posterior myelotomy and the intraoperative monitoring of SSEP, mMEP, and eMEP (d-wave) are considered the best predictors for the postoperative neurological outcome [9]. The myelotomy is the key step in the IMSCT surgery: this is the moment in which surgeons, technicals, and anesthesiologists should discuss every relevant change in IONM variables (amplitude and latency) in order to rule out eventual changes in medication and adjust anesthetic drugs, mean arterial pressure, body temperature, or oxygenation. SSEP is a highly sensitive parameter but it is limited in specificity: reduction of SSEP amplitude by 50% or an increase in latency by 10% of baseline after myelotomy indicates the need of a watchful waiting and a release of any local cord compression; a permanent intraoperative SSEP loss is considered in line with a new postoperative neurological deficit in 50–75% [10]. Different safe entry zones and myelotomy techniques have been proposed over the years: the standard posterior midline myelotomy—between the posterior columns—remains the mainstay for approaching IMSCTs. Mapping the posterior columns might help to identify the physiological midline sulcus for myelotomy in case of asymmetrically filocalized IMSCTs. To avoid the division of posterior large vessels coursing the mid surface of the spinal cord and the somatosensory related deficits, some authors proposed a discontinuous myelotomy technique in order to preserve the integrity of these main vessels (Fig. 67.1) [11]. A posterolateral sulcus (PLS) myelotomy might be feasible in cases of lateralized IMSCTs posterior to the level of the attachment of the dentate ligament (dorsal root entry zone), particularly if the lesion is associated with moderate to severe local pain symptoms (Fig. 67.2) [12].

Anterior or anterolateral approaches, even if very rare, were also described and should be considered in case of ventral intradural tumors with a paramedian anterior myelotomy [13].

The main goals of IMSTCs surgery are achieving the gross total resection with the best postoperative neurological outcome: the identification of tumor margins and the tumor-spinal cord interface should be mandatory in the first stages of the surgery in order to gain the early control of the tumor extension in all the lateral and craniocaudal directions. The decision of looking forward a gross total removal of the lesion should count mainly on two factors: the histology and the presence of a plan of dissection with the spinal cord are considered the most important predictors of neurological outcome in terms of "resectability" and recurrence [14]. This is particularly true for ependymomas and hemangioblastomas, whose benign histology and easy dissection plan encourage a GTR with a good neurological outcome. On the contrary, in case of malignant histologies (astrocytomas, higher-grade gliomas, or ependymomas) or the absence of a clear spinal cord-tumor interface leads to subtotal resections or biopsy in order to avoid neurological impairments (Case 5). Moreover, astrocytic tumors with an unclear cleavage plane were not associated with an improved progression free survival even after a gross total resection [15]. In summary, primary goals for astrocytic tumors, in case of an absent tumor-spinal cord interface, should be based also on patient's involvement, surgeon experience, and intraoperative changes of IONM variables (Figs. 67.3, 67.4, 67.5, 67.6, 67.7, 67.8, 67.9, 67.10, 67.11, 67.12, and 67.13).

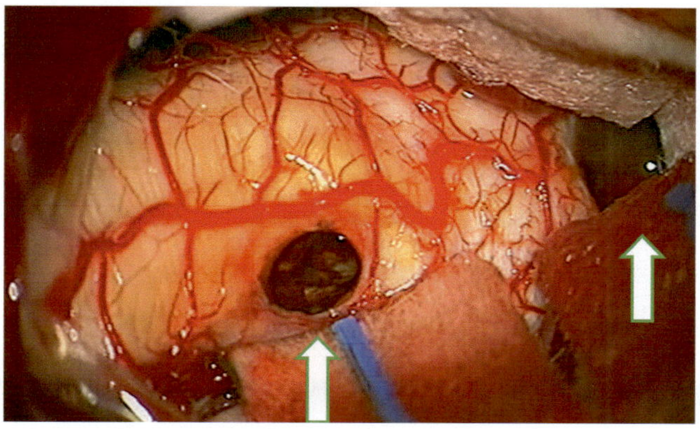

Fig. 67.1 Discontinuous myelotomy, personal case

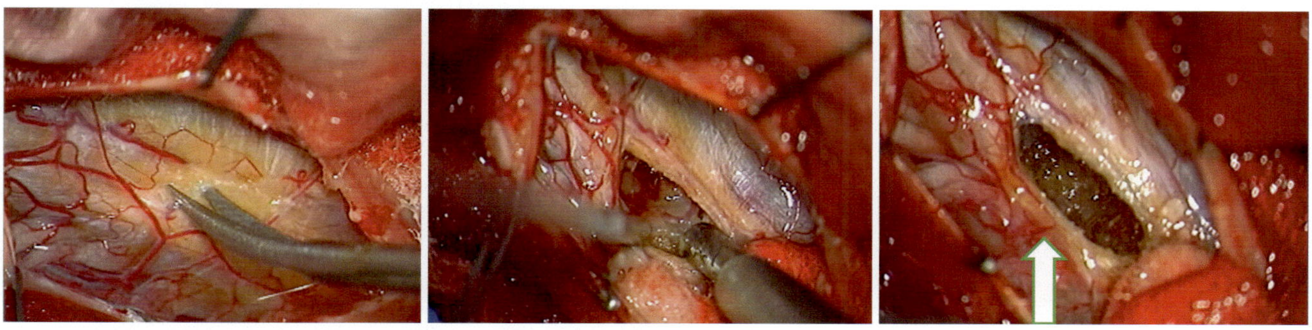

Fig. 67.2 Intraoperative pictures showing a dorsolateral myelotomy, at the root entry zone for lateral tumors

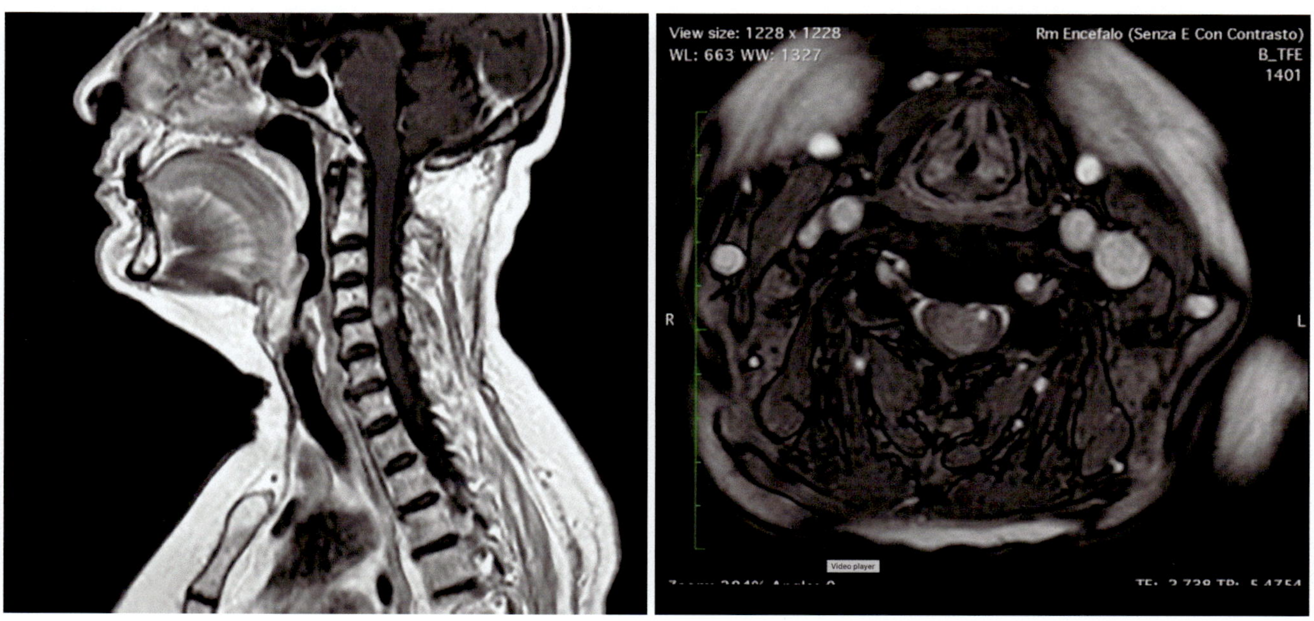

Fig. 67.3 Preoperative MRI showing the intramedullary nature of the lesion and the high intensity after contrast administration

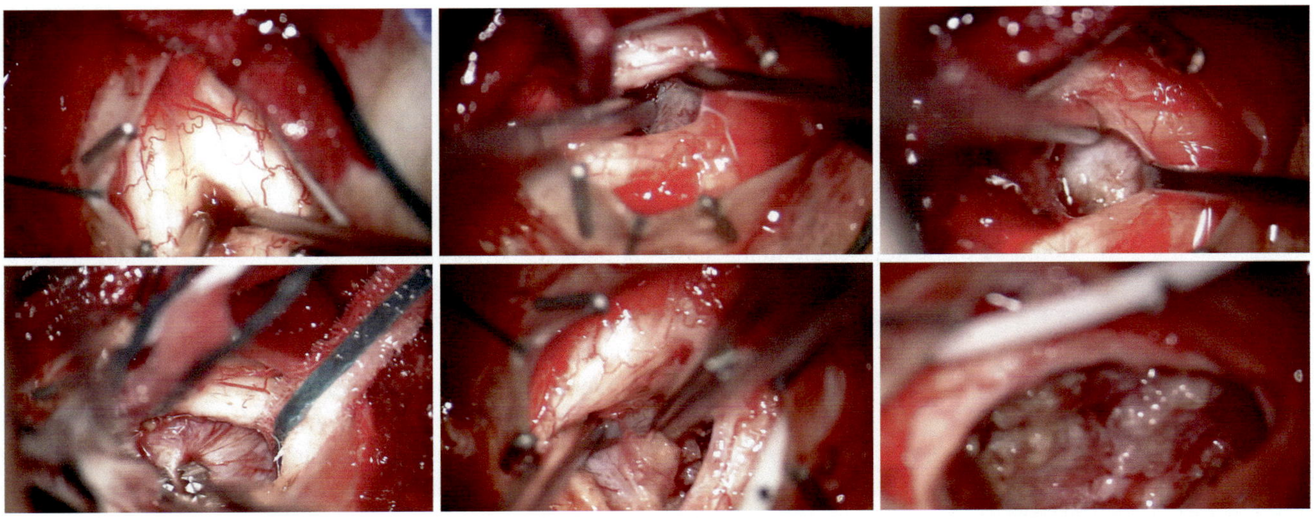

Fig. 67.4 Intraoperative pictures showing the removal of the lesion. The patient underwent a posterior bilateral C4–C5 hemilaminectomy and microscopic exeresis of the cervical lesion

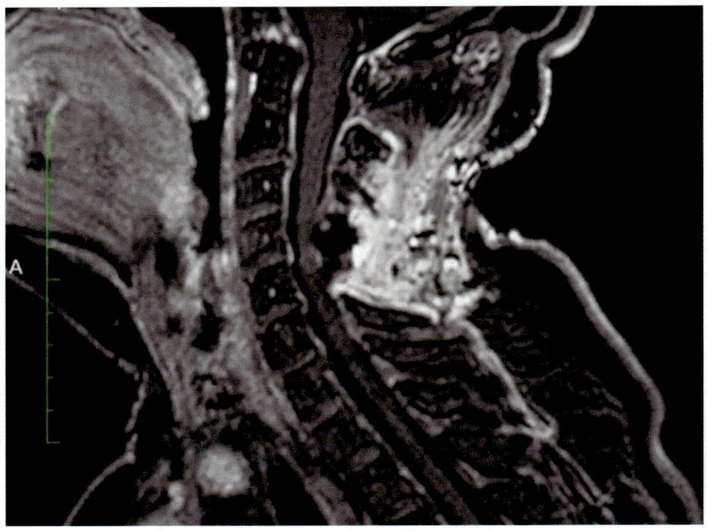

Fig. 67.5 Six months postoperative MRI

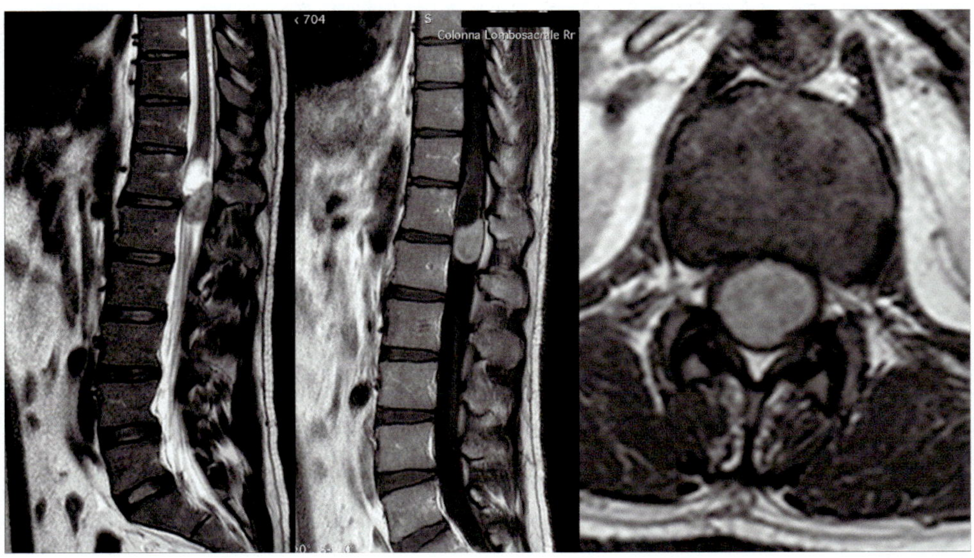

Fig. 67.6 Preoperative pictures showing the T12-L1 mass

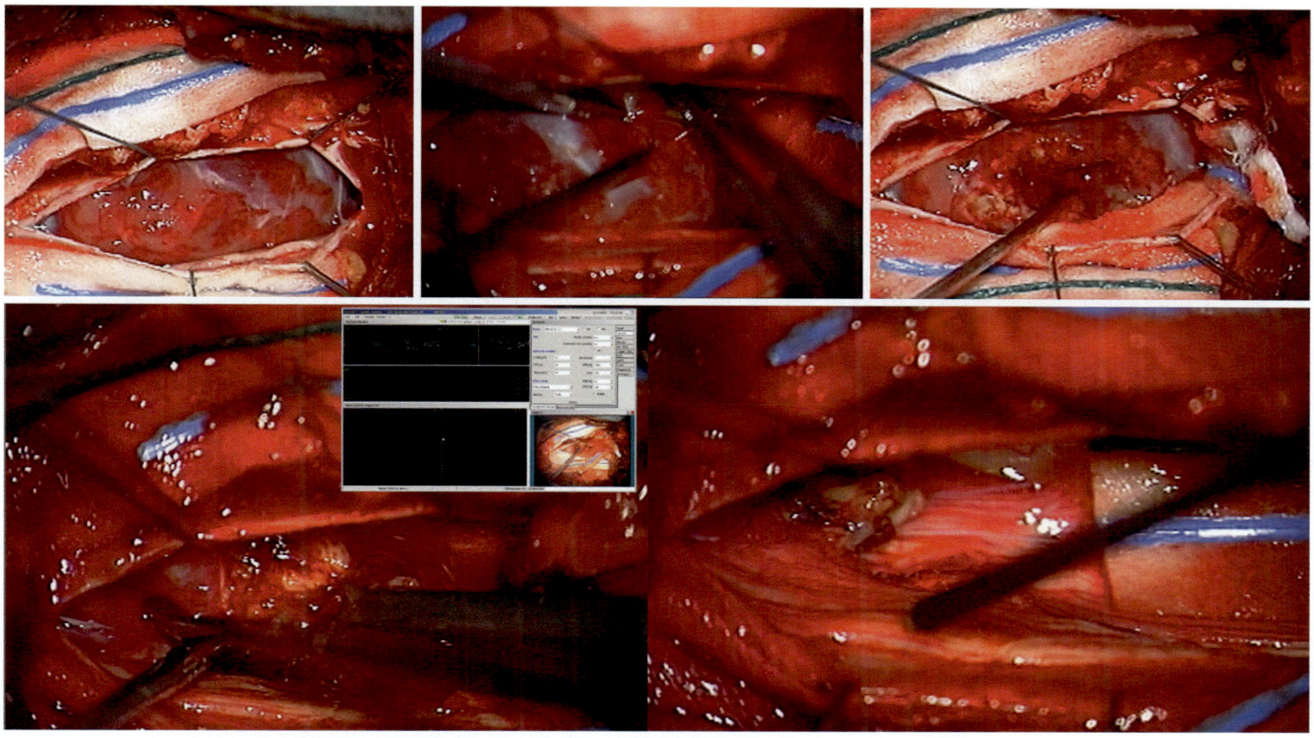

Fig. 67.7 Intraoperative pictures: removal of an intradural extramedullary lesion

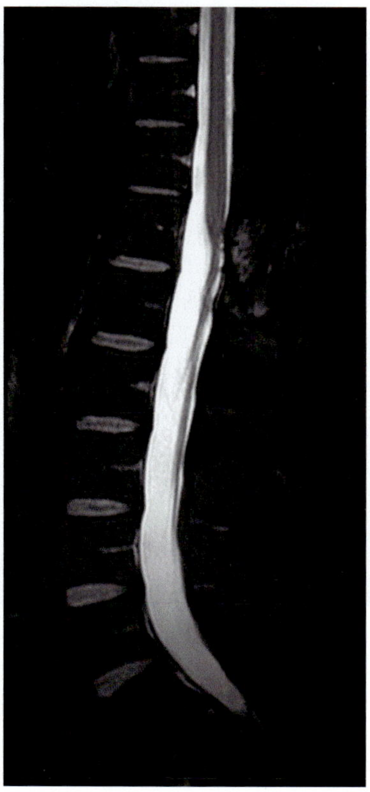

Fig. 67.8 Postoperative MRI

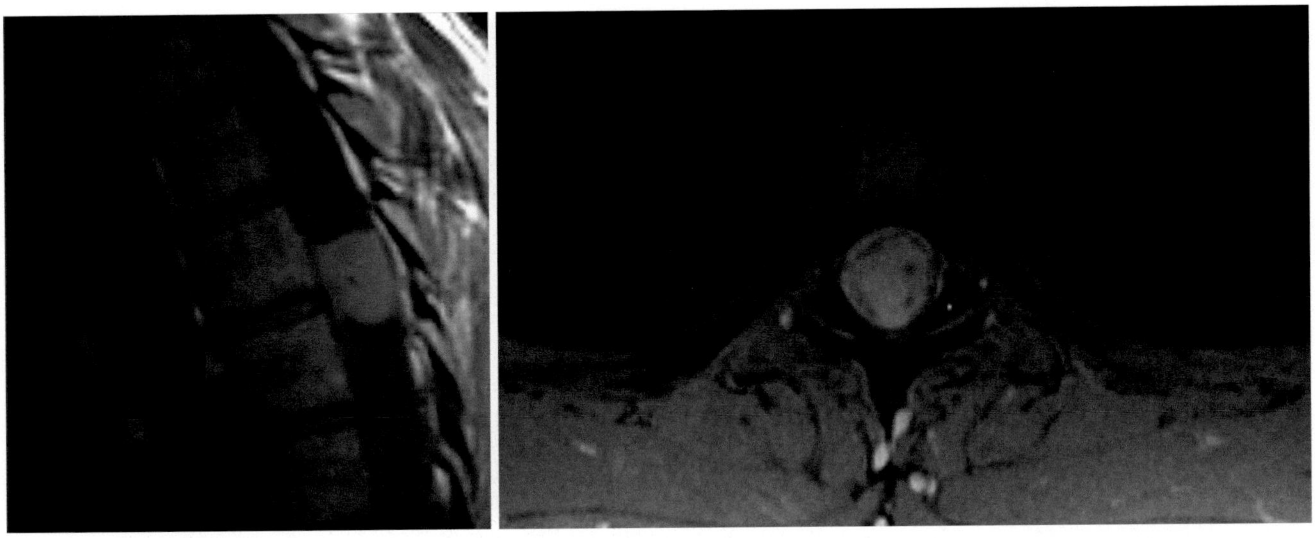

Fig. 67.9 Preoperative MRI

Fig. 67.10 Postoperative MRI and neurological outcome

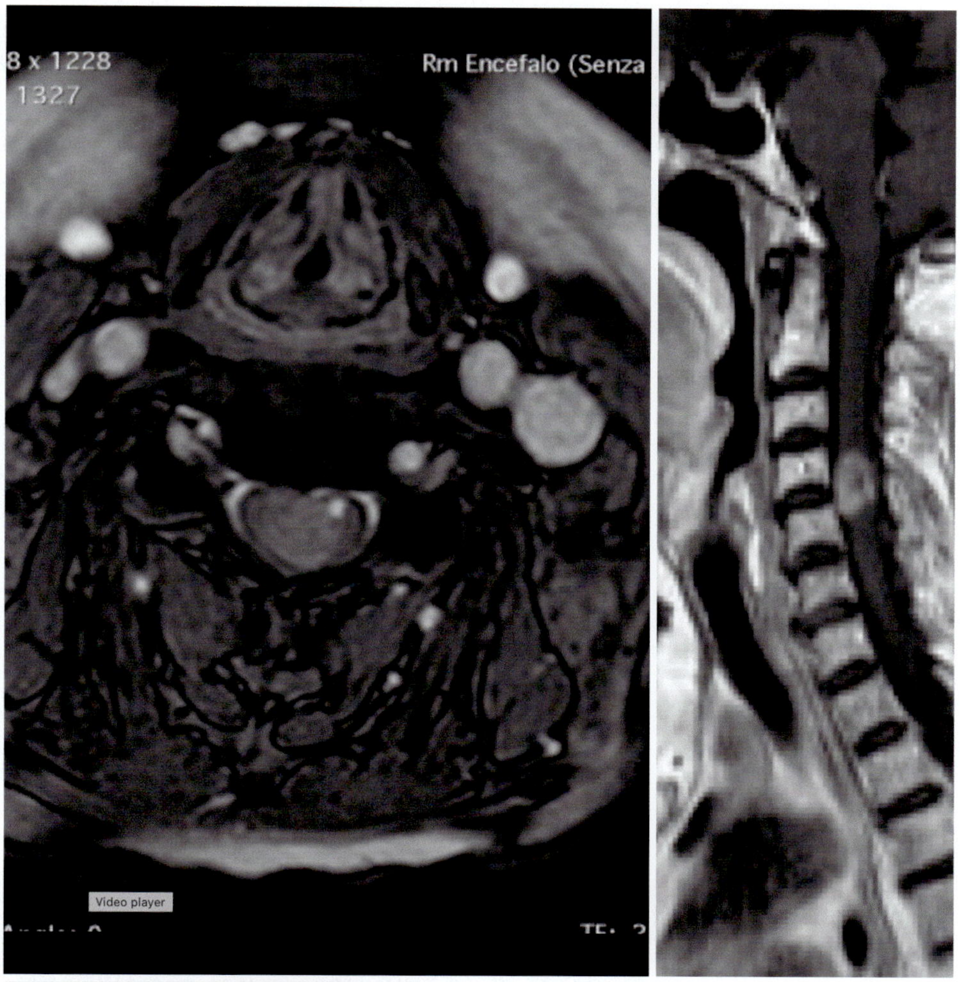

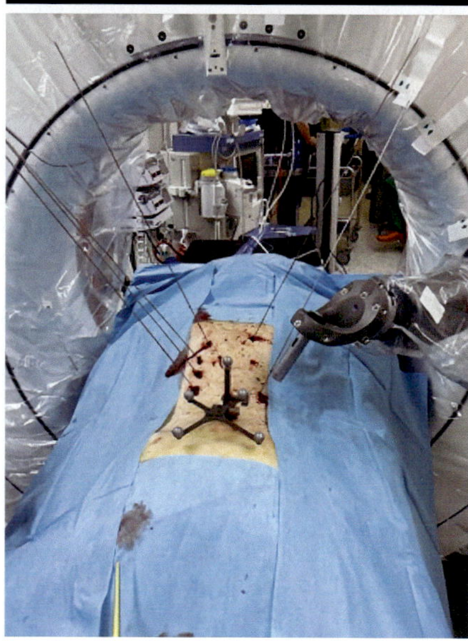

Fig. 67.11 Intraoperative picture showing multilevel skin incisions and Kirschner positioning after the robot-assisted procedure. Each level was treated from a robot-assisted monolateral approach, allowing the RFA procedure of vertebral bodies with a single incision

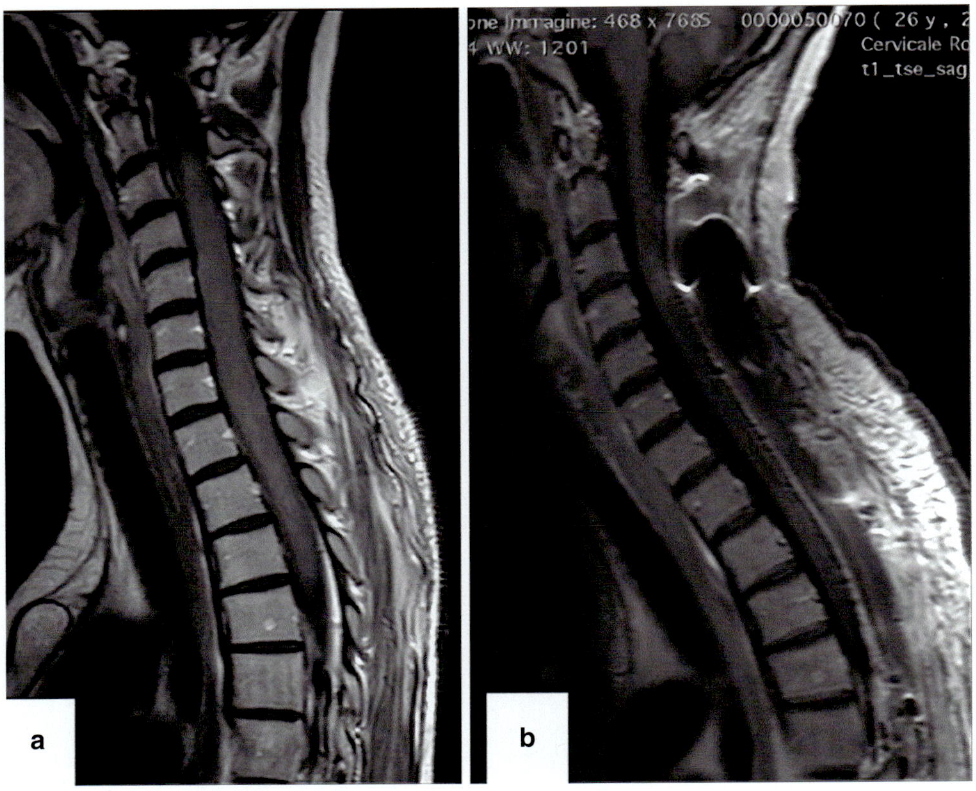

Fig. 67.12 (**a**) Preoperative sagittal image showing a huge intramedullary mass coursing along the cervicodorsal region. (**b**) Postoperative MRI

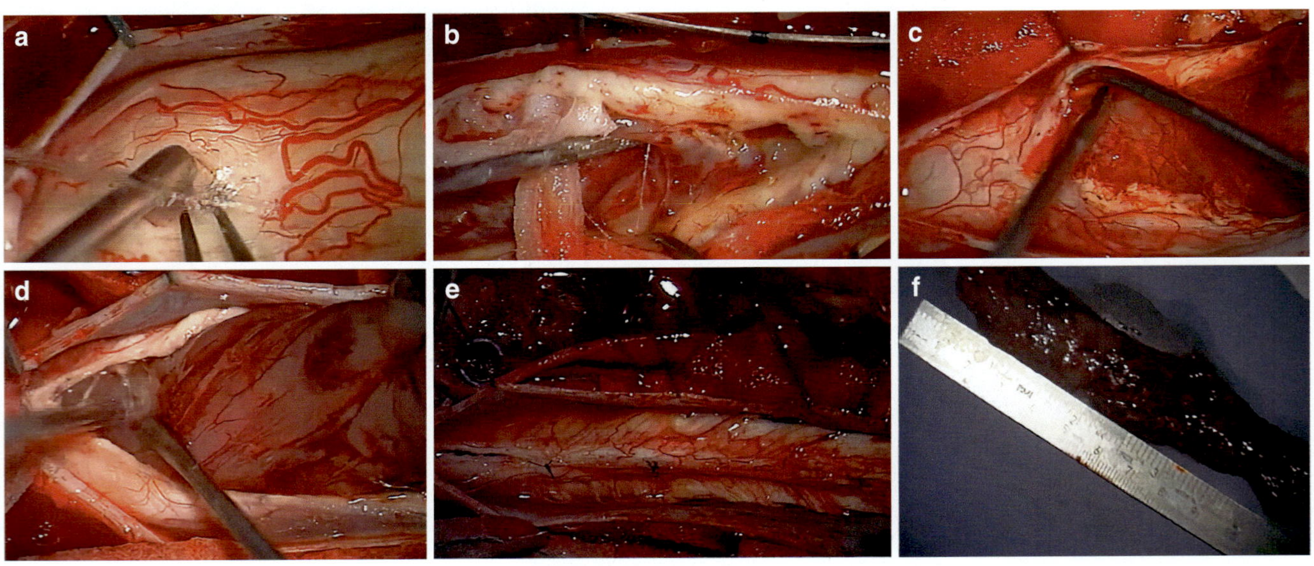

Fig. 67.13 Intraoperative pictures showing the removal of the lesion. (**a**) Midline myelotomy. (**b–d**) Microsurgical anatomical dissection. Identifying the interface tumor-intact shell of the spinal cord dissecting from the side of the tumor. (**e**) Suture of pia mater allowing anatomical reconstruction. (**f**) Tumor specimen after microsurgical removal (length 12 cm)

For laterally located EMSCT and IMSCT tumors, a hemilaminectomy can be performed to access the intradural space limiting postoperative pain and consequences of spine instability. The laminoplasty technique has the same goal and appears well indicated, especially in pediatric cases in order to reduce also delayed spine deformity.

67.4 Surgical Pearls

- Posterior midline myelotomy between the posterior columns.
- Discontinuous midline incision improves exposure of intramedullary spinal tumors.
- Identifying plane between tumor and spinal cord.
- Retraction of myelotomic borders by means of pial stitches.
- Pial suture after removal of intramedullary tumors.
- Hemilaminectomy and lateral dural opening for intradural extramedullary tumors to reduce instability in laterally located tumors.
- Laminoplasty reduces the risk of delayed spine deformity in pediatric cases.
- Intraoperative neuromonitoring as a fundamental tool for functional preservation.
- Debulking and mobilization of extramedullary tumors away from the spinal cord.

67.5 Discussion

Spinal tumors are classically classified as either extradural or intradural, depending on their location in relation to the spinal meninges. Intradural tumors can be further subcategorized into extramedullary or intramedullary spinal tumors. The treatment of these lesions is deeply changed in the years, thanks to the innovations in technology and the confidence of the neurosurgeons on such a tricky anatomical region and, moreover, is different from lesion to lesion. For intradural intramedullary and intradural extramedullary lesion, a neurophysiological monitoring is mandatory to avoid iatrogenic damage during the surgery; at the same time, extradural lesion which, most of the time, involves bony structure, the goal of the surgery is purely palliative or meant to recover a posttraumatic neurological damage; in those case, not only the tumor removal but also the decompression of the spinal cord and the spine arthrodesis are indicated.

However, surgical treatment of spinal cord lesions is considered nowadays the gold standard in most of the lesions diagnosed within the spinal cord: surgical treatment is the mainstay for intradural extramedullary lesions and for extradural extramedullary (mainly metastases) lesions which cause any neurological impairment, especially in those oncologic patients with high KPS and PF. On the contrary, a watchful waiting is considerable in patients with neurofibromatosis, when multiple lesions are diagnosed and there is no evidence of neurological deficits.

The anatomical morphology of the spinal cord and the dense distribution of eloquent fibers carry a significant risk of intraoperative neurological damage. In order to salvage function and to have the best surgical and neurological outcome, intraoperative neurophysiological monitoring (IONM) helps to prevent the functional integrity of essential neural pathways and optimize the surgical strategies. While SSEPs provide a constant monitoring of the dorsal column, MEPs monitor the cortico-spinal tract in discontinuous and very short intervals (2–10 s). The introduction of transcranial motor evoked potentials (MEPs) constituted a significant step forward for the intraoperative assessment of motor pathway function in spine surgery [16]. This technique could count on two different and complementary recording methods: the first consists in recording the D wave directly from the epidural space (e-MEPs), thanks to a single pulse; the second is to record MEPs along the peripheral muscles after a short train of electrical stimuli (m-MEPs; multipulse technique). The D-wave generated is the result of the direct activation of the axons of fast-conducting fibers of the cortico-spinal tract and is considered one of the most predictive variables of postoperative motor deficit. A decrease in the D-wave amplitude >50% is related to a postoperative motor deficit [17]. The literature reports that with the combined use of e-MEPs (d-WAVE) and m-MEPs, permanent motor deficit and paraplegia occur in less than 5% and 1%, respectively [18–20]. SSEP play a minor role in predicting the postoperative outcome: many authors share the idea that preserving SSEPs during surgery is considered a good sign but any intraoperative lowering or even loss of them has a small value in terms of outcome [21]. Finally, when single nerve roots need to be monitored, EMG is the best choice. EMG offers very high sensitivity, but limited prediction of functional changes.

For all those reasons, the use of IONM variations can contribute as key findings for the clinical decision-making process during spinal cord tumor resection.

Vertebral metastases with or without spontaneous pathological fractures affect 70% of cancer patients, and most patients with vertebral metastases present severe and chronic pain often resistant to opioids with remarkable impact on quality of life (QoL) [22]. Only for isolated vertebral metastases, at an early stage of disease, a radical surgical resection is suggested. In cases of multiple vertebral involvement, the preferred treatment is palliative, aimed to pain relief, spinal stabilization, and preservation of neurological function. Therapeutic options for patients with vertebral metastases have become more complex and require a multidisciplinary approach and, to date, there is no consensus on the best and QoL-centered practice. Vertebroplasty, kyphoplasty, radioablation with radiofrequency (RFA), and local biopsies are useful tools to achieve this goal. Robotics and neuronavigation help the surgeon to high the accuracies of the procedures, shorten the operation time, lower blood loss rates, and the radiation exposure which are valuable characteristics in fragile patients [23].

67.6 Illustrative Cases

Case 1 66-year-old woman with a C4–C5 lesion.

Symptoms at presentation: Gait disturbances, sensory deficits, four extremities weakness, and hyperreflexia.

Final diagnosis: Sclerosing meningioma (rare intramedullary variant).

Case 2 37-year-old male with a T12–L1 intradural tumor.

Symptoms at presentation: Progressive paraparesis.

Final diagnosis: Myxopapillary ependymoma WHO I.

Case 3 31-year-old male with a T5–T6 intradural extramedullary tumor.

Symptoms at presentation: Paraparesis.

Final diagnosis: Meningioma WHO I.

Case 4 78-year-old woman with breast cancer; KPS 90.

Symptoms at presentation: Dorsal and lumbar back pain.

Final diagnosis: Metastasis.

Case 5 17-years-old woman with an intradural intramedullary C3–T4 lesion.

Symptoms at presentation: Severe tetraparesis.

Final diagnosis: Intramedullary ependymoma.

67.7 Conclusion

The treatment of EMSCTs and IMSCTs is deeply changed in recent years, thanks to the innovations in surgical techniques, technology, and IONM. These advances have improved the extent of resection of spinal tumors and patient outcome in many cases. Knowledge of anatomy of spinal cord and its meningeal envelopment, especially arachnoidal and pial membranes, is crucial to identify the tumor-spinal cord interface. IONM is mandatory and should guide the resection maneuvers. In IMSCTS, the worst prognosis and resectability are related to astrocytomas, in which a surgical cleavage tumor-spinal cord plane is hardly identified. Minimally invasive techniques such as hemilaminectomy and lateral dural opening may reduce instability in laterally located tumors. Laminoplasty reduces the risk of delayed spine deformity in pediatric cases.

References

1. Tobin MK, et al. Intramedullary spinal cord tumors: a review of current and future treatment strategies. Neurosurg Focus. 2015;39(2):E14.
2. Rubinstein LJ: Tumors of the central nervous system, Second Series. Armed Forces Institute of Pathology, Washington, 1972.
3. el-Mahdy W, et al. Spinal intradural tumours: Part I—Extramedullary. Br J Neurosurg. 1999;13(6):550–7.
4. Ozawa H, et al. Spinal dumbbell tumors: an analysis of a series of 118 cases. J Neurosurg Spine. 2007;7(6):587–93.
5. Sonneland PR, Scheithauer BW, Onofrio BM. Myxopapillary ependymoma. A clinicopathologic and immunocytochemical study of 77 cases. Cancer. 1985;56(4):883–93.
6. Alafaci C, et al. Ossified spinal meningiomas: clinical and surgical features. Clin Neurol Neurosurg. 2016;142:93–7.
7. Kucia EJ, et al. Surgical technique and outcomes in the treatment of spinal cord ependymomas, part 1: intramedullary ependymomas. Neurosurgery. 2011;68(1 Suppl Operative):57–63; discussion 63
8. Ardeshiri A, et al. Intramedullary spinal cord astrocytomas: the influence of localization and tumor extension on resectability and functional outcome. Acta Neurochir. 2013;155(7):1203–7.
9. Nuwer MR, et al. Evidence-based guideline update: intraoperative spinal monitoring with somatosensory and transcranial electrical motor evoked potentials*. J Clin Neurophysiol. 2012;29(1):101–8.
10. Nuwer MR, et al. Somatosensory evoked potential spinal cord monitoring reduces neurologic deficits after scoliosis surgery: results of a large multicenter survey. Electroencephalogr Clin Neurophysiol. 1995;96(1):6–11.
11. Cantore G, et al. Discontinuous myelotomy: an alternative to standard myelotomy in the surgical treatment of intramedullary spinal cord tumours. Acta Neurochir. 2002;144(4):373–6.
12. Takami T, Yamagata T, Ohata K. Posterolateral sulcus approach for spinal intramedullary tumor of lateral location: technical note. Neurol Med Chir (Tokyo). 2013;53(12):920–7.
13. Angevine PD, et al. Surgical management of ventral intradural spinal lesions. J Neurosurg Spine. 2011;15(1):28–37.
14. Karikari IO, et al. Impact of tumor histology on resectability and neurological outcome in primary intramedullary spinal cord tumors: a single-center experience with 102 patients. Neurosurgery. 2015;76(Suppl 1):S4–S13; discussion S13
15. Garces-Ambrossi GL, et al. Factors associated with progression-free survival and long-term neurological outcome after resection of intramedullary spinal cord tumors: analysis of 101 consecutive cases. J Neurosurg Spine. 2009;11(5):591–9.
16. Costa P, Peretta P, Faccani G. Relevance of intraoperative D wave in spine and spinal cord surgeries. Eur Spine J. 2013;22(4):840–8.
17. Yamamoto T, et al. Intraoperative monitoring of the corticospinal motor evoked potential (D-wave): clinical index for postoperative motor function and functional recovery. Neurol Med Chir (Tokyo). 2004;44(4):170–80; discussion 181–2

18. Kothbauer KF, Deletis V, Epstein FJ. Motor-evoked potential monitoring for intramedullary spinal cord tumor surgery: correlation of clinical and neurophysiological data in a series of 100 consecutive procedures. Neurosurg Focus. 1998;4(5):e1.
19. Sala F, et al. Intraoperative neurophysiology in intramedullary spinal cord tumor surgery. Handb Clin Neurol. 2022;186:229–44.
20. Scibilia A, et al. Intraoperative neurophysiological monitoring in spine surgery: a significant tool for neuronal protection and functional restoration. Acta Neurochir Suppl. 2017;124:263–70.
21. Raco A, et al. Long-term follow-up of intramedullary spinal cord tumors: a series of 202 cases. Neurosurgery. 2005;56(5):972–81; discussion 972–81
22. Graham B, et al. The American academy of orthopaedic surgeons evidence-based clinical practice guideline on: management of carpal tunnel syndrome. J Bone Joint Surg Am. 2016;98(20):1750–4.
23. Wallace AN, Greenwood TJ, Jennings JW. Radiofrequency ablation and vertebral augmentation for palliation of painful spinal metastases. J Neuro-Oncol. 2015;124(1):111–8.

68

Anatomy of Brachial Plexus and Peripheral Nerves—Surgical Challenges

Lukas Rasulić and Miroslav Samardžić

Learning Objectives

- The reader should understand the topographic anatomy of the upper and lower extremities and identify the anatomical landmarks essential for adequate peripheral nerve exploration.
- The possible anatomical variations must be recognized to provide proper surgical exposure and prevent intraoperative complications.
- The general principles of surgical exposures should be adopted by all physicians that perform peripheral nerve surgery.

68.1 Introduction

The nerve identification and proper surgical exposure is one of the most important segments in treatment of patients with peripheral nerve injuries and disorders. The anatomical key points provide basic knowledge for understanding the topographic anatomy of the upper and lower extremity. Knowing the possible anatomical variations is of great importance for accurate diagnosis and adequate surgical exposure. The peripheral nerve surgical exposures are possible along their entire course; while in some cases, two nerves could be exposed through the same skin incision [1–5]. This chapter reviews the anatomical key points, important anatomical variations, surgical exposures, and pearls for peripheral nerve exploration in the upper and lower extremities, with inclusion of the original drawings and intraoperative case pictures.

L. Rasulić (✉) · M. Samardžić
Faculty of Medicine, University of Belgrade, Belgrade, Serbia

Department for Peripheral Nerve Surgery, Functional Neurosurgery, and Pain Management Surgery, Clinic for Neurosurgery, University Clinical Centre of Serbia, Belgrade, Serbia

68.2 Anatomical Highlights

68.2.1 Upper Extremity

- **The brachial plexus** is formed by the interconnections among the anterior rami of the lower cervical, C5–C8, and the first thoracic spinal nerves (Fig. 68.1). The main characteristics of the brachial plexus anatomy are (1) the nerves to the lower trunk exit from their foramina and have a shorter course along the bony groove between the anterior and posterior tubercules with loose connective tissue fixation, and consequently, they are prone to the avulsion injuries, (2) the components of the brachial plexus are (a) five anterior rami of the spinal nerves, (b) three trunks in order: the upper [C5–C6], the middle [C7] that is usually smaller in caliber and the lower [C8, T1] that is relatively short and lies behind the axillary artery and vein, (c) six divisions, three anterior, and three posterior, (d) three cords, the lateral formed by the anterior divisions of the upper and middle trunks, positioned superficially and the first to be revealed, the posterior formed from three posterior divisions, positioned posteriorly to the axillary artery and the medial which is simply a continuation of the anterior division of the lower trunk, and (e) the terminal nerves, (3) the anterior rami of the spinal nerves are located in the interscalene triangle bordered by the anterior and middle scalene muscles, and the first rib between the insertion points of these muscles, (4) the anterior rami form the trunks at the level of the posterior margin of the sternocleidomastoid muscle, (5) the trunks and their divisions are located over the base of the posterior neck triangle bordered by the sternocleidomastoid muscle anteriorly, the trapezius muscle posteriorly, and the middle third of the clavicle, (6) the neurovascular bundle which contains the brachial plexus elements and the subclavian vessel courses through the narrow spaces from the base of the neck toward the axilla including the most proximal interscalene triangle, then the costoclavic-

I. N. Kanaan, V. Beneš (eds.), *Neuroanatomy Guidance to Successful Neurosurgical Interventions*,
https://doi.org/10.1007/978-3-031-59838-8_68

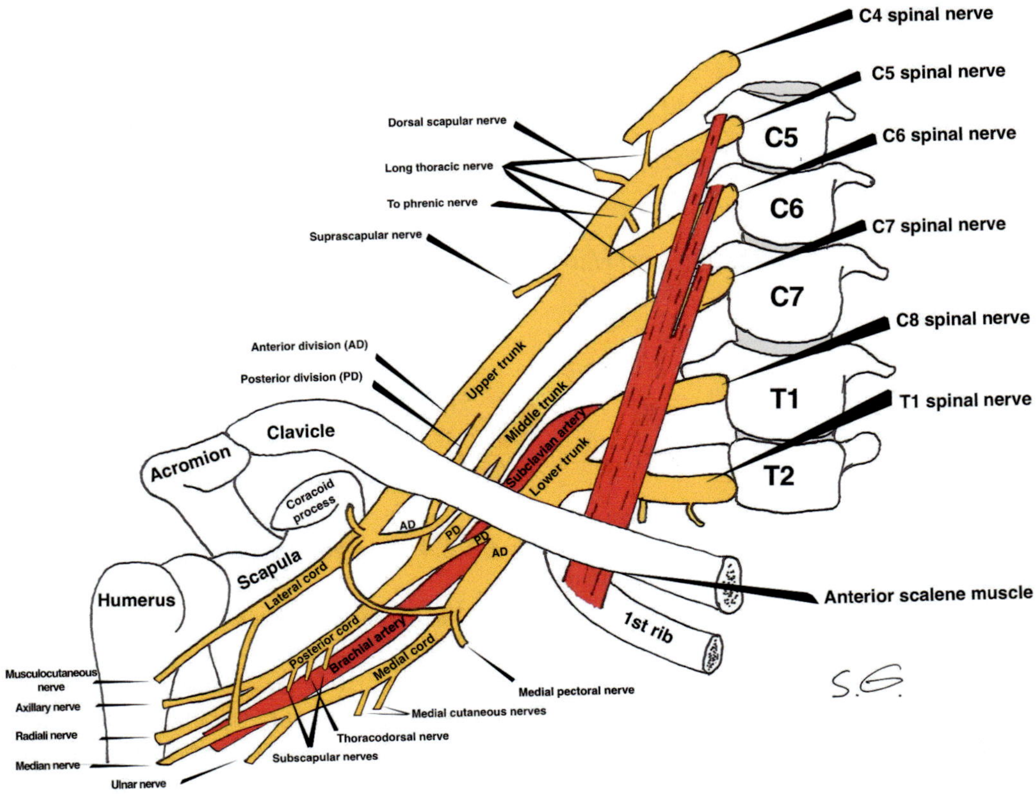

Fig. 68.1 Brachial plexus anatomy. With courtesy of Strahinja Gliorijević, BA (Faculty of Biology, University of Belgrade)

ular triangle and finally the subcoracoid space, (7) the collateral branches originate from the different components of the plexus in order (a) the dorsal scapular and long thoracic nerves from the anterior rami of the C5, C6, and C7 spinal nerves, (b) the suprascapular and subclavius nerves from the upper trunk, (c) the suprascapular nerve has its origin at the junction of the C5 and C6 spinal nerves that corresponds to the Erb's point located 2–3 cm above the clavicle, (d) the divisions have no branches, (e) the cords have seven collateral branches in order to the lateral pectoral nerve from the lateral cord, lower subscapular, and thoracodorsal nerves from the posterior cord, and the medial pectoral and the cutaneous nerves for the arm and forearm from the medial cord, and (8) the terminal branches of the plexus are the median nerve lateral component and the musculocutaneous nerve from the lateral cord, the axillary, and radial nerves from the posterior cord and the median nerve medial component and the ulnar nerve from the medial cord [6–8].

- **The axillary nerve** courses, together with the posterior circumflex artery and vein, along the inferior border of the shoulder capsule and travels through the quadrangular space, bordered by the humeral shaft laterally, the long head of the triceps medially, the teres minor muscle superiorly, and teres major muscle inferiorly. After passing this space, the nerve divides into the anterior and posterior branches [9, 10].
- **The radial nerve**, which is a continuation of the posterior cord, exits the axilla through the triangular space between the triceps medial and long heads and the teres major muscle, giving off the branches to the triceps muscle. The branch to the long head of the triceps is close to the axillary nerve, with an average distance of 2.5–5 cm. The nerve passes obliquely across the back of the humerus, in the spiral groove, from medial to lateral. By piercing the lateral intermuscular septum, it enters the anterior compartment. At the level of the lower arm, the radial nerve lies between the brachialis and brachioradialis muscles, where it is relatively superficial. At the level of the lateral epicondyle, 1–2 cm distally, the nerve divides into the superficial sensory and larger deep motor (posterior interosseus nerve) branches. The latter spirals around the radius between two heads of the supinator muscle and dives obliquely and deeply toward the posterior interosseus membrane passing through a series of fibromuscular tunnels (the arcade of Frohse, the superficial head of the supinator muscle and the tendinous origin of the extensor carpi radialis brevis muscle). However, upon exiting the distal border of the supinator muscle nerve divides into several muscular branches [11, 12].

- **The musculocutaneous nerve** runs laterally and obliquely toward the coracobrachialis and biceps muscles. The nerve pierces the first muscle and then continues on the medial aspect of the arm, between these two muscles. Midway of its course, the nerve gives off branches to the long and short heads of the biceps muscle, then within a few centimeters, it gives off the branch to the brachialis muscle. Beyond this point, it becomes a pure sensory nerve [13].
- **The median nerve** is formed from the lateral and medial cords contributions in the axilla and usually travels superficially to the brachial artery. Lower in the upper arm, it tends to lie deeply and medially to the artery, anterior to the brachialis muscle and enters the cubital fossa under the lacertus fibrosus. In the proximal forearm, the nerve lies between the superficial and deep heads of the pronator teres and gives off the branches as it passes through this muscle. The anterior interosseus nerve arises from the median nerve just proximal to the arch of the flexor digitorum superficialis muscle, and about 5–8 cm distally to the lateral epicondyle. This nerve passes between the flexor digitorum profundus and superficialis muscles, courses distally along the interosseus membrane, and gives no cutaneous innervation. The main trunk of the median nerve gives off the palmar cutaneous branch in the distal forearm, usually about 2 cm proximally to the transverse carpal ligament which begins 1 cm above the distal wrist crease. The ligament is approximately 4 cm wide and 3 mm thick. The nerve emerges from the carpal tunnel at the distal border of the ligament where it usually gives off the recurrent motor branch and divides into the proper and common palmar digital branches which further divide into the digital nerves [13–15].
- **The ulnar nerve** is an extension of the medial cord that continues medially to the brachial artery, between the biceps and triceps muscles, to the middle of the arm. It then pierces the intermuscular septum, 8 cm proximal to the medial epicondyle, and courses under the arcade of Struthers that presents an extension of the septum and exists in about 50% of cases, lying on the medial head of the triceps muscle. In the upper arm, the ulnar nerve is close to the median nerve and has no branches. At the elbow, it enters the postcondylar humeral groove, and then under the arcuate ligament of the cubital tunnel located between the medial epicondyle and the olecranon. Distally, the nerve courses deep to the aponeurosis between the two heads of the flexor carpi ulnaris muscle, which can be in about 75% of cases very thick and called the Osborne's band, then it continues into the forearm between this muscle and the flexor digitorum profundus muscle. The first muscular branch is to the flexor carpi ulnaris muscle. There are up to four branches that arise from 4 cm to 10 cm distal to the medial epicondyle. Near the wrist, the nerve runs medially to the ulnar artery, and laterally to the flexor carpi ulnaris tendon. Both the nerve and artery continue into the hand via the Guyon's tunnel, and distally, the nerve divides into the superficial and deep branches [16].

68.2.2 Lower Extremity

- **The femoral nerve** takes its origin from the L2, L3, and L4 spinal nerves and presents the largest branch of the lumbar plexus (Fig. 68.2). After their union, the formed nerve passes inferolaterally in the retroperitoneal location. Then, it pierces through the psoas muscle and travels in the groove between this and the iliacus muscle entering the thigh behind the inguinal ligament, peripherally to the vascular elements. The nerve enters the femoral triangle bordered medially by the sartorius muscle, laterally by the adductor longus muscle, and superiorly by the inguinal ligament. Approximately, 4 cm below the ligament, it divides into multiple motor and sensory branches. The saphenous nerve is the largest and longest sensory branch that arises in the femoral triangle and descends on the lateral side of the femoral vessels to enter the Hunter's adductor canal [17, 18].
- **The sciatic nerve** is the largest single nerve of the body, and it derives from the L4 to S3 spinal nerves (Fig. 68.2). Typically, the nerve runs between the piriformis and gemelli muscles emerging from the pelvis through the infrapiriform part of the greater sciatic foramen to enter the buttock and thereafter the lower extremity between the ischial tuberosity and the greater trochanter. Passing the lower border of the gluteus maximus muscle, it lies on the surface of the adductor magnus muscle, under the long head of the biceps femoris muscle. The nerve typically bifurcates in the distal thigh but may do so more proximally, including the pelvis [19].
- **The common peroneal nerve** is a smaller component of the sciatic nerve. After entering the popliteal fossa, the nerve crosses over the lateral head of the gastrocnemius muscle to run superficially, just posterior to the head of the fibula. Distally, the nerve dives between the two heads of the peroneus longus muscle and divides into the superficial and deep peroneal branches. The deep peroneal branch pierces the intermuscular septum and courses toward the anterior compartment [20].
- **The tibial nerve** has a relatively short course in the popliteal fossa where it is close to the popliteal artery and vein. The nerve accompanies these vessels through the leg where they are called the tibial artery and vein, medially to the tibia and posteriorly to the intermuscular septum that separates the anterior and posterior compartments. As it approaches the ankle, the nerve lies posteriorly to the tendons of the tibialis posterior

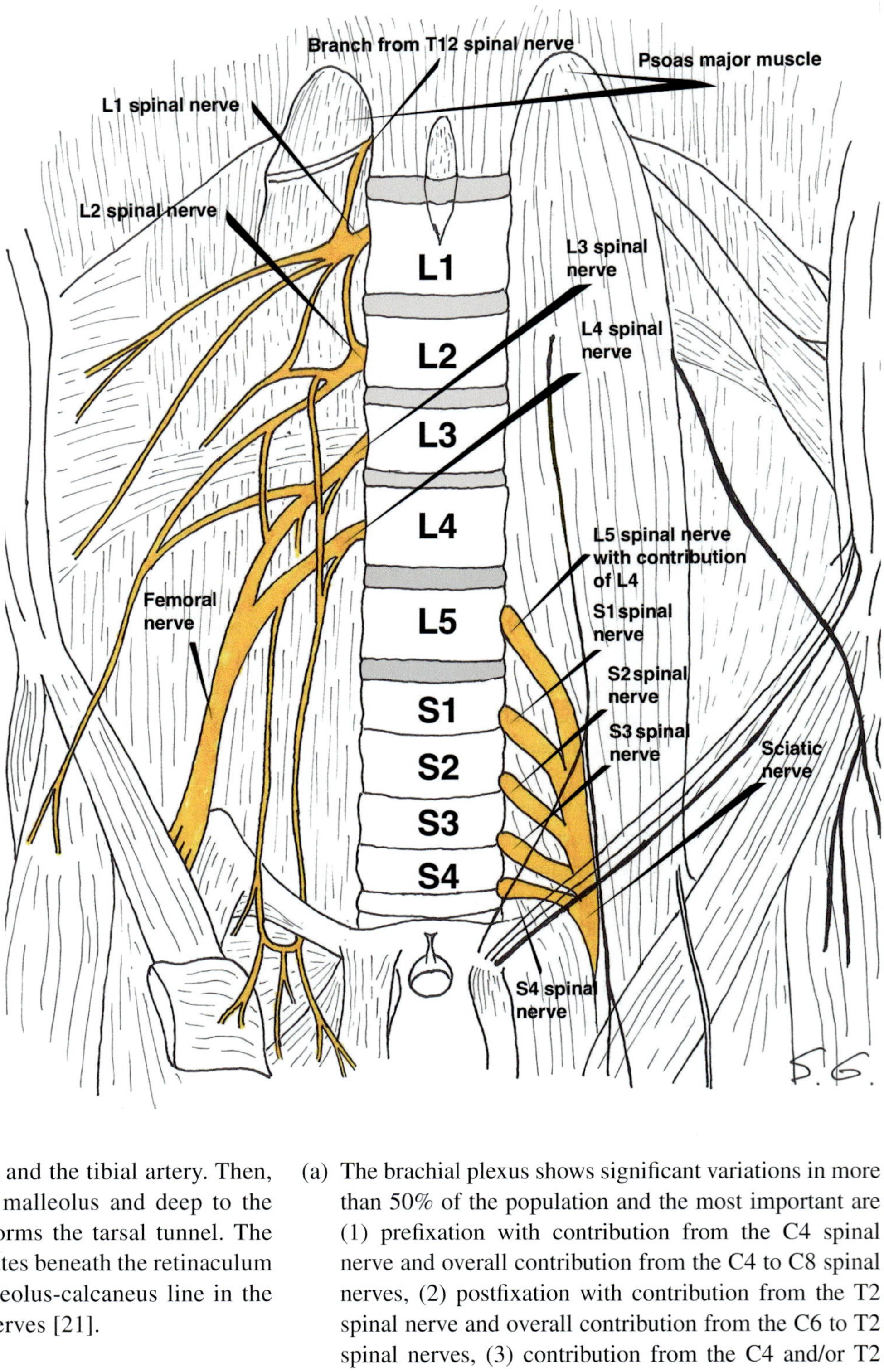

Fig. 68.2 Origin of femoral and sciatic nerve—lumbosacral plexus. With courtesy of Strahinja Gliorijević, BA (Faculty of Biology, University of Belgrade)

and flexor hallucis muscles, and the tibial artery. Then, it courses inferiorly to the malleolus and deep to the flexor retinaculum which forms the tarsal tunnel. The tibial nerve typically bifurcates beneath the retinaculum or 2 cm of the medial malleolus-calcaneus line in the medial and lateral plantar nerves [21].

68.3 Important Anatomical Variations

Among the numerous anatomical variations of the peripheral nerves, several of them should be emphasized, especially regarding their frequency and importance for accurate diagnosis and adequate surgical exposure [6].

(a) The brachial plexus shows significant variations in more than 50% of the population and the most important are (1) prefixation with contribution from the C4 spinal nerve and overall contribution from the C4 to C8 spinal nerves, (2) postfixation with contribution from the T2 spinal nerve and overall contribution from the C6 to T2 spinal nerves, (3) contribution from the C4 and/or T2 spinal nerves, (4) some trunks, divisions, or cords may be absent, and (5) variations in branching patterns of the cords [22].

(b) The median nerve has the largest number of variations including (1) the lateral cord may give off more than one branch which after fusion with the medial contribution gives the median nerve of the smaller caliber, (2) fusion of the medial and lateral components may be located at the different levels, as low as the elbow instead of the axilla, (3) fusion of two components under the axillary or brachial arteries, (4) small contribution from the lateral cord with additional communication between the musculocutaneous and median nerves, about halfway down to the upper arm, (5) Martin-Gruber anastomosis, which presents redistribution of the anterior interosseus motor nerve fibers toward the ulnar nerve, exists in 10–15% of cases, (6) Marinacci anastomosis which presents a reverse redistribution at the same level, (7) Riche-Cannieu anastomosis which presents communication between the recurrent motor branch of the median nerve and the deep motor branch of the ulnar nerve, and (8) origin of the recurrent motor branch within the carpal tunnel through the transverse ligament, proximal to the carpal tunnel or on the ulnar side of the median nerve [14].

(c) The ulnar nerve has also several variations including (1) the origin of the medial brachial and antebrachial cutaneous nerves from the ulnar nerve, (2) origin of the dorsal below the palmar ulnar cutaneous nerve, (3) early branching of the ulnar nerve proximally to the pisiform bone, and (4) bifurcation of the deep motor branch prior to entering the pisohammate hiatus [16].

(d) The radial nerve could have communication between the posterior interosseus nerve and the deep motor branch of the ulnar nerve, the so-called Froment-Rauber nerve.

(e) The lateral femoral cutaneous nerve may course over or through the inguinal ligament, and under or through the sartorius muscle.

(f) The sural nerve may be formed from one, usually the medial component.

68.4 Surgical Exposures

68.4.1 General Principles

Surgical exposures of all peripheral nerves are possible along their entire course. In some cases, two nerves could be exposed through the same skin incision, for example, the median and ulnar nerves in the arm or the tibial and peroneal nerves in the popliteal fossa. The surgical approaches could be combined, if necessary [23, 24]. There are several rules in nerve exposition.

(a) The skin incision characteristics (1) shouldn't be perpendicular to the flexion skin creases, (2) could include preexisting scars, (3) should avoid skin transplants, and (4) could be extended proximally and/or distally in cases of insufficient exposure.

(b) Nerve dissection should (1) be extensive and parallel to the anatomical structures, (2) take into consideration possible anatomical variations during the nerve identification, (3) be performed with minimal trauma to the neighboring structures and permanent hemostasis, and (4) avoid the sites of the tendon repair, whenever possible.

(c) External neurolysis is an obligatory procedure during exposure that include the removal of the adhesions, compressive scar tissue, and nerve adventitia. In some nerve lesions in continuity, this could be a definite procedure.

(d) The use of microsurgical techniques and neuromonitoring procedures are obligatory, and the latter is of special value for the nerve lesions in continuity.

68.4.2 Individual Key Steps—Upper Extremity

(a) Supraclavicular approach to the brachial plexus (Fig. 68.3) gives an exposition of the spinal nerves, trunks, and proximal divisions. Key steps are as follows: (1) the skin incision that starts at the jaw angle and courses along the posterior border of the sternocleidomastoid muscle, at the clavicle turns laterally along its superior border and ends at the midpoint of the clavicle, (2) division of the platysma muscle, (3) exposure of the posterior neck triangle with preservation of the spinal accessory nerve, ligation of the vessels crossing the operative field, detachment and medial retraction of the sternocleidomastoid muscle, and division and retraction of the omohyoid muscle, (4) dissection and lateral retraction of the supraclavicular fat tissue with exposure of the scalenus anterior muscle, and (5) iden-

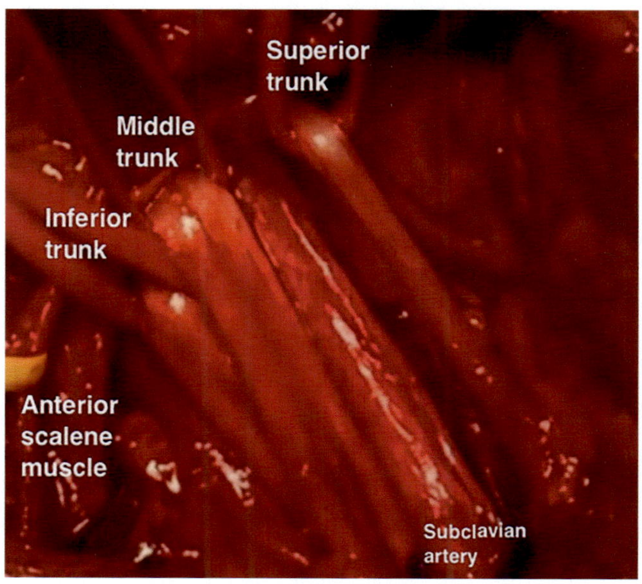

Fig. 68.3 Supraclavicular brachial plexus elements (intraoperative)

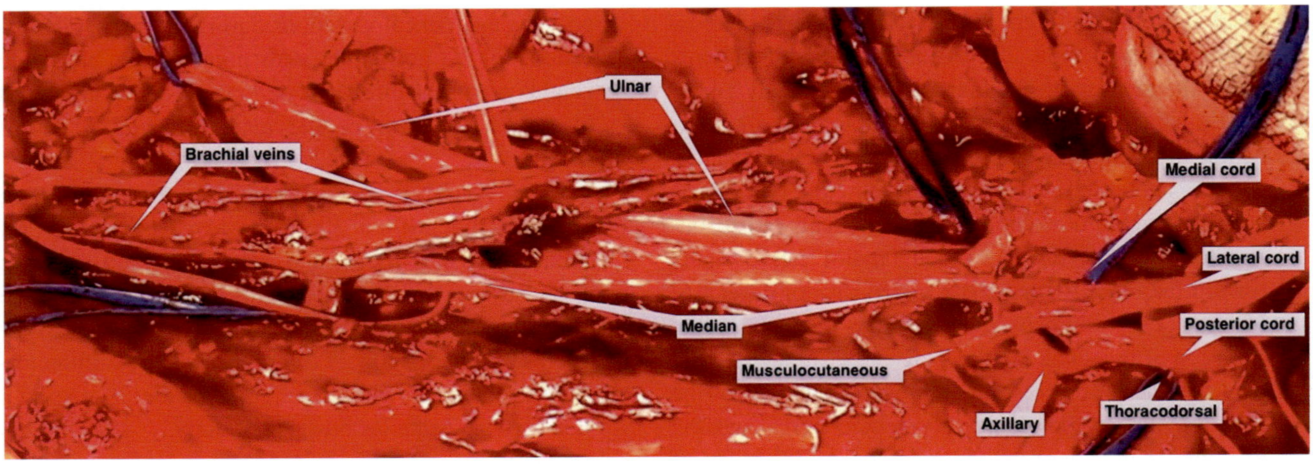

Fig. 68.4 Infraclavicular brachial plexus elements (intraoperative)

tification of the nerve elements including the phrenic nerve on the anterior surface of the scalenus anterior muscle, the C5 spinal nerve following the phrenic nerve superiorly, the C6 spinal nerve deep and slightly inferiorly, the C7 spinal nerve deep and inferior to the C6, three trunks with the C6 and T1 spinal nerves, and the divisions distally [25, 26].

(b) Infraclavicular approach to the brachial plexus (Fig. 68.4) gives exposition of the distal divisions, cords, and terminal branches. Key steps are as follows: (1) the skin incision that starts at the midpoint of the clavicle and courses inferiorly along the deltopectoral groove and anterior axillary crease, (2) ligation and resection of the cephalic vein, (3) division of the pectoralis minor tendon insertion with preservation of the muscle cuff for reapproximation, (4) incision of the clavipectoral fascia, (5) identification of the cords, (6) detachment of the pectoralis major muscle from the clavicle, ligation of the supraclavicular vessels, resection of the subclavius muscle, and retraction of the clavicle for exposure of the divisions proximally, and (7) identification of the terminal nerve branches [27].

(c) Exposure of the median nerve at the upper arm includes the following: (1) the skin incision that starts from the axillary fold, courses distally along the sulcus between the biceps and triceps brachii muscles, and ends about 4 cm above the medial epicondyle and curves laterally, (2) identification of the medial brachial cutaneous nerve before opening of the brachial fascia, (3) identification of the ulnar nerve which is located posteromedially to the median nerve and the basilic vein, and (4) in the proximal part identification of the median nerve that courses laterally to the ulnar nerve, and then crosses medially to the brachial artery.

(d) Exposure of the median and anterior interosseus nerves at the elbow includes the next common steps: (1) landmark is the biceps tendon, (2) the skin incision that starts 2–3 cm above the medial epicondyle, curves laterally to the biceps brachii tendon, and ends into the forearm between the flexor digitorum superficialis and brachioradialis muscles, and (3) identification of the median nerve between the biceps muscle and intermuscular septum (Fig. 68.5).

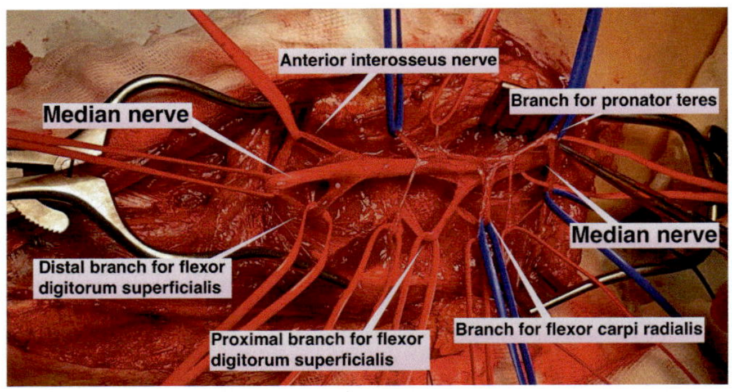

Fig. 68.5 Median nerve exposure at the elbow

(e) Exposure of the median nerve in the carpal tunnel includes the following steps: (1) the skin incision that begins at the distal wrist crease, courses along the ulnar side of the palmar crease, pointing to the ulnar edge of the middle finger up to 4 cm and ending at Kaplan's line (line parallel to the distal border of the extended thumb), (2) division of the subcutaneous fat and the palmar fascia, (3) exposition and gradual sharp incision of the transverse carpal ligament leading to the medial side of the nerve, (4) dissection of the nerve from the ligament, (5) potential extension of the incision distally until the palmar fat pad, and proximally 2 cm until the deep forearm fascia are identified, and (6) preservation of the recurrent motor and palmar cutaneous branches.

(f) Exposure of the ulnar nerve at the upper arm includes: (1) the skin incision that starts at the axillary fold over the pectoralis major muscle tendon, courses distally in the sulcus between the biceps and triceps brachii muscles and ends between the medial epicondyle and the olecranon, (2) identification and preservation of the medial brachial cutaneous nerve before opening of the brachial fascia, (3) identification of the ulnar nerve posteromedially to the median nerve and next to the basilic vein, and (4) incision of the intermuscular septum and identification of the ulnar nerve within the epimysium of the triceps muscle.

(g) Exposure of the ulnar nerve at the elbow involves the following steps: (1) the skin incision that begins 5 cm proximally to the medial epicondyle in sulcus between the biceps and triceps brachii muscles, curves between the medial epicondyle and olecranon, and ends 4–5 cm distally to the epicondyle between the flexor carpi ulnar and flexor digitorum profundus muscles, (2) identification and preservation of the medial antebrachial cutaneous nerve and distal tracing of the ulnar nerve, (3) dissection of the ulnar nerve under the arcade of Struthers, through the postcondylar humeral groove, the cubital tunnel, and release of the aponeurosis between two heads of the flexor carpi ulnaris muscle or the Osborne's band, and (4) distal dissection, if necessary, continues between the flexor profundus muscle medially and under and in the lower part of the forearm laterally to the flexor carpi ulnar muscle.

(h) Exposure of the ulnar nerve in the Guyon's tunnel involves the following steps: (1) Z-shaped skin incision that starts 4 cm proximally to the wrist over the flexor carpi ulnaris tendon, courses between the pisiform and hamate bones, and ends 4 cm distally to the wrist toward the midline, (2) identification and preservation of the palmar ulnar cutaneous nerve, (3) dissection and medial retraction of the flexor carpi ulnar tendon, (4) exposition of the ulnar nerve and artery medially to this tendon, (5) dissection and decompression of the nerve by dividing the palmar carpal ligament, palmaris brevis muscle fibers, and hypothenar fat tissue, (6) identification of the superficial and the deep ulnar nerve branches at the free border of the hypothenar muscles, and (7) dissection of the deep branch around the hamate bone (Fig. 68.6).

(i) Anterior exposure of the radial nerve at the upper arm includes the following: (1) initial steps as in the infraclavicular approach, (2) identification and dissection of the lateral cord and the axillary artery, (3) retraction of the lateral cord medially without excessive traction, (4) identification and dissection of the posterior cord and the axillary and radial nerves, and (5) identification and dissection of the radial nerve as it exits in the triangular space.

(j) Posterior approach for exposure of the radial nerve at the upper arm beyond the axilla consists of (1) the skin incision in the line connecting the acromion and the olecranon, (2) identification of the triceps muscle tendon, (3) proximal dissection for identification of the groove between the long and lateral triceps heads, (4) tracing the groove proximally until the teres major muscle and the triangular space, (5) distal dissection of the radial nerve with identification of the triceps muscle nerve branches and through the spiral groove until piercing the intermuscular septum, and (6) the separate anterior incision since the nerve is located in the anterior compartment.

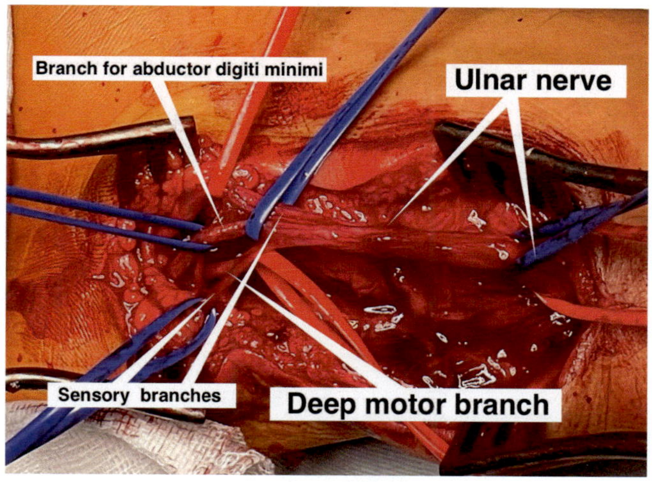

Fig. 68.6 Ulnar nerve exposure at the Guyon's tunnel

(k) Exposure of the radial nerve and posterior interosseus nerve at the elbow could be achieved through the anterior proximal incision between the brachioradialis and brachialis muscles and the posterior proximal incision between the brachioradialis and extensor carpi radialis longus muscles. Potential sites of compression should be explored and released including the radial recurrent vessels before the supinator muscle, the arcade of Frohse with complete section of the superficial supinator head and the extensor carpi radialis brevis muscle (Fig. 68.7).

68.4.3 Individual Key Steps—Lower Extremity

(a) Exposure of the femoral nerve is performed in a prone position and includes (1) the skin incision that begins at the spina iliaca anterior superior, continues parallel and 2 cm below the inguinal ligament, courses curvilinearly in the medial part of the anterior thigh, and ends at the junction of its middle and lower third crossing the sartorius muscle, (2) nerve dissection starts below the middle part of the inguinal ligament and courses into the femoral triangle where it branches, (3) the incision of the fascia lata parallelly to the medial border of the sartorius

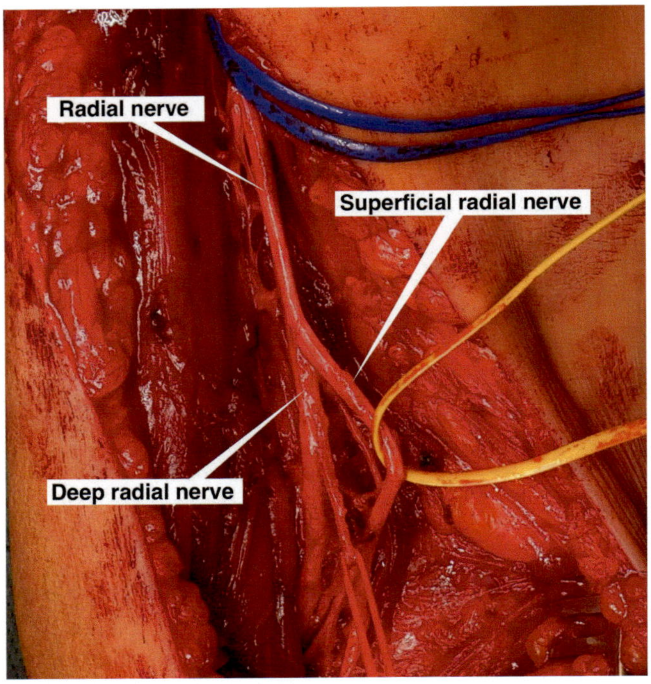

Fig. 68.7 Radial nerve exposure at the elbow

muscle and its lateral retraction, (4) incision of the fascia iliaca laterally to the femoral artery, in an upward direction, (5) identification of the femoral nerve, artery, and vein in order from laterally to medially, and (6) protection of the femoral artery and vein including the saphenous vein.

(b) Exposure of the sciatic nerve at the buttock includes the next steps: (1) curved skin incision that starts at the spina iliaca posterior superior, courses around the buttock along the gluteal crease, and ends in the midline of the posterior thigh, (2) division of the gluteus maximus muscle near the greater trochanter with preservation of the muscular cuff for reapproximation and its medial retraction, (3) identification and proximal dissection of the nerve up to the sciatic notch, (4) division of the piriformis muscle for better access, (5) preservation and protection of the cutaneous femoral posterior nerve, the inferior gluteal artery and nerve, and the branches of the sciatic nerve to the hamstring muscles, and (6) recommended separation of the peroneal and tibial nerve components for better evaluation and repair.

(c) Exposure of the sciatic nerve at the thigh includes (1) straight midline skin incision between the hamstring muscles, (2) identification of the long head of the biceps femoris muscle just inferior to the gluteus maximus muscle, (3) identification of the nerve in the fat pad laterally and deeply to the biceps long head, (4) distal nerve dissection under the bices long head that crosses nerve from medially to laterally, (5) splitting of the biceps femoris and semitendinosus/semimembranosus muscles for distal nerve identification, and (6) preservation and protection of the nerve elements as in previous exposure.

(d) Exposure of the common peroneal nerve at the knee includes the following steps: (1) identification of landmarks, namely, the fibular head, the biceps femoris tendon, and the midline popliteal fossa, (2) the skin incision that starts just superior to the popliteal fossa, courses along its midline, curves slowly laterally and cross obliquely posteroinferior to the fibular head, and ends over the peroneus longus muscle, (3) identification of the nerve in the fat pad, medially to the biceps femoris tendon, (4) nerve dissection over the lateral head of the gastrocnemius muscle and the fibular head, (5) incision of the peroneus longus fascia and muscle fibers until its division in the superficial and deep branches, and (6) identification of the deep branch piercing through the intermuscular septum on its way to the anterior compartment (Fig. 68.8).

(e) Exposure of the tibial nerve at the popliteal fossa is performed in a prone position and includes (1) the skin incision that starts just superior to the popliteal fossa, crosses it medially and slightly obliquely, and ends in the leg, medially to the midline, (2) nerve identification between two heads of the gastrocnemius muscle, and (3) protection proximally of the popliteal and distally the posterior tibial artery and vein, including the saphenous vein.

(f) Exposure of the tibial nerve at the lower leg is performed in a supine position with (1) the skin incision along the medial surface of the tibia, (2) nerve identification between the flexor digitorum longus and soleus muscles, and (3) protection of abovementioned vascular elements.

(g) Exposure of the tibial nerve in the tarsal tunnel consists of (1) the skin incision which in its central part curves around the posterior aspect of the medial malleolus and extends 5 cm proximally into the leg, and 5 cm distally along the medial border of the sole, (2) nerve identification anterior to the tendon of the flexor digitorum longus muscle along with vessels, and (3) division of the flexor retinaculum ligament, section of the abductor hallucis muscle fibers if needed, and dissection of all nerve branches.

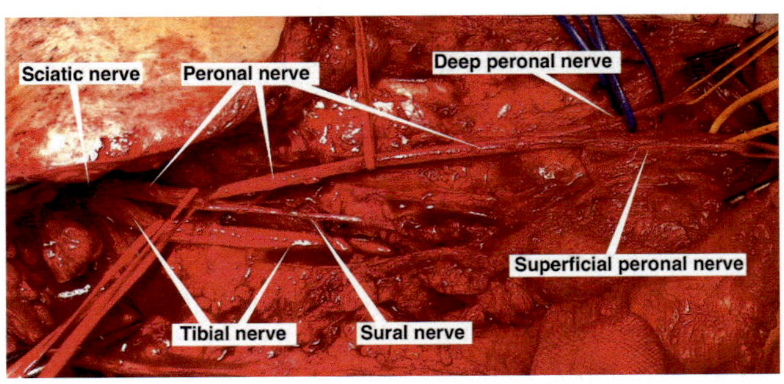

Fig. 68.8 Common peroneal nerve exposure at the knee

68.5 Surgical Pearls

- There is neither the same patient nor the same injury.
- Appropriate indications and timing of surgery are essential.
- The lesser indication may be a cause of greater complication.
- If you doubt, you look but don't cause additional damage.
- Don't look for things you don't want to find.
- Adequate and wide surgical exposure should be a rule.
- Gentle dissection because of significant diffuse scarring in the majority of cases.
- During dissection, don't be curious and don't look for things that you don't want to find.
- Always use microsurgical technique and neuromonitoring when necessary.
- In any complex situation, consider nerve transfer as a valuable option.

Acknowledgments The first author would like to express his gratitude to his Ph.D. students Aleksa Mićić and Stefan Radojević for their contribution to the preparation of this chapter.

References

1. Matloub HS, Yousif NJ. Peripheral nerve anatomy and innervation pattern. Hand Clin. 1992;8(2):201–14.
2. Catala M, Kubis N. Chapter 3—Gross anatomy and development of the peripheral nervous system. In: Said G, Krarup C, editors. Handbook of clinical neurology. 115: Elsevier; 2013. p. 29–41.
3. Stewart JD. Peripheral nerve fascicles: anatomy and clinical relevance. Muscle Nerve. 2003;28(5):525–41.
4. Mazurek MT, Shin AY. Upper extremity peripheral nerve anatomy: current concepts and applications. Clin Orthop Relat Res. 2001;383:7–20.
5. Toomayan GA, Robertson F, Major NM, Brigman BE. Upper extremity compartmental anatomy: clinical relevance to radiologists. Skeletal Radiol. 2006;35(4):195–201.
6. Orebaugh SL, Williams BA. Brachial plexus anatomy: normal and variant. ScientificWorldJournal. 2009;9:300–12.
7. Bayot ML, Nassereddin A, Varacallo M. Anatomy, shoulder and upper limb, brachial plexus. In: StatPearls. Treasure Island, FL: StatPearls Publishing LLC; 2023.
8. Polcaro L, Charlick M, Daly DT. Anatomy, head and neck: brachial plexus. In: StatPearls. Treasure Island, FL: StatPearls Publishing LLC; 2023.
9. Leechavengvongs S, Teerawutthichaikit T, Witoonchart K, Uerpairojkit C, Malungpaishrope K, Suppauksorn S, et al. Surgical anatomy of the axillary nerve branches to the deltoid muscle. Clin Anat. 2015;28(1):118–22.
10. Gurushantappa PK, Kuppasad S. Anatomy of axillary nerve and its clinical importance: a cadaveric study. J Clin Diagn Res. 2015;9(3):Ac13–C17.
11. Welle K, Prangenberg C, Hackenberg RK, Gathen M, Dehghani F, Kabir K. Surgical anatomy of the radial nerve at the dorsal region of the humerus: a cadaveric study. J Bone Joint Surg Am. 2022;104(13):1172–8.
12. Glover NM, Murphy PB. Anatomy, shoulder and upper limb, radial nerve. In: StatPearls. Treasure Island, FL: StatPearls Publishing LLC; 2023.
13. Kubwimana O, Ndata A, Ivang A, Ndahimana P, Nzayisenga A, Byiringiro JC, et al. Musculocutaneous and median nerve branching: anatomical variations. Case Series from UR clinical anatomy and literature review. Afr Health Sci. 2022;22(1):263–8.
14. Encarnacion M, Nurmukhametov R, Barrientos RE, Melchenko D, Goncharov E, Bernard E, et al. Anatomical variations of the median nerve: a cadaveric study. Neurol Int. 2022;14(3):664–72.
15. Savastano L, Yang LJS. Chapter 41—Anatomy of the median nerve and its branches. In: Tubbs RS, Rizk E, Shoja MM, Loukas M, Barbaro N, Spinner RJ, editors. Nerves and nerve injuries. San Diego, CA: Academic Press; 2015. p. 553–62.
16. Polatsch DB, Melone CP Jr, Beldner S, Incorvaia A. Ulnar nerve anatomy. Hand Clin. 2007;23(3):283–9. v
17. Chakravorty A, Jaeger M. Surgical anatomy of the lateral femoral cutaneous nerve for meralgia paraesthetica: a simple technical guide for surgeons and trainees alike. J Clin Neurosci. 2022;101:52–6.
18. Baker MD, Gullett JP. Ultrasound-guided femoral nerve blocks. Pediatr Emerg Care. 2015;31(12):864–8; quiz 9–71
19. Kanawati AJ. Variations of the sciatic nerve anatomy and blood supply in the gluteal region: a review of the literature. ANZ J Surg. 2014;84(11):816–9.
20. Capodici A, Hagert E, Darrach H, Curtin C. An overview of common peroneal nerve dysfunction and systematic assessment of its relation to falls. Int Orthop. 2022;46(12):2757–63.
21. Lau JT, Stavrou P. Posterior tibial nerve--primary. Foot Ankle Clin. 2004;9(2):271–85.
22. Patel NT, Smith HF. Clinically relevant anatomical variations in the brachial plexus. Diagnostics. 2023;13(5):830.
23. Lee SK, Wolfe SW. Peripheral nerve injury and repair. J Am Acad Orthop Surg. 2000;8(4):243–52.
24. Siemionow M, Brzezicki G. Chapter 8: current techniques and concepts in peripheral nerve repair. Int Rev Neurobiol. 2009;87:141–72.
25. Dengler J, Yee A, Mackinnon SE. Supraclavicular approach to the brachial plexus: how I teach it. Ann Thorac Surg. 2019;107(2):331–4.
26. Dawson SE, Gross JN, Berns JM, Weinzerl T, Adkinson JM, Borschel GH. Supraclavicular approach to the brachial plexus. Plast Reconstr Surg Glob Open. 2023;11(1):e4771.
27. Tender GC, Kline DG. The infraclavicular approach to the brachial plexus. Neurosurgery. 2008;62(3 Suppl 1):180–4; discussion 4–5

Part VII

Functional

White Fiber Anatomy Relevant to Epilepsy Surgery

69

Abhidha Shah, Dhiraj Parihar, and Atul Goel

Learning Objectives

- To study the normal anatomy of the white fiber tracts that are relevant for epilepsy surgery.
- To understand the epileptogenic zone that has to be tackled, the eloquent cortex that has to be preserved, and the white matter pathways that come into play during epilepsy surgery.
- To study the surgical steps involved in temporal lobectomy with amygdalohippocampectomy, corpus callosotomy, and hemispherotomy.

69.1 Introduction

69.1.1 History and Background

From ancient history, epilepsy has been an intriguing disease. It was attributed to be due to demonic possessions by some, believed to be a "sacred disease" by ancient Greeks, and considered to be a sign of genius. The oldest account of epilepsy can be found in a 4000-year-old Akkadian tablet from Mesopotamia which depicts the picture of a person having an epileptic attack. The Charaka Samhita accurately describes convulsions and also recognized the cause of the seizures to be the brain. The misconceptions around epilepsy remained largely unchallenged till Hippocrates hypothesized that the cause of epilepsy was the brain. This hypothesis was largely ignored till the seventeenth century and the Renaissance period when the science of epilepsy came into existence.

A. Shah (✉) · D. Parihar
Department of Neurosurgery, Seth G.S. Medical College and K.E.M Hospital, Parel, Mumbai, Maharashtra, India

A. Goel
Department of Neurosurgery, Seth G.S. Medical College and K.E.M Hospital, Parel, Mumbai, Maharashtra, India

Department of Neurosurgery, Lilavati Hospital and Research Center, Mumbai, Maharashtra, India

Epilepsy surgery has evolved dramatically and is now considered as the management of choice for drug-resistant epilepsy [1]. The goal of any epilepsy surgery is complete resection or disconnection of the cortical areas or networks (epileptogenic zone) that are responsible for the generation of seizures. The three categories of epilepsy surgery are resective surgery, disconnection surgery, and stimulation [1]. Resective surgeries include hemispherectomy (resection of the cerebral hemisphere), lobectomy (temporal or extratemporal), and topectomy (resection of focal areas of cortex). Disconnective surgeries include corpus callosotomy, multiple subpial transections, and hemispherotomy (disconnection of the cerebral hemisphere). Stimulation surgeries include vagal nerve stimulation, anterior thalamic stimulation, and responsive neurostimulation.

69.2 Anatomical Highlights

- The uncinate fasciculus, the inferior frontooccipital fasciculus (IFOF), and the Meyer's loop of the optic radiation are the white fibers at risk during surgery for mesial temporal sclerosis.
- A safe entry zone through the middle temporal gyrus and the anterior part of the temporal horn avoids injury to the fibers of the IFOF and the optic radiations.
- An anterior callosotomy disconnects the anterior and dorsal callosal radiations. A complete callosotomy disconnects all the callosal radiations.
- The central core of the brain is connected to the cerebral hemispheres by the cerebral isthmus and the brainstem by the cerebral peduncles.
- A disconnection procedure entails disconnecting the central core from the cerebral hemispheres and the brainstem by severing the cerebral isthmus and then disconnecting it from the opposite hemisphere by performing a corpus callosotomy.

I. N. Kanaan, V. Beneš (eds.), *Neuroanatomy Guidance to Successful Neurosurgical Interventions*,
https://doi.org/10.1007/978-3-031-59838-8_69

69.3 White Fiber Correlates of Temporal Lobectomy and Amygdalohippocampectomy

The temporal lobe with its unique mix of allocortex, mesocortex, and neocortex is considered the seat of epilepsy. Mesial temporal lobe epilepsy is one of the commonest forms of epilepsy. Controversy exists as to whether a temporal lobectomy should be performed along with amygdalohippocampectomy or a whether a selective amygdalohippocampectomy should be performed [2]. Various approaches, like the transsylvian approach, the middle temporal gyrus approach, the temporopolar approach, and the inferior temporal gyrus approach, have been described for performing the amygdalohippocampectomy (Fig. 69.1) [2]. Each of these approaches involves varying degrees of damage to the long association bundles in the vicinity of the temporal lobe and the temporal stem.

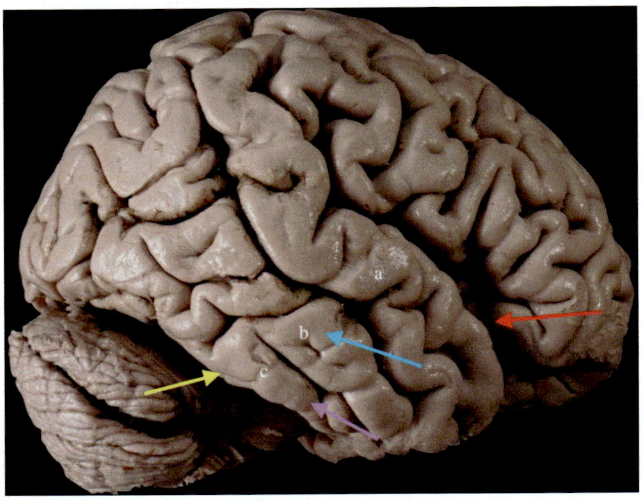

Fig. 69.1 Image showing the lateral surface of the temporal lobe and the various approaches used for temporal lobectomy and amygdalohippocampectomy. Red arrow indicates the transsylvian approach. Blue arrow indicates the approach through the middle temporal gyrus. Purple arrow indicates the approach through the inferior temporal gyrus. Yellow arrow indicates the supracerebellar infratentorial approach

69.4 Approach Via the Temporal Neocortex

The most common surgical technique used to perform a standard temporal lobectomy with amygdalohippocampectomy is via the middle or inferior temporal gyrus/sulcus. The lateral neocortex is initially removed, and an anterior temporal lobectomy is performed. The temporal horn is then entered, and an amygdalohippocampectomy is performed. Our approach to mesial temporal sclerosis involves entering the temporal horn through the middle temporal gyrus and then performing an en bloc resection of the amygdala, hippocampus, and the lateral neocortex. This approach was described by Goel et al. and initially presented as a university thesis in 1998 [2].

69.4.1 Relevant Surgical Anatomy

The long association fibers on the lateral aspect of the temporal lobe in the region of the posterior superior temporal gyrus are the fibers of the arcuate fasciculus (AF) superiorly and laterally and the fibers of the middle longitudinal fasciculus (MLF), the inferior frontooccipital fasciculus (IFOF) and the anterior commissure (AC) inferiorly and medially (Figs. 69.2 and 69.3). The fibers in the region of the anterior superior temporal gyrus are the temporal fibers of the uncinate fasciculus (UF) (Fig. 69.2). The fibers encountered in the region of the middle temporal gyrus are the MLF, the IFOF, temporal fibers of the AC, and the Meyer's loop of the optic radiation from lateral to medial (Figs. 69.2, 69.3, and 69.4). The fibers in the region of the inferior temporal gyrus are the fibers of the inferior longitudinal fasciculus (ILF) (Figs. 69.2, 69.3, and 69.4).

The IFOF connects the lateral orbital gyri, the inferior and middle frontal gyri to the temporal and occipital lobes. The fibers of the IFOF after exiting the temporal stem course posteriorly to form a major component of the sagittal stratum [2, 3]. The fibers of the IFOF, AC, and the optic radiation lie in the roof of the temporal horn in the middle portion of the middle temporal gyrus and in the lateral wall of the temporal horn in the posterior portion of the gyrus (Fig. 69.5). A safe

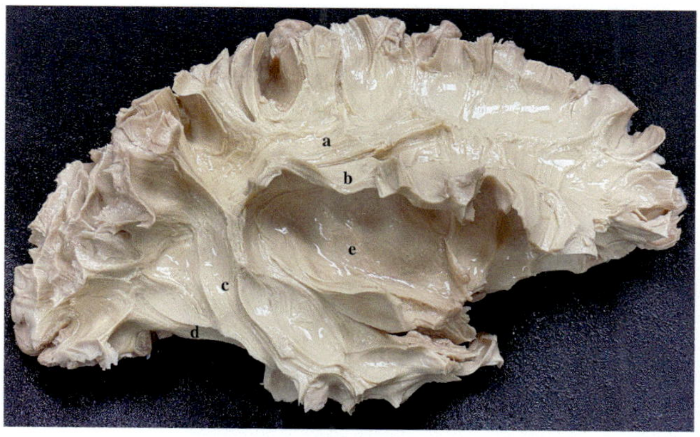

Fig. 69.2 Image showing the white fibers on the lateral surface of the hemisphere. (*a*) Superior longitudinal fasciculus. (*b*) Arcuate fasciculus. (*c*) Middle longitudinal fasciculus. (*d*) Inferior longitudinal fasciculus. (*e*) Extreme capsule

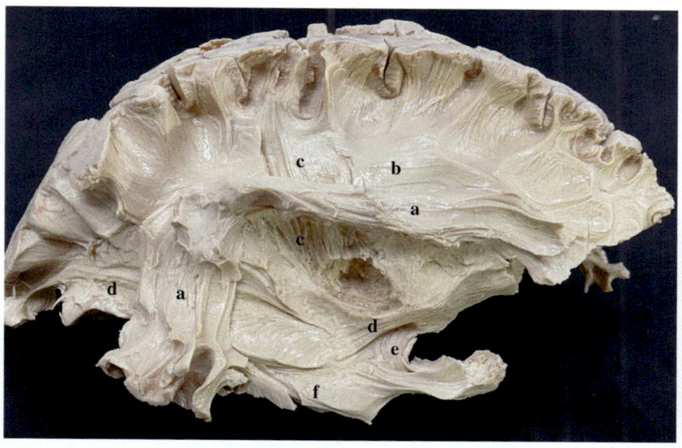

Fig. 69.3 Deeper dissection showing the white fibers on the lateral surface of the hemisphere. (*a*) Arcuate fasciculus. (*b*) Superior longitudinal fasciculus. (*c*) Dorsal external capsule. (*d*) Inferior frontooccipital fasciculus. (*e*) Uncinate fasciculus. (*f*) Inferior longitudinal fasciculus

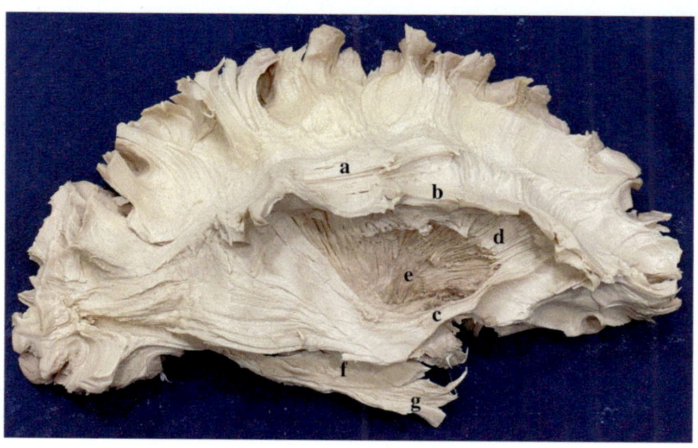

Fig. 69.4 The extreme capsule, parts of the dorsal external capsule, and the lentiform nucleus have been removed. (*a*) Superior longitudinal fasciculus. (*b*) Arcuate fasciculus. (*c*) Inferior frontooccipital fasciculus. (*d*) Dorsal external capsule. (*e*) Internal capsule. (*f*) Temporal horn. (*g*) Inferior longitudinal fasciculus

entry zone through the middle temporal gyrus and the anterior part of the temporal horn avoids injury to the fibers of the IFOF and the optic radiations (Fig. 69.6) [2]. The hippocampus occupies the medial aspect of the floor of the temporal horn of the ventricle and is divided into a head, body, and tail (Fig. 69.7). The tail of the hippocampus is seen to curve medially and then merges with the crura of the fornix. Superior to the hippocampus in the temporal horn is the inferior choroidal point, and it marks the junction of the head and body of the hippocampus. The inferior choroidal point is the point of emergence of the branches of the anterior choroidal artery as the choroid plexus into the temporal horn (Fig. 69.7). The amygdala can be recognized as a greyish structure, which lies at the junction of the roof and anterior wall of the temporal horn. The incision to resect the amygdala is made at the junction of the roof and anterior wall of the temporal horn, continued laterally beneath the roof of the temporal horn parallel to the lie of the hippocampus [4]. Extending the incision above this junction and deepening it will cause injury to the structures of the basal ganglia.

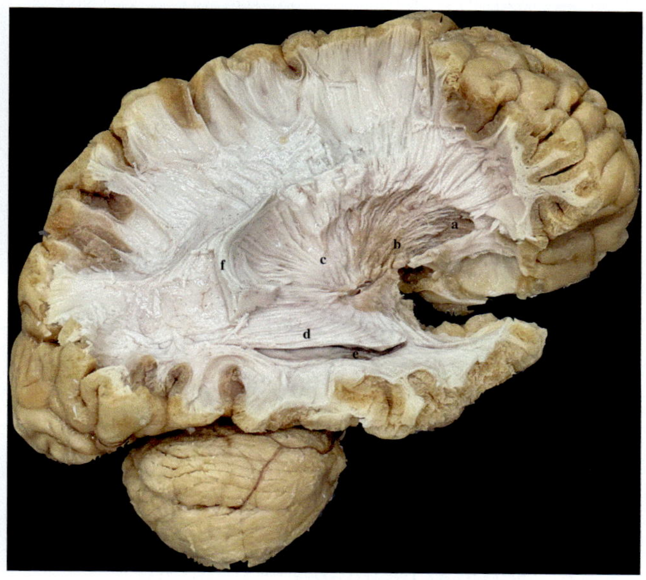

Fig. 69.5 The insular region has been dissected to show the entire extent of the internal capsule. The relation of the Meyer's loop to the temporal horn can be appreciated. (*a*) Anterior limb. (*b*) Genu. (*c*) Posterior limb. (*d*) Optic radiations. (*e*) Temporal horn. (*f*) Arcuate Fasciculus

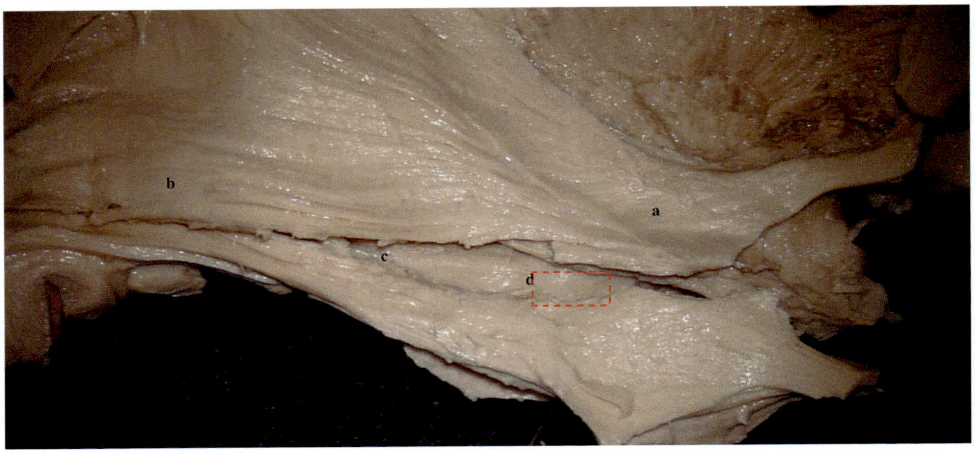

Fig. 69.6 The fibers of the IFOF, AC, and the optic radiation lie in the roof of the temporal horn in the anterior portion of the middle temporal gyrus and in the lateral wall of the temporal horn in the posterior portion of the gyrus. (*a*) IFOF. (*b*) Sagittal stratum. (*c*) Temporal horn. (*d*) Hippocampus. The rectangular area depicts a safe entry zone through the middle temporal gyrus and the anterior part of the temporal horn that avoids injury to the fibers of the IFOF and the optic radiations

Fig. 69.7 Image showing the hippocampus lying in the medial part of the floor of the temporal horn. (*a*) Head of hippocampus. (*b*) Body of hippocampus. (*c*) Tail of hippocampus. (*d*) Inferior choroidal point. (*e*) Temporal horn

69.4.2 Surgical Technique—Standard Temporal Lobectomy and Amygdalohippocampectomy Through the Middle Temporal Gyrus

A slightly anteriorly curving incision is taken starting 1 cm below the zygoma and extending superiorly about 1.5 cm anterior to the tragus. The temporalis fascia and muscle are dissected and reflected protecting the frontalis branches of the facial nerve and the neurovascular supply of the temporalis muscle. The temporalis muscle can be reflected anteriorly or inferiorly or split to obtain adequate exposure [5]. A temporal craniotomy is performed remaining below the superior temporal line. The basal extent of the exposure is till the root of the zygoma and till the floor of the middle cranial fossa. The dura is opened in a cruciate manner. The dural opening exposes the inferior portion of the superior temporal gyrus, the middle, and the inferior temporal gyri. The white fibers that can be damaged depend on the gyrus that is entered.

69.4.3 Safe Zone to Enter the Temporal Horn

The fibers of the IFOF, temporal extension of the AC, and the optic radiations form the major component of the sagittal stratum in the region of the temporal horn. As seen in the figure, the fibers initially course in the roof of the temporal horn and further posteriorly course in the lateral wall of the temporal horn. We described a safe area in the anterior portion of the lateral wall of the temporal horn (an average of

20 mm from the anterior wall of the temporal horn), which was devoid of any major fibers (Figs. 69.6 and 69.7) [2]. The amygdala, which lies at the junction of the anterior wall and the roof of the temporal horn, lies anterior to the sagittal stratum in this region and hence can be resected at this junction. We advocate that the temporal horn should be entered through this safe area (Fig. 69.6). Further disconnection and dissection should be carried out close to the hippocampus in the floor of the temporal horn to avoid injury to the sagittal stratum in the roof and lateral wall of the temporal horn.

69.4.4 Technique of *En Bloc* Amygdalohippocampectomy and Anterior Temporal Lobectomy

A corticectomy is performed in the middle temporal gyrus superior to the inferior temporal sulcus about 25 mm from the temporal pole. The corticectomy is deepened till the safe entry zone (25–30 mm) in the anterior portion of the temporal horn is entered, and the hippocampal head is visualized. From this point, all further dissection is carried out near the hippocampus inferior to the roof of the temporal horn [2]. The amygdala is disconnected first anteriorly and superiorly; then, the incision is extended into the floor of the temporal horn superior to the hippocampal head till the choroidal fissure. The incision is deepened across the entire extent of the hippocampus in the floor, including the parahippocampal gyrus till the arachnoid of the tentorial incisura is visualized. The arachnoid layer is not breached to protect the posterior cerebral artery, the third nerve and the cerebral peduncle. The incision is then extended inferiorly in the region where the tail of the hippocampus turns in the temporal horn and the tail is disconnected. This completes the disconnection of the amygdala and the hippocampus. Then, the whole lateral neocortex lateral to the disconnected amygdala and hippocampus is disconnected from the temporal lobe. The whole specimen consisting of the amygdala, hippocampus, uncus, parahippocampus, and the lateral neocortex is excised en bloc, thus completing the resection.

69.4.5 Alternative Approaches to Enter the Temporal Horn

The middle temporal gyrus approach to the temporal horn carries a risk of injury to the optic radiations and the IFOF. The technique described by Goel et al. of en bloc amygdalohippocampectomy with temporal lobectomy through the safe entry zone avoids injury to the fibers of the Meyer's loop and the IFOF [2]. The subtemporal approach avoids injury to the visual fibers as the temporal horn is entered through its floor. The minimally invasive subtemporal selective amygdalohippocampectomy entails a low temporal craniotomy and entry into the temporal horn via the collateral sulcus with the aid of image guidance. Even though the approach avoids injury to the sagittal stratum, a large recent study showed a significant change in word retrieval irrespective of side of surgery and a change in verbal memory more on the dominant side using this approach [6, 7]. The paramedian supracerebellar approach through the fusiform gyrus avoids injury to major subcortical fibers; however, it involves a long working trajectory to reach the most anterior portion of the amygdala [8].

Yasargil described the transsylvian approach to perform a selective amygdalohippocampectomy [9]. The procedure entailed wide opening of the proximal sylvian fissure, dissection of the middle cerebral artery branches, and sylvian veins and then making an incision in the limen insula and the inferior limiting sulcus to enter the roof of the temporal horn. In the region of the limen insulae, the UF and the IFOF run in close proximity to each other. This approach carries a risk of injury to the fibers of the IFOF and the UF as the incision traverses through the temporal stem. Subsequently, Yasargil modified the approach to the mesial temporal region by making an incision in the piriform cortex [10]. The amygdala was first removed, and the temporal horn was then entered to resect the hippocampus. This modification decreased the risk of injury to the IFOF but still carried a high risk of injury to the UF.

69.5 White Fiber Correlates of Corpus Callosotomy

The technique of surgical disconnection of the corpus callosum was first proposed by Van Wagenen and Herren with the aim of interruption of the interhemispheric transfer of impulses to control epilepsy [11]. The procedure disrupts the white fibers connecting the hemispheres and thus inhibits seizure propagation. Corpus callosotomy has been used to treat a wide variety of generalized seizures. The best results have been obtained in patients with atonic seizures or drop attacks. Pediatric conditions such as Lennox-Gastaut syndrome, Rasmussen's encephalitis, and infantile hemiplegia have been effectively treated with callosotomy.

69.5.1 Relevant Surgical Anatomy

The corpus callosum lies in between the two medial surfaces of the hemisphere and is divided into the rostrum, genu, body, isthmus, and splenium. The corpus callosum comprises transverse running fibers, extending from the rostrum to the splenium connecting both the cerebral hemispheres. The white fiber connections or radiations of the corpus callosum can be divided into four types such as superior/dorsal connections, ventral/inferior connections, anterior connections, and posterior connections (Fig. 69.8) [11]. Fibers emanate from the body of the corpus callosum curve either superiorly or inferiorly. The superior or the dorsal radiations are those fibers that traverse laterally and enter the medial surface of the hemisphere and then curve upwards. They traverse below the cingulum and the medial portion of the superior longitudinal fasciculus (SLF) and then turn upward lateral to the medial SLF (SLF—I) to join the fibers of the corona radiata. Fibers in the anterior part of the body connect the medial portion of the superior frontal gyrus of both the hemispheres. Fibers in the posterior part of the body connect the posterior portion of the medial superior frontal gyrus (supplementary motor area), precentral and postcentral gyri or the paracentral lobule of the two halves of the hemispheres. Fibers in the region of the isthmus, just anterior to the splenium, connect the region of the precuneus of the two hemispheres to each other.

The inferior or the ventral callosal radiations are those that traverse laterally from the genu and the body of the corpus callosum and curve inferiorly (Fig. 69.8). The ventral callosal radiations have an anterior and a posterior component. The anterior component of the ventral callosal radiations consists of fibers emanating from the genu and traversing laterally and inferiorly to terminate in the region of the caudate nucleus [11]. The posterior component of the ventral callosal radiations courses laterally, forming the roof of the body, atrium, and occipital horn of the lateral ventricle, and, then it turns inferiorly to form the innermost layer of the lateral wall of the temporal and occipital horns of the lateral ventricle. These fibers are better known as the tapetum [11]. The anterior radiations or the forceps minor originate from the genu and rostrum of the corpus callosum and curve anteriorly into the medial and basal frontal lobes (Fig. 69.8). The posterior radiations or the forceps major traverse posteriorly from the splenium of the corpus callosum and enter the medial aspect of the occipital lobes (Fig. 69.8).

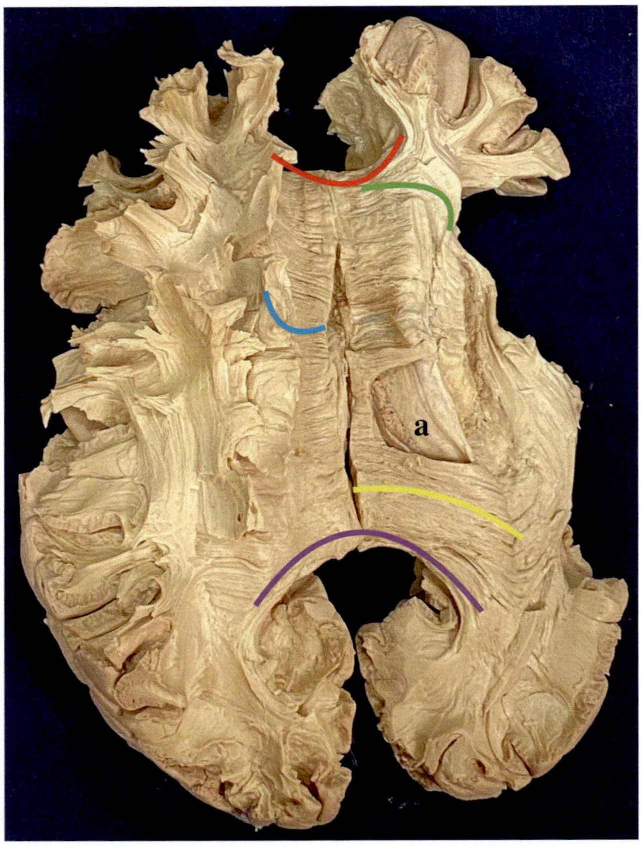

Fig. 69.8 Superior view of the corpus callosum and its callosal radiations. Red line indicates the anterior callosal radiations or forceps minor. (*a*) Lateral ventricle. Green line indicates the anterior ventral callosal radiations. Blue line indicates the dorsal callosal radiations. Yellow line indicates the posterior ventral callosal radiations. Purple line indicates the posterior callosal radiations or forceps major

Disconnection of the anterior commissure is also an essential component of surgery. The anterior commissure is shaped like a bicycle handle and has a frontal and a temporal component. The frontal component connects the olfactory regions on the basal frontal surface of the lobes to each other. The temporal component crosses beneath the internal capsule and traverses in the lateral wall of the temporal horn as a component of the sagittal stratum [3].

69.5.2 Surgical Technique of Callosotomy

A parasagittal craniotomy is performed with the midpoint just anterior to the coronal suture. If a complete callosotomy is being planned, then the craniotomy is larger. After performing the craniotomy, the dura is reflected based on the superior sagittal sinus. The interhemispheric fissure is then entered, taking care to avoid the bridging veins. The fissure is opened, taking care to avoid damage to the pericallosal and the callosomarginal arteries. The corpus callosum can be identified by its glistening white color. The corpus callosum must be distinguished from the cingulate gyri which often times are adherent to each other. The posterior end of the genu is identified. The callosotomy is commenced 3 cm posterior to the genu. The incision is deepened ventrally till the leaves of the septum pellucidum or the cavum septum pellucidum is identified. The anterior commissure is then identified and sectioned. For an anterior callosotomy, the incision extends posteriorly up to the isthmus of the corpus callosum, which lies at the junction of the body and the splenium. For a complete callosotomy, even the splenium has to be disconnected (Fig. 69.9). The landmark to indicate a complete disconnection of the splenium is the cavum velum interpositum. At the end of the disconnection, the vein of Galen should be visualized through the arachnoid membrane. Injury to the fornices should be avoided [12, 13].

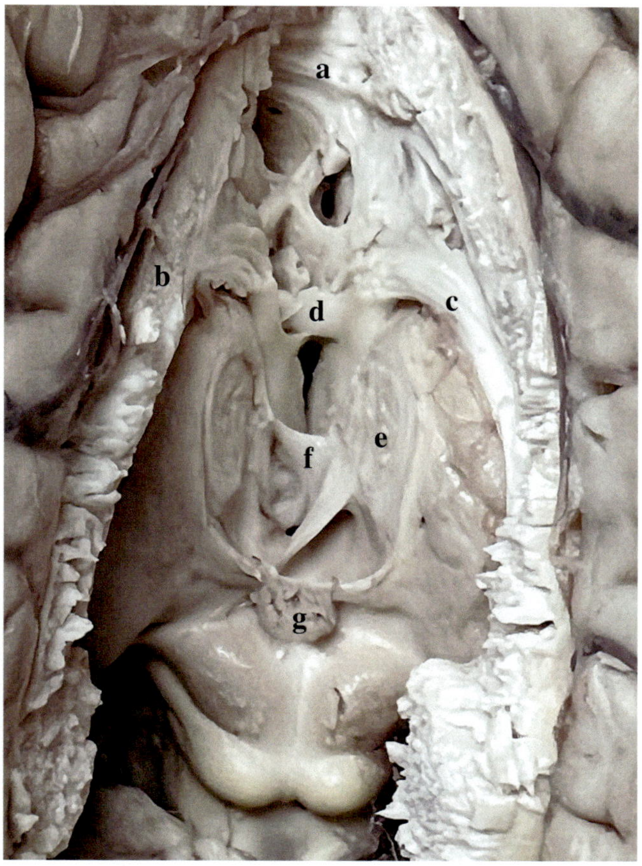

Fig. 69.9 Image showing the view after a complete callosotomy. (*a*) Rostrum of corpus callosum. (*b*) Callosotomy. (*c*) Fornix. (*d*) Anterior commissure. (*e*) Thalamus. (*f*) Massa intermedia. (*g*) Pineal body

69.6 White Fiber Correlates of Hemispherectomy and Hemispherotomy

The disconnection of one hemisphere from the contralateral hemisphere is performed for chronic refractory seizures that occur due to widespread hemispheric damage. Krynauw performed the first hemispherectomy for epilepsy in 1950 [14]. However, over the years, this procedure was largely abandoned due to a high incidence of complications related to blood loss, a large postresection subdural cavity, superficial cerebral hemosiderosis, occlusion of the foramen of Monro and aqueduct, and development of hydrocephalus. In view of the complications of the procedure, the concept of functional hemispherectomy was propagated which involved disconnection of the white matter tracts without removal of large portions of the hemisphere. Rasmussen's functional hemispherectomy consisted of removal of the temporal lobe, the supra sylvian central cortex combined with callosotomy, and disconnection of the frontal, parietal, and temporal lobes. The functional hemispherectomy has further evolved to a hemispherotomy, which involves mainly disconnection. Many variations of hemispherotomy procedures have been discussed in the literature [15].

The essential steps of the procedure include removal of mesial temporal lobe structures, interruption of the internal capsule and the corona radiata fibers, corpus callosotomy, disconnection of the frontal lobes, and the occipitoparietal disconnection.

The essential difference in the various described types of hemispherotomy is the avenue that is used to reach the ventricles: the lateral avenue—transylvian periinsular hemispherotomy or the medial avenue—the vertical parasagittal hemispherotomy.

69.6.1 Relevant Surgical Anatomy

The central core of the brain occupies a position on top of the brainstem and essentially lies between the insula laterally and the ventricular system medially (Fig. 69.10 and 69.11). It consists from lateral to medial of the extreme capsule, claustrum, external capsule, the putamen, the globus pallidus, the internal capsule, the caudate nucleus, ventral striatum, the stria terminalis, the septal region, and the thalamus [3, 16, 17]. The central core is connected to the cerebral hemispheres by the cerebral isthmus and to the brainstem by the cerebral peduncles. The cerebral isthmus consists of the terminations of the extreme, external and internal capsules, the temporal extension of the anterior commissure, and the ventral amygdalofugal pathway [3, 16, 17]. This isthmus is readily identifiable as a thick bundle of fibers in the region of the limiting sulci of the insula.

The insula can be viewed as an external shield for the central core of the brain. The insula is bounded by the circular sulcus of Reil, which can be divided into three parts: the anterior, superior, and inferior limiting or periinsular sulci.

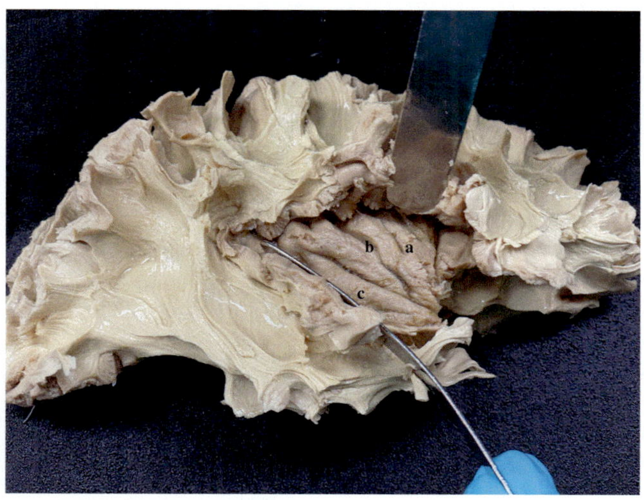

Fig. 69.10 Image showing the insula after the opercula have been lifted. (*a*) Anterior short insular gyri. (*b*) Central sulcus of insula. (*c*) Posterior long insular gyri

The lateral surface of the insula is divided by the central sulcus of the insula into an anterior and a posterior portion. The smaller anterior part of the insula is made up of the transverse, accessory and the anterior, and middle and posterior short gyri [3, 16]. The larger posterior part of the insula consists of the anterior and posterior long gyri separated by the postcentral insular sulcus.

Removal of the insular gyri reveals the extreme capsule (Fig. 69.2). The extreme capsule connects the adjacent insular gyri to each other and also consists of vertical fibers, which connect the insular gyri to the frontal parietal and temporal opercula. The claustrum lies beneath the extreme capsule, which is a thin layer of gray matter. Removing the gray matter of the claustrum reveals the external capsule (Fig. 69.3). The external capsule consists of dorsal and ventral parts (Fig. 69.3). The dorsal external capsule is composed of the claustrocortical fibers. This fiber system connects the claustrum to the frontoparietal cortex between the supplementary motor area and the posterior part of the parietal lobe. The ventral external capsule is composed of the IFOF superiorly and the UF inferiorly (Fig. 69.3). Removal of the external capsule leads to the gray matter of the lentiform nucleus. The lentiform nucleus is composed of the putamen laterally and the globus pallidus medially. Once the putamen is removed, one can see the temporal extension of the anterior commissure and the globus pallidus (Fig. 69.12). The globus pallidus is seen to lie just behind the anterior commissure and faces the genu of the internal capsule medially and the posterior limb of the external capsule posteriorly. Removing the lentiform nucleus exposes the entire extent of the internal capsule, which is discussed later in detail (Figs. 69.12 and 69.13). The external and internal capsules are capsules of the lentiform nucleus. The anterior, superior, and inferior limiting sulci are very closely related to the internal capsule medially. The anterior insular point (junction of the anterior and superior limiting sulci) is related to the anterior limb of the internal capsule. The posterior insular point (junction of the superior and inferior limiting sulci) is closely related to the posterior limb of the corticospinal tract and superior thalamic radiation and the

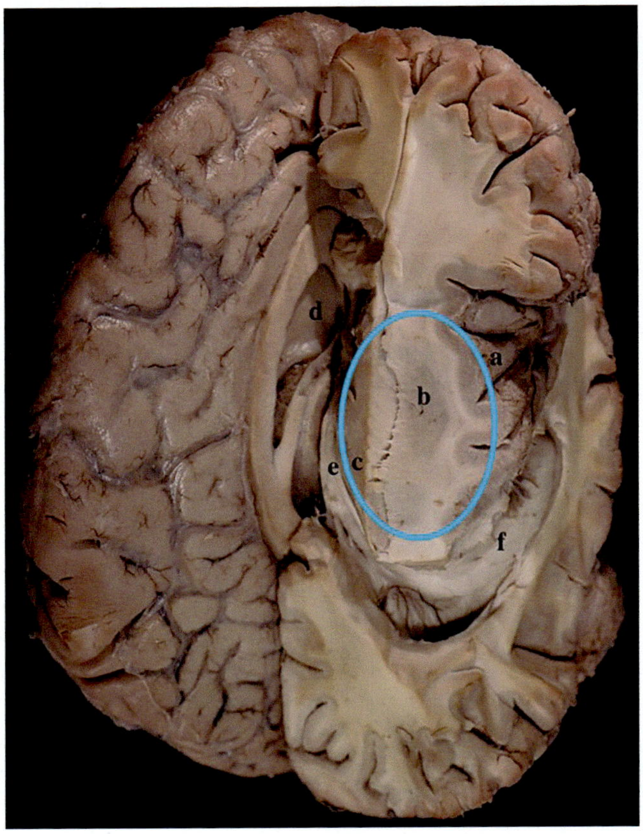

Fig. 69.11 Image showing the central core of the brain (blue circle). (*a*) Insula. (*b*) Basal ganglia. (*c*) Thalamus. (*d*) Caudate head. (*e*) Fornix. (*f*) Hippocampus

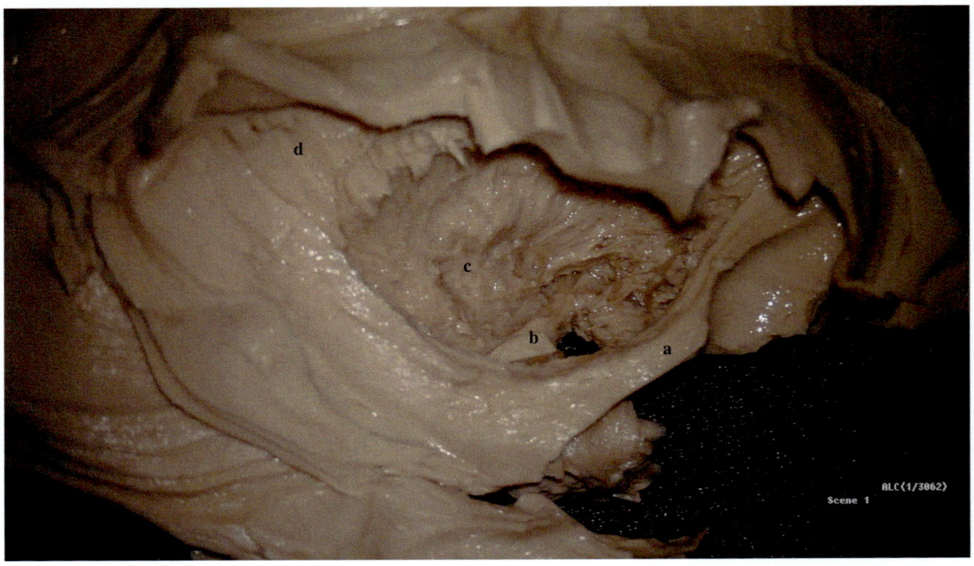

Fig. 69.12 Image showing white fiber dissection of the insular region. The putamen has been removed and the temporal extension of the anterior commissure can be visualized. (*a*) IFOF. (*b*) Temporal extension of the anterior commissure. (*c*) Globus pallidus. (*d*) Dorsal external capsule

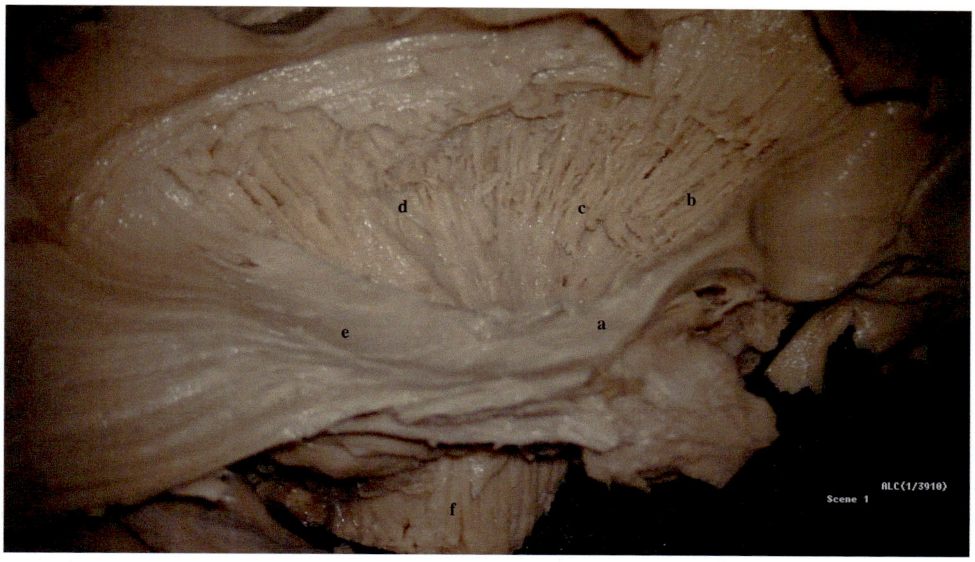

Fig. 69.13 Image showing the entire extent of the internal capsule. (*a*) IFOF. (*b*) Anterior limb. (*c*) Genu. (*d*) Posterior limb. (*e*) Optic radiations. (*f*) Cerebral peduncle

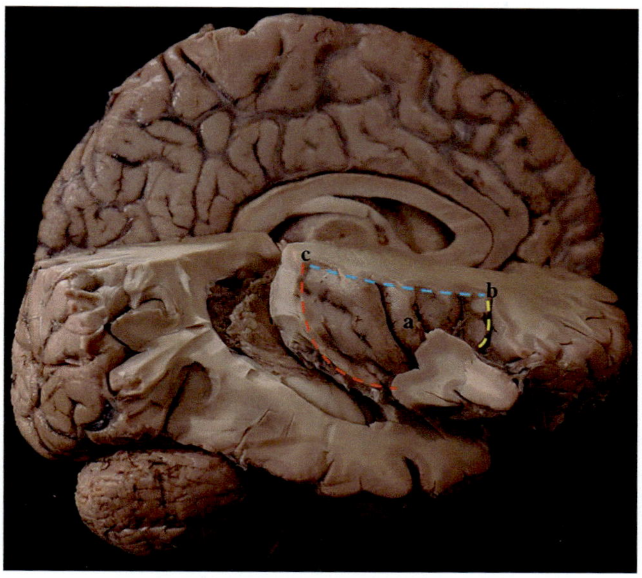

Fig. 69.14 Image showing the insula, central core of the brain, and the lateral ventricle. (*a*) Insular gyri. (*b*) Anterior insular point. (*c*) Posterior insular point. Blue line indicates the superior limiting sulcus. Red line indicates the inferior limiting sulcus. Yellow line indicates the anterior limiting sulcus

inferior limiting sulcus is medially related to the temporal stem (Fig. 69.14). The temporal stem is usually described as the area between the inferior limiting sulcus and the roof of the temporal horn. It is made up of the insuloopercular fibers, the ventral external capsule (UF and the IFOF), middle longitudinal fasciculus, anterior commissure, and the optic and auditory radiations [16].

69.6.2 Surgical Technique of Transylvian Hemispherotomy (Figs. 69.15, 69.16, and 69.17)

A frontotemporal craniotomy is performed, and the dura is opened. The sylvian fissure is opened widely till the three limiting sulci of the insula are identified. The disconnection is begun by incising the inferior limiting sulcus and is extended posteriorly till the atrium is visualized [14, 15]. This transects the insuloopercular fibers to the temporal and occipital lobes, the UF, the IFOF, the MLF, the temporal component of the anterior commissure, the optic radiations, the auditory radiations, tapetum, and the tail of the caudate nucleus. Next, the mesial temporal structures are removed which include the amygdala, hippocampus, uncus, and the parahippocampal gyrus. The next step is making an incision in the superior limiting sulcus. This step disconnects the extreme capsule, the dorsal external capsule, and the internal capsule. The first fibers to be transected are the insuloopercular fibers, connecting the insula to the frontoparietal operculum. The dorsal external capsule and the internal capsule merge above the superior limiting sulcus to form the corona radiata. The incision through the superior limiting sulcus transects the anterior limb, the genu and the posterior limbs of the internal capsule. Some fibers of the arcuate fasciculus and the SLF that run in the depth of the superior limiting sulcus are also transected. The incision is deepened from posterior to anterior till the body and the frontal horn of the ventricle are entered. The next step involves incising the anterior limiting sulcus and the frontobasal surface. This step

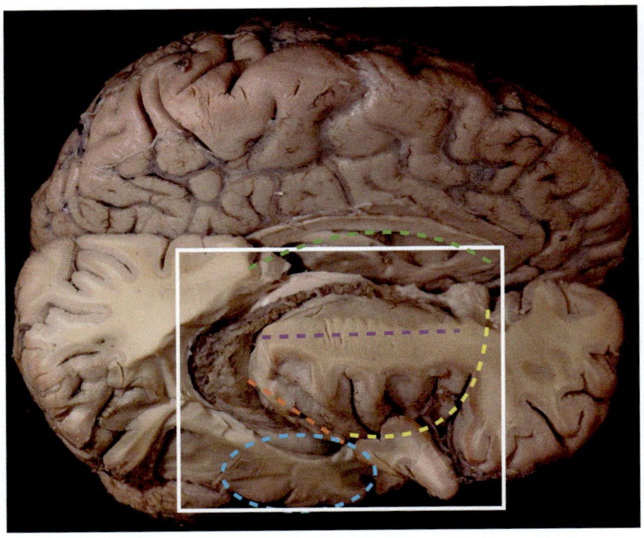

Fig. 69.15 Steps of a transylvian hemispherotomy. Step 1: Orange line indicating incising the inferior limiting sulcus. Step 2: Blue circle indicating removal of mesial temporal structures. Step 3: Purple line indicating incising the superior limiting sulcus. Step 4: Yellow line indicating incising the anterior limiting sulcus and the frontobasal surface. Step 5: Green line indicating intraventricular callosotomy. The white rectangle depicts the disconnected hemisphere

Fig. 69.16 The white fibers that are disconnected whilst performing the transylvian hemispherotomy. Step 1: Orange line indicating incising the inferior limiting sulcus. Step 2: Blue circle indicating removal of mesial temporal structures. Step 3: Purple line indicating incising the superior limiting sulcus. Step 4: Yellow line indicating incising the anterior limiting sulcus and the frontobasal surface

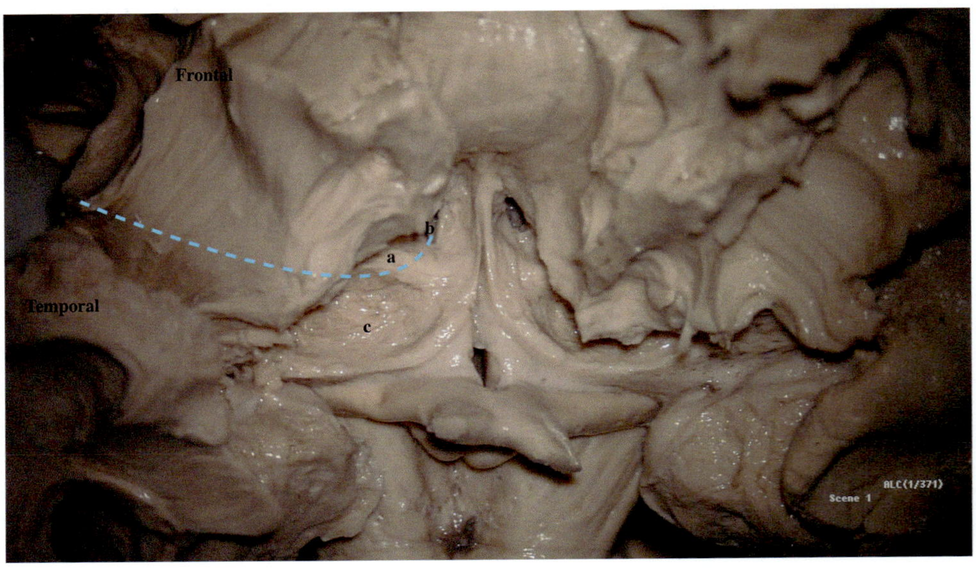

Fig. 69.17 Disconnection on the basal surface of the hemisphere. The blue line shows the incision that travels through the orbital gyri, the olfactory tract, and the gyrus rectus. This part of the incision transects the anterior or the frontal component of the anterior commissure. (*a*) Frontal component of the anterior commissure. (*b*). Frontal horn. (*c*). Substantia innominata

transects the insulooperacular fibers of the extreme capsule, which connect the insula to the frontal lobe, the frontal component of the UF and the IFOF and the most inferior portion of the anterior limb of the internal capsule. Deepening the incision will open the floor of the frontal horn of the lateral ventricle. The incision then continues in a lateral to medial direction across the floor of the frontal horn to reach the interhemispheric fissure. The incision travels through the orbital gyri, the olfactory tract and the gyrus rectus. This part of the incision transects the anterior or the frontal component of the anterior commissure. Care is taken to see that the incision remains anterior to the middle cerebral artery, the anterior perforated substance, and the proximal anterior cerebral artery. The last step of the surgery involves an intraventricular callosotomy. Starting anteriorly the floor, anterior wall, roof, body, and medial wall of the atrium are incised. The incision in the corpus callosum is deepened till the distal anterior cerebral arteries and the free edge of the falx is visualized. Posteriorly, the junction of the falx and the tentorium confirms the disconnection of the splenium. The incision is then turned into the medial wall of the atrium severing the forceps major and then continued inferiorly where it continues into the floor of the atrium and then merges with the earlier hippocampal removal, thus completing the hemispherotomy. Care must be taken to avoid injury to the internal cerebral veins and the vein of Galen.

Several modifications of the hemispherotomy procedure have been described. Schramm described a cortical incision extending from the superior temporal gyrus to the inferior frontal gyrus and then entering the ventricle and performing the disconnection. Villemure and Mascot described making suprasylvian and infrasylvian windows through the frontoparietal operculum and the superior temporal gyrus respectively followed by an insular resection and then entering the limiting sulci of the insula and performing the disconnection.

69.6.3 Surgical Technique of Vertical Parasagittal Hemispherotomy

The technique was described by Delalande et al. and maybe called as a dorsal transcortical subinsular hemispherotomy [18]. A parasagittal craniotomy one-third anterior and two-third posterior to the coronal suture is performed. A cortical window is made and the lateral ventricle is entered exposing the foramen of Monroe and the posterior thalamic region. A callosotomy is then performed extending posteriorly and the splenium disconnected (Fig. 69.18). The posterior column of the fornix is then disconnected at the level of the trigone. Then, a vertical incision is made lateral to the thalamus, which extends from the trigone to the anterior part of the temporal horn (Fig. 69.19). In the anterior part of the callosum, the genu is then disconnected. The incision is then continued laterally through the gyrus rectus, and the caudate nucleus till the anterior part of the temporal horn is reached. This completes the hemispherotomy [18, 19].

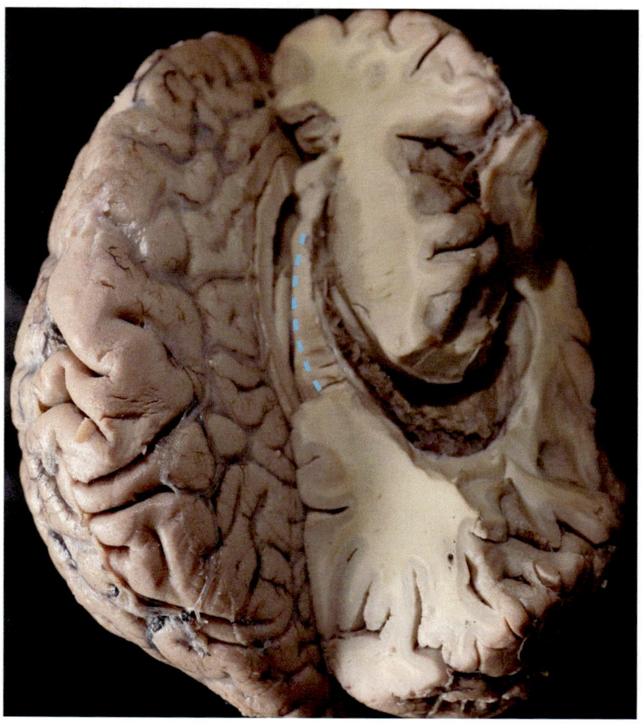

Fig. 69.18 Superior view of the cerebral hemisphere. The blue line depicts the callosotomy

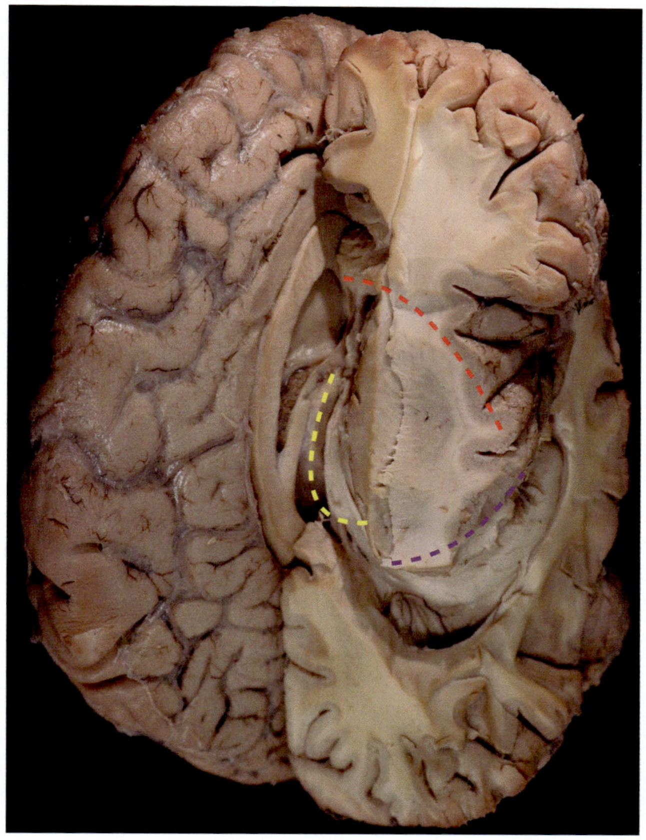

Fig. 69.19 Steps of vertical parasagittal hemispherotomy. The yellow line shows the disconnection of the fornix. The purple line shows the incision lateral to the thalamus, which extends from the trigone to the anterior part of the temporal horn. The red line shows the disconnection of the anterior part of the callosum continued laterally through the gyrus rectus, the caudate nucleus till the anterior part of the temporal horn is reached

69.7 Surgical Pearls

- Damage to the fibers of the IFOF and the optic radiations can be avoided by entering the temporal horn through the described safe entry zone through the anterior part of the temporal horn.
- Traditionally, an anterior callosotomy is preferred to decrease the morbidity associated with sectioning of the splenium but this often requires a second stage surgery and a complete callosotomy for persistent seizures and

drop attacks. A complete callosotomy is associated with better seizure outcomes [11].

- For an effective disconnection, we recommend that the callosotomy should include the posterior part of the body of the callosum to disconnect the supplementary motor, primary motor, and sensory areas.
- A selective posterior callosotomy as described by Paglioli et al. involves a disconnection of the posterior corpus callosum with sectioning of the premotor and motor fibers and sparing of prefrontal connectivity [20, 21].
- Incising the inferior limiting sulcus disconnects the temporal stem. Incising the superior limiting sulcus disconnects the internal capsule. Incising the anterior limiting sulcus and extending it medially disconnects the frontobasal surface.
- It is important to preserve the MCA and the ACA while performing the basal disconnection in a hemispherotomy.

69.8 Discussion

Surgery for epilepsy has evolved significantly and offers patients with drug-resistant epilepsy a chance of having seizure freedom and an improved quality of life. Newer imaging modalities have been effective in accurately localizing the epileptogenic zone and defining the specific areas to be resected or disconnected. Epilepsy is now being recognized as a network disorder where the occurrence of seizures is due to a disarray in the white matter connectivity. Understanding the white matter anatomy of the brain is a requisite for performing advanced epilepsy surgery. The propagation of seizure activity is through these white fiber bundles and disconnection of these fibers is essential for a successful outcome. Resective surgery involves removal of the epileptogenic focus and disconnecting it from the underlying white fibers, thus achieving seizure freedom. The most common resective procedures performed are for hippocampal sclerosis and focal cortical dysplasia. There has been a recent interest in surgery for resecting epileptogenic zones in the insula. Patients with nonlesional epilepsy and hemispheric epileptogenicity can be potential candidates for disconnective epilepsy surgery. The principles of hemispherotomy are to disconnect the vertical group of projection fibers, the commissural fibers, resection of mesial temporal structures, and orbitofrontal disconnection. This essentially disconnects the central core of the brain from the neocortex, thus halting seizure propagation. Surgery for epilepsy has a better outcome if offered early to patients with drug-resistant epilepsy. With advanced neuroimaging and anatomical understanding, surgery is beneficial in patients with drug-resistant epilepsy.

69.9 Conclusions

Epilepsy surgery has gratifying outcomes in patients suffering from intractable epilepsy. Anatomical understanding is a key in performing accurate resections and disconnections.

References

1. Mueller J, Hirsch L. Diagnosis and classification of seizures and epilepsy. In: Winn R, editor. Youmans and Winn's neurological surgery. 8th ed. Elsevier; 2023. p. 2414–31.
2. Shah A, Lunawat A, Jhawar SS, Goel A, Goel A. White fiber correlates of amygdalohippocampectomy through the middle temporal gyrus approach. World Neurosurg. 2022;157:e156–65.
3. Shah A, Goel A, Jhawar SS, Patil A, Rangnekar R, Goel A. Neural circuitry: architecture and function-a fiber dissection study. World Neurosurg. 2019;125:e620–38.
4. Shah A, Jhawar SS, Goel A. Analysis of the anatomy of the Papez circuit and adjoining limbic system by fiber dissection techniques. J Clin Neurosci. 2012;19(2):289–98.
5. Goel A. Splitting of temporalis muscle for basal exposure—technical note. Neurol Med Chir (Tokyo). 1997;37(1):59–62.
6. Little AS, Smith KA, Kirlin K, et al. Modifications to the subtemporal selective amygdalohippocampectomy using a minimal-access technique: seizure and neuropsychological outcomes. J Neurosurg. 2009;111(6):1263–74.
7. Whiting AC, Chen T, Swanson KI, Walker CT, Godzik J, Catapano JS, Smith KA. Seizure and neuropsychological outcomes in a large series of selective amygdalohippocampectomies with a minimally invasive subtemporal approach. J Neurosurg. 2020;12:1–9.
8. Harput MV, Türe U. The paramedian supracerebellar-transtentorial selective amygdalohippocampectomy for mediobasal temporal epilepsy. Oper Neurosurg (Hagerstown). 2018;14(4):462.
9. Yasargil MG, Türe U, Boop FA. Selective amygdalohippocampectomy. In: Kaye AH, Black PM, editors. Operative neurosurgery, vol. 2. London: Harcourt Publishers Limited; 2000. p. 1275–83.
10. Yasargil MG, Krayenbuhl N, Roth P, et al. The selective amygdalohippocampectomy for intractable temporal limbic seizures. J Neurosurg. 2010;112(1):168–85.
11. Shah A, Jhawar S, Goel A, Goel A. Corpus callosum and its connections: a fiber dissection study. World Neurosurg. 2021;151:e1024–35.
12. Shah A, Dandpat S, Goel A. Anatomic correlates and surgical experience with goel orbital cortical approach to caudate head tumors. World Neurosurg. 2021;148:e674–9.
13. Uda T, Kunihiro N, Umaba R, Koh S, Kawashima T, Ikeda S, Ishimoto K, Goto T. Surgical aspects of corpus callosotomy. Brain Sci. 2021;11(12):1608.
14. Schramm J. Hemispheric disconnection procedures. In: Winn R, editor. Youmans and Winn's neurological surgery. 8th ed. Elsevier; 2023. p. 2777–802.
15. Morino M, Shimizu H, Ohata K, Tanaka K, Hara M. Anatomical analysis of different hemispherotomy procedures based on dissection of cadaveric brains. J Neurosurg. 2002;97(2):423–31.
16. Ribas EC, Yağmurlu K, de Oliveira E, Ribas GC, Rhoton A. Microsurgical anatomy of the central core of the brain. J Neurosurg. 2018;129(3):752–69.
17. Kucukyuruk B, Yagmurlu K, Tanriover N, Uzan M, Rhoton AL Jr. Microsurgical anatomy of the white matter tracts in hemispherotomy. Neurosurgery. 2014;10(Suppl 2):305–24; discussion 324
18. Delalande O, Bulteau C, Dellatolas G, Fohlen M, Jalin C, Buret V, Viguier D, Dorfmüller G, Jambaqué I. Vertical parasagittal hemi-

spherotomy: surgical procedures and clinical long-term outcomes in a population of 83 children. Neurosurgery. 2007;60(2 Suppl 1):ONS19–32; discussion ONS32

19. Uda T, Tamrakar S, Tsuyuguchi N, Kawashima T, Goto H, Nakajo K, Umaba R, Sato H, Ohata K. Anatomic understanding of vertical hemispherotomy with cadaveric brains and intraoperative photographs. Oper Neurosurg (Hagerstown). 2016;12(4):374–82.
20. Paglioli E, Martins WA, Azambuja N, Portuguez M, Frigeri TM, Pinos L, Saute R, Salles C, Hoefel JR, Soder RB, da Costa JC, Hemb M, Theys T, Palmini A. Selective posterior callosotomy for drop attacks: a new approach sparing prefrontal connectivity. Neurology. 2016;87(19):1968–74.
21. Gerard C, Byrne R. Surgical techniques for non—temporal lobe epilepsy. In: Winn R, editor. Youmans and Winn's neurological surgery. 8th ed. Elsevier; 2023. p. 2683–722.

Functional Hemispherotomy and Variants for Treatment of Epilepsy

70

Roy Thomas Daniel and Giulia Cossu

Learning Objectives

- To understand and be able to reproduce the surgical techniques of disconnective procedures for the treatment of refractory hemispheric epilepsy
- To understand the functional neuroanatomy of the different surgical steps

70.1 Introduction

Epilepsy is a network disease, where afferent and efferent fibers connecting the epileptogenic foci with the rest of the cerebrum are responsible for the secondary spread of the epilepsy. Refractory epilepsy, particularly in childhood, can have a deleterious effect on brain development, making it a challenging disease to treat. Hemispherectomy and the evolution of techniques of disconnection thereafter stand out as highly effective surgical procedure for managing epilepsy in children with unilateral hemispheric epileptic syndromes.

Hemispherectomy, namely, the resection of the whole hemisphere, was first performed by Dandy in 1928 in a patient with hemispheric glioma, and then, McKenzie performed it for the treatment of a patient with infantile hemiplegia. This experience was later published by Krynauw in 1950, and then, the procedure became more popular and widely accepted [1]. In the 1970s, Rasmussen described the "functional hemispherectomy" [2], and then, a further progress was represented by the introduction of "hemispherotomy" techniques in 1990s [3–8], with the aim of further reducing the volume of brain parenchyma resected during surgery.

The ideal candidate is a patient with hemiplegia secondary to a damaged hemisphere and the most common causes are hemiconvulsion-hemiplegia-epilepsy (HHE) syndrome, Sturge-Weber syndrome, hemimegalencephaly, multilobar cortical dysplasia, congenital hemiplegia from a perinatal infarction, and Rasmussen's encephalitis. The final cause which is rarely acquired may led to intractable hemispheric epilepsy.

Surgical procedures can lead to seizure freedom and improved quality of life but should be considered early to prevent the negative effects of ongoing seizures on neuroplasticity and neurodevelopmental outcomes. Factors such as seizure semiology, disease duration, electroencephalogram, neurological and neuropsychological evaluations, educational level, professional plans, and hobbies should be considered before planning surgical treatment of intractable epilepsy. The precise location of the epileptogenic focus should be determined through a careful analysis of these phase I examinations (and eventually phase II) and through the correlation with the preoperative MRI. Resting state and task-based functional MRI can also be used to predict the location and modifications of eloquent cortices and associated pathways.

70.2 History and Background

Anatomical Hemispherectomy consists in the removal of one cerebral hemisphere, with or without sparing the basal ganglia. The use of this technique declines in the 1960s because of the occurrence of late complications and mortality, mainly due to hydrocephalus and chronic subdural collections. However, this technique has been again used in the 1990s, and an inferior rate of complications are reported when compared to older series [9–11].

R. T. Daniel (✉) · G. Cossu
Department of Neurosurgery, University Hospital of Lausanne, University of Lausanne, Lausanne, Switzerland
e-mail: roy.daniel@chuv.ch

I. N. Kanaan, V. Beneš (eds.), *Neuroanatomy Guidance to Successful Neurosurgical Interventions*,
https://doi.org/10.1007/978-3-031-59838-8_70

Hemidecortication was first described by Ignelzi and Bucy in 1968 [12], and it aims at removing the cortical gray matter with preservation of the ventricle and a layer of periventricular white matter. This can be of course particularly efficient when there is an epileptogenic abnormality in the cortical gray matter. The first surgical steps are represented by sylvian fissure opening with middle cerebral artery occlusion and then occlusion of anterior and posterior cerebral arteries. Then, the hemisphere is progressively removed while leaving the basal ganglia and thalamus. A variant of this technique was then described by Winston et al. [13], that left only a layer of white matter around the ventricles, while developing a dissection plane from the insula. Some complexity was however found in performing a cortical resection in the medial and basal surface of the lobes.

Functional hemispherectomy was described by Rasmussen [2], and it represents a partial anatomical resection with a complete disconnection of the remaining hemisphere. The resection is centered on the central lobe and the anterior two-thirds of the temporal lobe, including mesial structures, while the disconnection points at cutting the commissural fibers along with the projection fibers from thalamus and brainstem. The anterior frontal, posterior parietal, and occipital cortices are left intact. This aims at avoiding complications associated with large postsurgical cavities.

Following the description of functional hemispherectomy by Rasmussen, that allowed a disconnective procedure to obtain the same results as a resective one, Delalande and Villemure in the early 90s described the **vertical and lateral hemispherotomy**, respectively [3–8]. These procedures further reduce the brain resection while ensuring complete disconnection of the remainder of the hemisphere. However, for all these techniques, a rigorous and precise anatomical knowledge of the orientation and access points for the white matter tracts are necessary to achieve success in terms of epilepsy control and preservation of possible residual neurological functions.

70.3 Vertical Hemispherotomy

The transventricular vertical hemispherotomy was described by Delalande et al. in 1992 [3, 4].

The anatomic core is approached from a paramedian frontocentral window above the lateral ventricle, and then, the disconnection can be performed from the ventricle, with a minimal vascular exposure (Fig. 70.1). A paramedian frontoparietal craniotomy is performed: one-third anterior and two-thirds posterior to the coronal suture.

The key surgical steps are detailed as follows:

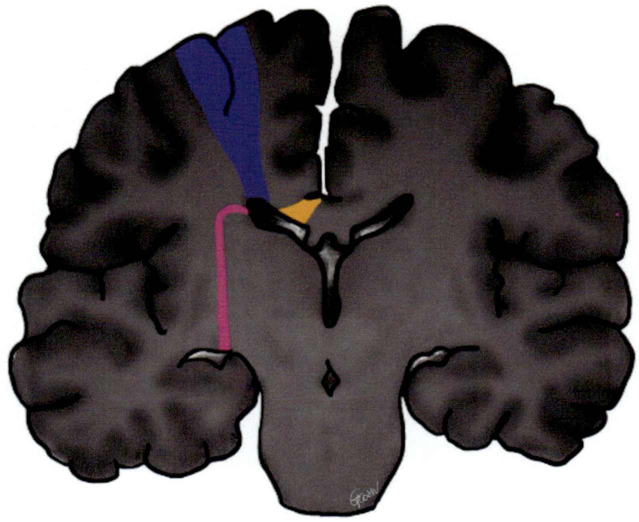

Fig. 70.1 Some surgical steps of the vertical hemispherotomy are illustrated in this coronal view: the frontal paramedian cut till the lateral ventricle is in blue, the perithalamic incision going from the frontal horn and body of the lateral ventricle till the temporal horn is in pink, and the parasagittal callostomy is in orange

1. A small cortical paramedian incision is performed parallel to the sagittal suture to enter the frontal horn of the ipsilateral ventricle. The surgical access can also be performed via an enlarged superior frontal sulcus.
2. From there, the posterior part of the body of the corpus callosum and splenium is incised along with the posterior column of the fornix at the trigone. The surface of the splenium is followed in a subpial fashion until the visualization of the deep veins confluences into the vein of Galen through the pia-arachnoid plane.
3. The crus of the fornix is cut at the level of the trigone.
4. The perithalamic incision is performed vertically starting at the trigone to reach the most anterior portion of the temporal horn, with the choroid plexus serving as guidance. Amygdalectomy and anterior hippocampotomy are also performed during this step.
5. Callosotomy was completed anteriorly through the anterior portion of the body, the genu, and the rostrum, following the subpially visualized pericallosal artery.
6. After subpial suction/resection of the subcallosal area and posterior gyrus rectus, the last cut is performed following the anterior cerebral artery (A1 portion), from the straight gyrus to the anterior temporal horn, through the caudate nucleus. The optic nerve, the chiasm, and the anterior cerebral artery can be visualized subpially and are left anteriorly.
7. During this last trajectory, the aspiration of the dorsomedial amygdala is performed, till the temporal horn is again reached.

70.4 Lateral Hemispherotomy Techniques

70.4.1 Peri-insular Hemispherotomy

Villemure and Mascott first described this technique in 1995 [5, 6, 8], with surgical transection of white matter tracts from a peri-insular opening of the cerebral hemisphere.

A modified question mark or a U-shaped scalp incision is suitable for the procedure. The inferior edge of the incision should reach the root of the zygoma, while the superior edge should be superior to the temporalis muscle. Most of the disconnective procedure is performed from within the ventricle, and access to the ventricle is performed through two windows: the supra- and the infrainsular window (Fig. 70.2).

The key surgical steps are detailed as follows:

1. Suprainsular cut performed at the level of the frontal and parietal opercula at some mm from the sylvian fissure and till the superior portion of the circular cistern.
   a. The frontal horn is reached and the corpus callosum cut towards the interhemispheric fissure along all its extension.
   b. A posterior hippocampotomy is performed.
   c. Frontobasal disconnection, performed in a coronal plane, and parallel to the sphenoid ridge in the direction of the basal pia of the straight and orbital gyri.
2. Infrainsular cut through T1 till the inferior portion of the circular sulcus. From the circular cistern, the cut proceeds downwards towards the temporal horn.
   a. The resection of amygdala and anterior hippocampus is performed, to reach the choroid fissure. A particular attention should be taken not to enter the basal ganglia superiorly, during the resection of the amygdala.
3. Insular cut at the level of the external capsule or claustrum, to avoid injury to M4 branches on the insular surface.

70.4.2 Transsylvian Keyhole Hemispherotomy

This technique represents a variant of the peri-insular hemispherotomy and was first described by Schramm et al. [13–15]. The craniotomy is smaller, and the insular cortex is reached after the opening of the sylvian fissure: the disconnection of the internal capsule is thus achieved through the sylvian fissure instead of through opercular widows (like in the peri-insular hemispherotomy). The rest of the surgical steps, namely, the posterior hippocampotomy, resection of amygdala and anterior hippocampus, the parasagittal callosotomy, and the frontobasal disconnection are similar to the peri-insular technique.

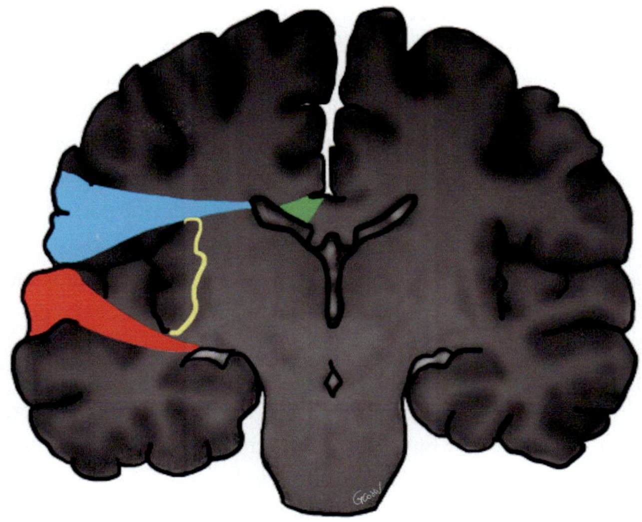

Fig. 70.2 The lateral peri-insular hemispherotomy is here illustrated in a coronal plane. The suprainsular window is in blue till the lateral ventricle. Then, a parasagittal total callosotomy is in green. The infrainsular window is represented in red, with the insular cut at the level of claustrum, and external capsule is in yellow

70.5 Endoscopic Hemispherotomy

Endoscopic hemispherotomy is a variant of the vertical approach. To be able to perform endoscopic procedures, cisternal and ventricular spaces should be primarily used to perform hemispheric disconnection. The exposure should allow the contemporary use of suction tubes and bipolar coagulation, through the use of very long instruments.

The cranial opening and the endoscopic entry point should be carefully planned: the disconnection can be performed using only an entry point (small frontal craniotomy anterior to the coronal suture) [16] or two bur holes pointing at accessing the frontal horn and the occipital horn [17]. With the small craniotomy, an interhemispheric approach can be used and the anatomy of bridging veins should be carefully studied. With the two bur holes techniques, a classical transparenchymal approach is performed to reach the ventricle.

A rigid endoscope holder can be a fundamental help in the performance of this procedure.

Most of the procedure may be performed with a 0° endoscope and the 30° scope may be useful during some particular steps.

The key surgical steps are detailed as follows:

1. The corpus callosum is exposed from the genu to the splenium. The callosum can thus be cut in a posterior-to-anterior direction. The homolateral ventricle is thus largely open. Anteriorly, the anterior commissure is also sectioned, and the anterior communicating complex is visualized through the basal pia.

2. Anterior disconnection: from the genu, the cut proceeds anterior to the head of the caudate nucleus and it reached the posterior part of the anterior cranial fossa laterally.
3. Middle disconnection: this cut connects the body of the lateral ventricle to the temporal horn and passes lateral to basal nuclei and thalamus, but remains medial to the insula. The ventral amygdala and the anterior part of the hippocampus should be removed during this step.
4. Posterior disconnection: the tail of the hippocampus and the fornix are cut during this step, staying under the choroid plexus.

This technique proved to be safe and effective in a few series [16, 18, 19] but a learning curve is necessary to gain the required expertise to perform this surgery. Neuronavigation is a valid support during these procedures to mark the cranial opening. In general, atrophic cases represent ideal cases for this technique.

70.6 Anatomical Highlights

Cortical anatomy is a key factor in guiding the surgical procedure. Neuronavigation systems can be useful but they cannot substitute for a good 3D appreciation of the cerebral neuroanatomy. For the peri-insular hemispherotomy, a large bone flap should allow the exposure of the per-Sylvian cortices in order to identify the sulcogyral anatomy. The sylvian fissure and the central sulcus are the two major landmarks to orient the early part of the surgery. Beside the sulcogyral anatomy, the vascular anatomy can also guide the surgeon using the main venous structures on the surface such as Labbe and Trolard's veins, which are additional landmarks to guide the procedure.

For the vertical hemispherotomy, the surgical exposure reveals the frontal parasagittal gyrus (F1). The coronal and sagittal sutures are the most important landmarks to plan the craniotomy. The anatomy of bridging veins should also be carefully studied.

The white matter tracts to be studied in any disconnective surgery are classified into projection, association, and commissural fiber systems. During hemispherotomy techniques, the essential fibers to disconnect are the projection and the commissural fibers, to isolate the epileptogenic focus and avoid secondary spreading. Some associative fibers can also be cut during a particular surgical step but their disconnection is not essential to obtain seizure control in hemispheric epilepsy. Table 70.1 summarizes the fibers cut during the different surgical steps.

During the vertical hemispherotomy:

- Some association fibers can be cut during the small cortical paramedian incision to reach the ipsilateral ventricle, such as the superior longitudinal fascicle (SLF) I and the fibers of the frontal aslant tract (FAT).
- Commissural fibers belonging to the callosum are cut during the callosotomy step, along with the crus of the fornix posteriorly and the anterior commissure (AC) anteriorly.
- During the perithalamic incision, the fibers of the anterior and posterior limb of the internal capsule are cut, namely, the thalamocortical fibers, the corticospinal tract and, the parieto-pontine and the occipitopontine fibers. The frontostriatal fibers are also cut. Furthermore, the retro- and sublentiform portions of the internal capsule are cut, along with the optic radiations when the temporal horn is reached.
- During the frontobasal incision, the frontal projections of AC, uncinate fasciculus (UF), and inferior frontooccipital fasciculus (IFOF) are cut.

During the lateral hemispherotomy:

- During the suprainsular window, some frontal projections of the arcuate fasciculus (AF), SLF III, FAT, and IFOF may be cut. In the trajectory from the superior portion of the circular cistern to the frontal horn, the corona radiata is cut, thus sectioning all the fibers entering into the internal capsule, including the thalamocortical radiation and the corticospinal tract. The FST fibers are also cut during this step.
  - A complete parasagittal callosotomy and a posterior hippocampotomy are then performed.
  - During the frontobasal disconnection, the projections of anterior commissure, IFOF, and uncinate fasciculus are cut.
- During the infrainsular window, the temporal projections of the arcuate fasciculus are cut. Furthermore, the optic radiations are sectioned when the temporal horn is reached, along with some fibers of the IFOF. A posterior hippocampotomy and an amygdalectomy and anterior hippocampectomy are also performed once in the temporal horn.
- During the insular cut, fibers of the external capsule are sectioned and the IFOF is again disconnected at this level.

Table 70.1 The key surgical steps are summarized along with the inherent white matter fibers disconnected during each step of vertical and lateral hemispherotomies

	Surgical steps	Projection fibers disconnected	Commissural fibers disconnected	Associative fibers disconnected
Vertical hemispherotomy 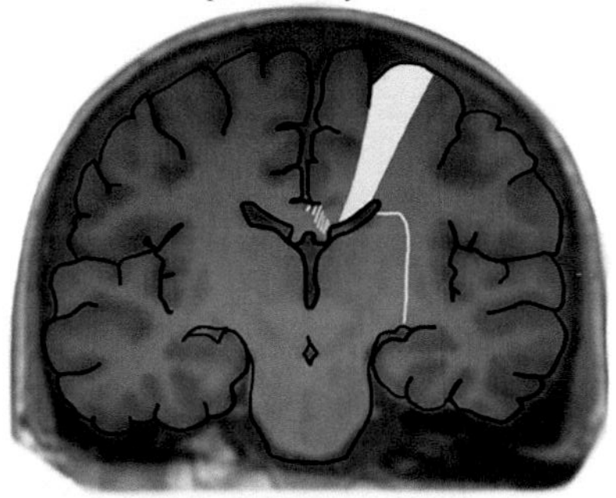	1. *Frontal paramedian incision till the frontal horn.*	Frontopontine fibers		SLF I FAT
	2. *Parasagittal total callosotomy* *Posterior hippocampotomy*		Corpus callosum Fornix	
	3. *Peri-thalamic incision* *Amygdalectomy and anterior hippocampectomy*	Anterior and posterior limbs, retro- and sublentiform parts of the internal capsule		Optic radiations FST
	4. *Frontobasal incision*		Anterior limb of AC	IFOF UF
Lateral hemispherotomy 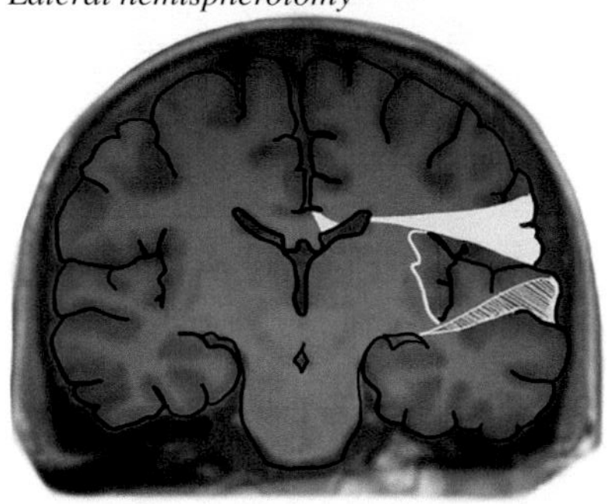	1. *Suprainsular window* *Parasagittal total Callosotomy* *Frontobasal disconnection*	Anterior and posterior limbs of the internal capsule at its junction with corona radiata	Corpus callosum Anterior limb of AC	Frontal projections of arcuate fasciculus SLF III IFOF FAT and FST UF
	2. *Infrainsular window* *Posterior hippocampotomy* *Amygdalectomy and anterior hippocampectomy*	Retro- and sublentiform parts of the internal capsule	Fornix	Temporal projections of Arcuate fasciculus Optic radiations IFOF
	3. *Insular disconnection*			Claustrum/external capsule

FAT frontal aslant tract, *FST* fronto-striatal tract, *IFOF* inferior fronto-occipital fasciculus, *SLF* superior longitudinal fasciculus, *UF* uncinate fasciculus

70.7 Surgical Pearls

- An intricate knowledge of the cerebral neuroanatomy is fundamental to perform epilepsy surgery, with a focus on sulcogyral and white matter anatomy.
- Quality of life preservation, as well as maintenance of neurological functions, are the two priorities of every surgical treatment.
- Disconnective surgery is based on the interruption of the epilepsy spreading network, while isolating the primary epileptogenic focus. Before considering surgery, the epileptogenic focus should be precisely localized through proper phase I or II investigations and surgery should be performed when an optimal electro-clinico-radiological correlation is found.
- Completeness of surgical disconnection is the main factor in determining epilepsy control and thus surgical outcome.
- During the many steps of the disconnective procedures, all efforts to preserve the cerebral vascularization (and therefore parenchymal viability) are made, essentially by using a subpial resection technique.

70.8 Discussion

The choice of the hemispherectomy technique is mainly dictated by the surgeon's experience. All hemispherectomy techniques are associated with a high-rate of seizure-free outcome, superior to 70%, with the best outcomes observed in patients with Rasmussen's encephalitis and Sturge Weber syndrome [20]. Pathology is thus an important determinant of surgical outcome, while incomplete disconnections are the most frequent reason of surgical failure. Disconnective techniques may thus be more challenging to perform in cases with large cortical dysplasia or hemimegalencephaly.

With a vertical hemispherotomy, the risk of sagittal sinus thrombosis should not be neglected when the midline is largely exposed. With lateral techniques instead, care should be taken in performing the insular cut as branches of the middle cerebral artery should be spared.

Hemispherotomy techniques present many advantages when compared to anatomical hemispherectomy as they were associated with a lower risk of infection, hydrocephalus, and trauma for the patients [21, 22]. The surgical complications are the same for the different techniques of hemispherotomy and include intraoperative blood loss with transfusion, postoperative fever, hydrocephalus, remote hemorrhage, infarction, and cerebral edema. The hydrocephalus rate seems however to be inferior with lateral hemispherotomies (less than 5%) [5, 7] than with vertical techniques (about 15%) [4].

70.9 Conclusions

Functional hemispherotomy and its variants represent surgical techniques for the treatment of hemispheric epilepsy, which can lead to seizure freedom and improved quality of life. Early surgery is necessary in children to prevent the negative impact on neuroplasticity and neurodevelopmental outcomes. A thorough understanding of the white matter anatomy is fundamental to successfully perform these procedures.

References

1. Krynauw RA. Infantile hemiplegia treated by removing one cerebral hemisphere. J Neurol Neurosurg Psychiatry. 1950;13(4):243–67. https://doi.org/10.1136/jnnp.13.4.243.
2. Rasmussen T. Hemispherectomy for seizures revisited. Can J Neurol Sci. 1983;10(2):71–8. https://doi.org/10.1017/s0317167100044668.
3. De Ribaupierre S, Delalande O. Hemispherotomy and other disconnective techniques. Neurosurg Focus. 2008;25(3):E14. https://doi.org/10.3171/FOC/2008/25/9/E14.
4. Delalande O, Bulteau C, Dellatolas G, et al. Vertical parasagittal hemispherotomy: surgical procedures and clinical long-term outcomes in a population of 83 children. Neurosurgery. 2007;60(2 Suppl 1):ONS19-32; discussion ONS32. https://doi.org/10.1227/01.NEU.0000249246.48299.12.
5. Daniel RT, Villemure JG. Peri-insular hemispherotomy: potential pitfalls and avoidance of complications. Stereotact Funct Neurosurg. 2003;80(1–4):22–7. https://doi.org/10.1159/000075155.
6. Daniel RT, Villemure JG. Hemispherotomy techniques. J Neurosurg. 2003;98(2):438–9; author reply 439. https://doi.org/10.3171/jns.2003.98.2.0438.
7. Villemure JG, Daniel RT. Peri-insular hemispherotomy in paediatric epilepsy. Childs Nerv Syst. 2006;22(8):967–81. https://doi.org/10.1007/s00381-006-0134-3.
8. Villemure JG, Mascott CR. Peri-insular hemispherotomy: surgical principles and anatomy. Neurosurgery. 1995;37(5):975–81. https://doi.org/10.1227/00006123-199511000-00018.
9. Gonzalez-Martinez JA, Gupta A, Kotagal P, et al. Hemispherectomy for catastrophic epilepsy in infants. Epilepsia. 2005;46(9):1518–25. https://doi.org/10.1111/j.1528-1167.2005.53704.x.
10. Davies KG, Maxwell RE, French LA. Hemispherectomy for intractable seizures: long-term results in 17 patients followed for up to 38 years. J Neurosurg. 1993;78(5):733–40. https://doi.org/10.3171/jns.1993.78.5.0733.
11. Di Rocco C, Iannelli A. Hemimegalencephaly and intractable epilepsy: complications of hemispherectomy and their correlations with the surgical technique. A report on 15 cases. Pediatr Neurosurg. 2000;33(4):198–207. https://doi.org/10.1159/000055953.
12. Ignelzi RJ, Bucy PC. Cerebral hemidecortication in the treatment of infantile cerebral hemiatrophy. J Nerv Ment Dis. 1968;147(1):14–30. https://doi.org/10.1097/00005053-196807000-00002.
13. Winston KR, Welch K, Adler JR, Erba G. Cerebral hemicorticectomy for epilepsy. J Neurosurg. 1992;77(6):889–95. https://doi.org/10.3171/jns.1992.77.6.0889.
14. Schramm J, Behrens E, Entzian W. Hemispherical deafferentation: an alternative to functional hemispherectomy. Neurosurgery. 1995;36(3):509–15; discussion 515–6. https://doi.org/10.1227/00006123-199503000-00010.

15. Schramm J, Kuczaty S, Sassen R, Elger CE, von Lehe M. Pediatric functional hemispherectomy: outcome in 92 patients. Acta Neurochir. 2012;154(11):2017–28. https://doi.org/10.1007/s00701-012-1481-3.
16. Sood S, Marupudi NI, Asano E, Haridas A, Ham SD. Endoscopic corpus callosotomy and hemispherotomy. J Neurosurg Pediatr. 2015;16(6):681–6. https://doi.org/10.3171/2015.5.PEDS1531.
17. Bahuleyan B, Manjila S, Robinson S, Cohen AR. Minimally invasive endoscopic transventricular hemispherotomy for medically intractable epilepsy: a new approach and cadaveric demonstration. J Neurosurg Pediatr. 2010;6(6):536–40. https://doi.org/10.3171/2010.9.PEDS10267.
18. Doddamani RS, Chandra PS, Samala R, et al. Endoscopic Hemispherotomy for nonatrophic Rasmussen's encephalopathy. Neurol India. 2021;69(4):837–41. https://doi.org/10.4103/0028-3886.325379.
19. Girishan S, Tripathi M, Garg A, et al. Enhancing outcomes of endoscopic vertical approach hemispherotomy: understanding the role of "temporal stem" residual connections causing recurrence of seizures. J Neurosurg Pediatr. 2019;8:1–9. https://doi.org/10.3171/2019.8.PEDS19148.
20. Griessenauer CJ, Salam S, Hendrix P, et al. Hemispherectomy for treatment of refractory epilepsy in the pediatric age group: a systematic review. J Neurosurg Pediatr. 2015;15(1):34–44. https://doi.org/10.3171/2014.10.PEDS14155.
21. Schramm J, Kral T, Clusmann H. Transsylvian keyhole functional hemispherectomy. Neurosurgery. 2001;49(4):891–900; discussion 900–1. https://doi.org/10.1097/00006123-200110000-00021.
22. Villemure JG, Vernet O, Delalande O. Hemispheric disconnection: callosotomy and hemispherotomy. Adv Tech Stand Neurosurg. 2000;26:25–78. https://doi.org/10.1007/978-3-7091-6323-8_2.

71 Evolution in Surgical Management for Movement Disorders

Giovanni Broggi

Learning Objectives

- Review history of movement disorders (MD).
- Provide indications and patient selection criteria for functional intervention.
- Outline the anatomical target for various DPS and alternative treatment options
- Refer to patient's preparation and planning for DBS or MRgFUS.
- Update on recent advancements in the management of MD.

71.1 Introduction

Deep brain stimulation (DBS) is today the most effective and well-known therapeutic adjunct in the treatment of drug-resistant movement disorders such as Parkinson's disease, essential tremor, and dystonia. DBS surgery has evolved considerably from its first inception to one of the most fast-evolving areas of neurosurgery. In addition to DBS, in the last decade, recent advances in imaging and lesioning technologies have rekindled interest in lesioning for the treatment of several movement disorders. Among these procedures, MRI-guided focused ultrasound (MRgFUS) ablation has emerged as the most promising one.

71.2 History of Movement Disorder Surgery

The history of the surgical treatment of movement disorders has been a story of experimental neurophysiology applied to the human brain. This history can be described in three key stages such as the prestereotactic era, the lesional stereotactic era, and the resurgence of functional surgery after the introduction of levodopa for Parkinson's disease, which opened the way for the modern era of implanted deep brain stimulators (DBS) and the newer ablation therapeutics, such as MRgFUS.

71.2.1 The Prestereotactic Era

Sir Victor Horsley was the first to perform cortical ablation for dyskinesia in 1890. He treated a patient with athetosis originating in the thumb by removal of the part of the cortex, representing the movements of that digit with cessation, for 2 weeks, of athetosis. The Horsley procedure was sporadically used by other surgeons for the next 20 years. Anschutz, Payr, and Polenow, all of them, reported therapeutic results on choreoathetosis by removing the corresponding area of the opposite precentral gyrus. Similar results were reported by Nazaroff in 1927 with intracortical alcoholic injections; in this case, an insufficient improvement in three of the four patients led to the abandonment of such a procedure. Other different types of interventions were tried in the period for movement disorders, such as ganglionectomy, dorsal cordotomy, posterior rhizotomies, and cerebellar nucleus dentate lesion with various results. For example, Delmas-Marsaleet tried electrical coagulation of a dentate nucleus in 1935, with improvement of rigidity in a parkinsonian patient but with worsening of tremor. In 1931, Bucy used cortical stimulation to reproduce athetoid movements in a patient with unilateral arm athetosis and decided to excise the cortex till the lateral ventricle (areas 6 and the anterior bank of area 4). The athetoid movements disappeared, but the patient had significant upper limb paresis. Bucy then concluded that athetoid movements originated in the premotor region, and so, he kept using this procedure during the decade, with the same results and drawbacks of the first intervention.

G. Broggi (✉)
Departement of Neurosurgery, Fondazione IRCCS Istituto Neurologico Carlo Besta, Milan, Italy

I. N. Kanaan, V. Beneš (eds.), *Neuroanatomy Guidance to Successful Neurosurgical Interventions*,
https://doi.org/10.1007/978-3-031-59838-8_71

Meyers in 1951 published his 10-year compilation of basal ganglia operations for movement disorders. After seeing a young woman who had arrived at an emergency room with an open skull fracture, through which he observed severe damage to bilateral frontal lobes and bilateral ventral striata and who, despite this, was awake and followed commands, Meyers thereafter performed a right transcortical transventricular resection of the anterior two-thirds of the caudate nucleus in the young woman with left hemiparkinsonism; postoperatively, the tremor was greatly reduced. Caudectomy and, later, caudectomy coupled with sectioning of the anterior limb of the internal capsule and ansa lenticularis (ansotomy) in 1940 seemed to be promising in abolishing tremor, but none of them provided lasting benefit in patients with choreoathetosis. The first patient who was offered the latter procedure died days later due to postoperative pneumonia, but Meyers noticed great tremor improvement and decided to employ the technique in his future tremor patients. Browder was a proponent of capsular operations for parkinsonism, too. In 1947, he reported a critical evaluation of his series of patients (15) who had undergone removal of the dorsal half of the caudate head and division of the anterior limb of the internal capsule to within a few millimeters of the capsular genu; of the 9 patients with unilateral tremor, complete remission was achieved in three patients. Anyway, the transventricular approach was burdened with high mortality, so Fènelon in Paris used a transfrontal approach with a small bone flap, through which an electrode was passed through the tip of the frontal lobe, 30 mm lateral to the midsagittal line, and at a depth of 75 mm; he coagulated the ansa lenticularis. His results were good in 70% of his 60 patients. Guiot and Brion reported a similar procedure of coagulation of globus pallidus and ansa lenticularis via a subfrontal approach. They passed an electrode into the anterior perforated substance 2 to 3 mm above the optic tract at the origin of the Sylvian fissure. Excellent results were reported in 42.5% of 40 patients. In 1952, during an intervention planned for a pedunculotomy, Cooper accidentally tore the anterior choroidal artery, which was coagulated. The intervention ended without the pedunculotomy, and the patient subsequently presented disappearance of tremor and rigidity without loss of strength. Cooper operated on other six patients with coagulation of the anterior choroidal artery, with similar results, and stated that the area susceptible to infarction was the globus pallidus and its afferent connections. Though after performing 55 anterior choroidal artery ligations for parkinsonism, with encouraging results, the mortality rate was high (10%) (maybe due to the variability in the territorial distribution of the artery), and the technique was abandoned [1].

71.2.2 The Stereotactic Era

In order to treat movement disorders (tremor) without producing severe side effects, Horsley and Clarke tried to target extracortical structures by introducing an apparatus with which it was possible to reach subcortical structures with a needle, a probe or a blade under control. Such apparatus relied on the Cartesian coordinate system. In a paper published in 1908, they reported making of an electrolytic lesion in the dentate nucleus of the cerebellum of the monkey. The Horsley and Clarke's apparatus was connected to the first stereotactic atlas, which consisted of a series of illustrations of sections of monkey brain, registered through a Cartesian coordinate system to the monkey's skull; in this way, the subcortical structures could be reached, considering their relationship to the skull. However, their pioneering work has never been applied to human surgery, probably due to the fact that Horsley and Clarke disagreed about other issues, while they were thinking of the idea of a human stereotactic apparatus still based on bony landmarks to find the landmarks from which Cartesian coordinates might be determined.

The establishment of internal (intracerebral) landmarks from which the structures of interest could be referred to is due to Ernest Spiegel (1895–1985) and Henry Wycis (1911–1972); they were also the first to perform, with their apparatus, stereotactic surgery in a patient (affected by Huntington's chorea). Their apparatus was essentially a Horsley-Clarke apparatus suspended over the head of the patient; Spiegel and Wycis used intraoperative X-rays for visualization of the internal landmarks (through an air-encephalogram, to appreciate the third ventricle); their method consisted of a spatial correlation of intracerebral nuclei with cerebral midline structures, such as the posterior boundary of the third ventricle, the posterior commissure, calcifications in the pineal gland, or the habenular commissure. They injected alcohol into the globus pallidus internus and the medial thalamus, with a temporary improvement of patient's symptoms. The choice of medial thalamus was justified by the fact that its lesioning decreased emotional reactivity, known to exacerbate abnormal movements: the fibers from mediodorsal thalamus project into globus pallidus. They described the case of a 53-year-old patient affected by Huntington's disease; they first performed a right pallidotomy and with alcohol injection and then a dorsomedial thalamotomy with an electrolytic lesion; later, they performed a right dorsomedial thalamotomy in the second stage 2 months later. Hand choreic movements improved after each thalamic lesion, and there was no morbidity but the improvement lasted only for 18 months.

They subsequently began to use electrolytic current to make lesions, because alcohol's spreading was unpredictable. Since the first procedures were performed with Spiegel and Wycis' apparatus, intraoperative electrophysiology began to be performed, in that, rather than confirming the exact target's localization, it is made possible to study the surrounding structures. Many neurosurgeons visited the two pioneers at Temple Medical School in Philadelphia, and since there were no commerce of such apparatus, they had to manufacture their own. Among them, Lars Leksell (1907–1986) in Sweden designed the first frame based on arc and center, and Jean Talairach (1911–2007) in Paris designed a fixed grid system. Traugott Riechert in Freiburg, Germany, developed an arc-guided stereotactic system in 1946. In the 1960s, many stereotactic centers arose throughout the world. By 1965, more than 25,000 stereotactic procedures had been made worldwide [1–3] (Figs. 71.1, 71.2 and 71.3).

Fig. 71.1 Dr. Victor Horsley

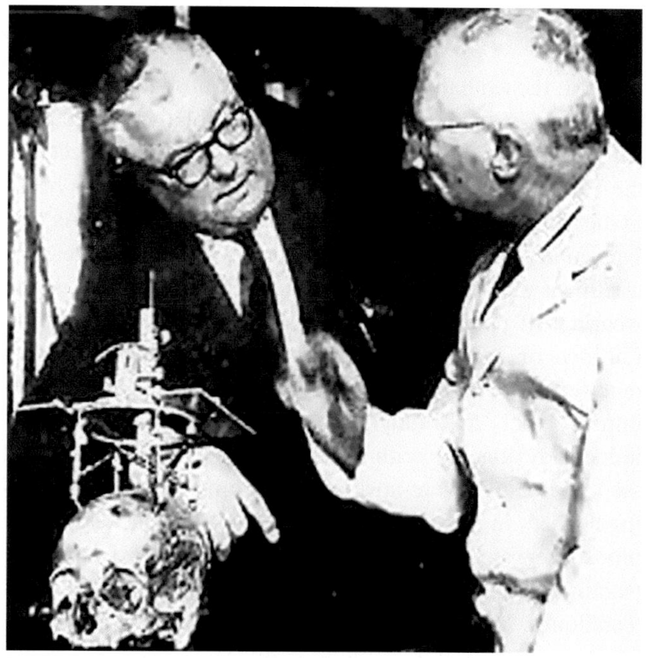

Fig. 71.2 Dr. Spiegel and Dr. Wycis

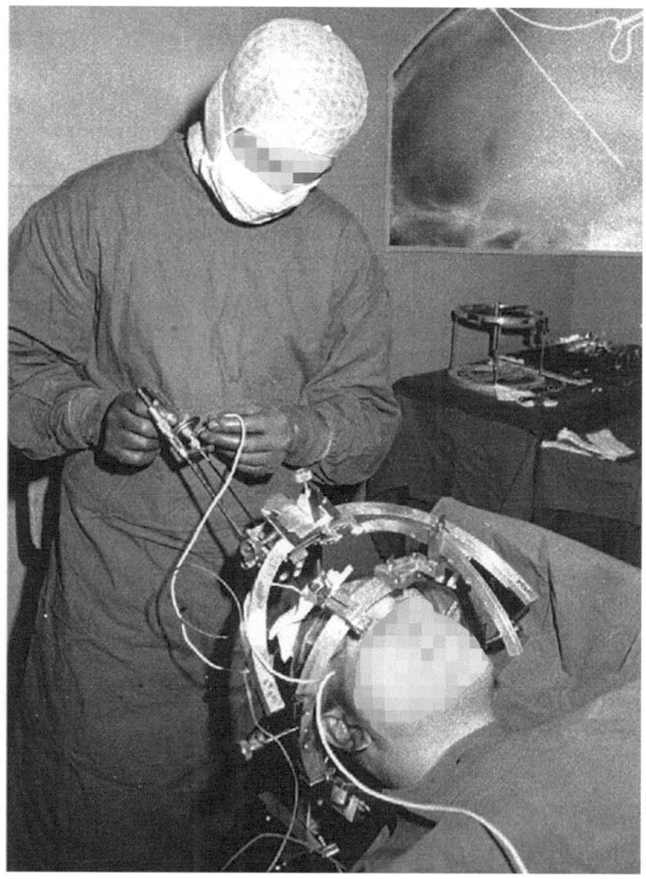

Fig. 71.3 Lesioning with Mundinger frame

71.2.3 The Resurgence of Functional Surgery

In 1968, the availability of L-Dopa and subsequently of botulinum toxin for dystonia reduced the ablative stereotactic surgery, mostly because of the potential adverse effect. By the 1980s, it was clear that the use of L-Dopa will not result in the miracle cure that it was initially thought to be. As parkinsonian patients started to develop resistance after years of treatment along with side effects such as dyskinesias and wearing off phenomenon (complicated PK D.), a report by Laitinen on pallidotomy for Parkinson's disease in 1992 renewed interest in stereotactic surgery. Laitinen reported an improvement in bradykinesia, rigidity, and L-Dopa—induced dyskinesias with a ventroposterior pallidotomy, and the good results were attributed to the inclusion of pallidofugal fibers in the context of the lesion. The success of the procedure was also to attribute to the introduction of imaging techniques and to the refinement of electrophysiological techniques. Michael Dogali in 1991 was the first surgeon to perform a new option of pallidotomy in the US, assisted by Laitinen and Tasker. Since then, many reports of pallidotomy were made, and many neurologists began to be involved in intraoperative neurophysiological studies, opening the way for the application of electrical stimulation to basal ganglia circuits.

Meanwhile, implantable medical device technology had been improving to the point that chronically implanted devices such as spinal cord stimulators and cardiac pacemakers were routine. At the same time, understanding the neurophysiological basis of movement disorders was improving rapidly. The most important theoretical advance was the development by Albin, DeLong, and colleagues of a coherent model of basal ganglia functional anatomy and its relation to movement.

In the past years, the stimulation of thalamic and other basal ganglia nuclei have been used for monitoring the area close to the target area before making a lesion at the target. The neurophysiologist Sem-Jacobsen was the first to give a precise account of this technique in 1965. For him, chronic stimulation allowed the identification of the optimal lesioning site, and the electrodes with the best stimulation results could be used for incremental lesions. Bechtereva (USSR) was the first to use deep stimulation as a chronic therapy. In her case, electrodes were implanted in the ventrolateral and centromedianus nucleus of the thalamus, and were used for "intermittent sessions of electrical stimulation with a high-rate pulses of suprathreshold current" before making a therapeutic lesion through the electrode, because of lack of availability of implantable stimulators.

Concurrently, the team in Grenoble led by Benabid and Pollack had been experimenting with the use of high-frequency DBS as an alternative to thalamic lesioning for patients with advanced PD, reporting their results in 1991 and demonstrating the first implementations of modern DBS. The early 1990s also saw publication of further primate data, supporting the value of the STN as a surgical target based on the MPTP nonhuman primate. This work was quickly translated into clinical practice when Pollak first reported the use of STN-DBS in humans in 1993, indicating efficacy in treating PD motor symptoms. Similarly, the internal globus pallidus (GPi) was identified as a viable target for DBS as an alternative to pallidal ablation for dystonia. By the end of the decade, clinical use of DBS was started to become widespread in the treatment of Parkinson's disease, essential tremor, and dystonia.

DBS targeted to the ventral intermediate nucleus of the thalamus (Vim) for use in essential tremor and in severe Parkinson's disease tremor obtained CE mark and FDA approval in 1993 and 1997, respectively. Since that period, DBS indications have expanded in parallel with new targets for a specific disease, as anterior nucleus of the thalamus (ANT) for epilepsy, and nucleus accumbens (Nacc) and anterior limb of the internal capsule (ALIC) for obsessive compulsive disorder, to name a few [2, 3].

71.3 Anatomical Highlights (Figs. 71.4, 71.5, 71.6 and 71.7)

- AC-PC line reference: Line between anterior and posterior commissure
- Globus pallidus I (GPi): Target for dystonia and Parkinson's disease
- Anterior nucleus of thalamus (ANT): Target for treatment of epilepsy
- Ventral intermedius nucleus (VIM): Target for essential tremor and Parkinson tremor
- Subthalamic nucleus (STN): Target for treatment of PD motor symptoms
- Caudal zone 1 (zona incerta): Potential target for Parkinson's disease
- Nucleus accumbens (Nacc): Target for treatment of obsessive compulsive disorder (OCD)
- Substantia nigra
- Red nucleus

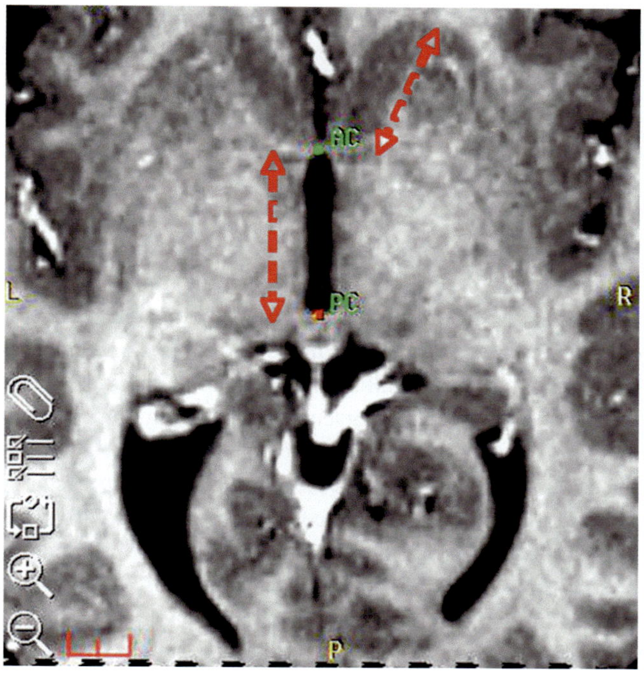

Fig. 71.4 Targeting for stereotactic lesioning, DBS, F. US

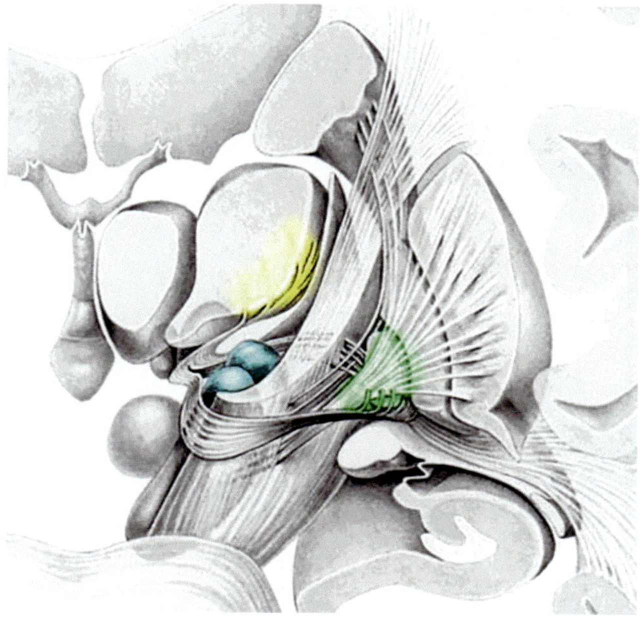

Fig. 71.5 Different structures for target: VIM, GPi, STN, and ansa lenticularis

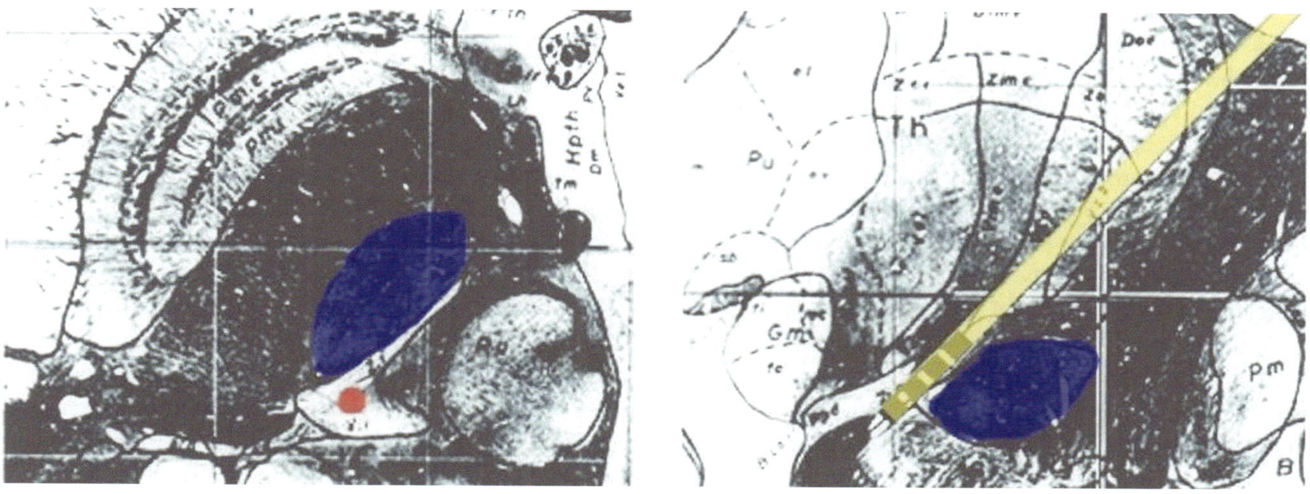

Fig. 71.6 Targets: Zona incerta Forel fields

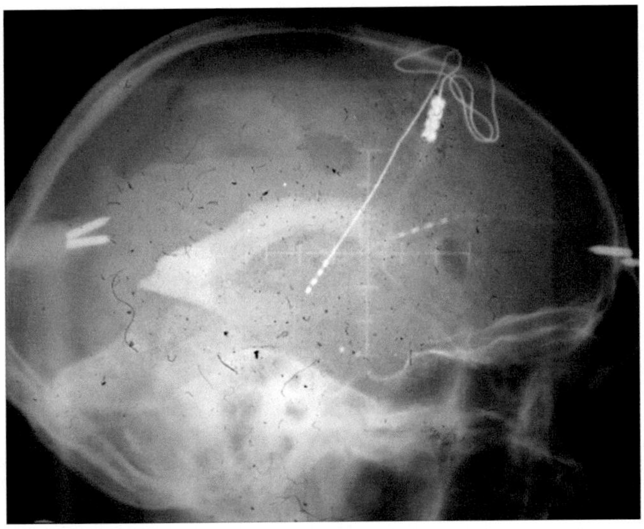

Fig. 71.7 Old-time DBS: Air and contrast medium ventriculography

71.4 Evolution in Movement Disorder Surgery

71.4.1 Indications

Choosing which patients represent ideal surgical candidates is still the first and most important decision of the DBS surgical team. Since the introduction in clinical practice more than 20 years ago of the criteria established by the "'Core Assessment Program for Surgical Interventional Therapies in Parkinson's Disease (CAPSIT-PD)," selection criteria for DBS have evolved over time. As DBS came into widespread use for PD in the 1990s, the median number of years following diagnosis of PD at time of surgery in two large, randomized studies was 11 and 13 years. These were explained by the fact that a patient considered for interventional surgery at the time should have a diagnosis of idiopathic PD and a minimum disease duration of 5 years, in order to exclude people with atypical parkinsonism, given the absence of benefits and the risk to harm patients with no idiopathic PD [4].

The cut for the indication is the age of the patient, i.e., 70 years, but comorbidity and biological age is more important that the real one.

Recently, the large RCT early-STIM trial has shown the superiority of subthalamic DBS compared to medical therapy alone in a group of patients included when having a PD diagnosis of ≥4 years, and fluctuations or dyskinesia present for 4 years or less. This has spawned a number of follow-up trials with even earlier interventional strategies—as early as 2 years following diagnosis—and before patients have developed significant medication-related side effects. A phase three randomized trial investigating this earlier point is still underway; however, each of these studies has brought into question the rationale of applying DBS for late-stage PD patients only. This development has had two significant effects. The first is that multidisciplinary teams caring for PD patients can apply less restrictive criteria in their patient selection. The second is that patients may be able to stay engaged in their work and social activities to a greater extent with early surgery than those who do not [5, 6].

The increasing observations that PD involves disturbances outside the primary motor domains is playing a role in the selection of patients as well. A large randomized trial (Veterans Administration Co-op Study) compared DBS outcomes of the two main targets in PD, the subthalamic nucleus (STN) and the globus pallidus internus (GPi). Neurocognitive features were found to worsen with STN stimulation, even though the motor scores were equivalent in both targets. These data opened the way to an ongoing controversy regarding which target should be selected in PD patients with some mild cognitive impairment (MCI-PD) at the baseline and if it was possible to predict potential worsening of neuropsychiatric features based on the nucleus targeted. Longer follow-up studies are now demonstrating though there is no clear-cut winner when it comes to STN versus GPi, and other factors including preference and experience of the surgical center, motor features (predominance of dyskinesias, for instance, might favor GPi), and the need to reduce levodopa requirements because of adverse effects (which instead might favor STN) should be considered in the preoperative target assignment [7].

A parallel argument on selection criteria can be made for dystonia, given that there is a growing appreciation of the genetic role in clinical outcomes of DBS for dystonia. In a series of 60 patients with generalized dystonia undergoing pallidal DBS, Fitzgerald et al. found significant greater motor outcomes in patients with genetic substrates (DYT1, DYT6, or DYT11). These data have been then confirmed in several other studies. Thus, the etiology of dystonia—and, perhaps, genetic subtype—may play a more important role than timing of surgery in this population. Experience with secondary dystonia is also increasing. There is good evidence for the improvement of symptoms in the context of tardive dystonia when the structural MRI appears largely intact, and there are also partial responses in dystonia of other genetic forms including pantothenate kinase-associated neurodegeneration as recently discovered [8, 9].

71.4.2 Imaging (Fig. 71.7)

DBS surgery is essentially a neuroimaging-based surgery. However, neuroimaging techniques have been advanced quite rapidly in the last 10 years, and their role in DBS surgery becomes much more critical. These advancements have questioned the role of microelectrode recording, the need for awake surgery, and other clinical controls during DBS procedures.

Ventriculography has been replaced over time by targeting on preoperative MRI, while the increasing field strength from 3 T to 7 T has made the indirect targeting of intracranial structures based on the AC-PC line a directed one. The targeted structures are now better visualized, and the surgeon may perform his or her preoperative planning directly seeing the STN on the T2 MRI sequences. Other special MRI sequences, such as the FAGTIR, opened the way to the possibility of delineating also the structure of the globus pallidus and subnuclei of the thalamus to a better extent. In addition, tractography can be combined with trajectory planning to influence the placement of contacts in reference to the position of white matter fiber tracts. For instance, there is an increasingly use of the diffusion tensor imaging (DTI) to target the dentatorubrothalamic tract (DRTT) to treat more effectively some tremor conditions, such as the rubral one. Similarly, the thalamocapsular border may be highlighted by seeding the thalamic tract and the thalamocortical fibers to make visualization of this structure possible [10, 11].

Another important innovation has been the intraoperative integration of CT or MRI scan (iCT and iMRI), which has allowed for real-time assessment of intracranial lead position, minimizing the risk of lead inaccurate placement. In case of iMRI, the surgeon works directly inside the bore of a 1.5-Tesla or 3-Tesla magnet. The procedure is performed under general anesthesia without microelectrode recording. Several studies are confirming the comparable accuracy of iMRI frameless versus frame-based DBS electrode insertion, and a recent report has shown clinical outcomes of iMRI that appear to be similar to frame-based STN DBS. [12]

71.4.3 Intervention

71.4.3.1 DBS Technological Advances (Figs. 71.5 and 71.6)

DBS hardware technological improvements have substantially contributed to the evolution of movement disorder surgery during the last years. One of the most important steps forward has been represented by the advent of directional electrodes. These electrodes allow current steering and may be able to optimize clinical benefits and avoid side effects in patients undergoing DBS. The directional electrodes currently on the market include electrodes, normally ring shaped, split into three components such that the current can be directed to one of the components rather than stimulating in a spherical fashion, as occurring with ring-type contacts. According to a provider survey, it is estimated that 20%–30% of patients could benefit from directional electrodes; however, further clinical outcome data are needed to assess whether most of the patients would benefit from having directional electrodes preferentially inserted. In the future, even higher-density designs with contacts that can be individually activated may allow additional opportunities for current manipulation. Tested mainly in silico at this point, there is relatively little clinical experience available, but computational modeling results suggest they might be quite effective in directing current and in improving therapeutic windows using complex electrode arrays in the near future [12]. Another recent DBS technological breakthrough has been the commercialization of sensing and closed-loop stimulation systems for DBS. The idea behind these systems is that of detecting a physiologic signal, such as beta oscillations from the subthalamic nucleus or phase-amplitude coupling from the sensorimotor cortex, which would then trigger the stimulation pulse, thus providing a feedback control for the stimulation. Adaptive DBS systems have in theory two possible advances: (1) the reduced need for current for stimulation duty because of stimulation only being applied when required rather than on a continuous basis and (2) the avoidance of stimulation when the neural activity is normal, thus not disrupting what is considered physiological activity. The evidence that these approaches may provide more benefits than continuous stimulation is growing, and RCT comparing conventional versus adaptative stimulation are currently underway.

Last but not least, the implementation of rechargeable batteries has in turn allowed the diminution in size of the implanted pulse generator (this aspect is fundamental in the pediatric population) and minimize the risk of IPG infection by avoiding multiple replacement over years. The rechargeable batteries are usually well tolerated by many patients, with some though who find challenging and difficult the way the IPG needs to be charged. Because of that, the choice between a rechargeable and a nonrechargeable device should be tailored for each case, depending most of all on the age, family compliance for pediatric patients, and personal preferences of the adult patient [11].

71.5 Magnetic Resonance-Guided Focused Ultrasound (MRgFUS) (Figs. 71.8, 71.9, 71.10 and 71.11)

MR-guided focused ultrasound (MRgFUS) involves noninvasive triangulation of 1024 transducer elements placed in a hemisphere-shaped transducer unit through the skull to a focal point in the brain under MR thermometry monitoring to lesion a target nucleus such as the ventral intermediate nucleus for essential tremor or tremor-dominant PD. Intraoperative MRI monitoring enables accurate intraprocedural targeting, while MR thermometry provides real-time feedback of the temperature at the target site during each sonication. Not all patients may undergo the procedure, depending on the eligibility on some skull parameters, such as the skull volume, thickness, and the skull density ratio

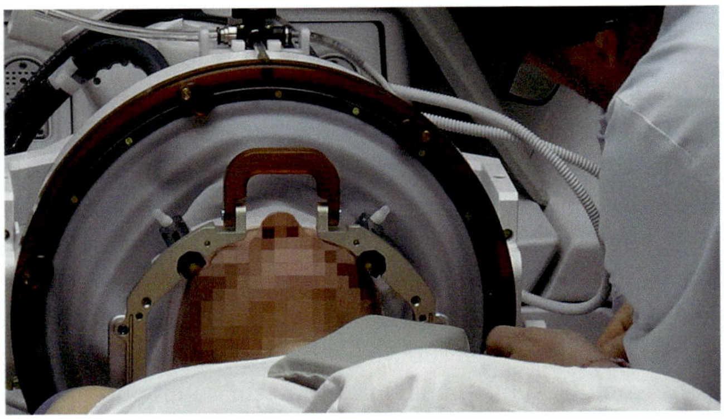

Fig. 71.8 MRI-guided focused ultrasound: The instrument

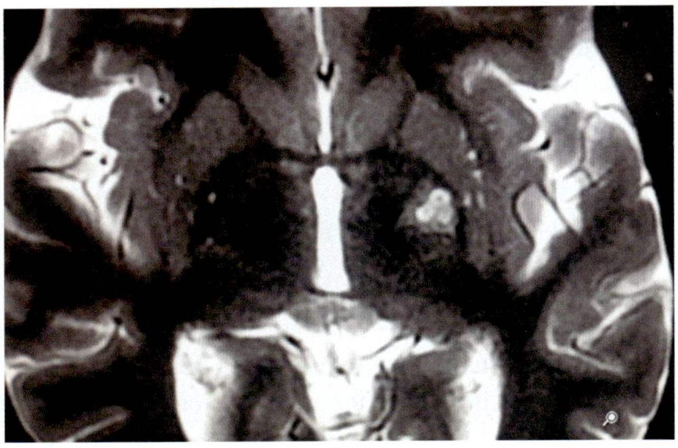

Fig. 71.9 FUS GPi lesion

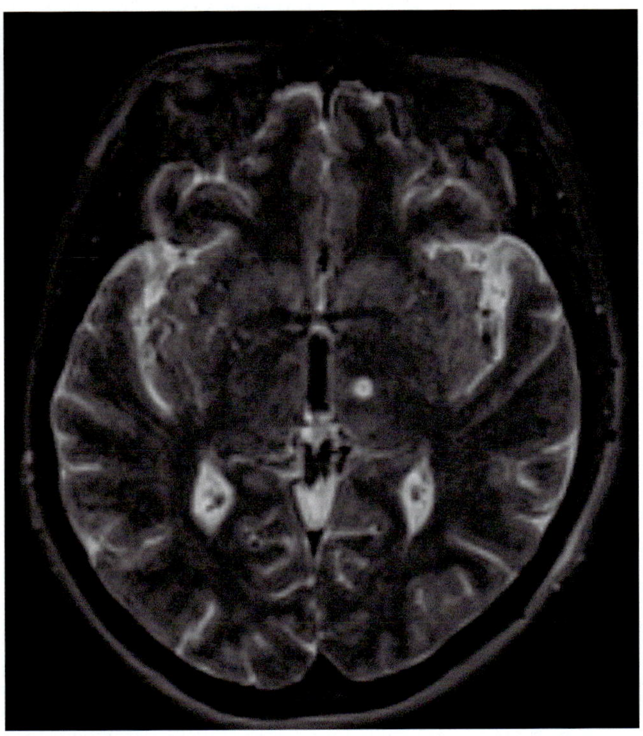

Fig. 71.10 FUS VIM lesion

(SDR), the ratio between the mean density values in Hounsfield units of bone marrow and cortical bone. In particular, patients with low SDR (<0.45) are prone to greater absorption and reflection of ultrasound and consequent difficulties in reaching high temperatures at the target site. Once a stereotactic frame is applied to the patient's head under local anesthesia, a silicone membrane is applied just above the head screws. The patient is placed in the supine position on the MR table, and the silicone membrane is attached to the transducer unit. Subsequently, water is filled between the silicone membrane and the transducer unit. Water is degassed to prevent cavitation and circulated to cool the scalp and skull to prevent heating. In the planning stage, multiplanar T2-weighted MR images are obtained, and the target is determined. In the treatment stage, the energy is increased in a stepwise fashion to gradually raise the temperature at the target site. Clinical evaluations are performed to observe therapeutic and adverse events. The treatment stage consists of three phases such as alignment, verification, and treatment. In the alignment phase, the location of the heated focus is checked in all the three directions (i.e., anteroposterior, mediolateral, and super inferior), referring to the MR thermometry with a mild temperature increase (approximately 40–45 °C). In the verification phase, temporary and reversible therapeutic and adverse effects, as well as the shape of the heated focus, are assessed with each moderate target temperature increase (around 50 °C) measured by MR thermometry. Once the location is confirmed with good clinical benefits and the absence of adverse effects, the target is ablated with high temperature (55 °C or higher) in the treatment phase.

The main advantages of this technique are that patients are awake during the procedure, obviating the need for general anesthesia, and that it allows for lesions that can be made incrementally while assessing clinical effects after each sonication. Second, the technique is completely noninvasive, in terms of cranial burr hole. A multi-center randomized controlled trial published in 2016 reported on 12-month outcome cranial of MRgFUS lesion for essential tremor (ET). [13] In this study, 56 patients with medication-refractory ET underwent unilateral VIM thalamotomy, while 20 patients underwent a sham procedure. At 3 months, tremor scores on the CRST significantly improved from 18.1 at baseline to 9.6 in the thalamotomy group, while the scores were 16.0 at baseline and 15.8 at 3 months in the sham group. Similarly, the mean disability (Part C) scores also significantly improved by 62% from 16.5 at baseline to 6.2 at 3 months in the thalamotomy group, while those in the sham group were 16.0 at baseline and 15.6 at 3 months. These improvements were maintained in the thalamotomy group at 12 months with a mean score of 10.9 and 6.3, respectively. Adverse events included paresthesia, taste disturbance, gait disturbance, dysmetria, and weakness. Eight patients (14%) had paresthesia or numbness, and five patients (9%) had gait disturbances at

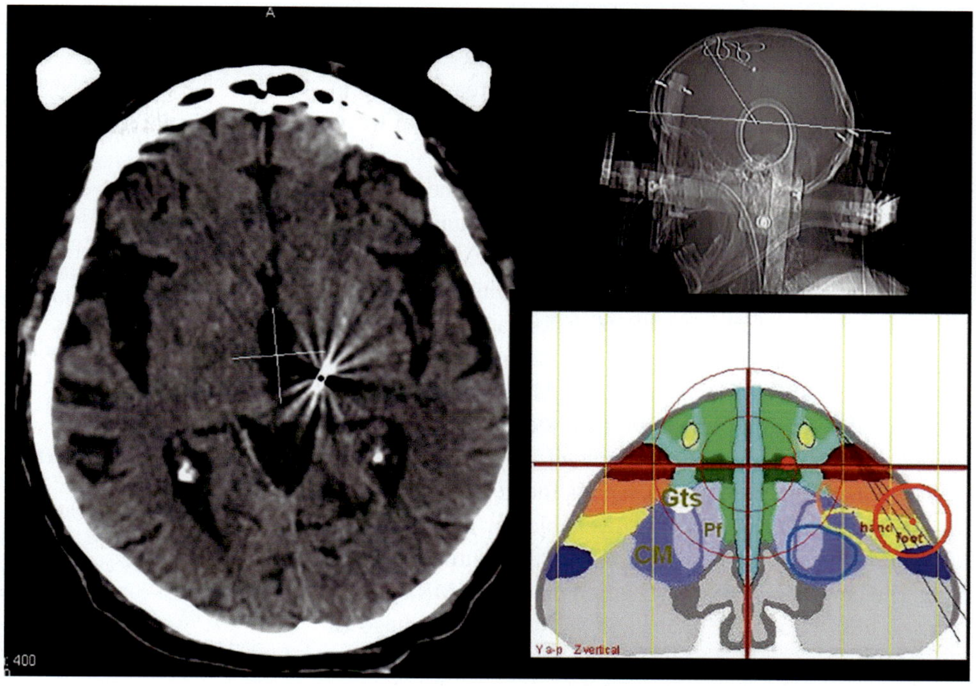

Fig. 71.11 Stereotactic lesioning of VIM

12 months. One patient had permanent hypoesthesia of the dominant thumb and index finger, categorized as a serious adverse event. A follow-up outcome of the same cohort published in 2018 showed that the improvements in the hand tremor and disability (Part C) scores were maintained at 2 years (56% and 60% on the CRST, respectively) [14].

A randomized controlled trial published in 2017 reported 20 patients with TDPD undergoing unilateral MRgFUS Vim thalamotomy while seven patients underwent a sham procedure. The contralateral hand tremor scores on the CRST in the on-medication state improved by 62% in the thalamotomy group at 3 months, while it improved by 22% in the sham group. Although the results indicated a placebo effect, the between-group difference was significant [15]. The improvement in the contralateral hand tremor was maintained at 1 year (70.6% from baseline on the CRST). Patients' quality of life measured using the 39-item Parkinson's Disease Questionnaire (PDQ-39) improved at 3 months by 63% and 30% in the thalamotomy and sham groups, respectively. On the other hand, the CRST Part C subscore improved by 73% from 13 at baseline to 3.5 at 3 months in the thalamotomy group, while it did not change substantially in the sham group (17 at baseline and 16 at 3 months). In the thalamotomy group, the PDQ-39 and CRST Part C both remained improved at 12 months by 65% and 81%, respectively. Thalamotomy-related adverse events included paresthesia, ataxia, hemiparesis, dysmetria, and vocal change, most of which were transient. Persistent adverse events included orofacial paresthesia (20%), finger paresthesia (5%), ataxia (5%), hemiparesis (10%), and mild vocal change (5%). One patient had serious adverse events of hemiparesis and associated ataxia. The hemiparesis gradually improved almost to the patient's baseline, but tone asymmetries and mild ataxia remained at last follow-up. This evidence indicates that VIM thalamotomy does significantly improve parkinsonian tremor, although it should be noted that it does not improve other parkinsonian symptoms such as bradykinesia and rigidity. Following these results, patients with Parkinson's disease for whom tremor reduction alone would considerably improve quality of life should be considered for this procedure [16, 17].

71.6 Surgical Pearls

- Treatment should be individualized.
- Team approach is mandatory for proper patient and procedure selection.
- Selection process should address response to L-Dopa, cognitive status, access to care, and patient's expectation.
- Presence of comorbidity and age of the patient are relevant in selection of the proposed surgical treatment, DBS vs MRI-F. US.
- DBS is a neuroimaging-based surgery. It is recommended to include high-quality MRI sequences, DTI, and iMRI and iCT images when possible.
- The use of directional electrodes for DBS should be considered. This may allow current steering, optimize clinical benefits, and minimize side effects.
- Select implementation of bidirectional rechargeable batteries (smaller oils generator) specially in pediatric age group has to be considered.

71.7 Discussion

Epidemiology of movement disorders is increasing due to the aging of population, with possible initial medical treatment of degenerative neurological diseases.

The stereotactic wax and later R.F. lesioning in the era when the only imaging has been the air or contrast medium ventriculography is over. The imaging of CT and more MRI, with the possibility of performing these examinations with the stereotactic frame on the patients, provide precision, and neurophysiological and clinical control during the electrode introduction through the skull and at brain target. Moreover, the use of "directional" electrodes improves the postoperative immediate follow-up, under the control of neurologist and neurosurgeons.

The dynamic of the DBS therapy is found on the hypothesis that this treatment may be "reversible" or better may be tailored on the single patient affected by the neurological disease. The new "lesional" era is carried out by the MRI-guided focused ultrasound. The main and substantial advantage of this stereotactic technique is the intraoperative MRI precision, and the reversibility of the initial thermic or sonification lesion, that could be modified following the clinical results in terms of adapting or enlarging the estimated target. This therapeutical possibility will be the future for movement disorders disease treatment. The possibility of repeating the treatment in the case of early or late recurrence is one of the major advantages. Since DBS or MRI F. US requires large economical investments, the "old" R.F. lessoning via stereotactically placed electrode may be considered in low-income countries.

71.8 Conclusion

Movement disorders surgery is about to face yet another revolution in the following decades, as a consequence of the technological innovation in imaging, DBS hardware, and customization of electrical therapy. Future challenges include the inception of more effective treatment for symptoms that are still resistant to DBS therapy, imaging allowing more precise targeting and expansion of "noninvasive" ablation techniques. Modification of the underlying disease remains an ambitious goal that could take the form of DBS in combination with delivery of stem cells or genes that restore dopamine neurotransmission. Finally, a very small number of patients who could benefit from movement disorders surgery are currently being treated. Understanding the impediments to such therapy whether educational, social, or economic that prevent patients from accessing surgery and developing strategies to overcome them is a crucial long-term objective.

Acknowledgements The authors thank Dr. Vincenzo Levi for helping in the preparation of the text and Fond I.E.N. for the scientific support.

References

1. Gildenberg PL. Evolution of basal ganglia surgery for movement disorders. Stereotact Funct Neurosurg. 2006;84(4):131–5. https://doi.org/10.1159/000094844.
2. Hariz MI, Blomstedt P, Zrinzo L. Deep brain stimulation between 1947 and 1987: the untold story. Neurosurg Focus. 2010;29(2):1–10. https://doi.org/10.3171/2010.4.FOCUS10106.
3. Speelman JD, Bosch DA. Resurgence of functional neurosurgery for Parkinson's disease: a historical perspective. Mov Disord. 1998;13(3):582–8. https://doi.org/10.1002/mds.870130336.
4. Artusi CA, Lopiano L, Morgante F. Clinical medicine deep brain stimulation selection criteria for Parkinson's disease: time to go beyond CAPSIT-PD. J Clin Med. 2020;9(12):3931. https://doi.org/10.3390/jcm9123931.
5. Charles D, Konrad PE, Neimat JS, et al. Subthalamic nucleus deep brain stimulation in early stage Parkinson's disease. J Parkinsons Dis. 2016;6(1):125–31. Published online 2014. https://doi.org/10.1016/j.parkreldis.2014.03.019.
6. Schuepbach WMM, Rau J, Knudsen K, et al. Neurostimulation for Parkinson's disease with early motor complications. N Engl J Med. 2013;7:610–32. https://doi.org/10.1056/NEJMoa1205158.
7. Wang JW, Zhang YQ, Zhang XH, Wang YP, Li JP, Li YJ. Cognitive and psychiatric effects of STN versus GPi deep brain stimulation in Parkinson's disease: a meta-analysis of randomized controlled trials. Published online 2016. https://doi.org/10.1371/journal.pone.0156721.
8. Fitzgerald JJ, Rosendal F, De Pennington N, et al. Long-term outcome of deep brain stimulation in generalised dystonia: a series of 60 cases. https://doi.org/10.1136/jnnp-2013-306833.
9. Artusi CA, Dwivedi A, Romagnolo A, et al. Differential response to pallidal deep brain stimulation among monogenic dystonias: systematic review and meta-analysis. J Neurol Neurosurg Psychiatry. 2020;91(4):426–33. https://doi.org/10.1136/JNNP-2019-322169.
10. Boutet A, Gramer R, Steele CJ, et al. Neuroimaging technological advancements for targeting in functional neurosurgery. Published online 1910. https://doi.org/10.1007/s11910-019-0961-8.
11. Schulder M, Mishra A, Mammis A, et al. Advances in technical aspects of DBS surgery HHS public access. Stereotact Funct Neurosurg. 2023;101(2):112–34. https://doi.org/10.1159/000529040.
12. Rowland NC, Sammartino F, Lozano AM. Advances in surgery for movement disorders. Mov Disord. 2017;32(1):5–10. https://doi.org/10.1002/mds.26636.
13. Elias WJ, Lipsman N, Ondo WG, et al. A randomized trial of focused ultrasound thalamotomy for essential tremor. N Engl J Med. 2016;375(8):730–9. https://doi.org/10.1056/NEJMOA1600159.
14. Chang JW, Park CK, Lipsman N, et al. A prospective trial of magnetic resonance-guided focused ultrasound thalamotomy for essential tremor: results at the 2-year follow-up. Ann Neurol. 2018;83(1):107–14. https://doi.org/10.1002/ANA.25126.
15. Bond AE, Shah BB, Huss DS, et al. Safety and efficacy of focused ultrasound thalamotomy for patients with medication-refractory, tremor-dominant Parkinson disease: a randomized clinical trial. JAMA Neurol. 2017;74(12):1412–8. https://doi.org/10.1001/JAMANEUROL.2017.3098.
16. Yamamoto K, Sarica C, Loh A, et al. Magnetic resonance-guided focused ultrasound for the treatment of tremor. Expert Rev Neurother. 2022;22(10):849–61. https://doi.org/10.1080/14737175.2022.2147826.
17. Ko TH, Lee YH, Chan L, Tsai KWK, Hong CT, Lo WL. Magnetic resonance–guided focused ultrasound surgery for Parkinson's disease: a mini-review and comparison between deep brain stimulation. Parkinsonism Relat Disord. Published online May 2023:105431. https://doi.org/10.1016/j.parkreldis.2023.105431.

The Role of Endoscopy in MVD for Treatment of Trigeminal Neuralgia and Hemifacial Spasm

72

Henry W. S. Schroeder and Ahmed Al Menabbawy

Learning Objectives

- Understand the historical background and evolution of neuroendoscopy in MVD surgery.
- Describe the advantages and limitations of using an endoscope in MVD surgery compared to traditional microscopic techniques.
- Identify the anatomical structures and cranial nerves that can be visualized using an endoscope during MVD surgery.
- Recognize the technical aspects of endoscopic MVD surgery, including the use of angled endoscopes and different surgical instruments.
- Evaluate the outcomes and complications associated with endoscopic MVD surgery compared to traditional MVD surgery.
- Understand the steps involved in an endoscopic MVD surgery, including patient positioning, craniotomy, arachnoid dissection, identification and separation of the offending blood vessel, and watertight closure.

72.1 Introduction

Neuroendoscopy has revolutionized neurosurgical practice in the last decades, making the endoscope an essential tool in almost all well-equipped neurosurgical operation theaters. Different nomenclature has been also used as endoscopic, endoscope-assisted, and endoscope-controlled procedures which is important to discuss and understand the difference between each one of them [1–3].

Endoscopic approaches have been adopted in microvascular decompression (MVD) for cranial nerve compression syndromes such as trigeminal neuralgia (TN) and hemifacial spasm (HFS). The main concept of MVD is relieving the pressure on the affected nerve by avoiding contact with the compressing blood vessel. This is usually achieved by repositioning or transpositioning the offending blood vessels rendering the root exit/entry zone (REZ), whenever achievable, free of any contact with any structure including the decompressing agent (e.g., Teflon) [4–7].

Our terminology is based on the optical device used (i.e., whether the endoscope was used alone or in conjunction with an operating microscope) and on the route of surgical manipulations (i.e., through or outside the endoscope). Accordingly, procedures are categorized into pure endoscopic neurosurgery (EN) and combined or endoscope-assisted techniques. Assisted techniques are further categorized into endoscope-assisted microneurosurgery (EAM), endoscope-controlled microneurosurgery (ECM), and endoscopic inspection. EN is used for all exclusively endoscopic procedures. Surgical manipulations are performed with special endoscopic instruments through the endoscope itself. EN is performed through a burr hole trepanation. EAM is used for procedures in which endoscopy is performed in addition to microsurgical manipulations under the operating microscope during the same operation. ECM is used for procedures in which microsurgical manipulations are performed with conventional instruments but are guided by the video image provided by the endoscope (video surgery). ECM is distinguished from EAM in that an operating microscope is not used, and ECM is distinguished from EN in that manipulations are performed outside the endoscope. Endoscopic inspection is used when the endoscope is needed as an adjunct during any neurosurgical procedure for inspection purposes only.

H. W. S. Schroeder (✉) · A. Al Menabbawy
Department of Neurosurgery, University Medicine Greifswald, Greifswald, Germany
e-mail: Henry.Schroeder@uni-greifswald.de; ahmed.almenabbawy@med.uni-greifswald.de; http://www.medizin.uni-greifswald.de/neuro_ch

I. N. Kanaan, V. Beneš (eds.), *Neuroanatomy Guidance to Successful Neurosurgical Interventions*,
https://doi.org/10.1007/978-3-031-59838-8_72

The most popular approach using the endoscope in MVD is the endoscope-assisted MVD which combines the benefits of endoscopic visualization (usually with 30° or 45° optics) as well as keeping the free mobility provided by using the microscope. In this technique, the endoscope is mainly used for visualization, and only some surgical manipulations are performed. In some cases, the neurosurgeon might prefer to use an instrument like a dissector to mobilize the offending vessel or to readjust the Teflon pledges while holding the endoscope. Many studies have shown the safety and effectiveness of using endoscopic techniques in treatment of TN and HFS. While some neurosurgeons perform MVD purely endoscopically [8], we prefer the combined endoscope-assisted technique because we do not see an advantage of using an endoscope at the beginning of the procedure [1, 9]. At first, we dissect all arachnoid, open the cisterns, and release CSF. Thereafter, the cerebellum falls back by gravity, and some space is created which is needed for endoscopic inspection. In young patients with HFS, there is sometimes very little space for dissection. If the endoscope would be inserted, it would block the way and hinder the bimanual dissection. In our institution, we prefer visualizing the whole course of the nerve from the REZ to the entrance into the internal auditory canal. With the microscope, that is not always possible, and the risk of overlooking an offending vessel remains. In TN, a prominent suprameatal tubercle may obscure the view to the Meckel's cave, where occasionally a small vein is strangulating the nerve. In HFS, a vessel down in the pontomedullary sulcus hidden behind the lower cranial nerve group may be easily missed with the microscope. However, some precautions are required when using the endoscope. The tip of the endoscope can become very hot and might damage neurovascular structures. Therefore, frequent irrigation is required, and a certain distance should be kept between the endoscope tip and the inspected region. Furthermore, one should keep in mind that only structures are visualized which are in front of the lens. Once one has passed a structure, there is no visual control and unintended damage may occur. Therefore, one has to be very cautious with instrument movements.

The use of endoscopy in MVD allows for better visualization of the surgical field, which can lead to improved accuracy and safety during the procedure and has shown promising results in the MVD for TN and HFS with low complication rates and short recovery times [10].

72.2 History and Background

Neuroendoscopy is a minimally invasive surgical technique that has been increasingly used in modern neurosurgery since the 1980s. In fact, the use of neuroendoscopy in MVD for TN and HFS is relatively new and has gained popularity in recent years. Historically speaking, the earliest recorded "minimally invasive" approach to the skull through the nasal cavity dates to more than 2000 BC by the ancient Egyptians [10].

The use of endoscopy in modern medicine dates to the early nineteenth century when the first endoscope was developed by the German physician Philipp Bozzini. Although and following him, Victor Lespinasse and Walter Dandy first reported using the endoscope intracranially, it was not until the 1980s that endoscopy became widely used in neurosurgery, thanks to the development of rod-lens endoscopes and video imaging technology [10].

In the late 1990s, the first reports of endoscopic-assisted MVD for TN and HFS were published, showing promising results in terms of safety and efficacy. The use of endoscopy in MVD allows for better visualization of the surgical field and identification of the offending blood vessel, leading to improved accuracy and safety during the procedure [11–13]. Since then, several studies have been conducted to evaluate the efficacy and safety of endoscopic-assisted MVD for TN and HFS. A systematic review published in 2019 declared that endoscopic-assisted MVD is a safe and effective alternative to traditional MVD, with lower complication rates and shorter recovery times [14].

Overall, the history of neuroendoscopy in MVD has been relatively short but promising and growing. As technology continues to advance and more studies are conducted, it is likely that using neuroendoscopy in MVD will increasingly confirm being the gold standard in the management of TN and HFS.

72.3 Anatomical Highlights

- An upper retrosigmoid craniotomy in the sinus knee between sigmoid and transverse sinus is performed for MVD in TN.
- Following the superior petrosal vein guides to the superior petrosal sinus between petrous bone and tentorium.
- The trigeminal nerve is running from the pons to the Meckel's cave. In most cases, the compression is caused by the superior cerebellar artery which runs above the trigeminal nerve under the tent.
- For HFS, a lower retrosigmoid craniotomy is done exposing the lower cranial nerve group.
- The facial nerve runs initially on the brainstem surface starting at the pontomedullary sulcus before turning into the free cisternal segment.
- The course of the posterior inferior cerebellar artery and anterior inferior cerebellar artery is extremely variable, causing a large variety of compression patterns.

- The proximal facial nerve at the pontomedullary sulcus is hidden behind the lower cranial neve group.
- Different anatomical types of arterial compressions are illustrated in Fig. 72.1. But in some cases, it may be a vein or a combination of both, and in very rare cases, compression can be caused by arachnoid bands, as shown in Figs. 72.1e and 72.2. The offending vessel can be located at different points along the nerve, including the root entry zone (REZ), the cisternal portion of the nerve, and the distal portion of the nerve. Distally located compressions are usually considered insignificant, but can be symptomatic, mostly in TN where a vein at the entrance to the Meckel's cave is squeezing the trigeminal nerve. In HFS, we have not seen distal symptomatic compressions so far.

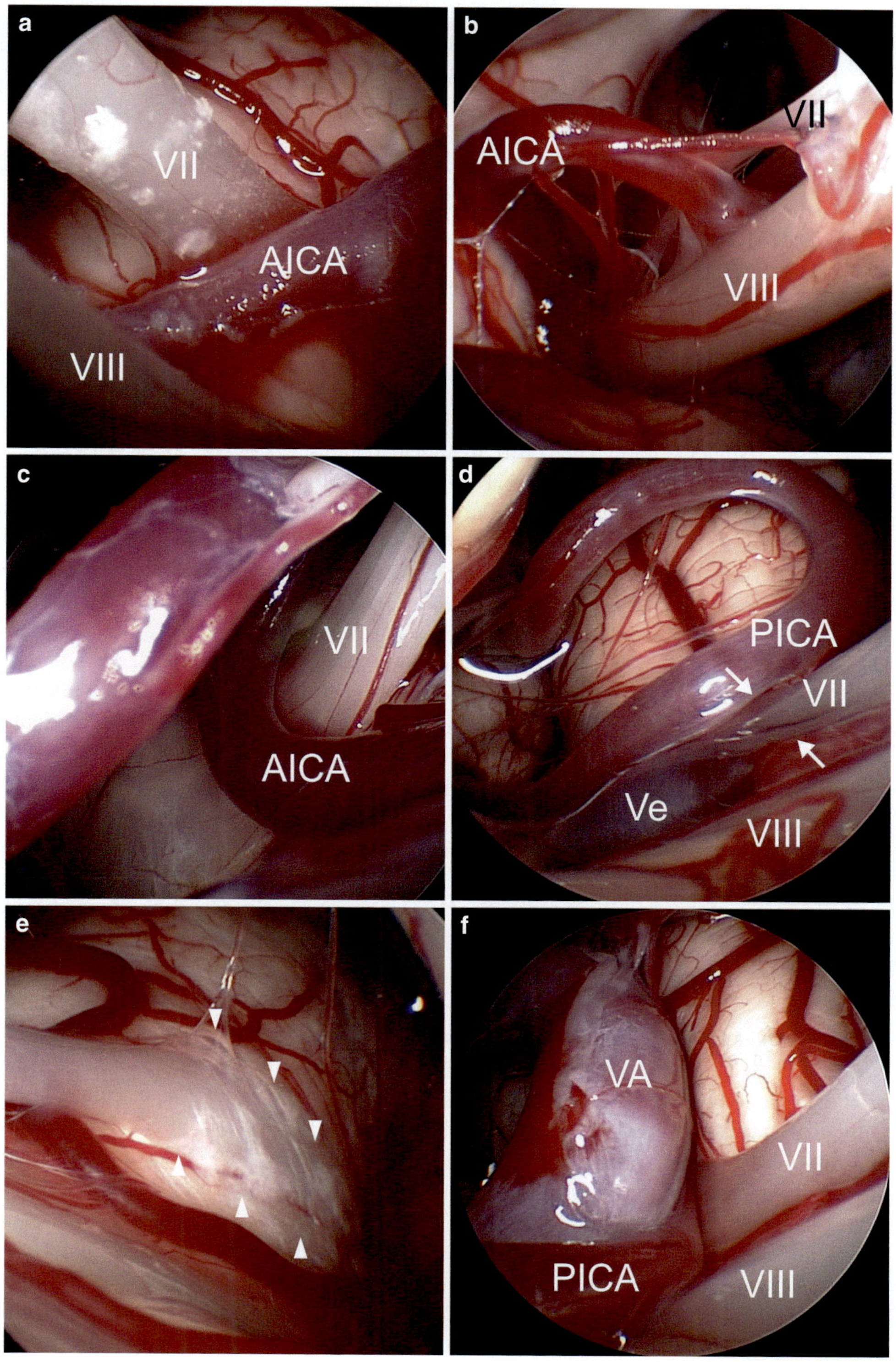

Fig. 72.1 Different anatomical types of arterial compressions visualized with a 45° endoscope. (**a**) Branch compression. (**b**) Perforator type. (**c**) Typical AICA loop type compression. (**d**) Sandwich compression with PICA loop and a vein (Ve). Arrows pointing to the sandwiched facial nerve. (**e**) Arachnoid type compression. Arrow heads pointing to the arachnoid bands. (**f**) Tandem compression with VA and PICA as it arises from it. Abbreviations: *AICA* anterior inferior cerebellar artery, *PICA* posterior inferior cerebellar artery, *VA* vertebral artery, *Ve* vein, *VII* facial nerve, *VIII* vestibulocochlear nerve

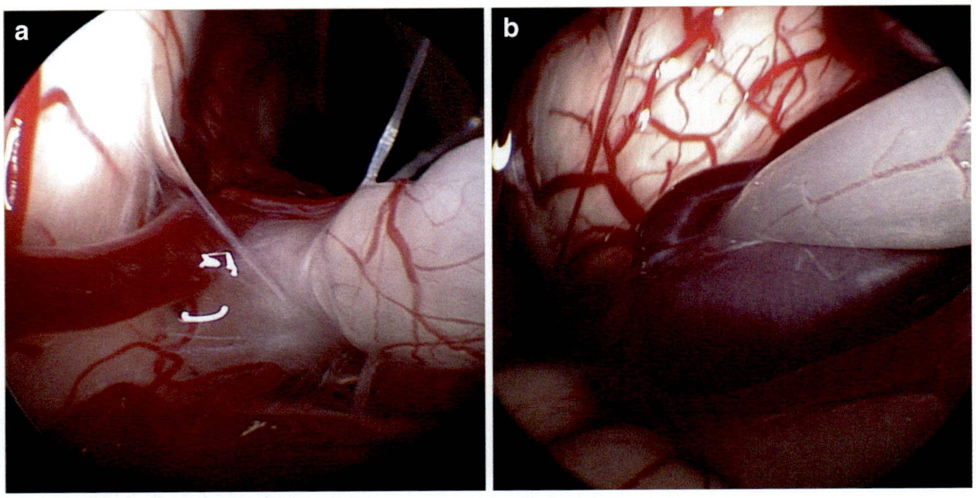

Fig. 72.2 Atypical compression etiologies visualized with a 45° endoscope. (**a**) Arachnoid strangulation of the facial nerve causing HFS. (**b**) Pure venous compression

72.4 Surgical Approach

72.4.1 Indication

Correct patient selection and establishing the right diagnosis is the first step for success of MVD for TN and HFS. The cure rates following MVD for TN and HFS exceed 90% in some series. After exploiting the other treatment options, MVD should be the procedure of choice. However, exclusion of other pathologies that might mimic the disease such as dystonia or facial tics or facial pain syndromes should be done first. In some cases, TN is just another presentation for multiple sclerosis (MS). For this reason, careful screening for MS from history and MRI images should be performed before surgery. In such cases MVD would not be indicated and if performed will not be successful. Other surgical and nonsurgical procedures could still be performed but these are out of the scope of this chapter.

The preoperative MR images with CISS and TOF sequences should be carefully examined for the previously mentioned anatomical consideration and specially to localize the compressing vessel [5, 15].

72.4.2 Anesthesia and Positioning

Following induction of general anesthesia (without muscle relaxation), the head is secured in a Mayfield head clamp to provide a stable base for the surgical procedure. The patient is typically positioned supine with the head slightly flexed and rotated 45° to the contralateral side, ensuring no compression is applied on the jugular veins. The contralateral side of the body is supported with side supports, and the ipsilateral arm is positioned near to the patient. Then, the table is additionally rotated 30° to improve the surgical corridor by cerebellar retraction due to gravity. Figure 72.3 shows the positioning of the patient. Now, the electrodes for neurophysiological monitoring are applied. The surgery is done under monitoring of facial EMG and brainstem auditory evoked potentials (BAEPs). The distribution of the staff and equipment should be done in a way that no obstruction or slowing of the workflow during the operation occurs.

72.4.3 Surgical Steps

An upper retrosigmoid craniotomy is placed for trigeminal nerve decompression, in contrast to HFS cases, where a lower retrosigmoid craniotomy is required which exposes the lower cranial nerves. Also studying the images preoperatively is very important to expect where to find the venous sinuses and large emissary veins. A linear retroauricular skin incision and a linear or slightly curvilinear dura opening near the venous sinus are sufficient to visualize the medial aspect of the CPA and the compression at the brainstem (Fig. 72.4) [16].

The bony landmarks for the craniotomy are the base of the mastoid process, asterion and the digastric groove. For TN, one burr hole is placed at the asterion, and for HFS, it is placed further down behind the digastric groove. Then, a small 2–2.5 cm craniotomy is made. The remaining bone covering the sinus is removed with a high-speed drill until the medial margin is exposed.

For TN, the sinus knee is exposed, and for HFS, the sigmoid sinus is exposed. The transverse sinus is usually, but not always, expected to be located along the line connecting the zygomatic arch and posterior occipital protuberance. In case of a large emissary vein, it can be skeletonized with high-speed drill and then coagulated and cut.

Noting any opening of the mastoid air cells is also crucial in order to seal them later. After localizing the sinus, a linear or curvilinear dura incision in line with the venous sinus and around 2–3 mm away from the sinus edge is performed (to allow for easier closure at the end of the procedure). Usually,

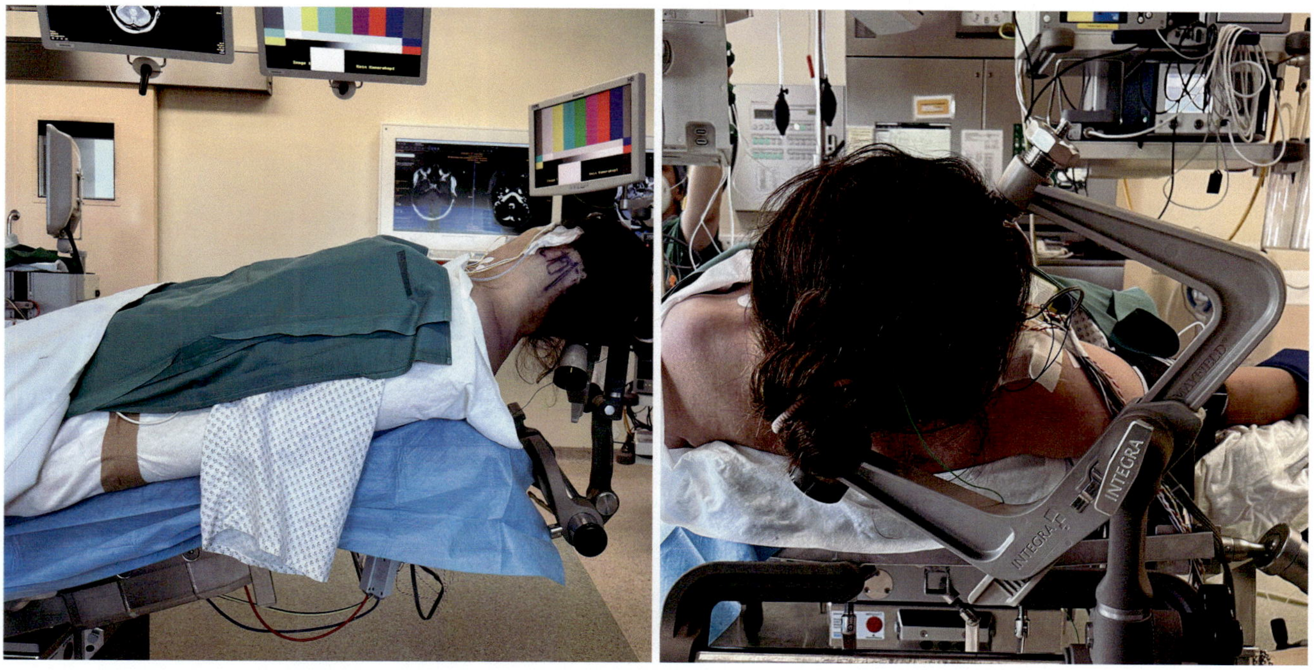

Fig. 72.3 Supine positioning of the patient with head rotation of 45° to the contralateral side, slight head flexion, and body side support on the opposite side

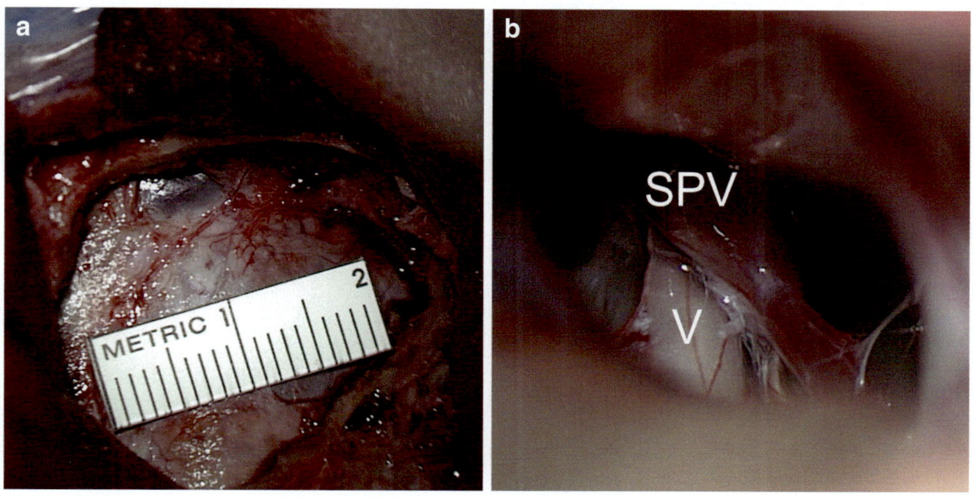

Fig. 72.4 (**a**) A small upper retrosigmoid craniotomy defect. (**b**) Microscopic view to the trigeminal nerve. Abbreviations: *SPV* superior petrosal vein, *V* trigeminal nerve

only a small incision is done in the beginning to avoid herniation of cerebellar tissue which might occur with larger incisions. However, when the initial incision shows that the cerebellum is relaxed, the entire dural incision is made from the beginning. Opening of the arachnoid over the lower cranial nerve group in HFS or around the superior petrosal vein/cerebellar artery releases CSF and gives more space for the dissection.

No fixed cerebellar retraction of the cerebellum is needed. We usually place a piece of cellulose and cottonoid on the cerebellar surface as a means of protection. Important landmarks in TN are the superior petrosal veins, trigeminal, trochlear, facial and vestibulocochlear nerve, and tentorium, while in HFS, important landmarks are the lower cranial nerves and facial and vestibulocochlear nerve. Following arachnoid dissection, the REZ of the nerves is inspected using a 45° endoscope in HFS and 30° in TN (Figs. 72.5 and 72.1).

Careful inspection of the course of the trigeminal or facial nerve should be done until the compressing vessel is located. Depending on the anatomical situation, the decompression can be done either using shredded Teflon, a Gortex or Teflon sling, or the bridge technique. Figure 72.6 illustrates MVD in a case of TN caused by a typical vascular loop and decompressed using shredded Teflon [17]. In Fig. 72.7, a case of MVD using shredded Teflon for HFS is presented.

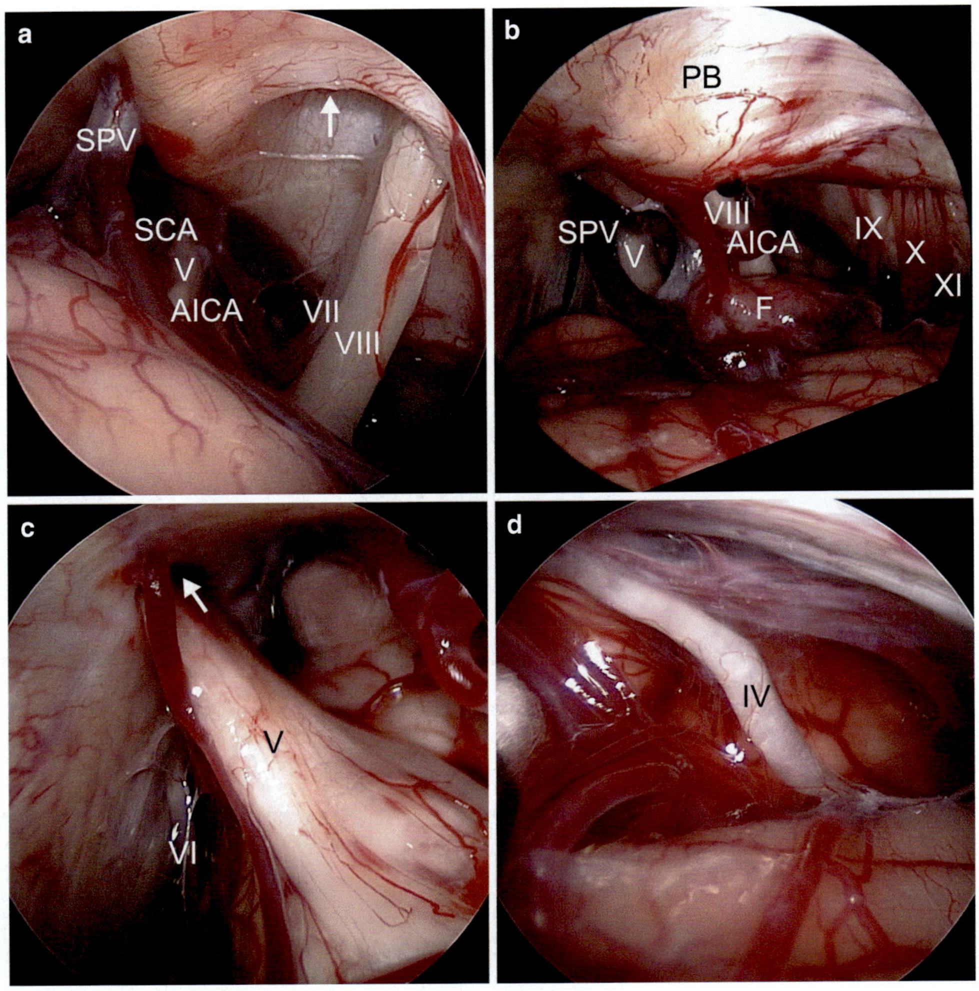

Fig. 72.5 Demonstration of the anatomy of the cerebellopontine using the endoscope during microvascular decompression (MVD). (**a**) Important contents of the right upper CPA including the superficial petrosal vein (SPV), superior cerebellar artery (SCA), trigeminal nerve (V), anterior inferior cerebellar artery (AICA), and facial and vestibulocochlear nerves (VII, VIII). Arrow pointing to internal auditory canal. (**b**) Overview of the entire CPA including cerebellar flocculus (F), lower cranial nerves (IX, X, XI), vestibulocochlear nerve (VIII), trigeminal nerve (V), SPV, AICA, and petrous bone (PB). (**c**) Trigeminal nerve (V) entering the Meckel's cave (arrow), abducent nerve (VI) can also be visualized as it enters the Dorello's canal. (**d**) The trochlear nerve (IV) as it courses near the tentorial edge

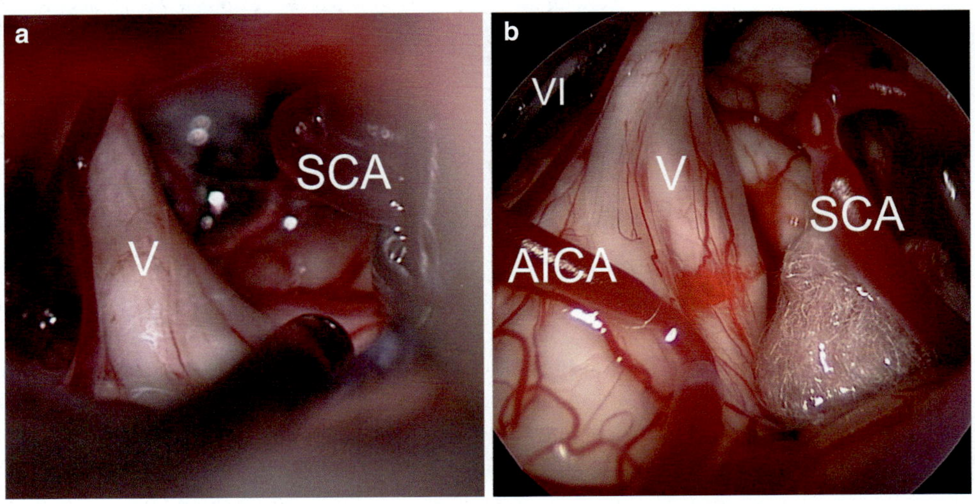

Fig. 72.6 Endoscopic inspection while performing an endoscope-assisted microvascular decompression MVD for trigeminal nerve (V). (**a**) Microscopic view of the trigeminal nerve (V) as it exits the brainstem. The compressing SCA loop is retracted using a dissector. (**b**) The endoscopic view shows the full decompression of the nerve. Abbreviations: *AICA* anterior inferior cerebellar artery, *SCA* superior cerebellar artery, *V* trigeminal nerve, *VI* abducent nerve

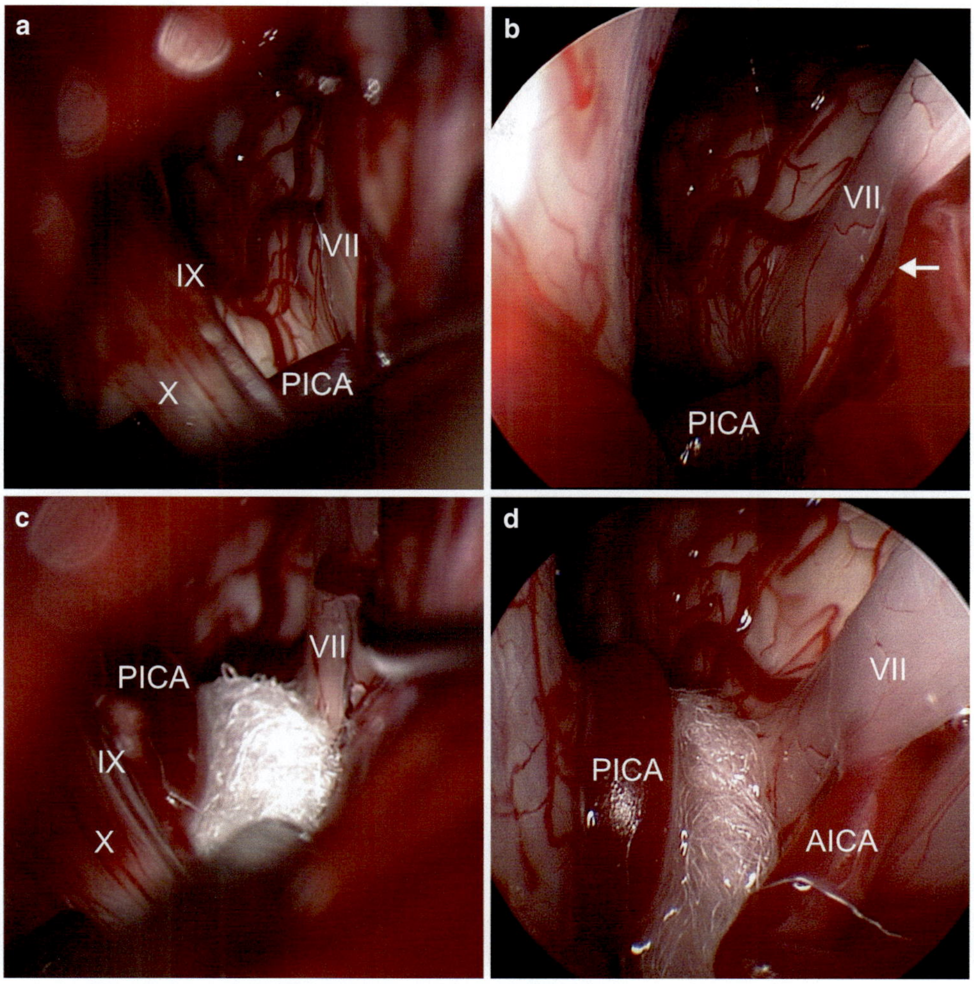

Fig. 72.7 Initial intraoperative microscopic (**a**) and endoscopic (**b**) images before performing the microvascular decompression (MV) for a case of HFS. The endoscopic view shows the vascular compression by PICA deep down in the pontomedullary sulcus more clearly. Following decompression using shredded Teflon, a microscopic view is shown (**c**) as well as the endoscopic view (**d**) which shows the root exit zone (REZ) clearer than the microscopic view. Abbreviations: *AICA* anterior inferior cerebellar artery, *PICA* posterior inferior cerebellar artery, *VII* facial nerve, *IX* glossopharyngeal nerve, *X* vagal nerve, arrow pointing to intermediate nerve

We demonstrate another case of MVD for TN, where a double sandwich compression by the superior cerebellar artery (SCA) (from above) and the anterior inferior cerebellar artery (AICA) (from below as shown in Fig. 72.8a) is witnessed. The main aim of decompression is to render the nerve completely free from contact with blood vessels, arachnoid and even the decompressing agent (Fig. 72.8b). Here, shredded Teflon and sling transposition were used to perform the decompression.

The idea of the sling technique is to encase the compressing vessel in a sling and transpose it away from the compressed nerve. Different variants of this technique have been introduced. We prefer fixation of the sling on the dura of the skull base after forming a pocket, as shown in Fig. 72.8c–f, using a microvascular clip. Another patient who underwent a sling transposition technique for HFS is shown in Fig. 72.9.

Another decompression technique is the "Bridge technique," where a bridge of Teflon lifts the compression artery without direct contact with the root exit zone or the facial nerve, as shown in Fig. 72.10.

In some cases, manipulations of the vessels and perforators can take more time and, even a visible spasm or kinking can be witnessed. In these cases, local papaverine can be used. In most cases, immediate visible dilation of the vessel can be witnessed. To check for adequate blood flow, a micro-Doppler probe or indocyanine green videoangiography can be used.

After decompression is achieved with the microscope, further visualization with the 45° endoscope is performed to make sure a sufficient decompression has been achieved. A dissector can be used to make some adjustments to the Teflon under endoscopic visualization.

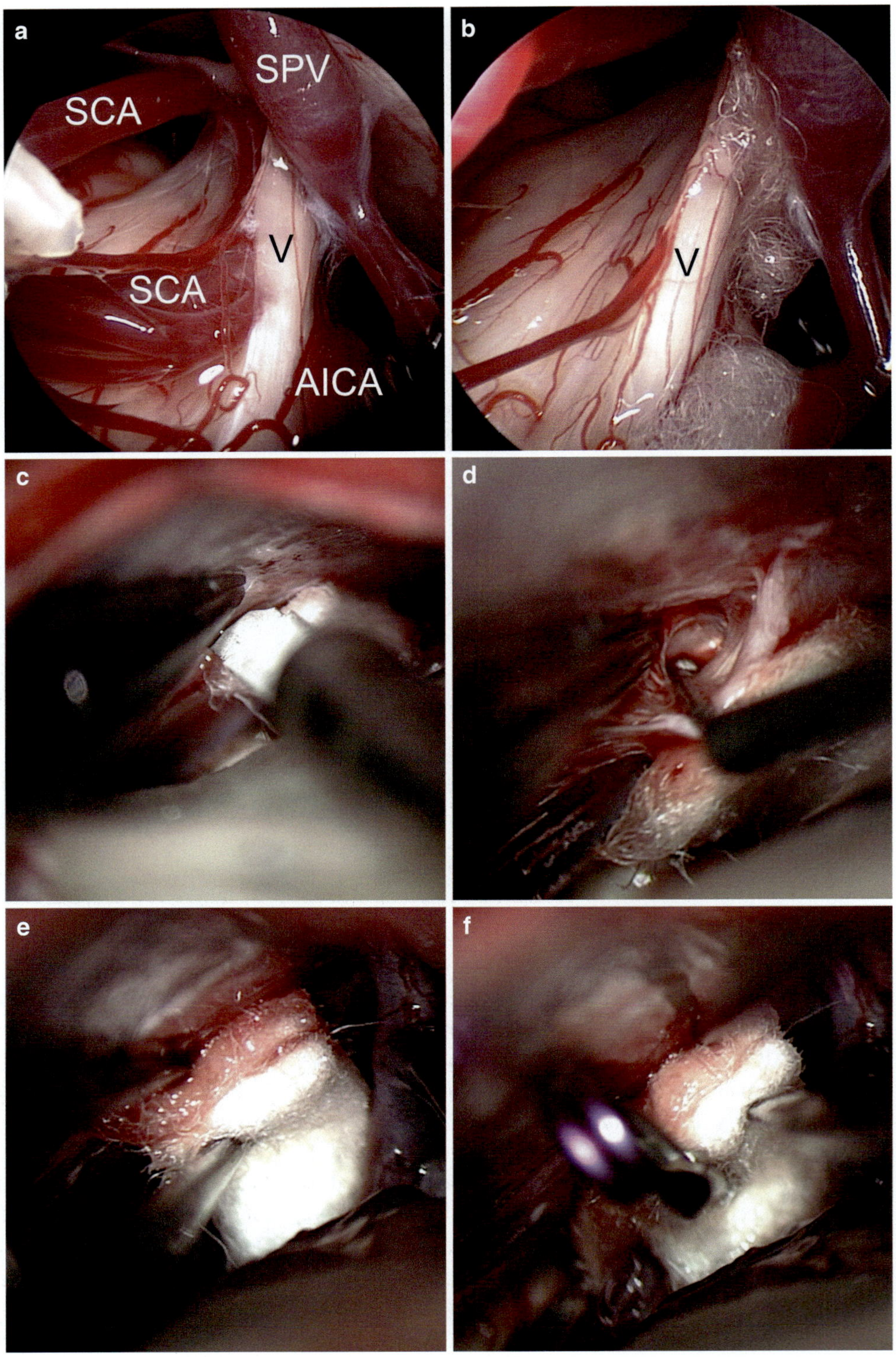

Fig. 72.8 Intraoperative images from a case of microvascular decompression (MVD) for trigeminal neuralgia (TN). (**a**) The double sandwich compression of the trigeminal nerve (V) with the superior cerebellar artery (SCA) and anterior inferior cerebellar artery (AICA). The superior petrosal vein (SPV) is also visualized. (**b**) No residual contact between the vessels and V after MVD via both shredded Teflon (from below) and sling technique (from above) is seen. (**c**) incision of the dura to create a pocket for the vascular clip that will be used to fix the sling is demonstrated. (**d**) The pocket. (**e**) Adjustment of the sling with great care not exerting too much pulling especially on the small perforators that might be emerging from this vessel. (**f**) Application of the vascular clip successfully

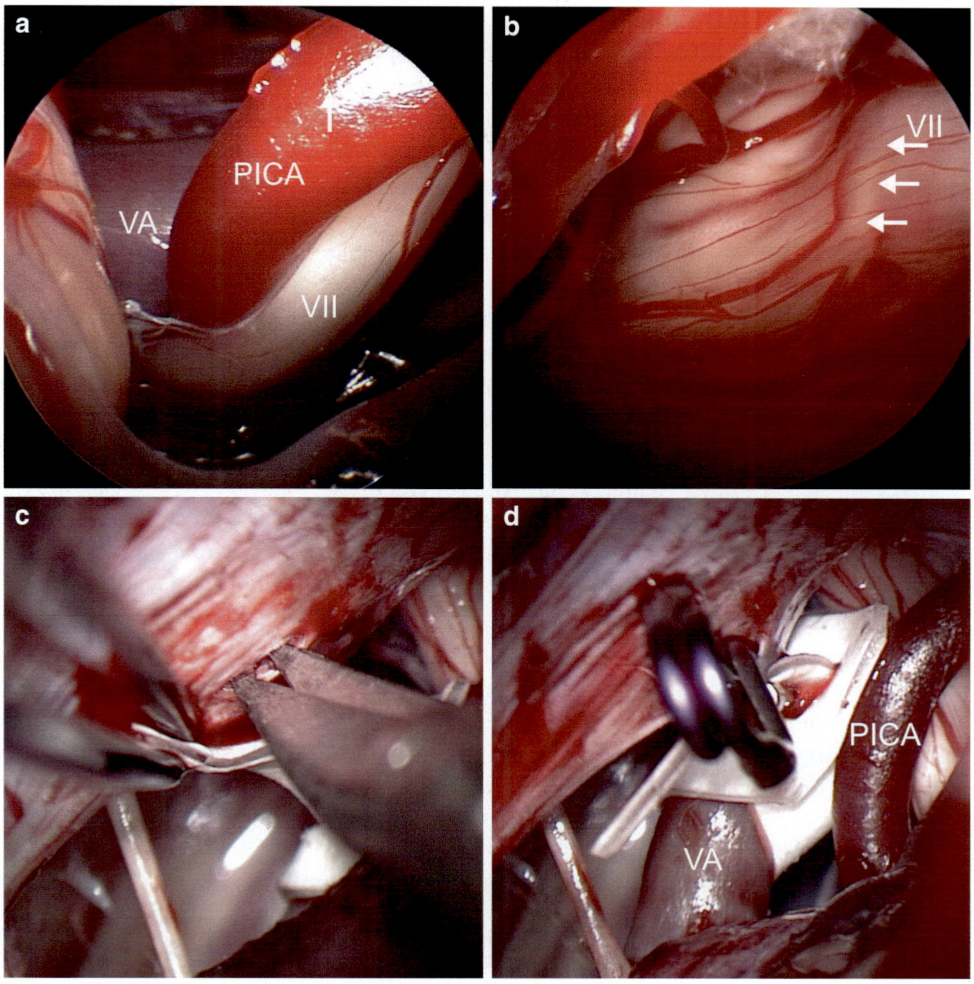

Fig. 72.9 Demonstration of a case of MVD for HFS using the sling technique. (**a**) An endoscopic view of the vascular compression caused by a tandem compression (with VA and its arising PICA). (**b**) The REZ free following decompression using the sling technique and severe grooving (arrows) of this part of the facial nerve due to prolonged compression. (**c**) A dural pouch for fixation of the sling at the skull base between jugular foramen and IAC is performed. (**d**) Fixation of the sling is done using a vascular aneurysm miniclip. Abbreviations: *IAC* internal auditory canal, *HFS* hemifacial spasm, *MVD* microvascular decompression, *REZ* root exit zone, *PICA* posterior inferior cerebellar artery, and *VA* vertebral artery

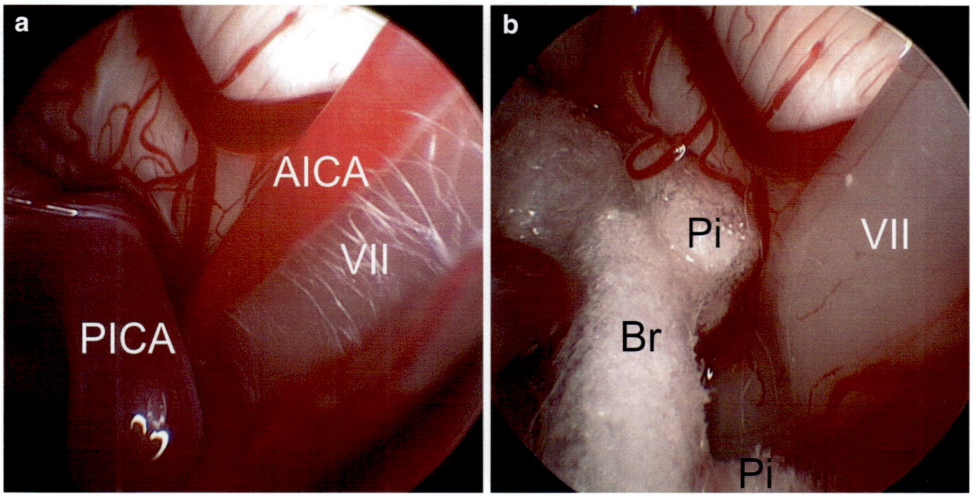

Fig. 72.10 Bridge technique of microvascular decompression (MVD) for hemifacial spasm (HFS). (**a**) Compression of the facial nerve (VII) at the root exit zone by a posterior inferior cerebellar artery (PICA) and anterior inferior cerebellar artery (AICA) loop is shown. (**b**) MVD successfully performed using the bridge technique where the two pillars (Pi) hold the Teflon bridge (Br) away from the facial nerve REZ. Apparent grooving of the nerve is visible due to prolonged compression

72.4.3.1 Role of Intraoperative Neuromonitoring

Intraoperative neuromonitoring (IONM) is an important adjunct to MVD surgery, as it allows for real-time monitoring of the cranial nerves' response to ensure the safety and efficacy of the procedure. Monitoring lateral spreads and BAEPs are especially important during MVD surgery. Lateral spreads refer to abnormal electrical discharges that can occur in the facial muscles. This is an indication for nerve irritation or compression, and usually after successful decompression, they should disappear and indicate an adequate decompression. Persistence of the lateral spreads following decompression must ring the bell for the neurosurgeon to look further for any missed compression.

BAEPs, on the other hand, refer to the electrical responses of the auditory nerve and brainstem in response to sound stimuli. By delivering a series of clicks or tones to the patient's ear, the surgeon can monitor the patient's auditory nerve and brainstem function in real time. Any changes in BAEP responses during surgery may indicate that the patient's brainstem or cochlear nerve is being compressed or irritated, which can be a serious complication of MVD surgery. By allowing the surgeon to make real-time adjustments based on the patient's neural responses, IONM can help reduce the risk of nerve damage or other complications during MVD surgery [18, 19].

72.4.3.2 Closure

Dura closure is done in a watertight fashion (Fig. 72.11). Any opened mastoid air cells should be adequately stuffed with muscle (preferably two layers) and fibrin glue. If any drilling of the skull base was needed, a careful check for any open-air cells must be done. A piece of TachoSil (Takeda, Linz, Austria) can be placed over the dural suture to assure that it is sealed. The small bone flap is fixed with titanium plates and screws. The wound is closed in layers.

72.4.3.3 Postoperative Management

The patient is usually monitored for 24 hours in the intermediate care unit. The postoperative continuous assessment for the outcome and screening for complications should be done within the hospitalization (usually 4–5 days after surgery) and then at different intervals, usually 3 months apart. The routine postoperative assessment by an otolaryngologist who performed tympanometry and pure tone audiometry should be done. Any swallowing problems or hoarseness should be noted, and confirmation should be done by the otolaryngologist using laryngoscopy and swallowing studies.

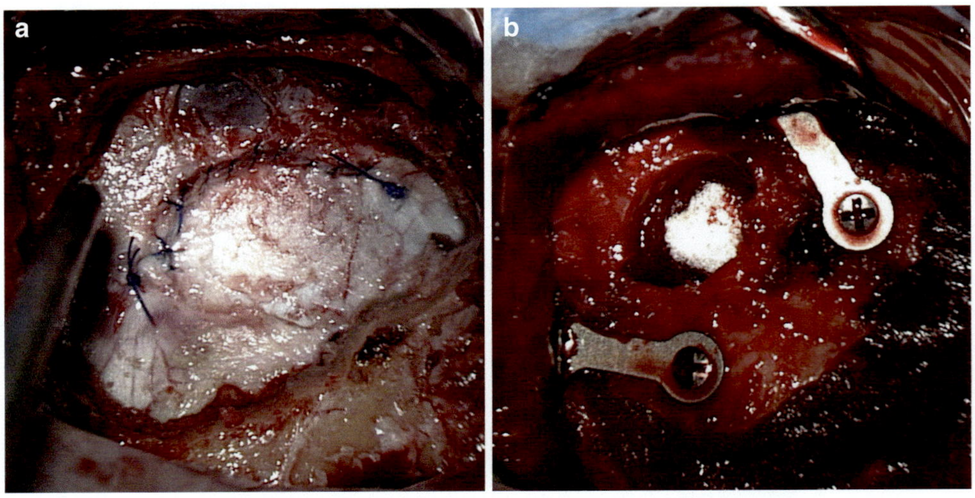

Fig. 72.11 (**a**) Dural closure in a watertight fashion using continuous prolene (4/0) monofilamentous sutures. (**b**) Reapplication of the bone fragment and fixation with two miniplates and screws (**b**)

72.5 Surgical Pearls

- Study the preoperative images in detail preoperatively to anticipate the intraoperative situation.
- Proper patient positioning: The patient should be placed in a supine position with the head rotated 45° to the opposite side.
- Sharp dissection along the plane of the neurovascular structures is the way for a safe MVD and bloodless field. Avoid any tension or traction on the sensitive neurovascular structures.
- Once the REZ is reached, better visualization can be achieved using an angled endoscope.
- Once determined, the compressing vessel can be transposed using shredded Teflon. In cases where there is not enough decompression with shredded Teflon, the sling or bridge techniques should be performed.
- During and after decompression, check the BAEP and lateral spreads.
- Before closing, make sure that any opened bony air cells are securely sealed to avoid CSF leak.

72.6 Discussion

MVD is a surgical procedure that has been shown to be an effective treatment for cranial nerves compression syndromes, especially TN and HFS [20, 21]. For instance, a metaanalysis from 2022 comparing MVD to Gamma knife surgery (GKS) involving 582 patients is treated with MVD and 607 patients GKS. All the patients had drug-resistant TN. Results demonstrated that MVD showed higher rates of pain relief and lower rates of pain recurrence compared to GKS, although GKS might be accompanied with less complications. Another systematic review also supported these results including 13 relevant studies with 683 patients treated by stereotactic radiosurgery compared with 670 patients treated by MVD [22, 23].

Regarding HFS, a systematic review from 2015 included 12 studies involving 2727 patients with HFS who underwent MVD showed that MVD had a mean postoperative success rate of 85.1% and in some studies, exceeded 90%. However, they pointed out that around 25% of the patients show delayed relief of the symptoms, rendering it the most effective treatment option for HFS [24]. Sindou et al. conducted a review in 2018 including 82 articles analyzing more than 10,000 cases of MVD for HSF. The success rates ranged between 85% and 90% and delayed relief was noted in more than 30% of the cases. Complications rates were as low as 1–2% for facial palsy, 2–3% for hearing deficits (nonfunctional), and 0.5–1% for lower cranial nerves deficits. Ischemia reached 0.1% and mortality at 0.1%. CSF leakage and its complications could be less than 2% in most series if careful closure was performed. They advise to wait 1 year before considering repeat surgery, as several patients may benefit from delayed effects. This is very important because reoperation is associated with more risks and complications [25]. Another metaanalysis also showed effectiveness of MVD for HFS even with complex cases involving vertebral artery compressions [20].

In summary, the literature supports the effectiveness of MVD for both TN and HFS, with success rates exceeding 90% in some series and low complication rates ranging from 1.6% to 5.5%. These findings demonstrate that MVD is a safe and effective surgical treatment option for patients with TN and HFS.

72.6.1 The Role of Endoscopy

Endoscopy has been increasingly used in MVD for TN and HFS due to its advantages of providing a wider and clearer view of the surgical field. This has been demonstrated in many series [26, 27].

Li et al. published a metaanalysis in 2019 which included 9 studies with a total of 1093 patients. In 543 patients, an endoscope was used; however, in 550 patients, purely microscopic decompression was done. Results showed remission rates of 92% vs. 86%. The offending vessel discovery rate was 99% with an endoscope compared to 95% with a microscope. Both remission rates and offending vessel discovery showed statistical significance and superiority of using the endoscope. Additionally, perioperative complications were significantly lower in patients where the endoscope was used [14].

72.6.2 Predictors for Outcome and Complications

Patient selection plays a very important role in deciding the outcome and complications rates, provided that the surgical technique is optimally performed. As we mentioned before, proper supine positioning, good examination of the preoperative MR images, careful sharp dissection of the arachnoid under IONM, and appropriate visualization of the compression using endoscopes with angulated view, performing the decompression appropriately without any residual contact with the nerve REZ or any harsh manipulations of the neurovascular structures including small arterial perforators and small cranial nerves fibers, dural closure in watertight fashion, and sealing of opened bony air cells are the cornerstones for an effective and safe MVD.

Familiarity with the time course of recovery following MVD is also important to know and communicate with the patients, that in some cases, an immediate cure of the symptoms might not occur, and symptoms might still last or recur for a couple of months, following the MVD, however, in a significantly lower intensities and frequencies.

But in some cases, and despite adopting all these measures, undesirable outcomes and complications might still occur. For instance, in HFS, intraoperative evident grooving of the facial nerve at the brainstem/pontomedullary sulcus has shown faster recovery in contrast to cases having distal grooving of the facial nerve. In our series, hearing deficits were more probable in cases with combined vascular compressions involving the vertebral artery and AICA. Lower cranial nerve affection was more seen in compressions involving the vertebral artery. Many tools can be used intraoperatively to assure that minimal affection of the neurovascular structures happens. For example, Doppler ultrasound can be used to assess the flow in larger arteries. However, the flow in smaller arteries like the small brainstem perforators can be assessed using intraoperative indocyanine green videoangiography [4, 17].

Intraoperative neuromonitoring is also important and has its value. One metaanalysis conducted in 2022 by Sprenghers et al. noted that a wave V latency prolongation of 1 ms or a wave V amplitude decrement of 50% are sensitive predictors for postoperative hearing loss. Other changes in the BAEP waves, such as complete loss of V wave, are more specific but mean irreversible damage and that is why they cannot be used as warning criteria. They also showed that lateral spreads (LSR) monitoring has high prognostic accuracy only at short-term follow-up. At long-term follow-up, its sensitivity decreases due to spasm relief in most patients, regardless of their LSR status [28].

72.6.3 Technological and Future Trends

In the era of technology and increasing use of augmented reality (AR), robotic-assisted surgery, and artificial intelligence (AI), the field of neurosurgery would surely witness some changes that might even lead to another revolution in the field in the near future resembling the revolutions of microscopy and endoscopy in the last five decades.

For example, AR-assisted surgery can provide a real-time overlay of the patient's anatomy during surgery, allowing the surgeon to see through different structures and anticipate deep unseen structures before encountering them. This can help improve precision and accuracy during surgery, reducing the time of the operation and the risk of complications [29]. AI algorithms are also used now to analyze medical images and assist with diagnosis and treatment planning. In MVD surgery, AI could be used preoperatively or intraoperatively to help identify the offending blood vessel and even predict the outcome, allowing for more precise and efficient surgery [30, 31]. Currently, robotic surgery has invaded many surgical disciplines. Although still not abundantly used in neurosurgery, in the near future, robots might assist with MVD surgery, providing a more stable platform for the surgeon to work on and improving precision and accuracy [32].

The use of AR and AI in MVD surgery is an exciting area of research that has the potential to improve outcomes and reduce complications. As these technologies continue to evolve, it is likely that they will become more widely used in neurosurgery, including MVD surgery.

72.7 Conclusion

In summary, our experience, data, and the literature support the use of endoscopy in MVD for both TN and HFS, with studies showing superior outcomes and lower complications compared to MVD without endoscopy. The advantage of the endoscope has been attributed to the wider and clearer view of the surgical field and the possibility to look around corners behind neurovascular structures. IONM is an important adjunct to the MVD procedure, and the goal of decompression is to avoid any residual contact with the REZ and usually is confirmed by disappearance of lateral spreads in HFS. Overall, endoscopy is a valuable tool for MVD in TN and HFS, improving surgical outcomes and minimizing complications.

References

1. Lee JYK, Pierce JT, Sandhu SK, Petrov D, Yang AI. Endoscopic versus microscopic microvascular decompression for trigeminal neuralgia: equivalent pain outcomes with possibly decreased postoperative headache after endoscopic surgery. J Neurosurg. 2017;126:1676–84.
2. Jallo GI, Conway JE, Bognár L. Neuroendoscopy of the central nervous system. Plural Publishing; 2008.
3. Kassam AB, Gardner PA. Endoscopic approaches to the skull base. Karger Medical and Scientific Publishers; 2012.
4. Al Menabbawy A, Refaee EE, Elwy R, Salem AA, Lehmann S, Vollmer M, Matthes M, Fleck S, Baldauf J, Schroeder HWS. A multivariable prediction model for recovery patterns and time course of symptoms improvement in hemifacial spasm following microvascular decompression. Acta Neurochir. 2022;164:833–44.
5. El Refaee E, Langner S, Baldauf J, Matthes M, Kirsch M, Schroeder HWS. Value of 3-dimensional high-resolution magnetic resonance imaging in detecting the offending vessel in hemifacial spasm: comparison with intraoperative high definition endoscopic visualization. Neurosurgery. 2013;73:58–67; discussion 67.
6. El Refaee E, Langner S, Marx S, Rosenstengel C, Baldauf J, Schroeder HWS. Endoscope-assisted microvascular decompression for the management of hemifacial spasm caused by vertebrobasilar dolichoectasia. World Neurosurg. 2019;121:e566–75.

7. Rosenstengel C, Matthes M, Baldauf J, Fleck S, Schroeder H. Hemifacial spasm: conservative and surgical treatment options. Dtsch Arztebl Int. 2012;109:667.
8. Jiang HT, Wang P, Zhou DW, Zeng LW, Lin B, Wu N. Fully endoscopic microvascular decompression for hemifacial spasm. Exp Ther Med. 2022;24:483.
9. Evans JJ, Kenning TJ, Farrell C, Kshettry VR. Endoscopic and keyhole Cranial Base surgery. Springer; 2019.
10. Abd-El-Barr MM, Cohen AR. The origin and evolution of neuroendoscopy. Childs Nerv Syst. 2013;29:727–37.
11. Magnan J, Chays A, Lepetre C, Pencroffi E, Locatelli P. Surgical perspectives of endoscopy of the cerebellopontine angle. Am J Otol. 1994;15:366–70.
12. Jarrahy R, Berci G, Shahinian HK. Endoscope-assisted microvascular decompression of the trigeminal nerve. Otolaryngol Head Neck Surg. 2000;123:218–23.
13. Caces F, Chays A, Locatelli P, Bruzzo M, Epron JP, Fiacre E, Magnan J. Neuro-vascular decompression in hemifacial spasm: anatomical, electrophysiological and therapeutic results apropos of 100 cases. Rev Laryngol Otol Rhinol. 1996;117:347–51.
14. Li Y, Mao F, Cheng F, Peng C, Guo D, Wang B. A meta-analysis of endoscopic microvascular decompression versus microscopic microvascular decompression for the treatment for cranial nerve syndrome caused by vascular compression. World Neurosurg. 2019;126:647–655.e7.
15. Chung SS, Chang JW, Kim SH, Chang JH, Park YG, Kim DI. Microvascular decompression of the facial nerve for the treatment of hemifacial spasm: preoperative magnetic resonance imaging related to clinical outcomes. Acta Neurochir. 2000;142:901–6; discussion 907.
16. Lehmann S, Al Menabbawy A, Matthes M, Schroeder HWS. Endoscope-assisted microsurgical resection of a large petroclival meningioma using a retrosigmoid approach: 2-dimensional operative video. Oper Neurosurg (Hagerstown). 2022;23:e325.
17. Al Menabbawy A, Refaee EE, Shoubash L, Matthes M, Schroeder HWS. The value of intraoperative indocyanine green angiography in microvascular decompression for hemifacial spasm to avoid brainstem ischemia. Acta Neurochir. 2022. https://doi.org/10.1007/s00701-022-05389-2
18. El Damaty A, Rosenstengel C, Matthes M, Baldauf J, Schroeder HWS. The value of lateral spread response monitoring in predicting the clinical outcome after microvascular decompression in hemifacial spasm: a prospective study on 100 patients. Neurosurg Rev. 2016;39:455–66.
19. El Damaty A, Rosenstengel C, Matthes M, Baldauf J, Dziemba O, Hosemann W, Schroeder HWS. A new score to predict the risk of hearing impairment after microvascular decompression for hemifacial spasm. Neurosurgery. 2017;81:834–43.
20. Li J, Lyu L, Chen C, Yin S, Jiang S, Zhou P. The outcome of microvascular decompression for hemifacial spasm: a systematic review and meta-analysis. Neurosurg Rev. 2022;45:2201–10.
21. Sharma R, Garg K, Agarwal S, Agarwal D, Chandra PS, Kale SS, Sharma BS, Mahapatra AK. Microvascular decompression for hemifacial spasm: a systematic review of vascular pathology, long term treatment efficacy and safety. Neurol India. 2017;65:493–505.
22. Alizadeh Y, Hayak H, Khoshnevisan A. Microvascular decompression versus gamma knife surgery in patients with drug-resistant trigeminal neuralgia: a systematic review and meta-analysis. World Neurosurg. 2022;167:67–73.
23. Lu VM, Duvall JB, Phan K, Jonker BP. First treatment and retreatment of medically refractive trigeminal neuralgia by stereotactic radiosurgery versus microvascular decompression: a systematic review and meta-analysis. Br J Neurosurg. 2018;32:355–64.
24. Jo KW, Kong D-S, Park K. Microvascular decompression for hemifacial spasm: long-term outcome and prognostic factors, with emphasis on delayed cure. Neurosurg Rev. 2013;36:297–301; discussion 301–2.
25. Sindou M, Mercier P. Microvascular decompression for hemifacial spasm: outcome on spasm and complications. A review. Neurochirurgie. 2018;64:106–16.
26. Zhao Z, Chai S, Xiao D, Zhou Y, Gan J, Jiang X, Zhao H. Microscopic versus endoscopic microvascular decompression for the treatment of hemifacial spasm in China: a meta-analysis and systematic review. J Clin Neurosci. 2021;91:23–31.
27. Refaee EE, Rosenstengel C, Baldauf J, Pillich DT, Matthes M, Schroeder HWS. Microvascular decompression for patients with hemifacial spasm associated with common trunk anomaly of the cerebellar arteries-case study and review of literature. Oper Neurosurg (Hagerstown). 2018;14:121–7.
28. Sprenghers L, Lemmens R, van Loon J. Usefulness of intraoperative monitoring in microvascular decompression for hemifacial spasm: a systematic review and meta-analysis. Br J Neurosurg. 2022;36:346–57.
29. Hou Y, Peng W, Tang J. Using augmented reality simulation for the surgery of trigeminal neuralgia caused by vertebrobasilar dolichoectasia: 2-dimensional operative video. Oper Neurosurg (Hagerstown). 2022;23:e289.
30. Bai R, Liu X, Jiang S, Sun H. Deep learning based real-time semantic segmentation of cerebral vessels and cranial nerves in microvascular decompression scenes. Cells. 2022. https://doi.org/10.3390/cells11111830
31. Hao W, Cong C, Yuanfeng D, Ding W, Li J, Yongfeng S, Shijun W, Wenhua Y. Multidata analysis based on an artificial neural network model for long-term pain outcome and key predictors of microvascular decompression in trigeminal neuralgia. World Neurosurg. 2022;164:e271–9.
32. Bhandari M, Zeffiro T, Reddiboina M. Artificial intelligence and robotic surgery: current perspective and future directions. Curr Opin Urol. 2020;30:48–54.

73 NPH Etiology and Updated CSF Management

Salman Sharif, Muhammad Yousuf Shaikh, and Uzair Ahmed Siddiqui

Learning Objectives

- Understanding the pathophysiology of NPH
- Diagnosing NPH and distinguishing it from other disorders
- Methods to assess or quantify features of NPH
- Invasive tests and radiological investigations for NPH
- Treatment options
- Different types of shunts and the role of ETV in NPH

73.1 Introduction

Idiopathic normal pressure hydrocephalus (INPH) is a common neurological disease that affects the elderly. This pathological disease distinguishes itself from other forms of hydrocephalus as it represents a potentially treatable cause of dementia. Studies estimate its prevalence to be about 22 per 100,000 individuals [1]. Hakim and Adams first described this syndrome in 1965, consisting of three classical symptoms, which are gait disturbance, dementia, and urinary incontinence without any significant signs of raised intracranial pressure [2]. Radiologically, the ventriculomegaly is out of proportion to the degree of brain atrophy, thus distinguishing it from age-related changes. The guidelines for the diagnosis of idiopathic NPH were first published in 2005, which consisted of four major topics such as clinical diagnosis of NPH, the role of supplementary diagnostic tests, surgical treatment, and outcomes with shunting [3]. Despite several decades since the formulation of diagnostic guidelines, the exact etiology remains unclear and poorly understood. Several possible theories are discussed here, which may explain the symptomatology of NPH.

S. Sharif (✉) · M. Y. Shaikh
Department of Neurosurgery, LNH, Karachi, Pakistan

U. A. Siddiqui
LNH, Karachi, Pakistan

73.2 Etiology

73.2.1 Abnormal Cerebrospinal Fluid Dynamics

The leading theory behind NPH is abnormal CSF dynamics, not CSF overproduction. The CSF pulsations theory is widely accepted [4]. These pulsations are linked to cardiac activity, causing small ICP changes to affect CSF pulsatility. Age and increased CSF volume amplify this impact, impairing CSF resorption and causing periventricular edema [5]. Studies on NPH's metabolic effects showed improved glucose metabolism post-CSF drainage [6]. Functional MRI revealed motor improvement in NPH patients, especially in gait, after lumbar drainage [7]. This highlights the significance of CSF dynamics in NPH pathogenesis and the potential for lumbar drainage in symptom improvement (Fig. 73.1).

73.2.2 Vascular Etiology

NPH has been associated with diabetes and hypertension, raising the possibility of vascular etiology [8]. Krauss initially blamed vascular risk factors in 1996 [9]. Momjian et al. observed lower cerebral blood flow (CBF) in the white matter near the lateral ventricles, but CBF normalized as one moved away from the ventricles. Periventricular edema in NPH might increase vascular resistance and cause ischemia, mainly in the periventricular area. This differs from the normal regulation of blood flow near the cerebral cortex [10]. Chang et al. pointed out that both CBF and cerebrovascular reactivity (CVR) were reduced in patients with NPH and reported a 90% improvement in patients after shunt placement [11]. Israelsson et al. [12] linked modifiable vascular risk factors like hyperlipidemia, obesity, diabetes, and psychosocial factors to NPH but found no direct connection to hypertension. However, managing these risk factors is crucial in iNPH.

I. N. Kanaan, V. Beneš (eds.), *Neuroanatomy Guidance to Successful Neurosurgical Interventions*,
https://doi.org/10.1007/978-3-031-59838-8_73

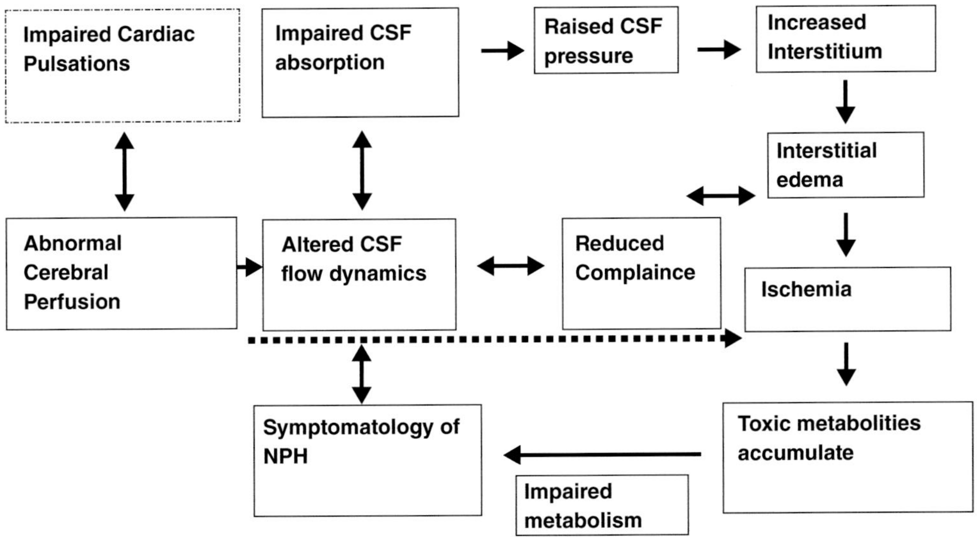

Fig. 73.1 Flow diagram showing pathways leading to NPH symptomatology

Table 73.1 Proposed theories of NPH

Various proposed pathophysiologies for NPH
Increased CSF pulsatility
Transcerebral mantle pressure gradient theory [13]
Bulk flow theory [14]
Venous congestion theory [15]
Osmotic gradient theory [16]
Reduced CSF drainage
Astrogliosis and neuroinflammation
Alterations of metabolism
Loss of blood-brain barrier integrity
Genetic and hereditary factors

73.2.3 Other Possible Etiologies

Over the past five decades, more than a dozen pathophysiological theories have been proposed which contribute to the development of NPH (Table 73.1).

73.3 Clinical Aspects

73.3.1 Gait Abnormalities

Gait abnormalities are the most frequent and troublesome symptom of NPH [17]. It is also the parameter that responds the most to CSF diversion procedures. The type of gait that has been associated with NPH is "broad-based gait" or "magnetic gait." Assessment methods include subjective descriptions, gait videos, and objective measures like step height and length. Williams [18] used the GAITRite system, showing significant improvement post-shunting. Other methods include three-dimensional trunk acceleration and measuring speed, stride length, and coefficient. Upper limbs may demonstrate motor issues in the form of bradykinesia, which is usually of mild severity. Strictly unilateral features in iNPH should raise a red flag for an alternative diagnosis.

73.3.2 Neuropsychiatric and Cognitive Issues

Patients with NPH often exhibit diverse cognitive decline, sometimes resembling Alzheimer's disease (AD). However, NPH's cognitive patterns differ, with subcortical dementia affecting subcortical structures, leading to slower processing and apathy. Frontal lobe symptoms dominate in NPH but not AD. Iddon used mini–mental state examination (MMSE) to highlight impaired spatial recognition in NPH [19]. Various tests like MMSE, Rey's auditory verbal learning test, controlled oral word association test, and Geriatric depression scale assess different aspects of NPH-related neuropsychiatric symptoms. No cognitive scale stands out for predicting NPH outcomes; new scales are emerging and need validation. Larger trials are necessary to identify a dominant scale.

73.3.3 Urinary Symptoms

Urinary incontinence is a common symptom in the elderly, affecting nearly up to 20% of males and 31% of females above 60 years of age in the US [20]. The etiology of urinary symptoms is diverse, but in cases of NPH, most patients present with initial symptoms of urinary urgency, which gradually progress to urinary incontinence [21]. Early shunting intervention can alleviate urgency but not incontinence. Assessing urinary symptoms is complex due to dementia, involving urodynamic studies and subjective methods. Overactive bladder symptoms may also result from frontal lobe lesions affecting voiding control.

73.4 Investigations

73.4.1 CSF Tap Test

The CSF tap test is one of the most frequently performed tests to predict the outcome of shunt surgery [22]. The procedure involves performing a lumbar puncture (LP), measuring the opening pressure and draining 40–60 ml of CSF. Removal of CSF lowers the ICP and transiently normalizes CSF hydrodynamics. The patient then undergoes repeat clinical examination to identify gait, urine, and cognition changes. Most studies show that gait improves the fastest after a CSF tap, with benefits seen as early as within 24 h [23]. Improvement in gait also serves as a favorable marker for good outcomes following shunt surgery [23].

73.4.2 Lumbar Infusion Test

The CSF infusion test, developed by Katzman [24], evaluates CSF dynamics by infusing artificial CSF into the lumbar subarachnoid space. It measures pressure changes at rest and during infusion, identifying plateau pressure when intracranial volume stabilizes. With a sensitivity of around 50–100% and specificity of 60–100%, it complements the CSF tap test in predicting shunt surgery responses.

73.4.3 Continuous CSF Drainage

Continuous/external lumbar drainage is an additional test routinely performed to evaluate patients with NPH. The patient is kept in a horizontal position while CSF is drained continuously at a rate of 5–10 ml/hr. over 3–5 days [25]. This procedure has one of the highest positive predictive values of 90–100% and patients showing no clinical improvement after the test are unlikely to improve with a shunt surgery [26]. Despite its high sensitivity, this invasive test is infrequently done due to the associated risks, such as infection, meningitis, and overdrainage.

73.4.4 CSF Biochemical Tests

CSF biomarkers, including amyloid beta-2 (AB42), t-tau, and p-tau, aid in diagnosing neurodegenerative diseases. Elevated tau and reduced AB42 levels are associated with both AD and NPH. Kudo et al. linked tau elevation to dementia severity in secondary NPH [27]; however, on the contrary, Lins et al. found a normal tau levels in his study for NPH patients [28]. Kapaki et al. identified another biomarker tp-182 to differentiate AD from NPH [29]. Coexistence of AD in NPH suggests a common pathogenesis involving impaired CSF protein clearance. Considering the various limitations of the studies available, such results still cannot be generalized and require much larger randomized studies to validate further.

73.4.5 Radiology

Computed tomography (CT) scan and magnetic resonance imaging (MRI) are used to evaluate patients with NPH. Although several features may predict the possibility of NPH, radiological investigations cannot independently diagnose iNPH as the size of the ventricles does not correlate with the symptomatology. The Evans index of more than 0.3 has been traditionally used to evaluate all patients with hydrocephalus (Fig. 73.2).

Radiological parameters including disproportionate enlarged subarachnoid spaces (DESH) and the callosal angle (Fig. 73.3) assist in diagnosing iNPH. Cerebral atrophy in elderly patients often widens subarachnoid spaces in neurodegenerative disorders, but NPH exhibits a unique DESH pattern with thinning near the cerebral convexity. A steep callosal angle (<90°) is also a frequently used marker with some special implications in the NPH. The proposed cause of which is an obstructed expansion of the corpus callosum at the falx and the continued rise of the roof of the lateral ventricles. Cagnin et al. suggested a 123° callosal angle cutoff, showing 95% sensitivity and 100% specificity for distinguishing AD from NPH [30].

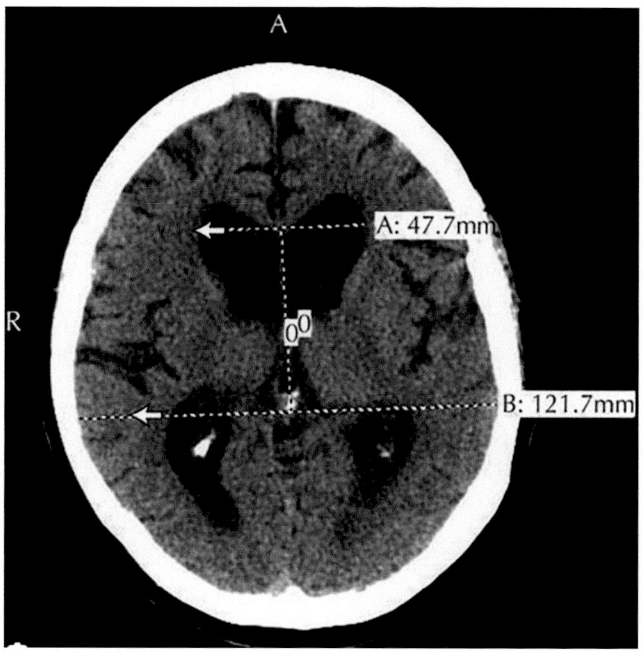

Fig. 73.2 Method to measure Evans index

Noninvasive radiological tests are now preferred for NPH diagnosis due to their cost and invasiveness compared to CSF tests. The Japanese Guidelines now emphasize DESH as a marker for shunt responsiveness [22] whereas some other systematic reviews evaluating the efficacy of the Evans index, callosal angle, and DESH [31, 32] reported that the callosal angle had the highest diagnostic accuracy and prediction to shunt response.

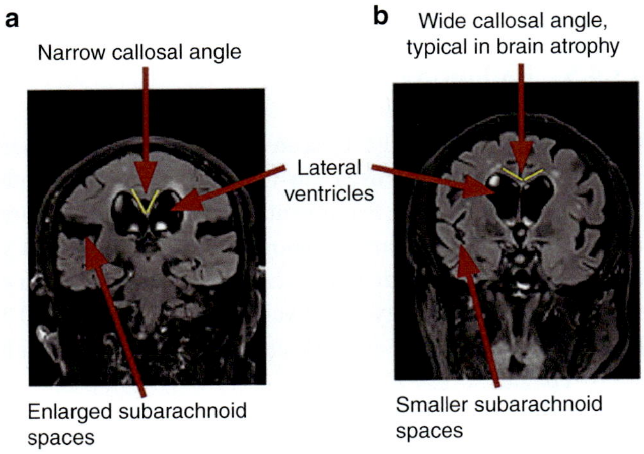

Fig. 73.3 DESH and steep callosal angle in NPH patients. (**a**) NPH. (**b**) Brain atrophy

73.5 Interventions

The chronic and progressive nature of iNPH complicates the diagnosis, which potentially results in treatment delays. Symptoms like gait, balance, and cognition worsen over time, and a delay of treatment by just 3 months after establishing a diagnosis can result in symptom worsening and only partial reversibility thereafter [33]. CSF diversion utilizing a shunt system remains the only effective treatment. Ventriculoperitoneal (VP), ventriculoatrial (VA), and lumboperitoneal (LP) shunts are proven effective for iNPH [22]. An emerging treatment is endoscopic third ventriculostomy, offering comparable results to shunting in patients with shorter symptom durations [34]. Table 73.2 summarizes landmark surgical intervention series.

Table 73.2 Surgical intervention series

Series	Type of study	Shunt type	Valve type	Comments	Complication
Zemack and Romner [35]	Retrospective mixed NPH	VP (90%) VA (10%)	Programmable differential pressure valve (DPV), no antisiphon device (ASD)	26% overdrainage, 54% overdrainage	Subdural hematoma (SDH) and hygroma 9.5%
Vanneste et al. [36]	Retrospective, multicenter, mixed NPH	VP, VA	Various DPV, no ASD	Improvement was noted in 31%	Severe complications 17%
Krause et al. [37]	Retrospective, all INPH	VA, VP	Medium DPV, programmable DPV, no ASD	90% improvement	5% over all complication rate: 2% SDH, subdural hygroma 7%, malfunction 10%
Takeuchi et al. [38]	Retrospective, all INPH	VP	Programmable DPV, no ASD	48% improvement rate, 8% required reprogramming because of under drainage	Not stated
Gangemi et al. [34]	Retrospective, all INPH		ETV	72% Improvement	4% Intracranial Hemorrhage (ICH)
Benzel et al. [39]	Retrospective, all INPH	VP, VPL	Various DPV, no ASD	Improvement was reported in 70% of cases	SDH 16% (all medium pressure valves)
Yamashita et al. [40]	Retrospective, mixed	VP	Hakim programmable DPV, no ASD	At follow-up, the mean valve setting was 100 mm H_2O. The valve was reprogrammed a mean of 1.25 times	Not stated
Kazui et al. [41]	Open-label randomized trial	LP	Codman Hakim programmable valve with a siphonguard	Benefit from shunt surgery in terms of functional status and triad symptoms	Adverse events in 15%, including SDH, shunt tube migration, postural headache

73.6 Shunt Systems for iNPH

The preference to use one shunt system over another varies globally. In North America and Europe, VP shunts are favored for iNPH [41–43], while Japan and Asia opt for LP shunts [41–43]. Although the results of VP and LP shunt surgeries seem equivalent, the decision to use one approach over the other appears to be influenced by cultural or regional factors [44].

73.7 Ventriculoperitoneal Shunt

VP shunt remains the mainstay of treatment for iNPH and has been commonly used worldwide. Various shunt valves are used, including fixed pressure and programmable valves. Fixed pressure valves are cost-effective but require replacement if the flow rate deviates, increasing overall costs. Programmable valves allow pressure adjustments via magnets. The 2021 Japanese iNPH Guidelines [22] recommend using a programmable shunt valve in iNPH patients. The programmable valves offer flexible pressure settings adjustable to match a patient's need where a physician can easily tailor the treatment without resorting to any revision surgery. More recently, shunts come with or without an antigravity/siphon device, an additional preventive mechanism, to avoid the overdrainage problem with the programmable valve set at a low pressure setting from the beginning [45].

The overall outcomes were comparable between patients with programmable valves and those with fixed-pressure valves. However, the incidence of subdural collections was much lower with programmable valves than with fixed valves. There was only a slight difference in the occurrence of subdural collections between programmable valves with antisiphon/antigravity components and programmable valves alone [46].

In a 12-year study, shunting with programmable valves demonstrated a significantly lower proportion of revisions than fixed-pressure valves. When comparing the outcomes between patients treated before 2005 (from 1990 to 2005) and those treated from 2005 to 2017, there were no significant differences in the proportion of patients with improved conditions (76% vs. 74%, respectively). However, there was a significant decrease in the need for revision surgeries over the past 12 years (12% vs. 31%) [46].

Case Presentation A 72-year-old male presented with a 3-year history of progressive gait disturbance, memory difficulties, and urinary incontinence. Neurological examination revealed a wide base while standing, difficulty moving while sitting, and slow walking. Neuropsychological testing showed cognitive slowing and impairments in executive function, learning, and memory, indicating mild dementia. MRI brain T2WI showed ventricular enlargement consistent with hydrocephalus, periventricular high signal intensity suggestive of normal pressure hydrocephalus, and extraventricular CSF collections indicating DESH (Fig. 73.4).

The patient underwent cognitive assessment using the MMSE and gait assessment involving the measurement of steps and time taken to reach a specific target. The initial MMSE score was 21, indicating mild cognitive impairment.

A pretap video recording of the patient's gait was obtained before the high-volume lumbar puncture procedure. The opening pressure measured 198 mm H_2O, while the CSF cell count, protein levels, and glucose levels were within the normal range.

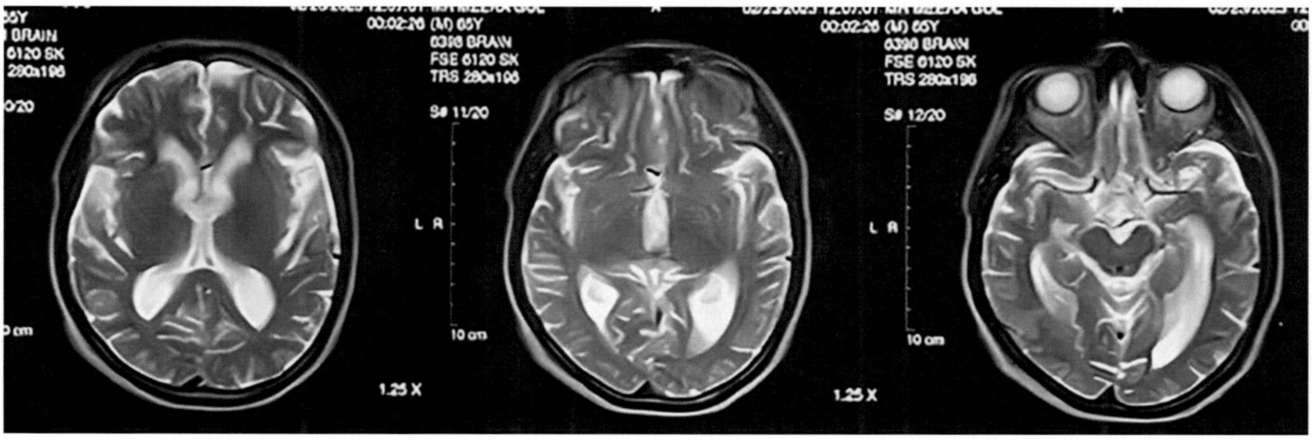

Fig. 73.4 MRI of the patient with clinical triad

Following a 30 ml CSF tap test, there was a significant improvement in the patient's condition. The MMSE score increased to 25, indicating a notable enhancement in cognitive function. Additionally, the gait assessment demonstrated remarkable improvements, with the number of steps taken to reach the target reducing to 15 and the time taken decreasing to 40 seconds, suggesting a substantial improvement in gait performance. The posttap video recording of the patient's gait further confirmed the qualitative enhancement in gait quality, turning ability, and start hesitation gait speed. These objective assessments, supported by the pretap and posttap video recordings, provided valuable evidence of the positive response to the CSF tap test.

Following the procedure, the patient was planned for a ventriculoperitoneal shunt placement with programmable settings. During the follow-up visit after 6 months, an MRI scan revealed only a subtle reduction in ventricular size and the presence of fluid in the sylvian and longitudinal fissures. However, clinically, the patient exhibited a much normal gait, resolution of urinary incontinence, and some cognitive function improvement.

73.8 Complications Avoidance in Treating iNPH with VP Shunt

iNPH surgery in geriatric patients requires special attention to the cardiopulmonary and thromboembolic risks. Proper screening reduces complications. The challenges associated with VP shunt intervention for iNPH are similar to general VP shunt complications seen in other conditions, like tube blockages and infections.

The malposition of the proximal ventricular catheter is a significant contributor to shunt malfunction. Hence, proper cerebral ventricle puncture and catheter placement are paramount [47]. Frontal and occipital horn/trigonum punctures are equally dependable for cerebral ventricular punctures. To ensure accuracy and avoid interference with structures like the ventricular wall and choroid plexus, it is crucial to base the puncture point on preoperative imaging. Utilizing tools like neuronavigation and stereotaxy aids in precise catheter placement [48].

The most frequent cause of VP shunt failure in patients with iNPH is distal peritoneal catheter issues, accounting for 90% of failures. Within the first 2–3 years after insertion, shunt failure is expected in 30–50% of patients overall [49]. Traditional VP shunt catheter insertion methods lack direct distal catheter visualization, raising risks like bowel perforation, catheter dislodgment, and shunt blockage. To reduce these issues, laparoscopic placement by a general surgeon is increasingly used for direct visualization and safer catheter positioning, reducing malfunction rates [50].

Infection is the most serious risk in VP shunt surgery, potentially leading to meningitis and peritonitis. Preventive measures include preincision antibiotics, double gloves, shunt device in antibiotic saline, and local wound antibiotics [51].

73.9 Lumboperitoneal Shunt

LP shunt has been widely used to treat iNPH by diverting CSF from the spinal canal to the peritoneal cavity [41]. LP shunt placement involves a small back incision for spinal catheter insertion, connecting to another catheter in the abdomen. A subcutaneous pressure valve regulates CSF flow. Possible complications include infection, CSF leak, headaches, spinal cord injury, catheter issues, and disconnection. While LP shunts have a lower risk of subdural hematoma and intracerebral hemorrhage than VP shunts, they are more prone to overdrainage and underdrainage, which can cause headaches and other issues. Despite a higher failure rate, LP shunts are preferred for their less invasive nature and lower risk of cortical injury [43].

73.10 Ventriculoatrial Shunt for iNPH

The ventriculoatrial (VA) shunt remains the most commonly used alternative for distal catheter placement when the peritoneal cavity cannot be used in shunting procedures for NPH [52]. Advanced surgical techniques, reduced shunt infection rates, and adjunctive tools like fluoroscopy and transesophageal echocardiography (TEE) have notably lowered complications and failures in ventriculoatrial (VA) shunt procedures, which were a major concern to limit its use, as mentioned in the earlier studies [53]. VA shunt intervention has a low incidence of severe complications now (less than 1%), making it suitable for cases with anticipated intraperitoneal adhesions but not advisable for patients with heart disease.

Although shunt surgery for iNPH helps to relieve the pressure in the brain's ventricles and improve neurological function, the size of the ventricles after the surgery usually does not change significantly. This is an important consideration for healthcare providers and patients when discussing the expected outcomes of iNPH shunt surgery [54].

73.11 Endoscopic Third Ventriculostomy for NPH

The third ventriculostomy was first performed by Dandy in 1922 [55]. With the technical advancements in neuroendoscopy, ETV has garnered significant recognition as a highly efficacious method for managing different hydrocephalus in carefully selected individuals. As a nonshunting alternative for iNPH [34, 56], it involves the formation of a shunt-free pathway connecting the ventricular system and the subarachnoid space. It induces a slight reduction in CSF pressure within the ventricular system, leading to an associated rise in cerebral blood flow and cerebral perfusion pressure. Moreover, it effectively eliminates transependymal resorption [57]. The fast movement of the pressure wave through the ventriculostomy to the basal cisterns can help restore cerebrospinal fluid (CSF) flow.

The ETV procedure involves creating a burr hole, inserting a fiberoptic endoscope into the lateral ventricle, and navigating it to the third ventricle. An opening is made in the floor of the third ventricle, gradually enlarged to ascertain the CSF flow, while visualizing the surrounding structures. Although ETV can be performed without image guidance, using image-guided systems can help locate the third ventricle more accurately and reduce the risk of complications [58]. These systems can also aid in identifying any obstacles or abnormalities in the procedure, leading to more effective treatment. A follow-up MRI can demonstrate a flow void in the floor of the third ventricle and has proven to be a useful noninvasive method to assess the functioning of ventriculostomy [59].

The overall rate of favorable outcomes after ETV was 63%. Among the different functional aspects assessed, gait improvement was observed in 79% of cases, cognitive function showed improvement in only 20% of cases, and urinary incontinence improved in 50% of patients following ETV. The occurrence of subdural collections among ETV patients was low, with only 1% experiencing this complication. Revision surgery was necessary in 16% of ETV cases during a follow-up period ranging from 12 to 80 months. Among patients requiring revision, more than 50% subsequently underwent VP shunt placement [46].

In conclusion, percutaneous ETV presents notable advantages compared to other methods. The risk of vascular and neural tissue damage is significantly reduced with the direct visualization provided by this technique [59].

73.12 Surgical Pearls

- Early intervention within 3 months → Better clinical outcomes.
- Preop assessment is the key to reduce anesthesia-related complications in this Geriatric age group.
- CSF diversion via VP shunt remains the mainstay of treatment.
- Programmable valves are a better alternative than fixed pressure valves regarding complication avoidance and redosurgery.
- Using neuronavigation and planning the ventricular placement unique to the patient's radiological findings minimizes the risk of malplacement.
- Infections can be minimized using the standardized protocol of infection control.
- LP shunt is a less invasive treatment used with proven efficacy but over−/underdrainage problems are more frequent.
- VA shunt is recommended for cases with expected intraperitoneal adhesions but not for patients with heart disease.
- TEE-guided insertion shows promising results in reducing catheter failures and complications in VA shunt.
- ETV has emerged as a recognized and effective nonshunting alternative for iNPH. Endoscopic visualization of the distinct anatomic structure is obligatory, whereas trajectory can be aided by using the IGS system.
- At present, VP shunt with a programmable valve has a well-established efficacy surpassing any other treatment modality.

73.13 Discussion

The efficacy and safety of ETV and VP shunts with fixed pressure valves were compared in patients with iNPH who responded to the tap test. A randomized controlled trial (RCT) included 42 Brazilian iNPH patients, and the results were analyzed as part of a systematic review from the Cochrane Library. The study showed that 50% of patients who underwent ETV experienced symptom improvement after 12 months, while 76.9% of patients who underwent VP shunt intervention showed improvement [60]. However, it is important to note that this study has limitations as it compared ETV to a nonstandard intervention using a fixed pressure valve instead of a programmable valve.

In a study conducted in the USA from 2007 to 2010, 652 patients who underwent ETV for iNPH were compared to 12,845 patients who underwent VP shunt intervention. The results showed that the mortality rate was 3.2% for ETV, compared to 0.5% for shunt intervention. In comparison, the surgical complication rates were 17.9% for ETV and 11.8% for shunt intervention, indicating that ETV has a higher risk than shunt intervention [61].

The available evidence on the efficacy of ETV for iNPH is currently limited and inconclusive, as it is based on a single RCT that used a nonstandard intervention. The quality of the evidence is considered very low, emphasizing the need for more rigorous research in this field. Clinicians should approach the results with caution and be aware of the limitations. Further studies are necessary to establish the effectiveness of ETV in treating patients with iNPH [56]. ETV

Table 73.3 Treatment options and efficacy

Treatment options	Efficacy: Pros and cons
Ventriculoperitoneal shunt	Definitive treatment; well-established role Shunt dependency; standard hardware-related complications: Obstruction, infection, malfunction
Fixed pressure valves	Cost-effective; ↑over/under drainage problem → ↑ redo-surgery
Programmable valves	Gold standard; expensive; ↓↓over/under drainage problem → ↓ redo-surgery
LP shunt	Lesser invasive; avoids cortical injury ↑↑ over/under drainage problem
VA shunt	Recommended in intraperitoneal adhesions ↑cardiac complications; not recommended with heart diseases
ETV	Avoids shunt dependency ↓efficacy as compared to shunt systems

treatment for communicating hydrocephalus with decreased CSF turnover was considered ineffective. In contrast to shunting, which has demonstrated consistent efficacy, the therapeutic outcomes of ETV have not been favorable in iNPH treatment [56, 62]. Consequently, shunting remains the solitary suggested remedy for individuals with iNPH [63] (Table 73.3).

73.14 Conclusion

Given the diverse etiology of NPH, identifying the sole culprit may still not be possible. However, with a better understanding of various factors contributing to its development, it may be possible to devise better treatment modalities in the future. Currently, amongst all CSF diversion procedures, the programmable VP shunt has shown a better outcome in patients with iNPH. An increasing number of surgeons are showing interest in ETV, and shunt remains the gold standard of care.

References

1. Brean A, Eide P. Prevalence of probable idiopathic normal pressure hydrocephalus in a Norwegian population. Acta Neurol Scand. 2008;118(1):48–53.
2. Hakim S, Adams R. The special clinical problem of symptomatic hydrocephalus with normal cerebrospinal fluid pressure: observations on cerebrospinal fluid hydrodynamics. J Neurol Sci. 1965;2(4):307–27.
3. Marmarou A, Bergsneider M, Relkin N, Klinge P, Black PM. Development of guidelines for idiopathic normal-pressure hydrocephalus: introduction. Neurosurgery. 2005;57(suppl_3):S2-1–3.
4. Raftopoulos C, Chaskis C, Delecluse F, Cantrainet F, Bidauti L, Brotchi J. Morphological quantitative analysis of intracranial pressure waves in normal pressure hydrocephalus. Neurol Res. 1992;14(5):389–96.
5. Chrysikopoulos H. Idiopathic normal pressure hydrocephalus: thoughts on etiology and pathophysiology. Med Hypotheses. 2009;73(5):718–24.
6. Ågren-Wilsson A, Roslin M, Eklund A, Koskinen LD, Bergenheim A, Malm J. Intracerebral microdialysis and CSF hydrodynamics in idiopathic adult hydrocephalus syndrome. J Neurol Neurosurg Psychiatry. 2003;74(2):217–21.
7. Lenfeldt N, Larsson A, Nyberg L, Andersson M, Birgander R, Eklund A, et al. Idiopathic normal pressure hydrocephalus: increased supplementary motor activity accounts for improvement after CSF drainage. Brain. 2008;131(11):2904–12.
8. Jaraj D, Agerskov S, Rabiei K, Marlow T, Jensen C, Guo X, et al. Vascular factors in suspected normal pressure hydrocephalus: a population-based study. Neurology. 2016;86(7):592–9.
9. Krauss JK, Regel JP, Vach W, Droste DW, Borremans JJ, Mergner T. Vascular risk factors and arteriosclerotic disease in idiopathic normal-pressure hydrocephalus of the elderly. Stroke. 1996;27(1):24–9.
10. Momjian S, Owler BK, Czosnyka Z, Czosnyka M, Pena A, Pickard JD. Pattern of white matter regional cerebral blood flow and autoregulation in normal pressure hydrocephalus. Brain. 2004;127(5):965–72.
11. Chang C-C, Asada H, Mimura T, Suzuki S. A prospective study of cerebral blood flow and cerebrovascular reactivity to acetazolamide in 162 patients with idiopathic normal-pressure hydrocephalus. J Neurosurg. 2009;111(3):610–7.
12. Israelsson H, Carlberg B, Wikkelsö C, Laurell K, Kahlon B, Leijon G, et al. Vascular risk factors in INPH: a prospective case-control study (the INPH-CRasH study). Neurology. 2017;88(6):577–85.
13. Hoff J, Barber R. Transcerebral mantle pressure in normal pressure hydrocephalus. Arch Neurol. 1974;31(2):101–5.
14. Rekate HL, Nadkarni TD, Wallace D. The importance of the cortical subarachnoid space in understanding hydrocephalus. J Neurosurg Pediatr. 2008;2(1):1–11.
15. Bateman GA. Idiopathic intracranial hypertension: priapism of the brain? Med Hypotheses. 2004;63(3):549–52.
16. Krishnamurthy S, Li J. New concepts in the pathogenesis of hydrocephalus. Transl Pediatr. 2014;3(3):185.
17. Kuriyama N, Miyajima M, Nakajima M, Kurosawa M, Fukushima W, Watanabe Y, et al. Nationwide hospital-based survey of idiopathic normal pressure hydrocephalus in Japan: epidemiological and clinical characteristics. Brain Behav. 2017;7(3):e00635.
18. Williams MA, Thomas G, De Lateur B, Imteyaz H, Rose JG, Shore WS, et al. Objective assessment of gait in normal-pressure hydrocephalus. Am J Phys Med Rehabil. 2008;87(1):39–45.
19. Iddon JL, Pickard JD, Cross JJ, Griffiths PD, Czosnyka M, Sahakian BJ. Specific patterns of cognitive impairment in patients with idiopathic normal pressure hydrocephalus and Alzheimer's disease: a pilot study. J Neurol Neurosurg Psychiatry. 1999;67(6):723–32.
20. Landefeld CS, Bowers BJ, Feld AD, Hartmann KE, Hoffman E, Ingber MJ, et al. National Institutes of Health state-of-the-science conference statement: prevention of fecal and urinary incontinence in adults. Ann Intern Med. 2008;148(6):449–58.
21. Sakakibara R, Kanda T, Sekido T, Uchiyama T, Awa Y, Ito T, et al. Mechanism of bladder dysfunction in idiopathic normal pressure hydrocephalus. Neurourol Urodyn. 2008;27(6):507–10.
22. Nakajima M, Yamada S, Miyajima M, Ishii K, Kuriyama N, Kazui H, et al. Guidelines for management of idiopathic normal pressure hydrocephalus: endorsed by the Japanese Society of Normal Pressure Hydrocephalus. Neurol Med Chir. 2021;61(2):63–97.
23. Bugalho P, Guimarães J. Gait disturbance in normal pressure hydrocephalus: a clinical study. Parkinsonism Relat Disord. 2007;13(7):434–7.

24. Katzman R, Hussey F. A simple constant-infusion manometric test for measurement of CSF absorption: I. Rationale and method. Neurology. 1970;20(6):534.
25. Marmarou A, Bergsneider M, Klinge P, Relkin N, Black PM. The value of supplemental prognostic tests for the preoperative assessment of idiopathic normal-pressure hydrocephalus. Neurosurgery. 2005;57(suppl_3):S2-17–28.
26. Nunn AC, Jones HE, Morosanu CO, Singleton WG, Williams MA, Nagel SJ, et al. Extended lumbar drainage in idiopathic normal pressure hydrocephalus: a systematic review and meta-analysis of diagnostic test accuracy. Br J Neurosurg. 2021;35(3):285–91.
27. Kudo T, Mima T, Hashimoto R, Nakao K, Morihara T, Tanimukai H, et al. Tau protein is a potential biological marker for normal pressure hydrocephalus. Psychiatry Clin Neurosci. 2000;54(2):199–202.
28. Lins H, Wichart I, Bancher C, Wallesch C-W, Jellinger K, Rösler N. Immunoreactivities of amyloid β peptide (1–42) and total τ protein in lumbar cerebrospinal fluid of patients with normal pressure hydrocephalus. J Neural Transm. 2004;111:273–80.
29. Kapaki E, Paraskevas G, Tzerakis N, Sfagos C, Seretis A, Kararizou E, et al. Cerebrospinal fluid tau, phospho-tau181 and β-amyloid1−42 in idiopathic normal pressure hydrocephalus: a discrimination from Alzheimer's disease. Eur J Neurol. 2007;14(2):168–73.
30. Cagnin A, Simioni M, Tagliapietra M, Citton V, Pompanin S, Della Puppa A, et al. A simplified callosal angle measure best differentiates idiopathic-normal pressure hydrocephalus from neurodegenerative dementia. J Alzheimers Dis. 2015;46(4):1033–8.
31. Park HY, Kim M, Suh CH, Lee DH, Shim WH, Kim SJ. Diagnostic performance and interobserver agreement of the callosal angle and Evans' index in idiopathic normal pressure hydrocephalus: a systematic review and meta-analysis. Eur Radiol. 2021;31:5300–11.
32. Park H, Park C, Suh C, Kim M, Shim W, Kim S. Prognostic utility of disproportionately enlarged subarachnoid space hydrocephalus in idiopathic normal pressure hydrocephalus treated with ventriculoperitoneal shunt surgery: a systematic review and meta-analysis. Am J Neuroradiol. 2021;42(8):1429–36.
33. Andrén K, Wikkelsø C, Tisell M, Hellström P. Natural course of idiopathic normal pressure hydrocephalus. J Neurol Neurosurg Psychiatry. 2014;85(7):806–10.
34. Gangemi M, Maiuri F, Buonamassa S, Colella G, de Divitiis E. Endoscopic third ventriculostomy in idiopathic normal pressure hydrocephalus. Neurosurgery. 2004;55(1):129–34.
35. Zemack G, Romner B. Adjustable valves in normal-pressure hydrocephalus: a retrospective study of 218 patients. Neurosurgery. 2002;51(6):1392–402.
36. Vanneste J, Augustijn P, Tan WF, Dirven C. Shunting normal pressure hydrocephalus: the predictive value of combined clinical and CT data. J Neurol Neurosurg Psychiatry. 1993;56(3):251–6.
37. Krauss JK, Droste DW, Vach W, Regel JP, Orszagh M, Borremans JJ, et al. Cerebrospinal fluid shunting in idiopathic normal-pressure hydrocephalus of the elderly: effect of periventricular and deep white matter lesions. Neurosurgery. 1996;39(2):292–300.
38. Takeuchi T, Kasahara E, Iwasaki M, Mima T, Mori K. Indications for shunting in patients with idiopathic normal pressure hydrocephalus presenting with dementia and brain atrophy (atypical idiopathic normal pressure hydrocephalus). Neurol Med Chir. 2000;40(1):38–47.
39. Benzel EC, Pelletier AL, Levy PG. Communicating hydrocephalus in adults: prediction of outcome after ventricular shunting procedures. Neurosurgery. 1990;26(4):655–60.
40. Yamashita N, Kamiya K, Yamada K. Experience with a programmable valve shunt system. J Neurosurg. 1999;91(1):26–31.
41. Kazui H, Miyajima M, Mori E, Ishikawa M. Lumboperitoneal shunt surgery for idiopathic normal pressure hydrocephalus (SINPHONI-2): an open-label randomised trial. Lancet Neurol. 2015;14(6):585–94.
42. Toma AK, Papadopoulos MC, Stapleton S, Kitchen ND, Watkins LD. Systematic review of the outcome of shunt surgery in idiopathic normal-pressure hydrocephalus. Acta Neurochir. 2013;155:1977–80.
43. Miyajima M, Kazui H, Mori E, Ishikawa M, Sinphoni-Investigators obot. One-year outcome in patients with idiopathic normal-pressure hydrocephalus: comparison of lumboperitoneal shunt to ventriculoperitoneal shunt. J Neurosurg. 2016;125(6):1483–92.
44. Bergsneider M, Black PM, Klinge P, Marmarou A, Relkin N. Surgical management of idiopathic normal-pressure hydrocephalus. Neurosurgery. 2005;57(suppl_3):S2-29–39.
45. Lemcke J, Meier U, Müller C, Fritsch MJ, Kehler U, Langer N, et al. Safety and efficacy of gravitational shunt valves in patients with idiopathic normal pressure hydrocephalus: a pragmatic, randomised, open label, multicentre trial (SVASONA). J Neurol Neurosurg Psychiatry. 2013;84(8):850–7.
46. Giordan E, Palandri G, Lanzino G, Murad MH, Elder BD. Outcomes and complications of different surgical treatments for idiopathic normal pressure hydrocephalus: a systematic review and meta-analysis. J Neurosurg. 2018;131(4):1024–36.
47. Wan KR, Toy JA, Wolfe R, Danks A. Factors affecting the accuracy of ventricular catheter placement. J Clin Neurosci. 2011;18(4):485–8.
48. Nesvick CL, Khan NR, Mehta GU, Klimo P Jr. Image guidance in ventricular cerebrospinal fluid shunt catheter placement: a systematic review and meta-analysis. Neurosurgery. 2015;77(3):321–31.
49. Isaacs AM, Williams MA, Hamilton MG. Current update on treatment strategies for idiopathic normal pressure hydrocephalus. Curr Treat Options Neurol. 2019;21:1–24.
50. Raysi Dehcordi S, De Tommasi C, Ricci A, Marzi S, Ruscitti C, Amicucci G, et al. Laparoscopy-assisted ventriculoperitoneal shunt surgery: personal experience and review of the literature. Neurosurg Rev. 2011;34:363–71.
51. Kestle JR, Riva-Cambrin J, Wellons JC, Kulkarni AV, Whitehead WE, Walker ML, et al. A standardized protocol to reduce cerebrospinal fluid shunt infection: the Hydrocephalus Clinical Research Network Quality Improvement Initiative. J Neurosurg Pediatr. 2011;8(1):22–9.
52. Hung AL, Vivas-Buitrago T, Adam A, Lu J, Robison J, Elder BD, et al. Ventriculoatrial versus ventriculoperitoneal shunt complications in idiopathic normal pressure hydrocephalus. Clin Neurol Neurosurg. 2017;157:1–6.
53. McGovern RA, Kelly KM, Chan AK, Morrissey NJ, McKhann GM. Should ventriculoatrial shunting be the procedure of choice for normal-pressure hydrocephalus? J Neurosurg. 2014;120(6):1458–64.
54. Meier U, Mutze S. Correlation between decreased ventricular size and positive clinical outcome following shunt placement in patients with normal-pressure hydrocephalus. J Neurosurg. 2004;100(6):1036–40.
55. Aschoff A, Kremer P, Hashemi B, Kunze S. The scientific history of hydrocephalus and its treatment. Neurosurg Rev. 1999;22:67–93.
56. Tudor KI, Tudor M, McCleery J, Car J. Endoscopic third ventriculostomy (ETV) for idiopathic normal pressure hydrocephalus (iNPH). Cochrane Database Syst Rev. 2015;2015(7):CD010033.
57. Hailong F, Guangfu H, Haibin T, Hong P, Yong C, Weidong L, et al. Endoscopic third ventriculostomy in the management of communicating hydrocephalus: a preliminary study. J Neurosurg. 2008;109(5):923–30.
58. Hayhurst C, Byrne P, Eldridge PR, Mallucci CL. Application of electromagnetic technology to neuronavigation: a revolution in image-guided neurosurgery. J Neurosurg. 2009;111(6):1179–84.
59. Jones R, Stening W, Brydon M. Endoscopic third ventriculostomy. Neurosurgery. 1990;26(1):86–92.
60. Pinto FCG, Saad F, Oliveira MF, Pereira RM, Miranda FL, Tornai JB, et al. Role of endoscopic third ventriculostomy and ventricu-

loperitoneal shunt in idiopathic normal pressure hydrocephalus: preliminary results of a randomized clinical trial. Neurosurgery. 2013;72(5):845–54.

61. Chan AK, McGovern RA, Zacharia BE, Mikell CB, Bruce SS, Sheehy JP, et al. Inferior short-term safety profile of endoscopic third ventriculostomy compared with ventriculoperitoneal shunt placement for idiopathic normal-pressure hydrocephalus: a population-based study. Neurosurgery. 2013;73(6):951–61.

62. Longatti P, Fiorindi A, Martinuzzi A. Failure of endoscopic third ventriculostomy in the treatment of idiopathic normal pressure hydrocephalus. Minim Invasive Neurosurg. 2004;47(06):342–5.

63. Halperin JJ, Kurlan R, Schwalb JM, Cusimano MD, Gronseth G, Gloss D. Practice guideline: idiopathic normal pressure hydrocephalus: response to shunting and predictors of response: report of the guideline development, dissemination, and implementation subcommittee of the American Academy of Neurology. Neurology. 2015;85(23):2063–71.

Part VIII

Complication in Neurosurgery

Complications in Neurosurgery: Ethics, Morals and Legalities

74

Keki E. Turel

Learning Objectives

- The two sides of the coin of complications: anticipation/prevention and recognition/management
- The World Health Organization's Surgical Safety Checklist: its importance in avoiding complications
- Ethical practice and moral principles: circumventing/defusing medicolegal liability
- Complications not due to surgical error: a patient's severe comorbidity or the natural history of the disease
- Root-cause analysis and how it helps analyse and objectively investigate the causes of problems
- Addressing moral issues and its challenges
- Measures to prevent medical errors (MEs) and adverse events
- The surgeon and their team: the 'second victim' of complications
- Steps to take when you are sued

74.1 Introduction

Medicine is an uncertain field, and there are no guarantees. Yet in medical practice, mistakes are not allowed and seldom forgiven. *Complication* is derived from the Latin word *complicare*, which means "fold together." It is a secondary disease or condition that makes an already-existing one worse: something more intricate, difficult or confusing. It is an unfavourable evolution or consequence of a disease, health condition or therapy. It also is an unfortunate part of surgery that makes the consequences worse than those of the original disease. It may bring great suffering to the patient and their family and even tarnish the reputation of the surgeon and the specialty at large.

K. E. Turel (✉)
Department of Neurosurgery, Bombay Hospital Institute of Medical Sciences, Mumbai, India

Surgery adds a new facet to a patient's life. But it violates nature (anatomically, physiologically, biochemically, etc.), and therefore, they can never be as nature made it, (with the exception of congenital malformations). So the surgical elimination of disease is a necessary invasion to help restore a patient's health. But it has its own inherent risks.

The human body suffers destruction through disease. By removing diseased organs or tissues, surgery compensates for what has been destroyed, but coping with that may entail further tissue destruction. Patients may encounter complications arising therefrom. Sound knowledge of anatomy, both normal and abnormal (caused by a disease), helps minimise these risks. Neurosurgery goes one step further because nervous tissue is delicate and injury to it is unforgiving. Complications may appear unexpectedly, complications such as difficulty and prolonged effort in achieving intubation, excessive bleeding, equipment failure or mishaps, and delays in blood transfusion, or the complications may be due to medical problems such as a myocardial infarct, pulmonary embolism, stroke, etc. We have also witnessed an anterior communicating (Acom) artery aneurysm re-rupture during the administration of general anesthesia (GA), resulting in fatality. Some complications may not be unexpected, because they may be related to comorbidities and the natural history of the disease or an inherent risk associated with a difficult disease process or procedure. Many years ago, one of our patients with a pacemaker was inadvertently subjected to MRI, resulting in pacemaker failure and cardiac arrest.

"It's easier to stay out of complications than get out of complications." In other words, prevention is better than cure. Emphasis, therefore, is on anticipation, on avoidance. Since complications are not uncommon in our practice, we must delve into the issues most relevant to their occurrence.

The practice of medicine over the centuries and across civilisations and the world has been fascinating and varied, and it has significantly evolved and differed from the way it was. Similarly, just as the scientific methods have evolved, so have the ethical, moral and legal aspects of science. In

I. N. Kanaan, V. Beneš (eds.), *Neuroanatomy Guidance to Successful Neurosurgical Interventions*,
https://doi.org/10.1007/978-3-031-59838-8_74

fact, history has shown that these issues originated much earlier than even modern medical diagnostics and therapy did.

Ethics is described as moral principles that govern a person's behaviour or professional conduct. In simple words, "do unto others as you would have them do unto you." Similarly, medical ethics refers to the moral principles that govern the practice of medicine. Generally, ethics varies from time to time, place to place and generation to generation, but medical ethics, as enshrined by Hippocrates (460–370 BC) in "primum non nocere" and the many others before and after him, has remained steadfast. Yet the practitioner has been frequently challenged by decision-making (doing the right thing) and their behaviour when faced with complications. Ethics and morals relate to "right" and "wrong" conduct. While they are used interchangeably, they can be defined differently: ethics refers to the rules provided by an external source (e.g. codes of conduct in workplaces or principles in religions), whereas morals refer to an individual's principles regarding right and wrong.

74.2 History

All societies are governed by ethics. In medicine, it assumes a prominent position.

In bygone centuries, life was often driven by superstition and witchcraft. Medicine was not a very distinct entity. Initially, herbs and other natural substances were used to treat diseases. They continued to be used down the generations through wise and experienced practitioners, and outcomes were often left to one's destiny. Hippocrates established a code of conduct, popularly known as the Hippocratic oath, which elaborates on a doctor's role, obligations and responsibilities. "By using experience, knowledge and insight, our ancestors recognised they could anticipate dangers and opportunities and take steps to exploit advantages and avoid hazards" (David Suzuki). There were no modern gadgets of investigation, only paying very careful attention to history and clinical examination, and a lot of experience and common sense were helpful in treating the majority of ailments. Over the centuries, several prominent anatomists and physicians have explored the human body and thus devised innovative methods of treatment. The credit for establishing surgery as a discipline goes to the great Indian surgeon Shushrut—who, 3000 years ago, described numerous surgical procedures in his famous treatise *Sushruta Samhita* [1] (Sushruta's doctrines should be dated on the basis of the composition date of Satapatha Brahmana). The composition date of the *Brahmana* is unclear, according to scholar Rudolf Hoernle (1841–1918), but he estimated it to be about the sixth century BCE). It is a compilation of numerous surgical procedures for the human body. Some of these operations and their principles are still being followed. On the other hand, the savagery of humankind destroyed many such invaluable handwritten documents on inventions and techniques. For example, Nalanda University in Bihar, India, operating from 427 until 1197 CE, with its 2000 teachers and 10,000 students, played a vital role in promoting the patronage of the arts and academics during the fifth and sixth century CE, a period that has been described by scholars as the Golden Age of India. Ransacked, burnt and destroyed by a goon, Muhammad Bakhtiyar Khalji, in ca. 1200 CE. The setup of guru (teacher), shishya (student) and Parampara (tradition) that Dr. Cushing's mentor, Dr. Halstead (1852–1922), later propagated (each of them taught another) originated during those ancient times: The student would completely surrender to the teacher and live with the teacher, where not only medical knowledge but also every other aspect of life was taught. The system has undergone a total transformation, becoming more and more impersonal and, in today's times, progressively electronic, but not without risks.

With the exception of the Hammurabi era (1792–1750 BCE), the medical profession has always been regarded as the most esteemed and noble; even today, the majority of the world's population on a survey [2] still trust doctors. Yet from this position of being regarded as the most trusted (of all professions), legislators have relegated doctors to being mere commodities or service providers. What is responsible for this change in attitude? The Consumer Protection Act (CPA) was introduced in India in 1986, and the medical profession has also been brought under the purview of this act since 1995, but the definition, interpretation and execution of it varies depending on the state.

74.3 Surgical Records

- Perform a detailed clinical examination and a review of every available image to plan the best approach to avoid complications.
- Check patient's preoperative comorbidity and consult the appropriate disciplines for clearance.
- Obtain informed consent from the patient and their family. Discuss their treatment options, the planned intervention and any potential risks or complications.
- Check sign-in, timeout and sign-out, and conduct a standard safety checklist.
- For microneuroanatomy, use neuronavigation for accurate localisation in both cranial and spinal surgeries.
- Use intraoperative monitoring methods to improve safety and surgical clinical outcomes.
- Practise safe intervention: Appropriate patient position, optimal exposure and minimal invasiveness in regional neurovascular structures are the highlights of safe intervention and should be complemented with excellent closure to achieve the best outcomes.

- Avoid experimental interventions and establish a multidisciplinary team to manage high-risk and complex pathologies.
- Recognise your complications, discuss them with the patient (Open Disclosure) and act swiftly to rectify them.
- Acknowledge your medical errors (MEs), and try to establish clinical pathways to avoid their recurrence in the future.

74.4 Discussion

74.4.1 Ethical Issues

Apart from facing the challenge of treating a complex disease or dealing with a "difficult" (highly emotional and/or doubting) patient and/or their family, doctors are also burdened with the demands of the corporate hospitals that expect them to churn out profitable results. If the ethical doctor makes a clinical decision that tries not to hurt the patient's purse by running fewer relevant tests, that tries to avoid unnecessary surgery or that promptly discharges the patient once the medical or surgical remedy has been proved effective (thus not withholding the patient for a prolonged hospital stay), then the doctor may be considered to be doing an injustice to their employer. At the other extreme, if the doctor does not carry out extensive tests and the patient's health deteriorates, the doctor faces flak, even litigation, for not doing enough. Overinvestigation has become the order of the day in this era of "defensive" medicine—a huge waste. So ethics are yet again challenged as we seek the middle path. However, a doctor is free to not accept a patient under the circumstances mentioned earlier but is still obliged to support/guide them by referring them to another service provider.

Many other ethical issues also come up, such as (1) the prophylactic use of antibiotics and anticonvulsants; (2) the postoperative use of the intensive care unit (ICU) for an otherwise-healthy patient; (3) operating on a disease that one has not seen before, without the help of an experienced colleague/superior (requires healthy work environment); (4) employing unproven therapies; (5) using drugs or devices not approved by a regulator agency or ethics committee; or (6) not accepting a patient's request for a second opinion (just to name a few).

In the neurosurgical setting, some complications are unavoidable because of the location or nature of the pathology. This differs from an **error**, which is the state or condition of being wrong in behaviour or judgement or the failure of a planned action to achieve its intended outcome—a deviation between what was actually done and what should have been done. A good and easy-to-remember definition of an error is "doing the wrong thing when meaning to do the right thing," resulting in an adverse event. A **surgical error** is a preventable mistake during the course of surgery, an unexpected complication and may occur because of insufficient preoperative planning or an improper work process. Sometimes even factors such as poor communication among practitioners or fatigue may be the cause. In some drastically bad cases, it may simply be caused by negligence or incompetence or, in the worst-case scenario, when under the influence of drugs. Not all errors are life-threatening, but many can severely affect quality of life. Shockingly, **medical error (ME)** is the third-leading cause of death, after heart disease and cancer [3]. According to an institute of medicine (IM) report, "preventable medical errors lead to the death of 100,000 Americans p.a. and cost about $17 billion per year." Of these known MEs, the top four are postoperative infection, device failure, failed back syndrome and operative haemorrhage. To what degree neurosurgical care contributes to these errors is not known [4]. MEs may be minor, serious or near misses: [5, 6] They may be errors in execution—i.e. the correct action doesn't produce the intended results—or errors in planning—i.e. the original action was not correct. Errors are problems not solely for young or less-experienced surgeons; even the most skilled and competent surgeons make mistakes and have complications.

Two main ethical issues crop up in discussions on errors: (1) the disclosure of errors to patients and their families and (2) the failure to embrace, discuss and study errors in an effort to improve patient safety and outcomes. At their occurrence, some mistakes may seem like the end of the world; however, below the surface, they present as perfect opportunities to learn, grow, improve and contribute to minimising errors in the future. "Mistakes are stepping stones to success." You can pick up a mistake and carry it as a burden, or you can set it down and use it as a stepping stone. A mistake increases your experience, and experience decreases your mistakes. So *experience is simply the name we give our mistakes.* It's folly to ignore even small mistakes; learning from them and correcting them prepares you for and prevents you from facing bigger ones. In my opinion, *the only real mistake is the one from which we learn nothing.*

Adverse events also occur, resulting in harm, though they may or may not be the result of errors. But they do result in undesired/unfavourable patient outcomes and need thorough introspection and analysis. In terms of numbers, 1 in 10 patients develops an adverse event, mostly surgery related; 7.4% of these are lethal [7]. More specifically, 43 million deaths occur worldwide because of complications/adverse events from surgery [8]. The causes of adverse events could be (1) human factors such as being underprepared or not trained for such a procedure, (2) environment factors or equipment failures, (3) administration policies or (4) the workings of local healthcare systems. One must not forget that "medicine remains a branch of probabilities and not of certainties." Complications of varying magnitude will happen.

Since surgeries involve an element of risk, labelling them is inappropriate and dangerous. **No surgery is minor.** In fact, we have seen more complications occurring from so-called minor surgeries (which are thus taken lightly and undeservingly given less attention) than from so-called major ones (which requires significant preparation, planning and teamwork). As part of our ethics, obtaining *detailed informed consent* pertinent to each individual case is vital. While doing so, one discusses various therapeutic options (optimal vs possible and even those unavailable to the treating physician), one's experience with that disease, the risks vs the benefits of being in one doctor's hands rather than in those of another doctor and the patient's safety and possible complications that may arise during the course of surgery or postoperatively with the suggested option that one would employ.

Another important ethical issue (though difficult) is to have an **open disclosure** of errors to the patient and their family when they occur. Open disclosure is an open discussion with the patient, their family and carers of the adverse events that result in harm to a patient while receiving healthcare. Who would know what really happened during this event more than the operating surgeon would? The elements of open disclosure are as follows: (1) an apology or expression of regret, which should include the phrasing "I am sorry" or "we are sorry"; (2) a factual explanation of what happened; (3) an opportunity for the patient, their family and carers to communicate their experience; (4) a discussion of the potential consequences of the adverse event; and (5) an explanation of the steps being taken to manage the adverse event and prevent its recurrence [9]. Such communication overcomes unwarranted suspicion, mitigates their angst and even minimises the likelihood of litigation. Here, when dealing with the patient and their family, the team must speak as one voice to avoid misrepresenting the events and plan, so the disclosure is best left to the team leader or their senior colleague.

As was said earlier, *studying complications* is important primarily to improve the outcomes of our patient care but also for science and research, thereby enhancing the overall healthcare system. One may look at complications as negative events, but studying complications and the factors leading to them helps us to understand them in a scientific way while aiming to prevent them in the future. **Root-cause analysis** helps analyse the cause of a problem—a technique developed by industries that *adopt a system approach* rather than a blame game that implicates individuals when an error occurs. This helps develop a plan of action *to avoid the repetition of errors in the future*. Additional skills and resources may be needed and be valuable in preventing such errors in the future. At an individual level, the surgeon/surgical team will ideally have objective evidence of what was performed during surgery. A postoperative review of the surgical videos, postoperative imaging information, and a review of the clinical videos before and after surgery play crucial roles in root-cause analysis. Additionally, they are useful for regular teaching purposes and also as backup evidence for mortality and morbidity (M&M) meetings, hospital inquiries and lawsuits. When a complication arises, honest and **objective introspection** may pave the way to understanding and overcoming a previously faced challenge. This is **solution-oriented thinking** (which depends on root-cause analysis).

74.4.2 Morals

The reactions of a physician after an error may vary, but generally, the first thing one may experience is a state of shock and denial. In a study, 93% intended to reveal the truth, but only 5% actually did [10]. Most physicians hold a false belief: What patients don't know can't hurt them [11]. Thus, after a complication, many doctors rely on both defensive strategies (denial, discounting and distancing) and constructive strategies [12]. After the initial shock and denial, the physician's next response, while disclosing, is to hide the details of mistakes in the complexities of medical science and procedures. They try to deliberately avoid using words such as *error* or *mistake* during counselling [11, 13]. A physician avoids confessing or shows of sympathy for multiple reasons, such as their fear of violent reactions from the patient's family, punishment, a besmirched reputation, losing their practice and lawsuits for malpractice liability [14]. Failure leads to fear, frustration, guilt, shame, remorse and anger at oneself. At the same time, human as the doctor is, they may also experience a loss of confidence, difficulty sleeping and marital disharmony. When the doctor develops a sinking feeling, disgust, anxiety, frustration and embarrassment, in their defence, they *choose their words very carefully* while explaining the mishap [15], describing the incident in a roundabout manner [16]. Accepting mistakes and sharing the bad news with the patient and their family is the right path [17, 18]. But news of physical attacks, even murder, scares doctors. In this condition, not many surgeons can analyse and devise corrective mechanisms; while they sulk in silence, they need succour.

On the other hand, **the initial response of patients and relatives** is one of shock, anger, frustration and sadness; they are often greatly pained and even feel betrayed, and they demand a full disclosure of the error [17, 18]. They want an admission that a mistake was made. The "choose your words carefully" approach makes them feel alienated, and they might start believing that the doctor is not remorseful [16]. They want an explanation of what actually happened, [19] how it happened [20] and the future implications of the error for the patient's health [21]. Discussing the factual details and offering a sincere apology are both necessary for the resolution process [22].

The plight of the patient and their family in the face of complications has been well appreciated and publicised, but what is not as well known is the other side of story, the story of the doctor. The surgeon is **the second victim** [23] of a complication. In a study by Deborah Sey [24], one out of seven patients is involved in an adverse event. The first priority after such an event is the patient and their family (the first victim). However, the involved healthcare professionals can also become victims in the sense that they are traumatised after the event (the second victim). They can experience significant personal and professional distress and may be traumatised by self-guilt and self-doubt about their own capabilities and clinical skills. Second victims use different coping strategies in the aftermath of an adverse event, which can have significant impacts on clinicians, their colleagues and subsequently their patients. Nearly half of healthcare providers experience the impacts of being a second victim at least once in their career. Because of this broad impact, they need to be offered support. Numerous supportive actions for second victims are described in the literature. Strategies include support that is organised at the individual, organisational, national and international levels.

Following their initial reaction, the physician goes *through a wave of repercussions*. They suffer decreased job satisfaction and avoid performing procedures and surgeries; consequently, they experience decreased professional efficiency and **burnout**. In fact, following the adverse event, the doctor (the second victim) often suffers a **post-traumatic stress disorder (PTSD)**. Indeed, failures seriously affect doctors professionally, where 12.2% experience an impaired ability to perform their jobs [25], 2% started avoiding certain procedures [25] and 6.3% had suicidal ideation [26]. Each year, **suicide rates among doctors** have been going up, where >400 doctors/year die by suicide in the United States alone [27]. This is 2.5 times more than that of the general population of the United Kingdom [28] or India [29]. In many other countries, suicide rates among doctors are a staggering five to seven times higher than those of the general population.

The **staff burnout** [30] concept is explored in terms of its physical signs and behavioural indicators and in terms of how cognitive, judgemental and emotional factors are intruded upon once the process is in motion. Burnout is highest among medical professionals. [31] An online survey showed that 80% of medical professionals felt burnt out, men more so than women, and 40% of them had mental problems. Indeed, burnout cannot be taken lightly as it leads to poor care, increased errors, isolation, a sense of failure, poor self-esteem, depression and even suicidal tendencies.

74.4.2.1 The Three Rs: Regret, Redemption and Restitution

A doctor–patient relationship (DPR) is considered to be the core element in the ethical principles of medicine. It is a sacred relationship, built on trust, respect, communication and a common understanding of both the doctor's and the patient's respective roles; any breach of confidence or faith by either person would lead to problems. Hence, in the event of an error, once all the facts are available, the mistake should be acknowledged as early as possible, without hesitation. "Mistakes are always forgivable if one has the courage to admit them" (Bruce Lee). The physician should proceed with disclosure in a systematic and timely manner. The basic problem-solving strategy of what, who, when, why and how can be used while the patient is provided with a *full disclosure of the mishap* [32]. As part of the human condition, patients will want to understand the circumstances that led to the error. The "choose your words carefully" approach may lead to an impression that the doctor is not remorseful and not open about the incident. Truth is central to all institutions. An ethos of transparency and frankness is the need of the hour. Failure to communicate honestly and clearly may leave patient and their family feeling deceived and angry, leading to a **malpractice suit**. [15] As a matter of fact, only 12% of suits are purely due to negligence [33]. Other causes for malpractice suits include the unavailability of the physician (31.5%), devaluing patients' or families' views (28.9%), dysfunctional communication (26.4%) and disrespecting patients' requests for information (13.1%). Remember that a sincere **apology** is a powerful tool to communicate regret and a necessary component of resolution process. While **regret** is an acknowledgement of the patient's suffering without an **acceptance** of responsibility, an **apology** is an acknowledgement of the patient's suffering with an **acceptance** of responsibility. To be fair, permitting and offering a **second opinion** are always appropriate [33].

To cope with guilt, the surgeon needs to be kind and compassionate to themself, forgive themself and move on. They don't have to *unduly* criticise themselves; there is a whole world to do that. They should deal with their guilt, both personally and professionally, without shame or feeling humiliated. How can the surgeon forget that they have saved thousands before this complication and will probably save thousands more hereafter? We have to accept that mistakes/complications are inevitable and unavoidable and part of the game, but so is learning from them. After all, we are only human.

Having analysed the unwanted effects of complications, both with the patient and their family (the first victim) and with the physician (the second victim), the creators of **the Michigan model** understood that the "deny and defend" model, as prevalent earlier, was counterproductive, and the desire to discover the truth was the most important reason for lawsuits, not financial gain. **Open disclosure and an apology** (based on the principles of accountability, honesty and consistency) were the most honourable solution. The model suggested compensating patients quickly and equitably if the care was not reasonable, but if the care was reasonable and didn't affect the outcome, it suggested fully supporting the healthcare provider and hospital.

74.4.3 Legal Implications

To avoid complications and subsequent legal issues, one needs to approach each procedure with respect. Besides the decided plan, it is good to have the **humility** to pay attention to cautionary warnings from one's anaesthetist, colleagues (junior though they may be), nurses and others. "I have never met a man so ignorant that I couldn't learn something from him" (Galileo Galilei, 1564–1642).

Having accepted the case for surgery, we are obliged to take responsibility for its results: That's **commitment**. One may have to spend one's own time and money, perhaps out of proportion to one's earnings. But that's what you have to prepare yourself for before wielding the knife. If you have performed a good surgery, so be it. That's expected out of you. But if you have made a mistake, you need to stand up and correct it. You may need help and guidance. A second opinion? Take it. You owe it to your patient. For not providing optimal care to the patient, you may be accused of negligence, which implies a deficiency in providing reasonable skill and care or implies harm caused by carelessness or neglect. That includes lapses in warning the patient and family about possible harm and/or ensuring the patient's safety—an act of commission or omission. Interestingly, negligence may or may not be punishable. **Negligence that is not punishable** applies to an error that occurs despite reasonable skill and care, harm from an unexpected accident, an unprovable medical deficiency in the surgeon. A statement from a respected colleague that there was no wrongdoing, that the complication is inherent to the procedure, and that you did everything possible to rectify matters helps to contradict claims of negligence. On the other hand, an error in judgement/performance implies that a competent surgeon with similar skills would *not* have made this mistake.

Never events are errors in medical care that are clearly identifiable, preventable, costly and serious in their consequences for patients, and they indicate real problems in the safety and credibility of a healthcare facility. Examples of never events include surgery performed on a wrong body part/side/level, a foreign body left in a patient during surgery, a mismatched blood transfusion or a major medication error. This is tantamount to **malpractice** of the highest order.

To summarise, doctors should adopt the following a **course of action after any complication**: Show concern and willingness to help through your actions, by making frequent visits to the patient; careful examining them on each occasion; and discussing the complication with your colleagues, resident doctors and nurses; and even bringing in other consultants who can help the patient. Briefing the patient and their relative(s) after each visit on the progress or lack thereof is necessary and augurs a healthy interaction. We also encourage relatives to visit the patient and interact with resident doctors, nurses and therapists.

A **sudden death** on the table in a healthy young patient poses a serious challenge to the operating surgeon even though they might not be directly involved. This may be experienced even by the most experienced surgeons. The psychological impact of such a death on the staff is profound. The behaviour of the neurosurgeon should try to restore calm. Steps to be taken by each member must be defined and performed without panic. All team members must be informed of the facts being placed before the family so that no discordant note is sounded. An accurate record of events—the anaesthetic, surgery and resuscitation—must be made. The time of crucial events must be noted. The patient records should be signed by the surgeon, anaesthetist and chief nurse. The original notes and charts should *not* be modified in any way. Subsequent amendments must be separately made and explanations attached. While dealing with the family, remain polite and answer carefully and truthfully, ensuring consistency with facts on record.

Breaking the news to the family needs honesty, empathy, politeness and patience. Hold the discussion in a quiet and comfortable room and explain the facts in a simple, understandable manner, avoiding esoteric medical jargon. The relevant information should be released gradually, and statements may need to be repeated as the shocked relatives absorb the sad news. All questions must be answered gently, without evidence of impatience.

The **steps** required after such a complication are as follows: (1) Immediately explain to the patient and their family why a complication occurred; (2) remain honest and truthful; (3) do not be evasive, shifty or vague; (4) do not shift the blame onto others, especially the junior members of the team; (5) provide patient and detailed responses to queries; and (6) emphasise that everything possible was done to help the patient.

As times have changed, **new advances in patient management** have been made. As the medical field went through a metamorphosis, **new terminologies** were being introduced, such as quality assurance, accountability, medical audits and accreditation. Any unfavourable outcome from a treatment hinted directly at negligence, a deficiency in service or something else. These entities made their way into our medical field **from the service industry**. Then they were amplified by an information explosion **through media**, including 24/7 TV channels, movies (that have often shown doctors in bad lights) and, to top it all, the Internet, which feeds diagnoses and treatments to untrained novices, who then may even challenge esteemed doctors.

Medicolegal liability is an increasingly challenging reality for doctors, especially those in surgical fields, like neurosurgery. Doctors need to know how to **handle allegations**. Allegations are likely to follow adverse or unfavourable outcomes from treatment, communication gaps or behaviour/ego problems, monetary issues (e.g. to avoid paying the doctor) and even provocations by colleagues and lawyers. Common allegations made by patients pertain to missed/delayed diagnosis, **negligence** in medical management, failure to provide a timely reference, failure to communicate, medication errors, failure to obtain informed consent and sexual misconduct. They may be oral or written, submitted via a hard copy by the patient or their legal representative or by a soft copy such as a text message or email, and are communicated to the individual doctor(s) or to hospital management. They may be valid and justified, merely perceived or even vexatious or malicious. When conveyed orally in person, one needs to give that person a patient hearing, answering the allegations to the best of one's ability without losing one's temper, even if the patient is agitated, and calmly promise to look into the matter. Always *document a summary of the conversation*, and note the name of the witness (if available). If the allegation is conveyed over the phone, request that they address their grievances in person. When confronting them personally, be aware and look for telltale signs of audio and/or video recording of the proceedings by the patient (one is within one's rights to refuse such a recording). When presented with written allegations, certain precautions need to be taken. *Do not reply without expert help*; inform your insurance company; approach the medicolegal cell of your medical association; and restrict your reply to the allegations only. Written allegations should not be ignored, as silence may be construed as consent or fear. Rebutting allegations sends the right signals, and a well-drafted reply often nips the matter in the bud. In court, not replying puts you in a bad light.

When you receive allegations accompanied by threats of physical harm, immediately file a police complaint and invoke legislation. If the allegations are followed by a demand for money to "settle" the matter, what should you do? This depends on the seriousness of the grievance and the circumstances of the aggrieved; hence, it will vary from case to case, and even here, proper legal counsel is needed.

Records are everything. In their practice, a doctor should have the full details of the patients; their histories; clinical and investigative findings; informed, written consents for surgery; meticulous surgical records (notes and videos); and documents on the postoperative events. The five Cs of a good patient record are as follows: correct, clear, comprehensive, chronological and contemporaneous. Good records are a good defence; poor records are a poor defence; and no records are no defence. Even legally, "what has not been documented never happened."

As quoted earlier, "prevention is better than cure." Try to pre-empt allegations; effective communication and empathy with patients are good antidotes to patient allegations. Sympathy, intelligence, tact and firmness (depending on the type of allegation) are all useful ingredients for *dealing with allegations*.

If you are sued, prepare your defence carefully, *consulting an experienced lawyer;* provide testimony of your qualifications and expertise, and emphasise all the steps that you took to provide reasonable care. Provide photocopies of consent forms, the patient's case sheets, operation notes/videos, etc., and obtain opinions from respected senior colleagues on your case. Judges pay special attention to opinions from experts at reputable teaching hospitals. It is advantageous to provide photocopies or reprints of statements on this complication from recent international and national textbooks and journals. These must show that the complication is well known to occur in such procedures and illnesses and that the measures described to prevent and treat the complication are the ones that you follow. *Never fear the judicial process.* If there are unhappy moments during any part of treatment process, try to have the matter go through the legal process rather than simply agreeing to any kind of settlement. Never hesitate to reach out to your colleagues for help, and take advice from those who are more experienced than you in dealing with such matters. Negligence can be both civil and criminal. Police tend to get involved even in civil matters over false pretexts allegedly to get kickbacks. Don't hesitate to look them back in the eye and stand your ground. Legal help is best given by lawyers; grab it and fight back. You can approach police, politicians and judges with your unity, knowledge, brotherhood, timely action, documentation and professional behaviour. So stop fretting and get prepared. Avoid medical jostling and jousting. *Stop criticising your fellow doctors.* Rise above a feudal mindset. Even though you are a specialist, give equal respect to others. There is no such thing as a plain Bachelor of Medicine, Bachelor of Surgery (MBBS) doctor. A doctor is a doctor qualified enough to treat and manage patients. Today, any trace of the famed doctor–patient relationship is long dead. The sooner we realise this, the better we'll be. *Now only a transactional relationship exists.* Charge your fees, stay professional, and keep documentations in place. One is well advised to obtain indemnity insurance and clinic/hospital insurance. Employ proper security at your hospital, have good CCTV cameras all around, and make an audiovisual counselling room and a health bulletin room. Spend some money from your earnings on contingencies like these. Count them as the cost of doing business. Realise that a patient is not loyal to any doctor. Gone are the days when patients thronged to one doctor for all their issues. Now the relationship is transactional. So in any case of even a minor dispute, *take your fellow doctor's side and not the patient's.* Even your relatives will not think twice before seeking a second opinion just to get the comfort they want. Give your time to your family, and foster a strong, value-based culture so that you have the emotional and psychological strength to face adversities like these. Last but not the least, actively participate in a local doctors association, and work to make it strong. Stay united!

74.5 Conclusion

Complications are part of the game. No neurosurgeon has no complications. Medical and surgical errors might be committed even by the most experienced practitioners. The question is not to demonise complications but rather to reduce their occurrence through anticipation, preparation and appropriate management. Surgeons may have profound knowledge and skill in their craft, but a wise and a mature surgeon is known by their prompt recognition and calm handling of a complication. They are quick to recognise the mishap, seek help when needed and do all that they can to remedy the error. The ethical course is to have a proper discussion of the error(s), and not only is the open disclosure to patients and their families necessary, but it may even absolve the surgeon of the mishap. Allegations cannot be ignored and need to be handled smartly, with a mix of apology and humility. When sued, obtaining a sound legal opinion and having well-documented records are essential for a strong defence. While the patient is the first victim of a complication, one rarely realises the second victim, the surgeon, who is also deeply affected and traumatised by self-guilt. They also need help.

References

1. Susruta-Ambika Dutta Shastri M, editor. 2008 – nlam.in.
2. Hawthorne M. Doctors still the most trusted profession. Aust Med. 2019;31(18):17. https://search.informit.org/doi/10.3316/ielapa.680724043792407.
3. National Center for Health Statistics (US) report. BMJ. 2016;353:i2139.
4. Rolston JD, Zygourakis CC, Han SJ, Lau CY, Berger MS, Parsa AT. Medical errors in neurosurgery. Surg Neurol Int. 2014;5(Suppl 10):S435–40.
5. Reason J. Cambridge University Press; 1990. https://doi.org/10.1017/CBO9781139062367
6. Ray Dorsey E, Rabbani A, Gallagher SA, Conti RM, Alexander GC. Impact of FDA black box advisory on antipsychotic medication use. Arch Intern Med. 2010;170(1):96–103.
7. de Vries EN, Ramrattan MA, Smorenburg SM, Gouma DJ, Boermeester MA. The incidence and nature of inhospital adverse events: a systematic review. Qual Saf Health Care. 2008;17(3):216–23. https://doi.org/10.1136/qshc.2007.023622. PMID: 18519629; PMCID: PMC2569153.
8. Jha AK, Larizgoitia I, Audera-Lopez C, Prasopa-Plaizier N, Waters H, Bates DW. The global burden of unsafe medical care: analytic modelling of observational studies. BMJ Qual Saf. 2013;22:798–801.
9. https://www.safetyandquality.gov.au/our-work/clinical-governance/open-disclosure
10. Kaldjian LC, Jones EW, Wu BJ, Forman-Hoffman VL, Levi BH, Rosenthal GE. Disclosing medical errors to patients: attitudes and practices of physicians and trainees. J Gen Intern Med. 2007;22(7):988–96.
11. Gallagher TH, Garbutt JM, Waterman AD, et al. Choosing your words carefully how physicians would disclose harmful medical errors to patients. Arch Intern Med. 2006;166(15):1585–93.
12. Chard R. How perioperative nurses define, attribute causes of, and react to intraoperative nursing errors. AORN J. 2010;91(1):132–45, ISSN 0001-2092. https://doi.org/10.1016/j.aorn.2009.06.028.
13. Mizrahi T. Managing medical mistakes: Ideology, insularity and accountability among internists-in-training. Soc Sci Med. 1984;19(2):135–46, ISSN 0277-9536. https://doi.org/10.1016/0277-9536(84)90280-6.
14. Singh L (personal Communication).
15. Hickson GB, Clayton EW, Githens PB, et al. Factors that prompted families to file medical malpractice claims following perinatal injuries. JAMA. 1992;267(10):1359–63.
16. Vincent C, Phillips A, Young M. Why do people sue doctors? A study of patients and relatives taking legal action. Lancet. 1994;343(8913):1609–13.
17. Vincent CA, Pincus T, Scurr JH. Patients' experience of surgical accidents. Qual Health Care. 1993;2:77–82.
18. Thomas E, Burroughs PD, Amy D, Waterman TH, Gallagher MD, Brian Waterman MPH, Jeffe DB, Dunagan WC, Garbutt J, Cohen MM, Cira J, Fraser VJ. Patients' concerns about medical errors during hospitalization. Jt Comm J Qual Patient Saf. 2007;33(1):5–14.
19. Mazor KM, Simon SR, Gurwitz JH. Communicating with patients about medical errors a review of the literature. Arch Intern Med. 2004;164(15):1690–7.
20. Bismark M, Dauer E, Paterson R, Studdert D. Accountability sought by patients following adverse events from medical care: the New Zealand experience. CMAJ. 2006;175(8):889–94.
21. Gallagher TH, Waterman AD, Ebers AG, Fraser VJ, Levinson W. Patients' and physicians' attitudes regarding the disclosure of medical errors. JAMA. 2003;289(8):1001–7.
22. Duclos CW, Eichler M, Taylor L, Quintela J, Main DS, Pace W, Staton EW. Patient perspectives of patient–provider communication after adverse events. Int J Qual Health Care. 2005;17(6):479–86.
23. Wu AW. Medical error: the second victim – the doctor who makes the mistake needs help too. BMJ. 2000;320(7237):726–7.
24. Seys D, Scott S, Wu A, Van Gerven E, Vleugels A, Euwema M, Panella M, Conway J, Sermeus W, Vanhaecht K. Supporting involved health care professionals (second victims) following an adverse health event: a literature review. Int J Nurs Stud. 2013;50(5):678–87.
25. Patel AM, Ingalls NK, Mansour MA, et al. Collateral damage: the effect of patient complications on the surgeon's psyche. Surgery. 2010;148:824–8; discussion 8–30. https://doi.org/10.1016/j.surg.2010.07.024.

26. Shanafelt TD, Balch CM, Dyrbye L, et al. Special Report: Suicidal Ideation Among American Surgeons. Arch Surg. 2011;146(1):54–62. https://doi.org/10.1001/archsurg.2010.292.
27. Stehman CR, Testo Z, Gershaw RS, Kellogg AR. Burnout, drop out, suicide: physician loss in emergency medicine, part I: west. J Emerg Med. 2019;20(3):485–94.
28. Gerada C. Doctors, suicide and mental illness. BJPsych Bull. 2018;42(4):165–8.
29. Kishor M, Chandran S, Vinay HR, Ram D. Suicide among Indian doctors. Indian J Psychiatry. 2021;63(3):279–84.
30. Freudenberger HJ. Staff Burn-Out. J Soc Issues. 1974;90(1):159.
31. Dyrbye LN, West CP, Satele D, Boone S, Tan L, Sloan J, Shanafelt TD. Burnout among U.S. medical students, residents, and early career physicians relative to the general U.S. population. Acad Med. 2014;89(3):443–51.
32. Burroughs TE. Patients' concerns about medical errors during hospitalizatio. Jt Comm J Qual Pt Saf. 2007;33(1):5–14.
33. Witman AB, Park DM, Hardin SB. How do patients want physicians to handle mistakes? A survey of internal medicine patients in an academic settin. Arch Intern Med. 1996;156(22): 2565–9.

Anticipation and Management of Complications in Pediatric Neurosurgery

75

Mubashir Mahmood Qureshi

Learning Objectives

- Children's diseases are best treated in an environment dedicated to their unique requirements.
- Recognizing the neurosurgical conditions that afflict the pediatric population, and those which are commonly managed within most pediatric neurosurgical units.
- The set of complications that may commonly occur when managing the pediatric neurosurgical patient.
- Developing best practice approaches that mitigate the risk of complications.
- Understanding the principles underlying the management of such complications.

75.1 Introduction

Pediatric neurosurgical patients represent up to 50% in many countries [1]. A large majority, especially in low- and middle-income (LMIC) countries, are managed by general neurosurgeons. They are best managed in an environment centered to their unique requirements, as the pathologies are often different to those of adults [2]. These pathologies include congenital and developmental disorders of the brain including hydrocephalus (Fig. 75.1), Chiari malformations, encephaloceles (Figs. 75.2 and 75.3), craniosynostosis (Figs. 75.4 and 75.5); congenital and developmental disorders of the spine including myelomeningocele (Fig. 75.11), craniovertebral anomalies and other spinal pathologies; accidental and non-accidental head injury, traumatic vascular injuries, brachial plexus injury, spinal column and spinal cord injuries; cerebrovascular conditions including cavernous and venous malformations, cerebral aneurysms, arteriovenous malformations (Fig. 75.14), spinal vascular malformations; infectious diseases, including shunt infections, epidural, subdural, and intracerebral infection, tuberculoma (Fig. 75.6), cerebral abscesses (Fig. 75.7), fungal and parasitic infections, spinal infectious disorders; neoplasms, including scalp and skull neoplasms, brain tumors, including gliomas and nonglial neoplasms. Functional disorders include management of epilepsy and spasticity disorders.

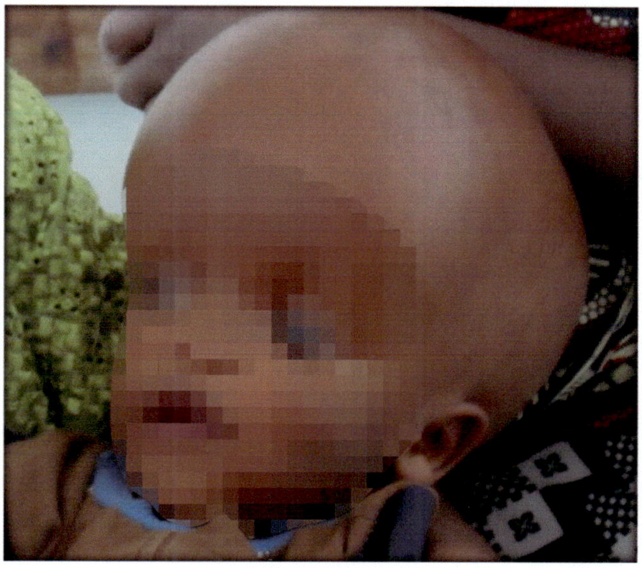

Fig. 75.1 Neonatal hydrocephalus

M. M. Qureshi (✉)
Aga Khan University Hospital, Nairobi, Kenya

I. N. Kanaan, V. Beneš (eds.), *Neuroanatomy Guidance to Successful Neurosurgical Interventions*,
https://doi.org/10.1007/978-3-031-59838-8_75

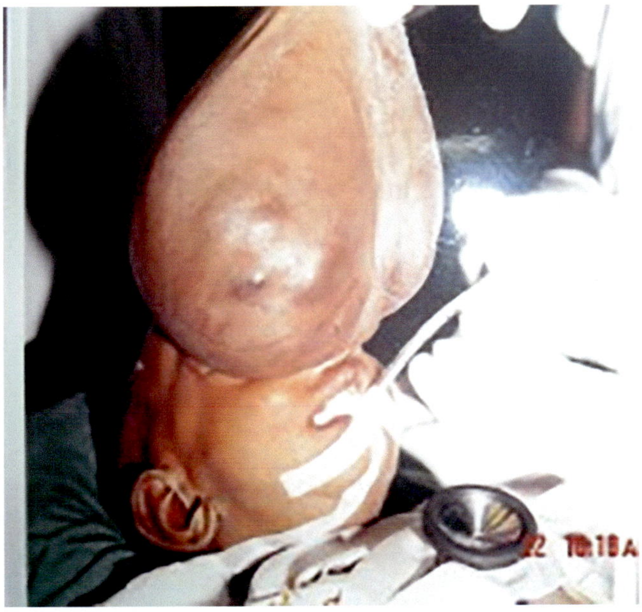

Fig. 75.2 Frontal encephalocele (preop)

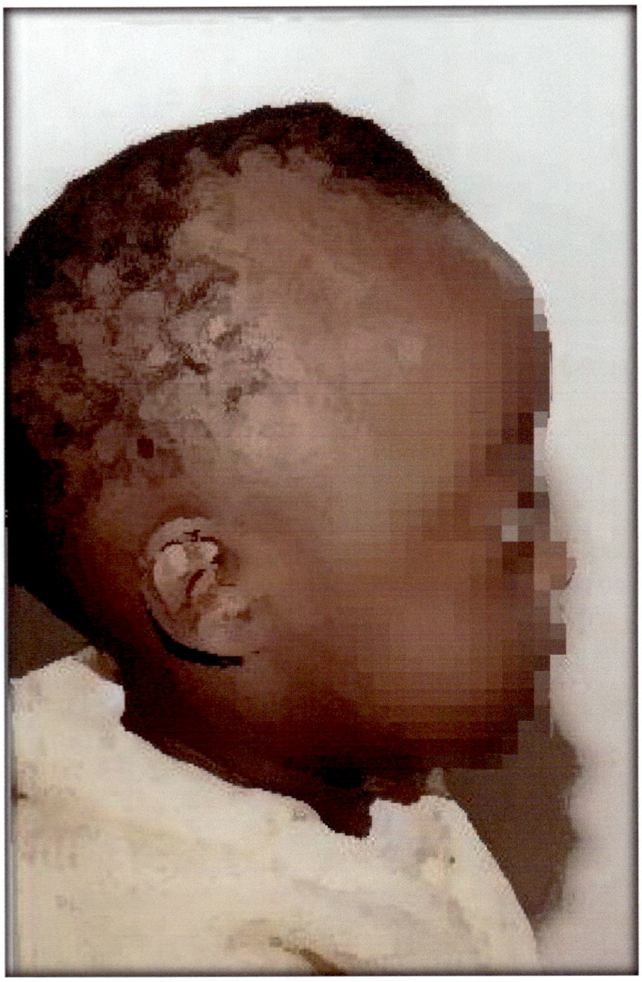

Fig. 75.3 Frontal encephalocele (postop–first stage)

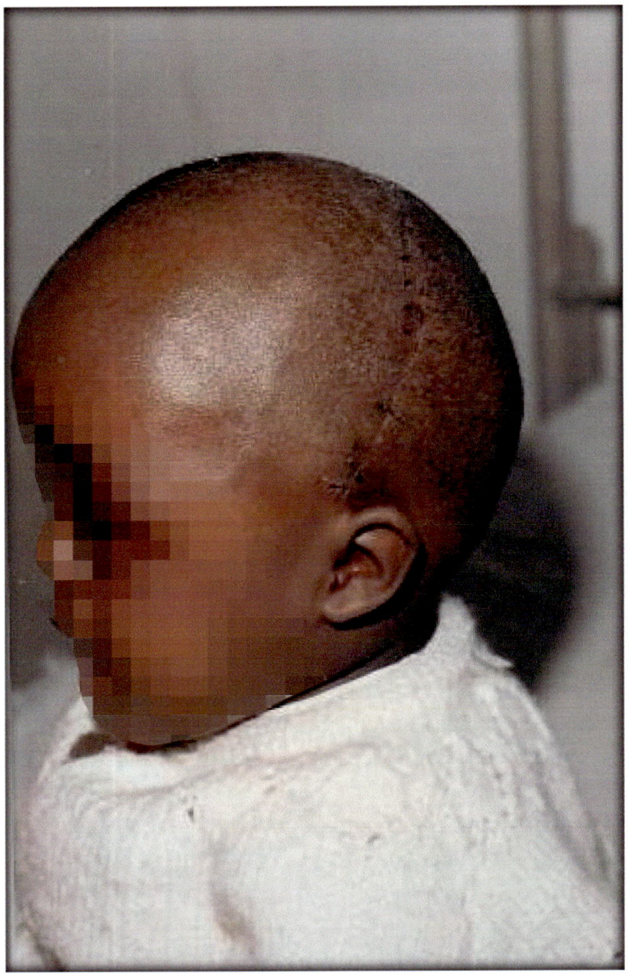

Fig. 75.4 Craniosynostosis (preop)

Children undergoing vascular surgery (41.7%) and brain tumor surgery (27.9%) are most prone to complication [3]. Other complications include CSF leakage, new onset neurological deficits, early shunt or endoscopic third ventriculostomy (ETV) obstruction, and shunt infection. Meningitis follows in 58% of shunt infections, 13% of CSF leaks, and 10% of wound infections [3]. These complications increase overall morbidity, often requiring a second procedure, such as an external ventricular drain (EVD) placement or shunt revision. Preventing these complications would significantly reduce morbidity and mortality risk [3].

Preoperative evaluation often indicates likelihood of complications, which are higher in neonates and infants [4]. Raised ICP signs require alteration in the anesthetic technique. Cerebral, cardiac, renal, and gastrointestinal anomalies are an important perioperative concern. Prematurity has a greater risk of respiratory distress syndrome, bradycardia, and apnea, along with retinopathy of prematurity. Calcium and glucose impairment is often seen in neonates due to

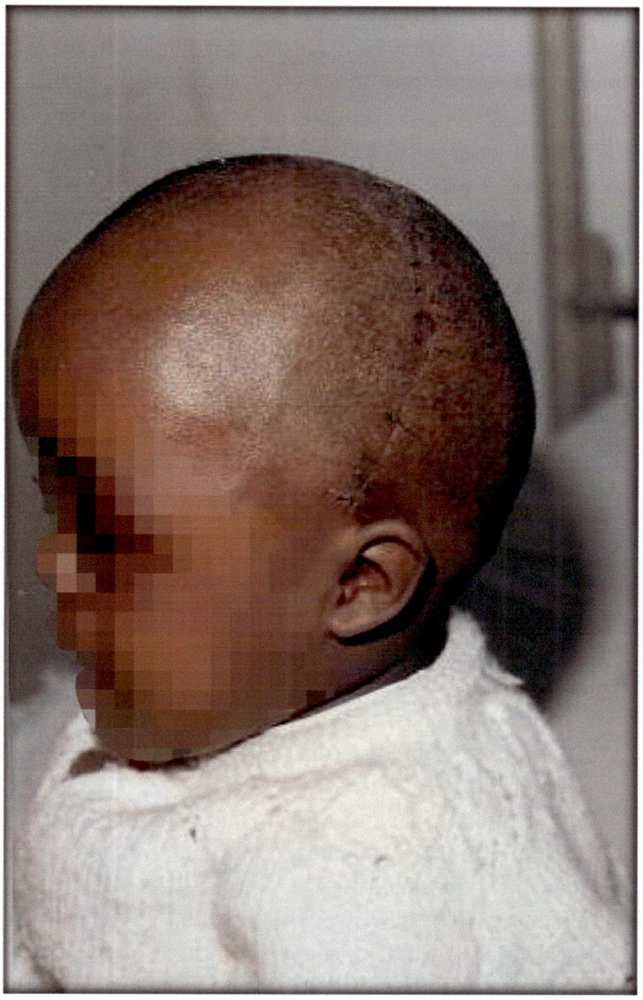

Fig. 75.5 Craniosynostosis (postop)

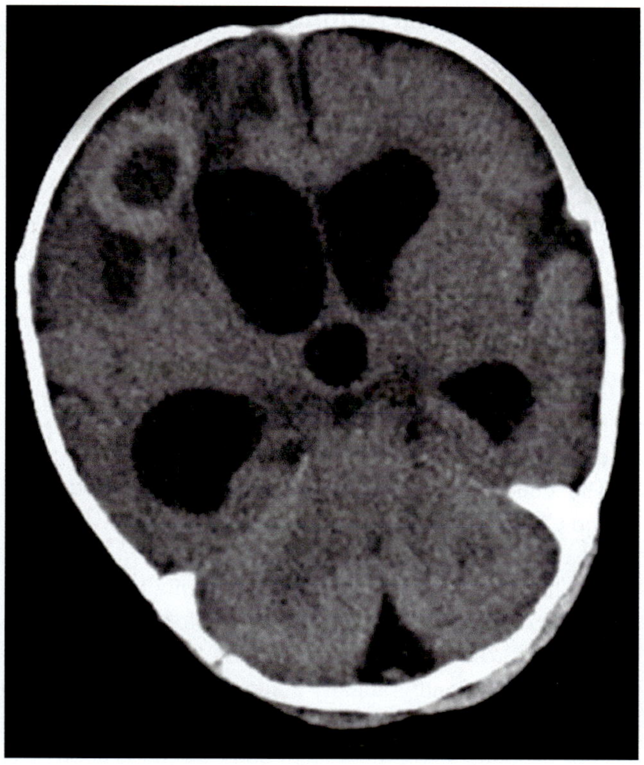

Fig. 75.6 Tuberculoma

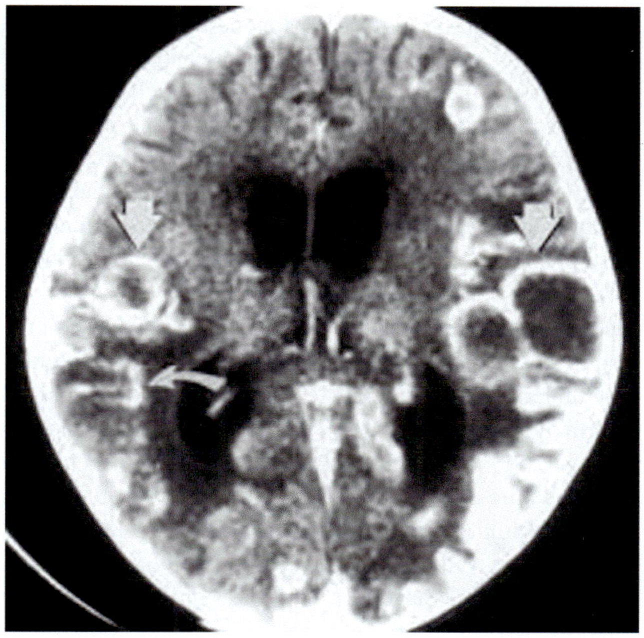

Fig. 75.7 Multiple cerebral abscesses

immature metabolic control [5]. Anemia presents risks of bradycardia and apnea. Neonates should be closely monitored postoperatively following a general anesthesia [6]. Settings such as prematurity, congenital cardiac disease, upper respiratory tract infection, craniofacial abnormality, arteriovenous malformations, neuromuscular disease, Chiari malformation, hypothalamic/seller and paraseller lesions, denervation injuries, and chronic anticonvulsant therapy, all place pediatric patients at a higher risk in the perioperative period [7].

75.2 Anatomical Highlights

- The child's head is disproportionally large. At birth, the brain is about 25% of the adult size even though body weight is about 5%.
- Given their narrow airway passages, children are at risk of airway obstruction.
- The tongue is large in relation to the oral cavity, as are the soft palate, soft tissues of the mouth, and the epiglottis.
- The middle meningeal artery is not as incorporated into bone as in adults.
- The calvarium is thin and pliable.
- Biomechanical maturation of the spine is a progressive process.
- The dentocentral synchondrosis between the odontoid and C2 body is yet to fuse.

The head, brain, and spine are fundamentally different in children compared to adults, both physiologically and anatomically [8]. In the newborn and infant, the head is disproportionally large and gradually assumes the head:body ratio of an adult over several years [8]. At birth, the brain is about 25% of the adult size even though body weight is about 5%. The disproportionally greater weight of the head also affects the movement of the head when a child falls or is struck by a moving object [9]. Young children are at particularly high risk of airway obstruction. Their tongues are relatively large for their oral cavity, as are the soft palate and soft tissues of the mouth and the epiglottis, which is relatively longer and stiff. They also have a larynx that is higher and more anterior, a cricoid ring that represents the narrowest point of the airway, and a shorter trachea that bifurcates higher [10]. Epidural hematomas in children are somewhat different to those in adults. The middle meningeal artery is not as incorporated into bone as in adults, but epidural bleeds from the edges of a fracture easily lead to hematomas. Because of the pliability of the skull, a linear fracture may represent significant underlying parenchymal injury sustained by marked deformation at the time of injury despite little evidence on the head CT scan. This makes growing skull fractures a unique feature of young children [11]. The calvarium is thin in young children; this, with the sutures and fontanelles, allows for easy deformation, with or without fracturing, under external pressure [12]. The thin skull may be a challenge for ICP monitoring—surgeons are often reluctant to use bolt systems or even measure ICP in the very young. The young age of a patient should not be a reason not to monitor ICP [13]. Biomechanical maturation of the spine is a progressive process that only starts to resemble the adult spine after the age of 8-9-years. Epiphyses fuse at different times and are easily mistaken for fractures. The pattern of injuries is determined by these progressive changes. Most spine injuries in children occur in the cervical region; in younger patients, they are more often subluxations or dislocations, often in the upper cervical spine, and more often associated with neurological injury [14]. The fulcrum of movement descends from the upper cervical spine in young children, where C1–C2 injuries predominate, to progressively lower in the subaxial spine as they grow, when mid to low cervical injuries become more common; paraspinal muscles are weaker, and the dentocentral synchondrosis between the odontoid and C2 body is yet to fuse [15].

75.3 Surgical Pearls

- Preoperative evaluation, including a preoperative anesthetic assessment, for associated anomalies, is invaluable.
- Intraoperative MRI and intraoperative ultrasound fusion imaging for real-time neuronavigation enhances tumor excision rates.
- Premature neonates run a higher risk of respiratory distress syndrome, bradycardia, apnea, and retinopathy of prematurity.
- Complications related to calcium and glucose commonly occur in premature neonates, due to an immature metabolic control mechanism.
- Prematurity, low birth weight, and multiple shunt revisions increase the risk of shunt infections. Shunt insertion should therefore be postponed, if possible, in premature and low-birth weight children.
- Early shunt or ETV failure and shunt infection are commonly seen, and meningitis may follow such occurrences.
- Infected VP shunts are best managed with antimicrobial treatment, removal of the shunt hardware, and placement of a temporary EVD.

75.4 Discussion

Any deviation from the optimal postoperative course should be recorded as a complication, even if asymptomatic and resolves spontaneously [16]. Every adverse event, whether surgical or clinical, should be recorded in order to have an exact understanding of the end result. Complications not directly connected to surgery or a surgical method should be identified in order to enhance overall multidisciplinary management [16]. A universally acknowledged classification is to be encouraged, as it would lead to the unification of outcome criteria, providing a comprehensive experience to enhance medical care and prevent adverse events [17].

Anticipation and mitigating the risk of complications.

(A) Complications in children undergoing treatment for hydrocephalus are more likely to occur in the following situations:
- Prematurity: Studies demonstrate prematurity as an independent risk factor for shunt infection [18–20].
- Number of shunt revisions: There is an independent relationship between the number of shunt revisions and shunt infection [21–23].
- Low birth weight: These infants have a higher rate of shunt infection due to their immature immune system. A significant difference is observed between the two groups in terms of birth weight ($p = 0.05$) [21, 24].

It is therefore advisable that shunt insertion be postponed as much as possible in premature and low birth weight patients. Given the increased risk in shunt revisions, it is recommended that shunt placement is carried out using strict surgical protocols. These include protocols regarding antibiotic prophylaxis, skin preparation, double gloving, handling of shunt components, and flow of operating room personnel [25]. Antibiotic-impregnated catheters have shown some benefit, although conclusive data for this are limited [24]. Infection rates improve with meticulous surgical technique, and adherence to wound-healing principles regarding incisions and wound closure. The management of an infected shunt (Fig. 75.8) generally involves antimicrobial treatment, removal of the shunt, and placement of a temporary external ventricular drain (EVD). Although there have been reports of effective management without removal of the shunt, the success rate is higher, and ultimate outcomes better, when the shunt is removed [26] (Figs. 75.9, 75.10 and 75.11).

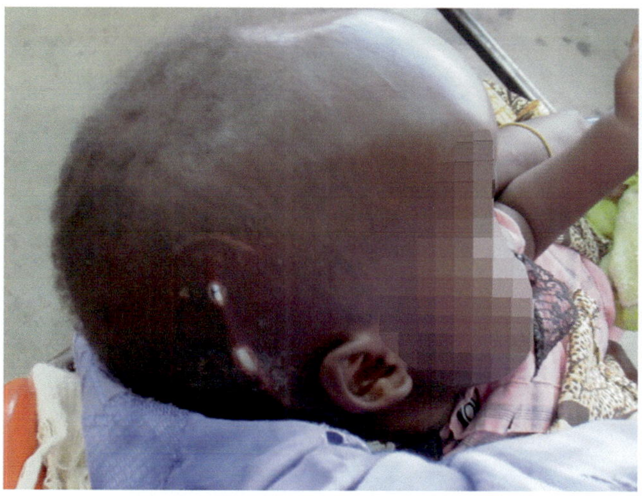

Fig. 75.8 Infected VP shunt

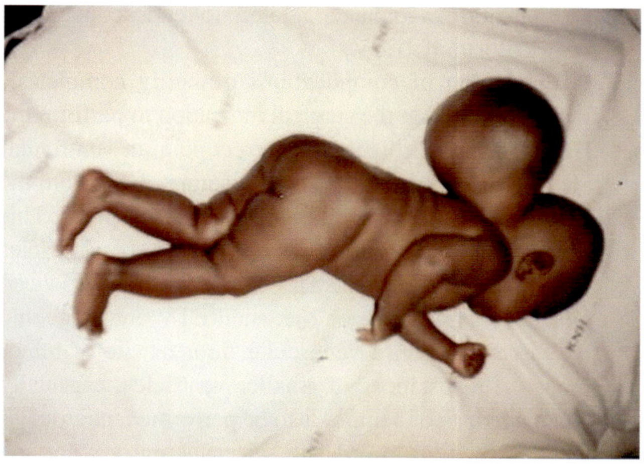

Fig. 75.9 Meningocele (preop)

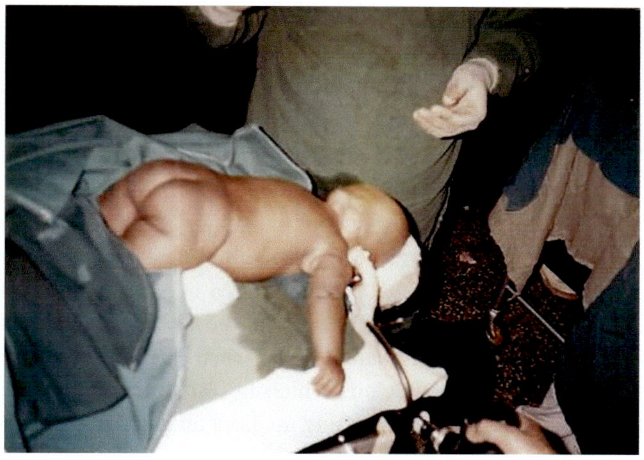

Fig. 75.10 Meningocele (postop)

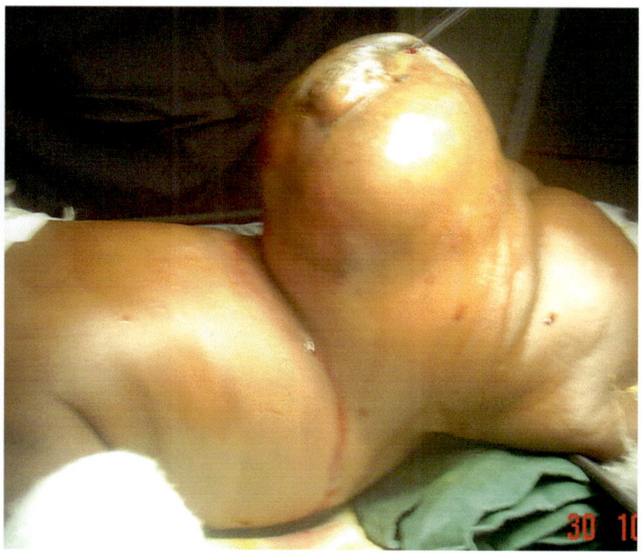

Fig. 75.11 Giant meningomyelocele

(B) Complication avoidance related to endoscopic third ventriculostomy (ETV):

Prevention of complications primarily commences with ensuring that the surgical indication to perform the ETV is sound. Utilizing the correct point of ventricular entry, usually around the coronal suture, 3 cm from the midline is the most commonly used entry point. This may be modified according to the shape of ventricles and age of the patient. The endoscope is introduced along the angle between the usual ETV entry site and the target site (third ventricular floor or site of tumor biopsy). In instances of smaller ventricles, especially when a biopsy of a tumor is to be performed, neuronavigation is helpful in order to select the entry point and trajectory. The key to successful ETV remains a thorough knowledge of endoscopic anatomy of ventricles.

Important anatomical landmarks (Fig. 75.12):

The choroid plexus, seen along the medial floor of the lateral ventricle, is a valuable landmark, and should be identified as it enters the third ventricle through the foramen of Monro, turning posteriorly to lie below the roof of the third ventricle. The thalamostriate vein, found in the groove between the caudate nucleus and the thalamus, meets the anterior cerebral vein. Together, they join to form the internal cerebral vein beneath the choroid plexus, just proximal to the foramen of Monro. The column and body of the fornix forms the anterior and medial borders of the foramen of Monro, respectively. The head of the Caudate nucleus which lies anteromedial to the foramen of Monro, forming the floor and wall of the ventricle. The thalamus forms the wall and floor of the ventricle posteromedially. The two lateral ventricles are separated by the septum pellucidum on their medial walls. Upon entry into the third ventricle, the following landmarks should be recognized: The optic chiasm is visualised anteriorly; the optic recess of the third ventricle lies above the optic chiasm, while the infundibular recess of the third ventricle lies below the optic chiasm. Two cream-colored structures, the mammillary bodies, are visualized posteriorly. Below the floor of the third ventricle lies a transparent membrane (of Lilliquist), between the optic recess anteriorly and the mamillary bodies posteriorly. The membranes of Lilliquist must be perforated in order to create a communication between the ventricular system and the basal subarachnoid spaces, allowing CSF flow, thus relieving the hydrocephalus. The basilar artery is visualized anterior to, and below, the mammillary bodies. Altered ventricular anatomy, occurring in many diseases, can interfere with the correct placement of the fenestration, leading to potential complication. Causes include postinfectious hydrocephalus, seen in up to 60% cases in sub-Saharan African studies [27], which may reduce the chance of a successful ETV. Developmental anomalies such as myelomeningocele and multilocular hydrocephalus make it difficult to identify the transparent membrane in front of the mammillary bodies [28]. Basal ganglia tumors and large thalamic tumors cause significant ventricular distortion [29]. A thickened third ventricle floor may prevent the basilar artery being identified in children who have experienced a previous cerebrospinal fluid (CSF) infection [30].

Risks in performing ETV and steps to mitigate them:

Minor bleeding occurs in up to 16.5% of cases [28]. The majority can be controlled through continuous irrigation with washout and gentle compression [31]. Major bleeding is reported following injury to the septal and thalamostriate veins, or branches of the basilar artery [32]. Such an intraoperative complication, occurring in 0.2–0.3% of cases, is associated with high mortality and morbidity. To reduce this risk, careful evaluation of the tortuosity and features of herniation of

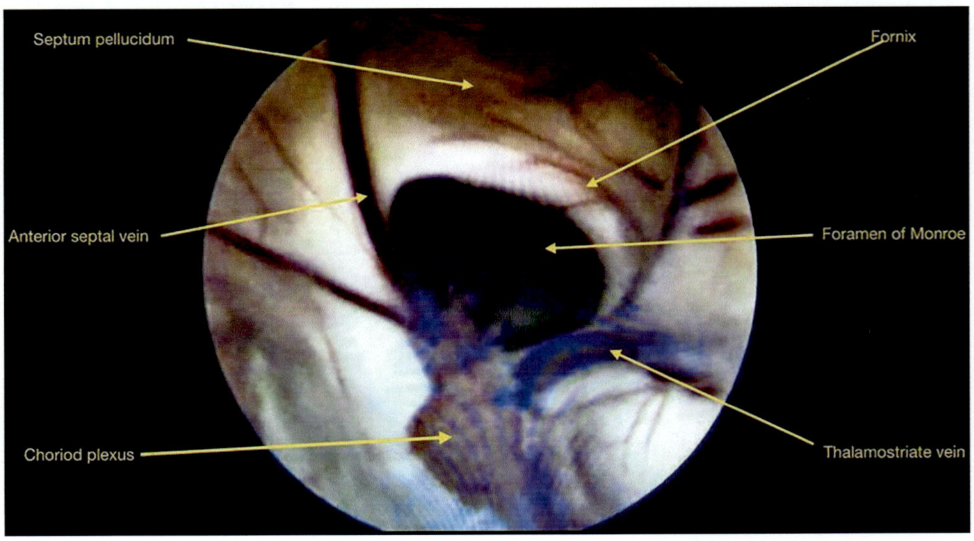

Fig. 75.12 Endoscopic anatomy—ventricle floor

the basilar complex against the floor of the third ventricle or an opaque neural floor should be carried out [32]. Cerebral injury may follow excessive irrigation, or raised pressure due to inadvertent obstruction of outflow channel during irrigation. Injury to the fornix during passage through the foramen of Monro, and to the thalamus by direct injury, and the hypothalamus are more likely due to eccentric location of the ostomy. The oculomotor nerve is at risk upon penetration of the third ventricle floor, causing transient ocular divergence and anisocoria from stretching of the midbrain and third or sixth cranial nerves palsies. These structures must be carefully identified during the procedure [32]. Hemodynamic alterations may occur from raised pressure and excessive irrigation, requiring fluid irrigation, and advancement of the endoscope being halted, until the alterations resolve. In rare occasions, cardiac arrest and uncontrollable postoperative high-frequency tachypnea have been reported [33]. The reported rate of bradycardia at the time of fenestration of the third ventricle floor is 26.8–43% [29].

Hormonal and metabolic derangement have been reported at 0.94% [32]. Transient diabetes insipidus of hypothalamic origin, remaining permanent in some cases, along with loss of thirst, is the most common. It can be avoided through placing the ventriculostomy as distant from the infundibular recess as possible [34]. Amenorrhea and partial growth hormone deficiency have been reported due to hypothalamic injury [35]. Electrolyte disorders, mainly hyponatremia, hypernatremia, and hyperkalemia can occur. These reverse spontaneously in most cases, and these may be explained by a transient hormonal disturbance related to the hypothalamus [28].

CSF leakage is often the first sign of early failure, while persistent CSF leakage, a sign of treatment failure. This is treated by resuturing of the skin incision, and serial lumbar CSF drainage if increased ICP is suspected. The leakage may lead to meningitis. Rates of CSF leaks between 1.7% overall, and as high as 5.2%, and meningitis or ventriculitis rates between 1.81% and 6.1% of cases in the immediate postoperative period, have been reported [30]. Postoperative fever, reported in 65.1% to 84.4% of cases, may be caused by hypothalamic irritation due to mechanical injury, small hemorrhages, or air in the ventricle. Fever frequently develops after procedures, peaking the day of surgery and/or rapidly subsiding, usually within 3 days [36]. However, if fever persists beyond 48 hours postoperatively, along with features of clinical illness, then meningitis, ventriculitis, or other infection should be suspected. Techniques recommended to prevent CSF leakage should be instituted [37].

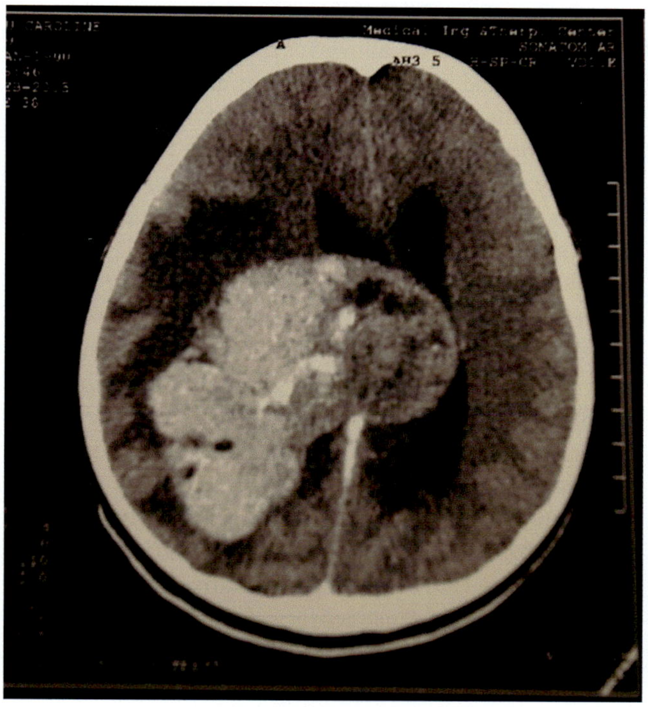

Fig. 75.13 Brain tumor

(C) Complication avoidance in patients undergoing surgery for brain tumors:

Brain tumors (Fig. 75.13) are the second most common malignancies in childhood, constituting approximately 35% of all childhood malignancies [38] and the leading cause of cancer-related deaths in children [39]. Mortality rates range between 0% and 20% in children aged 0-18 yrs. [40]. New-onset neurological deficits, presenting as weakness in one or more limbs, lower cranial nerve (LCN) palsies, oculomotor palsy, visual loss, and mutism are higher in children, ranging between 36% and 54% in infants [41]. Surgery carries a significant risk of morbidity and mortality and is related to tumor pathology, surgical approaches, intraoperative patient positioning, as well as the effect of anesthetics. Greater extent of excision improves survival in most children with brain tumors and obviates the need for adjuvant radiotherapy [42]. Measures to mitigate risk include image-guided neurosurgery, an essential tool for safe intracranial neurosurgery [43]. Accuracy is limited due to intraoperative changes such as “brain shift” [44]. Intraoperative magnetic resonance imaging (MRI) and intraoperative computed tomography (CT) are able to address this limitation [45]. CT is largely not acceptable for pediatric patients because of the ionizing radiation [46]. Advanced intraoperative ultrasonography (IOUS) is an effective way to localize and characterize juxtacortical lesions and to enable improved resectability, while reducing risk of cerebral injury. Real-time intraoperative fusion imaging between preoperative imaging and IOUS allows for real-time feedback during surgery, and it is less expensive and time-consuming than other intraoperative imaging techniques [47]. Diffusion magnetic resonance imaging (dMRI) tractography allows for noninvasive delineation and visualization of white matter (WM) tracts in vivo, and a major clinical application for presurgical planning and intraoperative image guidance, optimizing postsurgical outcomes [48]. However, high-quality prospective randomized clinical trials involving tractography intervention and a patient control group, similar to the study performed by Wu et al. [49], are desired in order to formally evaluate the clinical value of advanced dMRI tractography methods. Brain stem handling events occur during surgery in cerebello-pontine angle tumors (6%), skull base tumors (3%), and fourth ventricular tumors. These patients may need postoperative elective ventilation to allow gradual recovery of cardiorespiratory centers and protective reflexes over a period of time [50].

Control of massive blood loss is of concern, and it represents the second most commonly observed event. Bhatnagar et al. [51] reported that 31% of neurosurgical patients, including both children and adults, require allogeneic blood transfusion during neurosurgery, more so, with pediatric brain tumors. Pediatric craniotomies have been associated with massive blood loss, as high as 650% [52]. Brain swelling is a common complication, occurring in 13.6% of patients [50]. Additional doses of mannitol, withdrawal of volatile anesthetics, and total intravenous anesthesia (TIVA) are utilized, and hyperventilation instituted to maintain $PaCO_2$ of 30–32 mmHg. Despite these measures, some children will develop a tight brain at dural closure and require postoperative ventilation. Some children may require a decompressive craniotomy (DC) in the same sitting, or during the postoperative period [50]. Venous air embolism has been reported in 31% of cases, in one series, during surgery in the sitting position. It presents clinically, with a sudden fall of end-tidal CO_2 ($EtCO_2$) more than 2 mmHg, along with hypotension [50].

Prolonged postoperative elective ventilation is often required, especially when there has been significant brainstem handling, in the event of postoperative pulmonary complications (POPCs), sepsis, a second surgery for hydrocephalus, brain edema, venous infarct and hematoma, meningitis, and hypothalamic dysfunction. Postoperative lower cranial nerve (LCN) dysfunction often requires a tracheostomy, as does preoperative LCN palsy, especially in the setting of a ventilator-associated pneumonia (VAP). Patients having massive blood loss or transfusion should also be considered for elective ventilation as they may have hemodynamic instability and dilutional coagulopathy, increasing the risk of hematoma. Seizures often portend a poor outcome. The incidence in one study was about 2.4%, occurring almost equally in children who underwent either supra- or infratentorial surgeries [50]. Hyponatremia has been associated with increased morbidity; in one series, 55% hyponatremia was found in patients undergoing first surgery for intracranial neoplasms [51].

Meningitis is a dreaded complication, which remains at >5% in most pediatric centers despite antibiotic usage [53].

(D) Anticipation of complications in patients undergoing surgery for vascular lesions:

Neurovascular lesions in the pediatric age group include cranial and spinal cavernous malformations, venous malformations, intracerebral aneurysms, and arteriovenous malformations (AVMs). Their management involves a multimodality approach, utilizing microsurgical excision, radiosurgery, and endovascular embolization. Surgery may be considered for cerebral cavernous malformations (CCM), spinal cavernous malformations [54], and cerebral aneurysms. In cerebral aneurysms, surgery and endovascular treatment remain the definitive options, as in cerebral AVMs [55]. Wherever surgery is considered, risk of complication is present, and is similar to those outlined for surgery for brain tumors.

In cerebral cavernous malformations, location is an important factor that determines the natural history of CCM [56]. Microsurgical resection is the preferred treatment option for symptomatic CCM, as it is the only definitive treatment; however, the decision to operate remains challenging, as postoperative morbidity may exceed the complications of the untreated disease [57]. While good outcomes may be achieved in surgically resected brainstem lesions at high volume, specialized centers, complication rates are high, and new postoperative neurologic deficits are to be expected (53% of cases) [58]. In select cases, therefore, targeted radiotherapy may be used to treat lesions that are surgically inaccessible. There is no role for endovascular therapy in CCM [59]. In brain stem cavernomas, Ming-Guo Xie et al. [60] noted 43.4% of patients developed respiratory failure, while 30.2% developed dysphagia or cough. The meticulous microsurgical technique and postoperative airway management mitigate this risk. The independent factors for adverse long-term outcome were age $\geq$ 50 years and increased time of tracheal intubation postoperatively [60].

In spinal cavernous malformations (SCM), clinical presentation depends on the location and growth rate of the lesion. Microsurgery is the accepted method of treatment, especially for lesions which are greater than 5 mm and those that are located posteriorly [61]. Exact preoperative localization of the lesion is a critical factor in reducing morbidity. Preoperative imaging for localization may fail as the anatomy of the lesion may shift during surgical positioning and approach. Therefore, intraoperative ultrasound imaging is advised to localize the lesions precisely. Dural CSF leak is a recognized risk; its avoidance through watertight closure of the dura mater is an important surgical step [62]. Risk of deficit is reduced by using intraoperative neuromonitoring. Neurological complications are generally a result of ischemic damage due to sustained traction, manipulation, or rotation of the cord, or overheating of the spinal cord due to bipolar coagulation [61].

Pediatric cerebral aneurysms are rarely encountered, with estimates of 1–3 cases per million suggested [63]. Although traumatic cerebral aneurysms represent less than 0.5% of all intracranial aneurysms, 25% of reported traumatic aneurysms are found in children below age 20 years [64]. Due to their giant size and unusual location, e.g., vein of Galen aneurysms, surgical obliteration is often not straight forward. Review of the literature indicates that in only 29.5% of cases, clip ligation was possible [65]. A high percentage requires specialized and innovative procedures to obliterate the aneurysm [66]. In the modern era, the endovascular management of pediatric intracranial aneurysms is a viable alternative to surgical obliteration. However, this too is not without serious risk and potential complications, including vessel perforation, stroke, hemorrhage, and incomplete obliteration [67].

In pediatric arteriovenous malformations (AVMs), (Fig. 75.14) an autopsy-based series reveal a prevalence of AVMs estimated between 0.06% and 0.11% [68]. The annual incidence of symptomatic AVMs is noted at 1.1 per 100,000 [69]. The pediatric age group comprises 12–18% of all AVMs from major centers, and the overall prevalence in children is about 0.02% of the pediatric population [70]. Rupture of an AVM in a child is four times more likely to be a cause of brain hemorrhage than an aneurysmal rupture [71]. Surgical management of pediatric AVMs involves a microsurgery to occlude the vascular pedicle supply to the nidus. Endovascular embolization is considered an adjunct modality of treatment prior to surgical or radiation treatment for pediatric AVMs [55]. Biplane perioperative catheter angiography to confirm resection of the AVM, prior to the conclusion of surgery, improves obliteration rates. Bradley A Gross et al. [72] report no additional complications from the perioperative angiography, while the obliteration rates confirmed by angiography improved significantly, from 86% to 100% ($p = 0.01$) [72].

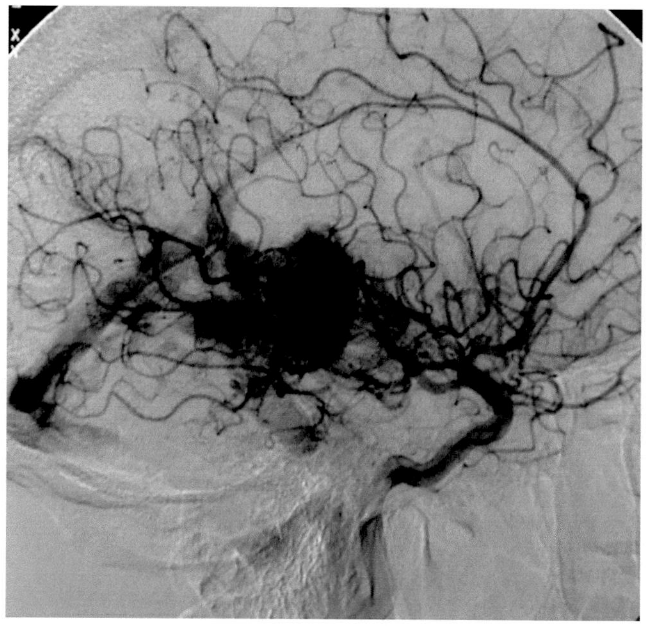

Fig. 75.14 Arteriovenous Malformation (AVM)

75.5 Conclusion

Pediatric patients represent up to 50% of patients in many countries, with the majority of cases, especially in LMIC countries, being managed by general neurosurgeons. Children's diseases, both congenital and acquired, are essentially different to those of adults, and are best managed in an environment dedicated to their unique requirements. Neurosurgical procedures in children are often complex. The most common complications occur in vascular and brain tumor surgery. Anticipating complications, ensuring appropriate preemptive measures, their timely early recognition, and management, enhance the safety profile of the procedure, thereby reducing the morbidity and mortality risk for the child.

References

1. Humphreys RP. Memories of ISPN. International Society for Pediatric Neurosurgery 30th Anniversary Members booklet. Tokyo: Narunia Ltd; 2002. p. 52–5.
2. Peter JC. Pediatric neurosurgery: an Odyssey of hope and discovery, Mauerberger Chair of Neurosurgery Inaugural lecture. New series. Cape Town: University of Cape Town; 1995.
3. Drake JM, et al. Prospective surveillance of complications in a pediatric neurosurgery unit. J Neurosurg Pediatr. 2010;5(6):544–8.
4. Cohen MM, et al. Pediatric anesthesia morbidity and mortality in the peri-operative period. Anesth Analg. 1990;70:160–7.
5. Raju TNK, Vidyasagar D, Torres C, et al. Intracranial pressure during intubation and anesthesia in infants. J Pediatr. 1980;96:861.
6. Welborn LG. Perioperative complications in the former premature infant. Curr Rev Clin Anesth. 1996;17:7–11.

7. Soriano SG, Rockoff MA. Principles and practice of pediatric neurosurgery. 2nd ed. New York: Thieme Medical Publishers. 2008. p. 1213–15.
8. Ludwig S, Loiselle J. Anatomy, growth, and development. In: Eichelberger MR, editor. Pediatric trauma: prevention, acute care, rehabilitation. St. Louis: Mosby-Year Book; 1993. p. 39–58.
9. Cory CZ, Jones MD, James DS, Leadbeatter S, Nokes LD. The potential and limitations of utilising head impact injury models to assess the likelihood of significant head injury in infants after a fall. Forensic Sci Int. 2001;123(2–3):89–106. https://doi.org/10.1016/S0379-0738(01)00523-0.
10. Adewale L. Anatomy and assessment of the pediatric airway. Paediatr Anaesth. 2009;19(Suppl 1):1–8. https://doi.org/10.1111/j.1460-9592.2009.03012.x.
11. Eckenhoff JE. Some anatomic considerations of the infant larynx influencing endotracheal anesthesia. Anesthesiology. 1951;12(4):401–10. https://doi.org/10.1097/00000542-195107000-00001.
12. Figaji AA. Anatomical and physiological differences between children and adults relevant to traumatic brain injury and the implications for clinical assessment and care. Front Neurol. 2017;8:685. https://doi.org/10.3389/fneur.2017.00685.
13. Bir SC, Kalakoti P, Notarianni C, Nanda A. John Howship (1781–1841) and growing skull fracture: historical perspective. J Neurosurg Pediatr. 2015;16(4):472–6. https://doi.org/10.3171/2014.12.PEDS14484.
14. Brown RL, Brunn MA, Garcia VF. Cervical spine injuries in children: a review of 103 patients treated consecutively at a level 1 pediatric trauma center. J Pediatr Surg. 2001;36(8):1107–14. https://doi.org/10.1053/jpsu.2001.25665.
15. Hickman ZL, McDowell M, Anderson RCE. Principles of pediatric spinal column trauma. In: Albright AL, Pollack IF, Adelson PD, editors. Principles and practice of pediatric neurosurgery. 3rd ed. New York: Thieme; 2015. p. 789–805.
16. Riabec P. Addressing problems in reporting and classification of complications in neurosurgery. Liet Chir. 2022;21(3–4).
17. Landriel Ibañez FA, Hem S, Ajler P, Vecchi E, Ciraolo C, Baccanelli M, Tramontano R, Knezevich F, Carrizo A. A new classification of complications in neurosurgery. World Neurosurg. 2011;75(5–6):709–15. https://doi.org/10.1016/j.wneu.2010.11.010. Complications in Neurosurgery. Elsevier, 2019.
18. Kulkarni AV, Drake JM, Lamberti-Pasculli M. Cerebrospinal fluid shunt infection: a prospective study of risk factors. J Neurosurg. 2001;94:195–201.
19. Moussa WM, Mohamed MA. Efficacy of postoperative antibiotic injection in and around ventriculoperitoneal shunt in reduction of shunt infection: a randomized controlled trial. Clin Neurol Neurosurg. 2016;143:144–9.
20. Spader HS, Hertzler DA, Kestle JR, Riva-Cambrin J. Risk factors for infection and the effect of an institutional shunt protocol on the incidence of ventricular access device infections in preterm infants. J Neurosurg Pediatr. 2015;15(2):156–60.
21. Simon TD, Whitlock KB, RivaCambrin J, Kestle JR, Rosenfeld M, Dean JM, et al. Revision surgeries are associated with significant increased risk of subsequent cerebrospinal fluid shunt infection. Pediatr Infect Dis J. 2012;31(6):551–6.
22. Rogers EA, KimiaA MJR, Nigrovic LE, Neuman MI. Predictors of ventricular shunt infection among children presenting to a pediatric emergency department. Pediatr Emerg Care. 2012;28(5):405–9.
23. Simon TD, Butler J, Whitlock KB, Browd SR, Holubkov R, Kestle JR. Risk factors for first cerebrospinal fluid shunt infection: findings from a multi-center prospective cohort study. J Pediatr. 2014;164(6):1462–8.
24. Bruinsma N, Stobberingh EE, Herpers MJ, Vles JS, Weber BJ, Gavilanes DA. Subcutaneous ventricular catheter reservoir and ventriculoperitoneal drain-related infections in preterm infants and young children. Clin Microbiol Infect. 2000;6(4):202–6.
25. Okamura Y, Maruyama K, et al. Detailed standardized protocol to prevent cerebrospinal fluid shunt infection. J Neurosurg. 2019;132(3):755–9.
26. Bonow RH, Browd SR. Principles of neurological surgery. 4th ed. Amsterdam: Elsevier; 2018.
27. Warf B. Pediatric surgery in the third world; Principles and practice of pediatric neurosurgery. 2nd ed. New York: Thieme Medical Publishers. 2008. p. 45.
28. Kawsar KA, Haque MR, Chowdhury FH. Avoidance and management of perioperative complications of endoscopic third ventriculostomy: the Dhaka experience. J Neurosurg. 2015;123:1414–9.
29. Roth J, Ram Z, Constantini S. Endoscopic considerations treating hydrocephalus caused by basal ganglia and large thalamic tumors. Surg Neurol Int. 2015;6:56.
30. Bouras T, Sgouros S. Complications of endoscopic third ventriculostomy. World Neurosurg. 2013;79(Suppl):S22.e9–12.
31. Wang KC, Cho BK, Kim CS, Kim SD. Control of intraoperative bleeding with hydrostatic pressure during endoscopic surgery. Childs Nerv Syst. 1998;14:280–4.
32. Kulkarni AV, Riva-Cambrin J, Holubkov R, Browd SR, Cochrane DD, Drake JM, et al. Endoscopic third ventriculostomy in children: prospective, multicenter results from the hydrocephalus clinical research network. J Neurosurg Pediatr. 2016;18:423–9.
33. Bernard R, Vallee F, Mateo J, Marsella M, George B, Payen D, et al. Uncontrollable high-frequency tachypnea: a rare and nearly fatal complication of endoscopic third ventriculostomy: case report and literature review. Minim Invas Neurosurg. 2010;53:270–2.
34. Schroeder HW, Niendorf WR, Gaab MR. Complications of endoscopic third ventriculostomy. J Neurosurg. 2002;96:1032–40.
35. Tafuri KS, Wilson TA. Growth hormone deficiency and diabetes insipidus as a complication of endoscopic third ventriculostomy. J Clin Res Pediatr Endocrinol. 2012;4:216–9.
36. de Kunder SL, Ter Laak-Poort MP, Nicolai J, Vles JS, Cornips EM. Fever after intraventricular neuroendoscopic procedures in children. Childs Nerv Syst. 2016;32:1049–55.
37. Jung T-Y, et al. Prevention of complications in endoscopic third ventriculostomy. J Korean Neurosurg Soc. 2017;60(3):282–8.
38. Rosemberg S, Fujiwara D. Epidemiology of pediatric tumors of the nervous system according to the WHO 2000 classification: a report of 1,195 cases from a single institution. Childs Nerv Syst. 2005;21:940–4.
39. Jemal A, Siegel R, Ward E, Murray T, Xu J, Smigal C, et al. Cancer statistics. Cancer J Clin. 2006;56:106–30.
40. O'Kane R, Mathew R, Kenny T, Stiller C, Chumas P. United Kingdom 30-day mortality rates after surgery for pediatric central nervous system tumors. J Neurosurg Pediatr. 2013;12:227–34. (Sandri A, Sardi N, Genitori L, Giordano F, Peretta P, Basso ME, et al. Diffuse and focal brain stem tumors in childhood: prognostic factors and surgical outcome. Childs Nerv Syst. 2006;22:1127–35.)
41. Sandri A, Sardi N, Genitori L, Giordano F, Peretta P, Basso ME, et al. Diffuse and focal brain stem tumors in childhood: prognostic factors and surgical outcome. Childs Nerv Syst. 2006;22:1127–35.
42. Sanford RA, Merchant TE, Zwienenberg-Lee M, Kun LE, Boop FA. Advances in surgical techniques for resection of childhood cerebellopontine angle ependymomas are key to survival. Childs Nerv Syst. 2009;25:1229–40.
43. Nimsky C, Ganslandt O, Kober H, Buchfelder M, Fahlbusch R. Intraoperative magnetic resonance imaging combined with neuronavigation: a new concept. Neurosurgery. 2001;48:1082–9.
44. Preuss M, Werner P, Barthel H, Nestler U, Christiansen H, Hirsch FW, et al. Integrated PET/MRI for planning navigated biopsies in pediatric brain tumors. Childs Nerv Syst. 2014;30:1399–403.
45. Jolesz FA. Future perspectives in intraoperative imaging. Acta Neurochir Suppl. 2003;85:7–13.
46. Ulrich NH, Burkhardt JK, Serra C, Bernays RL, Bozinov O. Resection of pediatric intracerebral tumors with the aid of intraoperative realtime 3-D ultrasound. Childs Nerv Syst. 2012;28:101–9.

47. Prada F, Del Bene M, Mattei L, Lodigiani L, DeBeni S, Kolev V, et al. Preoperative magnetic resonance and intraoperative ultrasound fusion imaging for real-time neuronavigation in brain tumor surgery. Ultraschall Med. 2015;36(2):174–86. [Epub 2014 Nov 27].
48. Yang JY-M, et al. Diffusion MRI tractography for neurosurgery: the basics, current state, technical reliability and challenges. Phys Med Biol. 2021;66:15TR01.
49. Wu JS, Zhou LF, Tang WJ, Mao Y, Hu J, Song YY, Hong XN, Du GH. Clinical evaluation and follow-up outcome of diffusion tensor imaging-based functional neuronavigation: a prospective, controlled study in patients with gliomas involving pyramidal tracts. Neurosurgery. 2007;61:935–49.
50. Mishra N, et al. Perioperative management of pediatric brain tumors: a retrospective analysis. Neurol India. 2022;70(3):1095–101.
51. Bhatnagar S, Udaya B, Rao GS. An audit of blood transfusion in elective neurosurgery. Indian J Anaesth. 2007;51:200–4.
52. Matuev KB, Gorelishev SK, Lubnin AY, Lemeneva NV, Sorokin VS. Blood loss in surgery of brain tumors in infants. Zh Vopr Neirokhir Im N N Burdenko. 2013;77:3–15.
53. Akay KM, Izci Y, Baysefer A, Atabey C, Kismet E, Timurkaynak E. Surgical outcomes of cerebellar tumors in children. Pediatr Neurosurg. 2004;40:220–5.
54. Scott RM, et al. Principles and practice of pediatric neurosurgery. 2nd ed. New York: Thieme Medical Publisher. 2008. p. 947–56.
55. Humphreys RP. Principles and practice of pediatric neurosurgery. 2nd ed. New York: Thieme Medical Publishers. 2008. p. 983–94.
56. Ellis JA, Barrow DL. Supratentorial cavernous malformations. Handb Clin Neurol. 2017;143:283–9.
57. Moultrie F, Horne MA, Josephson CB, Hall JM, Counsell CE, Bhattacharya JJ, Papanastassiou V, Sellar RJ, Warlow CP, Murray GD, Al-Shahi SR. Scottish Audit of Intracranial Vascular Malformations (SAIVMs) steering committee and collaborators. Outcome after surgical or conservative management of cerebral cavernous malformations. Neurology. 2014;83(7):582–9.
58. Abla AA, Lekovic GP, Turner JD, de Oliveira JG, Porter R, Spetzler RF. Advances in the treatment and outcome of brainstem cavernous malformation surgery: a single-center case series of 300 surgically treated patients. Neurosurgery. 2011;68(2):403–14; discussion 414–5.
59. Caton MT, Shenoy VS. Cerebral cavernous malformations. [Updated 2022 Oct 3]. In: StatPearls [Internet]. Treasure Island: StatPearls Publishing; 2023.
60. Xie M-G, et al. Surgical treatment of the medullary cavernous malformations: 53 cases. World Neurosurg. 2018;118:449–59.
61. Khalatbari MR, Abbassioun K, Amirjmshidi A. Solitary spinal epidural cavernous angioma: report of nine surgically treated cases and review of the literature. Eur Spine J. 2013;22(3):542–7. https://doi.org/10.1007/s00586-012-2526-2.
62. Ibrahim Sun M, Pamir N. Spinal Cavernomas: outcome of surgically treated 10 patients. Front Neurol. 2017;8:672.
63. Rosenorn J, et al. Clinical features and outcome in 1076 patients with ruptured intracranial saccular aneurysms: a prospective study. Br J Neurosurg. 1966;25:219–39.
64. Yazbak, et al. Pediatric traumatic intracranial aneurysms. Pediatr Neurosurg. 1995;22:15–9.
65. Raimondi AJ, Choux M, et al. Intracranial aneurysms in children. New York: Springer-Verlag; 1991.
66. Herman JM, Rekate HL, Spetzler RF. Pediatric intracranial aneurysms: simple and complex cases. Pediatr Neurosurg. 1991;17:66–73.
67. Kanaan I, et al. Spectrum of intracranial aneurysms in pediatrics. Minim Invas Neurosurg. 1995;38:1–9.
68. Olivecrona H, Riives J. Arteriovenous aneurysms of the brain: their diagnosis and treatment. Arch Neurol Psychiatr. 1948;59:567–603.
69. Jessurun GA, Kamphuis DJ, van der Zande FH, Nossent JC. Cerebral AVMs in The Netherlands Antilles. High prevalence of hereditory hemorrhagic telectasia-related single and multiple arteriovenous malformations. Clin Neurol Neurosurg. 1993;95:193–8.
70. Celli P, et al. Cerebral arteriovenous malformations in children. Clinical features and outcome of treatment in children and in adults. Surg Neurol. 1984;22(1):43–9.
71. Humphreys, et al. Pediatric cerebral aneurysms: a retrospective review of 56 cases. Childs Nerv Syst. 2002;18:557.
72. Gross BA, et al. Microsurgical treatment of arteriovenous malformations in pediatric patients: the Boston Children's Hospital experience. J Neurosurg Pediatr. 2015;15:71–7.

MIX
Papier aus verantwortungsvollen Quellen
Paper from responsible sources
FSC® C105338

If you have any concerns about our products,
you can contact us on
ProductSafety@springernature.com

In case Publisher is established outside the EU,
the EU authorized representative is:
Springer Nature Customer Service Center GmbH
Europaplatz 3, 69115 Heidelberg, Germany

Printed by Libri Plureos GmbH
in Hamburg, Germany

Neuroanatomy Guidance to Successful Neurosurgical Interventions

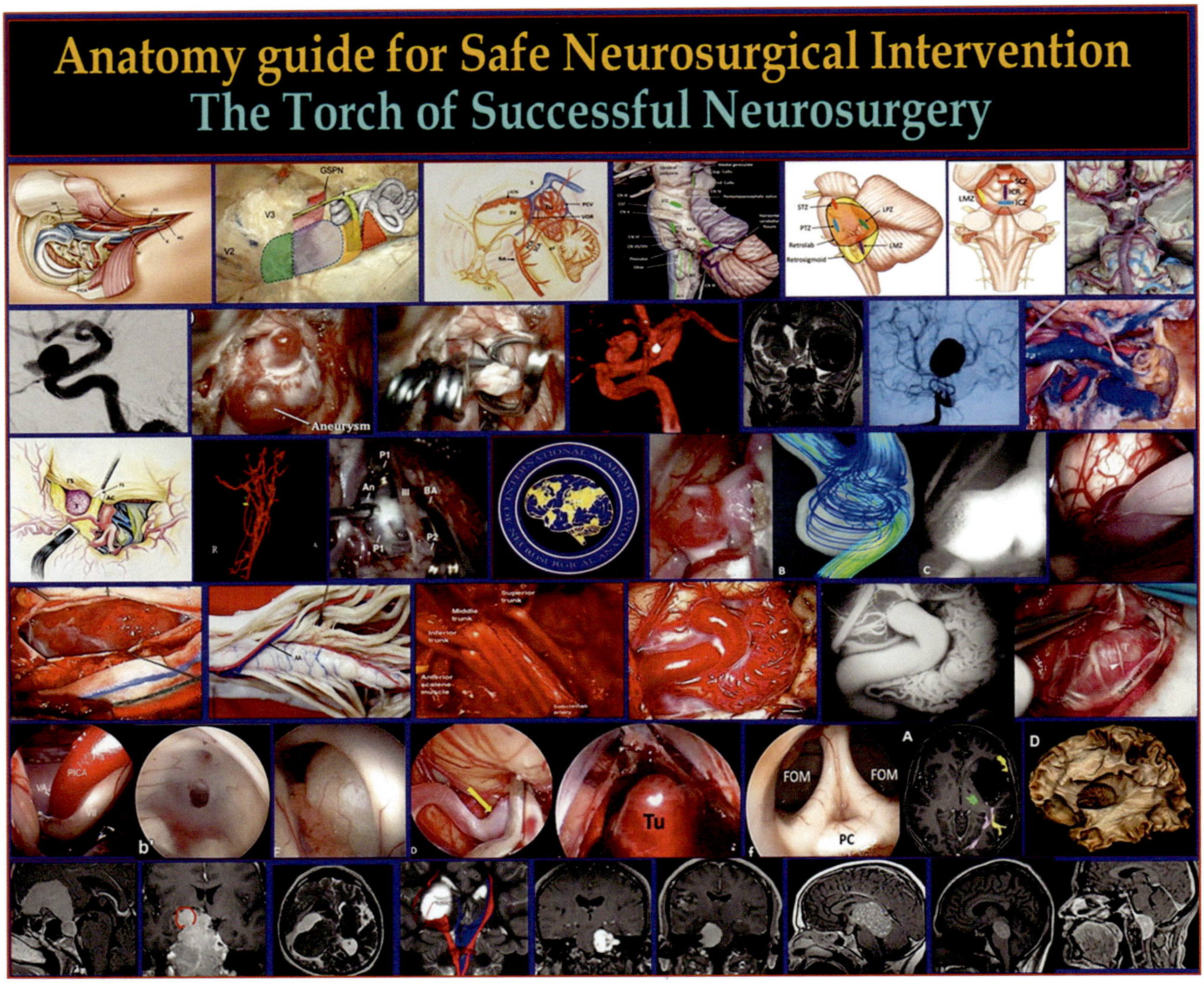
Anatomy guide for Safe Neurosurgical Intervention
The Torch of Successful Neurosurgery
GSPN
V3
V2
Aneurysm
P1
An
BA
P2
Middle trunk
Superior trunk
Inferior trunk
Anterior scalene muscle
PICA
Tu
FOM
FOM
PC

Imad N. Kanaan • Vladimír Beneš
Editors

Neuroanatomy Guidance to Successful Neurosurgical Interventions

The Torch of Neurosurgical Excellence

Volume I

Springer

Editors
Imad N. Kanaan
Neurosciences Center, Section of Neurosurgery
King Faisal Specialist Hospital & Research
Center Institution, Alfaisal University
College of Medicine
Riyadh, Kingdom of Saudi Arabia

Vladimír Beneš
Dept. of Neurosurgery
Central Military Hospital Prague
Praha 2, Czech Republic

ISBN 978-3-031-59837-1 ISBN 978-3-031-59838-8 (eBook)
https://doi.org/10.1007/978-3-031-59838-8

This Springer imprint is published by the registered company Springer Nature Switzerland AG
The registered company address is: Gewerbestrasse 11, 6330 Cham, Switzerland

Foreword

My personal congratulation to Professor Imad N. Kanaan and Professor Vladimír Beneš for publishing such a special neurosurgical book, which is the result of scientific programs of the last 2 years of webinars organized by both authors.

This was a very important neurosurgical educational program for many young and practicing neurosurgeons from all over the world, who had the opportunity to follow all these exciting lectures.

During the 2 years, the most outstanding and experienced neurosurgeons have delivered important insights from their expertise for the state-of-the-art of neurosurgery.

The publication of these important lectures on different aspects of neurosurgery such as "Management of intracranial skull base and intraspinal tumors," "Current technologies for treatment of neurovascular pathologies," as well as "Functional neurosurgery and peripheral nerve injury" is another step towards spreading this spirit of education to all other neurosurgeons who have not had the opportunity to attend live neurosurgical webinars in the last 2 years.

I hope that this important book will reach all neurosurgeons, particularly the new generation of neurosurgeons in every country.

My personal thanks go to both editors, Dr. Imad N. Kanaan and Dr. Vladimír Beneš, for promoting neurosurgical training and education worldwide and for publishing these important contributions to the world of neurosurgery.

Hannover, Germany

Madjid Samii

Preface 1

The successful global venture into neurosurgical education online had its genesis in navigating the constrictions imposed by the late stages of the coronavirus pandemic. This phenomenon is reminiscent of the proverbial maxim, "every cloud has a silver lining," underscoring positive outcomes born out of adversity. This laudable initiative owes its realization to the confluence of advancements in communication and digital technologies, spurred onward by the fervor of our eminent mentors and the aspirational zeal of international neurosurgeons and trainees.

This reference textbook of neuroanatomy guidance for successful neurosurgical interventions stands as a seminal academic publication, a "beacon to neurosurgical excellence," adeptly bridging the chasm between applied anatomy and the intricacies of neurosurgical intervention. Its 75 chapters authored by elite academic neurosurgeons cover the wide spectrum of neurosurgical practice. Adopting a stylistic and conceptual approach evocative of the visual artistry of Leonardo Da Vinci and the balance between art and science, the narrative intertwines vibrant illustrations with anatomical highlights and expertly selected surgical pearls, thereby offering profound guidance for the pursuit of safe neurosurgical interventions.

The expeditious responses of the devoted members of the "IANA" and our friends and esteemed colleagues within the international neurosurgical community were integral to the realization of this monumental endeavor. Their willingness to share their knowledge and experience with their younger international peers and trainees exemplifies the collaborative spirit within this field, and indeed was fundamental to accomplish this publication. Most importantly, I am really blessed to have the greatest support, inspiration, and understanding of my beloved family—my wife Huda and my two precious daughters Faten and Nadia—without whose encouragement this achievement would not have been possible.

Riyadh, Saudi Arabia — Imad N. Kanaan

Preface 2

Neuroanatomy Guidance to Successful Neurosurgical Interventions: The Torch of Neurosurgical Excellence is the book which is exactly what the title declares. The authors of 75 chapters are world known and respected leaders in their field. They have spent a considerable amount of time and effort to put together chapters which would be educational, very much pictorial and which would cover the recent state-of-the-art in respected areas. All areas of neurosurgery are covered with emphasis on educational value.

Starting with history with special emphasis on neuroanatomy and a few general chapters dealing with evolution, plasticity, and neurosurgical facilities requirements the following pages are devoted to individual subspecialties of neurosurgery. Part II of the book covers the whole vascular neurosurgery, aneurysms, AVMs, cavernomas, and ischemia. Part III is devoted to skull base tumors. All aspects are discussed, again with the special emphasis on neuroanatomy and surgical approaches. Part IV is dedicated to ventricular tumors which form the specific group and specific management requirements like hydrocephalus. Part V, gliomas, deals with the intrinsic tumors on the basis of white matter anatomy. One chapter deals with brainstem tumors and another with pediatric tumors which by all means form a specific and special group with problems different from that in adults. Part VI starts with an excellent anatomy of spinal cord vasculature followed by spinal tumors and peripheral nerves where the anatomical knowledge is of extreme importance. Part VII, functional, in its epilepsy chapters is again based on anatomy of white matter tracts. In this part, the movement disorders, trigeminal neuralgia, and normal pressure hydrocephalus are also included. Quite fittingly, the book is concluded by a special article on complications. Complications appear, despite the fact that generally all the preceding chapters do have one common feature—how to avoid the complications. The other common feature is the fact that all the authors have special interest in neuroanatomy and are either members or friends of International Academy of Neurosurgical Anatomy. The special interest of the individual authors, neuroanatomy, ensures the high number of high-quality images. These by themselves do have very high educational value.

We hope and believe the book will be of extreme help and reference, especially to young neurosurgeons, to those who want to excel in their chosen discipline. We believe the book will serve all neurosurgeons as well as others for a good orientation on what the recent neurosurgery may offer to our patients.

I would like to appreciate and thank all those who helped me put the book together, without such a help the task would be impossible. I also would like to thank my wife Eva for her understanding and never ending support.

Prague, Czech Republic Vladimír Beneš

Contents

Part VII Functional

Part VIII Complication in Neurosurgery

Contributors

Javier Abarca-Olivas Department of Neurosurgery, Hospital General Universitario de Alicante, Alicante, Spain

Hadie Adams Imperial College Healthcare NHS Trust, London, UK

Guilherme Agnoletto Department of Neurosurgery, Clinical Neurosciences Center, University of Utah, Salt Lake City, UT, USA

Mohamed Helmy Neurosurgery Department, Huashan Hospital Fudan University, China National Neurological & Neurosurgical Center, Shanghai, People's Republic of China

Neurosurgery Department, Shanghai Donglei Brain Hospital, Shanghai, People's Republic of China

Filippo Flavio Angileri Department of Neurosurgery, AOU Policlinico "G. Martino", University of Messina, Messina, Italy

Miguel A. Arraez Carlos Haya University Hospital, University of Malaga, Malaga, Spain

Cinta Arraez-Manrique Carlos Haya University Hospital, University of Malaga, Malaga, Spain

Heba Azooz Department of Neurosurgery, Children Cancer Hospital Egypt, Cairo, Egypt

Matias Baldoncini Laboratory of Microsurgical Neuroanatomy, Second Chair of Gross Anatomy, School of Medicine, University of Buenos Aires, Buenos Aires, Argentina

Elena Baño-Ruiz Department of Neurosurgery, Hospital General Universitario de Alicante, Alicante, Spain

Mustafa K. Başkaya Department of Neurological Surgery, University of Wisconsin-Madison, Madison, WI, USA

Jaafar Basma Arkansas Neuroscience Institute, CHI Saint-Vincent, Sherwood, AR, USA

Vladimír Beneš Institute of Clinical Neurodisciplines, Department of Neurosurgery and Neurooncology, First Faculty of Medicinefavo, Charles University and Military University Hospital, Prague, Czech Republic

Department of Neurosurgery and Neurooncology, First Faculty of Medicine, Charles University and Military University Hospital, Prague, Czech Republic

Giacomo Bertolini Neurosurgery Division, Head and Neck Department, Azienda Ospedaliero-Universitaria di Parma, Parma, Italy

Xu Bin Neurosurgery Department, Huashan Hospital Fudan University, China National Neurological & Neurosurgical Center, Shanghai, People's Republic of China

Luis A. B. Borba Department of Neurosurgery, Mackenzie Evangelical University Hospital, Curitiba, PR, Brazil

Department of Neurosurgery, Federal University of Paraná, Curitiba, PR, Brazil

Roman Bošnjak University Medical Center Ljubljana, Ljubljana, Slovenia

University of Ljubljana, Medical faculty, Ljubljana, Slovenia

Ondřej Bradáč Institute of Clinical Neurodisciplines, Department of Neurosurgery and Neurooncology, First Faculty of Medicine, Charles University and Military University Hospital, Prague, Czech Republic

Department of Neurosurgery and Neurooncology, First Faculty of Medicine, Charles University and Military University Hospital, Prague, Czech Republic

Department of Neurosurgery, Second Faculty of Medicine, Charles University and Motol University Hospital, Prague, Czech Republic

David P. Bray Department of Neurosurgery, Emory University School of Medicine, Atlanta, GA, USA

Giovanni Broggi Department of Neurosurgery, Fondazione IRCCS Istituto Neurologico Carlo Besta, Milan, Italy

Adéla Bubeníková Department of Neurosurgery and Neurooncology, First Faculty of Medicine, Charles University and Military University Hospital, Prague, Czech Republic

Institute of Clinical Neurodisciplines, Department of Neurosurgery and Neurooncology, First Faculty of Medicinefavo, Charles University and Military University Hospital, Prague, Czech Republic

Department of Neurosurgery, Second Faculty of Medicine, Charles University and Motol University Hospital, Prague, Czech Republic

Institute of Clinical Neurodisciplines, Department of Neurosurgery and Neurooncology, First Faculty of Medicine, Charles University and Military University Hospital, Prague, Czech Republic

Department of Neurosurgery and Neurooncology, First Faculty of Medicine, Charles University and Military University Hospital, Prague, Czech Republic

Department of Neurosurgery, Second Faculty of Medicine, Charles University and Motol University Hospital, Prague, Czech Republic

Alvaro Campero Department of Neurological Surgery, Hospital Padilla, Tucumán, Argentina

Alberto Campione Department of Neurosurgery, San Filippo Neri Hospital/ASLRoma1, Rome, Italy

Paolo Cappabianca Division of Neurosurgery, Department of Neurosciences, Reproductive and Odontostomatological Sciences, Università degli Studi di Napoli "Federico II", Naples, Italy

Salvatore Cardali Department of Neurosurgery, "Papardo" General Hospital, University of Messina, Messina, Italy

Andrea L. Castillo Department of Neurosurgery, Antonio Lenin Fonseca Hospital, Managua, Nicaragua

Guilherme H. W. Ceccato Department of Neurosurgery, Mackenzie Evangelical University Hospital, Curitiba, PR, Brazil

Fady T. Charbel Department of Neurosurgery, University of Illinois, Chicago, IL, USA

Dimitris Charitos Department of Neurosurgery, Lariboisière Hospital, APHP Nord, Assistance Publique – Hôpitaux de Paris, Paris Cité University, Paris, France

Experimental Neurosurgery Laboratory, Department of Neurosurgery, Lariboisière Hospital, APHP Nord, Assistance Publique – Hôpitaux de Paris, Paris Cité University, Paris, France

Iype Cherian Department of Neurosurgery, Krishna Vishwa Vidyapeeth, Karad, Maharashtra, India

Alfredo Conti Neurosurgery, IRCCS Istituto delle Scienze Neurologiche di Bologna, Bologna, Italy

Dipartimento di Scienze Biomediche e NeuroMotorie (DIBINEM), Alma Mater Studiorum University of Bologna, Bologna, Italy

Giulia Cossu Department of Neurosurgery, University Hospital of Lausanne, Lausanne, Switzerland

William T. Couldwell Department of Neurosurgery, Clinical Neurosciences Center, University of Utah, Salt Lake City, UT, USA

Antonello Curcio Department of Neurosurgery, AOU Policlinico "G. Martino", University of Messina, Messina, Italy

Roy T. Daniel Department of Neurosurgery, University-Hospital Lausanne, Lausanne, Switzerland

Chandrashekhar E. Deopujari Department of Neurosurgery, Bombay Hospital Institute of Medical Sciences, Mumbai, Maharashtra, India

Yücel Doğruel Department of Neurosurgery, Health Sciences University, Tepecik Training and Research Hospital, İzmir, Turkey

Víctor Fernández-Cornejo Department of Neurosurgery, Hospital General Universitario de Alicante, Alicante, Spain

Ioan Alexandru Florian Department of Neuroscience, University of Medicine and Pharmacy Iuliu Hatieganu, Cluj-Napoca, Romania

Department of Neurosurgery, Cluj County Emergency Clinical Hospital, Cluj-Napoca, Romania

Ioan Stefan Florian Department of Neuroscience, University of Medicine and Pharmacy Iuliu Hatieganu, Cluj-Napoca, Romania

Department of Neurosurgery, Cluj County Emergency Clinical Hospital, Cluj-Napoca, Romania

Sébastien Froelich Department of Neurosurgery, Lariboisière Hospital, APHP Nord, Assistance Publique – Hôpitaux de Paris, Paris Cité University, Paris, France

Experimental Neurosurgery Laboratory, Department of Neurosurgery, Lariboisière Hospital, APHP Nord, Assistance Publique – Hôpitaux de Paris, Paris Cité University, Paris, France

Giada Garufi Department of Neurosurgery, "Papardo" General Hospital, Messina, Italy

Turab Gasimov Department of Neurological Surgery, University of Wisconsin-Madison, Madison, WI, USA

Mario Giordano Department of Neurosurgery, INI – International Neuroscience Institute, Hannover, Germany

Atul Goel Department of Neurosurgery, Seth G.S. Medical College and K.E.M Hospital, Parel, Mumbai, Maharashtra, India

Department of Neurosurgery, Lilavati Hospital and Research Center, Mumbai, Maharashtra, India

Juan L. Gómez-Amador Department of Neurosurgery, National Institute of Neurology and Neurosurgery "Manuel Velasco Suárez", Mexico City, Mexico

Cristina Gómez-Revuelta Department of Neurosurgery, Hospital General Universitario de Alicante, Alicante, Spain

Pablo González-Lopéz Alicant General Hospital, University of Alicant, Alicant, Spain

Department of Neurosurgery, Hospital General Universitario de Alicante, Alicante, Spain

Takeo Goto Department of Neurosurgery, Osaka Metropolitan University Graduate School of Medicine, Osaka, Japan

J. André Grotenhuis Emeritus Professor of Neurosurgery, Radboud University Medical Centre, Nijmegen, The Netherlands

Michelle Grouls Department of Neurosurgery, Lariboisière Hospital, APHP Nord, Assistance Publique – Hôpitaux de Paris, Paris Cité University, Paris, France

Experimental Neurosurgery Laboratory, Department of Neurosurgery, Lariboisière Hospital, APHP Nord, Assistance Publique – Hôpitaux de Paris, Paris Cité University, Paris, France

Abuzer Güngör Faculty of Medicine Department of Neurosurgery, İstinye University, Istanbul, Turkey

Liv Hospital Vadistanbul, Neurosurgery Clinic, Istanbul, Turkey

Arthur Hosmann Department of Neurosurgery & Gamma Knife Surgery, P.D. Hinduja Hospital & MRC, Mumbai, Maharashtra, India

Department of Neurosurgery, Medical University of Vienna, Vienna, Austria

Timothée Jacquesson Skull Base and Pituitary Unit, Department of Neurosurgery B, Neurological Hospital Pierre Wertheimer, Lyon, France

Dragan Jankovic Department of Neurosurgery, University Medical Center, Johannes Gutenberg University Mainz, Mainz, Germany

DieuMerci Kantenga Kabulo Department of Neurosurgery, University of Zimbabwe, Harare, Zimbabwe

Kazadi K. N. Kalangu University of Zimbabwe, Department of Neurosurgery, Harare, Zimbabwe

Imad N. Kanaan Neurosciences Center, Section of Neurosurgery, King Faisal Specialist Hospital & Research Center Institution, Alfaisal University, College of Medicine, Riyadh, Kingdom of Saudi Arabia

Ekkehard M. Kasper Boston University School of Medicine, Boston, MA, USA

Department of Neurosurgery, St. Elizabeth Medical Center, Boston, MA, USA

Faculty of Health Sciences, McMaster University, Hamilton, ON, Canada

Yoko Kato Department of Neurosurgery, Fujita Health University Bantane Hospital, Nagoya, Japan

Abdel Raouf Kayssi Arkansas Neuroscience Institute, CHI Saint-Vincent, Sherwood, AR, USA

Mehdi Khaleghi Department of Neurosurgery, The James Cancer Hospital and Wexner Medical Center, The Ohio State University, Columbus, OH, USA

Ramez W. Kirollos Department of Neurosurgery, National Neuroscience Institute, Singapore, Singapore

Duke-NUS Graduate Medical School, Singapore, Singapore

Ali F. Krisht Arkansas Neuroscience Institute, CHI Saint-Vincent, Sherwood, AR, USA

Artem Kuptsov Department of Neurosurgery, Hospital General Universitario de Alicante, Alicante, Spain

José Alberto Landeiro Department of Neurosurgery, Antonio Pedro University Hospital, Fluminense Federal University, Niterói, Brazil

Yujun Liao Fudan University, Shanghai, China

James K. Liu, MD, FACS, FAANS Skull Base Institute of New Jersey, Neurosurgeons of New Jersey, NYU Langone Neurosurgery Network, Livingston, NJ, USA

Department of Neurosurgery, Cooperman Barnabas Medical Center, RWJ Barnabas Health, Livingston, NJ, USA

Martin Májovský Institute of Clinical Neurodisciplines, Department of Neurosurgery and Neurooncology, First Faculty of Medicine, Charles University and Military University Hospital, Prague, Czech Republic

Igor L. Maldonado CHRU de Tours, Tours, France

Laboratoire d'Anatomie, Faculté de Médecine de Tours, Tours, France

INSERM, Imaging Brain & Neuropsychiatry iBraiN U1253, University of Tours, Tours, France

Sai Mannam Perelman School of Medicine, University of Pennsylvania, Philadelphia, PA, USA

Mario Ammirati College of Science and Technology, Temple University, Philadelphia, PA, USA

Jaime L. Martínez Santos Department of Neurological Surgery, Medical University of South Carolina, Charleston, SC, USA

Luciano Mastronardi Department of Neurosurgery, San Filippo Neri Hospital/ASLRoma1, Rome, Italy

Laura Stone McGuire Department of Neurosurgery, University of Illinois, Chicago, IL, USA

José Orlando de Melo Junior Department of Neurosurgery, Paulo Niemeyer State Brain Institute, Rio de Janeiro, Brazil

Ahmed Al Menabbawy Department of Neurosurgery, University Medicine Greifswald, Greifswald, Germany

Ali Tayebi Meybodi Department of Neurological Surgery, New Jersey Medical School, Newark, NJ, USA

Basant K. Misra Department of Neurosurgery & Gamma Knife Surgery, P.D. Hinduja Hospital & MRC, Mumbai, Maharashtra, India

Jacques J. Morcos, MD Department of Neurosurgery, The University of Miami, Miami, FL, USA

Department of Neurosurgery, UT Houston, Houston, TX, USA

Hiroki Morisako Department of Neurosurgery, Osaka Metropolitan University Graduate School of Medicine, Osaka, Japan

Ramesh Nair Imperial College Healthcare NHS Trust, London, UK

Farshad Nassiri Department of Neurosurgery, Clinical Neurosciences Center, University of Utah, Salt Lake City, UT, USA

Juan Nieto-Navarro Department of Neurosurgery, Hospital General Universitario de Alicante, Alicante, Spain

Kenji Ohata Department of Neurosurgery, Osaka Metropolitan University Graduate School of Medicine, Osaka, Japan

Soichi Oya Department of Neurosurgery, Gunma University Graduate School of Medicine, Maebashi, Gunma, Japan

Nelson M. Oyesiku Department of Neurosurgery, University of North Carolina School of Medicine, Chapel Hill, NC, USA

Dhiraj Parihar Department of Neurosurgery, Seth G.S. Medical College and K.E.M Hospital, Parel, Mumbai, Maharashtra, India

Victor Hugo Pérez Forensic Sciences Institute, TSJ, Autonomous University of Mexico State, Mexico City, Mexico

Daniel M. Prevedello Department of Neurosurgery, The James Cancer Hospital and Wexner Medical Center, The Ohio State University, Columbus, OH, USA

Mubashir Mahmood Qureshi Aga Khan University Hospital, Nairobi, Kenya

Manuel de Jesús Encarnacion Ramírez Neurosurgery Department, RUDN University, Russian Federation, Moscow, Russia

Manuel Encarnacion Ramirez Laboratory of Microsurgical Neuroanatomy, Second Chair of Gross Anatomy, School of Medicine, University of Buenos Aires, Autónoma de Buenos Aires, Argentina

Lukas Rasulić Faculty of Medicine, University of Belgrade, Belgrade, Serbia

Department for Peripheral Nerve Surgery, Functional Neurosurgery, and Pain Management Surgery, Clinic for Neurosurgery, University Clinical Centre of Serbia, Belgrade, Serbia

Ehab El Refaee Department of Neurosurgery, University Medicine Greifswald, Greifswald, Germany

Department of Neurosurgery, Cairo University, Cairo, Egypt

Roberto Colasanti Department of Neurosurgery, Maurizio Bufalini Hospital, AUSL della Romagna, Cesena, Italy

Andrea De Rosa Division of Neurosurgery, Department of Neurosciences, Reproductive and Odontostomatological Sciences, Università degli Studi di Napoli "Federico II", Naples, Italy

Miroslav Samardžić Faculty of Medicine, University of Belgrade, Belgrade, Serbia

Department for Peripheral Nerve Surgery, Functional Neurosurgery, and Pain Management Surgery, Clinic for Neurosurgery, University Clinical Centre of Serbia, Belgrade, Serbia

Amir Samii Department of Neurosurgery, INI – International Neuroscience Institute, Hannover, Germany

Marcos V. Sangrador-Deitos Department of Neurosurgery, National Institute of Neurology and Neurosurgery "Manuel Velasco Suárez", Mexico City, Mexico

Thomas Santarius Department of Neurosurgery, Cambridge University Hospitals NHS Trust, Cambridge, UK

Department of Clinical Neurosciences, University of Cambridge, Cambridge, UK

Henry W. S. Schroeder Department of Neurosurgery, University Medicine Greifswald, Greifswald, Germany

Abhidha Shah Department of Neurosurgery, Seth G.S. Medical College and K.E.M Hospital, Parel, Mumbai, Maharashtra, India

Nishit J. Shah Department of ENT, Bombay Hospital Institute of Medical Sciences, Mumbai, Maharashtra, India

Muhammad Yousuf Shaikh Department of Neurosurgery, LNH, Karachi, Pakistan

Salman Sharif Department of Neurosurgery, LNH, Karachi, Pakistan

Puneet Sharma Imperial College Healthcare NHS Trust, London, UK

Rajeev Sharma Department of Neurosurgery, Osaka Metropolitan University Graduate School of Medicine, Osaka, Japan

All India Institute of Medical Sciences, New Delhi, India

Uzair Ahmed Siddiqui LNH, Karachi, Pakistan

Juan E. Small Lahey Hospital and Medical Center, Burlington, MA, USA

Anna Štekláčová Institute of Clinical Neurodisciplines, Department of Neurosurgery and Neurooncology, First Faculty of Medicinefavo, Charles University and Military University Hospital, Prague, Czech Republic

Zhangjie Su Department of Neurosurgery, Cambridge University Hospitals NHS Trust, Cambridge, UK

Department of Clinical Neurosciences, University of Cambridge, Cambridge, UK

Albert Sufianov Federal Centre of Neurosurgery (Tyumen) of Ministry of Health of Russian Federation, Tyumen, Russia

Rinat Sufianov I.M.Sechenov First Moscow State Medical University, Moscow, Russia

Matthew Z. Sun Department of Neurosurgery, UT Southwestern, Dallas, TX, USA

Francesco Tomasello Humanitas Clinical Institute of Catania, Misterbianco, Catania, Italy

Keki E. Turel Department of Neurosurgery, Bombay Hospital Institute of Medical Sciences, Mumbai, India

Sara Venturini Department of Neurosurgery, Cambridge University Hospitals NHS Trust, Cambridge, UK

Rodolfo Villalobos-Díaz Department of Neurosurgery, National Institute of Neurology and Neurosurgery "Manuel Velasco Suárez", Mexico City, Mexico

Robert Walker Department of Neurosurgery, St. Elizabeth Medical Center, Boston, MA, USA

Abhijit G. Warade Department of Neurosurgery & Gamma Knife Surgery, P.D. Hinduja Hospital & MRC, Mumbai, Maharashtra, India

Eva M. Wu, MD Department of Neurosurgery, The University of Miami, Miami, FL, USA

Part I

General

Phylogenetic Evolution of the Human Central Nervous System

1

Pablo González-Lopéz, Adéla Bubeníková, and Roy T. Daniel

Learning Objectives

- Understanding the central nervous system shape and morphology of different species and how the evolutionary changes have led to the development of the human brain.
- Introduce the parallelism between phylogeny and ontogeny.
- Study the human brain embryological development.
- Analyze the shape of the cortical and subcortical structures of the human brain.
- Adapt the phylogenetic concepts to surgical planning and disease understanding.

1.1 Introduction

The brain is a complex structure with a very specific architecture that serves as the network to provide humans with highly specialized abilities. Parallel, and because of the evolution of the species, the human brain has acquired a specific shape and external morphology based on fissures, sulci, and gyri, covered by layers of neurons and their underlying connections, all surrounded by cisterns containing the passing vessels. Neurosurgeons should master this organization from the surface to the depth and understand the way in which any neurosurgical disease is related to normal anatomy, aiming to optimize the surgical approaches and get better clinical results.

A nervous system might be defined as a group of neurons able to detect changes within the organisms themselves and in their environments, thanks to electrical signals that are carried out through a network of specialized cells, and finally react to create a response. The human central nervous system (CNS) should be understood as a continuum organ. However, for functional, practical, and educational purposes, some divisions and subdivisions have been established. Those divisions and subdivisions are based on the embryological, and therefore on the phylogenetic evolution (Fig. 1.1).

The human CNS has taken advantage of a tremendous cascade of evolutive changes for more than 3.800 million years, till getting its actual shape, size, organization, and functionality. Thanks to this sum of changes and to the needs marked by the different environments the species have had to face, the human CNS allows organisms to connect with the environment, stay safe, survive, and the most important function for the species, offspring. Despite the highly evolved brain areas with specific and unique functions humans have acquired, the final need of the CNS is to keep the human species alive.

Following the theory of evolution, these changes happened due to countless spontaneous mutations in the frame of different and specific needs and environments. Therefore, the different species suffering and benefiting from these breakthroughs have gained abilities and functions leading them to better interact with the environment and adapt to it in a beneficial way, thus, keeping those changes intact and somehow improving them during the next evolutive steps. During this long process of evolution, the different species' CNS have been exposed to changes which, indeed, created different evolutive branches leading to new species with common ancestors.

The *Homo sapiens* has appeared as a species occupying an ending position in one of those branches. Therefore, our CNS has acquired and stored many of those evolutive changes, and the story that summarizes the most relevant ones is defined as the phylogenetic evolution.

P. González-Lopéz (✉)
Department of Neurosurgery, Hospital General Universitario de Alicante, Alicante, Spain
e-mail: gonzalez_pab@gva.es

A. Bubeníková
Department of Neurosurgery and Neurooncology, First Faculty of Medicine, Charles University and Military University Hospital, Prague, Czech Republic

R. T. Daniel
Department of Neurosurgery, University-Hospital Lausanne, Lausanne, Switzerland
e-mail: roy.daniel@chuv.ch

I. N. Kanaan, V. Beneš (eds.), *Neuroanatomy Guidance to Successful Neurosurgical Interventions*,
https://doi.org/10.1007/978-3-031-59838-8_1

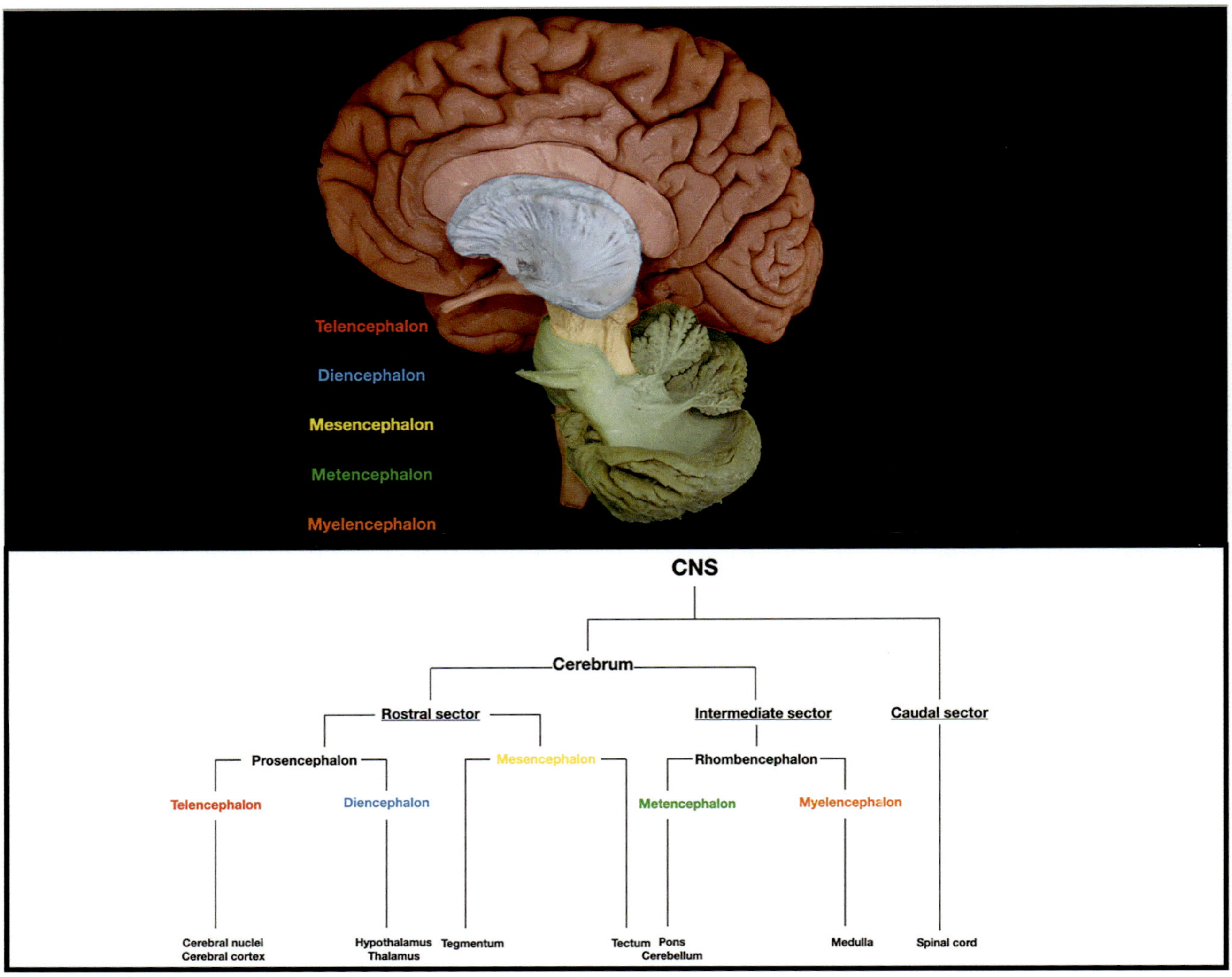

Fig. 1.1 The division of the human brain with the demonstration of different embryological origins

In this chapter, we will try to go through the most relevant changes, summarizing an outstanding story of millions of years of evolution in a few pages, with the final goal of discovering how and why the human brain is organized like this, and how physicians dealing with the human brain anatomy can take advantage of this knowledge to better understand the human brain anatomy, physiology and pathology, and finally optimize the treatment of brain diseases.

1.2 Background: The Brain Evolution Story

Neurosurgeons must assume the fact that a deep knowledge of the brain's architecture is essential to approach, explore, and remove brain lesions without disrupting its precisely organized nature. The delicate vascular system in the brain has been largely studied and analyzed with the aim of respecting its net when approaching brain lesions, especially when dealing with neurovascular diseases. However, it has not been as hardly assumed that neurosurgeons should also give such appreciable attention to cortical and white matter tract anatomy. While modern imaging techniques and surgical technology become increasingly precise, the surgeons' knowledge of the brain surface and deep pathways should improve hand-on-hand with this technical evolution. Modern radiological sequences such as DTI-based tractography, although extremely useful, should never replace the knowledge of the microanatomic relationships between the white matter structures. But something that should represent our study starting point is an understanding of how this cortical structure and underlying white matter network was created and organized.

1.3 An Evolutionary Perspective

This field is replete with theories about the evolution of species. The basis of the evolution is based on DNA changes in the species which can happen spontaneously or due to environmental exposure. There is no doubt that evolution is influenced by three aspects that feed into each other: mutations, survival stress, and environment. Mutations can result in a positive or a negative new character for a single organism. The key to classifying a new character as positive or negative is marked by its intrinsic property to help this organism survive and being able to create offspring. Moreover, and to close the circuit, this ability will, indeed, depend on the environment in which the organism inhabits.

There are arguments supporting the fact that the first bacteria appeared on the Earth about 3.800 million years back during the Precambrian time. These organisms possess the ability to interact with the environment in a very rudimentary but also practical way. Thus, even though they are unicellular organisms, this is considered to be the very first known nervous system. These prokaryotes respond directly to the surrounding environment (different organisms, nutritive substances, toxins) through many chemo- and mechanoreceptors located within the plasma membrane—such a communication with the surrounding habitat can be taken as a very simplistic "predecessor" of the nervous system (Table 1.1).

A big jump into evolution took place when prokaryotes suffered a symbiotic association, which derived into the eukaryote organisms [1], characterized by unicellular organization, such as paramecium or euglena. These organisms exhibit a much more profound control of behavior than prokaryotes and possess a complex cytoskeletal system. The response to the surrounding environment is driven by flagella, cilia, or pseudopodia. Mainly Ca^{2+} and K^{+} channels are found, facilitating the release of forward and backward movement. Many protozoans harbor specialized receptors, enabling them to react to various stimuli [2].

In the nineteenth century, Ernst Haeckel proposed a theory that facilitated the understanding of some of these changes in the different species over millions of years. Thus, in his work "General Morphology" of 1866, Haeckel proposed that the developing organism, in its growth from the egg to the adult stage, "recapitulates" the history of the race [3].

Mutations can occur in the DNA of a cell at any time during its replication. Obviously, only those mutations that occur during embryonic development will be inherited in subsequent organisms. When these mutations occur in the early periods of development, it is most likely that their gestation will be interrupted due to the impossibility of normal healthy development. However, those mutations that occur in the intermediate periods of embryogenesis and allow the normal development of the organism may be passed onto subsequent generations. From a strict point of view of the law of probability, it is most common that a mutation does not bring a positive ability to the organisms that undergo it. Therefore, only a minimum percentage of these will have the potential to change the intrinsic aspects of a species and therefore be inherited in the following generations [4].

Therefore, let us imagine a random mutation occurring during the intermediate stages of embryogenesis of an organism in question, and endowing this species with a morphological and/or functional change relevant enough to give rise to even a distinct subclass. The successive organisms derived from this new subclass will share the characteristics of the subclass from which they come in the first phases of embryogenesis, being different only in those phases from the moment in which the mutation appeared and began to take effect (Fig. 1.2).

This statement is summarized as "ontogeny recapitulates phylogeny." Thanks to this concept, the study of the embryological development of different species made it possible to understand and, in some way, visualize the macroscopic changes undergone by these organisms in their different evolutionary stages.

Table 1.1 Illustrative examples demonstrating evolutionary tendency among different species

Type of nervous system	Examples	Description
Predecessor of the nervous system	Procaryotes, simple eukaryotes	Intercell interaction via ion transportation
Diffuse	Cnidaria	The most primitive nervous system formed by interconnected neurons scattered irregularly throughout the body
Ganglial	Flatworm, Mollusca	Neurons grouped into nerve bands and ganglia
Ladder-like	Annelida, Arthropoda	Each body part contains a few ganglia, communicating with each other through nerve connections
Hollow tube	Chordata	The nervous system known in humans comprises peripheral nerves, white, and gray matter

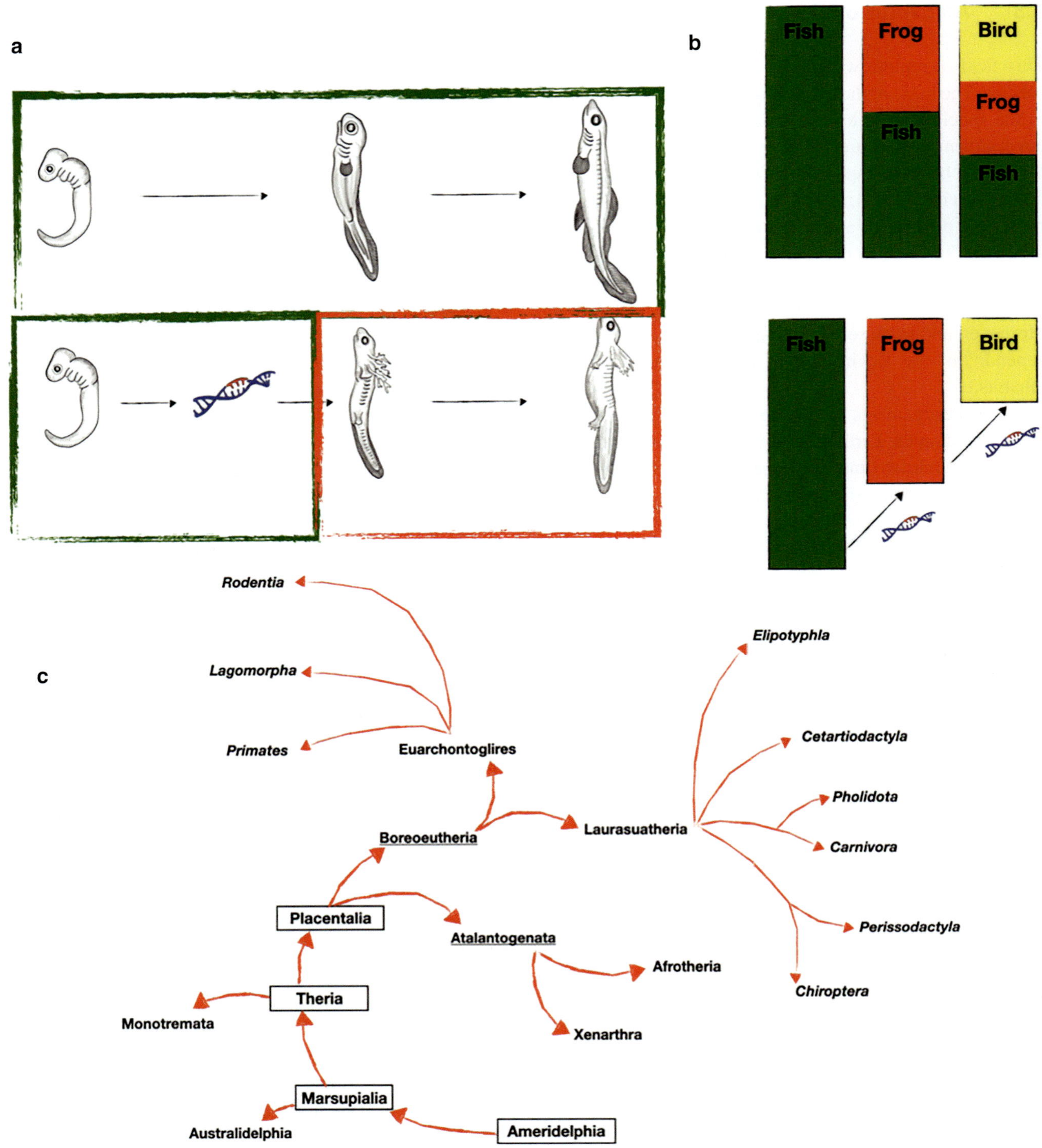

Fig. 1.2 (**a**) Example of a positive mutation affecting a fish embryo in the intermediate embryogenesis and causing the development of limbs instead of fins. The subsequent amphibian embryos will share the first stages as fish. (**b**) Developmental changes recapitulating evolutive changes. (**c**) Evolution *tree* in mammalian family. The idea that species change over time, give rise to new species, and share a common ancestor is closely connected to environmental changes, thus the extent of adaptability to these changes. Therefore, rather than being linear with the "most adaptable" species on the top of such a *tree*, the trend is more radial, specific to individual evolution branches and its species

1.4 Embryology of the Human CNS

All vertebrates, regardless of evolutionary subdivisions, share the same morphometric cornerstones of their CNS, and as it has just been stated, these structures' specific formation and final shape during the embryological development, somehow summarize this phylogenetic trip.

The basics are formed by a neural plate comprising thickened ectoderm at the beginning of the third week of development. During the fourth week the neural tube forms three primary brain vesicles at its cephalic end, namely, prosencephalon (forebrain), mesencephalon (middle brain), and rhombencephalon (hindbrain). Of note are developing two major bends: (1) cervical flexure, forming a border between the hindbrain and spinal cord, and (2) cephalic flexure, giving a formation of the midbrain region. Consequently, there are important modifications happening during the fifth week of development. The former is the division of the prosencephalon into two major parts including the telencephalon (also known as the terminal brain) and diencephalon (interbrain). The rhombencephalon also gives rise to two regions: metencephalon and myelencephalon. These embryonic structures form the main parts into which the adult CNS will be subdivided (Fig. 1.1).

Therefore, as early as in the fifth week, we can distinguish the following structures in the individual sections of the brain: a pontine flexure, just above the myelencephalon; a narrowing that separates the rhombencephalon and the mesencephalon; the lamina terminalis just above it (central part); a formation of the eye sacs in the diencephalon; and two lateral arches, the bases of the cerebral hemispheres in the telencephalon.

1.5 Evolution of the Telencephalon: What Makes Us Humans

At the embryological age of 8 weeks, both telencephalon sides grow caudally and warp in a rostral and ventral direction. This is classically described as a C-shaped shift, considering the occipital pole moving from cranial to dorsal, and the temporal pole from dorsal to ventrolateral as it must dodge the voluminous diencephalic–mesencephalic junction. This capricious movement is called the telencephalic flexure, and its anterolateral expression will hide the insular lobe into the depth of the Sylvian fissure around the 13th gestational week. This internalization of the future insular cortex will give rise to the fronto-orbital, frontoparietal, and temporal opercula, which is literally translated as "cover" from Latin. This opercularization process starts at the 20th week. Nevertheless, the lateral fissure will not be fully closed till the end of the first year.

All these movements that take place during the embryological development are supposed to summarize the phylogenetic changes of the CNS of different species as per Haeckel's recapitulation theory [5, 6]. Indeed, by pure observation of the different contemporary species, as well as analyzing fossils and endocasts, morphologically, the CNS have evolved from simple longitudinally oriented, to spherical, and finally flexed ones in primates [7].

During the development of the telencephalon, two essential parts are distinguished: the cerebral substance and the *pallium*. The former is eventually divided into deep gray and white matter. The latter, *pallium* (translated as *cloak* from Latin), refers to the gray matter ensheathing the surface of the vertebrates' cerebrum. Although its development is not entirely understood with many established theories along with its study, *pallium* has been historically divided into several parts: paleopallium (the oldest *pallium*), archipallium (superiority of this *pallium*—over the oldest one), subpallium (underlying *pallium*), and neopallium (the newest *pallium*). This concept of differentiating pallium into evolutionary older and younger parts has been also analogously interpreted by the concept of dividing pallium into medial, dorsal, lateral, and ventral (Fig. 1.3).

The paleopallium, the oldest part (in phylogenetic terms) of the *pallium*, developed in organisms 500–440 million years ago, consists of 1–3 layers of neurons into the telencephalon (missing two granular layers II and IV of the neocortex) [8]. These areas are described as the rhinencephalon, or primitive olfactory lobe, and include the olfactory bulb, tract, tubercle, lateral and medial striae, anterior olfactory nucleus, parts of the amygdaloid complex, and parts of the prepiriform cortex.

The paleopallium shows a strong functional and morphological relationship with the archipallium, which is indeed the next evolutive step appearing into the telencephalon development, together forming the structure named allocortex, which totally comprises 10% of the human brain. The archipallium develops from the paleopallium and acquires 1–2 extra layers of neurons, adding a needed complexity to the old rhinencephalon. This newly developed *pallium* appeared around 400 million years back, coinciding with the appearance of animals leaving the oceans. It is mainly composed of the primordial hippocampal-piriform formation, comprising the olfactory cortex and three cortical layers in the hippocampus in the human brain. Besides its indisputable role in olfaction, it facilitates free simple and directed memory, harboring memory cells with both excitatory and unmodifiable inhibitory synapses [9]. In humans, archipallium is present in the subcallosal area, fornix, mammillary bodies, cingulate gyrus, anterior part of the superior insular lobe, parahippocampal gyrus, dentate nucleus, or uncus. The most developed archipallial areas are also involved in the detection of threatening situations, showing a relevant role in the search for survival.

Almost 300 million years ago, the highly developed amphibia needed to perfectly coordinate movements with four limbs in their survival mission. Therefore, a compact group of neurons appeared ventrally into the telencephalon. This subpallium will finally migrate to the ventral area of the hemisphere and form the basal ganglia. The subpallium encompasses large parts of the basal ganglia, such as caudate nucleus, substantia innominata, striatum, or globus pallidus.

All these changes are adding complexity to the brains of those organisms that are accumulating all of them. But the highest evolutive change in the humans' evolutive branch

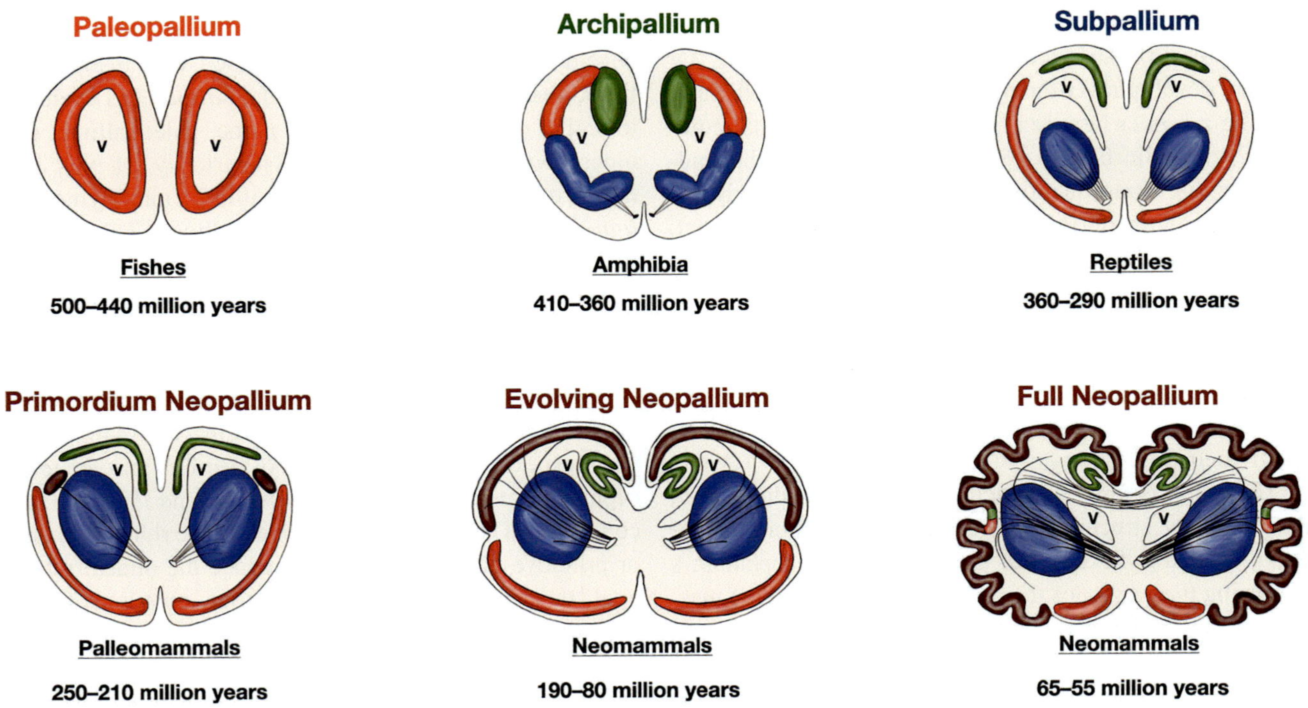

Fig. 1.3 The *pallium* concept and the evolution of *pallium* parts during the evolutionary development of more specialized organisms are illustrated. Abbreviation "V" represents the ventricular system; it is clear how the shape of the ventricle developed, related to the flexures of surrounding pallidal parts

happened around 250–210 million years ago when the first mammals appeared on the Earth. These paleomammals developed a very complex pallial area with up to six layers of neurons highly specialized and interconnected. This primordial neopallium appeared into the lateral convexity and started to expand into the brain surface of the different mammalian branches till pushing the paleo- and archipallial areas to the basal and medial aspects of the brain. The first primates are thought to have evolved prior to the Paleocene epoch, possibly as far as 90 million years back, during the Late Cretaceous period. These primates are also known as neomammals as they suffered a tremendous growth of the neopallium practically covering 90% of their brain surface. The paleo- and archipallial areas occupy a central position in their brains, receiving the name of *limbic* and *paralimbic* areas, which refer to its mediobasal location into the *hemisphere medial edge*. Therefore, the neopallium represents the most developed layers of neurons covering the human brain.

1.6 Brain Species Differences

1.6.1 Brain Size

It is generally accepted that the size of the brain correlates with intelligence and the level of cognitive capabilities [10, 11]. However, rather than explaining this relationship by a linear correlation, this fact is much more complex. The absolute brain size seems to be a good predictor of a higher level of cognition within primates. This is supported by the fact that in hominids, it has tripled since the Pliocene age: the average brain size of Australopithecus was approximately 450 cm^3, compared to 1345 cm^3 of the brain in *Homo sapiens* [9]. Supportive is also the fact that evolutionary, all closely related monkeys have smaller brains compared to humans [2]. On the contrary, large mammals having much larger brains (i.e., the sperm whale), do not scale in terms of cognitive capabilities as high as *Homo sapiens*. It has been shown that the absolute size of the brain does not increase across taxa uniformly in a linear pattern, thus reflecting the developmental patterns in evolution, being rather radial than linear (Fig. 1.1). The process of an increase in brain (or neocortex) size relative to body size, i.e., encephalization, can be quantitatively evaluated using the *encephalization quotient* [12].

1.6.2 Number of Neurons

Brain size as a predictor of cognitive level or intelligence has its limitations and is rather inadequate. This led to more detailed research dedicated to the estimation of the number, and the distribution of neurons in the CNS. *Homo sapiens*' CNS possess approximately 8.6 billion neurons, with the majority located in the brain (brainstem, cerebellum, and prosencephalon), while only around 20 million neurons are situated in the spinal cord [2]. The only organisms with a higher number of neurons in the brain are large aquatic placental marine mammals, namely, Risso's dolphin, short- and long-finned pilot whales, and finally the killer whale with over 43 billion cortical neurons [13–15]. Interestingly, Risso's dolphin has the highest number of neocortical neurons of all studied mammals so far, also possessing twice as many as humans [14]. When comparing the total number of neurons in the CNS, the only mammal foregoing *Homo sapiens* is the African elephant (20.57 billion neurons) [16]. Therefore, the quantitative explanation of a human brain's unique capacities may not be simply in its size (both absolute and relative), but also not in its number of neurons (Fig. 1.4). Instead, more nuanced components such as the diversity of neuronal types, macromolecular specifics, and more profound patterns of neuronal circuits are key to understanding the uniqueness of humankind.

1.6.3 Complexity of Neural Circuits

Connectome, a well-known term coined in many medical textbooks, refers to all neuronal cell types and their specific synaptic connections in neuronal tissue [21]. The exact number of synapses in the human CNS is approximated to be from several hundred trillion to over a quadrillion [21]. The expansion of the cerebral cortex particularly in human brains seems to be closely linked to the structural reorganization of the connectome, considering gyral anatomy changes in the brain of *Homo sapiens*. Although there is no definite evidence on human-specific neural circuits, changes in the morphology and abundance of both excitatory and inhibitory neurons have been identified [22, 23]. For example, human pyramidal neurons are larger, have more complex dendritic arborization, and have a higher density of spines compared to excitatory projection neurons of different species [23]. Such functional and morphological modification, specific to

a

	Homo sapiens	Chimpanzee	Dog	Goose	Crocodile	Frog
Brain size	1,300 g	368 g	115 g	13 g	5 g	0.3 g
Number of neurons	21 billion	7 billion	2.8 billion	482 million	80 million	16 million

b

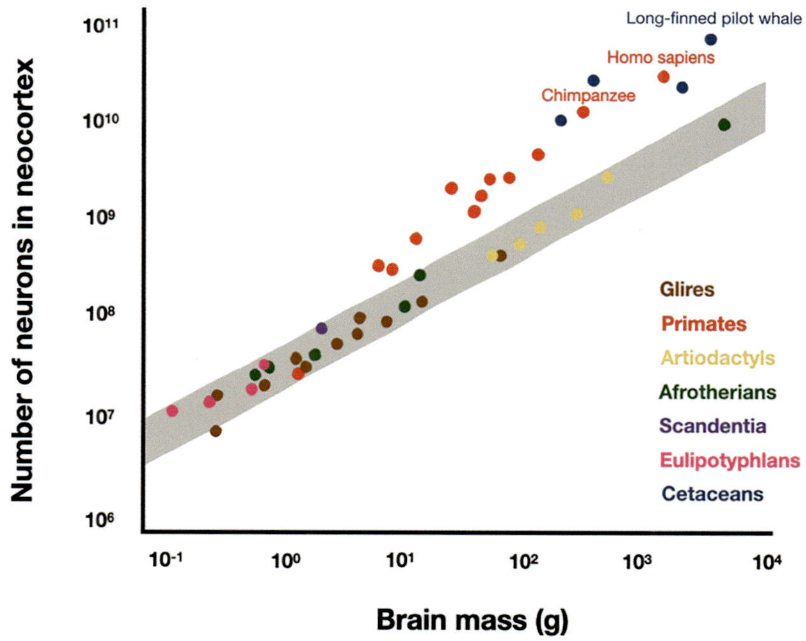

Fig. 1.4 (**a**) Comparison of the two first quantitative features, i.e., brain size of, number of neurons in the brain in several species distributed among taxa [2, 17–20]. (**b**) Graphical demonstration of how variable the correlation between the number of neurons and brain mass can be among species. (Adapted from Sousa et al. [2])

humankind, is suggestive of an increased potential for integrative connectivity. Other reported changes between human and nonhuman species brains were found in the regional localization and abundance of certain subgroups of inhibitory neurons from various neuromodulatory systems of the neocortex, biochemical properties of neural and glial cells, increased complexity of astrocyte behavior and morphology, or even differences in gene expression and signaling pathways—likely reflecting the improvement of glial role in synaptic modulation, thus possibly cognition level in the human brain.

The complexity and connectivity in the human brain during its evolution tend to increase in the number of cortical local short-range connections (such as short U-shaped fibers) [24]. These fibers connect neighboring areas, most likely involved in the generation of higher gyri complexity, one of the most distinguishing features of humans compared to other mammals [25]. Human association areas, especially in the prefrontal cortex, contain a higher fraction of neuropil compared to primary areas. In addition, neurons in association areas of the frontal and temporal cortices are more separated than in other areas of the brain in humans compared to other species [26].

Furthermore, the human brain seems to harbor specific characteristics of myelination in the neopallium (larger axonal lengths with less myelination), being fundamental for the characteristics of axonal conduction velocity [27]. Humans also have reorganized long-range corticopetal, intracortical, and corticofugal projection fibers, especially those associated with the prefrontal and temporal association cortices, i.e., regions involved in higher-order cognitive functions. These structural, molecular, and functional reorganizations likely happened during human evolution independently of brain enlargement.

1.7 Anatomy Highlights

A brief overview of some of the most relevant evolutive steps has provided us with an idea of how areas with different functions have reached their specific locations, from the most rudimentary based on survival to the most complex through which humans communicate and interact with our peers. Although each advance in the study of the evolution of the human brain brings more doubts than evidence, it seems plausible to accept that the human is one of the most developed systems, considering morphological changes that have a basis in molecular and cytoarchitectonic changes. Therefore, and considering perhaps an oversimplifying version, we can extrapolate that what makes our brain so special is the large number of neurons it has, and especially the interconnections that exist between them, especially between those known as association areas.

1.8 Blanket of Neurons: Sulcogyral Patterns

Unlike other more rudimentary and less-evolved CNS, the human brain and that of many primates show a pattern of sulci and gyri that mostly increases the useful surface area of the neo-, archi-, and even paleocortical regions. This capricious distribution of sulci and gyri presents several common characteristics in all encephalons apart from interindividual differences. Being aware of this sulcogyral architecture and how the telencephalon is organized remain fundamental for neurosurgeons dealing with intrinsic pathology. A careful study of the preoperative images as a part of the surgical planning should provide valuable information about the surface anatomical landmarks to decide the resection limits in the case of tumors (Fig. 1.5).

Fig. 1.5 Individual pallial parts in relation to their "age," i.e., time of development, during evolution in a human brain

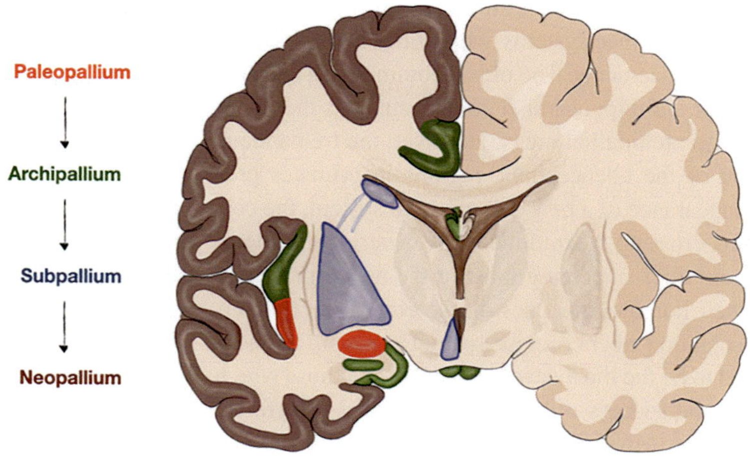

Paleopallium

1–3 layers of neurons

- Olfactory bulb
- Olfactory tract
- Olfactory tuberle
- Intermediate and mediate striae
- Anterior olfactory nucleus
- Parts of amygdaloid complex
- Parts of prepiriform cortex

Archipallium

3–5 layers of neurons

- Hippocampal rudiment
- Subcallosal area
- Body and columns of the fornix
- Mammillary bodies
- Cingulate gyrus
- Parahippocampal gyrus
- Dendate gyrus
- Uncus
- Superior part of the anterior insula

Subpallium

Ganglia

- Putamen
- Substantia innominata
- Striatum
- Globus pallidus
- Caudate nucleus

Neopallium

6 layers of neurons

- Neocortex

1.9 Cerebral Substance: White Matter Networks

The cerebral substance is defined as the underlying material to the cortical layers of neurons in the human brain. It is mainly composed of white matter fibers and also some gray matter structures such as the basal ganglia, thalamus, and other deep nuclei.

The white matter networks are classified as association, commissural, and projection fibers. Association fibers connect cortical regions into the same hemisphere, commissural ones cross the midline and connect the same or different areas on the opposite hemisphere, while the projection fibers (ascending or descending) connect the cortical neurons with deeper structures such as the basal ganglia, thalamus, cerebellum, brainstem, and the spinal cord.

Understanding the general organization patterns as well their relative location into the cerebral substance and direction of the fibers remain of great importance when approaching a single intrinsic brain lesion. Preoperative images such as anatomic sequences, functional MRI, and DTI tractography are currently in vogue, but nothing should surpass the anatomical study and understanding of how these structures have been developed.

1.9.1 Shape of the Fibers

The morphologic shift the human brain has suffered till getting its actual shape can provide us with an understanding of how the underlying fibers are arranged. Projection fibers have not changed too much during evolution as they are connecting areas which have kept a similar relative location among them. Commissural ones are special as there are three main commissural fiber systems with completely different evolutive commencements. The oldest commissural systems are the anterior and hippocampal commissures as they connect paleo- and archipallial areas. However, the tremendous development of the telencephalic regions in primates provoked the appearance of a new commissural system, the corpus callosum, crossing the midline while covering the lateral ventricles, and interconnecting most of the neocortical neurons with their homologous ones on the contralateral hemisphere. Association fibers require special mention, as these are supposed to be the most developed ones, and those which contribute to the success of the human brain as a perfectly organized device. Moreover, the tensional forces generated during phylogenetic development, and summarized in embryological development, have caused some of them to acquire particularly complex forms. Thus, although most of them are longitudinally oriented (superior longitudinal, inferior fronto-occipital, middle longitudinal, etc.), those connecting areas "affected" by the telencephalic flexure (arcuate, uncinate, cingulum, fornix, etc.) will show a curved or C-shaped skeleton, parallel to the development of the lateral ventricles.

The telencephalic flexure is responsible for the capricious shape of the central core, where the corona radiata occupies a central position with projection ascending and descending, as well as commissural fibers from the corpus callosum. Laterally and medially to these fibers, the telencephalic and limbic association systems are respectively represented. The diencephalon literally protrudes inside the telencephalon, while the temporal pole is directed anterolaterally dodging the same. The temporal horn of the lateral ventricle is pulled by the same force. Considering this flexion movement, all the association fibers related to the temporal lobe will suffer a rotational movement, providing their characteristic curved shape (Fig. 1.6).

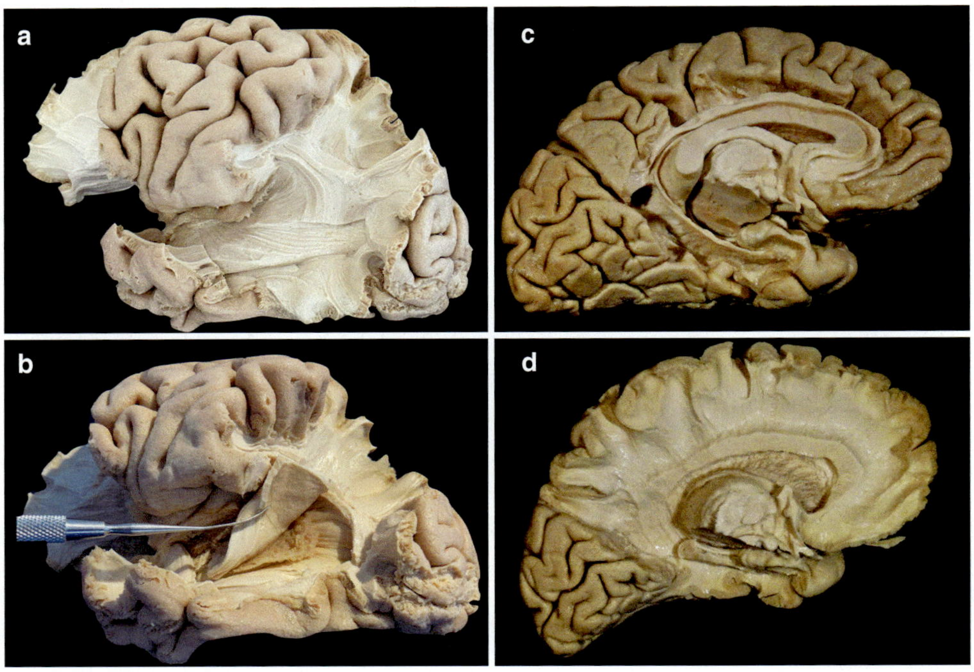

Fig. 1.6 Human brain specimen dissected from lateral to medial and medial to lateral. (**a**) The posterior frontal, central, and occipital lobes, as well as the temporal pole are preserved. The parietal hot spot is shown as a crossing fibers area where the curved arcuate covers the inferior fronto-occipital fascicle. (**b**) The atrium and temporal horn wall is opened showing its intrinsic anatomy from the lateral aspect. The hippocampal tail is covered by the choroid plexus. (**c**) The cingulum bundle is dissected. See how it is turning around the splenium of the corpus callosum and follows mediobasally as the parahippocampal fibers, which run into the temporal lobe, lateral to the mesencephalon. (**d**) The cingulum, parahippocampal and callosal fibers have been removed. The caudate nucleus gray matter has been scrapped away, and the hippocampus cut on its posterior body. The corona radiata medial aspect is exposed

1.10 Surgical Perspective and Future Directions

Illustrative case 1: Intrinsic tumors (Fig. 1.7)
Illustrative case 2: Epilepsy (Fig. 1.8)
Illustrative case 3: Blindness (Fig. 1.9)

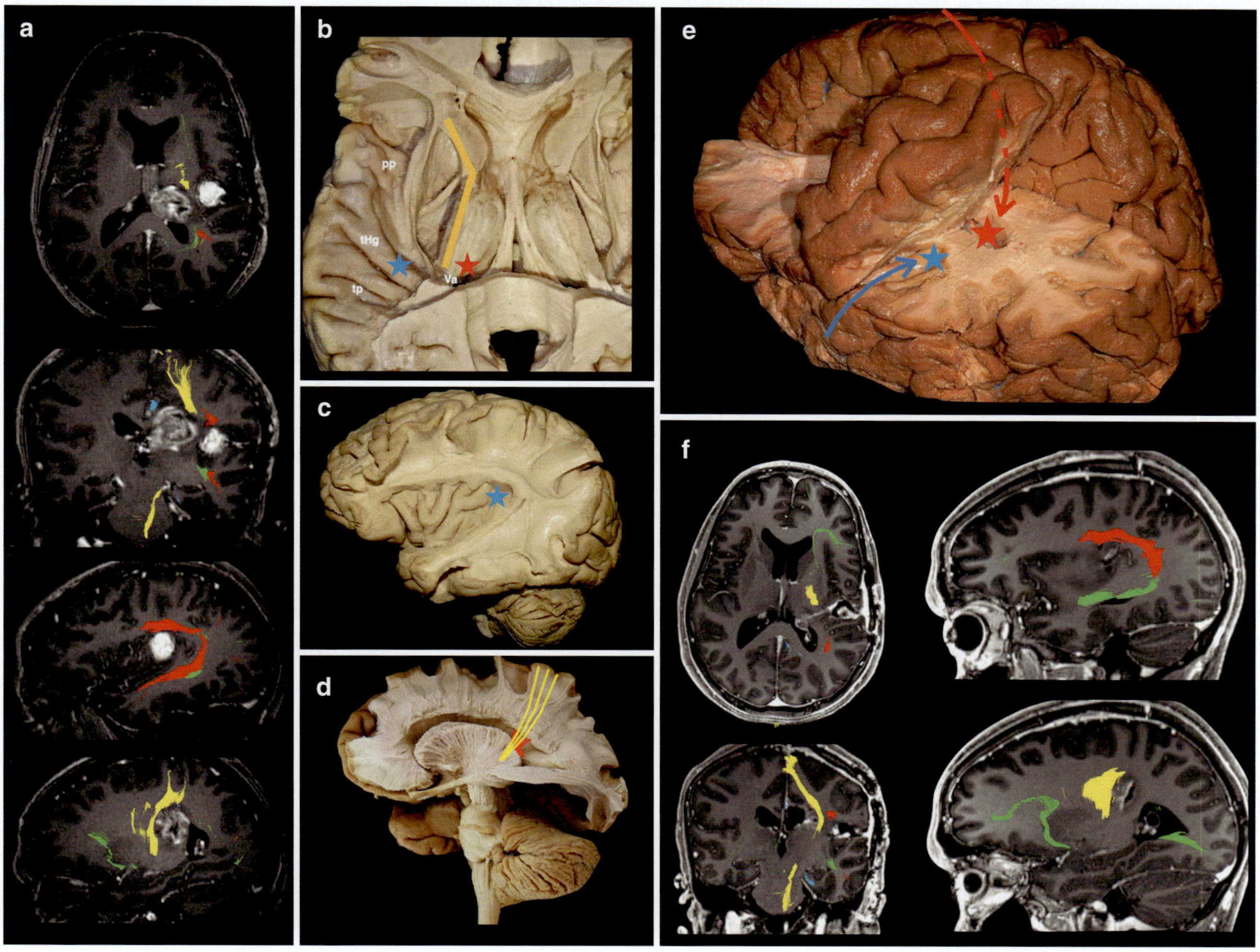

Fig. 1.7 A 35 years-old male presented with progressive mild right side hemiparesis. He was operated on 8 years before due to an insular pleomorphic xanthoastrocytoma. MRI shows a recurrence of the insular tumor with a new more medially located mass. DTI tractography shows how both lesions are separated by the corona radiata motor descending corticospinal fibers and superior thalamic radiations. The insular lesion is located lateral to the ventricular atrium and the inferior fronto-occipital fascicle and optic radiations, and in close relationship with the curve of the arcuate fascicle. The medial lesion seems to arise into the posterior thalamic surface where the ascending projection fibers to the primary somatosensory area arise. The preoperative DTI tractography was helpful in this case to clarify the location of the involved relevant white matter tracts. A combined transylvian and anterior interhemispheric approach was decided in two stages, with the aim of respecting as much as possible the healthy structures around the tumor. (**a**) Axial, coronal and sagittal preoperative T1-weighted MR images after embedding the DTI tractography reconstructions for the corticospinal (*yellow*), arcuate (*red*), inferior fronto-occipital (*green*) and cingulum (*blue*) fibers, related to a pair of tumors located in the posterior insular area and superior surface of the posterior thalamus. (**b–d**) Superior and lateral views of different human brain dissections where the location of the lateral (*blue star*) and medial (*red star*) tumors are shown in close relationship to the corticospinal tract (*yellow*). (**e**) Left human brain hemisphere dissection showing the central lobe and the location of both lesions (*blue and red stars*) and the representation of the surgical trajectories through the transylvian (*blue arrow*) and interhemispheric (*red arrow*) approaches. (**f**) Axial, coronal and sagittal postoperative T1-weighted MR images after embedding the DTI tractography reconstructions for the corticospinal (*yellow*), arcuate (*red*) and inferior fronto-occipital (*green*) fibers, related to the tumors removal

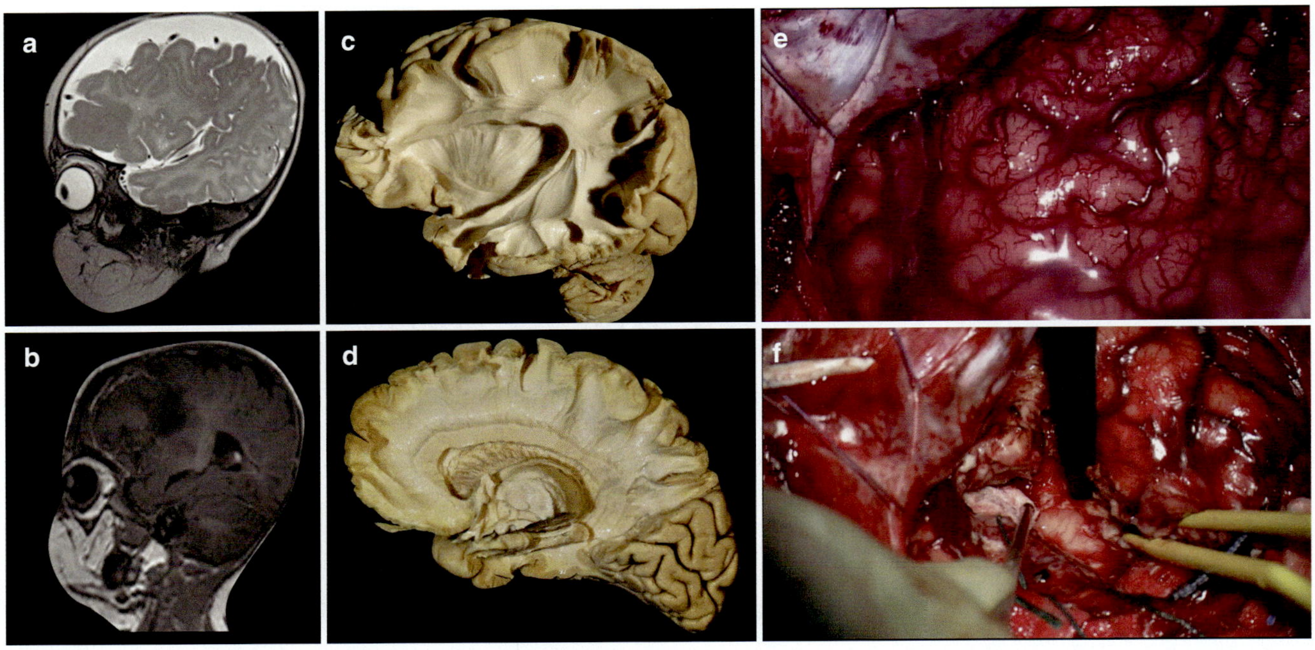

Fig. 1.8 A 5-month-old baby presented with drug-resistant epilepsy. He was presenting more than 25 seizures per day despite high doses of antiepileptic drugs. A large frontal lobe cortical dysplasia was found to be the main source of seizures. Due to the clinical presentation and age of the patient, an anterior quadrantotomy was offered and performed. This procedure requires an adequate understanding of the fibers connecting this area of the brain. The U fibers, superior longitudinal (SLF), uncinate, inferior fronto-occipital (IFOF) fasciculi, anterior limb of the internal capsule, anterior cingulum, anterior callosum, as well as the anterior limb of the anterior commissure, must be disconnected. To carry out this procedure, identifying the frontal horn of the lateral ventricle is of great importance. (**a**) Sagittal preoperative T2-weighted MR image showing a large frontal lobe lesion anterior to the pars opercularis. (**b**) Sagittal postoperative T1-weighted MR image after the anterior quadrantotomy. (**c**) Lateral dissection of a human brain specimen showing the superior longitudinal fasciculus, arcuate, inferior fronto-occipital fasciculus as well as the external capsule. (**d**) Medial dissection of a human brain specimen showing the corpus callosum, anterior limb of the internal capsule—thalamocortical projections, and the anterior commissure. (**e**, **f**) Pre- and post-disconnection intraoperative views

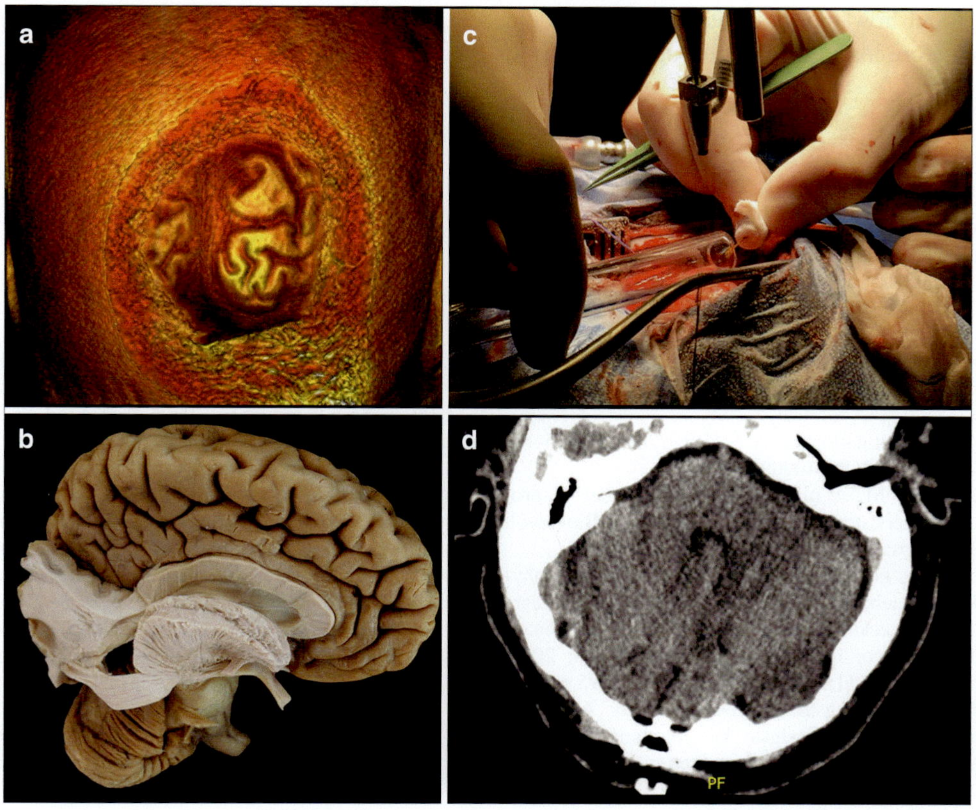

Fig. 1.9 A 61-year-old blind volunteer joined a clinical trial to implant a microelectrode array into the primary visual area. The optic radiations travel from the lateral geniculate nucleus toward the occipital lobe. On its way, they are found in close proximity to the lateral wall and roof of the temporal horn and atrium. A microelectrode array was placed on the occipital pole representing communication with the neocortical layer IV, which is highly developed in the visual occipital cortex. (**a**) 3D reconstruction from the blind volunteer brain MRI showing the right occipital pole. (**b**) Human brain specimen showing the internal capsule and its projection to the occipital lobe through the optic radiations. (**c**) Intraoperative view after tunneling the microelectrode array toward the occipital pole area. (**d**) Postoperative CT scan showing the electrode attached to the occipital pole

1.11 Surgical Pearls

- The human brain has adopted a specific shape due to the strong evolutionary needs and mutations that have happened during millions of years.
- The curved shape of some of the medial and lateral associative white matter tracts is the result of a complex evolutionary process.
- Understanding the evolutionary changes will allow us to understand how these white matter tracts are arranged and connect highly eloquent areas of the human brain.
- The study of the cortical and subcortical anatomy of the human brain from this phylogenetic perspective should allow us to better understand the behavior of some intrinsic lesions, and consequently optimize our surgical planning.

1.12 Conclusion

Understanding how our brain has been built over millions of years is a challenging task. But the effort to do so is worth it, as it may help us to understand some of its intrinsic and morphological features. Understanding the cortical surface (sulcogyral organization) and underlying cerebral substance (white matter fibers architecture) is the primary tool with which we count to optimize the surgical treatment of certain CNS diseases. This anatomical knowledge can help us to respect the healthy and eloquent neurons and fibers around a given glioma. Understanding the 3D orientation of the subcortical fibers, as well as their cortical connections, is mandatory to alleviate epileptic patients in the case of hemispheric disconnections. But even a deep understanding of the highly developed and functional associative areas and how they are interconnected could even drive us to restore lost functions through brain–compute interfaces.

References

1. Cooper G. The cell: a molecular approach. 2nd ed. Sunderland: Sinauer Associates; 2000. The origin and evolution of cells. https://www.ncbi.nlm.nih.gov/books/NBK9841/
2. Sousa AMM, Meyer KA, Santpere G, Gulden FO, Sestan N. Evolution of the human nervous system function, structure, and development. Cell. 2017;170:226–47. https://doi.org/10.1016/j.cell.2017.06.036.
3. MacBride EW. The recapitulation theory. Sci Prog. 2009;92:355–69. https://doi.org/10.1177/003685040909200310.
4. Tamuri AU, Dos Reis M. A mutation-selection model of protein evolution under persistent positive selection. Mol Biol Evol. 2022;39:msab309. https://doi.org/10.1093/molbev/msab309.
5. Hall B. Ontogeny does not recapitulate phylogeny, it creates phylogeny. In: Robert J, Richards ED, editors. A review of the tragic sense of life: Ernst Haeckel and the struggle over evolutionary thought, vol. 13. Chicago: University of Chicago Press; 2011. p. 401–4. https://doi.org/10.1111/j.1525-142X.2011.00495.x.
6. E.H. Ueber die Entwicklungstheorie Darwin's. In: Amtlicher Berich ueber die acht und dreissigste Versammlung Deutscher Naturforscher und aerzte. in Stettin. Stettin: Hessenland's Buchdruckerei; 1864. p. 17–30.
7. González-López P, Cossu G, Thomas CM, Marston JS, Gómez C, Pralong E, Messerer M, Daniel RT. Cadaveric white matter dissection study of the telencephalic flexure: surgical implications. Front Neurol. 2022;13:757757. https://doi.org/10.3389/fneur.2022.757757.
8. Butler AB, Reiner A, Karten HJ. Evolution of the amniote pallium and the origins of mammalian neocortex. Ann N Y Acad Sci. 2011;1225:14–27. https://doi.org/10.1111/j.1749-6632.2011.06006.x.
9. Valverde GF. The rhinencephalon; anatomical study & functional significance. Rev Clin Esp. 1959;73:91–106.
10. Marr D. Simple memory: a theory for achicortex. Philos Trans R Soc B Biol Sci. 1971;262:23–81. https://doi.org/10.1098/rstb.1971.0078.
11. Jerison H. Evolution of the brain and intelligence. New York: Academic; 1973.
12. Carel TZ, Bshary R, Sandra. A farewell to the encephalization quotient: a new brain size measure for comparative primate cognition. Brain Behav Evol. 2021;96(1):1–12. https://doi.org/10.1159/000517013.
13. Holloway R. Evolution of the human brain. In: Lock A, Peters CR, editors. Handbook of human symbolic evolution. Oxford: Oxford University Press; 1996. p. 74–125.
14. Ridgway SH, Brownson RH, Van Alstyne KR, Hauser RA. Higher neuron densities in the cerebral cortex and larger cerebellums may limit dive times of delphinids compared to deep-diving toothed whales. PLoS One. 2019;14:e0226206. https://doi.org/10.1371/journal.pone.0226206.
15. Mortensen HS, Pakkenberg B, Dam M, Dietz R, Sonne C, Mikkelsen B, Eriksen N. Quantitative relationships in delphinid neocortex. Front Neuroanat. 2014;8:132. https://doi.org/10.3389/fnana.2014.00132.
16. Herculano-Houzel S. Longevity and sexual maturity vary across species with number of cortical neurons, and humans are no exception. J Comp Neurol. 2019;527:1689–705. https://doi.org/10.1002/cne.24564.
17. Roth G, Dicke U. Evolution of the brain and intelligence in primates. Prog Brain Res. 2012;195:413–30. https://doi.org/10.1016/b978-0-444-53860-4.00020-9.
18. Collins CE, Turner EC, Sawyer EK, Reed JL, Young NA, Flaherty DK, Kaas JH. Cortical cell and neuron density estimates in one chimpanzee hemisphere. Proc Natl Acad Sci. 2016;113:740–5. https://doi.org/10.1073/pnas.1524208113.
19. Jardim-Messeder D, Lambert K, Noctor S, Pestana FM, de Castro Leal ME, Bertelsen MF, Alagaili AN, Mohammad OB, Manger PR, Herculano-Houzel S. Dogs have the most neurons, though not the largest brain: trade-off between body mass and number of neurons in the cerebral cortex of large carnivoran species. Front Neuroanat. 2017;11:118. https://doi.org/10.3389/fnana.2017.00118.
20. Kverkova K, Marhounova L, Polonyiova A, Kocourek M, Zhang Y, Olkowicz S, Strakova B. The evolution of brain neuron numbers in amniotes. PNAS. 2022;119:e2121624119. https://doi.org/10.1073/pnas.2121624119.
21. Silbereis JC, Pochareddy S, Zhu Y, Li M, Sestan N. The cellular and molecular landscapes of the developing human central nervous system. Neuron. 2016;89:248–68. https://doi.org/10.1016/j.neuron.2015.12.008.
22. Bianchi S, Stimpson CD, Bauernfeind AL, Schapiro SJ, Baze WB, McArthur MJ, Bronson E, Hopkins WD, Semendeferi K, Jacobs B, et al. Dendritic morphology of pyramidal neurons in the chimpanzee neocortex: regional specializations and comparison to humans. Cereb Cortex. 2013;23:2429–36. https://doi.org/10.1093/cercor/bhs239.
23. Elston GN, Benavides-Piccione R, Elston A, Manger PR, Defelipe J. Pyramidal cells in prefrontal cortex of primates: marked differences in neuronal structure among species. Front Neuroanat. 2011;5:2. https://doi.org/10.3389/fnana.2011.00002.
24. DeFelipe J. The evolution of the brain, the human nature of cortical circuits, and intellectual creativity. Front Neuroanat. 2011;5:29. https://doi.org/10.3389/fnana.2011.00029.
25. Van Essen DC. A tension-based theory of morphogenesis and compact wiring in the central nervous system. Nature. 1997;385:313–8. https://doi.org/10.1038/385313a0.
26. Schenker NM, Buxhoeveden DP, Blackmon WL, Amunts K, Zilles K, Semendeferi K. A comparative quantitative analysis of cytoarchitecture and minicolumnar organization in Broca's area in humans and great apes. J Comp Neurol. 2008;510:117–28. https://doi.org/10.1002/cne.21792.
27. Glasser MF, Goyal MS, Preuss TM, Raichle ME, Van Essen DC. Trends and properties of human cerebral cortex: correlations with cortical myelin content. NeuroImage. 2014;93(Pt 2):165–75. https://doi.org/10.1016/j.neuroimage.2013.03.060.

2 Brain Plasticity and Neurosurgery

Igor L. Maldonado

Learning Objectives

- Gain a comprehensive understanding of the dynamic changes occurring throughout the human brain's lifespan, driven by intricate interactions between both nearby and distant cortical areas
- Recognize the significance of lesions with gradual onset, appreciating how they promote the potential for medium-term recuperation after brain surgery when critical white matter pathways are preserved
- Familiarize yourself with the primary clinical scenarios within neurosurgical practice that prominently showcase brain plasticity, elucidating its manifestation and relevance
- Acquire insights into the fundamental mechanisms that underlie the network-based recruitment of alternative cortical sites employed when a region cannot fulfill a given task

2.1 Introduction

Always on the frontier of knowledge, cerebral neuroplasticity is a subject that integrates topics from different disciplines, such as neurosurgery, neuroanatomy, and rehabilitation. In this chapter, the spotlight falls on the tangible manifestations of neuroplasticity in clinical settings, followed by an exploration of surgical instances that underline their significance for both present and future neuroanatomy and neurosurgery endeavors. These topics, in a state of constant evolution, encapsulate the essence of modern neuroscience.

I. L. Maldonado (✉)
CHRU de Tours, Tours, France

Laboratoire d'Anatomie, Faculté de Médecine de Tours, Tours, France

INSERM, Imaging Brain & Neuropsychiatry iBraiN U1253, University of Tours, Tours, France
e-mail: limamaldonado@univ-tours.fr

The brain's remarkable ability to adapt and reorganize itself in response to injury and changing demands exhibits itself in diverse ways across distinct timeframes. Recognizing and harnessing these plasticity mechanisms are pivotal in the realm of therapeutic reasoning and surgical intervention, influencing techniques, resection aggressiveness, eligibility for future reoperation, and enhancing patient quality of life.

2.2 History and Background

The study of functional neuroanatomy is intrinsically connected to its historical development, shaped by key discoveries and breakthroughs. It is important for the reader to realize that crucial scientific discoveries that occurred in the nineteenth and twentieth centuries led to an understanding of brain function based on the association of brain topography to neural functions. For illustrative purposes, we will mention the main ones.

In the early nineteenth century, phrenology emerged from investigations by the German anatomist Franz Gall into relationships between cerebral areas, skull shape, and specific functions [1]. This approach, though considered a "pseudoscience" by today's standards, was a significant endeavor in the pursuit of comprehending brain localization [2].

In the mid-nineteenth century, the famous case of Phineas Gage emerged as one of the earliest well-documented works on frontal syndrome [3]. Following a traumatic brain injury when a metal bar penetrated both of his frontal lobes during an explosion, Gage's unique situation, documented by his physician, John Martyn Harlow, provided a meticulous description of a frontal syndrome. It offered comprehensive documentation on how damage to a specific brain area could lead to specific behavioral changes while preserving other cognitive capacities.

In 1861, Paul Broca, a prominent French physician at the Kremlin Bicêtre Hospital, made a groundbreaking discovery by examining the brain of a patient who had lost the ability

I. N. Kanaan, V. Beneš (eds.), *Neuroanatomy Guidance to Successful Neurosurgical Interventions*,
https://doi.org/10.1007/978-3-031-59838-8_2

to speak [4]. Broca identified a specific brain area that he claimed to be responsible for speech production. This region, known as Broca's area, remains a critical landmark in functional neuroanatomy.

In the early twentieth century, Wilder Penfield, a Canadian surgeon, revolutionized the study of functional neuroanatomy by using cortical electrical stimulation during epilepsy and tumor surgery. By stimulating specific areas of the cortex, Penfield identified functional brain regions and mapped their connections. The first cortical maps were published in 1937, and their methodological principles continue to be employed in contemporary neurosurgical procedures [5–9].

These seminal discoveries led to a shift in the physiological conception of the brain, transitioning from a generalized structure paradigm to a localization model comprising a set of specialized areas. This model posits that distinct brain regions serve specific functions, laying the foundation for research in functional neuroanatomy. These discoveries have transformed our understanding of the brain's organization.

The so-called *localizationist* model of the distribution of mental functions (cortical localization theory) has many virtues. It has been widely taught to medical students as an aid to thinking in terms of topographic diagnosis, i.e., deduction of the location of the lesion through the analysis of the patient's signs and symptoms. However, this model has several drawbacks [10, 11]. The first of these is its rigidity, failing to consider the fact that cortical areas, especially those that are not related to primary executive functions, may change during life. The second is that it underestimates the heterogeneity and dynamics of the participating cortical sites.

Within the landscape of modern neuroscience, intraoperative subcortical mapping has enabled the unveiling of the anatomical correlates of the connections underlying functional brain circuits. The increasing use of awake direct cerebral stimulation to map the cerebral white matter has ushered in a transformative era, enabling surgeons and researchers to chart neural circuits. This approach facilitates the exploration of the functional attributes of key fasciculi, as the mapping findings can be studied together with the results of neuroimaging and the latest data on white matter anatomy. As detailed further, structural connectivity has been identified as central to the viability of brain plasticity mechanisms.

Also propelled by advancements in neuroimaging and neural network analysis, this more recent *connectivist* model underscores the dynamic interactions between dispersed brain areas. This paradigm highlights how complex functions emerge from distributed computations, where communication among brain regions is paramount. The connectivist model is redefining our perception of brain function, offering a holistic framework that accommodates the brain's capacity for adaptability.

2.3 Anatomy Highlights

- In modern neuroanatomy, brain functions should not be regarded as outcomes solely derived from the activity of individual areas; rather, they should be understood as the consequence of the interplay among brain regions.
- Recognizing the critical interconnections among cortical sites is essential for optimizing postoperative functional outcomes. Within this context, the anatomy of white matter tracts and the circuits they comprise became essential topics in neurosurgery education.
- The concept of cortical remapping elucidates how cortical representations can shift in response to parenchymal damage. This process highlights the brain's adaptability in reallocating resources for improved function, emphasizing that the localization of brain functions does not follow a rigid pattern.

2.4 Dynamic Nature of the Human Brain: A Lifelong Journey of Adaptation and Learning

In humans, a task rarely arises solely from the activity of an isolated cortical area. Instead, it is the outcome of the interaction between cortical and subcortical sites, and their dynamic involvement, varying with the context and planning conditions. For instance, in an apparently simple motor task (e.g., kicking a ball), the final action results from a complex and bilateral dialogue involving the motor area and also the premotor cortex, prefrontal cortex, basal ganglia, spinal cord, sensory cortex, and other structures and networks. The nature and degree of participation of each of these elements depend on multiple factors, such as time, light, previous training, fatigue, ground characteristics, and body position. It can, therefore, be said that each action is unique.

A pure localizationist model undervalues the role of connective pathways [12]. As Gall suggested more than a century ago, no matter how specialized a cortical area may be, its functionality becomes futile if it cannot effectively communicate and interact with other areas of the brain. The localizationist model also underestimates cortical redundancy phenomena and compensation capacities. Understanding how the brain works across distributed and interrelated networks opens doors for rehabilitation if the appropriate stimuli are provided. Flexibility, dynamism, connectivity, and redundancy are pillars of human brain plasticity.

Indeed, the human brain undergoes continuous changes from birth until the final minutes of its lifespan. These transformations are closely intertwined with age, learning experiences, and environmental exposures, all of which shape the individual's development and capabilities. Our experiences and exposures significantly influence who we become and what we can achieve. These exposures encompass external environmental factors and how we engage and challenge our brain through new tasks, repetition, and exercises. Each individual's unique journey plays a role in the construction and development of the brain as a result of both stimulation and time.

Different periods emerge throughout life, each characterized by distinct responses and capacities. One acknowledged pivotal stage is childhood, a critical period for learning [13]. During this period, the brain is particularly sensitive to its environment and experiences, which shape cognitive abilities and form the foundation for future learning.

Learning does not cease in adulthood; it takes on a different character. The adult brain is characterized by the development of self-improvement and adaptation skills, which are facilitated by the capacity for metalinguistic and metacognitive processes [14, 15]. These higher-order abilities enable individuals to engage in self-modulation.

2.5 Network Models of Functional Neuroplasticity

When a cortical region can no longer perform a task, the recruitment of other sites always occurs as a network phenomenon. In this line, there are four primary mechanisms of functional neuroplasticity according to the Grafman (2000) classification (Fig. 2.1): homologous area adaptation, cross-modal reassignment, map expansion, and compensatory masquerade [16–18].

In homologous area adaptation, a region in the opposite hemisphere assumes a process previously carried out by the injured area, showing sensitivity to its loss. Moreover, the contralateral recruited region interacts with spared network elements on the same side as the injury, especially for functions with highly somatotopic organization. This observation highlights the physiological significance of heterotopic callosal fibers, which connect nonhomologous areas of the cerebral hemispheres.

Cross-modal reassignment involves a specialized cortical region becoming sensitive to a different input type and increasing its participation in the sensory processing network to which it is functionally and physically connected. Map expansion is facilitated by the presence of numerous short intergyral association fibers ("U fibers"), leading to the enlargement of a functional brain region. Finally, compensatory masquerade occurs when a new process allocation enables a subject to perform a task by circumventing or compensating for the missing process; this involves the recruitment and reorganization of preexisting nonlesioned functional networks.

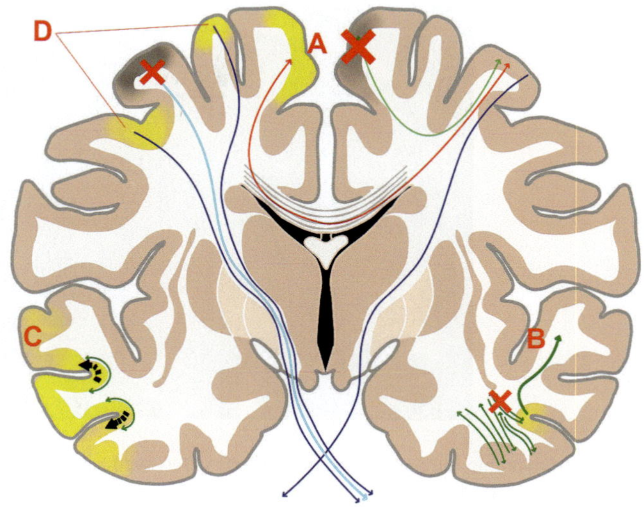

Fig. 2.1 Basic mechanisms of brain plasticity, based on the Grafman classification (2000). (**a**) Homologous Area Adaptation. In this mechanism, a region in the opposite hemisphere takes on a role previously performed by an injured area. An illustrative instance in adults is the recuperation of supplementary motor area syndrome, emphasizing the physiological significance of heterotopic callosal fibers. (**b**) Cross-modal Reassignment. This process involves a specialized cortical region that, following sensory input deprivation, becomes responsive to a different input type and integrates into an alternate functional network it is physically connected. A notable example is the recruitment of the visual cortex for sensory tasks in blind patients. (**c**) Map Expansion. This mechanism entails the enlargement of a cortical map dedicated to a specific information processing function, often due to repeated exposure to a stimulus. (**d**) Compensatory Masquerade. Here, an individual accomplishes a task by circumventing a missing process through the recruitment of pre-existing nonlesioned networks

Simultaneously with the reconfiguration of brain networks, it is postulated that plastic phenomena encompass biochemical and synaptic mechanisms, giving rise to both localized and widespread modifications in functional activities as well as focal volume expansions [19–24].

2.6 Neuroplasticity and Neurosurgery: A Multifaceted Phenomenon with Clinical Implications

Neuroplasticity may manifest clinically in various ways with distinct timeframes (Figs. 2.2, 2.3, 2.4, and 2.5). Short-term plasticity phenomena may involve rapid changes in the functional organization of brain networks in response to aggression. Medium-term plasticity occurs in response, for instance, to slowly evolving pathologies, such as slow-growing tumors. Lastly, long-term changes can be induced by established lesions over time, contingent upon rehabilitation efforts and the presence of adequate stimuli.

Evaluation of the potential for plasticity of a patient who has suffered brain damage or is going to have brain surgery should be part of therapeutic reasoning as this appraisal can influence

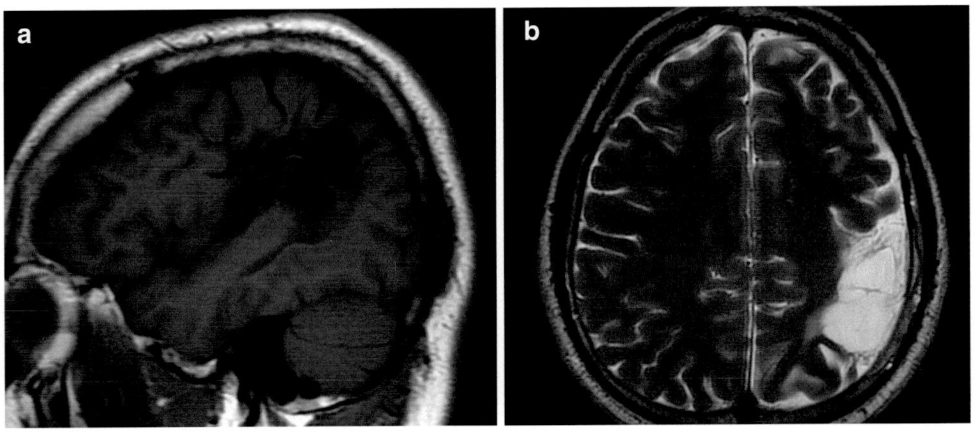

Fig. 2.2 Favorable long-term outcome despite extensive lesion in the left temporo-parietal junction. These images showcase a porencephalic and leptomeningeal lesion in the left temporo-parietal junction, initially identified during childhood due to seizures, likely of congenital origin. Despite requiring anticonvulsant therapy, this 29-year-old male patient enjoyed a normal adolescence and early adulthood. At the neurological examination, distal paresis of the right upper limb (weakly positive Mingazzini test) was observed and no language deficits. (**a**) Sagittal magnetic resonance T1-weighted image. (**b**) Axial T2-weighted image

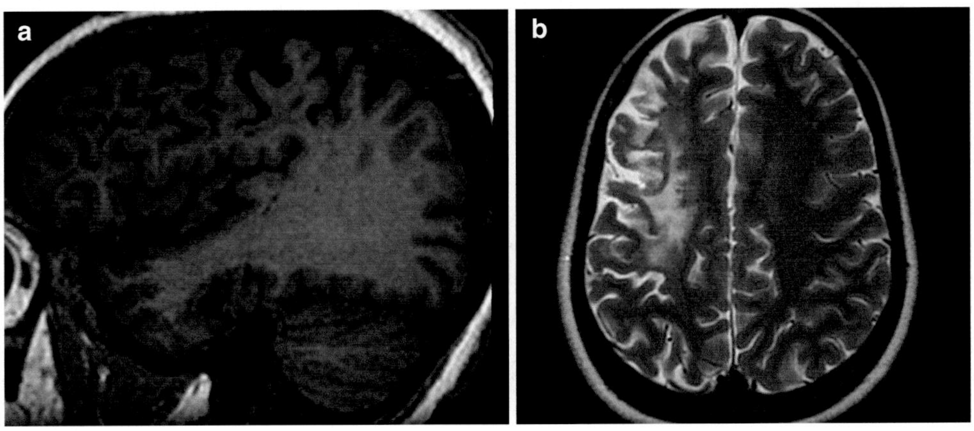

Fig. 2.3 Example of pathological change in handedness. Sagittal T1-weighted (**a**) and axial T2-weighted (**b**) magnetic resonance images show atrophy of vascular origin due to chronic ischemic insult. The patient was a left-handed 14-year-old girl with a history of sickle cell disease. At age six, she was diagnosed with a severe intracranial angiopathy in the right carotid vascular tree. The evolution was marked by the progressive shift in manual preference from left-handedness to right-handedness for writing and daily activities. There was no language impairment

the technique and aggressiveness of resection of intraparenchymal lesions. It also affects eligibility for future rehabilitation techniques or reoperation. In this context, recognizing the potential for compensation and recovery after oncological neurosurgery in eloquent areas, which is notably influenced by the patient's overall health status, becomes crucial for enhancing the patient's quality of life and should be taken into account when deciding on surgical intervention [24].

2.6.1 Short and Mid-Term Manifestations of Brain Plasticity

Short-term plasticity may be observed immediately or minutes after an injury or surgical intervention. Penfield documented phenomena of this nature, reporting modifications to the cortical map during resection procedures [9]. More recent reports in language mapping have shown that areas in the left temporal lobe outside the Wernicke's area (anterior and middle temporal lobe, anterior portions of the inferior longitudinal fasciculus) may respond to electrical stimulation and be resected without permanent post-operative language deficits when the critical long-range connectivity is preserved [25]. Similarly, other authors have reported language network reorganization evident with functional magnetic resonance imaging (fMRI) before and after temporal lobe epilepsy surgery [26].

Medium-term neuroplasticity phenomena may take days to months to manifest following resection or injury to specific brain regions. One of the most illustrative examples is the sur-

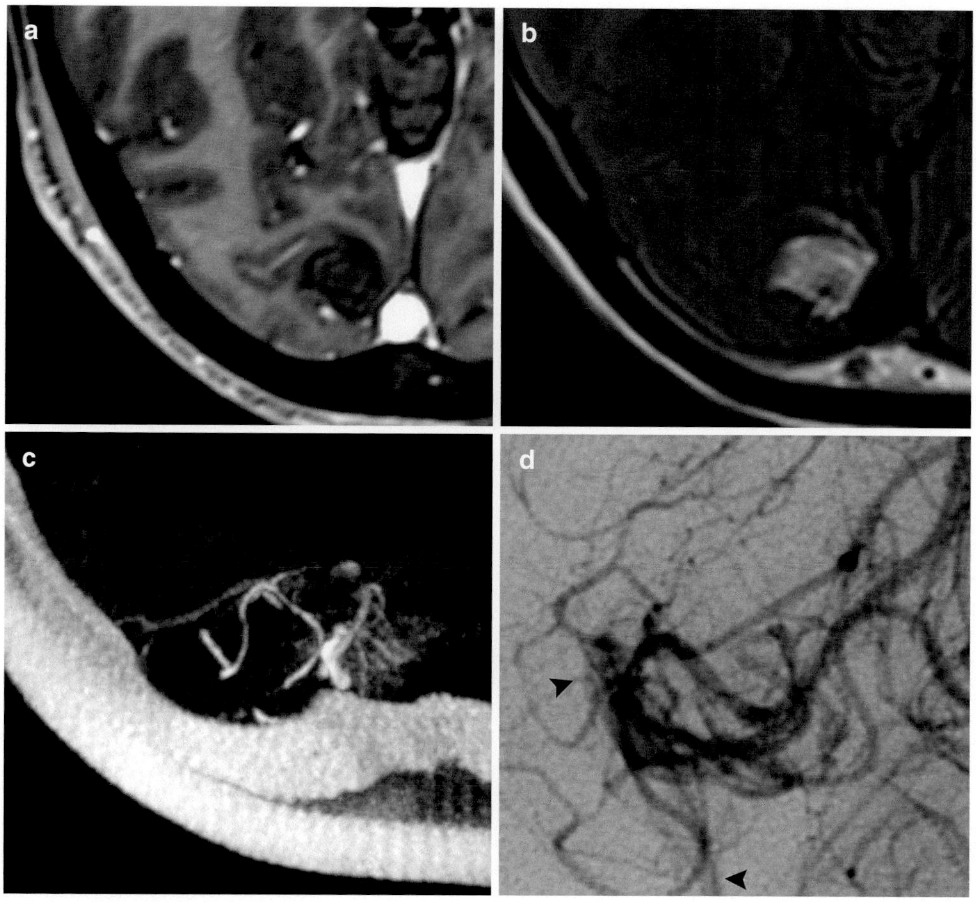

Fig. 2.4 Functional adaptation to paracentral scotoma. At the age of 19, this male previously healthy patient presented with a paracentral visual field defect due to the rupture and subsequent embolization treatment of a pial arteriovenous malformation in the posterior portion of the left lingula. The resulting scotoma proved debilitating, particularly for reading. Over time, the patient experienced progressive and complete recovery in their reading ability, despite the continued presence of the campimetry defect. Two years later, the patient, who initially described the defect as profoundly distracting, reported that he no longer noticed it unless they intentionally concentrated on it. (**a** and **b**) Initial aspect of the small intraparenchymal hematoma on magnetic resonance examination, axial enhanced T1-weighted (**a**) and FLAIR (**b**) images. (**c**) Results of rotational digital subtraction angiography (axial view reconstructed from 3D-DSA native images). (**d**) Lateral view, a 2D subtracted image. Arrowheads: early venous return

gery within the supplementary motor area (SMA) [16, 27]. It is worth noting that, for decades, patients with infiltrating tumors in eloquent areas were not considered for surgical treatment in most centers. Recent advances in functional mapping and a better understanding of neuroplasticity and connectivity mechanisms have challenged this conservative strategy [28].

The SMA is crucial in planning complex movements and is involved in movement initiation, including speech articulation. It is connected to the other frontal lobe areas including the primary motor cortex through locoregional and short-association fibers. The frontal aslant tract connects it with the frontal operculum (including Broca's area) [29]. Therefore, the SMA is functionally closely related to the adjacent motor cortex.

Resection of the SMA typically leads to SMA syndrome, a condition characterized by a kinetic mutism, and a lack of contralateral movement, a condition that may be mistaken for permanent hemiplegia. Nevertheless, in most cases, the SMA syndrome is reversible within days to a few weeks [30]. Remarkably, patients show impressive progress in large steps when they begin to recover. An interesting feature during this recovery is that patients frequently show motor responses to sufficiently intense stimuli (e.g., verbal). This responsiveness indicates that the underlying issue is a movement initiation disorder rather than a defect in the common final motor pathway.

fMRI studies using the blood-oxygen-level-dependent (BOLD) effect provide valuable insights into SMA plasticity. SMA syndrome is associated with a decrease in interhemispheric connectivity of the motor network hubs [27]. As recovery progresses, a shift in activation occurs, and the SMA on the opposite side starts to engage with the active primary motor cortex [27, 31], likely through commissural connections of the corpus callosum. This well-described recovery mechanism can also occur after acute injuries, such as trauma or stroke [32].

The progressive nature of brain damage appears to be a critical factor influencing neuroplasticity phenomena, afford-

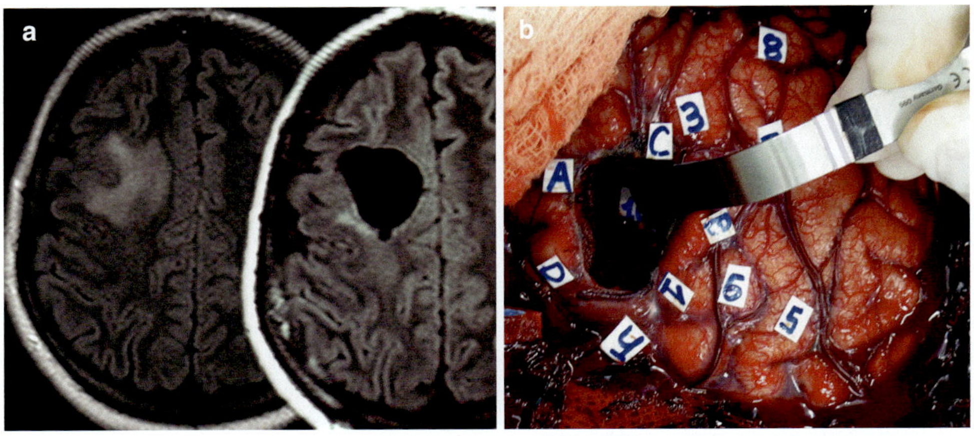

Fig. 2.5 Onset and recovery of motor apraxia after surgery near eloquent areas with preserved structural connectivity. A 44-year-old female patient with a left-sided low-grade glioma (diffuse astrocytoma) invading the pre-motor and motor cortices as well as the underlying white matter underwent an awake functional mapping-assisted surgical intervention. Tumor removal was carried out meticulously, ensuring contact with the primary motor area and safeguarding the corticospinal projection fibers through the aid of cortical and subcortical mapping. At the end of the procedure, the patient did not exhibit motor deficit but, as is customary, she experienced a transient moderate paresis on the second day attributed to the presence of moderate parenchymal edema around the operative cavity, which resolved within the first post-operative week. Following complete reabsorption and motor recovery, apraxia was noted and the patient complained of a noticeable loss of hand dexterity. This issue progressively improved through physical therapy efforts and the patient's dedication to concentrating on and performing her usual tasks over the course of some months. Ultimately, the patient achieved a full recovery of manual ability. (**a**) Axial FLAIR images from the pre- and post-operative magnetic resonance examinations. (**b**) Intraoperative view. (**a–d**) tumor boundaries according to the results of intraoperative ultrasound. Cortical map: 1 and 2 = articulatory troubles, 3 = face (motor), 4 = forearm and hand (motor), 5 = hand (sensory), 8 = tongue (sensory). Subcortical map: 10 = hand and forearm (motor)

ing the brain time for functional reorganization. This opportunity for adaptation is particularly evident in low-grade gliomas such as diffuse astrocytomas, where functional reorganization may occur over time. When such lesions are located in the so-called "dominant" hemisphere for language processing, this mechanism may involve language areas, such as the Broca's area in the frontal operculum, encompassing Brodmann's areas 44 and the posterior part of area 45 in the inferior frontal gyrus. This has allowed the partial or even total resection of this area, previously thought to be indispensable, in selected cases, without losing the ability to speak [33].

Functional mapping also has shown that brain plasticity mechanisms can develop between two sequential surgical procedures with preservation of function and extension of resection. Instances of this phenomenon have been documented in patients who underwent incomplete glioma removal and exhibited favorable outcome several years after the initial procedure [34]. In such scenarios, the prospect of a secondary surgery prior to anaplasia could be contemplated [35–37]. Failure to acknowledge the potential for neuroplasticity as the brain surgeon's ally would be a regrettable oversight.

As is evident from the earlier examples, the brain's capacity for functional reorganization following damage is influenced by several factors. The speed of lesion development plays a crucial role, as does the nature of the affected function. Recovery after injury to primary executive areas, such as the motor, sensory, or visual cortex, is considered more challenging due to these regions' high level of specialization.

Additionally, the composition of the injured area, particularly in terms of the eloquent cortex and white matter elements, also plays a significant role. A cortical area that sustains injury may rely on both short and long association pathways and commissural pathways to recruit cortical centers involved in specific functions. However, if these critical connecting bundles are damaged or lost, the potential for inducing compensation and functional recovery over time seems significantly reduced. This observation supports the prevailing notion that neuroplasticity phenomena are more pronounced in cases of pure cortical lesions compared to those affecting critical white matter fasciculi. Similarly, language and associative areas exhibit more robust neuroplasticity compared to the primary executive cortex [38–40].

Some authors have proposed probabilistic maps based on the observation of many patients, studying each brain region's recovery potential. The plastic potential of long association and projection pathways was shown to be lower than that of neocortical areas, except for primary unimodal areas and a few other centers [38–40]. As such, these findings emphasize varying degrees of plasticity between neocortical and white matter structures, which could hold significance for surgical strategies or the management of intraparenchymal pathological processes.

In light of the fact that neuroplasticity biomarkers are not taken into account in current neuro-oncological guidelines, Ng and Duffau have recently introduced an integrated workflow that involves therapeutic adaptations, contingent upon oncological characteristics and the evolution of clinical-radiological markers of neuroplasticity on low-grade glioma patients [24]. Within this workflow, the presence of a high neuroplastic potential and an estimated residual volume below 10 ± 5 ml assume pivotal importance when considering surgical intervention. The evaluation of plastic potential is grounded in the so-called 3Dtm system, which assesses three critical parameters: the tumor's location and the extent of white matter (3D) infiltration; the tumor's growth velocity ("*t*"); and the occurrence of meta-plastic cognitive compensation ("m") [24, 41].

2.6.2 Manifestations of Brain Plasticity in the Long Term

Long-term neuroplasticity phenomena may manifest in various tasks but are particularly evident in the rehabilitation of motor, sensory, and apraxia deficits. This form of plasticity involves harnessing the brain's adaptive mechanisms to overcome initial deficits strategically. It also enables the brain to cope with the absence of certain sensory inputs. Therefore, long-term plasticity plays a vital role in teaching the brain to adapt and compensate for deficits in motor and sensory functions [42].

An essential aspect of long-term plasticity is our ability to relearn complex tasks previously impaired due to injury or neurological conditions. In such instances, the brain undergoes extensive reorganization and remodeling, forming new neural pathways to facilitate task execution. Rehabilitation specialists, neurologists, physiotherapists, occupational therapists, and other healthcare professionals collaborate to tailor personalized rehabilitation strategies to optimize recovery. This ever-expanding knowledge base has the potential to revolutionize patient care and drive advancements in the field of neurorehabilitation.

Another crucial aspect that has progressively demanded attention is the domain of human–machine interaction. The human brain exhibits remarkable adaptability as it can learn to interpret electrical signals introduced into cortical sensory areas; provided certain essential conditions are met to adapt motor signals [43–45]. These conditions include sufficient time for learning, repeated stimulus exposure, feedback integration through alternative sensory modalities, and incorporation into broader cognitive tasks.

Moreover, with appropriate training, individuals can acquire the ability to transmit necessary signals through the motor cortex to achieve desired outcomes; this has immense potential for the treatment of motor deficits [46]. The brain's capacity to interpret and respond to electrical signals from invasive microelectrodes opens new avenues for developing brain–computer interfaces and neuroprosthetic devices. These innovations can potentially revolutionize the lives of individuals with sensory and motor impairments in the future, offering them enhanced control over external devices and empowering their daily activities.

2.7 Surgical Pearls

- Traditional "localizationist" model, while useful for diagnosis, overlooks the brain's adaptability and dynamic interactions. Compensation capacities are underestimated by a strict localizationist view.
- Considering the varying degrees of plasticity observed in the neocortex and white matter, a firm grasp of anatomical understanding and a meticulous approach to preserving crucial eloquent fiber pathways become imperative. This approach is vital to ensure optimal postoperative evolution following interventions in eloquent brain regions.
- Surgical intervention in SMA can lead to SMA syndrome, characterized by contralateral akinesia and mutism. SMA syndrome is typically reversible within weeks, showcasing the brain's capacity for recovery. Functional MRI studies reveal SMA plasticity during recovery, where the opposite-side SMA engages with the active primary motor cortex.
- Neuroplasticity exhibits its effects after short, medium, and long-term intervals. The impact of plasticity becomes noteworthy in individuals who have undergone prior surgeries for slow-growing lesions such as low-grade gliomas as it can significantly influence their suitability for reoperation.

2.8 Discussion

Neuroplasticity mechanisms and connectivity are thus inseparable concepts. A research endeavor important for neurosurgery is the identification of critical white matter structures for these processes. An assessment of the individual anatomy and how it relates to the pathology has become a necessary part of operative planning as it can influence the surgical approach and optimize the potential for further recovery [47]. This observation underscores the significance of understanding functional white matter anatomy and the circuits they integrate, making them vital subjects in modern neuroanatomy and neurosurgery education.

Throughout the various types of compensation mechanisms presented, the limited knowledge of neuroplasticity prompts us to study the brain from a more contemporary and likely more accurate perspective. The recognition of compensation processes highlights the crucial role of connectivity mechanisms. Indeed, neuroplasticity phenomena would be difficult to fathom without considering brain functions as the outcome of organized and distributed networks.

As neuroplasticity continues to captivate the attention of scientific exploration, these findings will undoubtedly contribute to a more thorough comprehension of the brain's functional organization and its role in cognition and behavior. Understanding neuroplasticity mechanisms may radically transform our understanding of surgical aggressiveness in the near future, leading to personalized treatment strategies tailored to individual patients' needs.

2.9 Conclusion

As our understanding of brain connectivity deepens, it highlights the intricacies of neural circuitry responsible for various cognitive functions. Recognizing these interconnections emphasizes the dynamic nature of the brain, constantly adapting and reorganizing to optimize performance and compensate for damage or functional deficits. This knowledge fuels advancements in medical research, brain–computer interfaces, and rehabilitative strategies, ultimately enhancing our ability to comprehend and harness the brain's remarkable plasticity. A profound understanding of these mechanisms equips neurosurgeons with the ability to incorporate the brain's remarkable plasticity into their surgical deliberations.

References

1. Gall FJ, Spurzheim JG. Anatomie et physiologie du système nerveux en général, et du cerveau en particulier, avec des observations sur la possibilité de reconnoître plusieurs dispositions intellectuelles et morales de l'homme et des animaux, par la configuration de leurs têtes. Paris: Schoell; 1810.
2. Zola-Morgan S. Localization of brain function: the legacy of Franz Joseph Gall (1758–1828). Annu Rev Neurosci. 1995;18:359–83.
3. Harlow JM. Recovery from the passage of an iron bar through the head. Boston: David Clapp & Son; 1869.
4. Broca P. Remarques sur le siége de la faculté du langage articulé, suivies d'une observation d'aphémie (perte de la parole). Bull Soc Anat. 1861;6:330–57.
5. Penfield W. Combined regional and general anesthesia for craniotomy and cortical exploration. I. Neurosurgical considerations. Curr Res Anesth Analg. 1954;33:145–55.
6. Penfield W, Boldrey E. Somatic motor and sensory representation in the cerebral cortex of man as studied by electrical stimulation. Brain. 1937;60:389–443.
7. Penfield W, Perot P. The brain's record of auditory and visual experience. A final summary and discussion. Brain. 1963;86:595–696.
8. Penfield W, Rasmussen T. The cerebral cortex of man. A clinical study of localization of function. New York: The Macmillan Company; 1950.
9. Penfield W, Welch K. Instability of response to stimulation of the sensorimotor cortex of man. J Physiol. 1949;109:358–65, illust.
10. Ffytche DH, Catani M. Beyond localization: from hodology to function. Philos Trans R Soc Lond B Biol Sci. 2005;360:767–79.
11. Rorden C, Karnath HO. Using human brain lesions to infer function: a relic from a past era in the fMRI age? Nat Rev Neurosci. 2004;5:813–9.
12. Duffau H. Hodotopy, neuroplasticity and diffuse gliomas. Neurochirurgie. 2017;63:259–65.
13. Sylva K. Critical periods in childhood learning. Br Med Bull. 1997;53:185–97.
14. Siegel ALM, Castel AD. Age-related differences in metacognition for memory capacity and selectivity. Memory. 2019;27:1236–49.
15. Overhoff H, Ko YH, Feuerriegel D, Fink GR, Stahl J, Weiss PH, et al. Neural correlates of metacognition across the adult lifespan. Neurobiol Aging. 2021;108:34–46.
16. Ho A, Khan Y, Fischberg G, Mahato D. Clinical application of brain plasticity in neurosurgery. World Neurosurg. 2021;146:31–9.
17. Maldonado IL. Letter to the editor regarding "Clinical application of brain plasticity in neurosurgery". World Neurosurg. 2021;146:399.
18. Grafman J. Conceptualizing functional neuroplasticity. J Commun Disord. 2000;33:345–56.
19. Almairac F, Deverdun J, Cochereau J, Coget A, Lemaitre A-L, Moritz-Gasser S, et al. Homotopic redistribution of functional connectivity in insula-centered diffuse low-grade glioma. NeuroImage Clin. 2021;29:102571.
20. Ng S, Deverdun J, Lemaitre A-L, Giampiccolo D, Bars EL, Moritz-Gasser S, et al. Precuneal gliomas promote behaviorally relevant remodeling of the functional connectome. J Neurosurg. 2022;138:1531–41.
21. Almairac F, Duffau H, Herbet G. Contralesional macrostructural plasticity of the insular cortex in patients with glioma: a VBM study. Neurology. 2018;91:e1902–8.
22. Yuan T, Ying J, Zuo Z, Gui S, Gao Z, Li G, et al. Structural plasticity of the bilateral hippocampus in glioma patients. Aging (Albany NY). 2020;12:10259–74.
23. Pasquini L, Jenabi M, Peck KK, Holodny AI. Language reorganization in patients with left-hemispheric gliomas is associated with increased cortical volume in language-related areas and in the default mode network. Cortex. 2022;157:245–55.
24. Ng S, Duffau H. Brain plasticity profiling as a key support to therapeutic decision-making in low-grade glioma oncological strategies. Cancers (Basel). 2023;15:3698.
25. Mandonnet E, Nouet A, Gatignol P, Capelle L, Duffau H. Does the left inferior longitudinal fasciculus play a role in language? A brain stimulation study. Brain. 2007;130:623–9.
26. Foesleitner O, Sigl B, Schmidbauer V, Nenning K-H, Pataraia E, Bartha-Doering L, et al. Language network reorganization before and after temporal lobe epilepsy surgery. J Neurosurg. 2021;134:1694–702.
27. Pinson H, Van Lerbeirghe J, Vanhauwaert D, Van Damme O, Hallaert G, Kalala J-P. The supplementary motor area syndrome: a neurosurgical review. Neurosurg Rev. 2022;45:81–90.
28. Duffau H, Taillandier L. New concepts in the management of diffuse low-grade glioma: proposal of a multistage and individualized therapeutic approach. Neuro Oncol. 2014;17:332.
29. La Corte E, Eldahaby D, Greco E, Aquino D, Bertolini G, Levi V, et al. The frontal aslant tract: a systematic review for neurosurgical applications. Front Neurol. 2021;12:641586.
30. Oda K, Yamaguchi F, Enomoto H, Higuchi T, Morita A. Prediction of recovery from supplementary motor area syndrome after brain tumor surgery: preoperative diffusion tensor tractography analysis and postoperative neurological clinical course. Neurosurg Focus. 2018;44:E3.
31. Krainik A, Duffau H, Capelle L, Cornu P, Boch A-L, Mangin J-F, et al. Role of the healthy hemisphere in recovery after resection of the supplementary motor area. Neurology. 2004;62:1323–32.

32. Boakye M. Implications of neuroplasticity for neurosurgeons. Surg Neurol. 2009;71:5–10.
33. Plaza M, Gatignol P, Leroy M, Duffau H. Speaking without Broca's area after tumor resection. Neurocase. 2009;15:294–310.
34. Robles SG, Gatignol P, Lehéricy S, Duffau H. Long-term brain plasticity allowing a multistage surgical approach to World Health Organization Grade II gliomas in eloquent areas: report of 2 cases. JNS. 2008;109:615–24.
35. Duffau H. Diffuse low-grade gliomas and neuroplasticity. Diagn Interv Imaging. 2014;95:945–55.
36. Gil Robles S, Duffau H. Surgical management of World Health Organization Grade II gliomas in eloquent areas: the necessity of preserving a margin around functional structures. Neurosurg Focus. 2010;28:E8.
37. Duffau H. Repeated awake surgical resection(s) for recurrent diffuse low-grade gliomas: why, when, and how to reoperate? Front Oncol. 2022;12:947933.
38. Bourdillon P, Apra C, Guénot M, Duffau H. Similarities and differences in neuroplasticity mechanisms between brain gliomas and nonlesional epilepsy. Epilepsia. 2017;58:2038–47.
39. Honey CJ, Sporns O. Dynamical consequences of lesions in cortical networks. Hum Brain Mapp. 2008;29:802–9.
40. Ius T, Angelini E, Thiebaut De Schotten M, Mandonnet E, Duffau H. Evidence for potentials and limitations of brain plasticity using an atlas of functional resectability of WHO grade II gliomas: Towards a "minimal common brain". NeuroImage. 2011;56:992–1000.
41. Duffau H. Dynamic interplay between lower-grade glioma instability and brain metaplasticity: proposal of an original model to guide the therapeutic strategy. Cancers. 2021;13:4759.
42. Maier M, Ballester BR, Verschure PFMJ. Principles of neurorehabilitation after stroke based on motor learning and brain plasticity mechanisms. Front Syst Neurosci. 2019;13:74.
43. Patil PG, Turner DA. The development of brain-machine interface neuroprosthetic devices. Neurotherapeutics. 2008;5:137–46.
44. Orsborn AL, Carmena JM. Creating new functional circuits for action via brain-machine interfaces. Front Hum Neurosci [Internet]. 2013 [cited 2023 Jul 17];7. http://journal.frontiersin.org/article/10.3389/fncom.2013.00157/abstract
45. Caria A, Da Rocha JLD, Gallitto G, Birbaumer N, Sitaram R, Murguialday AR. Brain–machine interface induced morpho-functional remodeling of the neural motor system in severe chronic stroke. Neurotherapeutics. 2020;17:635–50.
46. Martin AR, Sankar T, Lipsman N, Lozano AM. Brain-machine interfaces for motor control: a guide for neuroscience clinicians. Can J Neurol Sci. 2012;39:11–22.
47. Duffau H. The death of localizationism: the concepts of functional connectome and neuroplasticity deciphered by awake mapping, and their implications for best care of brain-damaged patients. Rev Neurol. 2021;177:1093–103.

History of Neurosurgical Anatomy

3

Ramesh Nair, Hadie Adams, and Puneet Sharma

Learning Objectives
- Understand the historical progression of knowledge and concepts in neuroanatomy from ancient times to the present day
- Appreciate the key figures and scientific advancements that have shaped our current understanding of neuroanatomy
- Recognise the impact of technological advancements on the field of neuroanatomy and their significance in shaping modern neuroscience

3.1 Introduction

The intricate structure and function of the human brain have captivated scholars for millennia. Unlocking the mysteries of this organ has been a gradual process, requiring diligent observation, experimentation, and innovation across disciplines and eras. This chapter will chart the key developments in the history of neuroanatomical studies, beginning in ancient times and continuing through the present day. Exploring how early insights emerged in ancient cultures like Egypt, Greece, and Rome, laying the foundations for future progress. The profound impact of pivotal figures including Herophilus, Galen, and Vesalius will be discussed. It will also examine how religious prohibitions stalled mediaeval anatomical progress, while the Renaissance and Enlightenment periods brought about a resurgence. The nineteenth century's critical theories and microscopic advances will be covered, followed by the landmark discoveries of the twentieth and twenty-first centuries enabled by new technologies. By tracing this long arc of discovery across the centuries, we will appreciate how each generation of anatomists built upon the work of their forebearers through meticulous observation, experimentation, and innovation. This spirit of scientific curiosity continues to propel the field forward today.

3.2 Anatomical Highlights

- The ancient Egyptians made early descriptions of the brain and spinal cord, recognising the cerebrospinal fluid and the spinal cord's role in transmitting neural signals.
- Herophilus provided detailed descriptions of the meninges, ventricles, cerebellum, cerebrum, grey/white matter, and cranial nerves.
- Galen accurately distinguished seven pairs of cranial nerves and their functions through dissection experiments.
- Vesalius created highly detailed and accurate illustrations of brain anatomy and structures like the cerebrum, cerebellum, and ventricles.
- Ramón y Cajal used microscopy techniques to demonstrate neurons as distinct entities and provided insights into the anatomy of brain regions like the cerebellum, retina, and cortex.

3.3 Early Studies of Human Neuro-Anatomy

In the annals of human history, our understanding of the brain and its functions has evolved over millennia. There is evidence to suggest that trepanation had been practised during prehistoric times, and many trephined skulls have been discovered belonging to the neolithic period. The ancient Egyptians believed that the heart held the secrets of thought and emotion, while the brain was merely in charge of the body. Imhotep (c. 2667–c. 2648 BC) was an Egyptian physician and architect credited with designing the first pyramid, and he is believed to be the first to describe the brain and its

R. Nair (✉) · H. Adams · P. Sharma
Imperial College Healthcare NHS Trust, London, UK
e-mail: hadie.adams@nhs.net; puneet.sharma2@nhs.net

I. N. Kanaan, V. Beneš (eds.), *Neuroanatomy Guidance to Successful Neurosurgical Interventions*,
https://doi.org/10.1007/978-3-031-59838-8_3

membranes [1]. The Egyptians were also the first to describe the cerebrospinal fluid and acknowledge the spinal cord's crucial role in transmitting information to the extremities. They also developed surgical procedures like transnasal excerebration to treat brain disorders and aid in mummification [2]. The ancient Mesoamericans also believed that the brain was the primary source of consciousness and thought, and they had a long tradition of trepanation to treat (mental) illness [3]. Their efforts required an understanding of neuro-anatomy, which they probably developed through close observation and experimentation. The fact that skulls with healed bones have been found following trepanation speaks volumes about the sophistication of their medical practices. In addition, the subsequent evidence found of cranioplasty use indicates that these ancient healers were not merely conducting crude experiments but working towards practical solutions to complex medical problems [4].

On the other hand, the ancient Greeks considered the brain the centre of our intelligence, consciousness, sensation, and movement [5]. They described the cerebellum as being responsible for coordination and balance. They were also pioneers in brain surgery and are believed to have employed opium and mandrake for anaesthesia during those procedures. These remarkable insights are probably achieved through intuition, experimentation, and an unwavering commitment to understanding the human brain. Although some of their contributions may have been lost forever, their pioneering efforts are part of human history and curiosity to comprehend and treat the brain with all its complexity, and their contributions are a reminder of this timeless pursuit.

In conversations about the history of neuro-anatomy studies, one cannot overlook the towering figure of Alcmaeon of Croton [6]. His pioneering work in anatomical dissection and invaluable contributions to studying the brain's anatomy have earned him a place of eminence in the annals of scientific history. He was one of the first to suggest that the brain is the seat of human intelligence and made detailed observations about the sensory nerves that transmit impulses to the brain. While his writings have not survived, his ideas were influential in developing ancient Greek and Roman medicine. Many of his insights influenced some of the great ancient Greek and Roman physicians, including Hippocrates, Herophilus, and Galen, who were also advocates of anatomical dissection [7]. Hippocrates is known to have performed dissections to understand the human body better. Still, it was not until the work of later anatomists, such as Herophilus and Galen, that more detailed knowledge of the brain's structure and function began to emerge [8]. Herophilus is credited with performing detailed dissections of the human body, including the brain. His work on the structure and function of the brain was highly influential, and some of their findings remain relevant to this day. Herophilus was the first to suggest that paralysis could be caused by damage to motor nerves. Herophilus was also responsible for the detailed naming of the meninges and ventricles of the brain, distinguishing between the cerebellum and cerebrum and between the grey and white matter of the brain. In addition, Herophilus is credited with identifying and describing the optic, oculomotor, motor division of the trigeminal, facial, vestibulocochlear, and hypoglossal nerves [9, 10]. Galen of Pergamon considered anatomy the basis of medical knowledge and extensively performed dissections and experiments on animals like pigs, sheep, goats, and the Barbary ape [11]. He emphasised the importance of dissection for research purposes and to enhance surgical skills. Galen accurately observed and distinguished seven pairs of cranial nerves and their functions. He conducted vivisection experiments, including the transection of the spinal cord to study the spinal nerves' functions and tying off the recurrent laryngeal nerve to demonstrate that the brain controls the voice [12]. His work on the brain's function was particularly groundbreaking, and his writings influenced the development of neuroscience for centuries to come.

3.4 Impact of Religious and Cultural Beliefs on Anatomy Studies

The study of anatomy, particularly the brain, has been shaped by various religious and cultural factors throughout history, influencing the methods used for studying the brain and the attitudes towards dissections, vivisections, and animal studies [13]. In ancient Egypt, for example, the religious emphasis on the heart led to the brain being considered less important and often discarded during mummification. Ancient Greeks and Romans believed that the brain was the seat of reason and intelligence, which led to a greater focus on its study. Galen dissected animals to study the brain and nervous system, but the dissection of human corpses remained taboo due to the perceived disrespect to the dead. Galen did, however, extensively dissect humans, including mainly gladiators and corpses from public executions [14].

During the early Christian era, the human body was considered sacred, and dissection was generally prohibited [13]. The Christian belief in the resurrection of the body contributed to the reluctance to perform dissections, causing the study of brain anatomy to stagnate during this period. Scholars often relied on the works of earlier Greek and Roman thinkers for their knowledge. In contrast, the Islamic Golden Era, spanning from the eighth to the thirteenth centuries, saw significant advancements in studying the brain. Islamic scholars, such as Ibn Sina (Avicenna) and Al-Razi (Rhazes), embraced human dissection as a means to acquire knowledge, as Islamic teachings emphasised the importance of learning for the betterment of society [15]. Their works, such as Avicenna's "The Canon of Medicine", laid the foundation for later advancements in brain anatomy.

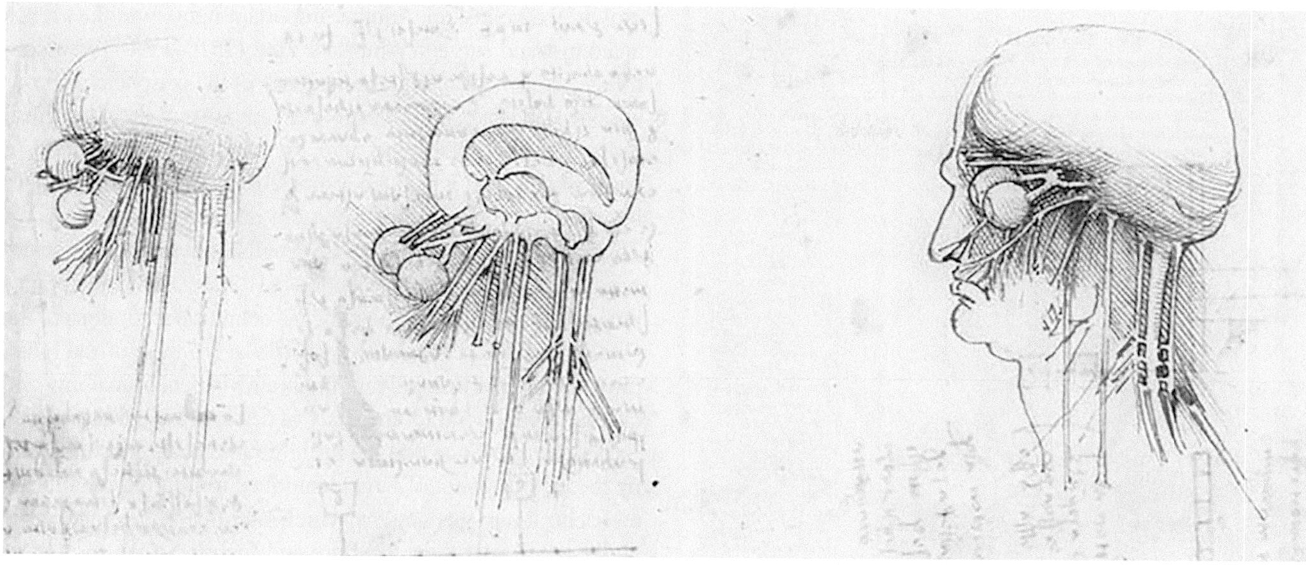

Fig. 3.1 Three drawings done around 1508, Leonardo da Vinci moves from the brain and nerve structures (left) adding the ventricular system (centre), and integrating the superimposed facial and bone structures (right). (From the Royal Collection Trust/© Her Majesty Queen Elizabeth II 2023)

The Catholic Church played a complex role in studying anatomy during mediaeval Europe. While it generally discouraged the dissection of human corpses, exceptions were sometimes made for medical purposes. Scholars such as Mondino de Luzzi conducted public dissections under the auspices of the Church, helping to advance the understanding of human anatomy [13]. The Renaissance, which began in the fourteenth century, marked a shift in attitudes towards the study of anatomy. Influenced by humanist ideals, artists and anatomists like Leonardo da Vinci and Andreas Vesalius began to perform detailed dissections of human cadavers [13]. Leonardo's detailed drawings of the brain (Fig. 3.1) and Vesalius's publication "De Humani Corporis Fabrica" are testaments to the advancements in brain anatomy during this period [16]. The Catholic Church was more lenient during this period, as the greater emphasis on humanism and studying the natural world aligned with their goals of promoting knowledge and understanding.

The Protestant Reformation in the sixteenth century brought about a renewed focus on studying anatomy [13]. Protestant scholars like John Calvin believed that learning about the human body was a way to understand the divine creation better, and this led to a more permissive attitude towards dissection and the study of the brain. In more recent times, the role of religion in the study of brain anatomy has continued to evolve. While religious beliefs no longer significantly hinder the study of the brain, ethical concerns have emerged, particularly in vivisections and animal studies. The rise of secularism and scientific ethics have led to more stringent regulations on the use of animals in research, with an increasing focus on humane treatment and the development of alternative methods.

3.5 Advancements During the Renaissance Era and the Enlightenment

The Renaissance Era (fourteenth to seventeenth centuries) and the Enlightenment (seventeenth to eighteenth centuries) were pivotal periods in science and medicine, marked by significant advancements in anatomical studies of the brain and central nervous system. At the forefront of these studies were groundbreaking figures such as Andreas Vesalius and other notable anatomists and physicians [16]. Their contributions not only revolutionised the understanding of the human brain but also laid the foundation for developing new techniques to study the brain and nervous system further.

Andreas Vesalius (1514–1564) was a Flemish physician and anatomist who played a critical role in the development of modern anatomy during the Renaissance. He is best known for his comprehensive work on human anatomy, De Humani Corporis Fabrica Libri Septem (On the Fabric of the Human Body in Seven Books), published in 1543 (Fig. 3.2). This work represented a significant departure from the prevailing reliance on ancient texts, such as those by the Greek physician Galen, who primarily studied animal anatomy and whose teachings had dominated the field for over a thousand years [16].

Vesalius's work on the brain and nervous system anatomy marked a significant departure from previous knowledge [17]. He was among the first to accurately describe the brain's structures, including the cerebrum, cerebellum, and ventricular system. His descriptions and illustrations of the cranial nerves and spinal cord were also groundbreaking. He is also credited with recognising the existence of the distinct white and grey matter in the brain and spinal cord [17].

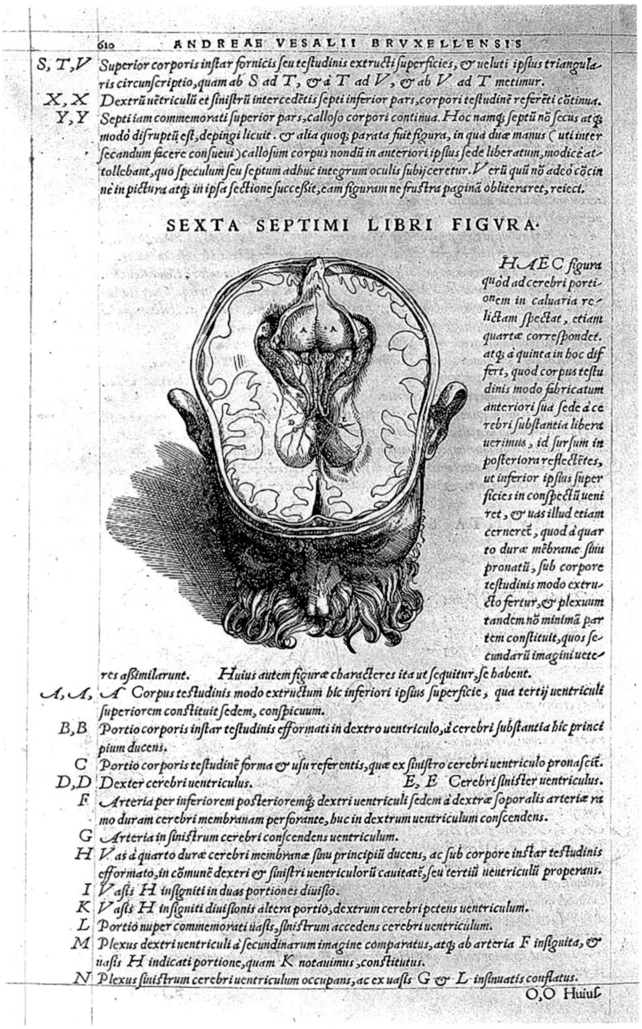

Fig. 3.2 Cross section of the head, by Vesalius. De humani corporis fabrica, Andreas Vesalius. (From the Wellcome Collection, Licence: Attribution 4.0 International (CC BY 4.0))

Vesalius's illustrations in De Humani Corporis Fabrica were highly detailed and accurate, and they profoundly impacted the study of neuroanatomy. Unlike previous anatomical drawings, which were often based on animal dissections and contained numerous inaccuracies, Vesalius's work provided a more accurate representation of human anatomy based on his dissections and observations. His illustrations were of such high quality that they were used as reference material for centuries.

While Vesalius is often considered the father of modern anatomy, several other notable anatomists and physicians made significant contributions to anatomical studies of the brain during the Renaissance and the Enlightenment. One such figure was the Italian anatomist and physician Costanzo Varolio (1543–1575), who made important observations about the structure of the brainstem [18]. Varolio was the first to describe the pons, a crucial part of the brainstem that regulates essential life-sustaining functions such as breathing, sleeping, and heart rate. Varolio's work laid the foundation for understanding the functional anatomy of the brainstem. Another important figure was the French anatomist and surgeon Ambroise Paré (1510–1590), who contributed to understanding the anatomy of the peripheral nervous system [19]. Paré's work on nerve injuries and their treatment provided valuable insights into the structure and function of peripheral nerves, which are responsible for transmitting signals between the central nervous system and the rest of the body. Thomas Willis (1621–1675), an English physician and anatomist, made significant contributions to understanding the brain's blood supply [20]. Most notably, his detailed descriptions of the Circle of Willis, a network of arteries at the base of the brain, provided crucial insights into the brain's vascular anatomy and laid the groundwork for the study of cerebrovascular disorders.

The Renaissance and the Enlightenment were also marked by the development of new techniques for studying the brain and central nervous system, which significantly expanded the understanding of their structure and function. One such technique was the use of injections and staining methods to trace neural pathways. Pioneered by the Italian physician and anatomist Marcello Malpighi (1628–1694), these techniques involved injecting coloured substances, such as wax or mercury, into blood vessels and nerves to visualise their structure and organisation [21]. Malpighi's work on the capillaries and the microscopic structure of the brain marked the beginning of histology, the study of tissues at the microscopic level.

Another significant development was the advent of electrophysiology, which allowed scientists to study neural activity directly. The groundwork for electrophysiology was laid by the Italian scientist Luigi Galvani (1737–1798) and the German physicist and physiologist Emil du Bois-Reymond (1818–1896). Galvani discovered that electrical stimulation could cause muscle contractions, and his work laid the foundation for understanding bioelectricity [22]. Du Bois-Reymond built upon Galvani's findings, developing techniques to measure the electrical activity of nerves and muscles, which provided a basis for studying the electrical properties of neurons [23].

3.6 Advancements During the Nineteenth Century

The nineteenth entury was a critical period for the study of brain and CNS anatomy as it saw numerous advancements, new techniques, and landmark moments driven by the work of several pioneering neuroscientists. The early decades of the century were dominated by the work of Franz Joseph Gall, an Austrian physician and anatomist who developed the controversial theory of phrenology [24]. Although phrenology's fundamental premise—that the brain's various faculties were localised in discrete regions on the skull—was ultimately discredited, Gall's work had the lasting impact of sparking interest in the brain's structural organisation and function.

In contrast to Gall's focus on localising specific faculties, Marie Jean Pierre Flourens, a French physiologist, emphasised the importance of the brain's interconnectedness [25]. Through experiments involving the ablation of various brain regions in animals, Flourens demonstrated that the brain operated as an integrated system, with different parts working in concert to produce behaviour. This work laid the foundation for a holistic view of the brain's organisation, which the discoveries of other researchers would later refine.

In the mid-nineteenth century, the localisation debate re-emerged with the work of Paul Broca, a French physician and anatomist [26]. Broca discovered that damage to a specific region in the frontal lobe of the brain's left hemisphere—now known as Broca's area—resulted in a loss of speech production ability while leaving comprehension intact. This finding provided compelling evidence for the localisation of function in the brain and was corroborated by subsequent research on the neurological basis of language. Broca's work has had a lasting impact on modern neuroscience, with Broca's area remaining a crucial concept in the study of language and the brain [26].

Another important figure in the localisation debate was Carl Wernicke, a German neurologist who, in the late nineteenth century, discovered a region in the left hemisphere's posterior temporal lobe—now known as Wernicke's area—that was responsible for language comprehension [27]. Wernicke's discovery complemented Broca's earlier work, together forming the cornerstone of the modern understanding of the neural basis of language. Wernicke also proposed a model of language processing, now known as the Wernicke–Geschwind model, that posited a connection between Broca's and Wernicke's areas via a neural pathway known as the arcuate fasciculus [28]. This model has since been refined and expanded upon but remains influential in contemporary language research.

The nineteenth century also witnessed groundbreaking advancements in the visualisation of the brain's microscopic structure, due in large part to the development of novel visualisation techniques. Two scientists, in particular, stand out for their contributions to this field: Santiago Ramón y Cajal and Camillo Golgi. Golgi, an Italian physician, developed a method of staining nervous tissue—now known as the Golgi stain or the black reaction—that allowed for the detailed visualisation of individual neurons, including their dendrites and axons [29]. This technique provided researchers with a means to study the brain's microscopic structure, revolutionising the field of neuroanatomy.

Ramón y Cajal, a Spanish histologist and neuroscientist, used Golgi's staining technique to make numerous groundbreaking discoveries about the nervous system's organisation [30]. He demonstrated that the nervous system was composed of distinct individual cells—neurons—rather than a continuous network, as previously believed. This concept, known as the neuron doctrine, laid the foundation for our current understanding of how the nervous system functions. Ramón y Cajal's work also provided crucial insights into the organisation of various brain regions, such as the cerebellum, the retina, and the cerebral cortex (Fig. 3.3) [31]. He is often considered the father of modern neuroscience due to his work's immense impact on the field.

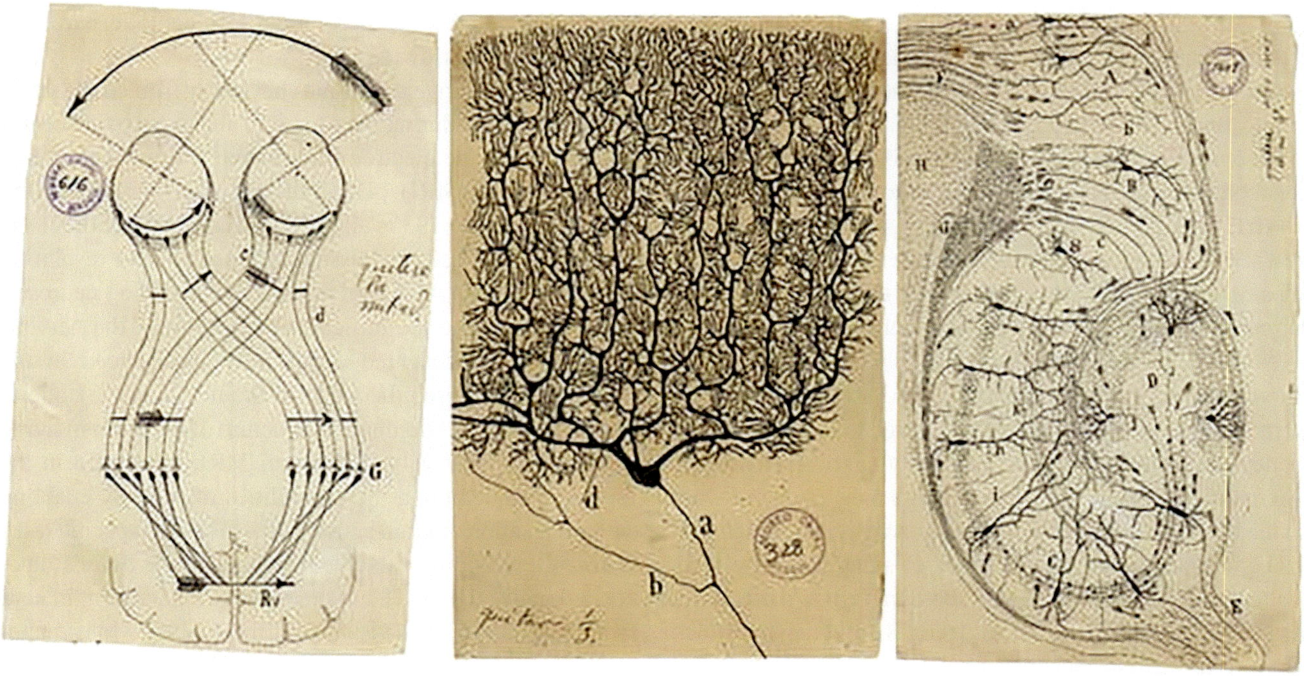

Fig. 3.3 Illustrations by Santiago Ramón y Cajal, Left: A diagram detailing the visual pathways; Middle: a purkinje neuron from the human cerebellum; Right: a diagram showing the tracts through the hippocampus. (Courtesy of the Cajal Institute and the Spanish National Research Council)

These pioneering scientists not only made significant discoveries about the brain's structure and function but also introduced new techniques and methods that have continued to shape the field in the centuries since. Their work has provided the foundation for countless subsequent discoveries, enabling researchers to delve ever deeper into the mysteries of the brain and the central nervous system [31]. This rich history of inquiry and innovation in the nineteenth century illustrates the power of scientific collaboration, curiosity, and determination to drive our understanding of the human brain forward, ultimately opening up new avenues of investigation and leading to remarkable advancements in neuroscience.

3.7 Advancements in the Twentieth and Twenty-first Century

Throughout the twentieth and twenty-first centuries, the study of brain and central nervous system anatomy has advanced tremendously, thanks to the tireless work of dedicated researchers, the development of new techniques, and a series of landmark moments that have shaped our understanding of these complex structures.

Wilder Penfield, a Canadian neurosurgeon, made groundbreaking discoveries in the 1930s and 1940s through his innovative cortical stimulation techniques [32]. By electrically stimulating the exposed cortex of conscious patients during brain surgery, Penfield created detailed maps of the brain's functional areas, which are still used as reference guides in neurosurgery today. German neurologist Korbinian Brodmann conducted seminal work on the cytoarchitecture of the cerebral cortex in the early twentieth century, classifying cortical areas based on their cellular organisation, which resulted in the Brodmann areas—a system still widely used in the study of the cerebral cortex [33]. Lastly, Brenda Milner, a British-Canadian neuropsychologist, made crucial contributions to understanding the brain's functional organisation through her work with a famous case study, who had large portions of his medial temporal lobes removed to treat severe epilepsy [34]. Her research on this patient's memory impairments revealed the critical role of the hippocampus and surrounding structures in forming new memories.

Notable surgical neuro-anatomy pioneers, including Harvey Cushing, Walter Dandy, Gazi Yasargil, and Albert Rhoton Jr. Harvey Cushing, often referred to as the "father of modern neurosurgery", developed new surgical techniques and improved patient outcomes through their extensive work on the brain, and most notably the pituitary gland, which ultimately led to the identification of Cushing's disease [35]. Walter Dandy introduced ventriculography, a method for visualising the brain's ventricles using X-ray imaging, which allowed for a more accurate diagnosis and localisation of brain tumours and other intracranial abnormalities, for which he was nominated for the Nobel Prize in 1933 [36]. Additionally, Dandy's work on the treatment of hydrocephalus led to a better understanding of cerebrospinal fluid dynamics and the anatomy of the choroid plexus, which was suggested to be responsible for producing the cerebrospinal fluid. Egas Moniz, a Portuguese neurologist and neurosurgeon, made significant contributions to the field of neuroanatomy through his development of cerebral angiography in the early twentieth century. This innovative technique, which involves the injection of a contrast medium into blood vessels to visualise the vascular structures of the brain, revolutionised the diagnosis and treatment of a wide range of cerebrovascular disorders. Additionally, Moniz is also recognised for developing the prefrontal leucotomy, a procedure also known as a lobotomy, which was intended to treat certain psychiatric conditions by severing connections in the brain's prefrontal lobe. Despite the controversial nature of this procedure in modern medicine, it was Moniz's work on the leucotomy, not cerebral angiography, that led to his receiving the Nobel Prize in Physiology or Medicine in 1949 [37]. Gazi Yasargil revolutionised neurosurgery by pioneering microsurgical techniques in the 1960s and 1970s, which enabled the safe removal of previously inoperable brain tumours and vascular malformations [38]. Yasargil's innovative use of the operating microscope and the development of specialised micro-instruments significantly impacted the field and allowed a unique appreciation of intra-operative neuro-anatomy. Albert Rhoton Jr., another influential neurosurgeon, made significant contributions to the field through his seminal work on microsurgical anatomy [39, 40]. His meticulous dissection and detailed brain and skull base anatomy documentation significantly improved the understanding of these structures. His extensive brain anatomy atlas has become an invaluable resource for neurosurgeons worldwide.

Several innovative techniques have transformed the study of neuroanatomy during the twentieth and twenty-first centuries, including the electron microscope and magnetic resonance imaging (MRI) [41]. The invention of the electron microscope in the 1930s allowed for the examination of the ultrastructure of cells and tissues at a much higher resolution than was possible with light microscopy, enabling neuroscientists to visualise the intricate architecture of the nervous system in unprecedented detail. MRI, introduced in the 1970s, revolutionised the study of brain anatomy by allowing researchers to visualise the structure of the brain in living subjects non-invasively. Functional MRI, developed in the 1990s, expanded upon the capabilities of MRI by enabling the visualisation of brain activity in response to specific tasks or stimuli, providing valuable insights into the brain's functional organisation. DTI, a more recent advancement in neuroimaging, allows for visualising the brain's white matter tracts, providing important information about the brain's structural connectivity. In recent years, integrating advanced imaging techniques with surgical planning and intraoperative guidance has further enhanced the field of surgical neuroanatomy. For example, neuronavigation systems, which use preoperative MRI or CT scans to create 3D models of the patient's brain, have become an essential tool for neurosurgeons [42]. These systems enable more precise surgical planning and intraoperative guidance, allowing for safer and more effective surgical interventions respecting the anatomy.

In the 1950s, Alan Hodgkin and Andrew Huxley developed the quantitative description of the action potential—the electrical signal propagating along nerve cells [43]. Their work revolutionised our understanding of how neurons communicate with one another and earned them the 1963 Nobel Prize in Physiology or Medicine. This laid the foundations for future techniques like optogenetics, introduced in the early 2000s, employing light-sensitive proteins to modulate neuronal activity, opening up new possibilities for studying neural circuits and their role in behaviour. The field of connectomics, which aims to create a detailed wiring diagram of the brain, has gained momentum in recent years, driven by advancements in imaging technologies and computational methods [44]. By mapping the complex network of connections between neurons, connectomics provides a comprehensive understanding of the brain's structure and function at a level of detail previously unattainable. The Human Connectome Project, launched in 2009, is an ambitious collaborative effort to map the human brain's connections at unprecedented detail and is expected to revolutionise our understanding of brain function [45].

It is important to mention the role of interdisciplinary collaboration in advancing brain and central nervous system anatomy research. Throughout the twentieth and twenty-first centuries, scientists from diverse fields, such as clinical disciplines, biology, psychology, physics, and computer science, have come together to tackle complex questions about the brain's structure and function. For instance, the field of computational neuroscience has emerged as a crucial component of modern neuroscience research as it combines experimental data with theoretical models to provide a more comprehensive understanding of the brain's inner workings.

3.8 Surgical Pearls

- Neuroanatomy's history is marked by technological advancements that have continually refined our surgical precision; being aware of these historical landmarks can help us appreciate the rationale behind current techniques and remain open to future innovations.
- Understanding the roots of neuroanatomy can provide valuable insight into surgical approaches: many traditional techniques have been influenced by historical understandings and constraints, and revisiting these can sometimes inspire new methods or refinements.
- Remember that much of our contemporary neuroanatomy knowledge is based on the diligent observation and documentation of our predecessors; similarly, detailed surgical anatomical papers, and documenting and sharing anatomical observations can contribute to further advancements in the field of neurosurgery.

3.9 Discussion

In ancient times, beliefs such as the heart being the centre of human emotions and intelligence shifted the focus away from a study of the brain. This, however, was later rectified by many prominent historical figures including philosophers which helped in further study of the brain.

There are several interesting anecdotes in the ancient and modern history of neuroanatomy that fall beyond the scope of the chapter. It is also interesting to learn about the role religious beliefs and churches played both in hindering and promoting advancement in such knowledge of human anatomy.

The major breakthrough in the study of the architecture of the nervous system occurred during nineteenth century with the development of staining techniques by Santiago Ramón y Cajal and Camillo Golgi.

From the pioneering work of Santiago Ramón y Cajal to the cutting-edge research being conducted today, the field of neuroscience continues to expand and deepen our knowledge of the brain, paving the way for new treatments and therapies for a wide range of neurological disorders. The integration of advanced imaging techniques, surgical innovations, and interdisciplinary collaboration have all played a role in shaping the current landscape of brain anatomy research, and these efforts are sure to continue propelling the field forward in the years to come.

3.10 Conclusion

An exquisite knowledge of neurosurgical anatomy is of paramount importance for safe and successful neurosurgical practice. We owe this knowledge to several key figures in history whose hard work, intense curiosity, persistence, and scientific temper, helped by the prevailing favourable societal environment and evolving technology over several centuries. A reflection of such endeavours helps us to put our current understanding of neuroanatomy into perspective and paves the way for further progress in an era of rapid technological advancements.

References

1. Wilkins RH. Neurosurgical classics. Park Ridge: American Association of Neurological Surgeons; 1992.
2. Fanous AA, Couldwell WT. Transnasal excerebration surgery in ancient Egypt. J Neurosurg. 2012;116(4):743–8.
3. Zuccaro G. The dawn of neurosurgery in pre-conquest Mesoamerican territories. Childs Nerv Syst. 2017;33(10):1621–9.
4. Kushner DS, Verano JW, Titelbaum AR. Trepanation procedures/outcomes: comparison of prehistoric Peru with other ancient, medieval, and American civil war cranial surgery. World Neurosurg. 2018;114:245–51.
5. Missios S. Hippocrates, Galen, and the uses of trepanation in the ancient classical world. Neurosurg Focus. 2007;23(1):E11.
6. Lloyd G. Alcmaeon and the early history of dissection. Sudhoffs Arch. 1975;59(2):113–47.
7. Bubb C. Dissection in classical antiquity: a social and medical history. New York: Cambridge University Press; 2023.
8. Rose FC. The neurology of ancient Greece – an overview. J Hist Neurosci. 1994 Oct;3(4):237–60.
9. Herophilus. Herophilus: the art of medicine in early Alexandria: edition, translation and essays. 1st ed. Cambridge: Cambridge University Press; 1989.
10. Wills A. Herophilus, Erasistratus, and the birth of neuroscience. Lancet. 1999;354(9191):1719–20.
11. Golder W. Between anatomy and pathology: The dissection reports of Galen of Pergamum. Pathologie (Heidelb). 2023;44(2):121–8.
12. Gross CG. Galen and the squealing pig. Neuroscientist. 1998;4(3):216–21.
13. Ghosh SK. Human cadaveric dissection: a historical account from ancient Greece to the modern era. Anat Cell Biol. 2015;48(3):153–69.
14. Robinson A. Galen: life lessons from gladiatorial contests. Lancet. 2013;382(9904):1548.
15. Alghamdi MA, Ziermann JM, Diogo R. An untold story: the important contributions of Muslim scholars for the understanding of human anatomy. Anat Rec (Hoboken). 2017;300(6):986–1008.
16. Zampieri F, ElMaghawry M, Zanatta A, Thiene G. Andreas Vesalius: celebrating 500 years of dissecting nature. Glob Cardiol Sci Pract. 2015;2015(5):66.
17. Abbott A. Neurophysiology: the man who bared the brain. Nature. 2015;521(7551):160.
18. Zago S, Meraviglia MV. Costanzo Varolio (1543–1575). J Neurol. 2009;256(7):1195–6.
19. Splavski B, Rotim K, Boop FA, Gienapp AJ, Arnautović KI. Ambroise paré: his contribution to the future advancement of neurosurgery and the hardships of his times affecting his life and brilliant career. World Neurosurg. 2020;134:233–9.
20. Dumitrescu AM, Costea CF, Cucu AI, Dumitrescu GF, Turliuc MD, Scripcariu DV, et al. The discovery of the circle of Willis as a result of using the scientific method in anatomical dissection. Romanian J Morphol Embryol. 2020;61(3):959–65.
21. Engelhardt E. Marcello Malpighi: the nervous system under a microscope. Arq Neuropsiquiatr. 2021;79(4):346–9.
22. Cajavilca C, Varon J, Sternbach GL. Resuscitation great. Luigi Galvani and the foundations of electrophysiology. Resuscitation. 2009;80(2):159–62.
23. Finkelstein G. Mechanical neuroscience: Emil du Bois-Reymond's innovations in theory and practice. Front Syst Neurosci. 2015;9:133.
24. Wijdicks E. The galling and gifted Franz Joseph Gall. Lancet Neurol. 2020;19(10):812–3.
25. Yildirim FB, Sarikcioglu L. Marie Jean Pierre Flourens (1794–1867): an extraordinary scientist of his time. J Neurol Neurosurg Psychiatry. 2007;78(8):852.
26. LaPointe LL. Paul Broca and the origins of language in the brain. San Diego: Plural Publishing; 2012.
27. Pillmann F. Carl Wernicke (1848–1905). J Neurol. 2003;250(11):1390–1.
28. Nasios G, Dardiotis E, Messinis L. From Broca and Wernicke to the neuromodulation era: insights of brain language networks for neurorehabilitation. Behav Neurol. 2019;2019:9894571.
29. Bentivoglio M, Cotrufo T, Ferrari S, Tesoriero C, Mariotto S, Bertini G, et al. The original histological slides of Camillo Golgi and his discoveries on neuronal structure. Front Neuroanat. 2019;13:3.
30. Garcia-Lopez P, Garcia-Marin V, Freire M. The histological slides and drawings of Cajal. Front Neuroanat. 2010;4:9.
31. Georgopoulos AP, Georgopoulos A. The beautiful brain and the influence of santiago ramón y cajal on medicine. JAMA. 2017;318(6):502–4.
32. Penfield W, Boldrey E. Somatic motor and sensory representation in the cerebral cortex of man as studied by electrical stimulation. Brain. 1937;60(4):389–443.
33. Zilles K. Brodmann: a pioneer of human brain mapping-his impact on concepts of cortical organization. Brain. 2018;141(11):3262–78.
34. Squire LR. The legacy of patient H.M. for neuroscience. Neuron. 2009;61(1):6–9.
35. Bliss M. Harvey Cushing: a life in surgery. Illustrated. New York: Oxford University Press; 2007.
36. Campbell E, Walter E. Dandy-surgeon, 1886–1946. J Neurosurg. 1951;8(3):249–62.
37. Tan SY, Yip A. António Egas Moniz (1874–1955): Lobotomy pioneer and Nobel laureate. Singapore Med J. 2014;55(4):175–6.
38. Rogers L. Yasargil: father of modern neurosurgery. Virginia Beach: Koehler Books; 2015.
39. Matsushima T, Matsushima K, Kobayashi S, Lister JR, Morcos JJ. The microneurosurgical anatomy legacy of Albert L. Rhoton Jr., MD: an analysis of transition and evolution over 50 years. J Neurosurg. 2018;129(5):1331–41.
40. Rhoton AL Jr. Rhoton: cranial anatomy and surgical approaches: neurosurgery. 1st ed. Philadelphia: Lippincott Williams & Wilkins; 2008.
41. Filler A. The history, development and impact of computed imaging in neurological diagnosis and neurosurgery: CT, MRI, and DTI. Nat Proc. 2009;10:3675–83.
42. Thomas NWD, Sinclair J. Image-guided neurosurgery: history and current clinical applications. J Med Imaging Radiat Sci. 2015;46(3):331–42.
43. Schwiening CJ. A brief historical perspective: Hodgkin and Huxley. J Physiol Lond. 2012;590(11):2571–5.
44. Emmons SW. The beginning of connectomics: a commentary on White et al. (1986) "The structure of the nervous system of the nematode *Caenorhabditis elegans*". Philos Trans R Soc Lond Ser B Biol Sci. 2015;370(1666):20140309.
45. Van Essen DC, Smith SM, Barch DM, Behrens TEJ, Yacoub E, Ugurbil K, et al. The WU-Minn human connectome project: an overview. NeuroImage. 2013;80:62–79.

Neurosurgery Theatres, Facilities, and Instruments

4

Albert Sufianov and Rinat Sufianov

Learning Objectives

1. Concepts of operating room and theatre complex design
2. Integrative role of the operating theatre design as an infection prevention protocol
3. Rationale for siting a neurosurgical theatre in a hospital (cul-de-sac, ancillary facilities)
4. The working set-up of neurosurgical theatres
5. Basic and some new neurosurgical equipment, instrumentation, and imaging technologies

4.1 Introduction

The surgical process involves implementing in practice the knowledge of anatomy, which is the basis of any neurosurgical intervention, and the neurosurgical skills of a neurosurgeon in combination with modern technological opportunities, tools, equipment, methods of intraoperative neuroimaging, and intraoperative control within the space of an operative theatre for ensuring the successful outcome of a surgical procedure.

Neurosurgical theatres are specialized operating rooms designed specifically for the performance of complex surgical procedures on the brain, spinal cord, and nerves. These theatres are equipped with advanced tools to provide optimal conditions for neurosurgeons to perform delicate procedures. The specialized equipment includes high-resolution imaging systems, microscopes, and robotic surgical instruments designed to enhance surgical precision and accuracy. Additionally, the theatres are staffed by highly trained and experienced surgical teams, who work together to ensure the best possible outcomes for patients undergoing neurosurgical procedures. Overall, neurosurgical theatres play a vital role in the diagnosis and treatment of complex neurological conditions and are an essential component of modern healthcare.

A. Sufianov (✉)
Federal Centre of Neurosurgery (Tyumen) of Ministry of Health of Russian Federation, Tyumen, Russia

R. Sufianov
I.M.Sechenov First Moscow State Medical University, Moscow, Russia

4.2 History

Cranial trephinations were reported in antiquity from different parts of the world. Their indications were clinical (headaches, seizures, and head injury) and non-clinical (bad spirits and 'bad-behaviours') [1].

The name 'operating theatre' goes back to its origins from the initial semi-circular amphitheatre designs, to enable students to watch the procedures [2].

The modern operating theatre as we know it was developed in the nineteenth century. In the early 1800s, surgeons began to recognize the importance of cleanliness and antiseptic techniques in surgery. This led to the creation of the first dedicated operating theatres, which were designed with features like ventilation systems and sterile surfaces to minimize the risk of infection.

Operating theatres continued to evolve throughout the nineteenth and twentieth centuries, with the introduction of new technology like anaesthesia and surgical tools such as forceps and scalpels. Today, operating theatres are highly specialized environments with advanced equipment and complex procedures that require specialized training and expertise [3].

4.3 Rationale for Neurosurgical Theatre Design

The contemporary neurosurgical theatre ensures the safe performance of procedures, improving the safety of patients (minimizing significantly the risks for surgical site infections

I. N. Kanaan, V. Beneš (eds.), *Neuroanatomy Guidance to Successful Neurosurgical Interventions*,
https://doi.org/10.1007/978-3-031-59838-8_4

and better visualization by the surgical team), and operating room personnel (improved capacity for surgeon-ergonomics amongst other occupational hazards, etc).

4.4 Types of Neurosurgical Theatres

There are various types of neurosurgical theatres used for different neurosurgical procedures. The most common types of neurosurgical theatres include:

1. Conventional operating rooms – These rooms are equipped with a basic set of neurosurgical and monitoring devices, and anaesthesia machines that are necessary for carrying out different types of neurosurgical procedures and ensuring the safety of a patient (Fig. 4.1).
2. Endovascular operating rooms – Some hospitals have separate operating rooms for endovascular surgeries, minimally invasive procedures that involve using micro-catheters and wires to navigate through the circulatory system to access and repair brain aneurysms, vascular malformations, and other abnormalities within the cranial cavity. Endovascular operating rooms are equipped with specialized imaging systems and machines for intraoperative monitoring and control.
3. Hybrid operating rooms – These are specialized operating rooms that allow for both conventional and minimally invasive procedures to be performed in the same setting. Hybrid operating rooms are equipped with advanced imaging systems such as CT and MRI scanners, and surgical equipment compatible with neuroimaging machines, and can be used interchangeably to allow neurosurgeons to perform a wide range of surgeries while monitoring the brain tissue in real-time (Fig. 4.2).

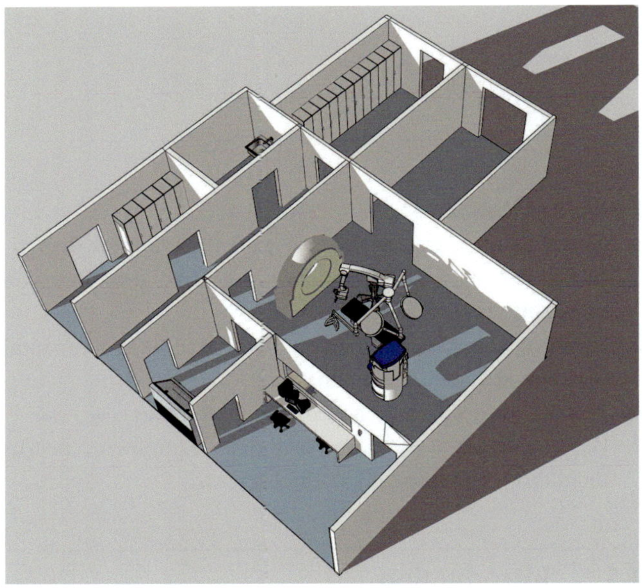

Fig. 4.2 3D scheme of the organization of the iCT hybrid operating room. (Federal Centre of Neurosurgery, Russia, Tyumen)

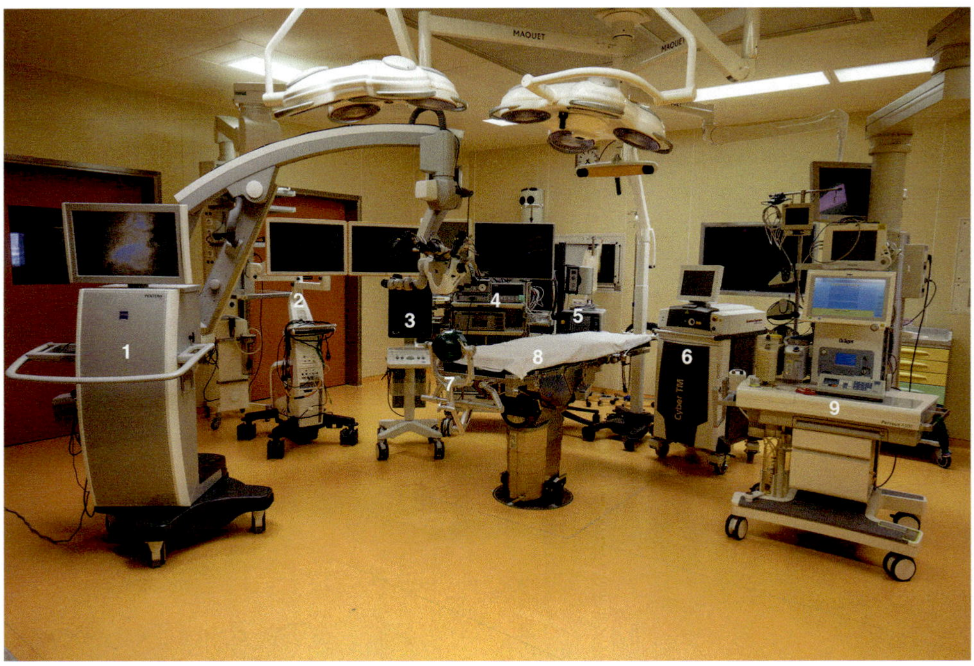

Fig. 4.1 Set-up of a standard operating theatre. (1) Operating microscope. (2) Neuronavigation system. (3) Intraoperative ultrasound. (4) Endoscopic stand. (5) Bipolar generator. (6) Thulium laser system. (7) Skull clamp. (8) Operating table. (9) Anesthesia machine. (Federal Centre of Neurosurgery, Russia, Tyumen)

4.5 Design of the Neurosurgical Theatre

There is adherence to the concepts of theatre design in the building of neurosurgical theatres. We present a summary of these principles below.

4.5.1 General Principles of Theatre Design

Site: It should be situated on the lower floors of a high-rise building [4]. It should be detached from the hospital's human traffic in a so-called cul-de-sac, where it is assessed from only 'semi-restricted' areas [5]. It needs to be in proximity (on the same floor or with efficient lifts if on another floor) to the neuro-intensive care unit, emergency department, central sterile services department, and ancillary units like the neurosurgical wards [4].

Compartments: The zones should be broadly categorized into dirty, clean, and sterile zones (Fig. 4.3). These are demarcated clearly (outside clothes are changed when entering the clean zone, also an additional layer of sterile personal protective equipment is necessary around the operating table) and are necessary when entering or exiting each zone. The 'sterile' zone being the cleanest part, traditionally serves as the operating room (and the anaesthesia induction room or ante-room); these are separated from the clean area by a 'white line'. The 'clean' areas will include the changing rooms (a transitional zone), the post-anaesthesia care unit, the ally from the theatre complex entrance and changing rooms to the operating rooms, and personnel lounges/offices within the theatre complex. This 'clean area' is separated from the 'dirty area' by a 'red-line' [4].

Ancillary Units: Facilities proximate and holding interdependency relationships with the theatre as already stated in the 'site' subsection.

4.5.2 Specific Operation Room Planning Principles

Size: Previous recommendations were 6.5 × 6.5 × 3.5 m; these are statutorily determined by the specific equipment and minimum number of personnel that would be routinely required for typical operations. [4, 5]. The average size is about 400 ft^2 and up to 600 ft^2 for a hybrid theatre because of additional equipment requirements [6].

Walls: The walls are traditionally made of materials that are not destructible by water, antiseptics, and disinfectants, as well as scrubbable. The wall edges should have no acute angles to minimize crevices for formites to collect and colonize [4–6].

Floor: The floor should be made of a water-impervious material, non-slippery, and stable with disinfectants and antiseptics; suggested materials include jointless conductive tiles, linoleum, and a mosaic of copper plates. It should have a light hue, to permit the identification of small objects [4–6].

Light: The operating room would consist of many surgical luminaires, most common are the overhead lights that directly illuminate the operating field, but also are the surrounding lightings in the suite. Other equipment also come with their lighting to amplify those from the overhead, and also some intra-cavitary illuminations during endoscopic procedures. The overhead light must provide a light field centre that delivers excellent central illuminance while providing adequate depth of illumination and shadow dilution to prevent obstructions in the operating field of view (about 50,000–100,000 lux centrally and a minimum of 15,000 lux peripherally) [4].

Operating Table: These should be easily malleable into a variety of shapes with hinges that are either manually or electronically controlled. They should be adjustable to different heights and angulation in multiple planes, to support surgeon ergonomics as well as preferences for the surgical approach [4].

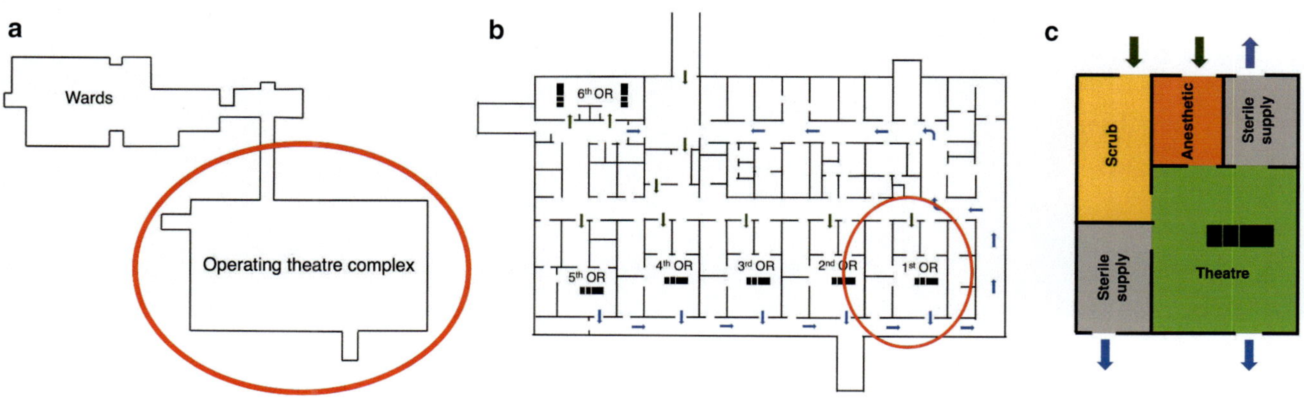

Fig. 4.3 Organization of operating theatres in the hospital (Federal Centre of Neurosurgery, Tyumen, Russia). (**a**) Organization of the operating theatre complex in relation to another department. (**b**) Operating theatre complex with interchange zone and sterile supply corridor. (**c**) Basic set-up of the theatre

Airflow System (types, positioning in relation to exit doors and table, minimum air exchanges per minute, and positive airflow pressure system): The traditional laminar (unidirectional), turbulent, and mixed airflow systems are utilized based on the preferences of the institution, availability, and accessibility [4, 7]. They should be positioned to create a constant airflow over and away from the operating table, making the table the point of highest airflow pressure in the operating room [4, 8, 9]. They should be directed to sweep towards the main entrance doors into the operating rooms. These ultimately create a positive pressure (5 cmH_2O) within the operating room and decrease pressure as the air exits from 'sterile', to clean and 'dirty' areas/zones; invariably creating a pressure gradient that drives aerosolized particulate matter away from the operating room [4]. There should be a minimum of 20–30 air changes per hour (ACH) to maintain the positive pressure relationship and recirculated air [4, 9]. These systems are typically fitted with HEPA (High-Efficiency Particulate Air) filters [or Ultra-Low Penetration Air filters (ULPA) with a filtration rate >99.99%], which are essentially a fine multi-layered, random, meshwork of polypropylene or micro-fiberglass; they ultimately prevent 99.97% of particulate matter between 0.15 and 0.2 μm (including aerosolized microbes), yielding an almost perfect clean air delivery rate (CADR) [7–9].

4.6 Hybrid Neurosurgical Theatre Concepts

All of the above concepts are combined with the greatest technological advances into a seamless work-space called the hybrid operating room, where there is the permanent integration of peri-operative imaging facilities and a combination of a conventional operating room and endovascular set-up.

The combination of additional imaging techniques and instruments with neuronavigation allows for organizing workflow eliminating complications and maintaining control over the operation at all stages [10]. High-quality neuroimaging with the use of iCT or iMRI ensures navigation in the operating field with altered anatomy and brain-shift conditions (Fig. 4.4).

C-arm and O-arm are the simplest intraoperative neuroimaging techniques used widely in spinal neurosurgery and can be compatible with the neuronavigation system. This combination contributes to the precise localization of spinal cord lesions, as well as during endoscopic skull base surgery in the case of O-arm (Fig. 4.5) [11, 12].

The next generation of imaging technologies in the operating room are iCT and iMRI. At present, iCT is used in all types of neurosurgical interventions. In combination with automatic registration, this is a straightforward, reli-

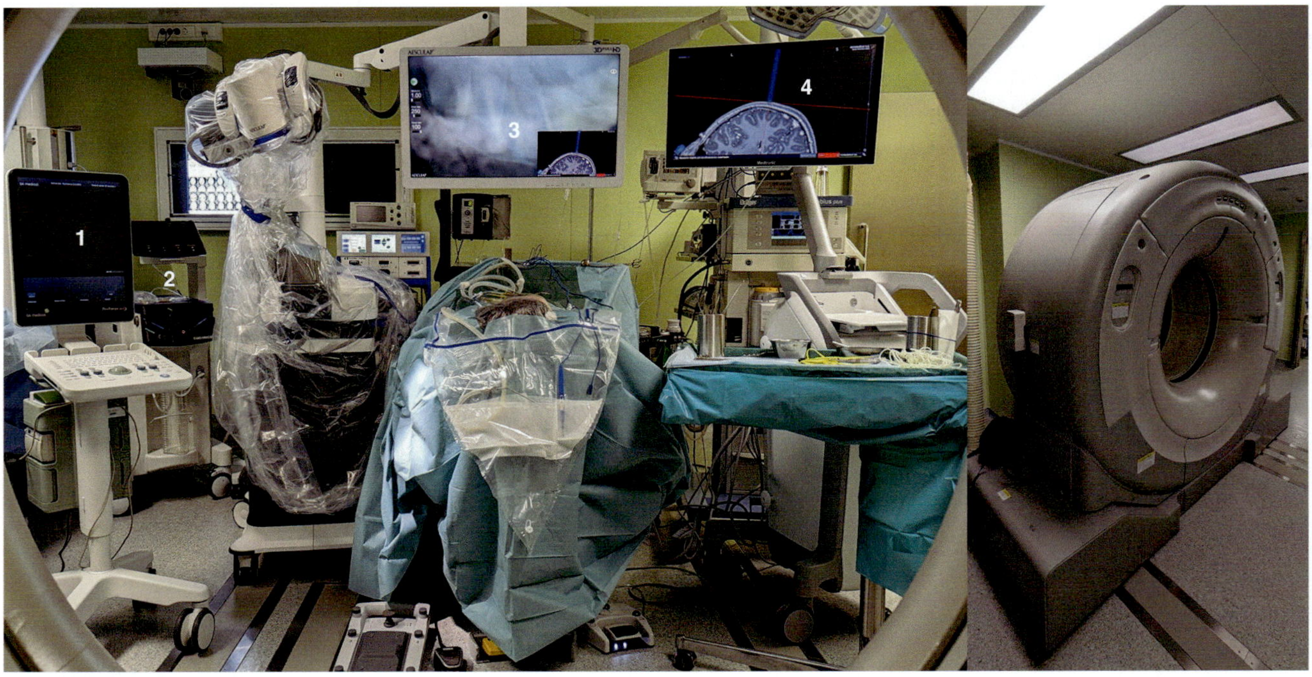

Fig 4.4 (Left) Organization of the iCT hybrid operating room (Federal Centre of Neurosurgery, Tyumen, Russia). (1) Intraoperative ultrasound. (2) Ultrasonic aspirator system. (3) Exoscope. (4) Endoscopic stand. (Right) Intraoperative CT. (Federal Centre of Neurosurgery, Russia, Tyumen)

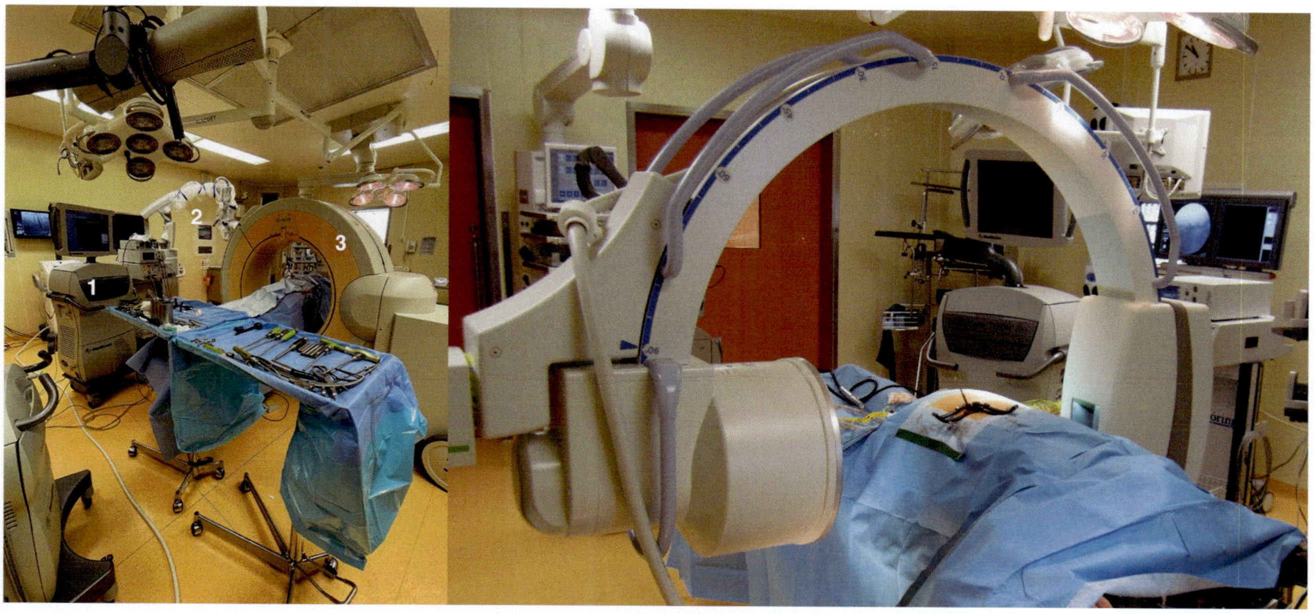

Fig. 4.5 (Left) Intraoperative view of the O-arm. (1) Neuronavigation system. (2) Operating microscope. (3) O-arm (right) intraoperative view of the C-arm. (Federal Centre of Neurosurgery, Tyumen, Russia)

able addition to cranial interventions with an accuracy of less than 1 mm [13]. For even greater precision, the main instruments can be navigated with the creation of a completely controlled surgical field in real time. Complementary modalities such as intraoperative CT perfusion and CT angiography are useful for managing challenged cases in vascular and oncology neurosurgery [14, 15].

Intraoperative MRI presents 'gold standard' state-of-the-art techniques for image-guided neurosurgery. Although this imaging technology has limitations conditioned by the use of magnetic fields, the organization of the operating room space allows overcoming these limitations. Head positioning beyond the edge of the 5 Gauss safety line makes it possible for the neurosurgical procedures to be carried out with standard instrumentation.

Another type of a hybrid operating room with a combination of standard and endovascular set-up useful for surgical and endovascular treatment of complex cerebrovascular diseases opens new possibilities to treat challenged vascular lesions in one session [16].

The design of hybrid ORs needs to be coordinated attentively by neurosurgeons to suit individual needs within the institution.

4.7 Neurosurgical Equipment and Instrumentation

Neurosurgical equipment and instrumentation refer to the various tools and devices that are used by neurosurgeons during surgical procedures. These tools and devices are specifically designed to help neurosurgeons perform intricate procedures with precision and accuracy, focusing on the brain, spinal cord, and nerves.

Modern neurosurgical equipment have advanced significantly in recent years, enabling neurosurgeons to perform more precise and minimally invasive procedures.

Some common types of neurosurgical equipment include:

1. **Microscopes and Exoscopes** to magnify the surgical area and visualize small structures in the brain and spine during open surgery.
2. **Neuroendoscopes** allow neurosurgeons to access and visualize difficult-to-reach areas of the brain.
3. **Image-Guided Systems** are accessory imaging techniques like intraoperative MRI, CT, and ultrasound; fluorescence-guided visualization techniques help surgeons orientate the brain and spine during surgery.
4. **Navigation Systems** guide to the precise location of the problem area in the brain or spinal cord and provide real-time feedback that helps the neurosurgeon stay on course during surgery, especially in difficult-to-reach areas or when dealing with complex anatomy.
5. **Intraoperative Neuromonitoring** in relation to a specific surgical intervention is aimed at evaluating the state of the reflex arc links, as well as mapping for the identification of the functional zones in the operating field, tracts, epileptogenic zones, and targets for neuromodulation.
6. **Ultrasonic Aspirators** are used to break down and remove tumours or other tissue from the brain or spine.
7. **Intraoperative Laser** utilizes high-energy light beams to perform precise incisions and tissue ablation during sur-

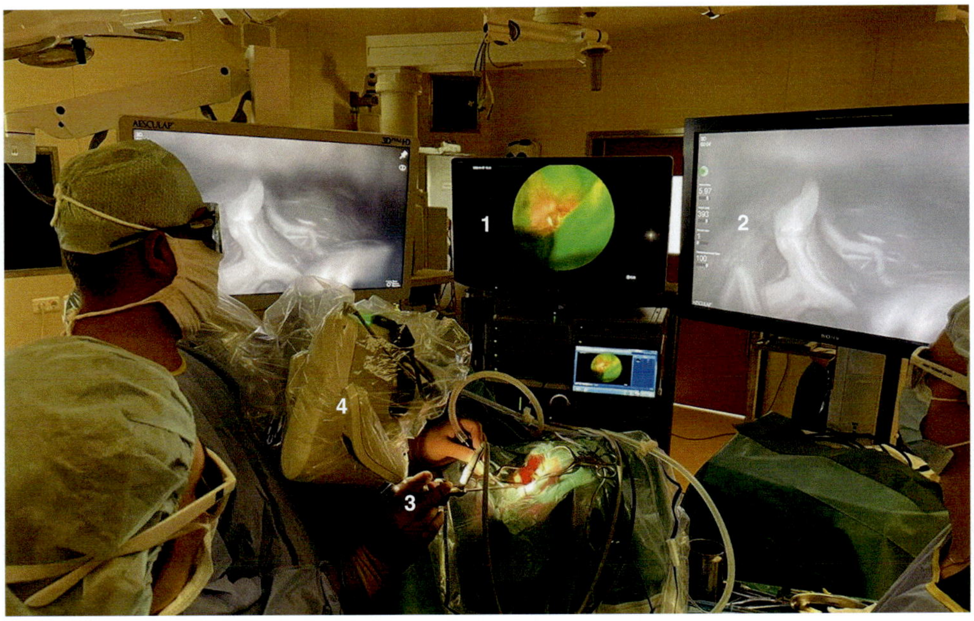

Fig. 4.6 Intraoperative view of usage of 3D technologies in neurosurgery. (1) Screen view from the endoscope. (2) 3D screen view from the exoscope. (3) Endoscope. (4) Exoscope. (Federal Centre of Neurosurgery, Tyumen, Russia)

gery. In neurosurgery, lasers are most often used for the treatment of brain tumours.

8. **Surgical Instruments** include a variety of scalpels, forceps, scissors, and retractors, which the neurosurgeon uses to manipulate the surgical area.

4.8 Microscopes and Exoscopes

Modern neurosurgery cannot be imagined without excellent magnification visualization equipment that allows performing high-precision manipulations. Nowadays, advanced types of visualization options are available, such as microscopy, exoscopy, endoscopy, and their combinations and variations [17].

In neurosurgery, microscopes are used to visualize the intricate structures of the brain and spinal cord during open surgeries. Since the advent of operating microscopes, their main principles have not changed significantly. Nevertheless, modern technologies allow better visualization and manoeuvrability, as well as adjusting assistant modules, such as fluorescence imaging, and compatibility with navigation systems, which improve surgical accuracy [18].

The use of 3D exoscopes in daily neurosurgical routines promotes an efficient workflow, thanks to the latest robotic interface and high-quality images. Automatic or manual positioning, excellent depth of field, and a wider field of view allow the surgeon to focus on the surgery itself rather than on setting up the machine. The advantages of an exoscope are the equal possibilities for the whole team to observe the surgical field in the monitor. Good ergonomics, and consequently less fatigue of the surgeon during surgery, also contribute to ensuring safety and successful outcome of the procedure (Fig. 4.6) [19].

4.9 Endoscopes

Endoscopy remains a relatively new but gaining popularity tool within the practice of many neurosurgical centres around the world. The endoscope offers a wider field of view than is generally possible with microscopy and is a helpful auxiliary tool for verifying the accuracy of microscopic resection. Used alone, the endoscope ensures minimally invasive character of the surgery.

The endoscope is an indispensable tool in neurosurgery, as it provides good visualization of the operative field, a wide view of anatomical landmarks, high image resolution, which helps in differentiating various structures, and also has the possibility of 3-D visualization. Angled scopes provide a better vision of corners and hidden areas in the deep-seated structures.

In principle, an endoscope for neuroendoscopic interventions consists of three parts: optical, irrigation and aspiration channels, and an instrument channel. Currently, the endoscope with the smallest diameter meeting these requirements is a miniature straight-needle endoscope (Fig. 4.7).

Among the advantages of the above endoscope, we can highlight the possibility of surgical manipulations in narrow and intricate anatomical spaces without any risk of damaging surrounding vital structures [20].

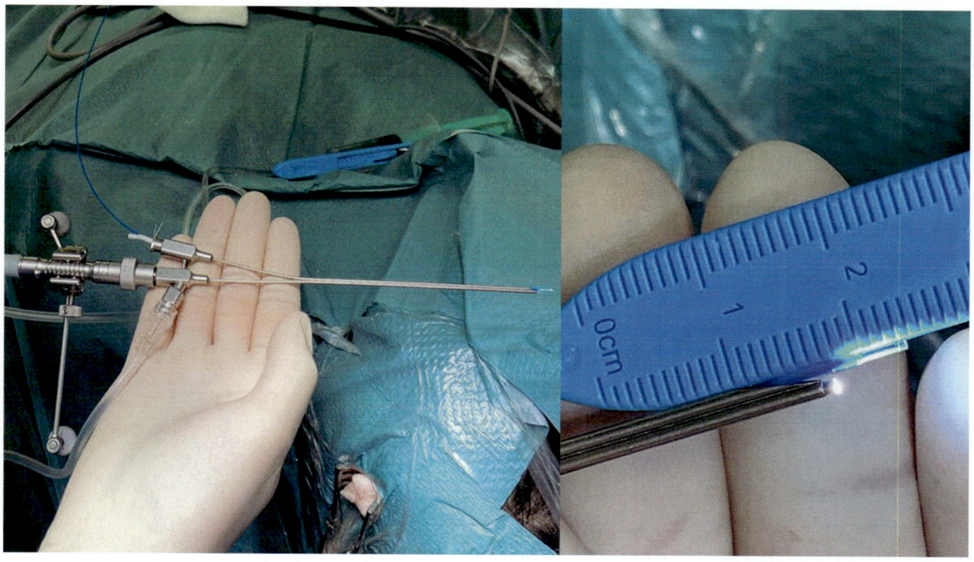

Fig. 4.7 (Left) Intraoperative view of the navigated needle endoscope with fibre-optic cable of Thulium laser. (Right) Tip of the operating sheath with the needle endoscope in the optic channel, and fibre-optic cable of Thulium laser in the working channel. (Federal Centre of Neurosurgery, Tyumen, Russia)

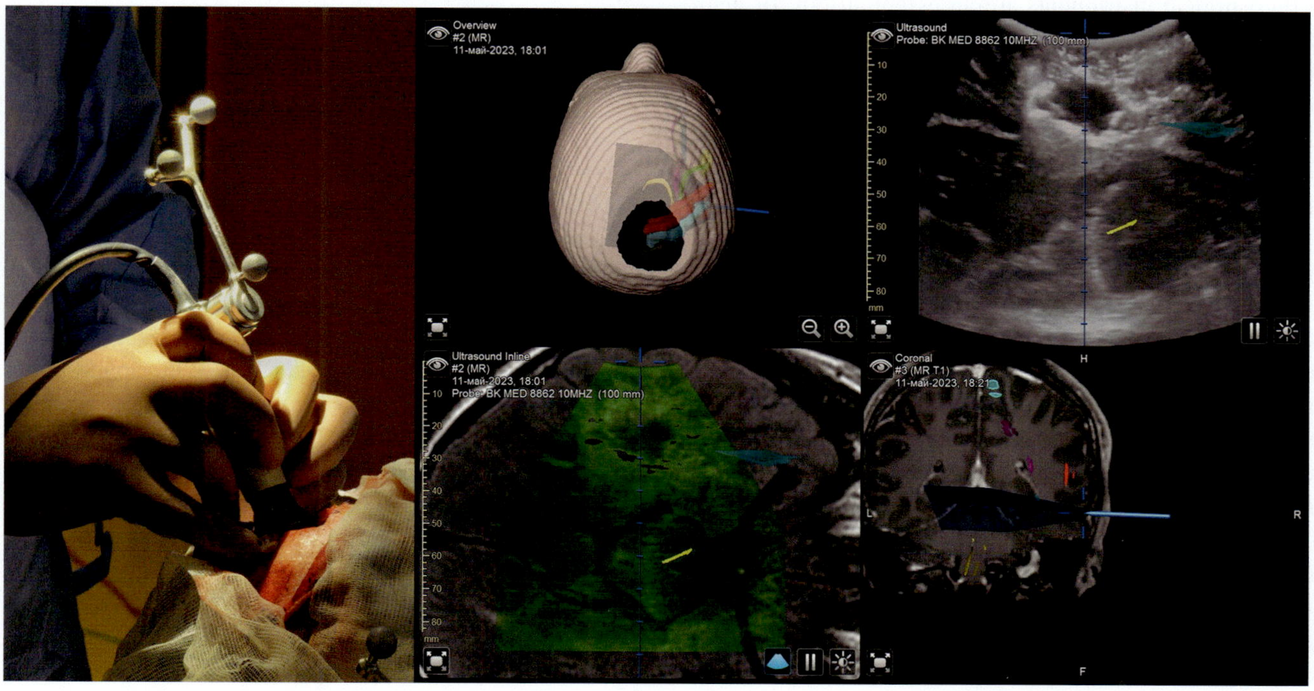

Fig. 4.8 (Left) Intraoperative view of the navigated iUS probe. (Right) Neuronavigation system screen view of usage iUS. (Federal Centre of Neurosurgery, Tyumen, Russia)

4.10 Ultrasonography

Ultrasonography is an imaging modality, which has been widely used in neurosurgery to visualize lesions and neurovascular structures/brain parenchyma.

Compared to other intraoperative neuronavigation techniques like MRI and CT, some of the key characteristics of iUS include the provision of real-time information on lesion location and extent even in the face of 'brain shift', significant reduction in surgical time, and cost-effectiveness (Fig.4.8) [21].

4.11 Neuronavigation

The use of neuronavigation technologies in the operating room allows many tasks to be solved at once. Examples of the most common ones are planning neurosurgical access at different stages, orientation in the surgical field during the main surgical phase, and so on. In addition, these technologies allow microscope-based augmented reality in which images from the neuronavigation station are transmitted into the eyepieces of the microscope or the monitor of the exoscope so that the neurosurgeon could clearly understand the situation at the moment. Neuronavigation is compatible with various instruments and an endoscope.

4.12 Instruments

Microsurgical instruments are an essential tool for neurosurgeons who work on structures in the brain or spinal cord that require precise, delicate surgical techniques. Basic microsurgical instruments such as scissors, forceps, spatulas, and elevators have been used since the dawn of microsurgery in the 1960s and still perfectly fulfil their function [22].

However, neurosurgery of the future will be defined by the development of solutions to new and emerging neurosurgical challenges. Newer devices being created, so called *de novo* (where new devices are created, either completely or as modifications of the already existing) and/or adaptive (where devices are adapted in a wide spectrum of combinations) innovation schemes. We present two devices, one each describing a *de novo* and *adaptive* devices.

A *de novo*, innovative bipolar diathermy forceps (Professor Albert Sufianov; patent for a utility model #187789) was designed for bipolar dissection for a wide range of surgical interventions [23]. The main features of these forceps are lightweight, wear resistance, and a non-stick coating that prevents the patient's biological tissue from sticking to the open tips of the forceps during surgery thus minimizing interruptions for cleaning (Fig. 4.9).

The *adaptive* use of Raabe's mapping suction probe (Inomed Medizintechnik GmbH, Germany) is integrated with an electromagnetic neuronavigation function of the shunt stylet. The mapping suction probe was designed for subcortical mapping of the corticospinal tract, while attempting to resect brain tumours deep into the motor eloquent areas. It is possible to combine adequate suctioning, while visualizing the intraoperative lesion, through the neuronavigation devices (Fig. 4.10).

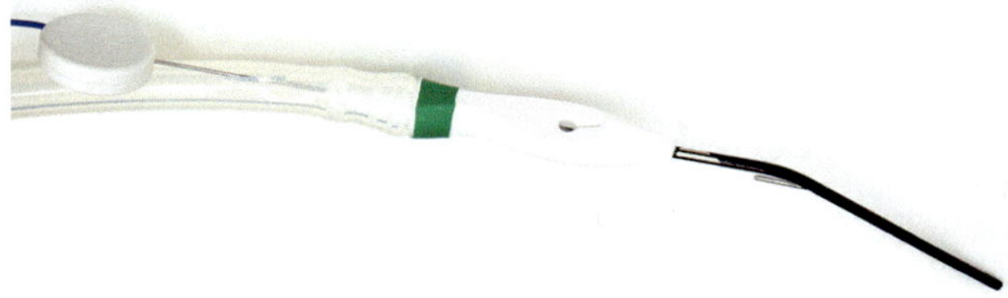

Fig. 4.10 Navigated mapping suction probe

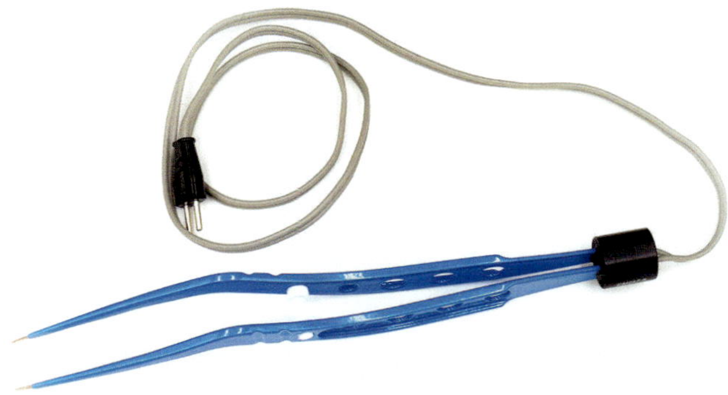

Fig. 4.9 Bipolar diathermy forceps

4.13 3D Printing for Neurosurgery

In neurosurgery, 3D printing technology can be used to create custom-made surgical instruments that are designed customized to a patient's specific needs and anatomy, and surgical needs. This can improve the accuracy and precision of surgical procedures, leading to better patient outcomes. In addition, 3D printed instruments can also be lighter, more durable, and less expensive than traditional surgical instruments. In modern practice, some custom-made 3D-printed instruments for endoscope-assisted surgery have already successfully implemented [24, 25].

4.14 Surgical Pearls

Ergonomic design and carefully thought out set-up of the operating room, as well as using modern neurosurgical equipment and instrumentation with their continuous upgrading and adapting to individual needs of the neurosurgeon and the surgical situation, make it possible to perform complex procedures with minimal risk and best outcomes for patients.

4.15 Conclusion

Post-operative outcomes of the neurosurgery patient are often dependent on the equipment in the theatre at the disposal of the neurosurgeon. The ideas put forth here provide a strong baseline of what an optimal neurosurgical theatre may encompass. In view of ever-evolving technology, more sophisticated equipment ranging from surgeon manual-dependent to robotic-assisted are yet to be developed. With the changing trends, we hope to update this information into novel neurosurgical theatres.

Acknowledgements The authors thank Dr. Ilshat Gaysin for writing assistance and preparing the images, Ms. Aliya Bukhtoyarova for writing assistance and language editing, and Dr. Promise Jaja for writing assistance.

References

1. Nikova A, Birbilis T. The basic steps of evolution of brain surgery. Maedica. 2017;12(4):297–305.
2. dos Souza FS, dos Silva RPS, Junior LSB, de Azevedo Filho HRC. The evolution of neurosurgery throughout the ages: from trepanations in prehistory to the robotic era. Arq Bras Neurocir. 2022;41(2):e153–8.
3. Adams LW, Aschenbrenner CA, Houle TT, Roy RC. Uncovering the history of operating room attire through photographs. Anesthesiology. 2016;124(1):19–24.
4. Harsoor SS, Bhaskar SB. Designing an ideal operating room complex. Indian J Anaesth. 2007;51(3):193–9.
5. Burlingame B. Operating room requirements for 2014 and beyond [Internet]. https://www.fgiguidelines.org/wp-content/uploads/2015/10/FGI_Update_ORs_140915.pdf
6. Langlands B, Ga L. The 2018 FGI guidelines proposed revisions to procedure, operating, and imaging room requirements [Internet]. https://www.fgiguidelines.org/wp-content/uploads/2017/08/SLS17_FGI_ExamProcedureOperatingImaging_170721.pdf
7. Sadrizadeh S, Aganovic A, Bogdan A, Wang C, Afshari A, Hartmann A, et al. A systematic review of operating room ventilation. J Build Eng. 2021;40:102693. https://doi.org/10.1016/j.jobe.2021.102693.
8. Christopherson DA, Yao WC, Lu M, Vijayakumar R, Sedaghat AR. High-efficiency particulate air filters in the era of COVID-19: function and efficacy. Otolaryngol Head Neck Surg. 2020;163(6):019459982094183.
9. Zhang Y, Cao G, Feng G, Xue K, Pedersen C, Mathisen HM, et al. The impact of air change rate on the air quality of surgical microenvironment in an operating room with mixing ventilation. J Build Eng. 2020;32:101770.
10. Uhl E, Zausinger S, Morhard D, Heigl T, Scheder B, Rachinger W, et al. Intraoperative computed tomography with integrated navigation system in a multidisciplinary operating suite. Neurosurgery. 2009;64(5):231–40.
11. Zhang P, Liu H, Sun Z, Wang J, Wang G. The application of O-arm and navigation system in precise localization of spinal cord lesions: a case series study. Clin Neurol Neurosurg. 2020;196:105922.
12. Václav Novák L, Hrabálek VJ, Jablonský J, Hoza J, Korčáková I, et al. The use of an O-arm in endonasal endoscopic operations of the skull base. BMC Surg. 2021;21(1):58.
13. Carl B, Bopp M, Saß B, Nimsky C. Intraoperative computed tomography as reliable navigation registration device in 200 cranial procedures. Acta Neurochir. 2018;160(9):1681–9.
14. Tonn JC, Schichor C, Schnell O, Zausinger S, Uhl E, Morhard D, et al. Intraoperative computed tomography. Acta Neurochir Suppl. 2011;109:163–7.
15. Thorsteinsdottir J, Sandner TA, Biczok A, Forbrig R, Siller S, Bernasconi P, et al. Detection of impending perfusion deficits by intraoperative computed tomography (iCT) in aneurysm surgery of the anterior circulation. Acta Neurochir (Wien). 2021;163:3501.
16. Schichor C, Rachinger W, Morhard D, Zausinger S, Heigl TJ, Reiser M, et al. Intraoperative computed tomography angiography with computed tomography perfusion imaging in vascular neurosurgery: feasibility of a new concept. J Neurosurg. 2010;112(4):722–8.
17. Uluç K, Kujoth GC, Başkaya MK. Operating microscopes: past, present, and future. Neurosurg Focus. 2009;27(3):E4.
18. Maurer S, Prinz V, Qasem LE, Lucia KE, Rösler J, Picht T, et al. Evaluation of a novel three-dimensional robotic digital microscope (Aeos) in neurosurgery. Cancers. 2021;13(17):4273.
19. Motov S, Bonk MN, Krauss P, Wolfert C, Steininger K, Picht T, et al. Implementation of a three-dimensional (3D) robotic digital microscope (AEOS) in spinal procedures. Sci Rep. 2022;12(1):22553.
20. Sufianov A, Kasper EM, Sufianov RA. An optimized technique of endoscopic third ventriculocisternostomy (ETV) for children with occlusive hydrocephalus. Neurosurg Rev. 2018;41(3):851–9.
21. Simfukwe K, Iakimov IA, Sufianov RA, Luis B, Mastronardi L, Shumadalova A. Application of intraoperative ultrasound navigation in neurosurgery. Front Surg. 2022;9:900986.
22. Spetzler RF, Samuelson RM. The evolution of neurosurgical instrumentation. World Neurosurg. 2011;75(1):20–1.
23. Sufianov A. Lightweight electrosurgical forceps for neurosurgery. Patent for utility model. 2019;RU 187789 U1.
24. Yang HS, Park JY. 3D printer application for endoscope-assisted spine surgery instrument development: from prototype instruments to patient-specific 3D models. Yonsei Med J. 2020;61(1):94.
25. Lador R, Regev G, Salame K, Khashan M, Lidar Z. Use of 3D printing technology in complex spine surgeries. World Neurosurg. 2019;133:e327.

Part II

Vascular

5 A Practical Guide to Cerebrovascular Anatomy

Igor L. Maldonado

Learning Objectives
- Origin of cerebral vascularization in the supra-aortic trunks
- Gross anatomy and subdivisions of the internal carotid artery
- Arterial branches of the supratentorial space and relationships with the human cerebrum
- Basis of the anatomical interpretation of cerebral angiography

5.1 Introduction

Cerebrovascular anatomy is of utmost importance to the neurosurgeon, as the vascular system plays a critical role in maintaining brain function. Detailed knowledge of the organization of intracranial arteries allows the neurosurgeon to understand the complexity of the brain's blood supply and identify possible anatomical variations. This understanding is fundamental for the planning of surgical procedures, such as the treatment of aneurysms, arteriovenous malformations, and the removal of intracranial tumors, as well as revascularization procedures. Furthermore, understanding the cerebrovascular anatomy is crucial to minimizing the risks of damage to blood vessels and surrounding neural tissues, ensuring better surgical results, avoiding complications, and preserving the patient's quality of life.

Cerebrovascular imaging plays an essential role in neurological evaluation and surgical planning. It provides information not only about the anatomy of vessels but also about the underlying brain structure. This wealth of information is because the vessels and the brain are harmoniously coupled. Therefore, imaging tests, such as cerebral angiography and magnetic resonance or computerized tomography angiography, offer a comprehensive view of the vasculature, allowing a detailed analysis of vessels and their connections with the surrounding brain tissue.

These examinations are exceptionally rich in detail, revealing useful information for diagnosing and treating both vascular and nonvascular pathologies. This is particularly the case for digital subtraction angiography (DSA), considered the golden standard study modality. The complexity of these images requires neurosurgeons to study them calmly and timely, ensuring an accurate and reasoned interpretation, resulting in safer and more effective treatment for patients.

In this context, this chapter aims to provide the neurosurgeon with a practical guide for understanding the anatomical basis of the supratentorial vasculature, interpreting arteriographic images, and recognizing the main cerebrovascular anatomical relationships.

5.2 Anatomy Highlights

- Two interconnected systems provide the arterial supply to the cerebral hemispheres: the anterior or carotid system and the posterior or vertebrobasilar system.
- When penetrating the skull, the internal carotid artery presents curvatures that give rise to its subdivision into cervical, petrous, lacerous, cavernous, clinoidal, and supraclinoidal segments. Different numbering systems exist without exact correspondence among them.
- The supraclinoidal carotid artery travels in the subarachnoid space and is subdivided into the ophthalmic, communicating, choroidal, and carotid bifurcation segments.
- The intracavernous internal carotid artery (ICA) supplies branches to the dura mater of the skull base, tentorium,

I. L. Maldonado (✉)
CHRU de Tours, Tours, France

Laboratoire d'Anatomie, Faculté de Médecine de Tours, Tours, France

INSERM, Imaging Brain & Neuropsychiatry iBraiN U1253, University of Tours, Tours, France
e-mail: limamaldonado@univ-tours.fr

I. N. Kanaan, V. Beneš (eds.), *Neuroanatomy Guidance to Successful Neurosurgical Interventions*,
https://doi.org/10.1007/978-3-031-59838-8_5

cranial nerve foramina, and pituitary gland through mainly the inferolateral and meningohypophyseal trunks.

- The anterior cerebral artery and its branches are subdivided into pre-communicating (A1), post-communicating (A2), pre-callosal (A3), supra-callosal (A4), and terminal (A5) segments. These segments and their branches indicate the relative position of the corpus callosum, which forms the roof of the lateral ventricle and structures on the medial aspect of the cerebral hemisphere.
- The middle cerebral artery and its branches are subdivided into sphenoid (M1, horizontal), insular (M2), opercular (M3), cortical (M4), and terminal (M5) segments. Their angiographic aspect indicates, among others, the relative position of structures of the central core, cerebral opercula, and the lateral aspect of the cerebral hemisphere.
- The posterior cerebral artery and its branches are subdivided into pre-communicating (P1), post-communicating-peduncular (P2a, anterior part of P2), post-communicating-ambient (P2p, posterior part of P2), quadrigeminal (P3), cortical (P4), and terminal (P5) segments. Their angiographic appearance indicates, among others, the relative position of the midbrain and posterior structures of the ventral and medial aspect of the cerebral hemisphere, such as the precuneus, calcarine fissure, and lingula.
- Several anatomical variants exist regarding the anatomy of the arterial circle of the brain, among which the most frequent ones involve hypoplasia of a communicating artery or a pre-communicating arterial segment.

5.3 Technical Aspects of Cerebral Angiography

Undoubtedly, the main name related to the development of cerebral angiography at the beginning of the twentieth century was António Caetano de Abreu Freire Egas Moniz, a prominent Portuguese physician. At that time and according to his developed protocols, cerebral angiography was performed through a direct puncture of the common carotid artery in the neck without any image post-processing for bone subtraction. Curiously, Egaz Moniz shared a Nobel prize in 1949 for the development of frontal leucotomy, considered one of the most controversial Nobel prizes ever awarded [1]. Regardless of this anecdote that continues to generate polemic today, he was responsible for major advances in vascular neurology years before, among which a major contribution was the development of cerebral angiography.

Cerebral angiography is rarely performed using direct carotid puncture nowadays. The most employed approach is the femoral artery puncture in the groin, though radial and brachial approaches are also possible in specific situations. Basically, the operator certifies that the lumen of the femoral artery (not the vein) is reached, given the characteristics of the reflux, such as color and pulsatility, and replaces the needle with a vascular sheath with the help of a guidewire. The sheath is equipped with an antireflux valve which allows the introduction of the angiographic catheter and navigation. Navigation is then performed until the neck for most diagnostic purposes or into the intracranial space for therapeutic procedures, mainly through the carotid or vertebral arteries. The catheter progression is followed through radioscopy in the angio suite. The vessels are opacified by the soluble contrast medium, and the two-dimensional native images are acquired.

In recent years, there have been notable advances in DSA image post-processing. Despite being an invasive examination, the operator can visualize cerebral blood vessels with high resolution. Technological developments have significantly improved the speed of image acquisition, reducing the amount of contrast required and, at the same time, maintaining the quality of the images obtained. This has resulted in safer and more comfortable examinations for patients, in addition to greater accuracy. The introduction of advanced image processing algorithms has improved the visualization of vascular structures, making images clearer and facilitating interpretation. Among the main advances are those able to perform three-dimensional reconstruction and 4D modalities that combine DSA with the temporal dimension, creating a dynamic visualization of cerebral blood vessels in real-time [2].

5.4 Origins of the Cerebral Arterial Supply

The vascularization of the encephalon can be subdivided into two interconnected systems: the anterior or carotid system and the posterior or vertebrobasilar system. An essential element connecting these systems is the arterial circle of the brain, commonly referred to as the anastomotic polygon of Willis. Understanding this network of arterial connections is crucial in comprehending the complexities of blood flow within the brain.

The vascularization of the encephalon originates from the supra-aortic trunks. Aortography provides a clear view of the emergence of supra-aortic trunks but is not mandatory for cerebral angiography and is usually not performed. From proximal to distal, the innominate artery or brachiocephalic artery is the first aorta's large aortic branch (after the coronary arteries) and the source artery for the right subclavian and common carotid arteries. The vertebral arteries are the first branches from the superior aspect of each subclavian artery. The second supra-aortic trunk is the

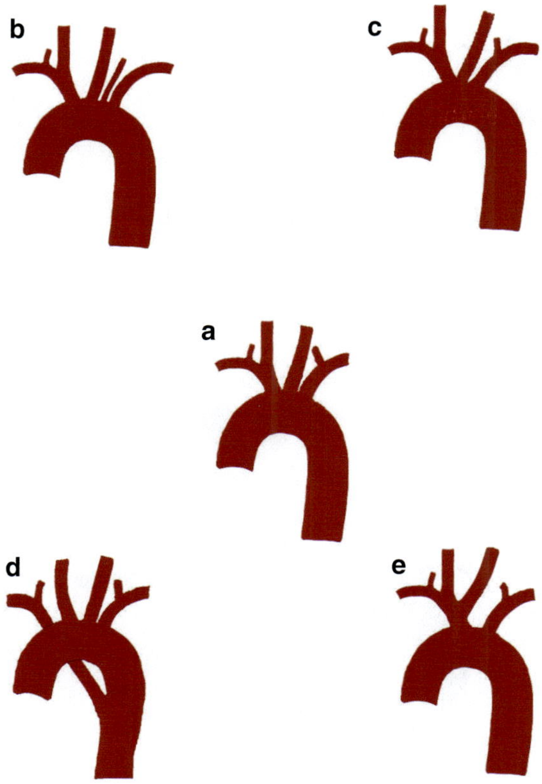

Fig. 5.1 Supra-aortic trunks and their main variations. (**a**) Typical pattern. (**b**) Left vertebral artery arising directly from the aortic arch. (**c**) Adjacent ostia of the brachiocephalic and left carotid arteries. (**d**) Aberrant right subclavian artery. (**e**) Common trunk for brachiocephalic and left carotid arteries ("bovine arch")

left common or primitive carotid artery, which emerges alone from the aortic arch in most cases. Finally, the third trunk is the left subclavian artery, which gives rise to the left vertebral artery.

Some variations can pose challenges during catheterization procedures, some of which are exceedingly rare while others are relatively common (Fig. 5.1). Among the more frequent ones are situations where the left common carotid artery emerges alongside the brachiocephalic artery or directly from the same ostium (the so-called "bovine" aortic arch variant). The vertebral artery may emerge directly from the aorta. Another less frequent situation is when a different order of appearance of the supra-aortic trunks is observed, with the right subclavian artery arising distally from the left subclavian (aberrant right subclavian artery, also known as *arteria lusoria*) [3–5].

The fact that the intricate vascularization of the encephalon is made up of connections of the aforementioned arterial axes holds significant clinical importance. Understanding the connections between the anterior and posterior systems and recognizing the variations are vital for a successful neurovascular procedure in either open or endovascular surgery.

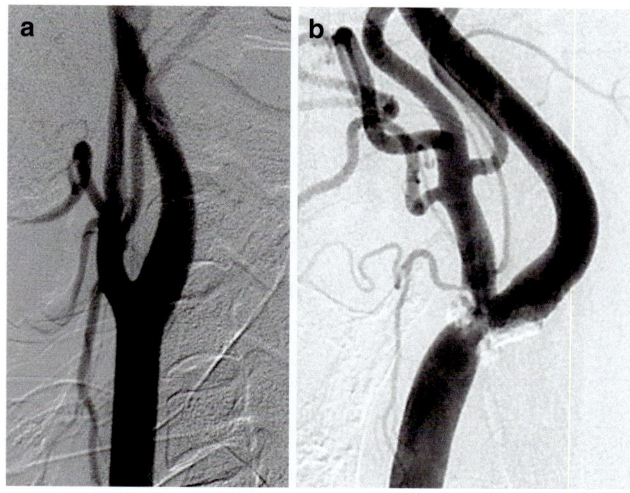

Fig. 5.2 Angiographic aspect of the internal carotid artery. (**a**) Healthy young subject (lateral view). (**b**) Elderly patient with atherosclerotic stenosis (front view)

The internal carotid artery exhibits distinct characteristics according to the patient's range of age. In young, healthy individuals, it typically assumes a posterior position relative to the external carotid artery and displays a slight lateral inclination as it enters the carotid canal. In lateral views, the ICA is easily distinguishable from the external carotid artery due to the absence of branches (Fig. 5.2). Conversely, the external carotid artery shows various branches, typically: i) the superior thyroidal, facial, and lingual arteries in its anterior aspect; ii) the ascending pharyngeal, occipital, and posterior auricular artery in its posterior aspect; and iii) the maxillary and superficial temporal artery in the terminal bifurcation. The ICA very rarely presents a thin cervical branch, an ascending pharyngeal artery. Generally, all cervical branches come from the external carotid artery, and the ICA remains branchless in the neck.

Another significant feature of the ICA is the carotid sinus or carotid bulb, a region where the lumen appears enlarged compared to the more distal ICA. This area is a major baroreceptor site in humans and is a frequent site of atherosclerotic stenosis. As individuals age, the ICA may elongate, and its diameter can naturally change over time, often resulting in curves, kinkings, or loops. Consequently, the coronal view (instead of a lateral view) may be more appropriate to depict the carotid bifurcation in many elderly patients.

5.5 Internal Carotid Artery in the Skull Base and Intracranial Space

As the ICA enters the skull, it is possible to recognize several segments. They have been named according to different proposed classification systems. The most used ones are those proposed by Fischer, Gibo, and Bouthillier [3, 6, 7] (Fig. 5.3).

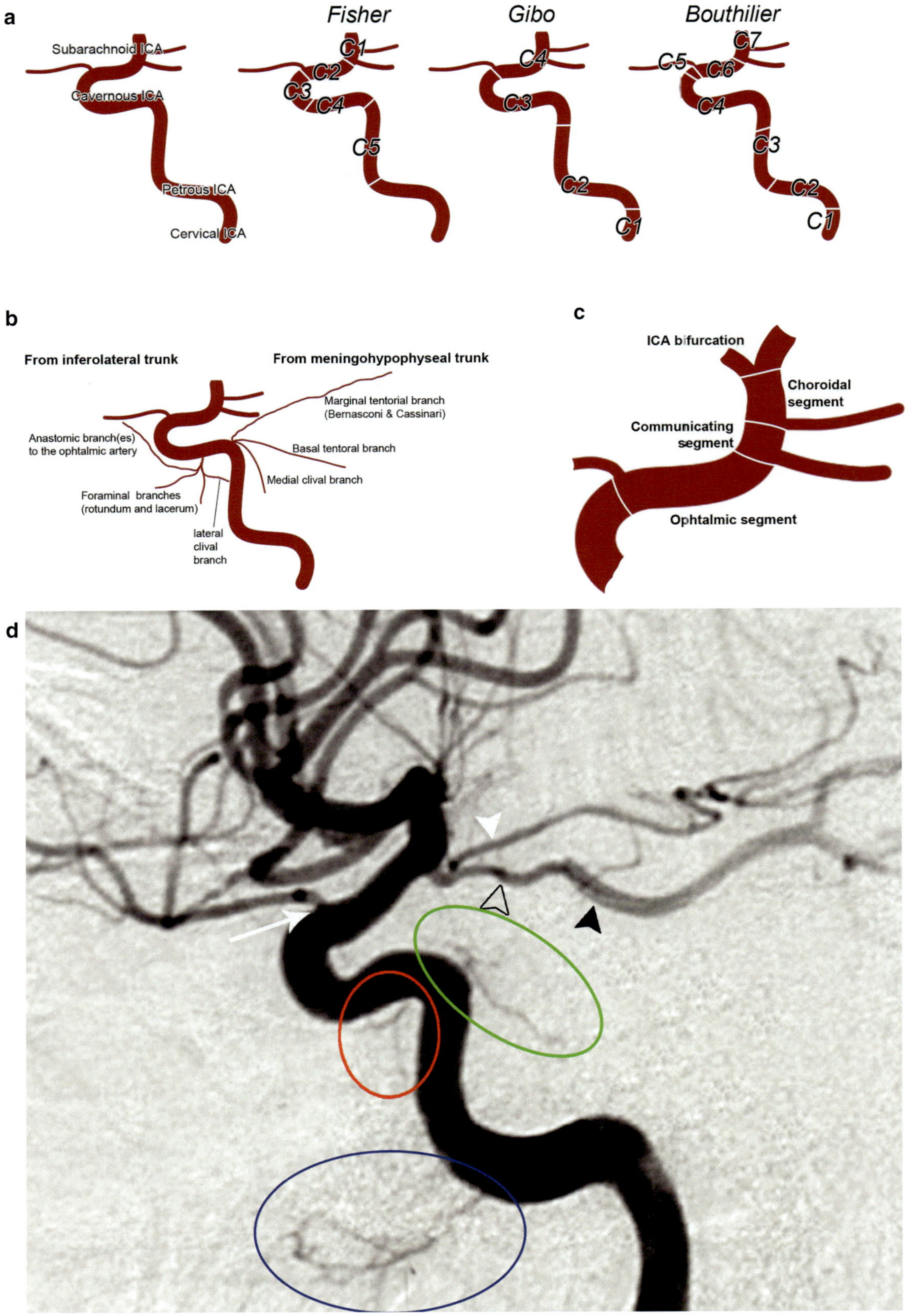

Fig. 5.3 Internal carotid artery segments, main classification systems. (**a**) Common terminology and correspondence among different numbering systems. (**b**) Intracavernous branches. (**c**) Subarachnoid segments (According to Rhoton et al.). (**d**) Aspect on digital subtraction angiography (a lateral view). White arrowhead, anterior choroidal artery; empty arrowhead, posterior communicating artery; black arrowhead, post-communicating segment of the posterior cerebral artery; full white arrow, emergence of the opthalmic artery; green ellipse, meningohypophyseal trunk and its branches; red ellipse, inferolateral trunk and its branches; and blue ellipse, prominent mandibulovidian artery (usually not seen in most angiograms) of the petrous ICA and its branches

It is worth noting that these classification methods differ in their subdivisions and numbering direction, which imposes the operator always to specify which system he/she uses for the sake of proper communication.

The petrous segment constitutes the portion of the ICA immediately after penetration into the carotid canal. Within the petrosal segment, tiny branches, such as the mandibulovidian, tympanic, or stapedious arteries, are challenging to see due to their inconstant character and very small size. The petrous segment is followed by the lacerum or pre-cavernous segment. The ICA then curves twice within the cavernous sinus forming the carotid siphon. Upon entering the subarachnoid space, the ICA gives rise to its first major branch, the ophthalmic artery. The ophthalmic artery's origin is a landmark for the superior limit of the cavernous sinus. The operator must be aware that this is a variable situation, as the origin of the ophthalmic artery can nonrarely be intracavernous [8].

The intracavernous branches of the ICA are grouped into two main groups: those originating from the meningohypophyseal trunk and those from the inferolateral trunk. The inferior hypophyseal artery originates directly from the cavernous ICA or as a meningohypophyseal trunk branch. The meningohypophyseal trunk gives rise to long, tiny branches that supply the tentorium and the dura of the clivus, including the medial clival branch. For the tentorium, one branch runs along its petrous attachment (the basal tentorial artery), while the other follows the free tentorial border and is referred to as the marginal tentorial branch or, still, the "artery of Bernasconi and Cassinari." The latter typically present small tortuosities along a predominantly ascendant trajectory. These two branches may seem similar in lateral views, but they can be effortlessly distinguished in front views, as only the marginal tentorial branch of Bernasconi and Cassinari outlines the tentorial margin. Less frequently, this artery arises from the inferolateral trunk.

The inferolateral trunk typically gives origin to branches to the clivus dura mater, as well as foramen rotundum and lacerum. Its distal branches are commonly observed to anastomose with the ophthalmic artery. This pattern represents the most prevalent arrangement in the intracavernous region. Such minor branches are typically challenging to visualize and often become evident under increased flow conditions, as observed in cases involving skull base tumors, arteriovenous malformations, and fistulas.

For a precise identification of subarachnoid segments, a subdivision system has been widely used and is presented in Fig. 5.3c [6]. The ophthalmic segment is defined from the proximal emergence of the ophthalmic artery to the proximal portion of the ostium of the posterior communicating artery (PCom). The communicating segment extends from the PCom to the proximal portion of the ostium of the anterior choroidal artery, while the choroidal segment spans from the anterior choroidal artery to the carotid bifurcation. These segments can be promptly recognized in angiograms, each exhibiting specific diameter changes and contrast flow as described below.

Understanding the ramifications of the internal carotid artery is important in diagnosing and treating various neurological conditions. The classification of segments, their unique characteristics, and the connections between arterial branches provide valuable insights for open and endovascular surgery alike. As we delve deeper into this intricate vascular network, further observations enhance our knowledge of the patient's individual anatomy and cerebral hemodynamics and contribute to more effective procedures.

5.6 Arterial Branches in the Supratentorial Space

In lateral and oblique views, the ophthalmic artery is the first large branch of the intracranial ICA and is prominently recognizable. Among its branches is the central artery of the retina. A choroidal blush is usually observed in later phases of contrast passage. Prior to embolization procedures involving the ophthalmic vascular territory, confirming the origin of the choroidal blush is crucial. This confirmation necessitates the selective catheterization of the internal and external carotid arteries. This precaution is particularly important due to the frequent presence of anastomoses between the ophthalmic and facial artery territories, which occur through the angular branch located at the medial canthus [9].

It is important to note that although the ophthalmic artery is often utilized as a landmark to denote the superior limit of the cavernous sinus, it may not be entirely precise. Variations may exist, and one must be mindful of the small extension of subarachnoid space along the medial aspect of the ICA, referred to as the carotid cave. As a result, the presence of an aneurysm in this region may be mistakenly identified as intracavernous when, in reality, it may have a partial or complete subarachnoid situation in the carotid cave. The superior hypophyseal arteries emerge from the ophthalmic or clinoid segment of the ICA [10].

More superiorly, the PCom and the anterior choroidal artery follow. The posterior communicating artery displays a distinct change in direction and diameter as it approaches the posterior cerebral artery on angiograms (Fig. 5.4). The posterior cerebral artery continues its path to the occipital lobe and further branches into the calcarine and parieto-occipital arteries.

Regarding more distal branches, the anterior cerebral artery is subdivided into five segments (in the direction of the blood flow): A1 or pre-communicating segment, A2 or post-communicating (infracallosal) segment, A3 or precallosal segment (around the genu of the corpus callosum, the bifur-

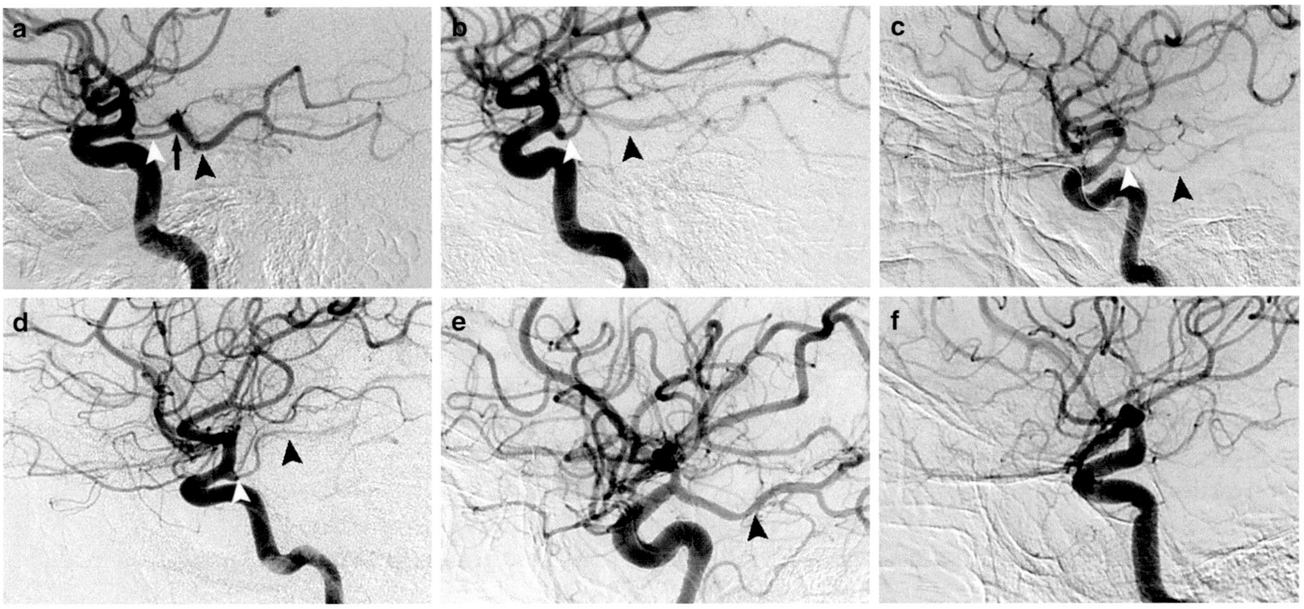

Fig. 5.4 Supraclinoidal branches of the internal carotid artery and variants of the PCom. Aspect on digital subtraction angiography. (**a**) Strong opacification of both PCom and post-communicating PCA. Note the angle that is typically seen between PCom and P2 (vertical black arrow). (**b**) Contrast washing of the P2 segment and distal PCA, highlighting the presence of a functional P1 segment (although not visible in the image). (**c**) Hypoplastic PCom. (**d**) PCom origin with infundibulum. (**e**) Fetal variant. (**f**) Absent PCom. White arrowheads, posterior communicating artery and black arrowheads, posterior cerebral artery

cates in pericallosal and callosomarginal branches are usually located in this segment), A4 or supracallosal, and A5 or the terminal postcallosal branch (that often anastomoses with the posterior cerebral artery territory). These segments play essential roles in supplying blood to specific brain regions (Fig. 5.5).

The perforators of the anterior cerebral artery supply part of the anterior commissure, the anterior pillars of the fornix, the anterior hypothalamus, and the anterior portion of the striatum. Few perforators originate from the anterior communicating complex, in proximity to the communicating artery, and they may be destined for the fornix, septal area, anterior cingulate gyrus, and corpus callosum. Those that arise directly from the anterior communicating artery are usually small. The anterior cerebral artery contributes to medial lenticuloestriate branches, one of which is particularly long and recurrent and penetrates the anterior perforate substance irrigating basal nuclei, especially the head of caudate: the recurrent artery of Heubner. This artery is a branch of the A2 segment in most cases.

In the medial cerebral aspect, the pericallosal artery courses along or very near the sulcus of the corpus callosum, while the callosomarginal artery reaches the cingulate sulcus. Two branches often reach the upper edge of the cerebral hemisphere. While the most posterior one crosses the precuneus, the anterior one, coming from the callosomarginal artery, penetrates the marginal branch of the cingulate sulcus. The operator then has a reference for an important structure in the medial aspect of the cerebral hemisphere, which is the paracentral lobule. It is situated above the cingulate sulcus, anterior to the marginal branch, and anterior to the precuneus. The posterior superior portion of the paracentral lobules has the superior extremity of the central sulcus in its superior portion, which contains a terminating branch of the central branch of the middle cerebral artery. Importantly, the paracentral lobule contains the somatotopic motor and sensory maps of the contralateral lower limb.

Similarly, the middle cerebral artery is subdivided into segments, as follows: M1 (horizontal or sphenoidal portion), M2 (insular), M3 (opercular), M4 (cortical), and M5 (terminal) branches. The curves of the middle cerebral artery serve notably as useful anatomical landmarks, with the transition from M1 to M2 indicating the location of the limen of the insula (in the subarachnoid space, sometimes called the "falciform fold"). DSA image gives us an overview of the arrangement of the cerebral hemisphere, especially in the late arterial or early parenchymal ("capillary") phase. The M2 branches frame the cortex of the insula, and we can promptly deduce the location of the basal ganglia (putamen and pallidum) and lateral lenticulostriate arteries in a front-view angiogram.

Along the M1 segment, on its upper aspect, we see the emergence of lenticulostriate arteries. The penetration zone is the anterior perforated space. This means that they penetrate behind the olfactory stria. In open surgery, the olfactory tract is a useful landmark. If one follows the direction of the

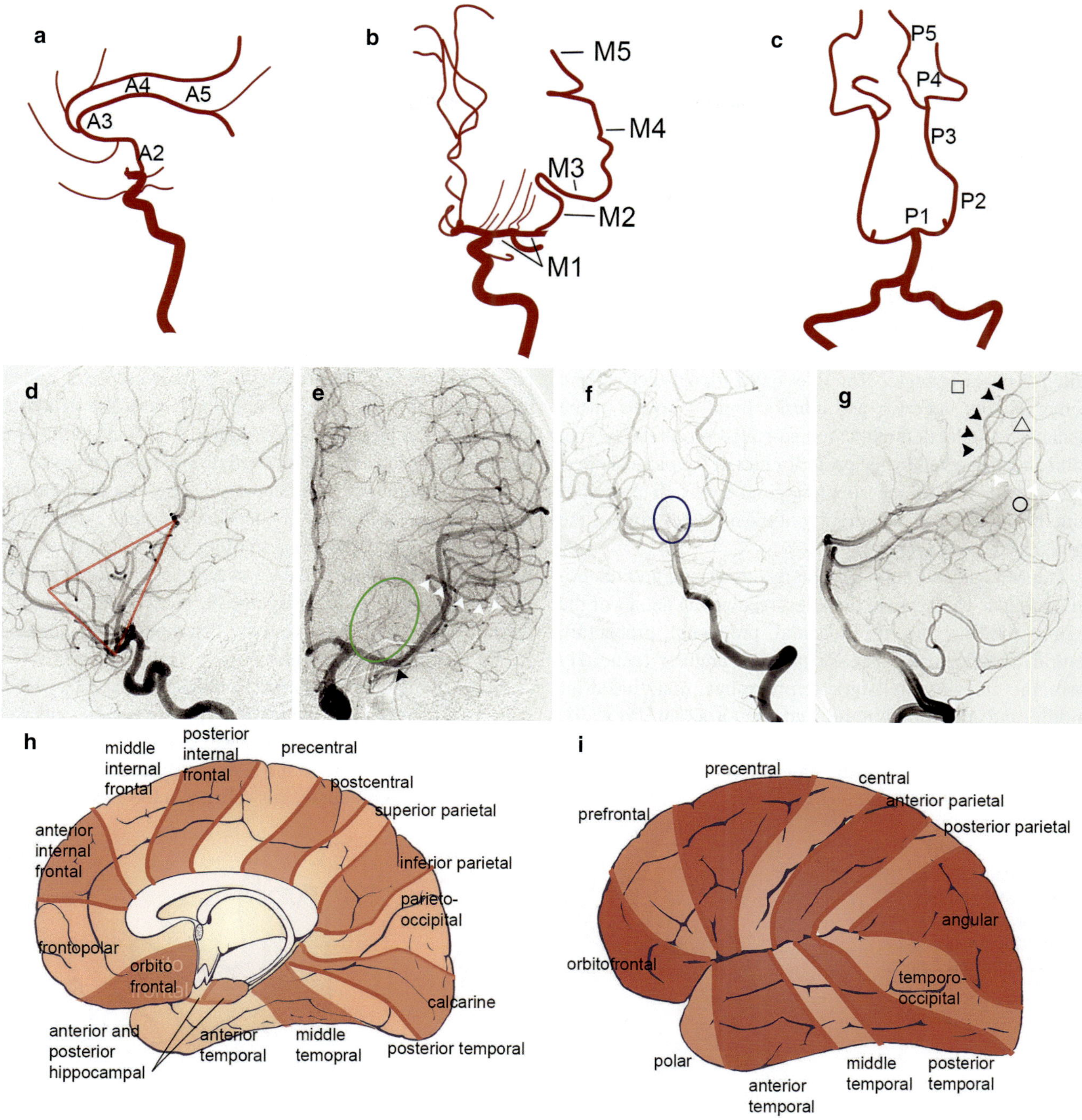

Fig. 5.5 Segments of the anterior (**a**), middle (**b**), and posterior (**c**) cerebral arteries, their appearance on digital subtraction angiography and examples of relevant cerebrovascular anatomical relationships. (**d**) Lateral view. The red triangle marks the so-called Sylvian triangle. Note the significant changes in the direction of the middle cerebral artery branches in the circular sulcus of the insula. (**e**) Frontal view. The M1–M2 transition is the main angiographic landmark for the limen of the insula (black arrowhead). The M2 branches and the lateral lenticulostriate arteries (green ellipse) allow the operator to visualize the situation of the central core of the cerebral hemisphere as they reveal the position of the lenticular nucleus and the cortex of the insula. The M3 segments draw the Sylvian opercula (white arrowheads). The distance between an M2–M3 transition and an M3–M4 transition reflects the thickness of the operculum at that site. (**f**) Vertebrobasilar system opacified after left vertebral artery injection, frontal view. Several important anatomical landmarks are found along the path of the posterior cerebral artery. P1 perforators reveal the position of the posterior perforated substance and the interpeduncular fossa. The P1 and P2 surround the cerebral peduncles of the midbrain before the more distal PCA progresses to the medial aspect of the cerebral hemisphere. (**g**) In the medial aspect of the occipital lobe, the calcarine (white arrowheads) and parieto-occipital (black arrowheads) branches reveal the position of the homonymous fissures, and consequently the relative position of the precuneus (□), cuneus (Δ), and lingula (○). (**h**) and (**i**). Branching pattern

olfactory tract in a straight line posteriorly, he/she will navigate directly to the olfactory stria and pass above the carotid bifurcation. Classically, the right lenticulostriate arteries assume an S-shape and the left ones have an inverted S-shape. The lenticulostriate arteries are bounded by neighboring structures such as the optic pathways medially and part of the uncus laterally. At the level of the optic tract and posteriorly, the anterior choroidal artery provides perforators that irrigate the posterior arm of the internal capsule.

The contours of the insula are identified by the deflection of insular branches in the circular sulcus of the insula, which may be seen in both views. In lateral projection, these deflections of the middle cerebral artery branches draw the so-called "Sylvian triangle." The sides of the triangle correspond to the superior, anterior, and inferior limiting sulcus of the insula. The deflection marks the so-called posterior Sylvian point and is covered by the posterior part of the parietal operculum, made up of the supramarginal gyrus. It coincides with the posteromedial extremity of the most posterior of the transverse temporal gyri.

The M3 and M4 delineate the frontoparietal and temporal opercula. The cortical branches receive the names of the regions they irrigate: orbitofrontal, prefrontal, precentral, central, anterior and posterior parietal, angular (generally form the inferior division), temporo-occipital, posterior middle, and anterior temporal, and temporopolar. Typically, the distal cortical branches provide pial anastomoses between the territories of the anterior and middle cerebral arteries.

The angular artery follows the general direction of the lateral sulcus, crosses the supramarginal gyrus, and reaches the angular gyrus. It is usually a large branch that deals with a number of higher functions, such as language circuits in the left hemisphere and visuospatial integration mechanisms in the right hemisphere of most individuals. It is thus an important anatomic reference for eloquent areas of the cortex. The three or four middle cerebral artery branches above and anterior to it are also destinated to the eloquent cortex, including primary motor and sensory cortices.

That is also the case of one or two temporal branches destined to Wenicke's area in the posterosuperior portions of the temporal lobe. More anterior temporal infarctions are generally better tolerated. Moreover, due to the presence of optical radiations, Meyer's loop, around the ventricular temporal horn, lesions of these branches are often associated with visual field defects, such as superior contralateral quadrantopsy.

Understanding the vascular distribution of the cerebral hemisphere allows us to understand vascular syndromes (Fig. 5.6). Given the disposition of the motor homunculus in which the somatotopy of the lower limb is located on the medial surface of the hemisphere, a middle cerebral artery infarction classically courses with disproportionate hemiparesis with brachial predominance.

In the dominant hemisphere, middle cerebral artery ischemia is also related to expression aphasia for the superior division and comprehension aphasia for the inferior division branch, classically, due to the distribution of the mentioned language areas. Of course, this scheme is too much simplis-

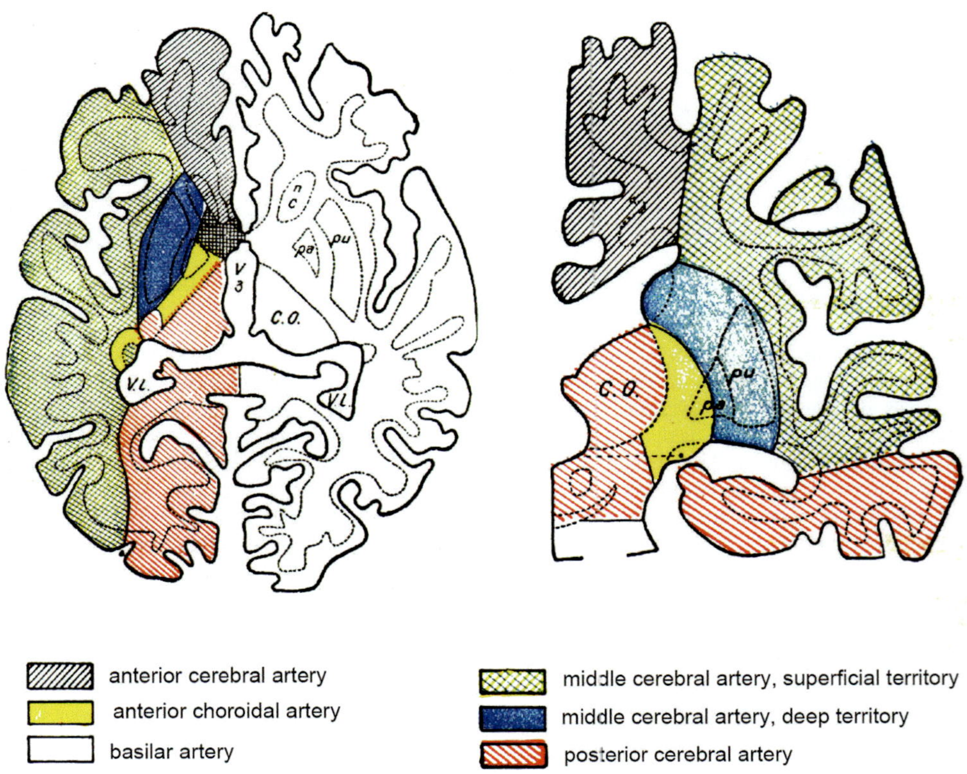

Fig. 5.6 Main vascular territories of the human cerebral hemisphere. According to the classical works of Testut and Latarjet (1899)

tic. It does not reflect at all the numerous combinations we can observe of superficial and deep infarction with transcortical aphasia, conduction aphasia, or even the anatomical variants we can have with the leptomeninges anastomoses, variations of these territories, or even functional interindividual specificities.

The inferior parietal lobule is often associated with visuospatial integration in the nondominant hemisphere. Damage to this region has been related to hemineglect, particularly difficult to rehabilitate, or to Gerstman syndrome with other associated manifestations such as acalculia, digital agnosia, or left–right confusion.

Regarding the posterior cerebral artery, four segments are usually described: P1 (pre-communicating), P2 (post-communicating with peduncular and ambient portions), P3 (quadrigeminal), and P4 (cortical/terminal, calcarine). The P2 segment can be subdivided into P2A (peduncular or crural) and P2P (ambient).

The precommunicating P1 segment relates to the interpeduncular fossa. Its perforators penetrate the posterior perforated space between the two cerebral peduncles, many of which are thalamoperforators. Sometimes a thalamic artery originates exclusively or predominantly on one side, the so-called Percheron's artery. The P1 segment is related inferiorly to the third nerve, which emerges above the superior cerebellar artery.

Among the cortical branches of the PCA are the perisplenial, parietooccipital, and calcarine arteries, and two or more temporal branches. The periesplenial artery nonrarely continues as a posterior pericallosal artery and provides a site of anastomosis with the ACA territory. In this situation, the operator may have the exact position of the splenius visualizable in an angiographic examination.

One horizontal branch enters the calcarine fissure (the calcarine artery) and a postero-superior one enters the parietal occipital sulcus (the parieto-occipital artery). Consequently, the lingula, forming the inferior lip of the calcarine fissure, and the boundaries of the cuneus and precuneus are promptly recognizable, thanks to this ramification pattern in a lateral view angiogram. Among the fine branches, a medial (more proximal) and a lateral (more distal) posterior choroidal artery arise from the P2 segment of the PCA as a single vessel in most cases. Less frequently, they may also arise as a duplicate of multiple rami from a nearby branch of the PCA.

5.7 Surgical Pearls

- The emergence of the ophthalmic artery serves as an approximate angiographic landmark for the start of the ICA's subarachnoid portion; the operator should be aware of frequent anatomical variations.
- A variable small extension of the subarachnoid space—the carotid cave—lies medial to the clinoid ICA; aneurysmal formations at this site may mimic an intracavernous location on lateral view angiograms.
- Two main branches of the meningiohypophyseal trunk contribute to the arterial irrigation of the tentorium (tentorial marginal artery and tentorial basal artery) and may increase in volume in cases of vascular (arteriovenous malformation or fistula) or tumoral (meningioma) pathology.
- During operative planning, the cerebral vasculature should always be studied considering the individual anatomy; comparing vascular and sulco-gyral anatomy is of great value for anatomical navigation.
- The pericallosal and marginal callosal arteries indicate the position of the corpus callosum and the cingulate gyrus in the lateral view.
- The M1–M2 transition of the middle cerebral artery in a face view reveals the position of the limen of the insula; together with the perforating and M2 branches, it helps understand the location of the anterior perforating space, lenticular nucleus, and insular cortex.
- The M2–M3 and M3–M4 transitions delineate the fronto-parietal and temporal opercula in face view; in profile view, the M2–M3 transitions reveal the boundaries of the insula (Sylvian triangle).
- In face view, the P1 and P2 segments reveal the cerebral peduncles; P1 perforators indicate the relative position of the posterior perforated substance.
- In lateral view, the posterior cerebral artery bifurcation reveals the position of the parieto-occipital fissure and calcarine fissure.

5.8 Discussion

Recognition of vascular anatomy is a fundamental step in many neurosurgical procedures. In the context of surgical planning, the study of the cerebral vasculature must take place on an individual scale (individual anatomy) in view of the—sometimes significant—anatomical variants that may exist.

The arterial circle of the brain exhibits numerous possible variations, including asymmetry of the PCom arteries due to unilateral hypoplasia, P1 hypoplasia, or fetal variants. Variations in this region are relatively frequent. Some cases show dominant flow from the ICA and excellent opacification of the PCom. In contrast, others exhibit better opacification of the PCom and poor opacification of the P2 segment when the ICA is injected, demonstrating the presence of a functional P1 segment and significant contrast medium dilution. The PCom may also be very small, almost absent, or be associated with an infundibular formation (enlarged origin, also nonrarely seen at the origin of the anterior choroidal artery). In an opposite arrangement, fetal variants of the posterior cerebral

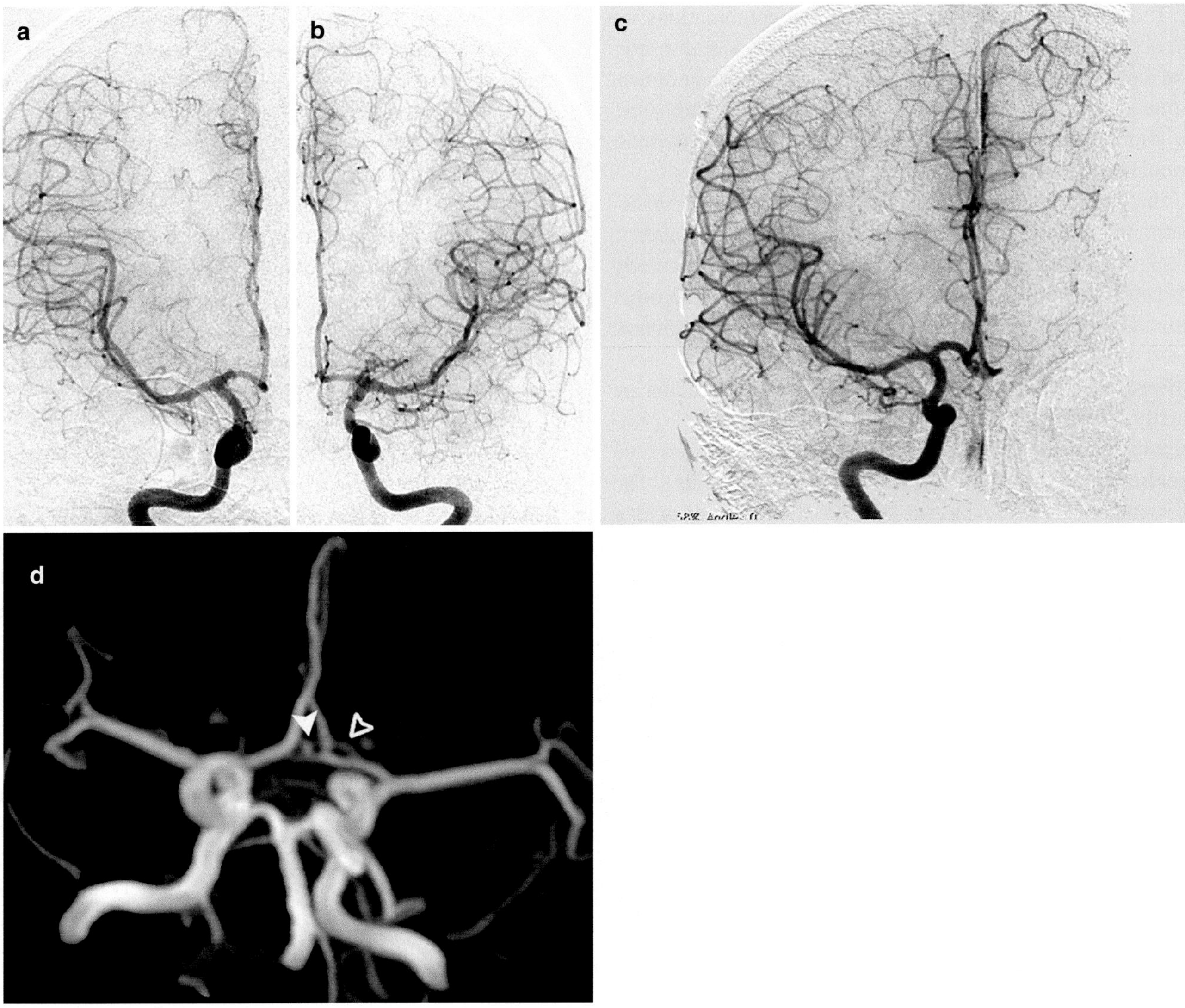

Fig. 5.7 Variants of the anterior cerebral and anterior communicating arteries. (**a** and **b**) Symmetric system. (**c**) Hypoplastic left anterior communicating artery with a large asymmetrical right A1 (both A2 are filling from the right A1). (**d**) Fenestrations of left A1 (empty arrowhead) and anterior communicating artery (solid white arrowhead) arteries

artery can also occur, where its flow originates from the ICA, and the PCom diameter is as large as the posterior cerebral artery, suggesting a hypoplastic or absent P1 segment.

Common variants of the anterior communicating complex are also observed, with varying diameters of the anterior communicating artery and asymmetric A1 segments, leading to a dominance phenomenon where one of the ICAs and A1 is responsible for filling both A2 segments. Surgical or endovascular disconnection of the anterior communicating artery, if needed, should be reserved for cases in which both A1 segments are functional without significant hypoplasia or asymmetry. Azygos A2 and duplications of certain elements are not uncommon. The so-called arteria termatica of Wilder appears as a third post-communicating segment from the anterior communicating artery. Interestingly, it is often directed to the paracentral lobule.

Figure 5.7 shows examples of some possible presentations: symmetric anterior system, each A1 filling one A2; large asymmetrical A1 (both A2 are filling from the right); and fenestrations in the A1 segment and anterior communicating artery. Figure 5.8 shows some frequent variants of the proximal middle cerebral artery. Intriguing, more proximal variants are persistent carotid-vertebrobasilar anastomoses as a remnant from embryonic development [11]. During carotid opacification, a persistent trigeminal artery produces opacification of both the ICA and the superior third of the basilar ramification tree, including the entire posterior cerebral artery. Other rare remnants are the otic, hypoglossal, and proatlantal arteries.

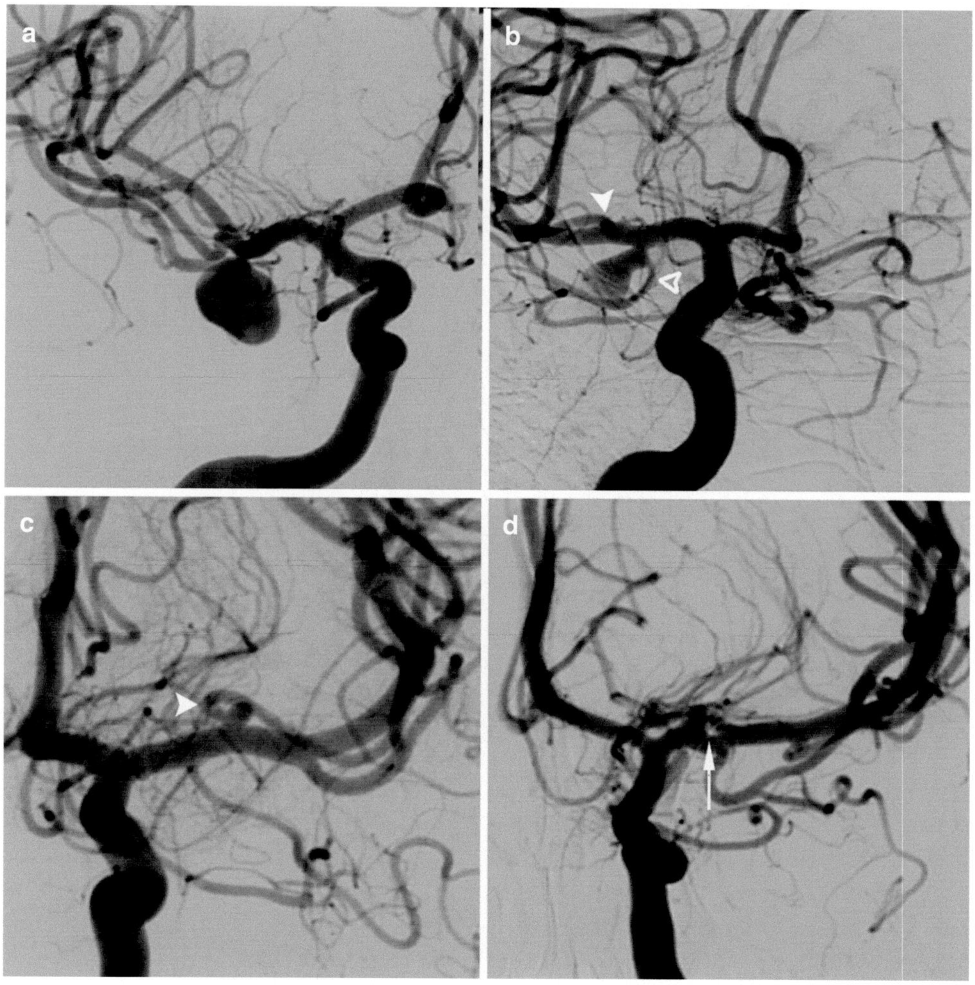

Fig. 5.8 Examples of anatomical variants of the middle cerebral artery. (**a**) Trifurcation with a saccular aneurysm. (**b**) An early temporal branch with an aneurysm (empty white arrowhead); note that an early frontal branch (solid white arrowhead) was also present in the same case. (**c**) An early frontal branch (white arrowhead) with an aneurysm. (**d**) An M1 fenestration (white thin vertical arrow) with an aneurysm

5.9 Conclusion

An in-depth understanding of cerebral arterial anatomy and its variations is crucial for accurate diagnosis and treatment of neurological conditions. Knowledge of the complex vascular network enables the neurosurgeon to navigate the intricacies of cerebral blood flow and identify potential anomalies that may impact patient care. The rich anatomy of cerebral arteries invites further research and exploration to enhance medical knowledge and improve patient outcomes in neurovascular medicine.

References

1. Artico M, Spoletini M, Fumagalli L, Biagioni F, Ryskalin L, Fornai F, et al. Egas Moniz: 90 years (1927–2017) from cerebral angiography. Front Neuroanat. 2017;11:81.
2. Ruedinger KL, Schafer S, Speidel MA, Strother CM. 4D-DSA: development and current neurovascular applications. Am J Neuroradiol. 2021;42:214–20.
3. Morris P. Practical neuroangiography. 2nd ed. Philadelphia: Lippincott Williams & Wilkins; 2007.
4. Hanneman K, Newman B, Chan F. Congenital variants and anomalies of the aortic arch. Radiographics. 2017;37:32–51.
5. Shapiro M, Raz E, Nelson PK. Aortic arch variants: a practical guide to safe and timely catheterization. Interv Neurol. 2018;7:544–55.
6. Gibo H, Lenkey C, Rhoton AL. Microsurgical anatomy of the supraclinoid portion of the internal carotid artery. J Neurosurg. 1981;55:560–74.
7. Bouthillier A, Van Loveren HR, Keller JT. Segments of the internal carotid artery: a new classification. Neurosurgery. 1996;38:425–33.
8. Matsumura Y, Nagashima M. Anatomical variations in the origin of the human ophthalmic artery with special reference to the cavernous sinus and surrounding meninges. Cells Tissues Organs. 1999;164:112–21.
9. Michalinos A, Zogana S, Kotsiomitis E, Mazarakis A, Troupis T. Anatomy of the ophthalmic artery: a review concerning its modern surgical and clinical applications. Anat Res Int. 2015;2015:1–8.
10. Cironi KA, Decater T, Iwanaga J, Dumont AS, Tubbs RS. Arterial supply to the pituitary gland: a comprehensive review. World Neurosurg. 2020;142:206–11.
11. Namba K. Carotid-vertebrobasilar anastomoses with reference to their segmental property. Neurol Med Chir (Tokyo). 2017;57:267–77.

6 Anatomy of the Vertebrobasilar System

Victor Hugo Pérez and Juan E. Small

Learning Objectives
- Label the major arterial blood vessels of the vertebrobasilar system
- Identify and describe the vertebral, basilar, cerebellar, and anterior spinal arteries as well as their branches
- Compare and contrast the differences and morphometric characteristics of different components of the vertebrobasilar system
- Discuss the course of the major branches of the vertebrobasilar system as they traverse the intracranial compartment

6.1 Introduction

A thorough understanding of the anatomy of the vertebrobasilar system is of crucial importance for the treatment of aneurysms and related tumors that affect and displace the vertebral arteries and their delicate branches such as the cerebellar arteries and the anterior and posterior spinal arteries. The details and intricacies of this vertebrobasilar system are therefore of great interest to the specialized personnel endeavoring in this field.

6.2 Anatomy Highlights

- The basilar artery originates from the union of the vertebral arteries at the pontobulbar junction.
- The vertebrobasilar system gives rise to the cerebellar arteries, known as the superior cerebellar artery (SCA), anteroinferior cerebellar artery (AICA). and posteroinferior cerebellar artery (PICA).
- The SCA is related to the oculomotor, trochlear, and trigeminal nerves, most of the times below the oculomotor nerve. The SCA originates from the basilar artery, near its bifurcation, and is divided into four segments.
- The AICA is related to the abducens, facial, and vestibulocochlear nerves. The AICA originates from the proximal part of the basilar artery and is divided into four segments.
- The PICA arises from the vertebral artery, is related to the glossopharyngeal, vagus, accessory, and hypoglossal nerves and is divided into five segments.

6.3 History and Background

The anatomy of the vertebrobasilar system is of crucial importance for the treatment of aneurysms and for avoiding vascular complications of posterior fossa surgery for tumors and neurovascular pathologies. The vertebrobasilar system gives rise to the cerebellar arteries and many perforating arteries that supply the midbrain, pons, medulla oblongata, and cerebellum. Surgeons should be well acquainted with this vaculature in order to avoid damage to vital vascular supply for the brain stem.

6.4 Basilar Artery

The basilar artery originates from the union of the vertebral arteries at the pontobulbar junction. It has an average length of 28 mm an average diameter of 3.5 mm, and 80% bifurcates at the pontomesencephalic junction [1–3]. The basilar gives rise to an average of five long circumferential arteries, predominantly in the distal third. The short circumflex arteries predominate in the proximal and middle thirds with an average of four vessels. The perforating branches predominate in the upper third with an average of six.

V. H. Pérez (✉)
Institute of Forensic Sciences, TSJ, Autonomous University of Mexico State, Mexico City, Mexico

J. E. Small
Lahey Hospital and Medical Center, Burlington, MA, USA
e-mail: Juan.E.Small@Lahey.org

I. N. Kanaan, V. Beneš (eds.), *Neuroanatomy Guidance to Successful Neurosurgical Interventions*,
https://doi.org/10.1007/978-3-031-59838-8_6

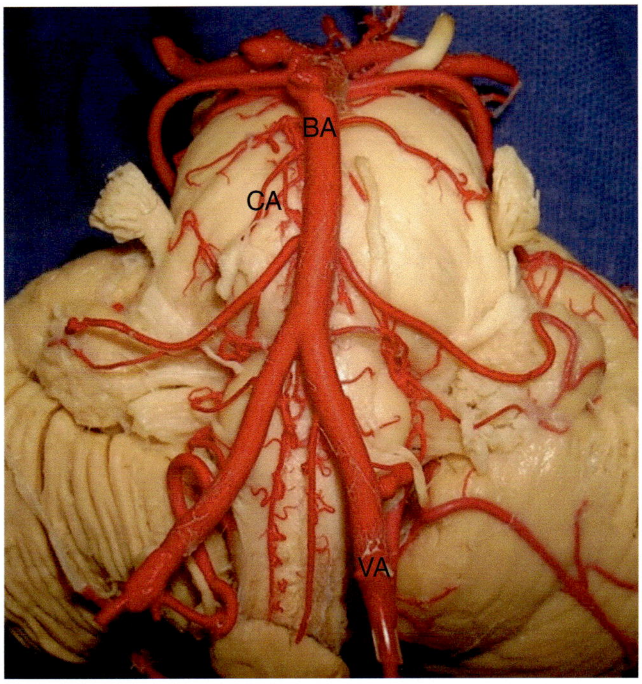

The vertebral arteries (VA) join together at the level of the potobulbar junction to form the basilar artery (BA). Note also the short circumflex arteries (CA) arising from the basilar artery

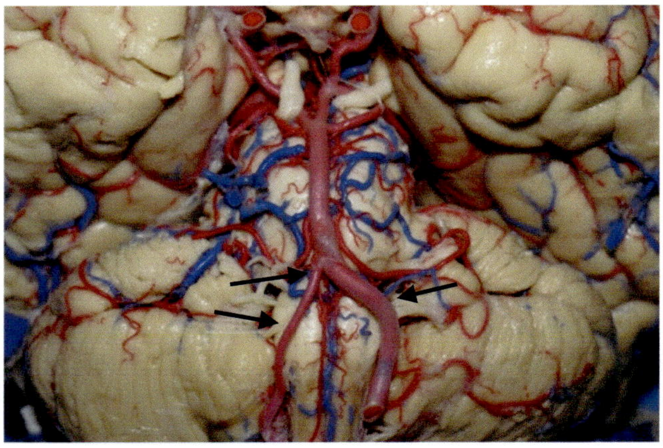

View of the ventral brainstem and the overlying posterior fossa vasculature, the asymmetry of a left dominant vertebral artery (arrow) and a hypoplastic right vertebral artery (double arrow) is evident

6.5 Superior Cerebellar Artery

The superior cerebellar artery (SCA) originates from both sides of the distal portion of the basilar artery, arising 1–2.5 mm below the basilar terminus [1–3]. A common origin of the SCA and the posterior cerebral artery can also be seen as an anatomic variant.

In our experience, the SCA on either side is single in 80% and duplicated in 20%. The SCA bifurcates into a rostral and a caudal trunk. The rostral trunk supplies the anterosuperior surface of the cerebellum and vermis. The caudal trunk supplies the surface of the cerebellar cortex, lateral to the area supplied by the rostral trunk. The SCA has been divided into four segments: anterior midbrain pontomesencephalic segment, lateral midbrain pontomesencephalic segment, cerebellar midbrain segment, and cortical segment.

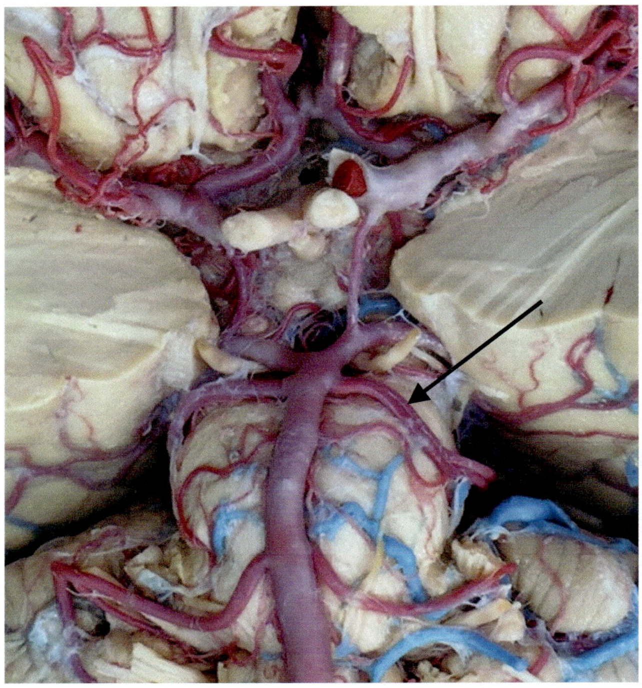

Duplicated left Superior Cerebellar Artery (arrow)

6.6 Anterior Pontomesencephalic Segment of the SCA

The anterior pontomesencephalic segment of the SCA originates from the basilar artery, just before its bifurcation at the level of the pontomesencephalic junction. It extends below the oculomotor nerve to reach the anterolateral margin of the brain stem. This segment is medial to the free edge of the tentorium.

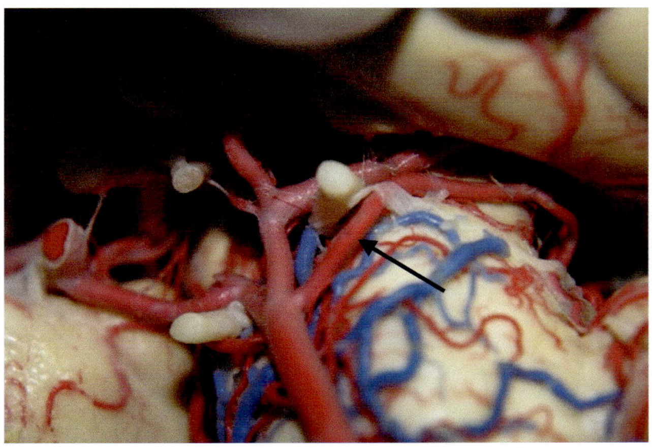

Anterior pontomesencephalic segment of the SCA (arrow)

6.7 Lateral Pontomesencephalic Segment of the SCA

The lateral pontomesencephalic segment of the SCA begins at the anterolateral margin of the brainstem, where the anterior portion is often visible above the free edge of the tentorium. It courses below the trochlear nerve and above the trigeminal nerve, curving along the lateral face of the superior portion of the pons. The caudal portion often reaches the trigeminal nerve root in the medial portion of the pons, terminating at the anterior limit of the cerebellar-mesencephalic fissure.

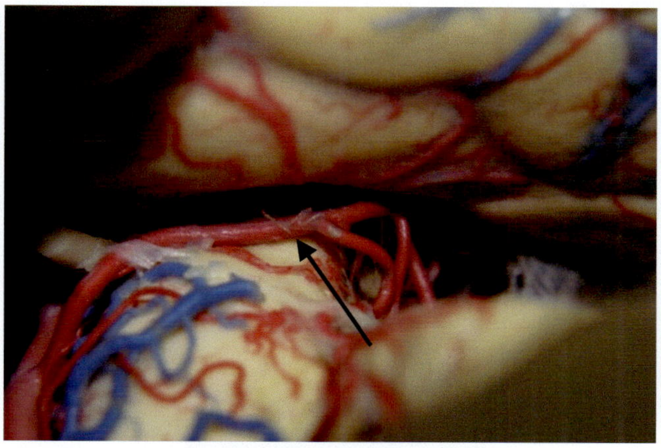

Lateral pontomesencephalic segment (arrow)

6.8 Midbrain Cerebellar Segment of the SCA

The midbrain cerebellar segment of the SCA courses within the cerebellar-mesencephalic fissure located between the cerebellum and the posterior face of the midbrain. This segment enters the posterosuperior fissure at the level of the trigeminal root entry zone, courses between the folia cerebellum and colliculi, over the superior cerebellar peduncle and anterior medullary veil, and ascends to reach the anterosuperior border of the cerebellum.

6.9 Cortical Segment of the SCA

The cortical segment of the SCA is formed by the hemispheric arteries, vermian arteries, and marginal arteries. The hemispheric arteries course toward the cerebellar-midbrain fissure and are distributed to the tentorial surface of the cerebellar hemispheres. Within the cerebellar-midbrain fissure, they send branches called precerebellar arteries to the deep nuclei of the cerebellum. The hemispheric arteries on the tentorial surface of the cerebellum divide into medial and medial–lateral branches. The vermian arteries derive from the rostral trunk and are distributed over the vermis. The vermian arteries give rise to the precerebral arteries that supply the inferior colliculi and the superior medullary veil.

The marginal arteries derive from the lateral pontomesencephalic segment of the SCA, derive from the caudal trunk, and supply structures in the horizontal fissure. The SCA gives rise to perforating branches toward the interpeduncular fossa at the junction of the superior and middle peduncles and the mesencephalic colliculi.

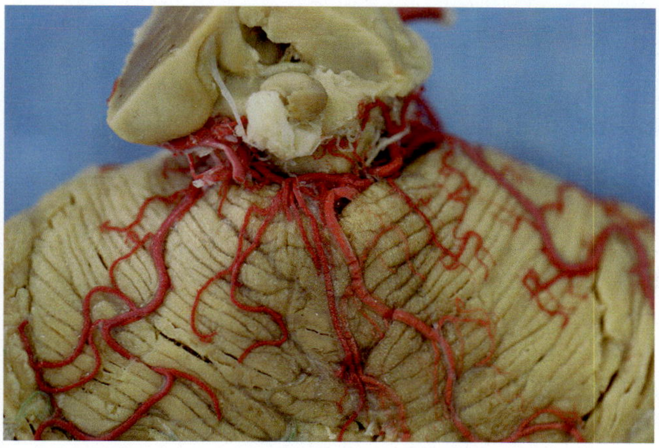

Cortical segments of the superior cerebellar arteries

6.10 Anterior Inferior Cerebellar Artery (AICA)

The anterior inferior cerebellar arteries arise from the basilar artery near the sulcus. This artery is connected to the facial nerve, Luschka's foramen, and flocculus, and passes around the brain stem as well as either above, below, or between the facial and acoustic nerves. The vessel then courses posteriorly over the middle cerebral peduncles above the flocculus. The main trunk of the vessel bifurcates before passing near the facial and acoustic nerve into two branches: rostral and caudal. The rostral branch courses laterally over the flocculus toward the middle cerebellar peduncle and into the horizontal fissure to supply the upper part of the petrous surface of the cerebellum. The caudal branch passes inferomedially below the flocculus and near the lateral recess of the fourth ventricle and supplies the inferior portion of the petrous surface of the cerebellum. The anterior inferior cerebellar artery has been divided into four segments: anterior pontine, lateral pontine, flocculopeduncular, and cortical segments. The lateral pontine segment can be further subdivided into premeatal, meatal segment, and postmeatal segments in relation to the vestibulocochlear nerve and the facial nerve.

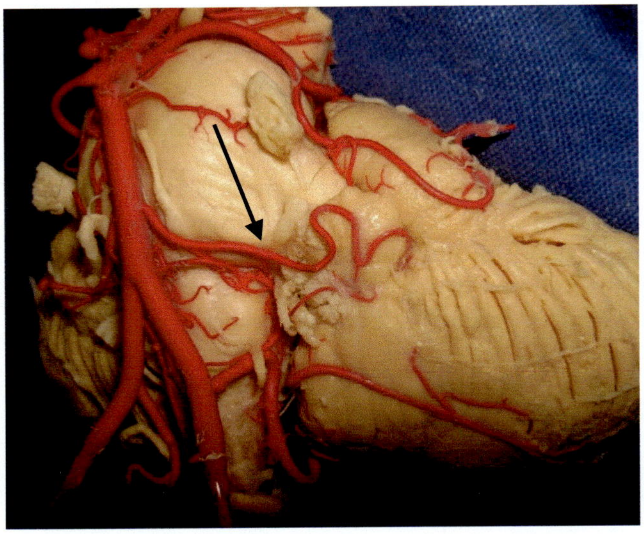

Anterior inferior cerebellar artery (arrow)

6.11 Posteroinferior Cerebellar Artery (PICA)

The posterior inferior cerebellar artery is the largest branch of the vertebral artery. Several variations of the PICA origin exist including hypoplasia, duplication, and a common origin of the PICA and AICA. The PICA courses around from the anterolateral surface of the medulla oblongata passing from rostral to caudal between the roots of the glossopharyngeal, vagus, spinal accessory, and hypoglossal nerves. In its course on the posterolateral surface it passes above, below, or between the roots of the glossopharyngeal, vagus, spinal accessory, and hypoglossal nerves. Subsequently, the vessel can ascend, descend, and pass medially or laterally to the cranial nerves. After reaching the dorsal glossopharyngeal, vagus, and spinal accessory nerves, the PICA passes around the cerebellar tonsils near the caudal half of the roof of the IV ventricle and bifurcates into lateral and medial trunks. The medial trunk supplies the vermis and part of the adjacent cerebellar hemispheres. The lateral trunk supplies the cortical surface of the tonsils and the cerebellar hemispheres.

The posteroinferior cerebellar artery is classified into five segments:

Anterior Medullar Segment. Located anterior to the medulla oblongata.

Lateral Medullar Segment. Courses lateral to the surface of the medulla oblongata.

Medullar Tonsilo Segment. Related to the caudal part of the tonsils and to the posterior surface of the medulla oblongata.

Telovelo Tonsilar Segment. Related to the superior pole of the cerebellar tonsil, the choroidal tissue, and the inferior medullary veil in the roof of the IV ventricle.

Cortical Segment, distributed over the vermis and cerebellar hemispheres.

Anterior Medullar Segment. Begins at the origin of the posterior inferior cerebellar artery, is located anterior to the medulla oblongata, and extends posteriorly passing through the roots of the hypoglossal, at the level of the inferior olive. Sometimes it can ascend, descend, and go laterally or medially before passing through the roots of the hypoglossal.

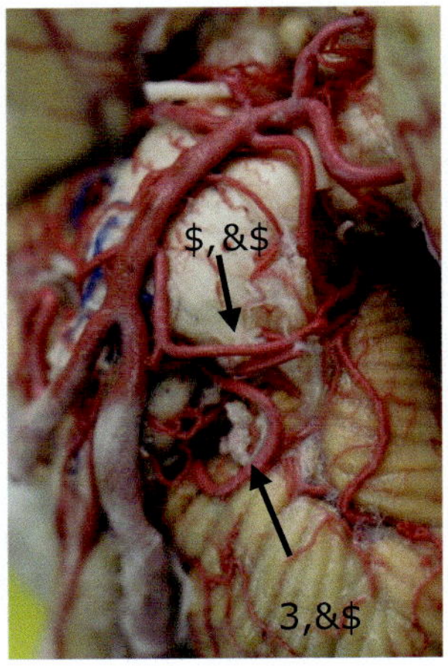

Common trunk of the anterior inferior cerebellar artery (AICA) and posterior inferior cerebellar artery (PICA) (labelled arrows)

Lateral Medullary Segment. It begins where the artery passes over the most prominent point of the olive and ends at the roots of the glossopharyngeal, vagus, and spinal accessory nerves.

Tonsilo-Medullary Segment. Begins at the level of the glossopharyngeal, vagus, and spinal accessory nerves; it extends across the posterior part of the medulla oblongata near the caudal half of the tonsils.

Tonsilar Telovelo Segment. Begins in the middle of the ascending portion of the PICA and in the middle portion of the tonsil; It ascends toward the roof of the fourth ventricle and ends in the fissure located between the vermis and the amygdala, and the cerebellar hemisphere to reach the cortical surface.

Cortical Segment. Begins in the groove located between the vermis medially and the tonsil and cerebellar hemisphere laterally, and proceeds to the lateral aspects of the tonsil, vermis, and cerebellar hemisphere. The posteroinferior cerebellar artery sends branches to the medulla oblongata, to the choroid plexus of the fourth ventricle and, in isolation, gives rise to meningeal branches.

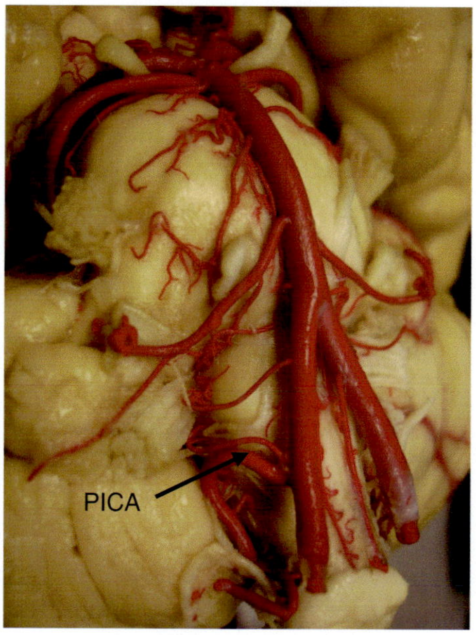

Posterior inferior cerebellar artery (PICA) (arrow)

6.12 Anterior Spinal Artery

The anterior spinal artery is formed by the union of the ventral spinal arteries, which arise from the anteromedial segment of the vertebral arteries, near the origin of the basilar artery. The junction of the anterior spinal arteries is located above the level of the foramen magnum and at the level of the inferior portion of the olives. The anterior spinal artery supplies the pyramids and their decussation, the median lemniscus, the interolivary fascicles, and the nerves and nuclei of the hypoglossal, and the posterior longitudinal fasciculus.

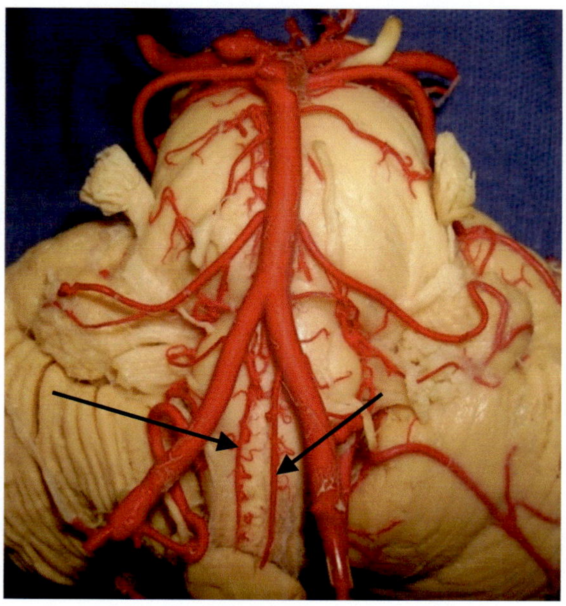

Paired anterior spinal arteries, one on each side of the anterior medulla oblongata (arrows)

6.13 Surgical Pearls

- Profound knowledge of the vertebrobasilar system is essential for safe surgical approaches or endovascular intervention.
- The posterior fossa is involved in several kinds of tumors and vascular malformations, for their management thorough knowledge is necessary of the neurovascular anatomy and its anatomical variants.
- Posterior giant fossa meningiomas are a surgical challenge, they displace neurovascular structures that could be damaged during sophisticated surgical approaches, for this reason, the comprehension of the vascular anatomy of the vertebrobasilar system is mandatory.

6.14 Discussion

It must be emphasized that a thorough understanding and comprehension of the normal vascular anatomy and anatomical variants of the vertebrobasilar system is of vital importance to avoid damage during surgical interventions to vascular malformations and posterior fossa tumors that still constitute a real surgical challenge in the current era.

6.15 Conclusion

The complex vertebrobasilar system provides the vital arterial vascular supply for the brain stem and cerebellum; therefore, it is important to know it in detail, including its delicate and fine branches whose damage during all types of surgical or endovascular interventions gives rise to neurological syndromes that could affect the quality of life of patients.

References

1. Rhoton AL Jr. The cerebellar arteries. Neurosurgery. 2000;47(3 suppl):S29.
2. Magklara EP, et al. Vertebral artery variations revised: origin, course, branches and embryonic development. Folia Morphol (Warsz). 2021;80(1):1.
3. Tsantili AR, et al. Anatomical variations of human vertebral and basilar arteries: a current review of the literature. Morphologie. 2023;107(357):169–75.

Advances in the Surgical Management of Cerebral Aneurysms

7

Yoko Kato and Dragan Jankovic

Learning Objectives

- Obtaining an endoscopic view of the relationship between the aneurysm and the parent artery is crucial for choosing the best clipping technique.
- Computational fluid dynamic (CFD) is useful when assessing the growth and rupture of intracranial aneurysms.
- An atherosclerotic plaque inside the aneurysm should not be mobilized because of the potential for migration and thromboembolic incidents.

7.1 Introduction

Intracranial aneurysms are pathological, localized dilatations of the arterial blood vessels in the brain [1].

Considering their complexity, aneurysm surgeries are demanding and challenging. A thorough knowledge of the nature of aneurysms and blood vessels is necessary to successfully treat aneurysms and prevent adverse damage and poor treatment results. Detailed preoperative planning and analysis to determine the best approach to treat the aneurysm and the most appropriate clipping technique are prerequisites for achieving a satisfactory postoperative result. In this educational chapter, we will discuss advances in the surgical treatment of cerebral aneurysms and evaluate the importance of choosing the best clipping technique for aneurysms with atherosclerotic plaque.

Y. Kato (✉)
Department of Neurosurgery, Fujita Health University Bantane Hospital, Nagoya, Japan
e-mail: kyoko@fujita-hu.ac.jp

D. Jankovic
Department of Neurosurgery, University Medical Center, Johannes Gutenberg University Mainz, Mainz, Germany

7.2 Background

The prevalence of unruptured intracranial aneurysms in the general population is 3.2%. However, only 1% of all intracranial aneurysms rupture [2, 3]. Multiple intracranial aneurysms were noted in 20–30% of patients [4]. Also, multiple cerebral aneurysms have a higher risk of rupture than a single aneurysm because patients with multiple aneurysms have a higher related death rate [5].

Neuroradiological imaging aims to reliably detect unruptured aneurysms or find the source of blood in ruptured intracerebral aneurysms.

Although, in most cases, treatment decisions can be made based on quality computed tomography angiography (CTA), digital subtraction angiography (DSA) represents the gold standard in aneurysm diagnosis in developed countries [6, 7]. The advantage of DSA is the potential to assess the direction and speed of cerebral blood flow, as well as vasospasms.

Although histological studies have shown that the primary cause of aneurysm formation is impaired integrity of the blood vessel wall, several factors can influence the development of aneurysms, such as age, gender, atherosclerosis, and high blood pressure [8, 9].

Furthermore, some genetic diseases increase the prevalence of aneurysms, such as autosomal dominant polycystic kidney disease, Ehlers-Danlos-Syndrome, and neurofibromatosis [10].

Proteolysis of arterial flow, metalloproteinase apoptosis, and chronic inflammation also contribute to aneurysm growth [11]. However, the primary mechanism of aneurysmal growth and rupture is still a matter of debate.

Morphology and hemodynamics are key determinants in the lifetime cycle of intracranial aneurysms. Morphological changes observed after aneurysmal rupture reflect the compound effect of time on growth and aneurysmal rupture [12]. The hemodynamic pattern of an aneurysm is often unclear because it is difficult to obtain data before rupture.

I. N. Kanaan, V. Beneš (eds.), *Neuroanatomy Guidance to Successful Neurosurgical Interventions*,
https://doi.org/10.1007/978-3-031-59838-8_7

Therefore, knowledge of morphological and hemodynamic patterns and changes within the blood vessel wall before aneurysm rupture may help neurosurgeons and researchers better understand aneurysm formation, growth, and rupture.

7.3 Anatomy Highlights

Cerebral blood vessels have three layers: tunica intima, tunica media, and tunica adventitia. Between the tunica intima and the tunica media is the internal elastic lamina, which is one of the key factors in forming intracranial aneurysms [8].

Increased mechanical stress can cause a cascade of events that result in endothelial dysfunction [13]. The subsequent inflammatory reaction leads to the swelling of cell adhesion molecules on the surface of endothelial cells and the expression of anti-inflammatory cytokines [14].

The internal elastic membrane is destroyed during the remodeling process. Therefore, the integrity of the blood vessel wall is often damaged, ultimately leading to the ballooning of the wall and the formation of an aneurysm [15].

7.4 Surgical Intervention

7.4.1 Case 1

A 69-year-old female patient with no family history of aneurysms was admitted to our hospital by the screening program. She has no medical history of hypertension or a smoking habit. A three-dimensional computed tomography (3DCT) with angiography revealed a small aneurysm on the internal carotid-posterior communicating (IC-PC) artery. The patient underwent surgical clipping of the aneurysm using pterional craniotomy. Figure 7.1a shows the microscopic view, presenting a limited operative field that does not give the surgeon much room for manipulation with the instruments. We used an adjunct intraoperative modality (indocyanine green video angiography (ICG-VA)) and endoscopic visualization during the operation. The endoscopic view allowed us to see the relationship between the neck of the aneurysm and the parent artery. In addition, with the help of an endoscope, we could see yellow atherosclerotic plaques in the parent artery. After clipping, an endoscope was used to confirm the occlusion of the aneurysm and to observe the fluorescence. Blood flow was identified, and no periprocedural complications occurred.

7.4.2 Case 2

A 56-year-old male presented with an aneurysm at the A2 segment of the anterior cerebral artery (ACA) on the right side (Fig. 7.2a). CFD analysis shows streamlined hitting with high velocity (Fig. 7.2b). The patient underwent microsurgical clipping through an anterior interhemispheric approach. After an aneurysm was found, ICG-VA was performed (Fig. 7.2c). Clipping is done using a straight clip in a simple technique.

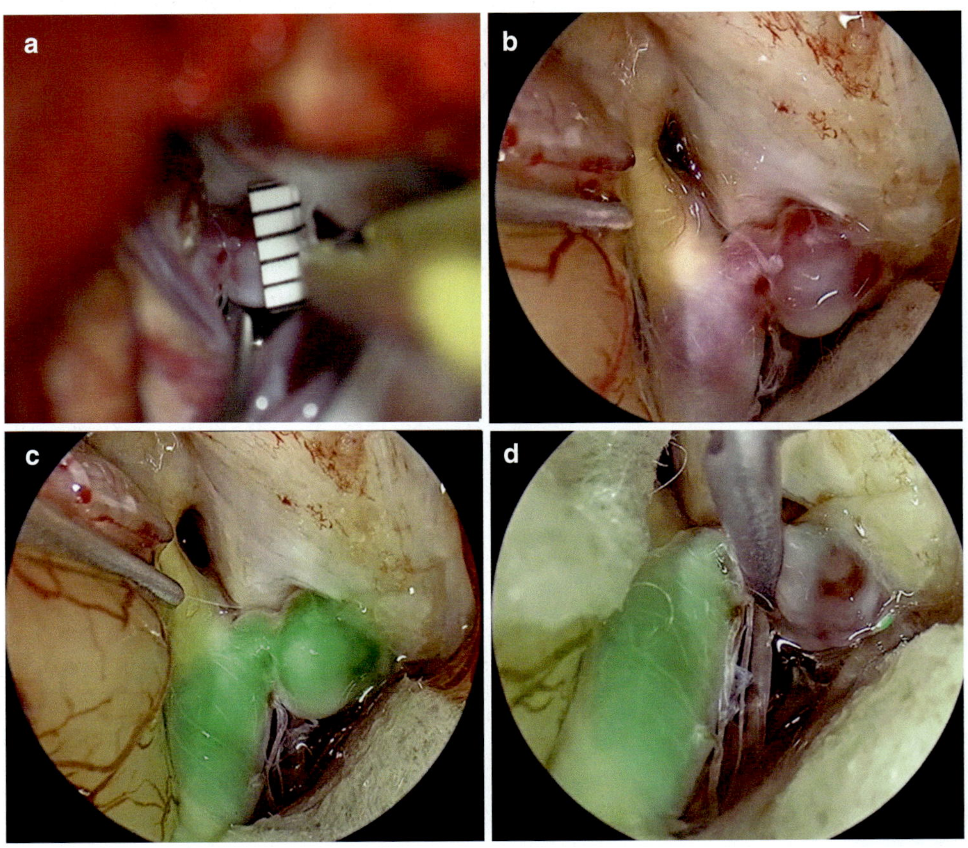

Fig. 7.1 Case of IC-PC aneurysm. Intraoperative view with the microscope (**a**) and endoscope (**b–d**)

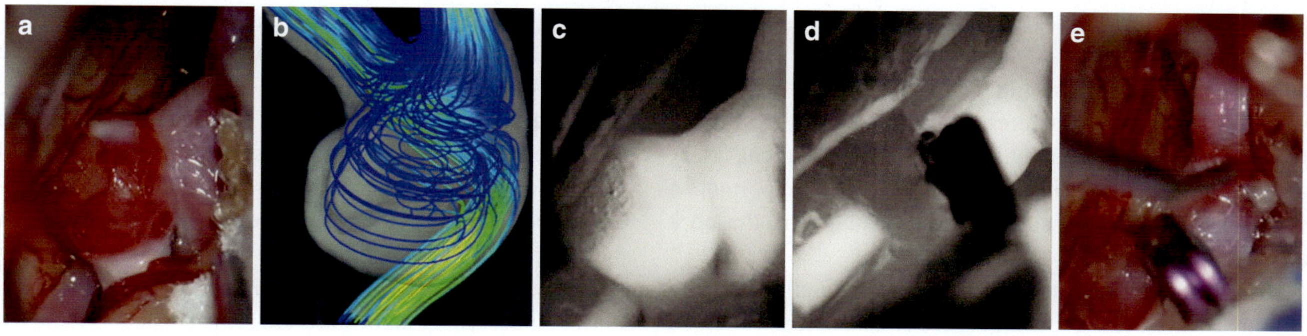

Fig. 7.2 Case of A2/3 aneurysm. Intraoperative view (**a**), CFD analysis (**b**), pre-clipping ICG-VA (**c**), post-clipping ICG-VA (**d**), and intraoperative view of small berry aneurysm (**e**)

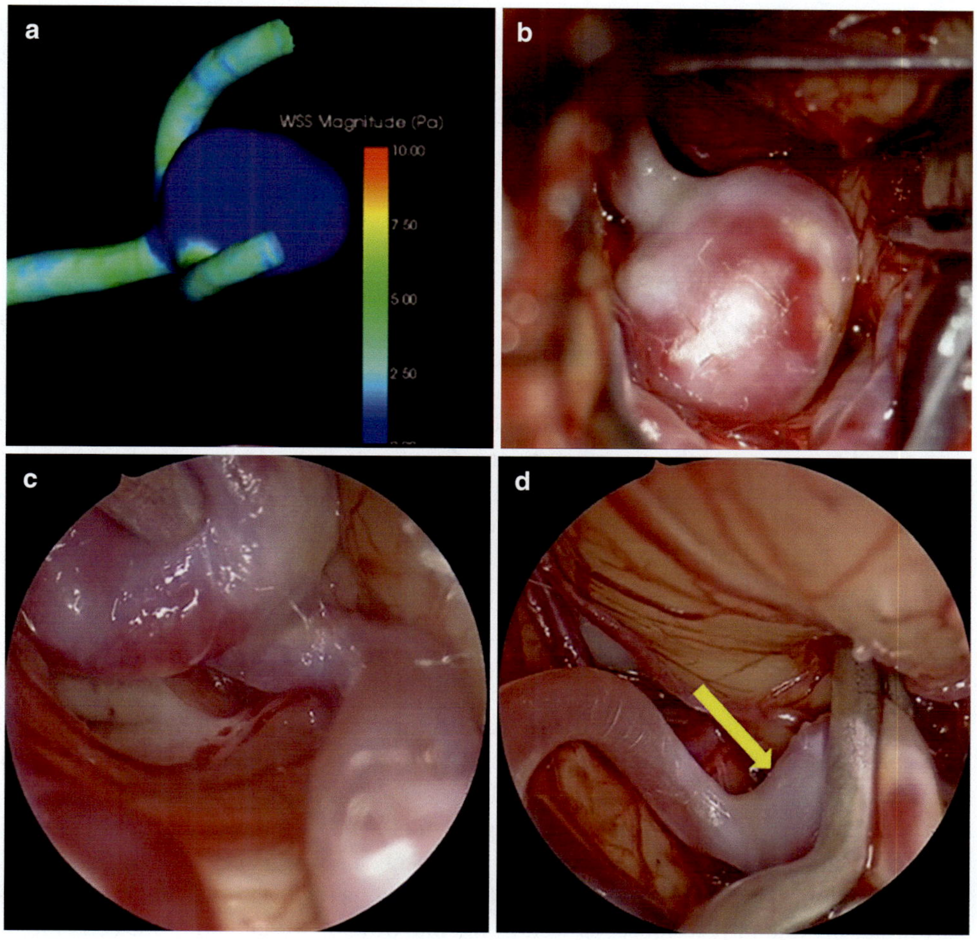

Fig. 7.3 Case of MCA bifurcation aneurysm. CFD of MCA aneurysm (**a**), intraoperative view with a microscope (**b**), and rigid endoscope (**c** and **d**). The arrow shows the remnant of the aneurysm

Post-clipping ICG-VA showed complete occlusion of the aneurysm (Fig. 7.2d). An endoscope was used to check all corners around the aneurysm, confirming the occlusion. However, the endoscope found a small berry aneurysm distal from the clipped aneurysm (Fig. 7.3e). Retrospectively, the second aneurysm was not seen on preoperative neuroradiological images. Intraoperatively, a decision was made to clip the second aneurysm. The berry aneurysm was also closed with a straight clip while preserving the surrounding perforators.

7.4.3 Case 3

A 78-year-old right-handed female was admitted to our hospital for observation after receiving a concussion during a traffic accident. In medical history, she had arterial hypertension and clipping of the right media cerebral artery (MCA) aneurysm 2 months ago. The patient presented no neurologic symptoms. CFD analysis was performed preoperatively to assess the hemodynamic characteristics of the aneurysm.

CFD showed low values of WSS in the aneurysm dome (Fig. 7.3a). Imaging studies showed an MCA bifurcation aneurysm on the left side (Fig. 7.3b). With a rigid endoscope, we illuminated the structures in the depth of the surgical field, especially the relationship between the parent artery and the aneurysm neck and dome. It is very important to verify the existence of atherosclerotic changes in the neck area.

After clipping the aneurysm with an angled fenestrated clip in an interlocking technique, we endoscopically found a small remnant of the aneurysm (Fig. 7.3d).

7.5 Surgical Pearls

- Preoperatively, it is necessary to examine the factors that can negatively affect the complete occlusion of the aneurysm.
- Computational fluid dynamics can evaluate the outer wall of a vessel both morphologically and hemodynamically.
- Using an endoscope for microsurgical treatment of cerebral aneurysms enabled the surgeon to “see all the corners.”
- Intraoperative fluorescence angiography enables the assessment of the aneurysm occlusion and the patency of the parent artery and the perforating arteries in real-time.
- Atherosclerotic plaque in the area around the aneurysm can lead to stenosis of the parent artery or a thromboembolic incident.
- Using temporary clips in the yellowish sclerotic area of the parent artery should be avoided.

7.6 Discussion

Due to the broad application of neuroradiological imaging, the accidental detection of unruptured intracranial aneurysms has increased over the last 20 years. In recent years, with the development of new neuroradiological tools and techniques, the endovascular method of treatment has become superior to surgical treatment. The ISAT study has shown that the rate of recurrence, as well as the rate of late rebleeding with endovascular therapy, is higher than with surgical clipping [16]. However, surgical treatment can be associated with severe complications. Therefore, choosing the best treatment method should be patient-specific to achieve the best results for each patient.

While ruptured aneurysms are treated immediately after detection, the decision to treat an unruptured aneurysm depends on the clinical symptoms, the size of the aneurysm, its morphology and complexity, and the risks for potential aneurysm rupture. There is no consensus on whether an unruptured aneurysm requires intervention.

Therefore, assessing the risk of aneurysm rupture is crucial to treat timely aneurysms and prevent aneurysm rupture and poor patient outcomes.

Although the PHASES score is the most well known in risk assessment; in recent years, introducing artificial intelligence (AI) into the risk assessment of aneurysm rupture has become more and more accessible [17, 18]. A recent study reported the superiority of the machine learning approach over the PHASES score [19].

Machine learning algorithms are sensitive and specific for assessing the rupture risk of intracranial aneurysms [20].

Preoperatively, examining the factors that can affect aneurysm clipping is necessary. Unexpected atherosclerosis of an aneurysm may present an additional challenge to neurosurgeons. Atherosclerotic plaque in the area of the aneurysm can lead to stenosis of the cerebral artery or atherosclerotic infarct [21]. Given that preoperative neuroradiological imaging cannot rule out the existence and severity of atherosclerotic plaque, the decision on the clipping technique is made intraoperatively.

Bae et al. reported that 27.8% of patients who underwent clipping had atherosclerotic changes in the aneurysm [22]. Atherosclerotic lesions occur mainly in large and proximal vessels [23]. However, the prevalence of atherosclerosis according to the location of the aneurysm was the posterior communicating artery, the bifurcation of the internal carotid artery, and the anterior communicating artery.

Treatment of atherosclerotic and calcified aneurysms is challenging, especially because surgical and endovascular treatments have limitations. The main limitation of surgical treatment is the development of postoperative infarction, while endovascular surgery has several limitations, including recurrence or thromboembolic events. Based on experience, our institute prefers the operation treatment for atherosclerotic aneurysms rather than the endovascular treatment. The surgical treatment of atherosclerotic aneurysms aims to achieve maximum occlusion while preventing consequent embolism and infarction.

Avoiding clip slippage near calcification or an atherosclerotic plaque area is crucial [21]. Atherosclerotic plaque of an aneurysm increases the risk of thrombus formation in the distal part of the cerebral flow [24].

This can be especially serious in the case of multiclipping when the atherosclerotic part of the aneurysm is manipulated, which can cause atherosclerotic plaque migration and lead to a consequent infarct in the distal area [21]. Ohno et al. reported an incidence of cerebral infarction of 50% after treatment of atherosclerotic aneurysm with conventional clipping [25]. In contrast, a study from Japan showed a very low incidence of embolic infarction (<3%) after clipping an atherosclerotic cerebral aneurysm [21]. The reasons for this difference are the development of new surgical techniques and tools, such as more modern microscopes,

ICG-VA, microdoppler, and electrophysiological monitoring during intracranial aneurysm surgery.

The stability of the plaque inside the blood vessel and aneurysm is almost impossible to assess. Unstable plaque may migrate easily due to compression force by the clip. In some aneurysms where the plaque is located both in the dome and the neck, it is necessary to place a dome clip to prevent aneurysm rupture in the future. In such a case, it is necessary to follow up with the patient for the potential growth of the not-clipped part. Manipulating or touching sclerotic areas should be avoided.

Intraoperatively, assessing the morphology of intracranial aneurysms under a microscope can be challenging. During operations, the surgeon may be surprised by unexpected aneurysms or parent artery atherosclerosis, which can significantly affect the safety of the surgical procedure.

At the beginning of the twentieth century, the usefulness of a rigid endoscope in the surgery of intracranial aneurysms was investigated [26]. The endoscope completes the microscopic view of the operative field, enabling additional visibility. In addition to enabling surgeons to see areas hidden from the microscope, it makes it possible to identify the relationship between the aneurysm and the parent artery and the presence of atherosclerotic plaque.

With a rigid endoscope, we can show the morphology of the clipped aneurysm from multiple angles through rotation [27]. The meta-analysis conducted by Shao and colleagues showed that endoscope use compensated for suboptimally or incorrectly positioned clips [28].

The endoscopic view helps visualize incomplete clipping and evaluate small berry aneurysms and occlusion on the parent or perforating arteries.

In addition, endoscopic assistance reduces intra- and postoperative complications. Improving orientation in the intraoperative field leads to less stimulation of blood vessels, which minimizes cerebral vasospasm postoperatively.

To understand the natural history of intracranial aneurysms, it is important to understand the mechanisms of blood flow. Over the last two decades, computational fluid dynamic (CFD) analysis has been used to investigate several aspects of intracranial aneurysms based on a combination of the computational domain and a patient-specific vascular model [29]. The ability of CFD to provide information on the hemodynamic characteristics of the aneurysm wall could be a valuable tool during surgery. CFD has improved our understanding of intraluminal flow patterns and intracranial aneurysms. Aneurysm initiation and growth is a long and complex process, and aneurysm architecture changes over time. Changing the aneurysm's morphology changes the fluid's dynamics in the aneurysm [30].

In recent years, several hemodynamic parameters have been proposed that could play a role in the growth and rupture of the aneurysm [31, 32]. However, the published results are contradictory and confusing in interpreting CFD parameters, especially regarding wall shear stress (WSS).

WSS is defined as "a frictional force exerted by the flowing blood tangentially on the vessel lumen." [33] WSS around the area of the aneurysm has a significant impact on vascular morphology. The spatial distribution of WSS is affected by a high flow of vortex formation. When blood enters the aneurysm, WSS patterns change qualitatively and quantitatively.

As a result of abnormal hemodynamics, there are destructive changes that make the aneurysm wall weaker and more prone to rupture [34].

Based on the hypothesis that WSS initiates different pathways of aneurysm growth and rupture, Meng et al. have proposed that the hemodynamics of both high and low pressure can initiate the process of growth and rupture [33].

- inflammatory-cell-mediated destructive remodeling caused by low WSS
- mural-cell-mediated destructive remodeling caused by high WSS

Which type of WSS will be dominant depends on the phenotype of the intracranial aneurysm. While high WSS is present in small, thin-wall aneurysms, low WSS is responsible mainly for the growth and rupture of large, thick-wall, thrombotic, and atherosclerotic aneurysms [33].

Analyzing the hemodynamic parameters obtained by CFD, it is possible to distinguish atherosclerotic, thick aneurysmatic areas from red, thin areas with a higher risk of rupture [30].

7.7 Conclusion

Understanding the neuroanatomy, hemodynamics, and morphology of intracranial aneurysms is crucial for the natural history of aneurysms and their rupture. Computational fluid dynamics can be used to assess rupture risk, while the combination of intraoperative fluorescent angiography and endoscopic-assisted microsurgery can improve the outcome of surgical treatment of intracranial aneurysms but also reduce surgical complications and increase surgical efficiency.

References

1. Xu Z, Rui YN, Hagan JP, Kim DH. Intracranial aneurysms: pathology, genetics, and molecular mechanisms. NeuroMolecular Med. 2019;21(4):325–43.
2. Pontes FGB, da Silva EM, Baptista-Silva JC, Vasconcelos V. Treatments for unruptured intracranial aneurysms. Cochrane Database Syst Rev. 2021;5(5):CD013312.

3. Williams LN, Brown RD. Management of unruptured intracranial aneurysms. Neurology. 2013;3(2):99–108.
4. Rinne J, Hernesniemi J, Puranen M, Saari T. Multiple intracranial aneurysms in a defined population: prospective angiographic and clinical study. Neurosurgery. 1994;35(5):803–8.
5. Deniwar MA. Management of multiple and unruptured cerebral aneurysms. Egypt J Neurosurg. 2022;37(1):26.
6. Burkhardt JK, Chua MH, Winkler EA, Rutledge WC, Lawton MT. Incidence, classification, and treatment of angiographically occult intracranial aneurysms found during microsurgical aneurysm clipping of known aneurysms. J Neurosurg. 2019;132(2):434–41.
7. Tanaka R, Liew BS, Yamada Y, Sasaki K, Miyatani K, Komatsu F, et al. Depiction of cerebral Aneurysm Wall by computational fluid dynamics (CFD) and preoperative illustration. Asian J Neurosurg. 2022;17(1):43–9.
8. Jung K-H. New pathophysiological considerations on cerebral aneurysms. Neurointervention. 2018;13(2):73–83.
9. Morel S, Diagbouga MR, Dupuy N, Sutter E, Braunersreuther V, Pelli G, et al. Correlating clinical risk factors and histological features in ruptured and unruptured human intracranial aneurysms: the Swiss AneuX study. J Neuropathol Exp Neurol. 2018;77(7):555–66.
10. Keedy A. An overview of intracranial aneurysms. McGill J Med. 2006;9(2):141–6.
11. Korde P. Aneurysm wall property. In: Recent progress in the management of cerebrovascular diseases: treatment strategies, techniques and complication avoidance. Singapore: Springer; 2021. p. 1–7.
12. Skodvin T, Johnsen LH, Gjertsen Ø, Isaksen JG, Sorteberg A. Cerebral aneurysm morphology before and after rupture: nationwide case series of 29 aneurysms. Stroke. 2017;48(4):880–6.
13. Theofilis P, Sagris M, Oikonomou E, Antonopoulos AS, Siasos G, Tsioufis C, et al. Inflammatory mechanisms contributing to endothelial dysfunction. Biomedicines. 2021;9(7):781.
14. Hudson JS, Hoyne DS, Hasan DM. Inflammation and human cerebral aneurysms: current and future treatment prospects. Future Neurol. 2013;8(6):663.
15. Turkmani AH, Edwards NJ, Chen PR. The role of inflammation in cerebral aneurysms. Neuroimmunol Neuroinflamm. 2015;2:102–6.
16. Molyneux AJ, Kerr RS, Yu LM, Clarke M, Sneade M, Yarnold JA, et al. International subarachnoid aneurysm trial (ISAT) of neurosurgical clipping versus endovascular coiling in 2143 patients with ruptured intracranial aneurysms: a randomised comparison of effects on survival, dependency, seizures, rebleeding, subgroups, and aneurysm occlusion. Lancet. 2005;366(9488):809–17.
17. Bijlenga P, Gondar R, Schilling S, Morel S, Hirsch S, Cuony J, et al. PHASES score for the management of intracranial aneurysm: a cross-sectional population-based retrospective study. Stroke. 2017;48(8):2105–12.
18. Shi Z, Hu B, Schoepf UJ, Savage RH, Dargis DM, Pan CW, et al. Artificial intelligence in the management of intracranial aneurysms: current status and future perspectives. AJNR Am J Neuroradiol. 2020;41(3):373–9.
19. Walther G, Martin C, Haase A, Nestler U, Schob S. Machine learning for rupture risk prediction of intracranial aneurysms: challenging the PHASES score in geographically constrained areas. Symmetry. 2022;14(5):943.
20. Shu Z, Chen S, Wang W, Qiu Y, Yu Y, Lyu N, et al. Machine learning algorithms for rupture risk assessment of intracranial aneurysms: a diagnostic meta-analysis. World Neurosurg. 2022;165:e137–47.
21. Sakarunchai I, Kato Y, Yamada Y, Inamasu J. Ischemic event and risk factors of embolic stroke in atherosclerotic cerebral aneurysm patients treated with a new clipping technique. J Stroke Cerebrovasc Dis. 2015;24(11):2497–507.
22. Bae H-J, Choi JH, Kim BS, Lee K-S, Shin YS. Predictors of atherosclerotic change in unruptured intracranial aneurysms and parent arteries during clipping. World Neurosurg. 2019;130:e338–43.
23. Aboyans V, Lacroix P, Criqui MH. Large and small vessels atherosclerosis: similarities and differences. Prog Cardiovasc Dis. 2007;50(2):112–25.
24. Sasaki T, Kodama N, Matsumoto M, Suzuki K, Konno Y, Sakuma J, et al. Blood flow disturbance in perforating arteries attributable to aneurysm surgery. J Neurosurg. 2007;107(1):60–7.
25. Ohno K, Arai T, Isotani E, Nariai T, Hirakawa K. Ischaemic complication following obliteration of unruptured cerebral aneurysms with atherosclerotic or calcified neck. Acta Neurochir. 1999;141(7):699–705; discussion-6.
26. Profeta G, De Falco R, Ambrosio G, Profeta L. Endoscope-assisted microneurosurgery for anterior circulation aneurysms using the angle-type rigid endoscope over a 3-year period. Childs Nerv Syst. 2004;20(11–12):811–5.
27. Nishiyama Y, Kinouchi H, Senbokuya N, Kato T, Kanemaru K, Yoshioka H, et al. Endoscopic indocyanine green video angiography in aneurysm surgery: an innovative method for intraoperative assessment of blood flow in vasculature hidden from microscopic view. J Neurosurg. 2012;117(2):302–8.
28. Shao D, Li Y, Zhang B, Wu J, Xie S, Zheng X, et al. Endoscope-assisted microneurosurgery for intracranial aneurysms: a systematic review and meta-analysis. J Clin Neurosci. 2022;103:62–71.
29. Staarmann B, Smith M, Prestigiacomo CJ. Shear stress and aneurysms: a review. Neurosurg Focus. 2019;47(1):E2.
30. Feletti A, Wang X, Talari S, Mewada T, Mamadaliev D, Tanaka R, et al. Computational fluid dynamics analysis and correlation with intraoperative aneurysm features. Acta Neurochir Suppl. 2018;129:3–9.
31. Murayama Y, Fujimura S, Suzuki T, Takao H. Computational fluid dynamics as a risk assessment tool for aneurysm rupture. Neurosurg Focus. 2019;47(1):E12.
32. Wang Y, Leng X, Zhou X, Li W, Siddiqui AH, Xiang J. Hemodynamics in a middle cerebral artery aneurysm before its growth and fatal rupture: case study and review of the literature. World Neurosurg. 2018;119:e395–402.
33. Meng H, Tutino VM, Xiang J, Siddiqui A. High WSS or low WSS? Complex interactions of hemodynamics with intracranial aneurysm initiation, growth, and rupture: toward a unifying hypothesis. AJNR Am J Neuroradiol. 2014;35(7):1254–62.
34. Qiu T, Jin G, Xing H, Lu H. Association between hemodynamics, morphology, and rupture risk of intracranial aneurysms: a computational fluid modeling study. Neurol Sci. 2017;38(6):1009–18.

8 Clipology, the Art of Aneurysm Clipping

Matias Baldoncini and Alvaro Campero

Learning Objectives

- Identify Indications for Aneurysm Clipping: Recognize the specific clinical scenarios and aneurysm characteristics that warrant surgical intervention over other treatment options, such as endovascular coiling.
- Comprehend Aneurysm Clipping Techniques: Explain the different approaches to aneurysm clipping, including microsurgical clipping, fenestrated clips, temporary clip usage, and specific considerations for different aneurysm morphologies.
- Evaluate Preoperative Planning and Patient Selection: Describe the critical steps involved in preoperative planning, including patient assessment, imaging evaluation, and the importance of individualized treatment plans based on aneurysm characteristics.
- Understanding the vascular neuroanatomy of the blood vessels and structures of the brain is essential for identifying the location of the aneurysm and planning the surgical approach.
- Assess Intraoperative Procedures and Challenges: xamine the surgical steps involved in aneurysm clipping, the nuances of clip placement, and strategies to overcome challenges, such as clip selection for complex aneurysms.

8.1 Introduction

Intracranial or brain aneurysms are pathological dilatations of intracranial arteries that often occur at a branching point of the arteries [1]. The four broad types of aneurysms are saccular, fusiform, dissecting, and mycotic, of which saccular aneurysms are the most common, accounting for around 90% [2]. The prevalence of intracranial aneurysms in the general population is approximately 3.2%, though most are asymptomatic [3–5] (Fig. 8.1).

The art of aneurysm clipping involves a combination of technical skill, experience, and knowledge of neuroanatomy. The neurosurgeon must have a thorough understanding of the blood vessels and structures of the brain to identify the location of the aneurysm and plan the surgical approach. Patient selection is also critical, and the decision to perform the procedure depends on several factors, including the size and location of the aneurysm, the age and overall health of the patient, and the risk of rupture [4].

"Microsurgical clipping of intracranial aneurysms remains a cornerstone in the treatment of cerebral aneurysms and still provides excellent outcomes in experienced hands." Surgical clipping remains the gold standard for certain aneurysms, we define *Clipology* as the set of systematic steps for clipping an intracranial aneurysm, encompassing knowledge of vascular anatomy, aneurysm morphology, surgical technique to be employed, as well the proper choice of aneurysm clip.

M. Baldoncini (✉)
Laboratory of Microsurgical Neuroanatomy, Second Chair of Gross Anatomy, School of Medicine, University of Buenos Aires, Buenos Aires, Argentina

A. Campero
Department of Neurological Surgery, Hospital Padilla, Tucumán, Argentina

I. N. Kanaan, V. Beneš (eds.), *Neuroanatomy Guidance to Successful Neurosurgical Interventions*,
https://doi.org/10.1007/978-3-031-59838-8_8

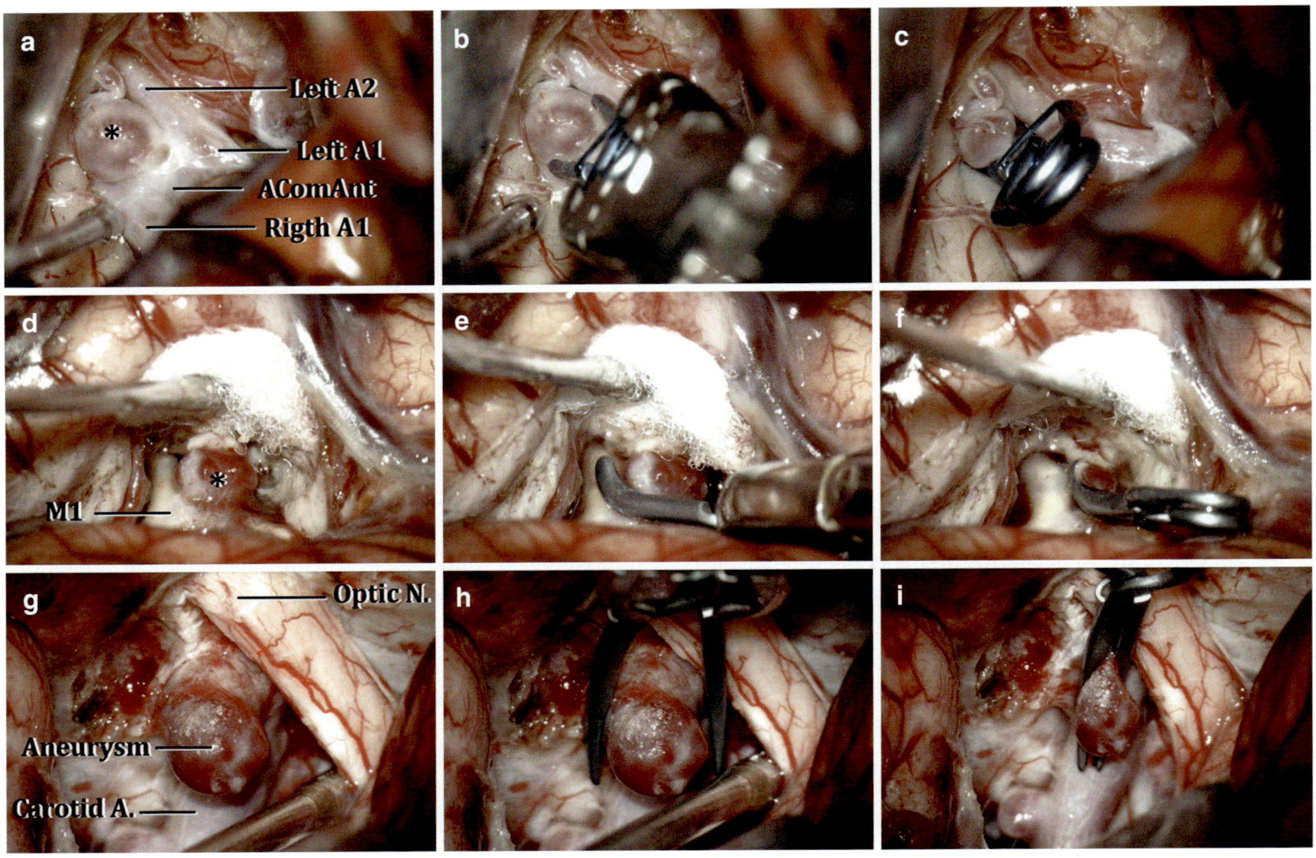

Fig. 8.1 (**a–c**) In this figure, we can observe a saccular aneurysm in the anterior communicating segment, where both A1 and A2 segments are exposed. After dissecting the aneurysm neck, a definitive straight 10 mm clip is placed. *Aneurysm. (**d–f**) After opening the Sylvian cistern, the medial cerebral artery is exposed, and this bifurcation aneurysm of the artery is exposed. A semicurved clip is placed parallel to the bifurcation edge. *Aneurysm (**g–i**) Left carotid-ophthalmic aneurysm, exposed after anterior intradural clinoidectomy. A curved clip is placed parallel to the paraclinoid internal carotid artery

8.2 Background

The treatment of cerebral aneurysms has evolved significantly over the past few decades, with endovascular techniques becoming more prominent. However, surgical clipping remains an essential tool in the management of complex aneurysms, particularly those with a wide neck, large size, or complex geometry [5]. Aneurysm clipping is a delicate and precise neurosurgical technique that requires a high level of technical skill and experience. In recent years, there has been a renewed interest in the art of aneurysm clipping, with a focus on achieving optimal surgical outcomes and minimizing complications.

8.3 Anatomy Highlights

- The art of aneurysm clipping requires a thorough understanding of neuroanatomy, including the blood vessels and structures of the brain.
- The neurosurgeon must then carefully dissect the cisternal space to access the aneurysm and place the aneurysm clip at the base of the aneurysm to prevent it from rupturing.
- Understanding the anatomy of the blood vessels and structures of the brain is essential for identifying the location of the aneurysm and planning the surgical approach.
- It is very important to know the arterial segment where the aneurysm originates, the morphology of the neck and the collateral vessels that are related to the anatomical sector to be operated on.

8.4 Surgical Intervention

Patient Selection: Patient selection is an important aspect of aneurysm clipping, as not all patients with brain aneurysms are good candidates for this procedure or endovascular treatment. In general, the decision to perform aneurysm clipping will depend on a number of factors, including the size and location of the aneurysm, the patient's age and overall health, and the potential risks and benefits of the procedure.

In addition, aneurysms that are located near critical structures in the brain, such as the optic nerve or major blood vessels, may also be more challenging to clip and may require careful surgery. The patient's age and overall health are also important factors in patient selection for aneurysm clipping. In general, younger patients with good overall health are better candidates for this procedure, as they are more likely to have a good outcome and fewer complications. However, even older patients or those with underlying health conditions may be candidates for aneurysm clipping if the potential benefits of the procedure outweigh the risks. Another important consideration in patient selection for aneurysm clipping is the potential risks and benefits of the procedure. Aneurysm clipping is a highly specialized and delicate surgery that carries certain risks, such as bleeding, infection, and damage to the surrounding brain tissue. However, the potential benefits of aneurysm clipping can be significant, including the prevention of a potentially life-threatening brain hemorrhage and the reduction of symptoms associated with the aneurysm.

In addition to patient selection, there are also certain preparations that are necessary before aneurysm clipping can be performed. These may include the correct imaging studies, such as MRI or CT scans and angiography to determine the size and location of the aneurysm, as well as tests to evaluate the patient's overall health and to identify any potential complications that may arise during or after the procedure.

After the procedure, patients will typically need to stay in the hospital for a period of time to be monitored for any complications or side effects of the surgery. In addition, patients may need to undergo rehabilitation or physical therapy to help them recover from the surgery and regain their strength and mobility.

Morphology of Aneurysm: Aneurysms most commonly occur at arterial bifurcations or where there is a change in the diameter of the artery. The most common location for aneurysms is the anterior communicating artery, followed by the middle cerebral artery, internal carotid artery, and posterior communicating artery. The size of the aneurysm is also an important consideration, as larger aneurysms may be more difficult to clip and may require more extensive dissection of the brain tissue or to perform a bypass. Saccular or "berry" aneurysms are the most common type, accounting for about 90% of all brain aneurysms. They typically have a rounded or sac-like shape that bulges out from the arterial wall. Saccular aneurysms are often associated with arterial bifurcations, where blood vessels branch out. Fusiform aneurysms are less common and have a spindle-shaped appearance. Unlike saccular aneurysms, which involve localized outpouchings, fusiform aneurysms involve a more generalized dilation of the artery. Giant aneurysms are generally larger than 2.5 cm (about 1 in.) in diameter. Their size makes them more prone to rupture and may present unique challenges in treatment. Some aneurysms have multiple lobes or outpouchings, creating a complex and irregular morphology. These aneurysms may require specialized treatment strategies due to their intricate shape (Fig. 8.2).

Aneurysm Clips: The development of aneurysm clips is a clear demonstration of the determination and creativity of vascular neurosurgeons in their pursuit of finding a treatment for a disease that was once deemed impossible to overcome. Aneurysm clips have undergone various modifications over time, both in the substances used to create them and in the intricate details of their mechanisms. Straight clips are one of the most basic and commonly used types of aneurysm clips. They have a simple straight shape and are available in various sizes to accommodate different aneurysm sizes and neck geometrically. Curved clips have a gentle curve in their design, which can be beneficial when the aneurysm's location or geometry requires a specific angle for optimal placement. Fenestrated clips have a small hole or window in the middle of the clip. This feature allows the clip to be positioned around the aneurysm neck while maintaining blood flow through the parent artery.

Several types of clips are worth mentioning, including booster clips, t-bar clips, and Sundt graft clips. A booster clip is utilized to reinforce a long clip placed across a giant aneurysm neck. Long clips may have insufficient closing force at the tip, and a booster clip can be placed over it to enhance closure and prevent slippage. T-bar clips aid in parent artery reconstruction by having a horizontal segment that is perpendicular to the direction of the clip itself. The horizontal segment occludes the neck while the parent artery is incorporated in a fenestration. Finally, Sundt graft clips are used to occlude the neck of an aneurysm while preserving a branch or perforator by using a fenestration to incorporate the branch or perforator (Fig. 8.3).

By selecting the ideal clip before surgery, several advantages can be gained. For one, it can help identify the most suitable clip for the occlusion of aneurysms, leading to better occlusion and avoiding any potential intraoperative modifications of less optimal clips.The surgeon needs to see both sides of the neck and the two clip blades at the time of application. The microscope needs to be angled to achieve this objective. However, this view does not usually show the tip of the blades once they are advanced and closed. After application, the microscope is angled obliquely to verify that the clip blades have passed the neck entirely. When the neck of the aneurysm is squeezed shut by the clip blades, the diameter of the neck increases by about 50%. Hence, the clip blades should be that much longer (1.5 times the neck diameter). Incomplete neck closure at the distal side of the aneurysm neck is the most common cause of persistent filling of the dome. In most cases, the clip can be advanced to close the neck completely.

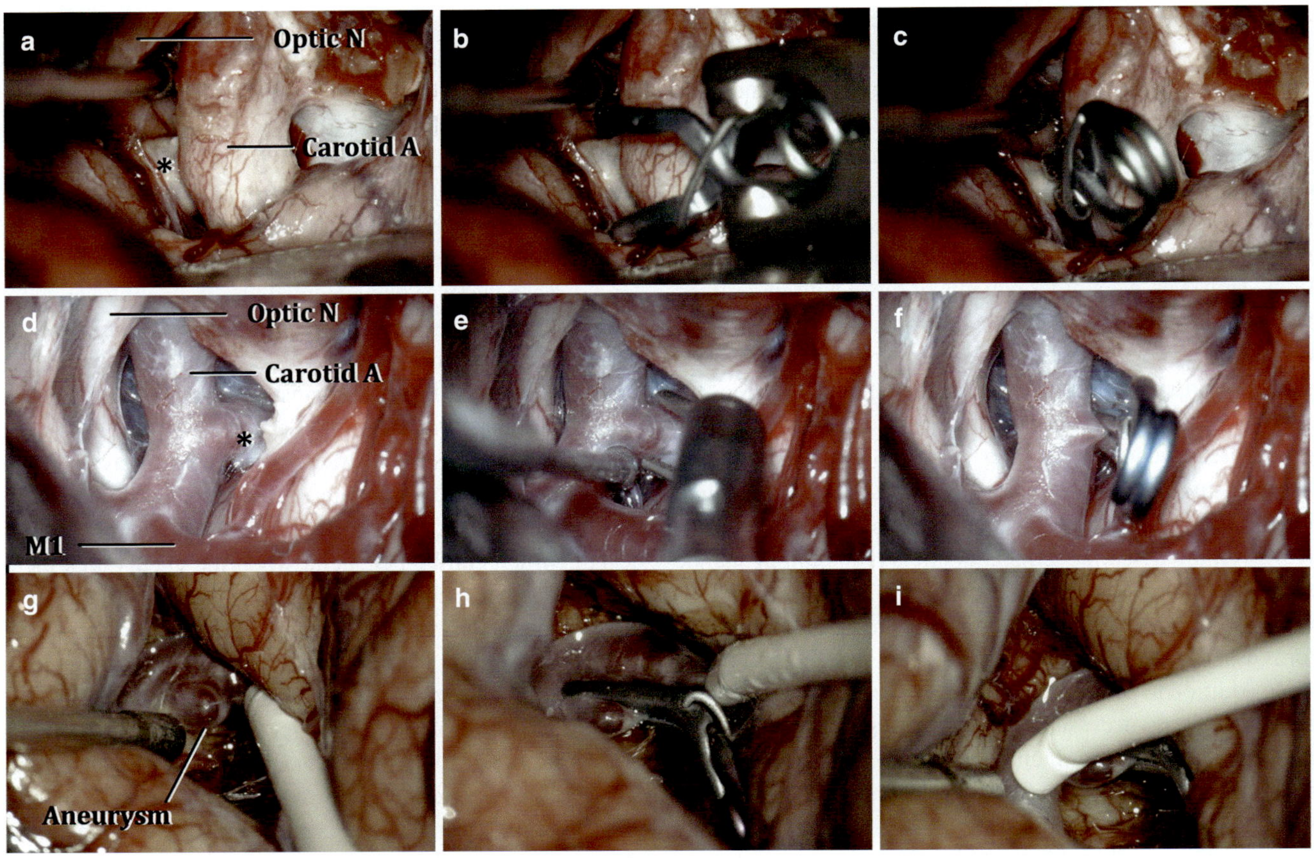

Fig. 8.2 (**a**–**c**) In this figure, a right internal carotid artery ophthalmic segment aneurysm with medial projection is observed. After dissecting the aneurysm neck and separating it from collateral branches, especially the posterior communicating artery and superior hypophyseal arteries, an angled clip at 45° parallel to the aneurysmal neck is placed. (**d**–**f**) In this case, a right posterior communicating aneurysm is observed, where it was necessary to cut the anterior petroclinoid ligament to expose the neck. Again, a 45-degree angled clip is used, but with a lateral to medial clipping trajectory. (**g**–**i**) In this case, after opening the left Sylvian fissure, the middle cerebral artery is exposed at its bifurcation with a small saccular aneurysm. Doppler control is performed before and after clipping. In this instance, a 5 mm mini-clip angled at 45° is used

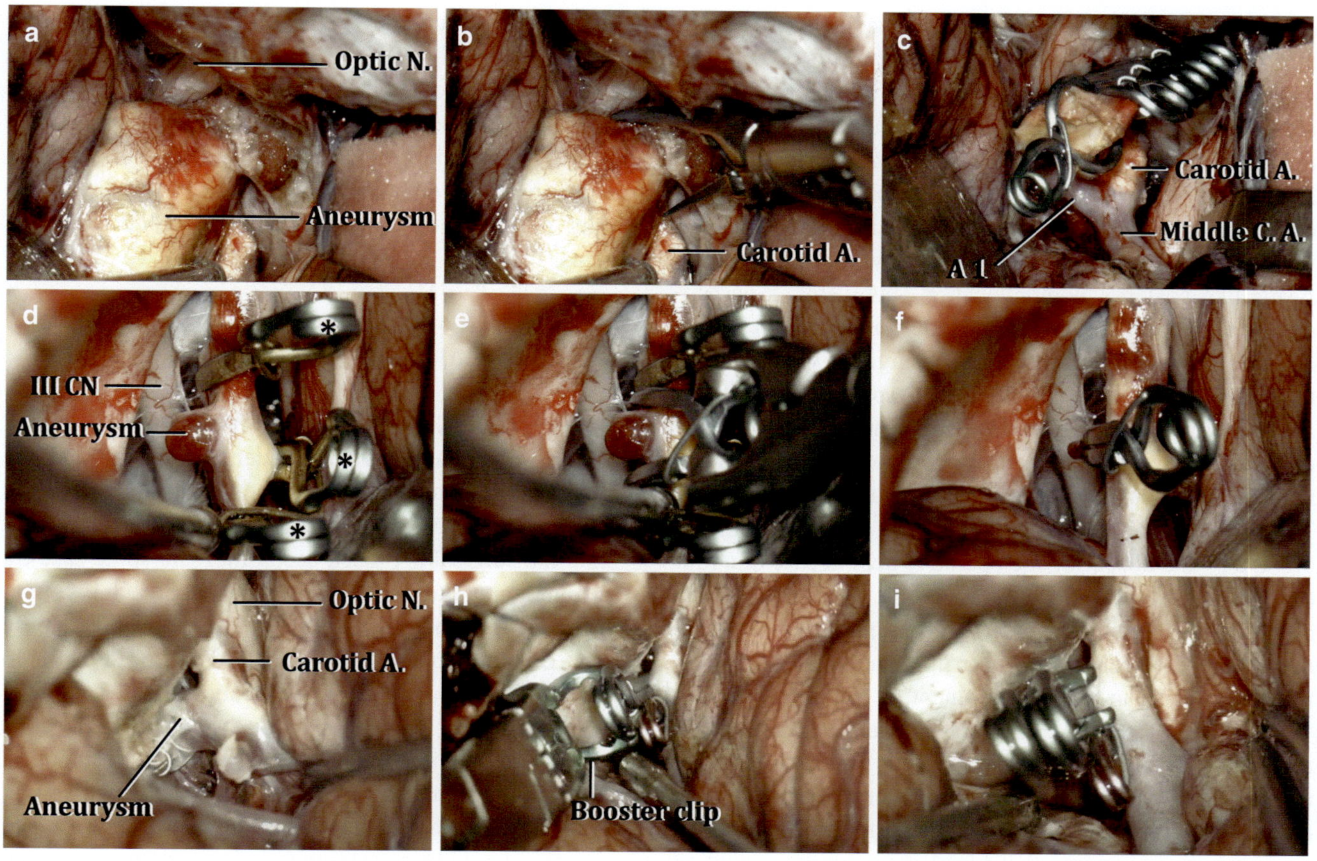

Fig. 8.3 (**a**–**c**) In this case, a right paraclinoid giant aneurysm is observed after exposure during an anterior intradural clinoidectomy. A first 20 mm semicurved clip is placed at the aneurysm's base, followed by a tandem clipping technique with 5 additional stacked clips to ensure sufficient closure strength. (**d**–**f**) In this case, a left anterior choroidal aneurysm is observed, where it was not possible to separate the base of the saccular aneurysm from the origin of the anterior choroidal artery. For this reason, temporary clips (*) are placed on the internal carotid, middle cerebral, and anterior cerebral arteries before proceeding with the definitive clip placement using a 7 mm semicurved clip. (**g**–**i**) In this case, a patient with two left-sided aneurysms is shown: a partially embolized posterior communicating aneurysm and an anterior choroidal aneurysm. A miniclip is placed on the anterior choroidal aneurysm, and a 10 mm straight clip is used on the posterior communicating aneurysm, which is reinforced with a clip booster due to flow detected during Doppler control

8.5 Surgical Technique

The patient's position during aneurysm clipping surgery is an important consideration. The surgical team will typically position the patient depending on the localization of the aneurysm, with their head secured in a special head holder that provides access to the surgical site. The head holder may be adjusted to allow for the precise positioning of the patient's head, and the patient's body may be tilted slightly to improve access to the surgical site. In some cases, the surgical team may also use specialized monitoring equipment to track the patient's brain function during the procedure. The surgical team will work carefully to avoid damaging the surrounding brain tissue or blood vessels and to ensure that the clip is securely in place to prevent bleeding or rupture of the aneurysm.

The use of mouth-operated controls for the microscope allows for delicate adjustments in the angle of visualization without the need to take hands off. Sharp arachnoidal dissection and gentle probing with a rounded dissector are used to reveal the anatomy of the aneurysm's neck, its associated branches, and perforators while avoiding the dome. High-risk procedures, such as freeing a vessel that adheres to the fragile dome of the aneurysm or advancing a permanent clip, should be done after temporary clipping to soften the aneurysm.Incomplete closure of the tips of the blade can also occur because the proximal parts of the blades are held open by voluminous or atheromatous aneurysmal tissue. Tandem clipping with a fenestrated clip can be done in such a situation. A fenestrated clip is placed first and then the proximal part of the neck is closed with additional straight clips.

To optimize the placement of aneurysm clips, it is crucial to align their direction parallel to the parent vessel and the neck. This alignment helps prevent kinking in the parent vessel and avoids catching perforators at the tip of the clip, reducing the risk of complications. Placing the clip perpendicular to the parent artery can increase the likelihood of rupture at the aneurysm's neck.

For specific locations such as the anterior communicating artery, middle cerebral artery, basilar bifurcation, and ICA-ophthalmic artery, parallel clip placement is of utmost importance. In cases where an ICA aneurysm points inferiorly, like superior hypophyseal aneurysms, a right-angled fenestrated clip is preferable. Conversely, for certain aneurysms like posterior communicating aneurysms, a perpendicular clip placement is often adequate and safe.

To facilitate the procedure, many surgeons suggest using temporary clips on the parent vessel, especially in ruptured aneurysms, before placing the permanent clip. Temporary clip placement softens the aneurysm and makes the neck more pliable, simplifying the dissection of adherent vessels. This proactive approach minimizes the risk of significant blood loss and obscuring the operative field in case of a rupture. However, in heavily atherosclerotic vessels, temporary clip placement may be avoided. Some surgeons reserve the use of temporary clips solely for intraoperative aneurysmal rupture, rather than routine placement.

Employing a tentative clip strategy can be advantageous in cases of suboptimal clip placement. Instead of hastily removing the initial clip, another clip can be stacked on top or below it. This approach is particularly beneficial when dealing with aneurysms that bleed prematurely, providing temporary flow control until proximal control is established.

Examples of situations where this technique proves valuable include giant aneurysms with thick and wide bases, aneurysms with thickened walls due to calcium or cholesterol plaques, and broad-based high-flow or high-pressure aneurysms, such as proximal carotid aneurysms near the artery's entry into the cranial cavity. Thickened aneurysm bases can lead to clip displacement onto the thinner-walled parent artery, potentially causing occlusion and subsequent ischemia.

When calcium or cholesterol plaques are present, they can alter proper clip placement by increasing the closing pressure required for complete occlusion and causing displacement toward the parent vessel or aneurysm fundus, resulting in suboptimal obliteration. This technique is especially advantageous in cases with limited surgical exposure or narrow corridors.

Using fewer clips for occlusion can prevent visual obstruction caused by overcrowding in the operative field. It is worth noting that the closing force of clip blades may vary along the blades, being strongest near the spring and weakest at the blade tips.

To reinforce and maintain the closing pressure of the initial clip, the additional booster clip must be placed as close to the spring as possible and must securely grip the proximal part of the clip blades. The closing force is also inversely proportional to its length; hence when facing giant aneurysms requiring longer clips, incomplete neck occlusion or clip slippage is at risk. In this sense, booster clips are designed to provide a higher closing pressure than standard permanent clips, boosting or supplementing jaw compression. For this reason, they should never be applied directly over an aneurysm or blood vessel. As with standard clips, booster clips can have straight, angled, or curved blade patterns.

8.6 Surgical Pearls

- Aneurysm clipping requires a tailored approach that takes into account the individual patient's anatomy and the location and size of the aneurysm. The surgeon should carefully assess the patient's neurovascular anatomy and plan the surgical approach accordingly. In some cases, a bifrontal craniotomy may be the best approach, while in other cases, a pterional or orbitozygomatic craniotomy may be more appropriate. A tailored approach can help minimize the risk of complications and optimize outcomes.
- Aneurysm clipping requires microsurgical techniques, including careful dissection and manipulation of blood vessels, meticulous hemostasis, and precise suturing. The surgeon should have a strong foundation in microsurgical techniques and should be comfortable working with a microscope and microsurgical instruments.
- Angiography is an important tool for ensuring the complete occlusion of the aneurysm and minimizing the risk of recurrence. The surgeon can use intraoperative angiography to visualize blood flow in real time and confirm that the aneurysm has been completely occluded. Intraoperative angiography can also be used to assess the patency of nearby vessels and detect any inadvertent injury to blood vessels.
- Intraoperative neuromonitoring techniques, such as motor evoked potentials (MEPs) and somatosensory evoked potentials (SSEPs), can help the surgeon assess the function of the brain and cranial nerves during surgery. The use of neuromonitoring can help prevent postoperative neurological deficits and improve outcomes. The surgeon should work closely with a neurophysiologist who is trained in intraoperative neuromonitoring techniques.
- Team-based approach to patient care is critical for ensuring the best possible outcomes for patients who undergo aneurysm clipping. A multidisciplinary team, including neurosurgeons, neurointensivists, neurologists, neuroradiologists, and nurses, should work together to optimize the patient's care before, during, and after surgery. Preoperative assessment and optimization, including blood pressure control, electrolyte management, and management of any comorbidities, are critical for mini-

mizing the risk of complications and ensuring a smooth recovery.

- Use of the proper microsurgical instruments: It is essential to work with appropriate microsurgical elements, non-stick bipolars of different sizes, 3 and 5 French microaspirators, use dynamic suction regulation, straight and curved tip microscissors, etc.
- Continuously improve your skills: aneurysm clipping is a complex and technically challenging procedure, and it requires ongoing education and training in the laboratory. As an airplane pilot must do his internship in flight simulators, the laboratory has the same significance for our training, but in these times when aneurysm microsurgery has decreased in many countries.

8.7 Discussion

Despite the best efforts of the neurosurgical team, there are several potential pitfalls that can occur during the procedure. Understanding these and taking steps to prevent them is crucial for successful outcomes. One of the most significant pitfalls during intracranial aneurysm clipping is the risk of intraoperative rupture. An aneurysm can rupture during the dissection and clipping process, which can lead to significant blood loss and a difficult surgical situation. Intraoperative rupture can occur due to several reasons, including inadequate temporary clipping, excessive manipulation of the aneurysm, and injury to the fragile dome of the aneurysm. To prevent intraoperative rupture, the surgical team must take several precautions.

First, temporary clips should be placed on the parent vessel before the final dissection and placement of the permanent clip. This softens the aneurysm and makes the neck more pliable, aiding in the dissection of adherent vessels. Planned temporary clip placement reduces the risk of massive blood loss and obscuration of the operative field in the event of a rupture. Additionally, gentle handling of the aneurysm and avoiding risky maneuvers like freeing a vessel adherent to the fragile dome of the aneurysm or advancing a permanent clip should only be carried out after temporary clipping to soften the aneurysm.

Another potential pitfall during intracranial aneurysm clipping is the inadequate occlusion of the aneurysm neck. The clip must be placed precisely to occlude the neck without causing damage to the parent vessel or adjacent structures. If the clip is not adequately placed, the aneurysm may continue to bleed and cause complications. Moreover, a poorly placed clip can also cause vessel kinking or obstruction of perforating vessels, leading to neurological deficits. To prevent inadequate occlusion, the surgical team must take several precautions, careful planning and selection of the ideal clip are crucial. The direction of the clip should be parallel to the direction of the parent vessel and the neck. That angle prevents kinking in the parent vessel and also avoids catching perforators at the tip of the clip. For aneurysms of the anterior communicating artery, middle cerebral artery (MCA), basilar bifurcation, ICA-ophthalmic artery, and many other aneurysms, parallel placement of clips is crucial.

Another potential pitfall during intracranial aneurysm clipping is injury to adjacent structures. The brain and adjacent blood vessels are delicate structures that must be protected during the procedure. Injury to these structures can lead to neurological deficits or even death. For example, injury to the optic nerve or chiasm during clipping of a paraclinoid aneurysm can result in vision loss. To prevent injury to adjacent structures, the surgical team must take several precautions. The surgeon should have a thorough understanding of the anatomy and the location of the aneurysm. Sharp arachnoidal dissection and gentle probing with a rounded dissector reveal the anatomy of the neck of the aneurysm, associated branches, and perforators while avoiding the dome. Additionally, the use of modern imaging techniques such as 3D angiography and neuronavigation can help the surgical team visualize the location of the aneurysm and adjacent structures more accurately.

8.8 Conclusion

In conclusion, brain aneurysm surgery represents a crucial and effective intervention in managing this potentially life-threatening condition. The surgical treatment of brain aneurysms aims to prevent rupture and reduce the risk of devastating consequences such as subarachnoid hemorrhage. Aneurysm clipping represents a profoundly specialized and intricate neurosurgical procedure, demanding meticulous planning, adept surgical expertise, and vigilant monitoring to achieve optimal patient outcomes. While all surgeries carry inherent risks, those related to aneurysm clipping can be mitigated through prudent patient selection, thorough preoperative preparation, and the utilization of cutting-edge tools and techniques designed to minimize potential complications.

Advancements in neurosurgical techniques, including microsurgical clipping and endovascular coiling, have significantly improved patient outcomes. The selection of the most suitable surgical approach depends on the aneurysm's size, location, shape, and the patient's overall health. Close collaboration among neurosurgeons, interventional neuroradiologists, and other healthcare specialists is essential in ensuring optimal patient care and successful surgical outcomes.

References

1. Lee KS, Zhang JJY, Nguyen V, Han J, Johnson JN, Kirollos R, Teo M. The evolution of intracranial aneurysm treatment techniques and future directions. Neurosurg Rev. 2022;45(1):1–25. https://doi.org/10.1007/s10143-021-01543-z.
2. Giotta Lucifero A, Baldoncini M, Bruno N, Tartaglia N, Ambrosi A, Marseglia GL, Galzio R, Campero A, Hernesniemi J, Luzzi S. Microsurgical neurovascular anatomy of the brain: the anterior circulation (part I). Acta Biomed. 2021;92(S4):e2021412. https://doi.org/10.23750/abm.v92iS4.12116. PMID: 34437363; PMCID: PMC9179062.
3. Giotta Lucifero A, Baldoncini M, Bruno N, Tartaglia N, Ambrosi A, Marseglia GL, Galzio R, Campero A, Hernesniemi J, Luzzi S. Microsurgical neurovascular anatomy of the brain: the posterior circulation (part II). Acta Biomed. 2021;92(S4):e2021413. https://doi.org/10.23750/abm.v92iS4.12119. PMID: 34437362; PMCID: PMC9179054.
4. Rhoton AL Jr. The supratentorial arteries. Neurosurgery. 2002;51(4 Suppl):S53–120.
5. Lawton MT. Seven aneurysms tenets and techniques for clipping. New York: Thieme; 2011.

Orbital-Meningeal Band Anatomy and Application for Exposure of Proximal Carotid Artery

9

Iype Cherian and Giacomo Bertolini

Learning Objectives

- Understand the microsurgical anatomy of the orbital-meningeal band (OMB) and its relationship with the surrounding structures.
- Apprehend the brain unlocking concept, leading to an extradural mobilization of the frontal and temporal lobe and gaining access to the skull base.
- Master the step-by-step microsurgical technique to dissect the OMB and its application to expose the pericavernous region and the proximal internal carotid artery (ICA).
- Discern tips and tricks from exemplificative cases to better appreciate the implications and advantages of OMB dissection in skull base surgery.

9.1 Introduction

The orbital-meningeal band (OMB) is a dural fold tethering the frontotemporal basal dura and the periorbital space, passing through and defining the superolateral and inferolateral aspect of the superior orbital fissure (SOF). Despite the OMB being not frequently appreciable in a standard pterional approach that not requires extradural dissection and/or skull base exposure, its knowledge and the surgical implication related to its detachment are paramount for anterolateral skull base approaches, requiring exposure of the anterior clinoid, proximal internal carotid artery (ICA), and pericavernous region.

In the following chapter, the microsurgical anatomy of the SOF and the surgical applications related to its dissection will be discussed along with operative consideration derived from representative cases that will help the reader to appreciate the surgical relevance of OMB.

9.2 History/Background

Although the first report of the OMB is dated back to the pioneering papers by Prof Vinko Dolenc in the description of extradural approaches to carotid artery aneurysms [1] and skull base tumors [2], the nomenclature of this dural fold changed over the years. Originally named solely as the "dural tent" of SOF [2], several authors over the years have described more in detail the microsurgical anatomy, the neurovascular structures embedded within the OMB, and the surgical benefits derived from wider corridors subsequent to OMB detachment. Coscarella and colleagues [3] reported an alternative extradural dissection method for the anterior clinoid process (ACP) exposure referring to the OMB as a "frontotemporal dural fold." Conversely, Froelich et al. [4], named the OMB as "orbitotemporal periosteal fold" in their paper, presented a refinement of the extradural anterior clinoidectomy technique, and described several surgical insights to minimize the morbidity associated with the sectioning of the OMB and elevation of the temporal dura near close to the SOF.

The first reference to the nomenclature "meningoorbital band" was introduced by Noguchi et al. [5] in 2005 and subsequently spread in the literature in the following years, becoming the most widely used term in referring to this anatomical structure [6–8]. The central role of the OMB in skull base surgery unequivocally emerges from the literature and requiring the extradural removal of the ACP and/or performing approaches in the paraclinoid or cavernous sinus (CS) region. In fact, the wider exposure granted by a proper OMB detachment allows for enhanced surgical orientation, access to otherwise not practicable surgical corridors, and increased surgical maneuvers safety related to an easier manipulation

I. Cherian (✉)
Department of Neurosurgery, Krishna Vishwa Vidyapeeth, Karad, Maharashtra, India

G. Bertolini
Neurosurgery Division, Head and Neck Department, Azienda Ospedaliero-Universitaria di Parma, Parma, Italy

I. N. Kanaan, V. Beneš (eds.), *Neuroanatomy Guidance to Successful Neurosurgical Interventions*, https://doi.org/10.1007/978-3-031-59838-8_9

preventing potential injuries to surrounding structures. A solid anatomical knowledge of the OMB appears; therefore, the sine qua nonbackground is necessary in a neurosurgeon's armamentarium in dealing with pathologies arising from or close to the proximal carotid artery.

9.3 Anatomy Highlights

- The OMB is composed of a dural reflection of frontal and temporal dura merging along the greater sphenoid wing and the sphenoid ridge, connecting the anterior and middle fossa dura.
- In order to perform a safe OMB dissection, it is mandatory to understand the three-dimensional relationship between the SOF and cavernous sinus.
- The superior orbital fissure (SOF) has a triangular shape, with the base directed laterally and the apex medially; the majority of the neurovascular structures are located in the medial portion of the SOF.
- The dura lateral to the CS is composed of two layers; one layer attached to the bone called periosteal dura—or endosteal layer—and one more superficial layer identified as dura propria—or meningeal layer. The identification of the loose tissue between those two layers is crucial to finding the right dissection plane and avoiding lesions to neurovascular structures related to the violation of the true lateral wall of the CS.

The appropriate understanding of the OMB anatomy requires a thoughtful knowledge of the SOF anatomy and the dural layers/sheets that constitute the middle fossa dura lateral to the CS. Schematically representable as a triangular-shape communication, with a medially oriented base and the apex laterally, interconnecting the middle cranial fossa with the orbit, the SOF is bounded by a roof or upper margin, represented by the small wing of the sphenoid that medializes in an anterolateral to posteromedial direction giving rise to the ACP bony prominence, a lower margin constituted by the large wing of the sphenoid, and a medial margin composite superiorly by the optic strut and inferiorly by the body of the sphenoid bone, in a strict relationship with the carotid sulcus immediately medial to it [9]. Numerous neurovascular structures are crossing through this bony fissure. Coursing from the posterior fossa into the CS, the oculomotor (IIIrd cranial nerve [CN]), trochlear (IVth CN), ophthalmic branch (V_1) of the trigeminal (Vth CN), and abducent (VIth CN) nerves approach the fissure. In its lateral part, delimitated by the lateral margin of the annular tendon, is located the trochlear nerve beside the lacrimal and frontal branches of V_1, and the superior ophthalmic vein with the lacrimal nerve residing in the most lateral part of the fissure. Contrarywise, the IIIrd CN, in its superior and inferior division, the nasociliary branch of V_1, and the abducent nerve pass through the medial aspect of the SOF. Immediately beneath the ACP and medial to the inner margin of the SOF course the proximal segments of the ICA. Along the carotid sulcus, a bony groove in the sphenoid bone directed in an anteromedial fashion from the ending of the carotid canal to the base of tuberculum sellae runs the horizontal segment of the cavernous carotid (C4), turning then upward and medial, defining an anterior vertical loop enclosed within the proximal and distal dural ring and sheltered by the ACP, into the clinoidal segment (C3). Emerging from the distal dural ring, infero-medial to the ACP and inferior to the optic canal, the ICA course intradural giving origin to the ophthalmic or intradural segment (C2). Finally, the ICA intradural segment, prior to the bifurcation (C1 segment) into its distal branches—the anterior and medial cerebral arteries –, gives rise to numerous arteries, including usually in a distal to proximal order, the ophthalmic, superior hypophyseal, posterior communicating, and the anterior choroidal artery [9–11].

Establishing the anatomy of the SOF and its relationships with the neurovascular structures that course through it, the next step to fully master the procedure of OMB dissection and avoid potential injuries to the neurovascular structures crossing the SOF is to understand the dural anatomy in this region. The dura lateral to the CS that covers also the middle fossa is composed of two layers; one layer attached to the bone called periosteal dura [6, 8] or endosteal layer [10], and one more superficial layer identified as dura propria [6] or meningeal layer [8, 10]. Those two layers regularly split in the lateral edge of the CS defining the "true" lateral wall of the CS, embedding the cranial nerves within it, as a thin extension of the periosteal dura, and a thicker layer of dura propria covering the brain, the "true" lateral wall of the CS sinus, and defining the dural margin of the SOF. Therefore, during the dissection of the OMB, a layer of loose tissue is found between the two dural layers, guiding the peeling of the dura propria of the temporal lobe from the lateral wall of the CS, allowing a safe dissection of the two dural folds along this cleavage plane and preserving the anatomical integrity of the CS and the anatomical structures held within it.

9.4 Approach/Surgical Intervention

Different surgical approaches can lead to the exposure of the OMB, from the standard pterional to its anterolateral variations (e.g., frontotemporal, frontotemporalorbitozygomatic, etc.). Once the craniotomy has been performed, the OMB is seldom directly appreciable, as composed of frontobasal and temporal dural reflections toward the SOF usually covered by the convexity frontal and temporal dura.

The first step entails the gentle retraction of the temporal and frontal dura in a latero-medial fashion from the temporal pole and sphenoid ridge, respectively. This procedure outlines the superior and inferior tent-shape margins of the OMB, along the borders of the SOF (Fig. 9.1a, b). Subsequently, partial drilling of the sphenoid ridge is required to better highlight the superior margin of the OMB and facilitate the detachment of the OMB from the roof of the SOF; this maneuver, which takes the name of "sagittal unlocking," allows a wider surgical exposure on the sagittal plane. Then a cut in the lateral aspect of the OMB is necessary to proceed with the release of the OMB and achieve the "axial unlocking" among the frontal and temporal lobes; the cut usually starts from the most lateral margin of the OMB (Fig. 9.1c, d), posing/paying attention to avoid injuries to the cranial nerves crossing the SOF. Of note, despite this region being deemed safe as it does not enclose any neural structure, the orbitomeningeal artery is frequently encountered within the lateral margin of the SOF but it can be frequently cauterized without detrimental consequences. Once the OMB has been sectioned in its lateral fold, the cleavage plane between the periosteal dura and dura propria is revealed (Fig. 9.1e, f), guiding the sharp dissection and progressive exposure of the lateral wall of the CS inferiorly and the ACP superiorly (Fig. 9.1g, h). After the ACP isolation from the surrounding dura, the extradural anterior clinoidectomy, up to a complete removal of the orbital roof exposing the optic nerve and opticocarotid window, can be performed to increase the exposure of the paraclinoid and pericavernous region (Fig. 9.1i, j). Finally, a further peeling of the lateral wall of the CS can be performed on an individual basis. Subsequently, we will present some illustrative cases to show the practical application of OMB dissection in cases requiring the exposure of proximal ICA.

9.4.1 Case 1 (Fig. 9.2)

A 43-year-old man presented with a thunderclap headache. Head CT and CT angiography revealed an aneurysmal subarachnoid hemorrhage from a left ophthalmic artery aneurysm. A left frontotemporal craniotomy was performed. Once the bone flap was removed the OMB is not appreciable, covered by the frontal and temporal dural fold over the Sylvian fissure (Fig. 9.2a). After the partial sphenoid ridge drilling and frontotemporal dural retraction, the OMB is visualized and incised in its lateral margin (Fig. 9.2b); then the dura propria is peeled away from the periosteal dura of the middle fossa (Fig. 9.2c), revealing the ACP. Subsequently, anterior clinoidectomy was performed (Fig. 9.2d) to gain a wider exposure over the aneurysm site that would otherwise have been covered. After the dura opening and proximal Sylvian fissure dissection, the II CN, the proximal ICA, and the aneurysm were identified (Fig. 9.2e). The aneurysm neck, after a gentle arachnoid dissection and aneurysm dome isolation, was closed with a straight clip (Fig. 9.2f); the post-clipping indocyanine green video angiography confirmed the aneurysm exclusion and the patency of the proximal ICA (Fig. 9.2g).

9.4.2 Case 2 (Fig. 9.3)

A 67-year-old woman came to medical attention for headache and vomiting. Head CT and CT angiography revealed an aneurysmal subarachnoid hemorrhage from a right posterior communicating artery (PComA) aneurysm. After a right pterional craniotomy, the sphenoid ridge was drilled (Fig. 9.3a) and, following the OMB section, the dura propria was detached from the periosteal dura, revealing the true lateral wall of the cavernous sinus and its structures passing through the SOF (Fig. 9.3b). Thereafter, the ACP was removed with a high-speed drill to increase the intradural working space for the aneurysm clipping (Fig. 9.3c). The proximal to distal ICA dissection showed the PComA aneurysm (Fig. 9.3d, e) secured with a slightly curved aneurysm clip (Fig. 9.3f), preserving the emergent PComA trunk.

9.4.3 Case 3 (Fig. 9.4)

A 71-year-old male suffered a traumatic brain injury after a road traffic accident (GCS 10). A head CT showed a left acute subdural hematoma and multiple parenchymal contusions; a cisternostomy was planned to revert the cerebrospinal fluid shift edema and prevent brain swelling. Subsequent to a fronto-temporo-parietal craniotomy (Fig. 9.4a), the sphenoid ridge was flattened with a high-speed drill (Fig. 9.4b). Then, the OMB was cut in the lateral part (Fig. 9.4c) and the dura propria was peeled away from the anterolateral floor of the middle fossa (Fig. 9.4d). Afterwards, the dura was opened, the subdural hematoma was removed, and the frontal lobe was gently retracted to visualize the optic nerve and the proximal carotid artery (Fig. 9.4e). Finally, the basal cisterns were opened through the optico-carotid (Fig. 9.4f) and carotico-oculomotor (Fig. 9.4g) windows and the cisternal drainage was placed (Fig. 9.4h).

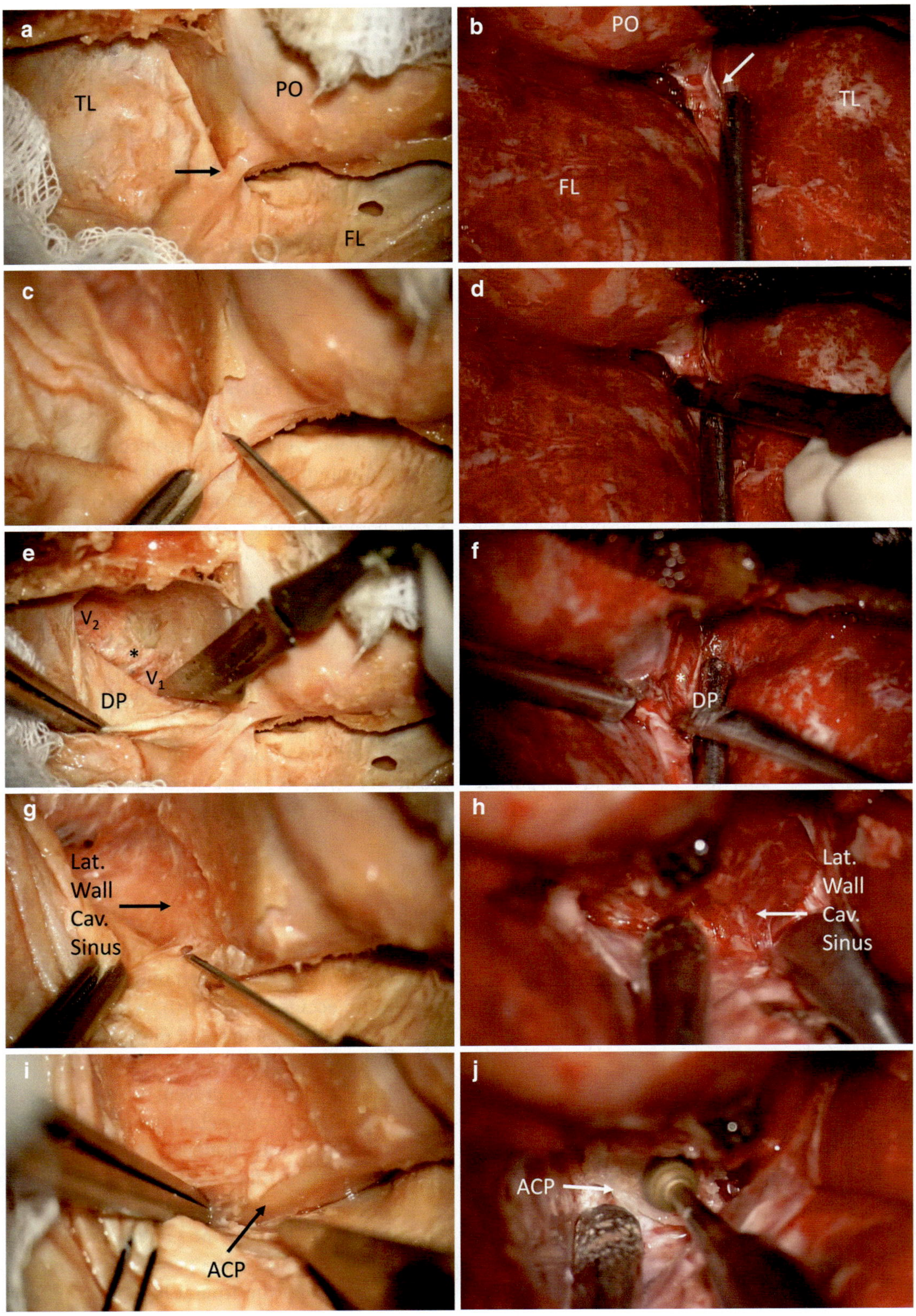

Fig. 9.1 Cadaveric (left) and intraoperative (right) stepwise anatomy dissection of the orbital-meningeal band (OMB). (**a** and **b**) dural exposure and identification of the frontal lobe (FL), OMB (arrow), periorbita (PO), and temporal lobe (TL). (**c** and **d**) magnification on the sharp cutting of the OMB in its lateral margin. (**e** and **f**) identification of the avascular plane and peeling of the dura propria (DP) from the periosteal dura (asterisk); the first (V_1) and second branch (V_2) of the left trigeminal nerve respectively crossing through the superior orbital fissure and foramen rotundum can be observed in the middle fossa floor in panel E. (**g** and **h**) the progressive detachment of the dura propria from the periosteal dura reveals the "true" lateral wall of the cavernous sinus (Lat. Wall Cav. Sinus). (**i** and **j**) the lateral to medial dissection along the sphenoid ridge uncovers the anterior clinoid process (ACP) in its most medial part (panel I); subsequently, the anterior clinoidectomy can be performed in an extradural fashion (panel J)

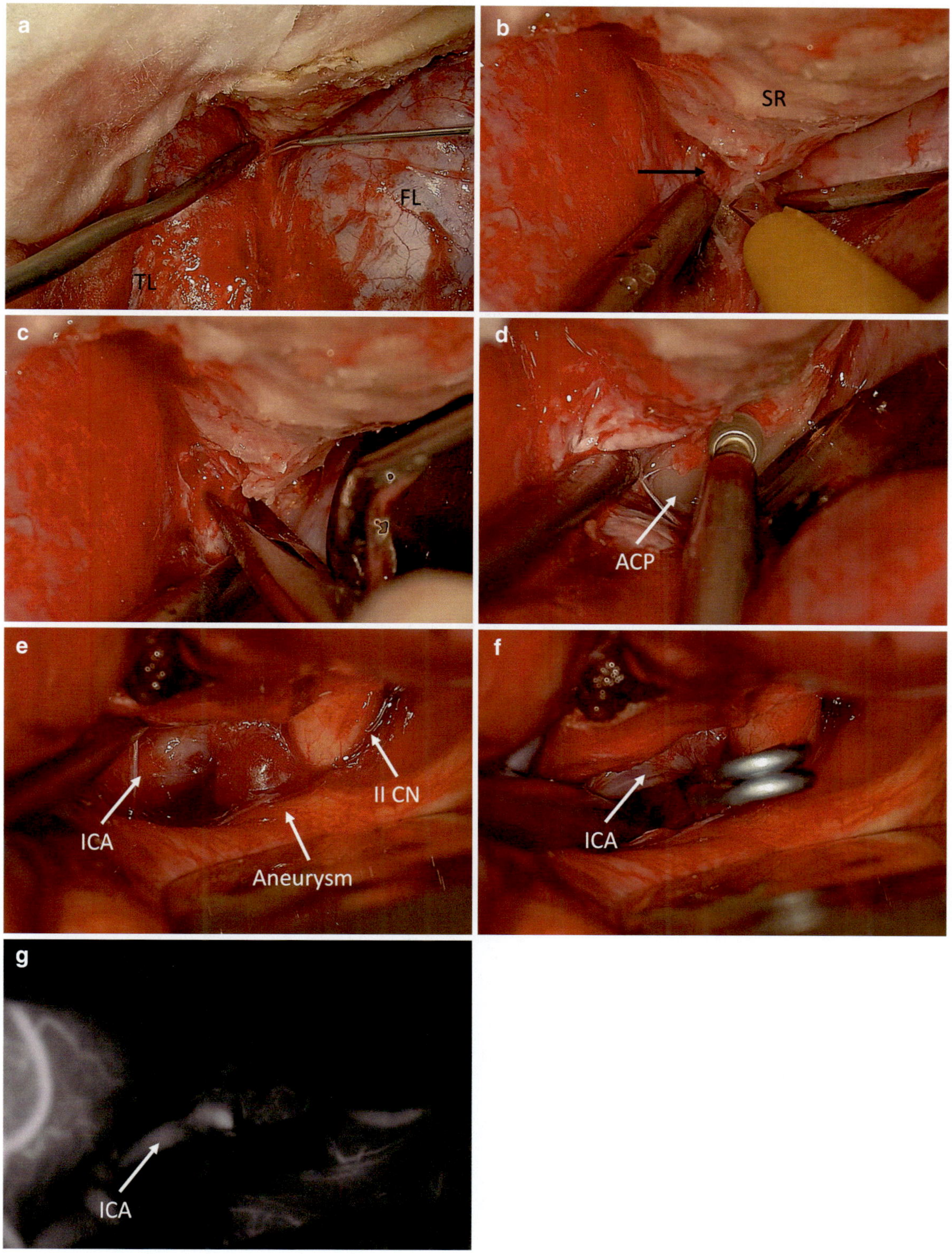

Fig. 9.2 Case one. (**a**) left pterional approach was performed, showing the left frontal and temporal lobe; the orbital-meningeal band (OMB) is not clearly visible covered by the frontal and temporal dural folds. (**b**) after a partial drilling of the sphenoid ridge (SR), the OMB is exposed (arrow) and cut along the lateral edge. (**c**) the peeling of the proper and periosteal dural folds follows the plane revealed after the OMB dissection. (**d**) exposure and drilling of the anterior clinoid process (ACP). (**e**) after the dural opening, the ophthalmic segment of the internal carotid artery (ICA), the optic nerve (II CN), and the aneurysm are visualized. (**f**) final intraoperative view following the aneurysm clipping. (**g**) Intraoperative indocyanine green videoangiography showing the patency of the ICA (arrow) after the aneurysm clipping

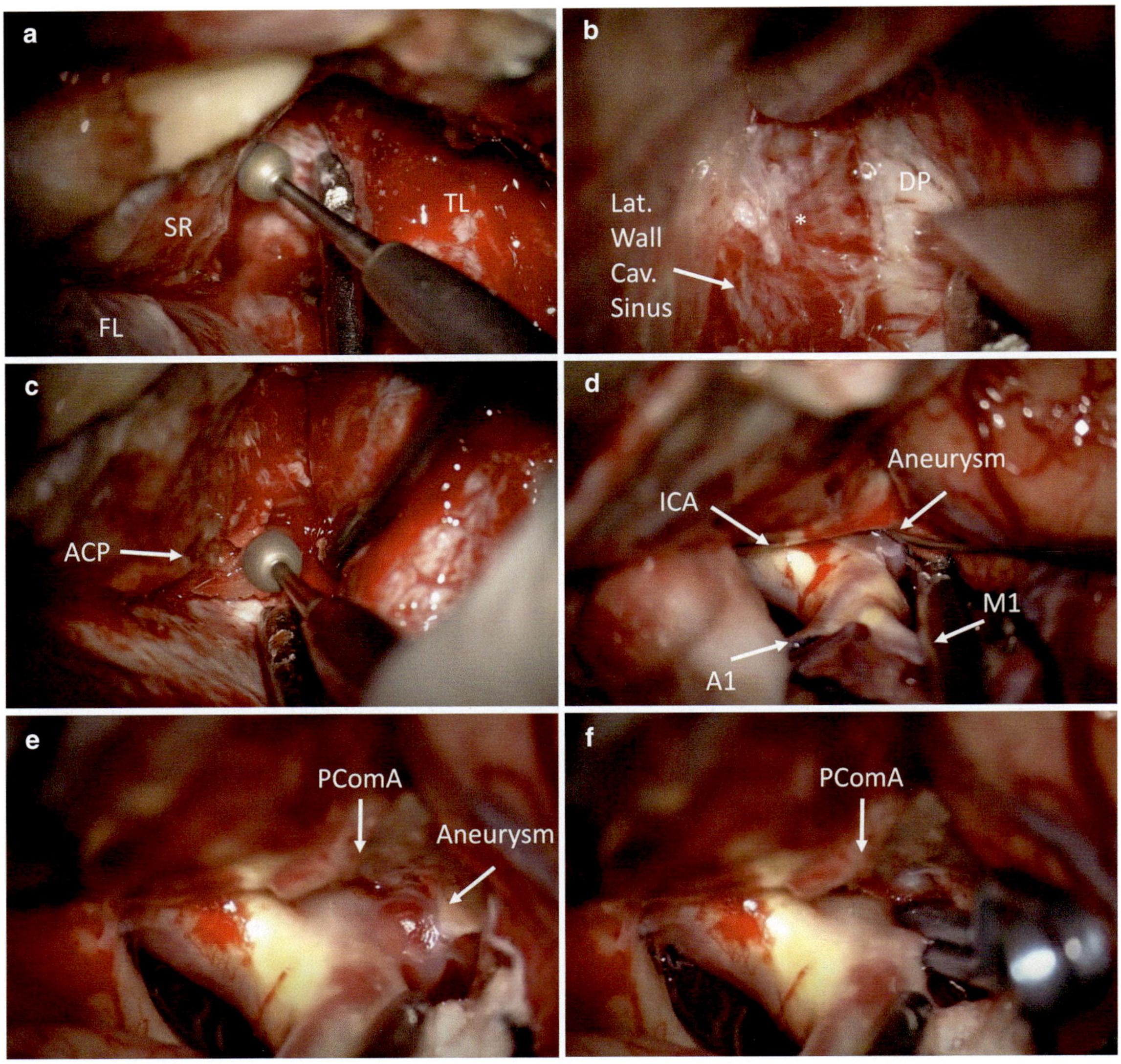

Fig. 9.3 Case two. (**a**) sagittal unlocking of the brain drilling the sphenoid bone (SR) next to the exposure of the frontal (FL) and temporal (TL) dura. (**b**) once the orbital-meningeal band is sectioned, the dura propria (DP) is peeled away from the periosteal dura (asterisk) revealing the lateral wall of the cavernous sinus (Lat. Wall Cav. Sinus). (**c**) the anterior clinoid process (ACP) is removed to gain a wider intradural working channel. (**d**) the proximal internal carotid artery (ICA) is visualized as well as the first segment of the anterior cerebral artery (A1) and middle cerebral artery (M1), and a partial vision of the aneurysm dome. (**e**) magnification showing the aneurysm and the origin of the posterior communicating artery (PComA). (**f**) final intraoperative view following the aneurysm clipping

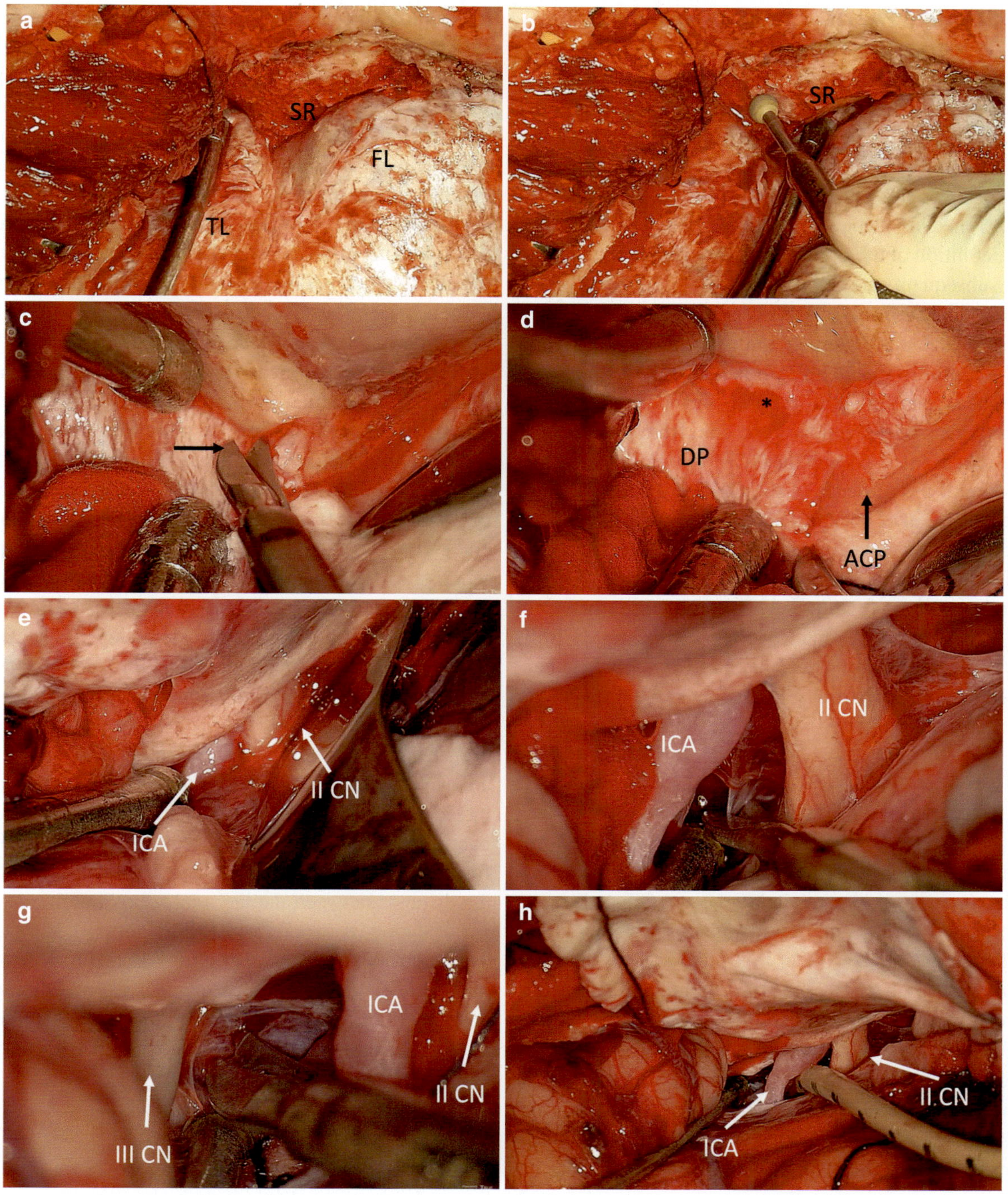

Fig. 9.4 Case three. (**a**) fronto-temporo-parietal craniotomy centered on the sphenoid bone (SR), and the dura of frontal (FL) and temporal (TL) lobes. (**b**) flattening of the sphenoid ridge (SR) with a high-speed drill. (**c**) initial cut of the orbital-meningeal band. (**d**) dissection of the dura propria (DP) from the periosteal dura (asterisk), revealing the anterior clinoid process (ACP). (**e**) after the dural opening and partial acute subdural hematoma evacuation the frontal lobe is gently retracted to expose the optico-carotid window between the optic nerve (II CN) and the internal carotid artery (ICA). (**f**) the optico-carotid window is open, dissecting the arachnoid folds including the membrane of Liliequist, and connecting the basal cisterns with the supratentorial cisterns. (**g**) the carotico-oculomotor window, between the ICA and oculomotor nerve (III CN), is also open with the same purpose. (**h**) the final vision of the cisternostomy procedure after the insertion of the ventricular catheter into the basal cisterns through the optico-carotid window

9.5 Surgical Pearls

- Once the bone flap has been removed, prior to the dissection of the OMB, we suggest initiating the gentle dural peeling from the greater sphenoid wing, forming the anterior wall of the middle fossa, and then detaching the dura of the anterior fossa covering the sphenoid ridge, both in a latero-medial fashion. This procedure will help in a clear definition of the OMB, its relationship with the SOF, and a subsequent safer incision.
- Perform the OMB dissection under a high magnification. This shrewdness is extremely important, as a better vision will be useful in the identification of the correct surgical plane between the dura propria and periosteal dura, permitting a safe dissection of the dura from the true wall of the cavernous sinus and the ACP.
- Be aware of the presence of the OMB artery. An aggressive coagulation of unexpected bleeding in the lateral portion of the OMB can cause an injury to the lacrimal nerve, usually close to this artery.
- Limit the drilling of the ACP whenever it is strictly necessary. This challenging maneuver is not exempted from risks and complications, and therefore it should be performed, entirely or partially, according to the needs on an individual basis.

9.6 Discussion

The peculiar anatomy of the OMB as well as the critical neurovascular structures intimately related to the CS made unfamiliar, especially for young neurosurgeons, the extradural approaches to the skull base in this region. Since the first descriptions reported by Dolenc [1, 2], anterolateral extradural approaches emerged as a useful and effective surgical corridor to treat suprasellar and parasellar lesions and cerebral aneurysms [11–13]. Indeed, the anterior clinoidectomy has been described as a crucial step to better expose the proximal ICA in case of aneurysm originating from the PComA or clinoidal/supraclinoidal segments of the ICA (e.g., ophthalmic segment) space otherwise covered by the ACP [5, 12, 14]. Numerous variations of the technique for the OMB dissection from the SOF and ACP were proposed over the years [2–9, 11], focused on minimizing the potential complications that this challenging procedure may cause. The lateral-to-medial cutting of the OMB, starting at the lateral edge of the SOF, has proved to be a safer maneuver [3–8], maximizing the exposure of the ACP before its removal and facilitating the preservation of the plane between the dura propria and periosteal dura and, consequently, the CS integrity. In this regard, a simple but effective expedient to identify the lateral edge of OMB, as proposed by Coscarella et al. [3], consists of a gentle and partial peeling of the dura along the sphenoid ridge and greater sphenoid wing. In fact, this procedure helps the surgeon to envision the OMB and the relationship with the SOF facilitating the establishment of the correct dissection plan and preventing neurovascular injuries. However, this procedure is not exempt from risk, as an excessive retraction of the temporal fossa dura from the SOF and CS has been associated with transient cranial nerve deficits, including diplopia and facial numbness; similarly, the aggressive coagulation of the orbitomeningeal artery may lead to lacrimal nerve injury because of the proximity between those two structures [4]. Therefore, a judicious dissection should be performed to avoid excessive traction along the inferior margin of the SOF together with a careful incision of the lateral margin of the OMB and prevent harmful maneuvers toward the trochlear nerve and the lacrimal and frontal branches of V_1, contained in the lateral part of the SOF. In this regard, some authors proposed to schematically divide the superior and inferior margin of the SOF into medial and lateral portions (line A—inferomedial region, line B—inferolateral region, line C—superolateral region, and line D –superomedial region), defining the OMB as the dural fold covering the superolateral and inferolateral regions [6, 8].

After the exposure of the OMB, a cut along the lines B and C was considered safe by Fukuda et al. [6] as well as an incision up to 4 mm in the OMB lateral margin according to Saenz et al. [8]. However, according to measurements performed on histological specimens, it has been reported that the distance between the lacrimal nerve and the lateral border of the OMB varied from 1.2 to 4 mm [4]. Refusing dogmatic concepts, we suggest limiting as much as possible the blind cut of the OMB and subsequently proceeding with the dural peeling to avoid potential injuries to the cranial nerves in the SOF lateral compartment.

9.7 Conclusion

Surgical access to the lesions harboring into paraclinoid, parasellar, and cavernous sinus regions continues to be a significant challenge. Extradural approaches have proven to be useful techniques in the neurosurgical armamentarium to treat those lesions. The knowledge of the OMB anatomy and the technical nuances to perform a proper and safe dissection are paramount for gaining adequate surgical exposure and working space. The step-by-step anatomical details coupled with illustrative cases provided in this chapter will help the reader to familiarize themself with this anatomical region and understand the potentiality of this technique.

References

1. Dolenc VV. A combined epi- and subdural direct approach to carotid-ophthalmic artery aneurysms. J Neurosurg. 1985;62(5):667–72. https://doi.org/10.3171/jns.1985.62.5.0667. PMID: 3989589.
2. Dolenc VV. Frontotemporal epidural approach to trigeminal neurinomas. Acta Neurochir. 1994;130(1–4):55–65. https://doi.org/10.1007/BF01405503. PMID: 7725943.
3. Coscarella E, Başkaya MK, Morcos JJ. An alternative extradural exposure to the anterior clinoid process: the superior orbital fissure as a surgical corridor. Neurosurgery. 2003;53(1):162–6; discussion 166–7. https://doi.org/10.1227/01.neu.0000068866.22176.07. PMID: 12823885.
4. Froelich SC, Aziz KM, Levine NB, Theodosopoulos PV, van Loveren HR, Keller JT. Refinement of the extradural anterior clinoidectomy: surgical anatomy of the orbitotemporal periosteal fold. Neurosurgery. 2007;61(5 Suppl 2):179–85; discussion 185–6. https://doi.org/10.1227/01.neu.0000303215.76477.cd. PMID: 18091231.
5. Noguchi A, Balasingam V, Shiokawa Y, McMenomey SO, Delashaw JB Jr. Extradural anterior clinoidectomy. Technical note. J Neurosurg. 2005;102(5):945–50. https://doi.org/10.3171/jns.2005.102.5.0945. PMID: 15926728.
6. Fukuda H, Evins AI, Burrell JC, Iwasaki K, Stieg PE, Bernardo A. The Meningo-orbital band: microsurgical anatomy and surgical detachment of the membranous structures through a frontotemporal craniotomy with removal of the anterior clinoid process. J Neurol Surg B Skull Base. 2014;75(2):125–32. https://doi.org/10.1055/s--0033-1359302. Epub 2013 Dec 11. PMID: 24719799; PMCID: PMC3969442.
7. Cherian I, Kasper EM, Agarwal A. The orbitomeningeal band as a way to bloodless transcavernous dissection and anterior clinoidectomy. Asian J Neurosurg. 2018;13(3):943–5. https://doi.org/10.4103/ajns.AJNS_198_16. PMID: 30283589; PMCID: PMC6159013.
8. Saenz A, Villalonga JF, Solari D, Baldoncini M, Mantese B, Lopez-Elizalde R, Campero A. Meningo-orbital band detachment: a key step for the extradural exposure of the cavernous sinus and anterior clinoid process. J Clin Neurosci. 2020;81:367–77. https://doi.org/10.1016/j.jocn.2020.09.055. Epub 2020 Oct 23. PMID: 33222945.
9. Natori Y, Rhoton AL Jr. Microsurgical anatomy of the superior orbital fissure. Neurosurgery. 1995;36(4):762–75. https://doi.org/10.1227/00006123-199504000-00018. PMID: 7596508.
10. Yasuda A, Campero A, Martins C, Rhoton AL Jr, de Oliveira E, Ribas GC. Microsurgical anatomy and approaches to the cavernous sinus. Neurosurgery. 2008;62(6 Suppl 3):1240–63. https://doi.org/10.1227/01.neu.0000333790.90972.59. PMID: 18695545.
11. Fukushima T, Day JD. Surgical management of tumors involving the cavernous sinus. In: Schmidek HH, Sweet WH, editors. Operative Neurosurgical Techniques. Philadelphia: W. B. Saunders; 1995. p. 493–510.
12. Yonekawa Y, Ogata N, Imhof HG, Olivecrona M, Strommer K, Kwak TE, Roth P, Groscurth P. Selective extradural anterior clinoidectomy for supra- and parasellar processes. Technical note. J Neurosurg. 1997;87(4):636–42. https://doi.org/10.3171/jns.1997.87.4.0636. PMID: 9322855.
13. Kawase T, van Loveren H, Keller JT, Tew JM. Meningeal architecture of the cavernous sinus: clinical and surgical implications. Neurosurgery. 1996;39(3):527–34; discussion 534–6. https://doi.org/10.1097/00006123-199609000-00019. PMID: 8875483.
14. Andaluz N, Beretta F, Bernucci C, Keller JT, Zuccarello M. Evidence for the improved exposure of the ophthalmic segment of the internal carotid artery after anterior clinoidectomy: morphometric analysis. Acta Neurochir (Wien). 2006;148(9):971–5; discussion 975–6. https://doi.org/10.1007/s00701-006-0862-x. Epub 2006 Aug 21. PMID: 16917665.

10 Surgical Approaches to Cerebral Aneurysms

Mustafa K. Başkaya and Turab Gasimov

Learning Objectives
- Short historical review of the development of aneurysm surgery
- Anatomy of the main approaches used in cerebral aneurysm surgery
- Pearls in patient positioning and the surgical technique

10.1 Introduction

The first direct surgical treatment of an intracranial aneurysm was performed by Norman Dott, who wrapped a ruptured aneurysm in 1933 [1]. The first clipping of an aneurysm was performed by Walter Dany in 1938, which was located on the internal carotid artery at the origin of the posterior communicating artery [2]. A brief history will also be provided in each section below that discusses a specific surgical approach.

The surgical treatment of aneurysms requires advanced knowledge in neuroanatomy, especially of the bony and dural skull base anatomy, and of the cisternal, parenchymal, and vascular anatomy, including their variations. In this chapter, we will only mention the names and locations of the subarachnoid cisterns; advanced cisternal anatomy may be reviewed from the reference list [3]. We believe that training in cadaveric dissection should be mandatory to strengthen theoretical anatomical knowledge and improve surgical skills.

10.2 Anatomy Highlights

- Skin anatomy and the anatomy of subcutaneous and muscle tissue along with nerves and vessels, facial nerve anatomy in aneurysm surgery
- Extradural bony anatomy, sphenoid ridge, anterior clinoid process, posterior clinoid process, optic canal, optic strut, and occipital condyle
- Dura anatomy, dural folds, relationship with vascular anatomy, distal dural ring, falciform ligament, and tentorium
- Cisternal anatomy
- Cerebrovascular anatomy and variations

10.3 Intervention

To simplify the choice of surgical exposure, we will discuss each approach in the following groups:

1. Approaches encountering the anterior and middle fossa
2. Approaches through the posterior fossa

10.3.1 Approaches Encountering the Anterior and Middle Fossa

10.3.1.1 Pterional Approach and Its Extensions

Pterional or frontotemporosphenoidal craniotomy was introduced by Yaşargil in the 1970s and is the most often used approach in aneurysm surgery. Indications for the pterional approach in aneurysm surgery include:

- Anterior circulation aneurysms other than distal pericallosal artery aneurysms
- Basilar tip and proximal superior cerebellar artery aneurysms

We will enlighten the main steps of this approach from the skin to the target artery.

Surgical positioning is an essential part of aneurysm surgery. The following steps should be completed while positioning the patient for the pterional craniotomy:

M. K. Başkaya (✉) · T. Gasimov
Department of Neurological Surgery, University of Wisconsin-Madison, Madison, WI, USA
e-mail: baskaya@neurosurgery.wisc.edu

I. N. Kanaan, V. Beneš (eds.), *Neuroanatomy Guidance to Successful Neurosurgical Interventions*,
https://doi.org/10.1007/978-3-031-59838-8_10

- Thorax is elevated approximately 20° to keep the head above the level of the heart to facilitate good venous return.
- Head is rotated to the opposite side approximately 20–30°, depending on the aneurysm location. For most aneurysms, rotations greater than 30° are not needed. For a more subtemporal exposure, additional head rotation may be required.
- Neck is slightly extended until the malar eminence becomes the highest point in the surgical field to allow frontal lobe retraction through gravity.
- Head is slightly tilted laterally to the opposite side; this brings the basal cisterns nearer to the surgical field and widens the angle for the maneuverability of the surgical microscope.

A curvilinear skin incision (Fig. 10.1) begins 5–10 mm anterior to the tragus and 5 mm below the inferior border of the zygomatic arc, which runs in a curvilinear fashion up to the midline behind the hairline. The inferior limb of the incision should not cross the zygomatic root inferiorly, and within 10 mm from the tragus anteriorly to avoid facial nerve injury. Care should be also taken to protect the main stem of the superficial temporal artery (STA). Having a closer incision to the tragus is an easy way to protect the STA. A line drawn between the inferior and superior end of the planned incision corresponds to the flap base, which is the most anterior part of surgical exposure. The curve may run more posteriorly or anteriorly depending on the patient's hairline and the exposure required for the location of the particular aneurysm.

The next step is the elevation of the scalp and the retraction of the temporal muscle flap, while preserving the frontotemporal branch of the facial nerve. The topographic anatomy of the superficial temporal fascia and the twigs (posterior, middle, and anterior) of the frontotemporal branch is of great importance. The frontotemporal branch runs between the superficial layer of the superficial temporal fascia and the scalp after the duplication of the fascia into two layers. The outer layer joins the scalp anteriorly and the zygoma inferiorly. A sickle-shaped fat pad (Fig. 10.2) between these layers helps the surgeon recognize the duplication. To avoid nerve injury, the dissection should proceed either interfascially or subfascially, or as a single myocutaneous flap. An interfacial technique was first described by Yaşargil. After the fat pad is recognized, the superficial temporal fascia's superficial layer is incised and the dissection proceeds along the fat tissue to keep the nerve within the scalp flap [4]. In this approach, an unusal anatomy, such as abberant nerve branch in the interfascial plane, makes the nerve susceptible to injury [5]. Therefore we prefer a single layer myocutaneous flap [6] for usual pterional craniotomies and a subfascial technique [7] (Fig. 10.3), if an orbitotomy for an orbitozygomatic or for an extradural anterior clinoidectomy is planned.

Inferiorly, the fascia attaches to the zygomatic arc. Opening this attachment exposes the subzygomatic subfascial fat compartment. This fat tissue can be harvested and used for skull base repair if needed [8].

In the next step, the temporal muscle is cut and reflected inferiorly while leaving a cuff at the attachment site on the superior temporal line for reapproximation at the end of the surgery. The keyhole, marginal tubercle (posterior extension of the zygomatic process of the frontal bone), and the most posterior part of the zygomatic root are exposed (Fig. 10.4). The places for the burr holes are marked over the keyhole (1), just superior to the zygomatic root (2), at the most posterior part of the Sylvian fissure (3), and if needed over the

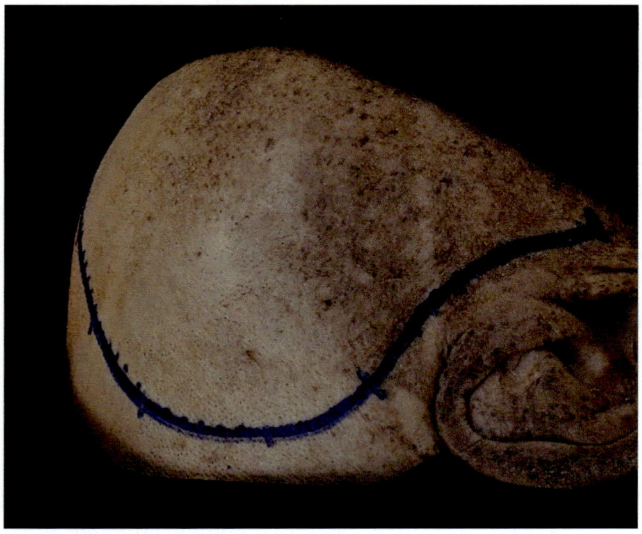

Fig. 10.1 Skin incision for the pterional craniotomy

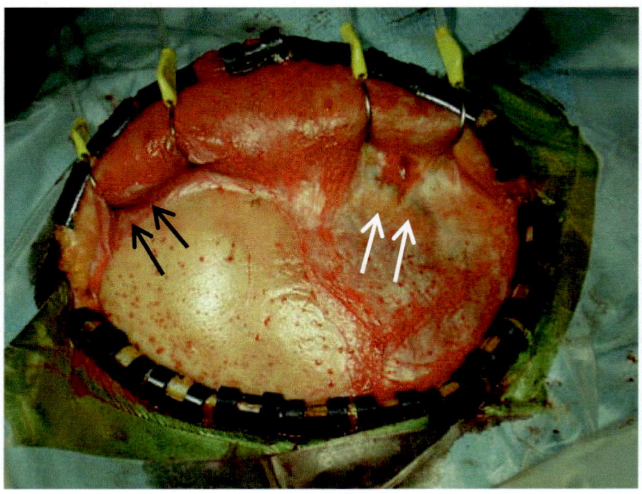

Fig. 10.2 Topography of the superficial temporal fascia. The interfacial fat pad is marked with white arrows. A vascularized pericranial flap (black arrow) is peeled and retracted anteriorly for closure of the frontal sinus if opened

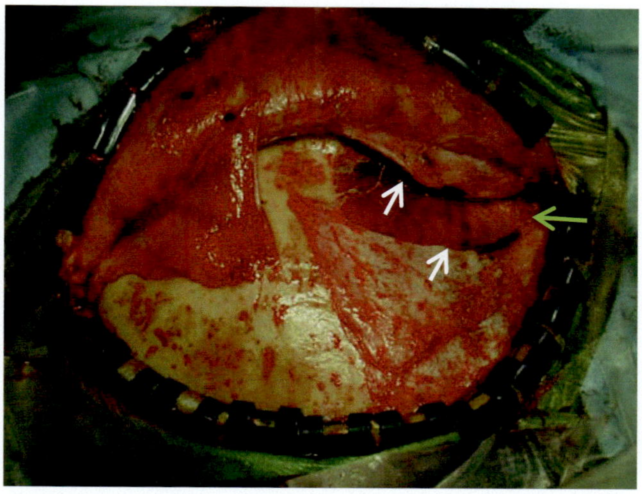

Fig. 10.3 Deep layer of the superficial temporal fascia is incised for a subfascial technique (white arrows). A green arrow indicates the attachment site of the fascia to the zygomatic arc

frontal dura (4). The dura underlying (3) is the thickest and (4) is the thinnest. After the burrholes are placed, dissection of the dura from the bone should be started at (3) to avoid dural laceration. The numbers and locations of burrholes depend on the surgeon's preference. The size and the shape of the bone flap depend on the aneurysm's location, shape, and size.

After the bone flap is elevated, the sphenoid ridge and the indentations of the orbital roof should be drilled until the meningo-orbital band (mob) is reached (Fig. 10.5). This provides a free working space for the surgeon, minimizes brain retraction, and enables the opticocarotid cistern to be reached before opening the Sylvian fissure [9]. The meningo-orbital band contains the meningo-orbital artery, which provides a rich vascularization to this area along with its branches (frontal and temporal). It should be coagulated and cut to expose the anterior clinoid process, if this is needed for access. The surgical anatomy of the anterior clinoid process will be reviewed separately. An extradural part of the pterional approach could be

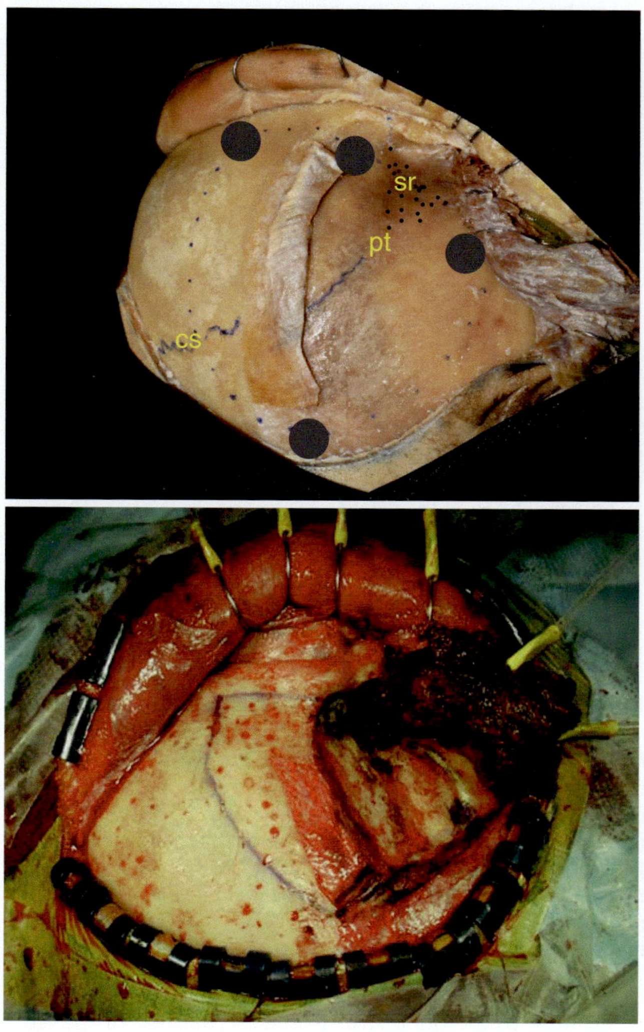

Fig. 10.4 *Bony landmarks for pterional approach. cs* coronal suture, *pt* pterion, *sr* shadow of the sphenoid ridge

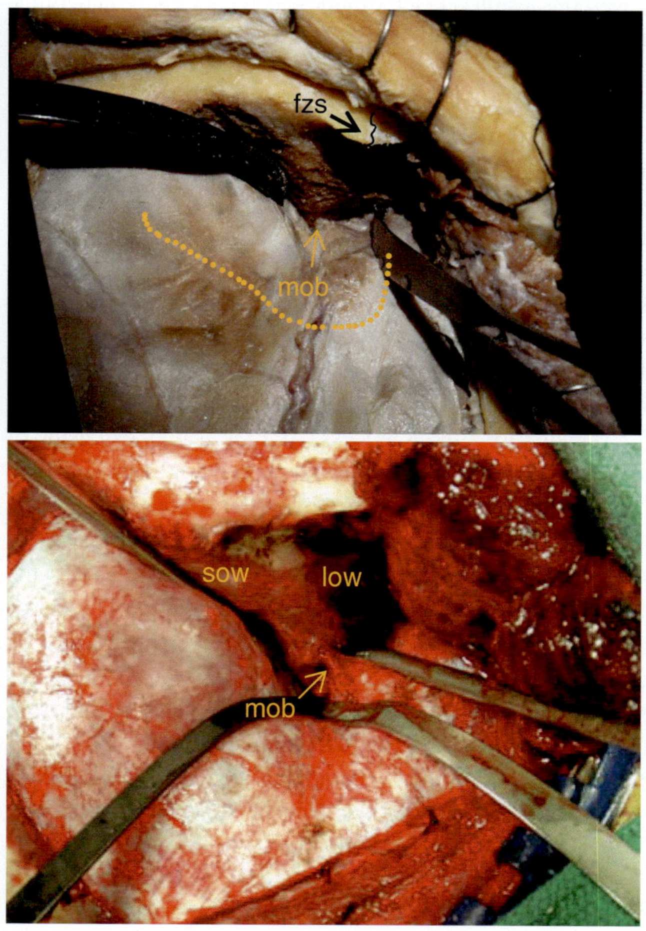

Fig. 10.5 Dotted yellow line indicates the borders of the osteotomy for the modified orbitozygomatic (OZ) approach. *mob* meningo-orbital band, *fzs* frontozygomatic suture, *sow* superior orbital wall, *low* lateral orbital wall

expanded by orbitotomy and by removing the supraorbital bar to reach the orbitozygomatic (OZ) approach. After the periorbita is dissected free from the superior and lateral orbital wall, a medial osteotomy is performed just lateral to the supraorbital nerve and the lateral osteotomy just lateral to the frontozygomatic suture. Finally, these cuts are connected through the superior orbital wall approximately 1.5–2 cm posterior to the orbital rim. This additional bone removal helps the surgeon work in the subfrontal area and interpeduncular cistern with minimal brain retraction, which is especially important when the brain is swollen after a subarachnoid hemorrhage. We use a modified OZ approach for clipping ruptured Acom and basilar tip aneurysms. This approach is often combined with an extradural anterior clinoidectomy.

10.3.1.2 Anterior Clinoidectomy

Anterior clinoidectomy is a crucial step for clipping paraclinoid and basilar tip aneurysms. An intradural technique was described first by Drake for clipping the carotid-ophthalmic aneurysms [10]. In our center, we prefer an extradural anterior clinoidectomy, which was pioneered by Dolenc for intracavernous lesions [11]. Extradural techniques have undergone modifications compared to that described by Dolenc.

Anterior clinoidectomy should begin with a careful examination of the anatomy of the anterior clinoid process in preoperative images. Care should be taken to exclude erosions in the clinoid process by subclinoid aneurysms, which can rupture intraoperatively and end fatally [12]. Possible ossification of carotico-clinoid ligaments should also be kept in mind. This connects the tip of the anterior clinoid process to the lateral border of the sella turcica and forms the carotico-clinoid canal. A thickened attachment site of the ossified ligament between the anterior and the posterior clinoid process is also called a middle clinoid process [13].

The anterior clinoid process (ACP) is a bulky extension of the sphenoid ridge and has three bony attachments: (1) Sphenoid wing forming the roof of the superior orbital fissure (SOF), (2) Planum sphenoidale forms the roof of the optic canal (OC), and (3) Optic strut, which separates the SOF and OC (Fig. 10.6a). To perform an extradural anterior clinoidectomy the following steps should be performed:

- Meningo-orbital band is coagulated, cut, and peeled from the SOF. The ACP is dissected free from the dura.
- Unroofing the SOF by removing the lateral orbital apex (sphenoid wing), and identifying the IV nerve and V1 in the superior orbital fissure. The III nerve hides under the lateral edge of the ACP.
- Expansion of the orbitotomy of the superior orbital wall, identifying the OC and optic strut from the intraorbital site (Fig. 10.6b).
- Unroofing the OC with a high-speed drill under copious irrigation to avoid thermal injury to the optic nerve.
- ACP is hollowed and partially removed in pieces.

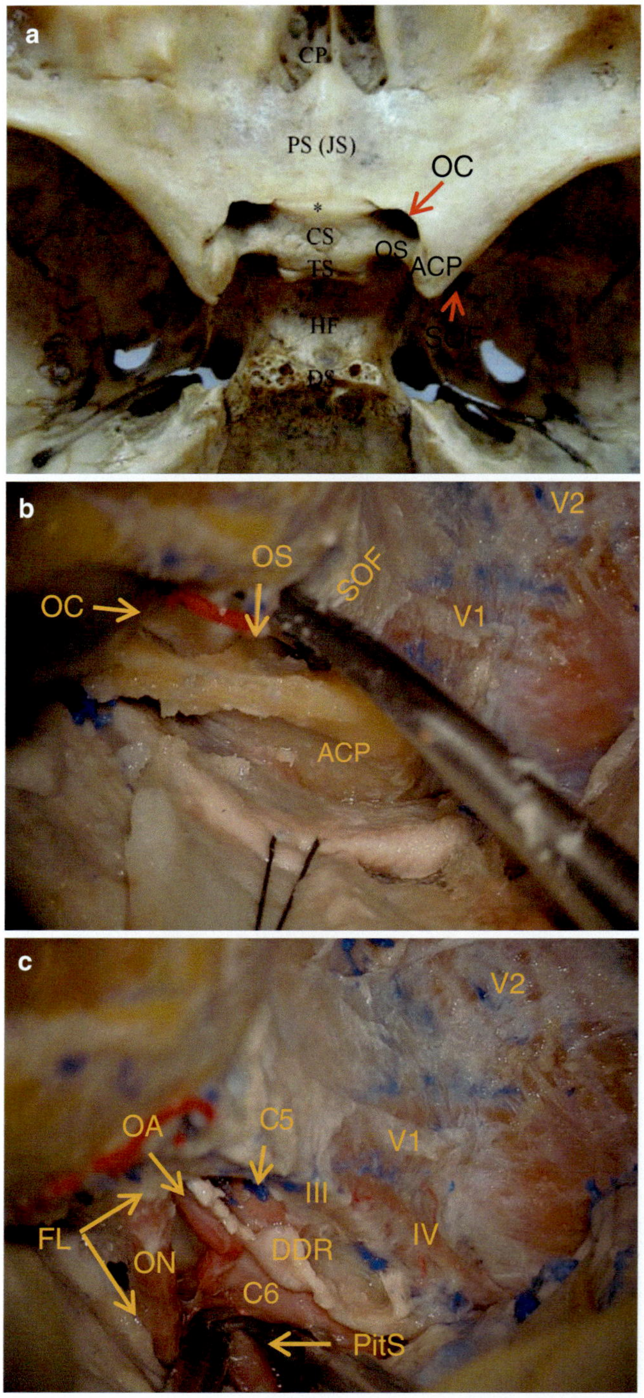

Fig. 10.6 (**a**) *CP* cribriform plate, *PS* planum sphenoidale, *CS* chiasmatic sulcus, *TS* tuberculum sellae, *DS* dorsum sellae, *ACP* anterior clinoid process, *OC* optic canal, *OS* optic strut, *SOF* superior orbital fissure. (**b**) Orbitotomy is performed, SOF unroofed, OC and OS identified. (**c**) ACP removed, III nerve identified, DDR and FL opened, OA identified. *DDR* distal dural ring, *FL* falciform ligament, *OA* ophthalmic artery, *C5* clinoid segment of the internal carotid artery, *C6* ophthalmic segment of the internal carotid artery, *PitS* pituitary stalk

- Drilling of the optic strut to release the rest of the ACP covering the clinoid segment of the internal carotid artery. This is dissected free and removed to expose the carotid-oculomotor membrane.

Complete removal of the optic strut would expose the sphenoid sinus mucosa. The clinoidal segment of the internal carotid artery (ICA) is covered by the carotid-oculomotor membrane, which continues laterally to cover the III nerve. Opening this membrane exposes the clinoid segment of ICA, which can be used as a proximal control for clipping paraclinoid aneurysms. As the optic nerve and ICA become intradural, they are covered by a dural fold, which holds these structures tightly. A dural fold over the optic nerve is called the falciform ligament (FL), which runs laterally to attach the ICA and is called the distal dural ring (DDR). Opening this dural fold releases the optic nerve and the ICA, which can be mobilized if needed.

10.3.1.3 Transcavernous and Subtemporal Approach

Dissecting and mobilizing the III nerve from the wall of the cavernous sinus allow the widening of the carotid-oculomotor triangle. This is used for clipping the basilar tip aneurysms by a transcavernous approach. The relationship of the anatomy of the basilar tip to the posterior clinoid process and dorsum sellae is important for choosing the correct approach for basilar tip aneurysms. For the low-lying basilar tip, a full transcavernous [14] or subtemporal approaches [15, 16] could be used.

10.3.2 Approaches Through the Posterior Fossa

10.3.2.1 Far Lateral Approach

The far lateral approach was first described by Heros for clipping vertebral and vertebrobasilar junction aneurysms while avoiding cerebellum and brain stem retraction [17]. Indications for far lateral approaches in aneurysm surgery are

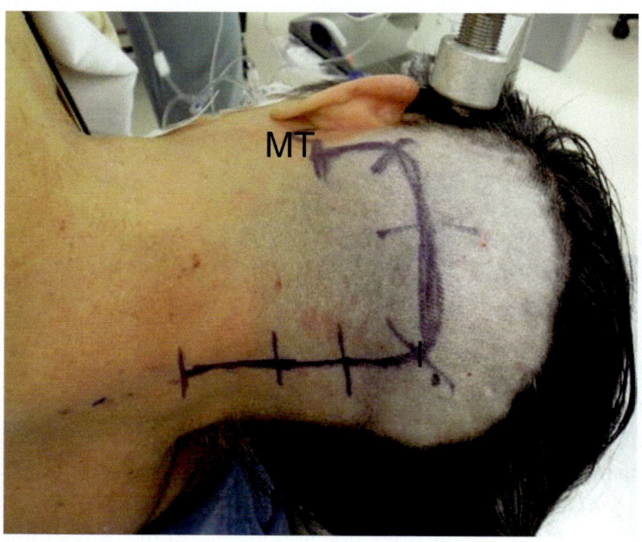

Fig. 10.7 *Positioning and skin incision for left-sided far lateral approach. MT* mastoid tip, *I* inion

- Vertebral and proximal posterior cerebellar artery (PICA) aneurysms
- Vertebrobasilar junction aneurysms

Later modifications to this approach include transcondylar and transjugular extensions to minimize brain retraction while enabling visualization of the anterior foramen magnum and clivus. The patient can be positioned in a semi-sitting or park bench position. We use the park bench position to allow the cerebellum to fall by gravitation and thus minimize the need for retraction. The following steps should be followed:

- Head is flexed, with a two-finger breadth between the chin and the chest.
- Head is tilted upwards (to the ipsilateral shoulder) by 20–30° to avoid venous engorgement and to allow CSF to drain by gravity.

We use the hockey stick incision (Fig. 10.7). The incision begins on the midline at the level of the C5 spinous process, turns laterally just cranial to the inion up to the mastoid base, and then turns caudally to reach the mastoid tip. An important landmark in this approach is to palpate the location of the transverse process of C1 caudal to the mastoid tip. A hockey stick incision has a lower chance for muscle atrophy compared to the transmuscular approach and allows for an avascular dissection plan through the nuchal ligament.

In the muscular stage of this approach the sternocleidomastoid, the splenius capitis, longissimus capitis, and the semispinalis capitis are detached from the mastoid in one layer while leaving a 1 cm cuff on the nuchal line for approximation during closure. The posterior belly of the digastric muscle should not be detached to protect the facial nerve exiting the stylomastoid foramen. In the final stage, muscles forming the suboccipital triangle (the superior and inferior oblique muscles, the m.rectus capitis posterior major, and the m.rectus capitis posterior minor) are dissected subperiosteally to expose the vertebral artery and the perivertebral venous plexus. A possible congenital C1-arc defect should be kept in mind while dissecting C1.

In the extradural stage, a lateral suboccipital craniectomy is performed by opening the foramen magnum. A C1 hemilaminectomy is then performed. The lateral part of the foramen magnum rim should be drilled until the occipital condyle is reached. In most cases, there is no need to drill the condyle and the jugular tubercle. If needed, the most medial part of the condyle could be drilled. A possible extradural origin of the PICA and posterior spinal artery should be kept in mind [18]. The dural incision starts from the midline at the C1 level, ascends toward the cerebellum, and turns laterally (Fig. 10.8b). The intradural structures in the surgical field are the vertebral artery, PICA, posterior spinal artery, IX-X-XI-XII nerves, C1 nerve rootlets, and the dentate ligament (Fig. 10.8c).

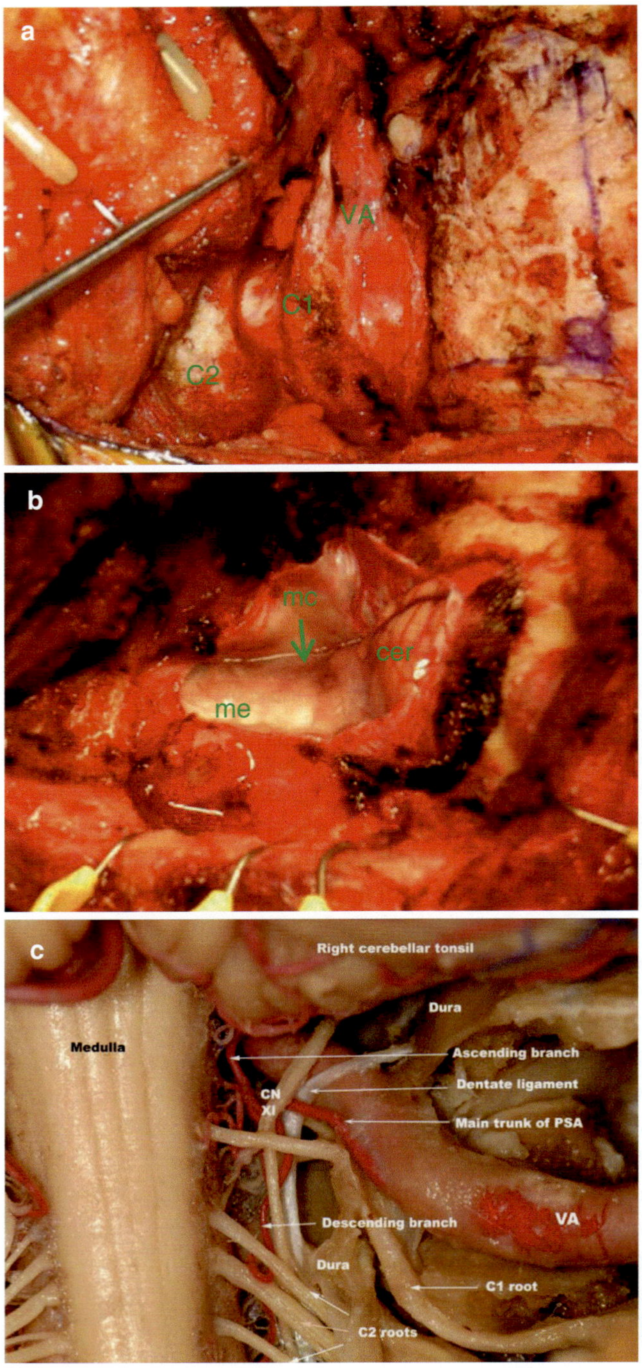

Fig. 10.8 (**a**) C1, posterior arc of atlas; C2, lamina of C2; *VA* vertebral artery. (**b**) craniectomy and C1 hemilaminectomy performed, dura opened and retracted laterally. *med* medulla oblongata, *cer* cerebellum, *mc* medullary cistern. (**c**) note the extradural origin of the posterior spinal artery

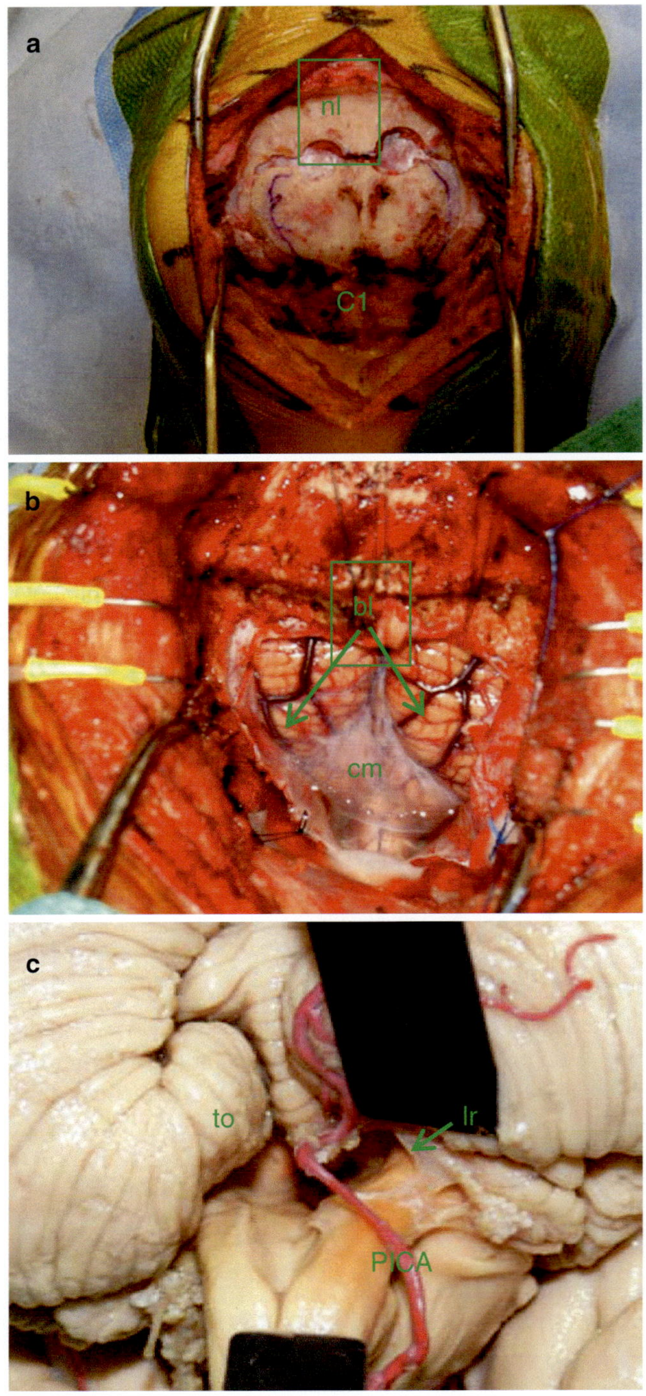

Fig. 10.9 Telovelar approach. (**a**) *nl* nucal line, C1, posterior arc of atlas. (**b**) *cm* cisterna magna, *bl* biventer lobule. (**c**) *to* tonsil, *lr* lateral recess

10.3.2.2 Telovelar (Transcerebellomedullary Fissure) Approach

This approach is used for distal PICA aneurysms occurring in the tonsillomedullar and telovelotonsillar segments of the PICA.

Patient positioning depends on the surgeon's preference for a semisitting, prone, or park bench position. The head is flexed and a median suboccipital craniotomy with an opening of the foramen magnum is performed (Fig. 10.9a). Before opening the dura, the possibility of an occipital and marginal sinus should be kept in mind.

After opening the dura, biventer lobules of the cerebellum and the cisterna magna are exposed (Fig. 10.9b). Opening the cisterna magna exposes the fourth ventricle and distal segments of the PICA. A thickened arachnoid wall of the cisterna magna could be opened linearly, fixated on the dural leaflets, and then sutured during closure with 10-0 sutures [19].

Opening the cerebellomedullary fissure and the inferior medullary velum exposes the intraventricular part of the PICA (Fig. 10.9c). With retraction of the tonsil lateral recess and with retraction of the nodulus, the cerebral aqueduct can be reached.

10.4 Surgical Pearls

- Appropriate surgical positioning should be chosen in order to use gravity to minimize direct brain retraction.
- Extradural bone removal (sphenoid ridge resection, anterior clinoidectomy, orbitotomy, partial resection of the occipital condyle) widens the surgical corridor, provides early proximal control, and minimizes brain retraction.
- Preoperative CT should be checked carefully to recognize the variations (middle clinoid porcess-clinoid carotid canal) and erosion of the anterior clinoid process.
- Anatomic variations of the ophthalmic artery origin (cavernous ICA, clinoid ICA) should be considered and checked in preoperative imaging.
- Opening the dural duplications (falciform ligament, distal dural ring) helps to release and mobilize the optic nerve and ICA.
- Clean arachnoid dissection should be performed to widen the surgical corridor and to expose the safe proximal and distal control before dissecting and clipping the aneurysm.
- CSF-release by opening the cisterns and lamina terminalis is a crucial step to widen the subarachnoid space.

10.5 Discussion

Although most saccular aneurysms can be treated by microsurgical clipping techniques, other treatment modalities such as microvascular bypass and endovascular techniques should be considered in selected cases. Blood-blister-like aneurysms are extremely fragile and thin-walled pseudoaneurysms located at nonbranching points over the ICA's superior, dorsal, or anterior surface. Because of the lack of a true wall, these aneurysms are not appropriate for microsurgical clipping. At our institution, trapping with bypass revascularization is the therapy of choice in cases with ruptured blister-like aneurysms. Trapping and microvascular bypass techniques are also widely used for giant aneurysms and tubular aneurysms. The advantage of microsurgery includes attaining an immediate and complete aneurysm occlusion, which eliminates the risk of re-bleeding, avoiding the use of dual antiplatelet therapy, and providing an extra conduit in the setting of severe vasospasm.

10.6 Conclusion

Despite the development of advanced endovascular techniques for aneurysm treatment, microneurosurgery continues to play an important role in the definitive cure of cerebral aneurysms. Advanced knowledge in surgical neuroanatomy is key in choosing and performing the best possible approach for cerebral aneurysm surgery. We believe that training in cadaveric dissection should be mandatory to strengthen anatomical knowledge and improve surgical skills. Correct surgical positioning and precise arachnoid dissection are important clues to widen and enlighten the surgical field and minimize brain retraction.

References

1. Dott NM. Intracranial aneurysms: cerebral Arterio-radiography: surgical treatment *read at a meeting of the medico-chirurgical society of Edinburgh, on 28th June 1933. Edinb Med J. 1933;40(12):T219–40.
2. Dandy WE. Intracranial aneurysm of the internal carotid artery. Ann Surg. 1938;107(5):654–9.
3. Yaşargil MG. Normal cisternal anatomy. In: Microneurosurgery. Stuttgart: Georg Thieme-Verlag; 1984. p. 25–53.
4. Yaşargil MG, Reichman MV, Kubik S. Preservation of the frontotemporal branch of the facial nerve using the interfascial temporalis flap for pterional craniotomy. Technical article. J Neurosurg. 1987;67(3):463–6.
5. Coscarella E, Vishteh AG, Spetzler RF, Seoane E, Zabramski JM. Subfascial and submuscular methods of temporal muscle dissection and their relationship to the frontalis branch of the facial nerve: technical note. J Neurosurg. 2000;92(5):877–80.
6. Spetzler RF, Lee KS. Reconstruction of the temporalis muscle for the pterional craniotomy: technical note. J Neurosurg. 1990;73(4):636–7.
7. Horimoto C, Toba T, Yamaga S, Tsujimura M. Subfascial temporalis dissection preserving the facial nerve in pterional craniotomy – technical note. Neurol Med Chir (Tokyo). 1992;32(1):36–7.
8. Lee P, Krisht KM, Cai L, Krisht AF. Autologous temporalis subfascial fat graft for skull base repair: a novel technique. Oper Neurosurg (Hagerstown). 2021;20(4):E274–8.
9. Yaşargil MG. Interfascial pterional craniotomy. In: Microneurosurgery. Stuttgart: Georg Thieme-Verlag; 1984. p. 222–4.
10. Drake CG, Vanderlinden RG, Amacher AL. Carotid-ophthalmic aneurysms. J Neurosurg. 1968;29(1):24–31.
11. Dolenc V. Direct microsurgical repair of intracavernous vascular lesions. J Neurosurg. 1983;58(6):824–31.
12. Korosue K, Heros RC. "Subclinoid" carotid aneurysm with erosion of the anterior clinoid process and fatal intraoperative rupture. Neurosurgery. 1992;31(2):356–9.
13. Rhoton AL. The cavernous sinus, the cavernous venous plexus and the carotid collar. In: Cranial anatomy and surgical approaches. Philadelphia: Lippincott Williams & Wilkins; 2002. p. 383–5.

14. Krisht AF, Krayenbühl N, Sercl D, Bikmaz K, Kadri PAS. Results of microsurgical clipping of 50 high complexity basilar apex aneurysms. Neurosurgery. 2007;60(2):242–50; discussion 250–2.
15. Hernesniemi J, Ishii K, Niemelä M, Kivipelto L, Fujiki M, Shen H. Subtemporal approach to basilar bifurcation aneurysms: advanced technique and clinical experience. Acta Neurochir Suppl. 2005;94:31–8.
16. Figueiredo EG, Tavares WM, Rhoton AL, de Oliveira E. Nuances and technique of the pretemporal transcavernous approach to treat low-lying basilar artery aneurysms. Neurosurg Rev. 2010;33(2):129–35; discussion 135.
17. Heros RC. Lateral suboccipital approach for vertebral and vertebrobasilar artery lesions. J Neurosurg. 1986;64(4):559–62.
18. Seçkin H, Ateş O, Bauer AM, Başkaya MK. Microsurgical anatomy of the posterior spinal artery via a far-lateral transcondylar approach. J Neurosurg Spine. 2009;10(3):228–33.
19. Keleş A, Harput MV, Türe U. Cisterna magna reconstruction with arachnoid suturing in brainstem surgery. Neurosurg Focus Video. 2019;1(2):V11.

11 Carotid-Ophthalmic Artery Aneurysms

Matias Baldoncini and Alvaro Campero

Learning Objectives

- Understanding the anatomy of the carotid-ophthalmic artery and its relation to the optic nerve and surrounding structures.
- Understanding the different treatment options available for carotid-ophthalmic artery aneurysms.
- Learning about the potential complications associated with carotid-ophthalmic artery aneurysms and their treatment.
- Understanding the role of imaging techniques such as CT angiography, MRI, and angiography in the diagnosis and management of carotid-ophthalmic artery aneurysms.
- Understanding the importance of timely diagnosis and treatment in preventing vision loss and other serious complications.
- Learning about the importance of exposure of the internal carotid artery in the neck and performing a clinoidectomy at the skull base in the management of carotid-ophthalmic artery aneurysms.

11.1 Introduction

Carotid-ophthalmic aneurysms arise from the internal carotid artery (ICA) between the distal dural ring and the origin of the posterior communicating artery at any point along its diameter. They are relatively rare and account for only 0.5–11% of all intracranial aneurysms. A higher incidence in women and a left-sided predominance have been reported. They are frequently large or giant and have a high association (21–64%) with multiple aneurysms. Aneurysms occurring at the origin of the ophthalmic artery usually point upward toward the optic nerve. Hence, they can cause visual deficits owing to optic nerve compression.

Symptoms of carotid-ophthalmic aneurysms may include headaches, eye pain, vision changes, double vision, or loss of vision in one or both eyes. However, many people with these aneurysms do not experience any symptoms until the aneurysm ruptures

The surgical treatment of these aneurysms will usually require optic canal unroofing, partial, or total clinoidectomy. However, clinoidectomy is not without complications, some of them severe.

11.2 Background

Ophthalmic aneurysms typically arise between the distal dural ring and the origin of the posterior communicating artery, with a superior orientation in the direction of hemodynamic flow. These aneurysms are commonly large or giant and multiple, which leads to superomedial compression of the optic nerve against the falciform ligament, resulting in visual morbidity. The incidence of ophthalmic aneurysms is relatively low, ranging between 0.5% and 11%, with a higher frequency in female individuals and on the left side.

Repairing ophthalmic aneurysms through direct methods is often considered challenging because of several factors. These include limited proximal control of the internal carotid artery (ICA), complex anatomy of the anterior clinoid process, difficult aneurysm exposure, anatomical variations of the ophthalmic artery and superior hypophyseal arteries, and the risk of further visual morbidity. Anterior clinoidectomy (AC), which requires a certain level of dexterity and detailed knowledge of the anatomy of the paraclinoid region, is often needed. However, the risks associated with this maneuver and its potential to worsen visual function remain significant challenges in treating ophthalmic aneurysms. Although optic

M. Baldoncini (✉)
Laboratory of Microsurgical Neuroanatomy, Second Chair of Gross Anatomy, School of Medicine, University of Buenos Aires, Buenos Aires, Argentina

A. Campero
Department of Neurological Surgery, Hospital Padilla, Tucumán, Argentina

I. N. Kanaan, V. Beneš (eds.), *Neuroanatomy Guidance to Successful Neurosurgical Interventions*,
https://doi.org/10.1007/978-3-031-59838-8_11

foraminotomy (OF) has been suggested as a less invasive alternative to AC for certain types of ophthalmic aneurysms, evidence of its efficacy and safety compared with AC is lacking in the current literature.

11.3 Anatomy Highlights

- Ophthalmic Artery Origin: The ophthalmic artery typically arises from the internal carotid artery within the cavernous sinus. Its proximity to the aneurysm site is crucial for surgical planning.
- Aneurysm Location: Carotid-ophthalmic artery aneurysms are located at the junction where the ophthalmic artery branches off from the internal carotid artery, often involving the arterial wall at this juncture.
- Aneurysm Size: The size of the aneurysm can vary significantly, from small saccular aneurysms to larger, more complex shapes.
- Branching Patterns: An understanding of the branching patterns of the ophthalmic artery and its relationship to the aneurysm is critical. Variations in these patterns can affect surgical approaches and repair techniques.
- Proximal and Distal Anatomy: Knowledge of the arterial segments proximal and distal to the aneurysm is essential for assessing the feasibility of repair and preventing occlusion of vital vessels.
- Adjacent Structures: Awareness of nearby structures like the optic nerve, cranial nerves (CN II and CN III), and the cavernous sinus is crucial to avoid damage during surgery. Analyze the relationship between the anterior clinoid process and the roof of the optic nerve in relation to this segment of the carotid artery.
- Anatomy for Proximal Control: Be familiar with the anatomy of the common carotid and internal cervical carotid arteries, to be able to access initially and have proximal control.

The paraclinoid carotid artery is a critical structure for neurosurgeons to be familiar with, as it is often involved in pathologies that require surgical intervention. These pathologies may include aneurysms, meningiomas, cavernous sinus tumors, and carotid artery stenosis, among others.

Neurosurgeons must also be aware of the anatomical variations that may be present in the paraclinoid carotid artery, such as the presence of fetal or persistent trigeminal artery. These variations can increase the complexity of surgical interventions and require careful preoperative planning to ensure optimal outcomes.

The most posterior bony projection connected to the lesser sphenoid wing and bridged to the sphenoid body by the optic strut is known as the anterior clinoid process (ACP). The ACP is typically round, but it may have sharp indentations. It forms the lateral wall of the optic canal and is in direct contact with the internal carotid artery on its medioinferior surface. In the intradural space, the ACP is in contact with the ophthalmic segment and is separated from the extradural paraclinoid segment by the dural ring. The superomedial border of the superior orbital fissure is formed by the lateroinferior surface of the ACP, which is in direct contact with the oculomotor and lacrimal nerves. The posterior part of the ACP is connected to the petroclinoid and interclinoid ligaments. The most lateral part of the superior orbital fissure contains the orbitotemporal periosteal fold, which is also called the meningo-orbital band or the fronto-temporal dural fold. This fold contains small dural veins and the orbitomeningeal artery. A transection of the periosteum medially to the orbitotemporal periosteal fold opens a cleavage plane between the temporal dura propria (inner layer of dura mater) and the inner layer of the lateral wall of the cavernous sinus. The lacrimal, oculomotor, trochlear, ophthalmic (V1), and abducens nerves are found in this cleavage plane, listed from medial to lateral discovery. In 5% of cases, a caroticoclinoid foramen with a bony bridge connecting the anterior with the middle clinoid process occurs.

11.4 Surgical Intervention

Microsurgical clipping of carotid-ophthalmic artery aneurysms (COAs) requires the exposure of the cervical internal carotid artery (ICA) to facilitate the placement of the aneurysm clip. The exposure of the cervical ICA can be achieved through various approaches, including the standard cervical approach, the retropharyngeal approach, the transcervical-transclival approach, and the extended retroauricular approach. The choice of the approach depends on the location and size of the aneurysm, as well as the surgeon's preference and experience. Another option to achieve proximal control in the treatment of these aneurysms is accessing the lacerum segment of the internal carotid artery. As a disadvantage of this approach, it is necessary to perform a more extended approach and a peeling of the middle fossa, increasing the risks of postoperative complications [1]. Regardless of the approach used, the goal of cervical ICA exposure is to provide adequate exposure of the vessel while minimizing the risk of injury to the surrounding structures. In all cases of paraclinoid aneurysms, we expose the cervical carotid artery with the aim of having it prepared for temporary clipping or for performing carotid puncture-aspiration to decrease pressure in giant aneurysms of this segment.

The standard cervical approach involves a skin incision along the anterior border of the sternocleidomastoid muscle, followed by dissection of the carotid sheath to expose the ICA. Each of these approaches has its advantages and disadvantages, and the choice of approach should be tailored to the individual patient.

The exposure of the cervical ICA requires careful dissection and identification of the surrounding structures, including the carotid sheath, the vagus nerve, the hypoglossal nerve, and the sympathetic trunk. Adequate exposure of the cervical ICA also requires the mobilization of the carotid artery from the surrounding tissues to provide adequate working space for the placement of an aneurysm clip. This mobilization is typically achieved by using vascular loops or shunts to temporarily occlude the ICA and facilitate the mobilization of the vessel.

Intraoperative monitoring is critical during cervical ICA exposure to ensure the safety of the procedure. This monitoring typically involves the use of electroencephalography (EEG), somatosensory evoked potentials (SSEPs), and motor-evoked potentials (MEPs) to detect any changes in cerebral perfusion or function during the procedure. The use of intraoperative angiography is also important to confirm the adequacy of the cervical ICA exposure and the placement of the aneurysm clip.

The approach used is the classic pterional approach with extradural dissection until reaching the meningo-orbital band to then drill the lesser wing of the sphenoid, flattening this sector, and generating a corridor between the anterior and middle fossa.

We make a C-shaped opening of the dura mater, with anchor points and with a microscope we open the Sylvian fissure from distal to proximal to separate the frontal lobe from the temporal lobe and improve the angle of vision. We believe that this last step is not mandatory since in some patients, after draining the cerebrospinal fluid from the carotid cistern, brain relaxation is achieved and is sufficient to begin to expose the aneurysm. As we mentioned at the beginning, in this type of aneurysm it is necessary to partially or totally unlock the bone structures that surround it, such as the roof of the optic canal and the anterior clinoid process. The intradural or extradural options are valid for this objective, although according to our philosophy we always perform the drilling intradurally. This modality is motivated by two situations: being able to have permanent visualization of the aneurysm and also being able to remove as much bone as necessary to expose the entire neck in its proximal and distal segments.

When exposing the skull base structures, we make an arcuate incision starting medial to the optic canal and ending lateral to the anterior clinoid process. The dura mater is dissected with a Penfield dissector and the dura mater is removed with microscissors, this prevents this dural fold from becoming entangled with the drill wick. With copious irrigation and a 5 mm diamond tip drill, we slowly remove the roof of the optic canal to avoid damaging the nerve that is covered by a special dural sheath. In many cases, only with the release and gentle mobilization of the optic nerve can these aneurysms be exposed and clipped. If the aneurysm cannot be adequately visualized up to this point, we continue to perform a partial clinoidectomy, working with a 3 mm diamond tip drill, the medial segment of the anterior clinoid process is removed [2, 3].

If we still cannot have adequate exposure for clip placement, we perform a total clinoidectomy with the possibility of exposing the proximal and distal dural rings and cutting them to fully expose the clinoid segment of the internal carotid artery and the aneurysm. With this last case, even in some patients, depending on the anatomy, it allows the placement of a temporary clip at the level of the internal carotid [2, 3].

In summary, we remove as much bone as necessary to effectively expose and occlude carotid-ophthalmic aneurysms. We can perform a total clinoidectomy as in the case of Fig. 11.1, partial clinoidectomy as shown in the case of Figs. 11.2 and 11.3, or just an optic nerve unroofing as shown in the case of Fig. 11.3.

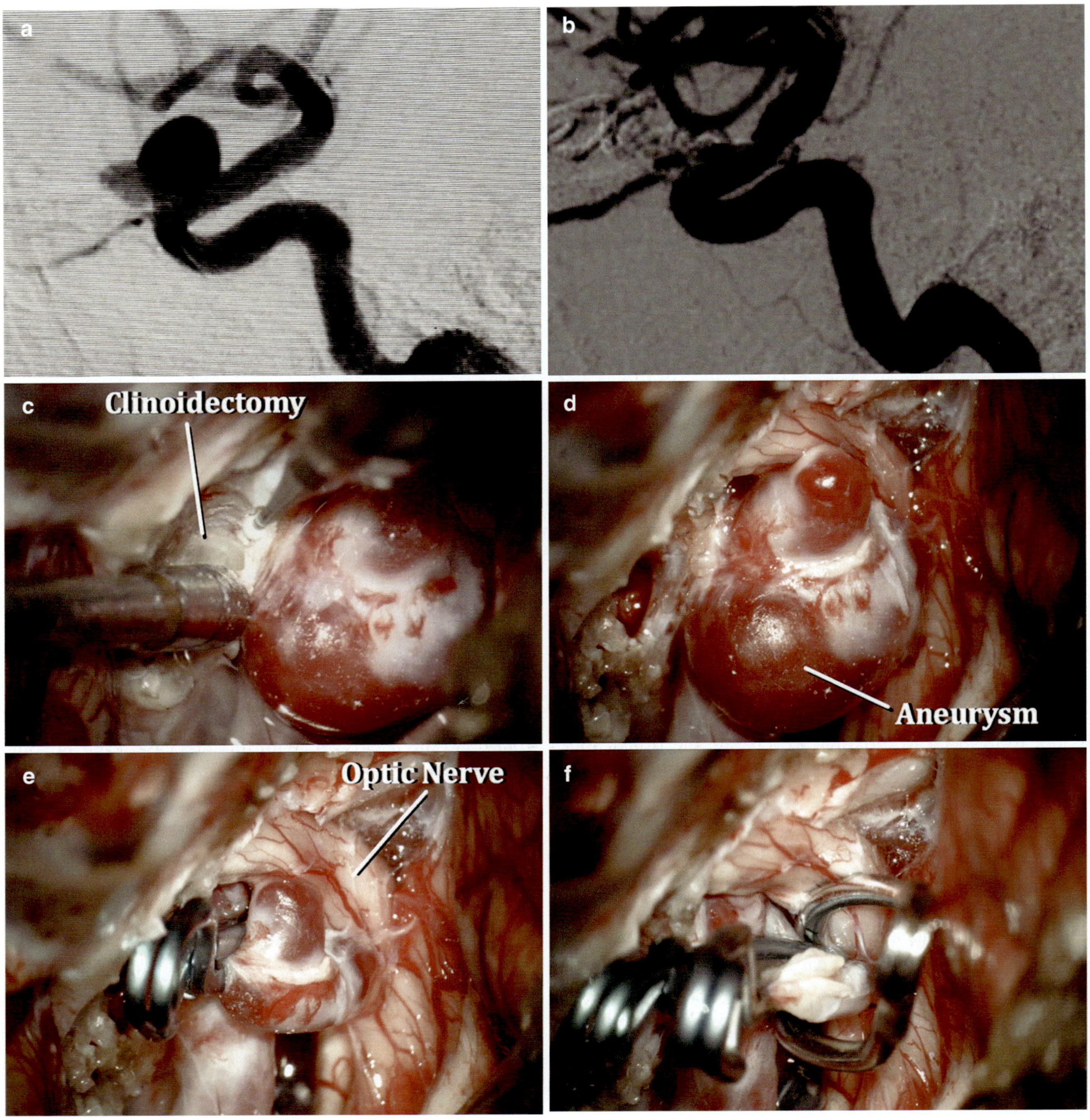

Fig. 11.1 (**a** and **b**) Preoperative and postoperative angiography. In images (**c** and **d**), the use of a drill to perform a total clinoidectomy can be observed, exposing the distal dural ring and aneurysm neck. In images (**e** and **f**), the aneurysm occlusion can be observed using multiple clips in a tandem clipping technique

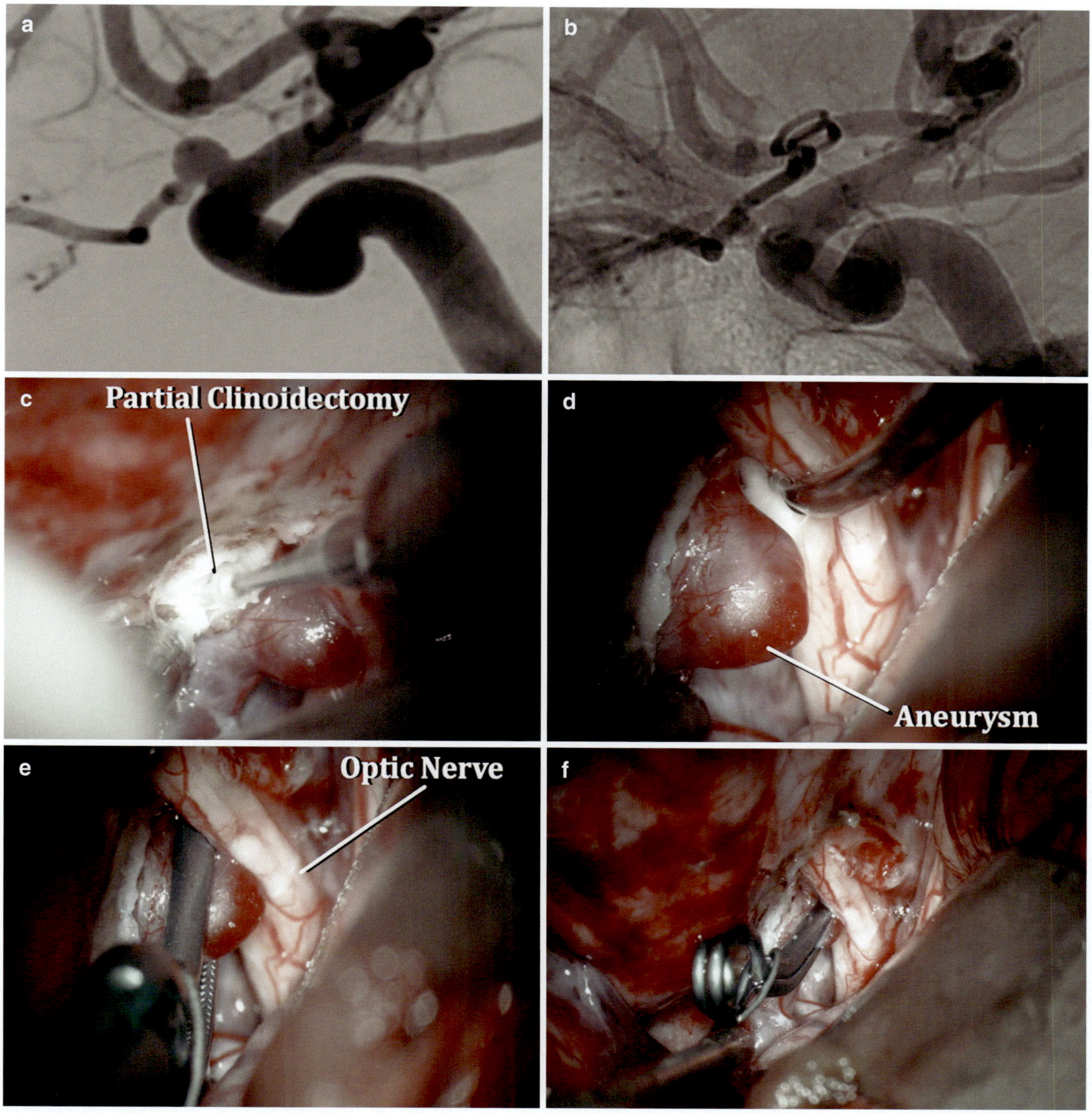

Fig. 11.2 (**a** and **b**) Preoperative and postoperative angiography. (**c**) The use of a drill to perform a partial clinoidectomy can be observed, removing only the medial aspect of the clinoid process. (**d**) Complete exposure of the proximal and distal aneurysm neck, observing the ophthalmic artery origin from the proximal segment of the aneurysm neck. In images (**e** and **f**), the aneurysm occlusion can be observed using a single 90° lateral angled clip

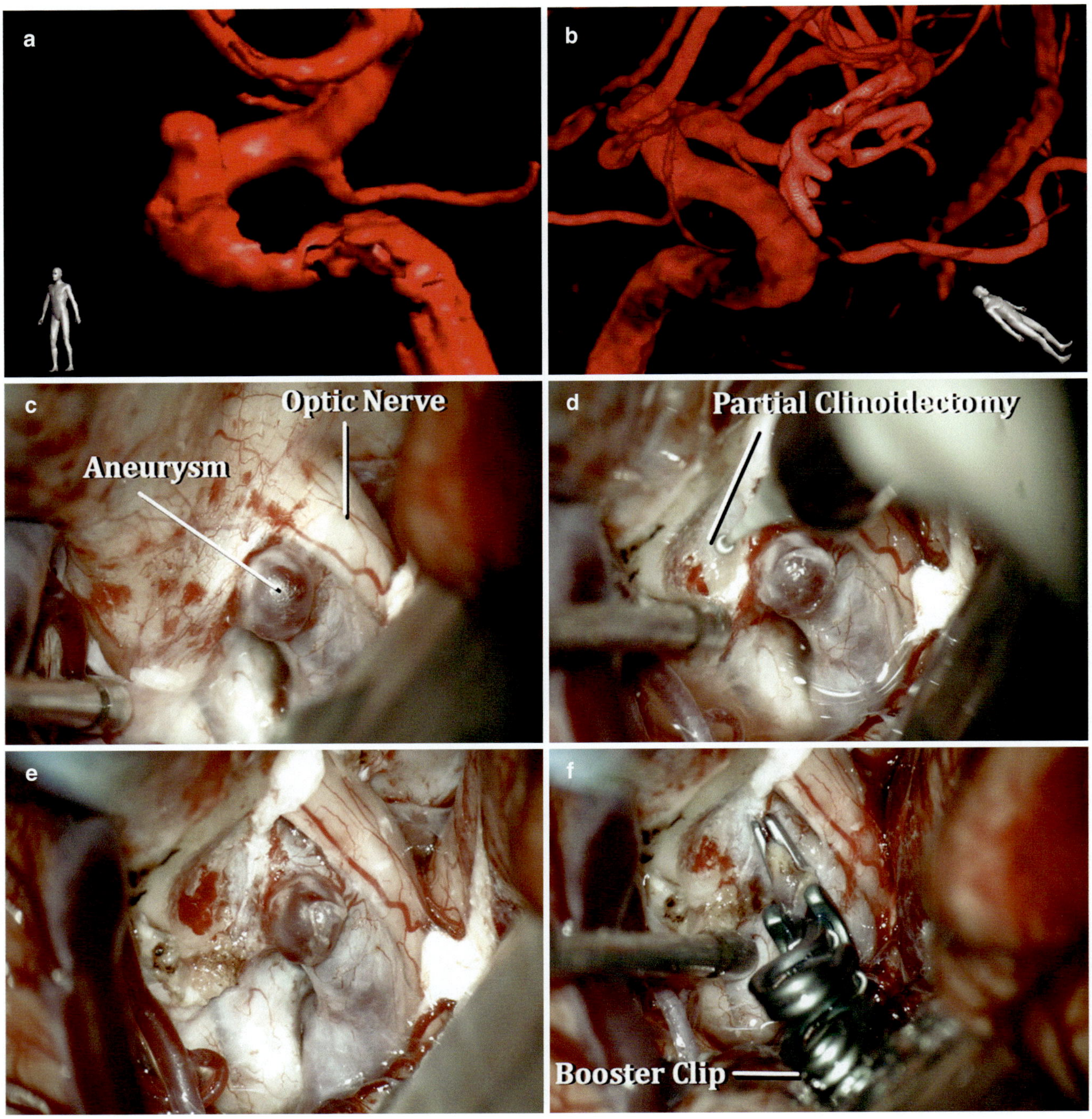

Fig. 11.3 (**a** and **b**) Preoperative angiography and postoperative AngioCT. In images (**c** and **d**), the use of a drill to perform a partial clinoidectomy can be observed. In images (**e** and **f**), the aneurysm occlusion can be observed using one curved clip and a booster clip to reinforce

11.5 Surgical Pearls

- Surgery for carotid ophthalmic aneurysms can be challenging due to their complex location and proximity to critical structures. Here are some tips and tricks that may be helpful:
- *Preoperative planning*: Comprehensive preoperative planning is essential for the success of the surgery. High-resolution imaging, such as CT angiography or MR angiography, can provide detailed information on the location, size, shape, and relationship of the aneurysm to the adjacent structures.
- *Experienced surgeon:* The surgery should be performed by an experienced cerebrovascular neurosurgeon who has a good understanding of the anatomy of the carotid ophthalmic segment and the necessary skills to perform the procedure safely and effectively.
- *Microsurgical techniques:* Microsurgical techniques such as high magnification, fine dissection, and meticulous hemostasis are crucial for successful surgery. The use of

intraoperative monitoring, such as electroencephalography and somatosensory evoked potentials, can also help to reduce the risk of neurological complications.

- *Temporary clipping:* Temporary occlusion of the parent vessel using a clip can help to reduce the blood flow to the aneurysm, facilitating its dissection and clipping. However, the duration of temporary occlusion should be minimized to avoid ischemic injury.
- *Postoperative management:* Postoperative management should include a careful monitoring of neurological function, blood pressure, and blood sugar levels to prevent complications such as stroke, bleeding, or infection.

11.6 Discussion

Aneurysms that arise from the internal carotid artery (ICA) between the cavernous sinus exit and the posterior communicating artery are referred to as paraclinoid aneurysms. Direct surgery for these types of aneurysms presents a challenge due to the small and intricate region they occupy, which includes several critical structures like the optic apparatus, oculomotor nerve, and branches and perforating arteries of the C5 and C6 segments of the ICA. In addition, these structures are concealed by dura and bone that must be removed to access them. An anterior clinoidectomy, optic strut removal, and distal dural ring dissection are required to safely clip paraclinoid aneurysms, but technical errors during these procedures can lead to serious complications like visual deficits, bleeding, and strokes. To become proficient in managing paraclinoid aneurysms, neurosurgeons need to gain an understanding of this complex region through study with expert neurosurgeons, cadaveric dissection, reviewing operative videos, and personal experience. However, the increasing use of endovascular technology like flow diverters has reduced the volume of cases needed for neurosurgeons to gain proficiency in microsurgical clipping of paraclinoid aneurysms. While endovascular treatment can be minimally invasive, it is associated with a higher rate of recurrence and may not address symptoms associated with optic nerve compression. Additionally, flow diverters require prolonged treatment with potent antiplatelet drugs that can cause complications like bleeding and side effects, limiting their use in acute aneurysm rupture. As a result, paraclinoid aneurysms and their management remain a controversial topic [4] (Fig. 11.4).

Cervical ICA exposure is a critical step in the microsurgical clipping of carotid-ophthalmic artery aneurysms. The choice of approach should be tailored to the individual patient, and the procedure should be performed with meticulous dissection and intraoperative monitoring to ensure the safety and success of the surgery.

In summary, surgery for carotid ophthalmic aneurysms requires careful preoperative planning, skilled microsurgical techniques, and careful postoperative management. The use of temporary clipping, intraoperative monitoring, and intraoperative angiography can help to reduce the risk of complications and improve the outcome of the surgery.

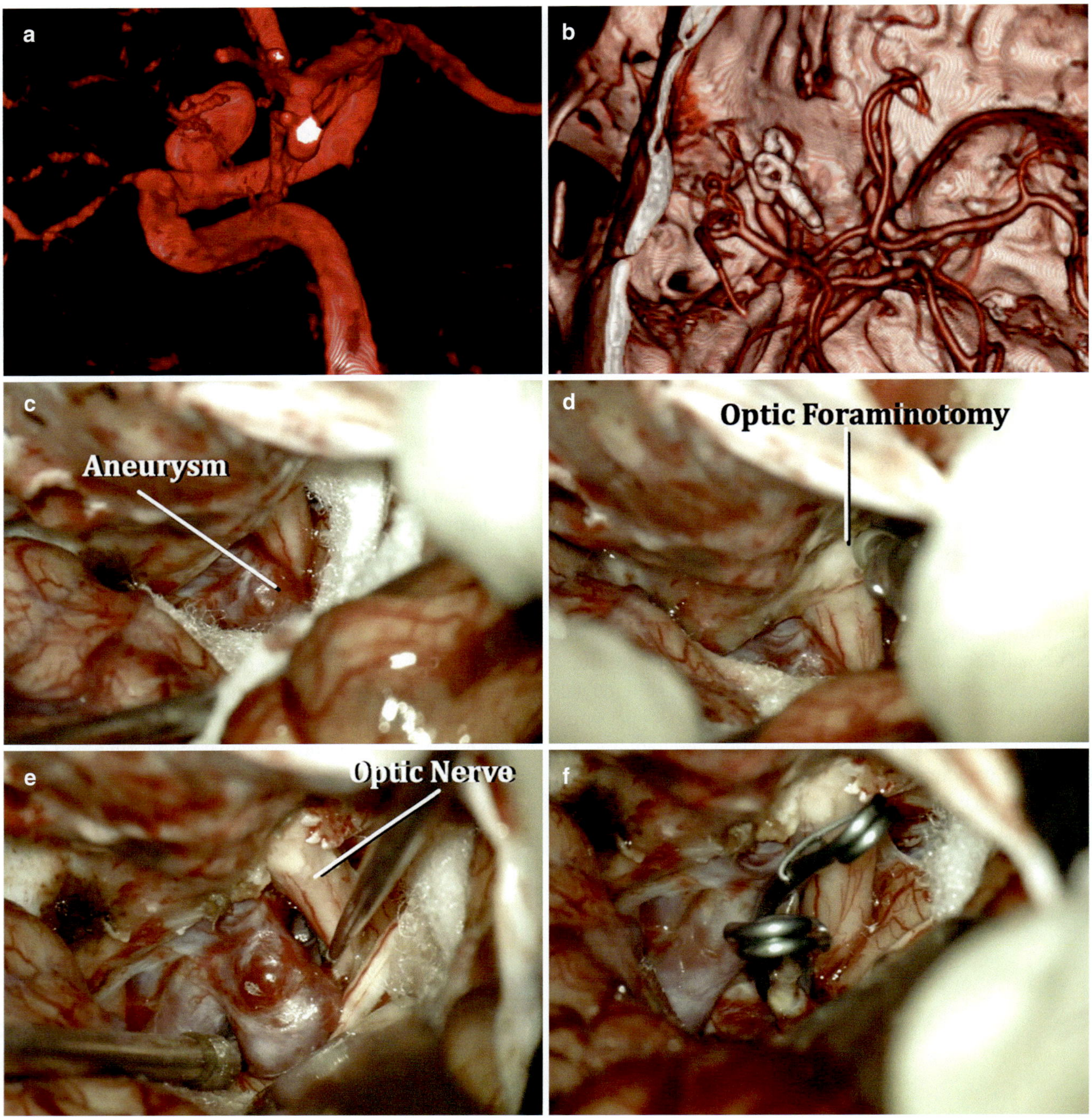

Fig. 11.4 (**a** and **b**) Preoperative angiography and postoperative AngioCT. In images (**c** and **d**), the use of a drill to perform a optic canal unroofing can be observed. In images (**e** and **f**), the aneurysm occlusion can be observed using two curved clips in a tandem clipping technique. In this case the patient presents a small anterior clinoid process and it was not necessary to remove it

11.7 Conclusions

The technical difficulty of surgical clipping for paraclinoid aneurysms is higher than that of other anterior circulation aneurysms due to their proximity to surrounding anatomical structures. Although surgical treatment has yielded positive results in the past, an increasing number of paraclinoid aneurysms are now being treated using endovascular therapy, which is associated with a lower rate of complete occlusion and a higher rate of recurrence than surgical treatment. Some studies have reported success using flow diverters, but these devices have also been linked to morbidity. Despite this, microsurgical clipping remains the preferred treatment for ruptured and wide-necked aneurysms and those compressing the optic nerve, as it can relieve the mass effect during the same surgery. It is also a good option for young patients with a long life expectancy as it has a lower risk of recurrence and retreatment than endovascular treatment. For very complex or giant aneurysms, an extracranial–intracranial bypass followed by microsurgical clipping or trapping may be performed [4]. The high aneurysm obliteration rates, low recurrence rates, lesser follow-up requirements, and avoidance of antithrombotic agents make microsurgical clipping a viable management option for paraclinoid aneurysms.

References

1. Liu JK, Fukushima T, Sameshima T, Al-Mefty O, Couldwell WT. Increasing exposure of the petrous internal carotid artery for revascularization using the transzygomatic extended middle fossa approach: a cadaveric morphometric study. Neurosurgery. 2006;59(4 Suppl 2):ONS309–19. https://doi.org/10.1227/01.NEU.0000232638.96933.A0.
2. Barrenechea IJ, Baldoncini M, González-López P, Campero Á. Optic nerve mobilization as an alternative to anterior clinoidectomy for superior carotid-ophthalmic aneurysms: operative technique. World Neurosurg. 2021;152:137–43. https://doi.org/10.1016/j.wneu.2021.06.008.
3. Luzzi S, Giotta Lucifero A, Baldoncini M, Campero A, Galzio R, Lawton MT. Optic foraminotomy versus anterior clinoidectomy for small superior-projecting paraclinoid aneurysms: visual and angiographic outcome evaluation. World Neurosurg. 2023;170:e612–21. https://doi.org/10.1016/j.wneu.2022.11.081.
4. Kamide T, Tabani H, Safaee MM, Burkhardt JK, Lawton MT. Microsurgical clipping of ophthalmic artery aneurysms: surgical results and visual outcomes with 208 aneurysms. J Neurosurg. 2018;129(6):1511–21. https://doi.org/10.3171/2017.7.JNS17673.

12 Surgical Treatment of MCA Aneurysms

Vladimír Beneš, Adéla Bubeníková, and Anna Šteklácová

Learning Objectives

- The "direct" surgical approach described in this chapter is adequate for all proximal and distal MCA aneurysms. More extensive exposure (30–35 mm long) of the Sylvian fissure is generally required for large and giant MCA aneurysms.
- Most fusiform MCA aneurysms are treated by different endovascular techniques; however, direct surgical management is also possible with good treatment results.
- Dissecting aneurysms were found at different MCA segments.
- MCA pseudoaneurysms are uniquely rare, but familiarity with their pathophysiology and possible presentation is required for adequate clinical decision-making.
- Good understanding and clinical experience with bypass techniques are required for the safe management of MCA aneurysms.

12.1 Introduction

Aneurysms found on the middle cerebral artery (MCA) account for approximately 14–43% of all intracranial aneurysms, thus representing one of the most common locations, along with aneurysms of the anterior communicating artery (AComA) and those found on the internal carotid artery (ICA) [1].

The surgical techniques for treating MCA aneurysms have been widely used since its introduction by Dandy [2], and the superiority of surgical treatment over endovascular treatment has been repeatedly documented [3–5]. This relatively uniform policy is supported by several factors: (1) straightforward surgical access, (2) minimal manipulation within the Sylvian fissure, and (3) minimal retraction of the frontal and temporal lobes. Also, in distal MCA aneurysms, the aneurysms are easy to reach with minimal dissection. In complex lesions, different bypass modes are available to ensure the flow into the distal MCA territory. MCA aneurysms generally pose a much higher risk for the patient than those in other locations. Once ruptured, the clinical outcome is much worse than in any other location, given that MCA aneurysms frequently present with intracerebral haemorrhage (ICH), undermining the eloquent cortical areas [6]. Up-to-date systematic reviews and meta-analyses have reported the microsurgical management of MCA aneurysms to render higher complete occlusion and lower morbidity/mortality rates than those treated by endovascular means [4, 6, 7].

Over the years, we have developed a minimally invasive surgical technique for MCA aneurysms. In the following paragraphs, the technique will be described in detail, documented with surgical images, and the introduction of relevant surgical anatomy will be provided. The approach is minimally invasive concerning the lesion (not the skin incision and craniotomy). In the authors' opinion, such an attitude is the core of neurosurgery.

12.2 Anatomy Highlights

As a part of the circle of the Willis anastomotic system, MCA is an extension of the ICA. It courses laterally into the Sylvian fissure, subsequently branching to supply the cere-

V. Beneš (✉) · A. Šteklácová
Institute of Clinical Neurodisciplines, Department of Neurosurgery and Neurooncology, First Faculty of Medicinefavo, Charles University and Military University Hospital, Prague, Czech Republic
e-mail: vladimir.benes@uvn.cz; Anna.steklacova@uvn.cz

A. Bubeníková
Institute of Clinical Neurodisciplines, Department of Neurosurgery and Neurooncology, First Faculty of Medicinefavo, Charles University and Military University Hospital, Prague, Czech Republic

Department of Neurosurgery, Second Faculty of Medicine, Charles University and Motol University Hospital, Prague, Czech Republic

I. N. Kanaan, V. Beneš (eds.), *Neuroanatomy Guidance to Successful Neurosurgical Interventions*,
https://doi.org/10.1007/978-3-031-59838-8_12

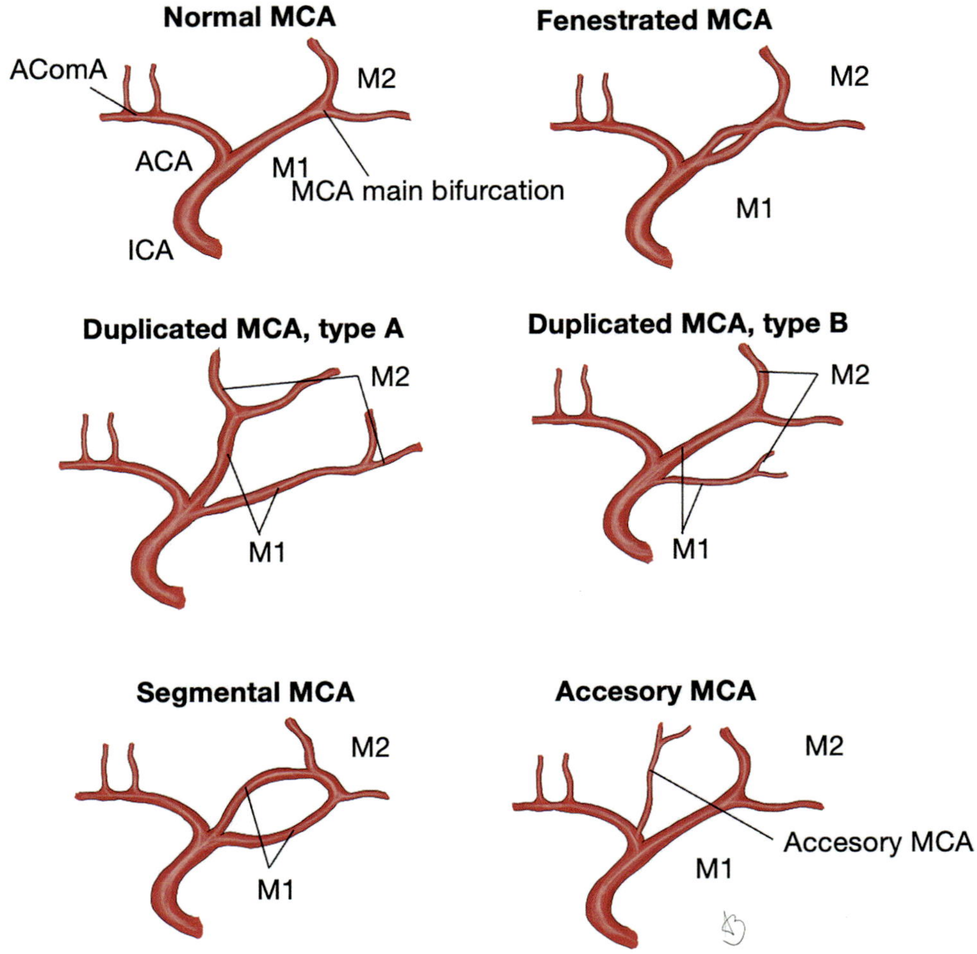

Fig. 12.1 Summary of the anatomical variations of MCA and its branches. Abbreviations: *ACA* anterior cerebral artery, *AComA* anterior communicating artery, M1/M2 segments of the MCA, *MCA* middle cerebral artery, *ICA* internal carotid artery

bral cortex. There are four major subdivisions (segments) of the MCA: (1) M1, sphenoidal or horizontal segment; (2) M2, insular segment; (3) M3, opercular segment; and (4) M4, cortical segment. M1 constitutes the MCA trunk since being one major vessel originating at the terminal bifurcation of the ICA, subsequently projecting laterally along the sphenoid ridge. Figure 12.1 illustrates the basic anatomy of MCA and anatomical variants of the MCA main trunk.

12.3 Characteristics and Classification Systems of MCA Aneurysms

One of the most comprehensive classifications of MCA aneurysms was published by Elsharkawy et al. [8] in 2013. The authors differentiated four categories: lenticulostriate, early cortical branch artery, bifurcation, and distal aneurysms (Fig. 12.2). The classification was built upon 1009 individuals harbouring 1309 aneurysms. Of these, 18% were early cortical branch artery aneurysms, comprising an important group to be anatomically distinguished from the other MCA aneurysms to prevent the risk of damage to lenticulostriate arteries. Because the lenticulostriate territory lacks a significant collateral blood supply, the occlusion may cause serious stroke syndromes.

Based on an exhaustive analysis of 561 patients harbouring 690 M1/2 or MCA bifurcation aneurysms, Rinne et al. [9] proposed that the vast majority of the aneurysms project laterally (45%), followed by inferiorly projecting (38%), superiorly projecting (15%), and only medially projecting accounting for the remaining 2% of described MCA aneurysm projections.

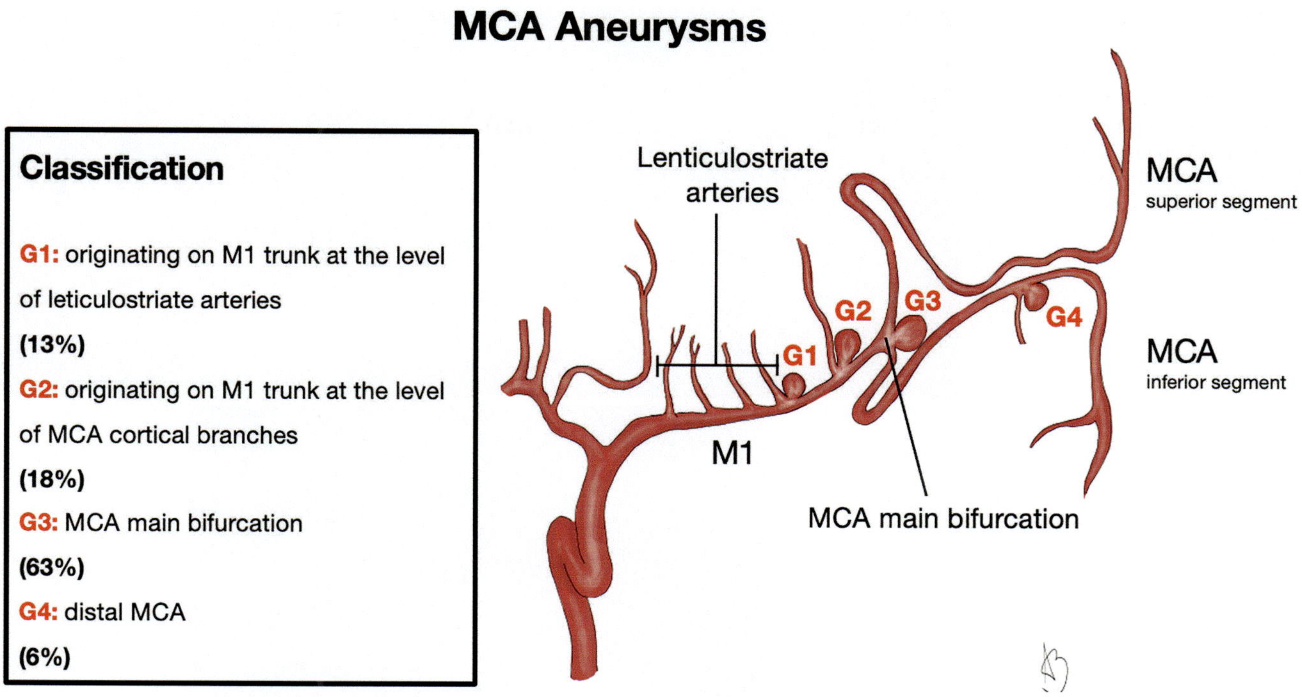

Fig. 12.2 Classification by Elsharkawy et al. [8] subdividing MCA aneurysms into four groups: G1 refers to aneurysms that arise on the M1 trunk between the bifurcation of the ICA and the MCA main bifurcation at the level of lenticulostriate arteries. G2 includes aneurysms originating on M1 at the level of early cortical branches. The most frequent are aneurysms within the third group (G3), consisting of those found at the main MCA bifurcation. The last group, G4, is composed of distal MCA aneurysms found on M2 segments of the MCA and their branches

12.4 Preoperative Considerations

12.4.1 Surgical Technique

The craniotomy is performed from the curvilinear skin incision with the bottom on the sphenoid wing. The craniotomy goes to the skull base; however, the resection of the sphenoid wing is unnecessary. The opening also reaches a little higher than the frontolateral or pterional craniotomy for other aneurysms. For distal MCA aneurysms, craniotomy is targeted over the Sylvian fissure where the aneurysm is located. In these cases, navigation is routinely used.

The dura is opened with the sphenoid wing in the middle of the flap base. Because the dura is reflected downwards, the Sylvian fissure is exposed from the bend skull base/convexity and for some 4 cm length. For typical M1/2 aneurysms, the arachnoid incision is some 20 mm long, usually from the base of the convexity (Fig. 12.3). The veins entering the sphenoparietal sinus are preserved, and the arachnoid cut is usually directed frontally to them. The Sylvian fissure is opened in the direction of the aneurysm. A sharp dissection combined with the waterjet technique is used.

Most frequently, the frontal M2 segment is exposed first, and this vessel is followed deeper. The aneurysm is first bypassed, and the dissection goes deeper to expose the distal M1 segment, which is dissected free to allow the placement of a temporary clip (Fig. 12.4). If the aneurysm projects frontally, the dissection is directed more temporally and vice versa. Should the aneurysm point downwards, the dissection goes deeper into the vessels. If an aneurysm points upwards or in the M1 axis, the dissection goes underneath the vascular tree. Exposure of only approximately 3–4 mm of distal M1 is required to obtain an adequate surgical corridor for further manoeuvrability and temporary clip placement.

The following steps are directed to the aneurysm itself. The sack, as well as both M2 segments, is entirely dissected. Thus, in the opened Sylvian fissure, the distal M1, its division, aneurysm, and both M2s are completely exposed. The veins are preserved. If any small artery, be it a perforator or cortical branch, adheres to the sack, it is dissected free or if it does not cover the aneurysm neck area, which would prevent clipping, it may be left attached to the aneurysm wall. Clipping the aneurysm is not always as easy as expected from the size and angiographic or CTAg appearance of the aneurysm. According to individual situations, all manoeuvres are employed, including temporary clipping, multiple clips, and neck modelling with bipolar coagulation (Figs. 12.4, 12.5, 12.6, and 12.7).

M1 division is frequently a part of the aneurysm base, and a perpendicularly placed clip does not sufficiently occlude the aneurysm. Several clips placed parallel and in the axis of the M1 solve this problem (Fig. 12.7).

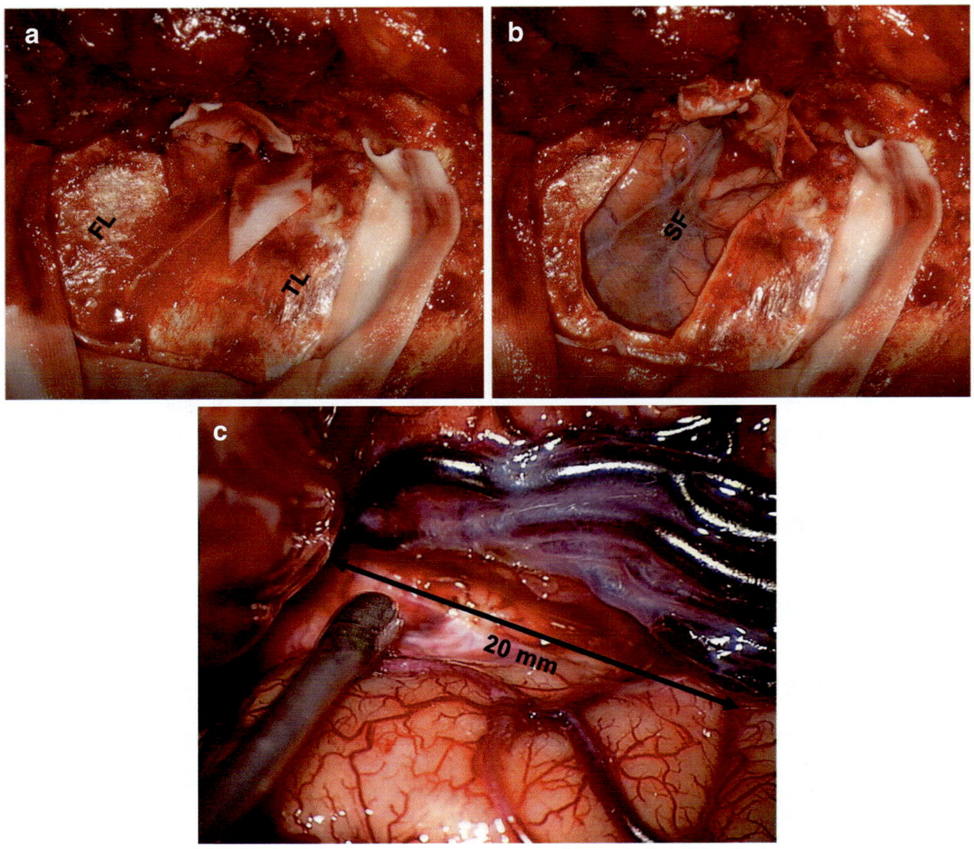

Fig. 12.3 (**a**) Pterional craniotomy. (**b**) Dural opening. (**c**) Minimal exposure of the Sylvian fissure, arachnoid opening is usually about 20 mm long. Abbreviations: *FT* frontal lobe, *SF* Sylvian fissure, *TL* temporal lobe

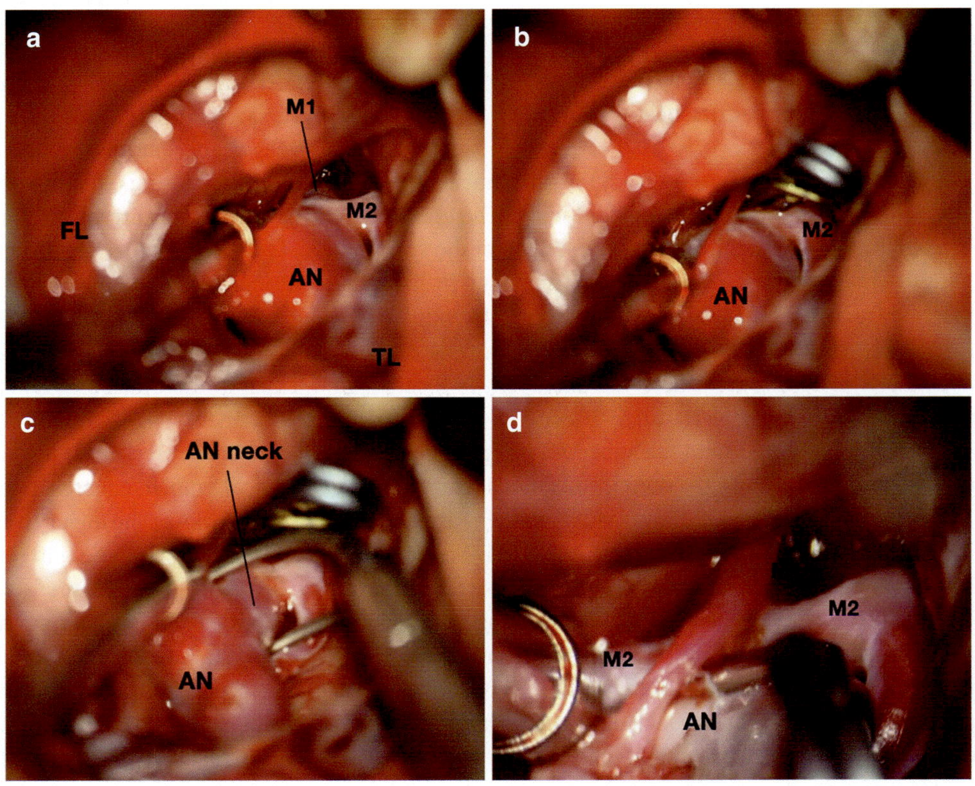

Fig. 12.4 (**a**) Exposure of the aneurysm at the MCA bifurcation. (**b**) A temporary clip is placed proximally to the bifurcation. (**c**) After proper check for any perforators, clip placement proceeds. (**d**) Decompression of the aneurysm, both M2 exposed. Temporary clip removed. Abbreviations: *AN* aneurysm, *FT* frontal lobe, *TL* temporal lobe

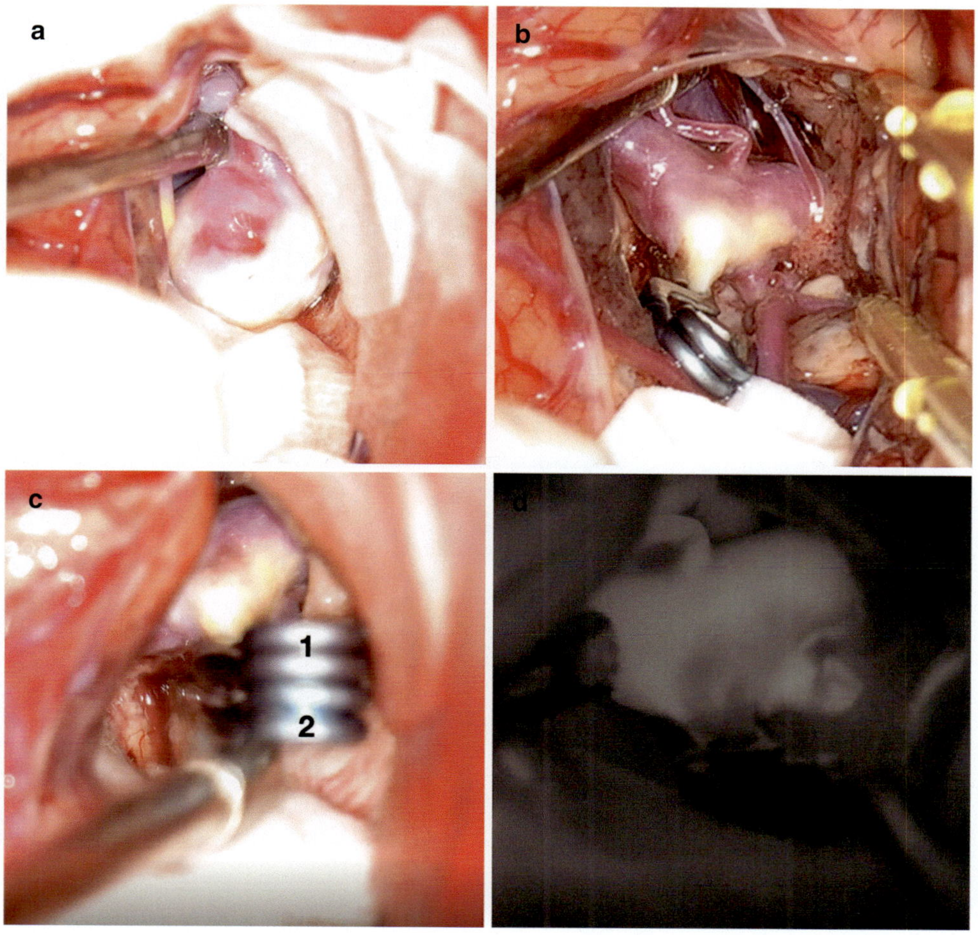

Fig. 12.5 MCA aneurysm. (**a**) Exposed aneurysm. (**b**) First clip applied and a little dog ear remains. (**c**) After clip adjustment and second clip introduction, the aneurysm is completely occluded. (**d**) ICG after clipping

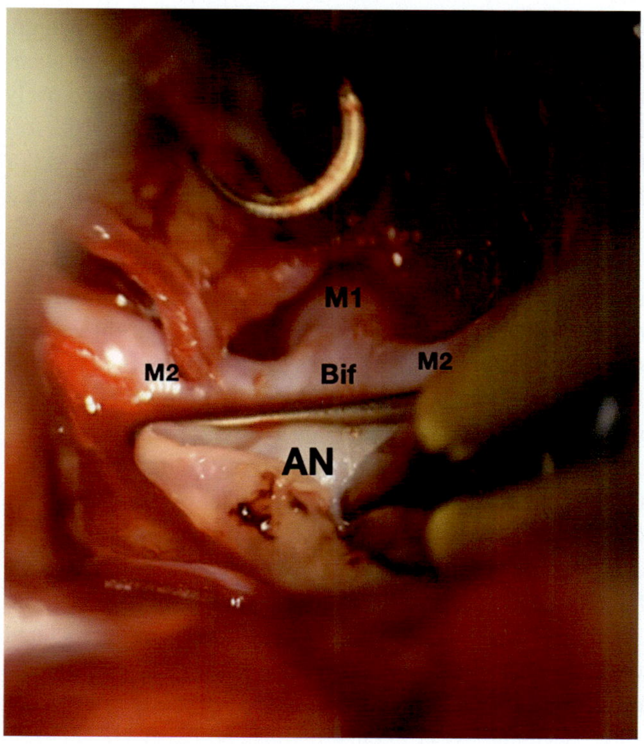

Fig. 12.6 (**a**) "Aneurysm killing"—opening the sack, coagulation, sack resection—after successful clipping, relevant surgical anatomy depicted. Abbreviations: *AN* aneurysm, *Bif* middle cerebral artery bifurcation

In general, it is not uncommon that an MCA aneurysm is not occluded with a single clip on the first attempt. More than one clip is often needed, and multiple manoeuvres are necessary (Fig. 12.5). However, these attempts should not be excessive, the more attempts, the more the danger of ischaemic complications.

Perioperative ruptures are not frequent in experienced hands. Still, ruptures occur. These ruptures usually arise when dealing directly with the aneurysm, i.e. at the point when M1 is secured and prepared for the temporary clip. Considering these factors, the rupture does not pose a serious threat when managed appropriately. In our hands, in MCA aneurysms adenosine is seldom used.

When the aneurysm is finally clipped, a thorough inspection is made for any eventual reddish spots at the MCA bifurcation. These are covered by small cottonoids. ICG is used to check the proper aneurysm occlusion and a good flow through M1, M2, and any eventual perforator (Figs. 12.5, 12.6, and 12.7). We usually open the sack, resect it for further studies, or shrink it to its complete absence with bipolar coagulation ("killing" the aneurysm in Juha Hernesniemi's words).

The Sylvian fissure is then flushed with saline, and the vessels are covered with small pieces of gelfoam soaked in Papaverine. The wound is then closed in layers.

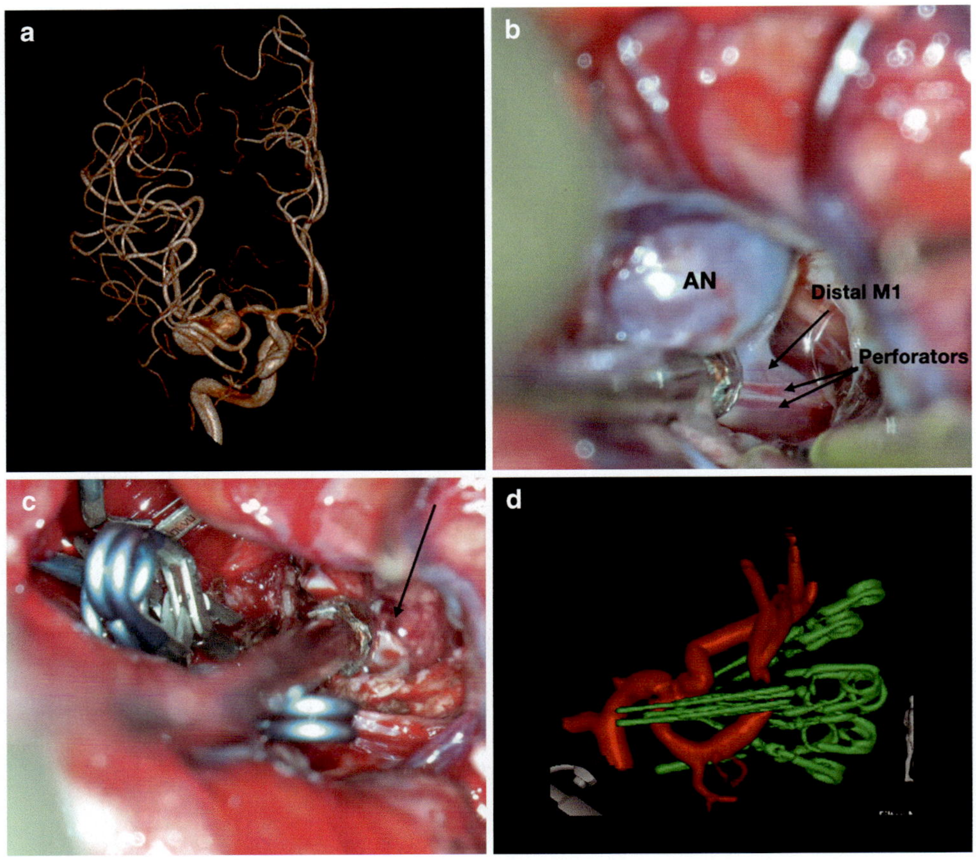

Fig. 12.7 MCA aneurysm 20 mm in long diameter occluded by multiple clips. (**a**) 3D angiography. (**b**) Distal M1 and aneurysm exposed. Note the two perforators along with the M1. (**c**) After the clipping and aneurysm sack resection, see the cut aneurysm (arrow). (**d**) Postoperative angiography

12.5 Specific Situations

12.5.1 Calcifications and Partial Thrombosis

In the bifurcation area, atherosclerotic plaques are frequently found in the neck area. Some may even prevent good clip closure from occluding the flow to the aneurysm. Performing too many manoeuvres to clip the aneurysm can dislodge some materials that may embolise the distal MCA territory. This possibility needs to be kept in mind and repeated manoeuvres to clip the aneurysm should be avoided. These plaques are almost always present in older adults with large aneurysms. Sometimes the clip slips to the mother arteries, and ultimately, some small parts of the aneurysm may be left outside the clip(s). These minimal remnants do not pose a serious risk in the future (in our material they remain stable).

12.5.2 Large and Giant Aneurysms

Large and giant MCA aneurysms most often appear at the MCA bifurcation, subsequently at the M1 segment and distal MCA branches [10]. In these aneurysms an analogous surgical approach from the convexity is used; however, exposure of the Sylvian fissure is usually more extensive (Figs. 12.8 and 12.9). The base of the sack is typically sclerotic and hard, and some remaining aneurysm base is often left behind. It is generally not considered a problem, as such an aneurysm base does not rupture, but the patients are more closely followed to diagnose eventual regrowth early during the follow-up. Preventive bypass, frequently double barrel to M3 or M4 on both frontal and temporal sides, is the standard first step of surgery (see below).

Large and giant MCA aneurysms are typical with asymmetric morphology, higher risk of intramural thrombus formation, calcification, and incorporation of adjacent vessels, resulting in more challenging surgical procedures [11, 12]. In some cases, an alternative treatment approach is required in which direct surgery (or an endovascular technique) cannot be implemented alone. These most commonly include trapping with an extracranial–intracranial bypass technique, clip reconstruction with aneurysm thrombectomy, aneurysm excision along with an in situ bypass, and mixed endovascular techniques with simultaneous bypass implementation. Therefore, large and giant MCA aneurysms are associated with a higher risk of residual aneurysms. A higher retreatment rate and procedure-related complications than “standard” saccular MCA aneurysms can be expected [11–13]. An individual surgical strategy is tailored to the individual patient and aneurysm.

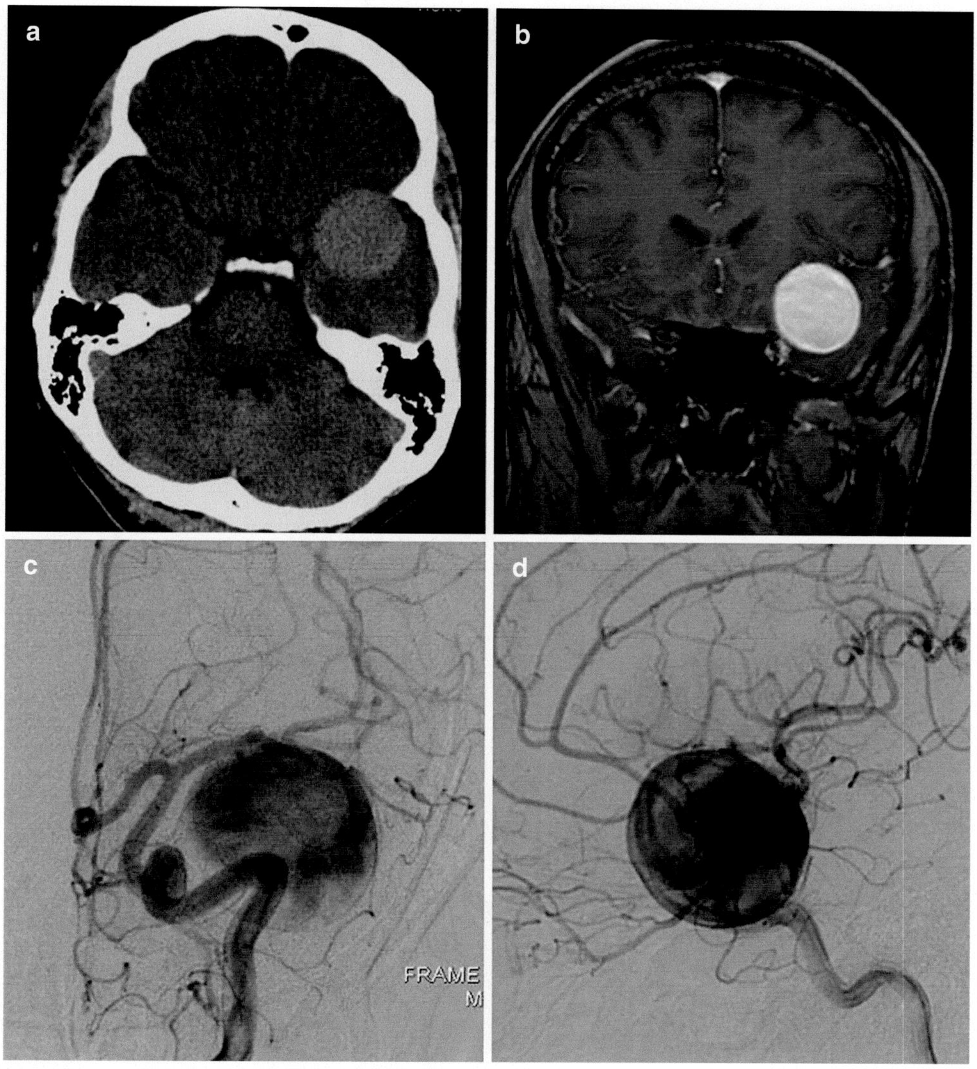

Fig. 12.8 Giant left MCA aneurysm

12.5.3 Fusiform Aneurysms

Fusiform aneurysms are a unique entity of vascular anomalies accounting for approximately 10% of all intracranial aneurysms [14]. They are characteristic of circumferential vessel dilatation with an absence of a definable neck, a pattern that may have resulted from various pathological mechanisms involving the whole artery. Considering the lack of a definable aneurysm neck, the surgical treatment of fusiform aneurysms notably differs from the saccular aneurysms we have discussed thus far. The vast majority of cases of fusiform MCA aneurysms are treated by endovascular techniques, although direct surgical management is possible. The treatment strategy is usually some kind of bypass surgery, followed by aneurysm trapping. Reconstruction by direct surgical clipping is seldom workable (Fig. 12.10). Quite a few distal small asymptomatic lesions are generally recommended to be left to observation unless the aneurysm dilates or worsens the patient's clinical status [15–17].

12.5.4 Dissecting Aneurysms

Because direct repair of the vessel is almost impossible, bypass techniques and aneurysm trapping (surgical or endovascular) are implemented in most cases.

12.5.5 Pseudoaneurysms

Pseudoaneurysms are rare lesions that require individual treatment (Figs. 12.11 and 12.12). Bypass and parent artery occlusion are often the treatment of choice.

12.5.6 Bypasses

All bypass techniques must be mastered to deal with diverse situations. At our institution, routine EC-IC, preferably double-barrel, is the treatment of choice. The flow in each M2 segment is some 25 ml (min), which is usually the same

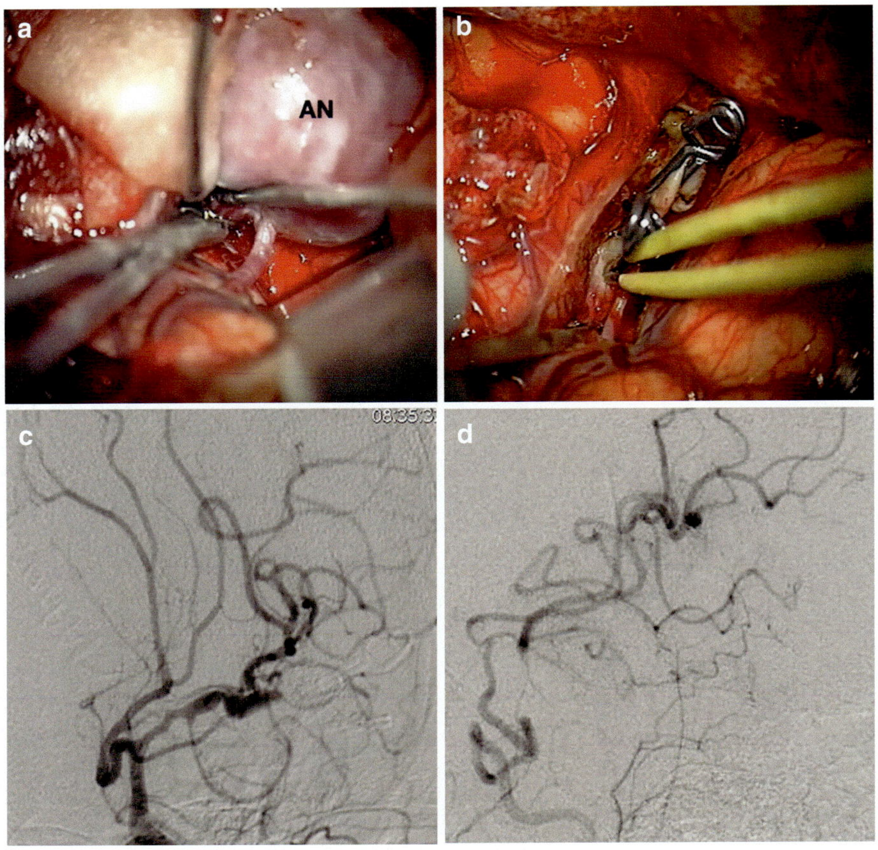

Fig. 12.9 (**a**) Temporary clips are placed on M1 and both M2 segments, aneurysm is punctured and decompressed. (**b**) After the clipping and sack coagulation. (**c**) Original M1 fills frontal MCA territory. (**d**) Temporal MCA territory is filled by EC-IC bypass

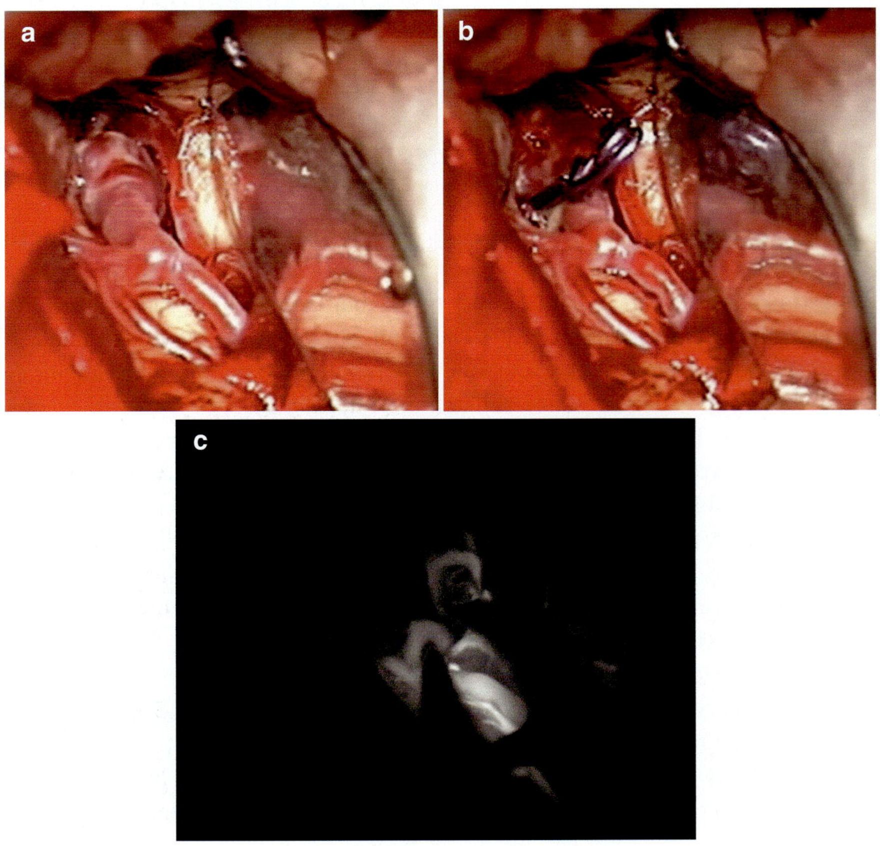

Fig. 12.10 Fusiform aneurysm of M2 segment. (**a**) Exposure of the MCA within the Sylvian fissure. (**b**) The clip is placed tangentially to the mother artery. (**c**) ICG

Fig. 12.11 Ruptured MCA pseudoaneurysm

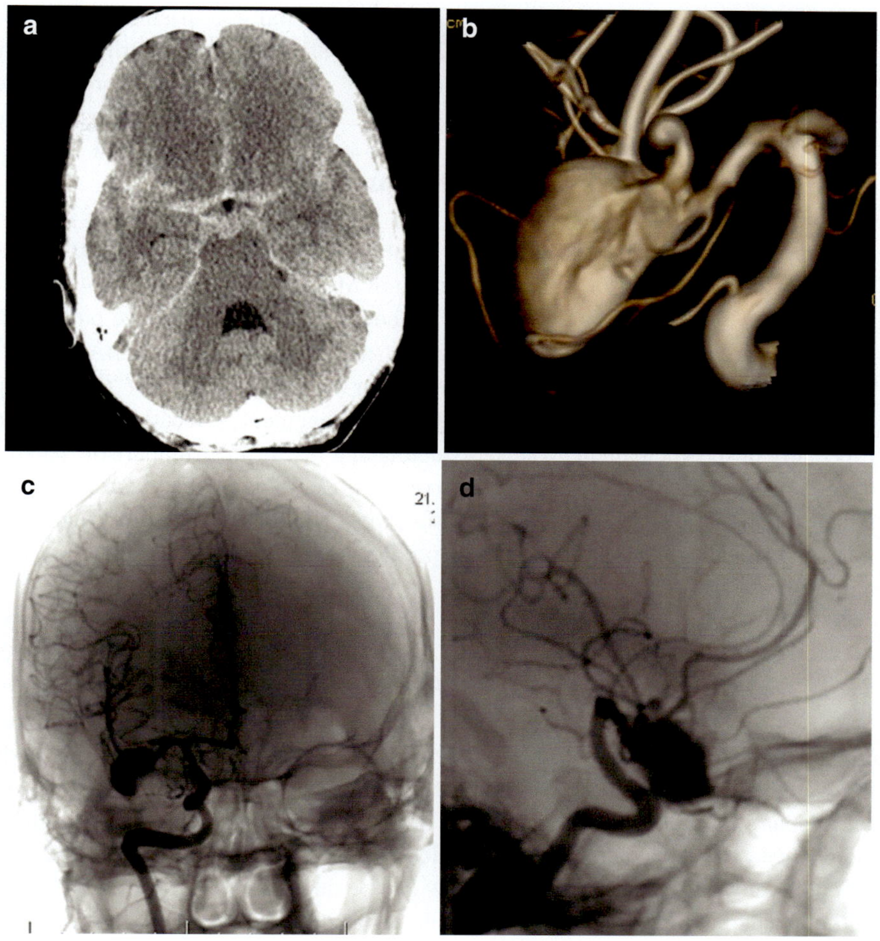

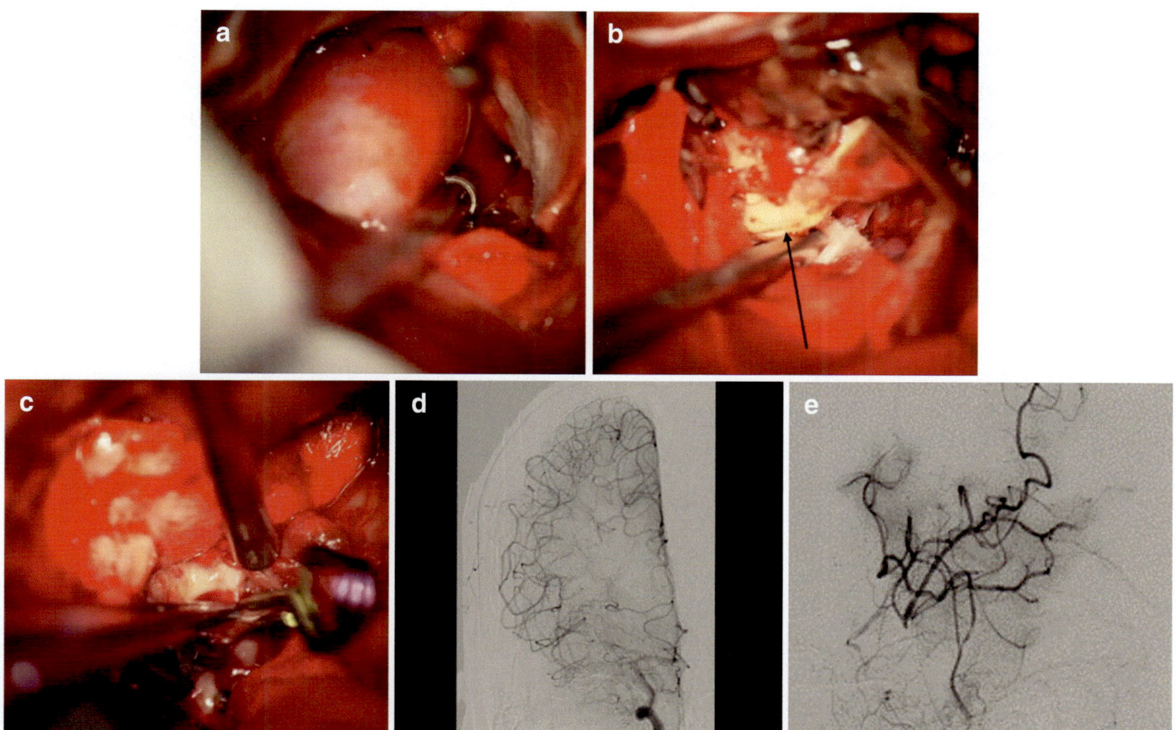

Fig. 12.12 (**a**) Exposed pseudoaneurysm. (**b**) Clot removal with the M1 temporarily clipped and the attempt at direct repair with suture (*arrow*). (**c**) Suture is enhanced by clip. Reconstruction of the MCA bifurcation did not succeed. (**d**) It seems like some minor flow through the MCA bifurcation if preserved. (**e**) However, the hemisphere is supplied predominantly by cortical anastomosis and bypass

as in the superficial temporal artery (STA) [18, 19]. Intra-intracranial bypasses can be implanted [20]. These are more physiological but technically more demanding.

12.6 Role of Endovascular Techniques

Most recent meta-analyses and comparative studies reported wider advantages of the surgical management of MCA aneurysms compared to coiling in terms of outcome variables such as the completeness of the occlusion, angiographic result, or clinical outcome [21–23]. Endovascular techniques play an important role in MCA aneurysms management, especially in specific patient groups, dissecting and fusiform aneurysms unsuitable for surgical intervention, i.e. older adult patients with significant comorbidities, those who primarily reject surgical treatment, or in particular types of aneurysms.

12.7 Complications and Their Avoidance

The most serious complications are ischaemic caused by inadvertent vessel occlusion. Proper dissection and clip placement allow for a careful inspection of all vessels. This is enhanced by indocyanine green injection (ICG) or direct Doppler study with microprobe. The flow can be quantified with the Charbel probe. Special care should be taken for M1 perforators; their patency is best confirmed by inspection and ICG. Ischaemic complications can also be caused by arterial hypotension. Thus, the anaesthesiologist should keep the patient normotonic throughout the whole surgical procedure. Another danger could be that the temporary clip application is in for too long. The anaesthesiologist or scrub nurse should call the time of occlusion.

Postoperative bleeding is prevented by meticulous intraoperative haemostasis. Vasospasms can occur on a regular basis in ruptured aneurysms. In unruptured aneurysms spasms can be caused by manipulation. Local application of Papaverine is routine. Later on, the vasospasms are dealt with accordingly. All other complications (e.g. wound infection) are generally surgical and can be treated best by prevention.

12.8 Outcome

We have published two articles [5, 24] on aneurysms found at AComA ("Coil first policy in AcomA aneurysms") and MCA ("Clip first policy in MCA aneurysms"). These two locations of intracranial aneurysms provide an appropriate comparison regarding the treatment algorithm. The active treatment policy in MCA aneurysms is mainly built on the superior effect of microsurgical clipping over endovascular techniques, i.e. the "clip first" policy should not be abandoned because it is supported by an extensive dataset of patients and a systematic review of the literature.

Over the past 22 years, 473 MCA aneurysms were treated at our institution, with the surgical approach proving to be a superior procedure (374 aneurysms in the surgical group vs. only 99 in the endovascular group). In the group of unruptured MCA aneurysms (203 aneurysms treated surgically and 42 treated by endovascular techniques), the overall morbidity–mortality rate in the neurosurgical group was 2.6% compared to 9.8% in the endovascular group. Most patients (90%) had a favourable outcome (GOS 5), regardless of the group. This outcome was reached in 78% of patients in the endovascular group. Mortality in the surgical group was zero and 5% in the endovascular group. On the other hand, morbidity and mortality in treated ruptured MCA aneurysms was 36% in the surgical group and 49% in the endovascular group. Fewer patients reached favourable outcomes (GOS 4–5 in 64% of clipped aneurysms and 51% managed endovascularly). In ruptured MCA aneurysms, the overall mortality was similar in the two groups (14% in the surgical group vs. 16% in the endovascular group) (12.13).

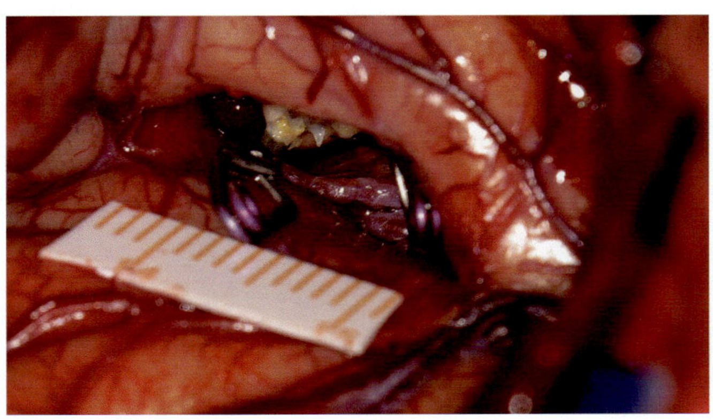

Fig. 12.13 Minimal invasive approach to MCA aneurysm. Aneurysm is clipped via some 20 mm long arachnoid opening and targeted dissection of M1-2 and aneurysm

12.9 Surgical Pearls

- Meticulous study of the length of the M1 segment preoperatively
- Check the direction of the aneurysm sack
- Opening Sylvian fissure arachnoid on the convexity some 20–25 mm long
- Follow any artery as far as distal M1
- Some aneurysms may appear at the surface of the brain. Adhesions between the sack and the sphenoid wing dura are rarely seen
- Check the aneurysm neck circumferentially
- Inspection of the aneurysm, usually the M1 tip, clip adjustment, multiple clips, and neck modelling

12.10 Conclusion

Considering easy surgical access, excellent results and minimal risk of complications, MCA aneurysms remain the dominant surgery-indicated entity. In addition to understanding direct surgical clipping, understanding of and clinical experience with bypass techniques are necessary for proper preparation for the microsurgical treatment of MCA aneurysms. Despite the apparent predominance of saccular aneurysms at the MCA, specific types (such as dissecting, fusiform, or pseudoaneurysms) are also a part of surgical indications. However, various endovascular methods are more commonly employed. Still, the final decision should be determined by a multidisciplinary team of well-experienced neurosurgeons and vascular neurointerventionalists.

The approach described in this paper fulfils the recent trend towards minimal invasivity: 20 mm arachnoid opening of the Sylvian fissure and targeted aneurysm dissection is minimally invasive in relation to the lesion and not towards the approach of standard craniotomy. In the author's opinion, such a trend is more logical than minimal invasivity in relation to the approach.

Supported by programme MO1012.

References

1. Rousseau O, Karakachoff M, Gaignard A, Bellanger L, Bijlenga P, Constant Dit Beaufils P, et al. Location of intracranial aneurysms is the main factor associated with rupture in the ICAN population. J Neurol Neurosurg Psychiatry. 2021;92(2):122–8. https://doi.org/10.1136/jnnp-2020-324371.
2. Dandy W. Intracranial arterial aneurysms. Ithaca: Comstock Publishing Company; 1969.
3. Bohnstedt BN, Nguyen HS, Kulwin CG, Shoja MM, Helbig GM, Leipzig TJ, et al. Outcomes for clip ligation and hematoma evacuation associated with 102 patients with ruptured middle cerebral artery aneurysms. World Neurosurg. 2013;80(3–4):335–41. https://doi.org/10.1016/j.wneu.2012.03.008.
4. Smith TR, Cote DJ, Dasenbrock HH, Hamade YJ, Zammar SG, El Tecle NE, et al. Comparison of the efficacy and safety of endovascular coiling versus microsurgical clipping for unruptured middle cerebral artery aneurysms: a systematic review and meta-analysis. World Neurosurg. 2015;84(4):942–53. https://doi.org/10.1016/j.wneu.2015.05.073.
5. Steklacova A, Bradac O, Charvat F, De Lacy P, Benes V. "Clip first" policy in management of intracranial MCA aneurysms: single-centre experience with a systematic review of literature. Acta Neurochir. 2016;158(3):533–46; discussion 546. https://doi.org/10.1007/s00701-015-2687-y. Epub 2016 Jan 6. PMID: 26733126.
6. Shimoda M, Oda S, Mamata Y, Tsugane R, Sato O. Surgical indications in patients with an intracerebral hemorrhage due to ruptured middle cerebral artery aneurysm. J Neurosurg. 1997;87(2):170–5. https://doi.org/10.3171/jns.1997.87.2.0170.
7. Alreshidi M, Cote DJ, Dasenbrock HH, Acosta M, Can A, Doucette J, et al. Coiling versus microsurgical clipping in the treatment of unruptured middle cerebral artery aneurysms: a meta-analysis. Neurosurgery. 2018;83(5):879–89. https://doi.org/10.1093/neuros/nyx623.
8. Elsharkawy A, Lehečka M, Niemelä M, Billon-Grand R, Lehto H, Kivisaari R, et al. A new, more accurate classification of middle cerebral artery aneurysms: computed tomography angiographic study of 1,009 consecutive cases with 1,309 middle cerebral artery aneurysms. Neurosurgery. 2013;73(1):94–102; discussion. https://doi.org/10.1227/01.neu.0000429842.61213.d5.
9. Rinne J, Hernesniemi J, Niskanen M, Vapalahti M. Analysis of 561 patients with 690 middle cerebral artery aneurysms: anatomic and clinical features as correlated to management outcome. Neurosurgery. 1996;38(1):2–11. https://doi.org/10.1097/00006123-199601000-00002.
10. Park W, Chung J, Ahn JS, Park JC, Kwun BD. Treatment of large and Giant middle cerebral artery aneurysms: risk factors for unfavorable outcomes. World Neurosurg. 2017;102:301–12. https://doi.org/10.1016/j.wneu.2017.03.028. Epub 2017 Mar 16. PMID: 28323182.
11. Zhu W, Liu P, Tian Y, Gu Y, Xu B, Chen L, Zhou L, Mao Y. Complex middle cerebral artery aneurysms: a new classification based on the angioarchitecture and surgical strategies. Acta Neurochir (Wien). 2013;155(8):1481–91. https://doi.org/10.1007/s00701-013-1751-8. Epub 2013 May 30. PMID: 23715946; PMCID: PMC3718994.
12. Shi ZS, Ziegler J, Duckwiler GR, Jahan R, Frazee J, Ausman JI, Martin NA, Viñuela F. Management of giant middle cerebral artery aneurysms with incorporated branches: partial endovascular coiling or combined extracranial-intracranial bypass--a team approach. Neurosurgery. 2009;65(6 Suppl):121–9; discussion 129–31. https://doi.org/10.1227/01.NEU.0000335173.80605.1D. PMID: 19934986.
13. Rodríguez-Hernández A, Sughrue ME, Akhavan S, Habdank-Kolaczkowski J, Lawton MT. Current management of middle cerebral artery aneurysms: surgical results with a "clip first" policy. Neurosurgery. 2013;72(3):415–27. https://doi.org/10.1227/NEU.0b013e3182804aa2. PMID: 23208060.
14. al-Yamany M, Ross IB. Giant fusiform aneurysm of the middle cerebral artery: successful Hunterian ligation without distal bypass. Br J Neurosurg. 1998;12(6):572–5. https://doi.org/10.1080/02688699844466.
15. Park SH, Yim MB, Lee CY, Kim E, Son EI. Intracranial fusiform aneurysms: it's pathogenesis, clinical characteristics and managements. J Korean Neurosurg Soc. 2008;44(3):116–23. https://doi.org/10.3340/jkns.2008.44.3.116.
16. Day AL, Gaposchkin CG, Yu CJ, Rivet DJ, Dacey RG Jr. Spontaneous fusiform middle cerebral artery aneurysms: characteristics and a proposed mechanism of formation. J Neurosurg. 2003;99(2):228–40. https://doi.org/10.3171/jns.2003.99.2.0228.

17. Lanzino G, Kaptain G, Kallmes DF, Dix JE, Kassell NF. Intracranial dissecting aneurysm causing subarachnoid hemorrhage: the role of computerised tomographic angiography and magnetic resonance angiography. Surg Neurol. 1997;48(5):477–81. https://doi.org/10.1016/s0090-3019(97)00178-x.
18. Amin-Hanjani S, Meglio G, Gatto R, Bauer A, Charbel FT. The utility of intraoperative blood flow measurement during aneurysm surgery using an ultrasonic perivascular flow probe. Neurosurgery. 2006;58(4 Suppl 2):ONS-305-12; discussion ONS-312. https://doi.org/10.1227/01.NEU.0000209339.47929.34. PMID: 16582654.
19. Amin-Hanjani S, Charbel FT. Flow-assisted surgical technique in cerebrovascular surgery. Surg Neurol. 2007;68(Suppl 1):S4–11. https://doi.org/10.1016/j.surneu.2007.07.051. PMID: 17963920.
20. Tayebi Meybodi A, Huang W, Benet A, Kola O, Lawton MT. Bypass surgery for complex middle cerebral artery aneurysms: an algorithmic approach to revascularisation. J Neurosurg. 2017;127(3):463–79. https://doi.org/10.3171/2016.7.JNS16772. Epub 2016 Nov 4. PMID: 27813463.
21. Zaidat OO, Castonguay AC, Teleb MS, Asif K, Gheith A, Southwood C, et al. Middle cerebral artery aneurysm endovascular and surgical therapies: comprehensive literature review and local experience. Neurosurg Clin N Am. 2014;25(3):455–69. https://doi.org/10.1016/j.nec.2014.04.005.
22. Diaz OM, Rangel-Castilla L, Barber S, Mayo RC, Klucznik R, Zhang YJ. Middle cerebral artery aneurysms: a single-center series comparing endovascular and surgical treatment. World Neurosurg. 2014;81(2):322–9. https://doi.org/10.1016/j.wneu.2012.12.011.
23. Blackburn SL, Abdelazim AM, Cutler AB, Brookins KT, Fargen KM, Hoh BL, et al. Endovascular and surgical treatment of unruptured MCA aneurysms: meta-analysis and review of the literature. Stroke Res Treatment. 2014;2014:1–11. https://doi.org/10.1155/2014/348147.
24. Steklacova A, Bradac O, de Lacy P, Lacman J, Charvat F, Benes V. "Coil mainly" policy in management of intracranial ACoA aneurysms: single-centre experience with the systematic review of literature and meta-analysis. Neurosurg Rev. 2018;41(3):825–39. https://doi.org/10.1007/s10143-017-0932-y. Epub 2017 Nov 28. PMID: 29181806.

13 Microsurgery of Giant Intracranial Aneurysm

Basant K. Misra, Arthur Hosmann, and Abhijit G. Warade

Learning Objectives

- Familiarize with the anterior and posterior circulation, including various anatomical variants.
- Understand the pathophysiology and identify the risk factors associated with Giant Intracranial Aneurysms (GIAs).
- Gain knowledge of the microsurgical techniques used in the management of GIA.
- Identify and comprehend the complications that may arise during GIA treatment, along with strategies for their effective management and prevention.

13.1 Introduction

Giant intracranial aneurysm (GIA) represents approximately 5% of all intracranial aneurysms [1]. In the International Study of Unruptured Intracranial Aneurysms (ISUIA) unruptured GIA accounted for 4.6% of all examined cases [2]. In contrast, in the Unruptured Cerebral Aneurysm Study (UCAS) of Japan GIA was present in only 0.5% of all patients [3].

GIAs have, by definition, a minimum diameter of 25 mm [4]. GIAs are treacherous lesions with grave prognosis and their management results in a combined surgical morbidity and mortality in the range of 20–30% [5]. Yet the GIAs need treatment, as these often have a downhill course without treatment, with a mortality rate at 2 and 5 years after diagnosis being 68% and 85%, respectively [6].

B. K. Misra (✉) · A. G. Warade
Department of Neurosurgery & Gamma Knife Surgery, P.D. Hinduja Hospital & MRC, Mumbai, Maharashtra, India
e-mail: dr_bmisra@hindujahospital.com

A. Hosmann
Department of Neurosurgery & Gamma Knife Surgery, P.D. Hinduja Hospital & MRC, Mumbai, Maharashtra, India

Department of Neurosurgery, Medical University of Vienna, Vienna, Austria

13.2 Anatomical Highlights

- Giant aneurysms are characterized as aneurysms measuring 25 mm or larger in size.
- The predominant location for GIAs is the internal carotid artery, followed by the posterior circulation.
- Aneurysms located in the posterior circulation are considered a predictor for rupture and are associated with poor outcomes.
- It is crucial to have precise knowledge of the vascular anatomy and its variations due to potential distortion caused by the mass effect of the GIA.
- Having a comprehensive understanding of the individual anatomical features of the GIA, such as the precise location of the aneurysm dome, neck, significant arterial branches, and perforators, is crucial for preventing complications.
- For anterior communicating artery (AcoA) aneurysms, the surgical approach is determined by the projection of the aneurysm, with the order of favorability for clipping being anterior > inferior > superior > posterior projection.
- The choice of surgical approach for basilar top aneurysms depends on the relation of the aneurysm neck to the posterior clinoidal process (above: sylvian/orbitozygomatic approach; below: subtemporal/transpetrosal approach).

13.3 Technical Nuances for Specific Aneurysm Subtypes

13.3.1 Anterior Circulation Aneurysms

13.3.1.1 Anterior Cerebral Artery (ACA)

Aneurysms mostly occur at the anterior communicating artery (AcoA) and are challenging to treat due to their close relationship to the perforating arteries, posterior projection, and usually small size. Several perforating branches arise

I. N. Kanaan, V. Beneš (eds.), *Neuroanatomy Guidance to Successful Neurosurgical Interventions*,
https://doi.org/10.1007/978-3-031-59838-8_13

from the A1/A2 segment. The most important and largest artery is the recurrent artery of Heubner. It originates most commonly from the A1/A2 junction (76%), followed by the A2 segment in 16%, and the A1 segment in 8% [7]. Another 2–15 perforating arteries (average 8) arise from the A1 segment, mostly from its lateral half [8]. The AcoA gives off one to four small perforating arteries from its superior and posterior surfaces [8]. Inadvertent occlusion or injury to these perforators can result in significant neurologic, endocrine, or cognitive deficits [9]. There is also a high risk of partial clipping because of inadequate neck visualization in superiorly and posteriorly pointing AcoA. Inferiorly pointing AcoA may interfere with securing proximal control of the contralateral A1 [10].

For the surgical exposure of AcoA aneurysms, the author prefers a pterional craniotomy, especially for anterior or inferior projecting aneurysms. The anteriorly projecting aneurysms are the most favorable for clipping from the neurosurgeon's perspective as the neck is well visible, both A2 are easily identified, the contralateral A1 can be exposed underneath the aneurysms' neck and the posterior projecting perforators are not encountered during the clipping procedure. Inferiorly projecting AcoA aneurysms are also favorable to clip but might adhere to the chiasm and the contralateral A1 is usually obscured from the aneurysm dome. Superiorly projecting aneurysms are less favorable as the contralateral A2 and perforators of the AcoA are displaced and obscured from the aneurysm sac. Identifying the perforators is even more difficult in the posterior projecting AcoA aneurysms, which makes these aneurysms the most challenging to clip. Partial resection of the gyrus rectus improves the exposure of the AcoA-complex and facilitates the clipping of especially superior and posterior projecting aneurysms [11].

Distal ACA aneurysms are rare and account for only 2–9% of intracranial aneurysms [12]. GIAs of the distal ACA are extremely rare and are reported in the literature as case reports only [13]. The majority of distal ACA aneurysms are clipped via an anterior frontal interhemispheric approach. In some proximal A2 segment aneurysms, a pterional or lateral supraorbital approach can be reasonable [12].

13.3.1.2 Middle Cerebral Artery (MCA) Aneurysms

The MCA is the largest branch of the ICA. In most cases, the main trunk of the MCA divides into a superior and inferior trunk. Following a pterional craniotomy and dural opening, the Sylvian fissure is split to expose the MCA branches. In ruptured MCA aneurysm cases, starting dissection from the basal cisterns might be reasonable for the early identification of the ICA for immediate proximal control.

From the M1 segment, multiple small perforating branches arise, which are called the lenticulostriate arteries [14]. In GIA these branches might be stretched and displaced. MCA aneurysms are commonly associated with atherosclerosis and calcification at the necks can interfere with the complete closure of clip blades [10]. Only 4% of MCA aneurysms are located distally to the M1 bifurcation and GIAs are very rare in these segments [15].

13.3.1.3 Internal Carotid Artery (ICA) aneurysms

The internal carotid artery (ICA) can be divided into the cervical, petrous, cavernous, and supraclinoid ICA [8].

The cervical ICA starts at the bifurcation of the common carotid artery and extends up to the carotid canal. Extracranial aneurysms of the ICA are rare and mostly asymptomatic.

The petrous ICA begins at the carotid canal and runs through the petrous bone of the skull. Aneurysms in this segment are very rare. These are mostly large and fusiform and present with intraluminal thrombus. Due to the limited surgical access to the petrous bone, these aneurysms are treated either endovascularly or surgically with bypass and parent artery occlusion [16] (Fig. 13.1).

The cavernous segment extends from the petrous segment to the point where the ICA enters the subarachnoid space. Aneurysms of the cavernous ICA segment account for 2–9% of all intracranial aneurysms [17]. Currently, these aneu-

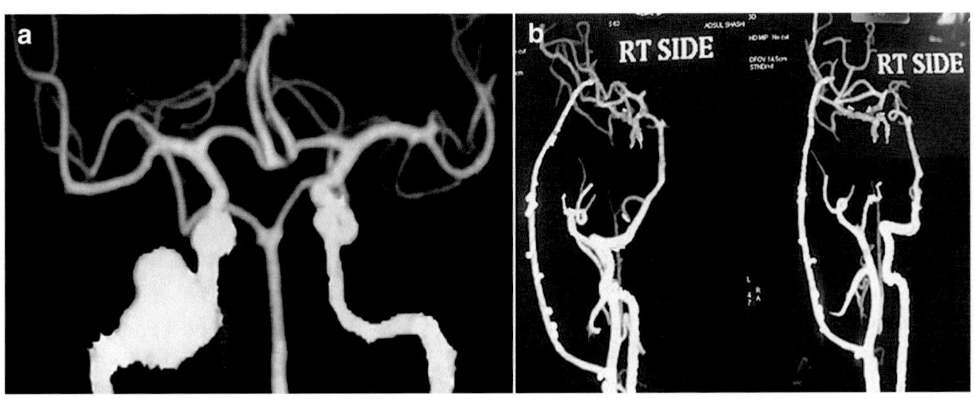

Fig. 13.1 (**a**, **b**) Petrous GIA: ECA-M2 radial artery graft, bypass and carotid ligation 20-yr follow-up shows patent graft

rysms are mostly treated by endovascular means using coiling, stent-assisted coiling, or flow diversion [18]. The microsurgical option is flow diversion (carotid ligation/trapping) with bypass. The microsurgical option may be more appropriate for GIA with thrombus and significant mass effect or occasionally in post-traumatic pseudoaneurysm [19].

The clinoid segment is a transitional segment of the ICA between the cavernous sinus and the subarachnoid space. It is the area between the proximal and distal dural rings of the ICA [20]. For the microsurgical exposure of clinoidal aneurysms, the clinoid process must be drilled. After exposure of the paraclinoidal ICA, the distal dural ring (sometimes the proximal ring needs to be sectioned as well) must be transacted allowing preparation along the clinoid segment without entering the cavernous sinus [21].

The supraclinoid ICA is the intradural portion of the ICA within the subarachnoid space after emerging from the dura mater of the cavernous sinus to the point where the ICA bifurcates into the anterior cerebral artery (ACA) and the middle cerebral artery (MCA). The supraclinoid ICA is approached via a pterional craniotomy and microsurgical dissection of the Sylvian fissure. The supraclinoid ICA is divided into the ophthalmic, the communicating, and the choroidal segment according to the major branches arising from each segment.

The ophthalmic segment extends from the cavernous sinus to the origin of the posterior communicating artery. During surgical exposure, the origin of the ophthalmic artery mostly cannot be visualized until the anterior clinoid is drilled [22].

Ophthalmic artery aneurysms arise immediately distal to and in relation to the origin of the ophthalmic artery and dural ring [22]. They project dorsally or dorsomedially and compress the lateral and superior surface of the optic nerve against the falciform ligament resulting in an inferior nasal field defect. The superior hypophyseal arteries emerge within 5 mm of the origin of the ophthalmic artery from the medial or posteromedial wall of the ICA. Superior hypophyseal artery aneurysms arise along the medial surface of ICA distal to the dural ring [22]. For surgical exposure of aneurysms of the ophthalmic segment, the anterior clinoid process must be drilled. Underneath, the ICA is fixed at its transdural passage by strong dural strands to the surrounding osseus structures, forming a fibrous ring around the ICA, also known as the distal dural ring [21]. Transection of this dural ring allows preparation along the clinoid segment of the extradural ICA but without entering the cavernous sinus or compromising cranial nerves [21]. This surgical approach provides access to the neck of ophthalmic and paraclinoid aneurysm and proximal control of the ICA. Otherwise, the cervical ICA needs to be exposed in the neck to achieve proximal control during aneurysm clipping. The exposure of the ICA in the neck can also facilitate clipping by suction-decompression (Dallas technique) of the aneurysm.

The communicating segment extends from the origin of the posterior communicating artery to the origin of the anterior choroidal artery. Aneurysms of the posterior communicating segment are usually thick-walled aneurysms and a perpendicular clip can cause incomplete closure of clip blades resulting in IOR [10]. The commonest projection is on the oculomotor nerve which gives the surgeon good accessibility to the neck on both sides between the tentorium and ICA. The aneurysm that projects the medial to oculomotor nerve abuts the posterior clinoid process, narrowing the space for the proximal clip blade application. One should be aware that the domes projecting above the tentorium can have adhesions to the temporal lobe and lead to rupture during dissection [23].

The anterior choroidal segment extends from the origin of the anterior choroidal artery to the ICA bifurcation. The anterior choroidal artery arises from the posterolateral or posterior aspect of the ICA. Multiple perforating branches (one to nine, average four) arise mostly from the posterior surface.

13.3.2 Posterior Circulation Aneurysms

Aneurysms of the posterior circulation account for around 10–15% of all intracranial aneurysms [24]. Saccular aneurysms of the posterior circulation most often occur at the basilar apex (BA) (33%) followed by the Vertebral artery-Posterior inferior cerebellar artery junction (BA-PICA) (31%) and the Basilar trunk (24%) [25].

The vertebral artery is divided into four segments, which are numbered from proximal to distal [26]. The V1 (preforaminal), the V2 (foraminal) segment, and the V3 (atlantic, extradural, or extraspinal) segment are located extradurally. The V4 (intradural or intracranial) segment is the intradural portion of the vertebral artery, ascending anterior to the hypoglossal nerve, runs along the lateral aspect of the medulla oblongata, and merges with the contralateral vertebral artery at the level of the pontomedullary junction to form the basilar artery. The most common site of aneurysms on the vertebral artery is at the origin of PICA. Therefore, most of the posterior inferior cerebellar artery (PICA) aneurysms are located in the lateral or posterior medullary cisterns in close relationship to the lower cranial nerves. Adherence and stretching of cranial nerves and arteries to the aneurysm dome might impede aneurysm preparation. Distal PICA aneurysms can be attached to the tonsils, hence excessive cerebellar retraction should be avoided. Anterior pointing aneurysms may be adherent to the dura of clivus, the manipulation of which can lead to tear at the fundus [10].

PICA aneurysms can be surgically exposed via a suboccipital approach. Depending on the distance to the origin from the vertebral artery, either bilateral suboccipital, lateral suboccipital, or even a far-lateral approach may be necessary [27]. Deep-seated PICA aneurysms in the middle portion of the posterior fossa can be exposed through a combined supra- and infratentorial sigmoid approach [28]. Infrequently, vertebral artery aneurysms appear at the junction of both vertebral arteries.

The basilar artery runs straight cranially and gives rise to the anterior inferior cerebellar (AICA) artery, the superior cerebellar artery (SCA), multiple pontine branches, and perforating arteries.

Aneurysms at the AICA are extremely rare, accounting for less than 1% of all intracranial aneurysms [29].

Aneurysms of the SCA account for less than 5% of all intracranial aneurysms. Most of them are surgically treated using a trans-Sylvian approach, either using a pterional or an orbitozygomatic craniotomy [30]. In case of a low origin of the SCA from the basilar artery, a subtemporal transtentorial approach is suggested [28].

Basilar top aneurysms can project superiorly, anteriorly, or posteriorly. The superior dome projection is most common and these aneurysms hide deep perforators behind the distal neck. Anterior dome projection is the least common and they hide the contralateral PCA and SCA. The most difficult are the posterior projections that hide the thalamoperforators by displacing them posteriorly and adhering them to the back wall [31]. For clipping of basilar top aneurysms, the relation of the aneurysm neck to the posterior clinoidal process must be considered. If the aneurysm neck is located lower than the posterior clinoidal process, a subtemporal approach may be used to improve visualization of the aneurysms' neck [32]. However, the contralateral PCA is obscured during this procedure and retraction of the temporal lobe is necessary, which might cause brain damage. A subtemporal approach might also be necessary in aneurysms, which are >5–10 mm posterior to the clivus [33]. In basilar top aneurysms with the neck >10 mm below the posterior clinoidal process, a transpetrosal approach might be necessary [33].

For basilar top aneurysms above the posterior clinoidal process, a trans-Sylvian approach is preferred. To visualize the aneurysms' neck during this approach extensive drilling of the anterior and posterior clinoid process and the dorsum sellae might be necessary. In case of a very high basilar tip (>15 mm above the posterior clinoidal process), an orbitozygomatic approach might be necessary [33].

PCA aneurysms are very rare and most of these occur in the proximal P1–P2 segment [28]. Up to 50% of PCA aneurysms are presented as large or giant aneurysms [34]. P1 and P1–P2 aneurysms are usually managed microsurgically via a pterional approach. This approach provides early proximal control and visualization of the basilar apex, posterior communicating artery, and both P1 segments. The subtemporal approach might also be used in proximal PCA aneurysms, which facilitate the identification of the thalamoperforating and perforating arteries of the P1 segment [35]. For more distally located PCA aneurysms of the P2 and P2–P3 aneurysms, a supracerebellar transtentorial approach may be chosen [34].

In complex or fusiform unclippable aneurysms, parent artery occlusion might be necessary. Although good collateralization of the PCA territory is present in some cases, combining these procedures with an extracranial–intracranial bypass is advisable to avoid ischemic complications [34].

13.4 Microsurgical Strategy

Microsurgery of aneurysms is the most definitive way of securing the aneurysm and has better long-term protection than endovascular treatment. However, it is a technically challenging operation and needs adequate experience by the neurosurgeon to have an optimal outcome and prevent complications [36]. Cantore et al. in their study concluded that surgical clipping remains the gold standard for GIAs and if this fails or is contraindicated the high-flow extracranial–intracranial bypass (ECIC) is a viable surgical option. Ensuring adequate flow in the vessels at risk and perforators is the key to success [37]. Sughrue et al. proposed indirect aneurysm occlusion (proximal occlusion, distal occlusion, or trapping) with or without a bypass as a more acceptable alternative than hypothermic cardiac arrest [5]. Campos et al. in their recent article recommended a multimodal approach [38]. Results of our experience with the microsurgery of GIAs justify its continuing use in GIAs [39].

13.4.1 Surgical Management

Decision-making depends on factors like aneurysm location, morphology, size, presenting clinical and neurological conditions, comorbidities, predicted treatment risk, preference of patient and family members, and the preference of the multidisciplinary team. Direct clipping is the preferred surgical technique for aneurysm occlusion. Novel clipping techniques should be employed to satisfactorily occlude most aneurysms (Fig. 13.2).

The pterional approach is the workhorse for anterior circulation aneurysms and an orbitozygomatic (OZ) extension should be added when required to reduce retraction. The preferred approaches for posterior circulation aneurysms are the OZ, trans-petrosal, and far-lateral for upper basilar, basilar trunk, and vertebral-posterior inferior cerebellar artery (PICA) aneurysms, respectively. Cervical carotid artery exposure for proximal control and clinoidectomy is

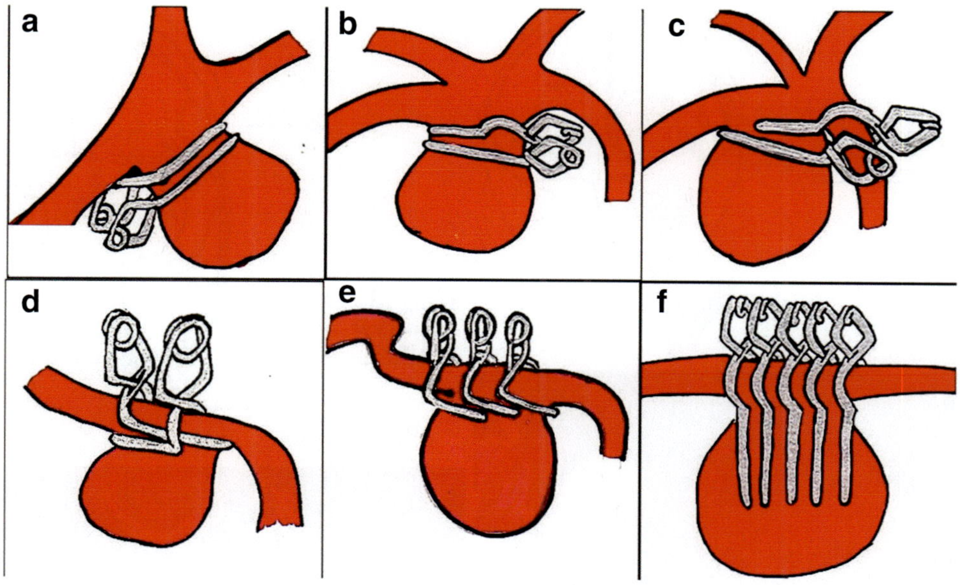

Fig. 13.2 Clipping techniques: (**a**) Clipping at the neck with two large clips, (**b**) clipping at the neck with a combination of fenestrated and nonfenestrated clips, (**c**) clipping with a branch in the fenestrated clip, (**d**) clipping with a fenestrated clip with clip blades parallel to the parent artery "face to face," (**e**) tandem stacking: clipping with fenestrated clips in series with clip blades on the aneurysm parallel to the parent artery and reconstruction of the parent vessel. (**f**) Vertical stacking: clipping with fenestrated clips with clip blades perpendicular to the parent artery

performed for ophthalmic segment aneurysms. High-flow bypass (saphenous vein/radial artery) is used in patients who failed balloon test occlusion (BTO) immediately, and low-flow bypass (STA-MCA) is usually used in patients who failed BTO after hypotensive challenge (lowering mean arterial pressure with nitroprusside drip by 20 mmHg or 20% of the mean arterial pressure, whichever was greater). A bypass helps in treating GIAs which otherwise were deemed impossible to clip. Post-operative angiography should be done routinely in post-clipping and post-flow-diversion procedures.

13.4.2 General Intraoperative Measures During Aneurysm Surgery

The useful intraoperative adjuncts routinely used by the authors are mild hypothermia, minimal or no brain retraction, temporary clipping, mild hypertension during temporary clipping, propofol-induced EEG burst suppression, and intraoperative ICG angiography [36, 40–42]. A lax brain reducing the need for a retractor can be obtained in most cases by employing skull base approaches and wide opening of the subarachnoid cisterns. Lumbar drain (especially in posterior circulation aneurysms) and ventricular drains (in case of hydrocephalus) are useful second lines of defence against a tight brain. The authors employ diuretics only if the above measures fail to achieve a lax brain.

Representative Case 1 A 25-year-old gentleman had presented with coma from an acute subdural haematoma (SDH) in an outside facility and underwent decompressive craniotomy. He was referred to us for definitive treatment after recovery. He was found to have a 77 mm partially thrombosed aneurysm involving the whole of right ICA from the petrous to the supraclinoid segment (the ruptured aneurysm) and a large wide-necked unruptured basilar apex aneurysm. He underwent decompression of the ICA aneurysm mass through the middle cranial fossa, had ECA-M2 bypass with RAG, and trapping of ICA (just after bifurcation in the neck and clipped intracranially proximal to the ophthalmic artery). The recovery was uneventful with good bypass patency and he was referred for endovascular therapy (EVT) of the basilar apex aneurysm. However, the EVT failed and in the second stage, after 3 months, he underwent fronto-temporo-orbito-zygomatic (FTOZ) craniotomy, extradural anterior clinoidectomy, intradural posterior clinoidectomy, and clipping of the basilar apex aneurysm. The patient had no postoperative deficit and DSA confirmed parent ECIC bypass and complete occlusion of both the aneurysms (Fig. 13.3).

Representative Case 2 A 35-year-old patient presented to us with progressive quadriparesis, gait ataxia of 6 months duration, and requiring support for daily activities. MRI/MRA brain and 3D cerebral DSA revealed a heavily thrombosed giant basilar top aneurysm. He underwent FTOZ craniotomy, extradural anterior clinoidectomy, intradural posterior clinoidectomy, and primary clipping of the aneurysm at the neck with two large multicurved Yasargil aneurysm clips. The aneurysm neck was free of thrombus and hence could be clipped without decompression. ICG angiography demonstrated complete occlusion of the aneurysm and preservation of normal branches and perforators. Postoperative third nerve paresis improved completely over a couple of months and the patient had no deficit at 3 months follow-up and check DSA showed the complete obliteration of the aneurysm and preservation of normal circulation (Fig. 13.4).

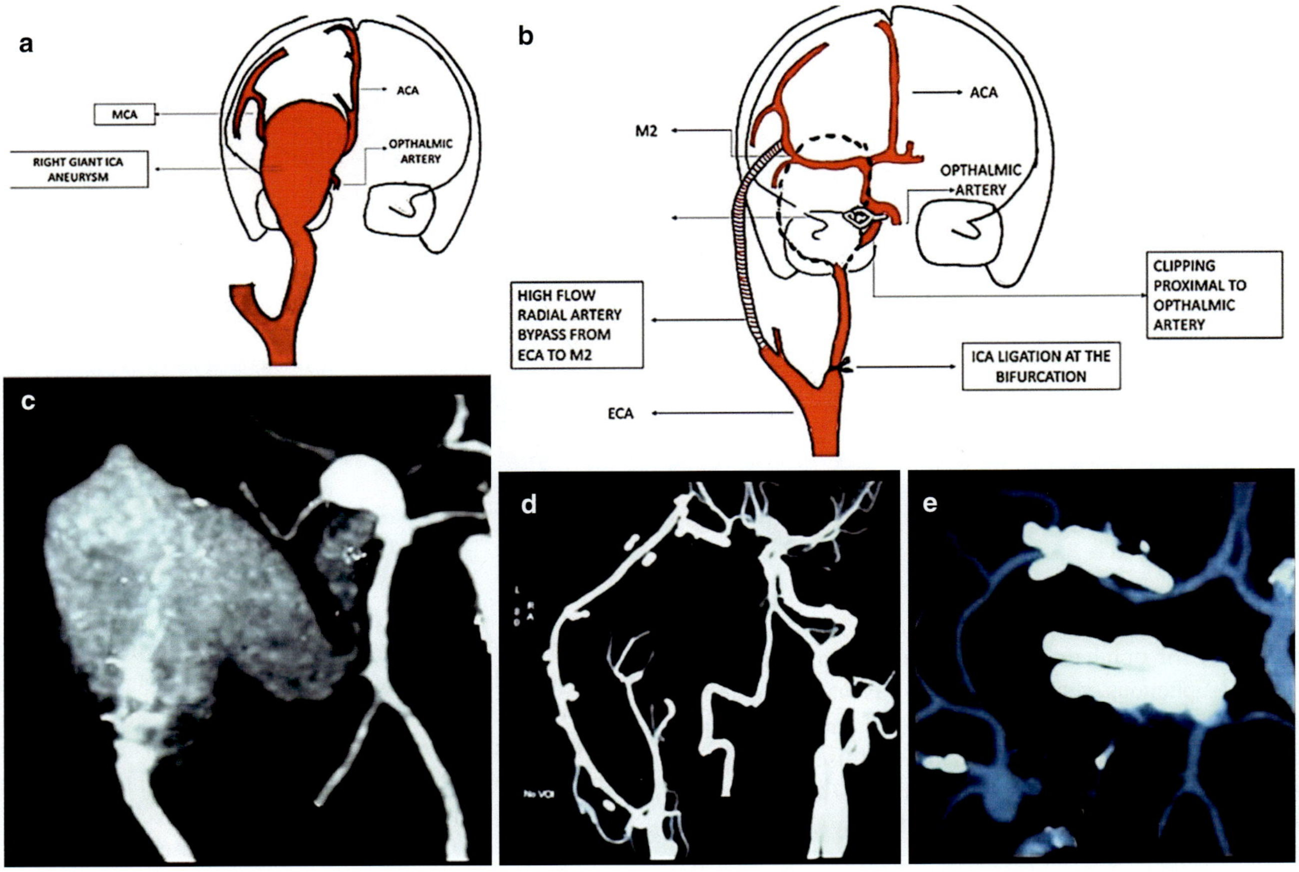

Fig. 13.3 (**a** and **c**) Diagram and CTA: a giant right ICA aneurysm and a large basilar apex aneurysm. (**d**) Postoperative CTA after the first stage. (**b** and **e**) Diagram and postoperative CTA post clipping basilar top aneurysm

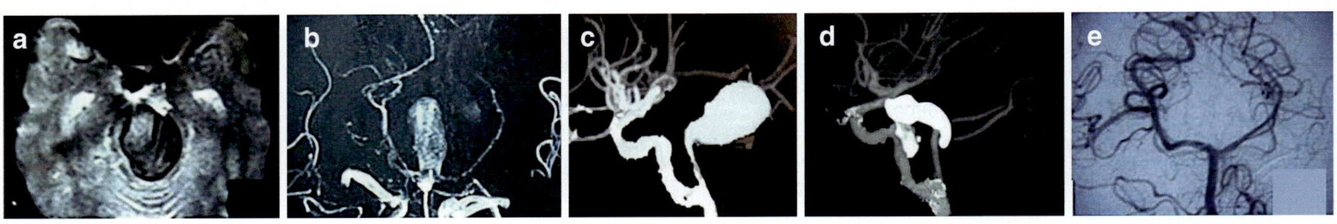

Fig. 13.4 (**a–c**) MRI, MRA, and CTA: A giant partially thrombosed basilar top aneurysm. (**d** and **e**) Post clipping CTA and DSA showing the complete occlusion of the aneurysm and preservation of parent vessels

13.5 Surgical Pearls

- The following intraoperative adjuncts are routinely used by surgeons to aid in aneurysm microsurgery: mild hypothermia, minimal or no brain retraction, temporary clipping, mild hypertension during temporary clipping, propofol-induced EEG burst suppression, and intraoperative ICG angiography.
- The pterional approach is the workhorse for anterior circulation aneurysms.
- Cervical carotid artery exposure for proximal control and clinoidectomy are performed for aneurysms in the ophthalmic segment.
- For upper basilar, basilar trunk, and vertebral-PICA aneurysms, surgical approaches such as the orbitozygomatic, trans-petrosal, and far-lateral approaches may be employed.
- In cases where parent artery occlusion or aneurysm trapping is necessary, high-flow bypass procedures utilizing saphenous vein or radial artery grafts may be considered if the balloon test occlusion (BTO) fails immediately. For patients who fail the BTO after a hypotensive challenge, low-flow bypass procedures using the STA-MCA bypass may be a preferable option.

13.6 Discussion

The goal of the treatment of GIAs is to exclude the lesion from the circulation while preserving function in the neural tissues. Treatment of GIA is a challenge given the giant size, partial thrombosis, calcification, and the typical presence of a wide neck. GIAs and large (20–24 mm) aneurysms have a dismal natural history, with neurological deterioration or death seen in 65–100% of cases over 2 years and hence need treatment [43, 44].

GIAs present many surgical challenges. The aneurysm sac of >15 mm in diameter in size blocks access to afferent and efferent vessels preventing dissection and gaining control of the parent artery [45]. Puncture followed by suction or suction decompression through cervical ICA softens the aneurysm facilitating access. The authors have also utilized adenosine-induced transient cardiac arrest to help dissection. Transient trapping of the parent artery and decompression of thrombosed aneurysm by ultrasonic surgical aspirator help, though not always. Protective low-flow bypasses were performed to prevent ischemic injury when prolonged trapping was anticipated. Previously coiled aneurysms prevent mobilization and remodeling of the sac by their resistance. Partial extraction of coils before clipping as well as novel multiclipping techniques help in complete occlusion. Excision of the residual neck close to the coils to completely isolate them and reconstruct the neck has also been advocated [46].

The postoperative outcome is directly proportional to the preoperative mRS score at presentation [47, 48]. Patients who become symptomatic with SAH and mass effect have the worst surgical outcomes [49]. Younger patients with GIAs seem to have a better prognosis [47]. In patients older than 70 years, the risk of surgery exceeded the lifetime risk of rupture [50]. Those patients presenting with TIAs, poor baseline condition, aneurysm ≥25 mm, and age ≥ 50 years have been shown to have a poorer outcome, 75% of poor outcomes were due to postoperative strokes [51]. The reported surgical morbidity and mortality rates are in the range of 20–30% [5].

13.7 Conclusion

The treatment of GIAs poses a formidable challenge. Yet, these need treatment because of their dismal natural history. Any decision to intervene should carefully consider the various options like microsurgery, EVT, and observation. The treatment should be individualized taking all the patient factors, pathological factors, and expertise available. Any intervention should be expected to better the natural history. EVT is an attractive option. Unfortunately, the high incidence of incomplete treatment, delayed complication, recurrence, and inadequate long-term follow-up data still make microsurgery the treatment of choice in most GIAs. However, the treatment of GIA is a major endeavor and should not be taken up by the occasional aneurysm surgeon. Committed groups in specialized centers with all available resources are best suited to tackle these life-threatening GIAs.

References

1. Lonjon M, Pennes F, Sedat J, Bataille B. Epidemiology, genetic, natural history and clinical presentation of giant cerebral aneurysms. Neurochirurgie. 2015;61(6):361–5.
2. Wiebers DO. Unruptured intracranial aneurysms: natural history, clinical outcome, and risks of surgical and endovascular treatment. Lancet. 2003;362(9378):103–10.
3. The UCAS Japan Investigators. The natural course of unruptured cerebral aneurysms in a Japanese Cohort. N Engl J Med. 2012;366(26):2474–82.
4. Bull J. Massive aneurysms at the base of the brain. Brain. 1969;92(3):535–70.
5. Sughrue ME, Saloner D, Rayz VL, Lawton MT. Giant intracranial aneurysms: evolution of management in a contemporary surgical series. Neurosurgery. 2011;69(6):1261–70. discussion 1270–1271
6. Vishteh A, David C, Spetzler R. Giant aneurysms. In: Sekhar LN, Fessler R, editors. Atlas of neurosurgical techniques, Atlas of neurosurgical techniques, vol. I. Thieme; 2016.
7. Matsuda W, Sonomura T, Honma S, Ohno S, Goto T, Hirai S, et al. Anatomical variations of the recurrent artery of Heubner: number, origin, and course. Anat Sci Int. 2018;93(3):317–22.
8. Rhoton AL. The supratentorial arteries. Neurosurgery. 2002;51(Suppl. 4):S1-53–S1-120.
9. Huang J, Germanwala A, Tamargo R. Anterior communicating artery aneurysms. In: Youmans JR, editor. Neurological surgery. 6th ed. Philadelphia, PA: Elsevier/Saunders; 2011. p. 3841–52.
10. Schuette AJ, Barrow DL, Cohen-Gadol AA. Strategies to minimize complications during intraoperative aneurysmal hemorrhage: a personal experience. World Neurosurg. 2015;83(4):620–6.
11. Valli D, Zhao X, Belykh E, Sun Q, Lawton MT, Preul MC. Partial Gyrus rectus resection as a technique to improve the exposure to the anterior communicating artery complex through the junctional triangle: a quantitative study. J Neurol Surg B Skull Base. 2021;82(Suppl. 3):e211–6.
12. Lehecka M, Lehto H, Niemelä M, Juvela S, Dashti R, Koivisto T, et al. Distal anterior cerebral artery aneurysms: treatment and outcome analysis of 501 patients. Neurosurgery. 2008;62(3):590–601. discussion 590–601
13. Lehecka M, Dashti R, Lehto H, Kivisaari R, Niemelä M, Hernesniemi J. Distal anterior cerebral artery aneurysms. Acta Neurochir Suppl. 2010;107:15–26.
14. Rosner SS, Rhoton AL, Ono M, Barry M. Microsurgical anatomy of the anterior perforating arteries. J Neurosurg. 1984;61(3):468–85.
15. Rinne J, Hernesniemi J, Niskanen M, Vapalahti M. Analysis of 561 patients with 690 middle cerebral artery aneurysms: anatomic and clinical features as correlated to management outcome. Neurosurgery. 1996;38(1):2–11.
16. Liu JK, Gottfried ON, Amini A, Couldwell WT. Aneurysms of the petrous internal carotid artery: anatomy, origins, and treatment. Neurosurg Focus. 2004;17(5):E13.
17. Eddleman CS, Hurley MC, Bendok BR, Batjer HH. Cavernous carotid aneurysms: to treat or not to treat? Neurosurg Focus. 2009;26(5):E4.
18. Fehrenbach MK, Dietel E, Wende T, Kasper J, Sander C, Wilhelmy F, et al. Management of cavernous carotid artery aneurysms: a retrospective single-center experience. Brain Sci. 2022;12(3):330.

19. Misra BK, Khandhar AV, Pradhan R. Bilateral blindness following flow diverter placement in a giant thrombosed cavernous carotid aneurysm: illustrative case. J Neurosurg Case Lessons. 2022;4(24):CASE22352.
20. Rhoton AL. The cavernous sinus, the cavernous venous plexus, and the carotid collar. Neurosurgery. 2002;51(Suppl. 4):S1-375–410.
21. Knosp E, Müller G, Perneczky A. The paraclinoid carotid artery: anatomical aspects of a microneurosurgical approach. Neurosurgery. 1988;22(5):896–901.
22. Day AL. Aneurysms of the ophthalmic segment. A clinical and anatomical analysis. J Neurosurg. 1990;72(5):677–91.
23. Lawton M. Posterior communicating artery aneurysms. In: Seven Aneurysms JR, editor. Tenets and techniques of clipping. New York, NY: Thieme; 2011. p. 45–64.
24. Peerless S, Drake C. Management of aneurysms of the posterior circulation. In: Youmans JR, editor. Neurological surgery: a comprehensive guide to the diagnosis and management of neurosurgical problems. Philadelphia: WB Saunders; 1990. p. 1764–806.
25. Yamaura A, Ise H, Makino H. Treatment of aneurysms arising from the terminal portion of the basilar artery—with special reference to the radiometric study and accessibility of trans-Sylvian approach. Neurol Med Chir (Tokyo). 1982;22(7):521–32.
26. Gailloud P. The segmentation of the vertebral artery: an ambiguous anatomical concept. Interv Neuroradiol. 2022;28(6):765–72.
27. Lister JR, Rhoton AL, Matsushima T, Peace DA. Microsurgical anatomy of the posterior inferior cerebellar artery. Neurosurgery. 1982;10(2):170–99.
28. Rhoton AL. Aneurysms Neurosurg. 2002;51(Suppl. 4):S121–58.
29. Gonzalez LF, Alexander MJ, McDougall CG, Spetzler RF. Anteroinferior cerebellar artery aneurysms: surgical approaches and outcomes—a review of 34 cases. Neurosurgery. 2004;55(5):1025–35.
30. Rodríguez-Hernández A, Walcott BP, Birk H, Lawton MT. The superior cerebellar artery aneurysm: a posterior circulation aneurysm with favorable microsurgical outcomes. Neurosurgery. 2017;80(6):908–16.
31. Lawton M. Basilar artery bifurcation aneurysms. In: Seven aneurysms: tenets and techniques of clipping. New York, NY: Thieme; 2011.
32. Hernesniemi J, Ishii K, Niemelä M, Kivipelto L, Fujiki M, Shen H. Subtemporal approach to basilar bifurcation aneurysms: advanced technique and clinical experience. Acta Neurochir Suppl. 2005;94:31–8.
33. Spiessberger A, Strange F, Fandino J, Marbacher S. Microsurgical Clipping of basilar apex aneurysms: a systematic historical review of approaches and their results. World Neurosurg. 2018;114:305–16.
34. Yonekawa Y, Roth P, Fandino J, Landolt H. Aneurysms of the posterior cerebral artery and approach selection in their microsurgical treatment: emphasis on the approaches: SAHEA and SCTTA. Acta Neurochir Suppl. 2011;112:85–92.
35. Taylor CL, Kopitnik TA, Samson DS, Purdy PD. Treatment and outcome in 30 patients with posterior cerebral artery aneurysms. J Neurosurg. 2003;99(1):15–22.
36. Misra B, Warade A. Complications of intracranial aneurysm microsurgery. In: Singh VP, Nair MD, editors. Progress in clinical neuroscience, vol. 30. Thieme; 2015.
37. Cantore G, Santoro A, Guidetti G, Delfinis CP, Colonnese C, Passacantilli E. Surgical treatment of giant intracranial aneurysms: current viewpoint. Neurosurgery. 2008;63(4 Suppl. 2):279–89. discussion 289–290
38. Campos JK, Ball BZ, Cheaney Ii B, Sweidan AJ, Hasjim BJ, Hsu FPK, et al. Multimodal management of giant cerebral aneurysms: review of literature and case presentation. Stroke Vasc Neurol. 2020;5(1):22–8.
39. Misra BK, Warade AG, Rohan R, Sarit S. Microsurgery of giant intracranial aneurysm: a single institution outcome study. Neurol India. 2021;69(4):984–90.
40. Misra B. Posterior circulation aneurysm: surgical approach. Prog Clin Neurosci. 2000;15:83–9.
41. Misra B, Thorat J, Ashok P, Udani V. Cerebral revascularization: our experience. Prog Clin Neurosci. 2005;20:133–52.
42. Misra B, Purandare H, Warade A. Does microscope integrated near-infrared indocyanine green videoangiography benefit in aneurysm surgery? An initial experience. Pan Arab J Neurosurg. 2013;17(2):54–67.
43. Drake CG. Progress in cerebrovascular disease. Management of cerebral aneurysm. Stroke. 1981;12(3):273–83.
44. Steinberg GK, Drake CG, Peerless SJ. Deliberate basilar or vertebral artery occlusion in the treatment of intracranial aneurysms. Immediate results and long-term outcome in 201 patients. J Neurosurg. 1993;79(2):161–73.
45. Lejeune JP, Thines L, Proust F, Riegel B, Koussa M, Decoene C. Selective microsurgical treatment of giant intracranial aneurysms. Neurochirurgie. 2016;62(1):30–7.
46. Lejeune JP, Thines L, Taschner C, Bourgeois P, Henon H, Leclerc X. Neurosurgical treatment for aneurysm remnants or recurrences after coil occlusion. Neurosurgery. 2008;63(4):684–91. discussion 691–692
47. Darsaut TE, Darsaut NM, Chang SD, Silverberg GD, Shuer LM, Tian L, et al. Predictors of clinical and angiographic outcome after surgical or endovascular therapy of very large and giant intracranial aneurysms. Neurosurgery. 2011;68(4):903–15. discussion 915
48. Hauck EF, Wohlfeld B, Welch BG, White JA, Samson D. Clipping of very large or giant unruptured intracranial aneurysms in the anterior circulation: an outcome study. J Neurosurg. 2008;109(6):1012–8.
49. Sorteberg A, Bakke SJ, Boysen M, Sorteberg W. Angiographic balloon test occlusion and therapeutic sacrifice of major arteries to the brain. Neurosurgery. 2008;63(4):651–60. discussion 660–661
50. Chang HS. Simulation of the natural history of cerebral aneurysms based on data from the International Study of Unruptured Intracranial Aneurysms. J Neurosurg. 2006;104(2):188–94.
51. Morley TP, Barr HW. Giant intracranial aneurysms: diagnosis, course, and management. Clin Neurosurg. 1969;16:73–94.

Surgical Nuances for Giant Cerebral Aneurysms

14

Soichi Oya

Learning Objectives

- Learn to recognize the differences between typical and giant cerebral aneurysms.
- Understand the effectiveness and limitations of bypass surgery, which plays a vital role in direct surgery for massive cerebral aneurysms.
- Understand the surgical techniques, instruments, and intraoperative monitoring required for giant cerebral aneurysm surgery.

14.1 Introduction

A "giant cerebral aneurysm" is defined as an aneurysm with a maximum diameter of 25 mm measured using angiography in the absence of thrombosis. However, in the presence of thrombosis, the diameter is usually measured using computed tomography (CT) or magnetic resonance (MR) imaging because angiography can only measure the diameter of the aneurysm lumen.

Giant cerebral aneurysm is a general term for heterogeneous lesions with respect to morphology and pathogenesis. Morphologically, aneurysms are classified into three types: (1) saccular, (2) fusiform, and (3) serpentine. Regarding the mechanism of occurrence, saccular aneurysms enlarge mainly due to hemodynamic mechanisms, whereas fusiform and serpentine aneurysms are associated with more complicated pathological conditions such as arteriosclerosis and thrombosis within the aneurysm. Giant cerebral aneurysms have a poor prognosis and an extremely high treatment risk. It is essential to understand the different mechanisms by which they grow and develop, and to address them with all the currently available surgical and medical treatment tools to ensure safe and effective management. Direct surgery for giant cerebral aneurysms includes the direct elimination of the aneurysm from normal hemodynamics and flow alteration in which spontaneous occlusion is expected without touching the aneurysm itself. This chapter focuses on the surgical treatment of giant cerebral aneurysms, particularly on the principles and surgical techniques for the direct elimination of aneurysms.

14.2 History/Background

Before 1970, giant aneurysms were often encountered as symptomatic intracranial masses in suspected neoplastic lesions, and their surgical outcomes were disastrous. However, the outcomes have improved with the development of skull base surgery and vascular surgery techniques, pioneered mostly by Drake and colleagues. In recent years, analyses of the etiology and hemodynamics of giant aneurysms and the development of endovascular treatment devices have made it possible to control most giant aneurysms.

In particular, treatment outcomes of giant cerebral aneurysms arising in the anterior circulation have improved dramatically. In addition to the improved surgical outcomes of craniotomies, the recent advent of flow divertor (FD) has enabled less invasive treatment of cavernous internal carotid artery (ICA) aneurysms and has become the mainstay treatment for this type of aneurysm. FD has also been widely used for intracranial ICA giant aneurysms, although the occlusion rate is lower than that of large or small aneurysms arising in this region. In addition, the development and widespread use of bypass techniques have increased the number of cases of giant aneurysms that can be safely

S. Oya (✉)
Department of Neurosurgery, Gunma University Graduate School of Medicine, Maebashi, Gunma, Japan
e-mail: sooya-gnm@gunma-u.ac.jp

I. N. Kanaan, V. Beneš (eds.), *Neuroanatomy Guidance to Successful Neurosurgical Interventions*,
https://doi.org/10.1007/978-3-031-59838-8_14

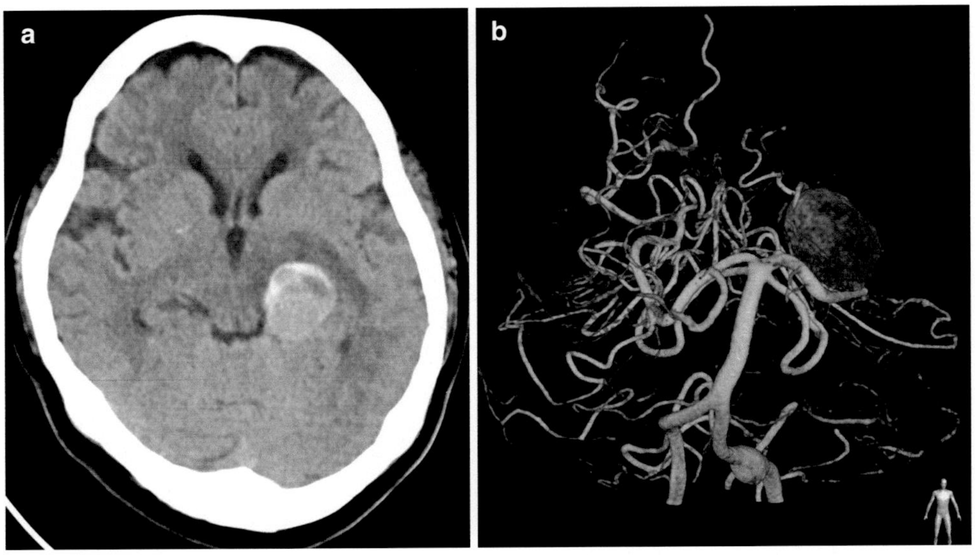

Fig. 14.1 (**a**) An axial CT scan showing a localized thin hemorrhage around the giant cerebral aneurysm. (**b**) A 3D rotational angiogram demonstrating a left giant P1P2 aneurysm

trapped. The development of instruments, such as intraoperative neuromonitoring and intraoperative blood-flow assessment methods, has supported this technological advancement in craniotomy. In contrast, giant aneurysms in the posterior circulation are frequently spindle-shaped and tortuous, unlike relatively simple saccular aneurysms at bifurcations. Although saccular aneurysms are thought to enlarge mainly due to hemodynamic stress, giant aneurysms in the posterior circulation increase due to the complex involvement of arteriosclerosis, arterial dissection, thrombus in the aneurysm, feeding vessels in the aneurysm wall, and other factors. In addition, because the perforating branches that feed the brain stem often branch from the aneurysm itself, occlusion of the aneurysm carries a high risk of ischemic complications. Recent advances in devices for endovascular treatment are expected to occlude the aneurysm and preserve the critical perforating branch. However, it remains unclear whether the effect of FD on the parent vessel prevents growth of the aneurysm. A previous study suggested that the vasa vasorum, a feeding vessel in the aneurysm wall, is involved in the growth of partially thrombosed giant cerebral aneurysms [1].

Giant cerebral aneurysms account for approximately 5% of the total and are slightly more common in women (60–70%) than in men. Aneurysms in the anterior circulation are often saccular; however, spindle-shaped and tortuous aneurysms have also been observed. By contrast, giant aneurysms of the posterior circulation often take the form of the vertebral or basilar artery itself, transforming into aneurysms. Therefore, surgical treatment of these aneurysms carries a high risk of perforating artery infarction around the aneurysm, even if the flow of the main arteries distal to the aneurysm is secured by vascular reconstruction. These anatomical features significantly increase the risks of surgical occlusion of the aneurysms.

As is the case with typical smaller aneurysms, giant cerebral aneurysms can cause subarachnoid or intracerebral hemorrhage due to rupture; however, the rupture rate of giant aneurysms is exceedingly high. The UCAS data from Japan show an annual rupture rate of 33% for giant cerebral aneurysms [2]. In giant aneurysms, the distribution of the hemorrhage may result in a localized thin hemorrhage around the giant cerebral aneurysm instead of a diffuse subarachnoid hemorrhage (Fig. 14.1). This condition may be related to narrowing of the cistern due to compression by the aneurysm or reduced perfusion pressure in the aneurysm. When such CT findings are observed, acute bleeding should be suspected and prompt treatment should be considered.

14.3 Anatomical Highlights

- For ICA giant aneurysms, preserve the anterior choroidal artery.
- For ICA giant aneurysms, p-com can be safely sectioned with preservation of the anterior thalamoperforating artery.
- For ICA giant aneurysms, manipulate the optic nerve and the oculomotor nerve carefully.
- For middle cerebral artery (MCA) giant aneurysms, beware of the origin of the lenticulostriate arteries.

- For anterior cerebral artery (ACA) giant aneurysms, preserve Heubner artery and hypothalamic artery.
- For giant aneurysms in the posterior circulation, respect the anterior spinal artery and perforators from the basilar trunk and basilar top.

It is well known that giant cerebral aneurysms can compress the surrounding brain and cranial nerves, causing symptoms of neurological deficits. These symptoms should be relieved quickly by blocking blood flow into the aneurysm and, if possible, by resection of the aneurysm. However, unlike small to large aneurysms that present with compression findings, adhesion of cranial nerves or perforating branches to the aneurysm can frequently be too severe to detach; therefore, it may be appropriate to block blood flow into the aneurysm without resection in some cases. This feature is particularly prominent in aneurysms of the posterior cranial fossa; perforating branches to critical areas such as the brainstem and thalamus often branch off from the aneurysm itself, making immediate closure of the aneurysm impossible. Thus, when considering treatment, it is necessary to understand the anatomy of the aneurysm and accurately assess the treatment risks.

The important blood vessels and nerves that should be identified in the preoperative imaging study and preserved intraoperatively are listed above. The anatomical and procedural aspects of the aneurysm are discussed in detail in the following section on approach/surgical intervention.

14.4 Approach/Surgical Intervention

14.4.1 General Concept of Approach and Direct Surgical Intervention

It must be recognized that the surgical treatment of giant aneurysms is quite different from that of small or large (<25 mm) aneurysms. First, a temporary blockage of the parent artery is required in almost all cases, even when aiming for clip closure of the aneurysm neck. Bypassing the blood flow distal to the aneurysm and intraoperative motor-evoked potential (MEP) and somatosensory evoked potential (SEP) monitoring are essential for a safe and sufficiently long temporary blockage of blood flow. If it is difficult to close the aneurysm wall directly using a clip, the feasibility and effectiveness of trapping the parent artery or proximal ligation should be considered. In surgeries for giant cerebral aneurysms, it is often optimal not to attempt to manipulate them.

The craniotomy should be as large as possible, and the operative field should be prepared for any hemorrhagic or ischemic problems that may occur intraoperatively. Even if a bypass is not planned, the superficial temporal artery (STA) and occipital artery should be preserved during skin incision and craniotomy for possible bypass. Another option is to establish an operative field to harvest the radial artery (RA).

In most cases, thrombus removal is required for thrombosed giant aneurysms. Various thrombectomy instruments, including the cavitron ultrasonic surgical aspirator (CUSA), should be prepared in advance to swiftly remove the thrombus for aneurysm obliteration. If thrombus removal or neck clipping is difficult, even if the thrombus is removed, proximal ligation or trapping after securing the blood flow distal to the aneurysm with a bypass may be a practical approach. However, trapping is preferred to ensure hemostasis in cases of ruptured aneurysms.

The surgical procedures for giant cerebral aneurysms differ greatly depending on the aneurysm site and are summarized in the next section.

14.4.2 Giant ICA Aneurysms

14.4.2.1 High-Flow Bypass Using RA Graft/Saphenous Vein Graft

Simple proximal ligation is often performed for giant ICA aneurysms but ischemic complications are problematic. Subsequently, proximal ligation is often performed with either extracranial (EC) or intracranial (IC) bypass using the RA graft/saphenous vein graft, or an STA-MCA bypass to reduce ischemic complications. In recent years, endovascular treatment of flow divertor (FD) has become the mainstay treatment for giant ICA aneurysms. However, it should be noted that the closure rate of FD decreases with the increase in aneurysm size. Although its minimally invasive nature has made FD treatment increasingly popular, direct surgery for giant ICA aneurysms remains safe and effective in expert hands.

14.4.2.2 Suction Decompression for ICA Aneurysms

Retrograde suction decompression (RSD) or Dallas RSD is effective for treating giant internal carotid artery aneurysms in the intradural space. In this method, the internal carotid artery distal to the aneurysm and carotid artery are trapped. The trapped internal carotid artery is punctured to aspirate blood from the lumen. By reducing the size of a giant cerebral aneurysm, clipping can be performed to form the lumen of the ICA without occluding the surrounding perforating branches. In a recent systematic review of 540 aneurysms in 26 previous reports, approximately 50% of the Dallas RSDs were used for giant cerebral aneurysms [3]. This technique plays an essential role in the treatment of giant cerebral aneurysms of the ICA. Recently, endovascular RSD has become more common. However, although selection bias exists, and simple comparisons are difficult, the efficacy and safety of RSD by direct puncture in the operative field are higher than those of endovascular RSD [3].

14.4.3 Giant MCA Aneurysms

Although giant MCA aneurysms are located in shallow areas that are relatively easy to access, manipulation of the aneurysm can damage the surrounding brain tissue, resulting in postoperative paralysis, aphasia, and epilepsy. Therefore, trapping the aneurysm by reconstructing the distal arteries without touching it may be effective. A previous analysis reported that the blood flow in the perfusion territory of the MCA peripheral to the aneurysm was 50 ± 25 mL/min. The measured blood flow in the STA after an STA-MCA bypass matched exactly 50 ± 25 mL/min [4]. Therefore, if the STA is sufficiently long, it is a safe and useful donor artery for trapping giant MCA aneurysms. However, the problem arises when the length is insufficient. In cases in which the aneurysm is huge (Fig. 14.2a, b), it may be necessary to use the RA (Fig. 14.2d red arrow) as an interposed graft to extend its length as much as possible (Fig. 14.2c, d). In any case, a technique that allows a stable bypass, when necessary, will increase the variety of surgical procedures appropriate for treating giant MCA aneurysms and lead to improved outcomes.

14.4.4 Giant ACA Aneurysms

Although revascularization techniques are essential in the surgery of giant ACA aneurysms, a higher level of skill in revascularization is required in comparison to ICA and MCA aneurysms. The anatomy of the a-com complex is complicated, differs from patient to patient, and requires access to the two proximal A1 and four distal vessels, namely, the bilateral A2 and Heubner arteries. It is difficult to approach these six proximal and distal vessels around a giant aneurysm in the same operative field. Furthermore, STA alone is rarely valid for the revascularization of ACAs, and complicated anastomoses, such as interposition grafts, are required.

14.4.5 Giant Posterior Circulation Aneurysms

The most challenging problem in treating giant aneurysms in the posterior circulation is the occlusion of perforators branching from the aneurysm dome or adjacent parent arteries. Historically, proximal ligation for complex clipping aneurysms of the vertebrobasilar artery system has been successful in approximately 70% of patients after an average follow-up of 9.5 years [5]. However, if the aneurysm is not completely occluded after this procedure, two-thirds of aneurysms rupture. The outcomes of FD are poor when limited to giant vertebrobasilar artery aneurysms, and no data

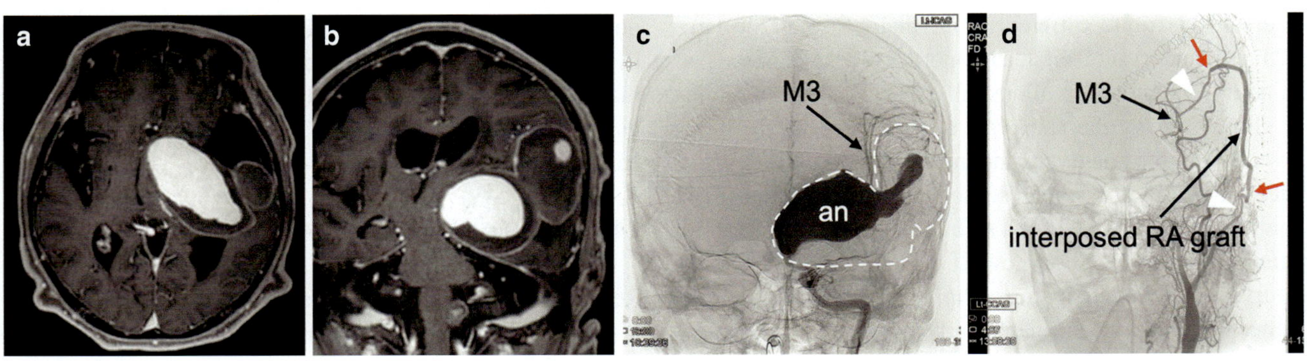

Fig. 14.2 (**a** and **b**): MR images showing a left giant partially thrombosed aneurysm. (**c**) Preoperative internal carotid angiogram revealing distal M3 branches (arrow) coursing behind this huge aneurysm. (**d**) Postoperative angiogram demonstrating successful obliteration of the aneurysm with preservation of distal M3 territory using STA-MCA bypass (arrowheads) with an interposed radial artery graft (red arrows). *an* aneurysm, *RA* radial artery. *Arrowheads* superficial temporal artery, *red arrows* the anastomotic sites of the superficial temporal artery and RA

are available on FD limited to giant aneurysms in the posterior circulation, even in some recent extensive outcome reports. Although flow alternation surgery with revascularization procedures, such as in the STA-superior cerebellar artery and STA-posterior cerebral artery, has been reported to be effective for these formidable aneurysms, methods to stop aneurysm growth and prevent rupture within an acceptable risk range remain to be established.

14.5 Representative Cases

14.5.1 Case 1: Giant MCA Aneurysm

14.5.1.1 History and Imaging Studies

This case illustrates fundamental guidances for clipping giant aneurysms. A 77-year-old female patient with a giant right MCA cerebral aneurysm (Fig. 14.3a, b) presented with

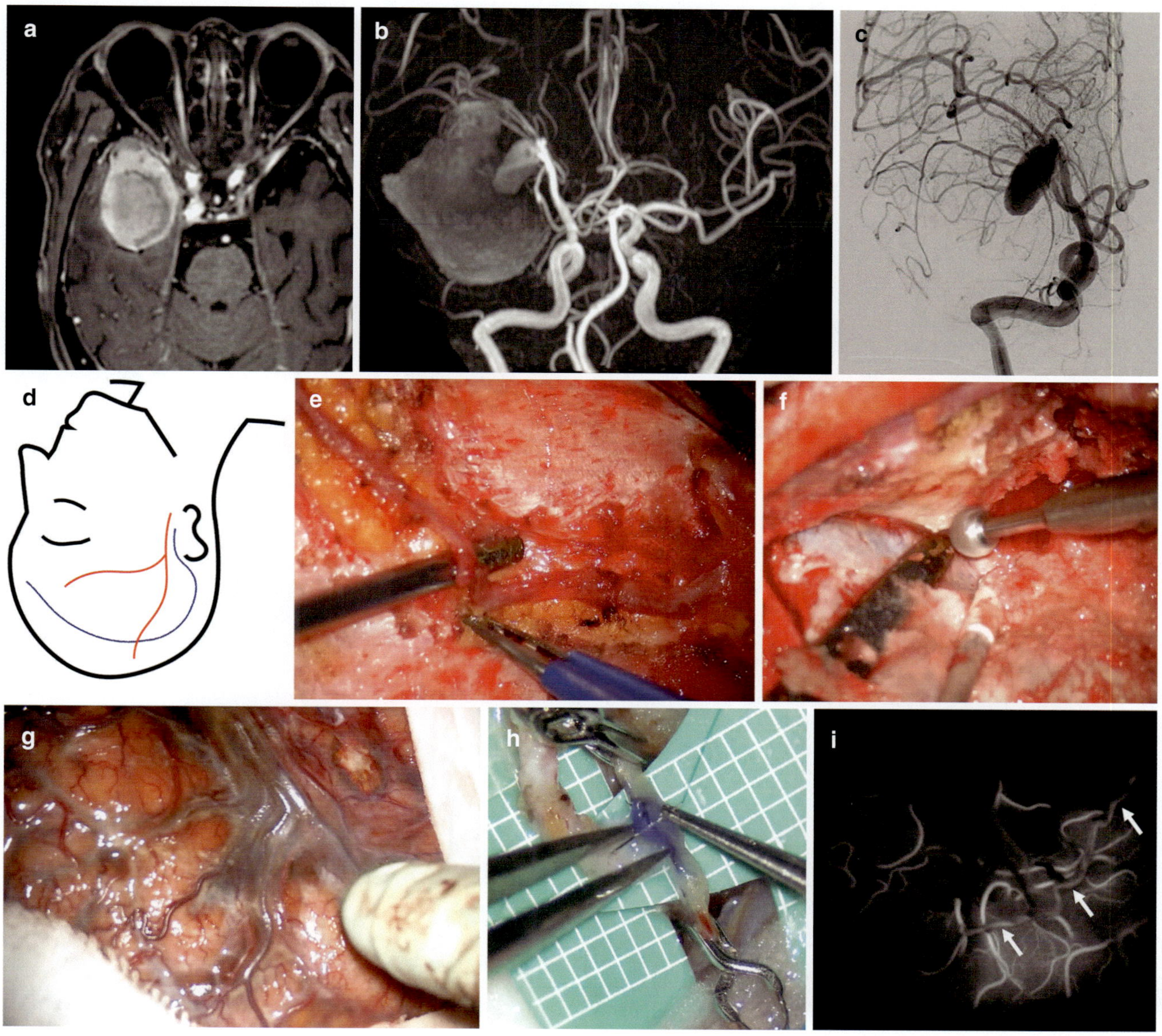

Fig. 14.3 An axial MR image (**a**) and MR angiogram (**b**) demonstrating a giant right MCA cerebral aneurysm with partial thrombosis (**c**). (**d**) A scheme showing a large right frontotemporal craniotomy. (**e–u**) Intraoperative photographs. (**e**) The STA was detached from the back of the skin for bypass. (**f**) The anterior cranial floor was flattened to provide a view of the aneurysm from below and to widen the operative field. (**g**) The central artery was selected as the recipient. (**h**) The STA-MCA anastomosis was conducted using a 10-0 nylon. (**i**) Adequate blood flow at the anastomotic site was confirmed using ICG video angiography. (**j**) The aneurysm was exposed. (**k**) A simulation was conducted to insert a temporary clip near the ICA to assess its feasibility. (**l**) The aneurysm was cut open using a No. 11 blade knife. (**m**) Tumor forceps were also used for thrombectomy. (**n**) After thrombectomy, the aneurysm became mobile. (**o**) The M1 segment and lenticulostriate arteries were identified. (**p**) The M2 superior trunk was identified (arrow) and safely clamped with a temporary clip. (**q**) Intense arterial bleeding started from M1 that opened into the aneurysm. (**r**) A temporary clip was applied to the M1 secured in advance. (**s**) The M2 inferior trunk was also identified behind the aneurysm. (**t**) The excess aneurysm wall was removed. A photograph (**u**) and scheme (**v**) demonstrating the picket fence clipping technique. (**w**) Postoperative angiography showing the complete clipping and good patency of the peripheral MCA. *an* aneurysm, *app* applier; *ica* internal carotid artery, *arrow* lenticulostriate arteries

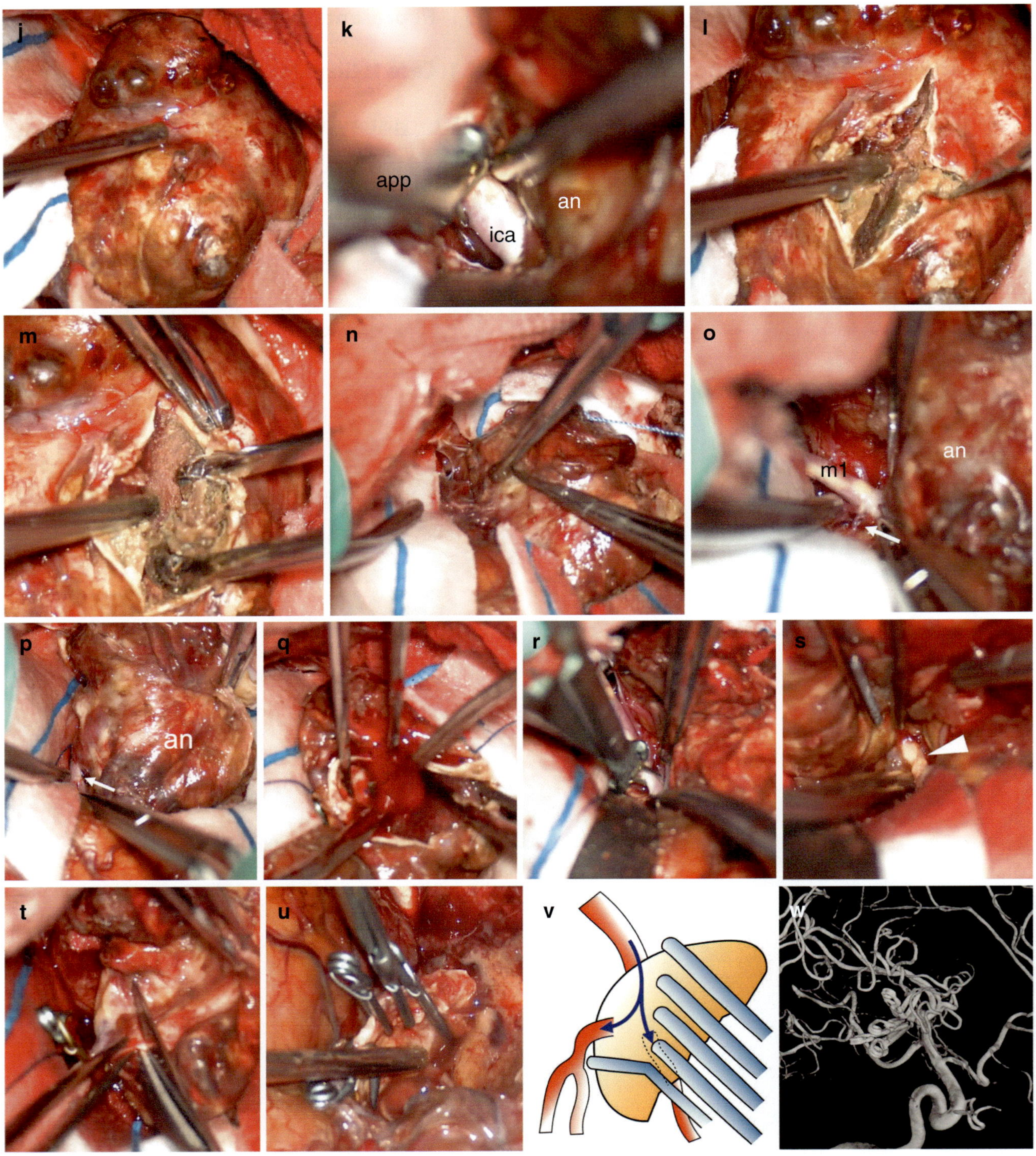

Fig. 14.3 (continued)

a chief complaint of cognitive decline. The aneurysm was 4.5 cm at its greatest diameter, and most of it was thrombosed (Fig. 14.3c). The wall of the aneurysm showed extensive calcification.

14.5.1.2 Strategy

We planned direct closure of the aneurysm with preservation of the M2 superior and inferior branches. A large frontotemporal craniotomy was performed to preserve the STA. To view the entire aneurysm from different angles, the temporal lobe must be widely exposed by reflecting the skin and muscles in two layers. Closure of the aneurysm requires temporary occlusion of the parent artery; however, it is difficult to visualize the proximal right ICA or M1 located behind the aneurysm. The craniotomy was designed to expose a wide frontal area to allow adequate observation of arteries proximal to the aneurysm. Next, an STA-MCA bypass was performed on the cortical branch of the M2 superior trunk to maintain peripheral blood flow to the motor cortex while the aneurysm was trapped.

The bypass patency was confirmed using Doppler ultrasonography and intraoperative indocyanine green-based video angiography (ICG-VA). Intraoperative transcranial MEP monitoring was performed. In addition, large scissors, various dissectors, tumor forceps, scalpels, and the CUSA were prepared for thrombus removal, and a preoperative CT scan showed that the aneurysm wall was thickened and stiffened. Therefore, we obtained multiple clips, including large clips of various lengths and fenestrated clips of various shapes, ready for use in the operating room.

14.5.1.3 Surgical Procedures

1. Craniotomy

   A large right frontotemporal craniotomy was performed (Fig. 14.3d). The STA was detached from the back of the skin for bypass (Fig. 14.3e) and moistened between cotton patties soaked in diluted papaverine hydrochloride. The anterior cranial floor was flattened to provide a view of the aneurysm from below and to widen the angle of insertion of instruments, such as a clip applier (Fig. 14.3f).
2. STA-MCA Bypass

   Ideally, the anastomosis should be placed in the cortical MCA branch, which perfuses the most eloquent area of the surgical field. The central artery was selected as the recipient (Fig. 14.3g). A 10-0 nylon was used for the anastomosis (Fig. 14.3h), and ICG and Doppler sonography confirmed adequate blood flow at the anastomotic site (Fig. 14.3i). We also confirmed that no decrease in MEP amplitude was evident.
3. Exposure of the Aneurysm

   The aneurysm was exposed via a partial corticotomy of the temporal lobe (Fig. 14.3j). The right ICA was identified behind the aneurysm, and a simulation was conducted to insert a temporary clip near the ICA to assess its feasibility. However, the aneurysm was so large that the wall required pressing down to visualize the ICA (Fig. 14.1k). A large anterior craniotomy was necessary to manipulate the aneurysm as gently as possible. At this point, the M1 segment proximal to the aneurysm and the M2 segment distal to the aneurysm were behind the giant aneurysm and could not be observed.
4. Removing the Intraaneurysmal Thrombus and Securing the Parent Arteries

   The parent artery could not be secured without removing the thrombus in the aneurysm; therefore, the aneurysm was cut open using a No. 11 blade knife in a cross-shaped incision (Fig. 14.3l). Tumor forceps were effective in removing the thrombi (Fig. 14.3m). After removing half of the thrombus, the aneurysm finally became mobile, and an anterior aneurysmal structure was approached (Fig. 14.3n).

   The M1 segment and lenticulostriate arteries were identified (Fig. 14.3o). The M2 superior trunk was located (Fig. 14.3p arrow) and safely clamped with a temporary clip because the STA-MCA bypass already perfused the peripheral portion of the M2 trunk. Furthermore, the patient's MEP amplitude was checked once every 3 min using a transcranial MEP. If the amplitude of the MEP decreased, we typically removed the temporary clip until it recovered. In this case, the MEP remained the same. To close the aneurysm, it was necessary to completely remove the thrombus. As further thrombus removal proceeded, intense arterial bleeding began from the orifice of M1 that opened into the aneurysm (Fig. 14.3q). However, when a temporary clip was applied to the previously secured M1 (Fig. 14.3r), the bleeding was well controlled. The M2 inferior trunk was also identified behind the aneurysm (Fig. 14.3s arrowhead), and the aneurysm was trapped by clamping the M2 inferior trunk.
5. *Angioplastic Clipping*

   The excess aneurysm wall was removed because it interfered with clipping (Fig. 14.3t). One tip for making clipping easier and safer is to leave the neck wall large enough to conduct the angioplastic clipping. This is useful because it avoids stenosis of the parent artery during clipping, and the aneurysm wall can also be pinched and held during clipping. To create an ideal closure line, picket fence clipping was performed using five clips (Fig. 14.3u, v). Postoperative angiography confirmed the complete clipping and good patency of the peripheral MCA (Fig. 14.3w). The patient was transferred to a rehabilitation hospital at mRS2 with no new neurological deficits.

14.5.2 Case 2: Giant ACA Aneurysm, the Importance of Various Bypass Techniques

14.5.2.1 History and Imaging Studies

This case emphasizes the importance of mastering different bypass techniques to treat giant ACA aneurysms. The patient was a 70-year-old man with progressive memory disturbances. MR imaging showed a 5-cm markedly thrombosed aneurysm in the left frontal lobe (Fig. 14.4a, b), which was determined to be symptomatic. Angiography revealed an aneurysm originating from the left A1–A2 junction (Fig. 14.4c).

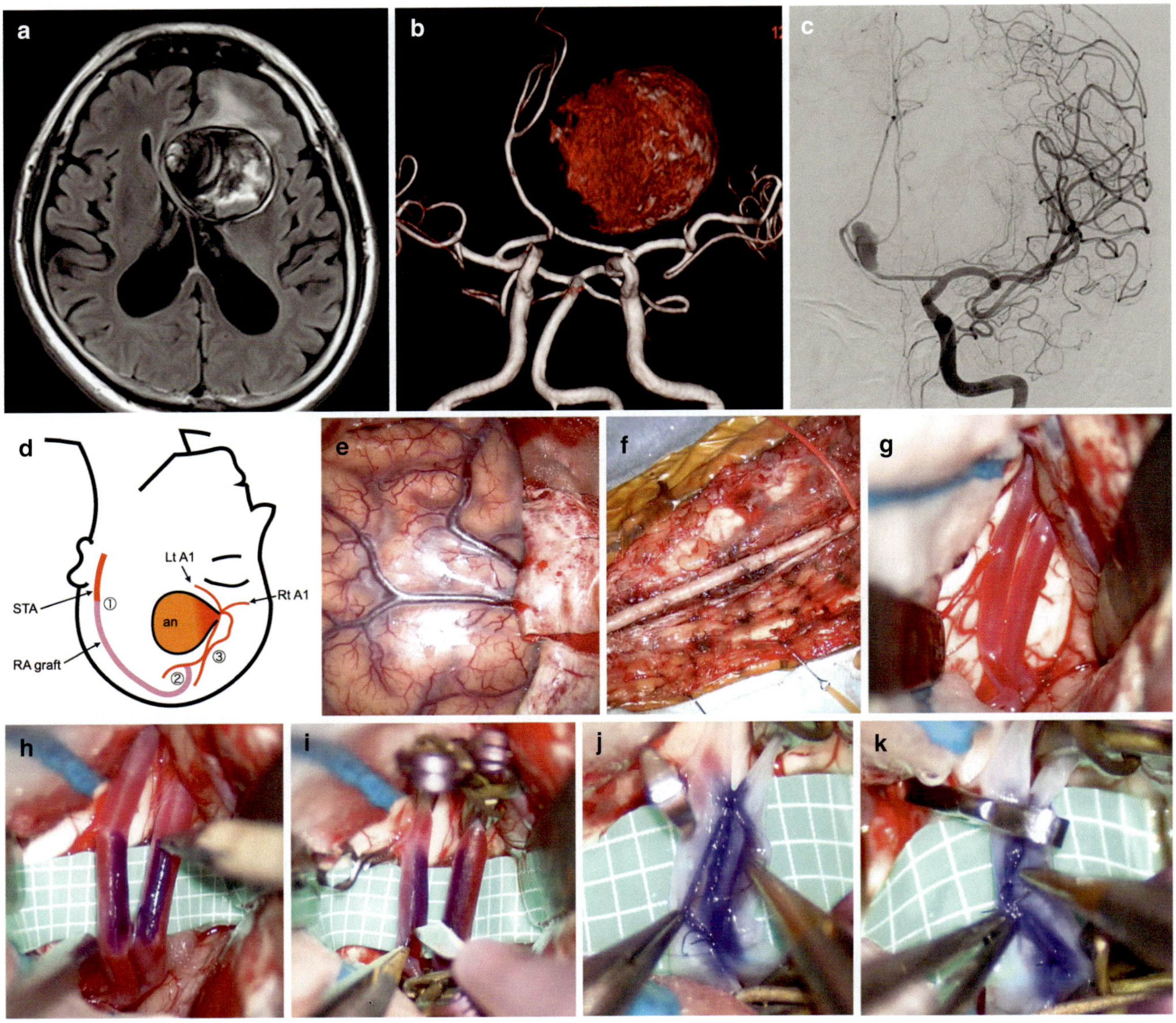

Fig. 14.4 (**a** and **b**) An axial MR image showing a partially thrombosed giant ACA aneurysm. (**c**) A left internal carotid angiogram demonstrating the aneurysm originating from the left A1–A2 junction. (**d**) A scheme depicting the surgical strategy consisting of clipping or trapping aneurysm with A3–A3 bypass and STA-RA-STA bypass. (**e**–**s**) Intraoperative photographs. (**e**) The left frontal lobe was exposed by reflecting the dura mater. (**f**) The radial artery was harvested from the left forearm. (**g**) The bilateral A3 trunks running along the surface of the corpus callosum were prepared. (**h**) The slightly medial surfaces of each artery were stained for better visualization. (**i**) Arteriotomy was made using microforceps and knife. (**j**) A running suture was placed on the side behind the anastomosis. (**k**) The anastomosis was conducted in a clean operative field with no blood contamination. (**l**) The anastomosis was completed. (**m**) Good anastomotic blood flow was confirmed using ICG-video angiography. (**n**) The RA graft was anastomosed to the left A3. (**o**) The distal end and the proximal end of the RA were anastomosed in an end-to-end fashion. (**p**) All anastomoses were completed. (**q**) ICG-video angiogram showing good intracranial blood flow via the RA. (**r**) Bilateral A1 segments were closed with clips. (**s**) The aneurysm was trapped by clipping the A1 and A2 segments. (**t** and **u**) Postoperative internal carotid angiograms demonstrating the occlusion of the RA graft and well-developed leptomeningeal anastomosis to the left distal ACA territory. (**v**) A postoperative MR image showing no ischemia. *an* aneurysm, *a-com* anterior communicating artery, *ICAG* internal carotid arteriogram, *RA* radial artery, *STA* superficial temporal artery, *VAG* vertebral angiogram. *Arrowheads* collateral flow to the left anterior cerebral artery territory

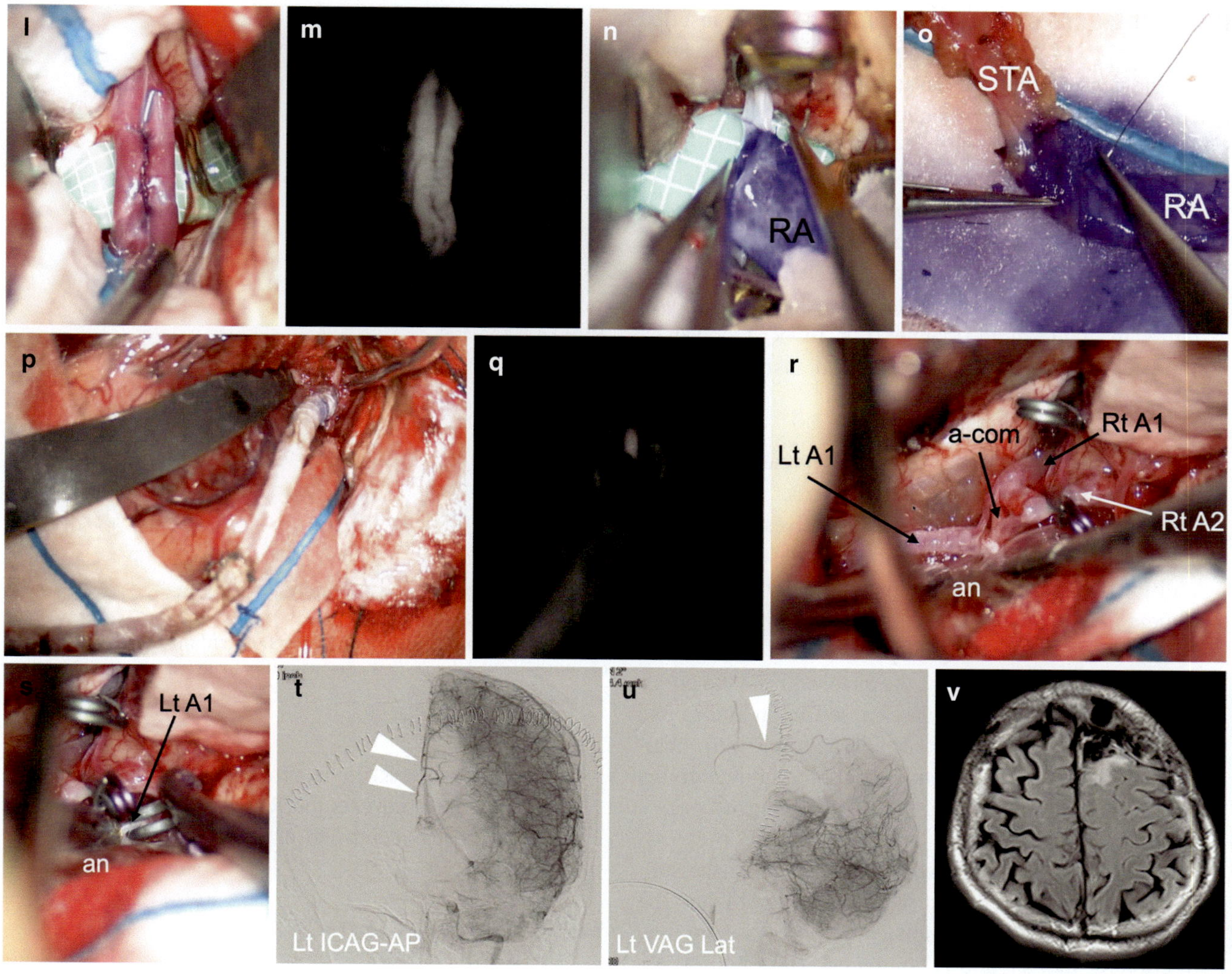

Fig. 14.4 (continued)

14.5.2.2 Strategy

Although the neck of the aneurysm itself appeared narrow, the vessel walls of the neck and parent arteries were considered thickened; therefore, standard clipping was expected to be impossible. In addition, removing the thrombus and dissecting the aneurysm from the brain surface would require a left frontal corticotomy. This procedure could cause further brain damage and worsen cognitive dysfunction. Therefore, we decided not to resect the thrombosed aneurysm but to trap it at the anterior communicating artery to eliminate blood flow into the aneurysm. Perfusion of the bilateral ACA territories was maintained by three bypasses consisting of an STA-RA-Lt A3 bypass with the RA graft and a further A3–A3 bypass (Fig. 14.4d).

14.5.2.3 Surgical Procedures

1. Craniotomy

   A left semicoronal skin incision was made, bilateral frontal craniotomy was performed across the superior sagittal sinus (SSS), and the dura mater was incised to expose the left frontal lobe (Fig. 14.4e). During this maneuver, the assistant simultaneously harvested the radial artery from the left forearm (Fig. 14.4f). The bone covering the SSS should be removed because limited maneuverability prevents the creation of an ample working space rendering anastomosis technically more difficult.
2. A3-A3 Bypass

   The bilateral A3 trunks running along the surface of the corpus callosum were exposed (Fig. 14.4g). The small arterial branches of the corpus callosum can be cauterized and cut, if necessary. A rubber sheet was placed under A3 and its surface was stained blue. The slightly medial surfaces of each artery were stained for better visualization (Fig. 14.4h). The arteriotomy is performed cautiously to avoid dissection, and an ophthalmic knife in ophthalmology is preferred. With gentle pinching of the recipient artery using microforceps, the knife blade can be inserted cleanly (Fig. 14.2i). A stay suture was made at the anterior and posterior ends of the anastomosis with one stitch at each end. A running suture was placed on the side behind the anastomosis (Fig. 14.4j). During anastomosis, a bloodless surgical field is essential to maintain excellent visibility and prevent clot formation. This requires adequate hemostasis using cotton patties to block blood from reaching the base of the operative field. After the suturing of the back side was completed, the shallow side was sutured in an interrupted manner (Fig. 14.4k). Good anastomotic blood flow was confirmed using ICG-VA (Fig. 14.4l, m).

3. STA-RA-A3 Bypass

   The RA graft obtained from the left forearm was anastomosed distally to the A3-A3 bypass. Anastomosis requires special attention because it involves the suturing of a large artery in a deep and narrow surgical field. Additionally, the RA running longitudinally in the surgical field may obstruct the view; therefore, the anastomosis site may not be clearly visible, making anastomosis difficult (Fig. 14.4n). Finally, the distal end and the proximal end of the RA were anastomosed in an end-to-end fashion (Fig. 14.4o). After all anastomoses were completed (Fig. 14.4p), good intracranial blood flow through the RA was confirmed using the ICA-VA (Fig. 14.4q).

4. Trapping of the Aneurysm

   Bilateral A1 segments were identified using the subfrontal approach and closed with clips (Fig. 14.4r). The aneurysm and a-com were isolated from the vascular flow by clipping the A1 and A2 segments (Fig. 14.4s). No problems with bypass patency or MEP potentials were encountered at closure.

5. Postoperative Course

   Postoperatively, no paralysis or disturbed consciousness was observed; however, postoperative angiography on day 1 showed that the RA graft was occluded. This may have resulted because the ACA originally developed a significant leptomeningeal anastomosis (Fig. 14.4t, u) due to chronic hypoperfusion. Another possible reason is that the RA was compressed between the skin and skull due to head bandaging. Fortunately, the patient avoided postoperative ischemia in the ACA region (Fig. 14.4v) and was discharged with no deterioration in cognitive function or independent activities of daily living.

6. Recent Change—Bone Flap Guttering

   Theoretically, delayed occlusion of the bypass can occur more frequently when the STA is extended, as in Case 2. To prevent the RA from delayed occlusion due to squeezing, we recently created a gutter in the bone flap to accommodate the RA in its groove, as shown in Fig. 14.5. This procedure prevents the RA from compression even when the head is bandaged.

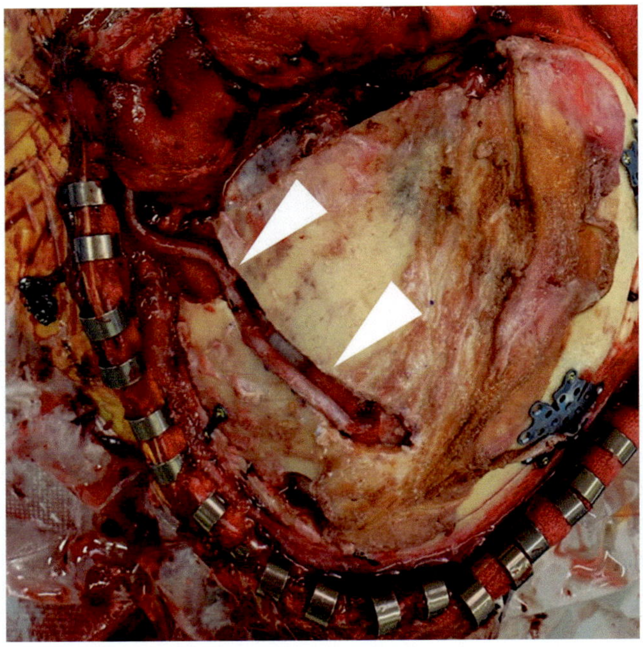

Fig. 14.5 An intraoperative photograph describing our recent method of guttering the bone flap (arrowheads) to avoid the postoperative occlusion of the bypass

14.5.3 Case 3: A Giant ICA Aneurysm

14.5.3.1 History and Imaging Studies

This case demonstrates the effectiveness of the combination of endovascular and direct surgical approaches. An 85-year-old woman with a large aneurysm with a maximum diameter of 20 mm (Fig. 14.6a) presented with right visual worsening over the previous year. A right EC-RA-M2 bypass with right ICA ligation was performed (Fig. 14.6b). The immediate surgical outcome was excellent without complications, and postoperative visual acuity improved. Three months after surgery, the patient had a contralateral lacunar infarction, and her family doctor initiated treatment with antiplatelet agents. One year after surgery, her vision deteriorated again, and MR imaging revealed partial thrombosis of the aneurysm with residual inflow, resulting in an increased size. Even after discontinuation of antiplatelet agents, the aneurysm continued to enlarge, reaching a diameter of 25 mm (Fig. 14.6c). The patient was considered to be at a high risk of rupture, and the decision was made to perform direct clipping surgery. Preoperatively, the patient lost vision.

14.5.3.2 Strategy

Traditionally, giant ICA cavernous aneurysms have been treated with proximal ICA ligation, with some form of revascularization as needed. Endovascular treatment using FD has been frequently performed in recent years. However, because the ICA had already been ligated in the previous surgery, FD could not be performed in this case. In addition, trapping the aneurysm was not possible without removing the thrombus inside the aneurysm to preserve the perforating branch from the posterior communicating artery (p-com) and the anterior choroidal artery near the aneurysm. Although the aneurysm was incised to remove the thrombus, blood flow via the p-com entered the aneurysm (Fig. 14.6d, white arrow). If the aneurysm was incised through a frontotemporal approach (Fig. 14.6d, blue arrow), the p-com might bleed profusely during thrombus removal. However, because the aneurysm was large and rigid, direct visualization of the p-com was not possible until after the thrombus was removed, causing difficulty in controlling the bleeding. Therefore, we preoperatively embolized the p-com at its entry into the aneurysm (Fig. 14.6e). Direct surgery was performed after the preparation described above to safely incise the aneurysm and remove the thrombus intraoperatively.

14.5.3.3 Surgical Procedures

The right frontotemporal craniotomy incision was reopened to expose the aneurysm. The intracranial blood flow was maintained through an RA graft. First, a space was created with a dissector between the anterior choroidal artery and aneurysm to insert a clip (Fig. 14.6f). The perforators branching from the top of the ICA and A1 were detached from the aneurysm wall to avoid obstruction when the aneurysm was closed with a clip (Fig. 14.6g). Next, the aneurysm was incised with a knife to remove the intra-aneurysmal thrombus (Fig. 14.6h). The thrombus inside the aneurysm was very stiff and various instruments were required to remove it, including an L-shaped dissecting hook, tumor forceps, and an ultrasonic aspirator (Fig. 14.6i). Intraoperative transcranial MEP repeatedly confirmed that blood flow to the anterior choroidal artery region via the bypass was maintained during thrombus removal. Fortunately, thrombus removal was safely performed despite weak retrograde bleeding from the distal ICA. Once the thrombus in the aneurysm was removed, a clip blade was inserted into the space between the aneurysm and the anterior choroidal artery to preserve blood flow to the anterior choroidal artery (Fig. 14.6j), and the clip was applied to the distal ICA. Intraoperative ICG-VA confirmed sufficient blood flow in the anterior choroidal artery (Fig. 14.6k and l). By moving the collapsing aneurysm medially, the p-com embolized by the coil outside the aneurysm was identified (Fig. 14.6m). Part of the coil penetrated the p-com. A clip was applied to the p-com to preserve the perforator (Fig. 14.6n). Trapping of the aneurysm and perfusion of the perforator were confirmed (Fig. 14.6o). A schematic of the surgical procedure is shown in Fig. 14.6p. Postoperative 3D-CT angiography confirmed the disappearance of the aneurysm (Fig. 14.6q). Postoperative MR imaging revealed no anterior choroidal artery infarction (Fig. 14.6r). However, thalamic infarction occurred due to the occlusion of the perforator from the terminalized p-com. Delayed occlusion of perforators from a terminalized artery during surgery for giant aneurysms remains an unresolved problem. In our experience, this is not predictable because these perforators remain patent during surgery. The patient was discharged to a rehabilitation hospital with a modified Rankin Scale 2.

14.6 Surgical Pearls

- Assessment of the characteristics of the aneurysm wall, intra-aneurysmal thrombus, and the branch site of the perforating arteries to determine the surgical strategy.
- Whether to aim for obliteration of the aneurysm or flow alteration.
- Variations in aneurysm closure with clips of various shapes and sizes.
- A variety of revascularization techniques.
- Limitation of vascular reconstruction regarding the preservation of the perforating branches.
- Stable intraoperative monitoring system.

14.7 Discussion

The treatment of giant cerebral aneurysms carries a high treatment-related risk. As the actual number of aneurysms is small, it is difficult for neurosurgeons to learn through experience in many cases. Consequently, surgeons should master excellent bypass techniques and understand the pathophysiology of aneurysm formation to treat this disease using creative treatment options. Intraoperatively, the surgeon must respond flexibly according to the results of intraoperative monitoring and hemodynamic evaluation using Doppler sonography and ICG-VA. This procedure requires a high level of skill and experience in vascular surgical treatment.

1. *Rationale for Performing Direct Surgery*

   Recently, most cerebral aneurysms have been treated using endovascular surgery instead of direct surgery. However, even in recent reports on FD treatment of giant cerebral aneurysms of the ICA, the aneurysm occlusion rate at 1 year is 60–75% [6, 7], and challenges remain. Therefore, direct surgery still plays a significant role in giant cerebral aneurysms, with high occlusion rates of 95%

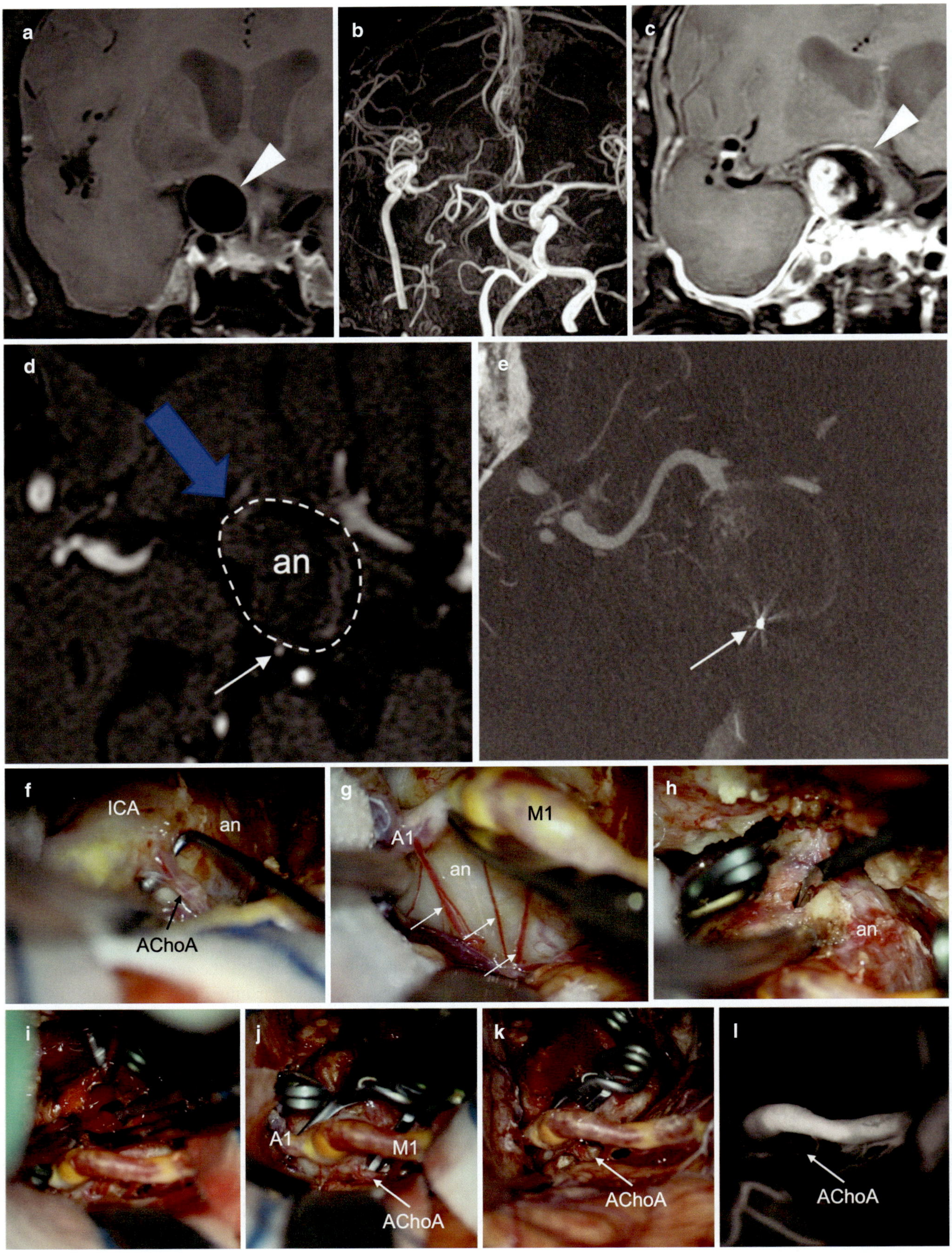
a
b
c
d
an
e
f
ICA
an
AChoA
g
A1
M1
an
h
an
i
j
A1
M1
AChoA
k
AChoA
l
AChoA

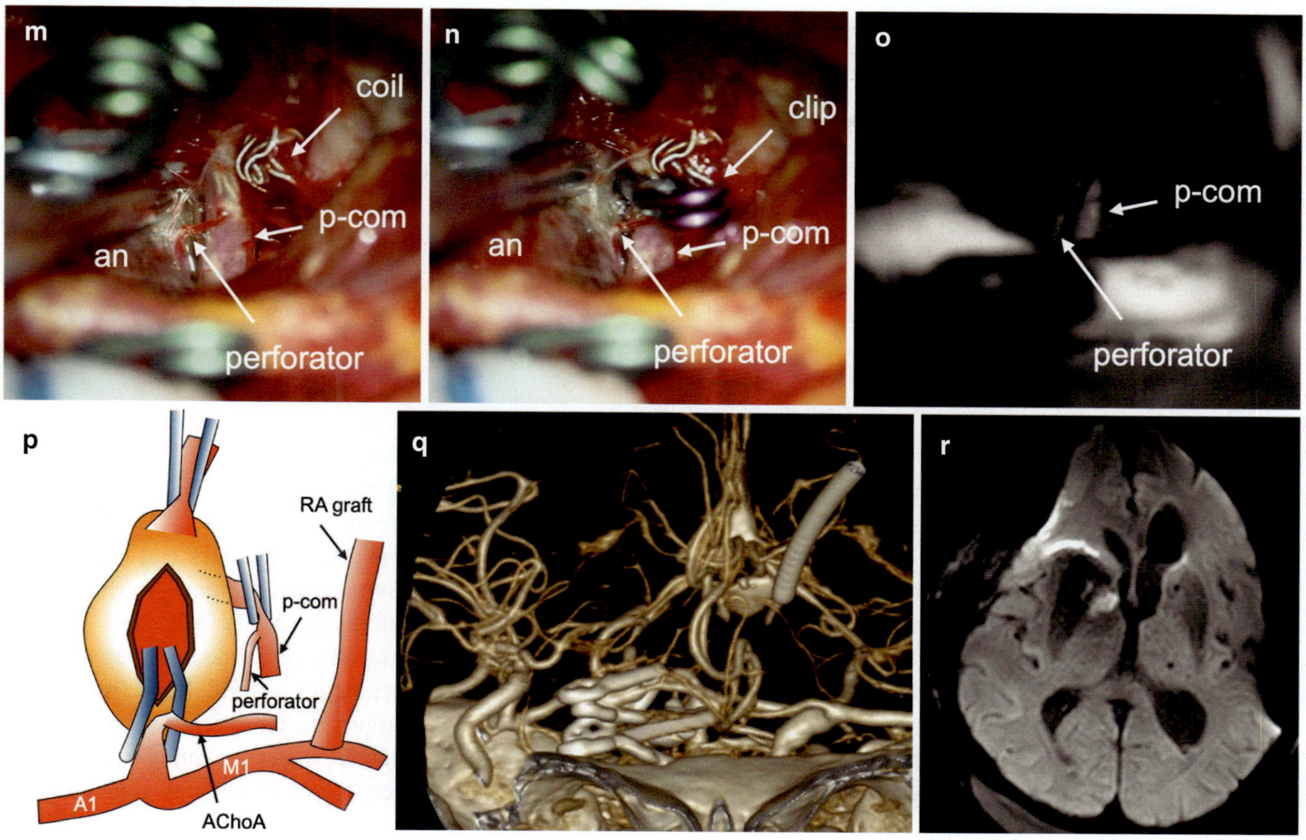

Fig. 14.6 (continued)

Fig. 14.6 (**a**) A coronal MR image showing a large right ICA aneurysm (arrowhead) causing the right visual deficit. (**b**) To improve her right vision and prevent bleeding, a right EC-RA-M2 bypass with right ICA ligation was performed. (**c**) However, the aneurysm (arrowhead) continued to enlarge to a diameter of 25 mm. (**d**) An MR T1 weighted image demonstrating a p-com (white arrow) behind the aneurysm (dotted line) which is invisible from the pterional approach (blue arrow). (**e**) A cone-beam angiogram showing the embolized p-com (arrow). (**f–o**) Intraoperative photographs. F: The anterior choroidal artery was dissected from the aneurysm. (**g**) The perforators branching from the top of the ICA and A1 were detached from the aneurysm wall. (**h**) The aneurysm was incised to remove the intraaneurysmal thrombus. (**i**) The hard thrombus inside the aneurysm was removed with various instruments. (**j**) After the thrombectomy, the clip was applied to the distal ICA. (**k** and **l**) Intraoperative ICG-VA confirmed sufficient blood flow in the anterior choroidal artery. (**m**) The p-com embolized by the coil outside the aneurysm was identified. (**n**) A clip was applied to the p-com to preserve the perforator. (**o**) Patency of the perforator was confirmed by ICG-video angiography. (**p**) A scheme describing the surgical procedure. (**q**) Postoperative 3D-CT angiography confirmed the disappearance of the aneurysm. (**r**) Postoperative MR imaging revealed no anterior choroidal artery infarction. *AchoA* anterior communicating artery, *an* aneurysm, *ICA* internal carotid artery, *p-com* posterior communicating artery, *RA* radial artery. *Arrowhead* aneurysm, *blue arrow* visual axis via the pterional approach; *white arrow* posterior communicating artery, *arrow* perforators from the top of the internal carotid artery and the proximal A1 and M1 segments

[3], 75% [8, 9], and 80% [10] for aneurysms arising at ICA, MCA, and ACA, respectively. Because incomplete occlusion of an aneurysm can lead to aneurysmal enlargement, worsening of neurological deficits, and even rupture, direct surgery remains an effective treatment option.

2. *Preoperative Radiological Examinations When Direct Surgery Is Contemplated*

When planning direct surgery for giant cerebral aneurysms, first, it must be determined whether direct occlusion of the aneurysm is feasible for preserving the flow of the parent artery, or whether trapping should be aimed at preserving distal blood flow using a bypass instead of manipulating the aneurysm itself. Temporary occlusion of the parent artery is necessary, even in cases where the parent artery is preserved. Therefore, preoperative STA, occipital artery, and RA angiography or 3D-CT angiography should be performed to confirm the availability of these arteries as donors. Preoperative imaging should also be evaluated for significant calcification of the aneurysm wall and perforating branches from the dome of the aneurysm, which are frequently not compatible with direct occlusion of the aneurysm.

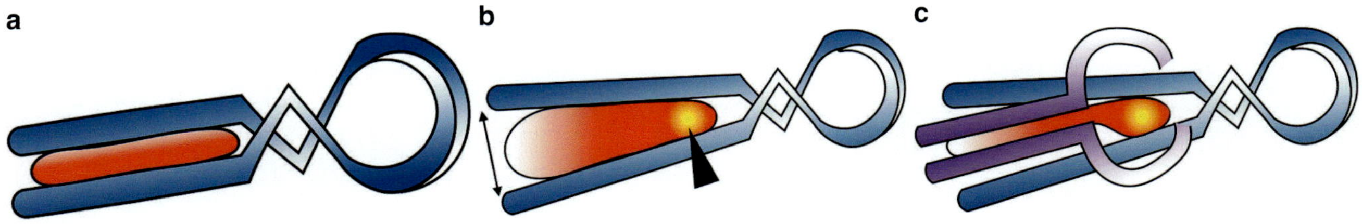

Fig. 14.7 Schemes demonstrating the clipping technique for giant aneurysms with thickened or calcified walls. (**a**) Aneurysms with soft walls are occluded using simple clipping. (**b**) The tip of the clip blade may not be able to close in cases of giant aneurysms due to their stiff or calcified wall (arrowhead). (**c**) A fenestrated clip applied over the stiff portion is useful to achieve complete closure

3. *Effectiveness and Limitation of Bypass*

   Bypass surgery plays a vital role in treating giant cerebral aneurysms. It is necessary to secure distal blood flow when trapping the aneurysm. In addition, even if preservation of the parent arteries is planned, aneurysmectomy is required in most cases to close the broad and stiff neck of giant aneurysms. Therefore, bypass must be performed before beginning to manipulate aneurysms. However, as shown in Case 3, no method has been established to reliably avoid the delayed obstruction that occurs when the perforator at the proximal end of the trapping site is terminated regardless of the bypass used. As intraoperative ICG-VA and neuromonitoring usually reveal normal findings, it is difficult to assess the risk of delayed occlusion precisely.
4. *Particularly Effective Methods for Clipping Giant Cerebral Aneurysms*

   Unlike normal-sized aneurysms, closing a giant cerebral aneurysm with one or two clips is rarely possible. Even if the aneurysm neck appears feasible for clipping on angiography or 3D-CT angiography, only the lumen is visible in these imaging studies, and the aneurysmal wall and parent artery are almost always adherent and difficult to dissect, making it impossible to insert a clip blade. Furthermore, the tip of a large clip may not close completely because of wall thickening or calcification near the aneurysm neck. Aneurysms with soft walls were occluded using simple clipping (Fig. 14.7a). However, if they had a stiff wall or calcification (Fig. 14.7b, arrowhead), the tip of the clip blade might not be able to close and the inflow could remain. Giant aneurysms frequently exhibit this type of neck wall. In these cases, a fenestrated clip was applied over the stiff portion to achieve complete closure (Fig. 14.7c). Therefore, a technique that combines multiple fenestrated clips is necessary. In addition, a unique clipping technique for giant aneurysms, such as the picket fence clip technique shown in Case 1, is useful. Clipping an aneurysm also requires the removal of the intraaneurysmal thrombus in most cases. Tumor forceps, large scissors, knives, and CUSAs should be available in the operating room for thrombus removal.
5. *Importance of Intraoperative Neuromonitoring and Vascular Flow Assessment*

   Surgery for giant cerebral aneurysms often requires the temporary closure of the parent artery. Consequently, intraoperative cerebral hemodynamics can change significantly after aneurysm obliteration. Intraoperative real-time monitoring using MEP, SEP, and visually evoked potentials should be performed. Therefore, these neuromonitoring techniques should be actively used during aneurysm surgery to familiarize surgeons with various types of potential issues including artifacts and to prepare for challenging cases such as giant cerebral aneurysms.

   Intraoperative evaluation of blood flow using ICG-VA and Doppler sonography is essential. However, evaluating the perforating branches on the dorsal surface of the aneurysm is impossible; therefore, it is preferable to treat patients in a hybrid operating room with intraoperative cerebral angiography, if possible.
6. *Postoperative Management*

   Because occlusion of a giant cerebral aneurysm can cause hyperperfusion syndrome due to a sudden increase in blood flow to the peripheral region of the parent artery, blood pressure elevation should be avoided in the acute postoperative period. Surgical occlusion of large cerebral aneurysms can cause postoperative hyperperfusion and neurological symptoms. In bypass cases, it should be also noted that the reconstructed vascular flow may be occluded by a drop in blood pressure. There is no consensus regarding the administration of antiplatelet or antiepileptic drugs.

14.8 Conclusion

For the surgical treatment of giant cerebral aneurysms, it is essential to understand the mechanism of aneurysmal growth and determine an appropriate treatment plan. Stable vascular surgical skills are also required to complete planned treatment and enable a flexible response to intraoperative findings. Although the small number of cases may preclude individual surgeons from experiencing a large number of cases, hopefully, this chapter may help neurosurgeons who encounter this formidable disease.

References

1. Iihara K, Murao K, Sakai N, et al. Continued growth of and increased symptoms from a thrombosed giant aneurysm of the vertebral artery after complete endovascular occlusion and trapping: the role of vasa vasorum: case report. J Neurosurg. 2003;98:407–13.
2. UCAS Japan Investigators; Morita A, Kirino T, et al. The natural course of unruptured cerebral aneurysms in a Japanese cohort. New Engl J Med. 2012;366:2474–82.
3. Flores BC, White JA, Batjer HH, et al. The 25th anniversary of the retrograde suction decompression technique (Dallas technique) for the surgical management of paraclinoid aneurysms: historical background, systematic review, and pooled analysis of the literature. J Neurosurg. 2019;130:902–16.
4. Amin-Hanjani S, Alaraj A, Charbel FT. Flow replacement bypass for aneurysms: decision-making using intraoperative blood flow measurements. Acta Neurochir. 2010;152:1021–32.
5. Steinberg GK, Drake CG, Peerless SJ. Deliberate basilar or vertebral artery occlusion in the treatment of intracranial aneurysms: immediate results and long-term outcome in 201 patients. J Neurosurg. 1993;79:161–73.
6. Colby GP, Lin LM, Caplan JM, et al. Flow diversion of large internal carotid artery aneurysms with the surpass device: impressions and technical nuance from the initial North American experience. J Neurointerv Surg. 2016;8:279.
7. Liu JM, Zhou Y, Li Y, et al. Parent artery reconstruction for large or giant cerebral aneurysms using the Tubridge flow diverter: a multicenter, randomized, controlled clinical trial (PARAT). Am J Neuroradiol. 2018;39:807–16.
8. Kalani MYS, Zabramski JM, Hu YC, et al. Extracranial-intracranial bypass and vessel occlusion for the treatment of unclippable giant middle cerebral artery aneurysms. Neurosurgery. 2013;72:428–36.
9. Kivipelto L, Niemelä M, Meling T, et al. Bypass surgery for complex middle cerebral artery aneurysms: impact of the exact location in the MCA tree: clinical article. J Neurosurg. 2014;120:398–408.
10. Abla AA, Lawton MT. Anterior cerebral artery bypass for complex aneurysms: an experience with intracranial-intracranial reconstruction and review of bypass options: clinical article. J Neurosurg. 2014;120:1364–77.

Microsurgical Treatment of Complex Aneurysms in the Current Endovascular Era

15

Mustafa K. Başkaya and Jaime L. Martínez Santos

Learning Objectives

- To identify the indications for an open microneurosurgical approach in the management of complex intracranial aneurysms in the "endovascular era."
- To understand the main perioperative considerations in patients undergoing microsurgery for complex aneurysms.
- To build a strong microneurosurgical strategy with multiple options to account for any intraoperative scenario including multiple revascularization options.

15.1 Introduction

Complex intracranial aneurysms (CIAs) are aneurysms where there are definitive aspects that make these challenging to treat via traditional open or endovascular routes. CIA treatment may be challenging due to their overall size (large or very small); the absence of a neck (e.g., fusiform and blister-type aneurysms); having necks that are severely calcified, atherosclerotic, or thrombosed; being located near branching arteries from the neck or the aneurysm dome; having a broad neck (dome-to-neck ratio < 1.5); and simply being located where they are difficult to access [1].

Treatment alternatives for CIAs include simple observation, if these appear innocuous and unruptured, and a range of open microsurgical and endovascular interventions. Such interventions include open microsurgical clip occlusion or reconstruction, variations of microsurgical trapping that include intracranial-to-intracranial (IC-IC) and extracranial-to-intracranial (EC-IC) bypasses for revascularizations, endovascular stent-coiling, flow diversion, and finally vessel sacrifice in those rare situations when this can be tolerated by collateral circulation.

In the current so-called "endovascular era" most CIAs are treated in the angiography suite, despite that open microneurosurgical strategies are more durable and versatile, with the ability to reconstruct and re-route blood flow in any situation. The rapid-pace evolution of neurointerventional radiologic techniques and the inception of innovative devices such as flow diverters and intrasaccular devices have made it possible to treat some CIAs effortlessly [2]. In fact, endovascular treatments might be the treatment of choice for some aneurysms in the basilar bifurcation and in the cavernous segment of the extradural internal carotid artery (ICA) with favorable parent artery anatomy [1]. Nonetheless, we believe that every patient and every aneurysm is different, and each must be discussed among an experienced multidisciplinary cerebrovascular team represented by a skillful complex skull base and open cerebrovascular neurosurgeon who continuously hones his/her microsurgical skills in the operating room and laboratory. The final selected game plan should be the best alternative for the individual patient and not be dictated by an individual physician's comfort level.

The pitfalls of endovascular therapies should always be considered, which include dual antiplatelet therapy, frequent follow-up with invasive diagnostic tests, frequent re-interventions, failures, and recurrences. These must be weighed against the microsurgical option, which can provide a definitive cure with much less burdensome follow-up, but also presents the risks of a craniotomy, which nonetheless is usually well tolerated. Another common mistake worth pointing out is the treatment of ruptured aneurysms that cannot be completely occluded endovascularly by selectively coiling, and thus are only partially treated at the presumed ruptured point of the aneurysm, with the hope and expectation to return later for the final occlusion after the vasospasm window. These partially coiled aneurysms represent a great

M. K. Başkaya (✉)
Department of Neurological Surgery, University of Wisconsin-Madison, Madison, WI, USA
e-mail: baskaya@neurosurgery.wisc.edu

J. L. Martínez Santos
Department of Neurological Surgery, Medical University of South Carolina, Charleston, SC, USA

I. N. Kanaan, V. Beneš (eds.), *Neuroanatomy Guidance to Successful Neurosurgical Interventions*,
https://doi.org/10.1007/978-3-031-59838-8_15

microsurgical challenge by preventing the closure of clip blades and by causing tears when clips are applied [3].

In this chapter, we define and illustrate using case examples, indications where an open microneurosurgical strategy should be considered as the treatment of choice for CIAs.

15.2 Anatomy Highlights

- The calvarial and skull base osteology is studied using computed tomography (CT) to fashion safe and effective approaches. Osseous anatomic highlights include calvarial hyperostosis, abnormal ossification of membranes and ligaments (i.e., arcuate foramen around the V3 segment of the vertebral artery), pneumatization of the frontal sinuses, erosion and/or pneumatization of the anterior clinoid processes, and petrous bone [4, 5].
- The extracranial vascular anatomy is studied using CT and catheter angiograms to assess collateralization from extracranial arterial branches and to potential arterial donors for revascularization. Extracranial vascular anatomic highlights include the superficial temporal artery (its size, level of bifurcation, location, and size of its anterior and posterior branches), the occipital artery, the maxillary artery, and the external carotid artery and its smaller branches (e.g., lingual artery) in case interposition grafts are needed. Lastly, it is always wise to study the limb's vasculature and feasibility (size and length) of harvesting venous (e.g., saphenous vein) or arterial (e.g., radial artery) grafts.
- The intracranial vascular anatomic highlights are particular to the individual patient and complex aneurysm, and in general, include a deep study and understanding of the angioarchitecture surrounding the aneurysm (projection of aneurysm dome, parent artery, associated branches and perforators, and irrigation territory), identifying any anatomic variations (in vessel origin and morphology), abnormal anastomoses, and collateral flow. Some examples, to name a few, of intracranial vascular anatomic highlights include the location of the proximal and distal carotid dural rings and length of the supraclinoid internal cerebral artery and the origin and course of the ophthalmic artery, anterior thalamic perforators, posterior communicating artery (PCOM), and anterior choroidal artery for aneurysms in these locations (paraclinoid carotid, PCOM, etc.); the recurrent artery of Heubner and location and number of lenticulostriate perforators for anterior communicating artery (ACOM), ICA bifurcation, and middle cerebral artery (MCA) aneurysms; the irrigation territory of individual branches of the anterior cerebral artery (ACA) accounting for any anatomic variation (e.g., azygous, duplications, bi-hemispheric, accessory, etc.) for distal ACA aneurysms; the branches the position of the basilar tip in relation to the posterior clinoid process and the origin, number and course of posterior thalamic perforators for basilar tip aneurysms; the origin (intradural or extradural), course and branches (including the posterior spinal artery and medullary perforators) of the posterior inferior cerebellar artery (PICA) for PICA; and vertebral artery aneurysms. Additionally, blood flow (antegrade or at least retrograde) to perforators (lenticulostriate, thalamic, and brainstem) and important branches (e.g., PCOM and AchA) must be preserved whenever the aneurysms are treated with trapping.
- Lastly, the patient's neuroanatomy should be carefully examined to identify any cavities from old hematomas, distortion of the brain parenchyma, cranial nerves in the aneurysm vicinity, location of relevant vessels and aneurysms in relation to cisternal spaces, and other neuroanatomical features that may impact the procedure.

15.3 Approach and Surgical Intervention

The basic tenets for cerebrovascular neurosurgery include: Making an only just sufficient bony exposure for safe microdissection while being prepared for any possible worst-case scenario; early access to the subarachnoid cisterns for cerebrospinal fluid drainage for brain relaxation; wide arachnoid microdissection that exposes normal and pathologic anatomy but also de-tethers and further relaxes the brain; being reluctant to brain retraction and only using fixed retractors for holding the untethered brain whenever the vector of gravity is not exactly what is needed for a brief microsurgical maneuver such as aneurysm dome microdissection or vessel micro suturing; and lastly securing early proximal and distal vascular control, which is the main tenet for cerebrovascular neurosurgery. Once the aneurysm and neighboring arteries are exposed, a simple clip occlusion (Plan A) or complex clip reconstructions (Plan B) are first considered. If the aneurysm cannot be clipped, the parent artery is "reconstructed" with combinations of clipping and revascularization techniques (Plan C). Finally, a parent artery "deconstruction" (or vessel sacrifice) is reserved as a last resort (Plan D). A complete occlusion is confirmed with intraoperative Doppler,

Table 15.1 Indications for microsurgical management of intracranial aneurysms

Aneurysm and patient characteristics	Rationale
1. Middle Cerebral Artery (MCA) Aneurysms	Microneurosurgery is a safe and durable alternative and achieves immediate aneurysm occlusion. The aneurysm's three-dimensional (3D) architecture can be visualized, and complete occlusion can be confirmed with real-time Doppler, indocyanine-green video angiography, and dome resection
2. Aneurysms located distal to the circle of Willis	Easier access via microneurosurgery with faster approaches and shorter surgical corridors. Improved real-time 3D visualization of the aneurysm architecture (see indication #1). Distal aneurysms are often associated with inflammatory processes or collagen diseases or infections (i.e., mycotic) and oftentimes microsurgery reveals and treats other smaller angiographically occult lesions. Navigating distal branches endovascularly with guidewires and catheters is more dangerous, and endovascular treatment often involves vessel sacrifice, which can be prevented with microneurosurgery
3. Aneurysms with simple microneurosurgical options and difficult endovascular options	Microneurosurgery reduces hospital costs, unnecessary invasive imaging for follow-up, unnecessary use of antiplatelet agents, and multiple endovascular re-interventions. Another important consideration is the patient's and surgeon's peace of mind about a fully treated aneurysm
4. Multiple aneurysms that are accessible with a single approach	Patients with multiple ipsilateral or paramedian aneurysms benefit from a single microsurgical intervention that attains definitive occlusion of all aneurysms in one setting
5. Small and blood blister-like aneurysms	Clipping small saccular aneurysms is simple, safe, and durable. Blood blister-like aneurysms have an extremely high intra-procedure rupture risk, lack normal vessel wall layers, and usually involve a large vessel circumference (>120°). In our experience, bypass and aneurysm trapping is the safest and most effective treatment modality
6. Giant aneurysms	Microsurgery is more versatile, with options that include clip reconstruction, aneurysmorrhaphy, aneurysm opening with the removal of any thrombotic or atherosclerotic material, identification and preservation of all branches including small perforators, vessel reconstruction with bypass and related methods that achieve immediate results including aneurysm decompression and occlusion
7. Aneurysms causing mass effect	Microsurgery can fully decompress any mass effects from any aneurysm and remove bony or ligamentous compression on cranial nerves
8. Ruptured aneurysms with intraparenchymal hematomas	Microsurgery allows for hematoma evacuation, a durable complete aneurysm occlusion, and a decompressive craniectomy if needed, in the same setting. Microsurgery eliminates the need for high-dose intra-procedure heparin and unnecessary post-procedure antiplatelet therapy if stents or flow diverters are used or if there is an unwanted coil loop. The subarachnoid cisterns are widely dissected and communicated, and the subarachnoid blood products are cleared away. Vasospasm can be further prevented by using sodium nitroprusside or Papaverine locally on the exposed arteries
9. Aneurysms in young patients	Microneurosurgery is the most durable treatment and the most reassuring throughout the lifetime of the patient (refer to indication 3)
10. Aneurysms with complex morphologies or at difficult parent artery bifurcations or 0branching angles with high hemodynamic stress	Microsurgery is a better option due to the high rate of failure of endovascular treatments for these indications
11. Failed endovascular treatments	Microneurosurgery is more versatile (refer to indication #6). Despite that treatment of failed endovascular lesions is dauting, there is always a feasible microneurosurgical solution for any aneurysm
12. Patients that would not comply with follow-up	Microneurosurgery is more durable and requires less follow-up and noninvasive imaging follow-ups
13. Patient Preference	With few exceptions (i.e., basilar tip and cavernous carotid aneurysms), all open and endovascular alternatives should be presented to the patient, and the patient should decide. This is the fundamental principle of patient autonomy

indocyanine-green (ICG) video angiography, and finally with puncturing the aneurysm, and bipolar remodeling and resecting of its dome.

In the following sections, we discuss and illustrate with case examples some of the indications for microsurgical management of intracranial aneurysms listed in Table 15.1 and highlight some of the principles for CIA microneurosurgery.

15.4 Representative Cases

Case 1. Large Aneurysm that Failed Endovascular Flow Diversion (Figs. 15.1 and 15.2)

A 48-year-old woman presented with progressive retro-orbital pain, diplopia, and right abducens (VI cranial nerve) palsy over the past few months. Workup including MRI (Fig. 15.1a, b) and digital subtraction angiography (DSA) (Fig. 15.1c–e) that

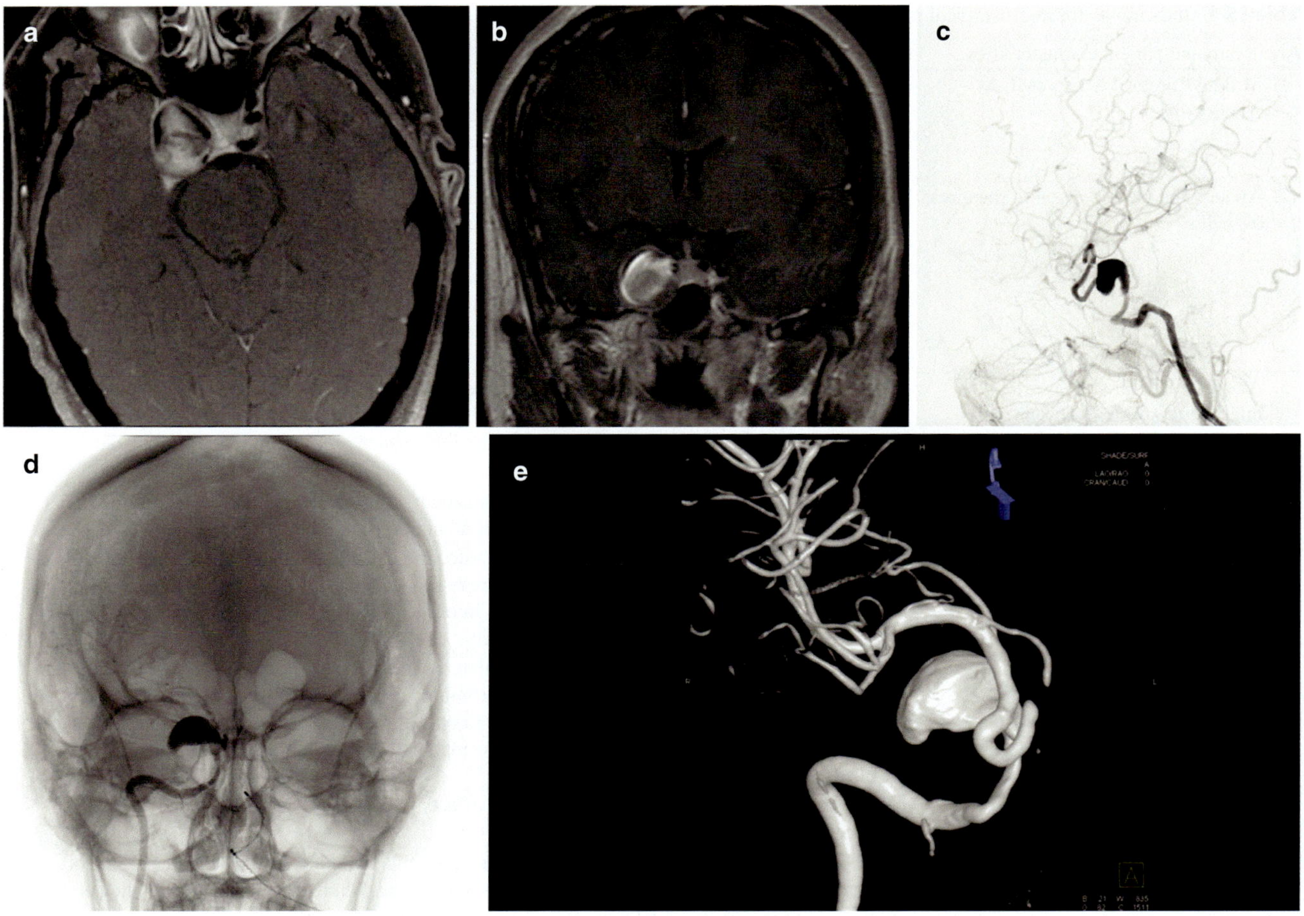

Fig. 15.1 Case 1 preoperative neuroimaging: Axial (**a**) and coronal (**b**) T1-weighted magnetic resonance imaging (MRI). Lateral (**c**) and anteroposterior (**d**) diagnostic cerebral angiogram. Three-dimensional reconstruction showing a right-sided, large partially thrombosed, fusiform aneurysm of the cavernous segment of the internal carotid artery

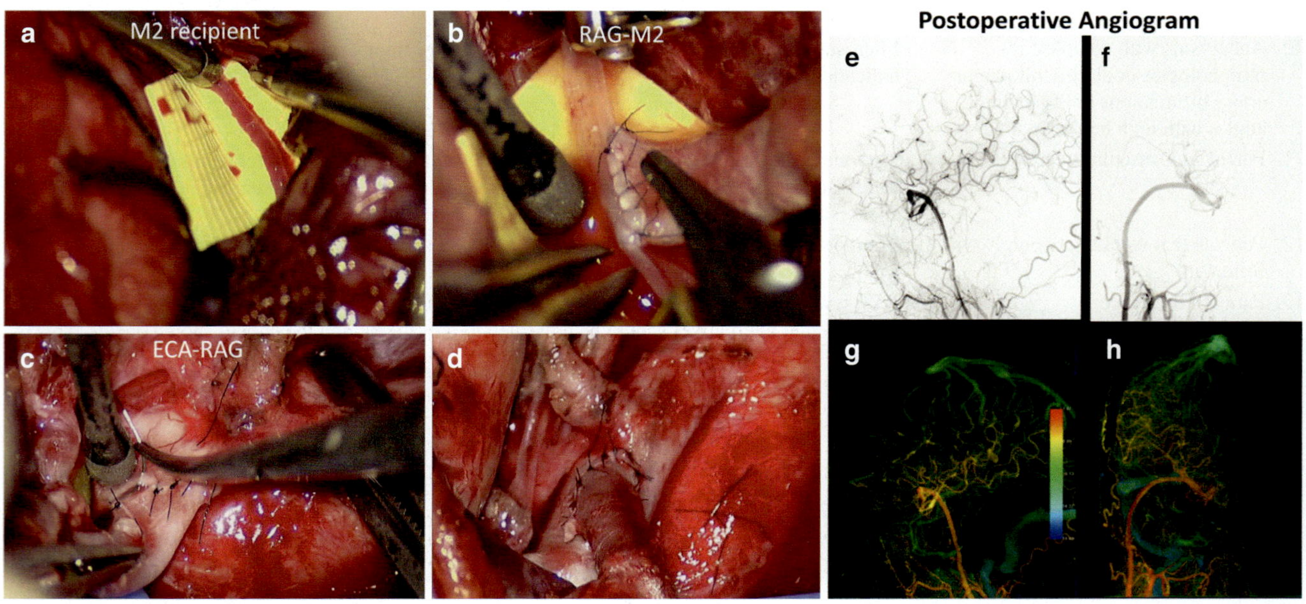

Fig. 15.2 Case 1 surgery and postoperative angiography: Intraoperative sequences (**a**–**d**). Preparing the right M2 recipient (**a**) after dissecting the Sylvian fissure, a distal radial graft (RAG) end-to-side anastomosis (**b**) using interrupted 10-0 suturing and a proximal radial graft end-to-side anastomosis to the right external carotid artery (ECA) (**c** and **d**). Postoperative angiograms (**e** and **f**) and postoperative i-flow angiograms (**g** and **h**) show patency of the bypass and the nonfilling of the aneurysm after trapping

revealed a large partially thrombosed, fusiform, right-sided aneurysm of the cavernous segment of the internal carotid artery. After discussing treatment options including endovascular flow diversion, bypass and trapping, clip reconstruction, and observation, the patient elected to undergo flow diversion. This endovascular treatment failed (Fig. 15.1). At this point, it was decided to treat this complex aneurysm definitively with microsurgery by performing a high-flow ECA-to-M2 bypass using a radial artery interposition graft, and aneurysm trapping (Fig. 15.2a–d). Surgery attained a dramatic, complete, and durable symptom resolution. The patient was discharged home shortly thereafter with a modified Rankin scale (mRS) of 0. Postoperative angiography showed patency of the bypass and nonfilling of the aneurysm (Fig. 15.2e–h).

Case 2. Giant Aneurysms Causing Mass Effect (Figs. 15.3, 15.4 and 15.5)

A 68-year-old male presented with aphasia and right hemiparesis. Neuroimaging revealed a giant, partially thrombosed, internal cerebral artery (ICA) bifurcation aneurysm causing significant mass effect and vasogenic edema (Fig. 15.3). Endovascular treatment of this aneurysm was deemed impossible. We planned for microneurosurgical aneurysm decompression and clip reconstruction. We performed a left cranio-orbital (modified orbito-zygomatic) approach and prepared the ipsilateral superficial temporal artery (STA) and left radial artery in the forearm for a possible bypass. The aneurysm was exposed superficially in the Sylvian fissure (Fig. 15.4a) and further dissection exposed the supraclinoid ICA for proximal control (Fig. 15.4b). Applying similar principles used for microsurgical resection of skull base tumors, it was necessary to debulk the aneurysm mass (Fig. 15.4b, c) to be able to visualize around the corner and expose the proximal middle cerebral artery, anterior cerebral artery, and lenticulostriate branches (Fig. 15.4d). The aneurysm wall was incised away from the live (filling) region (Fig. 15.4b) of the aneurysm to maximally decompress it before encountering bleeding. After we had secured

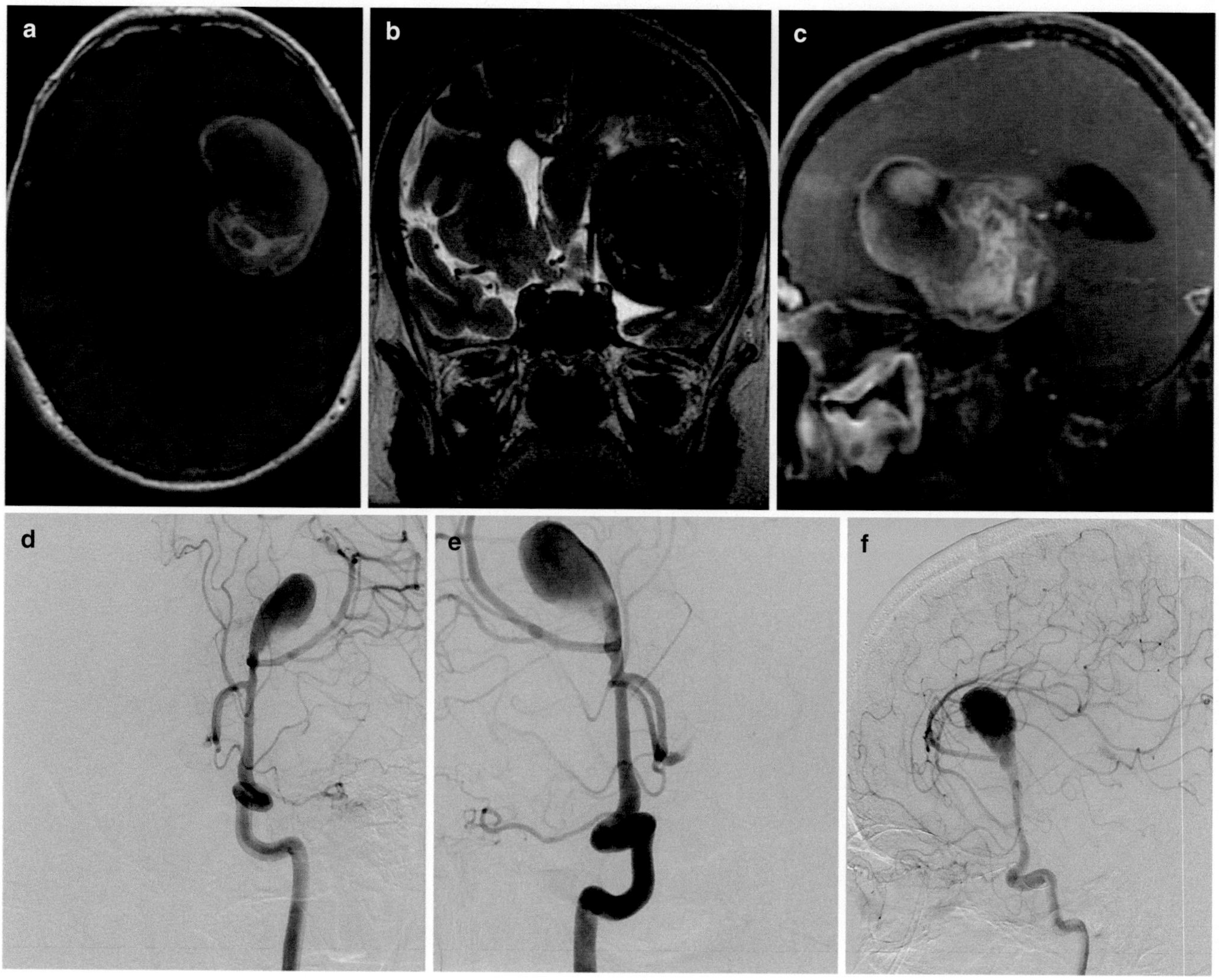

Fig. 15.3 Case 2 preoperative neuroimaging: Axial noncontrast T1 (**a**), coronal T2 (**b**), and sagittal post-contrast T1 (**c**) MRI. Rotated oblique (**d**) and lateral (**e**, **f**) diagnostic cerebral angiograms showing a giant (8 cm in size), partially thrombosed, internal cerebral artery bifurcation aneurysm causing significant mass effect and vasogenic edema

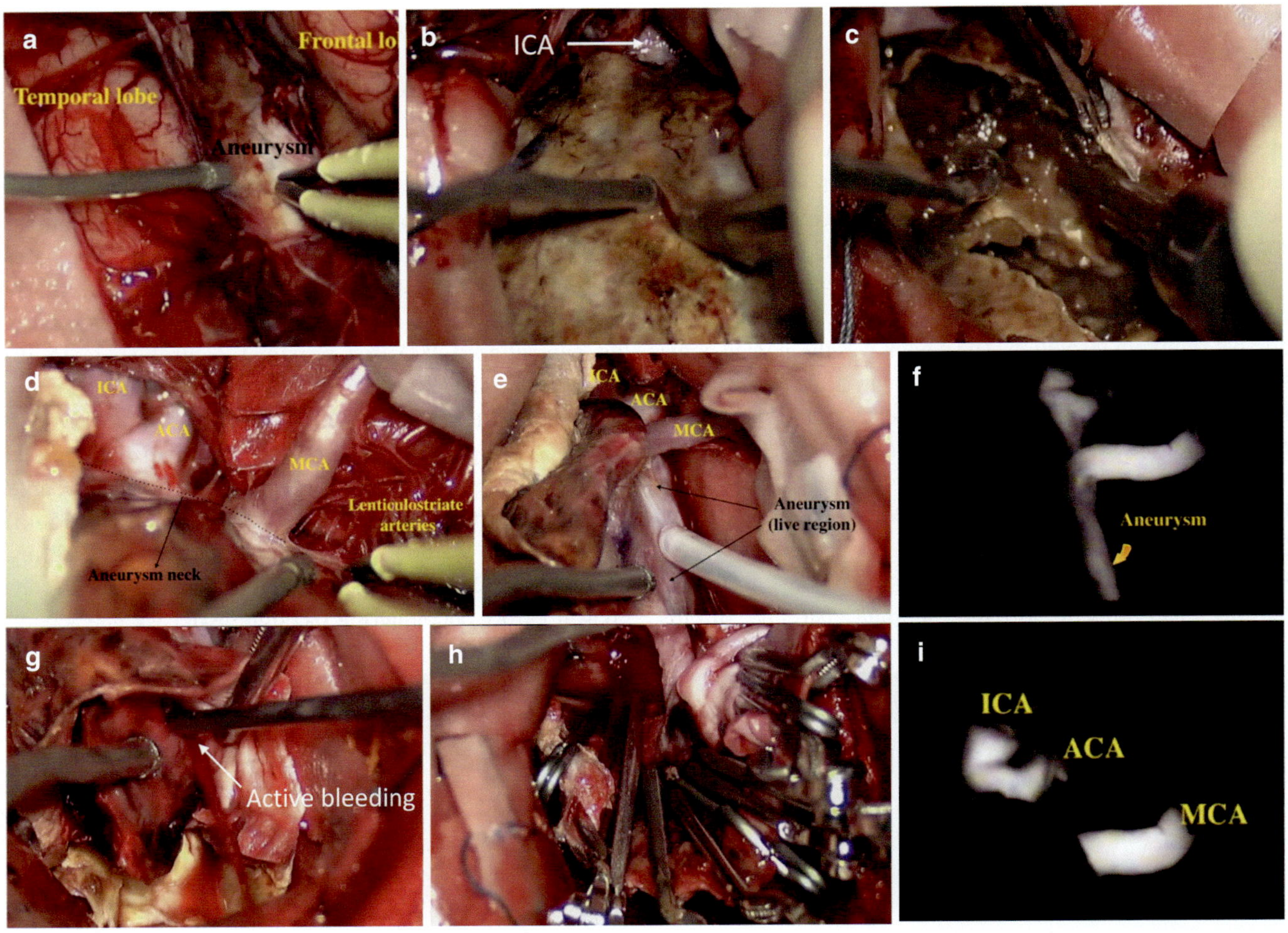

Fig. 15.4 Case 2 surgery and intraoperative angiography: Intraoperative sequence demonstrating exposure (**a**) of a giant aneurysm within the right Sylvian fissure. After obtaining proximal control of the internal cerebral artery (ICA), the aneurysm was internally decompressed (**b** and **c**). This decompression allowed us to expose the anterior cerebral artery (ACA), the middle cerebral artery (MCA), the lenticulostriate arteries, the aneurysm neck, and the live region of the aneurysm (**d** and **e**). Doppler (**e**) and ICG green video angiography confirmed flow within the live region of the aneurysm (**f**). Once better proximal control was obtained, the aneurysm was more aggressively decompressed and the clot near the live region of the aneurysm was removed until bleeding was encountered (**g**). Now that the aneurysm walls were soft, the aneurysm was clip reconstructed (**h**). ICG video angiography demonstrated good flow in the ICA, ACA, and MCA and nonfilling of the aneurysm (**i**)

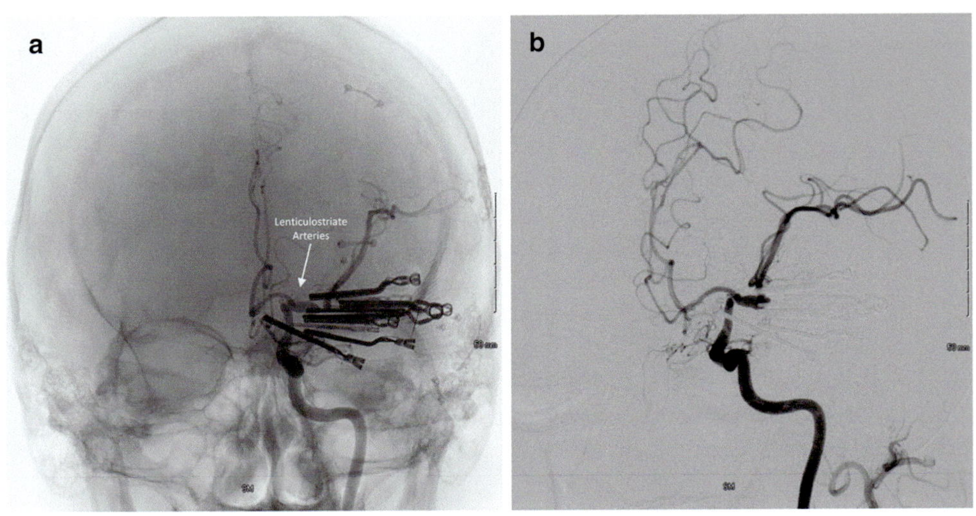

Fig. 15.5 Case 2 postoperative angiography: Anteroposterior diagnostic cerebral angiogram (**a** and **b**) shows clip construct, good flow in the neighboring branches including the lenticulostriate arteries, and nonfilling of the aneurysm

better proximal control and had identified the aneurysm neck and its live region (Fig. 15.4e, f), we proceeded to perform thromboendarterectomy until we encountered bleeding (Fig. 15.4g). At this point, we placed temporary clips in the ICA, M1, and A1 to stop the bleeding, and occluded the aneurysm using multiple large clips in a tandem fashion (Fig. 15.4h). ICG video angiography revealed good flow in the parent arteries and lenticulostriate branches (Fig. 15.4i). A postoperative angiogram confirmed complete occlusion of the aneurysm and good flow in all arterial branches (Fig. 15.5). The patient had significant improvement in his motor function and aphasia, with a modified Rankin scale (mRS) of 0 and full strength at his 1-year follow-up.

Case 3. Blood Blister-Like Aneurysms (Figs. 15.6, 15.7, and 15.8)

A 23-year-old male with no past medical history and no risk factors for intracranial aneurysms presented with a low-grade subarachnoid hemorrhage. The initial workup that included CTA did not reveal any obvious cause. A cerebral angiogram, however, showed a double-density dilatation in the superior surface of the left supraclinoid ICA (Fig. 15.6). These findings were highly suspicious for a blood blister-like aneurysm, which are extremely fragile, and thin-walled lesions that are usually located at nonbranching points over the superior, dorsal, or anterior surface of the ICA [6, 7]. There were no changes on EEG during a balloon test occlusion with a 15-minute hypotensive challenge. For treating this aneurysm, we first performed an STA-to-MCA bypass in anticipation of some ICA lumen narrowing when applying permanent clips occluding the aneurysm and incorporating all the involved ICA vessel walls (Fig. 15.7). Postoperative imaging, including a CTA and a diagnostic cerebral angiogram, revealed occlusion of the aneurysm and showed that despite the purposely narrowed ICA lumen, there was a good filling of the posterior communicating (PCOM), the anterior choroidal artery, MCA, and patency of the STA-to-MCA bypass (Fig. 15.8).

Case 4. Recanalized Posterior Inferior Cerebellar Artery Aneurysm after a Failed Endovascular Coiling (Fig. 15.9)

A 31-year-old female with a past medical history of Hunt-Hess grade II SAH secondary to a ruptured left Posterior Inferior Cerebellar Artery (PICA) origin aneurysm presented for evaluation of a growing residual left PICA aneurysm after coil embolization 2 years ago (Fig. 15.9a, b).

Due to the history of recurrence following endovascular treatment, microsurgical treatment was felt to be the most effective treatment modality. We performed a left far lateral transcondylar approach and clipped the aneurysm without difficulty (Fig. 15.9c–g). The surgical strategy involved early proximal control in the left vertebral artery; opening the lateral cerebellomedullary cistern for CSF drainage and

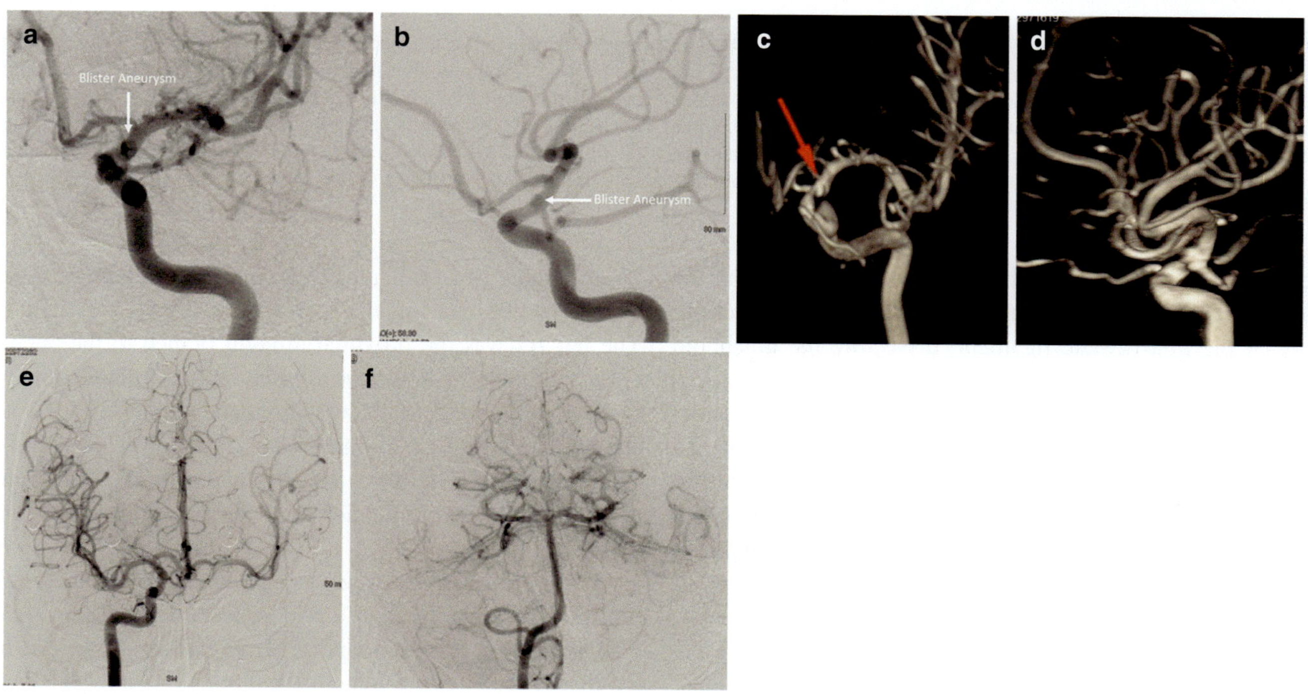

Fig. 15.6 Case 3 preoperative angiography: Diagnostic cerebral angiograms (**a**–**f**) and three-dimensional reconstructions (**c** and **d**) show a double-density dilatation in the superior surface of the left supraclinoid ICA suggestive of a blood blister-like aneurysm (arrows). A right ICA (**e**) and a right vertebral artery (**f**) injection showed good cross-filling, and the patient passed a balloon test occlusion with a 15-min hypotensive challenge

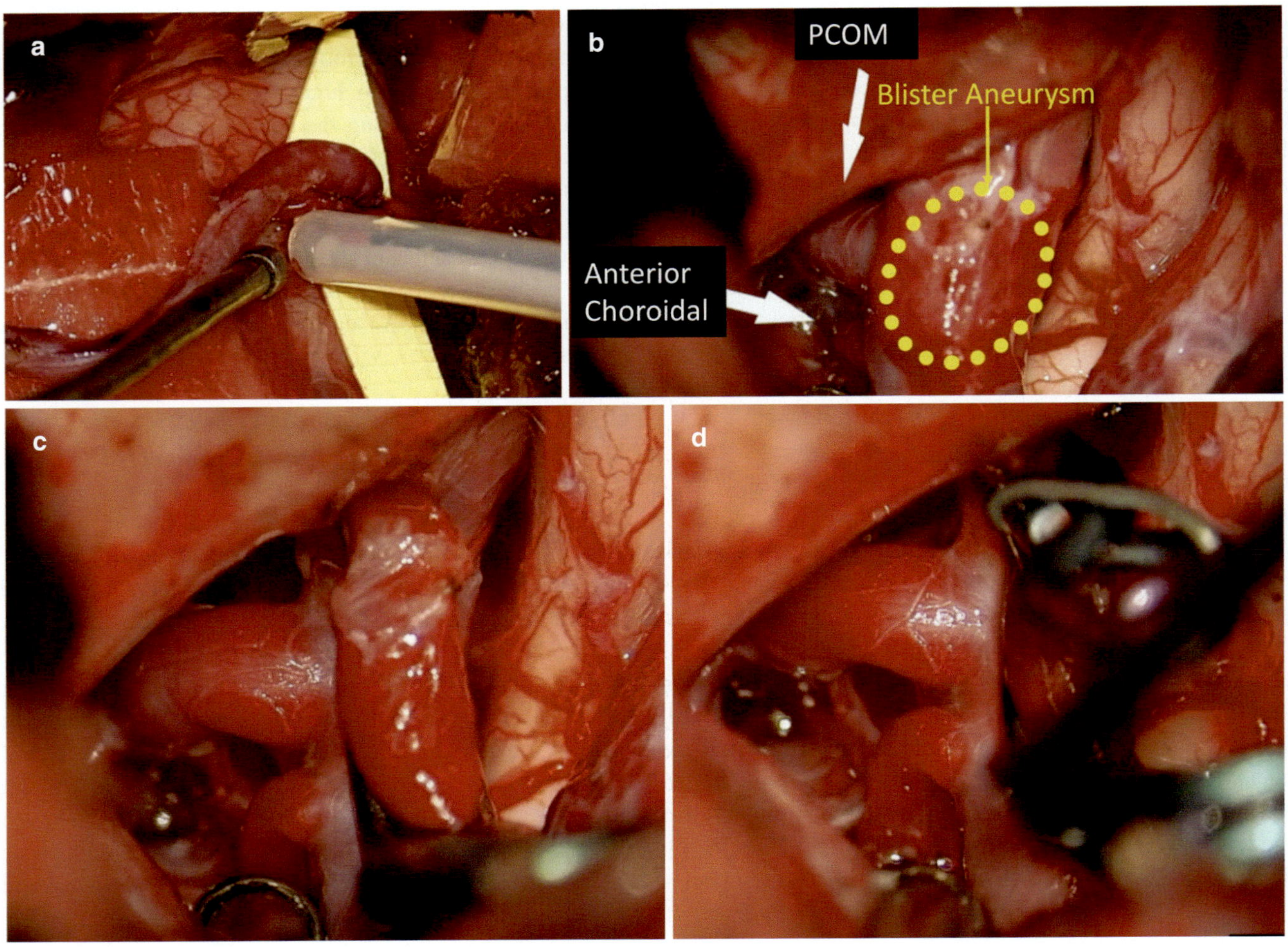

Fig. 15.7 Case 3 surgery: First, an STA-to-MCA end-to-side bypass was performed (**a**). Next, the supraclinoid ICA was exposed including the entirety of its diseased wall (dotted line) (**b**) along with the posterior communicating (PCOM) artery and the anterior choroidal artery. Since we had exposed the normal edges of the uninvolved ICA wall and we had secured good flow to the ipsilateral (left-sided) MCA after completing the bypass and a preoperative balloon test occlusion, we decided that instead of trapping the aneurysm proximal to the PCOM, it would be well tolerated to completely occlude the aneurysm using a right-angle clip and intentionally narrowing the ICA lumen while preserving flow into the PCOM and anterior choroidal branches (**c** and **d**), which may be needed if vasospasm were to occur

brain relaxation; wide arachnoidal dissection exposing the lower cranial nerves going to the jugular foramen; sectioning of the first dentate ligament; dissecting the arachnoid surrounding the cranial nerves IX, X, and XI to open up a corridor ventrally around the medulla to reach the aneurysm; final clip occlusion with a single curved clip; and aneurysm inspection and confirmation of complete occlusion. Post-operative cerebral angiography and 3D reconstructions showed complete occlusion and no recurrence to date (Fig. 15.9h, i).

Case 5. Aneurysms with Complex Morphologies: A Giant Fusiform Anterior Cerebral Artery Aneurysm in a Very Young Patient (Fig. 15.10)

A 10-year-old male presented with new-onset severe headaches. Neuroimaging (Fig. 15.10a–h) revealed a giant, partially thrombosed, fusiform anterior cerebral artery (ACA) aneurysm with a very complex morphology. The aneurysm was causing mass effect and vasogenic edema. The left A2 was the proximal afferent branch to the aneurysm and the left A3 was the afferent branch from the aneurysm distally. The right A2 was dominant, but the left A3 was supplying the left hemisphere (thus it was not a bihemispheric variant). There was no good endovascular option in this case and therefore microneurosurgical treatment was pursued.

As clip reconstruction was not possible, we planned for multiple revascularization options, including (1) A3-A3 side-to-side bypass and trapping, (2) STA-to-A3 bypass with a radial artery interposition graft and trapping, and (3) STA-to-A3 bypass with a saphenous vein interposition graft and trapping.

The surgical strategy involved mapping the left STA trunk, and frontal and parietal branches using Doppler, pre-

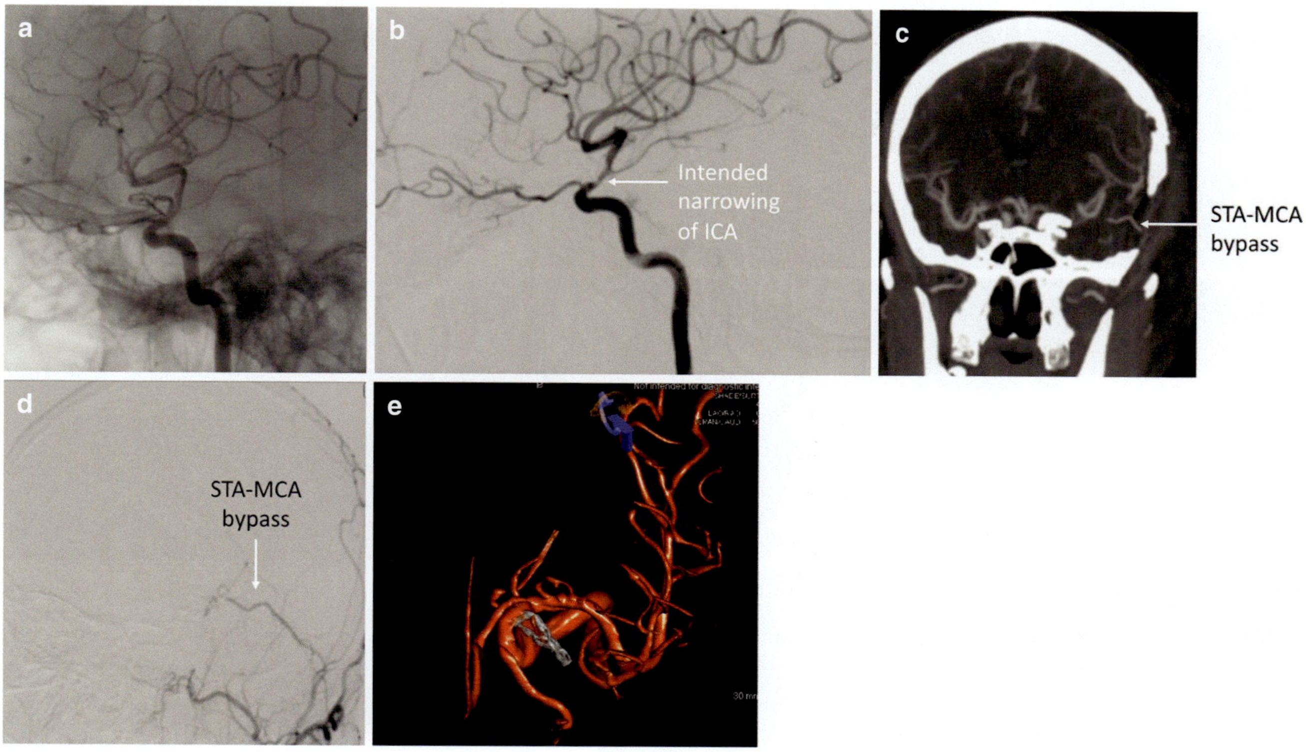

Fig. 15.8 Case 3 postoperative angiography: Lateral diagnostic cerebral angiography revealed aneurysm occlusion and good filling of the PCOM, anterior choroidal artery, and MCA (**a** and **b**) despite the purposely narrowed ICA lumen. CTA (**c**) and lateral diagnostic cerebral angiography (**d**) demonstrate patency of the STA-to-MCA bypass. Three-dimensional reconstruction (**e**) re-demonstrates the clip construct occluding the aneurysm and good filling of the neighboring arteries

paring the arm for a radial artery graft, and preparing for a saphenous vein graft in case that was needed. We planned a left frontotemporal incision extending across the midline for both an interhemispheric approach to the A3 distal to the aneurysm and a trans-Sylvian and sub-frontal approach to the A2 proximal to the aneurysm. After widely dissecting the interhemispheric fissure, we identified the A3 distal to the aneurysm (for distal trapping), which we temporarily occluded, and confirmed no good retrograde flow using ICG video angiography. This confirmed the need for revascularization to preserve flow in the left hemisphere. We also determined that the A3s were quite separated and gave up branches that tethered them to their ipsilateral hemisphere, and thus these could not be further mobilized and apposed for a side-to-sideA3-to-A3 anastomosis. Next, the Sylvian fissure was dissected, and we then went sub-frontally and then progressively opened the carotid, chiasmatic, lamina terminalis, and callosal cisterns. We routinely perform a third ventriculostomy by opening the *lamina terminalis*, and in addition to this, we also open Liliequist's membrane and communicate the aforementioned subarachnoid cisterns and third ventricle with the interpeduncular cistern. This dissection relaxed the brain and exposed the A2 proximal to the aneurysm for proximal trapping.

We ultimately executed our revascularization option #2: STA trunk-to-A3 bypass using a radial artery interposition graft and aneurysm trapping (Fig. 15.10i). We estimated we would need an 18 cm length of the radial artery. One obstacle that we encountered was that although a preoperative upper extremity MRA showed a long and healthy radial artery, the actual graft obtained was too short to bridge from the STA trunk to the A3. Therefore, we decided to anastomose the radial artery more distally to the frontal branch of the STA. We first proceeded with the radial artery-to-A3 end-to-side anastomosis and then performed the STA to radial artery end-to-end anastomosis. Another obstacle was the size mismatch between the radial artery graft and the frontal branch of the STA. To overcome this size mismatch, we used a tapering technique in which one side of the radial artery's "mouth" was cut obliquely and sutured to decrease the artery's diameter to better match the diameter of the smaller frontal branch of the STA (Fig. 15.10i, bubble c) [8]. Good flow was confirmed, and the aneurysm was trapped proximally and distally. Postoperative imaging revealed definitive and durable treatment of the aneurysm and good flow through the bypass (Fig. 15.10j–m). The patient was neurologically intact (mRS of 0), had an uneventful postoperative course, and his headaches resolved.

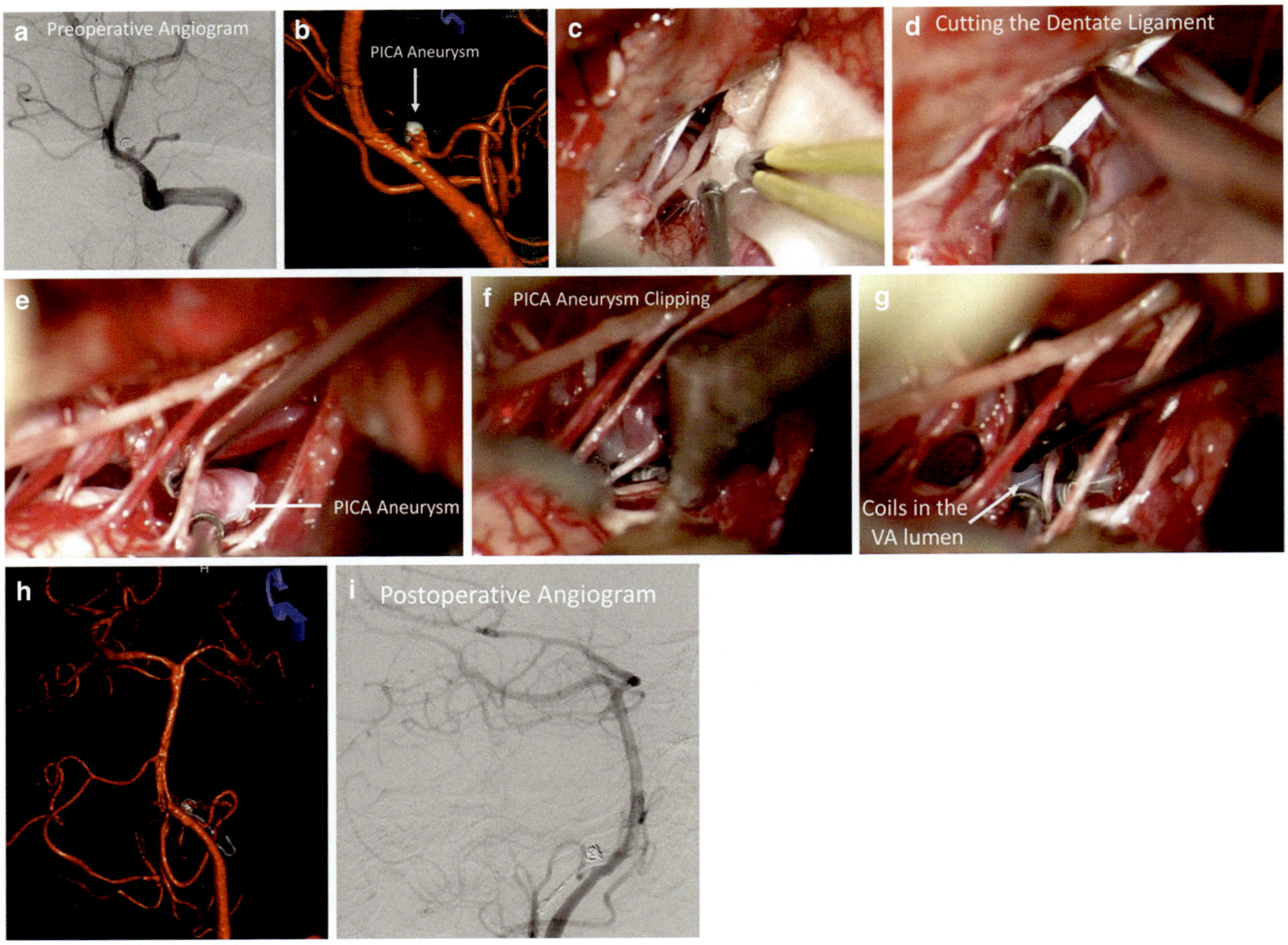

Fig. 15.9 Case 4 angiography and surgery: Preoperative angiography shows a recanalized posterior inferior cerebellar artery (PICA) aneurysm after failed endovascular coiling (**a** and **b**). For the microneurosurgical treatment, the patient was positioned in right lateral decubitus for a left-sided far lateral approach with minimal drilling of the occipital condyle. The dura was opened, and the cerebrospinal fluid was drained from the lateral cerebellomedullary cistern for brain relaxation (**c**). Cutting the first dentate ligament (**d**) opened more space and enabled attaining more proximal control on the left vertebral artery (VA). The arachnoid around the lower cranial nerves was dissected to carefully work between the rootlets of cranial nerves IX and X and expose (**e**) and clip the PICA aneurysm (**f**). Further inspection revealed coils within the vertebral artery (**g**) not revealed in the preoperative angiogram. Postoperative angiograms revealed good flow in the VA and PICA and nonfilling of the aneurysm (**h** and **i**)

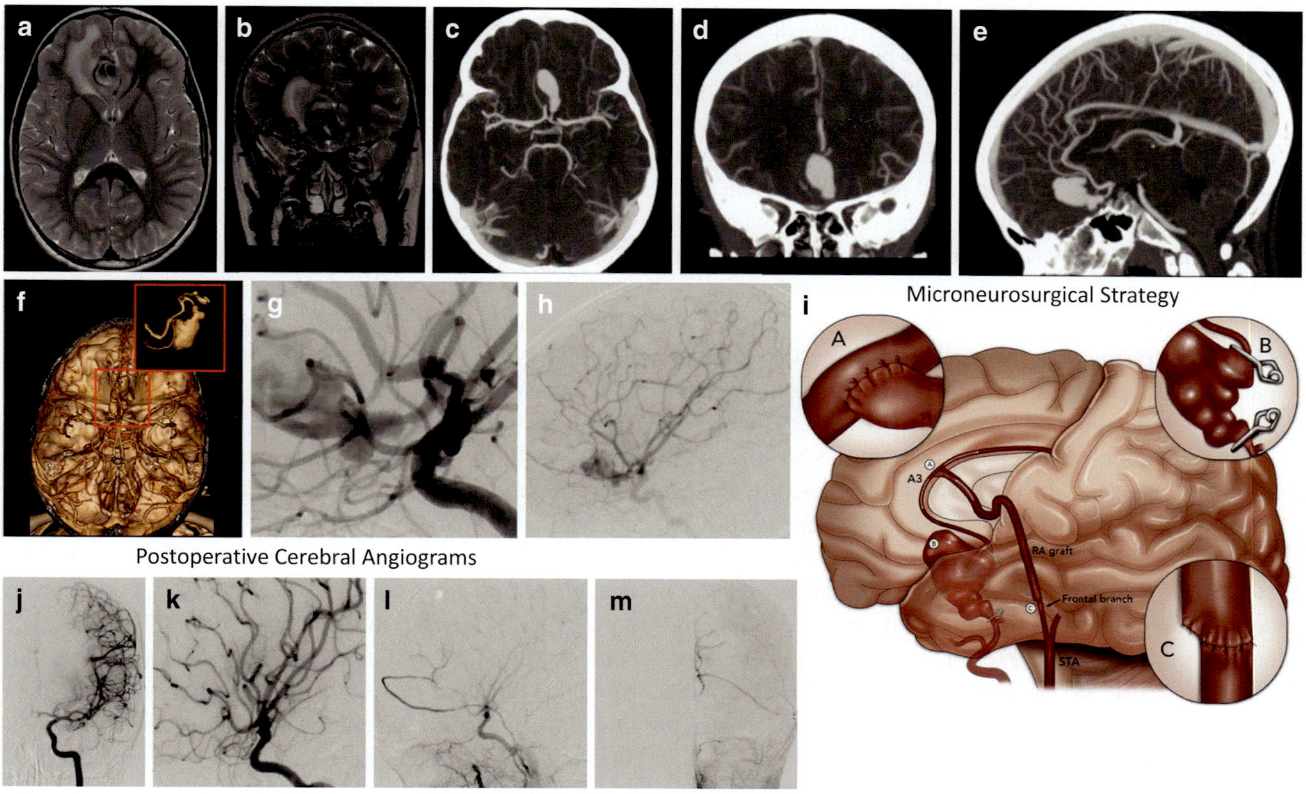

Fig. 15.10 Case 5 preoperative neuroimaging, angiography, and surgical plan: Preoperative MRI (**a** and **b**) showed a giant, partially thrombosed, fusiform distal ACA aneurysm causing mass effect and vasogenic edema. CTA (**c**–**e**) and diagnostic cerebral angiograms (**f**–**h**) further defined the angioarchitecture of this aneurysm with a very complex morphology arising from the left A2 (proximal or afferent branch) and A3 coming out of the aneurysm (distal or efferent branch). The microneurosurgical strategy (**i**) involved an STA-to-A3 bypass using an interposition radial artery graft (RAG) connecting the frontal division or branch of the STA to the afferent A3 (**bubble ia**) and aneurysm trapping with clips proximal and distal to the aneurysm (**bubble ib**). The size mismatch between the RAG and the frontal division of the STA was overcome using a tapering technique in which one side of the radial artery's "mouth" was cut obliquely and sutured to decrease the artery's diameter to better match the diameter of the smaller frontal branch of the STA (**bubble ic**). Postoperative diagnostic cerebral angiograms revealed nonfilling of the aneurysm (**j** and **k**) and patency of the bypass (**l** and **m**)

15.5 Surgical Pearls

- Indications for open microneurosurgical treatment of complex aneurysms include (1) large aneurysms causing mass effect, (2) aneurysms with complex morphologies, (3) aneurysms that failed endovascular treatment, (4) aneurysms with simple microsurgical treatment and difficult endovascular alternative, (5) aneurysm origin at a complex bifurcation angle with high hemodynamic stress, (6) aneurysms in young patients, and (7) aneurysms in distal locations (see complete list in Table 15.1). The endovascular treatment of aneurysms with complex fusiform morphologies, large sizes, and intraluminal clots that cause mass effects is not only challenging but also that partial aneurysm occlusion is associated with high rates of aneurysm recanalization.
- No matter how daunting the aneurysm is, there is always a good microsurgical option.
- Despite the size of the aneurysm, the microneurosurgical strategy follows the same basic principles of cerebrovascular neurosurgery: attaining proximal control, wide arachnoid dissection, exposure of the aneurysm, identification of the parent artery, identification of the proximal and distal aneurysm neck, identification and preservation of large and small arterial branches and perforators, temporary occlusion, aneurysm decompression with removal of thrombus and atheroma, clip reconstruction, and confirmation of complete occlusion and patency of important branches and perforators.
- Giant aneurysms usually obstruct the intraoperative visualization of their parent artery, their neck, and neighboring arterial branches and perforators. Therefore, the aneurysm must be decompressed either by suction decompression (if its content is mostly blood) or by incising the aneurysm dome and removing its contents (thrombus and atheroma) as much as possible until bleeding is encoun-

tered. At this point, the true angioarchitecture of the aneurysm and neighboring branches is better visualized and the aneurysm can be better occluded. A microsurgical treatment offers more versatility including clip occlusion or reconstruction, aneurysmorrhaphy (suturing the cut aneurysm wall), and trapping and revascularization. This offers an immediate and more robust occlusion whenever there is high hemodynamic stress at the aneurysm origin when compared to the range of endovascular options available currently.

- Large aneurysms that compress cranial nerves, such as paraclinoid aneurysms causing optic nerve compression, and posterior communicating (PCOM) aneurysms causing oculomotor compression, should be treated with microneurosurgery. In such cases, the cerebrovascular neurosurgeon, who should also be a skillful skull-base neurosurgeon, needs to apply the same basic principles of adequate bony exposure and early cranial nerve (CN) decompression, which include extradural anterior clinoidectomy and optic nerve unroofing for CN II and intradural dissection of the carotid's dural rings, cavernous sinus unroofing, and posterior clinoidectomy for oculomotor (CN III) decompression.
- Ruptured blister-like aneurysm should be treated microsurgically with trapping and bypass revascularization. The advantages of microsurgery include (1) attaining an immediate and complete aneurysm occlusion, which eliminates the risk of re-bleeding; (2) avoiding the use of dual antiplatelet therapy; and (3) providing an extra conduit in the setting of severe vasospasm.
- Whenever cerebral revascularization is considered, the microsurgeon must have multiple bypass strategies and always anticipate and prepare for any possible difficulty (i.e., short donor length, donor–recipient size mismatch, etc.).

15.6 Discussion

The microneurosurgical treatment of complex intracranial aneurysms requires meticulous planning, in-depth anatomic knowledge, readiness that comes with experience, practice and microsurgical dexterity and skills, and finally courage. The indications, principles, and microsurgical pearls of each representative case are discussed here.

Case 1. Large Aneurysm that Failed Endovascular Flow Diversion

This case highlights multiple indications for the open microneurosurgical treatment of complex aneurysms, including (1) large aneurysms causing mass effect, (2) aneurysms with complex morphologies, and (3) aneurysms that failed endovascular treatment.

Oftentimes, the endovascular treatment of complex large aneurysms achieves only partial occlusion and is associated with high rates of recanalization. No matter how daunting the aneurysm is, there is always a good microsurgical option to achieve complete occlusion.

Case 2. Giant Aneurysms Causing Mass Effect

This case highlights multiple indications for open microneurosurgical treatment of complex aneurysms: (1) the aneurysm origin at a complex bifurcation angle with high hemodynamic stress, (2) the aneurysm's giant size, and (3) the associated mass effect.

Despite the size of the aneurysm, the microneurosurgical strategy follows the same basic principles of cerebrovascular neurosurgery. Giant aneurysms usually obstruct the intraoperative visualization of their parent artery, their neck, and neighboring arterial branches and perforators. Therefore, the aneurysm must be decompressed either by suction decompression or by incising the aneurysm dome and removing its contents (thrombus and atheroma) as much as possible until bleeding is encountered. At this point, the true angioarchitecture of the aneurysm and neighboring branches is better visualized and the aneurysm can be better occluded. A microsurgical treatment offers more versatility including clip occlusion or reconstruction, aneurysmorrhaphy, and trapping and revascularization. Microsurgery offers an immediate and more robust occlusion whenever there is high hemodynamic stress at the aneurysm origin when compared to the range of endovascular options available currently. Large aneurysms that compress cranial nerves, such as paraclinoid aneurysms causing optic nerve compression, and posterior communicating (PCOM) aneurysms causing oculomotor compression, should be treated with microneurosurgery using skull base approaches.

Case 3. Blood Blister-Like Aneurysms

At our institution, trapping with bypass revascularization is the therapy of choice in cases with ruptured blister-like aneurysms. The advantages of microsurgery include (1) attaining an immediate and complete aneurysm occlusion, which eliminates the risk of re-bleeding; (2) avoiding the use of dual antiplatelet therapy; and (3) providing an extra conduit in the setting of severe vasospasm. In the present case, this included the flexibility to fully clip the aneurysm. While this led to a partial narrowing of the ICA lumen, this did not affect perfusion due to the STA-to-MCA bypass.

It must be emphasized that blister-like aneurysms are dangerous and friable lesions with only a thin layer of fibrous tissue holding them together. This places them at high risk for re-rupture, intraoperative rupture, and re-canalization if these are incompletely treated endovascularly. Attempting to microsurgically clip-ligate these lesions can be catastrophic because the clip blades very rarely incorporate the entirety of the diseased vessel wall that in most instances is more than 180 degrees of the parent vessel circumference.

In this case, we were able to place a clip occluding the diseased vessel wall while preserving the PCOM artery origin, knowing that this would cause stenosis of the parent artery. This is only feasible when the diseased vessel wall is readily visible and the distal arterial blood flow is secured via collateralization, as may be demonstrated with a favorable balloon test occlusion, a patent bypass, or both. Collateral flow in the PCOM can be confirmed by temporarily occluding the PCOM and observing simultaneous filling of the ICA and the PCOM perforators during ICG video angiography and by noticing no changes in the intraoperative neurophysiologic monitoring including motor and somatosensory-evoked potentials and EEG. In patients with high-grade subarachnoid hemorrhage and patients who failed a balloon test occlusion with cross-compression and hypotensive challenge, we recommend a high-flow, common carotid artery to M2 bypass with an interposition radial artery graft (see Case 1, Fig. 15.2) [9].

Case 4. Recanalized Posterior Inferior Cerebellar Artery Aneurysm after a Failed Endovascular Coiling
This case highlights the following indications for open microneurosurgical repair of intracranial aneurysms: (1) aneurysms in young patients, (2) aneurysms that failed endovascular treatment, (3) aneurysms at difficult branching angles, and (4) aneurysms with a simple microsurgical treatment and a difficult endovascular alternative. PICA-origin aneurysms are predominantly treated with microsurgical clipping at our institution. As in this case, the PICA angle is often acute, which makes it harder to navigate endovascularly, which risks aneurysm perforation and recurrence after partial coiling. If the aneurysm cannot be clip-reconstructed without occluding the PICA, there are three microneurosurgical options. Option one, the aneurysm can be removed and the PICA revascularized with an end-to-end anastomosis if there is a good proximal segment of the vessel and sufficient freedom to mobilize the distal PICA. Option two is to revascularize the PICA with a distal PICA-to-PICA side-to-side anastomosis and to trap the aneurysm. Lastly, option three, and the last resort, is to temporarily occlude the PICA over an extended period (>30 minutes) and sacrifice it proximal to the aneurysm if there is absolutely no change in neurophysiologic monitoring parameters.

Case 5. Aneurysms with Complex Morphologies: A Giant Fusiform Anterior Cerebral Artery Aneurysm in a Very Young Patient
This case highlights the following indications for open microneurosurgical repair of intracranial aneurysms: (1) giant size, (2) fusiform morphology, (3) very young patient, (4) distal location, and (5) mass effect. This case also highlights the need for careful and detailed preparation including having multiple bypass options and finding alternative techniques to overcome difficulties encountered.

15.7 Conclusions

Microneurosurgery continues to play an important role in the immediate and definitive cure of complex intracranial aneurysms. A microneurosurgical strategy should be entertained for any of the circumstances described in Table 15.1.

Despite the size and morphology of the aneurysm, the basic principles of cerebrovascular neurosurgery remain and include early proximal vascular control (prepare the internal carotid in the neck when needed), exposure of the proximal and distal neck of the aneurysm, identification, and protection of all neighboring branches (debulk the aneurysm mass when needed), and clip reconstruction of the aneurysm whenever possible. If clipping is not possible, consider trapping the aneurysm after securing good distal flow with a bypass, and be ready and able to use the STA, radial artery, saphenous vein, and carotid artery. Or, if clipping is not possible, consider using collateral flow, including checking with preoperative balloon test occlusion with hypotensive challenge and intraoperative temporary occlusion with ICG video angiography and neurophysiologic monitoring.

References

1. Straus DC, da Silva HB, McGrath L, Levitt MR, Kim LJ, Ghodke BV, Barber JK, Sekhar LN. Cerebral revascularization for aneurysms in the flow-diverter era. Neurosurgery. 2017;80(5):759–68. https://doi.org/10.1093/neuros/nyx064.
2. Rutledge C, Baranoski JF, Catapano JS, Lawton MT, Spetzler RF. Microsurgical treatment of cerebral aneurysms. World Neurosurg. 2022;159:250–8. https://doi.org/10.1016/j.wneu.2021.12.079.
3. Cohen-Gadol AA. Tackling technical challenges for microsurgical clip ligation of complex aneurysms during the endovascular era: a personal perspective. Neurosurgery. 2014;61(Suppl. 1):43–53. https://doi.org/10.1227/NEU.0000000000000374.
4. Korosue K, Heros RC. "Subclinoid" carotid aneurysm with erosion of the anterior clinoid process and fatal intraoperative rupture. Neurosurgery. 1992;31(2):356–359; discussion 359–360. https://doi.org/10.1227/00006123-199208000-00024.
5. Heros RC. Anterior clinoidectomy World Neurosurg. 2012;77(3–4):441–2. https://doi.org/10.1016/j.wneu.2011.08.007.
6. Richardson AM, Baskaya MK. Extracranial-intracranial bypass as primary treatment for internal carotid artery blood blister-like aneurysms, not just a last resort. World Neurosurg. 2021;145:320–2. https://doi.org/10.1016/j.wneu.2020.09.130.

7. Cıkla U, Baggott C, Başkaya MK. How I do it: treatment of blood blister-like aneurysms of the supraclinoid internal carotid artery by extracranial-to-intracranial bypass and trapping. Acta Neurochir. 2014;156(11):2071–7. https://doi.org/10.1007/s00701-014-2212-8.
8. Cikla U, Mukherjee D, Tumturk A, Baskaya MK. Overcoming end-to-end vessel mismatch during superficial temporal artery-radial artery-M2 interposition grafting for cerebral ischemia: tapering technique. World Neurosurg. 2018;110:85. https://doi.org/10.1016/j.wneu.2017.10.162.
9. Başkaya MK, Ahmed AS, Ateş O, Niemann D. Surgical treatment of blood blister-like aneurysms of the supraclinoid internal carotid artery with extracranial-intracranial bypass and trapping. Neurosurg Focus. 2008;24(2):E13. https://doi.org/10.3171/FOC/2008/24/2/E13.

Advanced Micro-Neurosurgery in Treating Basilar Aneurysms

16

Ali F. Krisht, Jaafar Basma, and Abdel Raouf Kayssi

Learning Objectives
- Conceive the basic anatomy of the transcavernous approach to the basilar apex.
- Discern the essential steps of the surgical approach to basilar aneurysms.
- Understand the main pitfalls and nuances of basilar aneurysm clipping.

16.1 Introduction

The treatment of basilar apex aneurysms has witnessed significant evolutions in the modern era. Contrary to what has almost become a common belief, such improvements were not made through endovascular means alone. Micro-neurosurgery continues to elevate the standard of care for these patients, as it provides advanced techniques and anatomical knowledge for a safer and more durable treatment. The historical surgical difficulties and increased risks, with the need for long and dedicated training, have widened the endovascular indications of these aneurysms to even include those which are not favorable for such treatments. These have resulted in higher rates of recurrences and the need for more complex and sometimes morbid interventions. In many centers throughout the world, endovascular therapy has become the only option, and all its drawbacks are accepted there as inexorable. However, mastering the anatomy of the surgical approach, and the surgical nuances and pitfalls, has simplified the microsurgical treatment and rendered it much safer nowadays, promising superior long-term results.

A. F. Krisht (✉) · J. Basma · A. R. Kayssi
Arkansas Neuroscience Institute, CHI Saint-Vincent, Sherwood, AR, USA
e-mail: ali.krisht@commonspirit.org

16.2 History/Background

Professors Olivecrona (1932) and Schwartz (1948) are probably to be credited with first clipping a basilar (or posterior fossa) aneurysm [1, 2]. However, such aneurysms remained associated with a high morbidity and mortality rate for a long time. Through a refined subtemporal corridor, which provides direct access to the basilar apex, Drake has then made significant advances to the treatment of basilar aneurysms, decreasing its mortality rate from 53% to 7% [3, 4].

Yasargil, who revolutionized micro-neurosurgery, advocated for a pterional approach to prevent excessive temporal lobe retraction, and to allow a view across the contralateral P1 and its perforators, through the Sylvian fissure and the optic-carotid-oculomotor windows [1, 5]. He admitted that the surgical space is "confining and requires advanced microsurgical experience and techniques." Yasargil described many tricks still used today to optimize the exposure of these difficult aneurysms, such as increased temporal exposure in the craniotomy, wide Sylvian fissure split, opening the optic-carotid and the carotid-oculomotor spaces, mobilizing the temporal lobe and any temporo-polar veins for safer retraction, excising and retracting the tentorial edge, dividing the circle of Willis (such as a hypoplastic posterior communicating or P1 arteries), and drilling the posterior clinoid process (PCP). He also advocated for aneurysm fundus coagulation to remodel it, permitting safer clipping and visualization of the surrounding perforators [1, 5].

Stimulated by Yasargil, Dolenc later developed a combined extradural–intradural trans-clinoid trans-cavernous approach to the basilar apex [6]. After the lateral wall of the cavernous is exposed, the clinoid is removed, and this immediately widens the optic-carotid and carotid-oculomotor spaces. Using this approach, the PCP is easier to remove, and lower access to the basilar trunk is obtained for proximal control. While Parkinson described a transcavernous approach to treat carotid cavernous fistulas, he had significant difficulties controlling cavernous sinus bleeding, often

I. N. Kanaan, V. Beneš (eds.), *Neuroanatomy Guidance to Successful Neurosurgical Interventions*,
https://doi.org/10.1007/978-3-031-59838-8_16

necessitating extra-corporeal circulation [7]. Nowadays, the cavernous sinus can be tamed through targeted injections of fibrin glue, guided by a detailed anatomical knowledge of its venous tributaries [8].

Professor de Oliveira published several reports on what he defined as the pretemporal approach, which added middle fossa and trans-cavernous extensions to Yasargil's pterional approach [9, 10]. As Drake has long thought of, it combines the advantages of both the pterional and the subtemporal approaches, with the adjoined trans-cavernous dissection. The trans-cavernous approach has continued to evolve, starting from the roof of the cavernous sinus through a clinoidectomy, to complete mobilization of cranial nerves, to even beyond the petro-clival junction and petrous apex down to the prepontine and premedullary spaces [11]. The cavernous sinus, instead of being the long-feared forbidden territory, has become an accessible route for the deep and confining basilar apex [12, 13]. Several steps have been added to optimize the deep surgical windows, proximal and distal control, and maneuverability around the basilar system.

16.3 Anatomy Highlights

16.3.1 Basilar Artery

The basilar artery divides in the interpeduncular cistern, into two superior cerebellar (SCA) and two posterior cerebral arteries (PCA), both of which flank the origin of the oculomotor nerve (III). As it joins the PCA, the posterior communicating (pcom) artery divides it into pre-communicating (P1) and post-communicating (P2) segments. The surgeon should always assess whether the patient has a fetal posterior communicating aretery, as this would make its sacrifice more problematic, and may require, if needed, applying a temporary clip to it for further control.

16.3.2 Perforators

The perforators in the interpeduncular region represent one of the key challenges of clipping basilar apex aneurysms, as these can be hindered by the aneurysm in a restricted space. These include the anterior thalamoperforating arteries arising from the pcom, the posterior thalamoperforating and posterior medial choroidal arteries from P1, and the paramedian and circumflex arteries from the SCA. These can have different configurations, such as instances where posterior thalamoperforating arteries have a common single origin on the ipsi- or contralateral P1 [1].

16.3.3 Oculomotor Nerve

The third nerve enters the cavernous sinus where the tentorial edge converges with its petroclinoid folds to attach to the anterior and PCP, forming the oculomotor trigone. It is then embodied within the lateral wall of the cavernous sinus, along with the trochlear nerve and the ophthalmic branch of the trigeminal nerve (V1) lateral to the cavernous internal carotid artery (ICA). When approaching the basilar apex from a pterional approach, the anterior and PCP, the dorsum sellae, the ICA and pcom arteries, and the neurovascular structures of the cavernous sinus represent natural obstacles that can either be removed or mobilized for enhanced exposure to the basilar apex.

The detailed anatomy of the sphenoid wing, clinoid process, middle fossa, petrous bone, cavernous sinus structures, and petro-clival region has been previously well described, and we will review pertinent key elements in the surgical approach [10, 11, 14].

16.4 Approach/Surgical intervention

The patient is positioned supine with the head rotated and tilted to the contralateral side and extended, to bring the malar eminence to the highest point of the field (Fig. 16.1). An interfascial temporalis muscle dissection is performed, and the muscle is detached from the zygomatic process of the frontal bone down to the temporal root inferiorly. This allows its full mobilization and its inferior and anterior retraction, permitting a full temporal exposure inferiorly, leveled with the middle fossa floor (Fig. 16.2). The pretemporal craniotomy must also extend anteriorly to the subfrontal region just posterior to the orbital roof (Fig. 16.3). This mobilizes the frontal dura and facilitates the epidural dissection.

Subfrontal and subtemporal dissection is carried out with further drilling of the squamous temporal and frontal bones until the middle fossa floor, frontal fossa floor and the foramen rotundum are well exposed (Fig. 16.4). The middle meningeal artery is coagulated and cut away from the foramen spinosum, and the foramen ovale is identified. The roof and lateral wall of the orbit are drilled while preserving the periorbita, from the superior orbital fissure to the optic canal (Fig. 16.5). The orbito-meningeal artery is coagulated and cut (Fig. 16.6). Cutting and dissecting the dura over the clinoid process and in line with the groove between the posterolateral orbit and the superior orbital fissure, opens the plane between the dura propria of the temporal lobe and the lateral wall of the cavernous sinus. During this dissection, the

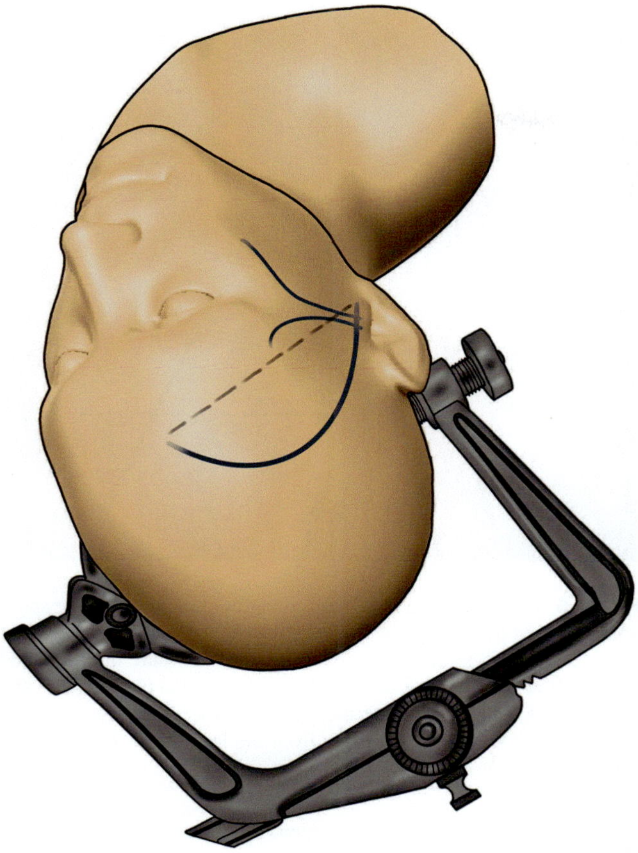

Fig. 16.1 Patient position and incision

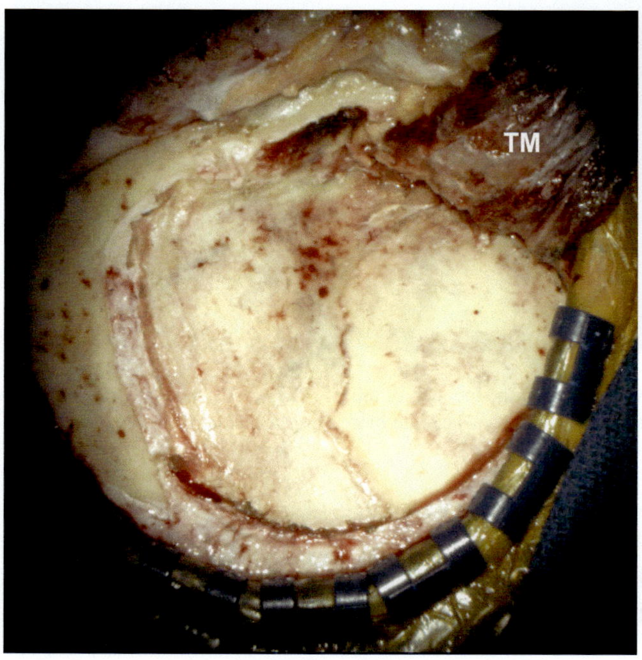

Fig. 16.2 Temporalis muscle and extent of bone exposure showing muscle flap and wide exposure of temporal bone and pre-temporal region

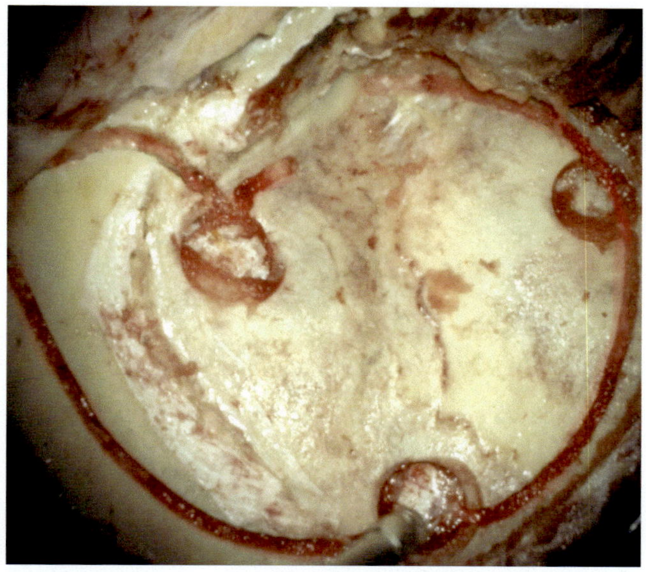

Fig. 16.3 Pretemporal craniotomy extending anteriorly to the subfrontal region

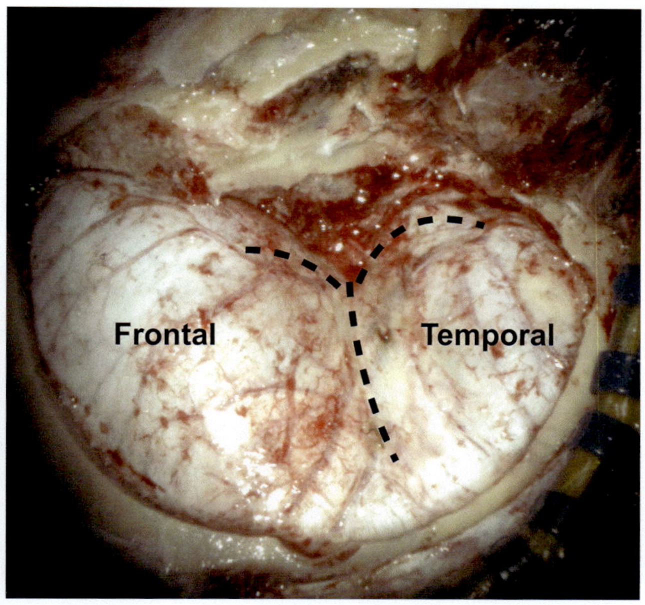

Fig. 16.4 Pterional craniotomy with wide temporal and pre-temporal exposure

spheno-parietal sinus (SPS) can be identified and preserved to avoid excessive premature bleeding (Fig. 16.7). The lateral aspect of the clinoid process will also start to surface on the surgical field. Further dissection will expose V1, V2 and V3, Meckel's cave, the greater superficial petrosal nerve (GSPN), and the petrous apex. As the temporal dura is elevated, the trochlear nerve becomes more visible medially in the lateral wall of the cavernous sinus [15].

Fig. 16.5 Removal of the posterior third of the orbital roof to the level of ACP after drilling of the sphenoid wing

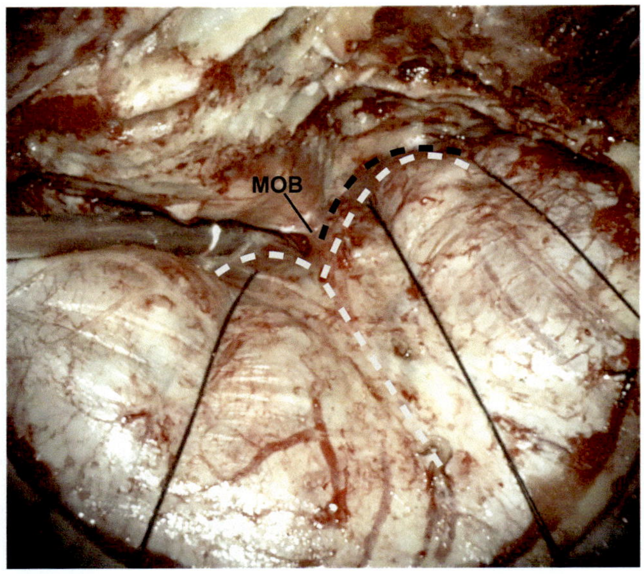

Fig. 16.6 Extent of extradural exposure after the removal of the sphenoid wing and posterior third of the roof of orbit. The white dashed line shows the line of dural opening over the Sylvian fissure in a T-shaped fashion. Black dashed line shows mthe eningo-orbital band (MOB) and artery (MOA) line of dissection that starts the process of dissecting dura propria of the temporal lobe away from the lateral wall of cavernous sinus

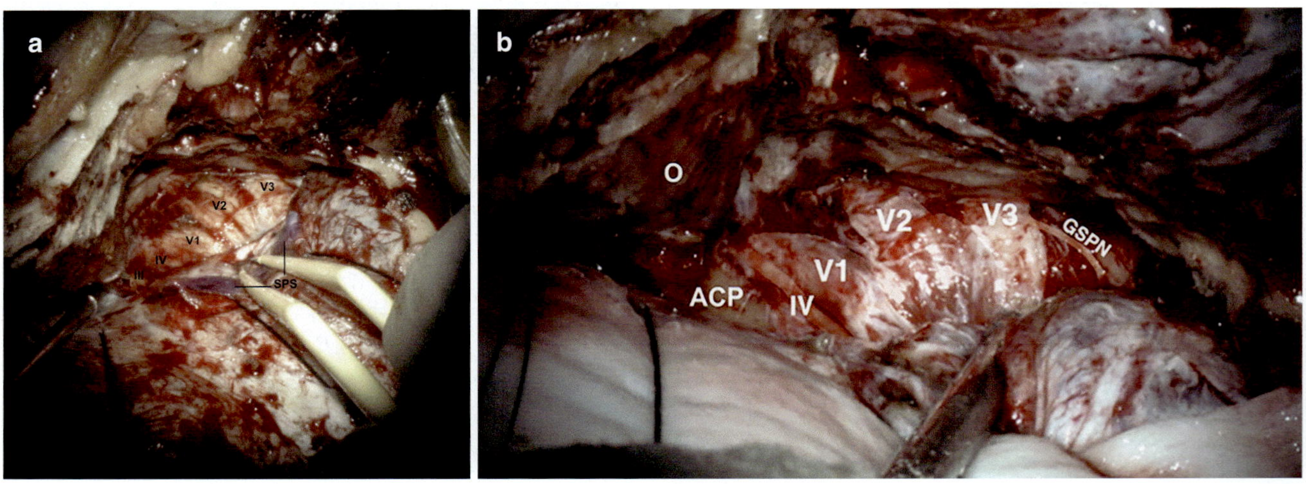

Fig. 16.7 (**a**) Exposure of lateral wall of the cavernous sinus after establishing hemostasis and peeling the middle fossa dura and showing how dissection went beyond the course of the SPS. (**b**) More lateral view showing the location of the GSPN laterally, and the tip of the ACP medially which has become more of a superficial structure

Venous hemostasis in the middle fossa and cavernous sinus, even before the anterior clinoid process (ACP) is removed, can be tricky. Understanding the regional venous anatomy is key to minimize any undue bleeding. The dissected SPS along the temporal dura is opened as it enters the cavernous sinus between V1 and V2. Fibrin glue is injected toward the cavernous sinus, while the proximal part is compressed to prevent reflux to the Sylvian vein. Fibrin glue is also injected between V1 and V2, and between V2 and V3. The superior petrosal sinus at the petrous edge on the other hand is corked with a piece of gelfoam, rather than injecting tisseel toward the veins of the posterior fossa [8, 15].

The ACP is then removed, by disconnecting its three attachments: the lesser sphenoid wing, the optic roof, and the optic strut (Fig. 16.8a). A small high-speed diamond drill bit is used with intermittent but frequent irrigation (Fig. 16.8b). It is preferable that the clinoid process be drilled all the way to the tip in order to be left with a small piece that can easily

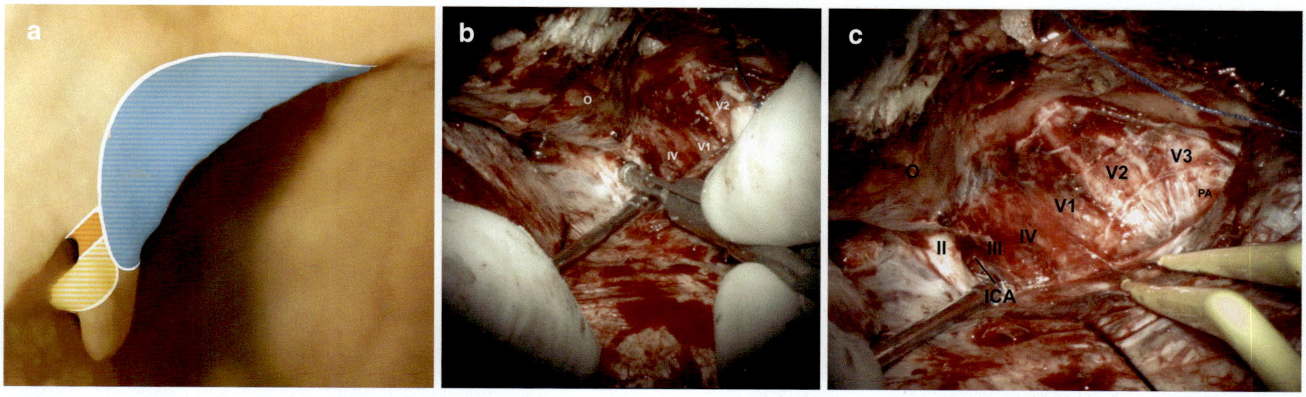

Fig. 16.8 (**a**) Anterior clinoid process and its three surgical attachments. (**b**) Drilling of the anterior clinoid using a high-speed diamond drill. (**c**) Shows the full extent of extradural exposure extending from the supra-optic region medially to the petrous apex laterally

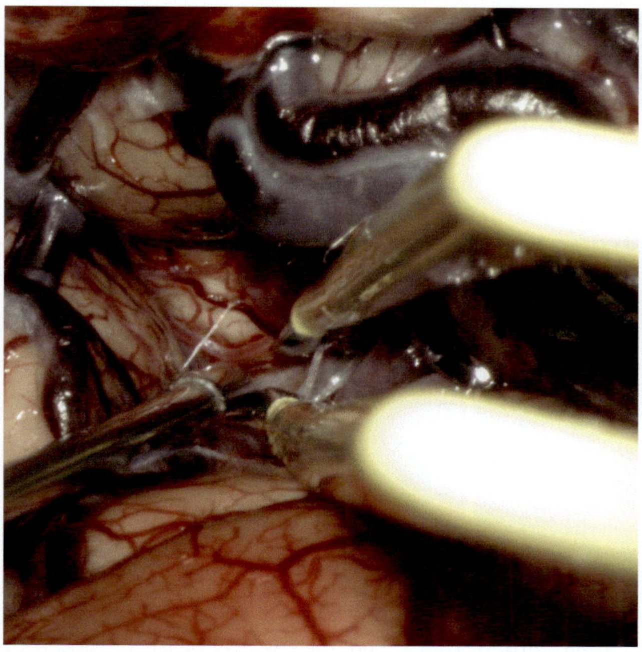

Fig. 16.9 Opening of Sylvian's fissure from inside-out

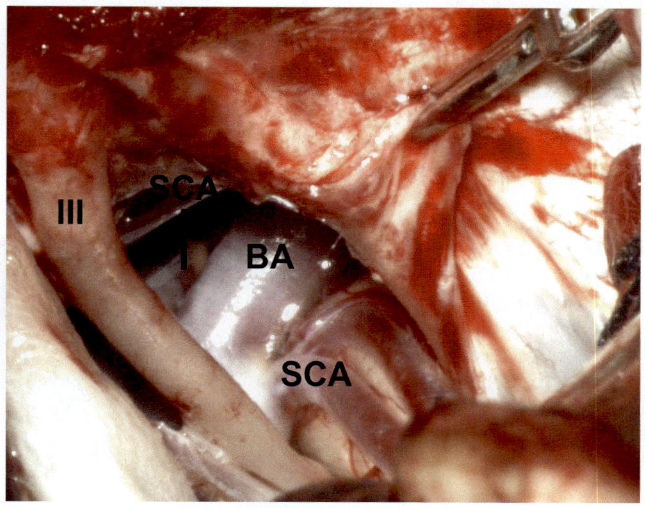

Fig. 16.10 The initial window to the inter-peduncular fossa after mobilizing the third nerve and reflecting its dural sheath laterally and holding it with a clip

be removed without any struggle to avoid injury to either the carotid artery or the adjacent oculomotor nerve. Removing the clinoid process unlocks the roof of the cavernous sinus, where fibrin glue can be injected, and where the path of the oculomotor nerve is exposed (Fig. 16.8c). The optic-oculomotor space is widened, and the course of the posterior communicating artery is better visualized [16, 17].

The dura is opened in a T-shaped fashion, starting over the Sylvian fissure, and then bifurcating toward the subfrontal and subtemporal regions (Fig. 16.6). To approach basilar aneurysms, the Sylvian fissure has to be opened widely from distal to proximal using professor Yasargil's inside-out technique [1] (Fig. 16.9). Splitting the Sylvian fissure and mobilizing the temporal lobe allow its gentle retraction to uncover the tentorial incisura down to the posterior fossa. The posterior communicating artery is dissected and the Liliquist membrane is perforated exposing the P1 and P2 junction of the posterior cerebral artery. The latter can be followed to the basilar apex; however, it is advised to complete the surgical approach prior to manipulating or disrupting the aneurysm. The origin of the oculomotor nerve and the ipsilateral SCA is easily identified, but the remaining structures are hindered, and the space is confining (Fig. 16.10). At this point, it is thus important to widen the deep surgical corridors to (1) optimize the proximal control at the basilar trunk inferiorly; (2) secure distal control at the contralateral P1 segment; (3) maximize the view to the aneurysm's dome, distal neck, and the posteriorly projecting basilar perforators; and (4) increase the deep working space and maneuverability for the clipping process. Several steps are available for these purposes and can be performed depending on the complexity of each case [15].

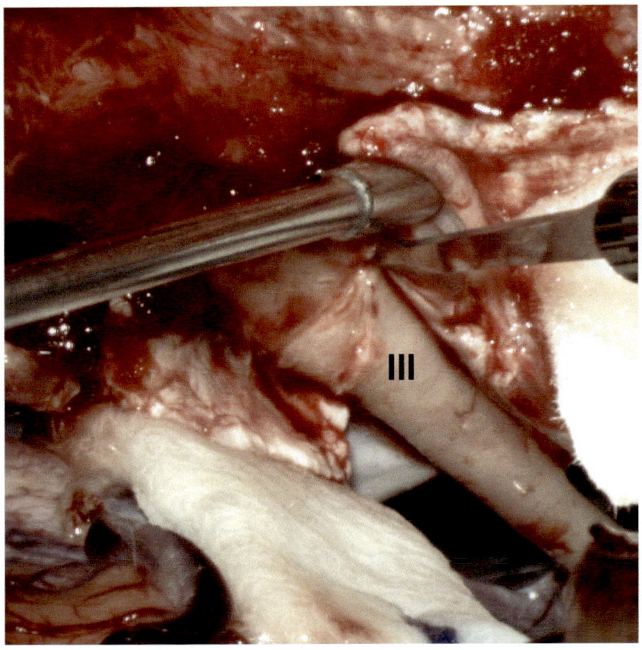

Fig. 16.11 Further dissection of the course of the third nerve toward the superior orbital fissure to widen the exposure at its depth

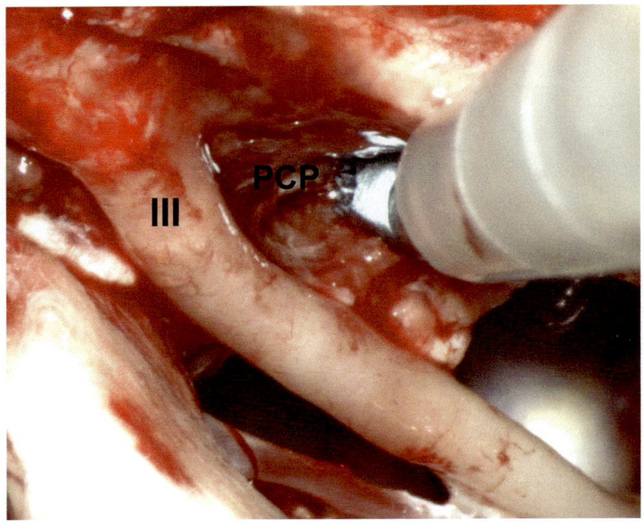

Fig. 16.12 The removal of the PCP with the ultrasonic aspirator which is done if needed to expose the basilar trunk at a perforator-free region

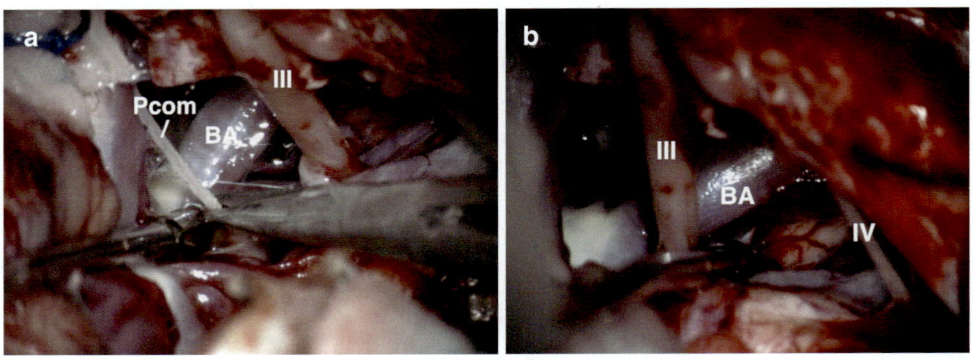

Fig. 16.13 (a) A wider window achieved by the previously made steps and cutting of the pcom (when hypoplastic) at a perforator-free zone. (b) Exposure of basilar trunk to allow access to the perforator-free zone below SCA

Cutting the clinoidal dura, the petro-clinoidal folds, and dissecting the oculomotor nerve off its dural sleeve along the oculomotor trigone will result in a complete mobilization of the third nerve from its brainstem origin to the superior orbital fissure (Fig. 16.11). This decreases its tension while increasing the exposure to the basilar trunk, and facilitating the removal of the PCP [18]. This can be drilled using an ultrasonic drill after gently retracting the mobilized third nerve (Fig. 16.12). The dura of the petroclival region can also be cut posterior to the oculomotor triangle after identifying the posterior genu of the intracavernous internal carotid artery using a Doppler ultrasound probe. The course of the posterior communicating artery as it relates to the view of the aneurysm and the region of the perforators is assessed. If it is deemed necessary, and if the posterior communicating artery is hypoplastic, then it can be coagulated and cut at a perforator-free zone, leading to a gentle mobilization of the internal carotid artery, and further access to the posterior aspect of the aneurysm's neck and the basilar perforators [19] (Fig. 16.13).

For low-lying aneurysms, a more complete trans-cavernous approach may be required. The trochlear nerve is then mobilized by releasing it from its leaflets in the lateral wall of the cavernous sinus. Its intra-dural course is inspected intra-durally through the extended subtemporal durotomy, and the tentorium is cut, thus leading to the nerve's full mobilization. Following this, the dura over Meckel's cave is also slit open and removed along with the pteroclival dura, exposing the junction between the petrous apex and the clivus [15]. Drilling bone at the dorsum sellae and petroclival junction extends the caudal approach to the prepontine cistern, providing an inferior view along the basilar trunk, almost to the level of the vertebro-basilar junction if necessary.

16.5 Surgical Pearls

- The process of clipping basilar apex aneurysms should not be restricted to one "hit or miss" step. The surgeon should take advantage of the 360-degree view procured by the transcavernous approach to ensure complete obliteration of the aneurysm while the basilar branches and perforators are well preserved beyond any doubt [20].
- Proximal and distal temporary clips deflate the aneurysm for safer dissection and clip application (Fig. 16.14). Note that the distal temporary clip may need to be placed on the posterior communicating artery when it is of fetal configuration.
- Basilar aneurysms usually project (1) anteriorly, (2) superiorly or supero-posteriorly, or (3) postero-inferiorly [1]. They may also have a combination of these projections. For instance, an aneurysm may project superiorly but may have a posterior bulge, which needs to be recognized and clipped. Anteriorly projecting aneurysms are more favorable because they are less likely to involve the perforators. Superiorly or supero-posterior aneurysms hide the contralateral perforators, but their neck does not usually tangle with the perforators. Postero-inferior aneurysms are embedded within the interpeduncular fossa, tightly surrounded by perforators. These are usually the most difficult to treat, especially when they are of a larger size (Fig. 16.15).
- For complex aneurysms, such as those that are ruptured, and/or larger in size, and/or posteriorly projecting, several strategies can be employed to maximize the safety of the clipping process. Short intermittent temporary clipping of the basilar trunk and bilateral P1 arteries at perforator-free zones is followed by the careful dissection of the perforators off the proximal neck of the aneurysm.
- Temporary clipping is usually maintained at short intervals not exceeding 2 min. A pilot clip can then be applied in the direction of the neck, through the dissected proximal window. This helps to visualize the perforators at the distal neck, and small pieces of cotton can be used to protect them.

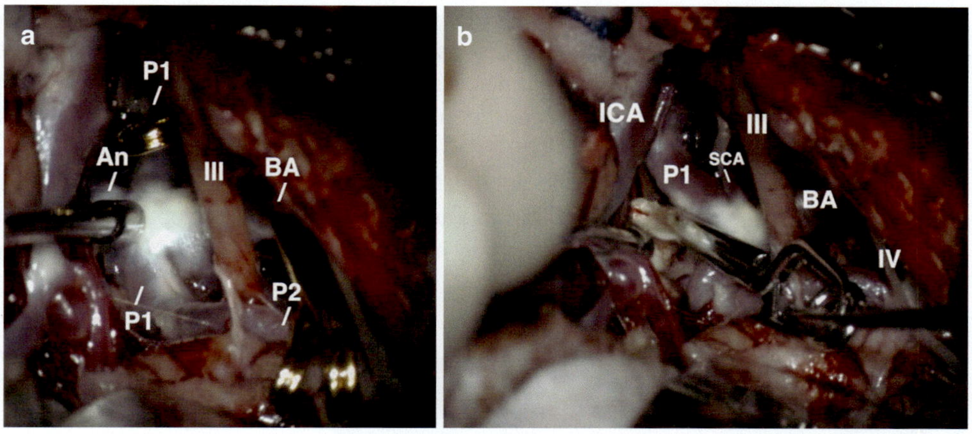

Fig. 16.14 (**a**) Temporary clipping of the basilar trunk and bilateral PCA allowing deflation of the system and full exposure of aneurysm base without obstruction. (**b**) Aneurysm clipped with dome deflated and opened

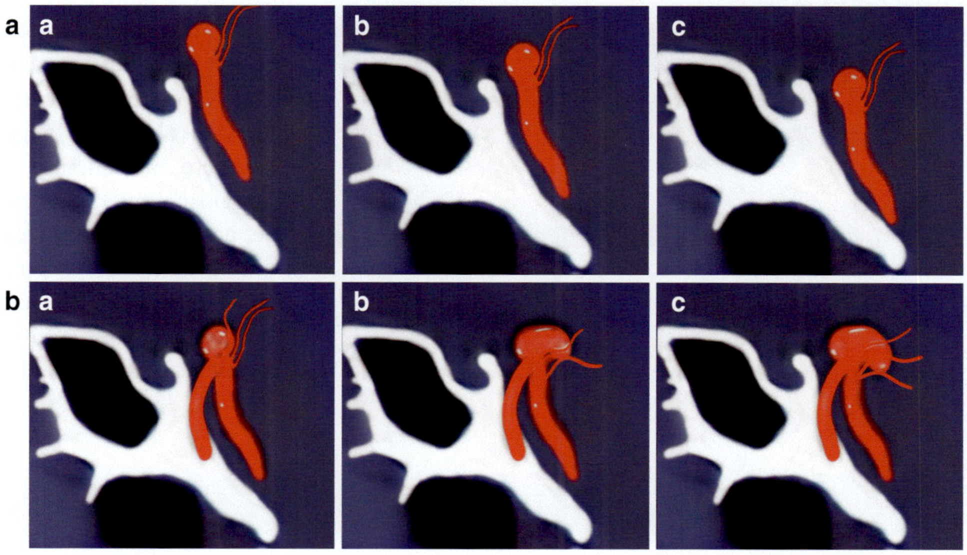

Fig. 16.15 Levels (**a**) and projections (**b**) of basilar tip aneurysms

- Depending on the aneurysm, it can then be deflated with a needle or reinforced with a second clip distal to the pilot clip, so that the pilot clip can be removed. This changes the initial configuration of a large and/or ruptured aneurysm, to a smaller and unruptured one, with the perforators at the distal neck now exposed and protected.
- Bipolar remodeling can also be used to shrink the aneurysm dome away from the perforators and expose them. At this stage, a permanent "smart" clip(s) can be applied, one (or more) which would completely occlude the aneurysm and avoid the perforators. Bayoneted clips can be helpful to avoid obstructing the view in the interpeduncular fossa. After the aneurysm is clipped, the dome is opened and deflated, ensuring that it has been completely occluded, and to inspect that the perforators are well preserved.
- For very low-seated basilar aneurysms, a full transcavernous approach may be necessary, requiring a full mobilization of the oculomotor and trochlear nerves; opening Meckel's cave; drilling the PCP, dorsum sellae, and petrous apex; and removing all the intervening dura in between (tentorial, petro-clival and clinoidal dura) [15] (Figs. 16.15a–c and 16.16). This leads to an extended view of the lower aspect of the prepontine cistern, where even the vertebral arteries could be seen in the premedullary space (Fig. 16.17).
- With very low-lying aneurysms, the approach is rather directed in the trajectory of the middle fossa and cavernous sinus and opening the Sylvian fissure may not even be mandatory.

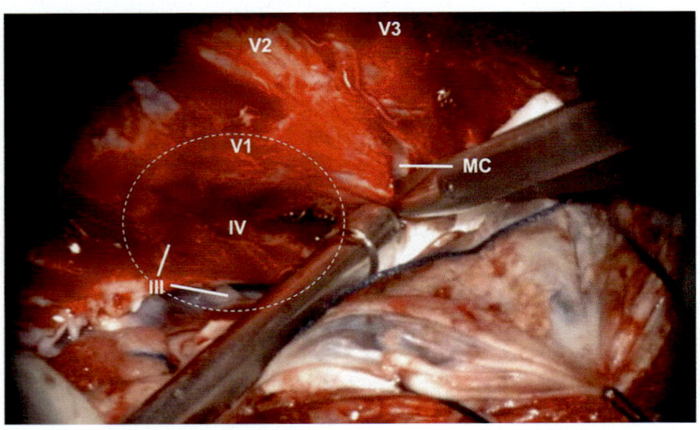

Fig. 16.16 In the very low ones, such as the aneurysm in A–C, a full transcavernous approach can be done in which third and fourth nerves are mobilized with the opening of Meckel's roof and the petroclival junction to unroof the prepontine cistern (dashed circle)

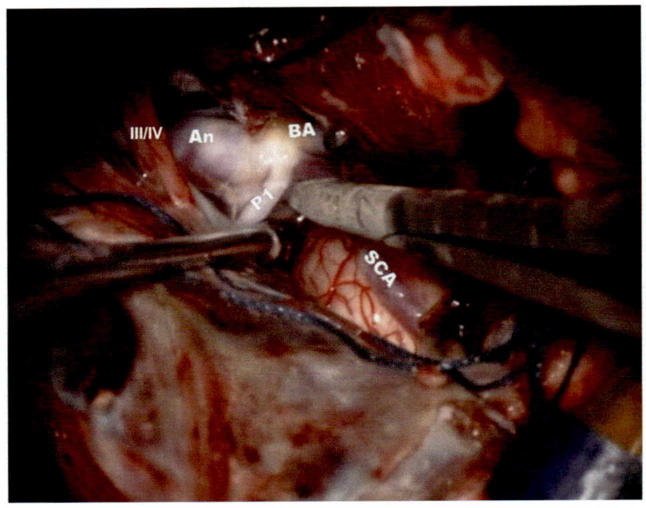

Fig. 16.17 View of a low-lying basilar artery after the full transcavernous exposure

16.6 Discussion

Microsurgical clipping of saccular aneurysms remains superior to endovascular treatment at least in cases that include larger size aneurysms or very small aneurysms, blister aneurysms, aneurysms with mass effect, high neck-to-dome ratio, hemodynamic stress locations, and in patients with kidney failure or difficult endovascular access. Basilar aneurysms are no exception to this list, and their microsurgical treatment should continue to be utilized as a vital, safe, and effective option.

In a recent metanalysis comparing different modalities of treating basilar apex aneurysms, endovascular therapy was found to have higher rates of retreatment and posttreatment hemorrhage, with lower rates of complete occlusion, compared to microsurgical clipping. However, the latter was found to be "more morbid" [21]. Clipping basilar aneurysms has historically been associated with a higher rate of difficulties and complications compared to other aneurysm locations. Nevertheless, clear progress was made, and basilar aneurysms should no longer be considered a dreaded pathology where microsurgery is to be avoided at all costs. Quite the opposite, microsurgical clipping can prove to achieve a safer and more definite treatment [20].

In his Microneurosurgery book, Yasargil reported that 10 out of 50 patients with basilar aneurysms were not amenable to surgical clipping. However, those occurred early in his experience, and as technical breakthroughs were made, his success rate continued to increase [1]. Introducing the transcavernous approach by Dolenc [6], then refined by de Oliveira [9], has paved the way for wider and even lower access to the basilar system. Since then, the surgical approach has been extended further and systematized to maximize its safety and advantages. The approach now encompasses a multitude of steps from which the surgeon can choose a tailored roadmap for each aneurysm, regardless of its complexity. The approach has also become more dynamic, where the cranial nerves can be mobilized and transposed. The spaces between them are no longer rigid triangles, but a continuous field with movable windows. The parts of the skull base that can be unlocked are no longer restricted to the parasellar region alone but go well beyond the clivus and petroclival junction.

While the transcavernous approach has been criticized for causing oculomotor palsy, this was previously demonstrated to be transient, and usually recovers completely in a few months [18]. Surgical risks in clipping basilar aneurysms, as in any other surgery or any endovascular procedure, cannot be made nil. Neurosurgeons, whether in the operating room or the angiogram suite, should continue to elevate the standard of their care. They should strive for better immediate and long-term results and decreased morbidity, notwithstanding the amount of training and effort required, as dictated by their ancestral oath.

16.7 Conclusion

The trans-cavernous approach to basilar aneurysms combines extradural and intradural dissections, with a wide variety of surgical steps, that can progressively deconstruct the space around the interpeduncular fossa and prepontine space. The approach is an anatomical exercise par excellence, thus demanding a profound anatomical apprehension of the region. Constant proximal and distal control is valuable to decrease the risks related to the procedure. The augmented maneuverability around the basilar artery and its branches, and the emerging perforators, permits a dynamic clipping process that can progressively occlude the whole aneurysm, while maintaining the integrity of the perforators.

Acknowledgment We would like to acknowledge the contribution of Li Cai, MD and Hitesh Inder Singh Rai, MCh.

References

1. Yasargil MG. Microneurosurgery: clinical considerations, surgery of the intra-cranial aneurysms and results, vol. 2. Stuttgart, Germany: Georg Thieme; 1984. p. 232–95.
2. Hook O, Norlen G, Guzman J. Saccular aneurysms of vertebrabasilar arterial system: report of 28 cases. Acta Neurol Scand. 1963;39:271–304.
3. Drake CG. Treatment of aneurysms of the posterior cranial fossa. In: Krayenbuhl H, Maspes PE, Sweet WH, editors. Progress in neurological surgery, vol. 9. Basel: Karger; 1978.
4. Drake CG, Peerless SJ, Hernesniemi JA. Surgery of vertebrobasilar aneurysms. New York, NY: Springer-Verlag; 1996.
5. Yasargil MG, Antic J, Laciga R, Jain KK, Hodosh RM, Smith RD. Microsurgical pterional approach to aneurysms of the basilar bifurcation. Surg Neurol. 1976;6(2):83–91.
6. Dolenc VV, Skrap M, Sustersic J, Skrbec M, Morina A. A transcavernous-transsellar approach to the basilar tip aneurysms. Br J Neurosurg. 1987;1(2):251–9.
7. Parkinson D. Carotid cavernous fistula: direct repair with preservation of the carotid artery: technical note. J Neurosurg. 1973;38(1):99–106.
8. Krayenbühl N, Hafez A, Hernesniemi JA, Krisht AF. Taming the cavernous sinus: technique of hemostasis using fibrin glue. Neurosurgery. 2007;61(Suppl. 3):E52.
9. de Oliveira E, Tedeschi H, Siqueira MG, Peace DA. The pretemporal approach to the interpeduncular and petroclival regions. Acta Neurochir. 1995;136(3–4):204–11.
10. Figueiredo EG, Tavares WM, Rhoton AL Jr, de Oliveira E. Nuances and technique of the pretemporal transcavernous approach to treat low-lying basilar artery aneurysms. Neurosurg Rev. 2010;33(2):129–35.
11. Krisht AF. Transcavernous approach to diseases of the anterior upper third of the posterior fossa. Neurosurg Focus. 2005;19(2):E2.
12. Basma J, Krisht KM, Lee P, Cai L, Krisht AF. Temporary clipping of the intracavernous internal carotid artery: a novel technique for proximal control. Oper Neurosurg (Hagerstown). 2021;20(2):E91–7.
13. Krisht AF, Kadri PA. Surgical clipping of complex basilar apex aneurysms: a strategy for successful outcome using the pretemporal transzygomatic trans- cavernous approach. Neurosurgery. 2005;56(Suppl. 2):261–73.
14. Seoane E, Tedeschi H, de Oliveira E, Wen HT, Rhoton AL Jr. The pretemporal transcavernous approach to the interpeduncular and prepontine cisterns: microsurgical anatomy and technique application. Neurosurgery. 2000;46(4):891–8.
15. Krisht AF, Basma J, Cai L. Transcavernous approach in vascular neurosurgery. Neurosurg Clin N Am. 2022;33(4S):e1–6.
16. Basma J, Moore KA, Krisht K, Abuelem T, Arnautovic K, Michael LM, Aboud E, Krisht AF. Morphometric comparison of the pterional trans-sylvian and the pretemporal trans-clinoidal approaches to the posterior communicating artery. Oper Neurosurg (Hagerstown). 2020;20(1):E22–30.
17. Basma J, Dacus MR, Kumar R, Spencer D, Arnautović KI. Cisternal, Falciform, and Optic canal decompression influencing optic nerve biomechanics: a microsurgical anatomic study. Oper Neurosurg (Hagerstown). 2023;24(2):e75–84.
18. Basma J, Ryttlefors M, Latini F, Pravdenkova S, Krisht A. Mobilization of the transcavernous oculomotor nerve during basilar aneurysm surgery: biomechanical bases for better outcome. Neurosurgery. 2014;10(Suppl. 1):106–14.
19. Beumer D, Delwel EJ, Kleinrensink GJ, Akouri S, Torres A, Krisht AF. The perforator-free zone of the posterior communicating artery and its relevance in approaches to the interpeduncular cistern, especially the transcavernous approach: an anatomic study. Neurosurgery. 2007;61(5 Suppl. 2):187–91.
20. Krisht AF, Krayenbühl N, Sercl D, Bikmaz K, Kadri PA. Results of microsurgical clipping of 50 high complexity basilar apex aneurysms. Neurosurgery. 2007;60(2):242–50.
21. Dandurand C, Prakash S, Sepehry AA, Tourigny K, Haw CS, Gooderham P, Moore J, Redekop G. Basilar apex aneurysm: case series, systematic review, and meta-analysis. World Neurosurg. 2020;138:e183–90.

Surgical Trajectories for Clipping of Different PICA Aneurysms

17

Ramez W. Kirollos

Learning Objectives
- Anatomy of the foramen magnum region and craniocervical junction.
- Anatomy of the posterior inferior cerebellar artery (PICA).
- Anatomical basis for the selection of surgical approaches to PICA aneurysms.
- Surgical nuances for approaches and microsurgical clipping of PICA aneurysms.

17.1 Introduction

Despite the feasibility of endovascular treatment of many posterior inferior cerebellar artery (PICA) aneurysms, there is a role for clipping of those aneurysms that are not suitable for coiling, those located distally, flow-related aneurysms associated with arterio-venous malformations or recurrence after endovascular intervention. Understanding the configuration of the aneurysm together with the 3D anatomical relationship with the cranio-cervical junction and lower brainstem is valuable to select the optimum approach and surgical trajectory to these aneurysms. Such knowledge would aid in rationalising the potential controversies regarding the choice between lateral and far lateral vs. midline suboccipital approaches and the extent of drilling of the occipital bone and its condyle if any.

R. W. Kirollos (✉)
Department of Neurosurgery, National Neuroscience Institute, Singapore, Singapore

Duke-NUS Graduate Medical School, Singapore, Singapore

17.2 Anatomical Highlights

- The PICA has five segments and is arranged by a medullary loop related to the medulla and a tonsillar loop before terminating into choroidal and cortical branches.
- The last two segments in general do not contribute significantly to the blood supply of the medulla.
- The PICA supplies the lateral medulla which is also supplied by direct branches from the VA while anteriorly the corticospinal tracts are supplied by the anterior spinal artery.
- The PICA may arise extradurally from the V3 segment of the VA.
- The shape of the posterior fossa is such that the lower 1/3 CPA can be reached by either a lateral or midline approach.
- The condylar part of the occipital bone is divided by the hypoglossal canal into an occipital condyle and a jugular tubercle.

17.3 Relevant Vascular Anatomy

The PICA is a branch of the V4 segment, which is the last segment of the vertebral artery (VA) extending from its intradural entrance to the vertebrobasilar junction (VBJ). At its entrance intracranially, the V4 segment is anterior to the first denticulate ligament which separates it from the spinal root of the accessory nerve being the only nerve root with motor fibres that is posterior to a denticulate ligament. At its origin, the PICA is in proximity to the XII nerve rootlets. During its course, it forms two loops, one related to the anterolateral aspect of the medulla and a second around the tonsil of the cerebellum following which it continues to supply the choroid plexus of the IV ventricle through the tela choroidea and gives cortical branches supplying part of the cerebellum (Fig. 17.1). The apex of the medullary loop reaches the cephalad at a level adjacent to the jugular foramen and occasion-

I. N. Kanaan, V. Beneš (eds.), *Neuroanatomy Guidance to Successful Neurosurgical Interventions*,
https://doi.org/10.1007/978-3-031-59838-8_17

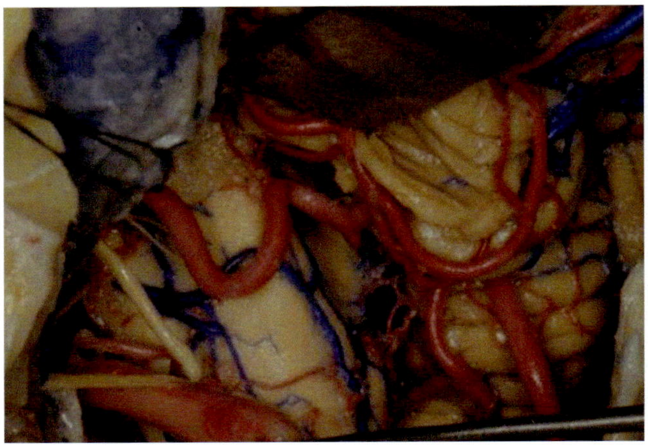

Fig. 17.1 Lateral view showing the course of the PICA forming loops related to the medulla comprised of an anterior and lateral medullary segment continuing to the tonsillo-medullary segment that is exposed by retracting the cerebellar tonsil and forming a tonsillar loop [from Rhoton. *Courtesy of the Rhoton Collection, American Association of Neurological Surgeons (AANS)/Neurosurgical Research and Education Foundation (NREF)*]

ally may be as high as the level of the internal auditory meatus (IAM), especially with the tortuosity of the VA. Anatomically the PICA is divided into five segments [1–3].

17.3.1 PICA Vessel Segments

Anterior medullary segment.
Lateral medullary segment.
Tonsillo-medullary segment.
Telovelo-tonsillar segment.
Cortical segment.

The general arrangement of the vascular supply of the brainstem follows a pattern akin to the spinal cord (Fig. 17.2). Perforating branches supply deep tracts and nuclei, reaching through the anterior median sulcus in the midline anteriorly and on either side postero-laterally in line with the dorsal root entry zones of the spinal cord, with motor fibres anteriorly and sensory/proprioception fibres postero-laterally. The midbrain is served by branches off the P1 segments of the posterior cerebral artery (PCA), proximal superior cerebellar artery (SCA) and directly off the basilar termination. Pons receives supply directly from the BA and proximal anterior inferior cerebellar arteries (AICAs). These arteries along their course also give off branches winding around the brainstem with short circumflex branches supplying the surface while the long circumflex branches also supply the surface with some perforating branches supplying the deep structures postero-laterally. In the medulla, the anterior midline

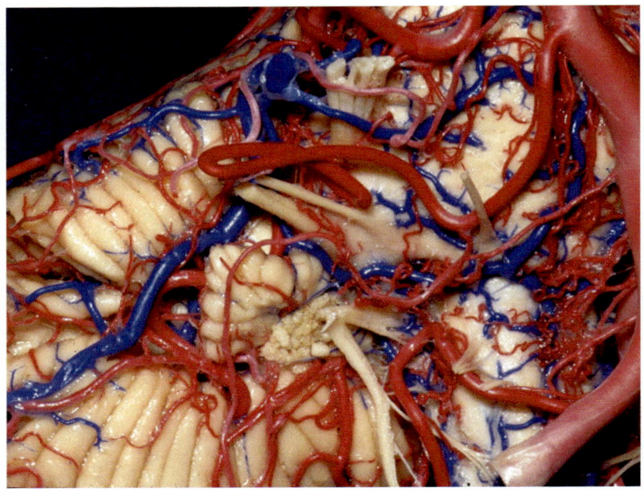

Fig. 17.2 Anterior view of the brainstem, middle cerebellar peduncle and petrosal surface of the cerebellum. The emerging cranial nerves and their relationship to the flocculus and the choroid plexus at the foramen of Luschka is demonstrated. This dissection depicts the arrangements of the blood supply to the brainstem from the vertebrobasilar system and its branches with anterior branches at the anterior median sulcus and the short and long circumflex arteries [from Rhoton. *Courtesy of the Rhoton Collection, American Association of Neurological Surgeons (AANS)/Neurosurgical Research and Education Foundation (NREF)*]

group is provided by the anterior spinal artery (ASA). The ASAs are paired arising from the VA proximal to the VBJ and may fuse into a single trunk but sometimes there is only one-sided ASA. Hence, the PICA supplies the posterolateral medulla through branches off the anterior and lateral medullary segments [1, 3]. These structures include the spinocerebellar, spinothalamic, spinal trigeminal tracts, vestibular nuclei and tractus solitarius. These are implicated in the neurological manifestations of the lateral medullary (Wallenberg) syndrome.

Beyond the tonsillo-medullary segments, the PICA does not significantly supply the medulla. The cortical branches to the cerebellum supply the tonsils, part of the sub-occipital surface (the rest is supplied by the SCA) and the vermis by prominent vermian branches reaching superiorly (Fig. 17.3). Some deep branches supply the dentate nucleus together with the SCA.

There are anatomical variations [1, 3]. The VAs are often asymmetrical with a more dominant side and occasionally the intradural distal V4 segment on one side is extremely hypoplastic with the main trunk of the VA ending its course as the PICA. In 8–15%, the PICA arises extradurally from the V3 segment of the VA, while less commonly it arises distally from the proximal BA (Fig. 17.4). In the latter arrangement, this could be independent of an existing AICA, while in some a prominent AICA will supply the absent PICA territory. Awareness of these relatively common variations helps with pitfall avoidance.

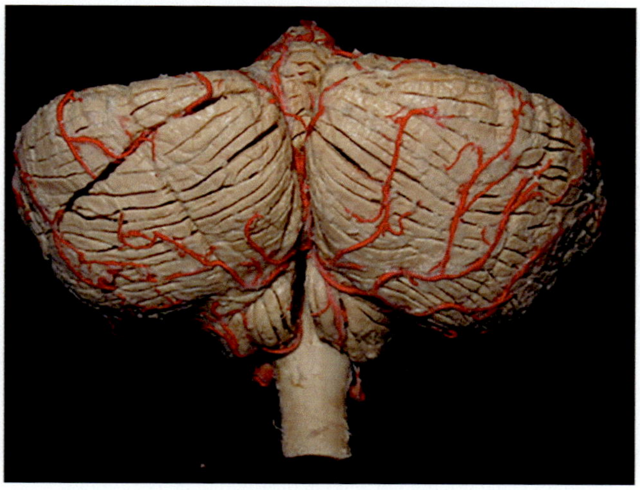

Fig. 17.3 The cortical branches of the PICA supplying the suboccipital surface of the cerebellum (together with terminal branches of the SCA), the tonsils and the vermis

17.3.2 Classification of Different PICA Aneurysms

17.3.2.1 Morphological Classification

Saccular.
Fusiform.
False/dissecting.

17.3.2.2 Haemodynamic Classification

"Berry" aneurysms
AVM-related flow aneurysms.

17.3.2.3 Classification by Location

VA/PICA junction.
Distal.

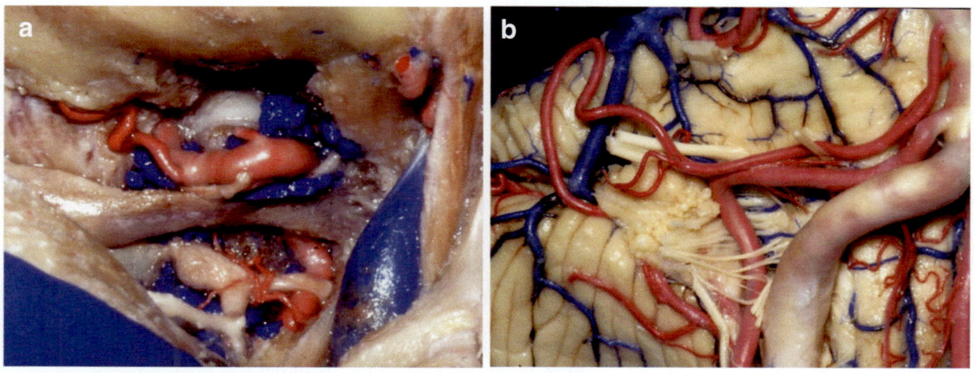

Fig. 17.4 Anatomical variations of PICA. In (**a**) an extradural origin of the PICA arising off the V3 segment of the VA. This may be confused with a dural branch. In (**b**) the PICA is originating from the BA near the VBJ [from Rhoton. *Courtesy of the Rhoton Collection, American Association of Neurological Surgeons (AANS)/Neurosurgical Research and Education Foundation (NREF)*]

17.4 Intervention

17.4.1 Relationship of the VA/PICA Segments and "3D Visualization" at the Region of the Cranio-Cervical Junction, Foramen Magnum and Posterior Fossa

17.4.1.1 Geometric Anatomical Considerations of the Far Lateral Approach

It is important to appreciate the 3D construction of the posterior fossa and the cranio-cervical junction. In the coronal (and sagittal) plane, the "funnel" concept depicts the posterior fossa shape with a sharp transition between the middle and lower thirds of the cerebellopontine angle (CPA) from laterally towards the midline (Fig. 17.5). At the apex of the angulation lies the superior tip of the jugular tubercle. The condylar part of the occipital bone is divided by the hypoglossal canal, which is two-thirds of the distance cephalad from the foramen magnum, forming the superior limit of the occipital condyle, while the remaining third between the hypoglossal canal and the jugular foramen is the jugular tubercle. The craniocaudal orientation of the cranio-cervical junction results in a relationship between the jugular tubercle and the lower third of the CPA, the occipital condyle and the foramen magnum and the C1 lateral mass and the upper end of the cervical spinal canal (Fig. 17.5, Table 17.1). The extent of "transcondylar" resection in approaches to the cranio-cervical junction should be tailored (Figs. 17.6 and 17.7). Therefore, for a lesion extending superiorly to the jugular

Table 17.1 The relationship between the different regions of the CCJ in a cranio-caudal orientation and the obstructing bony structure

Upper C canal	C1 lateral mass/tubercle
Foramen magnum	Occipital condyle
Jugular foramen	Jugular tubercle

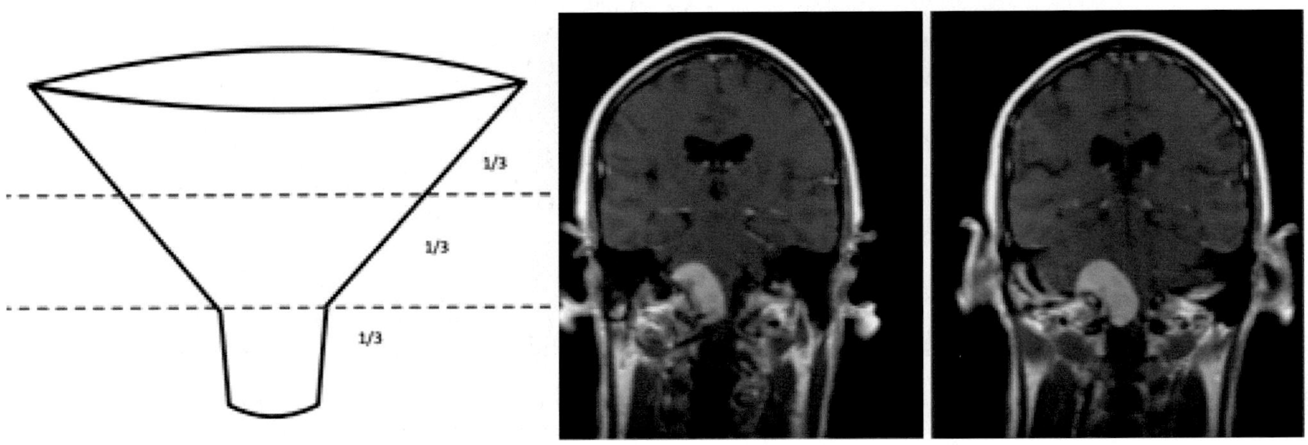

Fig. 17.5 The shape of the posterior fossa is similar to a funnel with the upper two-third of the CPA being more laterally located while the lower one-third near the midline with transition at the jugular foramen and superior tip of the jugular tubercle. A coronal view of an MRI, T1 weighted with Gadolinium, showing as an example a foramen magnum meningioma extending cephalad. From a far lateral approach, its upper pole is extending superior to the jugular foramen and would be obscured partially from view by the jugular tubercle

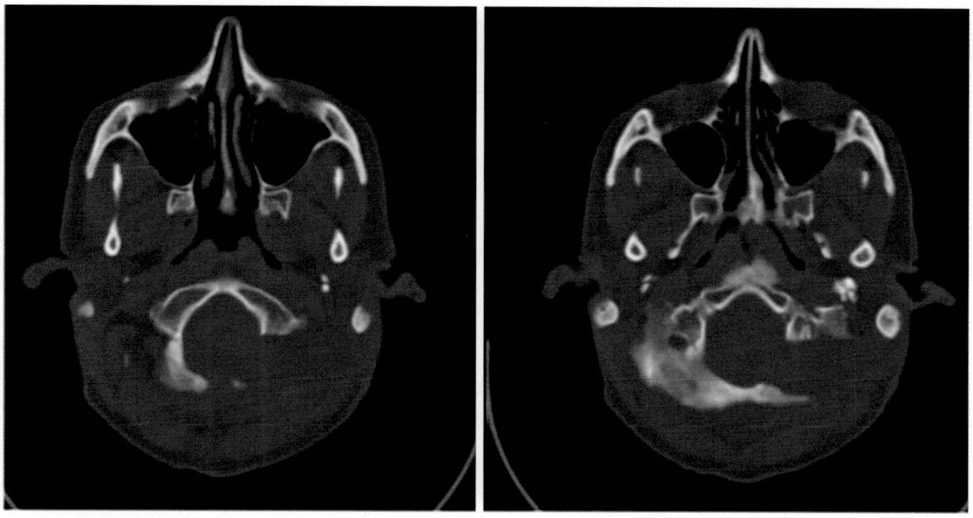

Fig. 17.6 A far lateral approach with minimal resection of the occipital condyle as depicted on the postoperative CT scan with bone windows

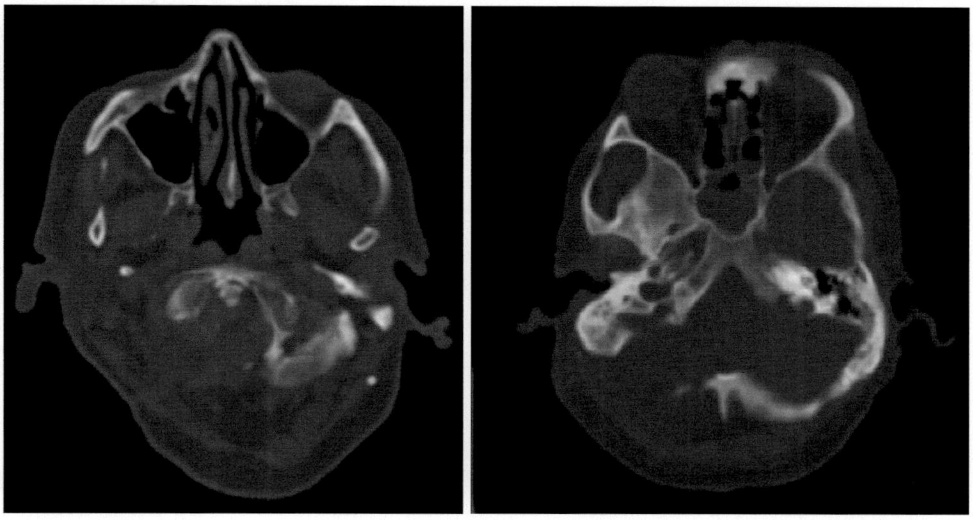

Fig. 17.7 An axial CT scan with bone windows setting showing a more extensive condylar resection

foramen, drilling and reduction of a prominent jugular tubercle would improve access to the anterior aspect of the superior pole of the lesion. This is often limited given the location of the jugular bulb and hence, this is frequently a potential "blind spot". Similarly, for a lesion just inferior to the foramen magnum and extending caudally drilling the occipital condyle would not provide additional exposure.

17.5 Factors to Be Considered for Selection of Safe Exposure and Trajectories for Clipping of PICA Aneurysms

17.5.1 Optimal Exposure/Visualisation of the "Neck" (+++)

As in any other intracranial aneurysm, exposure of the neck rather than the fundus is the primary aim. The exposure should visualise the presenting perimeter of the neck and need adequate space to allow inspection around the corner to identify the branches which in the case of PICA aneurysms either at the VA origin or on the medullary loop is likely to be an important perforator. Furthermore, the exposure should also provide the space to allow the freedom of manipulation of the clip applicator and the judicious placement of aneurysm clips of various configurations as required.

17.5.2 Proximal and Distal Control (++)

In particular with PICA aneurysms early access to distal as well as proximal control is strongly advised. In case of intraoperative rupture given the supply from the contralateral VA, not only that the bleeding will not be controlled by only applying a proximal temporary clip, but also in cases of excessive and continuous haemorrhage, there is a potential danger of brainstem ischaemia resulting from the "siphoning" blood off the basilar. Given the distal course and inclination of the VA anteromedially to the medulla then appropriate trajectory as described below is necessary.

17.5.3 Size (±)

Large aneurysm size would not per se necessitate modification of the operative approach as it is the exposure of the neck rather than the fundus that is important. Nevertheless, larger aneurysms tend to have wider necks and a large fundus may partially obstruct the access to distal control or the localisation of an ensuing perforator and hence adequate exposure will be necessary.

17.5.4 Proximity of Cranial Nerves (+)

In the majority of cases, the origin of the PICA off the VA which is the location of most "berry" aneurysms is related to the hypoglossal nerve rootlets while the medullary loop apex is in proximity to the IX, X and XI cranial nerves and sometimes even to VII and VIII nerves. The course of the distal VA is deep to IX and X nerves. Adequate exposure ensures providing space for safe manipulations by the clip applicator and for gaining distal control without traumatising these nerves.

17.5.5 Tortuosity/Off Midline VBJ (++)

Studying pre-operative vascular radiological images may reveal varying degrees of vertebrobasilar ectasia or tortuosity with an inclination of one VA horizontally and the other more vertically or vice versa and the VBJ could be one side or another off the midline (Fig. 17.8). Tailoring the approach to achieve the ideal exposure should be carefully planned.

17.5.6 Cranio-Caudal Location (+++)

The take-off of the PICA from the VA is not always within a short distance from its intradural entrance. A distal take-off along the course of the VA to the VBJ may be seen. Although the previous factors mostly dictate the size and adequacy of the exposure, the craniocaudal and 3D orientation of the aneurysm and its relationship to the brainstem is what determines the surgical approach, trajectory, head position and nuances of surgery. Hence, the influence of this factor on the selection of the approach is detailed below.

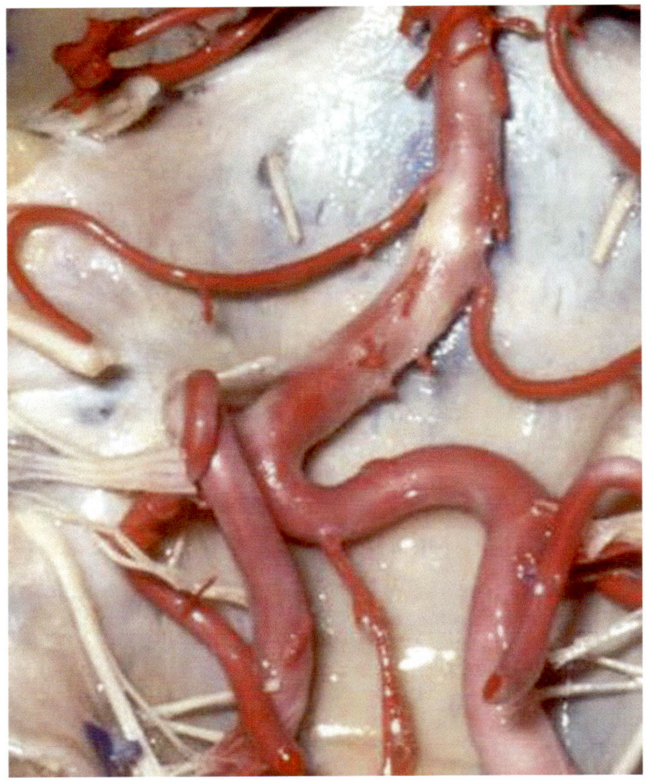

Fig. 17.8 Ectatic and tortous vertebral and basilar arteries with VBJ way lateral to the midline [from Rhoton. *Courtesy of the Rhoton Collection, American Association of Neurological Surgeons (AANS)/Neurosurgical Research and Education Foundation (NREF)*]

17.5.7 Anatomical Concepts for Tailored Approaches and Surgical Exposure for PICA Aneurysms

17.5.7.1 Dilemmas When Choosing the Best Exposure and Approaches for PICA Aneurysms

The common controversies regarding PICA aneurysm surgery [4–16] involves:

- Choosing a lateral vs. posterior midline approach.
- Placing the patient in a prone vs. lateral position.
- Drill the occipital condyle or not.

The anatomical depiction of the posterior fossa, lower third CPA and the cranio-cervical junction as detailed previously and the orientation, course and relationships of the VA and PICA vessels are taken into consideration in the selection of the different approaches and trajectories to clip a PICA aneurysm.

The extent and location of the craniotomy together with the head position and extent of its rotation provide the different trajectories and angles of view to approach PICA aneurysms at variable locations and configurations. Commonly, either a posterior midline craniotomy or a low suboccipital craniotomy with a far lateral approach combined with either a prone or lateral/park bench position, respectively, is what is usually applied. A far lateral approach, which is defined as opening the foramen magnum laterally up to the occipital condyle, is in reality a POSTEROLATERAL approach. If needed, then drilling of the posteromedial corner of the condyle widens the lateral angle of view but, as explained below, a trans-condylar approach is invariably not required. However, the jugular tubercle may pose some constraints to approaching certain PICA aneurysms.

17.5.7.2 Influence of the Location of the PICA Take-Off From VA on the Selection of Approach and Trajectory

The VA's course after entering the dura involves the VA-PICA junction moving supero-antero-medially with a more distal "take-off". The relationship of the VA to the medulla and pons starts posterolateral at its entry into the intradural space, then it inclines anteromedially and cephalad towards the vertebrobasilar junction, which is invariably anterior and at the level of the pontomedullary junction. Hence, the angle of view will dictate the trajectory achieving adequate exposure without manipulating neural structures, namely the medulla. For PICA aneurysms originating off the VA with a "proximal take-off", i.e. caudally at the foramen magnum, a trajectory providing a direct viewing angle is medial to lateral. This is readily provided through a midline suboccipital craniotomy widened ipsilaterally at the foramen magnum with the patient in the prone position. In fact, placing a patient in the lateral position with a far lateral craniotomy is a disadvantage as the relative blind spot is the deep surface of the V4 segment at its entrance through the dura and would potentially necessitate some drilling of the edge of the condyle and part of C1 tubercle for a better view.

Once the aneurysm is more cephalad due to a more distal PICA take-off, its neck is around the corner of the medulla and is obscured from a posterior midline view. For these, the trajectory with a direct viewing angle is transitioning from lateral to antero-medial and cephalad as the take-off approaches the VBJ. Therefore, it is best to approach these through a far lateral craniotomy with the patient in the lateral position with an operating table allowing some degrees of lateral tilt. For these, the adequate opening of the foramen magnum laterally should be insured; however, as the take-off of the PICA is now cephalad, neither the C1 tubercle nor the occipital condyle should be obscuring the view and hence do not require drilling apart perhaps from minor "shaving" to allow the space needed for the clip applicator. Challenges arise with more distal aneurysms being more medial and anterior to the medulla and also cephalad so that the trajectory through the far lateral approach may be obstructed by a prominent jugular tubercle. When prepared, the jugular tubercle may be drilled and flattened extradurally. However, the extent of bone removal may be limited to avoid injury to a prominent jugular bulb.

17.5.7.3 Influence of the Location of the Aneurysm on the PICA Segments on the Selection of Approach and Trajectory

Some aneurysms are located distally along the PICA rather than at the VA-PICA junction. These include some dissecting and flow-related aneurysms associated with cerebellar AVMs. For those located anterolateral to the medulla, i.e. on the medullary loop, being especially the anterior medullary and also the lateral medullary segments, a far lateral craniotomy with the patient in the lateral position is best (Fig. 17.9). Those aneurysms located on the tonsillar loop and beyond are simply approached through a midline suboccipital craniotomy in the prone position (Figs. 17.10 and 17.11).

The choices of the approach for the differently located PICA aneurysms are summarised in Table 17.2.

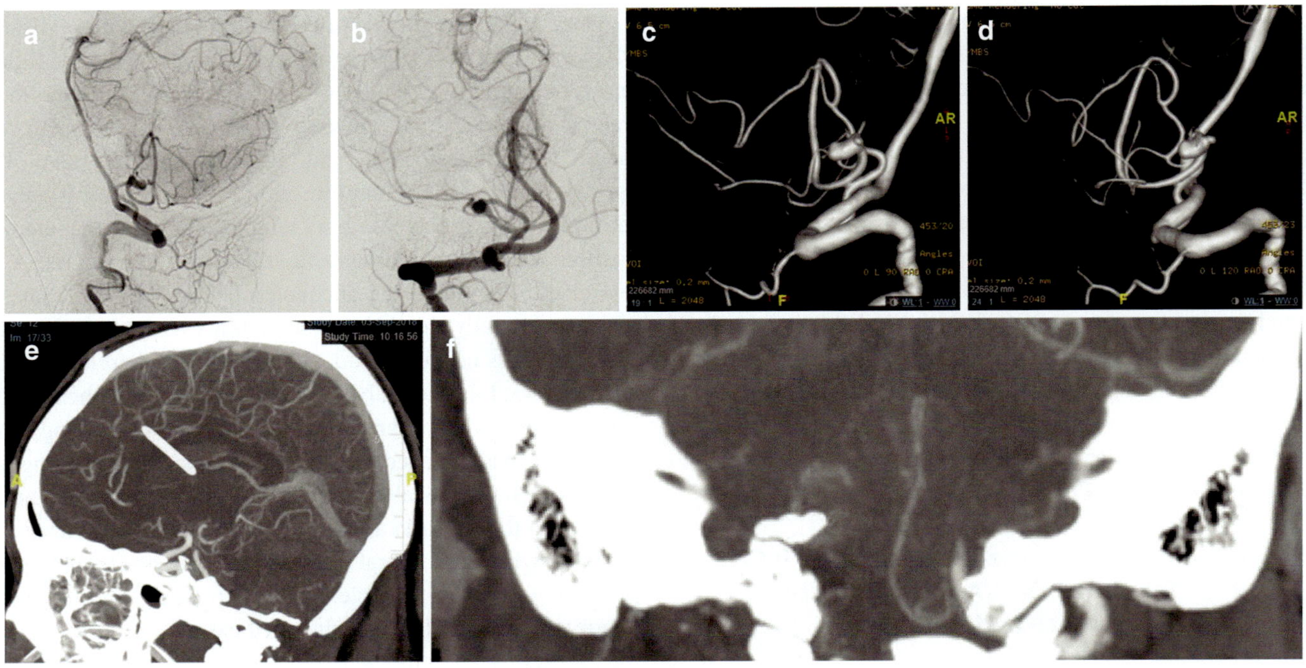

Fig. 17.9 Lateral (**a**), AP (**b**) and 3D reconstruction (**c** and **d**) views of an angiogram showing an aneurysm on the medullary loop of PICA. Post-operative CT angiogram (CTA), sagittal (**e**) and coronal (**f**) views were obtained and demonstrate the position of the clip anterior and cephalad from the FM and adjacent to the jugular tubercle. The surgery was performed by a far lateral approach and the jugular tubercle prominence was reduced by extradural drilling

Fig. 17.10 Lateral view (**a**) and 3D reconstruction (**b**) of an angiogram showing a distal PICA aneurysm on the tonsillar loop. This was operated upon through a midline suboccipital approach

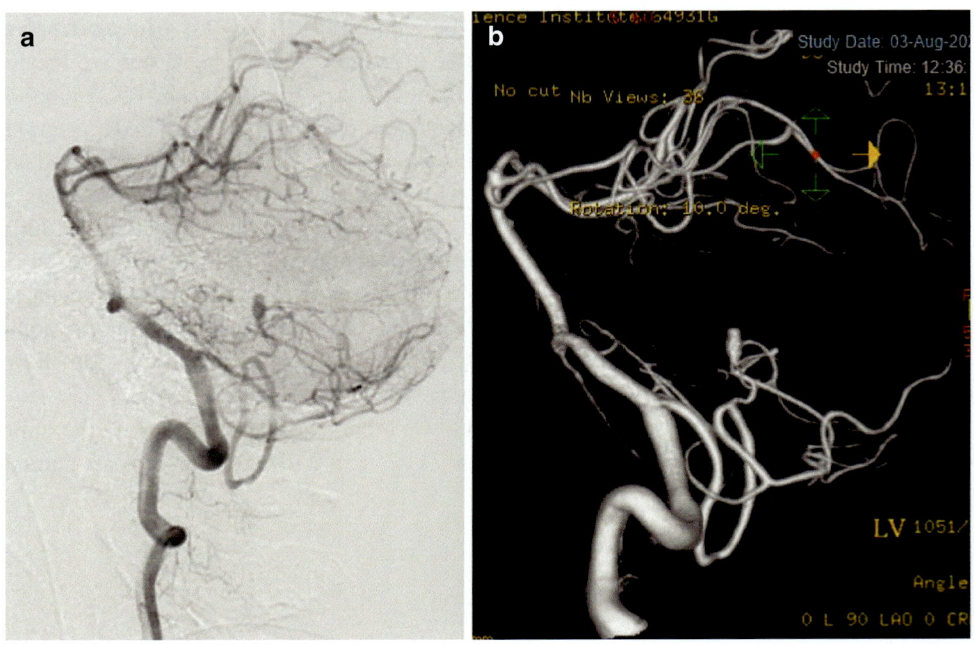

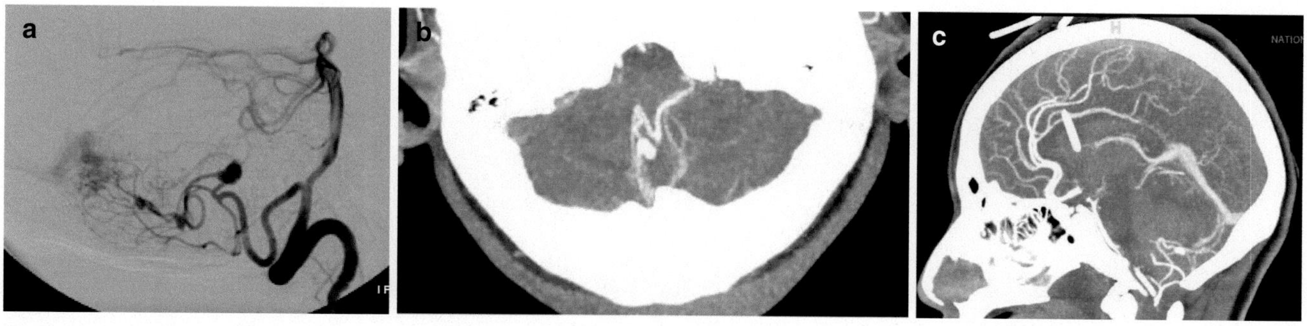

Fig. 17.11 An angiogram with a lateral view (**a**) and an axial (**b**) and sagittal (**c**) CTA from another case showing flow aneurysms on the distal PICA that is supplying a downstream cerebellar AVM

Table 17.2 The choice of the approach for the differently located PICA aneurysms

Midline sub-occipital approach	Lateral approach
• **VA-PICA aneurysms** • Proximal PICA origin off intra-dural VA	• **VA-PICA aneurysms** • Distal PICA origin off intra-dural VA
• **Distal PICA aneurysm** • Beyond "medullary" loop	• **Distal PICA aneurysm** • On "medullary" loop

17.6 Nuances of the Surgical Approaches Gaining Access to PICA Aneurysms at the CCJ

17.6.1 Midline Suboccipital Approach

For PICA aneurysms in these aforementioned locations, a bilateral limited suboccipital craniotomy incorporating the posterior lip of the foramen magnum and extending laterally on the ipsilateral side of the aneurysm is sufficient (Fig. 17.12). Removal of the posterior arch of C1 can be added as needed. The craniotomy need not be overly large or extend to the transverse sinus.

After opening of the dura, the tonsil is mobilised medially cautiously as it might be adherent to the aneurysm fundus, to gain exposure to those aneurysms arising from the V4 segment after its intradural entrance. Even in PICA aneurysms at the proximal V4 segment in the majority of cases, there is enough proximal intradural length of the VA to apply a proximal temporary clip if required. The advantage of this approach with a medial to lateral trajectory is the direct view to this segment providing proximal control. In truly low-lying aneurysms, gaining proximal control of the extradural VA as described below requires additional bone exposure from this approach.

This approach is also adopted for distal PICA aneurysms that are beyond the medullary loop with the tonsillo-medullary segment found lateral to the tonsil as a caudal continuation of the lateral medullary segment and winding around the tonsil. Preserving, in particular, branches supplying the medulla arising at the neck of the aneurysm is vital. The telovelo-tonsillar segment courses cephalad towards the foramen of Magendie to the IVth ventricular choroid plexus and is buried deep from the surface in the fissure between the vermis and tonsil in relation to the telovellar membrane formed by the telachoroidea and the inferior medullary velum (Fig. 17.13). For more distal aneurysms involving the distal cortical branches to the tonsils, suboccipital cerebellar surface and the vermis, these may need to be sacrificed for fusiform or dissecting aneurysms which are excised whenever a clip application is not readily feasible. Anatomical knowledge and awareness of the origin of important branches and perforators based on localisation of the aneurysm from imaging and anatomical landmark direct surgery, especially in cases of surrounding brain distortion from extensive haemorrhage (Fig. 17.13).

17.6.2 Far Lateral Approach

It is best to perform the craniotomy after full exposure of the occipital condyle, atlanto-occipital joint, C1 tubercle (lateral mass) and the course of V3 (Figs. 17.14, 17.15, 17.16, and 17.17). The size of the craniotomy is tailored to the pathology. Exceptionally, for a giant aneurysm, a retrosigmoid approach exposing the transverse and sigmoid sinus edges or even skeletonisation of the sigmoid sinus and jugular bulb and combining this retrosigmoid craniotomy with a trans- or retro-labyrinthine approach can be considered. Nevertheless, as elaborated before it is the exposure of the neck rather than the entire fundus that is required, and such large approaches are usually not necessary. For a far lateral approach, the foramen magnum is opened laterally to the edge of the condyle. For most intradural lesions, the required drilling of the condyle is minimal, if at all required (Fig. 17.16). This is in contrast to extradural or dumbbell lesions extending extradurally when more extensive drilling of the occipital condyle and possibly also the C1 tubercle (lateral mass) may be necessary. Similarly, resection of the posterior arch of C1 to the

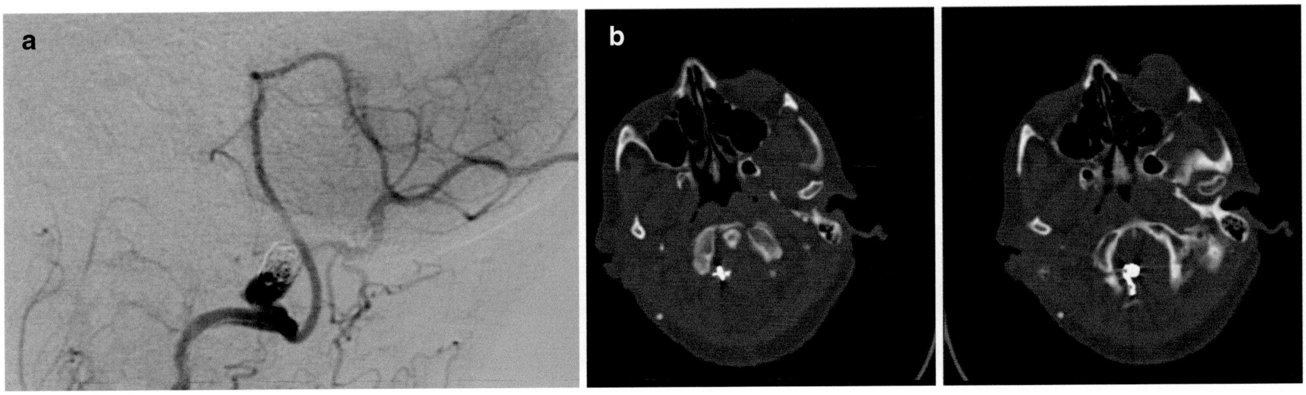

Fig. 17.12 A pre-operative angiogram (**a**) of a recurrent PICA aneurysm following previous endovascular coiling. Note that in this variation the right VA is ending in the PICA itself. Clipping was performed through a midline suboccipital craniotomy. The postoperative CT images (**b**) showing the relationship of the clip at the posterolateral lip of the FM and the extent of the craniectomy. In this case, the clip was placed on the aneurysm neck without the need to retrieve the coil mass

medial edge of the VA on the sulcus arteriae vertebralis is all that is required. The dislocation of the VA after drilling the posterior tubercle of the transverse process of C1 to open the foramen transversarium and mobilising the VA medially is not required.

The dura is opened vertically with caution particularly if the opening is too lateral so as not to injure the spinal XI nerve which may be pushed posteriorly and stuck onto the dura by an underlying large aneurysm. A horizontal dural slit laterally from the vertical incision just cephalad to the VA dural entrance and extending to the bony edge of the exposed condyle or if drilled beyond that point is most advantageous to improve the far lateral trajectory to the intradural compartment.

Anteriorly based foramen magnum aneurysms are limited by the first denticulate ligament at its bony attachment to the medial aspect of the foramen magnum which separates the spinal XI nerve posteriorly from the VA. Incising the fibres of this ligament untethers the cervico-medullary junction and prevents transmission of tension onto that area during dissection (Fig. 17.16). Also, this allows exposure of the proximal intradural course of the VA and the PICA origin, which is in the majority of cases related to the rootlets of the XII nerve (Fig. 17.17). Following the spinal XI nerve superiorly is a guide to the location of the jugular foramen and the X and IX nerves, especially if obscured by a lesion extending cephalad. It is important to be aware of the deep location of the anterior spinal artery (ASA) which can vary in origin off the VA or occasionally from the anterior medullary segment of the PICA these can be paired or single and in the case of the latter arrangement, compromise of the ASA may result in the typical anterior medullary spinal infarct appearing as a heart-shaped signal change on MRI and extensive distal neurological deficit (Fig. 17.17d).

Observing the configuration of the aneurysm on the pre-operative images dictates the approach and microsurgical manoeuvrers to allow the most appropriate access for its anatomical location (Fig. 17.18). Those originating distal to or even at the foramen magnum do not require additional drilling of the occipital condyle and ensure that the clip or its applicator does not compromise the related arterial branches or traumatise the cranial nerves (Fig. 17.19). Determining the craniocaudal location from the foramen magnum and the relationship to the occipital condyle and jugular foramen in the vertical plane, and in addition the lateral to medial location and relationship around the medulla in the axial and coronal plane, are the two important factors to plan the approach (Figs. 17.18 and 17.19).

17.6.3 Is There a Rationale for Occipital Condyle Resection to Gain Access Through the Far Lateral Approach to Its Anterior Location for Clipping of PICA Aneurysms?

In the axial plane, the trajectory provided by the far lateral approach to intradural lesions is still a posterolateral trajectory even with a transcondylar extension. Even in limited approaches for anteriorly placed foramen magnum lesions, exposure of the occipital condyle edge and the atlanto-occipital joint in these cases ensures opening the foramen magnum lateral enough. To maximise the view and to provide a direct trajectory to the retroclival space and towards the contralateral side for anteriorly located lesions, progressive drilling of the posterior lip of the occipital condyle—or the C1 tubercle (lateral mass)—from the lateral surface in an anteromedial direction (Fig. 17.20).

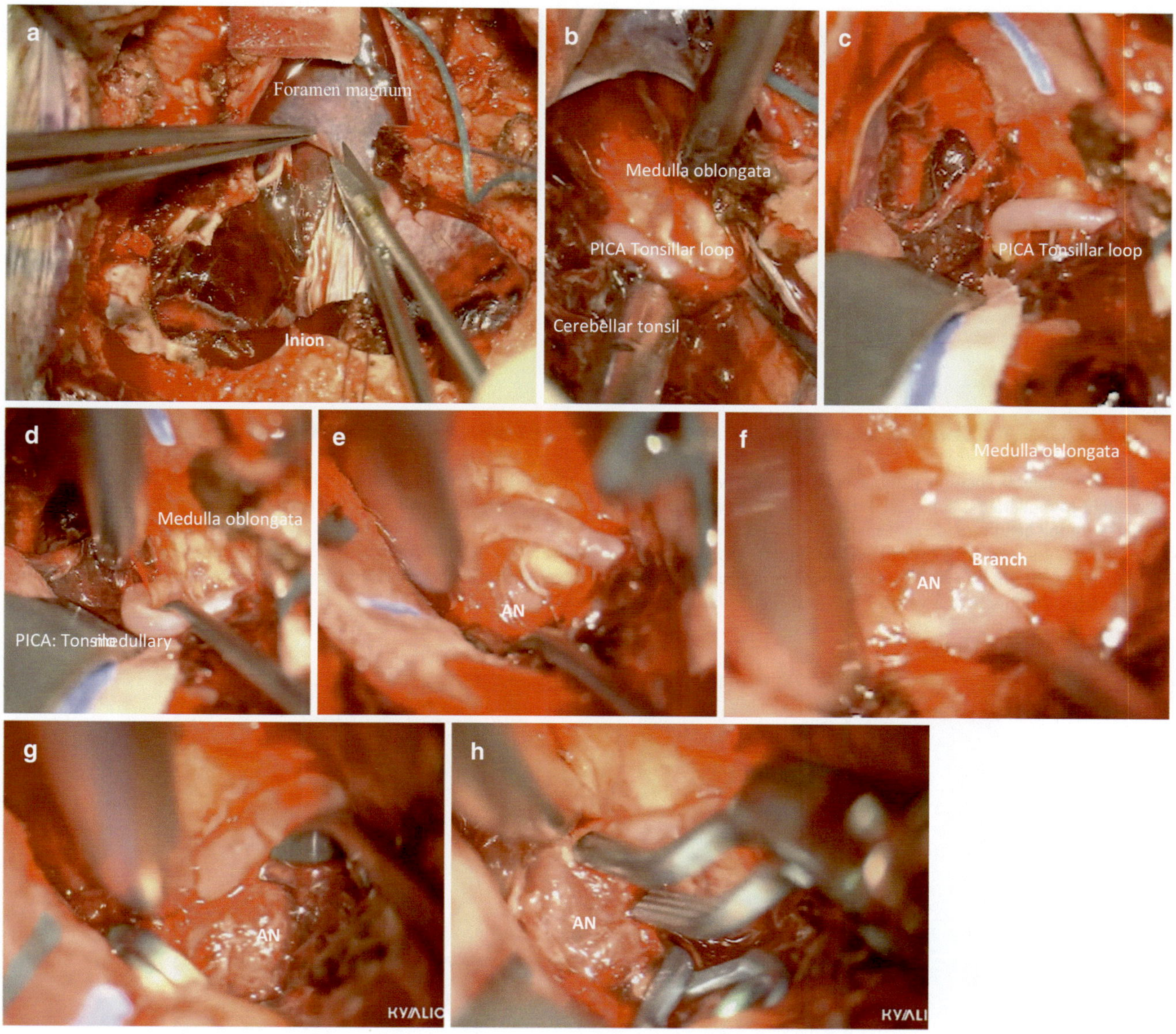

Fig. 17.13 Steps of dissection and clipping of a "distal" PICA aneurysm on the tonsillar loop involving the tonsillo-telovellar segment via a midline suboccipital craniotomy. The arachnoid of the cisterna magna is incised and opened on the right side of the midline showing extensive SAH with a thick blood clot filling the cisterna magna (**a**). Clearance of the blood identified the tonsillar loop of the PICA below a haemorrhagic cerebellar tonsil (**b**). This is dissected proximally towards the tonsillo-medullary segment postero-lateral to the medulla for proximal control (**c** and **d**). The aneurysm started to be exposed (**e**) with a distal small branch identified (**f**). The aneurysm was surrounded with adherent clots and fibrotic tissues, so a proximal temporary clip was placed to complete the dissection (**g**). A bayonet straight clip is applied to the main aneurysm, the blade negotiating the distal branch (**h**) before being rotated and applied to the neck in the final position. *AN* aneurysm

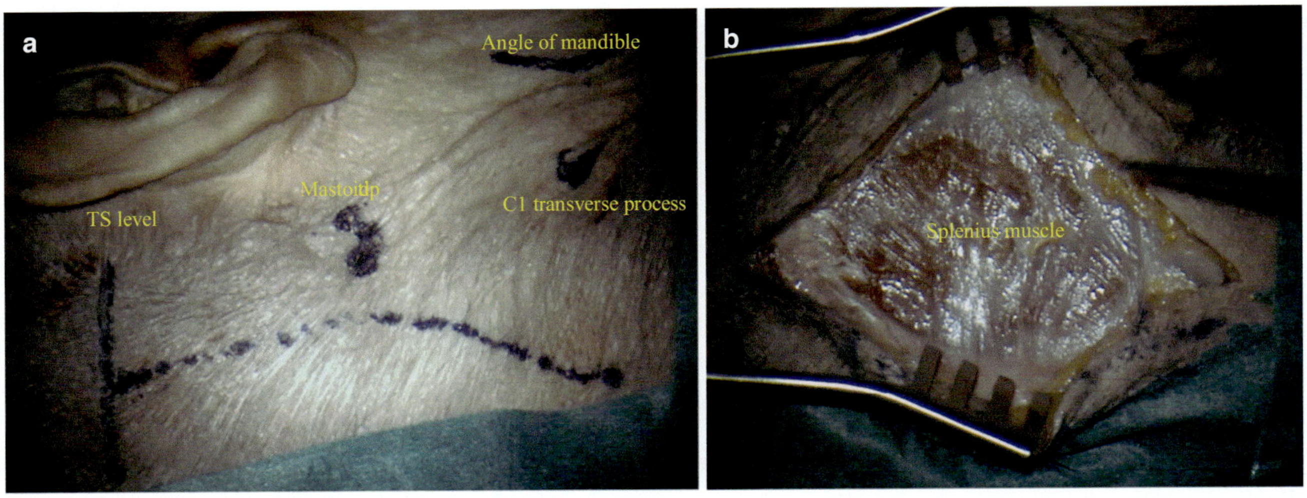

Fig. 17.14 A lateral position without head rotation (straight lateral) and a slight to moderate craniocaudal tilt may have an advantage over "Supine with the head turned" as it maintains the anatomical relationship of the atlanto-occipital joint and hence would be easier for orientation. The tip of the C1 transverse process is palpated behind the angle of the mandible in line with a vertical line dropped from the tip of the mastoid process and marked for orientation. The exposure through a long S-shaped retroauricular incision (**a**) and incising the sternomastoid and trapezius lateral to the line of the incision which is cosmetically fashioned behind the hairline after undermining the lateral skin edge exposes the deeper muscle layer which is the splenius muscle as shown in (**b**)

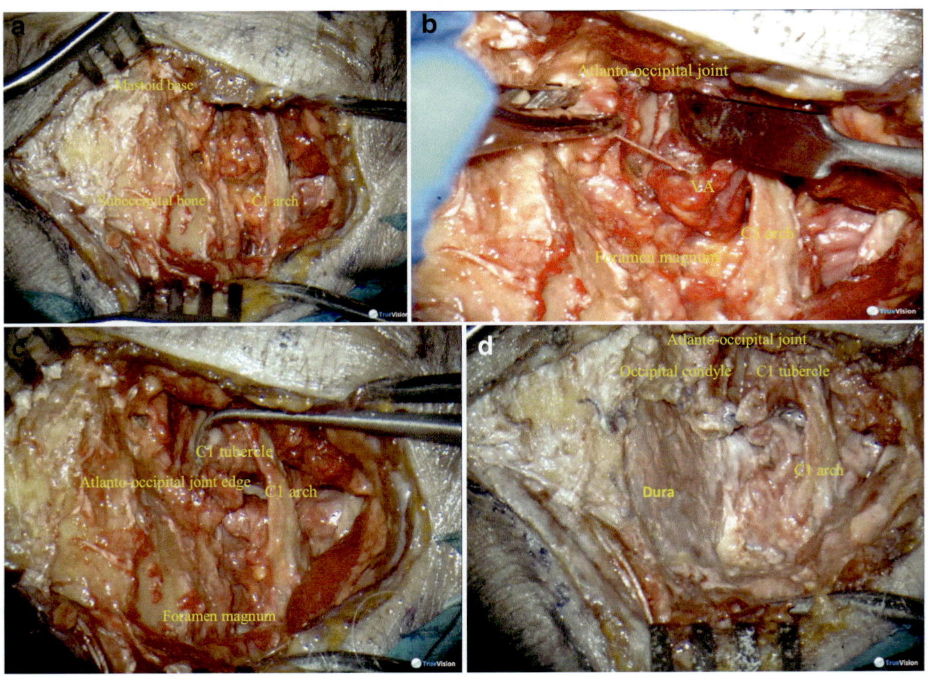

Fig. 17.15 A simple and safe modification is shown. A retrosigmoid exposure with a division of the muscles straight to bone to expose the suboccipital bone, asterion, mastoid base, start of digastric groove, foramen magnum, posterior ½ of the arch of C1 (**a**). The VA is identified by keeping the dissection on the bone of the mid-posterior C1 arch and identifying the rounded to flat transition inferior to the VA on the bone towards transverse process and clearing the bone edge supero medial to the VA area to C1 tubercle, i.e. follow the C1 from medial to lateral and identify the VA groove (round to flat) and follow the FM from medial to lateral and expose the atlanto-occipital joint (superolateral to VA) (**b**). To facilitate this, identify the condyle under the microscope BEFORE performing the craniotomy (**c**). The craniotomy extends as far laterally as the medial edge of the occipital condyle at least (**d**). This technical modification differs from the classic description of a retrosigmoid exposure and dissection of the muscle layers and exposure of the tip of the C1 transverse process to identify the VA in the vertebral triangle. This is followed by a craniotomy before the condyle exposure and then extending the craniotomy to the condyle

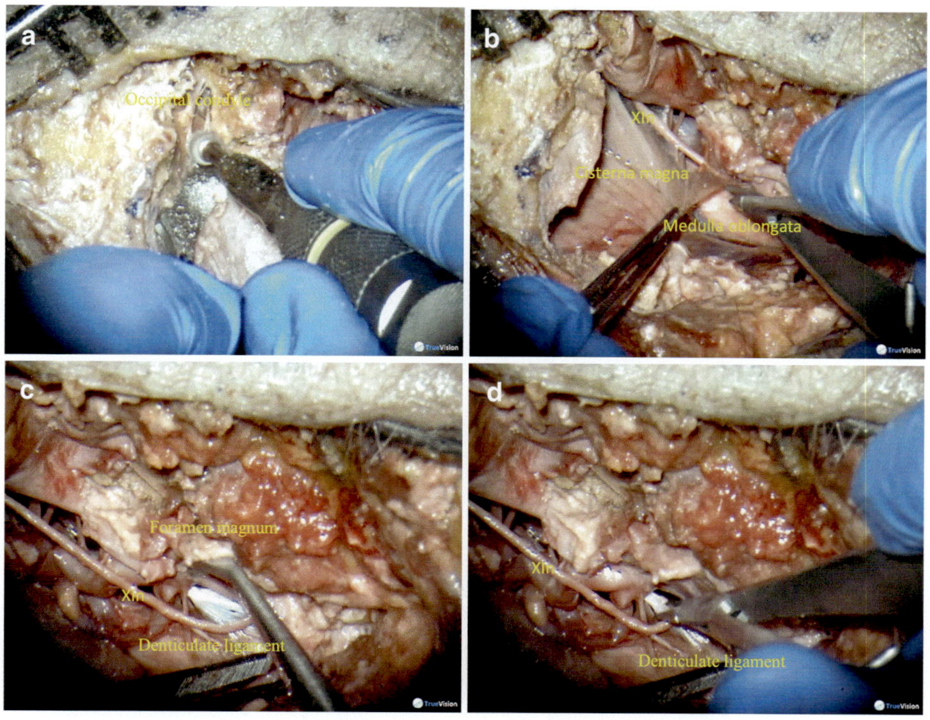

Fig. 17.16 If needed the medial edge may be flattened by a drill (**a**), the dura is opened (**b**) and the denticulate ligament is dissected (**c**) and divided (**d**)

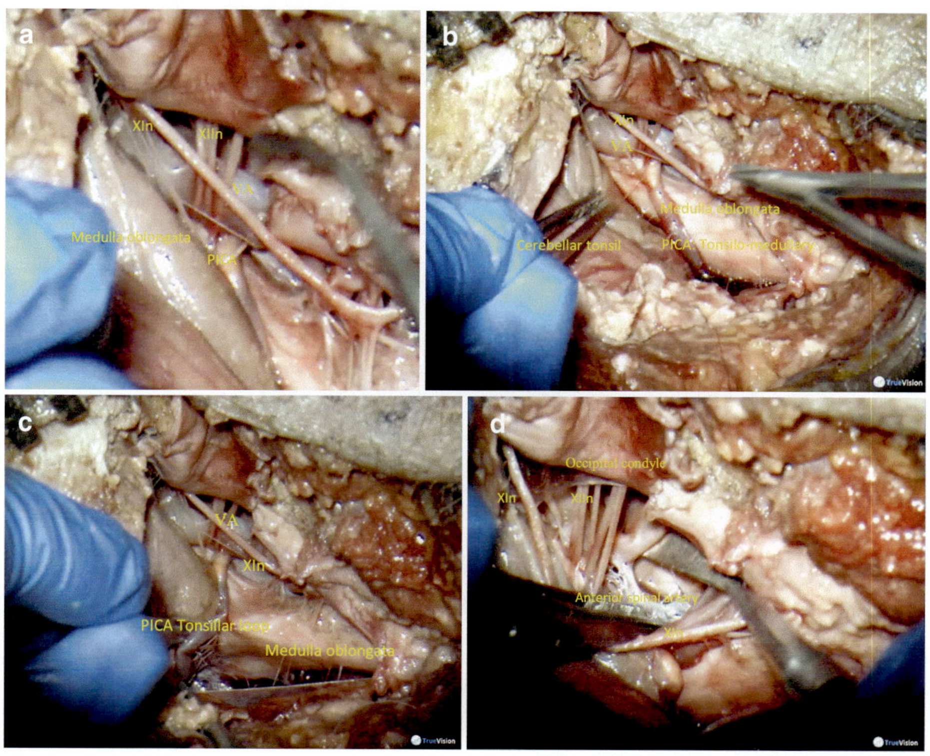

Fig. 17.17 The lateral exposure with the far lateral approach of C1 roots, Spinal XI nerve, Medulla, VA entrance and course, PICA origin and course, IX–XI cranial nerves, Jugular foramen, vertebral artery and PICA loops. Note the relation of the PICA to XII n and the lower cranial nerves (**a**–**d**). It is best suited for lesions at the upper cervical canal, FM and the pre-medullary space

Radical condylar resection–"extreme far lateral approach" is not required for purely intradural lesions including PICA aneurysms (Figs. 17.21 and 17.22). The nomenclatures describing the extent of far lateral approaches are confusing and it is best to describe the extent of condylar resection on individual cases.

17.6.4 Vertebral Artery and Exposure of the Cranio-Cervical Junction If Required

The VA emerges through the foramen transversarium of C1 and courses medially surrounded by an extensive venous

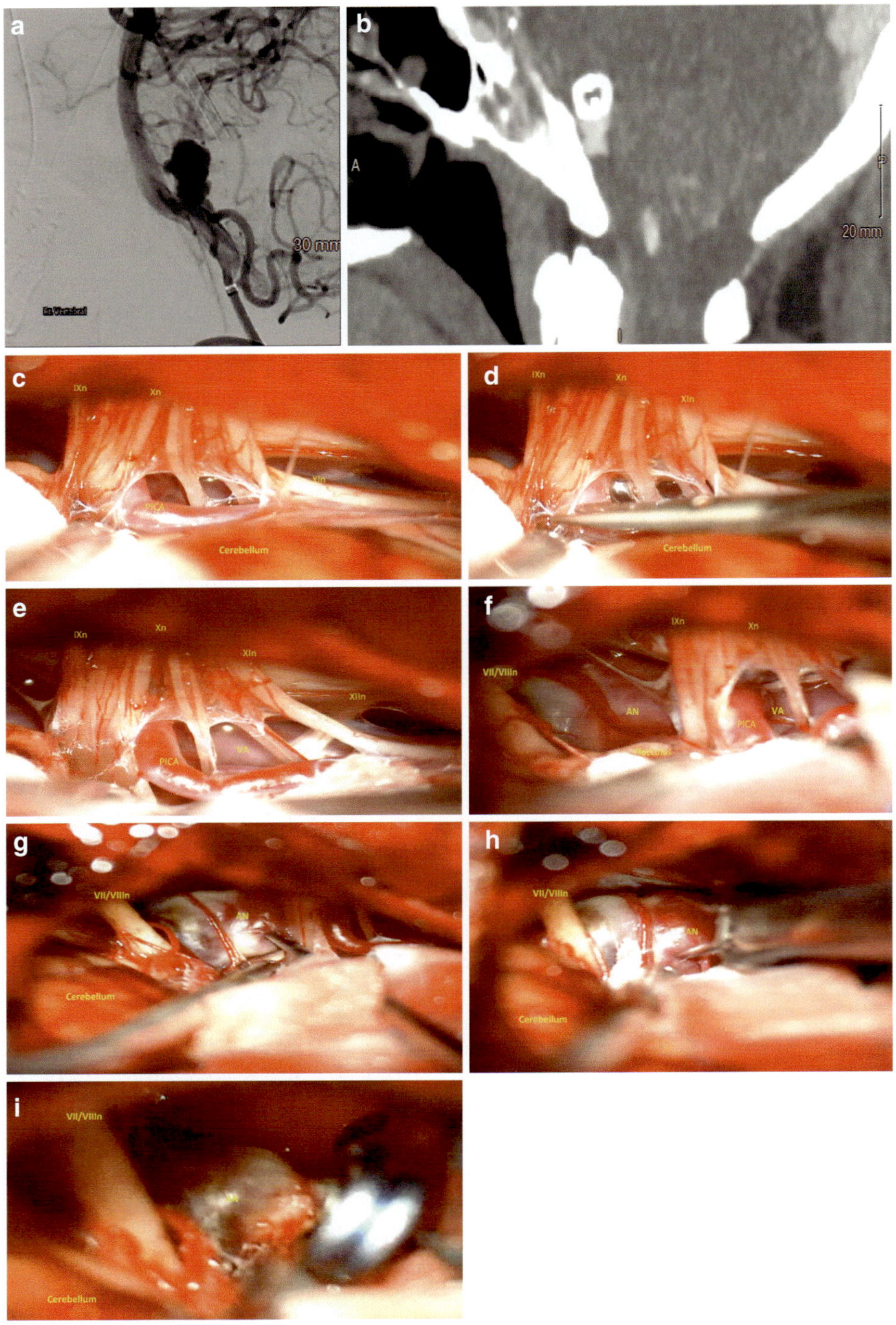

Fig. 17.18 A right PICA aneurysm arising more distally, i.e. "cranially" off the VA nearer to the VBJ as shown in the angiogram (**a**) and the fundus extends superiorly behind the clivus as seen on the sagittal CTA (**b**). Note the origin and course of the PICA travelling caudally leaving a clear space around the neck. Such aneurysm is exposed through a simple retrosigmoid craniotomy with the patient in the lateral position as the PICA is arising distally off the VA and the aneurysm is expected to be in juxtaposition and cephalad to the jugular foramen. The PICA is identified lateral to the tonsil and dissected cephalad towards its origin amongst the distorted fibres of the X nerve (**c**, **d**). The proximal VA is identified antero-lateral to the medulla with the rootlets of the XII nerve crossing it towards the relatively deeper hypoglossal canal by dissection through a window in between the fibres of the X nerve (**e**). The aneurysm sac is visible cephalad within the window in between the IX and VII/VIII cranial nerves (**f**). This allowed clearance of the neck as expected from the angiogram with PICA course caudally and the aneurysm fully exposed through a window between the IX and VII/VIII cranial nerves (**g**) to allow a smooth clip application and the blades sparing the perforator at the base (**h**). The final clip position is seen in (**i**). As such the planning and execution of the procedure were dictated by anatomical knowledge and predicting the configuration of the aneurysm from the pre-operative images to choose a relatively simple and safe approach. *AN* aneurysm

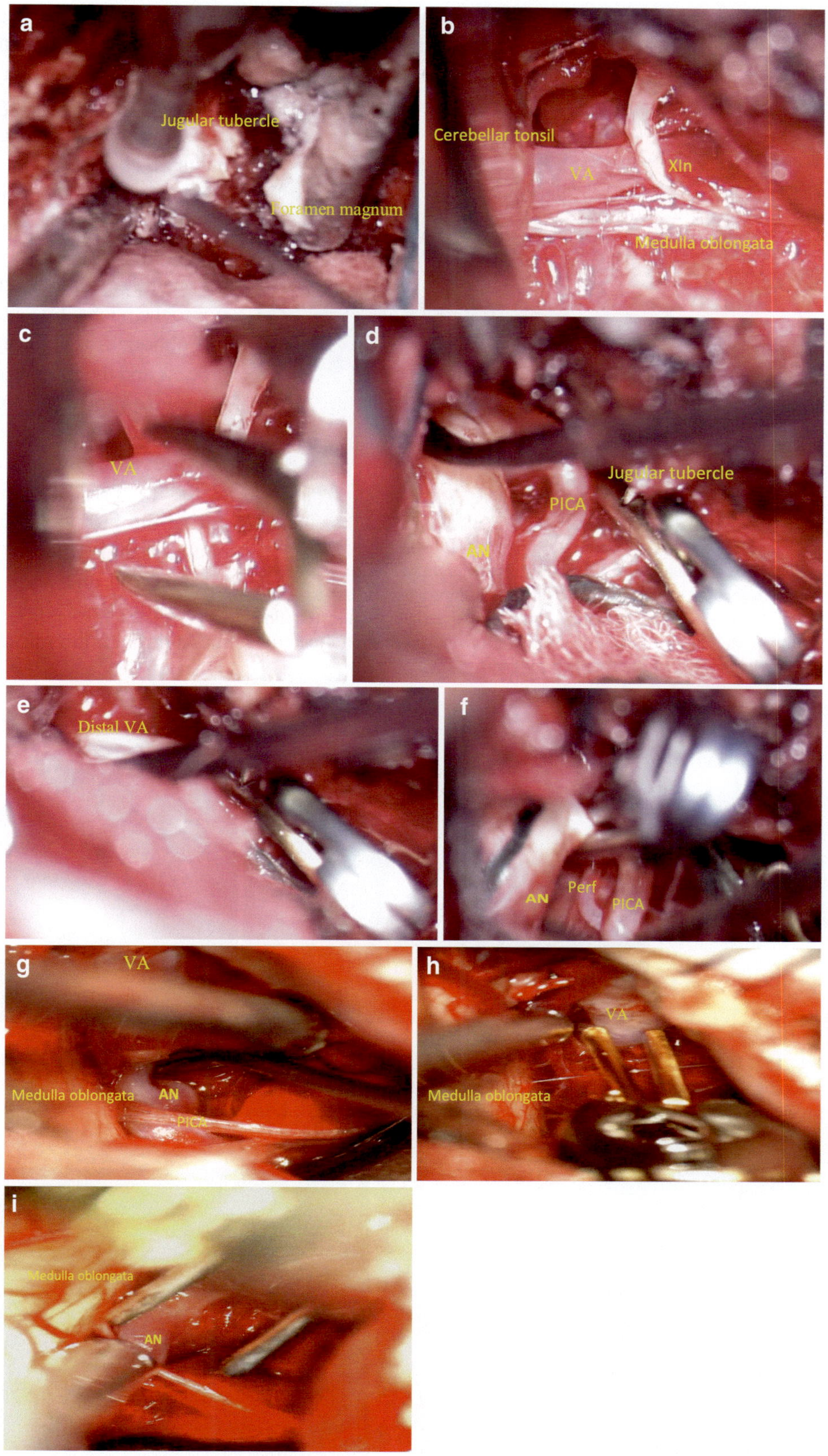

Fig. 17.19 Steps of clipping a large ruptured PICA aneurysm antero-lateral to the medulla with a "distal PICA take-off" from the VA and the neck adjacent to the jugular tubercle. A far lateral approach with the patient in the lateral position. The jugular tubercle is partially drilled extradurally (**a**) (see Fig. 17.21) and the VA proximal to the aneurysm identified just caudal to the jugular foramen with the XI nerve crossing it posteriorly (**b**). A temporary clip is applied (**c**) and the large aneurysm sac (AN) is dissected with clearance of the proximal neck from the PICA (**d**), followed by the identification of the distal VA (**e**). A clip is applied to the neck with care to spare a small perforating branch (Perf) near the neck (**f**). The occipital condyle was not drilled. The view is compared to another case approached through a midline suboccipital craniotomy in the prone position. In (**g**) the aneurysm (AN) is identified lateral to the medulla a short distance after the VA entrance intradurally. A viewing angle from medial to lateral readily exposed the proximal VA to apply a temporary clip in (**h**). The relatively wide opening of the foramen magnum allowed a straight view of the aneurysm to use a 45° angled clip in (**i**). As in the first case, a more distal aneurysm would require a far lateral approach

plexus over the relatively flat surface of the sulcus areriae vertebralis at the superior border of the lateral C1 posterior arch. It "snugs" the C1 lateral mass and passes in an antero-medial and cephalad direction through the dura at the foramen magnum and continues its intradural course. This V3 segment is related to the emerging C1 nerve root which passes inferior to it in a postero-lateral direction on the sulcus arteriosus of C1 (Fig. 17.23). The V3 gives off muscular branches and meningeal supply to the dura at the foramen magnum [1–3].

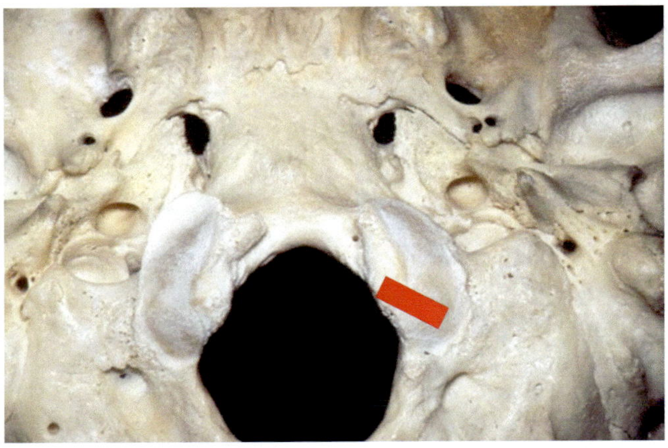

Fig. 17.20 A basal view of the foramen magnum. From a postero-lateral trajectory provided by the far lateral approach, drilling the posterior "corner" of the occipital condyle as marked by the red line improves the angle of view to the anterior and contralateral anterior paramedian region of the foramen magnum

Anatomical localisation of the V3 may be achieved by the identification of the suboccipital triangle. The boundaries of which are formed by the rectus capitis minor attached between the paramedian posterior suboccipital surface inferior to the nuchal line and the C1 posterior tubercle, the superior oblique from the suboccipital bone to the C1 transverse process and the inferior oblique with attachment between C1 transverse process and C2 spinous process. In the depth of the suboccipital triangle, the V3 and C1 nerve roots on the posterior arch of C1 are covered by soft tissue. Conventional textbooks classically describe surgical localisation of the VA during the far lateral approach by dissection of the suboccipital triangle muscles in layers through a skin flap in the retromastoid area such as a C-shaped or inverted U flap. Practically, however, a direct lazy S incision (Fig. 17.14) is made through the muscle layers, first exposing the postero-

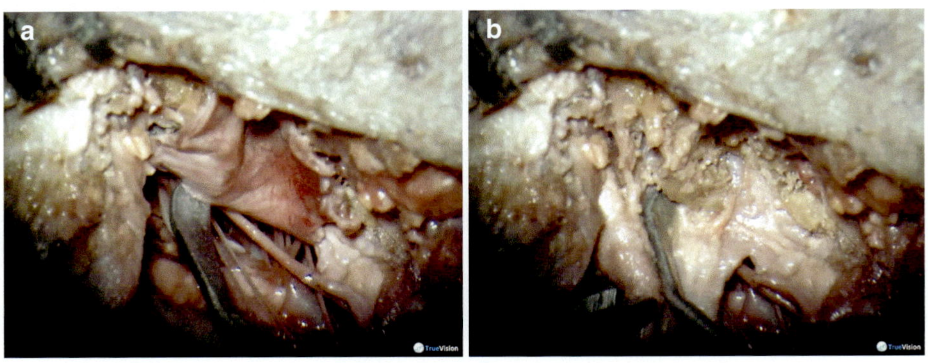

Fig. 17.21 In (**a**) the dura is reflected over the jugular tubercle indicated by the dissector to demonstrate the amount of obstruction by this process to the line of view to access the pre-pontine space cephalad. In (**b**) the dura is reflected back with the dissector delineating the jugular tubercle. Partial extradural drilling of its superomedial edge improves the access

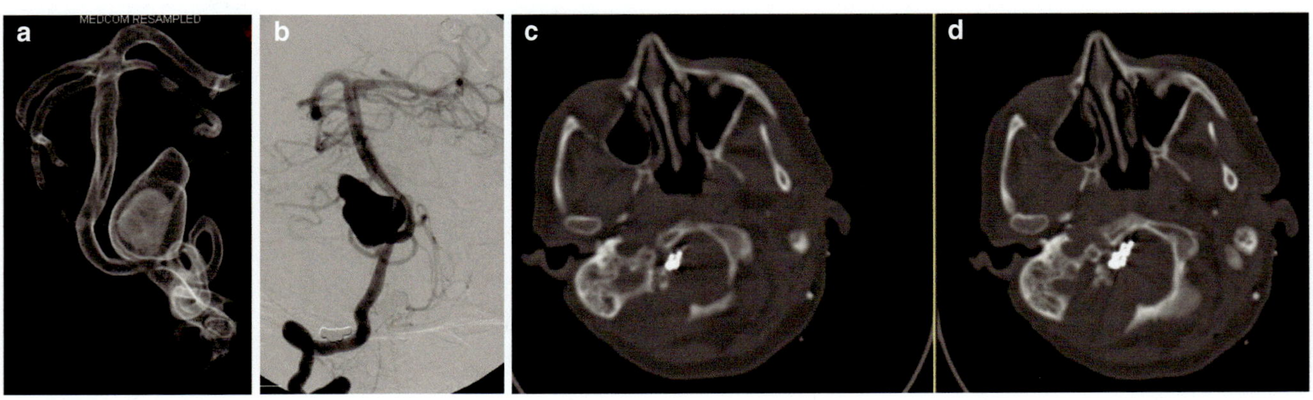

Fig. 17.22 Preoperative angiography (**a** and **b**) of a large vertebral aneurysm at the PICA origin. Postoperative CT scan with bone windows (**c**) after microsurgical clipping of the aneurysm via a far lateral approach not requiring drilling of the occipital condyle

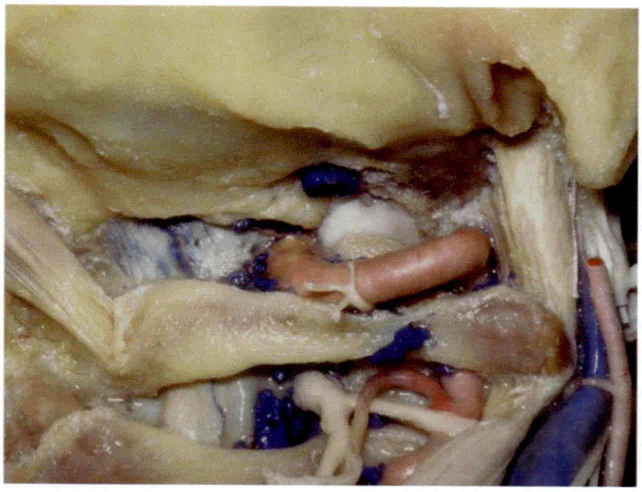

Fig. 17.23 The course of the extradural VA at the cranio-cervical junction from a postero-lateral view. The V3 segment is emerging from the C1 foramen transversarium on the sulcus arteriosus of the C1 lamina with the C1 nerve root emerging underneath. The VA is winding around the C1 tubercle (lateral mass) before its intradural entrance. A large condylar emissionary vein at the occipital condyle is frequently encountered [from Rhoton. *Courtesy of the Rhoton Collection, American Association of Neurological Surgeons (AANS)/Neurosurgical Research and Education Foundation (NREF)*]

lateral suboccipital bone and then through the muscles at the CCJ exposing the posterior lip of the foramen magnum and the C1 posterior arch. To avoid injury of the V3, the division of the tissues is best directed to expose the C1 posterior arch initially between the posterior tubercle and the flat surface of the sulcus arteriosus (Fig. 17.15). Guiding landmarks is by surface localisation of the position of the C1 transverse process which is deep on a straight vertical line extending inferiorly from the tip of the mastoid process just posterior to the angle and ascending ramus of the mandible (Fig. 17.14). Until there is a need to obtain extradural vascular control, the surrounding venous plexus can be kept intact, and the VA pulsations would be either visible, palpable or confirmed with a Doppler probe. During dissection in this area, brisk bleeding is usually from the condylar emissary vein and once identified can be controlled.

Appreciation of possible anatomical variations is important to avoid potential pitfalls and complications. C1 assimilation would have been easily identified on preoperative imaging. Ossification over the VA on the sulcus arteriosus of the C1 posterior arch converts this into a bony canal—"ponticulus posticus". Mistaking this as the normal rounded upper edge of the C1 arch can risk injuring the VA emerging from the foramen transversarium on proceeding with the dissection laterally. In about 10% or more the PICA origin from the VA is extradural before its dural entrance at the foramen magnum. Mistaking this for the muscular or meningeal branches may erroneously lead to its coagulation.

17.7 Surgical Pearls

- The main factors determining operative approaches and exposure of the PICA aneurysm neck are the size and location of the aneurysm, relationship with cranial nerves, provision to gain both proximal and distal control and the tortuosity of the parent vessel.
- The VA/PICA take-off along the cranial course of the VA from the foramen magnum to the VBJ is what determines the best trajectory that minimises the blind spot to approach an aneurysm at this location. Its relationship with the medulla and pons starts posterolateral at its intradural entry then it inclines anteromedially and cephalad becoming more anterior near the vertebrobasilar junction.
- A midline suboccipital craniotomy in the prone position is appropriate for a VA/PICA aneurysm at the foramen magnum or involving the distal segments of the PICA while a far lateral craniotomy with the patient in the lateral position provides the best trajectory to VA/PICA aneurysm arising more distally being anteromedial to the brainstem or for those arising on the proximal anterior and lateral medullary segments of the PICA.
- A far lateral approach is in reality a posterolateral approach.
- Exposure of the occipital condyle edge and the atlanto-occipital joint in these cases - which is best to identify under the microscope before performing the craniotomy - ensures opening the foramen magnum lateral enough.
- For most PICA aneurysms, there is no need to drill the occipital condyle.

17.8 Discussion

The success of surgical clipping for intracranial aneurysms depends upon planning based on anatomical considerations and the specific configuration and characteristics of each aneurysm [4–16]. The anatomical knowledge including the segments and branches of the PICA and its relationship in the posterior fossa to the neural structures and the craniocervical junction guides the best surgical trajectory with the best visualisation and minimal blind spots to allow safe clipping of different PICA aneurysms.

Variations in the craniocaudal "take-off" the PICA from the V4 segment of the VA affect the operative approach and patients' position, favouring midline suboccipital or far lateral approaches. Aneurysm neck's relation to the various components of the occipital condyle tailors the amount of bone drilling required. The far lateral approach identification and exposure of the occipital condyle ensure the opening of the foramen magnum is lateral enough to provide an adequate viewing angle.

With the current advances in endovascular treatment, the role of surgical clipping for PICA aneurysms shifted to those with locations and configurations—often complex—that are less ideal for such intervention [17, 18].

In a personal surgical series of 31 PICA aneurysms, 7 of these were flow aneurysms obliterated together with microsurgical resection of cerebellar AVMs. In fact, among 41 cerebellar AVM resections, 9 of these were associated with flow-related aneurysms that were obliterated at the same time and 7 out of the 9 were on a PICA. For the other 24 aneurysms clipped (out of over 750 clippings of aneurysms), 14 were approached through a midline suboccipital craniotomy (in the prone position), while 10 were approached by a far lateral approach. None of these required a transcondylar approach or extensive drilling of the occipital condyle. The jugular tubercle was drilled for approaching two of the high PICA aneurysms. It has to be noted that the excessive number of midline approaches reflects the fact that with endovascular intervention more of the aneurysms that were located distally on the PICA vessels were more favourable for microsurgical clipping.

Although further endovascular intervention is advocated to manage a proportion of those that recur following initial coiling, nevertheless in a number of these it is better to consider clipping for the recurrence. One particular situation is that of a recurrence, especially within a short period in relatively young patients, wherein there is coil "impaction" with a recurrent aneurysm neck that is circumferentially void of coils to accept a clip without the need to extract the coils. This may be regarded as a "window of opportunity" to perform definitive surgery in a potentially unstable situation. Extracting embedded coils from an aneurysm sac to allow coiling potentially increases the morbidity. In these situations, apart from the proper selection of cases, certain nuances should be observed. As stated regarding the anatomical considerations for clipping of PICA aneurysms, these in particular are anticipated to pose potential difficulties for mobilisation of a rigid sac to allow inspection for any small perforators arising at the neck and to access distal control if required. As such adequate exposure is required to allow a trajectory with minimal blind spots.

Certain considerations in performing the far lateral approach are aimed at selecting straightforward and simple technical steps even if the surgeon is capable of performing complex procedures. Some technical modifications may simplify the approach such as exposure of the occipital condyle prior to performing a craniectomy as the bony landmarks are still intact.

It has to be emphasised that clipping of PICA aneurysms should not be underestimated and it is not only the anatomical features described above but also the individual configuration of each aneurysm that need to be considered. The main priorities are adequate exposure to allow the safe application of the clip to the aneurysm neck without vessel compromise, especially important perforators, brainstem manipulation or disruption of the related cranial nerves. It is best to match the configuration of each aneurysm in combination with its location and anatomical relationships as described. A particular configuration may guide relatively straightforward microsurgical manoeuvrers even for those seemingly in less accessible anatomical locations. As mentioned above by far most PICA aneurysms originating near the foramen magnum do not require additional drilling of the occipital condyle and again the clip should be safely placed without compromising the arterial branches or traumatising the fibres of the XII nerve which is related to the PICA origin off the VA. Therefore, the simplest and safest procedure should be chosen with awareness of certain anatomical relationships and pathological configurations of different aneurysms that may dictate essential additional steps or manoeuvrers as required.

17.9 Conclusions

Planning for surgical clipping of different PICA aneurysms requires malleable thinking and tailored approaches using 3D imagination. Anatomical knowledge and careful planning based on pre-operative investigations aim to execute surgery through the least complex and safe manoeuvres as universal simplicity prevails. All operative steps should be tailored to the pathology and one must resist the temptation to perform manoeuvres just because one is capable/aware of doing them unless these are truly necessary.

By far in most cases of PICA aneurysms, there is **NO** need to drill the occipital condyle; circumvented simply through a posterior midline approach in the prone position. In some occasions, it is the jugular tubercle rather than the condyle that obscures the view and access to the aneurysm neck.

References

1. Lister JR, Rhoton AL Jr, Matsushima T, Peace DA. Microsurgical anatomy of the posterior inferior cerebellar artery. Neurosurgery. 1982;10(2):170–99.
2. Matsushima T, Natori Y, Katsuta T, Ikezaki K, Fukui M, Rhoton AL. Microsurgical anatomy for lateral approaches to the foramen magnum with special reference to transcondylar fossa (supracondylar transjugular tubercle) approach. Skull Base Surg. 1998;8(3):119–25.
3. Rhoton AL Jr. The cerebellar arteries. Neurosurgery. 2000;47(Suppl. 3):S29–68.
4. Bertalanffy H, Sure U, Petermeyer M, Becker R, Gilsbach JM. Management of aneurysms of the vertebral artery-posterior inferior cerebellar artery complex. Neurol Med Chir (Tokyo). 1998;(Suppl. 38):93–103.

5. D'Ambrosio AL, Kreiter KT, Bush CA, Sciacca RR, Mayer SA, Solomon RA, Connolly ES Jr. Far lateral suboccipital approach for the treatment of proximal posteroinferior cerebellar artery aneurysms: surgical results and long-term outcome. Neurosurgery. 2004;55(1):39–54.
6. Krayenbuhl N, Guerrero C, Krisht AF. Technical strategies to approach aneurysms of the vertebral and posterior inferior cerebellar arteries. Neurosurg Focus. 2005;19(2):E4.
7. Singh RK, Behari S, Kumar V, Jaiswal AK, Jain VK. Posterior inferior cerebellar artery aneurysms: anatomical variations and surgical strategies. Asian J Neurosurg. 2012;7(1):2–11.
8. Viswanathan GC, Menon G, Nair S, Abraham M. Posterior inferior cerebellar artery aneurysms: operative strategies based on a surgical series of 27 patients. Turk Neurosurg. 2014;24(1):30–7.
9. Seoane P, Kalb S, Clark JC, Rivas JC, Xu DS, Mendes GAC, Preul MC, Zabramski JM, Spetzler RF, Nakaji P. Far-lateral approach without drilling the occipital condyle for vertebral artery-posterior inferior cerebellar artery aneurysms. Neurosurgery. 2017;81(2):268–74.
10. Tjahjadi M, Rezai Jahromi B, Serrone J, Nurminen V, Choque-Velasquez J, Kivisaari R, Lehto H, Niemelä M, Hernesniemi J. Simple lateral suboccipital approach and modification for vertebral artery aneurysms: a study of 52 cases over 10 years. World Neurosurg. 2017;108:336–46.
11. Pilipenko Y, Eliava S, Okishev D, Okisheva E, Spyrou A. Vertebral artery and posterior inferior cerebellar artery aneurysms: results of microsurgical treatment of eighty patients. Surg Neurol Int. 2019;10:227.
12. Deora H, Nayak N, Dixit P, Vikas V, Rao KVLN, Pruthi N, Srinivas D, Shukla DP, Bhat DI, Malla BR, Devi BI, Somanna S. Surgical management and outcomes of aneurysms of posterior inferior cerebellar artery: location-based approaches with review of literature. J Neurosci Rural Pract. 2020;11(1):34–43.
13. Del Maestro M, Luzzi S, Galzio R. Microneurosurgical management of posterior inferior cerebellar artery aneurysms: results of a consecutive series. Acta Neurochir Suppl. 2021;132:33–8.
14. Murakami T, Toyota S, Nakagawa K, Hagioka T, Hoshikuma Y, Suematsu T, Shimizu T, Kobayashi M, Taki T. Midline suboccipital approach to a vertebral artery-posterior inferior cerebellar artery aneurysm from the rostral end of the patient using ORBEYE. Surg Neurol Int. 2022;13:87.
15. Yang Z, Song J, Quan K, Li P, An Q, Shi Y, Liu P, Yu G, Tian Y, Zhou L, Zhu W. Microsurgical treatment of posterior inferior cerebellar aneurysms based on angioarchitecture supplemented by high-resolution vessel wall MRI: a case series report. Stroke Vasc Neurol. 2022;7(4):337–44.
16. Sriamornrattanakul K, Akharathammachote N, Chonhenchob A, Mongkolratnan A, Niljianskul N, Phoominaonin IS, Ariyaprakai C. Far-lateral approach without C1 laminectomy for microsurgical treatment of vertebral artery and proximal posterior inferior cerebellar artery aneurysms: experience from 48 patients. World Neurosurg X. 2023;19:100216.
17. Song J, Park JE, Chung J, Lim YC, Shin YS. Treatment strategies of ruptured posterior inferior cerebellar artery aneurysm according to its segment. Surg Neurol Int. 2017;8:155.
18. Saal-Zapata G, Rodriguez-Calienes A, Malaga M, Velasquez-Rimachi V, Moran C, Bustamante-Paytan D, Pacheco-Barrios N, Pacheco-Barrios K, Alva-Diaz C, Walker M. Microsurgical and endovascular treatment of posterior inferior cerebellar artery aneurysms: a systematic review and meta-analysis. Neurosurg Sci. 2023;67(5):638–52.

Anatomy, Characteristics, and Management of Pediatric Aneurysms

18

Ioan Stefan Florian and Ioan Alexandru Florian

Learning Objectives
- Epidemiology, etiology, genetic factors, and distinguishing features of PIAs.
- Applied microsurgical anatomy for the most frequent locations of PIAs.
- Imaging diagnosis of PIAs.
- Decision making and therapeutical options.
  - Surgical treatment.
  - Endovascular treatment.
  - Conservative management.
- Prognosis of PIAs.

18.1 Introduction

Intracranial aneurysms (IAs) at the pediatric age are a rare occurrence, amounting between 0.5% and 4.6% out of the total [1–3]. Their rarity has been confirmed both in autopsy cases and on large angiographic studies, proving that their low incidence is not the result of underdiagnosis. The incidence is proportional to the age of the child, only 10–20% being diagnosed under 5 years of age, with the average age at diagnosis being at 10–14 years [4]. In our series of 1131 patients harboring a total of 1306 aneurysms, only 7 were diagnosed during childhood and a single patient was less than one year old, representing 0.6% of our retrospective series [5].

A similarity between pediatric IAs (PIAs) and the ones in the adult population is their predilection for the anterior circulation, although most resemblances end at this point [1, 6]. Common locations include the anterior communicating artery (ACoA), middle cerebral artery (MCA), and the bifurcation of the internal carotid artery [7]. In contrast to adult series, PIAs are more commonly encountered in males (between 55% and 70%), the male-to-female ratio changing from 1.1:1 in infants to 1.8:1 for children under 5 years [4]. Unlike adults, wherein the female sex is considered a risk factor, PIAs are more common in males. Another peculiarity of PIAs is the larger number of giant and dissecting aneurysms, suggesting congenital defects of the arterial wall, all the more evident when these lesions appear within the first year of life. Alongside genetic factors that contribute to these defects, infections and trauma also play a role in the development of PIAs. Most often, clinical diagnosis results either from the acute symptomology brought upon by intracranial hemorrhage, or a progressive symptomology in the context of mass effect from giant aneurysms. Their management raises a series of issues, especially in young children whose vessels have smaller diameters. If after 4 years the endovascular treatment is relatively safe before this age surgery seems the more feasible option. Nevertheless, treatment of giant or dissecting PIAs is made difficult by the small size of the parent vessels, as well as the lower experience of treating surgeons. Clinical and imagistic observation may be acceptable for selected cases. Considering all technical, surgical, and endovascular challenges, the results and prognostic of hemorrhagic PIAs are better than for ruptured aneurysms in adults, justifying the efforts for the complex treatment of this pathology.

18.2 History and Background

The first historical description of a subarachnoid haemorrhage (SAH) from a ruptured aneurysm dates from the eighteenth century, belonging to Biumi. Almost a century later, in 1871, Eppinger described the case of a 15-year-old boy who collapsed during some physical exercises and died 3 days later. At the autopsy, he described a ruptured saccular

I. S. Florian (✉) · I. A. Florian
Department of Neuroscience, University of Medicine and Pharmacy Iuliu Hatieganu, Cluj-Napoca, Romania

Department of Neurosurgery, Cluj County Emergency Clinical Hospital, Cluj-Napoca, Romania

I. N. Kanaan, V. Beneš (eds.), *Neuroanatomy Guidance to Successful Neurosurgical Interventions*, https://doi.org/10.1007/978-3-031-59838-8_18

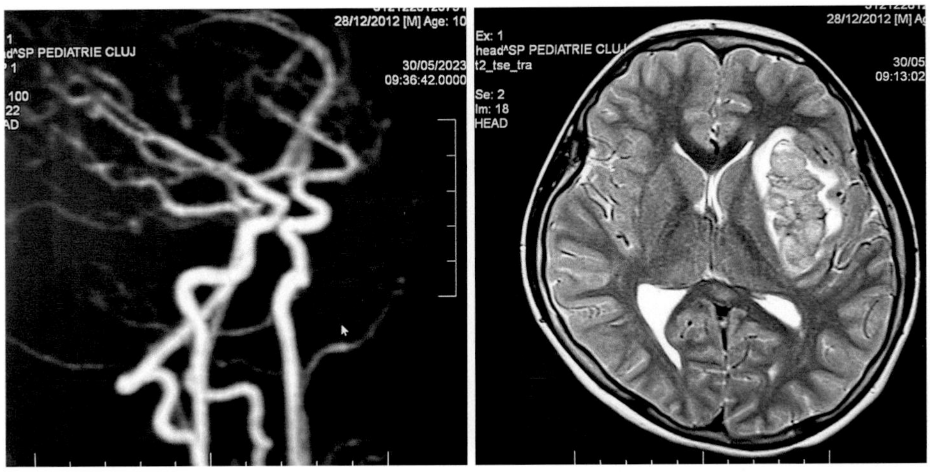

Fig. 18.1 Brain MRI in a 10-year-old comatose (GCS 4) boy in a septic state and kidney failure due to nephrotic syndrome. Time of Flight (TOF) MRI (left) shows an aneurysmal dilation on the distal M3 branch (*white arrow*), which led to the formation of a large insular hemorrhage in the left hemisphere

aneurysm on the right anterior cerebral artery (ACA) that caused an intracerebral hemorrhage. The boy also presented a stenosis of the aorta [4], coarctation of the aorta being nowadays recognized as a comorbidity associated with PIAs. This is considered to be the first PIA reported in the literature. Considering their comorbidities and peculiarities, especially in comparison to IA of the adults, a thorough analysis of their **background** is warranted.

The occurrence of aneurysms at such an early age, wherein etiologic factors ascribed to adult aneurysms (smoking, atherosclerosis, hypertension) cannot be applied, impels us to consider a genetic determinism and certain imperfections in the embryonal formation of the cerebral vasculature. It is beyond the scope of this chapter to delve into the genetic determinism and maldevelopment of primordial arterial vessels, these having been well characterized in the literature [8]. Some aspects however deserve discussion and will later be presented in this chapter.

As opposed to IAs of adults, which are mostly caused by the degeneration of arterial walls and enduring hemodynamic disturbances, those found in children typically possess a congenital, vasculopathy, traumatic, infectious, or even an idiopathic etiology [1, 9] On the etiological podium, hemodynamic stress stands at third place, superseded by traumatic and infectious causes. The higher incidence of infectious aneurysms (2–22%) (Fig. 18.1) or traumatic aneurysms (3–30%) at the pediatric ages suggests that the aneurysmal wall is much more susceptible to external factors than at the adult age. In children under 5 years, the histopathological analysis of the aneurysmal wall demonstrated the thinning, fragmentation, or even the absence of the internal elastic lamina or the tunica media, without a universally identifiable cause [4].

18.3 Anatomical Highlights

- The vascular development of the brain is completed at birth; therefore, the vascular anatomy of the child is quite similar to that of the adults.
- Knowing the normal vascular anatomy of the adults will make the recognition of vascular pathology of the child much more straightforward.
- Surgery of intracerebral aneurysm is mainly an *extracerebral* surgery.
- Knowing how to "navigate" the cisterns is an essential clue in aneurysmal surgery.
- Anatomical knowledge of cerebral vasculature through cadaveric dissection, anatomical atlases, and 3D video demonstrations is a mandatory step in understanding and memorizing cerebral vasculature (Fig. 18.2).
- Neurovascular surgical skills require many hours of microsurgical practice in an experimental laboratory and even many more hours of apprenticeship alongside surgical masters.
- Despite anatomical similarities, there are some peculiarities in infants and toddlers that should be underlined:
  - Overall parenchyma and vessel fragility (mainly those in the Sylvian fissure).
  - Size of the vessels—MCAs reach the diameter of adults after six months of life, whereas the ICA reaches diameters close to that of adults at 4 years.
  - In infants, due to a higher oxygenation of blood, it is sometimes difficult to discriminate between arteries and veins.
  - Generally narrower subarachnoid cisterns and spaces.
  - Coalescence of lobes.

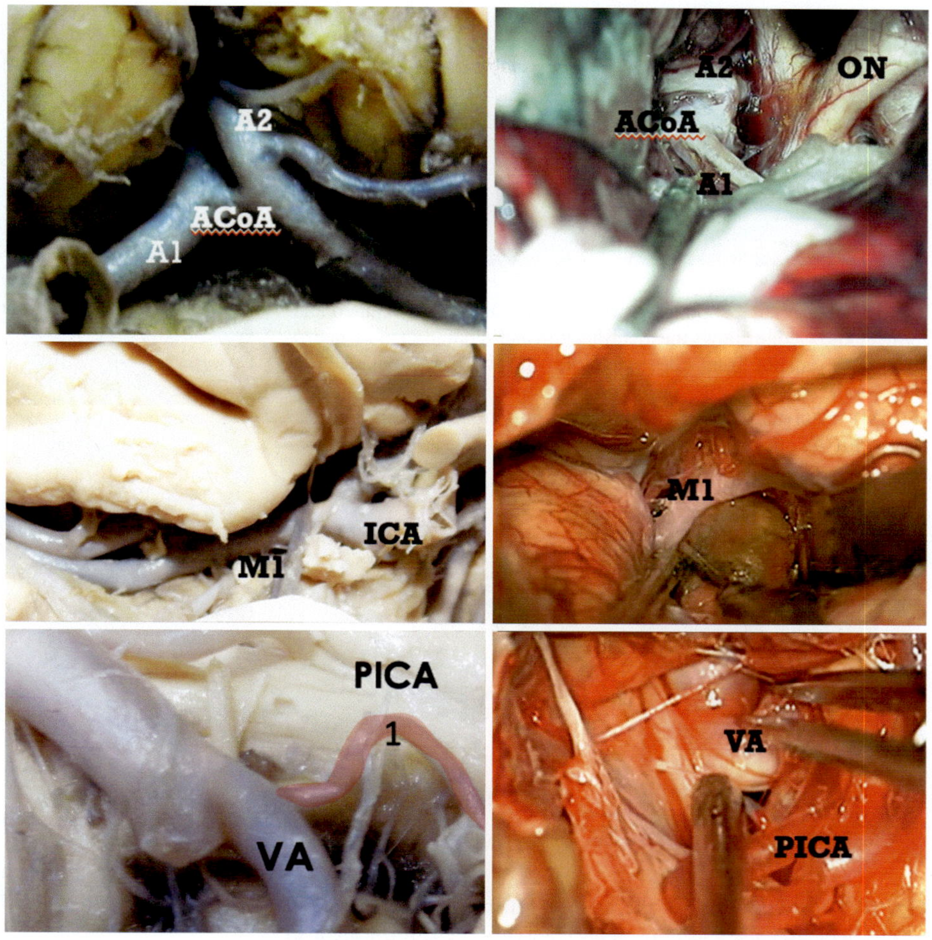

Fig. 18.2 Illustrative images of the vascular anatomy: left column—cadaveric specimens (images from the department archive) and Right column—intraoperative aspect of different aneurysmal locations: anterior communicating artery (ACoA) posteriorly oriented saccular aneurysm (top row), giant middle cerebral artery (MCA) M1 bifurcation aneurysm (middle row), and an aneurysm of the V4 segment of the vertebral artery (VA) (bottom row)

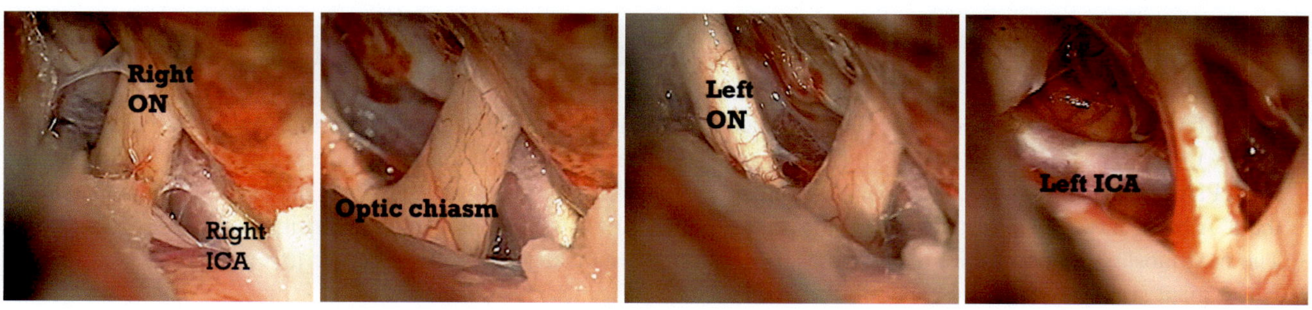

Fig. 18.3 Progressive dissection of arachnoid spaces starting above the right optic nerve (ON). The ON is the most constant landmark of anterior skull base surgery. Large sharp atraumatic dissection of the Sylvian fissure starting from the proximal end (preferred by the main author) or at the distal one (recommended by many authors) and basal cisterns is mandatory for the recognition of all regional vessels, definition of aneurysmal neck and proximal control; gentle "navigation" of the cisterns allows the exploration of even contralateral internal carotid artery (ICA) bifurcation and M1 bifurcation from one side

  – Incomplete definition of the cisterns and Sylvian fissure.
- As such, gestures should be minute and delicate during dissection in order to avoid additional parenchymal or vascular lesions.
- The same as for adults, the Sylvian fissures and basal cisterns represent the "highway" of aneurysmal surgery (Fig. 18.3).
- Delicate separation of adjacent arteries and nerves from the aneurysmal neck is difficult but mandatory for a good outcome (Fig. 18.4).

The location of PIAs has certain defining features:

– Anterior circulation PIAs are reported to be 70–75% of the total, 25–50% being in the posterior circulation (in

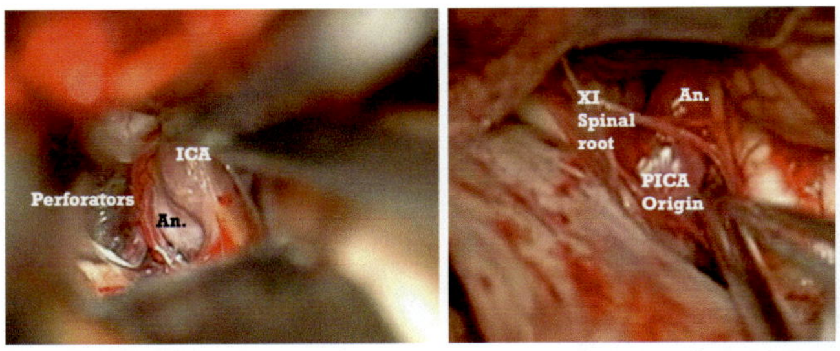

Fig. 18.4 Lage dissection of arachnoid spaces alongside detailed anatomical knowledge allows the identification and separation of all neurovascular elements from the place the clip will be applied. Left image—the intraoperative image of a posterior communicating artery (PCoM) aneurysm and Right image—a posterior inferior cerebellar artery (PICA) origin aneurysm

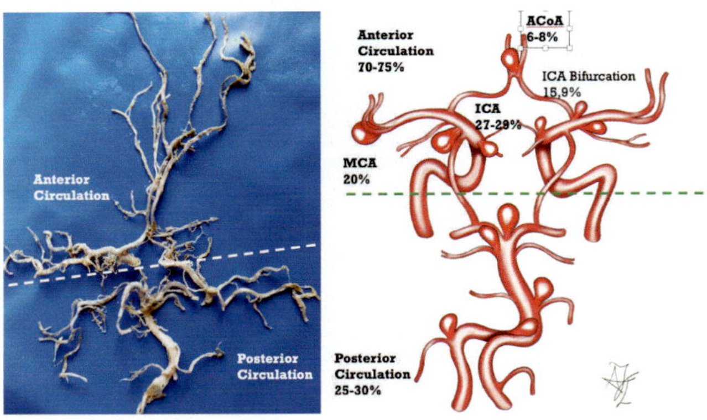

Fig. 18.5 "Arterial tree" of the brain: right image—the anatomical specimen of arterial arborization of the brain (from department archive) and Left image—a schematic representation of the major location of PIAs in anterior circulation (illustration by Alexandru I. Florian-coauthor)

contrast with adults where posterior lesions amount to 5–15%).

- Most PIAs arise on the ICAs (27–29%), the bifurcation of the ICA being a typical location (15.9%) [10].
- The next most common location is the MCA bifurcation where giant aneurysms are also generally encountered.
- The ACA-ACoA complex accounts for only 6–8% of cases (in adults is the most frequent location (Fig. 18.5)).
- PIAs of the posterior circulation are proportionately twice as more frequent [11].
- The most typical location is at the basilar trunk, followed by the vertebral arteries (VA) (Fig. 18.6).
- Giant, complex, fusiform, and dissecting aneurysms are the most common types of PIAs in the posterior circulation [4].

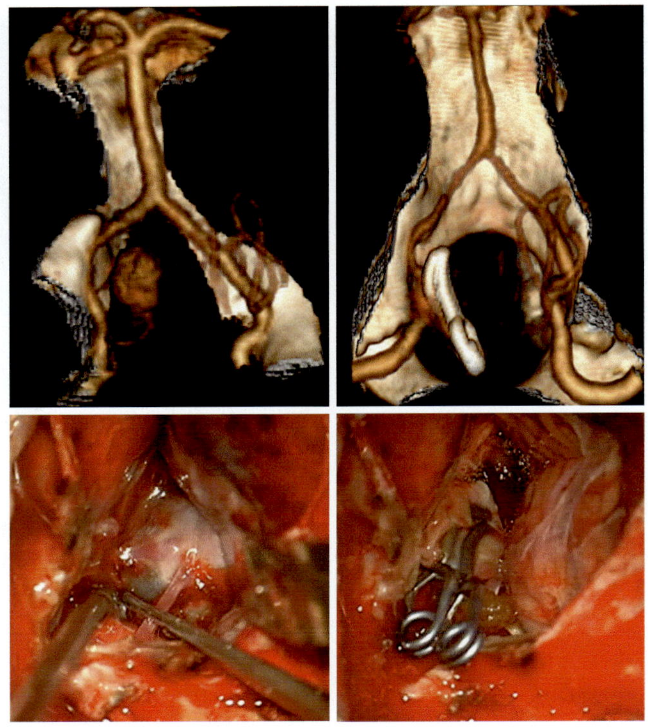

Fig. 18.6 Top row: three-dimensional reconstruction of a cerebral CT angiography (CTA) in a 17-year-old admitted with increased intracranial pressure and altered state of consciousness. The CTA reveals a large distal posterior-inferior cerebellar artery (PICA) aneurysm. Bottom row: the surgical clip was placed via a retrosigmoid approach, followed by the resection of the aneurysmal sac that was compressing the brainstem

18.4 Clinical Presentation

Most often, PIAs present through SAH, this being the case between 44 and 72% of the time [12] (Fig. 18.7). Aneurysmal SAH in children is much rarer than in adult patients (almost 90%), with PIAs being responsible for around 10–15% of hemorrhagic strokes within the first two decades of life [11] (Fig. 18.3).

This lower incidence of SAH is mainly due to the relatively higher number of cases that present with giant aneurysms. These typically manifest through a tumor mass syndrome, especially if belonging to the large or giant variants, however rupture and ischemia are also possibilities [1, 3, 6, 13]. Rebleeding is more often following PIA rupture due to a delay in diagnosis [12]. Cranial nerve involvement and seizures are more common than in adults [14] .

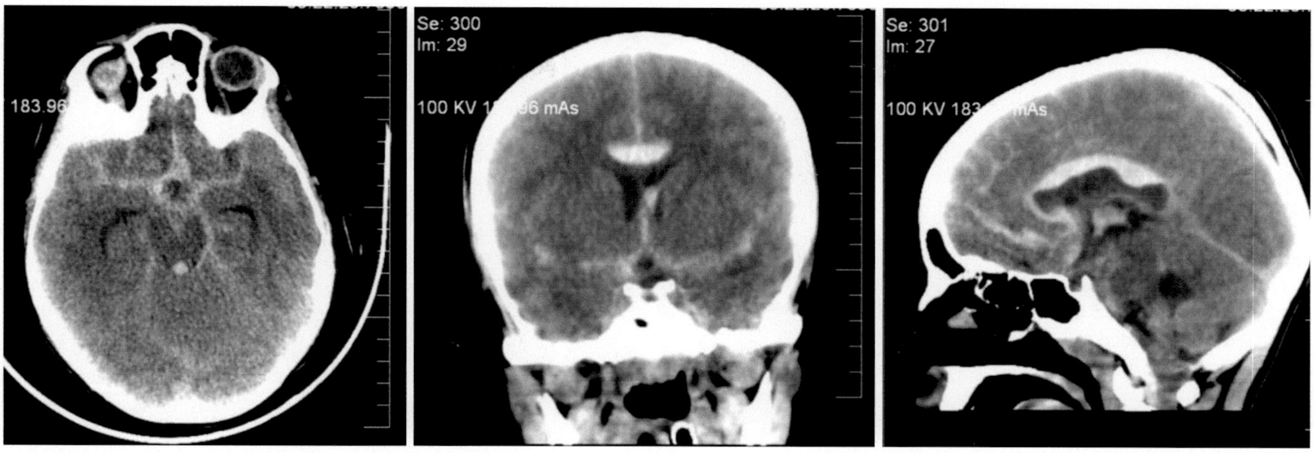

Fig. 18.7 Brain CT scan of a 10-year-old presenting in a coma (H&H 4), demonstrating hemorrhage in the subarachnoid spaces, inside the ventricles, and in the pericallosal cistern

18.5 Imagistic Diagnosis

In an emergency setting, CT and CT angiography (CT) are the mainstay diagnostic investigations. Identifying a spontaneous hemorrhage on a noncontrasted CT scan mandates the acquisition of contrast-enhanced CTA scans. Despite concerns regarding irradiation at the pediatric age, we consider that CTA accuracy is superior to Time of Flight (TOF) MRI, especially in acute settings (Fig. 18.8). Additionally, as MRI takes longer and is more difficult to perform in an emergency, requiring a series of supplementary logistic measures (tracheal intubation, MRI-compatible ventilator, ICU specialist on standby), irradiation-related concerns become secondary. Similarly, where available, digital subtraction angiography (DSA) is the elective diagnosis method, especially if an endovascular intervention is to follow (Fig. 18.9).

MRI is indicated primarily for giant aneurysms causing mass effects, wherein the differential diagnosis with a tumor requires contrast-enhanced MRI angiography. IRM is also the mainstay form of imaging follow-up in both conservatively managed and clipped cases (Fig. 18.10).

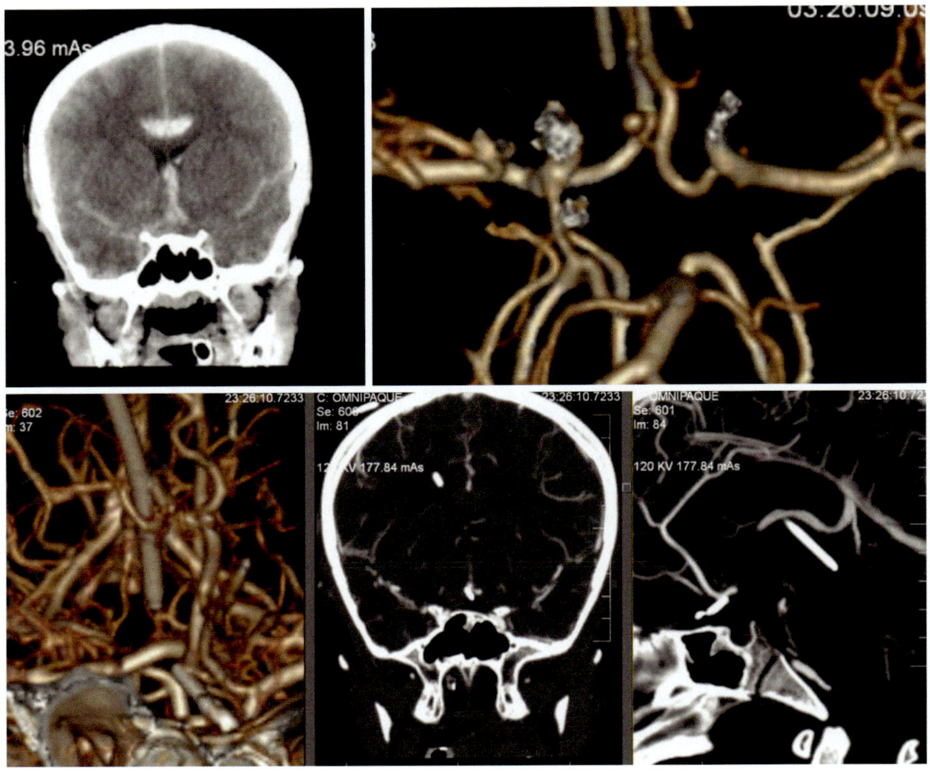

Fig. 18.8 Top row: coronal noncontrasted and 3D reconstruction of a CT angiography (CTA) in a 10-year-old boy admitted in a comatose state with subarachnoid hemorrhage, revealing a 3 mm ruptured ACoA aneurysm. Contrary to most guidelines that recommend follow-up for aneurysms of less than 5 mm in diameter, we decided to clip the aneurysm. Bottom row: control CTA demonstrating aneurysm occlusion by microclip. An external ventricular drainage (EVD) was placed after craniotomy, but before opening the dura mater, in order to obtain adequate brain relaxation. Placing an EVD before aneurysm clipping carries the risk of sudden intracranial pressure drop, theoretically increasing the chances of rebleeding

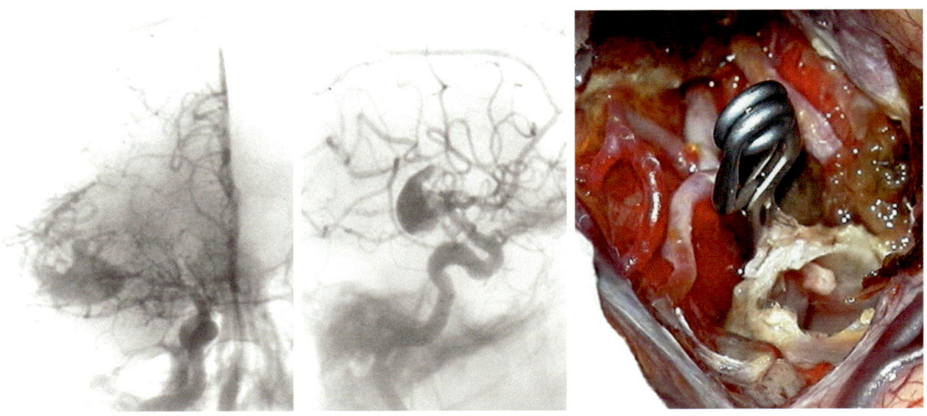

Fig. 18.9 Top row: DSA in a 15-year-old boy admitted for seizures and headache, revealing a giant right middle cerebral artery (MCA) aneurysm that was partially thrombosed. On the right: intraoperative aspect. Despite the large size of the aneurysm, the neck was small, allowing occlusion with a single straight Yasargil clip after internal debulking of the aneurysmal sac. No attempt was made to completely remove the residual sac because of the strong adherence with the adjacent brain and M2 segments

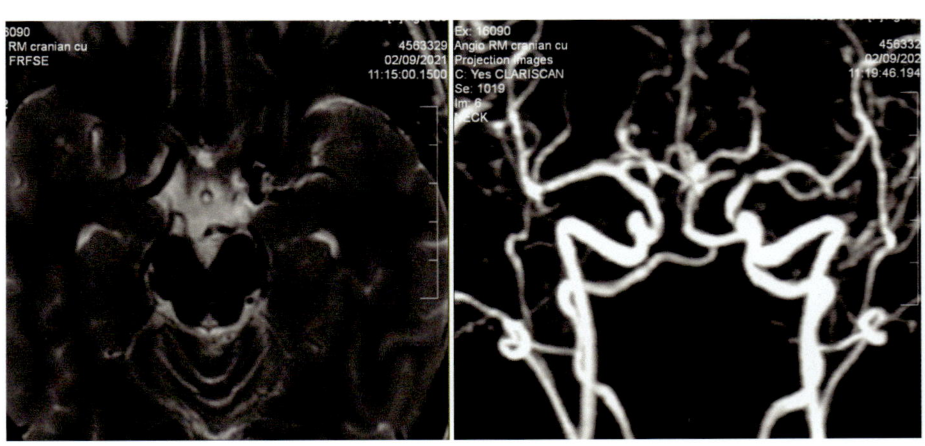

Fig. 18.10 Eight-year follow-up MRI in a girl operated at the age of 15 years for a ruptured left internal carotid artery (ICA) bifurcation aneurysm. Left: T2-weighted MRI. Right: MRI angiography showing no filling of the left ICA

18.6 Management of Pediatric Intracranial Aneurysms

This chapter, as well as the entire work, is addressed not only to the centers, regions, or countries equipped with all modern-day adjuvants but to all neurosurgeons across the globe. This is especially because PIAs may occur at any given place and appropriate treatment measures must be taken to ensure a favorable prognostic. Obviously, due to the rarity of PIAs, such cases should be directed to centers with the highest experience in pediatric and vascular neurosurgery. Nevertheless, in a severe emergency, the measures ensuring the survival of the child must be given within the first available center.

18.6.1 Surgical Treatment

To this day, microsurgical treatment remains the mainstay for the majority of PIAs [6, 14]. Extensive literature reviews demonstrate that microsurgical interventions were performed in 53–55% of cases, with a steady decline due to progress in endovascular treatment. Clipping shows excellent results across several series, with the majority of patients having a good recovery at follow-up [7, 15]. Microsurgical techniques are aplenty and should be tailored for each individual patient. In the event of saccular aneurysms of small and medium size, surgical clipping differs only in the size of the applied clip, with microclips or fenestrated clips being often sufficient for complete occlusion (Fig. 18.11).

Giant and fusiform PIAs are not always amenable to direct clipping. Therefore, familiarizing oneself with clip reconstruction techniques and the large pallet of available clips is a must for any vascular neurosurgeon. Given the complex nature of these aneurysms, special reinforcing techniques or clip reconstruction of vessels such as tandem overstacked clipping [16] or piggyback stacking and booster clipping [17], as well as clip reconstruction and sling-wrapping, have been successfully utilized (Fig. 18.12).

Certain details related to surgical techniques that are covered in separate chapters will not be discussed here. However, we deem it necessary to highlight some of the particularities pertaining to neurosurgical procedures at the pediatric age.

In children, especially toddlers and infants, a series of particular characteristics related to ***neuroanesthesia*** exists: the short and narrow airways, the low respiratory volumes, different ratios compared to adults pertaining to wight and body surface, hydration/dehydration characteristics, as well as anesthetic agent administration and metabolization (Fig. 18.13). Furthermore, the increased propensity for hypothermia requires a higher room temperature in the operating theater, the use of convective warming blankets, and the heating of perfusion substances. Considering that their total blood volume is low (70 ml/kg), special care must be given to intraoperative loss and especially to compensation. A compatible supply of blood equal in volume to the child's total volume should be made ready prior to the intervention. Plasma and platelets should also be supplemented appropriately [19].

Positioning the child on the operating table should be done with tremendous care to avoid producing limb luxation and decubitus ulcers. For anterior circulation aneurysms, the supine position is the obvious choice, with the head rotated and tilted accordingly to provide optimal visibility and brain relaxation. Considering the sitting position for posterior circulation PIAs, although there are no contraindications based on patient age [20], in our almost five-decade-long experience we cannot recommend using it in children under one year of age mainly because of issues related to fastening the head.

Incision and craniotomy. The thin and extremely well-vascularized scalp should be adequately infiltrated with a mixture of lidocaine and traces of adrenaline to avoid significant blood loss at the very start of the intervention. The incision should be as linear as possible and avoid sharp angles to preclude the risk of inappropriate healing and scalp necroses. Similarly, the thin bone implies shorter and narrower craniotome blades. Osteoplastic flaps are recommended, keeping a fragment of periosteum attached to the craniotomy for faster healing. Using titanium plates and screws in children under

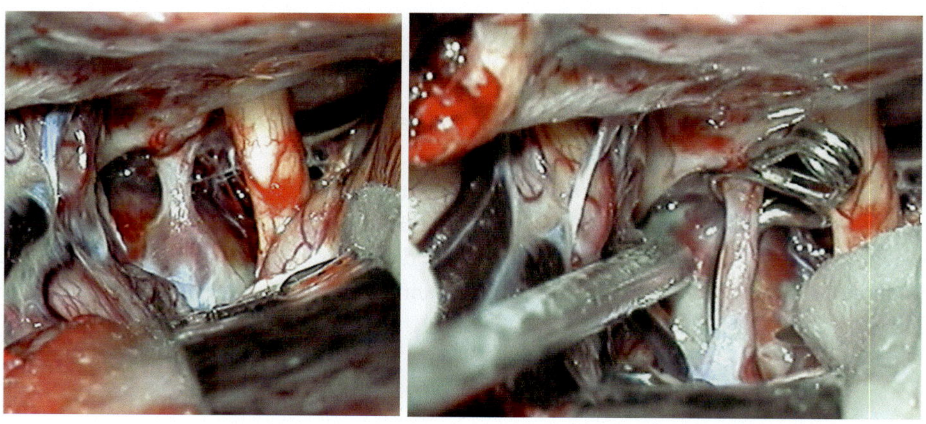

Fig. 18.11 Intraoperative aspect of a large globular left posterior communicating artery (PCoA) aneurysm occluded with a single fenestrated angled clip that was also used to reconstruct the left ICA

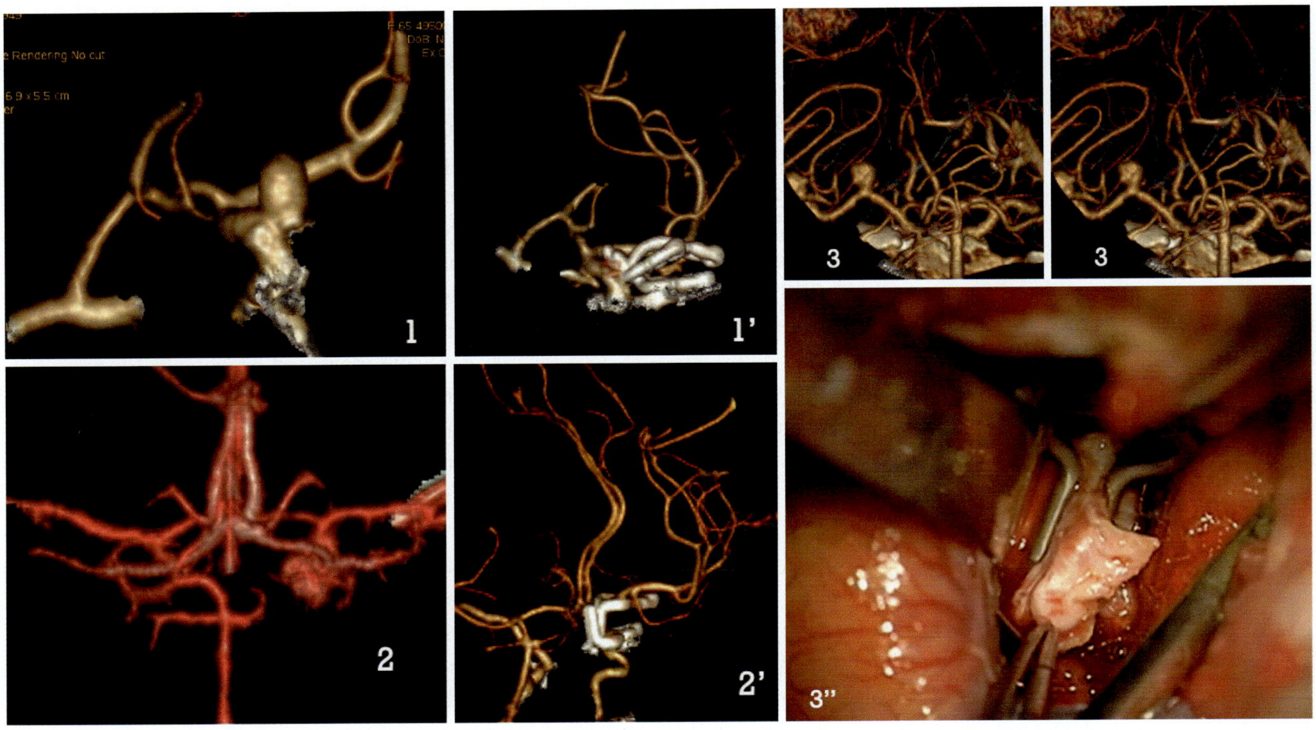

Fig. 18.12 Examples of the clip reconstruction technique. Images 1 and 1′ show a large ophthalmic segment aneurysm occluded through an interlocking tandem clipping technique. Images 2 and 2′ demonstrate a large posterior communicating artery (PCoA) aneurysm occluded through a tandem counter-clipping technique. Images 3, 3′, and 3″ reveal the sling wrapping technique used for the occlusion and prevention of further dilatation of an M1 segment fusiform aneurysm. 3″ Demonstrates the intraoperative aspect of the technique using dura mater instead of Gore-tex as originally described by Kim et al. [18]

5 years can also raise some issues, as a rigid fixation may hinder normal skull growth, leading to inaesthetic deformities. Ideally, resorbable implants could be utilized, however their availability and cost might be restrictive in certain regions. Consequently, fastening the flap with sutures remains a valid and effective option.

For anterior circulation aneurysms, we are in favor of the pterional flap, tailored to the exact location of the lesion. A fronto-lateral craniotomy is sufficient for older children with uncomplicated ACoA and ICA aneurysms. Whenever anterior clinoidectomy is anticipated for proximal control, a classic pterional craniotomy is recommended, as this allows the best access in both extradural and intradural compartments. Additionally, if an intracranial–extracranial bypass is to be expected, the skin incision must avoid and preserve the superficial temporal artery as much as possible.

Subarachnoid space dissection comes with a few challenges of its own, especially in smaller children. During this process, the brain might be damaged due to the smaller spaces, coalescence of lobes, overall parenchyma and vessel fragility (mainly those in the Sylvian fissure), and the incomplete definition of the cisterns and Sylvian fissure. As such, gestures should be minute and delicate during dissection. Using the operative microscope is a necessity.

Proximal control is mandatory, while temporary clipping is generally well tolerated in children. This allows faster aneurysm sac dissection and the prevention of a significant bleed. Microclips should be placed slowly and carefully so as not to detach the aneurysm at the site of the neck and the subsequent disastrous consequences. Moreover, when placing the clip, care must be given not to create parent or adjacent vessel strangulation or kinking. The clip should be placed parallel with the parent artery and as low as possible, otherwise there remains a risk for aneurysm recurrence. In our own experience, such an aneurysm recurred and rebled 18 years after the initial clipping.

Complications include intraoperative rupture, vasospasm, and cerebral ischemia due to arterial pinching. Vasospasms and angiographic disappearance of distal arteries such as the ACAs have been documented after the clipping of ruptured PIAs, with no discernible neurological deficit [5].

In cases with adequate collateral circulation demonstrated by balloon test occlusion, trapping or parent vessel occlusion can be performed. It is an older technique, but repeatedly proven to be effective [21]. It is not surprising that pediatric patients have a higher tolerance for parent vessel occlusion than adults [22].

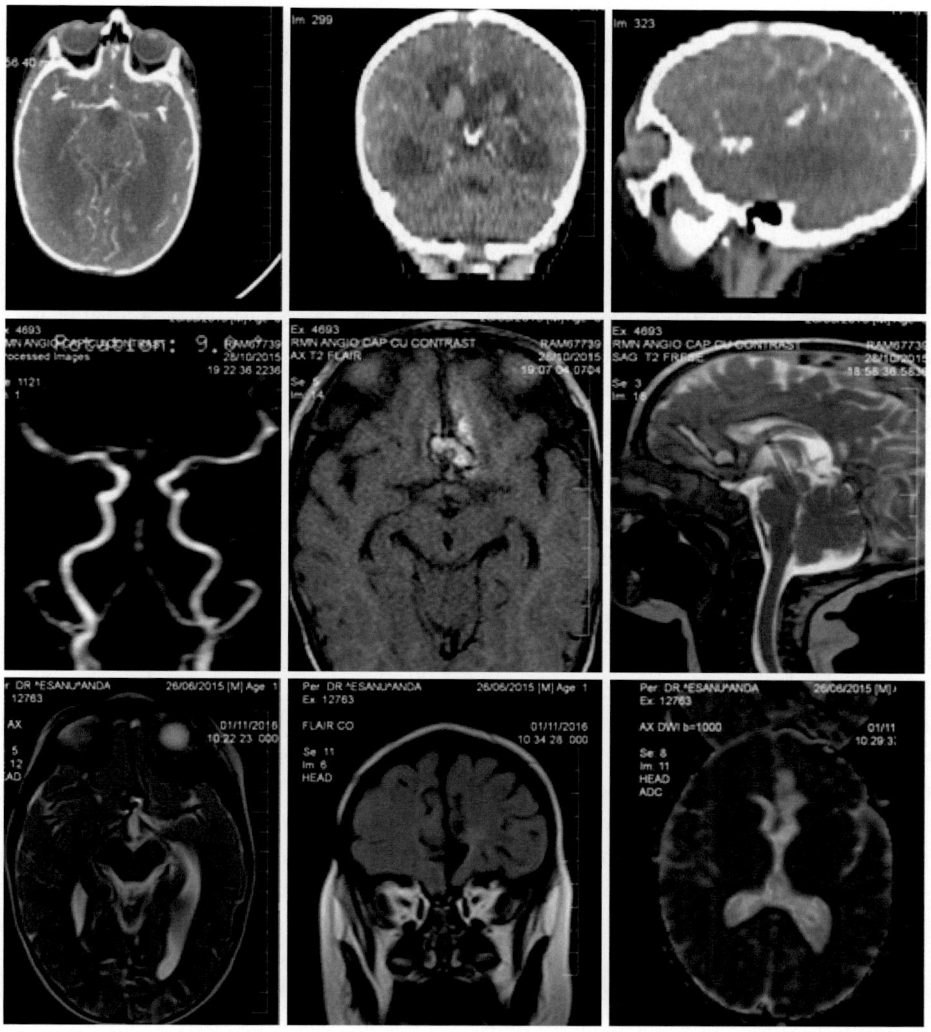

Fig. 18.13 A 4-month-old boy admitted for subarachnoid hemorrhage (SAH), right intracerebral hematoma, and intraventricular hemorrhage from a ruptured anterior communicating artery (ACoA) aneurysm. Symptoms began 4 days prior to admission, with intractable vomiting. Top row: CT angiography (CTA) demonstrates important intraventricular hemorrhage with a marked ventricular dilation and a large hematoma adjacent to the ACoA region. Also visible is the narrowing of the bilateral A2 segments (vasospasm). After placing an external ventricular drainage (EVD), the child's neurological state improved. Middle row: subsequent MRI angiography showed the failure of filling in bilateral A2 segments. Surgery was performed, clipping the aneurysm with a Yasargil microclip. Bottom row: postoperatively, the patient's evolution was very favorable, despite the lack of filling in the A2 segments, demonstrated by the control DWI and FLAIR sequences. This proves the extraordinary capacity of cerebral circulation in small children and justifies medical efforts for aneurysm occlusion at even the tenderest age [5]

Where low occlusion tolerance is expected or observed, another microsurgical option is extracranial-to-intracranial bypass for giant or complex PIAs, or those in which endovascular procedures did not yield a satisfactory occlusion [23–25]. This procedure is more rarely performed in the pediatric population than clipping and involves creating a bypass from an extracranial artery, such as the superficial temporal artery (STA) or the occipital artery, and a cerebral artery, for example, the MCA or the posterior cerebral artery. The saphenous vein or radial artery grafts are also a feasible option with long-lasting benefits and even an intact neurological status [24, 26, 27]. However, endovascular coiling might be needed in case of residual aneurysms after this type of procedure [27]. Postoperative complications include ischemic stroke and graft occlusion [25].

18.6.2 Endovascular Treatment

Endovascular interventions have already been established as a crucial component of PIA therapeutic arsenal [2]. One such technique is coiling, which is applicable for saccular PIAs. In cases of wide-necked (neck >4 mm or ASPECT ratio < 2) aneurysms, balloon-assistant coiling or stent-assistant approach is used in order to prevent migration of the coils [28, 29]. Based on the findings by Sanai et al., the major issue concerning endovascular coiling procedures are the lower rates of obliteration (82%) in comparison to surgery (94%) and the higher rate of recurrence, especially given the long life expectancy in these patients [30]. Another limitation is the low diameter of vessels in children under 4 years. From this age, ACA and ICA reach almost the diameter of adult vessels, whereas the MCA nears its adult diameter at 6 months of age [31]. Nevertheless, endovascular successes in extremely young patients have been published [32]. It should however be noted that the effects of radiation on tissue are cumulative, with younger patients being more vulnerable to radiation. As a consequence, the concept of the ALARA (As Low As Reasonably Achievable) concept is routinely used. We remind the high cost of angiographic equipment and devices as restrictive in some countries, making them not universally available at the time.

Despite the limited experience and data in the literature regarding PIAs, flow diverters seem to be an effective and safe treatment method in selected cases [1, 3, 9]. Endoluminal flow diverters such as the pipeline embolization device (PED) create a stagnation effect within the aneurysmal lumen, leading to the formation of a thrombus and the subsequent occlusion and shrinkage of the lesion [33, 34]. Complications of this procedure include a failure to deploy, as well as in-stent stenosis and distal occlusion of the parent arteries, although occlusion may sometimes be asymptomatic [1, 3].

Basilar aneurysms represent a particularly difficult challenge, especially in children. This is especially due to their increased size and fusiform morphology, and because microsurgical techniques and standard endovascular coiling are often inappropriate or ineffective. Mortality for this location of IA can be as high as 10% across treatment studies, with a poor outcome affecting 15% of patients [35]. The combined surgical and endovascular approach is rarely used in pediatric patients, but has shown promise for giant aneurysms that are not amenable to one or the other [36–38].

18.6.3 Conservative Treatment

Conservative management is based on the clinical observations in studies especially on dissecting aneurysms, which have been shown to spontaneously heal through intramural thrombosis and parent vessel occlusion. As for giant PIAs, over 50% present partial thrombosis that in time can completely obliterate the aneurysm. Thrombosis of a dissecting aneurysm takes between 3 weeks and 7 months, though the process itself is not completely understood [39]. As this takes a long time, it allows the innate plasticity of the cerebral vasculature to develop effective collateral circulation without significant or noticeable neurological consequences. However, spontaneous repermeabilization is also possible and should be treated aggressively due to the risk of hemorrhage. On the other hand, the progress of thrombosis and parent vessel stenosis can lead to an increased risk of ischemic stroke, mandating antiplatelet treatment. All things considered, conservative treatment should not be viewed as a primary option because of the poor long-term prognosis, with a 33–57% mortality and only a 7–44% long-term survival with a favorable neurological status [38].

18.7 Surgical Pearls

- Avoid pin fixation in children below 2 years of age.
- Sitting position is not recommendable in infants.
- Before skin incision infiltrate the scalp with a mixture of lidocaine and traces of adrenaline.
- Thin bone implies shorter and narrower craniotome blades.
- Osteoplastic flaps are recommended for supratentorial locations, whereas for posterior fossa craniotomy (not craniectomy) is the rule.
- Considering the incomplete definition of the cisterns and Sylvian fissure, use not only microscope and microsurgical instruments but microsurgical gestures as well.
- Use a combination of sharp and blunt dissection, moistened cotton patties for delicate retraction, and abundant lavage with saline during dissection.
- A large palette of clips and clip applicators should be at the surgeon's disposal.
- In complex, large, and giant aneurysms, different clip reconstruction techniques may be applied, as appropriate.
- After the clip occlusion always check the opposite side ("dark side of the moon").
- In complex aneurysms, always be prepared for plan B: Extra-Intracranial-Bypass.
- Fastening the flap with sutures remains a valid and effective option.

18.8 Discussions

Regarding the etiology of PIAs, the amplest study so far regarding genome-wide association studies (GWAS) [40] has identified 11 new risk loci and confirmed 6 previously described risk loci, thus elucidating over half of the heritability of intracranial aneurysms. The authors also identified a potential role for endothelial cells in the development of intracranial aneurysms and demonstrated that the genetic architecture of unruptured and ruptured aneurysms is analogous.

Analyzing the embryologic process is known as angiogenesis, it has been noted that a series of imperfections or alterations that do not follow the normal pattern of development may give rise to local conditions that favor the appearance of aneurysms. As such, the incomplete fusion of arterial precursors leads to a fenestration, the most common of which are found at the anterior cerebral artery (ACA)-ACoA system and the basilar trunk. The latter of the two is considered responsible for the occurrence of basilar artery (BA) aneurysms due to the circulation turbulences and tunica media defects that may appear [41, 42]. Persistent Vestigial Arteries (PVA), more often at the level of the Trigeminal Artery (TA) followed by the Hypophyseal Artery (HA), can result in alterations in flow dynamics, conferring a susceptibility to aneurysm formation. Similarly, an azygous A2 segment is considered a favoring factor for aneurysms distal to the fusion zone [8]. The incomplete fusion of the basilar tip, especially in its asymmetrical variant, is linked to the formation of aneurysms at the level of the inferior limb of the artery [43].

Cerebral endothelial cells (ECs) have a particular clinical relevance, given their involvement in certain diseases of the brain including stroke, arteriovenous malformations, and vascular dementia. Relevant to cerebrovascular diseases such as aneurysms, stroke, and neurodegeneration, blood flow influences the expression and epigenetics of ECs genes. Understanding the intricate mechanisms and actions of growth factors in cerebrovascular angiogenesis in both in vivo and in vitro models may yield more specific or tailored treatment methods for these pathologies, irrespective of age [44].

In the case of PIAs, one must first consider a pathology of the arterial wall [39], even though the cause may be identified in less than half of cases [45]. Among systemic pathologies more often associated with PIAs, we mention sickle cell disease, Ehlers–Danlos syndrome type IV, polycystic kidney disease, neurofibromatosis, type I, Marfan's syndrome, and hereditary hemorrhagic telangiectasia [46]. The Familial Intracranial Aneurysm (FIA) study revealed that familial IAs are indeed a distinct pathology, with a predilection for aneurysms at the MCA territory and at an earlier age [47].

The location of PIAs also displays a series of particularities when compared to adults. As such, saccular aneurysms are more frequently found in the anterior circulation, whereas fusiform and saccular aneurysms are met in the posterior circulation. Anterior circulation PIAs are reported to be 70–75% of the total, the rest being in the posterior circulation. This is in stark contrast with adult IAs, where posterior lesions amount to much less (5–15%). PIAs of the posterior circulation are proportionately twice as frequent [11], the most typical location being at the basilar trunk, followed by the vertebral arteries (VA). Concerning aneurysm morphology, giant, complex, fusiform, and dissecting aneurysms are the most common types of PIAs in the posterior circulation [4], being more often in children than in adults (11–41% vs. 5–9%), the proportion of giant aneurysms being all the higher with the younger age of the patient [14]. Moreover, the frequency of giant aneurysms in the anterior circulation is higher in children than in adults, many of these lesions being of the fusiform or dissecting type and raising serious management issues [2]. Generally, PIAs are saccular in 61% of cases, fusiform in 26%, 7% are traumatic, and 6% are infectious [4]. Among traumatic PIAs, those appearing after neurosurgical interventions (tumor resection, transsphenoidal procedures, shunt placement, or third ventriculostomy) may be included [48]. Radiotherapy for intracranial tumors has also been linked to the formation of aneurysms in children [49, 50].

From a clinical perspective, location within the distal arterial region or an age between 0 and 5 years was independently correlated with an increased risk of rupture in the study by Chen et al. [51]. Pediatric patients with IAs present a heightened risk for preoperative seizures [52]. Interestingly, thunderclap headache, which is associated with SAH in adults, is usually not associated with aneurysmal rupture and typically has a benign course [53]. Mood disorder has also been described as the presenting symptom in a child with an ICA aneurysm [54].

Regarding the treatment of PIAs, there is no unanimity. Management strategies are based on the application of algorithms and treatment criteria designed for the adult population, with specific adjustments based on the experience of the center and, ultimately, the surgeon. Both surgical and endovascular procedures have shown similar rates of good outcomes, regardless of rupture status in PIAs [55]. Conservative management may be recommended for a poor grade of the patient or in the event of spontaneous thrombosis [14].

The decision of when and how to treat is in itself a challenge. Pediatric patients present a unique set of difficulties such as a lower amount of blood, smaller blood vessels, and complexity of the lesions themselves [56]. For cases admitted with aneurysmal SAH, being a life-threatening condition, the surgical indication is absolute.

For giant, fusiform, or dissecting aneurysms, as well as the extremely rare case of multiple aneurysms, the therapeutic indication becomes more complex. Ideally, the decision to treat should be tailored to each patient and taken by a multidisciplinary team of neurosurgeons and neurointerventionists. Concurrently, a hybrid neurosurgeon can adapt the therapeutic indication depending on ensuing circumstances. However, these ideal scenarios are scarce on a global scale, thus decision-making depends on existent conditions. For complex aneurysms, a bypass may be required, demanding suitable expertise. Observation is an option in itself, yet it should be balanced against the child's long-life expectancy, implying the risk of rupture and ensuing complications [14]. For this reason, we are in favor of effective treatment with a long lifespan, preferably sooner rather than later.

Endovascular procedures represent an increasingly important alternative to open surgical obliteration, especially for posterior fossa aneurysms, giant aneurysms, and fusiform/dissecting aneurysms, which often require creative reconstructive or well-planned deconstructive endovascular approaches. Nevertheless, the low permanent complication rate and high percentage of patients with a favorable outcome make it an appealing treatment alternative [2]. In the series of pediatric aneurysms treated by Santos-Franco et al., the majority were located in the cavernous portion of the ICA, followed by the MCA [9].

The outcome of PIAs has shown improvement over the last years because of the progress made in microsurgical and endovascular techniques, as well as postinterventional care. Generally, the outcome of PIAs is better than those of adults, favored by a higher tolerance of SAH consequences, neuroplasticity, the rarity of vasospasm and delayed cerebral ischemia (DCI), and an increased capacity for recovery. When comparing the two main methods of treatment, they are basically equivalent in terms of the rate of a favorable outcome, complications, and mortality. Neurological status upon admission remains the decisive factor for the prognosis, with an 80% mortality for H&H IV and V. Aneurysm location also carries a significant influence, with PIAs of the anterior circulation having a better prognosis both in terms of survival and neurological complications [4].

18.9 Conclusions

PIAs represent a rare entity, being responsible for 10–15% of hemorrhagic stroke incidents in children. Unlike adults, there is a shift in the male-to-female ratio, favoring males. Anterior circulation PIAs are the most common, although in a smaller total proportion when compared to adults. Complex, giant, dissecting, and fusiform PIAs are more frequent at this age and have a predilection for the posterior fossa. Clinical diagnosis is often inconclusive and impels swift imaging studies. Management should be adjusted to the pathology, preferably striving for a lifelong solution, but also adapted to the center where the child is treated. All treatment methods presented are viable and effective. There are however limited possibilities regarding the effective comparison between management strategies, as randomized studies are rather unrealistic due to the rarity of these lesions and emotional nature of treatment selection. Children should not be viewed as downscaled adults, as their anatomy and physiology differ greatly. Pediatric neurosurgeons should be trained in vascular surgery, but so too should vascular neurosurgeons be capable of treating or assisting in the treatment of pediatric patients.

References

1. Vargas SA, Diaz C, Herrera DA, Dublin AB. Intracranial aneurysms in children: the role of stenting and flow-diversion. J Neuroimaging. 2016;26(1):41–5.
2. Beez T, Steiger HJ, Hänggi D. Evolution of management of intracranial aneurysms in children: a systematic review of the modern literature. J Child Neurol. 2016;31(6):773–83.
3. Shlobin NA, Raz E, Shapiro M, et al. Pipeline embolization of cerebral aneurysms in pediatric patients: combined systematic review of patient-level data and multicenter retrospective review. J Neurosurg Pediatr. 2021;27(6):668–76. Published 2021 Apr 23
4. Zeng G, Jiang L. Pediatric intracranial aneurysms. In: Di Rocco C, et al., editors. Textbook of pediatric neurosurgery. Springer Nature Switzerland AG; 2018. p. 2682–753.
5. Aldea CC, Florian IA, Timiş TL, Andraşoni Z, Florian IS. Ruptured AComA aneurysm and asymptomatic bilateral ACA vasospasm in an infant: a surgical case report. Childs Nerv Syst. 2022;38(3):633–41.
6. Pruvot AS, Curey S, Derrey S, Castel H, Proust F. Giant intracranial aneurysms in the paediatric population: suggested management and a review of the literature. Neurochirurgie. 2016;62(1):20–4.
7. Ciurea AV, Mohan A, Voicu A, Saceleanu V, Moreanu MS, Ples H. Characteristics and surgical outcomes of pediatric intracranial aneurysms in Romania. Turk Neurosurg. 2021;31(5):740–4.
8. Klostranec JM, Krings T. Cerebral neurovascular embryology, anatomic variations, and congenital brain arteriovenous lesions. J NeuroIntervent Surg. 2022;14:910–9.
9. Santos-Franco JA, Cruz-Argüelles CA, Agustin-Aguilar F, Abrego-Salinas AA, Casas-Martínez MR, Olivares-Peña JL. Intracranial aneurysms in pediatric population treated with flow diverters: a single-center experience. Surg Neurol Int. 2022;13:522. Published 2022 Nov 11
10. Ozgur BM, Khoo LT, Levy ML. Intracerebral aneurysms. In: Albright AL, Pollack IF, Adelson PD, editors. Principles and practice of pediatric neurosurgery. New York/Stuttgart: Thieme; 2008. p. 960–82.
11. Aeron G, Abruzzo TA, Jones BV. Clinical and imaging features of intracranial arterial aneurysms in the pediatric population. Radiographics. 2012;32(3):667–81.
12. Sorteberg A, Dahlberg D. Intracranial non-traumatic aneurysms in children and adolescents. Curr Pediatr Rev. 2013;9(4):343–52.
13. de Aguiar GB, Ozanne A, Elawady A, et al. Intracranial aneurysm in pediatric population: a single-center experience. Pediatr Neurosurg. 2022;57(4):270–8. https://doi.org/10.1159/000524523.
14. Garg K, Singh PK, Sharma BS, et al. Pediatric intracranial aneurysms--our experience and review of literature. Childs Nerv Syst. 2014;30(5):873–83.
15. Garrido E, Metayer T, Borha A, et al. Intracranial aneurysms in pediatric population: a two-center audit. Childs Nerv Syst. 2021;37(8):2567–75.
16. Campero A, Villalonga JF, Lucifero AG, Luzzi S, Baldoncini M. Tandem overstacked clipping for giant carotid ophthalmic aneurysm: 3-dimensional operative video. Oper Neurosurg (Hagerstown). 2022;23(1):e64.
17. Pipolo DO, Demichelis ME, Purves C, et al. Piggyback stacking and booster clipping technique: a technical note. World Neurosurg. 2022;164:306–9.
18. Kim LJ, Klopfenstein JD, Spetzler RF. Clip reconstruction and sling wrapping of a fusiform aneurysm: technical note. Neurosurgery. 2007;61(Suppl. 3):79–80.
19. Orliaguet GA, Hanafi M, Meyer PG, Blanot S, Jarreau MM, Bresson D, Zerah M, Carli PA. Is the sitting or the prone position best for surgery for posterior fossa tumours in children? Paediatr Anaesth. 2001;11(5):541–7.

20. Spennato P, Nicosia G, Quaglietta L, Donofrio V, Mirone G, Di Martino G, Guadagno E, del Basso de Caro ML, Cascone D, Cinalli G. Posterior fossa tumors in infants and neonates. Childs Nerv Syst. 2015;31(10):1751–72.
21. Chandela S, Langer DJ. Pediatric aneurysms. In: Jallo GI, Kothbauer KF, Pradilla G, editors. Controversies in pediatric neurosurgery. New York/Stuttgart: Thieme; 2010. p. 176–90.
22. Fulkerson DH, Voorhies JM, Payner TD, Leipzig TJ, Horner TG, Redelman K, Cohen-Gadol AA. Middle cerebral artery aneurysms in children: case series and review. J Neurosurg Pediatr. 2011;8(1):79–89.
23. Strickland BA, Attenello F, Russin JJ. Extracranial to intracranial bypass for the treatment of cerebral aneurysms in the pediatric population. J Clin Neurosci. 2016;34:6–10.
24. Abla AA, Zaidi HA, Crowley RW, et al. Optic chiasm compression from mass effect and thrombus formation following unsuccessful treatment of a giant supraclinoid ICA aneurysm with the Pipeline device: open surgical bailout with STA-MCA bypass and parent vessel occlusion. J Neurosurg Pediatr. 2014;14:31–7.
25. Kalani MY, Elhadi AM, Ramey W, et al. Revascularization and pediatric aneurysm surgery. J Neurosurg Pediatr. 2014;13:641–6.
26. Zhang YJ, Barrow DL, Day AL. Extracranial-intracranial vein graft bypass for giant intracranial aneurysm surgery for pediatric patients: two technical case reports. Neurosurgery. 2002;50:663–8.
27. Kim LJ, Tariq F, Sekhar LN. Pediatric bypasses for aneurysms and skull base tumors: short- and long-term outcomes. J Neurosurg Pediatr. 2013;11:533–42.
28. Boisseau W, Darsaut TE, Fahed R, et al. Stent-assisted coiling in the treatment of unruptured intracranial aneurysms: a randomized clinical trial. AJNR Am J Neuroradiol. 2023;44(4):381–9.
29. Alkhars H, Haq W, Al-Tayeb A, Sigounas D. Feasibility and safety of transradial aneurysm embolization: a systematic review and meta-analysis. World Neurosurg. 2022;165:e110–27.
30. Sanai N, Quinones-Hinojosa A, Gupta NM, et al. Pediatric intracranial aneurysms: durability of treatment following microsurgical and endovascular management. J Neurosurg. 2006;104(Suppl. 2):82–9.
31. Arat YO, Arat A, Aydin K. Angiographic morphometry of internal carotid artery circulation in Turkish children. Turk Neurosurg. 2015;25(4):608–16.
32. Colby GP, Jiang B, Bender MT, Beaty NB, Westbroek EM, Xu R, Lin LM, Campos JK, Tamargo RJ, Huang J, Cohen AR, Coon AL. Pipeline-assisted coil embolization of a large middle cerebral artery pseudoaneurysm in a 9-month-old infant: experience from the youngest flow diversion case. J Neurosurg Pediatr. 2018;22(5):532–40.
33. Taussky P, Tawk RG, Miller DA, Freeman WD, Hanel RA. New therapies for unruptured intracranial aneurysms. Neurol Clin. 2013;31(3):737–47.
34. Wong GK, Kwan MC, Ng RY, Yu SC, Poon WS. Flow diverters for treatment of intracranial aneurysms: current status and ongoing clinical trials. J Clin Neurosci. 2011;18(6):737–40.
35. Peters DR, Parish JM, Starnoni D, et al. Endovascular treatment of pediatric basilar artery aneurysms: case series and literature review. Childs Nerv Syst. 2023;39(1):25–34.
36. O'Neal CM, Ernst GL, Hughes KL, et al. Reported incidence and treatment modalities of giant cerebral aneurysms in the pediatric population: a systematic review and illustrative case report. J Clin Neurosci. 2021;90:14–20.
37. Heros RC. Pediatric intracranial aneurysms. J Neurosurg. 2006;104(Suppl. 2):77–81.
38. Koroknay-Pal P, Lehto H, Niemela M, Kivisaari R, Hernesniemi J. Long-term outcome of 114 children with cerebral aneurysms. J Neurosurg Pediatr. 2012;9(6):636–45.
39. Lasjaunias P, Wuppalapati S, Alvarez H, Rodesch G, Ozanne A. Intracranial aneurysms in children aged under15 years: review of 59 consecutive children with 75 aneurysms. Childs Nerv Syst. 2005;21(6):437–50.
40. Bakker MK, van der Spek RAA, van Rheenen W, Morel S, et al. Genome-wide association study of intracranial aneurysms identifies 17 risk loci and genetic overlap with clinical risk factors. Nat Genet. 2020;52(12):1303–13. https://doi.org/10.1038/s41588-020-00725-7. Epub 2020 Nov 16. Erratum in: Nat Genet. 2020 Dec 22
41. Tasker AD, Byrne JV. Basilar artery fenestration in association with aneurysms of the posterior cerebral circulation. Neuroradiology. 1997;39:185–9.
42. Krings T, Baccin CE, Alvarez H, et al. Segmental unfused basilar artery with kissing aneurysms: report of three cases and literature review. Acta Neurochir. 2007;149:567–74.
43. Kings T, Geibprasert S, Cruz JP. Neurovascular anatomy in interventional neuroradiology. New York, NY: Thieme Medical Publishers, Inc.; 2015. p. 231.
44. Chico TJA, Kugler EC. Cerebrovascular development: mechanisms and experimental approaches. Cell Mol Life Sci. 2021;78(9):4377–98. https://doi.org/10.1007/s00018-021-03790-1. Epub 2021 Mar 10
45. Krings T, Geibprasert S, terBrugge KG. Pathomechanisms and treatment of pediatric aneurysms. Childs Nerv Syst. 2010;26(10):1309–18.
46. Giotta Lucifero A, Baldoncini M, Bruno N, Galzio R, Hernesniemi J, Luzzi S. Shedding the light on the natural history of intracranial aneurysms: an updated overview. Medicina (Kaunas). 2021;57(8):742.
47. Broderick JP, Sauerbeck LR, Foroud T, Huston J 3rd., Pankratz N, Meissner I, Brown RD Jr. The familial intracranial aneurysm (FIA) study protocol. BMC Med Genet. 2005;6:17.
48. Egemen E, Massimi L, Di Rocco C. Iatrogenic intracranial aneurysms in childhood: casebased update. Childs Nerv Syst. 2012;28(12):1997–2004.
49. Wagner MW, Dewan MC, Dmytriw AA, Ramaswamy V, Taylor MD, Muthusami P. Radiation-induced intracranial aneurysm presenting with acute hemorrhage in a child treated for medulloblastoma. Childs Nerv Syst. 2021;37(4):1387–9.
50. Jensen FK, Wagner A. Intracranial aneurysm following radiation therapy for medulloblastoma. A case report and review of the literature. Acta Radiol. 1997;38(1):37–42. https://doi.org/10.1080/02841859709171239.
51. Chen R, Zhang S, Xiao A, Guo R, Ma J. Risk factors for intracranial aneurysm rupture in pediatric patients. Acta Neurochir. 2022;164(4):1145–52.
52. Chen R, Zan X, Xiao A, et al. Risk factors for preoperative seizures in pediatric patients with intracranial aneurysms. Clin Neurol Neurosurg. 2020;189:105616.
53. Levinsky Y, Eidlitz-Markus T. Thunderclap headache in children and adolescents. Curr Pain Headache Rep. 2022;26(3):235–9.
54. Madhusankha KHD, Rathnayaka D, Samaranayake M, Dharmasiri M, Wickramasingha R. Fourteen-year-old boy with intracranial internal carotid artery aneurysm presenting as mood disorder. Cureus. 2021;13(9):e18324. Published 2021 Sep 27
55. Yasin JT, Wallace AN, Madaelil TP, et al. Treatment of pediatric intracranial aneurysms: case series and meta-analysis. J Neurointerv Surg. 2019;11(3):257–64.
56. Ventureyra EC. Pediatric intracranial aneurysms: a different perspective. J Neurosurg. 2006;104(Suppl. 2):79–81.

Endoscopic Techniques Applied to Neurovascular Pathology

19

Juan L. Gómez-Amador, Rodolfo Villalobos-Díaz, and Marcos V. Sangrador-Deitos

Learning Objectives

- To describe the different endoscopic techniques applicable to cerebrovascular surgery, as well as their advantages and limitations.
- To describe the usefulness of endoscope-assisted microsurgery (EAM) for aneurysm clipping in different anatomical locations.
- To discuss the applicability of endoscopic-controlled aneurysm clipping through minimally invasive craniotomies, as a less invasive alternative to traditional microsurgical treatment.
- To describe the usefulness of extended endoscopic endonasal approaches (EEA) in aneurysm clipping and brainstem cavernomas resection, delineate which lesions are best addressed through these approaches, and discuss their limitations.

19.1 Introduction

The recent application of endoscopic techniques in vascular neurosurgery has been long preceded by the development and maturation of endoscopic techniques attained in the last decades, especially for the treatment of skull base and intraventricular tumors. As a result of this maturation, along with the recent technological advances in the field of neuroendoscopy, vascular neurosurgeons have been encouraged to implement these techniques for treating intracranial aneurysms and other vascular lesions, providing them with a new surgical armamentarium [1–3].

The endoscope in vascular neurosurgery may be used as a primary visualization tool to provide unique and direct corridors for treating vascular lesions (e.g., endoscopic keyhole and EEA) or as an adjunct to standard microscopic approaches (EAM) [2, 4]. These alternatives to classical vascular approaches have been found to reduce morbimortality and to improve functional and neurological outcomes [1, 4, 5].

In virtue of the available evidence and surgical experience in this field, this chapter is divided into three sections:

- Endoscope-Assisted Microneurosurgery (EAM) for intracranial aneurysm clipping.
- Endoscope-Controlled Microneurosurgery (ECM) for intracranial aneurysm clipping (Keyhole/ Endoscopic port).
- Endoscopic Endonasal Approaches for:
  - Aneurysm clipping.
  - Cavernoma resection.

19.2 History/Background

Although the first endoscopic intracranial procedure was performed at the beginning of the last century, it was not until the 1970s that Apuzzo et al. [6] described the utility of the endoscope for assessing the adequacy and accuracy of clip placement in aneurysm surgery. In 1994 Fischer and Mustafa [7] described its utility in improving the visualization of the anatomy of the aneurysms and surrounding structures. In 1994 Perneczky and Fries [8] elaborated on the general principles of EAM and described its utility as an adjunct to microsurgical clipping of aneurysms. The utilization of an EEA for aneurysmal surgery was first described in 2006 by Kassam et al. [9] which utilized an endoscopic transclival approach to treat a coiled vertebral artery aneurysm. Since these pioneering works, the use of the endoscope for clipping procedures has evolved from EAM to ECM ones, and more recently, into purely EEA. However, the utilization of the endoscope in neurovascular procedures has not been

J. L. Gómez-Amador (✉) · R. Villalobos-Díaz · M. V. Sangrador-Deitos
Department of Neurosurgery, National Institute of Neurology and Neurosurgery "Manuel Velasco Suárez", Mexico City, Mexico
e-mail: jlga@neurocirugia-innn.com

I. N. Kanaan, V. Beneš (eds.), *Neuroanatomy Guidance to Successful Neurosurgical Interventions*, https://doi.org/10.1007/978-3-031-59838-8_19

limited to aneurysmal surgery, and in recent years, techniques for resection of cavernomas [10], dural arteriovenous fistulas [11], and arteriovenous malformations [12] have been described.

19.3 Endoscope-Assisted Microneurosurgery for Intracranial Aneurysms

Microneurosurgical clipping of intracranial aneurysms has the purpose of excluding the aneurysm from the circulation while, simultaneously, maintaining the hemodynamic integrity of the parent artery, its branches, and perforators. To attain this, it is essential to have an optimal visualization of the regional anatomy, which is frequently impeded by the inherent limitations of the microscope, such as a restricted straight line of view and reduction of light intensity with increased zooming. In such cases, the better illumination, higher image quality, and extended viewing angles provided by the endoscope may be helpful to overcome these limitations [3, 13].

The endoscope has proven its usefulness in assisting microsurgical clipping of aneurysms in multiple ways, such as allowing to inspect and identify critical structures, such as branches or perforators hidden behind the aneurysm sac (and hence, blind to the microscope sight), and verification for the adequacy of the clipping from multiple angles through rotation [2, 3, 13, 14]. In a series by Fischer et al. [15] 180 aneurysms were clipped in 124 operations, performing an EAM procedure in 153 aneurysms (83%) before clipping to gain additional information, and in 130 (72%) after clipping to assess the adequacy of the clipping, being necessary the clip rearrangement in 27 (20%), reducing the complication rate to 3.8% instead of 18.9%.

In this section, the technique for EAM for aneurysmal surgery is described, as well as its utility in specific anatomical locations.

19.3.1 Operative Technique

To perform an EAM procedure, rigid endoscopes with a diameter of 2.7 mm or 4 mm and view angles of 0°, 30°, or 45° are preferred. It can be either handheld or fixed to a holding device, hence, allowing bimanual manipulation during the procedure [2, 13, 14]. The initial craniotomy, opening of the basal cisterns and aneurysm exposure are made in the usual microsurgical manner. At this stage, the use of the endoscope in assisting aneurysmal clipping can be made in three different steps:

- **Preclipping:** the endoscope is introduced free hand, using the same microsurgical route and under careful microscopic view. This initial inspection has the objective of obtaining additional topographic information about the aneurysm and its surrounding neurovascular structures. If the endoscopic view is found to be useful during the clipping procedure, then it should be fixed to the surgical field with a mechanical holder device [1, 16].
- **Clipping:** In cases where the endoscope is deemed to be useful during the clipping procedure and fixation of the endoscope is feasible, then the "true operative time" is made under simultaneous microscopic-endoscopic vision, taking advantage of the wider and clearer view provided by the endoscope to safely dissect the vessels attached to the aneurysm and achieve a better exposure of the aneurysmal sac and neck. Then, the temporal and permanent clipping are performed under simultaneous microscopic-endoscopic vision, employing the real-time endoscopic vision to avoid inadvertent occlusion of hidden perforating branches. However, in those selected cases, where the microscopic vision is restricted, the procedure may be made entirely under direct endoscopic vision (endoscopic controlled) [13, 14, 16].
- **Postclipping:** The last step consists of the final inspection, usually made by the freehand endoscopic view, which includes the assessment of the complete exclusion of the aneurysm, inadvertent occlusion of the parent vessels or perforators by the clip, and the integrity of the surrounding neurovascular structures. This step may be complemented by an endoscopic ICG video angiography, which has been reported to be useful in assessing the blood flow in dead angles under microscopic view. Lastly, if required by the visual information provided by the simultaneous microscopic-endoscopic view, changes or readjustments of the clip should be made [5, 6, 13–17].

19.3.2 Anatomic Highlights

- **Supraclinoid ICA**: EAM for aneurysm located in the posterior communicating artery (PcomA) and anterior choroidal artery (AchorA) is helpful due to the deep location of these aneurysms, and their surrounding by critical perforators. However, introducing the endoscope in the paracarotid cistern enables the visualization and dissection of perforators, especially those located at the medial aspect of the aneurysm and frequently blind under the microscope [13, 16].
- **ICA bifurcation**: The proximity of these aneurysms to ACA and MCA, along with their early branches and perforators (usually hidden behind the lesion itself), may imply great morbidity. In these cases, the endoscope is

helpful in identifying and preserving these vessels, mainly the perforators attached to the medial wall of the aneurysm [16].

- **Anterior communicating artery aneurysms (AcomA)**: These aneurysms may have an anterior, superior, inferior, or posterior projection, frequently hiding the vessels from the ACA complex and perforators during a pterional approach. In these cases, the introduction of the endoscope may be useful for assessing the vessels before and after clipping. However, because these lesions are not located in a true cisternal space, fixation of the endoscope is not recommended, and its use is limited to free-hand scopes [13, 14, 16].
- **Basilar Artery (BA tip, BA-SCA and proximal PCA)**: The deep localization of these aneurysms, in conjunction with the critical neurovascular structures and perforators surrounding them, makes the clipping very challenging. The endoscope may assist the microsurgical clipping by providing a clearer and wider angle of view and, therefore, bringing fundamental information for the safer dissection and preservation of vital branches and perforators, which may be attached to the lateral wall of the aneurysm and not visible under microscopic view. The fixation of the endoscope to the surgical field is also very feasible in these lesions because of their location in large cisternal spaces [13, 16].

19.4 Endoscope-Controlled Microneurosurgery for Intracranial Aneurysm Clipping (Keyhole/Endoscopic Port)

As originally defined by Hopf and Perneczky [2], in ECM procedures the microsurgical manipulations are performed with the guidance of the video image provided by the endoscope and with the employment of microsurgical instruments; therefore, in contrast to EAM procedures, there is no use of an operating microscope. Traditionally, microsurgical aneurysm clipping has been performed through large skin incisions, significant dissection of the temporal muscle and wide craniotomies, and is usually associated with extensive brain exposure and retraction; thus, increasing the surgical morbidity. To reduce these complications, Perneczky et al. [18] introduced the concept of keyhole surgery, based on a tailored and appropriately placed craniotomy, while maintaining minimal brain exposure and retraction. The supraorbital, subtemporal, retrosigmoid, suboccipital, interhemispheric, minipterional, and transcortical approaches are among the most representative [18–20]. These minimally invasive approaches have been demonstrated to improve postoperative results by minimizing tissue trauma, brain exposure, and manipulation; however, despite these advantages, there are some major drawbacks encountered, such as limited viewing and working angle and poor surgical field illumination [19]. In order to overcome these limitations, Fisher et al. [15] introduced the concept of endoscopic keyhole surgery, which combines the advantages provided by a minimally invasive craniotomy (keyhole) and endoscopic surgery.

Currently, there are a small number of case reports about endoscope-controlled aneurysms clipping through a keyhole approach. Perneczky et al. [21] reported seven aneurysms clipped through an EC standard. Hereinafter, the successful EC clipping of AcomA, MCA, ICA bifurcation, PcomA, and AICA aneurysms through a keyhole approach has been described in the literature [15, 22, 23].

More recently, the development of minimally invasive technologies, such as Endoport systems, has allowed us to effectively approach deeply located lesions through a minimally invasive craniotomy, while avoiding white matter injuries and brain retraction usually associated with conventional microsurgical approaches. The use of an Endoport distributes the retraction forces evenly over the surface of the sheath, thus minimizing the complications associated with the management of deep cerebral pathology [24]. The recent advances in endoscopic techniques and minimally invasive technologies, including Endoport or tubular retractor, have shown promising results in the treatment of deeply located cerebrovascular pathologies, such as periventricular [25] and posterior circulation aneurysms [26].

19.5 Endoscopic Endonasal Approaches for Aneurysm Clipping (Fig. 19.1)

Microsurgical clipping of intracranial aneurysms (IA) through conventional transcranial approaches, achieving a complete occlusion of the aneurysm, while preserving the parent, perforating, and branching vessels remains the most definitive treatment for IA [27]. However, transcranial approaches face some drawbacks, such as brain retraction and manipulation of neurovascular structures, which have led to the rapid development of endovascular techniques becoming the preferred strategy in the management of the majority of IA, especially those located in the posterior circulation [21, 28].

The recent advances in endoscopic techniques and technologies have encouraged some neurosurgeons to consider the EEA as a new tool for IA clipping, supported by the direct access route to the median ventral skull base provided by a natural sinonasal corridor. The EEA allows exposure and provides a panoramic view of aneurysms and associated vessels located in the ventral side of the skull base, while

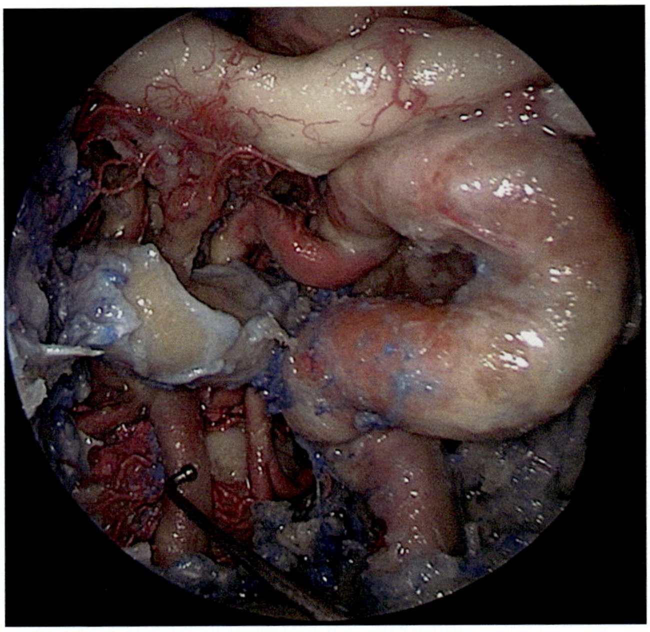

Fig. 19.1 Cadaveric dissection showing the scope of an extended endoscopic endonasal approach. The internal carotid artery can be visualized from the paraclival to the paraclinoid segment, including the posterior communicating artery through a transcavernous approach; and the posterior circulation can be accessed through either a superior or middle transclival approach, depending on the location of the lesion, with or without pituitary transposition. *Dissection by MVSD, at Weill Cornell NITC*

avoiding brain retraction and/or arachnoid dissection [21, 28–30]. However, there are still several limitations to this approach and should not be considered as a replacement for traditional transcranial or endovascular procedures. In addition, it is worth mentioning that EEA for cerebrovascular surgery is considered a "Level V" procedure and thus, should only be performed by teams with significant experience in microvascular surgery and extended EEA, and only in carefully selected cases, as part of the multimodality management of IA [31].

An extensive review of the advantages and limitations present in the current literature is provided in Table 19.1 [21, 28–30, 32]. The characteristics of the aneurysms potentially suitable for EEAC are exposed in Table 19.2.

19.5.1 Extended Endonasal Transplanum-Transtuberculum Approach (EETTA) for Anterior Circulation Aneurysms

Some anterior circulation aneurysms may be safely exposed through an EETTA such as small (<5 mm) and medium (5–10 mm) sized paraclinoid and AcomA complex aneurysms holding a favorable dome projection (Table 19.2) [21, 29].

Table 19.1 Advantages and disadvantages of endoscopic endonasal aneurysm clipping

Advantages	Disadvantages/limitations
• Provides direct anatomical access to the aneurysm located on the ventral skull base • Utilizes a natural sinonasal corridor • Avoidance of brain retraction, manipulation, or arachnoid dissection • Enhanced illumination and a panoramic view of the surgical field • Better visualization of clip blades, parent vessels, and perforating branches (e.g., angled lens) • Endoscope-integrated indocyanine green fluorescence (e-ICG) as a tool for assessing the adequacy of clipping • In selected cases, provides an easier proximal and distal control compared with a transcranial route	• Requires substantial familiarity and experience with extended endoscopic endonasal approaches • Lack of instruments and clips specially designed for EEAC • Limited maneuverability of instruments and angles of attack provided by the endonasal corridor • Impossibility to perform a cerebrovascular revascularization or anastomosis if required • Skull base reconstruction may be difficult because of the protrusion of the clip into the sphenoid sinus • Intraoperative bleeding (aneurysm rupture) can obscure the lens and limit proper visualization

Table 19.2 Characteristics of aneurysms suitable for endoscopic endonasal clipping

Location	Projection	Approach
AcoA	Anterosuperior, inferior, lateral	Extended transplanum-transtuberculum
Paraclinoid	Medial (superomedial, inferomedial)	Extended transplanum-transtuberculum *May require parasellar bone removal and optic canal decompression for adequate vascular control
Basilar apex	Ventral	Extended transclival *High-lying BA aneurysm may require pituitary transposition
Basilar trunk	Ventrolateral	Extended transclival

- **EETTA for AcomA Aneurysms**

  An EETTA provides a straightforward route to access the AcomA complex, including the pre- and post-communicating (A1 and A2) segments of the ACA, as well as AcoA, allowing for safe proximal and distal vascular control [21, 28]. However, regardless of the anatomical feasibility of clipping A1, A2, and AcomA aneurysms through this approach, it should only be attempted in those located on the AcomA and with a favorable dome projection. In this respect, the most suitable AcomA aneurysm for an EEAC is those having an anterosuperior projection, which allows for a direct exposure of the neck without requiring manipulation of the aneurysm dome, while permitting an adequate proximal and distal control as well as clear visualization of the perforators. Though aneurysms possessing an inferior and/or lateral dome projection may be clipped through this approach; conversely, those with a posterior projection are not well suitable due to the inadequate visualization of perforators and aneurysm neck provided by this corridor [21, 28, 33].

Technical nuances: There are several anatomical considerations that should be considered before proceeding to this approach. First, the size of the aneurysm plays a significant role in clipping feasibility since the proximal and distal control is easier for smaller aneurysms. Moreover, the AcomA complex is close to the optic chiasm, serving as an intraoperative obstacle and even requiring its manipulation while obtaining A1 control [28]. Additionally, this vicinity makes the optic apparatus prone to undue pressure exerted by the vascular clip, which may lead to visual deficits [21]. The position of the A2 also influences the difficulty of the clipping, being more amenable when the bilateral A2 are disposed in a horizontal position in the coronal plane [33].

- **EETTA for Paraclinoid Aneurysms**

  Paraclinoid aneurysms are always a challenge for microsurgical clipping and demand familiarity and experience with cranial base approaches due to their vicinity to the optic nerves, ICA, and anterior clinoid process (ACP). Though, in selected cases, some paraclinoid aneurysms, including those located in the ophthalmic and superior hypophyseal artery, arising from the medial wall of ICA, may be accessed through an EETTA [21, 29]. The rationale for using this approach is that it offers a straightforward route to the medial aspect of the ICA without requiring an anterior clinoidectomy and its associated morbidity. In addition, it has the theoretical advantage of easy proximal control of ICA in contrast to the transcranial route, which often requires exposure of the cervical ICA for proximal control [28, 29]. However, this approach carries the risk of injuring neurovascular structures during proximal and distal control, as well as during the dural opening. Moreover, working through the narrow corridor provided by an EEA yields little room for instrument maneuverability, a condition that may be improved by expanding the approach as much as possible [21, 28].

  Technical nuances: The ophthalmic artery, as seen through this approach is located beneath the optic nerve at the superomedial face of the parasellar carotid, at the level of the superior border of the lateral optic canal recess. Careful dissection of the parasellar ICA allows for lateral displacement of the ICA and exposes the medial wall up to the posterior turn of the ICA, where the superior hypophyseal artery is located [21]. Additionally, expanding the dural opening and bone removal over the optic nerve provide exposure to the distal dural ring and thus, the neck of paraophthalmic aneurysms [29]. The prompt achievement of proximal control is essential for clipping paraclinoid ICA aneurysms. The paraclival (both extra cavernous and intracavernous parts) and clinoidal segment of the ICA may be exposed through an EEA to secure proximal vascular control [21, 29]. In order to expose the paraclival ICA, it is necessary to make a bone window at the prominence of the ICA, followed by the drilling part of the quadrangular area lateral to the paraclival ICA (lingular recess). However, there is some risk of injuring the trigeminal and abducens nerves while drilling this area, thus, it should not extend superior to the level of the maxillary nerve [28, 34]. Moreover, obtaining proximal control from the clinoidal ICA requires significant bone removal from the cavernous sinus, medial opticocarotid recess, and middle clinoid area which may damage the oculomotor nerve. Conversely, obtaining distal control for paraclinoid ICA aneurysm through an EEA is more difficult, sometimes requiring bone removal around the optical canal and retraction of the optic nerve [34]. Therefore, this limitation must be considered prior to deciding the EEAC of paraclinoid aneurysms.

- **Illustrative Case** (Fig. 19.2)**:** An illustrative case of a medially projecting paraclinoid aneurysm that was successfully clipped through an endoscopic endonasal approach assisted by intraoperative angiography with roadmapping and proximal balloon control was recently published by our team [35].

19.5.2 Endoscopic Endonasal Transclival Approach (EETA) for Posterior Circulation Aneurysms (Fig. 19.3)

As a result of their deep location, neurovascular relationships, and their numerous perforating branches, posterior circulation aneurysms represent the most challenging group of aneurysms to be clipped through a transcranial route and the endovascular treatment has emerged as the preferred strategy in many centers [36–38]. However, some aneurysms may not be suitable for endovascular treatment due to their small size, distal location, neck-to-dome ratio, acute hemorrhage (contraindication for antiplatelet therapy) or significant mass effect, and compression of adjacent neurovascular structures [21, 29, 30]. In those cases, aneurysm clipping through an EETA may be an effective option by offering a direct route to the main components of the vertebrobasilar system, giving access to median vertebrobasilar aneurysms, their dome, neck, and associated vessels, while avoiding extensive brain exposure and retraction. Successful clipping of VA, AICA, SCA, P1, and BA aneurysms through this ventral approach has been described in a limited number of cases

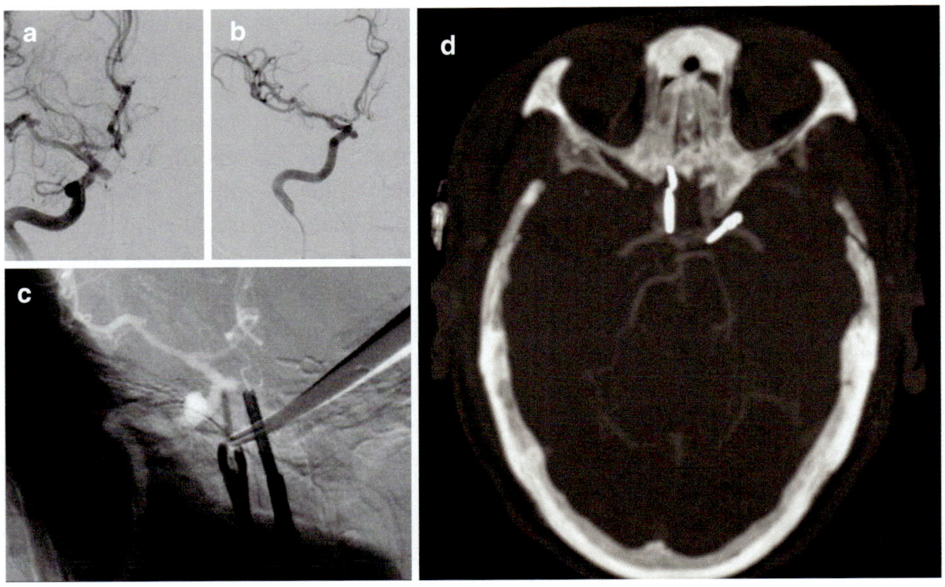

Fig. 19.2 Anteroposterior (**a**) and oblique (**b**) projection of preoperative DSA demonstrating a paraclinoid aneurysm with a medially projected dome, arising from the right superior hypophyseal artery. (**c**) Intraoperative image showing placement of a straight 7 mm vascular clip on the paraclinoid aneurysm through an EEA. (**d**) Axial projection of postoperative CTA showing aneurysms clips and no evidence of residuals

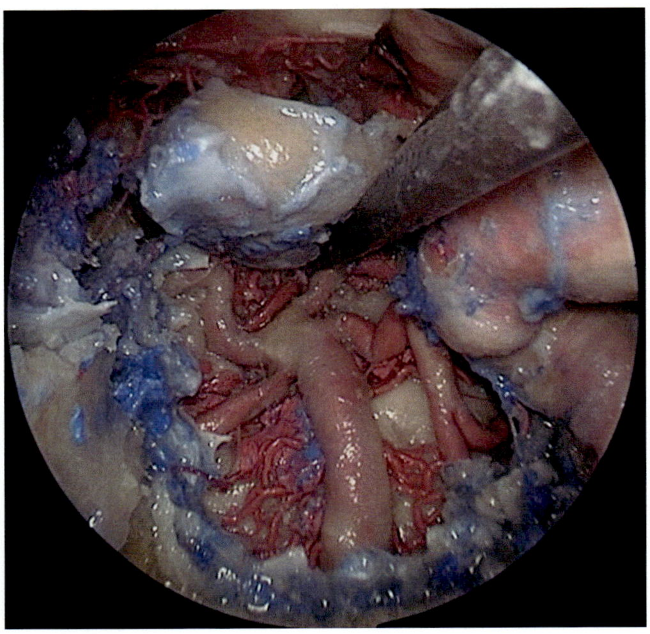

Fig. 19.3 Cadaveric dissection showing the endoscopic view of the vertebrobasilar system as obtained through an extended endonasal transclival approach. It is worth mentioning that the degree of exposition of the basilar artery may change in relation to anatomical variations, such as a low-lying/high-lying basilar artery. *Dissection by MVSD, at Weill Cornell NITC*

in the literature; consequently, its role remains undefined [21, 28–30]. The endoscopic view of the vertebrobasilar system as exposed through an EETA is shown in Fig. 19.3.

- **Basilar Apex (BA) Aneurysms**

  The exposure of the BA is optimal from the ventral perspective provided by the EETA [28]. However, the anatomical relationship of the BA and aneurysm neck with the clivus and posterior clinoid process (PCP) may lead to limitations in the sagittal exposure and must be assessed in the preoperative planning, along with the basilar bifurcation angle, the presence of fetal posterior circulation, size, and projection of the aneurysm [21, 30].

  Technical Nuances: BA aneurysms located below the dorsum sellae (low-lying) seem to be good candidates for aneurysms clipping through an EETA due to the facility to obtain proximal and distal control, in contrast to traditional transcranial approaches. Nonetheless, high-lying BA aneurysms (located above the PCP) are less accessible and the acquisition of distal control is restricted due to the anatomical barriers. However, these limitations can be overcome by the removal of the dorsum sellae and PCP, and performing a pituitary transposition, which significantly increases the exposed area, and allows for a more comfortable bimanual microdissection [21, 29, 30].
- **Illustrative case** (Fig. 19.4): We present the case of a 22-year-old female, with a long-term history of headaches, which presented to our emergency department due to the onset of gait disturbance and left peripheral facial palsy. A CTA was performed revealing a giant aneurysm arising from the distal third of the basilar artery with partial intra-aneurysmal thrombosis. The patient was initially treated with stent-assisted coil embolization of the aneurysm and discharged from the hospital. However, on follow-up appointments, the persistence of the neurologic signs and an increase in the aneurysmal diameter were noted. The patient's case was submitted to a multidisciplinary clinical session, and it was decided to treat this lesion by an endoscopic endonasal extended transclival approach in a hybrid operating room, and with the guidance of intermittent road mapping. The patient was positioned supine, with the head fixed with a three-pin holder,

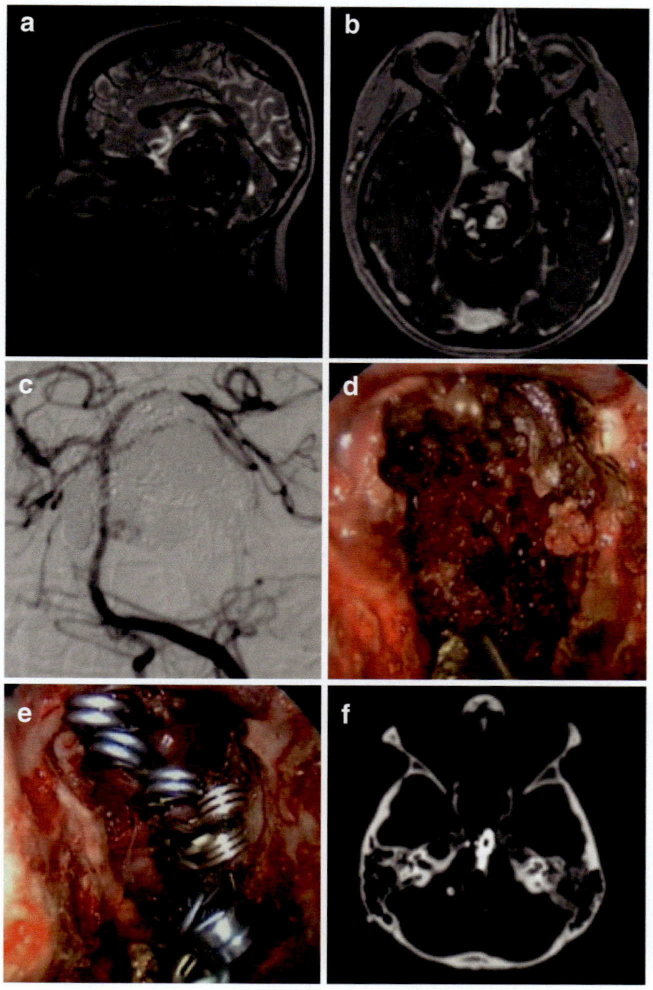

Fig. 19.4 (**a**) Sagittal T2-weighted MRI demonstrates a heterogeneous posterior fossa lesion, conditioning backward displacement of the brainstem. (**b**) Axial T1-post contrast weighted MRI showing a lesion with heterogeneous intensity on the pontine cistern and displacing the brainstem. (**c**) Digital subtraction cerebral angiography and anteroposterior view. Vertebrobasilar dolichoectasia can be observed, in addition to the superposed image of coils on the aneurysm wall. (**d** and **e**) Intraoperative endoscopic view of the extended transclival approach during the thro0mbectomy (**d**) and clipping (**e**). (**f**) Axial, noncontrast postoperative CT scan, bone window. Vascular clips are observed at the thrombectomy site

with 20° of flexion and turned 10° to the right. The procedure was performed with a 4 mm, 0°-degree endoscope (Braun-Aesculap AG, Tuttlingen, Germany). A bilateral nostril approach was used to allow a 2-surgeon technique and a right nasoseptal flap was harvested and placed in the right choana for later reconstruction. An extensive bone resection of the anterior wall and floor of the sphenoid sinus was performed to allow access to the middle and inferior third of the clivus. High-speed (diamond) drills were used to remove the middle clivus until the posterior fossa dura mater was identified with a lateral extension limited by the paraclival carotid arteries. A midline linear dural incision was performed to expose the basilar artery and ventral pons. The basilar artery could be directly identified, and arachnoid dissection was performed with micro dissectors under endoscopic view. The proximal control was obtained at the basilar trunk employing a Yasargil temporary titanium clip (Braun-Aesculap AG, Tuttingen, Germany. Then, a thrombectomy was performed under purely endoscopic view and finally, the aneurysm was clipped with a Picket fence technique, employing 5 straight clips of 7 mm. Skull base reconstruction was performed in a multilayered fashion, placing collagen matrix (DuraGen, Integra Neurosciences) above and under the dural defect, followed by autologous fascia lata graft, fat tissue, and the pedicled nasoseptal flap. Finally, fibrin tissue adhesive sealant (Beriplast, Aventis, Behring) was used to seal the surgical defect. No cerebrospinal fluid leak was observed in the postoperative period. The patient presented an uneventful postoperative course and in the follow-up appointment, a significant improvement of the neurological symptoms was observed.

19.6 Endoscopic Endonasal Approaches for Brainstem Cavernous Malformations (BSCM)

Brainstem cavernous malformations (BSCM) represent approximately 8–22% of all intracranial CM, nevertheless, they are associated with a higher annual incidence of hemorrhage (2–3%) compared to their supratentorial counterpart (0.25–1.1%) and when hemorrhage occurs, the rebleeding rate may raise to 5–35% per year. Repeated hemorrhage can be lethal, therefore surgical intervention must be considered in all patients with BSCM manifesting at the pial surface, and who have experienced a symptomatic hemorrhage [39, 40]. The ideal surgical approach to BSCM should provide adequate visualization of the surgical field, minimal manipulation of the neurovascular structures, and a good working angle. The surgical approach must also be determined by following the "2 point rule" which is defined as a point that connects the center of the lesion to the point where the lesion is closest to the brainstem surface. However, achieving these goals on ventromedially located BSCM is particularly difficult through the traditional transcranial approaches. As a result, expanded EEAs have been recently described as a safe and feasible alternative for selected mesencephalic and pontine BSCM resection [10, 36, 37, 39–41].

The rationale for using an EEA for BSCM resection is that this approach provides a direct route to the ventral brainstem with less brain retraction and cranial nerve manipulation, while simultaneously offering an enhanced view and illumination of the deeply located brainstem, allowing a prompt visualization of cranial nerves, arteries, and associated perforators. In addition, the displacement of the neural tissue caused by the CM makes theoretically possible the resection of the lesions presenting on the pial surface with minimal brain transection [36, 37, 39–41].

– **Mesencephalic BSCM**: Several transcranial approaches have been described to access the ventral midbrain (e.g., pterional/trans-sylvan approach); however, these approaches offer a narrow corridor between the cranial nerves and vascular structures, and limited access to the ventromedial mesencephalic area, in addition to requiring some degree of brain retraction. In this regard, the use of extended transtuberculum-transplanum and extended transclival approaches (with or without pituitary transposition) has been proven to provide a straightforward route to the ventral mesencephalon, allowing for a safe resection of BSCM located ventromedial in the midbrain [37, 41].
– **Pontine BSCM**: Lesions on this location have been traditionally resected through many surgical approaches, such as the transpeduncular, transpetrosal, or retrosigmoid. These approaches provide access to the anterolateral brainstem; conversely, the working angle provided is suboptimal when facing lesions located in the ventromedial pons. In addition, they require extensive bone drilling and considerable manipulation of neurovascular structures, thus, increasing morbidity. In these cases, the EETA provides a direct route to resect ventromedially located pons cavernomas [10, 36, 38, 40].

19.7 Surgical Pearls

19.7.1 Endoscope-Assisted Microneurosurgery for Intracranial Aneurysms

- The endoscope can be used in assisting aneurysmal clipping during the preclipping, clipping, and postclipping stages.
- During the preclipping stage, the endoscope provides additional topographic information about the aneurysm and its surrounding neurovascular structures.
- During the clipping stage a clearer view is provided by the endoscope to safely dissect the vessels attached to the aneurysm and achieve a better exposure of the aneurysmal sac and neck.
- During the postclipping stage, the assessment of the complete exclusion of the aneurysm, inadvertent occlusion of the parent vessels or perforators by the clip, and the integrity of the surrounding neurovascular structures are assessed.

19.7.2 Endoscope-Controlled Microneurosurgery for Intracranial Aneurysm Clipping (Keyhole/Endoscopic Port)

- Endoport systems allow to effectively approach deeply located lesions through a minimally invasive craniotomy while avoiding white matter injuries and brain retraction usually associated with conventional microsurgical approaches.
- However, the long narrow corridors provided by these approaches restrict the maneuverability of instruments and the presence of a major bleeding may compromise the endoscopic view.

19.7.3 Endoscopic Endonasal Approaches for Brainstem Cavernous Malformations (BSCM)

- The rationale for using an EEA for BSCM resection is that this approach provides a direct route to the ventral brainstem with less brain retraction and cranial nerve manipulation, while simultaneously offering an enhanced view and illumination of the deeply located brainstem, allowing a prompt visualization of cranial nerves, arteries, and associated perforators.

19.8 Discussion

The introduction of endoscopic techniques for treating cerebrovascular pathology is an expected result of the continuous advancements and maturation in endoscopic surgery. In the early stages, the endoscope was exclusively used as an adjunct to the microscope for aneurysm clipping and, in the last decades, it has proven its usefulness as a primary visualization tool to provide unique and direct corridors for treating vascular lesions. However, the use of the endoscope in cerebrovascular pathology has several considerations and limitations, which are discussed in this section.

The ECM clipping of aneurysms through a keyhole approach may represent a less invasive alternative to traditional microsurgical approaches; nevertheless, there are several aspects to be considered. First, the long narrow corridors provided by these approaches restrict the maneuverability of instruments. Additionally, the presence of major bleeding may compromise the endoscopic view and even requires transformation to an open approach, which may be deemed difficult in the context of a minimally invasive craniotomy.

Lastly, the two-dimensional view provided by the endoscope may compromise the depth perception, and thus, increase the complexity of the procedure in inexperienced hands [24].

On the other hand, the recently introduced EEAC may represent a new surgical armamentarium for treating well-selected aneurysms. However, some considerations to take into account are noted. First, it must be kept in mind that EEAC is not suitable for all IA and there is no actual consensus about their indications. However, it can be an effective alternative for well-selected IA difficult to access through a conventional transcranial route, and not suitable for endovascular treatment [21, 29, 32]. The meticulous selection of the cases must be based on the careful analysis of preoperative image studies, paying particular attention to the size of the aneurysm, projection, anatomic relationships, and sinonasal anatomy [27]. Moreover, the general principles of cerebrovascular surgery must be strictly applied in these approaches to achieve a safe aneurysm clipping: (1) gaining proximal and distal control, (2) adequate visualization of perforator vessels, (3) potentiality for a safe clip placement, and (4) adequate bony exposure to allow a bimanual microsurgical technique. If these principles cannot be safely achieved, EEAC should be discouraged [27, 29, 30, 32].

In addition, despite their feasibility, safety, and effectiveness being recognized under certain conditions, there is no randomized controlled trial assessing the outcomes of EEAC and comparing them with other treatment modalities. Therefore, there is no clear understanding about their role and applicability for treating cerebral aneurysms and further clinical studies are required to better define their role [27, 32]. Another concern is the capability for adequate vascular control in cases of major arterial bleeding in the scenario of a ruptured aneurysm, which may compromise visualization [21, 27]. Finally, the steep learning curve associated with endoscopic endonasal procedures must be considered, especially for cerebrovascular surgery which is considered a "Level V" procedure and therefore should only be performed by teams with extensive experience with expanded endoscopic endonasal approaches [31].

As for the EEA for BSCM, most of the technical and anatomical considerations aforementioned may be applied. Additionally, it must be reminded that most of the endoscopic instrumentation is designed for approaching the sellar region and, therefore, not long enough to properly reach the lesion located on the brain stem; thus, it is necessary to ensure that the length of the instruments is enough to safely reach the BSCM. Moreover, there are limited options for vascular control in cases of arterial injury, which may lead to catastrophic outcomes [38]. Therefore, significant surgical experience in both endoscopic and cerebrovascular surgery, as well as proper instruments and equipment, careful patient selection, and meticulous preoperative planning are mandatory for achieving good results with these approaches.

19.9 Conclusion

The combination of endoscopic techniques and minimally invasive approaches may represent a less invasive, safe, and effective alternative for treating selected intracranial aneurysms and BSCM. However, additional studies are needed to elucidate their role in cerebrovascular surgery and compare their efficacy with standard microsurgical approaches, and should not be considered as a replacement to the traditional microsurgical and endovascular treatment, but rather an option in limited cases and in the hands of an experienced surgeon, as part of the multidisciplinary management of this pathology.

References

1. Lai LT, O'Neill AH. History, evolution, and continuing innovations of intracranial aneurysm surgery. World Neurosurg. 2017;102:673–81.
2. Hopf NJ, Perneczky A. Endoscopic neurosurgery and endoscope-assisted microneurosurgery for the treatment of intracranial cysts. Neurosurgery. 1998;43(6):1330–6. discussion 1336–7
3. Perneczky A, Fries G. Endoscope-assisted brain surgery: part 1—evolution, basic concept, and current technique. Neurosurgery. 1998;42(2):219–24. discussion 224–5
4. Patel A, Abou-Al-Shaar H, Mallela AN, Algattas H, McDowell MM, Zenonos GA, Wang EW, Snyderman CH, Gardner PA. Endoscopic techniques in vascular neurosurgery. Neurosurg Clin N Am. 2022;33(4):491–503.
5. Shao D, Li Y, Zhang B, Wu J, Xie S, Zheng X, Jiang Z. Endoscope-assisted microneurosurgery for intracranial aneurysms: a systematic review and meta-analysis. J Clin Neurosci. 2022;103:62–71.
6. Apuzzo ML, Heifetz MD, Weiss MH, Kurze T. Neurosurgical endoscopy using the side-viewing telescope. J Neurosurg. 1977;46(3):398–400.
7. Fischer J, Mustafa H. Endoscopic-guided clipping of cerebral aneurysms. Br J Neurosurg. 1994;8(5):559–65.
8. Fries G, Perneczky A. Endoscope-assisted brain surgery: part 2—analysis of 380 procedures. Neurosurgery. 1998;42(2):226–31. discussion 231–2
9. Kassam AB, Mintz AH, Gardner PA, Horowitz MB, Carrau RL, Snyderman CH. The expanded endonasal approach for an endoscopic transnasal clipping and aneurysmorrhaphy of a large vertebral artery aneurysm: technical case report. Neurosurgery. 2006;59(1 Suppl. 1):ONSE162–5. discussion ONSE162–5
10. Gómez-Amador JL, Sangrador-Deitos MV, Guinto-Nishimura GY, Alvear-Quito NM. Middle transclival endoscopic endonasal resection of a ventral pontine cavernous malformation. World Neurosurg. 2022;158:34–5.
11. Hallaert GG, De Keukeleire KM, Vanhauwaert DJ, Defreyne L, Van Roost D. Intracranial dural arteriovenous fistula successfully treated by combined open-endovascular procedure. J Neurol Neurosurg Psychiatry. 2010;81(6):685–9. https://doi.org/10.1136/jnnp.2009.174771.
12. Kassam AB, Thomas AJ, Zimmer LA, Snyderman CH, Carrau RL, Mintz A, Horowitz M. Expanded endonasal approach: a fully endoscopic completely transnasal resection of a skull base arteriovenous malformation. Childs Nerv Syst. 2007;23(5):491–8.
13. Yoshioka H, Kinouchi H. The roles of endoscope in aneurysmal surgery. Neurol Med Chir (Tokyo). 2015;55(6):469–78.

14. Chavan VS, Yamada Y, Chandratej K, Gowtham D, Riccardo S, Firuz S, Yoko K. Intraoperative use of endoscope, a valuable, adjunctive tool for the surgical management of anterior communicating artery aneurysm surgery: our institutional experience. Asian J Neurosurg. 2020;15(2):338–43.
15. Fischer G, Oertel J, Perneczky A. Endoscopy in aneurysm surgery. Neurosurgery. 2012;70(2):184–90. discussion 190-1
16. Galzio RJ, Di Cola F, Raysi Dehcordi S, Ricci A, De Paulis D. Endoscope-assisted microneurosurgery for intracranial aneurysms. Front Neurol. 2013;4:201.
17. Fischer G, Rediker J, Oertel J. Endoscope- versus microscope-integrated near-infrared indocyanine green videoangiography in aneurysm surgery. J Neurosurg. 2018;1-10
18. Perneczky A, Muller-Forell W, van Lindert E, Fries G. Keyhole concept in neurosurgery. Stuttgart, Germany: Georg Thieme-Verlag; 1999.
19. Fischer G, Stadie A, Reisch R, Hopf NJ, Fries G, Böcher-Schwarz H, et al. The keyhole concept in aneurysm surgery: results of the past 20 years. Oper Neurosurg. 2011;68:ons45–51.
20. Nathal E, Gomez-Amador JL. Anatomic and surgical basis of the sphenoid ridge keyhole approach for cerebral aneurysms. Neurosurgery. 2005;56(Suppl. 1):178–85.
21. Martinez-Perez R, Hardesty DA, Silveira-Bertazzo G, Albonette-Felicio T, Carrau RL, Prevedello DM. Safety and effectiveness of endoscopic endonasal intracranial aneurysm clipping: a systematic review. Neurosurg Rev. 2021;44(2):889–96.
22. Wang JP, Wu ZY, Xu J, Dou YH. Fully Endoscope-controlled clipping bilateral middle cerebral artery aneurysm via unilateral supraorbital keyhole approach. J Craniofac Surg. 2016;27(8):2151–3.
23. Gao W, Liu P, Deng G, Li Z. Fully endoscope-controlled clipping ruptured intrameatal anterior inferior cerebellar artery aneurysm: report of 2 cases. J Craniofac Surg. 2021;32(5):e470–2.
24. Jo KW, Shin HJ, Nam DH, et al. Efficacy of endoport-guided endoscopic resection for deep-seated brain lesions. Neurosurg Rev. 2011;34(4):457–63.
25. Chen CJ, Caruso J, Starke RM, et al. Endoport-assisted microsurgical treatment of a ruptured periventricular aneurysm. Case Rep Neurol Med. 2016;2016:8654262.
26. Gómez-Amador JL, Sangrador-Deitos MV, Uribe-Pacheco R, Guinto-Nishimura GY, Mondragón-Soto MG. Endoport-assisted microsurgical treatment for a ruptured posterior cerebral artery aneurysm: a technical note. J Cerebrovasc Endovasc Neurosurg. 2022;24(1):73–8.
27. Xiao LM, Tang B, Xie SH, et al. Endoscopic endonasal clipping of anterior circulation aneurysm: surgical techniques and results. World Neurosurg. 2018;115:e33–44.
28. Shah VS, Martinez-Perez R, Kreatsoulas D, Carrau R, Hardesty DA, Prevedello DM. Anatomic feasibility of endoscopic endonasal intracranial aneurysm clipping: a systematic review of anatomical studies. Neurosurg Rev. 2021;44(5):2381–9.
29. Gardner PA, Vaz-Guimaraes F, Jankowitz B, et al. Endoscopic endonasal clipping of intracranial aneurysms: surgical technique and results. World Neurosurg. 2015;84(5):1380–93.
30. D'avella E, Fazzolari B, Cappelletti M, et al. Anatomical assessment of the endoscopic assisted lateral supraorbital approach and endoscopic endonasal transclival approach to basilar apex aneurysms. Turk Neurosurg. 2021;31(5):671–9.
31. Snyderman C, Kassam A, Carrau R, Mintz A, Gardner P, Prevedello DM. Acquisition of surgical skills for endonasal skull base surgery: a training program. Laryngoscope. 2007;117(4):699–705.
32. Gu Y, Zhong X, Gao Y, He L. Endoscopic endonasal approach for simultaneously treating a pituitary adenoma coexisting with a paraclinoid aneurysm: illustrative case. J Neurosurg Case Lessons. 2022;3(25):CASE22130. Published 2022 Jun 20
33. Qin K, Guo W, Zheng J, et al. Endoscopic endonasal approach for clipping anterior communicating artery aneurysms from cadaver studies and three-dimensional printed models to a clinical case. J Craniofac Surg. 2021;32(8):2854–8.
34. Tayebi Meybodi A, Borba Moreira L, Little AS, Lawton MT, Preul MC. Anatomical assessment of the endoscopic endonasal approach for the treatment of paraclinoid aneurysms. J Neurosurg. 2018;131(6):1734–174.
35. Sangrador-Deitos MV, Lopez-Valencia G, Guinto-Nishimura GY, Eguiluz-Melendez AG, Romano-Feinholz S, Gomez-Amador JL. Endoscopic endonasal clipping of a medial paraclinoid aneurysm with roadmapping assistance: two-dimensional operative video. Surg Neurol Int. 2023;14:31.
36. Gómez-Amador JL, Ortega-Porcayo LA, Palacios-Ortíz IJ, Perdomo-Pantoja A, Nares-López FE, Vega-Alarcón A. Endoscopic endonasal transclival resection of a ventral pontine cavernous malformation: technical case report. J Neurosurg. 2017;127(3):553–8.
37. Enseñat J, d'Avella E, Tercero A, Valero R, Alobid I. Endoscopic endonasal surgery for a mesencephalic cavernoma. Acta Neurochir. 2015;157(1):53–5.
38. Sanborn MR, Kramarz MJ, Storm PB, Adappa ND, Palmer JN, Lee JY. Endoscopic, endonasal, transclival resection of a pontine cavernoma: case report. Neurosurgery. 2012;71(1):198–203.
39. Fehrenbach MK, Kuzman P, Quaeschling U, Meixensberger J, Nestler U. Endoscopic resection of an intraventricular cavernoma: a case report. Int Med Case Rep J. 2019;12:249–52.
40. Erickson N, Siu A, Sherman JH, Gragnaniello C, Singh A, Litvack Z. Endoscopic transnasal transclival approach to a pontine cavernoma with associated developmental venous anomaly. World Neurosurg. 2018;118:212–8.
41. Topczewski TE, Di Somma A, Culebras D, et al. Endoscopic endonasal surgery to treat intrinsic brainstem lesions: correlation between anatomy and surgery. Rhinology. 2021;59(2):191–204.

Surgery for Brain Arteriovenous Malformations

20

Vladimír Beneš, Adéla Bubeníková, and Ondřej Bradáč

Learning Objectives

- AVMs S-M I–II are indicated for surgery regardless of presentation.
- AVMs S-M III are an individual, heterogeneous group. Any modality can be used depending on the lesion, patient's preference, or institutional experience.
- AVMs SM IV–V are most commonly given observation status. Only the bleeding calls for active treatment.
- Monotherapy is preferable. The role of endovascular treatment is questionable.
- Surgery following failed radiosurgical treatment has not been studied sufficiently.

20.1 Introduction

Arteriovenous malformations (AVMs) are abnormally developed fast-flow connections between arterial and venous circulation with the absence of a normal capillary bed occurring in various organ systems in the body [1]. The incidence of newly diagnosed brain AVMs is approximately 1–1.5 per 100,000 person-years [2–4]. Most AVMs are diagnosed without a previous history of rupture; thus, the incidence of haemorrhage is only approximately 0.5 per 100,000 person-years [2–4].

V. Beneš (✉)
Institute of Clinical Neurodisciplines, Department of Neurosurgery and Neurooncology, First Faculty of Medicine, Charles University and Military University Hospital, Prague, Czech Republic
e-mail: vladimir.benes@uvn.cz

A. Bubeníková · O. Bradáč
Institute of Clinical Neurodisciplines, Department of Neurosurgery and Neurooncology, First Faculty of Medicine, Charles University and Military University Hospital, Prague, Czech Republic

Department of Neurosurgery, Second Faculty of Medicine, Charles University and Motol University Hospital, Prague, Czech Republic

The indication for active AVM treatment is mostly the prevention of the rupture with all its consequences. Approximately 20–40% of patients with epileptic seizures have AVM presentation, and in some 10% of patients, a focal neurological deficit occurs. Rarely intractable headache is the only symptom; in rare cases, AVM causes a mass effect. Most AVMs are diagnosed during the second or third decade of life, but cases in children and adolescents have been reported [5]. In contrast, AVM presentation for the first time in the elderly is rare. With the widespread application of computed tomography (CT) and magnetic resonance imaging (MRI), many AVMs are diagnosed incidentally.

This chapter focuses on the surgical treatment of AVMs, which will be described in the essential points in the following paragraphs. However, first, some general remarks are necessary.

20.2 Background and AVM Anatomy Highlights

Anatomically, there are three morphological components of the AVM of utmost importance:

- Arterial feeders.
- Nidus (coming from the Latin word for "nest").
- Draining veins. Both arterial feeders and draining veins are very specific to the individual AVM.
- They can be both single or multiple, originating (feeders) and terminating (veins) at various anatomical locations (superficial versus deep venous drainage, noneloquence, eloquence, etc.).
- The nidus is the racemose centre of the AVM, composed of fragile and dysplastic vessels, with commonly present intervening brain tissue.

Despite historically earlier grading schemes, the most famous and most commonly used grading scheme was intro-

I. N. Kanaan, V. Beneš (eds.), *Neuroanatomy Guidance to Successful Neurosurgical Interventions*,
https://doi.org/10.1007/978-3-031-59838-8_20

duced by Spetzler and Martin (S-M grading) in 1986 [6]. This grading system is based on evaluating the diameter of the AVM nidus, the presence or absence of deep venous drainage, and the (non-)eloquence of adjacent brain tissue. Eloquent AVMs are generally considered in the brainstem, diencephalon, cerebellar peduncles, sensorimotor, language, or primary visual cortex. Later, Lawton and Young developed an extension of the S-M grading system, now known as the Lawton–Young grading scheme [7]. This grading system evaluates the age of patients, the compactness of the AVM nidus, and whether the patient has already suffered from AVM rupture. The combination of Spetzler–Martin and Lawton–Young schemes is called the supplemented Spetzler–Martin grading system, standing for the sum of the two grading together [8]. Also well known is the Spezler–Ponce 3-tier grading [9] classification of AVMs into three groups according to their S-M grade: Class A comprises S-M grade I and II AVMs, Class B of S-M grade III, and Class C of S-M grade IV and V AVMs.

20.2.1 Diagnostic Procedures

A gold standard diagnostic tool for AVMs is digital subtraction angiography (DSA) because it allows a detailed depiction of various parameters related to its anatomy: the nidus and its haemodynamic properties, anatomy of arterial feeders, and the pattern of venous drainage (deep or superficial), or diagnosis of AVM-related aneurysms. CT imaging is inevitable in the acute phase to investigate intracranial haemorrhage. CT is also useful in combination with CT angiography because of the possibility of investigating the presence of AVM during the acute phase of bleeding. Not surprisingly, MRI is an irreplaceable method for surgical planning. MRI is superior in several aspects when compared to the above-mentioned diagnostic tools. MRI makes it possible to differentiate between subacute or chronic haemorrhage, and evaluate the presence of oedema, mass effect, or perilesional gliosis. Although a valuable tool, functional fMRI (fMRI) is sometimes difficult to interpret because of haemodynamic changes caused by the AVM. The same applies to diffusion tensor imaging (DTI).

20.3 AVM Treatment

AVM treatment depends on patient presentation, age, general health, and preference. The other factor is AVM, its size, location, anatomical relations, angioarchitecture, accompanying aneurysms, etc. The biggest risk associated with brain AVMs is bleeding—exact numbers dedicated to overall case fatality and permanent disability rates following the AVM rupture vary, but they typically range from 5% up to 40% [10, 11]. The risk of unruptured AVM is approximately 2–4% per year, with an annual rehaemorrhage rate of 6–7% [10, 11].

One of the most important factors in AVM's active treatment policy is the institution's experience. AVMs are typical lesions treated at tertiary care hospitals with sufficient experience in cerebrovascular problems. The last factor is the surgeon; is his(her) experience and ability adequate to treat this specific lesion?

Recently, multimodal treatment has been available. Any active treatment should surpass the natural course in a long perspective, given that most patients are in their 20s and 30s. Thus, a 40-year perspective is logical. The most effective treatment is surgical, it is immediate, and its efficacy is close to 100% [12]. A valuable option is radiosurgery, with some 65% efficacy and patient friendliness [13]. The treatment delay of some 2–3 years is a negative factor. Endovascular treatment helps treat accompanying aneurysms and higher-grade lesions as a presurgical adjunct. We strongly believe that endovascular treatment alone with an efficacy of some 30% and risks of 7% is not beneficial to the patient. Medical treatment with beta-blockers is still experimental and should only be used as a last resort in otherwise untreatable lesions. Symptomatic treatment, usually by antiepileptic drugs, is obvious. Observation is a viable treatment option in some cases. We have nothing available in SM grades IV and V that could match the natural course.

In Fig. 20.1, all treatment options are compared to a natural course from a 40-year perspective [10, 11].

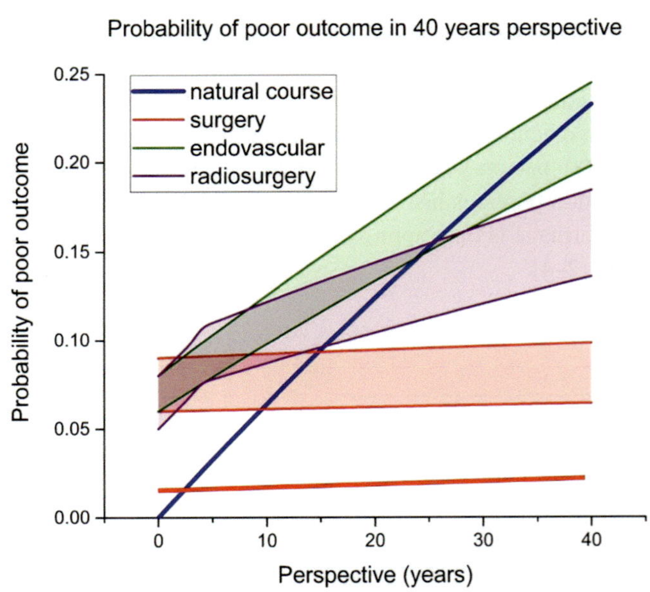

Fig. 20.1 Risks, effectivity, and long-term perspective of AVM treatment compared to the natural course [12]. The red line was calculated from our long-term surgical results of brain AVMs. The natural course represents 2.2% yearly risk of bleeding with 30%morbidity/mortality rate following any bleeding (ARUBA)

20.4 Surgery

At the end of the nineteenth century, Giordano and, later also, Pean performed presumably the first surgeries of intracranial AVMs, followed by such famous neurosurgeons as Cushing, Bailey, and Dandy [14]. Later on, in 1967, Yasargil was the first to use a microscope along with bipolar coagulation and automatic retractors [14]. The basic scheme of indications for "modern" surgical treatment of AVMs was introduced by Spetzler in 2011 [9] and remained relatively stable over the years. AVMs of S-M grade I or II are typically indicated for surgery. In contrast, S-M grade III should be managed by multimodal cooperation in relation to the AVM anatomy, localisation, and patient-related factors [15]. Patients with AVMs of S-M grade IV and V are, in most cases, not treated surgically because the risks of a surgical intervention exceed the risks of the natural course of the disease [16].

Surgery for AVMs is never an effortless task. Utmost care and concentration must be held throughout the procedure, which can be divided into two parts. The initial phase, the sulcal phase, is clean and elegant, while the second phase may be characterised as the white matter phase with the control of deep feeders. Routine navigation is nearly mandatory; intraoperative ultrasound or iMRI may be helpful. Electrophysiological monitoring in eloquent areas is routine. We never used awake craniotomy, and we avoided surgery in speech areas. Radiosurgery is the preferred treatment approach in these cases. Individual surgical steps that should be considered in AVM surgery are summarised in Tables 20.1 and 20.2.

In the following paragraphs, technical aspects of AVM surgery will be described and some graphically illustrated.

AVM surgery is not the subject of minimally invasive approaches. The approach and the patient positioning should be chosen and done to have an AVM on the surface and the long axis of the nidus vertical (Fig. 20.2, Points 1, 2, 3). The craniotomy should be large enough, well beyond the nidus margins. This ensures good orientation; the surface feeders and surface draining veins are more easily identified and the sulci delineated.

The dura should be opened widely and carefully. Adhesions between the AVM surface and dura are frequent, and these should be dealt with sharply while reflecting the dural flap carefully and gradually. Because these adhesions

Table 20.1 Summary of key steps in AVM surgery

AVM surgery step-by-step 1
1. Large enough craniotomy
2. AVM on the surface
3. Along the long axis, long axis vertical
4. Open the dura carefully
5. Correlation between the angiography and surgical field
6. Identification of borderline sulci
7. Dissect arachnoid circumferentially, respect sulci
8. Begin at the arterial side
9. Dissect en passage arteries
10. Coagulate patiently and thoroughly
11. Cut the coagulated vessels stepwise
12. Use miniclips
13. Non-stick and high-power coagulation seldom

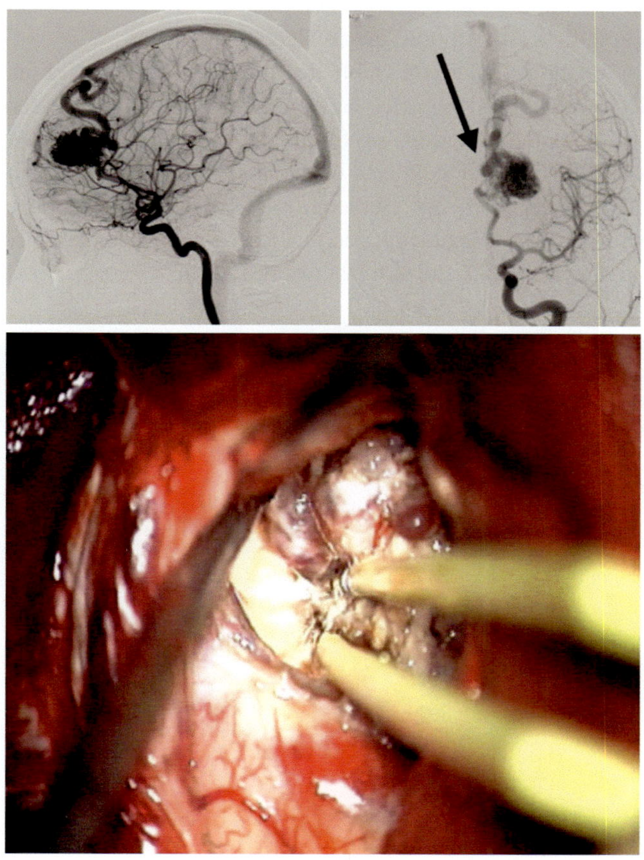

Fig. 20.2 Left frontal AVM. The AVM is approached from the right side. The falx is cut and reflected by stitches. Without retraction, the nidus is on the surface. The approach is in the axis of the nidus (*arrow*)

Table 20.2 Continued summary of key steps in AVM surgery

AVM surgery step-by-step 2
14. Keep the plane of dissection
15. Identify and deal with any bleeding before further resection. Keep the surgical field clean and transparent at all times
16. Don't turn too early underneath the nidus
17. Take extreme care with deep feeders
18. Have important structures constantly under control
19. Along the vein exceptionally
20. Choose the proper vein as the last target
21. Use ICG
22. Coagulate and cut the last draining vein
23. Thorough haemostasis before closure
24. High magnification/general overview
25. Immediate postoperative angiography under the same anaesthesia

are usually between the dura and the draining veins, it is important to be especially careful (Fig. 20.3, Point 4). Otherwise, the AVM resection may start with brisk, hardly controllable bleeding. Occluding the major draining vein at this stage would be disastrous.

As soon as the dura is opened, the correlation between the angiography and surgical field must be established, identifying the feeders, the draining veins, and the nidus margins. ICG, at this point, is helpful (Fig. 20.4, Point 5). The arachnoid is cut circumferentially, and the dissection continues in a sulcal fashion, keeping the proper dissection plane as long as white matter in depth is entered (Points 6, 7). This procedure does not apply to deep-seated AVMs; in these, the sulcus of approach is to be identified.

The nidus dissection should start at the arterial side, and the main feeders are identified and occluded, usually at this stage, by clips (Fig. 20.5, Point 8).

En pasage arteries must be identified and preserved. Generally, any artery in the surgical field is treated as en passage unless proven otherwise (Figs. 20.6 and 20.7). Any larger AVM vessel is coagulated patiently and thoroughly and then cut stepwise. First, opening the lumen with the vessel in continuity and only if the vessel does not bleed the rest is cut (Fig. 20.8, Points 9, 10). If the vessel is cut entirely and if it bleeds, the stump may retract into the brain tissue, making it difficult to follow such a bleeding source.

Regular bipolar is a viable instrument in AVM resection. High-power and nonstick bipolar are seldom used. In the white matter, mini clips are frequently used (Fig. 20.9, Point 12). These pathological and tortuous vessels with an insufficient vessel wall are challenging to coagulate, and once injured, it is extremely difficult to control the bleeding. Only on rare occasions and as a last resort, the "dirty" coagulation technique with high-power bipolar is employed to control bleeding.

Any bleeding during resection is dealt with immediately and permanently. If not, these bleeding points tend to multiply, making the haemostasis difficult and causing a worse orientation in the surgical field (Fig. 20.10, Point 14). The surgical field must be clean and anatomically transparent at all times. The plane of dissection is followed carefully; only if needed is the surrounding brain tissue entered (Fig. 20.2c, Point 15). It is important not to turn too early underneath the nidus, and there is no shame in measuring the nidus on preoperative MR and during the surgery to measure the depth of resection.

Many AVMs are conical, with the top at the ventricle wall. At this point, the feeder from the ventricle wall is invariably present. Such a vessel must be actively sought and identified. Occluding this vessel with a mini clip is appropriate (Fig. 20.11, Point 17).

It also is vital to have important structures under direct control at all times (Fig. 20.12, Point 18.). Rarely is it necessary to approach the nidus along the draining vein, but in some cases, this is the only way. Then the vein is carefully followed, and all the vessels entering the vein are dissected free and occluded (Fig. 20.13, Point 19).

An appropriate, usually the largest, vein is identified and occluded as last (Fig. 20.14, Points 14, 22). ICG is used to prove the proper resection (Point 21). However, any remnants obscured by the brain tissue escape ICG inspection. The high magnification of either microscope or exoscope is used throughout the surgery. However, it is frequently zoomed out to keep the general orientation and prevent the injury of the vessels, especially the nidus (Point 24).

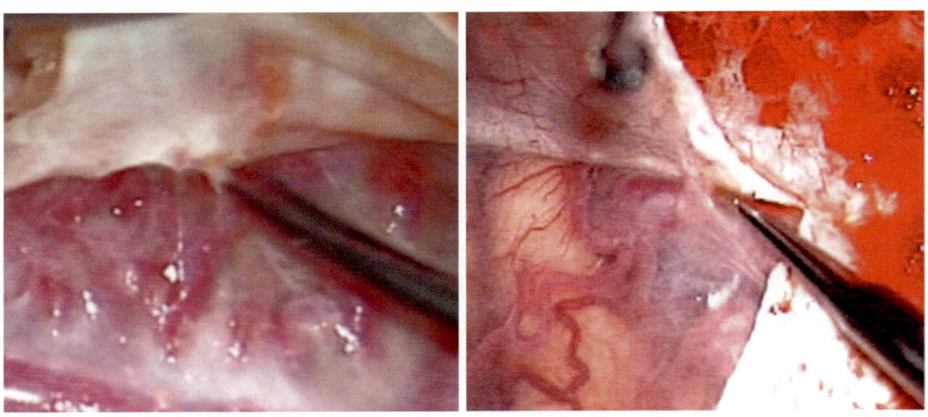

Fig. 20.3 Two AVMs. The adhesions between the dura and the nidus should be sharply divided

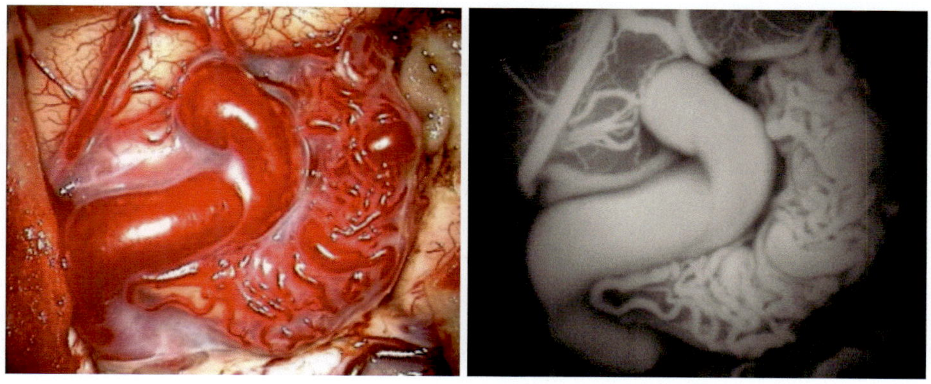

Fig. 20.4 The arachnoid on one side of the nidus is already dissected and the sulcus opened. ICG correlates with the surgical field, nicely depicting the margins and involved vessels

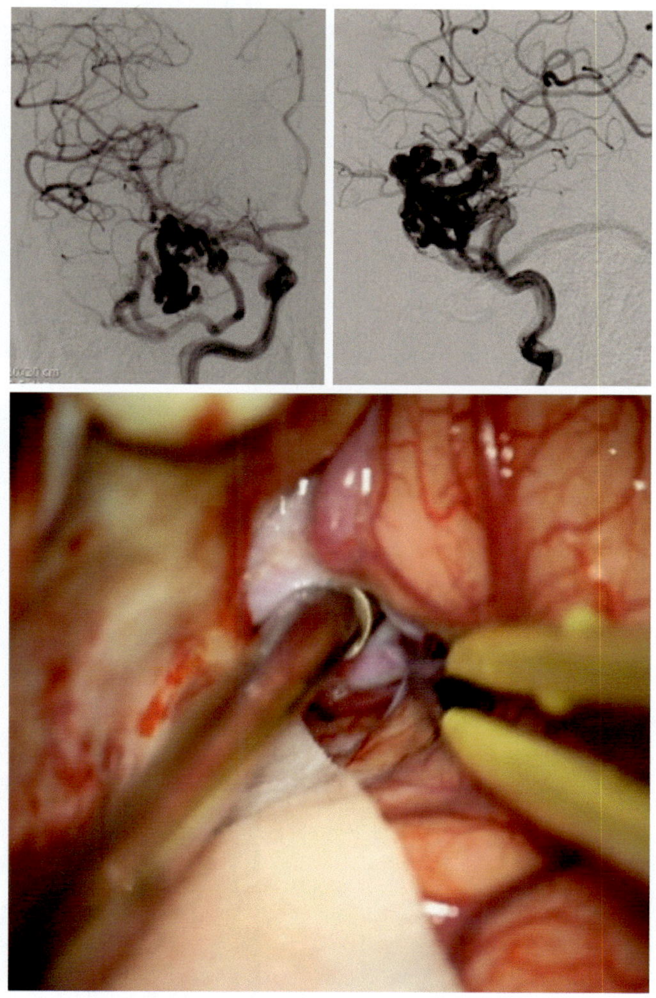

Fig. 20.5 Left temporal AVM. Feeders from the M1 and M2, draining veins more superficially. The dissection starts with Sylvian fissure dissection and identification of arterial feeders

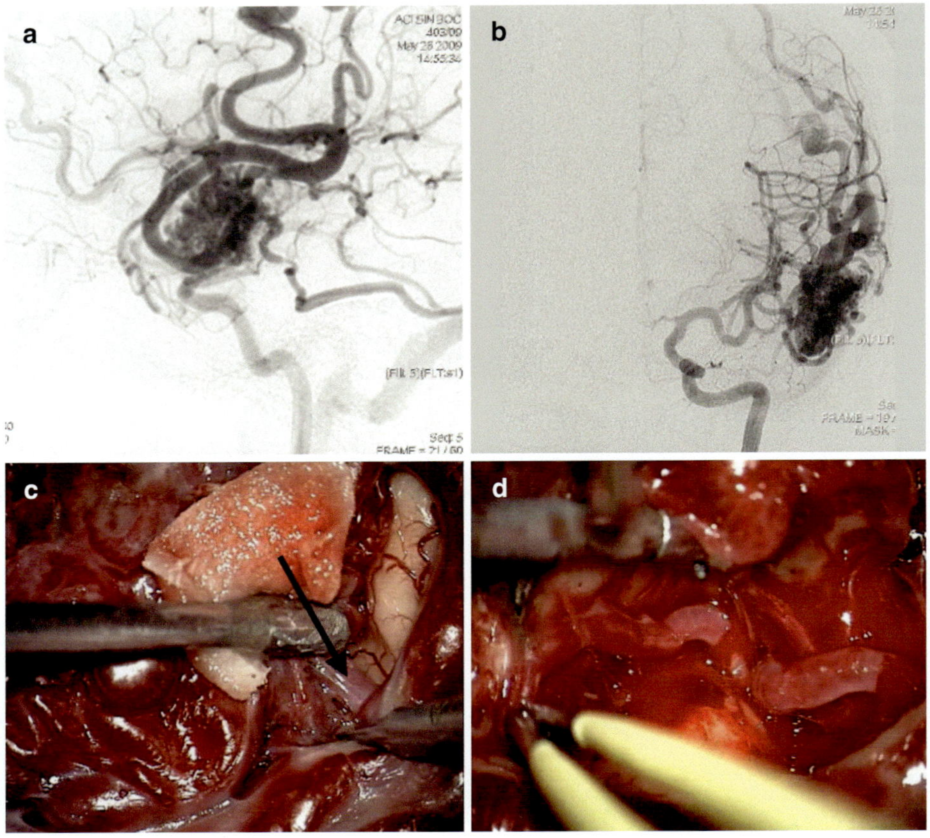

Fig. 20.6 Left temporal AVM. (**a**, **b**) Angiography. (**c**) The artery mimics the feeder (*arrow*). (**d**) It is apparent later that this artery is a major MCA branch

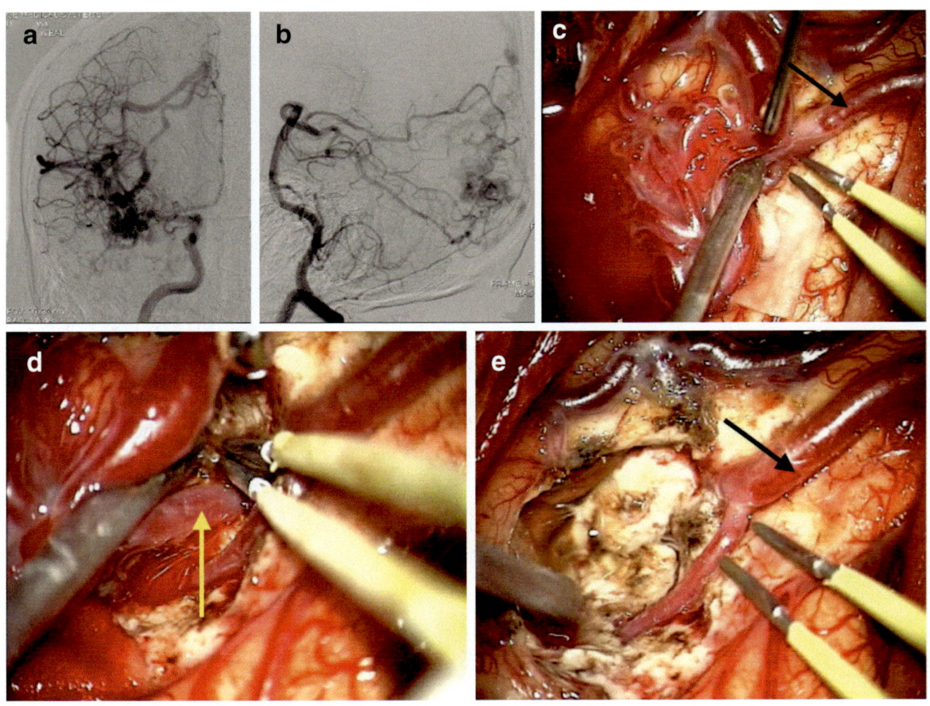

Fig. 20.7 Right occipital AVM. (**a**, **b**) Angiography. (**c**) The artery appears as an arterial feeder. (**d**) The true feeder is seen later on (*yellow arrow*). (**e**) The situation after the resection when the original "feeder-like"artery is obvious to have been an en passage artery (*black arrow*)

Thorough haemostasis before closure is mandatory; taking a 15–20-min break before dura closure may be reasonable. In our settings, control angiography is performed immediately after closure under the same anaesthesia (Point 25). The patient is then allowed to wake up slowly and spontaneously. Direct arterial line allows for continuous dealing with blood pressure changes. The patients are kept strictly normotonic or slightly hypotonic throughout the surgery and the next 24 h. Any blood pressure elevation is dealt with immediately.

Fig. 20.8 Same AVM as in Fig. 20.7. The main feeder is coagulated, and the first cut only opens the lumen (*arrow*)

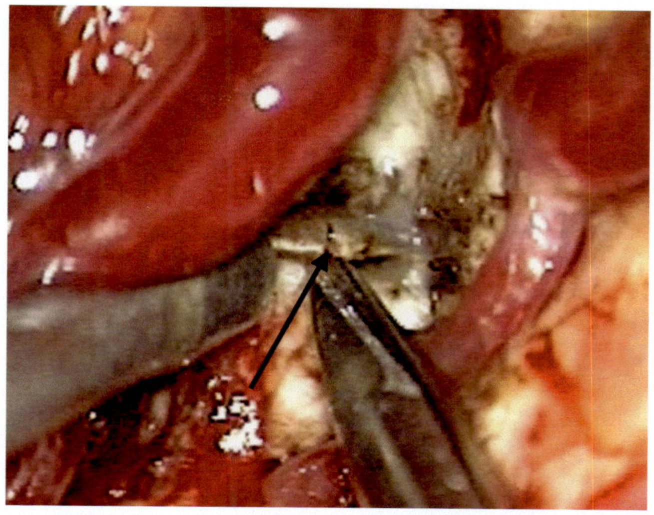

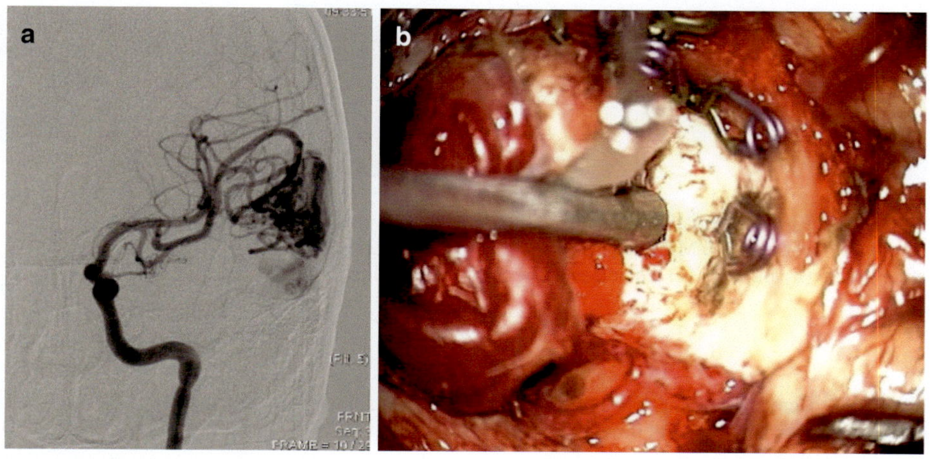

Fig. 20.9 Typical cone-shape AVM left temporal. Mini clips are used to control the bleeding from the deep feeders

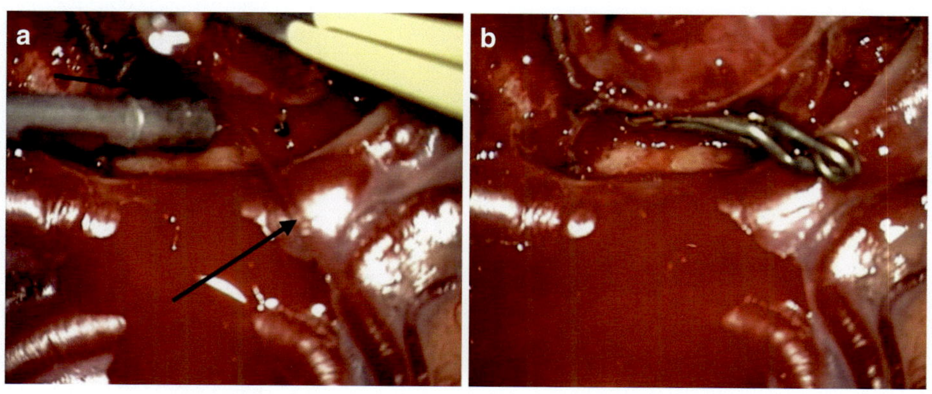

Fig. 20.10 The same case as in Fig. 20.6. (**a**) Shows the bleeding from the deeper draining vein with the jet of blood (*arrows*). (**b**) Tangentially applied clip to control the bleeding

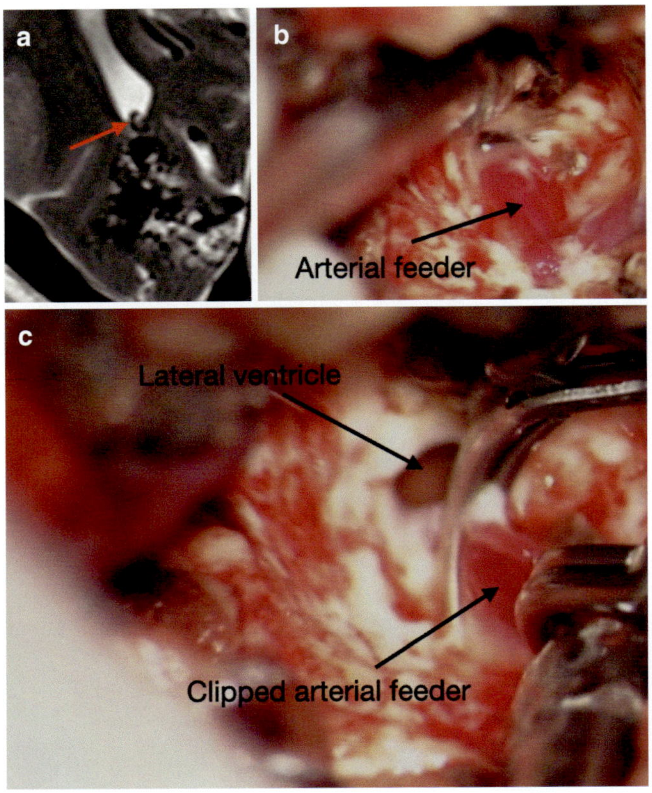

Fig. 20.11 Right occipital AVM. (**a**) Identification of the ventricular feeder (*arrow*) on the T2 MR image. (**a** and **b**) Depict the surgical situation before (**b**) and after (**c**) the occlusion by the two mini clips

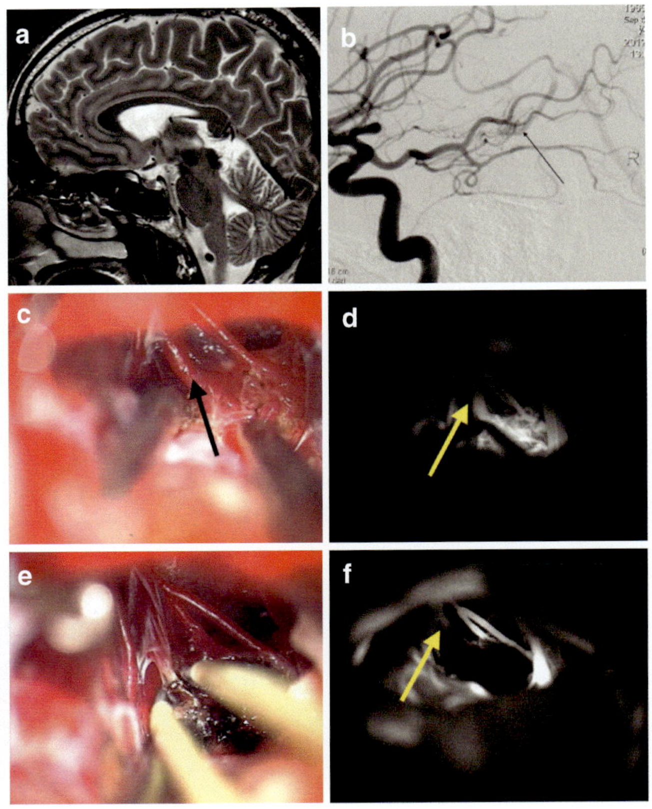

Fig. 20.13 Quadrigeminal AVM. (**a**, **b**) Preoperative MR and angiography. (**c**, **d**) Illustrate the draining vein seen through a microscope and ICG (*arrows*). It is impossible to distinguish the feeders and normal mesencephalic perforators. (**e**, **f**) The vein is coagulated last on ICG. The stump is seen (*arrow*), and the normal arteries are preserved

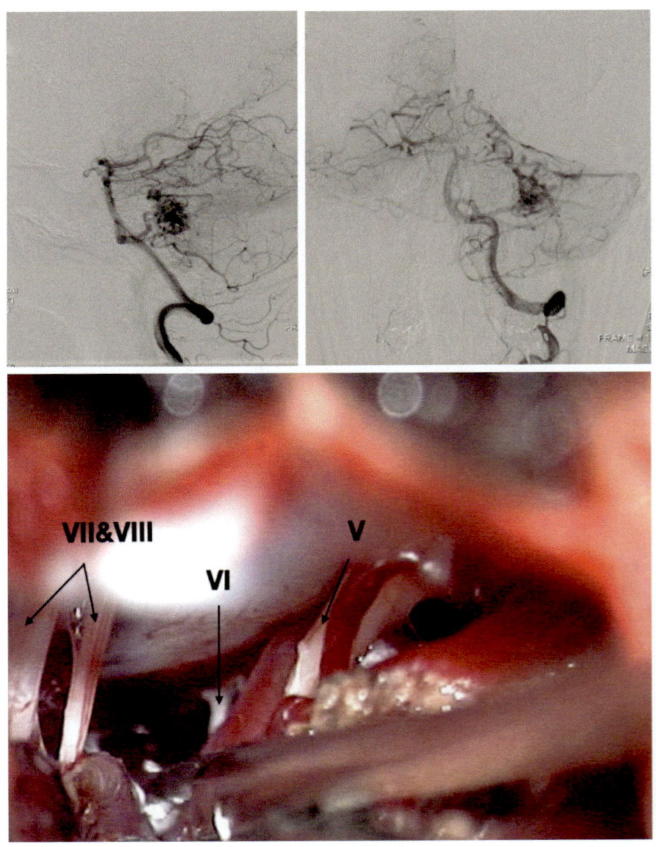

Fig. 20.12 Cerebellopontine AVM. Vth, VIth and VII/VIIIth cranial nerves are under visual control throughout the surgery

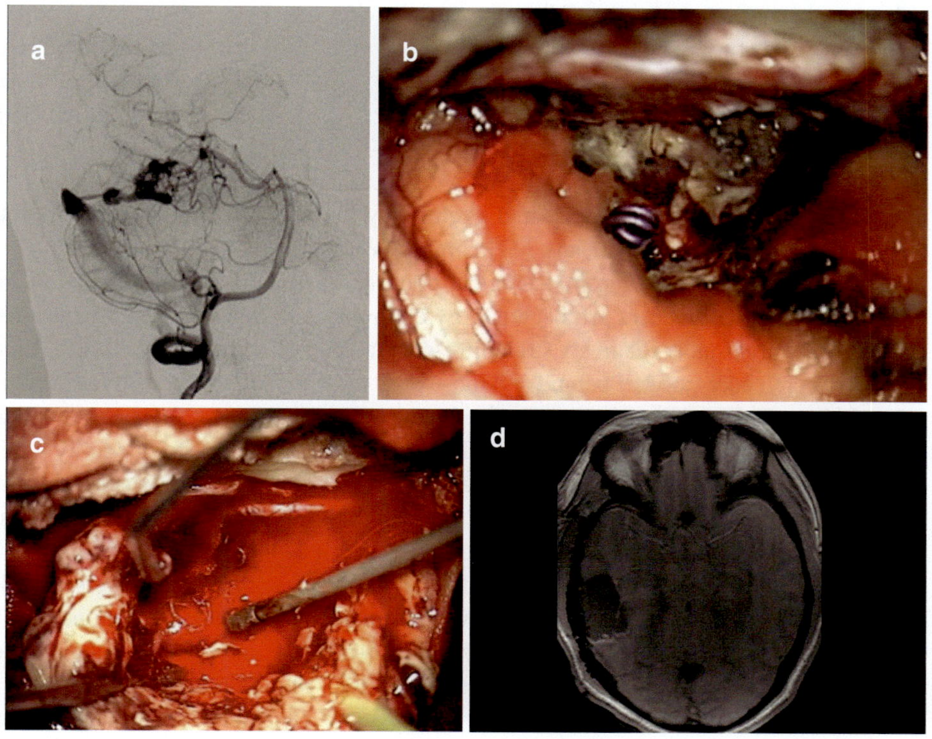

Fig. 20.14 Right temporal AVM. (**a**) Angiography. (**b**) Routine resection with a relaxed brain. The main AVM vein draining into the transverse-sigmoid sinus junction was injured and occluded. (**c**) Shows the brain prolapse and incessant bleeding; (**d**) illustrates the unnecessary final extent of resection. The patient is neurologically intact

20.5 Surgical Pearls

- Positioning and the extent of craniotomy are vital in any AVM surgery.
- Careful dura opening.
- Any artery in the vicinity of the nidus is to be treated as en passage unless proven otherwise.
- Patient coagulation and mini clips should be used for deep feeders.
- Thorough haemostasis throughout the whole surgery is vital.

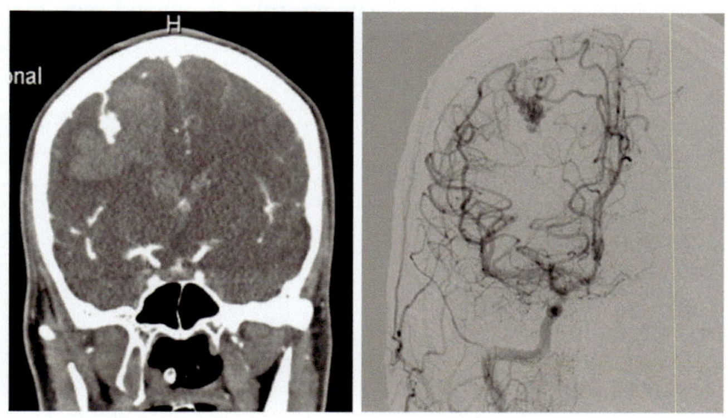

Fig. 20.15 Right frontal AVM with intracerebral hematoma. This case is suitable for immediate surgical intervention

20.6 Discussion

Most AVMs are hemispheric and reach the brain surface. However, AVMs are variable and the lesion may be located anywhere within the central nervous system. The literature is rich in describing the treatment at various locations. Somewhat more frequent are cerebellar and ventricular AVMs. Cerebellar AVMs are surgically treated the same way as hemispheric AVMs, with the provision that we do not have sulci to lead us during the resection (Fig. 20.12). Intraventricular AVMs are best approached through a routine approach, frontal to the anterior part of the ventricular system and parietal to the trigone regions. The resection should, as always, be started with the feeder occlusion, which in the case of ventricular AVMs, is some tributary of choroid arteries. The fragility of this feeder is similar to that of the pedicle of hemispheric AVMs, which reaches the ventricle.

20.6.1 Timing of Surgery

In ruptured AVMs subarachnoid haemorrhage is infrequent. The most frequent haemorrhage is represented by intracerebral hematoma. For AVM S-M grades 1 and 2, the surgery may be immediate in cases of raised intracranial pressure (Fig. 20.15). In all other cases, we prefer to wait some 2–3 weeks. Acute surgery means obscured anatomy, much worse orientation and more difficult control of bleeding in the malatic brain. In some cases of higher grades of AVMs only the hematoma can be carefully removed, leaving the AVM behind for further consideration.

20.6.2 Endovascular Treatment

Unlike dural AVMs, which are treated exclusively at our institution by endovascular means [17], brain AVMs are rarely embolised by Onyx. From a surgical point of view, predominantly, main feeders are occluded by endovascular means. These are usually relatively straightforward targets in surgery. Deep feeders are left behind and those pose the biggest problem. Unlike the soft non-embolised nidus, the bulk of vessels filled by Onyx is much more difficult to handle and manipulate. And the deep feeders are at a depth hidden by the hardly moveable mass of the nidus. Also, embolised vessels are difficult to coagulate, and even the vessels full of Onyx can bleed (Fig. 20.16). However, in the author's opinion, the reason not to use preoperative endovascular methods is given by the meta-analysis results [12]. The risk of surgery is 7% and the risk of each endovascular session is also 7%. Thus, the patient is exposed to the overall 14% risk, which importantly (and negatively) affects the long-term benefit (see Fig. 20.1). Although some authors published significantly higher proportions of successfully managed patients via presurgical embolisation followed by microsurgical resection [18], there is a lack of data to elucidate the benefits of preoperative embolisation and more extensive prospective studies are required [19].

20.6.3 Radiosurgical Treatment

Up-to-date, there are many radiosurgical techniques radiosurgical instruments that are used (Gamma Knife most commonly, Cyberknife, Linear Accelerator, etc.). Since the first case report of radiosurgically managed AVM by Steiner et al. in 1972 [xx], radiosurgery has become a well-established treatment modality for cerebral AVMs. Minimal invasivity is the main benefit of radiosurgical intervention, particularly in cases of eloquently seated AVMs where surgical risks are unacceptably high, but still harbour not negligible risk of bleeding. Moreover, older patients and/or patients with various comorbidities that would increase the risk of surgical intervention may be radiosurgical candidates as well. The obliteration rate of low S-M grade is high, up to 90% with minimal mortality and morbidity rates. Nevertheless, along with increasing S-M grade, the efficacy of radiosurgical treatment tended to decrease, with simultaneously increasing complication rates [85].

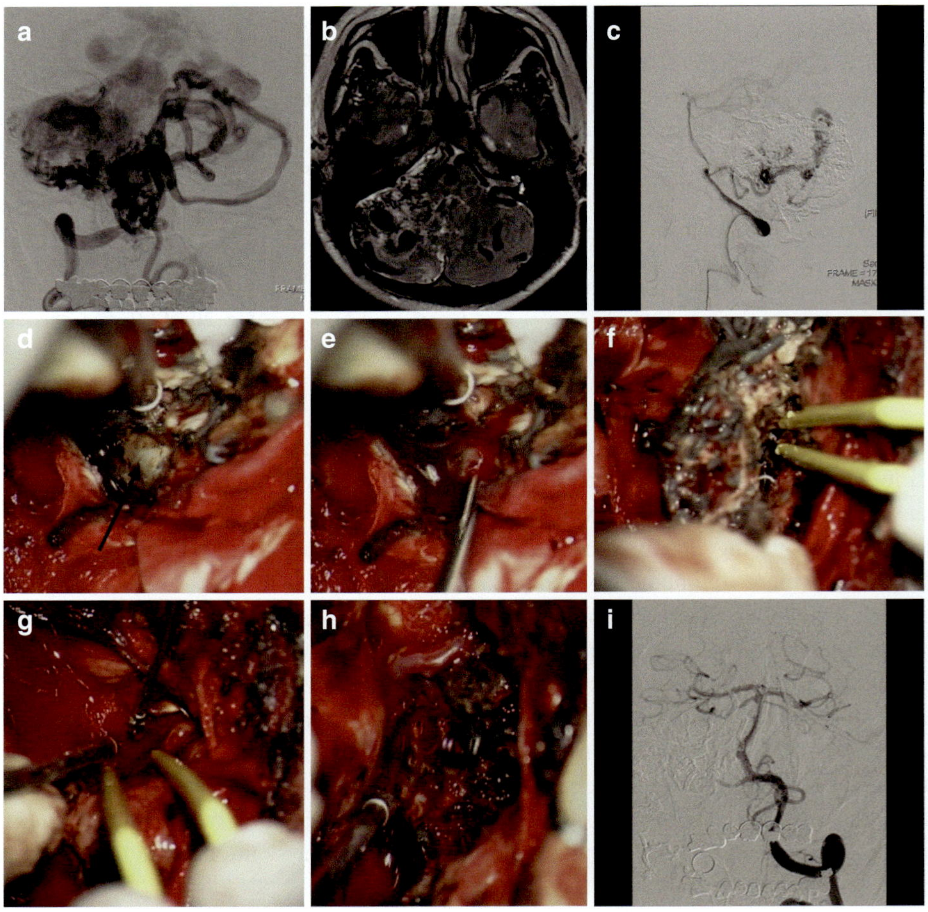

Fig. 20.16 (**a**, **b**) Large cerebellar AVM and, after the six embolisation attempts over 5 years, was finally (**c**) suitable for surgery. (**d**) The large embolised draining vein is seen. (**e**) After the vessel was coagulated and opened, it still bled profusely. (**f**) Manipulation of the embolised nidus is difficult and orientation is obscured. (**g**, **h**) In the craniocervical junction where the feeders from PICA are present, the bleeding was permanent, profuse, and hardly controllable. (**i**) The postoperative angiography. No new postoperative neurological deficits

20.6.4 Other Treatment Combinations

A promising combination would be surgical treatment after the failed radiosurgery. However, such an approach has never been studied, and there is only a subjective experience that radiosurgery before surgical resection of the AVM facilitates surgical outcomes and decreases operative morbidity. The author had an opportunity to treat several patients surgically after the failed radiosurgical treatment, and all these procedures were surprisingly straightforward. Generally, surgery should be recommended for irradiated AVMs with the following characteristics: (1) no signs of complete obliteration after the latency period of at least 3 years, (2) increasing in size, and/or (3) causing significant clinical deterioration. Radiosurgically treated patients with AVMs are exposed to a higher risk of haemorrhage during the latency period [12]. Moreover, the risk of AVM rupture may be as high as 8% within the first year after radiosurgical intervention [20]. Whether surgery should be promptly offered after radiosurgical intervention needs to be studied prospectively in more detail [21].

Our surgical experience with brain AVMs during the past 20 years has shown 97.3% efficacy in the completeness of AVM's radical resection with an overall 2.6% mortality/mortality rate. Of 188 surgically treated patients, three died and two developed de novo major neurological deficits (both groups comprised only AVMs of S-M grade III and IV). Three surgically treated AVMs partially resected were subsequently sent to Gamma Knife and two were post-surgically embolised. The surgical efficacy in solving S-M grade I and II AVMs has been 100%, with zero mortality and morbidity rates, clearly reflecting similar findings [10]. The efficacy of S-M grade III cases has been 95.5% (mortality/morbidity 6.3%) and of S-M grade IV/V cases 71.4% effectivity (mortality/morbidity 27.0%), respectively.

20.7 Conclusion

Well-established treatment algorithms play an essential role in the decision-making of active treatment policy in AVM management. The natural course of the disease is dynamic, and its detailed understanding is vital for surgical management. Overall mortality and morbidity rates after surgery have decreased thanks to advances in imaging modalities, treatment alternatives, and better experience with the disease. Surgery of AVMs is very complex. However, if performed with a high level of expertise, adequate indication (AVM characteristics, patient's demographic factors, institutional experience, etc.), and proper technical management, the risk of poor outcome is low while minimising the risk of bleeding, which might be fatal if left untreated. Every active treatment policy should be individualised to the needs of the patients and their AVM(s) while always bearing in mind that the risks of surgery should not exceed the benefits of the natural history of the disease or other treatment alternatives.

Supported by programme MO1012.

References

1. McCormick WF. The pathology of vascular ("arteriovenous") malformations. J Neurosurg. 1966;24(4):807–16.
2. Stapf C, Labovitz DL, Sciacca RR, Mast H, Mohr JP, Sacco RL. Incidence of adult brain arteriovenous malformation hemorrhage in a prospective population-based stroke survey. Cerebrovasc Dis. 2002;13(1):43–6. https://doi.org/10.1159/000047745.
3. Stapf C, Mast H, Sciacca RR, Berenstein A, Nelson PK, Gobin YP, et al. The New York Islands AVM Study: design, study progress, and initial results. Stroke. 2003;34(5):e29–33. https://doi.org/10.1161/01.Str.0000068784.36838.19.
4. Al-Shahi R, Bhattacharya JJ, Currie DG, Papanastassiou V, Ritchie V, Roberts RC, et al. Prospective, population-based detection of intracranial vascular malformations in adults: the Scottish Intracranial Vascular Malformation Study (SIVMS). Stroke. 2003;34(5):1163–9. https://doi.org/10.1161/01.Str.0000069018.90456.C9.
5. Brown RD Jr, Wiebers DO, Torner JC, O'Fallon WM. Frequency of intracranial hemorrhage as a presenting symptom and subtype analysis: a population-based study of intracranial vascular malformations in Olmsted Country, Minnesota. J Neurosurg. 1996;85(1):29–32. https://doi.org/10.3171/jns.1996.85.1.0029.
6. Spetzler RF, Martin NA. A proposed grading system for arteriovenous malformations. J Neurosurg. 1986;65(4):476–83. https://doi.org/10.3171/jns.1986.65.4.0476.
7. Lawton MT, Kim H, McCulloch CE, Mikhak B, Young WL. A supplementary grading scale for selecting patients with brain arteriovenous malformations for surgery. Neurosurgery. 2010;66(4):702–13. https://doi.org/10.1227/01.NEU.0000367555.16733.E1. discussion 713. PMID: 20190666; PMCID: PMC2847513
8. Kim H, Abla AA, Nelson J, McCulloch CE, Bervini D, Morgan MK, et al. Validation of the supplemented Spetzler-Martin grading system for brain arteriovenous malformations in a multicenter cohort of 1009 surgical patients. Neurosurgery. 2015;76(1):25–31.; discussion –2; quiz 2–3. https://doi.org/10.1227/neu.0000000000000556.
9. Spetzler RF, Ponce FA. A 3-tier classification of cerebral arteriovenous malformations. Clinical article. J Neurosurg. 2011;114(3):842–9. https://doi.org/10.3171/2010.8.Jns10663.
10. Beneš V, Bradáč O. Brain arteriovenous malformations: pathogenesis, epidemiology, diagnosis, treatment and outcome. Springer; 2017.
11. Mohr JP, Parides MK, Stapf C, Moquete E, Moy CS, Overbey JR, et al. Medical management with or without interventional therapy for unruptured brain arteriovenous malformations (ARUBA): a multicentre, non-blinded, randomised trial. Lancet. 2014;383(9917):614–21. https://doi.org/10.1016/s0140-6736(13)62302-8.
12. Davidson AS, Morgan MK. How safe is arteriovenous malformation surgery? A prospective, observational study of surgery as first-line treatment for brain arteriovenous malformations. Neurosurgery. 2010;66(3):498–504.; discussion –5. https://doi.org/10.1227/01.Neu.0000365518.47684.98.
13. Börcek A, Çeltikçi E, Aksoğan Y, Rousseau MJ. Clinical outcomes of stereotactic radiosurgery for cerebral arteriovenous malformations in pediatric patients: systematic review and meta-analysis. Neurosurgery. 2019;85(4):E629–e40. https://doi.org/10.1093/neuros/nyz146.

14. Yaşargil M. Microneurosurgery: AVM of the brain: clinical considerations, general and special operative techniques, surgical results, nonoperated cases, cavernous and venous angiomas, neuroanesthesia. Stuttgart: Thieme; 1988.
15. Schramm J, Schaller K, Esche J, Boström A. Microsurgery for cerebral arteriovenous malformations: subgroup outcomes in a consecutive series of 288 cases. J Neurosurg. 2017;126(4):1056–63. https://doi.org/10.3171/2016.4.jns153017.
16. Han PP, Ponce FA, Spetzler RF. Intention-to-treat analysis of Spetzler-Martin grades IV and V arteriovenous malformations: natural history and treatment paradigm. J Neurosurg. 2003;98(1):3–7. https://doi.org/10.3171/jns.2003.98.1.0003.
17. Voldřich R, Netuka D, Charvát F, Beneš V. Long-term stability of Onyx: is there any indication for repeated angiography after dural arteriovenous fistula embolisation? J Neurosurg. 2021;136(1):175–84. https://doi.org/10.3171/2020.12.JNS203811.
18. Saatci I, Geyik S, Yavuz K, Cekirge HS. Endovascular treatment of brain arteriovenous malformations with prolonged intranidal Onyx injection technique: long-term results in 350 consecutive patients with completed endovascular treatment course. J Neurosurg. 2011;115(1):78–88. https://doi.org/10.3171/2011.2.Jns09830.
19. Brosnan C, Amoo M, Javadpour M. Preoperative embolisation of brain arteriovenous malformations: a systematic review and meta-analysis. Neurosurg Rev. 2022; https://doi.org/10.1007/s10143-022-01766-8.
20. Zabel-du Bois A, Milker-Zabel S, Huber P, Schlegel W, Debus J. Risk of hemorrhage and obliteration rates of LINAC-based radiosurgery for cerebral arteriovenous malformations treated after prior partial embolisation. Int J Radiat Oncol Biol Phys. 2007;68(4):999–1003. https://doi.org/10.1016/j.ijrobp.2007.01.027.
21. Sanchez-Mejia RO, McDermott MW, Tan J, Kim H, Young WL, Lawton MT. Radiosurgery facilitates resection of brain arteriovenous malformations and reduces surgical morbidity. Neurosurgery. 2009;64(2):231–8.; discussion 8-40. https://doi.org/10.1227/01.Neu.0000338068.44060.Ea.

Arteriovenous Malformation Involving the Broca's Area and Precentral Gyrus

21

Matias Baldoncini, Alvaro Campero, and Manuel Encarnacion Ramirez

Learning Objectives

- The content delves into the anatomy and functions of the Broca's area and precentral gyrus, emphasizing their relevance when affected by AVM.
- Readers will be guided through the typical clinical manifestations, diagnostic modalities, and various therapeutic approaches. Special attention will be dedicated to surgical considerations and challenges in these regions.

21.1 Introduction

In the complex neural network of the human brain, one particular region of immense significance is the Broca's area, named after the French physician Paul Broca, who in the 1860s made the groundbreaking discovery linking this specific region to speech production [1]. Broca's area is located in the frontal lobe of the dominant cerebral hemisphere, which is typically the left hemisphere in most right-handed and about 70% of left-handed individuals. It encompasses two main regions in the frontal lobe: the pars opercularis and pars triangularis, identified as Brodmann areas 44 and 45, respectively. These areas are part of the prefrontal cortex, which is associated with higher cognitive functions, including problem-solving, decision-making, and social behavior. While the Broca's area is primarily associated with speech production, its role extends beyond just this function. The area is responsible for various aspects of language comprehension, articulation, word choice, syntax, and grammatical construction. Damage to this area can result in speech and language deficiencies, which are collectively referred to as Broca's aphasia [2].

Moreover, the region's involvement in sign language among the deaf population indicates its role in the general processing and production of language, whether it is spoken, written, or signed. This highlights the versatility and adaptability of the human brain.

The functionality of the Broca's area is not isolated. It works in close conjunction with other regions of the brain, particularly those associated with language processing. This includes Wernicke's area (involved in language comprehension), the primary motor cortex (which controls the physical aspects of speech), and the angular gyrus (involved in reading and writing). Together, these areas form a complex network that allows for the nuanced use of language, an essential aspect of human cognition and social interaction.

Symptoms of an AVM in Broca's area can range from subtle to severe, often depending on the size of the malformation and its impact on surrounding brain tissue. The primary clinical manifestation, when it occurs, is often a language disorder known as Broca's aphasia. This condition is characterized by nonfluent, effortful speech, impaired formation of complex grammatical sentences, and difficulty in language production. However, patients with Broca's aphasia typically retain their ability to comprehend language. Furthermore, patients may also experience seizures, headaches, or neurological deficits related to the location of the AVM. A high proportion of AVMs are discovered following a hemorrhagic event, such as an intracranial bleed, which can present as a sudden severe headache, neurological deficits, or in severe cases, loss of consciousness [2–4].

21.2 Anatomy Highlights

- AVMs involving the Boca's Area and Precentral Gyrus are located in the cerebral cortex of the brain. This area is associated with language processing, particularly expres-

M. Baldoncini (✉) · M. E. Ramirez
Laboratory of Microsurgical Neuroanatomy, Second Chair of Gross Anatomy, School of Medicine, University of Buenos Aires, Autónoma de Buenos Aires, Argentina

A. Campero
Department of Neurological Surgery, Hospital Padilla, Tucumán, Argentina

I. N. Kanaan, V. Beneš (eds.), *Neuroanatomy Guidance to Successful Neurosurgical Interventions*,
https://doi.org/10.1007/978-3-031-59838-8_21

sive language, while the Precentral Gyrus plays a crucial role in motor control.

- AVMs in this region are characterized by abnormal tangles of blood vessels (arteries and veins) that bypass the normal capillary network. These abnormal connections disrupt the usual blood flow and oxygen supply in the affected brain tissue.
- The symptoms of this AVMs' location can vary but often include seizures, motor deficits (such as weakness or paralysis in specific body parts), and language disturbances, which may manifest as speech difficulties or aphasia.
- A significant concern with AVMs is the risk of hemorrhage (bleeding) into the brain. If an AVM ruptures in this region, it can lead to a sudden and severe intracranial hemorrhage, potentially causing life-threatening complications.
- Emphasize with the patient the importance of early diagnosis and appropriate management of AVMs in this specific brain region due to their potential impact on language and motor functions, as well as the risk of hemorrhage.

21.3 Diagnosis of Arteriovenous Malformations in Broca's Area

The diagnosis of arteriovenous malformations (AVMs) in Broca's area, like other areas of the brain, can be a challenging process due to the highly variable clinical presentation. The diagnostic approach generally involves a combination of clinical evaluation, neuroimaging, and potentially invasive procedures such as cerebral angiography (Figs. 21.1, 21.2, and 21.3).

21.3.1 Clinical Evaluation

A detailed clinical evaluation is the first step in the diagnostic process of an AVM. Clinicians will obtain a thorough medical history, including the onset and progression of symptoms, the presence of seizures, headaches, or any neurological deficits. Particular attention is paid to symptoms suggestive of Broca's aphasia, such as nonfluent, effortful speech, and impaired sentence construction, given the AVM's location. The physical examination often includes a comprehensive neurological evaluation, assessing cognitive function, motor and sensory abilities, reflexes, balance and coordination, and vision. This evaluation will also include a detailed assessment of the patient's speech and language skills.

21.3.1.1 Neuroimaging

Neuroimaging is the cornerstone of AVM diagnosis and is used to confirm the presence of an AVM, evaluate its size, shape, and location, and assess its impact on the surrounding brain tissue.

Magnetic Resonance Imaging (MRI): MRI is an effective, noninvasive tool for identifying AVMs. It can provide detailed images of the brain's soft tissues, showing any abnormalities in the blood vessels and surrounding brain tissue. An MRI can also reveal any associated hemorrhage, brain edema, or infarction. Functional MRI (fMRI), a variant of the traditional MRI, can also be used to map the brain areas responsible for specific functions, such as speech and language in this case, providing critical information for planning any potential surgical intervention [5].

Computed Tomography (CT): A CT scan uses X-rays to produce cross-sectional images of the brain. It can quickly visualize a hemorrhage, making it especially valuable in emergency situations. However, it may be less effective than MRI at visualizing small AVMs or determining the exact location relative to other brain structures [6].

Cerebral Angiography: Also known as a cerebral arteriogram, this invasive test provides the most detailed images of the blood vessels in the brain. It involves the injection of a contrast dye into the blood vessels, followed by X-rays to visualize the blood flow through the arteries and veins. While it is highly accurate, it also carries a risk of complications such as stroke, making it generally reserved for situations where noninvasive imaging is inconclusive or in preparation for surgery or other interventions [5, 6].

21.3.2 Surgical Intervention

21.3.2.1 Treatment of Arteriovenous Malformations in Broca's Area

Patients with AVMs can present a range of symptoms, including headaches, seizures, or speech and language difficulties. Under conservative management, these symptoms are addressed individually:

Headaches: Analgesic drugs are typically prescribed for headache management. In some cases, prophylactic medications may be used to prevent headaches, especially migraines [4].

Seizures: Antiepileptic drugs (AEDs) are prescribed for patients presenting with seizures. AEDs help to reduce the frequency and severity of seizures, improving the patient's quality of life [4].

Speech and Language Difficulties: Since Broca's area is critical for speech and language, patients with an AVM in this location might experience related difficulties. Speech and language therapy may be recommended to help manage these issues. A trained therapist can provide exercises and strategies to improve communication abilities [3].

Close monitoring is a critical part of conservative management. This typically involves regular clinical evaluations

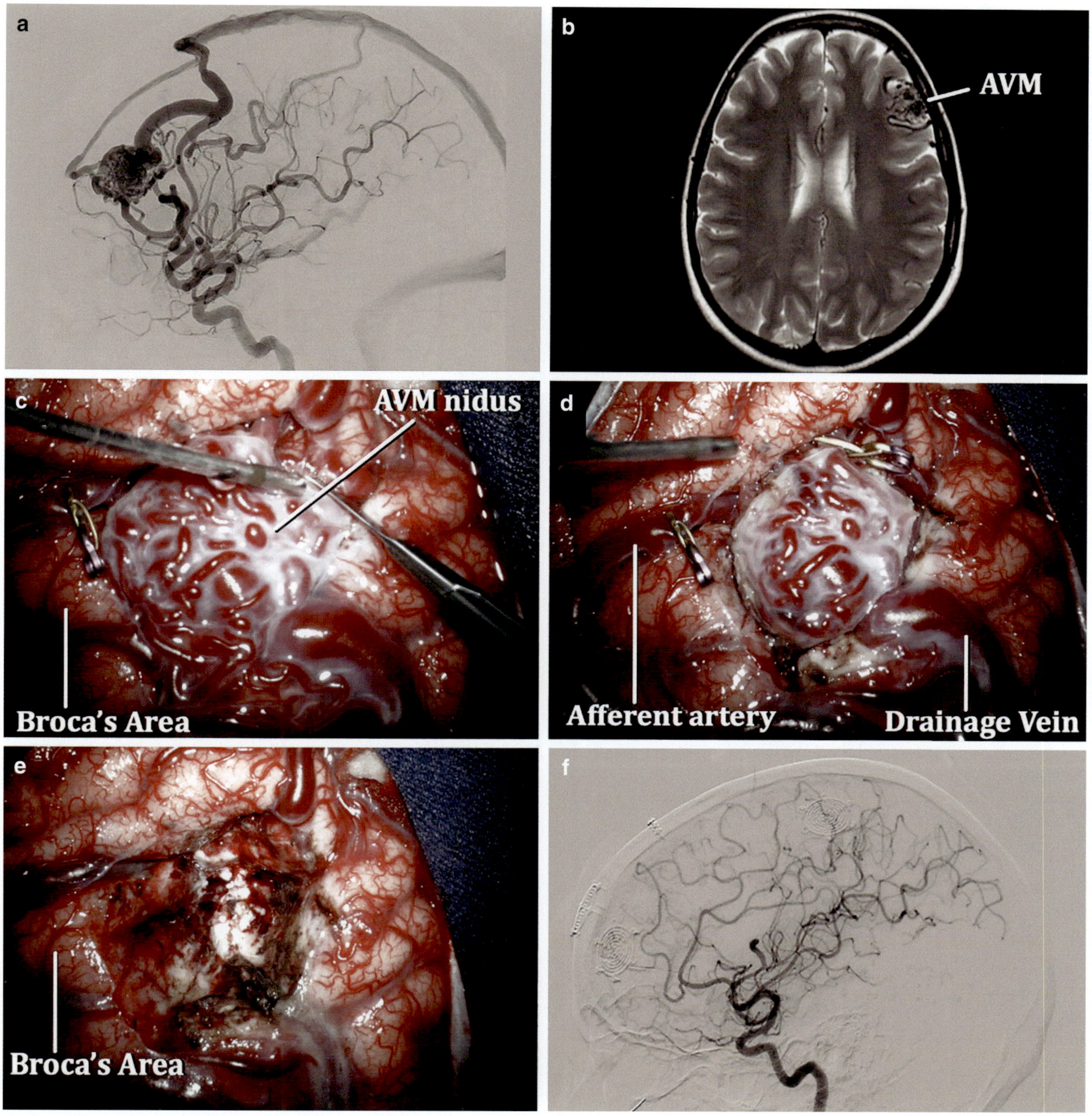

Fig. 21.1 (**a**) This depicts a left carotid artery angiography in profile view, showing an AVM located in the frontal area, with three feeders and two venous drainages. (**b**) In this MRI, the AVM is evident in the proximity of the Broca's area. (**c–e**) Highlighting the step-by-step process of microsurgical resection of the AVM, involving a 360-degree dissection, initial disconnection of feeders, and ultimately the venous drainage. (**f**) Postoperative angiography displaying complete resection

and repeated neuroimaging studies to check for any changes in the size, location, or characteristics of the AVM. Changes in symptoms, such as new onset or worsening of headaches, seizures, or neurological deficits, should also prompt further evaluation. Overall, conservative management is a strategy that aims to maximize the patient's quality of life while minimizing the risks associated with AVMs and their treatment. Each patient's situation must be individually assessed to determine the most suitable management approach [4].

Endovascular embolization is a minimally invasive procedure that has become an essential part of the treatment paradigm for brain arteriovenous malformations (AVMs). The primary goal of embolization is to reduce the size and blood flow of the AVM, thereby decreasing the risk of rupture and

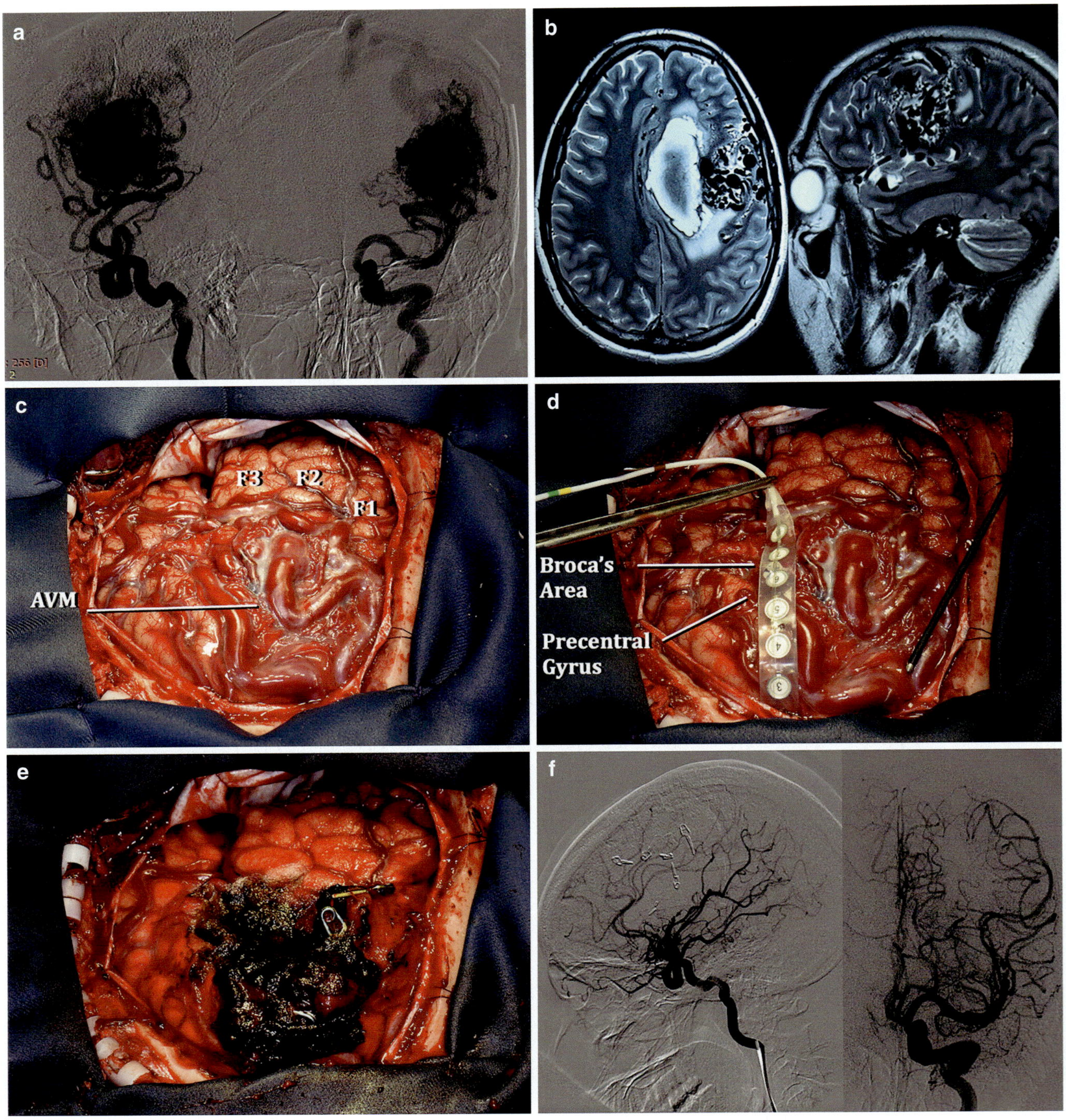

Fig. 21.2 (**a**) Profile and frontal angiography of the left internal carotid artery revealing a large 6 cm maximum diameter AVM. (**b**) MRI of the AVM located in the Broca's area and over the precentral gyrus, with an additional emphasis on medial hematoma due to prior bleeding. (**c–e**) Intraoperative images of the malformation, intraoperative motor area mapping, and the resection. (**f**) Postoperative angiography

facilitating subsequent interventions such as surgery or radiosurgery. Embolization can be used as a standalone treatment, particularly for smaller AVMs, or in conjunction with other treatments. When used in combination with surgical resection or stereotactic radiosurgery, embolization can reduce the size and complexity of the AVM, making the subsequent intervention safer and more effective. However, complete obliteration of the AVM solely with embolization is challenging and is achieved only in a small proportion of cases [7].

Microsurgical techniques allow for enhanced visualization, precision, and maneuverability when dealing with the minuscule and delicate structures within the brain. The Broca's area, being pivotal for speech production, and the precentral gyrus, playing a crucial role in voluntary motor functions, are areas where even a minor surgical mishap can

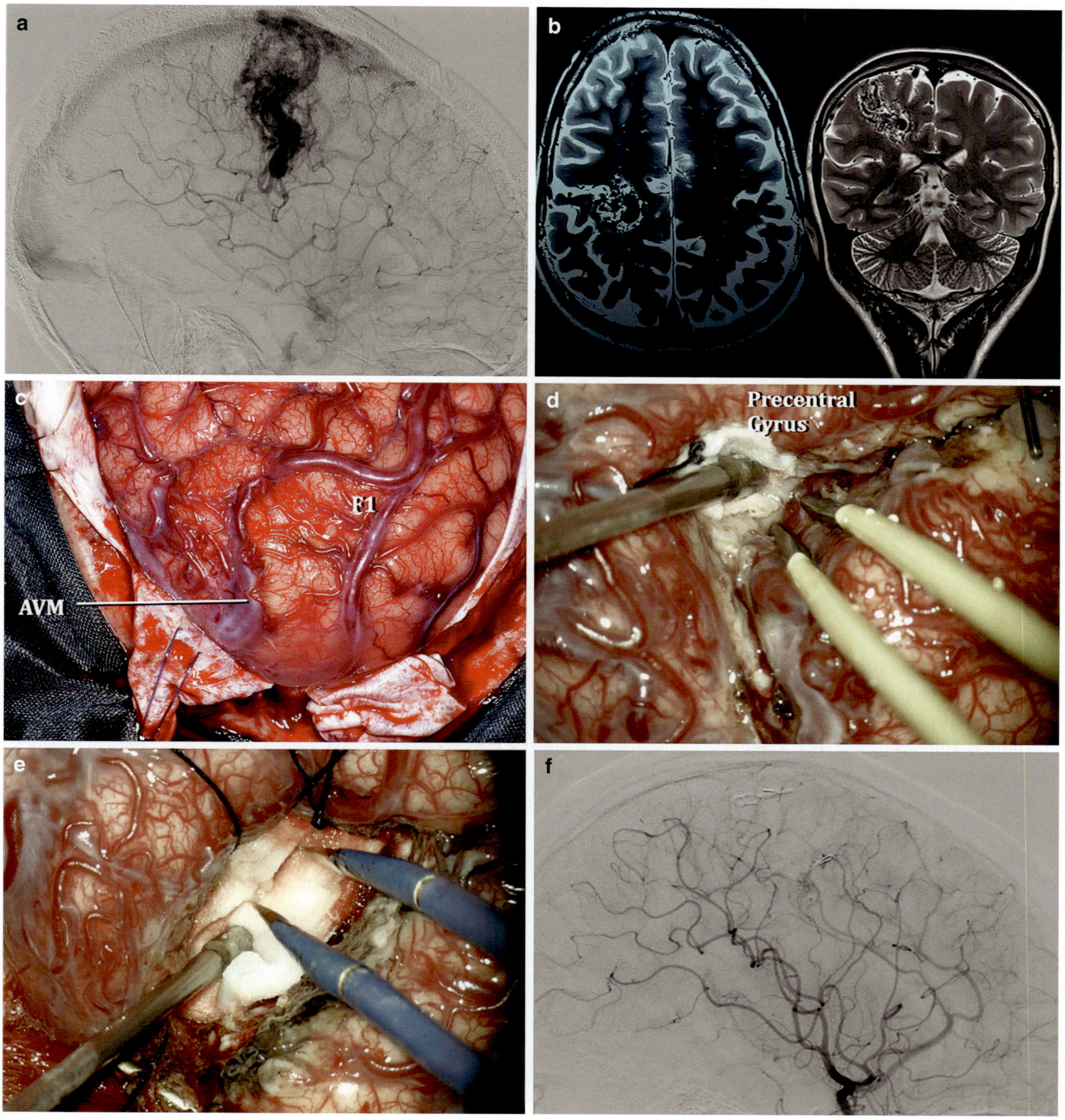

Fig. 21.3 (**a**) A right profile angiography showing a cone-shaped AVM. (**b**) MRI displaying a flow void of the AVM located in the right central sulcus. (**c–e**) Intraoperative images of the lesion resection, encompassing it on all margins. (**f**) Postoperative angiography confirming total resection

result in significant morbidity. Hence, the precision of microsurgery is invaluable.

Before delving into the surgical corridor, a robust preoperative plan is paramount. High-resolution MRI and CT angiography are essential to delineate the nidus, feeder arteries, and draining veins of the AVM. Often, digital subtraction angiography (DSA) is performed, which provides a dynamic picture of the AVM's flow dynamics and aids in visualizing any associated aneurysms. Functional MRI (fMRI) can be utilized to localize active functional areas around the lesion, providing a road map to avoid critical regions [4, 5].

The patient is positioned such that the surgical site is at the highest point, allowing gravity to aid in brain relaxation and minimizing the need for excessive retraction. An adequately sized craniotomy is performed to ensure good visualization and access. The bone flap is temporarily removed, and the dura is opened in a curvilinear or cruciate manner to expose the brain surface [4].

Under high magnification, the AVM's superficial component is identified. Microsurgical dissectors and bipolar cautery are used to isolate individual arterial feeders. These feeders are then either coagulated using bipolar diathermy or clipped using microvascular clips. This step is delicate; the surgeon must distinguish between the AVM's feeders and the normal brain vasculature.

The core of the AVM is the nidus, a tangle of abnormal blood vessels. Once the main feeders are controlled, dissection around the nidus begins. Using microscissors and fine-tipped forceps, the nidus is carefully separated from the surrounding brain tissue. This is a meticulous process; the goal is to preserve normal neural tissue and prevent rupture of the AVM. During this phase, intraoperative Doppler can be useful to assess the flow within the AVM and to confirm the complete obliteration of the nidus.

It is a cardinal rule in AVM microsurgery to deal with the draining veins last. This is because premature occlusion can lead to increased intranidal pressure, risking rupture. Once the nidus is completely excised, the draining veins are coagulated and divided.

Achieving meticulous hemostasis is critical to prevent postoperative hematoma. Bipolar coagulation, hemostatic agents, and surgicel are used to ensure the surgical bed is dry. The dura is sutured closed in a watertight fashion. The bone flap is then replaced and secured using titanium plates and screws. The scalp is sutured in layers to achieve a neat closure.

After the procedure, patients are typically transferred to the neurosurgical intensive care unit for close monitoring. Immediate postoperative imaging is done to ensure the complete removal of the AVM and to check for complications. Given the location near the Broca's area and the precentral gyrus, postoperative neurorehabilitation, including speech therapy and physiotherapy, may be beneficial to aid in recovery.

21.4 Surgical Pearls

- Always ensure thorough imaging. High-resolution MRI, CT angiography, and digital subtraction angiography (DSA) are invaluable.
- Use functional MRI (fMRI) or Diffusion Tensor Imaging (DTI) when operating near eloquent cortex to plan your surgical trajectory.
- *Elevate the head to promote venous drainage.*
- Optimal positioning minimizes the need for retraction and optimizes exposure. For example, in aneurysm surgery, position the head such that the aneurysm is at the highest point.
- Utilize intraoperative ICG angiography to confirm the completeness of AVM resection or to ensure an aneurysm is fully clipped and that surrounding vessels are patent.
- Intraoperative Doppler can quickly provide flow information in vessels obscured from view.
- Always use the bipolar coagulator on the lowest effective setting to minimize tissue injury.
- Keep the tips of your bipolar forceps clean and sharp. Regularly inspect and replace if they become eroded or blunt.
- Always have a wide variety of clips available. You may need a different shape, size, or angle than initially anticipated.
- When dissecting around large vessels, be constantly mindful of tiny perforating vessels. Their injury can lead to significant postoperative deficits.
- Liberal use of protective agents such as mannitol can help decrease edema and soften the brain.
- Warm saline irrigation can help to mitigate intraoperative brain swelling.
- Complications can occur. Staying calm, thinking clearly, and methodically addressing the situation are more beneficial than panicking.

21.5 Discussion

Arteriovenous malformations (AVMs) located in the Broca's area present a unique surgical challenge, given the region's critical role in language production. Addressing AVMs in this area requires a delicate balance: the need to mitigate hemorrhagic risks while preserving essential speech functions. Current imaging modalities, especially fMRI, are proving indispensable, providing surgeons the capability to accurately map eloquent cortex areas and determine a safer surgical trajectory.

As the medical landscape shifts toward less invasive interventions, the rise of the endovascular era has brought about transformative changes in the treatment of neurovascular anomalies. The allure of endovascular techniques, with their reduced invasiveness and promising early outcomes, has garnered significant attention. Nevertheless, while endovascular treatments offer remarkable advantages, they are not without limitations, especially when navigating complex vascular architectures or in scenarios where direct access is paramount.

This evolving dynamic highlights the enduring importance of microsurgery. The precision, flexibility, and direct approach afforded by microsurgical techniques remain irreplaceable in many clinical situations. As we forge ahead, it is crucial to recognize the symbiotic relationship between microsurgery and endovascular therapies. Rather than viewing them in isolation or in competition, an integrative approach promises the best outcomes, leveraging the strengths of each modality.

21.6 Conclusion

Looking to the future, we can anticipate a harmonious confluence of endovascular and microsurgical techniques, bolstered by the ongoing advancements in medical imaging, artificial intelligence, and augmented reality. Such innovations could lead to a new paradigm in vascular neurosurgery, where decisions are tailored more intricately to individual patient profiles, pathology, and anatomical considerations.

In essence, while the endovascular era brings forth a new dawn of possibilities, the art and science of microsurgery retain their pivotal role. Together, these modalities will define the future of vascular neurosurgery, ensuring that interventions are both cutting-edge and patient-centric, all the while holding the sanctity of the human brain at the forefront. Every patient, every case, is a new learning opportunity. While the above tips and tricks provide a foundation, it is the accumulation of personal experiences, learning from each surgery, that truly refines a neurosurgeon's expertise. Remember, in the delicate world of vascular neurosurgery, it is often the minute details that make the most significant difference.

References

1. Dronkers NF, Plaisant O, Iba-Zizen MT, Cabanis EA. Paul Broca's historic cases: high resolution MR imaging of the brains of Leborgne and Lelong. Brain. 2007;130(Pt 5):1432–41.
2. Amunts K, Schleicher A, Bürgel U, Mohlberg H, Uylings HB, Zilles K. Broca's region revisited: Cytoarchitecture and intersubject variability. J Comp Neurol. 1999;412(2):319–41.
3. Andrews JP, Cahn N, Speidel BA, Chung JE, Levy DF, Wilson SM, Berger MS, Chang EF. Dissociation of Broca's area from Broca's aphasia in patients undergoing neurosurgical resections. J Neurosurg. 2022;138(3):847–57. https://doi.org/10.3171/2022.6.JNS2297.
4. Lawton MT, Rutledge WC, Kim H, et al. Brain arteriovenous malformations. Nat Rev Dis Primers. 2015;1:15008.
5. Dobrocky T, Matzinger M, Piechowiak EI, Kaesmacher J, Pilgram-Pastor S, Goldberg J, Bervini D, Klail T, Pereira VM, Z'Graggen W, Raabe A, Mordasini P, Gralla J. Benefit of Advanced 3D DSA and MRI/CT fusion in neurovascular pathology. Clin Neuroradiol. 2023; https://doi.org/10.1007/s00062-022-01260-0. Epub ahead of print
6. Chang W, Wu Y, Johnson K, Loecher M, Wieben O, Edjlali M, Oppenheim C, Roca P, Hald J, Aagaard-Kienitz B, Niemann D, Mistretta C, Turski P. Fast contrast-enhanced 4D MRA and 4D flow MRI using constrained reconstruction (HYPRFlow): potential applications for brain arteriovenous malformations. AJNR Am J Neuroradiol. 2015;36(6):1049–55. https://doi.org/10.3174/ajnr.A4245. Epub 2015 Feb 19. PMID: 25698624; PMCID: PMC6278607
7. Brosnan C, Amoo M, Javadpour M. Preoperative embolisation of brain arteriovenous malformations: a systematic review and meta-analysis. Neurosurg Rev. 2022;45(3):2051–63. https://doi.org/10.1007/s10143-022-01766-8.

Surgical Strategy for Brainstem Cavernomas: Nexus of Anatomy, Pathology, and Ergonomics

22

Jacques J. Morcos and Eva M. Wu

Learning Objectives

- Understand the relevant anatomy of the brainstem.
- Understand the main "safe" entry zones of the brainstem.
- Understand the surgical subtleties needed for successful brainstem surgery.

22.1 Introduction

Brainstem cavernomas were once considered inoperable and were associated with significant surgical morbidity. However, the development of different microsurgical approaches and advancement in neuroimaging and neuronavigation have allowed for safer resection.

Due to the very dense composition of cranial nerve nuclei and tracts in a small cross-sectional area, operating within the brainstem can be associated with significant morbidity, particularly for lesions that do not reach a pial or ependymal surface. Knowledge of "safe entry zones" of the brainstem is critical to reduce morbidity in these scenarios. While no area of the brainstem is truly "safe," these regions merely represent entry points between eloquent neural structures through which a lesion can be accessed with the least possible damage.

This chapter discusses the 18 main entry zones that have been described in the literature. Various surgical approaches are reviewed with an emphasis on the area of exposure and entry zones provided by each approach. Surgical nuances and common pitfalls are reviewed at the end of the chapter.

J. J. Morcos (✉) · E. M. Wu
Department of Neurosurgery, The University of Miami, Miami, FL, USA
e-mail: jmorcos@med.miami.edu

22.2 History/Background

Surgery for brainstem lesions had once been discouraged and had a reported mortality rate of 60% [1]. The concept of safe entry zones first emerged in 1982, reported by Baghai et al. who reported an entry zone between the trigeminal nerve and facial nerve for use in biopsy of brainstem tumors in the ventrolateral pons [2]. This was followed by the introduction of two safe entry zones on the floor of the fourth ventricle by Kyoshima et al. in 1993 [3]. Two years later, Bricolo and Turazzi described five entry zones to the dorsal brainstem [4]. This was followed by the discovery of additional brainstem entry zones. To this date, at least 21 different safe entry zones for brainstem lesions have been described, some of which have been described only in a few case reports. This chapter discusses the 18 main entry zones that have been described in the literature.

22.3 Anatomy Highlights

- The brainstem can be divided from rostral to caudal into three structures: the midbrain (mesencephalon), the pons (metencephalon), and the medulla (myelencephalon).
- Each of these structures has a ventral, lateral, and dorsal aspect and contains a significant number of nuclei and white matter tracts.
- Figure 22.1 demonstrates the anatomy that all neurosurgeons should fully understand in order to safely operate on pathology arising in the brainstem.
- This section will discuss the most important anatomical landmarks in the midbrain, followed by the pons, the medulla, and lastly, the floor of the fourth ventricle.

The midbrain is the most rostral aspect of the brainstem between the diencephalon rostrally and pons caudally. It is separated from the pons by the pontomesencephalic sulcus (Fig. 22.1b). On the ventral aspect of the midbrain, the cere-

I. N. Kanaan, V. Beneš (eds.), *Neuroanatomy Guidance to Successful Neurosurgical Interventions*, https://doi.org/10.1007/978-3-031-59838-8_22

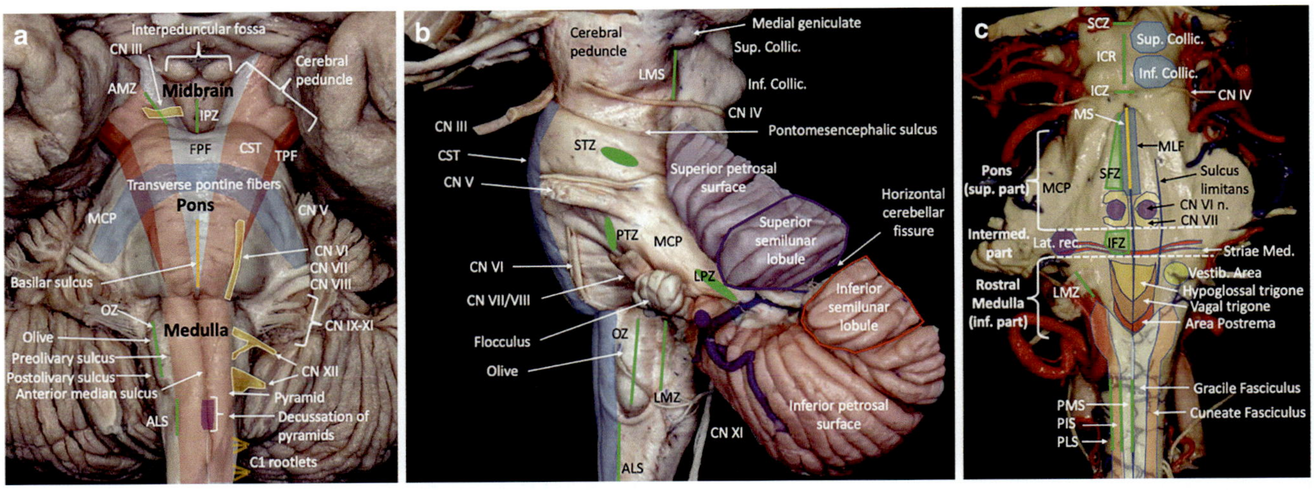

Fig. 22.1 Relevant anatomy of the brainstem. (**a**) Ventral view, (**b**) lateral view, (**c**) posterior view of the brainstem with relevant anatomical labels and the 18 entry zones that will be discussed in this chapter. The green solid lines and colored ovals are the main safe entry zones. The two orange solid lines are the theoretical entry zones that have been described but are not used due to significant morbidity (median sulcus) or are not practical due to the location of the basilar in this sulcus (basilar sulcus). Abbreviations: *ALS* anterolateral sulcus, *AMZ* anterior mesencephalic zone, *CN VI n.* cranial nerve VI nuclei, *CST* corticospinal tract, *FPF* frontopontine fibers, *ICR* intercollicular zone, *ICZ* infracollicular zone, *IFZ* infrafacial collicular zone, *Inf. Coll.* inferior colliculus, *Inf. Part* inferior part, *IPZ* interpeduncular zone, *Lat. Rec.* lateral recess, *LMZ* lateral medullary zone, *LPZ* lateral pontine zone, *MCP* middle cerebellar peduncle, *MLF* median longitudinal fasciculus, *MS* median sulcus, *OZ* olivary zone, *PIS* posterior intermediate sulcus, *PLS* posterior lateral sulcus, *PMS* posterior median sulcus, *PTZ* peritrigeminal zone, *SCZ* supracollicular zone, *SFZ* Suprafacial collicular zone, *Striae Med.* striae medullaris, *STZ* supratrigeminal zone, *Sup. Coll.* superior colliculus, *Sup. Part* superior part, *TPF* temporopontine fibers, *Vestib. area* vestibular area. (**a–c**): Dissections courtesy of the Rhoton Collection, American Association of Neurological Surgeons (AANS)/ Neurosurgical Research and Education Foundation (NREF)

bral peduncles are separated by a midline depression called the interpeduncular fossa (Fig. 22.1a). The interpeduncular fossa is the site from which the oculomotor nerve exits the midbrain (Fig. 22.1a). The cerebral peduncle is formed by three fiber tracts: the corticospinal tracts (CST) occupy the middle 3/5ths, the frontopontine fibers (FPF) occupy the medial 1/fifth, and the temporopontine fibers (TPF) occupy the lateral 1/fifth of the cerebral peduncle (Fig. 22.1a) [5]. The FPF and TPF are projections from the cerebral cortex to the pontine nuclei. The CST originates primarily from the precentral gyrus, descends through the posterior limb of the internal capsule into the middle 3/5ths of the cerebral peduncle, and continues down into the medulla forming the pyramids on the ventral surface of the medulla (Fig. 22.1a). At the caudal aspect of the medulla, 85–90% of these fibers decussate in the pyramidal decussation and continue as the lateral CST carrying information for voluntary movement of the limbs. The remaining uncrossed fibers form the anterior CST which controls movements of the trunk and proximal muscles. On the lateral aspect of the midbrain, the lateral mesencephalic sulcus (LMS) extends from the medial geniculate nucleus to the pontomesencephalic sulcus and separates the cerebral peduncle from the tegmentum (Fig. 22.1b). On the dorsal aspect of the midbrain, the tectum contains the superior and inferior colliculi which are important in the visual and auditory pathway, respectively (Fig. 22.1c). The trochlear nerve exits the midbrain below the inferior colliculus and courses laterally in the pontomesencephalic sulcus running between PCA and SCA (Fig. 22.1b, c). It is the only cranial nerve that exits from the dorsal brainstem.

The pons is the portion of the brainstem between the midbrain rostrally and medulla caudally. It is separated from the medulla by the pontomedullary sulcus. Anteriorly, the pontomedullary sulcus is the location where the abducens nerve, facial nerve, and vestibulocochlear nerve exit the brainstem in this medial to lateral sequence (Fig. 22.1a). The trigeminal nerve exits the ventral mid-pons adjacent to the middle cerebellar peduncle (MCP) (Fig. 22.1a). The transverse pontine fibers (pontocerebellar fibers) carry fibers from the contralateral pontine nuclei to the cerebellum forming the middle cerebellar peduncle which is important for coordinating and planning motor tasks. In the pons, the CST courses between the pontine nuclei and the transverse pontine fibers. The transverse pontine fibers also divide and travel anterior, posterior, and through the CST [6]. At the pontomedullary sulcus, the abducens nerve demarcates the lateral limit of the CST (Fig. 22.1a).

The medulla is the most caudal portion of the brainstem between the pons and upper cervical spinal cord. On the anterior aspect of the medulla, the anterior median sulcus divides the medullary pyramids which contain corticospinal tracts that descend into the spinal cord (Fig. 22.1a). Just lateral to the medullary pyramids is the medullary olive (Fig. 22.1a). The anterolateral sulcus (preolivary sulcus) divides the pyramids from the olive and is the location where the hypoglossal and C1 rootlets exit the brainstem

(Fig. 22.1a). The space between the hypoglossal rootlets and C1 rootlets corresponds to the pyramidal decussation (Fig. 22.1a *highlighted in purple*). The hypoglossal rootlets at the anterolateral sulcus demarcate the lateral limit of the CST (Fig. 22.1a). Posterior to the olive is the posterolateral sulcus (postolivary sulcus) which is the location where the glossopharyngeal, vagus, and accessory nerves exit in this rostro-caudal sequence (Fig. 22.1a). The dorsal surface of the medulla has an open part at the level of the fourth ventricle and a closed part below the fourth ventricle (Fig. 22.1c). The open part will be discussed separately in the following paragraph. The two main tracts that traverse the closed dorsal surface of the medulla are the fasciculi gracilis and cuneatus which carry proprioceptive sensory information (Fig. 22.1c). These fibers first synapse in the gracilis and cuneatus tubercle, then ascend and decussate in the caudal medulla before traveling along the medial lemniscus to the somatosensory cortex. The posterior median sulcus is located in the midline between the fasciculi gracilis (Fig. 22.1c). The posterior intermediate sulcus separates the fasciculi gracilis and cuneatus (Fig. 22.1c). The posterior lateral sulcus is lateral to the cuneate fasciculus (Fig. 22.1c).

The floor of the fourth ventricle (rhomboid fossa) is a diamond-shaped structure formed by a roof (superior and inferior medullary velum), sides (cerebellar peduncles), and floor (dorsal aspect of the pons and rostral medulla) (Fig. 22.1c). It is widest at the level of the lateral recess where the pons and medulla merge (Fig. 22.1c). The floor of the fourth ventricle is split into two triangular areas by the striae medullaris: the superior part (pontine) formed by the pons and the inferior part (medullary) formed by the rostral medulla (Fig. 22.1c *white brackets*). The intermediate part of the fourth ventricle is between the superior and inferior parts and corresponds to the upper and lower borders of the lateral recess (Fig. 22.1c *white dashed line*) [7]. The caudal portion of the fourth ventricle continues as the central canal. The median sulcus divides the rhomboid fossa longitudinally in the midline into symmetric halves (Fig. 22.1c). Parallel to and bordering the median sulcus is the medial longitudinal fasciculus (MLF) which connects the oculomotor and abducens nuclei (Fig. 22.1c). Injury to the MLF can cause internuclear ophthalmoplegia. The sulcus limitans further divides these two halves longitudinally and deepens lateral to the facial colliculus and hypoglossal trigone forming the superior and inferior fovea, respectively (*not labeled in* Fig. 22.1c). The facial colliculus is a raised area in the superior part of the rhomboid fossa lateral to the median sulcus. It is composed of fibers of the facial nerve wrapping around the nucleus of the abducens nerve before it exits the brainstem (Fig. 22.1c). This structure can be identified microscopically and by electrophysiologic monitoring. Further caudally, the hypoglossal nucleus and dorsal motor nucleus of the vagus form raised elevations in the inferior part of the rhomboid fossa termed the hypoglossal trigone (more cranial and medial) and vagal trigone (more lateral), respectively (Fig. 22.1c). These two trigones and the area postrema are stacked in the configuration of a pen nib; therefore, the area is commonly referred to as the calamus scriptorius. Injury to this area can cause dysphagia and cardiorespiratory disturbances. The vestibular area is medial to the lateral recess and overlies the vestibular nuclei (Fig. 22.1c). The dorsal cochlear nucleus is lateral to the vestibular area and is used as a target for auditory brainstem implants [8].

22.4 Approach/Surgical Intervention (Including Representative Own Cases)

22.4.1 Brainstem Entry Zones

Brainstem entry zones can be divided into midbrain, pons, and medulla. Further subdividing each of these structures into ventral, lateral, and dorsal entry zones can aid in understanding the different entry zones of the brainstem.

22.4.2 Midbrain Entry Zones

Lesions in the midbrain can be accessed using six entry zones via anterior, lateral, or posterior surgical approaches. The two entry zones that can be used to access ventral midbrain lesions are the interpeduncular zone (IPZ) and the anterior mesencephalic zone (AMZ) which is also known as the perioculomotor zone [9]. Both of these entry zones can be reached using an anterior surgical route via the transsylvian pterional, cranioorbital, or cranioorbitozygomatic approach (Fig. 22.2). The entry zone used to access lateral midbrain lesions is the lateral mesencephalic sulcus zone (LMS) and can be reached using a posterolateral route via the subtemporal, paramedian or extreme lateral supracerebellar infratentorial (SCIT) approach [10].

The pterional approach gives access to the ipsilateral cerebral peduncle. Adding an orbital osteotomy can be helpful to look more cranially for high lesions (Fig. 22.2). A unilateral or contralateral orbitozygomatic approach can be utilized depending on the anteroposterior extension of the lesion.

The subtemporal approach gives access to the entire lateral surface of the midbrain but carries the risk of temporal lobe retraction and injury to the vein of Labbe (Fig. 22.3). The extreme lateral SCIT provides a view of the posterolateral midbrain without these risks (Fig. 22.4). The long axis of the lesion in the midbrain tegmentum determines the approach used. Lesions closer to the lateral surface of the tegmentum should be accessed through the LMS, while lesions closer to the ventral surface should be accessed

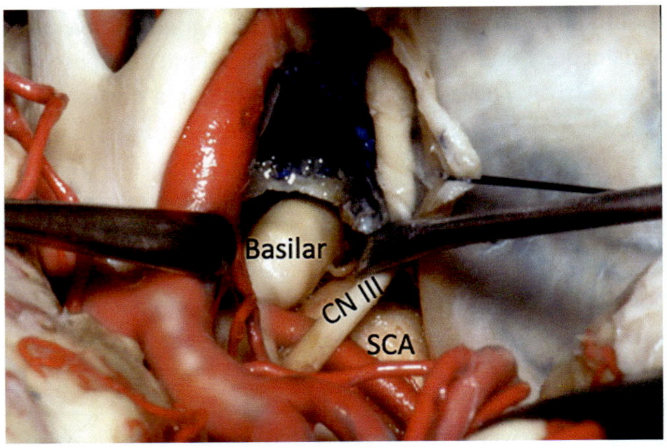

Fig. 22.2 Transsylvian approach. Rhoton dissection showing a right pterional approach with orbitozygomatic osteotomy, showing sylvian fissure is split. The dissection is carried proximally toward the carotid and chiasmatic cisterns. Rhoton dissection showing opening of the carotico-oculomotor triangle. The basilar is seen at the depth and SCA is seen below the oculomotor nerve. The carotico-oculomotor triangle is entered to reach the crural and interpeduncular cisterns where the AMZ and IPZ can be accessed, respectively. Mobilizing the oculomotor nerve medially allows access to the AMZ, whereas mobilizing the oculomotor nerve laterally allows access to the IPZ. Note the region of the interpeduncular fossa contains the basilar bifurcation, SCA, PCA, medial posterior choroidal, thalamoperforators, perforators from the SCA and the basilar and is therefore a challenging approach with little surgical freedom. A unilateral or contralateral orbitozygomatic approach can be utilized depending on the anteroposterior extension of the lesion. Abbreviations: *AMZ* anterior mesencephalic zone, *IPZ* interpeduncular zone, *PCA* posterior cerebellar artery, *SCA* superior cerebellar artery. Dissection courtesy of the Rhoton Collection, American Association of Neurological Surgeons (AANS)/Neurosurgical Research and Education Foundation (NREF)

through the AMZ using a pterional or orbitozygomatic approach. A paramedian SCIT approach is generally sufficient to expose dorsolateral lesions, but ventrolateral lesions may require an extreme lateral SCIT approach. The tentorium can be divided for lesions that extend more superiorly. The quadrangular lobule of the cerebellum can be cut to gain access to the pontomesencephalic fissure for lesions that extend more inferiorly.

The transsylvian pretemporal "half-and-half" approach is a combination of the transsylvian approach and subtemporal approach (Fig. 22.5). This half transsylvian and half subtemporal approach combines the anterior view from the transsylvian approach with the lateral view from the subtemporal approach by mobilizing the temporal lobe posterolaterally. It gives access to the lateral mesencephalic sulcus (LMS) and anterior mesencephalic (AMZ) entry zones.

Three entry zones used to access lesions in the midbrain dorsal to the aqueduct include the supracollicular zone (SCZ), infracollicular zone (ICZ), and intercollicular zone (ICR) (Fig. 22.1c). These entry zones can be reached using a posterior route via the median, paramedian, or extreme lateral SCIT (Fig. 22.4) or posterior interhemispheric transtentorial approach (Fig. 22.6). The choice of approach depends on where the lesion comes to the surface and the long axis of the lesion. The median SCIT is used for midline lesions or when the ICR needs to be visualized. The extreme lateral SCIT is used for lesions coming to the posterolateral surface of the midbrain and takes advantage of the less steep angle of the tentorium laterally (Fig. 22.4). The posterior interhemispheric transtentorial (PIT) can be used if the lesion is in the midline and the steepness of the tentorium is unfavorable (Fig. 22.6). The main disadvantage of the PIT is the need to mobilize the medial occipital surface and the poor visualization of the contralateral aspect, hindered by the Galenic complex.

22.4.3 Ventral Entry Zones

22.4.3.1 InterpeduncularZone (IPZ)

The IPZ utilizes the area between the mamillary bodies superiorly and the basilar apex/perforators inferiorly (Fig. 22.1a). The lateral borders are the intramesencephalic segment of the oculomotor nerve and the red nucleus which is located dorsal to this segment. These structures can be injured if the incision is not midline or if there is excessive retraction. Variability in basilar perforators can narrow the corridor making this approach more challenging. Lesions located ventral and medial to the oculomotor nerve can be accessed through this entry zone. In this approach, the oculomotor nerve is followed back to the brainstem and the medial arachnoid adhesions are dissected allowing the nerve to remain attached laterally.

22.4.3.2 Anterior Mesencephalic Zone (AMZ) aka Perioculomotor Zone

The AMZ is a narrow corridor in the medial 1/fifth of the cerebral peduncle that is bordered by the exit zone of the oculomotor nerve medially and the corticospinal and corticobulbar tracts laterally (in the middle 3/5ths of the cerebral peduncle) (Fig. 22.1a). The frontopontine fibers are located in the medial 1/fifth of the cerebral peduncle and are safe to enter. The red nucleus and the nigrostriatal fibers are located deep to the oculomotor nerve and corticospinal tracts, respectively. The superior and inferior limits of this entry zone are the posterior cerebral artery (PCA) and superior cerebellar artery (SCA), respectively. Lesions located in the ventral midbrain lateral to the oculomotor nerve can be accessed with this entry zone. In this approach, the arachnoid adhesions of the oculomotor nerve to the temporal lobe and tentorium are detached but the medial arachnoid adhesions are left intact so the nerve is not mobile laterally and injured during the approach.

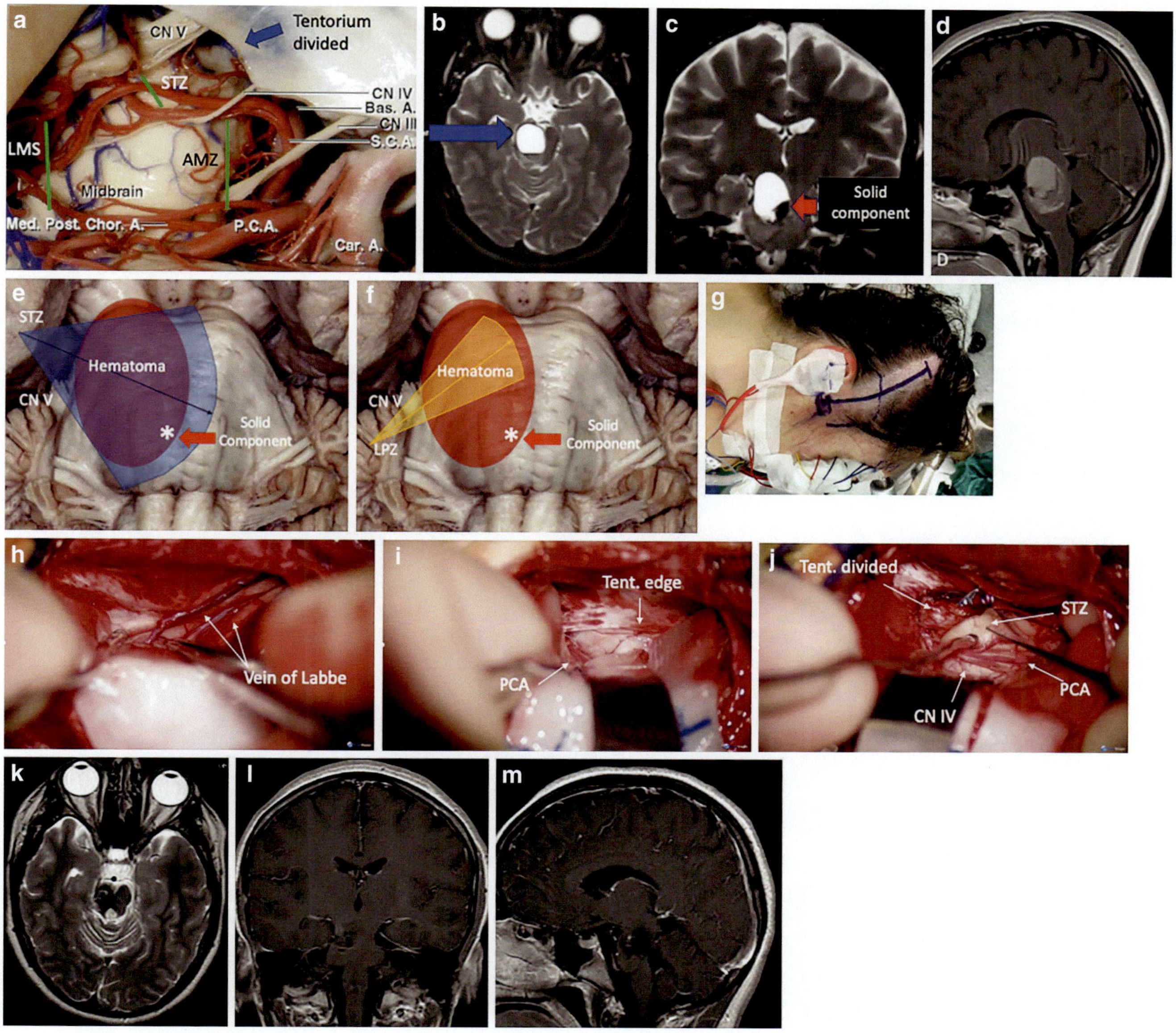

Fig. 22.3 Subtemporal approach. (**a**) Rhoton dissection showing view obtained through a left subtemporal transtentorial approach. The lateral midbrain down to the pontomesencephalic junction can be visualized in this approach. The PCA and medial posterior choroidal cross the operative field superiorly and SCA crosses inferiorly. This approach gives an oblique access to the AMZ and straight on access to the LMS. The tentorium is divided posterior to the trochlear nerve. Division of the tentorium provides improved visualization of the pontomesencephalic junction, gives a more anterolateral view of the rostral pons, and exposes the STZ entry zone. (**b–m**) 27-year-old-female presented with dysarthria, left central facial weakness, left facial hemianesthesia, left hemiparesis (left upper 1/5, left lower 2/5), and left hemianesthesia to all sensory modalities. (**b**) Axial T2-weighted, (**c**) coronal T2-weighted, (**d**) sagittal T1-weighted MRI show a pontomesencephalic cavernoma with varied signal intensity on T1 and T2 due to differing age of blood products. The lesion comes to the surface at the STZ (blue arrow) and there is a solid component at the inferior medial margin of the cavernoma (red arrow). (**e** and **f**) show the potential approaches that could be considered to remove this lesion. € Subtemporal transtentorial approach with entry through STZ (blue shaded area) was used because it provides optimal access to all components of the lesion and visualization of the inferior medial solid component (asterisks). (**f**) Retrosigmoid approach with entry through LPZ (orange shaded area) less ideal given it would be difficult to see the solid component of the lesion at the inferior medial margin (asterisks). (**g**) Patient is positioned supine, contralateral head rotation 90°, vertex tilted down, zygomatic arch should be the highest point of the field. Lumbar drain is used. (**h**) Microdissection is carried subtemporally until the tentorial edge is seen. Care is taken to preserve the vein of Labbe during this dissection. (**i**) Surgical view of the lateral surface of the midbrain after arachnoid dissection has been completed. A self-retaining retractor can be used once dissection is completed. (**j**) The tentorium is cut exposing the pontomesencephalic junction, rostral pons, and STZ. Immediately postop, patient had improvement in left hemiparesis to 4/5 strength. At 3-years follow-up, all preoperative symptoms resolved except for some difficulty with opening and closing of the left hand and left leg lag when ambulating. (**k–m**) T2-weighted (**k**) and T1 with gadolinium (**l** and **m**) MRI at 3-years postop shows no residual or recurrence. Abbreviations: *AMZ* anterior mesencephalic zone, *Ant. Chor. A.* anterior choroidal artery, *Bas. A.* basilar artery, *Car. A.* carotid artery, *LMS* lateral mesencephalic sulcus, *LPZ* lateral pontine zone, *Med. Post. Chor. A.* medial posterior choroidal artery, *PCA* posterior cerebellar artery, *Post. Comm. A.* posterior communicating artery, *SCA* superior cerebellar artery, *STZ* supratrigeminal zone, *Tent.* tentorium. (**a**) Dissections courtesy of the Rhoton Collection, American Association of Neurological Surgeons (AANS)/Neurosurgical Research and Education Foundation (NREF)

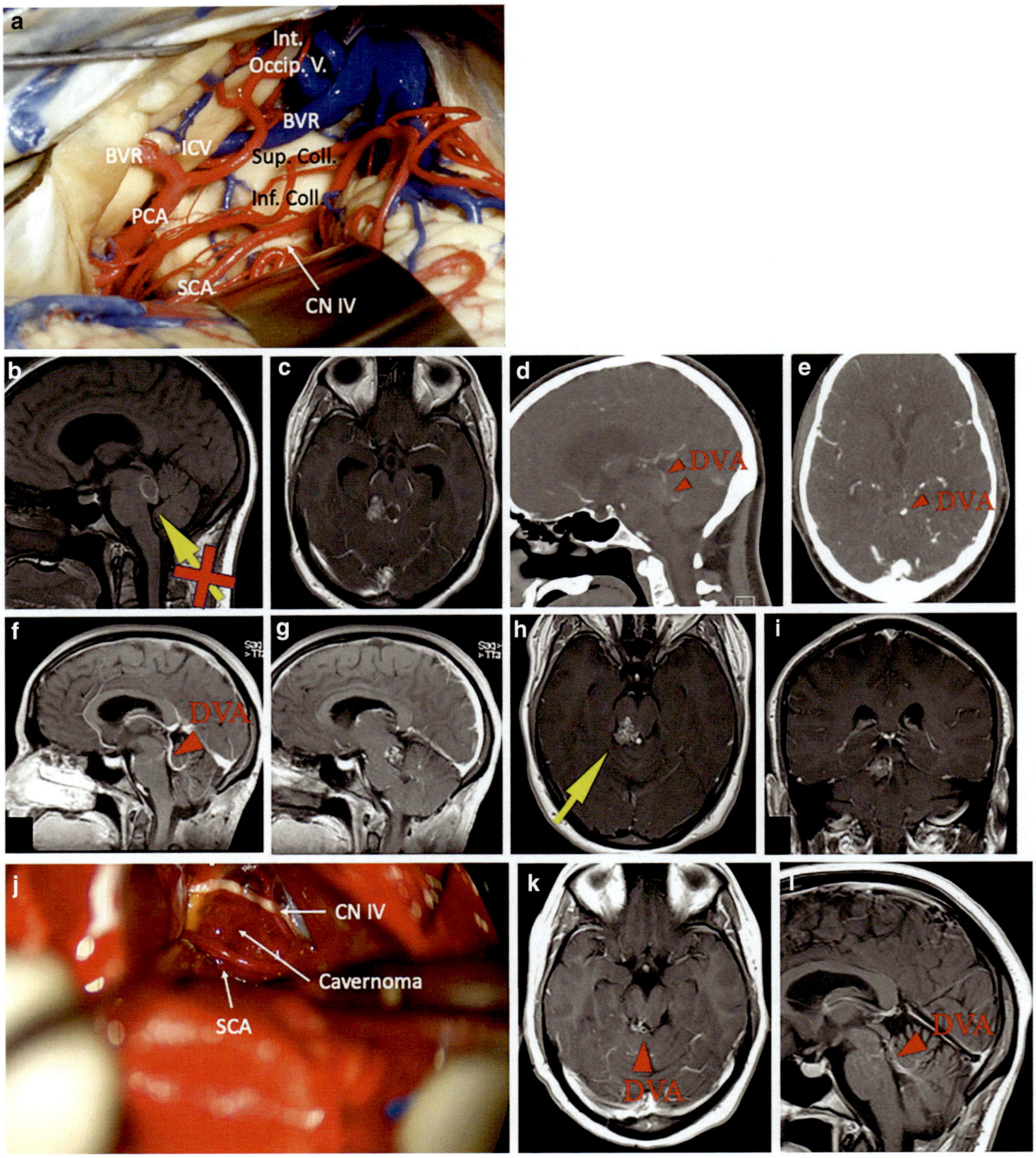

Fig. 22.4 Extreme lateral SCIT approach. (**a**) Rhoton dissection showing the view of extreme lateral SCIT approach. The extreme lateral SCIT takes advantage of the decreasing steepness of the tentorium as it goes out laterally. The extreme lateral SCIT provides an oblique view to the collicular plate and allows a more lateral view of the midbrain exposing the LMS. The disadvantage of this approach is that the midline cannot be crossed easily. The tentorium can be divided to gain a more superior view. (**b**–**l**) A 28-year-old female presented to an outside hospital with acute right CN IV palsy and left hemiparesis in 2019. (**b**) Sagittal T1-weighted and (**c**) axial T1 with gadolinium show a midbrain/superior cerebellar peduncle cavernoma. A telovelar approach was used (yellow arrow) but this was the incorrect approach given the DVA in the way and the long working distance (**d** and **e**). DVA prevented removal of the cavernoma and only a very partial resection was achieved. Patient recovered completely and presented to us in 2020. (**f**–**i**) T1 with gadolinium MRI when a patient presented to us shows hyperintense lesion in the right midbrain exophytic to the ambient cistern. An extreme lateral SCIT was used (yellow arrow). (**j**) Surgical view of the cerebellomesencephalic fissure. CN IV and SCA are pushed superiorly to allow for complete resection of the cavernoma (not shown here). Postoperatively, patient had transient CN IV palsy that resolved at 3 weeks postop. (**k** and **l**) Axial and sagittal T1 with gadolinium MRI at 1-year postop showing no residual or recurrence and preserved DVA. Abbreviations: *BVR* basal vein of Rosenthal, *DVA* developmental venous anomaly, *ICR* intercollicular zone, *ICV* internal cerebral vein, *ICZ* infracollicular zone, *Inf. Coll.* inferior colliculus, *Int. Occip. V.* internal occipital vein, *LMS* lateral mesencephalic sulcus, *med. Gen.* medial geniculate, *PCA* posterior cerebral artery, *SCA* superior cerebellar artery, *SCZ* supracollicular zone, *Sup. Coll.* superior colliculus. (**a**) Dissections courtesy of the Rhoton Collection, American Association of Neurological Surgeons (AANS)/Neurosurgical Research and Education Foundation (NREF)

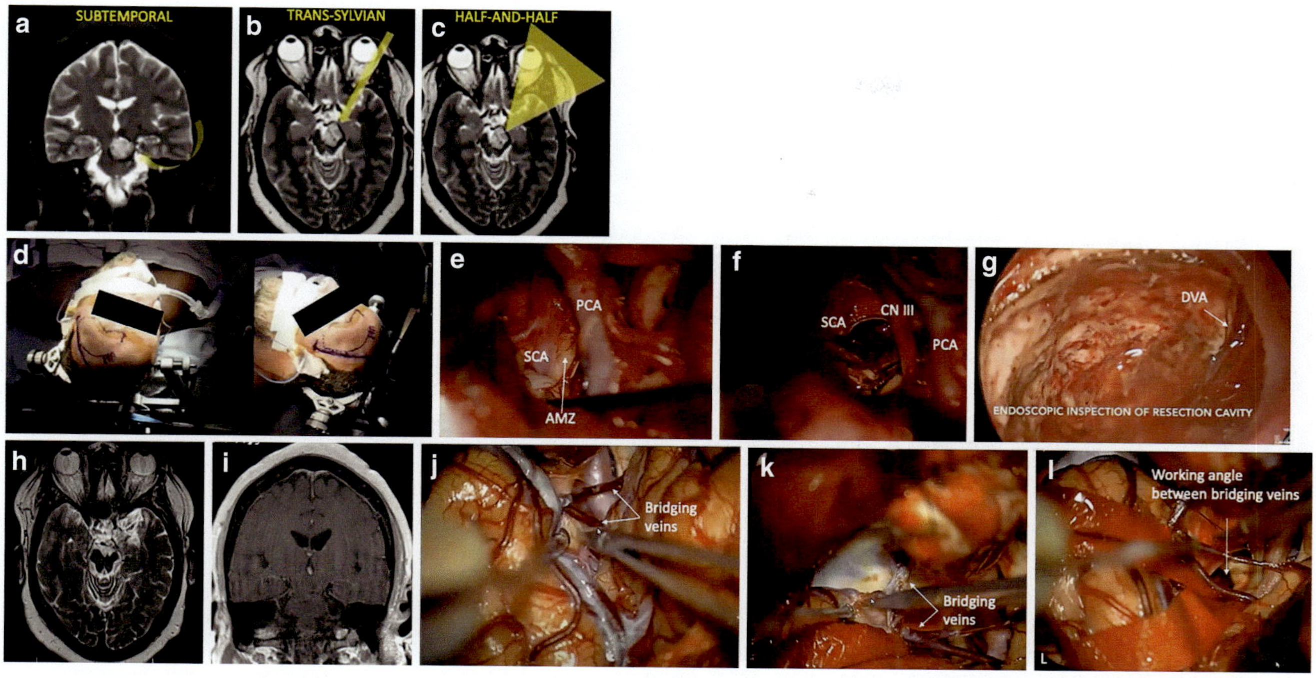

Fig. 22.5 Transsylvian pretemporal "half-and-half" approach. Cutting the tentorium improves visualization of the pontomesencephalic junction and rostral lateral pons and exposes STZ. (**a–l**) 51-year-old male presenting with multiple episodes of transient right hemiparesis. MRI shows a mesencephalo-pontine cavernoma reaching the surface of the right cerebral peduncle. This lesion could be resected using a subtemporal approach (**a**) or a transsylvian approach (**b**). (**c**) Demonstrates the corridor for the transsylvian pretemporal half-and-half" approach which is a combination of these two approaches. This half transsylvian and half subtemporal approach combines the anterior view from the transsylvian approach with the lateral view from the subtemporal approach by mobilizing the temporal lobe posterolaterally. This approach allows for less temporal lobe retraction and oculomotor nerve palsy. This was the approach used to resect this lesion. (**d**) Patient is positioned supine, head rotated 15–30° contralaterally and slightly extended.(**e**) Surgical view of the AMZ entry zone after wide sylvian fissure split. The PCA and SCA are seen. Fixed retractor is not used during sylvian fissure split but can be helpful at a later stage if a third hand is needed during cavernoma removal. (**f**) Surgical view showing entry zone with CN III medially, PCA superiorly, and SCA inferiorly. (**g**) Endoscopic assistance can be helpful to inspect blind spots. A 30-degree endoscope is used to inspect the resection cavity and shows complete resection and preservation of the DVA. Immediately postop, the patient had new transient CN III palsy that recovered fully and new right hemiparesis that improved from 2/5 to 4+/5 at 4 weeks postop. (**h** and **i**) Axial T2-weighted imaging and Coronal T1 with gadolinium at 5-years postop showing no residual or recurrence. (**j-l**) A case of a transsylvian pretemporal approach with preservation of bridging veins. (**j**) At first glance, it may be tempting to cut the bridging veins but it is not always necessary. (**k**) Dissecting arachnoid around bridging veins gives it more length and allows for a working angle in between the bridging veins. (**l**) Surgical view showing working angle between bridging veins. Abbreviations: *AMZ* anterior mesencephalic zone, *DVA* developmental venous anomaly, *LMS* lateral mesencephalic sulcus, *PCA* posterior cerebellar artery, *SCA* superior cerebellar artery, *STZ* supratrigeminal zone

22.4.4 Lateral Entry Zones

22.4.4.1 Lateral Mesencephalic Sulcus (LMS)

The LMS extends from the medial geniculate to the pontomesencephalic sulcus and separates the cerebral peduncle from the tegmentum (Fig. 22.1b). The lateral mesencephalic vein often overlies this sulcus. This entry zone is bordered by the PCA which crosses the LMS superiorly and the SCA, trochlear nerve, and tentorial edge which cross inferiorly. This entry zone is located between the substantia nigra anterolaterally and the medial lemniscus. Therefore, the incision must be angled 90° to the tectal surface to avoid hitting the medial lemniscus if the incision is angled 45° forward or the cerebral peduncle if the incision is angled even acutely forward [11]. The oculomotor fibers running from red nucleus to the substantia nigra form the anteromedial limit. This entry zone can be safely dissected up to a depth of 4.9–11.7 mm (mean 8.02 mm) [12]. Lesions located in the lateral midbrain can be accessed using this entry zone.

22.4.5 Dorsal Entry Zones

22.4.5.1 Supracollicular Zone (SCZ)

The SCZ is a transverse incision above the upper edge of the superior colliculus (Fig. 22.1c). The aqueduct is the landmark for the depth of the approach. Going deeper than the aqueduct would injure the intramesencephalic segment of the oculomotor nerve and MLF which are located in the midline ventral to the aqueduct. Extending the incision too far

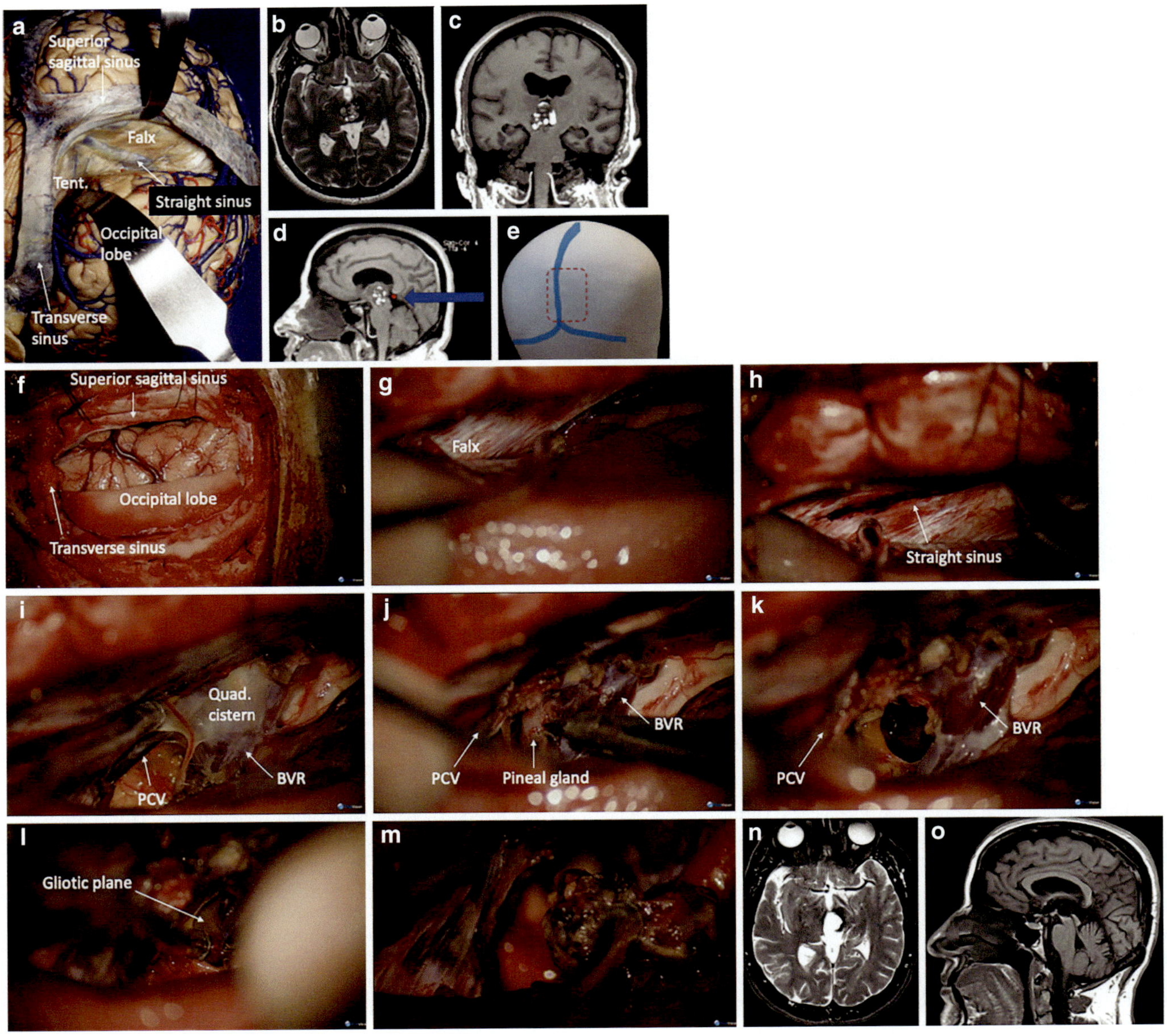

Fig. 22.6 Posterior interhemispheric transtentorial approach. (**a**) Rhoton dissection showing surgical corridor. (**b–o**) A 50-year-old female presented with sudden onset diplopia in 2016. Diagnosed with a thalamomesencephalic cavernoma and treated at an outside hospital with gamma knife. Symptoms gradually worsened and in 2019 she was referred to us with dysarthria, parinaud's syndrome, failure of inferior gaze, and right hemiparesis 4/5. Axial T2-weighted MRI (**b**) and coronal T1 with gadolinium (**c**) show the thalamomesencephalic cavernoma. (**d**) Sagittal T1 with gadolinium demonstrates the corridor for the posterior interhemispheric transtentorial supracollicular transpineal approach that was used. (**e**) The craniotomy is depicted in red dashed lines and extends past the superior sagittal sinus to allow for retraction of the falx and superior sagittal sinus superiorly. (**f**) Surgical view is shown. Patient is positioned lateral, lumbar drain is used to allow for the occipital lobe to fall away from the falx using gravity. A fixed retractor is not used. (**g**) Dissection is carried through the interhemispheric corridor and the falx is seen. The location of the straight sinus is not always obvious compared to a Rhoton dissection. (**h**) Neuronavigation is used to locate the straight sinus. The tentorium is cut starting from the middle, then working laterally toward the edge. (**i**) Surgical view of the quadrigeminal cistern, posterior cerebral vein (PCV), and Rosenthal. (**j**) The thick arachnoid over the quadrigeminal cistern is dissected and the pineal gland is revealed. The pineal gland is entered to expose the SCZ. PCV can be sacrificed but Rosenthal must be preserved. (**k**) Surgical view of the approach showing a perfect trajectory down the long axis of the cavernoma. (**l**) The dissection is carried in the gliotic yellowish plane. In this case, gamma knife created more scarring around the plane so it was not as pristine. (**m**) Cavernoma is removed in one piece whenever possible. (**n**) T2-weighted and (**o**) T1-weighted MRI at 4-years follow-up showing no residual or recurrence. At follow-up, patient continued to have superior and inferior gaze palsy but right hemiparesis and dysarthria resolved completely. Abbreviations: *BVR* basal vein of Rosenthal, *LMZ* lateral mesencephalic zone, *SCZ* supracollicular zone, *ICR* intercollicular zone, *ICZ* infracollicular zone, *Inf.* inferior, *PCV* precentral cerebellar vein, *Quad.* quadrigeminal, *Sup.* superior, *Tent.* tentorium. (**a**) Dissections courtesy of the Rhoton Collection, American Association of Neurological Surgeons (AANS)/Neurosurgical Research and Education Foundation (NREF)

laterally from midline would result in damage to the trigeminal mesencephalic tract, central tegmental tract, and red nucleus, respectively from medial to lateral.

22.4.5.2 Infracollicular Zone (ICZ)

The ICZ is a transverse incision below the lower edge of the inferior colliculus and above the exit zone of the trochlear nerve (Fig. 22.1c). The midline landmark is again the aqueduct. Traversing deeper than the aqueduct will result in damage to the trochlear nucleus, MLF, and decussation of the superior cerebellar peduncle in this order from superficial to deep.

22.4.5.3 Intercollicular Zone (ICR)

The ICZ is accessed through a longitudinal incision between the superior and inferior colliculi (Fig. 22.1c). It is a region sparse in fibers and can be used for lesions in the dorsal midbrain with relatively minimal morbidity compared to the SCZ and ICZ.

22.4.6 Pons Entry Zones

Pontine lesions can be accessed using anterolateral, posterolateral, and posterior surgical routes. Six main entry zones have been described: three on the lateral surface of the pons and three on the dorsal aspect of the pons (Fig. 22.1b, c). The lateral surface of the pons is a much more forgiving region for entering the pons than the floor of the fourth ventricle. For lesions in the dorsal pons, the eloquence of the floor of the fourth ventricle must be taken into account when selecting approaches using the two-point method and should be the primary consideration. A longer trajectory through the lateral pontine zone (LPZ) may be safer than entry through the fourth ventricle (Fig. 22.7). For exophytic fourth ventricular lesions, entry through the floor of the fourth ventricle may be considered (Fig. 22.8). See pearls/tips and tricks section for further discussion on this topic.

The three lateral entry zones to the pons are centered around the trigeminal nerve: supratrigeminal zone (STZ), peritrigeminal zone (PTZ), and lateral pontine zone (LPZ) (Fig. 22.1b). The subtemporal transtentorial approach is an anterolateral surgical approach that can be used to access the STZ (Fig. 22.3). Performing an anterior petrosectomy (Kawase) gives a more caudal exposure than the subtemporal approach, allowing for access to the PTZ (Fig. 22.9). The retrosigmoid (a posterolateral approach, Fig. 22.7) and retrolabyrinthine (an anterolateral approach, Fig. 22.10) can also be used to access the three lateral pontine entry zones. The retrolabyrinthine approach provides a more direct view to the lateral pons compared to retrosigmoid approach but at the expense of a narrower corridor (Fig. 22.10).

The three dorsal entry zones to the pons are the: median sulcus zone (MS), suprafacial collicular zone (SFZ), and infrafacial collicular zone (IFZ) (Fig. 22.1c). These entry zones are on the floor of the fourth ventricle and can be accessed through a posterior surgical route via the suboccipital telovelar approach (Fig. 22.8) with or without C1 laminectomy. For lesions located at the pontomesencephalic junction, maximum neck flexion and C1 laminectomy can be helpful to obtain a superior trajectory needed to reach the pathology.

22.4.7 Lateral Entry Zones

22.4.7.1 Supratrigeminal Zone (STZ)

The STZ is an oblique incision made in the middle cerebellar peduncle above the trigeminal root entry zone (Fig. 22.1b). The incision is made parallel to the transverse pontine fibers but must be posterior to the trajectory of the CST. This zone can be used for ventral lesions located just above the trigeminal root entry zone and can be reached through a subtemporal transtentorial (Fig. 22.3a), anterior petrosectomy (Kawase) (Fig. 22.9a), retrosigmoid (Fig. 22.7a), or translabyrinthine approach (Fig. 22.10).

22.4.7.2 Peritrigeminal Zone (PTZ)

The PTZ is a longitudinal incision between the trigeminal and facial-vestibulocochlear complex root entry zone (Fig. 22.1b). The superior limit is the intrapontine segment of the trigeminal nerve and the inferior limit is the intrapontine segment of the abducens and facial nerves. The CST borders it anteromedially and the trigeminal motor nucleus and trigeminal spinothalamic tract (TST) form the posteromedial boundary [13]. At this level, the distance from the trigeminal nerve to the CST is 4.6 mm (range 3.1–5.7 mm) and the depth to the trigeminal motor and sensory nuclei is 11.2 mm (range 9.5–13.1 mm) [12]. This entry zone can be used for lesions in the ventral pons. Note that the same surgical approaches used to access the STZ can be used to access the PTZ except for the subtemporal transtentorial approach as it does not provide enough caudal exposure to access this entry zone (Fig. 22.3a). Figure 22.9 gives an example of resection of a pontine cavernoma through the PTZ via a Kawase approach.

22.4.7.3 Lateral Pontine Zone (LPZ) aka Middle Cerebellar Peduncle Zone

The LPZ is an entry zone in the posterolateral middle cerebellar peduncle (MCP) used to access deep central or dorsolateral pontine lesions through a retrosigmoid approach with splitting of the horizontal cerebellar fissure (Fig. 22.7). The MCP is normally partially covered by the petrosal surface of the cerebellum. The horizontal cerebellar fissure (petrosal

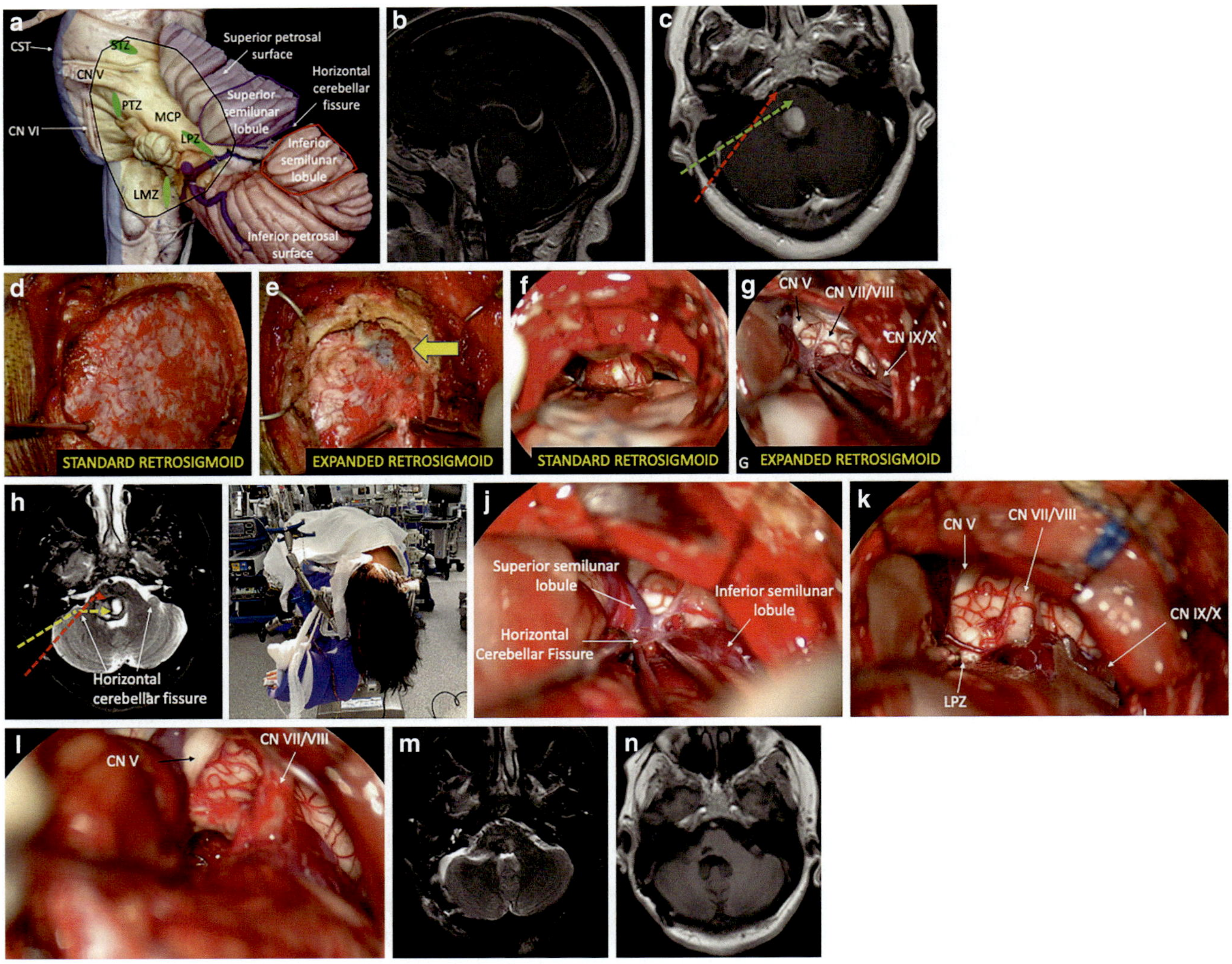

Fig. 22.7 Retrosigmoid approach. (**a**) Rhoton dissection showing relevant anatomy in relation to the brainstem entry zones of the pons. The MCP is normally partially covered by the petrosal surface of the cerebellum. The horizontal cerebellar fissure (petrosal fissure) divides the petrosal surface of the cerebellar hemisphere into superior and inferior parts. Splitting the petrosal fissure separates the superior and inferior petrosal surfaces and exposes the posterolateral middle cerebellar peduncle (posterior and lateral to the root entry zone of trigeminal nerve). The yellow shaded area shows the area of exposure and entry zones seen with a retrosigmoid approach: STZ, PTZ, LPZ, and LMZ. However, this approach provides a suboptimal angle of attack for the PTZ compared to the retrolabryinthine approach or Kawase approach. (D–N) This is a 63-year-old female with three separate episodes of sudden onset left hemianesthesia, left hemiparesis 4+/5, and right conjugate gaze center palsy. Sagittal (**b**) and Axial (**c**) T1 with gadolinium and T2-weighted (**h**) MRI show a pontine cavernoma. The lesion is eccentric to the right middle cerebellar peduncle and does not quite reach the surface of the floor of the fourth ventricle. Given this thin rim of tissue over the floor of the fourth ventricle, telovelar approach was not used due to the morbidity of traversing the floor of the fourth ventricle. Instead, an expanded retrosigmoid approach through the middle cerebellar peduncle was used. (**c**) Shows the difference in trajectory between a standard retrosigmoid corridor (red dashed arrow) and an expanded retrosigmoid corridor (green dashed arrow). In the standard retrosigmoid approach (**d**), the sigmoid sinus is blue lined. In the expanded retrosigmoid approach (**e**), bone over the sigmoid is drilled (yellow arrow). This additional 5–7 mm of anterior bony exposure allows for a more medial and posterior view of the brainstem (**g**) compared to the standard retrosigmoid approach (**f**). (**h**) Combining the expanded retrosigmoid approach with splitting of the horizontal cerebellar fissure allows for an even greater posterior view of the brainstem without need for cerebellar retraction or resection (yellow dashed arrow) which cannot be achieved with a standard retrosigmoid approach (red dashed arrow). (**i**) Patient is positioned lateral, head flexed and translated upwards, vertex tilted down. (**j**) Surgical view after an expanded retrosigmoid craniotomy has been performed. The horizontal cerebellar fissure is split between the superior and inferior semilunar lobules. (**k**) Surgical view after the horizontal cerebellar has been split exposing the LPZ on the posterolateral middle cerebellar peduncle (posterior and lateral to the root entry zone of trigeminal nerve). (**l**) Final surgical view after the cavernoma has been resected. Patient's conjugate gaze palsy improved immediately postoperatively and at 2-years follow-up she had complete resolution of her preoperative symptoms except for residual left hemianesthesia. (**m**) T2-weighted MRI on postoperative day 1 shows the tract used to resect the cavernoma through the middle cerebellar peduncle. (**n**) T1-weighted MRI at 2-years follow-up shows no residual or recurrence. Abbreviations: *CST* corticospinal tract, *LMZ* lateral medullary zone, *LPZ* lateral pontine zone, *MCP* middle cerebellar peduncle, *STZ* supratrigeminal zone, *PTZ* peritrigeminal zone. (**a**) Dissections courtesy of the Rhoton Collection, American Association of Neurological Surgeons (AANS)/Neurosurgical Research and Education Foundation (NREF)

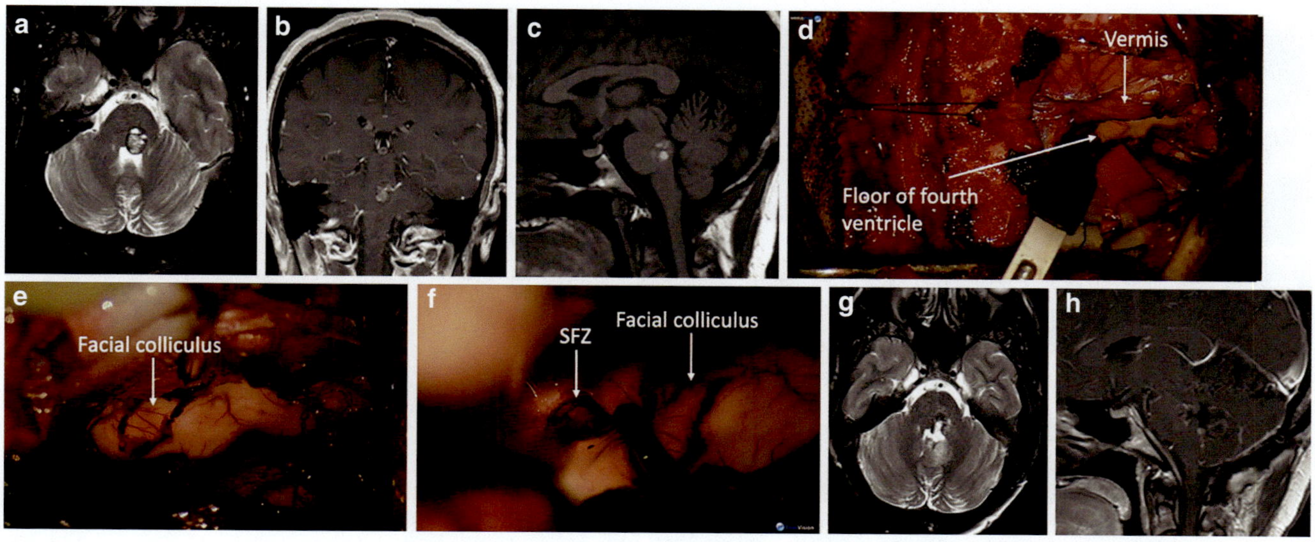

Fig. 22.8 Suboccipital approach. The cerebellar tonsils are retracted laterally exposing the inferior half of the roof of the fourth ventricle which is composed of the tela choroidea and inferior medullary velum. (A–H) This is a 41-year-old female with known progressively enlarging pontine cavernoma in the floor of fourth ventricle and progressive left hemibody numbness. Axial T2-weighted (**a**), coronal (**b**), and sagittal T1 (**c**) with gadolinium MRI show a pontine cavernoma exophytic to the floor of the fourth ventricle above the level of the facial colliculus. Given this, a telovelar suprafacial collicular approach was used. (**d**) Surgical view after the tela and inferior medullary velum are opened. A self-retaining retractor is used to hold the tonsils laterally after dissection has been completed. (**e**) Facial colliculus is mapped with the NIM stimulator and the borders are marked on the floor of the fourth ventricle. (**f**) Pial opening is made in the suprafacial collicular zone (SFZ) and cavernoma is removed. Facial nerve stimulation remained at baseline. Immediately postoperatively, the patient's left hemibody numbness remained at baseline but she had left House-Brackmann 5 facial weakness and left conjugate gaze palsy. At 2-weeks follow-up, she continued to have left conjugate gaze palsy but facial weakness improved to House-Brackmann 3. Patient was then lost to follow-up. T2-weighted (**g**) and (**h**) T1 with gadolinium MRI on postoperative day 2 show gross total resection. Abbreviations: *Facial coll.* Facial colliculus, *IFZ* infrafacial collicular zone, *Inf. Cere. Ped.* inferior cerebellar peduncle, *Inf. Med. Velum* inferior medullary velum, *LMZ* lateral medullary zone, *MS* median sulcus, *PIS* posterior intermediate sulcus, *PLS* posterior lateral sulcus, *PMS* posterior median sulcus, *SFZ* Suprafacial collicular zone. *Sup. Cere Ped.* superior cerebellar peduncle

fissure) divides the petrosal surface of the cerebellar hemisphere into superior and inferior parts (Fig. 22.7a). Splitting the petrosal fissure separates the superior and inferior petrosal surfaces and exposes the posterolateral middle cerebellar peduncle (posterior and lateral to the root entry zone of trigeminal nerve) and minimizes the need for cerebellar retraction or resection (Fig. 22.7) [14, 15]. The anterior limit for this entry zone is the CST. The retrolabyrinthine approach can also be used to access this entry zone (Fig. 22.10).

22.4.8 Dorsal Entry Zones

22.4.8.1 Median Sulcus Zone (MS)

This is a potential entry zone through the median sulcus in the midline of the floor of the fourth ventricle above the level of the facial colliculus (Fig. 22.1c) [4]. There is a paucity of crossing fibers in this region making it a potential "safe" entry zone. However, this zone is bordered laterally by the medial longitudinal fasciculus and even slight lateral retraction can cause INO making this approach less practical.

22.4.8.2 Suprafacial Collicular Zone (SFZ)

The SFZ is an entry zone in the floor of the fourth ventricle above the facial colliculus (Fig. 22.1c). It is the area bordered superiorly by the frenulum veli where the trochlear nerve passes and inferiorly by the upper margin of the facial colliculus and measures approximately 13.8 mm in length [16]. The medial border is the MLF and the lateral border is the sulcus limitans. The trigeminal mesencephalic tract (TMT), central tegmental tract (CTT), trigeminal motor, and sensory nuclei are located deep and lateral to the sulcus limitans and can be injured if the incision is made too far lateral. The trigeminal motor nucleus is located approximately 6.3 mm from midline [16]. This entry zone can be used to access lesions in the dorsal pons above the facial colliculus (Fig. 22.8).

22.4.8.3 Infrafacial Collicular Zone (IFZ)

The IFZ is an entry zone in the floor of the fourth ventricle below the facial colliculus (Fig. 22.1c). It is bordered superiorly by the lower margin of the facial colliculus and inferiorly by the upper edge of the hypoglossal trigone

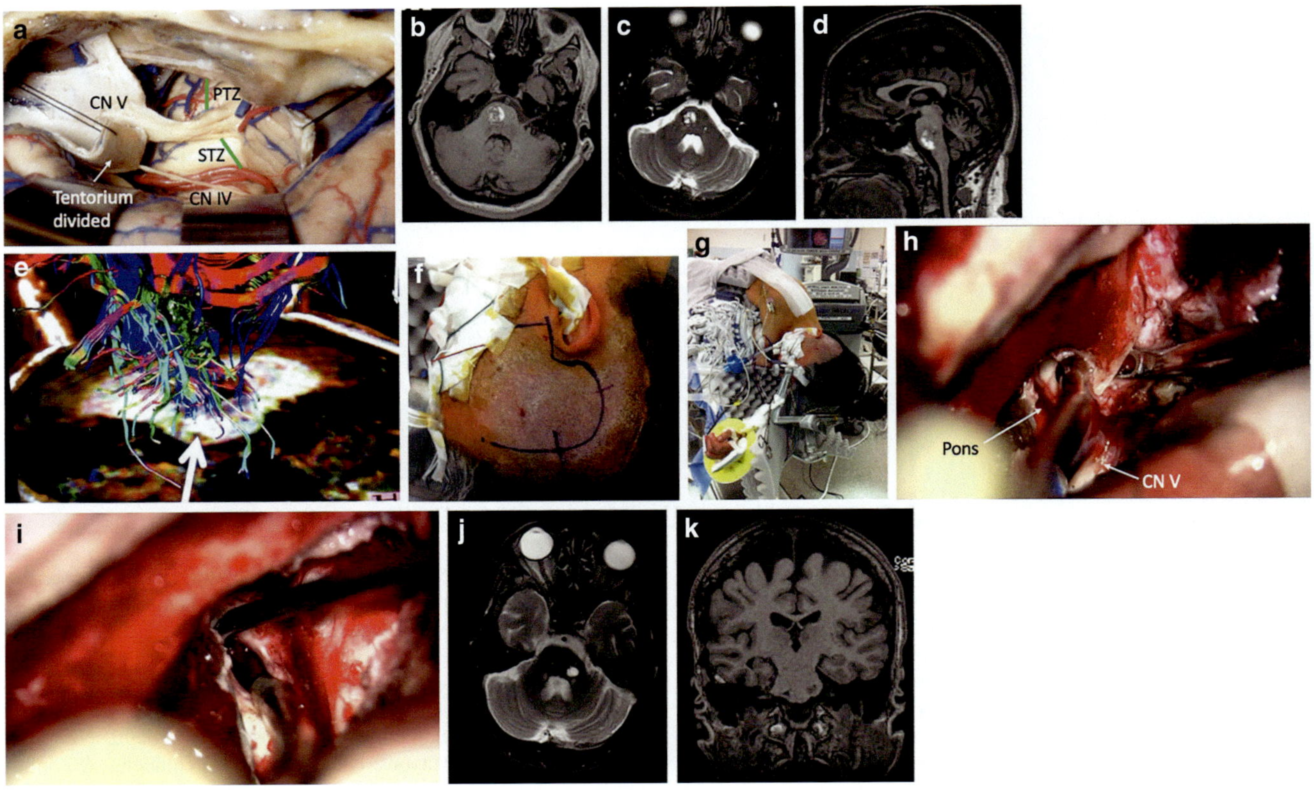

Fig. 22.9 Kawase approach. (**a**) Rhoton dissection showing the right temporal craniotomy performed for the Kawase approach. The Kawase rhomboid has been drilled. The Kawase approach can be performed alone or in addition to the subtemporal approach. Rhoton dissection showing the anterolateral view of the pons after the temporal lobe dura, tentorial dura, and posterior fossa dura have been divided. The Kawase approach allows a more caudal exposure than the subtemporal approach allowing for access to the peritrigeminal zone (PTZ). This approach provides the best direct anterior view of the PTZ. (**b–k**) A 61-year-old-female with familial cavernoma syndrome presented with dysarthria, left central facial weakness, left hemiparesis 3/5, and ataxia and progressive growth of the lower pontine cavernoma. (**b**) Axial T1-weighted, (**c**) axial T2-weighted, (**d**) sagittal T1-weighted MRI show a lower pontine cavernoma nearly reaching the surface of the pons laterally. (**e**) Diffusion tensor imaging shows corticospinal tracts pushed medially, therefore a lateral trajectory was planned. Kawase approach was used to resect this lesion and gives a perfect view down the long axis of the lesion. Retrosigmoid approach is less ideal as it would be difficult to visualize the posterior aspect of the lesion. (**f**) Question mark incision is used. (**g**) Patient is positioned lateral, head flexed, and translated upwards, vertex tilted down. Lumbar drain is used. If the neck is flexible, patient can also be positioned supine with a bump in the same position as done for the subtemporal approach. (**h**) Surgical view after the Kawase rhomboid has been drilled and temporal lobe dura, tentorial dura, and posterior fossa dura have been divided. The rostral pons and CN V are seen in this view. (**i**) The cavernoma is entered through the PTZ in between CN V and CN VII/III. Immediately postop, patient had new transient V1–2 hypoesthesia but was otherwise at neurologic baseline. In subsequent follow-up, patient had complete resolution of all preoperative and postoperative symptoms. (**j** and **k**) T2-weighted (**j**) and T1-weighted (**k**) MRI at 10-years follow-up show no residual or recurrence. There was growth in a separate left pontine cavernoma but the patient was asymptomatic so it is being observed with serial imaging. Abbreviations: *AMZ* anterior mesencephalic zone, *LMS* lateral mesencephalic sulcus, *STZ* supratrigeminal zone, *PTZ* peritrigeminal zone. (**a**) Dissections courtesy of the Rhoton Collection, American Association of Neurological Surgeons (AANS)/Neurosurgical Research and Education Foundation (NREF)

corresponding to the upper and lower borders of the lateral recess or intermediate part of the floor of the fourth ventricle [7]. The mean distance of this area is approximately 9.2 mm rostro-caudally [16]. The medial border is again MLF with the lateral border being nucleus ambiguous and the facial nucleus located deep to the rhomboid fossa. This entry zone can be used to access lesions in the dorsal pons below the facial colliculus.

22.4.9 Medulla Entry Zones

Medullary lesions can be accessed using posterolateral and posterior surgical routes. Six main entry zones have been described: three on the lateral surface of the medulla and three on the dorsal aspect of the medulla (Fig. 22.1b, c). The medullary portion of the floor of the fourth ventricle contains centers for respiratory function and vasomotor control. Care

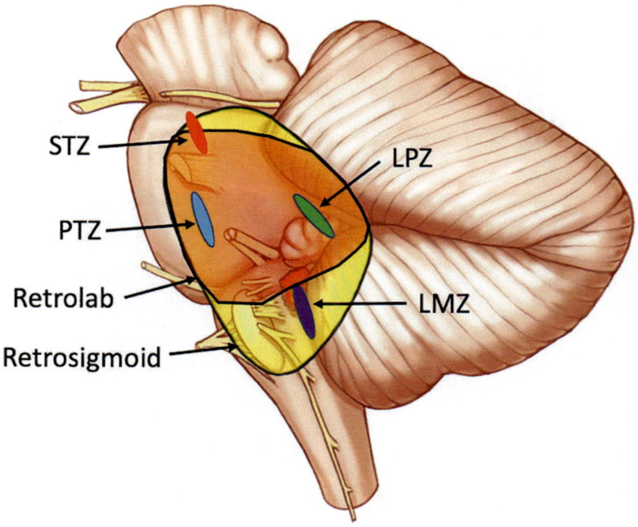

Fig. 22.10 Retrolabyrinthine versus retrosigmoid approach. Rhoton schematic showing the limited view seen in a retrolabyrinthine approach (orange shaded area) compared to the retrosigmoid approach (yellow shaded area). The retrolabyrinthine approach provides a more direct view to the lateral pons compared to retrosigmoid approach but has a more limited view and requires more time-consuming drilling. Due to the limited view, the LMZ cannot be seen with the retrolabyrinthine approach. Similar to a retrosigmoid approach, the STZ and LPZ can be accessed but the direct lateral view of the pons with a retrolabyrinthine approach enables a more optimal angle of attack for the PTZ. The retrosigmoid approach allows a much greater extent of exposure in the cranial-caudal axis but the view is tangential and does not allow for movement in the transverse plane of the brainstem. Whereas the retrolabyrinthine approach provides a more limited view but allows for a more medial and posterior angle of attack. Abbreviations: *LMZ* lateral medullary zone, *LPZ* lateral pontine zone, *STZ* supratrigeminal zone, *PTZ* peritrigeminal zone. Schematic courtesy of the Rhoton Collection, American Association of Neurological Surgeons (AANS)/Neurosurgical Research and Education Foundation (NREF)

must be taken when removing lesions in this location as damage to these structures can result in respiration and swallowing difficulty.

The three lateral entry zones to the medulla are the: olivary zone (OZ), anterolateral sulcus (ALS), and lateral medullary zone (LMZ) (Fig. 22.1b). An anterolateral approach via a far lateral (Fig. 22.11) or low retrosigmoid craniotomy can be used to access the OZ and ALS. The LMZ can be accessed using a posterolateral approach via a retrosigmoid craniotomy (Fig. 22.10) or a posterior approach via a midline suboccipital telovelar approach.

The three dorsal entry zones to the medulla are the: posterior median sulcus (PMS), posterior intermediate sulcus (PIS), and posterior lateral sulcus (PLS) (Fig. 22.1c). These entry zones correspond to the posterior medullary sulci and can be accessed through a posterior surgical route via the midline suboccipital approach with or without C1 laminectomy (Fig. 22.8). Dissection in the region of the calamus scriptorius should be avoided as it contains nuclei of the lower cranial nerves. Lesions located deep to the calamus scriptorius should be approached laterally to avoid traversing these nuclei.

22.4.10 Lateral Entry Zones

22.4.10.1 Olivary Zone (OZ)

The OZ is an entry zone in the olive located on the anterolateral surface of the medulla (Fig. 22.1b). It is bordered anteriorly by the preolivary sulcus and the pyramids, posteriorly by the postolivary sulcus, spinothalamic and tectospinal tracts, and medially by the hypoglossal nerve fibers and medial lemniscus. The safe depth of dissection ranges from 4.7 to 6.9 mm with a rostro-caudal length of 13.5 mm [12]. This entry zone can be used to access lesions in the anterolateral medulla.

22.4.10.2 Anterolateral Sulcus (ALS)

The ALS also known as the preolivary sulcus entry zone is located just lateral to the pyramids and anterior to the olive (Fig. 22.1b). The incision is made in the preolivary sulcus in the space between the caudal hypoglossal rootlets and rostral C1 rootlets and corresponds to the level of the decussation of pyramids [17]. It is bordered anteriorly by the pyramids, posteriorly by the olive, and medially by the hypoglossal nerve fibers and medial lemniscus. It is preferred for exophytic lesions in the anterior lower medulla given proximity to the pyramidal tracts.

22.4.10.3 Lateral Medullary Zone (LMZ)

The LMZ also known as the inferior cerebellar peduncle approach is a vertical incision made in the inferior cerebellar peduncle inferior to the facial nerve/vestibulocochlear complex and posterior to the origin of the glossopharyngeal and vagus nerve (Fig. 22.1b, c) [18]. This zone can be used to resect dorsolateral medullary lesions and is analogous to the lateral pontine zone (LPZ) in the pons.

22.4.11 Dorsal Entry Zones

22.4.11.1 Posterior Median Sulcus (PMS)

The PMS is an entry zone through the posterior median sulcus in the midline below the obex and is bordered laterally by the gracilis fasciculi (Fig. 22.1c).

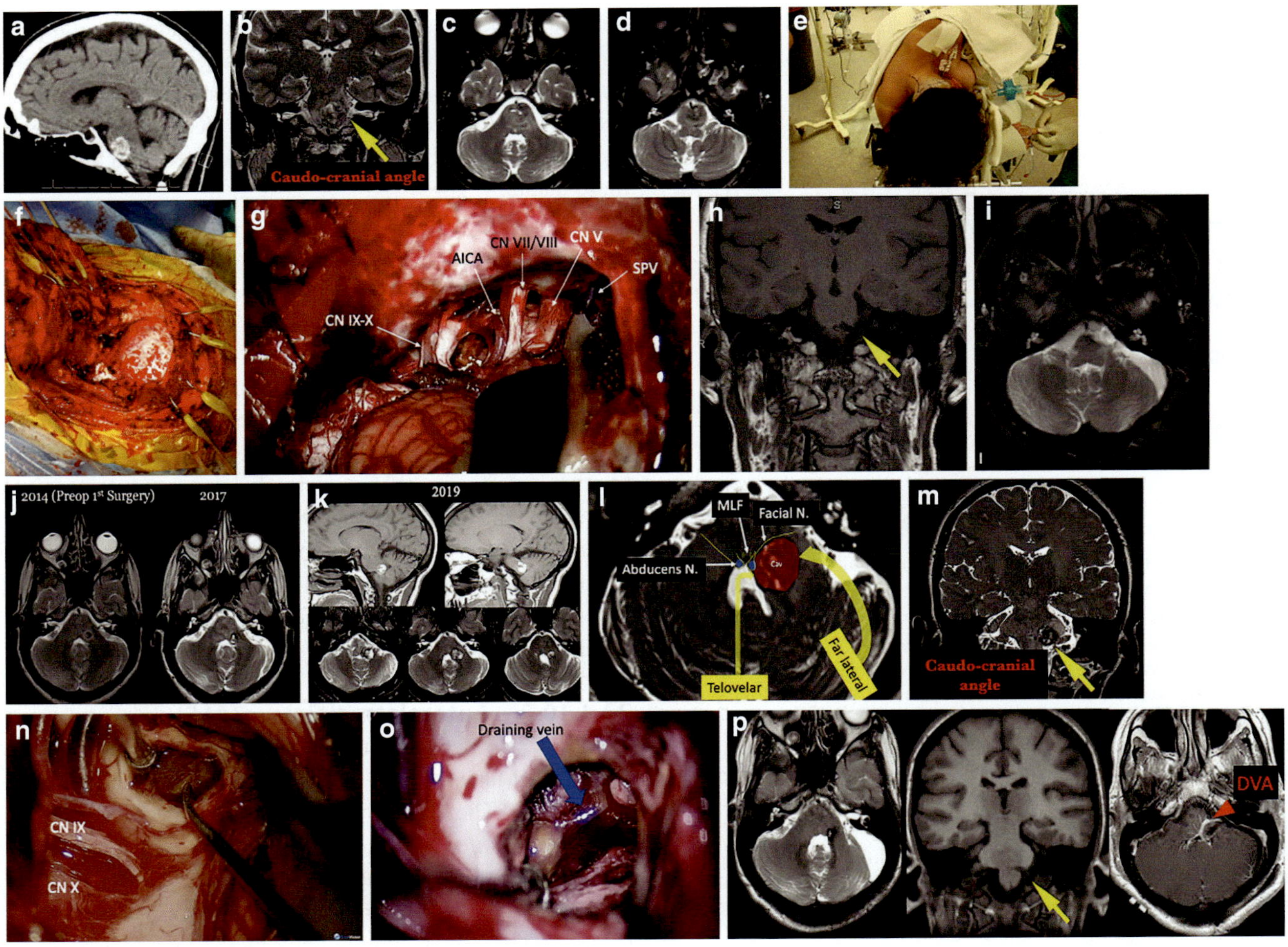

Fig. 22.11 Far lateral approach. (**a–i**) This is a 31-year-old female who presented with sudden onset right hemibody weakness (right upper 4/5, right lower 4+/5). (**a**) Sagittal CT shows lower pontine cavernoma with acute hemorrhage. (**b**) Coronal CISS sequence, axial T2-weighted MRI (C and D) shows a lower pontine that reaches the surface of the lateral pons at the level of the pontomedullary junction with the long axis of the lesion extending cranial-caudally from above the IAC to the pontomedullary junction. The far lateral approach was used as it allows a caudo-cranial ergonomic direction that is not allowed with the retrosigmoid approach (yellow arrow in E). (**e**) Patient is positioned 3/4 prone, neck is flexed, vertex tilted down, head rotated, and translated upwards. (**f**) Surgical view after the C1 laminectomy, occipital craniotomy, and paracondylar bone are drilled. The extent of the craniotomy, C1 arch, and paracondylar drilling is tailored based on the pathology. The angle of attack for anterolateral medullary lesions can be widened by drilling the posterior third of the occipital condyle. In this particular case, the far lateral approach was combined with a retrosigmoid craniotomy because the superior most extent of the pathology reached above the level of the IAC. (**g**) Final surgical view after the cavernoma has been removed. Note the caudo-cranial trajectory that is achieved with the far lateral approach. The patient remained at neurological baseline immediately postoperatively. She had complete resolution of her preoperative symptoms at 1-month follow-up. Coronal T1-weighted MRI (**h**) and axial T2-weighted (**i**) MRI at 3-years follow-up show no residual or recurrence. The yellow arrow demonstrates the caudo-cranial trajectory allowed by the far lateral approach. (**j–p**) This next case illustrates the concept that the most obvious entry point as suggested by the two-point method is not always the safest approach, especially for lesions in the floor of the fourth ventricle. This is a 27-year-old female who presented to another institution with left facial droop, numbness, diplopia, and imbalance in 2014 and was found to have a left middle cerebellar peduncle cavernoma with no surface representation (**j**, left image). She underwent a telovelar trans-fourth ventricular approach at that institution. Subtotal resection was achieved. Postoperatively, she required a two-month hospital stay, developed hemiparesis, House-Brackmann 6/6 facial palsy, and was ventilator dependent. Patient subsequently underwent facial reanimation procedure and a platinum implant in the left eye. The residual cavernoma continued to grow (**j**, right image). In 2019, she presented to our institution with gradual dizziness, imbalance, left V1/V2 hypoesthesia, House-Brackmann 3/6 facial weakness, left hypacusis, and nystagmus with left lateral gaze. (**k**) MRI at time of presentation to our institution shows a cavernoma extending from the medulla near the jugular foramen to the upper pons near the exit zone of the trigeminal nerve with the long axis in the cranial-caudal direction. (**l**) The lesion reaches the surface of the floor of the fourth ventricle but is not exophytic. Based on the two-point method, the telovelar approach would have been the most obvious approach. However, the far lateral approach through the middle cerebellar peduncle which traverses through more brain tissue was used in this case to avoid injury to structures of the floor of the fourth ventricle and allowed us to achieve a caudo-cranial trajectory along the long axis of the lesion (**m**). (**n**) An entry point is made in the middle cerebellar peduncle. (**o**) Surgical view showing the draining vein of the cavernoma that must be resected to prevent recurrence. Postoperatively, her dizziness, V1/V2 hypoesthesia, and hearing improved, her left facial palsy remained at preoperative baseline, and she has now returned to work fully. (**p**) MRI 2-years postop shows no residual or recurrence. The yellow arrow again demonstrates the caudo-cranial trajectory allowed by the far lateral approach. The DVA was preserved. Abbreviations: *Abducens N.* abducens nerve, *DVA* developmental venous anomaly, *Facial N.* facial nerve, *LMZ* lateral medullary zone, *MLF* medial longitudinal fasciculus, *OZ* olivary zone, *RCL* rectus capitis lateralis, *SPV* superior petrosal vein

22.4.11.2 Posterior Intermediate Sulcus (PIS)

The PIS is an entry zone between the gracile and cuneate fasciculi (Fig. 22.1c).

22.4.11.3 Posterior Lateral Sulcus (PLS)

The PLS is an entry zone lateral to the cuneate fasciculus (Fig. 22.1c). It is bordered medially by the cuneate fasciculus, laterally by the spinal trigeminal tract and nucleus. The nucleus ambiguus is typically 4 mm deep to the surface of this entry zone [7].

22.5 Surgical Pearls

Surgical principles for brainstem cavernoma resection are divided into three parts:

- Goals—conceptual reasoning of surgery.
- Strategies—analytical planning of surgery.
- Tactics—practical execution of those strategies.

These are further subdivided into four steps (Fig. 22.12) with the main goal of the surgery being (1) Get there, (2) Find the lesion, (3) Leave no trace, and (4) Resect it completely.

Step I: Get there

In order to get to the lesion, knowledge of *surface geometry* is important. Proper patient positioning and appropriately sized and placed craniotomies are necessary to optimize the surgical corridor. It is important to remember that the craniotomy only gives static exposure.

Step II: Find the lesion

To find the lesion, understanding *depth geometry* is critical. Wide arachnoidal dissection is important. During arachnoidal dissection, use of fixed self-retaining retractors is not necessary and even detrimental to surgical freedom. Instead, dynamic retraction with the shaft of instruments can be used and it minimizes trauma to the surrounding brain without sacrificing the surgical view. Self-retaining retractors can be used as a third hand at a later stage once arachnoidal dissection has been completed, if necessary. The two-point method is a simplified strategy that can be used to determine the most appropriate entry point in the brainstem. In this method, a point is placed at the center of the lesion (Point A). A second point is placed at a location where the lesion is most superficial (Point B). The trajectory of the line between Point A and Point B dictates the surgical approach [19]. However, this method is a general starting point and may need to be ignored if eloquent structures fall on the line. Intraoperative navigation can aid in selecting the appropriate entry point in cases

STEPS	GOALS	STRATEGIES	TACTICS
1	Get there	Surface geometry	- Positioning - Craniotomy choice (***static*** exposure)
2	Find it	Depth geometry	- Arachnoidal dissection - ***Dynamic*** retraction - Intraoperative navigation/MRI-DTI - Entry point in brainstem
3	Leave no trace	Optimal intra-axial neural path	- Minimal neural distortion - Avoidance of "cornering" - Preserve main DVA - Neurophysiological monitoring
4	Resect it completely	Microsurgical ergonomics	- Evacuate hematoma - Maintain gliotic plane - Bipolar shrinkage to gain space - Avoid piecemeal (if possible) - Endoscope assistance - Resect draining vein, NOT DVA

Fig. 22.12 Surgical principles of brainstem cavernoma resection are divided into goals, strategies, and tactics. These are further divided into four steps with the goals being (1) get there, (2) find it, (3) leave no trace, and (4) resect it completely. Strategies and tactics to achieve those goals are listed in the table above

where the lesion does not reach a pial or ependymal surface.

Step III: Leave no trace.

This means finding the optimal intra-axial path that minimizes distortion or damage to adjacent neural pathways. This is where the two-point method can be flawed and is an incomplete concept in planning approaches to brainstem cavernomas. There are two main caveats to the method. The first caveat is that the most obvious suggested entry point is not always the safest approach. The lateral surface of the pons is a more forgiving region for entering the pons than the floor of the fourth ventricle. The eloquence of the floor of the fourth ventricle must be taken into account when selecting approaches using the two-point method and should be the primary consideration. A longer trajectory through the lateral pontine zone (LPZ) may be safer than entry through the fourth ventricle floor. Figure 22.11j–p shows an example of this. In this case, the telovelar approach through the floor of the fourth ventricle would be the approach predicted by the two-point method. However, although the middle cerebellar peduncle entry zone traverses more brain tissue, it was used because it avoids the morbidity associated with traversing the structures in the floor of the fourth ventricle. Simply put, if steps 2 and 3 result in contradicting choices for an approach, always compromise in favor of step 3. For exophytic fourth ventricular lesions, entry through the floor of the fourth ventricle may be considered (Fig. 22.8).

The second caveat is that the surgical axis should approximate as much as possible the long axis of the lesion. Approaches that are perpendicular to the long axis of the lesion can create blind spots and increase the risk of retraction injury during resection. Blind spots can be identified preoperatively by the "right-angle method" [20]. A perpendicular line from the surgical axis to the point on the lesion farthest from the surgical axis is drawn. Any perpendicular line greater than 1 cm in length is a potential blind spot. Based on this concept, ledges of eloquent tissue in the near field can be blind spots making areas near corners more difficult to visualize than far corners [20]. It is in instances such as these that the use of an angled endoscope can be very useful to visualize and resect a residual cavernoma with angled instruments.

In addition to this, another tactic to leave no trace is to preserve the developmental venous anomaly (DVA) which drains normal brain parenchyma. Use of neurophysiologic monitoring also allows for more accurate and safer surgical resection of lesions. Neurophysiologic monitoring can identify the suspected location of the facial colliculus on the fourth ventricular floor (Fig. 22.8e, f). It can also alert the surgeon when the dissection is too rough or enters into tracts and nuclei. A sudden change in neurophysiologic signals can result in hesitation by the surgeon and lead to a remnant. Small decrements in signals may translate into only transient postoperative deficits or could be a warning signal that neural damage is actively taking place. A surgical pause should be taken and if the signal improves, then a slight modification of the manner of resection should be tried. There is no applicable general dictum in cases like these. It is during these moments that surgical experience will inspire wisdom and judgment, and each case should be analyzed individually. The decision to proceed or stop is a complex admixture of prior experience, an intimate understanding of surgical maneuvers being utilized, and the specific anatomopathologic features of the case at hand. There must be a balance between heeding and ignoring neuromonitoring to prevent recurrence and new postoperative deficits.

Step IV: Resect it completely.

Cavernomas must be resected completely to prevent recurrence. The cavernoma is first entered through a small opening and the liquified hematoma is drained. Intracapsular dissection follows to reduce the lesion's size, creating more working space within the brainstem. Care is taken to preserve the capsule while working intracapsularly in order to keep the margins of the lesion whole and protect the surrounding parenchyma during dissection. Working within the capsule, caverns are opened and interconnected internally. As this is done, the capsule begins to pull away from the surrounding parenchyma. Dissection then changes from intracapsular to extracapsular. Circumferential dissection is carried out while maintaining the gliotic plane using a combination of sharp dissection, traction, and countertraction. Most cavernomas are shaped like a "cauliflower," meaning that they have irregular protruding parts that bury into brainstem tissue and can therefore be easily amputated and missed, since they can be easily confused with the multicolored gliotic wall that harbors blood products of various ages. Bipolar shrinkage is used to collapse the capsule inward to gain more working space. Care must be taken not to coagulate the gliotic plane itself. The lesion is removed once circumferentially dissected. There is often a small draining vein leading to a developmental venous anomaly (DVA). The senior author used to avoid sacrificing any vein early in his practice but experienced recurrence in two such cases and reexplored them, re-resected the recurrences including a sacrifice of the small veins, with no further recurrence after several years. It is occurrences like these that give credence to the belief that cavernoma is a venous disease (Fig. 22.11o). The DVA itself, on the other hand, should be preserved as it drains normal brain parenchyma (Fig. 22.11p). Cavernomas are mulberry-like or cauliflower-like lesions so piecemeal removal should be avoided, if possible, to prevent residual.

In addition, they can be composed of several nodules separated by a thin layer of gliotic brain which may mask nodules so the resection cavity should be inspected thoroughly to ensure no residual remains. As mentioned earlier, the endoscope is a complementary tool that can be used to inspect anticipated blind spots (Fig. 22.5g).

22.6 Discussion

22.6.1 Common Pitfalls

Not unlike cavernous sinus lesions, brainstem cavernomas had once been considered inoperable owing to the rich concentration of cranial nerve nuclei and fibers in a small cross-sectional area imparting significant morbidity. Knowing the indications for surgery for patients with brainstem cavernomas is important but there is a lack of agreement on indications. In general, we recommend resection for (1) young symptomatic patients (high cumulative lifetime risk of hemorrhage); (2) progressive neurologic deficits; (3) cavernomas with significant mass effect causing neurological symptoms; and (4) after a hemorrhage if the lesion is surgically accessible. Patients with (1) incidentally discovered lesions; (2) asymptomatic; and (3) surgically inaccessible lesions are followed conservatively. It is very illuminating to see to what degree international experts agree/disagree with respect to various aspects of brainstem cavernoma management [21].

Although this chapter summarizes the approaches and "safe entry zones" to the brainstem, one must be aware that no area of the brainstem is truly "safe." These entry zones only represent regions between eloquent neural structures through which a lesion can be accessed with the least morbidity. It is also important to note that "anatomy" does not equate "pathology," but is modified by it. The concept of "safe entry zones" is meaningless when a lesion distorts the anatomy. A lesion can displace nuclei and tracts so that entry into a "safe entry zone" could result in significant morbidity. The displacement of tracts and nuclei must be taken into account when planning the surgical approach. This is indeed the crux of the matter with surgery of brainstem cavernoma.

Lastly, radiosurgery does not work for brainstem cavernomas. It often creates brainstem edema and can create more scarring in the plane between the brainstem and cavernoma making it more difficult to resect (Fig. 22.6b–o).

22.7 Conclusion

Neurosurgeons who wish to tackle brainstem cavernomas must have, among other attributes of technique, training, and character, a profound anatomical knowledge of brainstem anatomy and its dense concentration of tracts and nuclei. This chapter reviews the relevant anatomy, the main entry zones of the brainstem, and the surgical approaches utilized to access them with corresponding illustrative cases. Surgical tips and common pitfalls are also reviewed.

References

1. Bricolo A, Turazzi S, Cristofori L, Talacchi A. Direct surgery for brainstem tumours. Acta Neurochir Suppl (Wien). 1991;53:148–58. https://doi.org/10.1007/978-3-7091-9183-5_25.
2. Baghai P, Vries JK, Bechtel PC. Retromastoid approach for biopsy of brain stem tumors. Neurosurgery. 1982;10(5):574–9.
3. Kyoshima K, Kobayashi S, Gibo H, Kuroyanagi T. A study of safe entry zones via the floor of the fourth ventricle for brain-stem lesions: report of three cases. J Neurosurg. 1993;78(6):987–93. https://doi.org/10.3171/jns.1993.78.6.0987.
4. Bricolo A, Turazzi S. Surgery for gliomas and other mass lesions of the brainstem. Adv Tech Stand Neurosurg. 1995;22:261–341. https://doi.org/10.1007/978-3-7091-6898-1_5.
5. Ruchalski K, Hathout GM. A Medley of midbrain maladies: a brief review of midbrain anatomy and syndromology for radiologists. Radiol Res Pract. 2012;2012:258524. https://doi.org/10.1155/2012/258524.
6. Zenonos GA, Fernandes-Cabral D, Nunez M, Lieber S, Fernandez-Miranda JC, Friedlander RM. The epitrigeminal approach to the brainstem. J Neurosurg. 2018;128(5):1512–21. https://doi.org/10.3171/2016.12.Jns162561.
7. Spetzler RF, Kalani MYS, Lawton MT. 2 Anatomy of the Brainstem, Thalamus, Pineal Region, and cranial nerves. In: Surgery of the brainstem. Stuttgart: Georg Thieme Verlag KG; 2020. https://doi.org/10.1055/b-0039-173893.
8. Khan NR, Brown CS, Angeli S, Morcos JJ. Resection of a meningioma, vestibular schwannoma, and placement of auditory brainstem implant using translabyrinthine approach. Neurosurg Focus Video. 2021;5(2):V18. https://doi.org/10.3171/2021.7.Focvid2163.
9. Kalani MYS, Yağmurlu K, Spetzler RF. The interpeduncular fossa approach for resection of ventromedial midbrain lesions. J Neurosurg. 2018;128(3):834–9. https://doi.org/10.3171/2016.9.Jns161680.
10. Vishteh AG, David CA, Marciano FF, Coscarella E, Spetzler RF. Extreme lateral supracerebellar infratentorial approach to the posterolateral mesencephalon: technique and clinical experience. Neurosurgery. 2000;46(2):384–8.; discussion 8-9. https://doi.org/10.1097/00006123-200002000-00022.
11. Yagmurlu K, Rhoton AL Jr, Tanriover N, Bennett JA. Three-dimensional microsurgical anatomy and the safe entry zones of the brainstem. Neurosurgery. 2014;10(Suppl. 4):602–19.; discussion 19-20. https://doi.org/10.1227/neu.0000000000000466.
12. Recalde RJ, Figueiredo EG, de Oliveira E. Microsurgical anatomy of the safe entry zones on the anterolateral brainstem related to surgical approaches to cavernous malformations. Neurosurgery. 2008;62(3 Suppl. 1):9–15.; discussion–7. https://doi.org/10.1227/01.neu.0000317368.69523.40.
13. Cavalheiro S, Yagmurlu K, da Costa MD, Nicácio JM, Rodrigues TP, Chaddad-Neto F, et al. Surgical approaches for brainstem tumors in pediatric patients. Childs Nerv Syst. 2015;31(10):1815–40. https://doi.org/10.1007/s00381-015-2799-y.
14. Kalani MYS, Yagmurlu K, Martirosyan NL, Spetzler RF. The retrosigmoid petrosal fissure transpeduncular approach to central pontine lesions. World Neurosurg. 2016;87:235–41. https://doi.org/10.1016/j.wneu.2015.11.025.
15. Rutledge C, Tonetti DA, Raygor KP, Abla AA. How I do it: horizontal fissure approach to the middle cerebellar peduncle.

Acta Neurochir. 2022;164(3):763–6. https://doi.org/10.1007/s00701-021-05015-7.00.

16. Strauss C, Lütjen-Drecoll E, Fahlbusch R. Pericollicular surgical approaches to the rhomboid fossa. Part I. Anatomical basis. J Neurosurg. 1997;87(6):893–9. https://doi.org/10.3171/jns.1997.87.6.0893.
17. Cantore G, Missori P, Santoro A. Cavernous angiomas of the brain stem. Intra-axial anatomical pitfalls and surgical strategies. Surg Neurol. 1999;52(1):84–93. https://doi.org/10.1016/s0090-3019(99)00036-1. discussion-4
18. Deshmukh VR, Figueiredo EG, Deshmukh P, Crawford NR, Preul MC, Spetzler RF. Quantification and comparison of telovelar and transvermian approaches to the fourth ventricle. Neurosurgery. 2006;58(4 Suppl. 2):ONS-202-6. discussion ONS-6-7. https://doi.org/10.1227/01.Neu.0000207373.26614.Bf.
19. Brown A, Thompson B, Spetzler R. The two-point method: evaluating brain stem lesions. BNI Q. 1996;12(1):20–4.
20. Garcia RM, Oh T, Cole TS, Hendricks BK, Lawton MT. Recurrent brainstem cavernous malformations following primary resection: blind spots, fine lines, and the right-angle method. J Neurosurg. 2020;135(3):671–82. https://doi.org/10.3171/2020.6.Jns201555.
21. Dammann P, Abla AA, Al-Shahi Salman R, Andrade-Barazarte H, Benes V, Cenzato M, et al. Surgical treatment of brainstem cavernous malformations: an international Delphi consensus. J Neurosurg. 2021:1–11. https://doi.org/10.3171/2021.3.Jns2156.

Surgical Revascularization Strategies for Vertebral Artery Disease

23

Fady T. Charbel and Laura Stone McGuire

Learning Objectives

- Learn relevant skull base and neck anatomy related to vertebral artery bypass.
- Understand different surgical approaches to vertebral artery bypass.
- Establish flow-based paradigm for preoperative assessment and postoperative management.

23.1 Introduction

Posterior circulation strokes account for up to a third of ischemic strokes. Symptomatic vertebrobasilar disease carries a high annual recurrent stroke risk, averaging 10–15% per year. Narrowing or occlusion of the vertebral arteries may be caused by atherosclerosis, dissection, thrombosis, compression, and fibromuscular dysplasia. Vertebral artery (VA) disease can lead to reduced blood flow to the brainstem and cerebellum, resulting in a range of neurological symptoms, including vertigo, dizziness, dysarthria, and even stroke. While collateral circulation typically compensates for a flow deficit produced by VA stenosis, many patients have inadequate intracranial anastomoses due to incomplete circle of Willis, hypoplastic vessels, or diffuse atherosclerotic disease. Studies have indicated that stroke risk in vertebrobasilar disease is related to hemodynamic compromise [1–6]; in particular, stenosis of 80% or more generates the most significant lowering of cerebral blood flow, with tandem disease creating more compromise [7].

Revascularization of the VA should be considered in symptomatic patients after failure of medical management. In the age of growing endovascular devices and procedures, restoration of blood flow may involve transluminal balloon angioplasty or stent placement; however, certain cases may still require surgery. The choice of revascularization strategy depends on various factors, such as location and severity of the blockage as well as patient comorbidities and history of prior intervention. The complexity of surgical revascularization of the VA requires careful consideration of skull base and neck anatomy. While bypass procedures involving the VA may be performed for chronic steno-occlusive disease or for cerebral aneurysms, despite significantly different indications, several key principles of the surgical technique remain. This chapter will focus on bypass of the VA in the setting of ischemic pathology, highlighting technical nuances of the procedure, preoperative planning, and postoperative management.

23.2 History

Surgical techniques for bypass of the VA emerged in the 1970s. Berguer et al. pioneered the subclavian artery to VA bypass in their 1976 case series of four patients [8]. In 1978, Carney and Anderson described the surgical technique for a carotid-distal VA bypass for carotid occlusion, which supported the concept of improving blood flow to the anterior circulation via VA and subsequently the posterior communicating artery [9]. Nearly a decade later, in 1986, Carney et al.

F. T. Charbel (✉) · L. S. McGuire
Department of Neurosurgery, University of Illinois, Chicago, IL, USA
e-mail: fcharbel@uic.edu; lmcguir1@UIC.EDU

I. N. Kanaan, V. Beneš (eds.), *Neuroanatomy Guidance to Successful Neurosurgical Interventions*,
https://doi.org/10.1007/978-3-031-59838-8_23

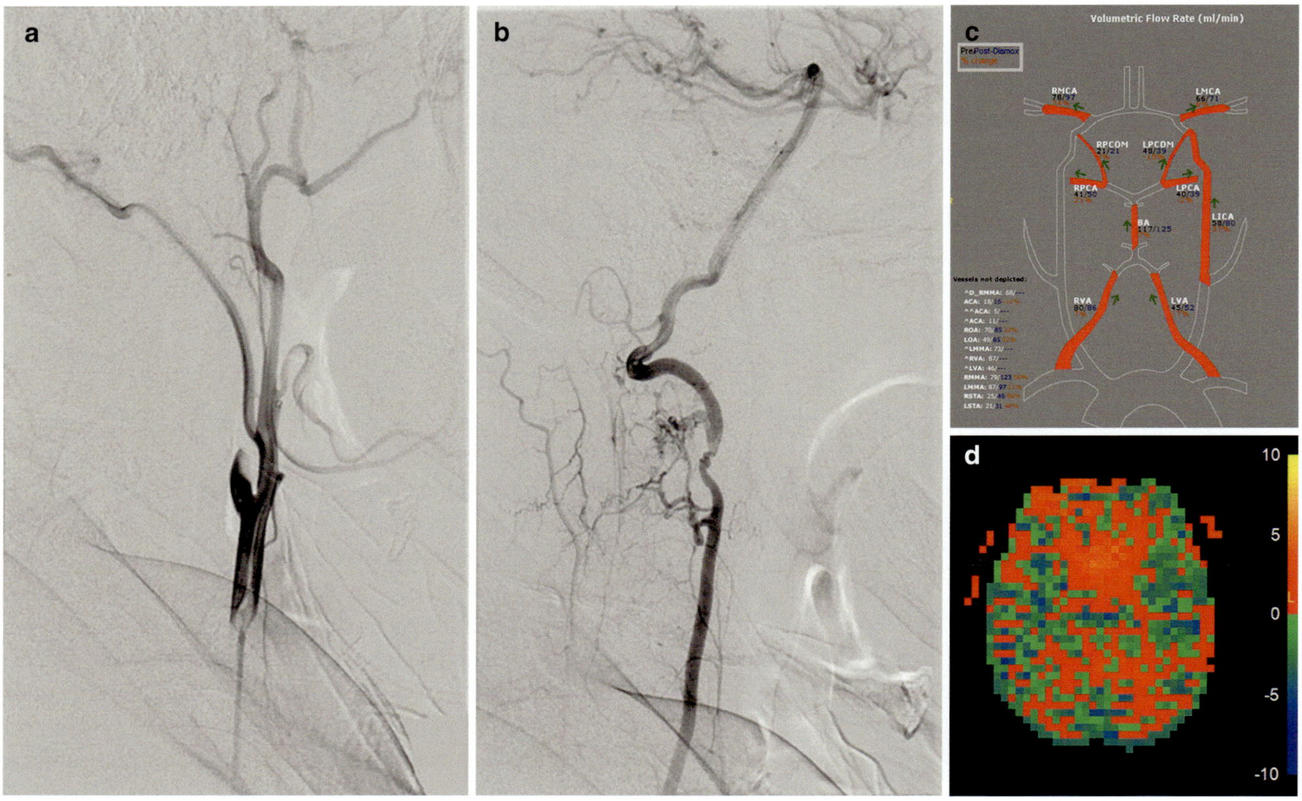

Fig. 23.1 Baseline imaging for patients with bilateral ICA occlusion, left VA occlusion, and progressive right VA stenosis. (**a**) Angiogram showing right common carotid artery (lateral) and (**b**) right VA (lateral). (**c**) Quantitative MRA with Diamox challenge and (**d**) global BOLD imaging showing impaired hemodynamic reserve

described their experience among 219 patients who underwent VA revascularization to increase the vessel's flow capacity in occlusive disease, including two intracranial, 118 skull base, and 99 neck base [10]. Blood supply for the bypasses was routed most commonly from the common carotid artery, followed by the external or internal carotid arteries, the subclavian artery, and the occipital artery. Of these surgeries, 91 completed the bypass between the transverse process of C1 and C2. Overall, they showed low perioperative mortality and stroke rate of 3% and long-term graft patency of 87%, featuring the general safety and efficacy of the procedure (Figs. 23.1, 23.2 and 23.3).

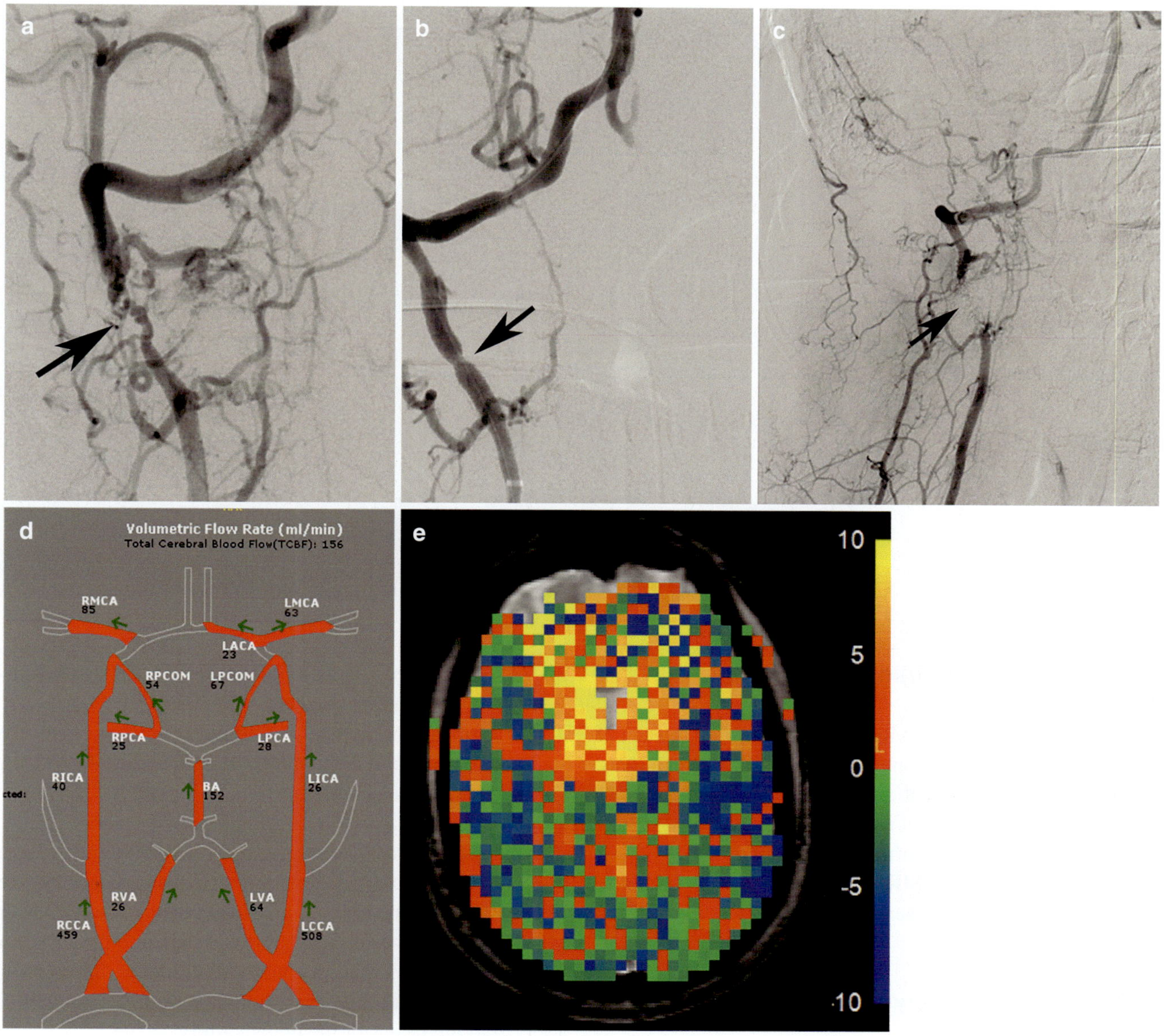

Fig. 23.2 Endovascular stenting and balloon angioplasty for patients with bilateral ICA occlusion, left VA occlusion, and progressive right VA stenosis. (**a**) Angiogram showing right VA stenosis preintervention (AP), (**b**) right VA postintervention (AP), and (**c**) right VA with occlusion (oblique). Black arrows point to regions of stenosis, intervention, and occlusion. (**d**) Quantitative MRA shows worsening cerebral blood flow and (**e**) global BOLD imaging indicates worsening hemodynamic reserve

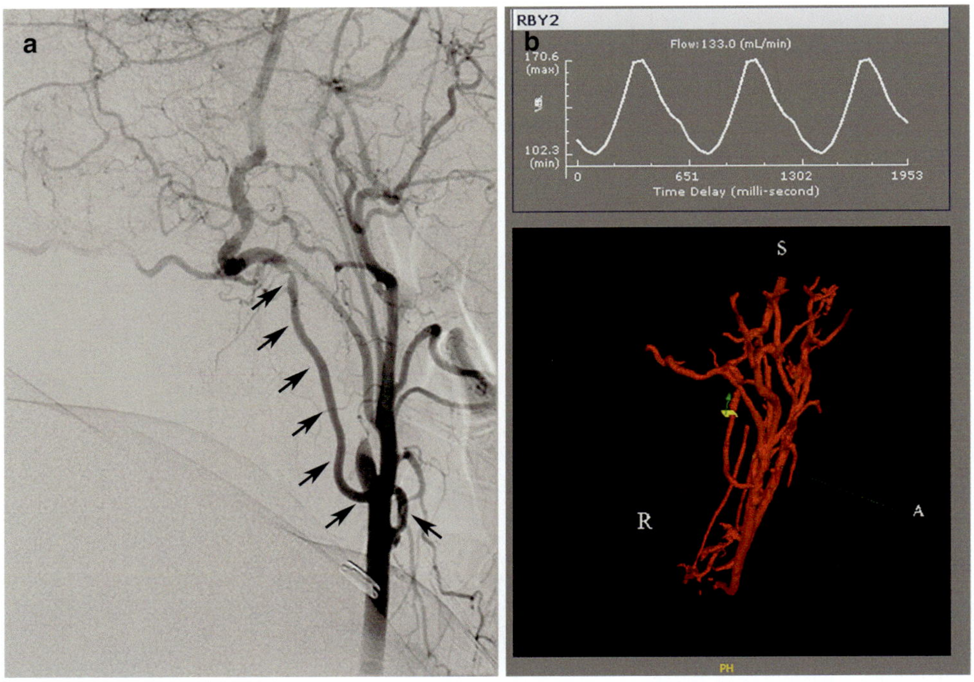

Fig. 23.3 Following occlusion of right VA, patients underwent right common carotid artery to V3 segment bypass. (**a**) Angiogram with patent right CCA-V3 bypass (lateral) and (**b**) quantitative MRA showing patent bypass with adequate blood flow. Black arrows follow the path of the bypass graft

23.3 Anatomy Highlights

- The vertebral artery is divided into four segments, each of which may require a different surgical approach to access.
- Preoperative assessment should include digital subtraction angiography as well as CT angiography to visualize surrounding bony and soft tissue anatomy.
- Consider quantitative MRA and MR-based protocols to assess hemodynamic reserve.

23.3.1 Vertebral Artery Anatomy

The VA typically originates from the subclavian artery, although in up to 5% it may originate at the aortic arch, and runs through the transverse foramina of the cervical spine, typically entering at C6. The VA then exits at C2 to loop around the sulcus arteriosus of C1 and enter the skull base through the foramen magnum. Occasionally, a bony bridge may form atop the sulcus arteriosus due to the ossification of the atlanto-occipital ligament of the atlas, creating a foramen and enclosing the VA and first cervical nerve root, called a *posticulus ponticus*. Awareness of this anatomical variant becomes important during surgical planning.

The VA is divided into four segments, essentially in relationship to the cervical spine:

1. V1 segment: from the origin at the subclavian artery to its entry point at the C6 transverse foramen.
2. V2 segment: from the transverse foramen of C6–C2.
3. V3 segment: from its exit point at the C2 transverse foramen to the foramen magnum.
4. V4 segment: intracranial, intradural component from the foramen magnum to its merge with the contralateral VA to form the basilar artery.

Atherosclerosis may occur at any point of the VA but most often at V1, while dissections more frequently occur at the V3 or V4 segments. The VA provides various branches as it ascends through the cervical spine and then into the foramen magnum, including multiple muscular branches, the anterior and posterior spinal arteries, the posterior inferior cerebellar artery, and the posterior meningeal artery.

23.3.2 Preoperative Assessment

Digital subtraction angiography (DSA) offers key insights into the anatomy of the VA, degree of vessel narrowing, collateral flow to the posterior circulation, and consideration for a possible graft and its implantation site. DSA will indicate the location of VA reconstitution, and thus, where the distal anastomosis may occur. Computed tomography angiography (CTA) also provides important information regarding surrounding bony anatomy, musculature, and soft tissue. The surgeon will note the path of the VA and any deviation from normal anatomy, including its entry into the cervical transverse foramina and its course around C1.

In addition, the assessment of hemodynamic reserve in steno-occlusive disease represents a crucial step in the decision-making process for cerebral revascularization. At our institution, this process entails a multimodal magnetic resonance imaging (MRI)-based protocol [11–13], including (1) local blood oxygenation level-dependent (BOLD) signal due to increased regional blood flow with task performance during functional MRI, (2) global BOLD signal response due to increased widespread cerebral blood flow following a vasodilatory challenge, and (3) volumetric flow measurements in the large vessels of the neck and head using phase-contrast quantitative MR angiography (QMRA) to assess flow before and after global vasodilation with intravenous acetazolamide (Noninvasive Optimal Vessel Analysis; VasSol, Inc., Illinois, USA). Absence of the expected responses in each sequence allows detection of compromised or absent hemodynamic reserve and may support the need for intervention.

Note that in ischemic disease, a balloon occlusion test (BOT) would not be required. Briefly, however, in the context of VA bypass for complex aneurysm surgery or skull base tumor resection, BOT should be considered in the preoperative evaluation. The BOT uses temporary balloon occlusion of the vessel of interest with an awake patient with constant electroencephalographic (EEG) monitoring and serial neurological exams. The patient is heparinized to avoid procedure-related thromboembolic complications. After 10–15 min of vessel occlusion under normal blood pressure, a hypotensive challenge ensues for an additional 10–15 min. While hypotensive, 99m technetium is administered, and a single photon emission computed tomography (SPECT) is performed immediately after the BOT to assess cerebral blood flow. The SPECT study may show asymmetries in distribution of the 99m technetium tracer if the patient experiences hypoperfusion of the affected vascular territory during hypotension. After completion of the BOT, the balloon is deflated and blood pressure normalized.

23.4 Surgical Approach

23.4.1 Approach by Segment

The affected VA segment dictates the surgical approach. Revascularization surgery tends to focus on either V1 to treat focal VA origin disease or on V2–3 to address stenosis or occlusion of the intraspinal segments. Below are possible bypass options based on the affected segment:

- V1 segment: vertebral artery transposition to common carotid artery.
- V2 or V3 segments: vertebral artery bypass with interposition graft to occipital artery, common carotid artery, or subclavian artery.

Because V2 ascends intraosseously, it is rarely accessed directly for surgery, as direct exposure would require resection of the transverse process of the cervical vertebrae. Therefore, the V1 and V3 segments, which are easily accessible by open surgery, may offer better options for control, and lesions in the V2 segment may be managed with a bypass to the V3 segment. When necessary, revascularization of the V4 segment is predominantly by endovascular intervention, and surgery of the V4 segment would require an intracranial approach, such as far lateral craniotomy to access.

23.4.2 Patient Positioning

Positioning of the patient relies strictly on the bypass planned, with the sites for proximal and distal anastomosis within view. Generally, the patient is positioned supine with the head in slight extension and turned away from the affected vessel. A shoulder roll may be necessary, depending on the flexibility of the patient. The use of head fixation depends on surgeon preference, but by placing the patient's head on a donut, the ability to rotate the neck during surgery is maintained.

23.4.3 Donor (Inflow) Selection

The selection of both sites for anastomosis depends on the anatomy of the patient and the precise location of vessel disease. For instance, if the stenosis or occlusion of the VA is focally located at its origin, then it may be disconnected from the subclavian artery, eliminating the diseased portion, and either reimplanted to the subclavian artery versus transposed to the carotid artery. In contrast, if the stenosis or occlusion of the VA is located more distally in the V2 or V3 segments, then an interposition graft may be required to connect the VA to its donor—the cervical internal carotid artery, external carotid artery, common carotid artery, subclavian artery, or occipital artery. Options for the interposition graft include radial artery and saphenous vein.

23.4.4 Vertebral Artery (V1) Transposition Technique

Proximal VA disease (V1 segment) yields itself to a VA transposition. The subclavian artery and common carotid artery have been described as options for the inflow artery, but either vessel must be free of disease. The patient is positioned supine with the head of the bed slightly elevated to reduce venous pressure. The head is turned away from the side of interest with the neck extended to optimize exposure of neck anatomy. A transverse incision should be planned 1–2 cm above the clavicle and directly over both heads of the sternocleidomastoid muscle. Subplatysmal flaps mobilize the skin, and dissection continues between the two heads of the sternocleidomastoid. The omohyoid muscle may be divided. The common carotid artery and vagus nerve are retracted medially, while the jugular vein is mobilized laterally. If operating on the left, the thoracic duct may be ligated and divided, and if operating on the right, accessory lymph ducts may be encountered and also may be ligated and divided. The prescalene fat pad represents an important anatomical landmark, as it overlays the anterior scalene muscle and phrenic nerve. The exposure should remain medial to this fat pad.

The vertebral vein and VA will be seen at the angle of the longus colli and anterior scalene muscle, and for added exposure and mobilization, the vertebral vein may be ligated and divided. Dissection of the VA should be carried from its origin at the subclavian artery to the tendon of the longus colli. The sympathetic chain should be identified and carefully managed. The VA must be mobilized from the sympathetic trunk, which rests on the anterior surface of the vessel.

After full exposure of the VA, the implantation site is selected, either on the common carotid artery or subclavian artery. At this point, the patient is systemically heparinized according to weight-based dosing. The distal edge of the exposed V1 segment is clamped, and the proximal edge is ligated and divided above the region of stenosis or occlusion at the origin. The carotid artery is clamped, and an aortic punch may be used to produce the arteriotomy on its lateral wall. A running 7-0 prolene suture is then used for the anastomosis between the V1 and common carotid artery. Care must be taken to avoid any tension in the VA, ensuring enough slack for neck rotation. Prior to cinching the final tie, the VA is flushed with heparinized saline, allowed to back bleed, and then reclamped, allowing any air, clot, or debris to exit the vessel. This process is repeated with the carotid artery, and then the final suture is tied. Clamps are then removed from the carotid artery and VA.

23.4.5 Vertebral Artery (V3) Bypass with Interposition Graft Technique

For distal vertebral artery bypass, a vertical skin incision is planned to incorporate the V3 segment at the superior most border of the incision and the carotid artery at the inferior most extent, extending from the angle of the jaw and coursing inferiorly along the anteromedial border of the sternocleidomastoid. Alternatively, if the inflow will be the subclavian artery, then two incisions can be made, with curvilinear or transverse incision at the upper neck for V3 exposure and a second transverse incision above the clavicle for visualization of the subclavian artery.

Prior to neck incision, the interposition graft should be harvested and can be a saphenous vein graft (autologous or cadaveric) or radial artery. The neck dissection then proceeds anterior to the sternocleidomastoid and posterior to the jugular vein until the spinal accessory nerve has been exposed. The spinal accessory nerve is then followed proximally to the transverse process of C1. The levator scapulae muscle is then exposed by dissecting overlying soft tissue, which will bring the anterior ramus of the C2 nerve root into view. The C2 anterior ramus represents an important anatomical landmark, as it guides the surgeon to the VA. A right-angle clamp then elevates the levator scapulae, and this muscle can be disconnected from its insertion on the C2 transverse process. In order to gain full exposure of V3, the C2 anterior ramus must be divided and should be disconnected before additional nerve branches. With the VA in view, it must be mobilized from the cervical venous plexus and then exposed circumferentially.

Next, the inflow vessel is prepared. A standard dissection for carotid endarterectomy can be used to free the common carotid artery. In the common carotid artery, attention should be paid to the presence of atherosclerotic disease, and plaques should be avoided at the proximal anastomosis. The distal anastomosis at V3 with the interposition graft should be completed first. The patient is systemically heparinized according to weight-based dosing. The proximal portion of the exposed VA is permanently clipped, and the distal portion is temporarily clamped. A longitudinal arteriotomy is produced, and an end-to-side anastomosis can then be performed with 8-0 nylon suture.

The proximal end of the graft is then brought to the lateral border of the common carotid artery. The common carotid artery is clamped, and an aortic punch can be used to create the arteriotomy. The proximal end-to-side anastomosis is then completed using running 7-0 prolene suture. Prior to cinching the final suture, the graft is flushed with heparinized saline, the clamp released allowing back-bleeding, prior to reclamping. Then, this flushing process is repeated for the carotid artery. The final suture is then tied, and clamps are sequentially released from the carotid and vertebral arteries.

23.5 Surgical Pearls

- *Intraoperative Flow Measurement.* Details of intraoperative direct flow measurement have been described fully within a different chapter in this text. To summarize, the intracranial Charbel micro-flowprobe (Charbel Micro-Flowprobe; Transonics Systems Inc., Ithaca, NY, USA) uses transit-time ultrasonic technology to measure blood flow, both quantifying it and providing audiovisual feedback on the data quality [14, 15]. Flow measurements can assess the baseline blood flow through the VA, the **cut**

flow through the donor graft following proximal anastomosis, and the final bypass flow at the distal anastomosis [16, 17]. Following completion of the distal anastomosis, the **cut flow index** (CFI) may be calculated: bypass flow (mL/min)/cut flow (mL/min). A CFI of ≥0.5 at the time of surgery is a significant predictor of bypass patency in ischemic cerebrovascular disease [18, 19].
- *Head Motion.* Neck and head mobility is an important consideration in the setting of bypass. As discussed earlier, the bypass must have enough slack to avoid tension along the VA or its interposition graft. Carney and Anderson suggested the graft should be parallel to the axis of the neck to minimize the kinking of the vessel with head movement [9].
- *Postoperative Management.* Following vertebral artery revascularization, the patient remains on aspirin 325 mg daily. Intraoperative or postoperative DSA then shows the bypass construct and can be followed to monitor patency. Our institution additionally implements a flow-based protocol using phase-contrast quantitative MRA with NOVA software (Noninvasive Optimal Vessel Analysis; VasSol, Inc., Illinois, USA). Using serial flow measurements monitors the patency of the bypass [19–23]. In addition, when available, the multimodal MR-based protocol described above may be implemented, using not only QMRA but also regional and global BOLD imaging with task performance and vasodilatory challenge to monitor hemodynamic reserve.

23.6 Conclusion

Surgical revascularization for vertebral artery disease may be considered in the symptomatic patient after failing maximum medical management. Although endovascular procedures have grown over the past few decades, understanding of the surgical indications and technique remains important and requires careful review of patient anatomy. Preoperative planning with assessment of hemodynamic reserve further supports the necessity of intervention, and a flow-based paradigm founded in the individual's vascular anatomy guides decision-making for the bypass construct.

References

1. Amin-Hanjani S, Stapleton CJ, Du X, Rose-Finnell L, Pandey DK, MSV E, Zipfel GJ, Liebeskind DS, Silver FL, Kasner SE, Caplan LR, Derdeyn CP, Gorelick PB, Charbel FT, VERiTAS Study Group. Hypoperfusion symptoms poorly predict hemodynamic compromise and stroke risk in vertebrobasilar disease. Stroke. 2019;50(2):495–7.
2. Amin-Hanjani S, See AP, Du X, Rose-Finnell L, Pandey DK, Chen YF, Elkind MSV, Zipfel GJ, Liebeskind DS, Silver FL, Kasner SE, Gorelick PB, Charbel FT, Derdeyn CP, VERiTAS Study Group. Natural history of hemodynamics in vertebrobasilar disease: temporal changes in the VERiTAS Study Cohort. Stroke. 2020;51(11):3295–301.
3. Amin-Hanjani S, Pandey DK, Rose-Finnell L, Du X, Richardson D, Thulborn KR, Elkind MS, Zipfel GJ, Liebeskind DS, Silver FL, Kasner SE, Aletich VA, Caplan LR, Derdeyn CP, Gorelick PB, Charbel FT. Vertebrobasilar flow evaluation and risk of transient ischemic attack and stroke study group. Effect of hemodynamics on stroke risk in symptomatic atherosclerotic vertebrobasilar occlusive disease. JAMA Neurol. 2016;73(2):178–85.
4. See AP, Pandey DK, Du X, Rose-Finnell L, Charbel FT, Derdeyn CP, Amin-Hanjani S, VERiTAS Study Group. Optimized hemodynamic assessment to predict stroke risk in vertebrobasilar disease: analysis from the VERiTAS study. J Am Heart Assoc. 2020;9(12):e016406.
5. Amin-Hanjani S, Rose-Finnell L, Richardson D, Ruland S, Pandey D, Thulborn KR, Liebeskind DS, Zipfel GJ, Elkind MS, Kramer J, Silver FL, Kasner SE, Caplan LR, Derdeyn CP, Gorelick PB, Charbel FT, VERiTAS Study Group. Vertebrobasilar Flow Evaluation and Risk of Transient Ischaemic Attack and Stroke study (VERiTAS): rationale and design. Int J Stroke. 2010;5(6):499–505.
6. Amin-Hanjani S, Turan TN, Du X, Pandey DK, Rose-Finnell L, Richardson D, Elkind MS, Zipfel GJ, Liebeskind DS, Silver FL, Kasner SE, Gorelick PB, Charbel FT, Derdeyn CP, VERiTAS Study Group. Higher stroke risk with lower blood pressure in hemodynamic vertebrobasilar disease: analysis from the VERiTAS Study. J Stroke Cerebrovasc Dis. 2017;26(2):403–10.
7. Amin-Hanjani S, Du X, Rose-Finnell L, Pandey DK, Richardson D, Thulborn KR, Elkind MS, Zipfel GJ, Liebeskind DS, Silver FL, Kasner SE, Aletich VA, Caplan LR, Derdeyn CP, Gorelick PB, Charbel FT, VERiTAS Study Group. Hemodynamic features of symptomatic vertebrobasilar disease. Stroke. 2015;46(7):1850–6.
8. Berguer R, Andaya LV, Bauer RB. Vertebral artery bypass. Arch Surg. 1976;111(9):976–9.
9. Carney AL, Anderson EM. Carotid distal vertebral bypass for carotid occlusion: case report and technique. Clin EEG Neurosci. 1978;9(3)
10. Carney AL, Anderson EM, Martinez DM. Advances in vertebral artery surgery at the skull base. Tex Heart Inst J. 1986;13(1):83–90.
11. Zhao M, Charbel FT, Alperin N, Loth F, Clark ME. Improved phase-contrast flow quantification by three-dimensional vessel localization. Magn Reson Imaging. 2000;18(6):697–706.
12. Charbel FT, Guppy KH, Zhao M, Clark ME. Computerized hemodynamic evaluation of the cerebral circulation for bypass. Neurosurg Clin N Am. 2001;12(3):499–508. viii
13. Thulborn K, McGuire LS, Charbel FT, Amin-Hanjani S. Magnetic resonance imaging methods for monitoring hemodynamic reserve in chronic steno-occlusive cerebrovascular disease. Cerebrovascular reactivity. Chapter 8. Berlin: Springer International Publishing; 2021.
14. Charbel FT, Hoffman WE, Misra M, Ostergren L. Ultrasonic perivascular flow probe: technique and application in neurosurgery. Neurol Res. 1998;20(5):439–42.
15. Charbel FT, editor. Flow assisted aneurysm surgery-cerebral aneurysm handbook. New York: Transonic Systems; 2004.
16. Amin-Hanjani S, Alaraj A, Charbel FT. Flow replacement bypass for aneurysms: decision-making using intraoperative blood flow measurements. Acta Neurochir. 2010;152(6):1021–32. discussion 1032
17. Amin-Hanjani S, Du X, Mlinarevich N, Meglio G, Zhao M, Charbel FT. The cut flow index: an intraoperative predictor of the success of extracranial-intracranial bypass for occlusive cerebrovascular disease. Neurosurgery. 2005;56(Suppl. 1):75–85. discussion 75–85

18. Stapleton CJ, Atwal GS, Hussein AE, Amin-Hanjani S, Charbel FT. The cut flow index revisited: utility of intraoperative blood flow measurements in extracranial-intracranial bypass surgery for ischemic cerebrovascular disease. J Neurosurg. 2019;133(5):1396–400.
19. Amin-Hanjani S, Shin JH, Zhao M, Du X, Charbel FT. Evaluation of extracranial-intracranial bypass using quantitative magnetic resonance angiography. J Neurosurg. 2007;106(2):291–8.
20. Kim T, Bang JS, Kwon OK, et al. Hemodynamic changes after unilateral revascularization for Moyamoya disease: serial assessment by quantitative magnetic resonance angiography. Neurosurgery. 2017;81(1):111–9.
21. Zhu F, Qian Y, Xu B, et al. Quantitative assessment of changes in hemodynamics of the internal carotid artery after bypass surgery for moyamoya disease. J Neurosurg. 2018;129(3):677–83.
22. Amin-Hanjani S, Singh A, Rifai H, et al. Combined direct and indirect bypass for moyamoya: quantitative assessment of direct bypass flow over time. Neurosurgery. 2013;73(6):962–7. discussion 967–8
23. Koch MJ, Stapleton CJ, Charbel FT, Russin J, Amin-Hanjani S. Intracranial-intracranial bypass for aneurysms: quantitative intraoperative assessment of flow preservation. Oper Neurosurg (Hagerstown). 2022;22(5):337–42.

24 Bypass Surgery in Moyamoya Disease and Ischemia

Jacques J. Morcos and Eva M. Wu

Learning Objectives

- Understand the indications for cerebral revascularization for cerebral ischemia and moyamoya disease/syndrome.
- Understand the main preoperative imaging tests needed to properly evaluate the need for cerebral revascularization.
- Understand the surgical subtleties needed for successful STA-MCA bypass as refined by the senior author.

24.1 Introduction

The first cerebral bypass surgery was performed by Yasargil in 1967. Since that time, techniques have evolved but controversy remains regarding its indications. The current indications for cerebral revascularization are (1) flow augmentation for cerebral ischemia and moyamoya disease/syndrome; (2) flow replacement for intracranial aneurysms and skull base tumors. We will only discuss flow augmentation in this chapter.

There has been much controversy on whether bypass should be performed for cerebral ischemia with different studies yielding conflicting results. Bypass for moyamoya is less controversial and in general symptomatic patients are thought to benefit. This chapter discusses the history, indications, relevant anatomy, and step-by-step surgical technique of bypass for cerebral ischemia and moyamoya disease/syndrome. Surgical nuances and common pitfalls are reviewed at the end of the chapter.

24.2 History

The history of cerebral revascularization began in 1939 with German and Taffel who performed the first documented encephalomyosynangiosis in dogs and primates [1]. In 1942, Kredel attempted this in humans but abandoned it due to a high rate of postoperative seizures [2]. The first documented extracranial to intracranial (EC-IC) bypass from the common carotid artery to the intracranial internal cerebral artery using a saphenous vein graft was performed by Woringer and Kunlin in 1963 [3]. It was not until 1967 that Yasargil performed the first superficial temporal artery (STA) to middle cerebral artery (MCA) bypass in humans [4]. Since that time, techniques for cerebral bypass have evolved but indications remained controversial.

Intracranial stenosis (caused by moyamoya or steno-occlusive disease) can cause ischemic cerebrovascular events through several processes: (1) hypoperfusion, (2) thrombosis at stenosis site, (3) thromboembolic event, (4) direct occlusion of small perforators [5]. Clinical presentation varies from an acute deficit due to an embolic or thrombotic event to intermittent neurological symptoms as a result of hypoperfusion. There are three stages of compensatory mechanisms in response to a decrease in cerebral perfusion pressure (CPP) [6]:

- Stage 0: Normal hemodynamics.
- Stage 1: Autoregulatory stage—reflex arteriolar vasodilation occurs in response to insufficient collaterals and decreased CPP. Normal regional cerebral blood flow (rCBF) is maintained as a result of arteriolar vasodilation but there is a resultant increase in regional mean transit time (rMTT) and regional cerebral blood volume (rCBV). Regional oxygen extraction fraction (rOEF) remains normal.
- Stage 2: Hemodynamic failure stage (misery perfusion)—occurs when all autoregulatory mechanisms available to overcome the decrease in CPP have been exhausted and rCBF drops. Sustaining oxygen requirements of the brain parenchyma are achieved by increasing rOEF. This increase in rOEF is a risk factor for stroke.

J. J. Morcos (✉) · E. M. Wu
Department of Neurosurgery, The University of Miami, Miami, FL, USA
e-mail: jmorcos@med.miami.edu

I. N. Kanaan, V. Beneš (eds.), *Neuroanatomy Guidance to Successful Neurosurgical Interventions*, https://doi.org/10.1007/978-3-031-59838-8_24

24.3 Background

24.3.1 Bypass for Cerebral Ischemia

Much controversy exists regarding indications for bypass in cerebrovascular occlusive diseases and many studies have been performed with inconsistent results. The 1985 EC-IC bypass trial, a large randomized controlled study, demonstrated a greater than 95% graft patency rate but concluded that bypass does not reduce the risk of ischemic stroke in patients with atherosclerotic disease in the carotid and middle cerebral arteries [7]. Many questioned the generalizability of these results due to several limitations of the study (failure to identify whether ischemic events were embolic or occlusive, lack of objective quantification of stenosis or cerebral perfusion and hemodynamics, majority were low-risk patients). Following this trial, the rate of bypasses performed for cerebrovascular occlusive disease decreased substantially and it became apparent that stricter preoperative selection criteria were needed. Hemodynamic factors were then considered important factors in patient selection. Patients with hemodynamic failure Stage II (misery perfusion) were considered to be the ideal candidates that would benefit from STA MCA bypass.

The 2006 Japanese EC-IC Bypass Trial (JET Trial) was conducted based on the criteria of hemodynamic failure Grade II and demonstrated improved two-year outcomes for patients who underwent bypass [8]. In 2011, a North American trial, the Carotid Occlusion Surgery Study (COSS), showed conflicting results. The study reported no reduction in risk of stroke at 2 years with bypass surgery compared to the best medical therapy alone and reported a 30-day risk of stroke of 14.4% [9]. The failure of COSS to show benefit with surgery was attributed to the less-than-expected stroke rate in the medical arm due to improved medical treatment over time, as well as a significant perioperative complication rate. However, in 2023, the Carotid or Middle Cerebral Artery Occlusion Surgery Study (CMOSS) trial, the largest EC-IC bypass trial in the Asian population showed similar results to the North American COSS trial which underrepresented Asian populations. The study concluded there was no benefit in stroke prevention compared to medical management [10]. In addition, advancement in endovascular techniques also led to the decline in bypass surgery for cerebral ischemia.

Despite these conflicting results, we certainly believe that, regardless of how well randomized controlled trials are designed and implemented, they necessarily favor internal validity over external validity. We had published some of our concerns regarding the COSS trial at the time of its publication [11]. We generally continue to offer bypass surgery to patients who have ALL the following (Fig. 24.1):

1. Angiographic findings of severe ICA/MCA stenosis or occlusion.
2. Recurrent ischemic symptoms despite maximum medical/antiplatelet therapy.
3. Evidence of misery perfusion on Transcranial Doppler-Vasomotor Reactivity study (TCD-VMR) and CT-Perfusion or MRI-Perfusion.

24.3.2 Bypass for Moyamoya Disease

Moyamoya disease is associated with considerable morbidity/mortality within 2 years of presentation in 75% of children and 20% of adults. Bypass has been associated with decreased morbidity and mortality compared to the natural history. The benefit of bypass surgery in preventing hemorrhage in MMD has been demonstrated through Level 1 evidence [12].

Revascularization is indicated for symptomatic moyamoya disease except in patients with strokes involving the entire MCA territory as there is little remaining salvageable brain (Fig. 24.2). For asymptomatic patients, lifestyle changes and medical management is key. In moyamoya, the controversy is not whether to perform a bypass, but rather which type, direct or indirect. For young children, the literature does support indirect bypass as the most appropriate, given their small-caliber vessels and great capacity to induce neovascularization. For adults, direct bypass is most appropriate because their vessels, in general, are large enough for anastomosis and the Japanese adult moyamoya trial showing better outcome with direct rather than indirect bypass [12]. We generally offer direct revascularization in hemorrhagic moyamoya-associated intracranial hemorrhage. In patients with bilateral disease, the most symptomatic side is revascularized first.

24.3.3 Preoperative Imaging

There are three main types of imaging needed to properly evaluate the need for revascularization. (1) MRI, (2) diagnostic cerebral angiography, and (3) perfusion/cerebrovascular reserve studies.

The DWI sequence on MRI is used to determine embolic/occlusive events. Multiple ischemic foci in a single vascular territory suggest an occlusive or stenotic cause. Ischemic foci in multiple different vascular territories suggest an embolic cause. Ischemic foci in watershed territories (between anterior cerebral and middle cerebral or middle cerebral and posterior cerebral) suggest hypoperfusion.

Diagnostic cerebral angiography is used to evaluate the laterality and severity of intracranial disease, availability of suitable donor for bypass (if there is a frontal and parietal

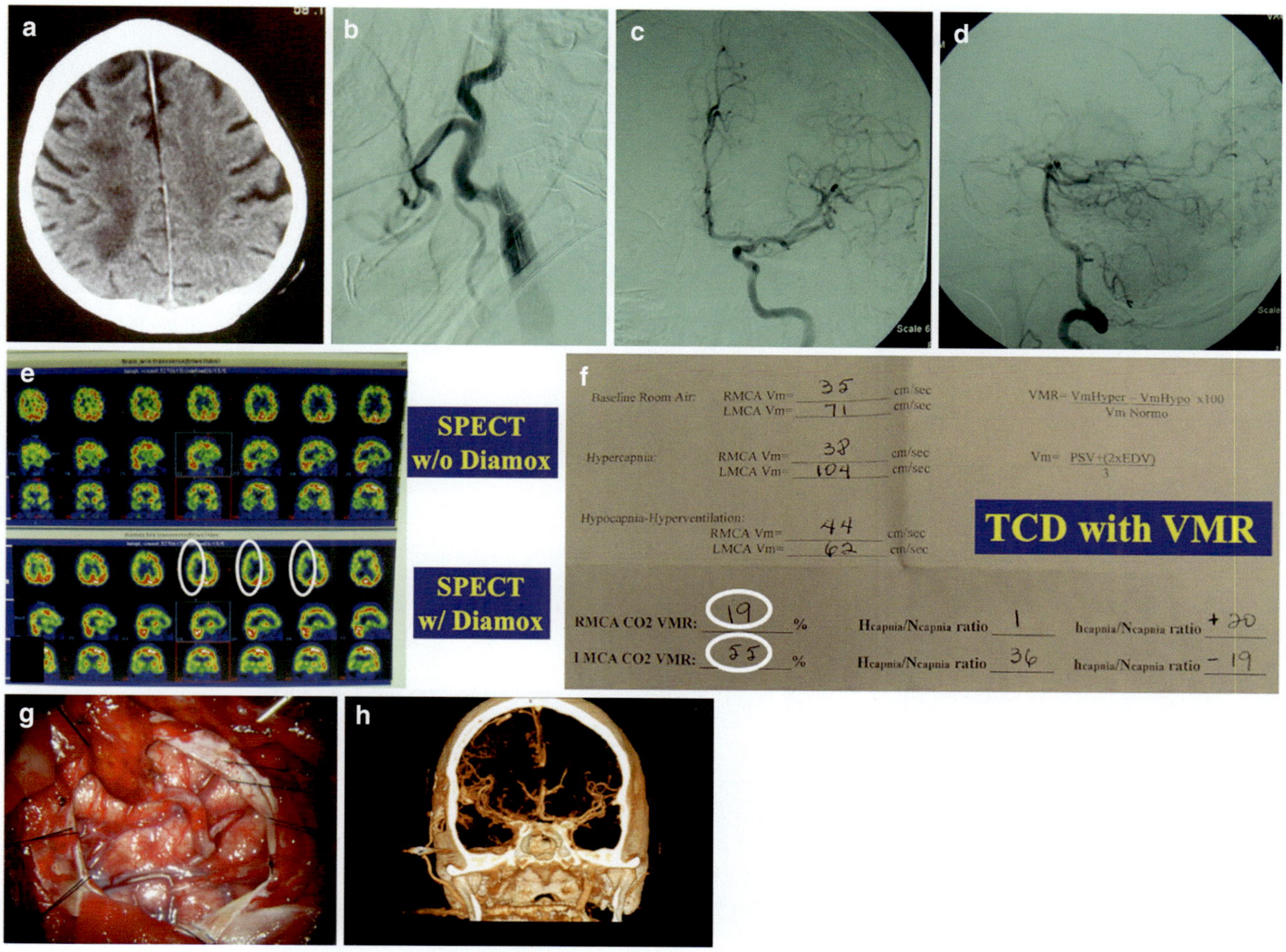

Fig. 24.1 STA-MCA bypass for steno-occlusive disease. 76-year-old man with multiple daily transient ischemic attacks for several weeks despite maximized medical treatment with antihypertensives and antiplatelets. (**a**) CT brain shows multiple watershed infarcts in the right hemisphere. (**b**) Right common carotid injection shows occluded cervical internal carotid artery. (**c**) Left internal carotid artery injection shows no crossover from Acomm. (**d**) Left vertebral artery injection shows poor collateral flow through Pcomm to MCA territory. (**e**) SPECT with Diamox shows misery perfusion (white circle) and paradoxical steal to the left hemisphere. (**f**) TCD with VMR showed severely impaired reserve in the right hemisphere. (**g**) Given angiographic findings of right carotid occlusion with recurrent ischemic symptoms despite maximum medical/antiplatelet therapy and evidence of misery perfusion on TCD and CT perfusion, the patient was taken for one-donor one-recipient (1D1R) bypass. (**h**) Patient did well postoperatively, postop CT Angiogram showed patent bypass. Patient had no further transient ischemic attacks for 10 years and died subsequently from a myocardial infarction. This case highlights that bypass in ischemic disease can be beneficial in appropriately selected patients and that the COSS study should not be generalized to the entire group. Abbreviations: Acomm = anterior communicating artery; COSS = carotid occlusion surgery study; CT = computed tomography; MCA = middle cerebral artery; TCD with VMR = transcranial doppler with vasomotor reactivity

branch), and degree of collateral compensation. Many classification systems have been described to characterize the different stages of imaging findings in moyamoya disease. Suzuki grade described in 1969 is the most common classification system [13]. There are six stages: stage 1 is characterized by stenosis of the terminal ICA. This is followed by the appearance of moyamoya collaterals (stage 2), intensification of moyamoya collaterals (stage 3), reduction of moyamoya collaterals (stages 4 and 5), and disappearance of moyamoya collaterals with development of collateralization from the ECA branches in stage 6 [13]. Three types of collateralization can form in moyamoya: (1) transdural collaterals through external carotid circulation; (2) leptomeningeal or pial-pial collaterals; (3) cross flow through Acom or Pcom (Fig. 24.3a–c). Angiographic collateralization can be determined by overlapping the internal carotid artery, vertebral artery, and external carotid artery injections and judged to be complete or incomplete. Angiographic collateralization is "incomplete" if there is an area of uncompensated perfusion deficit where flow is needed and a pressure gradient is likely to exist (Fig. 24.3d). In our study, we found that 36% of patients with complete collateralization will have asymptomatic delayed occlusion of their direct bypass. Interestingly, when there was incomplete collateralization preoperatively,

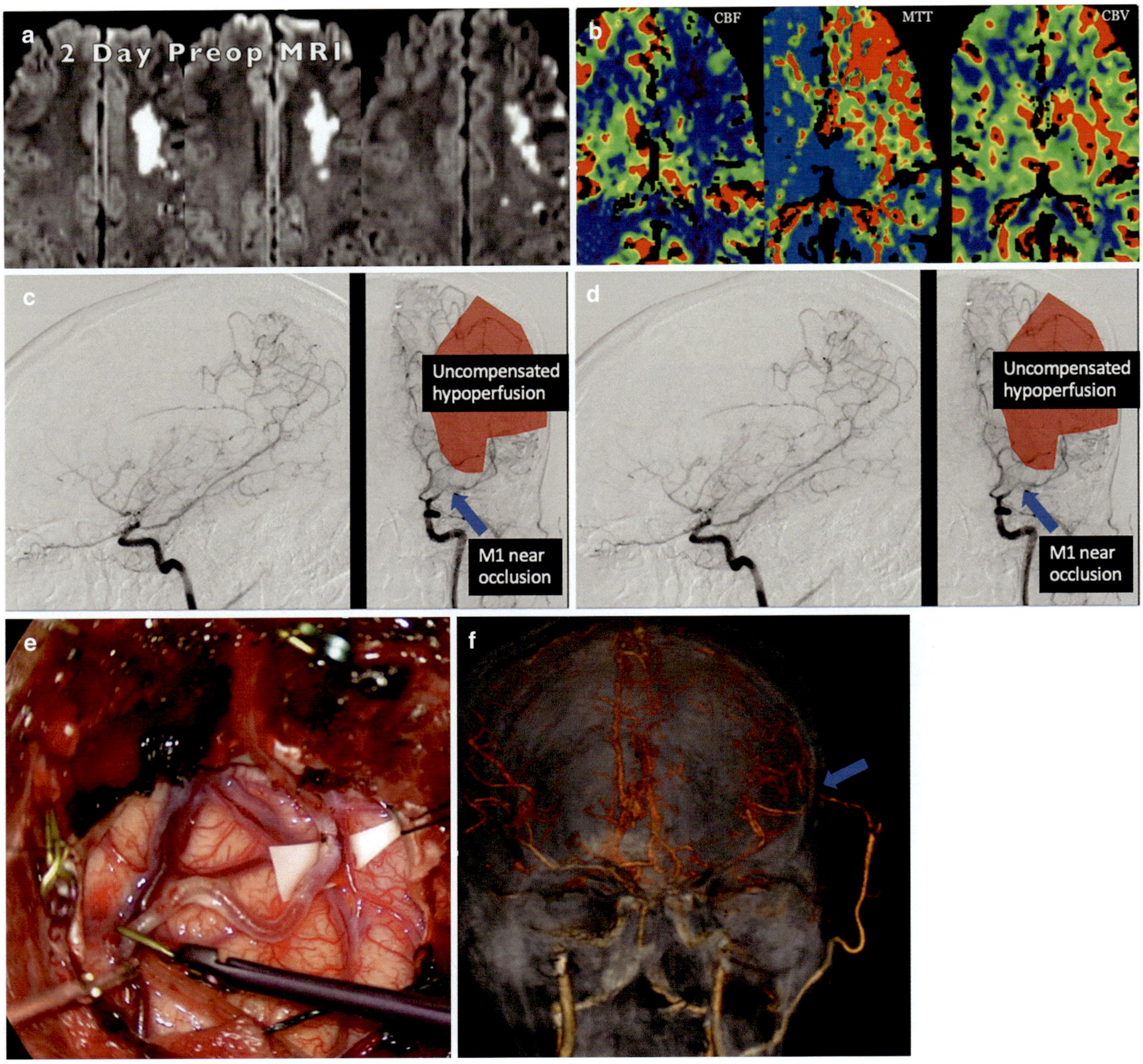

Fig. 24.2 STA-MCA bypass for moyamoya disease. 34-year-old male with HTN, DM, HLD presents with 4 months of limb shaking transient ischemic attacks in RUE, speech difficulty, and cognitive decline. (**a**) MRI revealed watershed infarcts in the left middle cerebral artery (MCA) territory. (**b**) CTP showed low cerebral blood flow (CBF) with increased mean transit time (MTT) and cerebral blood volume (CBV) in the left hemisphere. (**c**) Left internal carotid artery injection showed typical moyamoya changes with near M1 occlusion and an area of uncompensated hypoperfusion (shaded in red). (**d**) Left vertebral injection shows leptomeningeal posterior cerebral artery (PCA) to anterior cerebral artery (ACA) collaterals. (**e**) Patient was taken for 1D2R bypass given symptoms and imaging findings. (**f**) Patient did well postoperatively and speech improved at follow-up. CT Angiogram postop day 1 shows patent anastomosis

the bypass patency rate was much higher (96%). This should not be interpreted as an absence of indication for bypass surgery in patients with complete angiographic collateralization; but rather, that these patients had a better capacity for neovascularization, and postoperatively, the added development of further collaterals rendered the direct bypass conduit less necessary [14].

The use of angiographic findings alone to select patients for revascularization is inadequate. It is standard to use perfusion imaging to determine whether autoregulatory mechanisms are sufficient to prevent cerebral ischemia as a result of decrease in CPP. Perfusion studies include CT perfusion and MR perfusion, among many others. Parameters useful in assessing reserve mechanisms include: (1) regional cerebral blood flow (rCBF), (2) regional mean transit time (rMTT), (3) regional cerebral blood volume (rCBV), and (4) regional oxygen extraction fraction (rOEF). Stage I, autoregulatory stage, is characterized by normal rCBF and rOEF

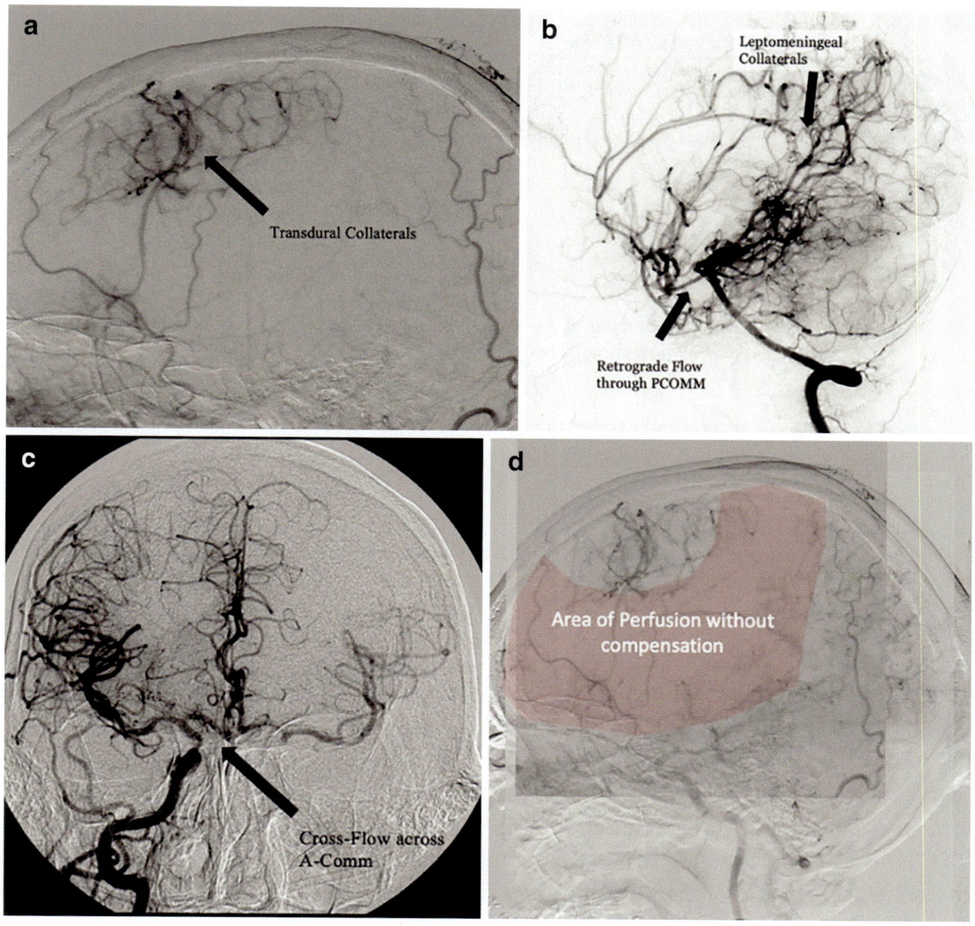

Fig. 24.3 Types of collateral formation. (**a**) External carotid injection showing transdural collaterals from middle meningeal artery branches. (**b**) Vertebral artery injection demonstrating retrograde flow through posterior communicating artery (Pcomm) and leptomeningeal collaterals. (**c**) Right internal carotid artery injection demonstrating filling of the left MCA territory through cross flow through the anterior communicating artery (Acomm). (**d**) Internal carotid artery, vertebral artery, and external carotid injection overlapped to show incomplete collateralization. There is an area of uncompensated perfusion deficit (pink shaded area) where flow is needed and a pressure gradient is likely to exist

and increased rMTT and rCBV. Stage 2, hemodynamic failure stage (misery perfusion) occurs when autoregulatory mechanisms have been exhausted and is characterized by increased rMTT and rOEF and decreased rCBF. Patients with misery perfusion are likely to benefit from bypass.

In addition to these studies, provocative testing such as the acetazolamide challenge test, transcranial doppler (TCD) with vasomotor reactivity (VMR) or breath holding index (BHI) can be used. In patients with misery perfusion due to critical stenosis or moyamoya changes, the ability to maintain normal rCBF by autoregulatory vasodilation is exhausted and acetazolamide (a vasodilator) causes a steal phenomenon and ischemia. In contrast, patients who have not reached maximal compensatory regional vasodilation show increased rCBF with acetazolamide. Results of the acetazolamide challenge test are stratified into three types. Type I patients have normal rCBF at baseline that increases with acetazolamide. Type II patients have decreased baseline rCBF that improves with acetazolamide. Type III patients have decreased baseline rCBF and a further decrease in rCBF with acetazolamide. The latter patients are thought to benefit from bypass. In our practice, TCD with VMR or BHI is used instead of acetazolamide. The VMR test uses inhaled CO_2 as a vasodilator to assess the percent change in MCA blood flow velocity. TCD is performed during inhalation of an O_2/CO_2 mixture before and after hyperventilation. The percent difference in mean MCA flow velocity during normocapnia is compared to velocity during hypocapnia. This value is added to the percent difference between MCA flow during normocapnia and hypercapnia. A relative change in mean MCA flow velocity from hypocapnia to hypercapnia of less than 70% is considered abnormal. BHI is calculated by having patients hold their breath for 30 s. The percent increase in mean MCA flow velocity is divided by seconds of breath holding. A BHI greater than 0.70 is considered normal. Patients with abnormal VMR or BHI are likely to benefit from bypass.

24.3.4 Types of Anastomoses

Although one can be quite creative in constructing EC-IC bypasses, three types of direct anastomoses can be utilized: one-donor one-recipient (1D1R), two-donor two-recipient (2D2R), or one-donor two-recipient (1D2R). Intraoperative flow measurements guide the choice of bypass type. In a *1D1R* configuration, a single donor and single recipient anastomosis is performed in an end-to-side fashion. A

Table 24.1 Advantages and disadvantages of types of STA anastomoses

	1D1R Single barrel (ES)	2D2R Double barrel (ES + ES)	1D2R (ES + SS)
Donors	STAa or STAp	STAa or STAp	STAp (less commonly)
Recipients	FM4 or TM4	FM4 or TM4	FM4 or TM4
Advantages	• Simplest	• Double territory revascularization • 2 independent donors • Provides higher flow than single barrel	• Double territory revascularization without extra dissection • Preserves STAa for skin perfusion & possible future use • Provides higher flow than single barrel
Disadvantages	• May create insufficient revascularization of entire MCA territory	• Tedious extra dissection of the STAa • Occasional poor wound healing	• Technical mistake jeopardizes both anastomoses
Caution	• None	• Avoid use in patients with poor skin healing (such as diabetes, redo surgery, radiation to head/neck)	• Not suitable if donor is small in caliber or STA bifurcation is at level of the craniotomy • May not be appropriate for inexperienced surgeons

Abbreviations: *STAa* superficial temporal artery anterior branch, *STAp* superficial temporal artery posterior branch, *FM4* frontal M4, *MCA* middle cerebral artery, *TM4* temporal M4, *ES* end-to-side; *SS* side-to-side

1D2R or 2D2R configuration can be used when additional flow is needed following the first anastomosis based on intraoperative flow measurements, or when it is believed that there is more than one isolated vascular recipient territory in need of flow augmentation. In a *2D2R* configuration, both STA branches are harvested and anastomosed to two separate recipient arteries in an end-to-side fashion. In a *1D2R* configuration, one donor is used with a side-to-side anastomosis to one recipient vessel and an end-to-side anastomosis to the other recipient vessel. The advantage of the 1D2R is that it can provide flow to multiple cerebral vascular territories and increase the cut flow index (CFI) while saving one branch of the STA to preserve scalp vascularity and for use in future bypass if needed. The advantages and disadvantages of each of the types of anastomoses are summarized in Table 24.1.

24.3.5 Modeling the Bypass

Flow hemodynamics surrounding bypass surgery are complex and difficult to model but some simplifications can help us understand the best bypass variation. Equations to help us understand the relevant flow hemodynamics are the Hagen-Poiseuille's equation, Ohm's law, and resistors in series and in parallel.

1. Hagen-Poiseuille's equation for laminar flow in cylinders: $Q = \Delta P \pi r^4 / 8\ \mu L$, in which Q = volumetric flow rate, ΔP = pressure difference between two ends, r = radius, μ = dynamic viscosity, and L = length.
2. Equation for resistors in series: $R_{\text{Total}} = R_1 + R_2 + R_3 \cdots + R_n$, in which R = resistance.
3. Equation for resistors in parallel: $1/R_{\text{Total}} = 1/R_1 + 1/R_2 + 1/R_3 \ldots + 1/R_n$, in which R = resistance.
4. Ohm's law has been applied to fluid dynamics where current is analogous to flow, and voltage is analogous to ΔP. Therefore, $V = I \times R \approx \Delta P = Q \times R$, in which: V = Voltage, I = Electric current, Q = volumetric flow rate, ΔP = pressure difference between two ends, R = resistance.

In our series of 162 bypass cases studied, 1D2R provides significantly higher total final flow and CFI compared to 1D1R due to the "lessening L" factor and "lessening R" factor [15]. In a *1D2R* configuration, one donor is used with a side-to-side anastomosis to one recipient vessel and an end-to-side anastomosis to the other recipient vessel. This side arteriotomy in the STA effectively shortens the length of the donor resulting in a decrease in overall resistance and therefore increase in flow. In addition, a 1D2R configuration is like an electric circuit in parallel and therefore the total resistance is less than each individual resistance in the frontal and temporal branches, resulting in increased flow for a given pressure gradient. As a result, 1D2R bypass provides significantly higher flow compared to 1D1R but has the benefit of saving one branch of the STA. The main disadvantage of this technique is that a technical error can jeopardize both anastomoses. For further explanation of the equations, refer to our publication [15].

24.4 Anatomy Highlights

The STA is a terminal branch of the external carotid artery. It passes superficial to the posterior root of the zygoma and divides into the frontal and parietal branches. It is important

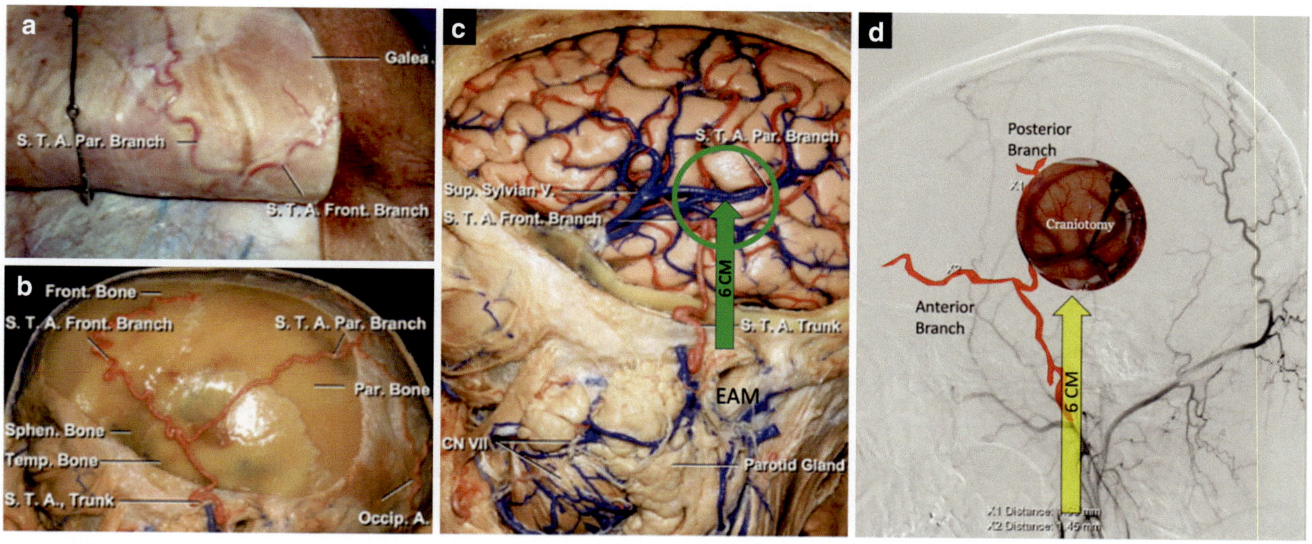

Fig. 24.4 (**a**) The STA is within the dermis located superficial to the galea. (**b**) The STA is a branch of the external carotid that passes posterior to the root of the zygoma and divides into the frontal and parietal branches. (**c**) A craniotomy centered 6 cm above the external acoustic meatus and perpendicular to the root of the zygoma exposes the distal aspect of the Sylvian fissure. On either side of the fissure, cortical frontal and temporal M4 branches can be seen. (**d**) Diagnostic cerebral angiogram showing anterior and posterior branches of the STA and the craniotomy centered 6 cm above the EAC. Abbreviations: EAM = external acoustic meatus; Front. Bone = frontal bone; Occip. A. = occipital artery; Par. Bone = parietal bone; Sphen. Bone = sphenoid bone; STA Front. Branch = STA frontal branch; STA Par Branch = STA parietal branch; Sup. Sylvian V. = superior Sylvian vein. (**a–c**) Dissections courtesy of Rhoton

to note that the STA is located within the dermis superficial to the galea (Fig. 24.4a, b).

A craniotomy centered 6 cm above the external acoustic meatus (EAM) and perpendicular to the root of the zygoma exposes the distal Sylvian fissure and exposes cortical frontal and temporal M4 branches (Fig. 24.4c, d).

24.5 Approach/Surgical Intervention

24.5.1 Patient Positioning

The patient is positioned supine with a bump under the ipsilateral shoulder, head rotated 90° contralaterally with a zygomatic arch, the highest point on the field (Fig. 24.5a). Patients have started on Aspirin 325 mg preoperatively. Systolic blood pressure is strictly maintained in the normal to hypertensive range based on the patient's baseline blood pressure values.

24.5.2 Incision Planning

The frontal and parietal branches of the STA are mapped using doppler ultrasonography. It is important to differentiate the venous signal of the superficial temporal vein from the classic triphasic pulsation of the STA. A 2–2.5 cm radius craniotomy centered 6 cm above the external acoustic meatus (EAM) and perpendicular to the root of the zygoma is marked in order to expose the distal Sylvian fissure. The frontal and parietal branches of the STA are traced a few centimeters past the edge of the craniotomy to ensure adequate length (Fig. 24.5b, c).

The incision is dependent on the location of the STA bifurcation and the presence and caliber of a frontal and/or parietal branch. The typical incision is linear over the parietal STA branch (Fig. 24.6a). If the parietal branch is absent, the incision is curved anteriorly in the shape of a 7 to avoid a scar on the forehead, and the frontal branch is dissected from within the flap (Fig. 24.6b). The 1D2R technique can only be used if the STA bifurcation is lower than the level of the planned craniotomy (Fig. 24.6a). A 2D2R technique is more appropriate if the STA bifurcation is within or higher than the planned craniotomy (Fig. 24.6c). The linear incision over the parietal STA branch can be used for the 1D2R technique, while the 7 incision is used for the 2D2R technique (Fig. 24.6a, c).

24.5.3 Harvesting the Superficial Temporal Artery

The STA donor vessel is harvested under the microscope to prevent inadvertent injury. The STA should be harvested from distally to proximally to the root of the zygoma (Fig. 24.7a). During dissection, it is important not to directly grasp the STA or avulse side branches. If a side branch is avulsed, the stump of the side branch can be bipolared at a low setting to

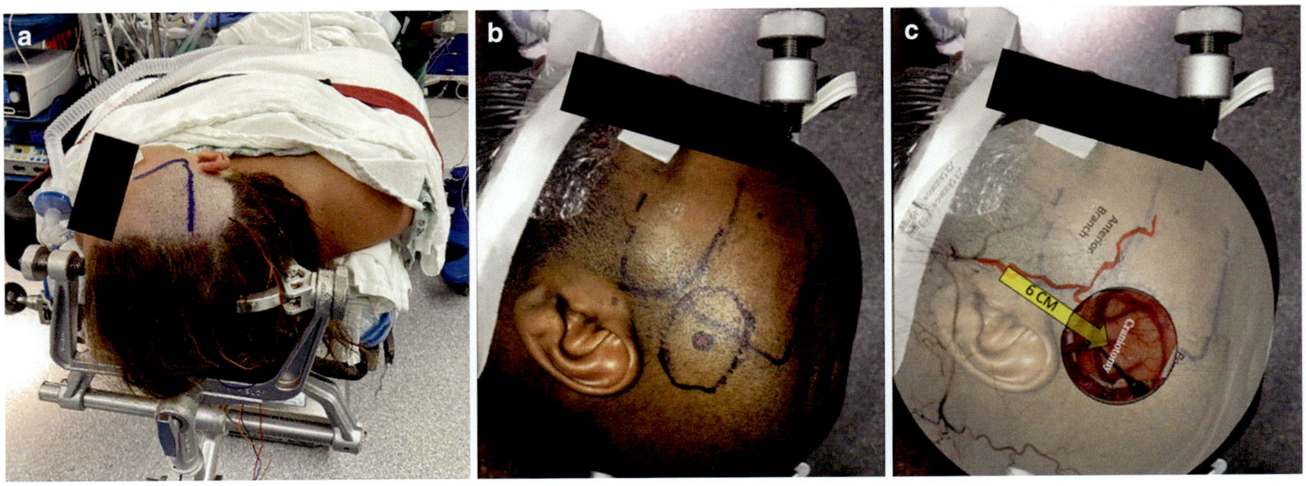

Fig. 24.5 (**a**) Patient is positioned supine, contralateral head rotation 90°, zygomatic arch should be the highest point of the field. (**b**) The frontal and parietal branches of the STA are mapped with Doppler ultrasonography and traced a few centimeters past the edge of the craniotomy to ensure adequate length. (**c**) A craniotomy centered 6 cm above the external acoustic meatus and perpendicular to the root of the zygoma is marked and exposes the distal aspect of the Sylvian fissure. *STA* superficial temporal artery

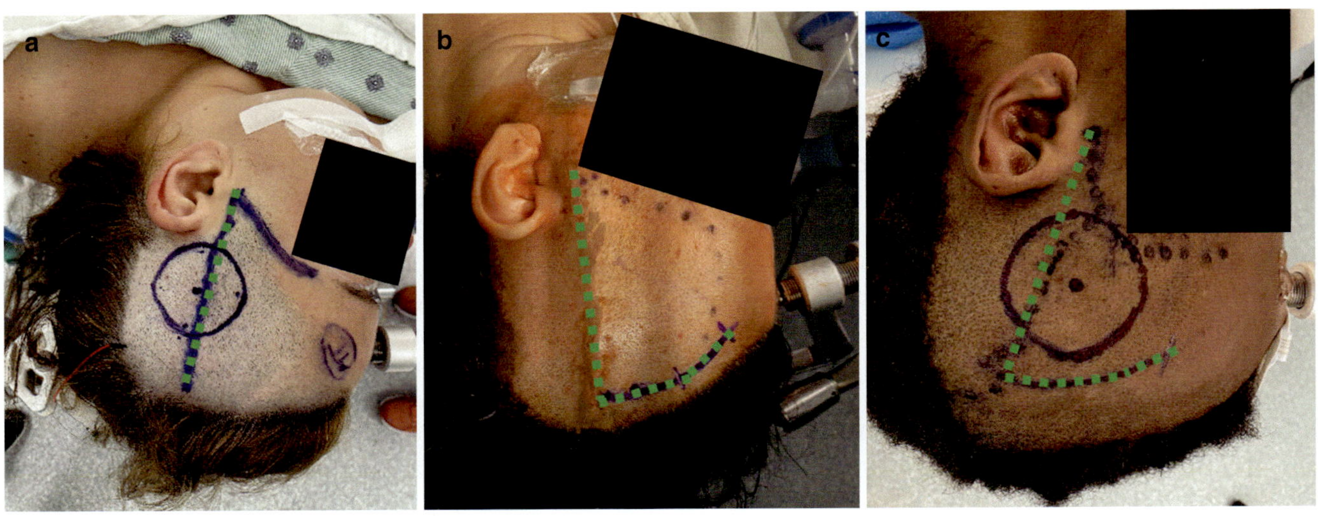

Fig. 24.6 Incision planning. The incision is dependent on the location of the STA bifurcation and the presence of a frontal and/or parietal branch. (**a**) The frontal and parietal branches of the STA are marked in purple. A craniotomy centered 6 cm above the external acoustic meatus is marked in purple. The typical incision is linear over the parietal STA (green dotted line). Note the STA bifurcation is lower than the level of the planned craniotomy. A 1D2R technique can only be used if the STA bifurcation is lower than the level of the planned craniotomy. (**b**) The frontal branch of the STA is marked in purple. The parietal branch is absent. The incision is curved anteriorly in the shape of a 7 to avoid a scar on the forehead and the frontal branch of the STA is dissected from within the flap. (**c**) The frontal and parietal branches of the STA and a craniotomy centered 6 cm above the external acoustic meatus are marked in purple. Notice that in this case, the STA bifurcation is within the craniotomy. A 2D2R technique is more appropriate if the STA bifurcation is within or higher than the planned craniotomy. Given both frontal and parietal branches will be used, the incision is curved anteriorly in the shape of a 7. Abbreviations: *STA* superficial temporal artery

avoid transmission to the main vessel. If the side branch is avulsed at its origin and there is a large defect in the STA, interrupted 10-0 sutures can be used to repair this. After the donor vessel is dissected circumferentially, the distal end is cut, and flow through the STA is confirmed. It is then flushed with heparinized saline to prevent clot formation and a temporary clip is placed proximally (Fig. 24.7c). The vessel is wrapped in papaverine-soaked telfa to prevent vasospasm and desiccation until the time of anastomosis (Fig. 24.7d).

24.5.4 Craniotomy

The temporalis muscle is opened in a T-shaped manner (in patients with thick temporalis muscle) or C-shaped manner based anteriorly (in patients with thinner temporalis muscle) (Fig. 24.7e). The craniotomy is performed with a single burr hole near the root of the zygoma and a 2–2.5 cm radius craniotomy centered 6 cm above the EAM and perpendicular to the root of the zygoma (Fig. 24.7f). This will center the craniotomy over the distal Sylvian fissure to allow for exposure of the cortical temporal M4 (TM4) and frontal M4 (FM4) branches emerging from the fissure. Care is taken to avoid injury to the STA during the craniotomy. Tack-ups are placed.

The dura is opened in a double Y-shaped manner based on the Sylvian fissure to maximize brain exposure. If the middle meningeal artery (MMA) is large on preop digital subtraction angiography (DSA) or if it is giving significant transdural collaterals, the dura should be opened around the MMA so that it is preserved. Meticulous hemostasis should be obtained to prevent blood influx into the operative field during anastomosis.

24.5.5 Harvesting the Recipients

The largest accessible recipient TM4 and FM4 arteries are identified. If possible, avoid choosing recipients at the edge of the craniotomy as this makes the anastomosis stage more challenging. Arachnoidal dissection is performed with microscissors and bypass forceps. Small branches on the recipient vessels can be coagulated and divided to create a perforator free zone. The bipolar should be turned to very low setting to avoid current transmission to the main vessel. A small triangle-shaped background material should be placed under the recipient vessels (Fig. 24.7g). Iatrogenically induced spasm can be reversed with topical papaverine applied to gelfoam. A 5Fr microsurgical suction device is placed in the most dependent corner of the field that will not interfere with the anastomosis (Fig. 24.7h).

24.5.6 Preparing for the Anastomosis

The adventitia is cleaned from the donor STA and the artery is shortened to appropriate length (Fig. 24.7i). The following flow measurements are made using a microvascular ultrasonic flow probe either 1.5 mm or 2 mm in size (Fig. 24.7g, h):

1. STA donor flow (Cut flow, CF)—max potential flow of donor STA prior to bypass.
2. Recipient flow (Frontal cortical, FM4).
3. Recipient flow (Temporal cortical, TM4).

The direction of flow in the recipients (+ for antegrade or – for retrograde) is documented.

For end-to-side anastomosis, the distal end of the STA is fish-mouthed to achieve a 30–45 degree angle in a direction that is parallel with the flow, if possible. To fish-mouth the donor, the distal end of the STA is cut usually at a 30-degree angle (Fig. 24.7i) along one edge, by a length approximately equal to the flattened vessel diameter (Fig. 24.7j). The arteriotomy in the recipient will therefore generally end up being 2–3× the length of the recipient diameter. The donor STA is always fish-mouthed as this increases the cross-sectional area. For an STA end cut at 90° and fish-mouthed by a diameter length, the cross-sectional area of the completed anastomosis is 4× larger than in the nonfish-mouthed case. Per Pythagoras theorem: d^2 versus $(d + d)^2 = 4d^2$. The only instances where it is not fish-mouthed are when there is a large size discrepancy between the donor and the recipient or when the donor orientation is unfavorable. In these cases, an end-to-side 90° to the recipient is constructed. The distal end of the prepared STA should be held up to the recipient MCA and the arteriotomy length on the recipient MCA should match this arteriotomy length (Fig. 24.7k).

The side-to-side arteriotomies are created by two linear arteriotomies of equal length at the 1:30 and 10:30 position (analogy to the face of a clock) on the donor and recipient arteries that are approximately twice the length of the recipient diameter (Fig. 24.11a).

24.5.7 Anastomosis

For the anastomosis, we like to use curved Castroviejo needle holders, diamond-dusted bypass forceps, and 10-0 Ethilon suture on a tapercut needle 3/8 circle. The suture is cut to the appropriate length. Before occluding the recipient, the needle is placed in the 12-o'clock position of the donor in an "out-in" fashion and readied to save on ischemia time (Fig. 24.7l). The recipient is occluded with temporary straight aneurysm "mini" clips with attention to the direction of flow of the recipient (Fig. 24.7m). The first clip is always placed distal to the direction of flow followed by the second clip placed proximal to the direction of flow to allow the trapped segment to fill with blood and be easier to incise. Clamp time is started when the first distal clip is placed. A disposable sharp sickle knife is used to make the arteriotomy on the recipient vessel taking care not to cut through to the other side (Fig. 24.7m). Heparinized saline solution is used to irrigate the recipient vessel to clear all blood products (Fig. 24.7n).

The recipient vessel is gently grasped with bypass forceps and everted and the needle is passed in an "in-out" fashion at the 12-o'clock position (Fig. 24.7o). The knot is tied and thread cut (Fig. 24.7p). Sutures near the apices should be tied with a surgeon's knot followed by two regular knots. Sutures not near the apices can be tied with three regular knots. This process is repeated (Fig. 24.7q). The transparency test should be done to ensure the back wall was not inadvertently grabbed. The needle should go through all three layers of the vessel including the intima. The bite size should be approximately the same thickness as the vessel. The suture interval should be approximately twice the thickness of the vessel. Both recipient and donor vessels should be everted, intima to intima, to prevent a clot from forming. The various techniques to suture end-to-side and side-to-side will be discussed below. After the "front wall" of the anastomosis is completed (Fig. 24.7r), a nerve hook is passed to ensure that the back wall was not inadvertently caught (Fig. 24.7s). The "back wall" is then completed (Fig. 24.7t).

Once the anastomosis is complete, the aneurysm clip is released in the distal direction of flow followed by the clip proximal to the direction of flow. Additional sutures are placed if a leak is seen but if there is only a small amount of oozing, it will generally stop with surgicel placed over the anastomosis.

24.5.8 Anastomosis Techniques

For end-to-side anastomosis, an interrupted (Fig. 24.8) or continuous (Fig. 24.9) technique can be utilized for the "front wall" and "back wall." For side-to-side anastomosis, a continuous endoluminal technique is utilized for the "back wall" (Fig. 24.10) and a continuous or interrupted technique can be utilized for the "front wall." If there is enough redundancy for the donor to be rolled sideways, a continuous or interrupted technique can be utilized for either wall of the side-to-side (Fig. 24.11b–d).

Interrupted technique—The needle is passed in an "out-in" fashion on the donor vessel followed by an "in-out" fashion on the recipient vessel. The knot is tied and thread cut.

Fig. 24.7 (A) The frontal and parietal branches of the STA and craniotomy are marked in purple. The parietal branch of the STA is harvested from distally to proximally to the root of the zygoma. Before identification of the STA, care must be taken to not incise too deeply to avoid injuring the STA. After the vessel is identified, a mosquito clamp can be used to dissect the vessel and elevate the soft tissue away from the vessel. A knife is used to cutover the mosquito clamp. This is repeated until the common STA is identified at the root of the zygoma. (**b**) The incision is then curved anteriorly, the scalp flap lifted, and the frontal branch of the STA is dissected from within the skin flap using the same method as discussed previously. During this stage, care must be taken not to inadvertently exit through the scalp. (**c**) After the frontal and parietal branches of the STA have been circumferentially dissected and freed from the surrounding soft tissue, they are cut distally, and flow through the STA is confirmed. The STA is then flushed with heparinized saline to prevent clot formation and a temporary clip is placed proximally. Direct manipulation of the STA is avoided to prevent injury and vasospasm. The STA should only be held by its fascial bands or the distal ends of the STA that will be sacrificed before the anastomosis. (**d**) The STA is wrapped in papaverine-soaked telfa to prevent vasospasm and desiccation until the time of the anastomosis. (**e**) The temporalis muscle is opened in a T-shaped manner (in patients with thick temporalis muscle) or C-shaped manner based anteriorly (in patients with thinner temporalis muscle). (**f**) A single burr hole is placed near the root of the zygoma and a 4–5 cm craniotomy is elevated in usual manner. The craniotomy should be centered 6 cm above the external acoustic meatus to maximize the view of the distal Sylvian fissure. Abbreviations: *STA* superficial temporal artery. (**g**) The largest TM4 and FM4 recipients are identified. Arachnoidal dissection is performed with bypass forceps and microscissors. Small perforating branches on the recipient vessel can be sacrificed to create a perforator-free zone but this should be done to a minimal extent. Small triangle-shaped background material should be placed under the recipient vessels. Flow measurements are made using a microvascular ultrasonic flow probe (Charbel probe) and the direction of flow is noted (+ for antegrade and – for retrograde). The flow in FM4 was 6 cc/min antegrade and 7 cc/min antegrade in TM4. (**h**) The STA donor is thoroughly cleaned of adventitia and any constricting fascial bands that could jeopardize the bypass. Once it is cleaned, it is shortened to appropriate length. STA cut flow is measured to be 68 cc/min. A 5Fr microsurgical suction device is placed in the most dependent corner of the field that will not interfere with the anastomosis. (**i**) The STA is fish-mouthed to increase the cross-sectional area by around 4×. The distal end of the STA is cut at a 30–45 degree angle, then the heel is fish-mouthed by the same length as the length of the trimmed distal end of the STA (**j**). (**k**) The length of the arteriotomy should be at least twice the length of the recipient diameter. The distal end of the prepared STA should be held up to the recipient MCA and the arteriotomy length on the recipient MCA should match this arteriotomy length. (**l**) Prior to occluding the recipient for the bypass, the needle is anchored in the 12-o'clock position of the donor in an "out-in" fashion to save on ischemia time. (**m**) The recipient is occluded with temporary straight aneurysm "mini" clips with attention to the direction of flow of the recipient. In this case, the flow was antegrade so the clip at the bottom right of the picture was placed first followed by the clip at the left side of the image. Note how the aneurysm clips were placed in a low-profile position minimizing the chance that the sutures are caught on the aneurysm clip during anastomosis. Clamp time starts after the first clip is placed. A disposable sharp knife is used to make the arteriotomy. Care is taken not to cut through the other side of the recipient vessel wall. (**n**) Heparinized saline is used to irrigate the vessel of blood products. Abbreviations: Fr = French; FM4 = frontal M4; TM4 = temporal M4; STA = superficial temporal artery. (**o**) The needle is passed in an "in-out" fashion near the 12-o'clock position of the recipient vessel. (**p**) The knot is tied and the thread is cut. Sutures near the apices should be tied with a surgeon's knot followed by two regular knots. Sutures not near the apices can be tied with three regular knots. (**q**) This process is repeated until the "front wall" of the anastomosis is closed (**r**). (**s**) Once the "front wall" is completed, a nerve hook is passed to ensure that the "back wall" was not inadvertently caught. (**t**) The "back wall" is completed. (**u**) In cases where a double bypass is planned à priori (1D2R or 2D2R), the interim flow (Q_i) should be measured after the first anastomosis and cut flow index calculated. CFI = bypass flow/cut flow, where bypass flow is flow in donor after anastomosis to the recipient and cut flow is the maximum potential flow of donor STA before bypass. The bypass flow post first anastomosis was 68 cc/min, for a CFI of 100%; it is important to position the donor artery in different positions to make sure it does not kink on itself. (**v**) Given the CFI was 100%, the second anastomosis was not performed for fear of creating hyperperfusion syndrome. Weck clips were placed at the origin of the frontal branch to occlude this vessel. (**w**) The inferior portion of the bone flap is craniectomized to accommodate the STA. Abbreviations: *CFI* cut flow index, *STA* superficial temporal artery

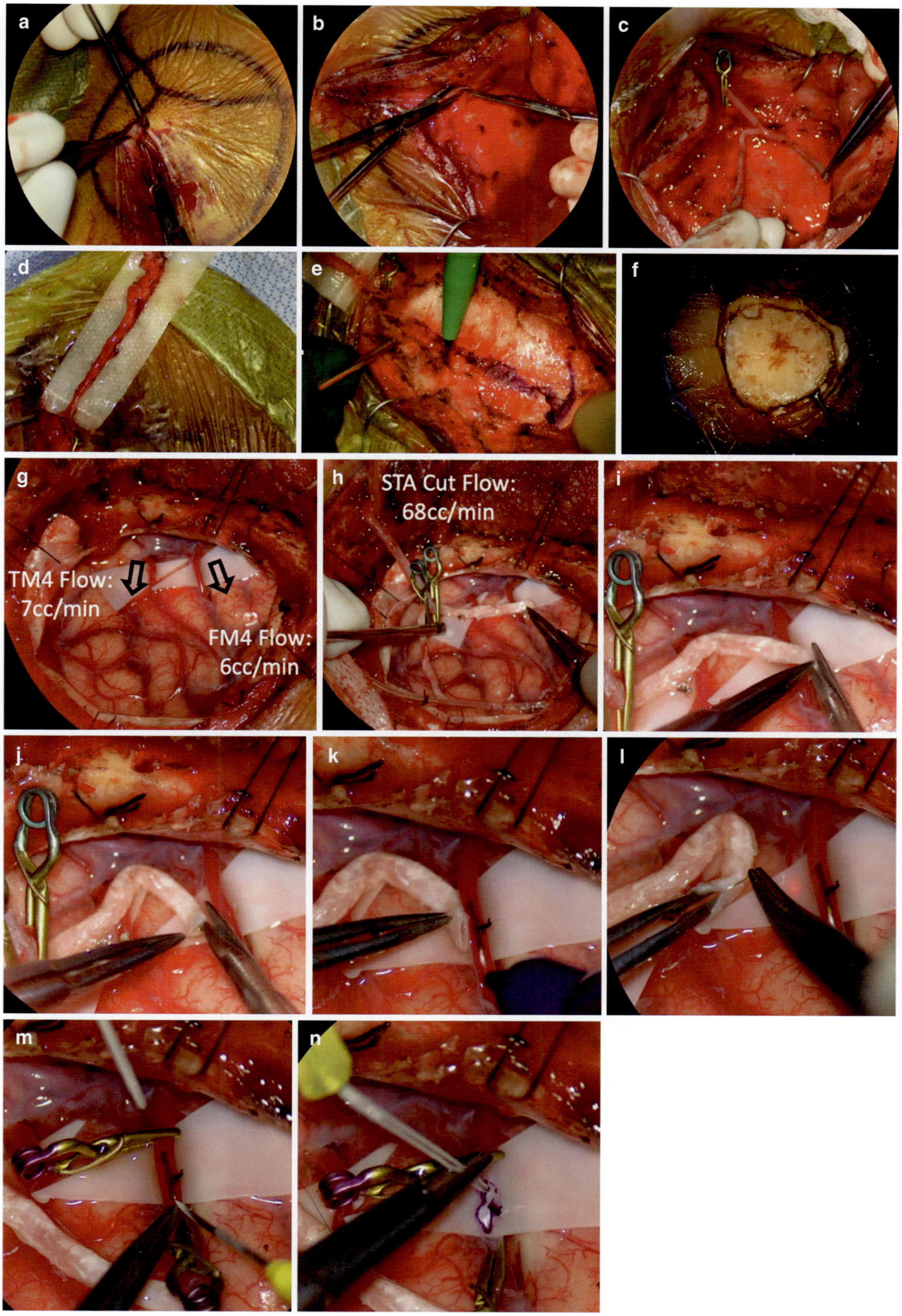
a
b
c
d
e
f
g
TM4 Flow:
7cc/min
FM4 Flow:
6cc/min
h
STA Cut Flow:
68cc/min
i
j
k
l
m
n

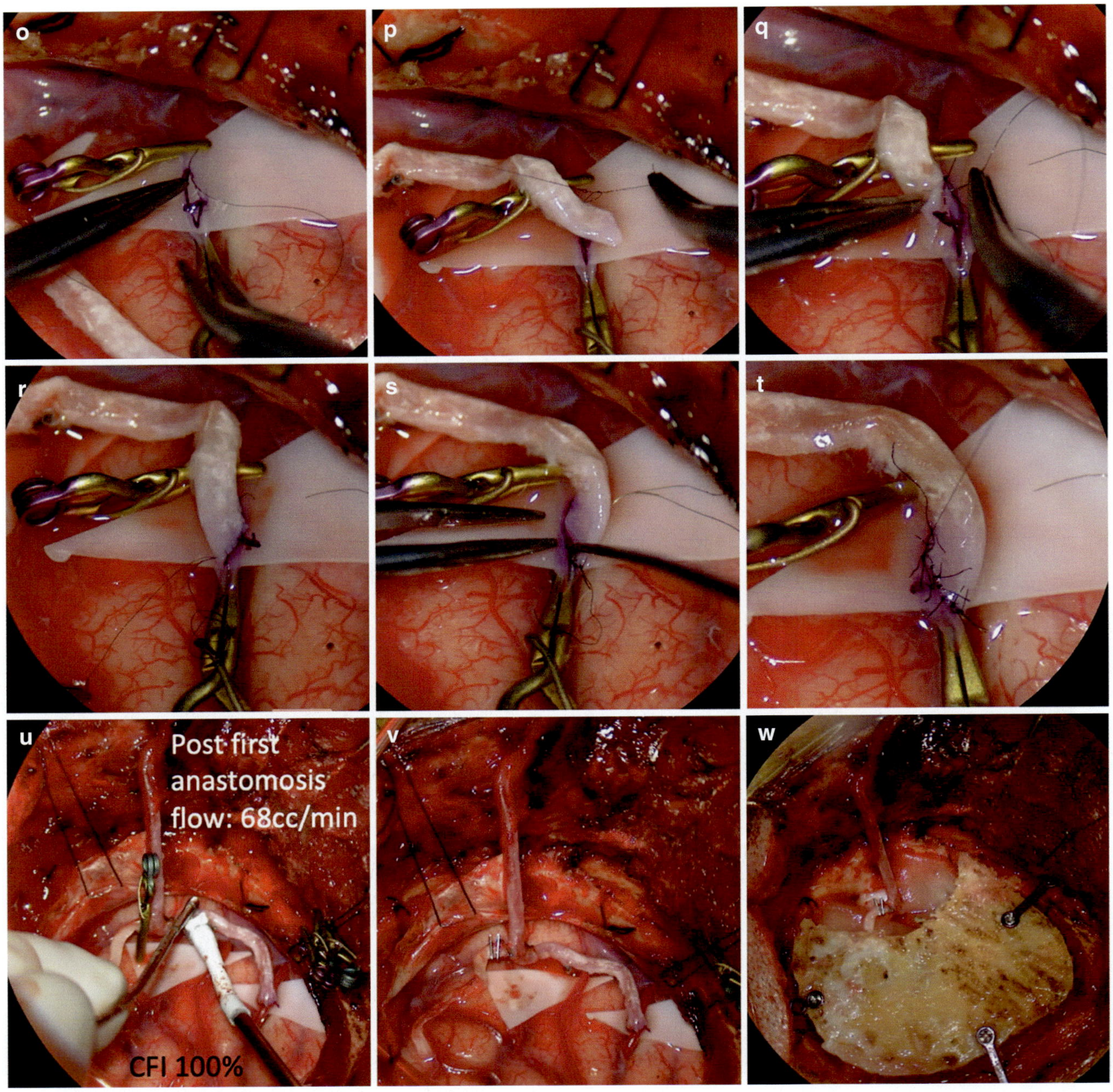

Fig. 24.7 (continued)

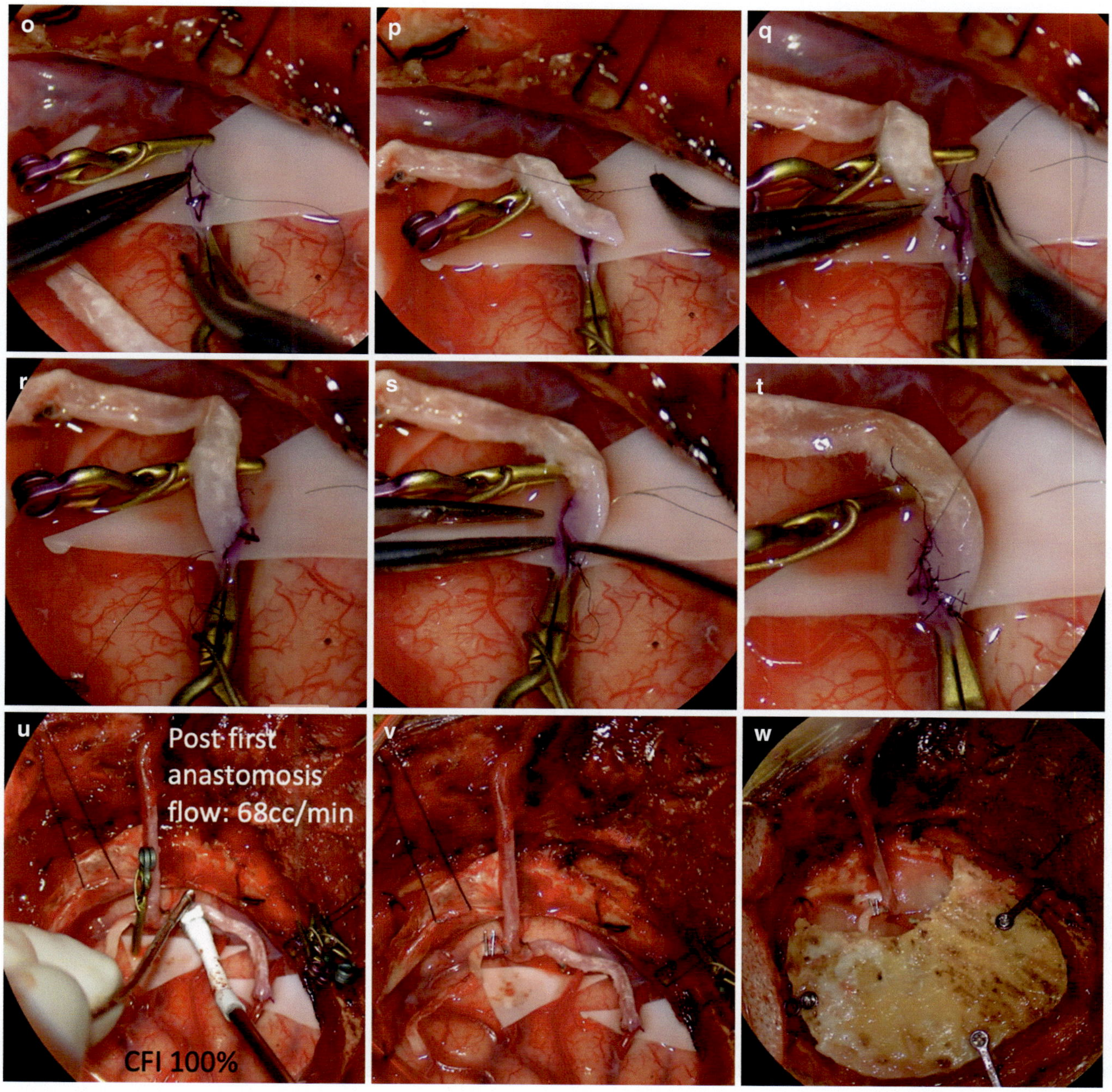

Fig. 24.7 (continued)

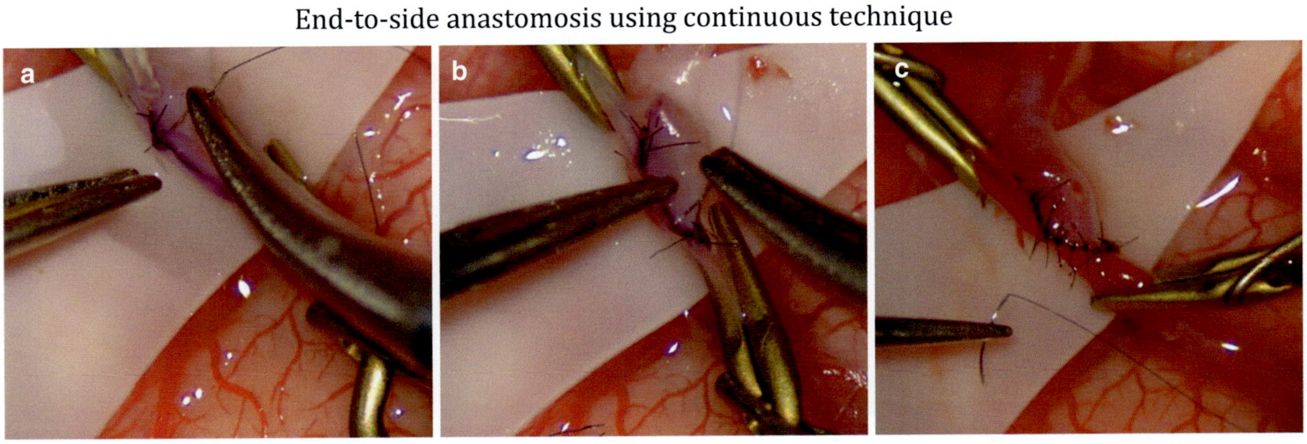

Fig. 24.8 End-to-side anastomosis using the interrupted technique. (**a**) The needle is passed in an "out-in" fashion on the donor vessel followed by an "in-out" fashion on the recipient vessel. The knot is tied and thread cut. (**b**) This "out-in-in-out" fashion is repeated until it is completed. (**c**) Shows end-to-side anastomosis using interrupted technique

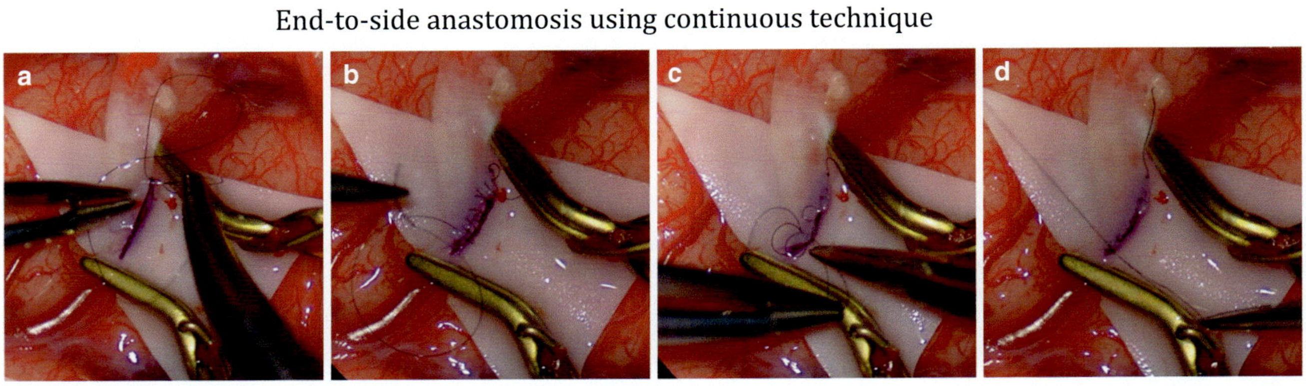

Fig. 24.9 End-to-side anastomosis using continuous technique. (**a**) The needle is passed in an "out-in" fashion on the donor vessel followed by an "in-out" fashion on the recipient vessel. The knot is tied but the thread is not cut. This "out-in-in-out" method is repeated. (**b**) The loops of the sutures are left loose until the last stitch is thrown. (**c**) Once the last stitch is thrown near the apex, bypass forceps are used to tighten each loop using a hand-over-hand technique. This will ensure no leaking will occur in the "back wall." (**d**) All loops are tightened and the knot is tied

This "out-in-in-out" fashion is repeated (Fig. 24.8). An anchor stitch can be thrown at the 6-o'clock position to help appose the two vessels and make it easier to suture.

Continuous technique—The needle is passed in an "out-in" fashion on the donor vessel followed by an "in-out" fashion on the recipient vessel. The knot is tied but the thread is not cut. This "out-in-in-out" method is repeated leaving the loops of the sutures loose. Once the last stitch is thrown near the apex, bypass forceps are used to tighten each loop using a hand-over-hand technique. The suture is tied using three regular knots (Fig. 24.9). Continuous technique is faster but is more prone to error so the interrupted technique is favored when conditions are suboptimal (small <0.5 mm or thin recipient, anastomosis under tension, awkward angle, novice). Caution must be taken when utilizing the continuous technique for both front and back walls as the adventitia may be pulled on both sides causing the vessel to kink and the bypass can occlude.

Continuous endoluminal technique—this technique is used for side-to-side anastomosis in which the vessel cannot be rolled sideways (Fig. 24.10). In this technique, the "back wall" needs to be anastomosed first. The first suture is placed in the 12-o'clock position. The needle is passed in an "out-in" fashion on the donor vessel followed by an "in-out" fashion on the recipient vessel. The knot is then tied in the usual fashion. This is done so that the knot will be on the outside of the lumen. If the knot is on the inside of the lumen, then a clot will form. The needle is then passed behind the knot. The needle is then passed "out-in" on the left edge then "in-out" on the right edge for right-handed surgeons. For left-handed surgeons, the reverse is done. This "out-in" on the left edge followed by "in-out" on the right edge should be repeated until the apex is reached leaving the loops of the sutures loose. Once the last stitch is thrown near the apex, bypass forceps are used to tighten each loop using a hand-over-hand tech-

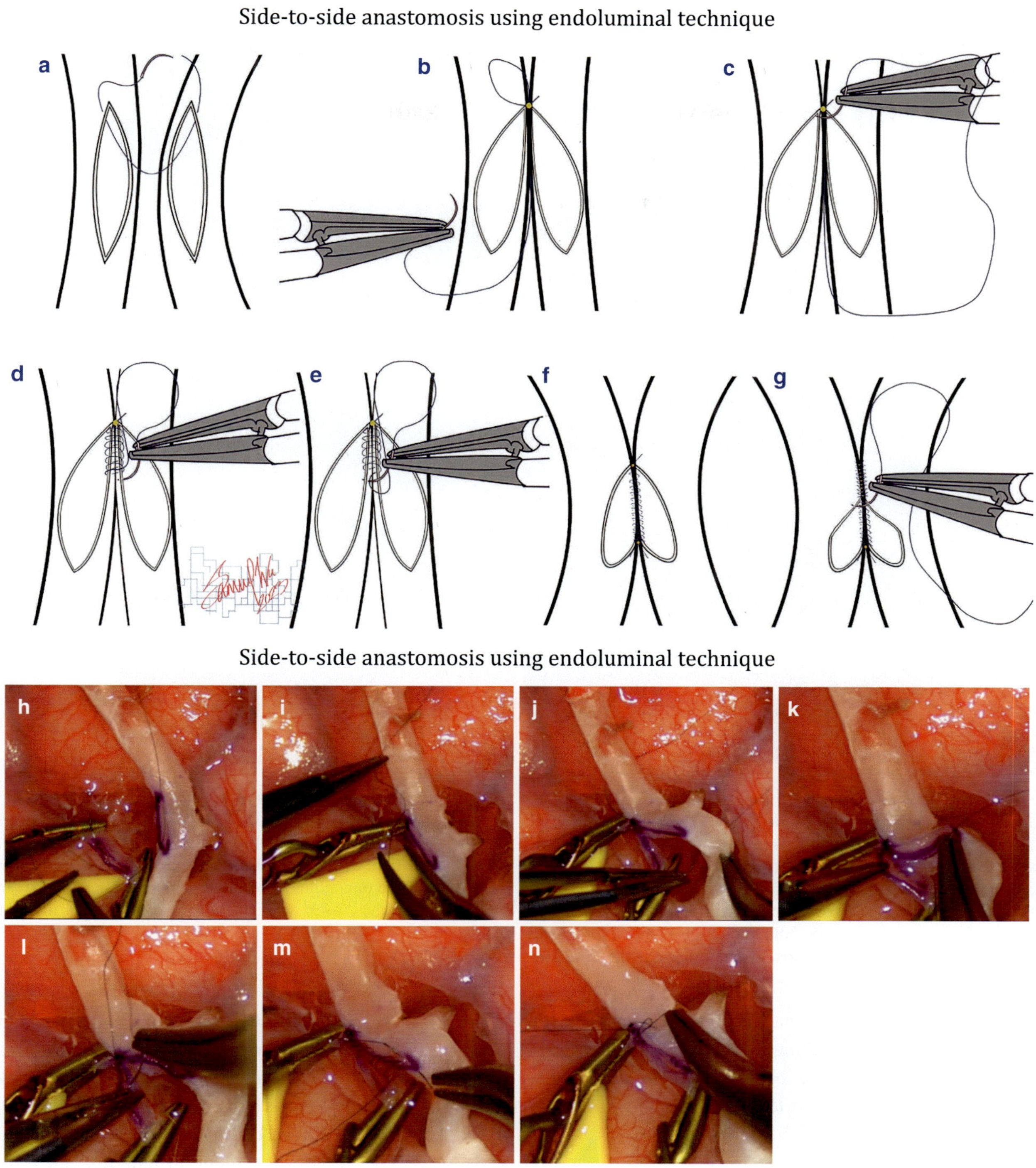

Fig. 24.10 Medical artist rendition of side-to-side anastomosis using endoluminal technique. This technique is used for side-to-side anastomosis in which the vessel cannot be rolled sideways and therefore the "back wall" must be done first. (**a**) The first suture is placed at the 12-o'clock position in an "out-in" fashion on the donor vessel followed by an "in-out" fashion on the recipient vessel. The knot is tied in usual fashion. Note that the knot must be on the outside of the lumen, otherwise a clot will form. (**b**) The needle is then passed behind the knot. (**c**) For right-handed surgeons, the needle must be passed "out-in" on the left edge followed by "in-out" on the right edge. (**d**) For left-handed surgeons, the reverse is done. (**e**) This "out-in" on the left edge followed by "in-out" on the right edge should be repeated until the apex is reached leaving the loops of the sutures loose. (**f**) Once the last stitch is thrown near the apex, bypass forceps are used to tighten each loop using a hand-over-hand technique. (**g**) The "front wall" of the anastomosis can be closed with interrupted or continuous technique. This illustration shows the "front wall" being closed with interrupted sutures. (**a–g**) Illustration courtesy of Samuel Wu (swuell@gmail.com). Intraoperative pictures showing side-to-side anastomosis using endoluminal technique. The technique is described in the previous fig. (**h**) Suture is passed in 12-o'clock position in "out-in-in-out" fashion. (**i**) Knot is tied. (**j**) Needle is passed behind the knot. (**k**) Surgeon is right-handed; therefore the needle is passed "out-in" on the left edge followed by "in-out" on the right edge (**l**). (**m**) This "out-in" on the left edge followed by "in-out" on the right edge is repeated until the apex is reached leaving the loops of the sutures loose. After the last stitch, a hand-over-hand technique is used to tighten the loops. (**n**) The "front wall" is closed with interrupted stitches in this case

Side-to-side using interrupted technique

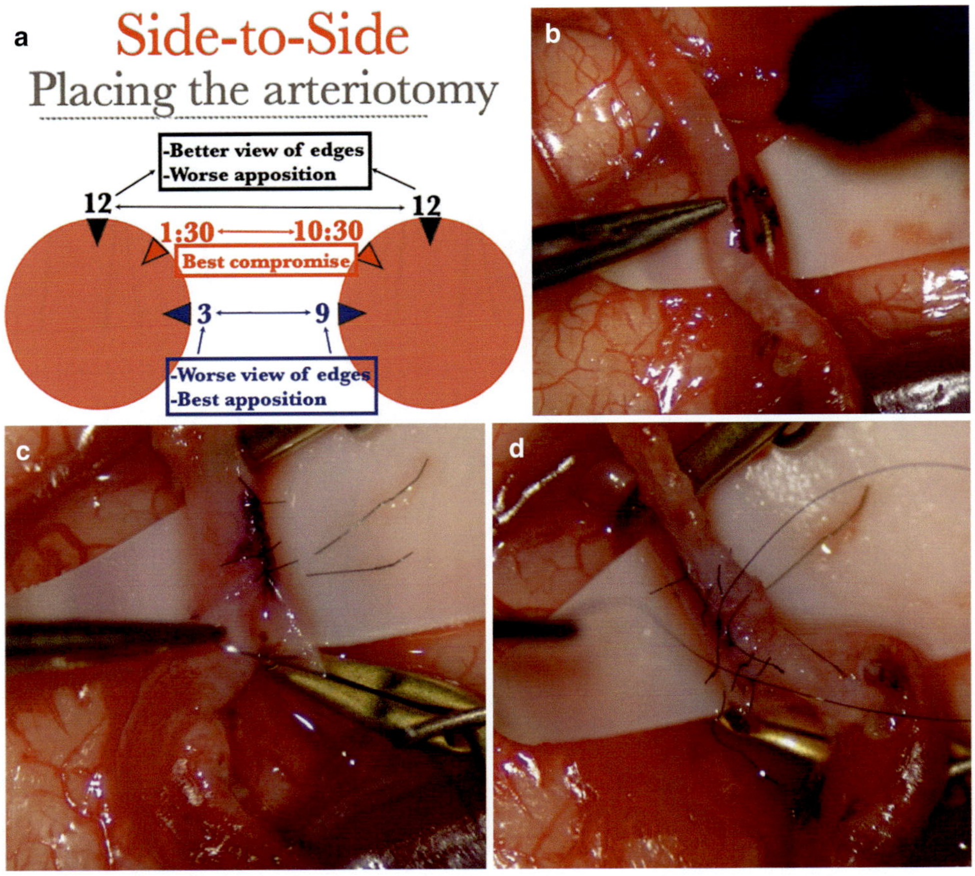

Fig. 24.11 Side-to-side anastomosis using interrupted technique. (**a**) The side-to-side arteriotomies are created by two linear arteriotomies of equal length on the donor and recipient arteries that are approximately twice the length of the recipient diameter. Placing the arteriotomies at the 3-o'clock and 9-o'clock positions gives the best apposition but a worse view of the edges making suturing difficult. Placing the arteriotomies at the 12-o'clock position on both vessels gives the best view of the edges but is worse for apposition. Arteriotomies at the 1:30 and 10:30 positions on the donor and recipient arteries is the best compromise. (**b**) Arteriotomies are equal in length on the donor and recipient arteries. Note the vessel can be flipped so the endoluminal technique does not have to be done here and instead both walls can be closed with either interrupted or continuous technique. In this case, both the "front wall" (**c**) and the "back wall" (**d**) were closed with interrupted technique

nique. The "front wall" of the anastomosis can be closed with interrupted or continuous technique. If the side-to-side anastomosis can be rolled sideways, both walls can be closed with either interrupted or continuous technique (Fig. 24.11c, d).

24.5.9 Flow Measurements/Analysis

In cases where a double bypass is planned *à priori* (1D2R or 2D2R), the interim flow (Q_i) should be measured after the first anastomosis and cut flow index (CFI) calculated (Fig. 24.7u). CFI is the ratio of bypass flow to cut flow, where bypass flow is flow in donor after anastomosis to recipient and cut flow is the maximum potential flow of donor STA prior to bypass [16]. If CFI is very high after the first anastomosis, the second anastomosis should not be performed for fear of creating hyperperfusion syndrome (Fig. 24.7v).

In cases where a double bypass is planned, the decision to perform the side-to-side or end-to-side bypass first is based on the geometry of the recipient and donor vessels. However, performing the side-to-side bypass first allows blood to flow through it while the second end-to-side anastomosis is being performed. The reverse technique would not allow flow while the second anastomosis is being performed. Flow measurements are made across each anastomosis site following the completion of the second anastomosis. For 2D2R, flow measurement is made at the common STA trunk and the frontal and parietal STA branches. For 1D2R, flow measurements are made as seen in Fig. 24.12. In 1D2R, as described above, since flow after both anastomoses ($CFI_{(f)}$) is usually greater than flow after the first anastomosis ($CFI_{(i)}$), we have introduced the ratio of $CFI_{(f)}$ / $CFI_{(i)}$ and defined it as "Second Anastomosis Relative Augmentation" (SARA), and it is usually >1. This indicates a demand for additional blood

Step	Perform	Measure	Calculate	Schematic
1	Dissect: -STA branch -FM4 -TM4	Initial Flow: -Q_{STA} (CF) -Q_{FM4}(0) -Q_{TM4}(0)	-	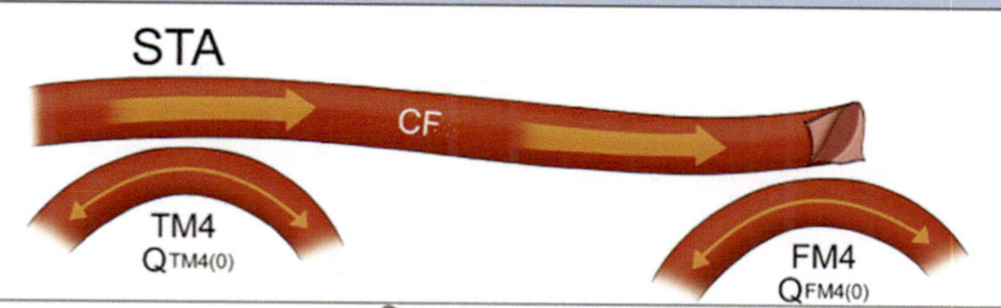
2	Anastomosis 1: STA-TM4 (SS)	Interim Flow: -Q_{TM4}(i)	Interim CFI: CFI(i)=Q_{TM4}(i) / CF	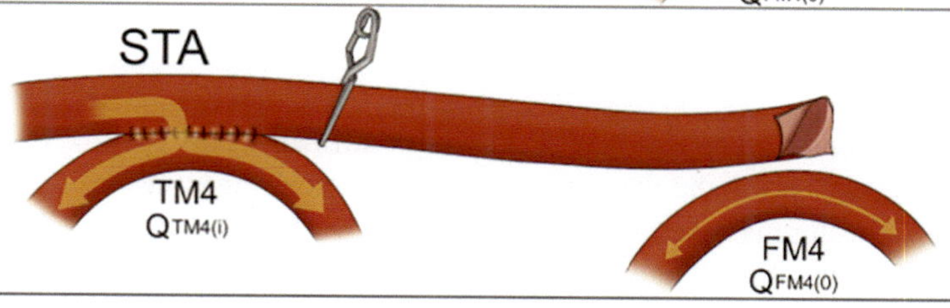
3	Anastomosis 2: STA-FM4 (ES)	Final Flow: -Q_{TM4}(f) -Q_{FM4}(f)	Total final bypass flow: Q_{Tot} = Q_{TM4}(f) + Q_{FM4}(f) Final CFI: CFI(f) = Q_{Tot} / CF Second Anastomosis Relative Augmentation (SARA): SARA = CFI(f) / CFI(i) Second Anastomosis Sink Index: (SASI) = Q_{FM4}(f) / Q_{Tot}	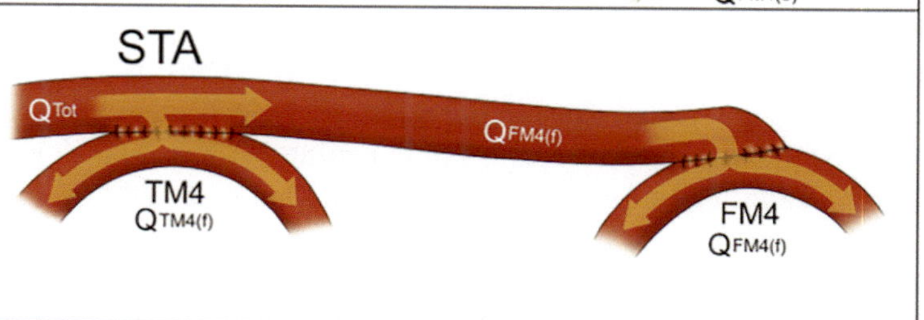

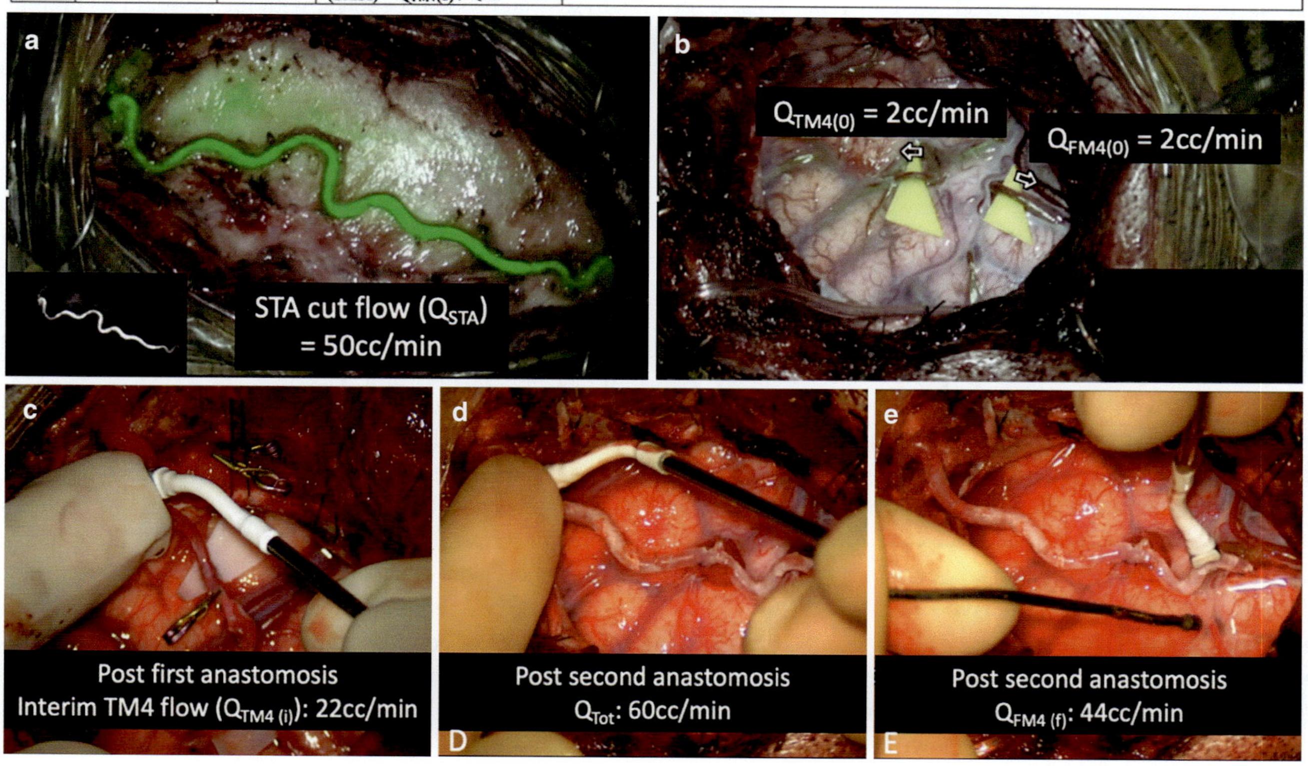

Fig. 24.12 The table shows the typical sequential surgical steps in a 1D2R bypass. Symbols used in this table apply to 1D2R when a side-to-side TM4 is performed first followed by an end-to-side to FM4. The same concept applies if the reverse is performed. (**a**) STA cut flow is measured to be 50 cc/min. (**b**) Prebypass TM4 flow ($Q_{TM4(0)}$) is measured to be 2 cc/min and Prebypass FM4 flow ($Q_{FM4(0)}$) is measured to be 4 cc/min. (**c**) Post first side-to-side bypass, interim flow in TM4 ($Q_{TM4(i)}$) is measured to be 22 cc/min. Interim CFI ($CFI_{(i)}$) is 44%. (**d**) Post second anastomosis, total final bypass flow (Q_{Tot}) is measured to be 60 cc/min. (**e**) Flow in FM4 ($Q_{FM4(f)}$) is measured to be 44 cc/min. Therefore, the ($Q_{TM4(f)}$) must be 16 cc/min (60-44 = 16). Final CFI is 60/50 = 120%. SARA = $CFI_{(F)}/CFI_{(i)}$ = 120%/44% = 2.7. SASI = $Q_{FM4(f)}$ /Q_{Tot} = 44/60 = 0.73. Abbreviations: Q_{STA} = Cut flow = max potential flow of donor STA prior to bypass; ($Q_{TM4(0)}$) = prebypass TM4 flow; $Q_{FM4(0)}$ = prebypass FM4 flow; $Q_{TM4(i)}$ = post first bypass TM4 flow; $CFI_{(i)}$ = cut flow index after first anastomosis; Q_{tot} = total final bypass flow post second anastomosis; $Q_{FM4(f)}$ = final flow through FM4 post second bypass; $Q_{TM4(f)}$ = final flow through TM4 post second bypass; $CFI_{(f)}$ = final cut flow index. *SARA* second anastomosis relative augmentation, *SASI* second anastomosis sink index

flow and that the patient was appropriately selected for bypass. Another measure that we have defined is the "Second Anastomosis Sink Index" (SASI) = $Q_{(f)(second)}/Q_{(f)(total)}$, where $Q_{(f)(second)}$ = final flow in second anastomosis and $Q_{(f)(total)}$ = total final flow. This ratio informs us how much sink effect the second anastomosis created [14, 15].

24.5.10 Closure

Indocyanine green (ICG) angiography is utilized to assess patency following the first and second anastomosis, if available. The dura is inverted to promote pial dural synangiosis and a dural substitute graft is placed as an onlay. A craniectomy is created in the inferior portion of the bone flap to accommodate the STA (Fig. 24.7w). The temporalis is loosely reapproximated. The inferior aspect of the incision is closed with interrupted vertical mattress until the STA is no longer visible. The superior aspect of the incision is closed with interrupted galeal stitches followed by running locking stitch for the skin. Care is taken to ensure there is no strangulation of the STA at the level of the skin, muscle or craniotomy.

24.5.11 Postoperative Considerations

The patient is monitored in the intensive care unit for one night with most patients discharged 48 h postoperatively. Computerized tomography angiography (CTA) is obtained prior to discharge. Aspirin 325 mg is continued the same day of surgery and taken for the rest of life. Patients are followed at 2 and 6 weeks without imaging, then 6 months, 1 year, 2 years, etc. with CTA or MRA. TCD with VMR never returns to normal values and is not an indication of bypass failure.

24.6 Surgical Pearls

- Always think STA first.
- Vessels adapt and remodel over time to achieve a constant wall shear stress. Using a medium or high capacitance wall conduit in situations with low-pressure gradient will result in low flow. Wall shear stress will cause vessels with low flow to shrink and potentially occlude.
- It is important to have as short L in the donor as possible to increase flow.
- The donor/recipient diameter ratio should be 1.5–2 ideally.
- Always fish-mouth the donor unless there is a large size discrepancy between the diameter of the donor and recipient. This increases cross-sectional area by at least 4x, minimizing the risk of stenosis.
- The ideal practical angle of end-to-side anastomosis is 30–45°. Smaller angle results in less turbulence, less stagnation, and more physiological wall shear stress.
- Sutures should be placed closer together at the toe and heel.
- Continuous technique is faster but interrupted technique is favored when conditions are suboptimal (small <0.5 mm or thin recipient, anastomosis under tension, awkward angle, novice).
- Continuous endoluminal technique is used for side-to-side anastomosis in which the vessel cannot be rolled sideways so the "back" wall must be done first. If the vessel can be flipped, both walls can be closed with either interrupted or continuous technique.
- Intraoperative flow measurements can help guide bypass technique (whether to do single or double barrel). Flow measurements should be made systematically after each bypass and should be made with the vessel in its optimal position.
- Bypass surgery in experienced hands has a high immediate patency rate; however, postoperative occlusion can occur and can be immediate or delayed. Delayed bypass occlusion must be recognized and distinguished from immediate bypass occlusion. Immediate bypass occlusion is usually due to technical errors and an urgent revision is usually indicated (Fig. 24.13a–c). Delayed asymptomatic bypass occlusion is not due to a technical issue and does not need to be revised, especially if the patient has complete collateralization on preoperative imaging as described above. Delayed bypass occlusion is typically due to the development of additional indirect collaterals, prompted by the surgical procedure, and as a result, there is less pressure gradient across the bypass, so less flow goes through the bypass and the bypass occludes through disuse (Fig. 24.13d–f). The extent of preoperative collateralization is a predictor of long-term patency. Complete collateralization preoperatively is associated with higher rates of delayed bypass occlusion due to the formation of indirect collaterals [14, 15].

24.7 Discussion

24.7.1 Alternate Options: Choice of Conduit

It is important to remember that the diameter of the conduit does not correlate with flow. The terms "low flow," "medium

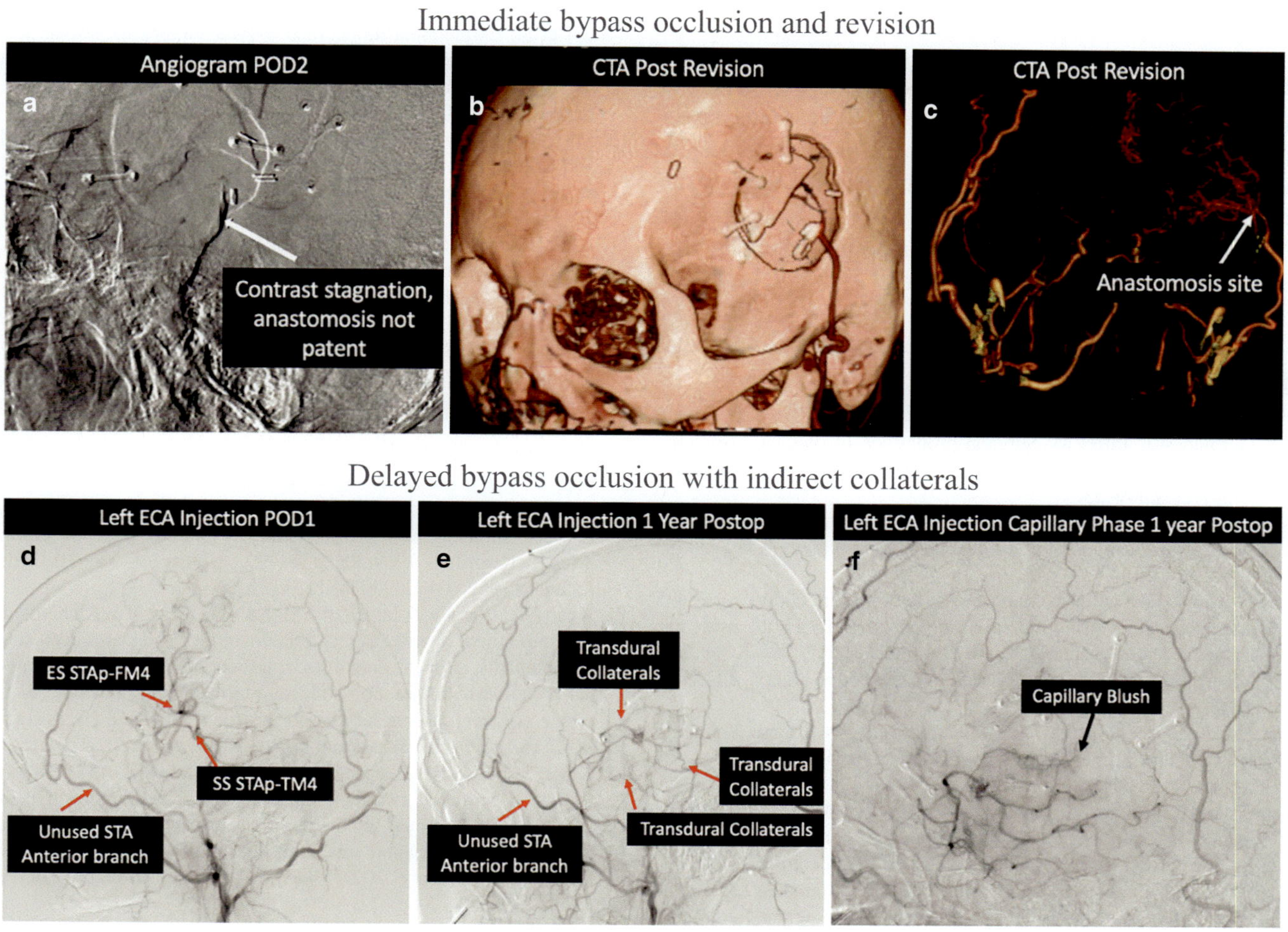

Fig. 24.13 Types of bypass occlusion—immediate (**a–c**) versus delayed (**d–f**). (**a**) Diagnostic cerebral angiogram left external carotid artery injection lateral view on postop day 2 showing contrast stagnation in the STA indicating occluded bypass. (**b**) Patient was taken immediately for revision of bypass using a different recipient vessel. 3D reconstruction post revision demonstrates the patent STA. (**c**) 3D reconstructed angiogram post revision shows patent bypass with filling of the MCA candelabra. Immediate bypass occlusion is due to a technical problem in either the donor or recipient and needs to be revised. This is in contrast to delayed asymptomatic bypass occlusion due to the gradual recruitment of indirect collaterals, which does not need to be revised. The development of indirect collaterals results in less pressure gradient across the STA and therefore less demand for flow through the bypass and subsequent occlusion of the bypass. (**d–f**) demonstrate the immediate postoperative and 1-year postoperative angiogram showing initially patent 1D2R anastomosis with end-to-side STAp to FM4 and side-to-side STAp to TM4. (**e** and **f**) show delayed bypass occlusion with the formation of transdural collaterals

flow," and "high flow" should not be used because the STA can have a cut flow of 160 cc/min or higher, as seen in our material and that of others. Instead, terms like low capacitance (STA), medium capacitance (radial artery), and high capacitance (saphenous vein) are more appropriate. For the majority of patients with anterior circulation moyamoya or atherosclerotic occlusive disease, the superficial temporal artery (STA) is the simplest conduit that can achieve the goal of revascularization. Always think STA first because the STA usually can provide enough flow but if flow is not enough, then others can be considered. Medium- and high-capacitance vessels should only be used if the STA is unable to supply adequate flow. The main driver of flow is pressure gradient, not diameter of donor vessel. Vessels adapt to achieve a constant wall shear stress. If pressure gradient is low, the flow will be low and wall shear stress will cause vessels with low flow to shrink and potentially occlude. In addition, high capacitance vessels should not be used for ischemia or moyamoya cases due to high risk of hyperperfusion syndrome.

In circumstances when the STA is unavailable, the occipital artery or the contralateral STA via a saphenous vein interposition bonnet bypass can be alternatives. The internal maxillary artery can also be used as a donor with an interposition radial artery graft. These alternatives are beyond the scope of this chapter. It is our practice to use direct bypass for all adults unless a donor or recipient artery cannot be identified (extremely rarely) in which case an indirect bypass can be used as a last resort.

24.8 Conclusion

Direct bypass surgery is effective in decreasing strokes in patients with symptomatic moyamoya disease and select patients with symptomatic steno-occlusive disease despite maximum medical management. This chapter reviews the history, indications, relevant anatomy, and surgical steps for bypass surgery as refined by the senior author. Surgical tips and common pitfalls are also reviewed.

References

1. German WJ, Taffel M. Surgical production of collateral intracranial circulation. An experimental study. Proc Soc Exp Biol Med. 1939;42(1):349–53.
2. Kredel FE. Collateral cerebral circulation by muscle graft: Technique of operation with report of 3 cases. South Surg. 1942;11:235–44.
3. Woringer E, Kunlin J. Anastomosis between the common carotid and the intracranial carotid or the sylvian artery by a graft, using the suspended suture technic. Neuro-Chirurgie. 1963;9:181–8.
4. Yasargil MG. Microsurgery: applied to neurosurgery. Elsevier; 2013.
5. Hwang S, Lee S, David C. Intracranial occlusive disease, Youmans neurological surgery. 6th ed. Philadelphia, PA: Saunders, An Imprint of Elsevier Inc.; 2011.
6. Grubb J, Robert L, Derdeyn CP, Fritsch SM, Carpenter DA, Yundt KD, Videen TO, et al. Importance of hemodynamic factors in the prognosis of symptomatic carotid occlusion. JAMA. 1998;280(12):1055–60. https://doi.org/10.1001/jama.280.12.1055.
7. EC/IC Bypass Study Group. Failure of extracranial–intracranial arterial bypass to reduce the risk of ischemic stroke. N Engl J Med. 1985;313(19):1191–200. https://doi.org/10.1056/nejm198511073131904.
8. Group JS. Japanese EC-IC Bypass Trial (JET Study). 脳卒中の外科. 2002;30(6):434–7.
9. Powers WJ, Clarke WR, Grubb RL, Videen TO, Adams HP, Derdeyn CP, et al. Extracranial-intracranial bypass surgery for stroke prevention in hemodynamic cerebral ischemia: the carotid occlusion surgery study randomized trial. JAMA. 2011;306(18):1983–92. https://doi.org/10.1001/jama.2011.1610.
10. Ma Y, Gu Y, Tong X, Wang J, Kuai D, Wang D, et al. The carotid and middle cerebral artery Occlusion Surgery Study (CMOSS): a study protocol for a randomised controlled trial. Trials. 2016;17(1):544. https://doi.org/10.1186/s13063-016-1600-1.
11. Amin-Hanjani S, Barker FGI, Charbel FT, Connolly ESJ, Morcos JJ, Thompson BG, et al. Extracranial-intracranial bypass for stroke—is this the end of the line or a bump in the road? Neurosurgery. 2012;71(3):557–61. https://doi.org/10.1227/NEU.0b013e3182621488.
12. Miyamoto S, Yoshimoto T, Hashimoto N, Okada Y, Tsuji I, Tominaga T, et al. Effects of extracranial-intracranial bypass for patients with hemorrhagic moyamoya disease: results of the Japan Adult Moyamoya Trial. Stroke. 2014;45(5):1415–21. https://doi.org/10.1161/strokeaha.113.004386.
13. Suzuki J, Takaku A. Cerebrovascular "moyamoya" disease. Disease showing abnormal net-like vessels in base of brain. Arch Neurol. 1969;20(3):288–99. https://doi.org/10.1001/archneur.1969.00480090076012.
14. Khan NR, Elarjani T, Jamshidi AM, Lu VM, Silva MA, Richardson A, et al. Direct bypass surgery for Moyamoya and steno-occlusive vasculopathy: clinical outcomes, intraoperative blood flow analysis, long-term follow-up, and long-term bypass patency in a single surgeon case series of 162 procedures. World Neurosurg. 2022;168:e500–e17. https://doi.org/10.1016/j.wneu.2022.10.015.
15. Khan NR, Lu VM, Elarjani T, Silva MA, Jamshidi AM, Cajigas I, et al. One-donor, two-recipient extracranial-intracranial bypass series for moyamoya and cerebral occlusive disease: rationale, clinical and angiographic outcomes, and intraoperative blood flow analysis. J Neurosurg. 2022;136(3):627–36. https://doi.org/10.3171/2021.2.Jns204333.
16. Amin-Hanjani S, Du X, Mlinarevich N, Meglio G, Zhao M, Charbel FT. The cut flow index: an intraoperative predictor of the success of extracranial-intracranial bypass for occlusive cerebrovascular disease. Neurosurgery. 2005;56(Suppl. 1):75–85. https://doi.org/10.1227/01.neu.0000143032.35416.41.

3D DSA-Guided Keyhole STA-MCA Bypass in Moyamoya Vasculopathy

25

Xu Bin, Yujun Liao, and Mohamed Helmy

Learning Objectives

- Demonstrate the ability to interpret angiographical data and neuroanatomical knowledge to identify suitable donor and recipient arteries for the keyhole STA-MCA bypass technique and to design the surgical approach accordingly.
- Understand the benefits of using 3D digital subtraction angiography (DSA) to select appropriate donor and recipient arteries with higher pressure differences (ΔP) before surgery.
- Describe the design of the skin incision and minicraniotomy based on the locations of the donor and recipient arteries, including when to use a linear or C-shaped incision.
- Evaluate the advantages and limitations of the keyhole STA-MCA bypass technique compared to conventional direct bypass with large incisions for moyamoya patients.
- Discuss the follow-up outcomes of keyhole STA-MCA bypass surgery in moyamoya patients.

Neuroanatomy knowledge is particularly important in STA-MCA bypass procedures, as these surgical interventions involve manipulating delicate structures within the brain to restore blood flow to the hypoperfused areas. In STA-MCA bypass-indicated for cerebrovascular diseases, the superficial temporal artery (STA) is harvested from the scalp and anastomosed to the middle cerebral artery (MCA) to create a new pathway for cerebral blood flow.

X. Bin (✉) · Y. Liao
Neurosurgery Department, Huashan Hospital Fudan University, China National Neurological & Neurosurgical Center, Shanghai, People's Republic of China

M. Helmy
Neurosurgery Department, Huashan Hospital Fudan University, China National Neurological & Neurosurgical Center, Shanghai, People's Republic of China

Neurosurgery Department, Shanghai Donglei Brain Hospital, Shanghai, People's Republic of China

A thorough understanding of the neuroanatomy surrounding the STA and MCA is essential for a successful bypass procedure. The neurosurgeon must be able to identify the STA, its branches, the MCA, and its tributaries. This includes knowledge of the origins, courses, and variations of these vessels, as well as the relationships between them and other structures in the brain.

Neuroanatomy knowledge is also crucial in the perioperative planning and follow-up during STA-MCA bypass surgery. The cerebrovascular surgeon also uses various radiological techniques to detect and monitor cerebral blood flow distribution and identify potential complications, such as magnetic resonance angiography (MRA), magnetic resonance venography (MRV), Computed tomography angiography (CTA), CT perfusion (CTP), Digital subtraction angiography (DSA), MR Arterial spin labeling (MR-ASL), etc. Understanding the underlying neuroanatomy is critical for interpreting these diagnostics and monitoring results, and making any adjustments to the surgical approach if necessary.

A comprehensive understanding of neuroanatomy is essential for a successful STA-MCA bypass procedure. Without this knowledge, the cerebrovascular surgeon risks damaging important structures and compromising the blood flow restoration the bypass aims to achieve. Therefore, neuroanatomy plays a fundamental role in the education and training of neurosurgeons performing STA-MCA bypass procedures. In this chapter, the authors will discuss their method of designing a three-dimensional (3D) keyhole STA-MCA bypass in patients with moyamoya vasculopathy based on neuroanatomical knowledge.

25.1 Anatomical Highlights

- The Superficial Temporal Artery (STA) is a commonly used vessel for surgical treatment of cerebral occlusive diseases with EC-IC bypass.
- Understanding the anatomy, origin, and course of the STA is critical for successful surgical outcomes (Fig. 25.1).

I. N. Kanaan, V. Beneš (eds.), *Neuroanatomy Guidance to Successful Neurosurgical Interventions*,
https://doi.org/10.1007/978-3-031-59838-8_25

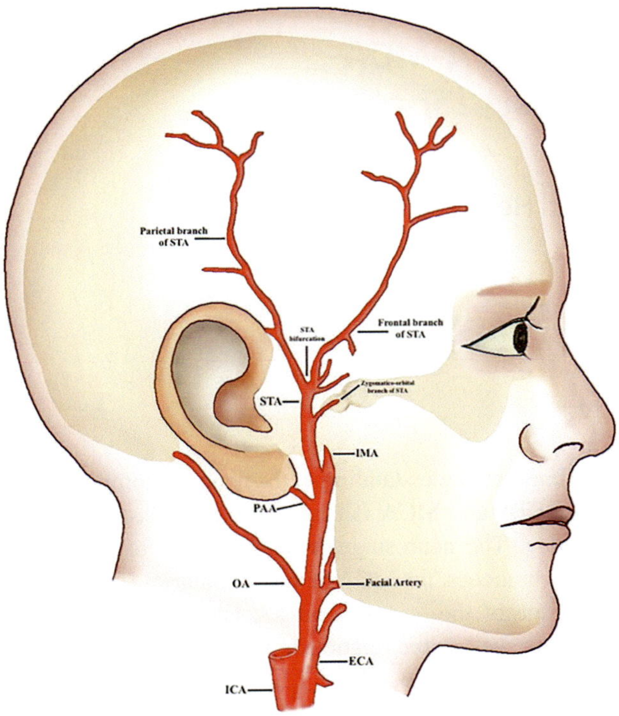

Fig. 25.1 Anatomy of the STA and other external carotid artery (ECA) branches. *ICA* internal carotid artery, *OA* occipital artery, *PAA* postauricular artery, *IMA* internal maxillary artery

- The STA originates from the external carotid artery (ECA) within the parotid gland at the neck of the mandible and follows an upward, tortuous course over the zygomatic process and across the tragus of the auricle, then runs anteriorly over the temporalis muscle accompanied by the corresponding vein and auriculotemporal nerve.
- At approximately 3–5 cm above the zygomatic process, the STA bifurcates into two terminal branches: the anterior (frontal) and posterior (parietal) branches, with the frontal branch supplying blood to the lateral frontal area's muscles, pericranium, and skin, and the posterior branch supplying the skin and pericranium of the parietal region of the scalp.
- The STA typically gives rise to several small branches, including the parotid branch, transverse facial artery, anterior auricular branches, zygomatico-orbital artery, middle temporal artery, frontal branch, and parietal branch.
- Although the STA typically has a typical course over the temporalis muscle, variations in its course and branches may complicate its accessibility as a donor vessel.
- A larger diameter of the STA may give rise to several branches usually derived from other arteries, such as the maxillary artery and the lateral nasal and labial arteries.
- Conversely, some branches of the STA can originate from adjacent arteries, such as the transverse facial artery, which can directly arise from the external carotid.
- The presence of these variations may complicate surgical procedures that utilize the STA as a donor vessel.
- A comprehensive understanding of the STA's anatomy and potential variations is crucial for minimizing possible complications during surgery and optimizing patient outcomes, particularly for those with minimal invasive keyhole craniotomy.

25.2 Introduction

Moyamoya vasculopathy (MMD/S) is a cerebral vascular ailment that manifests as bilateral steno-occlusive changes affecting the proximal divisions of the internal carotid arteries, encompassing Moyamoya disease (MMD) and Moyamoya syndrome (MMS) [1, 2]. Our clinical experience of more than 11,000 cases of bypasses has informed our management strategy, which entails employing STA-MCA bypass in conjunction with encephalo-duro-myo-synangiosis (EDMS) to treat Moyamoya vasculopathy. The measurements of the diameters and blood flow of the three corresponding branches of ECA, namely the superficial temporal artery (STA), middle meningeal artery (MMA), and deep temporal artery (DTA), have revealed that in patients with Moyamoya vasculopathy in Suzuki stages II or III, direct STA anastomosis constitutes the primary blood supply. Conversely, the contribution of MMA and DTA is relatively less significant. Consequently, we posit that a minimally invasive direct bypass may yield favorable outcomes in selected patients with early-stage Moyamoya vasculopathy.

The existing literature on this topic has primarily relied on the morphology-based selection of donor and recipient artery segments, with operators using neuronavigation and/or virtual reality systems to locate the recipient artery [3–6]. We contend that morphological criteria alone are inadequate. According to the Hagen-Poiseuille equation, the pressure difference (ΔP) between the donor and recipient arteries is the sole driving force of blood flow, and a significant positive ΔP is necessary for the STA-MCA bypass to be effective. In our practice, we adopted a novel practical approach to perform keyhole STA-MCA bypass in patients with Suzuki stage III Moyamoya vasculopathy, which includes a donor-recipient selection method that prioritizes ΔP as the primary consideration and is based on the vascular neuroanatomical presentation.

25.3 Surgical Strategy

25.3.1 Preoperative Preparation

As a hybrid cerebrovascular team, all patients underwent 3D DSA prior to surgery in our institution. The 3D rotational angiography was performed with SIEMENS ArtisZeego equipment. In moyamoya vasculopathy patients undergoing direct keyhole bypass, we adopt the following imaging setting: During the procedure, a diagnostic catheter was introduced to the common carotid artery (CCA), and a contrast agent was injected to fully display the internal carotid artery (ICA) and external carotid artery systems simultaneously. Iodinated contrast medium, with an injection rate of 3 ml/s and a total volume of 15 ml, was given at an injection pressure of 200 mmHg. Subsequently, a three-dimensional image was reconstructed. Depending on the reconstructed image result and/or the patient's vasculature conditions, the setting can be modified to 4 ml/s with a total volume of 20 ml.

On the coronal view of the 3D reconstructed image, all MCA cortical branches that were considered candidate recipients were identified and marked. The image was then viewed from a sagittal angle to determine the distance between the donor STA branch and the candidate MCA cortical branches. In selecting the candidate vessels, three factors were taken into consideration. Firstly, the distance between the donor and the recipient arteries, which is best to be closer to the donor artery in most cases. Secondly, the diameter of the recipient artery, where a thicker diameter is preferable. Thirdly, according to the theorem of Bernoulli, the time difference in the arterial phase development between the donor and recipient arteries [4], where a more extended time difference indicates a higher pressure difference (ΔP). If there were multiple candidate cortical vessels with similar diameters within the same spatial distance, the vessel with the larger ΔP was chosen. The more significant the pressure difference between the donor and recipient, the better the blood flow the donor's vessel will carry in the future. Therefore, it is crucial for the surgeon to carefully analyze the anatomical vascular structure on imaging to determine the donor and recipient sites.

Two situations were considered when designing the scalp incision and after selecting both the donor and recipient artery segments. The first was when the donor and the recipient were spatially close (≤2 cm) on the 3D DSA image, meaning the thickness of the bone flap only separated them (Fig. 25.2a, b). In this situation, the preselected candidate recipient segment was used as a reference to design the craniotomy, and an approximately 4 cm skin incision mark was proposed two-to-three millimeters lateral to the STA branch (Fig. 25.2c, d). The second situation was when there was a significant spatial distance between the donor and recipient arteries (>2 cm) on the sagittal view of the 3D DSA image (Fig. 25.3a). In this case, the STA branch was harvested and displaced. A curvilinear skin flap was required to expose the course of the donor with a sufficient length to meet the total exposure of the proposed bone window (Fig. 25.3b, c). The incision design referred to connecting three vertices of a triangle that could be marked by the operator on the 3D DSA images. The three vertices were composed of the bifurcation point of the STA, a 5 cm point on the donor artery distal to the bifurcation point, and the skin-correlated point to the expected bypass region based on the location of the proposed

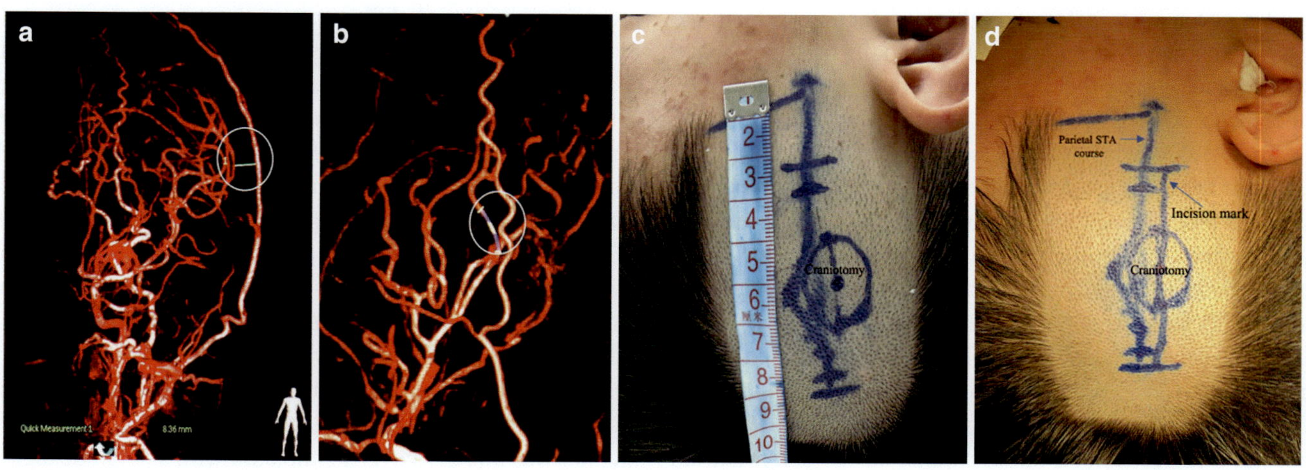

Fig. 25.2 The recipient artery was located just below the course of the parietal branch of the STA (**a**, **b**), and the spatial distance between the donor and recipient was 8.36 mm (**a**). A 5 cm skin incision was proposed in almost all cases (**c**). The craniotomy site, course of the STA branch, and incision mark were shown (**d**)

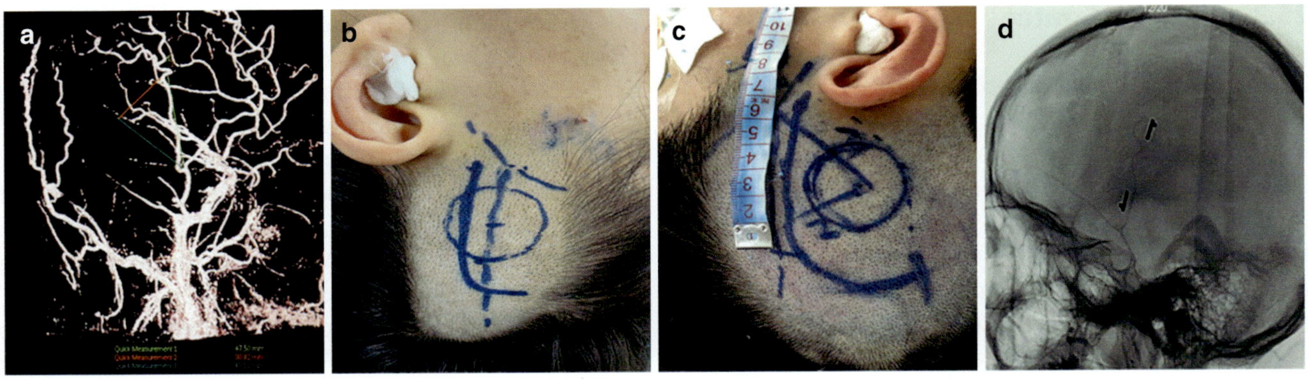

Fig. 25.3 Three vertices were connected to measure the distance between the bifurcation to the donor site and the bifurcation to the distal donor, which is usually kept longer than the distance between the bifurcation site to the donor as the donor needs to travel through the bone thickness to reach the recipient, and the donor's distal to the bifurcation (**a**). Curvilinear incisions were proposed for type-2 keyhole STA-MCA bypass (**b**, **c**). Skin stable secured by a medical tape can work as a guiding method to locate the vascular structures (**d**)

cortical branch. The three vertices could be correlated to taped scalp staples that worked as a radio-opaque surface marker (Fig. 25.3d). These anatomical data were crucial for the bypass surgery.

25.3.2 Intraoperative Strategy

In a hybrid operating room, with the patient in a supine position and under general anesthesia, the head was tilted contralateral to the operable side without using a head fixation clamp to avoid injury to the contralateral STA. A horse-shoe headrest was employed as an alternative. The course of the STA was commonly determined by finger palpation or, in difficult cases, by ultrasonography. As shown in previous figures, it is advisable to mark the course of the donor artery with a pen marker before anesthesia induction to facilitate easy palpation. After routine sterilization, the microscope was draped, and the entire procedure was performed microscopically.

To avoid donor injury, a scalp incision was initiated 2–3 mm lateral to the donor artery's distal part. The incision site was then grasped with toothed forceps, and the incision was carried on proximally. The STA runs within the dense connective tissue (galea aponeurosis) between the superficial temporalis fascia and the skin; thus, avoiding incising at this layer is recommended. After blunt dissection of the loose connective tissue space, it is critical to distinguish the STA from the superficial temporal vein, which runs parallel to it but has a larger caliber, thinner walls, darker color, and no tortuous morphology. By following this sequence, injury to the donor artery can be avoided. If the donor artery is accidentally injured at its distal end, the distal end can be ligated, and the remaining proximal length can be used to perform the anastomosis. The dissection of the donor trunk should be carried out as close as possible to the bifurcation so that the proximal blood flow force of the superficial temporal artery can be fully utilized. It is advisable to keep the recipient artery's depth in mind and make the incision length longer than the correlated marked point. Before harvesting the donor conduit, the temporalis muscle was divided linearly along the natural direction of its fibers until the periosteum was exposed (Fig. 25.4a). The range of bone flaps exposed by retraction of the temporalis muscle fibers is usually sufficient to make a craniotomy (Fig. 25.4b). The squamous suture on the skull can be used as an auxiliary skull surface anatomical landmark. A mini craniotomy was then completed (Fig. 25.4c).

After exposing and confirming the recipient segment, the donor artery under the skin flap was freed, and the vascular bypass was initiated (Fig. 25.4d). The surgical anastomosis process was not much different from those with conventional bypasses. The only difference was that cerebrospinal fluid (CSF) efflux during brain beating could cause visual blurring if the bone window became minimal. It was then necessary for the surgical assistant to promptly suck any excessive CSF in the surgical field with a 5-gauze mini suction tube. A small patch of a gelfoam or a neuro pattie can be used as a protective layer below the suction tip to avoid injury to the brain tissue. After the anastomosis, intraoperative indocyanine green (ICG) or DSA was used to verify the anastomosis patency. After drilling a bony gap for the donor artery or punching a hole with a rongeur forceps, the bone flap was fixed with titanium pins (Fig. 25.5a). Another alternative method was directly covering the dura with a small hexagonal titanium mesh (Fig. 25.5b). The skin was closed with a cosmetic subcuticular suture under the microscope (Fig. 25.5c, d). A follow-up DSA was applied to all patients 6 months after surgery. Matsushima's grade was evaluated for all postoperative cases.

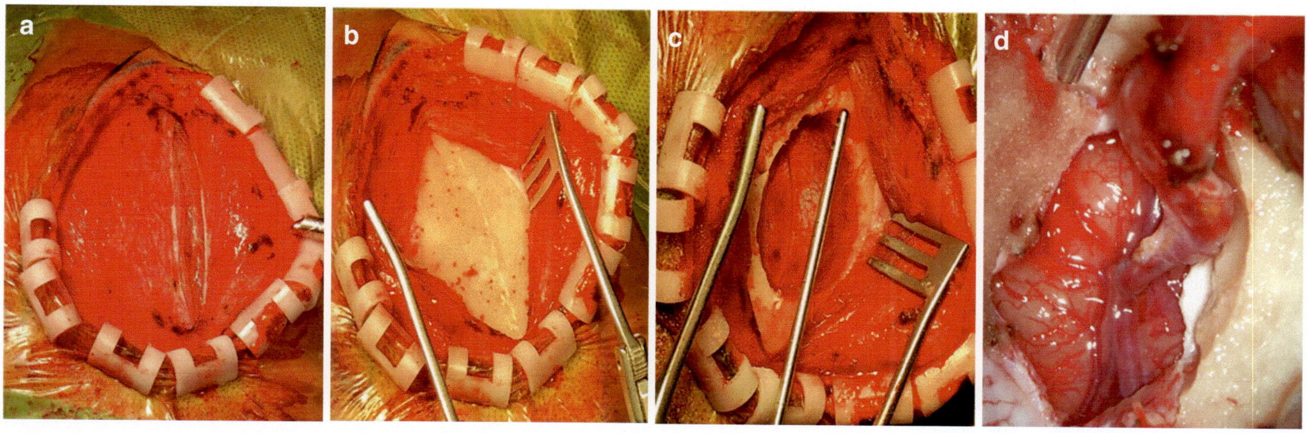

Fig. 25.4 The temporalis fibers were divided along their natural direction (**a**). After retraction (**b**), a minicraniotomy was performed (**c**), and the STA-MCA bypass was anastomosed (**d**)

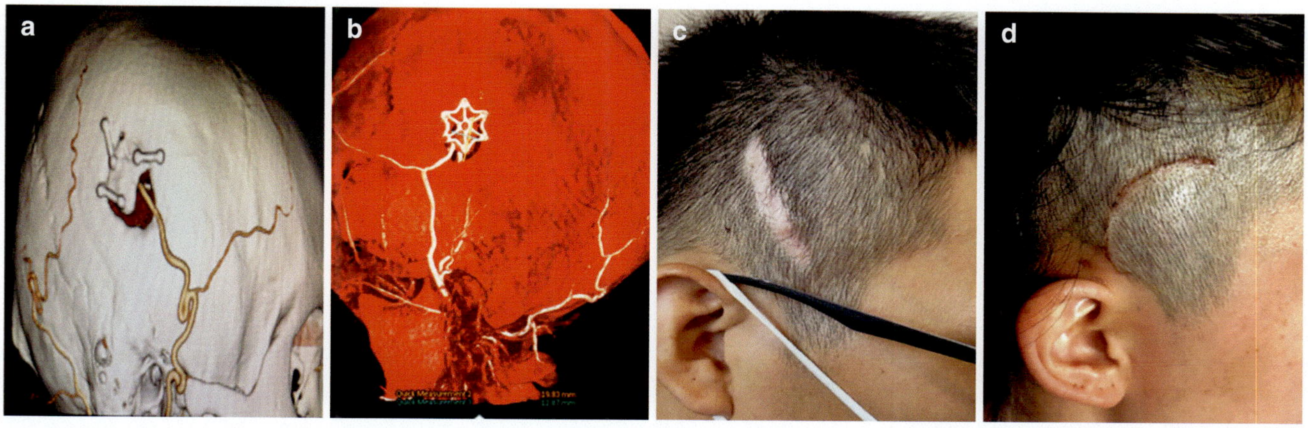

Fig. 25.5 The craniotomy could be secured by pins and screws (**a**) or covered directly by a titanium mesh (**b**). The skin incisions had better cosmetic results than the conventional large craniotomies (**c**, **d**)

25.4 Results

In our experience, all candidate cortical branches of the middle cerebral artery (MCA) were located at the center of the craniotomies, and the keyhole bypass strategy was applied to all included cases. The average length of skin incisions was 4.2 cm, and the average diameter of craniotomies was 2.05 cm (Fig. 25.6). At the 6-month postoperative follow-up with 3D DSA, eight out of ten (80%) patients achieved postoperative Matsushima grade A (Fig. 25.7). In comparison, one patient (10%) had Matsushima grade B, and another (10%) had Matsushima grade C. No significant neurological deficits were observed in any of the ten recruited patients during the follow-up period.

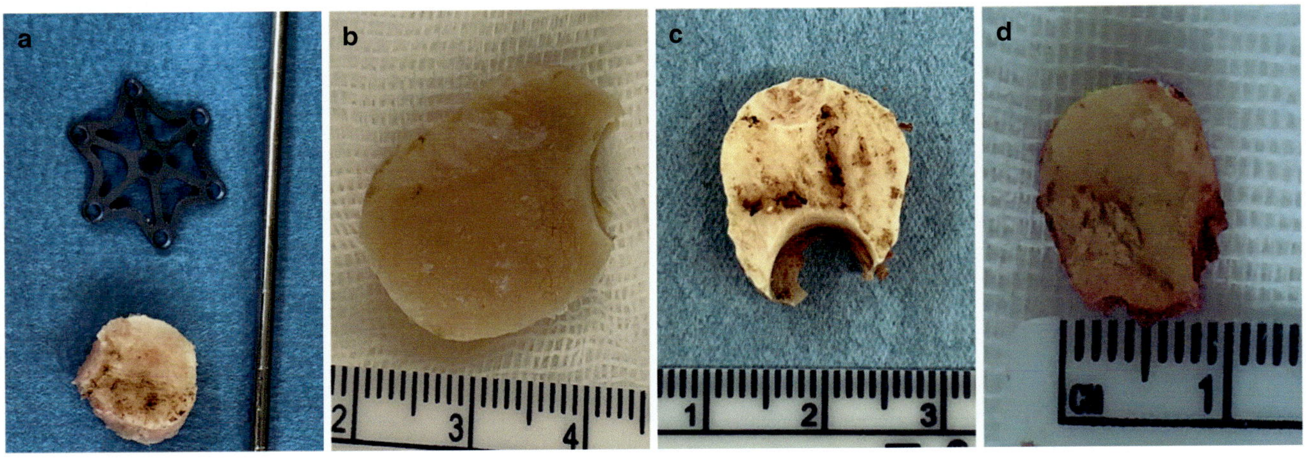

Fig. 25.6 Shows the average size of our keyhole STA-MCA bypass craniotomies

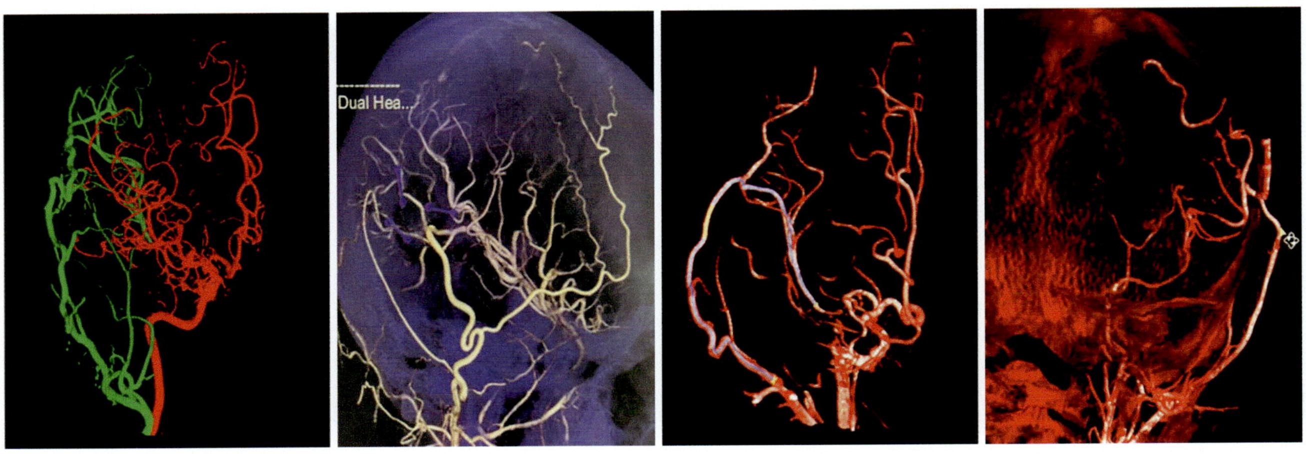

Fig. 25.7 Illustrates greater than 2/3 of MCA territories were perfused by the STA branch (Matsushima grade A) in most of our patients (80%)

25.5 Surgical Pearls

- In early stages of MMV, minimally invasive STA-MCA bypass surgery could offer advantages such as keyhole incisions, preserving blood supply, and reducing perioperative complications.
- Preoperative selection of suitable donor and recipient arteries based on angiographical data and neuroanatomical knowledge is crucial for a successful keyhole STA-MCA bypass.
- Visualization of the distal MCA network in bypass surgery can be enhanced through adjusting contrast agent injection time and 3D DSA reconstruction.
- 3D-DSA arterial phase development time difference can be used to select suitable recipient arteries based on pressure differences.
- The pressure difference (ΔP) between donor and recipient arteries is crucial for successful bypass surgery.
- The spatial relationship between the donor and recipient arteries guides the design of the skin incision and minicraniotomy.
- The length of skin incisions can be minimized, averaging 4.2 cm, with craniotomies averaging 2.05 cm which offer better cosmetic outcomes.
- Intraoperative indocyanine green test or postoperative DSA confirms anastomosis patency.
- Six months postoperative angiography follow-up shows satisfactory outcomes of our keyhole bypass technique, with 80% of patients achieving Matsushima Grade A.
- The "FAST" keyhole STA-MCA bypass technique emphasizes fast, adequate blood flow, simplicity, and tidy execution.

25.6 Discussion

As reported in previous studies, minimally invasive bypass surgery offers several unique advantages [3–10]. These include the use of a keyhole incision that minimizes dissection of the donor, maximizing preservation of blood supply to the flap through the temporalis muscle. Additionally, the procedure involves a smaller bone window, thereby reducing the risk of perioperative complications such as epidural and subdural hematomas. Patients also report high satisfaction with this technique, and the cosmetic benefits make it a preferred method for a wide range of patients. Considering these benefits, there is potential for the broader implementation of minimally invasive procedures in bypass surgery.

In patients with Moyamoya vasculopathy, the severity of M1 stenosis is milder in Suzuki stage II compared to stage III. Consequently, anterograde blood flow from M1 to the distal MCA network is more remarkable in stage II, resulting in a more significant blood flow in the Suzuki stage II hemisphere that competes with the blood flow from direct bypass. However, in stage III, the M1 segment's stenosis is more severe, and more moyamoya vessels develop, leading to delayed imaging time of distal vessels due to increased proximal stenosis.

The successful outcome of bypass surgery depends on identifying an appropriate recipient artery. The Hagen-Poiseuille equation states that the pressure difference (ΔP) between the donor and recipient is the sole driving force for blood flow. The efficacy of the STA-MCA bypass surgery is only possible when ΔP is significantly positive. Although prior studies have reported various technical approaches for the preoperative selection of proper vessels based on their morphology and performing minimally invasive STA-MCA bypass [6, 11], no mention was made regarding ΔP. Patients with stage III moyamoya often have more than two candidate recipient arteries with approximately 1 mm diameter at the cortex's surface. In this scenario, we had to rely on additional parameters to determine which ΔP was relatively significant and which segment was the more suitable recipient. We based our selection on the 3D-DSA arterial phase development time difference between the donor and the candidate recipients. According to the Bernoulli theorem and the Throttle principle, we can deduce that vessels with more significant time differences have higher pressure differences and, therefore, are preferred when all other morphological variables are the same.

Moreover, the proximal flow of the MCA in moyamoya patients has been reported to be significantly reduced, causing rapid contrast agent dilution upon entry into the distal vascular network and, subsequently, poor visualization of the distal MCA network. Therefore, bypass surgery often requires a large bone window diameter of over 3 cm to adequately expose cortical arteries for proper anastomosis selection. To enhance the visualization of Moyamoya collaterals, increasing the contrast agent injection time and adopting the 3D DSA reconstruction technology of the common carotid artery and its network is recommended. This approach enables the precise preoperative selection of the donor and recipient arteries, thereby reducing the bone flap size. Furthermore, the 3D-DSA technology provides sufficient angiographical information on the distal branches of the MCA during preoperative STA-MCA bypass planning and facilitates the acquisition of the spatial relationship between the donor, the recipient, and the skull surface marks.

Interestingly, in our cohort of Suzuki stage III patients, the Matsushima grade A rate was 80%, consistent with the recent conclusion of Han et al. [12] In cases where the recipient artery was the M4 segment of the MCA, indirect revascularization such as encephalo-duro-myo-synangiosis was unnecessary, as the conduit blood flow supplied up to the proximal M2 segment. However, one patient was Matsushima grade B, and only one case was Matsushima grade C in our series, with the entire recipient's blood flow being slow and almost stagnant when selected, which may have hindered the release of donor flow. This provides valuable experience for further exploration of this technology. Furthermore, we found that curvilinear and linear skin incisions resulted in better cosmetic outcomes than conventional large incisions [13], with all bypass vessels' diameters significantly increasing after surgery in stage III patients. Therefore, our findings suggest that stage III patients are the most suitable candidates for direct bypass.

In the hands of a skilled cerebrovascular surgeon, the STA-MCA bypass procedure can be performed expeditiously and with relative ease. Drawing from our extensive experience in the surgical management of patients with moyamoya vasculopathy, we have developed a "FAST" keyhole STA-MCA bypass technique, denoting the following attributes: F: Fast, A: Adequate "blood flow," S: Simple, T: Tidy & orderly. Our procedural data indicate an average duration of 120 min from skin incision to closure, with an average time of fewer than 40 min for donor harvesting and completion of the anastomosis.

Although the STA-MCA bypass approach we propose may have limitations, such as missing the opportunity to select a wider range of receptors or performing a double bypass, we speculate that finding a suitable recipient may still be difficult in advanced stages of moyamoya vasculopathy, despite wide cortical exposure. Furthermore, if either the donor or recipient sustains injuries during surgery, the option of switching to another blood vessel for the bypass may be restricted. Therefore, to mitigate these risks, the incision design should be comprehensive and allow for easy switching to traditional craniotomy if needed, while the surgical team should possess self-assurance, cooperation, and ample experience. As this technology evolves, conducting large comparative cohorts to refine the surgical indications for this innovative approach by comparing it with traditional STA-MCA bypass techniques will be critical.

25.7 Conclusion

A profound understanding of neuroanatomy can enable cerebrovascular surgeons to adopt novel surgical strategies for treating their patients. Our 3D DSA-based keyhole STA-MCA bypass technique has demonstrated satisfactory post-operative outcomes. The simplicity of our preoperative preparation, utilizing 3D DSA, offers a comprehensive and precise selection of suitable conduit and recipient arteries. For patients with Suzuki stage III Moyamoya vasculopathy and good integrity of the distal MCA network, minor bone window surgery may be a preferred alternative to traditional direct revascularization surgery.

References

1. Kim JS. Moyamoya disease: epidemiology, clinical features, and diagnosis. J Stroke. 2016;18(1):2–11.
2. Scott RM, Smith ER. Moyamoya disease and moyamoya syndrome. N Engl J Med. 2009;360(12):1226–37.
3. Cabrilo I, Schaller K, Bijlenga P. Augmented reality-assisted bypass surgery: embracing minimal invasiveness. World Neurosurg. 2015;83(4):596–602.
4. Charbel FT, Guppy KH, Ausman JI. Cerebral revascularization: superficial temporal middle cerebral artery anastomosis. In: Sekhar LN, Fessler RG, editors. Atlas of neurosurgical techniques. New York: Thieme; 2006.
5. Shaban S, Huasen B, Haridas A, Killingsworth M, Worthington J, Jabbour P, et al. Digital subtraction angiography in cerebrovascular disease: current practice and perspectives on diagnosis, acute treatment and prognosis. Acta Neurol Belg. 2021;
6. Fischer G, Senger S, Sharif S, Oertel J. Superficial temporal artery to middle cerebral artery bypass via a minimized approach: operative nuances and problem-solving aspects. World Neurosurg. 2016;88:97–103.
7. Coppens JR, Cantando JD, Abdulrauf SI. Minimally invasive superficial temporal artery to middle cerebral artery bypass through an enlarged bur hole: the use of computed tomography angiography neuronavigation in surgical planning. J Neurosurg. 2008;109(3):553–8.
8. Teo C. The concept of minimally invasive neurosurgery. Neurosurg Clin N Am. 2010;21(4):583–4. v
9. Welling LC, Figueiredo EG, Wen HT, Gomes MQT, Bor-Seng-Shu E, Casarolli C, et al. Prospective randomized study comparing clinical, functional, and aesthetic results of minipterional and classic pterional craniotomies. J Neurosurg. 2015;122(5):1012–9.
10. Tsunoda S, Inoue T, Ohwaki K, Akabane A, Saito N. Comparison of postoperative temporalis muscle atrophy between the muscle-preserving pterional approach and the mini-pterional approach in the treatment of unruptured intracranial aneurysms. Neurosurg Rev. 2021; https://doi.org/10.1007/s10143-021-01558-6.
11. Kadri PAS, Krisht AF, Gandhi GK. An anatomic mathematical measurement to find an adequate recipient M4 branch for superficial temporal artery to middle cerebral artery bypass surgery. Neurosurgery. 2007;61(Suppl. 3):74–8. discussion 78
12. Han Q, Yao F, Zhang Z, Huang Y. Evaluation of revascularization in different suzuki stages of ischemic Moyamoya disease by whole-brain CT perfusion. Front Neurol. 2021;12:683224.
13. Acker G, Schlinkmann N, Fekonja L, Grünwald L, Hardt J, Czabanka M, et al. Wound healing complications after revascularization for moyamoya vasculopathy with reference to different skin incisions. Neurosurg Focus. 2019;46(2):E12.

26 The Art and the Logic of Flow-Guided Bypass for Aneurysms

Fady T. Charbel and Laura Stone McGuire

Learning Objectives
- Overview the concept of flow replacement bypass for aneurysm surgery.
- Understand the principles of direct intraoperative flow measurement and how to use this measurement to select the appropriate donor graft and assess the bypass.
- Learn operative technique for the measurement of blood flow using the micro-flowprobe.
- Review flow-guided postoperative management with quantitative magnetic resonance angiography.

26.1 Introduction

The flow-guided approach to cerebrovascular surgery represents an important paradigm shift in bypass for aneurysms. The critically important conceptual starting point is that a bypass for an aneurysm is a "flow replacement bypass" [1]. Bypass in complex cerebral aneurysm surgery becomes necessary when blood flow must be replaced during clip ligation or reconstruction, such that a new conduit must be established to revascularize the efferent vessels of the aneurysm.

26.2 History

Classical teaching considered the terms "high flow" or "low flow" in reference to the donor vessel size, attempting to approximate the flow available and ranging from saphenous veins to radial artery to in situ pedicles. However, a supply-demand conceptual framework developed over the past two decades that highlighted the importance of direct intraoperative flow measurement to guide decision-making [2–4]. Ultimately, the best revascularization approach matches donor supply to recipient demand, and that assessment must be performed at an individual case level. Therefore, using real-time direct flow measurements intraoperatively during bypass surgery for cerebral aneurysms provides critical information for donor selection and about bypass patency. This chapter will review the preoperative assessment, surgical technique, and postoperative management in the flow-guided bypass for aneurysms.

F. T. Charbel (✉) · L. S. McGuire
Department of Neurosurgery, University of Illinois, Chicago, IL, USA
e-mail: fcharbel@uic.edu; lmcguir1@UIC.EDU

26.3 Anatomy Highlights

- Preoperative assessment should include digital subtraction angiography.
- Consider balloon occlusion testing for diseases that may require vessel sacrifice.
- Use quantitative flow measurements preoperatively, when available, including quantitative MRA.

26.3.1 Preoperative Assessment

Digital subtraction angiography (DSA) should capture adequate images not only of the aneurysm itself with three-dimensional reconstructions but also of extracranial circulation for surgical planning. The superficial temporal artery (STA) and occipital artery (OA) represent the most important potential extracranial in situ donor vessels. In addition, DSA aids in fully understanding any collateral flow from the anterior communicating artery and posterior communicating arteries.

Aneurysms of different parent vessels will have different considerations for potential bypass. Elective surgical planning for aneurysms of the internal carotid artery (ICA), otherwise not suitable for direct surgical repair or endovascular treatment, should include balloon occlusion testing (BOT). BOT involves temporary balloon occlusion of the ICA with

I. N. Kanaan, V. Beneš (eds.), *Neuroanatomy Guidance to Successful Neurosurgical Interventions*,
https://doi.org/10.1007/978-3-031-59838-8_26

the patient awake and under constant neurological assessment and electroencephalographic (EEG) monitoring. The initial vessel occlusion occurs under conditions of normal blood pressure for 10–15 min but followed by hypotension for an additional 10–15 min. During the hypotensive challenge, 99 m technetium is administered, such that single photon emission computed tomography (SPECT) may be performed immediately following the BOT to assess cerebral blood flow. Hypoperfusion during the hypotensive phase of the study will appear on the SPECT images as an asymmetric pattern of the 99 m technetium tracer when compared to baseline imaging. The balloon is then deflated, and blood pressure is normalized. During the procedure, the patient is heparinized according to weight-based dosing and reversed after the test.

In addition, the flow in the vessels at risk and which may be replaced can be assessed preoperatively with phase-contrast quantitative MRA (QMRA) using NOVA software (Noninvasive Optimal Vessel Analysis; VasSol, Inc., Illinois, USA) [5, 6]. As such, the flow demand from the donor vessel can be determined before surgery and the choice of donors further refined (Figs. 26.1, 26.2 and 26.3).

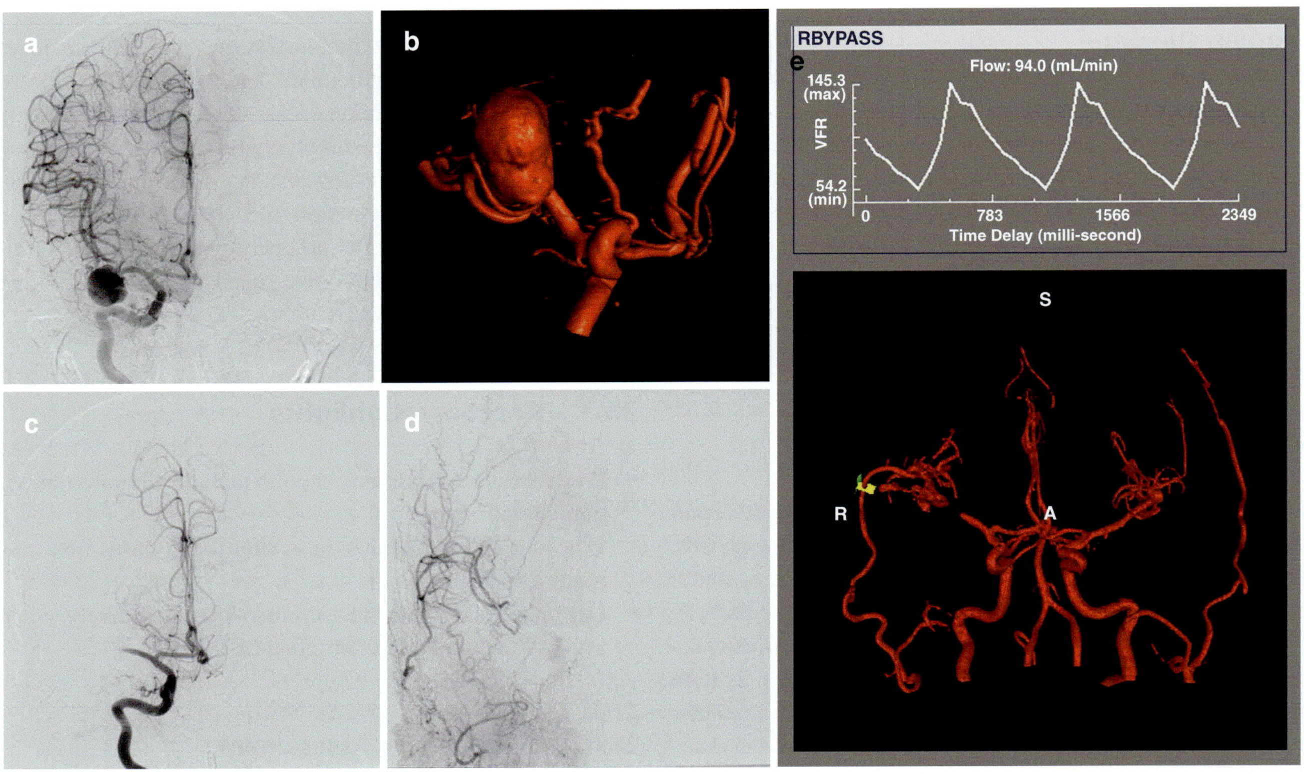

Fig. 26.1 Patient underwent right superficial temporal artery to MCA bypass for treatment of complex MCA aneurysm. (**a**, **b**) Preoperative digital subtraction angiography showing complex right MCA aneurysm. (**c**) Postoperative digital subtraction angiography showing clip occlusion and sacrifice of right M1 and (**d**) patent bypass. (**e**) Quantitative MRA used to follow the patency and adequacy of the bypass

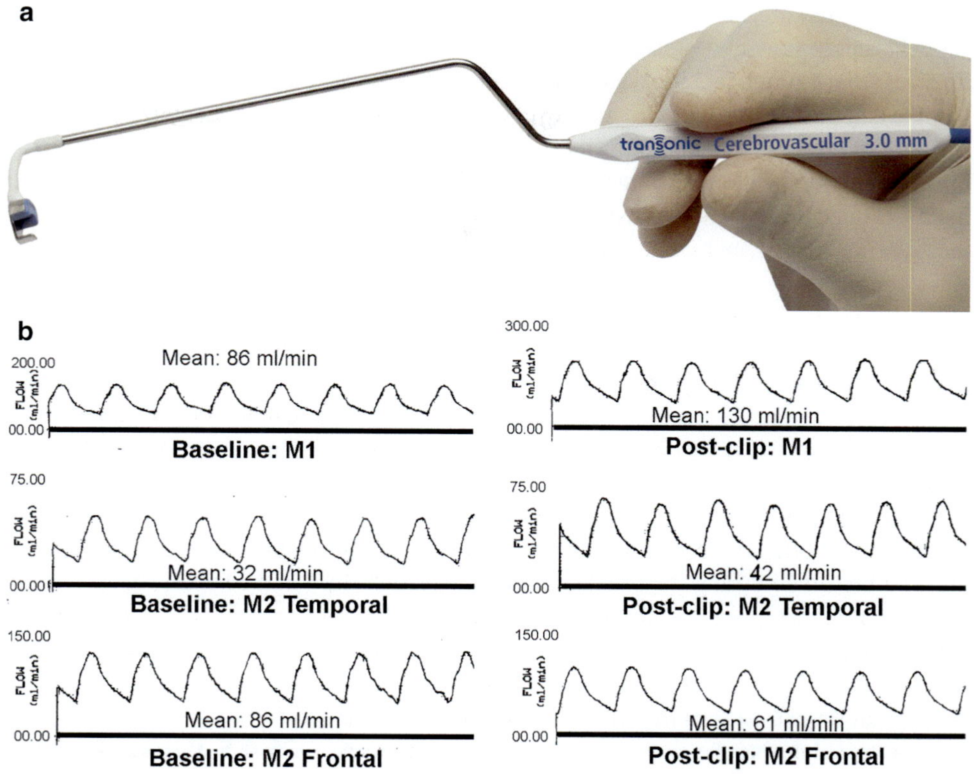

Fig. 26.2 (**a**) Intracranial Charbel micro-flowprobe for transit-time ultrasonic measurement of blood flow. (**b**) Sample tracing of micro-flowprobe measurement before and after bypass procedure

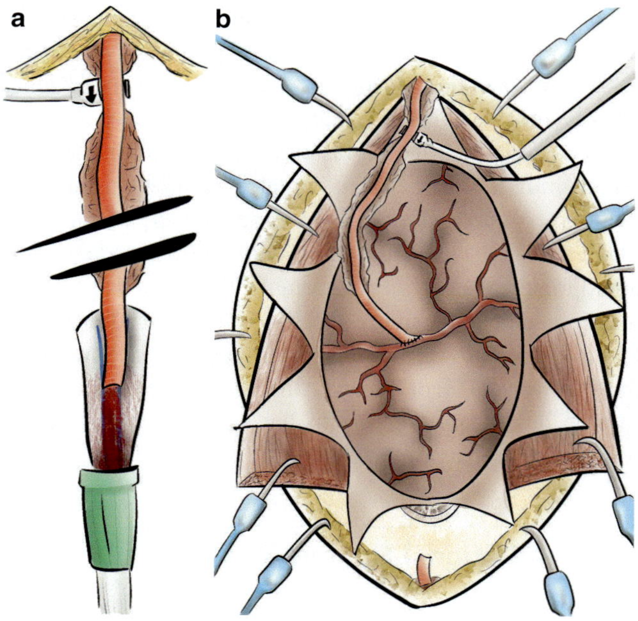

Fig. 26.3 Illustration of flowprobe measurement of donor vessel (**a**) before bypass to assess cut flow and (**b**) after bypass to measure flow and calculate cut flow index

26.4 Surgical Intervention

26.4.1 Preparation

When positioning for surgery, care must be taken to include the planned donor vessel in the preparation. Donor vessel mapping may be undertaken with a Doppler probe, using auditory feedback to plot the STA, OA, or radial artery, or with an ultrasound to visualize its path. The donor's course may then be marked and incorporated into the surgical draping. Aneurysms of the proximal intracranial ICA may additionally require inclusion of the neck in the field.

26.4.2 Intraoperative Blood Flow Measurement

Intracranially, once the aneurysm, its parent contributor, and its efferent vessels have been carefully dissected using microsurgical technique, intraoperative blood flow measurement may be taken. Assessing the flow of the vessels leading to and coming from the aneurysm before the bypass informs the surgeon about the demand required of the bypass and whether sufficient flow has been provided postanastomosis. The surgeon should note the physiological condition of the patient during flow measurements and ensure constant blood pressure, end-tidal carbon dioxide, and burst suppression on EEG throughout each phase of flow measurement.

The intracranial Charbel micro-flowprobe (Charbel Micro-Flowprobe; Transonics Systems Inc., Ithaca, NY, USA) utilizes transit-time ultrasonic technology to measure blood flow [7, 8]. It quantifies blood flow and also provides waveform data and acoustic feedback to assess the quality of the data capture and flow itself. Importantly, due to its use of transit-time ultrasound, the probe measures flow without the influence of turbulence, hematocrit, or flow velocity profile [9]. The micro-flowprobe features a bayoneted handle and tip sizes from 1.5 mm to 3 mm. The neck of the flowprobe is flexible and can thereby adjust to maneuver easily around the bypass. An arrow marks the tip of the probe to indicate the directionality of the blood flow measurement. When ready to measure blood flow, the flowprobe tip is gently placed around the vessel of interest, and saline irrigation is used as the ultrasonic couplant.

26.4.3 Operative Nuances

In aneurysms of the anterior circulation that might require bypass—the middle cerebral artery (MCA), anterior cerebral artery (ACA), and internal carotid artery (ICA) terminus—the STA may offer the best extracranial in situ donor and must be preserved during the standard cranial approach. In aneurysms of the posterior circulation that might require bypass—the posterior cerebral artery (PCA) and posterior inferior cerebellar artery (PICA) —the OA may present the best option for extracranial in situ donor and should be carefully protected during the surgery. In surgery for aneurysms of the proximal intracranial ICA—the cavernous, paraclinoid, and ophthalmic segments—the cervical ICA should be prepared. Following the initial cranial opening and subsequent microdissection of the aneurysm, its parent artery, and efferent branches, the microprobe may be brought into the field for assessing flow demand. Measuring flow in the parent artery may be difficult, and therefore, the summation of flow from the branches coming from the aneurysm should indicate the distal territory flow and bypass requirements.

26.4.4 Cut Flow

The **cut flow** refers to the maximum capacity of blood flow through the in situ graft, such as the STA or OA [3, 10]. Once the donor vessel is harvested and prepared for potential bypass, the surgeon uses the micro-flowprobe to determine the amount of flow that can be carried through this vessel in the absence of any distal resistance. For instance, when the STA or OA remains in continuity within the scalp, the flow usually ranges from 5 to 10 mL/min, which reflects the high resistance of the distal branches. In contrast, once the STA or OA has been dissected free and the distal end cut in preparation for the bypass, the flow measurement increases sharply. In the STA, the cut flow may rise to approximately 60 mL/min [11].

To prepare the in situ donor graft, the vessel is flushed with heparinized saline using a blunt syringe and then occluded distally with a temporary clip. The best practice to maximize the flow potential of the donor vessel is to wrap the graft in a papaverine-soaked cottonoid during the cranial approach to prevent vasospasm secondary to the dissection and tissue manipulation. After the recipient vessel has completed its preparation but before the graft has been trimmed to its final length, the micro-flowprobe is brought into the field to measure the cut flow. The donor graft is allowed to flow freely, usually onto a cottonoid with suction on the patty to keep a clean field, with the tip of the flowprobe around the vessel's distal end. Then, the vessel may be cut to its final size, removing the end manipulated by the temporary clip and syringe flush. Note that by shortening the donor, the cut flow will increase to a certain degree. At our institution, the cut flow should fall within 20% of that measured in the recipient, as prior studies have shown that reducing flow by 20–25% or more may result in ischemia [12–14]. If the cut flow falls beyond the 20% threshold, then an interposition graft, such as radial artery or saphenous vein, will be utilized in the bypass. In our institutional data, the STA may offer an average of approximately 60 mL/min, cadaveric vein graft 80 mL/min, and autologous vein graft 100 mL/min [11, 15].

26.4.5 Bypass Technique

The details of the bypass procedure will be described in other chapters of this book. The technique implemented for aneurysm surgery depends on the aneurysm itself, associated vessels, and location. In summary, however, the in situ donor graft is dissected and prepared as described above and then transected in a fish-mouth fashion using microscissors. The donor graft is sutured to its targeted recipient using 10-0 nylon interrupted sutures. When using an interposition graft, radial artery or saphenous graft is harvested as needed for the appropriate vessel caliber and length. The distal anastomosis may require a larger suture with 8-0 or 9-0 nylon for the distal anastomosis. The proximal graft may be anastomosed to the STA trunk [16], or it may be tunneled to the neck to anastomose its proximal end with the external carotid artery, using a 7-0 prolene running suture.

26.5 Additional Tips and Tricks

26.5.1 Cut Flow Index

The term **cut flow index** refers to the following calculation: bypass flow (mL/min) / cut flow (mL/min). The cut flow index (CFI) has been studied in the setting of bypass for ischemic cerebrovascular disease, and a CFI of ≥0.5 at the time of surgery is a significant predictor of bypass patency [10, 17]. Although not primarily studied in bypass for cerebral aneurysms, a poor CFI could indicate potential problems with the donor vessel, anastomosis, or recipient vessel during surgery.

26.5.2 Closing

During the surgical closure, the in situ donor draft may be measured using the micro-flowprobe to ensure no impingement on the vessel while suturing the tissue layers. If a significant drop in blood flow is encountered, the surgeon may first check to make sure physiologic parameters match those during the initial measurements, as those may change during emergence from anesthesia. Otherwise, the surgeon should address potential points of stenosis, including the temporalis muscle or the cranial edge.

26.5.3 Postoperative management

Following bypass surgery, the patient remains on aspirin 325 mg daily. Postoperative DSA establishes new baseline images to follow and allows for serial assessment of bypass patency over the long term. Our institution additionally implements a flow-based protocol using QMRA with NOVA software (Noninvasive Optimal Vessel Analysis; VasSol, Inc., Illinois, USA) to follow patients after bypass procedures to ensure not only patency but also adequate blood flow [18–22]. In a study of 22 patients with STA donor grafts for bypass for anterior circulation intracranial aneurysms, bypass flows were typically highest in the immediate postoperative period but generally remained stable across follow-up, and cerebral hemispheric flows were maintained, accommodating the recipient territory's long-term flow requirements [11].

26.6 Surgical Pearls

- Assess the intraoperative blood flow of the donor (cut flow), the recipient, and the bypass (cut flow index) to determine bypass adequacy.
- Check bypass blood flow before and during closure to ensure no impingement on the vessel.
- Follow quantitative blood flow measurements postoperatively using quantitative MRA to monitor bypass patency.

26.7 Conclusion

The flow-guided approach to bypass surgery for cerebral aneurysms offers important insights into the donor selection based on flow requirements in the recipient vessel territory, using direct intraoperative micro-flowprobe measurements. The cut flow indicates the maximum flow capacity of the donor graft and should approximate the recipient territory requirements. Furthermore, long-term bypass patency and adequacy of blood flow may be assessed with serial QMRA.

References

1. Charbel FT, Guppy KH, Ausman JI. Cerebral revascularization: superficial temporal middle cerebral artery anastomosis. In: Sekhar LN, Fessler RG, editors. Atlas of neurosurgical techniques. New York: Thieme; 2006.
2. Amin-Hanjani S, Charbel FT. Flow-assisted surgical technique in cerebrovascular surgery. Surg Neurol. 2007;68(Suppl. 1):S4–11.
3. Amin-Hanjani S, Alaraj A, Charbel FT. Flow replacement bypass for aneurysms: decision-making using intraoperative blood flow measurements. Acta Neurochir. 2010;152(6):1021–32. discussion 1032
4. Ashley WW, Amin-Hanjani S, Alaraj A, Shin JH, Charbel FT. Flow-assisted surgical cerebral revascularization. Neurosurg Focus. 2008;24(2):E20.
5. Zhao M, Charbel FT, Alperin N, Loth F, Clark ME. Improved phase-contrast flow quantification by three-dimensional vessel localization. Magn Reson Imaging. 2000;18(6):697–706.
6. Charbel FT, Guppy KH, Zhao M, Clark ME. Computerized hemodynamic evaluation of the cerebral circulation for bypass. Neurosurg Clin N Am. 2001;12(3):499–508. viii
7. Charbel FT, Hoffman WE, Misra M, Ostergren L. Ultrasonic perivascular flow probe: technique and application in neurosurgery. Neurol Res. 1998;20(5):439–42.
8. Charbel FT, editor. Flow assisted aneurysm surgery-cerebral aneurysm handbook. New York: Transonic Systems; 2004.
9. Lundell A, Bergqvist D, Mattsson E, Nilsson B. Volume blood flow measurements with a transit time flowmeter: an in vivo and in vitro variability and validation study. Clin Physiol. 1993;13(5):547–57.
10. Amin-Hanjani S, Du X, Mlinarevich N, Meglio G, Zhao M, Charbel FT. The cut flow index: an intraoperative predictor of the success of extracranial-intracranial bypass for occlusive cerebrovascular disease. Neurosurgery. 2005;56(Suppl. 1):75–85. discussion 75–85
11. Rustemi O, Amin-Hanjani S, Shakur SF, Du X, Charbel FT. Donor selection in flow replacement bypass surgery for cerebral aneurysms: quantitative analysis of long-term native donor flow sufficiency. Neurosurgery. 2016;78(3):332–41. discussion 341–2
12. Charbel FT, Zhao M, Amin-Hanjani S, Hoffman W, Du X, Clark ME. A patient-specific computer model to predict outcomes of the balloon occlusion test. J Neurosurg. 2004 Dec;101(6):977–88.
13. Eckert B, Thie A, Carvajal M, Groden C, Zeumer H. Predicting hemodynamic ischemia by transcranial Doppler monitoring during therapeutic balloon occlusion of the internal carotid artery. AJNR Am J Neuroradiol. 1998;19(3):577–82.
14. Jawad K, Miller D, Wyper DJ, Rowan JO. Measurement of CBF and carotid artery pressure compared with cerebral angiography in assessing collateral blood supply after carotid ligation. J Neurosurg. 1977;46(2):185–96.
15. Stapleton CJ, Hussein AE, Behbahani M, Alaraj A, Amin-Hanjani S, Charbel FT. Comparative efficacy of autologous versus cadaveric saphenous vein grafts in cerebral revascularization surgery. J Neurosurg. 2020;134(5):1562–8. https://doi.org/10.3171/2020.3.JNS192546. PMID: 32442978
16. Alaraj A, Ashley WW Jr, Charbel FT, Amin-Hanjani S. The superficial temporal artery trunk as a donor vessel in cerebral revascularization: benefits and pitfalls. Neurosurg Focus. 2008;24(2):E7.
17. Stapleton CJ, Atwal GS, Hussein AE, Amin-Hanjani S, Charbel FT. The cut flow index revisited: utility of intraoperative blood flow measurements in extracranial-intracranial bypass surgery for ischemic cerebrovascular disease. J Neurosurg. 2019;133(5):1396–400.
18. Amin-Hanjani S, Shin JH, Zhao M, Du X, Charbel FT. Evaluation of extracranial-intracranial bypass using quantitative magnetic resonance angiography. J Neurosurg. 2007;106(2):291–8.
19. Kim T, Bang JS, Kwon OK, et al. Hemodynamic changes after unilateral revascularization for Moyamoya disease: serial assessment

by quantitative magnetic resonance angiography. Neurosurgery. 2017;81(1):111–9.
20. Zhu F, Qian Y, Xu B, et al. Quantitative assessment of changes in hemodynamics of the internal carotid artery after bypass surgery for moyamoya disease. J Neurosurg. 2018;129(3):677–83.
21. Amin-Hanjani S, Singh A, Rifai H, et al. Combined direct and indirect bypass for moyamoya: quantitative assessment of direct bypass flow over time. Neurosurgery. 2013;73(6):962–7. discussion 967–8
22. Koch MJ, Stapleton CJ, Charbel FT, Russin J, Amin-Hanjani S. Intracranial-Intracranial Bypass for Aneurysms: Quantitative Intraoperative Assessment of Flow Preservation. Oper Neurosurg (Hagerstown). 2022;22(5):337–42. https://doi.org/10.1227/ons.0000000000000136.

Carotid Endarterectomy

27

Vladimír Beneš, Adéla Bubeníková, and Ondřej Bradáč

Learning Objectives

- Atherosclerotic plaque in the carotid artery can lead to atherosclerosis and stenosis of the artery, which may or may not be symptomatic clinically, increasing carotid artery risk for cerebrovascular disease and stroke.
- Carotid endarterectomy (CEA) is a surgery entailing removing plaque from the common carotid artery and/or internal carotid artery to improve blood flow.
- Techniques for CAE have been established over the years of experience with this disease along with individual indications to perform them. Detailed surgical technique and treatment algorithm are provided in this chapter.

27.1 Introduction

Carotid endarterectomy (CEA) is one of the most common procedures in vascular surgery to remove plaque build-up. In the 1980s, neurologists who had a good understanding of all the complications felt that surgery was overused and without proper indications [1]. Several randomised trials addressing this topic were organised, including the most world-recognised and foremost trials: The North American Symptomatic Carotid Endarterectomy Trial (NASCET) [2] and the European Carotid Surgery Trial (ECST) [3]. Based on these trials, the American Heart Association (AHA) proposed the indications for CEA [4]. These recommendations were updated several times over the years but the changes were minor. Later on, with the appearance of endovascular techniques, the AHA recommendations were applied for carotid stenting (CAS) as well.

In **symptomatic** patients with surgical morbidity/mortality (MM) <6%, recommended indications for CEA or CAS are one or more transient ischaemic attacks (TIAs) in the preceding 6 months and stenosis >50% [5]. In **asymptomatic** patients, the CEA or CAS is offered to those with stenosis >70% if the MM rate is <3% [6, 7].

27.2 Symptomatology

Patients are divided into two groups: asymptomatic patients, i.e. those in whom the stenosis has been diagnosed usually during screening or a preventive programme and (2) symptomatic patients, who have experienced either a TIA, single or repeated or completed stroke, typically of minor severity. In the large studies [2, 8, 9], patients with symptomatology relevant to the affected carotid artery were symptomatic >6 months before treatment. Patients with symptomatology from other territories were evaluated as asymptomatic [5].

27.3 Timing

Emergency procedures are rare [10]. In most patients, treatment is elective except for TIA; in these patients, treatment should be performed as soon as possible, given that TIAs are often followed by a major stroke [11]. A major stroke occurs within a week of a TIA event in approximately 20% of patients [11, 12].

V. Beneš (✉)
Department of Neurosurgery and Neurooncology, First Faculty of Medicine, Charles University and Military University Hospital, Prague, Czech Republic
e-mail: vladimir.benes@uvn.cz

A. Bubeníková · O. Bradáč
Department of Neurosurgery and Neurooncology, First Faculty of Medicine, Charles University and Military University Hospital, Prague, Czech Republic

Department of Neurosurgery, Second Faculty of Medicine, Charles University and Motol University Hospital, Prague, Czech Republic
e-mail: ondrej.bradac@uvn.cz

I. N. Kanaan, V. Beneš (eds.), *Neuroanatomy Guidance to Successful Neurosurgical Interventions*,
https://doi.org/10.1007/978-3-031-59838-8_27

27.4 Diagnostic Procedures

Because angiography offers valuable data, it was the gold standard in the past, which is the base of all randomised trials. Over the years, however, non-invasive techniques have become increasingly popular: ultrasound, magnetic resonance angiography (MRA) and, in particular, computed tomography angiography (CTAg). Axial slices of CTAg give the best correlation with angiography [8].

27.5 Endarterectomy Technique

Our CEA technique was developed over the past 40 years and has not changed significantly since the 1990s. Our monitoring technique has only changed a little in the past two decades. We have used the shunt in <2% of surgeries. The patch and the eversion technique have never been used [10]. Thus, CEA is a routine and uniform surgical procedure in our hands. In the following paragraphs, the anatomy and surgical technique will be described (Fig. 27.1).

The patient is put under general anaesthesia, an arterial line is introduced and the electrodes are attached for intraoperative monitoring. Next, the patient's head is inclined to the opposite side and downwards to ensure that the upper neck is stretched and kept at the highest point. A cushion is placed under the ipsilateral shoulder and the shoulder is pushed downwards by tape (Fig. 27.2). The surgical field is prepared and covered. Some 5000 units of Heparin are administered subcutaneously before surgery.

The incision runs along the anterior border of the sternocleidomastoid muscle (SCM). Haemostasis is achieved and maintained throughout the procedure. The platysma is incised and the anterior border of the SCM is exposed. The muscle is moved laterally, and the common carotid artery (CCA) is palpated and then exposed. The carotid sheath is opened, but the artery is not entirely dissected free. The dissection continues cranially. The only traversing structure is the facial vein, which is ligated and cut. The jugular vein is always lateral to the carotid artery and not dissected from the surrounding tissue. For endarterectomy, the microscope (or, more recently, the exoscope that provides high-resolution 3D imaging) and microtechniques are applied (Figs. 27.3 and 27.4).

The ansa nervi hypoglossi is moved laterally, and the vagal nerve, which runs along the CCA, is not exposed. Caudally, the omohyoid muscle is the caudal limit of approach, where the recurrent nerve is traversing the surgical field. Carotid bifurcation is identified and the superior thyroid and external carotid artery (ECA) are dissected free (10–20 mm). Then the internal carotid artery (ICA) is exposed and followed cranially to get as high as the plaque and beyond. In the event of high-positioned bifurcation or a long plaque, the lower head of the digastric muscle may be sectioned to gain more space. The hypoglossal nerve running directly underneath the digastric muscle is identified and slightly and gently retracted cranially. More superficially, superior to the digastric muscle runs the lower branch of the VII nerve. The branch is not exposed, but the dissection should not go too cranially to avoid injuring the nerve. Self-retaining retractors are introduced to open up the space and expose the bifurcation. After carotid bifurcation and enough length of ICA are dissected free, a tourniquet is introduced on the CCA for eventual shunt insertion. Manipulation with the arteries and their dissection should be minimal to prevent

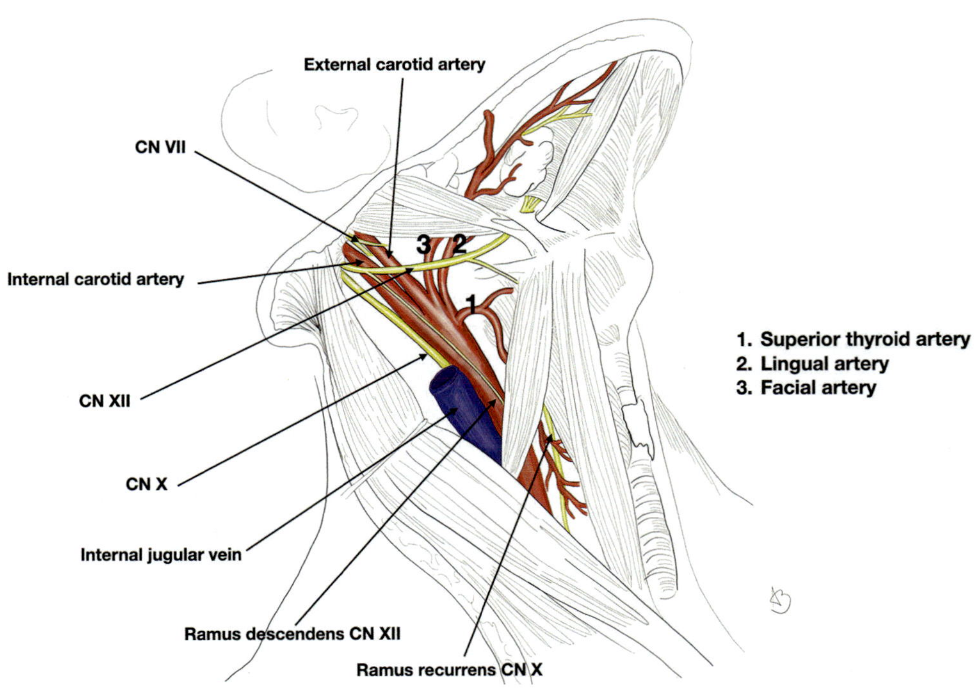

Fig. 27.1 Summary of the most important anatomical landmarks encountered during the carotid endarterectomy

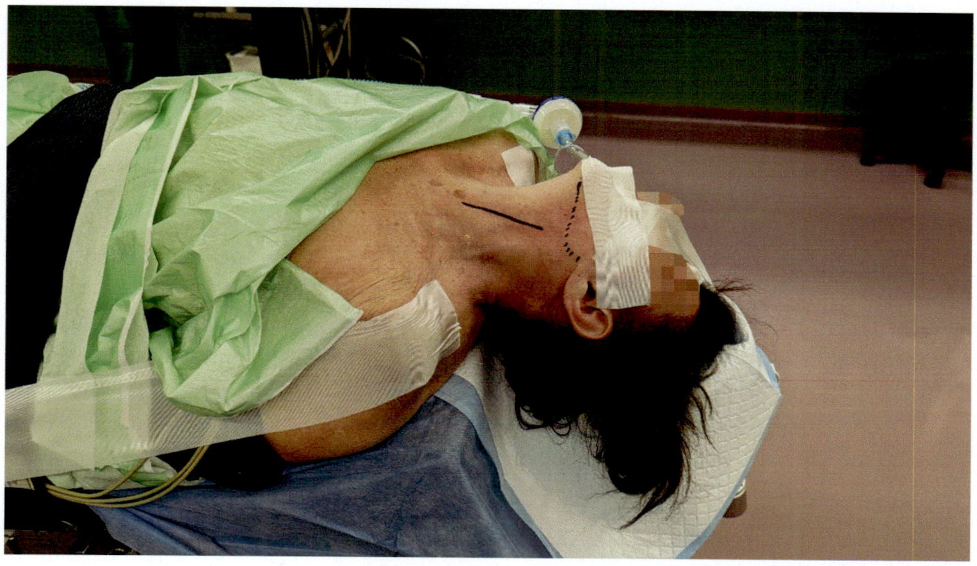

Fig. 27.2 Positioning for the carotid endarterectomy (CAE). The head is turned to the opposite side and inclined. Under the shoulder the paddy is put and the shoulder is pushed downwards by tape

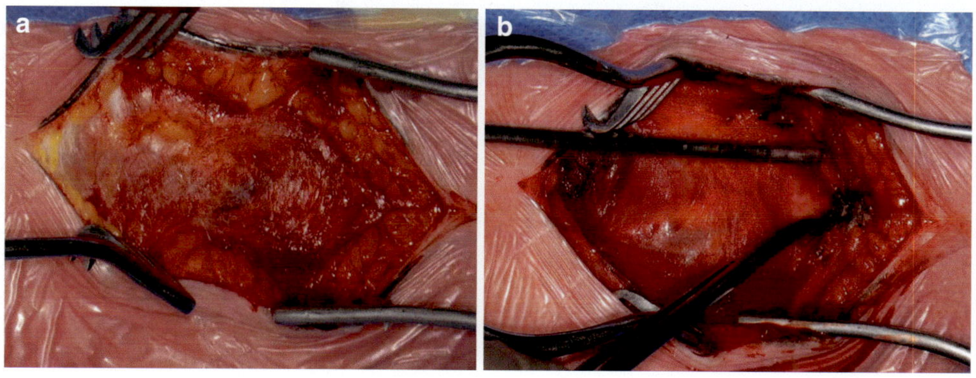

Fig. 27.3 (**a**) The skin is incised, and platysma exposed. (**b**) Platysma incision and movement of the sternocleidomastoid muscle laterally

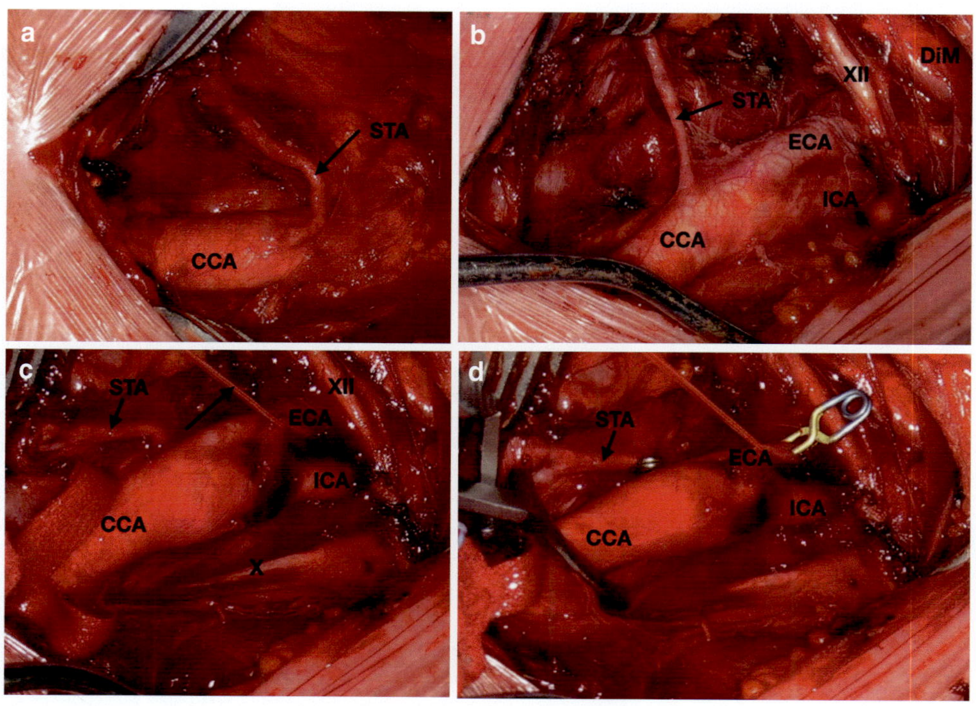

Fig. 27.4 (**a**) Common carotid artery and superior thyroid artery are exposed. (**b**) The dissection proceeds cranially. External carotid artery and XIIth nerve underneath the posterior venter of the digastric muscle (DiM) are exposed. (**c**) Internal carotid artery is dissected. In this case a stitch (arrow) is used to slightly rotate the bifurcation to expose the ICA better. The tourniquet around the CCA is introduced for eventual shunt. The Xth nerve is seen. (**d**) Superior thyroid artery and ECA are occluded by aneurysm clips, CCA is occluded by vascular clamp. Now we wait for 60 s for eventual deterioration of SEEPs which would call for shunt introduction

plaque material from dislodging and embolising into the intracranial circulation (Fig. 27.4).

Haemostasis is achieved; the most frequent points are the small veins running along the ICA cranially, forming something like a plexus. Now everything is prepared for endarterectomy. First, the ICA is closed to prevent dislodged material from being embolised distally. A small curved aneurysm clip is used, with the clip oriented so as not to hinder future manoeuvres, vessel wall suture in particular. Next, the ECA and the superior thyroid artery are closed with small aneurysm clips. Lastly, a vascular clip is applied to close the CCA.

We then wait 60 s for monitoring. If there is no alteration, we continue with the endarterectomy (Fig. 27.5). If alterations are not solvable by a slight increase in blood pressure (BP), we introduce a shunt after the arteriotomy.

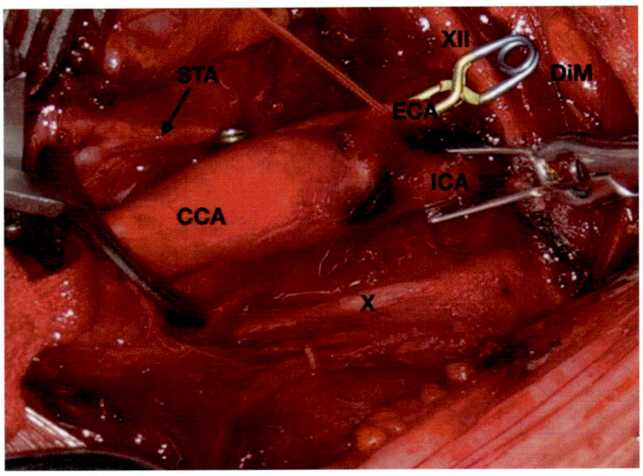

Fig. 27.5 Firstly, the ICA is shortly opened to flush back any loose material

The arterial incision running from the CCA to the ICA is performed. If we do not use the shunt (in 98% of the cases a shunt is not used), the arterial incision goes only as far as the plaque ends. After that, the proper plane of dissection in the CCA is found. The vessel wall should be smooth and a little pinkish compared to the atherosclerotic plaque. This plane is followed for the whole circumference of the CCA, and the plaque is subsequently sharply cut, usually some 8–10 mm proximal to the bifurcation. The plaque dissection continues to the superior thyroid artery orifice, and the plaque is sharply cut. Later, the ECA is freed from the plaque. It is usually divided by one haemostatic clamp closed as far as the first branch of the ECA. Using a second haemostatic clamp, the plaque is crushed and disrupted just proximally from the first pean and cleared from inside the ECA. The plaque within the ECA can also be divided sharply or bluntly from within the vessel (Fig. 27.6).

The plaque dissection continues to the ICA as far as the plaque ends. We remove the plaque without cutting through it, which is our usual and more straightforward technique. The entire plaque is used for further studies [8, 9, 13]. After plaque removal, a thorough inspection of the vessel wall was done. Any loose material and sclerotic remnants are removed, and the inside of the vessel is repeatedly flushed with heparinised saline. The water jet best shows any open remnants (Fig. 27.7).

The distal plaque end in the ICA is the crucial point. Any remnants of atherosclerotic material are removed, the intima frequently is cut sharply and if any intimal flap is apparent under the water jet, it is removed. If it is not technically pos-

Fig. 27.6 (**a**) The CCA-ICA is opened without plaque disruption. (b) The dissector is introduced underneath the plaque and is cut by the knife. (**c**) The plaque at CCA is cut. Note the colour of the proper plane of dissection. (**d**) The plaque is cut sharply at ECA and the dissection proceeds into the ICA

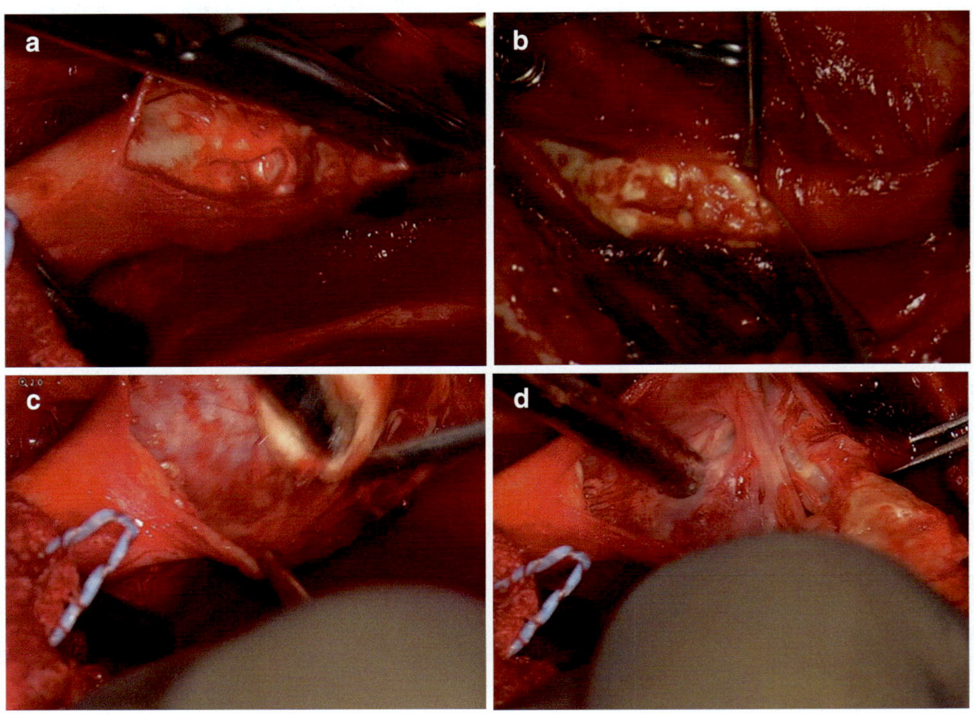

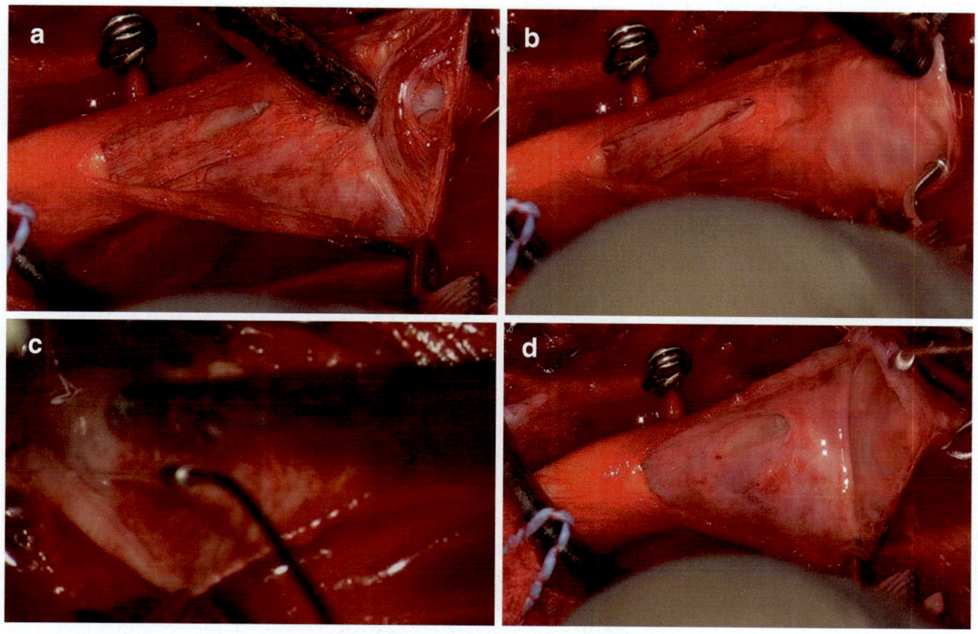

Fig. 27.7 (**a**) Finally, the whole plaque is removed. (**b**, **c**) Heparanised saline is used repeatedly to see any flaps at the distal end and any loose material within the whole endarterectomised segment. (**d**) The final image of the whole segment. Note the absence of any atherosclerotic material

sible in that the plaque may, at some point of vessel circumference, continue distally to the endarterectomy, it is fixed to the outer vessel wall. This is achieved by an 8/0 suture from outside in an inside-out technique. Depending on the individual situation, 1–2 stitches are sufficient. Tackling sutures were used to prevent potential post-operative dissection with all its consequences. After the plaque is removed, the clip on the ICA is slightly opened to achieve backflow, which flushes out any remnants from the clip area and distal ICA.

Arterial suture, after meticulous inspection, starts at the distal point of the ICA opening. The artery is closed by a 6/0 running suture. The intervals between individual stitches are approximately 1 mm. Before finalising the arterial suture, with 4–5 mm of arterial opening remaining, the backflow is again observed to flush out any loose material. Finally, the artery is flushed with heparinised saline. If shunt insertion is performed, this is when the shunt is removed. The suture is then concluded. A second running suture, which serves as a security against the rupture of the primary suture, is then performed. This suture is looser, with 3–4 mm intervals (Fig. 27.8).

After arteriotomy closure has been accomplished, the clips are removed. First, the ICA clip is shortly opened, followed by clip removal from the ECA and superior thyroid artery. The CCA is opened and the last branch to be opened is the ICA (Fig. 27.9).

Heparin is not reversed. At this time, any bleeding point in the entire wound is identified and managed. Platysma and subcutaneous sutures are introduced and the skin is closed using a plastic suture.

27.5.1 Shunt

Should the monitoring call for shunt insertion, the artery and plaque are first opened in the whole length. The shunt is first introduced into the ICA and secured by the aneurysm window clip of appropriate size. Thereafter, it is introduced into the CCA and secured by a tourniquet. The removal of the shunt follows the same sequence, i.e. from the ICA first and then from the CCA.

The patient is awakened at the operating theatre and then transferred to a semi-intensive unit for 24 h. During this time, the arterial BP is continuously monitored and any drop in BP is resolved. The patients are usually discharged on the second post-surgical day.

27.5.2 Patch

Patch technique is used to close the artery. With this manoeuvre, the diameter of the ICA is increased. In some series, the patch method is routinely used in all patients [14]. The aim is not only to increase the size of the vessel but also to prevent post-operative stenosis. With the use of the microscope and microtechniques, there is always an abundance of vessel walls and stenosis can hardly occur. The danger of stenosis is at the distal end of the arteriotomy, but the patch does not solve this danger. Two suture lines take twice the time to suture, which prolongs the occlusion time. It has never been shown that there are some differences in results between the primary suture and patch technique [14].

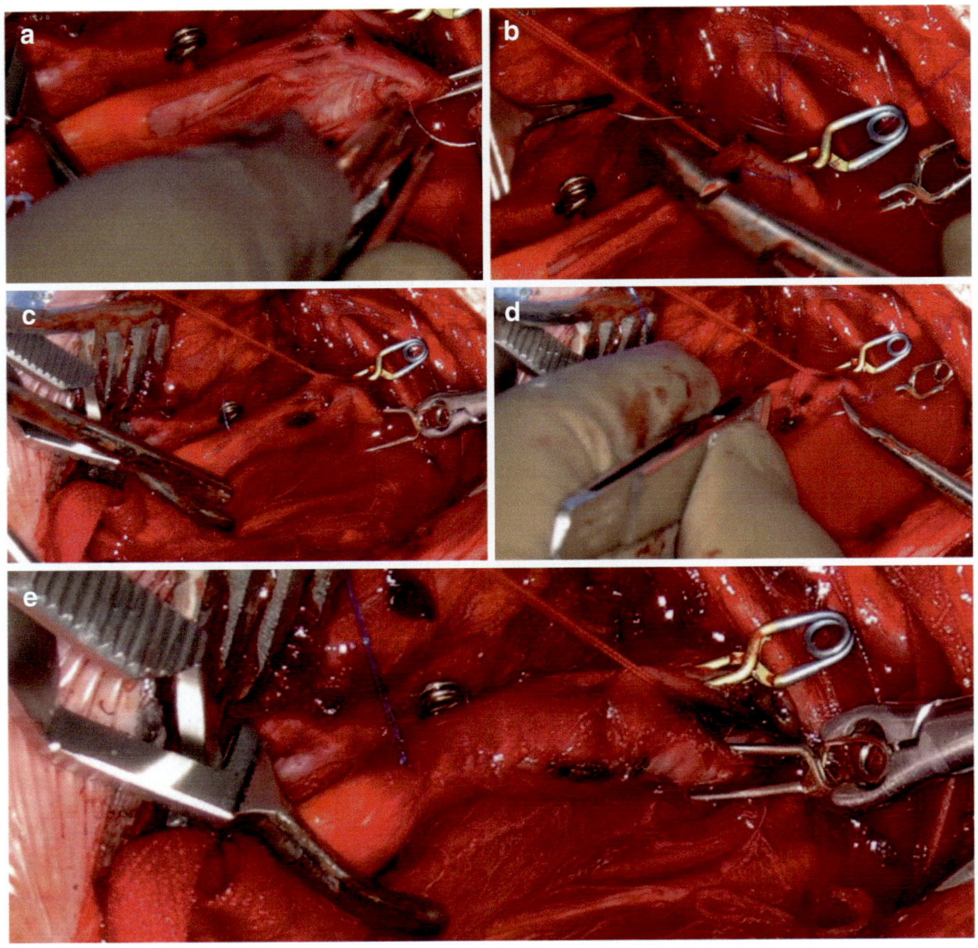

Fig. 27.8 (**a**) The suture starts at the distal end of arteriotomy. (**b**) With high magnification and microtechniques there always is abundance of vessel wall. No patch is needed. The critical point is distal arteriotomy end. (**c**) Before the final stitch the ICA is opened for a second to flush out any remaining loose material. (**d**) After the completing of the first suture, second, much looser suture, is used to prevent the rupture of the first one. (**e**) First, after the suture is completed, ICA clip is opened for a second

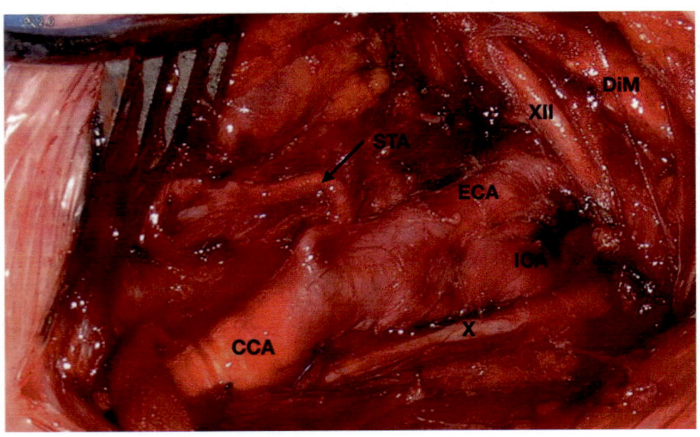

Fig. 27.9 ECA, superior thyroid artery, CCA and the ICA are released

27.5.3 Eversion Technique

The Eversion technique consists of sectioning the ICA at its origin from the CCA, and then the ICA wall is rolled down like a sleeve from the plague. The ICA is then reimplanted into the CCA in an end-to-side manner. Again, the suture is not as easy and fast as the linear suture of a standard arteriotomy approach. Like the patch technique, we have never used this technique. The disadvantage is that stenosis reaching the CCA necessitates additional longitunidal arteriotomy into CCA. Vascular surgeons favour the method, but no difference regarding linear opening has been documented.

27.5.4 Kink

Smooth asymptomatic kinks and coils are not addressed. We only correct those that are symptomatic. The symptomatology is typical repeated TIAs provoked by head rotation. We only correct those sharply bent like a bent garden hose. Smooth kinks and coils are subjected to surgery extremely rarely (Figs. 27.10 and 27.11). We have two solutions available—vessel straightening or vessel shortening. In the former, all bifurcation vessels are dissected entirely free and using 1–2 stitches between CCA adventitia and omohyoid muscle, the whole bifurcation is moved caudally [10]. The latter part of the vessel is resected and resutured in an end-to-end manner.

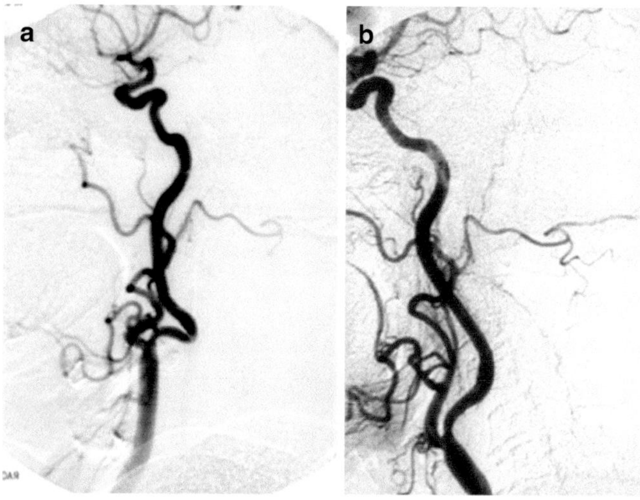

Fig. 27.10 Carotid kink before (**a**) and after surgery (**b**). The kink has only been straightened, no cross-clamping, no vessel opening

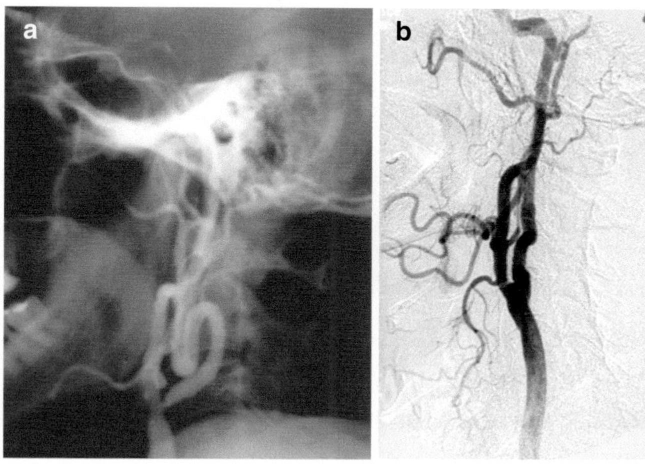

Fig. 27.11 (**a**) Carotid coil causing TIAs by head rotation. (**b**) After the coil resection and ICA reimplantation into the CCA

27.5.5 Pseudoocclusion

Pseuocclusion disappeared with the disappearance of angiography in the management of these patients (Fig. 27.12). The vessel lumen is not thrombosed but rather collapsed. At angiography, in some patients a little trace of contrast media could be traced within the ICA and the distal intracranial ICA is patent. In some of these patients, the vessel will reopen after CEA.

27.5.6 Carotid Occlusion

In some cases, particularly in those subjected to emergency surgery due to fluctuating neurological deficits, the vessel is usually occluded by the thrombus (Figs. 27.13 and 27.14). Thrombectomy with a Fogarty catheter can succeed in those patients where, on angiography, the backfilling of ICA is apparent as far as the foramen lacerum region.

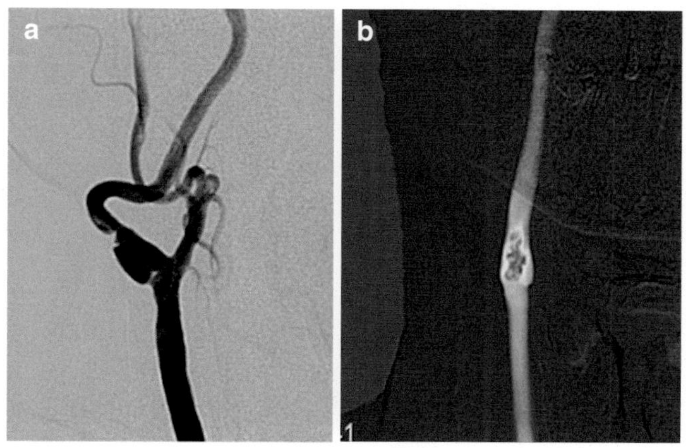

Fig. 27.13 Intraluminal thrombus and major ulcer. Measured by any NASCET or ECST methodology the stenosis would either be negative or 0%. Still, surgical indication is obvious. CEA in non-AHA recommendation

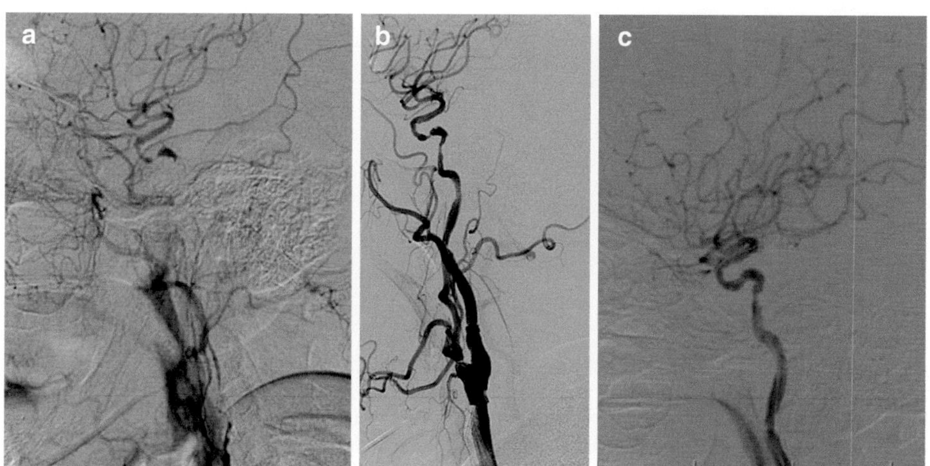

Fig. 27.12 (**a**) Carotid pseudoocclusion and tandem lesion. (**b**) Some traces of contrast media are apparent. After the CEA the tandem lesion is apparent. (**c**) Distal stenosis after the endovascular treatment. It is obvious the patient should have been treated by endovascular means

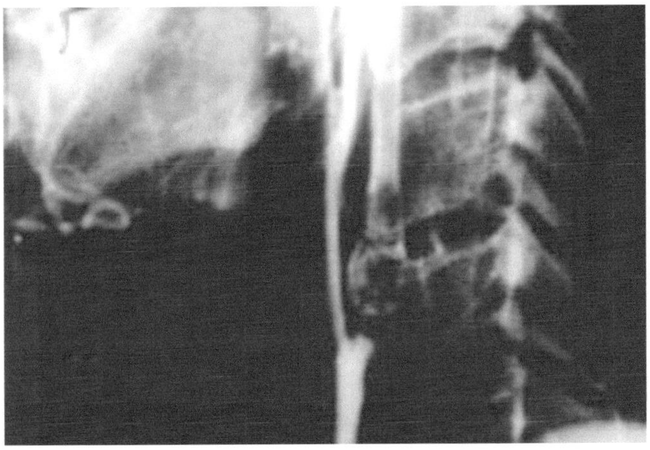

Fig. 27.14 Intraluminal thrombus. Emergency surgery is indicated despite the fact, that according to ECST measurements the stenosis is 0% and according to NASCET methodology the stenosis is in negative numbers

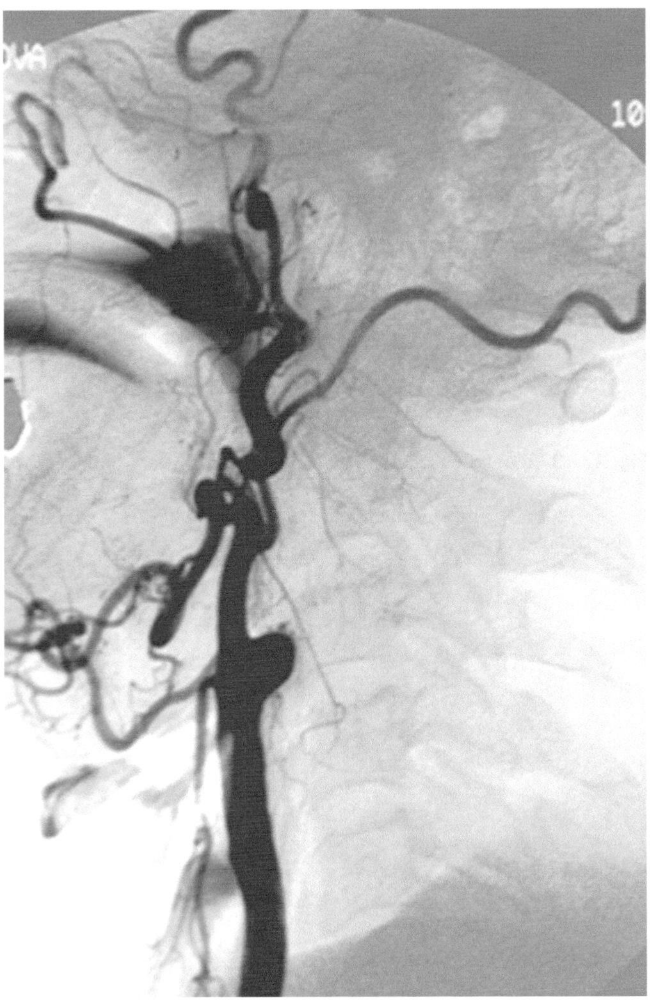

Fig. 27.15 Carotid stump after the carotid thrombosis. From the stump emboli may get into the brain via natural collaterals. Stumpectomy and CEA is the procedure of choice

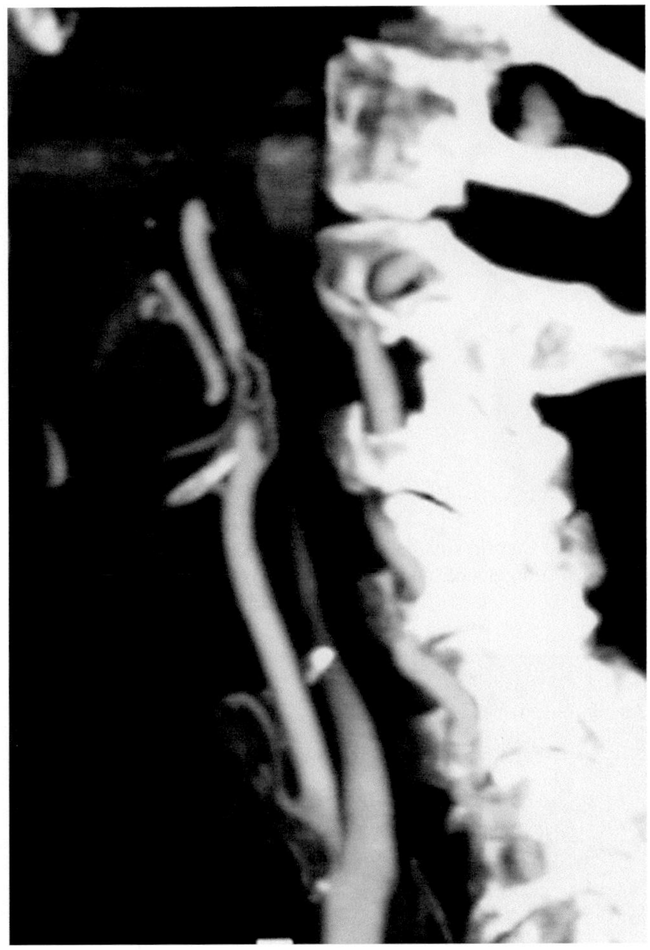

Fig. 27.16 Carotid dissection. Typical string sign

27.5.7 Carotid Stump

In rare cases, the patient has TIAs caused by embolisation via the ECA and ophthalmic artery from the stump of the occluded ICA. Seldom can stumpectomy and CCA-ECA endarterectomy be performed (Fig. 27.15).

27.5.8 Carotid Dissection

Carotid dissections are most commonly traumatic, although spontaneous ones may also occur (Fig. 27.16). Many heal spontaneously (as much as 80% [15]). If the dissection is symptomatic, stenting is the procedure of choice (Fig. 27.17).

27.5.9 Pseudoaneurysms

Pseudoaneurysms are closely related to dissection. The treatment is endovascular, using stents to treat the blocked area (Fig. 27.18). Surgery may be hazardous for some patients. Distal to the pseudoaneurysm, the ICA is usually affected as

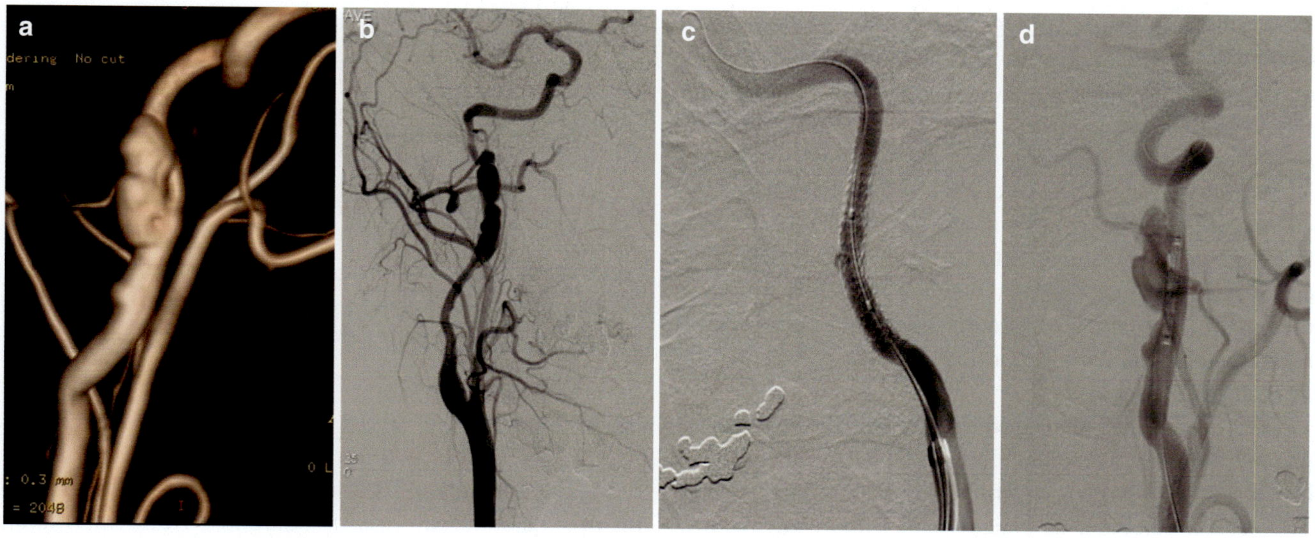

Fig. 27.17 Carotid dissection on both sides (**a**, **b**) before and (**c**, **d**) after the endovascular treatment

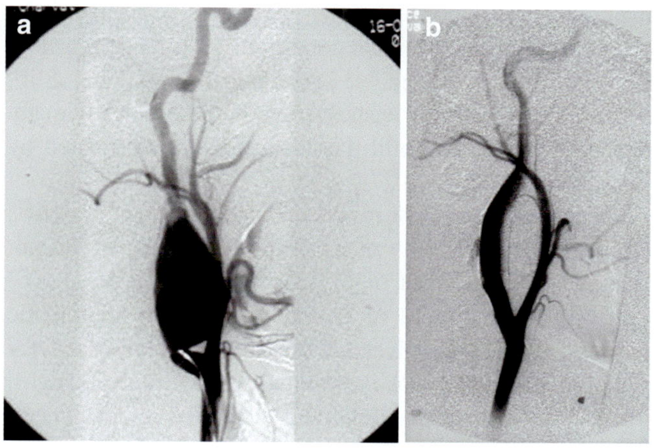

Fig. 27.18 Carotid pseudoaneurysm before (**a**) and after the stent introduction (**b**)

far as the foramen lacerum. Any reconstruction attempt commonly ends with ICA occlusion (personal experience from the 1980s, unpublished).

27.5.10 Other Non-Atherosclerotic Conditions Causing Cerebral Ischaemia

Fibromuscular dysplasia may cause ischaemic symptoms. Moyamoya disease can also affect the magistral vessels [16] and some other arterial wall diseases [17, 18]. These conditions are not surgical. If some form of active treatment is necessary, an endovascular procedure would be the most likely candidate [18].

27.6 Complications and Their Avoidance

Complications are usually flow-related or embolic. The most severe and feared complication is what we aim to prevent by surgery—a completed stroke. This type of stroke may be caused by endarterectomy site occlusion, embolisation of the dislodged material or too-long artery occlusion. The completed stroke can be avoided by meticulous technique and careful suture, especially at the distal end of the arteriotomy, frequent flushing with heparinised saline and distal flap, causing dissection, fixation. The vessels should be checked immediately after clip release by ICG, ultrasound or perop CTAg to be on caution. If any hesitation appears, the best option is to reopen the artery. Immediate angiography is indicated if the artery is patent and the patient awakes with a neurological deficit linked to the operated vessel. Because the thrombolysis is implausible, direct thrombectomy is indicated. Too long an occlusion can be prevented by the speed of surgery; however, it should not be hurried at the cost of quality, but it also cannot be slow. Monitoring provides good information if the occlusion is not too long. Shunt insertion is the solution. It is crucial to avoid any post-operative drop in BP after surgery. Such a decrease in BP may cause a stroke.

Hyperperfusion syndrome with headaches and confusion may appear post-operatively. These conditions are caused by sudden blood flow into previously depleted regions. Some decrease in blood pressure and careful monitoring almost always solve the problems, which soon disappear. In hyperperfusion cases, there is always a small risk of intracerebral bleeding. Thus, careful management of BP is mandatory to secure circulation to the brain.

Any distress and breathing difficulties after surgery can be dangerous. Either wound haematoma or wound swelling may cause such distress. Wound revision or reintubation should be done immediately.

A specific local complication in carotid surgery involving is the injury of one or more cranial nerves. In order of frequency, it is the hypoglossal nerve, the lower branch of the facial nerve, rarely recurrent nerve when dissection goes caudally more than the omohyoid muscle and the vagal nerve. All these potential complications can be avoided if meticulous dissection is performed. All these palsies are usually temporary only.

General surgical and medical complications occur in patients with atherosclerotic disease, which is reflected in a higher percentage of myocardial infarctions after surgery. The careful selection of patients and a thorough medical check are mandatory. The patients subjected to CEA are older and normally harbour several risk factors (e.g. coronary artery disease, hypertension, diabetes mellitus). Multidisciplinary medical management is mandatory.

The patient is usually discharged from hospital on the second post-surgical day.

27.7 Treatment Outcome and CAS Indications

Results on the efficacy of CEA and CAS in asymptomatic carotid stenosis typically indicate little or no differences in mortality, stroke or myocardial infarction. On the other hand, patients undergoing CAS generally have a higher probability of any stroke during the perioperative period, as well as an increased risk of non-disabling stroke. A meta-analysis of 9753 patients treated with CEA or CAS [19] demonstrated that the difference in the occurrence of the primary safety outcome between CEA and CAS was not statistically significant; however, there was a trend towards a higher risk of death or any stroke between randomisation and 30 days after treatment in the CAS group (odds ratio 1.72, 95% CI 1.00–2.97; $P = 0.05$). A recent meta-analysis [20] demonstrated in a long-term follow-up no difference in the composite outcome of any stroke, death and myocardial infarction between CEA and CAS groups.

Of note are the Asymptomatic Carotid Surgery Trial (ACST) and ASCT-2 [21], where investigators from our department participated. Overall, 1% of patients experienced a disabling stroke or death procedurally (15 allocated to CAS and 18 to CEA) and 2% had a non-disabling procedural stroke (48 assigned to CAS and 29 to CEA). Additionally, the risks of any non-procedural stroke in all CAS versus CEA were similar in symptomatic and asymptomatic patients. Similarly, ACST-2 proved comparable long-term effects of both procedures on fatal or disabling stroke [21, 22].

Table 27.1 CEA complications before and after the introduction of CAS in 2000

	1982–1999	2000–2022
Total number of patients	627	2067
TIA, *n* (%)	5 (0.79)	18 (0.87)
MS, *n* (%)	4 (0.63)	22 (1.06)
SD, *n* (%)	8 (1.27)	19 (0.91)
PN, *n* (%)	34 (5.5)	56 (2.7)
Shunt, *n* (%)	283 (45.10)	40 (1.93)

The rate of PN injuries dropped down significantly ($p < 0.001$) since extremely obese patients and patients with superiorly positioned bifurcation are treated by CAS

The results from both techniques did not differ substantially. However, some superiority of CEA was found in two major randomised trials: Carotid Revascularisation Endarterectomy versus Stenting Trial (CREST) [23] and the International Carotid Stenting Study (ICSS) [24]. CEA is the first line of treatment, with CAS reserved for specific conditions. Restenosis, stenosis after neck radiation and previous major surgery (opposite side IX-XII cranial nerve palsy) are all unfavourable anatomical conditions. The tandem lesion, however, is an acceptable anatomical condition. Still, because the patient's wishes should be respected, many patients who do not fulfil the above criteria are treated by CAS.

In our series of 2218 procedures over the past 20 years, the MM rate for CAS was significantly higher and did not change as a function of time. Because there was no significant difference in MM in symptomatic and asymptomatic patients, we are convinced CAS should be reserved for symptomatic patients. We also employ CAS in seriously ill patients given that all CEAs are performed under general anaesthesia [8]. Since 1982, our experience with CEA comprised 2694 cases. Of these, overal morbidity-mortality rate was 1.52%, while post-operative major stroke occured in 1.07% of treated patients. Small stroke was reported in 1% of cases, TIA in 0.89% and peripheral nerve injury in 3.26% of patients. The most common peripheral nerve injury was the facial nerve and/or hypoglossal nerve deficit (Table 27.1).

27.8 Surgical Pearls

- Careful dissection—XIIth, VIIth (lower branch), Xth (recurrent).
- Selective use of shunt according to monitoring.
- Linear arteriotomy and primary vessel wall suture.
- Micro- or exoscope and microtechniques.
- Thorough removal of any loose atherosclerotic material from the endarterectomised area.
- Special care to the distal end of arteriotomy (remaining material, intimal flap).

- Extreme care must be exercised with the first stitch in this area.
- Second loose suture to secure the eventual breakage of the first.
- Careful BP monitoring during and after surgery.

27.9 Conclusion

Carotid surgery is an established and effective technique for secondary stroke prevention in certain categories of patients. The surgery is routine, straightforward and technically not very demanding. Unfortunately, most cases are handled by vascular surgeons. Yet, because this is part of the surgical management of cerebral ischaemia, neurosurgeons are the best suited to perform this type of operation. The procedure should be performed in tertiary care hospitals where endovascular service is readily available and vice versa. It is the authors' opinion that this problem should not escape the hands of the neurosurgeon.

Supported by programme UK CU, Neuroscience

References

1. Barnett HJ, Taylor DW, Eliasziw M, Fox AJ, Ferguson GG, Haynes RB, Rankin RN, Clagett GP, Hachinski VC, Sackett DL, et al. Benefit of carotid endarterectomy in patients with symptomatic moderate or severe stenosis. North American Symptomatic Carotid Endarterectomy Trial Collaborators. N Engl J Med. 1998;339:1415–25. https://doi.org/10.1056/nejm199811123392002.
2. Ferguson GG, Eliasziw M, Barr HWK, Clagett GP, Barnes RW, Wallace MC, Taylor DW, Haynes RB, Finan JW, Hachinski VC, et al. The North American symptomatic carotid endarterectomy trial. Stroke. 1999;30:1751–8. https://doi.org/10.1161/01.STR.30.9.1751.
3. Warlow CP. Symptomatic patients: the European Carotid Surgery Trial (ECST). J Mal Vasc. 1993;18:198–201.
4. Biller J, Feinberg WM, Castaldo JE, Whittemore AD, Harbaugh RE, Dempsey RJ, Caplan LR, Kresowik TF, Matchar DB, Toole JF, et al. Guidelines for carotid endarterectomy. Stroke. 1998;29:554–62. https://doi.org/10.1161/01.STR.29.2.554.
5. Barnett HJM, Taylor DW, Haynes RB, Sackett DL, Peerless SJ, Ferguson GG, Fox AJ, Rankin RN, Hachinski VC, Wiebers DO, et al. Beneficial effect of carotid endarterectomy in symptomatic patients with high-grade carotid stenosis. N Engl J Med. 1991;325:445–53. https://doi.org/10.1056/nejm199108153250701.
6. Kernan WN, Ovbiagele B, Black HR, Bravata DM, Chimowitz MI, Ezekowitz MD, Fang MC, Fisher M, Furie KL, Heck DV, et al. Guidelines for the prevention of stroke in patients with stroke and transient ischemic attack: a guideline for healthcare professionals from the American Heart Association/American Stroke Association. Stroke. 2014;45:2160–236. https://doi.org/10.1161/str.0000000000000024.
7. Aboyans V, Ricco J-B, Bartelink M-LEL, Bjorck M, Brodmann M, Cohnert T, Collet J-P, Czerny M, De Carlo M, Debusa S, et al. The 2017 ESC Guidelines on the Diagnosis and Treatment of Peripheral Arterial Diseases, in collaboration with the European Society for Vascular Surgery (ESVS). Rev Española de Cardiol (English Edition). 2018;71:111. https://doi.org/10.1016/j.rec.2017.12.014.
8. Svoboda N, Bradac O, Mandys V, Netuka D, Benes V. Diagnostic accuracy of DSA in carotid artery stenosis: a comparison between stenosis measured on carotid endarterectomy specimens and DSA in 644 cases. Acta Neurochir (Wien). 2022;164:3197–202. https://doi.org/10.1007/s00701-022-05332-5.
9. Netuka D, Belšán T, Broulíková K, Mandys V, Charvát F, Malík J, Coufalová L, Bradáč O, Ostrý S, Beneš V. Detection of carotid artery stenosis using histological specimens: a comparison of CT angiography, magnetic resonance angiography, digital subtraction angiography and Doppler ultrasonography. Acta Neurochir (Wien). 2016;158:1505–14. https://doi.org/10.1007/s00701-016-2842-0.
10. Bradac O, Mohapl M, Kramar F, Netuka D, Ostry S, Charvat F, Lacman J, Benes V. Carotid endarterectomy and carotid artery stenting: changing paradigm during 10 years in a high-volume centre. Acta Neurochir (Wien). 2014;156:1705–12. https://doi.org/10.1007/s00701-014-2166-x.
11. Rothwell PM, Eliasziw M, Gutnikov SA, Warlow CP, Barnett HJ. Endarterectomy for symptomatic carotid stenosis in relation to clinical subgroups and timing of surgery. Lancet. 2004;363:915–24. https://doi.org/10.1016/s0140-6736(04)15785-1.
12. Coutts SB, Hill MD, Campos CR, Choi YB, Subramaniam S, Kosior JC, Demchuk AM. Recurrent events in transient ischemic attack and minor stroke: what events are happening and to which patients? Stroke. 2008;39:2461–6. https://doi.org/10.1161/strokeaha.107.513234.
13. Gasbarrino K, Hafiane A, Gianopoulos I, Zheng H, Mantzoros CS, Daskalopoulou SS. Relationship between circulating adipokines and cholesterol efflux in subjects with severe carotid atherosclerosis. Metabolism. 2022;155381 https://doi.org/10.1016/j.metabol.2022.155381.
14. Loftus CM. Carotid endarterectomy: principles and techniques. 2nd ed; 2007. https://doi.org/10.3109/9781420016277.
15. Schievink WI. Spontaneous dissection of the carotid and vertebral arteries. N Engl J Med. 2001;344:898–906. https://doi.org/10.1056/nejm200103223441206.
16. Hackel M, Benes V. Onemocnění moyamoya. Přehled a soubor 9 nemocných. Česká a slovenská neurologie a Neurochirurgie 1997;60/93:142–151.
17. Back T, Grünig S, Winter Y, Bodechtel U, Guthke K, Khati D, von Kummer R. Neuroborreliosis-associated cerebral vasculitis: long-term outcome and health-related quality of life. J Neurol. 2013;260:1569–75. https://doi.org/10.1007/s00415-013-6831-4.
18. Kalaria RN, Kalimo H. Introduction: non-atherosclerotic cerebrovascular disorders. Brain Pathol. 2002;12:337–42. https://doi.org/10.1111/j.1750-3639.2002.tb00448.x.
19. Müller MD, Lyrer P, Brown MM, Bonati LH. Carotid artery stenting versus endarterectomy for treatment of carotid artery stenosis. Cochrane Database Syst Rev. 2020;2:CD000515. https://doi.org/10.1002/14651858.CD000515.pub5.
20. Wang J, Bai X, Wang T, Dmytriw AA, Patel AB, Jiao L. Carotid stenting versus endarterectomy for asymptomatic carotid artery stenosis: a systematic review and meta-analysis. Stroke. 2022;53:3047–54. https://doi.org/10.1161/strokeaha.122.038994.
21. Halliday A, Bulbulia R, Bonati LH, Chester J, Cradduck-Bamford A, Peto R, Pan H. Second asymptomatic carotid surgery trial (ACST-2): a randomised comparison of carotid artery stenting versus carotid endarterectomy. Lancet. 2021;398:1065–73. https://doi.org/10.1016/s0140-6736(21)01910-3.
22. Halliday A, Harrison M, Hayter E, Kong X, Mansfield A, Marro J, Pan H, Peto R, Potter J, Rahimi K, et al. 10-year stroke prevention after successful carotid endarterectomy for asymptomatic stenosis (ACST-1): a multicentre randomised trial. Lancet. 2010;376:1074–84. https://doi.org/10.1016/s0140-6736(10)61197-x.

23. Mantese VA, Timaran CH, Chiu D, Begg RJ, Brott TG. The Carotid Revascularization Endarterectomy versus Stenting Trial (CREST): stenting versus carotid endarterectomy for carotid disease. Stroke. 2010;41:S31–4. https://doi.org/10.1161/strokeaha.110.595330.
24. Bonati LH, Dobson J, Featherstone RL, Ederle J, van der Worp HB, de Borst GJ, Mali WPTM, Beard JD, Cleveland T, Engelter ST, et al. Long-term outcomes after stenting versus endarterectomy for treatment of symptomatic carotid stenosis: the International Carotid Stenting Study (ICSS) randomised trial. Lancet. 2015;385:529–38. https://doi.org/10.1016/S0140-6736(14)61184-3.

Part III

Skull Base Tumors

Orbit Tumors: Anatomy and Surgical Approaches

28

Pablo González-Lopéz, Cristina Gómez-Revuelta, and Elena Baño-Ruiz

Learning Objectives

- Understand orbit cone components.
- Study the 3D representation of the intraorbital structures to understand the different orbit quadrants and anteroposterior regions.
- Plan the correct approach to an orbital lesion according to the involved compartments and the relationships with the different intraorbital structures.

28.1 Introduction

The orbit is a pyramidal-shaped structure confined in between the junction of several skull bones, localized into the facial skeleton and containing highly relevant anatomical structures that allow us to analyze, understand, and even somehow communicate with the environment. The orbit volume has been estimated to be around 24–28 mL [1], representing less than 2% of the whole intracranial volume. However, it is situated in a specific area of the skull where different anatomical regions are intimately related [2]. Moreover, in this relatively small volume, the orbit contains not only the eyeball but also many neural, muscular, and vascular structures that make this globe a unique organ with specific and highly developed features.

Understanding the orbit shape, limits, intrinsic anatomy, and the location of the different structures traversing its fissures and canals will provide physicians with the necessary tools to approach diseases inside this bony chamber, or even to use this region as a corridor to other different intracranial areas.

A common rule for any surgical approach is providing enough space to work with safety and efficacy while causing the least possible damage in terms of functionality and aesthetics. During the next few pages, we will give an effort to analyze and detail the microsurgical anatomy of the orbit and its contents, with the main goal of describing the most practical surgical approaches to this region, from a neurosurgical perspective.

28.2 History/Background

By simply looking at the medical terms related to ocular pathology, we will realize that ophthalmology is one of the oldest medical disciplines. While most medical terms derive from Latin roots, those related to the treatment and pathology of the eyes derive mainly from the Greek language. Thus, the prefix or the letter "*o*" that we find in most of those professionals in charge of eye care (opticians, optometrists, ophthalmologists, oculists...) derive from the Greek word "*opsis*," which means "view," and *ophthalmos* means "the eye."

Eye care dates back as far as the ancient Greek and Egypt. These were the first to perform and reflect surgeries in the lacrimal region, while the first sutures of wounds around the orbit date back to Babylonian times and ancient Indian culture. In tenth century Muslim culture, there are texts describing local treatments for exophthalmos, with little success and much trust in the hand of God. It was not until the end of the sixteenth century that the first enucleations by Bartish were described [3].

The most relevant advances came with the discovery of the effects of cocaine over the eye, which allowed surgery without pain or movements, and the introduction of the ophthalmoscope in the late nineteenth century [4]. Modern approaches to the orbit probably started after the publication from Krönlein in 1888, who was the first to report the use of a lateral orbitotomy to treat a dermoid cyst [5]. But it was undoubtedly Walter Dandy who revolutionized the treatment of intraorbital lesions with his transcranial approaches. By

P. González-Lopéz (✉) · C. Gómez-Revuelta · E. Baño-Ruiz
Department of Neurosurgery, Hospital General Universitario de Alicante, Alicante, Spain
e-mail: gonzalez_pab@gva.es

I. N. Kanaan, V. Beneš (eds.), *Neuroanatomy Guidance to Successful Neurosurgical Interventions*,
https://doi.org/10.1007/978-3-031-59838-8_28

approaching the orbit from its roof after performing an anterolateral craniotomy, he presented several advantages such as a panoramic view of the orbital contents from the globe to the apex, even reaching a complete view of the optic nerve from the chiasm to its entrance to the eyeball [6]. Short time after, Conley completed the orbital approaches with a transfacial vision though the maxillary sinus. This approach allowed the treatment of lesions inside the inferior and anterior orbital regions [7].

These first surgical descriptions allow treating the whole spectrum of orbital tumors and can be summarized in lateral, superior, and inferomedial surgical routes. Orbit surgery has evolved a lot in the last 50 years due to a deep knowledge of its intrinsic anatomy, the understanding of the biology of its tumors, and especially due to the introduction and improvement of operative technology, microsurgical, and recently endoscopic techniques. However, the way in which tumors are analyzed, planned to be treated, and finally resected is still based on a lateral, superior, and inferomedial understanding.

28.3 Anatomical Highlights

28.3.1 The Orbit Chamber

The orbit chamber can be described as a four-walled pyramidal structure limited by up to seven bones of the skull. This pyramid is usually divided into an anterior base, central axis, and posterior apex. While the base is covered by the orbital septum, tarsal plates, palpebral ligaments, and the globe, the posterior axis is opened to the anterior cranial fossa through the optic canal, middle cranial fossa through the superior orbital fissure, and to the pterygopalatine and infratemporal fossa through the inferior orbital fissure (Fig. 28.1).

The orbit, on its relatively small volume, hosts many muscles, nerves, ligaments, arteries, and veins. All these structures are somehow "floating" in a fatty content. This fat is aided by the periorbit, which is a thin membrane sealing the intraorbital structures, to create a special environment in which the muscles can move freely and provide the periorbital space with a specific pressure. This fat can be described as a "yellow sea" where surgeons must learn to navigate in a safe way.

The periorbit is a thin membrane that represents the continuum of the intracranial dura reaching the orbit through the optic canal and superior orbital fissure. All the orbit anatomic structures lie inside the periorbit, which remains attached to the orbit bones. The most voluminous intraorbital anatomical structures are the muscles, which are counted to seven (superior, middle, lateral, and inferior recti, superior levator palpebrae, inferior and superior oblique). The four recti and superior oblique are attached posteriorly to a fibrous ring located in the orbital apex known as the annulus tendon of Zinn. The trajectory of the recti muscles defines a kind of conal structure that follows the pyramidal shape of the bony orbit chamber. Finally, at the anterior pyramidal base, these muscles hold the eyeball in a quadrangular shape.

Understanding the pyramidal concept, as well as the periorbit and muscles disposition into the orbit, tumors can be divided into intra- and extraperiorbital. Those intraperiorbital can be furthermore divided into intra- and extraconal considering those which remain respectively inside and outside the conal area defined by the recti muscles. Being a bit more detailed, the extraperiorbital ones can be subdivided into tumors growing into the orbital wall (those growing into the bone) and those growing in between the bone and the periorbit. These extraperiorbital lesions should be named and defined regarding the bone(s) with which they are in contact. On the other hand, the intraperiorbital tumors can be better defined regarding the quadrant in which they are predominantly growing. These quadrants can be easily recognized on a coronal cut, where the optic nerve occupies a central position, and the recti muscles define the superior, inferior, medial, and lateral quadrants. This classification should mainly consider the relative location of the optic nerve to the tumor itself, as this will be a keypoint to define the surgical trajectory (superior, lateral, or inferomedial) (Fig. 28.2a–f).

28.3.2 The Pyramidal Concept: Surgical Routes to the Orbit Based on the Anatomy

As previously mentioned, the orbit is defined by a kind of messy joint of seven bones: frontal, zygomatic, maxillary, ethmoid, sphenoid, lacrimal, and palatine (Fig. 28.1). It might be difficult to properly correlate those bones into the different parts of the orbit pyramid (apex, axis, and base). However, some general concepts are relevant to understand these junctions and optimize our surgical approaches.

The base of the pyramid. The base of the orbit is known as the orbital rim from a purely bony perspective. The orbital rim hosts and protects the eyeball and serves as the physical support to keep it in a central and somehow fixed position. This orbital rim is delineated by the frontal, maxillary, and zygomatic bones. The frontal bone forms the whole superior quadrant, as well as the superolateral and superomedial ones. The superior orbital rim is in close relationship with the supraorbital nerve, which can cross it through a foramen or just a notch [8] (Fig. 28.1b–d).

Illustrative Case 1 A young woman presented with diplopia and proptosis, as well as some aesthetic complaints. She was presenting those symptoms for years. On CT scan and

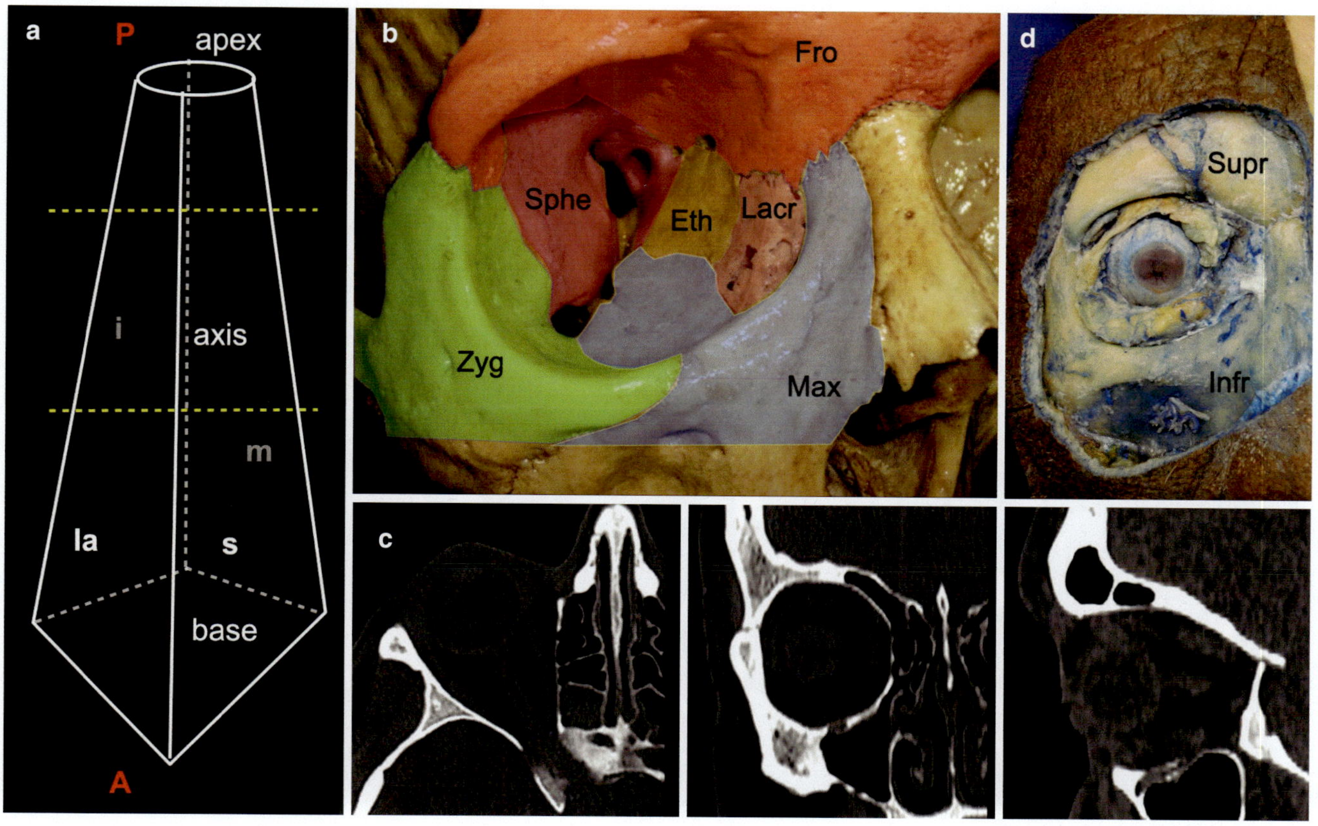

Fig. 28.1 (**a**) Pyramidal representation of the orbit region with an anterior "A" base, axis, and posterior "P" apex. Lateral (la), medial (m), inferior (i), and superior (s) walls. (**b**) Anterior view of the bony orbit camera: zygomatic (Zyg), maxillary (Max), frontal (Fro), sphenoidal (Sphe), lacrimal (Lacr), and ethmoidal (Eth) bones. (**c**) Axial, coronal, and sagittal computer tomography cuts at the mid-orbit level showing the close relationships of the orbit with the paranasal areas, middle and anterior intracranial fossae. (**d**) Anterior view of the orbit with the eyeball and periorbit fat around. The supraorbitary (Supr) and infraorbitary (Infr) nerves are visualized

MR images a frontal bone intradiploic hemangioma was present affecting the roof of the orbit and the superior orbital rim. The lesion was also invading the frontal paranasal sinus lateral wall and medial wall of the orbit. A bicoronal skin incision was performed, followed by a **unifrontal craniotomy and superior orbitotomy** including a safe margin of noninvaded bone. Then, the superior orbital rim was defined and drilled away from the frontozygomatic to the frontomaxillary sutures. A cranialization of the frontal paranasal sinus was performed. A key step consisted of identifying the supraorbital nerve around the frontal bone. The nerve was encased by bone that was carefully drilled and dissected away to preserve its function. A previously designed acrylic cranioplasty was designed and created to replace the resected bone. Patient's symptoms were completely released a few weeks after surgery when swelling disappeared and wound healing was completed. She only complained of a well-tolerated ipsilateral forehead hypoesthesia that fully disappeared when followed up 1 year after surgery (Fig. 28.2).

The frontal joins laterally the zygomatic bone joining through the frontozygomatic suture, which is a relevant landmark when performing a lateral orbitotomy, and medially with the maxillary bone on the frontomaxillary suture. Posteriorly on this medial wall, the maxillary and frontal bones join with the lacrimal one leading to a canal that communicates with the nasal space and hosts the lacrimal duct. Finally, on its superomedial quadrant the frontal bone shows a couple of tiny and sharp prominences to which the trochlea is attached. The trochlea is a pully-like cartilage through which the superior oblique muscle tendon turns posteriorly to reach the superomedial surface of the eyeball. Considering this anatomical detail is important when planning to approach lesions into the superomedial aspect of the orbit.

The inferior quadrant of the orbital rim is composed of the maxillary (medial) and zygomatic bones. These two bones expand laterally and medially to form the inferolateral and inferomedial quadrants of the base of the pyramid. The junction of the zygomatic and maxillary bones lies on the central part of the floor of the base and coincides in a sagittal plane with the infraorbital foramen, through which the infraorbital nerve enters the orbit and runs posteriorly into the infraorbital groove.

The axis of the pyramid. The axis is defined as the central part of the orbit and is located between the anterior base and

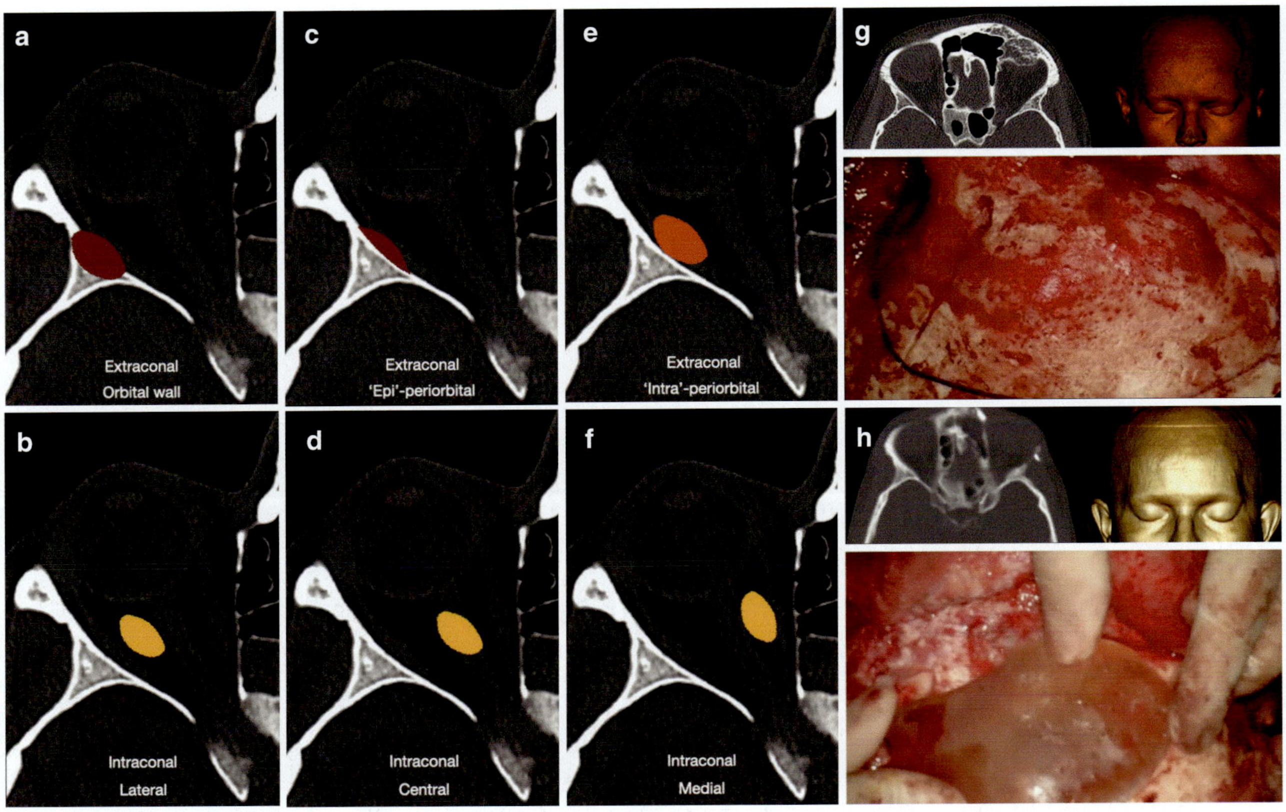

Fig. 28.2 (**a**–**f**) Orbit lesions classification based on their relative location to the periorbit and conic muscles net. Extraconal—extraperiorbital (red), Extraconal—intraperiorbital (orange), and intraconal (yellow). (**g**) Case illustration of an extraconal—extraperiorbital—orbital wall intradiploic hemangioma, before and (**h**) after resection

the posterior apex. Tracing well-defined boundaries for the axis is difficult and makes no sense from a surgical perspective. Therefore, we can define the roof as covered by the frontal bone; the lateral wall by the frontal, zygomatic, and greater wing of the sphenoid; the floor by the zygomatic and maxillary bones, and the medial wall formed by the lacrimal, ethmoid, and palatine bones.

Illustrative Case 2 A 37-year-old man presented with diplopia and left ocular pain especially when looking laterally. An intraconal cavernous hemangioma was discovered after performing a CT scan and MRI. The lesion was located into the lateral quadrant of the orbital axis, pushing and distorting the optic nerve medially and located in a tangential position relative to the superior and lateral recti muscles. This is an ideal lesion to be removed via a **lateral orbitotomy**. The patient was positioned supine, and head was slightly rotated to the contralateral side. Different incisions have been described: eyelid, Krönlein, and Stallard-Wright. A variation of the latter is our preferred and runs in the shape of an italic "s," with a path parallel to the external orbital rim from the tail of the eyebrow and a final stroke parallel to the zygomatic arch. This incision creates a wide space in the axial plane of the lateral orbit and protects the distal frontal branch of the facial nerve. The frontozygomatic suture is identified and the periorbit is dissected. A small part of the temporalis muscle is detached from the frontal, zygomatic, and greater wing of the sphenoid bones. A small hole is done with a diamond or cutting drill, and the orbitotomy is expanded anteriorly in a triangular shape. The greater sphenoid wing can be drilled posteriorly as much as needed. The periorbit is incised considering the relative location of the lesion (superior or inferior to the lateral rectus muscle). Blunt dissection of the fat tissue is carried out till the lesion is identified and carefully dissected. At this stage bipolar and microscissors should be avoided if possible. After tumor removal, the periorbit can be closed with a resorbable suture or even left opened. If the latter is the case, we usually cover the defect with a thin layer of collagen gelatine sponge. The bone flap is replaced with microplates and the temporalis muscle reattached to the lateral orbital rim. The skin incision is closed with an intradermal suture (Fig. 28.3).

The medial aspect of the central quadrant is always challenging in terms of surgical approaches as relevant structures are found on it, or even be found on the way to reach it. This region of the orbit is in close relationship with the trochlea superiorly and the bony attachment of the inferior oblique muscle inferiorly. The medial rectus muscle is situated in between the oblique muscles and innervated by lower divi-

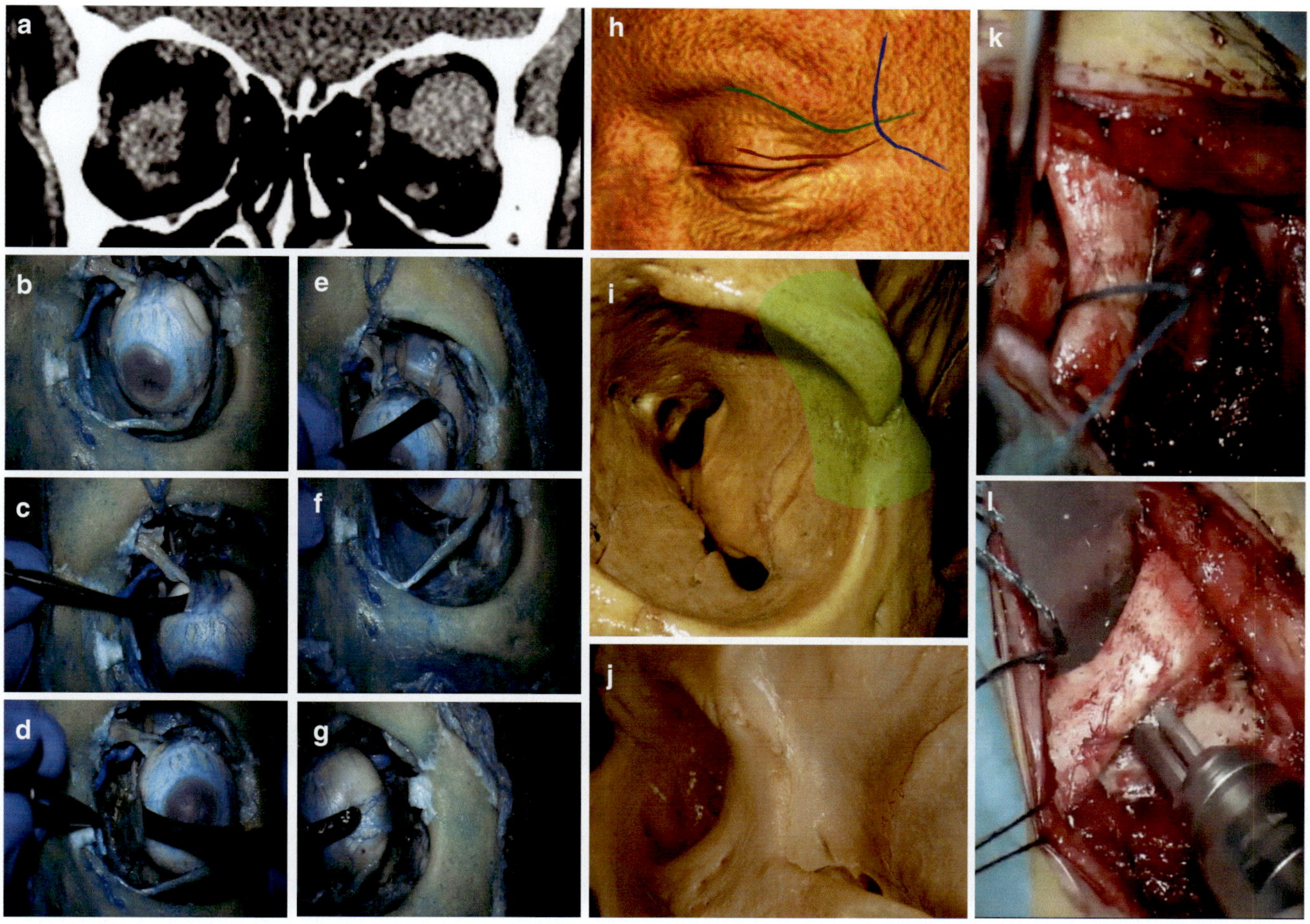

Fig. 28.3 (**a**) Coronal cut computer tomography slice at the level of the orbit axis. An intraconal lateral cavernous hemangioma is shown in the left orbit. (**b–g**) Different anterior views of the left orbit in a cadaveric specimen showing the different orbit muscles: (**b**) superior and inferior oblique, (**c**) superior oblique and superior rectus, (**d**) medial rectus, (**e**) superior levator palpebrae, (**f**) inferior oblique and inferior rectus, (**g**) lateral rectus. (**h**) Different skin incision styles to perform a lateral orbitotomy: eyelid (red), Krönlein (blue), and Stallard-Wright (green). (**i**) Lateral orbitotomy area (green) including the frontal and zygomatic bones, and partially the anterolateral extension of the sphenoid. (**j**) Lateral view of the orbit wall. (**k, l**) Intraoperative views of a lateral orbitotomy

sion of the oculomotor nerve. On the other hand, the medial wall on this central quadrant is traversed by the anterior and posterior ethmoidal arteries on their way to the ethmoidal sinuses and intracranial space to supply the sinus mucosa and the frontal pole dura and falx [9]. A general rule has been described to predict the position of the anterior and posterior ethmoidal canals. The anterior canal is located into the frontoethmoidal suture around 21 mm posterior to the junction of the lacrimal, frontal, and zygomatic bones. The posterior canal crosses this frontoethmoidal suture about 14 mm posterior to the anterior ethmoidal canal [10]. The posterior ethmoidal canal is located about 7 mm anterior to the optic canal entry into the orbit, a fact that defines the former as the medial limit between the axis and the orbital apex (Fig. 28.4a).

Illustrative Case 3 A 78-year-old man from a rural area consulted us due to a severe right orbit proptosis and blindness. A severe conjunctival chemosis was evidenced. The eyeball was pushed laterally and inferiorly and stuck in a fixed position. The MRI revealed a large lesion that came to be a solitary fibrous tumor into the medial quadrant, mainly located into the axis but invading anteriorly the base. It seemed to be an intraperiorbital extraconal lesion in origin but growing into the intraconal space. Due to the location of the tumor and the clinical presentation, a **transcranial superior orbitotomy** was performed (Fig. 28.4b–e).

The superior orbitotomy through a craniotomy requires to achieve enough medial exposure, especially in cases like this, where the lesion is in the medial quadrant. However, this approach might be also useful to approach the superior and superolateral quadrants. Therefore, this is probably the most versatile and efficient approach in terms of exposure. However, it requires a craniotomy, which increases the number and severity of potential postoperative complications.

Bicoronal or Frontotemporal question mark skin incisions are good options. Once the craniotomy is carried out, the

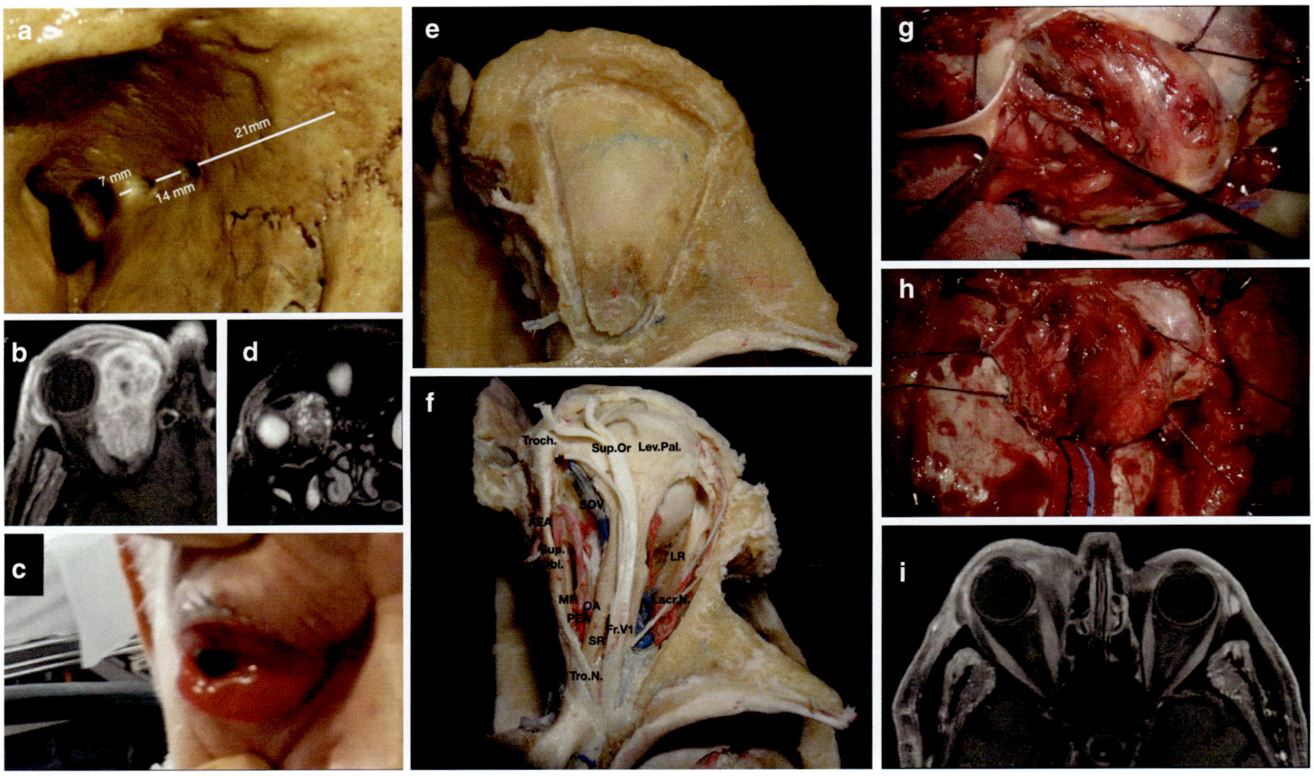

Fig. 28.4 (**a**) Anterior view of the bony orbit showing the distances between the middle orbital angle and anterior ethmoidal canal (21 mm), anterior and posterior ethmoidal canals (14 mm), and posterior ethmoidal canal and optic canal (7 mm). (**b–d**) Real case showing a fibrous solitary intraconal tumor. Axial (**b**) and coronal (**d**) orbit MRI showing a lesion located into the medial quadrant of the orbit occupying the orbit base and axis and even reaching the anterior apex. (**c**) Exophthalmos, ptosis, and conjunctival chemosis are evident. (**e**) Cadaveric dissection of a right orbit roof after removing the intracranial dura and drilling a triangular-shaped superior orbitotomy. (**f**) The orbit roof has been elevated, periorbit opened, and periorbit fat removed to show the intraorbit structures: trochlear nerve (Tro.N.), frontal branch of the first division of the trigeminal nerve (Fr.V1), superior rectus (SR), lacrimal nerve (Lacr.N.), lateral rectus (LR), ophthalmic artery (OA), posterior ethmoidal artery (PEA), medial rectus (MR), superior oblique muscle (Sup.Obl.), anterior ethmoidal artery (AEA), superior ophthalmic vein (SOV), trochlea (Troch.), supraorbitary nerve (Sup.Or), superior levator palpebrae muscle (Lev.Pal.). (**g–h**) Intraoperative views before and after resection of the fibrous solitary tumor through a superior orbitotomy. (**i**) Postoperative axial T1-weighted GD MRI view

anterior fossa dura is dissected enough to expose the frontal bone. Generally, there is no need to reach the ethmoid bone, as it could increase the chances of olfaction disturbances. The anteroposterior extension of the tumor will define how posterior to reach with the extradural dissection. Next, the superior roof is drilled. We generally use a thin diamond drill and take it into one piece to replace it posteriorly and diminish the chances of postoperative enophthalmos. The periorbit is then exposed and opened. Special care must be carried out to avoid damaging the frontal branch of the trigeminal nerve first division. This nerve lies just above the levator palpebral muscle. Sometimes this nerve can be visualized by transparency through the periorbit, especially when a high intraorbital pressure is present. This nerve is located on a central position at its emergence into the orbit apex, so opening the periorbit in a lateral and/or medial position seems a reasonably safe maneuver. This lateral or medial opening should be decided based on the location of the tumor. Therefore, we can perform lateral (between the superior and lateral recti muscles), medial (between the superior and medial recti muscles) explorations, or both through a superior orbitotomy. The direction of our blunt intraorbital dissections should be now based on the tumor position. The use of cottonoids® is of great help at this stage as due to the high pressure the fat will get herniated into our view and we must start gently navigating into this "yellow sea" minimizing or, even better, never using bipolar nor microscissors. In an oblique direction through the lateral space, the optic nerve will be found as our medial limit. This 45° trajectory will allow us to reach the optic nerve avoiding potential injuries on the abducens (inferolateral to our trajectory) and superior division of the oculomotor nerve (superomedial to our trajectory). Deep into this space and before reaching the optic nerve, the superior ophthalmic vein can be found on its posterior pathway to the superior orbital fissure. The ophthalmic artery runs parallel and lateral to the optic nerve. It then turns medially toward the medial quadrant above the optic nerve, giving rise to the posterior and anterior ethmoidal arteries (Fig. 28.4f).

An avascular lesion was found into the medial quadrant, just medial to the medial rectus posteriorly and pushing the superior oblique muscle anteriorly. Dissecting the tumor from its medial attachment was key and special care was taken in the anterior pole to minimize tractions over the trochlear and supratrochlear nerves. These are neural structures that generally we do not see during surgery, and this is why it is relevant to know their relative locations. When operating into this superomedial quadrant, special attention should be also given to the anterior and posterior ethmoidal branches, whose traction could damage the ophthalmic artery. The resection went uneventful, and after tumor resection, a thin layer of collagen was placed covering the periorbit defect and the bone flap was placed on site without the need to fix it with microplates as the intraorbital and intracranial pressure will fix it on site (Fig. 28.4g–i).

The apex of the pyramid. Even though the apex is the smallest of all the three regions in which the orbit is divided, it is probably the most important and challenging to deal with as all the neurovascular structures in the orbit are represented on its small volume.

The orbital apex is made by the confluence of the area where the orbit and craniofacial region merge. It is limited by the maxillary bone, and greater and lesser wings of the sphenoid bone. These bones do not fit in a perfect way, creating a series of couple fissures (superior and inferior) and a canal (optic canal). Moreover, it is relevant as it is in the orbital apex where the periorbit and tendons of four out of the seven extraocular muscles (recti muscles), as well as the dural layer crossing the superior orbital fissure and optic canal, join forming the so-called annular tendon of Zinn. The annular tendon of Zinn is located just anterior to the optic canal and the central (or oculomotor) space of the superior orbital fissure. The superior oblique muscle originates also in the orbital apex, not directly from the tendinous ring, but from the lesser sphenoid wing just above and medial to it (Fig. 28.5a–e).

Illustrative Case 4 A 61-year-old woman presented with right eye progressive loss of vision. Initially she was treated as an optic neuritis that seemed to partially respond to the administration of steroids. However, she was finally referred to our Neurosurgical Department with a complete loss of vision on her right eye, and an MRI showing the classic findings of an optic nerve sheath meningioma which had grown from the central to the apical compartments of the nerve. Generally, we treat these tumors surgically only when the nerve function is completely lost. In this case we decided to offer a resection of the optic nerve from the eyeball to the apical part of the nerve as posterior as the ophthalmic artery would let us go, but making sure that the tumor was completely confined in the resection piece. Furthermore, we offered a resection of the cisternal and intracranial part of the nerve to prevent potential remnants growing to the chiasm and contralateral nerve. Therefore, a **pterional craniotomy and a superolateral orbitotomy** were carried out. A classic frontotemporal question mark incision was performed behind the hairline. A pterional craniotomy was performed and the sphenoid wink was drilled till the lateral limit of the superior orbital fissure. Then, a superolateral orbitotomy was performed and the inferolateral window was dissected between the lateral and inferior recti muscles. A blunt dissection through the periorbit fat was carried out till discovering the severely thickened optic nerve which was completely removed in one piece. During this maneuver, special care must be paid to avoid any damage to the ophthalmic artery. Then, the dura was opened basally and the optic canal unroofed. After cutting the falciform ligament, the nerve was disconnected from the canal to the chiasm. The dura was closed and the periorbit reconstructed partially and covered with a thin layer of a collagen sponge. The bone flap was replaced and a titanium mesh was screwed inferiorly to prevent the temporalis muscle from getting into the sphenoid wing defect and keep an acceptable functional and aesthetic result (Fig. 28.5f–h).

The relevant anatomical structures crossing the orbital apex share complex anatomical relationships. The optic canal is positioned superomedially to the superior orbital fissure and is separated from it by the so-called optic strut, which is a pillar that fixes the anterior clinoid process to the body of the sphenoid bone. The optic canal runs anterolaterally in almost a 45° angle to the midline sagittal plane. It is crossed by the optic nerve and ophthalmic artery. These neural and vascular structures cross the annulus of Zinn when entering the apex.

The superior orbital fissure shows a "boomerang-shape" when seen through the orbit. It is divided in lateral, central, and inferior spaces. These spaces are separated by the annulus of Zinn, which surrounds the central space, and each one is crossed by different structures. The central or oculomotor space is crossed by the oculomotor, abducens, and nasociliary nerve. All these nerves pass through the tendinous ring. The superior or superolateral space is located between the lesser and greater sphenoid wings, and is traversed by the frontal, lacrimal, and trochlear nerves, as well as the superior ophthalmic vein on its way to the cavernous sinus. The inferior space is inconsistently traversed by the inferior ophthalmic vein.

The inferior orbital fissure is defined by the maxillary bone and the greater sphenoid wing and communicates the orbit with the infratemporal and pterygopalatine fossa, and is crossed by the infraorbital and zygomatic nerves, as well as the infraorbital artery and sometimes the inferior ophthalmic vein (Fig. 28.1).

Illustrative Case 5 A 53-year-old woman was referred to us complaining of a severe right retroocular pain, chemosis and a progressive ophthalmoparesis. Visual function was

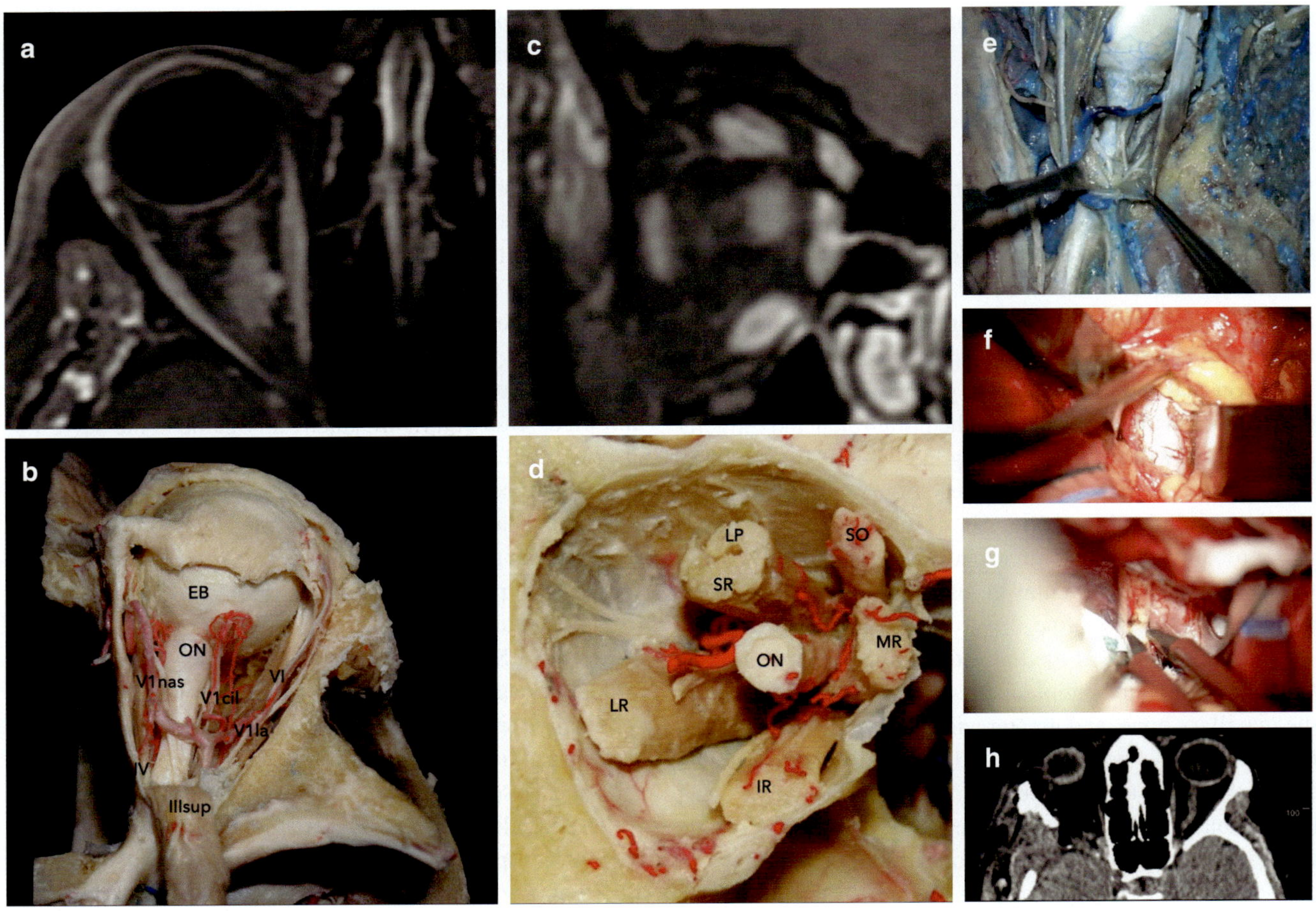

Fig. 28.5 (**a**) Axial T1-weighted GD MRI view of the right orbit showing an optic nerve sheath meningioma. (**b**) Superior view of the high orbit showing the central position of the optic nerve (ON) related to the eyeball (EB), and the trochlear nerve (IV), lacrimal (V1la), ciliary (V1cil) and nasociliary (V1nas) division of the first division of the trigeminal nerve, abducens (VI) and superior division of the oculomotor nerve (IIIsup). (**c**) Coronal T1-weighted GD MRI view of the right orbit showing an optic nerve sheath meningioma. (**d**) Coronal view of a cadaveric specimen showing the central position of the optic nerve (ON) surrounded by its sheath, and the conus defined by the lateral (LR), inferior (IR), medial (MR), superior (SR) recti muscles, the superior oblique (SO), and the superior levator palpebrae (LP) muscles. (**e**) Superolateral view of a cadaveric specimen's right orbit in which the periorbit fat has been dissected and a superolateral access to the optic nerve is shown. (**f**) Superolateral access to the intraconal optic nerve sheath meningioma before resection. (**g**) Intracranial view of the right optic nerve before resection. (**h**) Axial computed tomography slice of the case after resection where a lateral orbitotomy is seen as well as the lack of intraconal optic nerve that has been removed en bloc

surprisingly intact. The MRI showed an intraperiorbital—intraconal lesion into the orbital apex causing a severe compression of the optic nerve, which seemed to be completely engulfed by the tumor itself. Initial radiological diagnosis included two possibilities: optic nerve sheath meningioma vs. optic nerve glioma. Both the options and the clinical presentation obliged us to be conservative in surgical terms and offered the patient a bony decompression of the orbital apex with the aim of releasing the pressure over the extraocular muscles and nerves. The patient was positioned supine with very mild rotation to the contralateral side. A double approach **transcranial superolateral orbitotomy and endoscopic endonasal decompression** was performed. Two teams worked simultaneously with the aim of getting an almost 300° decompression of the orbital apex. The transcranial approach consisted of an extensive drilling of the sphenoid wings, starting lateral into the lesser and progressively also drilling the greater wing till reaching the superior orbital fissure laterally. Superiorly, the optic canal was unroofed till the ethmoid bone. The endonasal approach goal was to decompress the inferomedial quadrants of the orbital apex. A complete ethmoidectomy is performed to skeletonize the medial orbital wall. A wide sphenoidotomy is then performed. The last steps are performed with a diamond drill and under continuous irrigation to prevent heat injury. This decompression allows a range of about 150°. Finally, the periorbit was then opened on both the superolateral and inferomedial sides to release intraorbital pressure (Fig. 28.6).

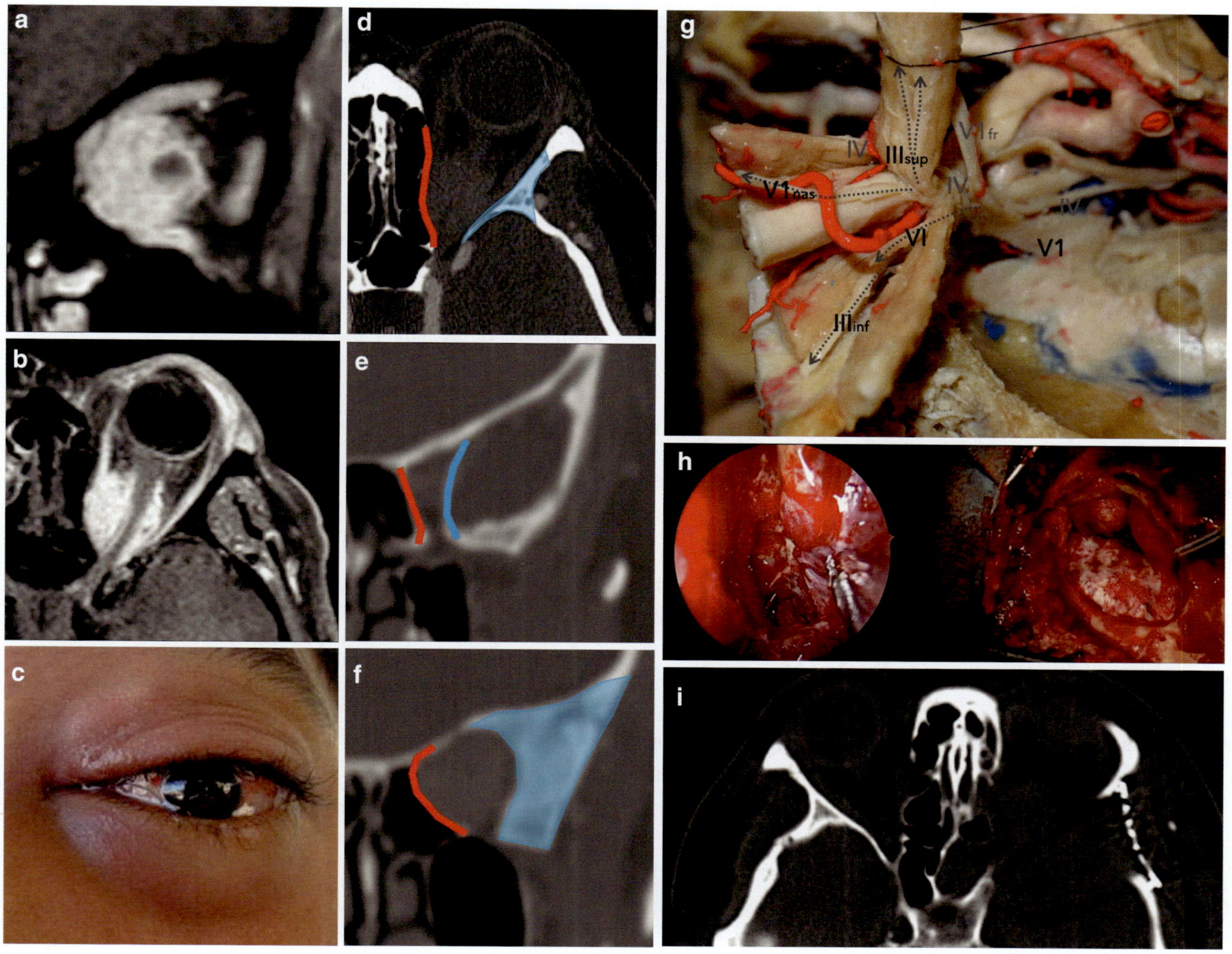

Fig. 28.6 (**a, b**) Coronal and axial T1-weighted GD MRI view of the left orbit showing an optic nerve sheath meningioma. (**c**) Conjunctival chemosis and exophthalmos in the same case. (**d–f**) Axial and coronal computed tomography slices showing the bony areas to decompress via an endonasal route (red) and transcranial approach (blue). (**g**) Superolateral view of the right orbit where the apical intraorbit structures are shown in relationship with the annulus of Zinn: trochlear nerve (IV), first division of the trigeminal nerve (V1), abducens (VI), inferior (IIIinf) and superior (IIIsup) divisions of the oculomotor nerve, frontal branch of the first division of the trigeminal nerve (V1fr), and nasociliary nerve of the trigeminal first division (V1nas). (**h**) Intraoperative endonasal and transcranial views after decompression. (**i**) Postoperative computed tomography axial scan after decompression

28.4 Surgical Interventions

There are many different routes to reach the orbit. The orbit contains the eye, its extraocular muscles, nerves, and vessels embedded in the orbit fat and enclosed by the periorbit. Each of these structures keeps a specific location in the intraperiorbital space, which is generally maintained by the skeleton and landmarks provided by the recti muscles. The four quadrants (medial, lateral, superior, and inferior) and their representation into each the three compartments of the orbit (apex, axis, and base), combined with the relative location of the extraocular recti muscles and bony structures protecting them, will define the different approaches we can use.

Summarizing these approaches from a neurosurgical perspective, we can enter all these chambers through a lateral, superior, or inferolateral orbitotomy. We can even combine some of these procedures and customize them depending on individual anatomy and our experience.

28.5 Surgical Pearls

- Choose the approach based on the location of the tumor into the different quadrants and parts of the orbit chamber.
- Drill gently trying to preserve the periorbit till the end of the bony work.

- Use a diamond drill with continuous irrigation when getting close to the apex to avoid heat neuropathies.
- Open the periorbit in a longitudinal way following the direction of the recti muscles.
- Consider the superomedial, superolateral, inferomedial, and inferolateral windows depending on the relative location of the lesion to the optic nerve.
- Once the periorbit is opened, use blunt dissection instruments and cottonoids to clean the field if bleeding.
- Try to avoid or minimize at maximum the use of bipolar coagulation and microscissors inside the orbit.
- Respect the "yellow sea" and give your best to keep it yellow, not red.
- Ligate the optic nerve anteriorly and posteriorly when you need to cut it.
- If the dura needs to be opened, try to reconstruct it as much as possible to avoid leaks.
- The periorbit is much gentle than the orbit in terms of healing, so do not "force" a watertight closure. Use thin layers of collagen sponges to fill potential periorbit defects.
- Replace the bone flap if possible, to prevent enophthalmos.
- A laterally placed titanium mesh separating the temporalis muscle from the periorbit can help preventing a functional and cosmetic defect.

28.6 Conclusion

The orbit is a fascinating area which represents less than 2% of the volume when compared to the intracranial space. Despite the small space in which it is confined, it hosts and protects one of the most important parts of the central nervous system, the eye. All the necessary elements to get a perfect equilibrium are located within the orbit walls, and slightly enclosed inside a watertight sac made of periorbit, which is a prolongation of the intracranial dura matter.

Its location among the craniofacial area and the anterior and middle cranial fossa, makes the orbit a limiting field for different specialists as ophthalmologists, otorhinolaryngologists, maxillofacial surgeons, and neurosurgeons.

A detailed understanding of the bony structures, periorbit, quadrants and spaces, muscles' relative location, and neurovascular structures inside the orbit remains mandatory to perform effective procedures and to alleviate our patients' symptoms in the safest and more efficient way.

References

1. Osaki TH, de Castro DK, Yabumoto C, Mingkwansook V, Ting E, Nallasamy N, et al. Comparison of methodologies in volumetric orbitometry. Ophthal Plastic Reconstruct Surg. 2013;29(6):431–6.
2. Hayek G, Mercier P, Fournier HD. Anatomy of the orbit and its surgical approach. Adv Tech Standards Neurosurg. 2006:35–71.
3. Newman SA. Orbital surgery: evolution and revolution. J Neurol Surg Part B: Skull Base. 2021;82(01):007–19.
4. Knapp H. On cocaine and its use in ophthalmic and general surgery. Arch Ophthalmol. 1884;13:402–8.
5. Krönlein RU. Zur Pathologie und operativen Berhandlung der Dermoidcystern der Orbita. Beitr Klin Chir. 1888;4:149–63.
6. Dandy WE. Orbital tumors: Results following the transcranial operative attack. Arch Neurol Psychiatr. 1941;46(3):568.
7. Conley JJ. Medial and inferior approach to orbital tumors. In: Troutman RC, Converse JM, Smith B, editors. Plastic and reconstructive surgery of the eye and adnexa. Butterworth; 1962. p. 85–9.
8. Haładaj R, Polguj M, Topol M. Anatomical variations of the supraorbital and supratrochlear nerves: their intraorbital course and relation to the supraorbital margin. Med Sci Monitor. 2019;25:5201–10.
9. Cankal F, Apaydin N, Acar HI, Elhan A, Tekdemir I, Yurdakul M, et al. Evaluation of the anterior and posterior ethmoidal canal by computed tomography. Clin Radiol. 2004;59(11):1034–40.
10. Martins C, Costa e Silva IE, Campero A, Yasuda A, Aguiar LR, Tatagiba M, et al. Microsurgical Anatomy of the orbit: the rule of seven. Anat Res Int. 2011;2011:1–14.

Anatomy of and Surgical Approaches to the Orbit for Neurosurgeons

29

William T. Couldwell, Farshad Nassiri, and Guilherme Agnoletto

Learning Objectives

- To understand the basic anatomy of the orbit
- To describe a topographical approach to classifying lesions of the orbit
- To understand the principles of commonly used transorbital and transcranial approaches to the orbit
- To understand the cranial corridors that may be accessed through an orbital approach

29.1 Introduction

Diverse pathologies can affect the orbit, each with a distinctive presentation and unique management options. The orbit may also become involved secondary to disease that is primarily intracranial (e.g., meningiomas). Symptomatic orbital tumors are often managed with surgery, and neurosurgeons and ophthalmologic surgeons have complementary expertise in accessing the orbit. This chapter will review critical surgical anatomy of the orbit and discuss the indications, contraindications, and risks of common surgical approaches to the orbit for neurosurgeons.

29.2 History and Background

The practice of orbital surgery represents the intersection of surgical expertise between neurosurgeons and ophthalmologists. Neurosurgeons typically approach intracranial pathologies that extend into the orbit from transcranial trajectories, whereas ophthalmologists traditionally adopt extracranial transorbital approaches to orbital pathologies that do not extend intracranially. The relationship between the orbit and the intracranial compartment is intricate, with a dense arrangement of interwoven nerves and vessels that traverse these two compartments. Thus, the interdisciplinary partnership offers significant benefits.

Although incidentally discovered small tumors may be managed conservatively with observation and serial imaging, surgery remains the cornerstone of treatment for symptomatic orbital tumors, the most common of which are meningiomas (either primary optic nerve sheath or invasion from periorbital tumors), cavernous hemangiomas, optic nerve gliomas, and lymphomas (either primary or metastatic) [1]. The most frequent symptoms of these tumors—orbital pain or headaches, proptosis, visual deficits, and ophthalmoparesis or complete ophthalmoplegia [1]—typically manifest secondary to local mass effect and, less commonly, neurovascular compromise.

Orbital tumors can be categorized based on their location relative to the extraocular muscle cone and optic canal. Tumors confined to the optic canal (intracanalicular) can lead to early vision loss, with more limited impact on extraocular movement limitations or proptosis. Extraconal tumors can present with proptosis, accompanied by delayed onset of visual and extraocular muscle movement impairments resulting from compression of individual muscles or the optic nerve. Intraconal tumors can present with early vision loss, proptosis, and ophthalmoparesis or ophthalmoplegia.

29.3 Anatomy Highlights

- The orbits are paired cone-shaped structures of the face with slight medial tilt and approximate volume of 30 cm^3.
- Figure 29.1 delineates the bony anatomy of the orbit, which is formed by articulations between seven bones that form four walls [2]. The roof is formed by articulations between the orbital surface of the frontal bone and the lesser wings of the sphenoid bone; the floor by articu-

W. T. Couldwell (✉) · F. Nassiri · G. Agnoletto
Department of Neurosurgery, Clinical Neurosciences Center, University of Utah, Salt Lake City, UT, USA
e-mail: neuropub@hsc.utah.edu; William.Couldwell@hsc.utah.edu; farshad.nassiri@hsc.utah.edu; guilherme.agnoletto@hsc.utah.edu

I. N. Kanaan, V. Beneš (eds.), *Neuroanatomy Guidance to Successful Neurosurgical Interventions*,
https://doi.org/10.1007/978-3-031-59838-8_29

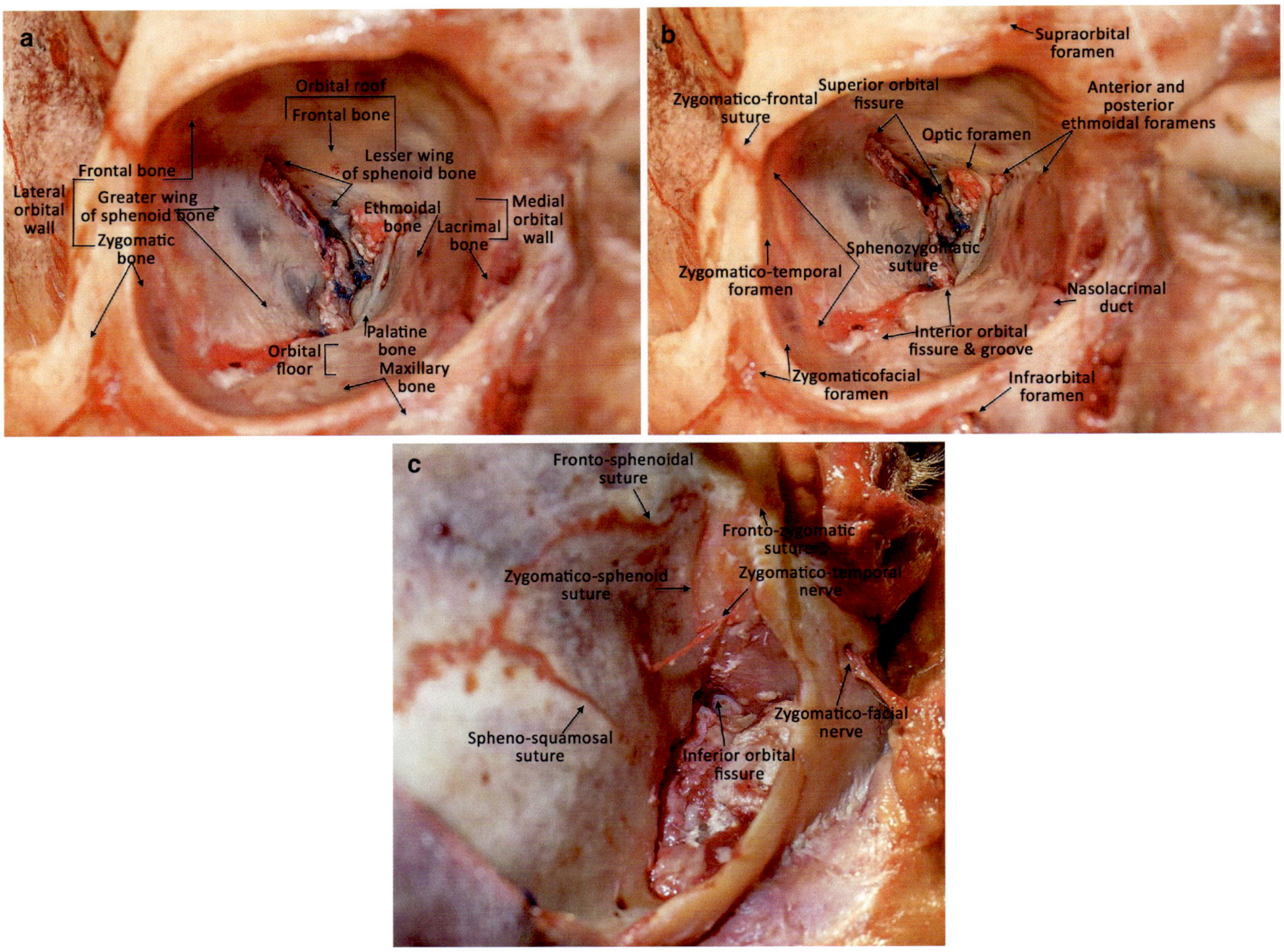

Fig. 29.1 Cadaveric dissection of the right orbit and cranio-orbital region. (**a**) Anterior view demonstrating the osteology of the orbit. (**b**) Anterior view of the orbit demonstrating the anatomically important orbital components. (**c**) Posterolateral view of the lateral orbital wall and the pterion. (Reproduced with permission from Alzhrani et al. [15]. https://journals.lww.com/onsonline/pages/default.aspx)

lations of the zygomatic, palatine, and maxillary bones; the lateral wall by articulations between the orbital surfaces of the greater wing of the sphenoid (GWS) and the zygomatic bone; and the medial wall by articulations of the ethmoid, lacrimal, and maxillary bones. Suture lines demarcate the bone-to-bone articulations and are important in understanding extent of dissections during surgery.

- All openings to the orbit are at the base (located anteriorly), at the apex (located posteriorly and formed by the optic canal and superior orbital fissure (SOF)), or between orbital walls.
- The anterior and posterior ethmoidal canals are formed by bony notches at articulations between the orbital plate of the ethmoid and frontal bones and transmit the ethmoidal terminals of the nasociliary branch of the ophthalmic division of the trigeminal nerve.
- The optic canal at the apex of the orbit transmits the optic nerve and ophthalmic artery and is approximately 5 mm in width, 5 mm in height, and 5–10 mm in length. The optic strut is a bony ridge that spans from the anterior clinoid process (ACP) of the lesser wing of the sphenoid to the body of the sphenoid, forming the lateral part of the optic canal and separating the optic canal from the SOF. The optic nerve is covered by a dural fold, known as the falciform ligament, as it enters the optic canal. The release of this ligament is a key step in optic nerve decompression.
- The SOF is located between the roof and lateral wall of the orbit. It separates the orbit from the cavernous sinus (CS) and transmits the nerves that control extraocular movement (oculomotor, trochlear, and abducens), branches of the ophthalmic division of the trigeminal nerve (frontal, lacrimal, and nasociliary), superior ophthalmic vein, and sympathetic fibers.

- The inferior orbital fissure (IOF) is located between the floor and lateral wall of the orbit and transmits the infraorbital and zygomatic branches of the maxillary nerve (which terminates as the zygomaticofacial and zygomaticotemporal nerves) and the inferior ophthalmic vein, infraorbital vein, and infraorbital artery. The IOF connects the orbit to the pterygopalatine fossa medially and the inferotemporal fossa laterally.
- There are seven muscles in the orbit: the four recti muscles (superior, lateral, medial, and inferior), the two oblique muscles (superior and inferior), and the levator palpebrae superioris. The four recti muscles (and rarely the superior oblique muscle) originate from a common tendinous ring known as the annulus of Zinn, which attaches to the orbital apex and encircles the optic canal and the medial portion of the SOF. Of the structures that traverse these canals, the optic nerve, ophthalmic artery, oculomotor nerve (superior and inferior divisions), nasociliary branch of the ophthalmic division of the trigeminal nerve, and abducens nerve pass through the tendinous ring. The superior ophthalmic vein, frontal and lacrimal branches of the ophthalmic division of the trigeminal nerve, and trochlear nerve pass outside the tendinous ring. The superior oblique muscle originates from the GWS along the superomedial wall of the orbit at the orbital apex, passes through the tendinous trochlea, and changes direction to attach to the superolateral surface of the globe. The inferior oblique muscle originates from the medial wall of the orbit lateral to the lacrimal sulcus and passes deep to the lateral rectus to insert on the lateral sclera. The levator palpebrae also originates near the orbital apex just above the optic canal and travels superficially to the superior rectus to reach the tarsal plate.
- There are also seven nerves in the orbit. The optic nerve courses through the optic canal, whereas the remaining nerves (oculomotor, trochlear, abducens, and frontal, lacrimal, and nasociliary branches of the ophthalmic division of the trigeminal nerve) course through the SOF. In general, the nerves that innervate muscles involved in extraocular movement do so at the midportion of the belly of these muscles on the medial surface. Approaching from the lateral aspect, the ophthalmic division of the trigeminal nerve and the trochlear nerve are the most superficial structures of the orbital apex. The trochlear nerve courses medially in the roof of the orbit to reach the superior oblique muscle. The frontal branch of the ophthalmic division of the trigeminal nerve courses medially and terminates as the supratrochlear and supraorbital nerves; the lacrimal branch courses laterally and superficial to the superior ophthalmic vein to supply secretory fibers and sensation to the gland; and the nasociliary branch divides into the nasal division that terminates as the ethmoidal nerves and the ciliary division that provides sensory innervation of the cornea and globe and carries parasympathetic autonomic fibers to the ciliary ganglion. The oculomotor nerve enters the orbit within the annulus of Zinn lateral to the optic nerve and immediately branches into its superior and inferior division. The inferior division has a branch that innervates the inferior rectus and inferior oblique muscles and a more medial branch that innervates the medial rectus muscle. The superior division courses medially over the optic nerve and innervates the superior rectus and levator palpebrae. Sympathetic nerve fibers that course along the internal carotid artery (ICA) are transmitted to their targets in the orbit along the ophthalmic artery and also along branches of the trigeminal nerve and abducens nerve.
- The intracranial and intracanalicular segments of the optic nerve are supplied by small perforating arteries from the supraclinoid ICA, whereas the intraorbital segment of the optic nerve is supplied by pial ciliary arteries. The ophthalmic artery is the major arterial supply to the orbit and, in most cases, is the first major branch of the supraclinoid ICA, medial to the ACP. The ophthalmic artery originates along the medial supraclinoid ICA and courses inferolaterally to the optic nerve as it enters the optic canal and then medially as it traverses the orbit with the central retinal artery as the first major branch ~10 mm posterior to the globe [2]. The central retinal arterial branch of the ophthalmic artery penetrates the optic nerve and provides the terminal arterial vascular supply of the retina. Aside from the critical central retinal artery, the ophthalmic artery also gives rise to a set of lateral orbital arterial branches including the lacrimal and recurrent meningeal arteries; a set of medial orbital arterial branches including the supraorbital, supratrochlear, and ethmoidal arteries; and the inferomedial trunk that supplies the inferior and medial recti muscles. The superior and inferior ophthalmic veins provide venous drainage of the orbit.

29.4 Approaches

Anatomical classification of lesions of the orbit based on their relationship to the periorbita, muscular cone, and optic nerve [2] can help with selection of the optimal surgical approach. Anterior to the optic canal and SOF, the orbit is covered by an extension of cranial dura—the periorbital or orbital dura—that supports the orbital structures. Lesions that are superficial to the periorbita are considered extradural, whereas those that involve or are located deep to the periorbita are considered intradural. Similarly, lesions that are outside the muscular cone are considered extraconal, whereas those that are within the muscular cone are considered intraconal. All intraconal lesions are intradural by definition, whereas extraconal lesions can be either intradural (between

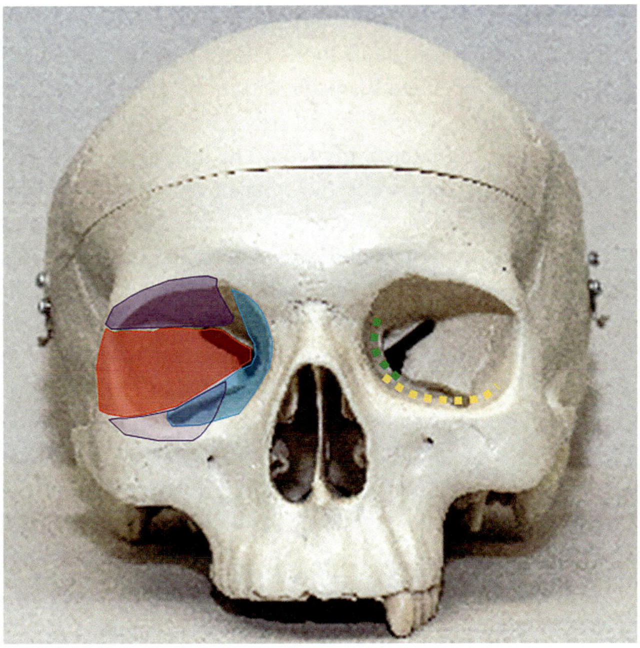

Fig. 29.2 Transorbital approaches to the orbit. Endonasal approaches (blue) can access the medial orbit and orbital apex. The lateral orbitotomy (red) can be used for lesions in the lateral orbital compartment. A modified lateral orbitotomy can be used to preserve the lateral orbital wall and decrease the need for retraction of orbital structures. The lateral orbitotomy can be expanded to total orbitotomy (purple) to remove the superior and inferior orbital rims. Anterior and basally located orbital tumors can be accessed via transconjunctival approach (yellow dashed line) that can be extended medially (anterior medial microorbitotomy, green dashed line)

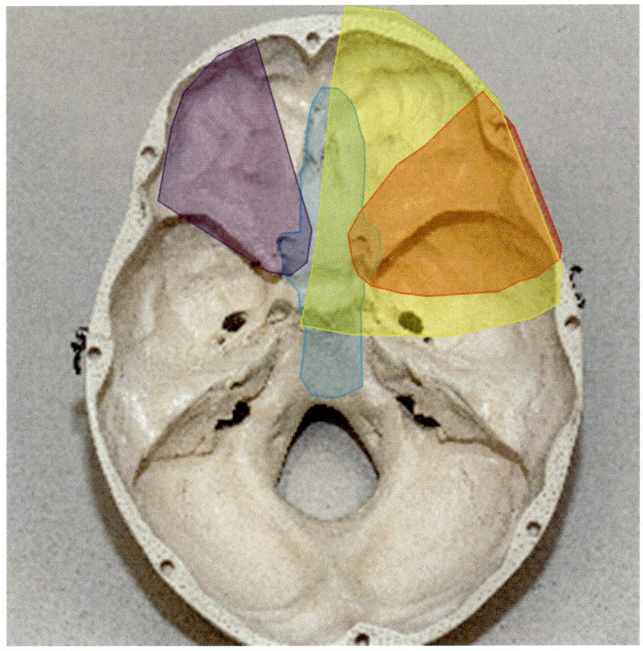

Fig. 29.3 Transcranial approaches to the orbit. These approaches are indicated for invasive intracranial tumors that expand into the orbit. The traditional frontotemporal pterional approach (red) is centered on the sphenoid ridge, allowing for access to the optic canal, anterior clinoid, and superior orbital fissure. By contrast, the lateral supraorbital approach (purple) centers the exposure along the subfrontal corridor. The orbitozygomatic approach (yellow) extends the standard frontotemporal flap with additional bony removal of the orbit and skull base that can include the lateral orbital rim, lateral wall, and roof of the orbit as well as the zygoma to allow multidimensional access and potentially shorter working distances to the orbit, orbital apex, anterior and middle fossa, and parasellar regions. Tumors within the medial orbit and orbital apex can be accessed by and endonasal approach (blue)

muscle and periorbita) or completely extradural (not involving the periorbita). Intraconal lesions can be described relative to their position of the optic nerve (medial, central, or lateral). In general, the orbit can be surgically accessed either using transorbital (orbitocranial) or transcranial (cranio-orbital) corridors (Figs. 29.2 and 29.3).

29.4.1 Transorbital Lateral Orbitotomy

Since its first description in 1889 [3], several modifications to the lateral orbitotomy approach have been proposed. This approach and its modifications can be used to resect lateral extradural lesions, as well as intraconal lesions that are lateral, dorsal, or ventral to the optic nerve, and can also provide access to the orbital apex, CS, and middle fossa [4–6]. Patients are positioned supine, with the head secured in 3-point fixation and extended and turned contralaterally 10–20° or more for lateral lesions. The lateral orbital rim is identified after lateral canthotomy with superior and inferior cantholysis (limiting the extent of canthotomy to improve cosmesis is also possible [7, 8]), and an extraorbital subperiosteal dissection is performed to elevate the temporalis muscle and then the periorbita off the lateral orbital wall. Pilot holes drilled in the lateral orbital wall facilitate reconstruction with titanium plates at the end of the case. Osteotomies are made at the level of the frontozygomatic suture superiorly and the zygomatic arch inferiorly. The orbital contents are carefully protected with malleable retractors. The remaining lateral orbital wall and GWS bone are then drilled using a diamond burr to expose the orbital apex, SOF, and ACP. At the end of the procedure, the lateral orbital rim is replaced in its native position, and reconstruction is completed with low-profile titanium craniofacial miniplates and screws. The lateral cantholysis is then repaired and the skin closed using absorbable sutures.

The lateral orbitotomy approach can be extended to access the anterior cranial fossa by dissecting the retro-orbicularis oculi fatty tissue and including the superior orbital rim, up to the supraorbital notch, in the orbitotomy. The inferior orbital rim can also be included in the orbitotomy to the level of the infraorbital nerve via a secondary infraorbital subconjunctival incision. Although this total lateral orbitotomy increases the extent of exposure, the necessary additional bony removal risks retraction- and pressure-induced injuries and ophthalmoparesis [9]. There have been recent reports describing the use of transorbital approaches to access select intracranial pathologies [10–12]. Opening lateral to the SOF and IOF from an endo-orbital route provides access to the middle and/or anterior cranial fossa, whereas opening medial to the SOF extends the opening to the opticocarotid area. Further work is being done to compare these approaches to more traditional transcranial approaches.

29.4.2 Transorbital Modified Lateral Orbitotomy

In the traditional lateral orbitotomy, handheld malleable retractors are used to protect the orbital contents while the lateral orbital wall is drilled [13, 14]. To minimize inadvertent retraction and pressure injuries, the senior author has described a minimally invasive extradural approach to access lesions of the SOF and CS by partially sparing the posterior lateral orbital wall as a protective barrier to orbital contents (Fig. 29.4) [15]. The skin incision along the lateral third of the upper eyelid is extended ~2.5 cm posteriorly from the lateral canthus, with a small fork in the lower eyelid for skin relaxation. The lateral canthal ligament is identified and protected. Dissection of the lateral orbital wall extends to the level of the sphenozygomatic suture. Osteotomies are made along the lateral orbital wall from the frontozygomatic suture and from above the zygoma to the sphenozygomatic suture, and then the orbital rim is outfractured. A 1-piece lateral and superior orbitotomy can be performed if access to ACP is needed by placing a burrhole at MacCarty's point to expose the frontal dura and periorbita. The lateral wall of the orbit is then drilled extraorbitally using a 4-mm diamond burr until a thin rim of cortical bone remains to protect the orbital contents. Frameless stereotactic navigation is used to confirm the trajectory toward the orbital apex. The GWS is drilled further extraorbitally until cancellous bone is reached, and the sphenoid ridge is also drilled flush to expose the meningo-orbital band, which can be sharply divided. A combination of gentle blunt peeling and sharp dissection of the dura can be used to expose the ACP, lateral wall of the CS, and middle fossa floor, if necessary. Closure is as described above.

29.4.3 Transorbital Transconjunctival Approach

This approach was first described as a means to perform lower lid blepharoplasty [16], but its use has been expanded to access basal lesions of the anterior orbit. The transconjunctival approach is optimally performed in collaboration with ophthalmologically trained orbital surgeons. A lateral canthal incision is started at the canthal angle and extended to the canthal skin crease, and lateral cantholysis is performed. The inferior eyelid is retracted, and the inferior fornix of the conjunctiva is exposed and incised laterally and then medially. Either a pre- or a postseptal dissection can be used to reach the orbital floor. A postseptal dissection is preferred and more commonly used to avoid ectropion deformities that are associated with injuries to the septum [17]. The capsulopalpebral fascia is opened over the inferior orbital rim and the periorbita identified. The periorbita can be incised to gain access to the intraconal space if needed. The limits of this exposure are the lateral rectus laterally and medial rectus medially. The medial, lateral, and inferior aspects of the optic nerve are accessible. The periorbita and capsulopalpebral fascia are not routinely closed. The conjunctiva is closed with absorbable 8–0 sutures, the canthal tendon insertion is repaired to the orbital rim, and the canthal incisions are closed with absorbable sutures. Complications are uncommon and are mainly related to issues of the eyelid (ectropion, entropion, granuloma) [18].

The transconjunctival approach can be expanded to access intraconal lesions within the anterior medial orbit or even medial extradural lesions (so-called anterior medial microorbitotomy approach), although endonasal approaches can also be used as minimally invasive alternative approaches [19]. To expand the approach, semilunar medial conjunctival incisions are made along the superior and inferior conjunctival fornices. The periorbita is identified and divided, and the medial rectus muscle can be hooked and tagged with a suture and disinserted from its attachment to the globe if necessary. Malleable retractors are used for gentle lateral retraction and protection of the globe. The conjunctiva is then repaired at the end of the procedure with absorbable sutures.

29.4.4 Transcranial Frontotemporal Approaches

The familiar transcranial frontotemporal approaches are versatile and provide excellent exposure of the superolateral surface of the orbit, including the optic canal and SOF (Fig. 29.5). The traditional pterional craniotomy is the workhorse of frontotemporal approaches, and the bony exposure is centered on the sphenoid ridge. The patient's head is placed in 3-point fixation, with the neck extended and head rotated contralaterally ~30°. A standard frontotemporal curvilinear incision 1 cm anterior to the tragus at the root of zygoma is carried forward above the superior temporal line toward the patient's hairline. A myocutaneous flap is raised, elevating the temporalis muscle off the temporal bone and orbital rim. A standard frontotemporal flap is raised with a burrhole at the MacCarty point to expose the dura of the frontal lobe, temporal lobe,

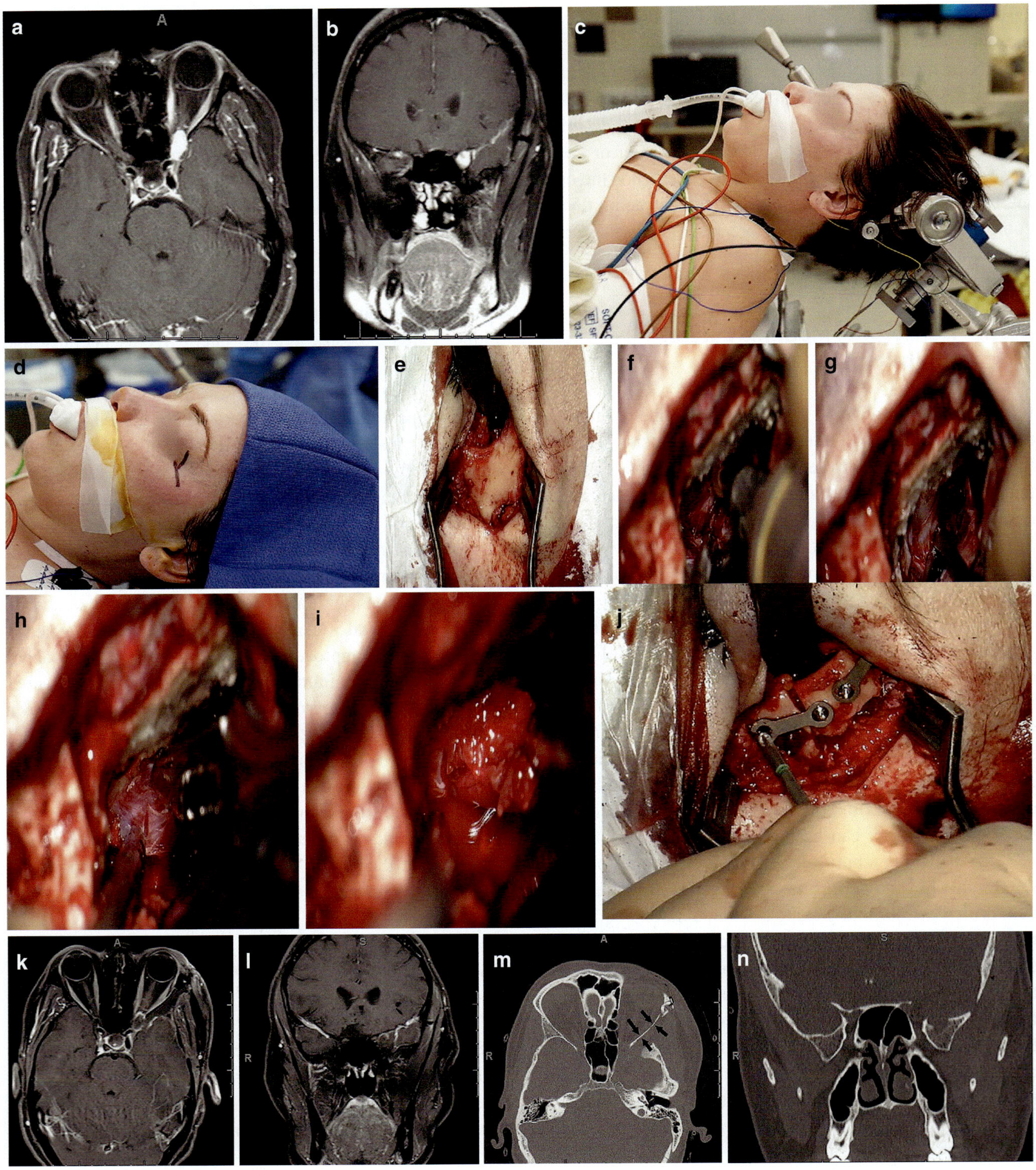

Fig. 29.4 Left lateral orbital approach for removal of superior orbital fissure lesion. Axial (**a**) and coronal (**b**) T1-weighted magnetic resonance imaging of the brain with contrast demonstrate an enhancing lesion in the superior orbital fissure. (**c**, **d**) The patient is positioned supine with the head held in a 3-point fixation head clamp and kept in neutral position. (**e**) Exposure of the lateral orbital rim and placement of the handheld malleable retractor to protect the orbital contents while making the bony cuts, not for retraction of the orbital contents. (**f**) Removal of the orbital rim and drilling the GWS to expose the dural of the temporal pole. In this case, the anterior portion of the lateral orbital wall was thinned using the drill and was not removed completely to help protecting the orbital content. (**g**) Peeling the temporal dura from the superior orbital fissure. (**h**) Tumor is evident at the superior orbital fissure. The tumor capsule is opened sharply using a vertical incision at its thinnest part. (**i**) Tumor is debulked and the tumor capsule is dissected free from the surrounding structures and removed completely. (**j**) The orbital rim is replaced using miniplates and screws. As above, the retractor seen in this image is positioned superficially to protect the globe during the removal of the lateral orbital rim and is not used for retraction of the orbital contents. Postoperative axial (**k**) and coronal (**l**) T1-weighted magnetic resonance imaging of the brain with contrast demonstrate gross total resection of the lesion. Postoperative axial (**m**) and coronal (**n**) computed tomography cuts (bone window) demonstrate the extent of the bone removal. Note the thinning of the lateral orbital wall without complete removal. (Reproduced with permission from Alzhrani et al. [15]. https://journals.lww.com/onsonline/pages/default.aspx)

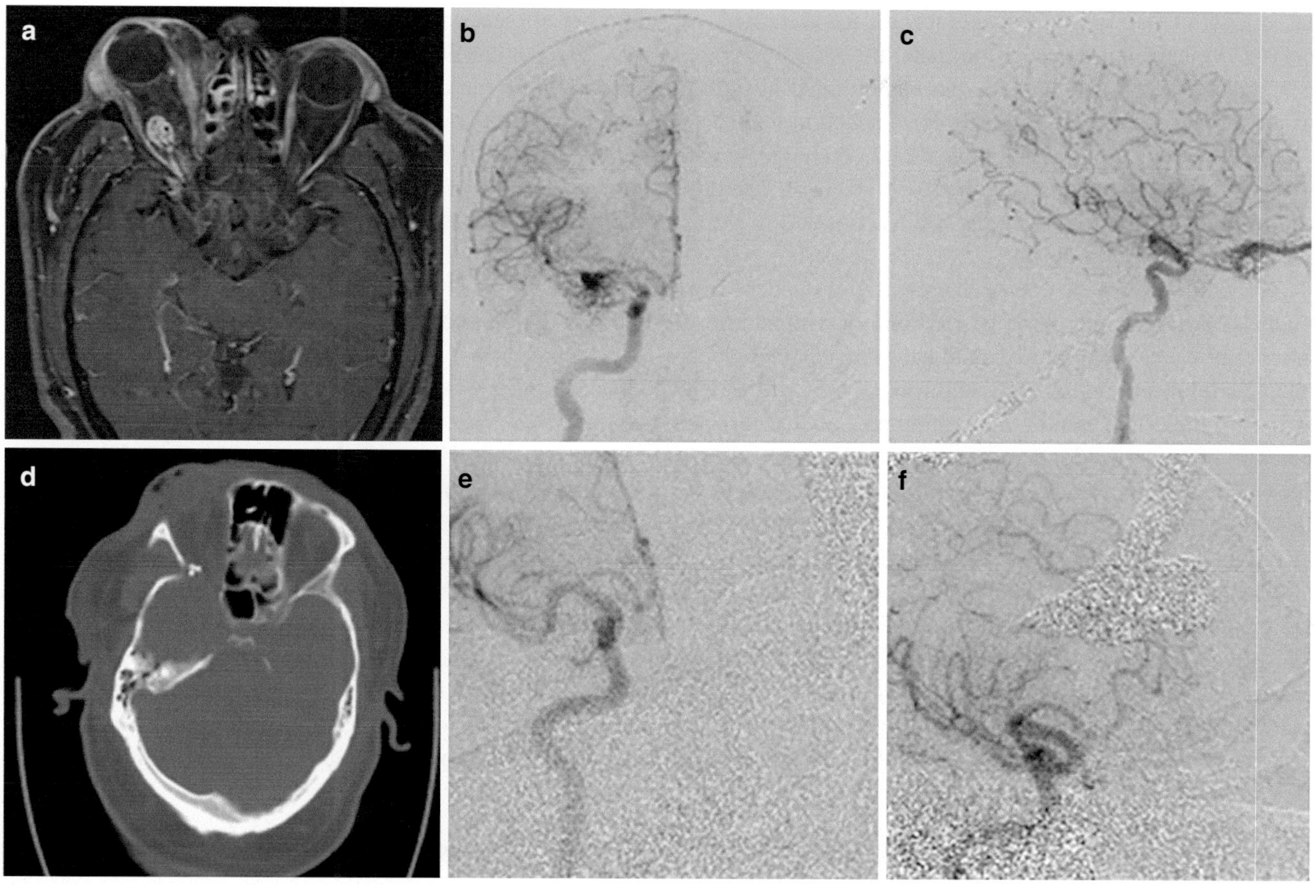

Fig. 29.5 A 45-year-old man presented with right eye chemosis, proptosis, and elevated intraocular pressures with progressive visual loss. (**a**) Magnetic resonance imaging demonstrated a right-sided intraconal mass located between the superior and lateral recti with prominent flow voids. Anteroposterior (**b**) and lateral (**c**) angiogram projections demonstrated an arteriovenous malformation of the orbit, fed predominantly by the ophthalmic artery and draining via the superior ophthalmic vein. (**d**) A frontotemporal craniotomy and lateral orbitotomy was performed to unroof the superior orbital fissure, remove the anterior clinoid, decompress the optic canal, and release the falciform ligament. The intracranial ophthalmic artery was identified so that an approximate distance to the lesion within the orbit can be measured. The intraconal lesion was accessed through a corridor between the superior and lateral rectus muscles. Feeders from the intraorbital ophthalmic artery were followed to a small nidus and the superior ophthalmic vein, which was ligated. The arteriovenous malformation was dissected from periorbital fat, which is unique to intraorbital arteriovenous malformations, and removed. (**e**,**f**) Postprocedural angiograms show disconnection and removal of the shunting lesion.

and orbit. A cutting or diamond burr is used to drill the sphenoid ridge flush to the level of the meningo-orbital band. This band is the connection of the middle fossa dura to the periorbita through the SOF and reflects the transition from the lesser wing of the sphenoid to the ACP. The lateral wall and roof of the orbit can be drilled until eggshell thin and then unroofed with Kerrison rongeurs. If optic nerve decompression is necessary, the meningo-orbital band can be divided and the ACP exposed. Generous decompression of the optic nerve can be achieved by unroofing the optic canal with a diamond burr, performing an anterior clinoidectomy and optic strut removal, and releasing the falciform ligament. Copious irrigation is necessary to avoid thermal injury to the nerve. The neurovascular contents of the optic canal and SOF can be accessed in the intracranial and orbital compartments in this approach. To gain access to superolateral intraconal lesions, the periorbita can be opened and a corridor between the superior and lateral rectus used. At the end of the procedure, the bone flap is fixed back to the cranium with miniplates and screws, and the temporalis muscle is reapproximated. A porous polyethylene implant is used to cover the bony defect in this region to avoid cosmetic deformity [20].

Several variations of the standard pterional approach have been described. The mini-pterional approach is very similar to a traditional pterional approach, but is a shorter and more focused curvilinear incision 1–2 cm above the zygoma to the midpupillary line that is centered on the sphenoid ridge and requires less dissection of the temporalis muscle. A smaller craniotomy flap is made centered over the sphenoid ridge with the same steps. Overall, the exposure of the orbit is similar to that of a standard pterional craniotomy, although access to the medial optic canal may be limited.

By contrast, the lateral supraorbital approach centers the exposure along the ACF floor [21]. The skin incision is similar in length to the mini-pterional incision and is located just behind the hairline; however, because the bony opening is centered along the frontal bone, there is minimal disruption of the temporalis muscle. A burrhole is drilled in the frontal pterional region, and a frontal bone flap is created flush with the ACF floor without entering the frontal sinuses. The orbit can be accessed by removing the orbital roof and, if necessary, the superior orbital rim.

The orbitozygomatic approach extends the standard frontotemporal flap from a pterional craniotomy with additional bony removal of the orbit and skull base that can include the lateral orbital rim, lateral wall, and roof of the orbit as well as the zygoma. This allows for multidimensional access and potentially shorter working distances to the orbit, orbital apex, anterior and middle fossa, and parasellar regions but requires more soft tissue dissection and exposure and increases operative time. Care should be taken to not injure the frontal branch of the facial nerve that runs superficial to the temporal fat pad in the superficial temporalis fascia. An interfascial or subfascial dissection with mobilization of the temporalis inferiorly can minimize injury to the frontal branch and also maximize exposure. Various technical descriptions of this approach offer slight variations in the number of pieces and cuts needed for complete osteotomies [22–27]. For lesions of the orbit and orbital apex, removal of the lateral orbital rim and roof (orbitopterional craniotomy) would be sufficient, but the addition of a zygomatic osteotomy expands access to the IOF if needed. Overall, the extent of exposure of the orbit and orbital apex is similar between the orbitozygomatic and pterional approaches.

29.4.5 Endoscopic Endonasal Approach

This approach provides excellent access to the medial aspect of the orbital apex as well as intraconal lesions that are inferior and medial to the optic nerve (Fig. 29.6). Because of the steep angle to the orbit, endoscopic visualization is preferred. The nasal turbinate of the ipsilateral side is lateralized and can be resected if it is large and bulky. An initial uncinectomy is followed by maxillary antrostomy and anterior and posterior ethmoidectomies on the ipsilateral side. The procedure can be completed through one or both nares. If a binarial approach is desired, a posterior septostomy is needed and a wide sphenoidotomy is performed. The anterior ethmoidal artery enters through the frontal recess, and the posterior eth-

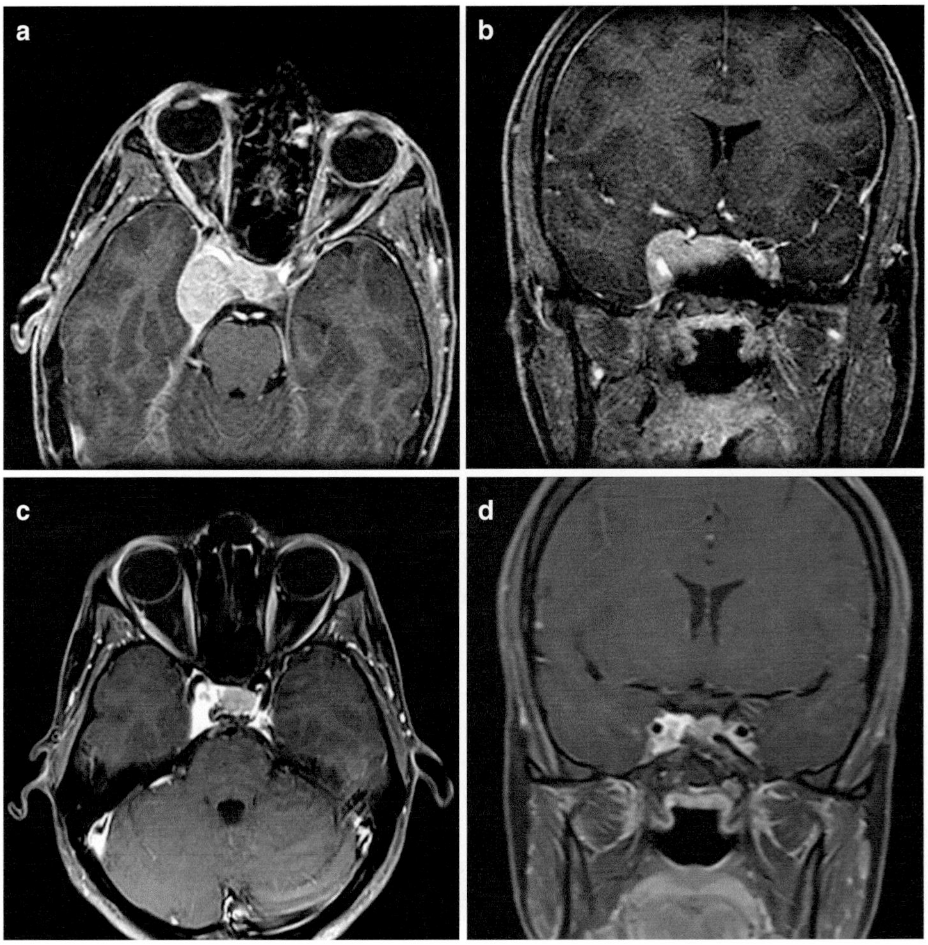

Fig. 29.6 T1-weighted magnetic resonance imaging with gadolinium enhancement of a 31-year-old woman who presented with a 3-month history of diplopia. Axial (**a**) and coronal (**b**) scans showed a right-sided paraclinoid meningioma with extension into the cavernous sinus, sella, and orbital apex. She had mild proptosis of the right eye. There was biochemical evidence of pituitary dysfunction, as well as disconjugate gaze. Given the significant sellar component and involvement of the medial orbital apex, an endoscopic endonasal approach was used to debulk the tumor. Postoperative axial (**c**) and coronal (**d**) scans show decrease in tumor volume, including the region of the orbital apex, with expected residual tumor in the right cavernous sinus. The patient received stereotactic radiosurgery for cavernous sinus residual. Her diplopia improved after surgery.

moidal artery will be identified adjacent to the lamina papyracea (the medial orbital wall); these are protected and then the lamina papyracea is outfractured. Rarely, this needs to be thinned first using a diamond burr. The degree of opening can be tailored to the pathology and can be maximally carried to the infraorbital foramen. The underlying periorbita should be protected to avoid unwanted herniation of orbital fat. There are two working corridors to access intraconal lesions. For inferomedial intraconal lesions, the surgical corridor is between the inferior and medial recti muscles. For superomedial intraconal lesions, the corridor is between the medial rectus and superior oblique muscles. The periorbita can be opened sharply in the region of interest to avoid excessive fat herniation. For reconstruction, either a pedicled and vascularized nasoseptal flap or a free mucosal graft can be used, followed by careful nasal packing to support the reconstruction without putting pressure on the globe. Nasoseptal flaps are sufficient to cover both the inferior and the medial limits of the orbit [28].

29.5 Surgical Pearls

- Electrophysiological neuromonitoring can be a helpful surgical adjunct to identify cranial nerves or branches in all approaches.
- In general, we reconstruct the orbit when there is no proptosis preoperatively to avoid enophthalmos. We typically use a biocompatible porous polyethylene implant or split thickness bone graft to complete the reconstruction. In cases of sphenoorbital meningiomas or other pathologies resulting in proptosis, such as fibrous dysplasias, reconstruction of the orbit is not routinely necessary because these patients do not get enophthalmos.
- Transorbital approaches

  The lateral orbital approaches provide excellent access to superficial and deep lesions of the lateral orbit, but injuries to the optic nerve and central retinal artery are possible. Although the ophthalmic artery may have significant anastomoses, injury or sacrifice of the central retinal artery invariably results in monocular vision loss. Because the ophthalmic artery courses lateral to the nerve and the central retinal artery typically branches from the ophthalmic artery just posterior to the globe, this injury can be minimized by avoiding blind coagulation of the lateral orbital space in lateral orbital approaches. Optimal reconstruction can be aided by drilling of pilot holes that allow replacement of the osteotomy in its native position using titanium plates and screws.
- Transcranial Approaches

  The main risks include the need for extradural brain retraction, which can be minimized by more extensive bony skull base removal, and cosmetic defects associated with soft tissue dissection, which can be minimized with careful reconstructive techniques or more focal approaches as outlined above. If optic nerve decompression is desired, the falciform ligament should be opened on the medial side of the ophthalmic artery after the anterior clinoidectomy, because this artery is lateral to the optic nerve in the optic canal.
- Endonasal Approaches

  This approach requires expanded dissection of the paranasal sinuses to reach the orbit. Thus, successful completion that minimizes morbidity benefits from collaboration between neurosurgeons and otolaryngologists. The steep angle required to reach the orbit necessitates the use of the endoscope (0, 30, and 45°) for visualization. Orbital fat herniation can be a significant challenge in this approach. Limiting the periorbital opening to the minimum extent needed, reduction with gentle bipolar coagulation, and dissection with cottonoid patties can help with the management of the orbital fat.

29.6 Discussion

The complex relationship between the cranium and orbit, the confluence of critical neurovascular structures that traverse these compartments, and the unforgiving nature of injuries in the orbit make surgery in this region challenging. The expertise of neurosurgeons and ophthalmologists, while overlapping in certain respects, are mainly complementary in treating pathologies that affect the orbit and for endonasal cases.

The selection of the optimal approach is governed by factors that include, but are not limited to, the location of the pathology, the intricacies of patient anatomy, the nature of the underlying disease, and the breadth of expertise and comfort of the surgeon. Given the complexity of these lesions, a multidisciplinary approach to surgical decision-making can be beneficial. Although this may not be necessary for all cases, development of a team-based framework can help when more challenging cases arise [29].

Transcranial frontotemporal approaches offer unparalleled panoramic visualization and remain the preferred surgical corridor for lesions of the orbital apex, particularly those superior to the optic nerve, which are generally less amenable to transorbital approaches [30]. Transcranial approaches also grant access to the anterior and middle cranial fossa and are the primary viable routes to approach invasive intracranial lesions that encroach or invade the orbit, such as sphenoorbital meningiomas [31, 32]. The modified lateral orbital approach can also be used as an alternative for select cases [15]. Additional bony removal of the skull beyond the standard frontotemporal flap decreases the need for extradural retraction. These approaches may entail cosmetic concerns such a temporal region defect; however, more focused incisions directed at the pathology and thoughtful reconstructive techniques with cranioplasty implants can mitigate these [20].

Lateral transorbital approaches provide excellent access to lateral extradural lesions as well as intradural intraconal lesions that are primarily lateral to the optic nerve. Lateral intraconal lesions that have inferior or superior extensions may require additional orbital or zygomatic osteotomies. Protection of the orbital contents during bony removal is paramount, and the retraction necessary to do this can risk inadvertent pressure- and retraction-related injuries [13, 14]. The modified lateral orbital approach allows for extensive access while minimizing the risk of retraction injuries by preserving a thin cortical rim of the lateral orbital wall to safeguard ocular structures [15]. Lesions positioned more anterior within the orbital compartment can be effectively managed using transconjunctival approaches that do not require bony removal, cutaneous incisions (with cosmetic scarring), or breach of the dura [33]; however, the surgical corridor is restricted, making these approaches better suited for small to moderately sized anterior lesions rather than those superior to the optic nerve or adjacent to the orbital apex.

The endonasal transethmoidal approach offers the most direct access to pathologies of the inferomedial orbit, with unmatched visualization of lesions in the medial orbital apex. The approach is unsuitable for lesions that are strictly lateral to the optic nerve, but lesions that are inferolateral may be accessible if the tumor is contained inferior to a plane between the long axis of the optic nerve and the contralateral naris. Optic canal decompression can also be achieved via an endonasal route. Advantages include the avoidance of brain manipulation or retraction, generally low rates of morbidity, and no external scarring [29]. Although access is direct, maneuverability is more restricted than with transcranial approaches. High-flow vascular lesions that may require direct vascular control may warrant alternative approaches (see Fig. 29.5).

29.7 Conclusion

The optimal surgical management of orbital lesions necessitates a judicious selection of the surgical corridor that accounts for lesion characteristics, anatomical subtleties, and surgical nuances. The range of transcranial, transorbital, and endonasal approaches, facilitated by multidisciplinary collaboration, allows versatility in patient management and enriches the armamentarium of orbital surgery that altogether provides panoramic access to the orbit.

References

1. Bonavolontà G, et al. An analysis of 2,480 space-occupying lesions of the orbit from 1976 to 2011. Ophthalmic Plast Reconstr Surg. 2013;29:79–86.
2. Martins C, et al. Microsurgical anatomy of the orbit: the rule of seven. Anat Res Int. 2011;2011:1–14.
3. Krönlein RU. Zur Pathologie und operativen Behandlung der Dermoidzysten der Orbita. Beitr Klin Chir. 1889;4:149–63.
4. McNab AA, Wright JE. Cavernous haemangiomas of the orbit. Aust N Z J Ophthalmol. 1989;17:337–45.
5. Carta F, Siccardi D, Cossu M, Viola C, Maiello M. Removal of tumours of the orbital apex via a postero-lateral orbitotomy. J Neurosurg Sci. 1998;42:185–8.
6. Altay T, Patel BCK, Couldwell WT. Lateral orbital wall approach to the cavernous sinus. J Neurosurg. 2012;116:755–63.
7. Nemet A, Martin P. The lateral triangle flap – a new approach for lateral orbitotomy. Orbit. 2007;26:89–95.
8. Moe KS, Jothi S, Stern R, Gassner HG. Lateral retrocanthal orbitotomy. Arch Facial Plast Surg. 2007;9:419–26.
9. Kim JW, Yates BS, Goldberg RA. Total lateral orbitotomy. Orbit. 2009;28:320–7.
10. Di Somma A, et al. Endoscopic transorbital route to the petrous apex: a feasibility anatomic study. Acta Neurochir. 2018;160:707–20.
11. Di Somma A, et al. Endoscopic transorbital superior eyelid approach: anatomical study from a neurosurgical perspective. J Neurosurg. 2018;129:1203–16.
12. Di Somma A, et al. Supraorbital vs endo-orbital routes to the lateral skull base: a quantitative and qualitative anatomic study. Oper Neurosurg. 2018;15:567–76.
13. Chabot JD, Gardner PA, Stefko ST, Zwagerman NT, Fernandez-Miranda JC. Lateral orbitotomy approach for lesions involving the middle fossa: a retrospective review of thirteen patients. Neurosurgery. 2017;80:309–22.
14. Mariniello G, et al. Lateral orbitotomy for removal of sphenoid wing meningiomas invading the orbit. Oper Neurosurg. 2010;66:ons287–ons292.
15. Alzhrani GA, Gozal YM, Sherrod BA, Couldwell WT. A modified lateral orbitotomy approach to the superior orbital fissure: a video case report and review of anatomy. Oper Neurosurg. 2019;16:685–91.
16. Bourquet J. Les hernies graisseuses de l'orbite: notre traitement chirurgical. Bull Acad Med Paris. 1924;92:1270–2.
17. Ilankovan V. Transconjunctival approach to the infraorbital region: a cadaveric and clinical study. Br J Oral Maxillofac Surg. 1991;29:169–73.
18. Mullins JB, Holds JB, Branham GH, Thomas JR. Complications of the transconjunctival approach. Arch Otolaryngol Neck Surg. 1997;123:385.
19. Abou-Al-Shaar H, et al. Cranio-orbital and orbitocranial approaches to orbital and intracranial disease: eye-opening approaches for neurosurgeons. Front Surg. 2020;7:1.
20. Couldwell WT, Chen TC, Weiss MH, Fukushima T, Dougherty W. Cranioplasty with the Medpor porous polyethylene Flexblock implant. J Neurosurg. 1994;81:483–6.
21. Hernesniemi J, et al. Lateral supraorbital approach as an alternative to the classical pterional approach. Acta Neurochir Suppl. 2005;94:17–21.
22. van Furth WR, Agur AMR, Woolridge N, Cusimano MD. The Orbitozygomatic approach. Oper Neurosurg. 2006;58:ONS-103–ONS-107
23. Zabramski JM, Kiriş T, Sankhla SK, Cabiol J, Spetzler RF. Orbitozygomatic craniotomy. J Neurosurg. 1998;89:336–41.

24. Seçkin H, Avci E, Uluç K, Niemann D, Başkaya MK. The work horse of skull base surgery: orbitozygomatic approach. Technique, modifications, and applications. Neurosurg Focus. 2008;25(6):E4.
25. Andaluz N, van Loveren HR, Keller JT, Zuccarello M. The one-piece orbitopterional approach. Skull Base. 2003;13:241–6.
26. Conway JE, Raza SM, Li K, McDermott MW, Quiñones-Hinojosa A. A surgical modification for performing orbitozygomatic osteotomies: technical note. Neurosurg Rev. 2010;33:491–500.
27. Lemole GM, Henn JS, Zabramski JM, Spetzler RF. Modifications to the orbitozygomatic approach. J Neurosurg. 2003;99:924–30.
28. Chhabra N, Healy DY, Freitag SK, Bleier BS. The nasoseptal flap for reconstruction of the medial and inferior orbit. Int Forum Allergy Rhinol. 2014;4:763–6.
29. Melder K, Zwagerman N, Gardner PA, Wang EW. Endoscopic endonasal approach for intra- and extraconal orbital pathologies. J Neurol Surg Part B Skull Base. 2020;81:442–9.
30. Lew H, Rootman DB, Nassiri N, Goh A, Goldberg RA. Transorbital approach without craniotomy to orbital tumors with extradural intracranial extension. Orbit. 2014;33:343–51.
31. Kim RB, Fredrickson VL, Couldwell WT. Visual outcomes in spheno-orbital meningioma: a 10-year experience. World Neurosurg. 2022;158:e726–34.
32. Ringel F, Cedzich C, Schramm J. Microsurgical technique and results of a series of 63 spheno-orbital meningiomas. Oper Neurosurg. 2007;60:214–22.
33. Kıratlı H, Bulur B, Bilgiç S. Transconjunctival approach for retrobulbar intraconal orbital cavernous hemangiomas. Orbital surgeon's perspective. Surg Neurol. 2005;64:71–4.

Olfactory Groove Meningiomas

30

Ramesh Nair, Puneet Sharma, and Hadie Adams

Learning Objectives

- To be able to describe the anatomy of the olfactory groove, both on gross section and on radiological studies (MRI and CT)
- To be able to name the commonly used surgical approaches to olfactory groove meningiomas
- To be able to discuss the relevant pros and cons of each approach, set against non-surgical alternatives
- Have, in particular, an understanding of anatomical considerations for each approach that will guide future operative endeavours

30.1 Introduction

Olfactory groove meningiomas are generally benign tumours of the anterior cranial fossa floor, related more specifically to the cribiform plate and frontosphenoid suture. They can grow to impressive sizes before the onset of symptoms, which can range from persistent headaches, seizures and vague personality changes (such as apathy and dysexecutive syndrome) to visual deficit and anosmia. Foster Kennedy Syndrome is a rare (<2% of all tumours) but oft-quoted presentation consisting of unilateral visual loss (due to optic atrophy from retro-orbital compression of the optic nerve) and contralateral papilloedema (from raised intracranial pressure), with anosmia. Meningiomas derive their blood supply principally from the anterior and posterior ethmoidal arteries (see Fig. 30.15d). The may also acquire small feeders from the anterior cerebral arteries, which often lie in close proximity to the tumour. The bicoronal subfrontal approach has been commonly utilised to tackle these tumours, although increasingly unilateral subfrontal, pterional and endoscopic approaches are being employed to minimise the surgical morbidity (Fig. 30.1).

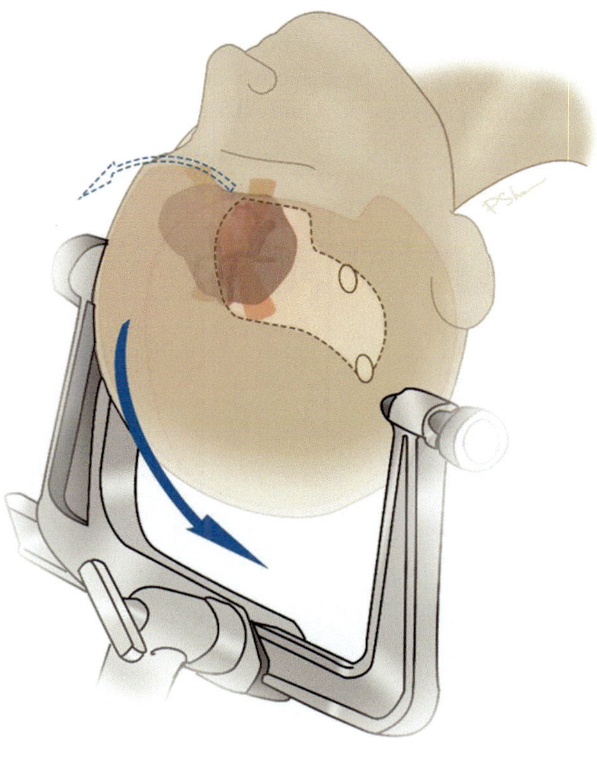

Fig. 30.1 Positioning and Craniotomy—The head is positioned in such a way that frontal lobe retraction is aided by gravity and is achieved by gentle neck extension tilting the vertex toward the floor (solid blue arrow below). Rotation of the head is decided by placement of craniotomy, i.e. lateral rotation is minimal when placement of craniotomy is more frontal (dotted arrow). (Illustration by P. Sharma)

R. Nair (✉) · P. Sharma · H. Adams
Imperial College Healthcare NHS Trust, London, UK
e-mail: puneet.sharma2@nhs.net; hadie.adams@nhs.net

I. N. Kanaan, V. Beneš (eds.), *Neuroanatomy Guidance to Successful Neurosurgical Interventions*,
https://doi.org/10.1007/978-3-031-59838-8_30

30.2 Background

Olfactory groove meningiomas represent approximately 10% of all meningiomas [1], all of which are said to arise from arachnoid cap cells. These cells are notable for their high density of calcified organelles known as psammoma bodies. Psammoma bodies are a hallmark of meningiomas, but also found in papillary thyroid carcinomas and papillary serous cystadenocarcinomas of the ovary [2]. Other histopathologic features of meningiomas are meningothelial 'whorls', cytoplasmic nuclear invaginations, and central chromatin clearing [3]. Meningiomas demonstrate positive immunohistochemical staining to epithelial membrane antigen, as well as progesterone and oestrogen receptors which have led to a postulated hormonal aetiology for their development, and may account for the 2:1 preponderance in females (which rises to 3:1 for pre-menopausal women [4]) (Figs. 30.2 and 30.3).

Studies relating to the association of multiple meningiomas in neurofibromatosis identified the NF2 gene, on 22q12, as being chiefly responsible for the effects of 22q chromosome deletion seen in the disease [5]. The NF2 gene was found to encode the tumour suppressor 'merlin' [3, 6] whose expression appears to be defective in at least half of all meninigiomas [7] (Figs. 30.4 and 30.5).

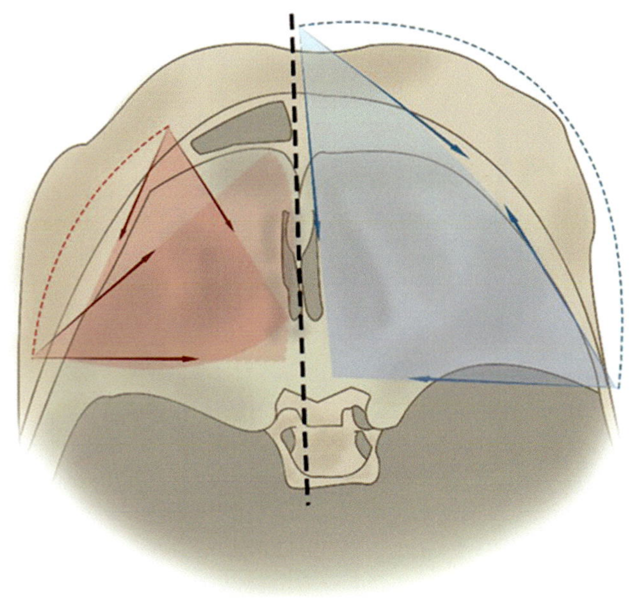

Fig. 30.2 The key factor in deciding the position of the craniotomy is the position of the frontal air sinuses. On the right, a poorly pneumatised frontal sinus permits a frontal craniotomy that extends to the midline and allows access to the entire anterior skull base (blue). On the left, the frontal air sinus is left unopened by a more laterally sited craniotomy. Much of the anterior skull base can still be accessed (red) by looking anteriorly or posteriorly through the operating microscope. In this way, a unilateral approach that avoids the sinus does not tend to restrict access. (Illustration by P. Sharma)

Other putative molecular mechanisms for meningioma development, away from NF2, include epigenetic modifier mutations (such as *SMARCB1*) and increased embryonic signalling activation through the Hedgehog pathway [3]. Perhaps more relevant for olfactory groove meningiomas are

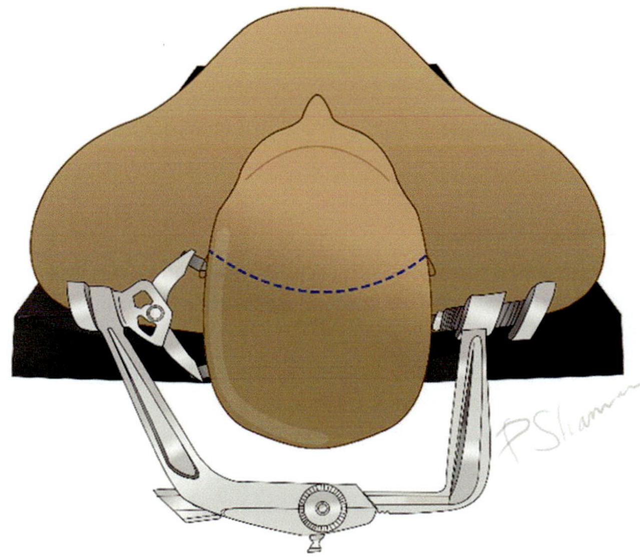

Fig. 30.3 The patient is positioned supine, with the head fixed in pins. As before, the patient may be sat slightly up, with the neck gently extended, to aid gravitational retraction of the frontal lobe. The incision extends from zygomatic arch to zygomatic arch so that the flap can be retracted forward and down sufficiently to access the anterior fossa floor. (Illustration by P. Sharma)

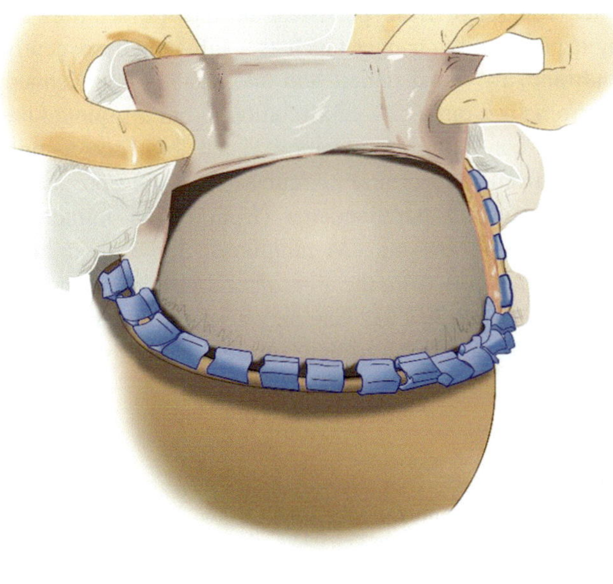

Fig. 30.4 Preservation of the pericranium is an important consideration which will require dissection of the skin flap above this layer. (Illustration by P. Sharma)

mutations in *SMO* ('Smoothened', a receptor closely associated with the PTCH1 tumour suppressor) and *AKT1-MTOR* (intracellular signalling pathway, important in regulation of the cell cycle) [7, 8] (Figs. 30.6 and 30.7).

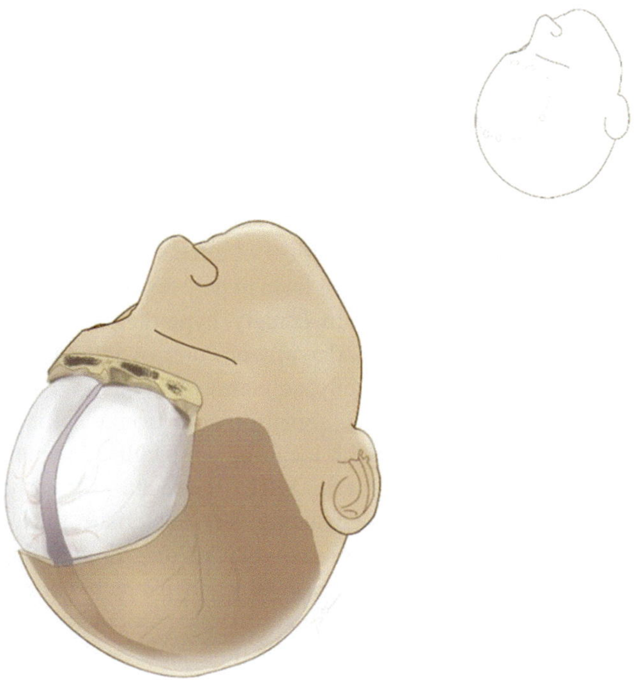

Fig. 30.5 A craniotomy is completed, crossing midline and exposing the superior sagittal sinus anteriorly. The craniotomy should be flush with the floor of the anterior fossa, generally leading to opening of the frontal air sinuses, which must then be cranialised. (Illustration by P. Sharma)

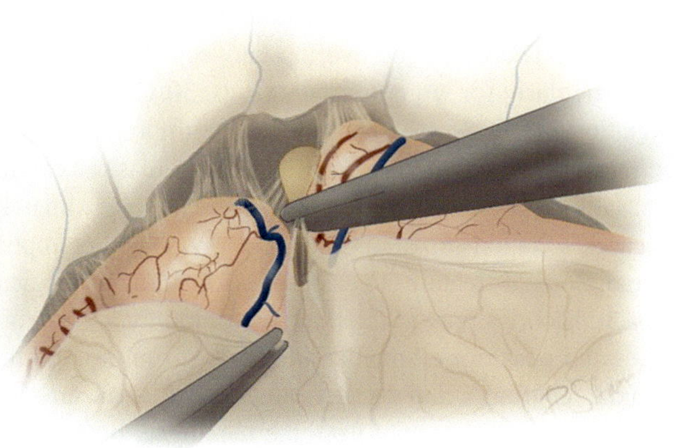

Fig. 30.6 The dura is itself widely opened parallel to the base and the falx freed from the crista galli (shown above). The frontal lobe is freed from its arachnoid attachments and gently elevated, using gravitational or dynamic retraction as much as possible. Brain relaxation is aided by CSF egress, pre-operative steroids and intra-operative mannitol or hypertonic (1.8%) saline. (Illustration by P. Sharma)

30.2.1 Tumour Grading

Meningiomas are typically considered benign entities with limited capacity for spread. The majority of meningiomas are indeed 'grade I' under the WHO 2016 guidelines, which additionally specify several variants based on staining pattern, including psammomatous, meningothelial, transitional and fibrous types. However, a minority demonstrate brain invasion and more aggressive growth patterns. The term 'grade II', under the WHO 2016 classification, specifies at least four mitoses be seen per ten high-power fields and/or direct invasion of meningioma cells into cortical tissue. Grade II tumours may also demonstrate additional features such as micronecrosis, and can exist as specific histopathological 'clear cell' or 'choroidal' types. Grade II tumours demonstrate higher rates of recurrence (~50% at 5 years vs. 0–10% of surgically excised Grade Is over this period) and shorter overall survival rates (~70% 10 year survival in Grade II tumours vs. >76% 10 year survival [9]. Grade III 'anaplastic' tumours represent the most aggressive form of brain-invasive meningiomas with mitotic rates above 20 per high power field (or else showing specific histophenotypes, such as rhabdoid meningiomas). Grade IIIs demonstrate a significantly reduced overall survival (0–23% at 10 years) and more frequent recurrence rate (90% at 5 years) (Figs. 30.7 and 30.8).

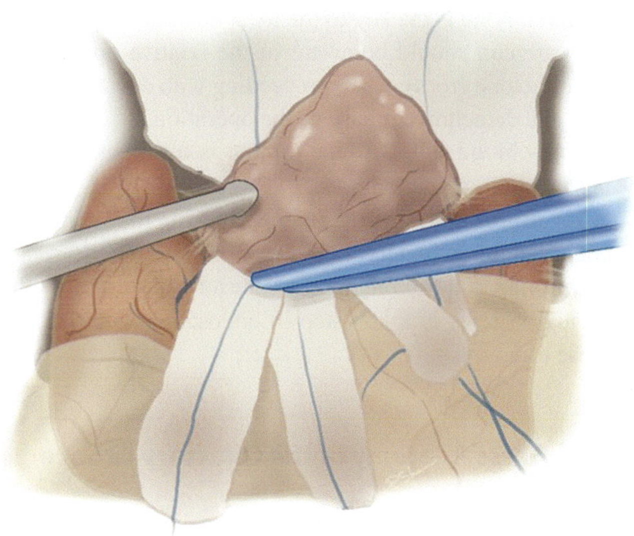

Fig. 30.7 The tumour capsular plane is explored upon brain retraction. Capsular vessels must be carefully inspected before coagulation, to ensure any en-passant vessels are preserved. The capsular plane is marked (and the brain protected) with cottonoids ('patties'). We caution against widespread electrocautery of the tumour capsule, as is often recommended, as a high proportion of vessels running over the capsule do not enter it. Meticulous microsurgical technique with systematic internal debulking (retaining haemostatic control and piecemeal resecting) is preferred and it leads to fewer complications. (Illustration by P. Sharma)

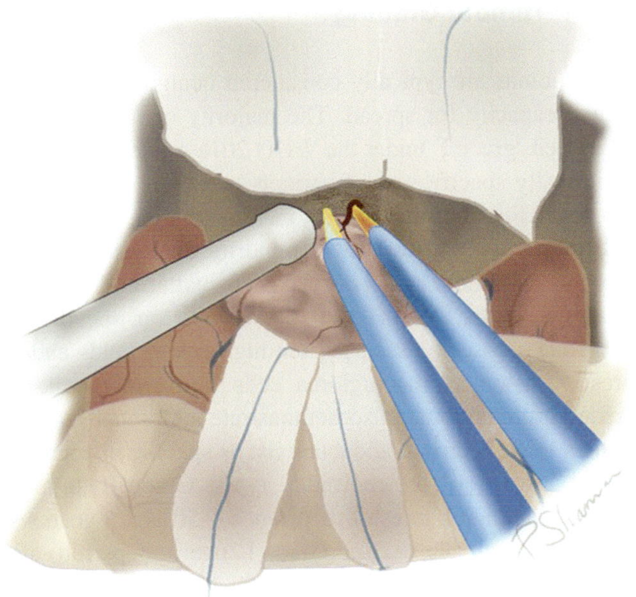

Fig. 30.8 Division of the ethmoidal feeding arteries will interrupt the majority of the tumour's blood supply. Posteriorly, care must be taken to identify and preserve the ACAs and chiasm. Perforators in this region (which supply in particular the chiasm and the hypothalamus) must not be mistaken for small feeding vessels—their entry into tumour must be identified before division. (Illustration by P. Sharma)

30.3 Anatomy Highlights

- Ethmoid bone is the key to the local anatomy.
- The surgeon should be aware of the relative position of the olfactory groove when proceeding with devascularisation and detachment phase of the operation.
- Primary blood supply to the tumour is from anterior and posterior ethmoidal arteries.
- A key aspect of the operation is identifying ipsilateral and contralateral optic nerves.
- Identifying and cauterising the anterior and posterior ethmoidal arteries is a key step in endoscopic approach to these tumours.

30.3.1 Gross Anatomy of the Olfactory Groove

The *olfactory grooves* are parasagittal depressions running in an anteroposterior direction above the cribiform plate of the ethmoid bone; the ethmoid bone being key to the local anatomy of this region. Its organisation can be simplified in coronal section to a cross-shaped structure consisting of a vertically oriented, midline *perpendicular plate* that crosses with the horizontally oriented *cribiform plate*. Above the level of the cribiform plate the perpendicular plate essentially becomes the vertical crista galli. The lateral laminae (or in its Latin diminutive form, lateral *lamellae*) are laterally sited upward projections on the ciribiform plate that define the lateral boundary of the olfactory grooves. The ethmoidal labyrinths hang down from the lateral aspects of the cribiform plate and contain the ethmoidal air cells, the most anterior of which communicate with the frontal sinus (Figs. 30.9 and 30.10).

30.3.2 Radiological Anatomy of Olfactory Groove Meningiomas

On computerised tomographic coronal section, the perpendicular plate, cribiform plate, crista galli, and lateral lamellae are easily seen (Fig. 30.15c). Meningiomas in this region are typically large as they often remain silent in the early stages of growth. They may have a variety of effects on the osseus anatomy of this region, ranging from hyperostosis to erosion. Meningiomas themselves are usually hyperdense on CT, enhancing homogenously and avidly on contrast administration. On MRI they are typically isointense on both T1WI and T2WI, but may demonstrate signal dropout from calcification which will correlate with localised or spiculated hyperdense regions on CT. On MRI, they are similarly homogenously and avidly enhancing. A characteristic, thickened and enhancing dural tail may indicate the origin of the tumour and is usually isointense to brain parenchyma on both T1WI and T1WI. Perilesional oedema, particularly evident as high signal on T2WI and FLAIR sequences, can indicate tumour activity and growth, and may correlate with acute clinical presentation. MRS demonstrates reduced N-acetyl aspartate and creatinine peaks, but increased choline and alanine peaks, in comparison with surrounding brain [3] (Figs. 30.11 and 30.12).

Olfactory groove meningiomas arise from either the cribiform plate, or the frontosphenoid suture. If arising more posteriorly along the planum sphenoidale or tuberculum sellae, they are referred to as planum or tuberculum meningiomas respectively. There may be overlap between these designations particularly for larger tumours (Figs. 30.13 and 30.14).

Relationship to osseus anatomy can be important in preoperative planning. Olfactory groove meningiomas can extend paranasally through the anterior fossa floor in approximately 15% cases. Relationship to vascular anatomy is another important radiological feature to assess prior to any surgical approach. Blood supply to the tumour commonly arises from anterior and posterior ethmoidal arteries that run through the ethmoidal labyrinths, and derive from the ophthalmic arteries. They may also draw supply from the meningeal branches of the ophthalmic artery. However, as these tumours enlarge, they demonstrate close relationships to the anterior cerebral arteries which may become: (a) displaced by tumour, (b) applied to its (posterior) capsule or, rarely, (c) encompassed within tumour itself. Branches from the ACA or ACOM occasionally contribute to tumour supply (Fig. 30.15d).

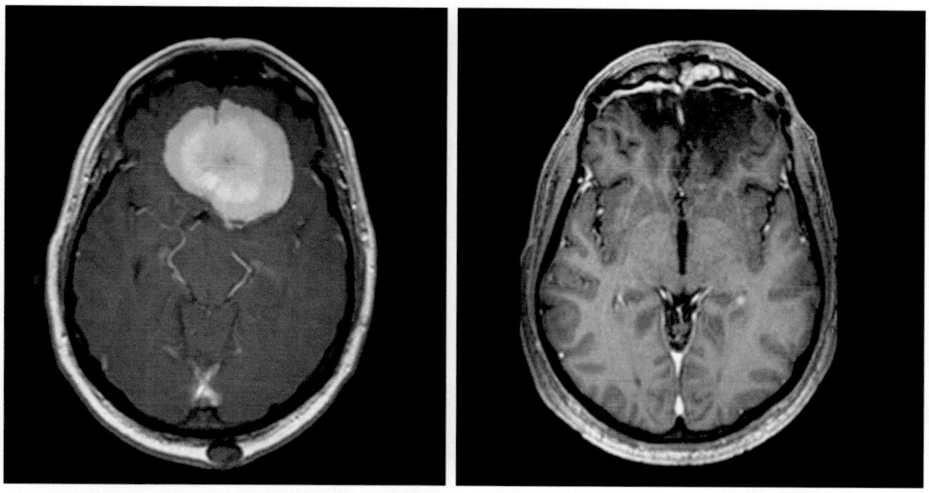

Fig. 30.9 The above (pre- and post-operative views) represents a somewhat historical case in our unit, treated at that time by bifrontal craniotomy. A similar sized tumour (below, pre- and post-op imaging) has been excised through a unilateral modified pterional approach, which has become the default technique. Even large, bilaterally extending tumours can be effectively excised through a unilateral approach. A similar sized tumour (pre- and post-contrast images) has been excised through a unilateral modified pterional approach, which has become the default technique. Even large, bilaterally extending tumours can be effectively excised through a unilateral approach

Fig. 30.10 A similar sized tumour (pre- and post contrast images) has been excised through a unilateral modified pterional approach, which has become the default technique. Even large, bilaterally extending tumours can be effectively excised through a unilateral approach

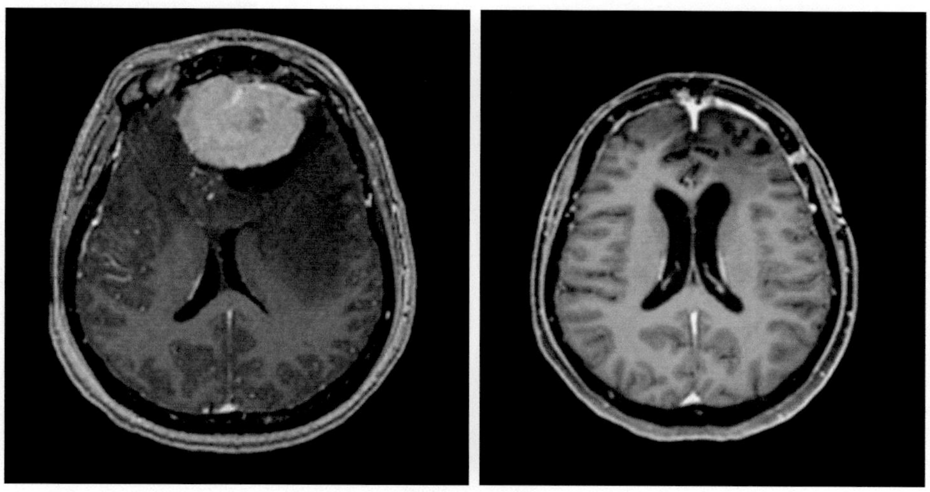

Fig. 30.11 One should try to foresee anatomical constraints to any approach. Here, despite the presence of a coincidental sellar tumour the choice was bifrontal approach. The concern was that tumour extension medial to the supraclinoid carotid may be difficult to excise from a unilateral exposure

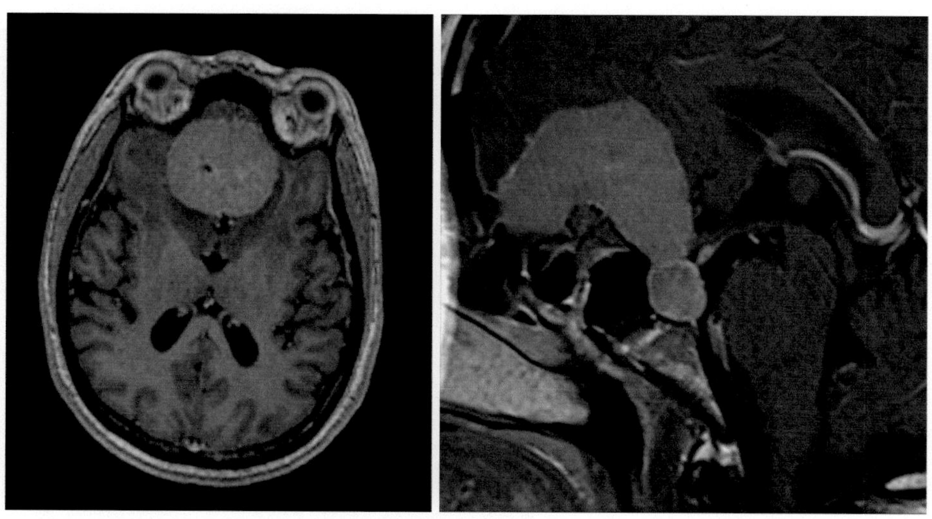

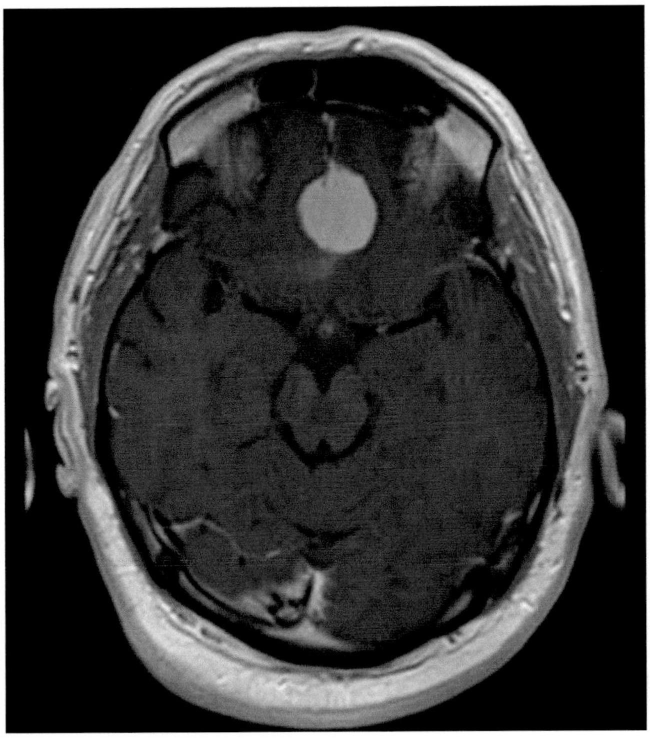

Fig. 30.12 Shows a small meningioma without significant lateral extension, which was amenable to endoscopic excision

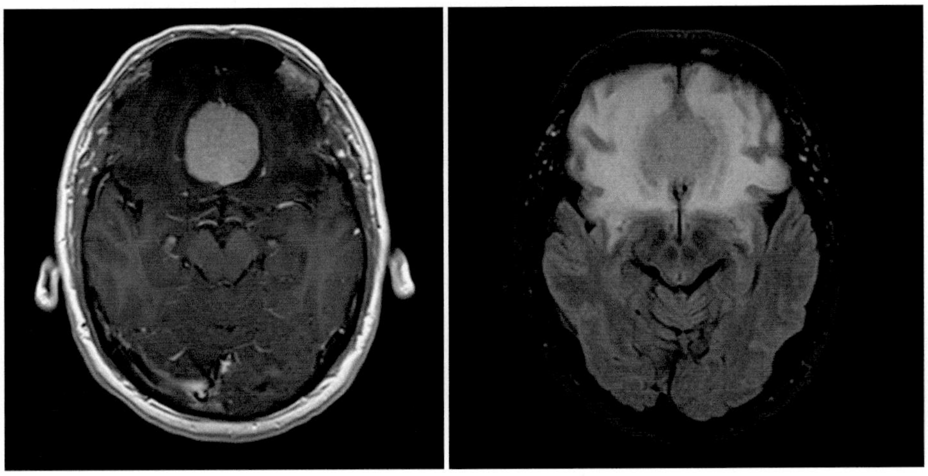

Fig. 30.13 Although the tumour is itself of modest size, the extent of oedema is large. This necessitated orbitotomy for safe tumour excision. In such cases, internal tumour debulking does not create sufficient space to mobilise the tumour. Early devascularisation is even more critical in such cases to maintain haemostatic control through the stages of tumour resection

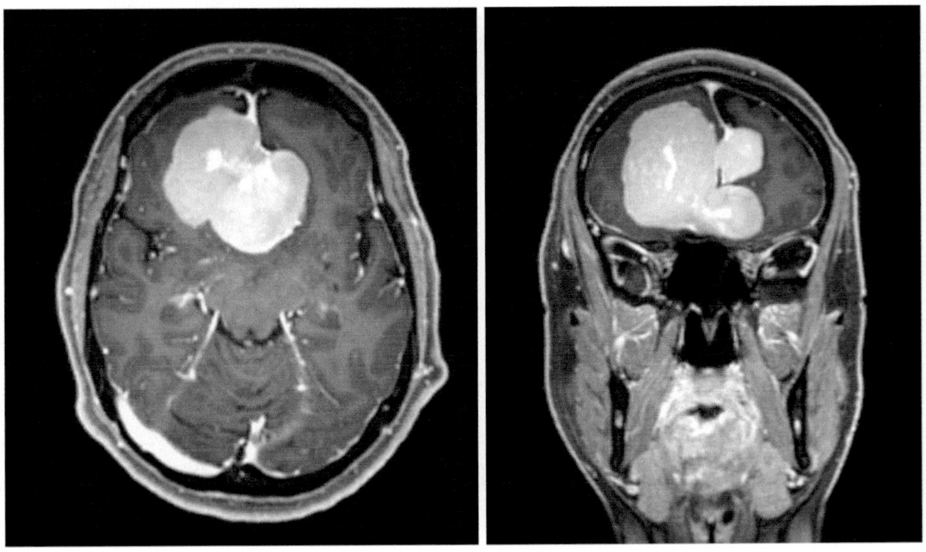

Fig. 30.14 Here a large frontal meningioma mimics an olfactory groove tumour but is in fact a parafalcine meningioma. This anatomical distinction is important as it determines the origin of the tumour's principal blood supply

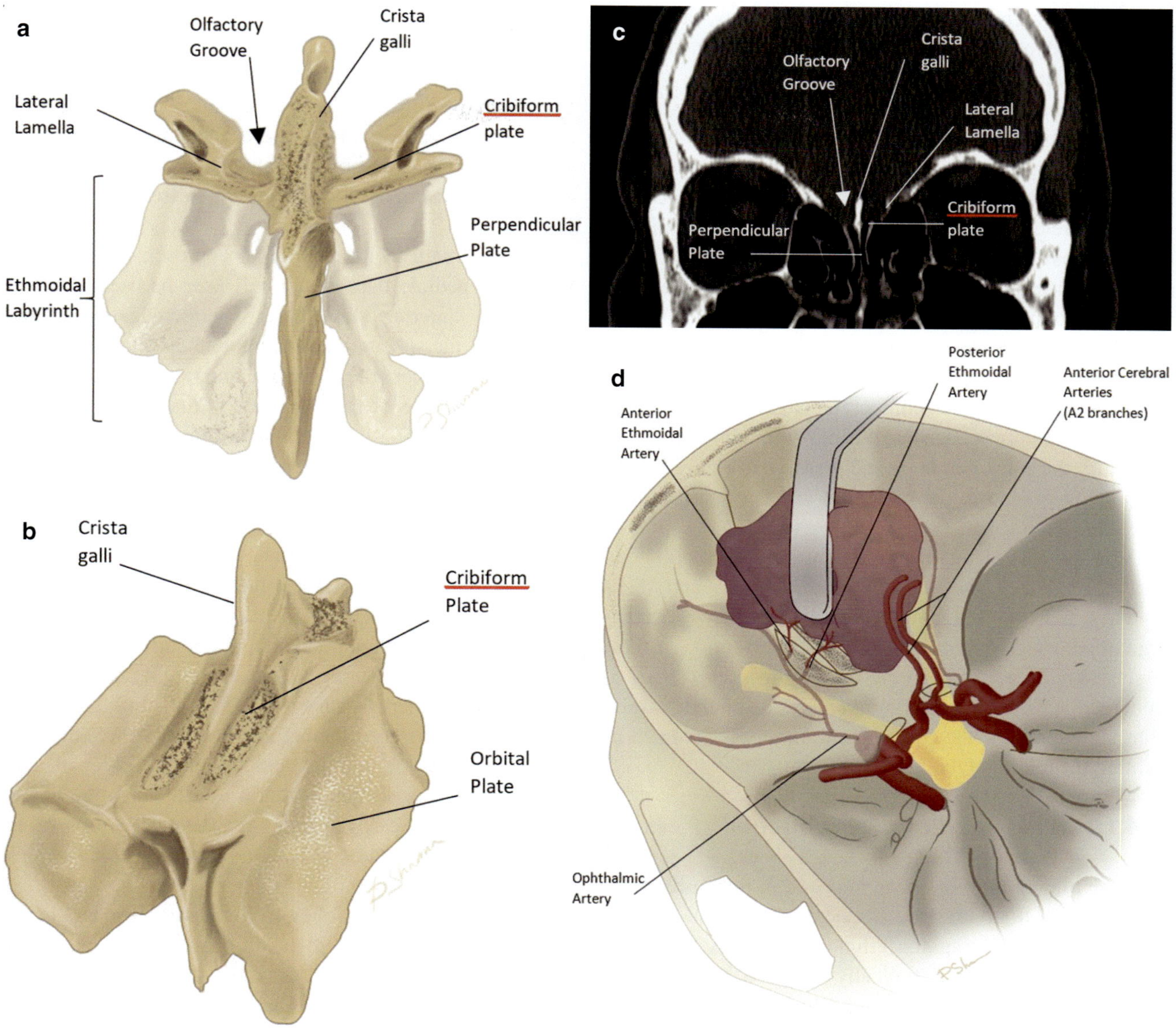

Fig. 30.15 (**a**) Anterior view of the ethmoid bone. The ethmoidal labyrinth has been faded to emphasise the cruciform appearance of the intersecting perpendicular and cribiform plates, completed by the cristal galli. The olfactory grooves are recesses which lie between the crista galli and the lateral lamellae. (**b**) Postero-superior view of the ethmoid (**c**) Coronal CT scan demonstrating key anatomy of the ethmoid. (**d**) demonstrates the typical blood supply to olfactory groove meningiomas—the posterior and anterior ethmoidal arteries derive from the ophthalmic artery. The ACAs pass in close proximity to the posterior tumour capsule and may send small feeding arteries to the tumour

30.4 Surgical Treatment of Olfactory Groove Meningiomas

For smaller tumours, or in patients with poor functional status or elevated surgical risk, a close clinical and/or radiological surveillance policy is adopted. Once a decision has been made for surgical intervention, however, several principles should be considered to guide the surgical approach.

30.4.1 Approaches

30.4.1.1 Unilateral: Frontal to Lateral Frontal to Modified Pterional

Incision

The more frontally the craniotomy is placed, the scalp incision extends to contralateral side. The more pterional the craniotomy is, scalp incision usually reaches and just crosses the midline

Following a conventional frontal craniotomy, the orbital rim may need to be removed to improve the subfrontal access if there is a large degree of brain oedema.

The dural incision is made parallel to the base.

Gentle subfrontal exposure of the tumour base is then performed. In a more laterally sited modified pterional craniotomy, there will be an option to access the optico-carotid cisterns to release CSF. In most olfactory groove meningiomas, it is possible to achieve substantial devascularisation by detaching most of the base as the primary step. However, in some large tumours with significant brain swelling, it will be necessary to perform some degree of internal debulking before devascularisation can be achieved.

Devascularisation, detachment, debulking and dissection are used in varying sequences. It is of paramount importance to identify the arachnoid layer and preserve it. Hence bipolar coagulation is used to the minimum and is often used in a targeted fashion.

Once sufficient internal debulking and dissection has been carried out, the ACAs, their proximal and distal branches, and the ACOM may come in to view. In many large olfactory groove meningiomas, the adventitia of these vessels may be adherent to the tumour. In such cases it is prudent to leave remnants of tumour capsule on the vessels.

In large tumours, the contralateral tumour is accessed by dividing the midline falx

30.4.1.2 Bilateral: Bicoronal Subfrontal ('Bifrontal')

Once the tumour bulk has been removed in this fashion, the anterior fossa floor must be inspected. Hyperostotic bone and tumour penetrating the ethmoidal sinuses should be removed. Dural repair is typically achieved by laying the pericranial flap over the floor, and suturing it to dural edges when possible. The repair can be reinforced by fascia lata graft and tissue or dural sealant. The bone is plated, a wound drain is sited and the bicoronal skin flap closed.

30.4.1.3 Extended Endoscopic Endonasal Approach

The use of a well-established endonasal corridor for removing olfactory groove meningioma is under debate. In the senior author's view, there is a subset of patients with olfactory groove meningiomas, who would benefit from this approach, based in part on anatomical considerations that include:

1. No olfactory function
2. Anterior extent of the tumour should not be beyond endoscope reach. This is to be ascertained pre-operatively by studying the CT sagittal anatomy. The anterior edge of the tumour should be posterior to a straight line drawn upwards from the anterior surface of maxilla in midline.
3. Small- to medium-size tumour
4. Coronal extent of the base not beyond the reach of endoscope: small- to medium-size olfactory groove meningiomas may still reach the optic canals and optic nerves to become a factor as in the case of planum meningiomas.
5. No significant cerebral oedema
6. No encasement of anterior cerebral vessels.

The first step in the endonasal approach is to raise an endonasal pedicle flap, which will later be used to reconstruct the anterior skull base. Neuronavigation-assisted opening of the sphenoid and ethmoidal air sinuses are then performed to access the extradural anterior fossa floor.

One advantage of this approach is that the hyperostotic base is drilled off at the beginning, immediately cutting off the anterior and posterior ethmoidal arterial supply.

The dura is opened around the tumour base. Tumour resection is carried out by systematic debulking, dissection, suction and retrieval of fragments.

No cerebral retraction is involved in this approach.

Navigation can help to identify position against bony landmarks, although intracranial brain-shift will naturally minimise its use to identify resection margins. Visualisation of the olfactory nerves, chiasm and ACAs is limited and late in the endoscopic corridor, but this approach offers the most direct route to the tumour and no visible external scar. However, wide resection of dural attachment may not be possible in many cases negating a Grade 1 Simpson resection which is often cited as an argument against endoscopic approach to these tumours.

Tumour resection is followed by haemostasis and multilayered reconstruction of the skull base, which can be reinforced with fascia lata graft, synthetic dural replacement, liquid dural sealant and/or other fat graft. The naso-septal flap is then rotated and applied to the reconstruction, which can be reinforced again by multilayer closure technique. Nasal splints are then left in situ and can be removed in 10–14 days (or sooner, depending on the surgeon's preference). CSF leak can complicate almost 6% of all anterior skull base procedure [10]. In many cases, use of a post-operative lumbar drain helps to prevent CSF leaks.

30.4.2 Other Approaches

Supraorbital approache has been described in the literature. This offers a minimal cosmetic defect, and potentially lesser rates of anosmia compared with the endonasal approach, and potentially superior resection rates. Detractors of the supraorbital approach cite that the rate of anosmia remains high (up to 50% in some series), and does not permit direct access to the tumour without some degree of brain retraction. The working corridor is the smallest of the three craniotomy options, increasing risk of vascular or nerve injury, and the difficulty of achieving gross total resection. Furthermore, when appropriately covered by hair, the extent of cosmetic defect in a pterional approach may be similar or even superior to that of a supraorbital approach, and so this advantage may not be so uniformly applicable. Finally, the supraorbital approach often requires use of the endoscope for larger tumours, and so technical facility with an endoscope is ultimately still required.

30.5 Surgical Pearls

- The approach is tailored in such a way that entire tumour base is accessible, irrespective of the total size of the tumour. In the senior author's experience, majority of olfactory grove meningiomas are operated through unilateral approach.
- Pneumatisation of frontal air sinus is a key factor deciding the placement of craniotomy. This means that in the case of a poorly pneumatised or non-existent frontal air sinus, the approach will be a frontal craniotomy. The more laterally the sinus extends, more posteriorly will be the placement of craniotomy making it resemble a pterional craniotomy. If breaching of frontal air sinus is likely to occur, harvest the pericranium.
- Side of approach is based on the asymmetrical extension of tumour and the proximity of tumour to the surface. In a symmetrical tumour, non-dominant side is chosen.
- In tumours with significant brain swelling, additional manoeuvres such as removal of the orbital rim or judicious use of a lumbar drain may be employed. However a lumbar drain may be avoided when the craniotomy is lateral and the basal cisterns can instead be accessed.
- **D**evascularisation and **D**etachment, **D**ebulking and **D**issection are key principles in surgery of meningiomas in general. The sequence of 4 'D's employed vary with each case of olfactory groove meningioma.
- When the olfaction is present, every attempt is made to preserve the olfactory system at least on one side. This is usually possible on the contralateral side. If olfaction is preserved pre-operatively, an open craniotomy is generally advocated as the endoscopic technique carries a higher risk for losing olfactory function.
- Preservation of arachnoid layer is a key step and hence indiscriminate use of bipolar coagulation of the surface of tumour is avoided.
- Most arterial vessels on the capsule are en-passant vessels and ACA trunks and branches make grooves on the tumour surface.
- Olfactory groove tumours tend to *displace* the chiasm inferiorly and posteriorly; vision is typically involved late once a tumour has grown very large, in contradistinction to tuberculum meningiomas (which compress the chiasm from below, earlier in their course).
- A bicoronal subfrontal approach will expose the sagittal sinus, which does not necessarily need to be taken, but may be at risk of injury through its exposure. The bicoronal approach, with its added exposure, can be useful if there is extensive disease traversing the ethmoids.

30.6 Discussion

An in-depth review of these tumours or detailed discussion regarding adjunct treatments is not intended in the scope of this chapter. An important concept when dealing especially with meningiomas is to think back in time and imagine the origin and subsequent growth of these tumours reaching to their current status. This gives the surgeon a unique understanding of the epicentre of the tumour and pattern of growth. This way of thinking helps to understand the way the local cisterns are displaced. Knowing the cisternal anatomy is crucial in most meningioma operations.

Olfactory groove meningiomas represent a distinct group of tumours with particular challenges, both technical and anatomical. No one approach should be considered superior to all others: the particular requirements of each case must be considered when selecting an approach. The following vignettes highlight the importance of maintaining critical judgement and flexibility of approach.

The above examples of Olfactory groove meningiomas highlight the importance of understanding relevant anatomical features of each case to have a successful surgical outcome. The goal is always radical resection without incurring complications. However, radical resection may not be possible nor necessary in many cases. Experienced surgeons make appropriate intra-operative decisions judiciously in such cases.

30.7 Conclusions

Olfactory groove meningiomas, though identical with regards to their point of origin, vary quite considerably when it comes to their clinical presentation, involvement of various anatomical regions, extent of aggressiveness, vascularity, management complexities, ease of surgical removal and post-surgical challenges.

References

1. Mirimanoff RO, Dosoretz DE, Linggood RM, Ojemann RG, Martuza RL. Meningioma: analysis of recurrence and progression following neurosurgical resection. J Neurosurg. 1985;62(1):18–24. https://doi.org/10.3171/jns.1985.62.1.0018. PMID: 3964853.

2. Das DK. Psammoma body: a product of dystrophic calcification or of a biologically active process that aims at limiting the growth and spread of tumor? Diagn Cytopathol. 2009;37(7):534–41. https://doi.org/10.1002/dc.21081. PMID: 19373908.
3. Buerki RA, Horbinski CM, Kruser T, Horowitz PM, James CD, Lukas RV. An overview of meningiomas. Future Oncol. 2018;14(21):2161–77. https://doi.org/10.2217/fon-2018-0006. Epub 2018 Aug 7. PMID: 30084265; PMCID: PMC6123887.
4. Hatch EE, Linet MS, Zhang J, Fine HA, Shapiro WR, Selker RG, Black PM, Inskip PD. Reproductive and hormonal factors and risk of brain tumors in adult females. Int J Cancer. 2005;114(5):797–805. https://doi.org/10.1002/ijc.20776. PMID: 15609304.
5. Rouleau GA, Merel P, Lutchman M, Sanson M, Zucman J, Marineau C, Hoang-Xuan K, Demczuk S, Desmaze C, Plougastel B, et al. Alteration in a new gene encoding a putative membrane-organizing protein causes neuro-fibromatosis type 2. Nature. 1993;363(6429):515–21. https://doi.org/10.1038/363515a0. PMID: 8379998.
6. Trofatter JA, MacCollin MM, Rutter JL, Murrell JR, Duyao MP, Parry DM, Eldridge R, Kley N, Menon AG, Pulaski K, et al. A novel moesin-, ezrin-, radixin-like gene is a candidate for the neurofibromatosis 2 tumor suppressor. Cell. 1993;72(5):791–800. https://doi.org/10.1016/0092-8674(93)90406-g. Erratum in: Cell. 1993;75(4):826. PMID: 8453669.
7. Brastianos PK, Horowitz PM, Santagata S, Jones RT, McKenna A, Getz G, Ligon KL, Palescandolo E, Van Hummelen P, Ducar MD, Raza A, Sunkavalli A, Macconaill LE, Stemmer-Rachamimov AO, Louis DN, Hahn WC, Dunn IF, Beroukhim R. Genomic sequencing of meningiomas identifies oncogenic SMO and AKT1 mutations. Nat Genet. 2013;45(3):285–9. https://doi.org/10.1038/ng.2526. Epub 2013 Jan 20. PMID: 23334667; PMCID: PMC3739288.
8. Strickland MR, Gill CM, Nayyar N, D'Andrea MR, Thiede C, Juratli TA, Schackert G, Borger DR, Santagata S, Frosch MP, Cahill DP, Brastianos PK, Barker FG 2nd. Targeted sequencing of SMO and AKT1 in anterior skull base meningiomas. J Neurosurg. 2017;127(2):438–44. https://doi.org/10.3171/2016.8.JNS161076. Epub 2016 Nov 25. PMID: 27885953.
9. Holleczek B, Zampella D, Urbschat S, Sahm F, von Deimling A, Oertel J, Ketter R. Incidence, mortality and outcome of meningiomas: A population-based study from Germany. Cancer Epidemiol. 2019;62:101562. https://doi.org/10.1016/j.canep.2019.07.001. Epub 2019 Jul 17. PMID: 31325769.
10. Coucke B, Van Gerven L, De Vleeschouwer S, Van Calenbergh F, van Loon J, Theys T. The incidence of postoperative cerebrospinal fluid leakage after elective cranial surgery: a systematic review. Neurosurg Rev. 2022;45(3):1827–45. https://doi.org/10.1007/s10143-021-01641-y. Epub 2021 Sep 9. Erratum in: Neurosurg Rev. 2022; PMID: 34499261.

Tailored Microsurgical Approach to Tuberculum Sella Meningioma

31

Imad N. Kanaan

Learning Objectives
- Review the distinct surgical approaches to the central skull base region.
- Tailor surgical approaches for optimal tumor exposure with least on-route damage.
- Highlight important anatomical landmarks to guide a safe TSM resection
- Maintain the integrity of regional neurovascular structures and provide excellent clinical outcomes.

31.1 Introduction

Tuberculum sella meningioma (TSM) is a distinctive neurosurgical entity, accounting for 10% of all intracranial meningiomas and 25% of anterior skull base meningiomas [1, 2]. Its origins can be traced back to 1899, when Steward first reported these tumors at autopsy, and the first complete resection performed by Cushing in 1916 [3–5]. However, the initial classification by Cushing and Eisenhardt in 1938 was imprecise, listing them incorrectly under the umbrella diagnosis of suprasellar tumors [6], and it was not until later that the true anatomical origin of TSM was recognized.

TSMs originate from the dura cap cells at the tuberculum sella, chiasmatic sulcus, limbus sphenoidale, and the diaphragm sella. They can extend in various directions, including anteriorly to implicate the planum sphenoidale, laterally to the optic nerves and internal carotid artery, posterosuperiorly to the optic chiasm, anterior cerebral artery complex, and hypothalamus, and inferiorly into the pituitary fossa. As a result, early and profound asymmetrical visual impairment is a hallmark of TSM clinical presentation, with frequent growth into both optic foramina [2, 7, 8].

In recent years, tailored surgical approaches have been developed to address the primary neurosurgical challenge of achieving complete resection of the tumor while preserving the integrity of the adjacent vital neurovascular structures with minimal brain retraction or on-route damage.

31.2 History and Background

Frontotemporal craniotomy was first described by Heuer in 1918 [9], and later modified by Dandy to clip an anterior communicating artery aneurysm in 1941 [10]. Hamby coined the term pterional approach in 1964 and Yaşargil refined and popularized it using a microscope in 1969 [11]. Perneczky pioneered a minimally invasive variant of the frontolateral approach in 1990 using the trans-eyebrow supraorbital keyhole approach assisted by endoscopy [12], which was limited to small anterior skull base lesions without major edema or vascular encasement. While some surgeons appreciated this approach for its less invasiveness and improved cosmetic outcomes, others criticized its confined exposure and lack of surgical maneuverability, leading to a higher percentage of incomplete resection.

The fronto-orbito-zygomatic (FOZ) approach, evolved from Krause's pioneering, yet ultimately unsuccessful, work in 1907. McArthur and Frazier were the first to remove the supraorbital ridge as part of the frontal craniotomy in 1912 and 1913 respectively [13, 14]. Jane et al. revived this approach to include the anterior orbital roof osteotomy in a single flap to approach orbital tumors and vascular lesions [15]. Pellerin reported the orbito-zygomatic-malar approach in 1969 [16] and Hakuba [17] described a newer complex version in the early 1980s, labelled as the orbitozygomatic

I. N. Kanaan (✉)
Neurosciences Center, Section of Neurosurgery, King Faisal Specialist Hospital & Research Center Institution, Alfaisal University, College of Medicine, Riyadh, Kingdom of Saudi Arabia

I. N. Kanaan, V. Beneš (eds.), *Neuroanatomy Guidance to Successful Neurosurgical Interventions*, https://doi.org/10.1007/978-3-031-59838-8_31

subtemporal approach to central skull base lesions. Eventually, Almefty and others modified and popularized this surgical exposure coined as the fronto-orbito-zygomatic approach (FOZ) [18–22]. Recently, several authors proposed a tailored version of FOZ to include the orbital rim, flattening of the orbital roof and avoiding the double sectioning of the zygomatic arch.

The bifrontal approach was derived from Cushing and Dandy's work in 1936 to resect a large frontobasal meningioma with involvement of the ethmoid sinus, but necessitated resection of the frontal lobe. Subsequently, this approach underwent several modifications and classification. Unterberger used it in 1958 to repair traumatic injuries of the anterior skull base, and Tessier adopted this approach to correct craniofacial anomalies [23, 24]. However, Derome was the first to coin this approach "the trans-basal approach" and proposed it for the surgical treatment of tumors involving the anterior midline skull base to permit craniofacial resection through a craniotomy-only approach [25]. Eventually, the interhemispheric access complemented the bi-frontal trans-basal approach (BBIA) to enhance exposure for the resection of giant midline anterior frontobasal tumors. This approach provides a true midline perspective and orientation, but it has fallen out of favor in recent years due to its associated shortcomings. The author concurred with other publications to limit its use for selected cases of giant anterior skull base tumors or en bloc resection of frontobasal malignancies [26–35].

The evolution of transsphenoidal surgery for treatment of pituitary tumors during the past decades and the resurgence of endoscopy in 1990 paved the way for a novel ventral skull base extended approach (EETSS) using the proposed sagittal (median) plane [36]. Weiss et al. reported the first description of the extended endoscopic transnasal removal of TSM meningioma in 1987 [37]. Recently, a wealth of publications advocate EETSS for small or medium size PSMs or TSMs, reporting its advantages whilst acknowledging its shortcomings and excluding its use for cases with intimate vascular structures encasement [38–46]. Opponents criticized these reports' failure to achieve an agreeable scientific consensus as a result of their selection bias, unbalanced comparison with limited cohorts, and short follow-ups [44, 47, 48].

31.3 Anatomical Highlights: (Fig. 31.1)

- Anterior clinoid process is an important anatomical landmark to indicate the location of optic nerve and the adjacent internal carotid artery (ICA) and origin of the ophthalmic A.
- Crista galli/falx insertion and the tuberculum sella are important midline anatomical landmarks of the anterior fossa.
- Blood supply to TSM arise from the posterior ethmoidal arteries, the superior meningio-hypophyseal arteries, and occasional twigs from ACA.
- The falciform ligament is an extension of the clinoid/tuberculum dura, forming a crescent-shaped veil over the ON as it enters the optic canal. Its length ranges from 2 to 3 mm and its width ranges from 7 to 9 mm.
- Anterior cerebral arteries at the anterior half of the circle of Willis give rise to important perforators including the hypothalamic artery and the recurrent artery of Heubner.
- Blood supply to optic nerves and optic chiasma includes superior hypophyseal A, ophthalmic A, and ACA.
- Liliequist membrane remains mostly free in TSM.
- Sphenoid rostrum (crest) is an important midline anatomical landmark in TSS or EETSS.
- Medial optic-carotid recess MOCR, optic strut, and both the carotid tubercles (prominences-CP) are important landmarks during EETSS approach to TSM.
- Beware of intercavernous sinuses during TSS.

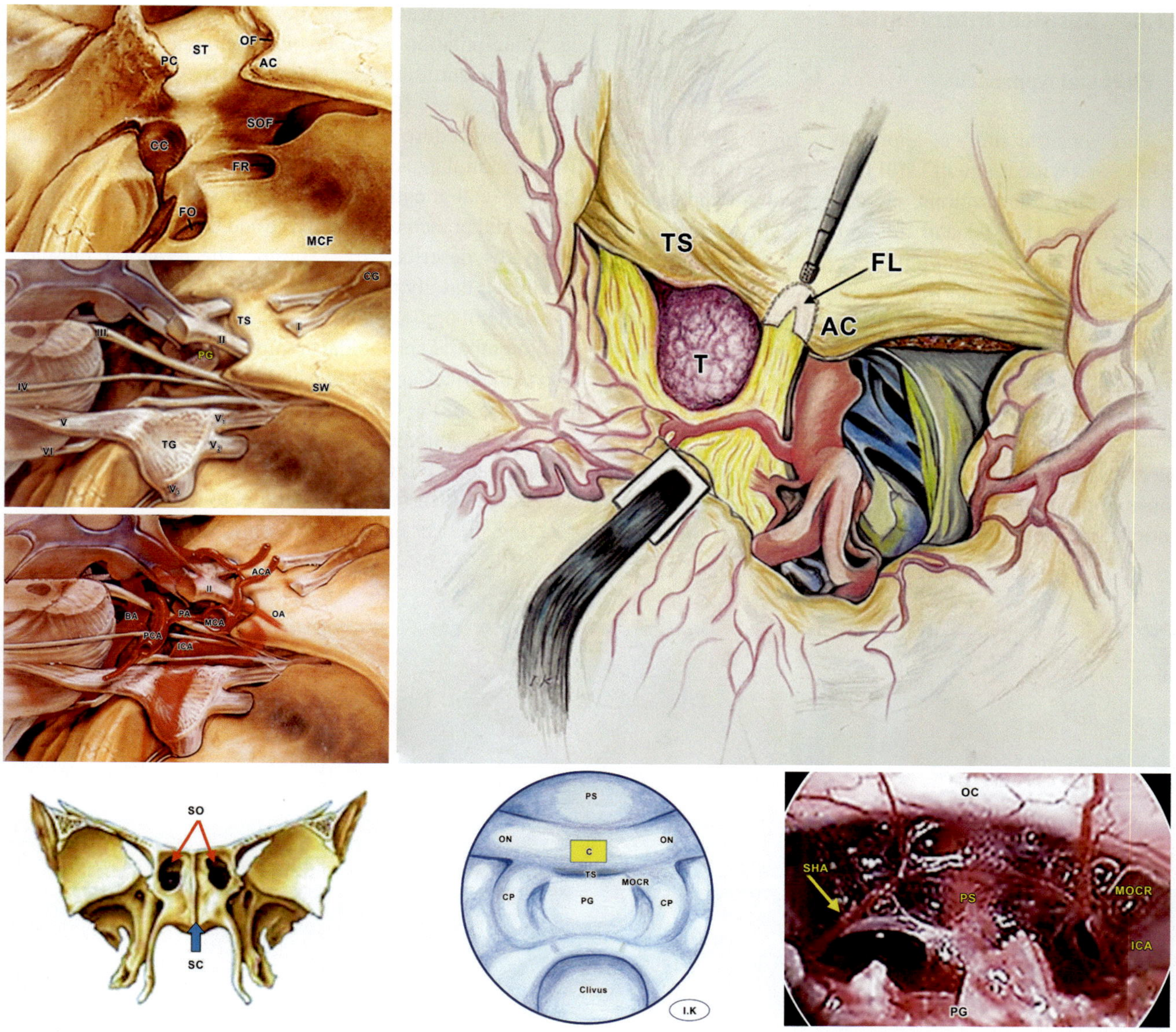

Fig. 31.1 *CG* crista gali, *PS* planum sphenoidale, *TS* tuberculum sella, *AC* anterior clinoid, *OF* optic foramina *ST* sella turcica, *PC* posterior clinoid, *ON* optic nerve, *C* chiasma, *FL* falciform ligament, *PG* pituitary gland, *PS* pituitary stalk, *SO* sphenoid ostium, *SC* sphenoid crest, *MOCR* medial opticocarotid recess, *CP* carotid prominent, *ICA* internal carotid artery, *OP* ophthalmic artery, *ACA* anterior cerebral artery, *SHA* superior hypophyseal artery

31.4 Surgical Intervention

- **Pterional Approach:**

This versatile surgical approach is highly effective for removing a wide range of skull base tumors in the subfrontal and suprasellar regions, including planum sphenoidale meningiomas (PSMs) and tuberculum sella meningiomas (TSMs), while also providing excellent functional exposure. The surgery is performed with the patient placed in supine position, with the ipsilateral shoulder elevated, and the trunk elevated by 25° to ensure gravity-assisted venous drainage. The head is secured in a Mayfield head clamp, extended, and rotated 25° to the contralateral side (Figs. 31.2 and 31.3). The author prefers to perform procedures using this approach on the non-dominant side, with consideration of contralateral intervention for left-sided optic nerve compromise, as recommended by others [48–51] The rationale for this approach is that it involves less manipulation of the involved optic nerve and results in more efficient tumor resection, with the potential for better patient recovery (Figs. 31.4 and 31.5).

The procedure begins with the elevation of a standard frontal skin flap using a curvilinear incision behind the hairline. Special attention is given to preserving the integrity of the frontal branch of facial nerve and the superficial temporal artery, as described by Yaşargil. The temporal muscle is mobilized basally to expose the pterion and the anterior part of the temporal squama. Bony exposure proceeds with a frontotemporal sphenoidal craniotomy, using two burr holes, one at the MacCarty point (thinking point) and the second at the bottom of the temporal squama. Flattening of the orbital roof and resection of the greater and lesser wing of the sphenoid enhance the exposure and minimize brain retraction. The dura is then opened in parallel to the orbital rim with a relaxing slit over the Sylvian fissure.

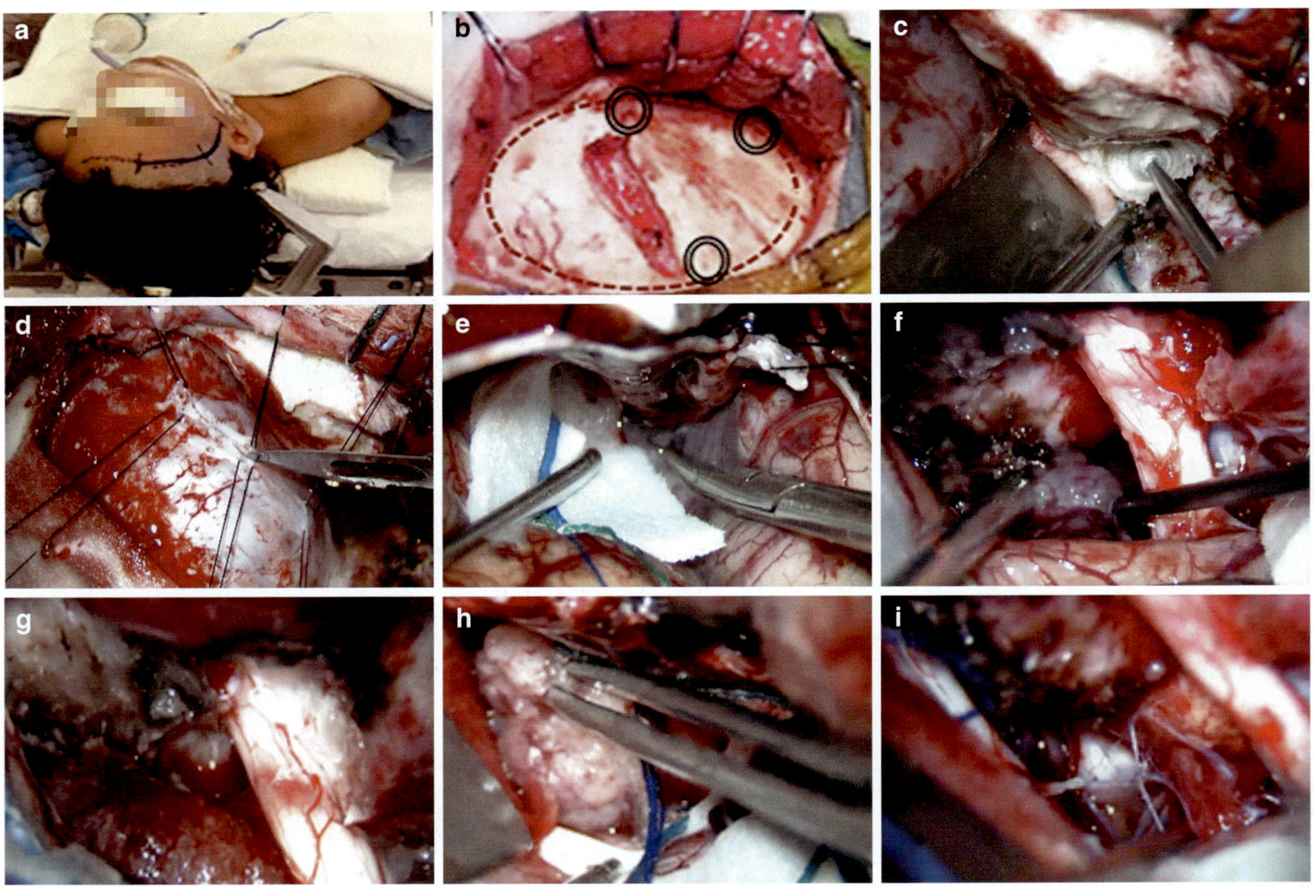

Fig. 31.2 Right pterional approach. (**a**) Frontal skin incision (solid line), (**b**) Fronto pterional craniotomy (dotted line) leaving a muscle cuff, (**c**) Extradural drilling of spenoidal wings and anterior clinoid process, (**d**) Fronto-basal opening of dura with relaxing incision over silvian fissure, (**e**) Proximal sharp opening of proximal sylvian cistern, (**f**) Unroofing of optic canal & incision of falciform ligament to retrieve a tumor extension, (**g**) Devascularization and progressive segmental resection of the meningioma, (**h**) Dissection of tumor capsule while maintaining the integrity of arachnoid membrane, (**i**) Total resection of meningioma preserving pituitary stalk

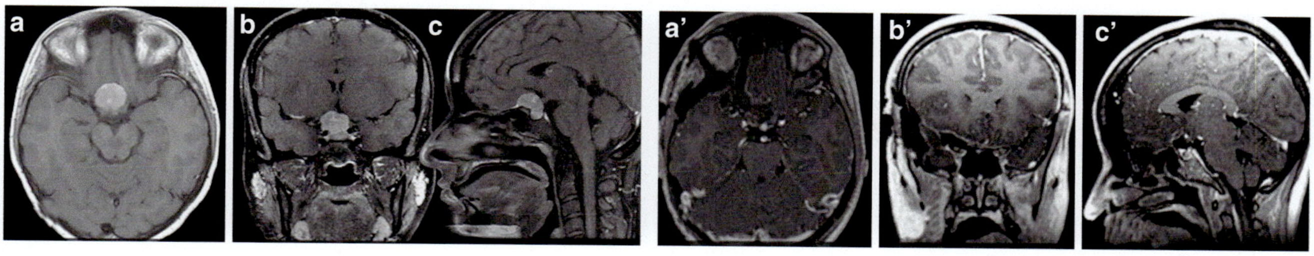

Fig. 31.3 TSM # 5232382. Adult patient with progressive history of headache and asymmetrical bitemporal visual impairment who underwent right pterional approach and gross microsurgical resection with early impressive visual recovery. Preoperative MRI + Contrast (**a**) axial, (**b**) coronal, and (**c**) sagittal views. Postoperative total resection of TSM (**a'**) axial, (**b'**) coronal, and (**c'**) sagittal views

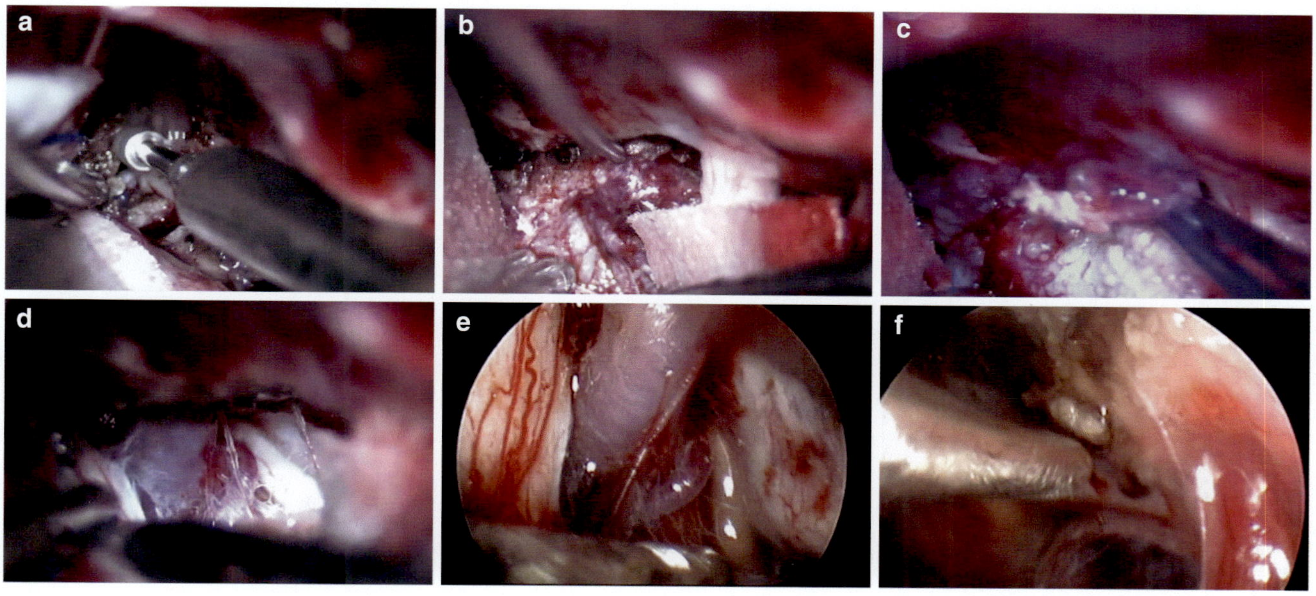

Fig. 31.4 The contralateral approach. Intraoperative images of a 37 years old female with headache and left visual deterioration, who underwent a contralateral pterional approach for gross total resection of a left-sided TSM with complete postoperative full visual recovery. (**a**) Right pterional approach and drilling the roof of left optic canal, (**b**) Devascularization and dissection of TSM, (**c**) Gross resection of TSM, (**d**) Microscopic view of left ICA, ON, SHA, III CN, and PC following the resection of TSM, (**e**) Endoscopic panoramic view of left ON, ICA, SHA, III CN, and PC. Post-tumor resection, (**f**) Pituitary stalk, diaphragm sella

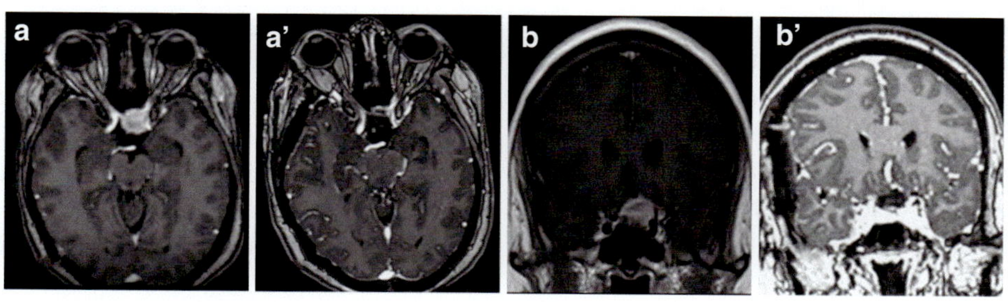

Fig. 31.5 MRI + C. Contralateral pterional approach for resection of left-sided TSM. Preoperative images in (**a**) axial, (**b**) coronal views. Postoperative images in (**a'**) axial, (**b'**) coronal projection confirmed the total resection

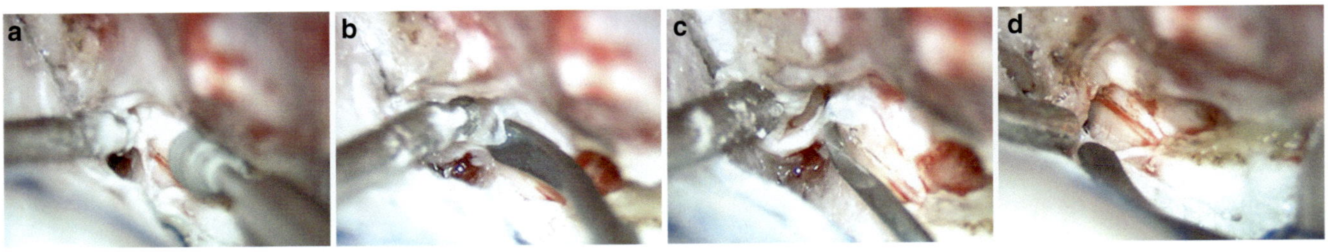

Fig. 31.6 Steps to decompress the right optic nerve at the optic canal. (**a**) Unroof the right optic canal using diamond drill, (**b**) Remove the residual thin bony shell after drilling, (**c**) Incision of the falciform ligament using sickle knife, (**d**) Showing the decompressed right optic nerve

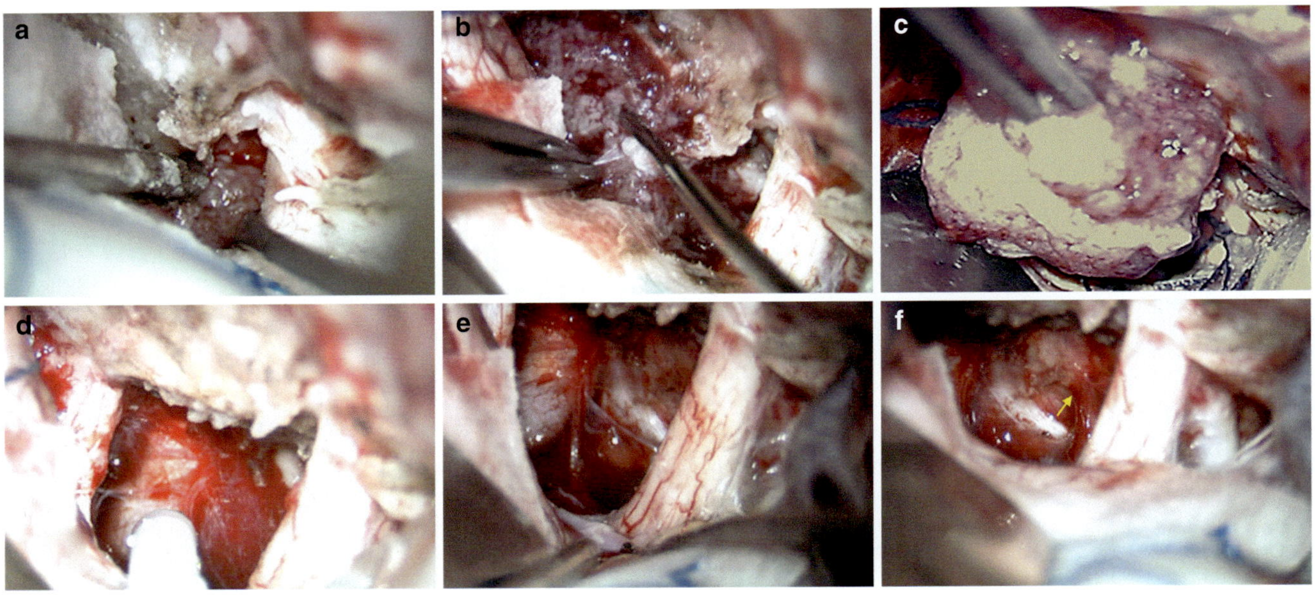

Fig. 31.7 Stepwise resection of TSM. (**a**) Devascularization and removal of the tumor tongue from underneath the Rt. optic nerve, (**b**) Decompression of tumor and peeling the arachnoid off the tumor capsule, (**c**) Dissection and resection of the tumor, (**d**) Value of using Doppler sonography during resection, (**e**) Gross resection with integrity of OC, ONs, ICA, SHA, A1 segment, (**f**) Showing intact pituitary stalk and diaphragm sella

Early introduction of the microscope is recommended prior to any retraction to open the proximal Sylvian cistern to release the cerebrospinal fluid (CSF) and identify the proximal internal carotid artery (ICA) and the entry of the optic nerve in its foramina, just medial to the anterior clinoid process. Early decompression of the optic canal using a 2–3 mm diamond drill and incision of the falciform ligament have been found to improve visual outcomes [52–54] (Figs. 31.2, 31.6, and 31.7). Tumor resection proceeds following the steps of the 5D's concept while maintaining the integrity of the arachnoid membrane as a natural voile protecting the brain and the neurovascular structures. The hybrid use of an endoscope is of value to inspect and retrieve any residual tumor in a blind spot (Figs. 31.4 and 31.5).

31.5 Trans-eyebrow Approach

The trans-eyebrow approach is a minimally invasive variation of the anterolateral approach. It involves either a keyhole supraorbital approach or a mini frontal craniotomy that may include the orbital rim (Fig. 31.8). These approaches are complemented by use of an endoscope, either alone or in a hybrid manner [12, 55]. Use of these variants is typically restricted to the removal of a very small tumors with no brain edema or involvement of adjacent neurovascular structures, as they provide good visualization of the subarachnoid space around the tumor. While the supraorbital keyhole approach may result in good cosmetic results, it may reduce the rate of good visual outcomes or gross total resection.

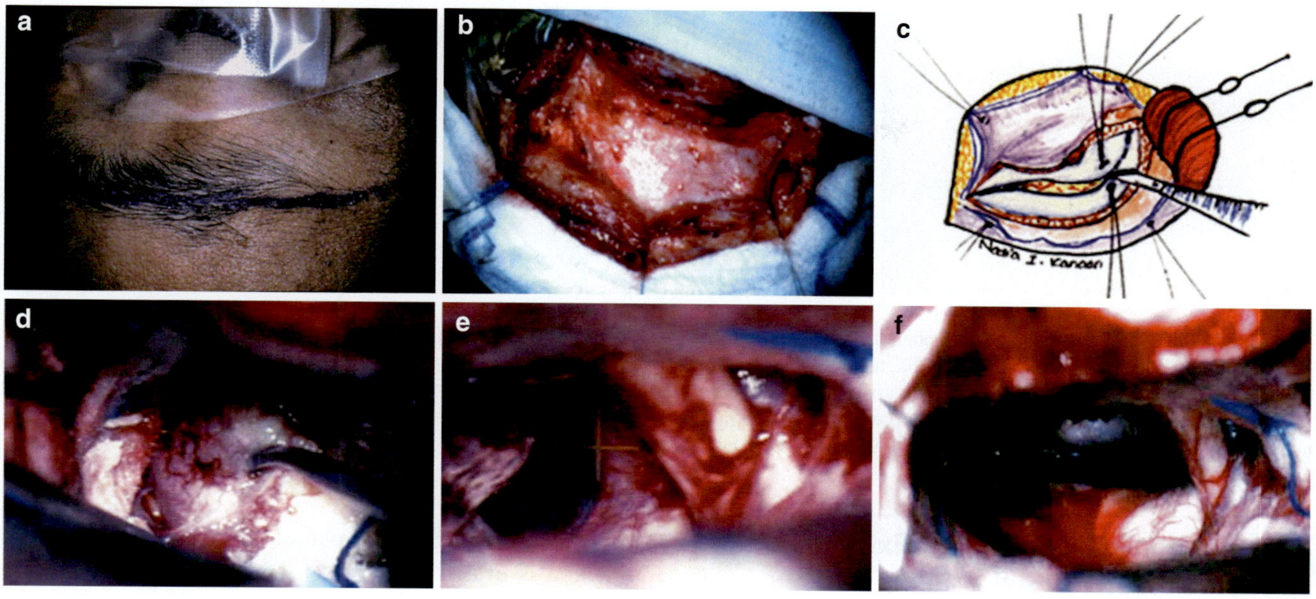

Fig. 31.8 Trans-eyebrow approach. (**a**) Right eyebrow incision, (**b**) Exposure of the supra-orbital fascia, (**c**) Artistic drawing of a small tailored right frontal craniotomy and dura basal opening, (**d**) Exposure of TSM. (**e**) Rt. optic nerve and pituitary stalk after gross tumor resection, (**f**) Illustrates both optic nerves and chiasm after complete removal of small TSM

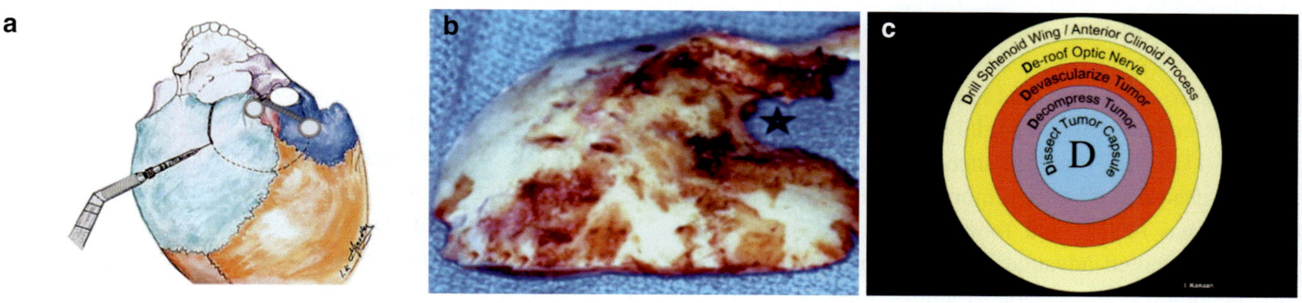

Fig. 31.9 (**a**) and (**b**) Modified Rt. Fronto–orbito-zygomatic approach, including the orbital rim with the pterional approach, (**c**) Scheme of 5D concept for meningioma resection

31.6 Fronto-Orbito-Zygomatic Approach and Variant

The fronto-orbito-zygomatic (FOZ) approach combines several surgical avenues to remove bone obstacles and improve exposure while minimizing brain retraction—very much in accordance with the philosophy of skull base surgery. In recent years, a less invasive version of the FOZ approach has been adopted: a simple tailored approach combining the standard pterional approach with occasional inclusion of the orbital rim, flattening of the orbital roof, sectioning of the orbito-meningeal fold, and tailored drilling of the anterior clinoid process, either extradurally or intradurally. However, double sectioning of the zygomatic arch is avoided (Fig. 31.9). The standard FOZ approach is reserved for management of giant central skull base tumors, complex basilar tip aneurysms, and cavernous sinus pathology, where more extensive exposure is required [17, 19, 21, 22].

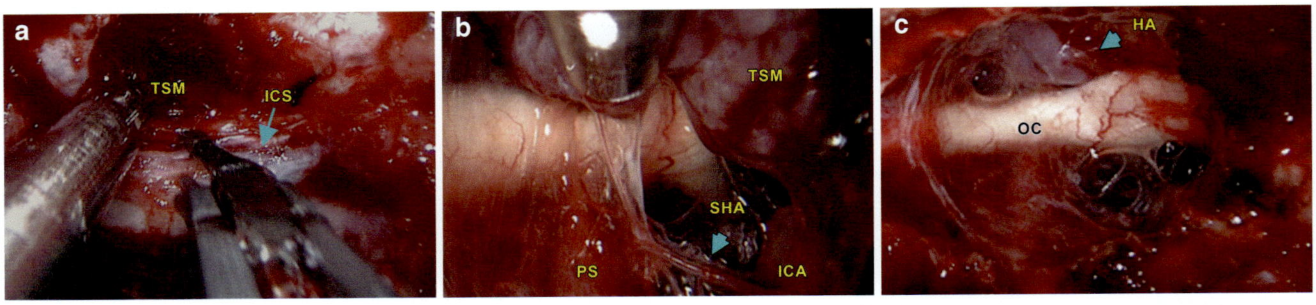

Fig. 31.10 Endoscopic Extended Transsphenoidal Surgery (EETSS). (**a**) Exposure of TSM after drilling of the sella, tuberculum, and part of the planum with coagulation of the intercavernous sinus (ICS), (**b**) Microsurgical resection of the meningioma while maintaining the integrity of the arachnoid membrane, pituitary stalk (PS), superior hypophyseal artery (SHA), and internal carotid artery (ICA), (**c**) Gross resection of TSM showing decompression of optic chiasm (OC), pituitary stalk (PS), anterior cerebral arteries and Hubner artery (HA). (Courtesy of H. Schroeder)

31.7 Endonasal Endoscopic Extended Transsphenoidal Surgery

The inferior trans-basal extended endoscopic approach to central skull base tumors is a novel ventral approach to central skull base tumors that complements the evolution of transphenoidal surgery and resurgence of endoscopic surgery. The Pittsburg group and the Napoli team, among other advocates, have been instrumental in guiding the progress of neuro-endoscopy and redefining the endonasal endoscopic extended transsphenoidal surgery (EETSS) approach [36, 39–43]. They reported successful results in removing selected central skull base tumors, including pituitary sellae meningiomas (PSM) and tuberculum sellae meningiomas (TSM), and noted advantages such as no brain retraction, early decompression of optic structures, maintenance of subchiasmatic perforators' integrity, and good visual outcomes. Shortcomings included diabetes insipidus and CSF leak. Critics expressed concerns about the validity of these reports, citing underreporting of pitfalls, selection bias, limited homogenous cohorts, and short follow-ups.

To perform EETSS, adequate sphenoidotomy is required for wide visualization of the ventral aspect of the skull base, including the planum sphenoidale, tuberculum sella, and sella (Fig. 31.10).

Neuronavigation is valueable in confirming anatomic landmarks, including optic canals, carotid protuberances, and medial and lateral opticocarotid recesses. The sella and the tuberculum are drilled between the medial opticocarotid recesses using a diamond bur under generous irrigation. The dura overlying the tuberculum and planum is sharply incised, and endoscopic bipolar and hemostatic agents are used to control any dural-intercavernous bleeding. Tumor resection is performed as in microscopic intervention, with devascularization, internal debulking of the tumor, followed by bimanual microsurgical dissection of the tumor capsule from the optic nerves and surrounding vital structures while respecting the arachnoidal plane. The extent of the tumor resection is evaluated using an angled endoscope, and the defect is properly reconstructed. This includes placement of dura substitute inlay, fat graft, and coverage with nasoseptal flap and fibrin glue [56, 57].

31.8 Surgical Pearls

- Review MRI images and anticipate potential surgical challenges and pitfalls.
- Simulate visual rehearsal of surgery preoperatively.
- Perform the surgical approach on the nondominant side and consider contralateral approach.
- Introduce the microscope for visual adaptation and early release of CSF from the basal cisterns.
- Unroof the optic canal and incise the falciform ligament for better visual outcomes.
- Implement the author's 5D strategy for meningioma resection.
- Respect the arachnoid membrane as a natural voile protecting the neurovascular structures.
- Consider solo or hybrid use of endoscopy as a visual extended arm to inspect blind corners.
- Utilize doppler sonography to identify vascular anatomy obscured by the tumor.
- Avoid EETSS approach for large TSMs with lateral extension or involvement of regional vessels.

31.9 Discussion

The merits of sophisticated skull base approaches are widely recognized in neurosurgical practice. However, their use has become limited to selective indications such as giant complex skull-base tumors, cavernous sinus interventions, or complex surgery for basilar artery aneurysms. Use of a bilateral frontal or interhemispheric approach for resection of such tumors is limited by the elongated learning curve to build command, time and cost consumption, and the resurgence of endoscopy for solo or hybrid use as an extended visual arm to reach hidden areas with least retraction. These shortcomings have led to the development of a less invasive version of the complex FOZ-approach.

Therefore, during resection of tumors such as TSM and PSM, drilling of the anterior clinoid process is rarely necessary except for lateralized tumors with extension around the proximal internal carotid artery or involvement of cavernous sinus. This recent trend suggests that a simpler approach is more effective for microsurgical resection of these tumors. This assumption is supported and validated by many publications that report excellent clinical outcomes related to visual outcomes, cognition, and extent of resection.

The tailored pterional approach, coupled with the aforementioned microsurgical strategies, remains the most popular and successful transcranial approach, achieving high scores of visual improvement exceeding 85% and gross total resection of tumors up to 90%. However, advocates of the emerging EETSS resection of PSMs and TSMs report good visual outcomes and resection rates, along with several advantages such as direct line of access to the tumor with no brain retraction and the preservation of blood supply to optic apparatus.

Nevertheless, the results of EETSS resection are relatively associated with a higher rate of CSF leak and the risk of diabetes insipidus. Furthermore, the published comparison studies to the transcranial approach failed to provide solid scientific proof for superiority of EETSS and were criticized by opponents for their biased selection, inhomogeneous limited cohort, and shorter follow-ups.

Therefore, EETSS approach to resect TSMs merits scrutiny, validation, and scientific long-term outcome studies to reach a mutual neurosurgical verdict and challenge the supremacy of a tailored pterional approach.

31.10 Conclusion

The tailored pterional approach or its variants provide a short, multi-channel avenue and are the leading and most-commonly used approach for successful resection of TSMs with superior visual outcomes and a high rate of gross total resection. The promising and attractive option of EETSS presents a possible alternative for managing smaller TSMs, but it is still under development and requires further technical refinement and long-term outcome studies. In order to achieve a successful treatment, it is crucial to have a comprehensive understanding of anatomical highlights of these approaches, proper selection of patients, and mastery of microsurgical techniques.

References

1. DeMonte F. Surgical treatment of anterior basal meningiomas. J Neuro-Oncol. 1996;29(3):239–47. https://doi.org/10.1007/BF00165654. PMID: 8858530
2. Schick U, Hassler W. Surgical management of tuberculum sellae meningiomas: involvement of the optic canal and visual outcome. J Neurol Neurosurg Psychiatry. 2005;76(7):977–83. https://doi.org/10.1136/jnnp.2004.039974. PMID: 15965205
3. Stewart J. The symptomatology of tumours involving the hypophysis cerebri. Trans Assoc Am Phys. 1899;14:282–9.
4. Cushing H, Eisenhardt L. Meningiomas arising from the tuberculum sellae: with the syndrome of primary optic atrophy and bitemporal field defects combined with a normal sella turcica in a middle-aged person. Arch Ophthalmol. 1929;1(2):168–206. https://doi.org/10.1001/archopht.1929.00810010177003.
5. Jallo GI, Benjamin V. Tuberculum sellae meningiomas: microsurgical anatomy and surgical technique. Neurosurgery 2002 51(6):1432–1439.; discussion 1439–40. https://doi.org/10.1227/01.neu.0000309119.55205.af. PMID: 12445348.
6. Cushing H, Eisenhardt L. Meningiomas, their classification, regional behaviour, life history, and surgical end results. Springfield: Charles C. Thomas; 1938. p. 228–43.
7. Sade B, Lee JH. High incidence of optic canal involvement in tuberculum sellae meningiomas: rationale for aggressive skull base approach. Surg Neurol 2009;72(2):118–123; discussion 123. https://doi.org/10.1016/j.surneu.2008.08.007. Epub 2009 Jan 14. PMID: 19147207.
8. Mahmoud M, Nader R, Al-Mefty O. Optic canal involvement in tuberculum sellae meningiomas: influence on approach, recurrence, and visual recovery. Neurosurgery 2010;67(3 Suppl Operative):ons108–18; discussion ons118–19. https://doi.org/10.1227/01.neu.0000383153.75695.24. PMID: 20679940.
9. Heuer GJ, Dandy WE. A new hypophysis operation. Johns Hopkins Hosp Bull. 1918;29:154–5.
10. Dandy WE. Intracranial arterial aneurysms. New York: Hafner; 1944, Reprinted 1969, 208, 415.
11. Yaşargil MG. Interfascial pterional (frontotemporosphenoidal) craniotomy. In: Yaşargil MG, editor. Microneurosurgery. Volume 1. Stuttgart: George Thieme Verlag; 1984. p. 215–20.
12. Perneczky A, Reisch R. Keyhole approaches in neurosurgery: volume 1: concept and surgical technique. Wien: Springer-Verlag; 2008. https://doi.org/10.1007/978-3-211-69501-2.
13. McArthur LL. An aseptic surgical access to the Pituitary body and its neighborhood. JAMA. 1912;58:2009–11. https://doi.org/10.1001/jama.1912.04260060362001.
14. Frazer CH. An approach to the hypophysis through the anterior cranial fossa. Ann Surg. 1913;57(2):145–50. https://doi.org/10.1097/00000658-191302000-00001.
15. Jane JA, Park TS, Pobereskin LH, et al. The supraorbital approach: technical note. Neurosurgery. 1982 Oct;11(4):537–42.
16. Pellerin P, Lesoin F, Dhellemmes P, et al. Usefulness of the orbitofrontomalar approach associated with bone reconstruction for frontotemporosphenoid meningiomas. Neurosurgery. 1984;15(5):715–8. https://doi.org/10.1227/00006123-198411000-00016.
17. Hakuba A, Liu SS, Shuro N. The orbitozygomatic infratemporal approach: a new surgical technique. Surg Neurol. 1986;26(3):271–6. https://doi.org/10.1016/0090-3019(86)90161-8.
18. Al-Mefty O. Supraorbital-pterional approach to skull base lesions. Neurosurgery. 1987;21:474–7. https://doi.org/10.1097/00006123-198710000-00006.
19. Andaluz N, Keller JT, van Loveren HR, Zuccarello M. Anatomical and clinical study of the orbitopterional approach to anterior communicating artery aneurysms. Neurosurgery. 2003;52(5):1140–9. https://doi.org/10.1227/01.NEU.0000057834.83222.9F.

20. McDermott MW, Durity FA, Rootman J, et al. Combined frontotemporal-orbitozygomatic approach for tumors of the sphenoid wing and orbit. Neurosurgery. 1991;28(1):69–74. https://doi.org/10.1097/00006123-199101000-00015.
21. Dolenc VV. A combined epi- and subdural direct approach to carotid-ophthalmic artery aneurysms. J Neurosurg. 1985;62(5):667–72. https://doi.org/10.3171/jns.1985.62.5.0667.
22. Zabramski JM, Kiris T, Sankhla SK, Cabiol J, Spetzler RF. Orbitozygomatic craniotomy: technical note. J Neurosurg. 1998;89(2):336–41. https://doi.org/10.3171/jns.1998.89.2.0336.
23. Unterberger S. Care of frontobasal wounds [in German]. Arch Ohren Nasen Kehlkopfheilkd. 1958;172(5):463–84. https://doi.org/10.1007/BF02103711.
24. Tessier P. Orbital hypertelorism: I. Successive surgical attempts. Material and methods. Causes and mechanisms. Scand J Plast Reconstr Surg. 1972;6(2):135–55. https://doi.org/10.3109/02844317209036714.
25. Derome PJ. The transbasal approach to tumors invading the base of the skull. In: Schmidek HH, Sweet WH, editors. Operative neurosurgical techniques: indications, methods, and results. Boston: Grune & Stratton; 1982. p. 357–79.
26. Feiz-Erfan I, Han PP, Spetzler RF, et al. The radical transbasal approach for resection of anterior and midline skull base lesions. J Neurosurg. 2005;103:485–90. https://doi.org/10.3171/jns.2005.103.3.0485.
27. Lang DA, Honeybul S, Neil-Dwyer G, Evans BT, Weller RO, Gill J. The extended transbasal approach: clinical applications and complications. Acta Neurochir. 1999;141(6):579–85. https://doi.org/10.1007/s007010050346.
28. Sekhar LN, Nanda A, Sen CN, Snyderman CN, Janecka IP. The extended frontal approach to tumors of the anterior, middle, and posterior skull base. J Neurosurg. 1992;76:198–206. https://doi.org/10.3171/jns.1992.76.2.0198.
29. Spetzler RF, Herman JM, Beals S, Joganic E, Milligan J. Preservation of olfaction in anterior craniofacial approaches. J Neurosurg. 1993;79:48–52. https://doi.org/10.3171/jns.1993.79.1.0048.
30. Chi JH, Parsa AT, Berger MS, Kunwar S, McDermott MW. Extended bifrontal craniotomy for midline anterior fossa meningiomas: minimization of retraction-related edema and surgical outcomes. Neurosurgery, 2006;59(4 Suppl 2):ONS270-8. https://doi.org/10.1227/01.NEU.0000223508.60923.91.
31. Nakamura M, Roser F, Struck M, Vorkapic P, Samii M. Tuberculum sellae meningiomas: clinical outcome considering different surgical approaches. Neurosurgery. 2006;59:1019–29. https://doi.org/10.1227/01.NEU.0000245600.92322.06. PMID: 17143236.
32. Chokyu I, Goto T, Ishibashi K, Nagata T, Ohata K. Bilateral subfrontal approach for tuberculum sellae meningiomas in long-term postoperative visual outcome. J Neurosurg. 2011;115(4):802–10. https://doi.org/10.3171/2011.5.JNS101812.
33. Ganna A, Dehdashti AR, Karabatsou K, Gentili F. Fronto-basal interhemispheric approach for tuberculum sellae meningiomas; long-term visual outcome. Br J Neurosurg. 2009;23(4):422–30. https://doi.org/10.1080/02688690902968836.
34. Terasaka S, Asaoka K, Kobayashi H, Yamaguchi S. Anterior interhemispheric approach for tuberculum sellae meningioma. Neurosurgery 2011;68(1 Suppl Operative):84–88; discussion 88–9. https://doi.org/10.1227/NEU.0b013e31820781ef.
35. Curey S, Derrey S, Hannequin P, Hannequin D, Freger P, Muraine M, Castel H, Proust F. Validation of the superior interhemispheric approach for tuberculum sellae meningioma: clinical article. J Neurosurg. 2012;117(6):1013–21. https://doi.org/10.3171/2012.9.JNS12167.
36. Kassam AB, Snyderman CH, Mintz A, Gardner P, Carrau RL. Expanded endonasal approach: the rostrocaudal axis. Part I. Crista galli to the sella turcica. Neurosurg Focus 2005;19(3):E3, 1.
37. Weiss MH. Transnasal transsphenoidal approach. In: Apuzzo MLJ, editor. Surgery of the third ventricle. Baltimore: Williams & Wilkins; 1987.
38. Cook SW, Smith Z, Kelly DF. Endonasal transsphenoidal removal of tuberculum sellae meningiomas: Technical note. Neurosurgery 2004;55(2):239–246. discussion 246–7. https://doi.org/10.1227/01.neu.0000126952.51782.4d.
39. Laufer I, Anand VK, Schwartz TH. Endoscopic, endonasal extended transsphenoidal, transplanum transtuberculum approach for resection of suprasellar lesions. J Neurosurg. 2007;106(3):400–6. https://doi.org/10.3171/jns.2007.106.3.400.
40. de Divitiis E, Cavallo LM, Esposito F, Stella L, Messina A. Extended endoscopic transsphenoidal approach for tuberculum sellae meningiomas. Neurosurgery 2007;61(5 Suppl 2):229–238. discussion 238–9. https://doi.org/10.1227/01.neu.0000303221.63016.f2.
41. Cappabianca P, Esposito F, Solari D. Endoscopic removal of tuberculum sellae meningiomas. In: Al-Mefty O, editor. Controversies in neurosurgery II. New York: Thieme; 2013. p. 234–48.
42. Kulwin C, Schwartz TH, Cohen-Gadol AA. Endoscopic extended transsphenoidal resection of tuberculum sellae meningiomas: Nuances of neurosurgical technique. Neurosurg Focus. 2013;35(6):E6. https://doi.org/10.3171/2013.8.FOCUS13338. PubMed PMID: 24289131
43. Cavallo LM, Andaluza N, Di Somma A, Solari D, Cappabianca P. Tuberculum sellae/planum meningiomas. In: Cappabianca P, Solari D, editors. Meningiomas of the skull base: treatment nuances in contemporary neurosurgery. New York: Thieme Medical Publishers; 2019. p. 127–37.
44. Gadgil N, Thomas JG, Takashima M, Yoshor D. Endoscopic resection of tuberculum sellae meningiomas. Neurol Surg B Skull Base. 2013;74(4):201–10. https://doi.org/10.1055/s-0033-1342922.
45. Frank G, Pasquini E. Tuberculum sellae meningioma: the extended transsphenoidal approach – for the virtuoso only? World Neurosurg. 2010;73(6):625–6. https://doi.org/10.1016/j.wneu.2010.05.031.
46. Van Gompel JJ, Frank G, Pasquini E, Zoli M, Hoover J, Lanzino G. Expanded endonasal endoscopic resection of anterior fossa meningiomas: report of 13 cases and meta-analysis of the literature. Neurosurg Focus. 2011;30(5):E15. https://doi.org/10.3171/2011.2.FOCUS118.
47. Schroeder HWS. Indications and limitations of the endoscopic endonasal approach for anterior cranial base meningiomas. World Neurosurg. 2014;82(6 Suppl):S81–5. https://doi.org/10.1016/j.wneu.2014.07.030.
48. Kachhara R, Nigam P, Nair S. Tuberculum sella meningioma: Surgical management and results with emphasis on visual outcome. J Neurosci Rural Pract. 2022;13(3):431–40. https://doi.org/10.1055/s-0042-1745817. Epub 2022 Mar 2
49. Fahlbusch R, Schott W. Pterional surgery of meningiomas of the tuberculum sellae and planum sphenoidale: Surgical results with special consideration of ophthalmological and endocrinological outcomes. J Neurosurg. 2002;96(2):235–43. https://doi.org/10.3171/jns.2002.96.2.0235. PubMed PMID: 11838796
50. Engelhardt J, Namaki H, Mollier O. Contralateral transcranial approach to tuberculum sellae meningiomas: long-term visual outcomes and recurrence rates. World Neurosurg. 2018;116:e1066–74. https://doi.org/10.1016/j.wneu.2018.05.166. Epub 2018 Jun 1
51. Jang WY, Jung S, Jung TY, Moon KS, Kim IY.The contralateral subfrontal approach can simplify surgery and provide favorable visual outcome in tuberculum sellae meningiomas Neurosurg Rev 2012;35(4):601–607;discussion 607–8. https://doi.org/10.1007/s10143-012-0397-y. Epub 2012 Jun 6.
52. Mathiesen T, Kihlström L. Visual outcome of tuberculum sellae meningiomas after extradural optic nerve decompression. Neurosurgery 2006 Sep;59(3):570–576; discussion 570–6. https://doi.org/10.1227/01.NEU.0000228683.79123.F9.

53. Kanaan IN. Tailored skull base approach to management of intracranial aneurysms. Adv Tech Stand Neurosurg. 2022;44:1–16. https://doi.org/10.1007/978-3-030-87649-4_1.
54. Jaafar B, Mallory D, Rahul K, David S, Kenan A. Cisternal, falciform, and optic canal decompression influencing optic nerve biomechanics: A microsurgical anatomic study. Oper Neurosurg (Hagerstown). 2023;24(2):e75–84. https://doi.org/10.1227/ons.0000000000000472. Epub 2022 Nov 4
55. Kanaan IN. Trans-eyebrow mini-orbitozygomatic pterional approach for minimally invasive skull base surgery. Minim Invasive Neurosurg. 2005;48(1):34–8. https://doi.org/10.1055/s-2004-830182.
56. Hadad G, Bassagasteguy L, Carrau RL, et al. A novel reconstructive technique after endoscopic expanded endonasal approaches: vascular pedicle nasoseptal flap. Laryngoscope. 2006;116(10):1882–6. https://doi.org/10.1097/01.mlg.0000234933.37779.e4.
57. Kassam AB, Thomas A, Carrau RL, et al. Endoscopic reconstruction of the cranial base using a pedicled nasoseptal flap. Neurosurgery 2008;63(Suppl 1):ONS44-52; discussion ONS52-3. https://doi.org/10.1227/01.neu.0000297074.13423.f5.

Management of Parasagittal Meningiomas

32

Kazadi K. N. Kalangu, Manuel de Jesús Encarnacion Ramírez, and DieuMerci Kantenga Kabulo

Learning Objectives

- Describe patients' workout including precise neurological examination, blood tests, neuroradiological investigations.
- Highlight the perfect positioning of patient and the head, the Precise location of the tumor and bone flap with neuronavigation
- Plan the preservation of cortical veins and know in advance the patency of the superior sagittal sinus (SSS) using MRV before surgery and ICGV during surgery.
- Be aware of the diploic veins which may function as collateral venous pathways.
- Use neuromonitoring during surgery and be prepared for the efficient control of hemostasis.
- Removal of the tumor to be done step by step using magnification with microscope or exoscope.

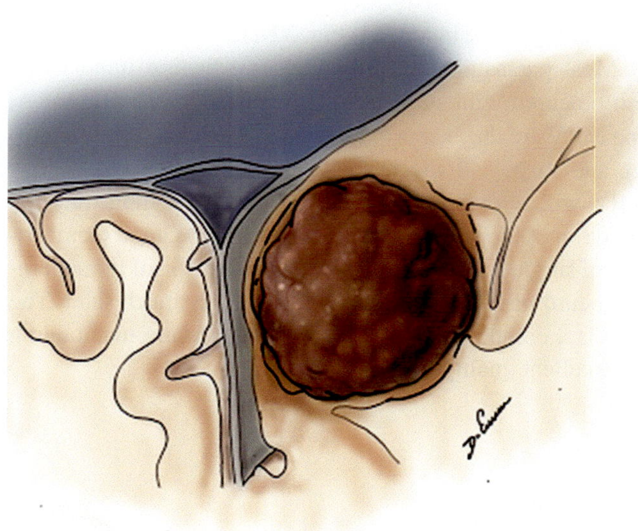

Fig. 32.1 Parasagittal meningioma

32.1 Introduction

Parasagittal region is considered to be the second most common origin of meningioma. They are challenging lesions, may invade the superior sagittal sinus (SSS), and/or encase/adhere to the bridge veins. Parasagittal meningiomas are defined as meningiomas that grow on one wall of the superior sagittal sinus [3, 5] (Fig. 32.1). Total microsurgical resection remains a challenge in terms of the postoperative neurological status of the patient. It is a surgery which starts from a thorough patient positioning and which requires a careful dissection and manipulation of venous structures [1, 7].

K. K. N. Kalangu (✉) · D. K. Kabulo
University of Zimbabwe, Department of Neurosurgery, Harare, Zimbabwe

M. d. J. E. Ramírez
Neurosurgery Department, RUDN University, Russian Federation, Moscow, Russia

The use of MRV and the ICGV play a valuable role in the preservation of collateral veins [29]. Despite the identification with indocyanine green video angiography, patients are still at risk of delayed cerebral venous infarction during the week following surgery [3]. The radical resection of the tumor without complications is the goal of every neurosurgeon. Long- and short-term morbidity is related to management of venous structures and flow [1, 3, 8]. Cortical functions may be affected because of compression by the tumor or aggravated during surgery if the tumor is removed without gentle manipulation of the adjacent brain. In general the majority of meningiomas are detected in females, but the gender distribution is more equal among parasagittal meningiomas than for meningiomas in other locations [8]. Parasagittal meningiomas can be challenging to be removed particularly when the SSS is invaded [9].

I. N. Kanaan, V. Beneš (eds.), *Neuroanatomy Guidance to Successful Neurosurgical Interventions*,
https://doi.org/10.1007/978-3-031-59838-8_32

32.2 History and Background

Cushing and Eisenhardt defined the parasagittal meningioma as one that fills the parasagittal angle, with no brain tissue between the tumor and the superior sagittal sinus [10]. Parasagittal meningiomas are tumors arising at the convexity of the hemisphere, just off the midline adjacent to SSS and falx, which may involve one, two, or three walls of the SSS without or with occlusion of its lumen [4].

They comprise 21–31% of intracranial meningiomas, and the distribution of the meningiomas along the SSS ranges from 14.8% to 33.9% in the anterior third, from 44.8% to 70.4% in the middle third, and from 9.2% to 29.6% in the posterior third of the sinus [1, 2].

The most known classifications are from Bonnal and Brotchi [4], and Sindou [5] (Fig. 32.1).

Most parasagittal meningiomas are benign tumors that can be cured with surgical treatment. To date, MRI, MRA, and MRV are the exams of choice [3, 8] to assess the diagnosis and to get all the needed information before surgery. MRV will also show if the SSS is patent, demonstrate the direction of the venous flow, thus helping the good planification of surgical technique [8]. The introduction of intraoperative neuromonitoring has made this surgery much safer. The goal is a complete removal of the tumor while maintaining an excellent quality of life [2, 7].

32.3 Anatomical Highlights

The superior sagittal sinus (SSS) is triangular in cross section and has right and left lateral angles at the junction with the dura. It begins in front of the crista galli and it is a caudally expanding dural structure (maximum at the coronal suture and minimum before draining to the confluens sinuum or Torcular Herophili) which collects venous blood from medial parts of the fronto-parieto-occipital cortex. The SSS is often deviated to the right side of the midline [11].

The superficial cortical veins are responsible of the vast majority of the volumetric contribution to the SSS [12]. They may pass directly into the sinus or may join the meningeal sinuses, which empty into the SSS.

The number of veins draining into the SSS are in general more numerous on the right side than the left, and anterior to coronal suture, followed by the area between the coronal suture and vertex, then between the vertex and lambdoid suture and lastly between the lambdoid suture and the confluens sinuum or torcular Herophili [13] (Fig. 32.1).

The diameter of veins draining into the SSS is larger between the coronal suture and the vertex. Nearly all the bridging veins enter the SSS at an angle opposite to the direction of blood flow [14] (Fig. 32.2).

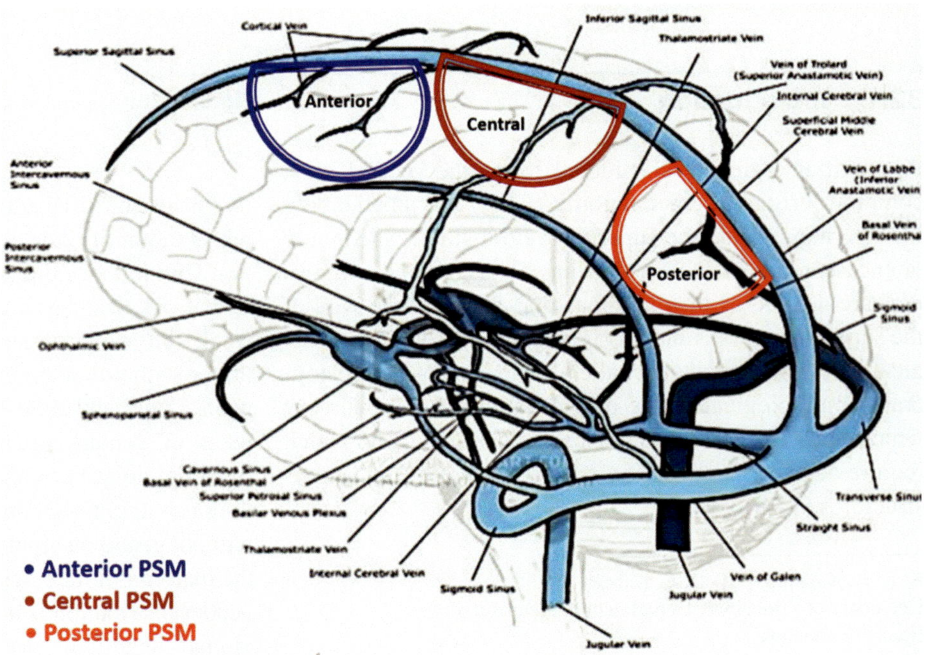

Fig. 32.2 Anatomy of the brain venous system

32.3.1 Arterial Vascularization of the Tumor

The tumor is generally supplied by the middle meningeal, anterior cerebral, or posterior cerebral artery (Fig. 32.3).

32.3.2 Clinical Presentation

Clinical signs are generalized seizures that are partial, headaches, and sometime focal deficit. The discovery of a parasagittal meningioma could be also be incidental or more rarely associated with visual loss, behavior disturbance, monoparesis, disorientation, and even apathy, dementia, or depression.

About 20% of the patients present manifestations of intracranial hypertension due to the occlusion of the posterior third of the SSS, and torcular or predominant lateral sinus [2, 3, 8] (Fig. 32.4).

32.3.3 Pathology

The histology is the same as for other locations of meningiomas. Histological classifications are performed according to the WHO recommendations with frequent benign grade I meningiomas (transitional, meningothelial, psammomatous, angiomatous, fibroblastic, secretory), more rarely grade II (atypical), and even more rarely grade III malignant meningiomas [5, 8].

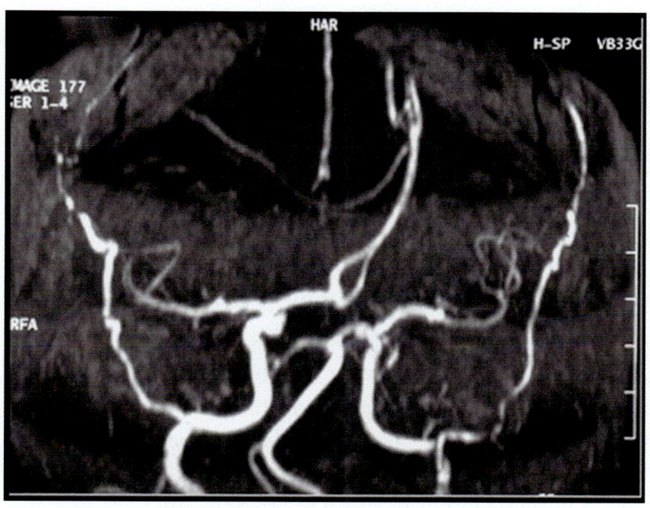

Fig. 32.3 Arterial supply of a parasagittal meningioma. (Courtesy of Professor Jacques Brotchi)

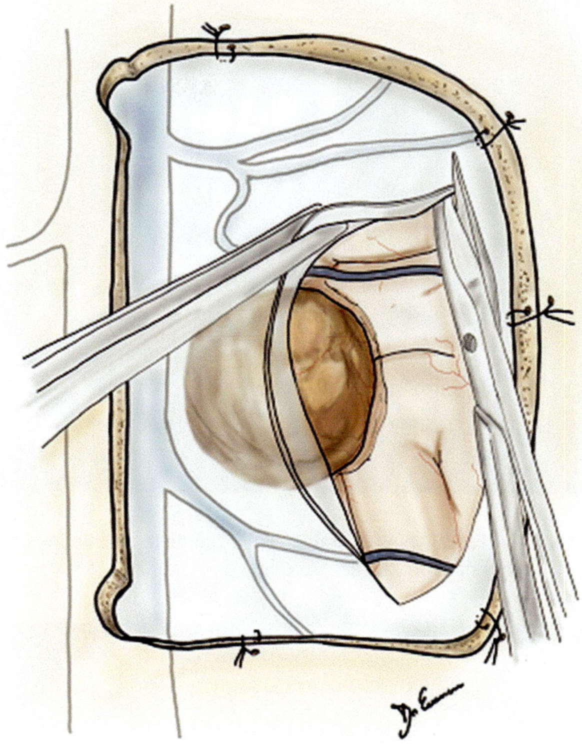

Fig. 32.4 Dura opening

32.3.4 Radiological Diagnosis

Parasagittal meningiomas may pose a difficult surgical challenge if venous patency and collateral anastomoses have not been clearly defined thus knowledge of the relevant venous anatomy would be extremely helpful during surgery (Fig. 32.5).

CT scan keeps its interest by visualizing the bone invasion, but to date, MRI is the exam of choice to assess the diagnosis and to get all the needed information before surgery, especially with angiographic sequences (MRA and MRV), allowing a precise study of the direction of the venous circulation without the invasive aspect of the digital substraction angiography (DSA) [29] (Fig. 32.6). The importance of DSA is obvious only to demonstrate the arterial supply of meningiomas or if preoperative embolization is being considered (Fig. 32.3). It is an invasive diagnostic procedure with a certain degree of risk, for well-known reasons inherent to the technique [1, 5, 8].

The diploic veins frequently communicate with the dural venous sinuses, they may function as collateral venous pathways in cases of meningioma with venous sinus invasion

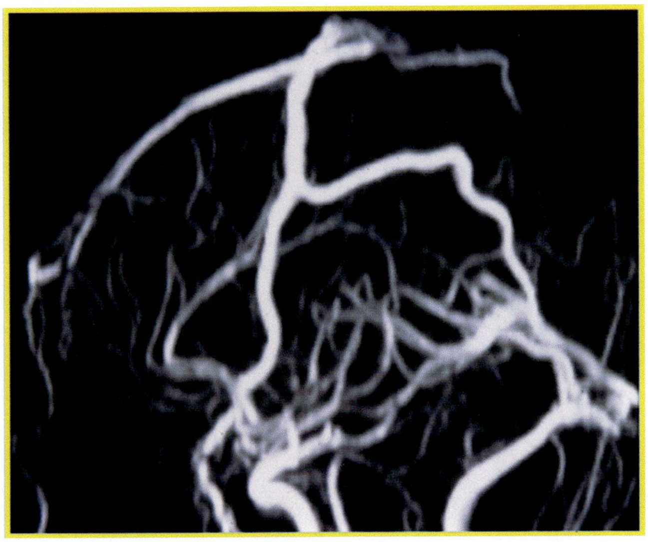

Fig. 32.5 Collateral veins of drainage in a complete obstruction of the SSS. (Courtesy of Prof. Jacques Brotchi)

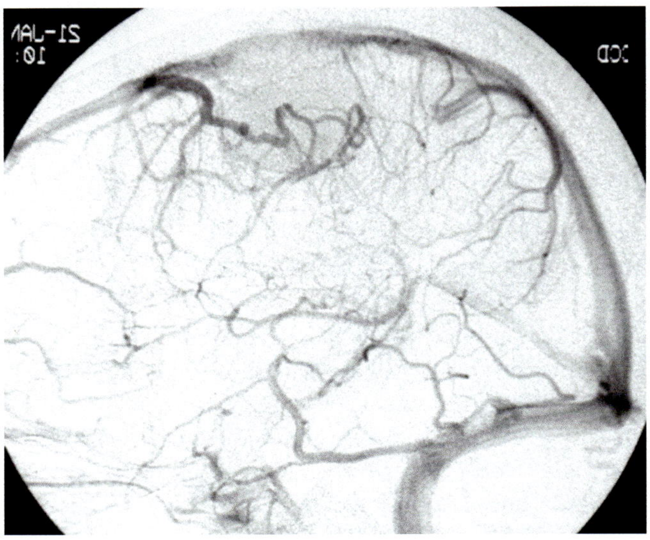

Fig. 32.6 MRV in a patient

[15], particularly in type V and VI of the Sindou classification. This can be determined by the use of the whole-head computed tomography angiography (4D-CTA) for assessing diploic vein blood flow and the positional relationship between typical craniotomy approaches and diploic veins. Diploic vein blood flow should be preoperatively evaluated to consider preserving the veins [16].

32.4 Surgical Intervention

32.4.1 Position

The head is supported by a three-point skeletal fixation. Patient is placed in a supine, lateral, or prone position according to the location of the meningioma (anterior, middle of posterior third of the SSS). Slight rotation of the head, 20–30° from the vertical plane to the side of the tumor to take advantage of gravity and allowing the brain to fall away from the midline and minimize brain retraction. If the attachment to the dural convexity is wide, it is important to position the head in such a way that the meningioma is located on the upper part of the operative field. The head position should provide excellent exposure of the tumor surface, provide control of its vascular supply, and maintain the integrity of the sagittal sinus and adjacent cortical veins [24]. For anterior third meningioma, it is recommended to have the patient in supine position with the head elevated and have the exposure over the uppermost central part ot the tumor dome, whereas, for the posterior third meningioma, it is advisable to place the patient in a three-quarter prone position with the tumor below the midline [3, 8].

32.4.2 Scalp Incision

A bicoronal skin incision is used for the anterior third of SSS to achieve better cosmetic results. For more posterior location, a horseshoe shape incision is used, extending approximately 2 cm across the midline with attempt to preserve a healthy pericranium for potential dural reconstruction [8].

32.4.3 Bone Flap

Cautious preoperative study of diploic veins [16] is of importance for the surgeon; it helps to preserve those important anastomic networks in designing and cutting the bone flap thus avoiding injuring those vital channels when SSS is occluded.

Based on the neuroradiological navigation, a free parasagittal bone flap [3, 23, 25], slightly larger than the meningioma of approximately 1.5–2 cm and crossing the midline by 2 cm, is performed. The SSS is exposed as needed to allow the visualization of both sides of the sinus [1, 5, 8]. This is performed by placing two or more burr holes, just lateral to the SSS and then connected with the craniotome (Fig. 32.7). The bone cuts across the midline are performed last, for the safe elevation of the bone flap. Pay special attention to possible tumor invasion of the dura and bone flap and the risk of injury to the underlying bridging veins during this process. Potential venous bleeding from the SSS may be controlled with Gelfoam, surgicel, or microsuture. Epidural tacking sutures are applied to prevent extradural bleeding. The intraoperative use of indocyanine green (ICG) has been recently considered in some centers, where the facility is available to evaluate the patency of the SSS and the adjacent bridging veins [3, 8].

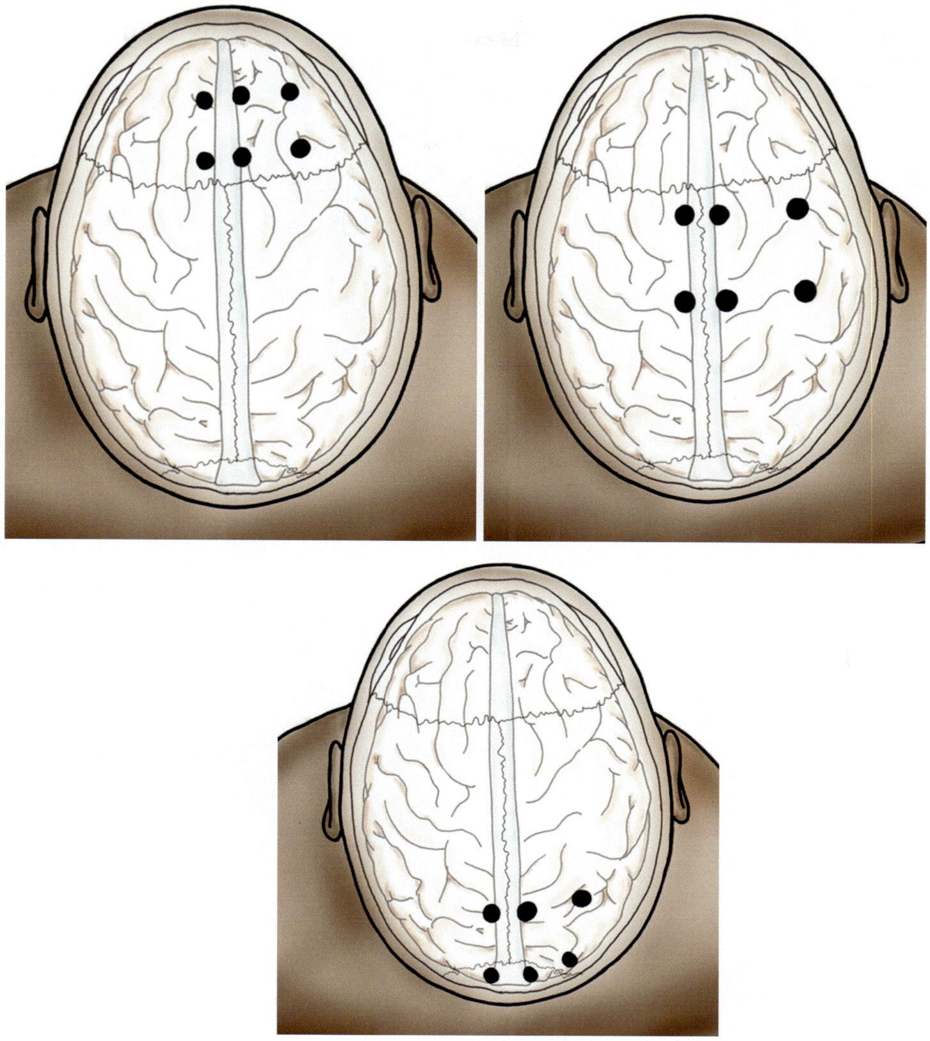

Fig. 32.7 Burr holes are made according to the tumor location

32.4.4 Dural Opening

The dura is opened in a horseshoe shape just lateral to the SSS and the tumor margin preserving the SSS and the draining cortical veins. The dura is not used as handle to retract the tumor as described in the convexity meningioma to prevent traction injury of the bridging draining veins in the vicinity [1]. There is no need to open the dura on the opposite side unless the tumor extends to the opposite area or in type V with an occluded SSS. Sagittal cutting of the dura is performed after most of the tumor has been dissected from the brain [1–3, 8].

32.4.5 Tumor Removal

This follows the standard 4 "D" concept of meningioma resection that includes define, devascularize, decompress, and dissect (Fig. 32.8).

Dissection of the meningioma from the surrounding brain tissue should be carefully carried out under the control of a microscope or exoscope respecting the integrity of the arachnoid membrane and the pial surface of the brain with special attention to preserve the adjacent bridging veins [26]. This is facilitated by the microsurgical technique and the use of cottonoid strip and fine forceps [1, 3]. When the lesion is subpial, it is important to define the plane of cleavage clearly with adjacent brain [1, 3, 8]. Intraoperative use of ICG may help to distinguish the normal collateral veins from the tumor-draining vein [17]. Some have reported the possible value of applying a temporary clip on the collateral veins to distinguish the direction of the blood flow and decide if the veins have to be preserved or coagulated [18]. It is recommended to decompress the tumor first by using miscrosurgical techniques and ultrasonic surgical aspirator. Some proposed the use of hydrogen peroxide intratumoral injection to reduce the amount of blood loss [19].

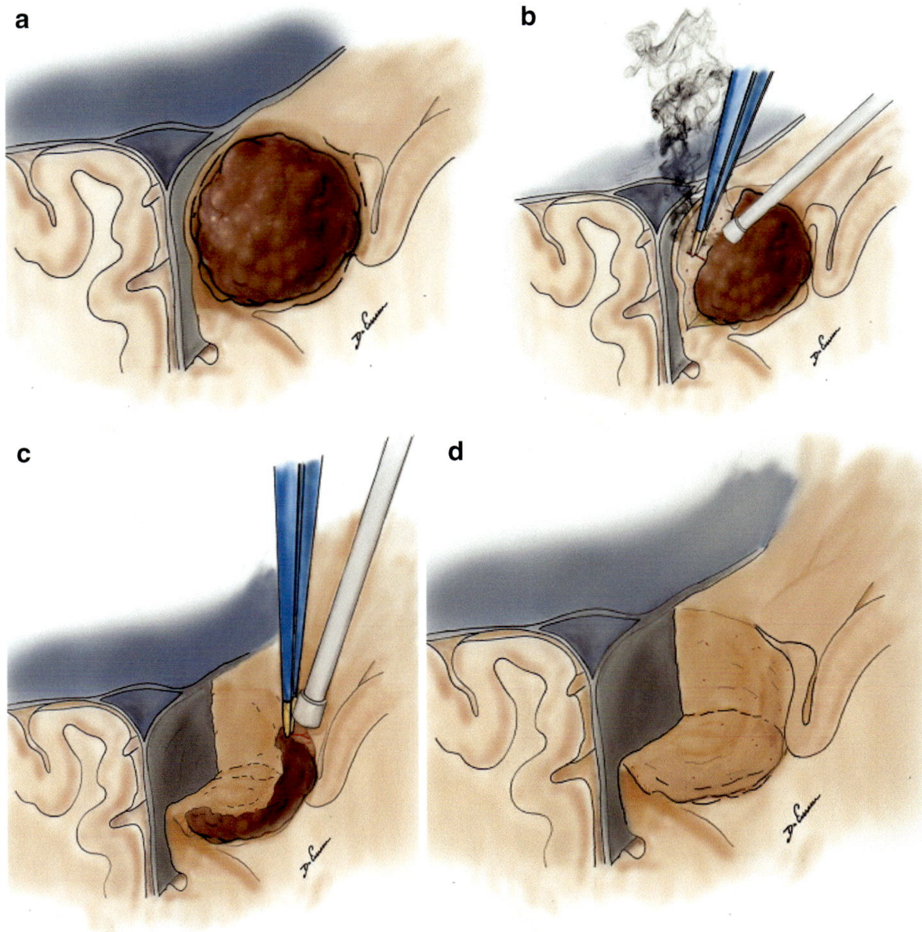

Fig. 32.8 (**a**) Define. (**b**) Devascularize. (**c**) Decompress. (**d**) Dissect

32.4.6 SSS Invasion

Based on Sindou's classification (Fig. 32.9) the surgery of the SSS is performed according to the scheme mentioned on Table 32.1. The SSS is opened with excision of the infiltrated wall. Both of the extremities of the SSS are filled with microsurgical patties to avoid air embolism or excessive blood loss. If an intrasinusal part of the tumor is present, it is removed using and ultrasonic surgical aspirator. Then the reconstruction may take place. At the end of the procedure, the patency of the SSS and bridging veins can be evidenced by intraoperative ICG. Whenever possible and to avoid the risk of a thrombotic occlusion, in type V and VI meningiomas, it is important to close the SSS at a distance of 1 cm from the bridging veins of the central group in the middle-third portion of the sagittal sinus [1, 3, 5].

This surgery requires also intraoperative neurophysiology as it facilitates the success of the procedure [8, 27]. If a blood vessel on the surface of the meningioma is encountered, it must first be identified, and if the vessel supplies the tumor (and the brain), it may be coagulated on the tumor site and cut [3, 8]. As little retraction as possible of the adja-

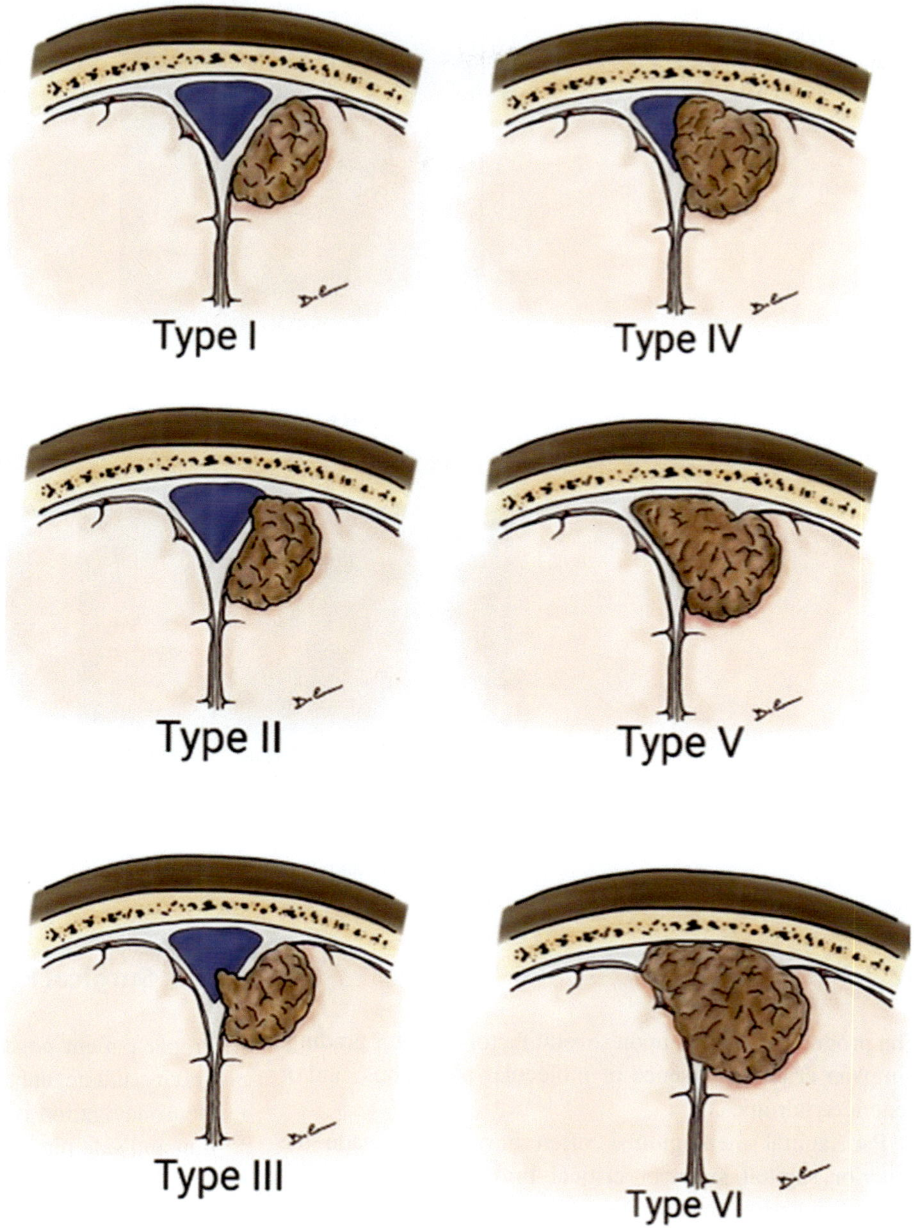

Fig. 32.9 Reproduction of the Sindou classification. Type I: Meningioma attached to the outer surface of the sinus. Type II: Lateral recess invaded. Type III: Lateral wall invaded. Type IV: Entire lateral wall and roof of the sinus both invaded. Type V: Sinus totally invaded with one wall being free. Type VI: Sinus totally invaded without any wall being free

Table 32.1 Sindou's classification and corresponding surgical strategies

Type I: Resection of the outer layer and coagulation of the inner layer
Type II: Resection of the invaded fragment of the lateral recess and suture of the recess defect.
Type III: Resection of the invaded lateral wall and repair with a patch of adjacent autologous periosteum
Type IV: Resection of the invaded lateral wall and roof of the sinus with a patch of reversal cerebral falx
Type V: Invaded sinus accompanied cortical draining veins, subtotal resection tumor, coagulation of wall of sinus; without cortical draining veins, the strategy and type IV were the same.
Type VI: Ligation of superior sagittal sinus and removal of the entire involved segment.

cent brain should be made, since most of those tumors have close relationship with the motor area. Nevertheless, it may be sometimes necessary to reconstruct a bridging vein. However, leaving a small tumor attached to a vein is better than injuring an important vein [1] even when the vein is totally embedded by the meningioma and the separation without wall injury risky. In the case of sinus repair, we take advantage of a venous graft collateral branch, which is end-to-end sutured to the stump of the bridging vein [7, 8]. When SSS is not concerned, surgery may end by cutting the dura parallel to the midline after separation of the parasagittal veins, and then the tumor may be completely removed (Fig. 32.10).

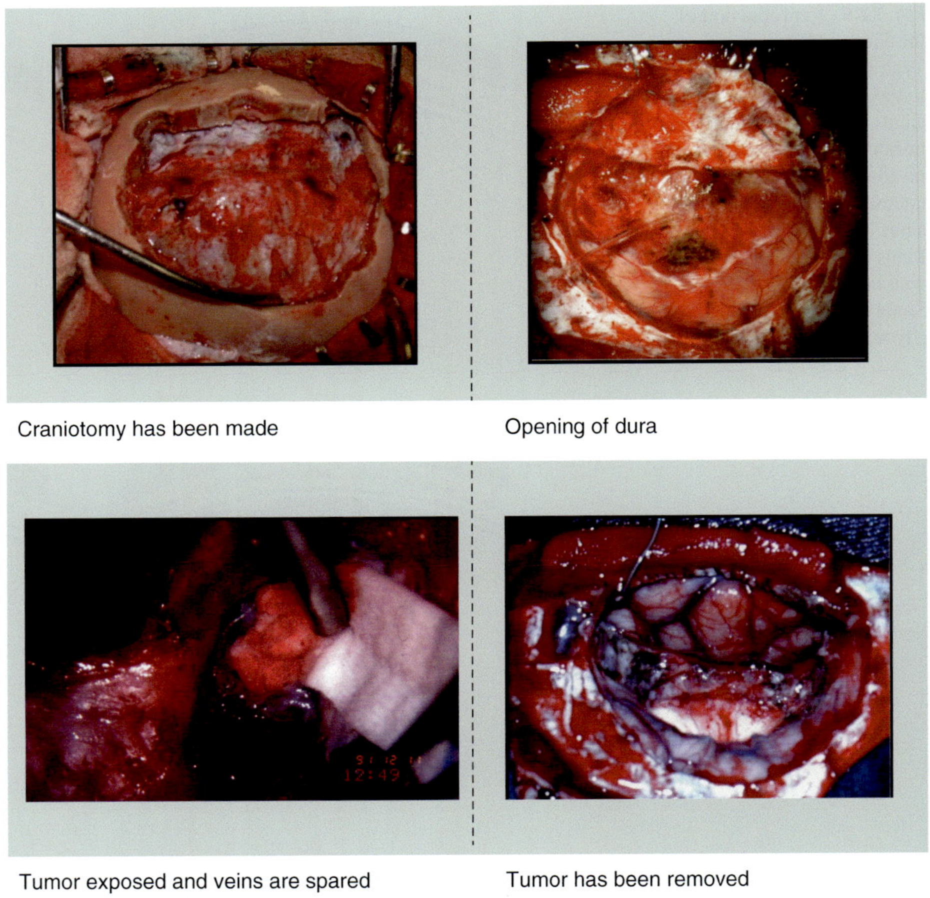

Fig. 32.10 (**a**) Craniotomy has been made, (**b**) Opening of dura, (**c**) Tumor exposed and veins are spared, (**d**) Tumor has been removed

32.4.7 Prognosis

The prognosis depends upon several factors: Tumor grading, Simpson criteria, presence of molecular alterations, and/or genetic syndromes.

Parasagittal meningiomas often intrude or invade the superior sagittal sinus or critical bridging veins such as Trolard superior anastomotic vein. Therefore, a complete resection is not always achievable. It was also stated that the risk of recurrence is considerable even with Simpson grade 1 resection. For these cases, radiosurgery (RS) or stereotactic radiotherapy (SRT) has been used as an adjuvant treatment option with satisfactory results in terms of efficacy and low incidence of toxicity [20, 28]. These options are valid for the treatment of the elderly and patients with high surgical risks and comorbidities. However, controversies remain regarding the timing, dose, and fractionation. It was also reported that radiation-induced brain edema was higher compared to other locations [21, 28].

32.5 Surgical Pearls

- Proper patient position for surgery to take advantage of gravity and negate the need for retraction.
- Neuronavigation is important for precise tumor localization and safe placement of craniotomy.
- The use of ICG has great value to identify the occluded segment of the SSS and recognize the regional cortical draining veins to prevent their injury.
- Consider the use of the microscope or exoscope early during the intervention to facilitate the microsurgical dissection of the tumor capsule while maintaining the integrity of the arachnoid membrane and neurovascular structures.
- Apply the standard concept of 4“D” for meningioma resection and maintain the integrity of the surrounding arachnoid membrane.
- Neurophysiology plays a key role to achieve good clinical outcome and avoid complications.

- Options for management of SSS infiltration are dictated by the extent and the location of the sinus obliteration and guided by IOM finding, those can be considered for excision and reconstruction but should never be on the cost of quality of life or resultant of severe neurological complications. Alternatively, near-total resection complemented by adjuvant radiosurgery or SRT is acceptable practice.

32.6 Closure

Dural closure is performed with the pericranial tissue that has carefully been preserved for dural reconstruction. It is important to do a watertight suture. If there has been a bone invasion by the meningioma, a piece is removed and replaced by an acrylic, titanium plate or mesh cranioplasty. Otherwise, the bone flap is replaced and secured with modern bone clips.

Dural closure must be watertight and taking into account the possible brain swelling.

32.7 Discussion

Cushing and Eisenhardt defined parasagittal meningiomas as tumors filling the parasagittal angle with no brain tissue between the tumors and the SSS. They may partially or completely invade the SSS [7, 8, 10]. Their correct management is still being discussed in the literature.

Nowadays, surgery remains the first treatment of choice [1, 3, 4], but in some well-chosen cases, radiosurgery may offer a valuable aid [20, 21]. The success of the surgery is also determined by the position of patient's head. If correctly applied, it eliminates the need for brain retraction. The supine or the prone position associated to an elevation of the head is preferred in order to control the intracranial pressure [2, 3].

Preparation of a galea capitis flap is recommended for the reconstruction of the SSS. Furthermore, the use of a pericranium patch avoids the heterologous material implant or the need for a levy of autologous veins [3, 5, 8].

Based on neuronavigation and diploic veins study, the craniotomy must be wider than the meningioma [1, 3, 8]. It is important that the bone flap extends at least 2 cm beyond the midline to allow the visualization of both sides of the sinus and to perform an easy tumor excision and sinus reconstruction [1, 2, 5], reducing the venous manipulation as a result of the wide exposure. Neuronavigation allows also to decrease the recurrence rate, the blood loss, the length of the operation, and finally to improve the recurrence-free period [3].

Before the dura opening, the ICGV is a safe investigation that allows the surgeon to protect the important parasagittal dural venous drainage. The role of the ICGV is very important during the surgical excision as it permits to distinguish the tumor's drainage veins from the collateral ones, owing to the use of a temporary clipping [2, 3, 16]. Moreover, intraoperative neuromonitoring/neurophysiology (EEG, VEPs, SSEPs, MEPs) facilitates the success of the operation [8, 27].

During the excision of the tumor, it is mandatory to preserve the bridging veins, especially those of the rolandic area through their dissection from the tumor and adjacent brain [1, 3], and edema should be treated aggressively with corticosteroid [8].

Some patients with meningiomas of the anterior third, who underwent a complete resection without reconstruction, developed neurological deficits [8].

Before the reconstruction of the SSS with a galea capitis, a temporary occlusion of the distal and proximal lumen of the sinus can be performed with surgical patties without the use of small pledgets of hemostatic material [1, 5]. It is also recommended not to use intraluminal balloons, because they may injure the sinus endothelium, nor vascular clamps, and not even aneurysm clips to avoid injuring the sinus wall and afferent veins [1, 3, 5].

Complete resection is not always achievable, and even if a Simpson grade I resection can be performed, the risk of recurrence is considerable. In these cases, radiosurgery (RS), in single- or multisession approach, and stereotactic radiotherapy (SRT) have been used as adjuvants with optimal results in terms of efficacy, and with a low incidence of toxicity [22, 28].

32.8 Conclusion

Treatment of parasagittal meningiomas involving the SSS is challenging and proved to be associated with higher risks compared to other brain locations [1, 9].

The success of this surgery is achieved through a management that starts from the patient's head positioning. It is a surgery that requires a limited and careful manipulation of the venous structures [1, 3]. The use of the ICGV plays a valuable aid in the preservation of collateral veins and it allows to avoid the use of DSA [3, 8].

Bonnal and Sindou's classifications provide a great help. It permits to establish an appropriate operative scheme during surgery. Furthermore, we also recommend the reconstruction of the anterior-third of the SSS to avoid changes in personality or mental disorders. In this manner, the respect of the patient's quality of life is prime [3–6].

Intraoperative neurophysiology monitoring and ICGV facilitate the success of surgery [8, 27].

Surgery in this pathology is the first option. In some well-selected cases, adjuvant radiotherapy or stereotactic surgery may be added [22, 28].

References

1. Brotchi J, Patay Z, Balériaux D. Surgery of the superior sagittal sinus and neighbouring veins. In: Hakuba A, editor. Surgery of the intracranial venous system. Tokyo: Springer-Verlag; 1986. p. 207–19.
2. Tomasello F, Conti A, et al. Venous preservation-guided resection a changing paradigm in parasagittal meningioma surgery. J Neurosurg. 2013;119:74–81.
3. Wang B, Zhang G-J, et al. Surgical outcomes and prognostic factors of parasagittal meningioma: a single-center experience 165 consecutive cases. Br J Neurosurg. 2022;36(6):756–61.
4. Bonnal J, Brotchi J. Surgery of the superior sagittal sinus in parasagittal meningiomas. J Neurosurg. 1978;48:935–45.
5. Sindou MP, Alvernia JE. Results of attempted radical tumor removal and venous repair in 100 consecutive meningiomas involving the major dural sinuses. J Neurosurg. 2006;105:514–25.
6. William C, Chen H, Perlow K, et al. Radiotherapy for meningiomas. J Neuro-Oncol. 2022;160(7):1–11.
7. Sindou M, Auque J. The intracranial venous system as a neurosurgeon's perspective. Adv Tech Stand Neurosurg. 2000;26:131–216.
8. Ricci A, Di Vitantonio H, et al. Parasagittal meningiomas: our surgical experience and reconstruction technique of the superior sagittal sinus. Surg Neurol Int. 2017;8:1.
9. Brotchi J. Should we pursue superior sagittal sinus grafting in parasagittal meningiomas in 2013? World Neurosurg. 2014;82:325–6.
10. Cushing H, Eisenhardt L. Meningiomas: their classification, regional behavior, life history, and surgical end results. J Am Med Assoc. 1978;112:185.
11. Cassius Vinicius C, Reis SNS, Grusmao, et al. Midline as a landmark for the position of the superior sagittal sinus on the cranial vault: an anatomical and imaging study. Surg Neurol Int. 2015;6:121.
12. Rothon AL Jr. The cerebral veins. Neurosurgery. 2002;51:S159–205.
13. Juskys R, Rocka S, et al. Anatomical variations of superior sagittal sinus and tributary bridging veins: a cadaveric study. Cureus. 2022;14(2):e 21979.
14. Sayhan S, Guvencer M, Ozer E, Arida MN. Morphometric evaluation of parasagittal venous anatomy for intracranial approaches: a cadaveric study. Turk Neurosurg. 2012;22:540–6.
15. Mathiesen T, Pettersson-Segerlind J, Kihlstrom L, Ulfarsson E. Meningiomas engaging major venous sinuses. World Neurosurg. 2014;81:116–24.
16. Yamashiro K, Muto J, Wakako A, et al. Diploic veins as collateral venous pathways in patients with dural venous sinus invasion by meningiomas. Acta Neurochir. 2021;163(6):1687–96.
17. He CJ, et al. Microsurgical treatment for parasagittal meningioma in the central gyrus region. Asian J Surg 2013. 2023;26:781–4.
18. Ferroli P, Nakaji P, Acerbi F, Albanese E, Broggi G. Indocyanine green (ICG) temporary clipping test to assess collateral circulation before venous sacrifice. World Neurosurg. 2011;75:122–5.
19. Lichtenbaum R, De Souza A, Jafar J. Intratumoral hydrogen peroxide injection. Oper Neurosurg. 2006;59(Supplement 2):470–3.
20. Pinzi V, Fariselli L, et al. Stereotactic radiotherapy for parasagittal and parafalcine meningiomas: patient selection and special considerations. Cancer Manag Res. 2019;11:10051–60.
21. Taylor et al. Reported an improved 10-year actuarial local control rate from 18% to 82%. Simila Taylor BW, Marcus RB, Friedman WA, Ballinger WE, Million RR. The meningioma controversy: postoperative radiation therapy. Int J Radiat Oncol Biol Phys. 1988;5:299–304.
22. Sindou MP, et al. Results of attempting radical tumor removal and venous repair in 100 consecutive meningioma involving the major dural sinuses, with relevant morbidity and mortality with relevant morbidity and mortality. J Neurosurg. 2006;105:514–25.
23. Mathiesen T, et al. Parasagittal meningiomas. Handb Clin Neurol. 2020;170:93–100.
24. Hancq S, Balériaux D, Brotchi J. Surgical treatment of parasagittal meningiomas. Semin Neurosurg. 2003;14:203–10.
25. Pettersson-Segerlind J, Mathiesen T. Long term 25-year follow-up of surgically treated parasagittal meningiomas. World Neurosurg. 2011;2015(76):564–71.
26. Tomasello F, et al. Venous preservation – guided resection: a changing paradigm in parasagittal meningioma surgery. J Neurosurg. 2013;119:78–81.
27. Paldor I, Doron O, Peso D, Jubran M, Sviri GE. Intraoperative neuromonitoring during resection of cranial meningiomas and its effect on the surgical workflow. Neurosurg Rev. 2022;45(2):1481–90.
28. Goldbrunner R, Stavrinou P, et al. EANO guideline on the diagnosis and management of meningiomas. Neuro-Oncology. 2021;23(11):1821–34.
29. Winestone JS, Baker Erdman H, Khweis M, Ben David E, Margalit N. Utility of MRI in surgical planning for parasagittal meningiomas. Acta Neurochir. 2023;165(7):1717–25.

Evolution and Revolution in Pituitary Medicine and Surgery

33

Nelson M. Oyesiku, Sai Mannam, and David P. Bray

Learning Objectives

- Understand the distinction between evolution and revolution in the context of change, particularly in relation to pituitary medicine and surgery.
- Examine the historical background and anatomical development of transsphenoidal (TS) surgery.
- Identify the key pioneers and their contributions to the evolution and revolution of pituitary surgery.
- Describe the various surgical approaches and interventions that have emerged over time.
- Explore the impact of technological and imaging advancements.
- Recognize the significance of medical advances in pituitary disease.
- Understand the limitations of traditional surgical approaches and the advantages of endoscopic techniques in pituitary medicine and surgery.
- Discuss the ongoing advancements and future directions in the field.

33.1 Evolution and Revolution

The terms "evolution" and "revolution" denote distinct modes of change. Evolution signifies gradual, continuous progress, characterized by incremental advancements and stepwise improvements, often resulting in lasting effects. In contrast, revolution denotes sudden, radical change, accompanied by upheaval and disruption, typically unfolding over shorter timespans. This chapter explores the pivotal role of ideas, discoveries, technologies, and individuals in the evolution and revolution of transsphenoidal (TS) surgery. It examines the historical backdrop, anatomical insights, surgical approaches, and techniques that have shaped the field of pituitary medicine and surgery over time. The chapter also assesses the significance of these changes and provides a comparison between traditional and modern approaches, offering a foundation for contemplating the future of pituitary medicine and surgery.

33.2 Anatomical Highlights for Transsphenoidal Surgery

- Sphenoid Bone: Understanding the anatomy of the sphenoid bone is crucial as it forms the surgical corridor for accessing the pituitary gland.
- Ethmoid Bone: Familiarity with the ethmoid bone and its relationship to the sphenoid sinus aids in the creation of a safe surgical pathway.
- Optic Nerve: Precise localization of the optic nerve is critical to avoid injury during surgery, as it lies close to the pituitary gland.
- Pituitary Gland: A comprehensive understanding of the pituitary gland's anatomy and its relationship to surrounding structures is fundamental for successful surgical intervention.
- Carotid Arteries: Awareness of the carotid arteries and their position relative to the pituitary gland is essential to prevent potential vascular complications.
- Cavernous Sinus: Knowledge of the cavernous sinus anatomy is crucial when dealing with invasive pituitary tumors that extend into this region.

N. M. Oyesiku (✉)
Department of Neurosurgery, University of North Carolina School of Medicine, Chapel Hill, NC, USA
e-mail: noyesik@med.unc.edu

S. Mannam
Perelman School of Medicine, University of Pennsylvania, Philadelphia, PA, USA
e-mail: mannams@pennmedicine.upenn.edu

D. P. Bray
Department of Neurosurgery, St. Luke's Health System, Boise, ID, USA
e-mail: brayd@slhs.org

I. N. Kanaan, V. Beneš (eds.), *Neuroanatomy Guidance to Successful Neurosurgical Interventions*,
https://doi.org/10.1007/978-3-031-59838-8_33

- Dura Mater: Proper identification and careful manipulation of the dura mater are essential steps during surgery to prevent cerebrospinal fluid leaks.

33.3 History/Background

33.3.1 Ancient Egyptian Transnasal Sphenoethmoidal Approach

The transnasal sphenoethmoidal approach, tracing its origins to ancient Egypt, involves accessing intracranial structures through the nose, possibly used in cranial mummification [1].

33.3.2 Endoscopic Techniques and Phillip Bozzini

Endoscopic techniques date back to the early 1800s, nearly a century before Schloffer's 1907 transsphenoidal surgery [2, 3]. Phillip Bozzini's early endoscope, known as the "Lichtleighter," though limited in visualization and patient comfort, revolutionized medicine [4].

33.3.3 Max Nitze and the Improved Endoscope

Max Nitze, a late nineteenth-century German urologist, played a vital role in advancing endoscopic techniques. His improvements, including a magnifying glass and an internal light source, significantly enhanced visualization during procedures [5]. Despite ongoing challenges with illumination, Thomas Edison's light bulb introduced further breakthroughs. In 1959, Harold Hopkins' rod-lens optical system revolutionized modern endoscopy, a milestone attributed to both Nitze and Hopkins [6].

33.3.4 Knowledge of Pituitary Disease and Pierre Marie and Minkowski

Our knowledge of pituitary disease can be traced back to the late nineteenth century, with Pierre Marie's publication on acromegaly [7] and Oscar Minkowski's discovery in 1887 linking the condition to pituitary hypertrophy [8]. In 1907, Nicolas Paulesco's hypophysectomy experiments on dogs further confirmed the importance of the pituitary gland for life [9]. Marie's work on acromegaly, which he coined as a distinct disease, and Minkowski's findings on the connection between acromegaly and pituitary enlargement were significant contributions to the field of endocrinology.

33.3.5 Sir Victor Horsley, Pioneer of Pituitary Surgery

Sir Victor Horsley (1857–1916) (Fig. 33.1) is widely regarded as the pioneer of pituitary surgery, having performed the first pituitary operation in 1889 [10]. Horsley published his surgical experience with pituitary tumors in ten cases, where he successfully removed a pituitary tumor employing a transcranial approach.

33.3.6 Davide Giordano and Francesco Durante, Pioneers of Endonasal Surgery

Davide Giordano (1864–1954) from Venice and Francesco Durante (1844–1934) were important pioneers of endonasal surgery. Giordano proposed a transglabellar-nasal approach to the pituitary, [11] which was based on cadaveric dissections and widely adopted by prominent surgeons such as

Fig. 33.1 Sir Victor Horsley. (Image licensed under Creative Commons Attribution 4.0 International license CC BY 4.0 (https://creativecommons.org/licenses/by/4.0/), Wellcome Library no. 12928i)

Herman Schloffer of Austria, Theodor Kocher of Switzerland, and Harvey Cushing, who modified the strategy to suit their own preferences. In parallel, Francesco Durante developed a transpalatine route to the gland [12]. Together, Giordano and Durante made significant contributions to the evolution of pituitary surgery.

33.3.7 Alfred Hirschmann and ENTs Applying Endoscope to Nasal Cavity

Alfred Hirschmann (1878–1961), an ENT specialist from Berlin, was another prominent figure who played a significant role in advancing the use of endoscopes in nasal surgery. In 1901, he modified a cystoscope for examining the maxillary sinuses, which paved the way for further developments in endoscopic techniques [13].

33.4 Approach/Surgical Interventions

33.4.1 Schloffer's Transsphenoidal Approach

In 1907, Schloffer, an Austrian surgeon, pioneered the TS approach for resection of pituitary tumors at the University of Innsbruck [3]. Guided by preoperative X-rays, Schloffer successfully removed a pituitary adenoma via a transfacial lateral rhinotomy incision, marking a revolutionary milestone in pituitary medicine and surgery. His achievement caught the attention of Harvey Cushing, who later introduced key concepts during his celebrated 1909 lecture, profoundly impacting the understanding and treatment of pituitary disorders [14]. Schloffer's TS procedure comprised three stages, involving an incision along the left nasolabial furrow, removal of the vomer and sphenoid rostrum, and the opening of the sphenoid sinus [15].

33.4.2 Theodor Kocher and the Submucosal Dissection Approach

During the early 1900s, there were rapid advancements in pituitary surgery as surgeons in both Europe and the United States developed various transnasal approaches to the sella. Theodor Kocher, from Berne, proposed a TS approach in 1909 that involved submucosal dissection of the nasal septum [16]. Kocher's approach addressed two significant challenges of TS surgery—reducing the risk of infection while providing better visualization within the sella and improving the cosmetic outcome.

33.4.3 Oscar Hirsch and the Endonasal Approach

In the same year that Halstead reported the first successful resection of a craniopharyngioma using an infranasal TS route, [17] Oskar Hirsch proposed an endonasal approach to the sella based on cadaveric studies [18]. Despite being dismissed by his peers, Hirsch performed the first direct transethmoidal operation without reflection of the nose (1910) and evolved the first endonasal, transseptal TS procedure shortly after [19]. He also introduced the nasal speculum.

33.4.4 Pituitary Milestones: Halstead, Hirsch, and Cushing's Innovations

In 1909, Halstead from Northwestern University in Chicago reported the first successful resection of a craniopharyngioma using an infranasal transsphenoidal (TS) route, following the guidance of Allen Kanavel [17]. Halstead's 1910 description of the sublabial approach to the sphenoid, preserving the septum and enhancing cosmetic outcomes, marked a significant advancement [20]. In the same year, Hirsch performed the first direct transethmoidal operation without nose reflection, progressing to the first endonasal, transseptal TS procedure [17–19]. Coincidentally, on June 4, 1910, Harvey Cushing (Fig. 33.2) combined a sublabial incision with submucosal septum resection. Hirsch and Cushing's exchange of ideas and Hirsch's demonstration of his endonasal approach to Cushing in 1911 further solidified the TS approach's importance [21]. Cushing's contributions included treating numerous pituitary tumors and defining Cushing's disease in 1932, while acknowledging the collective influence of innovators like Kanavel, Halstead, Hirsch, and others on his techniques.

Cushing was keenly aware of the collective influence of innovators like Kanavel, Halstead, Hirsch, and others on the evolution of his surgical techniques. He acknowledged their contributions and incorporated their modifications into his practice. This collaborative spirit and the exchange of ideas among these pioneering surgeons were instrumental in refining transsphenoidal (TS) surgery. The combined efforts of Halstead, Hirsch, and Cushing played a vital role

Fig. 33.2 Harvey Cushing. (Image licensed under Creative Commons Attribution-ShareAlike License CC BY-SA 4.0 (https://creativecommons.org/licenses/by-sa/4.0/), Wikimedia Commons file no. HarveyCushing.JPG)

in establishing TS surgery as the preferred approach for pituitary tumors, marking a significant turning point in the field of pituitary medicine and surgery [22].

33.5 Medical Treatments

33.5.1 Bromocriptine, Cabergoline, Octreotide, Pegvisomant, and Pasireotide

The discovery of genetic mutations, such as GAS and USP8, responsible for selective expansion of single-cell clones in GH adenomas and corticotropinomas led to the development of new medical treatments, including bromocriptine, cabergoline, octreotide, pegvisomant, and pasireotide, which have modified surgical indications, raised the bar for surgical interventions, and expanded the therapeutic repertoire [23, 24].

33.5.2 Development of Functional Endoscopic Sinus Surgery

In 1978, Wolfgang Draf and his colleagues revolutionized the diagnosis and treatment of sinus disease by introducing functional endoscopic sinus surgery (FESS) [25, 26]. The application of endoscopes to the paranasal sinuses allowed for minimally invasive and precise inspection and treatment of air-filled, empty, and dark sinus cavities with mucosa lining. With FESS, small incisions in the nose and a light source at the target allow for a thorough examination of multi-compartmental sinuses, even around corners.

33.5.3 Limits of the Microscope and Advantages of Endoscopy

The employment of microscopes in pituitary medicine and surgery comes with certain limitations that hinder complete and optimal visualization of the surgical field. Firstly, the intensity of light decreases as the field of view increases, thereby restricting the visual acuity and depth of field. Secondly, due to the parallel line of sight inherent to the microscope, asymmetric tumors and lateral extensions may remain obscured from view. In addition, the use of a corridor or exposure through a speculum is required, limiting maneuverability and flexibility during surgical procedures.

To address these limitations, endoscopic transsphenoidal approaches present themselves as a viable alternative. The advantages of endoscopy over traditional microscopy include superior illumination, enhanced resolution, and wider angle visualization capabilities. Additionally, the light source is in closer proximity to the target, providing greater clarity and contrast. Furthermore, the small entry point of endoscopy allows for a broader and more comprehensive view of the surgical field, without necessitating the use of a speculum.

33.5.4 Endoscopic Techniques in the Spotlight

During the 1970s, two independent groups almost simultaneously began utilizing the endoscope in conjunction with the microscope for the resection of tumors with extrasellar extension. In 1977, Apuzzo and his colleagues at the University of Southern California (USC) reported that the combination of techniques allowed for better visualization of anatomical landmarks and facilitated more effective tumor removal [27]. Similarly, Bushe and Halves in Germany published a comparable technique in 1978 [28].

The collaboration between neurological and ENT surgeons ultimately led to the development of the "pure endoscopic" TS technique in the early 1990s, in which the endoscope served as the sole visualization tool. The first use of the pure endoscopic sublabial technique for sellar lesions was reported by Roger Jankowski in France in 1992 [29].

In 1995, Sethi and Pillay from the Singapore General Hospital described a pure endoscopic endonasal TS approach for pituitary adenoma, and a few years later, Jho and Carrau used an endonasal, uninostril technique without the need for a nasal speculum [30, 31]. The indications for endoscopic TS surgery continued to expand with the introduction of additional technical adjuncts such as neuronavigation, microvascular Doppler, and other instruments. In 2003, Giorgio Frank and Ernesto Pasquini developed the trans-ethmoid-pterygoid-sphenoid (ePs) endoscopic approach for cavernous sinus lesions, which further expanded the potential applications of endoscopic techniques in pituitary medicine and surgery [32].

33.5.5 Advancements in Endoscopic Techniques for Anterior Skull Base Lesions and Overcoming Limitations

In the realm of anterior skull base lesions, the transtuberculum and transplanum approaches have been revolutionized by the advent of endoscopic techniques. The use of endoscopic TS surgery has opened up new avenues for treatment and has led to further advancements in the field. However, as with any new technology, these techniques have also presented their own set of challenges and limitations. But the medical community has been quick to respond to these challenges, and solutions are being developed to overcome these limitations. Among these solutions are the development of 3D endoscopes, as well as malleable endoscopes and instrumentation, which promise to address the difficulties encountered with standard endoscopic techniques.

33.5.6 The Nasoseptal Flap Technique

The introduction of the nasoseptal flap technique by Hadad et al. at the University of Pittsburgh Medical Center marked a significant turning point in the field of skull base surgery [33]. This technique facilitated access to areas of the brain that were previously difficult to reach, thus transforming the surgical approach to the anterior and ventral skull base. Prior to the implementation of the technique, patients with substantial dural defects resulting from endonasal skull base surgery faced a considerable risk of cerebrospinal fluid (CSF) leakage during the postoperative phase. The nasoseptal flap method has significantly improved patient outcomes by drastically reducing the incidence of postoperative CSF leaks. As a result, it has become the gold standard for reconstructing large dural defects, as well as in cases where postoperative radiation therapy is anticipated.

33.6 Medical Advances and Hormone Discoveries

In 1929, Cushing's support for the TS approach to pituitary surgery waned, leading to a 35-year stagnation due to safety concerns and improved outcomes with transcranial surgery. During this period, medical breakthroughs emerged as Herbert Evans and UC Berkeley colleagues isolated key pituitary hormones, propelling pituitary medicine forward [34–37].

The 1960s brought a significant endocrinological breakthrough with the development of radioimmunoassay (RIA). In particular, RIA's application in diagnosing acromegaly through somatomedin-C (IGF-1) in 1977 and the subsequent use of IGF-1 as a diagnostic marker in 1979 revolutionized the management of pituitary disorders [38, 39].

Concurrently, Professor GW Harris pioneered neuroendocrinology, uncovering the intricate connection between the nervous and endocrine systems. Harris's work led to the isolation of TRH in 1966, GnRH in 1971, and the discovery of somatostatin in 1973 and GHRH in 1982, significantly advancing our comprehension of pituitary function and therapeutic options [36, 40, 41].

33.7 Technological and Imaging Advances

33.7.1 Norman Dott's Lighted Nasal Speculum

Norman Dott, a Cushing's Rockefeller research fellow at the Brigham from November 1923 to June 1924, played a significant role in the evolution of pituitary surgery. After returning to Edinburgh, Dott continued to use the transsphenoidal approach and even co-authored Cushing's experience [42]. Dott also designed a groundbreaking lighted nasal speculum that revolutionized pituitary surgery by addressing the poor illumination of the deep surgical field.

33.7.2 Gerard Guiot's Intraoperative Fluoroscopy

Gerard Guiot, a French neurosurgeon, played a pivotal role in the revolution of pituitary surgery with the introduction of intraoperative fluoroscopy [43]. Guiot visited Norman Dott in Edinburgh in 1956 and was impressed by the efficacy and safety of the TS, which Dott had performed over 120 times with zero mortality in his final 80 cases. Guiot implemented the TS approach in Paris and introduced intraoperative fluoroscopy, which allowed for real-time visualization of surgical instruments' depth and position. Guiot convincingly demonstrated the superiority of the TS approach for pituitary adenomas compared to the transcranial approach, with lower morbidity and improved visual outcome in over 1000 cases. Guiot also used the TS approach for other tumors, such as craniopharyngiomas and chordomas. He was the first to report using an endoscope for resection of sellar and suprasellar lesions via the sublabial TS route in 1961. Despite occasional use of endoscopes to examine for residual tumors, Guiot abandoned endoscopic pituitary surgery due to inadequate visualization with Nitze's system-based endoscopes.

33.7.3 Jules Hardy with the Introduction of the Surgical Microscope and Selective Adenomectomy

Jules Hardy, who studied under Guiot and returned to Montreal in 1962, introduced the surgical microscope in 1965, marking the next significant technological revolution in TS surgery [43]. With superior illumination and binocular magnification, Hardy developed the selective adenectomy of pituitary microadenomas after coining the term "microadenoma" in 1968. This advance allowed for expanded indications for TS surgery, including functional microadenomas, but visualization remained limited to a straight and narrow view of the sella. To address this issue, neurosurgeons often used angled mirrors to view outside the line of sight. Ivan Ciric, who studied with Hardy, performed the first microscopic transsphenoidal surgery in the United States in 1968. Hardy's landmark technical paper in 1971 described gross total resection of pituitary adenomas and reliable separation from normal gland using the surgical microscope and selective adenectomy [44].

33.7.4 Hopkins-Storz Endoscope

The development of the Hopkins-Storz endoscope marked a major breakthrough in the evolution of endoscopy [6]. This advancement occurred in parallel with the development of a new lens system by Hopkins, which improved light transmission and image quality within a smaller diameter and provided a wider view. In collaboration with Karl Storz, a German businessman, Hopkins developed an endoscope that incorporated licensed fiberoptic external cold light transmission. The resulting Storz-Hopkins endoscopes were first produced in 1965. Other important improvements in endoscopy followed, including the introduction of the CCD camera, which allowed clinicians to document their endoscopic findings.

33.7.5 Advances in CT and MR Imaging

Imaging technology has brought about a significant revolution in pituitary medicine and surgery. CT and MRI were introduced as diagnostic tools, enabling surgeons to obtain unprecedented levels of insight into the anatomy and pathology of the pituitary gland [45, 46]. This improved the planning and execution of operations, leading to better outcomes for patients. The ability to visualize and accurately locate tumors has allowed for more precise and minimally invasive surgical approaches.

33.8 Molecular Biology and Pathological Diagnosis

In the early 1980s, the discovery of hormone-encoding genes marked a breakthrough in molecular biology and neuroendocrinology, providing crucial insights into pituitary disease and enabling targeted therapies. Pathological diagnosis also advanced through techniques like immunohistochemistry and electron microscopy, enhancing tumor differentiation. Diagnostic precision improved with dynamic assays and inferior petrosal sinus sampling (IPSS). This integration of molecular biology and pathology improved patient outcomes and shows promise in personalized treatments. Next-generation sequencing and liquid biopsy hold the potential to further transform the field. Recent techniques unveiled monoclonal origins and genetic mutations, like GAS and USP8, which has further deepened our understanding in the selective expansion of single-cell clones in these tumors [47, 48].

33.9 Surgical Pearls

- The transsphenoidal approach is the safest and most expeditious route to the pituitary region.
- Pituitary surgery has become safer as the operative microscope, endoscope, and improved instruments have improved surgical precision.
- Identification of the optic nerves and carotid arteries is important for the safe removal of pituitary tumors.
- Advances in imaging techniques allow for earlier diagnosis of pituitary lesions and identification of key anatomical landmarks.

33.10 Discussion

Pituitary medicine and surgery have evolved remarkably, propelled by pioneers, technology, and research. The shift to minimally invasive endoscopic approaches, alongside medical breakthroughs like bromocriptine and diagnostic advancements, has improved patient care. Innovations in surgical instruments and imaging, such as Norman Dott's lighted speculum, Gerard Guiot's fluoroscopy, and Jules Hardy's microscope, have enhanced precision. This journey underscores the value of embracing change for progress.

33.11 Conclusion

The history of pituitary medicine and surgery showcases continuous transformation. Pioneers, endoscopic techniques, medical advancements, and technological innovations have reshaped this field. Future prospects lie in personalized treatments and deeper insights through next-generation sequencing and liquid biopsy, emphasizing the role of transformative change in advancing pituitary medicine and surgery.

Disclosure of Funding Dr. David P. Bray is partly supported by the *Nell W. and William S. Elkin Research Fellowship in Oncology*, Winship Cancer Institute, Emory University Hospital, Atlanta, GA, and supported in part by the National Center for Advancing Translational Sciences of the National Institutes of Health under Award Number UL1TR002378 and TL1TR002382.

Conflict of Interest None.

References

1. Fanous AA, Couldwell WT. Transnasal excerebration surgery in ancient Egypt: Historical vignette. JNS [Internet]. 2012 Apr [cited 2023 Mar 29];116(4):743–8. Available from: https://thejns.org/view/journals/j-neurosurg/116/4/article-p743.xml
2. Spaner SJ, Warnock GL. A brief history of endoscopy, laparoscopy, and laparoscopic surgery. J Laparoendosc Adv Surg Tech [Internet]. 1997 Dec [cited 2023 Mar 29];7(6):369–73. Available from: http://www.liebertpub.com/doi/10.1089/lap.1997.7.369
3. Schmidt RF, Choudhry OJ, Takkellapati R, Eloy JA, Couldwell WT, Liu JK. Hermann Schloffer and the origin of transsphenoidal pituitary surgery. FOC [Internet]. 2012 Aug [cited 2023 Mar 29];33(2):E5. Available from: https://thejns.org/view/journals/neurosurg-focus/33/2/2012.5.focus12129.xml
4. Ramai D, Zakhia K, Etienne D, Reddy M. Philipp Bozzini (1773–1809): The earliest description of endoscopy. J Med Biogr [Internet]. 2018 May [cited 2023 Mar 29];26(2):137–41. Available from: http://journals.sagepub.com/doi/10.1177/0967772018755587
5. Engelsgjerd J, Deibert C. Cystoscopy [Internet]. StatPearls Publishing; 2022. Available from: https://www.ncbi.nlm.nih.gov/books/NBK493180/
6. Linder TE, Simmen D, Stool SE. Revolutionary inventions in the 20th century: the history of endoscopy. Arch Otolaryngol Head Neck Surg [Internet]. 1997 Nov 1 [cited 2023 Mar 29];123(11):1161–3. Available from: http://archotol.jamanetwork.com/article.aspx?articleid=624369
7. de Herder WW. Acromegaly and gigantism in the medical literature. Case descriptions in the era before and the early years after the initial publication of Pierre Marie (1886). Pituitary [Internet]. 2009 Sep [cited 2023 Mar 29];12(3):236–44. Available from: http://link.springer.com/10.1007/s11102-008-0138-y
8. Lindholm J, Laurberg P. Hypothyroidism and thyroid substitution: historical aspects. J Thyroid Res [Internet] 2011 [cited 2023 Mar 29];2011:1–10. Available from: http://www.hindawi.com/journals/jtr/2011/809341/
9. DeVito W. Neuroendocrine regulation of thyroid function. In: Neuroendocrinology in physiology and medicine [Internet]. Humana Press. https://doi.org/10.1007/978-1-59259-707-9_13.
10. Pascual JM, Prieto R, Mazzarello P. Sir Victor Horsley: pioneer craniopharyngioma surgeon. JNS [Internet]. 2015 Jul [cited 2023 Mar 29];123(1):39–51. Available from: https://thejns.org/view/journals/j-neurosurg/123/1/article-p39.xml
11. Artico M, Pastore FS, Fraioli B, Giuffrè R. The contribution of Davide Giordano (1864–1954) to pituitary surgery: the transglabellar-nasal approach. Neurosurgery. 1998;42(4):909–11.
12. Cavallo LM, Messina A, Esposito F, de Divitiis O, Fabbro MD, de Divitiis E, et al. Skull base reconstruction in the extended endoscopic transsphenoidal approach for suprasellar lesions. JNS [Internet] 2007 Oct [cited 2022 Jul 7];107(4):713–20. Available from: https://thejns.org/view/journals/j-neurosurg/107/4/article-p713.xml
13. Mudry A, Mlynski R, Kramp B. History of otorhinolaryngology in Germany before 1921. HNO [Internet]. 2021 May [cited 2023 Mar 29];69(5):338–65. Available from: https://link.springer.com/10.1007/s00106-021-01046-9
14. Cohen-Gadol AA, Liu JK, Laws ER. Cushing's first case of transsphenoidal surgery: the launch of the pituitary surgery era. J Neurosurg [Internet] 2005 Sep [cited 2023 Mar 30];103(3):570–4. Available from: http://thejns.org/doi/10.3171/jns.2005.103.3.0570
15. Liu JK, Das K, Weiss MH, Laws ER Jr, Couldwell WT. The history and evolution of transsphenoidal surgery. J Neurosurg. 2001;95(6):1083–96. http://thejns.org/doi/10.3171/jns.2001.95.6.1083. PMID: 11765830.
16. Kocher T. Ein Fall von hypophysis-tumor mit operativer Heilung. Deutsche Zeitschrift f Chirurgie [Internet] 1909 Jan [cited 2023 Mar 30];100(1):13–37. Available from: http://link.springer.com/10.1007/BF02819714
17. Barkhoudarian G, Laws ER. Craniopharyngioma: history. Pituitary [Internet]. 2013 Mar [cited 2023 Mar 30];16(1):1–8. Available from: http://link.springer.com/10.1007/s11102-012-0402-z
18. Hirsch. Eine neue Methode der endonasalen Operation von Hypophysentumoren. Wiener medizin. 1909:12.
19. Hirsch O. Endonasal method of removal of hypophyseal tumorswith report of two successful cases. JAMA [Internet]. 1910 Aug 27 [cited 2023 Mar 30];55(9):772. Available from: http://jama.jamanetwork.com/article.aspx?doi=10.1001/jama.1910.04330090046014
20. Halstead A. Remarks on the operative treatment of tumors of the hypophysis. With the report of two cases operated on by an oronasal method. Trans Am Surg Assoc. 1910:73–93.
21. Liu JK, Cohen-Gadol AA, Laws ER, Cole CD, Kan P, Couldwell WT. Harvey Cushing and Oskar Hirsch: early forefathers of modern transsphenoidal surgery. J Neurosurg [Internet]. 2005 Dec [cited 2023 Mar 31];103(6):1096–104. Available from: https://thejns.org/view/journals/j-neurosurg/103/6/article-p1096.xml
22. Cushing H. The Weir Mitchell lecture: surgical experiences with pituitary disorders. JAMA [Internet]. 1914 Oct 31 [cited 2023 Mar 31];LXIII(18):1515. Available from: http://jama.jamanetwork.com/article.aspx?doi=10.1001/jama.1914.02570180001001
23. Friesen H, Shiu R, Elsholtz H, Simpson S, Hughes J. Prolactin and growth hormone receptors. Ciba Found Symp. 1982:263–78.
24. Hwang P, Guyda H, Friesen H. A radioimmunoassay for human prolactin. Proc Natl Acad Sci USA [Internet]. 1971 Aug [cited 2023 Apr 1];68(8):1902–6. Available from: https://pnas.org/doi/full/10.1073/pnas.68.8.1902
25. Draf W. Therapeutic endoscopy of the paranasal sinuses. Endoscopy [Internet]. 1978 Nov [cited 2023 Apr 1];10(04):247–54. Available from: http://www.thieme-connect.de/DOI/DOI?10.1055/s-0028-1098303
26. Kennedy DW. Functional endoscopic sinus surgery: technique. archives of otolaryngology – head and neck surgery [Internet].

1985 Oct 1 [cited 2023 Apr 1];111(10):643–9. Available from: http://archotol.jamanetwork.com/article.aspx?articleid=612687
27. Apuzzo MLJ, Heifetz MD, Weiss MH, Kurze T. Neurosurgical endoscopy using the side-viewing telescope: Technical note. J Neurosurg [Internet]. 1977 Mar [cited 2023 Apr 1];46(3):398–400. Available from: https://thejns.org/view/journals/j-neurosurg/46/3/article-p398.xml
28. Bushe K, Halves E. Modifizierte Teehnik bei transnasaler Operation der Hypophysengeschwiilste. Acta Neurochir. 1978;41:163–75.
29. Jankowski R, Auque J, Simon C, Marchai JC, Hepner H, Wayoff M. Endoscopic pituitary tumor surgery: the laryngoscope [Internet]. 1992 Feb [cited 2023 Apr 1];102(2):198–202. Available from: http://doi.wiley.com/10.1288/00005537-199202000-00016
30. Sethi DS, Pillay PK. Endoscopic management of lesions of the sella turcica. J Laryngol Otol [Internet] 1995 Oct [cited 2023 Apr 1];109(10):956–962. Available from: https://www.cambridge.org/core/product/identifier/S0022215100131755/type/journal_article
31. Jho HD. Endoscopic endonasal transsphenoidal surgery: experience with 50 patients. J Neurosurg. 1997;87:44.
32. Pasquini E, Sciarretta V, Frank G, Cantaroni C, Modugno GC, Mazzatenta D, et al. Endoscopic treatment of benign tumors of the nose and paranasal sinuses. Otolaryngol Head Neck Surg [Internet]. 2004 Sep [cited 2023 Apr 1];131(3):180–6. Available from: https://onlinelibrary.wiley.com/doi/10.1016/j.otohns.2003.05.001
33. Hadad G, Bassagasteguy L, Carrau RL, Mataza JC, Kassam A, Snyderman CH, et al. A novel reconstructive technique after endoscopic expanded endonasal approaches: vascular pedicle nasoseptal flap. Laryngoscope [Internet]. 2006; Oct [cited 2023 Apr 15];116(10):1882–6. Available from: http://doi.wiley.com/10.1097/01.mlg.0000234933.37779.e4
34. Li CH, Simpson ME, Evans HM. Interstitial cell stimulating hormone. II. method of preparation and some physico-chemical studies. Endocrinology [Internet]. 1940 Nov [cited 2023 Mar 31];27(5):803–8. Available from: https://academic.oup.com/endo/article-lookup/doi/10.1210/endo-27-5-803
35. Li CH, Evans HM, Simpson ME. Adrenocorticotropic hormone. J Biol Chem [Internet]. 1943; Aug [cited 2023 Mar 31];149(2):413–24. Available from: https://linkinghub.elsevier.com/retrieve/pii/S0021925818721881
36. Li CH, Evans HM, Simpson ME. Isolation and properties of the anterior hypophyseal growth hormone. J Biol Chem [Internet] 1945 Jul [cited 2023 Mar 31];159(2):353–66. Available from: https://linkinghub.elsevier.com/retrieve/pii/S002192581952796X
37. Li C, Simpson M, Evans H. Isolation of pituitary Follicle-Stimulating Hormone (FSH). Science. 1949;109(2835):445–6.
38. Glick SM, Roth J, Yalow RS, Berson SA. Immunoassay of human growth hormone in plasma. Nature [Internet] 1963 Aug [cited 2023 Mar 31];199(4895):784–7. Available from: https://www.nature.com/articles/199784a0
39. Furlanetto RW, Underwood LE, Van Wyk JJ, D'Ercole AJ. Estimation of somatomedin-C levels in normals and patients with pituitary disease by radioimmunoassay. J Clin Invest [Internet]. 1977 Sep 1 [cited 2023 Mar 31];60(3):648–57. Available from: http://www.jci.org/articles/view/108816
40. Harris GW. Neural control of the pituitary gland. Physiol Rev. 1948;28(2):139–79.
41. Rivier J, Spiess J, Thorner M, Vale W. Characterization of a growth hormone-releasing factor from a human pancreatic islet tumour. Nature [Internet]. 1982 Nov [cited 2023 Mar 31];300(5889):276–8. Available from: http://www.nature.com/articles/300276a0
42. Lanzino G, Laws ER. Pioneers in the development of taranssphenoidal surgery: Theodor Kocher, Oskar Hirsch, and Norman Dott. J Neurosurg [Internet]. 2001 Dec [cited 2023 Mar 31];95(6):1097–103. Available from: https://thejns.org/view/journals/j-neurosurg/95/6/article-p1097.xml
43. Patel SK, Husain Q, Eloy JA, Couldwell WT, Liu JK, Key players in the resurrection and preservation of transsphenoidal surgery. FOC [Internet]. 2012 Aug [cited 2023 Mar 31];33(2):E6. Available from: https://thejns.org/view/journals/neurosurg-focus/33/2/2012.6.focus12125.xml
44. Hardy J. Transsphenoidal Hypophysectomy. J Neurosurg. 1971;34:582.
45. New PFJ, Schnur JA, Taveras JM. Computerized axial tomography with the EMI scanner. Radiology. 1974;110:109–23.
46. Araki T, Inouye T, Suzuki H, Machida T, Iio M. Magnetic resonance imaging of brain tumors: measurement of T1. Radiology. 1984;150:95–8.
47. Voellger B, Zhang Z, Benzel J, Wang J, Lei T, Nimsky C, et al. Targeting aggressive pituitary adenomas at the molecular level – a review. JCM [Internet]. 2021 Dec 27 [cited 2023 Apr 1];11(1):124. Available from: https://www.mdpi.com/2077-0383/11/1/124
48. Ballmann C, Thiel A, Korah HE, Reis AC, Saeger W, Stepanow S, et al. USP8 mutations in pituitary cushing adenomas – targeted analysis by next-generation sequencing. J Endcr Soc [Internet]. 2018 Mar 1 [cited 2023 Apr 1];2(3):266–78. Available from: https://academic.oup.com/jes/article/2/3/266/4874250

Functional Anatomy for Transsphenoidal Approach

34

Paolo Cappabianca, Andrea De Rosa, and Timothée Jacquesson

Learning Objectives
- To recognize the main anatomic landmarks of the nasal, sphenoid, and sellar steps of the standard transsphenoidal approach.
- To keep orientation while dealing with the anatomy as seen through an endoscopic perspective.
- To avoid complications during surgical procedure.

34.1 Introduction/History

The endoscopic transsphenoidal approach takes its roots from the evolution of microscopic approaches to reach the sellar region. Sellar region has been historically accessed by means of extensive craniotomies or through an extracranial transfacial transsphenoidal approach, with the endonasal corridor being introduced with the aim of providing a shorter and safer route to the sellar region and improving the esthetic outcomes: This approach was popularized by Harvey Cushing, who firstly reported 231 cases of pituitary tumors operated on [1]. The first application of the operative microscope during the transsphenoidal approach was attempted by Hardy; once the endoscope was introduced in the neurosurgical scenario, it was firstly applied as a pure visualizing instrument to enlighten the shady areas and hidden corners of the surgical field. At the end of the 1990s, a pure endoscopic approach to the sellar region, for the removal of a pituitary adenoma, was described by Jho and Carrau [2]. Since then, the endoscopic endonasal approach has become the mainstay for the treatment of tumors located in the sellar region [3, 4]. Thanks to the tremendous contribution of Amin Kassam, as a guide of the neurosurgical team of Pittsburg, the standard transsphenoidal approach became the starting point for the development of the so-called extended approaches. The anatomic and clinical contribution of the Pittsburg's team demonstrated the possibility of reaching multiple regions of the skull base, both in a sagittal plane, from anterior cranial fossa to the cranio-cervical junction, and in a coronal plane [5, 6]. Relying on their example, also our school gave its contribution to furtherly expand the knowledge of anatomy describing different anatomic regions as seen through an endoscopic exploration, such as the suprasellar, opto-carotid, and intraventricular regions [7–9].

Our aim is to describe all the anatomic landmarks, as seen through a ventral endoscopic perspective, which can serve as a guide for the neurosurgeon to maintain orientation during the whole surgical procedure thus avoiding eventual surgical complications.

34.2 Anatomy Highlights

- Nasal septum
- Middle turbinate
- Sphenoethmoidal recess
- Sphenoid sinus
- Sellar floor
- Optic protuberance
- Carotid protuberance
- Lateral opto-carotid recess
- Suprasellar notch

See Figs. 34.1 and 34.2.

A thorough knowledge of the anatomy of the nasal cavity and paranasal sinuses is mandatory: Many anatomic landmarks, which are already known from their "from above"

P. Cappabianca (✉) · A. De Rosa
Division of Neurosurgery, Department of Neurosciences, Reproductive and Odontostomatological Sciences, Università degli Studi di Napoli "Federico II", Naples, Italy
e-mail: paolo.cappabianca@unina.it

T. Jacquesson
Skull Base and Pituitary Unit, Department of Neurosurgery B, Neurological Hospital Pierre Wertheimer, Lyon, France
e-mail: Timothee.jacquesson@neurochirurgie.fr

I. N. Kanaan, V. Beneš (eds.), *Neuroanatomy Guidance to Successful Neurosurgical Interventions*,
https://doi.org/10.1007/978-3-031-59838-8_34

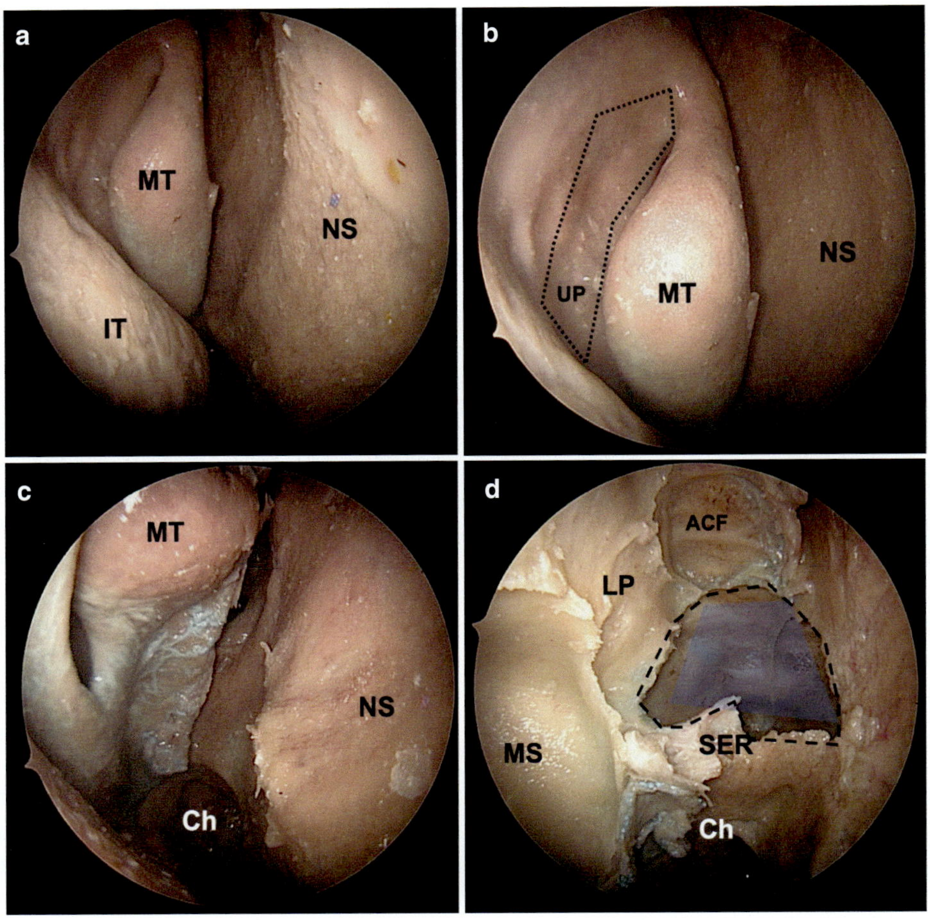

Fig. 34.1 Endoscopic endonasal anatomical picture disclosing the anatomical landmarks of the nasal phase of the transsphenoidal approach. The first landmarks to be encountered at the initial steps of the approach are the nasal septum, the inferior and middle turbinates (**a**). Lateral to the middle turbinate, the middle meatus (black dotted area) can be found. The bulla ethmoidalis is often covered by the uncinate process (**b**). The postero-inferior limit of the approach is the nasal choana, whose upper limit is an important landmark to identify the sphenoid ostium, which usually is located 1.5 cm above (black arrow) (**c**). In (**d**), an anterior sphenoidotomy (black interrupted line), a wide anterior and posterior ethmoidectomy together with removal of the lamina papyracea, has been conducted to highlight the relationship between orbit, the anterior cranial fossa, and the sellar region (shaded blue area). ACF anterior cranial fossa, Ch nasal choana, IT inferior turbinate, LP lamina papyracea, MS maxillary sinus, MT middle turbinate, NS nasal septum, SER sphenoethmoidal recess

Fig. 34.2 Endoscopic endonasal anatomical picture showing the principal landmarks of the sphenoid phase of the transsphenoidal approach. ** sphenoid septum removed, *C* clivus, *CP* carotid protuberances, *ICS* intercavernous sinus, *lOCR* lateral optocarotid recess, *mOCR* medial opto-carotid recess, *OP* optic protuberance, *pcICA* paraclival segment of the internal carotid artery, *SF* sellar floor, *SR* sphenoid rostrum, *TS* tuberculum sellae

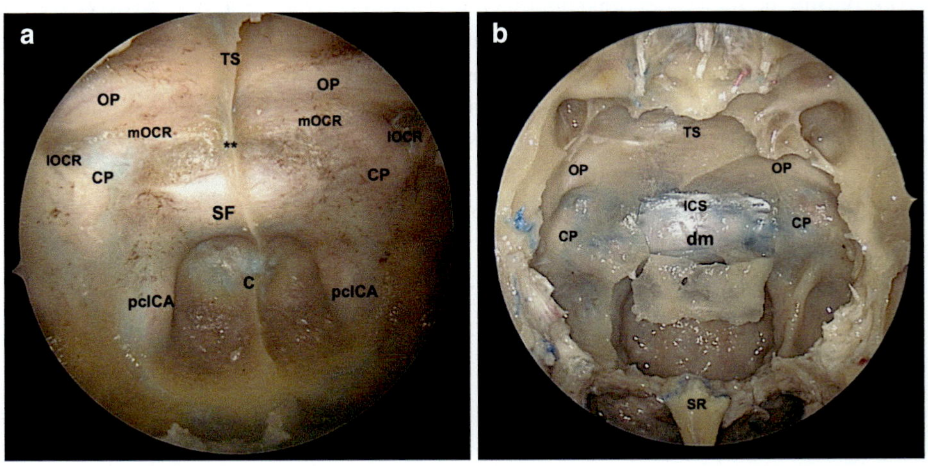

transcranial perspective, will be described as seen "from below."

The endoscopic endonasal trans-sphenoidal approach is divided into three stages, namely the nasal, sphenoidal, and sellar steps. The nasal step begins with the introduction of the endoscope through the nostril: The nasal septum forms the medial wall of the nasal cavity, and it is composed by cartilage, anteriorly, and bones in its posterior two-thirds, namely the perpendicular plate of the ethmoid, superiorly, and the vomer, inferiorly. Being a midline structure and forming the medial wall of the nasal cavity, the nasal septum is of utmost importance to keep orientation in a medial-to-lateral direction. The lateral wall of the nasal cavity is more complex, and it is composed of seven bones. The inferior turbinate is the first landmark that must be identified on the lateral wall, just above the floor of the nasal cavity. It is an autonomous bone which articulates with the conchal crest of the maxilla and ethmoid bones. The inferior turbinate covers the inferior nasal meatus, where the nasolacrimal duct drains. The medial aspect of the inferior nasal turbinate tail leads the surgeon to the posterior limit of the surgical corridor, the choana, which is the communication between the nasal cavities and the rhinopharynx. Angling the endoscope superiorly to the inferior turbinate, the head of the middle turbinate, which is a medial projection of the lateral wall of the ethmoid, comes into view; lateral to the middle turbinate, the middle nasal meatus with the ostium of the maxillary sinus can be observed. Superior to the maxillary ostium, a rounded projection could be identified, i.e., the bulla ethmoidalis, which corresponds to the middle ethmoidal air cells. During the transsphenoidal approach, the head of the middle turbinate can be luxated laterally to enlarge the surgical field. By pushing forward the endoscope along the body of the middle turbinate toward its tail, the superior portion of the choana is reached. This represents a key landmark because 1.5 cm superiorly to the upper limit of the choana, the sphenoethmoidal recess, a mucosal bridge which extend between the two bones, is identified. The sphenoethmoidal recess can cover partially or totally the sphenoid ostium, through which the sphenoid sinus can be entered. The tail of the middle is also important to identify the sphenopalatine foramen, which transmits the sphenopalatine artery, the main source of vascularization for the nasal cavity, which must be preserved if a pedicled flap is intended to be harvested for reconstruction [10, 11] (see Fig. 34.1).

The sphenoid phase of the transsphenoidal approach begins after entering the sphenoid ostium. The sphenoid sinus, depending on the degree of pneumatization, can be classified into three types: the sellar type, in which the air cavity is usually well-represented, the presellar type, where the air cavity is limited to a portion anterior to the sella turcica, and the conchal type, mostly present in children, that is the least pneumatized variation. The sphenoid septum or septa have to be removed to clearly expose the sellar floor over the posterior wall of the sphenoid sinus to identify landmarks and prepare the skull base for reconstruction at the end of the procedure. The bony landmarks in the posterior wall of the sphenoid sinus can be summarized into five bony protuberances and six grooves. The impaired, central prominence is represented by the sellar floor. Laterally to the sellar floor, on both sides, the parasellar segment of the internal carotid artery (corresponding to the distal portion of the cavernous segment of the carotid artery, following the classification of Bouthillier et al.) forms the carotid protuberances. Notably, the bone around the carotid artery may be dehiscent in some cases and the carotid appears as a reddish zone only covered by a thin rim of bone. This variation must be previewed and anticipated in order to avoid damage to the carotid artery. Superior to the carotid protuberances, the optic protuberances can be found, corresponding to the optic nerves in their canalicular segment. Between the carotid and optic protuberances, two recesses can be found: the medial and lateral opto-carotid recesses: The medial opto-carotid recess is a teardrop-shaped shallow groove which is usually less represented and corresponds intracranially to the lateral-most aspect of the *tuberculum sellae*, namely the lateral tubercular crest. The lateral opto-carotid recess, which is usually easily identified in a well-pneumatized sinus, corresponds to the pneumatization of the optic strut. The visualization of the lateral opto-carotid recess is of paramount importance as it represents the lateral limits of the bone removal during an extended transtuberculum-transplanum approach.

Superior and anterior to the sellar floor, the tuberculum sellae can be visualized as a groove, named for this reason "suprasellar notch" referred to its appearance from an endonasal perspective [12]. The clival notch is a groove located inferiorly to the sellar floor and corresponds to the sphenoidal segment of the clivus. It is interrupted by the hard palate which divides it from the rhinopharyngeal portion of the clivus. Lateral to the clival notch, the paraclival segment of the internal carotid arteries can be identified (see Fig. 34.2).

The sellar phase of the approach begins with bone removal at the sellar floor. Dura is opened to expose the pituitary gland. Extending the bone removal and dural opening superiorly, as during the transtuberculum-transplanum approach, access to the suprasellar region is gained and the chiasmatic and lamina terminalis cisterns can be exposed [13]. The medial portion of the optic nerves, the optic chiasm with the pituitary stalk surrounded by the branches of the superior hypophyseal arteries, the anterior cerebral artery with the anterior communicating artery, can all be visualized (see Fig. 34.3).

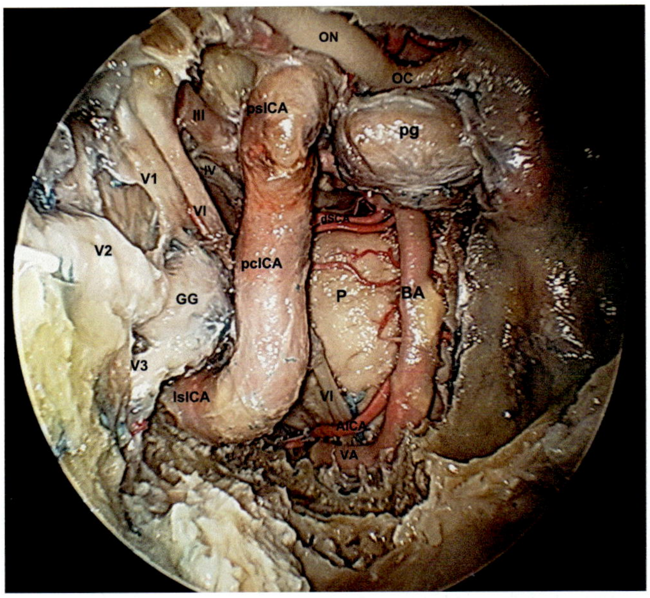

Fig. 34.3 Endoscopic endonasal anatomical picture showing the landmarks of the sellar phase of the transsphenoidal approach, depicting the principal structures of the sellar, suprasellar, parasellar, and clival regions. III oculomotor nerve, IV trochlear nerve, VI abducens nerve, V1 ophthalmic division of trigeminal nerve, V2 maxillary division of the trigeminal nerve, V3 mandibular division of the trigeminal nerve, AICA anterior inferior cerebellar artery, BA basilar artery, dSCA doubled superior cerebellar artery, GG Gasserian ganglion, lsICA lacerum segment of the internal carotid artery, OC optic chiasm, ON optic nerve, P pons, pcICA paraclival segment of the internal carotid artery, pg pituitary gland, psICA parasellar segment of the internal carotid artery, VA vertebral artery

34.3 Approach and Surgical Intervention/ Representative Cases

We herein report a case of a GH-secreting pituitary macroadenoma with suprasellar and right parasellar extension to show each step of the surgical procedure and underline relationships between the tumor and surrounding structures that must be recognized to gain the optimal surgical outcome, minimizing the risk of damage to these structures.

The anatomy of the sphenoid sinus could easily lead to disorientation due to the presence of a large right lateral recess and the presence of posterior ethmoid cells abutting into the anterior wall of the sphenoid. The high degree of pneumatization of the sphenoid sinus could also make at risk the right internal carotid artery which lies just beneath the thin lateral wall of the sinus (see Fig. 34.4).

The patient is positioned supine with the head in neutral position or slightly turned toward the operator, usually to the right. A slight extension is required if a transtuberculum-transplanum approach is planned, while a slight flexion is appropriate if a lower approach is required, i.e., to the clival region. Cottonoids soaked in a solution of lidocaine and adrenaline are inserted in both nostrils and left for about 5 min. This way, the local analgesia and vasoconstriction help in reducing bleeding during the nasal phase of the approach. The surgical field is then sterilized with iodopovidone. A small paraumbilical area is also prepped, so that a graft of autologous fat can be harvested during the reconstruction, if necessary.

The endoscope is inserted into the right nostril, parallel to the floor of the nasal cavity, usually at the superior apex of the triangle, while the other instruments are inserted through its base. The floor, inferiorly, the nasal septum, medially, and the inferior turbinate, laterally, are the first landmarks to be identified to gain orientation. Directing the endoscope superior to the inferior turbinate, the head of the middle turbinate must be identified. Lateralization of the middle turbinate is achieved by inserting a spatula along the long axis of the turbinate and gently pushing it laterally. The endoscope is advanced deeper so that the superior margin of the choana is reached. Angling the endoscope superiorly, the sphenoethmoidal recess is found 1.5 cm superior to the choana. After coagulation of the sphenoethmoidal recess and detachment of the mucosa, the sphenoid ostium is enlarged with Kerrison rongeurs or a high-speed drill. Attention must be paid not to extend the bony opening too laterally to avoid damage to the sphenopalatine artery. The endoscope is then inserted into the sphenoid sinus. As shown by our representative case, the sellar floor can be thinned and already opened by the tumor, so that the dura of the sellar floor already comes into view. Furthermore, the high degree of pneumatization of the sphenoid sinus can put in danger the internal carotid artery, which in our case was identified using the intraoperative doppler ultrasound. If tumor consistency is soft, as often happens with pituitary adenomas, it can be removed by suction and ring curettes (see Fig. 34.5).

Before addressing the suprasellar portion of the tumor, the right parasellar space was explored. This exact sequence of tumor removal—intrasellar, parasellar and, finally, the suprasellar component—is of particular importance to avoid an early decompression and fall of the suprasellar cistern, which could obstruct the control of the parasellar areas, thus improving the risk of leaving a residual tumor.

In our case, the right medial wall of the cavernous sinus was clearly infiltrated by the tumor. A 30° endoscopes and angled instruments are particularly useful when working into the parasellar spaces. We proceeded in removing the part of the tumor located superior to the parasellar carotid artery and then, by expanding the dural opening infero-laterally, by removing the portion located inferior to the carotid artery (see Fig. 34.6).

The procedure ended with the removal of the suprasellar portion of the adenoma. The fall of the suprasellar cistern represents a good intraoperative sign of the complete removal of the lesion. Intrasellar hemostasis was achieved by means of a sheet of sealant matrix (Tachosil ®) and reconstruction as there was no clear leakage aimed to close the sella by means of mucoperichondrial flap and fibrin glue.

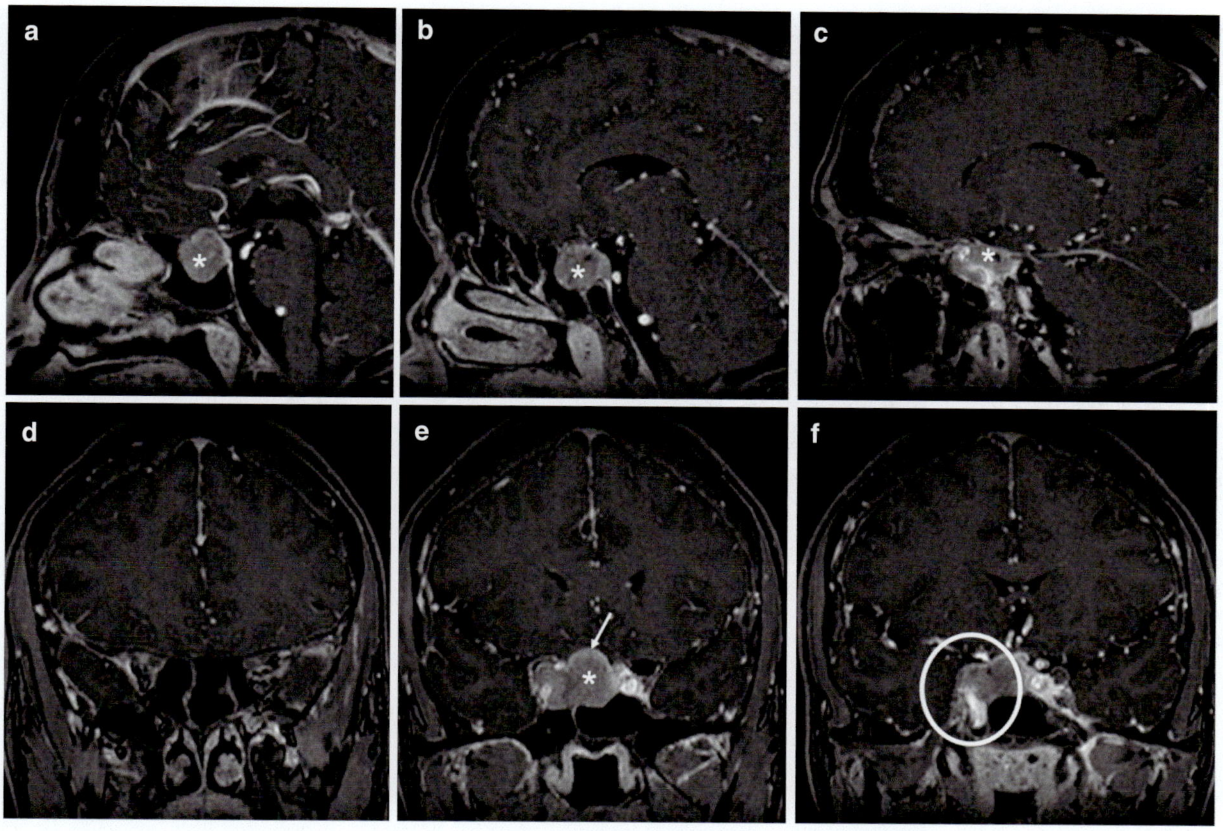

Fig. 34.4 Medial to right lateral sagittal (**a**, **b**, **c**) and anterior to posterior coronal (**d**, **e**, **f**) contrast-enhanced MRI scans depicting a pituitary macroadenoma (*) with suprasellar and right parasellar extension. Pituitary gland is pushed superiorly by the tumor (white arrow). Note the particular anatomy of the right lateral recess of the sphenoid sinus (**d**) which extends inferolaterally, and the strict relationship of the right internal carotid artery with the lateral wall of the sphenoid sinus (white circle) (**f**)

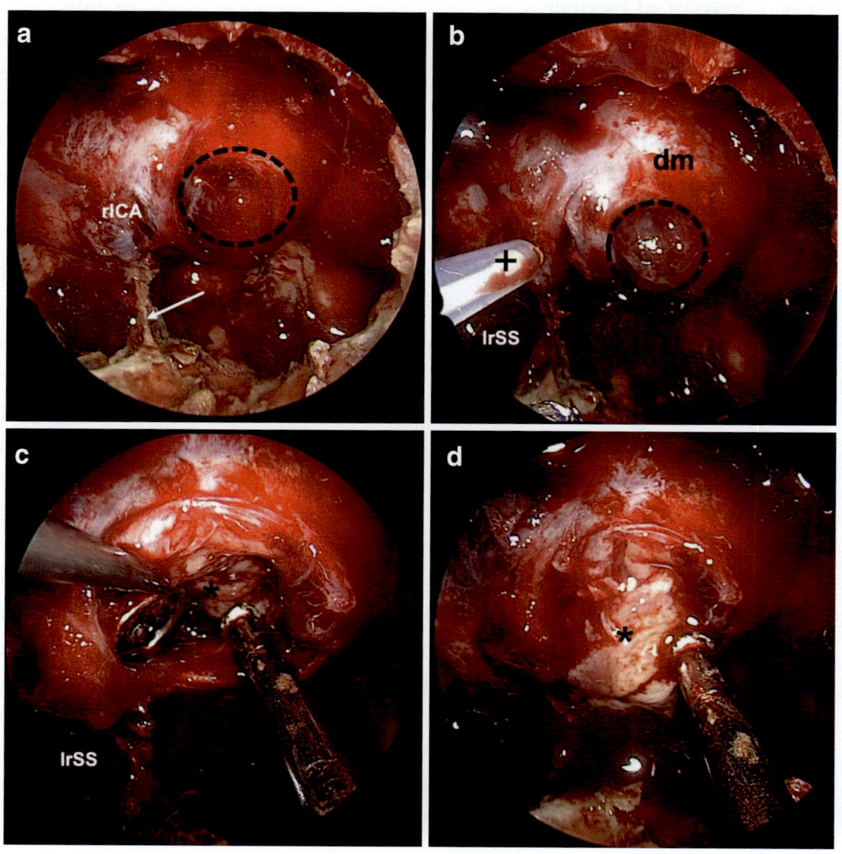

Fig. 34.5 Endoscopic endonasal intraoperative picture showing the exposure of the sellar floor and removal of the intrasellar portion of the tumor. After entering the sphenoid sinus, the sellar floor eroded by the tumor can be visualized (black dotted circle). Particular attention must be paid to the internal carotid artery and its relationship with the sphenoid septum (white arrow). Note that the bony covering of the internal carotid artery was dehiscent in this case (**a**). The correct position of the internal carotid artery can be confirmed by the Doppler ultrasound probe (+) (**b**). The intrasellar portion of the tumor (*) is firstly addressed with the aim of aspirator and ring curettes (**c**, **d**). Dm intact dura mater of the sellar floor, lrSS lateral recess of the sphenoid sinus, rICA right parasellar carotid artery

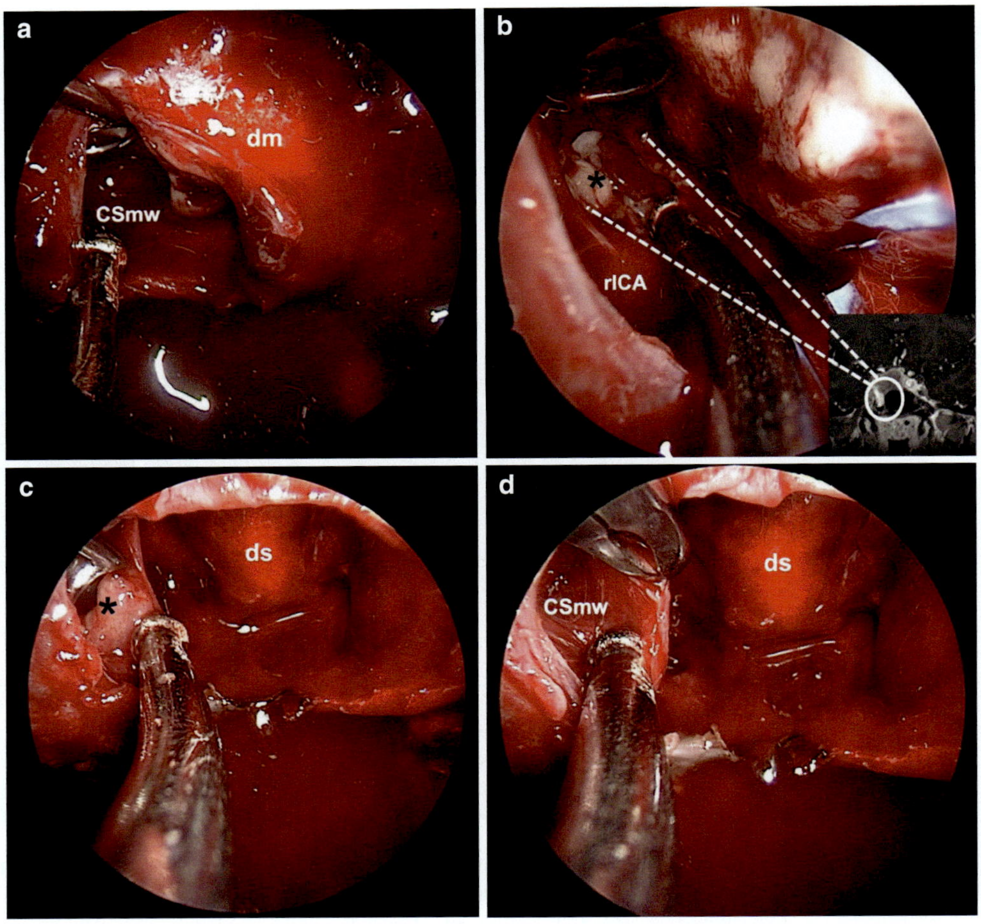

Fig. 34.6 Endoscopic endonasal intraoperative picture showing the removal of the right parasellar portion of the tumor. The medial wall of the cavernous sinus, invaded by the tumor, is explored (**a**). The superior parasellar portion of the tumor is exposed and removed (*). The insert demonstrates the correspondence between preoperative MRI and intraoperative findings (**b**). The parasellar intracarotid portion of the tumor is then identified (**c**) and removed (**d**). CSmw medial wall of the cavernous sinus, Dm dura mater of the sellar floor, ds diaphragma sellae, rICA right parasellar internal carotid artery

34.4 Surgical Pearls

- A deep preoperative examination of the neuroradiological data of the patient, focusing on both nasal and sellar anatomy, is mandatory.
- Check for anatomic individual variations, such as sphenoid sinus configuration, the presence of an Onodi cell, the distance between the internal carotid arteries.
- Minimize coagulation during the nasal step of the procedure.
- Avoid too lateral extension of the anterior sphenoidotomy to protect the sphenopalatine artery.
- Always stay in the midline! When loosing orientation, rely on median landmarks such as the nasal septum or the sphenoid rostrum.
- Open the sellar floor at its center and use the Doppler ultrasound to plan the dural opening if the inter-carotid distance is short.
- When dealing with pituitary macroadenomas with parasellar extension, the dissection should proceed sequentially: address the dissection firstly to the parasellar components and then to the suprasellar space. This will avoid an early fall of the suprasellar cistern that would hide the parasellar portions of the tumor.
- Incomplete fall of the suprasellar cistern may indicate the presence of residual tumor.
- Sellar reconstruction is the most important step of the procedure; plan preoperatively the reconstruction technique, tailoring it depending on the degree of the skull base opening.

34.5 Discussion

The transsphenoidal approach takes advantage of the nasal cavity to reach the median and paramedian skull base, with, the sellar region at the center. After the introduction of the endoscope as a unique visualization tool, the transsphenoidal approach has grown as the gold standard for the treatment of different lesions arising into and around the sellar lesions such as pituitary adenomas, craniopharyngiomas, Rathke's cleft cysts, tuberculum sellae meningiomas, and chordomas [14, 15]. The main advantage given by the endonasal route is to reach the central skull base avoiding brain retraction, which always represented one of the main drawbacks of the transcranial approaches.

A thorough knowledge of the anatomy of the nasal cavity, together with its vascular and nerve supplies, is of paramount importance: Violating structures as the nasolacrimal duct or the ostium of maxillary sinus can lead to ophthalmologic and rhinologic complications, such as lagophthalmos and sinusitis. Bipolar coagulation must be directed to the zones of interest and limited as much as possible to avoid unappropriated damages to the nasal mucosa and the formation of postoperative scarring and synechiae. Recognizing and respecting the sphenopalatine artery and its branches is fundamental to harvest a vascularized naso-septal flap which can be applied at the end of surgery for the reconstruction of the skull base defect [10, 16].

Another challenge for the neurosurgeon is to think about the anatomic landmarks in a different fashion as compared to renown to the transcranial perspective: Bony structures visualized as prominences during transcranial approaches can be recognized as bony depression from the endonasal route. One example is given by the tuberculum sellae, which represents the limit between the anterior and middle cranial fossa. It appears as a bony depression during the endonasal approach and, for this reason, our group proposed to rename it the "suprasellar notch" as referred to during endonasal surgery [12]. On the other hand, the bony prominences identified in the transsphenoidal approach correspond to the path of critical neurovascular structures, such as the optic nerve and internal carotid artery [17].

During the sellar step of the procedure, the anatomy of the sellar, suprasellar, and parasellar regions also must be deeply understood as seen from the endoscopic perspective. This is particularly true when tumors like pituitary adenomas distort the normal anatomy and alter the relationships between neurovascular structures and thus correct recognition of the landmarks can be challenging. One example is given by the relationships between the pituitary gland, the *diaphragma sellae*, and the suprasellar cistern after the removal of a pituitary adenoma. In normal conditions, the floor of the sella is formed by the *diaphragma sellae* which isolates the sellar region from the subdural space. A pituitary macroadenoma may push the normal gland against the *diaphragma sellae* and against the overlying arachnoid of the suprasellar cistern. Total removal of the lesion allows the roof of the sella to fall into the surgical cavity. At this point, care must be taken to identify the structures forming the roof of the sellar cavity: The pituitary gland often appears as a "butter-like" yellow tissue, the diaphragma sellae can be redundant and enlarged, and the arachnoid of the suprasellar cistern is seen as a bluish tissue through an enlarged *diaphragma sellae*. To underline the role of the sellar roof as an interface between the tumor and the arachnoid space, Campero et al. introduced the concept of "sellar barrier" which, depending on the relationship between the pituitary gland, the diaphragma sellae, and the suprasellar cistern, can be visualized in three possible anatomic configurations correlated to a different risk of postoperative CSF leak [18].

Nowadays, thanks to the evolution of a dedicated instrumentation and to the evolving experience in the field of endoscopic surgery, the transsphenoidal approach has been applied in the so-called extended approaches which provide access the to the entire midline skull base, from the anterior to posterior cranial fossa, and paramedian regions, such as cavernous sinus and Meckel cave [19, 20]. Mastering the anatomy of this approach would give the neurosurgeon the possibility of managing also complex skull base lesions avoiding surgical morbidities related to the transcranial routes.

34.6 Conclusions

The "natural route" given by the nasal cavities and the advantages in terms of surgical field illumination and visualization given by the endoscope make the endonasal transsphenoidal approach the gold standard for the management of lesions located in the sellar region. The complex anatomy of the nasal cavity and the ventral perspective of neurovascular structures represent a challenge to the neurosurgeons who firstly approach to this technique. A thorough knowledge and interpretation of the anatomy of this surgical route, which can be acquired only after an adequate training in the anatomic laboratory, are of paramount importance to keep the concept of "minimally invasiveness" of this technique, reducing the rate of postoperative complications and at the same time obtaining the best surgical outcome.

Acknowledgments We are grateful to **Prof. Domenico Solari** (Division of Neurosurgery, Department of Neurosciences, Reproductive and Odontostomatological Sciences, Università degli Studi di Napoli "Federico II," Naples, Italy), **Prof. Luigi Maria Cavallo** (Division of Neurosurgery, Department of Neurosciences, Reproductive and Odontostomatological Sciences, Università degli Studi di Napoli "Federico II," Naples, Italy), and **Prof. Emmanuel Jouanneau** (Skull Base and Pituitary Unit, Department of Neurosurgery B, Neurological Hospital Pierre Wertheimer, Bron, 69677 Lyon, France) for revising and drafting our manuscript.

All anatomical pictures were collected at the Laboratory of Anatomy, Université Claude Bernard Lyon I, Lyon, France. Courtesy of Prof. Timothée Jacquesson, Skull Base and Pituitary Unit, Department of Neurosurgery B, Neurological Hospital Pierre Wertheimer, Bron, 69677 Lyon, France.

References

1. Wang AJ, Zaidi HA, Laws ED. History of endonasal skull base surgery. J Neurosurg Sci. 2016;60(4):441–53.
2. Carrau RL, Jho HD, Ko Y. Transnasal-transsphenoidal endoscopic surgery of the pituitary gland. Laryngoscope. 1996;106(7):914–8.
3. Cappabianca P, Cavallo LM, Colao A, Del Basso De Caro M, Esposito F, Cirillo S, et al. Endoscopic endonasal transsphenoidal approach: outcome analysis of 100 consecutive procedures. Minim Invasive Neurosurg. 2002;45(4):193–200.

4. Dusick JR, Esposito F, Kelly DF, Cohan P, DeSalles A, Becker DP, et al. The extended direct endonasal transsphenoidal approach for nonadenomatous suprasellar tumors. J Neurosurg. 2005;102(5):832–41.
5. Kassam A, Snyderman CH, Mintz A, Gardner P, Carrau RL. Expanded endonasal approach: the rostrocaudal axis. Part I. Crista galli to the sella turcica. Neurosurg Focus. 2005;19(1):E3.
6. Kassam A, Snyderman CH, Mintz A, Gardner P, Carrau RL. Expanded endonasal approach: the rostrocaudal axis. Part II. Posterior clinoids to the foramen magnum. Neurosurg Focus. 2005;19(1):E4.
7. Cavallo LM, de Divitiis O, Aydin S, Messina A, Esposito F, Iaconetta G, et al. Extended endoscopic endonasal transsphenoidal approach to the suprasellar area: anatomic considerations – part 1. Neurosurgery. 2007;61(3 Suppl):24–33; discussion −4
8. Di Somma A, Torales J, Cavallo LM, Pineda J, Solari D, Gerardi RM, et al. Defining the lateral limits of the endoscopic endonasal transtuberculum transplanum approach: anatomical study with pertinent quantitative analysis. J Neurosurg. 2018;130(3):848–60.
9. Solari D, Chiaramonte C, Di Somma A, Dell'Aversana Orabona G, de Notaris M, Angileri FF, et al. Endoscopic anatomy of the skull base explored through the nose. World Neurosurg. 2014;82(6 Suppl):S164–70.
10. Hadad G, Bassagasteguy L, Carrau RL, Mataza JC, Kassam A, Snyderman CH, et al. A novel reconstructive technique after endoscopic expanded endonasal approaches: vascular pedicle nasoseptal flap. Laryngoscope. 2006;116(10):1882–6.
11. Cavallo LM, Solari D, Somma T, Cappabianca P. The 3F (fat, flap & flash) technique for skull base reconstruction after endoscopic endonasal suprasellar approach. World Neurosurg. 2019;126:439.
12. de Notaris M, Solari D, Cavallo LM, D'Enza AI, Enseñat J, Berenguer J, et al. The "suprasellar notch," or the tuberculum sellae as seen from below: definition, features, and clinical implications from an endoscopic endonasal perspective. J Neurosurg. 2012;116(3):622–9.
13. Aydin S, Cavallo LM, Messina A, Dal Fabbro M, Cappabianca P, Barlas O, et al. The endoscopic endonasal trans-sphenoidal approach to the sellar and suprasellar area. Anatomic Study J Neurosurg Sci. 2007;51(3):129–38.
14. Solari D, d'Avella E, Agresta G, Catapano D, D'Ecclesia A, Locatelli D, et al. Endoscopic endonasal approach for infradiaphragmatic craniopharyngiomas: a multicentric Italian study. J Neurosurg. 2023;138(2):522–32.
15. Laufer I, Anand VK, Schwartz TH. Endoscopic, endonasal extended transsphenoidal, transplanum transtuberculum approach for resection of suprasellar lesions. J Neurosurg. 2007;106(3):400–6.
16. Kassam AB, Thomas A, Carrau RL, Snyderman CH, Vescan A, Prevedello D, et al. Endoscopic reconstruction of the cranial base using a pedicled nasoseptal flap. Neurosurgery. 2008;63. (1 Suppl 1):ONS44-52; discussion ONS-3
17. Castelnuovo P, Dallan I, Tschabitscher M. Surgical anatomy of the internal carotid artery: an atlas for skull base surgeons. Berlin Heidelberg: Springer-Verlag; 2013.
18. Villalonga JF, Fuchssteiner C, Solari D, Campero A, Cavallo LM, Cappabianca P, et al. Endoscopic anatomy of the sellar barrier: From the anatomical model to the operating room. Clin Anat. 2020;33(3):468–74.
19. Jouanneau E, Simon E, Jacquesson T, Sindou M, Tringali S, Messerer M, et al. The endoscopic endonasal approach to the Meckel's cave tumors: surgical technique and indications. World Neurosurg. 2014;82(6 Suppl):S155–61.
20. Fernandez-Miranda JC, Zwagerman NT, Abhinav K, Lieber S, Wang EW, Snyderman CH, et al. Cavernous sinus compartments from the endoscopic endonasal approach: anatomical considerations and surgical relevance to adenoma surgery. J Neurosurg. 2018;129(2):430–41.

35 Approach to Giant Pituitary Adenomas

J. André Grotenhuis

Learning Objectives

- Understand the unique challenges that giant pituitary adenomas harbour compared to the more frequent smaller-sized micro- and macro-adenomas of the pituitary gland.
- Giant adenomas have different clinical features and anatomical specifications than the normal-sized adenomas.
- It is not only the size alone that poses a challenge for surgical treatment, but more so the often multi-compartmental, multi-lobular growth.
- Understand how to approach the different types of extrasellar extension, and to plan, based upon anatomical principles, the surgical treatment of giant pituitary adenomas, either in one stage, transnasal of transcranial or both routes simultaneously, or two (or more) sequential stages.

35.1 Introduction

Pituitary adenomas account for 10–15% of all intracranial tumours and have characteristic endocrinological functions affected by either hypopituitarism or hypersecretion of certain hormones. The most definitive initial treatment of any pituitary adenoma is surgical removal of the tumour by either the transsphenoidal or transcranial approach [1, 2]. Giant pituitary adenomas (GPAs) comprise about 6–14% of all pituitary tumours. The definition of giant pituitary adenomas in the literature has not always been uniform, but in the most recent published series, the definition used was pituitary adenomas with a maximum diameter in any direction of 40 mm or more [3, 4]. Some authors have named adenomas larger than 50 mm "huge adenomas" or "supergiant adenomas" but such an extension of the definition does not seem to be useful [5, 6].

The size, configuration and invasive tendency of giant pituitary adenomas make their management particularly challenging. Due to its large size, the GPA can confer a series of clinical symptoms on patients, such as headaches, dizziness, complete or partial visual loss and endocrine abnormalities, among others, depending on its specific extension.

They are mostly clinically non-functioning adenomas and occur predominantly in males. The presenting symptoms are usually secondary to compression of neighbouring structures, but also due to partial or total hypopituitarism. Functioning adenomas give rise to specific symptoms of hormonal hypersecretion. The use of dopamine agonists is considered a first-line treatment in patients with giant macroprolactinomas. Somatostatin analogues can also be used as primary treatment in cases of growth hormone and thyrotropin-producing giant adenomas, although remission of the disease is not achieved in most of these patients. Neurosurgical treatment continues to be the treatment of choice in most patients with giant pituitary adenomas. The intrinsic complexity of these tumours requires the use of different therapies in a combined or sequential way. A multimodal approach and a therapeutic strategy involving a multidisciplinary team of expert professionals form the basis of therapeutic success in these patients.

35.2 History/Background

Giant pituitary adenomas were first described in detail by Jefferson in his President's address before the Royal Society of Medicine in 1940. He described different pathways of extension of the adenoma outside the sella [7]. In Harvey Cushing's series of 338 adenomas (as reported by Henderson), some degree of unusual extension into the parasellar structures occurred in as high a ratio as 22%, while Jefferson described such extensions in 14% of his 128 cases [8].

J. A. Grotenhuis (✉)
Emeritus Professor of Neurosurgery, Radboud University Medical Centre, Nijmegen, The Netherlands

I. N. Kanaan, V. Beneš (eds.), *Neuroanatomy Guidance to Successful Neurosurgical Interventions*,
https://doi.org/10.1007/978-3-031-59838-8_35

It was already known then that large suprasellar extensions were associated with a high rate of complications and mortality. White and Warren have described a fatal case of an adenoma with very large extrasellar extension and express their frustration and concerns about such lesions clearly: "*These parasellar extensions not only produce profound alterations in the characteristic clinical syndrome but complicate the surgeon's task to a degree where his operative mortality may become prohibitive. If Jefferson's valuable paper had come to hand before this patient's operation, we should have realized from his greater experience that the best solution was to leave such a large growth respectfully alone. In it Jefferson stated that in most cases of pituitary tumours with unusual extensions operation is a course of doubtful wisdom. While his mortality was only 2 per cent for the usual small adenoma, 4 out of 12 patients with extrasellar extensions died. Finally, operation on long-standing massive tumours is not often profitable for the patient, as the severe changes in the chiasma brought about by prolonged and increasing compression are irremediable.*" [9]

This bleak outlook for patients with GPAs (operative mortality around 30% in the early 1950s and 18% in the late seventies) changed with the introduction of CT scan, and even more with MRI scan, and the operating microscope. Although the percentage of diagnosed GPAs did increase with better imaging techniques, the overall operative mortality came down to 5–10% with improved microneurosurgery and the start of endoscopic endonasal techniques, which is still a rather high number for a histologically benign tumour. At present, the main cause of mortality is the haemorrhagic apoplexy of larger tumour remnants with an incidence of around 6% but when it occurs, still a high mortality rate of 45% is observed.

Classification of GPAs starts, as with all pituitary adenomas, with an endocrine assessment, with measurement of levels of serum prolactin, insulin-like growth factor–I, growth hormone (oral glucose tolerance testing), cortisol, adrenocorticotrophic hormone, thyroid-stimulating hormone, follicle-stimulating hormone, and luteinizing hormone to evaluate pituitary insufficiency and to identify functioning adenoma.

The radiological growth pattern of GPAs on MRI can be classified as round, dumbbell-shaped, or multilobular according to Koutourousiou et al. [10]. Parasellar adenoma extension is mostly classified according to the Knosp grading system. Goel has proposed a grading system for GPAs based on the dural relationship [11]. Grade I is confined inferiorly to the sellar dura, is underneath the elevated diaphragma sellae and laterally is bordered by an intact medial wall of the cavernous sinus. The rounded superior wall or dome of the tumour is suggestive of the fact that the diaphragma sellae is intact.

The diaphragma sellae is elevated superiorly to varying extent (sometimes be elevated up to or even beyond the corpus callosum). The fact that diaphragma sellae is not transgressed is identified by the fact that the arteries of the circle of Willis are not encased by the tumour. This is important for the choice of surgical approach.

Grade II is when the tumour crossed the boundary formed by the medial wall of the cavernous sinus and invades into the confines of the cavernous sinus. Cavernous sinus invasion can be confirmed on MR imaging. Often the tumour extends on both sides of the carotid artery, but although bulged out laterally, a transgression of the lateral dural wall of the cavernous sinus is not observed, pointing to the relative firmness and the resistance offered by the lateral dural wall of the cavernous sinus.

Tumours, which selectively elevate the roof of the cavernous sinus, are labelled as Grade III, while giant pituitary tumours, which transgress the diaphragma sellae boundary and encase the arteries of the circle of Willis, are labelled as Grade IV. The exact site of dural dehiscence can often not be confirmed. This type of GPA spreads into the available spaces, engulfs or encircles the blood vessels and cranial nerves without invading these structures.

35.3 Anatomy Highlights

From an anatomical standpoint, one should study carefully the extension of the GPA. Sometimes, it is an extrasellar extension only in one direction, but more often, there are extensions into different intracranial compartments. It is important to look at all of them carefully because this will dictate the surgical approach.

- Extension into nasopharynx and sphenoid sinus.

Erosion of the sellar floor will lead to tumour extension into the sphenoid sinus and this will not cause any specific symptoms so it will remain unnoticed for a longer time unless a CSF rhinorrhoea occurs as can be seen in giant prolactinomas.

If the extension goes into the roof of the nasopharynx, patients will complain of nasal discharge, postnasal drip, and nasal obstruction. On inspection, this is often misdiagnosed as nasal polyposis and a biopsy will then finally lead to the correct diagnosis. Surgical debulking through an endonasal approach is the best for this type of extension.

- Lateral parasellar extension involving the cavernous sinus.

This is the type of extrasellar extension that has been studied and described most frequently. As we know now, there are openings in the medial wall of the cavernous sinus and the adenoma can grow through these available openings, so it does not mean invasive behaviour in an oncological sense. The same is true for the involvement of Meckel's cave. Typically, symptoms of cavernous sinus involvement will be extraocular palsies, sensory deficit of the ophthalmic first division of the trigeminal nerve and sometimes engorgement of orbital veins. The classification by Knosp is based upon the involvement or even complete encasement of the internal carotid artery inside the cavernous sinus and is very practical for deciding towards a surgical approach. Up to grade 3A from that classification can be approached through an endonasal route.

- Lateral (sub)temporal extension.

Some adenomas grow laterally between the superior part of the cavernous sinus and the overlying optic chiasm and then invade the middle cranial fossa, impinging upon the medial aspect of the temporal lobe and the optic tract. This can lead to homonymous hemianopia and to epileptic seizures, typically with an olfactory aura.

- Hypothalamic/intraventricular extension.

This is caused by projection of the tumour directly upwards into the third ventricle behind a prefixed chiasm, as is so frequently the case with craniopharyngiomas. Anatomically, it is important to remember that in such cases, the chiasm is prefixed and displaced anteriorly. De anterior cerebral arteries are usually stretched out over de already thinned-out chiasm, and hence, it is important to approach this from a more lateral side compared to a midline, interhemispheric approach. If the GPA is allowed to grow further, they have a great tendency to then extend more posteriorly since this is the route of least resistance.

- Frontal extension.

The adenoma can grow in an upward and forward direction in case of a post-fixed chiasm. It will extend into the anterior cranial fossa, compressing the olfactory bulb (often on both sides) leading to hyposmia, with further growth the adenoma will separate and compress the frontal lobe (this is often more on one side). Such an adenoma can grow to a considerable size before leading to distinct clinical symptoms, and quite a number of these extensions are cystic. During surgery, this is mostly shown as old hematoma, pointing to the diminished vascularisation to this specific region since vascular supply from both superior and inferior hypophyseal arteries does not run into frontal direction. Most of these patients present with a history of gradual visual loss and very gradual personality changes.

- Posterior, infratentorial extension.

The posterior and inferior extension of pituitary adenomas is the most uncommon type of extrasellar extension. The growth into the tentorial hiatus is usually on one side, and it then often extends into the cerebellopontine angle and can give symptoms of cranial nerve involvement (trochlear nerve, trigeminal nerve, facial nerve, and acoustic nerve!).

35.4 Approach/Surgical Intervention

Based upon the described anatomical details, the management will be described by a few representative own cases.

In a mono-compartmental GPA with large suprasellar extension (case 1), it should be clarified if the lesion stays medial from one or both optic nerves. In such a case, a straight-forward (endoscopic endonasal) transsphenoidal approach is the preferred surgical route. If there is an extension over de frontal skull base, that approach can be extended by removing the tuberculum sellae and/or planum sphenoidale.

In multicompartmental, multi-lobular GPAs (case 2 and case 3), the long axis of the lesion has to be determined. Often, there will be more than one. Then, those axis's lines must be extended to the surface, both towards the skull base/nasal part and cranial part. From those points, triangles are drawn that will cover the lesion but always considering any "obstructing" anatomical structure (e.g. in the transsphenoidal route bony structures and optic canals, or from the skull the internal carotid arteries, anterior communication artery complex, optic nerves and chiasm). By doing this, one can see how the large multi-lobular lesions can be best "covered" by one or by two different approaches. If the choice is for two approaches, one should then, finally, decide if this should be done simultaneously or in two separate sessions.

35.4.1 Case 1

A 26-year-old male patient had a check-up for "not feeling so well" and increasing fatigue. Hypopituitarism was found, MRI scan showed a giant pituitary adenoma. Macroprolactinoma was ruled out (Figs. 35.1, 35.2, 35.3, and 35.4).

35.4.2 Case 2

A 72-year-old male patient. Previous removal of a large adenoma more than a decade ago, patient did not want adjuvant

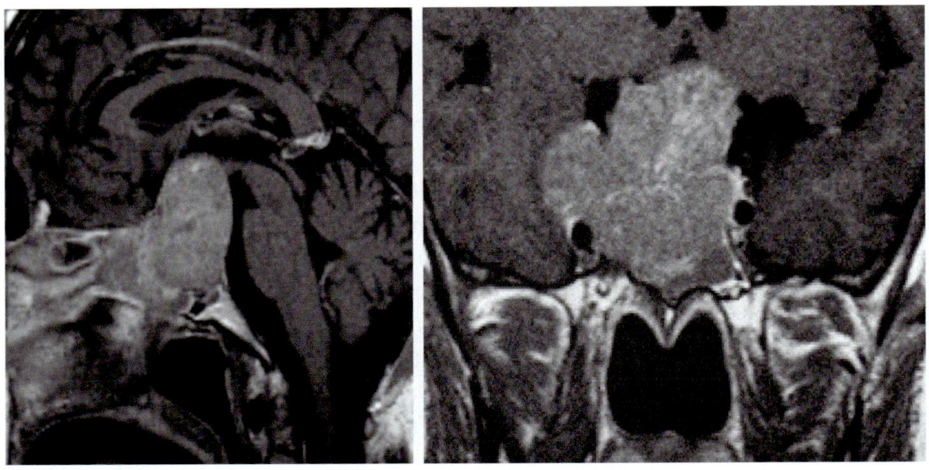

Fig. 35.1 MRI scan, sagittal and coronal T1 with Gadolinium, showing a giant pituitary adenoma with extension into the sphenoid sinus and into the third ventricle and on the right side some extension lateral to the internal carotid artery

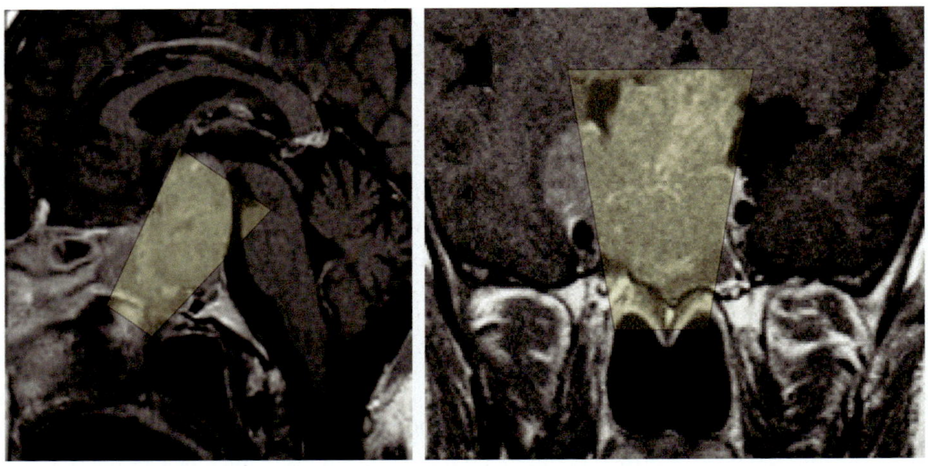

Fig. 35.2 An endoscopic endonasal approach will provide complete overview of the lesion and this route was chosen in this case (note that the area lateral of the internal carotid artery is not covered

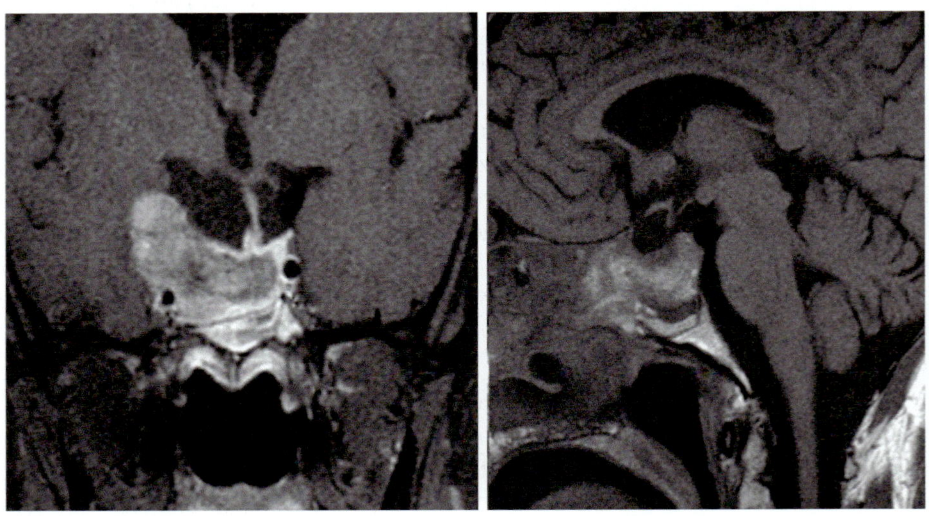

Fig. 35.3 MRI scan on second postoperative day, showing remnant in the right lateral part of the otherwise already nicely collapsing adenoma capsule

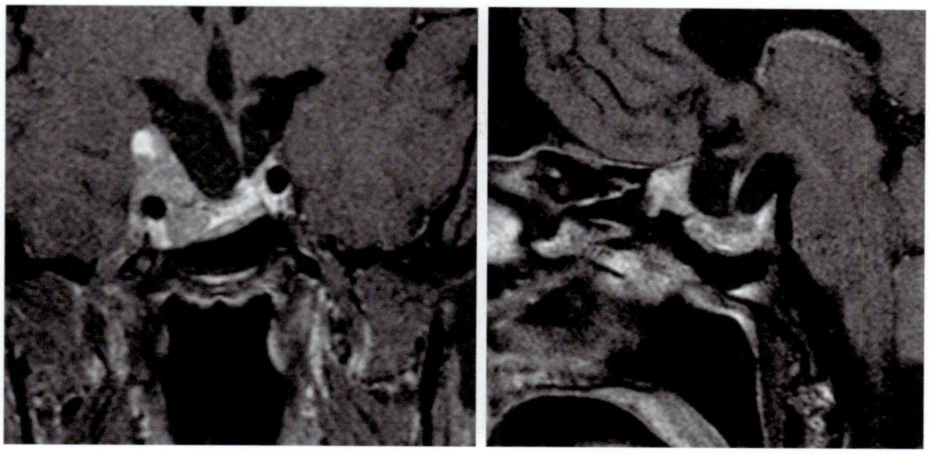

Fig. 35.4 MRI-scan 3 months postoperatively shows no residual sellar tumour but clearly the remnant inside the right cavernous sinus, initially watch by yearly MRI scans and when there was a slight increase in size, it was treated with surgical debulking of the cavernous tumour followed by stereotactic radiosurgery

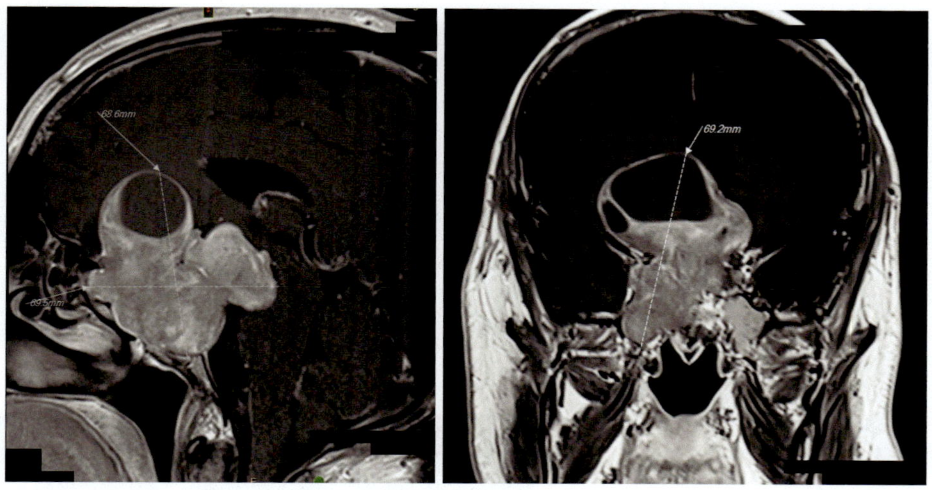

Fig. 35.5 MRI scan, sagittal and coronal T1 with Gadolinium, showing a giant adenoma with a cystic component. Planum sphenoidale is fully eroded, sphenoid sinus is filled up, and there is an irregularly shaped extension into the prepontine cistern. Optic nerve/chiasm is enveloped as is the right internal carotid artery, A1 and M1 segment

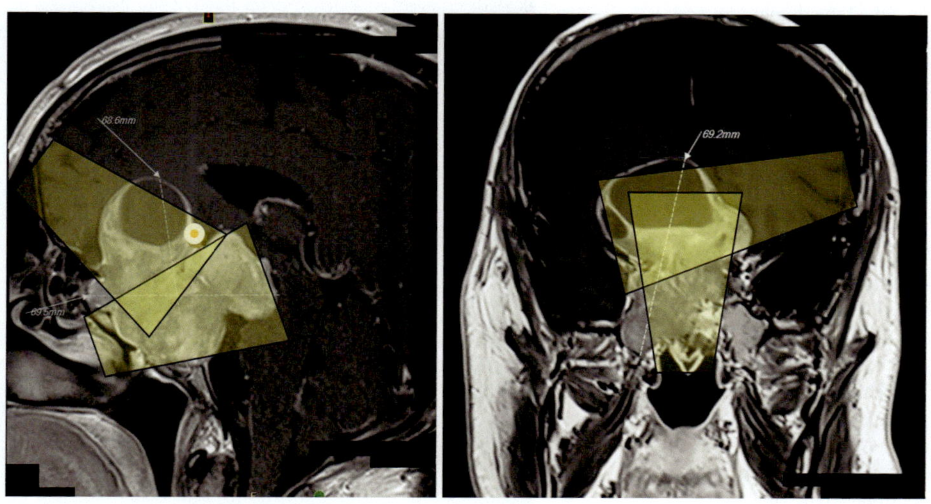

Fig. 35.6 Considering the position of the chiasm, indenting the tumour and, together with right A1 and M1 segments, dividing the lesion into two parts, the initial step should be an endoscopic endonasal transsplanum approach (the planum is also fully eroded by the tumour), but combined with a left-sided subfrontal/frontolateral craniotomy

treatment of the remnant (radiotherapy), not even when growth was evident. His wife and son urged him to undergo surgical treatment because of quick visual loss on his right eye and increasing symptoms of frontal syndrome (Figs. 35.5, 35.6, 35.7, and 35.8).

35.4.3 Case 3

A 58-year-old female patient with hypopituitarism and progressive difficulty in walking (wheelchair-bound), referred from abroad (Fig. 35.9).

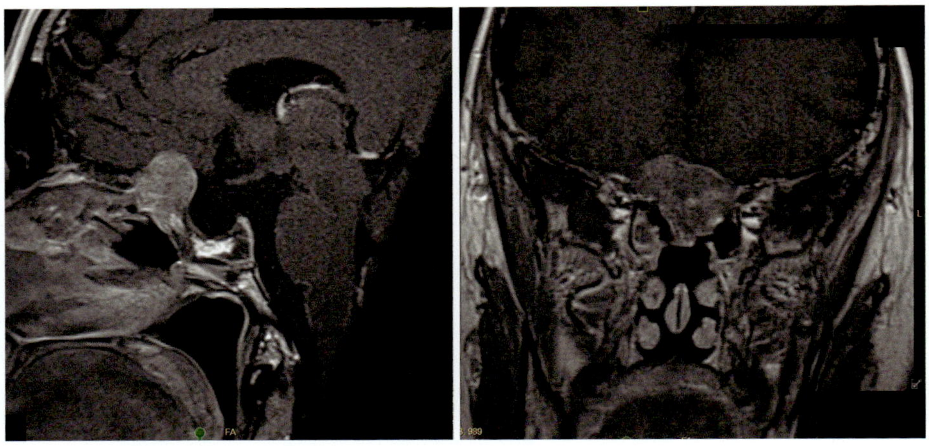

Fig. 35.7 Postoperative MRI scan showing remnant at the planum sphenoidale and the closure of the sellar approach from below and from above

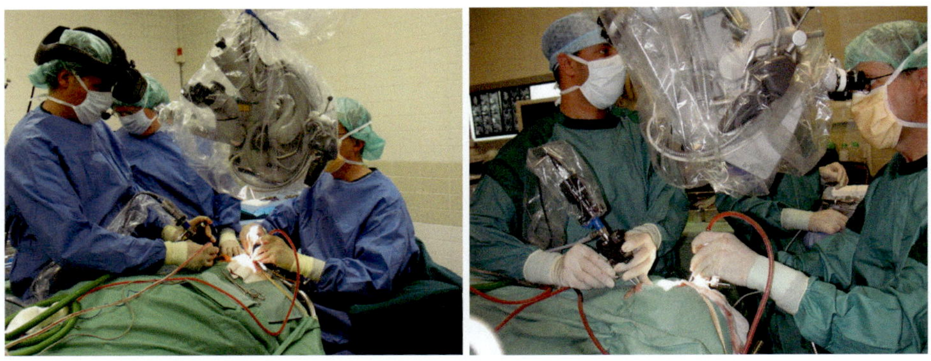

Fig. 35.8 A set-up for a combined endoscopic endonasal approach and left-sided (left image) and right-sided (right image) supraorbital subfrontal approach is shown here

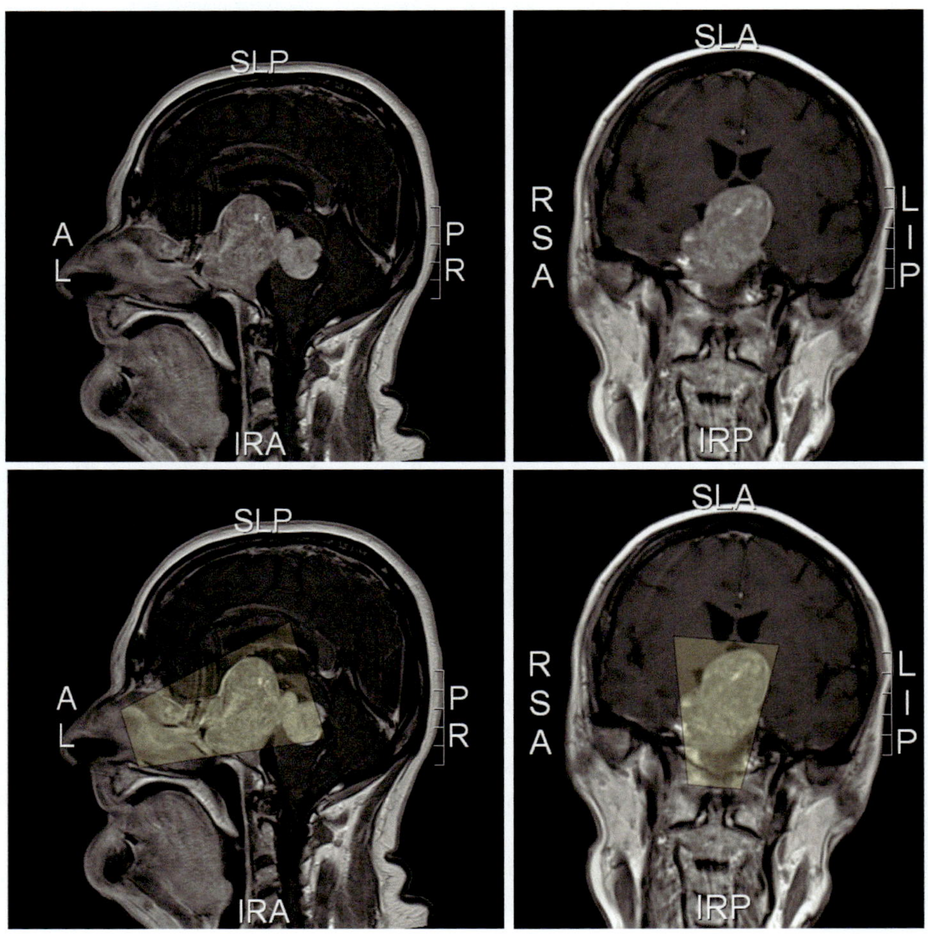

Fig. 35.9 The choice in this case is an endoscopic endonasal approach. The part of the adenoma that has breached the diaphragm and extends into the interpeduncular fossa and compressing mesencephalon and pons is difficult to reach because of the depth. It could only be debulked with the help of very long suction tubes and very long curettes. The patient improved clinically and could walk without assistance within a few days after surgery. She flew back home after a week. Unfortunately, no postoperative MRI scan was obtained there. The family has sent an annual update and she continues to do well and is still mobile

35.5 Surgical Pearls

- In most cases, the consistency of GPAs is soft and comparable to the more common micro- and macroadenomas.
- Irregularly shaped GPAs envelop neural and vascular structures rather than invading them.
- In GPAs with large midline suprasellar extension, the endoscopic endonasal/transsphenoidal approach is the approach of first choice.
- When a two-stage procedure has been chosen in irregularly shaped GPAs with extension away from the midline, the first stage should be a transcranial approach and a transsphenoidal approach as a second stage.
- Combined transsphenoidal and transcranial approach during the same procedure is technically and organizationally challenging but offers a higher rate of gross total resection (GTR), hence diminishing the risk of apoplexy in a larger adenoma remnant and allows a more secure and watertight closure of the skull base defect.

35.6 Discussion

Giant pituitary tumours usually have a meningeal cover and extend into well-defined anatomic pathways. Most enclosed adenomas even with large suprasellar extension provided that the tumour extends symmetrically in the midline above the sella turcica are an indication for endonasal transsphenoidal approach and transcranial approach is indicated only for irregular multinodular shaped tumour, or eccentric extensions into the subfrontal, temporal, or retrosellar (posterior fossa) regions that cannot be reached through the transsphenoidal route. Irrespective of the surgical approach, massive intracranial extension, an irregular configuration, and marked CS invasion are inherent factors that independently limit effective resection.

These high-risk tumours require an individualized therapeutic strategy. To achieve radical resection of the GPA, one should determine if a single stage with one approach, either endonasal-transsphenoidal or through a craniotomy, could achieve that goal. In most of the irregularly shaped multicompartmental GPA's, this will not be possible. Therefore, one should think of the possibility to combine endonasal-transsphenoidal approach with a craniotomy, either in two separate stages, or combined in a single surgery. The risk of a two-stage procedure is the apoplexy of the remnant which is associated with a high mortality rate. In my personal experience, this occurred more after using the endonasal-transsphenoidal approach as a first step, leaving a (sometimes still quite large) "non-collapsing" remnant intracranially that was scheduled to be removed through a craniotomy after e.g. 2 weeks. If apoplexy occurs in such a situation, it will happen rather soon postoperatively (within a few hours to 48 h postoperatively) and will quickly change the situation of an elective treatment for adenoma into a daring emergency. Resection of most of the tumour by first-stage surgery using the transcranial approach seems preferable in such cases of irregularly shaped GPAs with extensions away from the midline, followed by the total resection of the residual tumour by second-stage surgery using the endonasal transsphenoidal approach.

Previous studies have shown that single surgery with a single surgical approach (transsphenoidal or transcranial) for giant pituitary adenomas not only has a low GTR rate (44%) but also has a very high postoperative complication rate (11%) [3, 4, 11, 12]. Combining two approaches during one procedure (endonasal-transsphenoidal + transcranial or two different transcranial approaches) is challenging since it needs two neurosurgeons/neurosurgical teams at the same time and an experienced OR nurse who can ensure that the surgical instruments for the two approaches are strictly separated.

But the advantage is that the extent of tumour resection is much higher, with much smaller remnants and hence less risk of apoplexy. Furthermore, closing the skull base defect through the craniotomy opening using periosteum, combined with the nasoseptal flap from below and checking that it is watertight both from below and above, is another advantage. The risk of infection that is often quoted as an argument against combining an endonasal (contaminated) approach with (non-contaminated, clean) craniotomy is very low. In my personal series of 17 cases (over a 15-year period), there was not a single case of postoperative infection.

35.7 Conclusion

Management of GPAs remains challenging, despite all technological advances in imaging, medical treatment, stereotactic radiosurgery, and surgical techniques. Irrespective of the surgical approach, massive intracranial extension, an irregular configuration, and marked CS invasion are inherent factors that independently limit effective resection.

These high-risk tumours require a customized, personalized therapeutic strategy. Maximal surgical removal of giant adenomas aimed to relieve compression of the optic pathway and reduce tumour volume as much as possible through the transsphenoidal or transcranial approach, or both, used either simultaneously or subsequently, offers the best chance to control the tumour when followed with adjuvant medical and radiation therapies.

References

1. Iglesias P, Rodríguez Berrocal V, Díez JJ. Giant pituitary adenoma: histological types, clinical features and therapeutic approaches. Endocrine. 2018;61(3):407–21.

2. Micko A, Agam MS, Brunswick A, Strickland BA, Rutkowski MJ, Carmichael JD, Shiroishi MS, Zada G, Knosp E, Wolfsberger S. Treatment strategies for giant pituitary adenomas in the era of endoscopic transsphenoidal surgery: a multicenter series. J Neurosurg. 2022;136:776–85.
3. Goel A, Nadkarni T, Muzumdar D, Desai K, Phalke U, Sharma P. Giant pituitary tumours: a study based on surgical treatment of 118 cases. Surg Neurol. 2004;61(5):436–45.
4. Mortini P, Barzaghi R, Losa M, et al. Surgical treatment of giant pituitary adenomas: strategies and results in a series of 95 consecutive patients. Neurosurgery. 2007;60(6):993–1004.
5. Takakura K, Teramoto A. Management of huge pituitary adenomas. Acta Neurochir Suppl. 1996;65:13–5.
6. Penner F, Di Perna G, Baldassarre BM, Garbossa D, Zenga F. Endoscopic resection of supergiant pituitary adenoma. Surg Neurol Int. 2021 Oct;25(12):535.
7. Jefferson G. Extrasellar extensions of pituitary adenomas: President's address. Proc R Soc Med. 1940;33:433–58.
8. Henderson WR. The pituitary adenomata. A follow-up study of the surgical results in 338 cases. (Dr. Harvey Cushing's series). Br J Surg. 1939;26(104):811–921.
9. White JC, Warren S. Unusual size and extension of a pituitary adenoma; case report of a chromophobe tumour with unusually extensive compression of the base of the brain, and review of the literature on the pathways of extension of these tumours. J Neurosurg. 1945;2(2):126–39.
10. Koutourousiou M, Gardner PA, Fernandez-Miranda JC, Paluzzi A, Wang EW, Snyderman CH. Endoscopic endonasal surgery for giant pituitary adenomas: advantages and limitations. J Neurosurg. 2013;118:621–31.
11. Goel A, Nadkarni T. Surgical management of Giant pituitary tumours—a review of 30 cases. Acta Neurochir. 1996;138:1042–9.
12. Han S, Gao W, Jing Z, et al. How to deal with giant pituitary adenomas: transsphenoidal or transcranial, simultaneous or two-staged? J Neuro-Oncol. 2017;132(2):313–21.

Endoscopic Approach of Complex Pituitary Tumors with Involvement of Cavernous Sinus

36

Takeo Goto and Rajeev Sharma

Learning Objectives

- Learn the steps and technical nuances of the extended endoscopic endonasal approach to complex pituitary tumors involving cavernous sinus
- Assess the effectiveness of the extended endoscopic endonasal approach to complex pituitary tumors involving cavernous sinus
- Assess the safety of the extended endoscopic endonasal approach to complex pituitary tumors involving cavernous sinus
- Learn the likely complications of the extended endoscopic endonasal approach to complex pituitary tumors involving cavernous sinus

36.1 Introduction

Parasellar extension of pituitary macroadenomas into the cavernous sinus may be seen in 10–40% of surgical cases, thus making their complete excision challenging. Single dural layer composition of the medial wall of the cavernous sinus may be the factor responsible for these higher rates of parasellar extension into the cavernous sinus. Besides the common symptoms of vision impairment and hormonal dysfunction, diplopia is more commonly seen in these patients having cavernous sinus invasion [1].

Neuroimaging techniques (MRI, CT) are unable to delineate the medial CS wall [2]. So, medial cavernous sinus wall invasion cannot be identified preoperatively and is purely an intraoperative endoscopic finding. However, tumor extension into various CS compartments can be accurately delineated both on preoperative MRI and intraoperatively [3].

T. Goto (✉) · R. Sharma
Takeo Goto, Professor and Chairman, Department of Neurosurgery, Osaka Metropolitan University Graduate School of Medicine, Osaka, Japan
e-mail: gotot@omu.ac.jp

Safe entry zones to the various compartments of the cavernous sinus have been established by many authors [3]. Endoscopic endonasal route can assess cavernous sinus through its medial and anterior walls. So, pituitary tumors involving the CS can be removed endoscopically either by a midline approach (medial to the ICA) through the medial wall of the CS behind the ICA by using medial to the lateral trajectory or via a lateral approach (lateral to the ICA) through the anterior wall of the CS by using anterior to posterior trajectory [1, 3]. Ajlan et al [1] have suggested reserving the lateral approach for large tumors with symptomatic CS extension that cannot be appropriately excised by the medial approach. Transpterygoid approach is the lateral extension of the sellar approach used for the removal of pituitary invading the cavernous sinus [4].

36.2 History and Background

Bushe and Halves are among the first to use endoscope as an adjunct to an microscope for performing pituitary surgery in 1978. In the early 1990s, the collaboration of neurological and otorhinolaryngological surgeons developed the pure endoscopic transsphenoidal technique. Jho and Carrau of University of Pittsbergh Medical Center are recognized as the pioneers of pure endoscopic endonasal approach for pituitary adenoma treatment. In 2003, Frank and Pasquini reported the ethmoid-pterygoid-sphenoid (EPS) endoscopic approach for the treatment of cavernous sinus lesions [5].

Based on its spatial relationship with the cavernous ICA, Harris and Rhoton [6] divided the cavernous sinus into four venous spaces: posterosuperior, anteroinferior, medial, and lateral [6]. Recently, based on its spatial relationship with the cavernous ICA and with endoscopic perspective, Fernandez-Mirinda et al. [3] has divided cavernous sinus into four compartments: superior, posterior, inferior, and lateral. The later classification has completely excluded the clinical existence of the medial venous space because it is frequently obliter-

I. N. Kanaan, V. Beneš (eds.), *Neuroanatomy Guidance to Successful Neurosurgical Interventions*,
https://doi.org/10.1007/978-3-031-59838-8_36

ated by sellar tumors which grow either above, below, or behind the ICA [3].

36.3 Anatomy Highlights

- Main surgical corridor to cavernous sinus is through the medical wall of cavernous sinus. Inferior hypophysial artery should be carefully identified and coagulated and cut at the medial side of the cavernous carotid artery.
- Abducens nerve is the critical structure at the lateral side of cavernous carotid artery. Abducens nerve should be carefully differentiated from the sympathetic plexus around the carotid artery.
- Opening the oculomotor foramen with cutting the interclinoid ligament is a key step for the safe removal of intradural tumor through the superior wall of the cavernous sinus.

36.4 Approach (Surgical Intervention)

Case1: A 55-year-old post-menopausal lady presented with insidious onset gradually progressive painless vision impairment in both eyes for the last 1 year. Clinical examination showed acromegaly features and vision acuity of 6/12 in right and 6/60 in left eye, with bilateral temporal field cuts. CEMRI brain-pituitary showed a large sellar-suprasellar mass with modified Knosp right 2, left 4 compressing the optic chiasma from below with intracranial extension. Due to intracavernous tumor invasion, there was an expansion of the left superior compartment and elevation of the roof of left cavernous sinus with intracranial extension through the cavernous sinus roof into left parapeduncular space (Fig. 36.1a, b- red arc). Biochemical hormonal assessment showed raised growth hormone and IGF-1, and low cortisol levels, for which hormone replacement was started by the Endocrinology team. The patient underwent an extended endoscopic endonasal transoculomotor triangle approach. Tumor extension into the intracranial compartment through the left oculomotor triangle in the cavernous sinus roof provided a natural corridor (Fig. 36.1a- red arrow) which was enlarged by cutting interclinoid ligament anteromedially and posterior petroclinoid fold anterolaterally, and the tumor occupying left parapeduncular space was removed completely. Figure 36.2 shows how to open the oculomotor foramen. We cut the diaphragm sellae along the white dot line to open the oculomotor foramen. After opening the foramen, all subdural tumor was removed with preservation of the oculomotor nerve. The left third nerve was seen medially and intact, and left P1-PCA was seen (Fig. 36.2). Transient DI after surgery subsided spontaneously by day 7. Her vision improved to 6/9 right eye and 6/12 left eye, and her GH and IGF-1 levels showed biochemical remission without any significant residue at 3 months follow-up (Fig. 36.1c, d). Figure 36.3 shows a similar case of nonfunctional PitNET excised by the same approach.

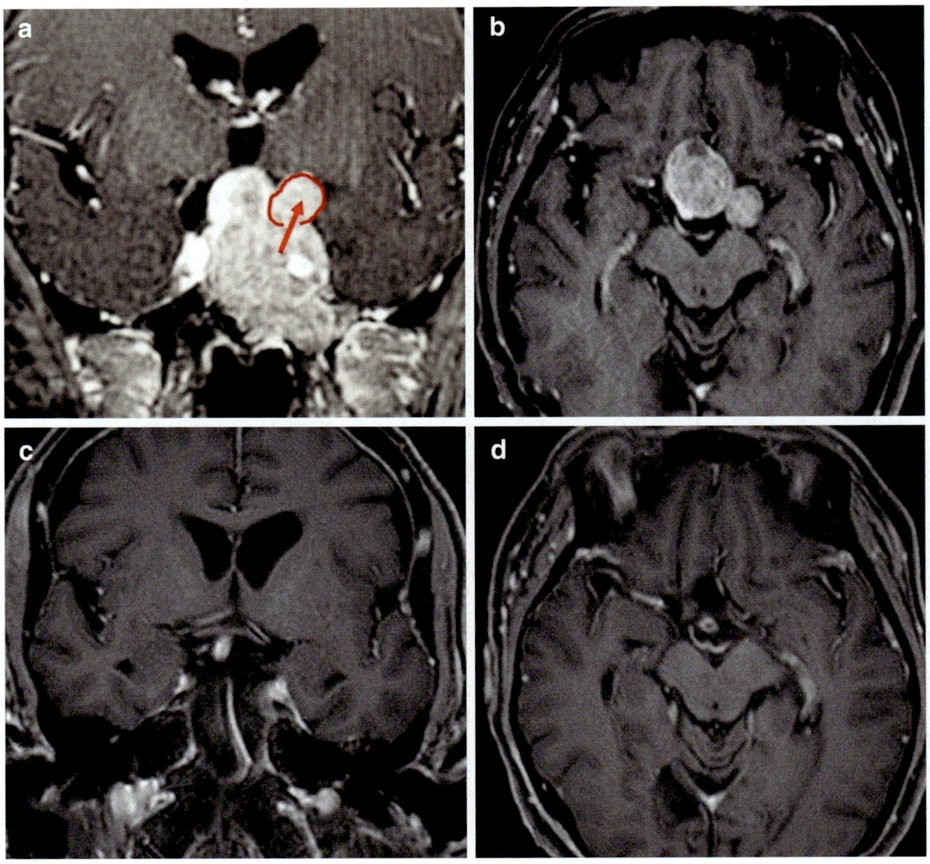

Fig. 36.1 (**a**, **b**) Preoperative CEMRI Brain showing a large sellar-suprasellar mass of modified Knosp right 2, left 4 compressing the optic chiasma from below with expansion of the left superior compartment and elevation of the roof of left cavernous sinus with intracranial extension through the cavernous sinus roof (red arrow) into left parapeduncular space (red arc); (**c**, **d**) postoperative CEMRI Brain showing no significant residue

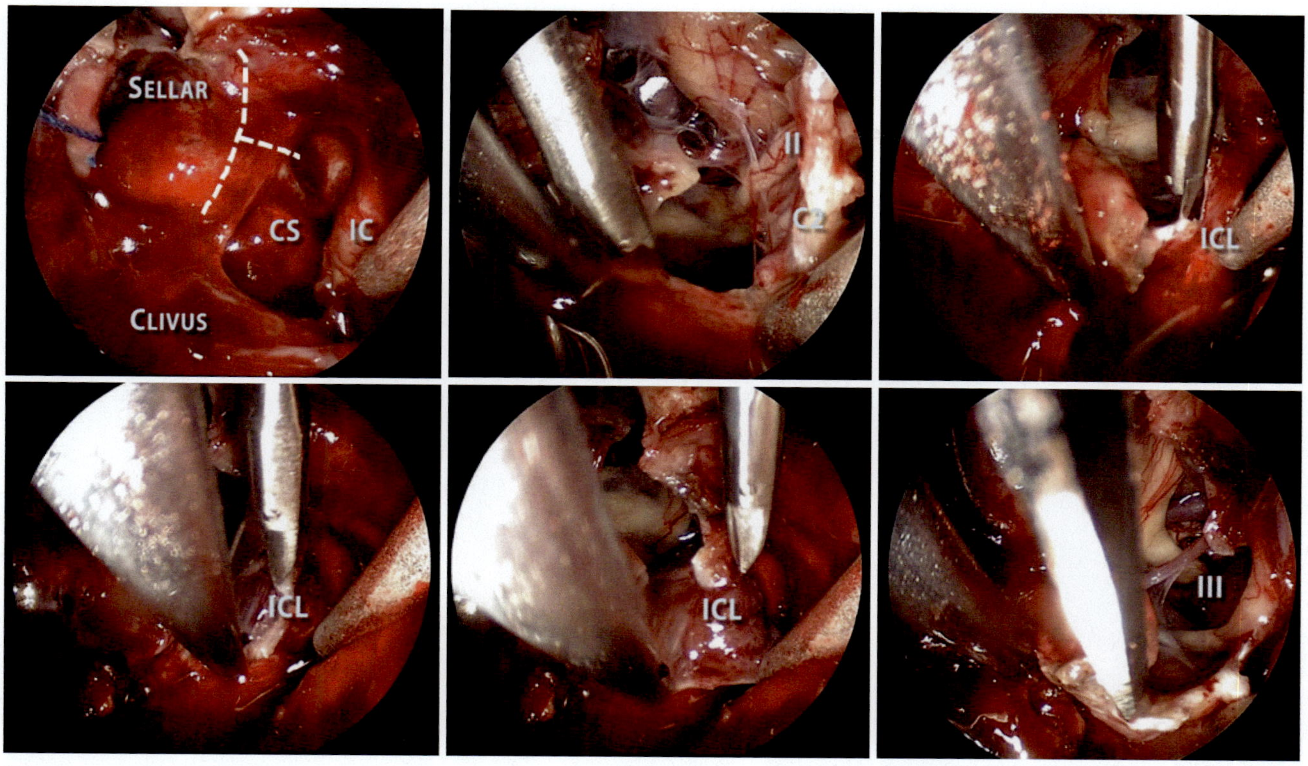

Fig. 36.2 Intraoperative photographs showing how to remove the subdural tumor after opening the oculomotor framen

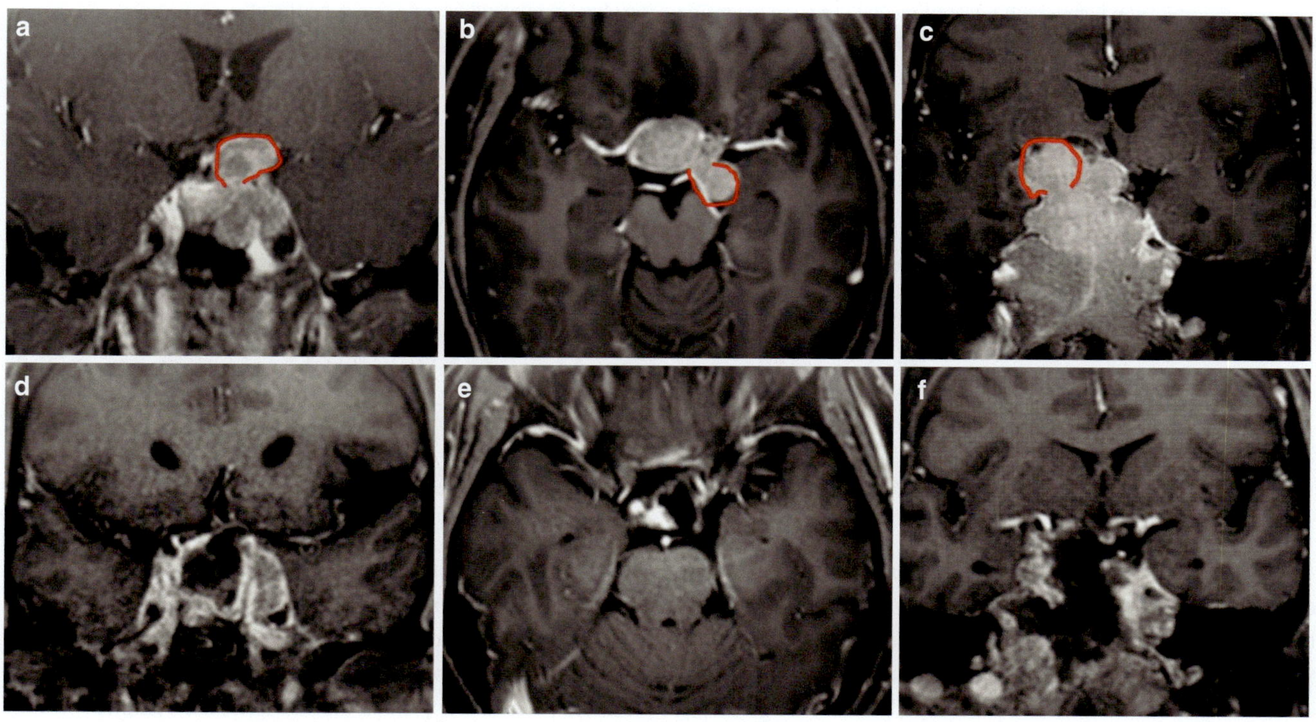

Fig. 36.3 (**a**, **b**, **c**) Preoperative CEMRI Brain showing a large sellar-suprasellar mass of modified Knosp right 2, left 3b compressing the optic chiasma from below with expansion of the bilateral superior compartments and elevation of the roof of bilateral cavernous sinuses with bilateral intracranial extension through the cavernous sinus roof (red arrow) into left parapeduncular space; (**d**, **e**, **f**) postoperative CEMRI Brain showing no significant residue

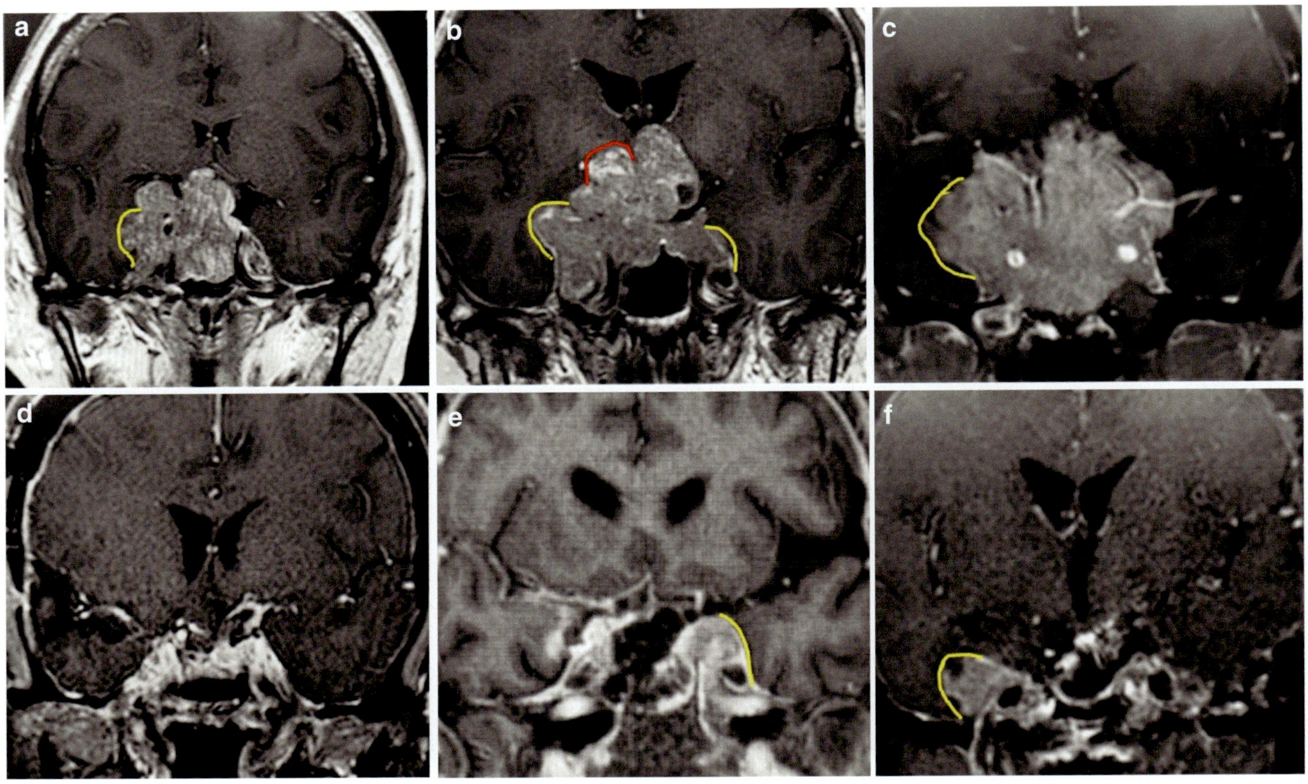

Fig. 36.4 (**a**, **b**, **c**) Preoperative CEMRI Brain showing a large multi-lobulated sellar-suprasellar mass of modified Knosp right 4, left 4 compressing chiasma from below with bilateral medial temporal extension through inferior and lateral compartments (yellow arc) and intracranial extension below the right basal ganglia through the roof of left cavernous sinus (red arc); (**d**, **e**, **f**) postoperative CEMRI Brain showing small residues in bilateral lateral CS compartments

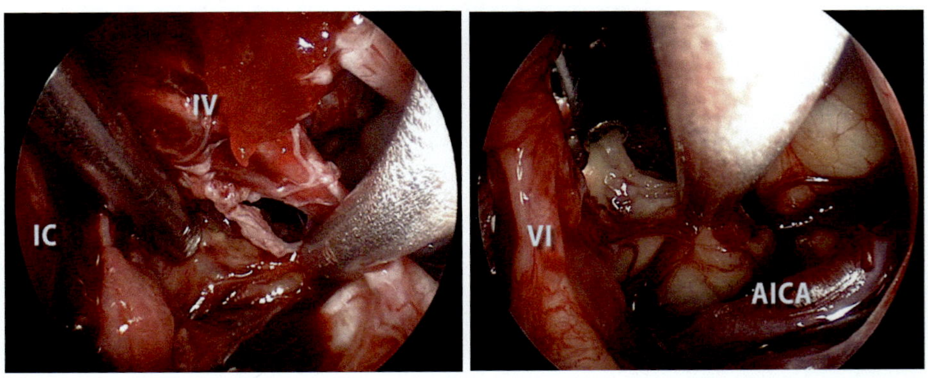

Fig. 36.5 Intraoperative photographs showing how to remove the multi-lobulate adenoma via combined surgical corridors

Case 2: A 40-year-old gentleman presented with insidious onset gradually progressive painless vision impairment in both eyes for the last 1 year. Clinical examination showed vision acuity of 6/60 in right and 6/48 in left eye, with bilateral field cuts. CEMRI brain-pituitary showed a large multi-lobulated sellar-suprasellar mass of modified Knosp right 4, left 4 compressing chiasma from below with bilateral medial temporal extension through inferior and lateral compartments (yellow arc) and intracranial extension below the right basal ganglia through the roof of left cavernous sinus (red arc) (Fig. 36.4a, b, c). Biochemical hormonal assessment showed panhypopituitarism, for which hormone replacement was started by the Endocrinology team. The patient underwent extended (transpterygoid) endoscopic endonasal and transoculomotor triangle approach combined with transcranial pterional approach on the right side. Anterior wall of the cavernous sinus was opened one by one on either side to remove tumor portion in inferior compartments. Tumor around the right abducens nerve was successfully removed via both surgical corridors from the endonasal and transcranial side (Fig. 36.5). Tumor portion in lateral compartments was left behind. His vision improved to 6/24 right eye and

6/12 left eye, and he continued to be on hormonal replacement with small residues in lateral compartments at 3-month follow-up (Fig. 36.4d, e, f). He is planned for gamma knife radiosurgery for the residual tumor portions.

36.5 Surgical Pearls

- Endoscopic CS surgery demands high expertize. So, it should be done by a team of skull-base surgeons well-versed with both the advanced endoscopic techniques and traditional transcranial techniques as some events may demand converting the endoscopic approach into an open transcranial approach. Also, it should preferably be performed in centers with endovascular support.
- Selection of the endoscopic approach is based on the specific CS compartments involved by the multidirectional tumor growth. The midline transsphenoidal approach is appropriate for pituitary tumors involving medial and posterosuperior CS compartments, and the ethmoido-pterygo-sphenoidal approach is appropriate for pituitary tumors involving lateral and anteroinferior CS compartments [7].
- The CS roof, lateral wall, and posterior wall of the CS are the potential intracranial breakthrough routes for PA invading CS [8].
- Posterior portion of the oculomotor triangle in the CS roof is single-layered, and thus a weak spot in the CS roof for intracranial tumor extension (into parapeduncular space) through the superior CS compartment.
- PAs with intracranial breakthrough through the superior and posterior CS compartments are best managed through EEA. PAs with lateral CS compartment breakthrough are the most inaccessible through EEA because of their lateral extension beyond the CNs and ICA; thus require transcranial approach [8].
- As ipsilateral sphenopalatine artery is highly manipulated during dissection in transpterygoid approach; a contralateral nasoseptal flap is harvested early during surgery for reconstruction [4]. Alternatively, the ipsilateral Hadad flap may be based on a wide pedicle of the posterior pharyngeal wall mucosal layer, thus allowing sphenopalatine artery sacrifice sometimes needed for more lateral exposure.

36.6 Discussion

Medial wall of the cavernous sinus has two parts: sellar part and sphenoidal part (Yasuda classification). Both the sphenoidal part (anterior wall of CS) formed by outer periosteal dural layer and sellar part (medial walls of the CS) formed by inner meningeal dural layer are single layered [9]. The sellar part of the medial wall of CS is formed by the single meningeal layer of dura and is called "weak spot" vulnerable to the lateral extension of pituitary tumors from sella into the cavernous sinus [10]. Medial wall of the cavernous sinus is anchored to the cavernous ICA via the caroticoclinoidal ligament [3] and various parasellar ligaments (superior, inferior, posterior). Excision of the medial wall requires cutting of these ligaments. Cavernous sinus invasion can be assessed most precisely by direct intraoperative visualization of medial wall (endoscopically) followed by its excision and histological analysis [11].

Based on its spatial relationship with the cavernous ICA, Harris and Rhoton [6] divided the cavernous sinus into four venous spaces: posterosuperior, anteroinferior, medial, and lateral [6]. Medial and posterosuperior venous spaces are in continuity with each other and do not contain nerves, thus representing a surgical corridor free from the risk of cranial nerve damage. The anteroinferior and lateral venous spaces contain the abducens nerve, thus representing the surgical corridors with the potential of cranial nerve injury [12].

Based on its spatial relationship with the cavernous ICA and with endoscopic perspective, Fernandez-Mirinda et al. [3] has divided the cavernous sinus into four compartments: superior, posterior, inferior, and lateral. These compartments may be either compressed or invaded by the tumor [3].

1. Superior compartment lies superior to the horizontal segment of intracavernous ICA and posterior to the anterior genu. Superior compartment contains oculomotor cistern and nerve in its lateral wall and interclinoid ligament in its roof. Endoscopic exploration of superior compartment is done through the medial wall of cavernous sinus using 0° and 30° endoscopes, and it requires the removal of bone covering paraclinoidal ICA and anterior wall of cavernous sinus. For accessing pituitary tumor portion extending intradurally through the superior compartment, the interclinoidal ligament needs to be cut. However, interclinoidal ligament may be mistaken for oculomotor nerve and vice versa [3].
2. Posterior compartment lies posterior to the initial short vertical segment of cavernous ICA and anterior to the lateral petroclival dura. The petroclival (gulfar) segment of the abducens nerve is located at the most inferior portion of this compartment. Endoscopic access to the posterior compartment requires extensive bone drilling to expose the anterior wall of the CS and to expose the paraclival-parasellar ICA junction.
3. Inferior compartment is located inferior to the horizontal segment and anterior genu of the intracavernous ICA and anterior to the short vertical segment of the intracavernous ICA. The anterior wall of this compartment is same as the anterior wall of the CS. It contains the sympathetic plexus and distal cavernous segment of abducens nerve.

Transpterygoid approach provides endoscopic access to this compartment.

4. Lateral compartment lies lateral to the horizontal segment and anterior genu of the intracavernous ICA. It relates to all the cavernous cranial nerves and the inferolateral arterial trunk. Proximal dural ring at the inferior surface of the optic strut marks its upper limit and the maxillary strut marks its lower limit. Thus, optic strut superiorly and maxillary strut inferiorly are the ideal landmarks for defining the complete extent of this exposure. Using a right-angled knife with a blunt tip, the exposed dura is opened in front of the anterior genu of the ICA starting from the inferior compartment and advancing superiorly and anteriorly. Medial mobilization of Cavernous and paraclinoidal ICA (from lateral to medial) requires partial sectioning of the proximal dural ring laterally. This may require coagulation and cutting of inferolateral arterial trunk and/or its branches. Electrostimulation helps in identifying and preserving cranial nerves [3].

Pituitary adenomas invading CS can extend intracranially through superior (oculomotor triangle, Dolenc triangle), lateral (Parkinson triangle), or posterior (Dorello canal) compartment penetration [13]. For pituitary tumors invading the specific compartment, an endoscopic approach focusing on providing access to that compartment is needed. For pituitary tumors invading the entire cavernous sinus, ipsilateral middle turbinectomy followed by ethmoido-pterygo-sphenoidal endoscopic approach provides simultaneous direct access to nearly all its compartments [12].

Traditionally, superior tumor extensions through oculomotor foramen are removed by transcranial approaches. However, this extended tumor portion provides a natural corridor to open the oculomotor triangle dura (through an extended endonasal approach) and resect tumor safely without causing any neurovascular injury. With the improvement in vision (advent of 4 K UHD endoscopic cameras) and technical expertize over the last decade, the cisternal portion of the oculomotor nerve (previously so-called endoscopic endonasal "blind spot") is now accessible by extended endonasal route (binostril 4-hands technique). Complete bone removal at the level of the paraclinoidal ICA and anterior wall of the cavernous sinus is a must as this facilitates the medial to lateral mobilization of the cavernous ICA for accessing the superior compartment of the cavernous sinus. After removing the sellar tumor portion, tumor portion invading inferior, posterior, and lateral compartments of the cavernous sinus is removed, as feasible. Then, the oculomotor triangle is identified and the interclinoidal ligament is delineated in the roof of the cavernous sinus. In the roof of the superior compartment of the cavernous sinus, the interclinoidal ligament separates the paraclinoidal ICA located medially and anteriorly from the oculomotor nerve, which runs lateral and posterior to the interclinoidal ligament. Dura forming the oculomotor triangle is then inspected to identify the tumor breakage point. Oculomotor nerve and PComA are then identified using electrostimulation and doppler ultrasonography respectively before cutting the interclinoidal ligament. The dura of the oculomotor triangle can be further opened laterally toward the anterior petroclinoidal ligament and inferoposteriorly toward the posterior petroclinoidal ligament. By opening the dura of the oculomotor triangle, the contents of the parapeduncular space superolaterally and the interpeduncular cistern superomedially get exposed widely. Then, the parapeduncular space tumor portion is removed by working below and lateral to the cisternal segment of the oculomotor nerve. Neuronavigation is a must during the whole procedure. Multilayered reconstruction is necessary [14].

36.7 Conclusion

The presence of lobulations from the superior surface of the pituitary tumors marks its extension into the intradural space. Single lobulation can be excised endoscopically, whereas multiple lobulations require a combined approach (endoscopic endonasal and transcranial). Endoscopic endonasal transoculomotor triangle approach is used to completely remove the tumor portion extending above the roof of the cavernous sinus. Tumor extension lateral to the intracavernous ICA segment represents a real surgical limitation. The role of radiosurgery may also be considered pivotal in recurrent or inoperable situations.

References

1. Ajlan A, Achrol AS, Albakr A, et al. Cavernous sinus involvement by pituitary adenomas: clinical implications and outcomes of endoscopic endonasal resection. J Neurol Surg B Skull Base. 2017;78(3):273–82. https://doi.org/10.1055/s-0036-1598022.
2. Balcerzak A, Tubbs RS, Zielinska N, Olewnik Ł. Clinical analysis of cavernous sinus anatomy, pathologies, diagnostics, surgical management and complications—comprehensive review. Ann Anat. 2023;245:152004. https://doi.org/10.1016/j.aanat.2022.152004.
3. Fernandez-Miranda JC, Zwagerman NT, Abhinav K, et al. Cavernous sinus compartments from the endoscopic endonasal approach: anatomical considerations and surgical relevance to adenoma surgery. J Neurosurg. 2018;129(2):430–41. https://doi.org/10.3171/2017.2.JNS162214.
4. Stamm AC, Vellutini E, Timperley D, Balsalobre L. Endoscopic transnasal pituitary surgery. In: Stamm AC, editor. Transnasal endoscopic skull base and brain surgery. Tips and pearls. 1st ed. New York: Thieme; 2011. p. 235–40. https://doi.org/10.1055/b-0034-81702.
5. Prevedello DM, Doglietto F, Jane JA, Jagannathan J, Han J, Laws ER. History of endoscopic skull base surgery: its evolution and current reality. J Neurosurg. 2007;107(1):206–13. https://doi.org/10.3171/JNS-07/07/0206.

6. Harris FS, Rhoton AL. Anatomy of the cavernous sinus: a microsurgical study. J Neurosurg. 1976;45(2):169–80. https://doi.org/10.3171/jns.1976.45.2.0169.
7. Frank G, Mazzatenta D, Sciarretta V, Zoli M, Farneti G, Pasquini E. Endoscopic cavernous sinus surgery. In: Stamm AC, editor. Transnasal endoscopic skull base and brain surgery. Tips and pearls. 1st ed. New York: Thieme; 2011. p. 246–55. https://doi.org/10.1055/b-0034-81704.
8. Xu Y, Mohyeldin A, Asmaro KP, et al. Intracranial breakthrough through cavernous sinus compartments: anatomic study and implications for pituitary adenoma surgery. Oper Neurosurg (Hagerstown). 2022;23(2):115–24. https://doi.org/10.1227/ons.0000000000000291.
9. Truong HQ, Lieber S, Najera E, Alves-Belo JT, Gardner PA, Fernandez-Miranda JC. The medial wall of the cavernous sinus. Part 1: surgical anatomy, ligaments, and surgical technique for its mobilization and/or resection. J Neurosurg. 2018;131(1):122–30. https://doi.org/10.3171/2018.3.JNS18596.
10. Raj PR, Raj PR, Smith V. Anatomical basis of clinical manifestations seen in cavernous sinus syndrome: a narrative review. Open Ophthalmol J. 2021;15:70–8. https://doi.org/10.2174/1874364102115010070.
11. Mohyeldin A, Katznelson LJ, Hoffman AR, et al. Prospective intraoperative and histologic evaluation of cavernous sinus medial wall invasion by pituitary adenomas and its implications for acromegaly remission outcomes. Sci Rep. 2022;12(1):9919. https://doi.org/10.1038/s41598-022-12980-1.
12. Frank G, Pasquini E. Cavernous sinus: endoscopic Endonasal approaches. In: Kassam AB, Gardner PA, editors. Progress in neurological surgery, vol. 26. Karger AG; 2012. p. 119–39. https://doi.org/10.1159/000329223.
13. Yang F, Bi Y, Zhou Q, et al. Pituitary adenoma with cavernous sinus compartment penetration and intracranial extension: surgical anatomy, approach, and outcomes. Front Oncol. 2023:13. https://www.frontiersin.org/articles/10.3389/fonc.2023.1169224. Accessed 1 July 2023.
14. Ferrareze Nunes C, Lieber S, Truong HQ, et al. Endoscopic endonasal transoculomotor triangle approach for adenomas invading the parapeduncular space: surgical anatomy, technical nuances, and case series. J Neurosurg. 2018;130(4):1304–14. https://doi.org/10.3171/2017.10.JNS17779.

Surgical Anatomy of the Intracavernous Carotid Artery

37

Victor Hugo Pérez and Juan E. Small

Learning Objectives
After reviewing this chapter, the reader will be able to:
- Identify the anatomical variations and origins of the meningohypophyseal trunk and the inferolateral trunk.
- Label the branches that originate from the meningohypophyseal trunk.
- Compare and contrast the differences and morphometric characteristics of the fine and complex anatomy of the inferolateral trunk to supply the ocular motor nerves in the lateral aspect of the cavernous sinus.

37.1 Introduction

A thorough understanding of the anatomy of the cavernous internal carotid artery is of crucial importance for both surgical and endovascular treatments. The details and intricacies of this segment of the internal carotid artery are therefore of great interest to the specialized personnel endeavoring in this field.

37.2 Anatomy Highlights

- The cavernous segment (C4) of the internal carotid artery is the only artery in the body that is surrounded completely by venous blood.
- The cavernous segment (C4) of the ICA spans from the petrolingual ligament and extends to the proximal dural ring.
- The meningohypophyseal trunk (MHT) and the inferolateral trunk (ILT) are the most consistent branches.
- MHT gives rise to three branches: 1. the tentorial (Bernasconi-Cassinari); 2. the inferior hypophyseal artery, which supplies the posterior lobe of the pituitary and the dura of the sellar floor; and 3. the clival (dorsal meningeal artery).
- Sympathetic plexus and sixth nerve have close proximity to the cavernous segment of ICA.
- The (MHT) is the largest and most proximal branch of the C4 segment of ICA. It originates from the medial loop and is always medial to the sixth cranial nerve as it crosses the vertical segment of the ICA.

V. H. Pérez (✉)
Forensic Sciences Institute, TSJ, Autonomous University of Mexico State, Mexico City, Mexico

J. E. Small
Lahey Hospital and Medical Center, Burlington, MA, USA

37.3 History and Background

The internal carotid artery (ICA) originates from the common carotid artery in the neck approximately at the level of the superior border of the thyroid cartilage or the fourth cervical vertebra. The ICA has been classified by Fisher, Giboi, Alain Bouthillier, and other authors [1–4]. Alain Bouthillier divides it into seven segments: cervical, petrous, lacerum, cavernous, clinoid, ophthalmic, and communicating. Alternatively, Gibo describes five segments: cervical, petrous, cavernous, infraclinoid, and supraclinoid. In turn, the supraclinoid segment is then described as dividing into three segments: ophthalmic, communicating, and choroidal.

37.4 Branches of the Cavernous Internal Carotid Artery

In the following discourse, we deal with the segments that are closely related to the cavernous segment and the directly related lacerum segment. The cavernous segment begins in the distal portion of the carotid channel, corresponding to a vertical posterolateral line to the exocranial margin of the foramen lacerum. The segment runs above the foramen lacerum, but not through it. In this portion, the lateral knee of the ICA is located and ascends in the vertical channel of the foramen lacerum toward the posterior portion of the cavern-

I. N. Kanaan, V. Beneš (eds.), *Neuroanatomy Guidance to Successful Neurosurgical Interventions*,
https://doi.org/10.1007/978-3-031-59838-8_37

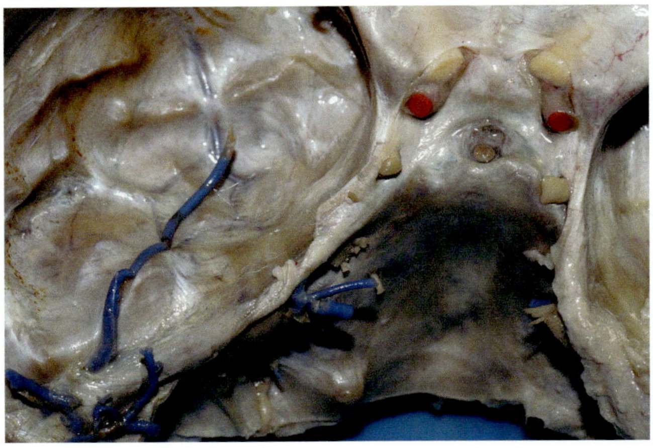

Fig. 37.1 Panoramic view of the upper portion of the sellar region, where we can see the two optic nerves in the front, followed by the carotid arteries in their supraclinoid portion, and toward the center the sellar diaphragm and the meningo-pituitary trunk. The clivus can be seen from behind, and we can also see both common ocular motor nerves, the trochlear nerves and the abducens nerves

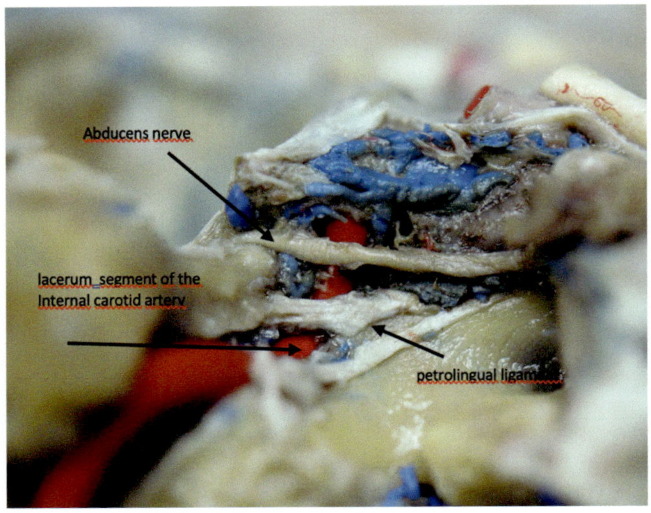

Fig. 37.2 The beginning of the cavernous segment on the right side is observed; just above the petrolingual ligament, the oculomotor, trochlear, and trigeminal nerves have been removed, and the wall of the cavernous sinus, lateral to the cavernous carotid, the abducens nerve, is observed. In blue, we can see the space occupied by venous blood within the cavernous sinus

ous sinus. The lacerum segment of the ICA is surrounded by periosteum. This layer of periosteum and fibrocartilaginous tissue covers the exocranial foramen below, which some authors call the foramen lacerum. The canalicular portion has an open roof where the ICA passes. The ICA enters the canalicular portion of the foramen lacerum of the carotid canal posterolaterally. The posterior, medial, and anterior walls of the vertical canalicular portion of the foramen lacerum are called the basisphenoid bone, which is covered by the periosteum and considered a posterior extension of the carotid sulcus (Fig. 37.1).

37.4.1 The Lacerum Segment

The lacerum segment ends at the level of the upper margin of the petrolingual ligament. The ligament runs between the sphenoid lingula anteriorly and the petrosal apex posteriorly. The petrolingual ligament is a continuation of the periosteum of the carotid channel. Distal to this ligament, the ICA enters the cavernous sinus. The lacerum segment is surrounded by fat, venous plexus, and postganglionic sympathetic nerves. The greater superficial petrosal nerve courses anteriorly, crossing the lateral foramen lacerum to the ICA, and leaves the foramen to enter the pterygoid canal (Fig. 37.2).

37.4.2 Cavernous Segment

It begins at the level of the superior margin of the petrolingual ligament and ends at the level of the proximal dural ring, which partially surrounds the ICA. The artery is surrounded by areolar tissue, fat, venous plexus, and postganglionic sympathetic nerves. This segment has two portions (the vertical portion and the horizontal portion) as well as two curves (the medial knee and the anterior knee). Several small branches originate from the cavernous segment including the meningohypophyseal trunk, which has been reported in the literature in up to 100% of cases, the inferolateral artery of the cavernous sinus in 84%, the capsular artery of McConnell in 28%, and the ophthalmic artery and the dorsal meningeal artery occasionally as a direct branch of the intracavernous carotid artery (Figs. 37.3 and 37.4).

The meningohypophyseal trunk originates mainly from the posterosuperior aspect of the medial knee of the ICA. It gives rise to three arteries: (1) the tentorial or Bernasconi-Cassinari artery, which in 100% of cases goes posterolateral to the roof of the cavernous sinus and to the free edge of the tentorium and supplies the cranial nerves III and IV; (2) the inferior pituitary artery, present in up to 80%, is directed medially joining with the contralateral inferior pituitary artery and supplies the neurohypophysis where it forms the pituitary portal system together with the superior pituitary arteries, branches of the supraclinoid internal carotid arteries; and (3) the dorsal meningeal artery pierces the dura mater of the posterior cavernous sinus and supplies the clival area and the cranial nerve VI. It originates in 90% of the meningohypophyseal trunk. 6% derive directly from the intracavernous carotid artery.

The inferolateral trunk of the cavernous sinus arises from 84% of the lateral portion of the horizontal segment of the intracavernous carotid artery and 6% of the meningohypophyseal trunk. It supplies the lateral wall of the cavernous sinus

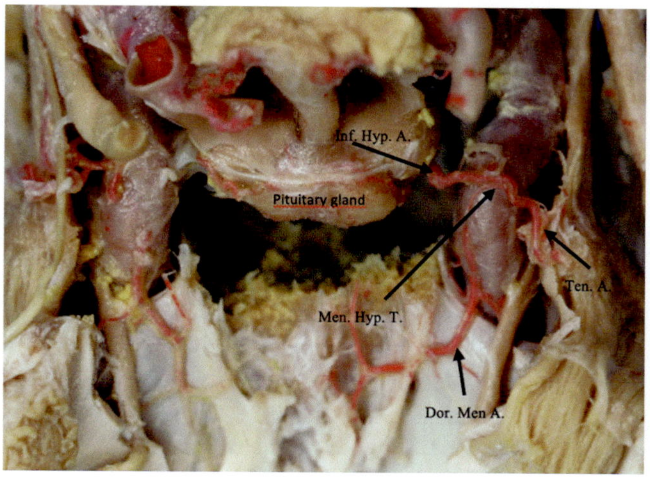

Fig. 37.3 Posterior portion of both cavernous sinuses. The clivus and the posterior portion of the sellar region have been removed. In the center of this image, the pituitary gland and pituitary stalk can be seen. The posterior curve of the right cavernous carotid artery is observed, giving rise to the meningohypophyseal trunk (Men. Hyp. T.), which in turn gives rise to the inferior hypophyseal artery (Inf. Hyp. A.), the tentorial artery (Ten. A.), and the dorsal meningeal artery (Dor. Men. A.), which in this case is not originating from the meningohypophyseal trunk but rather in isolation a few millimeters from it

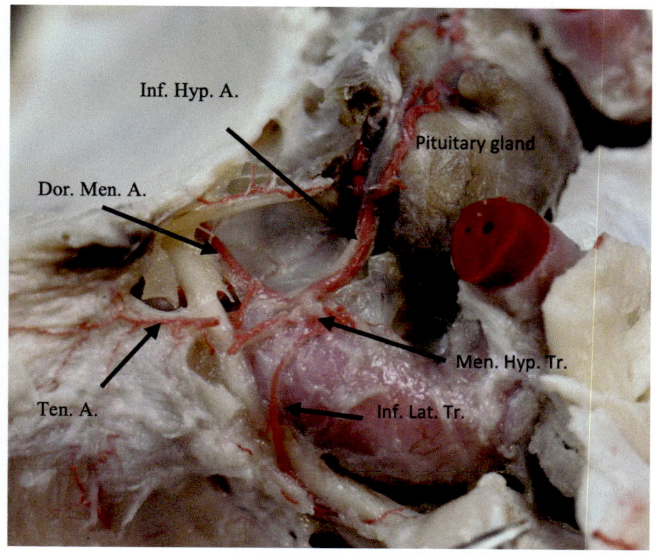

Fig. 37.5 Superior and lateral views of the right cavernous sinus showing the pituitary gland, the meningohypophyseal trunk (Men. Hyp. Tr), giving rise to the inferior hypophyseal artery (Inf. Hyp. A.), the dorsal meningeal artery (Dor. Men. A.), and the tentorial artery (Ten. A.) and as an anatomical variant also giving rise to the inferolateral artery or trunk of the cavernous sinus (Inf. Lat. Tr)

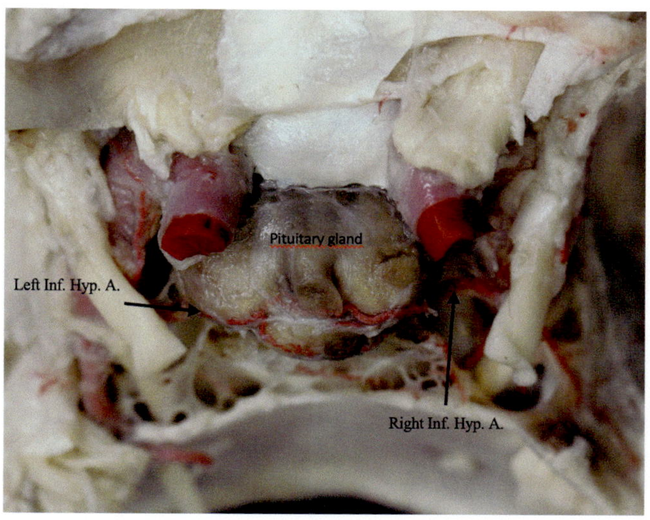

Fig. 37.4 Posterior portion of the pituitary gland and both cavernous sinuses. The two inferior pituitary arteries anastomose just posterior to the neurohypophysis

and makes an important contribution to the supply of cranial nerves III, IV, V, and VI, as well as the Gasserian ganglion. McConnell's capsular artery has been found in only 28%. It sends branches to the dura mater of the anterior sellar floor and anastomoses with the contralateral mater. In 8%, it has been reported that the ophthalmic artery arises from the intracavernous segment. Another artery, whose finding is incidental, is the persistent primitive trigeminal artery that originates proximal to the meningohypophyseal trunk, running posteriorly to join the basilar artery between the superior cerebellar artery and the anterior inferior cerebellar artery (Fig. 37.5).

The intracavernous carotid artery is intimately related to the pituitary gland. Therefore, during the trans-sphenoidal approach for the resection of pituitary tumors it has been a great challenge not to injure said artery. The challenge is even more difficult when the wall of the sphenoid sinus is very thin and can easily be pierced, injuring the carotid artery. The transsphenoidal approach was popularized by Harvey Cushing in the first decade of the last century, and currently, various endoscopic techniques are being used for the resection of pituitary tumors. Regardless of the approach and technique, the proximity of both cavernous carotid arteries to the pituitary gland continues to be a risk for the neurosurgeon (Figs. 37.6 and 37.7).

37.4.3 Venous Spaces of the Cavernous Sinus

The cavernous sinuses have multiple connections to different anatomical sites including the orbit, facial veins, and important connections through the inferior petrosal sinuses and superior petrosal sinuses. They communicate with each other through the superior intercavernous sinus and the posterior intercavernous sinus. The basilar sinus is located in the posterosuperior space of the sellar region; in this space, the meningo hypophyseal artery emerges (Figs. 37.8 and 37.9).

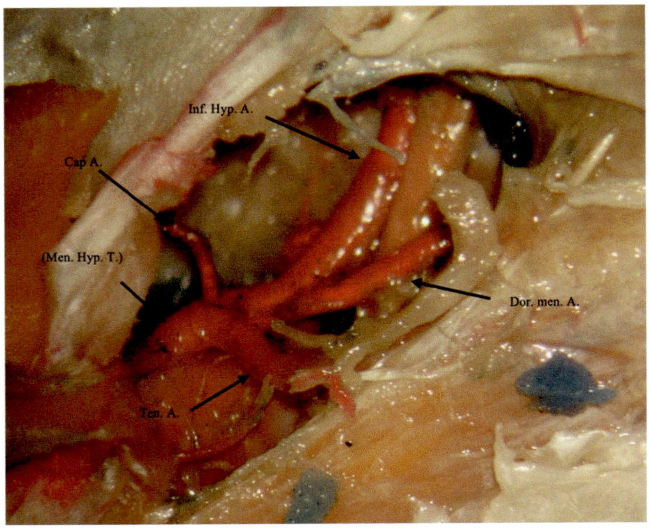

Fig. 37.6 View of the left cavernous sinus, showing the meningohypophyseal trunk (Men. Hyp. T.) giving rise to the inferior hypophyseal artery (Inf. Hyp. A.), the dorsal meningeal artery (Dor. Men. A.), the tentorial artery (Ten. a.), and the capsular artery of McConell (Chap. A.)

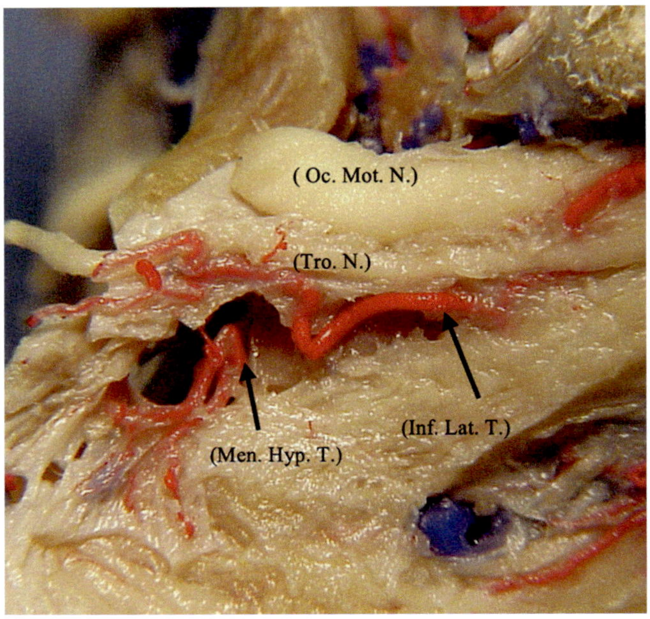

Fig. 37.7 Right lateral view of the infratrochlear triangle, also known as Parkinson's triangle, located between the lower margin of the trochlear nerve (Tro. N.) and the upper margin of the ophthalmic nerve, branches of the meningohypophyseal (Men. Hyp. T.), and the inferolateral artery or trunk (Inf. Lat. T.) of the cavernous sinus, slightly hyperplastic

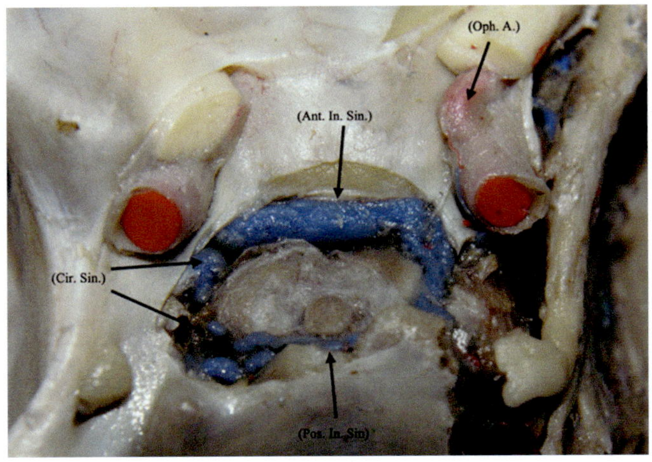

Fig. 37.8 Superior view of the intercavernous sinuses, where the anterior intercavernous sinus (Ant. In. Syn.) is larger than the posterior intercavernous sinus (Pos. In. Syn.), when both intercavernous sinuses coexist and form a venous ring around the pituitary gland, constitutes the circular sinus (Cir. Sin), and when a transsphenoidal surgical intervention is performed for resection of sellar tumors, there may be hemorrhage due to the opening of a hyperplastic anterior intercavernous sinus. The ophthalmic artery (Oft. A.) is also seen, originating from the right supraclinoid carotid artery

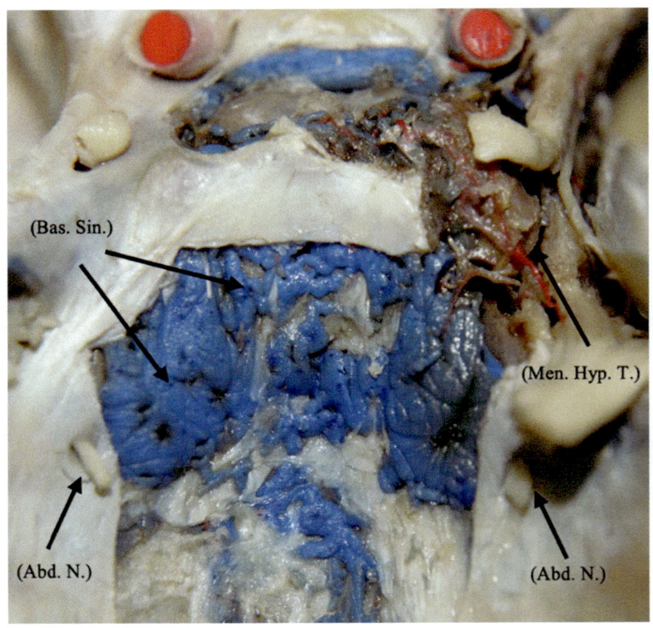

Fig. 37.9 Posterior view of the sellar region, showing the basilar sinus (Bas. Sin), the right cavernous sinus (R. Cav. Sin.) open, and the abducens nerves on both sides of the basilar sinus (Abd. N.)

37.4.4 Clinoid Segment

The clinoid segment follows the cavernous segment. Intimate knowledge of this segment is essential as it is common to drill the anterior clinoid process to have an approach with intracavernous structures or aneurysms such as those of the ophthalmic artery. It begins at the proximal dural ring and ends at the level of the distal dural ring where the internal carotid artery enters the subarachnoid space. The distal dural ring is a complete ring that surrounds the internal carotid artery and fuses laterally with the adventitia of the internal carotid artery. This ring is a continuation of the falciform

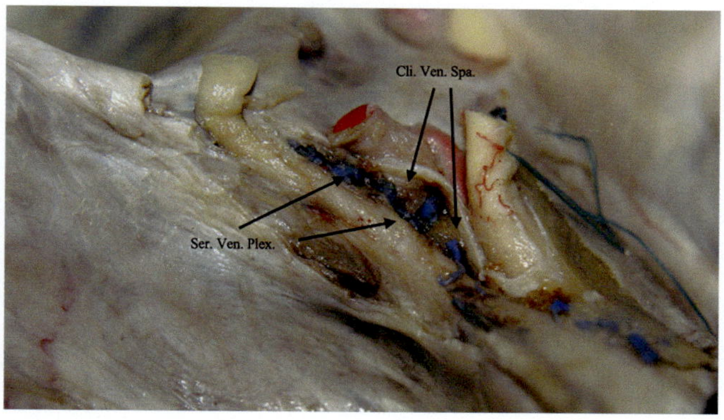

Fig. 37.10 Clinoid venous space (Cli. Ven. Spa.) corresponds to an anterosuperior extension of the cavernous sinus through the inferior dural ring. In its lower portion, it communicates with large channels venosus from the anterior portion of the cavernous sinus and superiorly tapers into a thin serpiginous venous plexus (Ser. Ven. plex.), indicating that the clinoid space is intracavernous as demonstrated by Dr. Rothon

ligament, the anterior clinoid process, and the roof of the cavernous sinus. The clinoid segment forms part of the anterior curve of the internal carotid artery. The clinoid segment is short and courses obliquely between the anterior clinoid process and the carotid groove; above and posteriorly, the clinoid segment is covered by dura mater, which is continuous with the roof of the cavernous sinus. Trabeculae from the cavernous sinus extend into this segment through an incomplete or incompetent dural ring (Fig. 37.10).

37.5 Surgical Pearls

- Profound knowledge of cavernous sinus anatomy, its vasculature, and its surrounding neural structures is essential for safe surgical approaches or endovascular intervention.
- MHT and ILT form important surgical landmarks, and trauma to these branches can generate pseudo-aneurysms or carotid-cavernous fistulae.
- Preoperative CTA and MRA play a pivotal role in imaging to select safe surgical corridors and anticipate the risk of injury to the cavernous segment (C4) during extended transsphenoidal surgery.
- The use of Doppler sonography is important during surgical intervention on cavernous sinus surgery with vascular encasement.
- Cavernous sinus aneurysms are most common in the elderly and carry limited morbidity, and many patients do not require intervention. However, when they are indicated, endovascular obliteration is an option.
- Type A carotid-cavernous sinus fistulas (high-flow, post-traumatic CCF) are symptomatic and need intervention to close the fistula. In contrast to type B, C, and D fistulas, they may close spontaneously.

37.6 Conclusion

The cavernous carotid arteries and their branches can be damaged during the resection of the most frequent sellar, parasellar, and suprasellar tumors, which include pituitary adenomas and meningiomas. Therefore, detailed knowledge of the surgical anatomy of the cavernous carotid artery, the arterial branches from which it originates and their distribution to the pituitary gland, the oculomotor nerves, and the optic chiasm is essential.

References

1. Rhoton AL. The Cavernous Sinus, The Cavernous Venous Plexus, and the Carotid Collar, Neurosurgery. Department of Neurosurgical Surgery, University of Florida, Gainesville, Florida. October 2002, Volume 51, Supplement 1.
2. Harris FS, Rhoton AL Jr. Anatomy of the cavernous sinus: a microsurgical study. J Neurosurg. 1976;45:169–80.
3. Umansky F, Nathan H. The lateral wall of the cavernous sinus: with special reference to the nerves related to it. J Neurosurg. 1982;56:228–34.
4. Zubair A, Das JM. June, 2022. https://www.ncbi.nlm.nih.gov/books/NBK556142/

38 Surgical Anatomy of the Cavernous Sinus and Refinement of Its Surgery

Jacques J. Morcos and Matthew Z. Sun

Learning Objectives
- Identify the relevant dural, arterial, venous, and nervous anatomy of the cavernous sinus
- Learn the various surgical approaches to the cavernous sinus
- Learn the relevant osseous anatomy around the cavernous sinus, including the attachments of the anterior clinoid process

38.1 Introduction

Surgery in and around the cavernous sinus was deemed too risky and dangerous only a few decades ago. Morbidity and even mortality were often expected, given the forbidding location, the neurovascular structures that enter and surround the cavernous sinus, and the complex dural and venous anatomy. With the development of modern techniques for safer approaches to the cavernous sinus and knowledge of various safe operative windows, nearly all lesions, including vascular, neoplastic, or infectious, can now be treated directly through open or endoscopic techniques. At the same time, other technologies/techniques have arisen over the last two decades that in many instances provide alternative and in some instances safer treatments for lesions within the cavernous sinus, namely endovascular techniques and radiosurgery. For example, the rise of endovascular techniques has supplanted open surgeries for some vascular lesions over the last two decades, particularly for most carotid-cavernous fistulas. Another example is that tumors that are only located within the cavernous sinus can be good targets for radiosurgery. Nevertheless, open surgery still plays a very significant role for most lesions in and around the cavernous sinus, as many cross anatomical boundaries and become multi-compartmentalized. The aim of this chapter was to review the history, anatomy, and representative surgeries in the cavernous sinus.

38.2 History and Background

The anatomical entity known as the cavernous sinus has been described since ancient times. Galen described it as a parasellar carotid retia bathed in venous blood [1]. Raymond de Vieussens appreciated the parasellar compartment as a well-defined space, which he termed the "receptacle," which received the cranial nerves alongside the "Turkish saddle" (seventeenth century) [2]. The term "cavernous sinus" itself was named by Sir Winslow in the eighteenth century as "reservatories of a very particular kind, containing not only blood but considerable vessels and nerves…and likewise a spongy or cavernous substance full of blood." [3, 4] Documented surgeries specifically in and around the cavernous sinus go back a little more than a century. The first attempt at cavernous sinus surgery was by Krogius in 1895 for a ganglionic trigeminal neuroma [5]. Frazier was the first to successfully remove a trigeminal neuroma in 1918 [5]. Browder in 1936 described muscle packing for direct carotid-cavernous fistula (CCF) [6].

Dwight Parkinson explored it further beginning in the 1960s and named it the "anatomical jewel box." [7] In several papers over the subsequent decades [7–11], he described treating various vascular lesions in the cavernous sinus and more specifically the repair of CCF through what is now known as Parkinson's triangle, between the fourth nerve and the ophthalmic nerve (V1). Then, in the 1980s, Hakuba and Dolenc separately described various epidural and intradural approaches for lesions in and around the cavernous sinus bringing forth the modern era of surgery around the cavernous sinus [12–16].

J. J. Morcos (✉)
Department of Neurosurgery, UT Houston, Houston, TX, USA
e-mail: jacques.morcos@uth.tmc.edu

M. Z. Sun
Department of Neurosurgery, UT Southwestern, Dallas, TX, USA

I. N. Kanaan, V. Beneš (eds.), *Neuroanatomy Guidance to Successful Neurosurgical Interventions*,
https://doi.org/10.1007/978-3-031-59838-8_38

Hakuba et al. described four approaches to the cavernous sinus [12]. Dolenc pioneered the extradural/intradural approach later modified by others to directly expose and operate within the cavernous sinus [14, 15, 17, 18]. The Dolenc approach, also known as the anteromedial transcavernous approach, was originally developed for exposure of intracavernous aneurysms. It uses a combined epidural and subdural frontotemporal approach, and subsequent modifications centered around the type of craniotomy and extent of skull base removal extradurally during exposure. The Hakuba approach involved the orbitozygomatic infratemporal approach [13]. Concurrently, surgical anatomical studies performed by many, starting in the late 1970s including Al Rhoton Jr. et al., also described the relevant anatomy of the cavernous sinus, the sella, and the sphenoid sinus from both cranial and endonasal approaches [19–21].

38.3 Anatomical Highlights

- The cavernous sinus is bound anteriorly by the edges of the superior orbital fissure, posteriorly by the petrous apex, medially by the sella, and laterally by the middle fossa.
- The posterior wall of the cavernous sinus is bound by the dorsum sella medially and the ostium of Meckel's cave laterally.
- The optic strut separates the optic canal and the superior orbital fissure.
- The clinoidal segment of the carotid artery is defined by the proximal and distal dural rings, which are exposed by removing the anterior clinoid process. This short but critical segment of the carotid is neither intracavernous nor subarachnoid.
- The clinoidal venous space is a separate venous space that is also an extension of the cavernous sinus through the lower dural ring within the carotid collar, surrounding the clinoidal segment of the carotid artery.
- Both the oculomotor nerve and the trochlear nerve are surrounded by a small amount of CSF within arachnoidal cuffs as they run through their respective cisterns just after their entrance through the dura.

The cavernous sinus is defined as the dural envelope containing the cavernous segment of the carotid artery. The paired cavernous sinuses are located on each side of the sella, with dural walls that surround a venous space that contains the carotid artery, its intracavernous branches, and extraocular nerves. The cavernous sinus is bound anteriorly by the edges of the superior orbital fissure, posteriorly by the petrous apex, medially by the sella, and laterally by the middle fossa. The posterior wall of the cavernous sinus is bound by the dorsum sella medially and the ostium of Meckel's cave laterally. The cavernous sinus is also a venous confluence that communicates with multiple venous structures that include the basilar sinus, the inferior and superior petrosal sinuses, the anterior and posterior intercavernous sinuses, the sylvian vein, the ophthalmic veins, the veins exiting the foramina rotundum, ovale, and spinosum, and the veins of the carotid canal and the sphenoidal emissary foramen.

38.3.1 The Four Walls of the Cavernous Sinus

The cavernous sinus has four walls: superior, medial, lateral, and posterior walls; these four walls shape the sinus into a boat-like shape with a narrow "keel" anteriorly, which is the superior orbital fissure (Fig. 38.1a). The superior wall is formed by the lower margin of the anterior clinoid process and the dura overlying the oculomotor triangle, which is lined by the interclinoid dural fold, the anterior petroclinoid fold, and the posterior petroclinoid fold. The medial margin of the posterior wall of the cavernous sinus is the lateral margin of the dorsum sella, where the cavernous sinus connects with the basilar sinus, the largest communication between the bilateral cavernous sinuses. The inferior margin of the posterior wall is at the petrosphenoidal ligament, under which the abducens nerve passes through the Dorello's canal to enter the cavernous sinus. The superior margin of the posterior wall is at the posterior petroclinoidal dural fold. The lateral margin of the posterior wall is the medial edge of the ostium of Meckel's cave.

The lateral wall of the cavernous sinus is bound anteriorly and posteriorly by the lateral margin of the superior orbital fissure and the medial edge of Meckel's cave, respectively. The lateral wall is bound superiorly and inferiorly by the anterior petroclinoidal dural fold and the carotid sulcus, respectively (Fig. 38.1a, b). The carotid sulcus is defined as the groove along the sphenoid body laterally, along which courses the carotid artery (Fig. 38.1c).

The medial wall of the cavernous sinus is the dura covering the lateral wall of the sella turcica that also covers the lateral surface of the sphenoid body. The anterior and posterior margins of the medial wall are defined by the medial edge of the superior orbital fissure and the lateral edge of the dorsum sella, respectively. The superior and inferior margins are defined by the interclinoidal dural fold and the carotid sulcus, respectively. The medial wall is more often viewed from the transsphenoidal approach (Fig. 38.1d). The lateral inferior margins of the cavernous sinus run just below the inferior edge of the ophthalmic (V1) division as it runs along the lateral wall of the cavernous sinus. The majority of Meckel's cave is located lateral and inferior to the cavernous sinus.

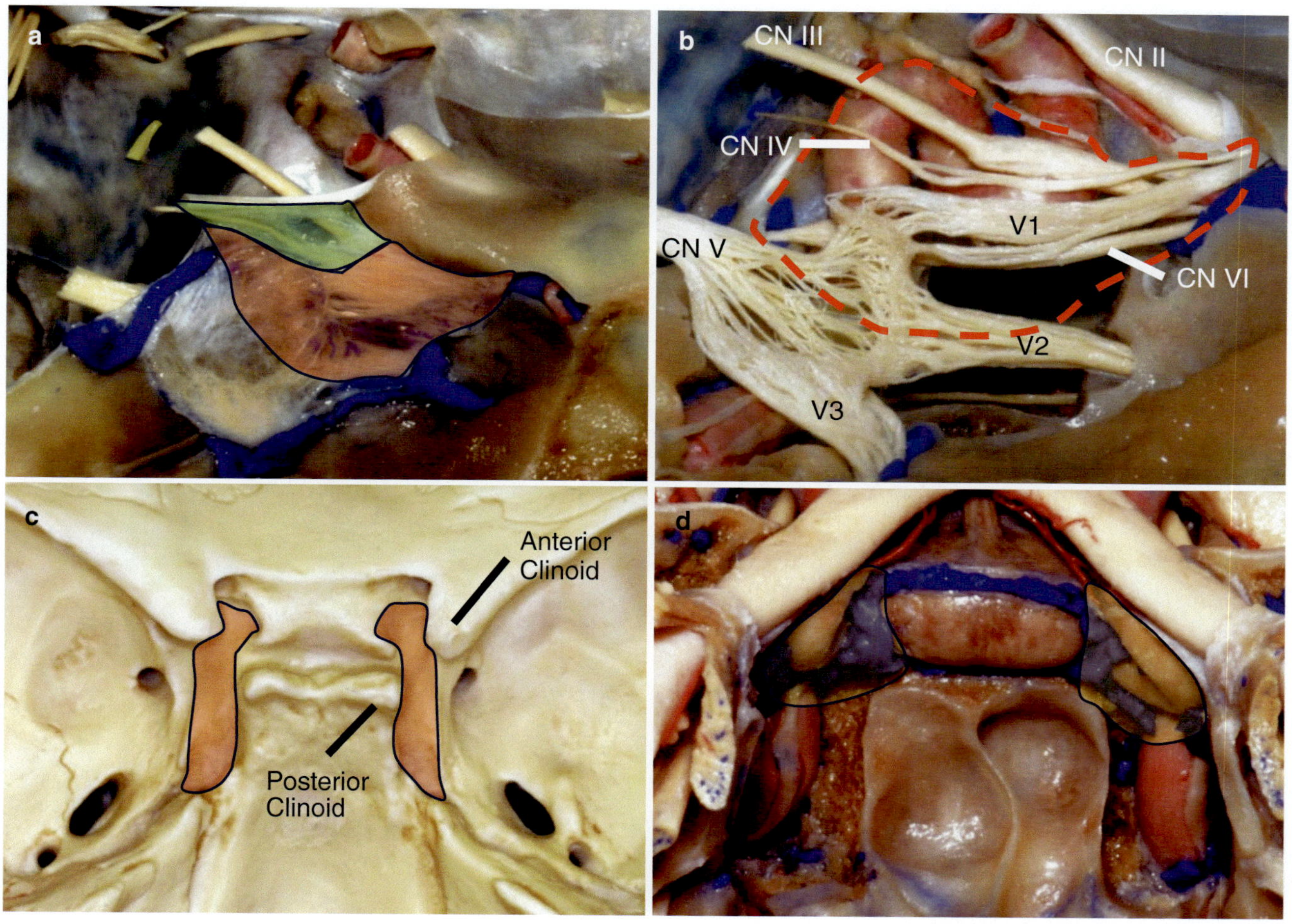

Fig. 38.1 The walls and boundaries of the cavernous sinus (Figures and images courtesy of the Rhoton Collection). (**a**) Lateral view of the right cavernous sinus after the dura has been peeled off of the lateral wall of the cavernous sinus. The green shaded area approximates the superior wall or the roof of the cavernous sinus, and the red shaded area approximates the lateral wall of the cavernous sinus, with the Meckel's cave posterolateral to it. (**b**) The lateral wall of the cavernous sinus has been peeled back, and the sphenoid bone between V1 and V2 (anteromedial triangle) as well as between V2 and V3 (anterolateral triangle) has been drilled to reveal the sphenoid sinus. The outline of the cavernous sinus is shown in the dashed line. (**c**) The carotid grooves on either side of the sella are shaded in red. (**d**) The anterior view of the sella revealing the pituitary gland in the center with cavernous sinuses on each side. The medial wall of the cavernous sinus has been removed, and its approximate outline is shaded in yellow

38.3.2 The Osseous Anatomy of the Cavernous Sinus

The cavernous sinus is located along the lateral aspect of the sphenoid body. The junction of the petrous apex and the sphenoid bone at the upper petroclival fissure demarcates the posterior inferior edge of the cavernous sinus. From the medial direction to the lateral direction, the cavernous sinus covers from the lateral border of the sella to the junction of the sphenoid body and greater wing, but excludes foramen rotundum (Fig. 38.2a, b).

38.3.2.1 The Anterior, Middle, and Posterior Clinoid Processes

The anterior clinoid process anatomy is important to understand as it forms the anterior roof of the cavernous sinus (Fig. 38.2b–d). The base of the anterior clinoid process forms the lateral optic canal and is attached to the medial edge of the sphenoid ridge. The other two attachments to the sphenoid are the anterior and posterior roots of the lesser wing, which form the roof of the optic canal and the optic strut, respectively. The optic strut separates the optic canal and the superior orbital fissure. The roof of the optic strut forms the floor of the optic canal, while the inferior aspect of the strut forms the roof of the superior orbital fissure and the anterior cavernous sinus roof. The air cells in the sphenoid sinus may extend through the optic strut into the anterior clinoid process. From the endoscopic endonasal perspective, the lateral opticocarotid recess (LOCR) is the medial aspect of the optic strut. In addition to being the site of attachment for the anterior petroclinoid and interclinoid dural folds, the base of the clinoid is the lateral attachment of the falciform ligament,

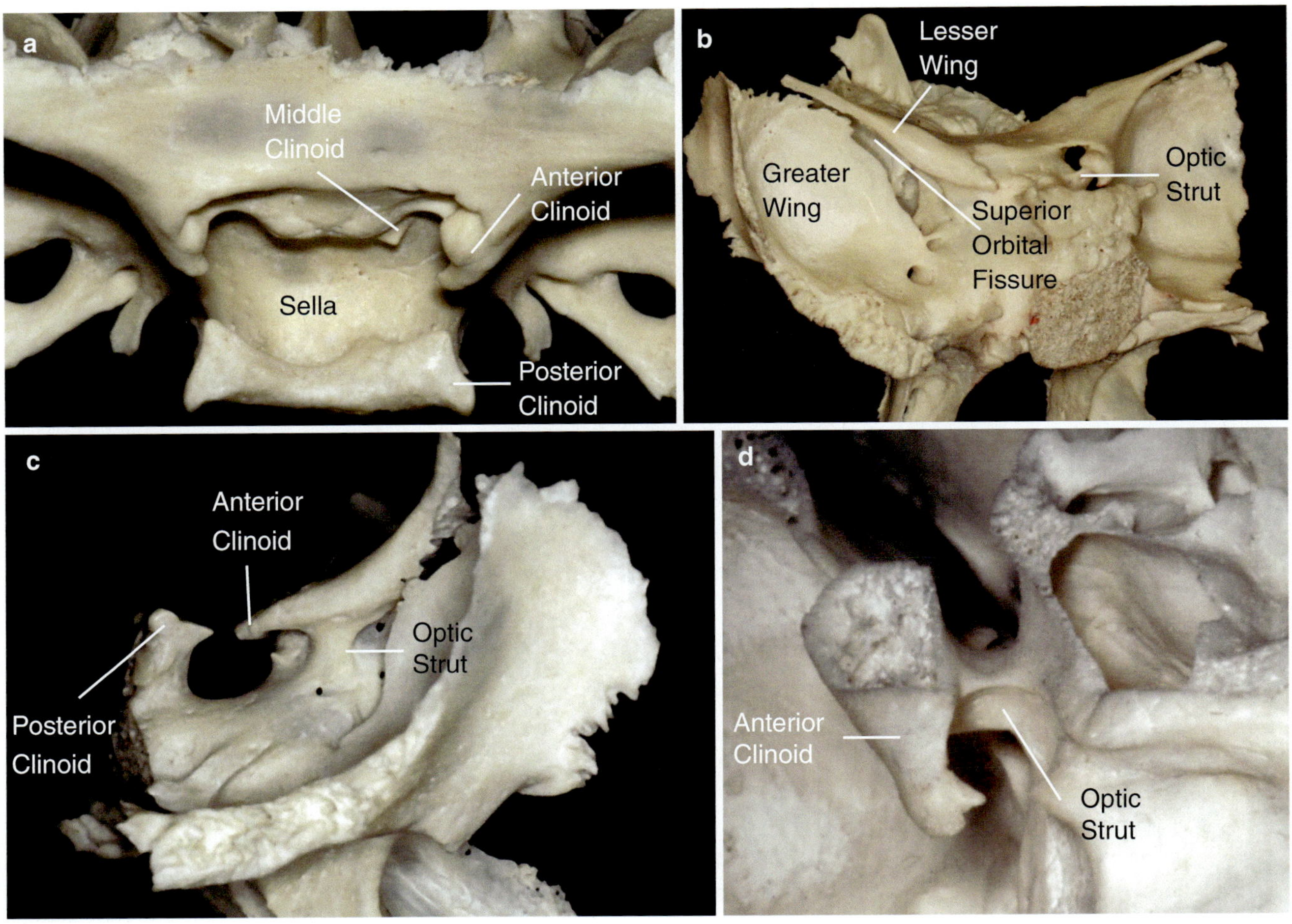

Fig. 38.2 The osseous anatomy of the cavernous sinus. (Photographs courtesy of the Rhoton Collection). (**a**) Superior view of the sphenoid bone with sella in the center. The anterior, middle, and posterior clinoids are seen. (**b**) Oblique view of the sphenoid bone from the left, and optic canal along with the optic strut can be seen. (**c**) Lateral view of the sphenoid from the right, and the optic strut is seen. (**d**) Inferior oblique view of the optic strut and the anterior clinoid process

which covers the roof of the optic canal and extends medially to the planum sphenoidale.

The middle clinoid process (Fig. 38.2a) projects superiorly from the medial side of the terminal carotid sulcus and may form an osseous bridge laterally with the tip of the anterior clinoid, creating a bony foramen known as the caroticoclinoidal foramen. Additionally, an interclinoidal osseous bridge may connect the anterior clinoid process to the posterior clinoid process. These bridges between the clinoid processes may increase the difficulty of removing the anterior clinoid process, increasing the risk to the carotid artery, and should be screened for preoperatively with a CT scan.

38.3.3 Arterial Anatomy

The three main branches of the cavernous segment of the carotid artery are, in order of appearance, the meningohypophyseal trunk, the inferolateral trunk, and the capsular arteries of McConnell (Fig. 38.3) [7, 22]. Anatomical variants of the trunks and their respective branches are well documented in the literature [22, 23]. The description here is representative of the typical anatomy encountered. The meningohypophyseal trunk is almost always present, and its branches typically include the inferior hypophyseal artery, the marginal tentorial artery of Bernasconi and Cassinari, the lateral tentorial artery, and the clival branches. The inferolateral trunk, also known as the artery to the inferior cavernous sinus, gives off branches that run along the cranial nerves within the cavernous sinus and arteries that enter the foramen rotundum, foramen ovale, and foramen lacerum. The capsular arteries arise from the medial wall of the cavernous ICA distal to the meningohypophyseal trunk and the inferolateral trunk and supply the sellar region including the sella dura and anterior pituitary gland.

The cavernous carotid artery enters the posterior cavernous sinus after exiting the carotid canal and passing under the petrolingual ligament before coursing superiorly. The clinoidal segment of the carotid artery is defined by the proximal

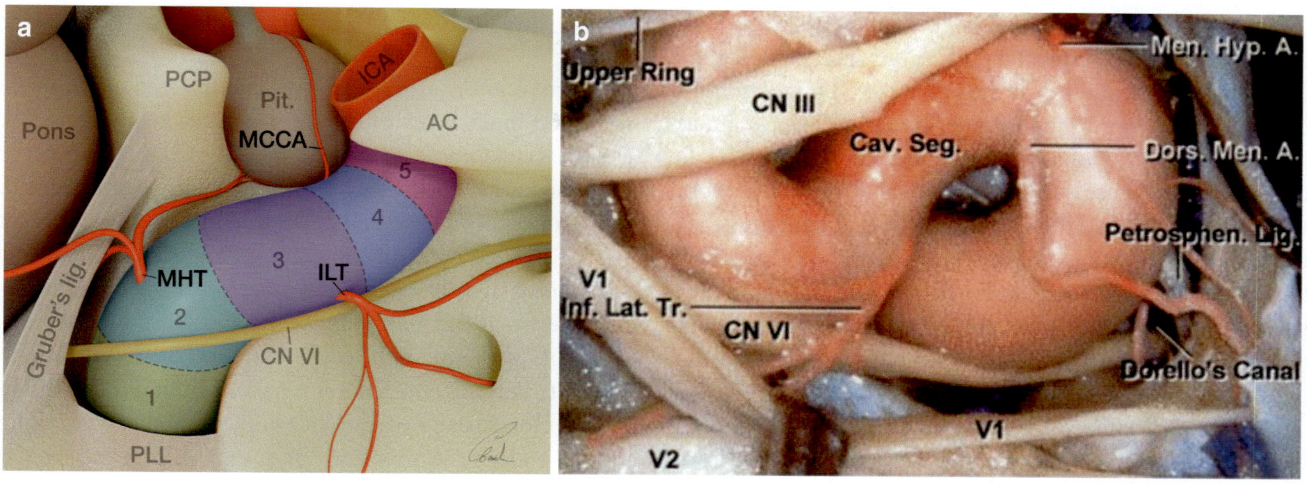

Fig. 38.3 Arterial branches of the cavernous carotid artery. (**a**) Schematic showing 3 of the major branches of the cavernous carotid artery: inferolateral trunk (ILT), meningohypophyseal artery(MHT), and McConnell's capsular arteries (MCCA). (1) Posterior vertical, (2) posterior bending, (3) horizontal, (4) anterior bending, and (5) anterior vertical. AC, anterior clinoid process; ILT, inferolateral trunk; MHT, meningohypophyseal trunk; PLL, petrolingual ligament; VI, abducens nerve. (©2024 The University of Texas Southwestern Medical Center). (**b**) Rhoton dissection showing inferolateral trunk (ILT), meningohypophyseal artery(MHT)

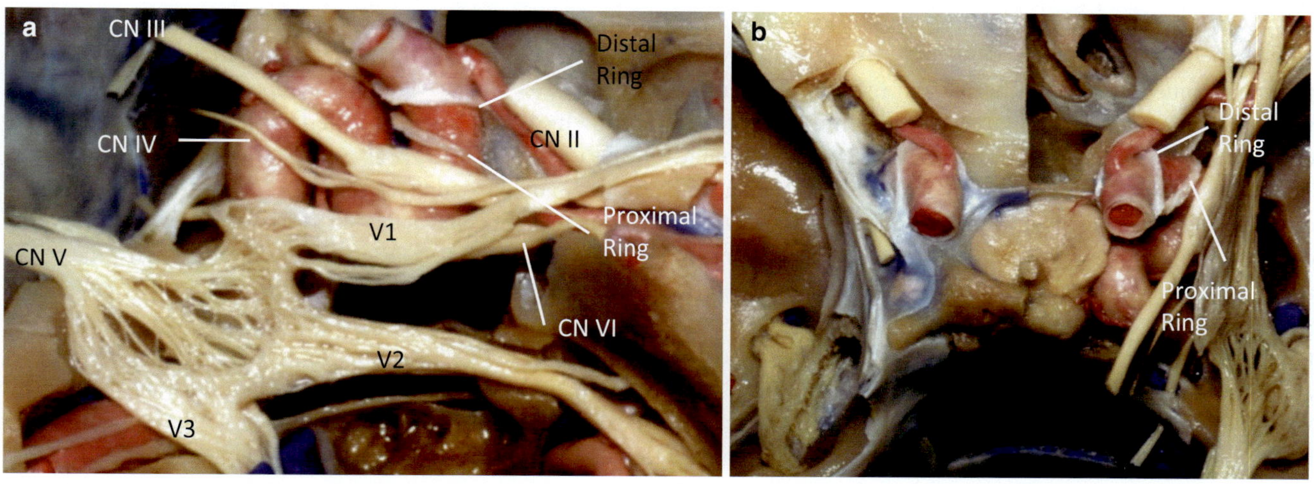

Fig. 38.4 Proximal and distal dural rings (Images courtesy of the Rhoton Collection). (**a**) Proximal and distal dural rings seen from a right lateral view. (**b**) Proximal and distal dural rings from a posterior view. These 2 images demonstrate clearly the inclined plane of the distal dural ring, with a slant that makes the medial edge lower (more caudal) than the lateral edge. Hence, the "carotid cave," that subarachnoid space immediately distal to the medial edge of the distal dural ring, is the anatomical consequence of this slanted orientation, and is sometimes a site for aneurysm formation

and distal dural rings (Fig. 38.4), which are exposed by removing the anterior clinoid process. This short but critical segment of the carotid is neither intracavernous nor subarachnoid. Cutting the dural rings is essential in controlling the carotid during surgery for superior hypophyseal or ophthalmic aneurysms. Incomplete understanding or exposure of this segment often results in suboptimal clipping and inadequate proximal control.

The carotid collar is a dural sleeve that wraps around the clinoidal segment of the carotid between the proximal and distal dural rings. The layer of dura medial to the anterior clinoid forms the lateral carotid collar. The anterior carotid collar is the dura covering the posterior surface of the optic strut. The clinoid venous plexus courses between the carotid collar and the clinoid segment, and it communicates with the cavernous sinus inferiorly. The lower ring is the connection point between the carotid collar and the carotid-oculomotor membrane, which forms the anterior roof of the cavernous sinus.

38.3.4 Venous Anatomy

The cavernous sinus is narrow anteriorly adjacent to the superior orbital fissure and is wide posteriorly near the dorsum sella, where it connects to the venous confluence at the junction of the basilar, cavernous, and superior and inferior petrosal sinuses. It has extensive connections. It connects to the orbit via the superior and inferior ophthalmic veins, to the retina via the central retinal vein, to the cerebral hemispheres via the middle and inferior cerebral veins, to the dura via the middle meningeal vein branches, to the transverse sinus via the superior petrosal sinus, to the jugular bulb via the inferior petrosal sinus, to the pterygoid venous plexus via the cranial (middle fossa) foraminae emissary veins, and to the facial veins via the ophthalmic veins.

Conceptually, there are three venous spaces within the cavernous sinus, defined by their relation to the carotid artery: the medial, anteroinferior, and posterosuperior compartments. The sizes of the various compartments can vary depending on the course of the cavernous carotid artery. The medial compartment sits between the cavernous carotid artery and the pituitary gland and contains the origin of the inferior hypophyseal artery. The anteroinferior compartment sits in the concavity inferior to the anterior genu of the cavernous carotid and is commonly the place of connection with superior and inferior ophthalmic veins. The abducens nerve passes lateral to the cavernous carotid before it enters the anteroinferior compartment. The posterosuperior compartment sits between the carotid artery and the posterior roof of the cavernous sinus and contains the origin of the meningohypophyseal artery. In addition to these three compartments, there is also a lateral space of the sinus between the lateral wall of the sinus and the carotid artery, through which the abducens runs, often adherent to the carotid artery medially and the lateral sinus wall laterally. This space is narrow in the majority of cases, and less than 10% of sinuses have this space extend lateral to the abducens nerve. The medial and posterosuperior compartments are the larger venous spaces in the cavernous sinus, and these spaces are better suited for entry into the sinus to minimize injury to the nerves.

Venous connections between the bilateral cavernous sinuses at the margins of the diaphragm and within the sella can be found both anteriorly and posteriorly and are named as such. If the anterior and posterior intercavernous sinuses coexist and connect, then an effective venous ring exists within the sella and forms the "circular sinus."

The clinoidal venous space is a separate venous space that is also an extension of the cavernous sinus through the lower dural ring within the carotid collar, surrounding the clinoidal segment of the carotid artery. It has connections with the diploic veins of the orbital roof through emissary veins that pass through the surface of the anterior clinoid process and the optic strut. These veins can be encountered during a typical anterior clinoidectomy.

38.3.5 Nerves

The oculomotor, trochlear, and ophthalmic nerves run along the inner layer of the lateral cavernous sinus wall. The abducens nerve runs medial to the ophthalmic nerve and just lateral to the cavernous segment of the carotid. The oculomotor nerve enters the cavernous sinus through its roof at the center of the oculomotor triangle, and this entrance is slightly lateral and anterior to the dorsum sellae. The trochlear nerve enters the roof of the cavernous sinus just posterolateral to the oculomotor nerve. It then runs between the oculomotor nerve and the dura covering the inferior surface of the anterior clinoid and the optic strut to end on the superior oblique muscle in the medial orbit. Both the oculomotor nerve and the trochlear nerve are surrounded by a small amount of CSF within arachnoidal cuffs as they run through their respective cisterns just after their entrance through the dura. As such, these dural cuffs can be opened to allow greater surgical freedom and reduce traction injury to these nerves.

The ophthalmic nerve is the first and smallest division of the trigeminal nerve. It courses just medial to the lateral wall of the sinus on its way to the superior orbital fissure and forms the lower border of the lateral wall. As it enters the superior orbital fissure, it further divides into the lacrimal, frontal, and nasociliary branches. The motor root of the trigeminal nerve courses medial to sensory fibers at the level of Meckel's cave. The superior petrosal sinus forms the upper margin of the ostium of Meckel's cave, which communicates with the subarachnoid space in the posterior fossa and contains the trigeminal ganglion.

The abducens nerve enters the dura forming the inferior posterior wall of the cavernous sinus, through Dorello's canal, below the petrosphenoid ligament (Gruber's ligament), which connects the inferior aspect of the dorsum sella to the petrous apex. Within the cavernous sinus, it first courses lateral to the proximal intracavernous carotid, before running just medial to the ophthalmic nerve. Of all the nerves in the cavernous sinus, the abducens nerve runs the deepest.

38.4 Surgical Relevance and Surgical Strategies

Approaches to the cavernous sinus allow exposure of the intra-, supra-, infra-, and parasellar areas and usually involve performing a frontotemporal craniotomy with optional additional bone removal to expose the roof and lateral wall of the cavernous sinus. These include but are not limited to orbitozygomatic osteotomy, anterior clinoidectomy, unroofing the optic canal, unlocking the superior orbital fissure, and posterior clinoidectomy.

While the lateral wall can also be exposed through a subtemporal approach with optional anterior petrosectomy, the angle of view from this approach alone is not optimal for exposing the roof of the sinus properly due to the amount of temporal lobe retraction required.

Both intradural and extradural approaches can be used. Extradurally, the peeling (Hakuba/Dolenc) starts laterally and anteriorly along the sphenoid ridge by elevating the dura from the greater and lesser wings until the fibrous dural band (meningo-orbital band) at the lateral edge of the superior orbital fissure, where the middle fossa dura folds to become the periorbita intra-orbitally. To continue peeling this dura to expose the lateral wall of the cavernous sinus requires a sharp cut of this dural fold, and the peeling of the thick outer layer will expose the thinner inner layer of the lateral wall encasing the oculomotor, trochlear, and ophthalmic nerves. We described a variation of the Dolenc technique that flips the sequence of the steps here, creating the middle fossa peel before cutting the meningo-orbital band [17]. The upper limit of this dural peeling is the anterior petroclinoid dural fold, and the posterior limit is the petrous ridge. At the end of this dural peeling from the floor of the middle fossa, the mandibular and greater superficial petrosal nerves will be seen and the lateral wall of Meckel's cave revealed at the postero-medial aspect. Occasionally, the petrous segment of the internal carotid artery is exposed through this dural peeling.

Intradurally, wide sylvian fissure dissection allows for optimal exposure of the optic nerves, chiasm, optic tract, carotid bifurcation, and oculomotor nerve. Intradural anterior clinoidectomy exposes the roof of the cavernous sinus. Posterior clinoidectomy, cutting of the posterior petroclinoidal ligament, and opening the dura around the oculomotor nerve within the oculomotor triangle further free up the nerve to allow for a larger working corridor between the carotid artery and the oculomotor nerve. The elevation and posterior retraction of the anterior pole of the temporal lobe provide a pretemporal exposure of the basal cisterns. This further allows for exposure of the basilar, posterior cerebral, superior cerebellar, and posterior communicating arteries, as well as the upper brainstem. Further dissection of the dura posterior to the oculomotor triangle can unlock the trochlear nerve as it enters the cavernous sinus.

All of the nerves of the cavernous sinus, except the abducens nerve, can be exposed by opening the outer layer of dura along the lateral cavernous sinus wall while leaving the inner layer intact, as these nerves run between the two layers of dura along the lateral cavernous sinus wall. Whenever possible, we limit the exposure to the extradural space without entering CSF space. This is only possible if the pathology is extradural and the surgeon understands the above layered anatomy perfectly.

38.5 Surgical Cases

Case 1: Transcavernous Transclinoidal Half-and-Half Approach for Clipping of Twice-Recurrent Stent-Coiled Giant Basilar Tip Aneurysm (Fig. 38.5)

For this giant twice-recurrent stent-coiled previously ruptured basilar tip aneurysm in a 60-year-old man, we needed sufficient exposure for both proximal control and room for dissection and clip placement. The widened exposure starts from the outside and continues on the inside. We started with a superolateral orbitotomy to enhance superior exposure, improve illumination, and refine the handling of the aneurysm. Then, anterior clinoidectomy is performed to enhance inferior exposure. This is followed by a transcavernous approach, cutting the dura around the third and fourth nerves as they enter the roof. When this is accomplished, we proceed with a half-and-half approach, also known as the half-transsylvian and half-pretemporal approach, initially described by Sano in 1980. The combination of these techniques ensures the widest corridor to the complex aneurysm. By then adding an intradural posterior clinoidectomy, a more caudal exposure of the proximal basilar artery became possible [24, 25]. Only then could we apply temporary clipping and temporary trapping, open the aneurysm, remove part of the coils and thrombus, and clip reconstruct the sac.

Case 2: Transcavernous Approach for Resection of a Large Cavernous Sinus Dermoid Cyst (Fig. 38.6)

For this large dermoid cyst in the lateral cavernous sinus wall in a 23-year-old woman, we used the Hakuba/Dolenc interdural peeling technique to widely expose the cyst wall and preserve the trigeminal nerve fibers. The dissection was carried out carefully to avoid premature rupture of the cyst until sufficient exposure had been achieved. The NIMS stimulator was used to map and avoid injuring the nerves that are draped over the tumor capsule but are not easily discernible. After cyst decompression, the tumor capsule is dissected away from the nerve fibers and Meckel's cave is reached posteriorly with free egress of CSF. The tumor capsule was eventually removed as one piece.

Case 3: Combined Transcavernous (Hakuba/Dolenc) and Anterior Petrosectomy (Kawase) Approach to a Trigeminal Schwannoma (Fig. 38.7)

For this large trigeminal schwannoma occupying both the middle and posterior fossa in a 25-year-old man who presented with blurry vision, headache, left VI palsy, and left V2-V3 hypoesthesia, we combined the transcavernous (Hakuba/Dolenc) and an anterior petrosectomy (Kawase) approach [26]. After the Hakuba/Dolenc peeling was performed, the middle meningeal artery was coagulated and divided, and extradural extracapsular dissection of the tumor was carried out. The trigeminal fibers were seen, and a small opening between V2 and V3 was created in order to decom-

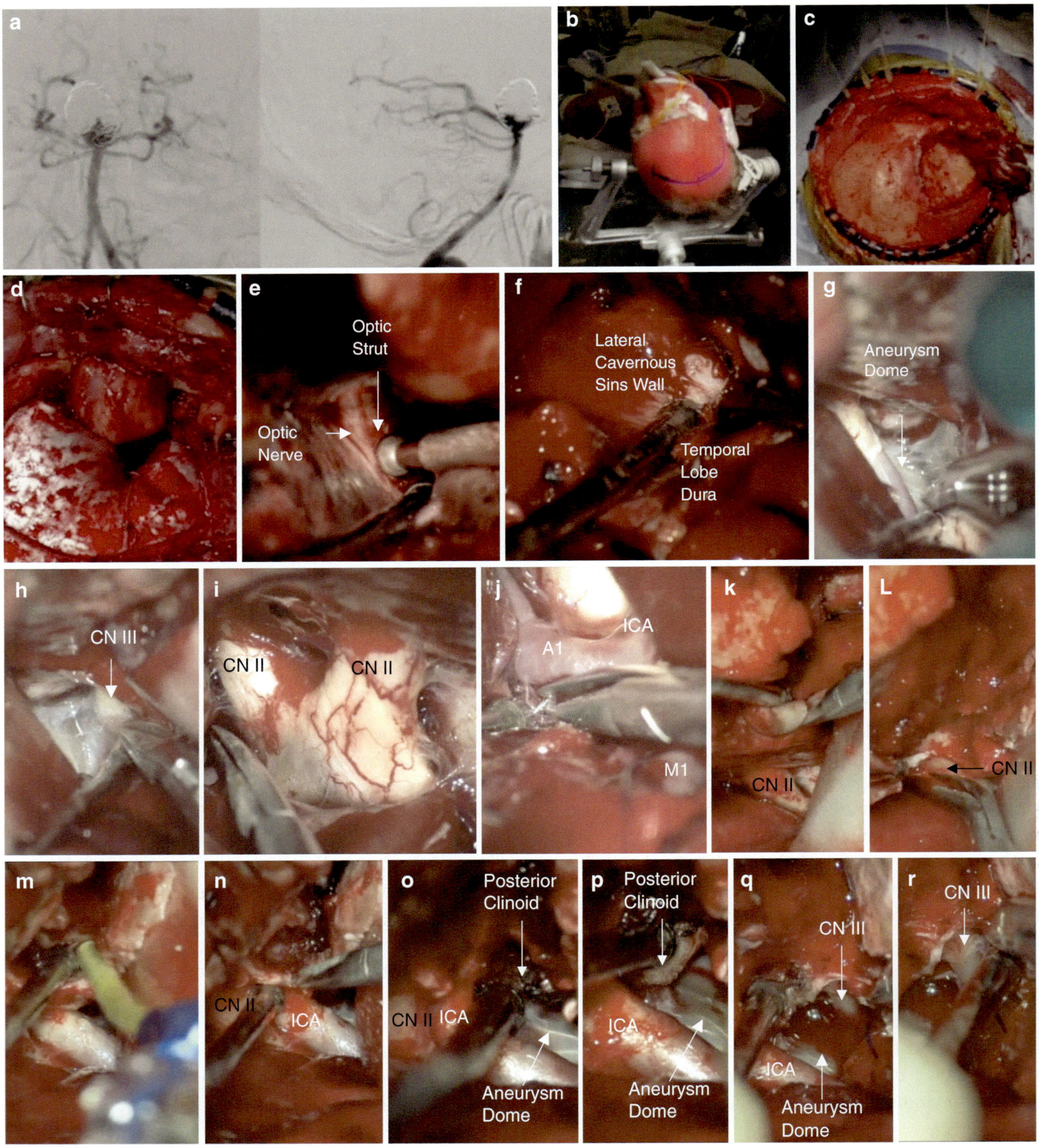
a
b
c
d
e
Optic
Strut
Optic
Nerve
f
Lateral
Cavernous
Sins Wall
Temporal
Lobe
Dura
g
Aneurysm
Dome
h
CN III
i
CN II
CN II
j
ICA
A1
M1
k
CN II
L
CN II
m
n
CN II
ICA
o
Posterior
Clinoid
CN II ICA
Aneurysm
Dome
p
Posterior
Clinoid
ICA
Aneurysm
Dome
q
CN III
ICA
Aneurysm
Dome
r
CN III

(continued)

Fig. 38.5 Case 1: Right transcavernous transclinoidal half-and-half (transsylvian pretemporal) approach for clipping of twice-recurrent stent-coiled large basilar tip aneurysm. (**a**) AP and lateral views of the angiogram of the recurrent large basilar tip aneurysm. (**b**) Patient was positioned supine with head in 30-degree contralateral tilt and 10-degree vertex down. (**c**) Subfascial dissection of skin flap. (**d**) Exposure after a fronto-temporal craniotomy with superolateral orbitotomy was performed. (**e**) Extradural anterior clinoidectomy is performed, and the drilling of optic strut is shown here after the optic canal has already been unroofed. (**f**) Hakuba technique of anterior to posterior peeling of temporal lobe dura from the lateral cavernous sinus wall. (**g**) After dural opening and a wide sylvian fissure dissection, the aneurysm dome is revealed after opening Liliequist membrane. (**h**) The third nerve (CN III) is dissected. (**i**) Dissection over the suprachiasmatic cistern to free the frontal lobe. (**j**) Dissection of ipsilateral A1. (**k**) Starting the dural cut down to the optic nerve (CN II). (**l**) Finishing the opening of the optic nerve sheath. (**m**) Packing the cavernous sinus with fibrin glue to minimize venous bleeding during dissection. (**n**) Cutting the anterior petroclinoidal dural fold and incising the distal dural ring behind the carotid to help mobilize it forward and medially. (**o**) Cutting the medial attachment of the coagulated posterior petroclinoidal fold. (**p**) Posterior clinoid exposed subperiosteally in preparation to drilling it. In this case, unless the posterior clinoid is drilled, proximal basilar artery control cannot be achieved. (**q** and **r**) Third nerve is dissected free from the dura forming the oculomotor cistern. (**s**) The trochlear nerve (CN IV) is dissected free at the posterior border of the cavernous sinus. (**t**) Drilling the posterior clinoid with a low profile, curved drill, with a 3mm diamond bit and shaft protective sleeve. (**u**) Finishing the posterior clinoidectomy. (**v**) Dissecting the basilar trunk and aneurysm neck. (**w**) After a retractor is placed, the SCA and PCA can be more easily identified. (**x**) The patient had passed preoperatively the Alcock's test, and the posterior communicating artery seen here is now being coagulated to widen the retrocarotid corridor to the basilar aneurysm. (**y**) Perforators have been dissected off of the aneurysm dome. Several unsuccessful attempts at direct clipping were made, even with adenosine administration, because of the presence of the unmovable coil mass. With the application of temporary partial trapping clips (to BA and left PCA), the aneurysm was opened, some of the coils were cut, and some thrombus evacuated. Coils are being cut here with microscissors for partial removal. The right PCA is labeled here as PCA. (**z**) Due to some expected backbleeding which can be seen here, a temporary clip needed to be placed on the right P1 segment as well. None of these maneuvers would have been possible without the extensive exposure provided by the thorough transcavernous dissection. (**ab**) After a first permanent curved clip was placed, a second tandem clip to the aneurysm neck is being placed here quickly after SSEP changes were detected. Note the amount of space required to place all of these clips. A smaller exposure would not have allowed this to be done comfortably. SSEP returned to normal at the end of the surgery. (**ac**) AP and lateral views of the postoperative angiogram showing complete occlusion of the aneurysm by the two clips, with patency of all other vessels

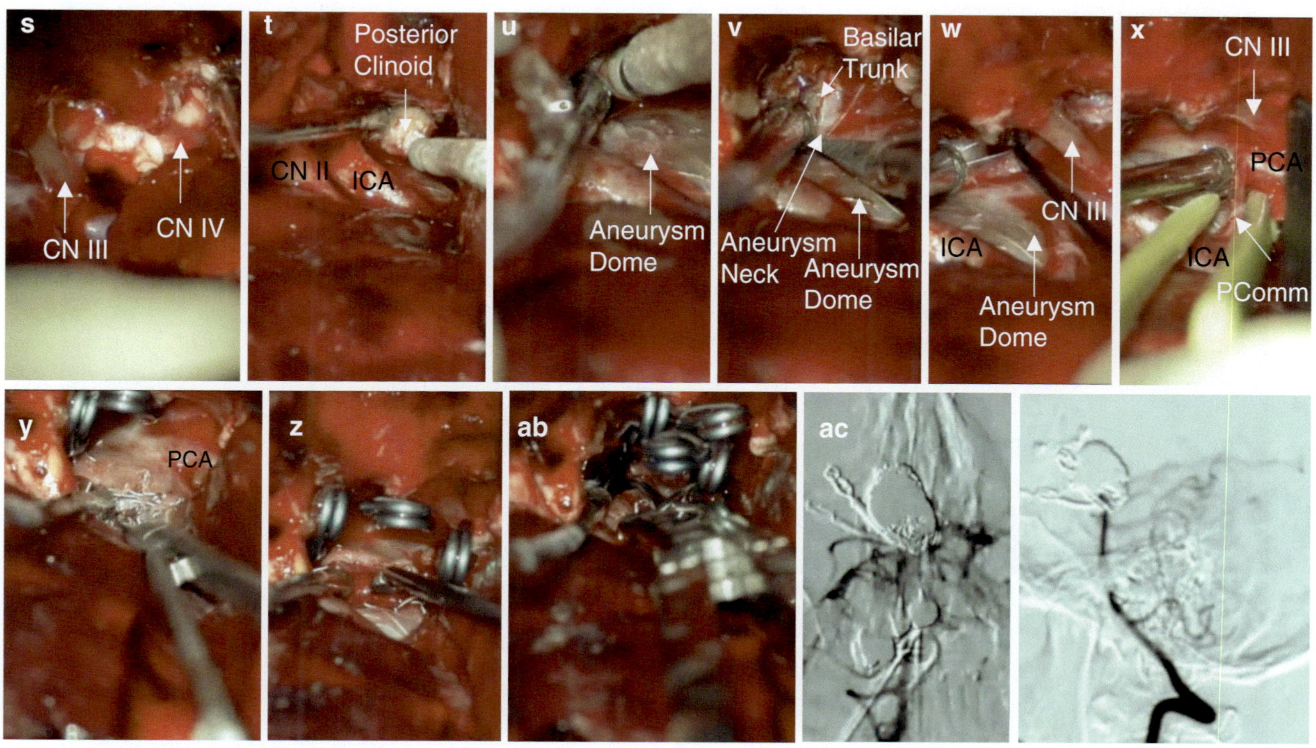

Fig. 38.5 (continued)

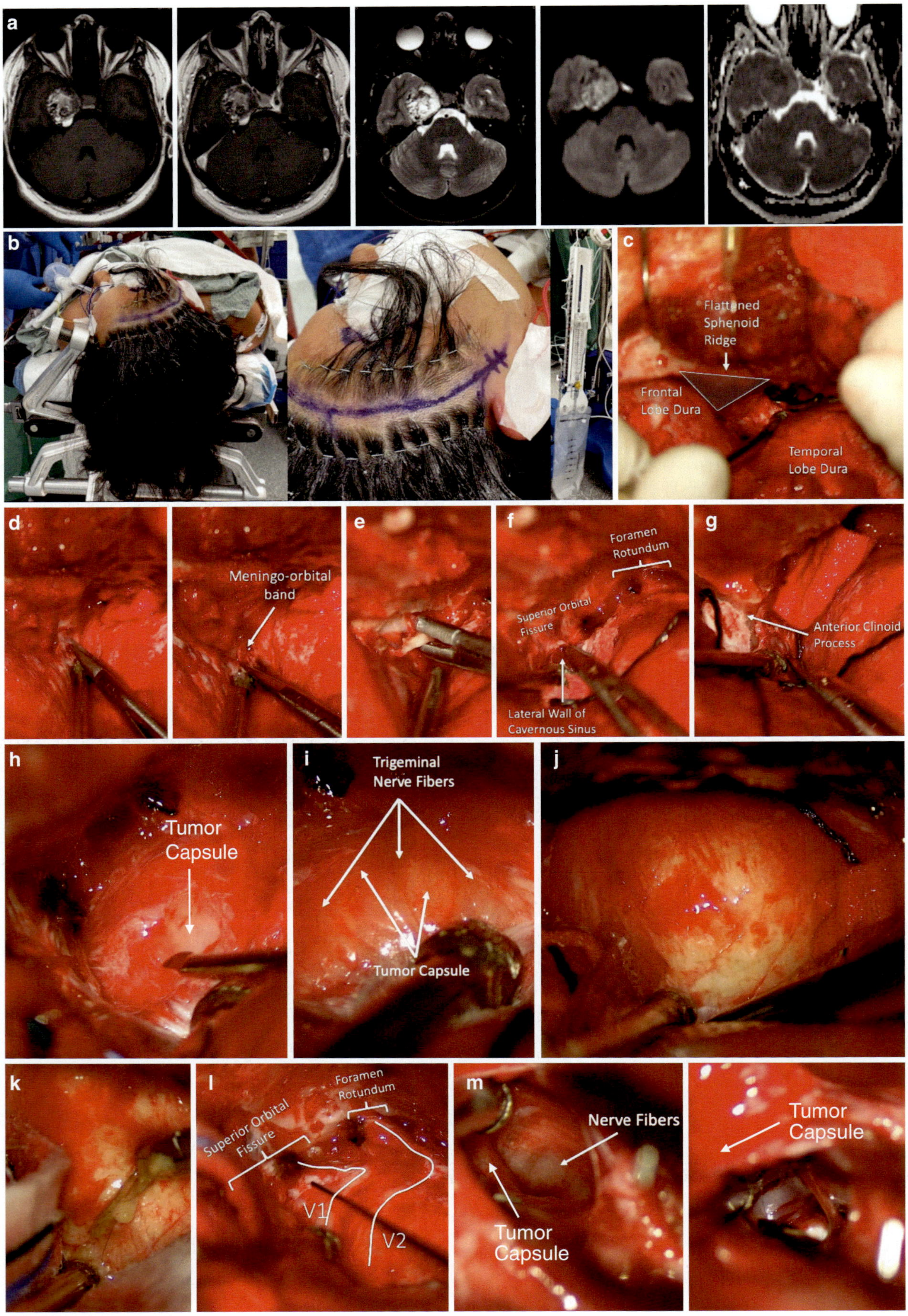
a
b
c
Flattened Sphenoid Ridge
Frontal Lobe Dura
Temporal Lobe Dura
d
Meningo-orbital band
e
f
Foramen Rotundum
Superior Orbital Fissure
Lateral Wall of Cavernous Sinus
g
Anterior Clinoid Process
h
Tumor Capsule
i
Trigeminal Nerve Fibers
Tumor Capsule
j
k
l
Foramen Rotundum
Superior Orbital Fissure
V1
V2
m
Nerve Fibers
Tumor Capsule
Tumor Capsule

(continued)

Fig. 38.6 Case 2: Hakuba-Dolenc Transcavernous Approach for Resection of a Large Cavernous Sinus Dermoid Cyst. (**a**) MRI of the dermoid cyst, occupying the lateral wall of the cavernous sinus with extension into Meckel's cave. Sequences from left to right: T1, T1 with contrast, T2, DWI, ADC. (**b**) Patient positioning and setup shown here. Head is turned 30 degrees contralaterally, neck extended, vertex down. Lumbar drain is placed to help with brain relaxation. (**c**) After performing the craniotomy and flattening the sphenoid ridge, we began the Hakuba interdural peeling of temporal lobe dura from anterior to posterior direction shown here. (**d**) Peeling of the lateral wall of the cavernous sinus begins with sharp dissection and cutting of the meningo-orbital band. (**e**) Unlocking the superior orbital fissure by drilling the lesser and greater wing of the sphenoid. Here a piece of greater wing is removed adjacent to the superior orbital fissure. (**f**) Interdural peeling continues sharply at the superior orbital fissure and reveals the lateral wall of the cavernous sinus. Foramen rotundum is seen here as well. (**g**) As the peeling progresses, the anterior clinoid process can be seen. (**h**) Continued peeling reveals the tumor capsule. (**i**) It is important to recognize the trigeminal nerve fibers draped over the tumor capsule and preserve them, stay within the correct dissection plane, while maintaining the integrity of the capsule during peeling. (**j**) Peeling continues on a wide front, and the goal is to avoid premature rupture of the cyst, because the counterpressure exerted by the cyst against the surgical instrument is helpful in maintaining the correct dissection plane. (**k**) Cyst decompression after cyst wall exposure. Hair and sebum can be seen within the lesion. (**l**) Direct nerve stimulation to minimize damage to cranial nerves. (**m**) Continued mobilization of deflated tumor capsule reveals nerve fibers on the other side of the capsule. (**n**) Traction and counter traction are used to pull the tumor capsule away from the nerves. Here the capsule is dissected away from V3 fibers. (**o**) Here the capsule is dissected away from V2 fibers. (**p**) As the dissection reaches its posterior end, Meckel's cave can be seen with CSF. (**q**) Tumor capsule removed as a single piece. (**r**) Relevant anatomy after tumor capsule is removed is shown here. (**s**) Post-Op MRI. T2, T1, and T1 with contrast sequences

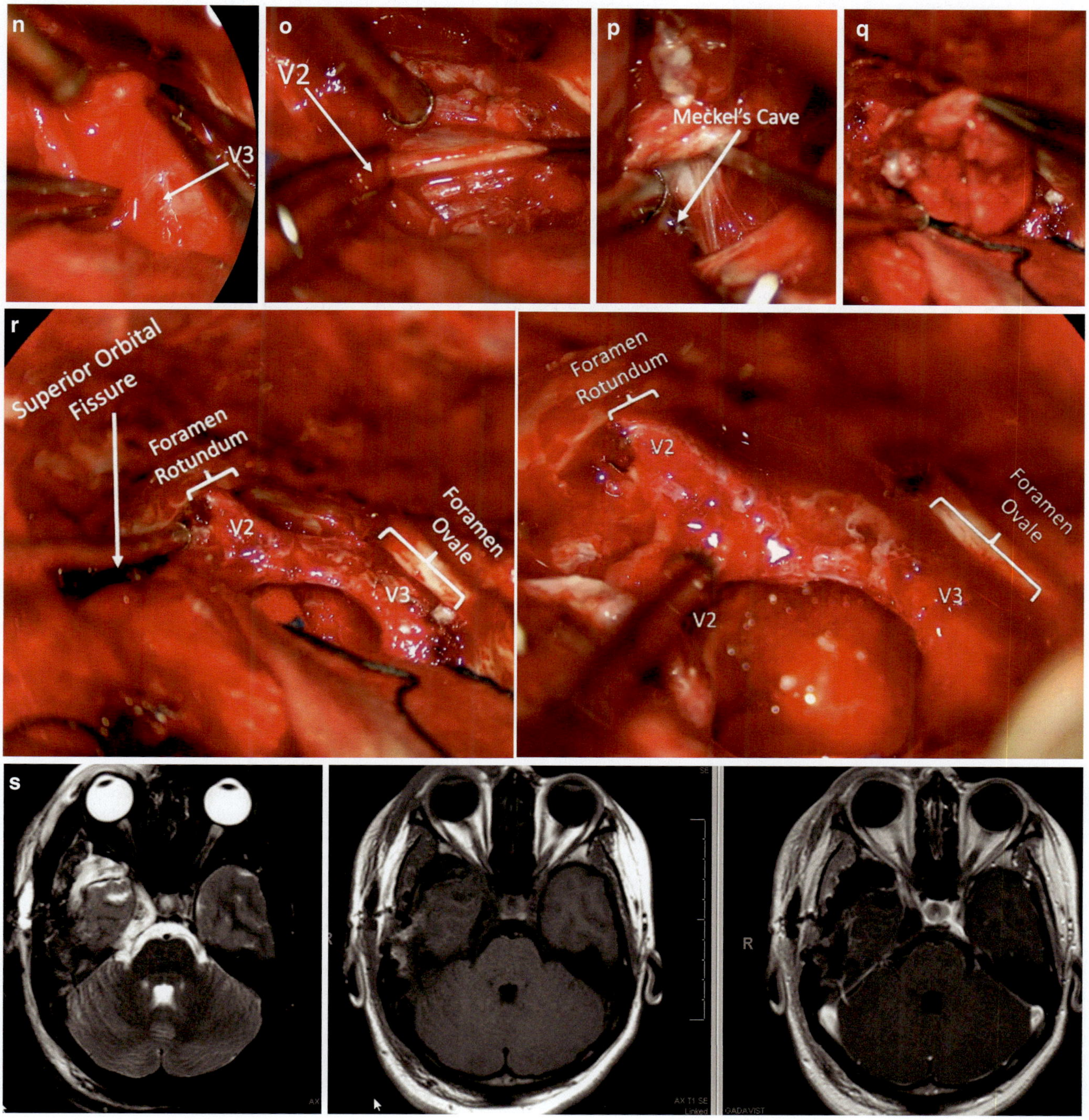

Fig. 38.6 (continued)

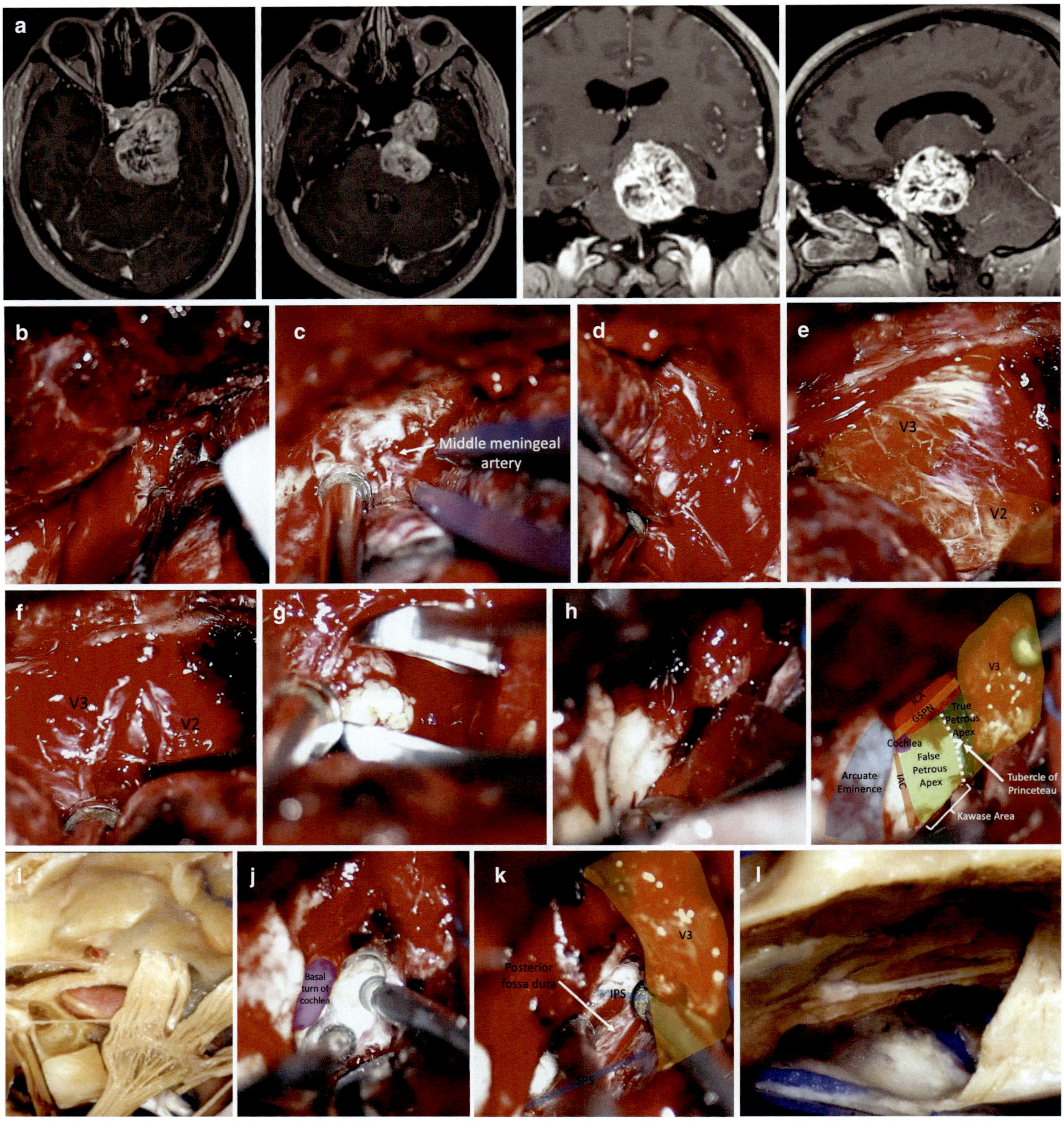

Fig. 38.7 Case 3: Combined transcavernous (Hakuba/Dolenc) and anterior petrosectomy (Kawase) approach to a large left trigeminal schwannoma. (**a**) Preop MRI T1 with contrast axial, coronal, and sagittal cuts. Typical dumbbell configuration of the tumor in the middle and posterior fossa, compressing brainstem and causing ventriculomegaly. (**b**) Peeling of temporal lobe dura from the lateral cavernous sinus wall. (**c**) Middle meningeal artery identified. (**d**) Extracapsular dissection carried out sharply. (**e**) The course of the very splayed V2 and V3 fibers draped over the tumor can be seen here. (**f**) Tumor capsule is opened between V2 and V3 parallel to the course of the trigeminal fibers. (**g**) Debulking the tumor in the middle fossa from the opening created in the tumor capsule. (**h**) As temporal lobe dura peeling continues further posteriorly, the middle fossa surgical view (left) and anatomic structure labels overlay (right) are shown here side by side. The exposure here is the view from the Kawase approach. (**i**) The Rhoton dissection of the middle fossa is shown here for reference. (**j**) Drilling the Kawase rhomboid area. The presumed location of the basal turn of the cochlea is shaded in purple. (**k**) The caudal limit of the drilling is reached once the inferior petrosal sinus (IPS) is uncovered. V3, IPS, superior petrosal sinus (SPS), and posterior fossa dura are labeled. (**l**) Rhoton dissection of the same area is shown here for reference. (**m**) The posterior fossa dura is opened sharply. (**n**) Superior petrosal sinus has been coagulated and divided. The trigeminal fibers can be seen in the background. (**o**) The tentorium is further divided. One needs to be cognizant to avoid cutting the trochlear nerve. (**p**) The tentorium has been divided. (**q**) Rhoton dissection view for reference. (**r**) Debulking the posterior fossa tumor. (**s**) Displaced rootlets of trigeminal nerve can be seen draped on the back of the tumor and preserved. (**t**) Basilar artery and its perforators can be seen after the tumor capsule is dissected away from the brainstem. The pontotrigeminal vein is seen here as well. (**u**) MRI 8 months post-op. Axial and coronal views of T1 with contrast

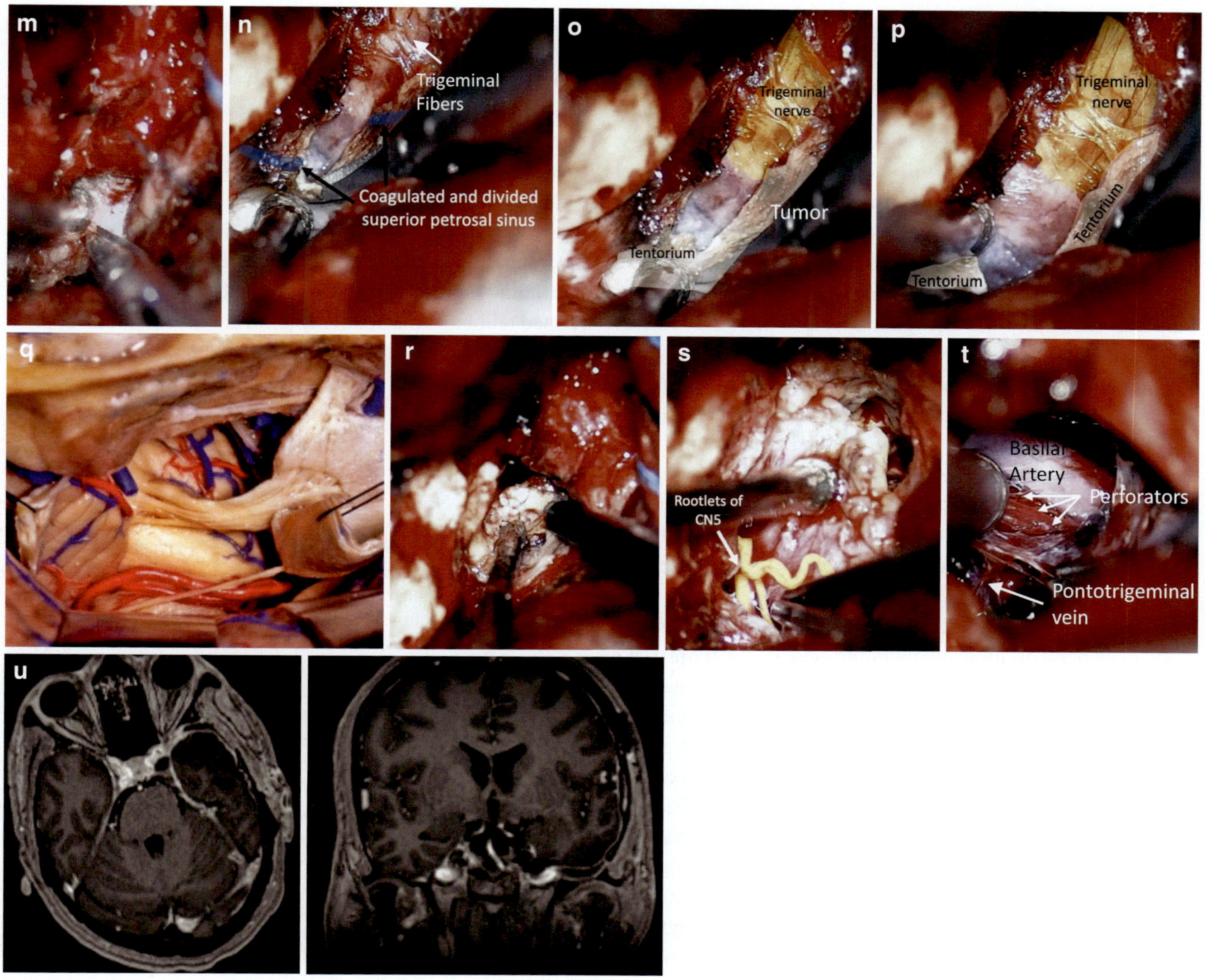

Fig. 38.7 (continued)

press the tumor. The deflated tumor capsule allows for continued peeling of the middle fossa until the Kawase rhomboid area is exposed. Anterior petrosectomy was performed, the posterior fossa dura was opened, and the superior petrosal sinus and the tentorium were divided. The posterior fossa tumor component was resected through that window. The basilar artery and its perforators were preserved. The trigeminal nerve roots were preserved.

38.6 Surgical Pearls

- The lateral wall of the cavernous sinus is not "intracavernous" per se; it is composed of two layers; and it is the inner one that contains CN 3/4/5, while CN 6 is deeper and closer to the ICA
- Dissection of the correct interdural plane can preserve the function of these cranial nerves, and surgical morbidity is minimized
- Lesions lateral to the cavernous carotid artery are best approached transcranially, while those medial to the cavernous carotid can be approached endonasally endoscopically
- Quality-of-life considerations should be included when weighing the risks and benefits of aggressive cavernous sinus surgery
- The texture, consistency, and adherence of lesions to critical structures will dictate the extent of resection. Pathology trumps location in this instance

38.7 Discussion

Given the various neuro-vascular structures passing through and around the cavernous sinus, surgeries of the cavernous sinus region can cause significant morbidity, including stroke, vision loss, diplopia, dry eye, and facial numbness. Therefore, quality-of-life considerations when weighing the risks and benefits of surgery are important. The approach for each lesion and for each patient must be individually tailored with quality-of-life considerations in mind. Data in the literature and our own experience have shown that not all cavernous sinus lesions are created equal. The location, pathology, and biology of the lesion will therefore dictate the optimal strategy and approach to the cavernous sinus. Some parts of the cavernous sinus are more easily accessible without significant associated morbidity compared to others. Based on our experience and what has been shown in the literature, surgical morbidity is not guaranteed when the correct approach is selected and when the tactical decisions, which are sometimes made intraoperatively, are not overly aggressive. The anatomical basis for at least a partial explanation for this is that the lateral wall of the cavernous sinus is not "intracavernous" per se, as described in the anatomy section of this chapter. There is yet another layer of separation between the cranial nerves running along the lateral wall and the intracavernous venous compartment, which encases the rest of the contents of the sinus.

This is not to overlook the fact that cavernous sinus surgery in general is more challenging and poses more risks than standard craniotomy approaches. Though the anatomy has been well described in the literature, an initial thorough 3-dimensional understanding of this anatomy requires the surgeon to have extensive first-hand experience, from cadaveric dissection training to operating around the cavernous sinus in patients, before a deeper understanding of what particular maneuvers are considered safe or unsafe is achieved. Additionally, a level of self-awareness or metacognition is needed for the individual surgeon to make the most appropriate preoperative and intraoperative decisions, based on each patient's priorities. Some intraoperative decisions are made in real time, and it is important to be clear about when it is appropriate to stop.

Our preference in general for lesions involving the cavernous sinus that are medial to the carotid and particularly those that involve the sella is to approach them through an endonasal endoscopic approach, which can expose that surface directly. In contrast, if the lesion is mostly lateral to the carotid, a transcranial approach is preferred. While holocavernous meningiomas can be managed conservatively or with radiosurgery, some lesions that are particularly aggressive but do involve the cavernous sinus may need cavernous sinus exenteration with or without carotid sacrifice that may or may not need cerebral revascularization via a bypass procedure. The texture of the lesion is another factor. If the consistency of the lesion is favorable, such as those that are soft and easily suckable (e.g., trigeminal schwannoma), a greater extent of resection can be achieved safely. If the lesion is firm, hard, or particularly calcified and adherent to nerves and the carotid, then a gross total resection cannot be safely achieved.

38.8 Conclusion

Cavernous sinus surgery has evolved over the last several decades with multiple different approaches to allow for tailored strategies for individual lesions, in order to maximize treatment effect and minimize morbidity. Our current anatomical understanding of the various walls and compartments of the sinus allows for very detailed risk and benefit analysis of each component of a modular type of surgical approach, depending on the goals of surgery. The surgeon needs to be a refined tailor and master of customization, rather than apply a cookie-cutter, one-size-fits-all approach to patients.

References

1. Parkinson D. Lateral sellar compartment O.T. (cavernous sinus): history, anatomy, terminology. Anat Rec. 1998;251(4):486–90. https://doi.org/10.1002/(sici)1097-0185(199808)251:4<486::aid-ar7>3.0.co;2-q.
2. de Divitiis O, Elefante A, de Divitiis E. Laterosellar compartment or cavernous sinus? What's the correct term? World Neurosurg. 2014;81:74–6. https://doi.org/10.1016/j.wneu.2012.10.047.
3. Winslow BJ. Exposition Anatomique de La Structure Du Corps Humain. London: N. Prevast; 1734.
4. Thakur JD, Sonig A, Khan IS, Connor DE, Pait TG, Nanda A. Jacques Bénigne Winslow (1669–1760) and the misnomer cavernous sinus. World Neurosurg. 2014;81(1):191–7. https://doi.org/10.1016/j.wneu.2012.06.030.
5. Thamburaj AV, Behari S, Chemate S. Comprehensive neurosurgery: (Volume 2). bohrpub; 2021. https://books.google.com/books?id=8LJoEAAAQBAJ
6. Browder J. Treatment of carotid artery-cavernous sinus fistula: report of a case. Arch Ophthalmol. 1937;18(1):95–102.
7. Parkinson D. A surgical approach to the cavernous portion of the carotid artery: anatomical studies and case report. J Neurosurg. 1965;23(5):474–83. https://doi.org/10.3171/jns.1965.23.5.0474.
8. Parkinson D. Arteries of the cavernous sinus. J Neurosurg. 1984;61(1):203. https://doi.org/10.3171/jns.1984.61.1.0203.
9. Parkinson D. The lateral sellar compartment (cavernous sinus) from Galen to Dolenc. 1994:443–446. https://doi.org/10.1159/000429888.
10. Parkinson D. Cavernous sinus or not? J Neurosurg. 1994;80(5):955–6. https://doi.org/10.3171/jns.1994.80.5.0955.
11. Parkinson D. Cavernous sinus exploration. J Neurosurg. 1992;76(2):341–2. https://doi.org/10.3171/jns.1992.76.2.0341.
12. Hakuba A, Nishimura S, Shirakata S, Tsukamoto M. Surgical approaches to the cavernous sinus. Report of 19 cases (author's transl). Neurol Med Chir (Tokyo). 1982;22:295–308. https://doi.org/10.2176/nmc.22.295.
13. Hakuba A, Liu S, Nishimura S. The orbitozygomatic infratemporal approach: a new surgical technique. Surg Neurol. 1986;26:271–6. https://doi.org/10.1016/0090-3019(86)90161-8.
14. Dolenc VV. A combined epi- and subdural direct approach to carotid-ophthalmic artery aneurysms. J Neurosurg. 1985;62:667–72. https://doi.org/10.3171/jns.1985.62.5.0667.
15. Dolenc VV, Škrap M, Šušteršič J, Škrbec M, Morina A. A transcavernous-transsellar approach to the basilar tip aneurysms. Br J Neurosurg. 1987;1(2):251–9. https://doi.org/10.3109/02688698709035309.
16. Dolenc V. Microsurgical removal of large sphenoidal bone meningiomas. Acta Neurochir Suppl. 1979;28(2):391–6.
17. Coscarella E, Başkaya MK, Morcos JJ. An alternative extradural exposure to the anterior clinoid process: the superior orbital fissure

as a surgical corridor. Neurosurgery. 2003;53(1):162–7. https://doi.org/10.1227/01.neu.0000068866.22176.07.
18. Dolenc V. Direct microsurgical repair of intracavernous vascular lesions. J Neurosurg. 1983;58:824–31. https://doi.org/10.3171/jns.1983.58.6.0824.
19. Rhoton ALJ, Hardy DG, Chambers SM. Microsurgical anatomy and dissection of the sphenoid bone, cavernous sinus and sellar region. Surg Neurol. 1979;12:63–104.
20. Harris F, Rhoton AL. Microsurgical anatomy of the cavernous sinus. Surg Forum. 1975;26:462–3.
21. Fujii K, Chambers SM, Rhoton AL. Neurovascular relationships of the sphenoid sinus: a microsurgical study. J Neurosurg. 1979;50(1):31–9. https://doi.org/10.3171/jns.1979.50.1.0031.
22. Banerjee AD, Ezer H, Nanda A. The artery of Bernasconi and Cassinari: a morphometric study for superselective catheterization. Am J Neuroradiol. 2011;32(9):1751–5. https://doi.org/10.3174/ajnr.a2552.
23. Tran-Dinh H. Cavernous branches of the internal carotid artery: anatomy and nomenclature. Neurosurgery. 1987;20(2):205–10. https://doi.org/10.1227/00006123-198702000-00001.
24. Sano K. Temporo-polar approach to aneurysms of the basilar artery at and around the distal bifurcation: technical note. Neurol Res. 1980;2:361–7. https://doi.org/10.1080/01616412.1980.11739588.
25. Fujimoto Y, Ikeda H, Yamamoto S. Pterional transcavernous approach for large basilar top aneurysm. Surg Cereb Stroke. 1992;20(3):191–5. https://doi.org/10.2335/scs1987.20.3_191.
26. Kawase T, Shiobara R, Toya S. Anterior transpetrosal-transtentorial approach for sphenopetroclival meningiomas: surgical method and results in 10 patients. Neurosurgery. 1991;28:869–75; discussion 875–6.

Surgical Treatment of Craniopharyngiomas

39

Francesco Tomasello, Filippo Flavio Angileri, and Antonello Curcio

Learning Objectives
- Epidemiology and diagnosis
- Knowledge of neurosurgical anatomy
- Anatomic classification of craniopharyngiomas
- Surgical approaches: open and related anatomy
- Surgical approach: endoscopic and related anatomy

39.1 Introduction

According to the WHO classification of 2021, craniopharyngiomas are benign (grade I), slow-growing tumors of dysembryogenetic origin. They are, however, characterized by local aggressiveness and a propensity to recur. Craniopharyngioma is a rare parasellar tumor representing 2–5% of all primary intracranial neoplasms, whose incidence is 0.18 per 100,000 inhabitants with a bimodal distribution with two characteristic peaks: the first in infancy with mean age between 5 and 9 years where the craniopharyngioma represents 5–9% of brain tumors and the second in the sixth-seventh decade of life with a mean age between 55 and 69 years without significant differences in distribution by sex [1]. If only sellar and parasellar tumors are taken into account, craniopharyngiomas are the most common tumor histotype in children (54%) while are less representative in adults (20%). The Rathke pouch epithelium, which gives rise to the craniopharyngeal duct, was the source of the cells that Mott and Barrett proposed as the origin of craniopharyngiomas in 1899. The latter joins the neural tube to the pharynx in the embryo, from where the tuber cinereum, the infundibulum, and the anterior lobe of the pituitary then emerge. As a result, craniopharyngiomas can appear anywhere along their course, from the pharynx to the hypothalamus, though the majority are found in the sellar and parasellar regions. Adamantinomatous and papillary are the two main subtypes according to histology; however, transitional variants and mixed forms have been described [2]. Uncertainty surrounds the etiopathogenesis of the lesion, and two different theories of its origin have been put forth: The first, known as embryogenetics, postulates that craniopharyngiomas result from the metaplastic transformation of pituitary cells within the gland, or p, rather than the neoplastic transformation of remaining embryonic squamous cells from the craniopharyngeal duct (the papillaries). The adamantinomatous histological type is the most common and can occur at any age, but it is more common in the first two decades of life. Macroscopically, this tumor variant can present solid, cystic, and calcified components, more common in children; in particular, the cysts can be multilobulated and contain fluid with characteristics ranging from the typical "motor oil" (Fig. 39.1) to the bright yellow fluid thick with cholesterol, which consists essentially of squamous epithelial cells. However, the papillary variant, with the exception of rare cases in the pediatric age, is prevalently diagnosed in adult patients. In this tumor subtype, calcifications are infrequent and the cystic content, unlike the adamantinomatous variety, is yellowish and viscous in consistency. Craniopharyngiomas frequently make tenacious contacts with neighboring neurovascular structures, optic nerves, chiasm, hypothalamus, tuber cinereum, carotid arteries, and their branches, making complete removal likely problematic. They are often surrounded by an abundant glial reaction, which, during surgical removal, can sometimes represent a useful cleavage plane for the surgeon. Many craniopharyngiomas have suprasellar extension, purely suprasellar are 20–41%, and intra- and suprasellar 53–75%; instead, intrasellar and intraventricular craniopharyngiomas are rarer, while those with sphenoidal, nasopharyngeal, and intracerebral localization are exceptional.

F. Tomasello (✉)
Humanitas Clinical Institute of Catania, Misterbianco, Catania, Italy
e-mail: ftomasel@unime.it

F. F. Angileri · A. Curcio
Department of Neurosurgery, AOU Policlinico "G. Martino", University of Messina, Messina, Italy
e-mail: fangileri@unime.it

I. N. Kanaan, V. Beneš (eds.), *Neuroanatomy Guidance to Successful Neurosurgical Interventions*,
https://doi.org/10.1007/978-3-031-59838-8_39

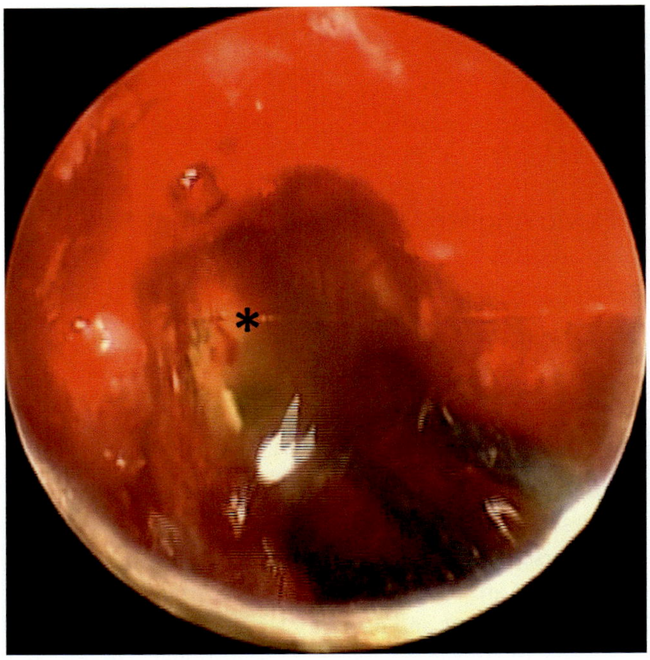

Fig. 39.1 Typical brownish color of "motor oil" cystic fluid

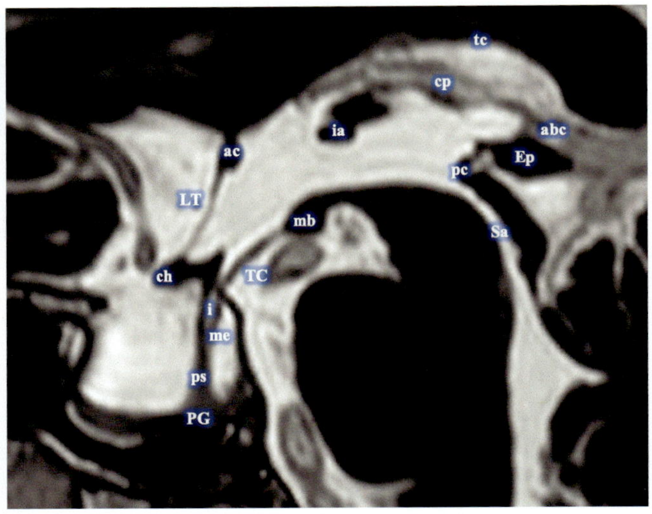

Fig. 39.2 3D sagittal view of LCR cavity of the third ventricle. 1. Optic recess; 2. infundibular recess; 3. infundibulum; 4. Sylvian aqueduct; 5. epiphyseal recess; 6. supraepiphyseal recess; 7. interthalamic adhesion; 8. Monro foramen

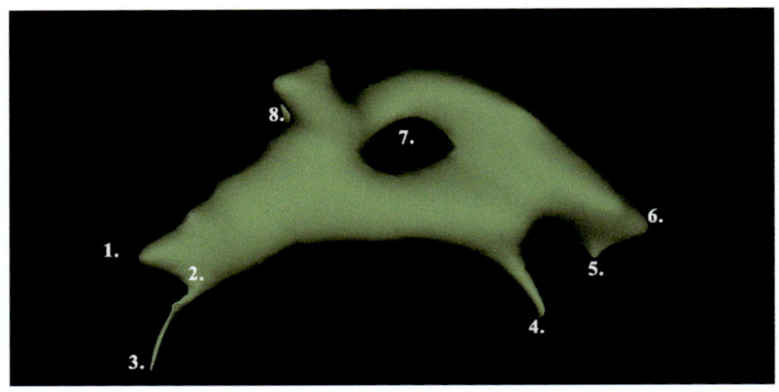

Fig. 39.3 MRI sagittal view of the third ventricle. *LT* Lamina terminalis, *ch* optic chiasm, *i* infundibulum, *me* median eminence, the posterior wall of infundibulum, connect neurohypophysis with hypothalamus, *ps* pituitary stalk, *PG* pineal gland, *TC* tuber cinereum, *mb* mammillary body, *Sa* Sylvian aqueduct, *pc* posterior commissure, *Ep* epiphysis, *abc* habenular commissure, *tc* tela choroidea, *cp* choroid plexus

Primary craniopharyngiomas in the temporal lobe, pineal gland, posterior cranial fossa, and cerebellar angle have been described in the literature.

39.2 Anatomical Highlights

- The pituitary gland, about 1 cm in diameter, forms the continuation of the infundibulum below. It is housed within the sella turcica of the sphenoid bone
- Pituitary stalk connects the hypothalamic region to the pituitary gland
- Microvasculature of pituitary gland and stalk arises from superior and inferior hypophyseal artery even if other less constant arteries may contribute
- The area immediately around the pituitary, the sellar region and parasellar region, is an anatomically complex area crowded of crucial neurovascular adjacent structures
- Optic nerves, chiasm, proximal tracts, oculomotor nerves, and post-part of olfactory tracts pass through suprasellar region and anterior incisural space
- Length of optic nerves (prefixed, normal, and post-fixed chiasm) significantly changes the surgical corridor accessing the sellar and parasellar areas
- The third ventricle is a median sagittal slit that connects the lateral ventricles with the fourth ventricle. The third ventricle extends forward to the lamina terminalis stretched between the optic chiasm and the rostrum of the corpus callosum (Figs. 39.2 and 39.3).

39.3 History/Background

Craniopharyngiomas are, by definition, benign tumors classified as WHO grade I. The peculiarity of these tumors is that in addition to being accompanied by neuroendocrine dysfunction, and they grow near vital neurovascular structures. Surgery for resection of craniopharyngiomas can be either standard or extensive transcranial or transsphenoidal surgery. Transcranial surgery is preferred for the resection of very large tumors offering the possibility of different corridors for the resection of the craniopharyngioma. In fact, it is possible to perform an interhemispheric approach, unilateral subfrontal approach, bilateral subfrontal approach, all the lateral frontotemporal approaches, i.e., the pterional and the orbitozygomatic in all its possible modifications, and also a posterior transpetrosal approach [3]. The transcranial approach, however, has some drawbacks. Transcranial approaches have been reported with a high recurrence of craniopharyngioma and with a high frequency of even severe adverse events, including perioperative death when compared with a less invasive transsphenoidal approach [4, 5].

The standard endoscopic transsphenoidal approach and the extended transsphenoidal approach are instead well-established microsurgical options for the resection of intra- and parasellar tumors through the sphenoid sinus access. The transsphenoidal corridor is especially indicated for intrasellar craniopharyngiomas in patients with an enlargement of the floor of the sella turcica [6]. The success rates of craniopharyngioma resections using standard or extended transsphenoidal approaches tend to be very high, ranging from 77.8% of complete tumor resection up to even 86% [7, 8]. Also in this case, as in transcranial surgery, postoperative complications with the onset or worsening of hypopituitarism and the new development of diabetes insipidus are frequent [7, 9]. Moreover, extended transsphenoidal approach carries a higher risk of postoperative CSF fistula rate compared to transcranial routes [8, 10, 11].

39.3.1 Pathogenesis

The adamantinomatous subtype of craniopharyngioma, which affects both adults and children, is the most prevalent type. The development of adamantinomatous craniopharyngiomas is explained by embryogenetic theory [12]. The buccopharyngeal membrane develops an ectodermal pouch called Rathke's diverticulum during embryogenesis, which gives rise to the anterior pituitary gland. Rathke's pocket develops as a result of this sac invaginating and ascending between the embryo's rhombencephalon and forebrain. The pharyngohypophyseal stalk serves as a temporary connection between Rathke's pouch cells and the embryo's roof of the oral cavity as they proliferate quickly during the fifth week of pregnancy. The craniopharyngeal duct, also known as Rathke's pouch extension, typically fully involutes at the seventh week of gestation [13]. The craniopharyngeal duct can occasionally fail to completely involute, which causes ectodermal cells that may be present along this path to persist and eventually give rise to the craniopharyngioma [12].

Instead, the papillary-type craniopharyngioma, which is mostly found in adults, is more frequently described by the metaplastic theory. As a result, the papillary develops from a metaplasia of the adenohypophyseal cells within the pituitary gland, and the formation of a cellular tissue results in differentiated squamous cells [13].

39.3.2 Classification

The growth trajectory and surgical approach of craniopharyngiomas have been classified by several authors over the years. All these classifications share the principle of subdividing lesions along the length of extension in the primary vertical axis, as related to the optic chiasm, diaphragma sellae, third ventricle, or, more recently, the infundibulum. Yasargil in 1990 [14] first described a classification in six levels depending on relationship between the craniopharyngioma and its vertical and lateral extension (Table 39.1). In contrast, Samii et al. [15] graded craniopharyngiomas according to the vertical projections they had in five levels (Table 39.2). A different classification that is useful during endonasal approaches was recently proposed by Kassam et al. [16], and it is based on the relationship between the craniopharyngioma and the infundibulum (Table 39.3) (Fig. 39.4). Because the adherence of the CP significantly affects the possibility of a safe and radical resection, Prieto et al. graded the predictors of CP adherence severity (Table 39.4). The grown pattern of CP is related anyway depending on its origin site. The origin site and the extension of the CP have been studied and classified by Fan et al., in 2021, and this is a useful tool for preoperative surgical planning (Table 39.5).

Table 39.1 Yasargil classification of craniopharyngioma based on its relationship with infundibulum, hypothalamus, and third ventricle

I	Purely intrasellar-infradiaphragmatic
II	Intra- and suprasellar, infra- and supradiaphragmatic
III	Supradiaphragmatic, parachiasmatic, extra-ventricular
IV	Intra- and extra-ventricular
V	Paraventricular
VI	Purely intraventricular

Table 39.2 Samii classification of craniopharyngioma based on its vertical extension

I	Intrasellar or infradiaphragmatic
II	Cisternal with or Without an intrasellar component
III	Lower half of the third ventricle
IV	Upper half of the third ventricle
V	Reaching the septum pellucidum or lateral ventricles

Table 39.3 Kassam classification of craniopharyngioma based on its relationship with infundibulum, for endoscopic trans-sphenoidal approach

I	Preinfundibular
II	Transinfundibular
III	Post- or retro-infundibular
IV	Isolated third ventricular

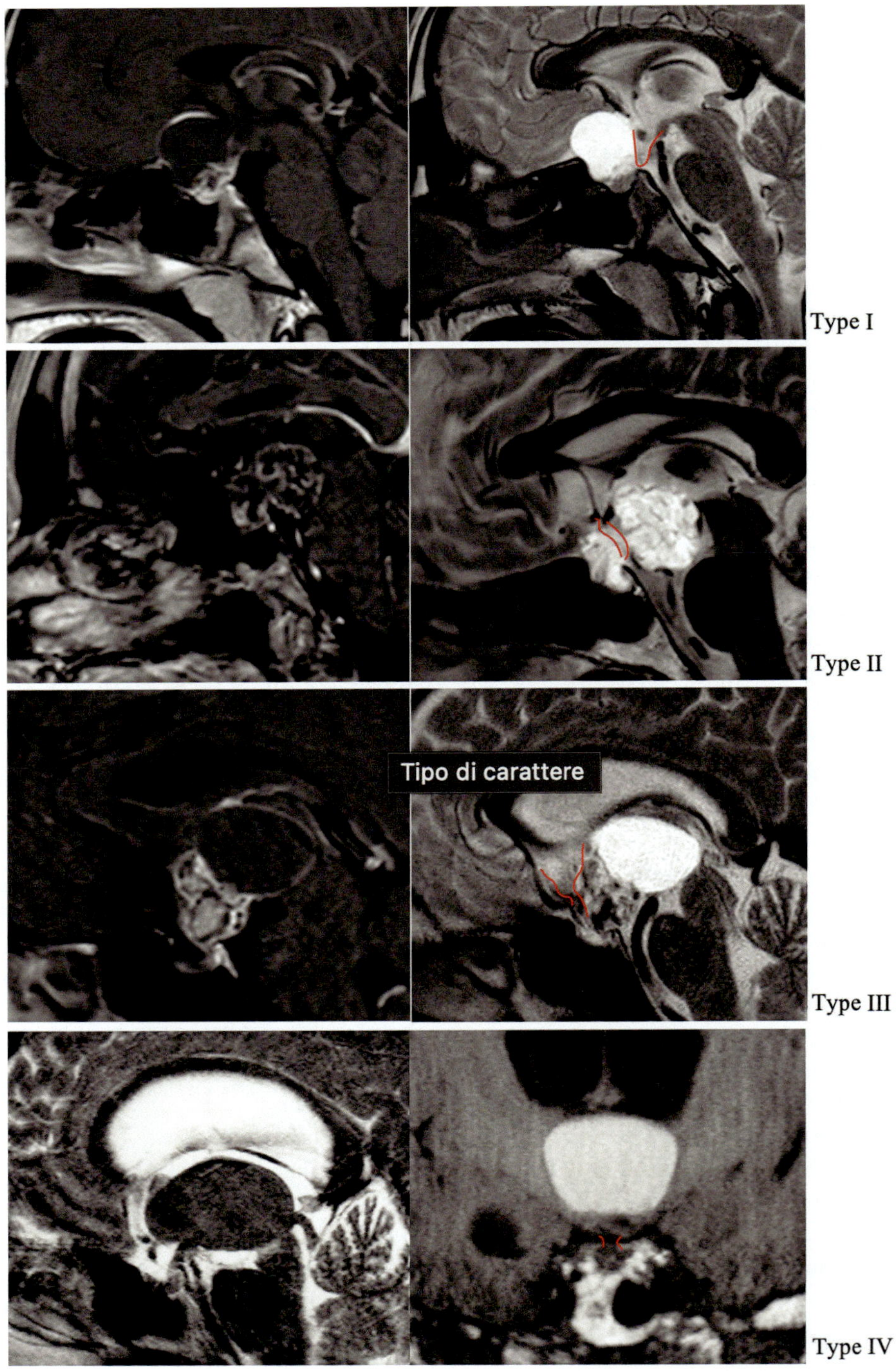

Fig. 39.4 Kassam classification system showed in MRI. Sometimes is quite difficult to differentiate subtypes, and its real relationship with infundibulum is really understood only during surgery. I and IV types are the easiest to define in MRI. Red line identifies infundibulum

Table 39.4 Prieto classification of adherence severity

I	Mild	The CP presents a loose or tight adhesion to the sellar region including the junction of the pituitary gland with the pituitary stalk
II	Moderate	The CP had developed within the third ventricle and presents a pedicle-like (fibrovascular stem) or a sessile (rectangular patch of tumor capsule) attachment to the inner lining of the third ventricle
III	Serious	The CP is tightly adhered or fused to the pituitary stalk and/or outer aspect of the infundibulum; this attachment may have either a sessile or cap-like (the adhesion covers the tumor's top portion) morphology
IV	Severe	The CP presents a tight attachment to, or is fused to, or replaces the third ventricle floor or both the floor and adjacent third ventricle walls; this attachment may have a bowl-like (the adhesion covers the tumor's bottom portion), a ring-like (the adhesion encircles the tumor's middle portion like a band), or a circumferential (the adhesion covers entirely the tumor surface like wrapping paper) morphology
V	Critical	The CP attachment involves all structures along the sella-third ventricle vertical axis; in this subgroup, the most severe adhesions occur at the level of the third ventricle floor, as this structure is usually invaded and replaced by tumor tissue

Table 39.5 Fan classification(QST) based on the location of the tumor and the nearby membranous and neurovascular structures

Q	CP emerge from the subdiaphragmatic intrasellar space and cling firmly to the surrounding residual gland
S	CP emerge from the middle or inferior segment of the stalk and typically extend between the subarachnoid cisterns
T	CP originate from remaining Rathke's pouch precursor cells at the top of the pars tuberalis, mostly extend upward, and have a close relationship with the third ventricular floors and hypothalamus

39.4 Surgical Approaches

39.4.1 Pterional Approach

Pterional approach is widely used to treat and remove many tumors in the sellar region. It is a very flexible approach that enables the development of various working channels through its various extensions, enabling access to the deepest regions of the brain. The patient is positioned supine, and the Mayfield clamp is fixed while the patient's head is raised hyperextended and rotated between 45 and 60 degrees. During the skin incision, it is important to remain 1 cm anterior to the tragus to avoid unintended injury to the superficial temporal artery. The temporalis muscle must still be cut out in order to expose the fronto-temporo-orbital area. The surgeon will take care to protect the frontal branch of the facial nerve when performing the interfascial, submuscular, or subfascial route for the muscle dissection. The middle cranial fossa and the anterior cranial fossa are simultaneously exposed during a craniotomy that is performed, from Dandy's keyhole, in a curvilinear shape superiorly. It is necessary to dissect the great wing of the sphenoid in its pterional process in order to better expose the middle cranial fossa and create a canal that is as straight as possible. Over the orbit and the sphenoid ridge, the dura mater is opened in a semicircular motion. The fissure at this location is divided in a proximal direction. Since the middle cerebral veins, given their anatomical course, tend to run more toward the temporal lobe, it is actually preferable to start from the more frontal area. As a result, the pars opercularis of the inferior frontal gyrus serves as the starting point for the dissection of the Sylvian fissure. Following the opening of the Sylvian fissure, it is possible to move deeper in the direction of the middle cerebral artery. The latter directs us to expose the carotid artery laterally. The entire Willis polygon, along with its efferents and anterior cerebral artery, can be seen when the Sylvian fissure is fully open. Additionally, if the approach has been properly executed, the sellar area, the parasellar area, the suprasellar area, and the retrosellar area can all be easily reached.

The surgeon may choose to continue his deep dissection in the parachiasmatic space after the prechiasmatic (interoptic), opticocarotid, carotid-tentorial triangle, in the supracarotid triangle (above the bifurcation of the carotid) (Fig. 39.5), or trans-lamina terminalis, after using the traditional pterional approach to reach the paraventricular paraclival space.

The medial margins of the optic nerves, the optic chiasm, and the limbus sphenoidale serve as the lateral, posterior, and anterior boundaries of the **interoptic triangle**, respectively. This triangle is simple to open, even without drilling the limbus sphenoidale. The suprasellar space, hypothalamus floor, and pituitary stalk are all visible through this window. Because it allows the surgeon to access the lesion's contralateral margin, the interoptic triangle is the best surgical technique for lesions that are bilaterally positioned between the optic nerves. The optic chiasm's relationship with the tuberculum sellae specifically, whether it is prefixed, normal, or postfixed, determines the usefulness of this window.

The optic apparatus (optic nerve chiasm) medially, the ICA laterally, and the A1 segment posteriorly serve as the landmarks for the **opticocarotid triangle**. This window can be used for lesions that originate anteriorly and extend posteriorly along the pituitary stalk. Two suggestions can increase the flexibility within this area: First, an anterior clinoidectomy allows for a greater mobilization of the optic structure to maximize the exposure of the triangle, allowing for more surgical trajectories; second, a gentle mobilization of the ICA can further optimize this trajectory.

The **carotid-tentorial triangle** is a wide opening between the two named structures. By the way, it contains the third cranial nerve, which represents an anatomical and stuck limit

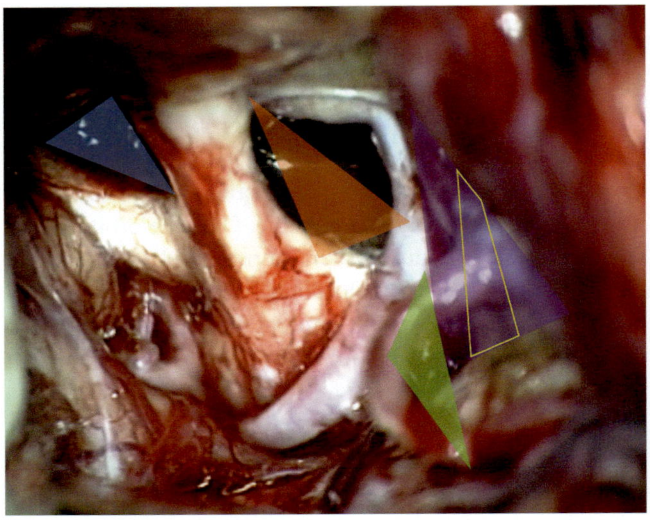

Fig. 39.5 Anatomical triangles useful for in skull base surgical corridor. *Blue* ▲: prechiasmatic triangle; *orange* ▲: opticocarotid triangle; *purple* ▲: carotid-tentorial triangle, divided by oculomotor nerve (*yellow trapeze* ⌂) in carotid-oculomotor triangle (on the left) and oculomotor-tentorial triangle (on the right); *green* ▲: supracarotid triangle

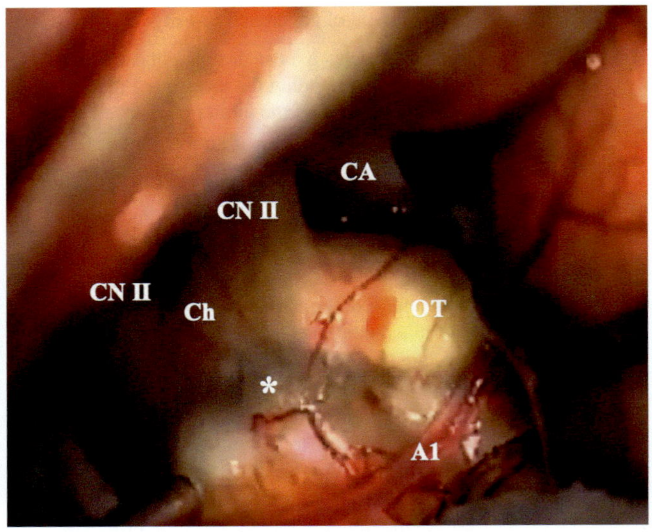

Fig. 39.6 Right pterional approach. *CN II* optic nerve, *CA* carotid artery, *OT* right optic tract, *Ch* chiasm, *A1* anterior cerebral artery, * lamina terminalis

for dissection. For this reason, this triangle is divided into carotid-oculomotor and oculomotor-tentorial triangles. The supraclinoid segment of ICA, a line from porus oculomotorius, where the nerve enters the roof of the cavernous sinus, to the anterior clinoid process, and a third line from the ICA bifurcation direct to the crossing oculomotor nerve over the tentorium all form the borders of the carotid-oculomotor triangle. The carotid-oculomotor triangle is an essential deep window for approach to the upper basilar artery, along with the opticocarotid triangle. For transsylvian or subtemporal approaches to the upper basilar region, the carotid-oculomotor triangle is typically the most advantageous trajectory because the opticocarotid triangle frequently includes eloquent perforator arteries. Even in this case, performing an anterior and a posterior clinoidectomy, and secondarily mobilizing the ICA, can be significantly widen this window. Add to this, the medial cisternal segment of the oculomotor nerve, the medial tentorial margin, and the posterior cerebral peduncle all encircle the oculomotor-tentorial triangle. For the purpose of treating retroinfundibular CP, this triangle provides access to the ventrolateral pontomesencephalic region.

The **supracarotid triangle**, which is bounded by the basal frontal lobe anteriorly, the M1 segment of the middle cerebral artery laterally, and the A1 segment medially, does not provide a straightforward trajectory. The surgery could aid the surgeon in extremely large CPs extending superiorly reaching the highest part of the third ventricle, as it was initially described for the treatment of lesion of the basal ganglia.

39.4.2 Translamina Terminalis Approach

Through anterior, antero-lateral, lateral, and trans-sphenoidal techniques, various surgical corridors have been described for the **translamina terminalis** approach to third ventricle. Once that the suprachiasmatic region has been exposed, anatomy is easily recognizable. The first step is to allow a greater mobilization of nervous structures. Following the visualization of the contralateral optic nerve, the arachnoid cisterns at the base of the skull, which surround the contralateral optic nerve and carotid artery, are opened to promote additional cerebral spinal fluid egress and brain relaxation. In order to gain access to the third ventricle, the lamina terminalis must be identified. It is a tiny grey sheet that extends between the optic chiasm's dorsal surface inferiorly and the anterior commissure postero-superiorly. The anterior communicating artery complex and anterior cerebral artery are located in the lamina terminalis cistern, which is anterior to the lamina terminalis itself. Lamina terminalis vascular organ, one of the seven circumventricular organs, is located there. To access lamina terminalis from anterior (subfrontal approach, anterior transcallosal translaminar terminalis approach) and antero-lateral (fronto-temporal approaches) routes, it is first of all fundamental the fenestration of lamina terminalis cistern and the identification of vascular and nervous structures that forms its edges (Fig. 39.6). Then, it is gently opened in midline gaining access to the anterior segment of third ventricle. Standard microsurgical methods are used for tumor dissection and debulking.

Because craniopharyngiomas are typically subarachnoid tumors, the surrounding structures can almost always be

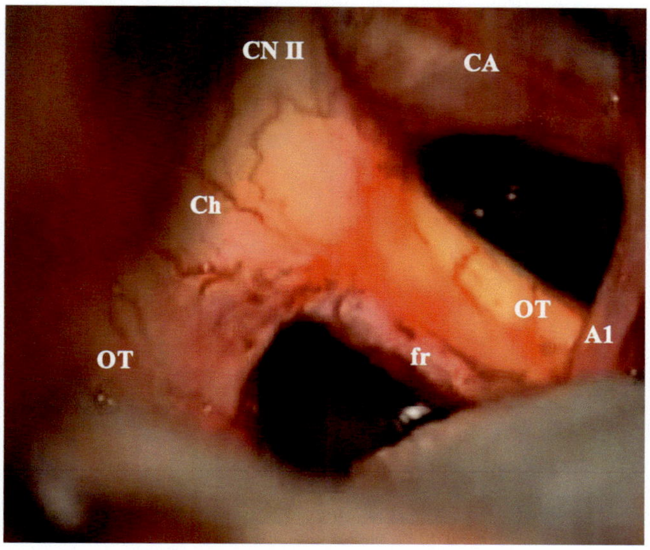

Fig. 39.7 Post-op surgical images showing intact nervous and vascular structures. Lamina terminalis has been widely opened. *CN II* optic nerve, *CA* carotid artery, *OT* right optic tract, *Ch* chiasm, *A1* anterior cerebral artery, *fr* residual arachnoid frustle of lamina terminalis

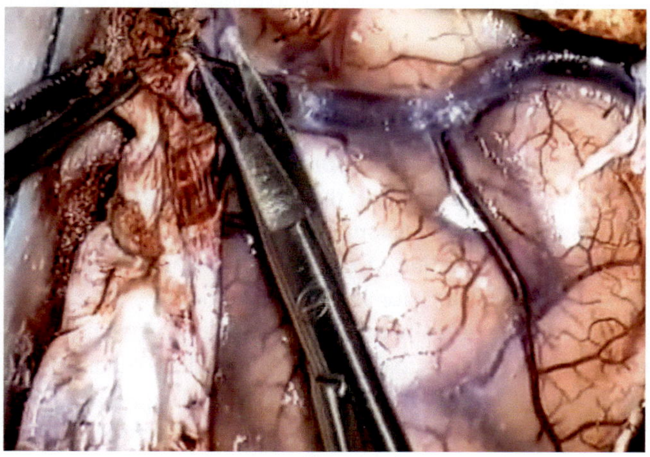

Fig. 39.8 Dissection of big superficial vein from dura adherences and arachnoid sheet, in latero-medial direction

removed easily. The main vascular and nervous structures must be kept intact during the lamina terminalis opening (Fig. 39.7), especially the hypothalamus, which may have been moved by the presence of an intraventricular mass.

39.4.3 Interhemispheric Approach

The procedure is carried out through a standard 6 × 4 cm right frontal paramedian craniotomy, which is made 2/3 anterior to the coronary suture and 1/3 posterior to it. Superficial draining vein must be cleaned of their arachnoid adherences until reaching the superior sagittal sinus (Fig. 39.8).

It is then important to find a space between veins where to proceed with interhemispheric approach (Fig. 39.9).

With all venous structures protected, the right frontal lobe is separated from the falx cerebri, allowing for a deeper procedure that may require the use of a retractor. To this purpose, it is useful to complete arachnoid interhemispheric dissection and right frontal lobe relaxation before gently positioning the retractor. Another strategy encompasses the use of cottonoids to maintain laterally retracted the frontal lobe avoiding the use of spatulas. The corpus callosum needs to be properly identified because it can be mistaken for the cingulate gyrus, and the pericallosal artery can be mistaken for the callosomarginal artery (Fig. 39.10).

Finding the central rafe of corpus callosum and the Lancisi nerves could aid in identifying the corpus callosum. The two pericallosal arteries can be lateralized to the left after being located and identified or the callosotomy can be performed between the two pericallosal arteries. After that, we perform a callosotomy between 1 and 2 cm deep in the latero-lateral direction, tracing an imaginary line between the ears that marks the foramen of Monro's projection. The intraventricular portion of the tumor can then be removed by first gaining access to the right lateral ventricle and then the third ventricle [17]. To allow for greater mobilization of the parenchyma and third ventricle, it may be necessary and completely safe to perform a pellucidotomy to reduce the size of the contralateral ventricle.

The interforniceal, interthalamic, suprachoroidal, and subchoroidal (transchoroidal) variations of the interhemispheric approach are occasionally used. To protect the brain's structures, the last three are strongly advised against. Protecting the fornix, anterior commissure, anterior choroidal artery, and deep cerebral veins actually requires top priority.

Anatomically, there are various surgical routes that can be used to access the third ventricle. A natural CSF outflow pathway, the foramen of Monroe, connects the lateral ventricle to the third ventricle physiologically. The **transforaminal approach**, which is also referred to as the parafornicial approach to distinguish it from the transfornicial approach, takes advantage of this exact route, which is advantageous because intraventricular lesions result in hydrocephalus, which widens the foramen of Monroe. The anatomy of the inferomedial wall of the lateral ventricle makes it easier to identify the foramen of Monroe during endoscopic microsurgical procedures. In fact, it is occasionally possible that the endoscope is discovered in the contralateral lateral ventricle when the access to the ipsilateral lateral ventricle is incorrect. The striated thalamic vein inferiorly, the choroid plexus posteriorly, and the anterior septal vein superiorly and medially must all be recognized in order to locate the foramen of Monroe (Fig. 39.11). We can determine the lateral ventricle's right or left identity based solely on the veins' medial and

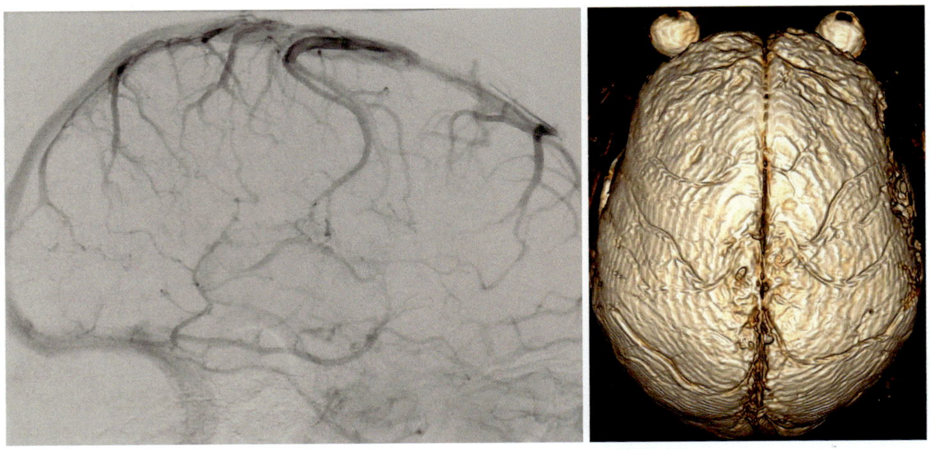

Fig. 39.9 On the left DSA in venous phase, useful for preoperative planning. It is important to evaluate the great venous trunk draining in SSS. On the right side, a 3D reconstruction of principal venous trunk, identified as sulci on the brain

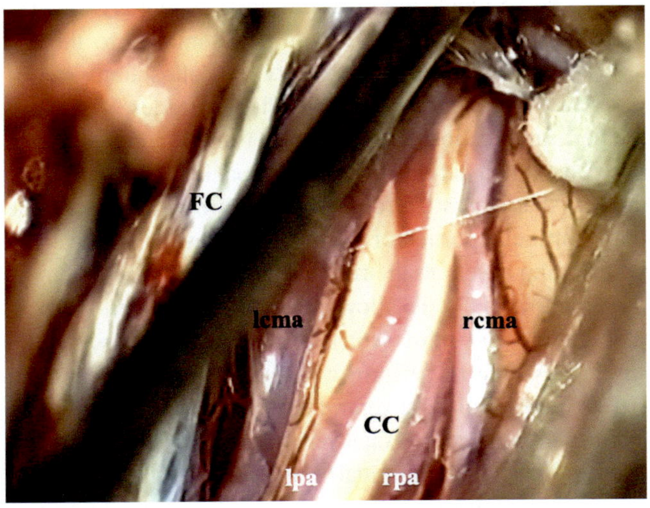

Fig. 39.10 Interhemispheric approach. On the right side, a gently positioned retractor avoids right frontal lobe collapse. On the left side, the falx cerebri represents a natural semi-rigid wall. *FC* falx cerebri, *CC* corpus callosum, *rpa* right pericallosal artery, *lpa* left pericallosal artery, *rcma* right callosomarginal artery, *lcma* left callosomarginal artery

lateral relationships. Contrary to what was previously revealed in the context of hydrocephalus, this can be helpful because, on occasion, the Morro foramen can be dislocated or completely blocked by the tumor lesion, making it challenging to locate.

The **interforniceal approach** (**transforniceal**) requires a midline dissection to separate the body of the fornix. The functionality of the fornix body can be preserved thanks to this expertly executed incision, but the psalterium fornicis should never be exposed. The fibers that make up this structure, which is situated about 2 cm posterior to the foramen of Monroe and is crucial to its proper functioning, crossing between the posterior columns. Once the incision has been made, the anterior columns of the fornix and the two half body of fornix can be gently lateralized while also protecting the deep intraventricular veins that are effectively hidden inferiorly medially to the hemifornix. Although an anterior interforniceal route, starting from the foramen of Monroe and moving forward, has been described, this is strongly discouraged because it necessitates cutting the anterior commissure. The interforniceal approach may be helpful because it provides a large access route to the third ventricle's higher and posterior region. Sometimes transforniceal approach has been meaning an incision inside an hemifornix and not between the two hemifornix.

In the **suprachoroidal approach**, the access to the third ventricle is gained between the choroid plexus and the fornix, resulting in the incision of the tenia fornicis, which symbolizes an extroversion of the fornix on which the choroid plexus rests. This incision is not harmful because the choroid plexus attaches laterally to another extroversion from the lateral ventricle's floor, called tenia choroidea (subchoroidal approach). Contrary to the interforniceal approach, this method requires the choroid plexus to be lateralized along with the deep veins in the medial wall of the lateral ventricle. Choroidal plexus represents an anatomical limit, in the form of a loop from the front, lowering the exposure of this strategy. However, by performing choroid plexus coagulation and separation, the suprachoroidal approach and the transforaminal approach can be combined.

Instead, the **subchoroidal approach** is carried out through the tenia choroidal, dislocating the choroid plexus, and the internal cerebral vein in the medial direction and the thalamostriate vein laterally. The latter, however, continues straight to the internal cerebral vein, posing an insurmountable anatomical barrier but also permitting only a very slight penetration.

An incision right in the floor of lateral ventricle represents an **interthalamic approach** to the third ventricle. This must be avoided to preserve thalamic fundamental functions.

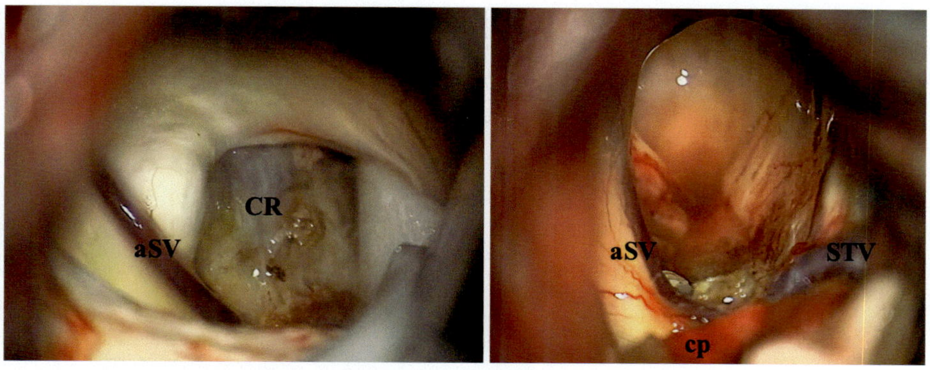

Fig. 39.11 On the left, craniopharyngioma seen through right Monroe foramen, occupying the third ventricle. On the right side, postoperative visualization of the floor of the third ventricle seen through Monroe foramen. aSV anterior septal vein, STV striated thalamic vein, cp choroid plexus, CR craniopharyngioma

Once entered in the third ventricle, the anatomical landmarks of the optic recess and infundibular recess could be lost as the tumor tissue must be carefully separated from the hypothalamic tissue, which can be dislocated posterosuperiorly, along with the mammillary bodies even more posteriorly, and the optic chiasm anteriorly. To obtain a safe tumor dissection, internal debulking should be pursued.

39.4.4 Trans-Sphenoidal Approach

Similarly, transsphenoidal exposure directed through the anterior sellar wall as for small pituitary tumors can be used to remove small craniopharyngiomas in the sella and surrounding suprasellar space. Larger tumors with suprasellar, parasellar, or presellar extensions, however, will call for a wide opening and a more difficult closure. The suprasellar space is typically affected by craniopharyngiomas, and their resection is frequently accompanied by a high-flow intraoperative cerebrospinal fluid leak. In order to avoid meningitis and postoperative cerebrospinal fluid leak, reconstruction with a vascularized nasoseptal flap is frequently required. The initial step entails locating the sphenoid sinus' natural ostium. The sphenoethmoidal recess, which is typically medial to the superior turbinate, is where the ostium is located. The ostium may be found between the superior and supreme turbinates if there is a supreme turbinate. It is infrequently discovered lateral to the superior turbinate. The ability to advance instruments inside the nasal cavity is improved by lateralizing, or pushing or shifting the inferior and middle turbinates laterally. Since the fracture line from the inferior turbinate lateralization can extend to the nasolacrimal duct and/or orbit, lateralizing the turbinates should be done carefully. To successfully view the sphenoethmoidal recess, the endoscope can be advanced and positioned between the middle turbinate and the nasal septum once more space has been made. The sphenoidotomy is carried out after the flap has been removed and placed in the nasopharynx. The superior turbinate's resection may be beneficial in some circumstances. If necessary, the planum sphenoidale and optic canals should be exposed by extending the sphenoidotomy and posterior ethmoidectomy superiorly up to the base of the skull. To enlarge the inferior sphenoidotomy, the rostrum is carefully drilled. It is crucial to keep the posterior septal artery, which supplies the septal mucosa, intact during the removal of the strip of mucosa from the sphenoid rostrum. About 9 mm below the sphenoid ostium, the artery crosses the sphenoid rostrum after emerging from the sphenopalatine foramen. To enhance instrumentation and bi-nostril visualization, the septal mucosa may be partially removed. The sphenoid septa are eliminated with the drill once a wide exposure of the sphenoid sinus has been achieved. The sphenoid sinus mucosa is then completely removed. The tuberculum sella is the primary bone that must be removed in order to access the suprasellar space. The suprasellar region is directly accessible through the tuberculum sella dura opening. The amount of bone removed is determined by the image guidance system based on the size, location, and characteristics of the lesion. Medial to the optic canals, the planum sphenoidale is drilled in a trapezoid pattern. The anterior osteotomy is enlarged as necessary, depending on the diagnostic images, and the posterior osteotomy is positioned just anterior to the tuberculum sella. The tuberculum is then meticulously removed. The optic canal serves as the lateral limit, but in cases of large craniopharyngiomas, a portion of the wall of the optic canal may need to be removed. To prevent heat from spreading to the optic canal during drilling, there must be extensive irrigation. The superior intercavernous sinus is bipolar coagulated after the dura mater has been thoroughly exposed, and the dura is then finally opened to reveal the suprasellar space. The pituitary stalk, optic chiasm (Fig. 39.12), gyrus rectus, and anterior communicating complex are all visible once the skull is opened. To view the intraventricular portion of the tumor, the lamina terminalis may be trespassed (Fig. 39.13). To drill the dorsum and nearby posterior clinoids, the pituitary gland can be transposed superiorly and medially then, and it is possible to approach the retroinfundibular part of the tumor in the interpeduncular fossa (Figs. 39.14 and 39.15).

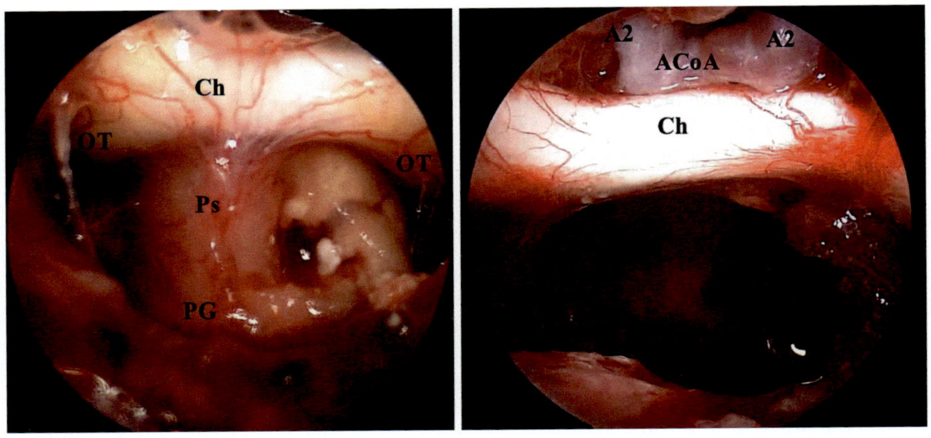

Fig. 39.12 Endoscopic view in extended transnasal transsphenoidal approach. Type III (Kassam) craniopharyngioma (*left image*). Type II (Kassam) postoperative. *Ch* chiasm, *OT* optic tract, *A2* anterior cerebral artery, *ACoA* anterior communicating artery, *Ps* pituitary stalk, *PG* pituitary gland

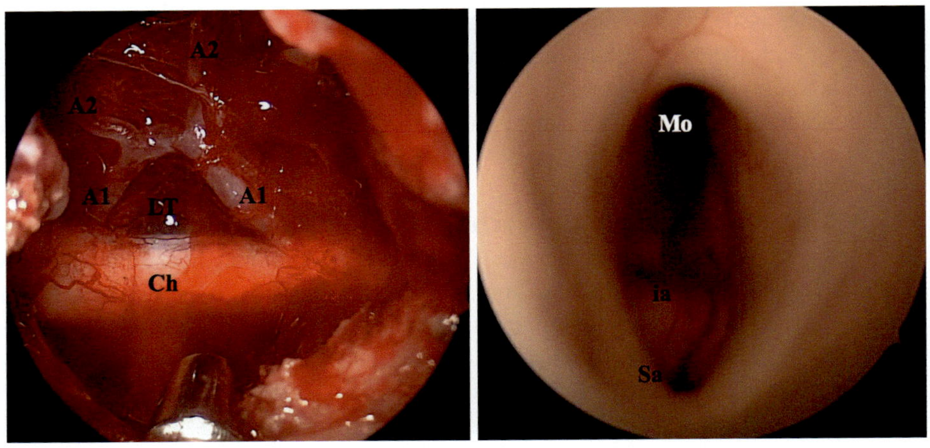

Fig. 39.13 Endoscopic view in extended transnasal transsphenoidal approach (*left image) Ch* chiasm, *A1* anterior cerebral artery, *A2* anterior cerebral artery, *ACoA* anterior communicating artery, *LT* lamina terminalis. Diving technique (right image). Mo Monro foramen, ia interthalamic adhesion, Sa Sylvian aqueduct

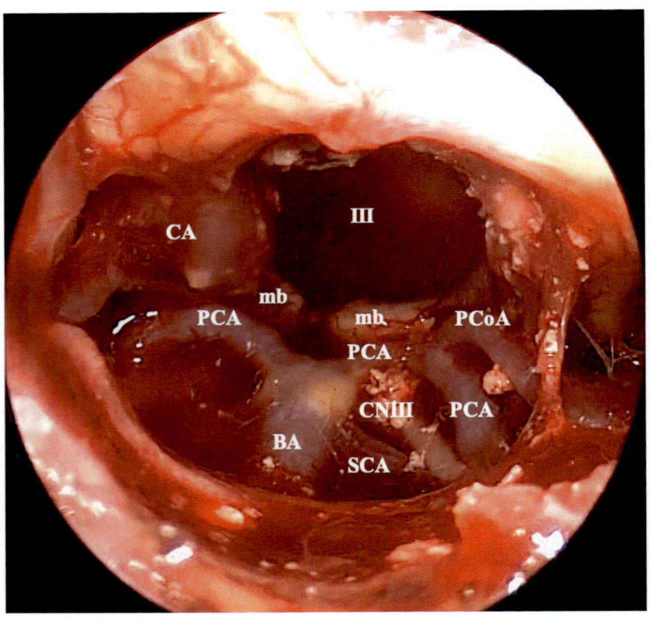

Fig. 39.14 Extended superior approach in a type II craniopharyngioma (Kassam). The pituitary gland has been transposed anteriorly (upper part of the image). BA Basilar artery, PCA posterior cerebral artery, PCoA posterior communicating artery, SCA superior cerebellar artery, CNIII oculomotor nerve, CA carotid artery, mb mammillary body, III third ventricle

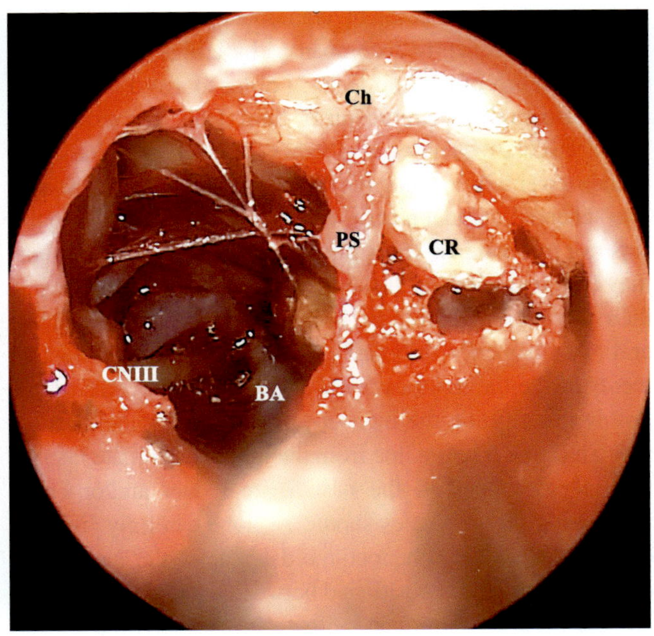

Fig. 39.15 Extended inferior approach in a type II craniopharyngioma (Kassam). *CR* craniopharyngioma, *Ps* pituitary stalk, *Ch* chiasm, *BA* basilar artery, *CNIII* oculomotor nerve

39.5 Surgical Pearls

- Craniopharyngiomas are benign tumors growing adherent to vital neurovascular structures
- Growth pattern dictates the surgical strategy
- Debulking as main step before dissection
- Identifying and possibly sparing the pituitary stalk and related microvasculature
- Transcallosal route limited to pure intraventricular tumors
- Hypothalamic sparing is the main goal
- Meticulous closure by means of vascularized mucoperiosteal flap in endoscopic endonasal approaches

39.6 Discussion

Localization in the suprasellar region can cause compression of the pituitary gland, hypothalamus, optic chiasm, and Monro foramen. The most common clinical manifestations include headache, endocrinological deficits, and visual field deficits [18, 19]. Histologically, they are epithelial neoplasms of the sellar, suprasellar, and parasellar regions. Adamantinomatous craniopharyngiomas is found in children and adults and contains differentiated epithelium, cystic cavities, and fibrosis and has a higher recurrence rate than papillary. Although it is associated with a high recurrence rate, complete resection via an open or endoscopic approach is the ideal treatment for patients with craniopharyngiomas. Figures 39.16, 39.17, and 39.18 show complete resection of different craniopharyngiomas, as classified by Kassam et al. gross total resection (GTR) is a challenging procedure to perform: Some authors describe resections of up to more than 95% of the initial tumor mass with calcified remnants or small tumor intruders detected only on postoperative imaging. Subtotal resection (STR) is instead defined as the persistence of a large portion of tumor in situ, as already estimated preoperatively, and as found postoperatively, and is certainly associated with a higher recurrence rate [6, 20, 21]. Adjuvant radiotherapy allows for residual tumor control many cases, but its application in the treatment of recurrent craniopharyngiomas has been found to be limited due to the dose that can be given to nearby structures [22, 23]. Craniopharyngiomas are known to recur after surgery with a recurrence rate of up to 90% with surgery alone and nowadays lower with surgery and adjuvant radiotherapy. Local recurrence may be related to the tight adherence to vital neurovascular structure limiting the potential radical resection. The best surgical approach must then be tailored on patient's and craniopharyngioma characteristics to obtain the better and more complete exposition of all anatomical structures implied, trying to get as much resection as possible.

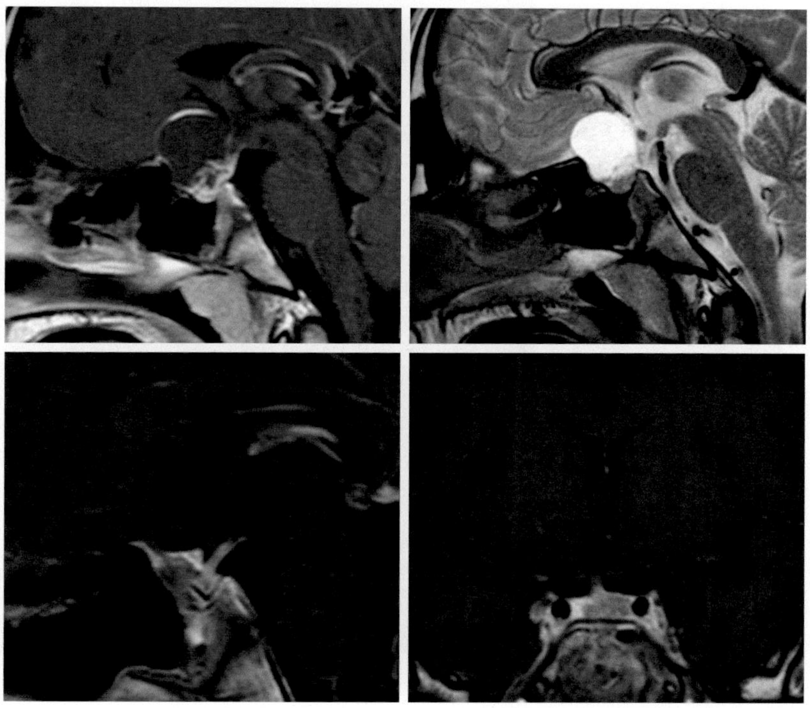

Fig. 39.16 Type I craniopharyngioma. Upper images show preoperative case. Bottom images show postoperative MRI

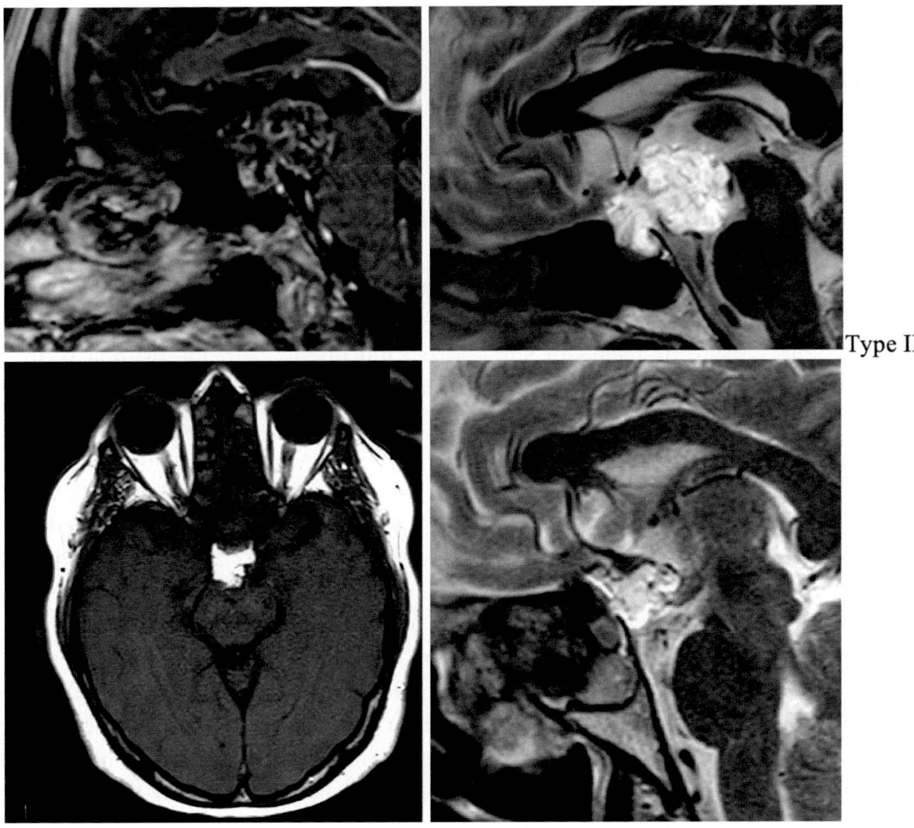

Fig. 39.17 Type II craniopharyngioma. Upper images show preoperative case. Bottom images show postoperative MRI

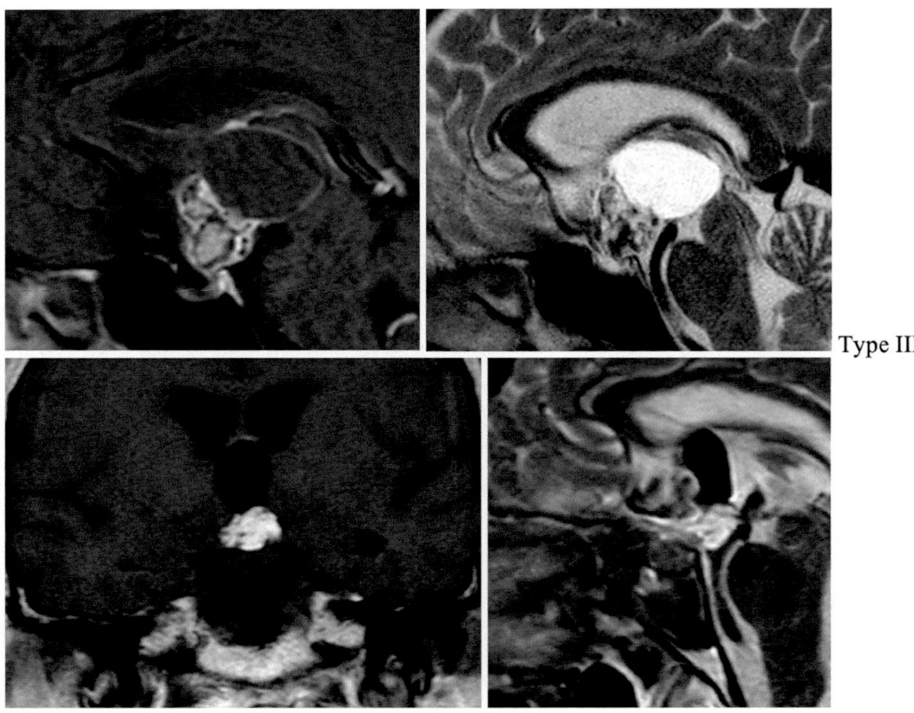

Fig. 39.18 Type III craniopharyngioma. Upper images show preoperative case. Bottom images show postoperative MRI

39.7 Conclusion

Craniopharyngiomas are benign tumors growing in the crowded sellar and parasellar area. The surgical excision is the cornerstone of the management. Standard or extensive transcranial or transsphenoidal surgery can be used to remove craniopharyngiomas. Even if gross total excision should be the main goal, it should be matched with the functional outcome mostly dependent by hypothalamic damage.

References

1. Momin AA, Recinos MA, Cioffi G, Patil N, Soni P, Almeida JP, Kruchko C, Barnholtz-Sloan JS, Recinos PF, Kshettry VR. Descriptive epidemiology of craniopharyngiomas in the United States. Pituitary. 2021;24(4):517–22.
2. Hölsken A, Sill M, Merkle J, Schweizer L, Buchfelder M, Flitsch J, Fahlbusch R, Metzler M, Kool M, Pfister SM, von Deimling A, Capper D, Jones DTW, Buslei R. Adamantinomatous and papillary craniopharyngiomas are characterized by distinct epigenomic as well as mutational and transcriptomic profiles. Acta Neuropathol Commun. 2016;4(1):20.
3. Alli S, Isik S, Rutka JT. Microsurgical removal of craniopharyngioma: endoscopic and transcranial techniques for complication avoidance. J Neurooncol. 2016;130(2):299–307.
4. Nie C, Ye Y, Wu J, Zhao H, Jiang X, Wang H. Clinical outcomes of transcranial and endoscopic endonasal surgery for craniopharyngiomas: a single-institution experience. Front Oncol [Internet]. 2022 [citato 13 aprile 2023];12. Disponibile su: https://www.frontiersin.org/articles/10.3389/fonc.2022.755342.
5. Mortini P, Losa M, Pozzobon G, Barzaghi R, Riva M, Acerno S, Angius D, Weber G, Chiumello G, Giovanelli M. Neurosurgical treatment of craniopharyngioma in adults and children: early and long-term results in a large case series. J Neurosurg. 2011;114(5):1350–9.
6. Flitsch J, Mueller H, Burkhardt T. Surgical Strategies in Childhood Craniopharyngioma. Front Endocrinol [Internet]. 2011 [citato 13 aprile 2023];2. Disponibile su: https://www.frontiersin.org/articles/10.3389/fendo.2011.00096.
7. Yamada S, Fukuhara N, Oyama K, Takeshita A, Takeuchi Y, Ito J, Inoshita N. Surgical outcome in 90 patients with craniopharyngioma: an evaluation of transsphenoidal surgery. World Neurosurg. 2010;74(2–3):320–30.
8. Kitano M, Taneda M. Extended transsphenoidal surgery for suprasellar craniopharyngiomas: infrachiasmatic radical resection combined with or without a suprachiasmatic trans-lamina terminalis approach. Surg Neurol. 2009;71(3):290–8, discussion 298.
9. Gallotti AL, Barzaghi LR, Albano L, Medone M, Gagliardi F, Losa M, Mortini P. Comparison between extended transsphenoidal and transcranial surgery for craniopharyngioma: focus on hypothalamic function and obesity. Pituitary. 2022;25(1):74–84.
10. Yamada S, Fukuhara N, Yamaguchi-Okada M, Nishioka H, Takeshita A, Takeuchi Y, Inoshita N, Ito J. Therapeutic outcomes of transsphenoidal surgery in pediatric patients with craniopharyngiomas: a single-center study. J Neurosurg Pediatr. 2018;21(6):549–62.
11. Zhang C, Ding X, Lu Y, Hu L, Hu G. Cerebrospinal fluid rhinorrhoea following transsphenoidal surgery for pituitary adenoma: experience in a Chinese Centre. Acta Otorhinolaryngol Ital. 2017;37(4):303–7.
12. Senthilvel HNP, Krishnan SS, Vasudevan MC. Extracranial infrasellar craniopharyngioma. Neurol India. 2014;62(1):100.
13. Ortiz Torres M, Shafiq I, Mesfin FB. Craniopharyngioma. In: StatPearls [Internet]. Treasure Island (FL): StatPearls Publishing; 2023 [citato 13 aprile 2023]. Disponibile su: http://www.ncbi.nlm.nih.gov/books/NBK459371/.
14. Yaşargil MG, Yaşargil MG. Microneurosurgery. Stuttgart: Thieme; 1984. 371 p. (Microneurosurgery/M. G. Yaşargil).
15. Samii M, Tatagiba M. Surgical Management of Craniopharyngiomas: a review. Neurol Med Chir (Tokyo). 1997;37(2):141–9.
16. Kassam AB, Gardner PA, Snyderman CH, Carrau RL, Mintz AH, Prevedello DM. Expanded endonasal approach, a fully endoscopic transnasal approach for the resection of midline suprasellar craniopharyngiomas: a new classification based on the infundibulum. J Neurosurg. 2008;108(4):715–28.
17. Tomasello F, Cardali S, Angileri FF, Conti A. Transcallosal approach to third ventricle tumors: how I do it. Acta Neurochir. 2013;155(6):1031–4.
18. Müller HL. The diagnosis and treatment of Craniopharyngioma. Neuroendocrinology. 2020;110(9–10):753–66.
19. Hoffman HJ, Silva MD, Humphreys RP, Drake JM, Smith ML, Blaser SI. Aggressive surgical management of craniopharyngiomas in children. J Neurosurg. 1992;76(1):47–52.
20. Mortini P, Gagliardi F, Boari N, Losa M. Surgical strategies and modern therapeutic options in the treatment of craniopharyngiomas. Crit Rev Oncol Hematol. 2013;88(3):514–29.
21. Weiner HL, Wisoff JH, Rosenberg ME, Kupersmith MJ, Cohen H, Zagzag D, Shiminski-Maher T, Flamm ES, Epstein FJ, Miller DC. Craniopharyngiomas: a clinicopathological analysis of factors predictive of recurrence and functional outcome. Neurosurgery. 1994;35(6):1001–11.
22. Ogino A, Niranjan A, Kano H, Flickinger JC, Lunsford LD. Optimizing stereotactic radiosurgery in patients with recurrent or residual craniopharyngiomas. J Neuro-Oncol. 2021;154(1):113–20.
23. Clark AJ, Cage TA, Aranda D, Parsa AT, Sun PP, Auguste KI, Gupta N. A systematic review of the results of surgery and radiotherapy on tumor control for pediatric craniopharyngioma. Childs Nerv Syst. 2013;29(2):231–8.

The Endoscopic Endonasal Approach to Suprasellar Craniopharyngioma

40

Daniel M. Prevedello and Mehdi Khaleghi

Learning Objectives

- Focusing on the extended endoscopic endonasal approach (EEA) for resecting suprasellar craniopharyngiomas (CP)
- Providing insights into anatomic consideration, surgical technique, and postoperative management of CP
- Discussing different types and their specific challenges from a surgical standpoint to enhance the understanding of neurosurgeons and otolaryngologists with experience in the EEA
- Covering critical aspects such as detailed anatomic landmarks, bone drilling, function-preserving tumor removal, and multilayer reconstruction
- Helping the reader to gain a clear perspective on patient selection, surgical intricacies, potential complications, and follow-up care, which contribute to improved surgical outcomes and patient management

40.1 Introduction

Coined by Harvey Cushing nearly a century ago, *craniopharyngiomas* (CP) stands as a rare benign sellar and suprasellar tumor, intimately situated near critical intracranial structures including the optic aperture, hypothalamic-pituitary axis, circle of Willis vasculature, and limbic system. It comprises 2–4% of intracranial neoplasms and follows a bimodal age incidence characterized by two peaks: one in childhood (age 5–14 years) and another in adulthood (50–74 years). CP derive from lingering epithelial cells of the Rathke's pouch, manifesting in two primary histologic variants: adamantinomatous and papillary. The adamantinomatous CP are the predominant type among both children and adults and is characterized histologically by wet keratin, necrotic debris, and associated infiltration of nearby brain structures. The presence of solid-cystic components and calcification in radiologic evaluation is almost uniquely associated with this type. However, the papillary type almost exclusively presents in adults, showing squamous epithelial proliferation around a fibrovascular core and the lack of cyst formation, necrosis, and calcification. From a molecular perspective, nearly 95% of papillary CP carry the BRAF V600E mutation, a promising target that has generated attention in numerous clinical trials. Conversely, the dysregulation of Wnt/β-catenin has been noted in the adamantinomatous type, yet its investigation remains less explored [1–3].

The surgical management of CP can be broadly categorized into two primary policies: The first strategy aims to achieve a gross total resection whenever it can be safely accomplished. This approach is grounded in the positive correlation between the risk of future recurrence and the degree of mass reduction in the first surgery. However, adopting this as the optimal surgical goal for all patients is challenging due to the tumor's potential adherence to the hypothalamus, optic chiasm, and critical perforators in the suprasellar region. The second strategy, which may seem more practical, involves performing a maximal safe resection followed by adjuvant therapies such as radiotherapy, chemotherapy, and emerging targeted therapy. This combined approach has demonstrated favorable outcomes comparable to the former in carefully selected patients [4, 5].

Historically, many surgical approaches have been proposed for the resection of the CP, necessitating careful individual evaluation tailored to the tumor characteristics, the growth pattern of the tumor, and the surgeon's experience and preferences. These approaches can be primarily categorized into two major groups: transcranial and transsphenoidal. Transcranial approaches through anterior (subfrontal) or anterolateral (pterional, frontolateral) corridors, often using the opticocarotid and carotid-oculomotor triangles to access

D. M. Prevedello (✉) · M. Khaleghi
Department of Neurosurgery, The James Cancer Hospital and Wexner Medical Center, The Ohio State University, Columbus, OH, USA
e-mail: daniel.prevedello@osumc.edu; mehdi.khaleghi@osumc.edu

I. N. Kanaan, V. Beneš (eds.), *Neuroanatomy Guidance to Successful Neurosurgical Interventions*, https://doi.org/10.1007/978-3-031-59838-8_40

the suprasellar region. However, transventricular approaches (through transcortical transcallosal or translamina-terminalis corridors) have been largely utilized in cases of purely intraventricular CP. Incorporating endoscopy as an adjunct to microsurgical techniques has notably expanded the scope of minimally invasive transcranial approaches. This integration provides a broader field of view and leverages the high illumination advantage of endoscopy, thereby enhancing the applicability of these approaches [6–11].

Transsphenoidal approaches, mainly through extended endoscopic endonasal approach (EEA), use a midline corridor to seamlessly access a wide area of the ventral skull base spanning from the C2 vertebra to the posterior table of the frontal sinus. Recent strides in surgical anatomic cadaveric studies and the advancements of the endoscopic endonasal equipment have propelled the evolution of these novel concepts, transcending its previous limitation to exclusively intrasellar tumors. The extended transplanum/transtuberculum technique, which adds more anterior bone drilling and dural exposure to the standard EEA, offers ideal access for reaching suprasellar tumors, including suprasellar CP. Employing tailored endoscopic visualization with varying angles offers an expansive high-definition panoramic view, thus guiding a meticulous arachnoid dissection and careful tumor removal within the suprasellar region to the region posterior to the pituitary stalk and interpeduncular fossa. The inherent advantage of this optimal visualization, coupled with enhanced access to infrachiasmatic and retroinfundibular regions, facilitates the meticulous isolation and preservation of critical perforators, frequently arising from the superior hypophyseal artery (SHA) and supraclinoid internal carotid artery (ICA). The preservation of these vital perforating branches that directly supply optic chiasm and infundibulum contributes to a superior visual and endocrine outcome when comparing EEA to transcranial counterparts in addressing suprasellar CP. Furthermore, the excellent exposure of the retrosellar region, the utilization of the inherent growth pathway of the CP along the hypothalamic-infundibular axis, and a shorter intradural working distance facilitate enhanced exposure of the posterior aspect of the circle of Willis, posterior thalamoperforating arteries, and third cranial nerve (oculomotor nerve). Collectively, these benefits culminate in achieving a greater extent of resection and ultimately result in improved neurological outcomes [12–20].

Several classification systems have been introduced to guide optimal surgical planning for reaching CP. These systems mainly focus on the spatial relationship of the tumor with adjacent structures, such as the infundibulum, optic chiasm, and diaphragm sella [21, 22]. Among these classifications, the one proposed by Kassam has gained widespread acceptance due to its technical utility. It categorized CP into four distinct types based on their relationship to infundibulum as a reference point:

Type 0: Intrasellar
Type I: Preinfundibular
Type II: Transinfundibular
Type II: Retroinfundibular, further subcategorizes based on rostral or caudal extension to:

- Type IIIa: Extension to the anterior third ventricle
- Type IIIb: Extension to the interpeduncular cistern

Type IV: Isolated third ventricular (not accessible via EEA) [19].

This chapter provides a comprehensive overview of the anatomic details and surgical nuances required to safely approach the suprasellar CP (referred collectively to as types I, II, and III) via the extended EEA, predominantly through the transplanum/transtuberculum corridor.

40.2 History and Background

Cushing, a pioneer in neurosurgery, described CP as "more easily diagnosed than treated." He laid the foundation for using the transsphenoidal route in the resection of CP. Guiot and his colleagues were among the first to use endoscopic-assisted transsphenoidal surgery for CP. Notably, their experiences were primarily centered on purely intrasellar CP accompanied by a large sella.

Expanding on Guiot's foundation, Laws advanced the field by adopting the transsphenoidal approach for intrasellar CP. He notably pushed the boundaries further by applying a transplanum approach in CP with suprasellar extension. This innovative "extended" version of the endoscopic-assisted technique found resonance among many neurosurgeons. In 1997, Carrau and their team embarked on an equally transformative journey by pioneering a purely endoscopic endonasal transsphenoidal approach to the sellar region. Still, this approach was primarily suited for the intrasellar CP during this period, requiring a large sella as a prerequisite for safe resection.

As technological advancements progressed and modern imaging techniques provided sharper endoscopic visualization of surrounding neurovascular structures, the landscape evolved further. Inspired by earlier concepts outlined by Weiss, Laws, and Couldwell, a purely endoscopic extended transsphenoidal approach for addressing suprasellar CP was introduced. This idea paved the way for Kassam and their colleagues, who shared their experiences in successfully resecting CP with retrochiasmatic extension [23–25].

40.3 Anatomy Highlights

- **The anatomical structures involved in accessing the suprasellar region and the general concept of endoscopic endonasal transplanum/transtuberculum approach closely mirror the standard EEA. However, specific anatomic landmarks set these approaches apart**

The EEA and its extended variants provide a wide exposure of the anterior skull base, sellar region, and posterior fossa from a midline ventral perspective, using a two-surgeon four-handed technique. These approaches delineate the same anatomic structures from a different viewpoint and sequence of exposure compared to their transcranial counterparts. The earliest structures to come into view when passing the endoscope into the nostrils are the nasal septum on the medial wall and the inferior and middle turbinates on the lateral wall, with the middle meatus in between. The nasal septum forms by the fusion of the cartilaginous part anteriorly and the bony part posteriorly. The bony septum is mainly comprised of the articulation of the vomer and perpendicular plate of the ethmoid bone anteriorly and the anterior aspect of the rostrum of the sphenoid bone posteriorly. For the EEA and its extended variants, the middle meatus commonly serves as the primary pathway for accessing the posterior aspect of the nasal cavity. As the endoscope advances in this space, the posterior wall of the nasal cavity is exposed, formed by the sphenoid rostrum superiorly and the choana inferiorly.

- **A pivotal early step in approaching this region is identifying the position of the sphenopalatine foramen before transgressing the mucosal lining, which is surgically relevant to extended EEA and final reconstruction**

The sphenopalatine foramen emerges between the perpendicular plate of the ethmoid bone and the medial pterygoid plate of the sphenoid bone in the posterior third of the middle meatus. This foramen contains an important artery with the same name (the sphenopalatine artery) that branches from the internal maxillary artery in pterygopalatine fossa. The sphenopalatine artery passes over the sphenoid rostrum approximately 5 mm superior to the choana and gives rise to the posterior septal artery, which its preservation is critical for a viable mucosal nasoseptal flap (NSF). This flap is almost always required as a safe and pliable barrier to secure the final closure of the endoscopic endonasal transplanum/transtuberculum approach. The base of the middle turbinate usually infers the location of the sphenopalatine artery and its posterior septal branch, aiding in elevating a customized NSF based on its pedicle.

- **A full opening of the sphenoid sinus is the common next step in all variants of the extended EEA and first begins with the identification of the ipsilateral sphenoid ostia**

The sphenoid ostia is located on the posteroinferior aspect of the superior turbinate, approximately 1.5 cm superior to the choana. It is imperative to consider the degree of the sphenoid sinus pneumatization during the preoperative planning, as the application of the EEA in pediatric CP might encounter challenges in patients with presellar and conchal types, making this approach possibly impractical.

- **The general principles of identifying intrasphenoid prominences, which infer the position of critical underlying structures such as the paraclival and cavernous segments of the ICA, optic nerves, and the sella follow the same rule as the standard EEA**

These prominences manifest on the posterior and superior walls depending on the degree of sphenoid sinus pneumatization, requiring careful navigation throughout the procedure. Likewise, various degrees of bony dehiscence may be encountered over these prominences, requiring extreme caution when removing the mucosal lining of the sinus. One or more sphenoid septa are often encountered upon entering the sphenoid sinus. It is important to not consider these septa as safe landmarks to infer the midline, as they usually lead to one of the internal carotid arteries.

- **When the sphenoid sinus is fully opened, four major bony depressions (recesses) need to be identified: the tubercular recess, the medial opticocarotid recess (MOCR), the lateral opticocarotid recess (LOCR), and the clival recess**

Only the tubercular recess and MOCR are relevant to the transplanum/transtuberculum approach and are briefly discussed here. The tubercular recess, located between the sella and planum sphenoidale, corresponds to the tuberculum sella intracranially. Upon drilling, it points to the diaphragm sella on the posterior half and the superior intercavernous sinus (SIS) on the anterior half. The MOCR is the shallowest recess and lies in the inferomedial angle created by the paraclinoid-supraclinoid ICA transition and the optic nerve just lateral to the tubercular recess. Notably, the closest part of the bilateral parasellar ICA typically resides in this area and should be cautiously investigated in preoperative imaging and during bone drilling. If drilled, the MOCR points to the opticocarotid cistern, providing access to the pituitary gland and suprasellar region. It also serves as a landmark for the distal dural ring (DDR) and the paraclinoid-supraclinoid

ICA transition where the paraclinoid ICA enters the intradural space. It is important not to confuse the MOCR with the middle clinoid process (MCP), a thick bony structure on the superolateral aspect of the sella, hence located more posteroinferiorly. The MCP infers the border between the paraclival and cavernous segments of the ICA, thereby effectively pointing to the proximal dural ring and the roof of the cavernous sinus. Although rare, the fusion of the middle and anterior clinoid processes around the paraclinoid ICA may give rise to the caroticoclinoid ring, necessitating preoperative radiologic assessment to avoid an ICA laceration during MCP removal. However, the drilling of the MCP is only needed when a complete exposure of the cavernous sinus and paraclinoid ICA is necessary and is therefore beyond the scope of this chapter.

- **After drilling the tubercular recess and MOCR and exposing the underlying dura, the dissection can proceed in a caudal or cranial direction, depending on the tumor's origin and growth pattern**

The key anatomic structures in selecting either of these directions are SIS and diaphragm sella. The SIS forms a single or a group of venous channels connecting both cavernous sinuses anterosuperiorly, although it can be completely absent. A careful coagulation and division of SIS may be needed to achieve optimal suprasellar access. Immediately posterior to the SIS, the diaphragm sella can be exposed, which usually needs to be opened along the direction of the pituitary stalk, particularly in cases of transinfundibular and retroinfundibular CP. This maneuver establishes a direct rostrocaudal axis of attack along the infundibulum and gives an optimal route to reach the interpeduncular fossa and third ventricular floor.

- **In general, achieving a full suprasellar exposure through the transplanum/transtuberculum EEA requires posterior septectomy, wide sphenoidotomy, and a posterior ethmoidectomy**

The anterior boundary of the planned posterior ethmoidectomy is defined by the posterior ethmoidal arteries. Early identification, coagulation, and division of this artery are paramount, as its inadvertent avulsion can result in significant intraorbital hemorrhage. Additionally, the presence of a sphenoethmoidal cell, also known as Onodi's cell, located superolateral to the sphenoid sinus should be ruled out. If present, the Onodi's cell should be drilled cautiously, as it may contain optic nerve prominence. The optic nerve lies in the posterolateral border of bone removal. Its cisternal, precanalicular, and canalicular (within the optic canal) segments are relevant in a planned inferomedial decompression of the optic canal, especially in anterior cranial base meningiomas. The optic strut forms the true floor of the optic canal. However, for optimal optic nerve decompression, the opening of the inferomedial part of the optic canal suffices, making the drilling of the LOCR unnecessary. A complete optic nerve decompression through the EEA is achieved by fully unroofing the canal and cutting the falciform ligament. However, this step becomes relevant in cases of optic canal involvement, especially in certain pathologies, such as tuberculum sella meningiomas. The precanalicular segment of the optic nerve is defined by the portion of the nerve located under the falciform ligament and can be released by cutting the lateral part of the ligament. The falciform ligament connects the superior border of the anterior clinoid process to the limbus sphenoidale, and its medial extension joins to the contralateral side to mark the limbus sphenoidale, on the intracranial side. The limbus sphenoidale marks the boundary between the planum sphenoidale anteriorly and the chiasmatic sulcus posteriorly and is relevant in both EEA and transcranial approaches to anterior skull base. The mediolateral extension of the falciform ligament can be estimated before dural opening by an assumed area between the two vertical lines drawing from the medial border of the LOCR and the most lateral part of the limbus dura (Fig. 40.1).

- **The intradural space of the suprasellar region is characterized by densely arachnoid membranes spanning between the prechiasmatic and interpeduncular cisterns**

The position of the chiasm can be identified through preoperative imaging and is crucial to accurate surgical planning. The optic chiasm most commonly lies in a normal position over the diaphragm sella. However, prefixed and postfixed chiasm have also been described, depending on whether the chiasm is positioned over the tuberculum sella or the dorsum sellae, respectively. Posteroinferior to the chiasm, the pituitary stalk originates from the tuber cinereum and pierces the diaphragm sella to join the pituitary gland. The diaphragm sella typically lacks a tight opening around the stalk, leaving a space through which the superior surface of the pituitary gland can be visualized.

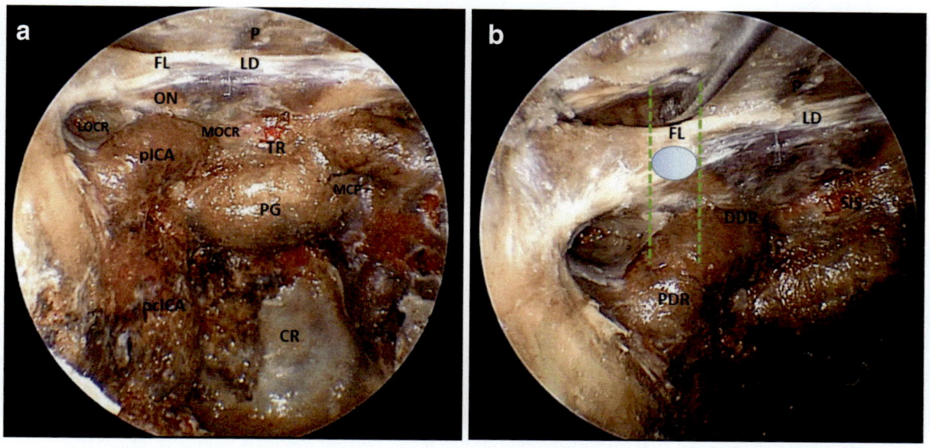

Fig. 40.1 Endoscopic endonasal transplanum/transtuberculum approach, bony and dural exposure(**a**, **b**) The tubercular recess marks the position of the tuberculum sella and superior intercavernous sinus. The medial opticocarotid recess, located just lateral to the tubercular recess and should be drilled for an optimal access to the opticocarotid recess. The middle clinoid process is located lateral to the sella and does not to be drilled in this approach. The pituitary gland lies within the sella and marks the posteroinferior border of the approach. The falciform ligament is the lateral extension of the limbus dura. The "blue oval" marks the precanalicular segment of the optic nerve under the falciform ligament, spanning between two imaginary vertical lines (dashed green lines) drawing from the medial border of the LOCR and the most lateral part of the limbus duraFL falciform ligament, LD limbus dura, P planum sphenoidale, ON optic nerve, LOCR lateral opticocarotid recess, MOCR medial opticocarotid recess, TR tubercular recess, PG pituitary gland, pICA paraclinoid ICA, pcICA paraclival ICA, CR clival recess, MCP middle clinoid process, DDR distal dural ring, and PDR proximal dural ring

- **From a surgical perspective, the suprasellar intradural space can be divided into four major compartments using intersecting horizontal and vertical planes drawing from the base of the chiasm and the pituitary stalk, respectively: the suprachiasmatic, third ventricular, infrachiasmatic/preinfundibular, and infrachiasmatic/retroinfundibular. Only the two latter compartments are relevant to this approach and are further discussed here.**

The infrachiasmatic/preinfundibular space is confined by the supraclinoid ICA laterally, the roof of the cavernous sinus and the diaphragm sella inferiorly, the ventral surface of the optic chiasm superiorly, and the pituitary stalk posteriorly. It is important to carefully open the underlying dura, as there exists a potential for one or more critical perforators to be in close proximity, particularly in cases of the preinfundibular CP. These perforators primarily arise from the SHA and the ICA, supplying the optic nerves, optic chiasm, and pituitary stalk. Therefore, meticulous isolation and preservation of these vessels are imperative to ensure optimal visual and endocrine outcomes. The infrachiasmatic/retroinfundibular compartment is bordered superiorly by the floor of the third ventricle and mammillary bodied, superolaterally by the optic tracts, laterally by the posterior communicating artery (PCOM), anteriorly by the stalk, inferiorly by the dorsum sellae, posterolaterally by the oculomotor nerve which crosses between the superior cerebellar artery (SCA) and the posterior cerebellar artery (PCA), and posteriorly by Lillequist's membrane, interpeduncular fossa, the upper part of the pons and basilar artery. The Lillequist's membrane is formed by a group of thick arachnoid layers that protect the posterior part of the circle of Willis, including important perforating branches arising from the posterior part of P1 and basilar apex. It originates from the dorsum sellae as a single bundle and then divides to insert into the SCA and mammillary bodies as two distinct leaflets. The Lillequist's membrane is of surgical relevance in retroinfundibular CP, as it is commonly pushed by the tumor posteroinferiorly (Fig. 40.2).

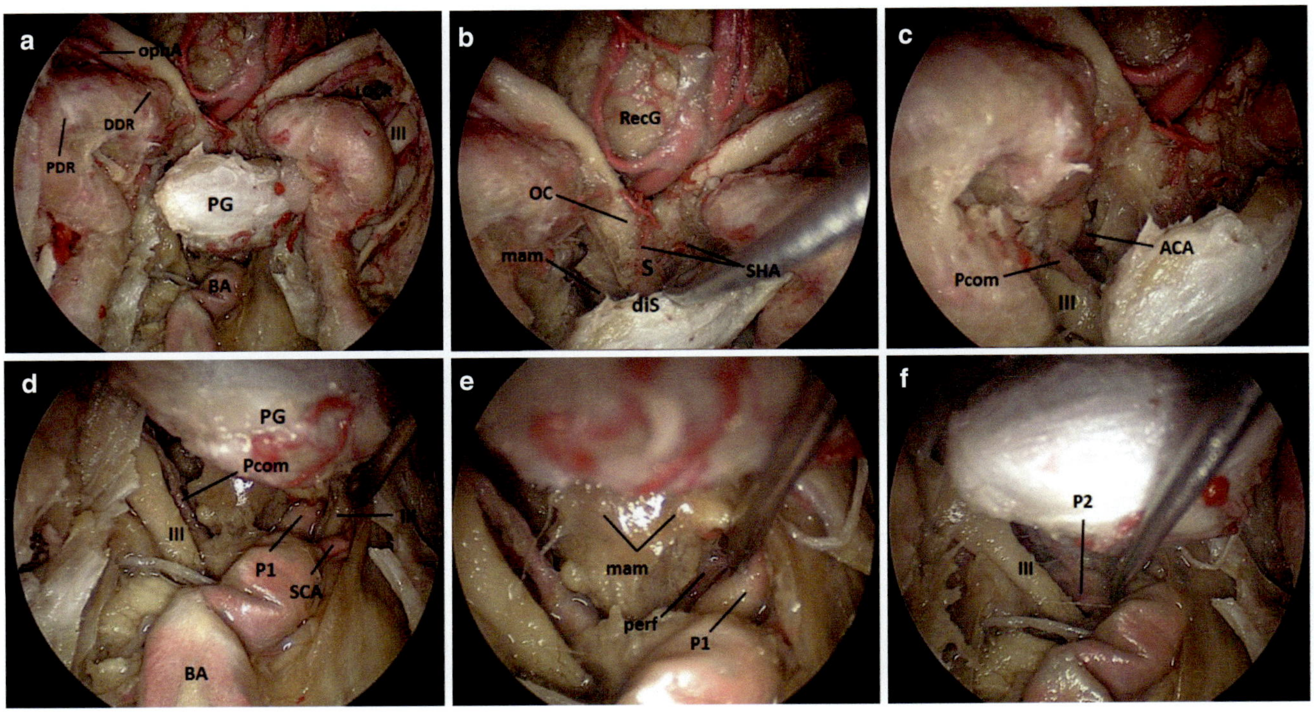

Fig. 40.2 Endoscopic endonasal transplanum/transtuberculum approach, preinfundibular and retroinfundibular spaces exposureThe pituitary gland (PG) hemitransposition requires the unilateral release of pituitary ligaments. The superior hypophyseal artery (SHA) supplies the pituitary stalk (S) and optic chiasm (OC). The mammillary bodies (mam) are located posterior to the stalk. The oculomotor nerve (III) travels between the first segment of the posterior cerebral artery (P1) and the superior cerebellar artery (SCA) and then enters the roof of the cavernous sinus through the oculomotor triangle. The basilar artery (BA) and P1 are retracted to show their important post-perforating branches (perf). Note the location of the posterior communicating artery (PCOM) in the lateral border of the retroinfundibular space. ACA anterior cerebral artery, P2 the second segment of the posterior cerebral artery, ophA ophthalmic artery, PDR proximal dural ring, DDR distal dural ring, RecG rectus gyrus, diS and diaphragm sella

40.4 Approach (Surgical Intervention)

The endoscopic endonasal transplanum/transtuberculum approach is a versatile approach to reach the suprasellar region, including CP that extend to the anterior part of the third ventricle. It presents a technical challenge and should ideally be performed by neurosurgeons and otolaryngologists experienced in the EEA at the end of their learning curves. The general principles of initial exposure are similar to the standard EEA, with more extensive bone drilling and demanding multilayered reconstruction at the end. It is basically a two-surgeon, four-handed technique requiring parallel working of both surgical teams until the final step of the procedure. High-definition endoscopic cameras with various-angled lenses and long-shaft bipolar and drills are necessary for optimal tumor dissection. Frameless stereotactic navigation by applying gadolinium-enhanced fine-cut T1-weighted MRI images fused with CT angiography is paramount to accurately delineate the tumor's vicinity and its relationship to sellar floor, dorsum sellae, and surrounding vasculatures during the procedure. The somatosensory evoked potential (SSEP) and electromyography monitoring of the oculomotor nerves are routinely applied in our center, especially for tumors with posterior or lateral extension.

As a general rule, the CP with significant lateral extension to the middle fossa or major vessel encasement are not good candidates for resecting solely through the EEA. In such cases, a staged surgical strategy may offer a better option. Similarly, an isolated third ventricular tumor with an intact third ventricular floor (type IV) or those displaying significant intraventricular extension accompanied by a small pituitary gland-optic chiasm corridor are not amenable to access through this approach and therefore the transcranial approaches may be a more reasonable choice.

After induction of general anesthesia, the patient's head is secured in a rigid three-pin head holder. The head is extended slightly and rotated to the right side, followed by a left vertex tilt. The chest is slightly elevated to a reverse Trendelenburg position to help venous return. Topical decongestant is applied by putting 0.05% oxymetazoline-soaked pledgets in bilateral nares and leaving them in place throughout the preparation process. Along the midface, the periumbilical region and right anterolateral thigh are also prepped and draped, should a fat graft, fascia lata, or muscle graft become necessary for the final reconstruction. Routine prophylactic antibiotic with fourth-generation cephalosporin is administered, but we do not use the prophylactic lumbar drain even in case of high-flow CSF leak.

After passing a zero-degree endoscope into the middle meatus of the wider nostril, the vicinity of the sphenoid rostrum, choana, and ipsilateral sphenoid ostia are defined and the trajectory of the posterior septal branch is estimated. A unilateral middle and superior turbinectomy is performed, followed by elevating an ipsilateral pedicled NSF. The NSF should be larger than the anticipated bony opening, as it may shrink over time. It subsequently is tucked in the nasopharynx for protection throughout the procedure. After disarticulation of the posterior nasal septum from the sphenoid bone, a small.

(<2 cm) posterior septectomy is performed, followed by contralateral middle turbinate lateralization to allow wide binostril surgical access. Once the contralateral sphenoid ostia is identified and widened, both sphenoid ostia are connected to achieve a wide sphenoidotomy. All sphenoid sinus septations are carefully drilled with diamond burr and under continuous irrigation to eliminate any potential dead space and maximize the bone-flap contact in the upcoming reconstruction step. Subsequently, a posterior ethmoidectomy is performed to convert the surgical corridor into a single unobstructed working space, which is necessary to safely navigate the underlying anatomic landmarks throughout the procedure.

The extent of bone removal from the posterior sphenoid sinus wall depends on factors such as the degree of sphenoid sinus pneumatization, the anterior extension of the tumor, and an estimated relationship between the solid portion of the tumor and the infundibulum (i.e., the type of CP), which are briefly discussed here. However, the definitive relationship of the tumor to the stalk cannot be determined until intraoperative inspection following initial tumor debulking. As a result, the surgeon should be prepared to extend or adapt the approach during the surgery.

40.4.1 Infrachiasmatic/Preinfundibular (i.e. Type I)

The boundaries of tumors in this region are the diaphragm sella inferiorly, the pituitary stalk posteriorly, the supraclinoid ICA laterally, and the optic chiasm superiorly. However, the anterior pole of the tumor may extend to the prechiasmatic region, abutting the anterior communicating artery complex. Bone removal starts from the upper third of the sellar floor and proceeds toward the tuberculum strut, bilateral MOCRs, and the posterior part of the planum sphenoidale, varying according to the anterior extension of the tumor. This should be performed using a diamond drill under continuous irrigation to prevent thermal injury to the optic nerves. Fully exposure of the sellar floor is unnecessary as it does not significantly enhance the anteroposterior access, already available for reaching the preinfundibular space. The bone removal exposes the underlying dura above and below the tubercular strut, leaving it as a single quadrangular bony process. The drilling continues over this trough until achieving an eggshell-like flattening, which is easier for a safe removal. Subsequently, the remaining bone is untethered by putting two cuts on both sides, followed by en bloc removal. A profuse epidural venous bleeding may occur after the exposure of the SIS and anterior attachment of the diaphragm sella, which should be controlled by injecting a collagen-based hemostatic agent (Floseal, Deerfield, IL). The dura is carefully opened by applying two separate small horizontal incisions above and below the SIS, with great care taken to avoid damage to the potential kissing perforating branches inside the underlying dura. The isolated SIS is then coagulated and divided to expose the diaphragm sella and suprasellar subarachnoid space. The two dural incisions are then connected at already divided SIS. The opening can be further extended posteriorly to expose the superior third of the pituitary gland or anteriorly depending on the extension of the tumor. Cutting the diaphragm sella may be performed in this type, if the pituitary stalk involvement is suspected in the posterior margin of the tumor. Following meticulous arachnoid dissection at the floor of the suprasellar cistern, the SHA and its perforating branches are identified and mobilized to expose the tumor capsule. Microsurgical principles of internal debulking and circumferential tumor isolation through sharp extracapsular dissection are generally applied to the EEA for all types of CP. In this type, the optic chiasm is typically pushed posterosuperiorly, leaving a relatively large working space anteriorly.

Internal debulking is carried out using a two-suction technique or a combination of suction microdebrider device (Myriad, Nico Corp., Indianapolis, IN) and ultrasonic aspirator (Sonopet, Stryker, Kalamazoo, MI) for the calcified tumor. This starts from the inferior pole of the tumor, progressing laterally to reach the right optic nerve. Mobilization of the tumor continues on the superior part along the antero-inferior surface of the optic chiasm and then advances to the left optic nerve in a clockwise fashion. Care should be taken to preserve any perforating branches that might appear on the posterosuperior pole of the tumor. This can be achieved by gentle mobilization of the tumor after a sufficient internal debulking, using a "countertraction" technique. This technique is performed by a slight traction on the closest part of the tumor to the arachnoid plane, followed by a systematic circumferential arachnoid release.

The final stage of tumor removal involves identifying the posteriorly located stalk. This is initiated by superior mobilization of the inferior pole from the anterior diaphragm sella, progressively moving toward the side "away" from the stalk. This strategy aids in reaching the largest remaining bulk of the tumor prior to the stalk. The tumor attachment to the stalk is finally identified and sharply dissected away. At this stage, a close inspection of this final attachment is necessary to determine if preservation of the stalk is possible.

40.4.2 Transinfundibular (Type II)

These tumors are characterized by the widening of the stalk, with or without an exophytic extension. They grow along the long axis of the stalk and may expand the diaphragm sellar opening or elevate the third ventricular floor. Therefore, a more vertical angle of attack along their axis of growth is necessary to optimally access these tumors. This can be achieved by exposing the entire sellar floor and removing the tuberculum strut and posterior planum. Additionally, a complete release of the diaphragm sella is the key step to expose these tumors by a steeper rostrocaudal angle of view. Therefore, the dural opening must extend posteriorly to include the anterior portion of the diaphragm sella and expose the widened pituitary stalk. Any exophytic part of the tumor should be resected first, followed by a vertical incision along the anterior surface of the stalk. The tumor within the stalk is subsequently removed in a piecemeal fashion using standard microsurgical techniques. Any extension of the tumor to the anterior third ventricle floor and the medial wall of the hypothalamus is traced toward the optic recess, always being cautious to avoid any traction force on these critical structures.

40.4.3 Infrachiasmatic/Retroinfundibular (Type III)

These tumors pose the most significant challenges among suprasellar CP, owing to their relationship with critical neurovascular structures and the need for extensive surgical exposure. Transcranial approaches to this region via either anterior or anterolateral corridors are limited by the optic chiasm and optic nerves, particularly in the uppermost part of the interpeduncular cistern. Although accessing through the carotid-oculomotor triangle in the orbitozygomatic modification of anterolateral approaches may enhance the exposure, the risk of causing ipsilateral oculomotor palsy remains a limiting factor.

The pituitary stalk in the anterior border, the Lillequist's membrane and basilar apex in the posterior border, the oculomotor nerve and PCOM in the lateral border, and the mammillary body and premammillary membrane in the superior border form the vicinity of these tumors. If the tumor penetrates the Lillequist's membrane and extends into the interpeduncular fossa (Type IIIb), the SCA, posterior perforated substance, and posterior thalamo perforating arteries may be encroached.

An extended transplanum/transtuberculum exposure followed by complete sellar floor drilling and division of both SIS and IIS is universally required to connect the sellar and suprasellar compartments. An intradural pituitary hemistransposition followed by the removal of dorsum sellae and bilateral PCP is usually necessary for an optimal access. This maneuver starts with the wide opening of the sellar floor dura and identifying the "true" capsule of the pituitary gland. Subsequently, a unilateral extracapsular release of pituitary ligaments from the medial wall of the cavernous sinus is performed. Once one side of the gland is released from its lateral anchors, the anterior part of the diaphragm sella is opened in the midline to expose the stalk. This maneuver releases the gland and aids in gently mobilizing it to superolaterally. This can be followed by applying fibrin glue to fix the gland in a lateral position and protect it from desiccation. However, we prefer to perform a dynamic retraction of the gland while drilling the posterior bony structures to minimize the risk of venous infarction. We have moved away from a complete pituitary transposition involving a circumferential dissection of the gland, given the potential risk of panhypopituitarism resulting from venous engorgement [26–28].

The upper part of the clivus and dorsum sellae is then drilled, and the ipsilateral PCP is mobilized, followed by the creation of a medial fracture line. The latter maneuver prevents the cavernous ICA laceration from an inadvertent force on the posterior carotid canal. Once desired posterior bony drilling is achieved, an intraoperative ultrasound is used to estimate the inferior extension of the tumor within the interpeduncular cistern. The dura is then opened accordingly, and the tumor is resected using standard microsurgical techniques. Efforts are made to preserve an intact Lillequist's membrane whenever possible, as it provides a protective barrier to the posterior part of the circle of Willis and the superior part of the brainstem. The abducent nerves should be identified and preserved in the inferior extent of the dural opening. A meticulous sharp dissection of the tumor from the posterior thalamo perforating arteries and direct mesencephallo-pontine perforating arteries arising from the posterior aspect of P1 and basilar apex is crucial in this step. Using a 45-degree endoscope is helpful when working in this space to optimally visualize the blind corners during dissection. Following adequate internal debulking, gentle anterosuperior mobilization of the tumor helps to individually isolate the posteriorly located perforators that might be attached to the tumor. All of these important microvessels should be preserved, as their damage could lead to catastrophic neurological consequences. The next step involves the dissection of the residual tumor from the oculomotor nerve and PCOM that often comes into contact with the lateral border of the tumor.

If a rostral extension of the tumor into the anterior part of the third ventricle (i.e., Type IIIa) is observed, the extraventricular part is resected first using the same principle, albeit with a less aggressive removal of the upper clivus.

Subsequently, by following the path of the tumor into the infundibular and optic recesses, the interior wall of the third ventricle is freed from the tumor by a sharp extracapsular dissection. Care should be taken to avoid aggressive resection of the tumor from the medial wall of the hypothalamus which may lead to significant morbidity. Hemostasis is then achieved using warm irrigation under close endoscopic inspection.

40.4.4 Reconstruction

A secure multilayer repair of a large intrasphenoid opening is equally paramount to the tumor resection, particularly in cases with wide dissection of the suprasellar cisterns [8, 29, 30]. The size of the defect is measured with the tip of the suction, and a customized graft is then designed accordingly. We routinely use synthetic collagen matrix (DuraGen: Integra LifeSciences, Plainsboro, NJ) as an inlay subdural layer. The graft should typically be larger than the defect to ensure complete coverage of the arachnoid layer. The lack of evidence of CSF blush while observing the regular CSF pulsation against the construct indicates an optimal positioning of the inlay graft. Subsequently, the construct is augmented with a pedicled NSF while ensuring maximal direct contact of the flap with the mucosa-free bony surface. We do not routinely apply tissue glue that may be used by some institutes to secure the flap. If the harvested NSF is insufficient to fully cover the hole, any remaining defect is augmented with a middle turbinate-free mucosal flap or fat. An absorbable nasal pack is inserted to hold the construct in place during the critical early healing phase. Following this, a removable sponge is placed and expanded with irrigation to buttress the whole system. Bilateral silicon septal splints are applied to prevent potential synechia and patient discomfort. Intravenous broad-spectrum antibiotic continues for 24 h postoperatively, then switched to an oral form for the next 5–7 days, while the removable nasal packing is in place.

40.4.5 Postoperative Care

A 30-degree head elevation is applied during the postoperative ICU stay. The pituitary function is assessed by measuring the early morning serum cortisol level the day after surgery while sparing the intraoperative steroid dose to prevent false results. Regular monitoring of urine output, urine osmolarity, urine specific gravity, and serum sodium is conducted to detect transient diabetes insipidus. Mild diabetes insipidus in awake patients can be managed with unrestricted access to clear water, while severe or prolonged diabetes insipidus may require desmopressin. An early endocrinology consultation is scheduled to establish the baseline pituitary profile. A brain CT is performed on the day of the surgery to rule out postoperative hemorrhages or significant pneumocephalus. A gadolinium-enhanced MRI is performed within the first 24 h to assess the potential residual tumor and the NSF health. Nasal saline spray is initiated every 6 h on the first postoperative day and continues after discharge. To prevent the displacement of the construct, the patient is recommended to use stool softeners and avoid straining, nose blowing, or positive pressure ventilation for 6 weeks. The patient is usually discharged on postoperative days 3–5.

Outpatient visits are typically scheduled on postoperative days 5–7. The nasal pack is removed, and the operative corridor is cleared under endoscopic guidance. Nasal saline spray and oral antibiotic are discontinued, and the patient is advised to irrigate the nose with 0.9% saline solution four times a day for 3–4 weeks. Ongoing biweekly otolaryngology visits are scheduled to debride possible crusts or clots. The nasal splint is usually removed 2 weeks after surgery.

40.4.6 Case 1

A 31-year-old man was referred to the clinic with bilateral peripheral vision loss (bitemporal hemianopsia). The brain MRI proved the presence of a heterogeneous enhancing suprasellar tumor pushing the optic apparatus superiorly and the pituitary stalk posteriorly, consistent with Type I CP. Endocrine tests did not show any abnormality. A gross total resection of the tumor and preservation of the pituitary stalk was achieved through the endoscopic endonasal transplanum/transtuberculum approach. The patient woke up without any neurologic and hormonal deficits. The histopathologic study demonstrated the adamantinomatous CP. The recent follow-up MRI showed no evidence of residual or recurrence (Fig. 40.3).

40.4.7 Case 2

The patient was a 32-year-old man who presented with acute onset bilateral peripheral vision loss and headache. The hormone study revealed a panhypopituitarism and partial diabetes insipidus. The brain MRI showed a large solid-cystic heterogeneous enhancing suprasellar tumor pushing the optic chiasm anterosuperiorly with some degrees of pituitary stalk expansion and third ventricular floor elevation, consistent with Type II CP. The detailed radiologic assessment also revealed evidence of early hydrocephalus. A near-total tumor resection and preservation of the pituitary stalk were achieved through the endoscopic endonasal transplanum/transtuberculum approach. The patient woke up with significant visual improvement and no new neurological deficits. The histopathologic study confirmed the presence of adamantinomatous CP. Subsequently, the patient underwent adjuvant radiation due to the age and the extent of the resection. The patient is on lifelong hormone replacement, and the recent follow-up MRI showed no evidence of recurrence or tumor residual (Fig. 40.4).

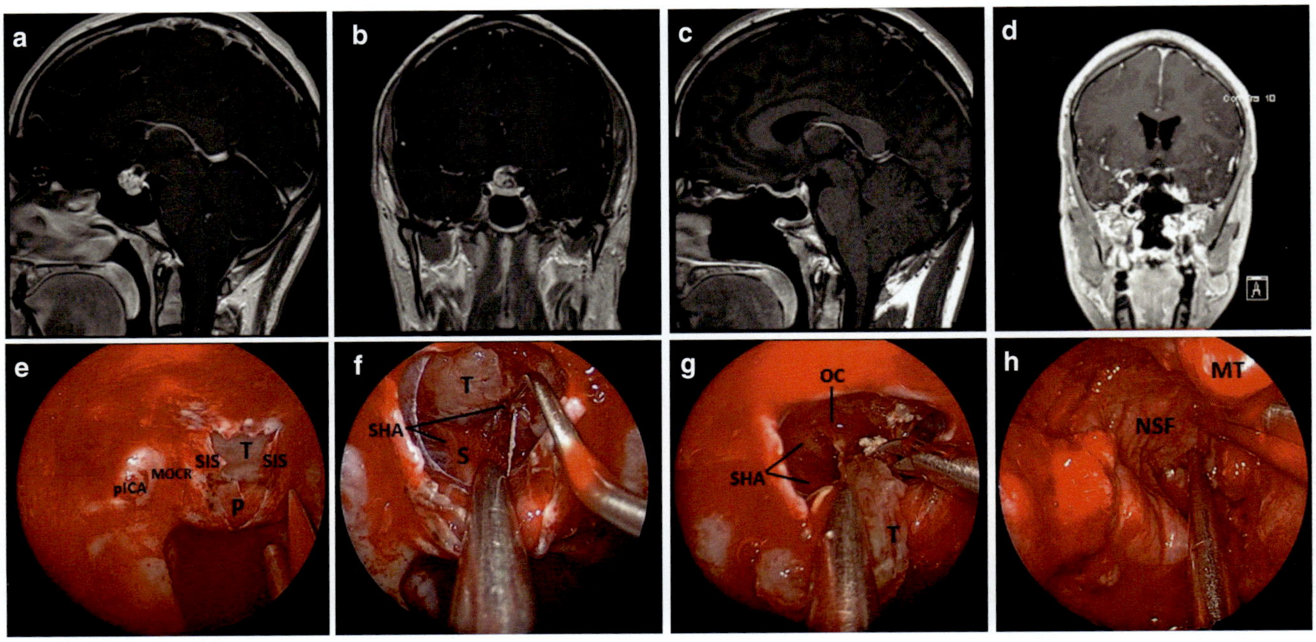

Fig. 40.3 Case 1: The endoscopic endonasal transplanum/transtuberculum approach, craniopharyngioma Type IUpper row: gadolinium-enhanced MRI in case 1. *Preoperative* (**a**) sagittal and (**b**) coronal views showing infrachiasmatic preinfundibular heterogeneous enhancing cystic-solid suprasellar mass and sellar sparing. *Postoperative* (**c**) sagittal and (**d**) coronal views demonstrated the gross total resection of the tumor. Lower row: intraoperative views of the same patient. (**e–h**) Stepwise access to the suprasellar region begins with the exposure of the tumor (T)-pituitary gland (P) interface following the division of the superior intercavernous sinus (SIS). Initial internal debulking and mobilization of the tumor revealed a healthy and posteriorly located pituitary stalk (S) and the preserved superior hypophyseal artery (SHA). Note the upward migration of the optic chiasm (OC) by the tumor. The final reconstruction should be augmented by a pedicled nasoseptal flap (NSF). The contralateral middle turbinate (MT) is usually preserved. pICA paraclinoid internal carotid artery and MOCR medial opticocarotid recess

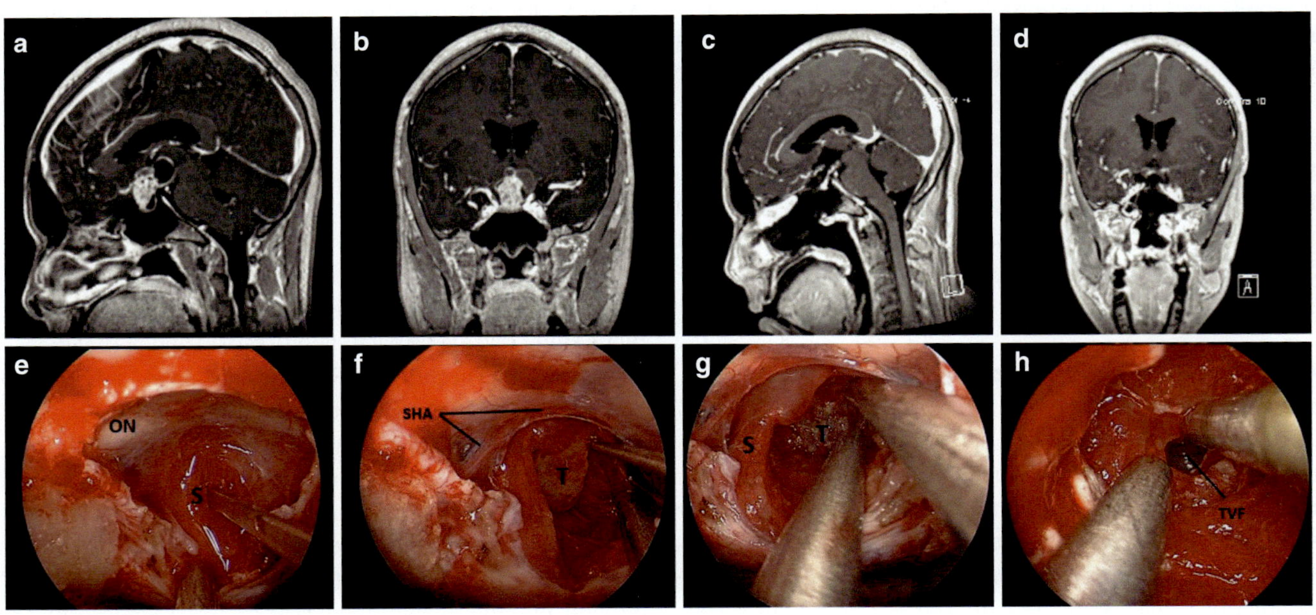

Fig. 40.4 Case 2: The endoscopic endonasal transplanum/transtuberculum approach, craniopharyngioma Type IIUpper row: gadolinium-enhanced MRI in case 2. *Preoperative* (**a**) sagittal and (**b**) coronal views showing transinfundibular heterogeneous enhancing cystic-solid suprasellar mass with a remarkable expansion of the pituitary stalk. *Postoperative* (**c**) sagittal and (**d**) coronal views demonstrated gross total resection of the tumor. Lower row: intraoperative views of the same patient. (**e–h**) Stepwise access to the tumor (T) begins with applying a vertical incision on the anterior surface of the pituitary stalk (S), followed by internal debulking and circumferential dissection of the tumor within the stalk. A tiny shell of the tumor was left behind due to firm adhesion to the anterior third ventricular floor (TVF). Note the absence of an exophytic tumor growth

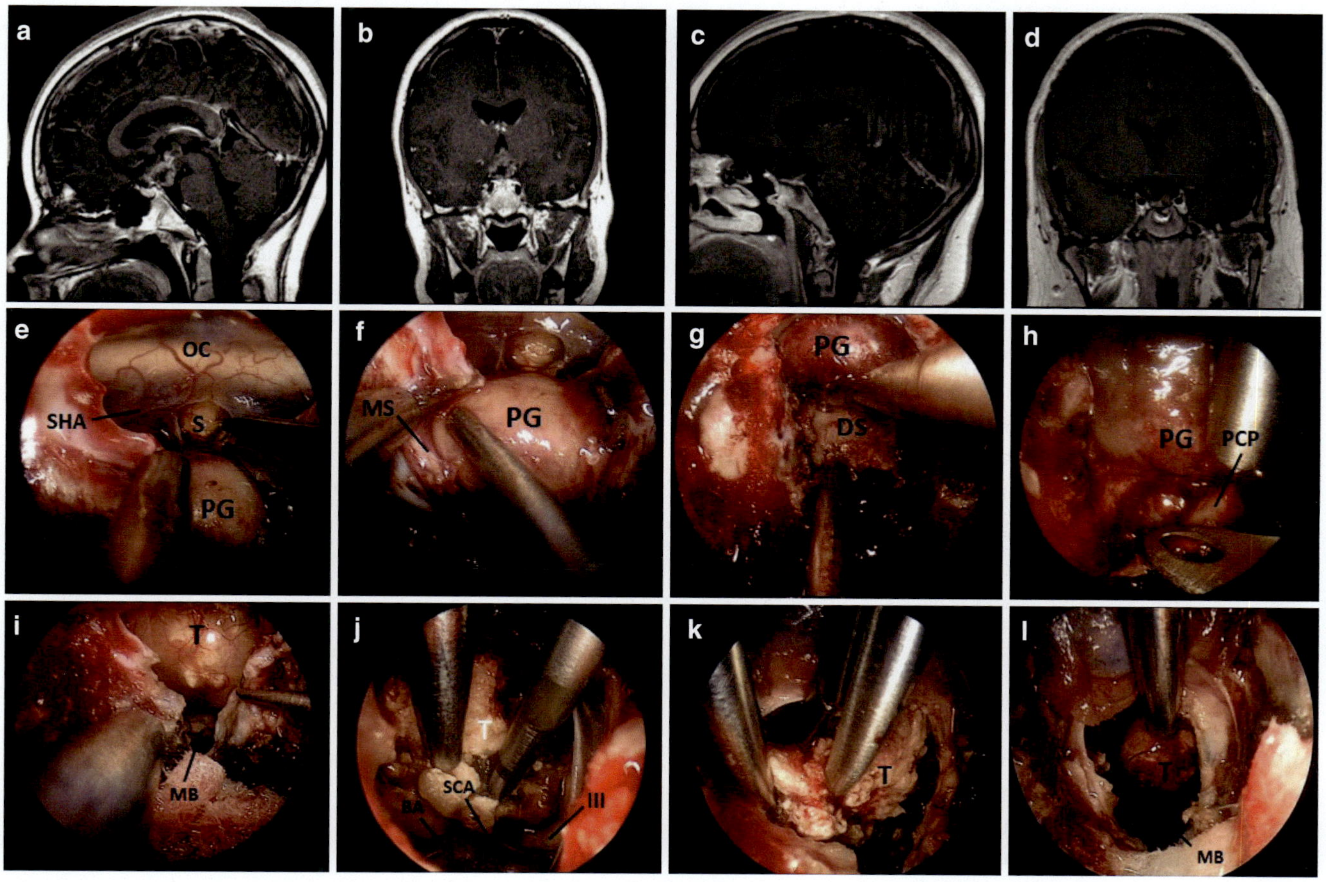

Fig. 40.5 Case 3: The endoscopic endonasal transplanum/transtuberculum-transclival approach, craniopharyngioma Type IIIUpper row: gadolinium-enhanced MRI in case 3. *Preoperative* (**a**) sagittal and (**b**) coronal views showing infrachiasmatic retroinfundibular heterogeneous enhancing cystic-solid suprasellar mass with a significant extension into the interpeduncular fossa. *Postoperative* (**c**) sagittal and (**d**) coronal views demonstrated gross total resection of the tumor. Middle row: intraoperative views of the same patient, dural exposure. (**e**–**h**) Stepwise access to the tumor begins with complete exposure of the sella and suprasellar regions. An intradural pituitary gland (PG) transposition is performed by releasing the pituitary ligaments from the medial wall of the cavernous sinus (MS), followed by the elevation of the gland. The dorsum sellae (DS) is then drilled, followed by an en bloc removal of the posterior clinoid processes (PCP). Lower row: intraoperative views of the same patient, tumor exposure. (**i**–**l**) The dural opening is extended inferiorly to expose the lower pole of the tumor (T) anterior to the midbrain (MB). The internal debulking is performed with a suction microdebrider device. The tumor is dissected from the posteriorly located basilar artery (BA), superior cerebellar artery (SCA), and oculomotor nerve (III). Note the heavy calcification of the tumor requiring a piecemeal resection

40.4.8 Case 3

The patient was a 17-year-old lady who presented with amenorrhea and the evidence of hyperprolactinemia and low growth hormone in the endocrine evaluation. The brain MRI showed the presence of a retroinfundibular heterogeneous enhancing tumor with inferior extension to the interpeduncular fossa and mild elevation of the mammillary bodies, consistent with type III CP. The patient underwent an endoscopic endonasal transplanum/transtuberculum-transclival approach with an intradural pituitary transposition. The significant intraoperative findings were the presence of a heavily calcified tumor and gradual pituitary gland engorgement following transposition. A gross total tumor resection was achieved, and the patient woke up without neurologic deficits. However, the postoperative course was complicated by the evidence of panhypopituitarism, requiring lifelong hormonal replacement. The histopathologic study demonstrated the adamantinomatous CP with HER2/Neu overexpression. The recent follow-up MRI showed no evidence of residual or recurrence (Fig. 40.5).

40.5 Surgical Pearls

- The extended endoscopic endonasal transplanum/transtuberculum approach provides excellent access to the midline structures, spanning from the sellar region to the posterior limit of the cribriform plate. It provides an optimal extent of resection of the suprasellar CP using enhanced illumination while minimizing the manipulation of the neurovascular structures lying in this area.

- Complete drilling of the tubercular recess and MOCRs is necessary to achieve a wide surgical corridor and free access to the suprasellar cistern using the opticocarotid cistern.
- The division of the SIS is required in all types of suprasellar CP, while the opening of the diaphragm sella is added in surgical resection of the transinfundibular and retroinfundibular CP.
- Intradural pituitary gland hemitransposition followed by removal of the dorsum sellae and posterior clinoid processes (i.e., transclival extension) provides optimal access to retroinfundibular CP while preserving the gland's function by establishing its venous return.
- If complete resection of the CP is not achievable, maximal safe subtotal resection followed by adjuvant therapy provides an optimal long-term outcome with similar progression-free survival.

40.6 Discussion

The surgical goal for CP should be tailored to each patient individually, considering factors such as tumor size, extension into the third ventricle and interpeduncular fossa, signal changes in surrounding neural tissue, and patient age. Whenever possible, a gross total resection must be achieved. However, the patient's quality of life should not be sacrificed to attain this goal. In cases of solid adherence to medial hypothalamus, infundibular recess, or posterior optic chiasm, a decision to leave the tumor behind and consider adjuvant therapy should be made, which can yield comparable outcomes to gross total resection.

Complications of the extended transplanum/transtuberculum EEA are similar to the standard approach, although slightly more frequent. This results from the higher extent of bone removal and dural opening and the need for generous subarachnoid dissection. The intraoperative carotid injury is always a concern, especially with such an extended form of bone removal over MOCR. However, with the advance in intraoperative navigation tools, the risk of vascular injury has been reduced significantly. The most common complication is CSF leak, even with the current novel reconstruction techniques. Early detection of CSF leak is paramount, given the high risk of meningitis and subsequent prolonged hospital stay. Serosanguinous nasal discharge is common and typically self-limiting. If there is continuous watery drainage, particularly worsening after leaning forward, a clinical diagnosis of CSF leak is made, and the patient should undergo early reexploration (<24 h) to secure the position of the reconstruction. Regular follow-ups with surveillance MRIs should be scheduled individually. Adjuvant radiotherapy with or without chemotherapy may be warranted, depending on the extent of resection or recurrence. The newer targeted therapy with BRAF inhibitors is recommended in cases of residual papillary CP, which can provide excellent tumor control.

40.7 Conclusion

Incorporating extended EEA for suprasellar CP offers a promising surgical strategy. The collaboration between neurosurgeons and otolaryngologists is essential for a comprehensive approach to complex cases. This is achieved through the synergy of neurosurgical intradural expertise and novel reconstruction techniques offered by otolaryngologists. Enhanced illumination, short surgical distance, tailored bony and dural opening, and preservation of critical neurovascular structures contribute to improved outcomes.

References

1. Brastianos PK, Taylor-Weiner A, Manley PE, Jones RT, Dias-Santagata D, Thorner AR, et al. Exome sequencing identifies BRAF mutations in papillary craniopharyngiomas. Nat Genet. 2014;46(2):161–5.
2. Karavitaki N, Wass JA. Craniopharyngiomas. Endocrinol Metab Clin N Am. 2008;37(1):173–93, ix-x.
3. Larkin S, Karavitaki N. Recent advances in molecular pathology of craniopharyngioma. F1000Res. 2017;6:1202.
4. Martinez-Gutierrez JC, D'Andrea MR, Cahill DP, Santagata S, Barker FG 2nd, Brastianos PK. Diagnosis and management of craniopharyngiomas in the era of genomics and targeted therapy. Neurosurg Focus. 2016;41(6):E2.
5. Yang I, Sughrue ME, Rutkowski MJ, Kaur R, Ivan ME, Aranda D, et al. Craniopharyngioma: a comparison of tumor control with various treatment strategies. Neurosurg Focus. 2010;28(4):E5.
6. de Divitiis E, Cappabianca P, Cavallo LM, Esposito F, de Divitiis O, Messina A. Extended endoscopic transsphenoidal approach for extrasellar craniopharyngiomas. Neurosurgery. 2007;61(5 Suppl 2):219–27. discussion 28
7. Fukushima T, Hirakawa K, Kimura M, Tomonaga M. Intraventricular craniopharyngioma: its characteristics in magnetic resonance imaging and successful total removal. Surg Neurol. 1990;33(1):22–7.
8. Gardner PA, Kassam AB, Snyderman CH, Carrau RL, Mintz AH, Grahovac S, et al. Outcomes following endoscopic, expanded endonasal resection of suprasellar craniopharyngiomas: a case series. J Neurosurg. 2008;109(1):6–16.
9. Jane JA Jr, Kiehna E, Payne SC, Early SV, Laws ER Jr. Early outcomes of endoscopic transsphenoidal surgery for adult craniopharyngiomas. Neurosurg Focus. 2010;28(4):E9.
10. Yaşargil MG, Curcic M, Kis M, Siegenthaler G, Teddy PJ, Roth P. Total removal of craniopharyngiomas. Approaches and long-term results in 144 patients. J Neurosurg. 1990;73(1):3–11.
11. Samii M, Tatagiba M. Surgical management of craniopharyngiomas: a review. Neurol Med Chir. 1997;37(2):141–9.
12. Jeswani S, Nuño M, Wu A, Bonert V, Carmichael JD, Black KL, et al. Comparative analysis of outcomes following craniotomy and expanded endoscopic endonasal transsphenoidal resection of craniopharyngioma and related tumors: a single-institution study. J Neurosurg. 2016;124(3):627–38.
13. Komotar RJ, Starke RM, Raper DM, Anand VK, Schwartz TH. Endoscopic endonasal compared with microscopic transs-

phenoidal and open transcranial resection of craniopharyngiomas. World Neurosurg. 2012;77(2):329–41.
14. Cappabianca P, Cavallo LM, Esposito F, De Divitiis E. Craniopharyngiomas. J Neurosurg. 2008;109(1):1–3; reply -5.
15. Cavallo LM, Messina A, Cappabianca P, Esposito F, de Divitiis E, Gardner P, et al. Endoscopic endonasal surgery of the midline skull base: anatomical study and clinical considerations. Neurosurg Focus. 2005;19(1):E2.
16. Cavallo LM, Prevedello DM, Solari D, Gardner PA, Esposito F, Snyderman CH, et al. Extended endoscopic endonasal transsphenoidal approach for residual or recurrent craniopharyngiomas. J Neurosurg. 2009;111(3):578–89.
17. Couldwell WT, Weiss MH, Rabb C, Liu JK, Apfelbaum RI, Fukushima T. Variations on the standard transsphenoidal approach to the sellar region, with emphasis on the extended approaches and parasellar approaches: surgical experience in 105 cases. Neurosurgery. 2004;55(3):539–47; discussion 47-50.
18. Kassam A, Snyderman CH, Mintz A, Gardner P, Carrau RL. Expanded endonasal approach: the rostrocaudal axis. Part I. Crista galli to the Sella turcica. Neurosurg Focus. 2005;19(1):E3.
19. Kassam AB, Gardner PA, Snyderman CH, Carrau RL, Mintz AH, Prevedello DM. Expanded endonasal approach, a fully endoscopic transnasal approach for the resection of midline suprasellar craniopharyngiomas: a new classification based on the infundibulum. J Neurosurg. 2008;108(4):715–28.
20. Cavallo LM, Frank G, Cappabianca P, Solari D, Mazzatenta D, Villa A, et al. The endoscopic endonasal approach for the management of craniopharyngiomas: a series of 103 patients. J Neurosurg. 2014;121(1):100–13.
21. Pascual JM, González-Llanos F, Barrios L, Roda JM. Intraventricular craniopharyngiomas: topographical classification and surgical approach selection based on an extensive overview. Acta Neurochir. 2004;146(8):785–802.
22. Puget S, Garnett M, Wray A, Grill J, Habrand JL, Bodaert N, et al. Pediatric craniopharyngiomas: classification and treatment according to the degree of hypothalamic involvement. J Neurosurg. 2007;106(1 Suppl):3–12.
23. Carrau RL, Jho HD, Ko Y. Transnasal-transsphenoidal endoscopic surgery of the pituitary gland. Laryngoscope. 1996;106(7):914–8.
24. Gardner PA, Prevedello DM, Kassam AB, Snyderman CH, Carrau RL, Mintz AH. The evolution of the endonasal approach for craniopharyngiomas. J Neurosurg. 2008;108(5):1043–7.
25. Prevedello DM, Doglietto F, Jane JA Jr, Jagannathan J, Han J, Laws ER Jr. History of endoscopic skull base surgery: its evolution and current reality. J Neurosurg. 2007;107(1):206–13.
26. Fernandez-Miranda JC, Gardner PA, Rastelli MM Jr, Peris-Celda M, Koutourousiou M, Peace D, et al. Endoscopic endonasal transcavernous posterior clinoidectomy with interdural pituitary transposition. J Neurosurg. 2014;121(1):91–9.
27. Kassam AB, Prevedello DM, Thomas A, Gardner P, Mintz A, Snyderman C, et al. Endoscopic endonasal pituitary transposition for a transdorsum sellae approach to the interpeduncular cistern. Neurosurgery. 2008;62(3 Suppl 1):57–72; discussion-4.
28. Truong HQ, Borghei-Razavi H, Najera E, Igami Nakassa AC, Wang EW, Snyderman CH, et al. Bilateral coagulation of inferior hypophyseal artery and pituitary transposition during endoscopic endonasal interdural posterior clinoidectomy: do they affect pituitary function? J Neurosurg. 2018;131(1):141–6.
29. Kassam AB, Thomas A, Carrau RL, Snyderman CH, Vescan A, Prevedello D, et al. Endoscopic reconstruction of the cranial base using a pedicled nasoseptal flap. Neurosurgery. 2008;63(1 Suppl 1):ONS44-52; discussion ONS-3.
30. Fernandez-Miranda JC, Gardner PA, Snyderman CH, Devaney KO, Strojan P, Suárez C, et al. Craniopharyngioma: a pathologic, clinical, and surgical review. Head Neck. 2012;34(7):1036–44.

MIX
Papier aus verantwortungsvollen Quellen
Paper from responsible sources
FSC® C105338

If you have any concerns about our products,
you can contact us on
ProductSafety@springernature.com

In case Publisher is established outside the EU,
the EU authorized representative is:
Springer Nature Customer Service Center GmbH
Europaplatz 3, 69115 Heidelberg, Germany

Printed by Libri Plureos GmbH
in Hamburg, Germany